COLLEGE BOTANY

[Vol. II]

COLLEGE BOTANY

[Vol. II]

Hirendra Chandra Gangulee MSc, DSc, FNA
Retired UGC Professor of Botany, Calcutta University

Asok Kumar Kar MSc, PhD [Illinois]
Retired Professor of Botany, Presidency College, Kolkata

AND

S C Santra MSc, PhD, PRS [SC], FAScT
Department of Environmental Sciences, University of Kalyani
Nadia, West Bengal

New Central Book Agency (P) Ltd

Co-Published by

Bijay Publications

E-mail : bijaypublications@gmail.com

Exclusively Distributed by

Bijay Pustak Bhandar (P) Ltd

90/6A M. G. Road, Bhawani Dutta Lane, Kolkata-700007

✆ : (033) 4605-7856, 8777898576, 8274916104

E-mail : bijaypustakbhandar@gmail.com

NCBA

REGD OFFICE
8/1 Chintamoni Das Lane, Kolkata 700 009, India
email: ncbakolkata@gmail.com

OVERSEAS
NCBA (UK) Ltd, 149 Park Avenue North, Northampton, NN3 2HY, UK
email: ncbauk@yahoo.com

CORPORATE OFFICE
12 Shahpur Jat, New Delhi 110 049
email: ncbaexports@ncbapvtltd.com

BIJAY PUBLICATIONS
90/6A M. G. Road, Bhawani Dutta Lane, Kolkata-7
E-mail : bijaypublications@gmail.com

BIJAY PUSTAK BHANDAR (P) LTD
90/6A M. G. Road, Bhawani Dutta Lane, Kolkata-7
E-mail : bijaypustakbhandar@gmail.com

COLLEGE BOTANY [Vol. II] • Hirendra Chandra Gangulee, Asok Kumar Kar and S C Santra

First Published: 1968
Fourth Edition: 1989
Reprinted: 2006, 2007, 2010, 2011, 2012, 2013
Thoroughly Revised and Updated Fifth Edition: 2015
Reprinted: 2016, 2017, 2018, 2019, 2020, 2023

PUBLISHER • TYPESETTER • PRINTER
New Central Book Agency (P) Ltd
8/1 Chintamoni Das Lane, Kolkata 700 009

CO-PUBLISHED BY
Bijay Publications
90/6A M G Road, Bhawani Dutta Lane, Kolkata 700007

PROJECT TEAM
Ashit Ghosh, Pradip Biswas and Prodip Baidya

ISBN: 978 81 7381 178 4

Price ₹ 1250.00

Contents

PART FOUR

Lichens

PART FIVE

The Myxomycota, Microbiology, Bacteria, Viruses

PART SIX

Plant Pathology

PART SEVEN

Bryophyta

Preface

PREFACE TO THE FIFTH EDITION

It is my pleasure and privilege to revise the entire book after the death of my revered teachers Prof. Hirendra Chandra Gangulee and Prof. Asok Kumar Kar. I hope the present edition will also fulfil the aspiration of students, researchers and other readers in the long run.

Independence Day
15th August, 2014

S. C. Santra

PREFACE TO THE FOURTH EDITION

It is a great pleasure to present to our students and fellow-teachers the thoroughly revised and enlarged fourth edition of the book. Part on Microbiology has been added newly to this edition. Other parts also have been revised and made up-to-date.

Our sincere thanks are due to Dr. (Miss) Manju Banerjee, Department of Botany, Calcutta University for her special article on the *Glossopuris* Flora.

We hope that the present edition also will be appreciated as before.

Janmastami
18th August, 1989

H. C. Gangulee
A. K. Kar

PREFACE TO THE THIRD EDITION

It is a pleasure to present to our students and fellow-teachers the revised third edition of our book. We hope that our patrons would realise how the present high prices are affecting the cost of book production. We can do nothing but only apologise for the consequential slight rise in price.

Republic Day
26th January, 1976

H. C. Gangulee
A. K. Kar

PREFACE TO THE SECOND EDITION

Our sincere thanks are due to our students and to our fellow-teachers all over India for the appreciation with which the first edition of this volume was received. In bringing out the second edition we tried to meet the requirements of our patrons whom we thank for their individual opinions expressed to us. We did not realise that the first edition would get exhausted so soon and we apologise for the consequent delay in bringing out the second edition. We hope that the present edition would satisfy the needs of our patrons.

Diwali
29th October, 1970

H. C. Gangulee
A. K. Kar

PREFACE TO THE FIRST EDITION

We owe an apology to our very appreciative patrons for not being able to bring out the Vol. II of our book before nine years had elapsed after publishing the first volume. The delay is because we endeavour to have every part written very carefully by teachers who really know their subjects. Numerous careful illustrations also consume much time. Our original aim was to publish the whole book in two volumes. But we find that a third volume is really necessary to finish it. The Vol. III is expected to come out shortly.

We expect this volume will be received with the same degree of enthusiasm by our students and fellow-teachers as was the case with the Volume I. We apologise for the printing mistakes which have crept into this volume and solicit from our patrons suggestions for improvement of the volume in future editions. Lastly, we thank all our friends, patrons and the publishers for the encouragement and support received from them.

Buddha Purnima
12th May, 1968

H. C. Gangulee
A. K. Kar

Part One

The Living World

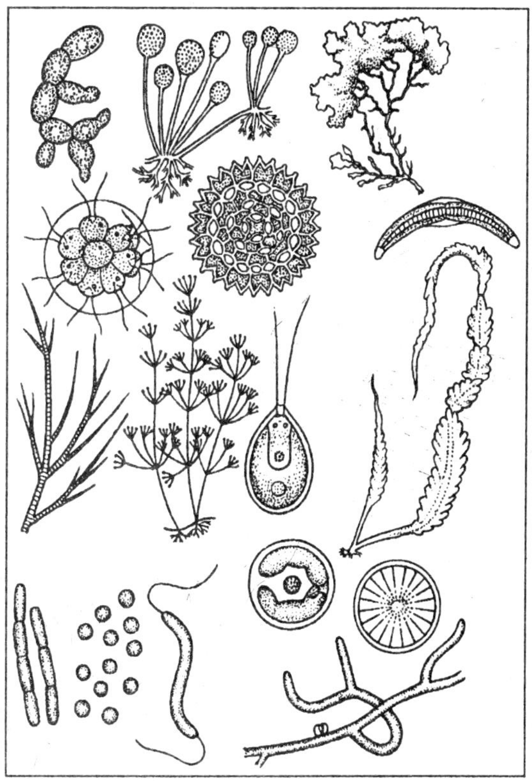

Chapter 1

The Living World

Our first impression of the living world is apt to be of bewilderment produced by its extraordinary diversity.

That the conditions under which life might be supposed to have arisen must have been quite different from those prevailing at the time, that need not be presumed to have been necessary, that amino acids can be produced from inorganic materials in an environment in which the attempt is made to approximate to such conditions that earliest living material need not have possessed capacity for photosynthesis — these are major view points held by Oparin (1957), Cloud (1965), Ponnam-peruna and Black (1965), and many others. These view points have brought about a change in our traditional concepts of the origin of life. The increasing emphasis on the biochemical approach to the understanding of living organisms has quite naturally had an impact on theories of relationship. The discovery of the remarkable similarity of the **deoxyribonucleic acid** molecule in widely different organisms is perhaps the most striking result of these studies. If this fundamental material is similar in all organisms as present evidence indicates, then it has been argued, all living things must be closely related. This is often extended to imply that all organisms have been derived from a single fortuitous chemical event under suitable environmental conditions, such as may have existed on the earth at a remote time. Such happenings must invariably have assumed by gradual stages, and certainly at a subcellular level, the characteristics which are now associated with life. The details of such a process are still speculative.

If one starts to examine the living world in terms of its chemical and microscopic organization, the initial impression of diversity among various organisms gives way to a profound realization of its fundamental unity. The great superficial diversity becomes understandable as the consequence of the operation of the evolutionary process on a kind of matter, **the living cell**, which probably had a single historical origin and which still preserves many common features. What are the common features of living organisms? *First*, all of them share a common **chemical compo-**

sition. The most fundamental feature of this common chemical composition is the invariable presence of three types of complex molecules: **protein, deoxyribonucleic acid** and **ribonucleic acid**. In addition they all perform certain **metabolic activities** known collectively as **metabolizm** or **metabolic activities**. There are, naturally, many differences in metabolic detail among the different kinds of organisms. However, all organisms are obliged to synthesize the universal constituents of living matter from external chemical building blocks and to generate the energy necessary for such synthetic activities by the formation of a characteristic kind of energy-bond at the expense of external energy sources. With respect to these fundamental features of metabolizm, resemblances, between different kinds of organisms are much more marked than other differences. Finally, all organisms share a **common physical structure**, being organized into microscopic subunits known as **cells**. As a consequence of their cellular organization, growth results in **cell division**, with an increase in the total number of cells.

The extraordinary diversity among the living organisms of the living world, then the scientists had to carefully workout and acceptable approach to the living organisms — their physical appearance, behaviour and suitable grouping together. Mention may be made about the work of Lamarck and his associate Treviranus who first started work on living organisms designating the branch of Science — **Biology**. But actually, Aristotle was also the first person to study the living organisms in detail.

Plant classification systems in particular, based on form or habit, were introduced by the Greeks and, for approximately 1000 years formed the basis of systems introduced by many subsequent workers and continued up to the time of Carl Linnaeus (1707-1778).

The publication of 'Genera Plantarum' by Linnaeus in 1737 introduced a second era of plant classification, characterized by the introduction of **artificial system**, more commonly known as **sexual system** because Linnaeus concentrated on he number of sex organs and presence or absence of flowers. Linnaeus grouped plants into 24 convenient classes. All flowering plants were subdivided among 23 classes. Under the twenty-fourth class, which he named as **Cryptogamia**, he included plants having no flowers and designated them as **Cryptogams**. Linnaeus's system was pioneer for it recognized for the first time the flowerless plants as a major taxonomic category.

During subsequent period of time there has been change in approach and a classification system of living world, established basically by Eichler (1886), was widely used for long time. This classification system is outlined below.

Kingdom: **Plantae**

- Division: **Thallophyta** — plant body not differentiated into stem, leaves and roots, i.e., thalloid
  - Subdivision: **Algae** — plant body without chlorophyll
  - Subdivision: **Fungi** — **true fungi, slime molds** and **bacteria**
- Division: **Bryophyta** — plant body with some differentiation into root-like structures
- Division: **Pteridophyta** — plant body differentiated into stem, leaves and roots
- Division: **Spermatophyta** — higher plants bearing flowers and seeds

Kingdom: **Animalia**

Traditionally, the living organisms were placed in two **kingdoms**: the plants (**Plantae**) and the animals (**Animalia**). Lecumenock's discovery of microorganisms and the gradual discovery of large number of diverse microorganisms made it impending to think of about creating a **third kingdom Protista** to accommodate these organisms, because they differed radically from both plants and animals. In 1866, Darwin's disciple Hauckel subsequently realized that simple organism like fungi, simple algae, protozoa and bacteria, were so much unlike, the higher plants and animals that a third kingdom — **Protista** be created for these simple organisms.

The protists include both photosynthetic and non-photosynthetic organisms. Some share many properties common to both plants and animals. What distinguishes all protists from higher plants and animals is their relatively **simple biological organization.** The protists are a large and very heterogenous group of organisms distinguished by their preservation, throughout the course of biological evolution, of a relatively simple and undifferentiated structure. Many protists are unicellular and even those which are multicellular do not show the differentiation into separate tissue regions characteristics of plants and animals.

It is now widely accepted that the two-kingdom system allows no place for organisms of simple structure which are neither plants nor animals, or which possess characters of both. In 1878 third kingdom was created, the living world can thus be subdivided into three kingdoms: **Protista, Plants and Animals.**

The component groups of the three kingdoms of living organisms are:

1. Unicellular or multicellular without tissue differentiation:

Protista: **Algae**
Protozoa
Fungi
Slime molds
Bacteria

2. Multicellular showing extensive tissue differentiation:

Plants: **Seed plants**
Ferns
Mosses and Liverworts

Animals: **Vertebrates**
Invertebrates

THE THALLOPHYTA

The term **Thallophyta** (Gk. **thallos**, a young shoot; **phyton**, a plant), was first introduced by Endlicher in 1836. It includes a vast and diverse assemblage of plants lowest in development with the least differentiation of vegetative bodies and having simple methods of reproduction. Hence it forms the lowest division of the plant kingdom comprizing of ancient members of the plant population.

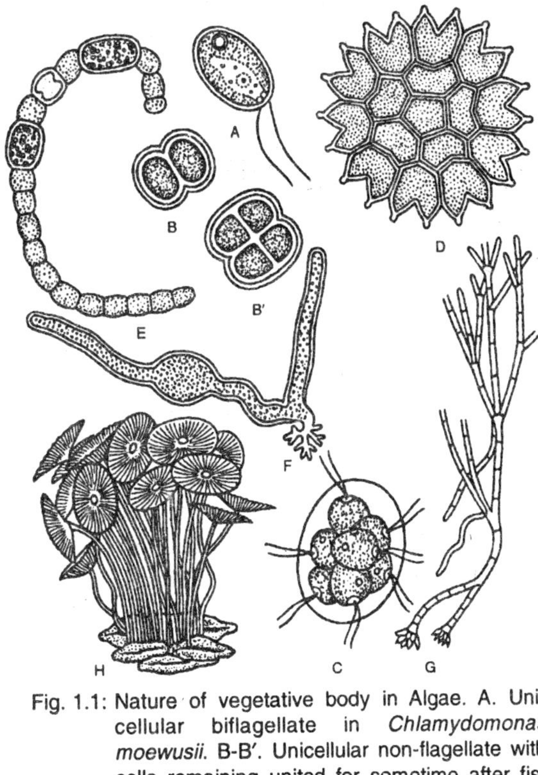

Fig. 1.1: Nature of vegetative body in Algae. A. Unicellular biflagellate in *Chlamydomonas moewusii*. B-B′. Unicellular non-flagellate with cells remaining united for sometime after fission in *Chroococcus* sp. C. Colony of cells embedded in a gelatinous matrix in *Pandorina* sp. D. Cells aggregated together forming a plate-like or star-shaped colony in *Pediastrum* sp. E. Filamentous unbranched in *Anabaena* sp. F. Branched coenocytic filament in *Vaucheria sessilis*. G. Filamentous branched with septa and rhizoids in *Cladophora glomerata* var. *simplicior*. H. Complex in form in *Acetabularia mediterranea*

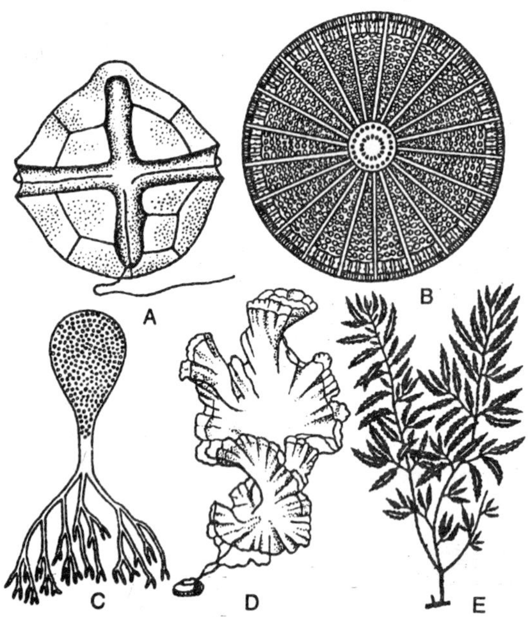

Fig. 1.2: Nature of vegetative body in Algae. A. Unicellular biflagellate in *Goniaulax* sp. B. Unicellular non-flagellate with attractive wall markings in a diatom. C. Coenocyte differentiated into overground vesicle and underground rhizoidal portion with division of labour, encountered in *Botrydium granulatum*. D. Parenchymatous thallus in *Ulva lactuca*. E. Elaborately developed branches bearing leaf-like expansions in *Sargassum longifolium*

Some of them are the border line between primitive plants and animals. In general, the thallophytes are aquatic, semiaquatic, and **epigeal** or hypogeal in habit. The vegetative **body** of a thallophyte is known as a **thallus (pl. thalli)**. Thallus is a relatively simple plant **body** which is not differentiated into roots, **stem** and leaves. A thallus may or may not be surrounded by a wall. The thalli may range from microscopic unicellular forms, free-living or otherwise (Figs. 1.1A, B and B′; 1.2A and B; and 1.3A and B), cells aggregated together forming colony (Figs. 1.1C and D); filamentous (branched, unbranched or false-branched) with or without transverse partition wall known as **septum** (pl. **septa**) (Figs. 1.1C, 1.1D, 1.1E, 1.1F) to macroscopic forms (Fig. 1.1H), some of which may be even two hundred feet or more in length. A filamentous branched thallus in which chlorophyll is absent is known as **mycelium** (pl. **mycelia**). Mycelium may be septate (Fig. 1.3C) or aseptate (Fig. 1.3D) with or without the presence of septa. It is designated as **rhizomycelium** when mycelium is poorly developed into a rhizoid-like structure (Fig. 1.3A). Thalli may also have considerable specialization of structure with corresponding specialization of function (Figs. 1.2C to E). There are thallophytes whose vegetative bodies may have structures resembling superficially as roots, stem and leaves. These are not true roots, stem and leaves, because they lack vascular tissues. A thallophyte having multi-nucleate unicellular (Figs. 1.2C and 1.3D) to multicellular (Fig. 1.2C) thallus is known as a **coenocyte** and the thallus is **coenocytic**. The thallophytes that are visible only under the high power magnification of a light microscope are commonly named as **microbes** (Figs. 1.4A to E). Again, the vegetative body of the certain thallophytes composed of multinucleate mass of protoplasm without any cell wall is known as

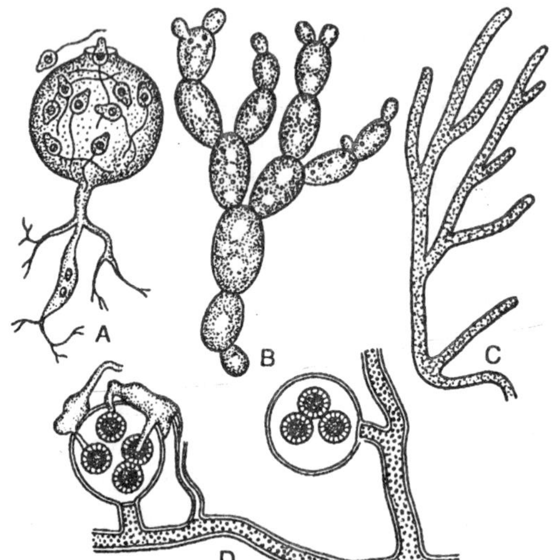

Fig. 1.3: Nature of vegetative body in Fungi. A. Rhizomycelium in *Nowakowskiella ramosa*. B. Chain of cells formed as a result of rapid budding and growth of unicellular uninucleate vegetative body of yeast (*Saccharomyces cerevisiae*). C. Septatebranched thallus (mycelium) of a fungus. D. Aseptate branched thallus (mycelium) bearing gametangia in *Saprolegnia* sp.

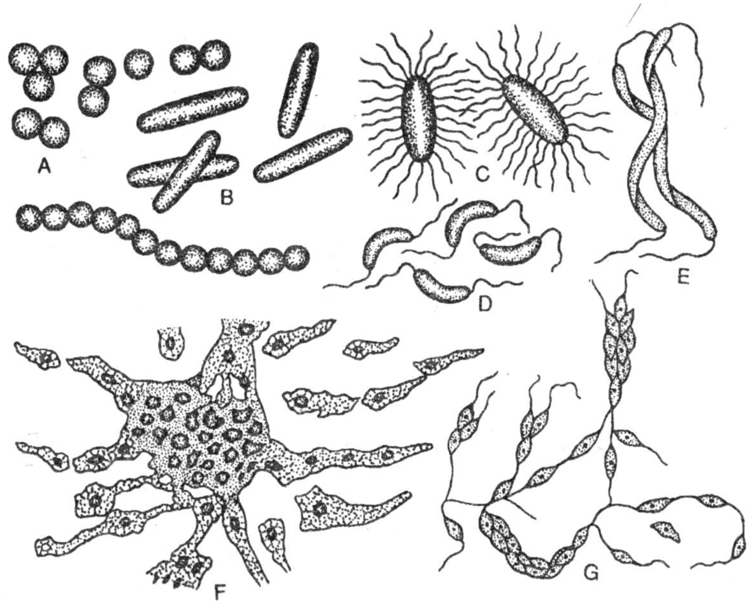

Fig. 1.4: Nature of vegetative body in bacteria and slime molds. A-E. Coccus, bacillus, and spirillum forms of bacteria. F. Pseudoplasmodium of *Dictyostelium discoideum*. G. 'Net plasmodium' of *Labyrinthula macrocystis* showing clumps of cels on the 'tracks' and cells sending out threads

a **plasmodium** (pl. **plasmodia**), which exhibits creeping movement (Fig. 1.5); or a **pseudoplasmodium**, a structure having resemblance with the plasmodium being developed as a result of aggregation of naked, uninucleate structures (Fig. 1.4F); or a **net plasmodium** characterized by naked, uninucleate spindle-shaped or oval cells which become interconnected by slime filaments along which the cells glide (Fig. 1.4G). The thallophytes may be with or without having the normal green pigment, chlorophyll in their thalli. They are **autophytes** (**autotrophic**) when their thalli possess chlorophyll and can prepare their own food from carbon dioxide and water with the help of chlorophyll, whereas those which lack chlorophyll in their thalli and live on food substances synthesized by other organizms are the **heterophytes (heterotrophic)**. Some of the heterophytes thrive as **parasites** on living organic matters; others receive their nutrition from dead and decaying organic matters, they are the **saprobes** (**saprophytes**). Again, there are other thallophytes which live in an intimate association with other living organizms deriving mutual benefit. They are known as **symbionts** and the phenomenon is **symbiosis**. But the thallophytes may also be **endophytes** or **epiphytes** when they live inside or outside plant parts respectively without receiving any nutrition from them. The endophytes are often

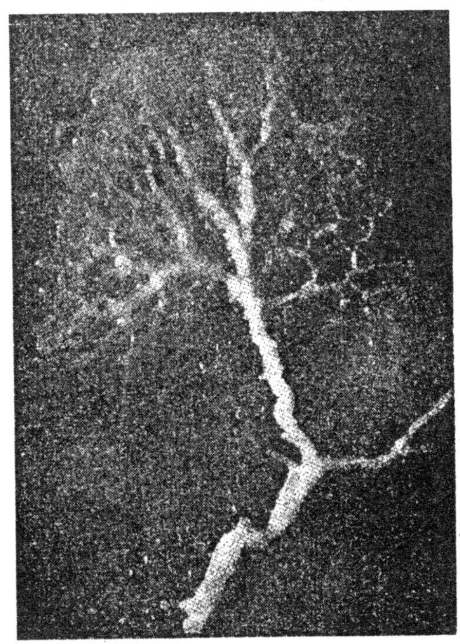

Fig. 1.5: Plasmodium of *Phycorum wingstense*

designated as **space parasites**. There are certain other thallophytes whose thalli possess a special kind of chlorophyll which is different from the chlorophyll that is present in green plants. With the help of this special kind of chlorophyll they can prepare their own food. Besides these, there are thallophytes which manufacture their food by the process of chemosynthesis. The nature of reserve food is extremely variable in the thallophytes, which may be carbohydrate of various forms, fats, oils and similar other substances.

The methods of reproduction in thallophytes may be: **vegetative, asexual** and **sexual**. In vegetative reproduction the thallus breaks up into many pieces and each piece subsequently develops into a new individual. The most common method of reproduction is by the development of minute one to many-celled propagative units, the **spores** which on germination give rise to new plants. The spores may be developed by processes other than sexual—**asexual spores** or as a result of sexual process followed by meiosis—**sexual spores**. They are usually developed in a structure known as a **sporangium** (pl. **sporangia**) which is normally a single-celled structure. Asexual reproduction takes place mainly in three ways : by **fission**, by **budding**, and by the development of **spores**. Fission usually is 'splitting easily' of the parent thallus into two daughter individuals which develop into new thalli. During budding the thallus produces one or more tiny outgrowths, the **buds** (Fig. 1.3B). These buds increase in size, each one receives a daughter nucleus from the parent thallus. All of them are ultimately cut off from the parent individual by a process of constriction and then carry on separate existence as new independent thalli. In cases where the asexual reproduction is by the development of spores, the spores may be flagellate or non-flagellate, that is, may or may not bear hair-like structures known as **flagella** (sing. **flagellum**). The flagellate spores are the **zoospores**. Sexual reproduction takes place by the union of **gametes**. The gametes are minute uninucleate sex cells which by themselves cannot normally produce new individuals. Hence they fuse in pair and the fusion product—**zygote** undergoes various changes leading to the development of new individual without the development of embryo. The gametes may or may not be developed in a gamete-producing structure known as a **gametangium** (pl. **gametangia**). The gametangium is simple and is usually unicellular without being surrounded by any protective tissue. There are few exceptional cases in which the gametangium is multicellular, but without any surrounding layer of protective tissue.

Many of the thallophytes are extremely beneficial to human beings and plants. They are of great use to human society either as food, as medicine or as manures and in similar other ways. Large number of thallophytes are used by fish as food. Again, many of the thallophytes are responsible for causing serious diseases of plants and animals including human beings. Thallophytes with beautiful markings on their cell wall are of great interest to the artists for their attractive designs (Figs. 1.2A and B).

The common thalophytes are : the **algae**, the **fungi**, the **slime molds**, and the **bacteria**. Unger (1838) included **algae, lichens** and **fungi** in the Thallophyta.

Validity of the Thallophyta. With the advancement of human knowledge about plants—the lower plants in particular, and their natural relationships, it was realized that Thallophyta, a designation for the lower plants is an artificial collection of unrelated plants having extremely diversified characteristics, have attained the same level of development. To accept the Thallophyta as a natural division of the plant kingdom implies acceptance of he view that both the Algae and the Fungi are each more or less homogeneous phylogenetic series.

It has become increasingly clear during the past quarter century that the morphology and the physiology of the individual cells are the fundamental bases upon which the Algae must be classified. This evidence shows that there are several series among the Algae, each of which has cells with certain distinctive physiological and morphological traits. Chief among the morphological characteristics is the structure of the motile cells, and for most of the groups among the Algae there is a striking constancy in its organization, especially with respect to the number, arrangement, and relative length of the flagella. The physiological distinction is in the plastids and a constancy in the chemical nature of the food reserves accumulating through photosynthetic activity. All these suggest very strongly that the various major groups of Algae have little in common with one another and as such the Subdivision Algae is an artificial group. It is actually made up of a number of groups which are not closely related.

Acceptance of the view that the various groups of Algae are more or less independent of one another means that both the Thallophyta and its Subdivision Algae must be abandoned as natural units, Algae being a correspondingly higher rank. Abandonment of the Algae as a Subdivision of the plant kingdom does not mean that the word alga must be abandoned. It is still of great service as a descriptive term for designating simple plants with an autotrophic mode of nutrition.

The taxonomic deposition of the Fungi, the other Subdivision of the Thallophyta depends upon the question of their origin. This is highly controversial and opinion is divided as to whether they had a monophyletic or polyphyletic origin. If they arose from the protozoa, they should be put in one or more divisions coordinate in rank with the various algal divisions; if they arose from Algae, they should be placed as classes of one or more of the algal divisions.

Adoption by many botanists of the conclusion that Fungi did not evolve from the Algae invalidated the use of the term Thallophyta as a Division and fostered the more recent system of dividing the Algae into several Divisions and the Fungi into several subdivisions and separating slime molds and bacteria as independent taxa, previously included under Fungi, ranking as Fungi. This aspect will be elaborated further later on.

Therefore, the Division Thallophyta is now regarded as a highly unnatural assemblage of unrelated forms of plants.

Some of the contrasting features of Algae and Fungi are incorporated below:

1. The thallus of Algae is characterized by the presence of the pigment chlorophyll which, in some species, is masked by other pigment or pigments producing various colourations.

Chlorophyll is completely lacking in Fungi.

2. Owing to the presence of chlorophyll the Algae are able to synthesize their own food from simple inorganic materials—**autotrophic**.

Most Fungi are incapable of manufacturing their own food and are dependent on food from external sources—**heterotrophic**. They are usually either **parasites** or **saprophytes**. But there are Fungi which can synthesize protein and other organic substances when grow under cultural condition.

3. Diffused or direct light is essential for the growth and nutrition of Algae.

The Fungi can grow both under light and darkness.

4. The reserve food in most Algae is a carbohydrate—**starch**, which turns blue with iodine solution.

The chief storage product in Fungi is a carbohydrate—**glycogen**, which may be distinguished from starch by a brown colouration produced by iodine solution.

5. The thallus of Algae ranges from unicellular to multicellular parenchymatous structure. Whereas, in Fungi, the thallus is unicellular-nonfilamentous or is made up of fine thread-like structure —**hypha** (pl. **hyphae**) interwoven in a mass—**mycelium** (pl. **mycelia**) which may be septate or aseptate.

6. In Algae, the cell walls are composed of cellulose.

The cell walls of Fungi are not of cellulose usually, but of **chitin**, a complex nitrogenous compound, but simpler than animal chitin.

7. In Algae, the rise in complexity of the vegetative body from a very simple form is correlated with the progressive evolution of sexuality from simple to complex condition.

Reverse is the condition in Fungi, where there is gradual degeneration of sexuality with the corresponding gradual progress in complexity of the vegetative body.

There are, naturally, many differences in metabolic detail among the different kinds of organisms. However, all organisms are obligated to synthesize the universal constituents of living matter from external chemical building blocks and to generate the energy necessary for such synthetic activities by the formation of a characteristic kind of energy-rich bond at the expense of external energy sources. With respect to these fundamental features of metabolism, resemblances between the different kinds of organisms are much more marked than are the differences. Finally, all organisms share a **common physical structure**, being organized into microscopic subunits known as **cells**. As a consequence of their cellular organization, growth results in **cell division**, with an increase in the total number of cells.

All the properties described above are common to plants, animals, and microorganisms, despite their gross differences of size, shape, and internal structure. There is, however, one group of biological objects, the **viruses**, which does not confirm to the list of common properties which have just been outlined. This is why, according to some, viruses are not microorganisms but are biological objects with a different set of fundamental properties. Others believe that viruses are microorganisms which have lost some of the principal properties of organisms. Again, others are of opinion that viruses are organisms that share properties of both living and non-living objects and represent the connecting link between them.

Present-day approach. With increase in knowledge about the living world, the traditional division of living world into the plant and animal kingdoms is thus no longer a reasonable approach to the living world. During the last century various solutions, involving the establishment of one or more new kingdoms have been offered.

In 1866, one of Darwin's disciples, the German Zoologist Haeckel, proposed the establishment of a third kingdom, the **Protista,** to include both photosynthetic and non-photosynthetic organsisms, some clearly 'plant-like', and some sharing many properties common to both kingdoms—plant and animal. What distinguishes all members of the Protista from higher plants and animals is their relatively **simple biological organization**. The protists are a large and very heterogeneous group of organisms, distinguished by their preservation, throughout the course of biological solution, of a relatively simple and undifferentiated structure. Many portists are unicellular and even those that are multicellular do not show the differentiation into separate tissue regions characteristic of plants and animals. Neither Haeckel's proposal, as first put forward or as later modified (1878, 1894), nor those of his contemporaries were generally adopted. In recent years renewed interest has been shown in this problem. It is now widely accepted that the two-kingdom system allows no place for organisms of simple structure which are neither 'plant' nor 'animal', or which possess characters of both. The living world can thus be subdivided most logically into three kingdoms: **Protista, Plants,** and **Animals.**

With the invention of electron microscopy, biologists studied the fine structure of cells. It was realized that biological cells could be distinguished basically into two types. One kind of cells possessed well organized double membrane-bound nucleus and several other membrane-bound organelles, while the other kind of cells, as those of bacteria, did not have these features. These organisms having first cells were designated as **Eucaryotic.** All animal organisms including the lower ones like protozoa, amoeba, as well as plants — both higher and lower ones; like bryophytes, algae and fungi possessing eucaryotic type of cells.

The eucaryotic cells shows innumerable specialization of form and of function having a cell wall composed of a variety of carbohydrate materials. Its nucleus has a well-defined nuclear membrane and hereditary information of the eucaryotic cell, coded in the DNA (deoxyribonucleic acid) of the nucleus, is carried on a number of different subunits, chromosomes. In all eucaryotic organisms gametic fusion is followed by nuclear fusion. A eucaryotic organism whose life cycle contains a sexual stage may exist almost executively in the diploid state, as to all higher animals and flowering plants. With a few exception, all eucaryotic cells contain mitochondria, and chloroplasts occur in eucaryotic photosynthetic cells. Nuclear division is both mitotic and meiotic.

The second type of cells, designated as **Procaryotic** forms the structural units like all **bacteria** and **blue-green algae** (Cyanophycae), now called **Cyanobacteria.**

Electron microscopy also revealed the structural details of **viruses** and it was also realized that viruses do not have a biological cell as unit of structure. Viruses are, therefore, **Acellular** and they are neither procaryotic nor eucaryotic. Strictly speaking, viruses are not organisms, though some biologists prefer calling them **Acellular organisms.**

To begin with, the classification of living organisms started with **2-kingdom**, the plants (Plantae) and the animals (Animalia).

The perfection of the electron microscope as an instrument for studying the fine details of cellular structure has revolutionized our knowledge about the internal organization of cells. One product of this revolution has been the demonstration that there are two quite **different kinds of cells** among living organisms. The more highly evolved type, the **eucaryotic cell**, is the unit of structure in all plants and animals and in several large groups of protists: **fungi, slime molds, protozoa,** and **most algae.** The eucaryotic cell shows innumerable specialization of form and of function having a cell wall composed of a variety of carbohydrate materials. Its nucleus has a well-defined nuclear membrane and the hereditary information of the eucaryotic cell, coded in the DNA (deoxyribonucleic acid) of the nucleus, is carried on a number of different structural subunits, the chromosomes. In all eucaryotic organisms gametic fusion is followed by nuclear fusion. A eucaryotic organism whose life cycle contains a sexual stage may exist almost exclusively in the diploid state, as do all higher animals and flowering plants. With a few exception, all eucaryotic cells contain mitochondria, and chloroplasts occur in eucaryotic photosynthetic cells. Nuclear division is both mitotic and meiotic. A much simpler kind of cell, the **procaryotic cells,** is the unit of structure in all **bacteria** and in **blue-green algae.** The procaryotic cell is small possessing a cell wall which contains a specific **mucopeptide** as its strengthening component. No mitochondria or chloroplasts have been detected in a procaryotic cell. Its nucleus is never separated from the cytoplasm by a nuclear membrane.

The nuclear region appears to be closely filled with a system of very fine fibrils which consist (at least in large part) of DNA. There is never any sign of organization of this material into individual chromosomes. Nuclear division does not take place by mitosis as in eucaryotic cell. Protists which are constructed of eucaryotic cells are **higher protists** and those which are constructed of procaryotic cells are **lower protists.**

The living world, based on the nature of internal organization of cells, can thus be subdivided into Kingdoms, Subkingdoms and Divisions in the following manner:

I. Kingdom **Protista**
 - Subkingdom **Lower Protista**
   - Division **Schizophyta** or **Bacteria**
   - Division **Cyanophyta** or **Blue-green Algae**
 - Subkingdom **Higher Protista**
   - Division **Chlorophyta**
   - Division **Euglenophyta**
   - Division **Xanthophyta**
   - Division **Bacillariophyta**
   - Division **Pyrophyta**
   - Division **Phaeophyta**
   - Division **Rhodophyta**
   - Division **Myxophyta** or **Slime molds**
   - Division **Mycophyta** or **Fungi**

II. Kingdom **Plants**

III. Kingdom **Animals**

Again one school of thought suggests that bacteria and blue-green algae should be termed the **Monera** (Dougherty and Allen, 1960, Whittaker, 1969); while another school terms them the **Procaryota** (Stainer and van Niel, 1962). This latter school includes all other plants in one major subdivision, the **Eucaryota.** Within this subdivision the simplest photosynthetic members are those plants known as the **Algae.** Those who recognize the Monera, they group the Algae, Protozoa, and Fungi as the **Protista** and sometime segregate the fungi into a separate subdivision, the **Mycota.** The terrestrial plants and aquatic higher plants with more complex organization are classed as the **Metaphyta.** Hence, an alternative classification system of the living world has been suggested to reflect our present knowledge of evolution based on the biochemical and ultrastructural studies of the living organisms. Each of these four categories has a taxonomic rank roughly equivalent to a kingdom, although it may not be desirable to use this rank so long as it is technically still reserved for 'plants' and 'animals'. The four basic categories are the **Monera,** the **Protista,** the **Metaphyta,** and **Metazoa.**

The Monera and Protista go back farthest in evolutionary history. Both groups are believed to be descended independently from the very first cells on earth. Monera include all those organisms in whch the cells do not possess true nuclei. Bacteria and blue-green algae are the modern representatives of this group. Protista possess true nuclei. Ancient protists were unicellular, and primitive-modern ones still are. Other modern protists are multicellular. They possess the ability to acquire food by both plant-like and animal-like methods. The major groups among modern protists are the **Algae,** the **Fungi,** the **Slime molds,** and the **Protozoa.**

The Metaphyta and Metazoa are both believed to have evolved from ancient protists. Certain ancient, green alga-like stocks were probably the ancestors of the Metaphyta, and other ancient algae or protozoa, or both may have been ancestors of the Metazoa. The derivation is far less certain in the latter case than in the former. The Metaphyta and Metazoa all are exclusively multicellular, their structural complexity reached the level of complicated organs and combinations of organs. Moreover, both Metaphyta and Metazoa have life cycles which include more or less distinct embryos. The

Metaphyta are almost exclusively photosynthetic and are unmistakably 'plants'. They include the Bryophyta and Tracheophyta (Vascular plants). The Metazoa are exclusively non-photosynthetic and they are unmistakably 'animals'.

This classification system is schematically presented below:

I. Category **Monera**

- Division **Schizophyta** (Bacteria)
- Division **Cyanophyta** (Blue-green Algae)

II. Category **Protista**

- Photosynthetic Protista—**Algae**
- Division **Chlorophyta**
- Division **Charophyta**
- Division **Euglenophyta**
- Division **Chrysophyta**
- Division **Pyrophyta**
- Division **Phaeophyta**
- Division **Rhodophyta**
- Non-photosynthetic Protista—**Fungi, Slime molds** and **Protozoa**
- Division **Mycophyta** (Fungi)
  - Class **Phycomycetes**
  - Class **Ascomycetes**
  - Class **Basidiomycetes**
  - Class **Fungi Imperfecti**
  - Class **Lichens**
- Division **Myxophyta** (Slime molds)
  - Class **Myxomycetes**
  - Class **Acrasieae**
  - Class **Labyrinthuleae**
  - Class **Plasmodiophoreae**

III. Category **Metaphyta**

- Division **Bryophyta** (Mosses and Liverworts)
- Division **Tracheophyta** (Vascular plants)

IV. Category **Metazoa—Animals**

In this four-part classification, every living creature has a proper place. The Monera and Protista include organisms which are plant-like, animal-like, or both, as well as some organisms traditionally regarded as 'true' plants (e.g., advanced algae) and 'true' animals (e.g., most protozoa). The Metaphyta and Metazoa include the remainder of the 'true' plants and 'true' animals.

A simplified classification system has been outlined in the following manner keeping the view points of both tranditional and modern approaches to the living world.

Kingdom **Plantae**

Division **Cyanophyta** or Blue-green Algae. Include unicellular to filamentous organisms; procaryotic cell; true nucleus and chromatophores lacking; pigmentation includes chlorophyll a, β-carotene, various xanthophylls, and two c-phycobilins; storage food cyanophycean starch, proteins; flagellation absent, reproduction by accidental breaking and by production of specialized cells; sexual system most inefficient.

Division **Chlorophyta** or Green Algae. Include both motile and non-motile organisms of diverse organization; eucaryotic cell; mainly chlorophylls a and b, α- and β-carotenes present; cell walls of cellulose; storage food starch; reproduction vegetative, sporulative and gametic.

Division **Euglenophyta**. One to three anterior tinsel-type flagella; eucaryotic cell; storage, food paramylum and lipids.

Division **Xanthophyta** or Yellow-green Algae. Eucaryotic cell; chlorophylls a and e present, carotinoids predominant; do not form starch; storage food oils; flagella unequal length.

Division **Bacillariophyta** or Diatoms. Largely unicellular; eucaryotic cell; chlorophylls a and c, fucoxanthin and special pigment diatomin present; walls conspicuously in two halves with silica deposition; storage food volutin, sexuality common; life cycles diplontic.

Division **Pyrophyta** or Fire Algae. Usually biflagellate; eucaryotic cell; chlorophylls a and c present; storage food lipids and carbohydrates including starch; cell walls when present of cellulose; sexuality rarely known.

Division **Phaeophyta** or Brown Algae. All sessile and multicellular; eucaryotic cell; chlorophylls a and c, carotene, xanthophyll and special pigment fucoxanthin present; storage food laminarin and mannitol; cell walls of cellulose and algin; exposed but multicellular reproductive structures; biflagellate reproductive cells.

Division **Rhodophyta** or Red Algae. Eucaryotic cell; pigments chlorophylls a and d, α- and β-carotenes, r-phycoerythrin, r-phycocyanin; storage food floridean starch; cell walls of cellulose and pectin; flagella entirely absent.

Division **Eumycota** or Fungi. Body unicellular uninucleate to filamentous branched mycelium; eucaryotic cell; cell walls of cellulose, chitin, or both; storage food glycogen; reproduction vegetative, sporulative, and gametic.

Division **Myxomycota** or Slime Molds. Diplohaplontic life cycles, with haploid phase often as independent solitary cells and diploid phase coenocytic plasmodium to uninucleate aggregated amoeboid mass; haploid phase flagellate cells; eucaryotic cell; reproductional sexual.

Division **Schizophyta** or Bacteria. Chlorophylls where present unique; procaryotic cell; flagella where present of unique structure; cell walls of polysaccharides, protein and lipid; storage food glycogen; reproduction usually by fission, sexual in some.

Division **Bryophyta** or Mosses and Liverworts. Vascular tissues absent; gametophyte dominant, sporophyte attached to gametophyte; homosporous.

Division **Tracheophyta or Vascular Plants.** Vascular tissues present in sporophyte; gametophyte independent or attached to sporophyte; homo- and hetero-sporous.

Kingdom **Animalia**

Even now some systematists prefer to place fungi and slime molds in a Division designating as **Mycota** or **Fungi** and subdivide Mycota as follows:

Division **Mycota**. These are the fungi having somatic structure ranging from a microscopic unicell to an extensive mycelium. Nuclei are eucaryotic. Reproduction is both asexual and sexual.

Subdivision **Myxomycotina**. These are slime molds. The somatic structure is a free-living plasmodium.

Subdivision **Eumycotina**. These are the true fungi. The somatic structure ranges from unicellular uninucleate to an extensive mycelium.

Class **Chytridiomycetes**. Fungi possessing variety of thalli whose motile cells bear a single posterior whiplash flagellum.

Class **Hyphochytridiomycetes**. A group of aquatic fungi whose motile cells possess a single anterior tinsel flagellum.

Class **Oomycetes**. Fungi possessing a usually well-developed coenocytic mycelium whose motile cells bear two flagella, one whiplash, the other tinsel. Product of sexual reproduction is an oospore.

Class **Plasmodiophoromycetes.** Parasitic fungi having plasmodial thalli, motile cells possess two anterior whiplash flagella of unequal size.

Class **Zygomycetes**. Saprobic or parasitic fungi with a well-developed coenocytic mycelium, no motile cells are produced. Product of sexual reproduction is a zygospore.

Class **Trichomycetes**. These are fungi with a simple or filamentous branched coenocytic mycelium attached to the digestive tract or the external cuticle of living arthropods. Asexual reproduction is by a variety of spores. Sexual reproduction is as in the Zygomycetes.

Class **Ascomycetes.** This group includes fungi which form sexual spores by karyogamy and meiosis in a sac-like structure called ascus.

Class **Basidiomycetes.** These are fungi which form sexual spores by karyogamy and meiosis on the surface of a special structure known as basidium.

Form-Class Deuteromycetes (Fungi Imperfecti). This is a group of fungi in which sexual stages have not been discovered.

Living organisms do not classify themselves. They are classified by systematists for convenience of reference. An ideal scheme of classification should reflect natural relationships among living organisms. But in considering natural relationships systematists differ in their approach by not attaching same weight to the criteria available resulting different schemes of classification of living organisms. Besides this, biological objects, like the viruses, which, according to some, do not conform to the common properties of living organisms are difficult to categorize. So also the **Rickettsiales** which bear a morphologic resemblance to bacteria but are biologically related to viruses. The situation becomes even more difficult with the discovery of very tiny organsisms which have similarities both to viruses and bacteria. Though they have been grouped by some as **Chlamydiales** and Mycoplasmatales, still they have to get a proper position in the scheme of classification of living organisms based on natural relationship.

The above facts indicate that it is extremely difficult to draw up an outline of classification of living organisms which will be acceptable to all systematists.

Since evolution in a living world is a continuous process, it is expected that the number of living organisms will increase. Hence grouping of organisms — classification will also be a continuous process.

We started with 2-kingdom classification. The process had to be followed by 3-kingdom, 4-kingdom. According to some biologists like Whittakar, Cavalier-Smith and others, the classification process may also be extended to 8-kingdom by dividing botanists into several well-defined kingdoms.

The details of evolution is still speculative.

NOMENCLATURE

Nomenclature refers to the correct naming of living organisms. The naming of living organisms started on the basis of their utility. Before the middle of the eighteenth century the names of plants commonly were **polynomials**, that is, being composed several words in a series, constituting more or less the description of a plant. In most cases it was rather arbitrary. The polynomial system of naming often became inconvenient to handle and it was felt that a better system of naming should be introduced.

A remarkable change came in the field of nomenclature when Carl Linnaeus in 1753 in his **Species Plantarum** put to practice the **binomial system of nomenclature** which was first introduced by G. Bauhin in 1623. The binomial system of nomenclature is naming of any plant or animal by two words usually Greek, Latin or of Latinform. The first word being that of the genus to which the organism belongs—**generic name**—and the second of the two words making up the name is called the **specific epithet**. The two words in combination form a **binomial**, and the binomial is a **binary epithet**. A particular specific epithet may be used only once within a given genus, but it may be used repeatedly in different genera. The first word is a noun or substantive designating the genus to which the organism belongs. The specific epithet is usually a Latin or Latinized adjective, which takes its grammatical gender from that of the genus, but it is sometimes a noun in apposition to the generic name. When the present system of naming plants was adopted, in the middle of the eighteenth century, Latin was the written language of scholars. Although Latin has diminished in importance, its use in scientific names is retained. Since it is not a spoken language it does not change, and it is intelligible to scientific workers of all nationalities. An important asset of the scientific name is its relative stability. Once a plant is named, the name remains, or if it is changed, the change is made according to established botanical rules. The scientific name is the name wherever the plant is found. But common names often vary with each locality, country, or other geographic subdivision. They differ, of course, from one language to another. It is customary to capitalize the initial letter to the generic name and not to capitalize specific epithets, although the rules of botanical nomenclature permit the optional capitalization of certain specific epithets. The binomials are, in general, descriptions of certain characters of the organisms to be named and are usually derived from internationally known language, Greek or Latin. When written they always should be underlined and printed in italics. The name of a plant is thus given on the basis of characters of a specimen, the **type specimen**. Again, the name (short names usually appear in full, e.g., Thom; Peck) or the abbreviated name (e.g., Fr. for Fries; L or Linn. for Linnaeus) of author who first described the name of the organism usually follows the binomial and is known as **author citation**. If the binomial is changed in part or whole in subsequent time, the name of the author who first described the particular organism should be put in parenthesis after which will follow the name of the author who consequently changed the binomial. The author citation is very helpful to find out the original and subsequent descriptions of the plant when found necessary. To make the situation even more clear, sometimes the year in which the organism was described in written after the author citation following the binomial.

With the gradual increase in human interest in plants there started to arize problems of nomenclature. For the settlement of these problems, first organized effort was made at the First International Botanical Congress, meeting in Paris in August 1867. There, attempts were made to standardise and to formulate certain principles of nomenclatural practices, the **laws of Botanical nomenclature** which in course of time with additions and alterations took the shape of **International Rules of Botanical Nomenclature**. The **International Rules of Botanical Nomenclature** has **Principles, Rules** and **Recommendations** which are printed as a series of Articles aiming at; fixity of plant names keeping out the use of forms and names that might lead to error or ambiguity and avoidance of useless creation of names.

A **legitimate** or **valid name** is the one that is published in accordance with the International Rules of Botanical Nomenclature. In general, the earliest name legitimately applied to a particular plant is the correct name. This is the **principle of priority** in nomenclature.

Part Two

Algae

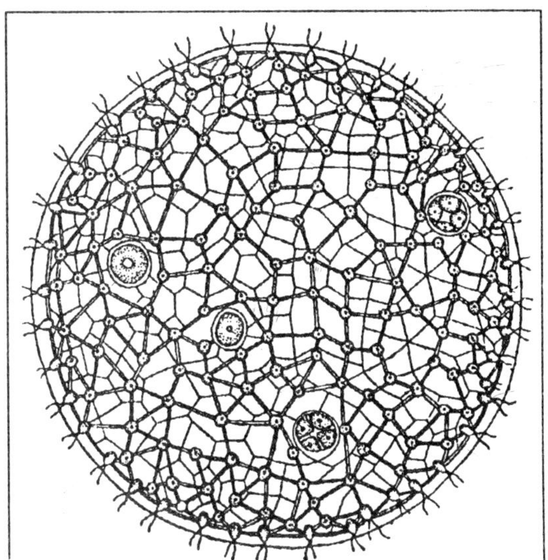

Chapter 2

Introduction To The Algae

Although Linnaeus (1753) first introduced the name **Algae**, but by Algae he meant the Hepaticae. The actual delimitation of a group of plants from their macroscopical features and naming as Algae was done by A. L. De Jussieu (1789). The term **algae**, (sing. alga) is applied to a vast assemblage of organisms which are highly diverse with respect to habitat, size, organization, physiology, biochemistry and reproduction. In these, organisms are found some of the oldest and most primitive types of plant life. The algae actually comprize of a large heterogenous primitive and simplest plant designated as **thallophytes** of the group **Thallophyta**, a primitive and simplest member of the **plant kingdom**. The study of Algae is phycology, and phycologists are those who study Algae.

What are Algae? The Algae are chlorophyll bearing lower plants which possess markedly different patterns of unicellular to multicellular organization, both prokaryotic and eukaryotic. Their reproduction is vegetative, by formation of spores, and gametic; gametes produced in unicellular body behaves as a gamete. During the latter part of the nineteenth century and the early decades of the twentieth, the term Algae was used both as a common name and as a formal taxonomic designation of a group of organisms. At the present time, however, the differences among these organisms are so strongly emphasised that the formal, collective taxon 'Algae' has been abandoned, and a polyphyletic treatment of these organisms is widely accepted. Accordingly, the old Subdivision Algae of the Division Thallophyta has been replaced by taxa of divisional rank: Cyanophyta, Chlorophyta, Phaeophyta, Rhodophyta, and so on.

Algae range from minute unicellular solitary or colonial organisms represented by some of the common pond scums which may or may not be motile and are entirely invisible as individuals to the naked eye, to large, diffuse and comparatively complex plants, such as seaweeds, usually growing submerged in water or in moist situations. They are largely aquatic, being common in both fresh and sea water although some are terrestrial, being found in moist soil as an important part of the soil flora essential in soil productivity, and on bark of trees. Generally speaking, the large and more complex forms are marine; the fresh-water and terrestrial representatives are both smaller and simpler. Even in apparently looking large and complex forms, however, the plant body shows very little differentiation. Some of the large sea-weeds which show considerable tissue differentiation with organs analogues to the roots, stem, and leaf of the higher plants, may still be termed a **thallus**, for being structurally simple. Algae are of various colours—blue-green, green, brown, red, etc. but majority of them contain the green chlorophyll pigments and hence are autotrophic. Although the mode of nutrition resembles that of higher plants, yet the method of absorption differs, as they do not possess specialized absorptive organs comparable with root hairs, but food material diffuses in, in solution through the cell wall over the whole surface. In some algae the chlorophyll is frequently masked by other pigments. These are mainly **phycocyanin**, a blue pigment; **fucoxanthin**, a yellowish-brown pigment; **phycoerythrin**, a red pigment. The principal food reserve is carbohydrate of some form or the other, in addition to which there are also fats and oils. Algae are very important as a source of food and serve as an early step in the food chain of larger aquatic animals especially fish. They are also of great use to human beings, being the source of food especially rich in vitamins A and E. Many marine algae are important source of food especially rich in vitamins A and E. Many marine algae are important sources of iodine, potassium (potash), and other minerals. There are certain algae which play a great role in enriching the soil fertility by fixing atmospheric nitrogen. In the more primitive members reproduction is often entirely asexual, while in advance forms it is sexual of varied kinds. Formerly, the classification of Algae was largely based on the basis of pigmentation. But the present tendency is to use a combination of other characters along with the nature of pigmentation of cells.

REVIEW OF ALGOLOGICAL STUDIES

I. ABROAD

The history of Algae is as old as that of botany. The first writings on Algae appeared in the ancient Chinese classics, where references to the collecting, eating, and to the medicinal uses of algae are found. There are also references in Roman and Greek literatures the use of algae for cosmetic purposes. Algae have been given many designations by different people. The Chinese named them as *Tsao* for having esthetic significance, whereas the Howaiians called *Limu* as they used algae as a food. References of several kinds of Tsao are available in the eighth century writings. It is interesting to note that the early century writings about Algae were restricted either to their use or else to their taxonomy. In the north coast of France, algae were being used for manual purposes as early as twelfth century. This practice spread to Great Britain, as mentioned in the sixteenth century reference. The use of brown sea-weeds for fertilizers climaxed in France during the seventeenth century. The utilization of certain algae for agar making introduced in China, slowly spread in Japan later on turned out to be a great industry which has its foundation in the seventeenth century. As already pointed out, although Linnaeus (1753) first introduced the name Algae, but the delimitation of the group Algae was made by A. L. De Jussieu (1789). Jussieu's characterization of Algae was not of much value since it was based entirely on macroscopic features. Actually the great pioneer in the study of Algae was C. A. Agardh (1824) who first started describing Algae considering almost all aspects of the life history. Revolutionary advances were made both in cryptogamic and phanerogamic systematics and a majority of genera of Algae were described before or after 1875. Before Agardh perhaps Vaucher (1803) should be credited for introducing life history studies and emphasising the importance of reproductive structures as basis of algal taxonomy. During the first part of the nineteenth century there were numerous workers in the marine Algae, and actually the knowledge of these Algae increased more rapidly than that of the fresh-water Algae. The works of Stackhouse and Lamouroux are important in the history on marine Algae. It was Blackman (1900) who postulated a flagellate ancestry for the Algae, expressed earlier by Klebs (1893), and suggested that there are three distinct tendencies the Volvocine, Tetrasporine, and Chlorococcine which could be recognized in the evolution of green algae from a *Chlamydomonas* prototype. The work of Adolph Pascher exerted a great influence in the modern concept of algal taxonomy and phylogeny. Parallel development of algal groups was his basic hypothesis and on the basis of this conception he presented an elaborate system of classification. His system did not receive wide acceptance, but provided helpful framework for the assignment of genera and families which were previously considered to be of uncertain position. The study and appreciation of the importance of soil algae were first introduced by Fritsch. In recent years algal plant physiology and biochemistry studies have also received a great impetus. Of late, Pringsheim has emphasised the importance of pure algal cultures as a means of establishing the taxonomic status of the smaller organisms.

II. IN INDIA

The earliest worker on Algae in India was Royle (1839), who first reported the occurrence of some members of the Lemaniaceae in the mountain streams of the Himalayas. Immediately following him were Decaisne and Montagne who described *Actinotrichea fragilliz* and *Calothrix indica* in 1842 and 1849. Griffin (1847) reported a number of species of *Chara* from Bengal. H. J. Carter (1858) collected *Eudorina elgans* from the pools of Bombay and recorded its fertilization process. Wallich (1860) and Hobson (1863) published an account of the desmids from Bengal and Bombay. For the period from the later part of 1800 to early 1900 mention may be made of Martens, Dickie, Lagerheim, Turner, who worked on the distribution of various groups of Algae in different parts of India. From 1919 onwards Indian workers got interested in Algae. During 1919 to 1932 Ghose carried on investigations on the blue-green algae of the Punjab. M. O. P. Iyengar is another pioneer investigator who started working from 1920 and published a good number of papers on algae of South India, both fresh-water and marine. Iyengar along with his students Balakrishnan, Desikachary, Kanthamma, Ramanathan, and Subramaniam described quite a few new species and genera of algae. Of which special mention may be made of the work on the life history of *Cylindrocapsa gemunella*, formation of gametes in a species of *Caulerpa*, morphology and cytology of *Polysiphonia platycarpa*, sexual reproduction in *Dictyosphaerium*, life history and cytology of *Microdictyon tenius*, reduction division and auxospore formation of *Cyclotella meneghiniana*. One of the outstanding contributions of Iyengar is the discovery of a very interesting terrestrial alga, *Fritschiella tuberosa*. For his very valuable voluminous contributions on Indian Algae, Iyengar is called the 'father of modern algology of India'.

Bruhl and Biswas (1922-26) and Biswas (1922-50) have made important contributions on the algal flora of Eastern India. Blue-green algae of Uttar Pradesh was worked out by Bharadwaja (1928-36). He established a school of algology at Benaras Hindu University. One of his students, R. N. Singh (1938-59) has made significant contributions on blue-green algae, Zygnemataceae, Oedogoniaceae, and Chaetophorales of Uttar Pradesh. He also worked on the life history of *Fritschiella tuberosa* and *Draparnaldiopsis indica*. C. B. Rao (1935-38) investigated on Myxophyceae, and Zygnemataceae of Uttar Pradesh and Madras. Randhawa (1932-59) made valuable contributions on Zygnemataceae, Oedogoniales and Vaucheriaceae of the Punjab and Uttar Pradesh. Indian fossil algae were worked out by S. R. Narayan Rao during 1941-49. Of the Indian workers who made contributions on the Characeae of India are Pal, Dixit, Kundu, and Sundaralingam. About the economic utilization of algae names of P. K. De (1939) and R. N. Singh (1938-59) are to be mentioned. De worked on the nitrogen fixation of the blue-green algae in the rice fields and Singh was able to indicate the possibility of reclamation of saline usar lands in India by the cultivation of blue-green algae. Singh and his associates (1961-73) have been doing significant work on the nuclear structure and sexual system of the blue-green algae.

ECONOMIC IMPORTANCE OF ALGAE

Except for the chemosynthetic bacteria, all organisms are directly or indirectly benefitted from algae. Some of the aspects of economic importance of algae are incorporated below:

1. **Algae constitute the link of food chain:** Both fresh and salt waters contain an enormous variety of algae which constitute the fundamental or primary link of many diverse food chains. Algae synthesize organic food stuffs, just as do the plants to the land. As the flesh of the land is dependent upon the activities of the green leaf, so the fish and other aquatic forms of animal life are dependent, directly or indirectly, upon algae, and fish in turn are important item in the daily diet of larger sea animals and man.

A number of aquatic algae form the food of fish either directly or indirectly. Diatoms, filamentous and some planktonic green algae, and a number of blue-green algae are very often found in the guts of various species of fresh and brackish water fish and they appear to be directly utilized as fish food. The reserve food materials in these algae, e.g., fats and volution in the diatoms; starch, often accompanied by oil in the green algae; sugars and glycogen in the blue-green algae; and polysaccharides in *Euglena* are utilized by fish.

2. **Algae useful in fish culture:** That algae are fruitfully utilized in fish culture can very well be indicated from the successful culture of the Siamese fish, *Tilapia mossambica* which is voracious feeder of filamentous algae. This particular fish has been successfully introduced in different parts of India. A culture of *Scenedesmus* is often exclusively used as daily dose of fish meal for the culture *T. mossambica.*

3. **Algae used for recreational purposes:** Certain selected algae are grown in recreational areas like—lakes and streams along with fish.

4. **Algae useful in sewage treatment plants:** Species of *Chlamydomonas, Scenedesmus, Chlorella* and *Euglena* are used in sewage treatment plants for providing through photosynthesis the oxygen necessary for rapid decomposition of the sewage by bacteria.

5. **Algae and water supplies:** In the summer months the phytoplankton in ponds, lakes and reservoirs may become so abundant as to be extremely conspicuous. The water becomes cloudy and may assume a yellowish or greenish tinge. A floating mat or scum may develop. These manifestations of algal growth are popularly termed 'water blooms'. Such concentrations of algae are extremely objectionable, not only in public water supplies but also in waters used for bathing, fishing and other recreational purposes. The blue-green algae are most frequently involved in the contamination of water

supplies, but the greens, the flagellated golden-brown, and the diatoms are also troublesome at times. Mention may be made of *Prymnesium parvum, Gymnodinium veneficum* and *Microcystis* spp. which cause mortality of fish and domestic animals that drink water infested with these algae. The living and the dead and decaying algae impart disagreeable oily or fishy odours to the water. Toxic protein decomposition products may be formed by blue-green algae and these are known to have caused death of cattle, sheep, and other animals which have drunk heavily infested water. In addition to the unsightly appearance and the unpleasant odours and tastes, the presence of algae in reservoirs requires a greater concentration of chlorine for bacterial control and causes difficulties in filtration.

6. **Algae as the origin of petroleum and gas**: The origin of oil and gas has been a matter of controversy, but it is now generally believed that, like coal, these fuels owe their energy to photosynthesis in ancient plants. Unlike coal, however, which was laid down in inland swamps, oil and gas were formed from organic matter in marine environments. The plankton of the seas was probably of the greatest importance as a source of this organic matter. Minute marine algae captured the energy of sunlight, which was in turn transferred to the animals that fed upon them. Organic compounds derived from the plankton, both plant and animal, accumulated in mud deposits in shallow waters of the ocean floor. In the source, materials were buried by sedimentary action, and in an oxygen-free environment, gradually converted into oil and gas. Natural gas is largely methane (CH_4), which can be produced by certain kinds of anaerobic bacteria. Gas is generally associated with oil and can result from the action of methane-producing bacteria upon organic compounds.

7. **Algae and limestone formation**: The most significant forms in this category are the blue-green and red, and certain green algae which withdraw calcium from water, both fresh and salt, and deposit it, in the form of calcium carbonate, in their cell walls or gelatinous sheaths. The blue-greens are chiefly responsible for the formation of extensive limestone deposits around hot springs and glaciers. The red algae are the most important calcareous algae of the seas; in particular, they play a significant role in the construction of coral reefs and islands. Although true coral results from the activities of minuet sedentary animals, it is recognised that lime-secreting red algae are almost as important in the formation of coral reefs as the coral organisms themselves. The calcareous red algae are best developed in the warmer seas, but certain species also flourish in temperate and polar regions, where they form extensive banks of limestone in coastal areas. The algae are not only important in the present age in the formation of calcareous deposits, both in the seas and fresh-waters, but also they have played a significant part in the production of beds of limestone rocks, which may be 1000 feet thick.

8. **Algae used in space research and other fundamental studies**: In recent years *Chlorella* is being used in space research. *Chlorella* has been found very suitable for keeping the air in space vehicles pure on long interplanetary flights. The stale air in which the carbon dioxide has been concentrated is fed into a flood-lit container containing a mixture of water and nutrient chemicals and *Chlorella.* The alga restores oxygen into the space vehicle by its photosynthesis. Again species of *Chlorella, Chlamydomonas,* and *Acetabularia* are used as tools for solving fundamental biochemical and genetics problems.

9. **Algae used as food**: Large number of algae have entered into the diets of human beings from ancient times. The earliest records are those of the Chinese, who mentioned such food plant as *Laminaria* and *Gracilaria* in their

'materia medica' several thousand years ago. The ancient inhabitants of Japan ate *Prophyra* as a healthful supplement to their rice diet. Its use became widespread, not only in Japan, but in China in course of time. **Kombu,** a Japanese food is prepared from stipes of species of *Laminaria.*

The most diversified dietary use of sea-weeds was developed by the Polynesians and reached its peak in Hawaii, where during the nineteenth century at least 75 species were separately named and used regularly as food in that island world. The Hawaiians called them 'limu' and considered them a necessary staple of their daily diet.

Perhaps the best known and most widely used food alga in Western Europe in recent centuries was Irish moss, or *Chondrus chrispus.* Sulfated galactan known as Carrangeenan is extracted from the walls of several red algae—species of *Chondrus, Gigartina,* and *Eucheuma* is cooked with milk, seasoned with vanilla or fruit, and made into a highly palatable dish known as **blancmanges**. The jellying qualities of Irish moss gave the alga an early food use.

Common Irish seaweeds *Porphyra linearis* and *Alaria esculenta* are used in Japanese foods. In 1994 Japan have imported almost half the national requirement of sea-weed consumed through Irish Seaweed Industry Organization and it is increasingly popular in Europe and the USA. Sea-weeds are popular as food for: low in fat, high in vitamin, and it helps longevity.

Man, thus obtains carbohydrates, vitamins (algae are especially rich in vitamins A and E, and they contain some C and D), and inorganic substances, e.g., iodine (goiter is unknown among the people who eat sea-weeds), not to mention the benefits of the mild laxative action of the ingested algae. Witsch (1959) stated that vitamin B value of young cultures of *Chlorella* equals that of lemon juice. In Japan, powdered *Chlorella ellipsoidea* has been used successfully mixing with green tea.

In Germany and in the United States considerable work is being carried out on the suitability of mass cultures of *Chlorella* as an alternative source of animal feed and of human vegetable food. Mass culture of **Spirulina** is successfully done in Mexico, Taiwan and India for its high protein content.

Irish farmers are already capitalizing on the surge in international demand for sea-weeds called "sea vegetables". Common Irish seaweeds Porphyra linearis and Alaria esculenta are used in Japanese foods, including misassup. Sea-weeds are preferable as food for: low in fat, high in vitamin and studies show that it helps longevity. Irish Sea-weed Industry Organization has found potential for success in seaweed market. In 1994, Japan has imported almost half the national requirement of sea consumed and it is increasingly popular in Europe and the USA.

10. **Algae used as fodder**: The orientals developed wide human uses for marine algae, but Europeans profited by extensive use of these plants for stock feed. In Iceland and Scandinavia, in the British Isles, and along the coast of France, stock has long been driven or allowed to wander to the seashore at low tide to feed on seaweeds. Some kinds of algae, such as *Rhodymenia palmata* and *Alaria esculenta,* are favourable food of goats, cows, and sheep, and in Scotland and Ireland the stock actively hunt the shores at low tide for particular algae, especially the former. The milk does not have any taste of algae, nor is the meat inferior because of the sea-weed diet. Such animals, that have for several generations been nourished on algae, show better ability to digest it than those not so habituated.

The shortage of grain in many parts of Europe during World War I led to considerable experimentation with the use of sea-weeds as food for cows and horses. Stock-feed factories were established in France, Norway, Denmark and Germany, and various methods of treating and reducing seaweeds to meal or powder were developed.

The favourable results in animal husbandry in Europe led to the industrial processing of the great Pacific-Coast kelp (*Macrocystic*) for animal rations. Sea-weed-meal factories have been operating in the United States for several decades, providing supplementary feeds for poultry, cattle and hogs. The high mineral and vitamin G content of kelp meal has made possible its use in various poultry and other animal rations.

11. **Algae used as fertilizers**: The value of sea-weeds in fertilising the soil was discovered early in the history of agriculture in coastal Asia, and by the ancient colonizers of the coasts and islands of North-western Europe. In some areas of Britain, and along the coast of North-West France, the cutting of rockweeds for manure has been so intensively practized that it became necessary to regulate it by laws that have now been in effect for nearly 100 years.

In the United States, long before the recognition of their potash content, sea-weeds were employed for fertilizers by the thrifty farmers. Not only the chemical fertilization, but also the water-holding capacity of fragments of the algae in the soil proved effective. These provided valuable small reservoirs of water in close contact with the roots of the cultivated plants. Furthermore, the bulky organic substances decay slowly in the soil and form humus. Again yield of paddy is increased substantially when paddy field is inoculated with nitrogen fixing blue-green algae. Some of them are: *Tolypothrix tenius, Aulosira fertlissima, Anabaena oryzae, Anabaenopsis arnoldii, Calothrix confervicola, Nostoc commune,* and *Cylindrospermum bengalense.*

12. **Algae used as medicine**: Medicinal applications of plants are almost as old as their food uses. From earliest times the Chinese used *Sargassum* and various Laminariales for treatment of goiter and other glandular troubles. *Gelidium* very early became employed for stomach disorders and for heat-induced illness. The gentle swelling of dried *Laminaria* stipes upon exposure to moisture make them surgical tool in the opening of wounds. Similarly, the orientals have employed the same technique in child-birth for expansion of the cervix. An **Antibiotic**, chlorellin is extracted from **Chlorella** is used to control Gram-negative bacteria.

Perhaps the algae used most widely and for the longest time for medicinal purposes, from which agar is extracted are the **agarphytes**, including *Gelidium, Pterocladia, Gracilaria,* and *Ahnfeltia.* The name 'agar-agar' is of Malay origin and means 'jelly'. This jelly was obtained by boiling up seaweeds and cooling the resulting liquid. Agar early became useful for stomach disorders and as a laxative, and was once employed as a dietetic. It was originally produced and marketed in China, but the Japanese took over production in about 1662 and maintained a world monopoly till 1940. The most significant date in the utilization of agarphytes was 1881, when Robert Koch proved the value of agar in the cultivation of bacteria. Since that time it has become essential to the work of hospitals and medical research laboratories throughout the world. Besides these, *Chlorella* is used for the preparation of antibiotic **Chlorellin.**

13. **Industrial utilization of algae**: The industrial utilization of algae may be outlined in the following manner:

(i) **Kelp industry**: Industrial utilization of sea-weeds in Europe had its principal early development in the production of 'kelp', a name that originally referred to the ash, rich in soda and potash, derived from burning marine plants. Kelp production started sometimes in the seventeenth century by French peasants and spread to other parts of North-West Europe. Driftweeds were first used, but cutting was later resorted to *Laminaria* and *Saccorhiza* in North Britain as of major importance. But *Fucus* and *Ascophyllum* were also widely used, and in some areas *Himanthalia* and *Chorda*. The kelp ash from these plants was widely brought by early industrialists for use in manufacture of soap, glass and alum. During the eighteenth and early nineteenth centuries the demands became considerable, and enormous quantities of sea-weeds were handled in areas of rich algal growth. Kelp extract contains a number of chemical elements, notably potassium and iodine. About 25 per cent of the dry weight of kelp is potassium chloride. Many species of kelp are used as food for man, especially in the Orient. In Northern Europe they also serve as food for domestic animals, such as sheep and cattle.

(ii) **Algin industry**: Algin is the general term designating the hydrophilic, or water-loving derivatives of alginic acid. The most commonly known algin is sodium alginate, but other commercially important compounds are the potassium, ammonium, calcium, and propylene glycol alginates, as well as alginic acid itself. With the exception of alginic acid and calcium alginate, the algin products offered commercially are soluble in water to form viscous solutions. Algin occurs generally throughout the brown algae (*Laminaria, Macrocystis, Sargassum* and *Fucus*) as a cell wall constituent. It has remarkable water-absorbing qualities that make it useful in numerous industries in which a thickening, suspending, stabilizing, emulsifying, gel-forming, or film-forming colloid is required. Thus, algin provides ice cream with a smooth texture by preventing the formation of ice crystals. In automobile polishes it suspends the abrasive; in paints, the pigments; also in pharmaceuticals, the drugs and antibiotics. As a stabilizing agent it serves in the processing of rubber latex and in the printing of textiles. As an emulsifier it is widely used in such products as water-base paints, French dressings, and cosmetics.

The algin industry has become so important to such a wide variety of industries that extensive survey of kelp-bed ecology is an effort to guard against loss of this important resource. Harvesting methods are now carefully regulated, and a huge amount of money is being spent on kelp-bed research throughout the world. Experimental studies are continuing on the relation of pollution to kelp survival and on kelp-bed grazing organisms.

(iii) **Agar industry**: The outstanding use of the red algae, however, is in the production of agar. This is a dried and bleached gelatinous extract obtained from red algae—*Gelidium nudifrons, G. pusillum, G. robustum,* and *Gracilaria verrucosa*. Agar is used extensively in medicine, chiefly as laxative, since it is not digested and increases greatly in bulk with the absorption of water. More important than this medicinal utilization is its use as an essential ingredient in the preparation of medium for the growth of bacteria and fungi. As such it is indispensable in acteriological laboratories because no adequate substitute for agar is known.

Since the introduction of agar into bacteriology in 1881, the agarphytes have become increasingly industrialized and the technical uses of agar enormously expanded. A few of the medical uses have been mentioned above, but modern industry has developed such a multitude of applications that only a fraction of them can be noted here. Large quantities of agar are used as a food adjunct. Agar serves widely as a substitute for gelatin, as an antidrying agent in breads and pastry, in improving the slicing quality of cheese, in the preparation of rapid-setting jellies and deserts, and in the manufacture of frozen dairy products. The use of agar in meat and fish canning has greatly expanded, and hundreds of tons are utilized annually. Agar has proved effective as a temporary preventive for meat and fish in tropical regions, due to the inability of most purifying bacteria to attack it.

Early industrial uses of agar in the Orient included sizing fabric, water-proofing paper and cloth, and making rice paper more durable. Modern industry has refined and expanded these uses to meet new needs in the manufacturing of such items as photographic film, shoe polish, dental impression molds, shaving soaps, and hand lotions. In the tanning industry agar imparts a gloss and stiffness to finished leather. In the manufacture of electric lamps, a lubricant of graphite and agar is used in drawing the hot tungsten wire.

The increasing applications have called for wide expansion of the collection of agarphytes, and since Japan supplied most of the world's markers before World War II, when those supplies were cut off, a great amount of hurried research was conducted in an attempt to develop domestic agar supplies not only in the United States, but in South Africa, Australia, Newzealand and Russia.

(iv) **Diatomaceous earth industry**: The Diatoms are equally important in comparizon with other algae that have industrial utilization. Most species of Diatoms are marine and when these minute plants die, they fall to the sea bottom and, because of their siliceous nature, the cell walls are preserved indefinitely. Great deposits of this material, known as diatomaceous earth, are found in many parts of the world. The largest beds in the United States, some 1400 feet thick, are in California. The beds are sedimentary deposits originally laid down on the floor of the ocean and later raised above the level of the sea.

Because diatomaceous earth is inert chemically and has unusual physical properties, it has become an important and valuable material in industry. It makes an excellent filtering agent, which is widely used to remove colouring matters from products as diverse as petrol and sugar. As a poor conductor of heat it is used in sound-proofing. It is used in the manufacture of paints and varnishes, of phonograph records, and as filler for battery boxes. Because of its hardness, it is used as an abrasive in scouring and polishing powders.

14. **Parasitic algae**: There are large number of parasitic algae of which special mention may be made of the genus *Cephaleuros*. Different species of the green alga *Cephaleuros* grow as parasites on and in the leaves of various angiosperms, such as *Magnolia, Rhododendron,* tea (*Thea sinensis*) and Pepper (*Piper nigrum*). Their plant body composed of one or more branched interwoven threads, produces rhizoids which penetrate the cells of the host and induce various abnormalities. *C. virescens* forms the red rust of tea plants causing economic loss. But the attack is only serious when the tea plants grow slowly, because during the period of rapid growth the alga is continually shed by exfoliation of the outer tissues. The disease can not be controlled by spraying with poisons. But the tea bushes can be made less susceptible to attack by treating the soil with potash. **Prototheca,** a colourless **Chlorella**-like organism which superficially resembles yeasts is a pathogen of humans. It has been found in systemic and subcutaneous infections, as well as inflammation of the joints.

15. **Biological control of algal growth**: The virus Cyanophage LPP-I and several other phycoviruses infect various blue-green algae and may be used to control their growth in water. These viruses also attack species of *Chlorella* and *Sirodotia*.

16. **Algal toxins**: Marine dinoflagellates belonging to the genera *Cymnodium* and *Cionyaulax* produce high molecular weight neurotoxin which causes death of aquatic animals.

17. **Biofertilizer**: *Nostoc, Anabaeno* and other members of blue green algae can **fix atmospheric nitrogen** to form nitrogenous compounds useful for soil fertility.

18. **Iodine production:** In Japan, sea-weeds belonging to **Rheophyceae** as used for the extraction of **Iodine**.

OCCURRENCE AND DISTRIBUTION

The algae are found to occur in moist situations which may be damp soil; rocks; old walls; tree-trunks; rain-pipes; still and running water, fresh or saline. They make up the great part of the vegetation of the oceans and lakes. Algae are, however, abundant in still waters. Some algae may be free-floating forming a surface scum on ponds, ditches, etc., collectively called **phytoplankton**. Phytoplankton furnishes food for most aquatic animals, particularly fish. Others are attached to submerged stones or to aquatic plants. A number of algae is associated with other organizms, being embedded in the tissue of plants either as **endophytes**, or as **symbionts** deriving mutual benefits, or as **epiphytes** without causing any injury to other plants, or as **parasites** causing damage to other plants and animals. Algae exist under wide range of temperature conditions. They thrive on long-persistent snows, by contrast there are some thermophilic algae that inhabit hot springs. Considerable number of algae are able to thrive in terrestrial situations and colonize on exposed rock surface and similar other wide range of substrata, like: **halophytes**, which grow in saline water; **cryophytic** growing in snow; **thermophytes** in hot springs; the algae which grow in sandy beaches are **psanaonas**; and those on moist rocks are **lithophytes**.

Algae also show different tolerance to pH. **Acidophilic** algae are able to live in water with a pH as low as 2.0. These algae include the red alga *Cyanidium caldarium* and the green algae *Dunaliella acidophila* and species of *Chlamydomonas*. Acidophilic algae have one or more of the following mechanizms:

(i) A cell-surface barrier that is extremely impermeable to protons (H^+), there by helping to keep the protoplasm at a more neutral pH.

(ii) A plasma membrane that has a higher concentration of saturated fatty acids, resulting in a decrease in the fluidity of the plasma membrane.

(iii) Accumulation of glycerol derivatives in the protoplasm in order to resist the osmotic imbalance caused by high acid concentrations. **Alkaliphiles** are algae that grow in water with a high pH. The cyanobacterium *Spirulina platensis* is an obligate alkaliphile that thrives in extreme alkaline habitats growing optimally at pH 9.0 to 10.0. However, the protoplasm pH is adjusted at 8.0 through sodium pump inside.

On the whole, algae group in terrestrial, and varied aquatic environment (fresh water, marine and estuarine environment). In antic and antertic environment there are selected algal species which profusely grow. The important environmental factors that govern the algae growth are light, temperature, and nutrients. Algae also developed suitable adaptive measures for protection from UV-damage. They contain high amount of flavonoids, carotenoids and also superoxide dismutase for UV-damage protection.

CELLULAR ORGANIZATION

It will not be wrong to generalize the algal characteristic of possessing a definite cell wall being made of cellulose; of course, this does not lose sight of many of the naked flagellate forms.

Algal cells are of two kinds: **prokaryotic** and **eukaryotic**. In a prokaryotic cell there is absence of internal membranes which separate the resting nucleus from the cytoplasm, and which isolate the enzymatic machinery of photosynthesis and respiration in specific organelles and as such the cell lacks organized nucleus, chromatophores, and mitochondria. Nuclear division does not take place by mitosis. In a prokaryotic cell, the nucleus is not bound by any membrane. Instead, the protoplast is differentiated into outer peripheral **chromoplasm** containing photosynthetic pigments and inner colourless **centroplasm** where the genetic material is not found within the membrane-bound nucleus and the DNA strands do not combine with histone to form chromosomes—the centroplasm the **incipient nucleus.** Whereas, a eukaryotic cell is characterized by the presence of well-organized nucleus, chromatophores and mitochondria. Nuclear division is by mitosis in a vegetative cell.

Cell wall: Generally, where a cell wall is present the surface layer of the protoplast is constituted by a plasma-membrane, but in numerous naked types it is developed as a more or less rigid **periplast.** The mucilage envelopes and mucilage accumulations found in some forms may be in part modifications of the cell wall. But in a number of cases the products of secretion of protoplast may perform diverse functions, like, holding the individuals of colonial forms together, as a protection against desiccation in many terrestrial algae, as a means of movement, etc.

When a wall is present, its chemical constituents vary from one group to another and are sometimes important indications of the taxonomic position of a particular alga. In general, algal cell wall is composed of a variety of carbohydrate materials which may be either soluble or insoluble in boiling water. The insoluble materials are usually considered as cell wall materials and the soluble carbohydrates as accessory matrix or sheaths which are outside the cell wall. The wall materials are composed of polymers of hexose and pentose sugars and of sugar acids, often a mixture of these. The major water-insoluble carbohydrates of algal cell wall include **cellulose I, mannan, xylan** and **alginic acid**. Cellulose I is widespread among the green algae and occurs also in certain red and brown algae; its presence in blue-green algae has not been confirmed. Mannan occurs in the cell wall of certain red algae and in a number of green algae, while water-insoluble xylans have been reported in the wall of certain marine green algae. Characteristic components of the cell wall include the **polyuronic acid, alginic acid, fucinic acid** and are found in the wall of brown algae; and a characteristic **mucopeptide component** is present in the cell wall of prokaryotic cell of blue-green algae. Alginic acid, a polymer of D-mannuronic and L-guluronic acids, may comprize up to 25 per cent of the dry weight of some brown algae. The extract of alginic acid in the form of sodium alginate is usually called **algin**. Certain algae, particularly the diatoms have a silicified wall. Water-soluble polysaccharides comprize the amorphous sheaths and matrix of algae. These include, among others, **agar, carrageenan** and **gelans (xylan)** of red algae; **pectin** and **ulvin** of the green algae; the **fucoidin** of brown algae; the mucilaginous compounds of blue-green algae and those of diatoms.

Septa: The septa may be complete or incomplete. The formers are mainly of five types: (i) **plane,** (ii) **replicate,** (iii) **semireplicate,** (iv) **colligate,** and (v) **unduliseptate.** The plane type has a homogeneous contour as in *Ulothrix.* The replicate septa are common in some species of *spirogyra* where middle lamella shows circular in foldings. A septum with two infolds in opposite directions is called replicate septum. When there is only a single infold in alternate positions, it is known as semireplicate septum. The colligate septum is an H-shaped cross wall, as in *Microspora.* In the unduliseptum the middle lamella is much expanded and has a wavy undulating margin, found in *Spirogyra undulisepta.* The incomplete septum found in members of the Rhodophyta and some Cyanophyta, has a central opening known as pit through which the cytoplasm of adjacent cells remains connected.

Nucleus: In majority of algae where the cells are eukaryotic, the cells are uninucleate

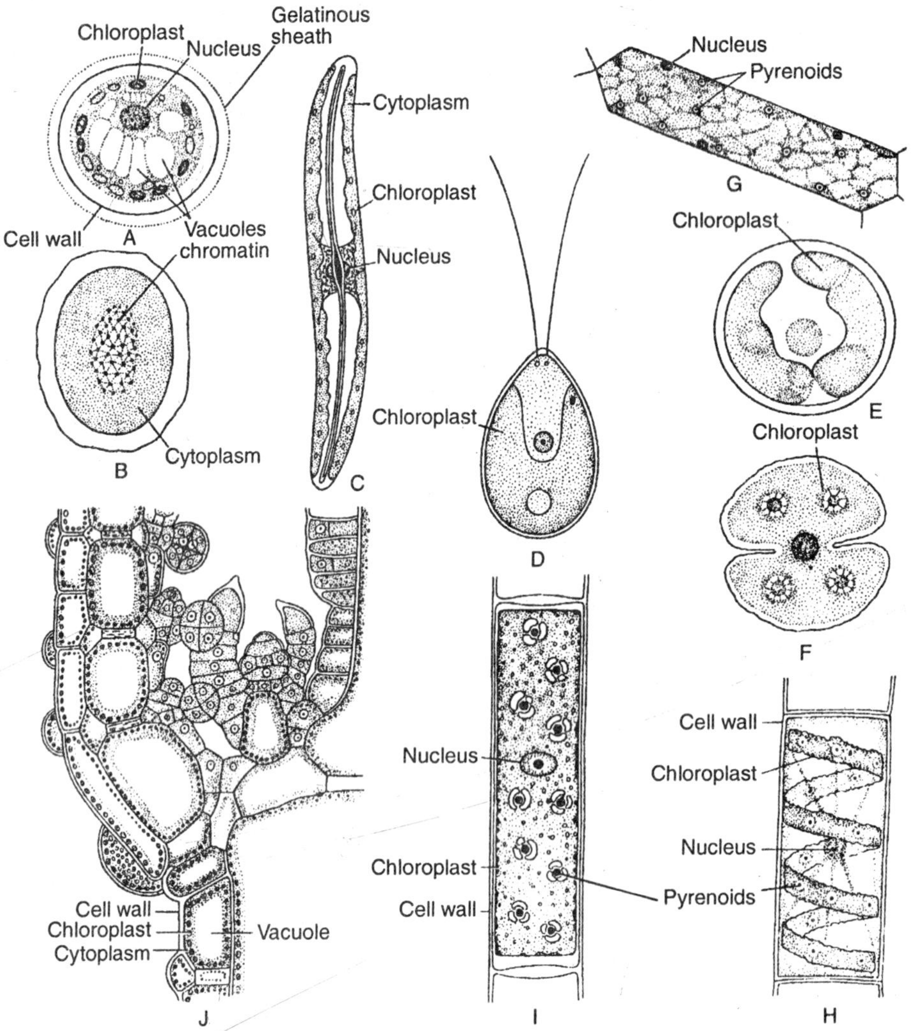

Fig. 2.1: Cell structure in Algae. A. Cross-section of a single cell of *Dinothrix* sp. showing structural details. B. A single cell of *Gloeocapsa* sp. showing cytoplasm and chromatin. C. Discoid chloroplast of *Pleurosigma spenceri*. D. Cup-shaped chloroplast of *Chlamydomonas* sp. E. Parietal chloroplast of *Pleurococcus naegelili*. F. Star-shaped chloroplast of *Cosmarium* sp. G. Cell of *Hydrodictyon reticulatum* showing nucleus and pyrenoids. H. Cell of *Spirogyra* sp. showing cell wall, nucleus and spiral chloroplast bearing pyrenoids. I. Cell of *Mougeotia scalaris* showing cell wall, nucleus, chloroplast and pyrenoids. J. Details of cell of *Chara* sp. showing cell wall, cytoplasm, chloroplast and vacuole

(Fig. 2.1A). But algae with multinucleate cells are also found in considerable number. The structure of nucleus in algae does not differ in any appreciate respect from that of higher plants. Again in algae (Cyanophyta) with prokaryotic cells where membrane-bounded nucleus is absent, the chromatin material of nucleus is located towards the centre of the cell and consists of DNA fibrils which are not associated with basic proteins (histones) so that no organized chromosomes are found (Fig. 2.1B).

Algal nuclear division may or may not be followed by wall formation. The chromosomes are small and chromosome number varies from 5 to 48.

Chloroplast, Pyrenoids and Pigments: There are three kinds of photosynthetic pigments in algae: chlorophylls, carotenoids, and biloproteins (also called phycobilins). They are lodged either in a well-organized photosynthetic apparatus, known as **chloroplast** which is composed of stroma and grana surrounded by a membrane; or in ill-organized photosynthetic apparatus, called **chromatophore.** The structure, number, and position of chloroplast in an algal cell is extremely variable depending on the species of algae. The simpler forms of algae lack chloroplast. It has long been accepted that in them, the pigment remains diffused in the peripheral region of the cytoplasm which is differentiated from a central colourless region having structural details resembling a nucleus. But recent researches have confirmatively indicated that in those algae, the pigment is located in lamellar structures (**thylakoids**) distributed in the cytoplasm, visible only under electron microscope. Apart from the nature of pigmentation which constitutes one of the basis of classification of algae, the distinctive features exhibited by the chloroplast are often of great value in generic diagnosis. Associated with the chloroplast are bodies spoken of as **pyrenoids** which consist of colourless masses of protein surrounded by starch plates (Figs. 2.1G to I). Again there are certain algae whose chloroplast do not bear any pyrenoids at all. Cells of most algae contain only one chloroplast each with the exception of a few species whose cells have more than one chloroplasts. Besides this, almost all chloroplasts bear one or more pyrenoids. There is a wide range of variation in the shape of the chloroplasts. Depending on species, the chloroplasts may be: **Cup-shaped, Parietal, Discoid, Lobed-discoid, Star-shaped, Spiral, Barrel-** or **Girdle-shaped with incised margins,** and **Reticulate** (Figs. 2.1 and 2.2).

The chloroplasts are genetically semi-autonomous systems containing histone-free DNA, messenger RNA (ribonucleic acid) transfer RNA and ribosomes. New chloroplasts arise from pre-existing ones.

The photosynthetic pigment chlorophyll may be of different kinds—**Chlorophyll a, Chlorophyll b, Chlorophyll c, Chlorophyll d** and **Chlorophyll e.** Chlorophyll a is present in all algae. Chlorophyll b is found only in the Chlorophyta and Euglenophyta. It is indispensible for the assimilation of true starch. Due to the presence of Chlorophyll b assimilation product in the Chlorophyta and Euglenophyta is true starch. Chlorophyll c is present in Bacillariophyta, Chrysophyta, Pyrophyta, Cryptophyta, Xanthophyta and Phaeophyta—

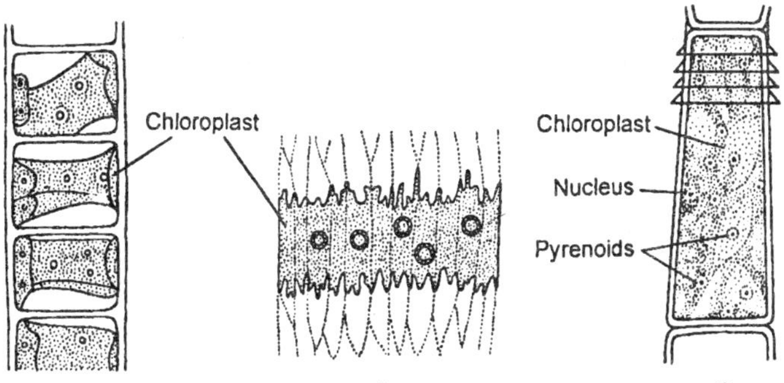

Fig. 2.2: Cel structure in Algae. A. Cells of *Ulothrix* sp. with girdle-shaped chloroplasts. B. Cell of *Draparnaldia glomerata* possessing girdle-shaped chloroplast with incised margins. C. Cell of *Oedogonium* sp. showing nucleus and reticulate chloroplast bearing pyrenoids

a feature which seems to link them together. Chlorophyll d is present the Rhodophyta. Chlorophyll e has been identified only in two genera: *Tribonema* and the zoospores of *Vaucheria*, belonging to the Xanthophyta. Besides Chlorophyll, two kinds of special pigments—**Carotene** and **Anthophylls**, collectively called **Carotenoids**, are present in algae. Carotene is fairly well distributed in algae except in the Siphonales. There are different Xanthophylls in algae which are important diagnostic features: **peridin** in the Pyrophyta, **myxoxanthin** in the Cyanophyta, **taraxanthin** in the Rhodophyta and **antheraxanthin** in the Euglenophyta. Besides these, there are various other pigments. Some of them are: **Phycobilin, Fucoxanthin, Diatoxamthin, Diadonoxanthin** and **Haematochrome**. The name of pigments Phycobilins has been changed to **Biloproteins**. Biloproteins are present in the Cyanophyta, Rhodophyta and Cryptophyta. Phycobilins include red coloured **phycoerythrins** and blue coloured **phycocyanins**. R. A. Lewin and N. W. Withers (1975) collected *Synechocystis didemni*, a unicellular marine alga with a prokaryotic cellular organization like that of a blue-green alga which possesses chlorophylls a and b, β-carotene, at least three xanthophylls and no demonstrable water-soluble phycobilin pigment.

It has been observed that certain algae completely lack pigments. Some of them are: colourless diatoms colourless dinoflagellates, colourless Cryptophyceae, colourless green algae (as in certain Volvocales), and colourless red algae that parasitize closely related hosts. Some colourless algae lack not only the pigments but also the chloroplast structure as well. Except few, almost all colourless algae lead a saprobic existence. They are often collectively known as **leucophytes**.

Nutrition: The algae, broadly speaking, like other chlorophyllous plants, require C, H, O, P, K, N, S, Ca, Fe and Mg and also traces of Mn, Bo, Zn, Cu, and Co. For certain organizms additional elements are required such as Si for diatoms and Mo for nitrate reduction in *Scenedesmus*. Algae do not differ markedly from other chlorophyllous plants with respect to the major and trace elements required for growth. With respect to the major elements they require, algae are able to use a wide range of inorganic and organic sources and probably do so in natural habitats. For example, the element nitrogen may be used by many algae in such alternate forms as nitrate, nitrite, ammonia, urea, certain amino acids, purines, and even as casein and gelatine. They are also able to use a wide variety of inorganic compounds as sources of the required major and trace elements. Those algae which can grow in an entirely inorganic medium in the presence of light are known as **photoautotrophic**. They synthesise protoplasm from exclusively inorganic sources using light energy. Again, algae which require in addition certain vitamins, usually B_{12}, thiamine or biotin, are said to be **photoauxotrophic**. In contrast to photoautotrophic algae a number of algae, are **heterotrophic**, either facultative or obligately so. Several algae, among them a species of *Ochromonas* and certain dinoflagellates may digest solid particles of food; such organizms are **phagotrophic**.

Storage food: The storage food in algae which is dependent on the nature of pigmentation, is extremely variable. In green algae it is starch, in other algae it may be carbohydrate of various forms. Some of the algal food reserves are : **Cyanophycean starch, Floridean starch, Laminarin, Mannitol, Diatomin, Leucosin, Oils, Fats, Paramylum**, etc.

Flagella: Whatever their number, the structure of algal flagella resembles that of cilia and flagella of other organizms (except bacteria) in showing the typical '9 + 2' pattern of component fibrils enclosed within a flagellar sheath, the latter continuous with the plasma membrane. Two fibrils forming the central core are surrounded by nine. Each fibril is again composed of subfibrils. The flagella may be equal (**isokontan**) or unequal (**heterokontan**) in length. For a long time algal flagella were thought to be of two kinds: **acronematic** or **whiplash** type being smooth and whip-like with stiff base and a flexible upper portion; and **pantonematic** or **tinsel** type is feathery in appearance having lateral rows of fine hairs arranged along the axis of the flagellum. For further details of the nature of the whiplash and tinsel types of flagella see chapter XI. Recent work with the electronic microscope has revealed another kind of flagella in which the flagellar surface is covered by minute hairs (different from those on the pantonematic flagella) and scales. Algae with such flagella are included in a new class, the **prasinophyceae** of uncertain systematic position. Flagella of algae, irrespective of kind, may be apical or lateral or all around the flagellate cell.

The motile cells of algae, like those of fungi, may be flagellate and/or amoeboid. Associated with motility are structures like **contractile vacuoles, flagella, stigmata** (sing. **stigma**) and **eye spots**. Contractile vacuoles range in number from one, two, four to many. They lie near the cell surface, often in close proximity to the flagelar bases. They apparently regulate osmotic balance in the cells. Contractile vacuoles are present in both flagellate and amoeboid cells, and in non-motile cells as well.

Other structures: Both **mitochondria** and **Golgi bodies** are present in algae possessing eukaryotic cells. A system of tubules and vesicles traversing the cytoplasm known as **endoplasmic reticulum** exists in algal except in the blue-green algae performing the function of synthesising proteins or enzymes. Except in the blue-green algae, the mature cells of algae possess one or more **vacuoles** bound by distinct membranes. Their function is osmoregulation of cell fluid. But the healthy cells of many planktonic blue-green algae contain numerous small bodies of irregular shapes known as **pseudovacuoles** or **gas vacuoles**.

PLANT BODY TYPES

Algae are plants of simple structure, simplest of which consists of a non-motile single-celled to filamentous forms with no well-organized nucleus and plastids. There exist many unicellular motile forms which may be solitary or colonial. Some are filamentous. Of the filamentous forms again some are unbranched with very little differentiation of base and apex having all the cells almost alike and every cell is capable of growth and division and formation of reproductive units, whereas others show diversified organization of plant body. Particularly, certain brown algae possessing long flexous stem-like structure and an expanded blade portion, have holdfasts attaching them to rocks. Some of these plants have considerable differentiation of tissue very similar to higher plants. But they lack vascular tissues. On the other hand, in the terrestrial forms there are distinct adaptations in the plant body to survive under constant inadequate supply of moisture. Again, there are **heterotrichous** forms where the plant body is highly developed being differentiated into prostrate and erect portions resembling prototype of simplest plant body of Bryophyte level with very clear division of labour. In general, the varied forms of algae may be outlined as follows:

1. **Unicellular motile form**: Single-celled plant body being spherical, oval or pear-shaped bearing two flagella in the anterior region (Fig. 2.3A).

2. **Unicellular non-motile (sedentary) form**: The cells are commonly small and spherical without any flagella and do not exhibit any movement. Some are solitary, others in groups being embedded in a gelatinous material (Fig. 2.3B). Also there are slightly elongated forms

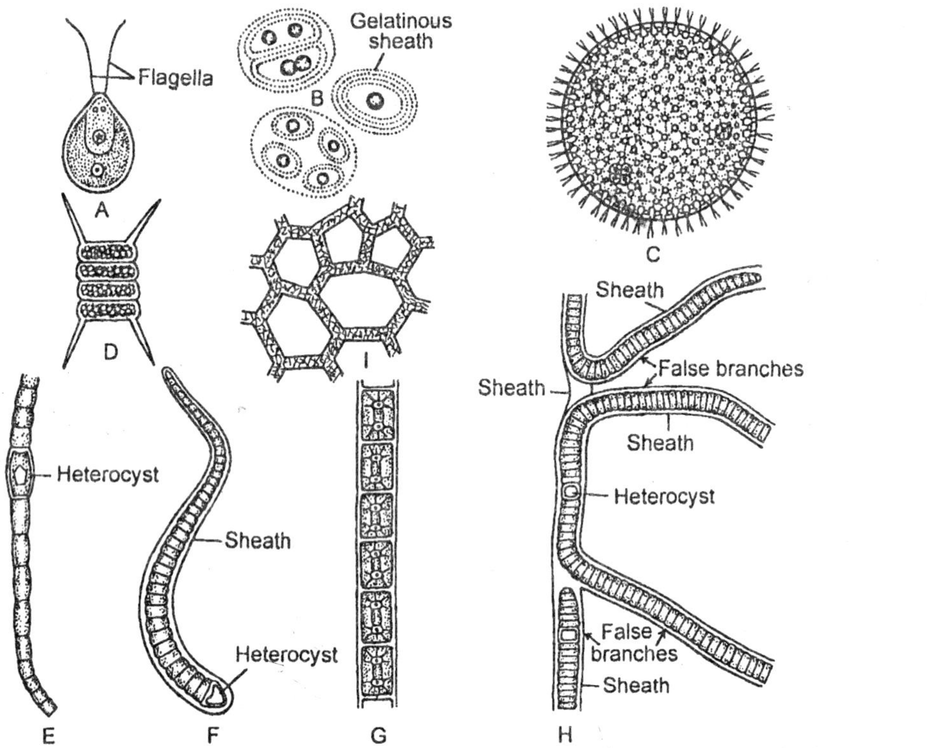

Fig. 2.3: Plant body types in Algae. A. Unicellular biflagellate in *Chlamydomonas* sp. B. Unicellular non-motile in *Gloeocapsa* sp. C. Motile coenobial in *Volvox* sp. D. Non-motile coenobial with cells arranged in a single layer and closely adpressed to each other along long axis in *Scenedesmus* sp. E. Unbranched filament in *Anabaena* sp. F. Unbranched filament in Calothrix parietina. G. A portion of the unbranched filament of *Zygnema* sp. with cells joined firmly end to end in a single file. H. False branching in *Scytonema arcangelili.* I. Meshes of the net of *Hydrodictyon* sp.

which can be differentiated into base and apex (Fig. 2.5E).

3. **Motile coenobial form**: Definite number of motile cells are embedded in a gelatinous matrix with their flagella protruded out, or are held together by cytoplasmic connection. The cells may be compact or loosely arranged. Thus a colony is formed of definite number of cells arranged in a specific manner forming what is called a **Coenobium** (pl. **coenobia**) (Fig. 2.3C).

4. **Non-motile coenobial form**: Coenobial may be composed of non-motile cells arranged in a single layer being closely adpressed to each other along the long axis (Fig. 2.3D) or may be attached end to end forming a network, the meshes of the net being pentagonal or hexagonal (Fig. 2.3I). The coenobium may also be star-shaped in appearance with a single central cell surrounded by peripheral cells of the coenobium (Fig. 1.1D).

5. **Filamentous form**: Filamentous thallus may be of indefinite length. Cell division in filamentous forms occurs in one plane so that a single row of cells is formed. The cells of the filament are usually joined firmly end to end either in a single file or having a tendency towards a multiseriate arrangement. The filament may or may not be differentiated into base and apex. All the cells of filament may or may not be alike. The filamentous form may be of various kinds—**unbranched, branched,** and having **false branches** (Figs. 2.3E to G), (Figs. 2.4B and D) and (Fig. 2.3H).

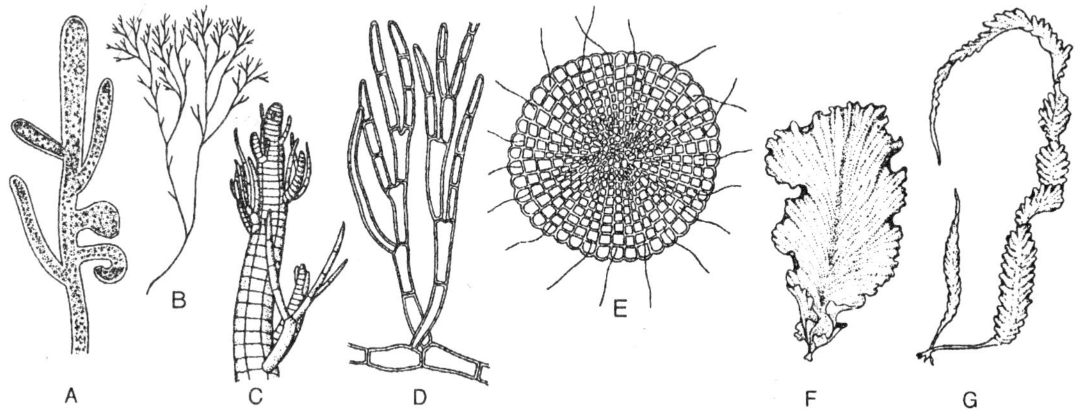

Fig. 2.4: Plant body types in Algae. A. Filamentous branched coenocyte bearing gametangia in *Vaucheria* sp. B. Filamentous branched thallus of *Cladophora glomerata*. C. Part of thallus of *Polysiphonia* sp. showing polysiphonous condition. D. Branching of *Cladophora* sp. E. Elaborately developed prostrate system of *Coleochaete scutata*. F. Thalloid form of *Ulva* sp. G. Complex thallus of *Laminaria agardhii*

6. **Thalloid form**: A parenchymatous thallus is resulted by the division of cells in more than one plane (Fig. 2.4F). Growth of parenchymatous thalli may be **diffuse** (all cells capable of division), **intercalary** (well-defined dividing regions not located terminally), **trichothallic** (a specialized intercalary meristem at the base of a terminal hair) or **apical** (one or more well-defined apical cells dividing to give remainder of the thallus).

7. **Siphonaceous form**: The thallus is non-septate multinucleate, a **coenocyte**. Depending on the organizm, a coenocyte may be simple branched or very elaborately developed with clear division of labour being differentiated into aerial and subterranean and in some cases into subaerial portions (Figs. 2.4A and Fig. 1.2C).

8. **Heterotrichous form**: This is a highly advanced type of thallus which is characterized by the differentiation of the vegetative body into a prostrate system growing along the substratum and a projecting system developing away from the substratum. In some algae both the systems are equally well developed, whereas in others one system is highly developed at the cost of the other. For example, the prostrate system may be very elaborately developed and the aerial portion being rudimentary (Fig. 2.4F) and in others, reverse is the case.

9. **Special forms**: The thallus in some algae is highly complex being differentiated into a perennial portion, an annual portion, and a meristematic region in between. The meristematic region produces the annual portion every year (Fig. 2.4G). In others, there is extreme degeneration of heterotrichous condition. The aerial portion is very elaborately developed being represented by a repeatedly branched erect system and the anchorage is performed by very poorly developed prostrate system. Again in others, the aerial portion of the thallus is highly developed into a multiseriate—**polysiphonous** branched condition with clearly visible cytoplasmic connections from cell to cell (Fig. 2.4C).

REPRODUCTION

All three processes of reproduction—**vegetative, asexual**, and **sexual** are encountered in Algae. The vegetative reproduction takes place by:

1. **Simple cell division**: This method is usually common in unicellular algae (Fig. 2.5A). In multicellular forms cell division leads

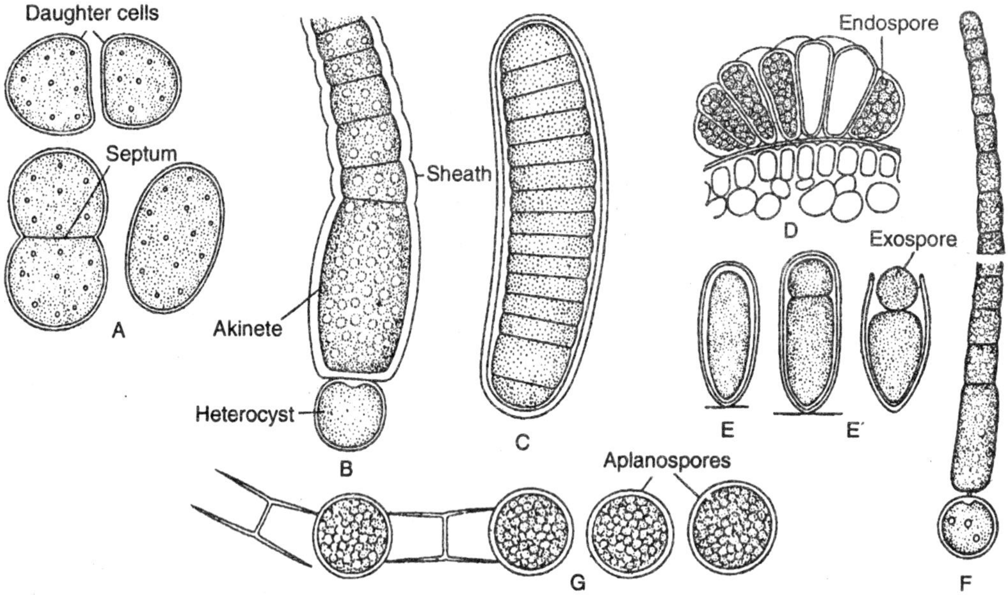

Fig. 2.5: Spore forms and vegetative reproduction in Algae. A. Simple cell division in unicellular alga *Synechococcus* sp. B. Filament of *Gloeotrichia natans* with akinete and heterocyst. C. Homospore of *Westiella* sp. D. Group of individuals with endospores of *Dermocarpa prasina*. E-E′. *Chamaesiphon incrustans* showing exospore formation. F. Fragmentation of filament of *Cylindrospermum muscicola*. G. Aplanospores of *Microspora wileana*

to growth. Growth of individual filament is usually intercalary, but in a few cases there is a definite apical cell growth.

2. **Fragmentation**: This process is common in the filamentous forms. Fragmentation of the filaments may be : (i) accidental (Fig. 2.5F); (ii) by the development of a special structure—**separation disc** (Figs. 3.4B and C), which facilitates the breaking down of the filaments into small fragments, or (iii) otherwise, resulting in the formation of multicellular fragments which grow into new filaments by the division of cells of the fragments. The process of splitting of mature colonies into small parts that takes place in many colonial forms may also be included under this category.

A large section of Algae reproduces asexually by the development of various kinds of spores and similar other structures. They may be:

1. **Alkinetes**: The vegetative cells of certain filamentous algae become elongated, thick-walled, and develop into structures known as **akinetes** (Fig. 2.5B). The akinetes can survive unfavourable conditions. With the return of favourable conditions they germinate into new individuals.

2. **Hypnospores**: These are thick-walled non-flagellate spores with abundant food reserve. They can stand unfavourable condition.

3. **Hormospores**: Fragments of some filamentous algae develop thick sheath, tide over unfavourable conditions, become functional and give rise to new filaments when conditions are favourable. These organs of perennation are also known as **hormocysts** (Fig. 2.5C).

4. **Endospores**: In some unicellular algae the entire protoplast of the vegetative cell divides to form internally a number of non-flagellate thin-walled spores known as **endospores** which develop into new individuals (Fig. 2.5D).

5. **Exospores**: These are thin-walled non-flagellate spores that are cut out externally from the protoplast of the vegetative cell of certain unicellular algae (Fig. 2.5E-E′).

6. **Heterocyst**: Certain vegetative cells of some filamentous algae become larger in size than the neighbouring cells, wall somewhat thickened being composed of two layers, outer of pectic substances and inner of cellulose, contents turn transparent and are transformed into **heterocysts**. The heterocysts develop button-like thickenings at the poles known as **polar nodules**, whose number, depending on the position of the heterocysts, may be one or two (Fig. 2.5B). A mature heterocyst is pale yellow due to the absence of photosynthetic pigments. During heterocyst formation a very young vegetative cell which has lost its capacity of division swells and becomes hyaline by the loss of photosynthetic pigments. The inner cellulose layer is added simultaneously keeping gap along the pores at the poles. By the pores the heterocyst retains connection with neighbouring cells. As the heterocyst matures these pores are plugged off with mucilage-forming polar nodules. Intensive enzymatic, structural, and functional changes take place during heterocyst formation. Functions of heterocysts are not clearly known. They are considered by some as dead cells serving the purpose of fragmentation of the filaments. Others are of opinion that they are spores. In certain species of algae, the heterocysts germinate into new individuals. Heterocysts are also considered as receptacles of enzymes which help formation of akinete. They may also take part in nitrogen fixation.

7. **Zoospores**: These are flagellate spores which may be formed either in a specialized structure, **zoosporangium** or directly from the vegetative cells. The zoospores may be bi- (Figs. 2.6A to C), quadri (Fig. 2.6D), or multiflagellate (Figs. 2.6E to F). In *Chlamydomonas*, zoospores, two whiplash type of flagella are present towards the anterior region of the cell. They are equal in length. The flagella may be very small, same in size of the cell (mother cell) or larger than the cell bearing skin, in most species.

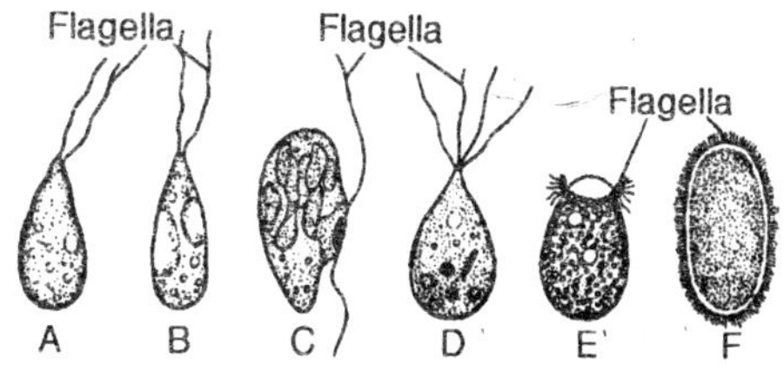

Fig. 2.6: Various types of zoospores among Algae. A-B. Pear-shaped biflagellate with anteriorly placed flagella in *Ulothrix zonata*. C. Biflagellate with laterally inserted unequal flagella in *Ectocarpus* sp. D. Quadriflagellate in *Ulothrix zonata*. E. Multiflagellate in *Oedogonium* sp. with flagella arranged in a ring surrounding a beak-like projection. F. Multiflagellate in *Vaucheria* sp. with flagella distributed throughout the entire body

The flagellation originates from the blepharoplast situated towards the anterior side of the cell. The flagella come out through very fine canals on the outer wall as described by Fritsoh (1935). Roberts (1972) and Hills (1973) studied the details of telepharoplasts by Electron Microspory. According to the blepharoplasts remain attached to the basal granules. The multiflagellate zoospores may again be of two kinds, with (i) flagella distributed throughout the entire body (Fig. 2.6F), and (ii) flagella arranged in a ring surrounding a beak-like projection (Fig. 2.6E). Each zoospore swarms around in water, comes to rest, sheds its flagella and germinates into new individual.

8. **Aplanospores**: These are non-flagellate spores that are formed in some aquatic algae by the failure of the development of flagella during zoospore formation due to certain unfavourable condition in the substratum (Fig. 2.5G). The aplanospores may also be formed in certain algae of semiaquatic habitat. When they appear identical to the parent cell they are referred to as **autospores**. Aplanospores with thickened walls and abundant food reserve are also called the **hypnospores**.

The occurrence of sexual reproduction probably marks a relatively advanced stage of evolution among Algae and is very rare in all

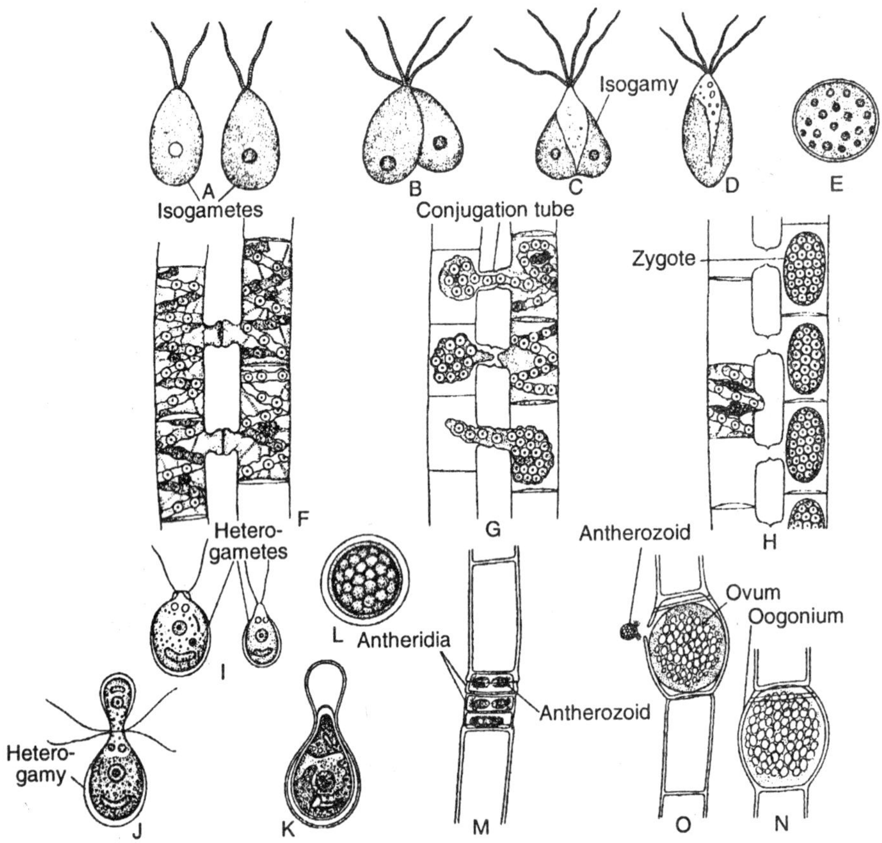

Fig. 2.7: Sexual reproduction among Algae. A-E. Stages in fusion of isogametes leading to zygote formation in *Ulothrix* sp. F-H. Successive stages in fusion between non-flagellate isogametes by conjugation tube development resulting in zygote formation in *Spirogyra* sp. I-L. Stages of heterogamy leading to zygote development in *Chlamydomonas braunii.* M-O. Structure of antheridia, antherozoid, oogonium and ovum in *Oedogonium* sp.

the types that are regarded as of relatively less advanced. There is a wide range of variations in the nature of the gametes and the mode of sexual reproduction. Any vegetative cell of the plant body may produce gamete and thus behaves as a gametangium or a specialized gametangium may be developed in which gametes are produced. The gametangia may be morphologically similar, **isogametangia** or dissimilar, **heterogametangia**. In heterogametangia, the smaller one is the male gametangium or **antheridium** (pl. **antheridia**) (Fig. 2.7M) which produces one or more male gametes which again may be flagellate or non-flagellate. The flagellate male gametes are the **antherozoids** or **spermatozoids** (Fig. 2.7O). The larger gametangium is the female gametangium or **oogonium** (pl. **oogonia**) (Fig. 2.7N), also known as **carpogonium** (pl. **carpogonia**) which produces non-flagellate female gamete. Such a non-flagellate female gamete is known as **ovum** (pl. **ova**) or **oosphere** or **egg** (Fig. 2.7O). The number of female gamete in a female gametangium is usually one, may also be more than one, but restricted in number. In general, gametes may be morphologically similar in appearance, **isogamous** or dissimilar, **anisogamous** or **heterogamous**. They may be flagellate, **planogamous** or non-flagellate, **aplanogamous**. The process of fusion between the two isogametes which may be planogamous (Fig. 2.7A-E) or aplanogamous (Fig. 2.7F-H) is known as **isogamy** or **conjugation** and that between the two dissimilar planogametes (Figs. 2.7I-L) or the smaller one planogamete and the larger one aplanogamete is **anisogamy** or **heterogamy**. The term conjugation is also applied to the process of fusion between two isogametangia. In heterogamy, when the smaller gamete is flagellate or non-flagellate and the larger one is non-flagellate, enclosed, and retained either permanently or up to maturity within the oogonium the process of fusion between these two gametes is known as **oogamy** or **fertilization**. During fertilization, the female gamete remains confined within the oogonium either permanently or until maturity or after being liberated out remains attached with some part of the oogonium. The antherozoids, when mature, are liberated out from the antheridium, swim around in water and enter the oogonium. But only one unites with the ovum. The fusion product of the two gametes, in general is called a **zygote**. The zygote bearing a nucleus with two chromosome complements (2n) is said to be **diploid** (Gk. **diploos**, double), one chromosome complement being contributed by each gametic nucleus. The gametic nucleus which possesses one chromosome complement (n), is designated as **haploid** (Gk. **haploos**, simple). The resulting zygote of isogamy is known as **zygospore**, and that of oogamy is called **oospore**. No new individual is produced so long the zygote remains enclosed within the oogonium. Hence, the zygote itself is liberated out of the parent individual and either directly develops into a new individual or the zygotic nucleus undergoes meiosis resulting in the development of spores from which new individuals are developed. All these three kinds of sexual reproduction—**isogamy, anisogamy**, and **oogamy** are encountered in Algae. The details of post-sexual stages exhibited by the members of the Algae are indicated below:

I. The diploid zygotic nucleus divides reductionally immediately after its formation resulting in the production of four haploid nuclei—**zygotic meiosis**. Further development may follow any one of the following ways:

(i) Of the four haploid nuclei, three abort and one survives to develop into a new haploid individual resembling the parent plant (Fig. 2.8).

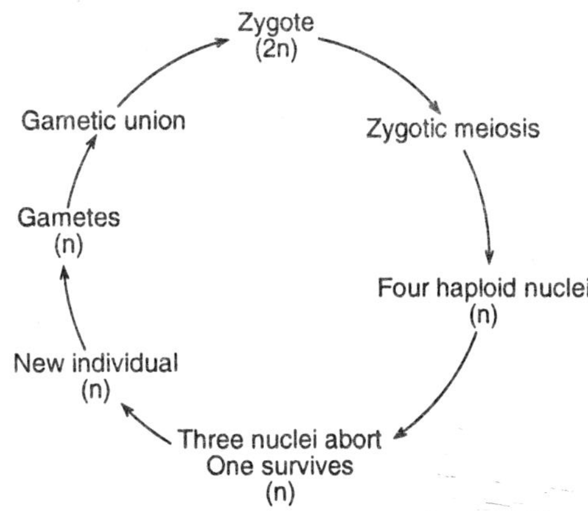

Fig. 2.8: Life cycle pattern in certain Algae

(ii) All the four haploid nuclei develop into flagellate spores—**zoospores**, which germinate into new individuals (Fig. 2.9).

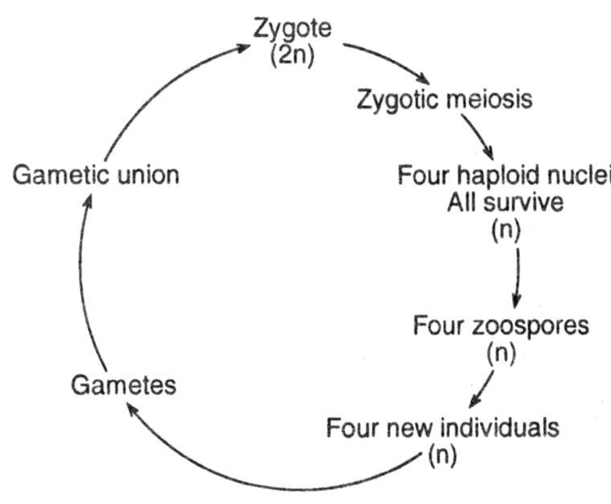

Fig. 2.9: Life cycle pattern in certain Algae

II. The zygote germinates directly without undergoing any meiotic division of the diploid zygotic nucleus. The subsequent stages may be:

(i) The zygote directly produces an elaborately developed diploid individual which bears gametangia. The haploid stage is very

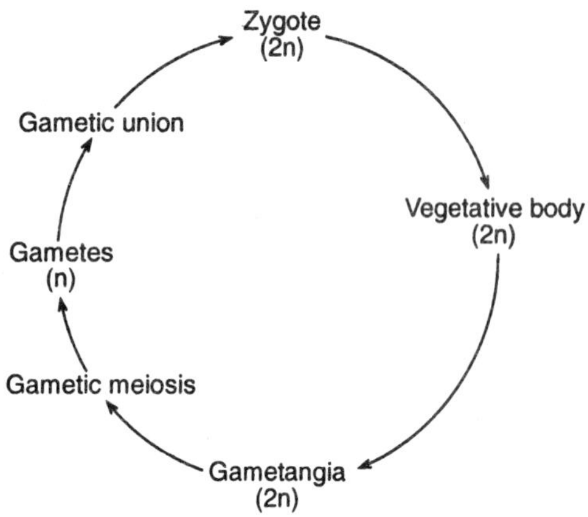

Fig. 2.10: Life cycle pattern in certain Algae

short being represented by the gametes only, which are again formed by meiosis just before the sexual reproduction—**gametic meiosis** (Fig. 2.10).

(ii) Due to delay in the reduction division of the diploid nucleus there is development of a new diploid phase—the **sporophytic generation** or the **sporophyte** which occurs regularly alternating with the haploid gamete producing phase—the **gametophytic generation** or the **gametophyte**. The sporophyte is the spore producing phase in which spores are produced by meiosis, **sporic meiosis** (Fig. 2.11). To complete their life cycle these algae pass through two phases of growth—the gametophyte and the sporophyte occurring alternately

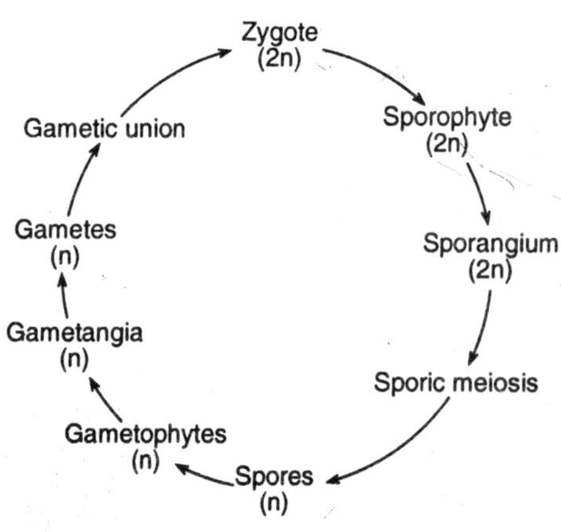

Fig. 2.11: Life cycle pattern in certain Algae

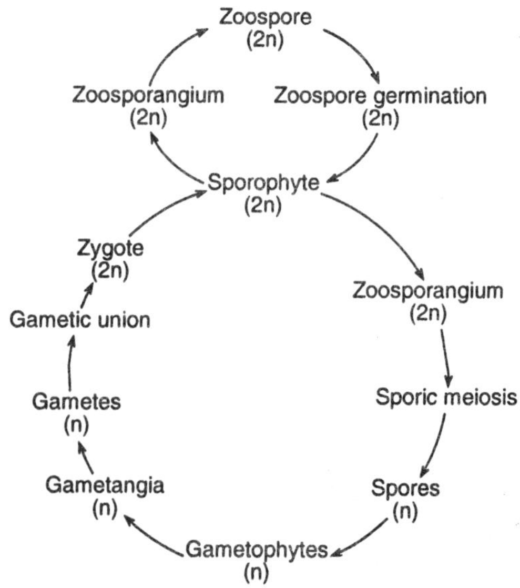

Fig. 2.12: Life cycle pattern in certain Algae

in regular sequence. This is known as **alternation of generations** (Fig. 2.12).

There are cases where during sexual reproduction gametic union may fail to take place. In such cases the gametes develop directly into spore-like structure, **parthenospores,** also known as **azygospores**. The process by which the parthenospores are formed is known as **parthenogenesis**. These parthenospores behave like zygospores without any meiotic division of the parthenospore nucleus.

LIFE CYCLE PATTERNS IN ALGAE

Members of the sexually reproducing algae exhibit various life cycle patterns which may be outlined as under:

I. An alternation of predominantly gametophytic generation or haploid phase with a very short and insignificant sporophytic generation or diploid phase which is the diploid zygote. This type of life cycle pattern is known as **haplontic or monognetic life cycle**.

II. In the second type of life cycle the haploid generation is very insignificant and is represented by the gametes only. It occurs alternating with a very predominant diploid generation. Life cycle of such pattern is called **diplontic life cycle.**

III. In the third kind of life cycle both the gametophytic and sporophytic generations are equally predominant, and is enumerated as **haplodiplontic** or **digenetic life cycle.** In such life cycle pattern, again, both the haploid and the diploid phases may be morphologically similar or dissimilar, that is, **isomorphic** or **heteromorphic** respectively.

IV. In fourth one both the gametophytic as well as the sporophytic generations are elaborately developed, the difference being, the diploid phase is long drawn, part of which is dependent on the gametophytic generation and later on gives rise to an independent diploid phase. Such a life cycle pattern is enumerated as **diplohaplontic** with **trigenetic life cycle.**

ALTERNATION OF GENERATIONS

Sexually reproducing algae, in general, pass through two phases of development to complete their life cycle. These are : (i) the spore producing phase, the **sporophyte** or **sporophytic generation** or **diploid generation** (2n), whose nucleus bears two chromosome complements (2n); and (ii) the gamete producing phase, the **gametophyte** or **gametophytic generation** or **haploid generation** (n) having only one chromosome complement (n) in each nucleus. These two phases or generations as they are commonly known, follow each other in regular sequence in the life cycle of algae. They are distinct from each other both functionally as well as for having the number of chromosome complements different. These two phases or generations which occur in regular succession, one alternating with the other is known as **alternation of generations.** The process of alternation of generations was discovered by Hofmeister in 1851.

As early as 1894, Strasburger made the generalization that, for plants showing alternation of generations, the generations differ in nuclear constitution. This difference lies in the fact that the nuclei of all cells of the sporophytic generation contain twice the number of chromosomes present in the nuclei of the cells of the gametophytic generation. The sporophytic generation is initiated by the fusion of two gametic nuclei, each possessing one chromosome complement (n). The fusing nuclei may be present in morphologically similar or dissimilar gametes. The two nuclei fuse to form a diploid nucleus having two chromosome complements (2n). Sooner or later, meiosis takes place whereby the chromosome number is reduced by half during the formation of spores. This marks the end of the sporophytic generation and the beginning of the gametophytic generation. The spore is thus the first cell of the gametophyte. The gametophyte produces gametes, which, on fusion, result in the formation of a new sporophyte.

A plant arizing either from a zygote or from a spore may give rise to an independent plant exactly like itself. Each member of such a series in a very natural use of the word constitutes a **generation.** Hence, the generation extending from zygote to a spore or from spore to gametes may consist of a series of **generations** of subordinate order. Furthermore, each generation, in either of these senses, may include hundreds or thousands of **cell generations.** A more adequate expression, however, is lacking for the two alternating arcs of the life cycle, one diploid and closing with meiosis; the other haploid and ending in the production and union of gametes, in what follows, each of these major arcs will be called a **generation,** however successive independent individuals it may include. The advantages resulting from alternation of generations are: (i) it provides for the production of many spores, so increasing the possibilities of reproduction; and (ii) meiosis of a zygote nucleus can give rise in consequence of new groupings of chromosomes and chromosome parts to at most four genetically different lines.

The two generations also differ in their relative conspicuousness in structure and duration. Majority of algae have a conspicuous gametophyte, and the sporophyte is relatively inconspicuous, short in duration, and more or less dependent for all food upon the gametophytic thallus. In such cases the diploid nucleus divides meiotically almost immediately after its formation. As such, there is alternation of an elaborate haploid or gametophytic generation and a very short single-celled diploid or sporophytic generation. A reverse condition is also encountered, where a well-developed diploid generation alternates with a very short haploid condition being represented by the gametes only, and meiosis takes place during the formation of gametes. Such conditions indicate that there is no clear alternation of phases having different functions and only alternation of chromosome numbers takes place in such life cycle.

Again, others possess the two alternating generations as independent plants that are often evenly balanced as to size and conspicuousness. In them, the zygote resulting from the gametic union directly develops into an elaborate diploid individual, the sporophyte. Meiosis takes place during the formation of spores. In such cases both the generations may be morphologically similar or dissimilar. An alternation of two morphologically similar generations is known as **isomorphic alternation of generations** and when the two alternating generations are morphologically dissimilar it is called **heteromorphic alternation of generations.**

There are some other algae which exhibit alternation of generations in their life cycle with little deviation from the typical one. Here the post-fertilization diploid condition is very much long drawn. The earlier part of the diploid condition is entirely dependent on the gametophyte, but subsequently an independent sporophyte is developed which bears sporangia where spores are formed as a result of meiosis. The spores thus produced give rise to the gametophytic generations. This is know as **triphasic alternation of generations.**

SEXUALITY IN ALGAE

In Algae, sexual reproduction generally takes place after the plant has had a certain amount of vegetative growth and a certain amount of food reserve has been built up. It involves two distinct processes—a fusion of protoplasts, **plasmogamy**, and a fusion of two nuclei, **karyogamy**. The occurrence of plasmogamy and karyogamy during the union of gametes to form a zygote is known as **syngamy**. But where sexual reproduction includes only karyogamy, that means union of two nuclei both of which are usually derived from the same parent cell, the process is called **autogamy**. Syngamy is of general occurrence among Chlorophyta, Euglenophyta, Chrysophyta, sparingly in Pyrophyta, Phaeophyta, and Rhodophyta. Autogamy has been reported in some Pytrophyta and Euglenophyta, whereas both syngamy and autogamy are lacking in Cyanophyta. The gametes taking part in the sexual process are of varied structure and in most cases their fusion takes place only when they are liberated out from the mother plant. But among the oogamous forms along with the enclosing of the ovum there is a distinct tendency for the retention of the fertilization product within the oogonium. This condition of advancement may very well be considered a case of incipient stage of embryo development. The most agreed kind of evolution is isogamy —most primitive, with anisogamy—a step in advance of isogamy, and oogamy—being the most advanced type of sexual reproduction.

ORIGIN AND EVOLUTION IN ALGAE

As already indicated, that members of Algae, like other plants, reproduced by classified methods. They are **sexual** and **asexual.** In cyanophyceae, takes place by fragmentation and specralial structures, homospores, heterocysts, akinities.

The topic **Origin of Sex** in Algae is treated separately.

Since true sexuality does not exist among the members of Cyanophyceae (blue-green algae), discussion will start from members of Chlorophyceae (green algae). Among the lower members of Chlorophyceae, asexual reproduction takes place by the development of asexual spores mainly flagellate of various kinds as already treated separately. These asexual spores developing sporous given and are liberated in equator at maturity, swim in water, and finally shed their flagella and develop into new algal filament with or without any resting period.

Now, during unfavourable atmospheric condition, these resting structures fail to develop flagella, undergo divisions and develop into smaller number of units. These smaller units at a later stage develop into gametes which fuse with each other producing **Zygoges.** This is how **Origin of Sex** takes place by starvation and hence designated as "Starvation theory" by Cholnoky.

The **evolution of sex in Algae** proceeds along the line **Isogamy—Aristogamy** and **Oogamy.** This line of evolution is encountered in Chlorophyceae and in almost all other algae groups. In Chlorophyceae (green algae) and in almost other algae, sexual reproduction takes place by different processes from simple **Isogamy**—between both the fusing gametes are morphologically and physiologically identical, hence both the fusing gametes cannot be differentiated as male and female gametes. The process is known as **Isogamy.** Again, in some algae gametes are morphologically alike, but differ in their physiological behaviour. The process is designated as **Physiological anisogamy.** In **true anisogamy,** both the fusing gametes are flagellated and hence motile and may or may not be similar in size. Again, in **Oogamy,** sexual fusion takes place between a smaller male gametes and a larger female gamete which is produced in a male gametangium and a female gametangium respectively. In some algae, both the fusing gametes may be flagellated, and in others, the smaller gamete is flagellated and the larger one is nonflagellate. They are produced in separate male female gametangia and fusion between the male and female gamete may or may not take place inside the algae filaments or outside in water.

This topic is further discussed in Generations of the Algae, in Chapter X.

GENERAL OBSERVATION

Names of many pioneer workers and their contributions in the topic: **Classification of Algae, Fungi, and Microorganizms,** are discussed in respective chapters as review in this book. Only reasonably recent workers involved in the above topic are mentioned.

CLASSIFICATION

The classification of Algae has been continually modified since 1753 when Linnaeus in Species Plantarum included Algae in the class Cryptogamia along with other organizms which are now recognized as Mosses, Vascular Cryptogams and Fungi.

Unger in 1838 grouped Algae, Lichens and Fungi under the taxonomic category Thallophyta which arrangement in spite of serious drawbacks is still being widely favoured. Although Unger separated Thallophyta from other plants because of their lack of differentiation into stem, leaves, and roots, this has created confusion, particularly for those algae whose body is differentiated into stem-like, leaf-like, and root-like fixing organs. Again to accept Thallophyta as a natural division of the plant kingdom means acceptance of the view that all the algae are more or less closely related to one another. Although limited in number, the members of the group Algae are of varied type and are sufficient to show that evolution has paralleled within the group.

Although it has become a customary practice to include Algae under the division Thallophyta, along with Fungi, in the late nineteenth century and first three decades of the twentieth, such classification has been abandoned in modern polyphyletic treatments of the plant kingdom. Wide range of classification system of Algae was introduced by A. Pascher (1927); F. E. Fritsch (1935, 1945); Pabenfuss (1946, 1951); G. M. Smith (1933, 1955); Christensen (1962); Chapman (1962); Bold and Wyne (1978); Lee (1989).

It has human age-old practice to classify Algae largely on the basis of the colour of pigments and to subdivide into major classes: Cyanophyceae, Chlorophyceae, Phaeophyceae and Rhodophyceae. But with the increase in knowledge, it has been realized that beside pigments, reserve foods, flagellations, and comparative chemistry of the several pigments should be considered in the classification of Algae. Christensen (1962) gave due consideration to micro-anatomy of protoplasmic structures and combined with characters like flagella present or lacking and chlorophyll to present or lacking and introduced the terminologies. **Prokaryote** and **Eukaryote** for algae possessing prokaryotic and eukaryotic cells respectively. Besides this, various classes of Algae are so distinct from each other each class should be placed in separate division adding the suffix–**phyta** to each division. **Cyanophyta, Chlorophyta, Euglenophyta, Xanthophyta, Bacillariophyta, Pyrophyta, Phaeophyta** and **Rhodophyta**. This arrangement was supported by Papenfuss, Bold and Wyne (1978) along with others. Papenfuss further suggested that in order to indicate the algal nature of different divisions of algae, the word **phyco** should be incorporated before **phyta**. Bold and Wyne (1978) prepared a new system of classification of algae as follows:

1. **Cyanophycophyta (Myxomycophyta)** or (**Blue-green algae**): The internal structure of the cell is extremely simple, true nucleus and chromatophores lacking. Cells are prokaryotic. The protoplast is differentiated into peripheral pigmented region together with oil drops and cyanophycean starch and non-pigmented central body. The pigmentation is characterized by chlorophyll α, β- carotene, myxoxanthin, myxo-xanthophyll, *c*-phycocyanin, and *c*-phycoerythrin. There is complete absence of motile cells. Reproduction is chiefly by accidental breaking and in some forms by the production of specialized cells. Inefficient sexual system has recently been reported.

   Members of this group are designated by some Microbiologists as cyanobacteria placing them under Bacteria with unusual properties under the section: Oxygenic Phototrophic Bacteria.

2. **Chlorophycophyta (Green algae)**: Cells are eukaryotic. Pigments are the same as those present in the higher plants and are in same proportions being located in definite plastids of variable nature. Starch is the customary form of storage food. The majority of the representatives exhibit sexuality ranging from isogamy, anisogamy to oogamy. Motile cells may be bi-, quadri- or mutiflagellate and are extremely variable in size.

3. **Euglenophycophyta**: They possess a definite nucleus and the pigmentation is chlorophyll associated with carotenoids localized in definite plastids. Cells are eukaryotic. The reserve food is **paramylum** (an insoluble

carbohydrates related to starch) and fats. Almost all are naked, unicellular flagellates having one, two or rarely three flagella. Reproduction generally occurs by longitudinal cell division with the exception of one genus which reproduces sexually.

4. **Xanthophycophyta (Yellow-green algae)**: This group of algae is mainly distributed in fresh-water with few marine forms. The vegetative body may be unicellular or multicellular. Cells are eukaryotic. The cell wall is composed chiefly of pectic compounds with very rare occurrence of cellulose. Cells contain numerous discoid chromatophores. The chromatophores are yellow-green in colour owing to the presence of excess yellow xanthophyll. Pyrenoids are usually absent. The food reserves are fats or leucosin. The formation of starch is completely absent. Motile cells bear two flagella of unequal length. Asexual reproduction may occur by the formation of flagellate or nonflagellate spores.

5. **Bacillariophycophyta (Diatoms)**: Members of this taxon of algae are commonly known as the **diatoms**. They occur in various habitats ranging from fresh and salt waters and in damp places including soil, bark of trees, rocky cliffs, etc. They have single-celled, usually diploid, vegetative body which may be solitary or colonial. The shape of diatom cells may be varied being composed of two nicely fitted halves. Cells are eukaryotic. The cell wall of pectic material with particles of silica impregnated on it is richly ornamented. The shape of the cells and their wall markings are very important characters of taxonomic interest. The rich deposit of diatom shells that is formed at the bottom of the water after the death of the diatom cells is known as **diatomaceous earth** which has great economic importance. The diatom cells are uninucleate. The pigments are located in the chromatophores whose shape and number are very variable. A special golden-brown pigment—**diatomin** masks the cholorophyll and other associated pigments. Fats are the main food reserves, besides which there may be present a special insoluble food of uncertain chemical composition—**volution**. Reproduction by cell division is rather very common. But when the cells become smaller and smaller with each succeeding division the original size is restored by the formation of rejuvenescent cells or renewal spores or **auxospores**. The formation of auxospores is associated with either sexual reproduction or parthenogenesis.

6. **Chrysophycophyta**: Members of this division constitute some of the most abundant kind of life on earth, being common components of phytoplankton. They cover a wide range of habitats, but are chiefly fresh-water. There is a range of vegetative body from the smallest flagellates to filamentous, and colonial forms. But vegetative body chiefly consists of unicellular forms, being uniflagellate and biflagellate (heterokont) having both flagella of the acronematic type or with a longer pantonematic type and the other of the acronematic type. The photosynthetic pigments include chlorophylls a and c, β-carotene, fucoxanthin and diadinoxanthin. The main reserve products are oil and chrysolaminarin (leucosin). The most common method of reproduction is by longitudinal division of the cell. A characteristic feature of the Chrysophyta is the formation of silica impregnated thick-walled **cysts** (statospores) from the vegetative cells. The contents of the cyst either give rise to a new vegetative cell or divide to produce 2-4 zoospores. Sexual reproduction may be **isogamy** (between two motile gametes), **hologamy** (fusion of protoplasts of two vegetative cells), or **autogamy** (fusion of two nuclei only within the cyst).

7. **Pyrrophycophyta (Fire algae)**: Most of the organisms are unicellular bearing two usually unequal flagella. A few are uncellular and colonial alga-like in form. Cells are eukary-

otic. Yellowish-green to golden-brown chromatophores are due to the presence of dinoxanthin, diadinoxanthin and peridinin. The reserve foods are starch or starch-like compounds or oils. Common mode of reproduction is by cell division, although some produce zoospores, while sexuality has been reported in a few.

8. **Cryptophycophyta (Cryptomonads):** The vegetative cells are naked, unicellular and motile with two equal (sometimes slightly unequal) usually anteriorly (in kidney-shaped cells laterally) located flagella. Members of this division occur in both fresh-water and marine habitats. Cells may be red, blue, olive-yellow, brown or green having photosynthetic pigments chlorophylls a and c, β-carotene, some distinctive xanthophylls and biloproteins which are different from those of blue-green and red algae. Nature of pigmentation suggests Cryptophyta as an intermediate stage in the evolution of eukaryotic forms from prokaryotic algae. Most forms are photosynthetic, but saprophytic nutrition also occurs. Again some reside in the tissues of host invertebrates. Reproduction is by longitudinal division of vegetative cells, thick-walled cyst formation has also been reported in some genera.

9. **Phaeophycophyta (Brown algae):** Cells of the brown algae are eukaryotic having definite nuclei, cytoplasm, and chloroplasts. The chloroplasts contain photosynthetic pigments masked by golden-brown pigments, called **fucoxanthin**. Food reserves are stored as simple sugars, the alcohol mannitol and complex polysaccharides such as **laminarin** or there may be fats in different forms. The motile reproductive cells are pear-shaped with two laterally inserted flagella. Reproduction is isogamy to oogamy.

10. **Rhodophycophyte (Red algae):** Cell structure of red algae is eukaryotic being of high organization with a single nucleus although some cells are multinucleate. One or several plastids floating in the cytoplasm possess chlorophyll associated with a red pigment—**phycoerythrin**, and often a blue pigment—phycocyanin. Food reserves may occur as alcohols, but are chiefly stored in the form of a polysaccharide as small granules free in the cytoplasm rather than in plastids and are known as **floridean starch**. Growth of the thallus may be apical as well as intercalary. One of the striking features of red algae is the conspicuous cytoplasmic connection from cell to cell. Another outstanding characteristic of this taxon is a total absence of any flagellate reproductive cells. Oogamous sexual reproduction is of general occurrence in red algae.

Robert Edward Lee (1989, 1999) classified algae into 4 evolutionary groups and 18 divisions, salient features of which are as follows:

[For further details, consult "Phycology" by R. E. Lee, 1999, 3rd Edn.]

Group I. **Prokaryotic Algae**
- Division 1. **Cyanophyta**
- Division 2. **Prochlorophyta**

Group II. **Eukaryotic Algae** with bimembrane chloroplast
- Division 1. **Glaucophyta**
- Division 2. **Rhodophyta**
- Division 3. **Chlorophyta**

Group III. **Eukaryotic Algae** with unimembrane chloroplast endoplasm reticulum
- Division 1. **Euglenophyta**
- Division 2. **Dinophyta**

Group IV. **Eukaryotic Algae** with bimembrane chloroplast endoplasmic reticulum
- Division 1. **Cryptophyta**
- Division 2. **Chrysophyta**
- Division 3. **Prymnesiophyta**
- Division 4. **Bacillariphyta**
- Division 5. **Xanthophyta**
- Division 6. **Eustigmatophyta**
- Division 7. **Raphidophyta**

Division 8. **Phaeophyta**
Division 9. **Synurophyta**
Division 10. **Dictyophophyta**
Division 11. **Pelagophyta**

In recent years, gene sequencing studies were also carried out in some algae which provide adequate information about the relationship of various algae groups.

ALGAL FOSSIL

The cyanobacteria are the oldest group of algae with definite fossil remains in the form of stromatolites, dating back almost 3000 million years under anaerobic environment. However most of the algal fossil are not well preserved due to its soft tissues and thus the exact time of appearance of each groups are not clear from the preserved fossil forms. However the appearance of fossil members of the algal classes in the geological timetable is presented in the table 2.1.

TOXIC ALGAE

Algae can be harmful to two basic ways:

(a) Producing large population in the aquatic environment and thus induce clog in other aquatic animals and also induce anoxic condition on decomposition and thus kill fish in particular. This is case of algal bloom in eutrophicated water bodies.

(b) Some algae produce toxins that sicken and till other organisms that prey on these algae. Many such algal toxins are persistant and bioaccumulated in sheelfish. The major toxin producing algae belongs to cyanophyta,

Table 2.1: First appearance of algae in the geological time scale

Era	Period	Epoch	Million of years ago	First appearance of algae fossil
Cenozoic	Quaternary	Holocene	1.8	
		Pleistocene		
	Tertiary	Pliocene	5.5	
		Miocene	25.0	Xanthophyta
		Oligocene	36.0	
		Eocene	53.5	Euglenophyta
		Palaeocene	65.0	
Mesozoic	Cretaceous		135	Chrysophyta
	Jurassic		191-205	
	Triassic		235-245	Bacillariophyta
Paleozoic	Permian		275-290	Prymnesiophyta
	Carboniferous		360-380	
	Devonian		405-430	
	Silurian		435-460	Dinophyta Charophyta
Proterozoic	Precambrian		3000	Cyanophyta Rhodophyta Chlorophyta

Source: Dill et al (1986); Loeblich (1974); Medlin et al (1997); Young et al (1994).

Raphidophyta, Dinophyta and Bacillariophyta. Cyanophyta produces neurotoxins like anatoxin, saxitoxin and hepatoxins like microcystin and nodularing Dinophyta produces toxins like okadaic acid, macrolide toxins and yessotoxins. They also produces ciguatoxin, maitotoxins, saxitoxins. Bacillariophyta produces neurotoxins like domoic acid. Rephidophyta produces some unidentified toxins too. There are selected algal species of each classes produce such toxins and thus poison the water bodies when they form algal bloom. However, most of the algal toxins are biodegradable in nature. The most promising strategy to control blooms of toxic marine phytoplankton is to spread flocculents, such as clay, that scavenge algal cells from seawater and carry them to bottom sediments. Clay is a non-toxic, naturally occurring material, that does not effect fish and bottom dwelling organisms.

Chapter 3

The Cyanophycophyta (Cyanophyta)

The division Cyanophycophyta (Cyanophyta) or Myxophyta, commonly known as blue-green algae, consists of a single class Cyanophyceae or Myxophyceae or Schizophyceae whose plants are extremely simple in several respects. Members of this division enjoy a wide range of distribution in all kinds of habitats, from the tropics to the polar regions, including marine and fresh-waters, and on tops of mountains. The fresh-water forms either inhabit damp terrestrial places or are aquatic, reaching their best development in stagnant waters to give the water a green or yellow-green colour, the familiar appearance 'water blooms'. A number of blue-green algae have a capacity to change their colour in relation to the wavelength (colour) of the incident light. This change in colour is often described as **complementary chromatic adaptation** or **Gaidukov phenomenon**. Several species grow in hot springs where the temperature of the water may be even about 85°C. Blue-green algae may also cause the precipitation of calcium carbonate from lake water. A considerable number of species live in association with other organisms; examples are afforded by their occurrence as algal constituents of Lichens, as symbionts in the cells of diverse unicellular animals and different groups of plants and as 'space-parasites' within the tissues of various higher plants. Whereas, a few species are mild parasites in the digestive tracts of human beings and other animals. Again others have been shown to be definitely able to fix atmospheric nitrogen. The capacity to thrive under such diverse circumstances indicates a great degree of adaptability.

CELL STRUCTURE

The cellular architecture is prokaryotic, internal structure of which is extremely simple. The cyanophycean cell does not contain golgi bodies, mitochondria, and endoplasmic reticulum. There are no membrane bounded chloroplast, organized nuclei, and aqueous vacuoles. Flagellate structures and true sexuality are completely absent in the blue-green algae.

Protoplast: The protoplast has a high degree of viscosity having jelly-like consistency filled with highly refractive granules of proteinaceous material, **cyanophycin.** Under light microscope the protoplast appears to possess two regions, a peripheral region, the **chromoplasm** and a colourless central region, the **centroplasm** or the **central body**. But the use of the electron microscope has shown that this idea is not valid. The protoplast is normally devoid of vacuoles, except in old cells. This is a major factor contributing to the great resistance of the plants to desiccation and of the cells to plasmolysis. In some planktonic species **pseudo-** or **gas-vacuoles** may be found which contribute towards their buoyancy by virtue of the gas that they are said to contain. The electron microscope study reveals a complex lamellar system is present in the protoplast. These lamellae are also known as **thylakoids** (Fig. 3.1). They are not enclosed in membrane-bounded groups to form chloroplasts. Instead, they lie free in the cytoplasm, in some species more or less restricted to the periphery. Depending on the intensity of the incident light and age of the cell, the thylakoids may be distributed irregularly throughout the protoplast or may be arranged in parallel stacks. But they are always separated by a layer of cytoplasm. Each thylakoid is an elongated flattened sac-like structure possessing small granules resembling ribosomes. The thylakoids are functionally analogous to the chloroplasts of other algae and higher plants.

Pigments: The thylakoids are the site of chlorophyll a, and the accessory pigments also occur on their surfaces in the form of small particles, the **phycobilisomes**. The accessory pigments are: c-phycocyanin, c-allophycocyanin, and c-phycoerythrin, the two former blue and the latter red. They are biliproteins and are composed of protein with chromo-

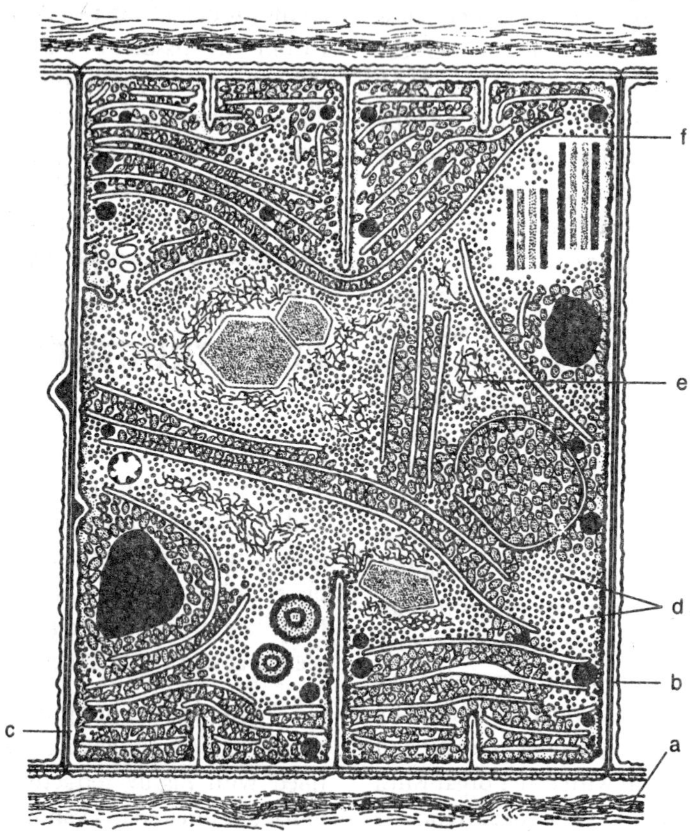

Fig. 3.1: Diagrammatic presentation of blue-green algal cell structure. (a) Gelatinous sheath, (b) Cell wall, (c) Plasmalemma, (d) Ribosomes, (e) Nuclear material, (f) Thylakoid

phoric groups. The accessory pigments transfer light energy they absorb to chlorophyll-a. They also have a protective shading function that prevents oxidation in intense light of the other photosynthetic pigments. Besides these, the other pigments present are: β-carotene, myxoxanthin, myxoxanthophyll, antheraxanthin, aphanicin, aphanizophyll, flavacin, lutein, oscilloxanthin, and zeaxanthin.

Nuclear material: Unlike other algae, the blue-green algal cell lacks an organized nucleus with a nuclear membrane and nucleolus. Fine fibrils of DNA are a feature of the nuclear material (Fig. 3.1). These are either distributed throughout the cell or concentrated in the central portion. The use of electron microscope has shown that the chromatin nuclear material, also designated as 'DNA-plasm', constituted of fine fibrils of DNA may occur as a single accumulation of nuclear material, or organized into a reticular structure or individual rod-like nuclear elements. The last type is found in many larger forms of blue-green algae and indicates the presence of several genetic complements or genomes per cell. The term 'chromatin apparatus' is generally used for the nuclear material of the blue-green algae. The DNA fibrils are not associated with basic proteins (histones) so that no organized chromosomes are found. They commonly appear as a reticulum (Figs. 2.1B and 3.4A).

Photosynthesis and **storage food:** Photosynthesis in blue-green algae occurs in the same manner as in other chlorophyllous plants. The storage food is **Cyanophycean starch** (probably amylopectin portion of starch) and proteins. Many of the blue-green algae also store reserve food in the form of oils. The proteins of some blue-green algae contain **diaminopimelic acid**, an amino acid also found in some bacteria but never found in higher plants or animals.

Nitrogen fixation: Certain blue-green algae can assimilate or fix elemental (gaseous) nitrogen of the atmosphere. They thus are independent of other combined nitrogen sources. The nitrogen fixed by blue-green algae can be assimilated by themselves as well as by other organisms. Three kinds of blue-green algae can fix nitrogen: (i) the filamentous heterocystous species, e.g., *Nostoc* and *Anabaena* (ii) certain nonheterocystous filamentous species, e.g., *Plectonema boryanum* only under microaerophilic conditions, and (iii) certain unicellular (nonheterocystous) species, e.g., *Gloeocapsa* and *Aphanothece.* The nitrogen-fixing enzyme complex **nitrogenase** is oxygen-sensitive, so that the highest rate of nitrogen fixation occurs under reduced oxygen tensions. In filamentous heterocystous blue-green algae, the site of nitrogen fixation is heterocyst. But microaerophilic nitrogen fixation may go on in the vegetative cells. Substantial amounts of soluble nitrogenous compounds are liberated from healthy cells of nitrogen-fixing blue-green algae either into the culture medium or in their natural habitats. Besides, when the algae die, their nitrogenous materials decay and form ammonia which may then be transformed into nitrates by nitrifying bacteria thus enriching the fertility of the habitat. Blue-green algae that contribute greatly to the fertility of rice fields are species of *Aulosira, Anabaena, Tolypothrix, Cylindrospermum, Nostoc* and *Mastigocladus.*

Cell wall: The cyanophycean cell possesses a surface layer of **gelatinous sheath** which is a constant feature, although it varies in thickness (Figs. 2.3B and 3.7C). The sheath consists of cellulose fibrils reticularly arranged within a matrix to give a homogenous appearance. The cell wall is inside the sheath. It is a multilayered structure lying external to the plasmalemma. Because of intricate relationship of plasmalemma with cell wall, the separation of plasmalemma from cell wall during plasmolysis fails. The cell wall is composed of various substances, some of them are: muramic acid, glucosamine, alanine, and glutamic and diaminopimelic acid. The inner layer of the cell wall contains a **mucopeptide** component comparable to that found in bacterial cell walls. In the filamentous forms the inner layer always completely envelopes the protoplast, but the cell sheath forms a cylindrical envelope which may be continuous (Figs. 3.5B, C) or interrupted between consecutive cells. The cell walls have a remarkable degree of elasticity. Division of a cyanophycean cell takes place by the ingrowth of a septum from the periphery and the gradual and consequent division of the chromoplasm. The septa composed only of the inner layer are exceedingly delicate. In the filamentous blue-green algae very delicate plasmodesmata or protoplasmic strands effect protoplasmic continuity across the transverse walls. Various pores and depressions in the longitudinal walls have been revealed by electron microscopy. These may be related to the gliding motility of some species.

ORGANIZATION OF PLANT BODY

The plants are diverse in structure, such as, unicellular, colonial to filamentous unbranched, falsely branched and branched forms. The single isolated cell with no polarity is the simplest condition, although by cell division the individuals remain combined to form **palmelloid** aggregates or colonies of different kinds. Cells are spherical to ovoid and are similar in structure being surrounded by a gelatinous sheath probably represent the most primitive types (Fig. 2.3B). There are also almost similar forms being unicellular but with marked polarity (Fig. 2.5E) which probably are the direct descendants from the previous forms which lack polarity. The simplest filamentous forms are made up of a long series of cells placed one upon the other to form a **thichome** (Fig. 3.4E). The trichome may be straight or spirally coiled (Figs. 3.6E and F). The trichome often secrete mucilaginous material or varying consistency which may be homogenous or lamelloid or striated (Figs. 2.3F and H) and (Figs. 3.5E and F). The trichome with the sheath together is termed a **filament.** In many filamentous forms cell division and growth are generally diffuse or spread over the entire length of the trichome. The most advanced forms have a **heterotrichous** condition of habit. The threads of many of the filamentous forms include structures known as **heterocysts** (Figs. 3.5B to E), specialized enlarged cells formed by differentiation of vegetative cells. One of the most conspicuous features of the heterocysts is the strong external thickenings of the cell wall. These thickenings are composed of three distinct layers, the outermost fibrous, the middle homogeneous and the innermost laminated. A mature heterocyst contains granular structures which are evenly distributed throughout the cytoplasm, giving a homogeneous appearance under the light microscope from which earlier workers concluded that heterocysts are 'empty' cells having no cytoplasmic content. The position of a heterocyst in a filament may be terminal (Fig. 2.5B), or intercalary (Fig. 3.4E) and depending on its position it bears one or two **polar nodules** which behave like plugs filling the pore channels of the heterocyst, thereby heterocyst loses its protoplasmic connections with the adjacent cells. There has been much speculation as to the nature and functions of the heterocysts. No doubt, in some cases they break-up the filament into **hormogonia** (sing. **hormogonium**) (Fig. 3.5E). It has been suggested that the heterocyst is a form of food storage organ or that it represents reproductive organ that is now functionless. In some species, however, it behaves as a reproductive body and is capable of germinating into a new filament. Fritsch (1951) suggested that the heterocysts, during vegetative period, secrete substances that stimulate growth and division of the adjacent cells. Present day workers are of opinion that heterocysts are involved in nitrogen fixation and may be the sites of effective actual nitrogenase activity. Besides heterocysts, a filament can also bear a structure much larger than the vegetative cells with appreciably thick wall enclosing accumulated food reserves in the form of cyanophycean granules. Such a thick-walled structure is known as an **akinete** (Figs. 3.4B, D and E) which being thick-allied is highly resistant to water shortage an unfavourable temperatures. Under favourable conditions the akinete germinates into a new filament. Akinetes may be formed in specific positions, especially in relation to the heterocyst (Fig. 2.5B). In others, they are formed in short or long chain. Only a few of the trichomatous forms of Cyanophyta exhibit various types of rapid movements, whereas, in filamentous genera movement is usually restricted in hormogonia. But the unicellular forms show only slow movements. The motion consists of creeping or gliding of the trichome backwards or forwards in the direction of its longitudinal axis accompanied by rotation either clockwise or anticlockwise depending on the species.

REPRODUCTION

The Cyanophycophyta are characterized by the complete absence of sexual reproduction; no sexual organs and no motile reproductive bodies have been observed. Propagation takes place by simple division, by spores (**akinetes, endospores,** and **exospores**) or else by fragmentation (**fission**). The multiplication of unicellular and colonial forms is brought about mainly by simple cell division. In all the filamentous forms multiplication is largely effected by the formation of **hormogonia**. The hormogonia are the fragments of the filament delimited by the occasional death of cells at intervals along the length of the filament. In certain forms, the hormogonia are enveloped by thick sheath and thus are modified as organs of perennation, known as **hormocysts** or **hormospores** (Fig. 2.5C), which on germination grow out into new filaments. Many of the filamentous forms produce heterocysts (Fig. 3.6D, E), which break up the filament into hormogonia. At times the heterocysts may behave as spores to germinate into new filaments. In some Cyanophycophyta, the **akinetes** germinate immediately after formation, while in others they remain dormant for long time behaving as resting spores. On germination the protoplast divides producing germling. In a number of genera, particularly in certain unicellular members, small spores, known as **endospores** are formed endogenously within a cell (Fig. 2.5D). They arise by successive division of the protoplast along three planes. The endospore forming cell behaves as a sporangium. The endospores are generally naked, but a thin wall is secreted after their liberation from the sporangium. In some epiphytic forms the delicate cell wall ruptures apically exposing the protoplasts from which spherical spores are abstricted successively from its tip, these are called **exospores** (Fig. 2.5E′). The abstricted spores are surrounded by a delicate membrane. The protoplast is continuously active. The exospores may germinate when already attached to the parent protoplast giving rise to new individuals.

True sexuality does not exist in the blue-green algae, but a kind of parasexual phenomenon designated as **genetic recombination** has been demonstrated in *Anacystis nidulans* by gene transfer and gene recombination through blue-green algal virus. Genetic recombination differs from true sexuality in that it is not through syngamy or meiosis, and yet the function of true sexuality is achieved. Genetic recombination has also been reported in *Cylindrospermum majus* and *Anabaena doliolum*. It is likely that genetic recombination is brought about by conjugation between donor and recipient cells, as in bacteria. It is also possible that gene recombination may be caused by transduction, a process in which the virus acts as a vector of certain genes transferring them from donor to recipient cells. Large number of blue-green algal viruses have already been discovered to play the above role, of which mention may be made of the blue-green algal virus, cyanophage LP-1 having host range (*Lyngbya, Plectonema* and *Phormidium*).

CLASSIFICATION

The Cyanophycophyta (also designated as Cyanophyceae, or Myxophyceae, or Schizophyceae) have been variously classified in recent decades, and there has been little agreement. Some authors have employed large numbers of orders and families. Geitler (1932) and Fritsch (1945) recognized 5 orders and 17 and 19 families respectively. Papenfuss (1955) included 3 orders and 22 families. Drouet (1964) preferred to treat Cyanophycophyta in a single order with 8 families. In as much as in 36 years Drouet examined more specimens of blue-green algae than any other botanist. He is inclined to follow a implified taxonomy by including eight families: Chroococcaceae, Chamaesiphonaceae, Clastidiaceae, Oscillatoriaceae, Nstocaceae, Rivulariaceae, Scytonemataceae and Stigonemataceae under a single order Chroococcales.

The Cyanophycophyta are also often grouped with bacteria in the **Procaryota** for having prokaryotic cell.

CONSIDERATION OF CYANOPHYCOPHYTA AS CYANOBACTERIA

Microbiologists designate cyanophycophyta (Blue-green algae) as **Cyanobacteria** placing under the Section. **Oxygenic Photosynthetic Bacteria** in the World of Bacteria. This arrangement does not receive support of Botanists like: Fritsch (1935 and 1945); Smith (1933, 1951, 1955); Chapman (1973); and Lac (1980, 1999); who still consider Cyanophycophyta closes Cyanobacteria—Division under the group Algae. Instead of entering into a prolonged contradictions, just a few salient features in favour of placing the Cyanophycophyta under the group Algae are incorporated here.

Members of the Cyanophycophyta (Blue-green algae) possess vegetative cells structurally much simpler than the bacterial cells. In them, flagellated structures are completely absent. In a blue-green algal cell the photosynthetic apparatus is a flattened membranous sac designated as the **thylakoid** in which (Chlorophylla, Carotinoid pigment, phytochemical reaction centres, and phytosynthetic electron transport chain) are contained. The surface of a thylakoid is studded with granules, called phycobilisomes containing phycobilin pigments. The thylakoids are thus analogous to the chloroplasts of other algae and higher plants. The pigment chlorophyla of blue-green algal cells is structurally and functionally different from the pigment—bacteriochlorophyll that is found in certain bacterial cells. Besides this, in a bacterial cell bacteriochlorophyll and auxiliary carotenoid pigments are located in the cytoplasmic membrane which may be greatly invaginated to form vesicles of folded layers (lamellae) or tubules. The photosynthetic system is encountered in a blue-green algal cell, which is similar to other groups of Algae, is quite different from that of a bacterial cell.

The nuclear material in a blue-green algal cell is either distributed through the cell or concentrated in a central portion. This is extremely different in a bacterial cell.

True sexuality does not exist among the numbers of the cyanopycophyta. But in their parasexual process worked out so far quite different from what is encountered in a bacterial cell.

Papenfuss treated blue-green algae as in Schizophyta along with bacteria. Whereas, Bold and Wynne placed blue-green algae in Algae designating as **Cyanochloranta** instead of **Cyanophycophyta.**

What are Cyanobacteria?

These organisms are placed under the Section: **Oxygenic Phototrophic Bacteria** in the **World of Bacteria**. Cyanobacteria constitute a large and morphologically diverse group of prokaryotic organisms that exhibit a great diversity of form. They may be unicellular, cells lung generally associated to/form colonies, or filamentous. cell mass generally has a blue-green colour due to the presence of blue phycocyanin and green chlorophyll pigments. But they may also appeared due to the presence of phycoerythrin. Both phycocyanin and phycoerythrin are phycobilins.

Cyanobacteria reproduce by budding, fragmentation of trichomes into hormones. Some cyanobacteria reproduce by **multiple fusion,** a process in which a cell undergoes internal divisions producing small cells called **bacocytes** within the mother cell. Balocytes are liberated by the rupture of the mother cell.

All cyanobacteria have a gram-negative type of cell wall. Many forms of cyanobacteria exhibit a oscillating movement like filaments of blue-green alga *Oscillatoria.* The photosynthetic, apparatus of cyanobacteria is contained in **thylakoids** which are flattened sac-like structures connected with the cytoplasmic membrane. The thylakoids contain chlorophylla, accessory carotinoids and the photosynthetic electron transport system as well as the photosynthetic reaction centres. On the surface of the thylokoids, numerous hemispherical bodies are orderly arranged. There are known as **phycobilisomes**, they are complex of proteins and hycobillin pigments. In some unicellular cyanobacterium thylakoids are absent. Cyanobacteria show a remarkable property of growing in a wide variety of habitats and tolerance to a wide range of environmental conditions. Many cyanobacteria can grow in symbiotic association with other organisms, like fungi, protozoa, bryophytes, gymnosporous and angiosperms. Classically, the cyanobacteria have been classified into five orders two of which are—Chroococcales and Chromosiphonales including the unicellular forms and the rest—Pleurocorpales, Nostocales and Stigonematales including filamentous genera.

How far Cyanobacteria differ from Cyanophycophyta?

Members of the Cyanophycophyta (Blue-green algae) possess vegetative cells structurally much simpler than the bacterial cells. In those, flagellated structures are completely absent. The nuclear material in a blue-green algal cell is either distributed throughout the cell or concentrated in the central portion. This is different in a bacterial cell. In a blue-green algal cell the photosynthetic apparatus (i.e., Chlorophylla, carotenoid pigments, phytochemical reaction centres, and phytosynthetic electron transport chains) is contained in flattened numerous sacs, the **thalakoids** which are different from the **thylakoids** of Cyanobacteria.

But Cyanobacteria and Cyanophyta have some similarities like: both are prokaryotic is both have Nitrogen fixing qualities, produce cysts and both can usefully be harnessed for the welfare of the society, blue-green alga *Oscillatoria* and *Beggiatoa*, a sulphur bacteria are similar in shape and movement and, both are sensitive to antibiotics. Genetic recombination has been reported in *Anacystis nidulans*, a member of blue-green algae, similar to bacteria.

True, sexuality does not exist among the member of Cyanophycophyta. But in terms the parasexual process worked out so far is quite different from what is encountered in a bacterial cell.

But Papenfuss treated blue-green algae as Schizophyta along with bacteria. Whereas, bold and Wynne placed blue-green algae in Algae designating as Cyanochloranta instead of Cyanophycophyta.

The Cyanophycophyta contain a single class Cyanophyceae which is grouped in five orders. These orders are mainly distinguished from each other by the nature of the plant body; presence or absence of gelatinous sheath, heterocyst; and mode of reproduction.

I. **Chroococcales:** Unicellular without polarity or colonial (commonly palmelloid); multiplication by cell division and by endospores.

Family: **Chroococeaceae**

Genus: *Chroococcus,* cells more or less solitary or forming very small colonies.

Genus: *Gloeocapsa,* cells forming large colonies, aggregated together to form gelatinous colonies of no definite shape, individual cell bears gelatinous sheath of more than one layer.

II. **Chamaesiphonales:** Unicellular or colonial epiphytes or lithophytes exhibiting marked polarity; multiplication by endospores or exospores.

Family: **Chamaesiphonaceae**

Genus: *Chamaesiphon*

III. **Pleurocapsales:** Heterotrichous filamentous types, devoid of heterocysts; multiplication by endospores.

IV. **Nostocales:** Non-heterotrichous filamentous types, often showing false (rarely true) branching; heterocysts commonly present; multiplication by hormogonia, hormocysts and akinetes.

Family: **Oscillatoriaceae,** simple filamentous unbranched forms, the threads occur singly or aggregated within a common gelatinous envelope; difference between base and apex may or may not be present, absence of heterocysts, reproduction by means of hormogonia, no akinetes or spores.

Genus: *Oscillatoria,* trichomes with insignificant sheath or sheath lacking, straight with curved extremities, differentiated into base and apex.

Genus: *Lygbya,* trichomes always singly within a sheath, differentiated into base and apex.

Family: **Nostocaceae,** gelatinous thallus of varied shape—solid or hollow, floating or attached to some object, numerous intertwined and contorted filaments with individual sheath may remain embedded in a gelatinous sheath, cells moniliform, presence of heterocysts, reproduction by hormogonia or spores.

Genus: *Nostoc,* trichomes contorted within a gelatinous investment formed in a firm colony.

Genus: *Anabaena,* trichomes more or less straight, nude or with a thin gelatinous sheath, no firm colony is formed.

Family: **Scytonemataceae,** filaments have basal and apical regions, forming little erect tufts, false branching.

Genus: *Scytonema.*

Family: **Rivulariaceae,** colonies form spherical, hemispherical or irregular masses attached to plants or stones or soil, containing numerous radiating filaments each of which terminating in a colourless narrow hair-like structure, the individual sheath may be near the base of the trichome or is diffluent further up.

Genus: *Rivularia,* filament partly within a gelatinous sheath, heterocysts terminal with spore produced next to it.

V. **Stigonematales:** Heterotrichous filamentous forms showing true branching, trichomes composed of several rows of cells, mostly with heterocysts, and usually showing clear pit connections between the cells; multiplication by hormogonia, hormocysts, more rarely by akinetes.

Family: **Stigonemataceae**

Genus: *Stigonema*

PHYLOGENY AND INTER-RELATIONSHIP

There is no positive evidence to indicate the phylogenetic origin of the division Cyanophycophyta. There are, however, certain intriguing characters which quite often give plenty of scope for speculation. The complete absence of flagellate stages might lead one to assume the origin of the Cyanophycophyta from unicellular forms that were devoid of flagella. This ancestral stock may or may not be related to the flagellate unicellular forms which apparently gave rise to the other groups of algae. But this assumption lacks evidence. Many authorities support the idea that their abundance of terrestrial forms suggests a terrestrial rather than an aquatic origin for the taxon as a whole which again needs further investigation.

Some have sought an affinity with the Rhodophycophyta owing to the presence of similar pigments (phycobilins), the absence of motile reproductive cells, and the occurrence of protoplasmic connections in some species of Cyanophycophyta. Such a view will not stand the test of a critical enquiry as it entirely overlooks quite apart from essential differences in metabolism, the fundamental contrast in cell structure in both the taxa and by the absence of any type of sexual in Cyanophycophyta. The simplicity of cell structure, mode of reproduction, and the varied morphological peculiarities, such as, presence of heterocysts, akinetes, hormogonia, etc., more or less eliminate any affinity with other algal taxa.

The Cyanophycophyta are more primitive than many other organisms, may be except the bacteria (division Schizophyta) with which there are similarities. Both groups possess prokaryotic cells and primitive nucleus which is different from all other organisms. Neither group has been demonstrated to have typical sexual process, although parasexual process is known in some bacteria. There is presence of similar cell wall components mucopeptide and storage food diaminopimelic acid in both the two groups. Both groups usually have very simple vegetative body which is again enclosed in a gelatinous sheath, although similar sheath occurs among some other algae as well. Certain blue-green algae and certain bacteria differ from all or nearly all other organisms in being able to fix atmospheric nitrogen. Again certain blue-green algae and bacteria thrive at temperatures far higher than that can be withstood by cells of other organisms. The fact that some blue-green algae may be infected with viruses (often designated as **cyanophages**) which resemble bacteriophages emphasizes further the similarity between blue-green algae and bacteria. In fact, in their method of genetic recombination the blue-green algae resemble bacteria. Some chemosynthetic bacteria are so similar in appearance to the blue-green algae that Stainer and van Niel (1941) have suggested that the organisms in question are more closely related to the Cyanophycophyta than the Schizophyta. Pringsheim (1949) questioned the significance of this statement. Again many botanists believe that the close relationship between the bacteria and the blue-green algae should be reflected by putting the two divisions together and the blue-green algae should be reflected by putting the two divisions together in a single division, the Schizophyta apart from other plants. Stainer and van Niel (1962) grouped blue-green algae with bacteria in the Prokaryota. Echlin and Morris (1965) prefer to emphasize the differences between the two groups, consider the blue-green algae to make up a division Cyanophycophyta, containing a single class, the Cyanophyceae.

Order: *CHROOCOCCALES*

The Chroococcales include the most primitive members of the Cyanophyceae. Almost all are fresh-water. Unicellular plant body solitary or colonia, multiplication is by cell division and by endospores.

Family: *CHROOCOCCACEAE*

The Chroococcaceae include algae having unicellular spherical (coccoid), ovoid, or cylindrical plant body either solitary or large number of cells united into spherical to flattened colonies. Each individual cell is surrounded by mucilage sheath and the colonies with homogeneous or lamellated sheaths. Reproduction of the cells takes place by means of fission in three planes. Generic distinctions are largely based upon a combination of cell shape and planes of division.

Genus: *Gloeocapsa*

This alga is abundantly found in wet rocks and other damp places. Its cells occur in small irregular groups within wide lamellate mucilage envelopes forming masses of jelly (Fig. 3.2A) which may be colourless or coloured yellow, brown, blue, violet, or red. It is an unicellular organism (Fig. 3.2B). Cells are rather spherical in shape without any polarity and are surrounded by well-defined stratified envelopes which, although distinct, are more or less confluent with one another (Fig. 3.2C). The sole method of reproduction is by simple cell division or fission (Fig. 3.2D to F). In this process, the chromatin material divides into two parts, and the cell constricts in the middle forming two new plants. In a group, however, cell division is along three planes which is rather a primitive condition, the products becoming irregularly arranged. Resting spores with firm thick walls are often formed in a number of species.

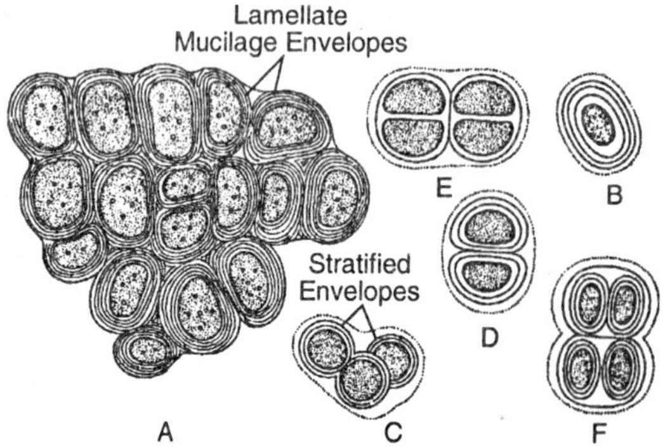

Fig. 3.2: *Gloeocapsa* sp. A. Cells in group possessing lamellate mucilage envelopes. B. A single cell with well-defined stratified envelope. C. Cells showing stratified envelopes. D-F. Stages in cell division

Some Indian Species: *Gloeocapsa atrata* (Turp.) Kutz.; *G. calcarea* Tilden; *G. decorticans* (A. Br.) Richter; *G. gelatinosa* Kutz; *G. punctata* Nag.

Special features

1. Unicellular thick-sheathed plant body.
2. Mostly cells are held together by a mucilage sheath.
3. Reproduction by fission.

Order: *NOSTOCALES*

This order includes filamentous unbranched algae often showing false branching. Majority of them are fresh-water. Heterocysts are present only in some genera. Reproduction is effected by hormogonia, hormocysts and akinetes.

Family: *OSCILLATORIACEAE*

The Oscillatoriaceae are by far the largest family in the blue-green algae and include all the relatively simple forms in which the trichomes are unrbanched and develop neither heterocysts nor hairs. Although in some genera such as *Oscillatoria,* the trichomes appear to be naked, a diffluent mucilage envelope formed during movement is probably present even though it is often indiscernible. Under favourable conditions, the sheath material is secreted copiously, sometimes in a well-hydrolyzed, sometimes in a solidified state. The most popular system of classification of the Oscillatoriaceae has been that of Gomont, which is based in large part on the structure of the sheath. Thus, in *Lyngbya* the sheath is discrete and contains a single trichome (Fig. 3.4D). *Symploca* is similar, but the filaments adhere in fascicles at the surface of the colony. In *Phormidium* the sheaths are commonly coalesced. *Hydrocoleum* has several trichomes within a single sheath; *Microcoleus* has filaments containing a single bundle of many trichomes (Fig. 3.4E). *Sirocoleum* has several such trichome bundles within a common sheath. Drouet's recent work, however, has shown that a single species may, under certain changing environmental conditions, modify its sheath characteristics to resemble species of *Oscillatoria, Lyngbya, Phormidium, Hydrocoleum,* and *Microcoleus.*

Trichome characters are also used in classification. *Spirulina* is distinguished by spirally coiled trichomes. In *Crinalium* the trichomes are flattened. Various species, especially of *Oscillatoria* and *Lyngbya* are distinguished by characteristics of the apical cells of mature trichomes. The apical cell may be attenuated or may possess a thickened hood, **calyptra**.

In *Oscillatoria* and *Lyngbya,* hormogonia are liberated soon after formation, while in *Hydrocoleum* and *Microcoleus* they germinate *in situ* to produce the numerous trichomes within a sheath.

False branching of filaments results in *Hydrocoleum* and *Schizothrix* from trichomes tending to grow out laterally and to continue to secrete a sheath as they extend.

Occasionally, *Lyngbya majuscula* occurs in abundance, notably in shallow waters in the tropics. In Hawaii it has been responsible for a severe toxic reaction of the skin known as "swimmers' itch". *Lyngbya* and other blue-green algae are the prime suspects in the chemical toxicity chain in fish poisoning in certain tropical Pacific areas.

Genus: *Oscillatoria*

Oscillatoria commonly occurs on the surface of damp earth or on the sides of flower pots and also on stones. The muddy banks of streams and ponds are frequently covered with extensive sheets of *Oscillatoria.* Patches of algal layer with adherent mud are often found floating on the surface of water. *Oscillatoria* can thrive in waters where other algae cannot survive, and it often grows luxuriantly in pools of stagnant putrescent water.

The plant body is of unbranched, cylindrical trichome being devoid of sheath (Fig. 3.3A), or with barely perceptible, cylindrical trichome being devoid of sheath (Fig. 3.3A), or with barely perceptible, evanescent sheath may be free or interwoven to form a thin stratum. Extremity of trichome is distinctly marked being straight (Fig. 3.3A) or curved like a sickle (Fig. 3.3B) or coiled more or less like a screw (Fig. 3.3D) and often attenuated (Fig. 3.3C); apical cell sometimes rounded (Fig. 3.3F), sometimes attenuated (Fig. 3.3C), or capitate (Fig. 3.3G), frequently furnished with a calyptra (slightly thickened hood) (Fig. 3.3E).

Trichomes of *Oscillatoria* can move from place to place, exhibiting a swaying back and forth or oscillating as well as a twisting or rotating movement. In narrow thrichomes, the cells are cylindrical with length greater than the breadth (Figs. 3.3D, G). Whereas, the broader trichomes have cells length less than the diameter (Figs. 3.3A, B, E and F). Depending on spices, the protoplasts are homogeneous (Fig. 3.3C), granulate (Figs. 3.3E to G), or with numerous pseudovacuoles. In *Oscillaoria princeps* and *O. tenuis,* the structure of central body has been worked out with the help of electron

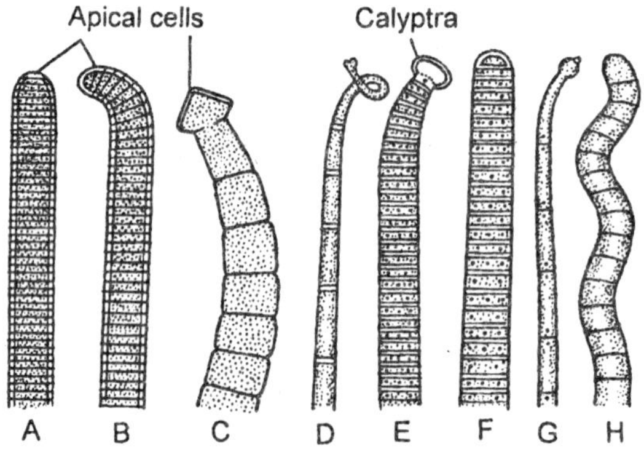

Fig. 3.3: Oscillatoriaceae. A. *Oscillatoria* sp. Unbranched straight plant body, B. *Oscillatoria* Sp. Unbranched curved plant body, C. *Oscillatoria* sp. Plant body with attenuated apical cell and cells having length less than diameter and homogeneous protoplasts, D. *Oscillatoria* sp. Coiled apical cell and cells with length greater than breadth, E. *O. proboscidea*. Apical cell furnished with a calyptra and cells having granulate protoplasts, F. *Oscillatoria* sp. Rounded apical cell, and cells having length less than diameter and granulate protoplasts, G. *O. splendida*. Capitate apical cell, and cells having length greater than breadth and granulate protoplasts, H. *Spirulina fenneri*. Coiled plant body

microscopy. This study has revealed that the central body is made up of a thin limiting membrane containing a homogeneous matrix in which fine chromatin threads are arranged (Fig. 3.4A).

Reproduction is by cell division, which again mainly effected by hormogonia. Hormogonia are formed by the death of cell or group of cells here and there in the trichome or by accidental breaking. At times double concave discs of gelatinous material—**separation discs** are formed between two adjoining vegetative cells in order to dilimit the hormogonia (Figs. 3.4B and C).

Oscillatoria differs from *Lyngbya* is not having a prominent gelatinous sheath surrounding the trichome (Fig. 3.4D). In *Phormidium*, however, numerous trichomes are embedded in an amorphous mucus formed by the gelatinization of the sheaths of the individual filaments.

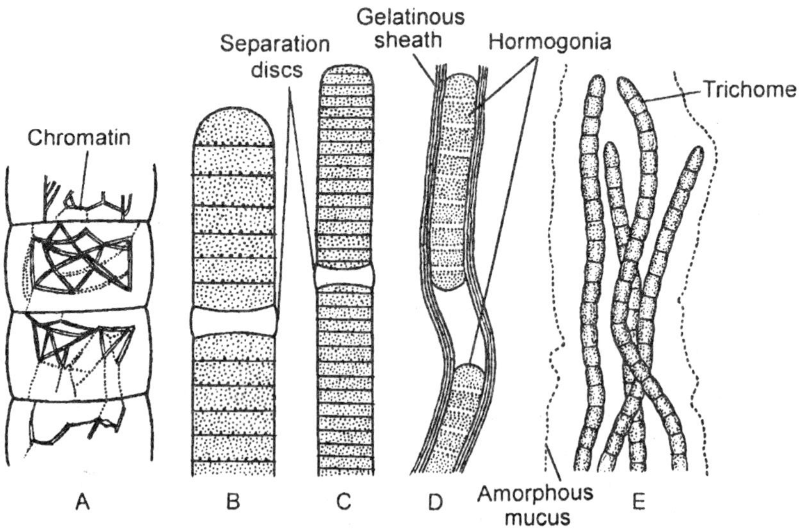

Fig. 3.4: Oscillatoriaceae. A. *Oscillatoria tenuis*. Cells containing chromatin. B-C. *Oscillatoria* sp. Dilimitation of hormogonia by separation discs. D. *Lyngbya majuscula*. Part of a filament showing prominent gelatinous sheath and dilimited hormogonia. E. *Microcoleus chthonoplastes*. Trichomes embedded in amorphous mucus

Some Indian Species: *Oscillatoria chilkensis* Biswas; *O. limosa* Ag. ex Gomont; *O. obscura* Bruhl et Biswas; *O. sancta* (Kutz) Gomont; *O. vizagapatensis* Rao, C. B.

Special features

1. Plant body exhibits oscillating, bending or gliding movement.
2. Filamentous unbranched plant body bearing extremely thin sheath.
3. Cells are cylindrical having length shorter or greater than the breadth.
4. Reproduction by accidental breaking and by separation disc.

Family: *NOSTOCACEAE*

The Nostocaceae resemble the Oscillatoriaceae in their diffuse growth and the unbranched structure of the trichomes, but are distinguished by the prevalent production of heterocysts and frequent reproduction by akinetes. The sheaths are well-hydrolyzed and do not function to restrict growth. The cells are uniseriate and generally ellipsoidal. Division may occur in al the vegetative cells of a trichome at the same time.

Hormothamnion is a common marine from of warm water seas. It has filaments resembling *Anabaena,* with subspherical vegetative cells and many heterocysts with discrete, cylindrical sheaths. It does not, however, produce spores. The sheaths are agglutinated to form superficial wefts on the substrate, and from these, *Symploca*-like erect turfs may later arise.

Genus: *Nostoc*

This is an alga of both terrestrial and habitats. Terrestrial species grow commonly either on bare soil or intermingled with plant parts. Sometimes aquatic species may be submerged lying on the bottom of the pools, or attached to a substratum, at other times floating freely, composed of a more or less firm jelly with denser limiting layer within which are innumerable contorted moniliform chains or strands of cells which resemble tiny beads in a necklace. Besides their association with fungi to form lichens, they behave as 'space-parasites' in the thalli of *Anthoceros.* There are certain subterranean species which can survive deep in the soil. *Nostoc* is ivory hardy and can be revived from dry herbarium sheet after prolonged drying.

The trichomes consist of a single series of uniform, often torulose ellipsoidal cells more or less depressed which are often contorted and sometimes from densely interwoven masses (Figs. 3.5A and C). The sheaths are thin and very mucilaginous, generally diffluent (Fig. 3.5B). Innumerable threads are aggregated together within mucilage, with a more or less firm boundary forming little bars, colonies of jelly—**Nostoc balls** (Fig. 3.5A). Here and there in a thread is a different kind of cell—heterocyst, distinguished by its thick walls, at times much larger in size tan its fellow having transparent contents with two polar nodules situated at two ends (Figs. 3.5B and C). The heterocysts may be terminal or intercalary. Their function is very obscure. But when intercalary they serve to break the filaments up into hormogonia. They may also occasionally function as spores.

Vegetative reproduction is effected by division of the cells, the threads thus formed remain surrounded by the sheath of the parent thread. Reproduction also takes place by hormogonia and spores. Spore formation does not take place until the colony is mature. Vegetative cells of the filament are metamorphosed into spores—akinetes (Fig. 3.5B). Akinetes are spherical or oblong much larger than the vegetative cells and are developed centrifugally in series between the heterocysts. Production of hormogonia and heterocysts generally takes place almost simultaneously throughout a colony. *Nostoc* shows a slight advancement in the formation of resistant spores from some of the ordinary vegetative

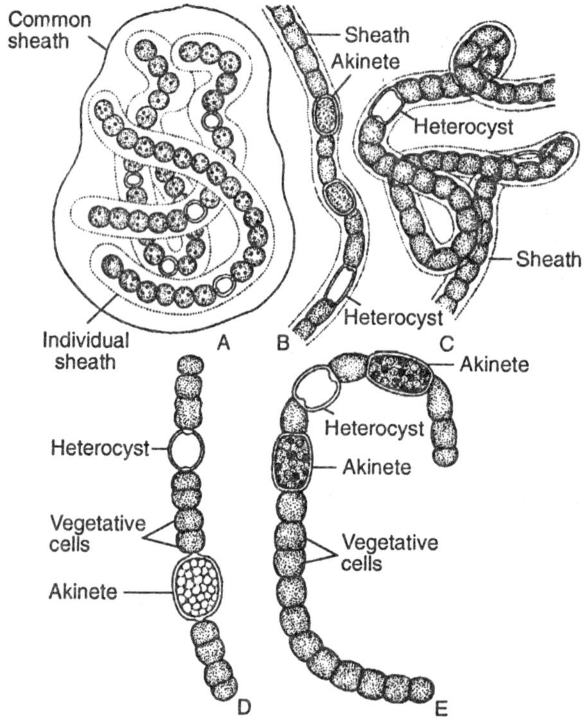

Fig. 3.5: Nostocaceae. A. *Nostoc* sp. A *Nostoc* ball of aggregated filaments. B. *Nostoc* sp. Portion of a filament showing sheath, akinete and heterocyst. C. *Nostoc* sp. A contorted filament with sheath and heterocyst. D. *Anabaena* sp. Portion of a trichome showing the nature of vegetative cells, heterocyst and akinete, E. *Anabaena* sp. A general view of the trichome consisting of vegetative cells, heterocyst and akinetes

cells. These spores germinate to form new filaments when the conditions become favourable.

Some Indian Species: *Nostoc commune* Vaucher ex Bron. et Flah.; *N. hatei* Dixit; *N. muscorum* Ag. ex Bron. et Flah; *N. paludosum* Kutz. ex Bron. et Flah.; *N. puntiforme* (Kutz.) Hariot.

Special features

1. Filamentous much contorted unbranched plant body.
2. Filaments sheathed and are held together by soft mucilage.
3. Cells rounded.
4. Presence of heterocysts in the filament.
5. Reproduction by hormogonia, akinetes and sometimes by heterocysts.

Genus: *Anabaena*

The closely related genus *Anabaena* differs from *Nostoc* in that no firm colony is formed. Trichomes of *Anabaena* either occur singly (Fig. 3.5D), as in many of the planktonic species, or form an ill-defined stratum; but never form colonies of definite form, several trichomes may then occur within a soft mucilaginous sheath. An individual trichome is never contorted and sheath is absent or more or less diffluent (Fig. 3.5E). The akinetes of *Anabaena* are generally much moe elongate than those of *Nostoc* (Fig. 3.5E). Some species are symbiotic. *Anabaena cycadearum* lives in the root tubercles of species of *Cycas,* and *Anabaena azollae* is found in hollows in the leaves of water fern (*Azolla*). Species of both *Nostoc* and *Anabaena* are capable of fixing atmospheric nitrogen.

Some Indian Species: *Anabaena ambigua* Rao, C. B.; *A. fertilissima* Rao, C. B.; *A. gelatinicola* Ghose; *A sphaerica* Bron. et Flah.; *A. spiroides* Kleb.

Family: *SCYTONEMATACEAE*

The Scytonemataceae are characterized by cylindrical, untapered, unbranched filaments of uniseriate cells, but with characteristic false branching of firmly sheathed filaments. Heterocysts occur here and there, tending by their shape to hold the trichome in the sheath and to favour apical growth. In the areas between heterocysts, however, cell division may occur which tends to cause pressures that burst the sheath. This allows one or both broken ends to protrude and to continue growth as a single branch or as twin branches.

Many of the Scytonemataceae suggest derivation from *Lyngbya*-like types by the acquisition of heterocysts and false branching. Reproduction may be by simple fragmentation of the trichomes or filaments, or by production of hormogonia.

The manner of branching is used in some generic separations. In *Scytonema* branching takes place well removed from heterocysts, so that twin branches are prevalent. In *Tolypothrix* the growth-break occurs at the heterocyst and results in a solitary branch.

Genus: *Scytonema*

Species of this genus usually prefer submaterial habitats. Some often occur on wet rocks as large felt-like masses even to the extent of an inch in thickness, others grow on damp soil, but very rarely submerged under water. The filaments interwoven into a felty mass (Fig. 3.6A). The individual filament with distinct basal and apical regions forming little turfs has a very wide hyaline or coloured sheath with prominently diverging strata of firm texture. The presence of heterocysts and the type of false branching are some of the characteristic features of this genus (Fig. 3.6C). False branching originates laterally as a result of the degeneration of an intercalary cell (Fig. 3.6D). The intercalary growth produces strong pressure to the sheath, which finally ruptures as a result of which the trichome forms a loop outside. Twin branches are produced out of this loop due to further growth (Figs. 3.6E-H). Degeneration of a heterocyst followed by growth of the two filaments on either side may also initiate production of false branching. Akinetes are of rare occurrence in a filament. The hormogonia usually constitute the only method of reproduction (Fig. 3.6B).

Some Indian Species: *Scytonema amplum* West et West.; *S. fritschii* Ghose; *S. hansgirgi*

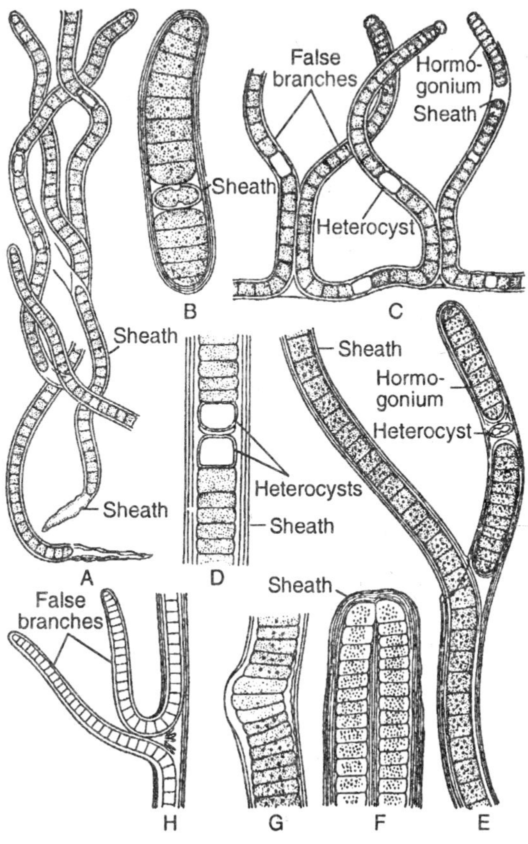

Fig. 3.6: *Scytonema* sp. A. Filaments as they occur in nature. B. A hormogonium. C. False branching, D. A degenerated intercalary cell in between the two heterocysts. E-H. Stages in the development of false branching

Schmidle; *S. iyengari* Bharadwaja; *S. pascheri* Bharadwaja.

Special features

1. Filamentous body with false branching.
2. Filaments bearing prominent sheath.
3. Presence of heterocysts in the filaments.
4. Reproduction by the development of hormogonia.

Family: *RIVULARIACEAE*

The Rivulariaceae are a well-circumscribed family including many marine forms that occur as gelatinous strata or hemispherical masses on rocks, plants, and marsh soil. The trichomes are unbranched and are oriented in their relation to the substrate to show a heterocyst terminating the basal end and a colourless hair terminating the distal. The cells are uniseriate, and transverse division occurs mainly in the region just above the heterocyst. The development of the marked difference between the base and apex of the trichome is secondary. Reproduction is commonly by hormogonia produced from portions of the trichome beneath the terminal hair. As the hormogonia germinate and grow, they become tapered at either extremity and ultimately break to form two or more trichomes, each with a basal heterocyst.

In the marine genus *Isactis* the filaments are erect and parallel, with their sheaths coalesced to form densely gelatinous colonies. *Calothrix*, which includes a number of species characteristic of the marine littoral, always has a single trichome in a sheath. *Gardnerula* has many trichomes packed inside a thin sheath.

Genus: *Gloeotrichia*

Gloeotrichia always occurs in colonies of spherical, hemispherical or irregular gelatinous masses being attached to plant parts or on wet rocks of cliffs or may grow directly on soil or free-floating in water (Fig. 3.7A and B). Each colony is exceedingly firm in consistency and contains numerous radiating filaments with repeated false branching, of which, each branch terminating in a multicellular colourless hair (Figs. 3.7D and E). The chief characteristic of this genus is the more or less gradual attenuation of the trichomes which terminate in a hair composed of elongate colourless cells (Fig. 3.7D). One or two heterocysts are located at the broad basal end of the trichome (Figs. 3.7C and D). Only the lower and broader portion of the filament is covered by a sheath (Figs. 3.7C, D and G). Growth of an individual trichome may be trichothallic and is initiated by a meristem of flat cells situated at the base of the hair-like structure (Fig. 3.7E). Many species exhibit false branching due to growth of the lower part of a trichome, mostly beneath an intercalary heterocyst. Hormo-gonia are formed from the upper part of the trichomes, the apical hair being shed. Reproduction is by hormogonia and akinetes. Number of akinetes may be one or more, and may be usually developed next to the heterocyst (Figs. 3.7G and H).

Some Indian Species: *Gloeotrichia ghosei* Singh, R. N.; *G. indica* Schmidle; *G. natans* Rabenhorst ex Bron. et Flah.; *G. pilgeri* Schmidle; *G. pisum* Thuret ex Bron. et Flah.

Special features

1. Filamentous unbranched vegetative body differentiated into base and apex.
2. Presence of a heterocyst at the broad basal end of the trichome.
3. The narrow end of the trichome is terminated by a hair of colourless cells.
4. The heterocyst is usually followed by an akinete.
5. The filament is partly covered by a sheath.

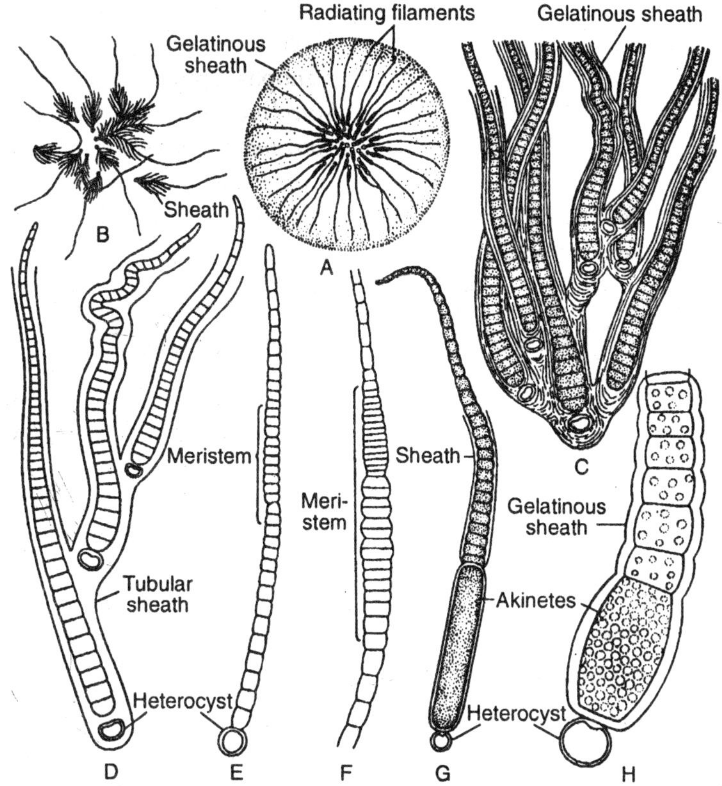

Fig. 3.7: Rivulariaceae. A-E. *Gloeotrichia prisum*. A. General view of colony. B. Portion of colony showing individual filaments. C. Closer view of the filaments. D. Group of filaments with attenuated ends. E. Trichome with meristematic zone. F. *Rivularia polyotis*. Trichome with meristematic zone. G. *Gloeotrichia pisum*. Filament showing attenuation with akinete and heterocyst. H. *Gloeotrichia natans*. Portion of filaments showing akinete and heterocyst

Closely related genus *Rivularia* differs from *Gloeotrichia* in its absence of formation of akinetes. In *Rivularia*, sheaths of the individual filaments are often diffluent as against *Gloeotrichia*, where sheaths only conspicuous near the bases of the trichomes, being gelatinous and confluent in the outer parts of the thallus.

Some Indian Species: *Rivularia beccriana* (De Not.) Born, et Flah.; *R. globiceps* West, G. S.; *R. hansgirgi* Schmidle; *R. manginii* Fremy.

CYANOPHYTAL SYMBIOSIS

The Cyanophycophyta (cyanobacteria) occur in basically two types of associations those in which the Cyanophycophyta (Cyanobacterium) is **extracellular** and those in which it is **intracellular.** The most common type of extracellular association is that with fungi to form lichens. In most lichens there is a single algal component (**phycobiont**) which is either cyanobacteria or green algae or xanthophycean algae. Cyanobacteria occur in about 8% of the species of lichens. Almost most of the fungal partners (**mycobionts**) are ascomycetes, but certain imperfect fungi and basidiomycetes also occur in lichens. The water ferm *Azolla* has cavities in the dorsal lobe of the leaf that are occupied by *Anabaena azollae,* a nitrogen fixing cyanobacteria. Thalloid bryophytes like *Anthoceros,* Cycads roots (Coralloid root) also contain *Anabaena.* Marine sponges generally contain symbiotic cyanophycophyta (cyanobacteria).

Intracellular associations involving cyanophycophyta (Cyanobacteria) are usually more specialized than extracellular associations, and it is not possible to culture cyanobacteria away from the host. The term **Cyanelle** for intracellular cyanobacteria in symbiosis, and the term **Cyanome** for the host cell. Such an association is noticed in some marine invertebrates.

ECOLOGY OF CYANOPHYCOPHYTA (CYANOPHYTA)

There is wide range of habitats for cyanophycophyta due to its unique adaptibility. Some of there specialized forms are discussed here:

- **Marine cyanophycophyta,**
- **Fresh-water cyanophycophyta,**
- **Hot spring cyanophycophyta** and
- **Terrestrial cyanophycophyta.**

Marine cyanophycophyta in the literal zone occur as a black encrustating film on rocks at the upper limit of the high tide mark. This zone is sandwiched between the lower barnacle zone and a higher zone of maritime lichens such as *Verrucaria* and *Lichina*. The dominant cyanophycophyta of this region are *Calothrix, Phormidium, Nodularia* and *Rivularia*. In the open sea, most of the picophytoplankton (2 μm diameter) is made up of coccoid cyanophycophyta. *Trichodesmium* is a major sea plankton which forms bloom in surface water, giving rise to the phenomenon known to sailors as "**sea sawdust**" or long orange or grey windows of algae.

Fresh-water cyanophycophyta is quite common. Many of them often furming bloom, viz. *Microcystis* bloom, *Anabaena* bloom, *Aphanizomenon* bloom or Oscillatoria bloom etc. Even arctic and antarctic region physhrophylic cyanophycophyta dominates. Cyanophycophyta are important in the colonization of non-acidic hot spring throughout the world. Many cyanophycophyta can also to grow upto 75-80°C temperature. The common hotspring cyanophycophyta are *Mastigocladus laminous, Synechococcus lividus, Oscillatoria terebrifornius, Aphanocapsa thermalis* and *Anabaena cylindrica.* In terrestrial environment like soil, rock surface and tree bark, Cyanophycophyta often colonized the commonest examples of species are *Microcoleus, Plectonema, Schizothrix* and *Scytonema.*

CYANOPHAGES

Cyanophages are viruses that infect selective Cyanophycophyta (Cyanobacteria) and thus kill them. In both fresh-water and estuarine water such cyanophages were frequently reported. They are very much host specific.

IMPORTANCE OF CYANOPHYCOPHYTA (BLUE-GREEN ALGAE)

Blue green algae are considered to be one of the major algal forms which are useful to mankinds in various ways. The major uses may be categories under the following heads:

- Utilization of blue-green algae as food.
- Utilization of blue-green algae in agriculture particularly with reference to nitrogen economy of soil.
- Utilization of blue-greens for conversion of solar energy through biophotolysis.
- Biogas and biomass production by the blue-green algae.
- Reclamation of saline and alkaline soil by the blue-green algae.

In addition, the abundant growth of blue-greens may be a source of considerable nuisance to man in some situations, as in the management of water reservoirs. Some blue-greens may also produce toxin that causes gastric disorders of farm animals and also dermal and other allergenic disorders of man.

Utilization of Blue-green Algae as Food

Blue-greens have recently been considered to be potentially excellent sources of food for animals and even for man. The introduction of blue-greens into the human diet would by no means constitute a genuine innovation. The Chinese consider *Nostoc commune* as a delicacy and *Aphanothece sacrum, Nostoc rucosum, N. commune* and *Brachytrichia* sp. have been used as side dishes since ancient time in Japan. In the northern parts of the African republic of Chad there are stretches of shallow alkaline lakes with high salt content which produce abundant growth of the planktonic *Spirulina platensis.* The planktonic material has been collected for food by the local people and dried in sun. The fried algal cake, called 'die' is being sold at the market throughout the country. It is used for the preparation of soup and sauce. It is likely that one of the basic food stuffs of the Aztees was of a similar nature, which they collected from shallow Mexican lakes. More recently, for the production of nutritions 'health food' *Nostoc* is utilized as food in Peru and in the Far East. *Microcystis* has also been grown on domestic wastes and used to feed fish in lagoons and fish ponds in India and Pakistan.

Chemical analysis of *Spiulina* showed that it contains 55-60% proteinaceous material, 16-18% carbohydrate, 2-3%, lipid as well as mineral salts. It also contains a number of essential amino acids, but the contents of sulphur-containing amino acids (cysteine and methionine) is quite low. Feeding experiments with animals indicate that the *Spirulina* diet is acceptable and well digested by farm animals.

Utilization of Blue-green Algae in Agriculture

Due to enormous rise of population on earth surface, the demand for food in the world is really unprecedented and thus malnutrition and starvation are already desperate problems. In view of the World's food crisis, greater attention has been turned to traditional farming practices. But the prospects of increased food-grain production are aggregated by the ever-increasing cost of fuel and chemical fertilizers.

In this context, the recent inventions which enable the more extensive and efficient utilization of biological nitrogen fertilizers in the field agriculture have greater significance. During past few decades, blue-greens are considered to be potent biological nitrogen fixer and thus recognized as 'nitrogenous biofertilizers'.

Many tropical paddy fields receive no chemical fertilizer nor natural manure, yet they remain productive and capable of supporting large populations with basic food. Here the fertility of paddy soil is maintained by the activities of blue greens which grow spontaneously and often luxuriantly in the water logged field. They provide fixed nitrogen to rice plants through both secretion of nitrogenous substances and on their decay and subsequent mineralization of organic substances in the soil.

Since De (1939) attributed the inherent fertility of the tropical rice field soils to the activity of nitrogen-fixing blue-green algae, considerable interest has been generated in exploiting the potential of these biological systems. In the south and south-east Asian regions, only about 4% of the 851 localities examined were found to harbour N_2-fixing blue-greens (Watanabe, 1959). The dominant forms in Japanese rice soils largely belong to the genera *Tolypothrix, Anabanea* and *Nostoc* (Okuda and Yamaguchi, 1952, 1956). In India, Singh was the first to observe a characteristic succession of blue-greens in the rice fields of Uttar Pradesh. However, it is interesting that only about 33% of the 2213 rice-field soil samples examined in India were found to harbour useful N_2-fixing blue-greens and their relative abundance showed a wide variation, with localization of specific forms in certain regions

(Venkataraman, 1975). Although blue-greens normally prefer warmer habitats they have been shown to play a major role as nitrogen fixers in temperate, polar and subpolar regions. The distribution and role of nitrogen-fixing blue-green in temperate regions have been investigated in Russia (Gollerbakh and Shtina, 1969; Shtina, 1969), northern and western Europe (Granhall, 1975; Henriksson et al., 1972) and Northern America (Jurgensen and Davey 1968; Shields and Dumell, 1964).

There is now positive evidence for the nitrogen-fixing ability of more than 100 species of blue-greens. A nitrogen accretion of 14 to 40 Kg/ha has been attributed to the activities of the N_2-fixing blue-greens in different localities (De and Mandal, 1958; Singh, 1961; Watanabe, 1962; Henriksoon, 1971; Sankaram, 1971; Yoshida and Ancaja, 1973; Alexander, 1975).

In India alone, the distribution of rice field blue-green algae in various states were carried for past two decades by various workers as stated in Table 3.1.

Table 3.1: Bibliography of Rice-Fields Blue-Green Algal Survey in India

States	References
Andhra Pradesh, Andaman and Nicobar, Bihar	Laloraya and Mitra, 1973; Prasad et al, 1987; Laloraya and Mitra, 1973; Jha et al, 1986.
Goa	Bongale, 1981.
Gujarat	Mahajan and Patel, 1983.
Himachal Pradesh	Kamat, 1970.
Jammu and Kashmir	Goyal et al, 1984; Khan, 1957.
Karnataka	Tiwari, 1972, 1975; Tiwari and Pandey, 1976; Kamat, 1979, Bongale and Bharati, 1980; Bongale, 1981, 1984; Shivaprakash and Shivappa Shetty, 1983; Bongale, 1987.
Kerala	Parukutty, 1940; Aier, 1965; Verma, 1965; Anand and Shanthakumar Hopper, 1987.
Madhya Pradesh	Tiwari, 1972, 1975; Mishra and Purohit, 1979.
Maharashtra	Gonzalves and Gangla, 1949; Marathe, 1963; Marathe and Navalkar, 1963; Sardeshpande and Foyal, 1981; Tarar and Shewale, 1984; Kolte and Goya 1985; Patil and Satav, 1986.
Meghalaya and Other North-Eastern States	Reddy et al. 1986.
Orissa	Rao, 1938; Laloraya and Mitra, 1973.
Panjab	Sarma and Swarnakanta, 1978; Grover and Pandhol, 1975; Sarma et al; 1979.
Rajasthan	Gupta and Kumar, 1968; Gupta, 1972.
Sikkim	Santra, 1987.
Tamil Nadu	Khurana and Venkataraman, 1968; Tiwari, 1972, 1975; Laloraya and Mitra, 1973; Tiwari and Pandey, 1976; Anand and Revathy, 1987, Anand, 1989.
Uttar Pradesh	Mitra, 1951; Gupta, 1957; Pandey 1965; Gupta, 1966; Single and Chaturvedi, 1970; Prasad and Mehrotra, 1980, Pal and Yadav, 1974; Khan and Mathur, 1976; Prasad and Srivastava, 1968.
West Bengal	Biswas, 1926; De, 1939; Laloraya and Mitra, 1973; Sinha and Mukherjee, 1975, 1984; Saha and Mandal, 1980; Mukhopadhyay and Chatterjee, 1981; Pal and Santra, 1982; Chatterjee and Chatterjee, 1983; Sinha and Mukherjee, 1984; Maity and Santra, 1985; Pal and Santra, 1985, Santra, 1991.

Recent advances in methodology have helped a great deal in our understanding of blue-green algal N_2-fixation. Nitrogenase is oxygen-sensitive and blue-green algae have evolved special devices such as heterocysts for the protection of their nitrogenone against oxygen damage (Fay et al., 1968; Stewart et al., 1969). On the whole, the basic mode of nitrogenase action appears to be similar with that of other N_2-fixing organisms. The mapping of nif* gene was also made in several N_2-fixing blue-green algae.

The role of N_2-fixing blue-green algae in the maintenance of the fertility of rice fields has been well substantiated and documented all over the world. In India alone, the beneficial effect of blue green algae on yield of many rice varieties has been demonstrated in a number of localities (Singh, 1961; Sankaram, 1971; Subrahmayam, 1972; Venkataraman, 1972, 1975, 1979, 1980, 1981; Saha and Mandal, 1981; Santra, 1991). In many areas, 10 to 20% increase in the grain yield has been observed as a result of algal application in the absence of any added chemical N-fertilizer (Table 3.2).

Beneficial effects of algal inoculation have also been reported on a number of other crops such as barley (Fuller et al., 1960), oats (Shtina, 1957), tomato (Kaushik and Venkataraman, 1979; Roger et al., 1979), radish (Roger et al., 1979), cotton (Kuchkarova and Maksudov, 1964), sugarcane and maize (Singh, 1961), Chilli and lettuce (Dadhich et al, 1969).

Considering these above stated facts, blue-green algal biofertilizer application in rice field has earned immense potentialities. Unlike *Rhizobium* and *Azospirillum* bacteria which are mainly produced in sterile conditions, bl ue-green algae, such as *Nostoc, Anabaena* and *Aulosira,* are grown in open air troughs, pits and tanks on the farm to raise a biomass. Upon sun drying, the algal flakes are collected and broadcast on the rice fields, one week after transplantation of rice seedlings at a rate of 10-15 kg/ha. The cost-benefit ratio of algal production in farmer's field indicates a net income of Rs. 2000 from half an acre of land (Venkataraman, 1980).

The beneficial effect of N_2-fixation for crop production are even more apparent in paddy fields which are populated by the water fern *Azolla* during the water-logged period. The importance of *Azolla* as an organic fertilizer in rice cultivation was first demonstrated in North Vietnam in 1957. Afterwards its potentiality was recognized in the USA, Indonesia, Japan, Philippines, China, India. *Azolla* is a tiny floating fresh-water fern and is ubiquitous in distribution. Each leaf has a dorsal and a ventral lobe. The dorsal lobe is green and has a blue green algal symbiont (*Anabaena azollae*) within a central cavity. The alga fixes atmospheric nitrogen, hence the importance of *Azolla* in rice cultivation. Estimates of the annual quantities of N_2-fixed by Azolla range between 120 and 312 kg nitrogen per hectare. The plant supplies substantial amounts of green manure between 150-300 ton per hectare per year), which supports the growth of soil microorganisms including heterotrophic N_2-fixers.

Table 3.2: Response of BGA-Application in Rice Cultivation (Santra, 1991)

Variety	Trial season	N + P + K	Gross yield (kg/ha) BGA + P + K	P + K	BGA + P + K	Additional yield over schedule doses of NPK (kg/ha)
High yield varieties	Aman	3620	3329	3015	3925	305 (10.5%)
(HYV) RATNA	Boro	3919	3050	2720	4520	601 (15.3%)

Solar Energy Conversion through Biophotolysis

Heterocystous blue-greens possess the unique ability to simultaneously evolve O_2 in photosynthesis (in vegetative cells) and H_2 by nitrogenase catalyzed electron transfer to H^+-ions (in heterocysts) in the absence of N_2 or other substrates of nitrogenase. This is the basis for the attempt of several workers to exploit this potential through the development of a 'biophotolytic system' for solar energy conversion. Following earlier laboratory studies successful outdoor systems were established in which *Anabaena cylindrica* was used to produce H_2 and O_2 in light when sponged continuously with a gas mixture containing argon and CO_2.

Among blue-green algae *Anabaena flos-aquae, A. cylindrica, Mastigocladus laminosus,* species of *Lyngbya, Plectonema, Nostoc, Tolypothrix, Scytonema, Gloeothece* etc. are known to photoproduce hydrogen.

Hydrogen has many advantages as a fuel energy source for future. It is the most abundant element in the universe and has maximum energy per unit weight (122 KJ/g). It has a grater energy conversion efficiency over petroleum. Hydrogen, when used as a fuel does not pollute the environment because its only conversion by-product is water. Besides as fuel, hydrogen is useful in industrial applications like iron one reduction and ammonia (fertilizer) production.

Biogas and Biomass Production by Blue-Green Algae

It is possible to make sewage productive by optimising the conditions of waste treatment by algal forms, particularly blue-green algae. The system developed long by Oswald et al. (1957) using a stabilization pond for waste water treatment coupled with algal production. This system is efficient, as it utilizes atmospheric carbon dioxide and sun light to obtain an yield of algal biomass. The oxygen released during photosynthesis favours the aerobic bacteria which degrade the organic material. In India, the work of BGA production from sewage was limited. At National Biological Research Institute, Lucknow, India extensive attempt has been made for large scale production of *Spirulina* from sewage with or without bicarbonate supplement. Then the harvested biomass either used for biogas production or as food substrate particularly for livestocks.

Reclamation of Saline and Alkaline Soil

Blue-green algae is one of the most well adopted algal forms in saline habitats. It has tremendous capacity to transform saline soil into productive neutral soil with greater amount of available nitrogen and carbon. Incubation of saline or alkaline soils with native blue green algae *in vitro* or *in vivo* resulted an improvement of physico-chemical properties of soils (Singh 1961, Subhashini and Kaushik, 1981, 1982, 1984). Blue-green algae are known to secrete substances like polysaccharides and lipids which have the capacity to transform the structure of saline and alkaline soil (Lewin, 1972; Hellebust, 1974). Again majority of the blue-green algae are endowed with capacity to fix atmospheric gaseous nitrogen and this fixed nitrogen is released during the growth and decay of algae along with effects of blue-green algae in improving the physical, chemical and nutritional properties of saline and alkaline soils, suggest the possibility of exploiting this biological system to ameliorate the salt affected soils.

It was revealed from the past study that the species of *Oscillatoria, Lyngbya, Phormidium, Microcoleus, Hydrocoleus* and *Calothrix* were most dominant in saline and alkaline habitats (Singh, 1950, Marathe 1965, Prasad and Srivastava, 1968; Santra *et al,* 1988, Kaushik, 1989). An example of soil quality improvement upto BGA application is depicted in the Table 3.3.

Table 3.3: Changes in the Physico-Chemical properties of Soil after Algalization and/or Gypsum Treatment for 7 days *in vitro* Experiment (Santra et al, 1987)

Treatment	Soil properties			
	pH	Organic carbon (%)	Exchangeable (mg/gms soil)	
			Ca^{++}	Mg^{++}
Soil (R)	6.3	0.17	0.20	0.20
Soil (R) + Algae (100 : 1)	6.5	0.77	0.14	0.20
Soil (R) + Gypsum (100 : 0.5)	6.4	0.31	0.20	0.20
Soil + Algae + Gypsum (100 : 1 : 0.05)	7.0	0.43	0.14	0.19
Soil (S)	6.5	1.27	0.21	0.16
Soil (S) + Algae (100 : 1)	7.0	3.33	0.15	0.16
Soil (S) + Gypsum (100 : 0.05)	6.8	2.67	0.19	0.16
Soil (S) + Algae + Gypsum (100 : 1 : 0.05)	7.2	3.24	0.15	0.15

R = River bank soil; S = Soil of cultivated land

The dominant blue-green algal forms in salt affected soils of different states of India, were presented separately in the Table 3.4. It was also further interesting to note the fact that limestone and phosphate rocks were biologically weathered by blue-green algae *in vitro* and *in vivo* studies. In this connection, it was reported earlier that some blue-green forms are capable of secreting acidic substances which in turn was able to dissolve the rock components. That is why they are also called as pioneer of soil formation on rock surface.

The ultimate success of bioameliorant cyanobacteria is reflected in the successful establishment of a crop with moderately high yields. The various package of practices recommended for amelioration of such soils have advocated the use of paddy as the first crop. The reclamation trials conducted for a period of 3 years in sodic soils of Delhi during summer season are given in Table 3.5.

In contrary to the above stated utility of blue-green forms there are a number of nuisance activities of blue-greens viz., nuisance in

Table 3.4: Dominant Blue-green Algal forms in Salt affected areas of Different States of India

States	Soil type	Dominant Algal genera
Andhra Pradesh	Sodic	*Nostoc, Plectonema*
Delhi	Sodic	*Anabaena, Oscillatoria, Plectonema*
Haryana	Sodic	*Anabaena, Calothrix, Nostoc, Plectonema, Westiellopsis*
Maharashtra	Saline	*Nostoc*
Punjab	Saline	*Calothrix, Hapalosiphon, Plectonema*
Rajasthan	Saline	*Nostoc, Scytonema, Plectonema*
Tamil Nadu	Saline-Sodic	*Nostoc, Tolypothrix, Plectonema*
Uttar Pradesh	Saline	*Calothrix, Plectonema, Hapalosiphon*
West Bengal	Saline	*Nostoc, Oscillatoria, Lyngbya, Hydrocoleus*

Table 3.5: Amelioration of Sodic soils and the Yield (Q/ha) of Paddy variety Pusa-33 due to algalization for three consecutive seasons (Kausik, 1989)

Treatment	1985		1986		1987	
	Paddy	Straw	Paddy	Straw	Paddy	Straw
Control	0	0	0	0	6.15	5.15
BGA	1.46	3.43	8.50	22.66	41.00	30.60
Gypsum	5.41	10.08	24.83	43.50	42.00	30.40
BGA + Gypsum	7.96	13.00	32.33	49.50	45.40	33.20

water supplies, water toxicity and as allergy causative agents. Extensive growth of algae can create severe problems in the maintenance of water supplies and in meeting the ever increasing demand for portable water. Numerous episodes of poisoning of farm animals (e.g. cattle, horse, pig) and of waterfowl and wildlife were described throughout the world and attributed to the ingestion by drinking animals of lethal doses of toxic blue-greens accumulating at the surface of water or along the shores of temperate lakes, ponds and reservoirs. Several species have been indicated but toxic strains of three common bloom-forming species—*Microcystis aeruginosa, Aphanizomenon flos-aquae* and *Anabaena flos-aquae,* appear to be responsible for most incidents of serious consequences. Several fresh-water and marine blue-greens were also implicated in outbreaks of human gastro-enteritis and in various allergic responses like dermal itching. The frequently occurring 'swimmers itch' is attributed to contact with *Lyngbya majuscula, Schizothrix calcicola* and *Oscillatoria nigroviridis,* which are commonly found in tropical and subtropical seawaters. The toxin responsible are lipid-soluble phenolic compounds.

Chapter 4

The Chlorophycophyta (Chlorophyta)

GENERAL FEATURES

This is the largest of the eight divisions of algae. Members of the Chlorophycophyta, or grass-green algae are similar to higher plants being characterized by a well-defined nucleus, photosynthetic pigments localized in chloroplastids in which usually pyrenoids are present, the food reserve is commonly stored as starch, and the possession of cell walls in which cellulose is usually a clearly recognizable ingredient. Most of the grass-green algae are of fresh-water forms with a few exception of marine species. They exhibit a considerable range of variation in the form and structure of the plant body. In short, the diversity of habit and habitat is very striking and in this respect the Chlorophyta surpass any other algal taxon. Motile cells both zoospores and gametes may be bi-, quadri-, or multiflagellate with the exception of forms where motility is lacking altogether. Gametic reproduction is common within the entire taxon. Moreover, clear division of labour indicated by some heterotrichous forms, and special adaptations of the terrestrial species to survive against external unfavourable conditions have been quite intriguing to some phycologists to trace the algal ancestry of land plants.

DISTRIBUTION

The green algae enjoy a wide range of distribution in aquatic (both fresh-water and marine) and terrestrial habitats. Some of the fresh-water plankton forms form greenish scum on the surface of quiet, stagnant water (*Volvox, Chlamydomonas, Spirogyra, Zygnema, Cosmarium)* or grow firmly attached to submerged rocks, plant parts, and similar objects in free-flowing streams (*Oedogonium, Cladophora, Chaetophora, Coleochaete*). Some of the marine green algae are: *Bryopsis, Caulerpa, Acetabularia, Codium*; and some of the common ones which prefer moist terrestrial habitats, such as damp soil and shaded sides of rocks and bark of trees are: *Trentepohlia, Pleurococcus.* Several are designated as snow or ice algae (*Scotiella* spp.) because of their habitats, whereas others are able to live in concentrations of salts many times higher than occurring in the oceans. Some of the green algae have lost their green pigments turning out to be parasites upon other plants causing serious diseases, for example, of tea and pepper plants (*Cephaleuros*), of *Sphagnum* (*Phyllobium sphagnicolum*), or *Ambrosia* (*Rhodochytrium spilanthidis*). Others merely grow upon other plants as **epiphytes.** Some are **epizooic forms** growing on small crustacians (*Chlorogonium* spp.). Again, some are **space parasites,** of which most common is *Chlorogonium* spp.). Again, some are **space parasites,** of which most common is *Chlorochytrium lemnae* growing inside the tissue of *Lemma.* Still others form algal component of most lichens (*Pleurococcus, Trentepohlia, Cladophora*). Again, *Zoochlorella* and *Carteria* live in association with *Hydra* and *Convoluta.*

CELL STRUCTURE

Shape and structure: The shape of the cell of green algae is exceedingly varied. Among the unicellular and simpler colonial forms, the cells may be spherical, oval, or pear-shaped with the exception of some specialized unicellular forms where the shape of an individual cell may be crescent-shaped or lobed. In the filamentous forms the cells are almost invariably cylindrical or barrel-shaped. Cells, irrespective of their shape and size, are eucaryotic in construction.

Cell wall: In great majority of cases protoplast of algal cell lies within a definite wall, which is a secretion product of the protoplast. It is a thin delicate structure, but in the larger forms it tends to be more robust being made of stratified layers. Usually the cell wall is composed of two layers, an inner layer immediately next to the protoplast consisting mainly of cellulose and an outer layer consisting of pectic substances. There is a continuous dissolution of the outer pectic layer as this layer consists of water-soluble substances. Again, with a continuous addition of fresh pectic material from the inside the constant thickness of the pectic layer is maintained. In some species there is a thin outer layer which is often sharply defined and more resistant to many reagents and is generally known as 'cuticle', though it is not identical with that of higher plants. In others, the cell wall may be incrusted with calcium carbonate. The mucilage envelopes are found in many forms and in part modifications of the cell wall, but in a large number of cases, represent secretions of the protoplast emitted through pores of the inner layer. This mucilage envelope at times helps movement of the individual plant.

Vacuole, pyrenoids and chloroplast: Depending on species and habitat, the protoplast may possess a conspicuous vacuole of extremely variable size. The vacuole when present is traversed by cytoplasmic strands. The most conspicuous part of the protoplast is the chloroplast which varies in number and shape from genus to genus. Usually the chloroplasts of green algae have a shape characteristic for the particular genus, elaboration of which is often related to increase in size of the cells. In some members, the chloroplasts exhibit movement in response to varying conditions of illumination. Associated with the chloroplast in the vast majority of the green algae are bodies known as **pyrenoids** which consist of a central core of colourless mass of proteinaceous material ensheathed by plates of starch. The pyrenoids are usually regarded as reserve protein, but they show a curious degree of permanence which, at times, is not in conformity with this view, for example, pyrenoids may or may not be present in closely related species and they disappear more or less completely during the formation of reproductive units. The shape of the chloroplast may be extremely varied. The chloroplast may be cup-shaped which is rather widespread and prevalent in unicellular forms, may indicate a sign of primitiveness. In some green algae the chloroplast may be stellate. In others it may be spiral, or girdle-shaped, or reticulate. Discoid chloroplasts of varying number are seen to occur in lining layer of cytoplasm.

Pigments and storage food: The pigments of green algae are lodged in the chloroplasts. The pigments are chlorophyll a and chlorophyll b in the same proportions as the higher plants and just about the proportion of α-, β-, γ-carotene, astaxanthin, lutein, neoxanthin, siphonein, siphonoxanthin, violaxanthin, zeaxanthin and lycopene. The amount of pigments and proportion of constituent parts present in an algal cell are extremely variable. A special pigment known as **haematochrome** may be present masking the chlorophyll in a terrestrial form whose function has not yet been clearly established. Certain Chlorophycophyta, however, may completely lack photosynthetic pigments. In great majority of the green algae the storage food accumulating from photosynthetic activity is usually starch and also oil in some green algae.

Nucleus: In the majority of cases, algal cells are uninucleate, but multinucleate condition of cells is not very infrequent and there are quite a few coenocytic forms too. The nucleus does not appear to differ in any appreciable respect from that of higher plants. Each nucleus possesses a definite nuclear membrane, one or more deeply staining nucleoli and a very poorly developed chromatin network. Usually the chromosomes are not very numerous. Nuclear division is mitotic, and the several stages of mitosis are quite similar to those of higher plants. Meiosis has also been reported in a number of algae resembling like that of higher plants.

Flagella: Both flagellate and non-flagellate cells are encountered in the green algae. Number of flagella in a flagellate cell may be two or four or more and flagella are equal in length. Flagella are whiplash type.

VARIATION RANGE OF PLANT BODY

Members of the green algae are well known for the extraordinary variety of forms and shapes and organization of plant body. They are, on the whole, simple algae and many of the unicellular forms belong here. But simple as they are in structure, some of them in their mode of development and in external complex form approach nearer to the higher plants than any other algae. Major forms of green algae are outlined below.

Unicellular motile forms: The simplest structure is seen under this category where unicellular spherical, oval or pear-shaped motile plant body may be isolated, individual or colonial form being not a mere assemblage of flagellated cells, but is a coordinated whole behaving as a well-organized unit with a definite number of cells arranged in a specific manner known as a **coenobium.**

Unicellular non-motile palmellate forms: A loose assemblage of cells which lie embedded in a common gelatinous matrix secreted by the several cells forming a colony. Colonies are generally amorphous, may often have definite shapes.

Unicellular non-motile coccoid forms: The plant body is small isolated unicellular non-motile spherical cell having sedentary habit.

Filamentous forms: The filamentous condition is an important advancement over the unicellular condition. In a large number of higher forms, the plant body is filamentous which may be unbranched or branched.

Parenchymatous thalloid forms: Plant body is parenchymatous thallus of simple to complex in form produced as a result of division of cells in more than one plane.

Heterotrichous forms: This is the most advanced type. Plant body is composed of two portions, a prostrate portion consisting of creeping threads attached to the substratum—the **prostrate system,** and a more or less richly branched erect portion growing from the prostrate portion—the **erect system.** In some forms both the systems are fairly well developed, whereas in others, one of these systems is developed better, the other being poorly developed, or very much reduced, or even completely suppressed.

Siphonaceous forms: In a number of green algae considerable enlargement of the plant body may take place without the occurrence of any septation producing a large multinucleate structure known as a **coenocyte.**

REPRODUCTION

The methods of reproduction encountered in the green algae show great diversity which may be considered under three headings—**vegetative, asexual** and **sexual.**

Vegetative reproduction: It is one of those processes in which portions of the plant body become separated off to give rise to new individuals without any obvious changes in the protoplasts. Common examples are afforded by the fragmentation into two or more pieces or through an accidental or natural separation of its parts. Each portion may then grow into an independent individual. Besides this, the process of multiplication by ordinary cell division is also characteristic of some unicellular algae.

Asexual reproduction: Most common method of asexual reproduction is by the formation of **zoospores.** The zoospores are flagellate spores that are formed singly or in numbers either in a vegetative cell or in a specialized part of the plant body known as a **sporangium.** The zoospores vary in size, shape, and nature of flagellation. The formation of zoospores is a complicated process and may take place in a variety of ways. Each zoospore, under favourable conditions, develops into a new plant. In some species non-motile spores are formed which are called **aplanospores,** but if these should, then secrete a thick wall they are known as **hypnospores.** All these spores produce new plants.

Sexual reproduction: Sexual reproduction is well established among the most primitive forms and persists in the highest members of the green algae with a few exception of autosporic forms. Gametic union ranges from isogamy to anisogamy, and even oogamy with a relatively advanced stage of evolution having a tendency of retention of the products of gametic fusion in the mother plant and to develop a mechanism to afford protection against unfavourable conditions. Depending on species, union of gametes both of which may be non-flagellate (**aplanogamy**) or flagellate (**planogamy**). Normally the two gametic nuclei fuse immediately after fusion of gametes, but in many cases they do not fuse until just before germination of the zygote. Germination of the zygote may take place soon after its formation, but in the great majority delayed as the time interval is necessary for the ripening of the zygote. Meiosis generally takes place during the germination of the zygote and as such practically in all fresh-water algae the diploid phase in the life cycle is restricted to the zygospore or oospore which is comparably with the sporophyte of the higher plants.

CLASSIFICATION

The Chlorophycophyta are subdivided into two classes. Chlorophyceae, and Charophyceae primarily on the basis of difference in their vegetative structure, reproductive organs and method of sexual reproduction. In these two classes of algae, nuclear and cell division are especially distinctive at telophase. In the Charophyceae, the telophasic nuclei are relatively far apart, and the microtubular spindle between them is persistent until the completion of telophase. By contrast, in the Chlorophyceae, the telophasic nuclei are in close proximity, and the microtubules of the spindle disappear before the completion of telophase.

Class: *CHLOROPHYCEAE*

The Chlorophyceae are characterized by algae having wide range of vegetative structure, greater than that of any other class. Despite the variability, they have main features: the photosynthetic pigments include chlorophylls a and b; starch accumulates as a reserve product; the cell wall invariably contains cellulose; flagella, when present, are 2 or 4 or more in number, and all are whiplash type; and methods of reproduction are diverse.

The Chlorophyceae are divided into the following orders based largely on the form of vegetative body, nature of chloroplast, and mode of reproduction.

I. **Volvocales:** Unicellular to colonial forms with definite number of cells in a colony; vegetative cells are usually motile throughout life; cup-shaped chloroplast; reproduction both asexual and sexual; sexual reproduction—isogamous, anisogamous and simple to advanced oogamous.

Family: **Chlamydomonadaceae:** Unicellular motile vegetative cells; asexual reproduction by zoospores and nonflagellae hypnospores; sexual reproduction ranging from isogamy to anisogamy and simplest type of oogamy.

Genus: *Chamydomonas.*

Family: **Sphaerellaceae:** Unicellular biflagellate plant body with strongly thickened and gelatinous cell wall; basin-shaped to more or less reticulate chloroplast; sexual reproduction isogamous.

Genus: *Sphaerella*

Family: **Volvocaceae:** Definite number of unicellular vegetative cells are either held together by cytoplasmic connection or embedded in a gelatinous sheath forming a coenobium; asexual reproduction either by the division of the ordinary vegetative cells or by the development of special vegetative cells; sexual reproduction isogamy, anisogamy to advanced oogamy by the development of antherozoids and egg.

Genera: *Gonium, Pandorina, Eudorina, Volvox*

II. **Tetrasporales:** Algae resembling *Chlamydomonas* which have their vegetative phase consisting an unspecialized aggregation of cells in a gelatinous matrix are included in this order. The vegetative cells sometimes acquire flagella and become motile. Asexual reproduction is by division of the contents of cells into 4 zoospores. Sexual reproduction occurs by isogamy, gametes are produced in the same way as zoospores.

Family: **Tetrasporaceae**

Genus: *Tetraspora*

III. **Chlorococcales:** Unicellular to colonial non-motile forms, the colonial forms having definite pattern during their development; parietal chloroplast; asexual reproduction by motile or non-motile spores; sexual reproduction isogamous.

Family: **Chlorellaceae:** Vegetative body unicellular non-motile; reproduction by non-motile spores.

Genus: *Chlorella*

Family: **Coelastraceae:** Cells of the coenobia grouped in one plane with the long axes of the cells parallel to one another; laminate chloroplast; reproduction by division of cells of the coenobium.

Genus: *Scenedesmus*

Family: **Hydrodictyaceae:** Occurring in colonies forming water-net; isogamous sexual reproduction.

Genera: *Pediastrum, Hydrodictyon*

IV. **Ulotrichales:** Forms filamentous unbranched or cellular expanse, cells having parietal or girdle-shaped or axile stellate chloroplasts; mostly isogamous sexual reproduction.

Family: **Ulotrichaceae:** Plant body consisting of a simple unbranched filament being differentiated into base and apex; asexual reproduction usually by the formation of zoospores and very rarely by aplanospores; sexual reproduction isogamous.

Genus: *Ulotrhix*

V. **Sphaeropleales:** Members of this order differ from the Ulotrichales having cells much longer than broad and containing several transverse cytoplasmic septa separated from each other by large vacuoles. Several nuclei and either a band-shaped chloroplast or a number of discoid chloropasts are present in each cytoplasmic septum. Sexual reproduction is advanced anisogamy to oogamy. Number of eggs is variable.

Family: **Sphaeropleaceae**

Genus: *Sphaeroplea*

VI. **Prasiolales:** The parenchymatous thallus of this group of algae is quite different from that of the order Ulvales. It is a small flattened blade, much simpler than the parenchymatous thallus of the order Ulvales and exhibits an intermediate form between the filamentous form of the Ulotrichales and the complex thallus of the Ulvales. The thallus is composed of flat sheet of cells which are small, polygonal and commonly arranged in groups of four. Each cell has a large axile stellate chloroplast with a central pyrenoid. Asexual reproduction is by the akinetes formed from marginal cells. The young vegetative thallus develops either into a spore-forming plant or a gamete forming plant exhibiting transitional condition of alternation of generations.

Family: **Prasiolaceae**

Genus: *Prasiola*

VII. **Ulvales:** Members of the order Ulvales are characterized by a parenchymatous form of thalloid vegetative body which is developed by cell division in more than one plane. Some authorities do not separate the Ulvales from the Ulotrichales since in the young stage the vegetative body is filamentous with uniseriate cells resembling *Ulothrix*. Vegetative propagation is by fragmentation of thallus. All genera produce quadriflagellate zoospores and biflagellate gametes. Sexual reproduction is isogamous. Isomorphic alternation of generations between a diploid sporophytic stage and a haploid gametophytic stage is characteristic feature of this order with meiosis during zoospore formation.

Family: **Ulvaceae:** Parenchymatous thallus body produced as a result of division of cells in more than one plane; asexual reproduction by zoospores; sexual reproduction usually isogamous; life cycle exhibiting isomorphic alternation of generations having resemblance with higher plants.

Genus: *Ulva*

VIII. **Cladophorales:** Filamentous branched or unbranched with multinucleate cells; reticulate or discoid chloroplast; reproduction usually isogamous.

Family: **Cladophoraceae:** Filamentous branched or unbranched plant body with multinucleate cells possessing stratified thick walls; mostly perennial; asexual reproduction by zoospores; mostly isogamous sexual reproduction with anisogamy in a few genera, sexually reproducing genera exhibiting isomorphic alternation of generations.

Genus: *Cladophora*

IX. **Acrosiponiales:** Members of this order are characterized by: (i) cell wall composed of cellulose II (mercerized cellulose), (ii) heteromorphic alternation of generations. These algae were previously grouped under the order Cladophorales.

Family: **Acrosiphoniaceae**

Genus: *Urospora*

X. **Dasycladales:** Algae of this group have juveline stage of an undifferentiated rhizoidal system with a small erect axis showing slight development of whorled structures at the apex. This vegetative stage is usually uninucleate. The giant nucleus of the uninucleate thallus remains situated in a basal rhizoid. When the thallus becomes reproductively mature this large nucleus undergoes fragmentation. The resulted numerous nuclei move upward in the whorls of branches near the apex. The whorls of branches which arise as terminal or lateral outgrowths behave as gametangia producing biflagellate isogametes. Zygote produced by isogamy germinates into the vegetative rhizoidal thallus. Members of this order frequently become encrusted with lime except the genus *Batopora.* This trait has helped a large number of genera to be fossilized. The fossil record extends back to the Ordovician. More than 50 fossil genera have already been described.

Family: **Dasycladaceae**

Genus: *Acetabularia*

XI. **Chaetophorales:** Filamentous simple branched to heterotrichous plant body, in several genera hair-like projections are developed from the projecting system; parietal or laminate or discoid or partial to complete girdle-shaped chloroplast; asexual reproduction by zoospores, sexual reproduction isogamy to oogamy.

Family: **Chaetophoraceae:** Heterotrichous plant body with branched projecting threads usually drawn out into long hyaline hairs, prostrate system ranging from a loosely branched thread of relatively few cells to a richy branched expanse or a compact one-layered disc of numerous coalescent branches or clusters of rounded cells, the development of the projecting system being in inverse ratio to that of the prostrate one; asexual reproduction by the development of zoospores; sexual reproduction isogamous.

Genera: *Chaetophora, Draparnaldia, Draparnaldiopsis, Fritschiella*

Family: **Coleochaetaceae:** A typical heterotrichous habit of numerous filaments of the projecting system combining to form a hemispherical cushion to only prostrate system being represented either as loosely arranged branching threads or as more or less circular discs; asexual reproduction being effected by large ovoid biflagellate zoospores; sexual reproduction by exhibiting the most specialized type of oogamy found among the green algae, specialized post-fertilization stages.

Genus: *Coleochaete*

Family: **Trentepohliaceae:** Majority of the genera possessing a thallus with both prostrate and projecting systems often combining to form dense cushion-like growths, plant body with specialized adaptations for protection against desiccation; swarmers, produced in sporangia or gametangia of varied dimension, take part in reproduction, reproduction both asexual and sexual—asexual by zoospores and sexual by the process of isogamy.

Genus: *Trentepohlia*

Family: **Pleurococcaceae:** Unicellular plant body usually found in groups of two or more owing to slow separation of the products of division, in moist places short filaments with a few branches often produced as a result of division mainly in one direction; the sole method of reproduction being by vegetative division in three directions followed by subsequent separation of the cells.

Genus: *Pleurococcus (Protococcus).*

XII. **Oedogoniales.** Filamentous branched or unbranched forms with multiflagellate swarm cells, swarm cells with flagella distributed at the anterior region only; reticulate chloroplast; sexual reproduction oogamous.

Family: **Oedogoniaceae:** Plant body consisting of branched and unbranched filaments being attached to substrata by specialized structures or of a branched creeping filament bearing rhizoids; cells of the filament with or without having hair-like structures, each cell having reticulate cylindrical chloroplast; one of the outstanding characteristics lies in the peculiar method of growth of the cell; asexual reproduction by multiflagellate zoospores; sexual reproduction of oogamous type.

Genus: *Oedogonium*

XIII. **Zygnematales (Conjugales):** Plant body unicellular or colonial or filamentous; cells with stellate, lobed to spiral chloroplasts; complete absence of any flagellate cells; sexual reproduction by conjugation.

Family: **Zygnemataceae:** Filamentous unbranched forms without any morphological differentiation between the cells; cells with one or more stellate or spiral chloroplasts; asexual reproduction by aplanospores or azygospores; sexual reproduction by conjugation—both scalariform and lateral.

Genus: *Spirogyra, Zygnema*

Family: **Desmidiaceae:** Commonly known as **desmids** having unicellular microscopic in form of great variety and beauty; each cell possessing two beautifully lobed or stellate chloroplasts; reproduction by cell division and sexually by conjugation.

Genera: *Cosmarium, Closterium*

XIV. **Siphonales:** Coenocytic elaborately developed plant body being differentiated into prostrate and aerial portions; discoid chloroplasts; isogamous or anisogamous sexual reproduction.

Family: **Caulerpaceae:** Coenocytic thallus consisting of rhizome-like creeping prostrate portion bearing rhizoids and upright shoot-like aerial portion with division of labour; sexual reproduction isogamous or anisogamous.

Genera: *Bryopsis, Caulerpa*

Class: ***CHAROPHYCEAE***

No group in the plant kingdom has presented greater difficulties in classification than the Charophyceae. Some authorities consider the group a distinct taxon; others give it only order status in the Chlorophyceae; still others maintain that it is a class within the green algae. Though the Charophyceae have a number of characteristics different from those of the Chlorophyceae, the body as a whole is still like that of a green alga and so they belong in the division Chlorophycophyta. This is the view taken in this book, in which the Charophyceae are considered a separate class co-ordinate with the Chlorophyceae.

This class differs from the Chlorophyceae in five important ways: the plant body consists of an erect axis differentiated into nodes and internodes, with laterals of limited growth at the nodes; the reproductive organs are complex and are surrounded by sterile structures; the motile male gamete, antherozoid, is unlike any found in the Chlorophyceae being an elongated biflagellate spiral structure; the zygote germinates into a protonemal stage, from which the adult plant develops later; and the nuclear and cell division.

This class includes all extant and extinct *Chara*-like plants. Its single order Charales has been subdivided into several families of which all but one (family Characeae) are extinct.

Order: *CHARLES*

Comparatively complex plant body being differentiated into root-like, stem-like, and leaf-like structures; discoid chloroplast; sexual reproduction oogamous with elaborate development of sex organs.

Family: **Characeae:** Elaborately developed plant body; prostrate portion being differentiated into root-like attaching organs; aerial portion distinguishable into notes, internodes and leaf-like structures; sexual reproduction oogamous; sex organs differ considerably from the corresponding organs of other green algae.

Genera: *Chara, Nitella*

ORIGIN AND EVOLUTIONARY TENDENCIES IN THE CHLOROPHYCOPHYTA

The Chlorophycophyta exhibit both primitive as well as advanced structural features and methods of reproduction, As such, opinions differ with regard to the origin of the Chlorophycophyta. According to some, the Chlorophycophyta are thought to be derived from the photosynthetic bacteria. Possibly, a single group of photosynthetic bacteria developed the capacity to use water as a source of hydrogen and release free oxygen in photosynthesis, and the Chlorophycophyta diverged from this proto-algal stock which contained chlorophyll as the principal chlorophyll and β-carotene as the principal carotene. Others believe that the Chlorophycophyta probably evolved from green flagellate forms resembling *Chlamydomonas.* From this primitive stock the following evolutionary lines or tendencies may be traced among the members of the Chlorophycophyta:

I. The first tendency involves in the progressive development of motile colonial forms from an ancestral motile unicellular condition. This line culminates in *Volvox.*

II. The second tendency is the development of non-motile unicellular types which may be solitary or colonial originating from a motile unicellular ancestry. This line is represented by *Pediastrum* and *Hydrodictyon.*

III. The third tendency is the development of an extensive tubular or siphonaceous type of coenocytic thallus (as in *Caulerpa*) starting from a motile unicellular condition without subdivision into cells.

IV. The unicellular motile condition has given rise to a fourth tendency which has developed into multicellular filamentous to thalloid types. They are: unbranched (as in *Ulothrix* and *Oedogonium*), branched of simple to complex in form (as in *Chaetophora, Draparnaldia, Coleochaete* and *Chara*), and parenchymatous thalloid condition (as in *Ulva*).

While the vegetative body evolved along these evolutionary tendencies, there was a progressive development of reproductive methods. In the more primitive forms, reproduction is asexual by cell division or later, by zoospores.

Then sexual reproduction developed (the sex cells, or gametes, probably having evolved from zoospores); first isogamy, then anisogamy, and finally oogamy. Those forms which reproduce sexually, evolved an alternation of generations in which the haploid phase is elaborate and dominant, while the diploid is short, being represented by one cell, the zygote. In *Ulva*, however, there is an alternation of a gametophyte with a sporophyte, the two being identical in appearance.

Order: *VOLVOCALES*

The order Volvocales is of interest, in that its members are primitive forms of aquatic plant life. The members of this order are widely distributed in fresh-water and in damp soil. The vegetative cell is rounded, ellipsoid, or pear-shaped in outline. Motility is secured by means of two flagella of equal length, which are longer than the length of the cell and attached to the anterior end of the body. Internally, a red eye-spot is present in addition to the nucleus and cytoplasm, with a cup-shaped chloroplast. A pyrenoid and one or more contractile vacuoles are present. All are motile unicellular or colonial plants, with the colonies often of definite shape but never filamentous. It is believed that they are very similar to the forms from which green plants arose early in the earth's history.

Asexual reproduction in the unicellular forms is by the longitudinal division of the parent cell into two, four, or eight daughter protoplasts which develop into new cells. Whereas in the colonial forms it takes place by the formation of asexual reproductive cells, the **gonidia**. A gonidium divides to produce a number of daughter cells, all of which are held together as a new colony. Sexual reproduction in the Volvocales ranges from isogamy, anisogamy to oogamy. Division of the zygotic nucleus, at the time of germination of zygote, is meiotic resulting in the formation of zoospores from which adult haploid plants are produced.

Family: *CHLAMYDOMONADACEAE*

This family includes algae possessing unicellular biflagellate vegetative cells. Asexual reproduction is by zoospores and hypnospores and sexual reproduction ranges from—isogamy, anisogamy to simple and advanced oogamy.

Genus: *Chlamydomonas*

The genus *Chlamydomonas* represents one of the most primitive members of the living green plants. It is widely distributed in fresh-water and damp soil, often giving a bright green colour. The vegetative cell is spherical to ellipsoid, or pear-shaped in outline (Fig. 4.1A). It passes its ordinary vegetative life in a state of active movement. Motility is secured by means of two anteriorly placed flagella of equal length, which are longer than the length of the cell. The cell is surrounded by a cell wall which is differentiated into firmer and mucilaginous portions. The cell contains a single large cup-shaped chloroplast, occupying most of the cell,

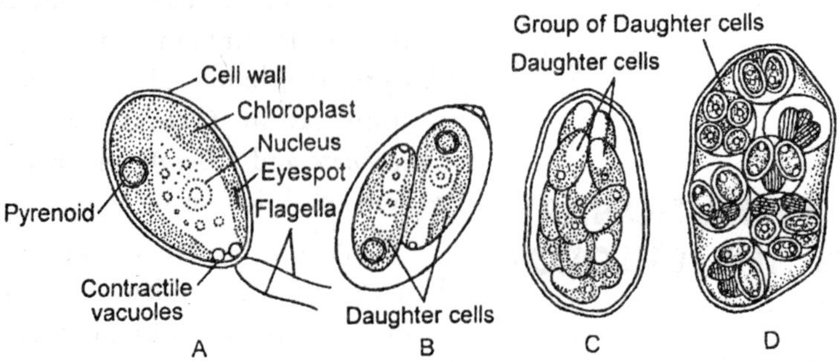

Fig. 4.1: *Chlamydomonas* sp. A. A single cell showing structural details. B-C. Successive stages of cell division. D. Palmelloid condition consisting of numerous groups of cells within a common wall

opening at the top and much thicker at the base than elsewhere. In the lower part of the chloroplast there is clear cytoplasm, and embedded in this is the nucleus. Two **contractile vacuoles**, which expand and contract, are generally situated in the anterior colourless cytoplasm, whilst a red **eyespot**, the most variable feature of the cell, is located at the side of the chloroplast. These cells swim about, through water with flagellate end foremost, rotating as they go and are positively phototrophic. During their movements the cells grow and ultimately divide. There may be produced within the cell two to eight or occasionally more daughter swarmers which resemble the parent cell, except being smaller (Fig. 4.1B and C). These swarmers are liberated out by the dissolution of the parent cell wall and form new plants. In forms growing on soil may result in the establishment of a **palmella**-stage by the failure of the daughter cells to escape or to become active, the result being a colony of many cells, commonly held together by a common mucilaginous investment (Fig. 4.1D). Ultimately, the cells acquire flagella and escape, again starting the active condition.

Most interesting of all is the sexual process, which takes place in three different ways, in different species. The simplest method is by the division of an ordinary cell into a larger number of parts than usual, so that the gametes are small and more slender in form compared with the vegetative cells. The gametes swim about as usual, and when two meet they entangle each other with their flagella, and the pair goes waltzing through the water together. Then they come in contact, and begin to fuse, either end to end or side to side, depending on species (Fig. 4.2A). After union the pair still go on swimming with their four flagella (Fig. 4.2B). The corresponding parts of the cell unite, and the two nuclei meet and fuse (Fig. 4.2C). The process of gametic union is called isogamy. A cell wall is formed around the whole producing a resting zygote (Fig. 4.2D). After some time, the diploid nucleus of the zygote undergoes reduction division with the production of four haploid nuclei which are incorporated into four zoospores which on being released from four *Chlamydomonas* plants (Figs. 4.2E and F and Fig. 4.4). This type of isogamy is exhibited by *C. moewusii.* Again isogamy between two amoeboid gametes is encountered in *C. eugametos*, in which the fusing gametes do not possess any flagella. It is interesting to note that the gametes are nothing but undersized vegetative cells which lack food and energy to produce new individuals directly and behave in a different way. This is known as 'hunger theory of sex'.

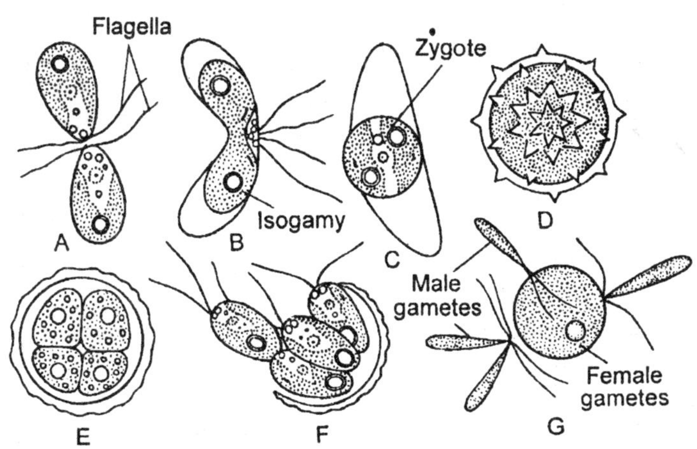

Fig. 4.2: Structure are reproduction of *Chlamydomonas moewusii* (A-F). A-C. Gametic pairing and union. D. Zygote. E-F. Stages of zoospore formation. G. *Chlamydomonas coccifera* showing sexual reproduction

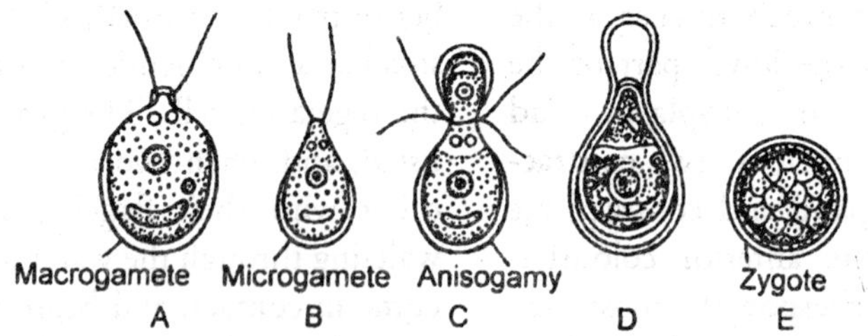

Fig. 4.3: *Chlamydomonas braunii.* A-E. Stages of anisogamy leading to the development of zygote

In the second type there is a marked and constant distinction between the two fusing gametes with a marked differentiation of sex. The microgametes are smaller and more active than the vegetative cells, and the macrogametes being much larger and less motile. Both kinds of gametes are flagellate (Figs. 4.3A and B). In fusion both the gametes come in contact with each other by their anterior extremities where the membranes coalesce forming an open canal and the contents of the smaller gamete drift over into those of the larger one (Figs. 4.3C and D). During the process of fusion the flagella disappear. A new and thick cell wall is developed around the fused contents consisting the zygote (Fig. 4.3E). This is a case of anisogamy which is more advanced than that of the previous isogamous condition, and is illustrated by *Chlamydomonas braunii.*

The third type represented by *Chlamydomonas coccifera* is more remarkable. Here the macrogamete is formed from the ordinary vegetative individual by the loss of flagella without undergoing any division, rather with a considerable enlargement accompanied by rounding off of the cell and may be designated as a female gamete. The male gametes are much smaller and elongated than the vegetative individual. The male gamete unites with the female and both of them fuse to form a zygote. Although the male gametes are not the typical spermatozoids, the sexual reproduction is certainly of a remarkably high level for an unicellular plant and almost amounts to oogamy (Fig. 4.2G).

In all the three cases the product of sexual fusion becomes enveloped in a thick membrane. The diploid zygote nucleus divides meiotically resulting in the formation of four haploid nuclei which is followed by the cleavage of the protoplast and formation of zoospores. The zoospores are liberated by the dissolution of the zygotic wall. *Chlamydomonas* possesses both homo- and heterothallic species (Fig. 4.4).

The various sexual processes exhibited by different species of *Chlamydomonas* illustrate all the stages in the evolution of sex in algae and can very well be arranged as isogamy → anisogamy → oogamy from the evolutionary

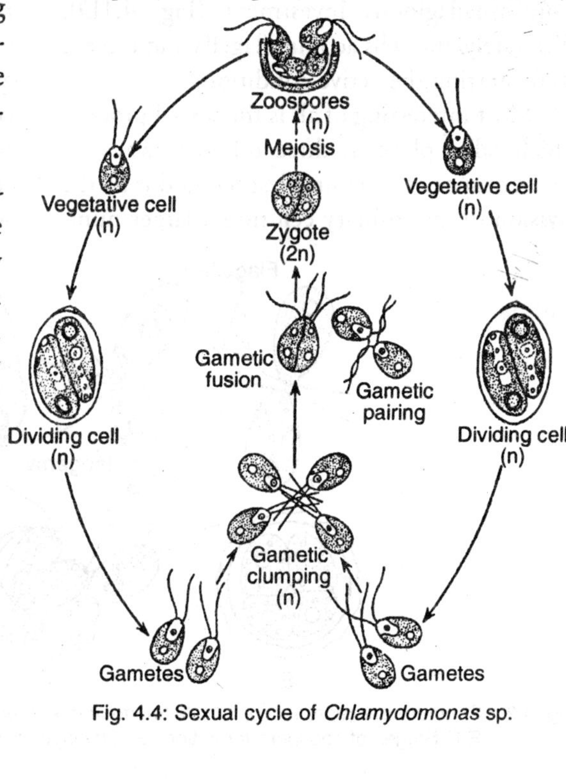

Fig. 4.4: Sexual cycle of *Chlamydomonas* sp.

point of view. The origin and differentiation of sex can also be traced within the different species of this genus. According to 'hunger theory of sex', both the zoospores and the gametes are remarkably similar and they start their development initially from the same material with their course of development being controlled by the food supply. The small individuals lose the power of independent development and behave like gametes. Further advancement takes place during this gamete development when, as a result of slowing down of the process of gamete development, there results gametes of dissimilar size. In the final stage of advancement, a vegetative cell instead of being divided starts enlarging losing its flagella and motility behaving like an ovum which is fertilized by an active male gamete and thus reaches the highest level of advancement in the sexuality that can be visualized.

Again, according to Smith the occasional occurrence of amoeboid gametes in *Chlamydomonas,* and the presence of a type of conjugation in *C. eugametos* suggest the possibility of a direct derivation of the Conjugales from motile, one-celled ancestors belonging to the order Volvocales.

Some Indian Species: *Chlamydomonas chrenbergii; C. eugametos; C. moewus; C. grandistigma.*

Special features

1. Unicellular biflagellate and pear-shaped plant body.
2. Cup-shaped chloroplast.
3. Presence of eyespot and two contractile vacuoles.
4. Asexual reproduction by biflagellate zoospores.
5. Presence of palmella-stage.
6. Sexual reproduction iso-, aniso- to simple cogamy.

Family: *SPHAERELLACEAE*

The members of this family possess unicellular thick gelatinous-walled biflagellate vegetative cells with basin-shaped to more or less reticulate chloroplast. Sexual reproduction is isogamous.

Genus: *Sphaerella*

It is a fresh-water alga being abundantly found in rain-water, pools, ponds and similar other places. In the Arctic and Alpine regions the alga is commonly known as 'red snow' as it imparts red colour to the snow. Structurally, the plant body is very similar to *Chlamydomonas* for being unicellular, uninucleate, anteriorly placed biflagellate, and broadly ellipsoid to ovoid in shape with several contractile vacuoles (Fig. 4.5A). It is characterized by the presence of a space between the protoplast and the cell wall. This space is filled with a watery gelatinous substance and is traversed by numerous delicate cytoplasmic strands extending from the protoplast to the cell wall. The chloroplast is basin-shaped to reticulate containing pyrenoids. Usually, the cells are coloured bright-red due to the presence of the pigment haematochrome in the cell sap. For long time the genus was named *Haematococcus* for having haematochrome in the protoplast. The cell wall is thick and is differentiated into a superficial strongly thickened layer and an extensive gelatinous zone.

The alga reproduces both asexually and sexually. Asexual reproduction may take place just by the division of the free-swimming cells into two to four daughter cells (Figs. 4.5B to E). But more frequently during asexual reproduction the cells withdraw their flagella and come to rest. Each cell increases in diameter, contents of which divide longitudinally into two to sixteen parts. All of them develop wall and are confined for some time within the mother cell. Ultimately they are liberated out by the rupture of the mother cell wall and develop into new individuals.

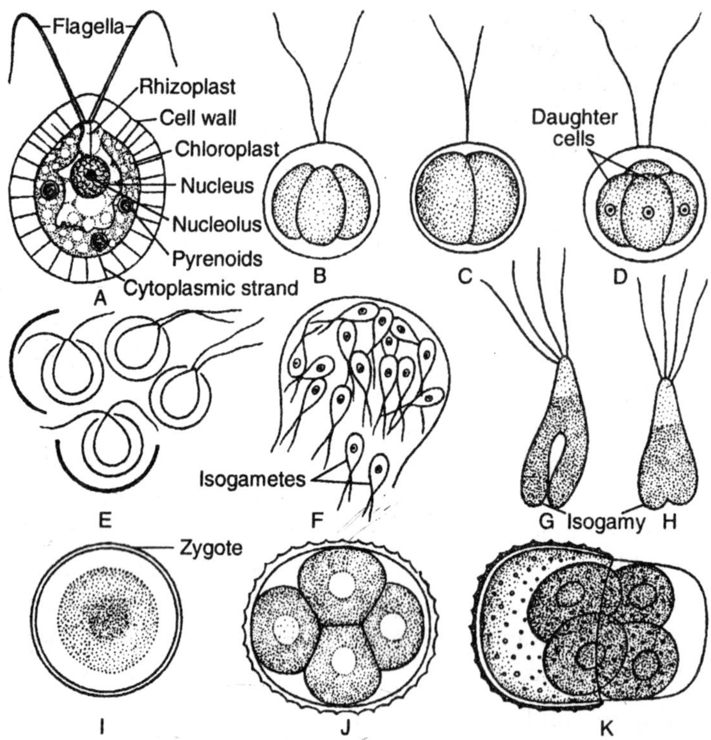

Fig. 4.5: *Sphaerella lacuistris*. A. Single individual showing structural details. B-E. Stages in asexual reproduction. F. Liberation of gametes. G-I. Stages in isogamy leading to zygote formation. J. Zygote producing haploid cells. K. Liberation of haploid cells by the rupture of zygote wall

Sexual reproduction is isogamous. The process is directly related with the prevailing unfavourable condition of the substratum. The flagella fall off and the cells come to rest, contents of which divide to form gametes much smaller in size and larger in number than the cells that are produced asexually. After remaining confined within the mother cell, the gametes are liberated out and ultimately unite in pairs resulting in the formation of a zygote (Figs. 4.5F to I). After a period of rest the diploid nucleus of the zygote, on return of favourable conditions divides twice, first division of which is reductional, to form four haploid cells (Fig. 4.5J). These haploid cells are set free by the rupture of the zygote wall, become flagellated and develop into new *Sphaerella* cells (Fig. 4.5K).

Depending on conditions of the substratum, the vegetative cells may form large akinetes which on germination can give rise to zoospores, hypnospores, or gametes.

Special features

1. Unicellular biflagellate plant body with characteristic space between the protoplast and the cell wall which is filled with a watery gelatinous substance.
2. Basin-shaped to reticulate chloroplast.
3. Cells with several contractile vacuoles.
4. Sexual reproduction isogamous.

Family: *Volvocaceae*

The Volvocaceae include forms in which flagellate cells are grouped into colonies (coenobia) of cells being multiple of two. The colonies may be flat, disc-shaped or spherical. All are fresh-water. Asexual reproduction is either by the division of ordinary vegetative cells or by the development of special vegetative cells—**gonidia**. Sexual reproduction ranges from isogamy, anisogamy to advanced oogamy with the development of antherozoids and egg.

Genus: *Pandorina*

This alga is abundant in fresh-water ponds and ditches. Typical chlamydomonas-cells are closely packed together to form spherical to ellipsoidal coenobium (Fig. 4.6A). The coenobium is composed of relatively definite number of cells ranging from 4 to 32, which are enclosed in a homogeneous gelatinous matrix with an outer watery sheath. In a coenobium cells are so compact that they are often angular as a result of mutual pressure. The coenobia are hollow with the cells forming a single layer in the peripheral gelatinous matrix. The two flagella of each cell pass out through specially differentiated canals of the gelatinous matrix. The coenobia exhibit a marked polarity with the constituent cells showing a definite orientation.

Asexual reproduction is by the forma ·on of daughter coenobia, prior to which the ɔarent coenobium ceases to be motile and sink to the bottom of the pool. Then each cell produces by several divisions, daughter coenobia (Figs. 4.6B to D). All the cells of the parent coenobium behave alike in reproduction. The daughter cells become inverted and assume a spherical shape and ultimately each cell produces two flagella. The daughter coenobia are ultimately liberated by the disintegration of the parent envelope (Fig. 4.6E). Sexual reproduction is anisogamous. Divisions leading to the formation of gametes are identical with those of asexual reproduction process. But the fusing gametes

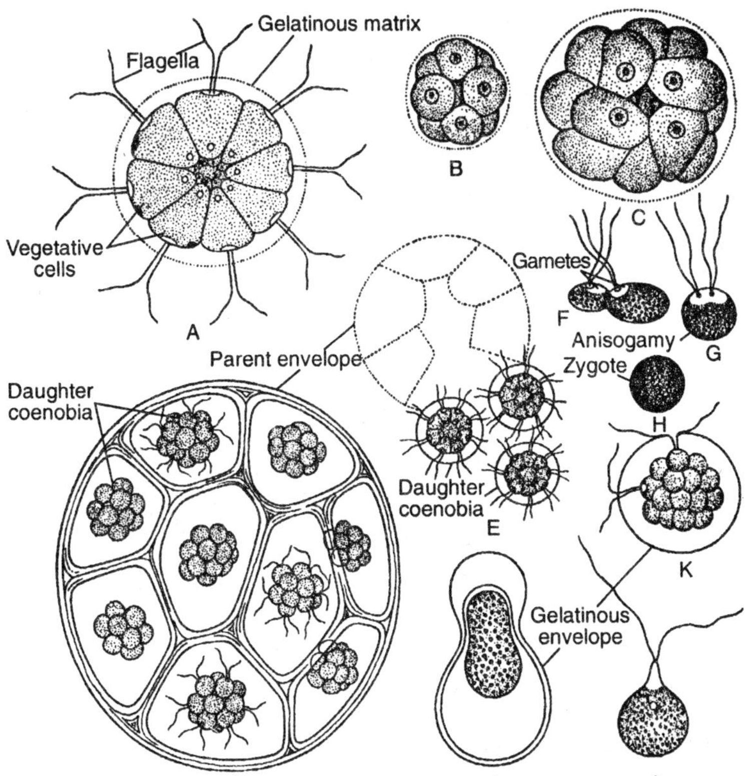

Fig. 4.6: *Pandorina* sp. A. Mature coenobium. B-D. Stages in coenobia development. E. Liberation of daugter coenobia breaking parent envelope. F-H. Stages in anisogamy resulting in zygote formation. I. Germination of zygote. J. Swarmer. K. New coenobium developed from the swarmer

are always unequal in size with the larger one more sluggish than the smaller counterpart. Gametes from different coenobia only fuse. Fusion may be terminal or lateral (Fig. 4.6F). The quadriflagellate zygote remains motile for a while before it loses its flagella and secretes a wall (Figs. 4.6G and H). Only one of the four resulting cells of the germinating zygote normally survives forming a swarmer (Figs. 4.6I and J). After swimming for a short time the swarmer withdraws its flagella and secretes a gelatinous envelope, and divides and redivides to form a new coenobium (Fig. 4.6K).

Some Indian Species: *Pandorina morum* (Mull.) Bory.; *P. morum f. major* Iyeng.

Special features

1. The coenobia are oblong to subspherical and are comparatively of small size.
2. Coenobium of typical Chlamydomonas cells.
3. Anisogamous sexual reproduction.

Genus: *Eudorina*

Commonly found in fresh-water pools and ponds, particularly during the rainy season. Spherical or ellipsoidal coenobia are distinguishable into anterior and posterior portions. There may be sixteen to sixty-four biflagellate, globose or somewhat pear-shaped cells present in the coenobium, which are not closely packed and are sometimes arranged in well-marked transverse series distributed in both anterior and posterior tiers (Fig. 4.7A). Cells with usual Chlamydomonad-structure are embedded in a mucus investment in the coenobium (Fig. 4.7B).

Almost all the cells of the coenobium give rise to new daughter coenobia with the excep-

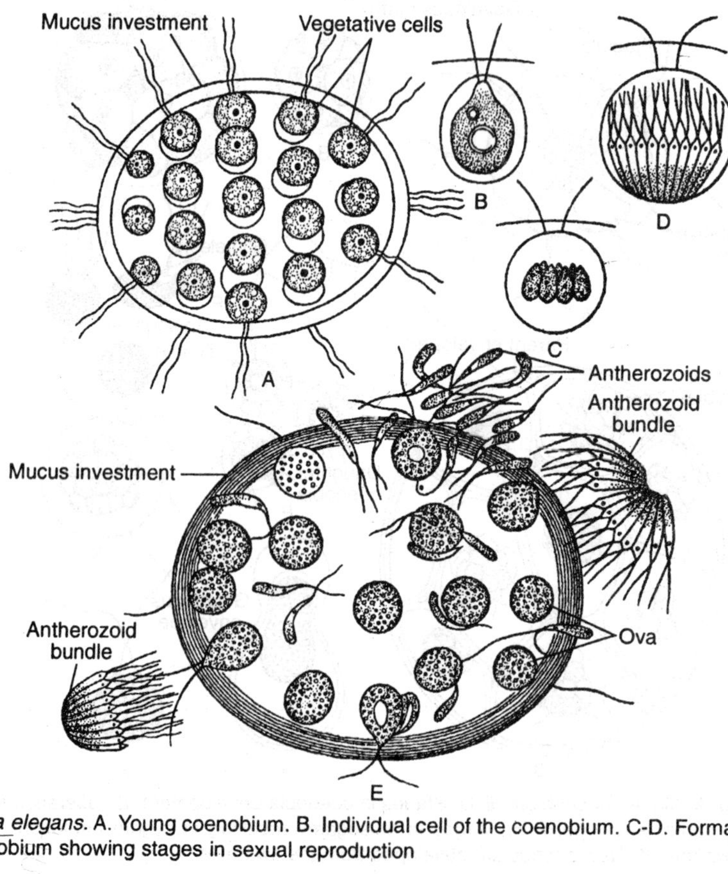

Fig. 4.7: *Endorina elegans*. A. Young coenobium. B. Individual cell of the coenobium. C-D. Formation of antherozoids. E. Coenobium showing stages in sexual reproduction

tion of a few anterior cells which are much larger and do not reproduce. The process of daughter coenobium formation is same as that of *Pandorina.* Sexual reproduction is advanced anisogamy indicating close approach to oogamy with the formation of typical anterozoids and non-motile female gamete—ovum (Fig. 4.7E). Coenobia are monoecious or dioecious. The anterior cells of a monoecious coenobium give rise to the anterozoids and the posterior ones form ova. But in heterothallic species there are distinct male and female coenobia which develop male and female gametes respectively. The vegetative cells of the coenobium divide and redivide forming anterozoids (Figs. 4.7C and D), whereas during female gamete formation, the vegetative cells are directly converted into female gametes—ova without undergoing any division. The zygote remains within the parent coenobium until it disintegrates. Zygote germinates to give rise to a single swarmer which swims about for a time and then, in the same manner as a vegetative cell, divides and redivides to form a new coenobium.

Some Indian Species: *Eudorina elegans* Ehrb.; *E. indica* Iyeng.

Special features

1. The coenobia are ellipsoidal and larger than in *Pandorina.*
2. Anisogamous sexual reproduction between an immobile large ovum and an elongated spindle-shaped biflagellate antherozoid exhibiting oogamous condition.

Genus: *Volvox*

This genus represents the most elaborate development of coenobial form, showing considerable advancement from the more simple and primitive Volvocales. The coenobia of *Volvox,* occurring in temporary and permanent fresh-water pools are spherical and are about the size of a small pinhead. Each coenobium is a hollow sphere of mucilage with a single layer of very numerous, even as many as fifty thousand cells (Fig. 4.8A), arranged in a single layer joined together by fine strands of cytoplasm (Fig. 4.8C). The whole coenobium moves and rotates by means of flagella of individual cells. Most of the cells in a coenobium are vegetative and do not have the capacity to form new individuals.

From the point of view of reproduction, *Volvox* differs essentially from *Pandorina* and *Eudorina;* in that, certain cells in a coenobium only take part in reproduction. Of the many cells of a *Volvox* coenobium, only a few differentiate as asexual reproductive cells growing larger with the disappearance of flagella, and thickening or with an increase in number of the cytoplasmic strands in the vicinity of them; these cells are called **gonidia** (sing. **gonidium**) (Figs. 4.8A and E). A gonidium divides and redivides producing a number of daughter cells (Figs. 4.9A to E), all of which are held together as a new colony. But since a number of gonidia divide at the same time several new colonies are formed simultaneously. After division ceases, the young colony turns itself inside out by inversion through a small pore, or opening—the **phialopore** (Figs. 4.9F to H). The individual cells then develop flagella. The new coenobium may or may not escape from the parent coenobium immediately after development (Fig. 4.8D).

Sexual reproduction is oogamous. There are both homothallic and heterothallic species of *Volvox.* Both kinds of gametes, in small numbers, are developed especially from cells of the posterior half of the coenobia. Cells producing female gametes or ova enlarge considerably without undergoing any division and they lose their flagella during development. Cells forming male gametes or antherozoids divide successively giving rise to numerous spindle-shaped, biflagellate anterozoids (Figs. 4.10A–C′).

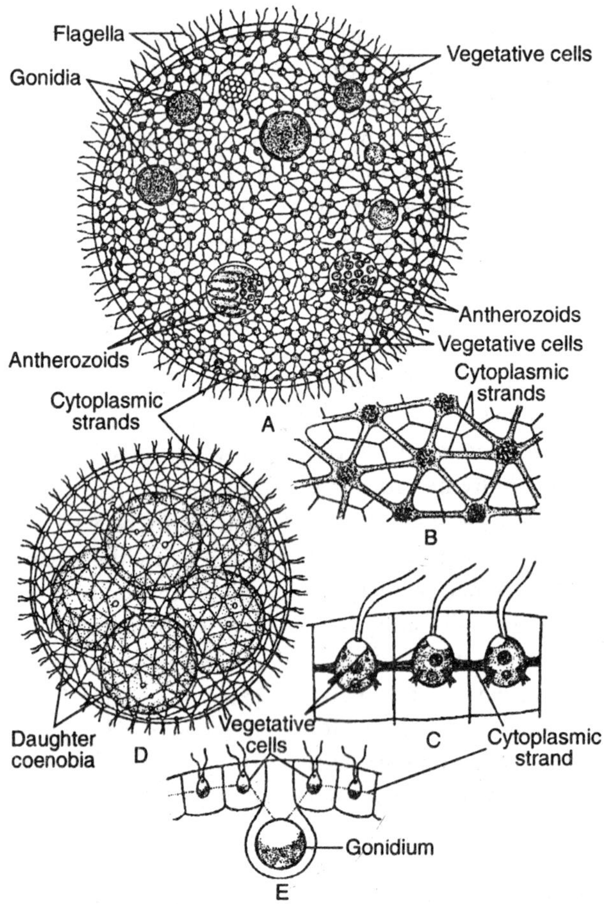

Fig. 4.8: *Volvox* sp. A. A mature coenobium bearing vegetative cells with cytoplasmic strands, gonidia, and antherozoids. B. Details of cytoplasmic strands existing between adjacent cells. C. Individual vegetative cells of the coenobium with cytoplasmic strand. D. A parent coenobium bearing daughter coenobia. E. A portion of the coenobium showing the position of vegetative cells, gonidium and cytoplasmic strads

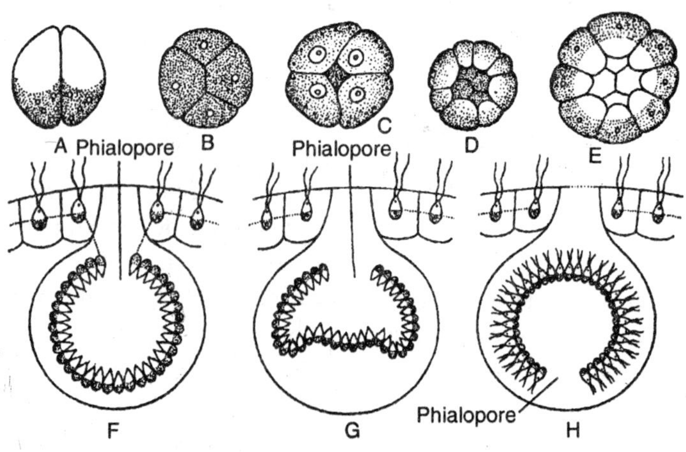

Fig. 4.9: *Volvox* sp. Stages in the development of a daughter coenobium. A-E. Two-, four-, eight-, and sixteen-celled stages. F. Just before inversion. G-H. Inversion and flagella development

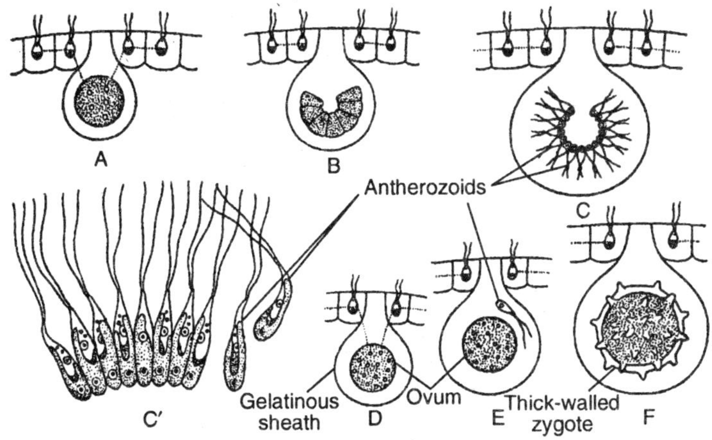

Fig. 4.10: *Volvox* sp. Stages in the development of antherozoids, ovum and zygote. A-C. Successive stages of antherozoid development. C′. Details of antherozoids. D. Portion of a coenobium showing ovum. E. A stage of fertilization. F. Portion of a coenobium with a thick-walled zygote

The female gametes are non-motile, much larger than the male gametes, and are surrounded by a gelatinous sheath (Fig. 4.10D). The antherozoids swim to the egg, enter through the gelatinous sheath, and finally fertilization is accomplished (Fig. 4.10E). The resulting zygote develops a thick, often stellate wall, and undergoes a resting period (Fig. 4.10F). The first division of the zygotic nucleus is meiotic. A single zoospore is produced which forms a new coenobium (Fig. 4.11).

Volvox represents the apex of this motile coenobial line for having an elaborate mechanism of coenobial development and oogamous sexual reproduction.

Some Indian Species: *Volvox africanus* West. *f. minor* Iyeng.; *V. aurens* Ehrb.; *V. globator* (L) Ehrb.; *V. merrille* Shaw; *V. prolificus* Iyeng.; *V. rousseleti* West var. *lucknowensis* Iyeng.

Special features

1. Biflagellate pear-shaped cells arranged in a coenobium of large number of cells.
2. Mature coenobium composed of vegetative cells, colony forming cells or gonidia and antherozoid mother cells and ovum mother cells.
3. Asexual reproduction by gonidia which produce daughter coenobia by an elaborate process.
4. Sexual reproduction oogamous.

Order: *CHLOROCOCCALES*

Unicellular plants vegetating in a motionless phase, free, less often attached, never filamentous, but are sometimes coenobic or endophytic. Cell wall is usually thin or at least not gelatinous, often of elaborate and distinctive form; protoplast with one to several chloroplasts, pyrenoids, and nuclei. Cell division is absent in the vegetative phase; asexual reproduction is by zoospores, aplanospores, autospores, daughter coenobia, or by other means. Sexual reproduction is by iso- or anisogametes.

Family: *CHLORELLACEAE*

This family includes the unicellular non-motile genus *Chlorella* which is a tool of many physiological experiments. It reproduces by the development of non-motile spores.

Genus: *Chlorella*

This unicellular alga has a very varied habitat. Cells are solitary or sometimes aggregated into groups being enveloped by mucilage. They may be free-living in stagnant pools. Species of

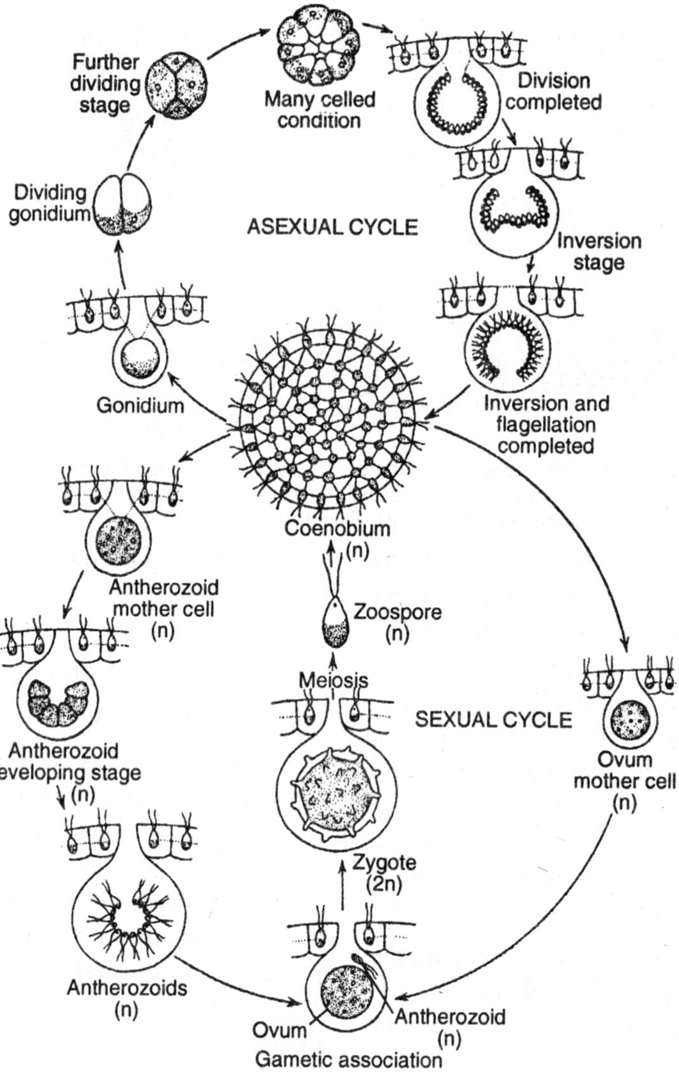

Fig. 4.11: Life cycle of *Volvox* sp.

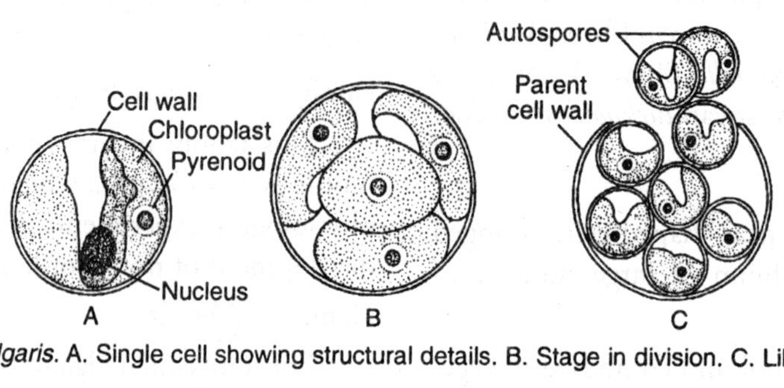

Fig. 4.12: *Chlorella vulgaris*. A. Single cell showing structural details. B. Stage in division. C. Liberation of autospores

Chlorella may also be terrestrial or may occur abundantly in habitat rich in organic matter (sewage waters and similar other places), within the cells of some invertebrates forming a symbiotic association, and in the exudation of trees. Being one of the simplest forms of algae,

Chlorella has been widely used in the study of respiration and photosynthesis. *Chlorella* is used for the purification of air in space capsule. Besides this, *Chlorella* has a great economic importance for it has been reported to yield an antibiotic, known as **Chlorellin**. The alga reacts remarkably under cultural condition with or without the supply of light. The cells of *Chlorella* are small, spherical, rarely ellipsoid. Each cell has a thin wall enclosing a single cup-shaped parietal chloroplast with or without pyrenoids, and a nucleus which is centrally placed (Fig. 4.12A).

Reproduction is effected by the production of **autospores** which are developed within a cell by the successive division of the protoplast (Fig. 4.12B). Two or even sixteen autospores may be formed in each cell which are liberated out by the rupture of the parent cell wall (Fig. 4.12C). These immobile spores form new individuals.

Some Indian Species: *Chlorella conglomerata; C. ellipsoidea; C. parasitica; C. vulgaris.*

Special features

1. Unicellular non-flagellate plant body—solitary or aggregated in groups.
2. Cup-shaped chloroplast.
3. Reproduction by autospores.

Family: *COELASTRACEAE*

In this family the vegetative cells of the coenobia are grouped in one plane with the long axes of the cells parallel to one another. Each cell possesses laminate chloroplast. Reproduction is by division of cells of the coenobium.

Genus: *Scenedesmus*

Coenobia occur abundantly in fresh-water pools as a part of planktonic flora. Each coenobium is a flat plate of usually four to eight elongated cells, arranged with their long axes parallel to each other. The two end cells of a coenobium often develop processes which are elaborations of the mucilaginous layer of the cell membrane. Cells of the coenobium contain a single nucleus and a single, longitudinal laminate chloroplast with a prominent pyrenoid (Fig. 4.13A). The only method of reproduction is by the division of a protoplast leading to form a daughter coenobium, which may or may not have the same number of cells as that in the parent coenobium (Figs. 4.13B to E). Global resting spores are known in a few species (Figs. 4.13F to I).

Some Indian Species: *Scenedesmus acuminatus* (Lagherheim) Chodat.; *S. bijugatus* (Turpin) Kutz.; *S. obliquous* (Turpin) Kutz.; *S. perforatus* Zemmerman.; *S. quadricauda* (Turpin) Brebisson.

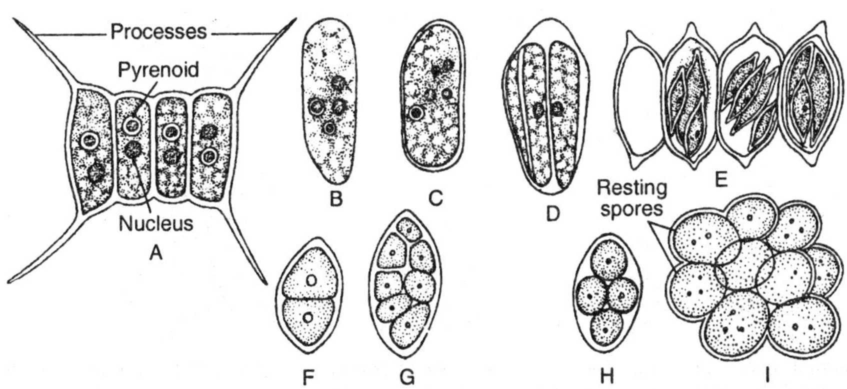

Fig. 4.13: *Scenedesmus* sp. A. Coenobium showing details of cells. B-E. Stages in formation of daughter cells. F-I. Development of resting spores

Special features

1. Flat plate-like coenobium of elongated cells arranged with their long axes parallel to each other.
2. The two end cells of a coenobium often develop processes.
3. Laminate chloroplast.
4. Reproduction by the division of a protoplast leading to form a daughter coenobium.

Family: *HYDRODICTYACEAE*

The members of this family occur in the form of colonies in fresh-water ponds and ditches. Sexual reproduction is isogamous.

Genus: *Pediastrum*

This jewel-like beautiful alga occurs as a free-floating disc-shaped coenobium in pools, ditches, and forms a component of the plankton flora of lakes. The alga sinks to the bottom of the pool when the water-level recedes. Its

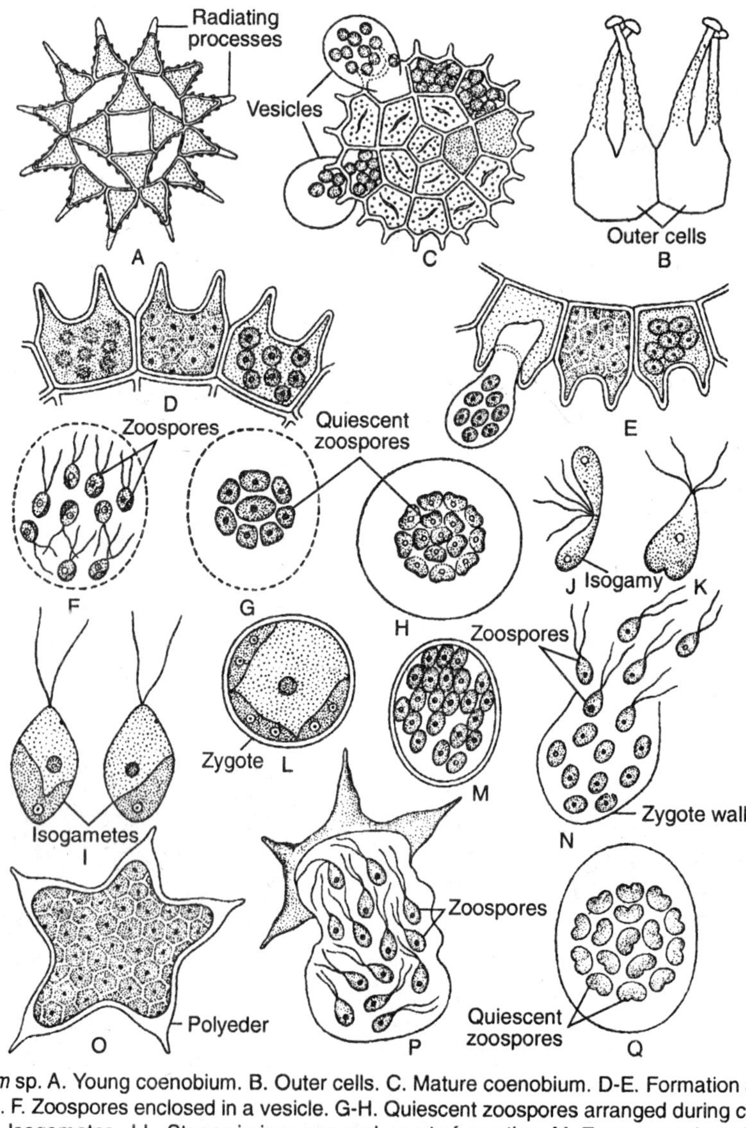

Fig. 4.14: *Pediastrum* sp. A. Young coenobium. B. Outer cells. C. Mature coenobium. D-E. Formation and liberation of zoospores. F. Zoospores enclosed in a vesicle. G-H. Quiescent zoospores arranged during coenobium development. I. Iscgametes. J-L. Stages in isogamy and zygote formation. M. Zoospore primordia. N. Liberation of zoospores. O. Polyeder. P. Zoospores in polyeder. Q. Quiscent zoospores before coenobium formation

abundance is very rare. The coenobia are flat and are formed of as many as 128 plane-faced or lobed cells arranged in a single layer producing a stellate plate-like in appearance (Fig. 4.14A). The outer cells differ from the internal ones for having radiating processes (Fig. 4.14B). Young cells have a single parietal chloroplast with one to four pyrenoids and are uninucleate, while the older ones are multinucleate.

Asexual reproduction is by biflagellate zoospores formed by successive nuclear division followed by progressive cleavage of the protoplast (Figs. 4.14D and E). Every cell in a coenobium is capable of giving rise to zoospores, but simultaneous division of the protoplast of the mother cells is rather rare. The zoospores produced by a cell are enclosed in a thin membrane and are suddenly liberated into an external vesicle through a slit in the wall (Fig. 4.13C). They swarm around inside the vesicle for a short time, then arrange themselves in a position like that of the cells of the parent coenobium, becoming motionless (Figs. 4.13F to H). The cells then acquire mature characteristics before liberation. Thus, this alga lends some support to the suggestion that this non-motile line of green alga descended from flagellated ancestors. *Pediastrum* reproduces usually by spindle-shaped biflagellate isogametes produced like zoospores (Fig. 4.14I). The isogametes fuse in pairs to form a zygote which increases remarkably in size (Figs. 4.14J to L). Zygote on germination produces a number of zoospores which when liberated swim in water (Figs. 4.14M and N). Each zoospore loses its flagella, comes to rest, becomes angular in shape, and subsequently turns into a thick-walled polyhedral cell, known as 'polyeder' (Fig. 4.14O). The polyeder now increases in size. Its protoplast divides to form a number of zoospores which ultimately form a new coenobium in the process described earlier (Figs. 4.14P and Q). The zygotic nuclear details have not yet been clearly worked out.

Some Indian Species: *Pediastrum duplex* Meyen var. *loktakense* Bruhl et Biswas; *P. incavatum* Turner.

Special features

1. Free-floating disc-shaped coenobium composed of single layer of cells of which the outer ones having radiating processes.
2. Parietal chloroplast.
3. Asexual reproduction by biflagellate zoospores.
4. Sexual reproduction isogamous.
5. The zygote gives rise to zoospores from which specialized structures—polyeders are developed which again produce zoospores. These zoospores ultimately give rise to new coenobium.

Genus: *Hydrodictyon*

This is a large macroscopic coenobic alga commonly known as 'water net' because of the fact that the vegetative plant body grows as a net-like structure usually floating on the surface of quiet fresh-water. The coenobium is composed of free-floating network of cylindrical to broadly ovoid cells that are united to form a sac-like reticulum (Fig. 4.15A). The meshes of the net, formed by five or six coenocytes, are of variable size being pentagonal or more usually hexagonal reaching a length of 8 to 20 cm. (Fig. 4.15B). The angles are formed by the junction of three of the cells (Fig. 4.15C). Young cells contain a single nucleus and a single pyrenoid, but older cells become multinucleate with cytoplasm containing a large reticulate chloroplast with many pyrenoids forming a lining layer around a large central vacuole (Fig. 4.15D).

Asexual reproduction is by means of biflagellate uninucleate zoospores that are formed by progressive irregular cleavage of the protoplast of a coenocyte (Fig. 4.15E). The zoos-

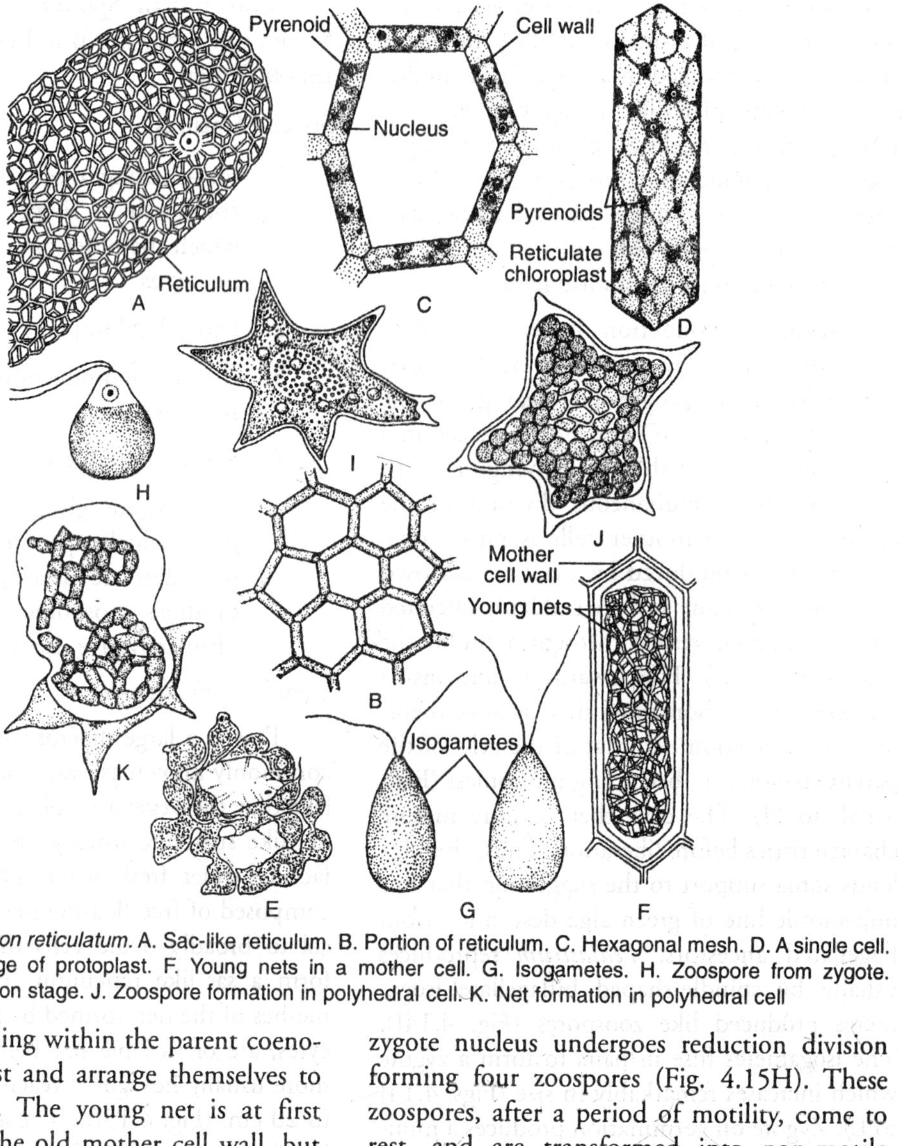

Fig. 4.15: *Hydrodictyon reticulatum*. A. Sac-like reticulum. B. Portion of reticulum. C. Hexagonal mesh. D. A single cell. E. Cleavage of protoplast. F. Young nets in a mother cell. G. Isogametes. H. Zoospore from zygote. I. Polyhedron stage. J. Zoospore formation in polyhedral cell. K. Net formation in polyhedral cell

pores after swarming within the parent coenocyte, come to rest and arrange themselves to form a new net. The young net is at first enclosed within the old mother cell wall, but this soon gelatinizes and the net liberates out (Fig. 4.15F).

Sexual reproduction is isogamous and the plants are monoecious. The biflagellate isogametes are formed in the same manner as the zoospores, but are larger in number and smaller in size (Fig. 4.15G). They are liberated out of the parent cell through a hole in the parent cell membrane. The fusion product of the isogametes is a thin-walled spherical zygote. After a short period of rest the resulting zygote nucleus undergoes reduction division forming four zoospores (Fig. 4.15H). These zoospores, after a period of motility, come to rest, and are transformed into non-motile aplanospores, each of which increases remarkably in size and develops into polyhedral cell also known as divide to form many zoospores. These zoospores come together and form a new net in the manner already described (Figs. 4.15J and K). Thus *Hydrodictyon* is essentially a collection of a number of individual plants, because it arises from a number of zoospores (Fig. 4.16).

Indian Species: *Hydrodictyon indicum* Iyeng.

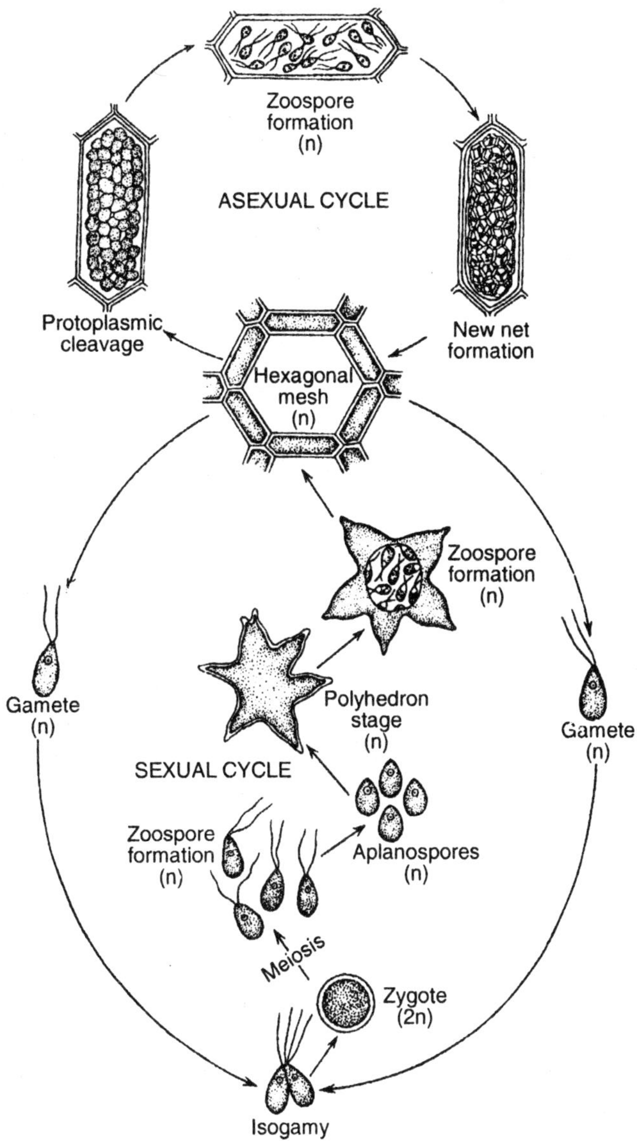

Fig. 4.16: Life cycle of *Hydrodictyon reticulatum*

Special features

1. Large net-like macroscopic coenobium.
2. Cells of the net coenocytic.
3. Reticulate chloroplast.
4. Asexual reproduction by biflagellate zoospores.
5. Sexual reproduction isogamous.
6. Development of polyhedral cell and post-sexual stages have similarity with *Pediastrum.*

Order: *ULOTRICHALES*

Plants are filamentous to foliaceous, simple or branched, with or without differentiation of attached base and free apex; in a few cases condensed toward a unicellular state or by irregular divisions becoming somewhat parenchymatous. Cells possess a distinct wall enclosing a single parietal chromatophore, pyrenoids, and a single nucleus.

Family: *ULOTRICHACEAE*

Plants are filamentous. Filaments are with or without an attaching base; typically unbranched, composed of cylindrical uniseriate cells with thin to gelatinous walls. Cells possess lateral plate-like or band-like chromatophores with or without pyrenoids.

Asexual reproduction is by bi- and quadriflagellate zoospores, although a few genera (e.g., *Stichococcus, Radiofilum,* and *Geminella*) fail to do so. Sexual reproduction is isogamous (*Ulothrix*), anisogamous, or oogamous (*Cylindrocapsa*).

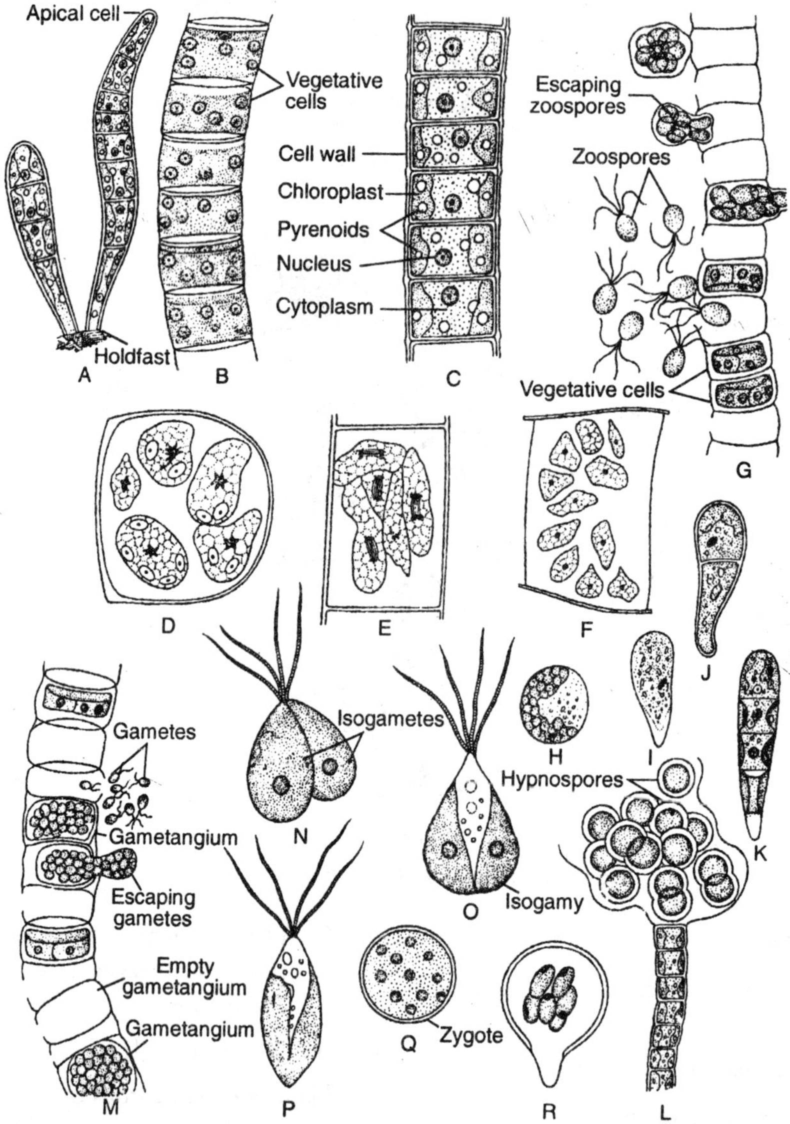

Fig. 4.17: *Ulothrix* sp. A. Filaments showing holdfast and apical cell. B-C. Structural detials of vegetative cells. D-F. Stages of development of zoospores. G. Portion of filament with vegetative cells and escaping of quadriflagellate zoospores. H-K. Stages of zoospore germination. L. Portion of filament with hypnospores. M. Formation of gametes and their escape. N-Q. Stages of isogamy leading to zygote development. R. Germinating zygote showig the division of its protoplast into a number of zoospore primordia

Genus: *Ulothrix*

This is a fresh-water alga of the filamentous or confervoid kind (in the days of Linnaeus such forms were included in the old genus *Conferva*, being simple filamentous), growing preferably in cold-flowing water. The unbranched slender filament consisting of a single row of cells, differentiated into base and apex, anchors to its substrate at one end by a specialized basal cell of the filament to form a **holdfast** (Fig. 4.17A). The filaments can, however, live equally well detached, floating in tangled masses on the surface of water. The filament is composed of numerous cylindrical cells placed end to end, each of which contains a single nucleus, and a band-shaped or plate-like parietal chloroplast resembling a slightly open ring curved to fit the rounded contour of the cell (Figs. 4.17B-C). Usually one or more pyrenoids are found in the chloroplast. Within the cell wall is the cytoplasmic layer, or primordial utricle, in which the nucleus is embedded (Fig. 4.17C). Walls surrounding the cells may be thick or thin, and homogeneous or stratified. The filament grows throughout its length, excepting the basal cell, and has no special growing point.

Apart from vegetative multiplication by accidental fragmentation of the filament *Ulothrix* reproduces by means of swarmers. All the cells but the holdfast are capable of producing swarmers. Three types of swarmers may be produced by the common species of *Ulothrix (U. zonata)*. Depending on their shape and behaviour they may be named as: 1. quadriflagellate **macrozoospores,** formed in small numbers from the mother cell, 2. quadri- or biflagellate **microzoospores,** formed in large numbers, and 3. biflagellate gametes, formed in large numbers. The macro- and micro-zoospores are formed in separate sporangia. The sporangia are nothing but the vegetative cells of the filament in which the zoospores are formed by the repeated division of the protoplast (Figs. 4.17D to F). The zoospores are of the type characteristic of the motile vegetative cells of the Volvocales, only difference being in size and number of flagella. The zoospores escape by the breaking down of the sporangial wall (Fig. 4.17G). After emerging, each zoospore swims rapidly for a short period, then comes to rest, loses its flagella, becomes attached to some convenient substrate, and grows into a new plant (Figs. 4.17H to K). In some species, the protoplast divides to form non-motile, round daughter protoplasts. These protoplasts instead of bearing flagella secrete a wall and become **aplanospores**. Under condition of drought the entire protoplast may round up to form a single, thick-walled aplanospore or the **hypnospore** (Fig. 4.17L). Occasionally wall of the sporangium producing the aplanospores gelatinizes followed by the gelatinization of the wall of the aplanospores as a result of which a number of green bodies become embedded in a mass of mucilage, this is known as **palmella-stage**, because the plant when in this state somewhat resembles the genus *Palmella*. The palmella-stage of *Ulothrix* is chiefly found when the filament is left by the receding water on the damp parts of the pool or stream in which it grows. Ultimately these cells reproduce the normal filaments, either by direct germination or by forming zoospores.

Sexual reproduction is isogamous and involves the formation of small biflagellate isogametes. Formation of gametes is like the zoospores, each gametangium producing to the extent of 64 gametes (Fig. 4.17M). After emerging from the gametangium, the gametes swarm around in water. The fusing gametes come from different filaments indicating **heterothallic** condition (Figs 4.17N to Q and Fig. 4.18). The zygote remains motile for a short time after its formation but soon becomes a resting cell, and later germinates, during which meiosis of the zygotic nucleus takes place. Four or more zoospores or

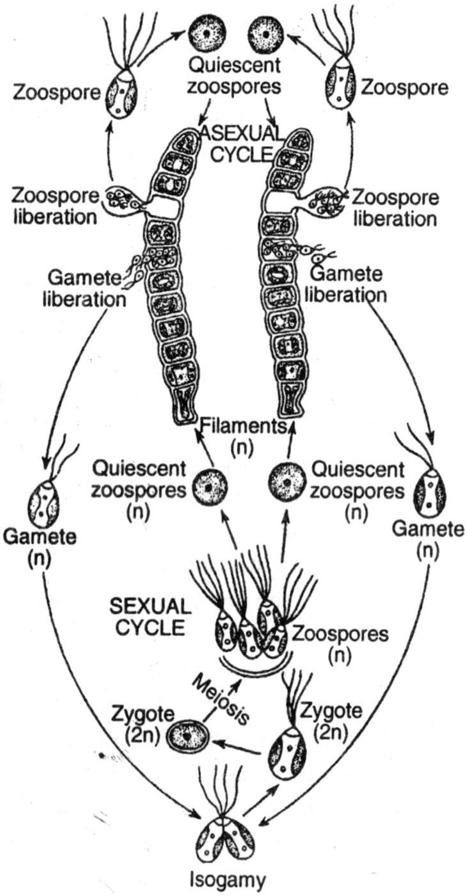

Fig. 4.18: Life cycle of *Ulothrix* sp.

aplanospores are formed, each of which produces a new filament (Fig. 4.17R).

Some Indian Species: *Ulothrix acqualis* Kutz.; *U. oscillarina* Kutz.; *U. tenerrima* (Kutz.) Kutz.; *U. tenuissima* Kutz.; *U. variabilis* (Kutz.) Kutz.; *U. zonata* (Web.) Kutz.

Special features

1. Filamentous unbranched plant body differentiated into base and apex.
2. Band-shaped chloroplast.
3. Asexual reproduction by bi- and quadriflagellate zoospores.
4. Sexual reproduction isogamous.

Order: *ULVALES*

The Ulvales contain a diverse assemblage of green algae, which are filamentous (biseriate or pluriseriate), tubular, or membranous (monostromatic or distromatic). The cells of all, except the rhizoidal cells of some genera are uninucleate. A few genera lack sexual reproduction, but the majority that have it exhibit a variety of life cycles including haplobiontic and haploid, as in *Monostroma*; diplobiontic and isomorphic, as in *Enteromorpha, Ulva,* and *Ulvaria.* Asexual reproduction is by biflagellate or quadriflagellate zoospores. Sexual reproduction is isogamous, anisogamous, or oogamous. Members of this order are largely marine.

Family: *ULVACEAE*

Plants are usually tubular or parenchymatous thalloid, capillary to broad, occasionally reduced to one or two rows of cells; attached or becoming free-floating; one or two cells in thickness. Cells possess one or two large lateral chromatophores with pyrenoids and single nucleus. Sexual and asexual plants are morphologically indistinguishable exhibiting isomorphic alternation of generations. Reproductive cells of the thallus are unaltered or slightly enlarged.

Genus: *Ulva*

Ulva, commonly known as **sea lettuce,** is rather a marine with the exception of a few species that grow in brackish water. The alga is a very interesting one for, besides having advanced parenchymatous thalloid type of vegetative body (Fig. 4.19A), it exhibits in its life cycle an **isomorphic alternation of generations** which has great resemblance with higher plants.

The expanded sheet of thalloid body is composed of two layers of cells which are developed from a single uniseriate filament expanding by lateral divisions. The basal part of the thallus is narrowed into a short stalk,

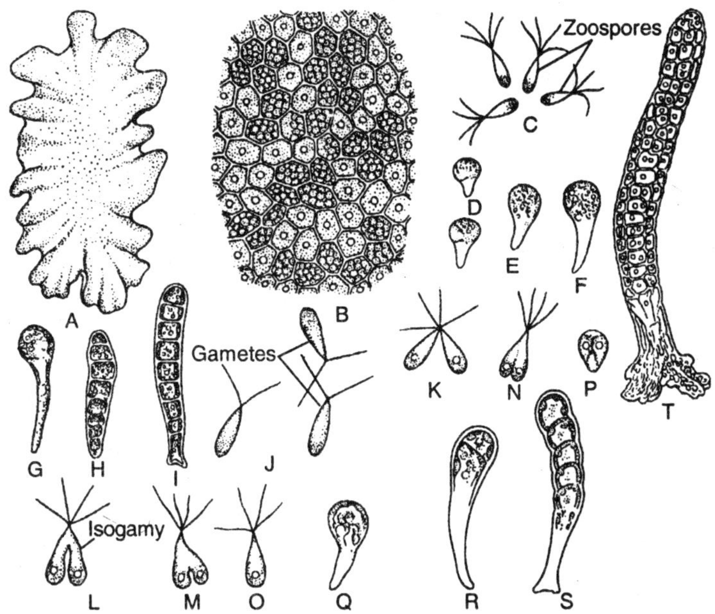

Fig. 4.19: *Ulva* sp. A. Parenchymatous thallus. B. Cells of thallus. C. Zoospores. D-I. Stages of zoospore germination leading to the development of young haploid thallus. J. Gametes. K-O. Stages of gametic union and zygote formation. P-T. Germination of the zygote and development of young diploid thallus

which ultimately develops into a holdfast of multinucleate rhizoids growing out from the uninucleate lower cells, with which the alga remains attached to its substrate. Each cell of the thallus is uninucleate, and contains a single laminate to cup-shaped chloroplast with a single pyrenoid (Fig. 4.19B). Cell division may take place at any part in the thallus.

Vegetative multiplication is practically unknown except in places where the alga grows in quiet water.

Asexual reproduction takes place by means of quadriflagellate zoospores (Fig. 4.19C) that are formed from the cells of a diploid thallus—sporophytic generation. These zoospores germinate to produce haploid thalli—sporophytic generation. These zoospores germinate to produce haploid thalli—gametophytic generation, whose cells again produce biflagellate gametes (Figs. 4.19D to J). The thalli are heterothallic. Both sporophytic and gametophytic generations are morphologically indistinguishable. The method of production of zoospores and gametes is identical except that during zoospore formation the division of the nucleus is reductional. The isogametes fuse to form a zygote which loses its flagella and germinates immediately and by division and redivision develops into a diploid thallus (Figs. 4.19K to T). Most species are isogamous, but a few are anisogamous.

The nature of alternation of generations in *Ulva* is isomorphic (Fig. 4.20).

Some Indian Species: *Ulva lactuca* Linn.; *U. lobata* Kg.; *U. reticulata* Forsk.

Special features

1. Advanced parenchymatous expanded sheet of thalloid vegetative body with holdfast.
2. Laminate to cup-shaped chloroplast.
3. Asexual reproduction by quadriflagellate zoospores formed from the cells of a diploid thallus (sporophyte).
4. Meiosis during zoospore formation.

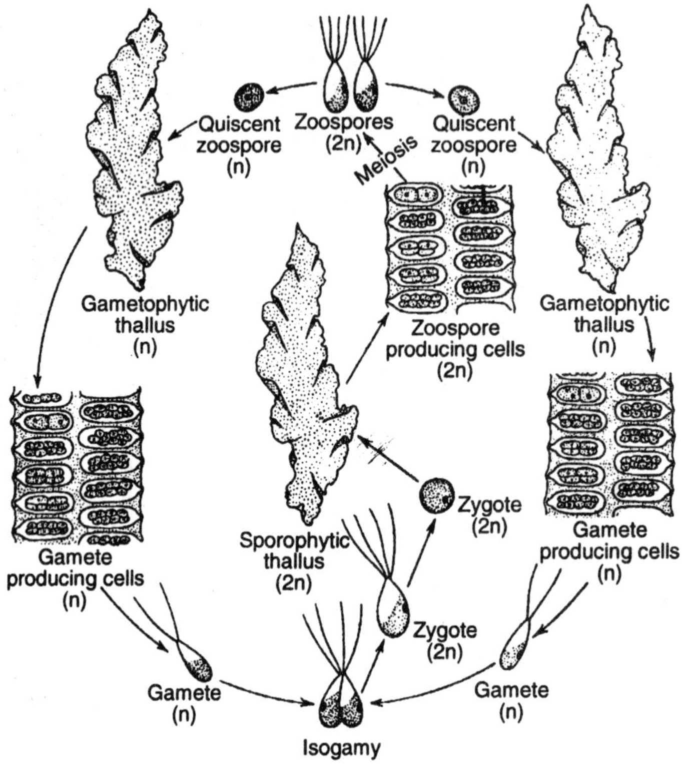

Fig. 4.20: Life cycle of *Ulva* sp.

5. Sexual reproduction is usually isogamous, in a few species anisogamous; gametes developed from the cells of the haploid thallus (gametophyte).
6. Exhibits an isomorphic alternation of generations which has great resemblance with higher plants.

Order: *CLADOPHORALES*

Plants are filamentous, usually with a distinct basal holdfast, simple or branched. The filaments are uniseriate having cells with a thin to thick firm wall, a large central vacuole, a generally multinucleate protoplast with a reticulate or perhaps ultimately fragmented chloroplast and one to many pyrenoids.

Family: *CLADOPHORACEAE*

This family contains many fresh-water forms. Some of the large genera, such as *Cladophora,* have both fresh-water and marine representatives, but there are several strictly marine genera. The plants are composed of multinucleate cells arranged in uniseriate fashion in simple or branched filaments. The chloroplast is parietal and reticulate. The filaments are usually attached by rhizoids. Generic separations have traditionally been made on vegetative characters such as the presence or absence of branches, the kind of branches, and the form of the basal cell and its attachment rhizoids.

Genus: *Cladophora*

Both fresh-water and marine species are quite common, growing abundantly in shallow water which may be stagnant or running. They inhabit mainly being attached to rocks and other objects under water including submerged water plants and Mollusca shells by fairly long rhizoidal branches. Filaments are typically branched with numerous cross-walls separating

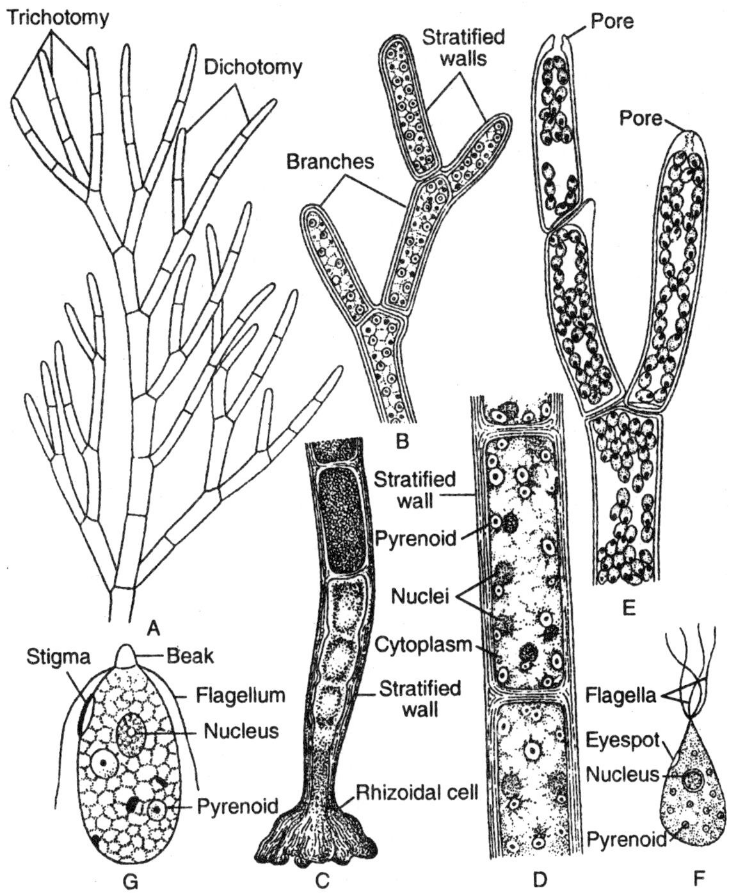

Fig. 4.21: *Cladophora* sp. A. Portion of a filament showing nature of branching. B. Filament cells magnified. C. Lower portion of the filament. D. Structural details of cells. E. Finer branches with zoospore primordia. F. Zoospore. G. Gamete

the coenocytic segments. Growth in length of the different branches is practically restricted to the apical cell of the filament. Branches nearly always arise just beneath a septum resulting the appearance of dichotomy or trichotomy (Fig. 4.21A). Elongated cells with thick and distinctly stratified walls, are sometimes encrusted with carbonate of lime. Walls are composed of an inner cellulose layer, an outer chitinous layer, and a pectic layer in between (Figs. 4.21B and C). Cytoplasmic strands traverse the large central vacuole of a mature cell. Each cell, contains a reticulate chloroplast bearing pyrenoids, or there may occur numerous discoid chloroplasts too. Internal to the chloroplast or chloroplasts as the case may be, usually numerous nuclei are embedded (Fig. 4.21D).

The perennial species of *Cladophore* are provided with a method of tiding over the unfavourable period of growth. The rhizoidal branches of the filaments undergo repeated divisions producing short branched filaments, the cells of which are filled with food material. These cells survive after the main parts of the filaments are dead and grow into new filaments in the next season. In some of the fresh-water species certain cells of the filament may become swollen to form akinetes which develop into new branches.

Asexual reproduction is accomplished by the production of numerous small, pear-shaped, diploid quadriflagellate zoospores (Fig. 4.21F). The zoospores arise by simultaneous division of the cell contents into numerous parts. Normally zoospores formation is confined to the finer branches (Fig. 4.21E) and takes place in basipetal succession, but there are cases where zoospores may be formed from any cell except the basal rhizoidal cell. The zoospores germinate directly into new diploid filaments.

The gametes are biflagellate and their formation takes place like the zoospores (Fig. 4.21G). The fusing gametes are of same size and thus the process of sexual reproduction is isogamous. The gametes from two different filaments fuse in pairs. Hence *Cladophora* is heterothallic. The zygote germinates directly into a diploid filament immediately after formation without undergoing any resting period or meiosis. The resulting diploid filament is known as the sporophyte and resembles the haploid gamete producing filament—the gametophyte. The diploid plant developing from the diploid zygote produces haploid biflagellate zoospores as a result of meiotic division. These zoospores germinate to form haploid filaments which resemble the diploid filaments morphologically. Hence morphologically identical hap-

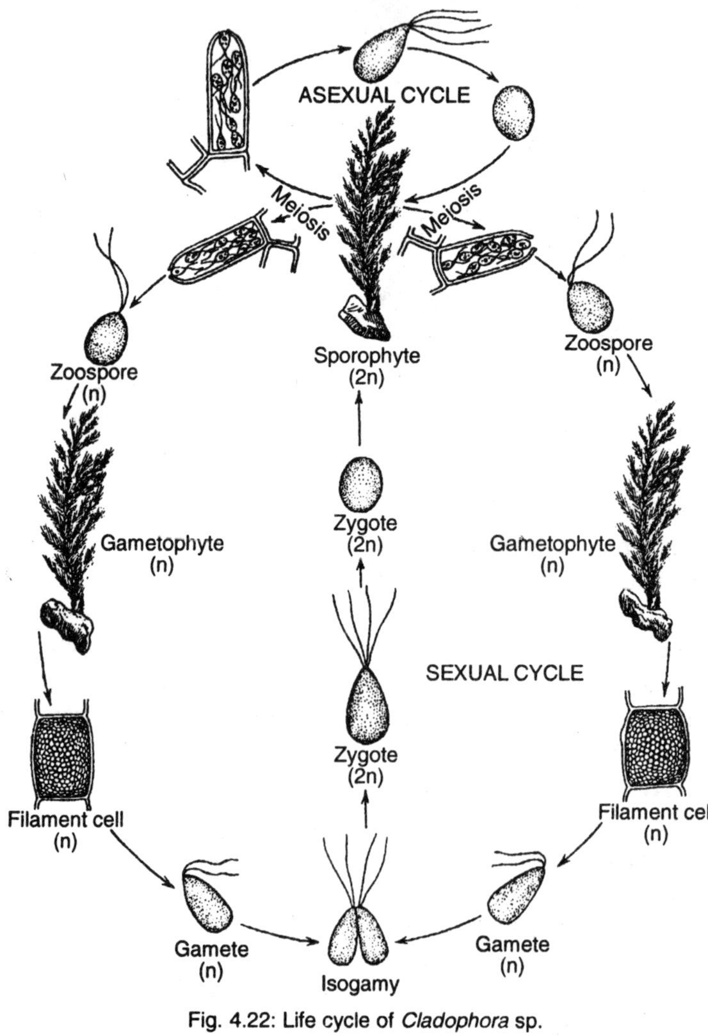

Fig. 4.22: Life cycle of *Cladophora* sp.

loid and diploid filaments alternate with each other to complete the life cycle indicating a distinct alternation of gene-rations which is again known as **isomorphic** or **homologous alternation of generations** (Fig. 4.22), the difference being in cytological details.

Cladophora glomerata, a fresh-water species present a wholly different type of life cycle. This difference may be compared with the various cycle found for *Ectocarpus siliculosus* under different conditions. In *C. glomerata* gametes and zoospores both are formed on diploid filaments and meiosis takes place only during gamete formation so that there is no haploid generation. The diploid filaments propagate by a long succession of diploid zoospores (Fig. 4.23). A sexual diploid plant in which only the gametes are haploid is known as **diplont.**

Some Indian Species: *Clodophora bengalensis* Mart.; *C. crispata* (Roth.) Kutz.; *C. furcata.; C. glomerata* (L.) Kutz.; *C. incurvata* West et West; *C. rottleri* Kutz.; *C. scitula* (Suhr.) Kutz.

Special features

1. Filamentous branched plant body.
2. Cell wall is thick and stratified.

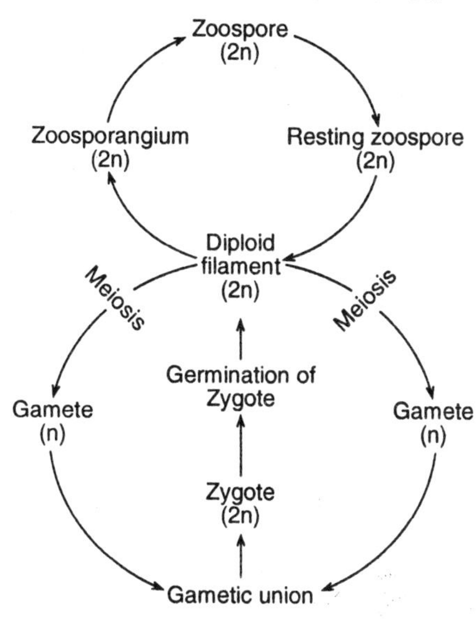

Fig. 4.23: Life cycle pattern of *Cladophora glomerata*

3. Cells multinucleate.
4. Asexual reproduction by diploid quadriflagellate zoospores.
5. Sexual reproduction by isogamy.
6. Exhibits isomorphic alternation of generations.

Order: *CHAETOPHORALES*

The Chaetophorales are characterized by a more or less complex organization of plant body being composed of heterotrichous filament. A distinct prostrate system gives rise to upright or erect threads constituting projecting system. In some cases the prostrate system is predominant over the projecting system; in others reverse is the condition. In majority of the genera hair-like projections—**setae** are developed. Asexual reproduction is by zoospores and sexual reproduction is from isogamy to oogamy.

Family: *CHAETOPHORACEAE*

Chaetophoralean algae are heterotrichous in habit frequently bearing terminal or lateral hair-like projections. In their heterotrichy both the erect and prostrate portions of the plant may be well developed (e.g., *Stigeoclonium*) or with an elaborate erect portion and feebly developed prostrate portion (e.g., *Chaetophora, Draparnaldia, Draparnaldiopsis, Fritschiella*). Cells are with him to gelatinous walls, usually uninucleate having a single more or less dissected plate-like or ring-like chromatophore with pyrenoids. Asexual reproduction is by biflagellate or quadriflagellate zoospores and sexual reproduction isogamous.

Genus: *Chaetophora*

This alga occurs abundantly in fresh-water being attached to some substratum preferably to submerged plant parts exhibiting a gelatinous macroscopic growth. Outstanding characteristics lie in the form of the vegetative body known as the **heterotrichous** filament being differentiated into feebly developed pros-

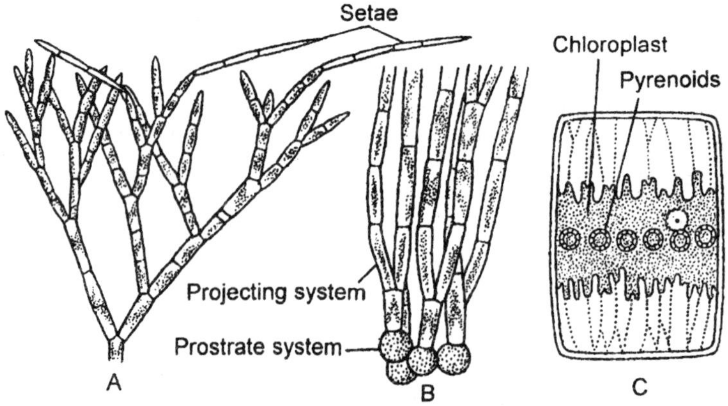

Fig. 4.24: *Chaetophora elegans*. A. Projecting system of branched filament. B. Prostrate system of rounded cells along with branched aerial filaments. C. A single cell showing chloroplast and pyrenoids

trate system often represented by loosely connected rounded cells, and a projecting system known as **setae** (Figs. 4.24A and B). The aerial branches are held together within a compact mucilage forming a tough envelope. The prostrate system may be impregnated with calcium carbonate. Cell structure and the mode of reproduction being very similar to *Ulothrix* may suggest common ancestry and development being along two divergent lines. Each cell contains a single banded chloroplast with several pyrenoids (Fig. 4.24C). Asexual reproduction is by quadriflagellate zoospores. Sexual reproduction is isogamous by the formation of biflagellate gametes.

Special features

1. Vegetative body heterotrichous being differentiated into feebly developed prostrate system and an elaborate projecting system of branched filaments, apices of which are terminated by long hyaline hair-like structures known as setae.
2. Banded chloroplast.
3. Isogamous sexual reproduction by the formation of biflagellate gametes.

Genus: *Draparnaldia*

Draparnaldia is a fresh-water alga. Its vegetative body indicates evolutionary tendency towards the development of elaborate aerial system as against a very insignificantly developed prostrate portion. The plant body is macroscopic and is pale-green in colour (Fig. 4.25A). The projecting part of the alga is composed of a main axis consisting of large, more or less barrel-shaped cells with barrel-shaped chloroplast having toothed edges forming a girdle in the equatorial region (Fig. 4.25B). The chloroplasts have several pyrenoids (Fig. 4.25D). The cells of the main axis are all alike and are uninucleate. From these cells arise short lateral branches in clusters, alternate, opposite, or verticillate, richly branched whose apices are drawn out into long hyaline setae. These branches may be borne on any cell of the main axis except the lower region from which arise numerous multicellular rhizoidal branches serving the function of anchorage of the alga to its substratum (Fig. 4.25C). Cells of the lateral branches are cylindrical, much smaller than those of the main axis, and are uninucleate with a more or less parietal chloroplast with a single pyrenoid. The main axis is completely covered by lateral branches which grow very profusely. Formation of elements connected with reproduction is confined to the short branches.

Asexual reproduction is by means of quadriflagellate macro- and microzoospores

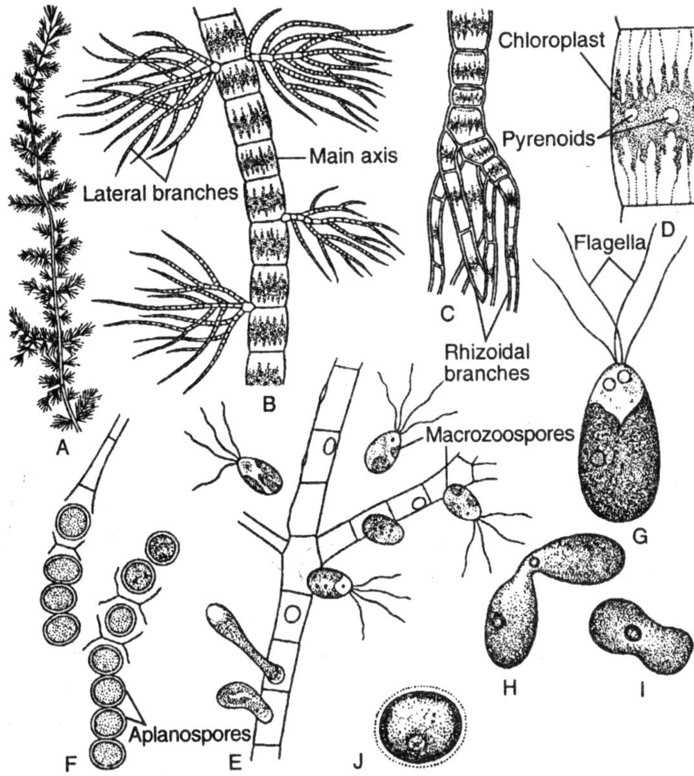

Fig. 4.25: *Draparnaldia* sp. A. General appearance of the plant body. B. Portion of plant body showing main axis and lateral branches. C. Prostrate portion of the plant body. D. A single cell showing barrel-shaped chloroplast with toothed edges and having several pyrenoids. E. Liberation of macrozoospores. F. Aplanospores. G. Gamete. H-I. Stages in isogamy. J. Zygote

and by aplanospores (Figs. 4.25E and F). The zoospores are pear-shaped. They are produced one to four in each cell. The zoospores are formed simultaneously in all the cells of the lateral branches and are liberated through an aperture in the lateral wall (Fig. 4.25E). The zoospores swim around in water, lose their flagella and develop into new plants. During unfavourable conditions cell contents of the lateral branches may develop into thick-walled aplanospores instead of producing zoospores. The contents of each cell recede from the cell wall and the whole structure comes to rest by developing a thick wall. On the return of favourable conditions the old wall is cast off and the new walls are developed. It then directly gives rise to a new individual.

Sexual reproduction is isogamous. The isogametes are quadriflagellate (Fig. 4.25G) like the zoospores, they are also produced in the cells of the lateral branches and are liberated out through an aperture of the lateral wall. Prior to union, the gametes lose their flagella and become amoeboid (Fig. 4.26). The amoeboid gametes fuse in pairs resulting in the formation of a zygote (Figs. 4.25H and I). The zygote develops a wall (Fig. 4.25J) and germinates into a new plant. Behaviour of the zygotic nucleus has not yet been confirmatively worked out. Cases of parthenospore development have been reported.

Some Indian Species: *Draparnaldia distans* Kg. var. *elongata; D. glomerata* Agardh; *D. groenlandica* Bachmann; *D. mutabilis* (Roth.) Bory.; *D. plumosa* (Vuch.) Ag.

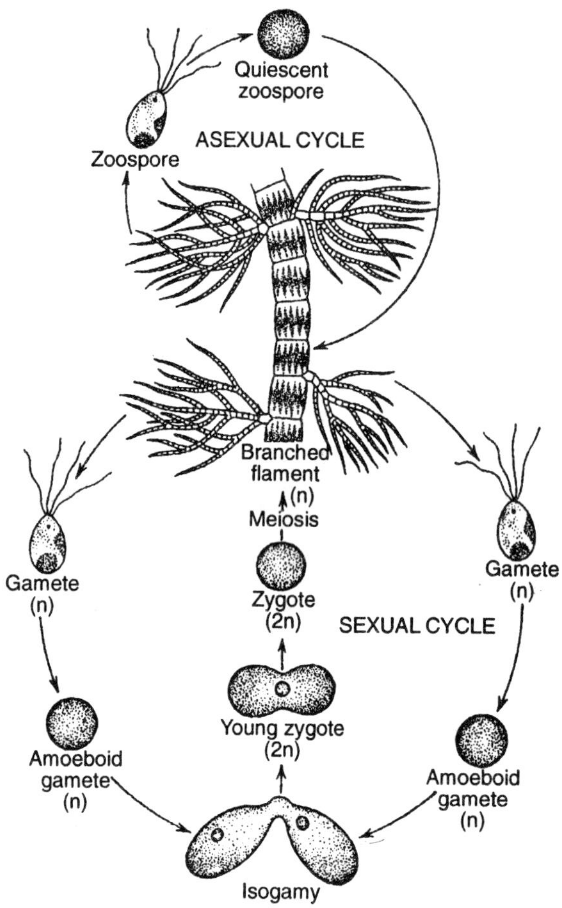

Fig. 4.26: Life cycle of *Draparnaldia* sp.

Special features

1. Heterotrichous macroscopic plant body having an elaborate aerial system and a very insignificantly developed prostrate system.
2. Barrel-shaped chloroplast with toothed edges.
3. Asexual reproduction by quadriflagellate macro- and microzoospores and by aplanospores.
4. Isogamous sexual reproduction by quadriflagellate gametes. Prior to union the gametes lose their flagella and become amoeboid and then fuse in pairs.

Genus: *Draparnaldiopsis*

Draparnaldiopsis is a fresh-water alga usually found in shallow pools and ponds. It grows affixed to aquatic angiosperms. The heterotrichous plant body has a remarkably elaborate aerial portion and an inconspicuous prostrate portion represented by rhizoidal branches with which the alga attaches itself to its substratum (Figs. 4.27C, D and F). In some species the branches have the tendency to coil around the main axis (Fig. 4.27A) giving a somewhat corticated appearance. The entire plant body is enveloped by a thick gelatinous sheath. Cells are uninucleate having zonate chloroplasts with several pyrenoids (Fig. 4.27B).

The alga reproduces asexually by three types of zoospores; quadriflagellate ellipsoidal to oval macrozoospores; quadriflagellate spherical microzoospores; and biflagellate spherical, ovoid to obovoid microzoospores (Figs. 4.28G to J). All the three types of zoospores are formed in the same plant and only from the cells of the lateral branches and never from the main axis. All the cells of the lateral branches except the basal one are capable of producing zoospores. Each cell produces a single zoospore (Figs. 4.28E and F). Both quadriflagellate macro- and microzoospores are developed out of the cells of the lateral branches of unlimited growth (Figs. 4.28A to F). They are diploid in nature and produce diploid plants—the **sporophytes**, whereas the biflagellate microzoospores are formed from the cells of the lateral branches of limited growth, during which meiosis takes place. These zoospores after being liberated swim around in water, lose their flagella, round up and produce new individuals resembling parent plant—the sporophyte. But these are haploid gamete producing plants—the **gametophytes.**

Sexual reproduction is usually isogamous, but cases of anisogamy have also been reported. Similar to the zoospores, biflagellate spherical to ovoid gametes (Figs. 4.27B and C) are pro-

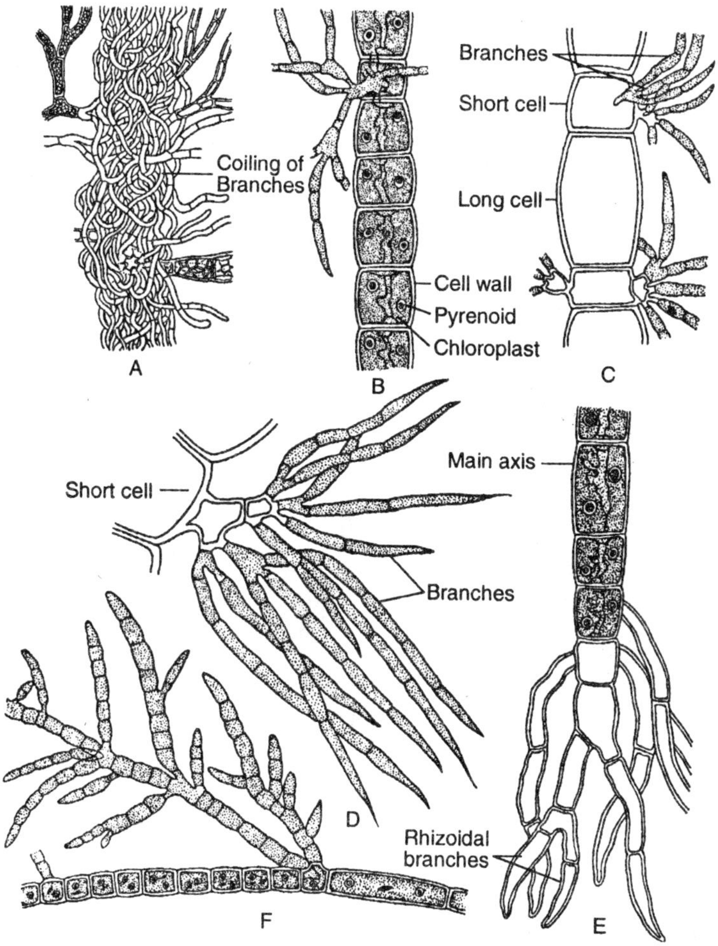

Fig. 4.27: *Draparnaldiopsis indica*. A. Portion of the plant body showing coiling of branches. B. Portion of the plant body showing cell structure. C. Main axis with long and short cells, short cells bearing branches. D. Short cell bearing branches. E. Main axis showing rhizoidal branches. F. Nature of branches

duced only from the cells of the lateral branches of limited growth borne on the gametophytic plants. Prior to the development of the gametes the cells divide into a row of daughter cells. The contents of each daughter cell metamorphose into one to several gametes (Fig. 4.29A) which on being liberated out by the gelatinization of the cell wall of the parent plant swim in water. The gametes from two different gametophytes pair together at the anterior end and finally fuse with each other resulting in the formation of a zygote (Figs. 4.29D to G). The zygote germinates immediately into new plant without going to rest (Figs. 4.29H to L) and produces a diploid plant—the **sporophyte** which resembles the gametophytic plant morphologically (Fig. 4.29). The alga *Draparnaldiopsis* thus exhibits **isomorphic alternation of generations** (Fig. 4.30).

Draparnaldiopsis has four main features which suggest affinity with the Charophyceae: (i) it is differentiated into nodes and internodes, (ii) it has a cortex-like covering derived from the basal cells of the laterals, (iii) the laterals have limited growth and, (iv) the reproductive organs are restricted to the laterals.

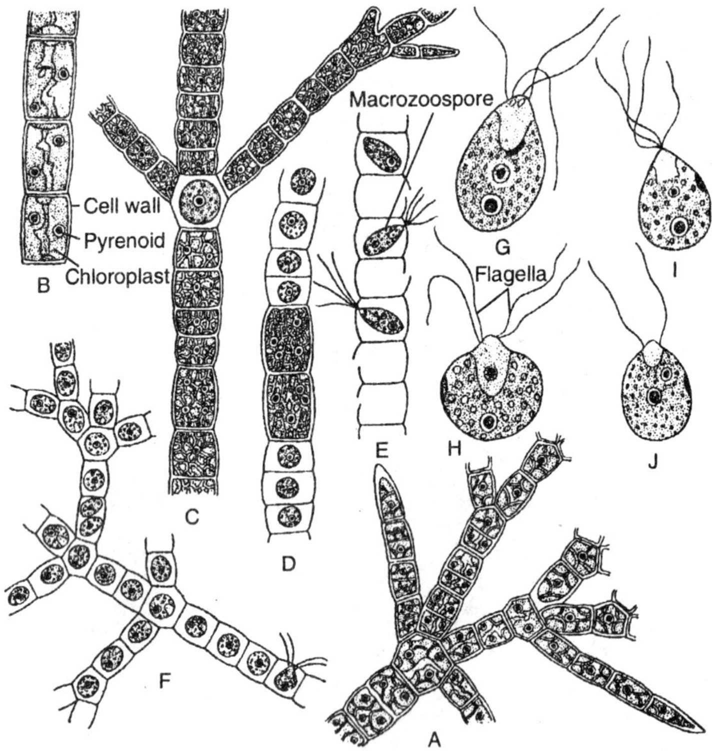

Fig. 4.28: *Draparnaldiopsis indica.* A. Portion of lateral branch. B. Cells magnified. C-F. Stages of development of macrozoospores in cells of lateral branches. G-J. Different types of zoospores

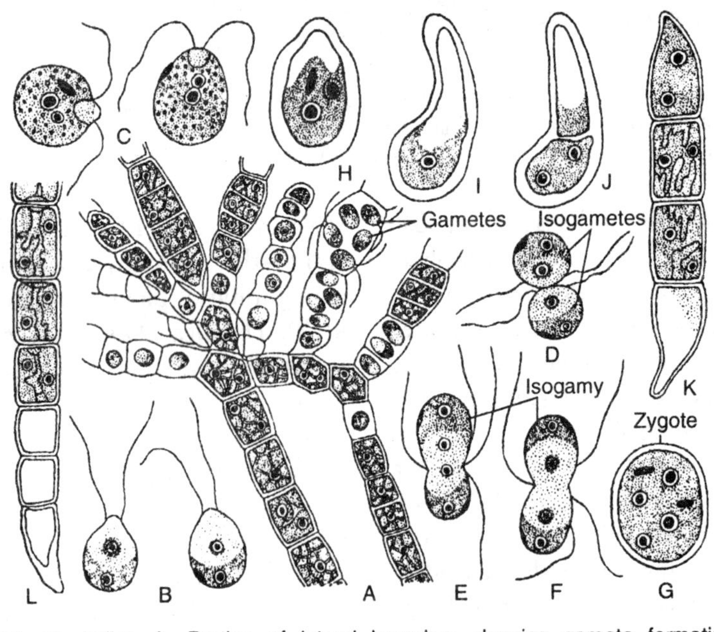

Fig. 4.29: *Draparnaldiopsis indica.* A. Portion of lateral branches showing gamete formation. B-C. Gametes. D-G. Stages in isogamy leading to zygote formation. H-L. Stages in germination of zygote

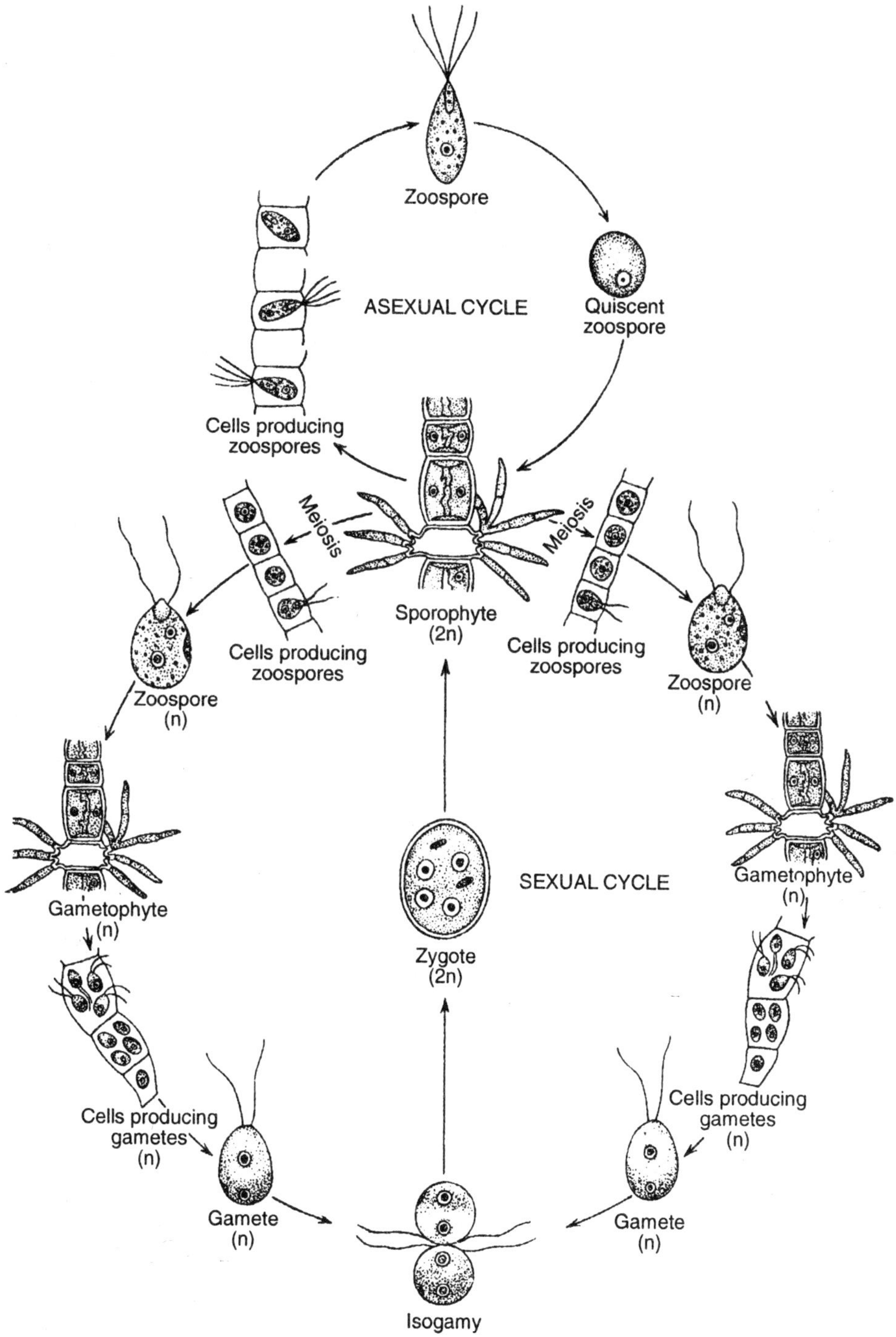

Fig. 4.30: Life cycle of *Draparnaldiopsis indica*

Some Indian Species: *Draparnaldiopsis indica* Bharadwaja; *D. indica* var. *robusta* Islam and Ahia.

Special features

1. Heterotrichous plant body composed of upright thread and poorly developed rhizoidal branch performing the function of anchorage.
2. The upright thread is differentiated into main axis and lateral branches. The main axis again, is composed of long and short cells arranged alternately, and the lateral branches are arranged in whorls originating from the short cells.
3. The alternate arrangement of short and long cells resembles nodes and internodes of higher plants.
4. Zonate chloroplast.
5. Asexual reproduction by quadriflagellate macro- and microzoospores and by biflagellate microzoospores.
6. All the zoospores are developed from the cells of lateral branches.
7. Sexual reproduction is isogamous, cases of anisogamy have also been reported.
8. Gametes are formed from the cells of lateral branches.
9. Isomorphic alternation of generations.

Genus: *Fritschiella*

Fritschiella grows gregariously on moist alkaline soil and on silt of drying pools and similar other habitats with a tendency of terrestrial habit. It has a highly developed heterotrichous plant body which is differentiated into a distinctly septate **rhizoidal system** consisting of downwardly directed rhizoid-like filament composed of elongated colourless cells, joined end to end; a **prostrate system** of rounded or irregularly swollen clusters of cells appearing to serve the purpose of perennation; a **primary projecting system** having short-celled branched threads; and **secondary projecting system** being composed of a tuft of slightly elongated branches with longer cells (Figs. 4.31A to C). In an immature plant, the prostrate system is very ill-developed in comparison with the projecting system. In a mature plant, only the secondary projecting system which arises as a tuft on the surface of the mud, is visible. The rest of the plant body remains buried in the damp soil. All the cells of the plant body are uninucleate. The cells of the rhizoidal system have scanty contents. Whereas, the secondary projecting system has cells with dense contents having curved plate-like chloroplast containing two to eight pyrenoids in each (Figs. 4.31D and E).

The alga produces three types of swarmers. The formation of swarmers is strictly confined to the prostrate system portion. The swarmers are: the quadriflagellate macrozoospores formed singly in each cell; the quadri-, biflagellate microzoospores developed two to four in each cell; and the biflagellate isogametes are formed in large numbers in each cell (Figs. 4.31F to J). Both macro- and microzoospores (Figs. 4.31K and L and Figs. 4.32A and B) are formed simultaneously by the same plants—the gametophytes. Both kinds of zoospores on liberation germinate to produce a new plants (Figs. 4.32C to N). The new plants so produced though morphologically indistinguishable, possess different nuclear constitution. The ones produced by the quadriflagellate zoospores are diploid—sporophytic plants, and those by biflagellate zoospores are diploid—sporophytic plants, and those by biflagellate zoospores are haploid—the gametophytic plants, since meiosis takes place during the formation of biflagellate zoospores.

Sexual reproduction is isogamous. The haploid gametophytic plants produce gametes. The gametes after being liberated from two different gametophytic plants, swim around in moisture for a while, meet in pairs (Figs. 4.32O and P). They then lose their flagella, gradually

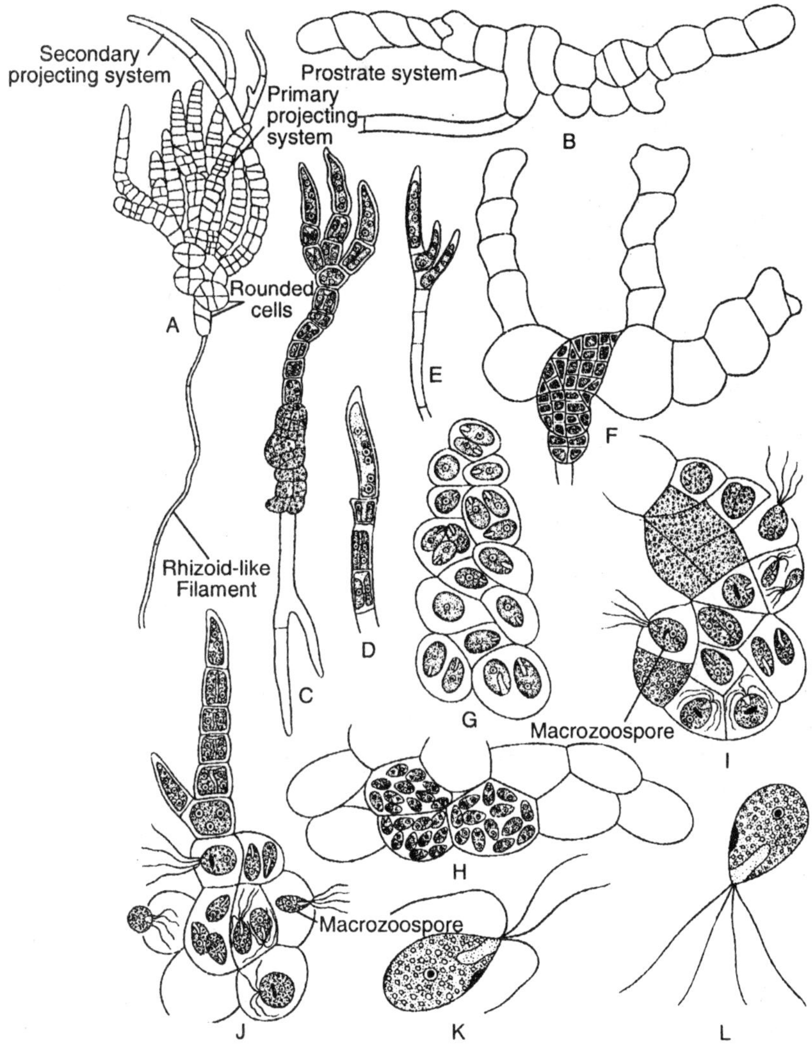

Fig. 4.31: *Fritschiella tuberosa*. A. Habit. B. Prostrate system magnified. C. Young thallus. D-E. Portions of secondary projecting system. F and H. Cells of prostrate system developing gametes. G, I-J. Cells of prostrate system developing macrozoospores. K-L. Macrozoospores

round off forming a zygote which germinates directly without any rest (Figs. 4.32Q to U) producing a diploid plant—the sporophyte which morphologically resembles the haploid gametophytic plants. The alga thus exhibits **isomorphic alternation of generations** (Fig. 4.33).

The life history of the alga exhibits certain points of interest. They are: the tendency of terrestrial habit accompanied with normal growth under highly alkaline condition; very clear distribution of labour indicated by photosynthetic subaerial portion, the prostrate system serving as an organ for the production of reproductive structures as well as perennation, and the underground rhizoidal portion performing the function of anchorage; the thallus resembles very well the gametophyte of some species of *Equisetum*. Besides all these, the alga exhibits isomorphic alternation of generations and the sexual reproduction is isogamous.

Indian Species: *Fritschiella tuberosa* Iyeng.

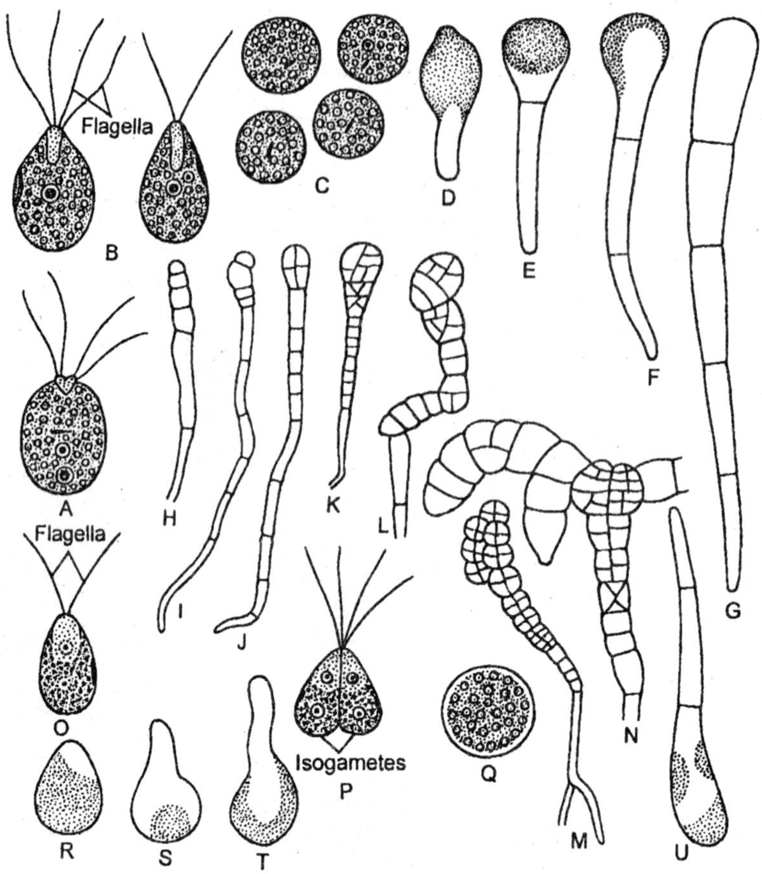

Fig. 4.32: *Fritschiella tuberosa.* A-B. Different types of zoospores. C-N. Stages in development of new individual from zoospores. O. Gamete. P. Pairing of gametes. Q. Zygote R-U. Stages in germination of zygote

Special features

1. Highly developed heterotrichous plant body differentiated into septate rhizoidal system, prostrate system of rounded or irregularly swollen clusters of cells, and erect system of a primary projecting system composed to short-celled branched threads and a secondary projecting system being composed of a tuft of slightly elongated branches with longer cells.
2. Curved plate-like chloroplast.
3. Asexual reproduction by quadriflagellate macro- and microzoospores and by biflagellate microzoospores.
4. Isogamous sexual reproduction by biflagellate gametes.
5. All swarm cells are developed from the cells of the prostrate system.
6. Isomorphic alternation of generations.

Family: *Coleochaetaceae*

The family Coleochaetaceae is represented by a single genus *Coleochaete.* This alga usually occurs as an epiphyte on fresh-water plants such as *Hydrilla* and *Potamogeton,* and also occurs attached to the fronds of *Lemna.* It has also been reported to grow endophytically in the cells of *Chara* and *Nitella.* Its plant body exhibits different degrees of heterotrichy-like— a definite prostrate system giving rise to the

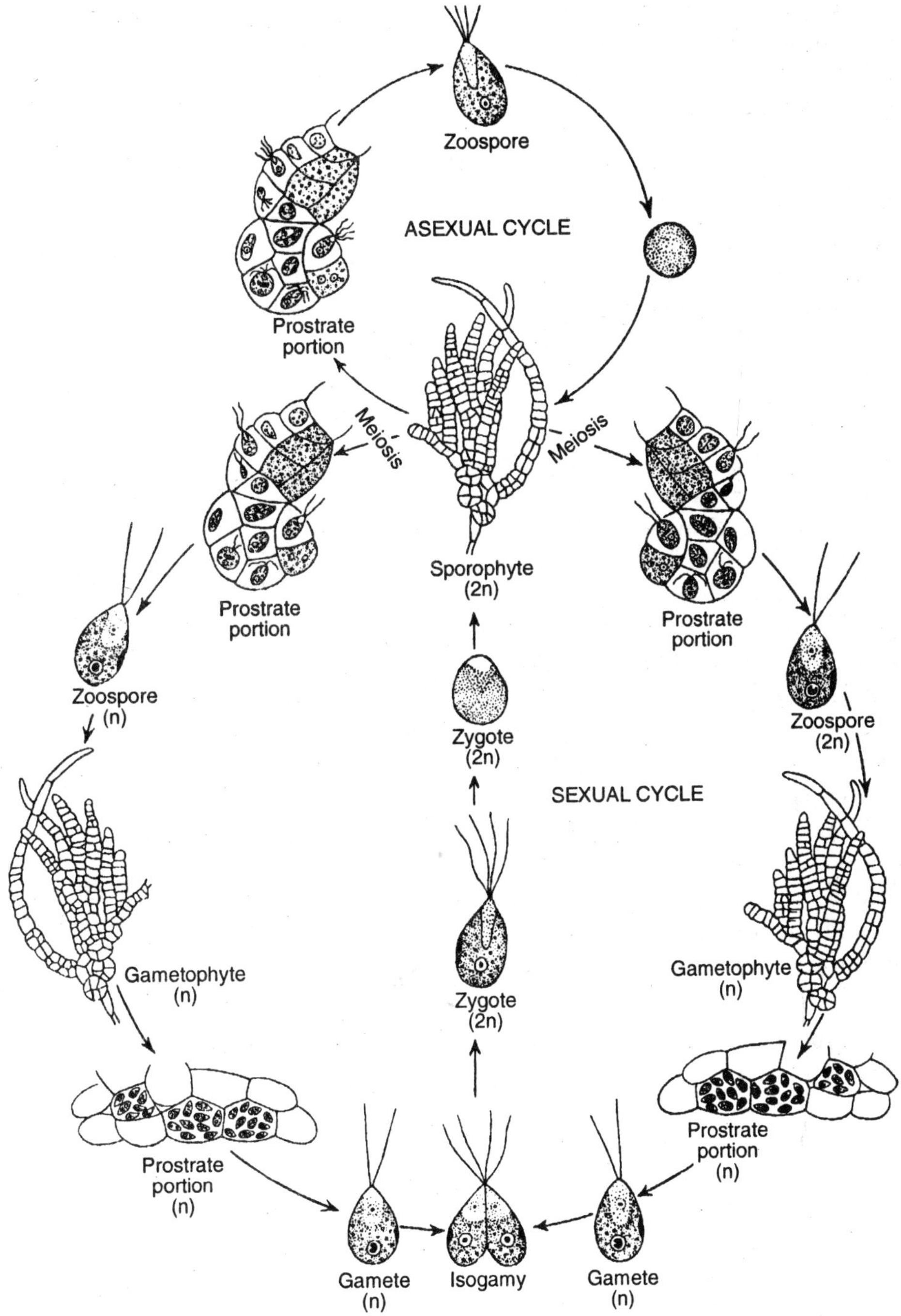

Fig. 4.33: Life cycle of *Fritschiella tuberosa*

projecting system, and only prostrate system present and the projecting system has disappeared. The projecting system grows by an apical cell and the prostrate system by means of marginal meristem. The vegetative cells are characterized by a laminate chloroplast and by the presence of sheathed bristles or hairs. Reproduction is both by asexual and sexual means. Asexual reproduction takes place by biflagellate zoospores. Sexual reproduction is oogamous. The ovum is borne in an oogonium having a long neck, the **trichogyne**. It is fertilized by a biflagellate antherozoid produced in an antheridium. The zygote is retained within the enlarged basal portion of the oogonium which is surrounded by pseudoparenchymatous investment. The first division of the zygotic nucleus is meiotic and by subsequent mitotic division 32 biflagellate zoospores are produced which give rise to new thalli.

Genus: *COLEOCHAETE*

This is an epiphytic alga occurring on fresh-water plants, for example, *Hydrilla, Potamogeton, Lemna,* etc. Some species may be also endophytic inhabiting in the cells of *Nitella* and *Chara.* There are quite a few heterotrichous species whilst others appearing either as hemispherical cushions composed of prostrate threads with branches distinct from one another exhibiting apical growth or as one-layered parenchymatous discs (Figs. 4.34A and B). The thallus is often enveloped in mucilage. In all these forms the thallus bears a single, long, unbranched cytoplasmic seta whose base is ensheathed by a gelatinous material (Fig. 4.34C). The cells are uninucleate with a large parietal chloroplast containing one or two pyrenoids.

Asexual reproduction is effected by large ovoid biflagellate zoospores (Fig. 4.34H) produced singly in the ordinary cells of the thallus. The zoospore escapes through a small pore formed at the apex of a short papilla. On germination the zoospore divides either horizontally or vertically into two daughter cells, one of which divides to form hair-like structure whilst the other develops into creeping thread.

Sexual reproduction is a type of advanced oogamy. There are both homothallic or heterothallic species. The oogonium is a flask-shaped structure with a swollen base and colourless neck, the **trichogyne** (Fig. 4.34I). Oogonia are borne terminally on short branches of the projecting system. In the discoid forms, their position is also terminal (Figs. 4.34E to G). The antheridia are developed in clusters at the ends of the projecting branches from which they are cut off as small colourless cells (Fig. 4.34D). In the discoid thallus, they are developed from marginal or midway cells. Each antheridium produces a single oval or spherical biflagellate antherozoid which is set free by the breaking down of the wall at the apex of the anteridium. The liberated antherozoid then swims towards the opened neck of the oogonium, enters through the neck, and finally fuses with the egg within the oogonium. After fertilization the neck of the oogonium is cut off by a septum, the basal part enlarges, the zygote secretes a heavy wall, and increases greatly in size (Figs. 4.34J and K). At the same time branches arise from the underlying and other adjacent cells and unite to form a complete pseudoparenchymatous envelope which more or less completely encloses the base the oogonium (Fig. 4.34L). The pseudoparenchymatous investment takes a red or reddish-brown colour. The entire structure is known as a **spermocarp** (Figs. 4.34B and M). The alga perennates in the spermocarp stage. With the advent of a favourable season, the zygote nucleus divides reductionally. With further division as many as 32 daughter nuclei are formed. Cytoplasmid division by the production of wall perpendicular to the long axis of the original oogonium continues simultaneously with the nuclear division ultimately forming 32 wedge-shaped cells. Each cell then gives rise to a zoospore by the bursting of the envelope. There is no difference in form and behaviour between these zoospores and the ordinary zoospores. All the zoospores are liberated out by the breaking of the spermocarpic wall into two halves (Fig. 4.34M). The zoospores then develop direction into new thalli after swarming for a short period of time (Fig. 4.35).

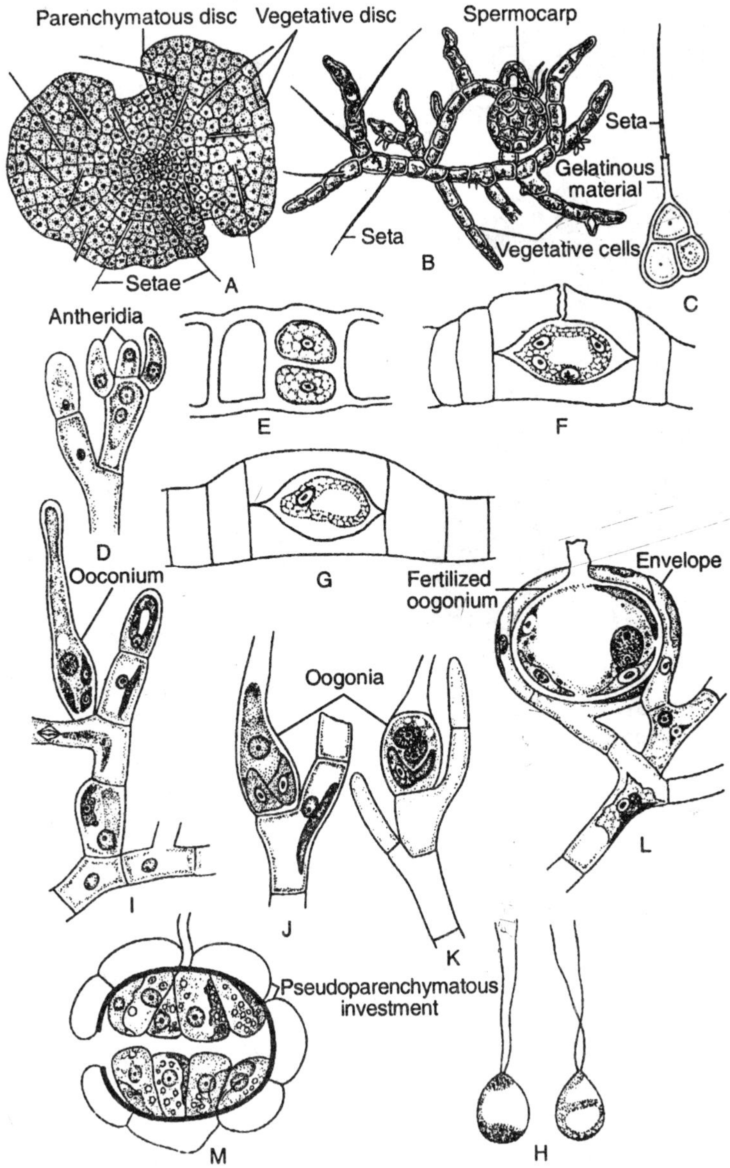

Fig. 4.34: *Coleochaete* sp. A. Parencymatous discoid thallus bearing setae. B. Portion of thallus showing seta and spermocarp. C. Very early stage of thallus showing seta with gelatinous material. D. Branch with antheridia. E-G. Stages in oogonial development in some discoid forms. H. Zoospores. I. Branch with oogonium. J-K. Stages in sexual fusion. L. Fertilized oogonium with an envelope. M. Breaking of spermocarpic wall of pseudoparenchymatous investment

The mode of development, structure, and behaviour of the spermocarp need particular discussion. During its development there is a clear mechanism to afford protection to the diploid oosporic nucleus against external adverse condition. That the first division of the diploid nucleus is meiotic has been clearly demonstrated by C.E. Allen in 1905. The ultimate production of zoospores by reduction division clearly indicates that the spermocarp which behaves as a sporophyte alternates with the thalloid gametophyte to complete the life

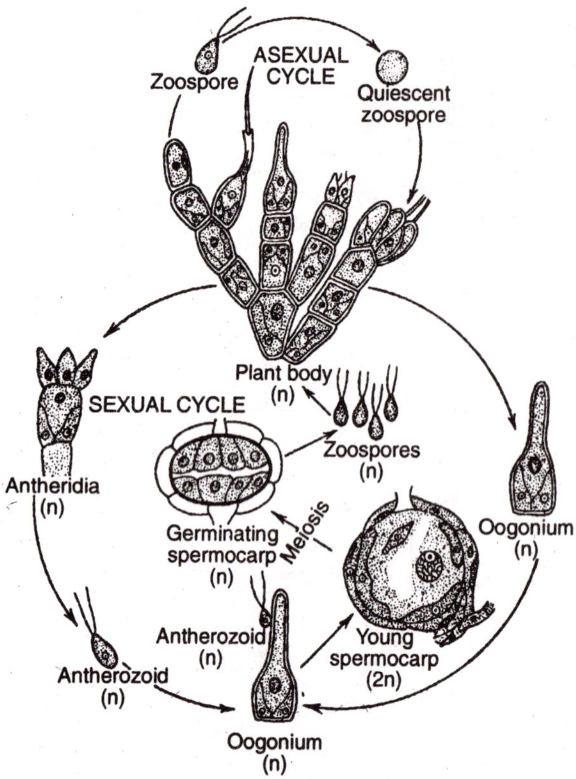

Fig. 4.35: Life cycle of *Coleochaete* sp.

cycle of *Coleochaete*. As such, there is somewhat indication of alternation of generations in the life cycle of this alga. Besides the development of spermocarp, the alga shows advancement in the thalloid gametophyte producing male and female gametes. The structure of the oogonium is also most advanced in the entire green algae for having long neck-like structure and a bulbous base.

Some Indian Species: *Coleochaete orbicularis* Pringsh.; *C. scutata* Breb.; *C. soluta* Pringsh.

Special features

1. Heterotrichous plant body with well-developed projecting system and a prostrate system represented by a thallus which appears to be a parenchymatous disc.
2. Asexual reproduction by biflagellate ovoid zoospores.
3. Sexual reproduction oogamous.
4. The oogonium is a flask-shaped structure composed of enlarged basal portion and a long neck—the trichogyne.
5. The male gametes are biflagellate oval.
6. Elaborate post-fertilization stages.
7. Formation of specialized structure—spermocarp during post-fertilization stages.
8. The spermocarp is composed of post-fertilization oogonium, enveloped completely by filamentous structures forming a pseudoparenchymatous investment.
9. Meiosis takes place in the spermocarp.
10. The spermocarp produces biflagellate swarmers which give rise to new individuals.

Family: *TRENTEPHOLIACEAE*

Members of this family grow as epiphytes and parasites on and in the leaves of various angiospermic plants. Some constitute the algal component of a number of lichens, *Cephaleuros virescens* causes the red rust of tea plant. The widely distributed genus *Trentepohlia* occurs attached to rocks, tree-trunks, etc. The plant body is composed of branched threads comprising of heterotrichous conditions. In the majority of cases the *Trentepohlia* appear as orange-yellow cushions due to the presence in the cells orange-red haematochrome dissolved in fat. Besides vegetative, reproduction is usually by the development of sessile, stalked, and funnel-shaped sporangia. The sessile sporangia often behave as gametangia and produce isogametes, the stalked sporangia bear bi- or quadri-flagellate swarmers, and the behaviour of funnel-shaped sporangia is still obscure.

Genus: *Trentepohlia*

It is an aerial alga growing epiphytically as orange-yellow cushions or felt-like layer on bark of trees and other damp situations. Plants can withstand prolonged desiccation without any appreciable change. Some species also constitute the algal component of a number of Lichens. The plant body is composed of heterotrichous filament. In the majority of the species, the thallus exhibits a differentiation of aerial as well as prostrate systems represented by richly branched threads. But there are some species where both prostrate and aerial systems are not equally well developed (Figs. 4.36, E and L). The branches of the aerial system are not equally well developed (Figs. 4.36D, E and L). The branches of the aerial system are opposite, alternate, or unilateral arising either from the tip of the parent cell or subterminally or even from the middle and are generally slightly attenuated (Figs. 4.36A to C). The cells are cylindrical or barrel-shaped and often length

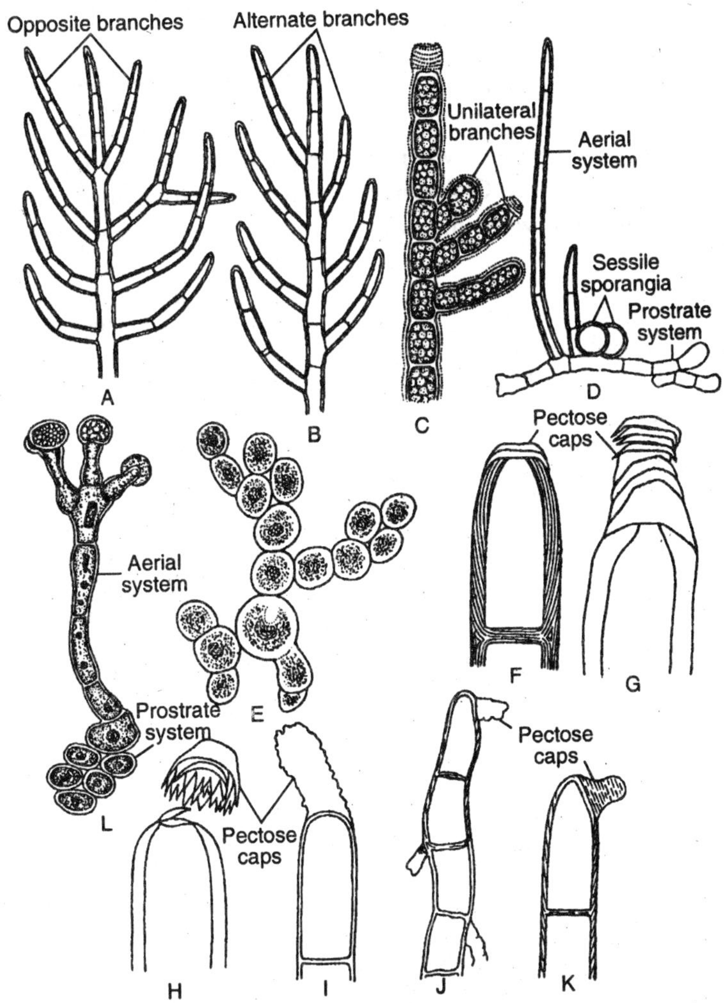

Fig. 4.36: *Trentepohlia* sp. A-C. Different forms of branches in the aerial system. D. Prostrate and aerial systems with sessile sporangia. E. Prostrate system. F-I. Structure and mode of formation of pectose caps on free ends of apical cells with stratified wall. J-K. Pectose caps pushed aside and remaining attached to the lateral walls with the growth of the apical cells. L. Pigmented cells of prostrate and aerial systems, aerial system cells having length greater than breadth

being twice the breadth (Fig. 4.36L). The cells have thick cellulose walls which are stratified with parallel or divergent layers (Figs. 4.36C and F). The free ends of the apical cells are often covered by **pectose caps** (Figs. 4.36F–I). In some species, during the growth of the apical cells, the pectose caps get pushed aside and remain attached to the lateral walls of the older cells (Figs. 4.36J and K). The cells have numerous band-shaped or discoid chloroplasts without pyrenoids. The characteristic orange-yellow colour of the cells is due to the presence of orange-red haematochrome dissolved in fat occurring as globules around the chloroplasts. The function of this special pigment is not yet clearly known. According to some, the haematochrome is a kind of reserve food, whereas others consider it as a light screen. Cells when young, are uninucleate and with maturity become multinucleate. Vegetative reproduction is through fragmentation. Bi- or quadriflagellate swarmers are developed in three different kinds of sporangia:

(i) **Sessile sporangia**—these may be terminal, lateral, intercalary, or rarely axillary in position (Figs. 4.36D and Figs. 4.37A to C). They are formed by enlargement of cells and liberate biflagellate swarmers without being detached from the mother plant. These swarmers behave as isogametes (Figs. 4.37N to R). Sessile sporangia are quite commonly produced on the prostrate portion of the plant. They behave as gametangia.

(ii) **Stalked sporangia** are formed terminally or laterally by cutting off from a tubular outgrowth developed from enlarged supporting cell. The apical portion of the outgrowth

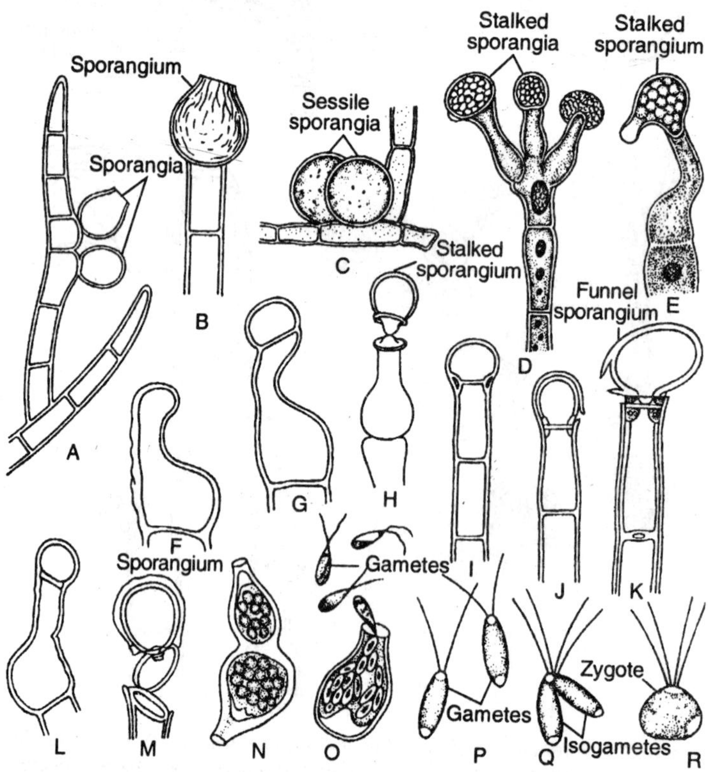

Fig. 4.37: *Trentepohlia* sp. A-C. Lateral, terminal and axillary sessile sporangia. D-E. Stalked sporangia. F-H. Formation of stalked sporangium. I-M. Development of funnel sporangium. N-O. Sporangia behaving as gametangia, liberation of gametes. P-R. Gametes, gametic union and zygote formation

swells to form sporangium and separates from the stalk cell underneath, which is often bent to assume a knee-shaped appearance (Figs. 4.37D and E). These sporangia produce bi- and quadriflagellate swarmers which behave as zoospores. Stalked sporangia are usually developed on the projecting system of the plant (Figs. 4.37F to H).

(iii) **Funnel-shaped sporangia** formed from the apices of the terminal cells, become detached at maturity (Figs. 4.37J-K). Their function has not yet been established.

Both sessile and stalked sporangia may be developed on the same plant or on separate plants. The development of sessile and stalked sporangia in the life cycle of the alga is suggestive of the possibility of alternation of generations.

Indian Species: *Trentepohlia monilia* de Wilderman.

Special features

1. Epiphytic alga forming orange-yellow cushions or felt-like layer on damp substrate.
2. Plant body with adaptations to withstand prolonged desiccation.
3. Heterotrichous plant body, aerial as well as prostrate systems with or without being equally well developed.
4. The branches of the aerial system are opposite, alternate, or unilateral.
5. Cells with stratified wall, special pigment—orange-red haematochrome and band-shaped or discoid chloroplasts.
6. Free ends of the apical cells are covered by pectose caps.
7. Bi- or quadri-flagellate swarmers are developed in specialized structures which according to the behaviour of the swarmers are gametangia or sporangia.
8. Biflagellate swarmers usually behave as isogametes and quadriflagellate one as zoospores.

Family: *Pleurococcaceae* (or *Protococcaceae*)

This family is best known for its genus *Pleurococcus*, also called *Protococcus*. It includes terrestrial algae occurring in all kinds of damp situations as a thin incrustation of green coat on the windward side of stones walls, tree-trunks, etc. The cells can withstand prolonged desiccation without change. The cells, which are globose in shape and occasionally branched, are single or else as many as four may be united into a group. Under certain cultural conditions branching may be copious. The sole method of reproduction is through vegetative division in three planes.

Genus: *Pleurococcus* (or *Protococcus*)

This is a simple alga growing in moist situations on the bark of trees, old walls, flower pots, and similar other places. It may occur singly as isolated individuals (Fig. 4.38A), or as small compact masses or groups of two, four or eight individuals (Figs. 4.38B to E). When

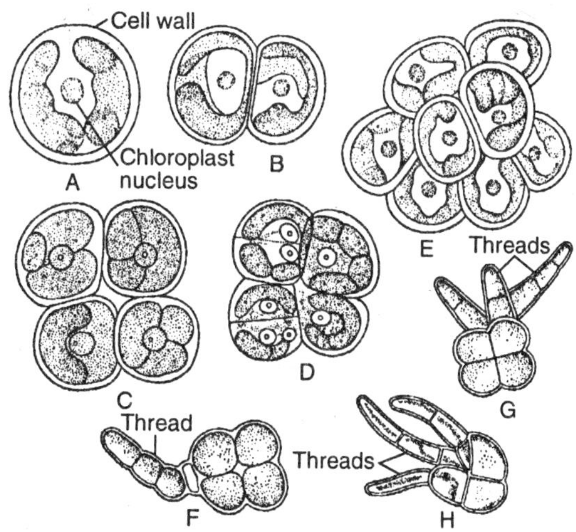

Fig. 4.38: *Pleurococcus* sp. A. A singe cell showing structural details. B-C. Cell division. D-E. Daughter cells adhered together. F-H. Thread formation

adhering together, the sides in contact are rather flattened. This unicellular alga carries on all essential life processes within the single cell. From the physiological point of view it has the same characteristics as higher plants. The cells can withstand prolonged desiccation. The resistance of the cells to desiccation is aided by a highly concentrated cell sap and a capacity to imbibe water directly from the air. Occasionally, when submerged under water, cell division may continue with the production of a colony or the daughter cells may grow out into short filaments or threads (Figs. 4.38F to H). For this character *Pleurococcus* is considered to be reduced and simplified form originating from a filamentous one in which both vegetative structures and reproductive functions have retrogressed. Each cell is spherical with a cellulose wall, and is densely filled with protoplasm containing a large lobed chloroplast occupying the peripheral portion of the cell, and a nucleus containing a nucleolus (Fig. 4.38A). The rest of the protoplasm is colourless. The only method of reproduction is by fission. The cells divide freely into two, with successive divisions taking place in three directions and at right angles to each other (Figs. 4.38B and C). The daughter cells may either round themselves off and separate immediately after each division, or may remain adhered together (Figs. 4.38D and E) for a few generations and later they fall apart.

Some Indian Species: *Pleurococcus cohaerens* Kutz.; *P. viridis* Ag.

Special features

1. Unicellular alga occurring singly or in compact masses.
2. The cells can withstand prolonged desiccation.
3. Large lobed peripheral chloroplast.
4. The only method of reproduction is by fission. The daughter cells either may remain adhered together for a long time or may produce a short filamentous structure.

There is probably no other algal genus about which there has been so much confusion. Phycologists are not even in agreement as to whether the genus should be called *Pleurococcus* or *Protococcus*. This confusion is partly due to the fact that any unicellular aerial green alga has been considered a species of *Pleurococcus*. Some authorities have even placed the genus *Pleurococcus* in the Chlorococcales, while other have treated it in a special group, the Pleurococcales.

In recent years it has been realized that the unicellular aerial green algae whose cells divide vegetatively and do not produce swarmers, belong to the pleurococcaceae. Besides this, as the alga *Pleurococcus* can occasionally develop branched filaments there would seem to be evidence for regarding it as a much reduced member of the Chaetophorales.

Order: *OEDOGONIALES*

Plants grow in quiet water, usually attached to other plants by a holdfast. Both branched and unbranched filamentous types are present in this order. Cells of the filament are uninucleate associates with a reticulate chloroplast containing pyrenoids. They undergo several successive divisions, each leaving a characteristic hemicellulosic **apical cap**. The caps, because of the repeated divisions of the protoplast, form a series of striations at the end of the cell. The outstanding characteristic feature in this order is a peculiar flagellation of the motile bodies. The numerous flagella of all the motile reproductive bodies, i.e., the zoospores, the androspores, and the male gametes, are attached in the form of a ring or 'crown' at the anterior end of the cell.

Asexual reproduction is by zoospores, and sexual reproduction is always oogamous, with certain species producing their male gametes in the terminal cell of special dwarf filament, the

nannandrium which develops from a special type of zoospore, the **androspore** produced in an **androsporangium**. Meiosis precedes or accompanies germination of the zygote, and four haploid zoospores are formed, each of which may develop into a mature filamentous adult.

Family: *OEDOGONIACEAE*

This family includes chiefly the fresh-water forms which prefer to grow in quieter situations and are often attached to submerged plant parts or any other object. The plant body may be of branched or unbranched filament. The distinguishing feature of the family is the presence of a large number of flagella at the anterior ends of the motile reproductive structures—zoospores, androspores, and the male gametes. Sexual reproduction may be heterogamous or even oogamous.

Genus: *Oedogonium*

It is a common fresh-water alga which grows in quiet water, usually attached to other plant parts and is abundant in ponds and ditches. Free-floating condition is not also very rare. Its occurrence in running water is rather less common. The outstanding characteristic features of this genus are, cell division resulting in cap formation, swarmers of varied size difference and function with a crown of flagella at the anterior end, oogamy with special dwarf

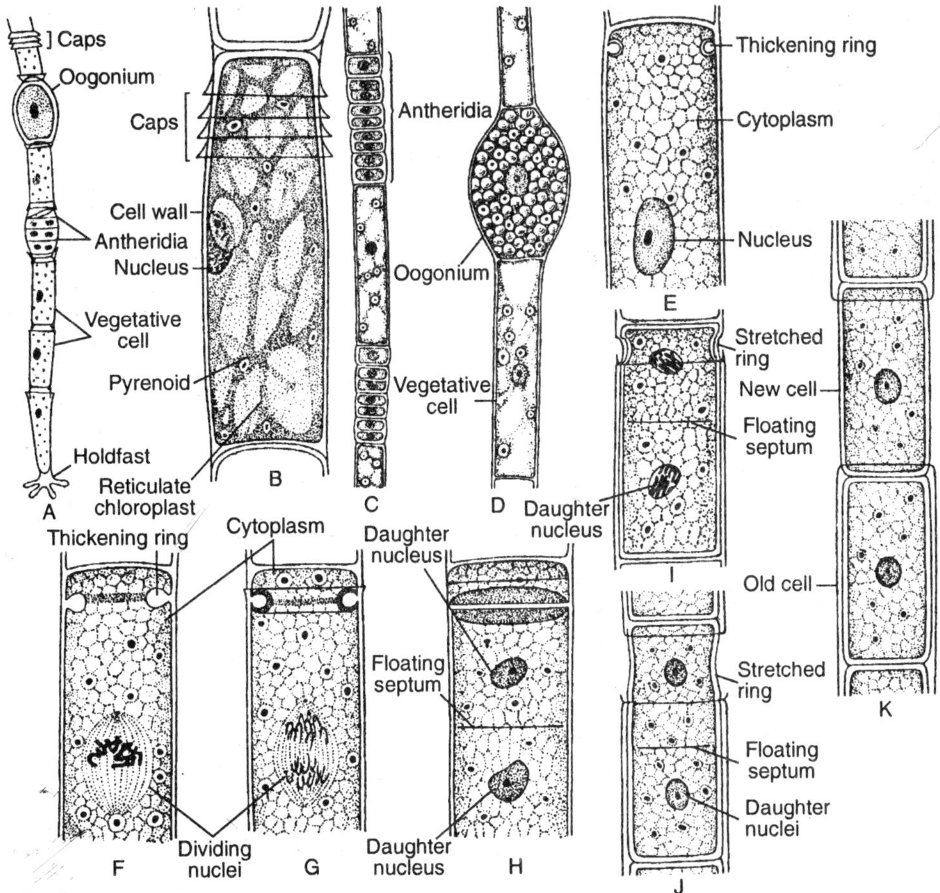

Fig. 4.39: *Oedogonium* sp. A. Portion of unbranched filament showing vegetative cells, caps, holdfast, antheridia and oogonium. B. A single cell showing structural details. C. Portion of filament wth antheridia. D. Portion of filament with oogonium and vegetative cell. E. Portion of a cell with thickening ring. F-K. Successive stages of cell division

male filament taking part in the sexual process, and clear sex segregation in the post-fertilization stages.

The plant body consists of filamentous unbranched thallus of cylindrical cells arranged end to end. The mature filament is differentiated into base and apex and consists of vegetative cells, antheridia, and one or more oogonia (Fig. 4.39A), with which the filament attaches itself with the substratum, whereas the apical cell is rounded. Each of the vegetative cells contains a single reticulate chloroplast with pyrenoids (Fig. 4.39B). Each cell contains a single large nucleus which lies near the middle of the cell remaining embedded in the cytoplasm just within the chloroplast. Certain cell in every filament has one or more ring-like markings which are known as caps (Figs. 4.39A and B). *Oedogonium* has no apical growing point. Cell division may take place in any cell of the filament except the basal cell. The mode of cell division resulting in the production of a cap is rather unique of its kind. With the migration of the nucleus towards the distal end of the cell, the cell division starts. The nucleus then divides mitotically, and simultaneously, there appears a ring of wall material (thickening ring) which encircles the inner face of the lateral wall just below the distal end of the cell (Figs. 4.39E to G). The division of the nucleus is followed by the appearance of a septum in between the two daughter nuclei (Fig. 4.39H). The septum remains floating for some time without being connected with the lateral wall (Figs. 4.39H to J). The protoplast soon elongates as a result of which the newly formed ring on the lateral wall is stretched resulting in the breaking of the lateral wall in the traverse direction at the level of the ring. The ring becomes stretched further to form a new wall (Figs. 4.39I and J), although a little of the original wall remains at the upper end of the cell and is known as cap. The elongation of the protoplast also causes the displacement of the floating septum upward until it takes up a permanent position near the lower end of the new wall developed from the stretched ring (Fig. 4.39K). Often, an individual cell may undergo several successive divisions, each leaving a characteristic cap, the number of which will denote the number of divisions the cell has undergone (Fig. 4.39B).

In addition to fragmentation of the filaments, reproduction may also be by the asexual and sexual methods. In the asexual method large multiflagellate zoospores are produced singly in the vegetative cells which are then transformed into zoosporangia. In this event, the whole content of the cell gradually withdraws itself from the cell wall contracting to form a pear-shaped body (Fig. 4.39A). At the same time a clear, colourless spot appears on one side of the contracted protoplasmic body around which is developed a characteristic ring of flagella (Fig. 4.40B). Ultimately a multiflagellate zoospore with a ring of flagella arranged around colourless spot in the anterior region is developed. The zoospore is liberated being enclosed in a mucilage sheath, by the breaking across of the cell which has functioned as a sporangium (Figs. 4.40C and D). A zoospore is almost spherical or pear-shaped and deep-green, but there is a well-marked colourless beak around the base of which short flagella arise (Fig. 4.40E). Such a type of zoospore is designated as **stephanokontan** zoospore. Eventually the zoospore comes to rest, attaches itself by its colourless end, withdraws its flagella, forms a cell wall, and divides to form a new filament (Figs. 4.40F and G).

Sexual reproduction is oogamous, with the production of large non-motile egg and small antherozoids. Some species are monoecious or homothallic (oogonia and antheridia develop on the same filament), others are dioecious or heterothallic (oogania and antheridia on different filaments). Depending on structure and distribution of antheridia, species of

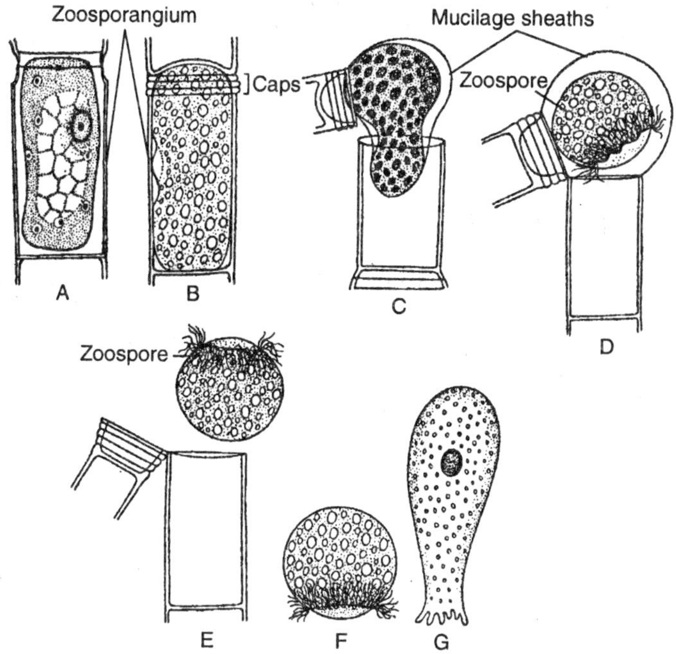

Fig. 4.40: *Oedogonium* sp. A-E. Stages in formation and liberation of zoospore. F-G. Germination of zoospore

Oedogonium may be grouped as **macrandrous type**, and **nannandrous type.** In **macrandous type**, antheridia are produced on regular-sized filaments, whereas in **nannandrous type** antheridia are produced on special **dwarf filaments** called 'dwarf male' plants or **nannandria** (sing. **nannandrium**). The macrandrous type may again he monoecious or dioecious. In macrandrous type antheridia are formed by successive transverse divisions of the vegetative cells of the filaments resulting in the formation of a row of antheridia (Fig. 4.41A). In every

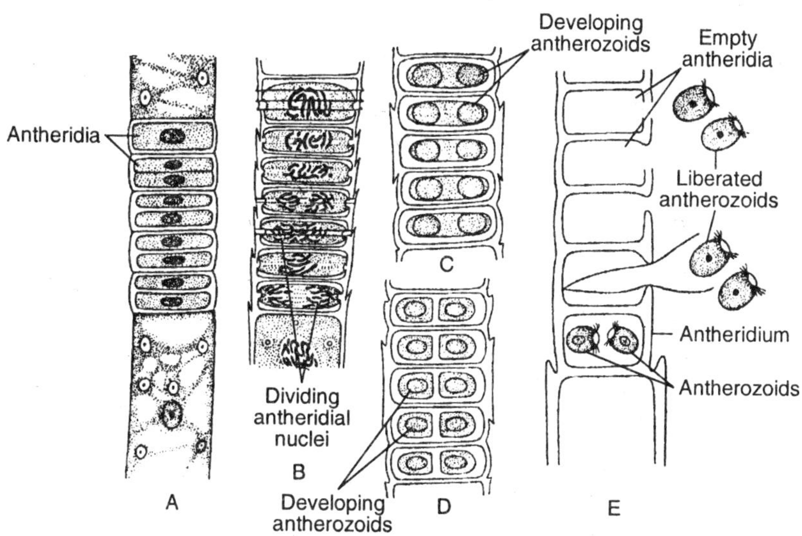

Fig. 4.41: *Oedogonium* sp. A-E. Successive stages in development of antheridia, antherozoids, and liberation of antherozoids from antheridia

antheridium the contents divide into two, and each mass develops into an antherozoid also known as spermatozoid (Figs. 4.41B to E). Except that they are much smaller and relatively poorer in chlorophyll content, the antherozoids resemble the zoospores, even to the characteristic crown of flagella (Fig. 4.41E). The structure and distributional pattern of oogonia are identical in both macrandrous type and nannandrous type of species of *Oedogonium.* The oogonia are highly differentiated structures, each one originating from a vegetative cell behaving as oogonium mother cell (Fig. 4.42B). This oogonium mother cell, by division forms two cells, the basal one becoming the supporting cell of the oogonium known as **suffulatory cell** and the upper one the oogonium proper (Fig. 4.42A). The oogonium swells out, assuming a round or oval in outline (Fig. 4.42C). The cell contents in the mean time withdraw themselves from the wall, and form a free, rounded protoplasmic body having a receptive spot—the ovum (Figs. 4.42D-E). Each oogonium bears one egg and opens, either by the formation of a pore or by the transverse splitting of the oogonial wall near the top corresponding to the receptive spot of the ovum, through which the swarming antherozoid reaches the immobile egg (Fig. 4.42E). The male gamete penetrates the egg, and fertilization is effected when the nucleus of the antherozoid fuses with the egg nucleus, the resulting zygote is an **oospore** (Fig. 4.42F). The oospore develops a heavy wall and gradually passes into resting state (Fig. 4.42G). After a period of rest, the oospore is set free either by the decay or rupturing of the oogonial wall (Fig. 4.42H). Meiosis precedes or accompanies

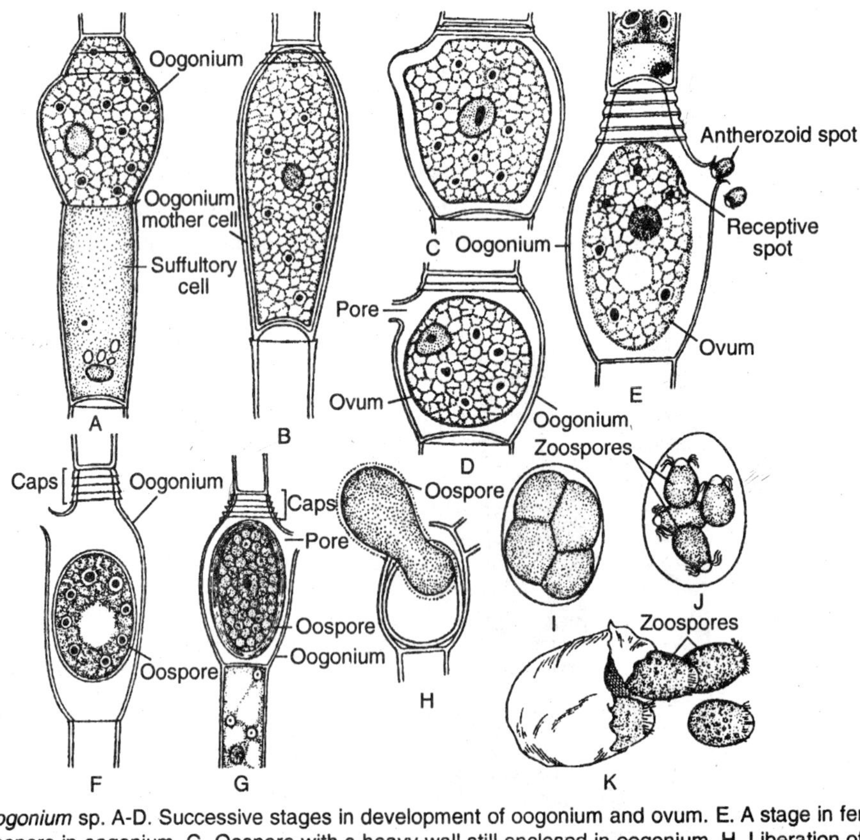

Fig. 4.42: *Oedogonium* sp. A-D. Successive stages in development of oogonium and ovum. E. A stage in fertilization. F. Oospore in oogonium. G. Oospore with a heavy wall still enclosed in oogonium. H. Liberation of oospore from oogonium. I-J. Germination of oospore and zoospore formation. K. Liberation of zoospores

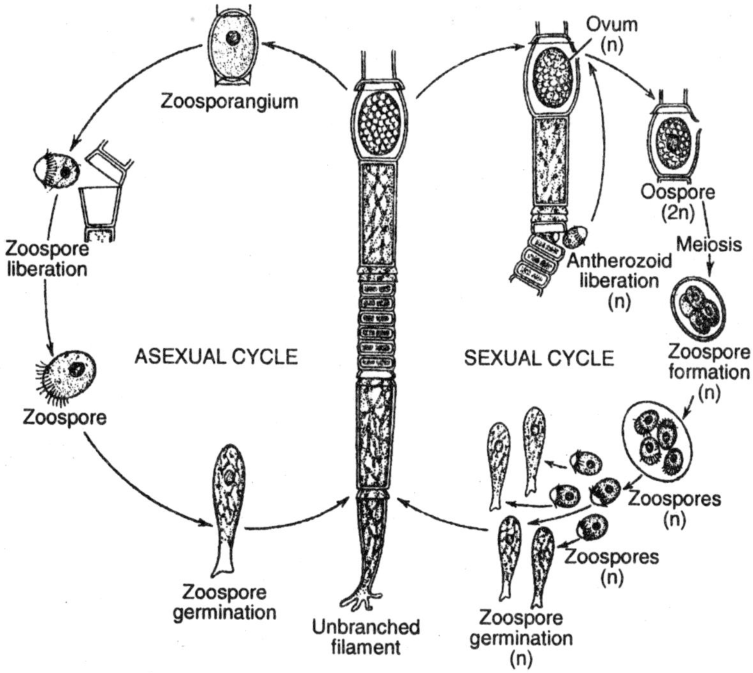

Fig. 4.43: Life cycle of a monoecious macrandrous type of *Oedogonium* sp.

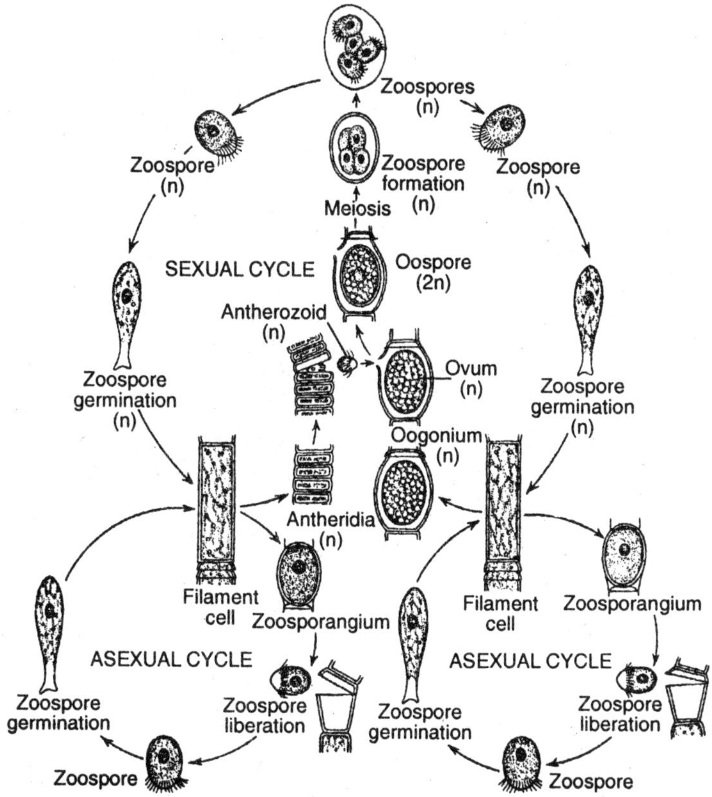

Fig. 4.44: Life cycle of a dioecious macrandrous type of *Oedogonium* sp.

germination of oospore with the formation of four haploid nuclei. The contents of the oospore (Figs. 4.42I and J). These four zoospores resemble the zoospores formed in the vegetative cells, except that they are much smaller than the latter (Fig. 4.42K). The resulting zoospores develop into new filaments. In monoecious macrandrous type (e.g., *Oedogonium fragile* and *O. nodulosum*), the new filaments are of same kind (monoecious) (Fig. 4.43). Whereas, in dioecious macrandrous type (e.g., *Oedogonium crassum*), two of the zoospores produce male filaments and the other two female filaments (Fig. 4.44).

In nannandrous type, the sexual process is a little different from that of the macrandrous one, where an interesting dwarf male plant or nannandrium plays a very important role in the entire process. The dwarf male plants are developed from special type of zoospores known as **androspores** which are smaller than the usually zoospores and are associated with the sexual reproduction. The androspores are produced in **androsporangia** which are formed by repeated transverse divisions of the ordinary vegetative cells of the filament (Fig. 4.45A). The contents of each of these androsporangia are transformed into a single zoospore (androspore) having a

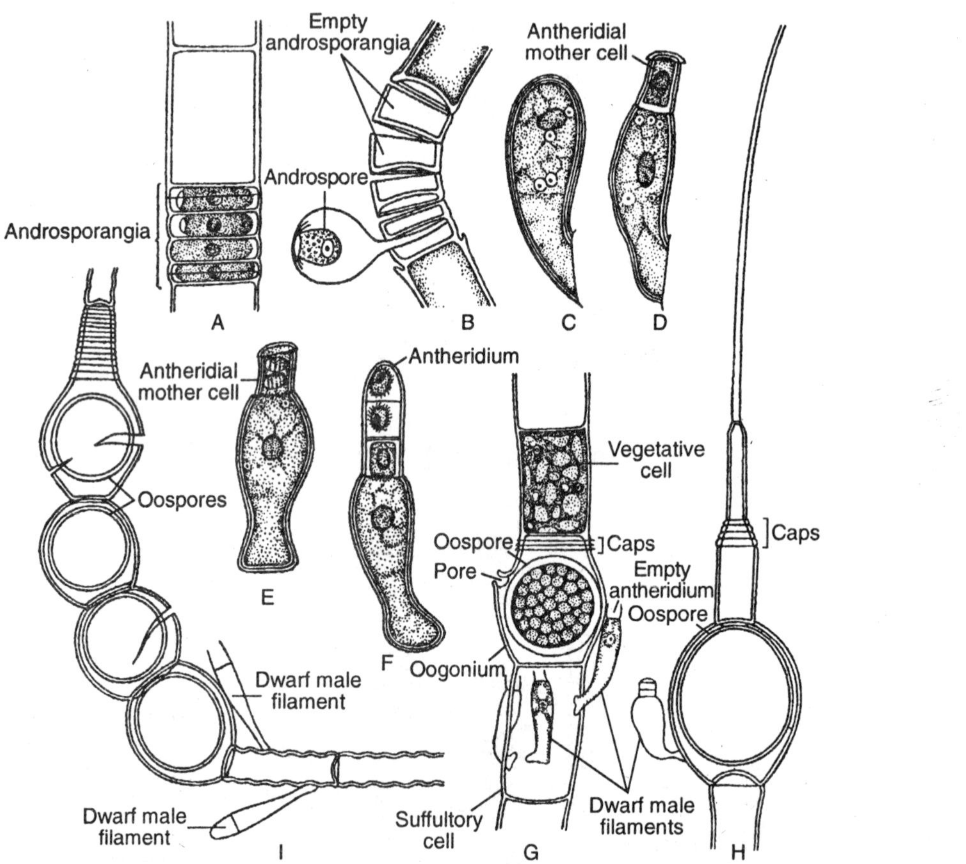

Fig. 4.45: *Oedogonium* nannandrous type (A-G and I) and *Bulbochaete* sp. (H). A. Portion of a filament bearing androsporangia. B. Liberation of androspore from an androsporangium. C. Young nannandrium (dwarf male filament). D-F. Stages in development of antheridium and antherozoid from nannandrium (dwarf male filament). G. Portion of a female filament bearing dwarf male filaments on suffultory cell and an oogonium with oospore. H. Portion of filament bearing oogonium with oospore and dwarf male filament developed on the oogonium wall. I. Oogonia in chain with oospores and dwarf male filaments on the suffultory cell

size intermediate between the usual zoospore and an antherozoid (Fig. 4.45B). Most commonly, the androsporangia are produced from the same filament which bears the oogonia—the **gynandrosporous** (e.g., *Oedogonium concatenatum*); more rarely they occur on a separate filament—the **idio-androsporous** (e.g., *Oedogonium confertum*). The androspore when liberated, after swarming, becomes attached either to the oogonial wall or to the wall of the suffolatory cell, surrounds itself with a cell wall, and germinates to form one-celled dwarf male filament (Figs. 4.45D to F). Each antheridial mother cell develops into an antheridium which produces two antherozoids each bearing the characteristic crown of apical flagella. The antherozoids escape by a lid-like opening of the antheridium and make their way to the oogonial aperture (Fig. 4.46). The act of fertilization and formation of oospore is exactly the same as in macrandrous type. The entire process of formation of new haploid filaments from the oospore is also quite similar to the macrandrous one.

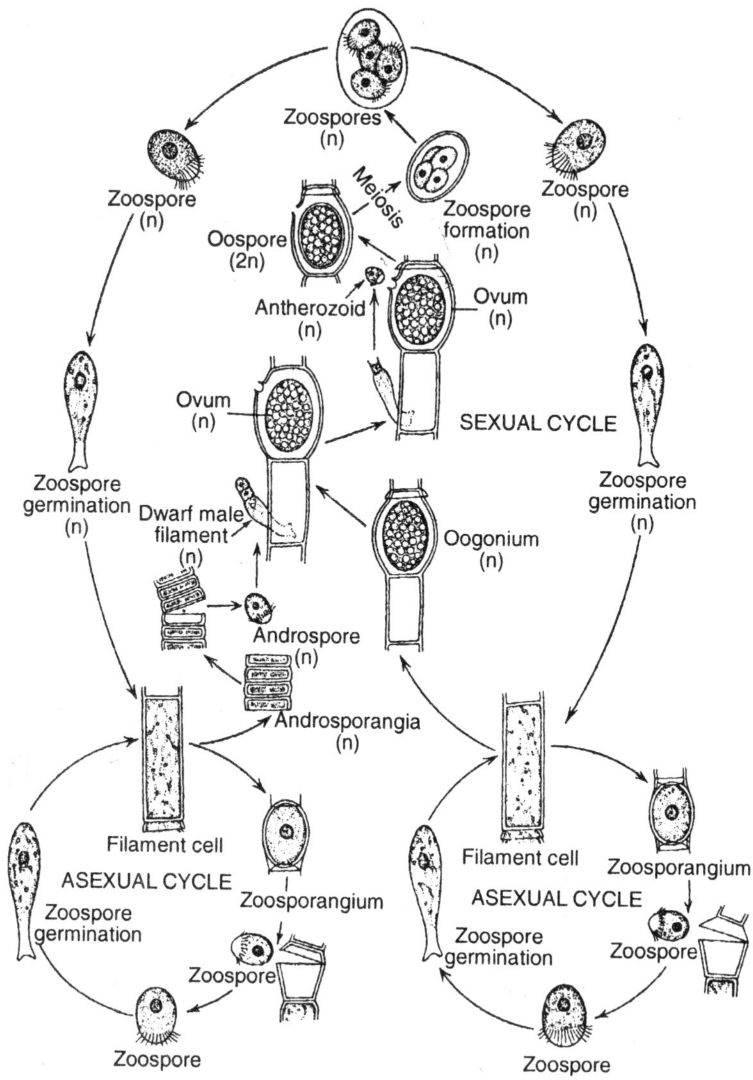

Fig. 4.46: Life cycle of a dioecious (idioandrosporous) species of *Oedogonium*—nannandrous type

The development of androspores which is directly linked up with the sexual process in a nannandrous type is rather a very interesting aspect in the life cycle of this alga. As to the origin of androspores, two hypothesis are considered. (i) The androspores are equivalent to smaller zoospores that are produced from the oospores by reduction division and subsequent other processes which, in the nannandrous type, instead of producing normal filaments give to dwarf male individuals. Whereas, in the macrandous type, there is no development of dwarf male filaments as the zoospores behave quite normally by producing normal filaments. As such, the macrandrous types are considered to be more advanced than the nannandrous ones. (ii) The androspores are nothing but prematurely liberated antheridial mother cells which at later stage undergo further development and produce antherozoids. According to this view the macrandous types are more primitive.

Some Indian Species: *Oedogonium areolatum* Lagarh.; *O. armigerum* Hirn. *f. tenuis* Singh; *O. cardiacum* Kutz.; *O. elegans* Kutz.; *O. oblongellum* Kirch.; *O. sociale* Wittr.

Special features

1. Filamentous unbranched plant body differentiated into base and apex.
2. Reticulate chloroplast.
3. Presence of caps in the dividing cells.
4. Very elaborate vegetative cell division.
5. Asexual reproduction by multiflagellate zoospores having flagella arranged in a ring around a beak-like area—stephanokontan zoospores.
6. Sexual reproduction oogamous.
7. The female gamete is produced singly in each oogonium. It is retained in the oogonium until fertilization is completed.
8. The male gamete is very much similar to zoospore but much smaller in size.
9. A curious dimorphism of sexual plants is exhibited by the nannandrous type.
10. Formation of dwarf male—nannandrium by germination of multiflagellate androspore.
11. The nanandrium produces antheridium which bears antherozoids.
12. Zygote produces four zoospores.

Order: *ZYGNEMATALES (CONJUGALES)*

The members of this order are strictly fresh-water plants of fairly wide distribution. The order contains both filamentous and unicellular forms with distinctive types of chloroplasts: spiral, stellate, or axile. No asexual reproductive bodies are formed by members of this order. The outstanding features of the order are a total lack of flagellated cells, and sexual reproduction isogamous by means of aplanogametes. Gametic union may take place through tubular connections between cells of two filaments, or between adjacent cells of the same filament. Meiotic division of the nucleus of the zygote results in the production of one, two, or four haploid nuclei. The two best known families are the Zygnemataceae and Desmidiaceae.

Family: *ZYGNEMATACEAE*

This family includes fresh-water multicellular filamentous forms of algae.

Genus: *Spirogyra*

It is one of the common green algae occurring chiefly in ponds, pools, ditches and similar other places. It prefers still, rather than running water and is one of the free-floating algae known as 'pond scum'. Quite often it is present in immense quantities, the filaments forming floating masses, sometimes several acres in extent, buoyed up by the bubbles of oxygen which their assimilation has produced. The

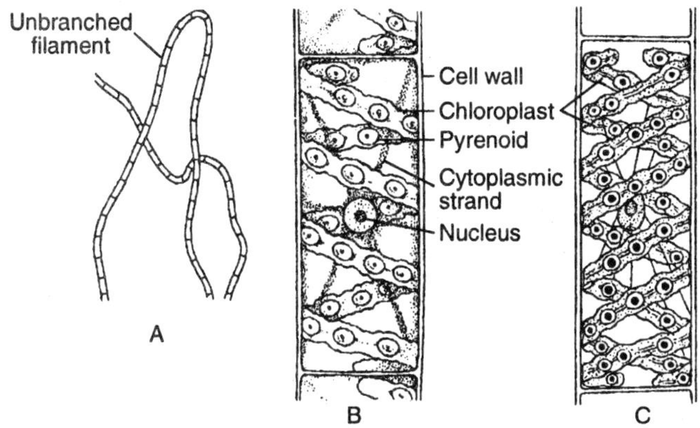

Fig. 4.47: *Spirogyra* sp. A. Unbranched filaments. B. A single cell showing structural details. C. A single cell with two spiral chloroplasts

plant is a multicellular, unbranched filament which is slippery to the touch because it is covered with a mucilaginous sheath of pectose (Fig. 4.47A). There is no distinction of base and apex and all the cells are alike throughout the filament. Within the stratified cell wall of the cylindrical cells composing the filaments there is the primordial utricle. Embedded in this are one or more spirally arranged ribbon-like chloroplasts with unsmooth edges, which form the most striking feature of the plant (Figs. 4.47B and C). Each chloroplast is studded at intervals with several pyrenoids. In every cell, there is one nucleus which is, either suspended in the centre by means of cytoplasmic strands, or embedded in the primordial utricle. The nucleus contains a large nucleolus. The cross-walls of the filaments are essentially of five types, for example, plane, replicate, semi-replicate, colligate and unduliseptate. In some species the middle lamella of the cross-wall develops on either side a ring-like ingrowth on which the subsequent apposition-layers are deposited, such a septum is known as **replicate septum** (Figs. 4.48A to C).

Strangely enough although *Spirogyra* is an aquatic alga, it produces neither zoospores, nor flagellate gametes. Except for the occasional breaking up of a filament into its constituent parts, *Spirogyra* possesses no asexual means of propagation. Its normal reproduction is always sexual, and is accomplished by **conjugation**— an aplanogametic isogamy. In this process, two vegetative cells combine their contents to form a zygote. Usually, conjugation takes place between the cells of two different filaments, but it may also take place between adjacent cells of two different filaments, but it may also

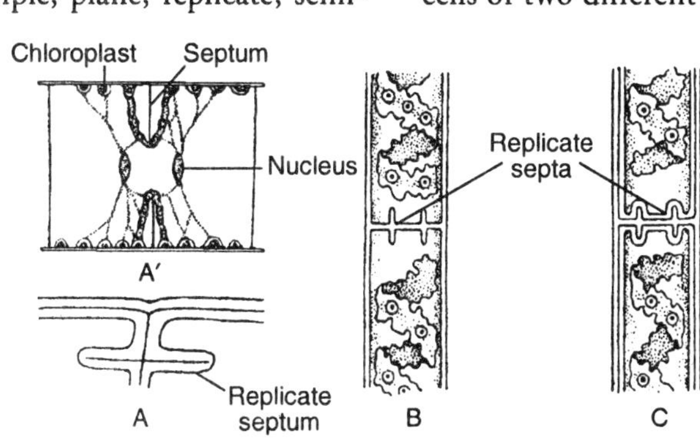

Fig. 4.48: *Spirogyra* sp. A-C. Stages in development of replicate septa

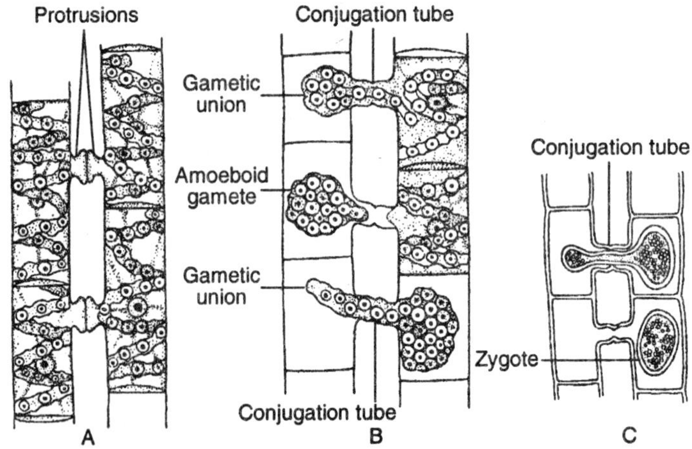

Fig. 4.49: *Spirogyra* sp. A-C. Successive stages in scalariform conjugation

take place between adjacent cells of the same filament. In the former type, the cells in two filaments which are in contact, form protrusions which lie opposite one another and as they elongate, the filaments are pushed apart (Fig. 4.49B). Meanwhile, the contents of the two participating cells contract. Fusion of the contracted cell-contents takes place within one of the two conjugating cells (Figs. 4.49B and C). This type of conjugation is known as **scalariform conjugation.** The active gamete may be designated as a male gamete and the passive as a female gamete though they are morphologically similar. In a few species where conjugation occurs between the adjacent cells of the same filament, small protrusions arise as a result of the growth of the lateral wall on either side of the cross-wall of the participating cells (Figs. 4.49A and B). With the growth of the protrusions the cross-wall also becomes stretched. Finally, as the cross-wall cannot keep pace with the growth of the protrusions, the latter lose their contact with the former, a side link is thus established between the two cells (Fig. 4.50C), through which the protoplast of one cell migrates into the other (Fig. 4.50D) fusing to form a zygote (Fig. 4.50E). This is called **lateral conjugation**. The fusion product

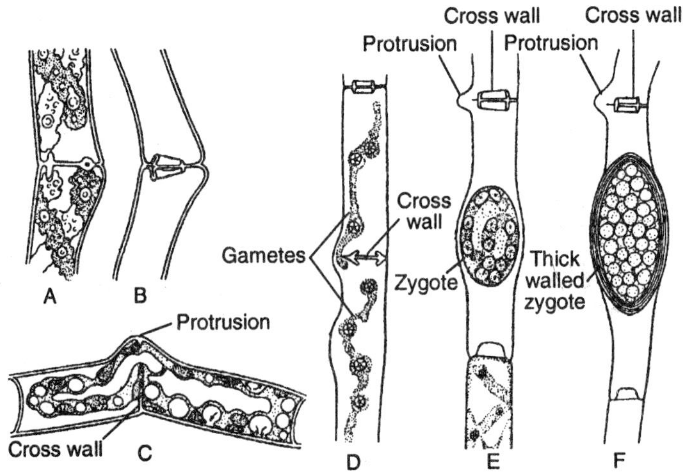

Fig. 4.50: *Spirogyra* sp. A-B. Canal formation in lateral conjugation. C-D. Gametic union in lateral conjugation. E-F. Zygote formation

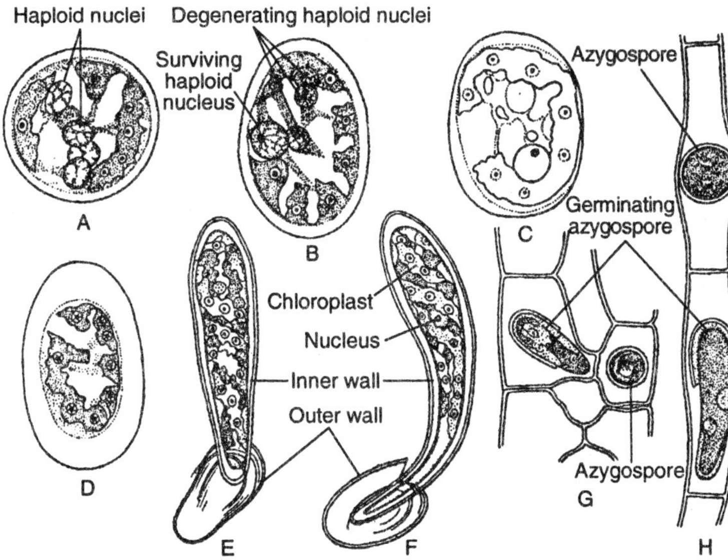

Fig. 4.51: *Spirogyra* sp. A-D. Stages of nuclear division during germination of zygote. E-F. Development of new filament during germination of zygote. G-H. Formation and germination of azygospores

of conjugation is a zygospore which develops three-layered thick wall and is highly resistant to extreme cold and drought (Fig. 4.50F). The zygospore generally sinks to the bottom of the pool where it is formed, and germinates after a period of rest. During germination the two outer layers burst open, the innermost cellulose wall elongates forming a cylindrical germ tube which divides transversely to form a new filament (Figs. 4.51E and F). Meiosis usually occurs during germination of the zygospore and thus the new filament is haploid (Fig. 4.52). Three of the four haploid nuclei formed in meiosis degenerate, so that each filament is derived from a haploid uninucleate cell (Fig. 4.51A to D). Sometimes the contents of certain cells of a filament round up and secrete walls to become spores parthenogenetically without any sexual fusion. Such spores are known as **parthenospores** or **Azygospores** (Figs. 4.51G and H).

The amoeboid gametes involved in the sexual process do not exhibit any morphological differences. Species in which only scalariform conjugation occurs are no doubt in part dioecious with genotypic sex determination. The morphologically similar gametes exhibit difference only in their behaviour. A situation where one of the fusing gametes is active and moves to the other awaiting gamete, indicates a case of physiological heterogamy. From the behaviour, the active gamete may be known as 'male' and the passive as 'female'. This differentiation of sex between the gametes is also accompanied by sexual differentiation between the filaments. The filaments bearing cells which produce active gametes behave as male and those having cells with passive gametes act as female. Thus such filaments are **unisexual** and the species having filaments of this nature as **dioecious.**

Some Indian Species: *Spirogyra maxima* (Hass.) Wittr.; *S. neglecta* (Hass.) Kutz.; *S. nitida* (Dillw.) Link.

Special features

1. Filamentous unbranched plant body without being differentiated into base and apex.
2. Presence of replicate septum.

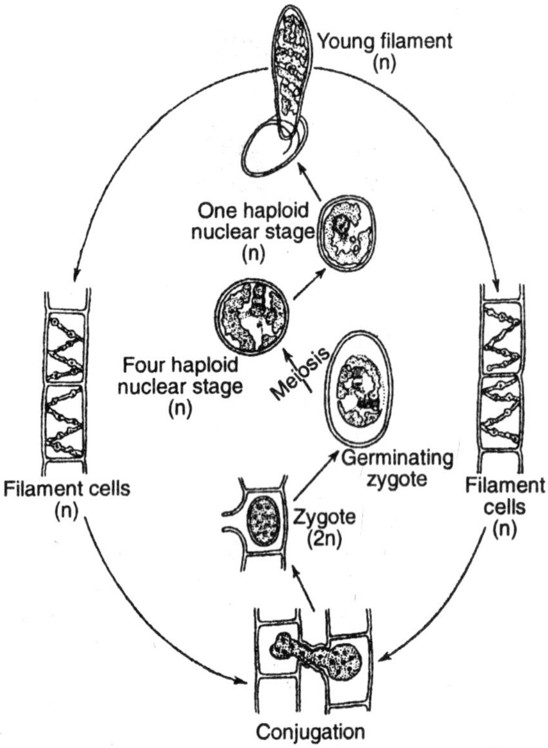

Fig. 4.52: Sexual cycle of *Spirogyra* sp.

3. Spiral chloroplast.
4. Absence of flagellate gametes.
5. Sexual reproduction by conjugation—morphological isogamy, but physiological heterogamy.
6. Conjugation both lateral and scalariform.

Sex differentiation. The conjugating cells of the filaments of *Spirogyra* behave as gametangia since the entire content of each cell metamorphoses into a gamete. The gametes produced are morphologically similar and are thus isogametes. But they exhibit difference in their behaviour. One of them is active and moves from the cell of the filament where it is developed to another, where its fusing partner waits remaining passive for fusing with it. This situation may be defined as primitive anisogamy or physiological heterogamy. The active gamete thus behaves as male and the passive as female.

This type of sex differentiation is also associated with sex differentiation between the filaments of whose cell contents are taking part in conjugation. The filaments whose cells produce active gametes are sexually distinct from that whose cells produce passive gametes. Both the filaments are thus unisexual. In spite of such sex differentiation the sexual reproduction in *Spirogyra* is isogamous since there is no other visible difference between the fusing gametes.

Genus: *Zygnema*

This alga is very commonly found to occur as yellowish free-floating masses in considerable quantities on the surface of slow-flowing as well as quiet fresh-water, particularly, in ponds and wayside ditches. The genus *Zygnema* is closely related to *Spirogyra* in many respects, differs only in having two stellate chloroplasts embedded in the cytoplasm, each with a central pyrenoid, with the nucleus embedded in a mass of cytoplasm suspended between them (Figs. 4.53A and B). The cell wall is composed of two cellulose layers. The characteristic pectose layer and the replication of cross-wall are practically absent. Asexual, reproduction is almost the same as that of *Spirogyra* with the exception of some species, where the formation of akinetes and aplanospores has been reported. Sexual reproduction is by conjugation, both lateral and scalariform (Figs. 4.53C to I), between two aplanogametes as encountered in *Spirogyra*. In scalariform conjugation, both the isogametes by amoeboid movement move towards each other and unite in the conjugation tube to form a zygospore (Fig. 4.53F). Again, one of the two fusing gametes may be active (male) and by amoeboid movement may pass through the conjugation tube, meet and fuse with the other passive (female) gamete lying in the cell where it is formed (Fig. 4.53H). The cell in which conjugation takes place is the **conjugating cell**. Here gametes, though morphologically similar, exhibit **physiological anisogamy**. Zygospore forma-

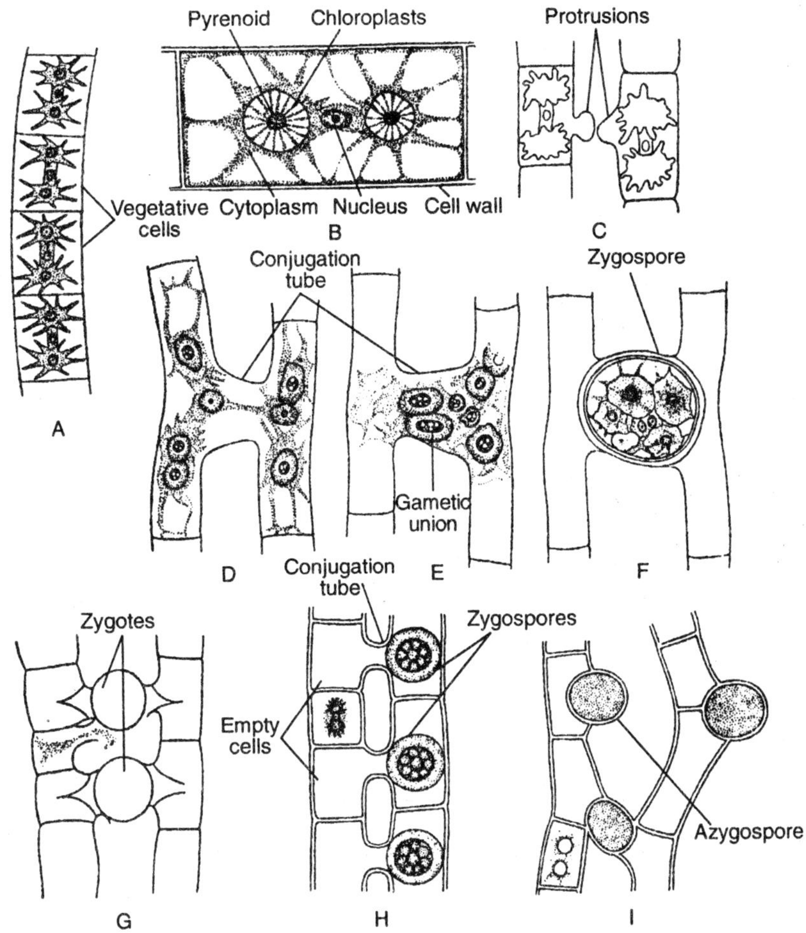

Fig. 4.53: *Zygnema* sp. A. Portion of a vegetative filament. B. Details of cell structure. C-I. Successive stages of scalariform and lateral conjugation

tion is more common in the conjugating cells indicating a tendency of differentiation of sex. Again, sexual dimorphism has been reported in certain species of *Zygnema*, where the cells producing passive gametes are longer and broader with larger chloroplasts and pyrenoids than the cells producing the active gametic partner. But since both isogamy and physiological anisogamy, as well as lateral and scalariform conjugation have been reported from the same species, the sex differentiation may not be very pronounced. The process of zygospore formation, and the behaviour of the zygospore and its diploid nucleus with ultimate development of a new haploid filament, all are very similar to those of *Spirogyra*. When conjugation fails, azygospore may be developed (Fig. 4.53I).

Some Indian Species: *Zygnema giganteum* Randhawa; *Z. gorakhporense* Singh; *Z. indicum* Misra; *Z. pectinatum* (Vauch.) Ag.; *Z. terrestre* Randhawa.

Special features

1. Filamentous unbranched plant body not differentiated into base and apex.
2. Stellate chloroplast.
3. Absence of flagellate gametes.
4. Conjugation both scalariform and lateral.

Family: *DESMIDIACEAE*

This family includes fresh-water unicellular forms of algae.

Genus: *Cosmarium*

The genus *Cosmarium* is commonly found free-floating in fresh-water small ponds and ditches, either solitary, or in amorphous colonies. Single-celled, non-motile plants of rare beauty with length rarely more than one and a half times the breadth, are characterized by the fact that the cells are usually divided into two symmetrical halves by a median constriction. The two halves are usually divided into two symmetrical halves by a median constriction. The two halves are referred to as **semicells;** the constriction is called the **sinus;** and the narrow connecting zone of the two halves is known as the **isthmus** (Fig. 4.54A). The cells may have smooth or ornamented wall (Fig.

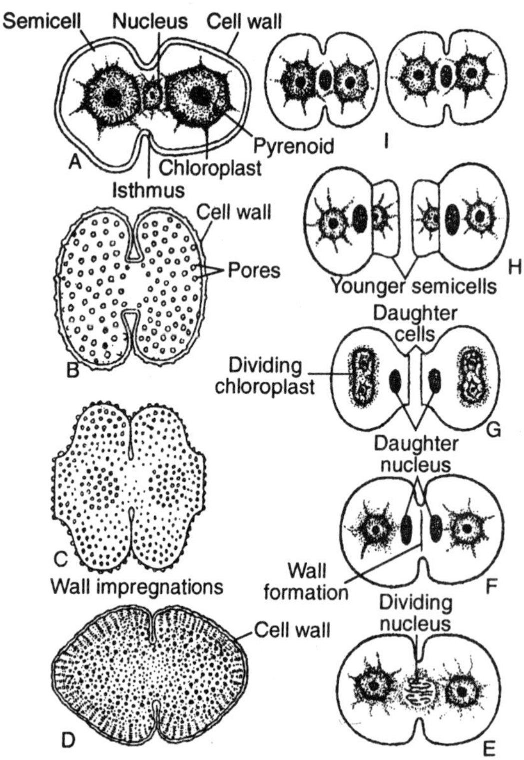

Fig. 4.54: *Cosmarium* sp. A. Details of cell structure. B-D. Nature of cell wall. E-I. Stages in the cell division

4.54C) which when mature is composed of two well-differentiated layers. The inner one of cellulose and the outer is firmer and thicker and is composed of cellulose impregnated with various substances including iron compounds (Fig. 4.54D). Numerous small pores occur in the cell wall except in the region of isthmus (Fig. 4.54B). Through these pores mucilage is secreted which sometimes resembles like a third layer of the cell wall. Most of the species have two relatively large chloroplasts with radiating plates having centrally placed pyrenoids, one located in each semicell. The single nucleus, embedded in a small mass of cytoplasm, is in the isthmus (Fig. 4.54A). Reproduction is by cell division and by the process of conjugation which resembles that in *Spirogyra* and *Zygnema.* As soon as the cell division starts the nucleus divides mitotically into two daughter nuclei (Figs. 4.54E and F) and is followed by the elongation of the isthmus region. Then there appears a constriction in the isthmus in between the two daughter nuclei dividing the parent cell into two daughter cells by the formation of a wall (Figs. 4.54F and G). The portion of the isthmus of each daughter cell later grows into a new semicell (Fig. 4.54H). A new chloroplast is formed in each of the new semicells by the division of the existing chloroplast and pyrenoid of the old semicell. Hence, the two semicells that constitute the *Cosmarium* plant, one is always younger than the other (Fig. 4.54H). The younger semicell is at first very small, but it gradually attains the same size as the other semicell (Fig. 4.54I).

Sexual reproduction is by conjugation, a process of aplanogametic isogamy (Fig. 4.56), in which the contents of the two conjugating cells behave as gametes. To start with the process of conjugation, the two conjugating cells come together, lie side by side, and are embedded in a common mucilaginous sheath (Fig. 4.55A). The cells split open at the isthmus and the semicells separate allowing the protoplasts to emerge (Fig. 4.55B). Each protoplast

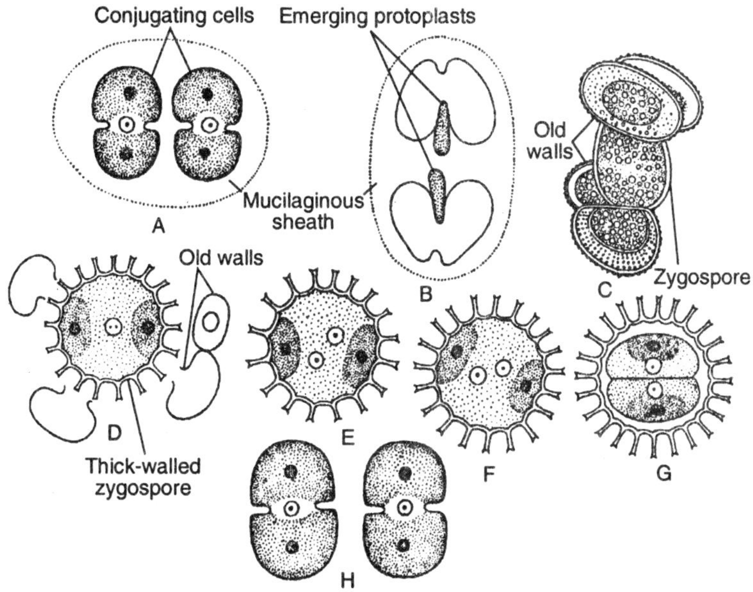

Fig. 4.55: *Cosmarium* sp. A-D. Stages showing sexual reproduction by conjugation. E-H. Germination of zygospore

behaves as a gamete. The gametes move towards each other by slow amoeboid movement and finally fuse together forming a zygote—zygospore which surrounds itself with a three-layered heavy wall (Fig. 4.54C), the outermost layer of which may be smooth or spiny (Fig. 4.55D). The zygospore sinks to the bottom of the pool and after a period of dormancy, the diploid zygote nucleus divides meiotically forming four haploid daughter nuclei, of which two are functional and the other two gradually disintegrate (Figs. 4.55E and F). The zygospore chloroplast then divides into two daughter chloroplasts (Fig. 4.55G). Each daughter chloroplast with one functional nucleus forms a new plant (Fig. 4.55H). The formation of **parthenospore** has also been reported in some species of *Cosmarium.*

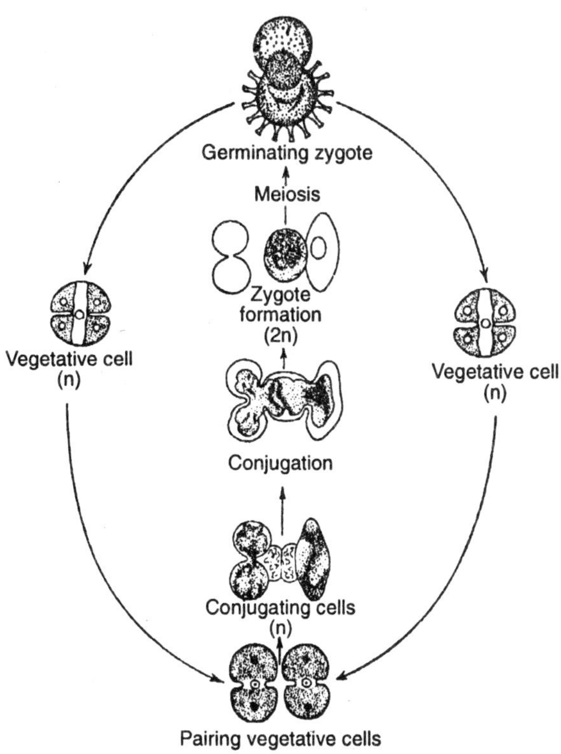

Fig. 4.56: Sexual cycle of *Cosmarium* sp.

Some Indian Species: *Cosmarium connatum* Breb.; *C. distichum* Nordst.; *C. granalum* Dreb.; *C. moniliforme* (Turp.) Ralfs.; *C. quadrilaterum* Bruhl et Biswas; *C. tenne* Arch.

Special features

1. Unicellular plant body differentiated into semicells and isthmus.
2. Star-shaped chloroplast.

3. Cell wall with characteristic pores.
4. Characteristic vegetative cell division.
5. Sexual reproduction by conjugation.
6. Gametes are non-flagellate.

Genus: *Closterium*

This is essentially a free-floating alga occurring abundantly in small ponds and ditches forming a component of the planktonic flora, may also be present at the bottom of pools. Vegetative cells are solitary, elongate with no median constriction. In most species the cells are curved with attenuated apices being markedly lunate or arcuate (Fig. 4.57A). The highly perforated cell wall is composed in two layers—an inner structureless cellulose layer surrounded by an outer firmer and thicker layer made of cellulose impregnated with pectic and iron compounds. In some species delicate ridges at intervals running from pole to pole appear as striations. In addition to cell wall, the cells are surrounded externally by a coat of mucilage which is secreted by the protoplast through the performations. An individual cell exhibits movement in constant with a substratum. This movement is facilitated by the secretion of mucilage by the cells. Each cell has a single nucleus which is embedded in the cytoplasmic mass and is median in position in the region of the isthmus (Fig. 4.57A). Axile chloroplast, one in each semicell bearing one to several pyrenoids arranged in linear series in the axis, is present in each cell. At the extremities there are vacuoles having one in each pole.

Cell division may continue indefinitely without conjugation. During cell division a slight internal thickening and a slight external constriction appear in the wall at the isthmus region (Fig. 4.57B). With the elongation of the cell this local thickening becomes drawn out into a cylindrical strip. The nucleus divides mitotically to form two daughter nuclei (Figs. 4.57C to E). Then there appears from inside the strip a septum between the two nuclei resulting in the division of the cell into two daughter cells (Figs. 4.57F and G). These daughter cells are cone-shaped in outline (Fig. 4.57G) and each one carries half of the isth-

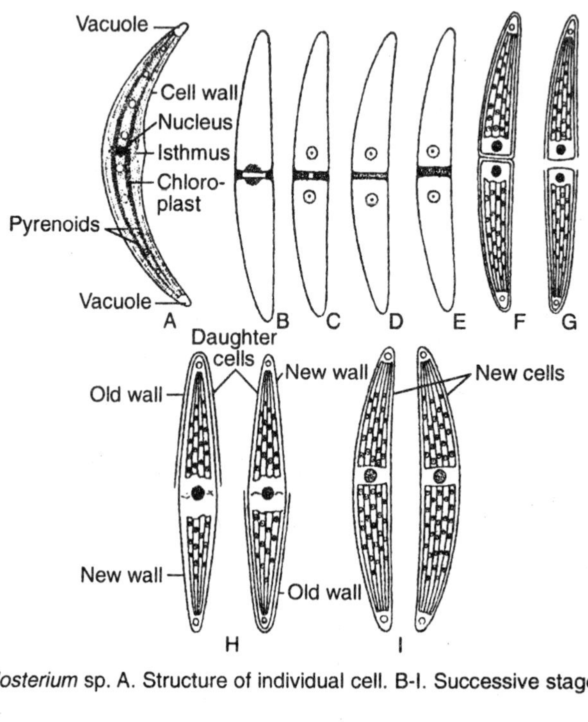

Fig. 4.57: *Closterium* sp. A. Structure of individual cell. B-I. Successive stages in cell division

mus. Each half of the isthmus then bulges out to form the new semicell. The old chloroplast divides into two daughter chloroplasts which are distributed into two semicells—the old semicell and the new semicell (Figs. 4.57H and I). This is how a *Closterium* cell is composed of one old semicell and a newly developed semicell (Fig. 4.57I).

During conjugation the two individuals become enveloped by mucilage sheath (Fig. 4.58A). They arrange themselves parallel to each other in pairs. Protrusions arising at the isthmus from the two paired cells meet and the two cells are pushed apart (Fig. 4.58B), as a result of which the cells break at the point of isthmus and the contents of the two cells are liberated out, behave as gametes fusing together to form a zygospore (Figs. 4.58C and D). In some species conjugation may also take place by the development of conjugation tubes. The zygospore goes to rest by developing a thick wall. On germination the zygospore nucleus divides twice of which one is reductional, forming four haploid nuclei. Of the four haploid nuclei, two degenerate. The two surviving nuclei along with the two daughter chloro-plasts take part in the formation of two new *Closterium* cells.

Special features

1. Free-floating solitary, elongate single-celled plant body with no median constriction.
2. Cells are usually curved with attenuated apices possessing lunate appearance.
3. Perforated cell wall.
4. Asexual reproduction by cell division.
5. Axile chloroplast.
6. Sexual reproduction by conjugation.

Order: *SIPHONALES*

Plants are coenocytic, either simple filamentous, or these filaments are branched or aggregated to more or less elaborate shapes, free or calcified. In the less simple forms, they usually show differentiation of rhizoidal and assimilative portions, and often of stoloniferous portions as well. The coenocyte walls are sometimes braced by extensions of the membrane across the cavity, and in some species at certain points with ring-like thickenings which almost divide the cavity into separate chambers. Nuclei and small chromatophores are very numerous, and pyrenoids absent or present. reproduction is both asexual and sexual.

This order includes the family Phyllosiphonaceae under which is the genus *Phyllosiphon*, an endophytic alga that occurs in the leaves and petioles of the Araceae, most of the species being confined to the tropics, although one is found in Europe. The thallus is composed of richly branched threads ramifying in the intercellular spaces of the host. As a

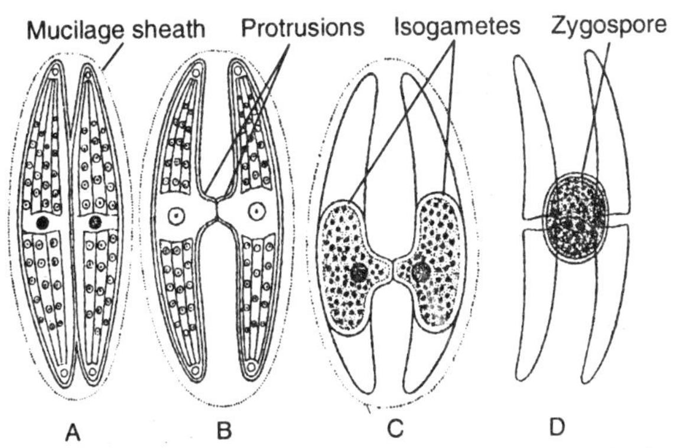

Fig. 4.58: *Closterium* sp. A-D. Stages in sexual reproduction

result of the presence of endophyte the chloroplasts of the host cells do not develop and yellow-green patches occur on the leaf, whilst at the same time the adjacent cells may be stimulated to active division resulting in gall formation, but later the affected cells die. Reproduction takes place by means of oval aplanospores.

Family: *CAULERPACEAE*

Plants are coenocytic, branched, slenderly filamentous, or quite large and then differentiated into rhizoidal, stoloniferous and erect portions, the latter assuming a great variety of forms. The coenocyte has firm wall braced internally by a system of trabeculae. Reproduction is by segregation of portions of the contents of a branch to form dimorphic swarmers, probably gametes, which are discharged through elevated papillae.

Genus: *Bryopsis*

This alga is found to be restricted particularly in warm seas with a few exceptional species which grow in cold waters.

The coenocytic plant body exhibits a very remarkably conspicuous development being differentiated into a comparatively inconspicuous prostrate rhizome-like portion creeping along the substratum (Fig. 4.59A) having anchored to it by the rhizoids and elaborately developed aerial erect axis with a dense and regular pinnate branching in acropetal succession. The erect branches are feather-like with the axis bearing numerous branchlets (Fig. 4.59B). The lower region of the axis is naked and the upper part bears pinnae of variable length. Each pinna is constricted at the point of origin. The rhizome also functions as an organ of perennation. The wall of the vegetative body is composed of cellulose and pectic substances. Internal to the wall is the cytoplasmic layer lining a continuous vacuole. The cytoplasm contains numerous nuclei and small discoid to spindle-shaped chloroplasts having one pyrenoid each (Fig. 4.59C).

Vegetative reproduction is by the growth and decay of new and old parts.

Sexual reproduction is anisogamous. The plants are dioecious. Markedly differentiated pear-shaped biflagellate anisogametes are formed in gametangia in two different plants (Figs. 4.59D and E). During reproduction the

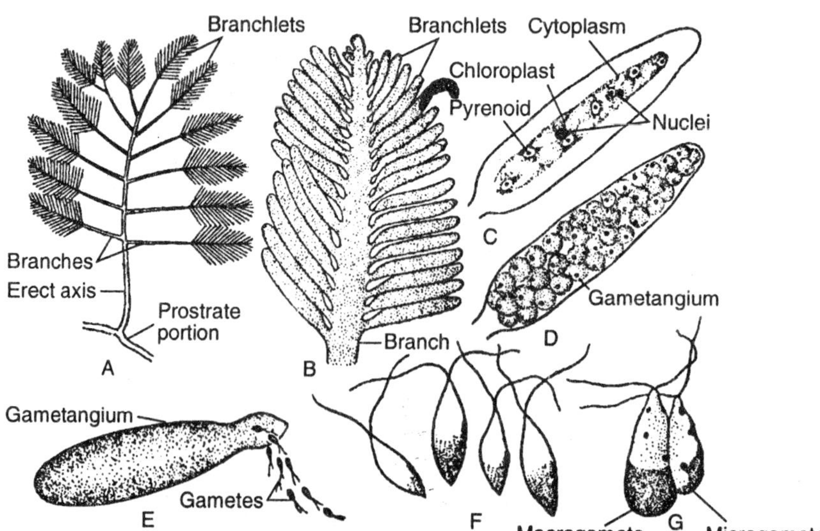

Fig. 4.59: *Bryopsis* sp. A. Habit. B. Branch with branchlets. C. Internal structure of a pinna. D. Gametangium. E. Liberation of gametes. F. Gametes. G. Gametic union

pinnae develop into gametangia being separated from the parent plant by means of septa. Macro- and micro-gametes are formed in different gametangia from the protoplasmic contents of the pinnae which behave as gametangia. The gametangia are distinguishable at maturity from the colour. The ones producing microgametes are yellow coloured, whereas, the others that produce macrogametes being green. The nuclei of the gametangia divide meiotically before taking part in the formation of gametes. The macrogametes are about three times the microgametes and have one to two chloroplasts, a pyrenoid and an eyespot (Fig. 4.59F), whereas the microgametes have a small yellow plastid without pyrenoid. The gametes are liberated by the gelatinization of the wall of the gametangia at the apex (Fig. 4.59E). They swim around for sometime and fuse together resulting in the formation of a zygote (Fig. 4.59G). The zygote remains quadriflagellate for a very short time, soon assumes round shape and germinates directly into new plant (Fig. 4.60).

The alga has an elaborate diploid phase. Meiosis being gametic the haploid phase is represented by the gametes only.

Some Indian Species: *Bryopsis corymbosa* J. Ag.; *B. hypuoides* Lamour.; *B pachynema* Martens.; *B. pennata* Lamour.; *B. plumosa* (Huds.) Ag.; *B. tenuissima* Notaris.

Special features

1. Very conspicuously developed coenocytic plant body being differentiated into a comparatively inconspicuous prostrate portion and an elaborately developed aerial erect axis with a dense and regular pinnate branching in acropetal succession.
2. Discoid to spindle-shaped chloroplast.
3. Vegetative reproduction by the growth and decay of the thallus.
4. Sexual reproduction anisogamous.
5. Elaborate diploid phase.
6. Gametic meiosis.

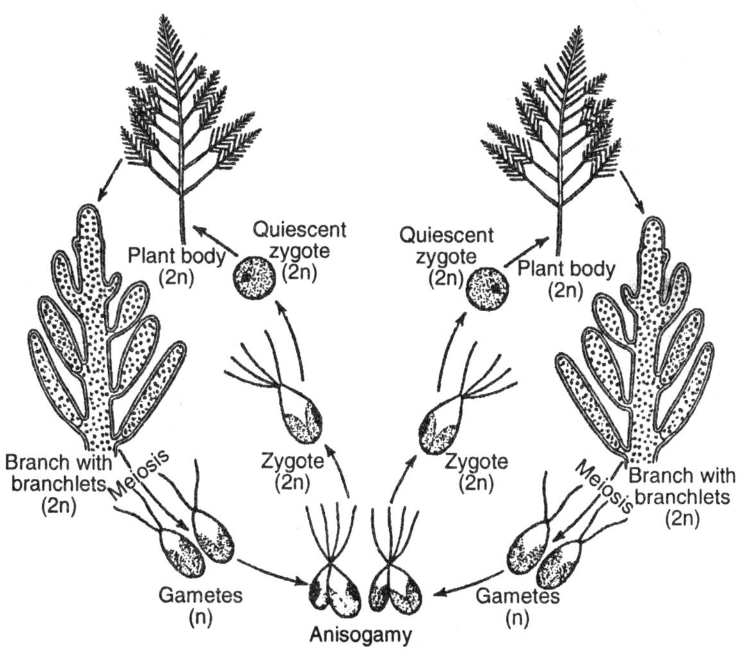

Fig. 4.60: Life cycle of *Bryopsis* sp.

Genus: *Caulerpa*

All species of *Caulerpa* are marine being frequent in the quiet shallow waters of the tropics although there are some Mediterranean species. They cover extensive tracts of the sea-floor and favour relatively shallow water. Several of them are often rooted in sand or mud with their prostrate rhizome bearing colourless rhizoids with which they attach themselves with the substratum. The production of deadly poisonous substance **Caulerpicin** has been recognized in Hawaii and in the Philippines, since freshly collected plants of *Caulerpa* eaten in salads causes health hazard. The plant body is elaborate in form being differentiated into prostrate portion represented by creeping rhizome with rhizoids (Fig. 4.61A) and an aerial portion performing the function of assimilatory shoot, or **assimilator** which may be very variable in structure being either simple flat blade-like, or lobed or a

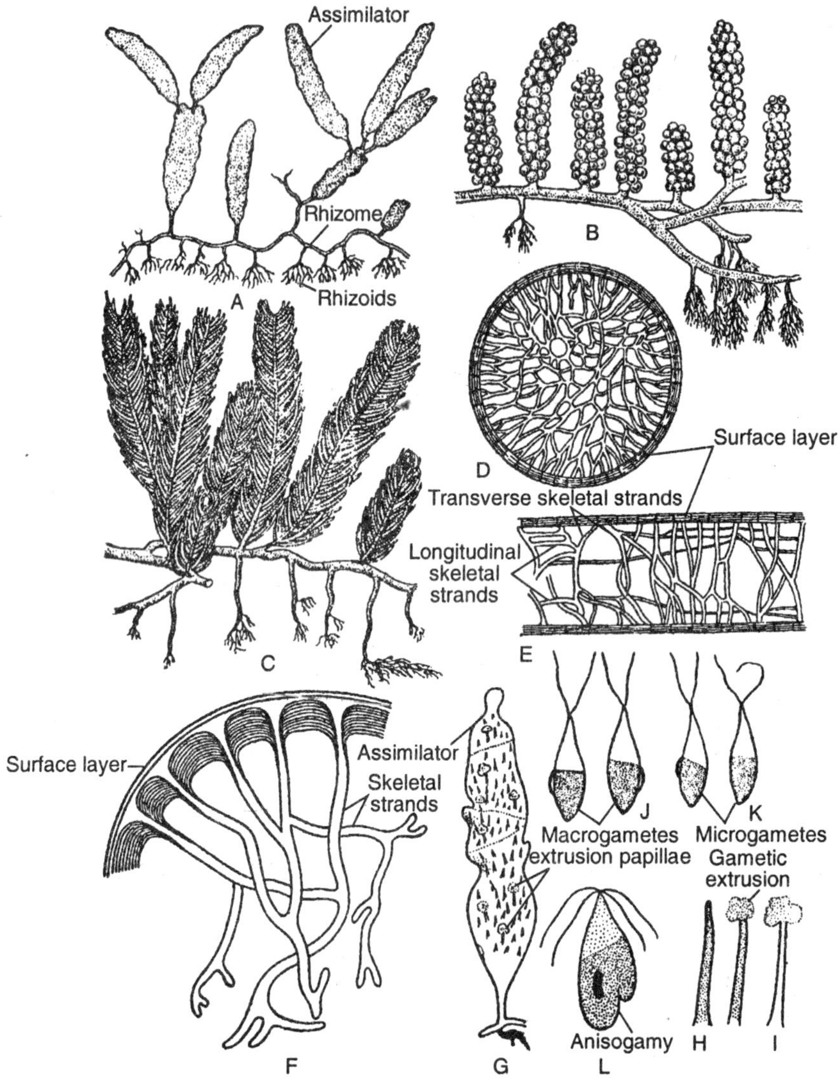

Fig. 4.61: *Caulerpa* sp. A-C. Habit and variations in the aerial portion. D. Section of rhizome showing skeletal strands. E. Longitudinal section of assimilator showing structural details. F. Internal structure of rhizome magnified. G. A fertile assimilator with extrusion papillae through which swarmers are liberated. H. Unopened extrusion papilla. I. Extrusion papillae showing gametic extrusion. J. Macrogametes. K. Microgametes. L. Anisogamy

highly complicated branched structure (Figs. 4.61B and C). Internally, plant body is characterized by the complete absence of septation and is composed of closely apposed and intertwined coenocytic threads traversed by longitudinal and transverse skeletal strands (Figs. 4.61D to F) perhaps to increase the rigidity of the plant body. These are very remarkably developed in the rhizome portion and are completely absent or poorly developed in the rhizoids. As such the entire plant body is a single multinucleate cell with a central vacuole and lining cytoplasm with numerous discoid chloroplasts. The longitudinal walls, as well as the wall of the skeletal strands are composed of callose, pectin, and similar other materials. But cellulose is completely absent.

Vegetative reproduction takes place by the growth and decay of new and old parts, resulting in the formation of innumerable plants.

Sexual reproduction, depending on species, may be iso- or anisogamous by the production of biflagellate swarmers which are more or less pear-shaped. They are developed in the entire body except the rhizoids. Prior to their liberation, papillate outgrowths are developed (Fig. 4.61G). Simultaneously with the development of papillae there starts the formation of gametes in the plant body. Before the formation of papillae the assimilators acquire a variegated appearance. The swarmers are liberated in a mass of mucilage through the apices of the extrusion papillae (Figs. 4.61H and I). In some species, soon after the liberation of swarmers the whole plant dies. In certain species the whole plant can produce swarmers, whilst in others the reproductive area is limited. The swarmers in *Caulerpa clavifera* are of two kinds—the macro- and microgametes (Figs. 4.61J and K). Gametic fusion in *Caulerpa* is commonly anisogamous (Fig. 4.61L).

Some Indian Species: *Caulerpa crassifolia* (Ag.) J. Ag.; *C. fastigiata* Mont.; *C. peltata* Lamour.; *C. racenosa* (Fursk.) Web. van. Bosse.; *C. sertulariodes* Gmel.; *C. taxifolia* (Vahl.) Ag.

Special features

1. Coenocytic plant body with remarkable differentiation of aerial and prostrate portions.
2. Development of internal skeletal strands for mechanical support of different parts of the plant body.
3. Sexual reproduction both iso- and anisogamous.
4. Fusing gametes are pyriform bearing two flagella.
5. Gametes escape from the plant body through the ends of extrusion papillae in a mass of mucilage.

Order: *CHARALES*

This order represents a very highly specialized group that must have diverged very early in the course of evolution from the rest of the green algae, the intermediate forms subsequently being lost. The group is very ancient because fossil members are found from almost the earliest strata. The living forms are widely distributed in quiet waters, fresh or saline, where they may descend to considerable depths so long as the bottom is either sandy or muddy. The plant body consists of a long, slender branched, jointed being differentiated into nodes and internodes attached to mud or sand by rhizoids. At each node arises a whorl of lateral branches. The young plants develop from a protonemal stage. The thallus is frequently lime encrusted. Sexual reproduction is oogamous. Antheridia and oogonia differ remarkably from the corresponding organs in other green algae. Zygotic nucleus divides reductionally resulting development of protonema and rhizoids from which new individual is produced.

Family: *CHARACEAE*

The Characeae are cosmopolitan, occurring in all continents and in many isolated oceanic islands. The plants are aquatic or sub-aquatic in habit with elaborately developed plant body having rhizoidal prostrate portion and aerial portion differentiated into nodes and internodes. The aerial portion is attached in the soil by hyaline rhizoids. The cells are with numerous chloroplasts and becoming multinucleate at maturity. Vegetative parts consist of an uniaxial series of alternating short nodal and elongated internodal cells. The nodal cells produce peripheral nodal cells from which arise a whorl of branches or 1-2 branches. In addition, 1-celled processes may arise at branchlet nodes or from gametangial laterals.

Sexual reproduction is oogamous and involves fertilization by biflagellate antherozoid which fuses with the egg while it is held in the oogonium surrounded by a sterile jacket of spirally twisted cells (convoluting cells). Such an oogonium should preferably be termed as **oocarp**. The antherozoids arise singly from cells in antheridial filaments enclosed in a complex **antheridiocarp** (taxonomically called the antheridium). The antherozoid gains access to the ovum through a terminal aperture produced between the convoluting cells. After fertilization the zygote develops into a thick-walled oospore. On germination of the oospore a horizontal filament (protonema) develops as a lateral branch. The normal, erect shoots arise from the protonemata. Meiosis is known to occur in the first division of the zygote, prior to initiation of the protonema. Vegetative propagation occurs by secondary protonemata which may develop from axial nodes, by bulbils, or by fragmentation.

Genus: *Chara*

It is a submerged aquatic alga growing attached by rhizoids to the muddy or sandy bottom of small pools, lakes, and slowly flowing streams. The plant body is slender and flexous, with individual plants attaining lengths 20 to 30 cm or more in height. The multicellular plant resembling the appearance of miniature horse-tail, has an erect main axis differentiated into a regular succession of nodes and internodes with whorls of lateral branches of limited growth, often spoken as 'leaves', arising from the nodes (Fig. 4.62A). Branches of limi-ted growth having nodes and internodes just like the main axis may arise axillary to the leaves. The internode is composed of a central cylindrical cell having length many times longer than breadth ensheathed by elongated narrow cells arising from basal nodes constitu-ting the **cortex** (Fig. 4.62D). The corticated nature of the stem is clear in cross-section. Incrustations of calcium carbonate make the plants coarser justifying the common name **stonewort**. The multicellular rhizoids with oblique septa deve-

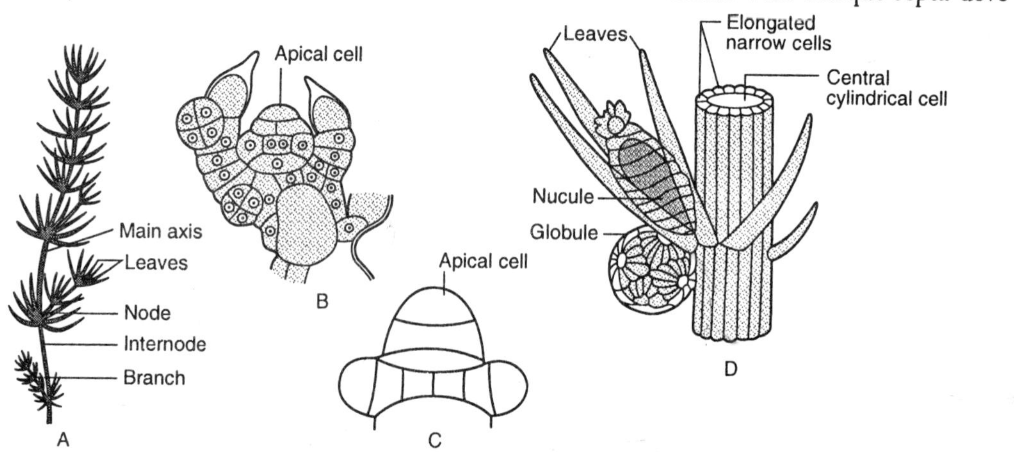

Fig. 4.62: *Chara* sp. A. Habit. B-C. Vertical section through apex of main axis showing methods of segmentation. D. Mature globule and nucule and leaves borne at the node, structure of main axis

loped from the lowest node are not differentiated into nodes and inter-nodes. Their functions are fixation and absorption. The apical and nodal cells are uninucleate with dense cytoplasm having no conspicuous vacuole, bear discoid chloroplasts without pyrenoids, whereas the elongated internodal cells are multinucleate with numerous small chloroplasts embedded in cytoplasmic layer lining a huge central vacuole. The inner layer of cytoplasm exhibits very conspicuous streaming movement.

The growth of the main axis and its branches is initiated by a single dome-shaped apical cell which cuts off segments parallel to its flat base (Fig. 4.62B). Each segment divides transversely into a biconcave upper nodal initial cell which divides and redivides to give rise to a node and a biconvex lower internodal initial cell which remains undivided but elongated considerably to form the internode (Fig. 4.62C).

Vegetative propagation usually occurs at all stages during the lifetime of the alga and is effected by the production of vegetative propagative structures which are: (i) star-shaped aggregated cells filled with starch that are developed about the lower nodes; (ii) tuber-like structures formed on the rhizoids; and (iii) protonema-like outgrowths originating from the rhizoids.

The alga exhibits a high degree of specialization in the nature of the gamete-producing structures which are named as **globule** and **nucule** instead of being called as antheridium and oogonium respectively. With a few exception, most species of *Chara* are homothallic having both globule and nucule always borne at the nodes with a definite orientation, the nucule always lying above the globule (Fig. 4.62D). A mature globule is large, stalked, spherical in structure and bright yellow to red in colour (Fig. 4.63A). Its wall is composed of eight curved plates, the **shield cells**, joined end to end giving the wall a pseudocellular appearance (Fig. 4.63B). Each shield cell bears in its centre a rod-shaped cell—the **manubrium** from whose distal end arise rounded cells—the **primary capitula**, each of which produces six **secondary capitula**. From each of these capitula arise two long, whip-like threads—the **antheridial filaments** (Figs. 4.63C and D).

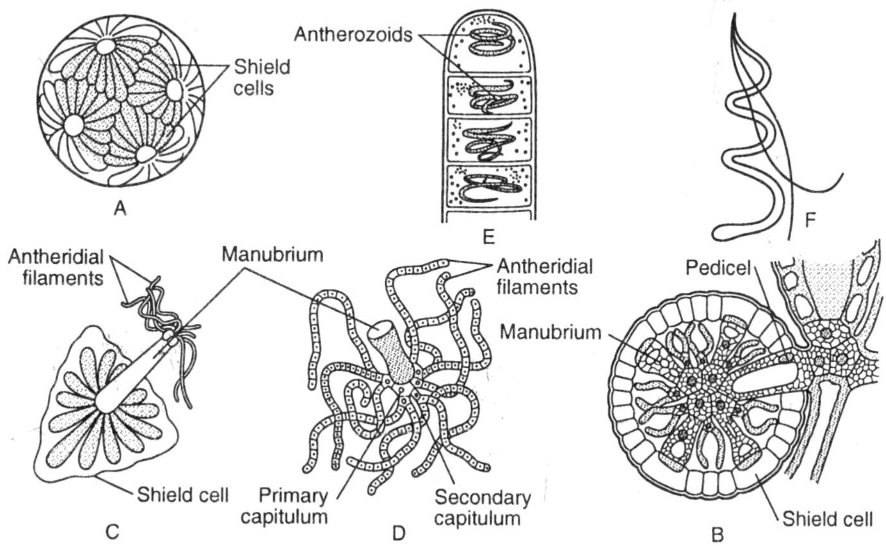

Fig. 4.63: *Chara* sp. A. A globule showing shield cells. B. A sectional view of globule and its attachment at the node. C. Shield cell with manubrium and antheridia filaments. D. Manubrium with capitula and antheridial filaments. E. Part of antheridial filament with cells bearing antherozoids. F. Mature antherozoid

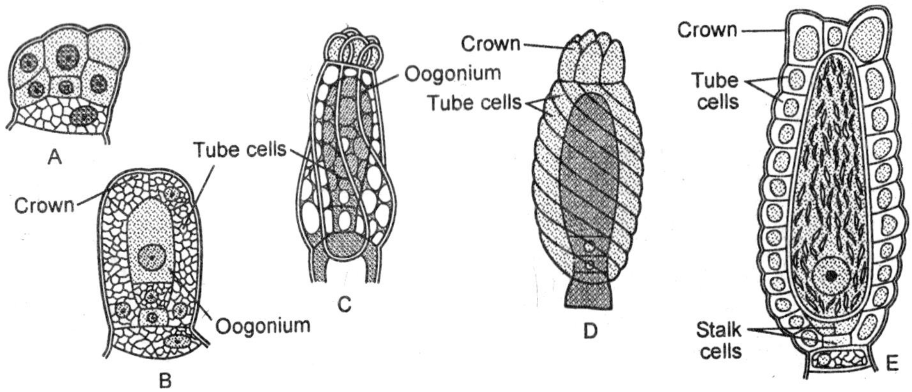

Fig. 4.64: *Chara* sp. A-C. Stages in development of nucule. D. Entire nucule. E. Nucule in section

Each antheridial filament is composed of single row of cells ranging between 100 and 200 in number, each of which produces one elongate biflagellate anthrozoid (Figs. 4.63E and F).

The nucule is a stalked oval structure being surrounded by an envelope of five long filaments placed side by side arranged spirally exhibiting a twisted appearance and affording protection to the enclosing oogonium. Each

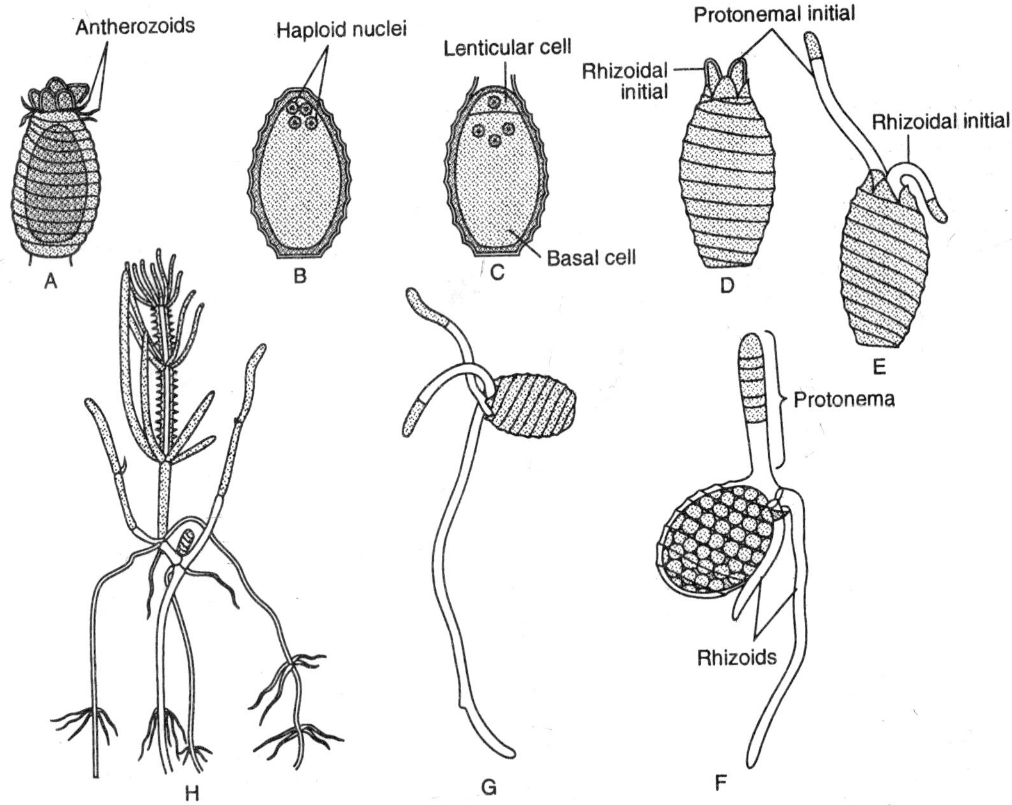

Fig. 4.65: *Chara* sp. A. Fertilization stage showing nucule with antherozoids. B-E. Successive stages in oospore germination leading to development of protonemal and rhizoidal initials. F-H. Developmental stages of protonema and rhizoids giving rise to young plant

filament terminates in a small erect cell. The five erect cells of the five filaments together form five-celled **crown** or **corona** capping a mature nucule (Fig. 4.64D and E). The oogonium produces a single large uninucleate ovum packed with starch grains and oil globules. The nucleus in the ovum lies at the base. The apex of the ovum remains clear and is occupied by finely granular cytoplasm which constitutes the receptive spot.

The development of the globule takes place from the apical cell of the lateral branch which cuts off one or two discoid cells at its base and then becomes spherical. As a result of longitudinal followed by transverse division, the spherical cell develops into **octants**, each of which undergoes further division and by enlargement, the inner segments become separated from one another. The eight peripheral cells develop into eight shield cells and gradually from the centre of each shield cell there arises a manubrium from which in turn develop primary and secondary capitula and antheridial filaments.

The adaxial cell of the basal node becomes the nucule initial which divides twice giving

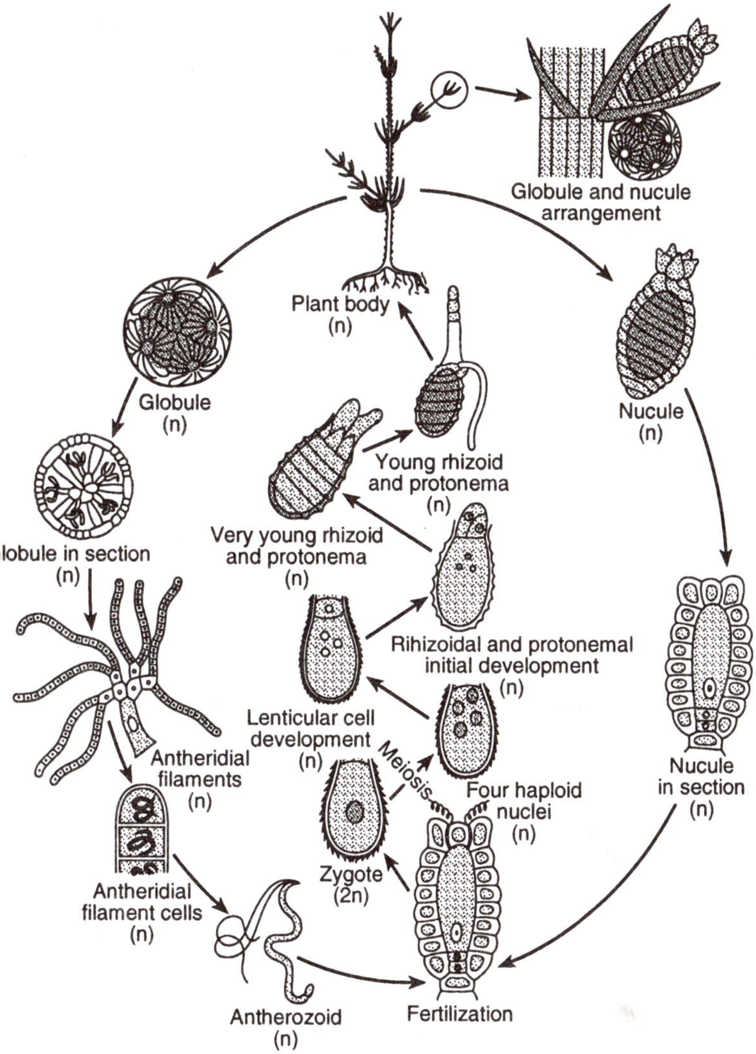

Fig. 4.66: Sexual cycle of *Chara* sp.

rise to a row of three cells, the uppermost of which develops into the oogonium proper, the lowest forms the stalk, and the middle cell cuts off five peripheral cells which elongate into filamentous structures forming **tube cells.** These tube cells surround the oogonium in a spiral fashion and finally each one cuts off coronal cell at the apex (Figs. 4.64A to C).

With maturity, the enveloping tube cells become dilated as a result of which there arises a narrow slit at the apex of the oogonium occupied by a mucilaginous liquid which permits easy entrance of antherozoids in the oogonium (Fig. 4.65A). The antherozoid nucleus traverses down in the base of the oogonium where the fusion of both the nuclei takes place. Immediately after fusion, the zygote nucleus travels to the apex of the oospore surrounding which develops a thick wall. Later, the thick-walled oospore with other structures of the nucule drops to the bottom of the pool. The surrounding envelope of the oospore hardens into nut-like structure due to deposition of calcium carbonate. Before germination, the oospore nucleus divides reductionally forming four haploid nuclei (Fig. 4.65B). One of these nuclei is cut off by a cell wall resulting two unequal cells. The upper or the distal cell is small and uninucleate, the **lenticular cell,** whereas the lower or **basal cell** is large containing three nuclei and is packed with reserve food (Fig. 4.65C). The three nuclei of the basal cell soon disintegrate. The small upper lenticular cell so formed later divides vertically into a **rhizoidal initial** and a **protonemal initial** which grow in opposite directions forming a colourless green and ultimately develops into the aerial portion of a mature *Chara* plant (Figs. 4.65G and H and Fig. 4.66).

POSITION OF CHARA

Chara exhibits marked advancement: (i) the vegetative body with highly developed aerial portion being differentiated into nodes and internodes bearing whorls of branches arising from the nodes and a less conspicuous prostrate portion resembling like root system, (ii) great elaboration of sexual reproductive organs, (iii) biflagellate elongated antherozoids, (iv) oogamous sexual reproduction, and (v) elaborate post-fertilization stages. All these features suggest to raise the status of *Chara* much higher than Algae and to place closer to the bryophytes. But there are certain weaknesses which retain *Chara* where it is at present. They are: (i) haploid plant body, (ii) the apparently complex nucule is actually a simple unicellular structure typical of any other alga, and (iii) the diploid stage being restricted in the zygote as the first division of the zygote nucleus is reductional. But the differences between *Chara* and the members of the green algae are again strong enough to put *Chara* under a special class the Charophyceae. To Church, *Chara* is a remnant of the many probable evolutionary tendencies that failed to attain land habit. The fossil evidence by no means gives any clue to determine its position. Perhaps *Chara* represents just a mere specialized sideline of the Chlorophyta.

Some Indian Species: *Chara benthamii* Braun.; *C. brachypus* Braun.; *C. braunii* Gmelin.; *C. corallina* Wild.; *C. fragilis* Desv.; *C. gymnopitys* Braun.; *C. zeylanica* Wild.

Special features

1. Plant body exhibits very elaborate complexity in structure and appearance.
2. The main axis and its branches are differentiated into a series of nodes and internodes. The nodes bear branches, leaves and sex organs.
3. The plant body is anchored with the substratum by rhizoids.
4. Vegetative propagation by means of specialized spore-like and protonema-like structures.

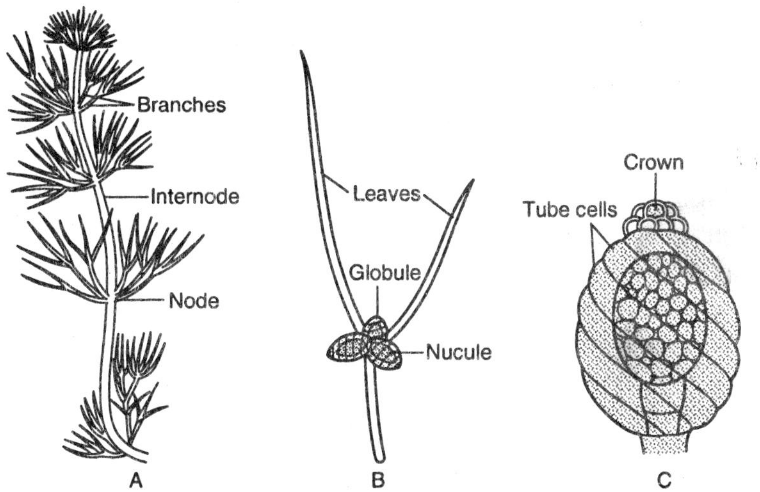

Fig. 4.67: *Nitella* sp. A. Habitat. B. Arrangement of globule and nucule. C. Nucule

5. Sexual reproduction oogamous. The sex organs are large and highly specialized and complicated in structure.
6. The female organ known as nucule is surrounded by a protective sheath of five long spirally wound filamentous structures ending in a five-celled crown or corona—a special mechanism by which the female organ receives protection.
7. The male organ or the globule is composed of eight plate-like cells which bear rod-like handle or manubrium, from whose inner end arise capitula ending in antheridial filaments. Each cell of the antheridial filament gives rise to a single biflagellate antherozoid.
8. Very elaborate post-fertilization stages.

Genus: *Nitella*

The genus *Nitella* has resemblance with *Chara* in aspects. Like *Chara, Nitella* also constitutes an extensive subaquatic vegetation in small ditches and similar other areas. But unlike *Chara, Nitella* grows in much deeper water. The vegetative body is somewhat shorter than *Chara* with bushy habit (Fig. 4.67A). It has the appearance of a miniature horse-tail (*Eqisetum*). The mode of sexual reproduction is very similar to *Chara.* There are both dioecious or monoecious species, in the latter case the globule and nucule are juxtaposed, the globule being directed upwards and the nucules downwards (Fig. 4.67B). Both organs appear on lateral branches of limited growth. The nucule is relatively less elongated than what is found in *Chara* and bears ten-celled crown which is again much shorter than that of *Chara* (Fig. 4.67C). Besides all these major differences, *Nitella* and *Chara* are very similar to each other.

Some Indian Species: *Nitella acuminata* Braun.; *N. flagellifera* Groves and Allen.; *N. furcata* Agardh.; *N. hyalina* (DC) Agardh.; *N. mucronata* (Br.) Miquel; *N. oligspira* Br.; *N. pseudoflabellata* Braun.

EVOLUTION IN SEXUALITY IN THE CHLOROPHYCOPHYTA

The Chlorophycophyta represent an evolutionary series in which the gametic union became most firmly established even in the most primitive forms and persists among the highest members of the series.

The evolution in sexuality has taken place from isogamy to oogamy through anisogamy. To start with, the fusing gametes are identical in size and shape but with gradual evolution become vastly differentiated from one another so that one becomes large, passive, non-motile and ultimately remains enclosed in a gametangium; while the other becomes small, active and motile.

Isogamy is evident in a large number of *Chlorophycophyta.* Special mention may be made of *Chlamydomonas moewusii,* where isogamy is between two flagellate isogametes. In the Conjugales, however, there is isogamy between aplanogametes, but the motility of the gametes is determined by physiological characters. In *Zygnema* and *Mougeotia* both the fusing gametes are equally active and fusion of isogametes takes place in the conjugation tube. But in *Spirogyra,* though the gametes are isogametes morphologically, they are distinguishable by their motility. Of the two fusing gametes, in some species, one remains passive and the other becomes actively amoeboid and moves to the passive gamete and fuses with it. This is a case where there is deviation from the typical isogamy. Here the gametes are isogametes morphologically but their behaviour is not similar, hence they are physiologically dissimilar. The passive as male, such a deviation from typical isogamy indicates a tendency which leads to anisogamy where the gametes are morphologically different. This is transitional stage between isogamy and true anisogamy.

Evolutionary trend may be traced from true anisogamy as follows:

(i) both the fusing gametes are anisogamous, flagellate, and active—exhibited by *Chlamydomonas braunii;*

(ii) both the fusing gametes are flagellate and active, but the larger one is sluggish and just before fertilization it loses its flagella, and becomes non-motile, as in *Aphanochaete repens,* which clearly indicates how anisogamy has been derived from isogamy;

(iii) of the fusing gametes, the larger one is non-flagellate and the smaller flagellate, such a condition is encountered in *Chlamydomonas coccifera.*

Oogamy is the most advanced method of sexual reproduction, where a large non-motile passive female gamete (egg) is fertilized by a small actively motile male gamete (antherozoid). Both male and female gametes are borne in male and female gametangia respectively. In majority of oogamous Chlorophycophyta the egg is retained in the oogonium and fertilization takes place there in the oogonium. But within the oogamous forms of Chlorophycophyta a less-advanced condition is exhibited particularly by *Chlorogonium oogamum,* where the female gametangium produces a single large egg which is liberated out from the female gametangium immediately after maturity and is fertilized by an elongated antherozoid. This is a case where fertilization takes place outside the female gametangium.

A most advanced type of oogamy is found in *Coleochaete, Chara* and *Nitella* where the female gamete is retained in the female reproductive organ and is fertilized by an antherozoid. The structure of female reproductive organ in these algae is very much advanced. In *Chara* and *Nitella* the female reproductive organ has a special protective device, but being unicellular, fails to rise much in the evolutionary scale. Again in *Coleochaete, Chara* and *Nitella* there is elaborate post-fertilization stage. Since meiosis takes place immediately after the formation of oospore this elaborate post-fertilization stage is not of much evolutionary importance.

EVOLUTIONARY TREND IN THE DEVELOPMENT OF PLANT BODY IN THE CHLOROPHYCOPHYTA

Members of the Chlorophycophyta exhibit a wide degree of variation in their external form, ranging from very simple unicellular motile and non-motile on the one hand to a highly developed complex structure with clear distribution of labour on the other. Considering all these various forms, trend of evolution can be traced broadly along four lines with a starting point from the Chlamydomonad type: (i) the motile coenobial line, (*Gonium-Eudorina, Andorina-Volvox*), (ii) the non-motile net-like line (*Chlorella-Scenedesmus-Pediastrum-Hydrodictyon*), (iii) the coenocytic line (*Caulerpa*), and (iv) the filamentous line (*Ulothrix, Oedogonium*). The filamentous line has again given rise to: (i) simple parenchymatous form (*Ulva*), (ii) a line terminating in the Conjugales, and (iii) the elaborately developed heterotrichous line. The heterotrichous line has further developed in two directions with the suppression of one form or the other. They are: (i) a line having the tendency to develop thalloid structure at the cost of the aerial portion which turns out to be very rudimentary,

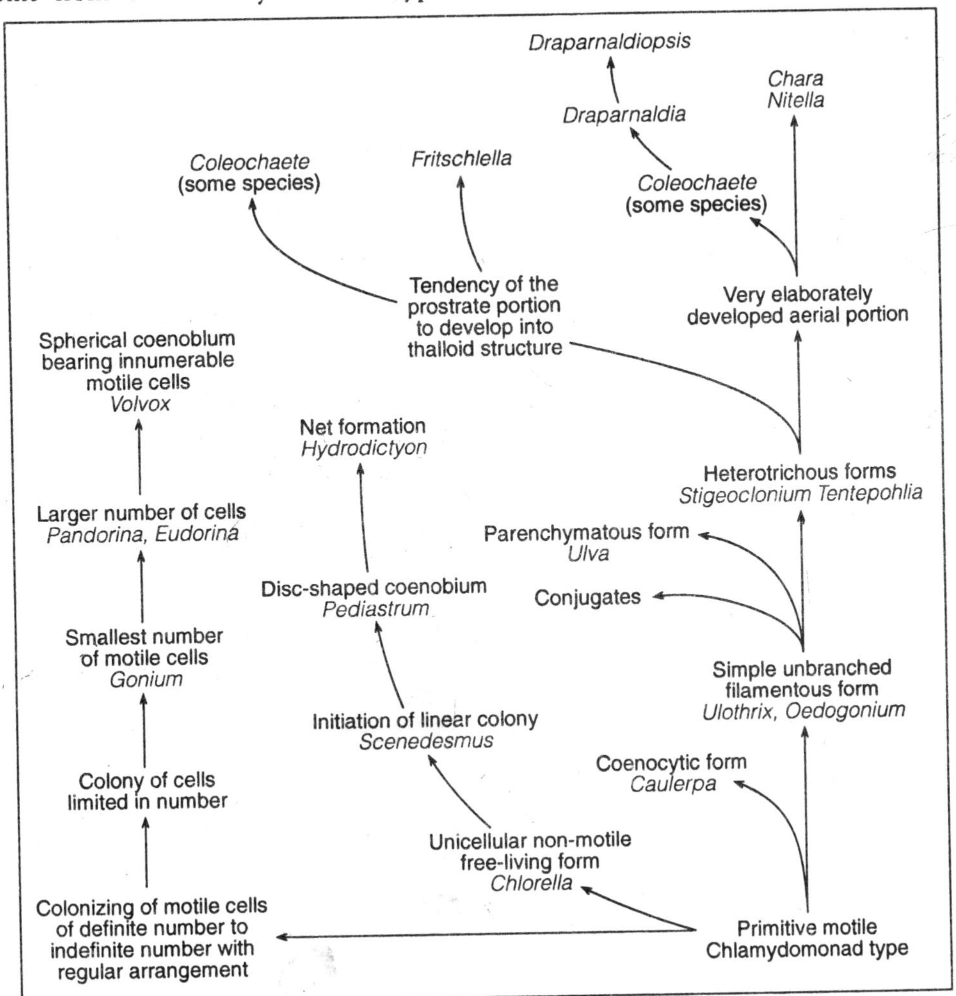

Fig. 4.68: Evolutionary trend in the development of plant body in the Chlorophycophyta

(*Coleochaete, Fritschiella*) and (ii) a very elaborately branched aerial portion with very insignificant prostrate portion (*Draparnaldia, Draparnaldiopsis, Chara, Nitella*). All these forms are schematically represented in Figure 4.68.

The evolutionary trend indicated in Fig. 4.68 is broadly based on the observation of E. F. Blackman, who is 1900 traced the evolution in the construction of plant body in the Chlorophycophyta from one-celled flagellated Chlamydomonad type of condition. He pointed out that evolution took place along the following lines, designated by him as **tendencies**:

(i) The **volvocine line** or **tendency** in which the flagellated vegetative cells divide to produce daughter cells which organize a motile colony.

(ii) The **tetrasporine line** or **tendency** in which motility is lost, except in reproductive stages; non-motile cells are produced as a result of vegetative division.

(iii) The **chlorococcine** or **endosporine line** or **tendency** in which just like the tetrasporine line, motility is lost, except in reproductive stages, but plant body has lost the capacity to divide vegetatively.

(iv) The **rhizopodal line** or **tendency** in which the trend of evolution is toward a naked amoeboid type of development.

Evolution along the volvocine line is very limited since size of motile colony and their cells are variable in number and perhaps *Volvox* is the culminating member of the line. But evolution along the chlorococcine line exhibited by the Chlorococcales, has taken place in different directions such as: unicellular to coenocytic condition, and to colonial habit with zoospores or aplanospores. Both of these conditions have limited scope of evolution. Again, the tetrasporine line has great scope for evolution. Here the tendency for the vegetative cells to become non-flagellated is a very dominant phase. This is accompanied by the tendency to organize filamentous condition in the Ulotrichaceae, heterotrichous condition in the Chaetoporales, and parenchymatous form by gaining the ability to divide both vertically and transversely seen in Ulvaceae. The other two lines that segregate from the tetrasporine line are, evolution of a coenocytic condition in cells of a filamentous branched condition in the Cladophorales and evolution of a distinctive type of zöoid as in the Oedogoniales or of apanogametes exhibited by the Conjugales.

EVOLUTIONARY TENDENCIES IN CHLOROPHYCOPHYTA TOWARDS A LAND HABIT

Land plants have advanced over the members of the Chlorophycophyta along the line of these adaptations:

(i) Increased structural complexity of the plant body.

(ii) Highly developed reproductive organs with special adaptation to protect the gametes.

(iii) Protection of zygote, embryo formation, and elaborate development of post-fertilization stages.

Though most of the members of the Chlorophycophyta are different from the plants which have become established on land, yet with a closer analysis it may be indicated that there is definite tendency among the members of the Chlorophycophyta to migrate towards land as can be exemplified by the following instances.

Terrestrial algae which grow on rocks, bark of trees, damp soil and other similar places show definite adaptations to protect themselves against desiccation. To elucidate this point, there may be cited as examples of various species of *Trentepohlia,* where the cell wall is stratified and cells develop a special pigment (haematochrome) to serve the purpose of a light screen. Besides, the delicate apical cell is protected by the pectose cap. *Cladophora calcicola,* a terrestrial alga which grows on rocks, develops cuticle on the cell wall. Similar adaptations may also be seen in different species of *hormidium, Zygnema, Characium, Pleurococcus, Oliveria* and *Fritschiella.*

As to the structural peculiarities of the plant body, members of the Orders Chaetophorales and Charales exhibit considerable morphological differentiation and show a wide range of construction from heterotrichous to a highly complex aerial portion with a capacity to grow indefinitely. In heterotrichous forms, the plant body consists of an expanse of dorsiventrally differentiated branched creeping threads spreading over the substratum and affording a firm anchorage, as well as a primary photosynthetic system is well exposed to the light. From this, sooner or later, branched filaments grow out at right angles into the surrounding water, or in subaerial form into the air, and take over most of the photosynthetic functions, as well as commonly bearing the reproductive organs. Both prostrate and projecting systems are well developed in *Stigeoclonium.* A further advancement is seen in *Draparnaldiopsis,* in which the aerial portion is composed of long and short cells, the latter alone bearing the laterals of limited and unlimited growth, commonly in whorls resembling more or less nodes and internodes. In *Draparnaldiopsis indica,* the branches have the tendency to coil around the main axis giving a somewhat corticated appearance.

In *Draparnaldia,* the aerial portion is strikingly differentiated into long and short branches, while the prostrate system may be rudimentary or sometimes may be represented by multicellular rhizoids. In all the more extreme forms the base shows a tendency to become pseudoparenchymatous, as in *Coleochaete.* As long as there is a properly developed upright system, the reproductive organs are usually confined to it, as in *Coleochaete.* In *Trentepohlia,* however, gametangia tend to arise from the base, whilst the sporangia are found on the upright threads.

The alga *Fritschiella tuberosa* found on relatively dry soils in India, Nepal, Sudan, and Africa has an extremely elaborate specialized heterotrichous habit, possessing a rhizoidal system comparable to the rhizoids of the bryophytes, a prostrate system with the potentiality to develop a parenchymatous structure from which arise projecting short-celled primary and long-celled secondary systems. Rhizoids serve for anchorage and absorption of nutrients. The prostrate system possesses the function of storage and reproduction. The projecting system is exclusively photosynthetic.

The maximum complexity in the structure of the aerial portion is found in *Chara,* where there is definite differentiation of nodes and internodes, the nodes bearing branches and leaf-like structures, and the internodes being corticated. Growth is by an apical cell with a definite mechanism for protection. Reproductive organs are borne at the nodes. The prostrate portion is represented by the rhizoids with oblique septa, resembling members of Bryophyta.

As far as the structure of the reproductive organ is concerned, in *Coleochaete* the oogonium has a long trichogyne which resembles the archegonial neck of the bryophytes. In *Chara,* the reproductive organs are invested with a cellular envelope which protects the gametes (male and female) against unfavourable external conditions.

The most remarkable feature in *Coleochaete* is the development of sterile filamentous structures surrounding the base of the oogonium immediately after fertilization; that is an adaptation to protect the zygote, and may very well be considered as a starting point for the gradual elaborate development of the post-fertilization stage towards further complexity.

In *Chara,* however, the oogonium is already well protected by a sterile covering which can very well serve the purpose of protecting the zygote. Elaborate post-fertilization stages and embryo development can very well be indicated in *Chara.* Though the so-called embryo is haploid, yet there is definite provision for food storage and clear separation of the protonemal and rhizoidal initials in the early stage.

Though many green algae represent 'dead ends' as far as subsequent evolutionary development is concerned, there is no doubt that they are extremely successful forms, since they are able to exist in a wide range of habitats all over the earth. Furthermore, the great diversity among the green algae would suggest that many forms among these plants would be able to adapt themselves to changing environmental conditions.

Chapter 5

The Euglenophycophyta

The Euglenophyriscophyta represent mostly unicellular naked motile individuals which usually occur in fresh-water. Some species form dendroid colony. The thallus possesses a definite nucleus and the grass-green colour is due to the presence of chlorophyll localized in chromatophores. Few species lack chromatophores and are colourless. Each euglenoid cell bears several chromatophores which may be discoid, band-shaped, or stellate. There are species whose cell may contain a red pigment—**haematochrome**. Cells contain pyrenoids which may be within or outside the chromatophores. The reserve food is **paramylum**—a carbohydrate related to starch, and fats. Besides assimilation of carbon with the help of chlorophyll the nutrition may also be holophytic, and holozoic. Number of flagella is one, two or rarely three being attached to the anterior end of the cell, close to which are the gullet and red eye-spot.

The euglenoid nucleus is of the mesokaryotic type, having chromosomes that are permanently condensed during the mitotic cycle, a nucleolus (endosome) that does not disperse during nuclear division, no microtubules from chromosomes to pole spindles, and a nuclear envelope that is intact during nuclear division. The chromosome number is usually high, and polyploidy probably occurs in some genera.

Reproduction is by cell division along the long axis. Many species form thick-walled resting cells—**cysts**, which germinate into new individuals. Sexual reproduction is practically unknown except in one genus only.

This division includes only one class Euglenophyceae.

PHYLOGENETIC RELATIONSHIPS OF THE EUGLENOPHYCOPHYTA

The Euglenophycophyta exhibit both plant as well as animal characters. The presence of gullet, holozoic mode of nutrition and the absence of chlorophyll in some species resemble animals and are classified as Protozoa by zoologists. On the other hand, they are classified by botanists under Algae, since they have simplest vegetative body and most of the forms possess chlorophyll with which they can carry on photosynthesis like green plants. The genus *Colacium* has true algal organization with immobile cells which have walls and *Colacium* forms colonies.

One may regard these euglenoids as animals which show some plant characters or may think of them as essentially plants, some of which have lost their chlorophyll and so have assumed as animal-like method of securing food.

They are possibly very primitive organisms, apparently closely related to the original stock from which probably both the plant and animal kingdoms have evolved. At least, some biologists hold this opinion.

Some avoid the difficulty of classifying these forms as plants or animals by placing them in a special group, the **Protista,** in which border-line organisms like the bacteria, the slime molds, and similar other flagellates are included.

Formerly, botanists placed the euglenoids along with all the other pigmented, flagellate unicellular organisms in a group known as the **Flagellata,** or the flagellates. There is a tendency to distribute them to the various taxa. They differ from the Cyanophycophyta for having definite chromatophores. They are not included under the Chlorophycophyta as their food reserve is not starch. They are thus classified as a separate taxon **Euglenophycophyta.** There are three orders:

Order I Heteronematales: Two emergent flagella, the longer flagellum directed anteriorly and shorter one directed posteriorly during swimming, special ingestion organeble present.

Order II Eutreptiales: Two emergent flagella, one directed anteriorly and the other laterally or posteriorly during swimming, no special ingestion organelles.

Order III Euglenales: Two flagella only one of which emerges from the canal, no special ingestion organelle.

Genus: *Euglena*

This is a single-celled uniflagellate free-swimming organism occurring exceedingly in fresh-water small ponds, especially rich in organic matter. A number of species inhabits damp mud and salt-marshes. All the species of *Euglena* are always solitary and never form colonies. The highly differentiated naked body shows a more definite trend in the direction of animal world than the pigmented plants. The shape of the individual cell is spindle-shaped, sometimes spirally twisted and at times forms a colourless tip at the posterior end, with anteriorly placed flagellum and the posterior end is more or less pointed. The exterior portion of the cytoplasm is differentiated into a soft more or less rigid periplast. The cell contains a centrally placed nucleus which is recognizable without staining and numerous discoid to band-shaped green chloroplasts which carry on photosynthesis (Fig. 5.1A). Pyrenoids may or may not be present; when present, may be naked or with a sheath of paramylum granules. The chloroplasts in most cases are bright-green in colour with the exception where the cells may also produce a red pigment (haematochrome). The products of assimilation are paramylum granules of characteristic shape originating in the cytoplasm. *Euglena* ingests solid foods. A few species lack chloroplasts, and these apparently are saprophytes absorbing organic matter by diffusion through cell surface. At the anterior end of a cell there is a flask-shaped gullet with a narrow neck and an enlarged posterior, the *reservoir,* adjoining to which are one or more contractile vacuoles (Fig. 5.1B). In the vicinity of the gullet, there is a red **eye-spot** which is sensitive to light.

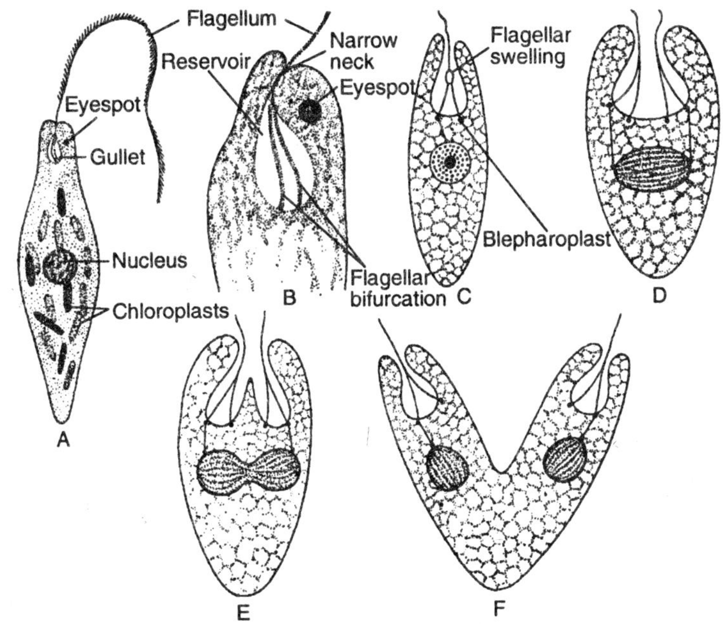

Fig. 5.1: *Euglena* sp. A. A single individual showing structural details. B. Anterior portion magnified. C. Details of flagellum. D-F. Stages in cell division

The single flagellum bifurcates as its lower end within the reservoir and each fork terminates in a blepharoplast (Fig. 5.1C). In reproduction the nucleus divides mitotically into two daughter nuclei which is followed by longitudinal division of the cell producing two new individuals (Figs. 5.1D to F). Sometimes the daughter cells are retained within an envelope and pass through pamelloid condition by further dividing and redividing. Formation of thick-walled cells to protect themselves against unfavourable condition is quite common in *Euglena.* Sexual reproduction is by conjugation of amoeboid isogametes.

Chapter 6

The Xanthophycophyta

The Xanthophycophyta or Heterokontae are commonly known as yellow-green algae including only one class Xanthophyceae. This division has close relationship with the Chlorophycophyta comprising both marine and fresh-water forms. Certain species grow on drying mud, on trunk of trees, on damp walls, and similar other habitat. There are varied forms of vegetative body ranging from unicellular coccoid, siphonaceous to filamentous condition. Each vegetative cell has a wall composed chiefly of pectic substances which, depending on species, may be impregnated with silica. The cell contains one to several discoid chromatophores. Except in the siphonaceous forms each vegetative cell is uninucleate. Due to the presence of excess of carotinoids, the colour of chromatophores is yellow-green. Chlorophyll a and chlorophyll c are present; rarely presence of chlorophyll e has been reported. But chlorophyll b is absent. The chromatophores lack pyrenoids. The usual food reserve is oil. One of the outstanding features of the Xanthophycophyta is the presence of motile cells bearing two flagella of unequal length.

Vegetative reproduction is by accidental breaking of the vegetative body. Asexual reproduction in most species is by zoospores with a few exceptional cases where aplanospores are formed for asexual reproduction. The zoospores are characterized for having two unequal flagella borne at the anterior end or are multiflagellate. Sexual reproduction is isogamous and oogamous.

Like the Chlorophycophyta three major evolutionary lines can be traced among the Xanthophycophyta. They are:

I. One line may have developed from an unicellular motile ancestry giving rise to non-motile unicells which may be solitary or colonial.

II. Another tendency is to produce a tubular, or siphonaceous form.

III. While the third one is leading to the formation of multicellular filamentous type.

Members of this division of algae were at one time included in the Chlorophycophyta. Because of their unequal flagella they were classed in the subgroup 'Heterokontae' and the others with equal flagella were classed in the subgroup 'Heterokontae' and the others with equal flagella were classified as 'Isokontae'. As an increasing number of species were examined, the separate division of Xanthophycophyta was created for the following characters they possess:

1. There is frequently an excess of carotinoids over chlorophylls; moreover, chlorophyll e appears to be restricted to this division.
2. The two flagella are normally of unequal length, being tinsel type and whiplash type.
3. The food storage products are oil and fat, but never starch.
4. A cell wall is frequently absent and when present it generally has a higher content of pectic material than the walls of Chlorophycophyta.

Three orders under the single class Xanthophyceae are: **Vaucheriales (Heterosiphonales)**.

Order I Vaucheriales

Order II Botrydiales

Order III Tribonematales

Order: **Vaucheriales (Heterosiphonales)**. Unicellular multinucleate—coenocytic plant body; asexual reproduction by bi- or multiflagellate zoospores; sexual reproduction iso- or oogamous.

Family: **Vaucheriaceae.** Coenocytic branched tubular thallus; asexual reproduction by multifagellate zoospores—flagella being distributed throughout the entire body of the zoospore, or by aplanospores; sexual reproduction oogamous.

Genus: *Vaucheria.*

Order: **Botrydiales.** Globose multinucleate thallus with colourless rhizoids.

Family: **Botrydiaceae**. Plant body of unicellular multinucleate vesicular, globose or dichotomously branched aerial portion and a colourless rhizoidal portion; discoid chromatophores; asexual reproduction by biflagellate zoospores, aplanospores or resting spores; sexual reproduction isogamous.

Genus: *Botrydium.*

Order: **Tribonematales.** Filamentous organisms, not coenocytic.

Family: **Tribonemataceae.**

Genus: *Tribonema.*

Family: *BOTRYDIACEAE*

This family comprises of algae whose plant body is characterized by unicellular multinucleate vesicular, globose or dichotomously branched aerial portion and a colourless rhizoidal portion. Asexual reproduction is by biflagellate zoospores, aplanospores or resting spores and sexual reproduction is isogamous.

Genus: *Botrydium*

It is a vesicular, coenocytic terrestrial alga growing extensively rooted either on drying damp muddy banks of streams and pools or on bare damp soil forming a dense green coating.

The plant body is differentiated into a pear-shaped or lobed aerial portion, also known as assimilator, may be one to two millimetres in diameter, and a branched, root-like underground rhizoidal portion (Figs. 6.1A and B). The shape of the aerial portion is considerably influenced by environmental conditions. The aerial vesicular or lobed portion is covered with a relatively tough wall chiefly of cellulose, within which there is a thin, peripheral layer of cytoplasm containing many nuclei; lenticular or fusiform plastids; oil droplets; and leucosin granules. The rhizoidal portion may be profusely or sparsely branched and has dense or

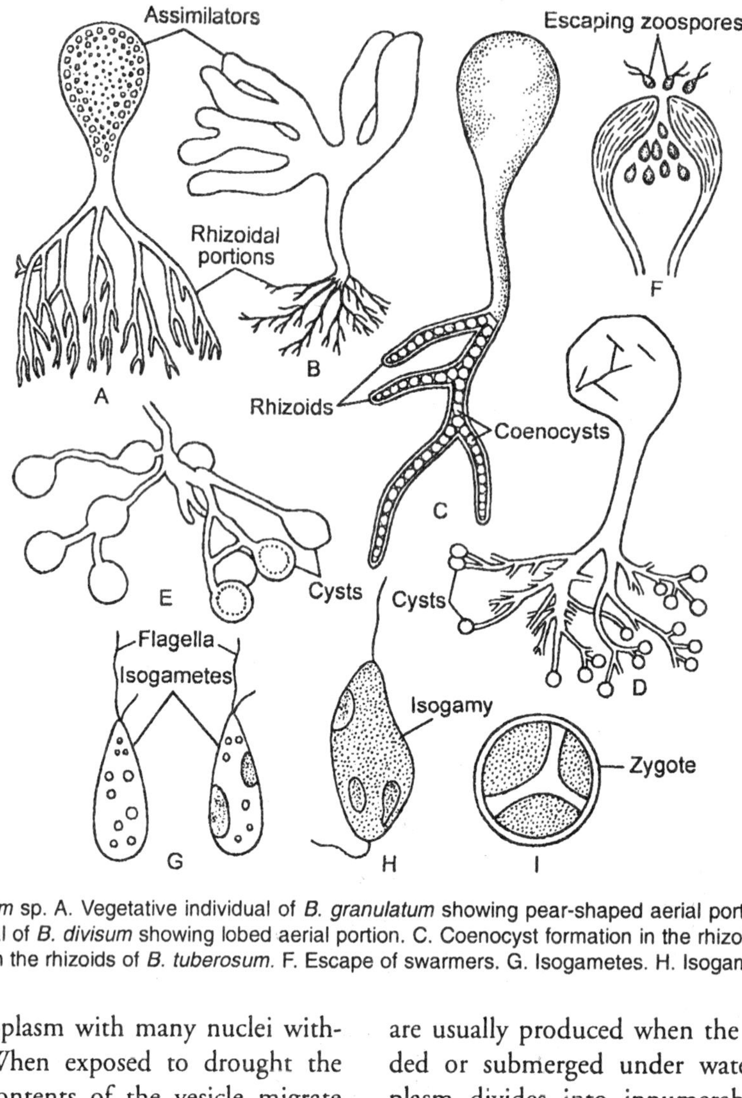

Fig. 6.1: *Botrydium* sp. A. Vegetative individual of *B. granulatum* showing pear-shaped aerial portion. B. Vegetative individual of *B. divisum* showing lobed aerial portion. C. Coenocyst formation in the rhizoids. D-E. Cyst formation in the rhizoids of *B. tuberosum*. F. Escape of swarmers. G. Isogametes. H. Isogamy. I. Zygote

vacuolated cytoplasm with many nuclei without plastids. When exposed to drought the protoplasmic contents of the vesicle migrate into the rhizoids and divide to form thick-walled globose to ellipsoid **coenocysts** also known as **cysts**, which either germinate directly or produce zoospores (Fig. 6.1C). In *Botrydium tuberosum* the protoplast of aerial vesicle migrates into the rhizoidal portion and collects at the ends of the rhizoids which become in flatted and swell to develop into rounded thick-walled **tubers** or **cysts** (Figs. 6.1D and E). The alga is incapable of vegetative division. The common method of reproduction is by the formation of zoospores which are usually produced when the vesicle is flooded or submerged under water. The protoplasm divides into innumerable uninucleate protoplasts. These protoplasts become metamorphosed into pear-shaped zoospores which bear two anteriorly placed unequal flagella. The method of liberation of zoospores from the mother cell is not definitely known, but it appears that it is accomplished by a gelatinization of apical portion in the vesicular wall (Fig. 6.1F). Occasional formation of many biflagellate isogametes or anisogametes which fuse in pairs being liberated from the same vesicle to form zygote indicates that *B. granulatum* is homothallic (Figs. 6.1G to I). The zygote

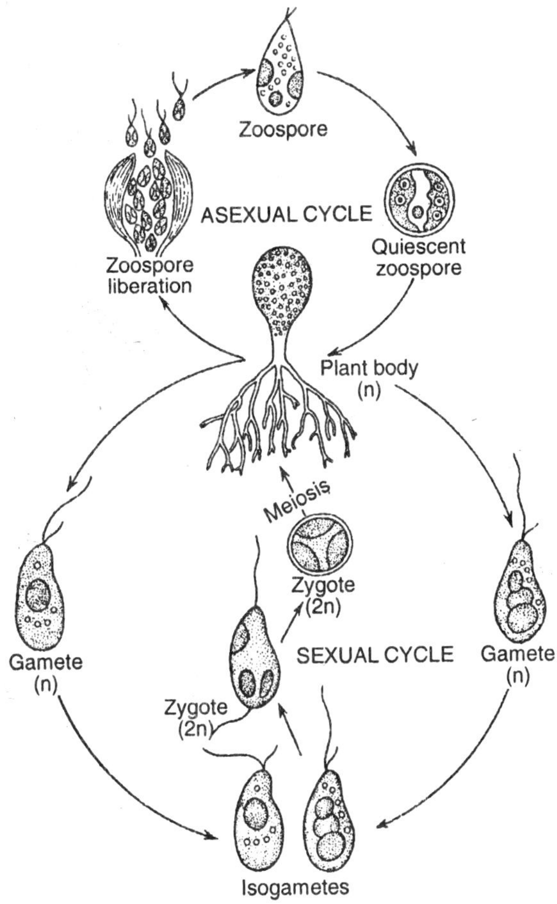

Fig. 6.2: Life cycle of *Botrydium* sp.

germinates immediately into new individual without undergoing any resting period when meiosis takes place (Fig. 6.2). Formation of parthenospore has also been reported in some species.

Special features

1. Coenocytic balloon-shaped plant body bearing rhizoidal structures.
2. Photosynthetic reserves are oils and leucosin granules.
3. Asexual reproduction by zoospores bearing two unequal flagella.
4. Asexual reproduction also takes place by resting spores.
5. Sexual reproduction isogamous or anisogamous.
6. Gametes biflagellate with unequal flagella.

Family: *VAUCHERIACEAE*

Plants are filamentous, irregularly or dichotomously branched coenocytic. Filaments are without regular wall formation except to segregate the reproductive organs. Asexual reproduction is by zoospores produced singly in sporangia formed at the tips of branches. The zoospores are multiflagellate, the flagella in pairs over the whole surface. Sexual reproduction is oogamous. Plants are dioecious or monoecious. The antheridia are cylindrical, generally curved, producing many small biflagellate anterozoids. The oogonia are single or in groups, producing single egg, which is fertilized in place and develops thick walls to form an oospore.

Genus: *Vaucheria*

This is an alga which has both terrestrial as well as aquatic species. The terrestrial species are often seen to form green felt on the surface of damp soil and on mud of drying pools, whereas the aquatic species occur in shallow fresh-water streams, while there are a few marine species. The thallus consists of a long, tubular, sparingly branched filament, which, in spite of its size, consists of one cell only as it lacks cross-walls. Cross-walls, however, occur only in the event of injury or in connection with the formation of reproductive organs. Attachment of the filament to the substratum is by a hapteron-like structure provided with colourless branched outgrowths known as **rhizoids** (Fig. 6.3A).

The filament wall is thin and is composed of an inner cellulose layer and an outer of pectose. The central portion of the filament is occupied by a large vacuole; the cytoplasm is in a thin, peripheral layer continuous along its entire length. Embedded in the cytoplasmic layer are numerous, small disk-shaped or ellip-

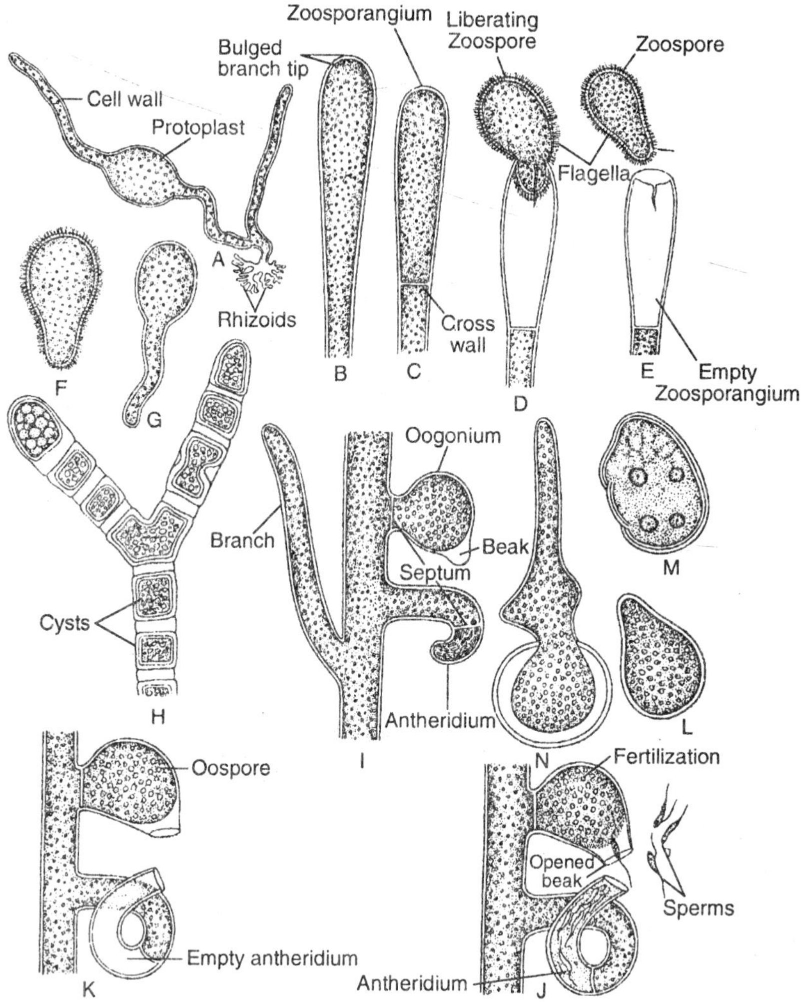

Fig. 6.3: *Vaucheria* sp. A. Young filament. B-E. Successive stages in the development and liberatiion of zoospore, F-G. Stages in germination of zoospore. H. *Gongrosira* stage with thick-walled cysts. I. Filament bearing mature antheridium and oogonium. J-K. Stages in fertilization and formation of oospore. L-N. Stages in the germination of oospore

tical chloroplasts containing a greater proportion of xanthophyll with no pyrenoids, many nuclei, and many oil globules, oil being the reserve food. Chlorophyll b is absent. But chlorophyll e has been detected in the zoospores. This undivided multinucleate, single-celled filament is known as a **coenocyte** which may attain a length of several inches. The elongation of the filament is accompanied by repeated division of nuclei without the formation of cross-walls. The filament and its branches grow by terminal elongation.

Vaucheria reproduces both asexually and sexually. The asexual method of reproduction is by means of large multiflagellate zoospores which are formed singly in **zoosporangia**, cut off at the ends of the tubular thallus by cross-walls. The branch tip bulges out forming a potential zoosporangium (Fig. 6.3B). A large number of nuclei and chloroplasts along with cytoplasm stream into swollen tip which is cut off by a cross-wall differentiating the zoosporangium from the rest of the filament (Fig. 6.3C). The protoplasmic vacuole diminishes in size as a result of which the protoplasm appears deep green in colour. The entire protoplast of the zoosporangium contracts to form an oval mass. The nuclei arrange themselves in the outer layer of the cytoplasm, and the chloroplasts in the central region. Opposite each nucleus arise a pair of flagella. The mature zoospore escapes through an opening formed by the gelatinization of the zoosporangial wall (Figs. 6.3D and E). Since this zoospore bears numerous chloroplasts and nuclei, in addition to the many flagella, it is considered a compound zoospore and is also known as a **coenozoospore**. After a short period of motility, followed by a subsequent loss of the flagella it germinates and forms one to several tubular outgrowths that may continue to grow indefinitely to produce an adult plant (Figs. 6.3F and G). The protoplasts of the sporangium may also divide to form many small immobile thin-walled aplanospores.

In some terrestrial species, if exposed to greater desiccation, the threads become septate and rows of cysts are formed giving rise to what is termed *Gongrosira* stage (Fig. 6.3H). Under favourable conditions these cysts germinate either into new filaments or into small amoeboid structures from which new filaments are produced.

Sexual reproduction is of frequent occurrence in all species of *Vaucheria* but less frequent in those species growing in flowing water. All species are oogamous. Both fresh-water and terrestrial species are homothallic, while there are quite a few heterothallic marine ones. Considering the simple structure of the plant body, the sexual reproduction is of fairly complicated nature. There is a great variation in the arrangement of sex organs in the different species and even in different individuals of the same species. In homothallic species antheridia and oogonia are generally produced on short branches close together on the filament. The antheridium is a slender, curled hook-like tubular structure which is walled off from the rest of the filament (Fig. 6.3I). The contents of the antheridium consist of cytoplasm, numerous nuclei, and chloroplasts. During development of sperms, also known as antherozoids or spermatozoids, small portions of cytoplasm surround each nucleus, thus making small pear-shaped bodies which acquire a pair of flagella. At maturity, numerous small male gametes, or antherozoids are formed. Each antherozoid bears two laterally inserted unequal flagella (one being whiplash, the other tinsel) and consists of a nucleus and a small amount of cytoplasm surrounded by a membrane. When the antherodium is mature, the antherozoids are set free through a terminal pore. The oogonium is a spherical or oval sessile or short-stalked body with a short, rounded beak which opens to receive the sperms (antherozoids) (Figs. 8.3I and J). The oogonium appears as a globular outgrowth, usually near an antheridium; and is separated from the rest of the filament by a septum. The oogonium consists of cytoplasm, a single rather large nucleus, numerous chloroplasts, and stored food in the form of oil. The contents of the oogonium form a single large spherical egg, or ovum laden with much food. The cytoplasm opposite to the beak is clear, while in the rest of the ovum chloroplasts are present. When the oogonium is mature, the end of the beak opens slightly and the sperms by chemotaxis, cluster round, swimming in a film of water. Eventually one passes in and fuses with the nucleus of the egg resulting in the formation of oospore (Figs. 6.3J and K). Actual fertilization, the fusion of nuclei, may be delayed. Following fertilization the oospore secretes a thick wall and passes through a resting stage, having within it reserve food in the form of oil. The oospore germinates and directly forms a new plant (Figs. 6.3L to N). Cytological behaviour at the reproductive stage has been insufficiently investigated. It is likely, however, that meiosis is zygotic and hence the vegetative filament is haploid (Fig. 6.4).

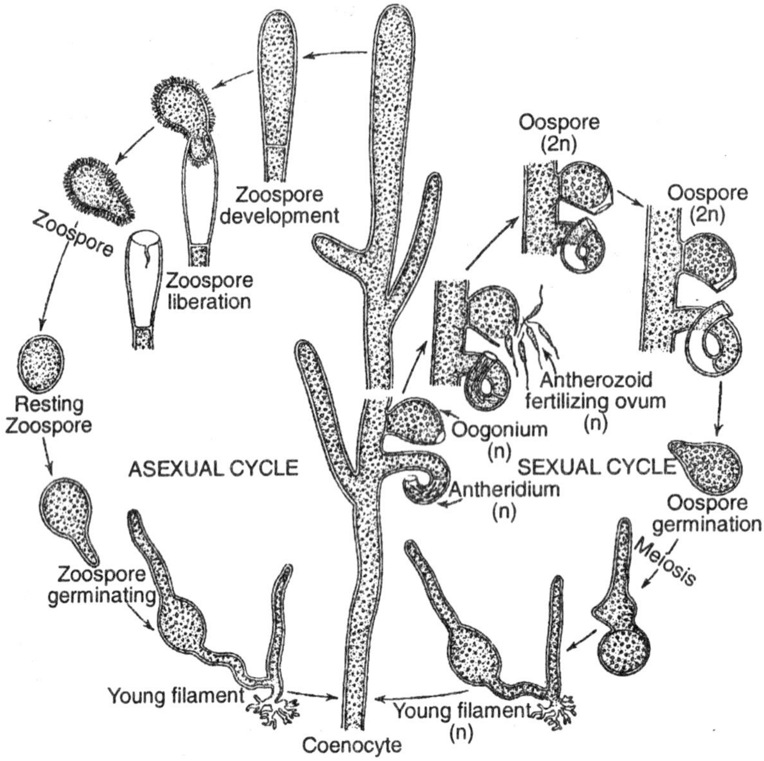

Fig. 6.4: Live cycle of *Vaucheria* sp.

Some Indian Species: *Vaucheria aversa* Hass.; *V. clavata* DC.; *V. geminata*; *V. globulifera* West et West; *V. orientalis* West et West; *V. piloboloides* Thur.; *V. polysperma*; *V. sessilis* (Vauch) DC.; *V. terrestris* Randh.; *V. uncinata* Kutz.

Special features

1. Coenocytic filamentous branched plant body attached to the substratum by specialized structure provided with rhizoids.
2. Reserve food fats and oils.
3. Asexual reproduction by means of multiflagellate zoospores whose flagella are distributed throughout the entire body—coenozoospores.
4. Sexual reproduction oogamous.
5. The oogonium bears single large, uninucleate oosphere, while the antheridium gives rise to numerous small biflagellate sperms bearing unequal flagella.

Genus: *Tribonema*

It is composed of barrel-shaped cells that are two to five times larger than their width. The cell wall composed of two H-shaped pieces overlapping in the middle of the cell. Protoplast uninucleate and contains number of discoid chloroplasts.

POSITION OF VAUCHERIA

Since long time the genus *Vaucheria* has been included in the Chlorophyceae. This arrangement was doubted by Bohin (1897), Printz (1927), Fritsch (1935), and Fldmann (1946). In spite of his own dislikeness, Fritsch placed *Vaucheria* in the family Vaucheriaceae under the order Siphonales belonging to the Chlorophyceae.

The main reasons for excluding *Vaucheria* from the Chlorophyceae are:

1. The carotinoids are in excess of the Chlorophylls.
2. Chlorophyll b is absent.
3. Chlorophyll e has been found to be present in the zoospores.
4. Two Xanthophylls—siphonein and siphonoxanthin are absent.
5. Starch never accumulates as food reserve.
6. Mercerized cellulose is not present in the cell wall.
7. Antherozoids with laterally inserted unequal flagella—one being whiplash, the other tinsel.

Although its inclusion in the Xanthophyceae appears well established, its elaborate oogamous sexual reproduction is unlike any process in the remainder of the Xanthophyceae, so that it is difficult to establish any affinity between Vaucheria and other genera of the class Xanthophyceae.

Smith (1950) has placed the genus *Vaucheria* in the Xanthophyceae among the Chrysophycophyta.

Chapter 7

The Bacillariophycophyta

The Bacillariophycophyta, commonly known as the **diatoms**, are ubiquitous algae of both fresh- and salt-water and of damp places including aerial habitats as old walls, rocky cliffs, bark of trees, and damp soils. Some are submerged in water growing attached to various substrata like aquatic algae and submerged objects. They play a very important

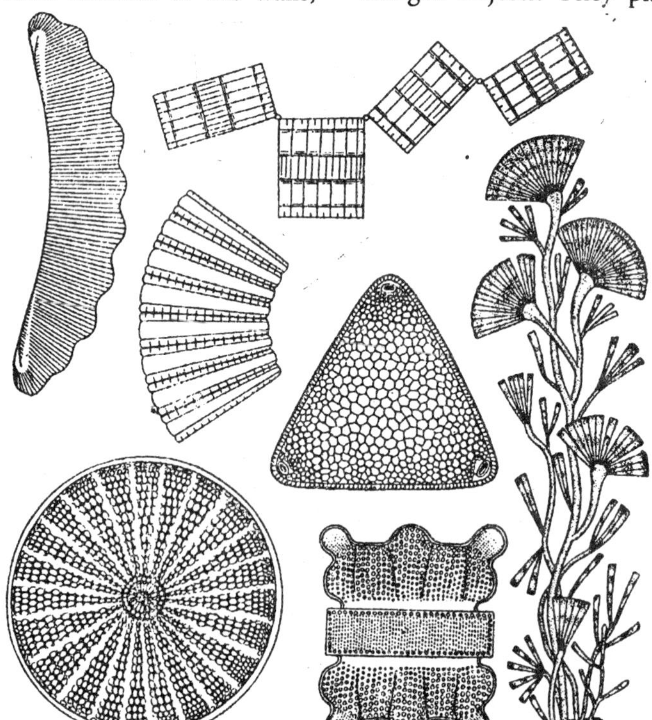

Fig. 7.1: Diverse forms of Diatoms and their wall ornamentations

role, particularly in the aquatic vegetation of the world forming an important part of the plankton, the basic food of aquatic fauna. They occur in countless individuals in even small quantities of water. Diatoms have been very appropriately named the 'jewels of the plant world', for their diverse forms and very delicately designed wall markings being the objects of great beauty (Fig. 7.1). These are unicellular, sometimes colonial microscopic algae. They are of innumerable shapes, some being boat-like in outline, while others are rods, disks, triangles, and wedges. A special branch of phycology known as **diatomology**, the science of diatoms, has developed dealing with the various aspects of the huge number of diatom species and their unique structure. The diatom cell, known as **frustule,** is composed of two overlapping halves, or **valves** very nicely fitted one upon the other like an old fashioned pill-box, or a pair of Petri dish. Of the two halves, the outer one is the **epitheca** and the inner the **hypotheca** (Fig. 7.2B). The edges of the two halves are more or less incurved, together forming the **girdle.** When one or the other valve is seen, it is known as **valve-view,** and in **girdle-view** the girdle is visible. The diatom cell wall is composed of pectic substances which are impreg-

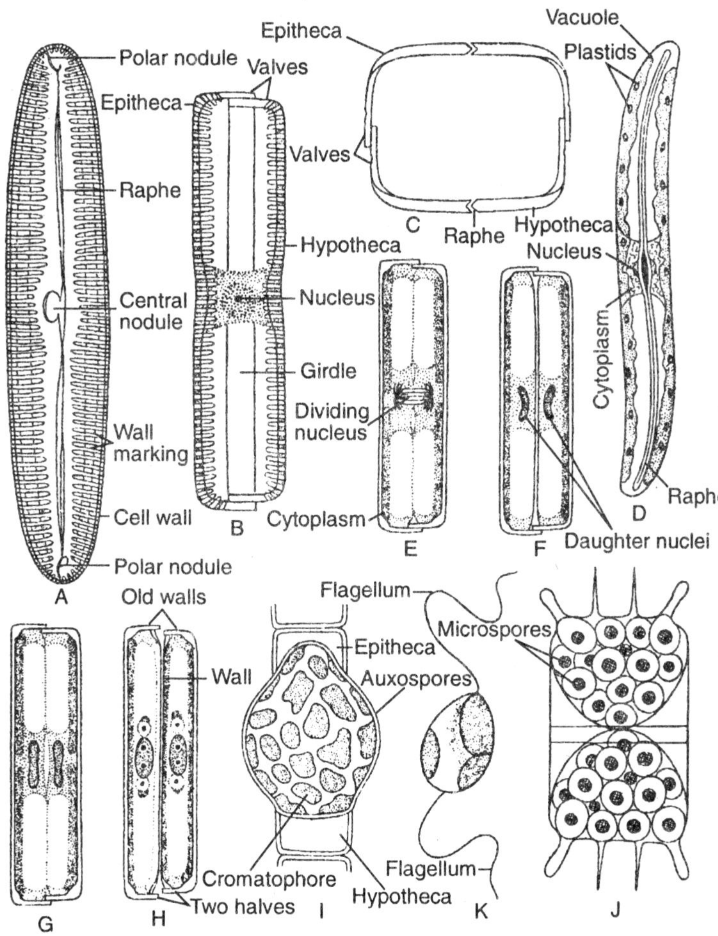

Fig. 7.2: Cell structure and cell division in the Pennales (A-H). Auxospore formation and microspore formation in the Centrales (I-K). A. *Pinnularia viridis* showing valve-view. B. *Pinnularia viridis* frustule in girdle-view. C. *Pinnularia viridis* frustule in transverse section. D. *Pleurosigma spenceri* showing cell structure. E-H. *Pinnularia viridis* showing successive stages in cell division. I. *Melosira varians* with auxospore. J. *Biddulphia mobiliensis* showing microspore formation. K. *Biddulphia mobiliensis* microspore

nated with silica. The siliceous material instead of being deposited as a smooth layer, forms different patterns which vary according to the genus and species. But usually the ornamentation follows two general patterns, for example, in the centric diatoms (Centrales) radially symmetrical about a central point, or drum-like, and in the pennate diatoms (Pennales) bilaterally symmetrical with respect to the long axis of the cell. On the back of each valve of most elongated types there are linear perforations forming longitudinal slit, called the **raphe** (Figs. 7.2A and C). But when without longitudinal slit is the **pseudoraphe**. There are also structural modifications of the cell wall forming thickenings which are known as **nodules,** appearing at both ends of raphe—**polar nodules**, and interrupting it near the mid-point—the **central nodule** (Fig. 7.2A). The markings of the wall, raphe and nodules constitute taxonomic criteria. The protoplast of a diatom cell forms a lining layer closely addressed to the cell wall and generally surrounding a large central vacuole. In species of elongated cells, the vacuole frequently bridges in the middle by a band of cytoplasm in which is embedded the large nucleus (Figs. 7.2B and D). Depending on species, the cell may contain one to many discoid or irregularly shaped plastids which are yellow to brown in colour. Pigments of a diatom cell are: chlorophyll a, chlorophyll c, fucoxanthin and a special pigment—**diatomin** casting a golden-brown hue to the diatom cell. Starch is never formed in diatoms; on the contrary, the reserve foods are fats and **volutin.** Diatom cells of some species exhibit movement which is slow and jerky and confined to forward and backward directions. Cause of movement has been a topic of considerable interest. The movement may be due to the flow of currents of water or by the circulation of cytoplasm.

After the death of diatom cells, their siliceous walls, or empty frustules drift to the bottom of the body of water in which they occur and under favourable conditions there accumulates a large deposit of diatom walls, known as **diatomaceous earth.** During tertiary and quarternary ages the formation of diatomaceous earth was so great that the great deposition was formed in ocean. Such deposition is known as **diatomite** or **kisegurh.** Diatomite because of its porosity, is light in weight and can stand high temperature. Its melting point is 175°C. Diatomite is used in various industries. Some of them are: used in polishes for silverware, metals, etc.; preparation of tooth powders; insulation of steam pipes and blast furnaces; to filter oils; as an absorbent for the liquid nitroglycerine in dynamite for safety in transportation; used in sugar-refineries; in brewing industry; and in manufacturing antibiotics to remove waste mycelium. At first Alfred Nobel used diatomite as an absorber of highly explosive nitroglycerine when he manufactured dynamite.

REPRODUCTION

Diatom cells multiply usually by a peculiar method of cell division. An increase in the volume of the frustule results in a slight separation of the valves which is followed by mitotic division of the nucleus with the axis of the spindle being parallel to the short axis of the cell to ensure longitudinal division of the protoplast (Fig. 7.2E). The division of the protoplast along the median line of the long axis follows immediately after the nuclear division resulting in the formation of two daughter protoplasts and ultimately the cell divides into two, with each half having half of the old wall and the side along the plane of cell division is bounded by a protoplasmic membrane, the surface of which is covered by a new half wall secreted by the protoplast (Figs. 7.2F to H). The new half wall fits inside the original old wall in both the newly formed cells. The old wall of each of the two newly produced diatom cells forms the epitheca no matter whether it was the epitheca or hypotheca of the parent cell and the new wall forms the hypotheca. Of the two cells that are formed one has the same size as the parent cell and the other slightly smaller. The process of cell division usually takes place during night. The size of some of the diatom cells continues to diminish with each succeeding division (Fig. 7.3). Theoretically it is possible that as new generations are produced, the cell will decrease more and more. This theoretical assumption was first laid down by Macdonald Pfitzer and was later known as **Macdonald Pfitzer's law.** This law, however, does not hold good, because, it has been experimentally demonstrated that the diatom cell wall being elastic, the progressive diminution in size is not a regular feature. But this process of diminution in size does not continue indefinitely. With the attainment of minimal size the cells die off. But this 'wasting away' process is prevented by the appearance of certain rejuvenescent cells, or **auxospores** by which restoration of normal size and physiological rejuvenescence follow.

The reestablishment of normal vegetative cell size is by the formation of auxospore. Until 1907 the auxospore formation was considered to be solely by asexual means. In recent decades it has been shown that auxospores result from a sexual process which may be autogamous, isogamous, anisogamous or oogamous. Auxospore formation by asexual means is by mere an enlargement of the vegetative cell. In the asexual process the protoplast escaping from the frustule enlarges several times its former size and surrounds itself with two siliceous valves.

The sexual process involved in the auxospore formation is accomplished in several

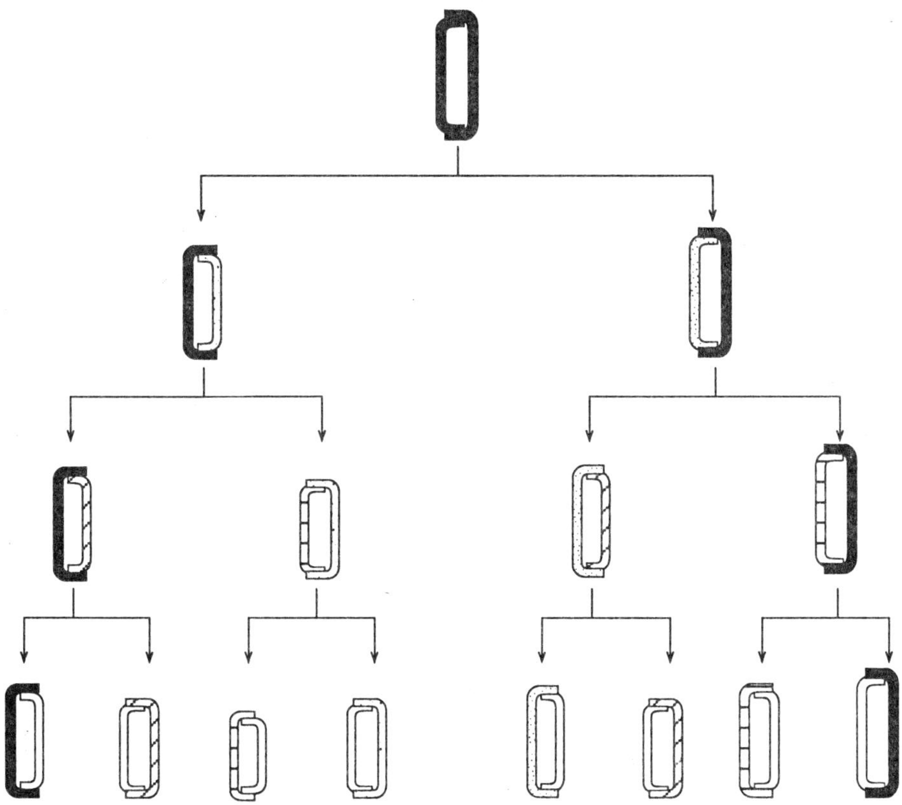

Fig. 7.3: Cell division in diatoms (diagrammatic) showing reduction in cell size in successive generations except one (extreme right)

different ways with considerable variation in the nuclear behaviour. In both pennate and centric diatoms the zygote begins to grow immediately after formation and develops into an auxospore. Two mitoses, at each of which only one daughter nucleus survives, ultimately give rise to the vegetative cell which has the largest dimension.

The process of sexual reproduction is, therefore, linked to reduction in cell size and, for a given species, there appears to be distinct size range (usually 30-40 per cent of maximum) within which the cells are capable of becoming sexual. In those diatoms which do not decrease in size sexual reproduction does not apparently occur. The vegetative cells of both Pennales and Centrales are diploid and gametes are formed directly following meiosis. However, the sexual process shows a striking difference between the centric and the pennate diatoms. In the Pennales the process is isogamous and the gametes are amoeboid, whereas in the Centrales the process is oogamous and there is presence of spermatozoids.

AUXOSPORE FORMATION IN THE PENNALES

In the Pennales, the auxospore is formed mainly by isogamy. Cases of anisogamy and oogamy are rather rare. During isogamy, cells aggregate in pairs (**gamontogamy**), nuclear division (meiotic) produces 4 nuclei, and of which only 1 or 2 normally develop into gametes. The parent cells fuse (**cystogamy**) and the fusion of gametes (generally derived from different cells) occurs in a populatory jelly. In most species, fusion is isogamous and the gametes move towards each other. However, some species show physiological anisogamy, in which, although the gametes are morphologically identical, one is stationary whereas the other is motile. There are four different ways of auxospore formation: (i) each cell produces two

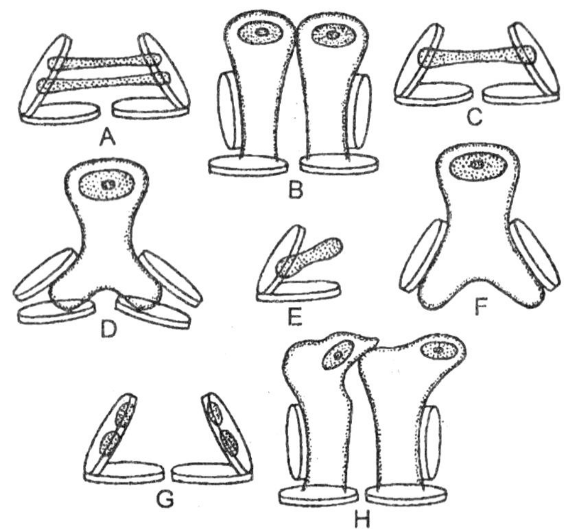

Fig. 7.4: Diagrammatic representation of auxospore formation in the Pennales

gametes but gametic fusion takes place between gametes of two different cells (Figs. 7.4A and B); (ii) one gamete formed by each cell and hence fusion of two gametes from two different cells (Figs. 7.4C and D); (iii) the gametes formed in each cell and they fuse with each other (Figs. 7.4G and H); (iv) the entire protoplast of a cell is transformed into an auxospore parthenogenetically (Figs. 7.4E and F).

The auxospore formed by any one of the above processes elongates considerably after which siliceous wall is formed producing a vegetative cell. In general, the vegetative cell of diatom is a diploid structure and meiosis is gametic (Fig. 7.5).

In anisogamy, one gamete is amoeboid while the other is non-motile. The amoeboid gamete (male) travels through a mucilage matrix, or a fusion tube, toward the female and nuclear fusion takes place there. Oogamy is rather very rare in the Pennales.

AUXOSPORE FORMATION IN THE CENTRALES

In the Centrales, the auxospore formation is usually by oogamous sexual process. Isogamy

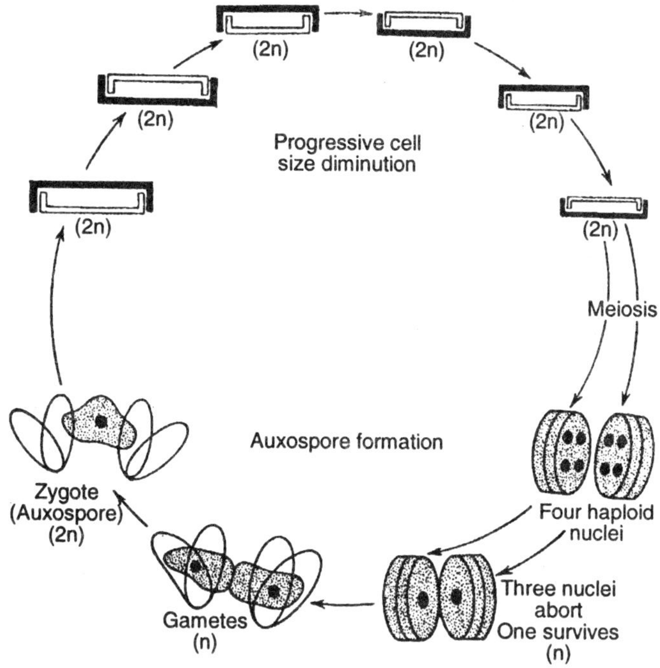

Fig. 7.5: Life cycle of the Pennales

has also been recorded in some species. Oogonia occur as slightly extended cells with an elongated nucleus. Normally, each oogonium contains only one egg nucleus. The egg cell normally remains enclosed in the oogonium or attached to it until fertilization. Four uniflagellate spermatozoids are formed by division of the conent of the male sex organ, the **spermatogonium,** also designated as a **spermatocyte.** However, the formation of the spermatogonium itself is variable. The vegetative cell either acts directly as a spermatogonium, or divides to form 4-8 spermatogonia. The spermatogonia are dispersed passively through water, before becoming

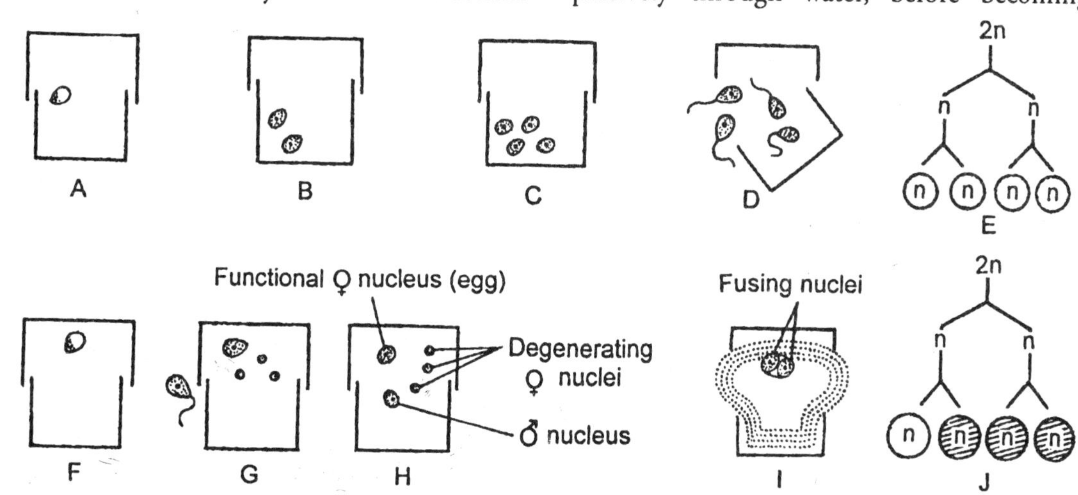

Fig. 7.6: Auxospore formation by Oogamy in Centrales: A-D. Formation of sperms, E. Nuclear behaviour during sperm formation, F-G. Formation of egg, (F. Single oogonium, G. Sperm approaching oogonium), H. Male nucleus entered inside the oogonium, I. Fusing male and female nuclei and J. Nuclear behaviour during Egg-formation

attached by mucilage pads to the oogonium. They then produce spermatozoids. At fertilization only the spermatozoid nucleus is injected into the egg cell (Fig. 7.6).

Besides the above process of oogamy an intracellular autogamy may also occur in some species. In isogamy, one to two gametes are produced in a cell and are freed from each mother cell. They fuse to form an auxospore. Asexual auxospore formation has also been reported in the Centrales.

Some marine diatoms produce flagellaed cells which are interpreted by some as **microspores** and others as zoogametes according to their behaviour (Fig. 7.2J and K).

STATOSPORE FORMATION IN THE CENTRALES

Many of the Centrales have been found to produce in their frustules, thick-walled resting spores known as **statospores.** The statospores are formed during unfavourable seasonal condition. The details of their development are not known. The statospores are very frequent in marine species, but are also found in freshwater ones (Fig. 7.7).

CLASSIFICATION

The Bacillariophyta include the single class Bacilariophyceae which has been subdivided into two orders:

I. Order **Centrales:** Valves circular, polygonal or irregular in outline and with an ornamentation on the wall; ornamentation radial or concentric about a central point; valves never have a raphe or pseudoraphe; protoplast with many chromatophores; statospores or microspores may be formed; auxospore formation mainly by oogamy.

II. Order **Pennales:** Valves bilaterally symmetrical or asymmetrical in surface view; wall ornamentation bilateral with respect to a line; valve always with a raphe or pseudoraphe; protoplast with one or two chromatophores; statospores or microspores are never formed; auxospore formation mainly by isogamy.

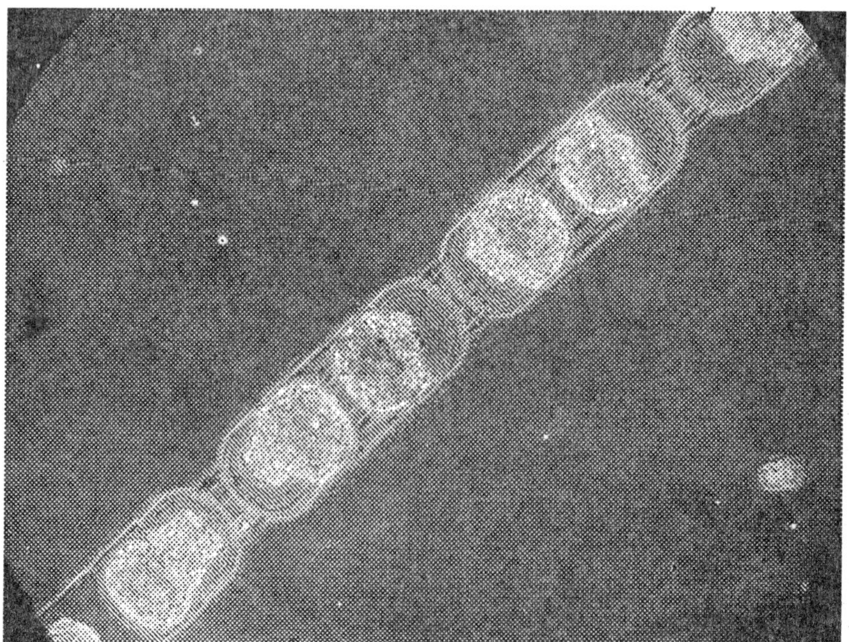

Fig. 7.7: Resting spores of Diatoms (Genus: *Stephanopyxis*, Order: Biddulphiales, Family: Melosiraceae)

Some Indian Species: *Cymbella affinis* Kutz.; *Eunotia pectinalis* var. *gibbulosus* Venkat.; *Eunotia pectinalis* var. *neglecta* Gandhi; *Navicula cryptocephaloides* Hust.; *Neidium gracile* Hustedt.; *Pinnularia graciloides* Hustedt.

Special features

1. Unicellular pill-box type of plant body composed of two valves.
2. Silica impregnated cell wall with beautiful markings.
3. Cell wall deposition produces diatomaceous earth which has great economic importance.
4. Specialized vegetative cell division.
5. Auxospore formation.

RELATIONSHIPS OF THE BACILLARIOPHYCOPHYTA

The cell structure suggests that the diatoms are associated with the Conjugales particularly with the desmids.

The presence of fucoxanthin and chlorophyll a and chlorophyll c suggests closer association with the Phaeophyta.

Pascher suggested a relationship with the Xanthophyta and Chrysophyta and classified diatoms under Chryosophyta for having golden-brown pigment and similar assimulation product.

Since diatoms are largely associated with the Peridineae forming plankton flora of the ocean, some relationship has been drawn by the phycologists between the Bacillariophyta and the Peridineae.

PHYLOGENY OF THE BACILLARIOPHYTA

Because of the siliceous skeleton an extensive fossil record of diatoms is known and a discussion of phylogeny based on the fossil evidence is possible with diatoms to a much larger extent than with other algae. Centric forms appear to be the most primitive appearing first in the Jurassic, whereas the pennate forms do not make their first appearance until the Tertiary period. The evidence also suggests that the pennate forms were derived from the centric forms.

FOSSIL DIATOM

After the death of diatom cells the frustules usually dissolve, but under certain circumstances they remain intact and accumulate at the bottom of any water bodies where the diatoms occur. Where conditions are exceptionally favourable and long continued, such accumulations may reach considerable thickness. Deposits of fossil diatoms, known as diatomaceous earth or kieselguhr, are found in various parts of the world. None of these deposits originated earlier than the cretaceous period. Various industrial uses of diatomaceous earth are known today.

ECONOMIC IMPORTANCE OF DIATOM

The deposit of dead cell wall of diatom often called **diatomite** or **diatomaceous earth.** These materials have a variety of uses:

(a) It is used as filter in different industries like sugar, oil and chemical industry.

(b) It is used as insulator in boilers and blast furnaces for its heat resistant ability.

(c) It is used as adsorbent of liquid nitroglycerine.

(d) It is also used as abrasive substance for the manufacture of metal paints, polish, varnish, toothpaste etc.

(e) Oils of diatoms have significant use value as biofuel.

(f) Diatom is one of the major producer in marine environment.

Chapter 8

The Phaeophycophyta

The Phaeophycophyta or brown algae, in general, are restricted to the sea—the seaweeds, with a few exceptional freshwater forms (species of *Heribaudiella, Sphacelaria, Lithoderma, Pseudobodanella* and *Pleurocladia lacustris,* A. Br.). Usually the marine members are abundant in cold waters. But warmwater brown algae are also fairly well represented. They are predominant in parts of the arctic and antarctic seas. Their colour varies from dark-brown to olive-green. Most marine forms grow attached to rocks or similar other submerged objects of the substratum by the development of certain mechanism for anchorage. Some members of this group grow either as epiphytes or as endophytes remaining in association with other algae. There is marked variation in the vegetative structure of brown algae, which range from minute unicellular to abundant and luxurious growth comprising 100 metres or more in length having morphological and anatomical differentiation. There are also various ranges of heterotrichous forms starting from simple heterotrichous individual with the gradual disappearance of heterotrichous condition being replaced by a very elaborate development of the aerial portion with the prostrate portion being reduced to rudimentary structure serving the function of anchorage. Growth of the thallus may be apical or by intercalary meristem exhibiting **trichothallic growth.**

Wall of each individual cell is composed of two layers: the inner cellulose layer, and the outer layer which is composed of a mixture of pectic material and **algin**—a colloidal substance which has commercial importance for being used in the preparation of adhesives and artificial silk. In general, the uninucleate vacuolate protoplast bears one to several chromatophores whose shape and position in a cell is extremely variable. The chromatophores contain Chlorophyll a, Chlorophyll c, Carotene and Xanthophyll which are masked by a special brown pigment—**fucoxanthin.** Due to the presence of this special pigment which is nothing but a mixture of two carotinoid pigments: Fucoxanthin a and Fucoxanthin b, the chromatophores are coloured brown. In some members, the pigment xanthophyll is also absent. Reserve food of most frequent occurrence is **laminarin**—a group of dextrin-like polysaccharides. Besides laminarin there may

be present a hexahydric alcohol—mannitol and some amount of fats or fat-like substances as reserve food. Certain brown algae are the sources of commercial iodine. Some are used as fertilizers.

Besides vegetative reproduction by fragmentation of the thallus, most brown algae, reproduce both asexually and sexually. Asexual reproduction is either by biflagellate zoospores or by non-flagellate tetraspores developed in sporangia. Each zoospore bears two unequal flagella which are inserted laterally and sometimes near the posterior end. Sexual reproduction ranges from isogamy to oogamy. In oogamy, fertilization may take place either after the female gamete or gametes are being liberated out from the oogonium or when the ovum is retained in the oogonium. The number of female gametes in an oogonium may be one or eight. The antherozoids are uni- or biflagellate.

Except a few, all the members of the group are characterized for exhibiting distinct alternation of generations which in most cases is **isomorphic,** and in others, **heteromorphic.** The zygote resulting from the fertilization without having the zygotic nucleus dividing reductionally, germinates producing a diploid individual, which depending on species, may or may not resemble the haploid individual morphologically. In such cases meiosis takes place at the sporogenesis in the sporangia. Moreover, there are other brown algae in which the plant body is diploid and the haploid condition is represented by the gametes only. In such case the reduction division takes place during the gametogenesis in gametangia.

The phaeophycophyta probably evolved from an organism in the Chrysomeridales of the Chrysophycophyta, which have motile cells similar to those in the Phaeophyceae. The algae in the Chrysomeridales, however, lack unilocular and plurilocular sporangia and plasmodesmata, and therefore cannot be classified in the phaeophycophyta.

CELL STRUCTURE

The cell structure is in many ways similar to that of the chrysophycophyta, Prymnesiophycophyta, Bacillariophycophyta and Xanthophycophyta, which are closely related to the Phaeophycophyta. The main difference lies in the large amounts of extracellular polysaccharides surrounding the protoplast.

CELL WALLS

Phaeophycophytean cell walls are generally composed of at least two layers, with **cellulose** making up the main structural skeleton. The amorphous component of the cell wall is made up of **alginic acid** and **fucoidin,** whereas the mucilage and cuticle are composed primarily of alginic acid. Alginic acid is basically made up of β-1, 4 linked mannuronic acid units that have a variable amount of guluronic acid units attached through C-1 and C-4 linkages (Fig. 8.1). Fucoidin is composed primarily of α-1, 2 linked sulfated fucose units, with a lesser amount of α-1, 4 linked sulfated fucose units (Fig. 8.1). The relative quantities of alginic acid and fucoidin vary between different species, different parts of the plant, and different environments.

Calcification of the wall occurs only in some species of *Padina* where calcium carbonate is deposited as needle-shaped crystals of aragonite in concentric bands on the surface of the fan-like thallus.

The parenchymatous Phaeophycophyta have plasmodesmata or pores between most of the cells. These pores are bounded by the plasmalemma, and protoplasm is continuous from one cell to the next through them. In the Laminariales, Fucales, and Dictyotales, the pores are grouped in primary pit areas, whereas in the more primitive parenchymatous Phaeophycophyta the plasmodesmata are scattered in the cell wall.

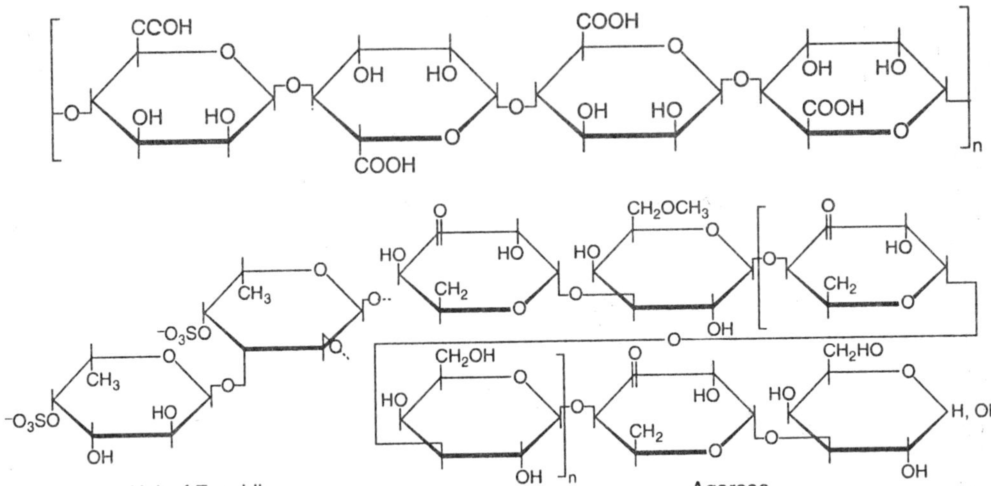

Fig. 8.1: Structural units of alginic acid, fucoidin, and agarose (after Percival and McDowell, 1967)

FLAGELLA AND EYESPOT

Generally the motile cells of the Phaeophycophyta (always zoospores or gametes, as there are no motile vegetative cells) have a long anterior tinsel flagellum with tripartite hairs and a shorter posteriorly directed whiplash flagellum (Fig. 8.2). The Fucales are an excep-

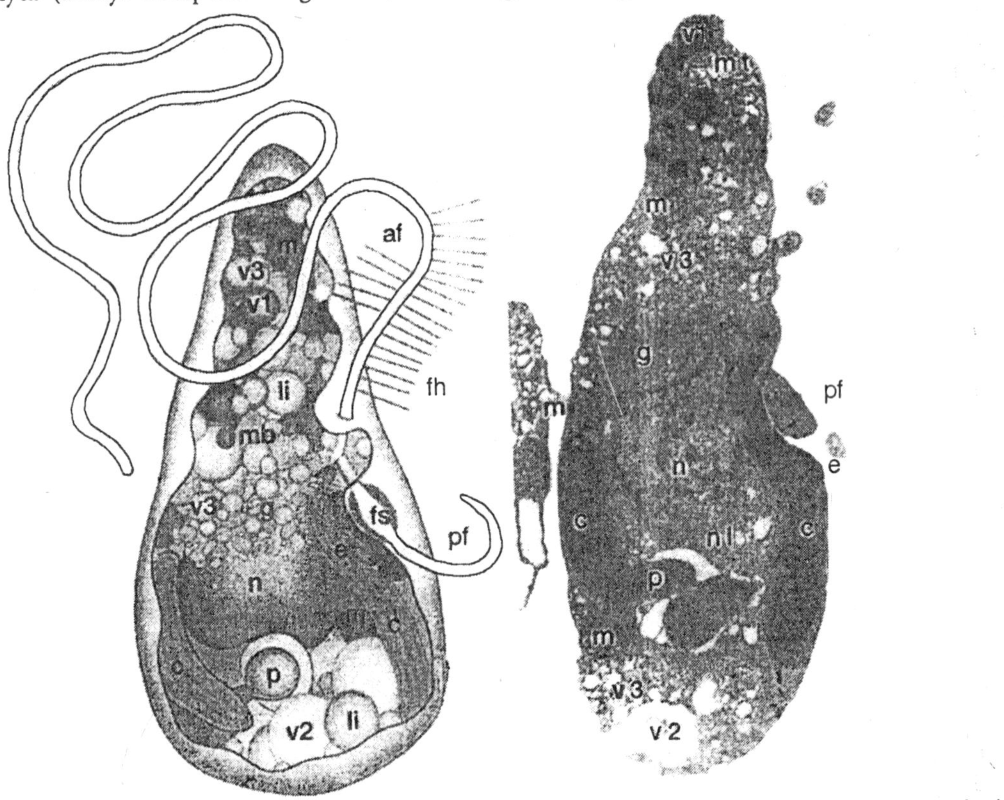

Fig. 8.2: *Left:* Diagrammatic representation of a male gamete of *Ectocarpus siliculosus* showing the distribution of cellular organelles. *Right:* Tranmission electron micrograph of a thin section of a male gamete of *E. siliculosus* (From Maier, 1997)

tion to this, with the posterior flagellum of the spermatozoid being longer than the anterior flagellum. The posterior flagellum usually has a swelling near the base, and this swelling fits into a depression of the cell immediately above the eyespot. The eyespot consists of 40 to 80 lipid globules arranged in a single layer between the outermost bands of the thylakoids, and the chloroplast envelope. The **eyespot (stigma)** acts as a concave mirror focusing light onto the flagellar swelling, which is the photoreceptor site for phototaxis in brown-algal flagellate cells. Light at 420 nm and 460 nm is most effective in phototaxis in the brown algae, and is probably detected by a flavin-like substance in the flagellar swelling of the posterior flagellum.

CHLOROPLASTS AND PHOTOSYNTHESIS

The chloroplasts of the Phaeophyceae have three thylakoids per band and are surrounded by the chloroplast envelope and two membranes of chloroplast E.R. (Fig. 8.3). The outer membrane of the chloroplast E. R. is generally continuous with the outer membrane of the nuclear envelope in the Ectocarpales but appears to be discontinuous in the Dictyotales, Laminariales, and Fucales. Membrane-bounded tubules are common in the area between the chloroplast E.R. and chloroplast envelope where the latter two are not closely appressed (Bouck, 1965; Evans, 1968). Microfibrils of DNA occur in the plastids, and in *Sphacelaria* sp. there is a ring-shaped genophore inside the outermost band of thylakoids (Bisalputra and Burton, 1969). The DNA microfibrils are both linear and circular and are attached to the thylakoid membranes. The plastids contain chlorophylls *a*, c_1, and c_2, with the major carotenoid being fucoxanthin.

All the phaeophycophytean orders have representatives with pyrenoids, but their presence, even in one species, can vary according to the stage of the plant. If the species is one

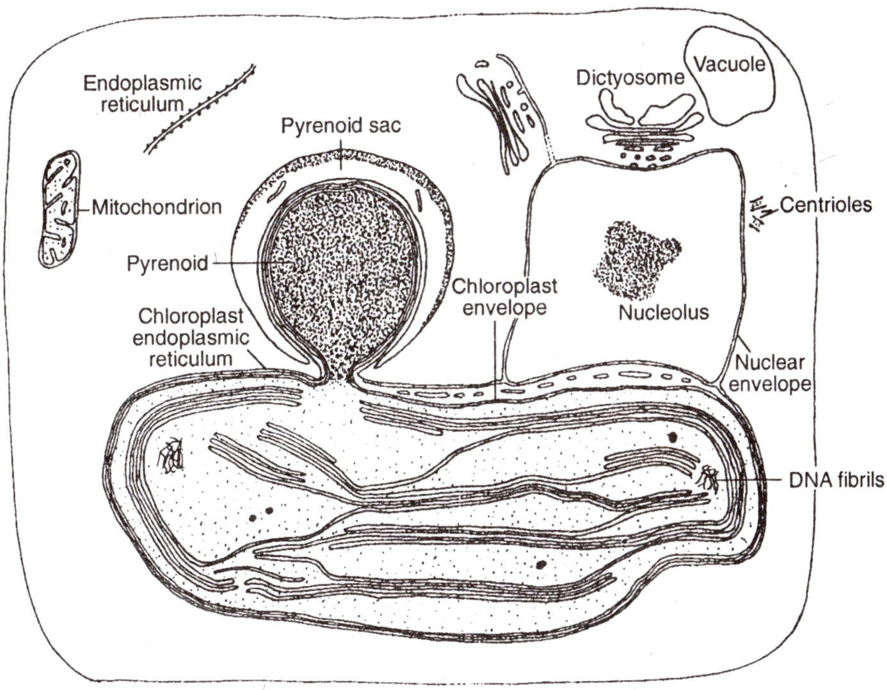

Fig. 8.3: Diagram of a hypothetical brown algal cell (From Bouck, 1965)

that has pyrenoids only in some stages, then a pyrenoid is usually present in the eggs and/or sporelings but absent in the macroscopic phase, spermatozoids, and/or zoospores. A pyrenoid in the Phaeophyceae is usually a stalk-like structure set off from the main body of the chloroplast and containing a granular substance not traversed by thylakoids. Surrounding the pyrenoid but outside the chloroplast E.R. is a membrane-bounded sac that presumably contains the reserve product. The long-term storage product is **laminarin,** a β-1, 3 linked glucan. The sugar alcohol, D-**mannitol** is, however, the accumulation product (up to about 25% of the dry weight of some *Laminaria* species in the autumn) of photosynthesis.

In a number of brown algae, the mannitol concentration in the cell increases or decreases as the salinity of the surrounding medium increases or decreases. This osmoregulatory mechanism prevents the cells from bursting in hypotonic media or shrinking in hypertonic media. The increase in mannitol concentration occurs in the dark as well as in the light, showing that photosynthesis is not involved in the process.

The brown algae are unique among the algae in having uptake of inorganic carbon, and therefore photosynthetic carbon fixation, stimulated by blue light. Most of the Phaeophyceae live in the littoral zone where they receive a generous amount of light. Photosynthesis is usually limited by the supply of inorganic carbon in this environment. The Phaeophyceae have evolved a mechanism to increase the amount of inorganic carbon uptake, but only when the cells are illuminated by blue light, thereby conserving the energy required for the process when the cells are in the dark. The only brown algae that do not have this mechanism are the flucoids, which appear to have evolved a separated carbon-concentrating mechanism.

PHYSODES AND PHAEOPHYCEAN TANNINS

Phaeophycean tannins (fucosan) are stored in **physodes (fucosan vesicles)** in the cytoplasm of many brown algae. These vesicles contain a colourless, highly refractive acidic fluid that stains red with vanillin and hydrochloric acid. The tannins are non-glycosidic (do not contain sugars), have strong reducing action, produce precipitates with lead acetate, and in aqueous solutions are astringent to the taste. Also they are readily oxidized in the air, resulting in the formation of a brown or black pigment, **phycophaein,** giving dried brown algae their characteristic black colour. Experiments have shown that the yellow solutions of tannins produced by *Fucus vesiculosus* will inhibit the growth of a number of unicellular algae at tannin concentrations of 25 to 150 $\mu g\ ml^{-1}$. Toxicity of phenols apparently depends on a denaturation of proteins, resulting in lysis of cells. The function of the tannins and phenols produced by algal cells is probably to regulate the occurrence and abundance of endo- and epiphytes on the host plants.

The physode vesicles originate in the plastids where the tannins are evidently produced, and then extruded from the plastid area into the cytoplasm in a vesicle. Once in the cytoplasm the vesicles have the ability to fuse and form larger vesicles.

Eggs of the Phaeophyceae contain phenolic vesicles just under the plasma membrane that are discharged outside the cells by exocytosis after fertilization. It has been postulated that the discharge of these phenolic vesicles has a toxic effect on spermatozoids and acts as a polyspermy block before the primary wall is secreted. These peripheral phenolic vesicles are distinguishable from physodes, which also contain phenolic compounds, but which are significantly larger and tend to be localized around the egg nucleus.

LIFE HISTORY

The unilocular sporangium is generally considered to be the site of meiosis, the haploid zoospores that are released forming the gametophyte generation. The gametophyte then produces the gametes, which fuse to form the zygote. Although meiotic divisions have been considered the rule in the unilocular sporangium, a disturbingly large number of investigations have not found this to be the case. In these investigations there is a 'direct' type of life history, with no meiosis or fusion occurring. More research is needed in this area to clarify the situation. In the phaeophycean life cycle, a **plethysmothallus** is a filamentous stage (or one composed of compacted filaments) that can multiply itself by spores (usually zoospores from plurilocular sporangia) (Papenfuss, 1951).

The thallus of many Phaeophycophyta is relatively large and complex with a number of different types of growth that include the following: (1) **diffuse**, with most of the cells of the plant capable of cell division; (2) **apical**, with a single cell at the apex giving rise to the cells beneath it; (3) **trichothallic**, in which a cell divides to form a hair above and the thallus below; (4) **promeristem**, with a non-dividing apical cell controlling a large number of smaller, dividing promeristematic cells beneath it; (5) **intercalary**, with a zone of meristematic cells forming tissue above and below the meristem; (6) **meristoderm**, with a layer of usually peripheral cells dividing **periclinally** (parallel to the surface of the thallus) to form a tissue below the meristoderm (usually cortex) and occasionally dividing **anticlinally** (perpendicular to the surface of the thallus) to add more cells to the meristoderm. Some Phaeophyceae have division of daughter cells without subsequent enlargement to give a **polysiphonous** structure, or one with parallel vertically elongate cells in rows. Daughter cells can form laterals or branches that branch in a **distichous** manner (branches disposed in two equal rows). **Monopodial** growth is a type of growth with one main axis.

Investigations on the sexual hormones of the brown algae in the 1970s and 1980s by Müller and his associates represent the major recent advancement in the knowledge of this group. A **sexual hormone (sirenine, phero-**

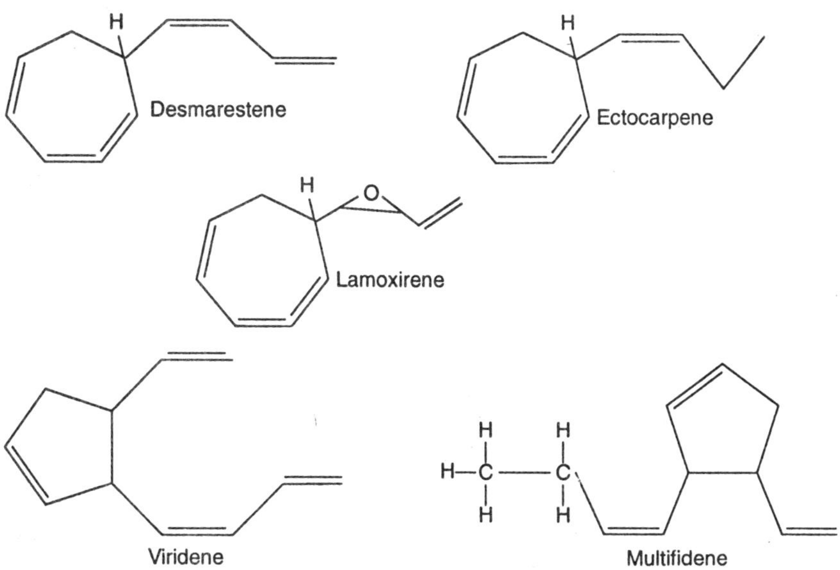

Fig. 8.4: The structure of some of the sexual hormones of the brown algae. (After Jaenicke *et al.*, 1974; Müller *et al.*, 1982)

mone) is a diffusible substance that coordinates cellular activities during sexual reproduction. Two types of biological effects mediated by sexual hormones occur in the brow algae: (1) the explosive discharge of spermatozoids from antheridia, and (2) the attraction of male gametes by female gametes or eggs. All sexual hormones in the brown algae are unsaturated hydrocarbons (have at least one double or triple bond) (Fig. 8.4). With the exception of the *Fucus* sperm attractant, all of the sexual hormones in the brown algae are C_8 to C_{11} olefins (unsaturated open-chain hydrocarbons containing at least one double bond), most of them incorporating a five- or seven-membered ring structure. All of the attractions in the brown algae identified so far are highly volatile and hydrophobic. The substances have a very strong tendency to leave the aqueous solution and escape into the air. This characteristic helps to avoid the build-up of chronic concentrations which would decrease the efficiency of the gradients around the female cells. The attractants probably do not function more than 0.5 mm away from the female cells, and thus attraction is clearly a short-distance phenomenon. The sexual hormones include **ectocarpene** from *Ectocarpus,* **desmarestene** from *Demarestia aculeata,* **lamoxirene** from *Laminaria,* **multifidene** from *Cutleria mutifida* (Jaenicke et al., 1974), **dictyopterene C′** from *Dictyota dichotoma,* and **fucoserratene** from *Fucus serratus* and *F. vesiculosus.*

CLASSIFICATION

Kylin (1933) divided Phaeophycophyta into three classes according to the type of life history. Further division of these classes into subclasses, and the arrangement of the orders, may best be understood from the key given below. This classification has received wide recognition for practical purposes although Papenfuss (1951) rejected the use of the classes and subclasses on phylogenetic grounds.

1. Plants with an alternation of generations 2
1. Plants without an alternation of generations
   Class **Cyclosporeae**
   Order **Fucales**
   2. With an alternation of macroscopic, usually isomorphic generations
      Class **Isogeneratae**
      Orders **Ectocarpales**
      Sphacelariales
      Cutleriales
      Tilopteridales
      Dictyotales
   2. With an alternation of heteromorphic generations 3
      Class **Heterogeneratae**
      3. Thallus composed of cells which adhere to one another to form a pseudoparenchymatous tissue
         Subclass **Haplostichineae**
         Orders **Chordariales**
         Sporochnales
         Desmarestiales
      3. Thallus composed of cells dividing by intercalary longitudinal walls to form a true parenchymatous tissue
         Subclass **Polystichineae**
         Orders **Dictyosiphonales**
         Laminariales

The Phaeophycophyta, as viewed by many phycologists, include a single class Phaeophyceae which again can be divided into eleven orders details of which are incorporated below:

I. Order: **Ectocarpales.** Thallus pseudoparenchymatous or true parenchymatous type or of branched filaments; growth usually intercalary, typical meristematic apical cells being rarely present; asexual reproduction by zoospores; sexual reproduction mostly isogamous, rarely anisogamous; usually alternation of isomorphic generations.

Family: **Ectocarpaceae.** Filamentous branched plant body; growth intercalary or trichothallic or apical without longitudinal subdivision of the cells of the filaments; gametangia or sporangia borne on all or part of an ultimate branch; gametes isogamous or anisogamous.

Genus: *Ectocarpus*

II. Order: **Laminariales.** Sporophytic generation represented by bulky parenchymatous forms of large dimensions with considerable morphological and anatomical differentiation having intercalary growth; gametophytic generation small and filamentous; asexual reproduction by zoospores formed in unilocular sporangia; sexual reproduction oogamous; heteromorphic alternation of generations.

Family: **Laminariaceae.** Sporophyte of large and coarse fronds composed of holdfast, stipe and lamina; lamina is renewed annually in many species by a meristematic tissue present at the junction point between the stipe and the lamina; internal differentiation into medullary tissue, cortical layer and assimilating layer; lamina with sori bearing unilocular sporangia with paraphyses; zoospores borne in sporangia germinating to form gametophytes; oogamous sexual reproduction, fertilized egg giving rise to sporophyte.

Genus: *Laminaria*

III. Order: **Fucales.** Very large diploid plant body of complex morphological and anatomical differentiation, true parenchymatous tissue being formed with apical growth; asexual reproduction absent; oogamous sexual reproduction with sex organs borne in conceptacles in the plant body, meiosis during the development of gametes; haploid phase very short being confined to the gametes only; oospore directly germinates into a diploid thallus; no evident alternation of generations.

Family: **Fucaceae.** Perennial monoecious or dioecious plant body having hold-fast and flattened dichotomously branched fronds, provided with air vesicles; conceptacles terminal or subterminal; conceptacles opening outwards by a narrow ostiole; antheridia on branching filaments; eight oospheres; fertilization outside the conceptacles.

Genus: *Fucus*

Family: **Sargassaceae.** Complex bilaterally symmetrical branched plant body; branches borne at the axil of leaf-like expansions; air bladders along the leaves or branches; monoecious or dioecious; oogonia liberated from the conceptacles are held by gelatinous stalks but fertilization outside the plant body; some species with cryptoblasts.

Genus: *Sargassum*

IV. Order: **Dictyotales.** Little differentiated expanded parenchymatous thallus with apical growth; asexual reproduction by non-motile tetraspores borne in a unilocular tetrasporangium; sexual reproduction oogamous; isomorphic alternation of generations.

Family: **Dictyotaceae.** Leafy olive-green thallus; asexual reproduction by tetraspores borne in tetrasporangia developed on the sporophyte; sexual reproduction by antherozoids and ova developed in sex organs on the surface of the gametophyte which may be monoecious or dioecious;

Genus: *Dictyota*

V. Order: **Desmarestiales.** Heteromorphic alternation of a large erect, essentially pinnately branched sporophyte with a microscope gametophyte; growth trichothallic initiated by a single filament; plants may be as large as many species of *Laminaria,* their structure simple in comparison with the Laminariales; the sporophytes bear unilocular sporangia; gametophyte plants, dioecious, and sexual reproduction oogamous.

VI. Order: **Sphacelariales.** Present in all seas, and a species of *Sphacelaria,* the largest genus, may be expected in almost any intertidal flora; usually short, densely branched, tufted plants up to a few centimetres in height; possess well-marked apical growth from a large apical cell, parenchymatous structure, and usually radial organization.

VII. Order: **Tilopteridales.** Live mostly in deep water and sporadic and ephemeral; plants generally resemble *Ectocarpus* in filamentous habit, and show marked trichothallic growth; lower parts may, however, become multiseriate somewhat as in *Sphacelaria;* the characteristic reproductive structures globose or oblong monosporangia whose entire contents are liberated as a monospore; monospores quadrinucleate, and some cytological investigations suggest that the four nuclei are the products of meiosis; monospores, thus, may be considered a kind of primitive, undivided tetrad of spores; sexual generation produces plurilocular gametangia, but fertilization and other details have not been described.

VIII. Order: **Cutleriales.** Both of the alternating generations are macroscopic but not always isomorphic; an advance toward strictly heteromorphic alternation of macroscopic and microscopic generations prevalent among the more highly developed brown algae.

IX. Order: **Sporochnales.** Mainly inhabitants of the Southern Hemisphere, but a few interesting species of a very wide distribution, occur along the North Pacific and North Atlantic; almost exclusively deep-water plants; all show an alternation of macroscopic sporophyte and microscopic, filamentous gametophyte; most distinctly characterized by the peculiar anatomy of the growing apices; the meristem at the branch tip sporophyte, consisting of a dome-shaped single layer of meristematic cells crowned by a tuft of simple hairs; detailed ontogeny of this mode of growth has been worked out only in the European genus *Nereia.*

X. Order: **Dictyosiphonales.** Structure of the thallus is parenchyma-like throughout, in which both longitudinal and transverse divisions of intercalary cells occur; characters of the subclass Polystichineae used by various authors to include the orders Dictyosiphonales (sometimes treated as Punctariales) and the Laminariales; in majority of Dictyosiphonales an alternation of a macroscopic sporophyte and a microscopic filamentous gametophyte (although there are several exceptions), but are, however, set off from the other orders by various characteristics: their solid, parenchymatous structure easily sets them apart from the Chordariales; their lack of trichothallic growth, from the Desmarestiales; their diffuse or apical growth, usually without a restricted intercalary meristem, from the Laminariales.

XI. Order: **Chordariales.** Trichothallic growth of the monoaxial or multiaxial sporophyte and isogamous gametic union (although gametes are still unknown for various genera and for most species); sporophyte plants range from minute, crustose epiphytes to erect, fleshy, or gelatinous plants up to 40 centimetres tall; gametophytes are microscopic, filamentous plants which have been observed only by culturing zoospores.

PHYLOGENETIC RELATIONSHIPS OF THE PHAEOPHYCOPHYTA

The Phaeophycophyta form a well-marked taxon not very closely related to the other algae. The nature of swarmers suggests that the Phaeophycophyta possibly arose from flagellate unicellular organisms. The Phaeophycophyta are often considered parallel to the Chlorophycophyta in the nature of plant body and in the method of reproduction, although the Phaeophycophyta have evolved multicellular reproductive organs and a higher type of vegetative body. The apically growing thallus of the Phaeophycophyta developing into an elaborate structure is often differentiated into epidermis, cortex, and medulla. Besides this, the development of sieve-tube cells and in some cases development of a cambium-like region are the features which can very well justify a very high position for the Phaeophycophyta in the evolutionary scale. Since they lack xylem for conduction and support, a cuticle for protection against evaporation, jacket layers around their multicellular reproductive organs for protection against desiccation, and the mode of sexual reproduction is not well advanced, the Phaeophycophyta did not receive any higher position although they possess high degree of specialization in their plant body.

The pigmentation of the Phaeophycophyta suggest a relationship to both the Chrysophycophyta and the Pyrrophycophyta. All these groups have an excess of carotinoid pigments over chlorophylls, as compared to the Chlorophycophyta. Chlorophyll c is known only in the Phaeophycophyta, Chrysophycophyta and Pyrrophycophyta. Fucoxanthin, the principal xanthophyll of the Phaeophycophyta occurs also in the Chryosophycophyta but is unknown among the Pyrrophycophyta. The sum of the evidences indicates that the Phaeophycophyta originated from a prechrysophytan stock after the pyrrophycophytan line had already diverged from the ancestors of this same stock.

A common ancestry of the Phaeophycophyta with the Chrysophycophyta has also been suggested considering the similarity of the flagellate cells in both these taxa.

The fossil record is hardly useful in determining relationships of the Phaeophycophyta. Fossils more or less resembling *Fucus* and other members of the Phaeophycophyta occur in rocks of the early Paleozoic. But they are not well preserved to be placed in any algal taxa, and hence they provide no indication of the ancestry of the Phaeophycophyta.

Order: *ECTOCARPALES*

Plants are generally filamentous and branched, uniseriate with apical, trichothallic, or dominantly intercalary growth. Sporophyte and gametophyte phases are of similar appearance, reproducing by zoospores and by iso- or anisogametes.

Family: *ECTOCARPACEAE*

Plants are filiform and more or less branched, or subsimple from a creeping, penetrating, or disciform base; generally uniseriate but occasionally some segments in the lower part with one or two longitudinal septa; reproductive organs lateral, replacing branchlets, or intercalary from transformed vegetative cells. In general, zoospores are produced in unilocular sporangia, followed by a generation producing gametes in plurilocular gametangia. Both sporangia and gametangia are occasionally found on the same plant.

Genus: *Ectocarpus*

This is one of the best-known brown algae containing many species. It is a marine alga of world-wide distribution being abundant particularly along the Atlantic Coast but rather scarce along the Pacific Coast. The plant body is composed of an irregularly prostrate portion and small sparsely or profusely branched filaments which grow upright in tufts (Fig. 8.5A). The ultimate branchlets of the erect portion are generally attenuated to a acute point (Fig. 8.5D). They may also taper into hairs. Plants are usually attached to other submerged plants, sometimes to stones or similar other objects. The individual cell uninucleate and contains plate-like or band-shaped chloroplast with or without pyrenoids. Growth of the prostrate system is apical, but that of erect threads shows considerably diversity. Apical growth is rather rare in the erect threads. In most species the erect threads exhibit diffuse growth. In a limited number of species growth is trichothallic, during which intercalary meristems are found at the base of the hairs. The cells of the meristem cut off segments both above and below, the former add to the length of the terminal hair, while the latter give rise to branches.

The alga has an isomorphic alternation of generations, the gametophyte and the sporophyte being essentially alike in appearance. The gametophyte produces **plurilocular gametangia**. The plurilocular gametangium is an elongated structure which develops from the termi-

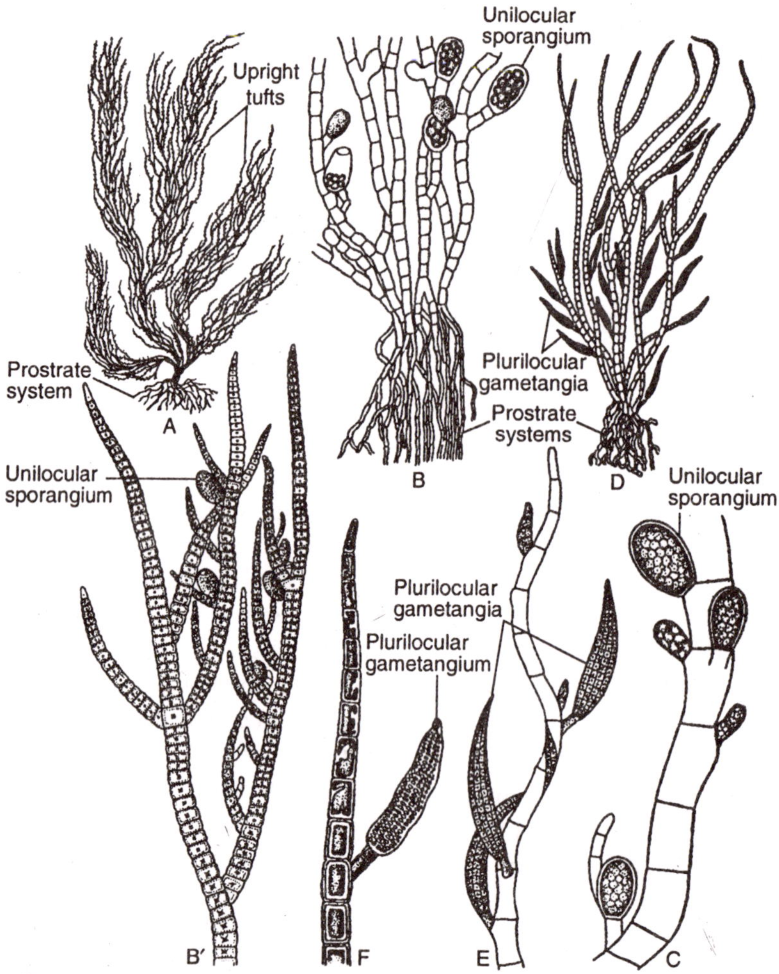

Fig. 8.5: *Ectocarpus* sp. A. Habit. B-C. Filaments bearing unilocular sporangia. D-E. Filaments bearing plurilocular gametangia. F. Portion of a filament showing plurilocular gametangium

nal cell of a lateral branchlet (Fig. 8.5D-F). The protoplast of each cell is metamorphosed into a single pear-shaped swarmer with two laterally inserted unequal flagella having the longer one directed forward and the shorter in the backward direction. Liberation of the swarmer is through a pore on the cell wall (Fig. 8.6A). The swarmers are of the same size and morphologically similar behaving as gametes. But some are less active and become passive after a short time. The more active ones cluster around the relatively passive one and fix themselves to it by their forwardly directed flagella. But fusion takes place between the passive one and the one of the more active gametes to form a zygote. This method of fusion and clustering of the active gametes around the passive one is known as the **clump formation** (Fig. 8.6B-F). The zygote so formed germinates directly producing a diploid plant—the sporophyte which resembles in every respect, the gametophyte that has produced plurilocular gametangia, only difference being the diploid plant bears **plurilocular sporangia** also known as **neutral sporangia** (Fig. 8.7), and more or less oval **unilocular sporangia** (Fig. 8.5B-C) that are developed with an enlargement of the terminal cells of short lateral branchlets. The single

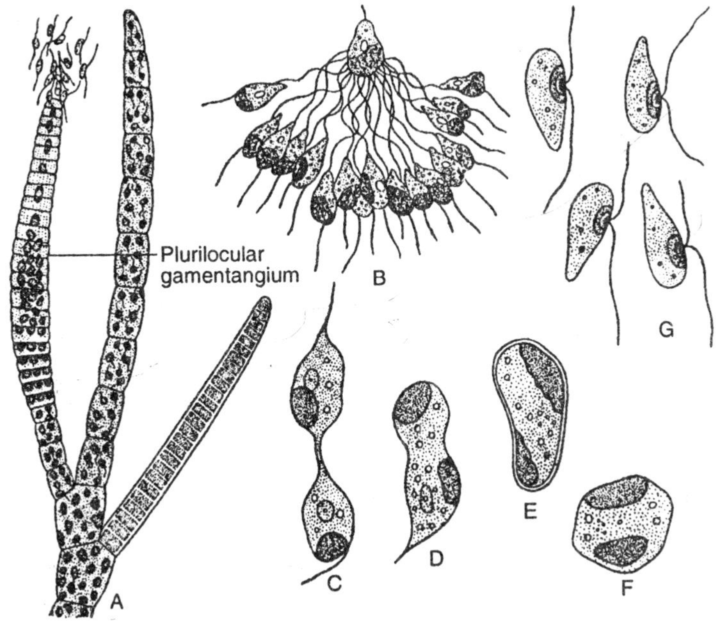

Fig. 8.6: *Ectocarpus* sp. A. Liberation of gametes from a plurilocular gametangium. B. Clump-formation of gametes during sexual reproduction. C-F. Progressive stages in fusion of gametes leading to zygote formation. G. Zoospore

nucleus of the young unilocular sporangium divides and redivides producing 64 nuclei. The first division of the nucleus is reductional and the rest being equational. With the completion of the nuclear division there is a cleavage into uninucleate protoplasts. Each protoplast then metamorphoses into a pear-shaped zoospore. The zoospores (Fig. 8.6G) resemble gametes, but differ in their behaviour by producing haploid individuals, the gametophytes. Again, the plurilocular sporangia borne on the sporophytic plant resemble morphologically the plurilocular gametangia, but they produce diploid zoospores and not gametes (Fig. 8.7). Each mature plurilocular sporangium consists of vertical rows of many small cubical cells or compartments. The protoplast of each compartment is metamorphosed into a single biflagellate zoospores. No reduction division takes place during zoospore formation, hence all the zoospores are diploid. These diploid zoospores and the haploid zoospores produced in the unilocular sporangia are morphologically indistinguishable. The diploid zoospores germinate to give rise to diploid individuals which bear plurilocular and unilocular sporangia. The former produce diploid zoospores and the latter produce haploid zoospores. The haploid zoospores produce the haploid individuals—the gametophytes, whereas the diploid zoospores serve to multiply the diploid individuals—the sporophytes. This is how the life cycle is completed (Fig. 8.7).

This typical life cycle of *Ectocarpus* exhibits morphologically identical filaments representing sporophyte and gametophyte—isomorphic alternation of generations. The filaments can only be differentiated from each other through cytological studies and the nature of reproductive structure, a particular filament bears. The filament produced by the germination of haploid zoospore bearing pluriolocular reproductive structure which produces gametes, is the gametophyte, and its cells are haploid. While the one formed by the direct germination of the zygote bearing unilocular sporangia and plurilocular sporangia is the sporophyte having diploid cells. Reduction division takes place

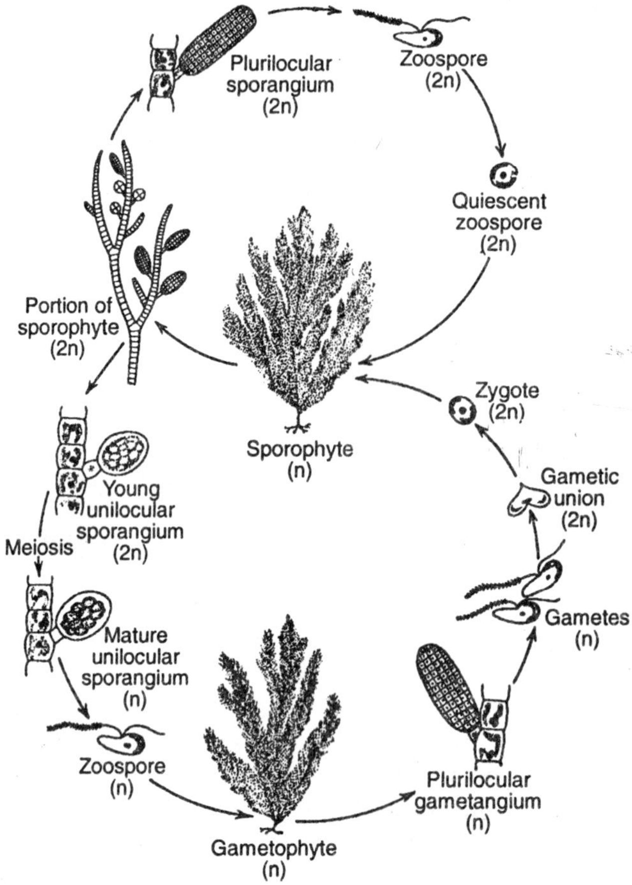

Fig. 8.7: Life cycle of *Ectocarpus* sp.

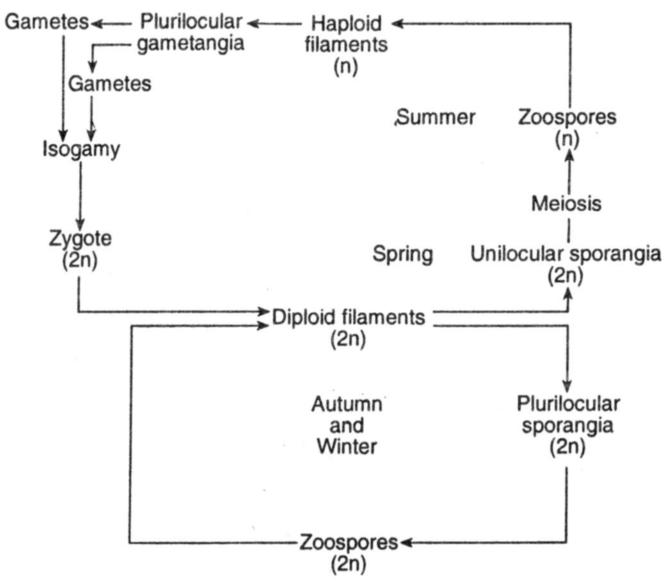

Fig. 8.8: Life cycle pattern of *Ectocarpus siliculosus*

during the production of zoospores in the unilocular sporangia which behave as zoosporangia. This type of life cycle has a great resemblance with the life cycle of the green alga *Ulva.* Besides this, the nature and development of swarmers in *Ectocarpus* possibly throw light on the origin of sex in algae. The gametes and zoospores are similar in structure, only difference being in their behaviour. It has been thoroughly investigated in *Ectocarpus siliculosus* that environmental conditions, probably temperature, commonly inhibit the development of one of the two generations. During warmer part of the year haploid filaments producing gametangia and gametes are developed. While during colder part diploid filaments bearing unilocular and plurilocular sporangia are produced. Knight, Kylin and others working on *Ectocarpus siliculosus* got some interesting results. They found diploid plants bearing both unilocular and plurilocular reproductive structures having a reduction division in the former and nothing in the latter. They named the plurilocular reproductive structures as neutral sporangia. The neutral sporangia produce diploid swarmers which germinate to form sporophytes, whereas the unilocular sporangia produce haploid zoospores. Zoospores from unilocular sporangia develop into gametophytes (Fig. 8.8).

Some Indian Species: *Ectocarpus arabicus* Fig. et De Not.; *E. breviarticulatus* J. Ag.; *E. dermonematis* Boergs.; *E. enhali* Boergs.; *E. filifer* Boergs.; *E. geminifructus* Boergs.; *E. thyrsoideus* Boergs.

Order: *LAMINARIALES*

This order contains the largest and most elaborately organized plant body of all algae. The members of this order, better known as **kelps,** are predominant plants of cold water.

With the exception of the genus *Chorda,* the all kelps show a large, coarse sporophyte generation differentiated into holdfast, stipe, and one or more blades. Growth in length results from an intercalary meristem between stipe and blade which adds tissue to both. Secondary growth in diameter of the stipe occurs in some annual and in many perennial species. The plants have a highly developed internal structure, showing clear tissue differentiation. The sporophyte is the evident plant, the life history always is completed through an alternate, microscopic filamentous gametophyte whose ecological requirements probably are more often the determining factors in the distribution of a given species.

The sporophytes of Laminariales produce superficial unilocular sporangia either on vegetative blades or on special sporophylls. The sporangia are sometimes in remarkably definitive sori. These mature on succession and fall out of the blade at maturity as an irregular plate, leaving a large hole that rapidly erodes. The sporangia are associated with unicellular paraphyses, both arising from the same basal cells. Meiosis occurs in the developing unilocular sporangia, which usually release 16 to 64 zoospores (32 in some cases). The zoospores attach and form a wall after brief motility and germinate within a few hours to grow into haploid gametophytes. The gametophytes, which occur in 1 : 1 ratio, are always dioecious, and this has now been determined in a large number of species in culture. The life history of Laminariales was first worked out by Sauvageau in 1915. The general appearance of the gametophytes is remarkably uniform throughout the order. All are small, filamentous thalli of a relatively few cells in uniseriate arrangement. The male plants have smaller cells and are more abundantly branched than the female ones. Although these gametophytes have almost always been studied in culture, it is possible to observe them in nature.

Sexual reproduction in Laminariales is oogamous. The oogonia are terminal or intercalary in position. Each forms a single egg which, though extruded, usually remains attached to the ruptured oogonium. The antheridia occur at the tips of branches or as outgrowths from intercalary cells of male gametophytes. Each releases a single biflagellate antherozoid, and the plant degenerates after the gametes are shed. The fertilized egg secretes a wall and soon undergoes division to begin the development of the sporophyte, but the female gametophyte persists during the early phases of this growth.

Family: *LAMINARIACEAE*

Unbranched stipe terminates in a single blade; blade smooth to perforate or imperforate, with or without ribs, often with tufts of hairs on the surface, blade may be fan-shaped spirally twisted and rolled up at the base. The sporangial sori are produced on the vegetative blades and not on special sporophylls; holdfast may be rhizomatous to discoid or conical, usually holdfast of branched haptera; stipe is solid or hollow and with or without mucilage ducts.

Genus: *Laminaria*

This is an alga that has attained the largest size forming an important element of perennial marine flora. There are many species of *Laminaria,* all of which have a very wide distribution being conspicuous in colder as well as temperate and tropical waters. They are popularly known as 'devil's aprons'. Most species

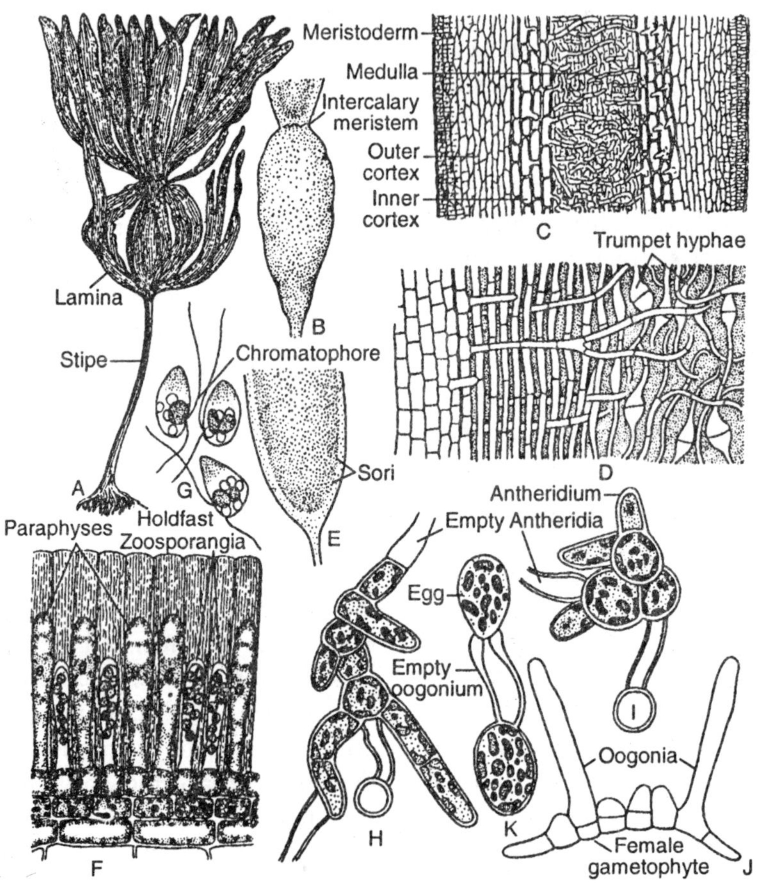

Fig. 8.9: *Laminaria* sp. A. Thallus with lamina, stipe and holdfast. B. Portion of the thallus showing intercalary meristem. C. Cross-section through lamina showing internal structural details. D. Portion of medulla in section indicating the arrangement of trumpet hyphae. E. Distribution of sori on lamina. F. Cross-section through a sorus. G. Zoospores. H-I. Male gametophytes. J. Young female gametophyte. K. Mature female gametophyte

favour quieter water. To the industrialists they are better known as kelps and are generally employed as a source of food, iodine and fertilizers. The alga is characterized by a conspicuous sporophyte, which may be even six feet or more in length consisting of a basal discoid or branched holdfast, a stipe, and a lamina which is simple or digitate (Fig. 8.9A). Growth occurs at intercalary meristem, usually located between the lamina and the stipe (Fig. 8.9B). Anatomically both lamina and stipe can be differentiated into different regions, the **meristoderm**—composed of one or more layers of similar small cells covered externally by mucilage, the **outer cortex**—much wider and elongated cells with pointed ends, the **inner cortex**—longer cells with square ends, and the **medulla** of a mass of tangled threads (Fig. 8.9C). Basically lamina shows the same structure as the stipe, only the elements of the medullary region are drawn out owing to the great surface enlargement. The medulla is characterized by the presence of cells modified into elongate hypha-like structures 'trumpet hyphae' (Fig. 8.9D), the transverse cell wall ends of both cells swell out to form bulbs, of which the upper bulb always being larger. The transverse wall is perforated resembling and behaving like sieve-plate. The **trumpet hyphae** are still objects of speculation as being suggested to be a storage or conducting tissue, or support. There is a system of anastomosing mucilage ducts which are particularly confined to the stipe.

The sporophyte bears patches of unilocular zoosporangia and paraphyses in sori on both surfaces of the lamina (Figs. 8.9E and F). The sporangia develop successively in considerable numbers maturing and liberating their contents at the same time. The zoospores are formed by the repeated division of the diploid sporangial nucleus one of which being meiosis, followed by typical cleavage of the protoplast of the sporangium with ultimate production of uninucleate bits which metamorphose into zoospores. The zoospores are biflagellate with laterally inserted unequal flagella (Fig. 8.9G). They, on germination, produce two types of gametophytes of microscopic size.

The male gametophytes are simple filamentous small-celled copiously branched structures (Figs. 8.9H and I). The antheridia arise singly or in groups at the terminal cells of the projecting threads or as lateral outgrowths from the upper side of the creeping threads. Each antheridium produces a single pear-shaped laterally inserted biflagellate antherozoid. The male gametophyte disintegrates after the liberation of the antherozoid. The female gametophyte is sparingly branched filamentous structure (Fig. 8.9J), any cell of which can function as an oogonium bearing one egg. Prior to fertilization the egg comes out through a narrow apical aperture and remains attached to the wall for a considerable time (Fig. 8.9K). The egg is fertilized by an antherozoid when already attached with the oogonial wall and is never set free. The zygote secretes a thin membrane and develops immediately into the sporophyte without undergoing any resting period (Figs. 8.10A-F).

Laminaria exhibits a heteromorphic alternation of generations, the sporophyte being dominant whilst the gametophyte is much

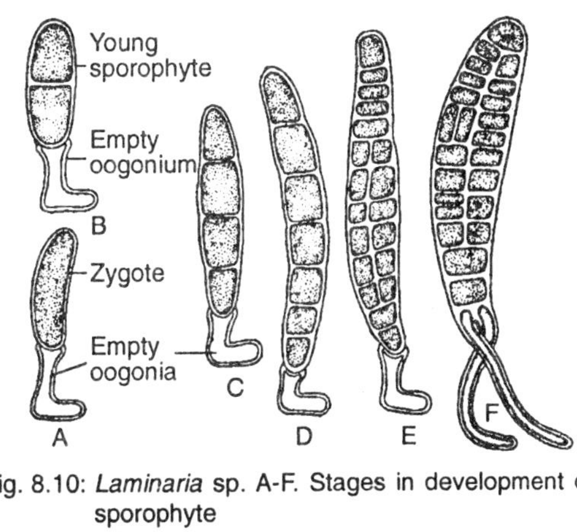

Fig. 8.10: *Laminaria* sp. A-F. Stages in development of sporophyte

reduced. Meiosis is delayed and takes place during the formation of zoospores in the zoosporangia. The entire process of sexual reproduction is unique of its kind. Though the egg is not retained within the oogonium it still remains attached with oogonial wall, which indicates a trasitional stage between primitiveness and advancement. The spontaneous germination of the zygote when already remaining attached with the female gametophyte without going to rest is also an interesting feature in the life cycle of *Laminaria* (Fig. 8.11).

Special features:

1. Gigantic size sporophytic plant body.
2. The plant body differentiated into a distinct holdfast, stipe and lamina.
3. Annual replacement of blade by the growth of meristematic tissue lying between stipe and blade.
4. Complex internal structure of the sporophytic plant body.
5. Zoosporangia in sori developing on the blade.
6. Reduction division during zoospore development in the zoosporangia.
7. Heteromorphic alternation of generations.

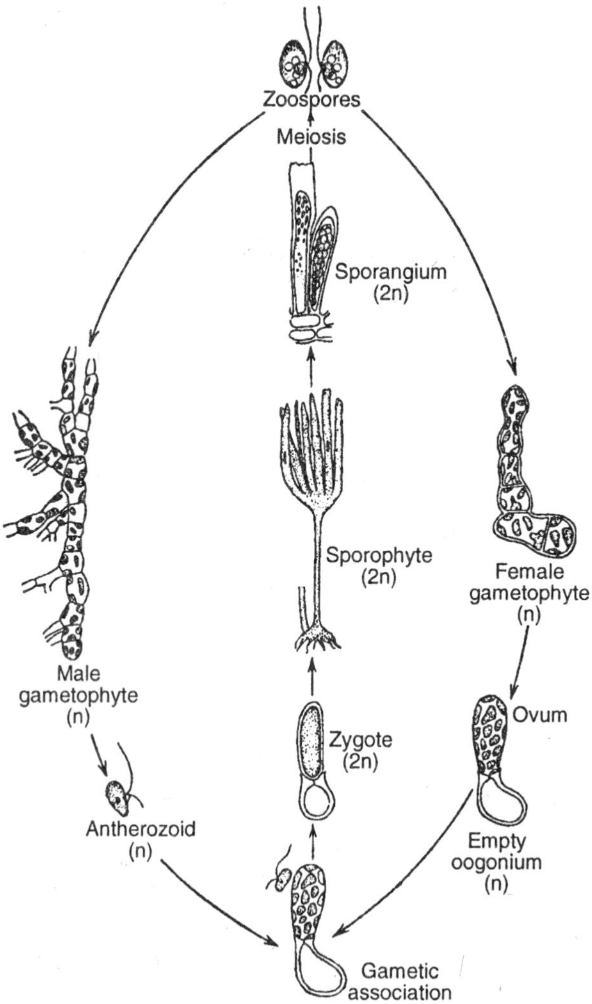

Fig. 8.11: Life cycle of *Laminaria* sp.

8. Predominant sporophyte alternating with much reduced gametophyte.
9. Gametophytes heterothallic.
10. Sexual reproduction oogamous.
11. Fertilization of ovum outside the oogonium but remaining attached to its apex.

Order: *FUCALES*

Plants are leathery dichotomously branched differentiated into holdfast, stipe and blade; growth from an apical cell group in the adult branches; sporophytic. The gametangia are developed in conceptacles bearing branched paraphyses. Meiosis takes place during the development of gametes which represent cytological gametophyte phase. When matured, the gametangia discharge their gametangial contents in coherent masses from the conceptacles; these, disintegrating in water. The liberated gametes—the antherozoids fertilize the eggs in a free state or still confined within the oogonia.

Family: *FUCACEAE*

Plants are dichotomous or pinnate strap-shaped simple or with costate branches, in some cases with piliferous cryptostomata, often with buoyant air bladders. Receptacular portions of the thallus terminate main branches or form stalked lateral branchlets, heterogamous to oogamous. Each oogonium forms several eggs.

Genus: *Fucus*

This is a common marine alga containing a number of species that are widely distributed in the sea coasts of temperate and Arctic regions. Most species are found attached to rocks between low and high tide marks and are commonly known as **rock-weeds**. The plant body of *Fucus* consists of a leathery, parenchymatous, dichotomously branched ribbon-like frond, stem-like stipe and a basal disc-like holdfast or hapteron by which it is attached to the substratum (Fig. 8.12A). The plants may be attached to completely or partly submerged rocks. The thallus is buoyed up in water by **air vesicles**, or bladder-like structures or floats. The swollen tips of the thalli, the receptacles, which lack midrib, are covered with small scattered pimple-like projections with small openings which lead into cavities, known as **conceptacles** (Fig. 8.12B). The thallus is diploid, may be monoecious or dioecious and is characterized by anatomical complexity. The thallus has a peripheral layer, which is known as the **limiting layer,** composed of small cells contain-

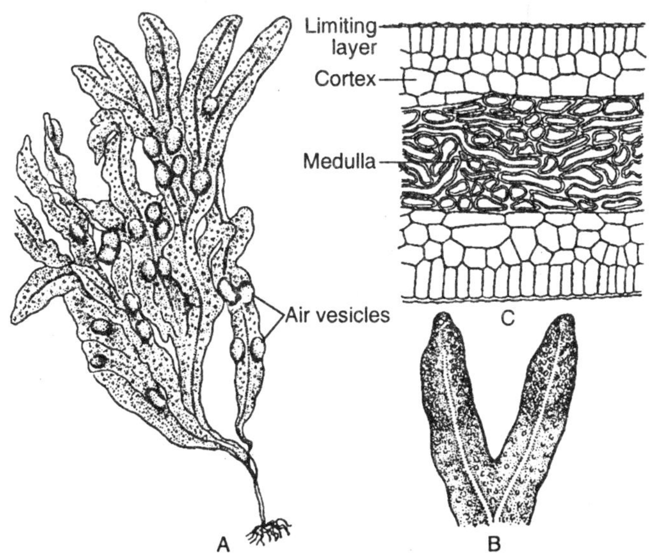

Fig. 8.12: *Fucus* sp. A. Habit. B. Apex of thallus. C. Portion of the thalius in section

ing abundant plastids and performing the function of assimilation (Fig. 8.12C). Below this is the **cortex** of several layers of elongated mucilaginous parenchymatous cells probably forming the storage system. The central tissue is the **medulla,** which is composed of hypha-like elongated cells very similar to *Laminaria,* probably performing the function of conduction of food material. Rudimentary sieve-plates are present in *Fucus.* Growth in length of the thallus takes place by means of an apical cell which lies in the depression at the tip of each branch.

The only method of asexual reproduction in *Fucus* is by fragmentation.

The method of sexual reproduction is heterogamous. The gametes are borne in oogonia and antheridia, which are produced in the conceptacles. In some species each conceptacle contains both oogonia and antheridia, in others there are distinct male and female conceptacles borne on two different thalli (Fig. 8.13A and D). Within each conceptacle are many multicellular hairs, the **paraphyses,** which are mixed with antheridia and oogonia. The oval oogonium is produced from a single cell and is borne on a short stalk (Fig. 8.13E). By cell division eight egg-cells are produced being enclosed in a cell wall of two or three layers (Fig. 8.13F-I). All eight eggs or ova when mature, are liberated out in water. The antheridia are smaller and more numerous than the oogonia, and are borne in groups on short, much branched hairs (Fig. 8.13B). Like oogonium antheridial wall is also

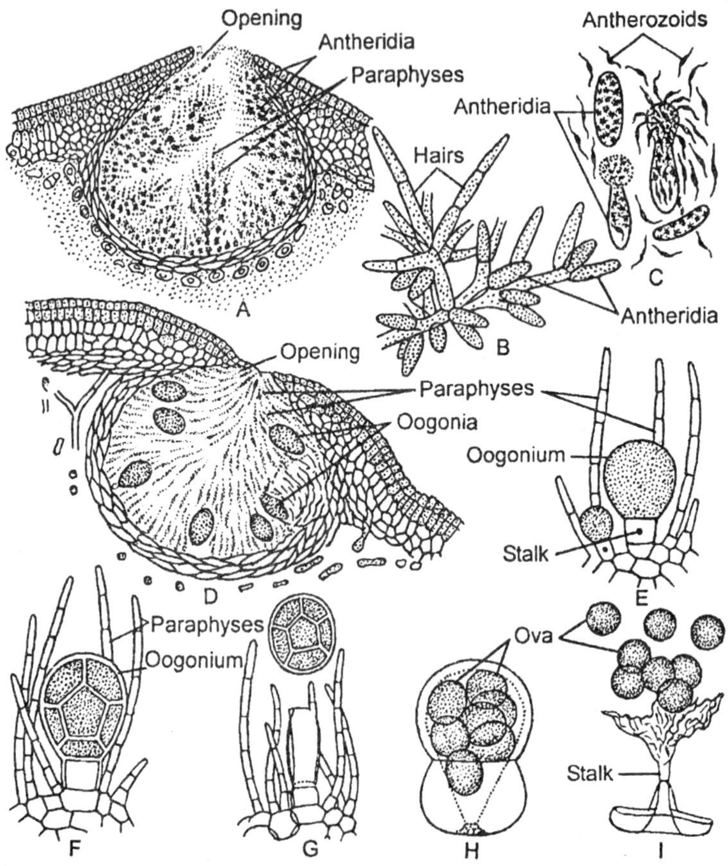

Fig. 8.13: *Fucus* sp. A. Section of a male conceptacle. B. Antheridia and hairs. C. Liberation of antherozoids from antheridia. D. Section of a female conceptacle. E-F. Stages in oogonium development. G. Liberation of packet of ova. H-I. Stages in liberation of ova

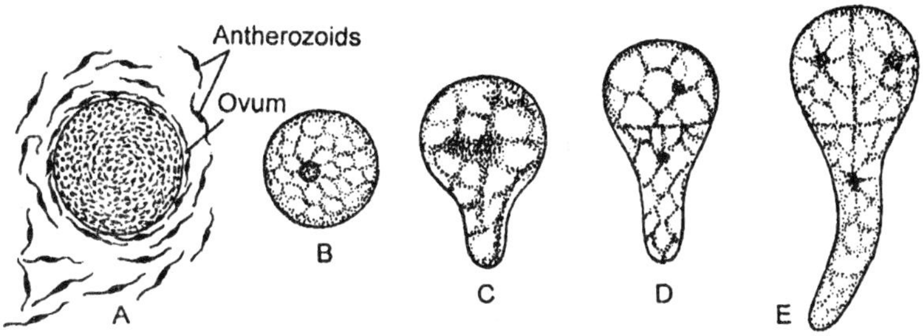

Fig. 8.14: *Fucus* sp. A. Ovum surrounded by antherozoids during fertilization. B. Oospore. C-E. Stages in oospore germination

composed of two or three layers. Numerous pear-shaped antherozoids with two laterally inserted unequal flagella are produced in the antheridium (Fig. 8.13C). Eggs floating free in water are soon surrounded by innumerable antherozoids which set the non-motile eggs in rotation and may continue until an antherozoid penetrates the egg and fertilizes it resulting in an oospore (Fig. 8.14A and B). The oospore forms a thin wall and at once divides to produce new *Fucus* plant without undergoing a resting period, the lower cell growing into the holdfast and the upper one the thallus (Fig. 8.14C-E). Meiosis in the life cycle of *Fucus* takes place during gametogenesis, as such the diploid phase is very predominating as against

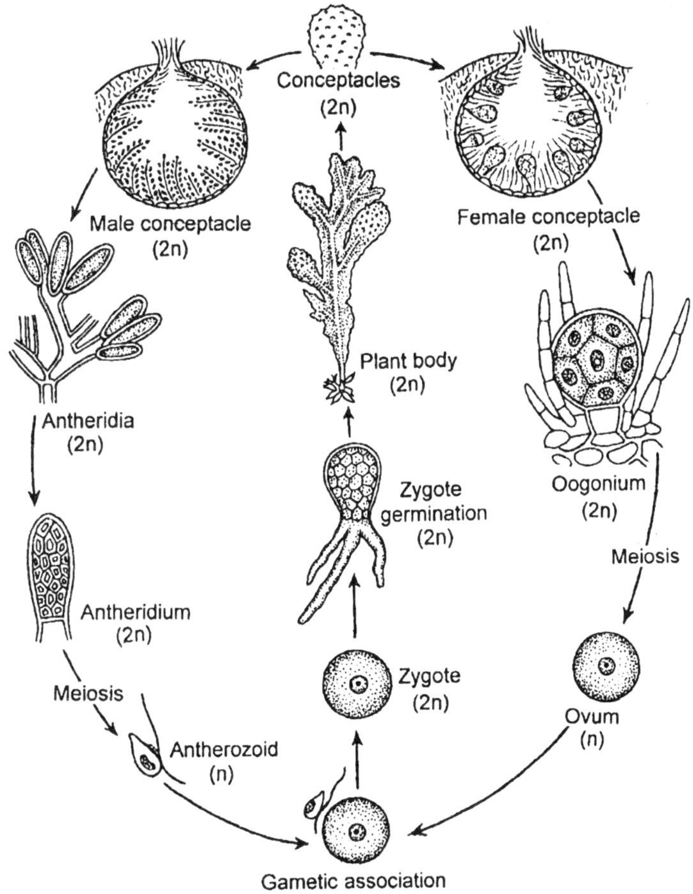

Fig. 8.15: Life cycle of *Fucus* sp.

haploid phase which is reduced merely to the gametes (Fig. 8.15).

Fucus exhibits a unique feature in its life cycle pattern. It possesses a remarkably elaborate diploid phase which bears the male and female sex organs. Reduction division occurs during the development of gametes (antherozoids and eggs) only. Hence the antherozoids and the eggs represent the haploid gametophytic phase in the life cycle of *Fucus*. The gametophytic phase instead of being distinctly organized has thus been appreciably reduced to the gametes alone. The diploid phase, on the other hand, does not produce any sporangia or spores. Instead, it poduces gametangia and gametes, which is rather very uncommon in Algae. In *Fucus*, therefore, there is only an alternation of chromosome numbers from diploid to haploid and back to diploid state, whereas the regular alternation of corresponding spore-producing and gamete-producing plants is lacking. There is thus no morphological alternation of generations in the life cycle of *Fucus*.

Besides above, it has also been suggested that the so-called antheridia and oogonia are micro- and macrosporangia. In the microsporangium four haploid cells are formed by meiosis which are potential microspores. Each potential microspore by mitosis produces gametophytes from which antherozoids are produced. Again, the macrosporangia by meiosis produce four haploids, each one is a potential macrospore. Each macrospore by mitosis produces female gametophyte from which ovum is produced. According to this view there is heteromorphic alternation of generations in *Fucus*. If this interpretation is correct, the order Fucales should be placed in the class Heterogeneratae.

Special features:

1. The diploid plant body differentiated into holdfast, stipe and dichotomously branched ribbon-like frond.
2. Presence of floating device by bladders.
3. Complex internal structure of the plant body.
4. Absence of asexual reproduction by means of spores or similar other structures.
5. Sexual reproduction heterogamous.
6. Sex organs in the same or different conceptacles.
7. Reduction division during the development of gametes in the antheridium and oogonium.
8. The development of eight eggs in each oogonium.
9. Fertilization of ova outside the plant body in the open water.
10. Absence of morphological alternation of generations.
11. Presence of chromosomal alternation of generations.

Plants possess flat costate branches or transitional stages to cylindrical branches bearing numerous macroscopic spine-like projections, turbinate foliar organs, or costate leaves, usually with cryptostomata and buoyant airbladders. Conceptacles are in ordinary branches or in special receptacular branchlet systems. Sexual repoduction is oogamous. Each oogonium produces a single egg.

Genus: *Sargassum*

It is principally confined to warmer seas in the tropical zone, and is commonly known as 'gulf weed'. It is one of the most conspicuous and common of the sea-weeds being very abundant along the Indian coast growing attached to rocks below low-tide level along the East and West Coasts of India. The so-called Sargasso weed is found as large floating mass in the Sargasso Sea near the West Indies. The bushy habit of the plant body has a great resemblance with the seed plants as its body is composed of stem-like stipe and leaf-like blades. The predominant sporophyte is attached to the substratum by means of a more or less irregular, warty, solid, parenchymatous base from which arises a stipe-like structure which grows into the main axis bearing alternate branches. The branches bear short, phyllode-like structures or leaf-like expansions commonly known as 'leaves', with or without midrib depending on species (Fig. 8.16A). Besides this, the secondary branches are also sterile and reduced to air-bladders (Fig. 8.16B-D). There occur tertiary branches which are fertile finger-like in appea-rance, called **receptacles** (Fig. 8.16C) whose development varies with species. In these receptacles are embedded flask-shaped **conceptacles**. In certain species of *Sargassum,* the thallus is free-floating, where it perpetuates itself by fragmentation. Structurally, the thallus is very similar to that of *Fucus* being differentiated into a peripheral layer of compact, columnar cells containing abundant plastids performing the function of photosynthesis. Inner to it is the **cortex** of many cells whose walls are gelatinized (Fig. 8.16E). These cells probably perform the function of storage. The central zone is the **medulla** of loosely arranged cells. Medulla is perhaps the food-conducting region. Growth of the

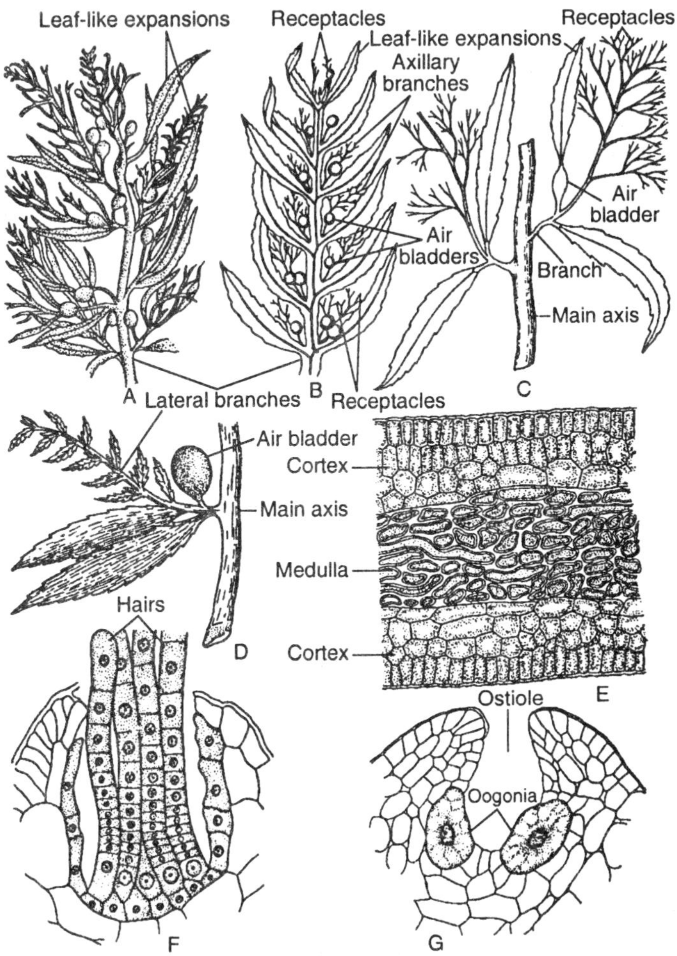

Fig. 8.16: *Sargassum* sp. A-B. Branches with leaf-like expansions, air-bladders and receptacles. C. Small portion of the main axis showing arrangement of branches, leaf-like expansions and receptacles. D. Lower portion of the main axis with air-bladder and branch. E. Portion of thallus in section. F. Section of a cryptoblast. G. Section of a young female conceptacle

thallus is carried by a three-sided apical cell situated at a depression at the apex of the thallus.

Sargassum reproduces vegetatively by fragmentation and by the growth and decay of the thallus.

Antheridia and oogonia are borne in flask-shaped conceptacles that are embedded in the receptacles. Some species are monoecious and others dioecious. The monoecious species may have both antheridia and oogonia developed in the same conceptacle or in different conceptacles on the same thallus, whereas the dioecious plants are either male or female. The two types of plants are distinguishable from the nature of the receptacles, the male receptacle being smooth and the female one spinous.

In the early stages, the development of male and female conceptacles is very identical as it initiates from a single initial cell which by repeated division and enlargement gives rise to a conceptacle.

The oogonia are sessile and borne directly on the wall of the conceptacle (Fig. 8.16G). The uninucleate young oogonium is filled with dense cytoplasm. Its nucleus has a very prominent nucleolus. The nucleus enlarges and undergoes three successive divisions with the first division meiotic, resulting eight haploid nuclei, only one survives and others degenerate. The entire content of the oogonium rounds up

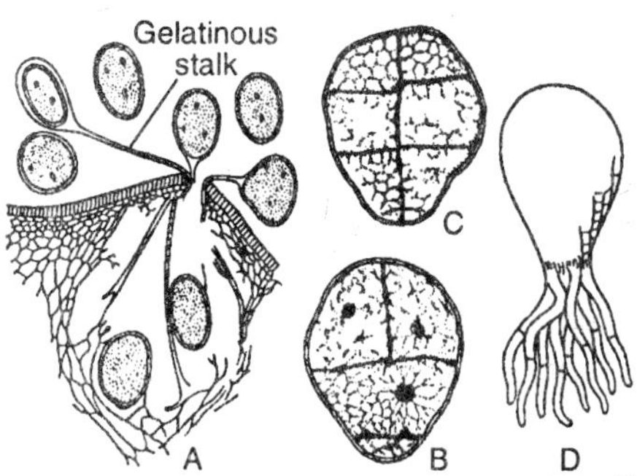

Fig. 8.17: *Sargassum* sp. A. Female receptacle with exuded oogonia, one oogonium remaining attached to the conceptacle wall by gelatinous stalk. B-D. Stages in the development of new individual

to form an uninucleate ovum. Each mature oogonium after emergence through the ostiole, remains attached to the conceptacle wall by means of a 'long gelatinous stalk which is formed by the elongation of the thickened apex of the oogonium (Fig. 8.17A).

The antheridia occur in large numbers and are crowded in each conceptacle. They are borne in clusters on the lower branches of the branched paraphyses which again arise from the wall of the conceptacle. The mature antheridium is a small ovoid structure with a thick mucilaginous wall. It produces 64 pear-shaped biflagellate antherozoids. Meiosis takes place during the development of antherozoids. The mature antherozoids are liberated in sea-water.

The liberated antherozoids swim towards the female conceptacle and several of them surround each oogonium. The antherozoids get attached to the oogonial wall by their flagella. One of the attached antherozoids after penetrating through the oogonial wall enters the ovum. Ultimately, fusion of male and female nuclei takes place. Thus, the fertilization is effected while the ovum is still confined within the oogonium which again is attached to the conceptacle wall by a gelatinous stalk.

The zygote starts germinating immediately after formation and while already enclosed in the oogonium, without undergoing any reduction division of the diploid nucleus. The oogonium also remains attached to the wall of the conceptacle. Ultimately, the zygote is liberated out by the gelatinizaion of the oogonial wall, rests on some solid object and then divides transversely. The upper cell gives rise to the erect portion and the lower to the rhizoids of the diploid plant (Fig. 8.19B-D).

In certain species of *Sargassum* there are found sterile flask-shaped structures borne on primary branches and on leaf-like expansions. These structures look like the conceptacles and are known as **cryptoblasts** (Fig. 8.16F). They

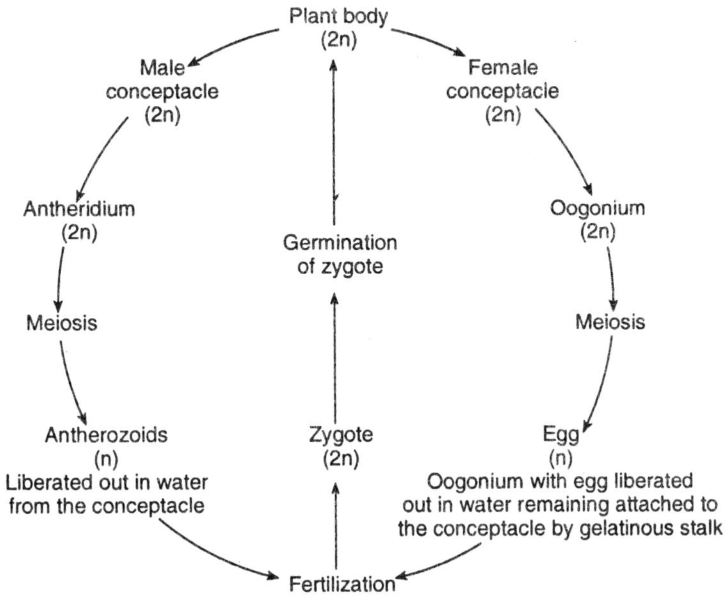

Fig. 8.18: Life cycle pattern of *Sargassum* sp.

have their early stages of development very similar to those of conceptacles. These cryptoblasts do not bear any sex organs, colourless unbranched hairs arise from the floor of the cryptoblasts. These hairs secrete mucilage and project in tufts through the openings of the cryptoblasts. These openings are called **cryptoblasmata.**

Sargassum exhibits certain special features of advancement. The complex diploid thallus is very predominant being differentiated into main axis with branches on which leaf-like structures are developed. There is great internal differentiation of tissue with fairly well division of labour. Meiosis is gametic. Haploid generation is restricted only in the gametes. Oogamous sexual reproduction is accompanied with retention of ovum within the oogonium even up to the early stage of development of embryo (Fig. 8.18).

Some Indian Species: *Sargassum caryophyllum* J. Ag.; *S. duplicatum* J. Ag.; *S. swartzii* (Turner) C. Ag.

Special features:

1. Diploid plant body of bushy habit having a main axis producing branches bearing leaf-like structures and air-bladders, a special floating device.
2. Internal differentiation of tissues of the plant body with division of labour.
3. Absence of any kind of spores.
4. Sexual reproduction oogamous.
5. Sex organs (antheridia and oogonia) developing in conceptacles borne on the same or different plants.
6. Reduction division during the development of gametes in the antheridia and oogonia.
7. Formation of a single egg in an eight-nucleate oogonium, the rest of nuclei degenerate.
8. Fertilization taking place while the egg is still enclosed in the oogonium which is liberated out in water but remains attached to the female conceptacle by means of a gelatinous stalk—a condition much more advanced than *Fucus*.
9. Absence of any alternation of morphological generations correspondingly

with chromosomal alternation of generations.

10. The diploid plant body instead of producing sporangia and spores produces gametangia and gametes—a condition encountered also in *Fucus.*

Order: *DICTYOTALES*

Plants are thalloid, with alternation of similar generations; growth apical. Asexual reproduction is by aplanospores and sexual reproduction by eggs and antherozoids.

Family: *DICTYOTACEAE*

Plants are of moderate to large size, simple or branched, the divisions growing from apical cells or marginal rows of apical cells, forming fan-like foliaceous, strap-shaped branching thalli, usually composed of cortical and medullary layers of cells. Sexual and asexual generations are distinct, the plants of the two phases similar in form. Asexual reproductive elements consist of sporangia in which four or eight aplanospores (commonly called tetraspores) are produced. Sexual reproductive elements, eggs are borne singly in the superficial oogonia; and uniflagellate antherozoids, borne in the superficial antheridia.

Genus: *Dictyota*

This genus is characterized by an isomorphic alternation of generations, with the growth of thalli initiated by a single apical cell. It is an annual of the warm seas and is common along the Indian coast, usually in rocky pools between the tide marks. The prostrate portion of the thallus is represented by cylindrical rhizome with rhizoids which affords anchorage of the flat dichotomously branched, fan-shaped upright portion of the thallus to the substratum. The upright portion has a lower cylindrical stalk-like structure gradually expanding

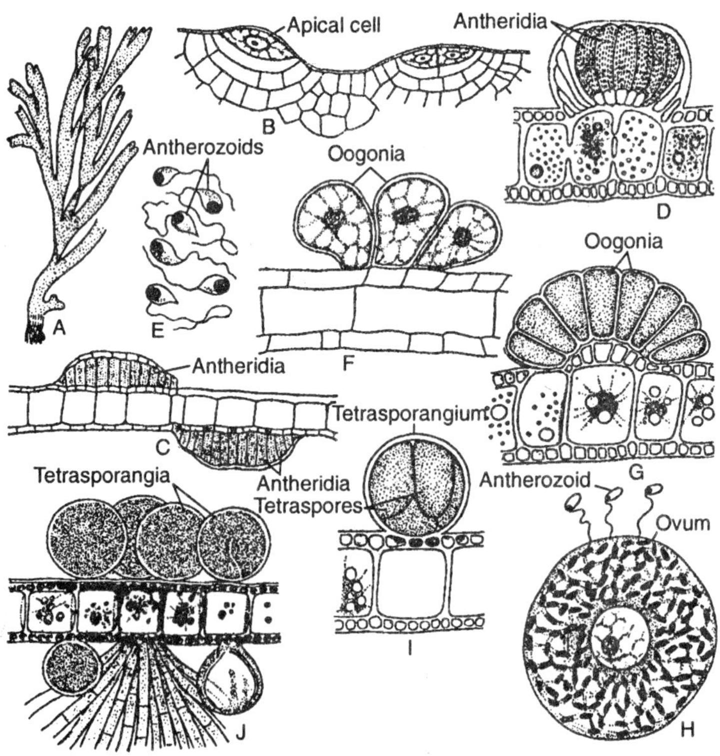

Fig. 8.19: *Dictyota* sp. A. Habit. B. Apex of thallus in surface view. C-D. Antheridial sori in vertical section. E. Antherozoids. F-G. Oogonial sori in vertical section. H. Fertilization. I. Section of thallus bearing tetrasporangium with tetraspores. J. Section of thallus showing tetrasporangia and hairs

into dichotomously branched ribbon-like in form (Fig. 8.19A). A mature thallus consists of a middle layer of large colourless cells with few or no chromatophores bounded by layers of small assimilating cells, densely packed with chromatophores. On both surfaces of the thallus there are colourless multicellular hairs which are shed during reproductive phase. Dichotomy of the upright portion is initiated by the longitudinal division of a single apical cell (Fig. 8.19B). The two daughter cells so formed give rise to dichotomous branching by continued independent division.

Asexual reproduction is brought about by means of tetraspores produced in superficial tetrasporangia which are scattered on both surfaces of the sporophyte, occurring singly or in small groups (Fig. 8.19I and J). Each tetrasporangium has a basal stalk cell and a spherical body. Four tetraspores are formed in the tetrasporangium as a result of meiotic division of the large diploid nucleus followed by the cleavage of the cytoplasm into four bits corresponding to each nucleus. All the four spores germinate into new *Dictyota* plants—the gametophytes (male and female).

Sexual reproduction is oogamous. Both male and female gametophytes morphologically resemble the sporophyte, only difference being that the gametophytes bear sexual reproductive organs, whereas the sporophytes bear sporangia. Both antheridia and oogonia are borne in sori on separate plants. The male gametophyte bears antheridia in small groups of sori (Fig. 8.19C and D). The number of antheridia in a sorus ranges from 100 to 200.

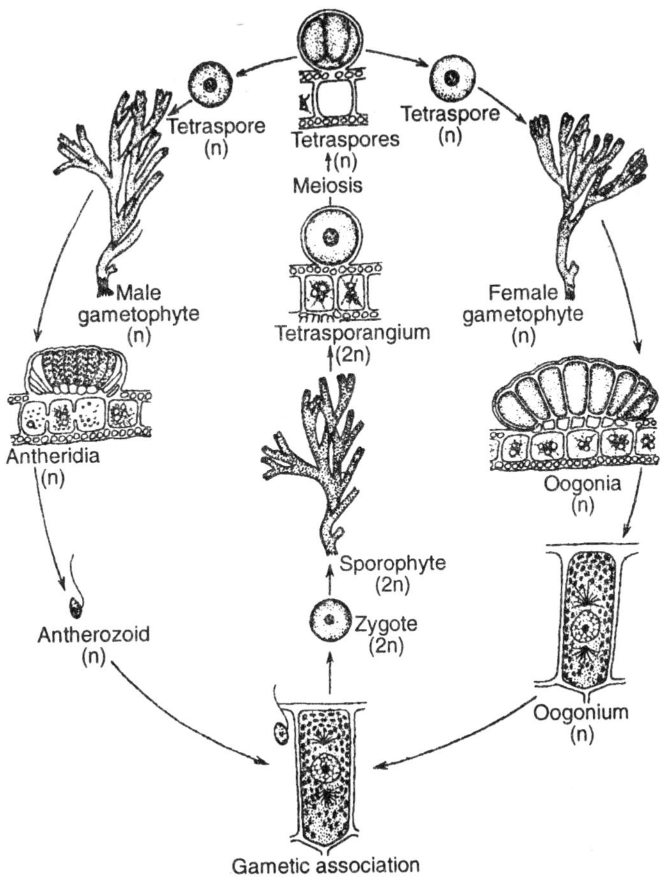

Fig. 8.20: Life cycle of *Dictyota* sp.

Each antheridium is formed from a single epidermal cell of the thallus which by subsequent divisions forms antheridia. A mature antheridium is a large colourless structure, the contents of which are metamorphosed into a single uniflagellate pear-shaped antherozoid (Fig. 8.19E), which is liberated by the gelatinization of the antheridial wall. The oogonial sorus is very similar to the antheridial sorus (Fig. 8.19F and G). The number of oogonia in a sorus goes even up to 100. The oogonial sorus is developed from a superficial cell of the thallus which elongates vertically and by transverse division is differentiated into a stalk cell and a primary oogonial cell. The primary oogonial cell develops into an oogonium, contents of which develop into a single egg. Both antherozoids and eggs are liberated out in the surrounding water and a large number of antherozoids are attracted by each ovum (Fig. 8.19H). Fertilization is entirely external. A single antherozoid penetrates the egg, and fertilization is effected. The zygote secretes a wall and without undergoing resting period develops into a sporophytic plant which morphologically resembles the gametophyte. In *Dictyota* meiosis takes place in the tetrasporangia during spore formation (Fig 8.20).

Some Indian Species: *Dictyota atomaria* Hauck.; *D. bartayresiana* Lamour.; *D. dichotoma* (Huds.) Lamour.; *D. dumosa* Boergs.; *D. maxima* Zan.; *D. pinnatifida* Kutz.

Special features:

1. Plant body differentiated into prostrate rhizome and rhizoids, and aerial dichotomously branched ribbon-like structure.
2. Internal differentiation of tissue of the plant body for the division of labour.
3. Isomorphic alternation of generations.
4. Reduction division during the development of tetraspores in the tetrasporangia borne on the sporophyte.
5. Gametophytic plant body heterothallic.
6. Sexual reproduction oogamous.
7. Fertilization of ovum outside the plant body in the open water.

Chapter 9

The Rhodophycophyta

The Rhodophycophyta or red algae are almost entirely confined to the sea. Both fresh- and brackish-water members of this group are comparatively few in number. The freshwater forms prefer well-aerated water. The marine forms are of worldwide distribution being predominantly represented in the warm waters. They generally grow attached mostly to rocks. Some grow on or in other plants either as epiphytes or as endophytes, whereas others thrive as parasites. The red algae exhibit various colourations from bright red and purple to dark brownish red, brownish green, blue-green, and black. The vegetative body, in the majority of cases, is multicellular which ranges from simple filamentous to branched, heterotrichous, to often expanded macroscopic form of apparently appearing parenchymatous structure having more than a metre or so in length. Some members have very simple unicellular thalli.

The thallus of red algae is characterized by the presence of gelatinous material which, in some species, is utilized for the preparation of agar, a gelatinous material of great use as a medium of culturing fungi and bacteria. In some of the Oriental countries red algae are extensively used as food. The thalli of red algae are composed of highly organized cells whose wall is again composed of a mixture of cellulose and pectic material. Depending on species and maturity, cells may be uni- or multinucleate. Each cell bears one to many chromatophores of varied shape. One of the striking features of most red algae is the presence of a pore-like opening in the cell wall through which conspicuous, broad cytoplasmic connection is maintained from cell to cell, known as **pit-connection.** The chromatophores contain **chlorophylls a** and **d,** α- and **β-carotenes, lutein, taraxanthin, allophycocyanin,** a special red water-soluble pigment **r-phycoerythrin,** and a blue pigment **r-phycocyanin.** The colour of the vegetative body depends upon the quantities and kinds of pigments present in the chromatophores. The food reserves may be alcohols, oils, but the most important one is—**floridean starch,** an insoluble carbohydrate, a compound intermediate between true starch and dextrin. One of the outstanding characteristics of red algae is complete absence of flagellate reproductive structures.

In general, red algae reproduce sexually. Vegetative reproduction by the fragmentation of the vegetative body is not very common. Besides sexual and vegetative reproduction of the vegetative body is not very common. Besides sexual and vegetative reproduction, most red algae reproduce asexually by the formation of **monospore, bispore, polyspore,** or **paraspore.** All these are non-motile single-celled spores and are formed inside respective sporangia. Sexual reproduction is oogamous with or without elaborate post-fertilization stages which, of course, depend whether meiosis of the diploid zygotic nucleus is delayed or takes place immediately after its formation. The male reproductive organ—**spermatangium** is a very simple structure which bears a single uninucleate single-celled male gamete—**spermatium,** whereas the female one—**carpogonium** has a long neck-like structure—**trichogyne** and a bulbous base—the **carpogonial base** where remains the female nucleus. The red algae which have a very elaborate post-fertilization stage develop an elaborate diploid phase, the earlier part of which is entirely dependent on the gametophytic individual and the latter part is represented by an independent sporophyte which may or may not be different from the gametophyte morphologically. The sporophyte bears **tetrasporangia** and **tetraspores.** Reduction division takes place during the sporogenesis in the sporangium of the independent sporophyte. Such forms exhibit isomorphic to heteromorphic alternation of generations.

CELL STRUCTURE

The major features of a red algal cell (Fig. 9.1) are a chloroplast with one thylakoid per band and no chloroplast E.R., floridean starch grains in the cytoplasm outside the chloroplast, no flagella, pit connections between cells in filamentous genera, and a eukaryotic type of nucleus (Scott et al., 1980).

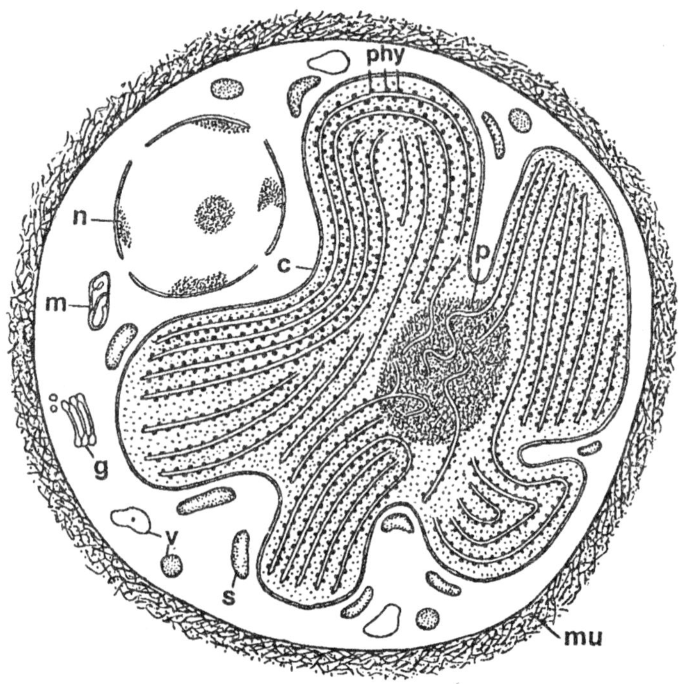

Fig. 9.1: Semi-diagrammatic drawing of a cell of *Porphyridium cruentum.* (c) Chloroplast; (g) Golgi; (m) Mitochondrion; (mu) Mucilage; (n) Nucleus; (p) Pyrenoid; (phy) Phycobilisomes; (s) Starch; (v) Vesicle (Adapted from Fantt and Conti, 1965)

CELL WALLS

In general, cellulose forms the microfibrillar framework in most rhodophycean cell walls, although in the haploid phase of the Bangiales (*Bangia* and *Porphyra*) a β-1, 3 linked xylan (polysaccharide composed of xylose residues) performs this function. Unicellular red algae have an amorphous matrix of sulphated polysaccharides without cellulose surrounding the cells (Arad et al., 1993). The amorphous polysaccharides or mucilages occur between the cellulose microfibrils in the rest of the red algae. The two largest groups of amorphous mucilages are the **agars** (Fig. 9.1) and the **carrageenans.** These mucilage may constitute up to 70% of the dry weight of the cell wall. Cuticles, composed mostly of protein, can occur outside the cell wall.

CHLOROPLASTS AND STORAGE PRODUCTS

Chloroplasts are usually stellate with a central pyrenoid in the morphologically simple Rhodophyceae (Fig. 9.1), whereas in the remainder of the Rhodophyceae they are commonly discoid (Fig. 9.2). In the Rhodophyceae with apical growth, the chloroplasts usually originate from small, colourless proplastids with a few thylakoids in the apical cell (Lichtlé and Giraud, 1969). Chloroplasts are surrounded by the two membranes of chloroplast envelope with no chloroplast E.R. present (Figs. 9.1 and 9.2). Inside the chloroplasts thylakoids occur singly. DNA occurs as microfibrils, and the phycobilin pigments are localized into phycobilisomes on the surfaces of the thylakoids, a situation similar to that in the Cyanophyceae.

Chlorophylls *a* and *d* are in the chloroplasts. Chlorophyll *d* has been considered as an altered form of chlorophyll *a* in the past, but it is now accepted as a separate molecule. Sagromsky (1964) claimed that it was present in plants in strongly illuminated habitats only. Of the carotenoids, zeaxanthin is found in greatest quantities.

The phycobiliproteins include R-phycocyanin, allophycocyanin, and three forms of phycoerythrin, the latter being present in the greatest amount, giving the algae their pinkish colour. B-Phycoerythrin is present in the more primitive Rhodophyceae and has been found in *Porphyridium, Rhodosorus, Rhodochorton,* and *Smithora.* R-Phycoerythrin is found in almost all higher Rhodophyceae and C-*phycoerythrin* has been found in *Porphyridium, Porphyra, Smithora, Grateloupia,* and *Polysiphonia.* The phycobiliproteins are in phycobilisomes on the surface of thylakoids (Fig. 9.1). If both phycoerythrin and phycocyanin are present, the phy-

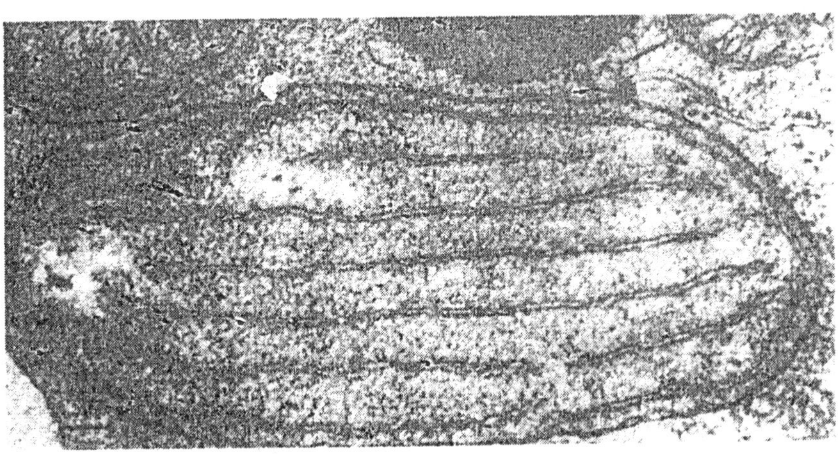

Fig. 9.2: Chloroplast in a carpospore of *Polysiphonia*

cobilisomes are spherical. If, however, only phycocyanin is present, then the phycobilisomes are discoid.

Complementary chromatic adaptation occurs in the red algae. Orange and red light stimulates the production of long-wavelength absorbing phycocyanin, while green light stimulates the formation of short-wavelength absorbing phycocyanin. The colour will vary, depending on where the plant grows; a species of red alga may be yellow-green when growing in the intertidal zone, whereas it is deep purple-red when growing 10 to 20m beneath the water surface. The ratio of phycoerythrin to chlorophyll is much greater at low light intensities. As the intensity of light increases, the ratio of phycoerythrin to chlorophyll drops. This is reflected in the colour of the plant and is the reason why many red algae turn green at higher light intensities. The greatest increase in the phycoerythrin: chlorophyll ratio occurs when the light intensity is dropped from 3000 to 400 lux. The ratio of phycoerythrin to phycocyanin to allophycocyanin stays about the same at different light intensities. There is little change in the thylakoid structure of the chloroplasts in different light regimes, but the number of phycobilisomes is much greater at low light intensities although the size of the individual phycobilisome remains constant. This adaptation to greater amounts of phycoerythrin under low light intensities allows the Rhodophyceae to grow at greater depths than other algae because of the ability to utilize better the green and blue light that penetrates farthest into water (pink-red pigments absorb blue-green and reflect pink-red).

Floridoside (O-α-D-galactopyranosyl-(1-2)-glycerol) Fig. 9.3) is the major product of photosynthesis in the red algae. An isomeric form of floridoside, isofloridoside, occurs along with floridoside in the Bangiales. The concentration of floridoside increases in red algal cells as the salinity of the medium increases. This change in floridoside concentration is thought to compensate, at least in part, for the changes in external osmolarity, thereby preventing water from leaving the algal cells as the salinity increases. The levels of floridoside can be as high as 10% of the tissue dry-weight in some marine red-algal thalli. The first observable product of photosynthesis is phosphoglyceric acid, as is the case in higher plants. Floridoside appears after 30 seconds of illumination and after 2 hours floridoside is the major product of photosynthesis. Floridoside apparently has the same function as sucrose, the common product of photosynthesis in green algae and higher plants.

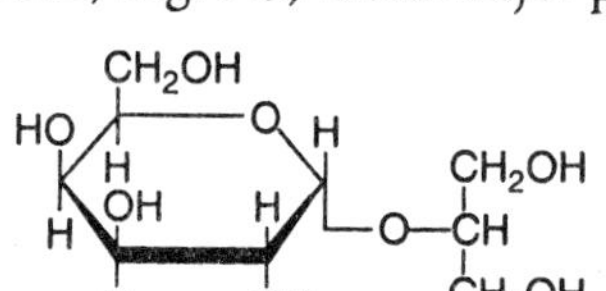

Fig. 9.3: The structure of floridoside

Floridean starch (Fig. 9.3) is the long-term storage product, occurring as grains in the cytoplasm outside of the chloroplast (Fig. 9.1). Floridean starch is similar to the amylopectin of higher plants, staining red-violet with iodine. In the more primitive Rhodophyceae the starch grains are clustered as a sheath around the pyrenoid of the chloroplast, whereas in the more advanced Rhodophyceae the starch grains are scattered in the cytoplasm.

PIT CONNECTIONS

Pit connections occur between the cells in all of the orders except the Porphyrdiales, and the haploid phase of the Bangiales. It has been pointed out that the term 'pit connection' is inappropriate because the structure is neither a 'pit' nor a 'connection'; however, because the term has been used for so long, it is probably best to retain it. A pit connection consists of a proteinaceous plug core in between two thallus cells (Figs. 9.4 and 9.5). Cap membranes separate the plug core from the adjacent cytoplasm. The cap membrane is continuous with the

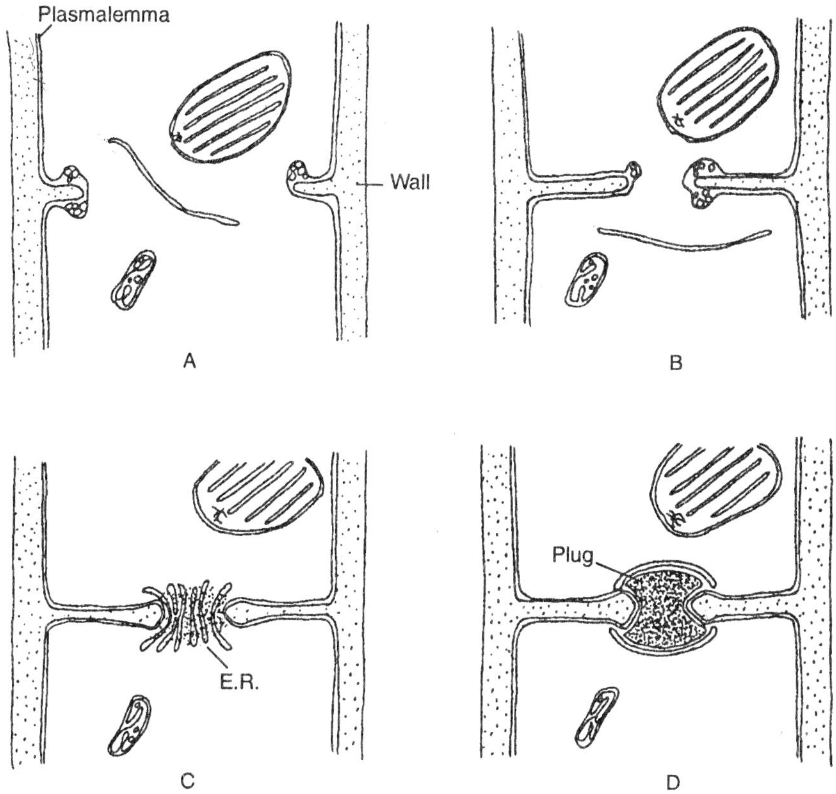

Fig. 9.4: Semi-diagrammatic drawing of the formation of a pit connection in a red alga; A. The cross wall begins forming inward with the wall precursors found in vesicles derived from the cytoplasm; B. The cross wall septum is complete, leaving an opening (aperture) in the centre; C. Endoplasmic reticulum lies across the opening in the wall, and electron-dense material condenses in this area; D. The pit connection is formed, consisting of a plug with the plasmalemma continuous from cell to cell (Adapted from Ramus, 1969, 1971; Lee, 1971)

plasma membrane, which in turn is continuous from one cell to the next. On the inside of the cap membrane can be an inner layer, while on the outside of the cap membrane can be an outer cap layer (Pueschel, 1987). The structure of the pit connection can vary. The more primitive red algae, such as *Rhodochaete* and *Compsopogon,* lack cap membranes and cap layers, with only a plug core present. It has been postulated that this represents the ancestral condition.

There are two types of pit connections. **Primary pit connections** are formed between two cells during cell division. **Secondary pit connections** result when two cells fuse. Both types of pit connections have the same structure. **Primary pit connections** are formed as follows (Fig. 9.4): Soon after nuclear division, the cross wall grows inward from the lateral wall. When the cross wall is complete, there remains a hole (aperture) in the centre through which the protoplasm of the two cells is continuous. A number of parallel vesicles traverse the hole, with electron-dense material condensing around the vesicles. Eventually the vesicles disappear, and the electron-dense material fills the hole. A membrane is formed around this material, producing a plug in the hole. The pit connection has been reported to contain proteins and polysaccharides. The pit connection may function as a site of structural strength on the thallus (Kugrens and West, 1973). In some algae the plugs of the pit connections become dislodged from between the cells of a developing gonimo-

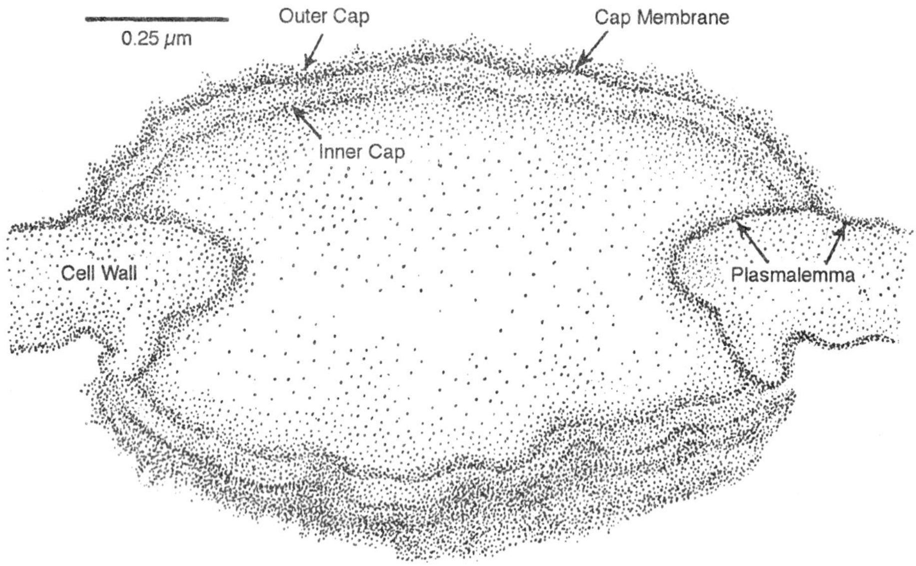

Fig. 9.5: A pit connection of *Palmaria mollis*. The plasma membrane is continuous from cell to cell. The cap membrane is continuous with the plasma membrane. The inner and outer cap layers are on each side of the cap membrane (From Pueschel, 1987)

blast, leaving the protoplasm continuous between the cells and allowing the passage of metabolites to the developing reproductive cells.

CALCIFICATION

All members of the Corallinales and some of the Nemaliales (*Liagora; Galaxaura*) deposit $CaCO_3$ extracellularly in the cell walls. Anhydrous calcium carbonate occurs in two crystalline forms, **calcite** (rhomboidal) and **aragonite** (orthorhombic). The two forms differ markedly in specific gravity, hardness, and solubility. The Corallinales deposit $CaCO_3$ as calcite, whereas the calcified members of the Nemaliales deposit $CaCO_3$ as aragonite. In *Liagora cenomyce* (Nemaliales) the aragonite occurs as needle-like crystals in the wall, whereas in the Corallinales the calcite occurs as massive deposits. Magnesium carbonate and strontium carbonate are commonly deposited along with calcium carbonate, there being less magnesium carbonate in aragonite than in calcite (Dixon, 1973). *Liagora* aragonite has about 1% magnesium carbonate, whereas in the Corallinaceae the calcite can contain 7% to 30% magnesium carbonate. Calcified walls of living cells probably have a mucilaginous component that shows the loss of Ca^{2+} into the medium. If a calcified thallus is killed, the dispersal of the calcified wall is greatly accelerated.

The coralline algae thrive in rock pools and on rocky shores exposed to very strong wave action and swift tidal currents. The red algae that have the highest rates of calcification also have the highest rates of photosynthesis and are usually found in waters less than 20 m deep (Goreau, 1963). Calcification of the thallus occurs about two to three times more rapidly in the light than in the dark, although significant calcification does occur in the dark. The above observations have led to the theory that calcification may be linked to photosynthesis (Pearse, 1972). The most quoted theory on calcification is that calcium salts are precipitated from seawater by the alkalinity brought about by the extraction of carbon dioxide during photosynthesis, calcium carbonate being less soluble in alkaline waters than acid. The obvious and often-mentioned drawback to this theory is that because all algae carry out photosynthesis, it is difficult to understand why they do not all calcify.

Sea-water is more or less saturated with respect to calcium carbonate, and the addition of either calcium or carbonate will cause the carbonate to precipitate. The concentration of CO_3^{2-} is related through a complex series of equilibria:

$$CO_2 + H_2O \rightleftharpoons H_2CO_3 \rightleftharpoons H^+ + HCO_3^- \rightleftharpoons 2H^+ + CO_3^{2-}$$

then

$$CO_3^{2-} + Ca^{2+} \rightleftharpoons CaCO_3 \text{ (ppt.)}.$$

The addition of acid will drive the reactions to the left and cause carbonate to dissolve, whereas the addition of base will drive the reactions to the right and form more carbonate. At the pH of seawater (8.4), almost all of the CO_2 in the water is in the form of bicarbonate ion, HCO_3^-. The addition of one equivalent of hydroxyl ions to sea-water saturated with respect to calcium carbonate will precipitate one equivalent of calcium carbonate:

$$Ca^{2+} + HCO_3^- + OH^- \rightleftharpoons CaCO_3 \text{ (ppt.)} + H_2O$$

The fact that sea-water is nearly saturated with calcium carbonate was demonstrated with sea-water from the coast of Maine by Digby (1977a). By raising the pH of this sea-water to 9.6, he caused precipitation of carbonates. Calcium carbonate precipitated first, being less soluble, followed by carbonate richer in magnesium.

The oxygen then diffuses out of the cell. As mentioned above, in the sea most of the carbon dioxide is in the form of bicarbonate ions; these ions diffuse into the cells and receive the electrons freed initially by photosynthesis. The bicarbonate ions are then converted into carbonate ions and hydrogen according to the following reaction:

$$2HCO_3^- + 2e^- \rightleftharpoons 2H^+ + 2CO_3^{2-}$$

The carbonate ions diffuse out of the cell where they partially dissociate, forming bicarbonate and hydroxyl ions and thereby raising the pH:

$$2CO_3^{2-} + H_2O \rightleftharpoons 2HCO_3^- + 2OH^-$$

When saturation with regard to calcium and carbonate is reached by a rise in pH outside the cells, calcium carbonate precipitates on the walls:

$$2Ca^{2+} + 2CO_3^{2-} = 2CaCO_3 \text{ (ppt.)}$$

REPRODUCTIVE STRUCTURES

The Rhodophycophyta have no flagellated cells or cells with any vestigial structure of flagellation such as basal bodies. In sexual reproduction **spermatia** are produced which are carried passively by water currents to the female organ, the **carpogonium** (Fig. 9.6A, C). The fertilized carpogonium produces **gonimoblast filaments** that form **carposporangia** and diploid **carpospores** (Figs. 9.6B and 9.7) at the tips. The carpospores produce the diploid **tetrasporophyte**, which subsequently gives rise to haploid tetraspores (Figs. 9.6D, E and 9.7). Advanced red algae form chiefly tetrahedral tetrasporangia (Fig. 9.6D) with large spores, whereas less advanced groups generally form cruciate or zonate tetrasporangia (Fig. 9.6D, E) with smaller spores (Ngan and Price, 1979). Tetraspores are generally larger than carpospores. The tetraspores complete the life cycle by germinating to form the gametophyte. Although this is the general life cycle of most Rhodophyceae, there are a number of modifications of it.

The post-fertilization events vary from one order to another. The more advanced orders have **auxiliary cells** with which the fertilized carpogonium fuses to form a multinucleate fusion cell. Papenfuss (1966) recognizes two types of auxiliary cells, nutritive and generative. **Nutritive auxiliary cells** provide nutrients for

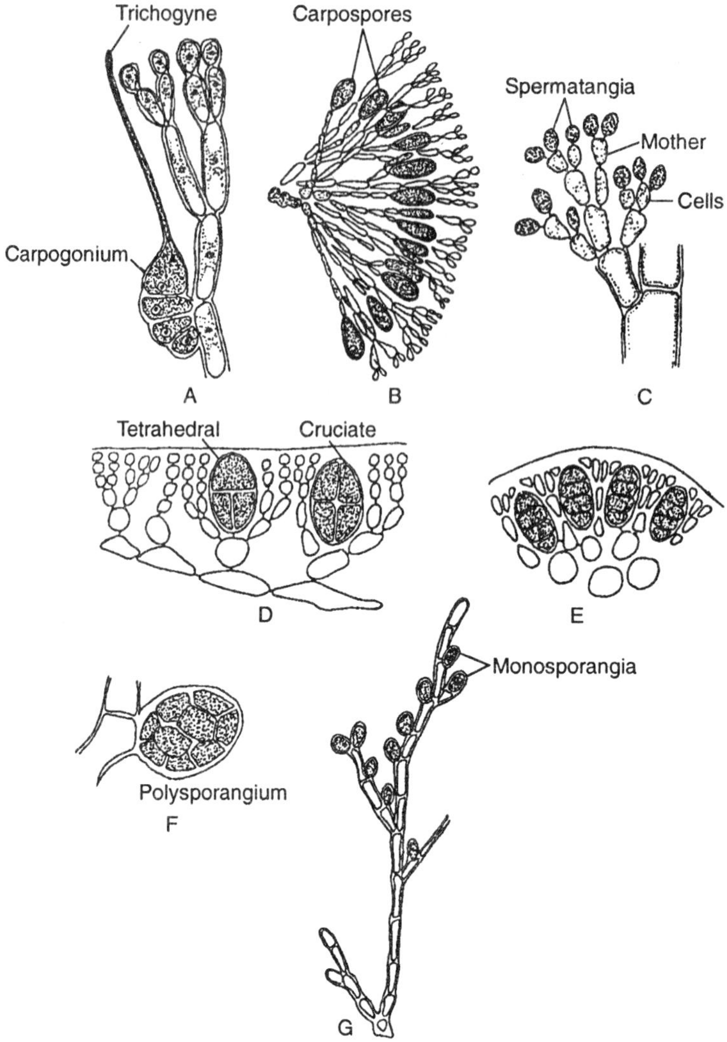

Fig. 9.6: A. A filament of *Liagora viscida* with a carpogonial branch. B. Gonimoblast filaments of *L. viscida* with carpospores. C. A spermatangial branch of *Acrochaetium corymbiferum* with spermatangia and spermatangial mother cells. D. Tetrahedral and cruciate tetrasporangia of *Nemastoma laingii.* E. Zonate tetrasporangia of *Hypnea musciformis.* F. Polysporangium of *Pleonosporium vancouverianum.* G. Monosporangia of *Kylinia rhipidandra.* A. B. C. E after Kylin, 1930; F. after Kylin, 1924; G. after Kylin, 1928)

the developing carposporophyte, whereas **generative auxiliary cells** give rise to gonimoblast filaments. The diploid tissue formed from the fertilized carpogonium forms the gonimoblast filaments. The gonimoblast filaments produce terminal carposporangia, which in turn form the carpospores. The carposporangia enlarge considerably during their maturation because of the development of the chloroplasts and the vesicles containing wall precursors. The pit connection between the carposporangium and the gonimoblast breaks before release of the carpospore. Also during the development of the gonimoblast filaments, the pit connections between the older gonimoblast cells usually dissolve.

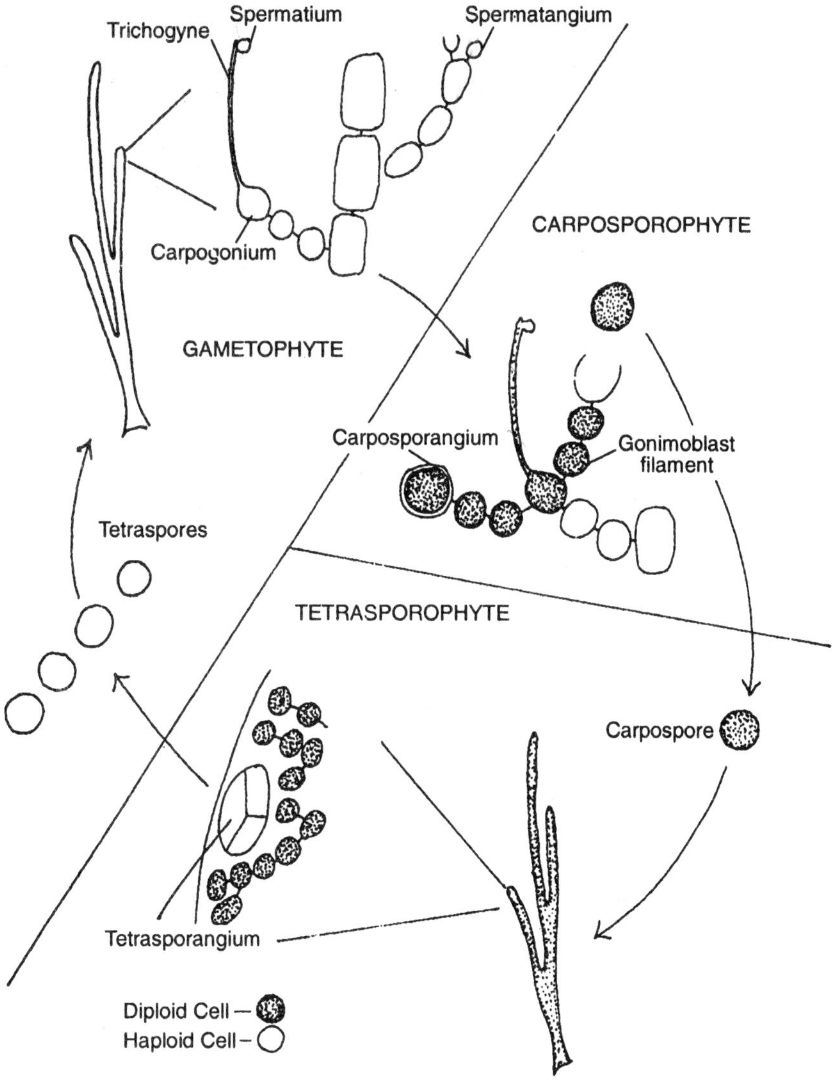

Fig. 9.7: Simplified life cycle of a typical red alga

CARPOGONIUM

The female organ, or carpogonium, consists of a dilated basal portion and a usually narrow gelatinous elongate tip, the **trichogyne,** which receives the male cells (Fig. 9.6A). Usually there are two nuclei in a carpogonium, one in the trichogyne, which degenerates soon after the *Carpogonium matures,* and one in the basal part of the carpogonium, which functions as the female gamate nucleus. In most Rhodophyceae the carpogonium terminates a short, often branched, three- to four-celled lateral called the *carpogonial branch.* The cell from which the carpogonial branch arises is the **supporting cell.** The carpogonium and carpogonial branch are commonly colourless, although in some Nemaliales this is not true.

SPERMATIUM

The spermatia of the Rhodophyceae are spherical or oblong cells produced in **spermatangia,** a single spermatium being formed in a

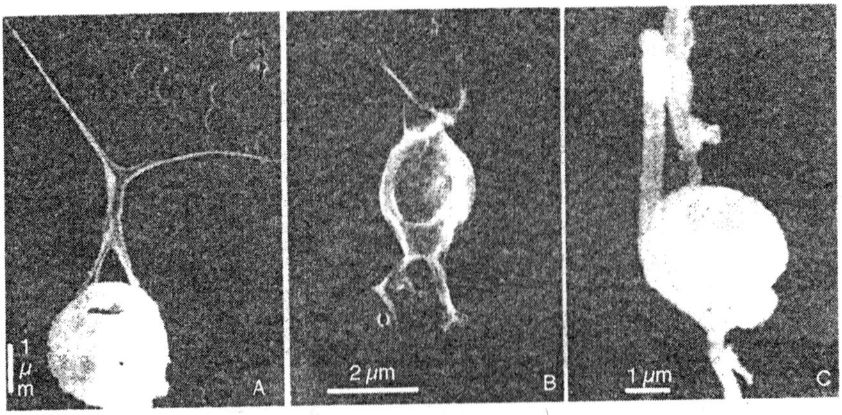

Fig. 9.8: Examples of spermatangial appendages. A. Appendage of *Antithamnion nipponicum*. B. Fimbriate cone-shaped appendages at each end of *Aglaothamnion neglectum*. C. Fibrous strands attached to spermatia of *Tiffaniella snyderae*. [A. from Kim and Fritz, 1993; B. from Magruder, 1984; C. from Fetter and Neushul, 1981]

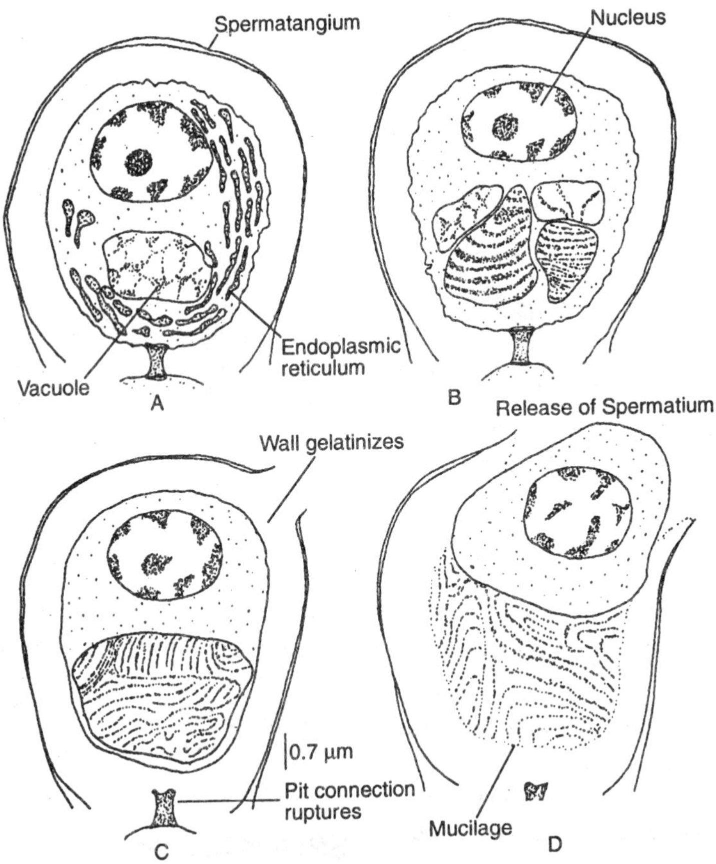

Fig. 9.9: A semi-diagrammatic drawing of the formation of a spermatium in the Rhodophyceae. A. Young spermatium. B. Formation of vacuoles containing fibrous material. C. Fusion of vacuoles, breaking of the pit connection, and gelatinization of the wall of the spermatangium. D. Extrusion of mucilage and release of the spermatium. (Adapted from Scott and Dixon, 1973; Kugrens, 1974)

spermatangium and then released, leaving the empty spermatangium (Fig. 9.9). The spermatangia (Fig. 9.8) are formed on spermatangial mother cells (Fig. 9.6C). The young spermatangia frequently have a pronounced polar orientation, with the nucleus in the apical portion and one or more vacuoles in the basal portion (Scott and Dixon, 1973A). As the spermatangium ages, vacuoles form in the basal area. These vacuoles contain fibrous materials (probably mucopolysaccharides) and make up half the volume of the spermatangium. Subsequently the vacuoles fuse to form one large vacuole. The spermatium is released by the gelatinization of the spermatangial wall near the apex and the concurrent release of the fibrous material in the basal vacuole. The fibrous material presumably swells and pushes the spermatium out of the spermatangium (Fig. 9.9). The fibrous material is sticky, and some of it adheres to the spermatium, thereby facilitating attachment to the trichogyne (Fig. 9.8C). During the development of the spermatium the pit connection with the spermatangium mother cell is severed. The mature spermatium is uninucleate, and wall-less but surrounded by mucilage, and may or may not contain functional chloroplasts.

FERTILIZATION

The spermatium usually is carried passively by water currents to the trichogyne of the carpgonium, although some spermatia can glide in a manner similar to the gliding in *Porphyridium*. At the point of contact the wall of the spermatium and the trichogyne dissolve, and the male nucleus moves into the carpogonium. Fusion of the male and female nuclei occurs in the basal portion of the carpogonium. The trichogyne will usually continue to grow until contact is made with a spermatium. After fertilization has occurred, the trichogyne becomes separated at its base from the rest of the carpogonium by the progressive thickening of the cell wall.

There is a relatively low statistical probability that the non-motile spermatia will be carried to receptive trichogynes of carpogonia. Species of red algae have evolved two mechanisms to increase this probability:

1. Some species have appendages on the spermatia that extend the reach of the spermatia five- to tenfold, resulting in a better chance of attaching to a receptive carpogonium. The appendages are initially contained within vesicles with the spermatia and unfold when the spermatia are released (Fig. 9.8).
2. The outer cell walls of the spermatia differ in their carbohydrate composition from the cell walls of the vegetative cells. The spermatia have α-D-methyl mannose residues on their surface and these residues bind specifically to the trichogynes of the carpogonia. Thus, vegetative cells may be contacted by spermatia as the spermatia are moved by water currents, but the spermatia will only bind to the trichogynes of the carpogonia.

MEIOSPORANGIA AND MEIOSPORES

Tetrasporangia, polysporangia, and bisporangia formed on diploid plants are generally regarded as being the seat of meiosis although there are exceptions to this. The **tetrasporangia** form four tetraspores either in a row (zonate), crosswise (cruciate), or most commonly in a tetrad (tetrahedral). In the formation of tetraspores, a wall is laid down inside the tetrasporangia by the protoplast, which has prominent dictyosomes (each dictyosome associated with a mitochondrion as is common in the Rhodophyceae. The tetraspores are not joined by pit connections.

In some of the red algae it is possible to control tetrasporogenesis by varying the light period, but there is no general rule that can be applied to the response. In *Rhodochorton purpureum* and *Acrochaetium asparagopsis*, tetrasporophytes produce tetrasporangia under short-day conditions. According to the physiological clock hypothesis, photoperiodism is controlled by an endogenous free-running **circadian** (*approximately 24-hour*) oscillation of some biochemical change. Each oscillation involves a regular alternation of two phases (each lasting approximately 12 hours) with a different sensitivity to light. The two phases are a **photophile** (light-loving) phase and a **skotophile** (dark-loving) phase. The initiation of a particular event depends on initiation or inhibition of metabolic changes of short-day or long-day organisms, respectively, by exposure to light at a particular point in the skotophile phase. In the case of *Rhodochorton purpureum* and *Acrochaetium asparagopsis,* light-breaks in the dark (skotophile) phase result in inhibition of tetrasporogenesis, whereas dark-breaks in the light (photophile) phase result in stimulation of tetrasporogenesis. Tetrasporogenesis in these algae is, therefore, a short-day phenomenon.

Polysporangia contain more than four spores, usually in multiples of four (Fig. 9.6f). Polysporangia probably evolve from tetrasporangia because polysporangia occur predominantly in the most advanced order, the Ceramiales. **Bisporangia** are probably reduced tetrasporangia in which cell division has resulted in two spores after meiosis instead of four.

ASEXUAL SPORES

Monosporangia (one spore per sporangium) and **parasporangia** (more than one spore per sporangium) produce asexual spores that reform the parent thallus. Monosporangia can be formed by the release of a vegetative cell from the thallus (*Goniotrichum, Asterocystis*), or by the formation of sessile one-celled branches (*Kylinia,* Fig. 9.6G).

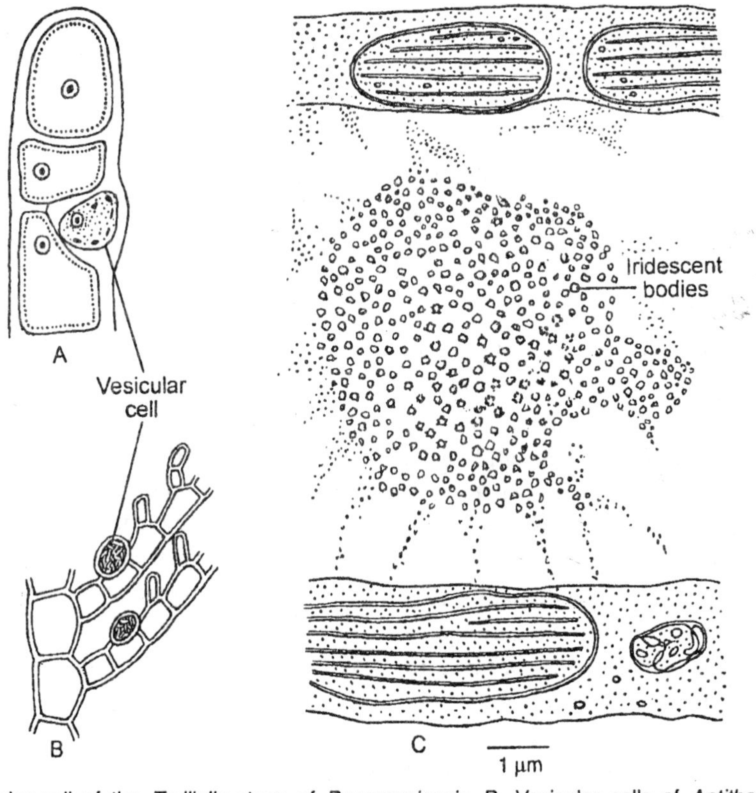

Fig. 9.10: A. Vesicular cell of the *Trailiella* stage of *Bonnemaisonia*. B. Vesicular cells of *Antithamnion plumula*. C. Iridescent bodies in the vacuole of a cell of *Chondria caerulescens*. [A. after Kylin, 1956; B. after Kylin, 1930; C. Adapted from Feldmann, 1970]

SECRETORY CELLS

Secretory cells (vesicular cells) occur in some Rhodophyceae (Fig. 9.10A, B). These cells are colourless at maturity and commonly have a large central vacuole. The secretory cells in *Bonnemaisonia* are prominent and associated with high concentrations of iodine (Fig. 9.10A). The concentration of iodine can be high enough to produce a blue colour in herbarium paper with starch as a filler (the chemical test for starch). Wolk (1968) has shown that if these algae are grown in an absence of bromine, then the secretory cells are vestigial, lacking the normal large vacuole with its refractile contents. Bromine can also occur as granular deposits in mucilage, such as in the mucilage of the thallus medulla in *Thysanocladia densa* (Pallaghy et al., 1983) or in the cuticle as in *Polysiphonia nigrescens* (Peders'en et al., 1981).

Other types of secretory cells not associated with the formation of halogens occur. The cells are often called secretory cells, even though they are apparently not involved in secretion. In *Antithamnion* (Figs. 9.10B and 9.11A), these cells have a large central vacuole containing sulfated acidic polysaccharide. In *Opuntiella californica,* there are "gland cells" with a large vacuole containing a homogeneous proteinaceous material (Young, 1979) (Fig. 9.11B). These "secretory cells" and "gland cells" may have compounds that act as deterrents to grazing, or they may accumulate special reserves for metabolic use.

IRIDESCENCE

The thalli of some Rhodophycophyta show a marked blue or green iridescence when observed in reflected light. Iridescence is solely a physical interference and is not related to any light-producing phenomena such as phosphorescence or bioluminescence. It results from the interference of light waves reflected from the surfaces of very thin multiple laminations separated by equally thin or thinner layers of material with a contrasting refractive index; the layers are uniform and produced by periodic

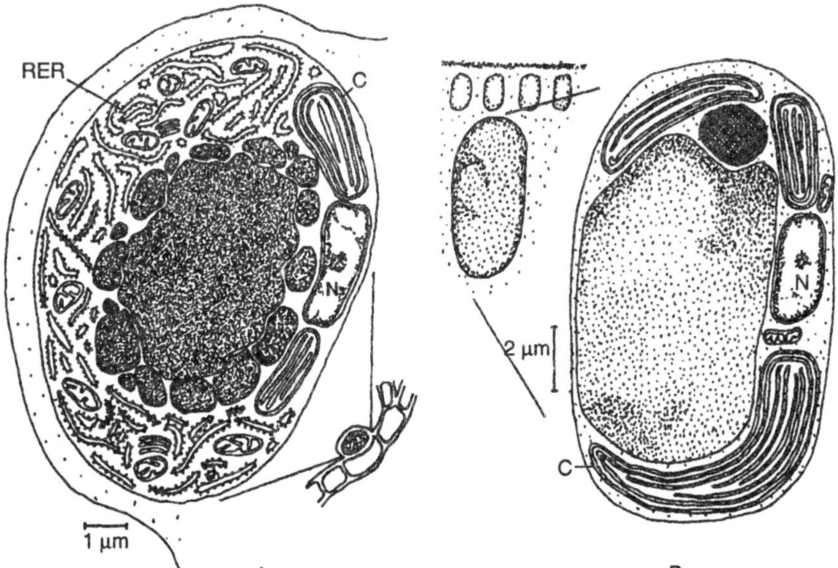

A B

Fig. 9.11: A. Semidiagrammatic drawing of the fine structure of a vesicular cell of *Antithamnion*. The cell has a large central vacuole surrounded by protoplasm containing rough endoplasmic reticulum (RER), mitochondria, chloroplasts (C), and a nucleus (N), B. Semidiagrammatic drawing of the fine structure of a gland cell of *Opuntiella californica*. The cell has a large central vacuole containing chloroplasts (C), a nucleus (N), and mitochondria, [A. after Young and West, 1979; B. after Young, 1979]

secretion and deposition. Iridescence in the Rhodophyceae has been attributed to different causes by different investigators.

EPIPHYTES AND PARASITES

Rhodophycean organisms range from autotrophic, independent plants to complete heterotrophic parasites. The spectrum includes non-obligate epiphytes (in the *Acrochaetium-Rhodochorton* complex), obligate epiphytes (*Polysiphonia lanosa* on *Ascophyllum*), semi-parasites that have some photosynthetic pigments (*Ceratocolax, Gonimophyllum*), and parasites with no colouration (*Harveyella, Holmsella*).

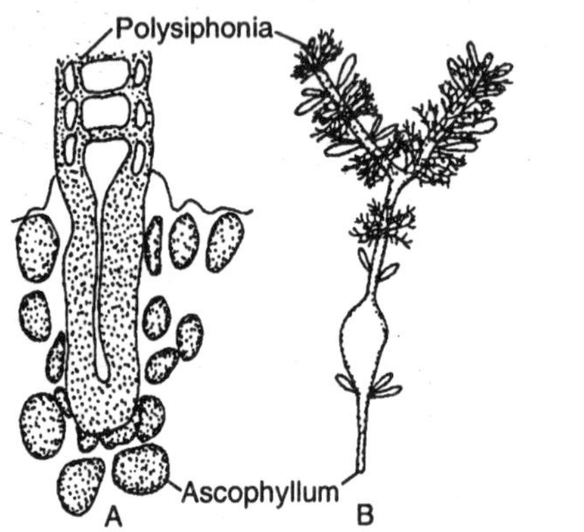

Fig. 9.12: A. The rhizoid of *Polysiphonia lanosa* penetrating tissue of *Ascophyllum nodosum*. B. *Polysiphonia* epiphytic on *Ascophyllum*. [A. after Rawlence, 1972]

The association between the obligate epiphyte red alga *P. lanosa* and its brown alga host *Ascophyllum* has been well studied. After the spore of *P. lanosa* germinates on the host, the red alga sends down a rhizoid that digests its way into the host tissue by means of enzymatic digestion of the host tissues. The enzymes are discharged from vesicles at the tip of the rhizoid. Once the rhizoid has established itself, intrusive cells form the basal parietal cells of the thallus (Fig. 9.12). Although *P. lanosa* is an obligate epiphyte, there is no transfer of metabolites from the host to the epiphyte, the epiphyte manufacturing all of its own requirements through photosynthesis.

Parasitic red algae can be either adelphoparasites (*adelpho* = brother) or alloparasites (*allo* = other). **Adelphoparasites** are closely related to, or belong to the same family as their hosts, and constitute 90% of parasitic red algae. **Alloparasites** are not closely related to their hosts. The parasitic habit apparently has been adapted more easily when the host is closely related to the parasite (adelphoparasites) than when it is not (alloparasites), partially because

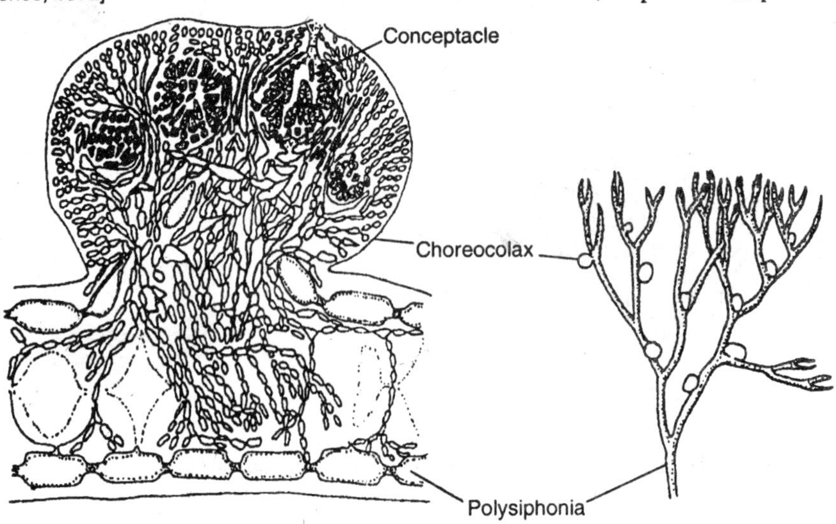

Fig. 9.13: Drawing of a section of *Choreocolax polysiphoniae* on *Polysiphonia*. The parasitic *Choreocolax* has carposporophytes in various stages of development (After Sturch, 1926)

it is easier for the parasite to establish secondary pit connections with the host (and therefore, transfer nutrients), if the host and parasite are related.

Choreocolax polysiphoniae is an example of a rhodophycean parasite (Fig. 9.13). The alga is a complete parasite and is interesting in that it is parasitic on *Polysiphonia fastigata,* which is itself epiphytic on *Ascophyllum.* Because *Choreocolax* is in the Gigartinales, and *Polysiphonia* is in the Ceramiales, this is a case of alloparasitism. *Choreocolax* consists of a more or less hemispherical while external portion is made up of subdichotomously branched filaments enclosed and surrounded by gelatinous matter, and a mass of haustorial cells growing inside the host (Sturch, 1926). When the haustorical cells have reached a certain distance from the original point of infection, they frequently give rise to a second external cushion. Secondary pit connections are established between the haustorial internal filaments of the parasite and the larger cells of the host. Apically dividing filaments of the parasite *Choreocolax* cells produce small conjunctor cells (Fig. 9.13) by the asymmetrical division of a cell. A conjunctor cell contains a highly condensed, small nucleus. The conjunctor cell fuses with an adjacent host *Polysiphonia* cell. The nucleus and cytoplasm of the conjunctor cell are incorporated into the host cell. The pit connection that has originally connected the conjunctor cell and the sister *Choreocolax* cell now connects the host *Polysiphonia* cell and the parasite. This is often called a "secondary pit connection", even though it is actually an ordinary pit connection. Up to several hundred *Choreocolax* cells can fuse with a single host cell. The infected host *Polysiphonia* cell enlarges and the vacuole decreases in size with a concomitant increase in cytoplasmic contents. The chloroplasts become scattered throughout the cytoplasm, instead of lying in the peripheral cytoplasm as they do in the non-infected cells. Although the parasite cell are colourless, they do contain very reduced plastids. In the association between the colourless parasite *Holmsella pachyderma* and the host *Gracilaria verrucosa,* both of which are Rhodophyceae, the main product of photosynthesis, floridoside, is transferred from the host to the parasite where it is accumulated as floridoside, mannitol, and starch.

COMMERCIAL UTILIZATION OF RED ALGAL MUCILAGES

The two most important polysaccharides derived from the Rhodophyceae are agar and carrageenin. **Agar** is defined pharmaceutically as a phycocolloid of red algal original that is insoluble in cold water but readily soluble in hot water; a 1.5% solution is clear and forms a solid and elastic gel on cooling to 32 to 39°C, not dissolving again at a temperature below 85°C. Agar is composed of two polysaccharides, agarose (Fig. 9.14) and agaropectin.

Agar is obtained commercially from species of *Gelidium* and *Pterocladia* as well as from various other algae, such as *Acanthopeltis, Ahnfeltia,* and *Gracilaria* (Therkelsen, 1993). These algae are often loosely referred to as **agarophytes.** Commercial production of agar was a world monopoly of the Japanese for many years, and even in 1939 Japan was still the major producer. Wartime demands in areas deprived of Japanese agar led to the development of agar industries in many of the Allied countries, some of which have continued and prospered while others have declined or disappeared. The agarophytes are collected by diving, dragging, or raking them offshore at low tide. In the traditional processing procedure the plants are then cleaned and bleached in the sun, with several washings in freshwater used to facilitate bleaching. The material is boiled for several hours, and the extract is acidified. This extract is then frozen and thawed. On thawing, water flows from the agar, carrying impurities

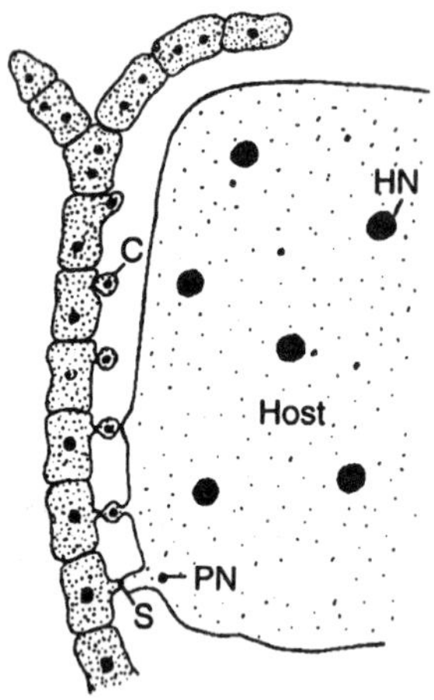

Fig. 9.14: A drawing illustrating the progressive formation of conjunctor cells by a filament of the parasitic red alga *Choreocolax* living on a host *Polysiphonia*. The parasite conjunctor cell fuses with the host *Polysiphonia* cell, leaving the pit connection of the conjunctor cell as a secondary pit connection. (C) Conjunctor cell; (HN) *Polysiphonia* host nucleus; (PN) *Choreocolax* parasite nucleus introduced into the host cell after fusion with the conjunctor cell of the oarasite; (S) Secondary pit connection (After Goff and Coleman, 1984)

with it. The agar that remains is dried and marketed as flakes or cakes. The more modern method extracts the agar under pressure in autoclaves. The agar is decolourized and deodorized with activated charcoal, filtered under pressure, and evaporated under reduced pressure. Further purification by freezing is then undertaken.

The greatest use of agar is in association with food preparation and technology, and in the pharmaceutical industry. It is used for gelling and thickening purposes, particularly in the canning of fish and meat, reducing the undesirable effects of the can and providing some protection against shaking of the product in transit. It is also used in the manufacture of processed cheese, mayonnaise, puddings, creams, and jellies. Pharmaceutically, agar is used as a laxative, but more frequently it serves as an inert carrier for drug products where slow release of the drug is required, as a stabilizer for emulsions, and as a constituent of cosmetic skin preparations, ointments, and lotions. The use of agar as a stiffening agent for growth media in bacteriology and mycology, which was its main use almost a century ago, is still responsible for a very considerable part of the demand.

Carrageenin is a phycocolloid similar to agar but with a higher ash content and requiring higher concentrations to form gels. It is composed of varying amounts of the principal components, κ-carrageenin and λ-carrageenans, both negatively charged high-molecular-weight polymers (Chiovitti et al., 1995; Therkelsen, 1993). κ-carrageenin is distributed throughout the wall while λ-carrageenin is localized to the cuticle (Vreeland et al., 1992). κ-carrageenin precipitates selectively from a cold, dilute solution in the presence of potassium ions. It forms a gel when heated and cooled with potassium ions and is, therefore, the gelling component. λ-carrageenin is the non-gelling component and is not precipitated or gelled by potassium. λ-carrageenin contains galactose-2, 6-disulfate, whereas κ-carrageenin contains 3, 6-anhydro-D-galactose. It has been shown in *Chondrus cripsus* and *Gigartina stellata* that the proportion of κ- and λ-carrageenin in the cell wall varies according to the ploidy of the plant. In the tetrasporophyte the amount of λ-carrageenan present is high as compared to the amount of κ-carrageenin, whereas just the opposite is true in the gametophyte. Such results may prove valuable in determining the ploidy of Rhodophyceae that have unknown life cycles.

Carrageenin is usually obtained from wild populations of Irish moss, the name for a mixture of *C. crispus* and the various species of *Gigartina*, particularly *G. stellata*. In the Philippines, *Eucheuma*, and in Vietnam, *Kappophycus* are extensively cultured as a source of carrageenan. Commercial extraction is similar to that for agar although carrageenin cannot be purified by freezing. The dried alga is washed with freshwater to reduce the salt content and then boiled with 2 to 4 parts of alga to 100 parts of water. The soluble carrageenin is separated from the insoluble residue in a centrifuge. Following filtration and some evaporation under vacuum, the carrageenin is dried on a rotary drier.

Carrageenans are used extensively for many of the same purposes as agar; however, because of their lower gel strength, carrageenans are used less for stiffening purposes than is agar, although for stabilization of emulsions in paints, cosmetics, and other pharmaceutical preparations carrageenans are preferred to agar. Also, for the stiffening of milk and dairy products, such as ice cream, carrageenans have supplanted agar completely in recent years, and it is in this area that demands for these products are the greatest. One particular use is for instant puddings, sauces, and creams, made possible by the gelling action, which does not require refrigeration.

Carrageenans inhibit human immunodeficiency virus (HIV) replication and reverse transcriptionase *in vitro* (in the test tube) (Bourgougnon et al., 1996). Replication of the HIV virus depends on interaction of a glycoprotein on the HIV virus envelope with a receptor on the target cells in the human body. The sulphated carrageenans prevent attachment of the HIV virus to the target cells. This occurs by the stronger negative R-O-SO_3^- groups on the carrageenin binding to a loop on the HIV molecule.

CLASSIFICATION

The presently accepted classification of the Rhodophyta having the single class Rhodophyceae is the system first devised by Schmitz (1883) and refined through years of painstaking work by Kylin (1932) and others to a form finally adopted and presented by Kylin in his monograph of 1956. Modifications have been made in this system in recent years and continue to be made as life histories are worked out and new relationships of the reproductive phases are recognized. On the other hand, some authors (Dixon, 1963) consider the system inadequate, unstable, and in need of complete revision.

The following is an outline of the distinctive characters of the classification of the class Rhodophyceae presently recognized by a majority of phycologists.

1. Cells with cytoplasmic connections; cell division rarely intercalary; carpospores produced in carposporangia borne on gonimoblast filaments 3

   Subclass **Florideophycidae**

   3. Life history usually without a free-living tetrasporophyte; or, if such present, then sexual and sporophytic generations usually heteromorphic

      Order **Nemalionales**

   3. Life history usually with free-living tetrasporophyte; sexual and sporophytic generations isomorphic 4

      4. Gonimoblasts growing directly from a carpogonium fusion cell

         Order **Gelidiales**

      4. Gonimoblasts growing mostly from an auxiliary cell 5

         5. Auxiliary cell an intercalary vegetative cell

            Order **Gigartinales**

         5. Auxiliary cell not an intercalary vegetative cell 6

            6. The auxiliary cell formed before fertilization 7

            6. The auxiliary cell formed after fertilization and cut off directly supporting cell of carpogonial filament

               Order **Ceramiales**

               7. Auxiliary cell in a special filament and usually not borne on the supporting cell of the carpogonial filament

                  Order **Cryptonemiales**

7. Auxiliary cell the terminal cell of a usually two-celled filament borne on the supporting cell of the carpogonial filament

Order **Rhodymeniales**

1. Cells usually without cytoplasmic connections; cell division intercalary; carpospores formed by direct division of the zygote

Subclass **Bangiophycidae**

2. Spore formation usually preceded by cell division; sexual reproduction present in most genera

Order **Bangiales**

2. Asexual reproduction only by monospores without cell division; sexual reproduction unknown

Order **Goniotrichales**

The Class Rhodophyceae is also conveniently divided as follows:

Subclass: **Florideae.** Well-developed thalli of filamentous branched to pseudoparenchymatous forms; growth of the thallus by apical cell; pit-connections invariably present between the cells; sex organs highly differentiated; carposporangia developing on filamentous gonimoblasts derived directly or indirectly from the zygote; meiosis immediately after the formation of diploid nucleus or delayed; chiefly marine. Include six orders.

Order: **Nemalionales.** Plant body heterotrichous with uni- or multiaxial construction; cells with a single chromatophore; asexual reproduction by monospores or tetraspores; male and female gametophytes present; sporophyte represented by zygote (fertilized carpogonial nucleus); zygotic meiosis; the fertilized carpogonium giving rise to branches which form the gonimoblasts; carposporangia are haploid and bear haploid carposopres; usually marine, but a few forms are freshwater.

Family: **Batrachospermaceae.** Soft, thick, gelatinous branched thallus being differentiated into structures resembling nodes and internodes; branches arise in whorls from the nodes; reproduction by monospores, carpogonia and spermatangia; fertilized carpogonial nucleus divides reductionally producing short gonimoblast filaments from which carposporangia and haploid carpospores are developed.

Genus: *Batrachospermum*

Order: **Gelidiales.** Small order with only a few genera in two families, but some of the species are common and very widely distributed, some of considerable economic importance; show an alternation of portion of sporophyte and gametophyte plants; sexual plants dioecious and distinguishable only when fertile; most members of the order plants of moderate size, of firm, cartilaginous consistency, of uniaxial construction, and of monopodial branching, carpogonial filaments single-celled and deeply embedded in the cortex of very young thallus parts; gonimoblasts developing directly from a multinucleate fusion cell formed from the carpogonium alone or as a result of the fusion of the carpogonium and certain neighbouring cells.

Order: **Gigartinales.** Diversified having an assemblage of algae as the Cryptonemiales, but, with a few exceptions, the members being fleshy and non-calcareous forms; plants of monoaxial or multiaxial construction and of cylindrical to foliaceous forms; only a few prostrate or crustose, and none filamentous or microscopic.

According to Kylin's system, the order is technically distinguished by the auxiliary cell being an intercalary vegetative cell and not a cell borne in a special filament. It should be pointed out, however, that in some Gigartinales the supporting cell may function as an auxiliary cell and that this presents a seeming contradiction. A parallel situation obtains in the Cryptonemiales and makes these ordinal distinctions extremely obscure.

Order: **Ceramiales.** Uni- to multi-axial construction of thallus; male and female gametophytes alternating with sporophyte; sporophyte producing tetrasporangia and tetraspores; carpogonial branch with procarp; presence of auxiliary cells being cut off from the basal cell of the carpogonial branch after fertilization; diploid carpospores; meiosis during tetraspore development.

Family: **Ceramiaceae.** Thallus formed of monosiphonous filaments, naked or corticated by secondary filaments developed at the nodes and adhering to the main axis; procarp exterior to the thallus; cystocarp often formed of gonimoblast; tetrasporangia usually tetrahedral and exterior.

Genera: *Ceramium, Callithamnion*

Family: **Rhodomelaceae.** Polysiphonous thallus with basal filaments attached to the substrate by small flattened disks; laterally or dichotomously, branched thallus; spermatangia and procarps developed exterior to the main axis; cystocarps urn-shaped with a pore, usually shortly pedicellate; tetrasporangia with tetraspores.

Genus: *Polysiphonia.*

Order: **Cryptonemiales.** Diversified in forms, structure, and substance than any other group; gelatinous, fleshy, cartilaginous and calcareous types in remarkable array; all of these having the auxiliary cell formed before fertilization and borne on a special filament of the gametophyte, the auxiliary cell filament usually resembling the carpogonial filament in its dense cytoplasm and may be distinguished before fertilization, it may be borne far removed from the carpogonial filament, or both may rise from a common cell.

Order: **Rhodymeniales.** Characteristic multiaxial growth of thallus; a three- or four-celled carpogonial branch; the presence of a procarp with one or occasionally two auxiliary cells not part of the vegetative system; auxiliary cell branch (produced by the supporting cell usually composed of two cells) initiated before fertilization; the mature tetrasporangia either cruciately or tetrahedrally divided; the life cycle almost always diplobiontic.

Family: **Champiaceae.** Most of the thallus, or at least the reproductive part hollow and with longitudinal filaments bordering the cavity; tetrasporangia generally tetrahedrally divided, but some genera with polysporangia.

Family: **Rhodymeniaceae.** Thallus solid or hollow, but if hollow, longitudinal filaments lacking; tetrasporangia generally cruciately divided.

Subclass: **Bangioideae.** Simple thallus with diffuse growth never exhibiting aggregation of filaments; pit-connections between cells of the thallus lacking; plants either entirely asexual or consisting of monoecious or dioecious gametophytes; the zygote representing the sporophyte; sex organs little specialized; carposporangia formed by direct division of the zygote; terrestrial, freshwater, and marine. Include one order.

Order: **Bangiales.** Cells forming the thallus isolated or united into filamentous or membranous colonies, of one or two cell layers; cells usually similar in form, with intercalary growth, obvious cytoplasmic continuity never established; red, violet, yellowish or occasionally bluish-green; plastids usually stellate. Vegetative reproduction by cell division in unicellular forms, by fragmentation in multicellular forms; asexual reproduction by non-motile spores, produced in an undivided vegetative cell or in a portion cut off by a curved wall; sexual reproduction by spermatangia and carpogonia, formed in the cells of the thallus; monoecious or dioecious, fertilization resulting in the producing of carpospores.

Family: **Erythrotrichiaceae.** Plant body filamentous branched or unbranched or laminate to cylindrical thalloid structure; reproduction by the development of large and small spores formed by the division of vegetative cells.

Genus: *Compsopogon.*

Order: **Goniotrichales.** Microscopic, epiphytic representative of this order common; it occurs on a wide variety of hosts and almost cosmopolitan; readily recognizable by its branched filaments of seriate cells with very thick, gelatinous walls; only the simplest kind of asexual reproduction known; the vegetative cells directly transformed into monospores freed by dissolution or rupture of the adjoining gelatinous walls.

PHYLOGENETIC RELATIONSHIPS OF THE RHODOPHYCOPHYTA

The Rhodophycophyta are unique among the algae for having specialized pigments (phycoerythrin), reserve food (floridean starch), characteristic female reproductive organ (carpogonium), absence of any flagellate structure, and the sexual process. These algae with fossil record extending back to the Ordovician, possibly arose independently from non-flagellate unicellular organisms.

The Rhodophycophyta resemble no other algae, except the Cyanophycophyta which also lack flagellate structure and possess phycobilin pigments. Furthermore, floridean starch, the characteristic carbohydrate reserve of the red algae, is chemically allied to cyanophycean starch, the characteristic carbohydrate reserve of the blue-green algae. As such many of the phycologists including Kylin (1943) suggested affinity of the Rhodophycophyta with the Cyanophycophyta. In spite of these points of similarity there are very many fundamental points of difference between the Cyanophycophyta and the Rhodophycophyta. They are: the presence of a nucleus, in having definite plastids, in having chlorophyll *d* in addition to chlorophyll *a*, in the usually more complex structure of the thallus, and sexual reproduction with often very complex life cycle. All these features of the Rhodophycophyta may be regarded as evolutionary advances over the condition in the Cyanophycophyta. Again within the Rhodophycophyta, the Bangoideae are less highly modified than the Florideae, and some of the genera of Bangio-ideae which have cyanophycean cell structure and lack sexual processes are very suggestive for being closely allied to the Cyanophycophyta. It is possibly justifiable to consider both Cyanophycophyta and Rhodophycophyta as living representatives of evolutionary lines of descent which have diverged considerably from the same common ancestor in the extremely distant past (Fig. 9.15).

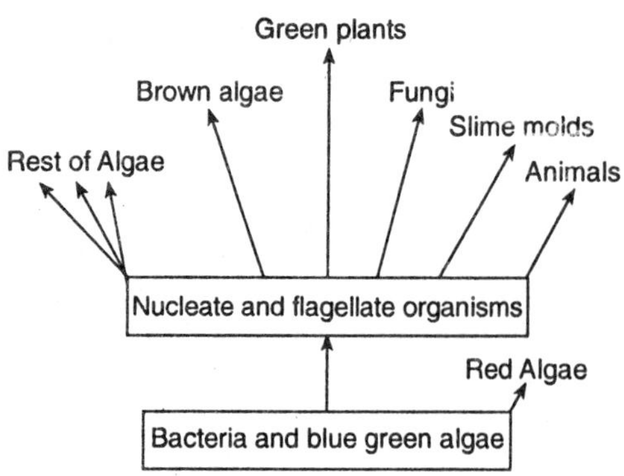

Fig. 9.15: Diagram showing the possible origin of and relationships between the Algae, Fungi, Bacteria and Slime molds

Some consider Rhodophycophyta to have derived from a member of the Chlorophycophyta. But the structure of the plant body and pigment composition in both these taxa are so different from each other that it is difficult to justify such an evolutionary line. But in *Prasiola,* the structure of chloroplast and the similarity in morphological structure and reproduction make it possible to suggest as an ancestor of the Rhodophycophyta.

Two alternatives have been suggested by Klein and Cronquist (1967) about the ancestry of the Rhodophycophyta: (i) they originated from some simple procaryotic blue-green algae, and (ii) they were derived from some archaic eucaryotic algae which themselves originated from blue-green algae. But the fundamental differences between the eucaryotic and procaryotic cellular organizations rule out the existence of any direct link between the red algae and the blue-green algae. Further, the biliproteins of Rhodophycophyta differ in spectral properties from those of Cyanophycophyta. It may be speculated that the red algae probably originated from some primitive non-flagellate eucaryotic ancestor possessing biliproteins.

Order: *NEMALIONALES*

Plants are filamentous, the free uniseriate filaments creeping or erect, or the plants much larger and corticated, developing relatively soft branches with either uniaxial filament or the multiaxial type of structure. Cells are uninucleate, with axial or lateral chromatophores. Asexual reproduction takes place by monosporangia, very exceptionally by bi- or tetrasporangia; sexual reproduction by spermatia in spermatangia, formed (in the larger species) from the surface vegetative cells, and carpogonia borne on few-celled carpogonial branches. The zygote itself produces the carpospores either directly or after establishing connections with nutritive cells on the carpogonial branch.

Family: *BATRACHOSPERMACEAE*

This family includes algae whose plant body is of soft, thick, gelatinous branched thallus differentiated into structures resembling nodes and internodes. Branches arise in whorls from the nodes. Asexual reproduction is by monospores and sexual reproduction oogamous by the development of carpogonia and spermatangia. Zygotic nucleus divides meiotically immediately after development. This is followed by the formation of gonimoblast filaments from which carposporangia and carpospores are developed in a cystocarp.

Genus: *Batrachospermum*

Batrachospermum, commonly known as 'frog-spawn' alga, has a very wide range of distribution both in temperate and tropical regions. Generally it prefers cool, shady and well-aerated situations in comparatively slowly flowing streams and margins of lakes. There are species which occur in rapidly flowing streams being attached to stones and similar other submerged objects. Most of the species are annuals but a few are perennial. The plant body is of gelatinous texture appearing as chains of delicate beads (Fig. 9.16A). The primary axis which grows from an apical cell consists of a large number of elongated cylindrical cells placed end to end in a single row exhibiting an example of simple uniaxial construction. Numerous branches of both limited and unlimited growth arise in whorls from the upper ends of the cells just below the septa. This peculiar arrangement of the branches has resulted the main axis to be differentiated into nodes and internodes (Fig. 9.16B). The basal cells of the lateral branches grow into narrow threads (Fig. 9.16F). These threads grow downward forming an envelope around the main axis which is completely concealed assuming a corticated appearance. Cells are uninucleate containing a number of pariental, lobed chromatophores with pyrenoids, *Batrachospermum* possesses **haplobiontic** life cycle.

Batrachospermum reproduces both asexually and sexually. Flagellate cells are completely absent in the life cycle of this alga. Asexual reproduction is brought about by means of non-flagellate uninucleate spores known as **monospores**, developed singly within **monosporangia** which are again developed on the erect portion of the heterotrichous filament called **Chantransia stage** that is produced during the post-fertilization stage of the sexual reproduction (Fig. 9.16C-E).

Sexual reproduction is a case of advanced oogamous condition. Thallus may be monoecious or dioecious depending on species. The **carpogonia** (female sex organs) are borne terminally on small branches of the thallus.

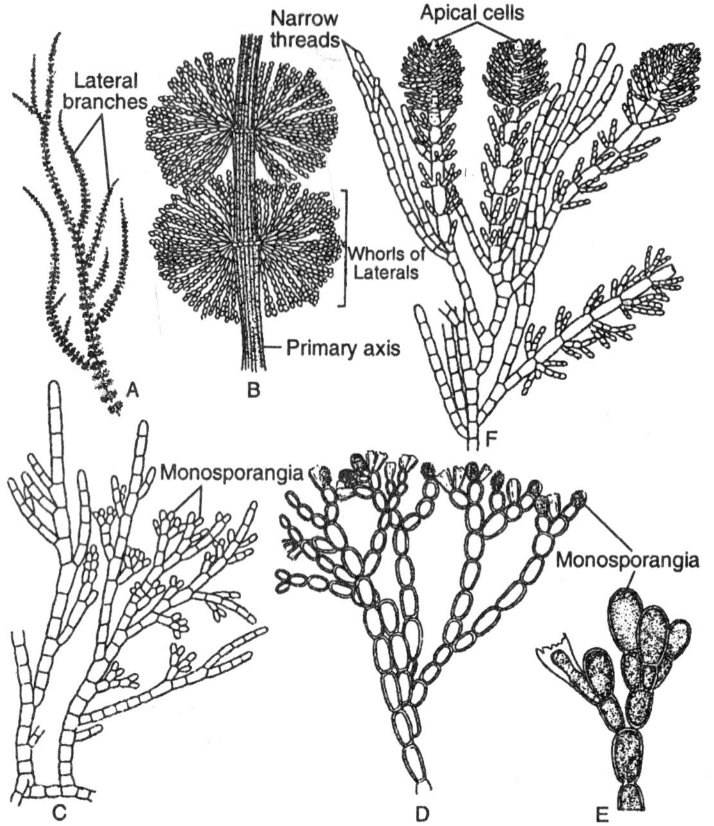

Fig. 9.16: *Batrachospermum* sp. A. Habit, B. Two whorls enlarged, C. Adult shoots with monosporangia. D-E. Monosporangia. F. Adult shoots

The carpogonium at its upper end is prolonged into a **trichogyne** which shrivels away after fertilization (Fig. 9.17A-F). The nucleus, chromatophores and the reserve food material are located in the lower swollen portion of the carpogonium.

Antheridia or **spermatangia** (male sex organs) are single-celled spherical structures borne in clusters at the tip of the lateral branches of the thallus (Fig. 9.17B). Contents of each antheridium become metamorphosed into a non-motile, spherical male gamete or **spermatium,** which is liberated through a narrow apical slit of the antheridial wall. The spermatium thus liberated floats in water and is finally carried away by water current to come in contact with the trichogyne of the carpogonium. The walls dissolve at the point of contact of the spermatium and the trichogyne and the contents of the speranatium pass through the open passage and move down the trichogyme into the base of the carpogonium where both male and female nuclei fuse together (Fig. 9.17C-E). After fertilization the trichogyne is separated from the carpogonium by a mucilage plug and finally shrivels and disappears (Fig. 9.17F). The diploid carpogonial nucleus divides meiotically into two haploid nuclei. Simultaneously with the nuclear division a lateral protuberance is developed from the carpogonium. One of the two daughter nuclei moves into this protuberance, the other remaining in the carpogonium. The protuberance is cut off from the carpogonium by a wall and is known as **gonimoblast initial** (Fig. 9.17G). Then the daughter nucleus of the carpogonium divides mitotically along with the formation of another protuberance on the other side of the carpogonium in which the daughter nucleus migrates and ultimately a second gonimoblast initial is produced (Fig. 9.17H) and like this several gonimoblast initials are produced which by repeated divisions form branched or unbranched filaments known as **gonimoblast filaments** (Fig. 9.17I and J). The terminal cells of the gonimoblast filaments become enlarged, each one developing into a **carposporangium**, the contents of which become metamorphosed into a single, non-motile **carpospore.** Along with the development of the carposporangium and carpospore numerous sterile threads are developed from the cells below the carpogonium which ultimately envelope the gonimoblast filaments. These threads are known as the **enveloping threads** (Fig. 9.17I). The structure so formed with gonimoblast filaments sur-

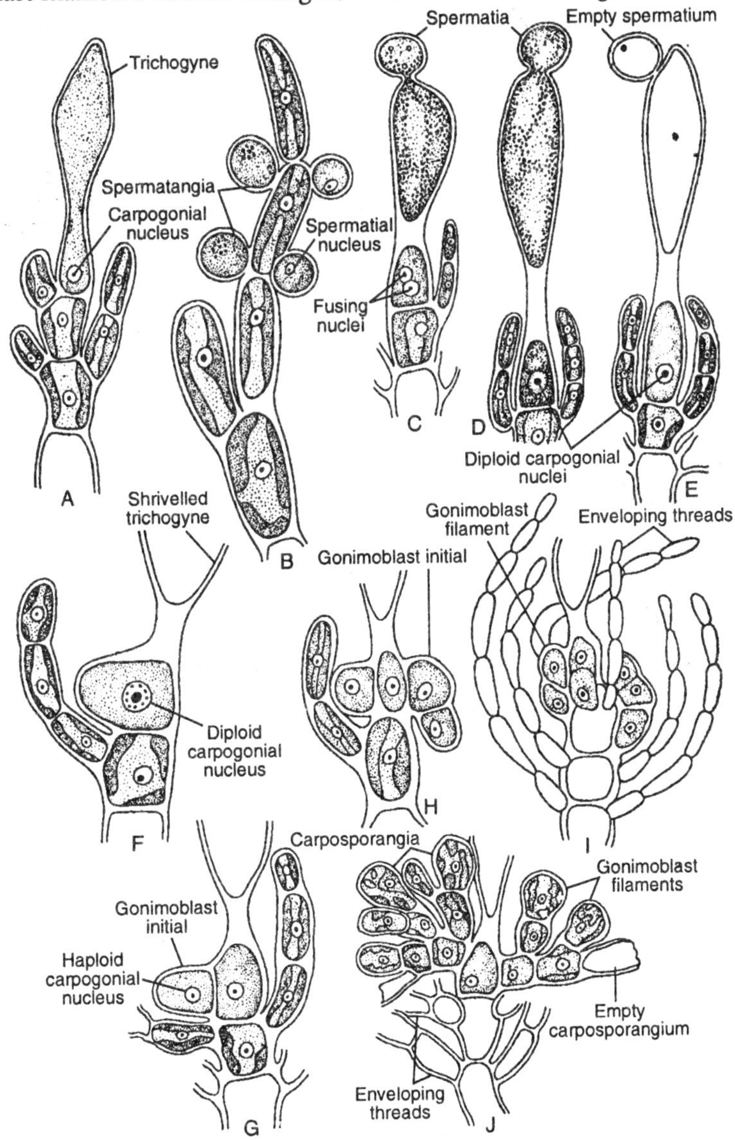

Fig. 9.17: *Batrachospermum* sp. A. Mature carpogonium. B. Spermatangial branch bearing spermatangia. C-F. Stages in fertilization. G-H. Development of gonimoblast initial. I. Development of gonimoblast filament and enveloping threads. J. Mature gonimoblast filaments, carposporangia and enveloping threads

rounded by the enveloping threads is the **cystocarp.** The carpospore is liberated from the carposporangium (Fig. 9.17J) and germinates into a heterotrichous filament which morphologically is quite different from the main thallus. The heterotrichous filament is known as **Chantransia stage** or **juvenile stage.** For long time this heterotrichous filament was considered as a new genus *Chantransia.* But later on it was found that the apical cells of the lower branches of the erect threads develop into new *Batrachospermum* thallus and the so-called *Chantransia* is nothing but a stage in the life cycle of the alga *Batrachospermum.* Finally it was linked up in the life cycle of *Batrachospermum* naming as **Chantransia stage** (Fig. 9.18).

The life cycle of *Batrachospermum* consists of two gametophytic phases alternating with one sporophytic phase which is, however, confined in the zygotic stage. The early gametophytic phase—Chantransia stage, is formed by the germination of the carpospore which is the product of post-fertilization stages. The late gametophytic phase, the main *Batrachosper-*

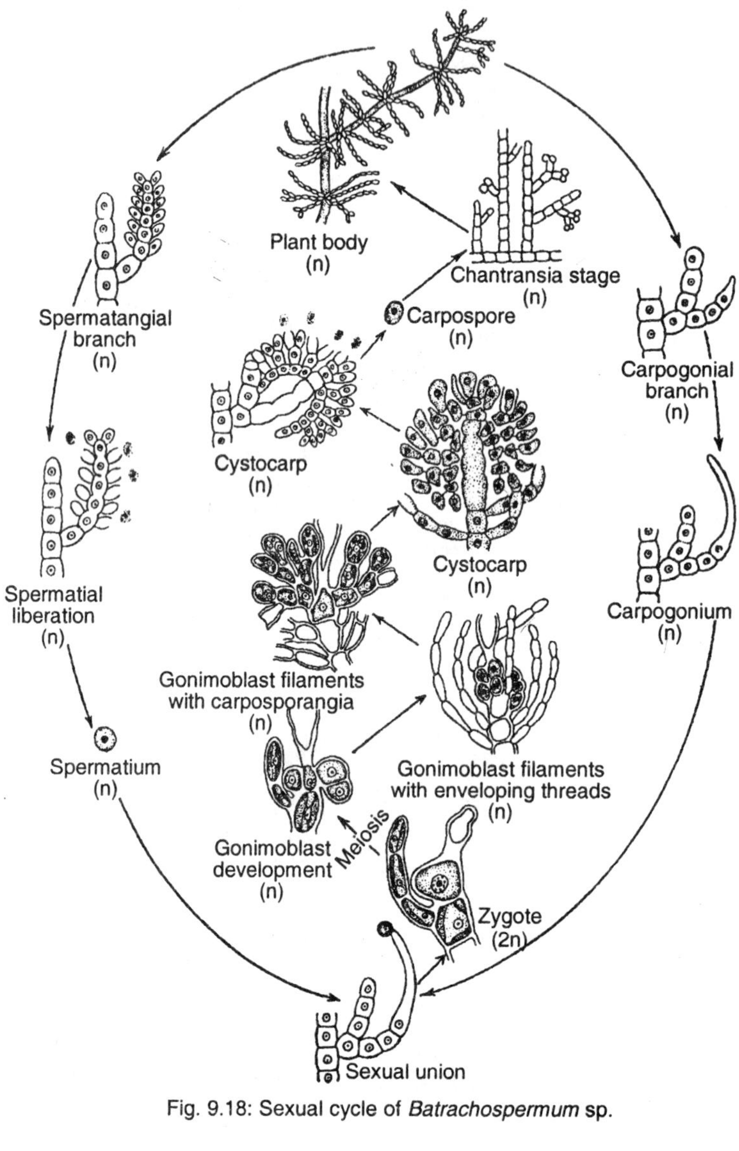

Fig. 9.18: Sexual cycle of *Batrachospermum* sp.

mum plant is developed from the Chantransia stage.

Special features:

1. Complete absence of any flagellate cell.
2. Structure of the carpogonium, the process of fertilization.
3. Advanced oogamy followed by elaborate post-fertilization stages with the production of carposprangium and carpospore in a special structure called *cystocarp*. All these structures are haploid.
4. The alga has attained great advancement but the zygotic meiosis has at the same time kept it far down in the evolutionary level.

Order: *CERAMIALES*

Plants are generally slenderly filamentous and branched, sometimes coarse, strap-shaped, or membranous; naked or corticated, with the central-filament type of structure developing from an apical cell. Asexual reproduction is by tetraspores formed in sporangia external or more or less covered by the cortex of ordinary branches, or grouped in specialized sporiferous branchlets or **stichidia.** Sexual reproduction takes place by spermatangia borne on the axial filaments, often in masses, or covering areas of the flat blade-like plants, or covering special colourless branchlets (spermatangial clusters or antheridia), and by carpogonia on carpogenic branches borne, together with sterile cells, on the axial or the pericentral cells; auxiliary cells formed after fertilization from the cell supporting the carpogenic branch; gonimoblasts developed from the auxiliaries to form a mass of carposporangia, which may be naked, partly enveloped by branchlets, or covered by a **pericarp.**

Family: *CERAMIACEAE*

Plants are usually bushy; branches usually terete, in some genera uniseriate, in others corticated; growth from the tip producing an axial row of cells, true colourless hairs often present, also hyaline hair-like extensions from the tips of branches; cortication, if present, usually developed first about the nodes, consisting of single ring of cells, or spreading over the internodes in more or less filamentous fashion, later appearing parenchymatous, or else consisting of an investment of rhizoidal filaments. Sporangia are superficial or stalked on the branches, single or whorled at the nodes, or in corticated species in the cortex, ultimately dividing in tetrapartite or tetrahedral fashion. Spermatangia are developed on special determinate branchlets, forming small colourless clusters, or covering the cortex portions of larger species. Carpogenic branches are of four cells, borne on supporting cells, which may also give rise to sterile cells and after fertilization to one or two auxiliaries near the carpogonium. Cystocarp is composed of groups of gonimoblast filaments, the outer cells of which produce the carposporangia; cystocarps are naked, enveloped in jelly, or partly enclosed by subtending filaments from below

Genus: *Ceramium*

Ceramium occurs very commonly between tide levels and in deeper water being especially abundant along the Mediterranean coast. Some species may also thrive well in river water. The

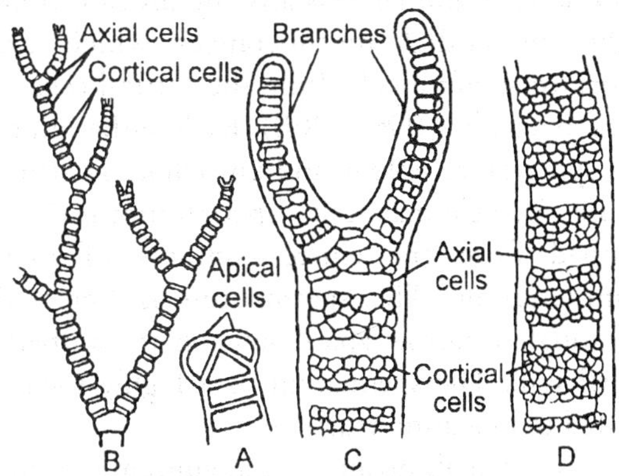

Fig. 9.19: *Ceramium* sp. A. Apex of a branch system. B. Branch system showing tong-like forking of the apical region. C. Thallus-branch showing dichotomy. D. Structure of a mature thallus-branch

richly branched uniaxial plant body is recognizable from the banded appearance and the tong-like forking of the apical region of the branches. These are considered very valuable diagnostic characters (Figs. 9.19B and 9.20A, C and F). The thallus, when young, is attached to the substratum by a cushion-like structure, which with maturity develops into rhizoids. The typical dichotomous branching is initiated by the forking of the dome-shaped apical cell (Figs. 9.19A and C). The plant body is made up of a single row of large cylindrical or barrel-shaped cells arranged one upon the other. The banded appearance of the filaments is developed as a result of the formation of an envelop of small-celled cortical cells cut out from the central axial cell (Fig. 9.19D). This development of the envelope is not continuous throughout the entire surface for which there results the banded appearance of the filaments. The colour of the plant body is extremely variable, may be red, yellowish-green, or reddish-brown, depending on whether a particular filament is inhabiting in deep or shallow water. *Ceramium* possesses specialized **diplobiontic** forms in the life cycle.

Ceramium has both monoecious and dioecious species. The antheridia occur in clusters on the upper side of the lateral branchlets. The antheridial mother cells arise by division from the antheridial protuberances which are formed subterminally from the cortical cells of the branchlets (Fig. 9.20B). Each antheridium is spherical to oblong and unicellular in structure. A single male cell—**spermatium** is developed from the entire contents of each antheridium. The spermatium is liberated through a narrow apical split on the antheridial wall. The spermatium is a highly vacuolated uninucleate structure.

The female organ or **carpogonium** is borne terminally on a short lateral branch known as **carpogonial branch.** The carpogonium consists of a swollen basal portion in which lies the female nucleus and a narrow elongated structure, the **trichogyne** which serves as a receptive organ for the spermatium. An open passage for the male nucleus is developed due to dissolution of the walls at the point of contact of the spermatium and the trichogyne. The male nucleus travels down to the female nucleus in the basal portion of the carpogonium. In the mean time the trichogyne becomes separated from the rest of the carpogonium by the development of a membrane at its base. After fertilization an **auxiliary cell** with dense content is formed. Connection between the carpogonial base and auxiliary cell is maintained by the development of a connecting cell which is very short. The entire structure is known as **procarp.** The diploid nucleus now passes in the auxiliary cell. A gonimoblast initially originates from the upper side of the auxilary cell which gives rise to several tufts of threads in which all the cells, except the basal one, develop into **carposporangia.** The carposporangia are surrounded by a loose envelope of sterile branches, the entire structure developing into a cystocarp (Figs. 9.20C-E). A carpospore on liberation from a carposporangium germinates to produce a diploid **tetrasporic plant** which morphologically resembles the gametophytic individual except that it bears **tetrasporangia** forming **tetraspores** (Fig. 9.20F). In the tetrasporic plants, tetrasporangia are developed in clusters from the enlarged cells of the cortical bands within which they remain partly embedded (Figs. 9.20G-J). In each tetrasporangium four tetraspores are formed as a result of the meiotic division of the diploid nucleus of the tetraspore mother cell (Figs. 9.20K-M). The tetraspores germinate to produce gametophytes.

The life cycle of *Ceramium* includes a haploid gametophytic phase which bears sex organs and two diploid asexual phases—the carposporangia producing carpospores and the tetrasporic plant. These are the three successive phases that follow in regular sequence comprising isomorphic alteration of generations exhibited by the sexual and the tetrasporic individuals.

In some species the development of special spores—**polyspores** inside **polysporangia** has been recorded. They bear polyspores numbering mostly four or multiples of four. The polysporangia appear as large spherical structures projecting from the cortical bands.

Some Indian Species: *Ceramium cruciatum* Collins et Harv.; *C. elegans* Ducl.; *C. maryae* Web.; *C. strictum* Grev. et Harv.; *C. subdichotomum* Web.; *C. trasversale* Collins et Harv.; *C. truncatum* H. & P.

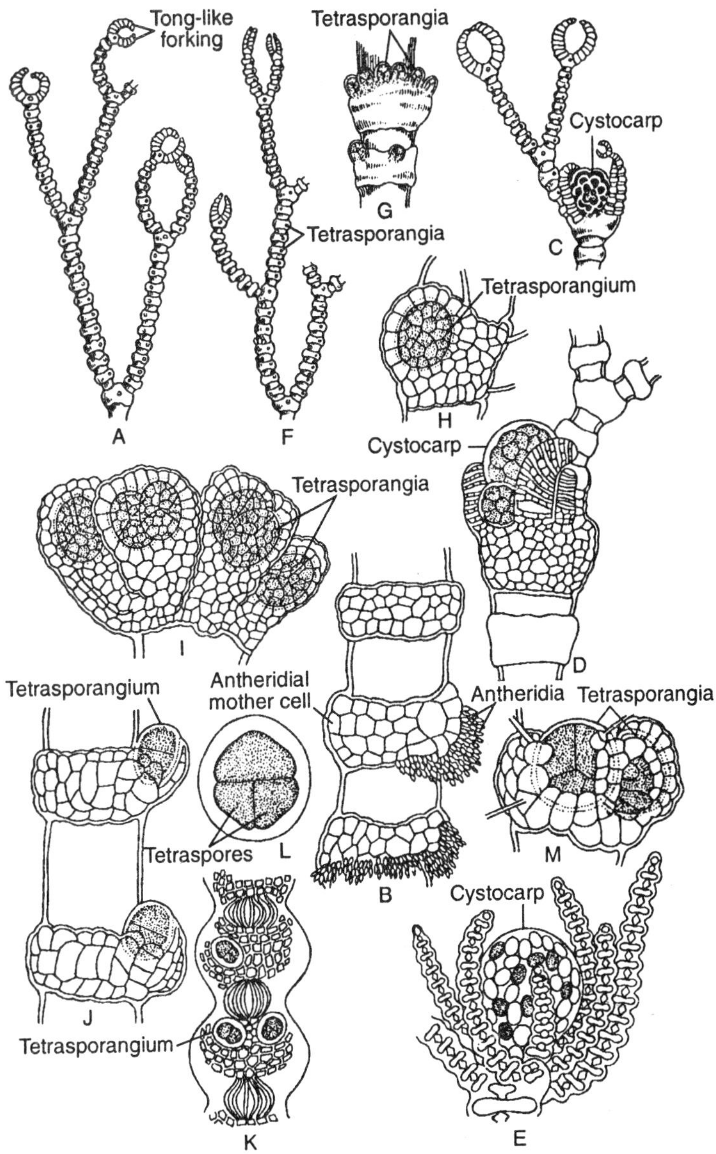

Fig. 9.20: *Ceramium* sp. A. Branch-system showing tong-like forking of the apical region. B. Part of the thallus showing antherídal mother cell and antheridia. C. Dichotomous branching of young threads bearing cystocarp. D. Portion of thallus with cystocarp. E. Cystocarp. F. Distribution of tetrasporangia on tetrasporic plant. G. Mature tetrasporangia. H-J. Development of tetrasporangia. K. Tetrasporangium with tetraspores. L. Tetraspores. M. Tetrasporangia

Special features:

1. Richly branched uniaxial plant body possessing banded appearance.
2. Tong-like forking of apical region of the branches.
3. Isomorphic alternation of generations.
4. Sexual reproduction advanced oogamy.
5. Elaborate post-fertilization stages.
6. Carposporangium and carpospores diploid.
7. Meiosis during the development of tetraspores in the tetrasporangium borne on sporophytic tetrasporic plant.
8. Triphasic life cycle.

Genus: *Callithamnion*

This is a marine alga possessing a delicate body of richly branched tufts of filaments (Fig. 9.21A). It attaches itself to the mud-covered rocks near the low water-level with the help of multicellular rhizoidal filaments emerging from the lower part of the main axis. A similar condition is encountered in epiphytic species. There are species which persist more than one season, others are mostly annual. The main axis of the plant body is composed of thick-walled elongate cells bearing branches (Fig. 9.21B). The branches develop from just below the septa of the cells of the main axis. In a mature plant the development of the branches shows somewhat pseudo-dichotomous arrangement which may be due to some irregularity in the development of the cells of the main axis. Certain species at times produce a profusion of long slender hairs on the ultimate brunches (Fig. 9.21B). The filaments of the plant body are monosiphonous or at times may be corticated by the rhizoidal filaments in the mature region. The cells of the vegetative body may be uninucleate or multinucleate having ribbon-shaped chromatophores. *Callithamnion* exhibits isomorphic alternation of generations.

Both antheridia and carpogonia are borne on separate plants. Antheridia form hemispherical or ellipsoidal tufts arising as lateral appendages on the branches of the male plant (Fig. 9.21C). The lateral appendages may produce crops of antheridia two or three times successively from the same place (Figs. 9.21D and

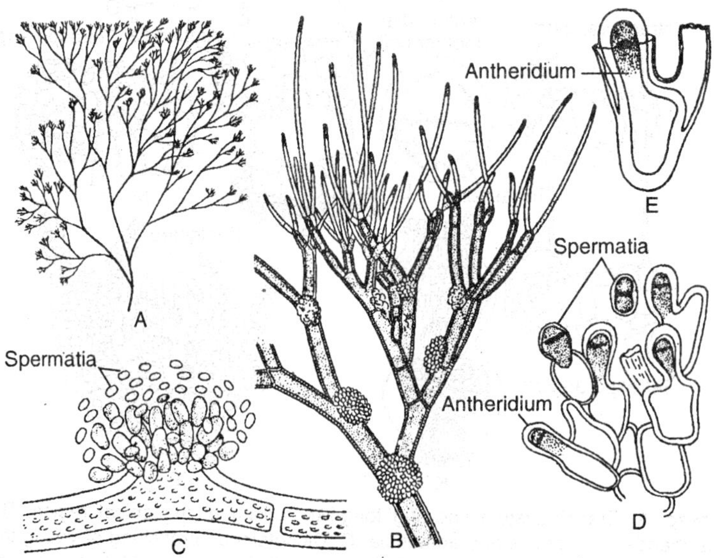

Fig. 9.21: *Callithamnion* sp. A. Habit. B. Male plant bearing antheridia, branches terminating into hairs. C. Portion of the male plant showing liberation of spermatia. D. Liberation of spermatia from antheridia. E. Antheridium

9.21E). The entire content of each antheridium develops into a spermatium (Fig. 9.21D).

The carpogonia are borne on female plants in combination with auxiliary cells, exhibit a primitiveness in character. The combination of a carpogonial branch and the two auxiliary mother cells is known as **procarp.** Of the two auxiliary mother cells only one bears the carpogonial branch. The carpogonial branch is four-celled, the apical cell of which is the carpogonium with long hair-like trichogyne (Figs. 9.22A and B). A floating spermatium in water comes in contact with the trichogyne (Fig. 9.22A) and rests for a while when the walls between dissolve and the spermatial nucleus moves down the trichogyme and fuses with the carpogonial nucleus. Simultaneously, with the process of fertilization the two auxiliary mother cells divide, each one forming a basal cell and an auxiliary cell. Along with the development of the auxiliary cells a cytoplasmic connection is maintained from cell to cell. The carpogonial base along with the diploid nucleus divides producing two daughter cells,

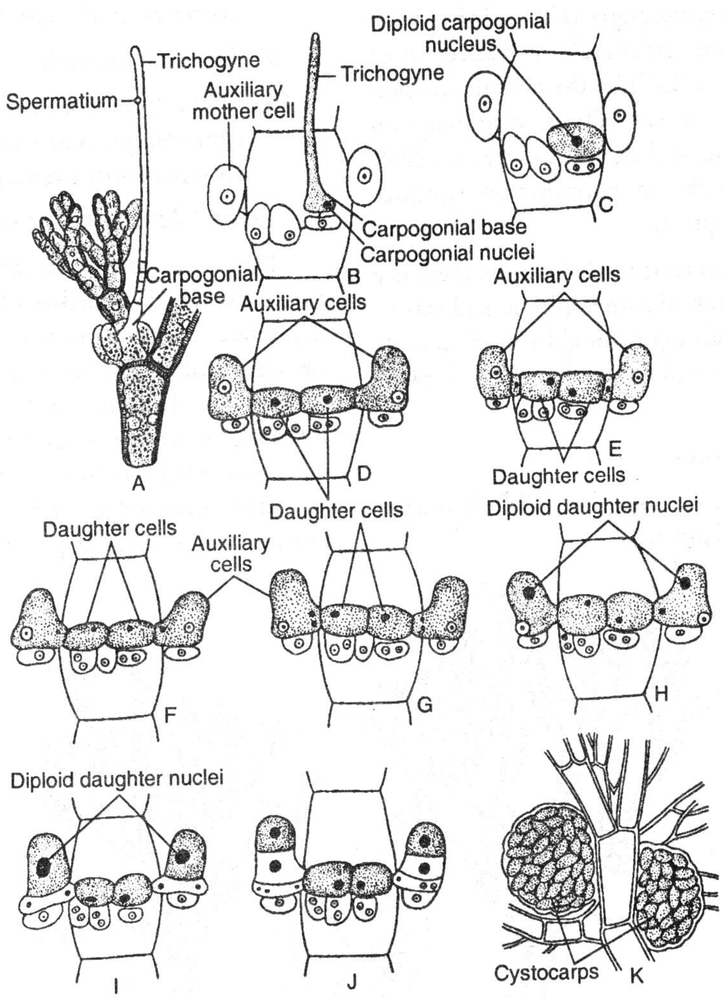

Fig. 9.22: *Callithamnion* sp. A. Portiion of female plant showing carpogonium with spermatium attached to trichogyne. B. Carpogonium with auxiliary mother cell. C-G. Early post-fertilization stages showing development of auxiliary cells. H-J. Late post-fertilization stages showing distribution of daughter diploid nuclei in the auxiliary cells. K. Cystocarps

each one bearing one diploid daughter nucleus (Fig. 9.22C-G). The two daughter cells extend laterally and ultimately fuse with the adjacent auxiliary cells, and thus the two diploid daughter nuclei are distributed in the auxiliary cells (Figs. 9.22H-J). Ultimately, the two auxiliary cells give rise to two gonimoblasts of the cystocarp. Cystocarps are usually developed in pairs (Figs. 9.22K and 9.23A) on branches of the female plant and are enveloped by mucilage. Carposporangia are spherical bearing one carpospore each, which germinates into a tetrasporophyte. The tetrasporophyte bears tetrasporangia (Fig. 9.23B). Each tetrasporangium ultimately produces four tetraspores (Fig. 9.23C) by the meiotic division of the diploid nucleus. The tetrasporangium dehisces by a lid and the tetraspores are liberated out, which on germination produce haploid gametophyes.

In the life cycle of *Callithamnion* there is a regular alternation of gametophytic and sporophytic individuals represented by haploid male and female gametophytes and tetrasporic plant respectively.

Special features:

1. Plant body comprising richly branched tufts of filaments.
2. Development of branches from just below the septa of the cells of the main axis.
3. Pseudodichotomous development of branches.
4. Isomorphic alternation of generations.
5. Sexual reproduction advanced oogamy.
6. Development of cystocarp.
7. Cystocarp, carposporangia and carpospores—all are diploid.
8. Meiosis takes place during the development of tetraspores in the tetrasporangia borne in the tetrasporic plant.
9. Triphasic life cycle.
10. Diploid phase partly dependent on the gametophyte and partly represented by an independent tetrasporic plant.

Family: *RHODOMELACEAE*

Plants are usually bushy, sometimes sparingly branched; branches often delicate, usually terete, occasionally flat; growth from persisting apical cells producing an axial cell row, the principal branches of the thallus developed from successive segments; branched colourless hairs (trichoblasts) often present. Axial cells are generally surrounded, at least in the fruiting portions, by a series of pericentral cells cut off

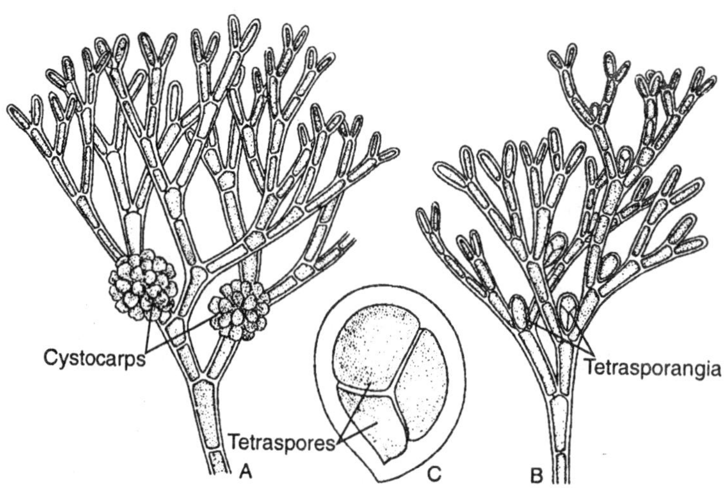

Fig. 9.23: *Callithamnion* sp. A. Female plant bearing cystocarps. B. Tetrasporic plant bearing tetrasporangia. C. Tetrasporangium with tetraspores

from them by longitudinal walls, producing a typically polysiphonous structure, and sometimes further corticated either by subsequent divisions of these to several degrees or by appressed rhizoidal down growths; tetrahedral sporangia formed from internal segments of the pericentral cells, the branchlets bearing them usually little modified, but in extreme cases like stichidia. Spermatangial clusters are developed from trichoblast rudiments in the form of colourless tufts, cones, or plates of spermatangia. Procarps are developed from polysiphonous basal trichoblast segments, the fertile pericentral cell as a supporting cell producing sterile cells and the four-celled carpogenic branch; from the supporting cell beside the carpogonium an auxiliary is cut off after fertilization, from which, by means of a fusion cell, the sympodially branched gonimoblasts are produced; outer cells of the gonimoblasts alone forming carpospores. Cystocarps become enclosed by an ostiolate pericarp.

Genus: *Polysiphonia*

This marine alga is relatively common along the coasts of the United States and British seas. In India, it has been reported from the West coast. Most species prefer quiet waters. Its plant body is attached to the substrate, may be submerged stones, rocks, shells, wood, or other algal body, by means of flattened discoid holdfasts. The plant body is only a few to several inches in length (Fig. 9.24A). The plant body is filamentous, branched laterally or dichotomously with brownish-red to purple-red in colour. There are species of thick

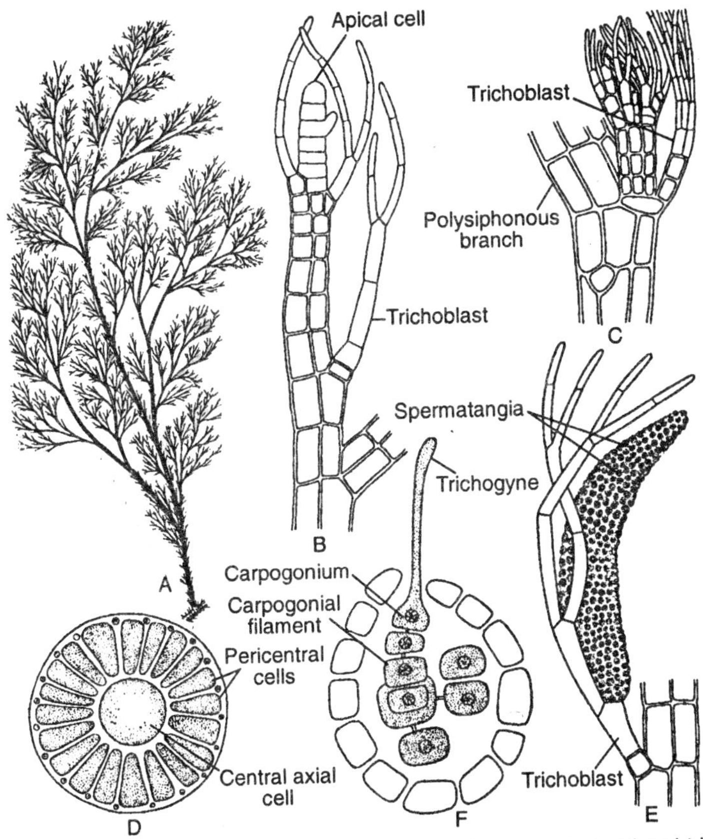

Fig. 9.24: *Polysiphonia* sp. A. Habit. B. Portiion of a young thallus showing apical cell and trichoblast. C. Portion of older thallus showing polysiphonous branch. D. Transverse section through mature thallus showing arrangement of cells. E. Spermatangial branch with spermatangia. F. Carpogonium with trichogyne

bushy habit. Both annual as well as perennial forms are present with regular shedding and regeneration of branches which is influenced by the seasonal variation. The alga is readily recognizable from the arrangement of its cells placed end to end in tiers. The main axis and its branches possess a polysiphonous appearance as the central axial cell (siphon) is surrounded by pericentral cells (siphon) of variable number (Fig. 9.24C). The cells have prominent cytoplasmic connections between cell to cell and each cell has one nucleus and many discoid plastids embedded in dense cytoplasm. The central axial cells are comparatively large, elongated and uninucleate. They are surrounded by pericentral cells (Fig. 9.24D). Growth of the thallus is initiated by a dome-shaped apical cell (Fig. 9.24B). In many species the main axis produces two kinds of branches, for example, some are polysiphonous like the parent axis and others uniseriate structures of unlimited growth. These uniseriate structures are known as **trichoblasts** (Fig. 9.24B).

There is well-developed alternation of generations in the life cycle of *Polysiphonia.* The gametophytic generation is represented by morphologically identical male and female filaments with which alternately occurs asexual or **tetrasporic plant** which produces **tetrasporangia** and **tetraspores.** The lateral branches of the male plant bear antheridia known as **spermatangia,** in dense clusters (Fig. 9.24E). The spermatangia are short-stalked colourless, spherical to oval structures. Each spermatangium produces a single male cell or a **spermatium** from its entire contents. The wall of the mature antheridium splits open through which the spermatium is liberated.

The carpogonium is developed terminally on a lateral carpogonial branch of the female filament. The carpogonium has a swollen base in which lies the female nucleus, and a narrow elongated process, the **trichogyne** (Fig. 9.24F) constituting the receptive organ for the spermatia. The spermatia are carried by water currents to the carpogonium and finally one of them adheres to the trichogyne. Both the membrane of the spermatium and the wall of the trichogyne dissolve at the point of contact producing an open passage for the male nucleus which travels down the trichogyne to the female nucleus, where fusion between the two nulclei takes place.

The sporophytic generation has its beginning in the zygote. After fertilization the supporting cell of the carpogonial filament cuts off an **auxiliary cell** from its apex (Fig. 9.25A). Soon a tubular connection is established between the auxiliary cell and the carpogonium. The thread connecting the auxiliary cell with the carpogonium is known as a **procarp.** The diploid nucleus of the carpogonium divides mitotically and one of the two daughter nuclei passes in the auxiliary cell through the connecting filament. The carpogonium gradually shrivels and finally disappears. A small lateral outgrowth, known as **gonimoblast initial,** arises from the auxiliary cell (Fig. 9.25B). This is repeated with the ultimate production of a set of gonimoblast initials. The diploid nucleus of the auxiliary cell further divides and each gonimoblast initial receives a single nucleus. Now each gonimoblast initial gives rise to a short **gonimoblast filament,** with its cells bearing a diploid nucleus. The apical cell of each gonimoblast filament develops into a **carposporangium,** the contents of which are transformed into a single **carpospore.** Simultaneously, with the development of carposporangia and carpospores lateral sterile cells closer to the carpogonium grow out to form sterile filaments. These sterile filaments form one- to two-layered walls surrounding the carposporangia forming an urn-shaped structure open at its tip through which the carpospores are liberated. This wall affords protection to the carposporangia. This entire urn-shaped structure is known as **cystocarp** (Figs. 9.25C and D).

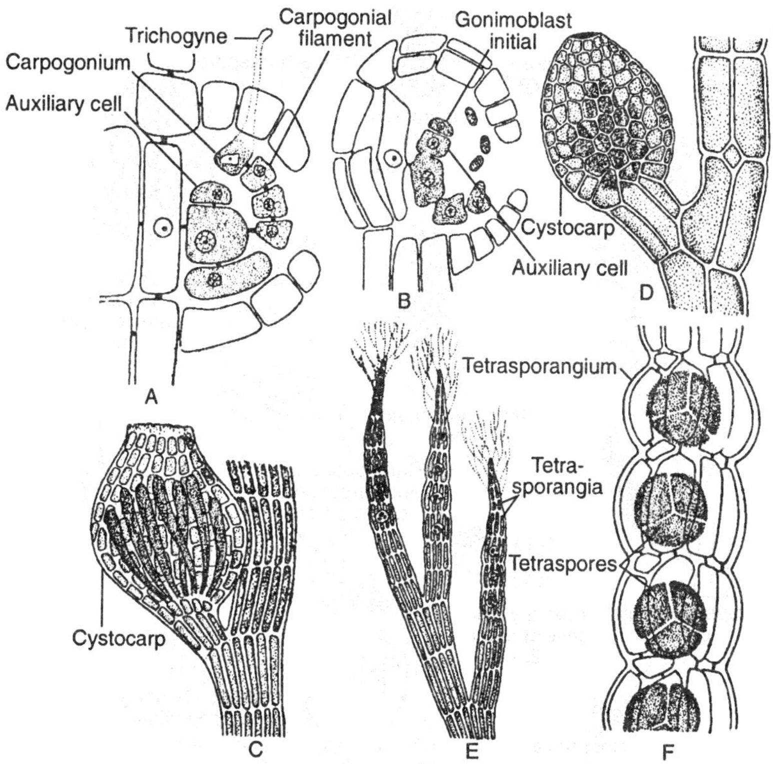

Fig. 9.25: *Polysiphonia* sp. A-B. Post-fertilization stages of auxiliary cell and gonimoblast initial development. C. Young cystocarp. D. Mature cystocarp. E. Portion of tetrasporic plant bearing tetrasporangia. F. Tetrasporangia with tetraspores

The diploid carpospores, after being released, form new plants—**tetrasporophytic plants,** which look like the gametophytic individuals except that a few cells of which become sporangia—**tetrasporangia** (Fig. 9.25E). The tetrasporangia are borne on the central cells of the axis (Fig. 9.25F). The diploid nucleus of a tetrasporangium divides meiotically to form four haploid nuclei around which four tetraspores are developed. After liberation, the tetraspores germinate to form male and female gametophytes.

The life cycle of *Polysiphonia* is a complex one, it comprises of three phases. There is a haploid sexual phase or the gametophytic generation. It is represented by gametophytic filaments which bear sex organs and gametes. After fertilization, the female gametophyte bears carposporangia and carpospores in a cystocarp. The carposporangia, carpospores and cystocarp all are diploid structures. The cystocarp is entirely dependent on the female gametophyte. This phase is followed by a diploid independent tetrasporophyte bearing tetrasporangia and tetraspores. It starts along with the germination of carpospore and terminates with the formation of haploid tetraspores by meiosis. This phase is concerned with asexual reproduction. Thus, *Polysiphonia* exhibits **triphasic** life cycle. It also exhibits distinct alternation of generations represented by independently alternating well-developed sexual (gametophytic) and asexual (sporophytic) plants (Fig. 9.26).

Some Indian Species: *Polysiphonia platycarpa* Boergs.; *P. sertularioides* (Gvat.) J, Ag.; *P. suotilissima* Mont.; *P. tuticorinensis* Boergs.; *P. unguiformis* Boergs.

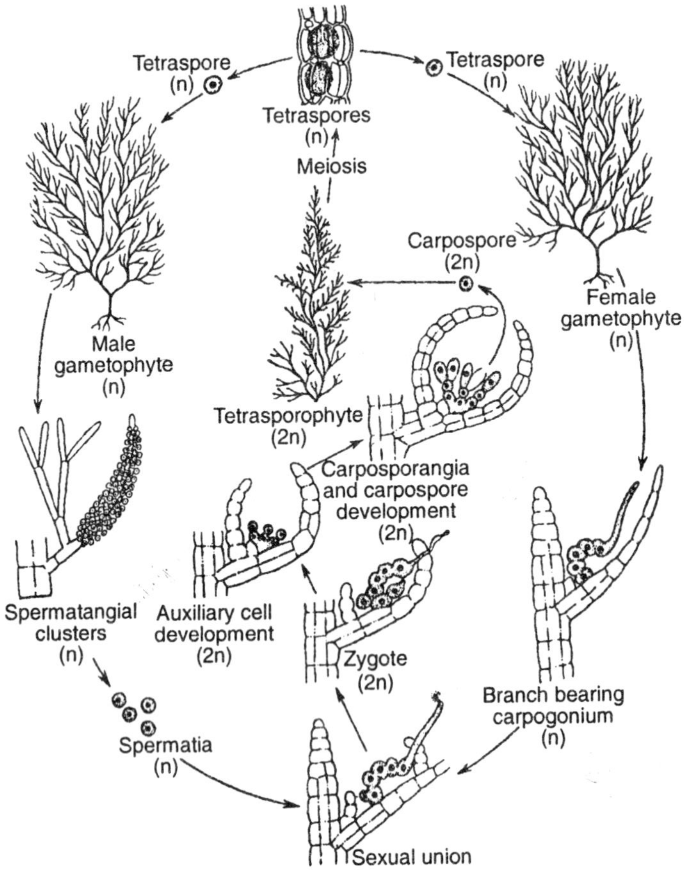

Fig. 9.26: Life cycle of *Polysiphonia* sp.

Special features:

1. Polysiphonous plant body.
2. Growth in length by a single dome-shaped apical cell.
3. Cytoplasmic connections between adjoining cells.
4. Isomorphic alternation of generations.
5. Sexual reproduction advanced oogamy.
6. Elaborate post-fertilization stages.
7. Presence of well-developed cystocarp.
8. Cystocarp, carposporangium and carpospores are diploid.
9. Meiosis taking place during the development of tetraspores in the tetrasporangium borne on tetrasporic plant.
10. Triphasic life cycle.
11. Sporophyte comprising of two phases, an early phase dependent on the gametophyte represented by cystocarp which is followed by an independent phase exhibited by the tetrasporic plant.

Order: *BANGIALES*

Plants are filamentous or membranous, simple or branched. Cells are usually with stellate chromatophores. Asexual reproduction is by naked monospores; sexual reproduction, if present, by spermatia and carpogonia; the fertilized carpogonium developing several carpospores.

Family: *ERYTHROTRICHIACEAE*

Algae belonging to this family possess filamentous branched, unbranched to thalloid plant body. Reproduction is by the development of large and small spore produced by the division of vegetative cells of the plant body.

Genus: *Compsopogon*

Compsopogon is chiefly confined in the tropical and subtropical freshwater streams. It occasionally occurs in bunches along the edge of the ponds being attached to submerged aquatic plants, other plant parts (Figs. 9.27A and B), bricks, etc., also in mud below water-level. The alga when growing in still water may remain unattached. In West Bengal, it appears in January and disappears with the advent of

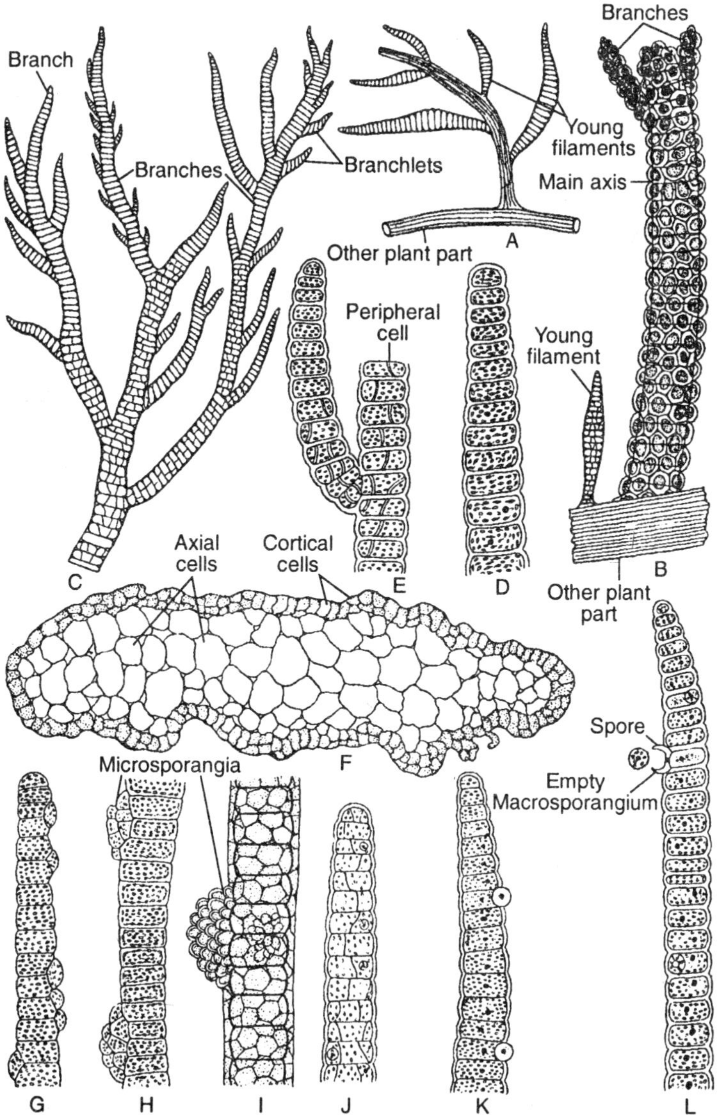

Fig. 9.27: *Compsopogon* sp. A. Young filaments attached to other plant part. B. Mature filament with branches and young filament attached to other plant part. C. Profusely branched filament. D. Uniseriate discoid cells of younger part of a filament. E. Separation of peripheral cells from central axial cells by septa. F. Transverse section through mature filament showing pigmented cortical cells around almost colourless axial cells. G-I. Stages in development of microsporangia. J-L. Stages in development of macrosporangium and liberation of its spore

hot season. Plant body is filamentous and always more or less copiously branched (Fig. 9.27C). In the coarse mature regions, the filament is blue-green in colour. The younger parts of the filament are uniseriate with discoid cells (Figs. 9.27A and D) and the mature parts are with an axial row of cells which become greatly enlarged being barrel-shaped and are surrounded by one or more layers of cortical cells (Fig. 9.27B). The outermost cortical cells are smaller with inner cortical cells being progressively larger (Fig. 9.27F). The axial cells in very old filaments break down and are replaced by the enlarged cells of the inner cortex.

The filaments increase in length through intercalary divisions, and at no great distance from the terminal cell longitudinal septa begin to make their appearance, irregularly at first, but succeeding one another in such a fashion that central axial cell is separated from a layer of deeply pigmented small peripheral cells (Fig. 9.27E) that completely surround it and continue to undergo anticlinal divisions without regularity, while the axial cell constantly enlarges assuming a flattened barrel-shaped form (Fig. 9.28M). These pigmented cells constitute the cortex around the large and almost colourless cell which is filled with watery contents and contains a few chloroplasts. The originally single cells of the younger filaments are thus converted into corticated segments indicated by slight constrictions in the older filaments. The monstrously developed axial cells form a continuous axial series (Fig. 9.28F). This cortical layer is lacking in the basal part of the thallus. Here the cells are commonly overgrown by septate rhizoids which perform the function of anchorage. Young individual after attaining a certain length begins to give rise to a subdichotomous branch, and the main axial portion becomes somewhat arched. Subsequently a second subdichotomous branch arises on the same side of the axial portion from a cell nearer to the apex. This branch behaves in a similar manner and a further arching of the main axis towards the substrate takes place. With repeated branching several branches and branchlets are formed resulting in the formation of richly branched bluish-green or violet-green coloured macroscopic thallus. The primary filament and the branchlets are distinctly constricted at the joints. Cells of the thallus are uninucleate containing numerous spherical chromatophores which are densely aggregated to the cell wall. There are no pit-connections between the adjacent cells.

The method of reproduction so far reported is asexual by **monospores.** The monospores are non-flagellate spherical spores with a large nucleus embedded in dense cytoplasm. They are of two kinds being borne in sori of **microsporangia** (Figs. 9.27H and I) and some are developed in solitary **macrosporangia** (Figs. 9.27K and L). Both kinds of sporangia are developed either from the cortical cells or from cells cut off from the uncorticated cells (Figs. 9.27G and J).

The microsporangial sori are more or less hemispherical to oblong in shape and consist of 15-25 or even more microsporangia. The microsporangia are normally spherical, but very commonly polyhedral due to mutual pressure. They may arise from any cell of the filament appearing as scattered minute heaps of cells (Figs. 9.26H and I). During the formation of spores the contents of each microsporangium contract and become spherical. The spores are borne singly in each microsporangium. The microsporangial wall opens by a slit and the spore is discharged with a jerk. The spore thus liberated secretes a wall and directly develops into a new plant.

The macrosporangia are developed in rather restricted areas of the filament. The formation of each macrosporangium is initiated by the appearance of a septum in the cell from the daughter cell of which the macrosporangium will be ultimately produced. The septum separates the cell into two daughter cells. The daughter cell which ultimately matures into a macrosporangium has denser contents and the nucleus larger and more conspicuous (Fig. 9.28A) than the other daughter cell. The separation of daughter cells takes place during night. As the day advances the protoplast of the cell and its wall assume a subtriangular outline (Fig. 9.28B). One of its rounded angles projects in the form of a papilla. The cell then becomes a macrosporangium (Fig. 9.28C). The basal septum of the macrosporangium which originally separated it from the other daughter cell bulges more and more strongly towards the macrosporangium, and finally the discharge of the spore is accomplished by the dehiscence of the papilla, slowly at first and then more rapidly (Figs. 9.28D and E). The basal septum then reverses to its previous position. The position of the sporangium is merely indicated by the slightly projecting margins of the orifice through which the spore escaped (Fig. 9.27L). The spore which is discharged with a force is sufficient to carry to a very short distance from the filament. Each macrosporangium produces a single spore which is spherical, completely filled with chloroplast, except for a small area (Fig. 9.28E). The liberated spore directly germinates into a new filament.

The spore on coming into contact with the substratum becomes somewhat flattened at the surface of contact and then divides to form a short-creeping filament of three or four cells (Figs. 9.28F-H). Then one to two of these cells give rise to a short lateral branch (Figs. 9.28I-K). Subsequently, a cell in a median position elongates vertically and forms the initial cell of an erect filament (Fig. 9.28L). It divides to form a dome-shaped apical cell which gives rise to discoid segments. In the latter stages, the growth is by the division, largely, of these segments with the gradual differentiation of central axial cells and peripheral cells (Fig. 9.28M).

The initiation of the erect portion of the filament curtails further extension of the creeping portion which, therefore, remains small. In fact, the attachment of the thallus is subsequently augmented by a rhizoidal development. The erect filament is at first uniseriate, but in the later stages the cells, a little distance from the apex, enlarge and divide, forming a

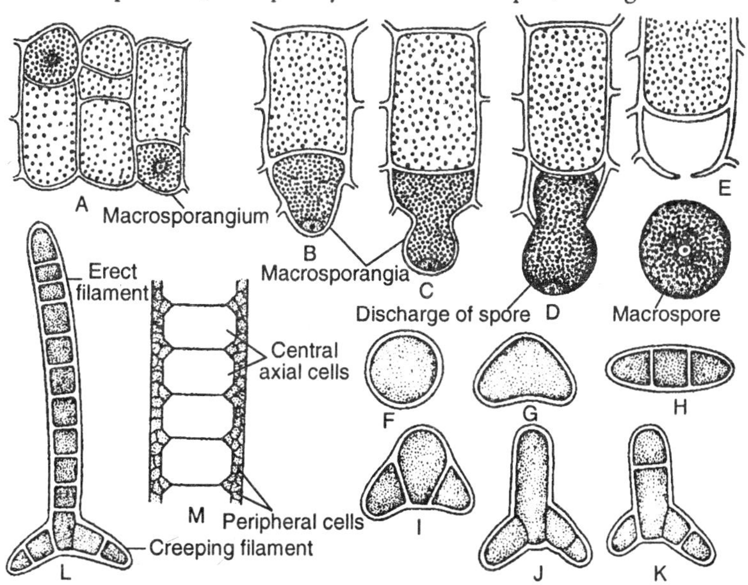

Fig. 9.28: *Compsopogon* sp. A-E. Successive stages in development of macrosporangium and discharge of macrospore. F-K. Stages in germination of macrospore. L. A very young filament. M. Portion of mature filament in vertical section

peripheral layer of small cortical cells. This involves a lateral expansion of the thallus. The distal part of the filament remains uniseriate and the cells of this region contribute to the growth in length of the filament. As any or all of the cells in this region are capable of division in a transverse plane, it constitutes a growing region. The lowermost few cells of the erect filament do not form a cortical layer but only bear rhizoids as tubular extensions at the lower ends. These rhizoid-bearing cells are delimited early in the development of the erect portion before the formation of the corticated portion of the filament. Thus, the filament consists of three regions: (i) the growing region, (ii) the region differentiated into cortical and axial cells, and (iii) the rhizoidal region. A striking analogous condition to higher plants thus exists so far as the general developmental pattern of the alga is concerned.

Some Indian Species: *Compsopogon coerulens* (Balbis) Mont.; *C. hookeri* Mont.; *C. indica* Das; *C. iyengarii* Krishnamurthy; *C. lividus* (Hooker) Detoni.

Special features:

1. Copiously branched filamentous macroscopic thallus.
2. Thallus differentiated into growing region, region differentiated into cortical and axial cells, and the rhizoidal region.
3. Asexual reproduction by non-flagellate spores—the monospores.
4. Monospores borne singly in microsporangia and macrosporangia.
5. Microsporangia are developed in sori, whereas the macrosporangia are solitary.

Chapter 10

General Considerations of The Algae

A. ORIGIN OF SEX IN THE ALGAE

Sexual reproduction is one of the means of perpetuation of species in algae with ultimate increase in the population of the individuals. The process involves the fusion of two gametes or sex cells which may originate from the same thallus (monoecious) or from different thalli (dioecious). The gametes may be flagellate or non-flagellate. In algae, the mode of sexual reproduction ranges from isogamy, anisogamy to oogamy. In general, sexual reproduction takes place after the alga had attained a certain amount of vegetative growth and having certain amount of reserve food built up. In algae, an isogamous fusion of two zoogametes is the most primitive type of sexual reproduction. The evolutionary tendency is isogamy, anisogamy to oogamy. Anisogamy is an advanced condition than isogamy. Among anisogamy again there are two conditions prevailing: (i) both the fusing pair of gametes may be flagellate, and (ii) the larger one is non-flagellate as against the smaller which is flagellate. It is evident that the development of a large-sized gamete permits the greater accumulation of reserve food for the use of the zygote during the post-sexual stages. At the same time increase in size means greater surface area which facilitates the chance of meeting of gametes. Hence, anisogamy has distinct advantage over isogamy. Within oogamy also there is wide range of variations including incipient oogamy to highly advanced stage where the female gamete is permanently retained within the female gametangium. A progressive evolution from isogamy to oogamy has taken place along independent phyletic lines in the Chlorophycophyta and the Phaeophycophyta. The development and differentiation of gamete producing structures and the gametes are extremely variable. The simplest condition is exhibited by filamentous algae *Spirogyra* and *Zygnema* where all the cells of the thallus have the potentiality to behave as gamete producing structures (gametangia) and can bear gametes. The vegetative cell of the filament and the gametangium resemble each other morphologically except in the fact that in the former there is no change in the protoplasm of the cell, whereas in the latter the entire protoplasm of the cell is metamorphosed into a gamete. As such no specialized gametangia are developed and any vegetative cell is a potential gametangium.

But in filamentous alga *Ulothrix* any cell of the filament except the apical and basal one has the potentiality to behave as a gametangium without undergoing any morphological change except the protoplasm which divides and redivides to develop into gametes. Some of the gametes are morphologically very similar to the zoospores, but difference is only in their behaviour. In unicellular alga *Chlamydomonas* also the vegetative cell divides and redivides to form either daughter vegetative cells or gametes depending on the conditions of the habitat. In both *Ulothrix* and *Chlamydomonas* difference in the gamete or zoospore formation lies in the number of division of the protoplasm and the behaviour of the daughter cells. The number of division during the formation of gametes is much larger than that of the formation of daughter vegetative cells.

In *Oedogonium, Vaucheria, Ectocarpus* there is distinguishable gametangia development but the gametangia may be formed anywhere in the thallus without having any specificity for any particular region of the thallus. It is much more interesting in *Echocorpus,* where the plurilocular structures may behave as gametangia as well as sporangia. In some species the controlling factors are the conditions of the substratum in which the alga is growing.

In *Volvox, Coleochaete, Chara, Nitella, Laminaria, Fucus, Polysiphonia,* there are particular regions or branches or particular cell or cells where after a certain period of vegetative growth there is development of specialized gamete producing structures of varied names.

Usually, when the sexual fusion between the gametes fails to occur the gametes degenerate. But there are exceptions to this process. Particular mention may be made in case of *Spirogyra* and *Zygnema,* where in absence of sexual fusion the gametes behave as spores.

What induces the formation of gametes? What factor or factors control the sexual reaction? It is broadly agreed that conditions of the substratum play a very important role in the entire process. In case of certain marine algae the formation of gametes is very much pronounced during the warmer months of the year. It has also been experimentally demonstrated that the gametes secrete certain substances in the surrounding medium. The secretion products may have some effect to envigourate the sexual reaction. The sexual reaction inducing substances in some cases have the capacity to change zoospores into gametes. It is also believed that certain algae reproduce vegetatively under favourable conditions, whereas the same algae will start reproducing sexually by the development of gametes when the prevailing conditions of the substratum are unfavourable, which suggests that the origin of sexual reaction is directly correlated with the environmental conditions of the surrounding medium. Hence the origin of sex is dependent upon the microclimate which again accelerates the division of the protoplasm.

In case of gametes behaving as spores after the failure of sexual act, there is a possibility that with the failure of sexual act the sexual reaction inducing substance of the gametes are destroyed due to which the gametes behave as spores. This situation suggests that a gamete has basically the property to behave as a spore. But this quality is suppressed by the development of so-called sexual reaction inducing substance. Similar interpretation is tenable in case of *Ulothrix* where the development of gametes or zoospores is dependent on the number of division of protoplasm. With more division of protoplasm the result will be development of gametes. It is quite possible that there is some relation between the number of divisions of protoplasm and the development of sexual reaction inducing substance.

That gametes secrete sexual substances into the surrounding water was demonstrated by Jollos (1926), Geitler (1931), Pascher (1931). Moewus (1933, 1940), and many others. These substances can be utilized for increasing sexual intensities of weak gametes. Moewus was able to extract substances from fertile sexual plants of the members of the Ulvaceae and species of *Chlamydomonas* and with these substances he made zoospores function as gametes. It has been ascertained by spectroscopic analysis that these sexual substances are carotinoids in nature and the first-formed substance is **protocrocin** which breaks down into two molecules of **picrocrocin** and one of **crocin.** Each molecule of carotinoid picrocrocin breaks down into a molecule of glucose and molecule of **carotinoid safranol.** Each molecule of carotinoid crocin again breaks down into two molecules of **sugar gentiobiose** and one molecule of **ciscrocetin dimethyl ester** and ultimately becomes **trans crocetin dimethyl ester.** Each step in the degradation of protocrocin is brought about by different genes. Crocin is the chief substance that induces motility in gametes, and cis and trans crocetin dimethyl esters are primarily concerned with the mutual attraction of gametes. The carotinoids picrocrocin and safranol are sex determining substances.

Recent researches indicate that although the sexual process is a manifestation of profound physiological changes within organisms, it may be evoked or modified by a manipulation of certain external factors such as duration of illumination, temperature, the amount of nitrogen available, etc.

B. ALTERNATION OF GENERATIONS IN THE ALGAE

Alternation of generations in algae has been manifested in the manner indicated below.

An elaborately developed gametophyte followed by a very short sporophyte represented by the zygote nucleus only, is exhibited by *Chlamydomonas, Volvox, Ulothrix, Spirogyra, Zygnema, Cosmarium, Oedogonium,* and *Vaucheria.* In them meiosis is zygotic. There is no organized sporophytic phase with a different function. In the life cycle of these algae there is alternation of a well-developed gametophyte with a single-celled sporophyte. This results in definite alternation of chromosome numbers from haploid to diploid and back to haploid stage. There is, however, no alternation of individuals with different functions. Such a condition is the starting point of the origin of the diploid body or the sporophyte in the life cycle.

A deviation from above is found in *Coleochaete,* where an elaborately developed gametophyte followed by a post-zygotic spore-producing phase is present. It thus indicates a tendency to develop an elaborate post-zygotic phase, but could not go further due to zygotic meiosis. If the zygote were to divide without meiosis, the natural consequence would be a transfer of the post-zygotic phase from haplont to diplont. Meiosis would be delayed until spore formation. The two generations would be unlike, as they are throughout the bryophytes.

The development of an organized sporophytic phase is encountered in *Ulva, Enteromorpha, Cladophora,* and *Chaetomorpha.* These algae exhibit isomorphic alternation of generations in their life cycle. Both the gametophytic and sporophytic plants are independent and are alike except for the reproductive cells they bear. The delayed meiosis has given rise to the sporophytic plant where meiosis takes place during the development of spores which on germination give rise to gametophytic individuals.

A further advancement along this line exists in *Ectocarpus* and *Dictyota,* where evolution of gametangia and sporangia has taken place on an isomorphic alternation of generations.

In *Cladophora glomerata* there exists only a diploid generation, meiosis results in the production of gametes. Haploidy, except for the gametes, has disappeared in consequence of a mutation. There are indications of a similar history in certain races of *Ectocarpus*. This condition of diploidy except for the gametes appears also in the Bacillariophyta.

Steps in the gradual dominance of diplont over haplont are illustrated by *Laminaria, Fucus*, and *Sargassum*. In *Laminaria* the diplont is large, often perennial; the haplont is a distinct but very small plant and thus exhibits heteromorphic alternation of generations. In *Fucus* and *Sargassum* the haplont, no longer independent, has almost disappeared, being represented by the gametes only. Similar condition of reduction of the haplont appears in *Acetabularia, Bryopsis*, and *Codium*. The haplont has disappeared, gamete nuclei now being produced by meiotic division of the diploid nuclei.

A somewhat different method of alternation of generations is found in red algae. In *Batrachospermum* a post-zygotic phase (zygotic meiosis) is dependent upon the gametophyte, gives rise to haploid carpospores. It increases the possibilities of multiplication, because through it a single zygote gives rise to many spores. A condition of delayed meiosis has been exemplified by *Ceramium, Callithamnion* and *Polysiphonia*, where an elaborately developed post-zygotic diploid phase results in the development of diploid carposporangium and carpospores. The diploid carpospores on germination produce independent sporophytic individuals which are functionally different from the gametophytic ones. These sporophytic individuals bear sporangia in which spores are produced by meiosis. The spores again on germination give rise to gametophytic plants exhibiting a distinct alternation of generations. Since part of the diploid phase is dependent upon the haploid phase and the rest is entirely independent, and both these phases alternate with gametophytic plants, such a condition is also known as triphasic alternation of generations.

C. ORIGIN AND EVOLUTION IN THE ALGAE

As different authors differ widely in their lines of thinking with regard to the origin and evolution in algae, it is felt necessary to review those possible lines instead of expressing one's personal opinion in this aspect of algae. This will give everyone a wide scope for independent thinking after giving due consideration of everybody's line of thought. This discussion is primarily based on living forms of algae. The difficulties associated with the interpretation are mainly, wide variation in the morphology of the thallus and the diversity in the mode of sexual reproduction in all the taxa of algae. Cases of parallelism are very common which need quite a careful consideration. Distinguishing differences among the major groups of algae are not primarily morphological, but biochemical concerning types of pigments and metabolic processes. Structure and placement of flagella are also significant. The principal algal groups: Chlorophycophyta, Phaeophycophyta and Rhodophycophyta have probably originated from a common primitive motile chlamydomonad type of ancestry which followed a single line of evolution (Fig. 10.1). The parallelism existing within various groups may be due to the plasticity of primitive organism. The separation of Rhodophycophyta was possible by a sideline from an intermediate simple filamentous condition and that of Phaeophycophyta much higher up from the heterotrichous condition. Considering the morphological features and the mode of reproduction, *Coleochaete* appears to be also another point

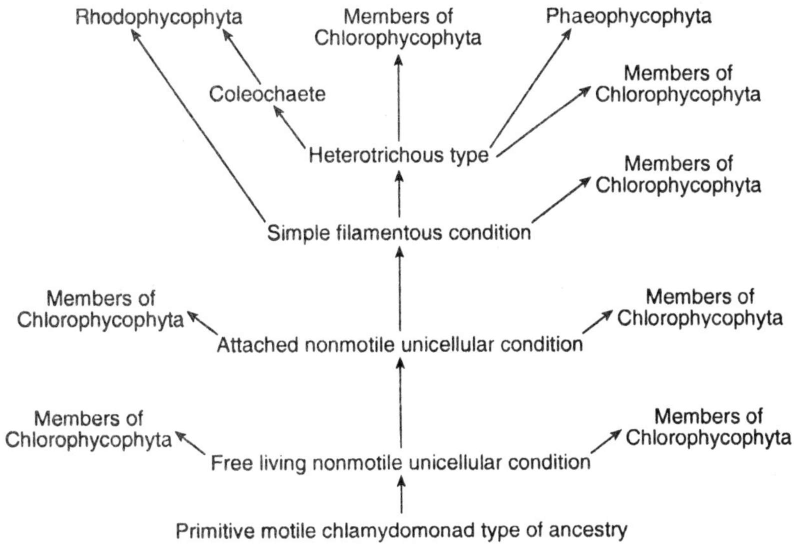

Fig. 10.1: Schematic representation of the possible origin and evolution in the Algae

from which some members of the Rhodophycophyta might have evolved either directly or from a point very close to it.

But from the phylogenetic standpoint both Chlorophycophyta and Rhodophycophyta follow parallel lines of development as in both the taxa the haploid condition is primitive and the diploid generation is interpolated in the life cycle due to delay in the occurrence of meiosis of the diploid zygotic nucleus. The evolution of heteromorphic alternation of generations has taken place due to certain modification from the isomorphic alternation of generations which starts earlier. Again there are Chlorophycophyta which have maintained the simple form of life cycle along with very elaborate morphological differentiation of the thallus forming divergent lines of evolution. One other interesting feature is the development of completely dominant diploid generation indicating an extreme case of interpolation and elaborate development of doploid generation by the gradual suppression of the haploid generation. As to the unicellular red algae, it is probable that they are reduced forms than being primitive. There is a possibility that the Rhodophycophyta have evolved from the Cyanophycophyta, as both the classes have similarities in the pigments, but chemically they are different; and complete absence of any motile structure. The main drawback of this line of interrelationship is the complete absence of sexual reproduction in any form in Cyanophycophyta, whereas, in most red algae the development of sexual reproductive organs and reproduction by sexual means are quite common.

Cyanophycophyta and Chlorophycophyta are considered by some as most primitive. But if the Phaeophycophyta and Rhodophycophyta had a flagellate ancestry, in that case the four taxa: Cyanophycophyta, Chlorophycophyta, Phaeophycophyta and Rhodophycophyta may have originated from the same stalk. Again, there are workers who are of the opinion that all the three taxa: Chlorophycophyta, Phaeophycophyta and Rhodophycophyta have originated independently from different stalks.

In 1952 Feldmann expressed his opinion in considering algae having isomorphic alternation of generations as primitive. This may be true only in case of Phaeophycophyta and

Rhodophycophyta and not in Chlorophycophyta, where the primitive orders have simple life cycle. Elaborately developed thallus, associated with isomorphic alternation of generations is usually encountered only in advanced types of Chlorophycophyta, whereas according to Feldmann, this is the possible starting point in Phaeophycophyta and Rhodophycophyta. Hence it does not seem tenable to have the same characteristic features of primitiveness in all the taxa of algae.

The possible source of origin of the algae has been suggested from an ancestral flagellate Chlorophycophyta. If the chlorophycean ancestry is accepted of the Phaeophycophyta, then the diploid condition has evolved by the interpolation subsequently in the haploid state. The Phaeophycophyta have their origin probably among the Chaetophoraceae or Cladophoraceae. Similarities are in the morphology of the thallus and the nature of alternation of generations.

Besides these, on the basis of pigments it has also been suggested that Phaeophycophyta have their origin from the Chrysophycophyta according to the Scheme outlined by Senn (Fig. 10.2).

D. FOSSIL ALGAE

The soft delicate bodies of algae do not ordinarily leave much fossil impressions or petrifications as after death they decompose very rapidly. Thus, quite often fossil records of most algae are imperfect. In spite of these difficulties quite a good number of fossil representatives of algae have been found in various era of the earth's crust. One should not restrict himself only to the study of the present-day living forms of algae, since a large number of fossil algae have also been worked out. Since invertebrates existed in pre-Cambrian times it is quite obvious that algae for long ages, before animals began eating each other, have been required as food of animals were in existence even earlier than animals, may be in the Archeozoic era.

It is fairly well accepted that the blue-green algae existed in the early Archeozoic era. Many of them secreted lime or silica during their lifetime. Careful examination of the sedimentary rocks of the Proterozoic era has revealed the presence of fossil algae in great abundance. The blue-green algae may be among the earliest algae to appear, since a considerable number of them have been found in ancient rocks. The Proterozoic limestone beds were possibly resulted due to accumulation of algae. Besides these, well-preserved fossils of both marine and freshwater blue-green algae are also known. *Girvanella*, a cyanophycean fossil form-genus is one of the calcarious algae. Some other fossil forms of the blue-green algae are: *Cryptozoon*, *Collenia* and *Newlandia*. The vast warm sea that covered much of the earth of the Paleozoic era was very suitable for the growth of algae. Thus, the Paleozoic era has been quite often called the 'Age of Seaweeds'. Many algae thrived in the earliest period of the Paleozoic

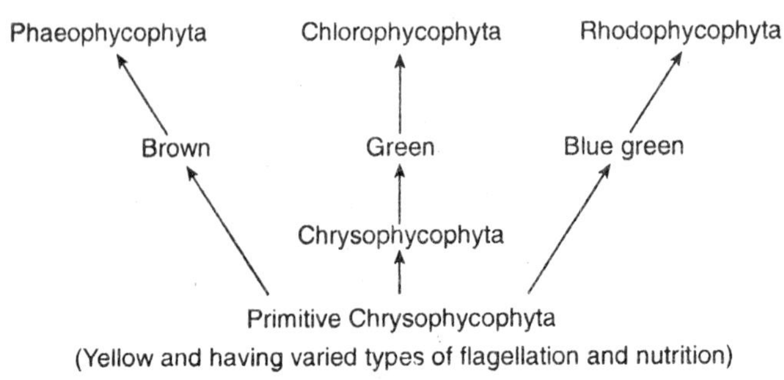

Fig. 10.2: Diagram showing the origin of the Phaeophycophyta from the Chrysophycophyta as suggested by Senn

era. They are not only of simple pattern, such as blue-green, and green, but also more highly differentiated types similar to our modern brown algae.

Among green algae, the Characeous remains have been found from the Paleozoic era. Kidston and Lang have reported from the Paleozoic era the two form-genera, *Lagynophora* and *Palaeonitella* which have very close similarity with the present-day Characeae. Among the Chrysophycophyceae, *Scyphosphaera* and *Thorosphaera* have been found in early Paleozoic time. Chrysophycophycean fossil algae are rather abundant in Cretaceous onwards. Several genera of the family Codiaceae under Siphonales have been found to be common on the Lower Silurian. *Ovulites,* a representative of Eocene has been considered to be allied to the alga *Penicillus.* The form-genus, *Microcodium,* a member of the Miocene may have some affinities with the members of the Codiaceae under Siphonales. About sixty well-preserved fossil genera of the family Dasycladaceae (Siphonales) have already been discovered from the Ordovician. They serve link with living members of the Dasycladaceae. It is quite possible that the incrustation with carbonate of lime on the various parts of these algae have been advantageous for the preservation of the numerous fossil forms of this family. Mention may also be made of various other genera which are of wide interest, for example, *Mizzia* from Permian, *Gyroporella* from Triassic, *Triploporella* and *Tetraploporella* from Jurassic and Cretaceous periods, and *Dactylopora* from Eocene.

Almost all imperishable siliceous shells of Diatoms are laid down in various habitats to form thick deposits. This large accumulation of Diatom shells is known as **diatomaceous earth.** The earliest reported diatomaceous deposit is during late Mesozoic (Cretaceous). But immensely thick deposits of diatomaceous earth have been found to be associated with rocks of the Tertiary age. In the Diatoms, the Centrales are older than the Pennales for having only a few living representatives.

The brown algae which probably had their origin in the early Proterozoic can very well be traced in many geological ages. One of the fossil types of Phaeophycophyta that is most striking of the Silurian and Devonian in *Prototaxites.*

Though some red algae are traceable in the later Archeozoic actually their fossil forms are found in the early Cambrian. Certain fossil forms of Rhodophycophyta have resemblance with the living species with regard to external appearance. It is quite probable that highly organized red algae have attained their maximum development in the middle Cambrian. Walcott named these Cambrian algae as *Waputika, Dalvia, Wahpia* and *Bosworthia,* and compared them with the corresponding living forms, such as *Dasya, Ceramium* and *Griffithsia.* Limestone rocks consisting largely of *Lithothamnion* are well known from various parts of the world being predominant in Miocene. Besides this, *Solenopora* of Ordovician-Jurassic is an allied genus.

E. ALGAE CAUSING BIOLOGICAL DISTURBANCES

Nowhere can there be found more striking examples of interactions between and, interdependence of organisms in nature than in the ramifying effects produced by algal populations. The roles which these plants play in aquatic biology and in limnology are clearly recognized but often incompletely understood. Whereas the place of algae in nature may involve some benefits to man and to other organisms, some of the effects produced are radically offensively disturbing, sometimes costly and even lethal. The putrefying masses of algae in lakes and in water-supply reservoirs are aesthetically disturbing and economically aggravating. But of greater importance are the indirect and far-reaching effects algae have on other aquatic organisms, including their ability to kill.

It is from populations of algae in standing water, of course, that effects are most obvious and most disturbing because of the concentration of numbers of individuals. It is well known that of the countless species of freshwater algae, only a few produce disturbances which attract our attention. This may be exemplified by the common 'water bloom'. Some of the so-called water bloom producing species of algae are *Microcystis aeruginosa, M. toxica, Oscillatoria rubescens, O, lacustris, Anabaena circinalis,* and *A. flos-aquae.*

The disturbances attributable to algae are both directly and indirectly produced. The direct effects are those arising from substances given off by certain objectionable algae, or through their physiology wherein nutrients necessary for the growth of less offensive forms are taken up; clogging of fish gills because of their concentration in shallow water; producing disagreeable tastes and odours in drinking water, etc. Indirect effects arise as a result of a chain reaction in modification of ecological conditions leading oxygen depletion, and formation of toxic substances.

Most of the disturbances of algae are related directly to their physiology, it may be noted in the instances of cyanophycean trouble-makers especially, that their effects are related to morphology and to their habits. For example, the sticky sheath possessed by blue-green algae makes it possible for plants to adhere to one another, thus forming dense mats. Many blue-green algae, especially the trouble-makers, possess pseudovacuoles which permit or force the plants to float high in the water. Because of their tremendous numbers, the mat is actually elevated above the water-level. Thus, in intense illumination and in warm water layers, a decaying blanket of blue-green algae forms.

The indirect effects produced by dense algal populations is found in fish deaths and in human poisonings from eating fish. Because blue-green algae especially are high in proteins, when they decompose proteinaceous by-products are formed, some of which are poisonous. It has been demonstrated that when algal blooms are dense in shallow water, the decomposition products may be concentrated enough to kill fish.

Another indirect effect chargeable to algae is that of fish-kills by oxygen depletion. As the oxygen decreases below the threshold necessary for them, various groups of organisms including fish are suffocated to death.

The incompatibility of two algal species and the inhibition of growth of one plant in presence of another are often due to the antibiotic substances that are produced by algae and are liberated out in water in which they survive. Chlorellin from *Chlorella vulgaris* is a well-known antibiotic which inhibits the growth of other algae. The antibiotics may even have inhibitory effects on fish. Pandorine from *Pandorina morum* produces marked effects and abnormal shapes in *Scenedesmus.* Both *Microcystis* and *Chlorella* secrete substances active against *Closterium.*

A serious disease of sheep is reported from Africa, Canada, and the United States. Death from blue-green algal poisoning had experienced lung haemorrhages and liver lesions, where horses, cattle, and dogs were killed by drinking water infested with *Microcystis aeruginosa* and *Anabaena lemmermannii.*

The blue-green alga *Microcystis toxica* produces 'soupy' blooms in the shallow pans and reservoirs. Cattle drinking these waters died by the thousands in Central South Africa.

Part Three

Eumycota (Fungi)

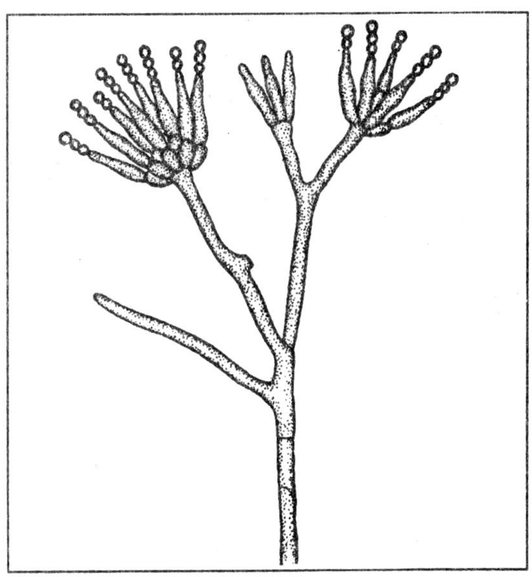

Chapter 11

Introduction To The Eumycota (Fungi)

The **fungi** (sing. **fungus;** L., a mushroom) are heterotrophic organisms of very diverse form, size, physiology and mode of reproduction. They may, however, be defined as, achlorophyllous organisms whose nucleated somatic bodies are usually surrounded by cell walls containing cellulose or chitin or both, and which reproduce asexually and sexually. The study of fungi is known as **mycology** (Gr. **mykes,** a mushroom and **logos,** discourse). Most familiar fungi are mushrooms, toadstools, the molds that grow on foodstuff, on wet shoes, the yeast that is used in brewery and bakery industries, etc., but thousands of others are so minute or so evanescent being only slightly developed and soon disappearing, or grow in such obscure situations that they are to be found by trained persons. Most fungi have basic similarities with algae in the simplest nature of the vegetative body possessing definite cell wall and their mode of reproduction being by means of spores. In general, fungi have cell wall composed of chitin and the reserve food is glycogen. In both fungi and algae the spores may be formed either directly from the vegetative body through certain asexual processes or as a result of sexual act. But unlike algae, the majority of fungi are to depend on external sources for their nutrition since they lack the synthesizing mechanism of food from the simpler substances. Furthermore, in fungi with the progressive complexity of the vegetative body there is a gradual degeneration of visible sexuality, whereas in algae reserve is the case.

REVIEW TO THE STUDY OF FUNGI

I. Abroad

Just like the study of other living organisms the study of fungi—mycology, had its beginning centred around the use of fungi for food and medicine without being concerned with the knowledge of their nature. The writings of the Greeks and Romans indicate that they could distinguish various edible and poisonous mushrooms. For centuries the Chinese had used certain fungi as food and others as medicine.

With the invention of printing, there started a period of renewed interest in plants when herbalists collected, described and made illustrations of many fungi along with seed plants. Of the early herbalists Gaspard Bauhin (1560-1624) who first initiated 'binomial nomenclature', listed 100 species of fungi and lichens and grouped them under the names **Fungus** and **Muscus Saxatilis vel Lichen** respectively. Joseph Pitton Tournefort (1656-1708) in his **Elements de Botanique** arranged fungi in six groups: (i) Fungus, comprising all centrally stalked agarics, boletes, polypores; (ii) Boletus, comprising Morchella, Phallus, and Clathrus; (iii) Agaricus, comprising Auricularia, and all polypores laterally attached to trees, logs, etc.; (iv) Lycoperdon, comprising various Lycoperdaceae and also some larger Myxomycetes; (v) Coralloids, comprising coral fungi and other branched fungi; (vi) Tubera, comprising subterranean fungi. Mention may be made of Dillenius (1687-1747) and Sebastein Vaillant (1669-1722) who, besides adding many genera and species of fungi with their illustrations, made valuable changes of Tournefort's naming and grouping of fungi. Dillenius named the cup-shaped or saucer-shaped fungi as *Peziza*.

Pier' Antonio Micheli (1679-1737), an Italian botanist made outstanding contributions to the knowledge of fungi. He in his classical publication **Nova Plantarum Genera** gave accurate descriptions and illustrations of various genera and species of fungi. Micheli paid special attention to the larger fungi and grouped them into **Fungi lamellati** (Agaricaceae), **Fungi porosi** (Polyporaceae), **Fungi ramosi** (Clavariaceae) and **Fungi pulverentes** (Lycoperdaceae and similar others). He by his simple technique of inoculating Squash pieces with *Mucor, Aspergillus* and *Botrytis* and keeping some of them covered and others exposed demonstrated that: (i) the spores of various moulds were in the air, (ii) each fungus produces its own kind. In fact, Micheli is the founder of the science of mycology. During the period from 1707-78 Carl Linnacus worked on different groups of plants. He in his 'Species Plantarum' placed fungi along with algae in his twenty-fourth class 'Cryptogamia'. Linnaeus put Lichens among algae. His treatment of fungi was very superficial and different from that of Micheli and his predecessors. After Linnaeus mention should be made of the works of Bulliard (1791) which contain descriptions and accurate illustrations of many micro-fungi, special mention may be made of Mucorales, Discomycetes, Pyrenomycetes, etc.

Most significant contribution in the classification of fungi was made by Christian Hendrik Persoon (1755-1838). The knowledge of higher fungi, particularly the 'Hymenomycetes' was enriched through very valuable contributions of Elias Magnus Fries (1794-1878) in Sweden. Fries published his work in three volumes of **Systema Mycologicum**, on the basis of which the present-day system of classification is drawn and it serves as the starting point of classification of fungi. Contemporary to Fries was Lewis David von Schweinitz (1780-1834), who is often called the 'Father of American Mycology'. He collected thousands of species of fungi from North Carolina and Pennsylvania, described and published them. Many monographic studies that were made in subsequent years had the foundation on Schweinitz's collections and publications. Beautiful illustrations of fungi by Tulasne brothers, published during 1861-65 have still remained unequalled and unsurpassed through years and years. Italian mycologist Pier' Andrea Saccardo (1845-1920) became famous due to his 'Sylloge Fungorum', in which he published the descriptions of ail recognized fungi. Sylloge Fungorum first appeared in one volume in 1882 and it came out in twenty-five volumes in 1931, and contained descriptions of all known fungi in Latin.

The first outstanding work of Anton De Bary was to workout the life history of the

Myxomycetes (1859). Besides this, he made more fundamental contributions of which special mention may be made of his work in which he determined the life cycle and heteroecious nature of *Puccinia graminis* Pers. He inspired many young people to be interested in fungi. Among his students the name of Oscar Brefeld may be mentioned, who for the first time introduced the study of fungi growing under pure culture by which it was possible to study the developmental stages as well as their physiology. De Bary and his students thus opened up a new line of developmental morphology. Thus, De Bary is universally known as the father of modern mycology.

The contributions of Fontana (1767) on rust (*Puccinia graminis*) and Prevost (1807) on bunt of wheat (*Tilletia*) have clearly indicated that fungi produce spores and through these spores they produce plant disease. Their evidence was the starting point from which the ancient dogma of spontaneous generation started to fall apart. It was Louis Pasteur (1882-95) through his classical experiments demonstrated that the microbes are airborne, induce fermentation, and reproduce their own kind. Pasteur's revolutionary experiments shook off the foundation of the spontaneous generation. Researches of Berkeley (1837), Kuhn (1858), De Bary (1866), Hartig (1874), Woronin (1878) and Brefeld (1884-1912) established the fact that fungi cause some of the plant diseases. Their work gave incentive to the wide appreciation of the economic importance of fungi as a result of which ultimately opened up the scope of Mycology and Plant Pathology.

Mention should be made on the mycoplasm hypothesis of Eriksson (1894) which was unique of its kind by which Eriksson tried to explain the annual recurrence of rust disease of cereals. P. A. Dangeard (1894) and R. A. Harper (1896-97) were the pioneers to introduce and develop the cytological studies of fungi, whereas A. F. Blakeslee (1904) worked out in details the sexual phenomena in the Mucorales and introduced the names homothallism and heterothallism. Brief mention is made of some of the names of fairly recent workers and their lines of investigations: Mlle Bensaude (1918), Hans Kniep (1913-17), Craigie (1927) and others who worked on the sexuality of the various members of the Basidiomycetes. A. H. R. Buller (1924-34) through his publication 'Researches on Fungi' clearly indicated how principles of physics may be applied to interprete the various functions of fungi. Mention is also made of the work of Stakman (1927) on the races of various parasitic fungi. B. O. Dodge (1927), the Lindegrens (1933-43), and others made intensive studies on the genetics of the Ascomycetes with special reference to *Neurospora* and yeast. Dodge pointed out the properties that made *Neurospora* an almost ideal organism for the study of the laws of heredity. In a series of publications he laid the foundations of a new branch of science known as haploid genetics. Dodge's discoveries stimulated geneticists and biochemists to start using *Neurospora* as an experimental tool.

Mention may also be made about the valuable contributions of the following workers in the different areas of Mycology. G. W. Martin on different groups of fungi and Myxomycetes; J. S. Karling on chytrids; F. K. Sparrow on aquatic fungi; S. D. Garrett on soil fungi and soil fertility; E. S. Salmon on the powdery mildews; L. Olive on fungal cytology; C. T Ingold on spore liberation in fungi; E. S. Luttrell on the Pyrenomycetes; J. A. Nannfeldt, M. Le Gal, R. P. Korf, R. W. G. Dennis on the Discomycetes; E. J. K. Corner on the Clavariaceae; M. K. Nobles, B. Lowy, D. N. Pegler on different aspects of the Polyporaceae; R. Singer and A. K. Smith on the agarics; M. A. Donk on the Hymenomycetes; and C. W. Hesseltine on fungi and food fermentation.

II. In India

In the nineteenth century some foreign visitors made stray collections of fungi from different parts of India with which started the collection and study of Indian fungi. They used to send their collections to various European Laboratories for the study of characters and identification. Hooker and Thompson were the pioneers who started a more comprehensive and systematic survey of Indian fungi. They made extensive collection of fungi in India. Their collection included a large number of Agarics from the Eastern Himalayas which were described at later time by Berkeley. Sulpis Kurz, the then curator, Royal Botanic Garden, Calcutta, made similar contribution, whose collections of fungi from Bengal and Burma were described by Currey. Special mention may be made of the names of D. D. Cunningham (1875-97) and A. Barclay (1886-92) who for the first time initiated the investigation of Indian fungi in India, thereby they made a very valuable change in the practice of sending abroad the fungi collected from India for study and identification. Cunningham studied the members of the Mucorales, Ustilaginales, and Uredinales; whereas Barclay carried on researches on the rusts occurring in the neighbourhood of Simla.

During the last decade of the nineteenth century Indian workers started to be interested in the study of fungi, among which was K. P. Kirtikar who collected and studied Indian fungi, particularly the fleshy fungi. But it was E. J. Butler, who, for the first time, initiated, organized and finally made a tradition on a large-scale mycological and phytopathological researches in India. After making voluminous contributions over a period of twenty years Butler left India in 1921. His research publications on various aspects of the different groups of fungi and textbooks entitled 'Fungi and Diseases in Plants' and 'Plant Pathology' are being widely used by different teaching and research institutions in India. Butler also trained many Indian mycologists and plant pathologists who are now working in the various research institutions in India. The scientific monograph, 'Fungi of India' written by Butler and G. R. Bisby published in 1931 provides a list of Indian fungi and host index of pathogenic fungi. W. McRee (1911-34), almost a contemporary of Butler, contributed on the South Indian fungi and also worked on the various aspects of the different species of *Phytophthora* occurring in India.

K. C. Mehta (1924-41) made outstanding contributions on the annual recurrence of rusts on wheat and on the wheat rust problem in India. The fungus flora of Allahabad, Nainital, and Mussorie was worked out by J. H. Mitter (1929-38). Beside this, he also worked on *Fusarium, Sclerospora* and heterothallism in fungi. Mycological contributions of H. P. Chaudhuri (1929-47) comprise a wide range of various groups of fungi. His primary interest was the study of the Indian aquatic fungi, in which area he made very valuable contributions. He, with his associates, collected, described, and studied the behaviour of water moulds in culture, their growth characters and the factors influencing sexual and asexual reproduction; and published a very useful book, 'A handbook of Indian Water Moulds'. Chaudhuri discovered a genus and named it after the name of one of his collaborators, A. Hamid as *Hamidia* of the family Saprolegniaceae. He, with his group of workers, described various species of *Myzocytium, Saprolegnia, Protoachlya, Isoachlya, Achlya* and *Rhizidiomyces.* Under his supervision the description and illustration of a number of Myxomycetes were made by S. Ahmad in 1934, which was later on published in the form of an illustrated monograph entitled 'Indian Slime Moulds'. The name of M. Mitra (1931) is associated with the history of bunts in India as because he recorded a new bunt on wheat in India—*Tilletia indica* which was later renamed by C. S. Holton (1949) as *Neovossia indica* (Mitra) Mundkur taking into consideration of the fundamental differences and similarities between it and *Tiletia caries* which causes bunt of wheat.

Working at the Forest Research Institute, Dehra Dun, K. D. Bagchee (1939-51) made very remarkable contributions in the field of forest mycology and pathology. His researches are of applied nature through which he has been able to work out many new fungi causing damage to the timber yielding plants both in nursery and field in India. Mention may be made of G. W. Padwick's (1939-45) publications, particularly on *Fusarium* and *Helminthosporium.* His publication, on the use of field observation in classifying Fungi Imperfect, is an interesting piece of work. Most of his contributions are plant pathological in nature which will be treated in a separate place. Many important contributions to the taxonomic and morphologic studies of Indian smut and rust fungi were made by B. B. Mundkur (1939-46) and M. J. Thirumalachar (1938-50) and their collaborators. Mundkur and his associates, in a series of publications, prepared a historical account of the collections of smuts in India and studied various species of Indian smuts. Their studies contributed a great deal towards the knowledge of Indian Ustilaginales. B. P. Pal and Mundkur (1939) published very valuable papers on cereal smuts and their control with special reference to the development of resistant varieties of cereals. Thirumalachar described a new genus of smut fungus which has interesting features of heterosporous sori, structure of promycelium, etc. Being quite different from other genera of Ustilaginaceae, this new genus of smut fungus was named by him as *Mundkurella.* Thirumalachar also studied the germination of chlamydospores of *Ustilago striaeformis* and cultured it in the laboratory. In 1952 Mundkur and Thirumalachar presented a consolidated account of Indian smuts in an excellently well-illustrated monograph entitled 'Ustilaginales of India'. Mundkur reported the occurrence of eight species of *Phakopora* and two species of *Bubakia* from India. Mundkur and N. Prasad (1938) described a new species of *Ravenelia* occurring on *Acacia modesta.* Thirumalachar also made extensive studies on the various aspects of several genera of rusts, and a number of new species, such as *Puccina leiocarpum, Melampsora mundkuri* and many others. Thirumalachar and Mundkur jointly published a number of papers on the genera or rusts, in which they put together information regarding 120 rust genera of the world.

In course of last three decades S. R. Bose has devoted his research activities on the Polyporaceae and other fleshy fungi of Bengal, subsequently West Bengal, and worked out the history and geographical distribution of Bengal polypores and their high altitude occurrence. Through a series of publications he suggested the use of anatomical characters for the identification of species of these fungi along with the characters of basidia and basidiospores. He also studied the production of fruiting bodies of the polypores under cultural condition and demonstrated that their basidial transformations, to a great extent, are controlled by the water-relations of the external environment. Bose is one of those pioneers who searched for antibiotics in higher fungi. T. S. Sadasivan (1938-39) working on the competition in fungi studied the succession of fungi decomposing wheat straw in different soils. Valuable contributions on the fungi of Bengal and subsequently West Bengal have been made by S. N. Banerjee. Special mention may be made of his work on the Thelephoraceae of Bengal.

In 1927, P. Bruhl and J. C. Sengupta compiled a check list of Indian Myxomyetes recording 161 species. During 1962 to 1964, A. K. Kar collected, preserved, identified different genera of Myxomycetes, and isolated plasmodium it, a species of *Licea,* genus of

Myxomycetes and studied its plasmodium under cultural condition.

During 1970 to 1989, A. K Kar and others surveyed distribution; collected, described and published papers on different groups of fungi of West Bengal.

ECONOMIC IMPORTANCE OF FUNGI

The first acquaintance of human beings with fungi started knowingly or unknowingly from their beneficial as well as harmful characters. There are numerous ways in daily life that people are directly or indirectly concerned with fungi. Scarcely a day passes during which all of us are not benefited or harmed by them in some way or the other. Some fungi are prized for their usefulness directly or indirectly to the society whilst others are shunned for causing great harm to both plant and human communities.

Since the fungi lack the chlorophyll pigment essential for independent existence, hence they must get their food from other living organisms or from dead or decaying organic matters. Some of the fungi cause serious diseases of human beings and of other animals, and plants. Many, however, are of great benefit to mankind and green plants. Thus, mushrooms, morels, truffles, etc., are used as food; yeasts are highly important in the baking, brewing and many other industries; and many soil fungi are important in crop production. Antibiotic substances like penicillin is obtained from fungi, and certain alkaloids extracted from fungi have become extremely important in medicine. Again, fungi are responsible for much of the disintegration of organic matter. As such they destroy food, fabrics, leather and other consumers' goods manufactured from raw materials which are subject to fungi attack. They also cause the majority of known plant diseases.

The fungi may be formidable enemies to mankind. They may destroy the stable food supply of a human society. Example may be cited from the damage caused by the fungus *Phytophthora infestans* which was responsible for the Irish famine of 1845. This fungus was responsible for the complete destruction of the entire potato crop in Ireland by inducing a disease known as potato blight which causes severe damage to the potato tubers in humid weather during the growing season. In the years from 1845 to 1860 a million people died in Ireland alone as a direct consequence of the famine, and one and a half million emigrated from Ireland to the United States. It also unlocked the doors of England on the nature of plant disease and called into being a new branch of applied science, Plant Pathology. Following are some of the economic aspects of fungi:

1. **Useful for the maintenance of soil fertility:** The majority of the known fungi along with bacterial live on dead organic matters, decompose them performing a very useful service in returning to the soil the nutrients originally extracted by the plants and also prevent many inorganic substances from being lost by leaching. Thereby they play an essential role in maintaining soil balance as well as in recovering the loss of soil nutrients useful for the green plants. Thus they are very important in preserving soil fertility. Some of the soil fungi are more active in producing ammonia from proteins than are the ammonifying bacteria.

2. **Essential for various industries:** The fungi form the most useful basis in industries like: baking and brewing (*Saccharomyces cerevisiae*), preparation of cheese (*Penicillium roqueforti* and *P. camemberti*), commercial production of organic acid—citric acid with the help of *Aspergillus niger* and of some vitamin preparations. Yeast played a very important role in the nutrition of German people during World Wars I and II. The high vitamin content

of yeast cells is very useful as food. Some fungi are grown on commercial basis for the extraction of pigments which are used for the preparation of various paints. *Thamnidium elegans* is used industrially in the U.S.A. for improving the taste and tenderness of beef.

3. **Utilization of antibiotic substances and alkaloids:** The fungi produce a variety of substances as their metabolic products which include enzymes, alkaloids, toxins, organic compounds, and similar other substances. Most of these substances are of great economic importance. Hence a number of fungi are cultured on a commercial scale for the utilization of their metabolic products.

I. **Antibiotic substances:** The metabolic products produced by certain fungi which inhibit the growth of or destroy bacteria and other microorganisms are called **antibiotics.** Hence these are antibacterial, antifungal, and to some extent antiviral substances. The antibiotic substances can also accelerate the growth of certain smaller animals and birds. The production of antibiotic substances by fungi is common. Whether the antibiotic substances are produced in sufficient amounts to inhibit the life processes of pathogenic organisms under natural conditions is not certain. That fungi produce antibiotic substances was first established by Sir Alexander Fleming in 1929.

(i) **Used as drugs:** Mention may be made of the wonder drug **penicillin,** a boon to the sufferers, is an antibiotic substance extracted from the fungus *Penicillium notatum.* Since 1941 penicillin has been widely used as a therapeutic agent against infections induced by certain bacteria. The use of various antibiotics has far transcended their original utility in curing human and animal diseases.

(ii) **Used for other purposes:** Non-therapeutic uses of antibiotics are growing into importance day by day. Recent developments have taken place in their stimulatory concentrations to chicks, pigs, and calves increasing growth rate and decreasing infant mortality of these young animals. As a result, the antibiotics are generally mixed with animals feeds. It is not clearly known how these antibiotics act to promote increased growth of animals but evidence is in favour of the view that they eliminate harmful bacteria and reduce the number of intestinal microorganisms which competitively eat up some of the essential nutrients. The effect of antibiotics on the growth and protection of plants from diseases is coming into importance day by day. Recently various types of antibiotics are being extensively used to increase the storage life of perishable foods, particularly poultry, meat, fish and egg. Main advantages include saving of foods, increased availability and distribution capacity of such easily perishable foods in areas where cold storage facilities are inadequate or totally lacking, the use of antibiotics within certain limits in food preservation is very effective. It has been found that antibiotics at low concentrations (5 to 7 parts per million) may be advantageously used for poultry and meat before storage. For fish preservation, the above concentration of antibiotics may be mixed with crushed ice or with salt in which the catch is covered, particularly after sea-fishing.

Antibiotics are also used as fungicides, details of which have been treated separately.

II. **Alkaloids:** The sclerotia of the fungus *Claviceps purpurea* contain a number of alkaloids which are utilized in medicine for the preparation of abortifacients and are also useful in controlling haemorrhage during child-birth.

4. **Used as food:** Certain fungi like, mushrooms, morels, and truffles are eagerly used, particularly in the western countries, as food. Mushrooms have been used by man as food since ancient times, but not until the seventeenth century they were cultivated as a crop and then not extensively. The early cultivation of mushrooms took place in France. Soon many other European countries started to

grow them. In the United States they were not grown before the latter part of the nineteenth century, and the method of growing them was a guarded secret for a long time. Now they are produced in many places all over the world. They are grown in caves, cellers, and green houses, but especially in houses built for the purpose. In the western countries mushrooms are extensively cultivated with great profit but with very little cost of production. There are on the market of the large cities as many as 50 or 60 different species of mushrooms for sale in their season. *Agaricus bisporus* is the one which is widely cultivated at a nominal cost of production and is most often seen on the menu of the restaurants and public eating places. This is also sold as fresh mushroom or is canned and made into soups, sauces, and other food products. Mushrooms have some value for its protein and vitamin content. People who like to eat them do so for their flavour and need no other inducement. The fact that a great mushroom growing industry has developed and is constantly growing in the United States is a proof that a great many people appreciate mushroom as food.

The genus *Morchella* is rather well-known as an edible fungus under the names morel or sponge musroom. The morels are highly prized for being delicious.

Truffles are the species of the genus *Tuber.* Besides just being edible, they have already earned a fame of special importance for their delicacy of flavour. In France, Spain and Italy they form an important article of commerce. When mature, their fruiting bodies emit a powerful odour. The subterranean fruiting bodies are gathered usually with the assistance of trained dogs or pigs. These animals locate them by their odour. In Italy and France they are canned for export. Fruiting bodies of the genus *Lycoperdon* commonly known as puff-ball fungus are also edible when young. The young maize cob tumours formed as a result of localized infection induced by the fungus *Ustilago maydis* when fried forms a popular side dish.

Most of the Clavarias are tender and may be used as food. In fact, several of them have been reported as edible such as *Ramaria aurea, R. botrytis, R. botrytoides, R. flava, R. flavobrunnescens, R. obtusissima, R. rufescens, R. strasseri, Clavaria vermicularis, Clavariadelphus pistillaris, Clavariadelphus truncatus* and many more. In India, too, several of the large-sized Clavarias would be quite edible though little is known on this aspect as yet. Uptill now only *Ramaria apiculata* and a species of *Sparassis* are widely used as food in Simla and Dalhousie Hills respectively. The latter is very large-sized and forms a favourite dish. It grows profusely at the base of coniferous trees during the rainy season. The food value of Clavarias would apparently be quite similar to that of other well-known edible fungi such as edible mushrooms, morels and truffles.

5. **Induce diseases of animals and human beings:** It is widely known that fungi (species of *Aspergillus, Cercospora, Cryptococcus,* and others) infest animal and human bodies and thereby cause a disease known as **mycosis** (pl. **mycoses**). Depending on the fungi causing the disease and the host infected mycoses may affect only the skin or may cause generalized infection of a more serious nature, examples of which are body ringworm, aspergillosis, cryptococcosis, etc.

6. **Induce disease of plants:** Fungi have great economic importance as they cause the majority of known plant diseases. Of all parasitic diseases of plants, these caused by fungi are most common. The diseases in plants have been recognized since the time of Aristotle, Theophrastus, and others. But during that time no clear explanation was put forward for the incidence of the diseases. There is hardly a plant that is not subject to the attack of one or more fungi. They are responsible for the

damage of agricultural crops and crop plants (*Phytophthora infestans* causes late blight of potato disease, *Puccinia graminis tritici* causes black stem rust of wheat and *Helminthosporium oryzae* causes brown spot of rice); Ornamentals (*Puccinia malvacearum* causes hollyhock rust disease of *Althaea rosea* and *Malva rotundifolia*); and the forest and shade trees (*Polyporus shoreae* causes root rot of *Shorea robusta*) whose annual yield is crores of rupees. Again a considerable amount of money is spent every year for sprays, dusts, and their plant protection work to combat fungous diseases, and almost the same amount is spent in breeding plants resistant to fungous attack. The fungi are more dangerous to plants than bacteria to us. Since most of the plant diseases caused by fungi spread very rapidly, the damage is ofen very serious in nature which may even be more serious and widespread if prompt control measures are not undertaken. Instances are not infrequent to indicate clearly how in extreme cases due to the severe attack of fungal diseases the entire plantation has been wiped out from the country. Some of them are: coffee rust of Ceylon, chestnut blight of the U.S.A.

7. **Causes loss of food and other essential commodities:** The fungi cause spoilage of fruits, vegetables, all kinds of foodstuff and other consumers' goods, destruction of fabrics, leather, paper, optical equipment, and cause rotting of lumber, railroad ties, wooden beams and joints, etc.

8. **Useful for academic studies:** Certain fungi are being used as essential medium for important academic studies.

(i) **Cytological, genetical and biochemical studies:** Of late, fungi are being widely used by cytologists, geneticists, and biochemists as important research tools in the studies of fundamental genetical and biochemical processes. These studies have contributed immensely to our knowledge of the action of genes in the inheritance of characters including the biochemical pathways and the manner in which genes control enzyme action.

Such studies are possible because some fungi grow and reproduce rapidly and consequently a number of generations can be easily obtained with a very little effort within a much shorter time than of higher plants or animals. Furthermore, they occupy very little space and all the stages involved during sexual reproduction can be easily followed under cultural conditions with minimum expenses. Fungi used are: *Neurospora, Saccharomyces, Ascobolus* and *Sordaria.*

(ii) **Physiological studies:** It is interesting that the fungus *Gibberella fujikuroi* causing serious rice and maize diseases serves as the source of gibberellic acid which is being extensively used as a growth promoting substance in the studies relating to growth of various plants.

9. **Useful for biological control of soil borne pathogens:** Certain fungi parasitize amoebae, nematodes and other microscopic terrestrial or aquatic animals. These fungi are known as **predaceous fungi.** Some of these fungi are profitably utilized for controlling soil borne pathogens like nematodes as a part of biological control of plant diseases. These fungi possess a special trapping mechanism by which they catch and destroy the soil borne pathogens. For details see Part: Plant Pathology.

MORPHOLOGY OF SOMATIC STRUCTURES

Thallus. The assimilative body, or somatic body, or vegetative body of a fungus is a thallus. It ranges from unicellular uninucleate being rounded, elongated (Fig. 11.1A to B′), or lobed in shape with or without having cell wall developed around it; a filamentous branched structure called **mycelium** (pl. **mycelia**) to a poorly developed rhizoid-like system resem-

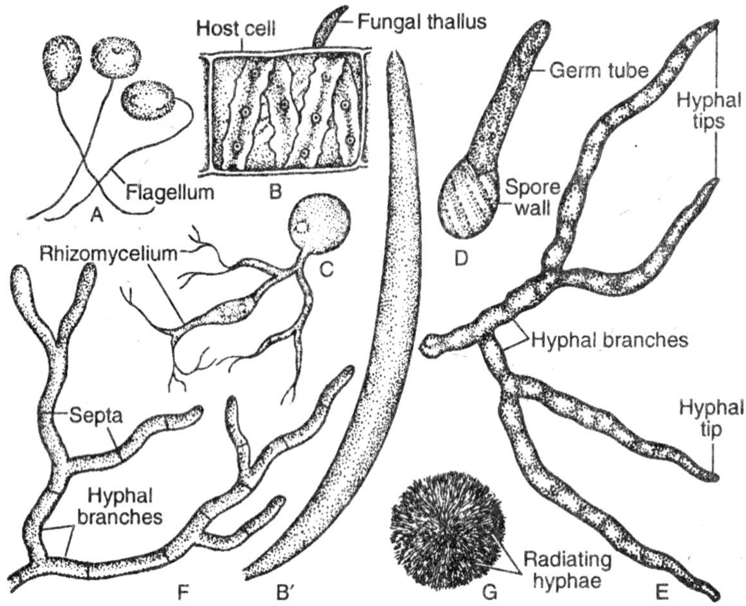

Fig. 11.1: Variations in the somatic structures of the Fungi. A. Unicellular uninucleate rounded with a single flagellum in *Synchytrium* sp. B-B′. Unicellular uninucleate elongated in *Harpochytrium* sp. C. Rhizomycelium in *Nowakowskiella* sp. D. Spore germination by germ tube. E. Aseptate hyphae. F. Septate hyphae. G. Spherical hyphal colony showing radiating hyphae

bling mycelium and is known as **rhizomycelium** (Fig. 11.1C). Despite this apparent simplicity, fungi show great diversity in size, metabolic activity and organization of specialized structures.

When unicellular, thalli produce bud cells in succession, these may remain attached to one another in a chain known as a **pseudomycelium.** The occurrence of a pseudomycelium is especially notable in yeast. Some of the pathogenic fungi have a mycelial thallus in the host but a yeast-like thallus in culture, e.g., *Taphrina* and members of the Ustilaginales. These fungi thus possess **dimorphic thalli.**

Hypha. Each individual filament of the **mycelium** is a **hypha (pl. hyphae)** which is usually a branched tube-like structure having protoplasm with reserve food bounded by a wall of a transparent pliable material (Fig. 11.2). The hyphae may be hyaline or variously coloured. Each hypha may vary enormously in over all length and diameter, the latter ranging from 0.5 μ to 1 mm. The apex of a hypha is a thin-walled region where growth materials are added, differentiation takes place, elongation occurs in a zone behind the tip. Further back from the apex the wall sets and becomes rigid in time with the deposition of wall substance. Branching **dichotomous** when the apex of a hypha ceases elongating and forks into two equal branches. More often branching is subapical and lateral, leaving the leading hyphal

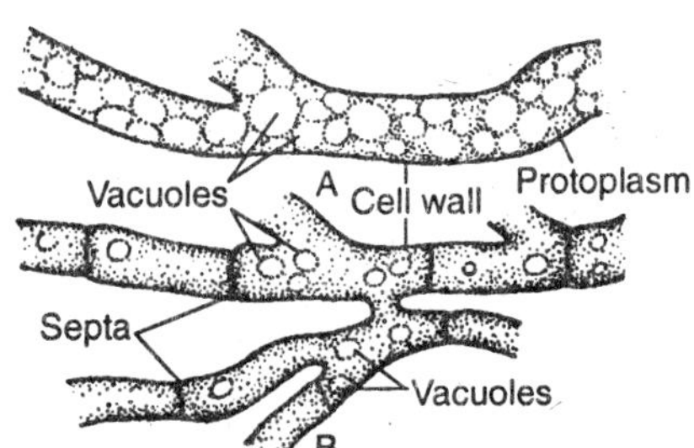

Fig. 11.2: Structural details of somatic hyphae. A. Portion of an aseptate hypha showing cell wall and protoplasm containing vacuoles. B. Portion of a septate hypha showing cell wall, protoplasm containing vacuoles and septa

apex free to continue its growth. The type of branching may be: **dichotomous, verticillate, cymose** and **racemose.**

The protoplasm of the hypha may be continuous without being interfered by cross-wall or transverse partition or **septum** (pl. **septa**). **The septa** may be **primary** and **adventitious.** Primary septa are formed in association with mitotic or meiotic nuclear division, and they separate the daughter nuclei. Adventitious septa are formed in the absence of mitosis or meiosis and occur especially in association with change in the local concentration of cytoplasm. A hypha is called **aseptate** (Figs. 11.1E and 11.2A) when septa are absent in a hypha, and thus the protoplasm is continuous, and that which is divided into cells by septa is known as **septate hypha** (Figs. 11.1F and 11.2B). An **aseptate hypha** is nothing but a multinucleate **tube-like** structure, a **coenocyte** and the vegetative **body** is thus **coenocytic.**

In an aseptate hypha, septa are produced when reproductive structures are delimited or when a wall is developed to seal off a damaged region of the hypha. In septate hyphae the septa, in majority of cases, are found to possess a central pore, and the septa are thus **porous** or **perforated septa** (Fig. 11.3B, C). More rarely the septa possess more than one pore. Again pore may be absent in a septum which is then known as **non-porous septum.** In certain fungi the hypha may form **pseudosepta** which are plugs or septum-like partitions of cellulin or other substance deposited at intervals and have chemical composition different from that of the hyphal wall. The widespread occurrence of perforated septa, through which cytoplasm, cytoplasmic organelles or even nuclei can regularly pass (Fig. 11.3A), indicates that the distinction between coenocytes and septate fungi is not so profound as has sometimes been imagined. The septa can, in fact, be regarded primarily as strengthening features of the hyphae.

All septa appear to originate in the same way as rim of material which develops centripetally either to cut off a segment completely or, more usually, to stop development before this happens, thus leaving the characteristic central pore. Septum formation by centripetal growth is rather rapid. The septal pore may be simple or complex. A simple septal pore is without swollen pore rim, but a complex septal pore possesses a typical swollen pore rim. Around the rim of the central pore amorphous material develops as a swelling. This swollen region thickens. A septum and its swollen pore rim are invested closely by the cytoplasmic reticulum on either side of the pore giving rise to a cup-shaped, perforated pore cap as soon as the septum has formed. A septum bearing a complex pore is designated as a **dolipore septum.** Again a complex septal pore may break down and get disorganized resulting in what is known as a **secondarily simple pore.** Both light

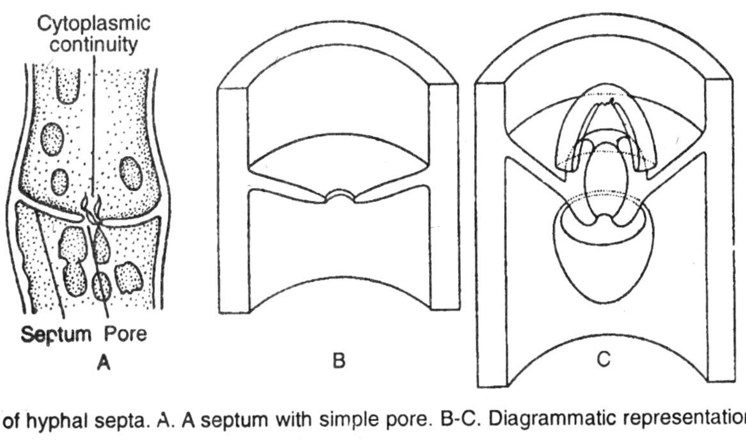

Fig. 11.3: Nature of hyphal septa. A. A septum with simple pore. B-C. Diagrammatic representation of hyphal septa

microscopy and electron microscopy studies reveal that septa of large number of fungi possess micropores, possibly in groups having narrowest in the centre.

The features of the septal pore characterize distinct groups of higher fungi. The Ascomycetes have a pore represented by a simple opening that is usully plugged by or closely associated with electron dense, membrane-bounded Woronin bodies. With the exception of the Uredinales and Ustilaginales (which have an ascomycetous type of septum), the Basidiomycetes have a highly elaborate dolipore septum.

Depending on the number of nuclei in each cell, the septate hyphae may be mainly of three kinds: with multinucleate cells; with binucleate cells known as **dikaryotic hyphe**, each nucleus usually derived from a different parent cell, the corresponding mycelium is **dikaryotic mycelium;** and having only one nucleus in each cell, called **monokaryotic hyphae**, the mycelium being **monokaryotic mycelium.** The mycelium may be very narrow or wide-angled. The hypha, in most fungi, is developed by the rapid elongation accompanied by branching of a tube-like or filament-like structure known as a **germ tube** which again originates through the germination of a spore (Fig. 11.1D). The hypha ramifies spreading over or within a **substratum,** the medium from which it receives its nutrition producing a hyphal colony. The hyphal colony is composed of radiating hyphae which have a tendency to grow in all directions starting from a central point and ultimately to develop a spherial colony (Fig. 11.1G).

Fungal cell structure. Although minor differences in composition and arrangement do exist, the fine structure of fungal cells is broadly the same in unicellular as well as filamentous thalli of fungi.

Cell wall. The development of electron microscopy since 1950 has provided clear evidence for the occurrence of microfibrillar, structural elements in the walls of fungal cells. There is clear evidence that fungal cell walls include microfibrillar components associated with non-fibrillar material. The composition of hyphal wall is, however, extremely variable. The hyphal wall is laminated, usually being composed of two to several layers of microfibrils arranged in various ways in an amorphous matrix. This type of construction is also found in cell walls of higher plants, but with difference in chemical composition. The chemical heterogeneity of hyphal wall has been recognized for several decades, since De Bary (1866, 1887) distinguished 'true cellulose' from 'fungal cellulose'. Three classes of compound: **cellulose, chitin** and insoluble **glucan** can occur as microfibrillar elements and are presumably the basis of the structural rigidity of fugal cell walls. The microfibrils of cellulose and chitin are of indefinite length and similar diameter. The fungal cell wall is generally composed of a complex nitrogenous substance known as **chitin** which is often impregnated with some salts or similar other substances. The chitin that is present in the wall of fungi is not strictly indentical with animal chitin; the formula $(C_{22}H_{54}N_4O_{21})_n$ has been suggested for the fungus chitin. In some fungi true cellulose has been detected. But in most fungi cellulose is lacking. **Callose,** a complex carbohydrate, **lignin**-like substances and other **organic materials** have also been detected. Cellulose and glucan may well be the predominant material in the walls of some lower fungi. Chitin, or perhaps, **chitosan** predominates in some other lower fungi. In the higher fungi, however, it looks as though insoluble glucans may be predominant structural material although chitin may be present in quite appreciable amounts. In certain fungi (*Neurospora crassa*) the presence of a chitin core covered by an outer layer of glucan has been demonstrated.

Cell inclusions. The fungal cell inclusions are enclosed within a cytoplasmic membrane—

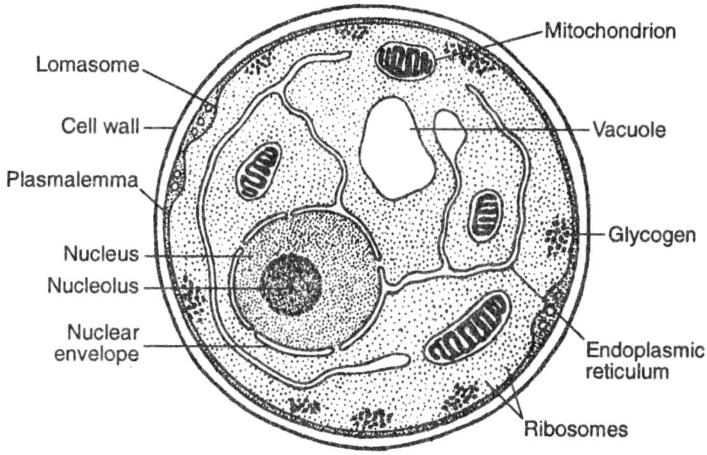

Fig. 11.4: Structural details of a fungus cell

plasmalemma. Within the cytoplasmic membrane, endoplasmic reticulum (Fig. 11.4) has been demonstrated, which seems to be looser and more irregular than in the cells of green plants and animals. In several fungi the endoplasmic reticulum is composed of microtubular structures. In many fungi in the apical growing region, the endoplasmic reticulum is highly vesicular. A further remarkable feature is the occurrence of aggregations of complex concentric lamellae associated with an invagination of the cytoplasmic membrane (**plasmalemma**) or free in the cytoplasm. The small pockets adjacent to the wall, formed by invagination of the plasmalemma are kown as **lomasomes** (Fig. 11.4). A Golgi apparatus, similar to that found in higher plants and animals, has been detected in some fungi. The principal cell inclusions are **dictyosomes, mitochondria, ribosomes, vacuoles, centrioles, nuclei,** and **storage product.**

(i) **Dictyosomes.** The dictyosomes consist of a 'stack' of plate-like membrane bound sacs with the edges of which are associated small, more or less spherical vesicles.

(ii) **Mitochondria** (sing. **mitochondrion**) (Fig. 11.4). Mitrochondia are remarkable amongst the filamentous fungi both for their size and for their ability to change shape. They may be small spherical structures at, or below, the level of optical resolution, which may elongate to 30 μ, form unequally thickened structures resembling a row of beads on a thread, or even branch quite considerably and may then apparently break up into a mass of discrete, smaller bodies. In growing hyphae they exhibit the most active motion. Studies with the electron microscope have not revealed any obvious fundamental structural differences between fungal mitochondria and those of other organisms, although some workers are of opinion that fungal mitochondria are fewer, flatter than those of green plants.

(iii) **Ribosomes.** Ribosomes, similar to those found in higher plants and animals occur in the cytoplasm of fungal cell (Fig. 11.4).

(iv) **Vacuoles.** Vacuoles are not apparent in the apices of hyphae but can readily be seen to arise further back, to enlarge and to show a tendency to coalesce. In most fungi the vacuoles remain as distinct structures often apprently located near to the cell walls, their number increasing with the age or degree or degeneration of the cells. The vacuoles are bounded by a distinct membrane, the tonoplast (Figs. 11.2, 11.4). Vacuoles may contain pigments or inclusions of various kinds, both crystalline and amorphous.

(v) **Centrioles.** Centrioles are not common in hyphae. They are usually observed during divisions of nuclei being located one in each pole of the nuclear spindle particularly during ascospore formation (Fig. 11.7F-H). More than one kind of centriole can exist in fungi.

(vi) **Nucleus.** Fungal nuclei are usually small; the majorty range from 2-3 μ in diameter but in some they are much larger, e.g., 25 μ. In living cells, as seen by phase contrast, they comprise a central dense area surrounded by a clear area, around which some observers believe they can detect a definite nuclear membrane (envelope). The nuclear membrane is in continuity with endoplasmic reticulum (Fig. 11.4). Unlike the endoplasmic reticulum in higher plants and animals, the endoplasmic reticulum in fungi is usually comparatively sparse and is not closely packed. The nuclei do not, in general, stain readily apart from the central body which stains heavily with iron haematoxylin (Fig. 11.5A-C). The central body is usually Feulgen-negative and appears as an amorphous, or granular mass in electron micrographs. It is usually described as the nucleolus (Fig. 11.4) but whether it is really comparable to that structure in higher green plants is uncertain. This body has been claimed to contain RNA. The central body behaves differently in different fungi during nuclear division. Electron microscopy has provided clear evidence for a typical three-layered nuclear membrane. The membrane is perforated (Figs. 11.4).

In life, fungal nuclei can vary greatly in size and shape and can move, possibly at great rates, through hyphae. There may be as many as 20 to 30 in a cell, or there may be one or two nuclei per cell, up to even 100 nuclei have been reported in *Neurospora crassa.*

In somatic structures of fungi there may coexist, as a natural phenomenon, in the cytoplasm, nuclei of different genetic constitution and size variations. These nuclei are known as **heterokaryons,** as against nuclei having similar genetic constitution **homokaryons.** Such association of genetically different nuclei is designated as **heterokaryosis.** These nuclei are characterized by variability in the proportions of the sets of

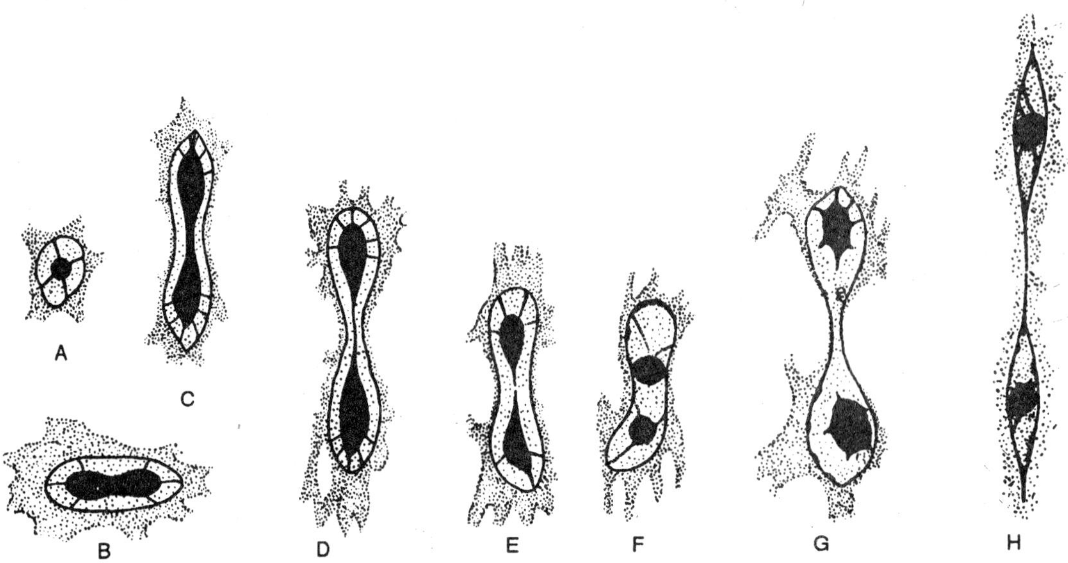

Fig. 11.5: Stages in mitotic division of fungus nucleus showing intact nuclear membrane and heavily stained central body

chromosomes existing side by side and provide a unique system of somatic variation. Heterokaryons can originate by mutation in a homokaryon or by the migration *via* hyphal anastomoses of nuclei of one mycelium into another.

After fixation the chromatin of resting nuclei appears finely or coarsely granular or filamentous. In most, but not all, instances it is not organized in the form of identifiable chromosomes. It is generally taken for granted that mitosis in fungi follows the same course as in higher orgnisms, but cytologically acceptable evidence for this belief is hard to find. Recent work has already indicated that fungal mitoses are of several different kinds and have peculiarities that set them off from the mitotic division encountered in higher plants.

The study of mitotic division in fungi is at present in an active almost explosive, state. There are three documented cases of mitosis in fungi. The first is a more or less typical mitosis, as generally understood; the second is a most unnatural type of division which is well documented for a number of fungi; and the last is somewhat intermediate in its features. Mitotic division of nuclei in *Macrophomina phaseoli* illustrates the first case, *Mucor hiemalis* second case, and *Basidiobolus ranarum* is the third case.

In general, metaphase in mitosis is usually short and really well-condensed metaphase plates are unusual in mitosis of fungal nuclei. In some, like *Macrophomina phaseoli* at metaphase a top-shaped spindle with spindle fibres and terminal centrioles develops. No distinct metaphase-plate arrangement of the chromosomes has ever been recognized. During anaphase chromosomal material separates into two chromosomal groups and pass in two opposite poles of the spindle (Fig. 11.5). In others, like *Basidiobolus ranarum* closely packed minute chromosomes form on the equator of the spindle a 'Saturn's ring' which separates during anaphase (Fig. 11.6). Again, in others, like *Mucor hiemalis* the nuclear material separates as two half-cups or shells which separate during anaphase. In a number of

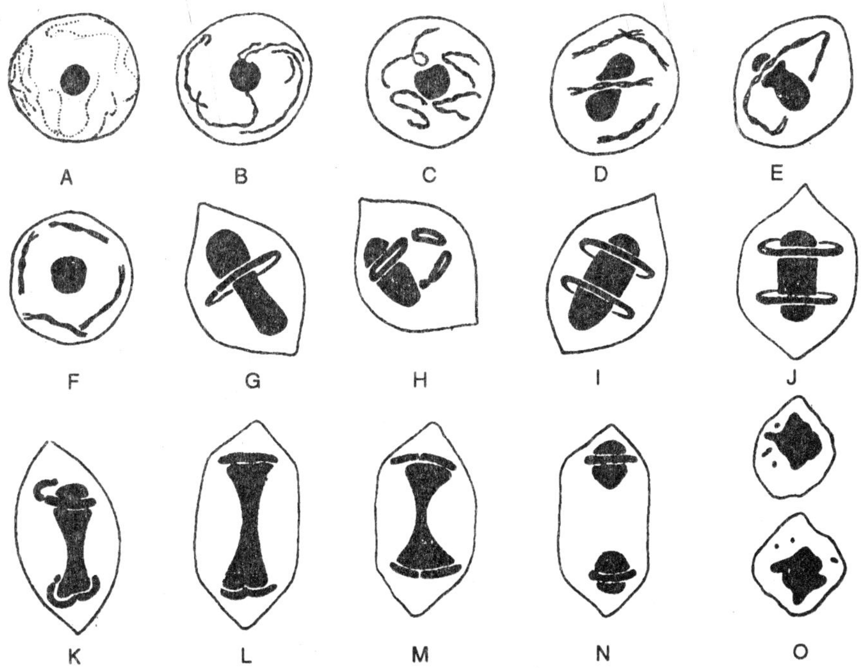

Fig. 11.6: Stages in mitotic division of fungus nucleus showing intact nuclear membrane, deeply stained central body and 'Saturn's ring configurations' of dividing chromosomes

fungi, there is persistence of the nuclear membrane which finally constricts between the two poles of the spindle in the telophase (Figs. 11.5G, H and 11.6M, N). Structures like intranuclear fibres have also been observed during mitosis in other fungi. A great deal of controversy exists concerning the presence of a mitotic spindle apparatus during division of fungal nuclei. According to Olive, the spindle mechanism is intranuclear in the fungi. This would imply that it takes its origin within an intact nuclear membrane. Recent electron microscopic study supports Olive's hypothesis.

The best authentic account of meiosis in fungi indicates that in fungi meiosis follows a normal course as found in higher plants, except in prophase. In many fungi synapsis, at zygotene, occurs between greatly elongated chromosomes but, in *Neurospora,* this is not so and the chromosomes are considerably contracted when pairing begins. They elongate as synapsis proceeds and, by pachytene structural differentiation can be clearly recognized. After attaining their maximum length and degree of linear differentiation at late pachytene, chromosomes condense and contract until they reach

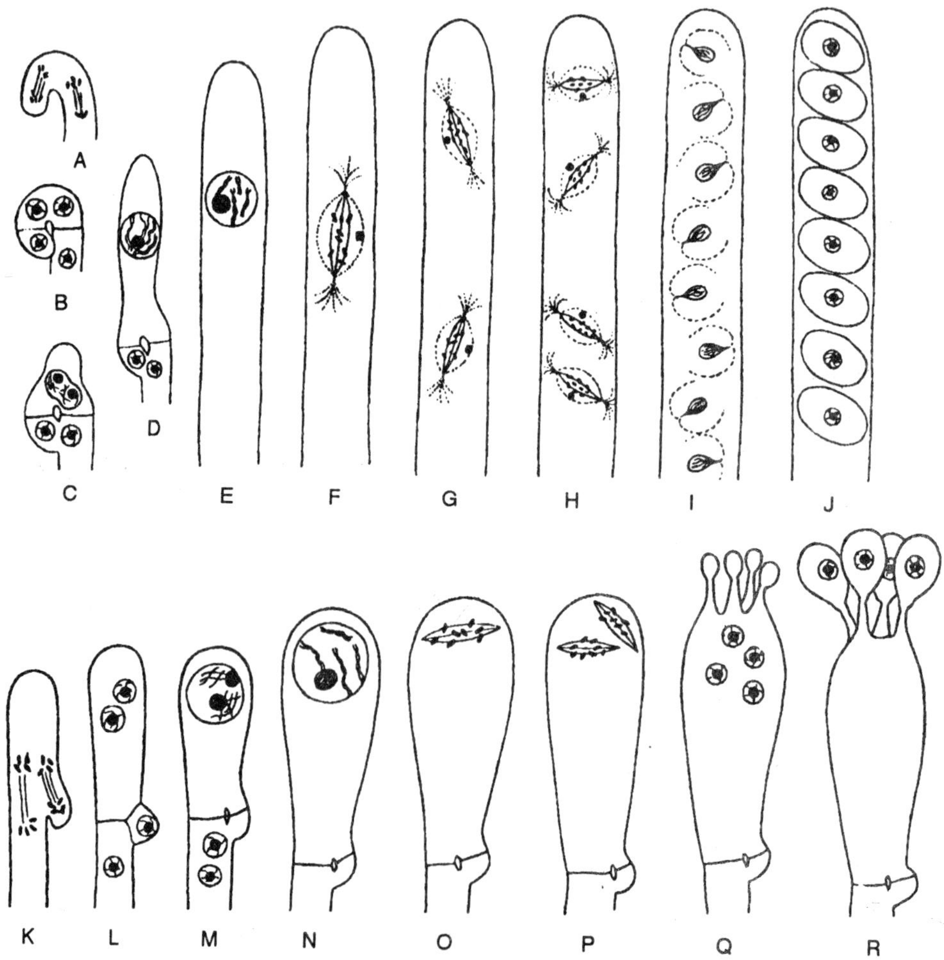

Fig. 11.7: Stages in meiotic and mitotic divisions of fungus nuclei. A-J. In an Ascomycete, K-R. In a Basidiomycete. A-B. Conjugate mitoses associated with crozier formation. C. Karyogamy. D. Synapsis. E. Diplotene, F. Anaphase I. G. Anaphase II. H. Anaphase III. I. Ascospore delimitation. J. 8-spored ascus. K-L. Conjugate mitoses associated with clamp formation. M. Karyogamy. N. Diplotene, O-P. Meiotic divisions. Q. Basidium with four haploid nuclei. R. 4-spored basidium

their smallest size at metaphase I. Diplotene nuclei are difficult to observe because of a characteristic fuzzy appearance. Bivalents are readily recognized at diakinesis as are chiasmata. An intranuclear spindle now develops although centrioles have not been observed at this stage. The bivalents become arranged at the periphery of the equator of the spindle and this is followed by disjunction and spindle elongation (Fig. 11.7F, O). The nucleolus usually becomes detached during anaphase I and is lost although it may persist until anaphase II (Fig. 11.7G, H). Nuclei are reconstituted after telophase. In division II, spindles are formed and the chromosomes contract rapidly in prophase. In telophase II the spindles elongate much more rapidly than in telophase I. The third division follows rapidly. Centrioles usually become clearly visible at interphase II. The spindle fibres are attached at one pole of the plate. The increase in size of the centrioles from metaphase I to interphase II as shown by their visibility seems to be associated with their function in spore delimitation which now occurs. A group of fibres radiates out from each centriole and surrounds a fusiform mass of cytoplasm including the nucleolus (Fig. 11.7F-H). A hyaline wall develops around the delimited mass and the nucleus undergoes a final division to produce a spore (Fig. 11.7I, J). Divisions III and IV are in all respects apparently normal mitosis. After further differentiation, the wall appears and eventually eight spores lie freely in the ascus (Fig. 11.7J). A general outline of the stages in meiosis in the basidial nucleus in the Basidiomycetes is given (Fig. 11.7M-R).

(vii) **Storage product.** The chief storage product in a fungal cell is a carbohydrate, **glycogen** (Fig. 11.4) which turns brown with iodine solution. The alcohol **mannitol** is also frequently found in fungi. Many fungi store food as minute droplets of **oil.**

Fungal tissue. In certain fungi due to more or less localized mycelial growth an accumulated mass of mycelial structure is formed which is known as **mycelial mat.** By further growth this mycelial mat may develop into a compact structure called **stroma** (pl. **stromata**). Again, at certain stages in the life history of some other fungi the mycelium grows together, intertwines and becomes organized forming a thick-woven tissue which is called **plectenchyma.** Plectenchyma may be: (i) a rather loosely woven tissue whose hyphal components lie more or less parallel to each other and are recognizable, is known as **prosenchyma** or **prosoplectenchyma** (Fig. 11.8); or (ii) hyphae lying adpressed to one another forming a compactly woven tissue whose hyphal components have lost their entities and appear isodiametric and continuous in sections resembling the parenchyma of higher plants, it is called **pseudoparenchyma** or **paraplectenchyma** (Fig. 11.8). In some fungi the mycelium may pass into a dormant or resting stage by the formation of hard resting bodies resistant to unfavourable conditions, called **sclerotia** (sing. **sclerotium**). The sclerotia may remain dormant for long periods of time and germinate upon the return of favourable conditions. Each sclerotium is composed of central prosenchymatous and peripheral pseudoparenchymatous tissues which are again surrounded by a rind of pigmented hyphae (Fig. 11.8).

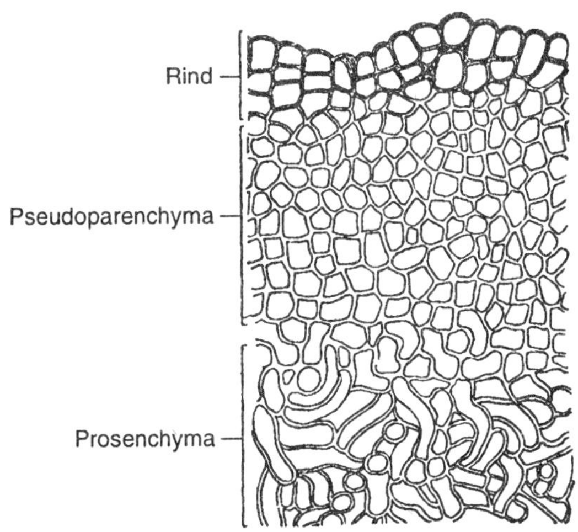

Fig. 11.8: Section through sclerotium of *Claviceps purpurea* showing structural details

Reserve carbohydrates and lipids are usually translocated to, and stored in, cells of developing sclerotia. The development of a compact, thick-walled outer rind results in a restriction of water loss and sclerotia can retain their viability for long periods.

Specialized somatic structures. In fungi with a mycelial thallus, the somatic hyphae are modified into specialized structures.

(i) **Sclerotia** and **pseudosclerotia.** As already indicated sclerotia are hard resting-bodies formed by aggregation of somatic hyphae, but pseudosclerotia are completely different structures. Pseudosclerotia are sclerotium-like masses composed of materials of the substratum (remnants of plant or animal tissue of the host or otherwise) bound together in a hard mass of mycelium. Thus they are not composed purely of fungal material.

(ii) **Rhizoids.** Many fungi produce relatively short thin hyphae which penetrate the substratum and serve the purpose of anchorage as well as food absorbers of the mycelium that remains externally on the surface of the substratum. Such hyphae superficially resembling roots are known as **rhizoids** (Fig. 11.11H) though they are structurally very simple.

(iii) **Rhizomorphs.** There are fungi whose hyphae aggregate together, behave as an organized unit to form a root-like strand in a thick, hard cortex and develop a growing tip somewhat resembling that of a root tip. Such a structure is known as a **rhizomorph** (Fig. 11.11I) which mainly serves the function of absorption. Rhizomorphic structures are mainly limited to the higher fungi. They are often blackish and hard on the outside and resemble a bootlace reaching length of 9 m or so. The rhizomorph can also survive unfavourable condition and its growth may be renewed with the return of favourable condition.

(iv) **Pseudorhizas.** The mycelium of some fungi originates from materials buried a considerable distance underground. These materials are the **pseydorhizas.** Fruit bodies of some members of the Agaricaceae often remain connected with underground pseudorhizas. The pseudorhizas of species of *Termitomyces* commonly originate from the comb of living termite nests.

(v) **Hyphopodia.** A hyphopodium is a short branch, one or two cells in length, of an external hypha in certain leaf-inhabiting parasitic fungi. The terminal cell of a hyphopodium may be expanded and rounded, or lobed, or pointed.

(vi) **Mycelial setae.** These are stiff, straight, septate, simple and acute or 2-4 dentate structures that are produced erect on the surface of the mycelium of leaf-inhabiting parasitc fungi.

(vii) **Appressoria** (sing. **appresorium**). An appressorium is a simple or lobed mucilaginous swelling on a germ tube or hypha, attaching to the surface of the host or other substratum. Appressoria are formed by some parasitic fungi.

(viii) **Haustoria** (sing. **haustorium**). A haustorium is an organ that is developed from a hypha usually performing the function of absorption (Fig. 11.11C-G).

(ix) **Laticiferous hyphae.** These are hyphae which exude drops of latex when they are damaged. The latex may be watery or milky or coloured, and its colour may change on exposure to air.

Flagellation. The somatic body of certain lower fungi bears hair-like structure known as **flagellum** (pl. **flagella**) (Fig. 11.1A). Again certain fungi, in course of their development, produce flagellate cells (swarmers) which according to their behaviour are either zoospores or gametes and may be uni- or bi-flagellate. Couch (1938, 1941) and Vlk (1939) clearly demonstrated the presence of two kinds of flagella in fungi. One narrows more or less abruptly towards the tip, which forms a thin extension of

variable length: this is the 'whiplash' flagellum or Peitchgeisel (Fig. 11.13I-K). The second does not obviously narrow and has fine lateral hairs. This is the 'tinsel' flagellum or Flimmergeisel (Fig. 11.13J and K). Details of whiplash type and tinsel type of flagella have already been discussed in previous chapter.

FRUCTIFICATION

In the life history of a large number of fungi, after an extensive growth, or with or without the stimulus of sexual reproduction, the mycelium undergoes change in form resulting in the loss of individuality and takes part

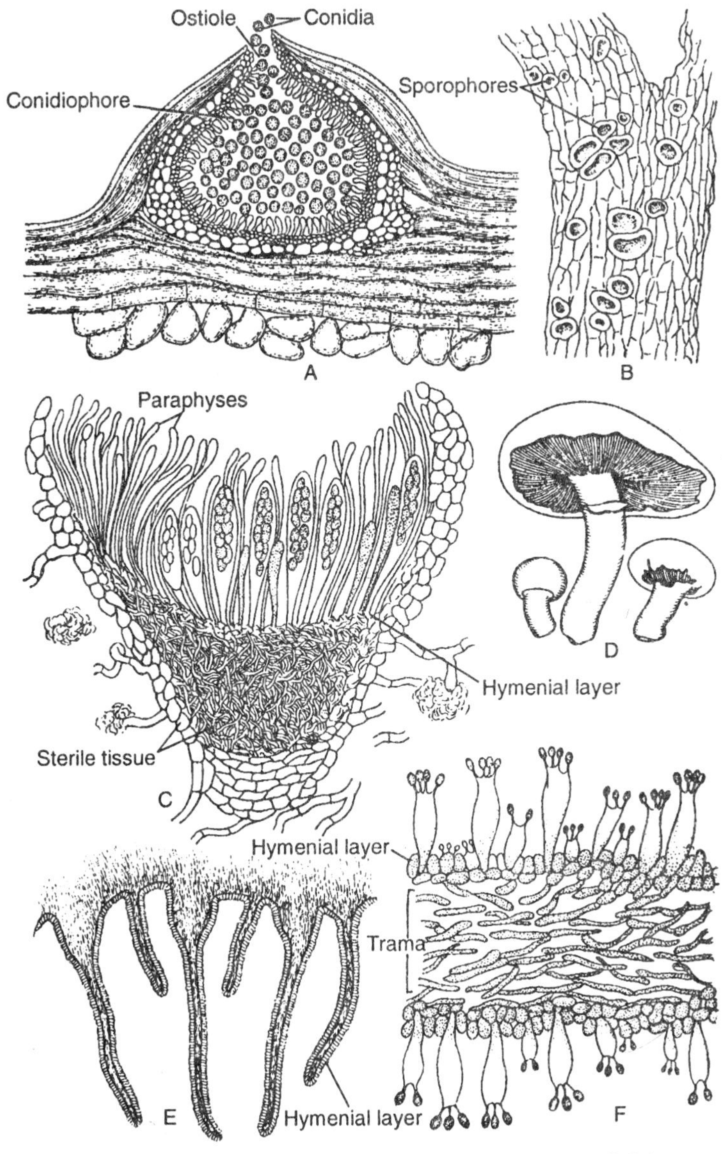

Fig. 11.9: Nature of some of the fructifications. A. Section through a pycnidium of *Sphaeropsis* sp. showing ostiole, conidiophores and conidia, B. Sporophores of *Dasyscypha* sp. C. Section through a fructification of *Diplocarpon* sp. showing hymenial layer, sterile tissue and paraphyses. D. Fructifications of *Agaricus* sp. E. Portion of a section of fructification of *Coprinus* sp. showing hymenial layer. F. Portion of a section of fructification of *Agaricus* sp. showing trama and hymenial layer

in the development of a structure known as **fructification** or **fruit body**, also called **fruiting body** or **carpophore**, also known as **sprophore** (Figs. 11.9 and 11.10). A fructification is a structure which may be simple to complex and minute to very large consisting of fertile portion composed of spore-producing elements and sterile portion of various tissues and accessory organs for purposes of maintaining mechanical rigidity, protection of fructification and for liberation of spores. The sterile portion is usually composed of pseudoparenchymatous tissue (Fig. 11.9C). Depending on the nature of fungi, the sterile portion is known as **hypothecium** and **excipulum** (Fig. 11.9D and E), **context,** or **trama** (Fig. 11.9F), whereas the fertile portion, in a fructification is known as **hymenium** or **hymenial layer** (Fig. 11.9C, E and F). The hymenium is usually enclosed, supported by, or carried aloft on sterile tissue and it is the range and form of the sterile tissue that gives the diversity of form to the fructification (Fig. 11.10). The hymenium may be well developed being a continuous layer or may be irregularly scattered at various levels within the fructification and is usually composed of spore-producing structures like **asci** (sing. **ascus**) (Fig. 11.9C) or **basidia** (sing. **basidium**) (Fig. 11.9F) borne interspersed with certain sterile hyphae called **paraphyses** (sing. **paraphysis**) (Figs. 11.9C and 11.9F) or similar other structures known as **pseudoparaphyses** which usually assist in the liberation of spores. The hymenium may also bear in addition to these, certain other sterile structures which are of great taxonomical importance for being diagnostic anatomical features of certain fungi. The fructifications differ in details of their structure (Figs. 11.10) and development. Again, the development of fructification is not universal in fungi. There are large number of fungi which do not develop any fructification at all.

OCCURRENCE AND MODE OF NUTRITION

Like all living beings the fungi also flourish in any environment which provides suitable food, moisture, and temperature for their existence though their requirements are extremely variable. For ideal growth, the fungi need a combination of warmth and humidity. In general, they prefer an acid medium and grow in the temperature range between 0°C and 35°C, the optimum temperatures being 20°C-30°C. There are fungi which can withstand extremely low temperature even up to –195°C. Light is not required for the growth of fungi. But in most fungi light is essential for sporulation and spore dispersal. Respiration in fungi is just like all other plants, the intake and consumption of oxygen accompanied by the giving off of carbon dioxide. None of the fungi appear to be obligate anaerobes. Many are facultative anaerobes, while some appear to be strict aerobes. Free oxygen is used by the fungi in respiration, chiefly as an acceptor of hydrogen. The facultative anaerobes have another mechanism of oxidation which does not involve free oxygen. This is called anaerobic respiration, or fermentation. The rate and amount of growth, sporulation, and the formation of metabolic by-products of a given fungus are affected by the oxygen supply.

Occurrence. Broadly speaking, the fungi are heterophytes. Due to the absence of chlorophyll in their body, they are unable to synthesize their food from CO_2 and water. Hence their mode of nutrition is quite different from that of the green plants. Naturally, they are to depend on various external sources for the supply of their nutrition which may again be in the form of prepared food or simpler desynthesized organic substances. It is due to this inherent character in them the fungi have varied mode of nutrition and are thus very widely distributed in various situations ranging from soil,

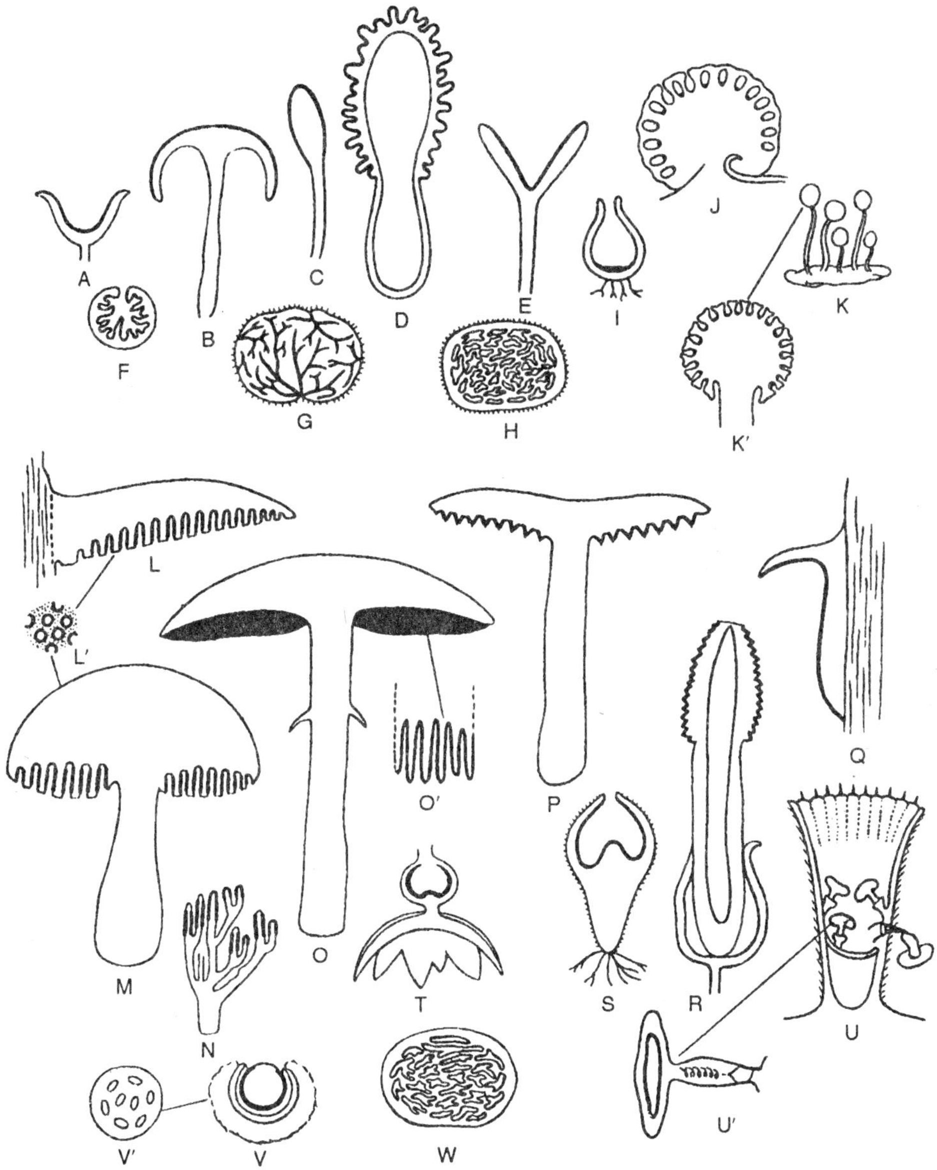

Fig. 11.10: Diagrammatic presentation of various types of fructifications in sections. The hymenial layer is represented by a darker line or area in each case. A-K′. Fructifications in the Ascomycetes. L-W. Fructifications in the Basidiomycetes. A. *Peziza*. B. *Helvella*. C. *Mitrula*. D. *Morchella*. E. *Sarcoscypha*. F. *Genea*. G. *Tuber*. H. *Balsamia*. I. *Sordaria*. J. *Daldinia*. K-K′. Sclerotium and stromata of *Claviceps*. L. *Polyporus*. M. *Boletus*. L′. Pores of *Polyporus* and *Boletus* N. *Clavaria*. O. *Agaricus*. O′. Gills of *Agaricus*. P. *Hydnum*. Q. *Stereum*. R. *Phallus*. S. *Lycoperdon*. T. *Geaster*. U. *Cyathus*. U′. Peridiolum of *Cyathus* magnified in section. V. *Sphaerobolus*. V′. Projectile of *Sphaerobolus* containing basidiospores. W. *Hymenogaster*

both hypogean and epigean; rocks; water; dead and decayed organic matter; animal dungs, clothes, feathers, hides, skin of animals, food-stuff; living green plants, living animals, living and dead insects and flies, sometimes even in acids and laboratory equipment and similar other substrates. The fungi growing on animal dung are called **coprophilous fungi.**

Mode of nutrition. Some of the fungi are omnivorous and subsist on anything containing organic matter and moisture. In general, they prefer to have organic matters in some form or the other. But there are fungi which are specific with respect to certain objects except which they fail to occur in any other substrate. Those which receive nutrition from dead or decaying organic matters are known as **saprophytes** or **saprobes**, and fungi surviving on living protoplasm of plants or animals are known as **parasites.** A living organism, plant or animal which entertains a parasite is called a **host.** Usually food material is absorbed over the whole hyphal surface from the substratum on which the fungus grows.

The fungi produce **enzymes**, which are nothing but the products of their metabolic processes. Some enzymes formed by fungi are excreted out from their body. These enzymes normally perform their functions outside the fungal cell that produces them. These are termed **exoenzymes** (extracellular enzymes), such as cellulase, amylase, and pectinase. Different kinds of fungi produce different exoenzymes. The fungi release exoenzymes in the surrounding substrate. Food materials, particularly organic substances—complex food materials, are broken down into simpler substances by the exoenzymes. These simpler substances are absorbed by the fungi. After entering the fungal cell these substances are acted upon by another set of enzymes which are formed in the fungal cell. These are **endoenzymes** (intracellular enzymes) which never come out from fungal cell. These endoenzymes make the absorbed simpler substances acceptable for the growth and nutrition of fungal cell. The enzymes are the specific organic catalysts which form an unstable union with the substrate on which they act, accelerate and often initiate chemical change. The life processes of fungi are controlled and directed by a complicated and interrelated series of enzymes or enzyme systems. Some fungi produce certain enzymes only in response to particular environmental conditions. Such enzymes are called **adaptive enzymes.**

(i) **Parasitic fungi.** In parasitic fungi, the hyphae usually grow inter- or intra-cellularly in the host tissue (Figs. 11.11A and B). The intercellular hyphae send out short-absorbing organs, **haustoria** (sing. **haustorium**) in the adjacent cells (Fig. 11.11C). With the help of these haustoria the hyphae receive nutrients from the cells of the host tissue. The haustoria may be knob-like in shape, short or long, or branched resembling a miniature root system (Fig. 11.11C-G). They are developed from the somatic hyphae probably due to a response to the contact stimulus along with the stimulus of host protoplasm. The haustoria usually enter into the host cells though minute pores punctured in the cell wall. Again, there are parasitic fungi whose mycelium does not ener in the host tissue and is entirely superficial. These fungi secure anchorage by numerous haustoria which sink in the epidermal cells of the host tissue and obtain nourishment from their protoplasts. Some other parasitic fungi produce flattened, hyphal pressing organs, from which minute infection pegs usually grow and enter the epidermal cells of the host. These organs are known as **appressoria** (sing. **appressorium**). Besides these, there are fungi which grow epiphytically on the surface of the green leaves. These fungi do not cause any major harm to the plant on which they grow except that their dark coloured superficial mycelium produces shedding effect and unwidely appearance. These are **epiphytic fungi.** But fungi whose mycelia ramify only on the host surface, develop appressoria as organs of attachment and send haustoria into the epidermal layer for deriving nutrition are known as **ectoparasites,** whereas those which penetrate the host and develop their mycelia within the host tissues are the **endoparasites.** The development of haustoria by the mycelium of the fungi parasitizing animals is unknown.

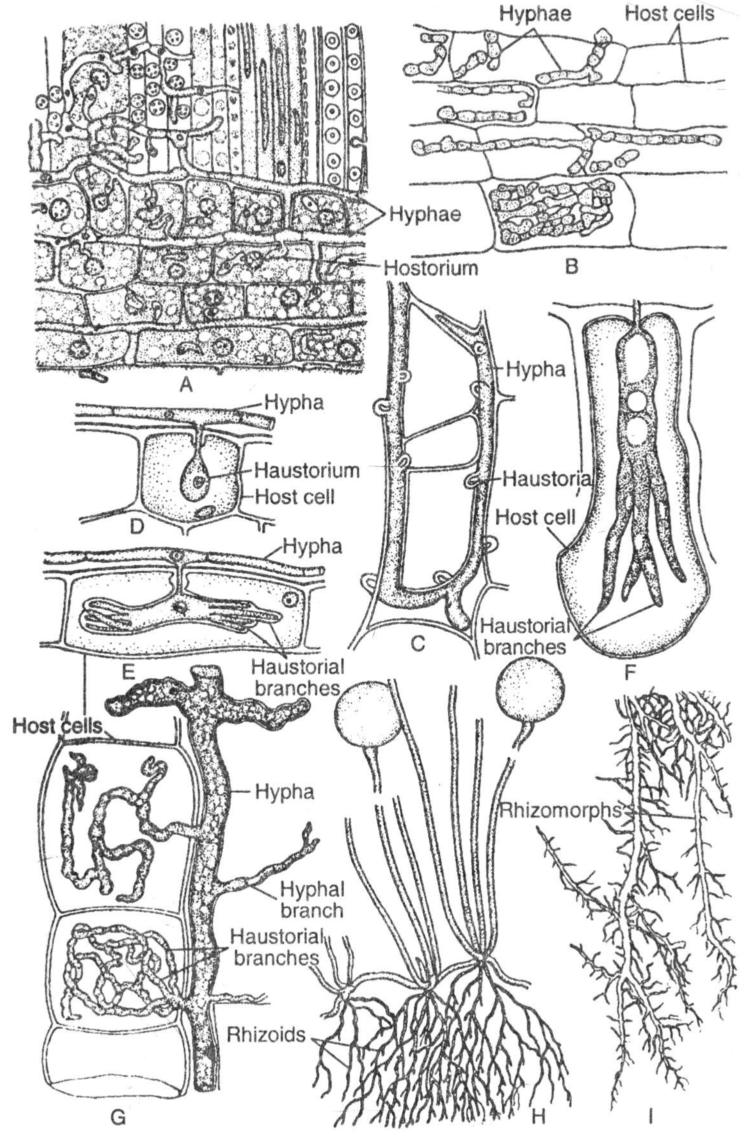

Fig. 11.11: Behaviour of fungal hyphae. A. Intercellular hyphae of *Cronartium* sp. in stem of *Pinus* sp. B. Intracellular hyphae of *Helminthosporium avenae* in the cells of *Avena sativa.* C-G. Various types of haustoria. H. Rhizoids penetrating substratum in *Rhizopus stolonifer.* I. Rhizomorphs of *Merulius* sp.

The penetration of mycelium in the host tissue is effected through natural openings: stomata, lenticels, etc., wounds or by direct penetration by the secretion of various enzymes which dissolve the cell wall. In 1944 Barghoorn first suggested that some fungi are capable of enzymatic hydrolysis of lignin. This has been confirmed experimentally by Johnson, who demonstrated the penetration of wood cells by fungi and their destructive cellulolytic activity in the absence of any contaminating organisms such as bacteria. The enzymes are capable of softening and dissolving either the middle lamella or the entire cellulose wall and thus help penetration of mycelium or haustoria in the host tissue. Again some of the enzymes are so toxic that they kill the cells of the host tissue.

Parasitism, a rather common form of life among many groups of heterotrophic organisms basically involves a nutritional relationship that is favourable to the parasite, and usually unfavourable to the host. The parasitism of one fungus on another is known as **mycoparasitism**, and the parasitizing fungus is the **mycoparasite**. This phenomenon has been recognized by mycologists for nearly one hundred years. The different forms of parasitism among fungi testify to the continuous struggle between organisms occupying the same habitat. Parasitic fungi have beeen classified by Gaumann in 1946 as either **biotrophic parasites**, those which obtain nutrients from living host regardless of whether they can grow on artificial media; or **necrotrophic parasites**, which have the ability to kill the host cells and use the nutrients thereof. Quite often the mycoparasites upon contact with a host hypha develop an appressorium-like structure. This is usually followed immediately by formation and penetration of a haustorium. A favourable nutritional relationship is quickly established with the susceptible host and rapid development of the parasite occurs. In others, contact with nearby host stimulated the hyphal branches to grow directly towards the nearby parasite. This is an adaptation for survival. The degree of parasitism, estimated visually by the amount of parasite growth or amount of host inhibition, varies with the host and parasitic species. Those fungi which complete their life history parasitically on certain animals or plants and thus cannot survive except on living protoplasm are known as **obligate parasites**. The parasitic fungi which pass their life history in or on the tissues of a single host are called **autoecious** in contrary to the **heteroecious** ones which require two hosts to complete their life history. These terms are particularly used for rust fungi.

Fungal races. Certain parasitic fungi which are capable of attacking different species of host plants may develop into distinct entities known as **races**. Such races though morphologically similar to each other have adapted themselves to those particular hosts and can be differentiated only on the basis of their reaction on the host plants. Hence the difference between them lies in their physiological characters because they produce differnt effects on various hosts. Thus, due to splitting of parasitic fungi specialized races are produced. The specificity of the races to different host plants is termed **specialization of parasitism** and the races have been called by various names, e.g., **specialized forms, specialized races, biological forms, biologic races, biologic species, physiologic forms**, and **physiologic races**. The use of the term physiologic race is a recommendation of the International Botanical Congress at Amsterdam in 1935. Many parasitic fungi can be grown on culture media and it is possible to distinguish races in them by differences in growth, consistency, topography of the surface, and other characteristics. Such races are termed **cultural races** to distinguish them from physiologic races which are differentiated only by their restriction of host choice. There is no direct or satisfactory evidence to correlate cultural races with physiologic ones. The physiologic races, in turn, may be subdivided into the ultimate unit, **biotypes**. A biotype is a population of individuals which gives consistent pathological reactions on a standard set of differential varieties of a species of a host genus. Differential varieties are chosen from among a large number of varieties tested for various disease reactions (resistant, susceptible, etc.), produced by a large number of isolates. Now, by the process of permutation and combination, those varieties showing similar reactions with different isolates are discarded and representatives from only those showing distinct reactions are chosen. The reactions on these varieties are so distinct as to distinguish a particular physiologic race or biotype by means of specific combinations of disease reactions on them. As for example, physiologic races and biotypes of *Puccinia graminis* can be identified by means of combination of disease reactions on a standard set of differential varieties.

(ii) **Saprophytic (saprobic) fungi.** There are also fungi which live only on dead or decaying organic matter, they are called obligate saprophytes, whereas according to the circumstance prevailing, fungi which are capable of growing on living protoplasm or on dead or decaying organic matter, are called **facultative parasites** or **facultative saprophytes.**

Many saprobic fungi produce rhizoids or rhizomorphs which perform the function of absorbing organs for the mycelium that grow superficially on the surface of the substrate.

(iii) **Mycorrhiza.** There are fungi which live in intimate association with some other unlike living organisms deriving mutual benefit, they are known as **symbionts,** and the process is **symbiosis.** Such association between the two unlike living organisms is **symbiotic association.** A symbiotic, non-pathogenic or weakly pathogenic association of various fungi and bryophytes, pteridophytes, or flowering plants is known as **mycorrhiza,** a condition of balanced parasitism. The fungi live in the **rhizosphere** (a region in the vicinity of the roots or the rhizoids and where physiological interaction of roots or rhizoids may still take place), on the surface of the roots, or in the roots or similar other organs of the bryophytes, pteridophytes, and flowering plants. They play a role in the nutrition and hormone exchange of the bryophytes, pteridophytes, and flowering plants using up some carbohydrates produced by the latter groups of organisms. The mycorrhiza may be **ectotrophic** (Figs. 11.12B and C), when the fungal hyphae remain restricted largely on the surface of the roots and rootlets forming **a**

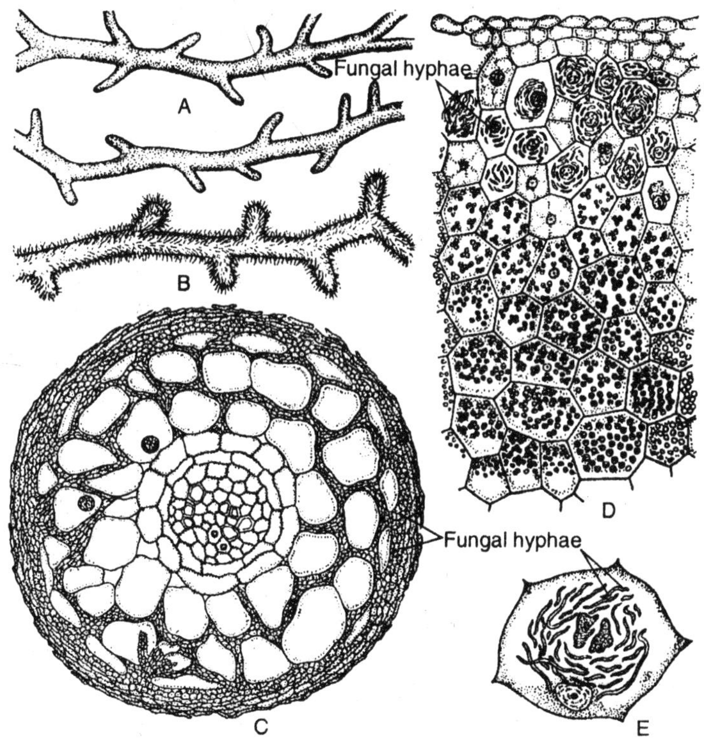

Fig. 11.12: Types of mycorrhiza. A. Root of *Pinus* sp. without mycorrhiza. B. Root of *Pinus* sp. covered with mycorrhiza. C. Section through a root of *Pinus* sp. showing ectotrophic mycorrhiza forming a compact mantle covering the root, with some portion of the fungal hyphae in between the cortical cells. D. Section through a root of orchid showing endotrophic mycorrhiza. E. A single cell of orchid root containing fungal hyphae

mantle or sheath of light pseudoparenchymatous fungal tissue which sends branches inward between the cortical cells of the roots and outward in the soil; and **endotrophic** (Figs. 11.12D and E), in which the fungus enters into the roots, where it is frequently kept restricted to well-marked layers—here the external hyphal mantle or sheath is lacking. There may also be **pseudomycorrhiza,** where the fungal hyphae, though superficially appear to be symbiotic, are definitely parasitic.

(iv) **Utilization of nutrients.** True, the majority of the fungi require already elaborated food for their survival, and are incapable of preparing their own food. But it has also been experimentally established that some fungi are capable of synthesizing their own proteins when grown in a medium containing carbohydrats preferably in the form of glucose, sucrose, or maltose; nitrogen in inorganic or organic form and; various mineral elements essential for their growth. There are also fungi which can synthesize vitamins required for their growth and reproduction. Some are also able to make up their deficiency of thiamine or biotin, or both utilizing from the substratum. Hence it is obvious that all fungi are not really incapable of synthesizing at least to some extent, their own food in the strictest sense.

Almost all the organic compounds used by or present in fungi contain hydrogen. One of the most important hydrogen-containing compounds is water. This compound is associated with proteins in the form of bound water, and it functions as a solvent in which most, if not all, biochemical reactions take place. Water enters into many reactions, particularly in the hydrolytic processes of 'digestion'. Apparently fungi do not utilize free hydrogen.

Most fungi utilize sulphate sulphur, but some require reduced sulphur. Other species are unable to synthesize specific sulphur-containing amino acids, especially methionine. Sulphur enters into the composition of enzymes and other proteins, peptides, and at least two vitamins.

The fungi utilize phosphorus in the form of phosphate salts and esters. Some specificity in the different sources of phosphate has been found. Phosphate esters enter into a wide variety of enzymatic reactions, and many coenzymes are phosphate esters. It is thought that certain phosphate esters act to transfer chemical energy to certain enzymatic reactions. Phosphorus enters into the composition of proteins, especially the nucleoproteins, which are found in the nucleus or cytoplasm of every cell.

Fungi differ in ability to utilize different froms of nitrogen. A few utilize atmospheric nitrogen; many utilize nitrate nitrogen; and a still greater number utilize ammonium nitrogen. All species are able to utilize some form of organic nitrogen. Other constituents in media, especially the four-carbon dicarboxylic acids, modify the availability of ammonium nitrogen and certain amino acids. Not all amino acids are of equal value in fungus nutrition. The primary amino acids are those which enter directly metabolic pathways, while secondary amino acids are deaminated before the nitrogen is used. Most of the nitrogen utilized by fungi enters into the synthesis of proteins. The primary amino acids are formed first, and the secondary amino acids are formed from primary amino acids.

REPRODUCTION

The common method of reproduction in fungi is by means of **spores** which may be developed with or without any sexual act, that means by sexual or asexual mehods respectively. The spores, in general, are unicellular; they may also be multicellular. A fungus, during its life history, may produce only one or more than one kind of spores. A spore becomes a **resting spore** when it can withstand extreme variations of climatic may not be developed inside a sporangium or a sporophore. Usually a fungus during its reproductive phase produces spores both by asexual and sexual methods. The part of the reproductive phase utilized for the development of asexually produced spores is known as **asexual stage** or **imperfect stage** and that during which spores are formed by sexual method, is the **sexual stage** or **perfect stage.** The life history of a fungus is complete when both the stages are developed. There are fungi which propagate themselves by imperfect st.ge only. It is quite likely that their perfect stage is not yet known. These fungi with their incompletely known life cycles are grouped under **Fungi Imperfect** or **Deuteromycetes.** As soon as their perfect stages are worked out they are transferred to their respective places depending on the spores they produce by sexual method and related characters. The asexually produced spores, in general, are very delicate and may be both flagellate and non-flagellate, whereas the sexually produced spores are comparatively thick-walled and do not bear any flagella. The asexually produced spores are formed profusely and are much larger in number than the sexually produced spores. They are repeatedly formed several times under favourable conditions. These spores are responsible for the rapid increase in number of new individuals and their distribution in various localities. But they are not normally very useful for perennation as most of them are very delicate and are incapable to stand extreme climatic changes, whereas the sexually produced spores are formed only once in the life history of a fungus. During the process of their development an interchange of characters takes place which leads to the origin of genetically new individuals. Besides this, these spores are very useful organs through which a fungus can tide over unfavourable conditions. When the entire somatic body of a fungus is converted into one or more reproductive structures the fungus is called **holocarpic,** and **eucarpic** where only certain portions of the thallus take part in the development of reproductive structures and the rest of the thallus performs its somatic function. Certain fungi reproduce by a simple process of budding, whereas in others the vegetative mycelium may break into small fragments, each of which under favourable conditions develops to form a new individual. Usually there are two broadly recognizable types of reproduction: **asexual** and **sexual.**

Asexual Reproduction: Asexual reproduction is a method of production of new individuals by non-sexual means, that is, without involving any sexual process. Common methods of asexual reproduction are as follows:

(i) division of a vegetative body into one or more fragments, each of which grows into a new indivdual;

(ii) simple splitting of vegetative cells into two daughter cells by constriction;

(iii) development of small outgrowths, called **buds** from the vegetative body, each bud grows further to form a new individual;

(iv) production of spores from the vegetative body by various processes and each spore grows into a new inividual.

The fragmentation of the vegetative body may be accidental and the fragments under favourable conditions develop into new individuals. In some fungi fragmentation is a normal means of reproduction. In them the hyphae break up into numerous small fragments known as **oidia** (sing. **oidium**) (Fig. 11.13A-B″) or **arthrospores** which behave as spores and produce new individuals. Oidia may also behave as male gametes (spermatia) and take part in sexual reproduction. The fragments of hyphae are called **chlamydspores** (Fig. 11.13C-F) when they are separated from each other and become thick-walled. The chlamydospores are resting spores, being thick-walled they can stand unfavourable conditions. They

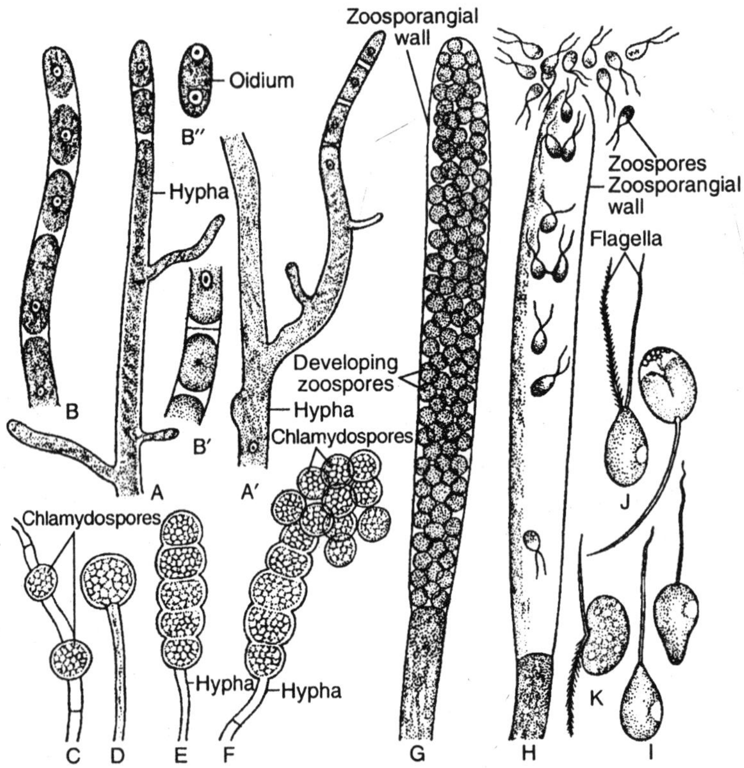

Fig. 11.13: Various types of fungal spores. A-B″. Stages of oidium development. C-F. Chlamydospores. G. A zoosporangium with developing zoospores. H. Zoospores liberating from a zoosporangium. I. Uniflagellate zoospores with whiplash flagellum. J. Biflagellate zoospores with whiplash and tinsel flagella anteriorly placed. K. Biflagellate zoospore with whiplash and tinsel flagella laterally placed

may remain viable for considerale period, up to several years. With the return of favourable conditions they germinate to produce new individuals. As such the chlamydospores are the organs of perennation. They may be terminal or intercalary in origin bearing either one or two nuclei, that is, monokaryotic or dikaryotic which again depends whether the fragments are of monokaryotic hyphae or of dikaryotic hyphae.

The simple process of splitting of vegetative cells into daughters cells is known as fission rather common in bacteria, is also present in some fungi. The splitting takes place by constriction and the daugher cells so formed develop into new vegetative cells.

There are certain fungi which produce small outgrowths, buds from their vegetative body. These buds receive one daughter nucleus each as a result of division of the parent nucleus. Eventually the buds are cut off from the parent cell and mature to form new individuals. The buds may also remain attached to the parent cell for a long time resulting in the formation of chains of buds (Fig. 11.13B).

The most common method of asexual reproduction in fungi is by means of spores. These spores are formed from the vegetative body by various processes other than sexual and are known as **asexual spores.** They are extremely variable in structure and development and may be flagellate or non-flagellate. The flagellated spores are called **zoospores** (Fig. 11.13H) which may be developed in a sac-like structure known as a **zoosporangium** (Fig. 11.13G and H). The zoospores may be uniflagellate or bi-flagellate when they bear one or two flagella (Fig. 11.13I-K). In biflagellate zoospores, the flagella are equal in length. The flagellum of uniflagellate zoospore may be of whiplash type (Fig. 11.13I) or of tinsel type, whereas, the biflagellat zoospores bear flagella one of whiplash type and the other of tinsel type. They are of two kinds: pear-shaped or pyriform with two flagella placed at the anterior end (Fig. 11.13J); and kidney-shaped or bean-shaped bearing two oppositely directed flagella inserted laterally in the furrow on the concave side (Fig. 11.13K).

The non-flagellate spores produced asexually are in general of two kinds: **sporangiospores** and **conidia** (sing. **conidium**). The sporangiospores are developed in a sac-like structure called **sporangium** (Figs. 11.14A and B). Stalk of the sporangium is **sporangiophore.** A sporangium may or may not have a dome-shaped sterile structure known as **columella** (Fig. 11.14A). Some mycologists have defined **sporangiospores** as spores that are produced asexually in a sporangium. According to them, flagellate sporangiospores are **zoospores** and non-flagellate sporangiospores are **aplanospores.** Again according to others, sporangiospores, may also be formed in a sporangium as a result of sexual act and meiosis of diploid nuclei. Such spores may be termed **meiospores;** in contrast with spores produced asexually and solely by mitotic division of haploid nuclei; these spores are the **mitospores.** A sporangium may thus be defined as a spore producing strcture in which spores are produced either by the mitotic division of haploid nuclei, e.g., sporangium of *Mucor;* or by meiotic and mitotic divisions of diploid nucleus produced by sexual act, e.g., an ascus. As against these spores, conidia are not formed in any kind of sporangium. They are developed singly or in groups or in chains either from an ordinary vegetative hypha or from a specialized hypha called **conidiophore** (Fig. 11.14C). The conidiophore is also known as **sporophore** since this is a spore-producing or supporting organ. When a conidium is formed endogenously inside a conidiophore and subsequently extruded, such a conidium is known as **endoconidium** (Fig. 11.14E-H), whereas, a conidium formed exogenously on the surface of a conidiophore is **exoconidium** (Figs. 11.14C and D). The conidiophores may be free from each other or may group together to form a fascicle. When the conidiophores join together to produce an elongated upright fascicle of spore-producing structures it is known as **synnema** (pl. **synnemata**) (Fig. 11.14I). A synonymous term **coremium** is frequently used for a looser fascicle. Again, a mycelial mat may give rise to short conidiophores which are closely packed together forming a bed-like mass from which conidia are produced. Such a structure is known as an **acervulus** (pl. **acervuli**) (Fig. 11.14D). Acervuli are often found in parasitic fungi and may be small or large, subepidermal, subcortical or superficial and may not be provided with hair-like structure **seta** (pl. **setae**) (Fig. 11.14D). An acervulus with a well-developed basal stroma is known as a **sporodochium** (pl. **sporodochia**) (Fig. 11.14D). Sporodochia may form crusts, discs or cushions of fungal tissue and may be fleshy, woody or carbonaceous in texture. The conidiophores and conidia may also be formed in a hollow fruiting body known as a **pycnidium** (pl. **pycnidia**) (Fig. 11.9A). The pycnidium is composed of hyphae joined in groups of plectenchymatous structure. The inside of which is lined by conidiophores from which conidia are cut off. The pycnidium is usually provided with an opening called **ostiole** (Fig. 11.9A). It may be provided with a small papilla or with a long neck leading to the opening. The pycnidia are extremely variable in size, shape, color, and consistency of wall; and may be superficial or embedded in the substratum. They may be uniloculate, simple or labyrinthiform and may be formed directly by loose mycelium or may be embedded in a stroma.

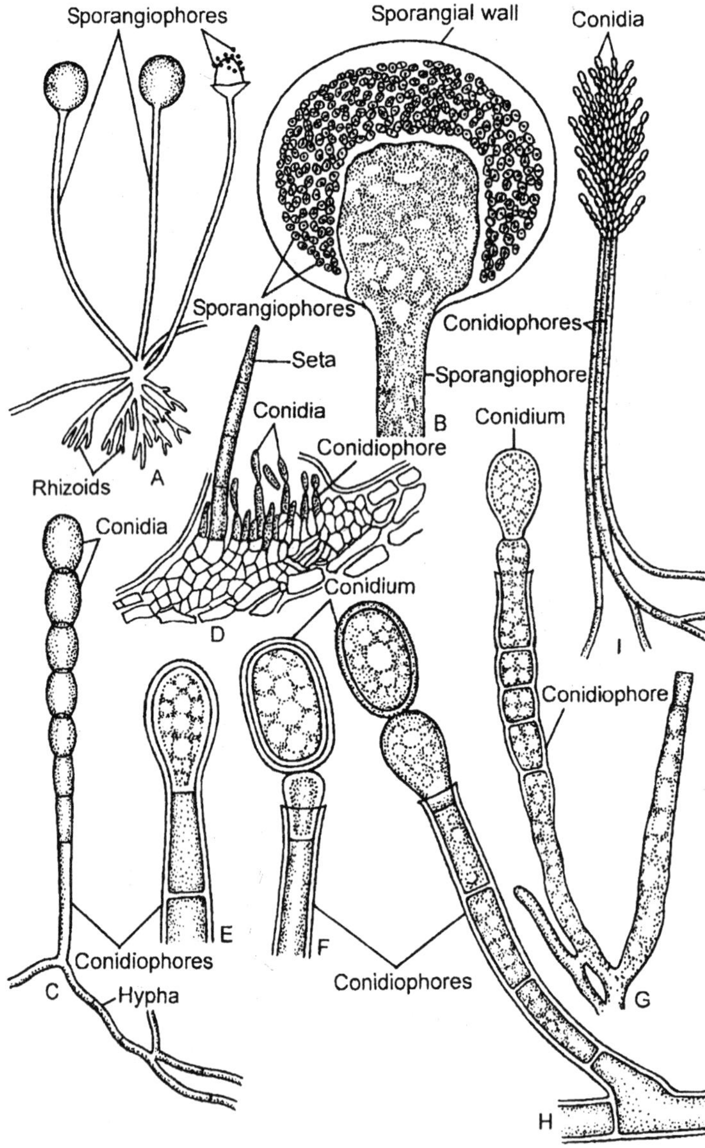

Fig. 11.14: Various types of fungal spores. A. Sporangia, sporagiophores and dispersal of sporangiospores in *Rhizopus stolonifer.* B. Sporangiospores borne in a sporangium of *Rhizopus stolonifer.* C. Conidia in chain borne on a conidiophore in *Erysiphe* sp. D. Section of acervulus of *Colletotrichum* sp. showing seta and conidia borne on conidiophores. E-H. Development of endoconidium from conidiophore in *Ceratostomella* sp. I. Synnema of *Stysanus* sp.

Fungi reproducing asexually by spores, usually produce only one form of spore. But those having more than one independent form or spore stage in the life cycle are **pleomorphic** or **polymorphic** fungi. This may be a normal and regular phenomenon or such changes may be due to degeneration in culture. The polymorphic condition of a fungus is known as **polymorphism.**

Sexual Reproduction: Sexual reproduction in fungi involves the union of two nuclei of different parentage. The nuclei, in some fungi are contributed by two well-organized gametes which may or may not be developed in

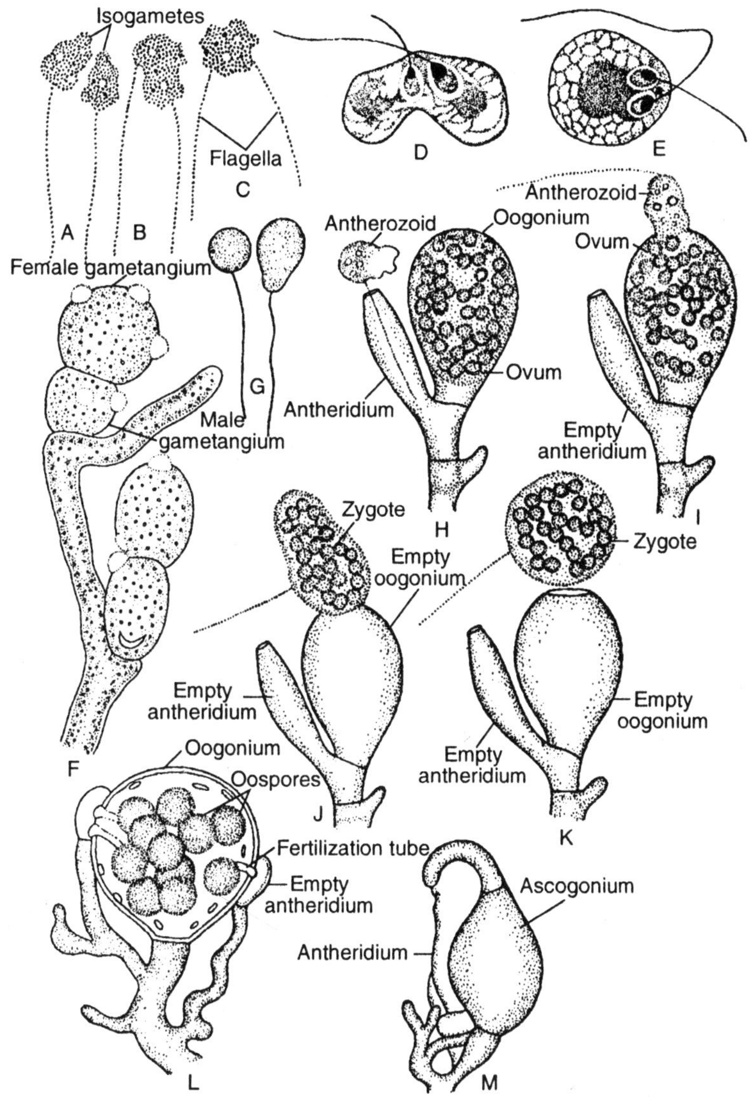

Fig. 11.15: Sexual reproduction. A. Uniflagellate isogametes of *Olpidium* sp. B-E. Stages of isogamy. F. Tip of hypha of gametangial plant of *Allomyces arbuscula* showing terminal female gametangium and subterminal male gametangium. G. Anisogametes. H-K. Stages of sexual reproduction in *Monoblepharella* sp. showing union between a smaller planogamete and a larger aplanogamete. L. Gametangial contact in *Saprolegnia* sp. showing fertilization tube formation. M. Gametangial contact in *Pyronema* sp. without fertilization tube

gametangia, whereas in others, they are present in an undifferentiated mass of protoplast. Sexual reproduction comprises of distinct processes which occur in a regular sequence.

(i) **Plasmogamy,** i.e., a fusion of two nuclei which are brought together by plasmogamy. There may or may not be any interval between plasmogamy and karyogamy. In some fungi karyogamy follows immediately after plasmogamy, whereas in fungi, where karyogamy is somewhat delayed, each pair of nuclei of different parentage is known as **dikaryon,** and when the paired nuclei divide simultaneously, it is called **conjugate division.** The dikaryotic condition, also known as **dikarophase** may be prevalent occupying greater part and performing very useful functions in the life history of certain fungi. The fusion between nuclei of different parentage results in the establishment of **diploid** condition. The diploid condition, also known as **diplophase,** established by karyogamy, is very short.

(ii) **Meiosis** follows immediately after karyogamy resulting in the reestablishment of the number of chromosomes to the haploid condition.

The process by which the dikaryotic condition is established is known as **dikaryotization.** Such a process by which a haploid cell or a mycelium or similar structure becomes diploid (according to Buller) having conjugate nuclei is designated by Buller as **diploidization.** In sexually reproducing and fructification producing fungi, the establishment of dikaryotic condition is a prerequisite to the development of fructification.

The gametangia when developed may be either morphologically indistinguishable as male and female and are known as **isogametangia** (Fig. 11.16C), or **heterogametangia** where the gametangia are morphologically different and can be distinguished as male and female (Fig. 11.15L). In the later case the male gametangium which is the smaller one is known as **antheridium** (pl. **antheridia**) and the female gametangium is called an **oogonium** (pl. **oognia**) or **ascogonium** (pl. **ascogonia**) (Fig. 11.15H-M). An oogonium contains one or more well-organized female gametes, whereas no well-organized gametes are formed in an ascogonium. An ascogonium contains one or more female nuclei embedded in cytoplasm. When both male and female gametangia are developed in the same thallus, the thallus is called hermaphrodite or bisexual or **monoecious.** Where the male and the female gametangia are borne in two different thalli, the thalli are unisexual or **dioecious.**

The gametes in fungi may be flagellate (**planogametes**) or non-flagellate (**aplanogametes**). Morphologically they may be similar (**isogametes**) (Fig. 11.15A) and dissimilar (**heterogametes** or **anisogametes**) (Fig. 11.15G). These are all well-organized gametes. Each well-organized gamete has a nucleus surrounded by cytoplasm which gives a gamete a definite shape. Well-organized gamete formation is usually encountered in lower fungi. But in large number of fungi gametes are represented by nuclei embedded in a mass of cytoplasm contained in a gametangium. In them the development of well-organized gametes either male or female is completey lacking. In majority of fungi there is absence of visible sexuality exhibited by the development of well-organized gametes and gametangia. In such cases the term **compatibility** which is closely related to sex is of common use.

The common methods by which nuclei of different parentage are brought close together through plasmogamy are the following:

1. **Gametic Copulation:** This is the simplest type of sexual reproduction which involves the fusion of two well-organized gametes. When the fusion is between the two isogametes, both of which being either planogamous or aplanogamous, the process is called **isogamy** (Fig. 11.15A-E). Where the gametes taking part in the copulation differ in size it is known as **anisogamy** or **heterogamy.** In the heterogamous copulation both the gametes taking part in the process may be planogamous or the smaller one planogamous and the large one being aplanogamous (Fig. 11.15H-K).

2. **Gametangial Contact:** In this process the two gametangia come in contact with each other (Fig. 11.15L-M). Either a passage is formed by the dissolution of the common walls at the point of contact (Fig. 11.15M), or fertilization tubes (Fig. 11.15L) are developed to facilitate the migration of the entire content of the male gametangium into the female

gametangium. Both the gametangia neither fuse together nor lose their identities. But either only the male gametangium (Fig. 11.16A) or both male and female gamentangia may collapse during the post-fertilization stage. In some fungi one or more female gametes are developed in the female gametangium and in others the content of the female gametangium is not organized into gametes. But in all cases, with a few exceptions, the entire content of the male gametangium does not develop into well-organized male gametes rather remains as a mass of undifferentiated multinucleate protoplast.

3. **Gamentangial Copulation:** This is a process in which the fusion of the entire contents of two gametangia is accomplished in any one of the two ways: (i) The two gametangia fuse with each other by the dissolution of their common walls resulting in the formation of a single cell in which the protoplasts of the two gametangia mix with each other (Fig. 11.16B-E). The entities of the two gametangia are completely lost. (ii) Transfer of the entire protoplast of one gametangium into the other takes place through a pore developed in the gametangial walls at the point of contact of the two gametangia. The basic difference between this

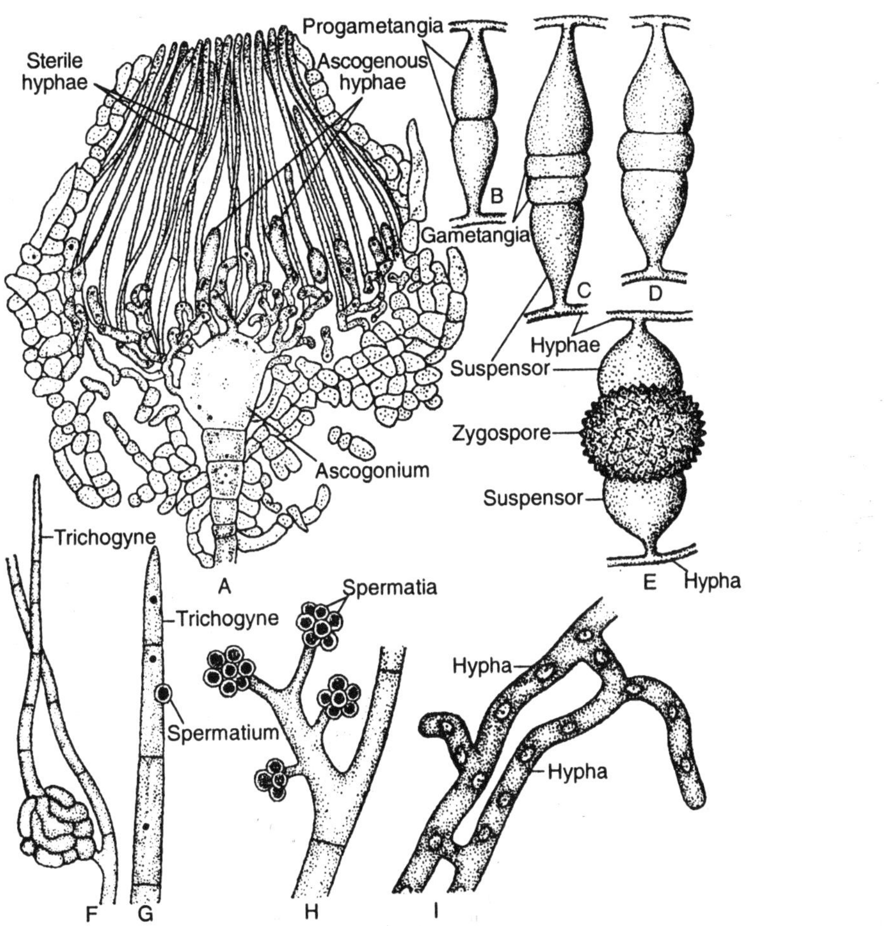

Fig. 11.16: Sexual reproduction. A. Stage of sexual reproduction in *Lachnea* sp. after plasmogamy, antheridium collapsed. B-E. Stages in isogametangial copulation leading to zygospore formation in *Mucor* sp. F. Ascogonium with trichogyne in *Pleurage anserina*. G. Tip of trichogyne with spermatium attached. H. Clusters of spermatia. I. Somatogamy in *Ascophanus* sp.

and the process of gametangial contact is that, in gametangial contact the two gametangia are developed on somatic hyphae, whereas here the two gametangia originate from two zoospores which come in contact with each other.

4. **Spermatization:** Certain fungi in various ways, produce numerous spore-like minute, single-celled, uninucleate structures which are known as **spermatia** (sing. **spermatium**) (Fig. 11.16H) or **microconidia** which behave as non-motile gametes. Oidia also may behave as gametes. Depending on fungi, these structures are transferred through the agencies like insects, wind, water, etc., either to special receptive hyphae or to the trichogyne (Fig. 11.16F and G) of the ascogonium or even to somatic hyphae to which they remain attached. Their contents migrate into those receptive structures through passages formed by the dissolution of the walls at the points of contact. Thus the dikaryotic condition is established. This process is known as **spermatization.**

5. **Somatogamy:** In many fungi the development of gametes and gamete-producing structures is completely lacking. In such fungi anastomosis takes place between the somatic structures bearing nuclei of different parentage, whereby the nucleus of one passes into the corresponding somatic cell by dissolving the walls between them leading to the establishment of a dikaryotic condition (Fig. 11.16I). In such cases the somatic structures take part in the process of sexual repoduction and the process is designated as **somatogamy.**

PARASEXUALITY

In a conventional method of sexual reproduction and normal sexual cycle the three processes: plasmogamy, karyogamy, and meiosis occur in a regular sequence and usually at specified points. But some fungi do not possess conventional method of sexual reproduction and normal sexual cycle. In them, plasmogamy, karyogamy, and meiosis take place in a regular sequence, but not at specified time or points in the thallus or in the life cycle. Such a cycle is known as **parasexual cycle** and the mechanism is **parasexuality.** Parasexuality was first discovered by Pontecorvo and Roper (1952). The sequence of events in a parasexual cycle is: establishment of heterokaryotic condition during which nuclei of differnt genotypes coexist side by side, karyogamy, multiplication of diploid nuclei accompanied with occasional mitotic crossing-over and simultaneous multiplication of haploid nuclei, occasional meiosis of diploid nuclei, and distribution of both haploid and diploid nuclei. But there are fungi in which both normal sexual cycle and parasexual cycle have been detected.

Qualitatively parasexual cycle is equivalent to the normal sexual cycle, but differs in the absence of a precise time sequence for its stages.

Further details of this aspect have been incorporated in Chapter 14.

SEXUAL COMPATIBILITY

In sexually reproducing fungi where sexual reproduction takes place by the development of morphologically distinguishable male and female gametangia and gametes the condition may be described as the existence of **visible sexuality.** The situation becomes difficult in cases where there is absence of visible sexuality, i.e., complete absence of the development of morphologically distinguishable gametangia and gametes. In such cases sexuality is in the nuclear level only. According to some, there is **degeneration of sexuality** which actually means, visible sexuality. There may be a third condition when male and female organs are borne on the same thallus but being self-sterile there is absence of sexual fusion.

Sexual act basically means, the fusion between two nuclei which are contributed, in cases where there is visible sexuality, by the male and female gametes. Actually the fusion is between the two nuclei which have certain affinity to unite with each other and this affinity is due to the presence of certain factors which help them to unite, and the fusing nuclei may be known as **compatible.** In absence of such factors they fail to do so, i.e., the nuclei are **incompatible.** These factors may be described as **factors controlling compatibility** or **factors of compatibility group.**

In cases where visible sexuality is lacking the term compatibility is very appropriate and it often becomes very easy to explain the whole process of sexuality in the light of **factors of compatibility group.**

Similar is the condition where though male and female organs are developed in the same thallus, the organs are self-sterile due to the absence of factors controlling compatibility. Here sexual act will take place only when the nuclei from thalli having factors controlling compatibility present in them are brought together, otherwise not.

Sexual reproducing fungi may thus be grouped into following categories:

1. **Monoecious** or **hermaphrodite,** where both male and female gametangia are borne in the same thallus.
2. **Dioecious** or **sexually dimorphic,** where male and female gametangia are borne in two different thalli.
3. **Sexually undifferentiated,** where gametangia are not morphologically distinguishable as male and female.

Again on the basis of compatibility, fungi of above categories belong to any one of the following groups:

(i) **Homothallic fungi** are those in which every thallus is sexually self-fertile and sexual reproduction takes place without the help of any other thallus.

(ii) **Heterothallic fungi** are those in which every thallus is sexually self-sterile and sexual reproduction takes place with the help of another compatible thallus.

Heterothallic fungi again belong to either of the following groups:

(a) **Bipolar heterothallic.** Fungi in this group are of two mating types, which differ in their genetic makeup with respect to the compatibility factor. Sexual compatibility is governed by one pair of factors Aa located on the same locus on different chromosomes.

(b) **Tetrapolar heterothallic.** Fungi in this group consist of four mating types of individuals. Sexual compatibility is governed by two pairs of factors AaBb located on different chromosomes.

(iii) **Secondarily homothallic fungi.** In some of the bipolar heterothallic fungi during spore development, two nuclei of opposite mating type are invariably present in each spore which on germination, produces a thallus containing both A and a nuclei and naturally behaves as though it where homothallic.

BEHAVIOUR OF COMPATIBLE NUCLEI

After the compatible nuclei are brought together by the methods disussed above, they may fuse with each other immediately without any delay, or their fusion may be delayed.

Where the fusion between the compatible nuclei takes place immediately after they are brought together without any delay, a diploid cell (zygote) is formed. Subsequent stages of development of the zygote follow any one of the following ways:

1. The zygote remains active for a short period, then develops a thick wall, goes to rest

and germinates to produce zoospores or gametes, during which meiosis takes place.

2. The zygote becomes active to produce a diploid thallus bearing two types of zoosporangia. In some of these zoosporangia, diploid zoospores are formed which give rise to diploid thalli; in others, haploid zoospores are developed by meiosis. These haploid zoospores produce haploid thalli exhibiting an alternation of generations.

3. The zygote develops thick wall and goes to rest. Under favourable conditions depending on the nature of fungi, it germinates in various ways, during which meiosis takes place. Such a thick-walled zygote is **oospore** when it is formed from a female gamete through fertilization, whereas a zygote which results from the fusion of two isogametangia or isogametes is a **zygospore.**

But where the fusion between the compatible nuclei is delayed there is establishment of a **dikaryophase,** when the compatible nuclei without fusing, either pair or remain close to each other. This dikaryophase is usually very much prolonged. Ultimately dikaryophase is followed by **diplophase,** when karyogamy of the compatible nuclei takes place. Karyogamy is immediately followed by meiosis resulting in the formation of haploid nuclei around which spores are developed. Thereby, haploid condition or **haplophase** is reestablished. Of these spores, that are formed endogenously in a saclike structure known as **ascus** (pl. **asci**), are the **ascospores,** whereas others borne exogenously on a more or less club-like structure called **basidium** (pl. **basidia**), are the **basidiospores.**

LIFE CYCLE PATTERN

Although the life cycles of individual fungi vary in their details, the general pattern is the same in the group as a whole. The deviations which exist, will be discussed as and when the respective fungi will be considered.

The occurrence, sequence of processes and types of products involved in asexual and sexual reproduction have profound effects upon the life cycles and hence the whole biology of the fungi.

Asexual cycle. Somatic structue (vegetative body) after a certain period of growth takes part in the asexual reproduction, hence there initiates the asexual cycle either directly or by the formation of special reproductive structures. There are certain fungi which reproduce only by asexual method since their sexual process has not been discovered. Again in others, the process of asexual reproduction continues for some time and is followed by the sexual phase or the sexual cycle. Some of the last two kinds of fungi may produce only sterile hyphae without producing any reproductive structures, the **sterile fungi.** They produce sterile hyphae only, whereas others produce asexual spares (conidia) and reproduce with them, the **imperfect fungi.**

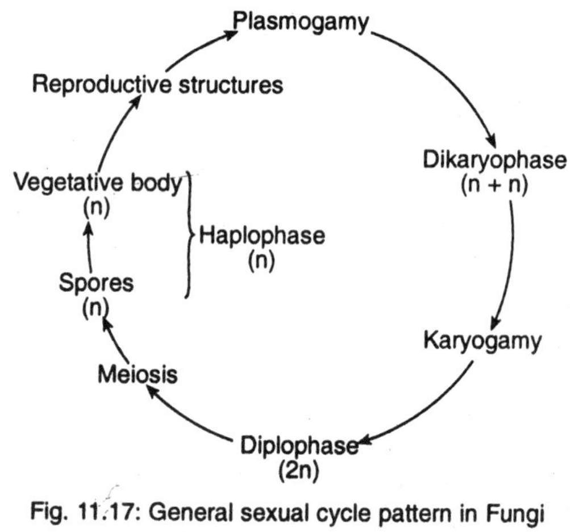

Fig. 11.17: General sexual cycle pattern in Fungi

Sexual cycle. In course of sexual cycle compatible nuclei are brought together by different methods. Tracing the behaviour of compatible nuclei, the sexual phase comprises of (i) **Haplophase,** the haploid phase bearing haploid nuclei (n), when the compatible nuclei remain separated from each other, may be in different thalli or in the same thallus. (ii) Subsequently the compatible nuclei are brought together as a result of which dikaryotic condition—**dikaryophase** with two haploid compatible nuclei remaining in pairs (n + n), is established. During which the compatible nuclei either remain in pairs or close together but do not fuse. This phase may be completely absent or may be very much prolonged. (iii) The dikaryophase is followed by **diplophase** with diploid nuclei (2n), when fusion between the compatible nuclei takes place giving rise to diploid condition. The diplophase is very short and is immediately followed by meiosis, thereby haplophasic condition is reestablished. Ultimately spores are developed, each of which again gives rise to somatic structure. All these three phases follow in regular sequence. The general sexual cycle pattern in fungi is presented in Fig. 11.17.

The above general sexual cycle pattern in fungi can, however, be elaborated in the following manner.

1. **Haploid cycle,** in which meiosis immediately follows nuclear fusion and the meiotic products are then dispersed. The diploid phase is, therefore, of minimal duration. This cycle is shown by many Mastigomycotina and Zygomycotina and some Ascomycotina.

2. **Haploid cycle with restricted dikaryon,** similar to the preceding except that paired, compatible nuclei persist in close physical association in the same hyphal segment (hence dikaryon) and divide synchronously, for a greater or lesser period. At one extreme the association may be for a few cell generations only, e.g., in many Ascomycetes, binucleate ascogenous hyphae are developed just prior to ascus development and such a dikaryon cannot apparently exist independently of the haploid phase—'haploid cycle with restricted dikaryon'.

3. **Haploid-dikaryotic cycle,** at the other extreme is the condition where the dikaryotic phase persists almost throughout the life cycle associated with a short-haplophase. This can occur in yeasts (Endomycetales, also known as Saccharomycetales), but is more frequent in many Basidiomycotina. A highly characteristic type of life cycle is shown by the majority of the Basidiomycotina. Here the mycelium derived from germination of a basidiospore may persist in the haploid condition as a monokaryon, but once a dikaryon is formed, through hyphal fusion for example, it shows potentially unre-

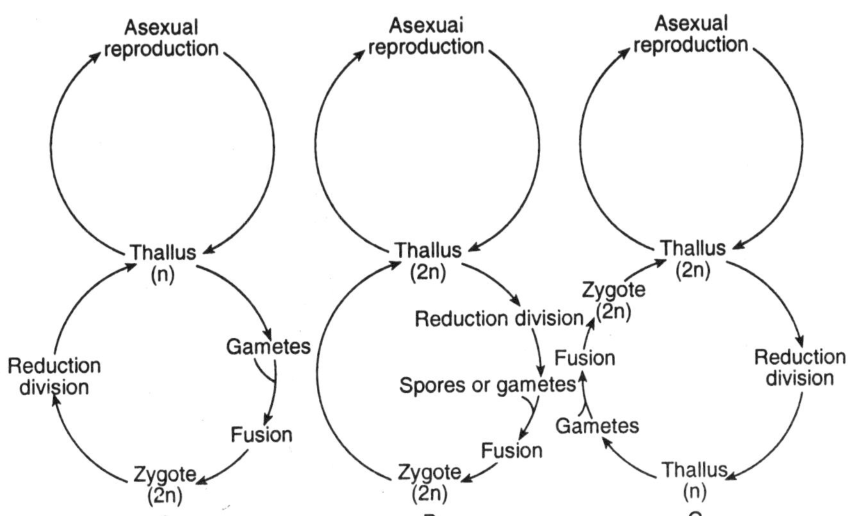

Fig. 11.18: A-C. Life cycles of sexually reproducing fungi. A. A-haplobiontic, B. B-haplobiontic

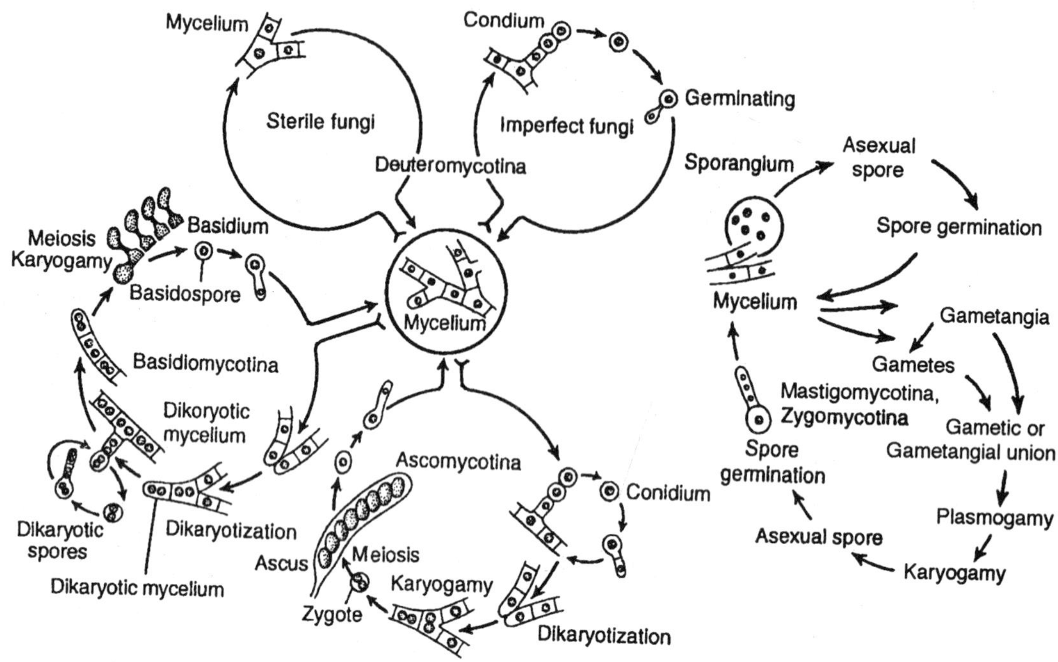

Fig. 11.19: Schematic presentation of Sexual and Asexual Life cycles combined together encountered in Fungi

stricted and independent growth so that it may well comprise the most long-lived phase of the life cycle—'haploid-dikaryotic cycle'. For example, the dikaryotic phase of the members of the Agaricaceae pro-bably persists for several centuries.

4. **Haploid-diploid cycle,** in which these phases (haplophase and diplophase) alternate regularly exhibiting alternation of generations. This condition is encountered in species of *Allomyces* and *Saccharomyces cerevisiae.*

5. **Diploid cycle,** in which the haploid phase restricted to the gametes or gametangial phase is followed by an elaborate diploid phase. This may be exemplified by *Saccharomycodes ludwigii.*

SPORE DISSEMINATION

Survival of species is very much conditioned by dispersal of spores into habitats where food is available. In most species of fungi special mechanisms for spore dissemination are lacking. In them mature spores remaining on the conidiophores or in the ruptured fruit body are removed by air currents or other agency. Many fungi, however, exhibit special and often intricate mechanisms for the more or less violent discharge of the mature spores, very interestingly exhibited by the genus *Sphaerobolus.* Turgour often provides the necessary energy as in the sporangial discharge of the genus *Pilobolus* or the ascospore discharge of many Ascomycotina. Other mechanisms include the fluid drop-excretion mechanism of the Basidiomycotina and those in which the kinetic energy of rain drops is utilized by the 'splash cup' of the Nidulariaceae, or by the bellows effect of rain drops falling on the mature fruit body of the genus *Lycoperdon.*

In many aquatic Mastigomycotina spore dissemination is effected by self-motile zoospores.

In terrestrial fungi the most important agencies for spore dissemination are air currents, insects and animals.

The dissemination of spores of *Puccinia graminis* may be cited as one of the important examples of air currents as a factor in dispersal of spores. The transport of *Ceratocystis ulmi,* pathogen of the Dutch elm disease by bark beetles may be cited as a classical example of insects as agencies for dissemination of spores. A number of fungi, particularly the coprophilous ones, such as the species of *Pilobolus, Ascobolus, Sordaria,* and *Coprinus,* their spores are disseminated by herbivorous animals like horses, cattle, sheep and goats. These animals swallow the spores and herbage together. After having been eaten, the spores pass through the alimentary tract of the animals and by the time they pass out with the feces, the animals may transport them for miles. After being discharged from the animal body along with the feces the spores develop into new fungi.

These and other mechanisms of spore dissemination will be discussed further in details in Chapter 16.

PIGMENTS OF FUNGI

Many fungi are beautifully pigmented. This character has often been used in classification of certain groups of fungi. Again certain wood-attacking fungi cause staining in wood. The pigments that are produced in the fungi themselves or in the wood that is attacked by the fungi are linked up with the metabolic activities of the fungi. Their specific role in the fungal cell is still answerable. Pigments of variable chemical nature have been isolated from different fungi, some of them are indicated below:

Phycomyces blakesleeanus, Mucor hiemalis, Allomyces javanicus and *Piobolus* sp. contain—carotine. Erythroglaucin, auroglaucin and flavoglaucin have been obtained from *Aspergillus glaucus.* Citromycetin and citrinin have been extracted from *Penicillium glabrum* and *P. citrinum. Helminthosporium gramineum, H. cynodontis,* and *H. catenarium* contain helminthosporin.

Ceratocystis ulmi causes blue stain of elm wood. *Verticillium* sp. produces green discolouration of maple wood. Stored wood may be stained by the growth of the fungi like *Mucor, Rhizopus, Aspergillus, Penicillium, Alternaria* and many others by growing on the surface of the wood.

LUMINESCENT FUNGI

Some fungi give out light, sometimes cause the attacked wood or leaves to become luminous. Some of them are: *Panus incandescens, Pleurotus incandescens, P. igneus*—pilei are luminescent; *Clitocybe illudens*—fruit body and mycelium are luminescent; the rhizomorphs of *Armillaria mellea* and the sclerotia of *Collybia tuberosa* and *C. cirrhata* give out light; *Omphalia flavida* and *Xylaria hypoxylon*—mycellium is luminous.

Luminosity in these fungi is due to the action of an enzyme luciferase on a substance known as luciferin present in the fungal cells in presence of oxygen.

FOSSIL FUNGI

The constitution of the fungi has not given much scope for being fossilized. The most common kind of fossil fungi is by incrustation with calcium carbonate. Sometimes fungi inhabiting leaves and stems remain as impressions. These impressions are formed when the plant parts along with fungi are deposited in the siliceous or calcareous matrix. Gradually the matrix hardens and sets, and the impressions often clearly portray the details of fungi. Besides these, fungi may also be preserved in masses of Baltic amber.

According to James (1893), fossil fungi may be traced back in the Devonian period. This was partly supported by Seward (1933), who felt that the existence of fungi may be looked from the Devonian period onwards and representatives of some of the parasitic and saprophytic fungi may be anticipated to have existed even in a more remote age.

The generic names of the fossil fungi are expressed with the addition of the suffix **ites** to the present-day genera of fungi with which their resemblance is indicated. Saccardo in his Sylloge Fungorum (1882) included 350 names of fossil fungi, their fruit bodies (very infrequently with spores) on or with fossil leaves or stems. Some of the fossil fungi are: *Peronosporites gracilis* Renault, *Mucorites cambrensis* Renault, *Phycomycites frodinghamii* Ellis, *Erysiphites metilli* Pamp., *Penicillites curtipes* Berk., *Pezizites candidus* Gopp et Ber., *Polyporites bowmanii* Lindley et Hutton, *Agaricites wardianus* Mesch., *Puccinites cretaceous* Velen.

CLASSIFICATION

The classification of the fungi is a very difficult task since it presents various problems which originate from the differences of opinion of different workers. Such differences are due to the differences in the approach and the interpretation of the data collected during the investigations through years and years. The situation becomes even more difficult as we learn new facts about the fungi which often demand the change of concept of their relationships, reclassifiction and even alteration of their names.

Since classification includes the grouping of different fungi and ultimately naming them according to the internationally accepted system, it assumes a great responsibility which can be neatly served only when the various aspects are taken into consideration and put together, failing which serious discrepencies may appear leading to even more confusion. The wide diversity in form and structure of the fungi led mycologists to attempt to arrange them into different groups for the convenience of study and discussion. Different systems of classification have been used, but the ones most generally accepted are based upon fundamental similarities or relationships and are called **natural systems.** The grouping of fungi into various taxonomic categories was necessary along with the study of the various aspects of them. Hence to facilitate methodical study of the fungi the following taxonomic categories were used: Kingdom—Division—Subdivision—Class—Subclass—Order—Family—Genus—Species. Kingdom is the largest and species being the smallest. According to convenience species is sometimes broken down into even more smaller categories, like variety, biological strain or physiological strain or cultural strain. But in general, species is the unit of classification. These categories are used as an aid to identification of fungi. It will be rather impractical to consider the various taxonomic categories as concrete and stable units since the living organisms are constantly evolving and any attempt to put strict lines of demarkation between the taxonomic categories is rather an impossible proposition. These taxonomic categories are laid down on the basis of characters of the various organisms which are liable to change from time to time by hybridization, mutation, and similar other natural processes.

The history of classification of fungi can be traced back from the time of herbalists. Bauhin, an early herbalist in 1623, listed in **Pinax Theatri Botanici** nearly 100 species of fungi.

Tournefort (1694) in his **Elements de Botanique** arranged fungi in six groups: (i) Fungus, including all centrally stalked agarics, boletes, and polypores; (ii) Boletus, including *Clathrus, Morchella,* and *Phallus;* (iii) Agaricus, including *Auricularia* and all laterally attached polypores; (iv) *Lycoperdon,* including the various Lycoperdaceae and certain slime moulds; (v) Coralloides, including the coral fungi and other branched fungi; and (vi) Tubera, including subterranean fungi.

Linnaeus in 1753 in his **Species Plantarum** described fungi under the twenty-fourth class Cryptogamia Fungi. He included in Cryptogamia Fungi—27 species of *Agaricus,* 12 of *Boletus,* 4 of *Hydum,* 2 of *Phallus,* 3 of *Clathrus,* 2 of *Elvela,* 8 of *Peziza,* 8 of *Clavaria,* 9 of *Lycoperdon,* and 11 of *Mucor.*

Then followed the taxonomic treaties of Persoon. His **Synopsis—Methodica Fungorum,** in 1801, and his comprehensive **Mycologia Europaea,** a three-volume work appeared between 1822 and 1828. Persoon divided the fungi into 2 classes, 6 orders and 71 genera and was the first to establish a usable system for the classification of fungi.

Elias Fries's **Systema Mycologicum** consisting of three volumes published between 1821 and 1832, is a fundamental contribution to the knowledge of fungi. The **Systema Mycologicum** is the basis of our present-day system of classification and constitutes the starting point of classification of many major groups of fungi. In **Systema,** Fries grouped fungi into four classes: Coniomycetes, Hyphomycetes, Gastromycetes, and Hymenomycetes, and then further dividing these classes.

Saccardo (1882) in his **Sylloge Fungorum** classified fungi into the following classes: Schizomycetes, Myxomycetes, Phycomycetes, Ascomycetes, Basidiomycetes and the form-class Deuteromycetes or Fungi Imperfect.

Gwynne-Vaughan and Barnes (1926) divided fungi into three main classes (Phycomycetes, Ascomycetes and Basidiomycets) according to the septation of the mycelium and on the nature of the principal spores. Their fourth class was Fungi Imperfect whose members lack sexual process and they possess mycelia and spores very much similar to those of the Ascomycetes.

Fitzpatrick (1930) separated the then Thalophyta into two subdivisions: Myxothallophyta and Euthallophyta. Under the Euthallophyta he included Bacteria (Schizomycetes), Fungi (Eumycetes), Lichenes and Algae.

Clements and Shear (1931) have emphasized on three basic principles: (i) The fungus is a physiological adjustment to the environment. (ii) The fungi do not constitute a natural group, and that all the phyletic lines lead sooner or later to holophytic origins. (iii) The ecological approach to the morphology and development of the fungi constitute the best attack upon their evolution and phylogeny. They also subdivided fungi into Phycomycetes, Ascomycetes and Deuteromycetes (Fungi Imperfect). On the basis of the three principles stated above Clements and Shear have made further subdivisions which are different from the view points of other mycologists.

Martin (1936, 1941) subdivided fungi into three classes: Phycomycetes, Ascomycetes and Basidiomycetes; and two form-classes: Fungi Imperfect and Lichenes.

Bessey (1950) included Mycetozoa under the Fungi separating the former as a Subclass Mycetozoa from the latter what he designated the True Fungi. Under the True Fungi he placed four classes: Phycomycetes, Ascomycetes, Basidiomycetes and Fungi Imperfect.

Martin in 1961 designated Fungi (Eumycetes) as Subdivision Eumycotina (Eumycetes) which include three classes (Phycomycetes, Ascomycetes and Basidiomycetes and two form-classes (Lichenes and Deuteromycetes).

According to the rules of International Code of Botanical Nomenclature (ICBN), the fungal taxa of different taxonomic ranks are as follows:

Division :	mycota
Subdivision :	mycotina
Class :	mycetes
Order :	ales
Family :	aceae

Alexopoulos (1962) followed Martin and classified Fungi and Fungus-like organisms devoid of chlorophyll in the followig manner:

A. Somatic phase consisting entirely of simple, uninucleate amoebae
 B. Amoebae aggregating into a pseudoplasmodium
 Order: **Acrasiales**
 BB. Amoebae sliding on a network of slimy filaments
 Order: **Labyrinthulales**
AA. Somatic phase not of simple, uninucleate amoebae
 B. Nuclei devoid of a nucleolus and of a nuclear membrane; somatic phase of minute single cells or filaments rarely exceeding 2 μ in diameter
 Class: **Schizomycetes**
 BB. Nuclei with one or more nucleoli and provided with a nuclear membrane; somatic phase filamentous, unicellular (the filaments or cells usually over 2 μ in diameter), or plasmodial
 Division: **Mycota**
 C. Somatic phase a free-living plasmodium
 Subdivision: **Mycota**
 Class: **Schizomycetes**
 CC. Somatic phase not a free-living plasmodium
 Subdivision: **Eumycotina** (True Fungi)
 D. Flagellated cells characteristically produced
 E. Motile cells uniflagellate
 F. Flagellum posterior, whiplash
 Class: **Chytridiomycetes**
 FF. Flagellum anterior, tinsel
 Class: **Hypochytridiomycetes**
 EE. Motile cells biflagellate
 F. Flagella of unequal size, both whiplash
 Class: **Plasmodiophoromycetes**
 FF. Flagella nearly equal, one whiplash, the other tinsel
 Class: **Oomycetes**
 DD. Flagellated cells lacking or rarely produced
 E. Hyphae aseptate, or if septate then sexual reproduction resulting in the formation of a resting spore
 F. Some filamentous, attached to digestive tract or external cuticle of living arthropods, mycelium not immersed in host tissues
 Class: **Trichomycetes**
 FF. Mainly saprobic; if attacking living arthropods, then mycelium developing in host tissues before sporulating
 Class: **Zygomycetes**
 EE. Hyphae septate, sexual reproduction not resulting in the formation of a resting spore
 F. Sexual reproduction resulting in the formation of ascospores or basidiospores
 G. Spores resulting from karyogamy and meiosis borne in asci
 Class: **Ascomycetes**
 GG. Spores resulting from karyogamy and meiosis borne on basidia
 Class: **Basidiomycetes**
 FF. Sexual reproduction lacking; a parasexual cycle may be present
 Form-Class: **Deuteromycetes**

On the basis of fruit body development, the phycomycetes are often designated as **Lower Fungi** as because they do not develop any fruit body. Since Ascomycetes and Basidiomycetes produce fruit bodies, they are called **Higher Fungi** or **Carpomycetes.**

Much earlier than Martin and Alexopoulos, Sparrow (1958) classified lower fungi based on flagellation in the following manner.

Elizabeth-Moore-Landecker (1972) classified Fungi separately for lower fungi and higher fungi.

Classification of lower fungi based on flagellation was adopted from Sparrow (1958).

Uniflagellates

Class: **Chytridiomycetes**

Class: **Hyphochytridiomycetes**

Biflagellates

Class: **Pasmodiophromycetes**

Class: **Oomycetes**

Non-flagellates

Class: **Zygomycetes**

The somatic phase of the plasmodiophoromycetes is so different from that of the true fungi with which they are often classified based on the nature of flagellate cells, needs careful consideration.

Again Talbot (1969) subdivided Fungi and Fungus-like organisms as follows:

Fungi with a plasmodial thallus

Division: **Myxomycota** (Slime Moulds)

Non-colonial slime moulds

Sporangiate fructification present, not endoparasites

Class: **Myxomycetes**

Sporangiate fructification lacking, endoparasites

Class: **Plasmodiophoromycetes**

Colonial slime moulds

Net slime moulds, mostly parasitic on marine plants

Order: **Labyrinthulales**

Cellular slime moulds, saprobes

Order: **Acrasiales**

Fungi with a non-plasmodial thallus

Division **Eumycota** (True Fungi)

Thallus unicellular uninucleate, poorly to well-developed aseptate mycelium with a tendency to develop septa; asexual spores motile or non-motile sporangiospores; sexual reproduction results in the formation of oospores or zygospores; absence of fruit bodies

Sporangia with zoospores, predominantly oogamous

Subdivision: **Mastigomycotina**

Sporangia with aplanospores, entirely zygogamous

Subdivision: **Zygomycotina**

Members of the Subdivisions: Mastigomycotina and Zygomycotina, together formerly known as 'Phycomycetes', include what are usually thought to be the most primitive types of Eumycota, even though the simplest show much diversity of form and complexity of life cycle.

Thallus with septate mycelium; asexual spores other than sporangiospores; sexual reproduction results in the formation of asci and ascospores or basidia and basidiospores usually in fruit bodies.

Presence of asci and ascospores

Subdivision: **Ascomycotina**

Presence of basidia and basidiospores

Subdivision: **Basidiomycotina**

Thallus structure and asexual spores are similar to Ascomycotina and Basidiomycotina; sexual reproduction lacking

Subdivision: **Deuteromycotina**

TREND OF CLASSIFICATION SYSTEM IN FUNGI

All is not settled in the classification of fungi as because differences of opinion are very numerous. Divergent viruses arise because of our incomplete knowledge on the structure, development, and physiology of fungi. Therefore, as our knowledge of fungi increases our classification schemes are bound to change.

If we trace back we find that Copeland (1950) followed by Ingold (1959) favoured the division of living world into four kingdoms: Plant, Animal, Fungi, and Bacterial. It is now agreed that the origin of the fungi from the algae presents more difficulties than does a theory of protozoan origin. Mycologists believe that the fungi had a very common ancestry with the protozoa which split off at a very early stage of evolution. Hence fungi are neither plants nor animals, they belong to a separate kingdom—Fungi. This virus was expressed earlier by Martin (1932, 1955).

Both Martin (1961) and Alexopoulos (1962) outlined their classification systems as: Kingdom-Plantae and Division-Mycota (chlorophyll lacking organisms) with two sub-divisions. Eumycotina (The Fungi) and Myxomycotina (Slime moulds), whereas Ainsworth (1966) treated the Fungi (Mycota) as a subkingdom of the Plant Kingdom. He recognized two divisions of Fungi (Chlorophyll lacking organisms)—Eumycota (True Fungi) and Myxomycota (Slime Moulds).

Ainsworth (1971, 1973) elaborated his classification of Fungi and Fungus-like (Chlorophyll-lacking) organisms in further details in the following manner:

Kingdom: Fungi
Division: Myxomycota (Slime moulds)
Division: Eumycota (True Fungi)
Subdivision: Mastigomycotina
Zygomycotina
Ascomycotina
Basidiomycotina
Form-Subdivision: Deuteromycotina

This approach of Ainsworth has replaced the use of the age old terminologies. Myxomycetes, Phycomycetes, Ascomycetes, Basidiomycetes, and Deuteromycetes (Fungi Imperfecti).

The classification system outlined is based to a large extent on the particular life cycle involved and on morphology of the reproductive structures and spores produced. To a lesser extent, morphology of tissues is used. But in recent years, data based on physiology have been used. In general, the classification systems which are currently used are based on the classic morphological studies of past centuries. These systems are being constantly modified as our knowledge and understanding of the fungi increase. But no system of classification has yet been devised which is based on physiological and genetic information alone, though such modern information supplements taxonomic systems based on morphological studies. As such there is ample scope for work to make an all-round improvement so that microanatomy supported by physiological and genetical studies can produce fruitful results to make a widely acceptable classification system.

Following is a simplified key to the subdivisions and Form-subdivision of the Division Eumycota (True Fungi), as suggested by Ainsworth (1975) with slight modification.

A. Mycelium, when present, aseptate; asexually produced spores in sporangia; sexual reproduction by gametic union or, by gametangial contact or copulation resulting in the formation of oospore or zygospore.

B. Sporangia with zoospores; sexual reproduction by gametic unions, by gametangial contact; sexual reproduction resulting in the formation of oospore
Subdivision **Mastigomycotina**

BB. Sporangia with aplanospores; sexual reproduction by gametangial copulation resulting in the formation of zygospore
Subdivision **Zygomycotina**

AA. Mycelium, when present, septate; asexual reproduction, when present, by spores not borne in sporangia; both oospore and zygospore lacking; sexual reproduction, when present, resulting in the formation of spores produced by karyogamy and meiosis, borne in or on specialized structures—asci or basidia

C. Spores resulting from karyogamy and meiosis borne in asci

Subdivision **Ascomycotina**

CC. Spores resulting from karyogamy and meiosis borne on basidia

Subdivision **Basidiomycotina**

CCC. Sexual reproduction unknown; usually possessing spores produced by method other than sexual

Subdivision **Deuteromycotina**

Members of the Subdivision: Mastigomycotina with other fungi put together were formerly designated as the Class **Phycomycetes** (Lower Fungi) to include what are usually thought to be the most primitive fungi, even though the simplest show much diversity of form and complexity of life cycle with evolutionary tendency from an aquatic habitat to complete adaptation to land habit as saprocus or parasites in plants and animals. The changes have involved considerable modification of the life cycles. With increase in knowledge it has been realised that the above grouping is an artificial conglomeration of unrelated forms of fungi. This has lead to discontinue the use of the taxonomic category: Class Phycomycetes.

There are still anomalies in this classification system: (1) designating as subdivisions Zygomycotina, Ascomycotina, Basidiomycotina and Deuteromycotina. (2) Placing Zygomycotina, Ascomycotina, Basidiomycotina and the Deuteromycotina under Subdivision Mastigoycotina.

Members of Mastigomycotina, have adapted aquatic mode of life and include those fungi which produce zoospores in their cycle. From among the five kinds of Zoospores found in fungal world, three types are most common in these fungi. These are—anteriorly uniflagellate, posteriorly uniflagellate and laterally biflagellate. These differences in Zoospores, suggest that they may be polyphyletic.

This also suggests Mastigomycotina is not a natural group, but simply an assemblage of unrelated organisms.

Ainsworth (1973) placed all such Mastigomycotina under a class **Comycetes.** But earlier Oomycetes were placed under a Class **Phycomycetes.**

Much earlier, Saccardo (1866) separated mycelial—**Eumycota** forms of Fungi into four classes as: (1) **Pycomycetes** (zoospore or zygospore producing), (2) **Ascomycetes** (ascospore producing), (3) **Basidiomycetes** (basidiospore producing), and (4) **Deuteromycetes** (asexual spore producing).

For a long time mycologists found it convenient to use the classification of Fungi laid down by Ainsworth (1973). Later on Hawkorth and others (1983 and 1995) revised Ainsworth's classification of Fungi on the basis of sequence of nucleic acid structural details, according to which they stated that Fungi are polyphylatic assemblage of unrelated organisms. Based on the nucleic acid study, Hawkorth and others separated organs into their kingdoms: **Fungi, Straminopila (Chromista)**, and **Protozoa.** The structure of animal male gametes and zoospores in fungi have close similarity than two other straminopila and Protozoa. All the above groups are placed under **Eukaryota.**

On the basis of above findings Hawksworth and others outlined the following classification:

Kingdom: **Fungi**
Phylum: Ascomycotina (no classes)
Phylum: Basidiomycotina
Class: Basidiomycetes

Class: Teliomycetes
Class: Ustomycetes
Phylum: Chytridiomycetes
Class: Chytridiomycetes
Phylum: Zygomycetes
Class: Trichomycetes
Class: Zygomycetes
Kingdom: **Straminopila (Chromista)**
Phylum: Hyphochytridiomycota
Phylum: Labyrinthulomycota
Phylum: Oomycota
Kingdom: **Protozoa**
Phylum: Plasmodiopharmycota
Phylum: Dictyostelismycota
Phylum: Acrasiomycota
Phylum: Myxomycota
Class: Myxomycota
Class: Protostediomycetes

The replacement of the use of **division** by **phylum** is a deviation from the International code of Botanical Nomenclature. Further elaborations of the different phyton are available in relevant reference books.

The discussion on the topic—classification of Fungi—is a continuous one, as because workers in this line will always take part in it.

NOMENCLATURE OF THE FUNGI AND MYXOMYCETES

The nomenclature of Fungi, Myxomycetes and Lichens is controlled by the International Code of Botanical Nomenclature, printed by authority of an Inernational Botanical Congress, at which all changes in the Code have to be made. The purpose of these Rules is the fixing of plant names and the keeping out of use of uncertain or wrong names, or those which may be a cause of error. They are made up of Principles, Rules, and Recommendations which are printed as a series of Articles. Some of the more important aspects are:

1. The start of nomenclature of most Fungi is Fries, **Systema Mycologicum**, 1821; for Uredinales, Ustilaginales, and Gasteromycetes, Persoon's **Synopsis Methodica Fungorum**, 1801.

2. The start for Fossil Fungi is the year 1820 and that for Myxomycetes and Lichens the **Species Plantarum** of Linnaeus, 1753.

3. The name of a taxonomic group is based, e.g., a Family name on a genus, a generic name on a species, a specific name generally on a specimen, sometimes on a culture, a Figure, or a description.

4. Taxonomists in all their writings should make clear on which types their accounts are based. In addition, the name of the herbarium (or other place) having the type material on which the account of a new species is based is to be given.

5. Names of new groups first published after January 1, 1935 are valid only when there is a Latin diagnosis; new transfers or new combinations are valid only when the name or epithet-bringing synonym is given with its author and date and place of publication.

6. Names of species to begin with small letters and that names be not long, hard to pronounce, and so on.

7. A valid name should be in binomial composed of two letters—a generic name and a specific epithet.

8. The authority for a name, generally abbreviated comes after the binomial. With a future change, the name of the first author must be in round brackets before that of the author making the change. A name in use before the starting point for nomenclature, if later taken into use again, is legitimate under the Rules only from the time of its use by the

later worker who is given as the authority, often preceded by the earlier name.

9. The valid generic name or specific epithet is the earliest.

10. In Ascomycotina and Basidiomycotina with pleomorphic life cycles the first valid name or epithet applied to the perfect stage takes precedence, but the names of the imperfect stages may still be used.

PHYLOGENY OF THE FUNGI

Views on the phylogeny of the fungi have been based on comparative morphology, cytology, serodiagnosis, and study of fossil records of fungi. Since evolution in plants and animals has produced from simple to complex structures, it will be logical to search for simple organisms as the probable primitive types from which plants and animals have evolved. But at the same time retrogression from complex to simpler structures that has taken place should be given proper attention. The fact that most of the simpler animals, green plants and fungi possess flagellate condition, it will not be unreasonable to search for their ancestry in the unicellular motile organisms.

Did fungi arise directly from the unicellular motile organisms, or through the Algae or otherwise? It is a big question.

I. **Algal origin**. The existence of thread-like filaments, the antheridial branches and oogonia of the water moulds and particularly the zoospores produced by these forms of fungi naturally suggested algae as their nearest relatives. Alexander Braun (1847) declared that the fungi were no more than collateral groups of algae, and Pringsheim (1858) put the great weight of his authority behind this view. The wise acceptance of evolution following Darwin's proposal of natural selection in 1858 and 1859 stimulated this line of thought and was supported by Cohn (1872), Sachs (1875), Brefeld (1889), C. E. Bessey (1895), E. A. Bessey (1935, 1942, 1955), and others.

According to them, the Chytridiales have evolved from unicellular algae, the Saprolegniales from the Siphonales, the Zygomycetes from the Conjugales and the higher fungi from the red algae.

As a matter of fact, Brefeld's derivation of the higher fungi in two lines, one through the Zygomycotina and the other through the Masigomycotina, both regarded as unquestionably offshoots from distinct groups of algae is of special importance because of the wide acceptance it received for many years. But in course of time it failed to stand tests of critical observers. Through physiological and cytological studies it has been established that the Algae and the Fungi are parallel lines of evolution rather one originating from the other, both of them have unicellular motile organisms as their ancestors.

Actually it was De Bary (1881) who first protested against the algal ancestry of fungi. He stated that the arrangement for classification of the Thallophyta by putting green and non-green organisms together is not a natural one. Besides this, the idea of algal ancestry of fungi is based on some speculative assumptions than on sound arguments. The physiological processes of both algae and fungi are different and he did not believe that evolutionary change in metabolism could take place in such a way that algae, by change of metabolism could have given rise to fungi. According to De Bary in the remote past algae and fungi originated from unicellular green and non-green ancestors independently and their lines of evolution paralleled somewhat with each other. Atkinson (1909) supported De Bary.

Hence the belief that fungi are basically no more than algae whch in the course of their evolution have dispensed with chlorophyll, while still held, has in recent years been less favoured. The idea that the earliest organisms must have been photosynthetic has been shown to be unnecessary in the light of current theories as to the origin of life. It is becoming common to regard the Fungi as an essentially homogeneous group, at least conceivably monophyletic if, as is often the case, certain aberrant forms are excluded. Many would exclude the Myxomycetes; nearly as many, those water moulds as to what is meant by fungi. Flowering plants, red algae and a few apparent green algae have long been known to lack chlorophyll. It is possible that some groups still generally incuded in the Fungi, the Mastigomycotina for example, may have to be removed to algal groups. But on the whole, the bulk of the fungi may be regarded, in the light of present information, as constituting a very large, extremely variable but on the whole remarkably coherent group of organisms, which may reasonably be treated as a discrete major taxonomic unit.

II. Non-chlorophyllous unicellular origin. Fischer (1892) and Atkinson (1909) suggested that the source of origin might possibly be sought in non-chlorophyllous unicellular forms.

Copeland (1950) placed the great majority of the fungi along with red and brown algae, all pigmented flagellates, algae other than the green algae, and the **Protozoa** in his phylum **Inophyta** but excluded the water moulds and the Myxomycetes. He restricted the **Kingdom Plantae** to those plant whose pigments are restricted to chlorophyll a and b, carotine, and xanthophyll, that is, the green algae, bryophytes, and vascular plants; and the **Kingdom Animalia** to multicellular animals. Zuck (1953) elaborated this idea and argued that three nutritional roads, the **autotrophic,** the **Lysotrophic,** and the **phagotrophic,** may well have been followed by plants, fungi, and animals from very early time, hence we should recognize the three kingdoms: **Phyta, Myketes** and **Zoa.**

Ingold (1959) remarked that many present-day biologists would favour the division of living world into four kingdoms: Plant, Animal, Fungal, and Bacterial which is essentially a simplification of Copeland's treatment.

III. Protozoan origin. Gobi (1884) was first to propose an independent derivation of the fungi from the **Protozoa.** Dangeard (1886) supported Gobi's view. Others who have favoured it include Scherffel (1901, 1925), Cavers (1915), Cook (1928), Martin (1932, 1955), Ramsbottom (1941), Langeron (1945), Langeron and Vanbreuseghem (1952), Heim (1952), Moreau (1954), and Ingold (1959). All agreed that the derivation of the fungi from the algae presents more difficulties than does a theory of protozoan origin. Atkinson also considered that if the product of metabolism and type of flagella are taken into consideration fungi should have originated from the protozoa rather than the algae.

Whatever may have been the origin, evolution and phylogeny of Fungi, it must be kept in mind tha they have been on the earth for a long period of time. Seward traced the existence of parasitic and saprophytic fungi from the Devonian period onwards and even from a more remote age. As such fungi are the most ancient members of the plant kingdom.

Chapter 12

Subdivisions of Mastigomycotina and Zygomycotina

INTRODUCTION

The Mastigomycotina and Zygomycotina are ordinarily considered to be most primitive fungi. They constitute a very diversified assemblage of individuals ranging from a microscopic cell to those with a conspicuous, filamentous branched form. Some show relationships with the flagellates, others closely resemble algae and still others being true molds. Because of their morphological similarity with algae they were for long time known as 'algal-fungi' and were regarded as degenerate algae by loss of chlorophyll and consequent loss of ability to elaborate food have become fungi. The supporters of this view emphasized only on the morphological resemblances between algae and fungi and ignored the most important physiological differences between them.

OCCURRENCE AND ECONOMIC IMPORTANCE

The members of these subdivisions of fungi occur on a wide variety of substrate. Some are aquatic (both freshwater and marine) being parasitic on plants and animals including fish and insects, or are saprophytic on submerged vegetable and animal debris. Very few are mycorrhizal. The aquatic ones are known as **water molds**, whereas others are amphibious, and still others terrestrial. Many of the advanced species attack leaves and shoots of angiospermous plants and induce various plant diseases. Some of the terrestrial members are the most destructive parasites of crop plants and certain insects. Again others often cause spoilage of foodstuff, various fruits and vegetables. Some others are utilized in industrial fermentations. There are also species which inhabit damp soil. Many of these soil-inhabiting species cause serious root troubles of economic plants and others destroy seedlings in the seed bed. Most of the aquatic species are very susceptible to hydrogen-ion concentration of the substratum. They decrease in number with the decrease in hydrogen-ion concentration along with the rise in temperature and falling of water-level. Again some of the aquatic fungi are reported to play the role of natural purification of water.

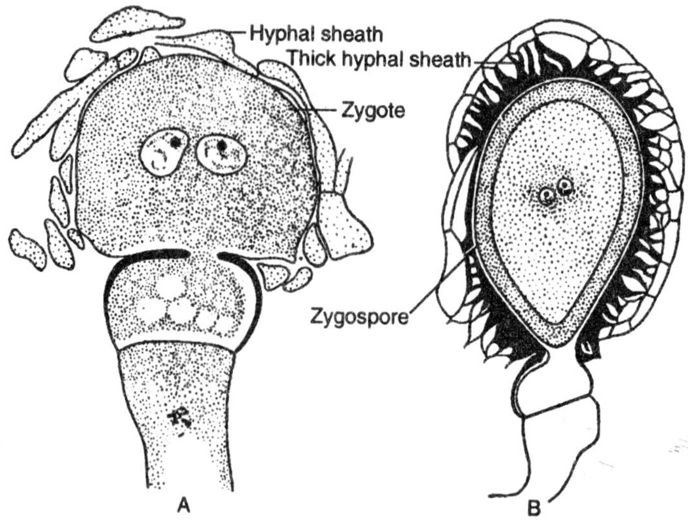

Fig. 12.1: *Endogone* sp. Tendency of fruit body development. A. Formation of hyphal sheath around zygote B. Development of thick hyphal sheath around zygospore

SOMATIC STRUCTURES

The most primitive ones have unicellular uninucleate uniflagellate holocarpic thalli with or without any surrounding membrane. In somewhat advanced species the somatic phase is a rhizomycelium, and in more advanced forms it is represented by a scanty mycelium consisting of a few short hyphal branches. The majority of these fungi, however, have a profusely branched well-developed, filmentous aseptate mycelial thallus. Most of them are eucarpic. Even in the aseptate mycelium the septa appear in old hyphae of some species, and during the development of reproductive structures, septum being formed at the base of each reproductive organ. The septa of these fungi are different from other fungi being solid and are not perforated in the centre. In a profusely developed mycelium the reproductive bodies are less conspicuous. In the group as a whole the development of fruiting bodies, comparable to those of other groups of fungi is absent, but in a few, a tendency toward the formation of such structures is present, encountered in the genus *Endogone* (Fig. 12.1).

ASEXUAL REPRODUCTION

The mode of reproduction in the Mastigomycotina and Zygomycotina is asexual and sexual. Asexual reproduction is by the development of spores which may or may not be borne in a sporangium. In certain simpler forms the vegetative body itself behaves as a spore. But a majority of cases after the mycelium has attained sufficient growth, apices of hyphae undergo certain changes to develop into sporangia with their contents being metamorphosed into spores. These spores are single-celled and may be flagellate or non-flagellate. The flagellate spores or zoospores are of two kinds: uniflagellate and biflagellate. The uniflagellate zoospores may have anteriorly or posteriorly placed whiplash type of flagellum, whereas the biflagellate zoospores whose one flagellum is whiplash type and the other of tinsel type may be of two kinds: (i) pear-shaped or pyriform with anteriorly placed flagella, and (ii) kidney- or bean-shaped having flagella inserted in the groove on the lateral side. A zoospore usually swims for a while in water, then loses its flagellum or flagella as the case

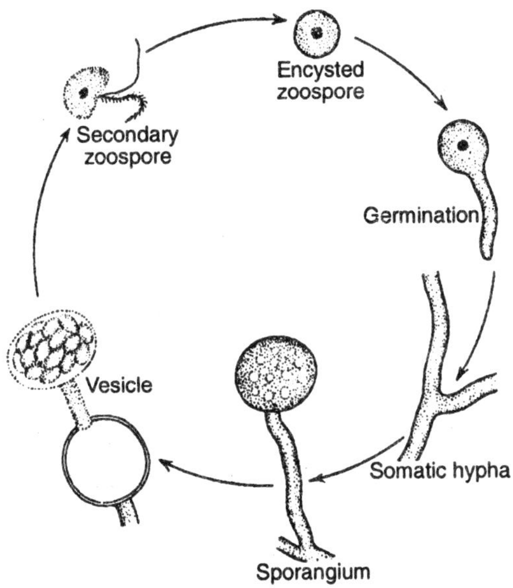

Fig. 12.2: Monoplanetism in *Pythium* sp.

may be, then gives rise to new individual. The biflagellate zoospores may behave in a different manner. (i) A zoospore on being liberated out from a vesicle of the zoosporangium swims around in water, after some time sheds its flagella, comes to rest, becomes rounded, encysts, and germinates by a germ tube which then develops into hypha of a new individual. This type of zoospore is known as **monoplanetic zoospore** and the phenomenon is **monoplanetism,** exhibited by species of *Phythium* (Fig. 12.2). (ii) Zoospore behaves initially like a monoplanetic zoospore and goes to rest by forming a rounded structure which instead of producing a new individual develops into a zoospore. The zoospore so formed swims for sometime and goes to rest by the loss of flagella, encysts, and eventually germinates by a germ tube which becomes the hypha of a new individual. Here, zoospore alternates with two resting stages. Such a zoospore is called **displanetic zoospore** and the phenomenon exhibited by it is **diplanetism,** found in *Saprolegnia* (Fig. 12.3). (iii) Zoospore after swimming goes to rest in a manner as stated above. After a period of rest, again produces a secondary zoospore, which again goes to rest and finally develops into a new individual. The process of formation of zoospore and resting condition may be repeated several times before a new individual is formed. Such a process is known as **repeated emergence,** encountered in *Dictyuchus.* It is

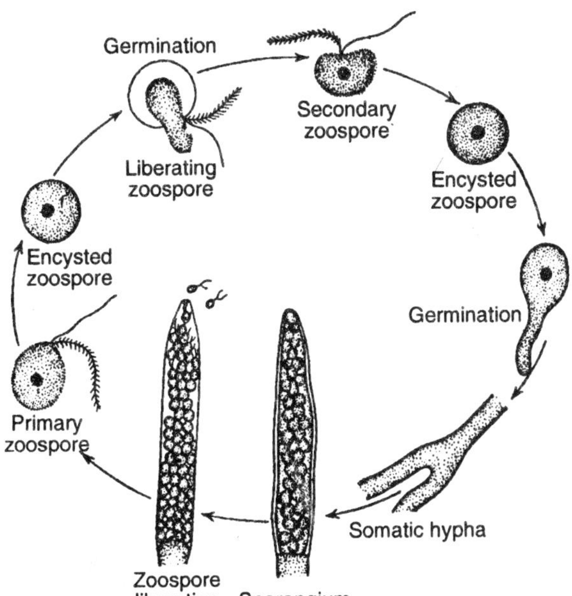

Fig. 12.3: Diplanetism in *Saprolegnia* sp.

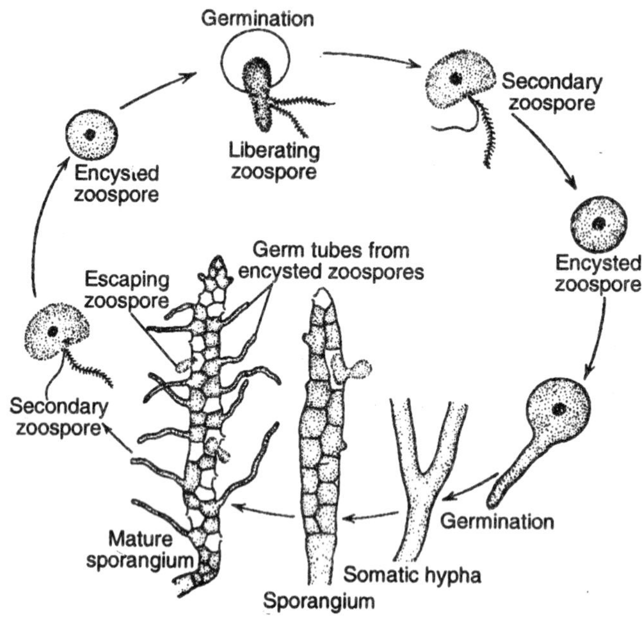

Fig. 12.4: Repeated emergence in *Dictyuchus* sp.

also known as **polyplanetism** (Fig. 12.4). In some terrestrial forms due to certain unfavourable atmospheric conditions, the zoospores fail to develop flagella and become non-motile and are known as **aplanospores** which behave like non-flagellate spores.

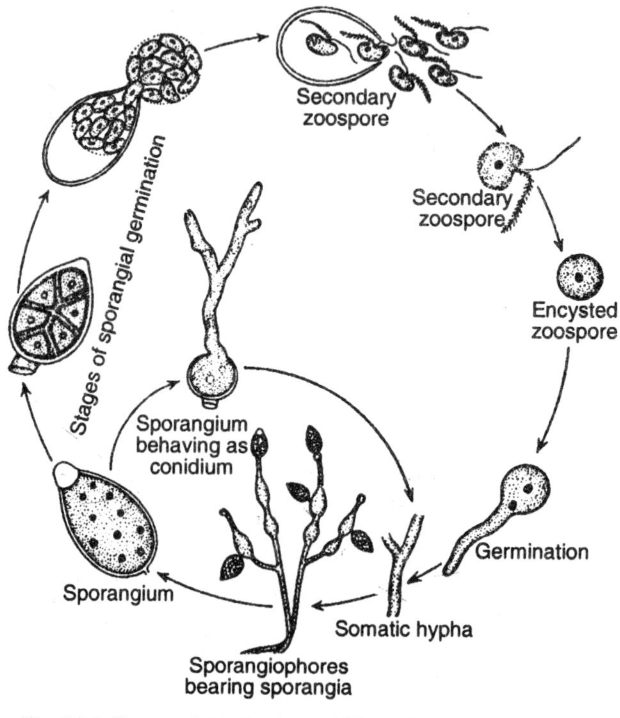

Fig. 12.5: Sporangial behaviour of *Phytophthora infestans*

The non-flagellate spores are: (i) sporangiospores (some designate them as aplanospores) developed in a sporangium, and (ii) conidia. A sporangium contains a large number of spores and may or may not have **columella** in it. Besides this, few-spored sporangia of much smaller size known as **sporangiola** (sing. **sporangiolum**) (Fig. 12.25) are also developed. A sporangiolum may be with or without columella and may be even monosporous. The monosporous sporangiolum has evolutionary significance with regard to the origin of conidia from the sporangial condition. A cylindrical sporangiolum designated as **merosporangium** (pl. **merosporangia**) (Fig. 12.21), represents as advanced evolutionary condition. Both sporangiospore and conidium germinate by germ tube which ultimately develops into new mycelium. Members of the Entomophthorales reproduce asexually by conidia. Here each conidium is developed at the apex of the conidiophore. The conidiophore is again produced from **hyphal body** which is nothing but one of the segments of the segmented mycelium. But Thaxter showed that here conidium is a one-spored sporangiolum which constitutes important evidence of the origin of conidium from a sporangiolum.

In certain fungi, for example, *Phytophthora infestans,* the behaviour of sporangia is very much controlled by conditions of moisture and temperature, depending on which a sporangium either produces zoospores or germinates directly by a germ tube behaving like a conidium. Usually a sporangium at an optimum temperature range of 18°C-22°C and at an optimum relative humidity of 100 per cent produces zoospores. But at an optimum temperature of 24°C and under less humid condition the sporangium germinates directly by a germ tube to form a new individual (Fig. 12.5). The sporangium thus behaves as a conidium. This type of sporangium with a dual function is called a **conidiosporangium.**

Besides the development of different types of spores and sporangia, in some Phycomycetes, the hyphae may directly produce thick-walled resting spores known as **chalmydospores** which can stand unfavourable condition and with the return of favourable condition the chlamydospores develop into hyphae of new individuals.

SEXUAL REPRODUCTION

There is a great diversity in the method of sexual reproduction and, very similar to the green algae, all gradations, from isogamous, anisogamous to oogamous types occur in the Mastigomycotina and Zygomycotina. Again sexual reproduction may involve union of gametes which may or may not be well organized and may or may not be borne in gametangia. Different methods of sexual reproduction are the following:

1. **Gametic Union.** Union takes place between the two well-organized gametes. The fusing pair of gametes may be isogamous as in *Olpidium* (Fig. 12.15A); anisogamous, (i) with two free anisogametes found in *Allomyces* (Fig. 12.8) or (ii) between an antherozoid developed in an antheridium and an ovum which remains enclosed in an oogonium exhibited by *Monoblepharella* (Figs. 12.15H to K).

2. **Gametangial Contact.** Two gametangia taking part in the process are distinguishable as antheridium and oogonium. Contact between their contents is established through a tube-like structure developed from the antheridium which pierces the wall of the oogonium, such a tube-like structure is known as the **fertilization tube**. The oogonium develops well-organized female gamete whose number may be more than one, as in the genus *Saprolegnia* (Fig. 12.11C) or just one, in the genus *Pythium* (Fig. 12.13N). But the antheridium does not bear well-organized male

gametes. The male gametes are represented by nuclei embedded in a mass of cytoplasm. The nuclei and cytoplasm are discharged in the oogonium through the fertilization tube and come in contact with the female gamete (Fig. 12.13N).

The absence of the development of well-organized male gametes indicates the tendency of gradual degeneration of visible sexuality. The entities of the two gametangia are not lost at any stage of the sexual process. This type of sexual reproduction is encountered in the general *Saprolegnia, Pythium, Phytophthora, Albugo,* and others.

3. **Gametangial Copulation.** Union between the two gametangia takes place in such a manner that they lose their entities during the process. The fusing pair of gametangia are multinucleate, nuclei being embedded in a mass of cytoplasm. The gametangia do not bear well-organized gametes. The gametes are nothing but a large number of nuclei embedded in a mass of cytoplasm. This again indicates a further degeneration of visible sexuality. The fusing gametangia may be isogametangia, found in *Mucor* sp. (Fig. 11.16B to E) or heterogametangia, as in *Zygorhynchus heterogamus* (Figs. 12.30D and E). In the latter case of heterogametangial condition the two gametangia differ only in size but there is no difference in their contents and behaviour during the sexual process.

The gametangial copulation exhibited by species of *Rhizophidium* is different from that of *Mucor* sp. an *Zygorhynchus* sp. In *Rhizophidium* sp. the two gametangia are distinguishable as male and female from ther external morphology and behaviour. The male gametangium empties its contents into the female game-tangium through a pore or short fertilization tube. The two gametangia originate from two equal-sized zoospores which behave as holocarpic thalli.

BEHAVIOUR OF COMPATIBLE NUCLEI

In fungi belonging to the Mastigomycotina and Zygomycotina there is no interval between plasmogamy and karyogamy which occur almost simultaneously. But in most of them are post-sexual stages are not clearly known. In general, meiosis takes place before the development of new individuals. In a small number of these fungi, for example, species of *Allomyces,* however, meiosis of the zygotic nucleus is delayed resulting in the development of a diploid structure which may be equally well developed like the haploid vegetative body indicating the presence of an alternation of generations. The post-sexual stages so far known are extremely variable in this group of fungi. In general, they follow along the lines as indicated below.

1. The fusion product of two well-organized free gametes is usually a flagellae zygote. The flagella are shed and the zygote then forms diploid structure which, depending on the nature of fungi, behaves in a different manner as outlined below:

(a) In case of parasitic fungi, e.g., *Synchytrium endobioticum, Olpidium viciae,* the zygote infects the host tissue and ultimately develops into sporangia or gametangia as the case may be, and meiosis takes place during the formation of spores and gametes.

(b) In some water molds like the genera *Allomyces, Blastocladiella,* the zygote directly germinates to form a conspicuous diploid structure called **sporothallus** which bears two types of sporangia. Some of them produce diploid zoospores which on germination give rise to sporothalli, whereas others bear haploid zoospores, meiosis takes place during their development. These haploid zoospores give rise to the somatic body.

2. The oospore formed as a result of gametangial contact or otherwise behaves in any one of the following ways:

(a) The oospore directly germinates to produce germ tube or hypha of the new individual, exhibited by the genera *Monoblepharis, Saprolegnia, Pythium.* Meiosis takes place during the germination of oospore.

(b) The oospore germinates by a germ tube which terminates in a sporangium. Meiosis takes place in the oospore at the time of germination. The sporangium produces secondary zoospores which encyst and ultimately germinate to produce germ tubes. The germ tubes give rise to the hyphae of new individuals. Examples are the species of *Phytophthora* and *Plasmopara.*

(c) The entire content of the oospore, on germination is transformed directly into secondary zoospores from which new individuals are formed and found in the species of *Albugo.* Meiosis takes place during the germination of oospore.

3. Where gametangia are not distinguishable as male and female, the zygospore that is formed by gametangial copulation germinates producing a sporangiophore bearing a germ sporangium at its tip. The germ sporangium produces sporangiospores which give rise to hyphae of new individuals. Meiosis takes place during the zygospore germination, exhibited by the genera *Mucor* and *Rhizopus.*

LIFE CYCLE PATTERN

The life cycle pattern in the Mastigomycotina and Zygomycotina is rather simple in comparison with that of the Ascomycotina and the Basidiomycotina. It consists of well-differentiated asexual cycle and sexual cycle.

The somatic body may directly take part in the asexual reproduction or may produce sporangium from which spores are developed asexually. Each of these spores on germination produces somatic body. Thereby asexual cycle is completed. This process is repeated to enable the development of new individuals.

The sexual cycle is rather less elaborate than what is encountered in the Ascomycotina and the Basidiomycotina. Here also the somatic body, depending on surrounding conditions, may directly take part in the sexual reproduction or may produce gametangia from which gametes are developed and by gametic union zygote is formed. Again the gametangia without producing well-organized gametes may also take part directly in the sexual process. As soon as the compatible nuclei are brought together by plasmogamy, almost immediately after that zygote is formed by karyogamy resulting in the establishment of diplophase. The diplophase is followed by meiosis leading to the development of spores or germ tube from which somatic body is developed. The zygote may also be thick-walled and becomes active only after a period of rest. Meiosis may thus be delayed.

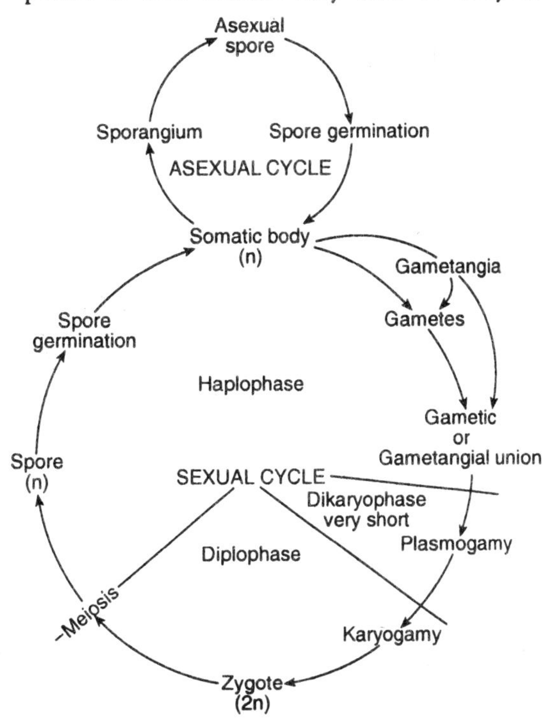

Fig. 12.6: Life cycle pattern in the Mastigomycotina

The major portion of the sexual cycle in the Phycomycetes comprises of haplophase, whereas the dikaryophase is particularly very short or may often be absent. life cycle pattern is indicated in Fig. 12.6.

CLASSIFICATION: SUBDIVISION MASTIGOMYCOTINA

Members of this subdivision possess a wide range of somatic body ranging from simple unicellular and holocarpic to complex forms that develop a system of rhizoids or a rhizomycelium, and finally to eucarpic predominant mycelium. Some are entirely equatic as saprobes or as parasites of algae or fish. Others inhabit soil moisture films causing root-rots or damping-off diseases of seedlings. Again others are adapted to land life as saprobes, or as parasites of higher plants.

Asexual reproduction is by zoospores in sporangia. Zoospores may be posteriorly uniflagellate (whiplash type) or anteriorly uniflagellate (tinsel type) or biflagellate with one tinsel and another whiplash flagellum.

Sexual reproduction in Mastigomycotina is extremely variable—plasmogamy by isogamous, anisogamous or heterogamous planogametic conjugation or by gametangial contact, resulting in a resting sporangium or an oospore.

The following is an outline of the key to the classes, their respective orders and families:

A. Mycelium lacking or poorly developed (a rhizomycelium), thallus holocarpic or eucarpic, posteriorly uniflagellate (whiplash type) zoospore; sexual reproduction by various methods—planogametic copulation (isogamy to anisogamy), fertilization of a non-motile female gamete (egg) by a motile male gamete (antherozoid), or gametangial copulation

Class: CHYTRIDIOMYCETES

B. Mycelium lacking, rhizomycelium present, thallus holocarpic or eucarpic

Order: **Chytridiales**

Family: ***Chytridiaceae***

Genus: *Synchytrium*

BB. Mycelium present, scanty or profuse; thallus eucarpic

C. Sexual reproduction by fusion of planogametes, thick-walled resistant sporangia always present

Order: **Blastocladiales**

Family: ***Blastocladiaceae***

Genus: *Allomyces*

CC. Sexual reproduction by fusion of a free-swimming male planogamete with a female apalanogamete, resistant sporangia lacking

Order: **Monoblepharidales**

Family: ***Monoblepharidaceae***

Genus: *Monoblepharis*

AA. Freshwater or marine chytrid-like fungi, parasitic on algae and fungi or saprobicon plant and insect debris; holocarpic or eucarpic thallus with anteriorly uniflagellate tinsel type flagellum; on germination thallus producing rhizoids in the substratum and converted into a zoosporangium; zoospores releasing through discharge tube reproduce the thallus; sexual reproduction or resting spores not observed.

Class: HYPHOCHYTRIDIOMYCETES

Order: **Hyphochytriales**

Family: ***Rhizidiomycetaceae***

Genus: *Rhizidiomyces*

AAA. Mycelium well-developed, thallus eucarpic, reproductive parts separated from assimilatory part by cross-walls

B. Gametangia morphologically distinguishable as male and female, sexual reproduction by gametangial contact with oospore as end product, asexual reproduction by biflagellate zoospores

Class: OOMYCETES

C. Oospheres one to several in each oogonium; periplasm absent; zoospores biflagellate, diplanetic to monoplanetic

Order: **Saproleginiales**

Family: ***Saprolegniaceae***

Genus: *Saprolegnia*

CC. Oosphere one in each oogonium; periplasm present; sporangia germinating by zoospores or germ tube, zoospores biflagellate, monoplanetic, formed within the sporangia or within an evanescent vesicle arising from the sporangium.

Order: **Peronosporales**

D. Sporangiophores mostly similar to the somatic hyphae or, when different, with indeterminate growth

Family: ***Pythiaceae***

Genus: *Pythium*

Genus: *Phytophthora*

DD. Sporangiophores clearly different from the somatic hyphae with determinate growth; sporangia borne on branched or unbranched sporangiophores, or on various types of sporangiophores.

E. Sporangia in chains on clavate sporangiophores borne in dense soti

Family: ***Albuginaceae***

Genus: *Albugo*

EE. Sporangia borne singly or in clusters at the tips of branched or unbranched clavate sporangiophores.

Family: ***Peronosporaceae***

Genus: *Peronospora*

Genus: *Plasmopara*

PHYLOGENY

Fascinatingly speculative accounts have appeared from time to time with regard to the origin and phylogeny of the members of the Mastigomycotina without leading to any agreement. More critical and comprehensive work will have to be taken up on the various groups of lower organisms before drawing any widely acceptable conclusion. Widely diversified views have been put up by various workers which may be indicated in the following lines:

1. These fungi are degenerate algae (C. E. Bessey, 1895, and others). The statement was made only taking into consideration the superficial resemblances between the two groups—Algae and Fungi, but completely overlooking the fact that their physiological processes are quite different. The Algae and Fungi probably arose as separate groups from unicellular chlorophyll-bearing and non-chlorophyll-bearing organisms respectively.

2. The two groups have evolved as two parallel series was suggested by De Bary and his student Vuillemin 1912 and was given prominence by Atkinson 1909, 1915 and later by Fitzpatrick. In a number of points of morphology and behaviour, the Fungi differ from Algae, e.g., zoospores, structure and mode of sexual reproduction, etc. The fungi have no connection with the algal line and are derived from primitive colourless organisms below the level of the Chytridiales.

3. The origin of the Mastigomycotina is to be sought among non-pigmented flagellates (Dangeard, 1886; J. H. Ashworth, 1923) with which there are many points of likeness. It is probable that the Fungi have evolved from lower forms along lines independent of those leading to green plants and to animals.

4. The Phycomycetes have a protozoal ancestry and should be placed in a position distinct from the Algae.

5. The aquatic members of the Mastigomycotina were considered by A. Scherffel, 1925, to be made up of two distinct series, the Chytridineen and the Saprolegniineen-Peronosporineen. These he believed to have arisen from different monad ancestors, the first from an uniflagellate type, and the second from a biflagellate, and to have undergone independent development. Scherffel's idea was supported by Sparrow.

Sparrow (1958), however, was of opinion that these fungi are not a homogeneous monophyletic group, but rather an artificial category of coenocytic zoosporic fungi, consisting of four groups which may have come from radically different progenitors. The evolution of the lower fungi has become of current interest through Koch's (1956) discovery of the presence of what he called a second, non-functional, vestigial blepharoplast in the chytrid motile cell, suggesting a biflaggellate ancestry. This supposition was supported by conclusions from observations on the movement of chytrid zoospores (Koch, 1959, 1961). Koch observed, that the double image of the flagellum formed when the flagellum is undulating in a plane horizontal to the observer, may be asymmetrical, with one side more 'humped' than the other. He suggested that this asymmetrical double image originates phylogenetically in a motile cell in which the posteriorly directed flagellum is laterally inserted on the body, so that the primitive or ancestral condition is not the posteriorly attached whiplash flagellum, but rather the laterally attached whiplash flagellum. This, and the second non-functional blepharoplast suggest that the chytrids, and perhaps all posteriorly uniflagellated fungi, have evolved either from biflagellated organisms or from organisms with biflagellated swimming cells.

Order: **Chytridiales**

The members of this order, commonly known as chytrids, are parasites or saprophytes usually in water, rarely on land. They are regarded as the lowest of all the fungi. Some of the chytrids induce serious diseases of plants of economic importance, for example, (i) *Synchytrium endobioticum* causes wart disease of potato, (ii) *Physoderma zeaemaydis* causes brown spot of maize, (iii) *Urophlyctis alfalfae* causes crown wart of alfalfa, and (iv) *Olpidium brassicae* causes diseases of roots of seedlings of cabbage, and tobacco.

The thallus is unicellular either naked or walled, may be rounded, elongated or of various other shape. It is either holocarpic or eucarpic being differentiated into a vegetative and fertile portion. In some chytrids the thallus may develop into a rhizomycelium.

Family: *Synchytriaceae*

In the family Synchytriaceae the thallus is posteriorly uniflagellate and holocarpic. The thallus is at first naked, in course of growth develops a thin membrane and ultimately forms a thick-walled structure whose protoplast divides by cleavage to form a sorus of either zoosporangia or gametangia. Zoospores on being released from the zoosporangium repeat the cycle, whereas the gametes by planogametic isogamy result in the development of zygote which ultimately becomes thick-walled resting sporangium. The resting sporangium on germination produces zoospores. The zoospores and the gametes are morphologically similar but their behaviour is different.

Genus: *Synchytrium*

Synchytrium is a very widely known chytrid. It is world-wide in distribution and several of its species are of considerable economic importance. All species are parasitic on both higher and lower plants. They have been reported to occur in abundance on algae, mosses, ferns, and flowering plants and are rather more common on flowering plants growing in a moist environment since water is essential for their zoospores to mature, disseminate, and also to infect host cells. The intensity of infection is often correlated with environment in which the host plants are growing and usually seedlings are very susceptible. The species which occur on flowering plants cause the development of galls on leaves, petioles, flower buds, and stems, whereas those which parasitize algae may cause marked local enlargement as well as elongation of the infected cell. Most terrestrial species occur on the leaves, petioles, and stems of the hosts, but some infect the roots and underground stems only.

The life cycles of the different species of *Synchytrium* vary considerably in length. Some are described as short-cycled and others referred to as long-cycled depending on the sequence of development of different stages and their number. Most widely known and thoroughly investigated species is *Synchytrium endobioticum* which causes the destructive wart disease of potato. The first classical study of the life history and cytology of *Synchytrium endobioticum* was that of Miss K. M. Curtis, published in 1921. The minute endobiotic, holocarpic thallus represented by a naked posteriorly uniflagellate zoospore with whiplash flagellum swarms in water, comes to rest on the epidermis of a young host cell at or near ground level. It withdraws its flagellum and penetrates the host cell by a very thin naked penetration tube which pierces the cuticle and epidermal wall making a narrow channel. There is no further inward migration of the parasite in the host body. The subsequent development of the parasite is carried on in the epidermal cell originally invaded in which it remains confined. The parasite is carried to the bottom of the epidermal cell by the streaming of protoplasm. In the mean time under the influence of the stimulus produced due to the presence of the parasite, the infected cell swells and becomes pyriform. The neighbouring cells start dividing repeatedly (hyperplasia) followed by swelling (hypertrophy) resulting in the formation of a tumour and the surrounding epidermal cells divide similarly forming a rosette with the infected cell being placed in the middle. The parasite along with its nucleus enlarges markedly, rounds off and forms a wall of two layers, a thick golden-brown exospore and a thin hyaline endospore, developing into a **prosorus** or an **initial cell.**

At first the prosorus is uninucleate. Soon after maturing, the prosorus germinates within the host cell. The exospore ruptures and the endospore extrudes to produce a vesicle. The cytoplasm and nucleus pass out into the vesicle consisting of extruded endospore, in this vesicle nuclear division occurs. At this stage 5 to 7, even up to 9 multinucleate segments are delimited by the development of hyaline walls forming a sorus of segments whose identity will be determined in the later stages. The number of nuclei in each segment increases to as many as 300 resulting from repeated nuclear division. The entire protoplast of each segment divides into as many portions as there are nuclei and each portion with its nucleus and cytoplasm develops into an uniflagellate swarmer. Whether a segment will be a sporangium or a gametangium depends on its environmental condition. When water is in abudance the swarmers will behave as zoospores, whereas with shortage of water in the substratum the swarmers when released fuse with each other behaving as gametes. The gametes originating from the same gametangium fail to fuse with each other indicating heterothallic condition of the gametangia. The enlarging sporangia or gametangia, as the case may be, exert pressure on all sides and are forced out of the sorus and the disorganized host cell. The swarmers are set free through one or two ruptured papillae which are developed on each sporangium or gametangium. The zoospores on being liberated out from the sporangium complete the asexual cycle by inducing new infection in the

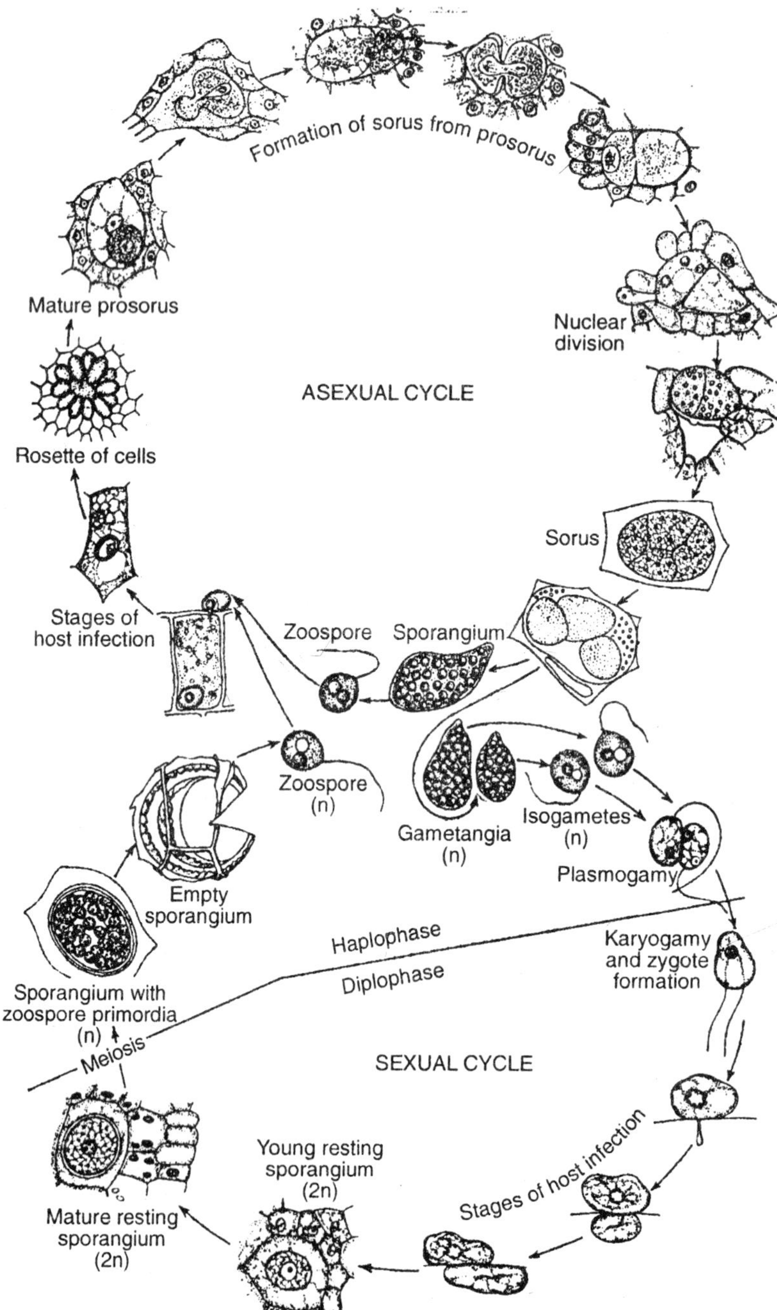

Fig. 12.7: Life cycle of *Synchytrium endobioticum*

host cells. The asexual cycle repeats itself several times initiating series of new infections during the same season. In *Synchytrium endobioticum* with such a rapid method of reproduction taking place at a time when the potato tubers are making active growth, fresh infections are constantly taking place, and as every infection involves the formation and proliferation of rosette cells there is so much hypertrophy around infected cells that the tuber soon becomes more or less covered with the warted excrescences which are so characteristic of the wart disease of potato.

Gametic union takes place on the surface of the host or in a film of water. Karyogamy follows immediately after plasmogamy resulting in a biflagellate zygote which swarms around for a short time comes to rest on the host cell, and penetrates the host epidermal cell wall in a process similar to infection by zoospore. The zygote sinks to the bottom of the infected cell and simultaneously with this process the surrounding cells of the host tissue divide repeatedly as a result of which a pressure is developed resulting the infected cell being buried deeply in the host tissue.

The zygote enlarges considerably, develops a three-layered thick wall and is converted into a resting sporangium. With the completion of resting period the diploid nucleus divides reductionally. Several hundred free haploid nuclei are formed by the repeated division of the haploid nuclei that are resulted from meiosis of the diploid nucleus. These free nuclei with cytoplasm subsequently form innumerable zoospores. The thick wall of the sporangium breaks open from pressure set up by the swelling of the inner wall due to imbibition of water resulting in the bursting of the wall and the zoospores are set free. The zoospores again cause new infection on the host cells and thereby the life cycle is completed. The life cycle of the genus *Synchytrium* is illustrated from the species *S. endobioticum* (Fig. 12.7).

Some Indian Species: *Synchytrium ajrekari* Payak and Thirum; *S. brownii* Karling; *S. collapsum* Syd.; *S. cookii* Lingappa; *S. endobioticum* (Schilb.) Percival.

Order: **Blastocladiales**

The order Blastocladiales includes both aquatic and terrestrial fungi. Some are parasitic on aquatic insects or fungi or saprophytes on dead twigs submerged in water or in soil. All are filamentous fungi with more or less well-developed mycelium. The fungi under this order exhibit fairly well-developed alternation of generations. The vegetative body is the gametophyte which produces planogametic anisometes. Planogametic anisogamy leads to the development of diploid sporophytic individual which bears sessile zoosporangia. Both haploid and diploid zoospores are formed in different zoosporangia. The haploid and diploid zoospores germinate into haploid and diploid thalli respectively.

Family: ***Blastocladiaceae***

In the family Blastocladiaceae the thallus is attached to the substratum by rhizoids. Usually the thallus is large enough to be visible with unaided eye. The true hyphae of the vegetative body besides performing metabolic function also bear reproductive organs on their branches. Uniflagellate gametes of two kinds, large and small, are developed in different gametangia. Gametic union takes place outside the thallus, karyogamy taking place immediately after plasmogamy resulting in the development of a diploid zygote which germinates giving rise to a diploid thallus known as **sporothallus.** The diploid sporothallus bears two kinds of sporangia some of which produce haploid zoospores and others diploid zoospores. The haploid zoospores germinate to produce new vegetative body, whereas the development of diploid sporothallus is repeated by the germination of the diploid zoospores, thereby the life cycle is repeated.

Genus: *Allomyces*

It has world-wide distribution occurring in soil and on dead animal material in water. Investigations by Ralph Emerson on the isolates of *Allomyces* from different parts of the world have revealed valuable information on the life cycle of the different species of this fungus. The somatic body consists of a stout trunk-like structure with the basal portion attached to the substratum by rhizoids. Numerous slender hyphal branches arise dichotomously from the distal end of the hyphal trunk. The hyphal branches are separated into segments by pseudosepta in the form of thickened rings of cellulin at fairly regular intervals. Typical sieve-like pseudosepta resembling true cross-walls also occur in some cases. The nature of cytoplasm is changeable, differing markedly in the same plant under altered condition of environment and at various ages.

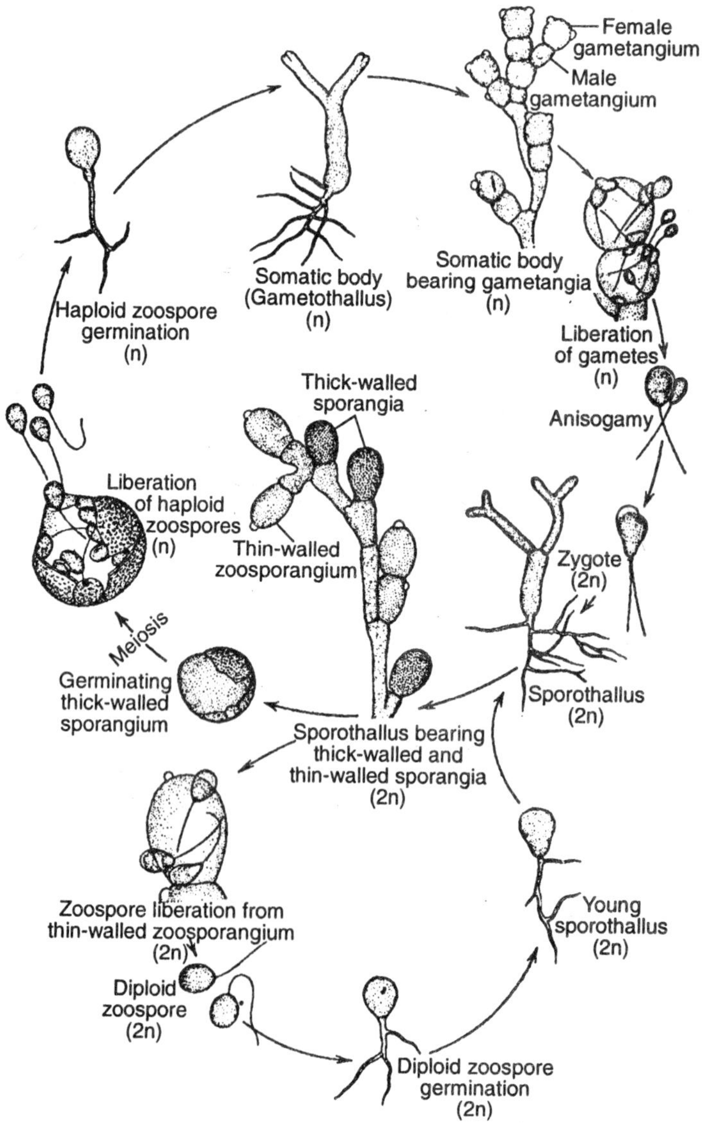

Fig. 12.8: Life cycle of *Allomyces* sp.

The cytoplasm is colourless except in the male gametangia being orange in colour. *Allomyces* is homothallic. The gametangia appear in pairs, a male and a female, at the tips of the hyphae. The male gametangia are smaller than the female gametangia and are orange in colour due to the presence of carotene. Posteriorly uniflagellate gametes are liberated out from the gametangia by the formation of papillae on the gamentangial surface. The male gametes are smaller than the female gametes and possess conspicuous orange globule one in each. The anisogametes on being liberated out in water copulate to form a motile biflagellate zygote without having any interval between plasmogamy and karyogamy. The motile zygote soon comes to rest and develops into a diploid thallus which is known as sporothallus. The sporothallus bears two types of sporangia; colourless thin-walled zoosporangia and reddish-brown thick-walled resistant sporangia. Both of them produce uniflagellate zoospores. The zoospores that are set free from the thick-walled resistant sporangia develop into haploid thalli bearing paired male and female gametangia. Meiosis takes place during the development of these zoospores, whereas the thin-walled zoosporangia produce diploid zoospores which germinate giving rise to sporothalli bearing thin-walled zoosporangia and resistant sporangia. This process may continue indefinitely. Usually the haploid thallus occurs alternating with the diploid sporothallus exhibiting an alternation of generations.

The vegetative body of *Allomyces* is composed of true hyphae. It exhibits considerable differentiation of parts performing both metabolic function and bearing reproductive organs. The genus *Allomyces* exhibits a definite alternation of generations manifesting ansogamous sexual reproduction in its life cycle. Thus *Allomyces* shows advancement over the chytrids. Life cycle is presented in Fig. 12.8.

Some Indian Species: *Allomyces anomalus* Emerson; *A. arbusculus* Butler; *A. javanicus* Kniep; *A. javanicus* Knicp var. *macrogynus* Emerson.

Order: **Monoblepharidales**

The order Monoblepharidales includes only a few species of fungi, most of them are aquatic. This order is closely related to the Blastocladiales. The fungi belonging to this order have outstanding characteristic of the method of sexual reproduction in which a large non-motile egg is fertilized by a small uniflagellate antherozoid. This type of sexual reproduction is not encountered in any other fungi.

Family: ***Monoblepharidaceae***

The mycelium is saprophytic, occurring usually on decaying twigs, sometimes on animal remains lying in cool clear water. It is fixed to the substratum by rhizoids. The cytoplasm of the mycelium is characterized by the presence of vacuoles. These vacuoles produce foamy appearance of the mycelium. Asexual reproduction is by uniflagellate zoospores developed in an elongated zoosporangium. Sexual reproduction is oogamous. Well organized male and female gametes are developed in antheridia and oogonia respectively. More than one antherozoids are developed in an antheridium, whereas the oogonium produces only one oosphere. The mode of fertilization is an outstanding feature of the family. A single antherozoid effects fertilization. In some cases, the fertilized oosphere pushes through the mouth of the oogonium and rounds up outside before it is transformed into an oospore. In others, it matures within the oogonium. Germination of the oospore occurs by a germ tube during which meiosis takes place.

Genus: *Monoblepharis*

The genus *Monoblepharis* is saprophytic primarily on submerged aquatic vegetable debris, particularly twigs, and on animal remains. The thallus consists of delicate non-septate hyphae, colourless or with a slightly brownish tinge usually protruding from the lenticels of the twigs submerged under water. The thallus has a rhizoidal system which serves the function of anchorage to the substratum. The protoplasmic contents of the hyphae are highly vacuolated containing oil globules, which impart a characteristic foamy appearance to the hyphae. Asexual reproduction is by posteriorly unifagellate zoospores developed in a zoosporangium borne terminally on actively growing somatic hypha. During the development of the sporangium cytoplasm and nuclei accumulate in the end of a branch which is cut off from the rest of hyphae by a cross-wall. The sporangial protoplast then divides into many uninucleate portions, all of which ultimately develop into posteriorly uniflagellate zoospores. At maturity a pore is formed at the tip of the sporangium, the zoospores are released through the pore. They swim for some time, settle to a suitable substratum, become rounded, and each germinates by two germ tubes, one of which is narrowly cylindrical or somewhat irregular in shape. Many lateral sporangia may grow out behind each other giving the impression of sympodial branching. The sporangia may also develop by the process of **internal proliferation** in which a new sporangium arises inside an emptied old one and grows through it by the growth of the septum through its empty membrane which is again filled with protoplast. The sporangium is cut off by a septum and the zoospores are developed from the protoplast. The same thallus under higher temperature produces gametangia. Both antheridium and oogonium are developed in the same thallus. The antheridium is narrow-elongated structure below which is the oogonium which is larger and rounded. The antheridium, like a zoosporangium, is cut off from the tip of a hypha by a septum. The development of antheridium and antherozoids occurs in a manner very similar to that of zoosporangium and zoospores. The antheridium produces small number of antherozoids. The antherozoids are smaller than the zoospores, but otherwise they resemble them in appearance. The oogonium is developed from a lateral outgrowth arising from below the septum of the antheridium. A mature oogonium bears a single oosphere and has a well-defined receptive papilla on its wall. During fertilization the receptive papilla forms the point of entrance of the antherozoid. Plasmogamy is followed immediately by karyogamy. Immediately after fertilization the fertilized oosphere may or may not be extruded from the oogonium. Whatever may be the case, after fertilization the oospore becomes thick-walled, which may be either smooth or bullate when mature. In exogenous species of *Monoblepharis* where the oosphere emerges out from the oogonium, immediately after fertilization the oospore moves, with revolution and amoeboid alterations of form, to the bottom of the oogonium, and then toward the tip when it escapes. It swims around in water by means of the flagellum. Possibly this motility is caused by the flagellum derived from the antherozoid which is thrown off rather late. It exhibits amoeboid movement, becomes spherical, and surrounds itself with a membrane which gradually thickens to form a thick wall. In the endogenous species the oospore remains in the oogonium, and development of thick wall takes place there. If the oosphere is not fertilized it is surrounded by a membrane and develops parthenogenetically to a thick-walled resting spore. During the germination of the oospore, nuclear divisions occur, in course of which meiosis takes place. At first a hypha-like structure is produced from which ultimately a new thallus is developed. Details of the life cycle are indicated from which ultimately a new thallus is developed. Details of the life cycle are indicated in Fig. 12.9.

The genus *Monoblepharis* possesses very interesting characters for which its position in the evolutionary line of fungi is rather parallel to none.

It has resemblance with water molds for having mycelial soma, elongated hypha-like proliferating zoosporangium, and heterogametangial condition. Again, characters like development of well-organized male and female gametes, specialized nature of fertilization, the behaviour of fertilized egg, and germination of oospore by the formation of a tube-like structure instead of producing zoospores are of special interest in this genus.

Order: **Saprolegniales**

The fungi included under this order are mostly inhabitants of water. Some are soil-inhabiting. They are best and widely known among the aquatic Phycomycetes and are commonly known as water molds and fish molds. They occur abundantly in all freshwaters and usually prefer those which are clear and relatively pure. Most of the species are saprophytes on plant and animal debris in freshwater or in soil. A few are parasites of fish and amphibians causing their serious diseases. Practically all the species may be readily grown in pure culture, either on solid media or in water cultures.

Some are parasitic on freshwater and marine algae, especially diatoms, and on the

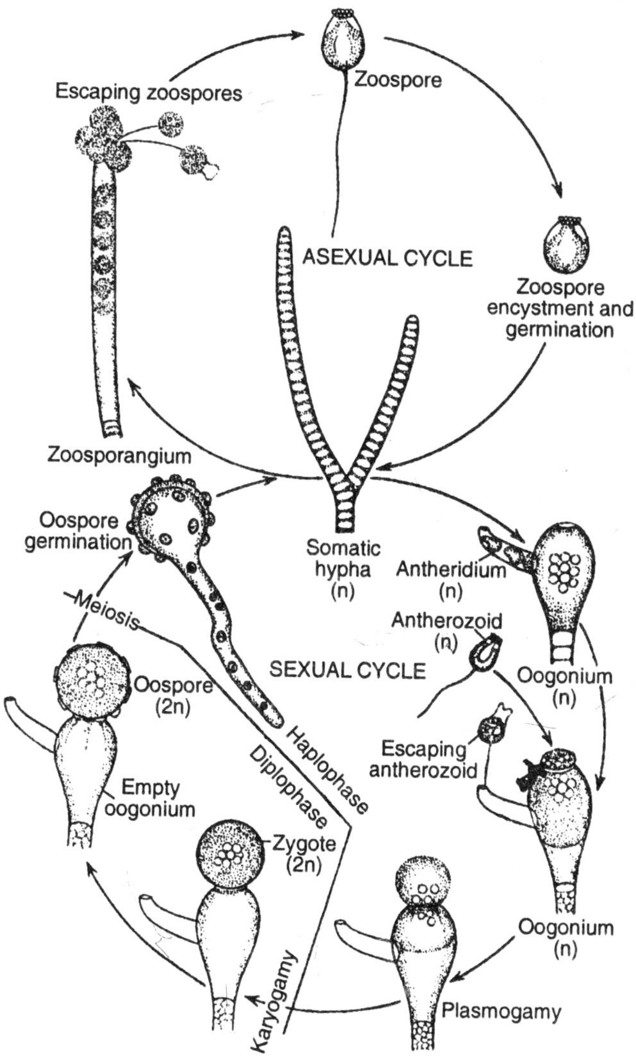

Fig. 12.9: Life cycle of *Monoblepharis* sp.

roots of terrestrial flowerig plants. All members of this order are characterized by well-developed mycelium. Biflagellate zoospores are produced in zoosporangia of various types. Sexual reproduction occurs by heterogametangial contact. The gametangia are known as antheridia and oogonia. The oogonia containing one to several oospheres and periplasm is lacking. The entire mycelium does not take part in the development of reproductive organs. The reproductive organs are mostly terminal or subterminal. The development of abundant mycelium and the formation of morphologically distinct oogonia and antheridia show a marked advancement of the Saprolegniales from the Chytridiales. Again, the Saprolegniales are quite distinguishable from its neighbouring order Peronosporales by the aquatic habit and persistent sporangia, whereas the Peronosporales are characterized by a terrestrial habit and deciduous sporangia which are mostly terminal.

Some relationship can, however, be drawn between the Saprolegniales and the Peronosporales as both groups have walls containing cellulose, similar zoospore structure and oogamous reproduction. The Leptomitales is another group of aquatic fungi showing clear similarities with the Saprolegniales.

Family: ***Saprolegniaceae***

In members of the Saprolegniaceae the thallus is a richly branched coenocytic aggregation of eucarpic hyphae, usually with unlimited capacity for growth. This aggregation may form a dense turf, easily visible with unaided eye on the substratum. The profusely developed mycelium consists of two portions, the rhizoidal for anchorage with the greatest diameter near the point of attachment being known as intramatrical hyphae which decrease slowly giving rise to extramatrical hyphae or the vegetative mycelium bearing reproductive organs. The nearly cylindrical hyphae with vacuolate protoplasm lying next the wall have branching usually of the monopodial type, though rarely dichotomous. The hyphal wall unlike that of higher fungi, gives test for cellulose which turns blue with chloro-zinc-iodine. Septa are formed in the mycelium just below the reproductive organs, separating them from the somatic hyphae, which generally remain aseptate.

When the extramatrical hyphase have attained a certain age, their ends become clavate or spherical, filled with thick cytoplasm and numerous nuclei which ultimately take part in the development of biflagellate zoospores. In the mean time the clavate or spherical ends of the extramatrical hyphae are cut off by transverse septa. The zoospores are discharged by the internal pressure and usually through the softened apex of the zoosporangium (Figs. 12.10B and C). Often undifferentiated masses of spores are poured out of the sporangium. The zoosporangia are typically terminal segments of hyphae, but sometimes several are formed one behind the other. Occasionally the sporangia are formed intermingled with oogonia. Usually the zoosporangia are cylindrical, fusiform, or clavate in shape (Figs. 12.10A, D and E) with the exception of *Pythiopsis* where the sporangia are spherical to pyriform in shape. In *Aplanes* sporangia are rarely produced; but when they are developed, the spores germinate *in situ* within the sporangium, and the germ tubes protrude through the sporangial wall. Sometimes sporangia are cut off in basipetal succession, as in *Dictyuchus*. Rarely intercalary sporangia are formed. In general, the zoosporangia of the Saprolegniaceae remain permanently attached to the somatic hyphae, and even after they have discharged their zoospores. But exception is the genus *Dictyuchus* where the sporangia with maturity usually fall off from the somatic hyphae. Occasionally they may remain attached to the somatic hyphae. Again, the sporangial wall may be persistent or evanescent. When the sporangium has emptied its contents of spores, the basal septum grows through the empty sporangium and forms a second sporangium in the emptied space of the first sporangium. This process may be repeated several times as a result of which several sporangia may thus be formed one within the other and each maturing and shedding its spores before the next one is formed. This phenomenon is known as **sporangial proliferation** (Fig. 12.10A). The sporangial proliferation is commonly encountered in *Saprolegnia* and *Leptolegnia*. In *Achlya, Aphanomyces, Pythiopsis,* and *Thraustotheca,* the sporangia are renewed by lateral outgrowth of a portion of the hypha next to the empty sporangium and proliferation is absent. In such cases the procss may be repeated forming sympodially divided, verticillately or spirally arranged sporangia. In *Isoachlya* and *Protoachlya* cymose branching and proliferatioin may both occur together.

The structure and behaviour of zoospores are very variable in different genera of the Saprolegniaceae. In *Saprolegnia, Isoachlya, Leptolegnia,* and *Leptolegniella,* the zoospores

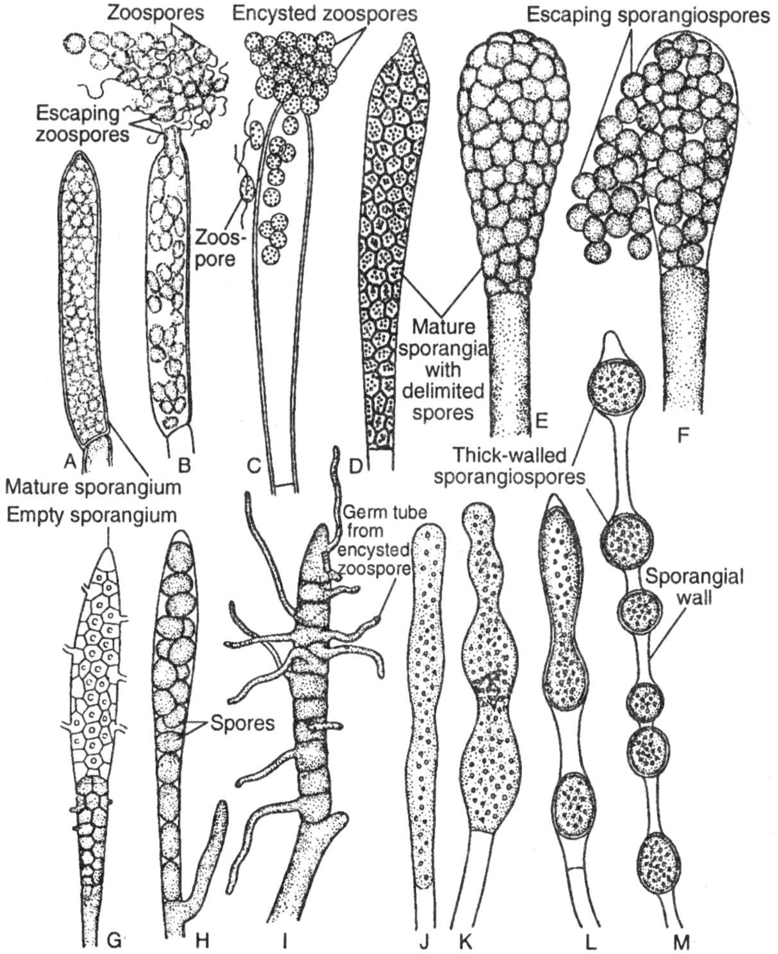

Fig. 12.10: Types of sporangia and spores of different genera of the Saprolegniaceae. A-B. *Saprolegnia*. A. Mature sporangium. B. Escape of zoospors through a terminal pore. C-D. *Achlya*. C. Sporangium freeing its spores through an apical pore. D. Mature sporangium with delimited spores. E-F. *Thraustotheca*. E. Mature sporangium with delimited spores. F. Escape of sporangiospores. G-I. *Dictyuchus*. G. Empty sporangium. H. A sporangium with spores. I. Germ tubes from encysted zoospores. J-M. *Geolegnia*. Stages in formation of sporangia and thick-walled sporangiospores

are diplanetic, that is, have two motile phases (Fig. 12.3).

In *Pythiopsis,* only the primary zoospores are formd which come to rest immediately after they are liberated, shed off flagella and germinate with a hypha, the zoospores are monoplanetic. Monoplanetism is also exhibited by *Thraustotheca.* The zoospores of *Achlya* and *Aphanomyces* are also diplanetic, but the first swarming period is very brief being confined to the sporangium and serves the purpose of carrying the spores through the terminal pore of the sporangium. Then encyst just outside the mouth of the zoosporangium (Fig. 12.10C), and eventually the secondary zoospores are released from the encystments. Thus in *Achlya* and *Aphanomyces* there is a strong tendency to suppress the first swarming period of the zoospores. In *Dictyuchus,* no primary zoospores are liberated, instead, they encyst in the sporangium and form secondary zoospores. The secondary zoospores escape through short exit papillae, swarm for a time and encyst (Figs. 12.10G and H). After a period of rest those encystments release another crop of secondary zoospores. Following the second swarming, encystment and germ tube formation occur. Thus exhibiting a phenomenon of **repeated zoospore emergence** or **polyplanetism** (Fig. 12.4). Again, some zoospores may fail to escape from the sporangium, encyst within it, form germ tubes there, which pass through the openings in the sporangial wall (Fig. 12.10I), whereas in case of a sporangium having evanescent sporangial wall, the primary zoospores are rounded and remain aggregated together and ultimately form secondary zoospores repeatedly. In *Aplanes,* the zoospore initials, while still in the sporangium, surround themselves with a wall and germinate there piercing the sporangial wall with their germ tubes. Very similar is also in *Geolegnia,* where the zoospores initials remain multinucleate and surround themselves with a thick wall (Figs. 12.10J to M) and await the decay of the sporangial wall before germinating directly to mycelium. The sporangium is rather functionless. Both these two genera are aplanetic and they exhibit the phenomenon called **aplanetism.** With the suppression of the process of zoospore formation, these genera represent the culmination of an evolutionary series where degenerate asexual reproduction is possibly in favour of replacement by sexual reproduction. It is also noteworthy that this may be the point of origin of the Subdivision Zygomycotina.

At times the sporangia, after being cut off from the hyphae, may remain quiescent for a considerable period, and later discharge their spores just like normal sporangia. Such sporangia are known as **resting sporangia.** At the tips of certain hyphae unicellular, globoid to ovoid or pyriform, thick-walled resting spores are developed in chains which fall apart easily at the dividing septa. After a resting period they germinate by germ tubes which either grow into hyphae or develop into short-stalked sporangia. They are termed differently by various authors as **resting spores, conidia gemmae,** or **chlamydospores.**

Sexual reproduction is oogamous. The sexual organs are developed on extramatrical hyphae, and are usually terminal in position, though occasionally intercalary. The antheridia and oogonia may be formed on the same hypha, homothallic or on different hyphae, heterothallic. But majority of the Saprolegniaceae are homothallic. The oogonia are spherical bodies containing one or more eggs inside them. The oogonia may have smooth or pitted walls. The multinucleate antheridial branches are slender, cylindrical or clavate in form, and somewhat thicker than the hyphal branches. They are usually applied to the oogonium by their side rather than tips. Usually one or more antheridia are applied to each oogonium. Fertilization occurs with the help of fertilization tubes, which originating from antheridia pierce through the oogonial wall and reach the oospheres. The young oogonium is multinucleate. After nuclear division there occur degeneration of many nuclei and reorganization of the oogonial contents into a number of uninucleate eggs. In *Leptolegnia, Aphanomyces, Dictyuchus,* and *Geolegnia,* only a single egg is formed within the oogonium. In most of the remaining genera more than one eggs are produced whose number again varies from species to species. As many as thirty egg may be formed in certain species. During fertilization one nucleus of each antheridium passes through each fertilization tube into each egg and fuses with the nucleus therein. The fertilized eggs develop thick walls and are converted into oospores. Each oospore, after a period of rest, germinates by a germ tube developed by the inner membrane of the oospore which pushes out through a crack in the outer wall. The germ tube develops directly into a new plant. It has been reported by Schrader in 1938 that meiosis occurs during oospore germination.

Genus: *Saprolegnia*

Saprolegnia lives in all sorts of water both saprophytically on vegetable and animal remains and parasitically on living fish, and may cause epidemic by infecting fish eggs. *Saprolegnia parasitica* causes disease of fish. Usually the infection of fish is preceded by a weakening or injury of the individuals infected. The somatic hyphae consist of: (i) branched rhizoidal portion embedded in the nutritive matrix serving the purposes of anchorage and absorption of nourishment, and (ii) profusely branched hyphae lying outside the substratum forming cottony growth being visible even in the naked eye, some portions of which take part in the development of reproductive organs. The vegetative hyphae are coenocytic with tapering branch tips. Septa appear only at the base of the reproductive organs.

Under suitable environmental conditions the hyphal branches give rise to zoosporangia. The zoosporangia are terminal. They are elongated tapering, clavate, often slender to thicker

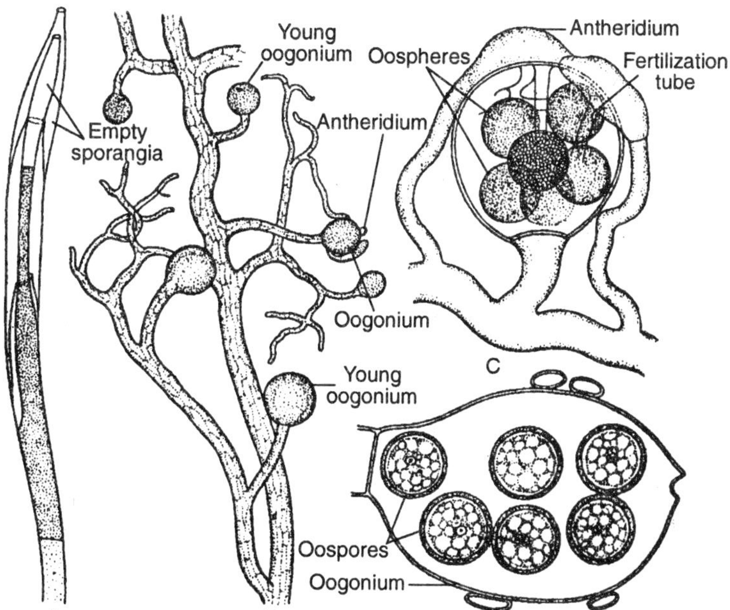

Fig. 12.11: *Saprolegnia* sp. A. Sporangial proliferation. B. Portion of mycelium with young sex organs. C. Mature sex organs, oogonium with oospheres and fertilization tubes entering from the antheridia. D. Oogonium with oospores

toward the distal end. Each sporangium is separated from the rest of the hyphae by a septum laid down at the base. The sporangium is clearly distinguishable from the rest of the somatic hyphae from its dense protoplasm. With the initiation of the sporangial development, a large number of nuclei are embedded in cytoplasm stream in the hyphal branch, which is the potential sporangium, from the rest of the hyphae. This is followed by the gradual development of a septum at the base of the sporangium by which it is separated from the rest of the hyphae. The dense mass of sporangial protoplast is then divided into as many portions as there are nuclei and each portion with cytoplasm and a nucleus develops into a spore. The whole protoplast is utilized for the development of spores. Thus, numerous primary zoospores are developed in a sporangium. The zoospores are diplanetic. They squeeze out one by one through the terminal opening of the sporangium (Fig. 12.10B), swim in the surrounding water, come to rest, and encyst at a distance from the sporangium. After a short period, a thin papilla develops on the encystment, its tip dissolves, and a secondary zoospore creeps out. The secondary zoospores after a short swarming period, encysts and the encystment germinates by a germ tube which ultimately develops into a hypha of the new individual. Several asexual generations follow one another by sporangial proliferations (Fig. 12.11A). Chlamydospores or gemmae often in chains of different shapes and sizes are formed, which after a period of rest produce zoospores. These zoospores on germination by germ tubes produce new individuals. Sexual reproduction is oogamous. When conditions are favourable, the somatic hyphae give rise to antheridia and oogonia. The oogonia are terminal on main hyphae or on lateral branches or intercalary singly or in chains. They are spherical, oval, pyriform or sometimes fusiform bearing one to many oospheres. Each mature oosphere bears a single nucleus. The oogonial wall is smooth, papillate or often pitted and much thicker than the walls of the somatic hyphae.

The antheidia are developed on the same hypha which bears the oogonium, and arise immediately below it or on different hyphae (Fig. 12.11B). They are multi-nucleate, slender structures borne terminally on slender branches of the somatic hyphae. The antheridia may be short or long, simple or branched. They grow toward the oogonia and become attached to them. More than one antheridia may be attached to an oogonium. During fertilization the antheridium sends out fertilization tubes which penetrate the oogonial wall and reach the oospheres (Fig. 12.11C). Upon entering the oogonium a fertilization tube often produces branches sending one branch to each oosphere. In cases where more than one antheridia contact the oogonium, an oosphere, is usually approached by one fertilization tube only. The antheridium does not bear well-organized male gametes and contains a large number of nuclei embedded in a mass of cytoplasm. The antheridial nuclei with cytoplasm are discharged from the antheridium to the oospheres through the fertilization tubes. One antheridial nucleus enters each oosphere, approaches its nucleus, and fuses with it resulting in the formation of a diploid nucleus. Immediately after fertilization the diploid nucleus with cytoplasm is surrounded by a thick wall and an oospore is formed (Fig. 12.11D). The oospore is smooth-walled. After a prolonged resting period, the oospores are liberated from the oogonium by the disintegration of its wall. The oospore germinates to produce hypha of the new indiviual. Meioss of its diploid nucleus takes place during the germination of the oospore. Life cycle of the genus *Saprolegnia* is presented in Fig. 12.12.

Some Indian Species: *Saprolegnia ferax* (Gruith.) Thuret; *S. monoica* (Pringsh.) De Bary.

Order: **Peronosporales**

The order Peronosporales has reached highest point of advancement in the Oomycetes. The members of this order lead a terrestrial existence as parasites in the tissue of higher plants. Many of them induce destructive diseases of plants of economic importance and cause tremendous loss to crops. Among them, mention may be made of: *Pythium debaryanum* causes damping off of seedling disease, *Phytophthora infestans* causes late blight of potato disease, *Albugo candida* causes white rust disease of crucifers, *Peronospora pisi* causes downy mildew disease of peas, *Plasmopara viticola* causes downy mildew disease of grape, *Sclerospora graminicola* causes green ear disease of Bajra, and *Bremia lactucae* causes downy mildew disease of lettuce.

The coenocytic profusely developed mycelium is composed of freely branched hyphae which grow intercellularly, receive nutrition from the host cells by the formation of haustoria. The hyphae may also be intracellular.

Asexual reproduction is usually accomplished by the development of biflagellate secondary zoospores borne in a sporangium.

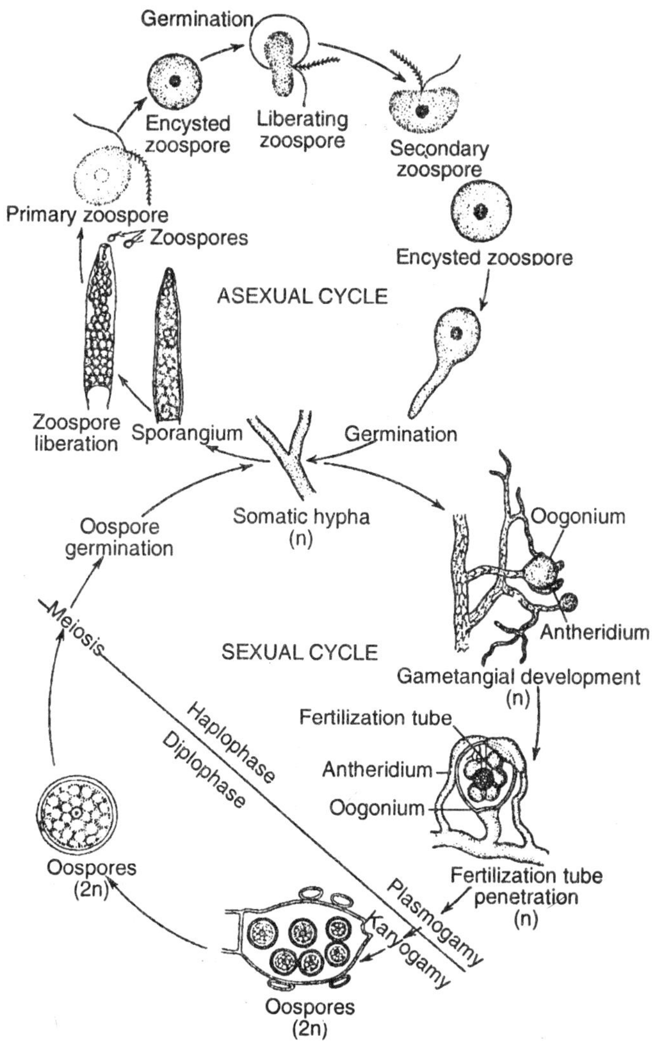

Fig. 12.12: Life cycle of *Saprolengia* sp.

The sporangia are developed terminally or laterally on sporangiophores which may or may not be distiguishable from the somatic hyphae. The sporangiophores may again be branched or unbranched. Depending on the environmental conditions, especially on temperature, the sporangia may also behave as conidia by directly germinating by germ tube without the development of zoospores. The production of zoospores indicates the aquatic ancestry of these fungi.

Sexual reproduction takes place by the development of oogonia and antheridia which may be borne in the same thallus or on different thalli. The oogonium contains in the centre an uninucleate oosphere and cytoplasm in the periphery which is known as **periplasm.** There is absence of well-organized male gametes in the antheridium. The antheridium contains one or more antheridial nuclei embedded in cytoplasm. Fertilization is accomplished by the development of a fertilization tube through which the antheridial nucleus or nuclei as the case may be, with cytoplasm pass into the oogonium. One antheridial nucleus fertilizes the oosphere resulting in the formation of an oospore around which smooth or variously sculptured or ornamented thick wall is developed. The periplasm probably serves as nourishment for the developing oospore. The nature of oospore germination is either by the production of a germ tube or zoospores during which meiosis takes place.

Family: *Pythiaceae*

The Pythiaceae comprise fungi of wide range of habitat. The lower members are aquatic, the higher ones terrestrial, and some are amphibious intermediate species. Many of these fungi are destructive plant pathogens. They induce serious diseases of plants of economic importance by attacking both herbaceous and woody plants. Some of them are: *Pythium debaryanum* causes damping off of seedlings, *Pythium aphanidermatum* causes a seedling and root-rot disease of sugar beets and radishes, also causes foot rot of papaya; *Pythium ultimum* causes tip blight of pea; *Pythium arrhenomanes* is associated with root of maize, wheat, and sugarcane; *Phytophthora infestans* causes late blight of potato and tomato; *Phytophthora parsitica* causes wilt of betel vine, *Phytophthora palmivora* causes bud rot of cocoanut, and fruit rot of arecanut, and *Phytophthora citrophthora* causes brown rot and gummosis of lemons. Some members of the Pythiaceae attack fungi, algae and insects. There are others which form mycorrhizal component in liverworts and ferns. Many inhabit in soil. In general, members of the Pythiaceae have less parasitic tendency than their neighbouring fungi included under the family Peronosporaceae. Rather they possess saprophytic tendency than being obligate parasites.

In general, sporangia are developed at the tips of the sporangiophores represented by a sympodially branched system of hyphae which are very little differentiated from the somatic hyphae. In some species well-defined sporangiophores are developed. The sporangia may remain permanently attached to the hyphae or may fall off as soon as they are mature, deciduous. The species with permanently attached sporangia are more primitive than those having detactable sporangia. Liberation of zoospores takes place either through a pore in the sporangial wall or by the development of a thin-walled vesicle in which imperfectly differentiated zoospores are discharged before they mature and are liberated out.

The mode of sexual reproduction in the members of the Pythiaceae follows the general pattern as indicated under the Peronosporales.

Genus: *Pythium*

The species of *Pythium* are usually encountered in soil or water, living as saprophytes on dead organic matter or as parasites upon plants causing rot of the tissues or damping off of seedlings. In *Pythium debaryanum* the slender mycelium composed of branching coenocytic hyphae may be both intercellular and intracellular generally does not form special haustoria penetrating the host. Septa are formed during the development of reproductive structures.

Asexual reproduction takes place by means of biflagellate secondary zoospores produced in a sporangium which may be terminal or intercalary on the somatic hypha (Figs. 12.13A and B). Depending on species, the sporangium may be filamentous and scarcely distinguishable from the somatic mycelium. It may also appear as more or less inflated mycelium with or without any definite shape, or may be distinctly spherical to oval in form (Figs. 12.13C and D).

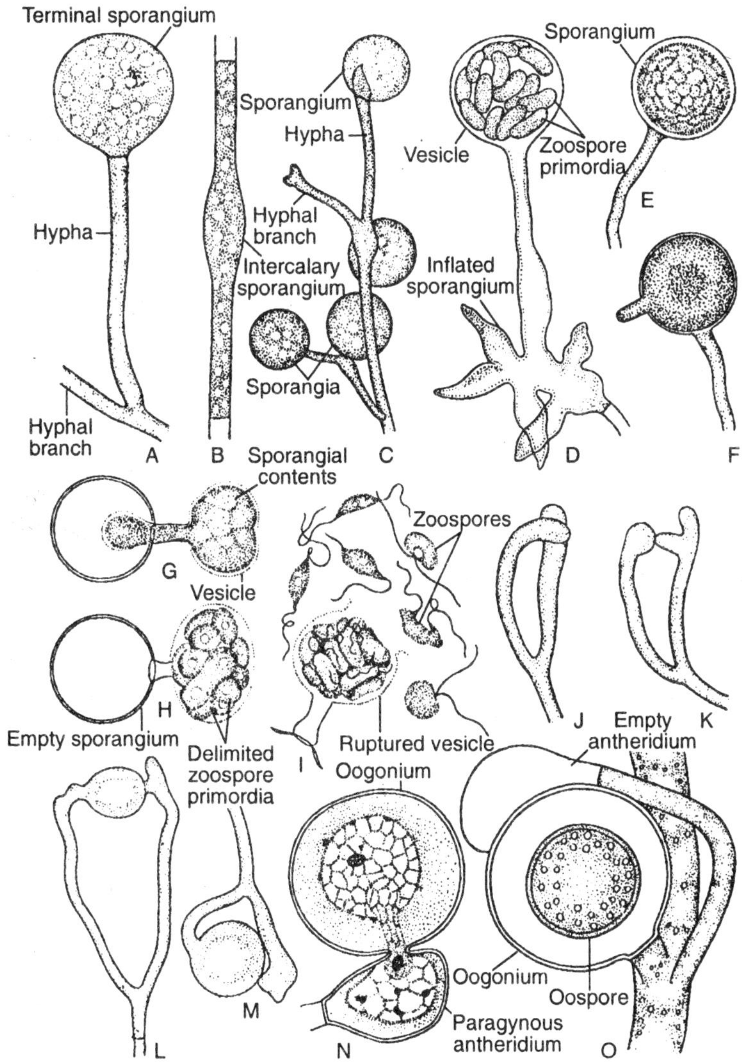

Fig. 12.13: *Pythium* sp. A. Terminal sporangium. B. Intercalary sporangium. C. Sporangia formed successively and becoming lateral by further elongation of the hypha. D. Inflated filamentous sporangium with vesicle containing zoospore primordia. E. Young sporangium. F. Sporangium with emission tube. G. Vesicle containing sporangial contents. H. Vesicle containing delimited zoospore primordia. I. Escape of zoospores by bursting vesicle wall. J-M. Stages of development of antheridium and oogonium. N. Arrangement of antheridium and oogonium, antheridium discharging male nucleus through the fertilization tube. O. Oospore still enclosed in the oogonium and the empty antheridium still attached with the oogonial wall

Usually the tips of the somatic hyphae swell, in them protoplasm flows, and ultimately are cut off by septa from the rest of hyphae as multinucleate structures. Each sporangium then forms a beak or emission tube which is usually short but sometimes several times the diameter of the sporangium (Figs. 12.13E and F). The tip of the beak or emission tube softens and develops into a bubble-like vesicle in which passes the protoplasm of the sporangium in an undifferentiated state (Fig. 12.13G). The protoplasm in the vesicle becomes differentiated into biflagellate secondary zoospores (Figs. 12.13D and H). The zoospores move about within the vesicle and escape singly or several in a clump by bursting of the vesicle wall (Fig. 12.13I). After swimming for while each zoospore comes to rest, encysts, and then germinates by a germ tube from which somatic hypha is developed. In some species the sporangia germinate by the production of one to six germ tubes, e.g., *Pythium ultimum,* exhibiting an adaptation to a subaerial habit where there may be shortage of sufficient water for the successful distribution of zoospores.

Sporangial proliferation is absent in the genus *Pythium,* only exception is *Pythium proliferum* where sporangia are renewed by proliferation. Again, sporangia are developed in basipetal chains in *Pythium intermedium* indicating similarity with the Albuginaceae.

Sexual reproduction is typically by the development of antheridia and oogonia which appear when the host has been killed and the fungus lives saprophyticallly. The oogonia and antheridia are developed on branch tips of the same hypha or of different hyphae of the same thallus (Figs. 12.13J and K), with some exceptions where the antheridium may be intercalary, e.g., *Pythium aphanidermatum.* The hyphal tips first become inflated, multinucleate, and are set off by septa (Figs. 12.13L and M). As the oogonium enlarges to form a globose structure all nuclei but one migrate into the peripheral layer where they degenerate. The centrally placed nucleus with a portion of cytoplasm forms the oosphere and the cytoplasmic layer in the periphery is the periplasm. The number of oosphere is usually one, in some species two to six oospheres are produced, e.g., *Pythium polysporum.* The antheridium is also multinucleate at first but normally only one nucleus survives. The antheridium is applied at the side of the oogonium, such an antheridium is known as **parazynous** (Fig. 12.13N). During fertilization the antheridium produces a fertilization tube which penetrates through the oogonial wall and periplasm. The male nucleus with cytoplasm passes through the fertilization tube and fertilizes the oosphere. The fertilized oosphere then forms a thick wall and becomes an oospore (Fig. 12.13O). The oospore wall may be smooth or reticulate. The oospore then goes to rest, and germinates by the production of a long germ tube during which meiosis occurs (Fig. 12.14). At a temperature range of 10°C-17°C, however, the germ tube after growing up to certain length stops growing. The protoplast of the oospore migrates through the germ tube and pushes out through the tip forming a vesicle in which secondary zoospores are developed. Life cycle of the genus *Pythium* is illustrated in Fig. 12.14.

Some Indian Species: *Pythium aphanidermatum* (Edson) Fitzp.; *P. butleri* Subramaniam; *P. debaryanum* Hesse.; *P. graminicolum* Subramaniam; *P. vexans* De Bary.

Genus: *Phytophthora*

The genera *Pythium* and *Phytophthora* are so closely related that though the genus *Phytophthora* was established by De Bary (1876) later accepted by Butler (1907) and others, did not receive full support of Fitzpatrick (1923). The main differences are: the method of sporangial germination, and absence of vesicle development. These characters do not hold good with respect to many of the species of *Phytophthora* indicating an

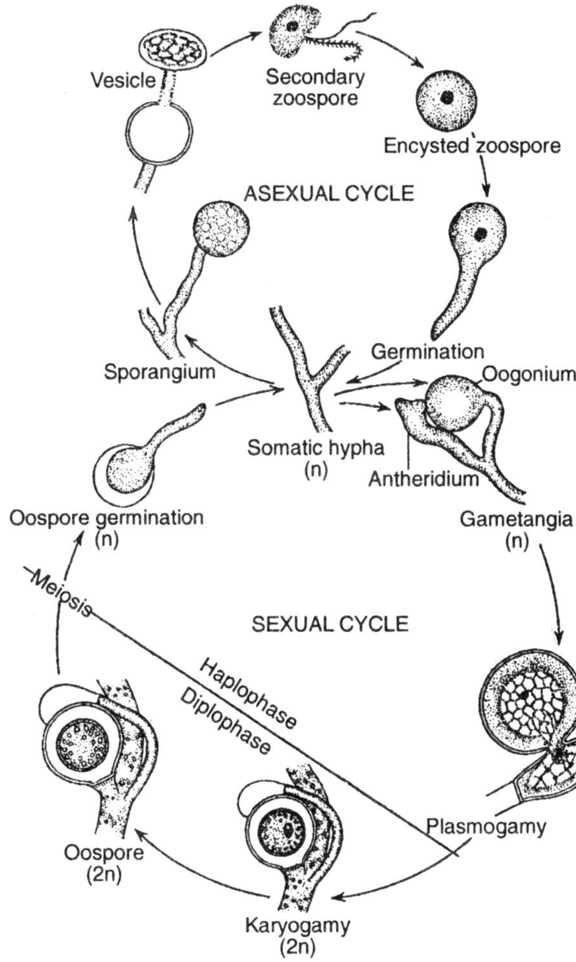

Fig. 12.14: Life cycle of *Pythium* sp.

unsatisfactory separation of the two genera. But is is also true that the genus *Phytophthora* exhibits preference for terrestrial flowering plants as hosts, and possesses looser branching of the hyphae than *Pythium*. The hyphae when old, become irregularly separate.

This genus is of special interest on account of its one exceedingly destructive representative, *Phytophthora infestans,* which occupies a historic position in Phytopathology. It is one of the earliest of parasitic fungi to receive study in any way complete or adequate, which did much to turn attention and interest toward Plant Pathology. The genus *Phytophthora* contains species some of which live as saprophytes in the soil, developing as parasites in the presence of suitable host plants. Other species are usually found as parasites on higher plants. *Phytophthora infestans* which causes the late blight of potato, tomato, and rot of potato tubers was first described in 1845 as a *Botrytis.* Since then various investigators have studied different aspects of the fungus and it was fully described by De Bary in 1876.

The somatic mycelium which is composed of branched, non-septate, hyaline hyphae is both intracellular and intercellular forming rudimentary haustoria. In damp air the soma-

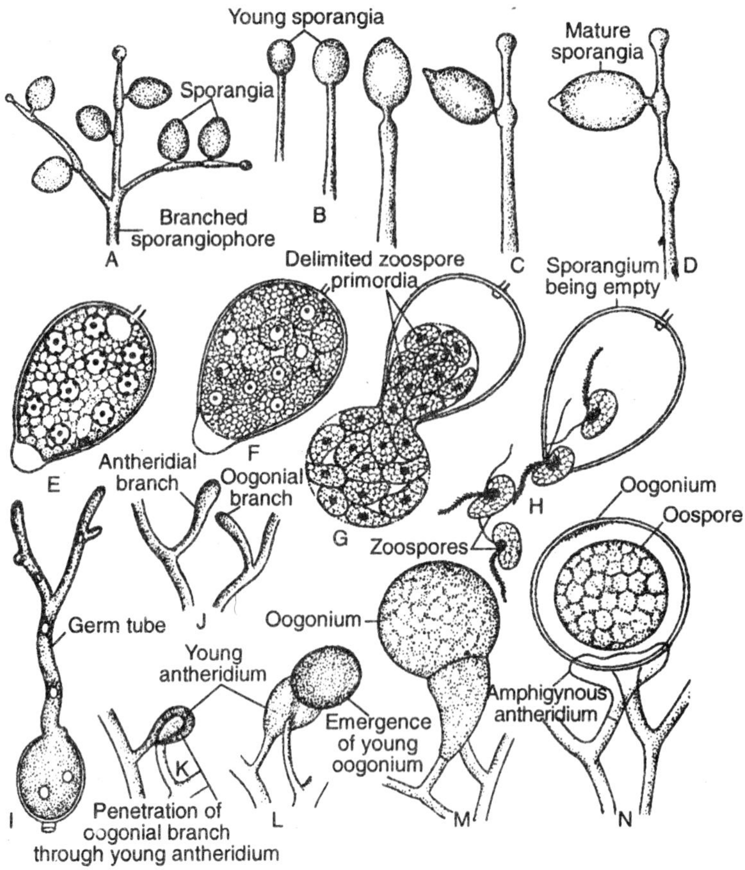

Fig. 12.15: *Phytophthora* sp. A. Branched sporangiophore bearing sporangia. B-D. Stages of formation of sporangium. E. Lemon-shaped sporangium with an apical papilla. F-H. Stages of formation of zoospores from a sporangium. I Germination of sporangium by germ tube. J-M. Stages of development of antheridium and oogonium. N. Oospore enclosed in the oogonium and funnel-like empty antheridium at the oogonial stalk

tic hyphae produce abundant sporangiophores (Fig. 12.15A). The sporangiophore is at first simple and bears a single apical sporangium after the production of which a lateral branch arises below the sporangium and grows on in such a way as to give the first sporangium a lateral appearance producing a sympodial type of branching with characteristic swellings at the nodes (Figs. 12.15B to D). The sporangia are readily detachable from the sporangiophore and may be dispersed by rain-splash or air currents. The sporangium is thin-walled, lemon-shaped with an apical papilla (Fig. 12.15E). It may germinate directly by a germ tube (Fig. 12.15I) to produce somatic hypha. But most commonly the contents of the sporangium cleave to form biflagellate zoospores which escape throgh an apical pore (Figs. 12.15F to H). After a few minutes the zoospores come to rest and each germinates by a germ tube producing somatic hypha. The nature of germination of sporangia is determined by temperature and moisture. A lower temperature is favourable for zoospore formation, whereas germination by germ tube occurs at higher temperature (Fig. 12.5). The genus *Phytophthora* differs from *Pythium* in the formation of zoospores in the sporangium and the discharge of zoospores through an opening at the apical papilla without the formation of a vesicle. Exception being *Phytophthora palmivora,* where a vesicle may at time be formed.

Phytophthora infestans when grown on culture media, the mycelium often forms chlamydospores.

Sexual reproduction takes place by the development of antheridia and oogonia. In species of *Phytophthora,* e.g., *P. cactorum* and *P. nicotianae,* the nature of development and arrangement of antheridium and oogonium are essentially like *Pythium debaryanum,* that is, paragynous. But in others, e.g., *Phytophthora infestans, P. erythroseptica, P. phaseoli, P. parasitica, P. arecae, P. himalayensis* the oogonium pierces through the antheridium which takes a funnel-like appearance at the oogonial stalk. This type of arrangement of antheridium and oogonium is known as **amphigynous** (Fig. 12.15N). In *P. infestans* the antheridium and oogonium always arise from the lateral branches of somatic hyphae. They grow, and come close to each other. The young oogonium penetrates the young antheridial wall, passes into the young antheridium, grows through it, projects out, and swells to form a mature oogonium in which an oosphere is developed (Figs. 12.15J to M). Both antheridia and oogonia are multinucleate when young. Ultimately with maturity only one nucleus survives in each. A mature antheridium contains only one male nucleus with cytoplasm and a mature oogonium bears an oosphere surrounded by periplasm. During fertilization due to bulging of antheridial wall, a large receptive papilla is developed. As soon as the receptive papilla is withdrawn the antheridium produces a fertilization tube through which the antheridial nucleus with cytoplasm is brought in contact with the oosphere where karyogamy takes place resulting in formation of an oospore

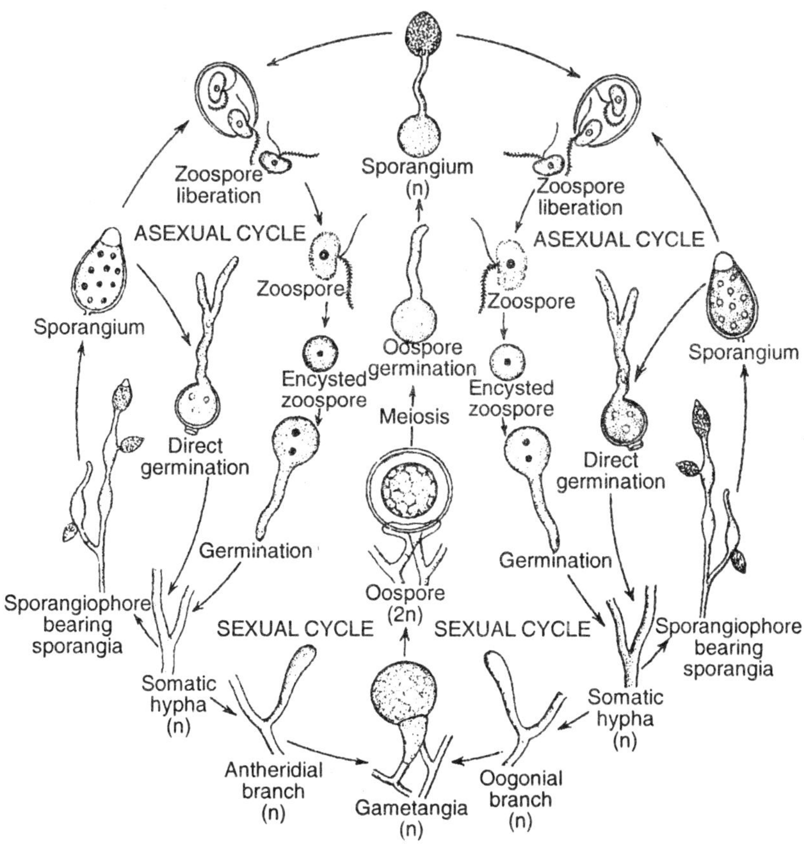

Fig. 12.16: Life cycle of *Phytophthora* sp.

(Fig. 12.15N). The oospore becomes thick-walled, goes to rest, and germinates to produce a short hyphal system which terminates in a sporangium (Fig. 12.16). The sporangium produces biflagellate secondary zoospores which on germination produce somatic hyphae. Meiosis takes place during the germination of oospore. Life cycle of *Phytophthora* is presented in Fig. 12.16.

The earlier workers were unable to find any oospore for which the existence of oospore was a much controversial point for years. Even De Bary created the genus *Phytophthora* on the basis of a type species in which the oospores were unknown. Clinton (1911) first obtained perfect oogonia, antheridia, and oospores of *Phytophthora infestans* in pure cultures. The first report of oospore formation in nature was made by Murphy (1927) from Ireland. Barrett (1948) reported that in *Phytophthora infestans* oospores are developed only when two compatible strains are brought together. Oospores have recently been found in Mexico where compatible strains of the fungus co-exist.

Some Indian Species: *Phytophthora arecae* (Coelem.) Pethybridge; *P. cactorum* (Leb. and Cohn) Schroet; *P. colocasiae* Racile; *P. infestans* De Bary; *P. parasitica* Dast. var. *sesami* Kale and Prasad.

Family: ***Albuginaceae***

The members of this family are obligate parasites upon flowering plants. They induce white rust disease. The only genus in this family, *Albugo* has several species. Each species is capable of infecting a certain group of closely related host plants. For example, *Albugo candida* attacks practically all cultivated crucifers, including cabbage, radish, turnip, horse-radish, etc.; *A. ipomoeae-pandurance* attacks various species of the Convolvulaceae, very common on sweet potato; *A. tragopogonis* occurs on the Compositae; *A. portulaceae* on *Portulaca;* and *A. bliti* on members of the Amarantaceae. Wilson (1907) reported that *Albugo candida* infects only the crucifers, but no plants outside the family Cruciferae. In spite of the fact that *Albugo candida* attacks so many different host plants, it exhibits remarkable stability of essential characters, i.e., size of sporangia, sculpturing of oospore wall, etc. Hence *Albugo candida* is a good morphological species. But cross-inoculation from one species of host to another does not always result in infection which indicates that the species *Albugo candida* is composed of a number of **biological forms** or **strains.** These biological forms are morphologically identical, but differ in their pathological characters since their differences are detectable only in their ability to infect different cruciferous plants. The specialization of parasitism of these biological forms is known as **biological specialization.** The genus *Albugo* is characterized by sporangia borne in chains at the tips of short, clavate and unbranched sporangiophores which are formed in clusters in a sorus.

Genus: *Albugo*

The genus *Albugo* is also commonly known as *Cystopus.* The name *Albugo* was given in 1821 and *Cystopus* in 1874. In view of the recommendation of the International Botanical Nomenclature the earlier name *Albugo* is retained and the latter name *Cystopus* abandoned.

Species of *Albugo* possess very fine coenocytic mycelium which is strictly intercellular. Small globular haustoria, developed by the hyphae pierce the host cell wall. The hyphae ramify along the intercellular spaces and collect beneath the epidermis, branch profusely and after maturity produce numerous short, club-shaped sporangiophores arranged in a palisade-like layer at right angles with the host epidermis (Fig. 12.17A). Sporangia are abstricted from the tips of the sporangiophores in basipetal succession and remain attached in rather long chains unless disturbed. During the development of the sporangium cytoplasm accumulates in the apex of the sporangiophore, the apex becomes multinucleate and is cut off by the appearance of a septum below (Figs. 12.17B and C). The septum is gelatinous in nature and as the sporangium matures the gelatinous septum shrinks and forms a disjunctor whose dissolution permits the sporangium to fall off from the sporangiophore (Fig. 12.17D). The entire process is repeated simultaneously with the elongation of the sporangiophore producing a chain of sporangia and disjunctors arranged alternately (Fig. 12.17E). Large number of sporangiophores are developed in clusters (Fig. 12.17A). The sporangia are produced simultaneously in chains. Both the growth of the fungus and the production of sporangia exert a pressure from below on the host epidermis causing it to bulge producing a white blister-like appearance. The blister-like structure eventually ruptures exposing the sorus bearing sporangia and sporngiophores. The sporangia are released from the sorus. They are disseminated by wind, water, and by other agencies as well. The sporangia are normally globose. They may appear cuboid or polyhedral by pressure during their formation. The sporangia are thin-walled, multi-nucleate and are filled with cytoplasm. Depending on the temperature, the sporangia germinate by germ tubes or by the formation of biflagellate zoospores (Figs. 12.17F to J), which develop into hyphae of new individuals.

Antheridia and oogonia are formed after the completion of sporangial development, when the mycelium has already invaded deep in the host tissue. They are buried in the host tissue. The details of sexual reproduction differ very little from those of *Pythium debaryanum.* The club-shaped multinucleate antheridium is applied to the side of the globose multinucleate oogonium (Fig. 12.18A). During fertilization

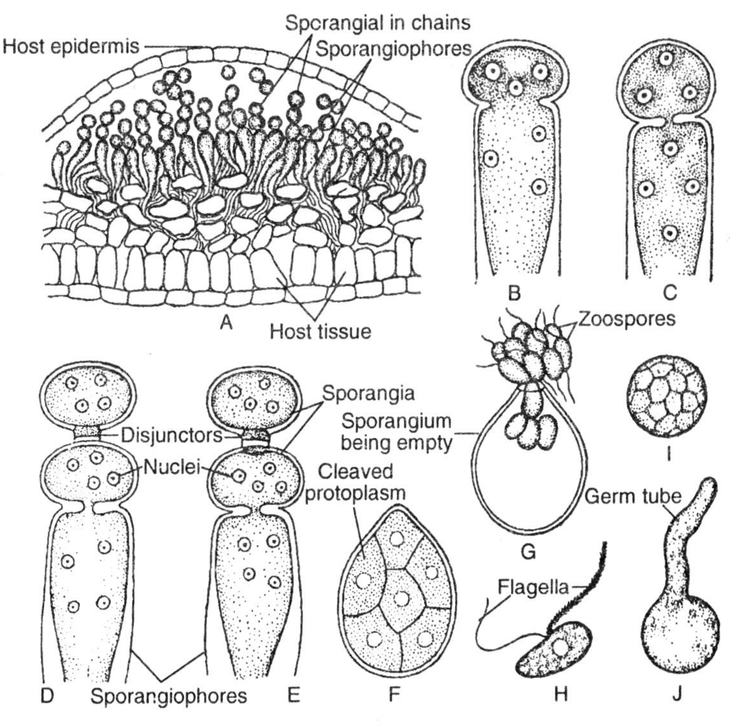

Fig. 12.17: *Albugo candida.* A. Section through sporangial sorus showing club-snaped sporangiophores producing chains of sporangia, host epidermis and host tissue. B–E. Stages in sporangial development and separation. F. Sporangium with cleaved contents, G. Discharge of biflagellate zoospores. H. Biflagellate secondary zoospore. I. Encysted zoospore. J. Germination of encysted zoospore by germ tube

the oogonial wall at the point of contact of the antheridium becomes thin and a papilla develops which protrudes in the antheridium. With the disappearance of the papilla the antheridium produces a fertilization tube which passes through the thin spot of the oogonial wall and periplasm reaching the multinucleate oosphere (Fig. 12.18B). A deeply granular body bearing single egg nucleus arises at the centre of the multinucleate oosphere which is known as **coenocentrum** (Fig. 12.18B). The fertilization tube enters the coenocentrum, ruptures and discharges a single antheridial nucleus which fuses with the egg nucleus forming an oospore (Fig. 12.18C). Immediately after fertilization the fertilization tube collapses and the coenocentrum disappears. The oospore becomes thick-walled and variously sculptured (Fig. 12.19D). The pattern of sculpturing is of taxonomic interest for the identification of species. Each species has a distinctive pattern of sculpturing (Figs. 12.18F to H). The oospore germinates by the production of biflagellate zoospores and thereby behaves like a resting sporangium. Meiosis takes place during the germination of oospore. The rough outer wall of the oospore cracks open and the inner wall grows out to form a vesicle bearing the zoospores (Fig. 12.18E). The zoospores disperse by the breaking of the vesicle ultimately giving rise to new individuals. Life cycle of the genus *Albugo* is presented in Fig. 12.19.

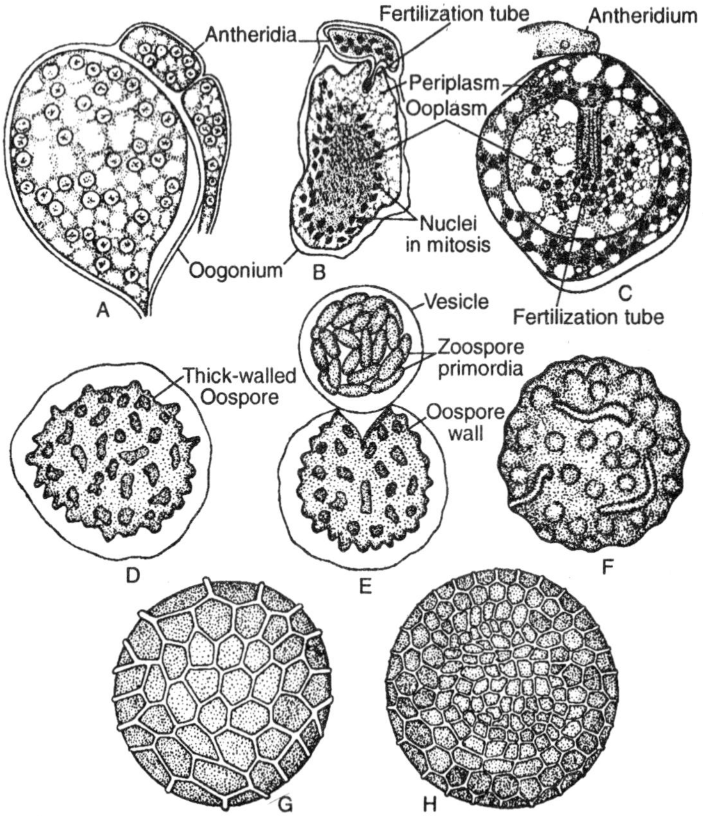

Fig. 12.18: *Albugo* sp. A. Club-shaped multinucleate antheridium applied to the side of globose multinucleate oogonium. B. Penetration of fertilization tube through periplasm, presence of deeply granular central region of the oosphere surrounded by multinucleate layer. C. Fertilization tube discharging antheridial nucleus. D. Mature thick-walled oospore. E. Germinating oospore with vesicle containing zoospore primordia. F. Oospore of *A. candida*. G. Oospore of *A. bliti*. H. Oospore of *A. platensis*

Some Indian Species: *Albugo bliti* (Biv.) Kuntze; *A. candida* (Pers.) Kuntze; *A. ipomoeae-panduranae* (Schw.) Swingle; *A. platensis* (Speg.) Swingle; *A. portulacae* (DC) Kuntze.

Family: ***Peronosporaceae***

The members of this family, commonly known as downy mildews, are obligate parasites. They induce disease of vascular plants called downy mildew. Many of them cause destruction of crop plants. Some of the important plant pathogens of this family are: *Sclerospora graminicola* on cereals, *Plasmopara viticola* attacks grapes, *Peronospora destructor* on onion, *Pseudoperonospora cubensis* on cucurbits, *Bremia lactucae* on lettuce, *Peronospora tabacina* on tobacco, *Peronospora pisi* attacks *Pisum sativum,* and *Basidiophora entospora* attacks weeds. They possess intercellular somatic mycelium from which haustoria of various

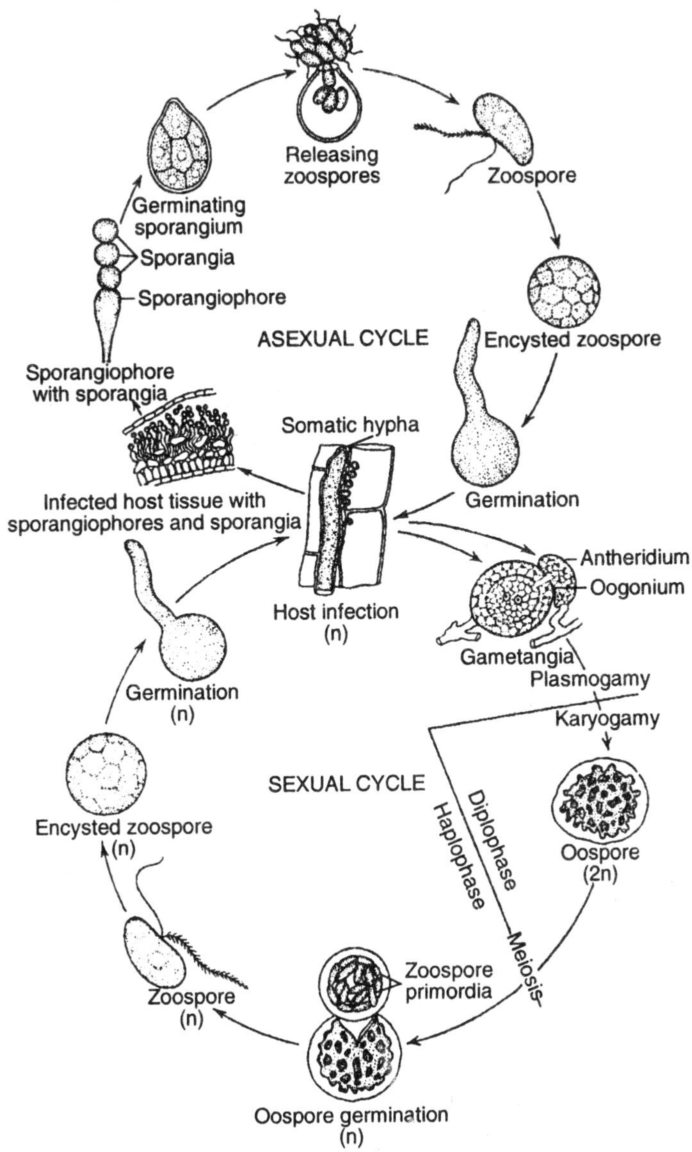

Fig. 12.19: Life cycle of *Albugo* sp.

shapes are produced. The sporangiophores are distinct from the somatic hyphae and constitute the most important identifying character of the different genera. They are branched and unbranched. When branched, the sporangia are borne singly at the tips of the branchlets. The sporangia are deciduous and are disseminated by wind. They germinate either by germ tube or by the formation of biflagellate secondary zoospores. Sexual reproduction takes place by the formation of antheridia and oogonia arising on closely adjoining hyphal tips. Mature antheridium bears a single antheridial nucleus with cytoplasm and the mature oogonium contains one uninucleate oosphere with periplasm. Fertilization takes place by the formation of a fertilization tube resulting the formation of an oospore which becomes thick-walled and may germinate by a germ tube or by biflagellate zoospores.

This family has suffered much from renaming of genera. There are five or more genera which are distinguished on the basis of their asexual characters. In *Basidiophora* the sporangiophores are unbranched, club-shaped, having apex covered with numerous short sterigmata each of which cuts off a single nearly spherical sporangium (Fig. 12.20F). In all other genera the sporangiophores are branched. The genus *Bremia* has a long sporangiophore which is dichotomously or trichotomously branched having branches branching again. The branchlets possess saucer-shaped terminal enlargements with a few sterigmata round the margin each of which bears a single sporangium (Fig. 12.20B). The sporangiophores of *Peronospora* are also branched like *Bremia* but the tip of each branch tapers to a point from which the sporangium is cut off (Fig. 12.20C). In *Sclerospora* the sporangio-

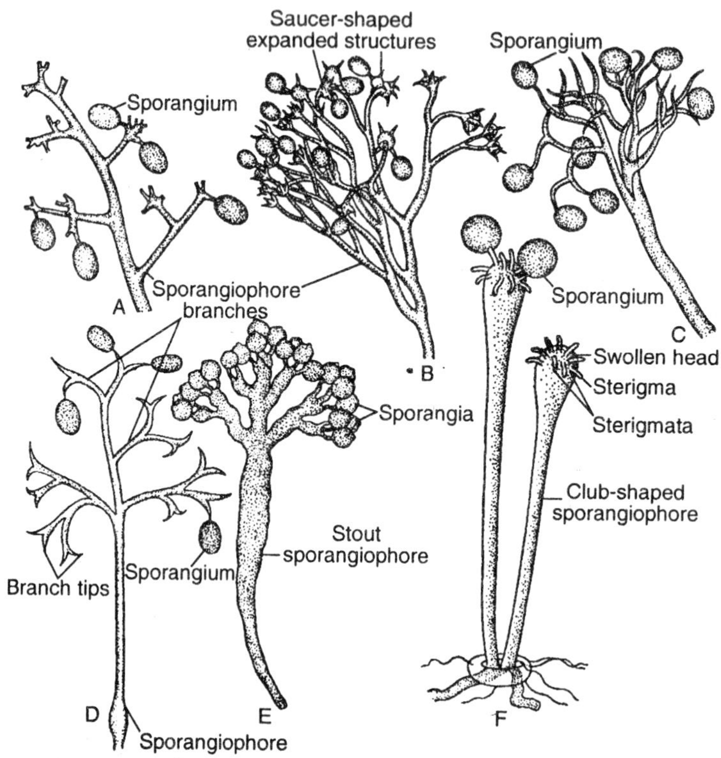

Fig. 12.20: Nature of sporangiophores of six genera of the Peronosporaceae. A. *Plasmopara viticola*. B. *Bremia lactucae*. C. *Peronospora effusa*. D. *Pseudoperonospora humuli*. E. *Sclerospora graminicola*. F. *Basidiophora entospora*

phore is thick, stout and sparingly branched. The branches are short and thickened producing a bushy appearance. The sporangia are borne at the tips of the branches (Fig. 12.20E). In *Plasmopara* the sporangiophores are irregularly branched. The branches arise nearly at right angles to the main axis and the secondary branches are at right angles with the primary. The tips of the branches are truncate which bear globose to ovoid sporangia (Fig. 12.20A). The genus *Pseudoperonospora* shows characters intermediate between *Peronospora* and *Plasmopara* and is also known as *Peronoplasmopara.* It possesses pseudomonopodially branched sporangiophores. The primary branch arises at acute angles with the main axis. The ultimate branchlets are acute from whose tips elliptical sporangia are developed (Fig. 12.20D).

Genus: *Peronospora*

The coenocytic branched mycelium develops along the intercellular space of the host tissue producing haustoria, which penetrate the host cells. The haustoria, in a few species, are short and knob-like, but in general they are filamentous and more or less branched. The members of this genus are aggressive parasites. They cause serious damge to the host. The mycelium can survive from season to season in the tissues of perennating organs of the host. Repeated dichotomously branched sporangiophores project from the host tissue usually through stomata covering the green parts of the host with a dense white growth giving a downy appearance from which the name downy mildew has been given. The sporangia are borne singly at the acute more or less reflexed tips of the branchlets (Fig. 12.21A). The sporangia are broadly elliptic to globose, blunt without any apical papilla. They are hyaline or very light-coloured. The sporangia, when mature, fall and germinate directly by lateral germ tubes (Fig. 12.21B). Such sporangia are often termed conidia.

The antheridia and oogonia arise from closely adjoining hyphal tips. Both of them are multinucleate when young, but with matu-

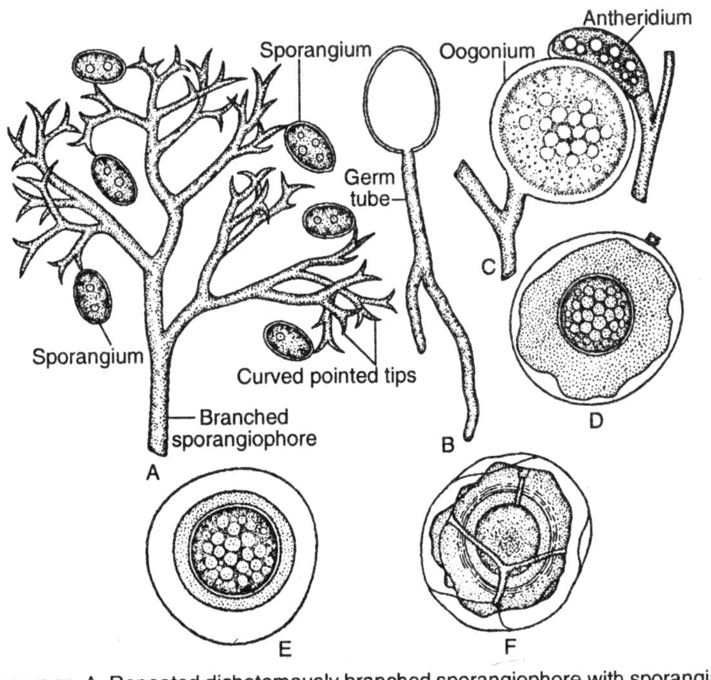

Fig. 12.21: *Peronospora* sp. A. Repeated dichotomously branched sporangiophore with sporangia borne at the curved pointed tips of the branchlets. B. Sporangium germinating by germ tube. C. Arrangement of antheridium and oogonium. D–F. Stages in formation and maturation of thick-walled oospore

rity become uninucleate. A mature oogonium bears an uninucleate oosphere with periplasm. So also a mature antheridium contains a single male nucleus embedded in cytoplasm (Fig. 12.21C). Fertilization takes place by the development of a fertilization tube through which the antheridial nucleus comes in contact with the oosphere. Plasmogamy and karyogamy lead to the development of an oospore which becomes thick-walled (Figs. 12.21D to F), goes to rest and germinates by the formation of a germ tube. *Peronospora tabacina* is an exception, where oospore produces sporangial vesicle which releases biflagellate secondary zoospores. Meiosis takes place during oospore germination. The nature of life cycle of *Peronospora* is illustrated in Fig. 12.22.

Some Indian Species: *Peronospora aestivalis* Syd. ex Gaum; *P. brassicae* Gaum; *P. gaumannii* Mundk.; *P. pisi* Syd.; *P. viciae* (Berk.) Gaum.

Genus: *Plasmopara*

The genus *Plasmopara* is widely known for its species *P. viticola* which, besides being the causal organism of downy mildew disease of grape, is associated with the discovery of the fungicide Bordeaux mixture. Bordeaux mixture was discovered by Professor Millardet, University of Bordeaux, France during 1882, and in 1885 Professor Millardet got the clue of his discovery from the vineyards of Saint-Julien in Medoc, France. While he was strolling through the vineyards he was surprised to notice that the vines growing along the path

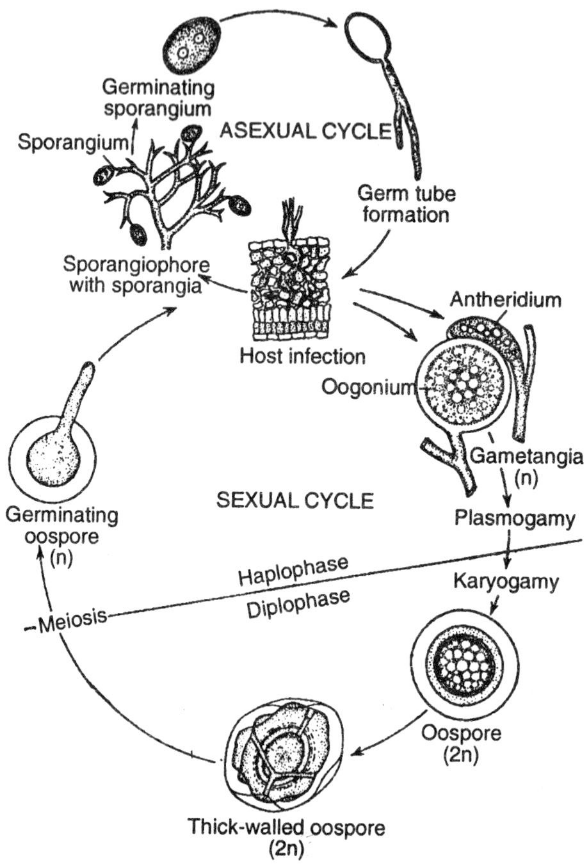

Fig. 12.22: Life cycle of *Peronospora* sp.

had leaves with bluish-white deposit on them and were quite healthy, whereas the vines growing deep in the vineyards were severely destroyed by downy mildew. He learnt from the people of the locality that it was a custom of the vinegrowers of that locality to bespatter the vines beside the paths with a conspicuous poisonous-looking substance. This was a practice to discourage the passers-by from pilfering the grapes. They used either verdigris—the bluestone, salts of copper or a mixture of copper sulphate and lime. Millardet was actually looking for some substance which could be applied as a protective coating on the leaves, so that fungus spores falling on them would be destroyed before they can cause any infection to the leaves. Through experimental trials Professor Millardet discovered Bordeaux mixture, which is the most inexpensive and effective fungicide.

Plasmopara produces intercellular branched coenocytic mycelium which receives nutrition from the host cells by the production of a large number of simple globose haustoria. Erect solitary or fasciculate branched sporangiophores emerge through stomata of the host (Fig. 12.23A). The branches arise at right angles to the main axis, so also the secondary branches. The ultimate branches are apically obtuse which bear globose to ovoid hyaline to smoky papillate sporangia (Figs. 12.23A and B). The sporangia germinate either by the production of biflagellate zoospores (Figs. 12.23C to G), or the entire protoplasmic mass escapes and then sends out a germ tube.

The oogonia and antheridia are developed rather late. Globose oogonia are formed terminally on the branches of the mycelium. A club-shaped antheridium is attached to each oogonium (Fig. 12.23H). During fertilization a fertilization tube is developed by the antheridium. The details of the process of

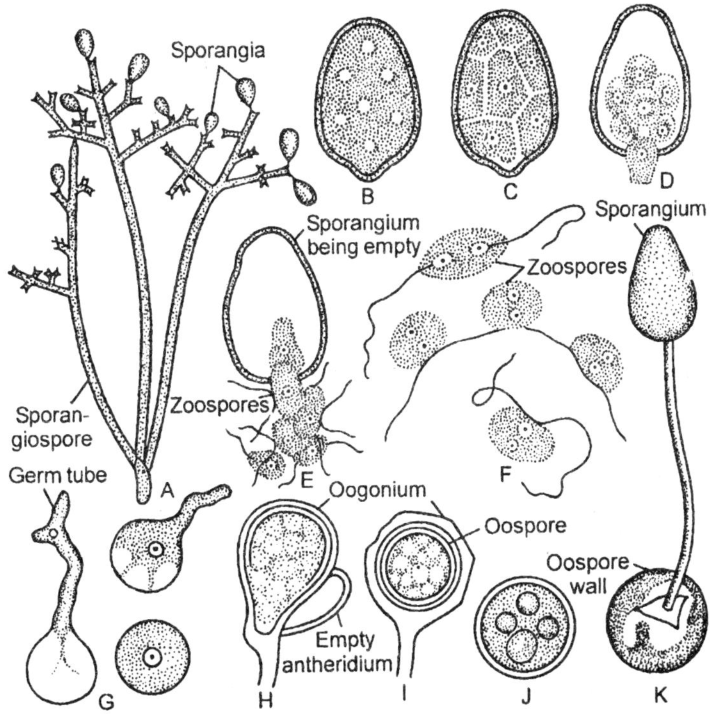

Fig. 12.23: *Plasmopara viticola.* A. Fasciculate branched sporangiosphores with branches at right angles to the main axis. B. Papillate sporangium. C–E. Stages in germination of sporangium and liberation of Zoospores. F. Biflagellate zoospores. G. Germination of zoospores. H. Oogonium with empty antheridium. I. Oospore in oogonium. J. Oospore. K. Germination of Oospore

fertilization have great similarity with those of *Peronospora.* The oospore is thick-walled (Figs. 12.23I and J). The oospore wall is composed of a smooth endospore surrounded by variously wrinkled, sometimes appearing somewhat reticulate epispore.

The oospore germinates after a period of rest by producing a hypha-like germ tube which protrudes through a small crack in the oospore wall. A sporangium is produced at the tip of the germ tube which gives rise to biflagellate secondary zoospores (Fig. 12.23K). Meiosis takes place during the germination of oospore. Life cycle of *Plasmopara viticola* is presented in Fig. 12.24.

Some Indian Species: *Plasmopara obducens* Schroet.; *P. vernoniaechinensis* Sawada; *P. viticola* Berk. and Curt.; *P. wildemaiana* P. Henn; *P. wildemaniana* var. *macrospora* Sawada.

ZYGOMYCOTINA

The great majority of the members of this subdivision are saprobes on dung and decaying plant or animal matter. A few are weak parasites growing on fruits causing diseases in transit and in storage, others are parasitic on other fungi. Again, others are parasitic on insects or saprobic on insect skins. The somatic body of the members of this group is a well-developed

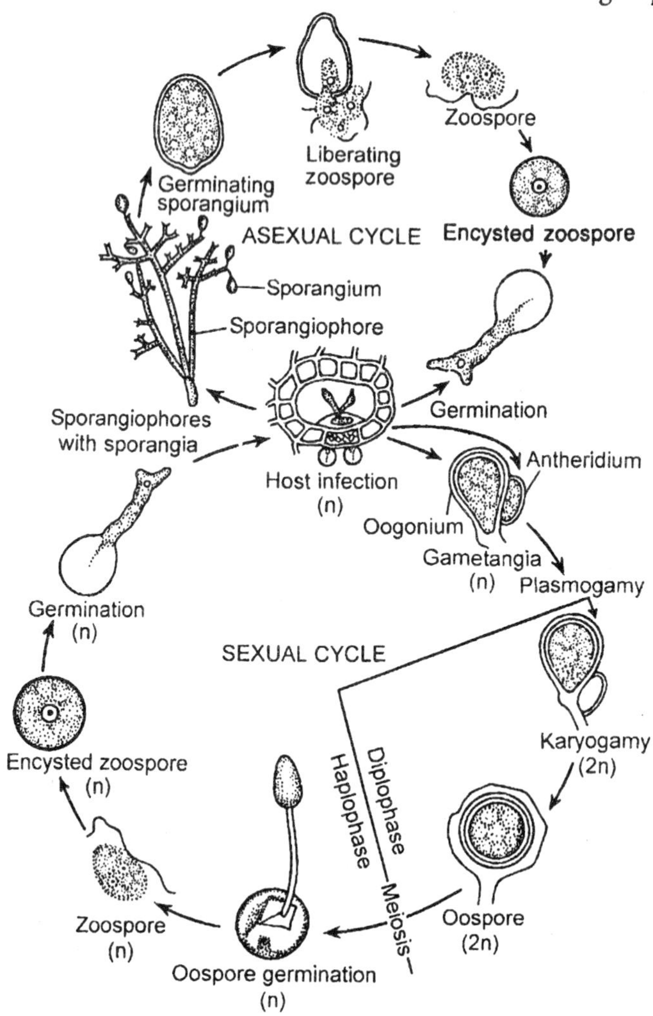

Fig. 12.24: Life cycle of *Plasmopara viticola*

mycelium having coenocytic hyphae, producing septa at the base of the repoductive organs, and elsewhere with age. In some members, the mycelium produces rhizoids at points of contact with hard surface. In othes, the somatic body consists of a long, slender, unbranched, straight or spirally curved, coenocytic hypha with cellulose walls.

Members of this group are characterized by lacking motile cells. Their mode of asexual reproduction is by one or more aplanospores contained in sporangia developed on simple or branched sporangiophores, or by the development of violently discharging conidia. In some members, asexual reproduction is by spores variously formed in the hyphal tips.

Sexual reproduction is by the copulation of two multinucleate gametangia which may or may not differ in size. Gametangial copulation results in the formation of zygospore.

The following is simple key to the classes, their respective orders and families:

A. Chiefly saprobic, some weakly parasitic on plants or specialized parasites on animals (both higher and lower); asexual reproduction by sporangiospores formed inside a sporangium, sometimes replaced by sporangiolum or by conidia produced by modified sporangia; sexual reproduction by gametangial copulation with zygospore as end production.

Class: Zygomycetes

B. Chiefly saprobic, some weak parasites on plant parts, on fungi, green plants, or animals including humans causing human diseases; asexual reproduction by sporangiospores formed inside sporangia, sometimes replaced by sporangiola to conidia; sexual reproduction by zygospore formation.

Order: Mucorales

C. Sporangial wall not cutinized, sporangiospores liberated by breaking or dissolution of sporangial wall, suspensors rarely tong-like, mycelium grows along the substration.

Family: Mucoraceae

D. Hyphae are usually prostrate, but sometime penetrate the substrate and serve the function both anchorage and absorption of nutrients from the substratum.

Genus: *Mucor*

DD. Hyphae after growing along the substratum for sometime produces long aerially project sporangiophore with sporangia at the tip, hyphae also develop root-like branches—stolons serving the function of anchorage.

Genus: *Rhizopus*

CC. Sporangial wall cutinized, sporangium violently discharged spore, or passively separated as a unit from sporangiophore, suspensor always tong-like.

Family: Pilobolaceae

DDD. Coprophilous fungi, sporangiophores and sporangia bent towards the incident light.

Genus: *Pilobolus*

CC. Sporangium many to few spored, or replaced by unicellular conidium.

Family: Choanephoraceae

DDDD. Monosporous sporangiola which are hyaline and are borne on swollen, globose vesicles produced at the tips of branched and unbranched sporangiophores.

Genus: *Cunninghamella*

BB. Parasitic on insects (entomogenous) or spaphobic on insect skins that have been cast, or animal excreta, reproduction asexually by means of violently discharged conidia; sexual reproduction by means of the development of zygospores.

Order: Entomophtherales

Parasitic on arthropod and insects; conda formed through break of hyphal body, smooth and produced outside the body of host.

Genus: *Entomophthora*

BBB. Parasitic on small animals—amoebae, rhizopods and nematodes; reproduce asexually by true conidia which are not forcibly discharged; sexual reproduction by the formation of zygospores.

Order: Zoopagales

CCCC. Parasitic on small animals with non-septate mycelium produced outside host warty zygospores are borne on sprirally twisted suspensors.

Family: Zoopagaceae

DDDDD. The haustoria are inflated or lobed, the external mycelium bears conidia in chain on short profections.

Genus: *Zoopage*

AA. Parasitic with a simple or branched filamentous somatic hyphae attached by a basal cell to the digestive tract or external cuticle of living arthropods; coenocytic hypha long, slender, unbranched, straight or curved; asexual reproduction by sporangiophores formed signly in linearly arranged sporangia separated by cross wall; sexual reproduction not confirmatively reported.

Class: Trichomycetes

BB. Trichospore are present, sexual spore zygospore type.

Order: Harpellales (with two families)

BB. Arthrospores are present.

Order: Asellariales (with single family)

BB. Sporangiospores are present.

Order: Eccrinales (with three families)

BB. Amoeboid spores are present.

Order: Amoebidiales (with single family)

CC. Amoebord spore with rigid wall, thallus attached outside host, or attached to the hindgut of host.

Family: Amoebidiaceae

DD. Spore are amoeboid type with rigid wall, derived from negetative mycelium, thallus is attached outside the host.

Genus: *Amoebidium*

Order: MUCORALES

The Mucorales, commonly known as 'black-molds' or 'pin-molds' grow on damp organic materials, such as bread, overripe fruits, exposed foodstuffs, and animal dung. Although most species are terrestrial saprophytes, a few species are pathogenic to man and other animals, causing diseases like **bronchomycosis,** and ailments of the cornea and skin; and others are of considerable importance in the decay of fruits and vegetables. Others are parasitic on other fungi, e.g., *Piptocephalis* and *Syncephalis.* Again, several species are utilized in industrial processes, such as for the manufacture of alcohols, fumaric acid, lactic acid, and certain enzymes.

The vegetative body is composed of profusely developed filamentous and richly

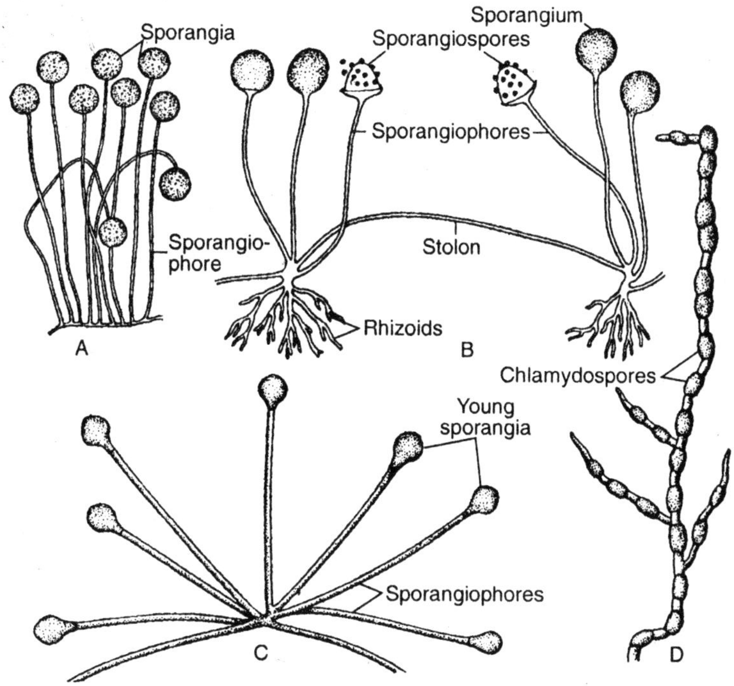

Fig. 12.25: Mucorales, Family Mucoraceae. A. *Mucor* sp. Sporangiophores arising from the prostrate hypha. B. *Rhizopus stolonifer.* Sporangiophores arising in fascicles from the nodes of the stolon opposite the tufts of rhizoids. C. *Absidia* sp. Sporangiophores in fascicle not opposite the rhizoids. D. *Mucor* sp. Hyphae with chlamydospores.

branched mycelium. The mycelium is white in the early stage of growth and turns grey or brownish with age. It is encountered lying on and in the substratum. The mycelium is differentiated into aerial hyphae and prostrate hyphae. The prostrate hyphae serve the function of absorption (Fig. 12.25A), in some species are buried in the substratum, whereas in others remain on its surface and produce **rhizoids** (Fig. 12.25B) at certain intervals. These rhizoids perform the functions of absorption and anchorage. A connecting hypha of the two groups of rhizoids is called a **stolon** (Fig. 12.25B) which usually grows over the surface of the substratum.

ASEXUAL REPRODUCTION

Reproductive bodies are usually borne on the aerial hyphae. In liquid media, particularly in presence of high concentration of sugar the hyphae may be segmented and the segments bud in yeast-like fashion. This is known as the **Torula** condition. Under adverse conditions, whether drought or excess of water the hyphae become septate, and within each segment the protoplasm becomes rounded off, develops a thick wall around itself, and forms very resistant spore. Such a spore is designated as a **chlamydospore** (Fig. 12.25D). Such a chalamydospore with the return of favourable conditions develops directly into new mycelium. The usual mode of asexual reproduction is by aplanospores which are also known as sporangiospores. These spores are produced in a very large number in a sporangium (Fig. 12.26A). The sporangium develops at the tip of aerial hypha, the lower part of which is known as **sporangiophore** (Figs. 12.25B and C). The sporangiophore supports the sporangium. The sporangiophores of many Mucorales are positively phototropic. Within the Mucorales there are varied forms of spores and spore-producing structures. In the Mucorales, in general, innumerable minute spores are developed in a sporangium, whereas in general, innumerable minute spores are developed in a sporangium, and in some, the spore-producing structure is much smaller in size than a typical sporangium and it produces much less number of spores. The spores produced in it are also larger in size than those of sporangium. Such a spore-producing structure is known as a **sporangiolum** (pl. **sporangiola**) or **sporangiole** (pl. **sporangioles**). Again, members of the Entomophthorales reproduce asexually by conidium produced singly at the tip of the conidiophore arising from hyphal body produced as a result of segmentation of mycelium. Here the conidium is structurally a one-spored sporangiolum (sporangium).

A. Evolutinary Tendencies in the Sporangial Development

The evolutionary tendencies in the sporangial development in the Mucorales may be traced along the following lines:

(i) **Series with sporangia:** Thc simplest and probably the most primitive type of sporangium is encounerd in the genus *Mortierella.* Here the sporangium is the spherical apex of the sporangiophore delimited by a septum at the base (Fig. 12.26C). The next step in the evolutionary line is the condition that is seen in *Mucor,* in whose sporangium there is a dome-shaped sterile structure in the centre known as **columella** (Fig. 12.26A), which is nothing but an arching up of the septum by further growth. A similar type of sporangium is found in *Sporodinia grandis.* But here the sporangiophore is dichotomously branched (Fig. 12.28A). In *Pilobolus* there is modification of the *Mucor* type. In it, the columella is comparatively small and the sporangium is rather bean-shaped with its groove placed on the columella. The sporangiophore is with an enlarged subsporangial vesicle (Fig. 12.26B). In *Absidia,* the sporangiophore widens gradually and forms a wedge-shaped base to the columella, known as **apophysis.**

(ii) **Series with sporangiola:** Gradual development of sporangiola from sporangial condition is encountered in different members

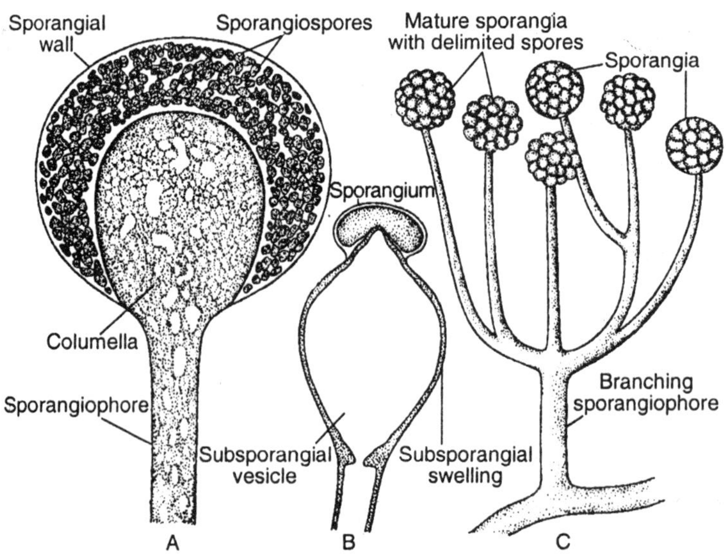

Fig. 12.26: Nature of sporangiophores and sporangia. A. *Mycor* sp. Sporangiophore bearing sporangium at the apex, sporangium with columella and a large number of sporangiospores. B. *Pilobolus* sp. Sporangiophore wtth an enlarged subsporangial vesicle, bean-shaped sporangium with small columella. C. *Mortierella* sp. Branching sporangiophore bearing spherical sporangia without columella.

of the Mucorales. In *Mucor proliferus* and in some others where the sporangiophore is branched, the terminal sporangium is larger than the laterals. In *Mucor proliferus* the spores that are produced in the lateral sporangia are smaller in number than those produced in the terminal one. In *Thamnidium* the terminal sporangium possesses a well-developed columella, whereas the lateral sporangia are much smaller and few-spored without any columella they are best known as sporangiola (Fig. 12.27A). In some species of *Thamnidium,* under controlled conditions the terminal sporangium does not develop at all, only the spo-

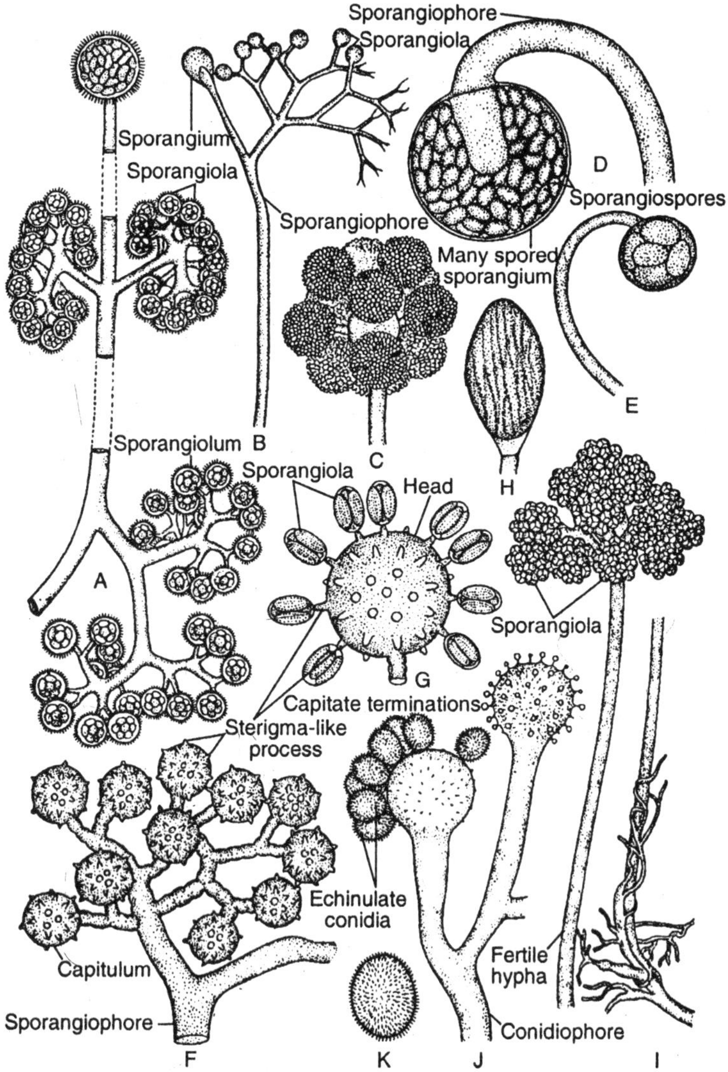

Fig. 12.27: Variations in the sporangial development. A. *Thamnidium* sp. Terminal sporangium and lateral sporangiola. B. *Dicranophora* sp. Sporangiophore bearing one sporangium and several sporangiola. C. *Choanephora cucurbitarum.* Heads of monosporous sporangiola (conidia). D. *Blakeslea* sp. Sporangium bearing sporangiospores and columella. E. *Blakeslea* sp. Sporangiolum. F. *Choanephora trispora.* Portion of sporangiophore from which the sporangiola have been removed from the capituli. G. *C. trispora.* Three-spored sporangiola borne on sterigma-like processes over the surface of the capitulum (head). H. *Choanephora cucurbitarum.* Detached monosporous sporangiolum (conidium). I. *Choanephora trispora.* Typical sporangiola borne over globose heads at the apex of a fertile hypha. J. *Cunninghamella echinulata.* Echinulate conidia on one of the capitate terminations of the conidiophore. K. *C. echinulata.* Detached echinulate conidium.

rangiola on the lateral branches are formed. The sporangiola that are formed by the genus *Dicranophora* on the lateral branches are only one-to-two-spored and their spores are much larger than those of terminal spornagium which often fails to develop (Fig. 12.27B), whereas the terminal sporangium is entirely lacking in *Chaetocladium,* suggesting a step advanced in the evolution. A further advancement is exhibited by the development of one-spored sporangiola, which in some species are dehiscent and in others indehiscent. These indehiscent monosporous sporangiola are considered as conidia. In the genus *Blakeslea* a large columella is developed in the sporagium (Fig. 12.27D) when it is well nourished, but the same fungus with shortage of nourishment produces smaller and few-spored sporangia without any columella—the sporangiola (Fig. 12.27E). In *Choanephora* also both sporangia and sporangiola are formed. Again, in *Choanephora trispora* (=*Blakeslea trispora*) three spores are formed in a sporangiolum (Figs. 12.27F, G and I), whereas in *Choanephora cucurbitarum* the sporangiola are monosporous and indehiscent behaving as conidia (Figs. 12.27C and H). The genus *Cunninghamella* is much more advanced than the genera already mentioned. No sporangia are developed in this genus, only indehiscent sporangiola are formed which function as conidia (Figs. 12.27J and K). It is evident that along with the decrease in the number of spores there is increase in the size of spores in a spore-producing structure leading to the development of one-spored indehiscent condition resembling conidial development of the Ascomycetes.

(iii) **Series with merosporangia:** Another line of evolution very much parallel to the above, exhibited by the genera *Syncephalis* and *Syncephalastrum* is in the formation of numerous cylindrical sporangi—merosporangia,

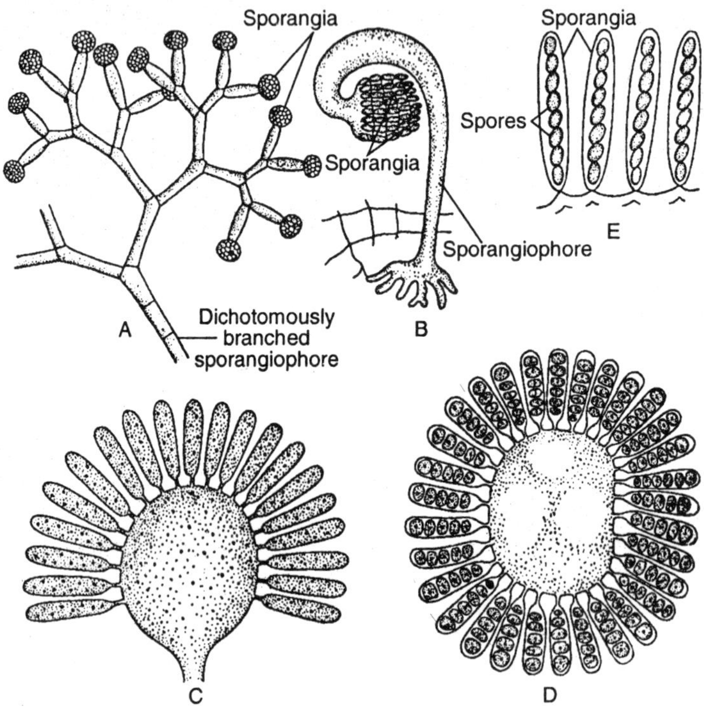

Fig. 12.28: Variation in the sporangial development. A. *Sporodinia grandis.* Dichotomously branched sporangiophore with branches terminated by sporangia. B. *Syncephalis cornu.* Sporangiophore bearing sporangia. C–D. *Syncephalastrum cinereum.* Stages of development of merosporangia on swollen sporangiophore head. E. Sporangia (merosporangia) containing spores.

which are again borne either on swollen sporangiophore or terminally on branchlets of the sporangiophore (Figs. 12.28B to D). The spores are uniseriate in the sporangium with varying number (Fig. 12.28E). The sporangial wall may be constricted showing chain of spores. Eventually, the merosporangial peridium degenerates and the aplanospores then look like a chain of exogenous spores, or conidia. The number of aplanospores may, in some instances, be reduced to only one. A monosporous merosporangium is, in effect, a conidium. Monosporous sporangial condition is encountered in the genum *Conidiobolus* (member of the Entomophthorales), where hyphal bodies produce conidiophores at the apex of each of which a conidium is produced. That the conidium represents monosporous sporangial condition can be shown by soaking conidium in water, when an outer peridial wall and a single 'spore' within it will be visible. These instances have some resemblance with conidial development of the Ascomycetes.

B. Development of Spores

In the Mucorales, there are three types of development of spores. In one type, the sporangial conent is usually cleaved into uninucleate portions along the lines of the vacuoles using up all the protoplasm. These portions round up and after several nuclear divisions become invested with spore membranes. This condition is illustrated by species of *Pilobolus*. A second type is found in *Rhizopus nigricans, Phycomyces nitens, Sporodinia grandis,* and others, where cleavage directly results in the formation of multi-nucleate masses which without undergoing nuclear division, round up, form a membrane, and spores are formed. The third type is exhibited by species of *Cunninghamella*. Here the tip of the conidiophore enlarges into a capitate structure on the surface of which short protuberances—**sterigmata** appear. Small globular enlargements develop at the tips of the sterigmata in which protoplasm enters and ultimately these enlargements develop into **indehiscent monosporous sporangiola (conidia)**. In *Syncephalis* and *Syncephalastrum* the tips of the sterigmata instead of being globular form long tube-like structures in which spores are arranged in a linear series. The sporangia, whether globular or elongated, may bear one or more spores. In *Blakeslea trispora* the sporangia contain three spores, whereas in *Chaetocladium jonsii* and *C. brefeldii* the sporangia are single-spored. Again, in *Cunninghamella echinulata* sterigmate conidia are formed which is cosidered to be the highest in the line of evolution from sporangium to conidium through sporangiolum.

SEXUAL REPRODUCTION

Sexual reproduction in the Mucorales is effected by the copulation of two multinucleate gametangia which, in general, are morphologically similar—isogametangia, with a few exceptional cases in which the copulating gametangia are dissimilar—heterogametangia. The copulating gametangia may be produced by the same thallus or by different thalli. Members of the Mucorales in which the copulating gametangia are developed on the same thallus are known as **homothallic** and when produced by the interaction of two different thalli they are called **heterothallic**. Homothallic member is *Sporodinia grandis* and heterothallic members are *Mucor mucedo, Rhizopus nigricans, Mucor hiemalis, Absidia spinosa, Phycomyces nitens.* During the process of sexual reproduction the tips of hyphae or hyphal branches of the same thallus or different but complementary thalli on coming in contact with each other swell, and constitute **progamentangia** (Fig. 12.29A). A transverse septum develops in each progametangium to delimit it into two portions, the terminal cell thus delimited in each function as a gametangium and the basal portion of the progametangium is known as **suspensor** (Fig. 12.29B). The walls between the two conjugating gametangia dissolve and the contents mix

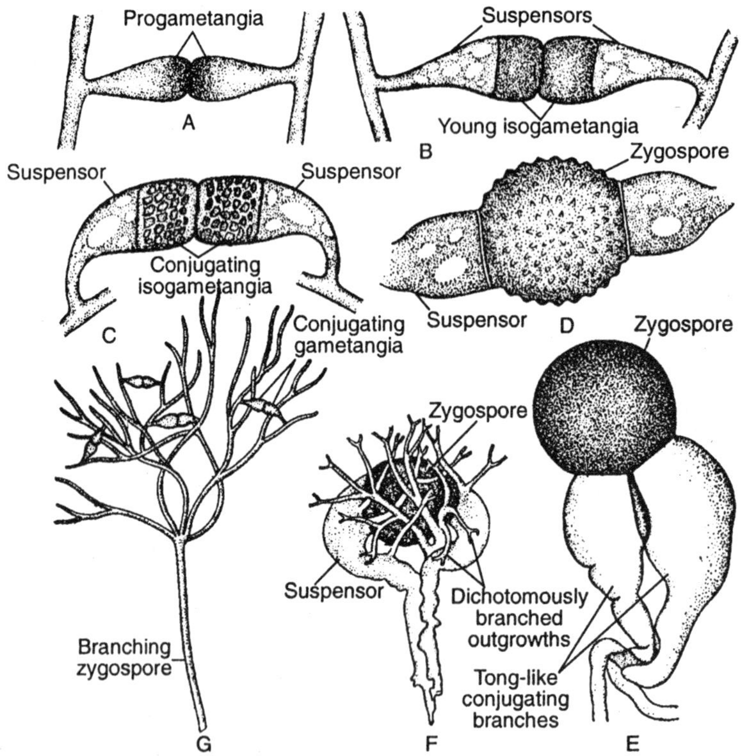

Fig. 12.29. Gametangia and and zygospores of Mucorales. A–D. *Mucor* sp. Stages in development of gametangia leading to zygospore formation. E. *Pilobolus* sp. Zygospore with tong-like conjugating branches (suspensors). F. *Phycomyces nitens*. Zygospore with dichotomously branched outgrowths from the suspensors. G. *Sporodinia grandis*. Zygophore showing conjugating gametangia

resulting in the formation of a fusion cell termed **coenozygote** which develops to form a thick-walled structure with warty surface called **zygospore** (Figs. 12.29C and D). When both the conjugating gametangia are morphologicaly similar and their entities lost as a result of conjugation the entire process is known as **isogametangial copulation.** But in some members, a distinct tendency of heterogametangial condition is exhibited by the unequal size of the copulating gametangia and supporting suspensors. The product of their copulation is, however, known as zygospore since their mode of origin is very much simialr to what is encountered in isogametangial condition. Heterogametangial condition is found in *Zygorhynchus heterogamus* (Figs. 12.30D and E), species of *Dicranophora* (Figs. 12.21B and C), *Absidia orchidis* and others. In general, the zygospore remains naked. But exceptions are *Phycomyces nitens* (Fig. 12.29F) and *Absidia glauca* (Fig. 12.31A), where long outgrowths grow out from the suspensors forming a protective covering around the zygospore. Much more advanced condition is seen in *Mortierella.* Here a large number of short hyphae form a solid felt having its outer surface cutinized turning brown in colour, within which lies the zygospore (Figs. 12.30A and B). This, Brefeld considered a true fructification. Very similar condition is found in *Endogone,* where the zygospore is surrounded by a tight weft of hyphae indicating a tendency to develop a fructification (Fig. 12.30C).

The zygospore during its development remains along the substratum. Exceptional condition is in *Phycomyces nitens,* where the suspensors resemble clippers or tong-like in

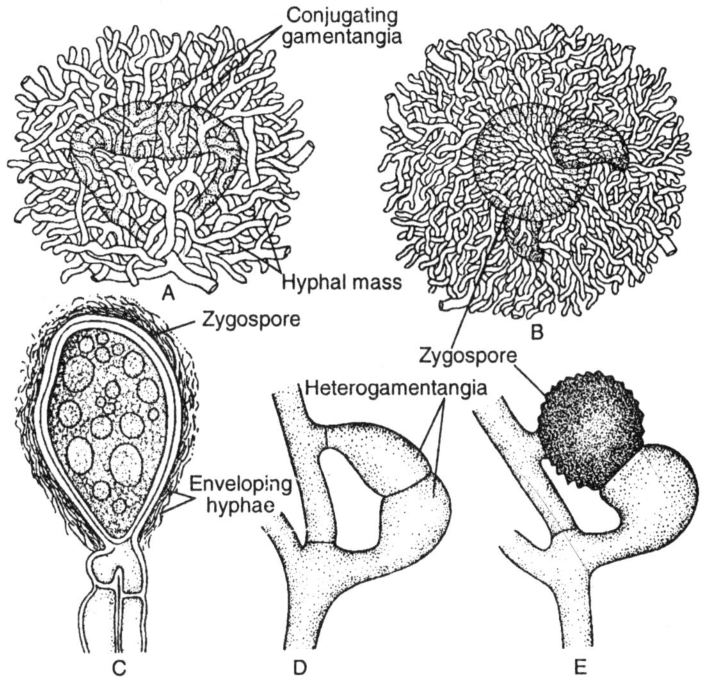

Fig. 12.30: Zygospores of Mucorales. A–B. *Morticrella rostafinskii.* Stages in formation of zygospore, hyphal mass forming a solid weft covering around the conjugating gametangia and zygospore. C. *Endogone* sp. Zygospore surrounded by a tight weft of enveloping hyphae. D–E. *Zygorhynchus heterogamus.* Stages in heterogametangial copulation leading to zygospore formation.

appearance (Fig. 12.29F) holding a zygospore between their tips pushing it out to a height of even 6 to 12 inches or more above the substratum. A similar condition is also exhibited by *Pilobolus* (Fig. 12.29E). Again in *Sporodinia grandis* special hyphal branches arise from ordinary hyphae on which conjugating gametangia are borne. These special hyphal branches are known as **zygophores** (Fig. 12.29G). When a gametangium fails to find its copulation mate, and in some cases is surrounded by a many-layered wall, it is called **azygospore.**

The zygospore wall usually consists of five layers. The innermost is thin granular. The next

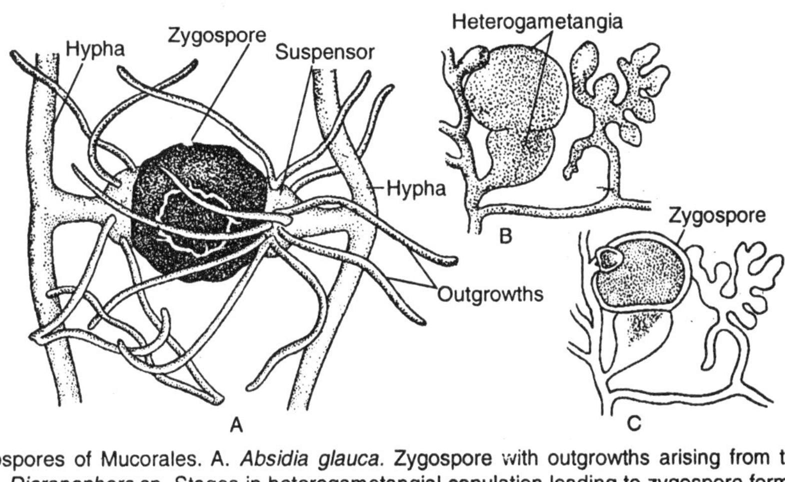

Fig. 12.31: Zygospores of Mucorales. A. *Absidia glauca.* Zygospore with outgrowths arising from the suspensors. B–C. *Dicranophora* sp. Stages in heterogametangial copulation leading to zygospore formation.

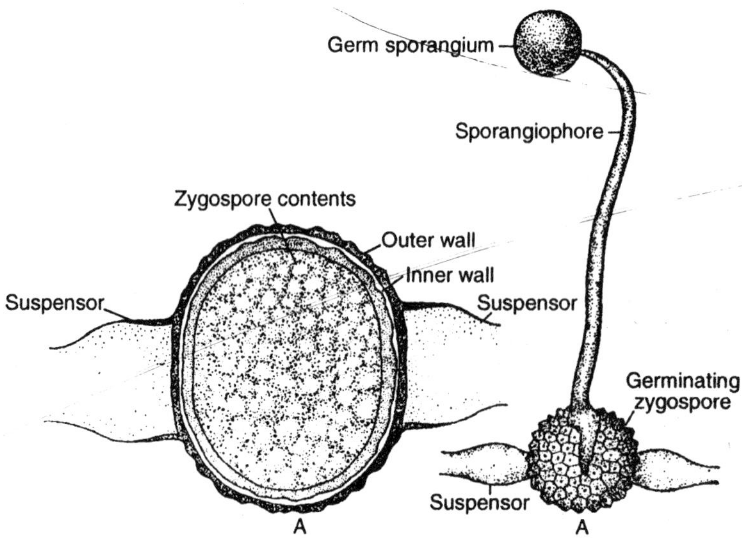

Fig. 12.32: Structure and germination of zygospore in *Mucor* sp. A. Structural details of zygospore, B. Germinating zygospore

one is elastic known as cartilaginous layer which is covered by a thin sheath. The fourth one is fragile, brown or black and is the carbonaceous layer surrounding which is the outermost layer which is either pale and elastic, or dark and fragile. The three inner layers are grouped as endospore and the two outer as exospore (Fig. 12.32A). The zygospore germinates only after a long resting period. With the rupturing of the exospore, the endospore grows out to form a germ tube which develops to a mycelium-like structure known as **sporangiophore,** whose tip swells to form a sporangium, **zygosporangium** or **germ sporangium** (Fig. 12.32B). But with the shortage of nourishment a sporangium is developed directly by the endospore.

HETEROTHALLISM

In some members of the Mucorales zygospores occur commonly in nature, whereas in others the formation of zygospore is rather infrequent, if not rare. The formation of zygospores in the Mucorales was for the first time observed by Ehrenberg (1820) and De Bary (1864) in *Sporodinia grandis.* It was assumed by earlier investigators that the zygospores are formed only under special condition of environment. But they were unable to offer any clear explanation for the process, as such the problem remained unsolved for a long time. The mystery behind the process of zygospore formation was revealed by A. F. Blakeslee (1904) through his classical studies. He was successful in producing zygospores of *Sporodinia grandis* in pure culture from a single spore. But under similar conditions he failed to obtain zygospores in *Rhizopus nigricans.* That really was the starting point of Blakeslee's studies. He carried on his experiments with different members of the Mucorales like *Mucor mucedo, Absidia spinosa, Phycomyces blakesleeanus,* and many others.

Blakeslee, through his intensive studies showed that the production of zygospores is possible in the laboratory under cultural condition. According to him, the formation of zygospores is primarily dependent on the inherent nature of the individual and secondarily on external factors. He found out that in some special hyphal branches of the mycelium originating from a single sporangiospore interact to form zygospores. Blakeslee called them

homothallic species and the phenomenon **homothallism.** But the majority of species tested by him did not produce zygospores unless hyphae arising from sporangiospores of different potentialities are interacted. To him, they are **heterothallic species** and the phenomenon is **heterothallism.** On the basis of the mode of zygospore formation, Blakeslee subdivided the members of the Mucorales into two groups: those which produce zygospores from the same thallus or mycelium obtained from a single sporangiospore, **homothallic**; and others where the zygospore formation is dependent on thalli or mycelia of two different potentialities, and zygospores can never be obtained from the mycelium arising from a single sporangiospore, **heterothallic.** To the homothallic group belong: *Sporodinia grandis, Dicranophora fulva, Rhizopus sexualis, Mucor genevensis, Zygorhynchus dangerdi,* whereas *Absidia caerulea, Blakeslea trispora, Choanephora cucurbitarum, Mucor mucedo, Phycomyces nitens,* and *Rhizopus nigricans* are heterothallic. He further explained that in heterothallic individuals production of zygospores is brought about through the interaction of mycelia of two distinct individuals which are morphologically indistinguishable, though a physiological difference evidently exists. Blakeslee labelled them (+) **strain** and (–) **strain.** In a very few exceptional cases, however, visible differences exist between the two strains, one being rather more luxuriant than the other. To the relatively well-developed strain sometimes the term (+) is applied, and the term (–) to the strain with which zygospores are produced. During the study of sexual reproduction in Mucorales, Blakeslee (1904) observed that, in some species zygospores are produced independently, but in others two mycelia of different strains are needed. Based on this fact Blakeslee divided Mucorales into two groups: (a) Homothallic and (b) Heterothallic. Such mycelia did not show any morphological differentiation. Based on this fact Whitehouse (1945) recognized two types of heterothallism as 1. **Morphological heterothallism** and 2. **Physiological heterothllism.** According to Blakeslee, heterothallism is a condition when zygospore formation takes place by the interaction of two thalli or mycelia of different potentialities, whereas under homothallic condition zygospore development is completed by the same thallus and without the union of two different thalli. He considered the terms homothallism and heterothallism corresponding to monoecism and dioecism used in higher plants. He stated that in heterothallic species, mycelium of one sex comes in contact with that of the opposite sex resulting in the development of zygospores, segregation of sex having been taken place in two different thalli. Since there is no morphological difference between the fusing gametangia in the members of the Mucorales in which Blakeslee carried on his investigations, the interpretation of the process of zygospore development on the basis of sex as is being used in higher plants where the sexual organs are clearly distinguishable, was questioned by Blackman (1906). According to Blackman, the application of the term sex in cases where the fusing cells are not morphologically distinguishable is not appropriate. In such cases he prefers to use the term **syngamy (syngametic)**. Blakeslee (1906) further emphasized that the morphological similarity of the gametangia does not rule out the possibility of differentiating (+) and (–) strains on the basis of sex. He pointed out that the mutual indifference of two mycelia of the same sex and the active sexual reaction between mycelia of opposite sex which leads to the formation of zygospores, indicate that the strains are fundamentally different though morphologically identical. The progametangia arise only from the stimulus of actual contact between hyphe which touch each other, and are from the very beginning adherent. To establish the sexual identity of the (+) and the (–) strains, Blakeslee performed experiments with species having heterogametangia (*Zygorhynchus hetero-*

gamus, Z. macrosporus, and *Dicranophora fulva*) whose larger gametangium may be considered female and the smaller as male having some analogy with the Oomycetes. The strain reacting with the larger gametangium is to be considered male and the one reacting with the smaller gametangium female. The results were up to his expectation only with *Absidia spinosa* and *Mucor hiemalis*, both of which are heterothallic. But his findings could not be repeated in any other heterothallic species for which the explanation offered by him did not receive wide support. That the difference in size of the gametangia is not constant and is sometimes controlled by the conditions of the substratum has been indicated in the heterothallic species *Rhizopus nigricans*, where the gametangia often differ in size indiscriminately in response to nutritional changes in the medium in which the fungus is growing. As such the heterogametagial conditions of the Mucorales and the Oomycetes are not analogous.

Previous view: Satina and Blakeslee (1926, 1927) attempted to indicate the difference between the two strains (+) and (–) by biochemical studies. The results they obtained were, however, not of very encouraging nature. Their observations were based on the response of (+) and (–) strains separately to what they named **Gosio reaction**, which involves the capacity of the living cell to absorb and reduce salts of tellurium and selenium. The (+) strains of *Absidia blakesleana* and *Circinella spinosa* show great capacity to reduce tellurium salts and also grow better on maltose than the (–) strains. Similarly, they observed that (+) strains of *Phycomyces nitens* and *Rhizopus nigricans* reduce potassium permanganate of **Manolov's reagent** more vigorously than the (–) strains.

It will be more widely acceptable if the use of the terms 'homothallic' and 'heterothallic' and interpretation of the mechanism of zygospore formation on the basis of sex are avoided. Instead, the terms compatability groups or mating types, self-sterility or interfertility are used, since these terms are not ambiguous.

On the basis of compatibility, heterothallism may be explained as the phenomenon of sexual incompatibility. In a heterothallic species where the factors controlling compatibility are distributed in different thalli, no development of zygospores unless there is mating between thalli possessing factors controlling compatibility, or in other words, the thalli belonging to two compatibility groups. The two compatible strains which are physiologically distinct are not distinguishable morphologically, whereas in the homothallic species zygospores are produced on a single thallus which bears the factors controlling compatibility in the same thallus.

In 1930 Ling-Young showed that when two similar strians of *Phycomyces nitens* are grown in culture medium poor in nutritive substances, after a short period of growth the hyphae of the two strains fail to proceed to each other. Whereas, hyphae of two different potentialities grown under similar condition produce zygospores. This has a resemblance with the phenomenon 'barrage sexual' exhibited by some Basidiomycetes between mycelia of incompatible sexual phases. This phenomenon was discovered by Vandendries and Brodie (1933).

Among the heterothallic species there are also strains in which (+) and (–) strains fail to react with each other, and zygospores are not formed. These strains are called 'neutral strains'. Similar neutrality may be induced in *Mucor mucedo* by prolonged cultivation of (+) and (–) strains under unfavourable conditions causing the loss of conjugation ability and developing into **neutral races**.

Blakeslee working with the Mucorales, reported a very interesting phenomenon of **imperfect hybridization**, that is, development of immature zygospores which ultimately do not mature. Immature zygospores are formed when mating occurs (i) between unlike strains of different heterothallic species of the same or different genera or (ii) between homothallic species and both strains (+) and (–) of a heterothallic species. This special feature of the members of the Mucorales can be utilized to label the strains of the heterothallic forms in which only one strain has been obtained for use.

According to Whitehouse (1949a) two fundamentally different types of heterothallism exist.

1. **Morphological heterothallism (haplodioecism):** In this type morphologically dissimilar sexual organs are produced by different thalli and cross-breeding becomes obligatory (if mating between differentiated gametangia occurs in the same thallus, the fungus is homothallic). Morphological heterothallism is present in many members of lower fungi.
2. **Physiological heterothallism (haploid incompatibility):** This type of heterothallism is entirely independent of morphological differences between male and female and depends on genetic factors conferring compatibility or incompatibility. Physiological heterothallism may be determined by either two allelomorphs at one locus or by multiple allelomorphs at one or two loci as discussed below.

Bipolar (unifactorial) heterothallism: Fungi in this category consist of two groups (mating types) of individuals that differ in their genetic make up for their compatibility factor. In a unifactorial system, compatibility is controlled by alleles of a single factor (gene). Since only a single factor is involved the genetic basis for the bipolar condition is described as unifactorial. Segregation of the two alleles during meiosis ensures that a single spore carries only one allele. They are generally designated as + and –, or A and a, or a and A, or A_1 and A_2. Here all mycelia with the same mating type are self-sterile or self-incompatible (+ x +; – x –). Fertilization occurs only between different mating types (+ x –). Since each mating type can act as donor and receptor of nuclei, there are two reciprocal crosses between + and – mycelia. It can be controlled by two alleles at one locus as in members of Mucorales, such as *Mucor, Rhizopus* and *Phycomyces;* most species of the Pyrenomycete *Neurospora;* the Discomycetes *Ascobolus magnificus* and *Sclerotinia gladioli;* the rust *Puccinia graminis;* and the smut *Ustilago levis* (Whitehouse, 1949). Or it can be controlled by a multiple allelic series that occurs at a single locus. Therefore, in these species more than one mating type exist. Their mating behaviour obeys the same rule; genetically like mating types are incompatible; and genetically unlike mating types are compatible as illustrated in Fig. 12.33A. Bipolar multiple allele heterothallism occurs in about 35 per cent of the Basidiomycetes (Whitehouse, 1949b). It is also found in a few Gasteromycetes and the mushroom *Coprinus.*

Tetrapolar (bifactorial) heterothallism: When monosporous mycelia derived from a single fruit body of *Coprinus cinereus* are intercrossed, fertile dikaryons are the product of one-quarter of the matings. The genetic explanation for this situation is that compatibility is controlled by two factors, with two alleles at each locus (i.e., with two loci). These loci segregate independently at meiosis. Since two separate factors are involved, the genetic basis is said to be bifactorial. The two factors can be designated as A and B and their two alleles as A_1, A_2 and B_1, B_2 respectively. Fungi in this category consist of four mating types of individuals: A_1B_1, A_2B_2, A_1B_2 and A_2B_1. The sexual behaviour among these is determined by the Kniep rule (Kniep, 1920). Neither pair of factors can undergo karyogamy in the homozygous state. Each locus is multiple-allelic with at least 100 alleles at each locus (Raper, 1953

Raper et al. (1958) found 96 alleles for the A factor and 56 for the B factor in *Schizophyllum commune.* Later on Raper and Flexer (1971) reported 450A factors and 90B factors in the same fungus. In *Coprinus cinereus,* 164A factors and 79B factors have been reported (Day, 1963).

Octopolar (trifactorial) heterothallism: There is only one report of the occurrence of octopolar trifactorial heterothallism in a basidiomycetous fungus *Psathyrella coprobia.* In this fungus, incompatibility is controlled by three factors A, B and C which are thought to be inherited independently (Jurand and Kemp, 1973). Since three factors are involved, the genetic basis is said to be trifactorial. All the three factors must be heterozygous for nuclear migration to occur and mature fruit bodies to form.

PSEUDOCOMPATIBILITY (SECONDARY HOMOTHALLISM)

In some monoecious higher fungi, at the time of spore formation, two nuclei of opposite mating type are incorporated regularly in each spore. Each spore, upon germination, therefore, gives rise to a thallus that contains both A_1 and A_2 nuclei and consequently behaves as if it was homothallic. This phenomenon is called secondary homothallism (also pseudocompatibility) (Esser, 1966) different mating (compatible) type for sexual reproduction. If the self-sterile thalli differ in respect of sex (dioecism), they are referred to by some authors as cases of morphological heterothallism. Such authors termed the cases of true heterothallism as physiological heterothallism (self-sterile thalli differing in respect of compatibility and not morphologically in sex). These are species in which difference in strains of their thalli are in respect of both, i.e., in sex (morphological heterothallism or dioecism) as well as in **compatibility** (true or physiological heterothallism).

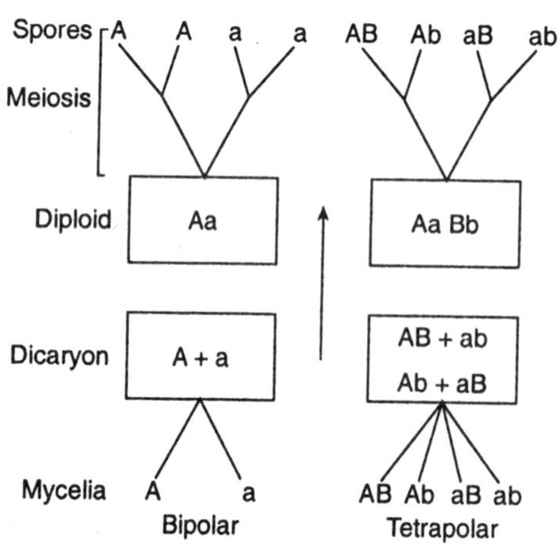

Fig. 12.33: Diagram to show segregation of compatibility factors and polarity in heterothallic fungi

Heterothallic fungi may have the following two general groups:

1. **Bipolar heterothallic.** In these species, there are two mating types of individuals which differ in their genetic make-up for the compatibility factor (Fig. 12.33). The compatibility is governed by one pair of factors (alleles), **A** and **a** located at single same locus of chromosome. The appropriate chromosome of one strain of thallus carries the allele **A** and that of the other thallus allele **a.** Only thalli carrying complementary alleles of this pair are compatible and interfertile giving an **Aa** zygote. There are 50% chances of interfertility between opposed mycelia. This form of heterothallism was originally discovered by Blakeslee (1904) in the mucorales.
2. **Tetrapolar heterothallic.** In these species, there are four basic mating types of thalli. The compatibility is governed by two pairs of factors, **Aa** and **Bb**. Thus it involves two compatibility loci each with a pair of alleles, A and **a** or **B** and **b.** The four strains of thalli with the respective allelic constitution would be: **AB, Ab, aB** and **ab**; only those thalli which combine to give an **AaBb** zygote are compatible and fertile. There are 25% chances of interfertility between opposed mycelia of the

	A_1	A_2	A_3	A_4
A_1	–	+	+	+
A_2	+	–	+	+
A_3	+	+	–	+
A_4	+	+	+	–

A

	A_1B_1	A_1B_2	A_2B_1	A_2B_2
A_1B_1	–	FL	B	+
A_1B_2	FL	–	+	B
A_2B_1	B	+	–	FL
A_2B_2	+	B	FL	–

B

Fig. 12.34: Tetrapolar heterothallism

species. Both bipolar and tetrapolar species occur in higher fungi, particularly Basidiomycotina.

In higher fungi with bipolar and tetrapolar heterothallism, a great number of alleles (not simple one or two pairs) may occur at one or two compatibility loci respectively, responsible for **multiple allelomorph heterothallism.** As in (Fig. 12.34) mating reactions of four possible isolates of bipolar and tetrapolar heterothallic fungi can also be presented in the following manner where two allelic bipolar fungus-A and tetropolar fungus-B expressed in a Cheker Boards in the folllowing manner.

Sex regulation mechanism in the Mucorales

Information obtained so far regarding the mechanism of sexual development in the Mucorales points towards a hormonal basis of sex regulation.

Working on *M. mucedo, M. hiemalis,* and *Phycomyces blakesleeanus* Burgeff in 1924 reported for the first time that each strain (+) or (–) produces a diffusible or volatile substance which induces and directs the gametangial development. Burgeff's observation was confirmed by Kohler (1935).

The most extensive and successful studies of the regulation of the reaction have been made with *M. mucedo* by Plempel and his associate (1957 to 1961). Plempel postulated the existence of the two progametangia-inducing substances produced by the respective plus and minus strains of the fungus. Each of the two substances were found to be active towards the opposite strain. But he failed to separate them, despite repeated attempts. These sexual factors were characterized after isolation from the mycelium of the mating strains of *M. mucedo* and *Blakeslea trispora* by Austin and others (1969). The compounds have been found to be identical with trisporic acid B and trisporic acid C characterized earlier by Caglioti and others (1967); and Cainelli, Grasselli and Selva (1967) who isolated them from culture liquids of *B. trispora.* Both trisporic acid B and trisporic acid C and their methyl esters were equally effective in inducing progametangia in the plus and minus mating types (Ende and others 1970). Initiation of trisporic acid production, however, occurs only when mycelia of either mating types are grown together in a culture medium. Evidently, some initiator substance(s) is(are) believed to be responsible for primary interaction between the mating types resulting in trisporic acid formation. These initiator substance(s) may be very labile or rapidly metabolised (van den Ende and Stegwee, 1971). However, nothing is known definitely uptil now regarding the nature of these initiator substance(s). But it has been speculated earlier that the substance(s) concerned may be proteinaceous in nature (Ende, 1970). It is also known that the plus strain is the major producer of trisporic acid in *M. mucedo* and the same strain might be the controlling partner in respect of trisporic acid production (Ende, 1971).

The above findings have been verified by Marsman using radioactive mycelia for the detection of trisporic acid. Sutter reported that a compound extracted from a plus culture of *B. trispora* is converted into trisporic acid C when applied to the minus culture of the same fungus. The source of the trisporic acids and their role in sexuality of some members of the Mucorales are now known. But factors affecting the biosynthesis of these trisporic acids and the individual role of each on the development of the mating systems need further investigation.

The biosynthetic pathway of trisporic acid has been elucidated by Bu'lock and Austin during 1971 to 1974. It is suggested that trisporic acid could be synthesized in unmated cultures of **Mucor muceds,** as compared to mixed cultures. It is now believed that very low levels of some trisporic acid precursors are present in single cultures. They have been identified as methyl dehydrotrisporate B in the (+) strain and Trispor in the (–) strain, whch were earlier designated as Progamone (P^+) and Progamone (P^-) respectively by Plempel.

Germination of Zygospore: The zygospore germinates only after a long resting period. The germination of a mucoraceous zygospore usually culminates in the formation of a terminal sporangium. Such a sporangium is known as a **germ sporangium,** also as a **zygosporangium** and is produced at the tip of a germ tube which develops to a mycelium-like structure, also designated as **germ sporangiophore.** In the germ sporangium, sporangiospores also designated as **germ sporangiospores,** are produced which on germination give rise to new mycelia. The germ tube which ultimately produces germ sporangiophore and germ sporangium is developed by the growth of the endospore being exposed by the rupturing of the exospore of the zygospore during germination. But with the shortage of nourishment, a germ sporangium may be developed directly by the endospore. The contents of the germ sporangium and the germ sporangiophore presumably represent the surviving meiotic products of zygospore germination. In majority of cases a single unbranched germ sporangiophore is produced. In *Mucor hiemalis* occasionally the germ sporangiophore may be branched and two germ sporangia are formed. Rarely more than one germ sporangiophore arises from a single zygospores. Zygospore may also germinate to produce vegetative mycelia.

The results obtained so far in the study of the germination of mucoraceous zygospores may be summarized as follows:

Blakeslee (1906) reported that a homothallic strain which he studied and which he designated *Mucor I,* gives rise to germ sporangiospores which, when isolated singly, gives rise to zygospore-producing mycelium. He designated this type of zygospore germination as **pure homothallic.**

Kohler (1935) in his studies on *Mucor mucedo* found that all of the sporangiosphores obtained from a single germ sporangium give rise to mycelia which are of the same mating type. A zygospore germiantion of this type is known as **pure heterothallic.**

The germination of zygospore in *Mucor hiemalis* has been studied by Cutter (1942), Sjowall (1945), and Gauger (1965). They reported that the zygospores of *Mucor hiemalis* germinate as pure heterothallic with deviation from what is encountered in *Mucor mucedo.*

Here the zygospores on germination produce vegetative mycelia before any sporangiophores are produced. Such a type of germination is referred to as a **mycelial germnation.**

Blakeslee working on *Sporodinia gradis, Mucor mucedo, Phycomyces nitens,* and *Rhizopus nigricans* classified the possible sexual reactions exhibited by sporangiospores obtained from such germinations as being of four types. With reference to plus (+) and minus (–) characters of the spores the four types of germiantion may be indicated as follows:

(i) In *Sporodinia grandis* which is homothallic, all the spores of the germ sporangium on germination give rise to homothallic mycelia, that is, all are (±).

(ii) In the heterothallic species *Mucor mucedo,* all the spores that are developed in a germ sporangium are of one strain, either all (+) or all (–). In this case the segregation of sex is completed before the germination of zygospore.

(iii) In another heterothallic species *Phycomyces nitens,* the germ sporangium contains three kinds of spores. They are (+), (–) and (±). This is a case of partial segregation of sex which results, besides the spores of (+) and (–) strains, in the development of some of the spores (±) which on germination give rise to homothallic mycelia. It is most likely that some of the diploid nuclei take part in the formation of spores without undergoing meiotic division.

(iv) Blakeslee was unsuccessful in classifying the organism, he called *Rhizopus nigicans,* because of the difficulty he encountered in germinating the zygospores. But he hypothesized that a mixed type having (+) and (–) sporangiospores will be found in the same germ sporangium. This was designated by Blakeslee as the fourth type.

But Gauger (1961) was able to germinate zygospores of *Rhizopus stolonifer* (*R. nigricans*). He found that a single zygospore produces sporangiospores which, when cultured, react as either all (+), all (–) or as a mixture of the two (+) and (–).

Blakeslee's findings indicate that in the members of the Mucorales segregation of sex may take place at any stage immediately after the zygospore formation to the development of spores which varies from species to species.

Nuclear behaviour during the Formation and Germination of Zygospore: Cytological features associated with the formation and maturation of zygospores are not yet clearly understood due to hardness of zygospore wall, density of zygospore protoplast, presence of a large amount of oil as well as granular structures in the protoplast and the nuclei being very small in size. However, the nuclei in course of zygospore development can possibly behave in any of the following ways:

(i) Survival of only one pair of compatible nuclei which pair and ultimately fuse producing a diploid nucleus.

(ii) Pairing of all the compatible nuclei, survival of only one pair which ultimately fuse forming a diploid nucleus.

(iii) Pairing of all the compatible nuclei, conjugate division of paired nuclei, and survival of only one pair of nuclei which ultimately fuse resulting a diploid nucleus.

(iv) Pairing and fusion of all compatible nuclei producing many diploid nuclei. In this case either all the diploid nuclei survive or just one diploid nucleus survives and others degenerate.

Dangeard (1906), Moreau (1911, 1912, 1913) and Keene (1914, 1919) reported that the nuclei in the zygospore remain in pairs and ultimately fusion between the paired nuclei takes place. Lendner (1908) in *Sporodinia grandis* and McCormick (1912) indicated that only one pair of nuclei survives which fuse together and the rest of nuclei degenerate. Cutter (1942) made extensive studies on the Mucorales and encountered widely different conditions. In *Mucor hiemalis, Blakeslea trispo-*

ra, and *Absidia spinosa,* within a few days of conjugation there is pairing of nuclei. Later on fusion of nuclei and meiosis of diploid nuclei take place as a result of which the dormant zygote contains haploid nuclei. In *Rhizopus nigricans, Absidia glauca* and *Absidia coerulea,* some nuclei degenerate, others fuse in twos, and the meiosis of the dormant zygote having nuclei takes place immediately before the germination of the zygospore.

Family: *Mucoraceae*

Mycelial threads are all alike or of two kinds being differentiated into aerial and some buried in the substratum. Sporangiospores are borne in a globose to pyriform sporangium. The sporangium is typically with a columella bearing many spores. In some genera the sporangia are few-spored and without columella. The sporangial wall is thin and not cutilized. Zygospores are variable, smooth or spiny.

Genera included under the family Mucoraceae are: *Pirella, Circinella, Absidia, Mycocladus, Rhizopus, Spinellus, Sporodinia, Phycomyces, Zygorhynchus* and *Mucor.*

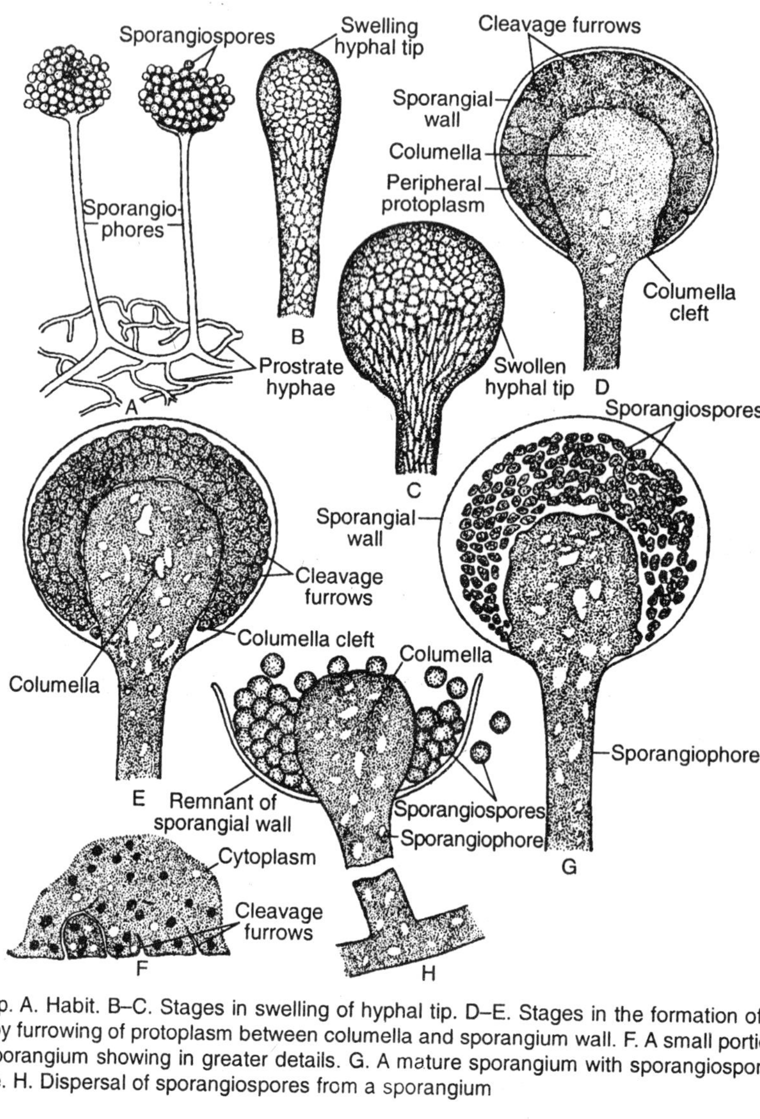

Fig. 12.35: *Mucor* sp. A. Habit. B–C. Stages in swelling of hyphal tip. D–E. Stages in the formation of columella, and spores by furrowing of protoplasm between columella and sporangium wall. F. A small portion of cytoplasm of the sporangium showing in greater details. G. A mature sporangium with sporangiospores and sporangiophore. H. Dispersal of sporangiospores from a sporangium

Genus: *Mucor*

This is the largest genus of the order Mucorales under the family Mucoraceae. It includes a number of species which are of widespread occurrence and considerable importance being saprophytic and parasitic on variety of substrata. It is found on almost every kind of damp material. Some species inhabit soil and hence are found on numerous soil-contaminated objects. Others are associated with rotting fruits. A number of species are used for the manufacture of alcohol. Again the occurrence of species of *Mucor* on horse-dung is not uncommon.

Stout, well-developed, freely branched coenocytic mycelium grows loosely on substratum forming typical colonies, white in the early stages of growth and becoming coloured with the production of reproductive structures. The hyphae generally spread out evenly within and upon the substrate lacking in rhizoids and stolons. These are nutritive hyphae. Besides these, some aerial hyphae arise from the nutritive hyphae as branches standing erect on them (Fig. 12.35A).

The tip of each aerial hypha swells in which cytoplasm with nuclei migrates (Fig. 12.35B). The tip swells considerably and the nuclei divide repeatedly. The contents of the young swollen tip divide into a central zone mainly filled with sap surrounded by a rich peripheral zone containing cytoplasm and nuclei. The two zones (outer layer of nucleated cytoplasm, **sporeplasm**, and inner dome-shaped area of cytoplasm, **columellaplasm**) are delimited by a foamy protoplasmic layer containing narrow flattened vacuoles lined one after the other (Figs. 12.35C and D). These vacuoles ultimately form a continuous vacuolate layer by fusing laterally with one after the other which ultimately develops into a dome-shaped septum and the dome is known as a **columella** (Fig. 12.35E). In the mean time cleavage of cytoplasm takes place (Fig. 12.35F) resulting into innumerable small uninucleate portions which round up, become invested with spore membranes, and develop into non-flagellate spores, the sporangiospores. The swollen hyphal tip ultimately develops into a sporangium in which the sporangiospores are formed (Fig. 12.35G). The sporangium is large, globose and many-spored. It is supported by the rest of the aerial hypha which is called a sporangiophore (Fig. 12.35G). The sporangial wall is evanescent (Fig. 12.35H). The sporangiospores are globose to ellipsoidal with a thin smooth wall. On liberation, they germinate into fresh mycelia. Spores may also be dispersed by bursting of the thin wall of the sporangium due to pressure that is set up in the columella exerted by the quantity of fluid that increased in the sporangiophore and columella during maturity of the sporangium.

Sexual reproduction in a heterothallic species of *Mucor* is effected by the copulation of two multinucleate isogametangia arising from two compatible hyphae, (+) strain and (–) strain. The gametangia may be developed either as terminal swellings of other. The process is initiated by the stimulus produced by their contact. They swell and constitute the progametangia (Fig. 12.29A). Each progametangium delimits a terminal cell termed as the gametangium, by a wall laid down from the edge inwards. The remainder of the progametangium is known as the **suspensor** (Fig. 12.29B). The walls separating the two gametangia gradually dissolve from the middle toward the edge resulting a fusion cell, **coenozygote**, in which fusion of contents of both gametangia takes place. The entities of the two copulating gametangia are completely lost. A thickened, several-layered wall is developed around the fusion cell which is then called a zygospore (Figs. 12.29C and D), in which karyogamy takes place. The zygospore germinates after remaining dormant for several months. The thick exospore is ruptured and the endospore puts forth a germ tube, the tip of which develops into a sporangium known as a **germ sporangium** or **zygosporangium** (Figs. 12.32A and B). In *Mucor mucedo* the sporangium produces only one kind of spores (+) or (–). Occasionally the copulation between the gametangia does not take place and the zygospore fails to develop. The gametangia are then surrounded by a many-layered wall and are known as **azygospores.** Life cycle of a heterothallic species of *Mucor* is presented in Fig. 12.36

Some Indian Species: *Mucor hiemalis.* Wehmer; *M. indicus* Lendner; *M. mucedo* (L.) Fr.; *M. praini* Chodat and Nechitsch; *M. silvaticus* Hagem.

Genus: *Rhizopus*

Rhizopus exists everywhere. Usually as a saprophyte and sometimes as a weak parasite on stored organs of plants. The aseptate mycelium of the fungus produces long, aerial sporangiophores at the tips of which black spherical sporangia develop. The sporangia contain thousands of spherical sporangiospores. When the mycelium grows on a surface of the substratum, it produces stolons, that is, hyphae that arch over the surface and at the next point of contact with the surface produce both root-like hyphae, called rhizoids, which grow toward the surface, and aerial sporangiophores bearing sporangia. From each point of contact more stolons are produced in all directions. Adjacent hyphae produce short branches called progametangia, which grow toward one another. When they come in contact, the tip of each hyphae is separated from the progametangium by a cross wall. The terminal cells are gametangia. These fuse and their nuclei

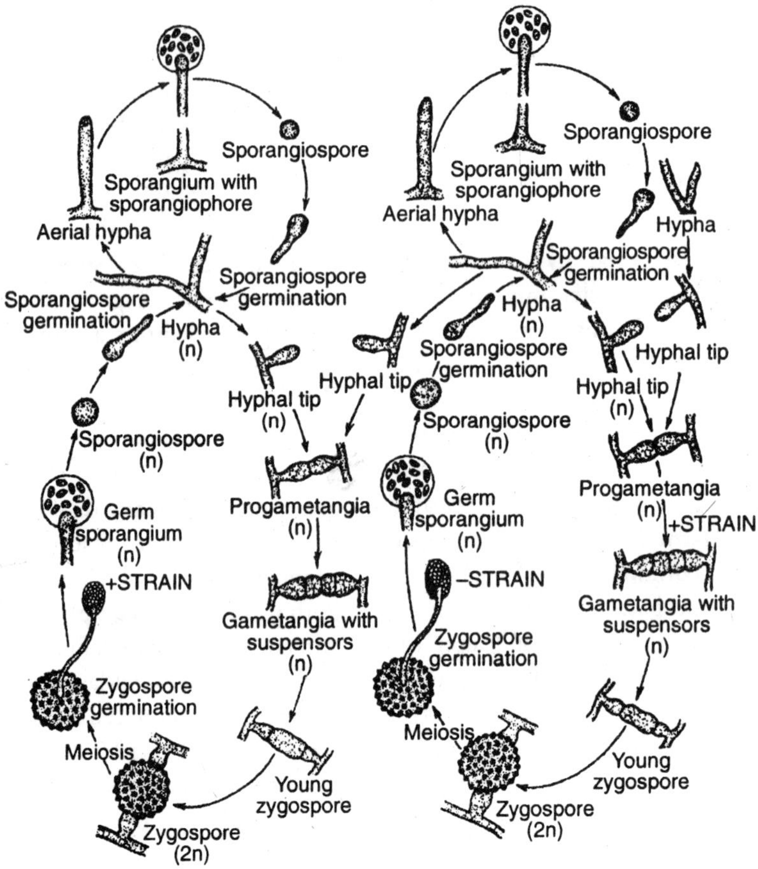

Fig. 12.36: Life cycle of a heterothallic species of *Mucor*

pair. The cell formed by the fusion enlarges and develops a thick, black, and warty cell wall in which compatible nuclear fusion takes place. This sexually produced structure is called a zygospore and is the overwintering or resting stage of the fungus. Where it germinates it produces a sporangiophore bearing a sporangium at the tip full of sporangiospores resulting from meiosis of diploid nuclei (Fig. 12.37).

Different species of **Rhizopus** cause **Rhizopus** soft-rot of fruits and vegetables throughout the world and on harvested fleshy organs of vegetable crops during storage, transit and marketing of these products.

Common species is *Rhizopus stolonifer.* It causes bread molds and soft rot of fruits and vegetables when moisture conditions are favourable.

Family: *Pilobolaceae*

Sporangiospores are developed in a sporangium which is discoid to lenticular and many-spored. The sporangial wall is black and heavily cutinized in the upper portion. The sporangium, in most species, is provided with a subsporangial vesicle and is usually lifted up from the substratum by tong-like hyphal growth.

Genera included under the family Pilobolaceae are: *Pilobolus* and *Pilaira.*

Genus: *Pilobolus*

Pilobolus grows on the fresh dung of herbivorous animals, especially on horse-dung. In spite of its disagreeable habitat, *Pilobolus* has been studied with great interest by many investigators since it exhibits phototropic reactions. The mycelium of *Pilobolus* is composed of repeatedly branched coenocytic hyphae which spread in all directions through the substratum. The part of the fungus which rises above the surface of the substratum consists of the stalk bearing the sporangium. Water drops remain adhering on the stalk glitter in the early mor-

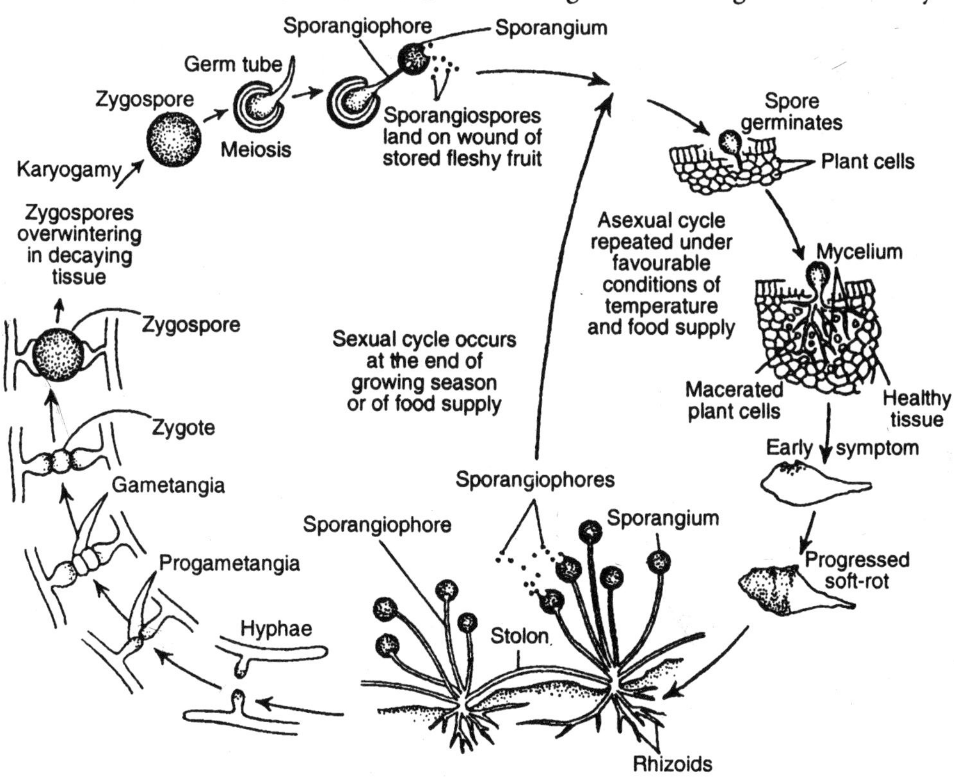

Fig. 12.37: Life cycle of *Rhizopus* sp.

ning sun (Fig. 12.38F). The stalk is about a quarter of an inch high, and swells up near the tip into a little globule, surrounded by a black cap, which is the sporangium itself.

Asexual Reproduction: The fungus reproduces asexually by the development of sporangia and sporangiospores. At first the spherical sporangiophore primordium is visible to the naked eye as a minute orange bulb protruding slightly above the general level of the substratum. This bulb is cut off from the greatly enlarged end of the hypha by a transverse septum. This is the primary bulb or the primordium from which sporangiophore and sporangium are developed (Fig. 12.38A). The spherical primordium is packed with minute globules of yellow oil. Then an intensely yellow branch grows up from the bulb into the air assuming upright position (Fig. 12.38B). This is the young sporangiophore whose tip is markedly phototropic. The tip of growing sporangiophore then swells up into a minute spherical bulb which is destined to become the sporangium (Fig. 12.38C). The swollen tip enlarges more and more to form a nearly spherical sac-like structure which is filled with protoplasm containing oil drops giving yellow colour and the sporangiophore region becomes more or less colourless. The contents of the sac-like structure which ultimately develops into a sporangium divide to form a large number of protoplasmic fragments, each of which sur-

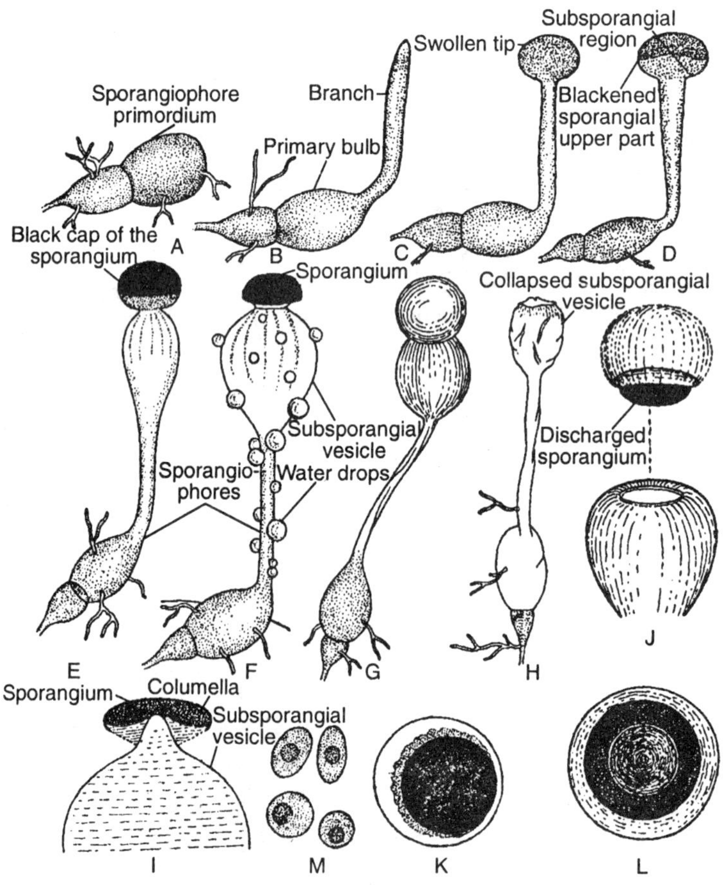

Fig. 12.38: *Pilobolus* sp.— A–F. Stages in the development of sporangium, sporangiophore and subsporangial vesicle. G, H and J. Stages in the discharge of sporangium. I. Details of sporangium, columella and subsporangial vesicle, K–L. Closer view of sporangia. M. Sporangiospores.

rounds itself with a cell wall and becomes a spore (Fig. 12.38M). At the same time the upper part of the sporangial wall begins to blacken and the subsporangial region starts to swell into a bladder-like structure which is much larger than the sporangium itself, and is known as a **subsporangial vesicle** (Figs. 12.38D to F). Often a second bladder is formed in the lower part of the sporangiophore which is anchored to the substratum by rhizoids (Figs. 12.38E to G). The entire portion of the sporangiophore between the two transverse septa constitutes a water-reservoir, in which a high hydrostatic pressure is set up so that drops of liquid often exude through the lateral wall of the sporangiophore (Fig. 12.38F). Owing to this pressure, the septum lying at the base of the sporangium bulges into it forming what is known as **columella** (Fig. 12.38I). The sporangium is solitary, discoid to lenticular and many-spored, (Fig. 12.38F, K and L). The wall of the sporangium itself is not uniform throughout; the upper part is thickened dark and cutinized bearing bristles with crystals of calcium oxalate, while the lower part adjoining the columella remains thin and is light-coloured (Fig. 12.38I). A mature sporangium is discharged with violence from the end of the sporangiophore. Inside the lower part of the sporangium there is a mucilaginous layer which coming in contact with water swells bursting the delicate wall so that the sporangium becomes free from the columella. The balanced hydrostatic pressure of the columella is no longer maintained as soon as the sporangium is set free from the columella, consequently the columella bursts which results in the squirting out of a jet of fluid from the sporangiophore which has been behaving as a reservoir, and ultimately projecting the sporangium violently to a great distance (Figs. 12.38G, H and J) and (Fig. 12.38K, L and M). Hence the fungus behaves as a thrower of missile. The sporangium so hurled sticks to any object owing to the mucilage which still clings to and also due to adhesive sporangial wall. The sporangiophores are positively phototropic. The sporangiophore including the subsporangial vesicle contains clear watery liquid except at the base of the subsporangial vesicle where a conspicuous zone is invariably found due to accumulation of carotinoid orange pigment. This zone appears to the naked eye as a minute orange spot at the base of the subsporangial vesicle and has been interpreted by Buller (1921) as an eye-spot, the subsporangial vesicle being regraded as a lens focussing the light on the sensitive spot (Fig. 12.26B). The subsporangial vesicle functions as a lens in much the same way as does a flask filled with water. If the rays of the sun strike on one side of the vesicle they are refracted through it and converge on the opposite side forming a spot of light (Fig. 12.39). The protoplasm at that point thus receives a phototropic stimulus; growth and elongation occur on that side of the sporangiophore. In consequence, the sporangium is turned toward the light until the ray strikes it head-on. A condition of physiological equilibrium is then reached and the turning movement ends. Due to phototropic movement of the sporangiophore, the sporangium tends to be directed towards open space, and in nature it is often shot forth to grass surrounding the dung. The spores are eaten along with grass by herbivorous animals and become disseminated with the dung. Passage of the spores through the alimentary canal of the animals facilitates germination producing new individuals.

Discharge of the Sporangium: The fact that the sporangium of *Pilobolus* is discharged in response to light has been thoroughly studied by many investigators like Allen and Jolivette (1914), Parr (1918), Pringsheim and Czurda (1927) van der Wey (1929), and Buller (1934). The sporangium exhibits remarkable accuracy when light is admitted through a pin-hole. The highest degree of accuracy is obtained with white or blue light, comparatively less with yellow light, and very inaccurate with red light. The process was explained in greater details by Buller (1934) through his series of experiments. According to him, the subsporangial vesicle acts as a simple eye with its lens below. When light strikes the upper surface of the sporangium the rays falling on the wall of the subsporangial vesicle bulging around the sporangium are refracted through it and converge on the opposite side below on the protoplasm and the wall at the base of the subsporangial vesicle acts as a region for light perception (Fig. 12.39). At this region a photochemical change is produced including growth and elongation of the subsporangial wall. In response to differential light intensity, differential growth and elongation of the subsporangial wall take place which ultimately induce the movement of the subsporangial vesicle to the direction of higher light intensity resulting in the development of a curvature and a positively phototropic movement of the subspornagial vesicle.

Buller further demonstrated that the release of the sporangium from the sporangiophore takes place due to photochemical reactions on the protoplasmic contents of the subsporangial vesicle and result in the increase of osmotic pressure stimulating the separation of the thin sporangial wall from the columella followed by bursting of the latter with ultimate dispersal of the sporangium from the sporangiophore.

The pressure inside the sporangiophore finally becomes so great that an explosion results rupture occurring along a very definite line of weakness where the columella joins the subsporangial vesicle. The elastic wall of the sporagiophore contracts almost instantaneously with a very audible click. The sporangiophore immediately after explosion shows that the line of dehiscence is a very perfect circle. The explosion takes place so quickly that it is impossibe to follow the exact details at this stage. Water ejected from the sporangiophore tears off and carries away the sporangium from the sporangiophore. The amount of water is about twice the volume of the sporangium itself. The sporangium on leaving the sporangiophore, and before it hits on object, turns upside down.

Immediately on the discharge of the sporangium the stalk region of the sporangiophore loses its rigidity and the subsporangial vesicle, which has contracted to about half its original volume, is brought down to the level of the substratum. The subsporangial vesicle is, however, still full of liquid, part of which is soon squeezed out. After a short time the old sporangiophore becomes completely deflated.

Buller has found using *Pilobolus kleinii* and *P. longipes* that the sporangia are shot to a maximum height of six feet and a horizontal distance of eight feet. With *P. kleinii*, Ingold has found that sporangia may be shot to a height of 118 cm.

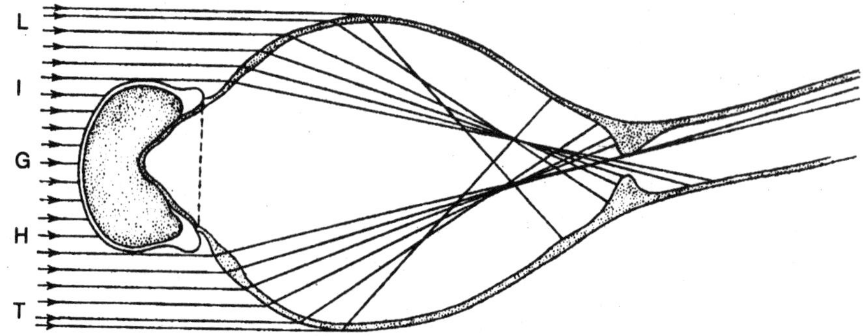

Fig. 12.39: Sectional view of a sporangium and subsporangial vesicle of *Pilobolus kleinii* directed towards the source of light

Sexual Reproduction: Sexual reproduction takes place by the process of isogametangial copulation. The development of gametangia takes place only when there is some hindrance in the formation of sporangia. During sexual reproduction two neighbouring hyphae of the mycelium enlarge, and become club-shaped. The swollen portions grow upright placing themselves together side by side. At the same time a large quantity of protoplasm accumulates in them. The tips of the conjugating hyphae lie in such a position as to give the appearance of a pair of tongs (Fig. 12.29E). The zygospore grows in size with a very tough and thick wall. After a long period of rest the zygospore germinates by the production of a germ tube which grows into a new mycelium. But more frequently the germ tube produces at its tip a germ sporangium, the spores of which are dispersed after maturity and germinate to produce new individuals. Meiosis is during zygospore germination.

It has been experimentally demonstrated that in *P. crystallinus* definite sexual strains are in existence which will alone conjugate with one another and produce zygospore, whereas the hyphae of the same strain will not fuse sexually among themselves and fail to form zygospores. The results have indicated that *P. crystallinus* is heterothallic.

Some Indian Species: *Pilobolus crystallinus* (Wiggers) Tode; *P. kleinii* van Tieghem; *P. longipes* van Tieghem; *P. minutus* Speg.; *P. nanus* van Tieghem.

Family: ***Choanephoraceae***

Sporangium is present in some forms, but absent in others. Sporangium when present is globose to pyriform. In some genera the sporangium is many-spored and in others it is replaced by few-spored sporangiolum or unicellular conidium covering terminal capitate enlargement of the branch of the sporangiophore or conidiophore. Zygospore is formed by gametangial copulation.

Genera included under the family Choanephoraceae are: *Blakeslea, Choanephora* and *Cunninghamella.*

Genus: *Cunninghamella*

The vegetative hyphae are coenocytic, branched and are differentiated into erect and prostrate portions. Sporangia are never observed in this genus. Asexual reproduction, as far as known, takes place exclusively by conidia. Conidia are borne on conidiophores which arise from vegetative hyphae. The erect vegetative hyphae are transformed into more or less branched, sometimes septate conidiophores. Each conidiophore is terminated in a capitate vesicle covered with sterigmata bearing small, echinulate, unicellular, globose to oval or pyriform deciduous conidia (indehscent monosporous sporangiola) (Figs. 12.27J and K). The nature of branching of the conidiophores is extremely variable—cymose to racemose. There may be a main hypha bearing a subterminal whorl of branches each of which is apically swollen. Often there is development of intercalary globose chlamydospores in the mycelium.

Almost all species are heterothallic having gametangia of approximately equal size. The formation of zygospore is very much like that in *Mucor* (Figs. 12.29A to D). The zygospore wall is rough but without appendages.

There are both parasitic and saprobic species.

Some Indian Species: *Cunninghamella echinulata* Thaxter; *C. elegans Lend.; C. intermedia* Despande and Mantri; *C. verticillata* Pain; *C. vesculosa* Misra.

Chapter 13

Subdivision of Ascomycotina

INTRODUCTION

The Ascomycotina and the Baridiomycotina are commonly known as higher fungi. They include fungi which possess much complexity of structure. The Ascomycotina, or sac fungi, as they are commonly known, are characterized by the development of spores, the **ascospores** in a sac-like structure called **ascus.** The ascus is distinguished from the sporangium of the Zygomycotina in the origin and method of formation of spores. A more obvious difference is that the sporangium contains an indefinite number of spores and that too produces without any sexual act. Whilst in the Ascomycotina, the ascus invariably contains 8 spores produced by karyogamy and meiosis. In a few exceptional cases the number is a higher multiple of 8. In a few primitive members of the Ascomycotina the number of spores in an ascus is less than 8. The Ascomycotina constitute the largest of the three classes of sexually reproducing fungi. As may be expected in such a large group there is considerable diversity in form and structure associated with their wide range of distribution and mode of nutrition. At one end are the unicellular organisms commonly known as yeasts, and, at the other, members having their bodies composed of branched hyphae which are always septate and they produce large and elaborate fruiting structures known as **ascocarps.**

OCCURRENCE AND ECONOMIC IMPORTANCE

The fungi included in this group occur throughout all seasons of the year. Some are saprophytes (growing largely on decayed wood—**lignicolous** or on soil—**terricolous**) and others parasites. Again, there are others which are coprophilous. A few are **hypogean,** growing under soil level. Their ascocarps are underground, they are closed, and their spores remain imprisoned until some animal digs them up for food and scatters the ascospores while devouring the ascocarps. The Ascomycotina have the highest economic importance among fungi. A large number of Ascomycotina are utilized profitably for the welfare of the society in the brewery, bakery, cheese-making, organic acid manufacturing industries, preparation of medicine, etc. The parasitic Ascomycotina induce various diseases of plants and animals. Some of the plant pathogenic Ascomycetes are virulent enough to wipe off cultivation of plants of economic importance in the Western countries, whereas the Ascomycotina inducing diseases of animals including human beings are also quite well-known. Again, some of the edible Ascomycotina form delicacy of dishes particularly in the Western countries. But the Ascomycotina causing destruction of food grains in storage, clothes, leather, and similar other essential articles are also not uncommon. Particulars of the respective fungi of economic importance have already been discussed in Chapter 11 and more will be discussed along with the details of different genera of fungi.

SOMATIC STRUCTURES

In the Ascomycotina the somatic body is generally composed of septate hyphae having a tendency to develop into an aggregate mass of mycelium (Fig. 13.1) with a few exceptions where the vegetative body is unicellular uninucleate. In others, a series of cells which remain adhering end to end to form a chain, is known as **pseudomycelium.** Again, some other Ascomycotina are unicellular under certain conditions, but mycelial under others. Cells of the hyphae are usually uninucleate. But there are also members whose hyphae possess multinucleate cells. In some Ascomycotina the hyphae become compactly aggregated to form a sclerotium, in which stage they tide over unfavourable climatic conditions. A short branch of one or two cells known as **hyphopodium** (pl. **hyphopodia**) may be developed on epiphytic mycelium. Hyphopodia may be **mucronate** (Fig. 13.1B) with pointed end or **capitate** (Fig. 13.1C) having end rounded. They are of great taxonomic importance, so also the mycelial setae (Fig. 13.1D).

ASEXUAL REPRODUCTION

The Ascomycotina reproduce asexually by **fission, budding, fragmentation, arthrospores, chlamydospores** or **conidia.** A new individual may be produced directly by budding or by

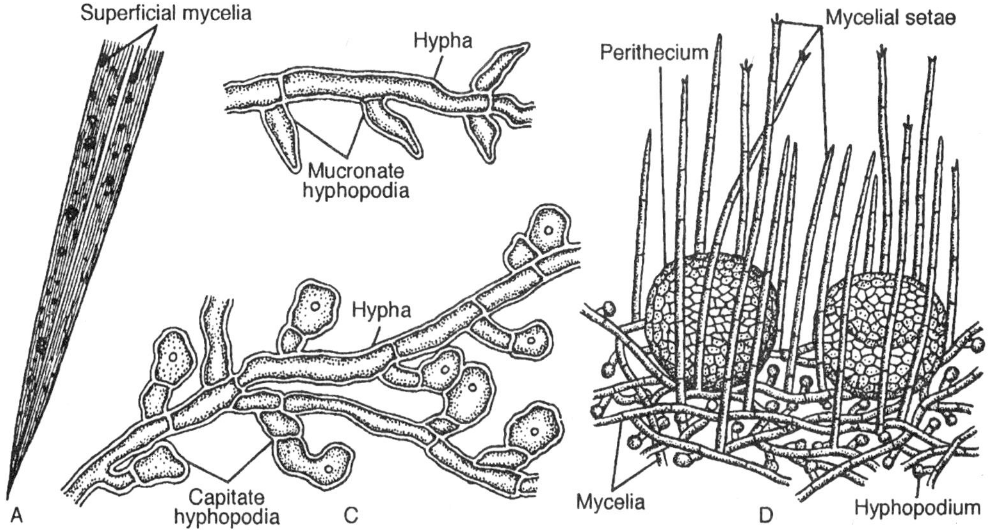

Fig. 13.1: *Meliola palmicola* var. *africana*. A. Superficial dense patches of mycelia on leaf, B. Septate hypha with mucronate hyphopodia, C. Septate hypha with capitate hyphopodia, D. Mycelia with hyphopodia, mycelial setae and perithecia

budding spores known as **blastospores** which on germination give rise to new individuals. The development of conidia is rather common, especially among pathogenic ones. The conidia are usually produced on conidiophores which may be branched or unbranched and are developed from the somatic hyphae. The conidiophores may be solitary or in groups forming an elongated spore-bearing structure, synnema or structures, synnemata. They may also be developed in a pycnidium. Inside the pycnicium is lined with conidiophores from which conidia are produced. Again, the conidiophores may be developed in the peripheral layer of an erumpent, cushion-like mass of hyphae, the acervulus. Conidia and conidiophores may also be borne around a sporodochium. The conidia developed profusely, are disseminated by wind or insects or by some other agencies suitable for quick and wide spreading of the organism. They are usually thin-walled and cannot stand extreme climatic conditions. Sometimes the conidia can also survive seasonal variations.

SEXUAL REPRODUCTION

In comparison with the Mastigomycotina the mode of sexual reproduction in the Ascomycotina is much more elaborate. In the Mastigomycotina and Zygomycotina there is almost no interval between plasmogamy and karyogamy, one follows immediately the other, whereas in majority of the Ascomycotina there is an interval between plasmogamy and karyogamy. The duration of this interval is again extremely variable. During this interval the fusing nuclei pair forming one or more dikaryons, where the dikaryotic condition is very much prolonged, the paired nuclei divide by conjugate division. In the, Ascomytina the dikaryotic phase is dependent on the monokaryotic mycelium for nutrition, whereas in the Basidiomycotina it is completely independent. The three phases clearly encountered in the sexual cycle of the Ascomycotina are: **haplophase, dikaryophase,** and **diplophase.** Just like all fungi the diplophasic condition in the Ascomycotina is very brief. Meiosis follows immediately after karyogamy producing haploid nuclei and reinstating haploid condition. There is wide diversity in the methods by which the compatible nuclei are brought together leading to the development of ascus and ascospores in the members of the Ascomycotina. The most common methods are indicated below.

1. **Gametangial Copulation:** This method has great similarity with what is encountered in some members of the Ascomycotina. The copulating gametangia are morphologically similar and are developed either from mycelial soma or unicelular uni-nucleate somatic bodies which behave directly as gametangia. Their entities are lost during the sexual process producing an unicellular structure in which karyogamy follows immediately after plasmogamy and dikryotic condition is almost absent. An unicellular zygote is formed. The unicelular zygote so formed is transformed directly into an ascus in which ascospores are developed. The variations in the method of gametangial copulation are as follows:

(i) In the genus *Dipodascus* gametangia are separated from the branches of the same or different hyphae by septa. The gametangia soon come into contact and fuse with each other by the dissolution of the intervening walls (Fig. 13.2A). The protoplasts of the two gametangia intermingle. Two nuclei, one from each gametangium, fuse (Fig. 13.2B). The zygote nucleus then undergoes several divisions of which the first one is reductional. The zygote cell elongates into a tapering saccate ascus. Numerous ascospores are developed in the ascus by free cell formation (Figs. 13.2C to E).

(ii) A similar condition is also found in the genus *Eremascus.* But here the ascospore number is strictly eight. In *Eremascus albidus,* two copulating gametangia arise simultaneously,

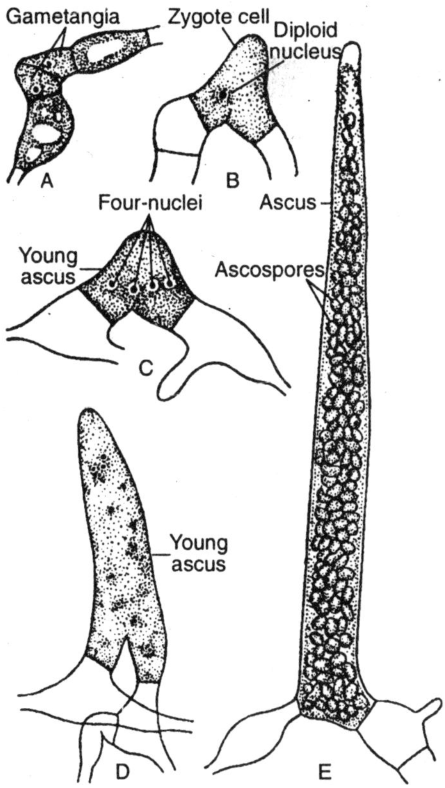

Fig. 13.2: *Dipodascus uninucleatus*. A–B. Stages in gametangial copulation leading to zygote cell formation. C. Four-nucleate young ascus. D. Multinucleate young ascus. E. Ascus bearing numerous ascospores

wound spirally, their tips touch, the walls at the point of contact dissolve and there arises a globular ascus, in which eight ascospores are developed (Fig. 13.3).

(iii) In *Schizosaccharomyces octosporus* any somatic cell is a potential gametangium. During sexual reproduction, cells copulate in pairs by short tubes which ultimately develop into a copulation canal by the dissolution of the common walls. Two nuclei migrate into the copulation canal and fuse (Figs. 13.4A to E). The copulation canal then broadens to form a barrel-shaped structure. Here three nuclear divisions take place, of which the first one is reductional resulting in the formation of eight nuclei. The barrel-shaped structure then becomes an ascus in which eight ascospores are formed (Figs. 13.4F to I).

2. **Gametangial Contact:** In this method two morphologically distinguishable gametangia known as antheridium and ascogonium come in contact with each other and their entities are not lost during the sexual process. This method has variations as shown below.

(i) In *Sphaerotheca humuli* the uninucleate antheridium is applied to the uninucleate ascogonium and at the point of contact a broad pore is developed through which the male nucleus with cytoplasm passes into the ascogonium (Figs. 13.5A to D). The antheridium then collapses. The two nuclei in the ascogonium then divide mitotically, and the ascogonium grows giving rise to a row of cells, the penultimate of which is binucleate and behaves as an **ascogenous cell** (Fig. 13.5E) producing

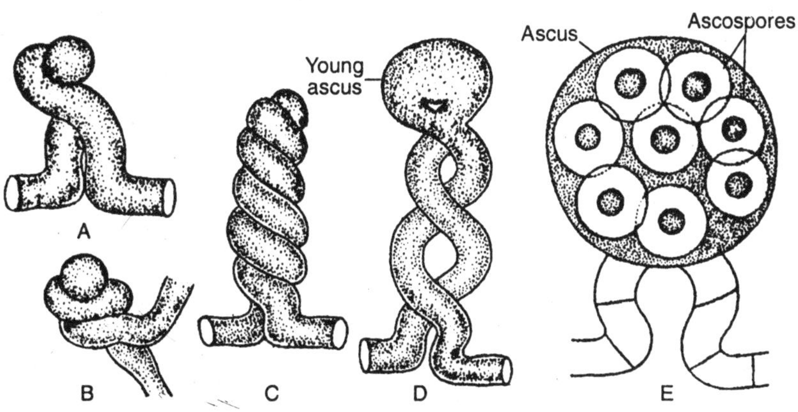

Fig. 13.3: *Eremascus albidus*. A–C. Stages in gametangial copulation. D. Young ascus. E. Eight-spored ascus

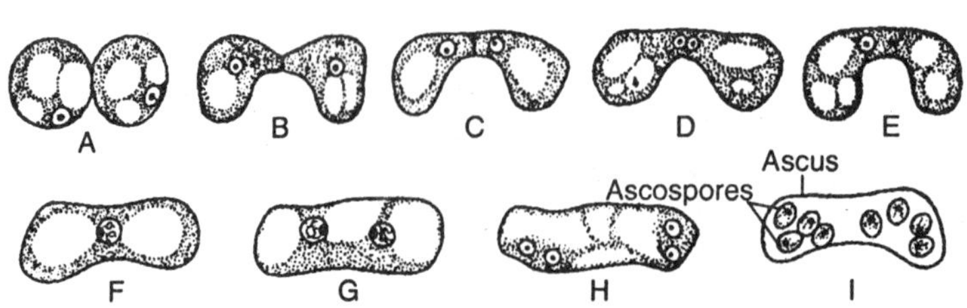

Fig. 13.4: *Schizosaccharomyces octosporus.* A–I. Stages in gametangial copulation leading to the development of ascus and ascospores

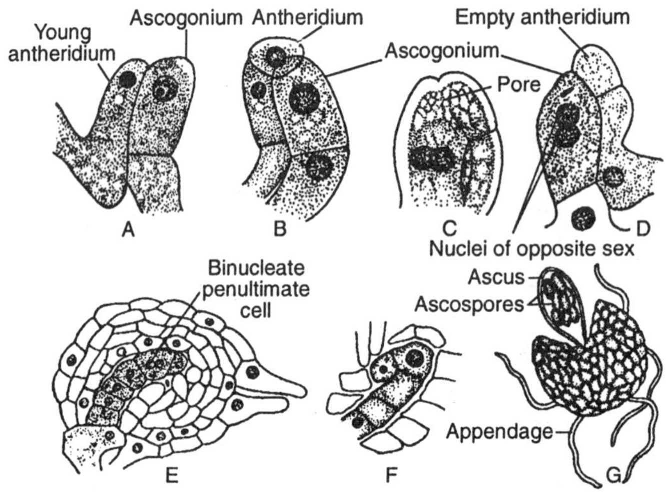

Fig. 13.5: *Sphaerotheca humuli.* A–G. Stages in gametangial contact leading to the development of ascocarp, ascus and ascospores

the single ascus in which ascospores are developed (Figs. 13.5F and G).

(ii) In *Pyronema confluens* club-shaped multinucleate antheridium arises near the multinucleate ascogonium surmounted by a curved tube-like structure, the **trichogyne** (Fig. 13.6A), which affects the union between the ascogonium and the antheridium. During sexual reproduction the trichogyne grows on towards a neighbouring antheridium, and presses firmly against it. The walls at the point of contact are dissolved forming an aperture through which the multinucleate protoplast of the antheridium passes, by way of the trichogyne, into the ascogonium; plasmogamy takes place. Plasmogamy is followed by an increase in size of the ascogonium, and by the formation of protuberances whch ultimately develop into thick short-celled **ascogenous hyphae** (Fig. 13.6B) from the penultimate cells of whose recurved tips the asci are developed (Fig. 13.6C).

3. **Aplanogametic Union:** In *Spermophthora gossypii,* two non-motile fusiform gametes come in contact with each other and unite by a conjugation tube within which two nuclei fuse with each other (Figs. 13.7A and B). From the conjugation tube rather limited branched septate hyphae of uninucleate cells are developed which behave as ascogenous hyphae. The tips of the ascogenous hyphae become spherical asci containing eight ascospores (Figs. 13.7C and D).

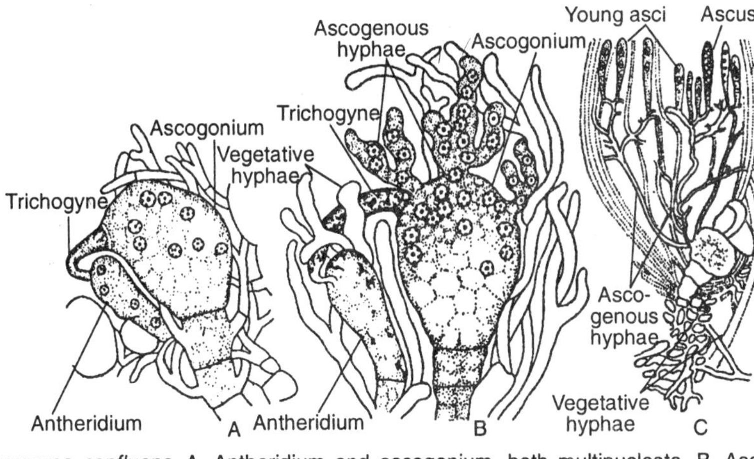

Fig. 13.6: *Pyronema confluens*. A. Antheridium and ascogonium, both multinucleate. B. Ascogonium giving rise to forked ascogenous hyphae. C Young and mature asci developing from ascogenous hyphae

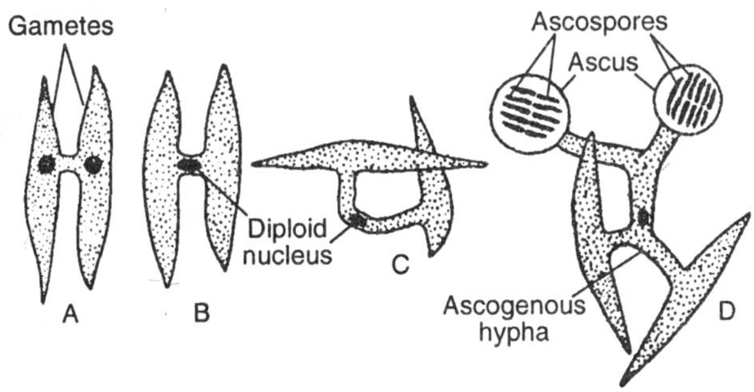

Fig. 13.7: *Spermophthora gossypii*. A–C. Stages in union of gametes. D. Ascogenous hypha and asci bearing ascospores

4. **Spermatization:** In some Ascomycotina (*Pleurage anserina* and *Neurospora* sp.) antheridium is not developed. Specialized sex cells known as **spermatia** (sing. **spermatium**) are developed from hyphae. The spermatia are brought in contact with the trichogyne of the ascogonium or with hyphae where there is no ascogonium, usually through some external agency like insects, water, wind, etc. Dikaryotic condition is achieved by the dissolution of common walls which is followed by the development of asci and ascospores (Figs. 11.6F to H).

5. **Somatogamy:** Here the dikaryotic condition is attained by the union of somatic hyphae of opposite strains which ultimately leads to the development of asci and ascospores (Figs. 13.8F-H), encountered in *Ascophanus* sp. and *Sordaria fimicola*.

DEVELOPMENT OF ASCUS AND ASCOSPORES

The process leading to the development of ascus and ascospores is extremely variable in the Ascomycotina. It is very simple in the genera *Dipodascus, Eremascus, Schizosaccharomyces,* and *Saccharomyces,* where the zygotic cell directly develops into an ascus, whereas in *Taphrina deformans* the dikaryotic cells of the mycelium behave as **ascogenous cells** (Fig. 13.8A). The cells elongate breadth-wise and the nuclei fuse (Fig. 13.8B). The diploid nucleus divides mitotically into two daughter

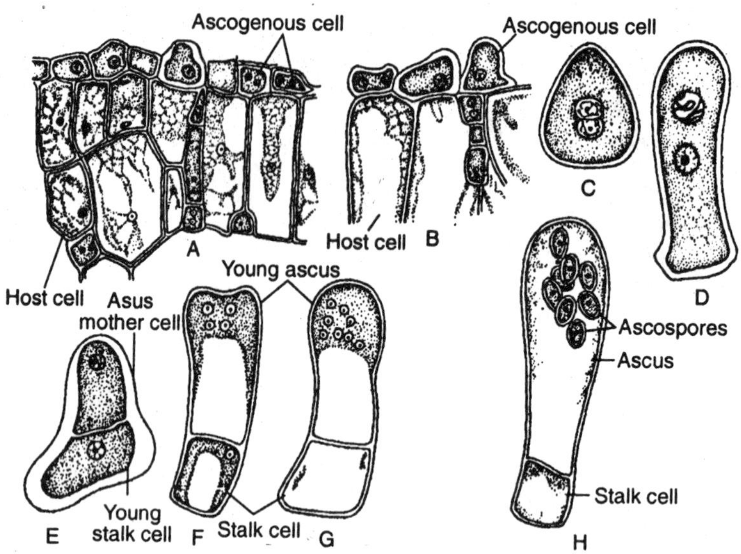

Fig. 13.8: *Taphrina deformans*. A. Dikaryotic ascogenous cells with host tissue. B. Ascogenous cell with diploid nucleus. C–E. Stages in the development of ascus mother cell and young stalk cell. F–H. Stages in the development of stalk cell, ascus and ascospores

nuclei and the ascogenous cell also divides into two uninucleate cells. The lower one of which develops into a stalk cell and the upper one behaves as an ascus mother cell (Figs. 13.8C to E). The diploid nucleus of the ascus mother cell divides meiotically and then mitotically to form eight daughter nuclei (Figs. 13.8F and G). Soon eight ascospores are delimited centred around the nuclei in the ascus which is again developed from the ascus mother cell (Fig. 13.8H). A slightly complex situation is exhibited by the genus *Sphermophthora* where a short ascogenous hypha is formed from the top of which the ascus is developed. Again, in *Sphaerotheca humuli* there is development of an ascogenous cell resulting from the gametangial contact. Most complex condition is exhibited by *Pyronema confluens* which has attracted attention of investigators for many years. It is one of the few Ascomycetes which has been very thoroughly worked out. The study of development of ascus and ascospores in *Pyronema confluens* has produced many interesting points of discussion.

The study of the development of sex organs in the Ascomycotina was started by De Bary and his associates during the latter half of the nineteenth century. Of the earlier workers, mention may be made of Woronin (1866) who working on *Ascophanus pulcherrimus* (*Ascobolus pulcherrimus* Cr.) described the **archicarp** (ascogonium as a row of swollen cells which he termed 'scolecite'). It becomes enclosed in a sheath of mycelial branches, some of which he regarded as probably antheridial. In 1871 Janezewski published a similar account of *Ascobolus ferfuraceus* Pers. He said that the scolecite arises as a curved multicellular branch and becomes surrounded by dense sheath of hyphae in which antheridia are present. One of the cells near the apex of the scolecite increases in size and gives rise to branches upon which asci are borne. But they neither made critical studies of the development of asci nor paid attention to the nuclear behaviour during the process. Details of nuclear behaviour during the development of asci and ascospores in the Ascomycotina were observed for the first time by Dangeard in 1894. In subsequent years Harper (1895), Blackman and Fraser (1906),

Welsford (1908), Brown (1909), Claussen (1912), Dodge (1912), Bagchee (1925), and Gwynne-Vaughan and Williamson (1931), and many others carried on investigations in this line.

The sex organs in *Pyronema confluens* as described by De Bery consists of **archicarp** and **antheridium** produced side by side. The archicarp consists of a large globular cell known as ascogonium having a receptive neck called **trichogyne** at the base of which a septum in present. The antheridium is about the same length as the ascogonium but much more slender (Fig. 13.6A). Both ascogonium and antheridium are multinucleate, as well as the trichogyne. The trichogyne at first grows upright but later curves about and partially surrounds the top of the antheridium and its tip presses hard against the tip of the antheridium. The common walls at that point and the septum at the base of the trichogyne dissolve resulting in the formation of a continuous passage through which the contents of the antheridium pass into the ascogonium and plasmogamy takes place in the ascogonium. The ascogonium now swells up and produces innumerable outgrowths on its surface which develop into hyphal structures and are known as **ascogenous hyphae** (Fig. 13.6B). There is some difference of opinion on the behaviour of the antheridial and ascogonial nuclei and with regard to the existence of the septum at the base of the trichogyne, the details of which will be considered in due course under cytology of ascus develcpment. Immediately after plasmogamy the antheridial and ascogonial nuclei pair up within the ascogonium resulting in the formation of dikaryons. The dikaryons then pass into the ascogenous hyphae (Fig. 13.9A). Opinions differ as to whether they fuse in the ascogonium or merely pass in pairs in the ascogenous hyphae. The ascogenous hyphae arising from the ascogonium are aseptate when young, develop by repeated forking, and subsequently become septate so that cells neighbouring the ascogonium bearing many nuclei and cells further away have only two (Figs. 13.6B and 13.9A). The tip of each ascogenous hypha bends forming a hook and is known as **crozier** (Fig. 13.9B). The nuclei of the crozier divide by conjugate division with spindles parallel to each other (Fig. 13.9C). During the division of the nuclei two septa appear between the two daughter nuclei. These two septa divide the crozier into three cells which are designated as **basal** or **stalk cell, loop** or **dome cell** or **subterminal cell** or the **penultimate cell,** and **ultimate** or **tip cell** (Fig. 13.9D). The stalk and tip cells are uninucleate, whereas the loop cell is binucleate. The loop cell or the penultimate cell is the **ascus mother cell.** The two nuclei fuse to form a diploid nucleus in the ascus mother cell which elongates into an ascus (Figs. 13.9F to J). The diploid nucleus divides three nucleus surrounding itself with a part of the cytoplasm of the ascus and develops into an ascospore which is surrounded by a rather firm wall (Figs. 13.9O to Q). The development of ascospores is by **free cell formation.** The entire mass of cytoplasm of the ascus is not used up in the formation of ascospores. The small portion of cytoplasm that is left in the ascus is known as the **epiplasm** which serves to nourish the young ascospores and also furnishes material for the sculpturing of the ascospore wall, particularly in those species which have sculptured spores.

Most Ascomycotina have eight-spored asci. But asci with more than eight spores are not very rare. In such cases more than eight ascospores are developed in two ways: (i) More than eight nuclei may be present as a result of successive nuclear divisions before any ascospores are delimited. In that case the final number of ascospores in each ascus is likely to approximate to some multiple of eight. (ii) Alternatively eight ascospores are differentiated but they proceed to bud off secondary spores or conidia.

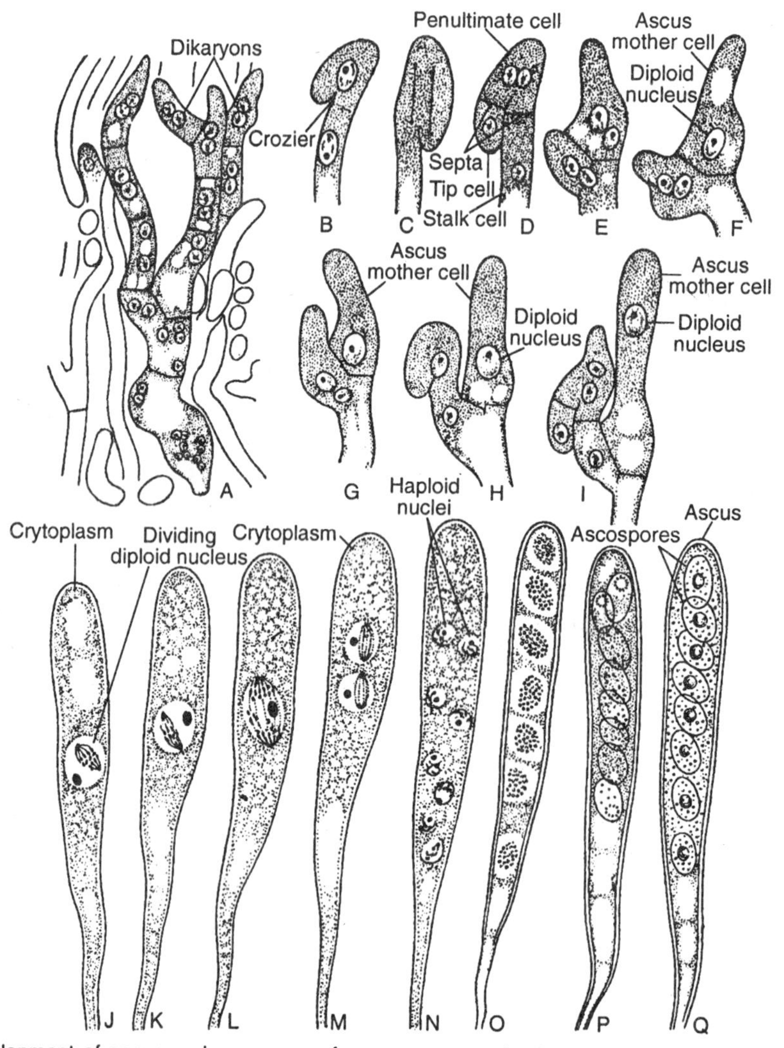

Fig. 13.9: Development of ascus and ascospores from ascogenous hypha as exemplified by *Pyronema confluens*. A. Forking of ascogenous hyphae with dikaryotic cells. B. Bending of tip of ascogenous hypha and crozier formation. C. Nuclei of crozier dividing by conjugate division having spindles parallel to each other. D. Binucleate penultimate cell, tip and stalk cells are uninucleate. E. Elongated binucleate penultimate cell, walls between tip and stalk cells dissolved forming a dikaryotic cell below the penultimate cell. F. Ascus mother cell with a diploid nucleus and a dikaryotic cell below. G–I. Subsequent stages of growth of the ascus mother cell and further growth of the dikaryotic cell below with a tendency to develop into an ascus mother cell. J–Q. Successive stages in the development of ascus and ascospores from diploid nuclear condition.

Mating behaviour. Fungi of this group may be homothallic or heterothallic. The basis of heterothallism is typically a single gene with two alleles, 'A' and 'a'. Because of segregation during the meiotic division which precedes ascospore formation, the eight ascospores normally present in an ascus will include four of one mating type and four of the other.

Other processes of ascus formation. By the time ascus is being developed from the loop cell, the apical cell bends and touches the wall of the stalk cell. The walls at the point of contact between the tip and stalk cells may dissolve to form a dikaryotic cell which develops into an ascus (Figs. 13.9E to I). This is rather common in *Pyronema confluens* and is known as **curvascous type** of ascus formation.

In another case, the loop cell elongates to develop into a new crozier. The process may be continued with several times, only the last terminal loop cell develops into an ascus.

Whatever has been described above, is known as **hook type** or **Pyronema type** of ascus development. Besides this, there are different other processes of ascus formation from ascogenous hyphae or similar other structures. They are as follows:

(i) In *Geopyxis catinus*, ascus development is not preceded by hook formation. Here the terminal cell of the ascogenous hypha is uninucleate, but the subterminal cell is binucleate. The binucleate subterminal cell grows out laterally to form an ascus mother cell which directly develops into an ascus, known as **Geopyxis type.**

(ii) In *Plicaria succosa*, the ascogenous hyphae are terminated by binucleate cells which directly take part in the development of asci. This is known as **Plicaria type.**

(iii) A synthesis between Pyronema type and Plicaria type is exhibited by *Pustularia vesiculosa*, in which there is development of crozier, but the loop cell of the crozier instead of developing into an ascus grows further to form a hypha whose terminal cell develops into an ascus like the Plicaria type.

(iv) Again there are Ascomycetes, where asci are developed from all the binucleate cells of the ascogenous hyphae producing a chain of asci and is known as the **Rectascous** type.

(v) In the Laboulbeniales dikaryotic condition is achieved by the process of spermatization. The female reproductive organ itself develops into an ascogenous cell and ascogenous hyphae are absent. The ascogenous cell directly develops into an ascus. This type of ascus development is known as **Laboulbenia type.**

NATURE OF ASCI AND ASCOSPORES

The members of the Ascomycotina are characterized by the development of asci and ascospores. In the large majority of them the asci are produced in an ascocarp. There are also Ascomycotina whose asci remain naked and are not enclosed in ascocarp. In general, they remain intermingled with paraphyses forming a well-developed hymenium in the ascocarp. The asci may also remain scattered at various levels within the ascocarp and may as well be loosely clustered even when they are enclosed in an ascocarp. They are extremely variable in shape being cylindrical to rectangular, globose,

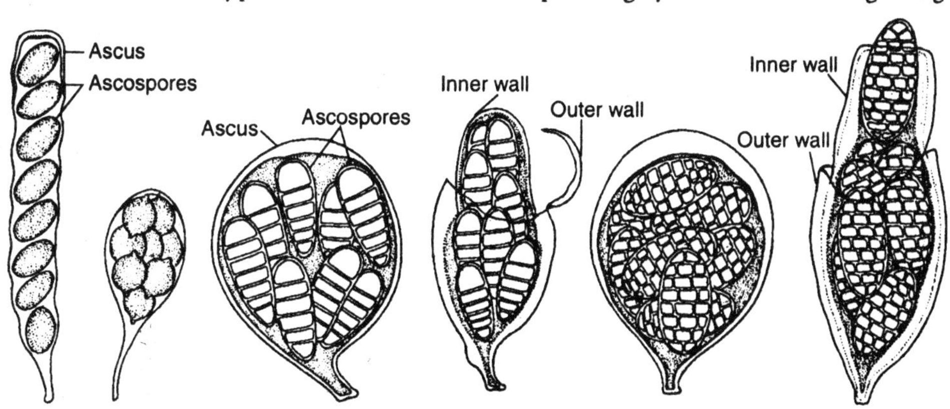

Fig. 13.10: Nature of unitunicate and bitunicate asci. A–B. Unitunicate asci. C–F. Bitunicate asci. C and E. Intact bitunicate asci. D. Expanding inner wall splitting the outer wall. F. Escape of ascospores through a pore at the tip of the expanded inner wall

subglobose, or ovoid. The asci may be stalked or sessile (Figs. 13.10 to 13.12). Usually the asci are single-chambered but in exceptional cases each one is divided into a many-chambered structure by the development of transverse septa. A mature ascus may be **persistent** or **evanescent** depending on whether the ascus wall is persistent or of short existence respectively. A mature ascus with persistent wall is known as **persistent ascus.** But when the ascus wall dissolves away soon after the formation of ascospores releasing them in the ascocarp in a mass of mucilage formed by the gelatinization of the dissolved ascus wall, this type of ascus is known as **evanescent ascus,** as in *Eurotium repens.*

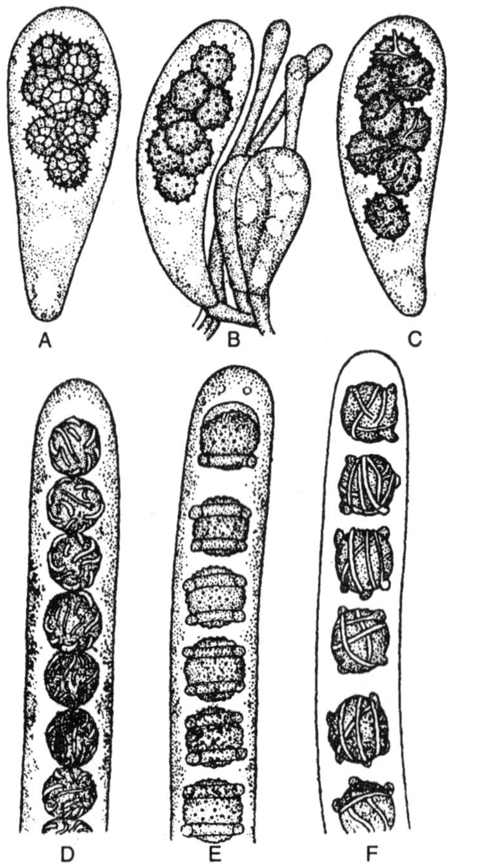

Fig. 13.11:A–C. Broad-clavate to ovoid asci bearing eight sculptured ascospores encountered in different species of *Ascodesmis.* D–F. Clavate to subcylindric asci with sculptured ascospores of different species of *Lamprospora*

Nannfeldt (1932) suggested that the asci may be double-walled as well as single-walled. But Chadefaud (1942) said that all asci have two walls—an outer thick wall (**ectoascus**) and an inner thin wall (**endoascus**). But it was subsequently demonstrated that both the walls may be extremely variable in texture. An ascus is designated as **unitunicate** when both the inner and outer walls are more or less rigid and do not separate during spore ejection (Figs. 13.10A and B). Again, it is known as **bitunicate** when the inner wall of the ascus is elastic and expands beyond the thick outer wall at the spore liberation (Figs. 13.10C to F).

Unitunicate asci are of several kinds. (i) **Operculate**—when they open by a **lid** or **operculum** at the tip (Fig. 13.13B) which is pushed open when the ascospores are shot into the air, as in *Ascobolus scatigenus.* Operculate asci may or may not be blued in iodine. They usually contain symmetrical ascospores which are not septate and are usually large and globose to broad-ellipsoid but often have roughened or warted walls. (ii) **Inoperculate**—where there is no operculum and the ascospores escape through an apical pore or **ascostome** (Figs. 13.13D and E), as in *Bulgaria inquinans.* In others there is a ring-like thickening on the inner surface of the ascus wall at the tip. This is visible as two minute highly refractive dots in optical section and sometimes also stains blue with iodine. Inoperculate asci have ascospores that are slightly asymmetrical—the upper half of the ascospore is slightly broader than the lower half. The ascospores of inoperculate asci are rather small, narrow and often very long and slender or filiform. They may be septate but do not have warted or otherwise ornamented walls. (iii) **Suboperculate asci** are the asci which possess thick apical ring and the apical pore is blocked usually with some substance, **pore plug** different in its staining pro-

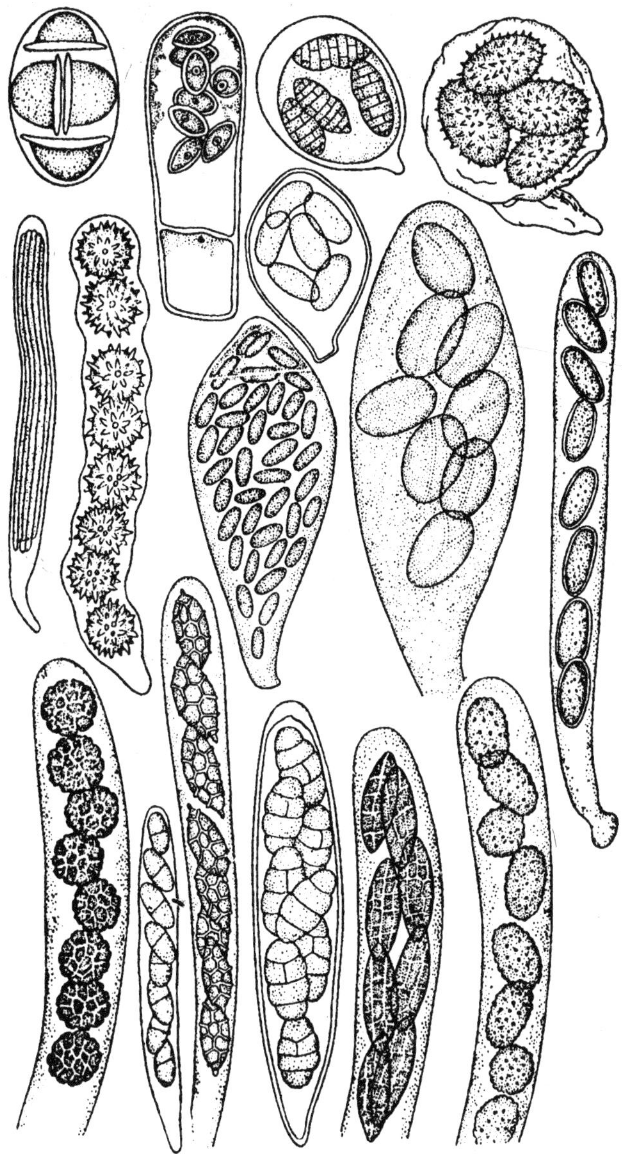

Fig. 13.12: Various types of asci and ascospores

perties from the rest of the wall. This pore plug stains blue with iodine, either as a whole or round its edges, as in *Xylaria filiformis*. (iv) **Bilabiate** are the asci which dehisce by means of a slit across the tip (Fig. 13.13A). (v) **Indehiscent asci** liberate their ascospores either by deliquescing or by becoming turgid and bursting.

Most unitunicate asci are accompanied by sterile hyphae, called **paraphyses,** which occur between them and apparently afford some protection to the developing asci, at least in species with hymenium wide open.

Bitunicate asci possess two distinct, separable walls—outer thick and rigid and inner thin and elastic. A bitunicate ascus is distinctly thick-walled toward the tip and very commonly a well-defined dimple in the inner wall can be seen at the tip of the ascus. The apex is never coloured blue with Melzer's chloral-hydrate iodine solution, i.e., it is **non-amyloid**—and it lacks an operculum or a plugged pore or peculiar ring-like thickenings which are common in unitunicate asci. A bitunicate ascus is usually very short-stalked with the stalk rather sharply delimited from the body of the ascus containing the spores. There is a tendency for the spores in bitunicate asci to be multicellular and coloured (Figs. 13.10C to F). Bitunicate asci usually occur in a hymenium within closed ascocarps superficially resembling perithecia but without true paraphyses. Instead the hyphae between the asci are commonly attached to the ascocarp wall at top and bottom and are often branched or form a network among the asci. These interascal threads are called **pseudoparaphyses** or **paraphysoids** and the ascocarp is a **pseudothecium**, commonly encountered in *Pleospora herbarum.*

In general, a mature ascus contains eight ascospores (Figs. 13.11A to C). But the number of ascospores may be less than eight or may be as many as sixteen, thirty-two, or even more (Fig. 13.12). Similarly, the ascospores are extremely variable in size, shape, colour, nature of wall surface, and other characters. Ordinarily they are unicellular uninucleate (Figs. 13.11D to F). Single-celled binucleate ascospores are also not uncommon. There are also cases where the ascospores are two to many-celled (Fig. 13.12). The ascospores may be colourless to extremely variable in colour. The spore wall may be smooth or variously sculptured (Figs. 13.11 and 13.12). The ascospores may be of any shape from globose to narrow thread-like (Fig. 13.12). They may be arranged irregularly within the ascus, in a single row—**uniseriate,** or a double row—**biseriate,** or parallel with one another, if they are long and narrow. Each ascospore may have one or more oil globules—**guttules.**

DEHISCENCE OF ASCI AND ASCOSPORE LIBERATION

The mechanism of dehiscence of asci and the liberation of ascospores are extremely variable in the Ascomycotina. They are dependent upon the nature of ascocarp, whether an ascocarp has hymenium wide open or the ascocarp has a pore or the ascocarp is completely closed. Again, in cases where the asci are naked and not borne in an ascocarp the mechanism of ascus dehiscence and spore liberation are also different. In a large number of fungi of this group the ascospores are forcibly shot out from the ascus which may rupture or open by a lid at the top to permit the ascospores to escape.

In fungi possessing ascocarp with hymenium wide open having unitunicate asci, there exist the following different types of dehiscence of asci:

(i) In the case of operculate ascus, the ascus opens at the top by means of a circular lid of variable size. The **lid** or **operculum** (Figs. 13.13B and C) remains attached at right angles or obliquely to the long axis of the ascus after the discharge of the spores. The opening formed at the top of the ascus is known as the **ascostome** (Figs. 13.13B and C). When these asci are borne in a wide open ascocarp—**apothecium,** the ascospores are discharged simultaneously (Figs. 13.19D to F).

(ii) But in the inoperculate form the ascus instead of developing an operculum ruptures forming an ascostome with toothed margin produced due to discharge of spores from the ascus (Figs. 13.13D and E).

(iii) In case of suboperculate ascus, the ascospores are liberated through the apical pore by the displacement of pore plug.

(iv) Dehiscence of an ascus may also be brought out by means of a slit across the top and is known as the **bilabiate** method (Fig. 13.13A).

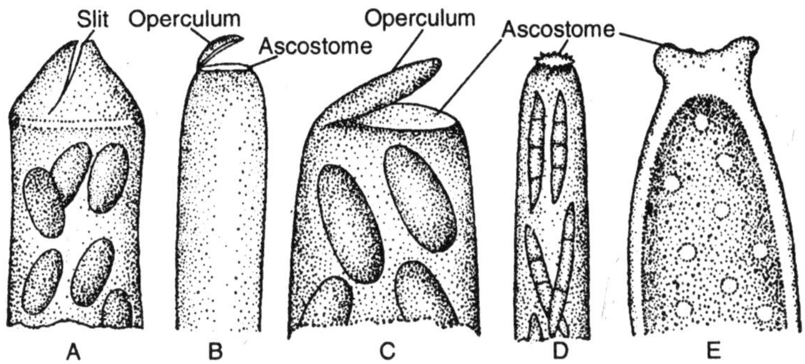

Fig. 13.13: Methods of dehiscence of the ascus. A. Bilabiate. B–C. Operculate. D–E. Inoperculate

(v) In indehiscent asci the ascospores are liberated from the asci either by deliquescing or by the bursting of the ascus wall by becoming turgid due to hydrostatic pressure from inside.

(vi) In an ascocarp with a small opening—**perithecium** expulsion of ascospores may be brought about by the elongation of asci so that their tips reach the opening—**ostiole** one by one and shoot their contents (Fig. 13.20A). Asci also break down, in some fungi, into a slimy mass in the perithecial cavity. The slimy mass containing ascospores then swells and oozes out of the ostiole and is dispersed either by insects or splashed by raindrops.

(vii) Where the asci are developed in a closed ascocarp, the ascospores are liberated by the disintegration of the ascocarp and ascus wall. Again, in the hypogean Ascomycetes, the ascocarps are broken by animals and they feed upon them and the spores are dispersed through their dung.

(viii) In fungi producing naked asci, the dehiscence of asci and liberation of ascospores are to a great extent controlled by diurnal conditions.

A bitunicate ascus normally opens by a transverse split in the outer wall, a short distance below the tip, which is then pushed off like a thimble by the rapidly expanding elastic inner wall. The ascospores then escape through a pore at the tip of the inner wall (Figs. 13.10F and 13.14).

CYTOLOGY OF ASCUS DEVELOPMENT

Opinions and interpretation are so conflicting that it is rather difficult to take a clear view with regard to the cytology of ascus development. In spite of everything an attempt is being made to give a picture of the whole situation as clearly as possible taking into consideration of the view points of most of the workers. The phenomenon of nuclear fusion in the Ascomycotina was reported for the first time by

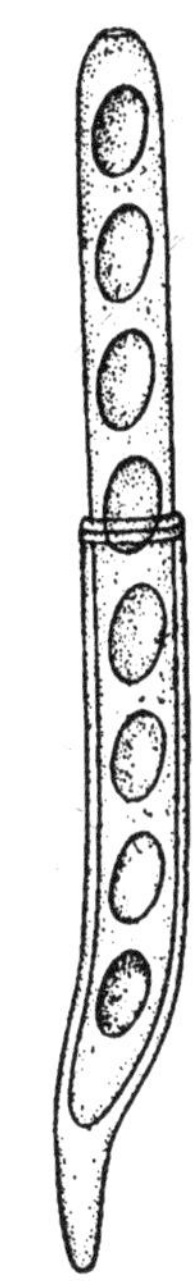

Fig. 13.14: Dehiscence of bitunicate ascus with pore at the tip of the expanded inner wall

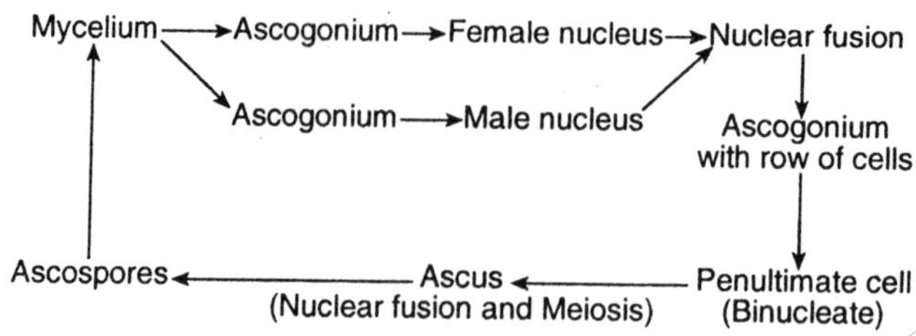

Fig. 13.15: Cytology of ascus development as observed by Harper in *Sphaerotheca humuli*

Dangeard in 1984. He observed the formation of a crozier at the tip of ascogenous hypha in *Peziza vesiculosa.* The two nuclei of the subterminal cell or loop cell of the crozier fuse and the subterminal cell elongates forming an ascus. Dangeard's findings have been widely confirmed by many investigators. In 1895 Harper reported that in *Sphaerotheca humuli* sexual reproduction is initiated by gametangial contact when a passage is developed at the point of contact of the gametangia through which the male nucleus passes into the ascogonium where it fuses with the female nucleus. Immediately after nuclear fusion septa appear in the ascogonium resulting in the development of a row of cells, the binucleate penultimate cell of which develops into an ascus. Harper stated that there are two nuclear fusions, one between the male and female nuclei in the ascogonium and the other in the ascus, hence the nucleus is tetraploid. Harper's account indicates the two-chromosome reductions are necessary during the nuclear divisions in the ascus (Fig. 13.15).

In 1906 Blackman and Fraser (*nee* Gwynne-Vaughan) described a case of reduced sexual process in *Humaria granulata,* a self-sterile species. Here antheridium is not developed. A multinucleate ascogonium without any trichogyne is present. The female nuclei fuse in pairs in the ascogonium and the diploid nuclei so formed enter in the ascogenous hyphae that are developed from the ascogonium. A second nuclear fusion takes place in the ascus.

In 1907 Dangeard and his followers working on *Pyronema confluens* reported that there is development of antheridium and ascogonium bearing trichogyne. But antheridium and trichogyne do not fuse and the septum at the base of the trichogyne remains intact as a result of which the antheridial nuclei fail to reach ascogonium. The trichogyne and the antheridium ultimately disintegrate. The ascogonial nuclei arrange themselves in pairs in the ascogonium. The paired nuclei then pass into the ascogenous hyphae and eventually ascus and ascospores are formed by crozier formation. Dangeard denied the occurrence of a nuclear fusion in the ascogonium, but admitted one nuclear fusion and one reduction division that too are in the ascus (Fig. 13.16).

Gwynne-Vaughan (*nee* Fraser) in 1908 stated that in *Humaria rutilans* there are two nuclear fusions and two reduction divisions in the ascus.

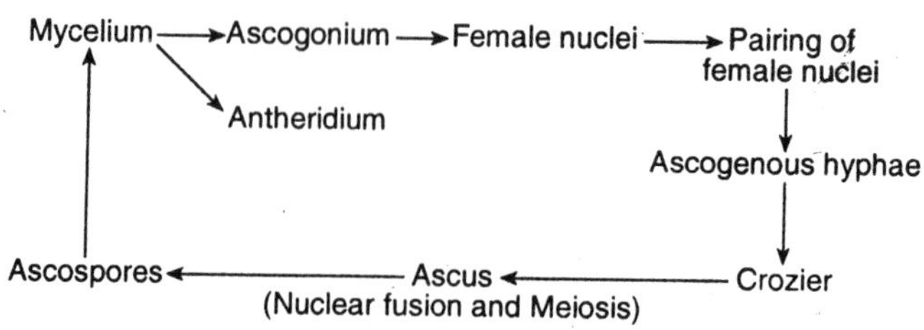

Fig. 13.16: Cytology of ascus development as observed by Dangeard in *Pyronema confluens*

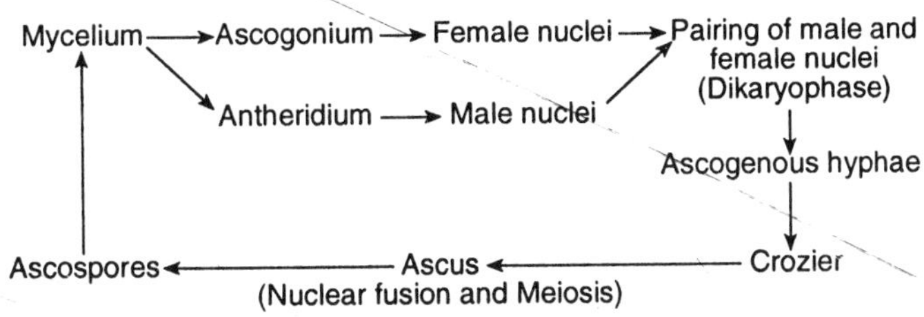

Fig. 13.17: Cytology of ascus development as observed by Claussen in *Pyronema confluens*

Claussen working on *Pyronema confluens* in 1912 said that the male and the female nuclei pair closely in the ascogonium but they do not fuse. The paired nuclei then enter in the ascogenous hyphae where the ascus is developed by crozier formation, and nuclear fusion takes place in the ascus. Meiosis of the diploid nucleus takes place immediately after nuclear fusion. According to him, there is one nuclear fusion and only one meiosis in the entire process of sexual reproduction leading to the development of ascospores and that too in the ascus (Fig. 13.17).

Dodge in 1920 described the life history of *Ascobolus magnificus.* According to him, a large septate ascogonium wraps about a club-shaped antheridium. The antheridial tip fuses with the tip of the trichogyne. Ascogenous hyphae develop from the ascogonium. Simultaneously numerous sterile hyphae grow from the neighbouring cells producing a closed structure. Dodge did not enter into nuclear details.

Bagchee (1925) working on *Pustularia bolarioides* reported that the number of chromosomes in the third telophase in the ascus is the same as the first.

Schweizer in 1931 described the development of ascus and ascospores in *Ascobolus strobilinus.* Here each fructification has sex organs. The male nuclei pass through the unicellular trichogyne to the ascogonium, where they pair with the female nuclei, but do not fuse. The male nuclei can be recognized by their small size. He said that the ascogenous hyphae also contain paired nuclei, one small male nucleus and one large female nucleus. But he did not go any further.

Gwynne-Vaughan and Williamson in 1931 found that in *Pyronema confluens* the trichogyne and the antheridium make contact. The nuclei of the trichogyne degenerate and the wall at the base of the trichogyne ruptures. Hence a continuous passage is developed from antheridium to ascogonium through trichogyne by the dissolution of the walls at the point of contact of the trichogyne and the antheridium. The male nuclei pass from the antheridium through the trichogyne to the ascogonium and mix with the female nuclei there. The male and female nuclei fuse in pairs in the ascogonium and the diploid nuclei so formed then pass into the ascogenous hyphae. Asci are formed by typical crozier formation and a second nuclear fusion takes place in the ascus resulting a tetraploid nucleus. During the development of ascospores the tetraploid nucleus undergoes three successive divisions to produce eight haploid nuclei. The first and the third divisions are both reductional and the second one is mitotic (Fig. 13.18). The second reduction division that takes place in the ascus has been termed **brachymeiosis.** Their statements were supported by chromosomal counts at every stage of nuclear division. They reported the chromosome number of the tetraploid nucleus as 24 which reduces to 12 immediately after first division, whereas in the second division the chromosome number remains 12 which again is reduced to a haploid number of 6 as a result of the third nuclear division which is also reductional.

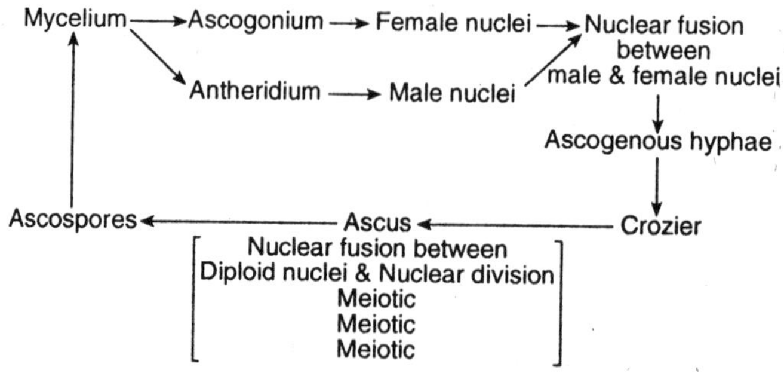

Fig. 13.18: Cytology of ascus development as observed by Gwynne-Vaughan and Williamson in *Pyronema confluens*

Gwynne-Vaughan and Williamson in 1932 investigated on *Ascobolus magnificus* and confirmed the findings of Dodge. They were the first to report that a continuous passage is developed from antheridium through trichogyne to the ascogonium along which antheridial nuclei pass into the ascogonium. The nuclei of the antheridium possess a large chromatin body and a small stainable granule, whereas the ascogonial nuclei show a small rounded chromatin body. The nuclei bear 4 chromosomes. Fusion of male and female nuclei takes place in the ascogonium. The diploid nuclei then pass into ascogenous hyphae that are developed from the ascogonium. They identified the diploid nature of the nuclei by chromsome counts. The chromosome number of the diploid nuclei is 8; along with the chromsomes, the large and the small chromatin bodies were also present in the diploid nuclei. The binucleate apical cells of the ascogenous hyphae produce asci by crozier formation, when fusion of the two diploid nuclei results a tetraploid nucleus in the ascus having 16 chromosomes. During the first division of the ascus nucleus 8 bivalents are visible and 8 chromosomes travel to each pole. In the second division also 8 chromosomes are seen. In the third division prophase 4 recurved and elongated bodies are seen. 4 chromosomes pass to each pole, so that the ascospore nuclei are haploid with 4 chromosomes. The first division in the ascus is reductional, second equational, and the third division is also reductional which they called brachymeiosis.

In 1937 Gwynne-Vaughan reported two nuclear fusions and two reduction divisions in *Patella* (*Lachnea*) *melaloma* (Alb. and Schw.) Seaver.

The whole idea of two nuclear fusion and two reduction divisions was strongly criticized by many workers resulting into difference of opinion. To a large number of investigators it appeared very unusual to have two successive fusions and two reduction divisions in the same life cycle. From time to time there has been debate concerning the number of nuclear fusions and reductional divisions in the life cycle of certain members of the Ascomycotina. The solution of problem of this kind is necessary before the investigators can arrive at a true insight into the interrelationships and evolutionary tendencies of the group.

Olive in 1949, 1950 using propinocarmine staining technique demonstrated that there was no evidence of brachymeiosis in *Patella melaloma* which has previously been reported by Gwynne-Vaughan in 1937. He clearly worked out that there is one nuclear fusion in the young ascus giving rise to a diploid nucleus which is followed by a single reduction division.

Hirch in 1950 using techniques especially devised for staining chromosomes stated that there is no evidence that brachymeiosis occurs in *Pyronema confluens*, as was claimed by Gwynne-Vaughan and Williamson in 1931. He stated that the young ascus nucleus is diploid and not tetraploid with 24 chromosomes as claimed by them. A single reduction division occurs, in the young ascus, and the haploid number of chromosome is 12.

In 1953, working on the cytology of ascus development in *Ascobolus magnificus*, Wood reported that asci arise in the usual fashion from the binucleate penultimate cells of the croziers. There is a conjugate division in the ascogenous hyphae. The darkly staining spots or granules present in the ascus nucleus may be heterochromatin in nature. The number of bivalents in the first division metaphase is eight. The diploid number of chromosome in the fusion nucleus of the ascus, therefore, is sixteen. Following first division metaphase the homologous chromosomes separate. During first division anaphase eight univalent chromosomes pass to each spindle pole. Two daughter nuclei with their nucleoli are formed. At prophase and metaphse of the second division there appears to be 8 chromosomes in each nucleus. During second division anaphase 8 chromosomes pass to each pole. At the end of the second division four nuclei with four prominent nucleoli are present in the ascus. It would appear from Wood's work that the haploid number of chromosomes in *Ascobolus magnificus* is eight. No evidence of brachymeiosis is found. The evidence contradicts the claims of Gwynne-Vaughan and Williamson that the haploid number is four and that the fusion nucleus is tetraploid. The deeply staining bodies claimed by Gwynne-Vaughan and Williamson in the fusion nucleus of the ascus may be two heteropyknotic regions of a pair of chromosomes.

Wood's observations were mainly on the nuclear behaviour in the ascus nuclei. In the ascus, the nucleus is a prominently vacuolate body. The number of bivalent chromosomes at metaphase I is eight. The diploid number of chromosomes in the fusion nucleus is sixteen. Following metaphase I the homologous chromosomes separate. During anaphase I eight univelent chromosomes pass to each pole. The two daughter nuclei with their nucleoli are formed.

Hence no evidence of brachymeiosis has ben found in *Patella melaloma* (Olive, 1949, 1950), *Pyronema consfluens* (Hirch, 1950), or in *Ascobolus magnificus* (Wood, 1953). The whole idea of brachymeiosis would now be unacceptable.

In general, Claussen's findings have received wider support being more reasonably acceptable than those of other workers.

Those who still hold that there occur two successive nuclear fusions followed by a double reduction division could very well settle this question by a genetic study of the species *Ascobolus*. With only a single nuclear fusion and that one in the ascus, there cannot be normally more than four genotypically different kinds of ascospores in an ascus. If the spores are arranged in one line they will be found to alternate 4 and 4; and 2 or 2, 4, 2 for any pair of genetic factors. With double fusion and double reduction division, one should find many asci with eight different kinds of spores. If the spores are in linear arrangement, they should very frequently alternate 1 and 1 for members of a pair of factors, such as, for example, mating type factors.

RANGE AND DEVELOPMENT OF ASCOCARPS

The fructifications or fruit bodies or **ascocarps** of the Ascomycotina are structures containing asci surrounded by, or enclosed within, protective tissue—**peridium**. Their general structure, conditions of growth and shape are extremely variable. Ascocarps may occur singly

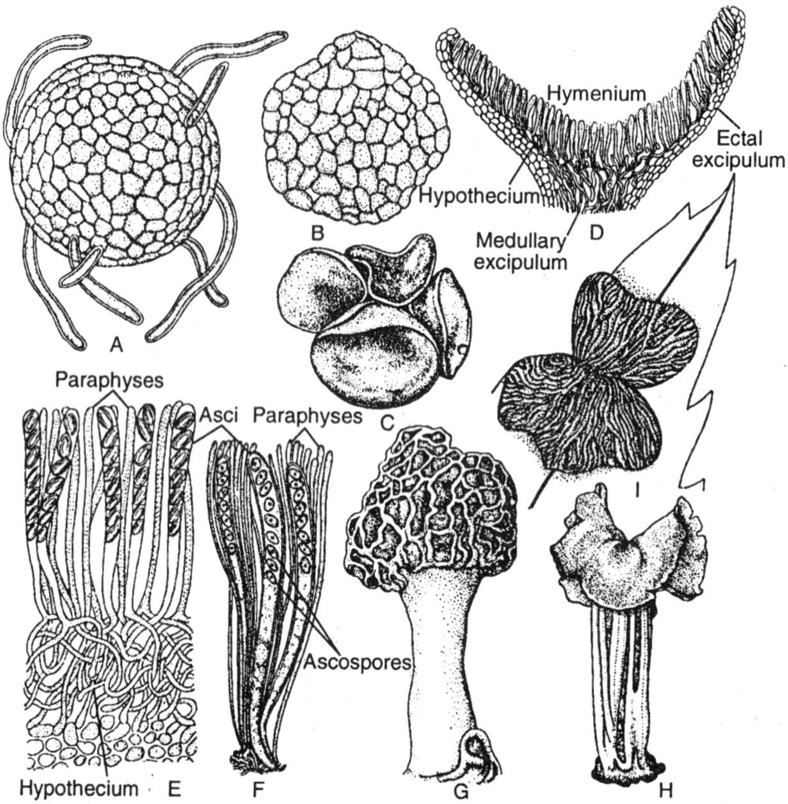

Fig. 13.19: Various types of ascocarps. A. *Erysiphe* sp. Cleistothecium with appendages. B. *Eurotium* sp. Cleistothecium without appendages. C. *Peziza* sp. Apothecia. D. Diagram of a section through an apothecium showing structural details. E. Portion of a section through an apothecium showing hypothecium, asci and paraphyses. F. Portion of hymenium with asci, ascospores and paraphyses. G. *Morchella* sp. Stipitate apothecium with sponge-like pileus. H. *Helvella* sp. Stipitate apothecium with saddle-shaped pileus. I. *Rnytisma* sp. Tar-like stroma with radiate wrinkles under which apothecia are formed

or in groups, and may be superficial, erumpent (bursting through the substratum) or embedded in the substratum or host tissues. The origin and distribution of hymenium are also different in them. The hymenium may consist of asci only or the asci may be interspersed with sterile hyphal elements, **paraphyses** (sing. **paraphysis**) (Figs. 13.19E, F and 13.20B) arising in the ascogenous layer. The paraphyses may be simple or branched, sometimes coloured, sometimes inflated at the apex (**capitate**), or at intervals along their length (**moniliform**), and with or without septation. Parapyses are said to space out the asci and to secrete mucilage which is concerned in spore discharge mechanisms. The fructifications may be flat expansions, crusts, rounded, cup-like, flask-like, alveolar, or erect club-like, etc. But the most common and widely encountered forms are: (i) a more or less rounded structure completely closed, and having hymenium enclosed within the wall, a **cleistothecium** (pl. **cleistothecia**) (Figs. 13.19A, B); (ii) a saucer- or cup-shaped ascocarp with wide open hymenium, an **apothecium** (pl. **apothecia**) (Fig. 13.19C); (iii) a flask-like ascocarp designated as a **perithecium** (pl. **perithecia**) (Figs. 13.20A to C) perforated at the apex by a pore known as **ostiole.** Sometimes the ostiole is prolonged into a beak—**rostrum** and usually it is lined with hair-like hyphal structures, **periphyses** (sing. **periphysis**) (Fig. 13.20B) directed towards the opening and assisting in expelling asci or ascospores. A perithecium has its own distinct peridium of specialized tissue and may or may not be embedded in a stroma and is thus designated as **stromatic** or **non-stromatic perithecium** (Fig. 13.20A). In addition to these forms of ascocarp, there are others which are not so widely encountered. They may be black, carbonous or leathery, boat-shaped or branched, etc. Some of them are designated as: (i) a **hysterothecium** (pl. **hysterothecia**) (Fig. 13.20F), an elongated ascocarp closed when young, but opens at maturity by a long slit following a line of dehiscence parallel to the long axis extending almost the entire length of the ascocarp; and (ii) a **thyriothecium** (pl. **thyriothecia**), also known as **catathecium** or **catothecium** (Fig. 13.20E), is an inverted saucer-shaped ascocarp having wall more or less radial in structure. The hysterothecium is often considered to be an intermediate form between an apothecium and a perithecium.

Again, there are ascocarps that are developed as **locules** or cavities, especially ascigerous cavities, in a stroma where asci are directly borne. These locules are also designated as **pseudothecia** (sing. **pseudothecium**) or **pseudoperithecia** or as **ascostromata** (sing. **ascostroma**) (Fig. 13.20D). A pseudothecium may be uni- or multilocular in which asci are borne in the base of the locule. The asci may or may not be mixed with hyphal structures having no free ends are known as **pseudoparaphyses** (sing. **pseudoparaphysis**), also designated as

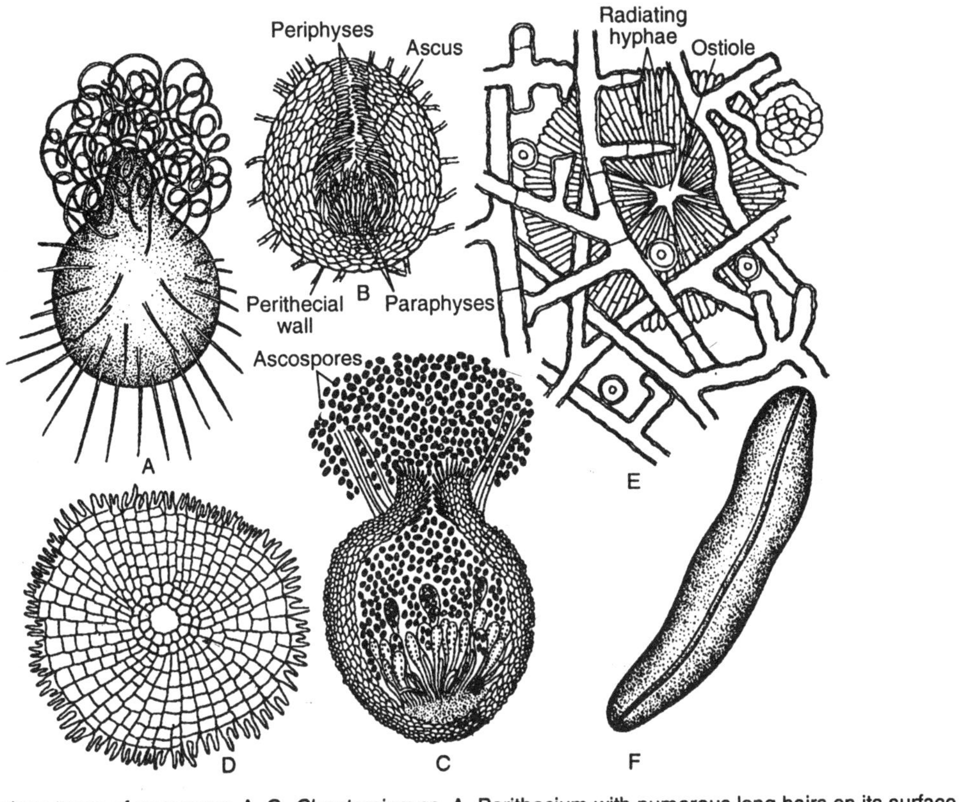

Fig. 13.20: Various types of ascocarps. A–C. *Chaetomium* sp. A. Perithecium with numerous long hairs on its surface. B. Section through the perithecium showing structural details. C. Section through the perithecium showing asci and dispersal of ascospores. D. *Microthyrium microascopicum*. Shield-shaped, dimidiate ascostroma with an opening. E. *Asterina* sp. Thyriothecium composed of radiating hyphae. F. *Glonium* sp. Boat-shaped ascocarp (hysterothecium) with a long slit-like opening parallel to the long axis and extending almost through the entire length of the fruit body

paraphysoids. The cavity of a pseudothecium may or may not be developed before the formation of asci and paraphysoids. In a pseudothecium, there is no special wall similar to what is present in a perithecium. Only a cavity is present within the stroma in which asci are located. The development of a regular ostiole is absent in a pseudothecium, instead a **pore** or **slit** is formed at a point above the asci where the stromatic tissue is dissolved. Through this pore the ascospores escape.

Besides the above forms, the fructifications may be **hypogean** and closed, as in the genus *Tuber* (Fig. 13.21A). Ascospores are liberated by the decay or by the breaking of the fructifications by animals.

The development of fructifications in the Ascomycotina, up to the present time, has been closely studied in a comparatively few species. In general, the process is initiated by the stimulus produced during sexual reproduction. In some members of the Ascomycotina the ascocarp tissues are differentiated around the sex organs after plasmogamy, while in others the ascocarp tissues form first, forming a stroma within which the sex organs appear later. In a fructification the hymenium and sterile tissue originate from different hyphal elements which in some cases, may be very well differentiated from the very inception or at best in a very early stage of frutification, whereas in others they are not clearly distinguishable. Usually the development of hymenium and the sterile tissue is simultaneous. Immediately after plasmogamy, the ascogonium gives rise to structures known as **ascogenous hyphae** from which asci are developed. The ascogenous hyphae are generally broader than the associated vegetative hyphae and have thinner walls, and dense easily stained contents. They grow out radially to form a plate or hollow disk in the lower part of the fructification, but they may also ramify throughout the central tissue of the ascocarp. Along with this process, profuse growth of sterile hyphae takes place from the adjacent hyphal cells. In a mature fructification the elements producing asci and the sterile hyphae become so intimately united and interwoven with one another that it is often difficult to separate or distinguish them. But most fructifications differentiate into two, and in more highly developed ones three, hyphal systems: (i) tissues derived from the ascogonium comprising of ascogenous hyphae and asci; (ii) vegetative protective tissue developed from the surrounding mycelium; and (iii) secondary protective tissue formed under stimulus from the ascogonium. All the tissues and structures enclosed by the peridium or other boundary layer of the ascocarp are known collectively as the **centrum.** Thus the centrum includes sex organs, ascogenous hyphae, asci and sterile tissues. The structural details and development

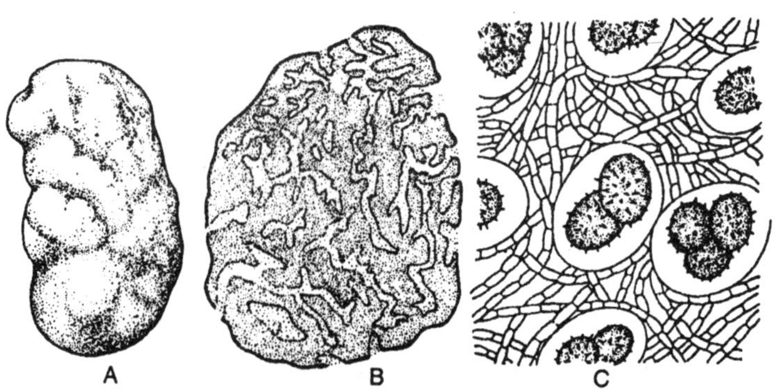

Fig. 13.21: Nature of ascocarp of *Tuber* sp. A. An entire ascocarp. B. Section through an ascocarp. C. Asci containing ascospores

of three widely encountered forms of fructification are discussed separately.

I. **Cleistothecium.** A cleistothecium is a rounded, completely closed ascocarp which has no natural opening but brusts irregularly at maturity, or along sutures in the peridium. The cleistothecial wall may (Fig. 13.21A) or may not (Fig. 13.21B) be provided with outgrowths on its external surface. These outgrowths are known as **appendages.** The appendages when present serve character of taxonomic interest. In a cleistothecium the hymenium may be distributed at different levels or at a particular area within the ascocarp. The number of asci in a cleistothecium may be one or more (Fig. 13.22F). Again, the asci may be persistent or evanescent. The number of asci in a cleistothecium and whether the asci are persistent or evanescent are also characters on the basis of which genera of certain fungi are separated. The mature ascospores are released from a cleistothecium by external influences which cause the rupture or decay of its wall.

In *Podosphaera,* the process of cleistothecium development is very simple. Two hyphal branches put out short protuberances at the same time and are delimited by septation (Fig. 13.22). The one which becomes elongated ellipsoid cell is the ascogonium, whereas the other which remains cylindrical in shape and narrow is the antheridial branch. The antheridial branch remains closely applied to the ascogonium and its upper extremity bends over and covers the apex of the ascogonium, and is delimited by a transverse wall, forming a short nearly isodiametric cell, the antheridium. After plasmogamy the ascogonium divides by transverse wall into two cells, an upper which becomes the solitary ascus and subsequently produces eight spores, and a lower which is a stalk cell (Fig. 13.22E) bears the ascus. The wall of the cleistothecium also begins to develop at the same time with the ascus. Tubular outgrowths (Fig. 13.22B) appear close round the base of the ascogonium on the hyphae which bear it and the antheridial branch. These tubular outgrowths grow up round the ascogonium and in close contact with it and in close lateral contact with one another and with the antheridial branch, till they all meet together above the apex of the ascogonium. Each one of the tubular outgrowths then divides by one or two transverse walls, so that the ascogonium is surrounded by an envelope formed of a single layer of many cells (Figs. 13.22C and D). The

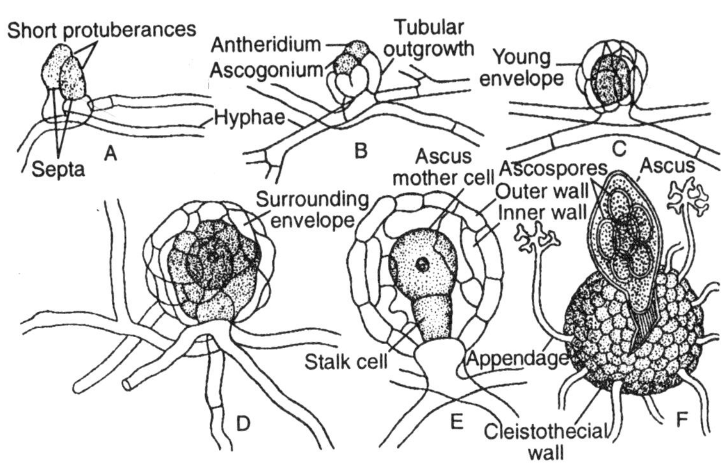

Fig. 13.22: Development of cleistothecium with ascus and ascospores in *Podosphaera* sp. A. Short protuberances cut off by septa. B. Antheridium and ascogonium surrounded by tubular outgrowths. C–D. Stages in development of surrounding envelope. E. Differentiation of ascus mother cell and stalk cell. F. Mature cleistothecium with appendages and ascus bearing ascospores

cells then increase in size in the surface-direction, their walls thicken and assume a dark-brown colour and they thus form the outer wall of the cleistothecium from the outer surface of which number of hyphae with delicate ramifications are developed which ultimately mature into appendages (Fig. 13.22F). Branches from the inner surface of the cells of the outer wall ramify and develop into a dense parenchyma-like weft resulting in the formation of the inner wall of the cleistothecium.

II. Apothecium. An Apothecium is a cup- or saucer-shaped ascocarp in which the hymenium remains wide open. The hymenium is uniformly continuous and is developed lining the wide open surface of the apothecium (Fig. 13.19D). It is made up of asci and paraphyses (Figs. 13.19E and F). The paraphyses may be as long as the asci, longer or somewhat shorter. They are extremely variable in shape and may be septate or aseptate. The tips of the paraphyses are usually unbranched. In cases where the tips are branched, the tips of the branches may unite above the asci forming a layer known as **epithecium** (pl. **epithecia**). The rest of the tissue of the apothecium is composed of interwoven hyphae. The thin layer of interwoven hyphae immediately below the hymenium is the **hypothecium** (Figs. 13.19D, E and 13.23) whose hyphal elements are rather less dense than the rest of the tissue of the apothecium. The fleshy part of the apothecium which supports the hypothecium and the hymenium is the **excipulum** (Fig. 13.23). The excipulum consists of two parts: (i) The **ectal excipulum** is the outer layer of the apothecium, and (ii) the **medullary excipulum** forms the inner portion (Figs. 13.19D and 13.23).

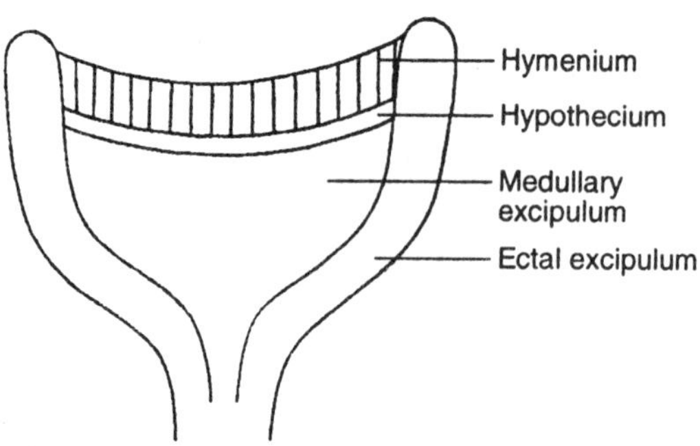

Fig. 13.23: Diagram of an apothecium in section showing arrangements of tissues

Korf (1958) defined in details the types of tissue found in the excipulum (Fig. 13.23). According to him, the ectal excipulum is short-celled, in whch the individual hyphae are not recognizable; and the medullary excipulum is with long cells, in which the component hyphae are visible (Fig. 13.19D). The short-celled tissue he subdivided into three types as follows: (a) cells globose with intercellular spaces—**textura globulosa;** (b) cells polyhedral by mutual pressure, no intercellular spaces—**textura angularis;** (c) cells rectangular in section—**textura prismatica** (Fig. 13.24). The long-celled tissue he divided into four subgroups. Long-celled tissue of medullary excipulum may have: (i) hyphae running in all directions: (a) hyphal walls not united, usually with distinct interhyphal spaces—**textura intricata;** (b) hyphal walls united, with distinct interhyphal spaces—**textura epidermoidea;** and (ii) hyphae are more or less parallel: (c) hyphae with strongly thickened walls, cohering—**textura oblita;** (d) hyphae without thickened walls, not cohering—**textura porrecta** (Fig. 13.24).

Since the hymenium is wide open, the discharge of ascospores from the asci is rather simple and often clouds of spores are seen on the surface of mature apothecium.

The apothecia are white or brilliantly coloured red, yellow, or orange. Some may be brown. A few are black. The apothecia may be **stipitate** or **non-stipitate,** that is, with or without any stalk or **stipe** respectively. Besides the typical cups or saucers, there are various other forms of apothecia: (i) apothecium with a thick stalk and a cap or **pileus** on which the pitted or ridged hymenium looks like a sponge (Fig. 13.19G); (ii) apothecium having heavily ridged short or long stalk and the pileus may be quite large being saddle-shaped (Fig. 13.19H) or convoluted and brain-like; (iii) apothecium may be tongue-, club-, or fan-shaped with long stalk (Figs. 13.53A to D); (iv) flat, circular, black, tar-like stroma which bear radiate wrinkles under which apothecia are formed (Fig. 13.19I).

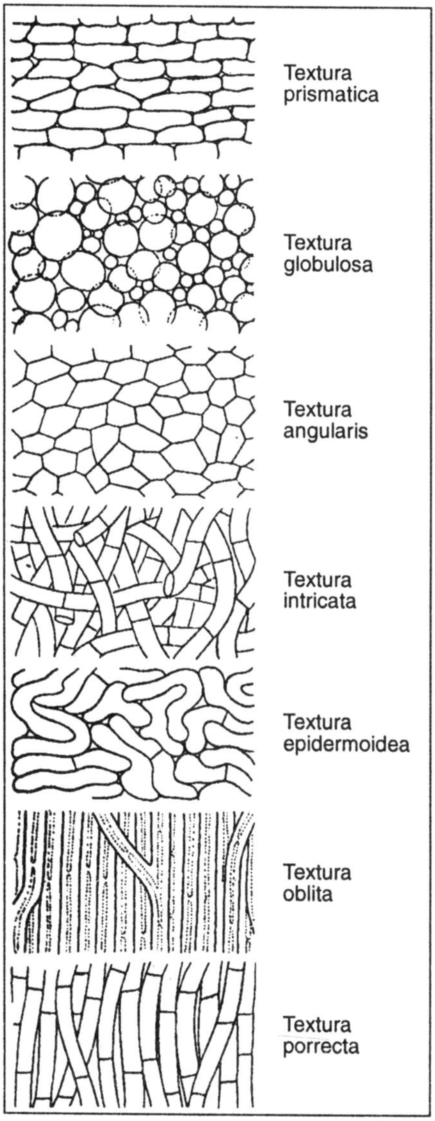

Fig. 13.24: Types of tissue in the excipulum

The development of apothecium in *Pyronema confluens* was imperfectly described by De Bary for the first time in 1863. Tulasne also added some information after De Bary. Later on Kihlmann studied in further details. The details of gametangial contact, plasmogamy, development of dikaryons and ascogenous hyphae, crozier formation, karyogamy, meiosis, and formation of asci and ascospores have already been included under the development of asci and ascospores (Figs. 13.6 and 13.9). Simultaneous with all these processes from gametangial contact to the development of ascospores, the vegetative cells at the base of the ascogonium and antheridium put out a growth of vegetative mycelium which surrounds the ascogenous hyphae and ultimately the asci (Fig. 13.25). The two elements form a compact structure to constitute the apothecium.

Corner (1929) defined three types of development of apothecia: (i) **angiocarpic** development, in which the tissues of the excipulum initially grow over and protect the developing hymenial layer; (ii) **gymnocarpic** development, in which the hymenium is exposed throughout its development; and (iii) **hemiangiocarpic** development, an intermediate condition in which the hymenium is partially covered. However, Corner did not attribute any significance to these types of development.

III. Perithecium. The perithecia vary in shape from globose to pear-shaped or elongated to flask-shaped with long or short neck. They may be associated with stroma and are known as **stromatic perithecia** (Fig. 13.60C) or free of any stromatic connection, **non-stromtic perithecia** (Fig. 13.63B). Again, the stroma of stromatic perithecia may be only of fungal hyphae or it may be composed of a mixture of fungal hyphae and host tissue. The stroma may not necessarily surround the perithecia, but hyphae and host tissue. The stroma may not necessarily surround the

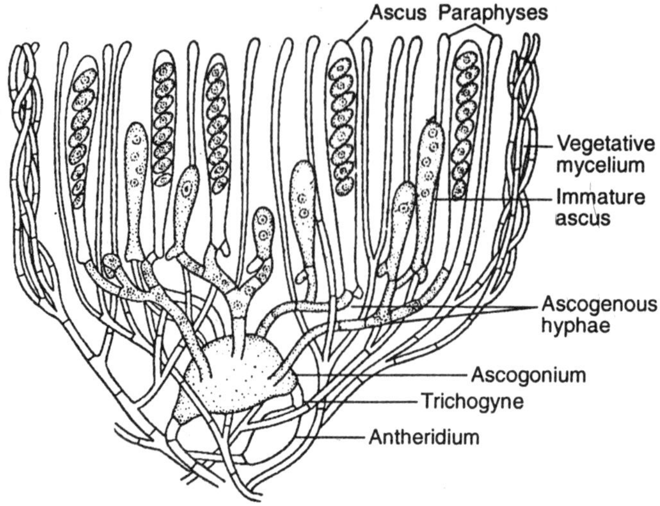

Fig. 13.25: *Pyronema confluens*. Diagrammatic representation of the development of different parts of the apothecium

perithecia, but may instead be reduced to a shield-like **clypeus** around the ostiole, or to a **subiculum** of stromatic interwoven hyphae below the perithecium. A perithecium is furnished in the full-grown state with a schizogenous pore, ostiole through which asci or ascospores are discharged (Figs. 13.20C and E, 13.63B). It is a conspicuous feature of the perithecium. The ostiolar canal is lined by short hair-like growths known as **periphyses** (sing. **periphysis**) (Figs. 13.20B and C, 13.63B). A perithecium, stromatic or non-stromatic has a definite wall of its own with a true ostiole. In a stromatic perithecium, the differentiation of wall may not be sometimes very distinct. The perithecial wall is usually formed of a dense hyphal weft or pseudoparenchyma (Fig. 13.20B). The outer surface of a non-stromatic perithecium may (Fig. 13.20A) or may not (Fig. 13.63A and B) bear numerous long hairs which when present have taxonomic importance. In a perithecium, the hymenium either occupies a narrow bit of surface opposite the ostiole (basal portion of the inner surface of the wall) on which asci grow as a small tuft parallel and eret towards the ostiole, or is spread over the entire inner surface of the wall, and the asci then converge radially towards the middle line of the perithecium. The asci are inserted in the delicate tissue lying inside the perithecium. They fill the inner space of the perithecium or at least the largest part of it, excepting the neck. All the space not occupied by them is filled with branches of the hyphae which grow out from the inner layer of the wall. Some of these branches which are basally attached to hymenium and lie between the asci are termed **paraphyses** (sing. **paraphysis**) (Fig. 13.20B).

Depending on fungi, the asci of perithecium may be evanescent or persistent. In case of evanescent asci the ascus wall with maturity of asci breaks down to form a mass of gelatinous material in which lie embedded the ascospores. This mass of gelatinous material with ascospores is discharged from the perithecium by hydrostatic pressure and remains at the mouth of the ostiole. The ascospores are then transferred to distant places by the visiting insects which are attracted by the gelatinous material, whereas in case of perithecia bearing persistent asci, the ascospores when mature are forcibly expelled from them. The ascospores borne in the perithecial ascocarp are one to more than one-celled and are extremely variable in shape and size.

The perithecium of *Venturia inaequalis* is initiated by the development of a coiled hyphae (Figs. 13.26A and B) arising within a stroma. The hyphal cells at the periphery of the stroma are uninucleate, and their walls become thickened, whereas the cells of the inner hyphae remain thin-walled and multinucleate. One of these thin-walled cells produces a chain of cells, the ascogonium, each cell of which is bi- to quadri-nucleate. The apical cell of the ascogonium becomes clavate giving rise to a well-defined trichogyne (Fig. 13.26D). Meanwhile, near the developing ascogonium another hyphal tip thickens and

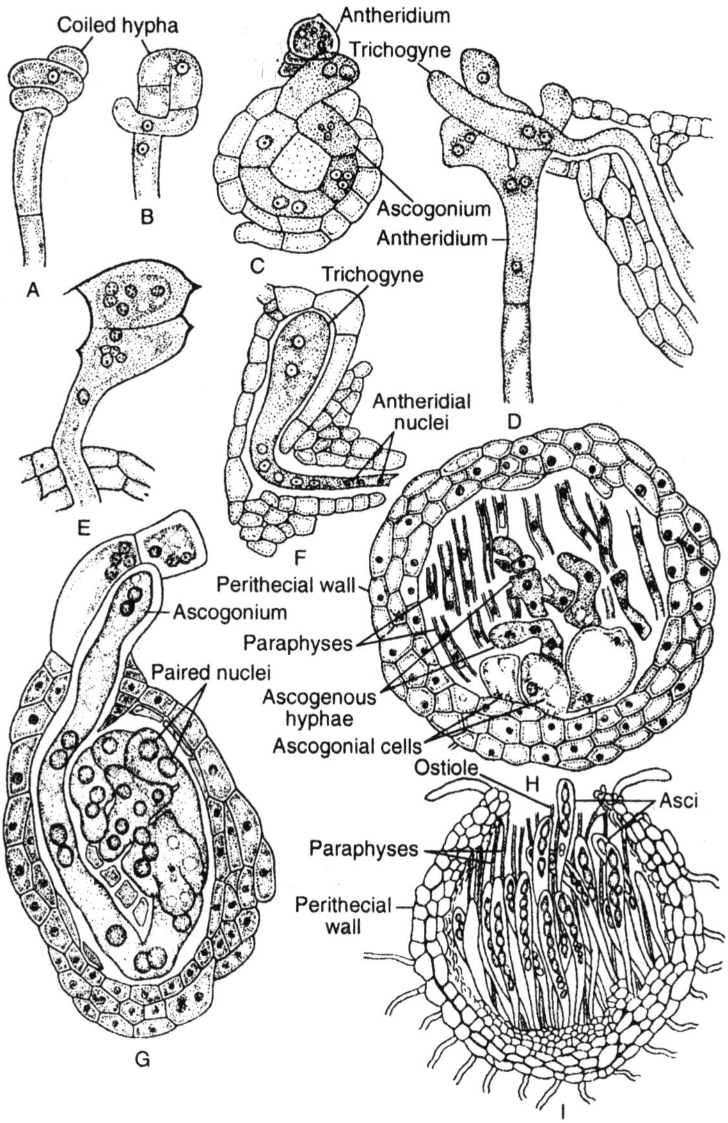

Fig. 13.26: Development of perithecium in *Venturia inaequalis*. A–B. Stages of coiled hypha development. C. Antheridium and ascogonium developed in stroma. D. The antheridium is in intimate association with the trichogyne. E. Plasmogamy. F. Trichogyne after nuclear migration, the antheridial nuclei are already at the lower end of the trichogyne. G. Ascogonium with paired nuclei. H. Section through perithecium with ascogenous hyphae and paraphyses. I. Section of a mature perithecium showing the nature of perithecial wall, asci with ascospores and paraphyses

becomes lobate, and multinucleate. This is the antheridium (Fig. 13.26D). Its growth continues until the lobes contact the trichogyne and become closely applied to it (Fig. 13.26C). A pore in the adjoining walls then forms, and the antheridial content empties into the trichogyne. Septa in the ascogonial chain are then dissolved, whereupon the antheridial nuclei migrate to the ascogonium and become associated in pairs with the ascogonial nuclei (Figs. 13.26E to G). Immediately after this, ascogenous hyphae arise as outgrowths from the ascogonium. Croziers are formed at the tips of the ascogenous hyphae (Fig. 13.26H). A binucleate penultimate cell forms, and the two nuclei fuse promptly to form the diploid stage. The ascus elongates, and three successive divisions of the diploid nucleus occur producing 8 haploid nuclei, the first of which is the reduction division. The cytoplasm is delimited around each of the eight nuclei, and eight uninucleate haploid ascospores are formed. Meanwhile, the uninucleate cells of the hyphae multiply to form the wall of the developing perithecium and the nurse tissue for the developing asci and paraphyses. Ultimately a perithecium with well-developed perithecial wall bearing ostiole and containing asci and paraphyses is developed (Fig. 13.26I).

LIFE CYCLE PATTERN

Despite their diversity in many features, the Ascomycotina possess certain common unifying characteristics, namely, the somatic body composed of a loose, indefinite mass of septate mycelium; the mode of asexual reproduction; and sexual reproduction. In general, they have two reproductive phases; the asexual stage and the sexual stage. Besides sexual process, some members of this class reproduce by budding and others by the development of spores, the conidia. The conidia are produced profusely either directly from the somatic mycelium or from conidiophores after the organism has grown vigorously for quite some time. They are dispersed by external agencies, like, wind, insects, and simialr others. Under favourable conditions each conidium germinates by germ tube which ultimately grows into somatic mycelium of the new individual. This is how the fungus reproduces asexually. During sexual reproduction the changes involved in the process occur in regular sequence in cyclic order. The stage during which a fungus reproduces asexually is known as **asexual stage** or **asexual cycle** or **conidial stage** or **imperfect stage.**

Under favourable climatic conditions the asexual stage may be repeated resulting in the production of conidia in profuse quantities. This is a means how a large number of fungi of this group propagate very quickly and disseminate themselves in distant localities. The conidia are very much susceptible to the alternation of climatic conditions. In some cases, however, the conidia can stand unfavourable conditions and perennate. In course of development, the same mycelium from which conidiophores and conidia were being developed now takes part in the formation of gametangia. Gametangial union is followed by plasmogamy. Ultimately asci are formed, within which ascospores are produced by karyogamy and meiosis. The haplophase and diplophase are intervened by dikaryophase which may be short or long in duration. Depending on the organism, the asci may or may not be enclose in an ascocarp. Members of the Ascomycetes may be homothallic or heterothallic. In a homothallic species asci are formed in the same thallus, whereas in heterothallic species asci are developed by the interaction of two compatible thalli. The basis for heterothallism is typically a single gene with two alleles 'A' and 'a' and because of segregation during the meiotic division which precedes ascospores formation, the 8 ascospores normally present in an ascus will

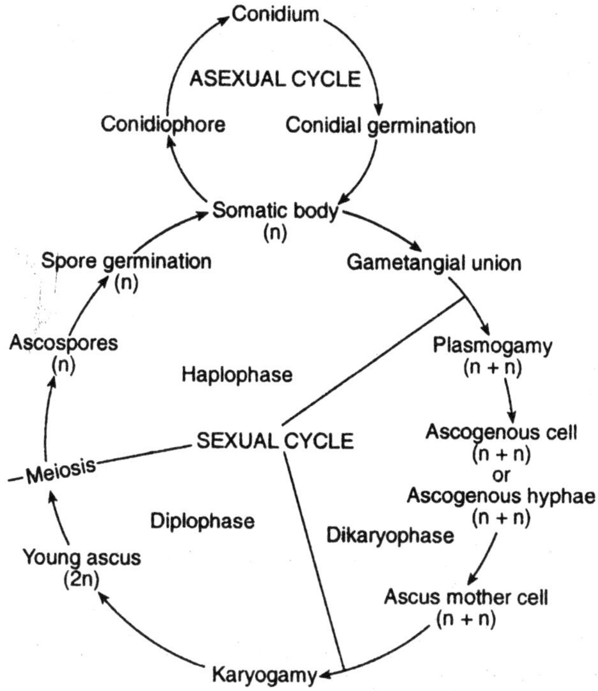

Fig. 13.27: Life cycle pattern in the Ascomycotina

include 4 of one mating type and 4 of the other. In certain 4-spored ascomycetous fungi, e.g., *Neurospora tetrasperma* and *Podospora anserina,* the ascospores are binucleate, and commonly contain nuclei of both mating types. Such ascospores on germination would give rise to fully fertile mycelia, and it would appear that the fungi are homothallic. Occasionally uninucleate ascospores are also formed which on germination do not produce fertile myceli and ascocarps are formed by interaction between mating types. Since basically these fungi are heterothallic in their mating behaviour, the term homothallism is used to describe the behaviour of their binucleate ascospores. The ascospores are the organs through which most of the fungi of this group perennate. The germination of ascospores takes place under favourable conditions by the production of one or more germ tubes which grow into somatic mycelium. The stage during which asci and ascospores are developed is known as **sexual stage** or **sexual cycle** or **ascigerous stage** or **perfect stage.** A large number of Ascomycotina have both conidial and ascigerous stages by means of which they propagate and disseminate themselves. But many Ascomycotina are known only in the ascigerous stage, their asexual stages either do not exist or have not been found. Usually both asexual and sexual stages are completed either parasitically or saprophytically depending on whether an organism is a parasite or a saprophyte. But in many of the parasitic ascomytenous fungi the asexual stage is completed on the living tissue of the host and the sexual stage is formed only on the dead host tissue. Life cycle pattern in the Ascomycotina is presented in Fig. 13.27.

CLASSIFICATION

The classification of the Subdivision Ascomycotina constitutes perhaps the most difficult taxonomic problem in mycology.

The large number of species and the morphological complexities and variations in these fungi have often led to an understandable hesitancy on the part of the majority of mycologists to accept revisions of the group which attempt to incorporate the data that have been obtained into the system of classification.

The system of Lindau presented in the Naturlichen Pflanzenfamilien in 1897, with various modifications is still in general use today despite the fact that the knowledge of the morphology, cytology and genetics of the fungi of this group has developed since its publication.

In the widely accepted age-old classification system, the Ascomycotina were subdivided into three major groups: **Plectomycetes, Pyrenomycetes,** and **Discomycetes,** based on the gross morphological structure of the ascocarp being **cleistothecium, perithecium,** and **apothecium** respectively.

This above classification system has, however, been modified by different mycologists. One of such modified classification systems is presented below. Here the Ascomycotina (designated as Class Ascomycetes) are classified into two subclasses which are again subdivided into orders and families. In one of the subclasses the taxonomic category **series** is introduced. The names—form-genera *Aspergillus* and *Penicillium* are introduced for convenience, instead of introducing the names of their perfect stages.

AA. Asci are naked and are formed singly or in loose clusters, absence of ascogenous hyphae or ascocarp

Subclass: **Protoascomycetes**

B. Assimilatory phase uninucleate cells or mycelioid, ascus formed directly from zygote which is the product of copulation of two cells or part of hyphae

Order: **Endomycetales**

C. Asci bearing more than eight ascospores

Family: **Ascoideaceae**

CC. Asci bearing one to eight ascospores

D. Mycelium scanty or lacking

Family: **Saccharomycetaceae**

Genus: *Saccharomyces*

DD. Mycelium abundant

BB. Assimilatory phase mycelioid, asci arising from ascogenous cells derived from the binucleate cells of septate mycelium

Order: **Taphrinales**

Family: **Taphrinaceae**

Genus: *Typhrina*

AA. Asci produced from ascogenous hyphae and enclosed in well-developed ascocarps

Subclass: **Euascomycetes**

B. Asci scattered at various levels within the ascocarp, ascocarp lacking any opening—a cleistothecium

Series: **Plectomycetes**

C. Interior of ascocarp filled with evanescent asci, ascocarp without appendages.

Order: **Eurotiles**

Family: **Eurotiaceae**

Form—Genus *Aspergillus*

Form—Genus *Penicillium*

BB. Ascocarp wide open—an apothecium or a modified form thereof

Series: **Discomycetes**

C. Ascocarps above ground

D. Asci operculate or suboperculate

Order: **Pezizales**

E. Receptacle stalked, ascospores elliptical

Family: **Morchellaceae**

Genus: *Morchella*

EE. Receptacle usually sessile, cup-shaped or button-shaped

F. Asci narrowly cylindrical; turn blue, at least at the tip with Melzer's reagent; asci not protruding beyond the general level of the hymenium at maturity; ascospores uniseriate

Family: **Pezizaceae**

Genus: *Peziza*

FF. Asci broad, not blued at the tip by Melzer's reagent, asci protruding beyond the general level of the hymenium at maturity, ascospores lie in two to three irregular rows.

Family: **Ascobolaceae**

Genus: *Ascobolus*

DD. Asci inoperculate

Order: **Helotiales**

Family: **Geoglossaceae**

Genus: *Geoglossum*

CC. Ascocarps below the ground

BBB. Ascocarp with differentiated wall and with or without a regular ostiole, asci arising as a fascicle at a common level in the ascocarp

Series: **Pyrenomycetes**

C. Ascocarp without an ostiole

Order: **Erysiphales**

Family: **Erysiphaceae**

Genus: *Erysiphe*

CC. Ascocarp with an ostiole—a perithecium

D. Ascospores thread-like

Order: **Clavicipitales**

Family: **Clavicipitaceae**

Genus: *Claviceps*

DD. Ascospors not thread-like

E. Ascocarp borne singly or in a stroma, dark membranous or carbonous.

Order: **Saphaeriales**

F. Perithecia free, ostiole without long hairs

Family: **Sordariaceae**

Genus: *Neurospora*

FF. Perithecia embedded in stroma, stroma free from substratum

Family: **Xylariaceae**

Genus: *Xylaria*

BBBB. Ascocarp with differentiated wall lacking, asci borne in stromatic locules

Series: **Loculoascomycetes**

Luttrell in 1955 put more emphasis on the structure of asci—**unitunicate** or **bitunicate**, and combined this character with that of ascocarp and similar other ascus-bearing structure. He recognized two subclasses: **Euascomycetes** and **Loculoascomycetes.**

In recent years there has been a tendency to follow the line of approach indicated by Luttrell. The main subdivisions of this group of fungi are based on the structure of the asci themselves and of the ascocarps within or upon which they assemble. The primary division depends on the nature of the ascus wall, which may be composed either of one or two layers, i.e., ascus may be **unitunicate** or **bitunicate.** The Ascomycotina (designated as Class Ascomycetes) may thus be classified as follows:

A. Asci unitunicate, arranged in hymenium borne in an ascocarp or otherwise

Subclass: **Euascomycetes**

B. Asci produced singly

C. Developed from thick-walled chlamydospores produced within the tissue of vascular plants.

Order: **Protomycetales**

CC. Developed directly from gametangial copulation

Order: **Endomycetales**

Family: **Endomycetaceae (Saccharomycetaceae)**

Genus: *Saccharomyces*

BB. Asci in continuous layer but not in ascocarp

Order: **Taphrinales**

Family: **Taphrinaceae**

Genus: *Taphirna*

BBB. Asci in hymenium of an ascocarp

C. Asci operculate

Order: **Pezizales**

D. Receptacle stalked, ascospores elliptical

Family: **Morchellaceae**

Genus: *Morchella*

DD. Receptacle usually sessile, cup-shaped or buttom-shaped

E. Asci narrowly cylindrical; turn blue, at least the tip with Melzer's reagent; asci not protruding beyond the general level of the hymenium at maturity; ascospores uniseriate

Family: **Pezizaceae**

Genus: *Peziza*

EE. Asci broad, not blued at the tip by Melzer's reagent, asci protruding beyond the general level of the hymenium at maturity, ascospores lie in two to three irregular rows

Family: **Ascobolaceae**

Genus: *Ascobolus*

CC. Asci inoperculate

D. Asci indehiscent or at least without a definite apical dehiscence mechanism

E. Hymenium continuous, ascocarps subterranean

Order: **Tuberales**

EE. Hymenium not continuous

Order: **Plectascales**

F. Ascocarp more or less globose, without ostiole, opening by splitting

G. Saprophytes, mostly

H. Ascocarp sessile

Ascocarp globose or ellipsoidal, minute fungi on plant debris above ground

Family: **Eurotiaceae**

Form–Genus *Aspergillus*

Form–Genus *Penicillium*

HH. Ascocarp stalked

GG. Parasites, especially on leaves and other green parts of flowering plants

H. Mycelium without hyphopodia, conidia abundant

Family: **Erysiphaceae**

Genus: *Erysiphe*

HH. Mycelium bearing hyphopodia, conidia absent

Family: **Meliolaceae**

Genus: *Meliola*

FF. Ascocarp with a long beak, asci diffluent and ascospores emerging in a drop of mucilage at the apex of the beak

DD. Asci with an apical pore

E. Asci very long, narrowly cylindrical with thread-like ascospores

F. Asci in perithecia borne in stroma

Order: **Clavicipitales**

Family: **Clavicipitaceae**

Genus: Claviceps

FF. Asci in perithecia without stroma

EE. Asci not cylindrical, ascospores narrow but not thread-like

F. Asci in apothecia

Order: **Helotiales**

Family: **Geoglossaceae**

Genus: *Geoglossum*

FF. Asci in perithecia

G. Paraphyses usually present, perithecia with or without stroma

Order: **Saphaeriales**

H. Perithecia free, ostiole without long hairs

Family: **Sordariaceae**

Genus: *Neurospora*

HH. Perithecia embedded in stroma, stroma free from substratum

Family: **Xylariaceae**

Genus: *Xylaria*

AA. Asci bitunicate, borne in locule

Subclass: **Loculoascomycetes**

Talbot (1969) and Webester (1970) categorized formerly designated class Ascomycetes as Subdivision Ascomycotina. The chief distinguishing features of the classes of Ascomycotina as outlined by them are shown below. The classification adopted by them is a compromise between older systems in which the nature of the ascocarp was emphasized to the virtual exclusion of other characters, and newer ones in which microscopic characters are prominent.

Asci unituncate, or if bitunicate then in an exposed hymenium of an apothecium

Asci naked, i.e., formed as discrete free cells or in a hymenium of indefinite extent, not bounded by a stroma or by ascocarp tissue; asci indehiscent

Class: **Hemiascomycetes**

Order: **Endomycetales**

Asci arising from a common cell formed by gametangial conjugation

Order: **Taphrinales**

Asci arsing from binucleate ascogenous cells of the mycolium

Asci formed in ascocarps

Asci scattered at various levels within a cleistothecium or a beaked perithecium; asci indehiscent

Class: **Plectomycetes**

Order: **Eurotiales**

Asci persistent to evanescent formed at various levels in the clustothecial ascocarp lacking paraphyses but containing ascogenous hyphae

Asci forming a hymenium or arising as a fascicle at a common level in the ascocarp, or rarely single

Ascocarp an apothecium or its hypogean derivative; asci operculate, inoperculate or indehiscent

Class: **Discomycetes**

Order: **Pezizales**

Asci opercualte or suboperculate, apothecial ascocarp epigean

Order: **Helotiales**

Asci inoperculate, dehiscing by an apical pore or slit, apothecial ascocarp epigean

Ascocarp usually a perithecium, less often a cleistothecium with fasciculate asci or an ascostroma with unitunicate asci; asci inoperculate, with an apical pore or slit; not minute external parasites of insects and arachnida

Class: **Pyrenomycetes**

Order: **Erysiphales**

Asci persistent forming a single layer or tuft within a cleistothecium, develop at the expense of pseudoparenchymatous cells of the centrum and eventually fill the cleistothecial cavity

Order: **Clavicipitales**

Asci long cylindrical persistent bearing thread-like ascospores which escape perforating the thick ascus cap, paraphyses are formed on the lateral walls of the perithecium with a well-developed stroma

Order: **Sphaeriales**

Asci persistent, perithecial wall with or without discernible stroma, paraphyses persistent to evanescent, perithecium with conspicuous ostiole

Ascocarp a perithecium with inoperculate asci whose walls soon disintegrate; minute external parasites of insects and arachnida

Class: **Laboulbeniomycetes**

Asci bitunicate, formed in an ascostroma but not in an apothecium

Class: **Loculoascomycetes**

PHYLOGENY

There are two widely divergent lines of thought regarding the phylogeny of the Ascomycotina. The one proposed by Sachs in 1875 suggests Floridean ancestry of the Ascomycotina. Sachs pointed out similarities between the Ascomycotina and the red algae in the following aspects:

1. Spermatia produced in both the groups perform identical function.
2. The development of receptive organ called trichogyne by the female reproductive structure.
3. Elaborate sexual process in both the groups.
4. Ascocarp and cystocarp are identical organs.
5. Continuity of cell to cell by cytoplasmic connection in the thallus.
6. Growth in length of the somatic body of both the groups by the elongation and division of terminal cells.

This line of thought failed to consider the physiology of the two groups which is so different to suggest that one cannot evolve from the other. The resemblances are due to parallel development.

The other line of evolution proposed by De Bary in 1887, indicated the possibility of the derivation of the Ascomycotina from the lower fungi. The arguments in favour of his hypothesis are:

1. Asci developed directly from the zygote.
2. Evolution of conidium from sporangium.

These evidences are not strong enough to eliminate the possibility of a separate flagellate ancestry of the Ascomycotina.

COMPARISON BETWEEN THE LOWER FUNGI AND THE ASCOMYCOTINA

Habitat. Most lower fungi are aquatic and even the terrestrial ones have aquatic tendency. They are largely saprophytic than parasitic. Reverse is the case with the Ascomycotina where very few species are true aquatic. They are mostly parasites of varying degree.

Somatic Body. In the lower fungi it is very simple being composed of unicellular uninucleate, ill-developed mycelium to well-developed aseptate mycelium. In the Ascomycotina the mycelium in septate and may be developed to form plectenchymatous structures being differentiated into prosenchyma and pseudoparenchyma.

Asexual Reproduction. In a large number of asexual reproduction takes place by the development of flagellate spores which are borne in sporangia. But in the such spores are non-flagellate and are not developed in sporangia.

Sexual Reproduction. In many of the Ascomycotina sexual reproduction taken place by the formation of well-organized gametes developed in gametangia. But in the lower fungi there is usually absence of any such structure. Here the gametangia instead of bearing well-organized gametes contain nuclei embedded in a mass of cytoplasm. In the lower fungi karyogamy follows immediately after plasmogamy, whereas in the Ascomycotina there is interval between plasmogamy and karyogamy which is occupied by a phase of dikaryotic condition. Here sexual reproduction leads to the development of a fruiting structure composed of fertile and sterile tissue which is altogether absent in the lower fungi.

Order: **Endomycetales**

This order includes both saprophytic and parasitic forms. Only a few forms possess hyphae but a large number have unicellular plant body. In them sexual union may be either between two vegetative cells, two gametangia or two ascospores. Sexual fusion involves immediate plasmogamy followed by karyogamy leading to the formation of a zygote which directly develops into an ascus. These fungi bear asci singly and never in ascocarps. The fungi included in this order are considered by many as primitively simple possessing greatly reduced plant body.

Family: ***Saccharomycetaceae***

The Saccharomycetaceae are commonly known as the yeasts. Yeasts were first detected by Anton van Leeuwenhoek as early as 1680. They were compared with globular, oval bodies of albuminoids. After Pasteur (1875) had demonstrated that alcoholic fermentation results from the activity of yeasts in grap juice, these microorganisms, previously regarded as lifeless, attracted attention of botanists and were the subjects of morphological and systematic investigations. What are yeasts? In the botanical sense, yeasts are unicellular fungi of biochemical interest which when placed in sugar solutions decompose them into alcohol and carbon dioxide causing alcoholic fermentation. Yeasts may be: (i) fungi with special morphology, spherical or oval in shape, multiply by fission or budding and produce ascospores at certain stages in their life cycles—these are known as true yeasts; and (ii) fungi living normally with a mycelium are able to reproduce by budding of their spores to form cells which have the shapes of yeasts, the basidiospores of some Basidiomycetes (*Ustilago, Coprinus lagopus*) and ascospores of certain Ascomycotina (*Taphrina*) multiply by budding, and hyphae of *Mucor* when submerged in sugar solution dissociate into round structures, but in a similar manner as the yeasts—all these are yeast-like fungi. Physiologically true yeasts differ from the yeast-like fungi by their resistance to anaerobic conditions and high fermenting capacity. Again, some yeasts may be deprived of fermenting function. Yeasts are typically single-celled. Under certain conditions of the medium some yeasts may develop short hyphae. Yeasts are widespread in their distribution. Mostly the yeasts are saprophytic fungi. They are commonly found in organic substrates like soil, decaying vegetables, ripe fruits and grains, sugary exudates of trees, and in the nector of flowers. Yeasts are of considerable economic importance. They are capable of fermenting carbohydrate solutions producing ethyl alcohol and carbon dioxide. The beer, alcohol, and wine yeasts are included in the genus *Saccharomyces,* which has a number of species. *Saccharomyces cerevisiae* is by far the most important species being the one employed in beer-brewing and bread-making. Both the biological interest and the economic importance of yeast centre round its metabolism, which is closely connected with its enzyme content. The yeast cells absorb certain sugar solution which is decomposed within the protoplast forming ethyl alcohol and carbon dioxide. This change is brought about by an enzyme, **zymase**, which is a product of the protoplasm. The enzyme zymase was discovered by Buchnar (1900). It is not always necessary to add yeast to the solutions containing the fruit sugars since wild yeasts are present in nature on the surface of the fruits. Another enzyme present is invertase, which hydrolyses cane sugar into glucose and laevulose. Many other enzymes, too, are present, such as lipases and proteases, which split fats and proteins respectively, so that they may be absorbed in the yeast protoplasm. In bread-making, yeast plants are mixed with flour and water. The flour contains some sugar in addition to the starch. At a suitable temperature, the yeast cells multiply rapidly. Zymase is formed, which aids in the breaking down of the sugar into alcohol and carbon dioxide. The bubbles of carbon dioxide gas are prevented from escaping by the dough, which, however, is caused to rise by the expansion of the bubbles. At baking temperatures there is further expansion of the gas bubbles, and also a driving off alcohol and water. The carbon dioxide released in fermenting dough raises the bread in making its escape. Yeasts are employed to impart flavour to cacao beans. Yeast cells are rich in protein and certain strains afford considerable fat. They are, therefore, mingled with other foodstuffs to increase their food value. Yeasts are also important as a commercial source of vitamins. Fresh yeast cells are excellent source of vitamins B and G.

A few species of yeasts are parasites of higher plants. They occasionally induce diseases of tomato fruits and pods of beans. Several species of yeast are pathogenic to man, causing diseases, of which **blastomycosis,** or **torulosis** is common. They attack the central nervous system and skin of man. Often they also cause diseases of ears, liver, spleen, and lymph glands.

Genus: *Saccharomyces*

Somatic Structures. The yeasts are unicellular fungi. Cells may remain attached in short chains forming a **pseudomycelium,** but they do not produce true mycelium. The cells are extremely variable in shape being globose, oval, elongated, or rectangular. The yeast cells are very polymorphic and are capable of assuming different forms depending upon the medium

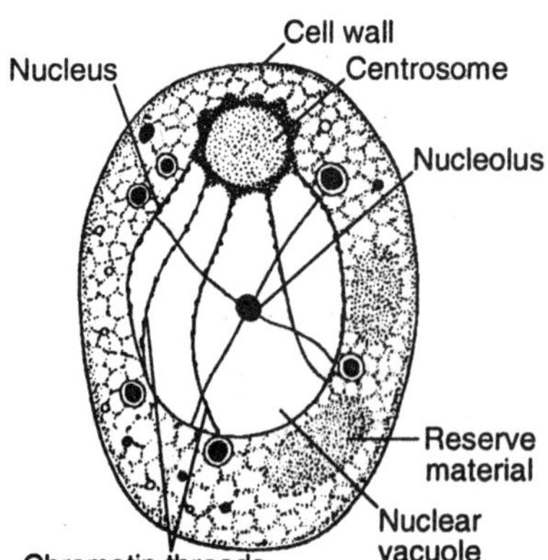

Fig. 13.28: Diagrammatic representation of parts of a yeast cell

in which they grow and their age. Individually yeast cells are hyaline but in colonies they appear white, cream-coloured or slightly brownish. The structure of yeast cell has been very thoroughly worked out by a large number of investigators who differ in their interpretations. Each yeast cell has a distinct cell wall enclosing granular cytoplasm, within which can be seen a large vacoule and a nucleus (Fig. 13.28). The vacuole varies much in size according to the state of activity of the cell. It may at times become much contracted, but it does not disappear completely except during spore formation. The wall is thin delicate and is composed of chitin in combination with other compounds. Reserve materials are present in the cytoplasm in the form of oil globules, glycogen, and volutin. Glycogen accumulation increases with the decrease in fermentation. The volutin content is also very much linked with the metabolic conditions of the yeast cells. Difference of opinion exists whether the vacuole is a part of the nucleus or the vacuole and the nucleus are separate entities. Despite the fact that *Saccharomyces cerevisiae* has been subject of cytological study for over 50 years, investigators at the present time are still not in agreement. According to old idea of Wager and Peniston the nucleus has a large vacuole and that the body to one side of it is the nucleolus. But Wager during later time described the nucleus as a homogeneous body having its nucleolus at one end. His explanation was supported by Lindegren, who stated that the central area of the yeast nucleus is not a vacuole but a true nucleus which is tranversed by chromatin threads and the body which Wager described as nucleolus, is **centrosome.** But Guillermond is of opinion, there is a central vacuole which does not belong to the nucleus. Apposed to the vacuole is the nucleus which according to Wager is the nucleolus and to Lindegren is centrosome. But great majority of investigators have concluded that the vacuole is a cytoplasmic constituent and that

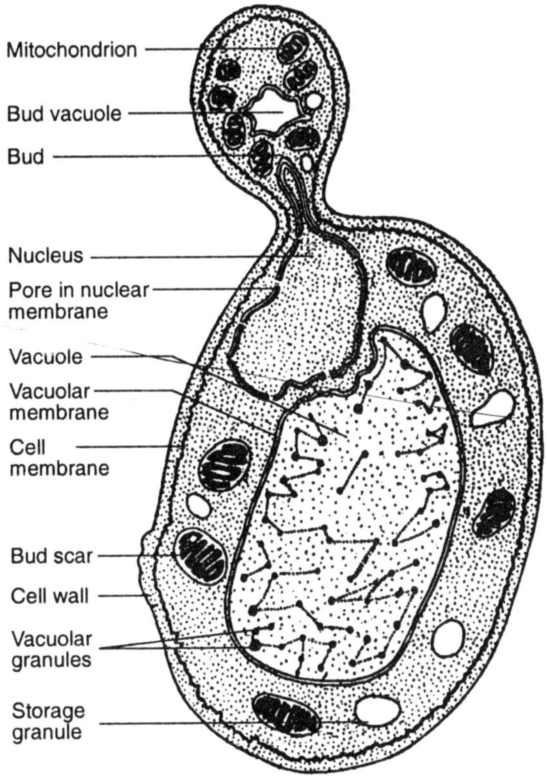

Fig. 13.29: *Saccharomyces cerevisiae.* Sectional view of a budding cell

the distinctive chromatic body referred to by Wager and Peniston is actually the nucleus.

The small size of the yeast cell renders observation of its contents under the light microscope difficult, and there has previously been controversy about the interpretation of its structure. Studies of thin sections of yeast cells (*Saccharomyces cerevisiae*) under the electron microscope have clarified our knowledge. The cell wall contains protein, lipid, and at least two polysaccharides (a mannan and a glucan). Chitin has also been reported. Immediately below the cell wall is the cytoplasmic membrane. The membrane is made up of particles penetrated by fibrils which probably correspond to the glucan fibrils of the cell wall. Within the cytoplasm are mitochondria, endoplasmic reticulum and reserves of fat and glycogen (Fig. 13.29). A large vacue surrounded by a single unit membrane contains strands and granules of dense material sometimes linked into a network. The larger granules are probably volutin, i.e., polymetaphosphate. The nucleus is surrounded by a double unit membrane (nuclear membrane) perforated by pores. The nuclear membrane is distinct from the vacuolar membrane which surrounds the vacuole.

The problem in *Saccharomyces* has been both to identify the nucleus and to describe its division. Nagel (1946), in a valuable and painstaking comparative study, employed a wide range of stains and fixatives as well as living cells and decided that the nucleus is Feulgen-positive, small body lying to the side of a central vacuole. This view was confirmed much later by electron microscopy (Moor and Muhlethaler, 1963). Moor and Muhlethaler

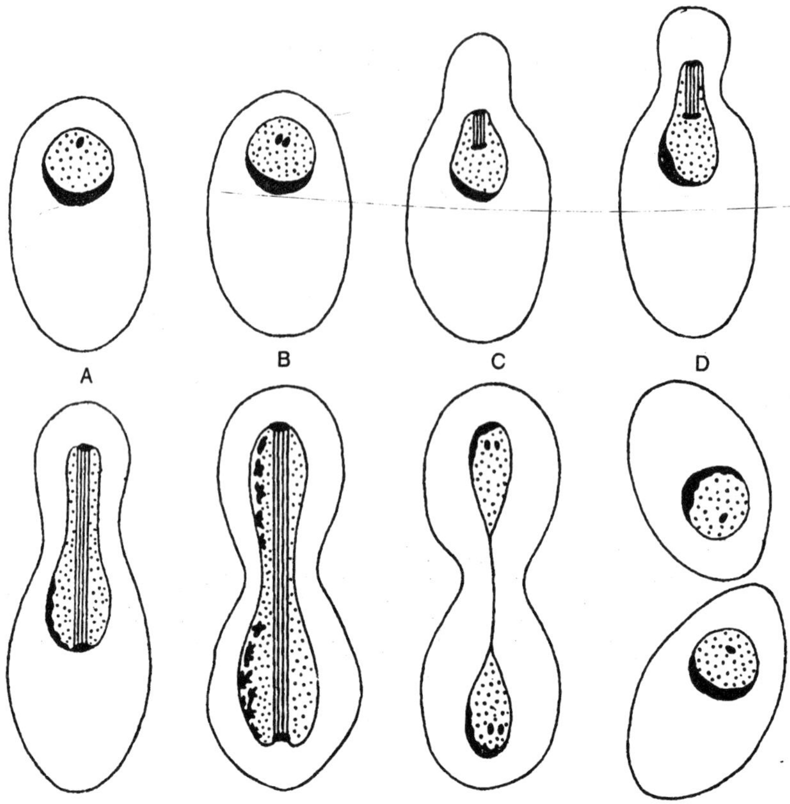

Fig. 13.30: Diagram showing mitotic division of nucleus in *Saccharomyces.* Dots represent Feulgen-positive part and continuous black indicates the nature and behaviour of basiphilic crescent

made it clear that there is a large, more or less central vacuole with which is closely associated a much smaller nucleus. There does not seem to be a structure comparable to the 'nucleolus' or 'central body' of other fungi. The nucleus is bipartite. The larger part is Feulgen-positive and during division becomes elongated, constricted in the centre and eventually separates into two halves (Fig. 13.30). Individual chromosomes are not usually recognizable. A second, crescentic, smaller part is strongly basiphilic but Feulgen- and Giemsa-negative (Figs. 13.30A-D). This body becomes attenuate, sometimes breaking into small, irregular pieces and eventually is more or less evenly divided between the daughter nuclei (Figs. 13.30E-H). Robinow and Marak (1966) provided evidence that there is a single, intranuclear fibre. Opposite the crescentic, basiphilic body within the Feulgen-positive material is a small bead-like structure (Fig. 13.30A). This bead of fibre intial, expands into a long, straight fibre which becomes anchored at each end to the nuclear membrane (Figs. 13.30C-F). It is, perhaps, not unreasonable to apply the terms nucleolus and centriole to the basiphilic crescent and fibre initial respectively. Other yeasts, notably *Lipomyces* (Robinow, 1961), differ from *Saccharomyces* in possessing distinct, countable chromosomes.

Recent investigtors, regardless of their diverse opinions on the mechanics of nuclear division and the number of chromosomes present, have agreed that the nucleus is the relatively small body and is distinct from the large vacuole in the yeast cell. The nucleus is sur-

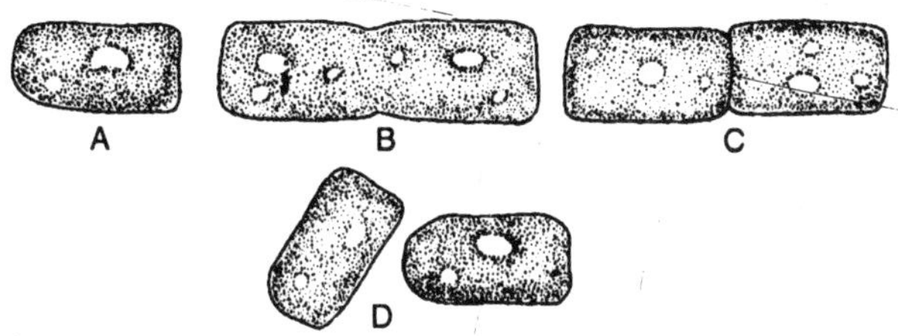

Fig. 13.31: *Schizosaccharomyces octosporous.* Stages in cell multiplication by fission

rounded by a membrane. The nuclear membrane remains intact during mitosis and the spindle is intranuclear as it is in all fungi. Mitotic division of the yeast nucleus was for the first time demonstrated by Guillermond (1901) and was subsequently persued by many other investigators. He also pointed out that the yeast cells have typical nuclei similar to other fungi.

Asexual Reproduction. Yeasts reproduce asexually either by fission or by budding. Depending on this character they are grouped as fission yeasts, *Schizosaccharomyces* and budding yeasts, *Zygosaccharomyces.*

During reproduction of fission yeasts the parent cell elongates (Figs. 13.31A and B), the nucleus divides into two daughter nuclei, and gradually a transverse partition wall is laid down somewhat near the middle starting from periphery to the centre dividing the mother cell into two daughter cells (Figs. 13.31C and D). The two daughter cells so formed may remain together for some time and begin to divide again or they may separate soon and then divide.

Budding yeasts are rather common than the fission yeasts. At the commencement of budding a small portion of the cell wall, usually near the end, softens. The protoplast of the mother cell, covered by a thin membrane, bulges out in the form of bud which ultimately develops into a daughter cell (Fig. 13.32A and B). Meanwhile, the nucleus of the mother cell, according to some, divides mitotically. One of the two daughter nuclei migrates into the enlarging bud (Figs. 13.32C and D). The bud grows until it attains the size of the mother cell. The daughter cell then becomes separated from the mother cell and process may be repeated indefinitely (Fig. 13.32E). Others suggest that when the yeast cell buds its nucleus appears to divide by constriction and the nuclear envelope does not break down. The cytoplasmic connection is closed by the lying down of wall material. Eventually the bud separates from the parent cell leaving a bud scar (Fig. 13.29). Quite often the daughter cell also starts producing bud before being abstricted from the mother cell and the process may be repeated giving rise to chains or groups of yeast cells. In this way a large number of buds are

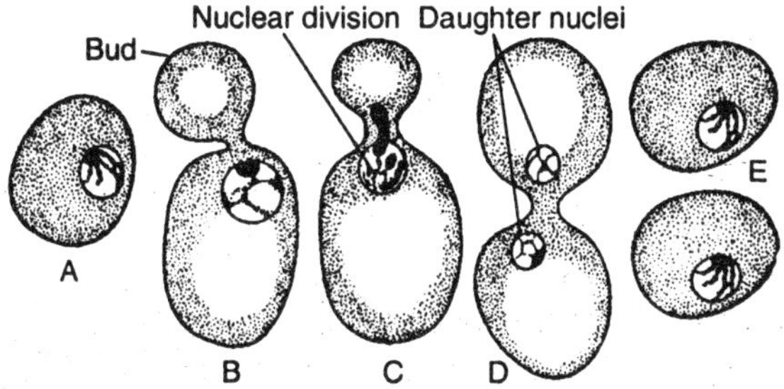

Fig. 13.32: *Schizosaccharomyces cerevisiae.* Stages in cell multiplication by budding

developed without being detached from one another resulting in the formation of branched or unbranched chains of cells constituting **pseudomycelium.** The cells in chains or pseudomycelium are loosely joined together. Sooner or later, however, the chains break into their constituent cells.

In some yeasts when the food in the surrounding medium is exhausted, or when there is danger of desiccation, spores are formed from the mother cells. In the process of spore formation the protoplasm divides, usually into four portions, each of which becomes surrounded with a comparatively thick wall. Thus four thick-walled spores known as **endospores** are formed. After some time the spores are liberated. By virtue of their thick walls the spores can remain alive under adverse conditions. On germination the spores bud and give rise to chains of cells. Welten (1914) challenged the view that starvation condition is essential for sporulation. Working with *Saccharomyces cerevisiae,* he found that on prune extract agar, yeasts sporulate well. Even colonies embedded in the agar sporulated as well as those on the surface.

Sexual Reproduction. It takes place by the union of two cells more often similar in size but sometimes they may be dissimilar in appearance, and by the development of short protuberances which unite to form a conjugation tube. This is followed by the dissolution of intervening walls and nuclear fusion which takes place in the conjugation tube. The subsequent stages are extremely variable and are discussed separately. The copulating pair of cells may be vegetative cells or ascospores. Often copulation occurs between a mother cell and its bud. This is known as **pedogamy** and is observed in *Zygosaccharomyces chevalieri.* Yeasts may be homothallic or heterothallic.

Sexual reproduction of yeasts was first clearly recognized by Guillermond (1901-1902). He demonstrated copulation of yeast

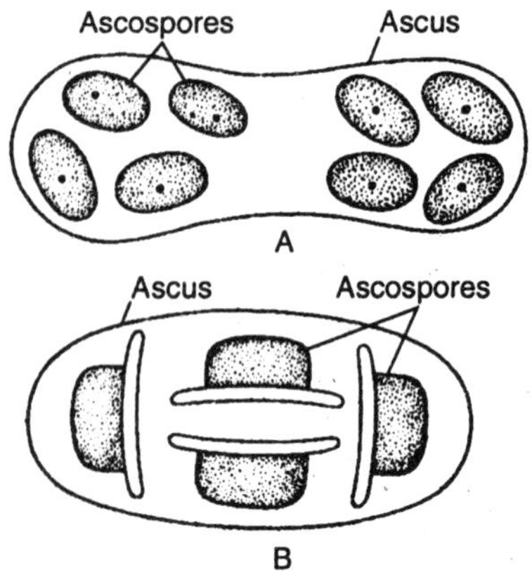

Fig. 13.33: Various types of yeast ascospores. A. *Schizosaccharomyces octosporous.* B. *Hansenula* sp.

nuclei and the subsequent stages leading to the ascospore formation. The number and shape of ascospores are variable (Fig. 13.33). In 1940 Guillermond showed that three life cycle patterns are distinguishable among yeasts. They are:

I. **Haplobiontic life cycle.** This is exhibited by *Schizosaccharomyces octosporus* which is homothallic. Here the haploid stage (haplophase) is very elaborate, whereas the diploid stage (diplophase) is very short being confined to the zygote cell only. Meiosis of the diploid zygotic nucleus takes place immediately after karyogamy. The somatic cells are haploid and elongated. They divide by fission forming daughter cells. Any somatic cell is a potential gametangium. During sexual reproduction two cells come in contact. A beak-like protuberance develops from each conjugating cell at the point of contact. A continuous passage is developed by the dissolution of intervening walls at the point of contact where the two nuclei migrate. The passage between the two cells enlarges forming a conjugation tube, where karyogamy takes place. Gradually the two cells along with the conjugation tube form the zygote cell. The zygote cell ultimately develops into an ascus. The diploid zygotic nucleus undergoes three divisions, of which the first

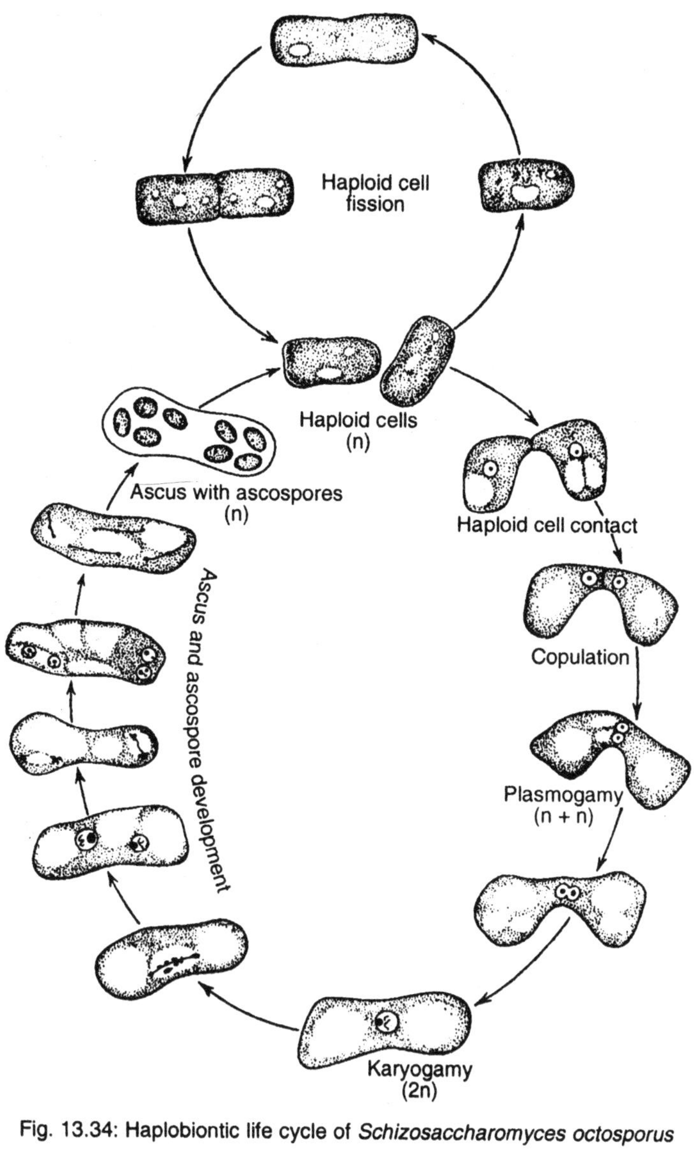

Fig. 13.34: Haplobiontic life cycle of *Schizosaccharomyces octosporus*

one is meiotic, producing eight haploid nuclei. Each nucleus with cytoplasm develops into an ascospore and the ascus contains eight ascospores. The ascospores liberate by the breaking down of the ascus wall. They now behave as somatic cells. Life cycle is presented in Fig. 13.34.

II. **Diplibiontic life cycle.** This is exemplified by *Saccharomycodes ludwigii*. Here the diploid somatic stage is long and the haploid stage is very short. The diploid somatic cells produce buds which eventually enlarge to function as asci. The diploid nucleus divides meiotically formaing four haploid nuclei around which four ascospores are developed. The ascospores remain confined in the ascus and copulate there forming two diploid cells. Each diploid cell germinates by a germ tube which pushes out through the ascus wall ultimately forming a tubular structure. This tublar structure behaves as a **sprout mycelium** from which diploid cells are produced by budding. Thus,

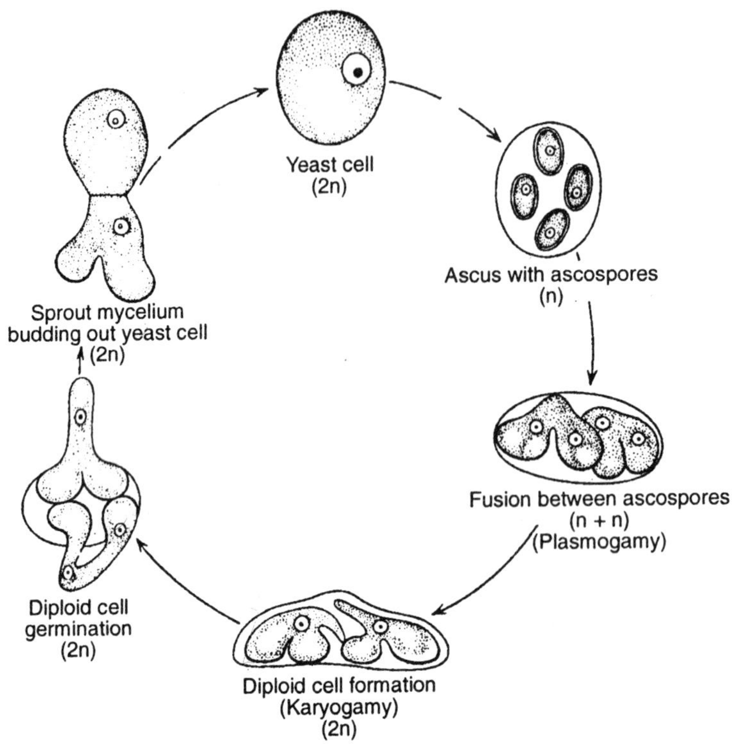

Fig. 13.35: Diplobiontic life cycle of *Saccharomycodes ludwigii*

the haploid stage is represented by the ascospores only. Life cycle is presented in Fig. 13.35.

III. **Haplo-diplobiontic life cycle.** This is exhibited by *Saccharomyces cerevisiae.* In this type of life cycle both haploid and diploid phases are equally well represented constituting somewhat an alternation of generations. Two haploid cells copulate forming a diploid cell. The diploid cell multiplies by budding producing large number of diploid cells. Eventually, each diploid cell behaves as an ascus bearing four ascospores and meiosis takes place during the development of ascospores. The ascospores on being liberated from the ascus multiply by budding proucing haploid cells. Thanks to the investigations of Winge who worked out in detail the life cycle of *Saccharomyces cerevisiae.* It is evident that in this life cycle there are two distinct stages, the diploid stage and the haploid stage which alternate in cyclic order. The vegetative cells may be both haploid and diploid reproducing asexually by budding. Nuclear cycle of *Saccharomyces cerevisiae* was worked out in detail by Guillermond. He observed that karyogamy does not always immediately follow plasmogamy. A dikaryotic condition may sometime be interpolated between plasmogamy and karyogamy. Dikaryotic buds are formed from the dikaryotic cells simultaneously along with the conjugate nuclear divison. In these dikaryotic buds, nuclear fusion occurs. Guillermond has compared this with the dikaryophase of other fungi. Renaud confirmed Guillermond's findings and, in addition, pointed out that some dikaryotic cells produce not only dikaryotic buds, but also uninucleate buds which may later copulate and produce diploid cells. Life cycle is presented in Fig. 13.36.

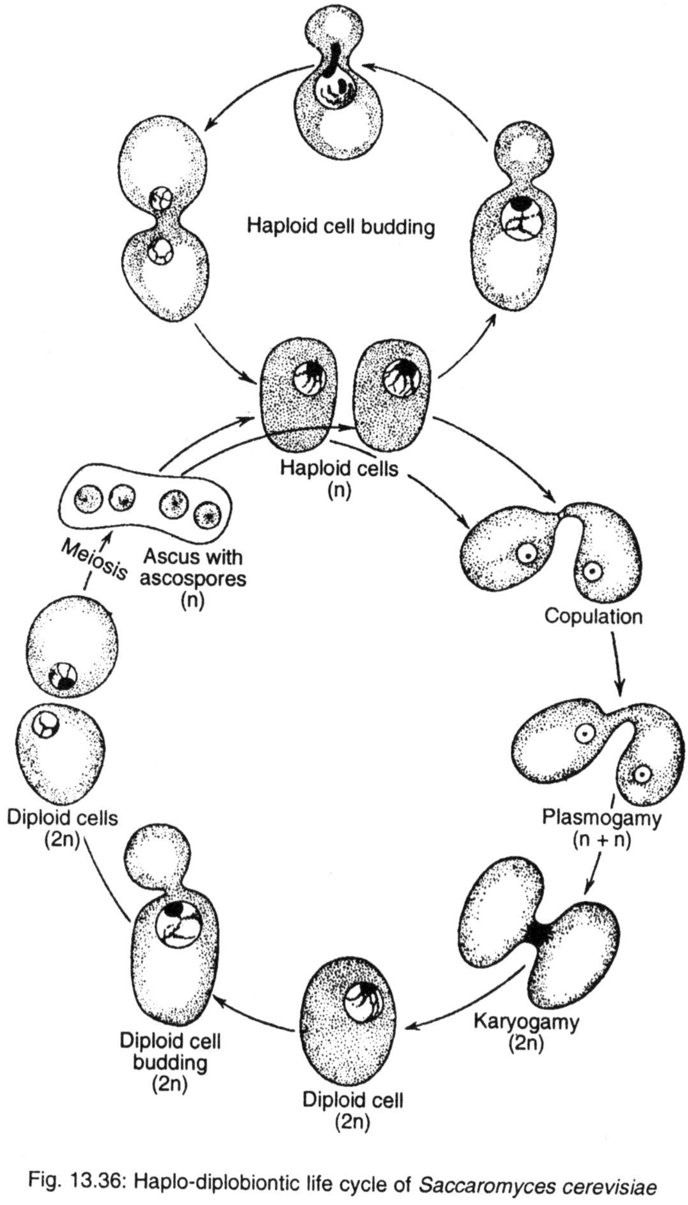

Fig. 13.36: Haplo-diplobiontic life cycle of *Saccaromyces cerevisiae*

Yeast Genetics. Beginning in 1935 Winge and Lausten extended our knowledge of sexuality in the yeasts. Later, the Lindegrens (1943) has demonstrated how yeast may be used as tool for genetical studies. Some of the aspects are as under:

1. ascospores can be cultured to study segregation of characters;
2. dominant, recessive, and lethal characters can very well be studied by simple cultural technique;
3. new strains can be developed by making hybrids of commercial importance;
4. inbreeding degeneration, legitimate and illegitimate pairing, instability of

the mating type alleles and similar other fundamental studies can be performed within a short period of time applying various cultural techniques.

Some Indian Species: *Saccharomyces apiculatus* Rees: *S. cerevisiae* Meyen; *S. fructuum* Lodder and Kreger van Rij; *S. steineri* Lodder and Kreger van Rij; *Schizosaccharomyces octosporus* Beijerinck.

Order: **Taphrinales**

The Taphrinales include fungi which are parasitic on seed plants and ferns. They are characterized by typical parallelly arranged cylindrical asci which are not borne in an ascocarp. These fungi do not produce sex organs. In them, asci are developed directly from dikaryotic cells of the mycelium. The ascospores of the Taphrinales, like yeast cells, multiply by budding. The Taphrinales include a single family, the Taphrinaceae, and a single genus *Taphrina.*

Genus: *Taphrina*

The genus *Taphrina* (old generic name *Exoascus* still in use by many authors) contains several species which are very important pathogens. They induce hypertrophic malformations of buds, leaves, twigs, flowers and fruits producing diseases known as leaf curl, blister and fasciation. In woody twigs often unnatural, profuse, tufted branching "witches' broom" is developed. *Taphrina deformans* causes peach leaf curl disease and *T. cerasi* causes witches' broom of cherries.

The somatic mycelium grows intercellularly and forms a network under the epidermis, or the cuticle of the host tissue. Its cells are irregular in size and shape and are dikaryotic. The mycelium in most species of *Taphrina* is annual, but in some species it is perennial.

Asexual reproduction takes place by uninucleate, thin-walled spores which are referred to as conidia. The conidia are developed from the ascospores. The ascospores produce conidia by budding. The conidia themselves bud indefinitely producing secondary, tertiary, etc., conidia. They germinate by germ tubes which penetrate through cuticle of young leaf and cause infection in the host tissue.

Sexual reproductioin is accomplished by the development of palisade-like layer of rectangular asci which are produced from the dikaryotic cells of a compact mycelial layer. These cells are the ascogenous cells. The mycelial layer is one cell thick and is formed subcuticularly. The ascogenous cells are ovoid, pyriform, or dome-shaped. During the development of an ascus the ascogenous cell elongated perpendicularly to the host surface. Its nucli fuse forming a diploid nucleus. The diploid nucleus then divides mitotically into two daughter nuclei of which one moves to the distal end of the elongated ascogenous cell and the other remains at the base. The elongated ascogenous cell now divides into two unequal cells by a transverse septum. The upper larger cell is the ascus mother cell and the lower smaller cell is the stalk cell. The ascus mother cell now develops into an ascus. The protoplasmic contents of the ascus mother cell crowd the tip where the diploid nucleus divides reductionally into daughter nuclei which again divide mitotically to form eight haploid nuclei. Ultimately eight ascospores are formed.

There is no development of ascocarp. Mature asci are exposed by the rupture of the cuticle or epidermis of the host tissue, when palisade-like asci become visible. The ascospores, soon after they are formed when already in the ascus, produce small, round or ovoid uninucleate blastospores (also known as conidia) by budding. Copulation of conidia takes place establishing dikaryotic condition. The ascospores with adhering conidia forming spore balls are ejected forcibly from the asci. They may be carried by wind or splashed in raindrops. On reaching host surface, the dikaryotic conidia germinate by germ tubes

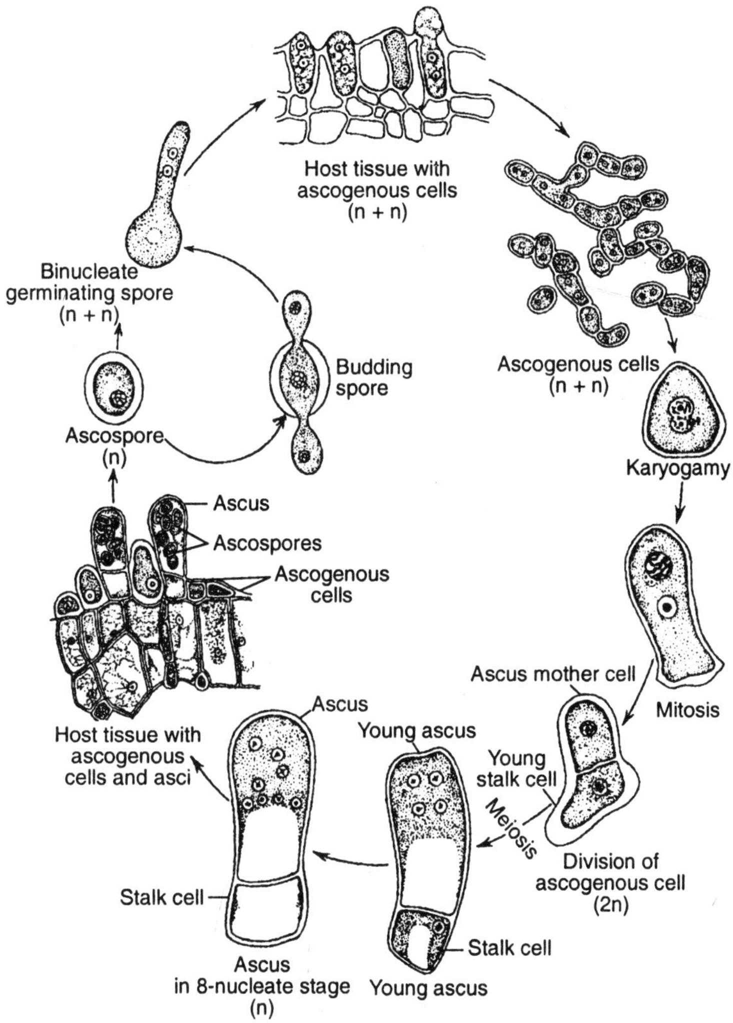

Fig. 13.37: Life cycle of *Taphrina deformans*

which infect the host and produce hyphae with dikaryotic cells. The hyphae grow intercellularly and conjugate division of the nuclei perpetuates the dikaryotic condition of the hyphal cells. Life cycle of the genus *Taphrina* is illustrated by *T. deformans* in Fig. 13.37.

Some Indian Species: *Taphrina deformans* (Berk.) Tul.; *T. maculans* Butler; *T. purni* Tul. *T. rhomboidalis* Syd. and Butler; *T. tubiforme* (Rabenh.) Lagerh.

Order: **Eurotiales**

The Euratiales, also called Aspergillales (Bessey, 1950) and Plectascales (Gaumann, 1952) include fungi most of which are saprophytes. Only a few are parasites upon plants and animals. The saprophytic ones are familiar under the names of black mold, blue mold, and green mold. They are of great economic importance. The chief means of reproduction of the Eurotiales is by asexual means by conidia which are developed in enormous numbers. The Eurotiales are characterized by the development of cleistothecia which are produced without the formation of a stroma. The asci are scattered at various levels within the cleistothecia with definite hymenial layer representing primitive condition.

Family: **Eurotiaceae**

The Eurotiaceae are very widely distributed. They have been intensively studied mainly because of their great economic importance. Members of this family possess cleistothecia with well-developed peridia without appendages.

Many of the Eurotiaceae are best known in their asexual stages comprising the form-genera: **Aspergillus** and **Penicillium**, which are ubiquitous and highly important molds. Different form-species of term-genus **Aspergillus** are placed in perfect-stage genera: **Eurotium, Emericella,** and **Sartorya.** So also is in form-genus **Penicillium,** whose form-species are recognized as perfect-stage genera: **Talaromyces** and **Carpenteles.** Changes in the names of order, family and genus are often found essential by replacing Aspergillales with Eurotiales, Aspergillaceae with Eurotiaceae, *Aspergillus* with *Eurotium* as the name applied to the ascosporic form determines the generic name. Since the non-ascosporic forms vastly outnumber the ascosporic ones it is felt more convenient to use the name *Aspergillus* than *Eurotium* though technically it may not be sound. Hence the names of conidial stages of the form-genera *Aspergillus* and *Penicillium,* so widely well-known, will be rather more convenient to describe them on the basis of their imperfect stages instead of introducing the names of their perfect stages at the beginning.

Of late, some mycologists do not use the prefix 'Form' and designate Form-Genus *Aspergillus* as just *Aspergillus.*

Genus: *Aspergillus*

The species of *Aspergillus* form a very large proportion of all the molds encountered in nature. They are found almost everywhere on every conceivable type of substratum. Indeed, the aspergilli can occur even in a slight trace of moist organic matter. Most common are the black aspergilli or *Asergillus niger* group being widely known as black mold. They quite often cause trouble as contaminants of cultures in the laboratory. Molds being typically aerobic, normal colonies develop only on the surface of the substratum (Figs. 13.38A and B). The activities of the aspergilli, are, however, largely destructive but are also utilized for the welfare of our society. They affect our welfare by causing decay of exposed foodstuffs, leather and fabrics. There are certain species which induce diseases (mycoses) of animals and human beings, particularly of birds. These diseases are known as **aspergilloses** (sing. **aspergillosis**). They are mostly caused by *Aspergillus fumigatus* and *A. flavus.* Again, *A. fumigatus, A. flavus, A. niger,* and *A. nidulans* are responsible for disease of human ear, **otomycosis.** The activities of the aspergilli, however, are not entirely destructive in nature, several species have had their fermentative powers harnessed for com-

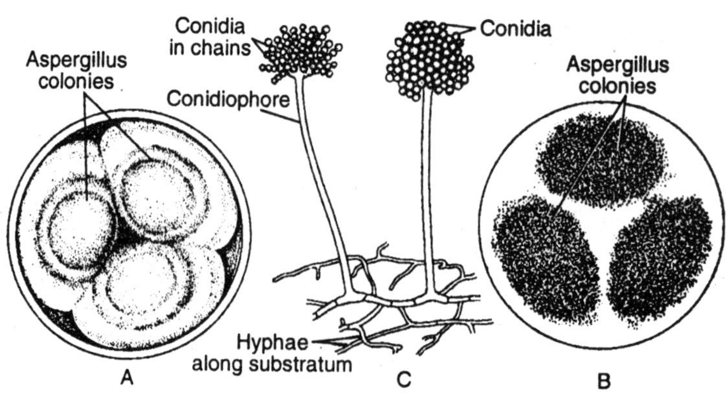

Fig. 13.38: *Aspergillus* sp. A and B. Colonies on the surface of substratum. C. Hyphae and conidiophores with conidia in chains

mercial purposes. In Japan, strains of *A. oryzae* have long been used for the saccharification of rice starch in the production of sak's and similar portable liquors. Strains belonging to the same group are used in the manufacture of soy sauce and of the enzymic mixtures which are sold under the names 'Takadiastase' and 'Polyzime'. The black aspergilli, long known to be capable of producing considerable amounts of oxalic acid from sugar, are now used for the successful commercial production of citric acid. The species *A. niger* is so sensitive that sometimes it is used to detect the prsence of copper in soil and other substances. *A. niger* also causes decay of stored tobaccos and cigars.

The mycelium of aspergilli is colourless or bright or pale coloured, or bearing surface concentrations of colouring matter. It is composed of profusely branched septate hyphae, partly aerial (Fig. 13.38C). Cells of the hyphae are multinucleate. Many species and strains of *Aspergillus* often produce conspicuously zonate colonies on the substratum (Fig. 13.38A). Most commonly, these take the form of fairly regular concentric zones and result from an increased production of conidial structures periodically during the development of the colony. The most striking character of an *Aspergillus* colony is usually its colour production. This takes two general forms: (i) colour in the aerial parts, including hyphae, conidiophores, heads and conidia; and (ii) colours appearing in the substratum, and representing the specific response and effect of the organism upon particular media.

Asexual reproduction takes place by conidia. The first step toward conidium formation in the aspergilli is the differentiation of certain cells, the **foot cells** in the mycelium for propagative purposes (Fig. 13.39A). These cells become larger thick-walled and each usually bears a single fertile hyphae, perpendicular to the long axis of the cell and usually about midway between the ends of the cell. In age these cells frequetly become curved and twisted with their connection to vegetative hyphae, incospicuous but usually still determinable. These foot cells are commonly submerged in the substratum. The erect, perpendicular hypha arising from the foot cell constitutes the **conidiophore** (Fig. 13.39B). The apex of the conidiophore enlarges forming a swelling, the **vesicle** which may be variously globose, subglobose, club-shaped, or hemispherical (Figs. 13.39C and D). The lumen of the vesicle is continuous with the rest of the conidiophore; a septum near the base of the head is occasionally, but only rarely, seen, and has not been regularly found in any species. The conidia-bearing surface, represented by the fertile area of the vesicle, is closely covered by the simultaneous development of a layer of cells the **sterigmata** (sing. **sterigma**) (Figs. 13.39E to G), each in a general way perpendicular to a point on the fertile surface of the vesicle. The sterigmata often take bottle-shaped appearance. Each sterigma produces an unbranched chain of conidia (Figs. 13.39I and L). The first series of cells, or **primary sterigmata** may also bear two to several cells, the **secondary sterigmata,** forming a crown at the apex. Each of the secondary sterigmata bears one chain of conidia (Figs. 13.39H, J and K). In cases where there is a sngle layer of cells each of which produces a chain of spores, in the strict sense these cells ar the sterigmata. Where there are both primary and secondary series, the primary sterigmata are essentially supporting cells and vary much more in size and shape than do the secondary sterigmata (Figs. 13.39H, J and K). The actual spore-producing cell, or sterigma, is definitely specialized. It ordinarily consists of an essentially cylindrical body, which, after reaching certain length, narrows into a spore-producing tube whose diameter is fairly uniform within the species. Elongation is thenceforth confined to this spore-producing tube. The nuclei in the sterigma divide and one of each pair of daughter nuclei passes into the tube; cell division follows. Parallel with the repeated division of the sterigma nucleus, the tube continues to elongate rapidly successively cutting off new sections and pushing the older cells outward. Each such chain of spores typically consists then of series of equal sections borne on tip of a sterigma. Each section carries a daughter nucleus derived directly from the

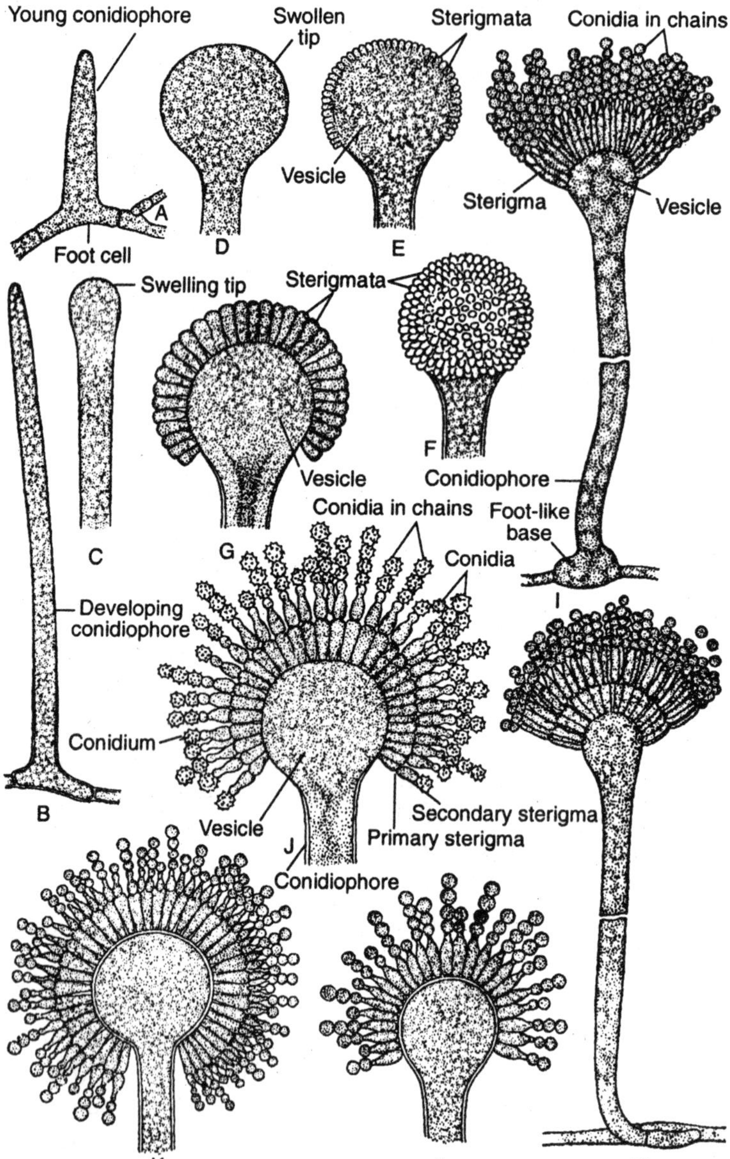

Fig. 13.39: Conidiophores and conidia of different species of *Aspergillus*. *A niger* (A–G and J). A. Foot cell bearing young conidiophore. B. Developing conidiophore. C–D. Development of the vesicle by swelling of the terminal portion of the conidiophore. E–G. Development of sterigmata. H. *A. nidulans.* Typical head showing sterigmata in two series and conidia in chains. I. *A. fumigatus.* Shortened conidiophore with foot-like base and vesicle with one series of sterigmata producing conidia in chains. J. *A. niger.* Portion of conidiophore with vesicle bearing primary and secondary sterigmata and conidia in chains. K. *A. carbonarius.* Vesicle showing two series of sterigmata bearing chains of conidia. L. *A. violaceofuscus.* Small head with one series of sterigmata producing conidia in chains

active nucleus of the sterigmatic cell at the base of the chain. No further divisions occur among the cells in the chains. Such chains often contain several hundreds of conidia. After the conidium is separated by a septum from the parent cell, it assumes the size and shape characteristic of the species, then it lays down on its original, or primary wall, a secondary wall whose colour, texture, and marking are those of the species. The secondary wall completely separates this spore from the parent cell. The conidia are typically globose and unicellular with externally roughened walls (Fig. 13.39J). They are at first uninucleate, soon become multinucleate in many species. Differences in size, colour, and wall markings are evident from species to species.

Aspergillus is commonly encountered as a conidial fungus. In most of its species the perfect stages have not been reported. Maybe due to the fact that such species have lost their ability to reproduce sexually or else their perfect stages have not been discovered. Actually the perfect stages of some of the species of *Aspergillus* have been observed long time ago, but they were described and figured the essential features of cleistothecial development of the *Aspergillus glaucus* group. He also established genetical connection between *Aspergillus glaucus* and *Eurotium herbariorum*. Again the general morphology of the cleistothecium of *Aspergillus nidulans* was described by Eidam (1883). The findings of De Bary were verified later by Fraser and Chambers (1907) and Dale (1907). It has also been reported that *Aspergillus herbariorum* forms cleistothecia even on bread, the substratum on which it very frequently occurs. The stages leading to the

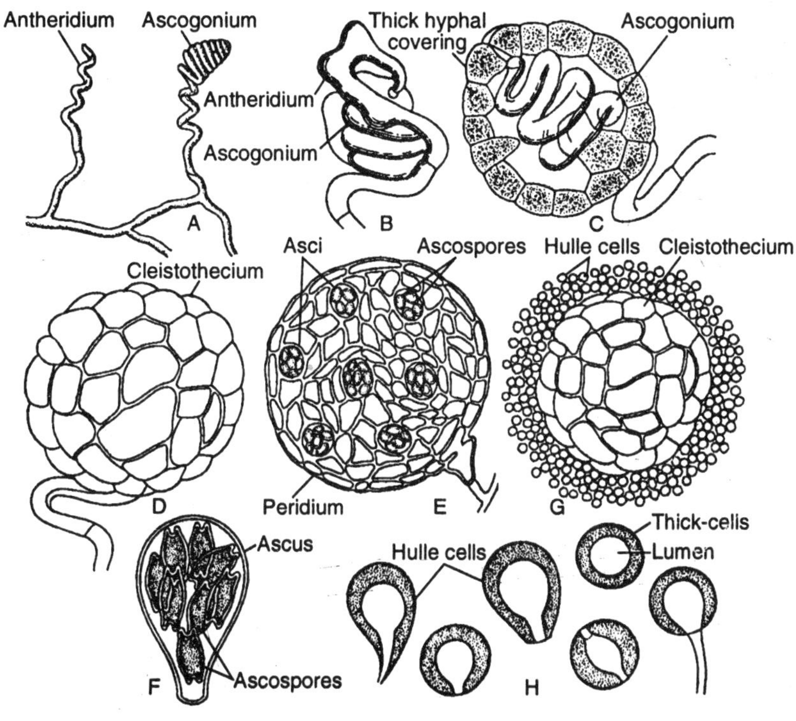

Fig. 13.40: Stages in development of cleistothecium in *Eurotium* sp. (A–F). *Aspergillus nidulans* (G–H). A. Coiled antheridium and ascogonium. B. Antheridium and ascogonium coiled round each other. C. Ascogonium surrounded by thick hyphal covering. D. Cleistothecium. E. Cross section of a cleistothecium showing peridium, asci and ascospores. F. Ascus bearing ascospores. G. Cleistothecium with hulle cells. H. Details of hulle cells

development of cleistothecia were studied in detail by Dangeard (1907) in *Aspergillus flavus* and *A. fischeri, A. fumigatus;* by Fraser and Chambers in *A. herbariorum;* and by Dale (1907) in *A. repens.* In 1946, E. Gaumann gave a clear picture of the sexual pattern of the Eurotiaceae. Again, C. R. Benjamin (1955) discussed the genera possessing ascocarp along with Aspergillus type of conidia. He suggested reviving the use of the name *Eurotium* for the *Aspergillus glaucus* series. In *Eurotium* both antheridium and ascogonium are developed close to each other on the hyphal branches of the somatic mycelium (Fig. 13.40A). They are multinucleate structures. The ascogonium has in its terminal portion a receptive organ, the trichogyne. During their growth, both antheridium and ascogonium approach each other and wind spirally around each other (Fig. 13.40B). The spiral is then divided by transverse septa into a many-celled structure. Meanwhile, sterile hyphae develop from the lower hyphal cells of the spiral forming a dense mass enclosing the spiral (Fig. 13.40C). This dense mass of sterile hyphae becomes thick and resistant and ultimately forms the **peridium,** the wall of the ascocarp. Since the spiral is invested by the dense mass of sterile hyphae, the subsequent stages of development is rather difficult to follow. This has resulted in controversial view points. The antheridium may fuse with the trichogyne which is followed by plasmogamy and pairing of antheridial and ascogonial nuclei. It has also been reported that the spiral puts out numerous branches, the ascogenous hyphae, which thrust themselves in between the inner wall cells of the peridium in every direction. The many extremities of their numerous ramification become ovoid eight-spored asci. The continuity of the ascogenous hyphae is more and more lost as the asci are formed, so that as the spores begin to ripen the peridium encloses only asci and the remains of the hyphae. Finally the walls of the asci themselves disappear and the ascocarp contains scarcely anything but ripe ascospores. The antheridium may also fail to fuse with the trichogyne and ultimately get disorganized. In such case the ascogonial nuclei become paired. It has been agreed that the ascogenous hyphae grow out from the ascogonium and eventually bear asci. But it has not been confirmatively proved whether or not the antheridium is functional in the spiral. The ascocarp is more or less rounded smooth-walled without any opening and is known as **cleistothecium.** The cleistothecium is without appendages (Fig. 13.40D). The asci lie irregularly distributed in the cleistothecium. They are globose, ovoid or pear-shaped. Eight ascospores are formed in each ascus (Figs. 13.40E and F) by three successive divisions of the diploid nucleus, of

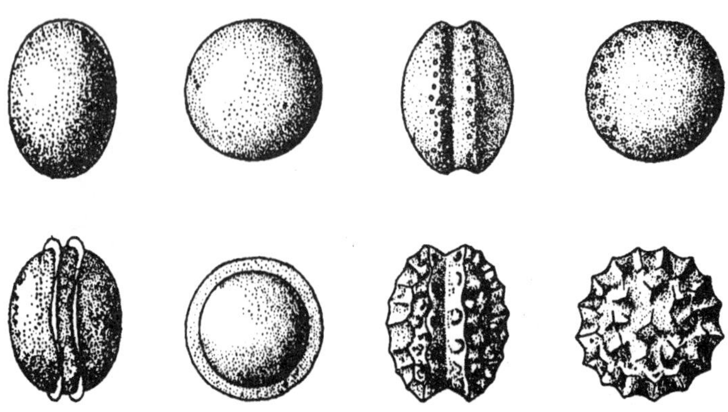

Fig. 13.41: Ascospores of different species of *Eurotium*

which the first one is reductional. The asci are evanescent and the ascospores remain free in the cleistothecium as soon as the ascus wall is dissolved immediately after maturity of the asci. The ascospores are finally set free by the rupture of the cleistothecial wall, they germinate and form new mycelia. The ripe ascospores are multinucleate. They may have variously sculptured to double-convex wall resembling pulley wheels (Fig. 13.40F), which may vary from species to species (Fig. 13.41). Mature cleistothecia of certain species of *Aspergillus* bear loose network of hyphae containing large number of terminal or intercalary cells which are vesiculose, elliptical or almost globose, thick-walled with almost obliterated cell lumen. They are described as **hulle cells** (Figs. 13.40G and H). The appearance of these cells along with the cleistothecia of *Aspergillus nidulans* was reported for the first time by Eidam (1883). Origin and function of these cells are still obscure. Dangeard (1907) designated them as chlamydospores. But he failed to present any evidence in support of his statement. Life cycle of *Eurotium* sp. (*Aspergillus glaucus* group) is presented in Fig. 13.42.

Some Indian Species: *Eurotium herbariorum* Link, reported earlier under *Aspergillus herbariorum* Wiggers; *Eurotium repens* De Bary, reported earlier under *Aspergillus repens* (Corda) De Dary and Woron.

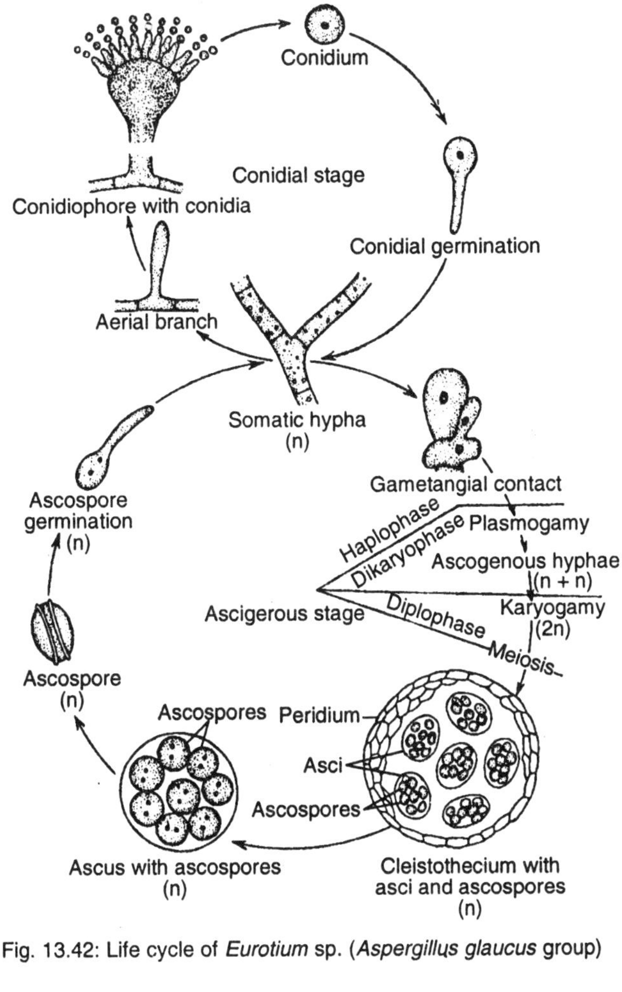

Fig. 13.42: Life cycle of *Eurotium* sp. (*Aspergillus glaucus* group)

Form-Genus: *Penicillium*

Present designation just *Penicillum.*

The penicillia are commonly known as 'green mold' or 'blue mold' and are universally present on decaying organic matter. They are closely related to the aspergilli and are just as widespread and omnivorous. Like the aspergilli, many of them are common and serious agents of destruction, but on the other hand, there are several species which are best known as the means of conducting industrial fermentations. They frequently occur as contaminants in the Biological laboratory. Usually, *Penicillium* grows as a saprophyte on decaying fruits and vegetables. Decaying oranges, lemons and other fruits usually exhibit a bluish growth on them. This is due to the blue-coloured spores of *Penicillium.* It also forms the common blue-green mold on damp fruit, leather, stale bread and other exposed food-stuffs. Some species of *Penicillium* have been reported to grow on the forest floors, a few of them grow on cultivated and manured ground.

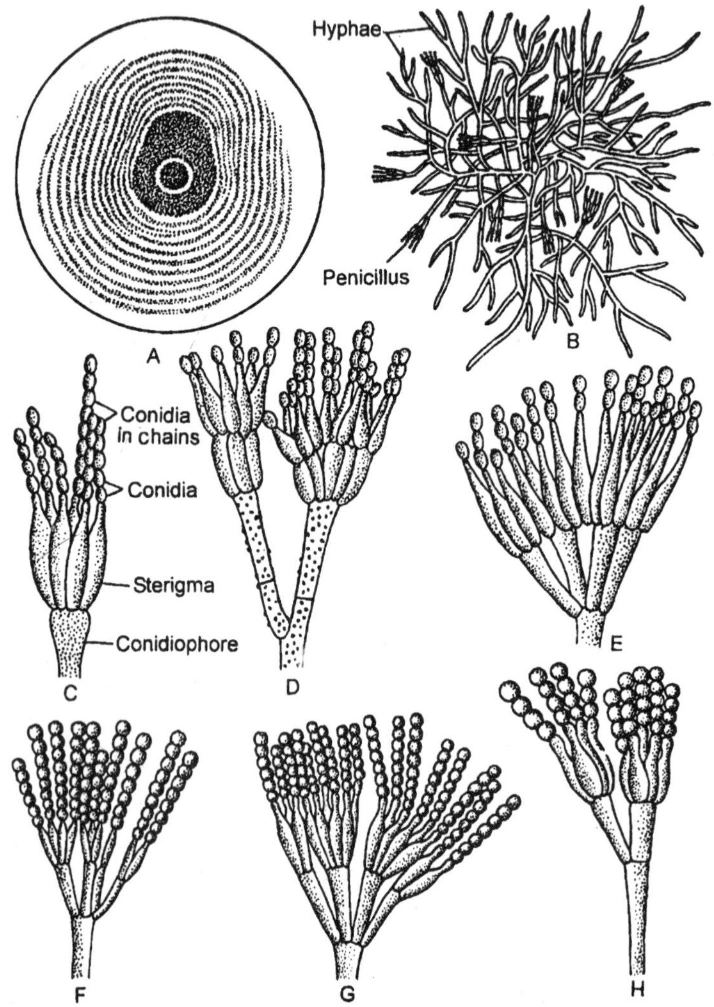

Fig. 13.43: Conidiophores and conidia of different species of *Penicillium*. A. A colony of *Penicillium* sp. B. Mycelium with conidiophores of *P. gloucum*. C. Conidiophore with sterigmata bearing conidia in chains. D. Conidiophore of *P. lanoso-caerulium*. E. Conidiophore of *P. vermiculatum*. F. Conidiophore of *P. glaucum*. G. Conidiophore of *P. herquei*. H. Conidiophore of *P. chrysogenum*

Most of the species of *Penicillium* cause economic loss like, *P. expansum* is responsible for the spoilage of fruits, e.g., apples, pear and grapes in storage, while *P. italicum* and *P. digitatum* cause blue and green molds of citrus fruits, whereas *P. purpurogenum* spots print paper, books and engravings. Again, some of the penicillia have been reported to be associated with animal and human diseases, but they are less virulent than the aspergilli. Besides these, the penicillia are also used on large-scale fermentations in the industries for ripening of cheeses and for producing organic acids. Most important of the cheese molds are the *P. roqueforti* and *P. camemberti* which are involved in the ripening of Roquefort and Camembert cheeses by means of products of hydrolysis of casein thereby lending them a characteristic aroma and consistency. A member of the *P. purpurogenum* series is used for the large-scale manufacture of gluconic acid. The wonder drug, **penicillin,** is prepared from *P. notatum* and *P. chrysogenum*. Recent researches have shown that the penicillia exhibit the most varied metabolic activity, different species being capable of synthesizing from glucose a bewildering variety of substances of complex chemical constitution. The production of ergosterol, the parent substance of vitamin D, and of a number of substances closely related to ascorbic acid (vitamin C) has indicated the potentiality of employing these molds in future commercial processes.

The mycelium, composed of septate branched hyphae, remifies on the surface of, and in, the substrate on which the fungus survives (Fig. 13.43B). The portion of the mycelium remaining on the surface of the substratum becomes closely matted, loosely floccose, or partially as popes of hyphae (Fig. 13.43A). The mycelium is colourless or pale or brightly coloured. The cells of the hyphae are multinucleate.

In the asexual reproduction hyphae grow erect from the mass of mycelium, and function as **conidiophores.** From the tip of the conidiophore several smaller branches (**metulae**) project which divide again to form sterigmata. The sterigmata, because of their bottle-shape, are termed **phialides.** The sterigmata bear conidia in long chains (Fig. 13.43C). The branching of conidiophores may be symmetrical, asymmetrical, or broom-like (Figs. 13.43D to H). The terminal portion of conidiophore along with its branches and sterigmata, gives the appearance of a broom and is known as **penicillus** (Fig. 13.43B). The branches supporting the metulae if comparatively short and obviously part of the penicillus are known as **rami** (sing. **ramus**). Sometimes the conidiophores become aggregated into fascicles or compacted into definite **coremia.** The conidia are produced by abscission forming unbranched chains. They are globose, ovoid, elliptical, or pyriform, smooth or rough, in most cases green during the growing period but sometimes colourless or in others pale colour. The conidia become detached, and are carried to different substrata, where they germinate to give new mycelia. The mechanism of development of conidia, their dispersal and germination are the same as in *Aspergillus.*

The sexual stage of most species of *Penicillium* has not yet been clearly worked out. The development of ascocarp (**cleistothecium**) in *Penicillium* was reported for the first time by Brefeld (1874) in *Penicillium glaucum.* The sexual process has been described by Dangeard (1907) in *Penicillium vermiculatum,* by Dodge (1933) in *Penicillium stipitatum.*

Langeron (1922) proposed the generic name *Carpenteles* for the species *Penicillium glaucum,* described by Brefeld in 1874. Its ascocarp originates at the initial stage as a sclerotium-like body and gradually asci are formed from the centre outwards having hard wall. Raper and Thom (1949) also adopted the name *Carpenteles* for a series of penicillia which produce ascocarp. Subsequently C. R.

Benjamin (1955) introduced the use of the name *Carpenteles.* He has also erected the genus *Talaromyces,* to include the species of *Penicillium* which form soft, mostly yellow ascocarps. Raper (1957) strongly disapproved the suggestions put forward by Benjamin, and argued that the needs of taxonomy are best met by retaining the name *Penicillium* for all the species, whether ascocarpic or entirely conidial.

Of the species known to produce asci, some are heterothallic. The development of ascocarp of *Penicillium crustaceum* begins with the appearance of a pair of short hyphae spirally coiled about each other, the antheridium and ascogonium (Fig. 13.44A). The spirally twisted structure is invested by three or four layers of felted mycelium which arise close to one another and are spirally twisted round one another in one or two turns. Then from the spirally twisted body, short ascogenous hyphae grow out as branches in every direction, and at the same time numerous branches begin to form on the neighbouring hyphae which grow rapidly round the others, and enclose them in compact envelope (Figs. 13.44B and C). The elements of this enveloping weft thrust themselves everywhere in between the ascogenous hyphal branches. The ascogenous hyphae gradually elongate and force themselves in between the growing tissue of the envelope (Fig. 13.44B). The ascogenous hyphae are of two kinds; comparatively slender one which penetrate between the cells of the tissue of the envelope, ramify copiously in it and form asci; secondly, thicker much curved froms with short branches of their own, from the ramifications of which the asci are ultimately produced (Fig. 13.44E). The asci bear eight ascospores. As the asci mature the ascospores are set free by the dissolution of the wall of asci (Figs. 13.44G and H). The ascospors are finally liberated out

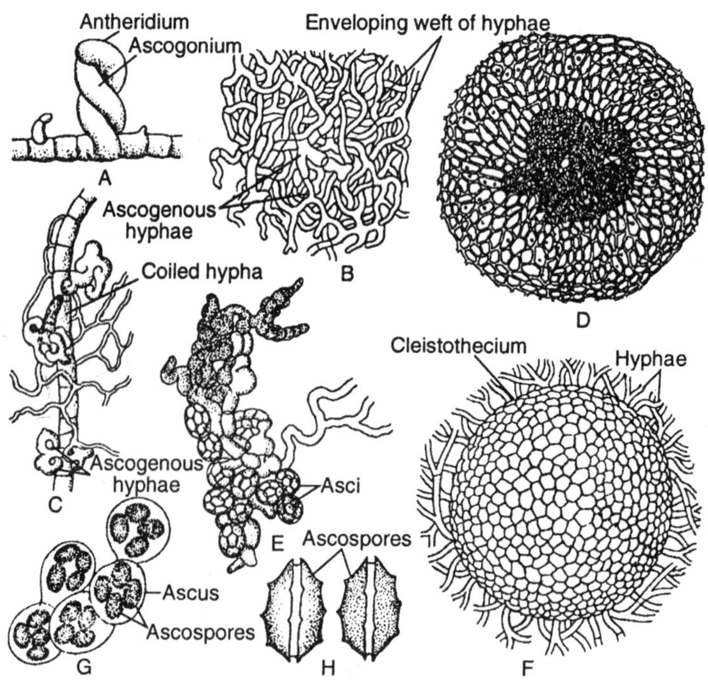

Fig. 13.44: *Penicillium crustaceum.* Stages in development of cleistothecium, asci and ascospores. A. Coiled antheridium and ascogonium. B. Growth of ascogenous hyphae along with enveloping weft of hyphae. C. and E. Development of asci from ascogenous hyphae. D. Young cleistothecium. F. Mature cleistothecium. G. Asci with ascospores. H. Ascospores

by the rupture of the ascocarp wall (Figs. 13.44D to F).

In *Talaromyces vermiculatus* (*Penicillium vermiculatum*) both antheridium and ascogonium are differentiated morphologically as well as dynamically. The antheridium appears a long time after the formation of the ascogonium. The ascogonium which is rich in protoplasm and uninucleate at the initial stage appears as a branch of any hyphae. It elongates rapidly and by repeated nuclear division becomes 16, even up to 64-nucleate. When the ascogonium has already become multinucleate, the young antheridium winds several times around the ascogonium (Fig. 13.45A). It then cuts off an uninucleate apical cell which swells slightly. The tip of the apical cell of the antheridium then touches the ascogonium (Fig. 13.45B), at which point the intervening walls dissolve forming a continuus passage through which plasmogamy takes place (Fig. 13.45C). But the actual migration of the contents of the antheridium to the ascogonium followed by plasmogamy has not been observed. The ascogonium then divides into numerous binucleate cells (Fig. 13.45D). Each of these binucleate cells sends out one or more branched ascogenous hyphae whose cells are also binucleate. Simultaneously with the development of ascogenous hyphae, entangled sterile hyphae grow up from the adjacent hyphal cells of the antheridium and the ascogonium. The sterile hyphae ultimately develop into wall of the ascocarp (Fig. 13.45E). Further growth of the ascogenous hyphae leading to the development of asci and ascospores is unknown. The wall of the ascocarp is composed of two layers, the inner serving as the nourishing layer. The mature asci are irregularly distributed within the ascocarp. The asci are globose or pear-shaped bearing four to six ascospores (Fig. 13.45F). The ascospores are shaped like pulley wheels. Sooner or later the wall of the asci disintegrates so that the ascospores are set free inside the cavity of the ascocarp. The ascospores are liberated with the decay of the wall of the ascocarp. The ascocarp (cleistothecium) is without appendages. Life cycle of *Talaromyces vermiculatus* is presented in Fig. 13.46.

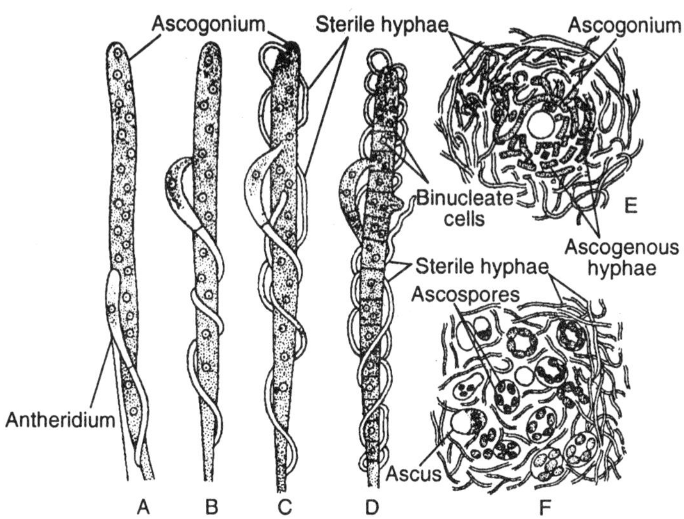

Fig. 13.45: *Talaromyces vermiculatus* (*Penicillium vermiculatum*). Stages in development of asci and ascospores. A. Multinucleate ascogonium and young antheridium. B. Tip of the apical cell of the antheridium touching the ascogonium. C. Antheridium in open communication with the ascogonium. D. Ascogonium surrounded by sterile hypha is divided into binucleate cell. E. Sectional view of ascogonium, ascogenous hyphae and sterile hyphae. F. Sectional view of asci, ascospores and sterile hyphae

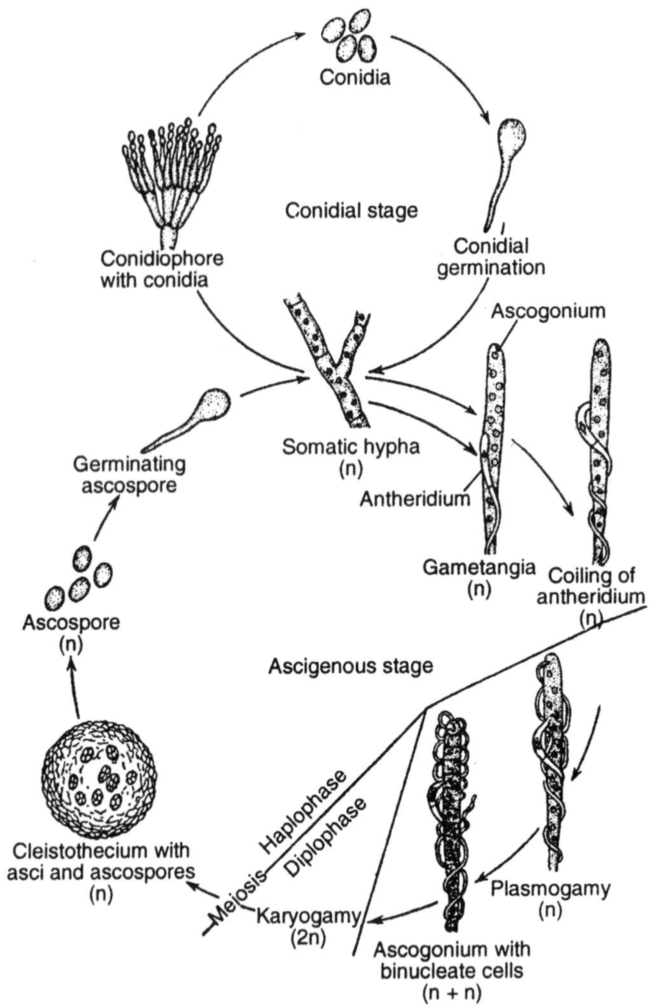

Fig. 13.46: Life cycle of *Talaromyces vermiculatus* (*Penicillium vermiculatum*)

Some Indian Species: *Talaromyces vermiculatus* (Dang.) Benjamin, earlier reported under *Penicillium vermiculatum* Dang.; *Talaromyces wortmanni* (Kloecker) Benjamin, earlier reported under *Penicillium wortmanni* Kloecker.

Order: **Pezizales**

The fungi making up this order are mostly saprophytes along with a number of parasitic ones (*Sclerotinia, Pseudopeziza,* and others) which cause serious diseases of cultivated plants. These fungi, in general, reproduce by both asexual and sexual methods. Asexual reproduction is by conidia, but not so widely encountered in these fungi in comparison with other Ascomycotina. Sexual reproduction is by the development of distinguishable gametangia. The product of sexual reproduction is the formation of an ascocarp in which asci and ascospores are produced. The ascocarps may be from cup-shaped or discoid being sessile or stipitate (Figs. 13.19C and 13.49); tongue-, club-, fan-shaped or discoid being sessile or stipitate (Figs. 13.19G, H). They may be extremely variable in size from very minute microscopic to large macroscopic, some are even very gigantic in size. In texture also there may be very wide variation in the ascocarps from fleshy, leathery, horny to various other gradations; so also in colour, being hyaline to bright-coloured of various shades.

Asci are unitunicate and operculate; usually large with symmetrical elliptical or spherical ascospores. They dehisce by an apical lid, the operculum (Figs. 13.13B, C). In the comparatively larger ascocarps the discharge of ascospores is simulataneous with asci being emergent or non-emergent from the hymenial layer (Fig. 13.51).

The ascospore walls are smooth or ornamented with external warts, spines or ridges, the shape and distribution of which afford valuable specific characters.

Family: **Morchellaceae**

The apothecia of this family are large and vary greatly in shape. Their receptacle may be honeycomb-like, companulate—in both cases distinctly stipitate—or sometimes pezizoid and substipitate.

This family is characterized by its smooth elliptical ascospores, without prominent internal oil drops but bearing a number of small external guttules at their polar regions.

Genus: *Morchella*

The species of this genus are known as Morels or Sponge Mushrooms and are among the most delicious edible fungi known. The common edible species is *Morchella esculanta* having stalked apothecia about 10–15 cm high. Unfortunately, no one has yet succeeded in cultivating any species of *Morchella* although cultures of mycelium are not difficult to establish on agar.

Species of *Morchella* usually occur under trees in open woods, orchards or in open grassland areas. They are common on moist soil rich in humus in the forests of Kashmir, Punjab, Uttar Pradesh, Nepal, and in Darjeeling in West Bengal.

Apothecia are variable in shape and size. They are scattered or rarely gregarious, erect, distinctly stipitate, morchelloid, namely with modified disc and receptacle (Fig. 13.19G). Disc yellowish brown to brown, covering numerous shallow depressions of sponge-like appearance on the surface of the subglobose, ovoid to narrowly conical receptacle which has either no distinct margin or with well-developed and free margin. The depressions are separated from one another by anastomosing, sterile or rarely fertile-edged ridges, which are either honeycombed or irregular and paler than the disc, or the ridges may run more or less longitudinally in which case their edges are often darker than the disc or even almost black. The hymenium lines the shallow depressions, whilst the ridges between lack asci (Fig. 143D). The ridge which is also known as hymenophoral trama is composed of mostly horizontally running, septate, occasionally branched and rather large-celled hyphae. The tips of asci are curved so that the spores are projected outwards, and do not impinge on the opposite side of the depressions in which the asci are formed. In section it can be seen that each cavity represents an apothecium and that these are compounded at the surface of the rather large, hollow pileus.

The stipe or stalk is hollow (Fig. 143D), brittle, subcylindrical, sometimes distinctly bulbous at the base, occasionally furrowed, minutely scurfy, yellowish or whitish cream.

Asci are rather subcylindrical, narrower at the rounded base, unitunicate and operculate; 8-spored, and pores are not blued in Melzer's reagent.

Ascospores are broad elliptical to oblong elliptical, smooth-walled, hyaline to subhyaline when viewed singly but appearing cream coloured in mass, without oil globules but during their development the polar regions are crowned with numerous oil globules.

Paraphyses are stout, multiseptate near the base, branched, enlarged at the clavate apices.

Ascospores on germination produce septate hyphae which under favourable conditions give rise to ascocarp.

Some Indian Species: *Morchella conica* Pers.; *M. esculanta* (L.) Pers.; *M. hybrida* (Sow.) Pers.; *M. rotunda* (Pers.) Boud.

Family: **Pezizaceae**

The apothecia are mostly cup-, disc-, or lentil-shaped (Fig. 13.19C). They may be sessile or stipitate, minute to very large, bright-coloured to dark-brown. The apothecia are fleshy, waxy, leathery, cartilaginous or corneous with ectal excipulum smooth, velvety, hairy, or bristly. The asci are narrowly cylindrical, turn blue at least at the tip, with Melzer's reagent; unitunicate and operculate or more rarely opening by a slit at the apex giving the open ascus a bilabiate appearance (Figs. 13.13A–C), occasionally marked by a thickened ring or collar near the apex. They do not protrude beyond the general level of the hymenium at maturity. The asci bear two to many ascospores which are globose, ellipsoid or fusiform; hyaline or coloured of various shades. The ascospores are smooth or variously sculptured. The asci are mixed with filiform to clavate, hyaline or variously coloured paraphyses. The mycelium of the members of this family is septate and branched.

This family is divided into two tribes: Pezizeae and Otideae.

Genus: *Peziza*

Species of *Peziza* grow on or beside rotting wood; on the ground, even on burnt soil; on animal dung; some partly buried in sand and others mostly on soil in woods, fields and gardens. They form more frequently saucer-shaped than cup-shaped apothecia up to 5 cm or more in diameter (Fig 13.19C). Apothecia are sessile or very short-stalked, receptacle usually thin, brittle-fleshed and often more or less scurfy externally. Some are brilliantly coloured, especially in shades of red and orange. The ascocarps are composed of tightly compacted hyphae and are typically fleshy. The hymenial layer of asci and paraphyses line the cup on the inside (Fig. 13.19D). The excipular tissue is either textura globulosa or textura intricata (Figs. 13.19D and 13.24).

Asci are unitunicate and operculate; cylindric or club-shaped and nearly always contain uniseriate eight ascospores. They are blued at the tip by Melzer's reagent. This character distinguishes *Peziza* from *Aleuria. Aleuria aurantia* is sometimes called the orange-peel *Peziza.* The asci do not protrude beyond the general level of the hymenium at maturity.

Ascospores are large, usually elliptical, occasionally somewhat spindle-shaped, hyaline or subhyaline to brownish with or without oil drops. The species of *Peziza* fall into two series, one with smooth spores without oil drops, the other with spores ornamented by warts or ridges and usually containing two large oil drops. The second series includes a few species whose flesh yields a coloured fluid and was formerly given generic rank under the name *Galactinia.* The two groups are not completely clear cut, however, and it is generally agreed that they cannot be separated genetically. The spore ornament is a valuable specific character which can be demonstrated by staining with cotton blue in lactophenol.

Paraphyses are slender, or stout, septate, unbranched, apex subclavate or clavate, often containing coloured granules.

Some Indian Species: *Peziza multiguttulata* Kar and Pal; *P. repanda* Pers.; *P. subcupularis* Rohm; *P. succosa* Berk.; *P. vesiculosa* Bull.

Family: **Ascobolaceae**

The Ascobolaceae are characterized mainly by relatively broad asci which protrude above the general level of the hymenium as they mature. Ascospores commonly lie in two or three irregular rows in the ascus instead of a single vertical row. They are dark-coloured at maturity, elliptical or, if globose, there is a purplish outer layer to the spore wall soluble in alkali. The ascospores have a double wall, a hyaline persistent inner layer and a dark brown or purple outer layer, the epispore, which commonly cracks into a characteristic pattern. The epispore swells up and disappears in KOH solution and dried specimens of the Ascobolaceae should never be soaked in this reagent to soften them for microscopic examination. The spores are not cohering within the ascus but are ejected singly. The apothecia are short-stalked and occur mainly on dung, a few species on burnt ground, soil or debris. They may be incospicuously hairy to furfuraceous.

Genus: *Ascobolus*

The genus *Ascobolus* was established by Persoon to include Discomycetes in which mature asci protrude from the hymenium.

Plants are generally found on dung of animals (*Ascobolus scatigenus,* common on dung of cow; *A. minutus,* on dung of cow and buffalo; *A. viridis,* on dung of buffalo, goat and rabbit); but often occurring on decaying plant materials (*A. striisporus);* on fabrics (*A. saccoboloides*) and similar other substrata. Although saprophytic in habit, same species of *Ascobolus* like, *Ascobolus scatigenus, A. carbonarius, A. stercorarius,* and *A. immersus* have been grown in pure culture to the fruiting stage.

The mycelium of the different species of *Ascobolus* resembles that of many other fungi. The hyphae are well-developed, profusely branched and septate. They ramify along the surface of the substratum producing a cottony appearance. When young, they are white and with age turn creamish due to accumulation of fat globules in the protoplasm.

While asexual reproduction has been reported for a few species of the genus *Ascobolus,* such as *Ascobolus scatigenus, A stercorarius, A denudatus,* and *A lignatilis;* on the whole this type of reproduction is conspicuous by its absence in many of the species of the genus *Ascobolus,* or at least, has not been observed.

In 1905 Claussen observed that, in *Ascobolus stercorarius,* great number of oidia are formed in chains from the mycelium derived from an ascospore. These oidia on germination produce mycelium which again produces oidia, but the mycelium never produces any apothecia. This has also been observed by E. S. Dowding (1931). That the oidia are produced in several species of *Ascobolus* has been reported earlier by Brefeld (1891) in *Ascobolus denudatus* and by Falck (1902) in *A. lignatilis.*

J. W. Hotson (1912), B. O. Dodge (1920) and E. M. Betts (1926) have found that single-ascospore cultures of *Ascobolus scatigenus* (*Ascobolus magnificus*) never give rise to apothecia, but produce certain specialized sclerotium-like structures, the **papulospores** (Fig. 13.47). A papulospore has one or two large central storage cell surrounded by a covering of hyphae which develop blister-like outgrowths of the storage cell. Papulospores arise in about ten days in all cultures. But they fail to produce ascocarp. They are the spores developed by asexual method. Hotson described it as the asexual stage of *Ascobolus scatigenus* (*Ascobolus magnificus*) and named it as *Papulaspora magnifica* Hotson. This has been confirmed by Dodge (1920), and H. C. I. Gwynne-Vaughan and H. S. Williamson (1932).

Sexual reproduction takes place by the development of antheridium and ascogonium bearing a trichogyne (Fig. 13.48E). The eventual product of sexual reproduction is an apothecium.

The development of an apothecium of *Ascobolus scatigenus,* as described by Dodge (1912, 1920) and Gwynne-Vaughan and Williamson (1932) is initiated only when two

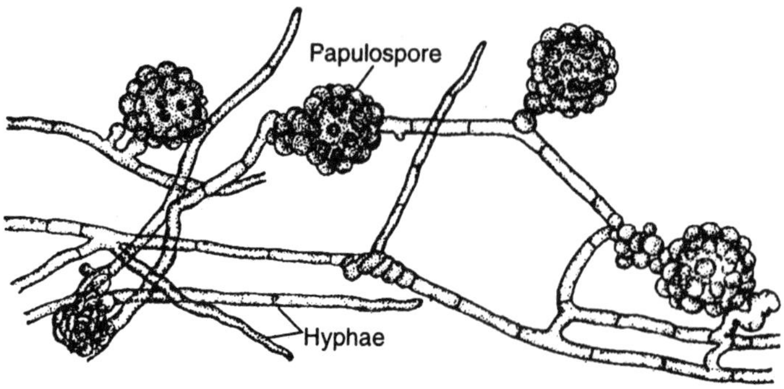

Fig. 13.47: *Papulospora* stage of *Ascobolus satigenus* (*Ascobolus magnificus*)

complementary mycelia are intermingled. The thallus of *A. scatigenus* is monoecious, that is, capable of bearing both male and female organs, but is not homothallic. The young sexual branches are multinucleate and divide into short club-shaped multinucleate one or two celled structures, their nuclei being larger and clearer than those of vegetative hyphae. They pass beyond rudimentary stage only when the two complementary mycelia are intermingled. These sexual branches grow in an erect or oblique position at the surface of the medium (Fig. 13.48A). They may be scattered singly or associated in pairs or in groups of 3 or 4. Very frequently both structures arise at a short distance apart. The paired branches elongate, one which can be recognized as the female branch elongates more than the other which remains erect as the male branch. Soon both the branches become septate, the female especially, develops many septa and is differentiated into a long terminal, septate trichogyne, a large unicellular ascogonium, and a multicellular stalk. The trichogyne elongates rapidly, grows out and circles widely about the antheridium (Figs. 13.48B to E). The trichogyne and the antheridium fuse with each other at or near their apices. The antheridium is plump and almost cylindrical with granular contents having large number of nuclei. The stalk of the antheridium is made up of two to four cells. The ascogonium also contains a large number of nuclei which are arranged close to the ascogonial wall, leaving a central space of reticulate or finely granular cytoplasm.

When the sex organs are fully grown, continuity is established between the contents of the antheridium and of the end cell of the trichogyne by the dissolution of walls at the point of fusion of the antheridium and the trichogyne. The nuclei begin to leave the antheridium. They move from cell to cell through the trichogyne, lying in each cell against the proximal wall till an aperture is formed and they travel further. The opening of the wall is due to the dissolution of the central area. After the passage of the male nuclei, the apical cell of the trichogyne becomes crushed, proximal cells retain their form, the breach in the wall is closed after the male nuclei have passed. Later the nuclei of these cells become swollen and their contents finally disintegrate. During the passage of the male nuclei through the trichogyne, the nuclei of the ascogonium divide. The male nuclei crowd agianst the final wall, it is penetrated and they pass into the ascogonium. The male nuclei reach and pair with the female nuclei which are lying in the periphery of the ascogonium. The nuclei remaining unpaired in the ascogonium and those left behind in the antheridium and nuclei of the cells of the trichogyne, all of them disintegrate. Gywnne-Vaughan and Williamson (1932) were the first to report the passage of antheridial nuclei

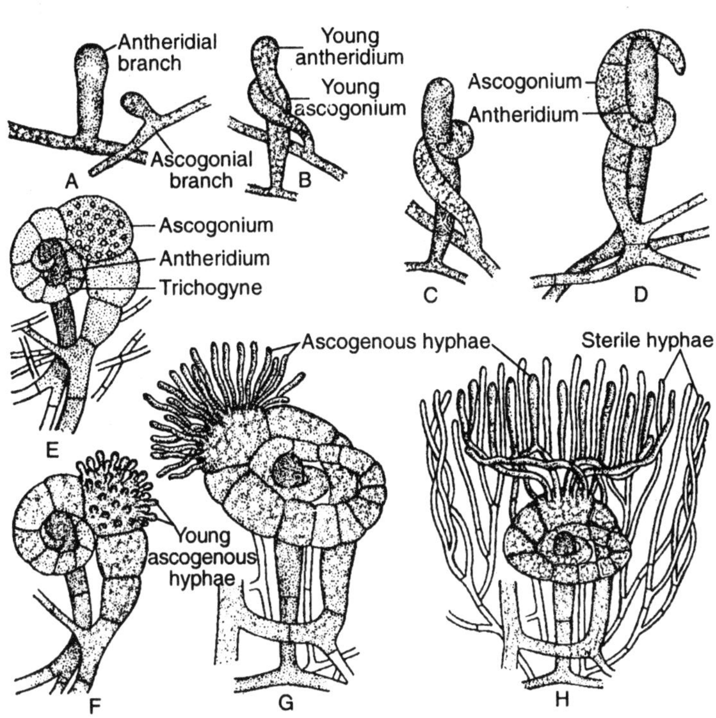

Fig. 13.48: *Ascobolus scatigenus*. Stages in development of sex organs and ascogenous hyphae. A. Antheridial and ascogonial branches being paired. B-D. Formation of antheridium and ascogonium. E. Trichogyne circling around erect antheridium. F-G. Ascogenous hyphae formation. H. Further development of ascogenous hyphae along with formation of sterile hyphae.

taking place through the cells of the trichogyne into the ascogonium. The ascogonium increases in size and buds out ascogenous hyphae (Figs. 13.48F and G). Simultaneously, with the development of the ascogenous hyphae, sterile hyphae begin to develop from the stalk of the antheridium and ascogonial stalk to form a sheath around the antheridium, ascogonium, and the growing ascogenous hyphae (Fig. 13.48H). The hyphal structures which compose the sheath may develop still below the antheridium and the ascogonium so that they insinuate themselves among the vegetative hyphae from the very beginning. In all cases the sheath sooner or later completely encloses the ascogonium and ultimately a closed globose structure is formed. The paired nuclei pass in the ascogenous hyphae which may remain unbranched or may become dichotomously branched. The ascogenous hyphae then become septate. The terminal cell of the ascogenous hypha forms a crozier and along with division of the paired nuclei three cells are formed, of which the binucleate penultimate cell elongates and becomes the ascus in which eight ascospores are developed, when karyogamy and meiosis take place. Meanwhile, some of the sterile hyphae grow along with the growing ascogenous hyphae giving rise to paraphyses which remain mixed with the asci in the

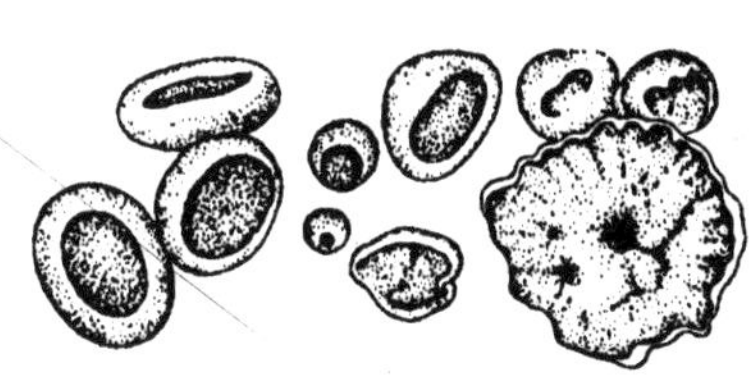

Fig. 13.49: *Ascobolus scatigenus*. Apothecia at different stages of maturity.

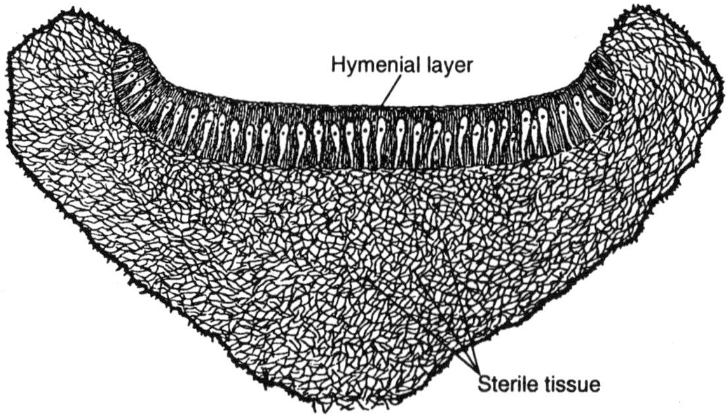

Fig. 13.50: *Ascobolus* sp. Diagram of apothecium in section showing hymenial layer and sterile tissue

hymenial layer of the apothecium. During the early stages of development the apothecium remains closed globose, later more or less cup-shaped and a fully mature apothecium bcomes saucer-shaped (Fig. 13.49). The major portion of the sterile hyphae takes part in the development of different layers of sterile tissue of the apothecium (Fig. 13.50).

The apothecia are sessile or substipitate, grow superficially or partially immersed in the substratum. They are soft, fleshy or waxy, usually greenish. The excipulum is of textura globulosa to angularis (Figs. 13.19D and 13.24). The ectal excipulum is smooth, furfuraceous or clothed with soft hairs. The hypothecium is pseudoparenchymatous. The hymenium is concave, plane or convex; dotted with the ends of the asci (Fig. 13.50). Asci are cylindrical or clavate, unitunicate and operculate, protruding at maturity (Fig. 13.51), and blued by Melzer's reagent. The emergent character of the asci is of very important diagnostic feature. 4-8 ascospores are borne in each ascus. The ascospores are at first hyaline becoming purple, fading to brown or blackish. They are free in the ascus. The ascospores are ellipsoid to subglobose smooth or becoming sculptured. Spore-sculpturing is very variable, consisting of warts, ridges or crevices or longitudinally striated. Paraphyses are septate, hyaline, slender and usually adhering together, and scarcely enlarged upwards (Fig. 13.51). Life cycle of *Ascobolus scatigenus* is presented in Fig. 13.52.

Although heterothallism has become an established fact in the Zygomycetes and higher Hymenomycetes, the acceptance of the idea of its occurrence in the Ascomycetes has not been

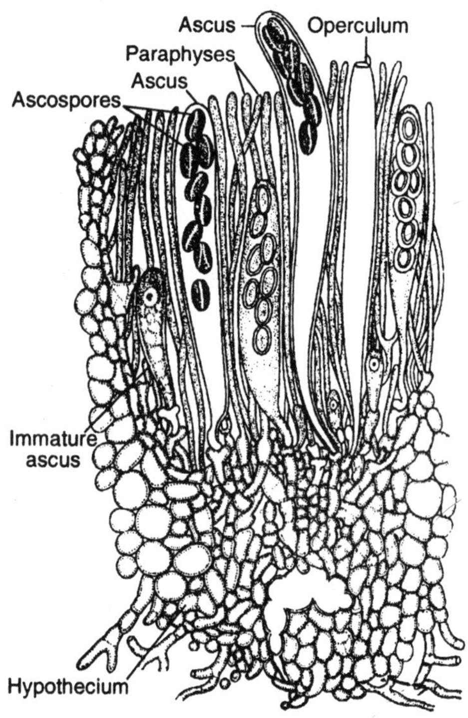

Fig. 13.51: *Ascobolus* sp. Portion of a section of apothecium showing emergent and operculate asci, ascospores and paraphyses

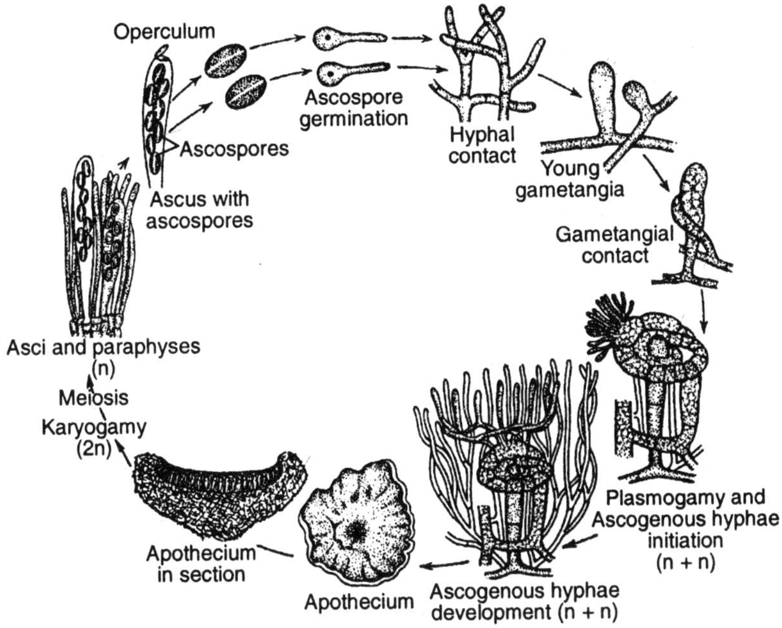

Fig. 13.52: Life cycle of *Ascobolus scatigenus*

well-established in a large number of Ascomycotina. The heterothallic nature of *Ascobolus scatigenus, A. carbonarius, A. stercorarius* and *A. immersus* has been established by Dodge (1920), Betts (1926), Ames (1930) and Rizet (1939) respectively.

Certain principles of inheritance are much more clearly expressed by these haploid organisms than they are in diploid plants and animals. For this reason as is abundantly evident from the work of the Lindegrens, Beadle and others that such haploid organisms like the species of the genus *Ascobolus* specially the heterothallic ones possess immense potentialities to be used as tools for genetical studies.

Some Indian Species: *Ascobolus carbonarius* Karst.; *A. immersus* Pers.; *A. indicus* Sanwal; *A. scatigenus* (Berk.) Brumm.

Order: **Helotiales**

This is probably the largest and most diverse order of the Ascomycotina. Most of the Helotiales are saprobes, a few are serious pathogens of plants. Ascocarps are discoid to cupulate; long-staked, club-shaped, fan-shaped or pileate. Hymenium is of limited extent, may be borne in or on some receptacle formed of fungus tissue. Asci are unitunicate, inoperculate with an apical pore (Figs. 13.13D-E). Ascospores are sometimes needle-shaped but not thread-like.

Family: **Geoglossaceae**

This family was formerly associated with the Helvellaceae to form the order Helvellales, before the importance of the type of dihiscence of the asci became apparent. The fruiting bodies are mostly found on rotton wood, decaying leaves, moss or soil, usually where plenty of moisture is available. They are long-stalked, culb-shaped, fan-shaped or pileate, more rarely discoid or turbinate varying from a few millimetres to five or more centimetres in height. Several genera have bright-coloured fruiting bodies, while those of others are black.

The hymenium either forms a closely adherent layer to the somewhat thickened upper part of the stipe or forms the upper surface of a head borne at the apex of the stipe. The fruiting body may be entirely made up of hyphae or the cortex may be pseudoparenchymatous. The hymenium usually covers only a part of the fruiting body but in some cases extending over the whole of it, or the hymenium may occupy the upper surface only. The fruiting bodies are usually elastic or fibrous rather than brittle, a few leathery, others are distinctly gelatinous.

Asci are with distinct paraphyses. The ascus contains eight ascospores and opens by the ejection of a plug. The ascospores are ellipsoid and one- or two-celled or long elliptical to filiform or narrow several to many septate. They are hyaline to smoky to dark brown.

Genus: *Geoglossum*

This is the largest genus in the family Geoglossaceae. It grows on rotten wood, log, stumps, on wet ground, or among leaves or rich humus. Fruiting bodies are fleshy, erect, stipitate, clavate (Figs. 13.53A-D), hymenium covering only the upper portion, black or brownish black, commonly known as "earth-tongues". The terminal fertile region becomes club-shaped, laterally compressed, or forming a cup, or a pileus. In some of the simpler forms, as in *Geoglossum hirsutum,* there is no clear line

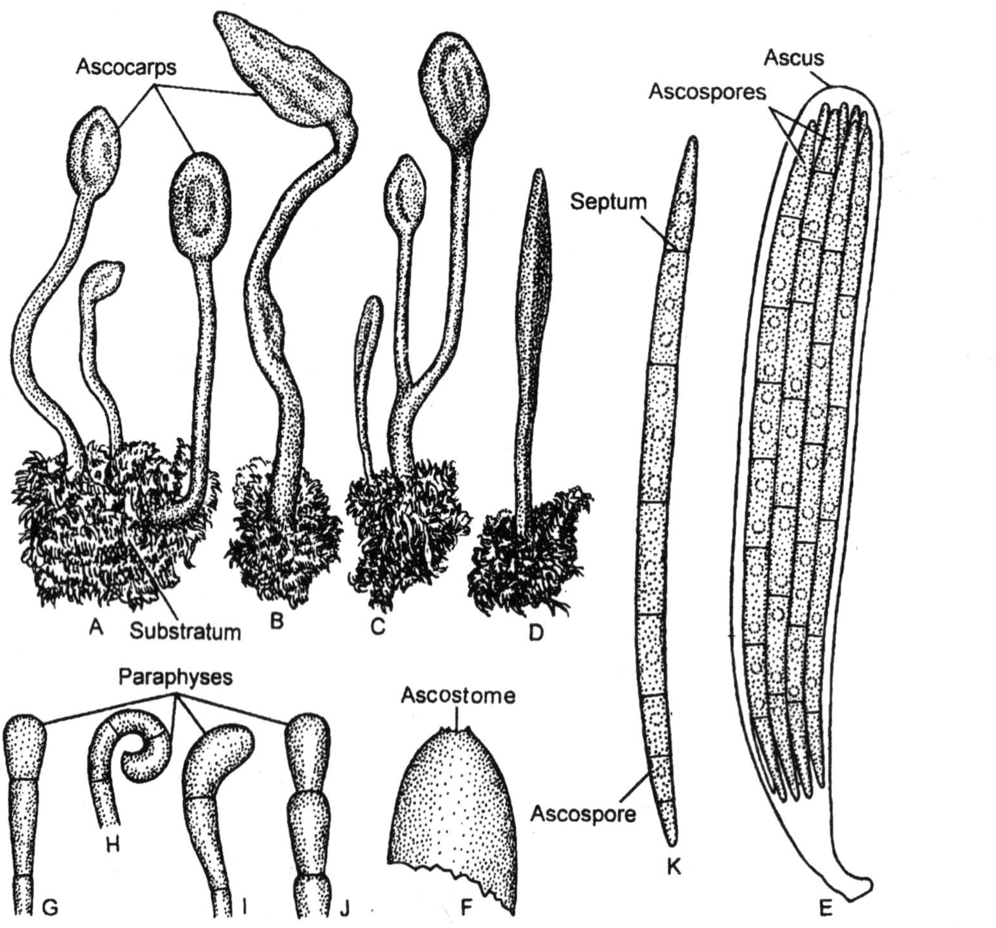

Fig. 13.53: *Geoglossum* sp. A-D. Ascocarps of varying shape and size. E. An ascus bearing septate ascospores. F. Portion of an ascus showing ascostome. G-J. Different types of paraphyses. K. An ascospore showing nature of septation

of demarcation between the fertile and sterile regions. The genus *Geoglossum* is separated from its closest genus *Trichoglossum* by the absence of setae from the hymenium.

The fruiting bodies arise as a tangle of closely twined hyphae which are elongated at the periphery and entirely surrounded the tangle. In the middle of the base, however, they are more rounded or polyhedral. The middle cells elongate upwards and thus cause the elongation of the whole fruiting body, while the basal cells remain irregularly angular with thickened walls. The ascogenous hyphae push from the ground tissue and form asci. The sexual organs undergo a peculiar degeneration. Antheridia are lacking; the nuclei in the ascogonia are arranged in pairs and migrate into the ascogenous hyphae. Thus the sexual processes are autogamous.

Asci are unitunicate and inoperculate; cylindric-clavate, pore turns blue with iodine, and 8-spored (Figs. 13.53E and F). Ascospores are cylindric-clavate, brown, multiseptate lying parallel in the ascus (Figs. 13.53E and K). The ascospores are apt to protrude from the ripe asci giving the hymenium a downy appearance. Paraphyses are numerous, slender, septate, slightly longer than the asci, may be abruptly swollen at the tips or cells may be swollen to give a beaded appearance (Figs. 13.53G and J).

Some Indian Species: *Geoglossum alveolatum* Durand; *G. hirsutum* Pers.; *G. ophioglossoides* (L.) Sacc.

Order: **Erysiphales**

The Erysiphales are essentially plant-inhabiting obligate parasites being limited in habitat to the surface of green plants. They fail to grow on artificial media. Their septate mycelium without penetrating through the host epidermis remains superficially on the surface of the host. The mycelium, however, sends out haustoria which penetrate the host epidermal cells. Through these haustoria the mycelium draws nutrition from the host cells. The mycelium is frequently quite conspicuous. Asexual reproduction takes place by conidia. Their ascocarps are cleistothecia with appendages. The cleistothecia are developed from a superficial mycelium without the development of a stroma. Large, globose or ovoid asci are arranged in a single basal layer or forming a tuft in a cleistothecium. The asci are persistent. The cleistothecium possesses pseudoparenchymatous dark wall. Different authors have subdivided the order into more than one families. Here only one family Erysiphaceae is recognized.

In the Erysiphales the ascocarp is technically a cleistothecium because it has no natural opening, but the asci arise in a fascicle at a single level near the base of the cleistothecium and thrust into the mass of pseudoparenchyma which at first occupies most of the centrum. Rupture of the cleistothecial peridium is by swelling of the asci as they mature, and either the asci themselves, or the ascospores, are dispersed. Sometimes the asci are thrown out violently by sudden splitting open of the peridium. These features closely connect the Erysiphales with perithecia producing Ascomycotina—the Pyrenomycetes rather than with the Plectomycetes.

Family: **Erysiphaceae**

The family is very well-known on account of its abundance everywhere, its simplicity of structure, and its possession of typical ascigerous and conidial stages. Members of the family are easily recognizable since they form a coating of white conidia, conidiophores and mycelium upon the surface of the host producing a powdery appearance by infecting leaves (Fig. 13.54A), stem, flowers and fruits. They are largely known as powdery mildews. The powdery mildews, except *Erysiphe graminis* which infects grasses, are usually limited to dicotyledonous angiosperms. They produce powdery mildew disease. Except perhaps the powdery mildew of grape-vine caused by *Uncinula necator,* the powdery mildews induce very mild infection and cause little damage to the hosts. Most of the genera have a limited host range with a large number of physiological races. Some of the important members of the Erysiphaceae are noted here: *Erysiphe graminis* is wide-spread on wheat, oat, barley, rye and grasses; *Erysiphe cichoracearum* seriously attacks cucurbits, *Sphaerotheca humuli* causes powdery mildew of hops and *Sphaerothec pannosa* attacks roses; *Phyllactinia corylea* has world-wide distribution and has very wide host range including plants of economic importance. Geographically the Erysiphaceae are widely distributed, practically of world distribution,

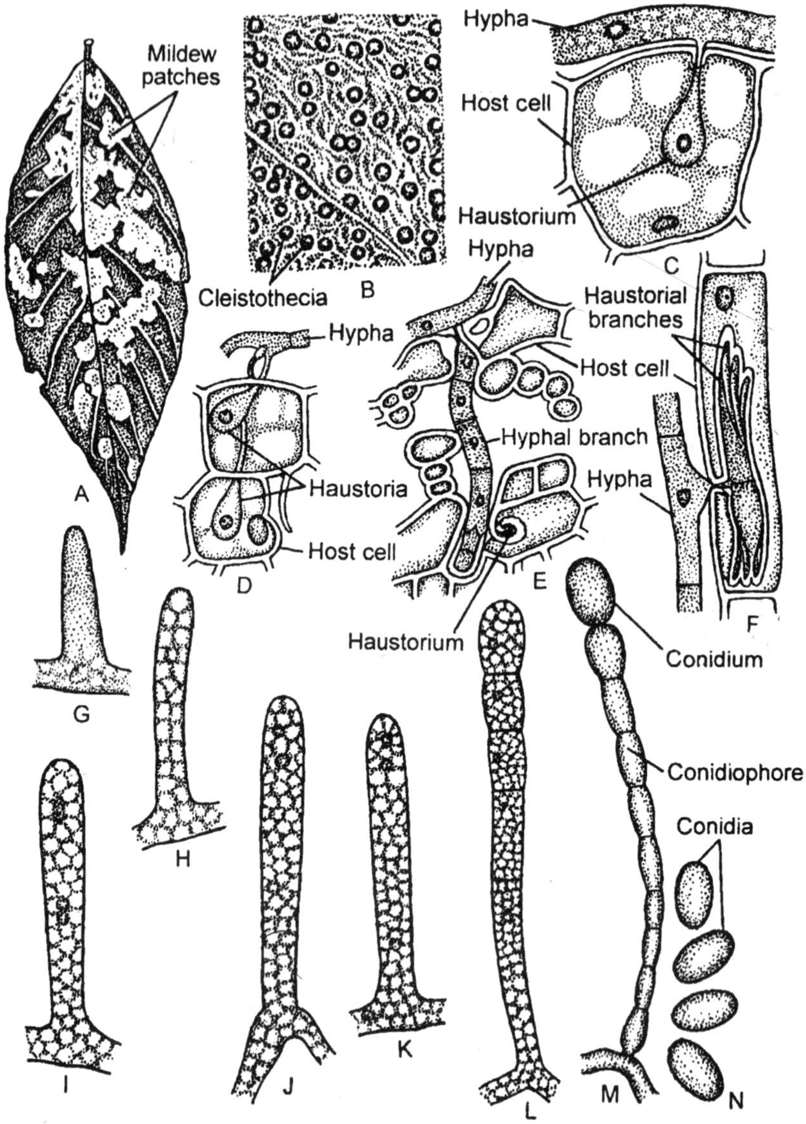

Fig. 13.54: Structural features of the Erysiphaceae. A. White mildew patches on the leaf surface. B. Cleistothecia on host surface. C. Bulbous haustorium, a type common among the Erysiphaceae. D. *Uncinula salicis.* Haustoria penetrated the hypodermal cells. E. *Phyllactinia corylea.* Hyphal branch penetrating the mesophyll sending haustoria. F. *Erysiphe* sp. Digitately branched haustorium. G-M. *Sphaerotheca* sp. Stages in development of conidiophore and conidia. N. Nature of conidia in the Erysiphaceae

but they are more abundant in the temperate zones than elsewhere. Working with *Sphaerotheca humuli,* Harper (1895) traced for the first time the development of antheridia and ascogonia, and the cytological features that accompany nuclear fusion and nuclear division during the development of asci and ascospores.

The mycelium of the Erysiphaceae is usually hyaline and is branched, septate and its cells are uninucleate. The mycelium is hyaline when young and greyish-brown when old. It is entirely external to the host tissues except for that of *Phyllactinia.* The mycelium fastens to the host epidermal cells by haustoria. The haustoria may be grouped in three general categories: (i) those arising directly from the lower surface of the mycelium; (ii) those arising at the side of the mycelium as small semicircular processes; and (iii) arising from more or less deeply-lobed lateral swellings of the mycelium. In most species the haustoria are bulbous and uninucleate (Fig. 13.54C). In certain species of *Erysiphe* they are ellipsoid, with digitat processes at the ends (Fig. 13.54F). In *Phyllactinia corylea* some hyphae of the mycelium remain external and do not develop any haustoria, and the rest of the hyphal branches enter into the host through the stomata. When they come in contact with the mesophyll cells they send haustoria in them to obtain nourishment from the host cells (Fig. 13.54E). This is how the entire mycelium receives nutrition. Again, in *Uncinula salicis* the haustoria may even penetrate the cells of the hypodermis (Fig. 13.54D). Some members of this family perennate as mycelium.

A few days after infection, several of the hyphae grow erect, perpendicular to the host surface. The cells are short, at first cylindrical, then become barrel-shaped and finally, when ready for dissemination become elliptical and are then known as conidia. The conidia arise in basipetal succession on simple unbranched, scattered conidiophores. They are produced in chains (Figs. 13.54G-M). The conidia are hyaline smooth and thin-walled, and one-celled (Fig. 13.54N). They vary greatly in size with nutrition and other conditions. The conidia are short-lived summer spores. Conidial production goes on at a quick rate and they are produced in profuse quantity which is responsible for giving the powdery appearance on the infected plant parts. The name powdery mildew very appropriately describes the white, mealy appearance of affected plant parts like leaves, stems, flowers, and fruits. The whiteness being imparted by the profuse growth of external mycelium and conidia. With favourable conditions each conidium germinates by germ tube which ultimately gives rise to new mycelium.

Sexual reproduction occurs at the end of summer, when conidial production slows down and eventually ceases. The white powdery appearance of the host surface now changes to greyish or brown shade, and the hyphae prepare to form ascocarps which are cleistothecia. The cleistothecia are large enough to be visible to the naked eye as black dots on the infected host surface (Fig. 13.54B). Sexual reproduction involves the production of recognizable ascogonium and antheridium. At their point of contact, a passage is developed by the dissolution of walls. The antheridial nucleus passes through this passage into the ascogonium where plasmogamy takes place which is followed by the development of ascogenous hyphae and ultimately asci and ascospores are developed within the cleistothecium. Some species are homothallic, others heterothallic. The cleistothecium is a small more or less spherical structure bearing appendages which are commonly simple, curled or branched at the tip, in a manner characteristic of each genus. The function of these appendages is rather dubious, but their structure furnishes a useful taxonomic character. They serve by hygroscopic movements to aid in the distribution of the fungus. The cleistothecium contains one to several asci which arise in one or more tufts from the base. The asci are globose to ovoid, and may have short stalk. The genera *Podosphaera* and *Sphaerotheca* produce only one ascus in each cleistothecium. The asci of the Erysiphaceae bear two to eight ascospores. Members of the Erysiphaceae survive the winter as ascospores in the asci developed in the cleistothecium, and in the spring both cleistothecium and asci absorb water and swell. The cleistothecium cracks open, and the asci discharge the ascospores which on germination on a suitable host produce new mycelia. The origin and structure of cleistothecia were, for the first time, traced and described by De Bary in 1863.

In *Sphaerotheca humuli* sexual organs originate at the points where two hyphae cross or come into contact with each other. Each of the adjoining hyphae sends out an upright branch; the one which enlarges and becomes club-shaped forms the ascogonium, the other remains more slender is the antheridium, both are cut off by the septa. The antheridium comes into close contact with the ascogonium, and grows up with it. The antheridium soon overtops the ascogonium and bends over its apex. Both the organs are uninucleate structures. The ascogonium enlarges and the antheridium lengthens. The antheridial nucleus divides into two daughter nuclei and simultaneously a septum appears which divides the antheridium into an uninucleate stalk cell and

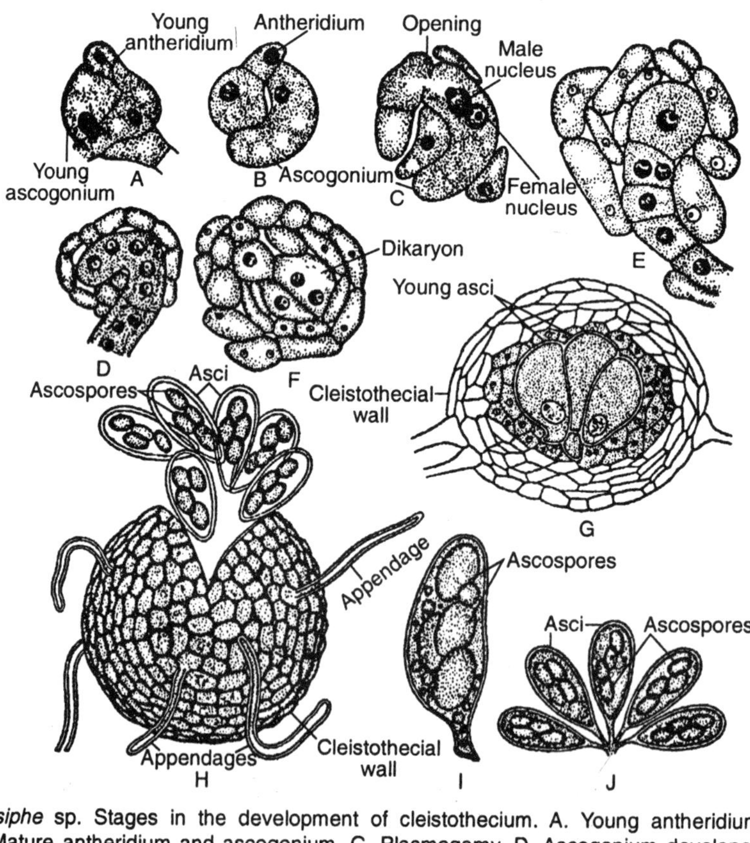

Fig. 13.55: *Erysiphe* sp. Stages in the development of cleistothecium. A. Young antheridium and ascogonium. B. Mature antheridium and ascogonium. C. Plasmogamy. D. Ascogonium developed into multinucleate tube-like structure. E and F. Subterminal binucleate cell of the tube-like structure. G. Section of young cleistothecium. H. Liberation of asci by rupturing of cleistothecial wall. I. Ascus with ascospores. J. Asci with ascospores

an uninucleate terminal cell which is the antheridium proper. At the point of contact of the antheridium and ascogonium, the common walls dissolve resulting an opening through which the contents of the antheridium pass into the ascogonium. This is followed by plasmogamy between the protoplasm of antheridium and that of ascogonium. Harper (1895) believed that karyogamy of the antheridial and ascogonial nuclei takes place in the ascogonium. But researches of Bergmann (1941) reveal that there is no such nuclear fusion in the ascogonium. The two nuclei form a dikaryon in the ascogonium and undergo mitotic division there. The ascogonium then elongates and becomes divided by transverse septa into a 3- to 5-celled structure, whose binucleate penultimate cell becomes the ascus mother cell. The ascus mother cell enlarges to develop into an ascus. Its two nuclei fuse to form a diploid nucleus which by three divisions, of which one is reductional, poduces eight haploid nuclei, around each of which an ascospore is developed. While the ascus development is going

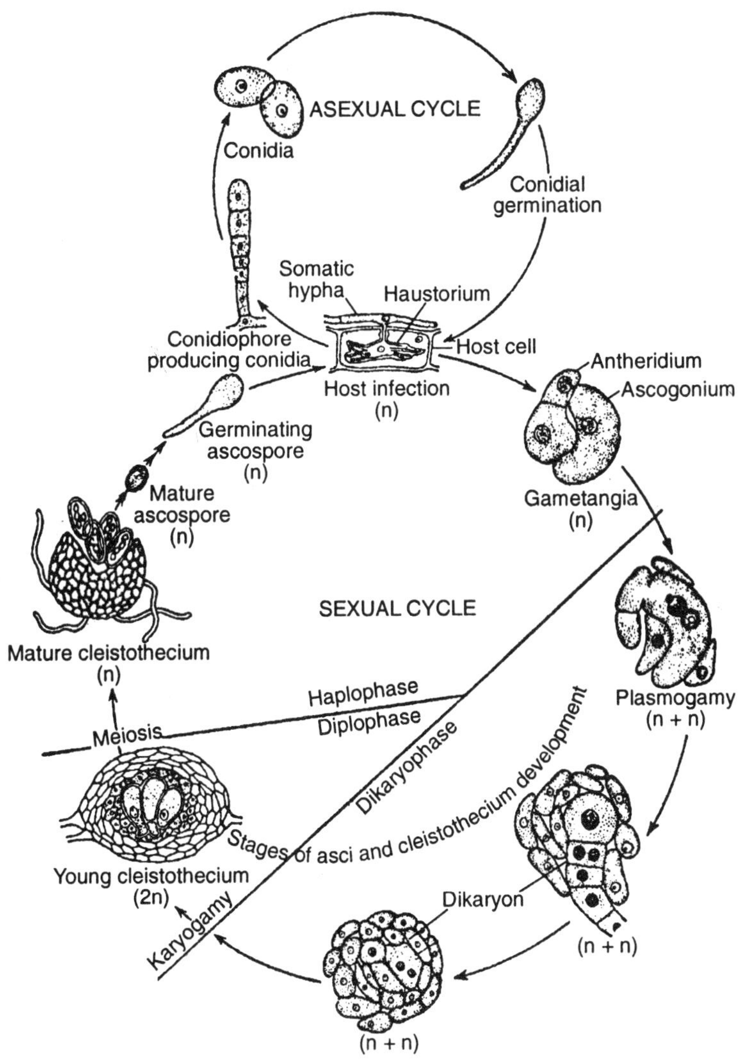

Fig. 13.56: Life cycle of *Erysiphe* sp.

on, the hyphae next below the ascogonium send out several branches which grow rapidly around the ascogonium, completely envelope it forming a thick wall, the peridium. From the inner cells of the peridium thus formed, short branches filled with very dense protoplasm grow inwards and apply themselves closely to the ascus, probably supplying it with food. The cells supplying nourishment to the ascus are the **nurse cells.** The outer cells of the peridium become thick-walled, and form a dense protective layer. Some of its superficial cells grow out into long filamentous appendages. The cleistothecium is subglobose and bears mycelioid appendages. At first its colour is white but turns black at maturity. The ascus absorbs water, swells up, and bursts the cleistothecium, whereupon its own membrane dehisces at the top, and the ascospores are violently shot into the air. Nature and stages of development of cleistothecium, ascus and ascospores have been indicated in Fig. 13.5.

In *Erysiphe,* sexual organs are constricted from the erect branch-tips of two adjacent but distinct hyphae. The ascogonium is large, ovoid in shape, while the antheridium is smaller and more slender, but each contains one nucleus only (Figs. 13.56A & B). The tips of these organs come together and plasmogamy takes place in the ascogonium when the antheridial protoplasm comes in contact with that of the ascogonium through an opening that is developed by the dissolution of walls at the point of contact of the antheridium and the ascogonium (Fig. 13.57C). The male and female nuclei then divide mitotically. The ascogonium develops to a multinucleate tube-like structure which divides simultaneously into several cells, the subterminal one of which is binucleate (Figs. 13.57D-F). This binucleate cell gives rise to one to several ascogenous hyphae, the cells of which give rise to asci. Karyogamy and the development of ascospores take place in a process similar to that of *Sphaerotheca humuli.* By the time the asci were being developed, sterile branches have arisen from the base of the ascogonium surrounding it, closely intertwining to form the cleistothecial peridium. The outer surface of the peridium eventually becomes dark-brown, and develops long filamentous mycelioid appendages. The inner layer of tissue forms a source of food for the asci which are developing in the cleistothecium (Fig. 13.55G). The cleistothecium bears more than one asci, in each of which 2 to 8 ascospores are developed. The asci and ascospores are liberated out by the rupture of the cleistothecial wall (Figs. 13.55H-J). Life cycle of *Erysiphe* is presented in Fig. 13.56.

The family Erysiphaceae is subdivided into six genera depending on the nature of appendages and the number of asci in the cleistothecium. The genera are: *Sphaerotheca* with mycelioid appendages and one ascus in the cleistothecium (Fig. 13.57B), *Erysiphe* bears mycelioid appendages having several asci in the cleistothecium (Fig. 13.57A), *Podosphaera* possesses appendages having tips dichotomously branched and one ascus in the cleistothecium (Fig. 13.57E), *Phyllactinia* having appendages with bulbous base and several asci in the cleistothecium (Fig. 13.57D), *Microsphaera* possesses appendage tips dichotomously branched and several asci in the cleistothecium (Fig. 13.57F), and in *Uncinula* appendage tips are coiled and cleistothecium bears several asci (Fig. 13.57C).

Some Indian Species: *Erysiphe acaciae* Blumer; *E. communis* Wallr. ex. Fr.; *E. galeopsidis* DC.; *E. graminis* DC.; *E. polygoni* DC.; *Microsphaera alni* (Wallr.) Salm.; *Phyllactinia guttata* (Fr.) Lev.; *P. heterophragmitis* Ramakr.; *P. leucotricha* (Ellis and Everh.) Salm.; *P. terminaliae* Ramakr.; *P. tridactyla* (Wallr.) De Bary; *Sphaerotheca euphorbiae* (Cast). Salm.; *S. macularis* (Wallr. ex Fr.) Cooke; *S. pannosa* (Wallr.) Lev.; *Unicinula necator* (Schw.) Burr.; *U. salicis* (DC.) Wint.; *U. tectonae* Salm.

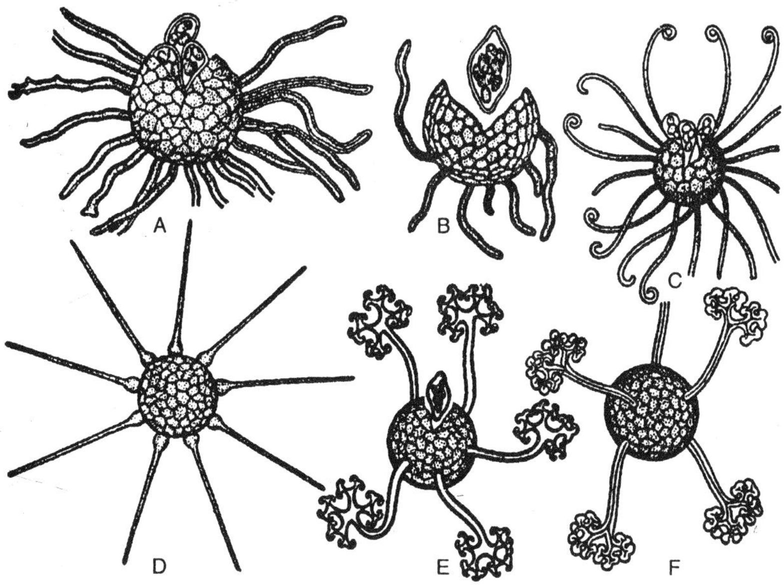

Fig. 13.57: Taxonomic characteristics of the different genera of the Erysiphaceae. A. *Erysiphe*. Cleistothecium with several asci and mycelioid appendages. C. *Uncinula*. Cleistothecium with several asci and appendage tips coiled. D. *Phyllactinia*. Cleistothecium with several asci and appendages with a bulbous base. E. *Podosphaera*. Cleistothecium with one ascus and appendages with tips dichotomously branched. F. *Microsphaera*. Cleistothecium with several asci and appendage tips dichotomously branched

Order: **Clavicipitales**

The order Clavicipitales includes fungi which subsist on plant tissue either as parasites or as saprophytes. Certain are entomogenous. Again, few grow on fruit bodies of various Basidiomycetes. Many of the Clavicipitales reproduce asexually by conidia or chlamydospores. They bear ascocarps which are perithecia. The perithecia are either partially or completely immersed within a stroma with some exceptions where stroma is absent. The perithecium has a wall of its own. Nearly all perithecia possess soft, brightly coloured wall. The colour of perithecial wall may be bright red or yellow purple and various pale shades. Asci originate from the base and the side of the inner wall of the perithecium. The asci are clavate or cylindric in shape. They bear usually eight, in some species four and in others many ascospores. The ascospores are hyaline to dark-coloured, elliptical or filiform and may be one- or more-celled. Ostiole of the perithecium is lined with periphyses.

The single family Clavicipitaceae has been classified by Bessey (1950) and Martin (1961) under the order Hypocreales, and Miller (1949) an Luttrell (1967) under the order Sphaeriales. Alexopoulos (1962) has recognized the order Clavicipitles following Nannfeldt (1932), Gaumann (1952), and Dennis (1960).

Family: **Clavicipitaceae**

The family Clavicipitaceae includes fungi parasitic on angiospermic plants, insects, and on the fruit bodies of some fungi. They form conspicuous stroma in which perithecia are embedded. Asci are cylindrical each of which has a cap with perforation through which thread-like ascospores are released.

Genus: *Claviceps*

The genus *Claviceps* is parasitic on rye and other grasses. It is characterized by filiform ascospores which germinate on the flowers of the host. The mycelium ramifies in the ovary and forms an acervulus from peripheral region of which conidia are produced. The conidia are dispersed by insects to the ovaries and thereby the infection is spread. The mycelium of an infected ovary is ultimately transformed in a hard pink or purplish sclerotium. The sclerotium on germination produces numerous stromata with violet stalks and rose-pink heads in which perithecia are immersed. Each mature perithecium bears several elongated cylindrical asci, each containing eight filiform ascospores.

Some Indian Species: *Claviceps digitarie* Hansf.; *C. paspali* Stev. and Haller; *C. purpurea* (Fr.) Tul.; *C. pusillla* Ces.; C. *viridis* Padwick and Azmatullah Khan.

Claviceps Purpurea (Fr.) Tul.

The species of *Claviceps* are parasitic on grasses, including rye and barley and occasionally wheat. *Claviceps purpurea* is a parasitic fungus. It causes ergot disease of rye. The fungus attacks only the inflorescence of its host. Grains of rye are replaced by blunt trilateral horn-shaped black sclerotia known by the name of **ergots** (Fig. 13.58A). These sclerotia are resting stages. In the mature state they are composed of a dirty-white medullary tissue surrounded by a violet-brown rind. The medulla or the central core consists of a prosenchyma formed of looser cylindric prismatic cells of pale-brown colour which are about one to four times as long as breadth (Fig. 13.55). The cells are arranged in straight or sinuous longitudinal rows possessing the characters of the fungal hyphae. The medulla has usually shorter and broader cells towards the periphery of the sclerotium than in its centre. The cells contain large colourless drops of oil. The surrounding rind which is at every point firmly connected with the medulla, is composed of one or two layers of cells with contents showing no oil and has a dark-brown colour. In between the rind and the medulla there are few layers of pseudoparenchyma of longitudinally arranged or irregular branched rows of isodiametric cells resembling parenchyma of higher plants. The cells are narrow and pale violet-brown in colour. The sclerotia contain valuable therapeutic properties. They yield about a dozen different alkaloids. These alkaloids have different pharmacological properties. Some of these alkaloids are chiefly used in gynaecology and obstetrics. Most ergot of trade (the sclerotia of *Claivceps purpurea*) comes from naturally diseased rye, but lately ergot has been produced by artificial infection from the fungus in culture. Some of the alkaloids contained in the sclerotia are: **ergometrine** causes rapid contraction of the uterus; **ergotoxine, ergotamine** have very similar action but much slower with prolonged effect and **ergotoline** has physiological importance. Ergotamine tartrate which centres its activity on constricting the dilated blood vessels in the head, is frequently used to relieve headache. For these properties the sclerotia are widely used during childbirth and to control haemorrhage after delivery. The sclerotia are poisonous to live-stock and human beings, often causing disease known as **ergotism**—ergot poisoning, a disease also known as St. Anthony's fire. This is the effect of poisonous alkaloids that are also present in the sclerotia. *C. purpurea* has been cultured under laboratory conditions but forms neither sclerotia nor alkaloids under these conditions.

The sclerotium on germination usually produces several spherical stalked **stromata** (sing. **stroma**), also known as **sphaeridia** (sing. **sphaeridium**) (Figs. 13.58B and C). The sphaeridium is covered with numerous closely set **papillae** (Fig. 13.60A), each papilla surrounding an ostiole leading down to a perithecium which is sunk halfway beneath the surface. The perithecia form a single layer of numerous flask-shaped cavities which follow

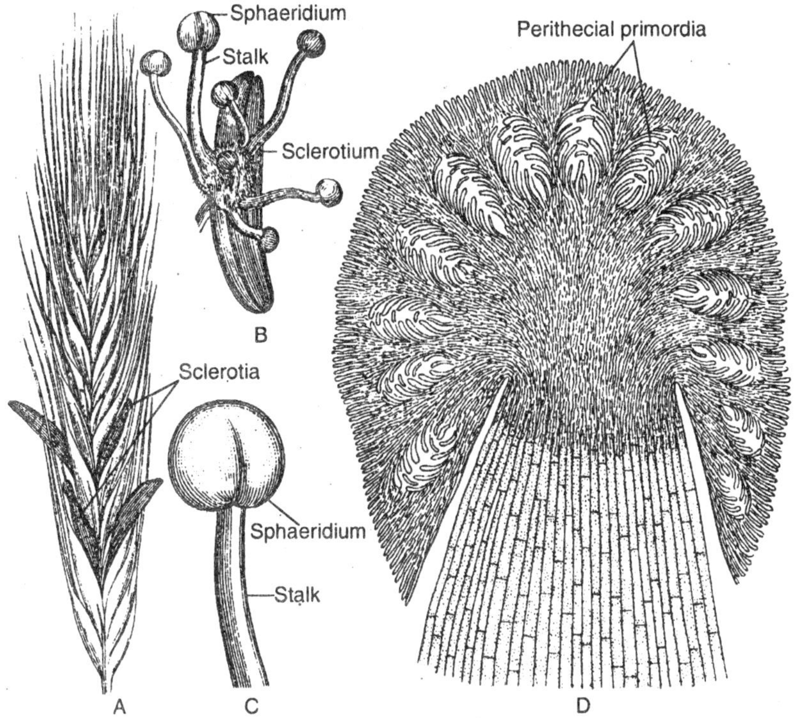

Fig. 13.58: *Claviceps purpurea*. A. Sclerotia on inflorescence of rye. B. Germinating sclerotium showing stalked sphaeridia. C. A single stalked sphaeridium. D. Section through a sphaeridium showing primordia of perithecia.

closely the contour of the sphaeridium (Fig. 13.60B). The actual mechanism of formation of the cavities is not clearly known. In all possibility it is affected by the mutual separation of the individual hyphae. The roof of the cavity ultimately develops into the wall of the perithecium, and its floor becomes the hymenium (Fig. 13.58D).

Details of the formation of perithecia were first elucidated by Killian (1919). According to him, the sex organs are produced from the terminal cell of a hypha. The hypha is usually unbranched, elongated composed of multinucleate cells. It is easily recognizable from the adjacent stromatic hyphae. The terminal cell of the hypha swells (Fig. 13.59A), the nuclei divide and arrange themselves in pairs. It becomes broader and thicker and then develops into an ascogonium from whose base branch out one or two antheridia (Figs. 13.59B-D). The antheridia elongate considerably in which nuclei divide repeatedly. They are longer and more slender than the ascogonium. One of these antheridia comes in contact with the ascogonium at a point where the ascogonium forms a small papilla on the side toward the antheridium (Fig. 13.59E). The walls at the point of contact dissolve and opening is formed through which the nuclei of the antheridium migrate into the ascogonium (Fig. 13.59F). The emptied antheridium collapses and disappears completely. Ascogenous hyphae then grow from the ascogonium. The ascogenous hyphae are composed of binucleate cells. Asci are formed at the tips of ascogenous hyphae by typical crozier formation. The young perithecium soon acquires the form of an elongated cone by growth and forms a canal, the ostiole

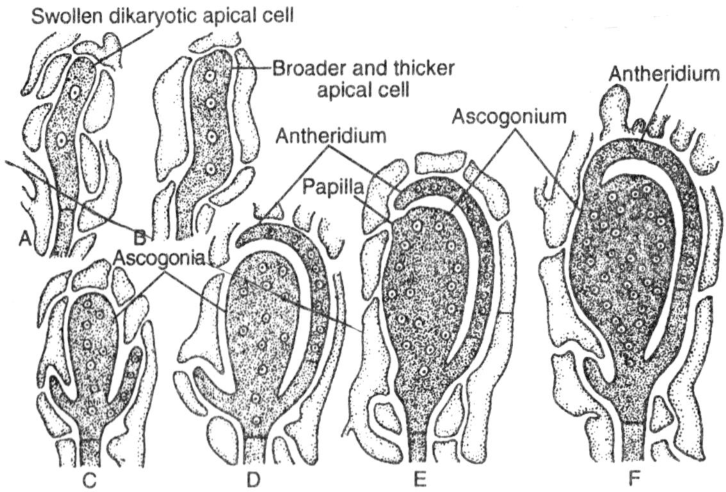

Fig. 13.59: *Claviceps purpurea*. A-D. Successive stages of development of antheridium and ascogonium. E. Ascogonium with papilla. F. Plasmogamy.

(Fig. 13.60C) which is beset all round from below upwards with periphyses. The perithecia contain long, curved hyaline asci, narrowed at both ends (Fig. 13.60D) and surrounded by paraphyses which are thickest at the apical end. An ascus contains eight long cylindrical filiform, slightly curved, hyaline ascospores (Fig. 13.60E). The ascospores under the influence of moisture, are extruded and collect at the ostioles of the perithecia for dispersal. They are either forcibly ejected into the air by raindrops or dispersed by insects. This constitutes the **sphaeridial stage** or **ascigerous stage** or **perfect stage.** The ascospores during germination under favourable conditions put out germ tubes at several points.

Hyphal branches developed from the germ tubes infect the ovary only at the time of flowering and the infection remains restricted to the pistil only. Gradually the hyphal branches spread through the entire ovary replacing the ovarial tissue by mycelial mat. The ovary is thus changed into a white fungus-body of nearly its own shape. The mycelium at first occupies the base of the young ovary and then completely replaces the entire ovary. During the early stage of infection a sweet sticky yellow fluid commonly known as 'honey-dew' is exuded from the infected flowers. In this fluid remain embedded minute oval uninucleate conidia cut off from the conidiophores that are developed in the peripheral portion of the soft cottony mycelial mat, an **acervulus-like** structure, which has partly replaced the ovary tissue (Fig. 13.60). This is known as **honey-dew** or **Sphacelia stage** and was originally known as *Sphacelia segetum.* Before the establishment of connection between this stage and the ascigerous stage, *Claviceps purpurea* was known by its conidial stage *Sphacelia segetum.* The full life history of the fungus was established by Kuhn in 1863. The sweet exudate which accumulates on the surface of the inflorescence like dewdrop is also known as honey-dew. It is mixed with conidia. Insects on being attracted by the honey-dews visit diseased inflorescences and invariably eat the honey-dews. At the same time the conidia are carried by the insects from infected to healthy plants and thereby the infection is spread. On the completion of the conidial stage the mycelial mat develops considerably even beyond the ovary and is ultimately metamorphosed into a compact tissue which becomes hard purple or bluish-black

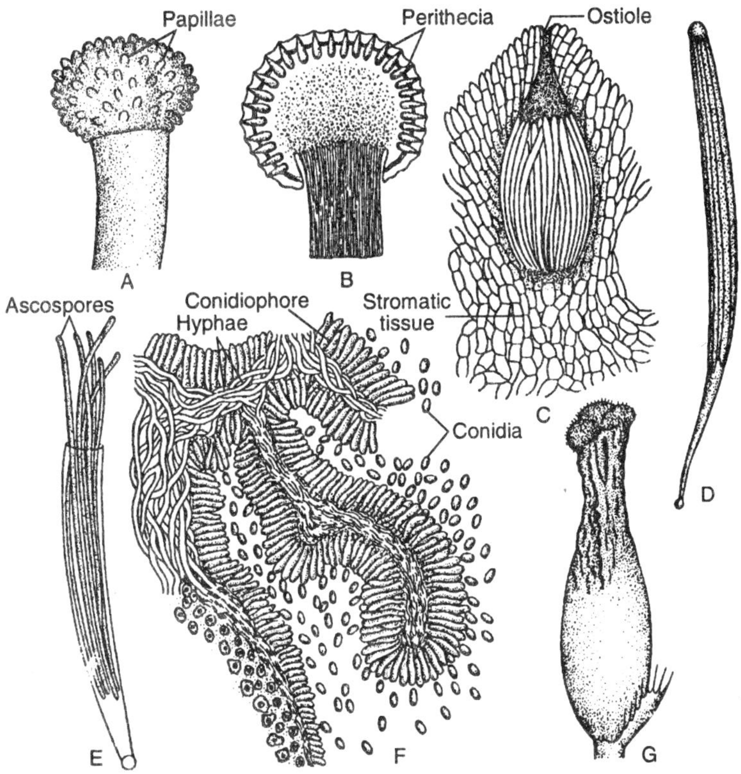

Fig. 13.60: *Claviceps purpurea*. A. Papillae on the surface of the sphaeridium. B. Longitudinal section of a sphaeridium showing single layer of perithecia. C. A perithecium. D. An ascus. E. Ascus discharging ascospores. F. Conidial, 'honey-dew' stage on the ovary. G. Immature sclerotium.

sclerotium, the **sclerotial stage** (Figs. 13.60G and 13.58A). A sclerotium is several times larger than the normal grain of the host plant. After a period of rest, usually lasting till the following season the sclerotium germinates. *Claviceps purpurea* is differentiated into a number of physiologic races. Life cylcle of *Claviceps purpurea* is presented in Fig. 13.61.

The Sphaeriales include fungi, mostly saprophytic, e.g., *Neurospora;* but many are parasitic, e.g., *Phyllachora,* especially on seed plants. Others are coprophilous, e.g., *Sordaria.* Among the Sphaeriales, some members produce conidia which may be of the same type or of a wide variety of conidial types, whereas others have lost their ability to produce conidia completely. In the conidia producing Sphaeriales, the conidia may behave as spores or as spermatia and take part in the sexual reproduction. The majority of the Sphaeriales have either lost the ability to produce antheridia or even if the antheridia are developed they are non-functional. Sexual reproduction usually takes place by spermatization.

The Sphaeriales have an ascocarp, that is, a perithecium which varies in shape from globose to pear-shaped, or elongated with or without a long neck. A perithecium is either formed directly from the loose mass of mycelium, **non-stromatic** or is associated with a well-developed strong stroma standing on top or sunken within it, **stromatic.** The wall of the perithecium is definite, whether the latter is stromatic or non-stromatic. The asci are club-shaped or cylindrical. Many of the Sphaeriales lack paraphyses, in others the paraphyses are commonly inter-

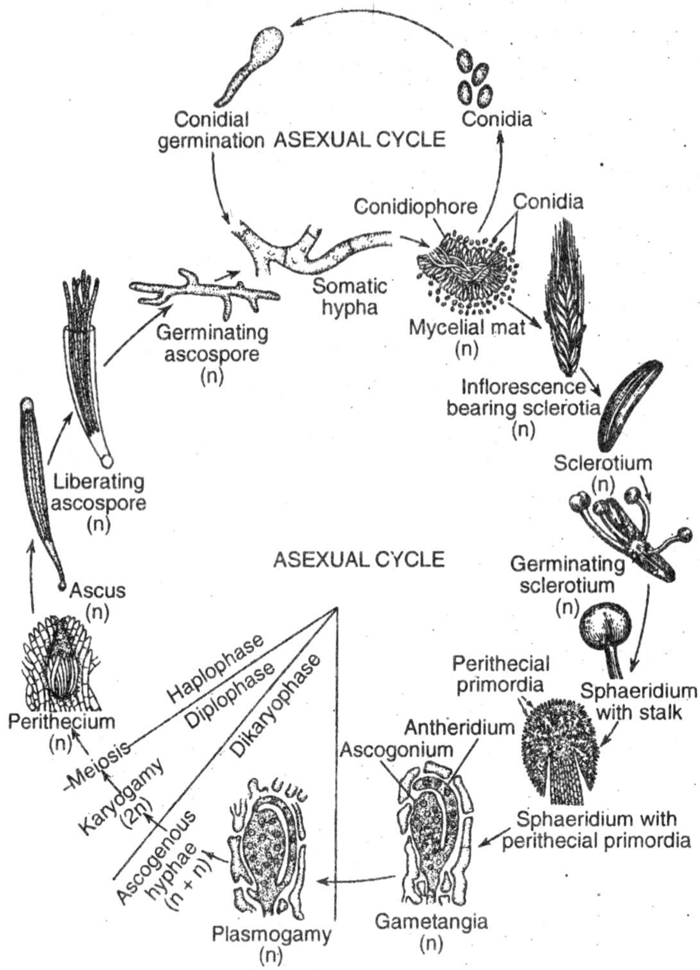

Fig. 13.61: Life cycle of *Claviceps purpurea*

mingled with the asci. The ascospores, in general, are forcibly expelled through an apical pore of the ascus. But in cases where the asci are evanescent, the ascospores are extruded through the ostiolar neck in mucoid droplets. The number of ascospores in an ascus is mostly 8, but occasionally 4, or even 2. Again, there are cases where large number of ascospores are formed in an ascus. In certain cases the number of ascospores may even be more than thousand in each ascus.

Family: **Sordariaceae**

The Sordariaceae are saprophytes, mostly occurring on dung or on decaying plant parts. The distinguishing feature of this family is the somewhat beaked, superficial, separate perithecia. The perithecia are dark-brown or black with membranous or carbonous wall. Their long, club-shaped or cylindric asci are often interspersed with paraphyses, in most of the species the paraphyses are evanescent. The ascospores are dark-brown to black and variously ornamented. Conidial stages are infrequent. Sexual reproduction may be by somatogamy or spermatization. The two genera *Sordaria* and *Neurospora* are very well-known for their great value as tools for the genetical studies.

Genus: *Neurospora*

Neurospora is one of the best known and most intensively studied genera in the Ascomycotina. The species of *Neurospora*, particularly *Neurospora sitophila*, the red bread mold or commonly known as bakery mold, frequently infest bakeries causing severe damage to the baking industry. It also causes considerable trouble as a contaminant of cultures in bacteriological and mycological laboratories. The genus *Neurospora* for the last few decades, is of considerable interest to a large and diverse group of geneticists and biochemists who have established, Neurospora Genetics. Some of the pioneer workers are: Dodge, Lindegren, Beadle and Tatum. *Neurospora sitophila, N. crassa*, and *N. tetrasperma* have been widely used as tools for fundamental biological studies.

In the early 1840's the occurrence of a pink mold on bread was detected in France and in 1841 a French pharmacist brought the mold to the attention of student of the fungi, who published a description, a name, and some detailed microscopic drawings. Some time later, the mold caused such extensive spoilage in army bakeries, that a commission was appointed to examine the matter. Subsequently, the mold was noted not only in Western Europe, but as far away as Brazil and Japan, and in the United States.

The mold was known as *Monilia sitophila*—'monilia' from a word necklace, because it bears spores in series like strings of beads, and 'sitophila', meaning bread-loving; and included under the Fungi Imperfecti.

In 1927, perfect stage of the mold was collected in Brazil, Japan, Surinam, and in Louisiana. All of these, and other strains lacking the perfect stage, were gathered together by B. O. Dodge and C. L. Shear and assigned to a new genus *Neurospora* and to four distinct species.

When Shear and Dodge (1927) described the genus *Neurospora*, four species were included: *Neurospora sitophila* Shear and Dodge, *N. crassa* Shear and Dodge, *N. tetrasperma* Shear and Dodge, and *N. erythraea* (Moll.) Shear and Dodge. But up to 1969 the total number of recognized species of Neurospora is twelve. Following is a key to the twelve species of *Neurospora* now recognized:

A. Asci four-spored
- B. Mature perithecia mostly 250-300 μ; conidial masses pale salmon — *N. tetrasperma* Shear and Dodge
- BB. Mature perithecia larger
  - C. Asci about 250 μ long; conidial masses orange — *N. erythraea* (Moll.) Shear and Dodge
  - CC. Asci about 190 μ long; conidial masses pink — *N. toroi* Tai

AA. Asci eight-spored
- B. Perithecia embedded in a stroma — *N. phoenix* (Kze.) Dennis
- BB. Perithecia not embedded in an expanded stroma
  - C. Heterothallic with imperfect stage
    - D. Perithecia 200-300 μ — *N. sitophia* Shear and Dodge
    - DD. Perithecia larger than 300 μ
      - E. Ascospores mostly 23 × 13 μ — *N. intermedia* Tai
      - EE. Ascospores mostly 27-30 × 14-15 μ — *N. crassa* Shear and Dodge

CC. Homothallic without imperfect stage

D. Ascospores with one germ pore *N. terricola* Gocheanaur and Backus

DD. Ascospores with one germ pore at each end

E. Spores larger, mostly 30-33 × 14-18 μ, ribs and intercostal veins prominent

F. Spores broad, mostly with continuous, conspicuous ribs *N. dodgei* Nelson and Novak

FF. Spores narrow, occasionally with branched or anastomosed ribs *N. galapagosensis* Mahoney and Backus

EE. Spores smaller, mostly less than 25 μ long, intercostal veins distinct or incospicuous

F. Intercostal veins distinct; spores broad, 14-17 μ in width *N. africana* Huang and Backus

FF. Intercostal veins indistinct; spores narrow, 10-15 μ in width. *N. lineolata* Frederick and Uecker

The species of *Neurospora* besides being pest in bakeries also live saprophytically on many decaying substrates of both animal and plant origin. Again, some species have been obtained from soil.

The somatic mycelium is composed of rapidly growing pigmented hyphae. The colouration of hyphae varying with the substratum may be red, orange dark-brown, or pinkish-grey. The hyphal cells are multinucleate.

Some of the species of *Neurospora* reproduce asexually by conidia. Again in others, conidial stage is absent. The vegetative hyphae produce two types of conidia borne on conidiophores. The two types of conidia are: large multinucleate **macroconidia** produced abundantly on **macroconidiophores** having moniliform in arrangement (Fig. 13.62A), and uninucleate **microconidia** growing singly or in short chains on **microconidiophores** (Fig. 13.62B). Both types of conidia germinate to form mycelium. The process may be repeated and the fungus can propagate itself indefinitely.

There are both homothallic and heterothallic species of *Neurospora*. The development of ascocarp is initiated by the formation of a short, stout, specialized hypha arising from an ordinary vegetative hypha growing up at approximately a right angle to it. This hypha, which at first is without cross-walls, coils at the tip to form an ascogonium. A septum soon develops in the stalk, and one or two branches arising below the septum grow out to form an envelope around the fertile central structure.

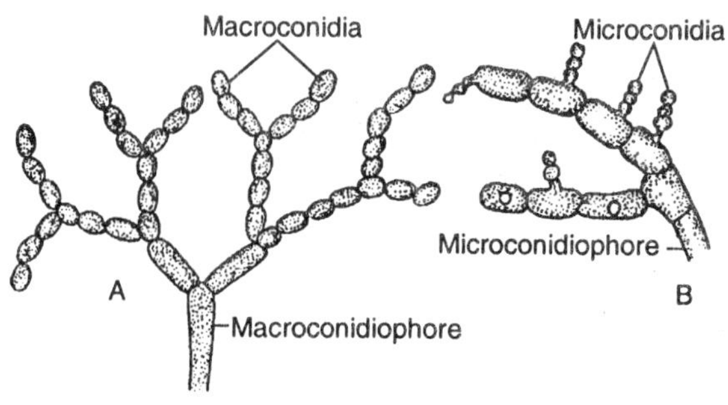

Fig. 13.62: *Neurospora* sp. A. Macroconidia borne on macroconidiophore. B. Microconidia borne on microconidiophore

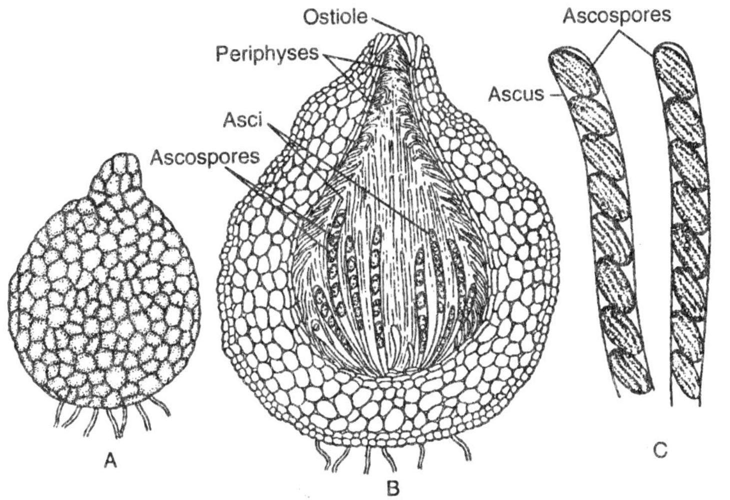

Fig. 13.63: *Neurospora* sp. A. An entire perithecium. B. Sectional view of perithecium. C. Asci with ascospores

The entire structure is the primordium of perithecium, also known as **protoperithecium.** The ascogonium develops into a multinucleate structure bearing a long hyphal branch which functions as a trichogyne. The antheridium is not developed in *Neurospora.* During fertilization the conidium behaves as a spermatium either directly coming in contact with the trichogyne or germinates producing a germ tube and provides with the male nucleus. In the heterothallic species the performance of sexual functions is regulated by compatibility factors which prevent self-fertilization. With the disappearance of antheridium development, *Neurospora* exhibits partial degeneration of sex. Once initiated, growth of the ascocarp is typically uninterrupted. The perithecial wall is composed of several cell layers, the outer portion of which becomes firm and dark-coloured. An ostiole lined with well-defined periphyses develops in a short beak at the apex of the ascocarp (Fig. 13.63B). The asci arise from croziers. The young asci grow upward among paraphyses but the latter structures disintegrate as the asci mature and are not evident in mature fruiting bodies. The mature perithecia are dark-coloured, pyriform and beaked (Fig. 13.63A). They may or may not be associated with stroma. Asci bear eight ascospores (Fig. 13.63C). There are both eight-spored and four-spored species of *Neurospora.* The ascospores of *Neurospora* are dark-brown or black and possess characteristic wall sculpturing in the form of ribs or veins running roughly parallel to the long axis of the spore (Fig. 13.63C). Nelson and Backus (1968) studied the development of ascus in *N. crassa* and reported that the cytology of ascus development is similar to that of higher Ascomycetes.

Life cycle of a heterothallic species of *Neurospora* is presented in Fig. 13.64.

Some Indian Species: *Neurospora dodgei* Nelson and Novak; *N. tetrasperma* Shear and Dodge.

Family: ***Xylariaceae***

The Xylariaceae occur chiefly on wood; they constitute the highest development of the Sphaeriales and are characterized by the free, superficial stroma, though a few, including certain species of *Hypoxylon,* have stromata partly sunk in the substratum. The stromata show every variety of form, from a spreading crust, to the almost spherical cushions of *Daldinia, Hypoxylon,* and the erect, simple branched expansions of *Xylaria.* The perithecia are arranged just below and at right angles to the surface of the stroma; their development may

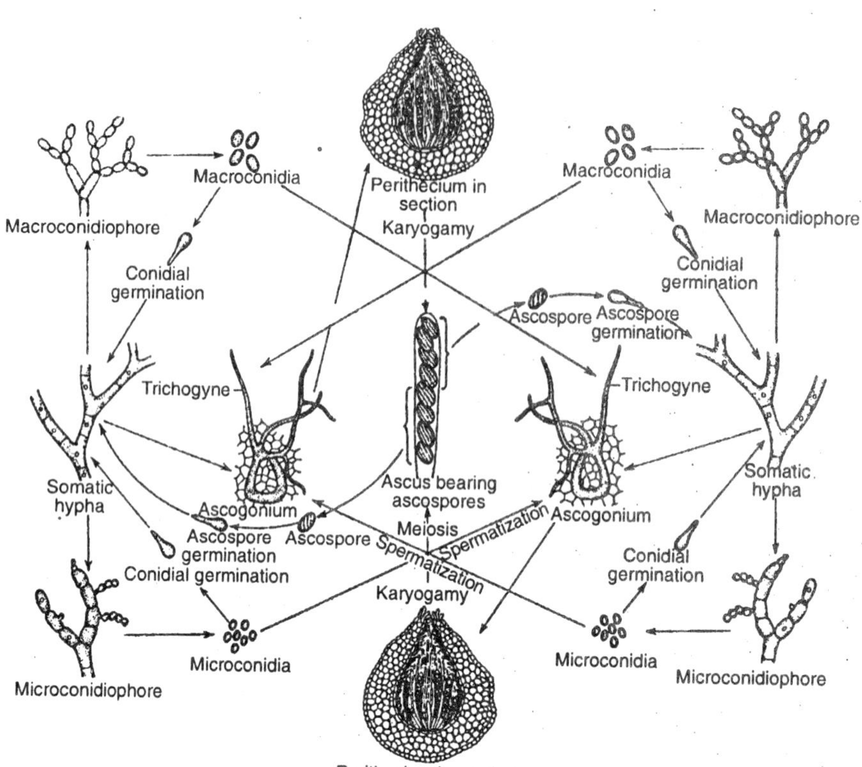

Fig. 13.64: Life cycle of a heterothallic species of *Neurospora*

be preceded by the formation of conidia, which often cover young stromata with a white or brown powder. Asci are 8-spored. Ascospores are non-septate, dark-brown to black.

Genus: *Xylosphaera* (= *Xylaria*)

It is a cosmopolitan genus, but especially well represented in the tropics. It generally inhabits dead wood, rarely dung or dry fruits in wet and shady places.

All the species vary somewhat in shape and size. However, the general habit is fairly characteristic. The stromata are erect, more or less stalked, simple or branched and sometimes forked, more or less club-shaped, cylindrical or fusiform (Figs. 13.65A-C), flesh white or pale buff to black. *Xylaria polymorpha* is commonly known as dead man's finger for its typical stromata. Stromata are sometimes partly covered with light-coloured conidia when immature. In some species the stroma becomes hollow at maturity, apex apiculate, free from perithecia (Figs. 13.65A, C). Perithecia embedded in the stromatic tissue (Figs. 13.65E, F), are in a single layer usually inserted beneath a black crust of stromatic tissue (Figs. 13.65C-F). In a mature stroma, ostioles of the perithecia are visible from outside with unaided eye (Figs. 13.65C, D).

Perithecia are with protruding tips (Figs. 13.65C-F). Asci are cylindrical 8-spored and are mixed with paraphyses (Figs. 13.65F, G). The ascus tip contains a blue-staining apical apparatus pierced by a narrow pore. Ascospores are uniseriate, fusiform, inequilateral (i.e., with one side more strongly curved than the other) and often with a hyaline equatorial germ pore, non-septate, dark-brown to black at maturity (Figs. 13.65G, H).

From the ascospores develops an extended mycelium. Mycelial hyphae unite into thick strands; these grow tall and show an intense

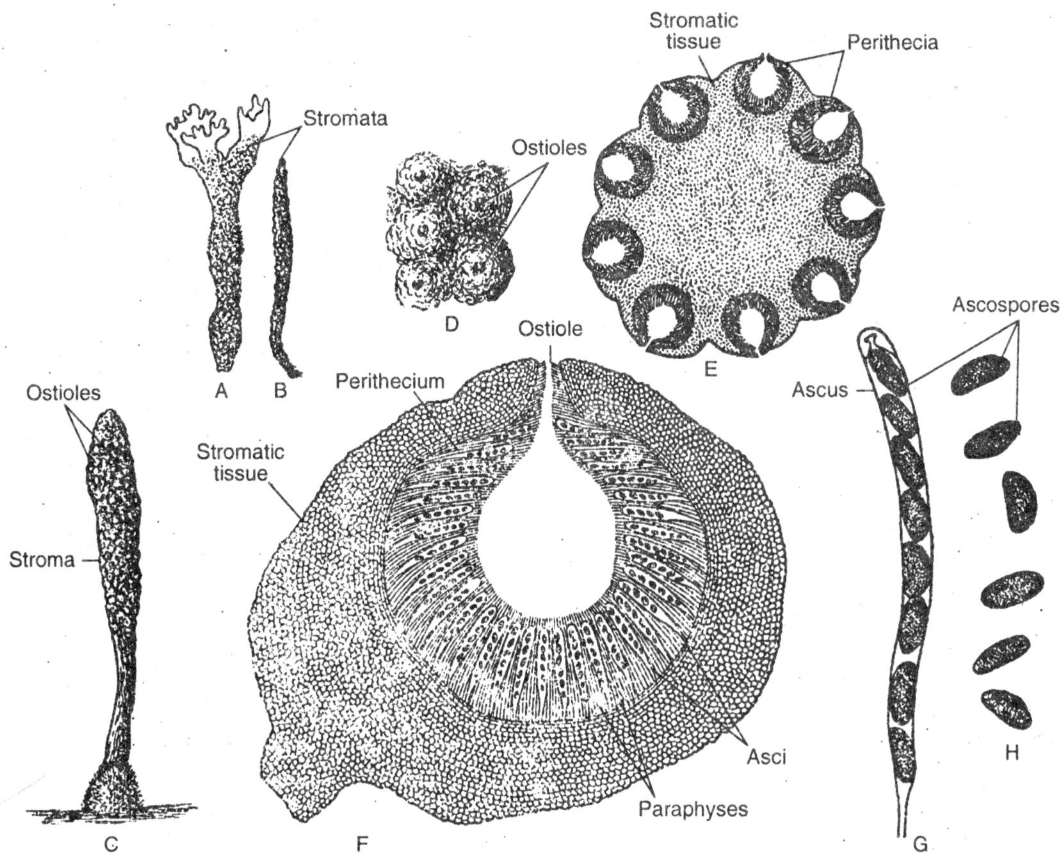

Fig. 13.65: *Xylosphaera* sp. (*Xylaria* sp.) A-B. Nature of stromata. C. Stroma showing ostioles. D. Surface view of ostioles. E. Stroma in section showing perithecia in stromatic tissue. F. A perithecium in section showing asci, paraphyses, ostiole and associated stromatic tissue. G. An ascus with uniseriate eight ascospores. H. Ascospores.

heliotropism so that even when under bark or tree trunks they easily come to the outer surface. They are first differentiated into a black pseudoparenchymatic rind and a light fibrous core and then gradually develop to the cylindrical, clavate or branched black fructifications. The growing tip of the stroma remains white for a long time and is covered with a strikingly regular hymenium of palisade-like conidiophores which, if unicellular, cut off ovoid conidia; if they are multicellular, however, at any position they cut off fusiform conidia at their tips. The conidia form a white coating in marked contrast to the exposed portions of the black stroma, and justify the name "candle-snuff fungus". After the disappearance of conidia, the stromatal branches swell clavately in the upper part and proceed to form perithecia. If the stroma is sectioned during the conidial stage, nests of small hyphae are found, and form the first indication of perithecia.

Note: Dennis (1958) claimed that the name *Xylosphaera* should have priority over *Xylaria*. But the proposal has not been welcomed by Petrak (1962), Holm and Muller (1965).

Some Indian Species: *Xylaria aemulans* Starb; *X. aspera* Massee; *X. fimbriata* Lloyd; *X. obovata* Berk; *X. tricolor* Fr.

Chapter 14

Subdivision of Basidiomycotina

GENERAL CHARACTERISTICS

The Basidiomycotina comprise the highest group of fungi which are often noticed conspicuously occurring in the fields and forest areas. Most of them possess highly developed fructifications, known as **basidiocarps** which may be fleshy or woody. The fungi, commonly known as mushrooms, toadstools, puff-balls, and bracket fungi, are the Basidiomycotina. To this group also belong the rusts and smuts, responsible for diseases of plants of economic importance. There are both parasitic and saprophytic fungi in this group. Many used as food are cultivated in a large-scale for marketing. Again, there are highly poisonous ones which should be carefully avoided. Some cause destruction of forest trees and ornamentals, whereas others are responsible for the decay of fallen logs of wood. The Basidiomycotina resemble the Ascomycotina in possessing septate hyphae and production of non-motile spores and fructifications. The fungi of this group are characterized by the production of spores known as **basidiospores** resulting from karyogamy and meiosis, and are borne externally on slender protuberances, the **sterigmata** (sing. **sterigma**). The sterigmata are developed from a more or less club-like structure, called a **basidium** (pl. **basidia**) for which the Basidiomycetes have been frequently named as *club fungi*. The basidium was discovered by Levelle in 1837.

SOMATIC STRUCTURES AND OCCURRENCE

The somatic mycelium is composed of septate branched hyphae, a portion of which usually serves the function of absorption of nutriment from the substratum and the rest performing different other functions. It may be white, bright yellow or orange, deep-brown to charcoal-black. The mycelium usually grows in a wide range of substrata like damp shady places containing decayed vegetative matters, rotten logs, grassy land, sometimes on animal dung, and similar other places. In parasitic forms the mycelium lives in or along the host tissue receiving nutrition from the host protoplasm. In most Basidiomycetes there is profuse development of hyphae which may even become interwoven into macroscopic structures like **sclerotia, rhizomorphs,** and **basidiocarps.** The basidiocarps are usually of definite form showing considerable internal differentiation of tissues. Again, there are fungi included under the Basidiomycotina which possess poorly developed hyphae. The somatic mycelium in course of development passes through three stages. When young, the mycelium is multinucleate, a stage that follows immediately after the production of a germ tube during the germination of a basidiospore. This multinucleate condition of the mycelium is very short and is followed by the appearance of septa (dolipore septa) resulting in the formation of septate mycelium having uninucleate cells, **monokaryotic,** which is known as the **primary mycelium.** The hyphae of the primary mycelium are usually wide-angled. The primary mycelium gives rise to **secondary mycelium** in which the cells are binucleate, **dikaryotic,** the two nuclei are derived from different parent cells. But Buller used the term **haploid mycelium** to the mycelium possessing uninucleate cells and **diploid mycelium** to that with cells each containing a pair of **conjugate nuclei.** Since nuclei in a primary mycelium (haploid mycelium) are derived from the original single nucleus of the basidiospore, they are all idential—**homokaryotic,** and the mycelium is **homokaryotic mycelium.** In the majority of the Basidiomycetes the homokryotic mycelium does not produce basidiocarp. The dikaryotic condition is attained from the monokaryotic condition in a variety of ways, a process known as **dikaryotization** or **diploidization.** The details of the process will be discussed later. The dikaryotic mycelium is the **secondary mycelium** whose hyphae are narrow-angled. The secondary mycelium is responsible for the development of basidia and basidiospores. The basidiocarp of the Basidiomycotima is developed from the dikaryotic mycelium. In a basidiocarp, the portion of the dikaryotic mycelium producing sterile tissue is the **tertiary mycelium** and that producing fertile layer is the secondary mycelium. In majority of the Basidiomycetes the secondary mycelium is characterized by the presence of bridge-like hyphal connections known as **clamp connections.** The clamp connections are more often present at every septum particularly in narrow hyphae. These structures are usually absent in broader hyphae. Clamp connections were discovered by Hoffmann in 1856 and were observed by Bail (1856), De Bary (1859, 1866, 1884), Tulasne (1861), Brefeld (1877, 1888, 1889), and many others. But the details of nuclear behaviour associated with clamp connection were worked out by Kniep (1915) and Mlle. Bensaude (1918).

The formation of clamp connection starts when the binucleate terminal cell of a dikaryotic hypha is ready to divide (Fig. 14.1A). A small pocket known as **clamp** is formed on the lateral wall usually in the middle of the cell and between the two nuclei (Fig. 14.1B). The clamp grows backwards away from the apex of the cell. The two nuclei of the terminal cell then change their position. One of them moves into the newly formed clamp and the other remains behind its base (Fig. 14.1C). They then divide simultaneously, **conjugate division** (Fig. 14.1D). During division the spindle axis of the nucleus in the clamp lies obliquely and that of the other remains more or less parallel to the main axis. Towards the end of the division of the nuclei, the four daughter nuclei which are about to be formed, exhibit remarkable change of position. One daughter nucleus from the clamp and one from the main axis pair toward the end of the hypha. Of the other two daughter nuclei, one remains in the clamp and the other moves down in the main axis. In the mean time the clamp grows further and bends down touching the wall of the main axis. Now two septa appear. One at the base of the clamp along the main axis which cuts off a basal cell from the original apical hyphal cell,

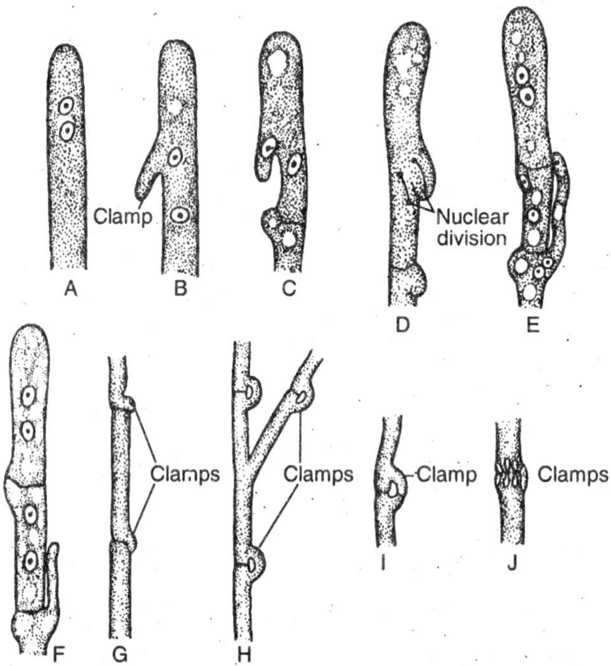

Fig. 14.1: A–F. Successive stages in the development of a clamp connection in a hypha. G–J. Different types of clamp connection

whereas the other septum is formed directly under the base of the clamp being arranged obliquely with the first septum. In the mean time one of the four daughter nuclei is imprisoned in the clamp until the clamp grows further and its end comes in contact with the lateral wall (Fig. 14.1E). The walls at the point of contact between the clamp and the lateral wall dissolve and a passage is formed. Through this passage the nucleus of the clamp migrates in the basal cell and meets its conjugate mate. Thus a short semicircular hyphal outgrowth is developed just behind a septum during cell division in such a way as to connect the two cells together (Fig. 14.1F). This short semicircular hypha is known as a **clamp connection.** Functionally, however, it is what may be called a by-pass hypha through which the contents of the two adjacent cells are in communication with each other. Each of the septa has a minute pore in its centre, and through the two pores of the two septa protoplasm flows. Clamp connections ensure separation of the sister nuclei of the dikaryon, at each division, into two daughter cells. Since they occur often on the dikaryotic mycelium, it was at one time believed that they are connected with the diploidizing process. But it has now been shown that they merely help in the flow of protoplasm and food material from cell to cell through the pores in the septa. They are, however, a visible sign of the presence of conjugate nuclei. But again though clamp connections are always associated with conjugate nuclear division the converse is not true, for such division may take place without their formation. Depending on the thickness of hypha the nuclei can pass from cell to cell through the central pores of the septa. In the opinion of some, the development of clamp connection and crozier formation of an ascogenous hypha are homologous processes. Their contention is based on the fact that a clamp generally occurs at the base of a basidium which again arises from the terminal cell of a dikaryotic hypha, a condition very similar to the development of ascus from an ascogenous hypha whose terminal cell is also dikaryotic, whereas others express grave doubt and rather consider them as analogous processes.

In general, clamp connections arise singly (Figs. 14.2G-I). But in some cases they may develop in two or more (Fig. 14.2J). Again, clamp connections are entirely lacking in certain Basidiomycotina and interestingly enough they have been reported in some Tuberales belonging to the Ascomycotina. But in almost all Basidiomycotina clamp connection is present and this is a character of taxonomic importance of the group. Although clamp connections are made with dikaryotic mycelium, they appear in monokaryotic mycelium also, e.g., *Stereum hirsutum* and in multinucleate primary mycelium, e.g., *Coprinus narcoticus*.

Normally the two nuclei of a conjugate pair attract one another and divide simultaneously, thus providing pairs of conjugte nuclei for new cells. Yet, under certain conditions in some Basidiomycotina and Ascomycotina, the two nuclei of a conjugate pair may become separated in different cells and these cells may develop into monokaryotic hyphae. Thus a dikaryotic mycelium may produce monokaryotic cells or monokaryotic branches. This is a process which is just the reverse of diploidization and is known as **de-diploidization**. De-diploidization is thus the production of monokaryotic cells or hyphae by a dikarytic mycelium.

ASEXUAL REPRODUCTION

Asexual reproduction in the Basidiomycetes takes place by fragmentation of the mycelium, and by the development of spores, such as **oidia, conidia** or **arthrospores**. The oidia may be uninucleate or binucleate depending on whether they are produced from primary or secondary mycelium. They are produced from the tip of special, short, lateral hyphal branches called **oidiophores**. They arise mostly from the primary mycelium. The oidia perform a dual function. They may either germinate to form primary mycelium or may behave as spermatia and bring about diploidization by uniting with somatic hyphae. The production of conidia among the Basidiomycotina is not so common as among the Ascomycotina. They are produced by basidiospores and mycelium. The conidia are disseminated by the wind and serve to propagate the mycelium. Besides these, some Basidiomycotina reproduce asexually by the production of small unicellular mycelial fragments. They are the arthrospores. They may round up forming thick-walled chlamydospores.

SEXUAL REPRODUCTION

It is rather interesting that in the Basidiomycotina sexual reproduction takes place without the differentiation of characeristic gametangia as is encountered in other fungi. In this group, similar to other fungi, the sexual process comprises fundamentally of the establishment of dikaryophase by plasmogamy, diplophase by karyogamy, and meiosis. But unlike members of the Ascomycotina there is elaborate and physiologically independent dikaryotic condition. There are three clear and distinguishable phases in the life cycle of the Basidiomycotina. They are: **haplophase** represented by the monokaryotic condition, **dikaryophase** exhibited by the association of two compatible nuclei, and the **diplophase** which is very short and is attained by karyogamy. The **establishment of dikaryotic condition** or in other words, the compatible nuclei are brought together by plasmogamy which is accomplished either by somatogamy or by spermatization. In somatogamy two monokaryotic hyphae bearing compatible nuclei come in contact, the walls between the adjacent cells at the point of contact dissolve, and the nucleus of one cell migrates into the other resulting a dikaryotic condition. The dikaryotic condition is perpetuated by the development of a dikaryotic mycelium. But when a dikaryotic mycelium meets a monokaryotic mycelium, the two mycelia form hyphal fusions with one another, and then the dikaryotic mycelium diploidizes the monokaryotic mycelium. The diploidization of monokaryotic mycelia by dikaryotic mycelia has been called by Quintanilha (1937–1939) the **Buller phenomenon.** This phenomenon is found to occur in the Uredinales as well as in the Hymenomycetes and also, in principle, in certain Ascomycetes. Again, during spermatization an oidium comes in contact with a cell of the monokaryotic hypha. At the point of contact the walls dissolve and a pore is formed through which contents of the oidium migrate into the cell of the hyphae initiating a dikaryotic condition. In some fungi spermatization takes place between a single-celled uninucleate spermatium and a receptive hypha. Both spermatium and the receptive hypha are borne in structures known as **spermogonia** (Fig. 14.18A). They are of two compatible groups. The union of spermatium and receptive hypha takes place between the compatible ones only. In smuts, however, the dikaryotic condition is established in various other ways which will be discussed later.

In most Basidiomycotina the terminal cell of the dikaryotic hypha of the secondary mycelium develops into a basidium from which basidiospores are produced exogenously (Fig. 14.5). In some, the entire dikaryotic hypha develops into a mass of dikaryotic thick-walled resting spores. Again, there are others, where only the terminal cells of the dikaryotic hyphae are transformed into dikaryotic thick-walled resting spores. Such dikaryotic thick-walled resting spores are known as **chlamydospores** or **teliospores** or **teleutospores.** Each resting spore on germination develops a short mycelial structure behaving as a basidium which is known as **promycelium** from which basidiospores are developed. The basidiospores so formed are commonly designated as **sporidia** (sing. **sporidium**) (Figs. 14.14A and 14.23). The basidiospores constitute the chief and most efficient organs of reproduction.

THE BASIDIOCARP

In a large number of Basidiomycotina the dikaryotic somatic mycelia, both seconday and tertiary intertwine producing highly organized sporocarps or fruiting bodies or fructifications or sporophores which are known as **basidiocarps,** whereas in a relatively few Basidiomycotina there is total absence of the development of basidiocarps. The basidiocarps may be epigeous or hypogeous, or there may be many transitional forms between these conditions. They are extremely variable in size, texture, colour and shape ranging from very large macroscopic to even minute microscopic in size. The basidiocarps besides being thin crust-like cartilaginous, or fleshy, spongy, gelatinous, corky, woody, may be almost of any other texture. So also in colour they may be bright orange-red, yellow, brown, dark, white and different grades of these and some other shades of colour. The basidiocarps are often so highly developed and attract attention so easily that the main body of the fungus which is the extensive mycelium, usually goes unnoticed. Most common ones are crust-like, partly or completely shelved without any stalk; or stalked umbrella-like, fan-shaped, coralloid, round to oval and of various other configurations. Structurally the basidiocarps may also vary enormously. But in general they are composed of fertile tissue, the **hymenium** or the **hymenial layer** produced by the secondary mycelium and sterile tissue, the **trama** or **context** (Figs. 14.2C and E) of tertiary mycelium.

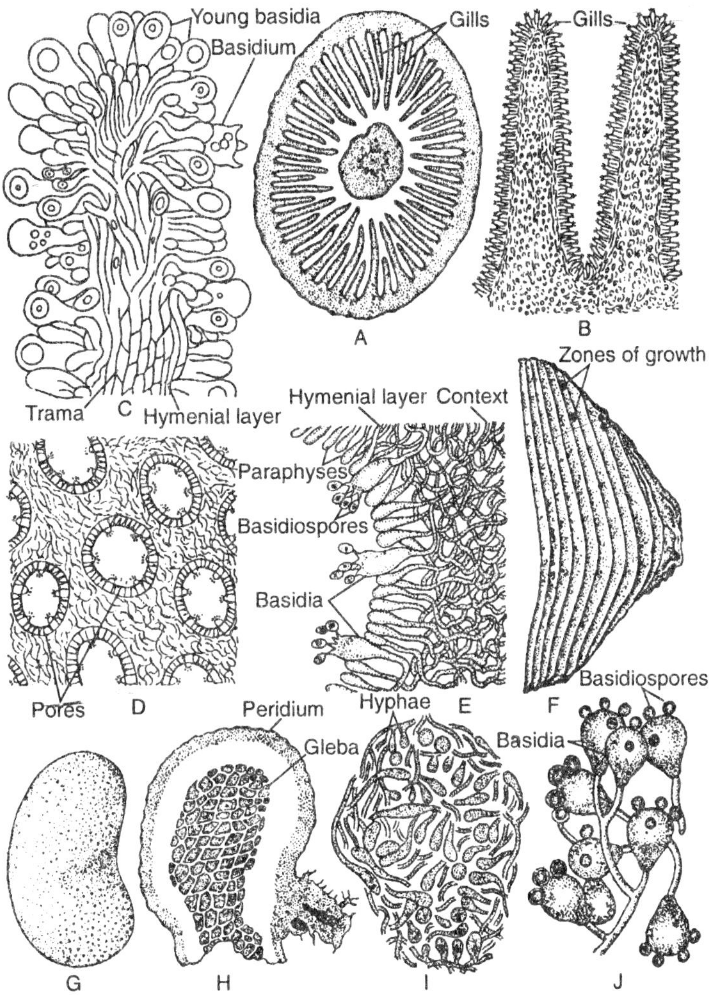

Fig. 14.2: Structural details of basidiocarps. A. Complete section across the pileus of a basidiocarp of *Coprinus* sp. showing the radiating gills covered with blackened hymenium. B. Closer view of the gills. C. Details of a gill showing trama and hymenial layer. D. Portion of a section through a basidiocarp of *Polyporus* sp. showing pores lined with hymenium. E. Details of a portion of a pore showing hymenial layer composed of basidia bearing basidiospores and paraphyses, and context, F. Zones of growth of hymenial layers in the basidiocarp of *Fomes* sp. G. *Scleroderma* sp. An entire basidiocarp. H. Sectional view of a basidiocarp of *Scleroderma* sp. I. Immature gleba. J. Basidia with asterigmate basidiospore.

The hymenial layer may be smooth or spread along plate-like structures known as **gills** or **lamellae.** The part of the basidiocarp bearing the hymenium is known as **hymenophore.** The basidiocarp is often covered by a covering known as **peridium** (Fig. 14.2H). The basidiocarp may be: (i) open without being covered by a peridium from the very beginning of its development exposing the hymenium—**gymnocarpous,** (ii) the hymenium is enclosed by basidiocarp tissue at first but later becomes exposed to the open air before the spores are mature—**hemiangiocarpous,** (iii) the fertile portion (**gleba**) of the basidiocarp is enclosed in basidiocarp tissue until well after the spores are matured—**angiocarpous,** and (iv) when the mature hymenium remains covered—**endocarpous** (Figs. 14.2G and H). The hymenium may be borne on all sides of the basidiocarp—**amphigenous,** or on one side only—**unilateral.** In cases where the basidiocarps are closed or closed up to maturity of spores, enclosed within the peridium are the tramal plates lined with hymenium, known as **gleba** (Fig. 14.2H). The tramal plates and the hymenium are ultimately transformed into a characteristic mass, it is called *glebal mass.* Where the basidiocarps are open from the very beginning of their development, the spores are dispersed mostly by a special mechanism known as **fluid drop** or **drop-excretion mechanism of dispersal of spores,** whereas in basidiocarps opening at a later stage, the spores are dispersed by agencies like wind, raindrops, animals and insects. Again in others, where the basidiocarps remain closed, the spores are liberated by the disintegration or accidental breaking of the basidiocarps.

The hymenium of a basidiocarp consists of basidia and sterile structures interspersed with them. The sterile structures are often regarded as immature basidia. But they are generally designated as **paraphyses** (sing. **paraphysis**) (Fig. 14.2E) and have been compared with those of the Ascomycotina. This has been questioned by Leutz (1954) and others. According to them, the sterile structures of the hymenium are nothing but the sterile hyphal ends which are basically, just like basidia, nothing but the components of the dikaryotic secondary mycelium, whereas the asci and paraphyses in the Ascomycotina originate from different hyphal elements. As such the similarities between the paraphyses of the Basidiomycotina and the Ascomycotina are superficial and they are not homologous structures. Hedwig (1789) was the first person to use the term 'paraphysis' in relation to fungi. Functionally the paraphyses are considered as spacing organs.

Besides paraphyses, the hymenium may also have certain other sterile structures whose taxonomic interest has been emphasized largely by Bourdot and Galzin (1928), and Overholts (1929). Some of them are:

1. **Acanthophyses** (sing. **acanthophysis**): These are sterile hymenial hyphal ends having a number of short pin-like outgrowths on their surface. Burt (1914) designated them as 'bottle-brush paraphyses', present in certains species of *Aleurodiscus* (Fig. 14.3F).

2. **Dendrophyses** (sing. **dendrophysis**): These are hyphal ends of the hymenium or context having irregular tree-like branching. They are thick- or thin-walled and covered throughout with spines of variable length, also encountered in certain species of *Aleurodiscus* (Fig. 14.3C).

3. **Dichophyses** (sing. **dichophysis**): These sterile hyphal ends arising from the hymenium or subhymenium are antler-like in appearance because of their successively dichotomous, wide-angled branching and the prong-like terminal branchlets (Fig. 14.4A). They are present in some tropical species of *Hymenochaete* and species of *Vararia.*

4. **Pseudophyses** (sing. **pseudophysis**): These are smooth thin-walled, unbranced hymenial structures which are knobbed, knotted or nodose or moniliform (Figs. 14.3F and 14.4B). They are present in species of *Aleurodiscus* and *Coprinus.*

5. **Cystidia** (sing. **cystidium**): These are hymenial or subhymenial organs, often projecting beyond basidia. They are elongated subcylindrical globose to sub-globose (Figs. 14.3A and B); hyaline to uniformly coloured or wall coloured and sap not coloured never dark-brown or black even in KOH. They are present in species of *Stereum, Poria, Armillariella, Coprinus, Lentinus,* etc. The cystidium has more or less rounded tip which may be smooth or provided with spines. The wall of the cystidium may have deposition of Ca-oxalate crystals, found in species of *Peniophora.* The

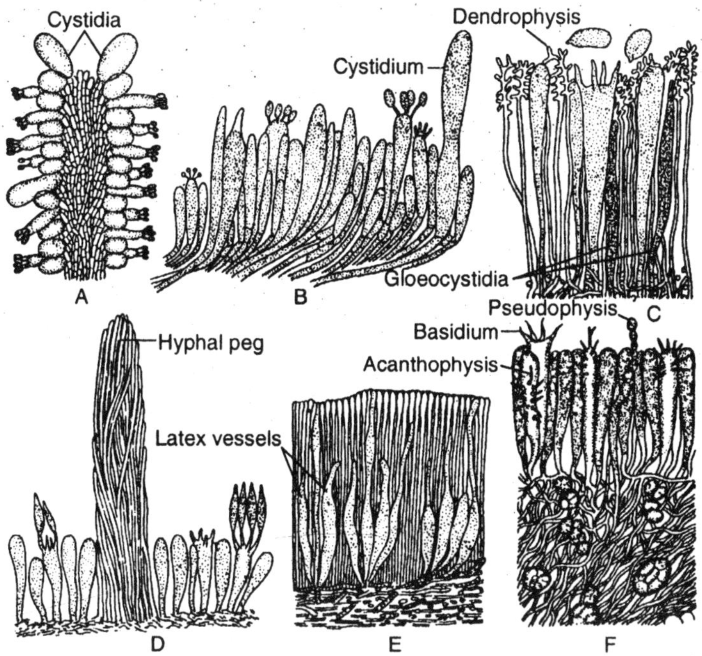

Fig. 14.3: Various types of sterile structures of the hymenial layer. A. Cystidia in *Armillariella mellea*. B. Cystidium in *Stereum purpureum*. C. Dendrophysis and gloeocystidia in *Aleurodiscus sparsus*. D. Hyphal peg in *Epithele* sp. E. Latex vessels in *Corticium seriale*. F. Acanthophysis and pseudophysis in *Aleurodiscus oakesii*

existence of cystidia was first demonstrated by Micheli (1729) and was put to wider use by Hedwig (1789). The taxonomic importance of cystidia in the study of the family Thelephoraceae has been introduced by Massee (1889) and later on adopted by Burt (1914-1926) and many others.

In some fungi, e.g., *Vararia* sp., cystidioid prolongations occur in the hymenium. These structures do not possess staining qualities of true cystidia and are known as **pseudocystidia** (sing. **pseudocystidium**) (Fig. 14.4A).

6. **Gloeocystidia** (sing. **gloeocystidium**): These are subhymenial organs extending up to the hymenium, may or may not remain associated with cystidia. They are of varied form, some of them being nearly indistinguishable from submerged, non-incrusted cystidia, others being long, flexuous, refractile structures. A typical gloeocystidium is an elongated, usually rather slender, tortuous object, sometimes slightly enlarged toward the base and with a long slender apical neck, sometimes slender over the entire length and then often tapering very gradually toward the apex (Fig. 14.3C). The walls are thin and colourless, very rarely with true septa. The walls of gloeocystidia are never incrusted with crystalline particles. The gloeocystidium contains sap which on mounting in glycerine divides into several opaque blocks, orange or brown in colour containing oily substances which do not exude when the gloeocystidium is broken. The gloeocystidia are especially prominent in the genera *Aleurodiscus, Stereum,* and *Merulius.*

7. **Setae** (sing. **seta**): These are hymenial and subhymenial organs, mostly subhymenial and extended up to hymenium. The setae may be broadly ovate in shape and may extend beyond hymenial level. They are usually bristle or bristle-shaped bodies, typically deep yellow or brown in colour and dark-brown or black when treated with KOH. Setae contain brown-coloured sap. Their tips may be pointed

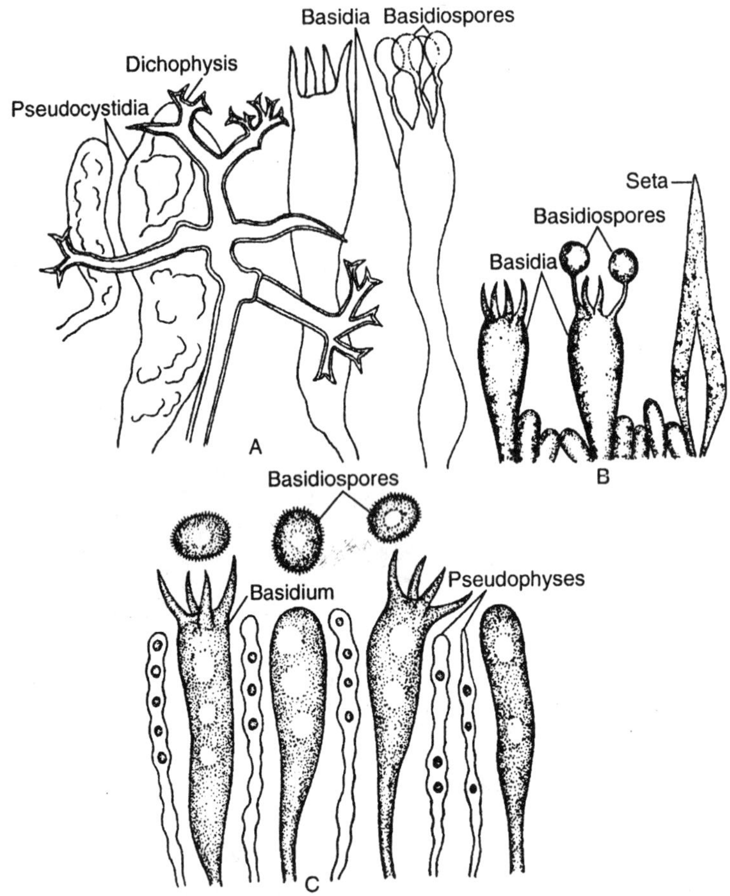

Fig. 14.4: Various types of sterile structures of the hymenial layer. A. Dichophysis and pseudocystidia in *Vararia* sp. B. Pseudophyses in *Aleurodiscus amorphus*. C. Seta in *Fomes pini*

(Fig. 14.4C), blunt, curved or twisted. The genus *Hymenochaete* is characterized by having setae. Setae differ from most cystidia by their brown colour. When the setae have several radiating spines they are known as **stellate setae**. A stalked stellate seta is designated as **asterophysis.**

8. **Hyphal pegs:** Compound hyphal fasciculate projections derived from undifferentiated hyphae extend beyond the general level of the hymenium forming the hyphal pegs. Each hyphal peg consists of two or more parallel or interwoven hyphae forming a soft column (Fig. 14.3D). It may be pyramidal or cylindrical or conical in outline. The hyphal pegs are found in the members of the families Thelephoraceae, Polyporaceae and Agaricaceae.

9. **Latex vessels:** These are elongated, branched, non-septate, labyrinthiform, anastomosing hyphae of variable thickness. They rise from the subhymenium, penetrate the hymenium growing between the basidia and end on the upper surface of the hymenium (Fig. 14.3E). The latex vessels are multinucleate, vacuolate structures which contain milky or coloured or a hyaline sap which becomes coloured upon contact with air. They are present in *Corticium seriale.*

DEVELOPMENT OF BASIDIUM AND BASIDIOSPORES

A basidium develops from a dikaryotic terminal cell of a hypha of the secondary mycelium. The dikaryotic terminal cell is separated from the rest of the hypha by a septum over which a clamp connection is usually found. During the development of the basidium the terminal cell enlarges to form young basidium, also known as **basidiole** (Figs. 14.5A-D). The two nuclei of the basidiole fuse to form a diploid nucleus (Figs. 14.5B and C). The diploid nucleus then undergoes meiosis producing four haploid nuclei. The growing basidium then produces four protuberances at the top. These protuberances are known as **sterigmata** (sing. **sterigma**) (Fig. 14.5D). They may be lacking in some Basidiomycetes. When present, the tips of the sterigmata enlarge to produce **basidiospore initials.** The four haploid nuclei then squeeze through sterigmatal passage into each basidiospore initial. The basidiospore initial then becomes walled off to form a **basidiospore** with a single nucleus (Fig. 14.5E). The portion of the basidiospore in contact with the sterigma has been designated by Heim (1931) as **hilum;** and **hilar appendix** or **apiculus** to the short, often sharp protrusion near the hilum at the basal end of the basidiospore. Each basidium typically bears four bsidiospores on four sterigmata, a **tetrasterigmate basidium** (Fig. 14.5E). The basidiospores may be borne symmetrically or asymmetrically on the sterigmata. Sometimes a basidium may produce only two sterigmata and two basidiospores. Such a basidium is called **bisterigmate basidium** (Fig. 14.6C). In such case each of the two spores may receive two nuclei resulting in the development of binucleate basidiospores. In case of

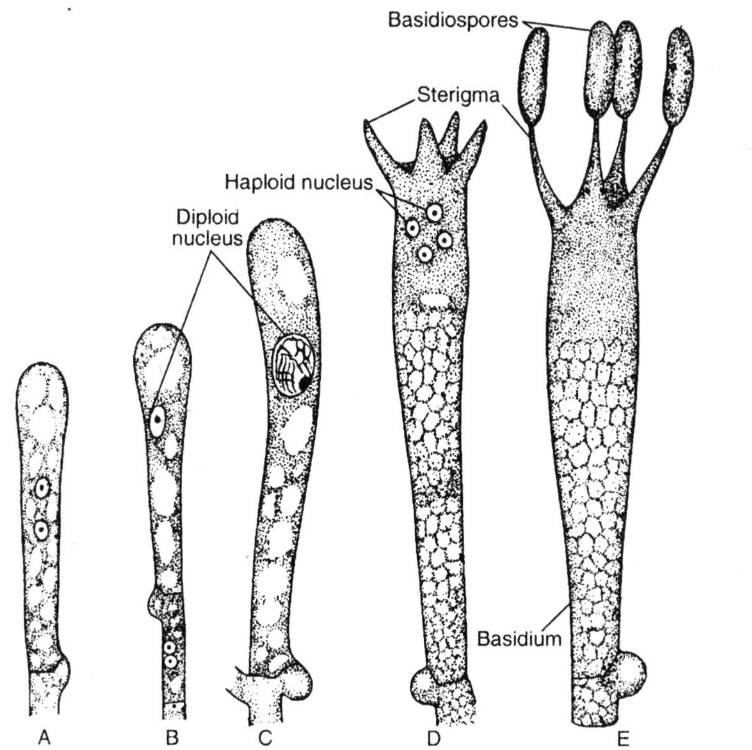

Fig. 14.5: Successive stages in the development of a basidium and basidiospores. A. Dikaryotic terminal cell. B-C. Karyogamy. D Young basidium with four haploid nuclei and four sterigmata. E. Mature basidium with four basidiospores borne on sterigmata

bisterigmatic basidia the basidiospores may also be uninucleate. Here, the other two nuclei remain unused within the basidium and eventually disintegrate. The number of basidiospores may also be more than four.

The basidia are usually slender, clavate to broadly clavate. Spherical to elongate basidia are also not uncommon. They may be septate or aseptate. Septation is usually transverse or vertical, may also be oblique. During nuclear division the orientation of the nuclear spindles may also vary in different basidia. When the nuclear spindles are oriented transversely to the basidium, such basidia are regarded as of the **chiastobasidial type**. Again, basidia having nuclear spindles oriented longitudinally or obliquely are the **stichobasidial type.**

Following are the main types of basidia:

1. **Homobasidia:** The basidia are slender, clavate to broadly clavate, or nearly globose and aseptate. The number of basidiospores may be four or more, with (Figs. 14.6A, B, F and H) or without sterigmata (Figs. 14.6E and G). They are also known as **autobasidia** or **holobasidia.** The homobasidia may be chiasto-, or stichobasidial type. When the basidia are developed inside a fruit body, they are the **endobasidia.** Again, basidia with basidiospores symmetrical on sterigmata are the **apobasidia** and those with basidiospores asymmetrical are the **autobasidia.**

2. **Heterobasidia:** The basidia are either septate or deeply lobed. They are also known as **phagmobasidia** or **metabasidia.**

In heterobasidia, the enlarged basal portion of the mature basidium in which nuclear fusion takes place has been designated by Neuhoff (1924) as **hypobasidium** and which bears upon it the **epibasidium,** but not the sterigmata directly. According to Neuhoff the

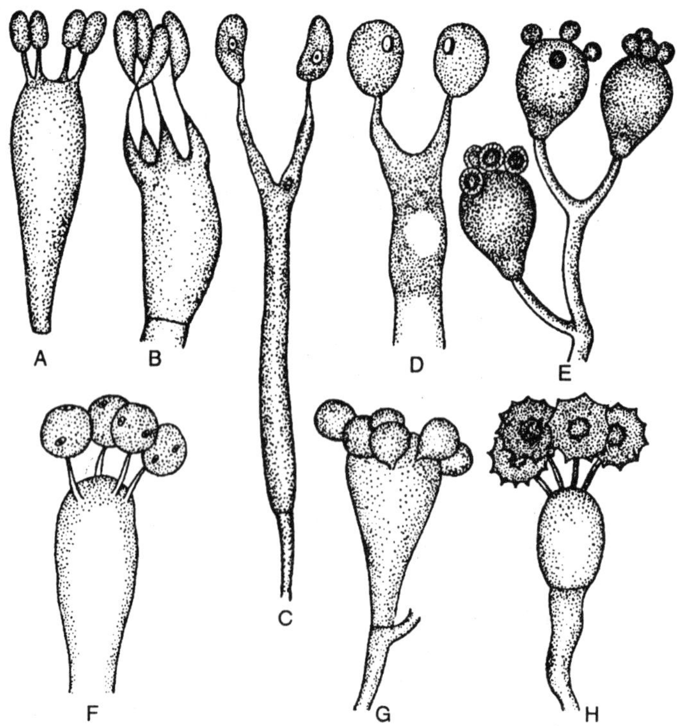

Fig. 14.6: Various types of basidia, A, B, E and F-H. Homobasidia. C and D. Deeply-lobed heterobasidia—tuning fork type of basidia. A, B, F and H. Homobasidia bearing sterigmate basidiospores. E and G. Homobasidia bearing asterigmate basidiospores

hypobasidium and the epibasidium make up a **heterobasidium.** Linder (1940) discarded this terminology and preferred **probasidium** for hypobasidium and **basidium** for epibasidium. The heterobasidia may again be: (i) transversely septate, (ii) longitudinally septate, and (iii) deeply lobed.

The transversely septate basidia may be differentiated into the following:

(a) With thin-walled hypobasidium and the epibasidium is traversed by three septa. Each cell of the epibasidium produces a long tube at whose apex a sterigmatic structure is formed on which arises the one-celled uninucleate basidiospore (Fig. 14.7A). The basidiospores are at the same level, as in the genus *Auricularia.*

(b) The hypobasidium produces hooked epibasidium which again produces rather long sterigmata (Fig. 14.7B), seen in the genus *Helicobasidium.*

(c) The hypobasidium is firm-walled from which grows out the straight or curved epibasidium which becomes four-celled. The

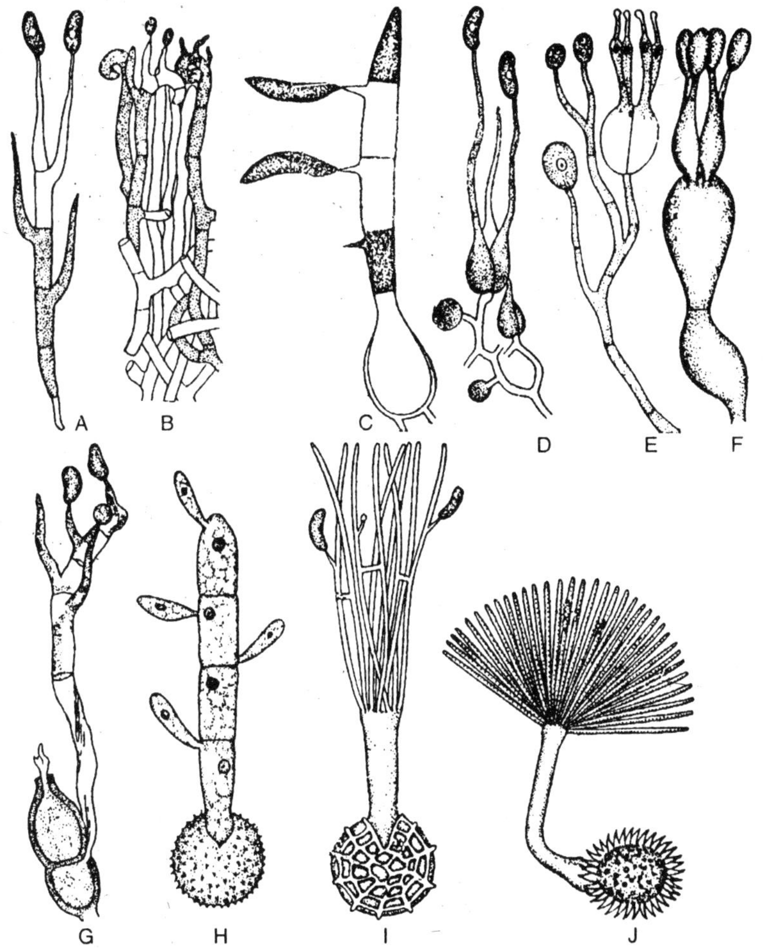

Fig. 14.7: Various types of basidia. A-C. Transversely septate heterobasidia. A Thin-walled hypobasidium in *Auricularia* sp. B. Hooked epibasidium in *Helicobasidium* sp. C. Firm-walled hypobasidium in *Septobasidium* sp. D-E. Longitudinal to obliquely septate heterobasidia. F. Heterobsidium with four epibasidia separated from hypobasidium by a septum at the base of each. G-J. Thick-walled probasidia with tubular basidia producing basidiospores. G. Basidiospores borne on sterigmata in *Puccinia* sp. H-J. Basidiospores not borne on sterigmata. H. *Ustilago* sp. I. *Tilletia* sp. J. *Neovossia* sp.

basidiospores are borne on well-developed sterigmata (Fig. 14.7C), found in the genus *Septobasidium.*

(d) In the Uredinales and Ustilaginales the mature teleutospore, a uninucleate but diploid thick-walled structure, is the **probasidium.** On germination it gives rise to a tubular basidium, designated as a **promycelium.** By meiotic division of the diploid nucleus and by transverse septation, the promycelium becomes four-celled. Each cell of the promycelium contains a single haploid nucleus which takes part in the development of a basidiospore. In the Uredinales the basidiospores are borne on sterigmata and number of basidiospores is four (Fig. 14.7C), whereas in the Ustilaginales there is absence of sterigmata and more than four basidiospores are formed (Figs. 14.7H-J).

The longitudinally septate basidia are the so-called **cruciate type.** The hypobasidium is longitudinally septate. The nuclear spindle is chiastobasidially oriented. Two vertical or oblique septa are formed at right angles to one another producing a four-celled structure from each of which arises a prolongation—the epibasidium which is terminated by a sterigma, bearing the basidiospore (Figs. 14.7D and E). The various genera of the Tremellaceae exhibit this form of basidial character.

In the members of the Tulasnellaceae the basidium is not divided by vertical septa. The hypobasidium is subglobose, pyriform or broadly clavate. From the upper portion of the hypobasidium arise usually four stout cells—the epibasidia, each of which is narrowed at the tip to form a sterigma upon which a single basidiospore is formed. The epibasidia are separated from the hypobasidium by a septum at the base of each (Fig. 14.7F).

In the Dacrymycetaceae the basidia, when young, are long cylindrical or somewhat clavate. The basidial nucleus divides stichobasidially. The broadened apex of the basidium becomes deeply lobed to produce two epibasidia and almost of the same length as the hypobasidium. Each epibasidium is terminated by a sterigma at whose apex the basidiospore is produced. The basidium takes a 'tuning fork' in appearance, from which it is known as **tuning fork type of basidium** (Figs. 14.6C and D and Figs. 14.26C-J).

THE BASIDIOSPORES

The basidiospore is a haploid structure. Basidiospores are generally unicellular, uninucleate and thin-walled structures with the exception of some septate and some thick-walled ones. They may be globose oval, elongated, or sausage-shaped or apically **truncated** with cut top. Their colour en masse, may be hyaline, pink yellow, green, orange, brown, or black. The basidiospore wall is almost universally smooth, except orange, brown, or black. The basidiospore wall is almost universally smooth, except a few cases where wall markings are present. Some basidiospores give a positive **amyloid** reaction by being coloured blue with Melzer's reagent. Again, some thick-walled spores absorb aniline blue dye strongly to the inner-wall surface. Such spores are said to be **cyanophilous.** Basidiospores are **asterigmate** (Figs. 14.6E and G) when the sterigmata are lacking. But usually they are **sterigmate** (Figs. 14.6A and B), i.e., borne on sterigmata. The sterigmate basidiospores may be symmetrical or asymmetrical depending on whether they are arranged symmetricaly or asymmetrically on the sterigmata. In majority of the Basidiomycotina, the basidiospores are arranged asymmetrically on the sterigmata.

In most of the Basidiomycotina where the basidiospore are produced in the open air there is usually a special provision for the discharge of these spores from their points of attachment. The spores are usually discharged with considerable violence. The term **ballistospore** has been applied to violently projected spores

(Derx, 1948). Whilst most basidiospores are ballistospores, some, e.g., those of the Gasteromycetes, are not. Basidiospores on a four-spored basidium are discharged in a regular succession and not simultaneously. The interval between discharge of the first and second and third, or the third and fourth spore may be somewhat longer. Since the maturity of basidia is usually a prolonged process, the period of basidiospore discharge is also prolonged which may last for hours, days or even weeks. The present-day knowledge of spore discharge among the Basidiomycotima is from the researches of Buller (1909-1933). Buller observed that in cases where the basidiospores are attached asymmetrically on the sterigmata, the spores are discharged violently by the mechanism which is termed the **drop-excretion mechanism**, also known as fluid drop mechanism (Fig. 14.8A). A minute or so before the spore discharge, a small droplet of liquid exudes at the hilum (apiculus) (Figs. 14.8B-D), it is often referred to as 'Buller's drop'. Within 5 to 20 seconds this liquid drop increases in volume to about one-fifth the size of the spore and then both spore and liquid drops suddenly shot off from the sterigma (Fig. 14.8E). The sterigma may or may not collapse after the spore is discharged.

Buller (1909) initially postulated that basidiospore discharge was due to the sudden rounding-off of the contacting walls of the spore and sterigma. He subsequently (1922) rejected this hypothesis and suggested that a pressure developing in the vicinity of the hilum was the source of the force for discharge. Buller also suggested that the force for discharge is caused by surface tension energy. Ingold (1939) explained that by the time the liquid drop flow is at the hilum, the spore also will tend to move in the opposite direction exerting pressure on the end of the sterigma. This pressure causes the discharge of spore. Corner (1948) stated that the liquid drop formed at the hilar appendix was surrounded by a membrane, which, he suggested, was an extension of the wall of the sterigma. Olive (1964) suggested that the discharge was due to the explosion of a gas bubble in the vicinity of the hilar appendix and to the residual gas between the inner wall and the outer membrane of the spore and sterigma. Wells (1965) with the electron microscope studied the ultrastructure of the basidia and basidiospres in *Schizophyllum commune*. He observed that a thin fragile extension of the wall of the sterigma is present around the base of the basidiospore. The liquid drop is formed within that portion of the extension of the sterigma wall covering the hilar appendix. The enlargement of the drop contributes to the rupturing of the contacting surfaces between the sterigma and the spore, and to the force of discharge. Besides this, the

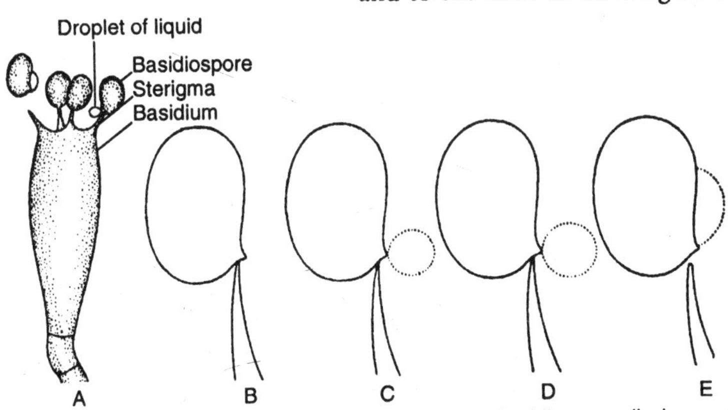

Fig. 14.8: Discharge of basidiospores by drop-execretion mechanism. A. Basidiospore discharge. B-E. Successive stages in the fluid drop excretion and basidiospore discharge

turgour pressure within the basidium also contributes to the force of basidiospore discharge.

The basidiospores germinate in presence of moisture by germ tubes and produce primary mycelium. The uninucleate basidiospores produce monokaryotic primary mycelium. But in case of binucleate basidiospores the condition is different. A basidiospore may be uninucleate in origin but subsequently becomes binucleate by the mitotic division of the single nucleus. Such a binucleate basidiospore behaves like a uninucleate basidiospore. However, it may so happen, two of the four haploid nuclei of the basidium may pass into a single basidiospore and the other two in another basidiospore. Here the binucleate basidiospores possess two nuclei which are genetically different. Such binucleate basidiospores give rise to dikaryotic secondary mycelium.

Again, basidiospores instead of germinating by germ tubes may bud out large number of conidia from which the mycelium is produced.

In the Basidiomycotina there are homothallic and heterothallic species. Since the term heterothallic as originally applied refers to mycelia representing different sexes, the use of the terms heterothallic and homothallic in the Basidiomycotina should be avoided, if not otherwise qualified. It is advisable to use self-compatible or self-fertile and self-incompatible or interfertile for homothallic and heterothallic respectively. The large majority of the Basidiomycotina are self-incompatible and rather few are self-compatible. Among the self-incompatible Basidiomycotina, the compatibility may be governed by one allelomorphic pair of factors, Aa, located on different chromosomes, or by two pairs of allelomorphic factors Aa Bb located on differnt chromosomes and segregating independently. The former condition is known as **bipolarity**, the corresponding species **bipolar species**; and the latter is designated as **tetrapolarity**, the corresponding species **tetrapolar species.** The common explanation of these phenomena is that in bipolar species fruit body development is dependent on two factors and in tetrapolar species on four. Basidium of a tetrapolar species again will produce two kinds or four kinds of basidiospores depending on the method of chromosome segregation during meiosis and on corssing-over. When segregation of the factors controlling compatibility takes place in the first division, only two kinds of nuclei will result; AB, AB and ab, ab or Ab, Ab and aB, aB. In such case, when mycelium bearing nuclei containing AB or Ab unites with mycelium having nuclei possessing ab or aB, a dikaryotic condition bearing all factors Aa Bb in each cell of the dikaryotic mycelium will be established to produce basidia and basidiospores.

If, however, segregation takes place at the second division of meiosis, a basidium will give rise to four types of spore: AB, Ab, aB, ab. Basidia and basidiospores are produced only when the secondary mycelium has the combination of compatibility factors of Aa, Bb.

Mating systems of the Basidiomycotina

About 10 per cent of Basidiomycotina are homothallic. Three types of homothallic behaviour may be distinguished:

(i) **Primary homothallism.** In *Coprinus sterquilinus,* a single basidiospore germinates to form a mycelium which soon becomes organized into binucleate segments bearing clamp connections at the septa. There is no genetical distinction between the two nuclei in each cell, and mycelium is capable of forming fruit bodies.

(ii) **Secondary homothallism.** In *Coprinus ephemerus* f. *bisporus,* the basidia bear two spores, but the spores are heterokaryotic. After meiosis two nuclei enter each spore and a mitotic division may follow. On germination, a single spore germinates to form a dikaryotic mycelium capable of producing fruit bodies.

(iii) **Unclassified homothallism.** The four-spored wild mushroom *Agaricus campestris* and the two-spored cultivated mushroom *Agaricus bisporus* are homothallic in the sense that a mycelium derived from a single spore is capable of producing fruit bodies. In both species there is nuclear fusion in the basidium, followed by two nuclear divisions, presumably meiotic. In *A. bisporus* two nuclei enter each spore, but in *A. campestris* only one. However, paired nuclei, conjugate nuclear divisions and clamp connections have not been observed.

Amongst the remaining 90 per cent of Basidiomycotina reported to be heterothallic, can be distinguished as follows:

(i) **Bipolar.** About 25 per cent of Basidiomycotina examined have been shown to be bipolar. Besides the Uredinales and most Ustilaginales, *Coprinus comatus, Fomes roseus,* and *Polyporus betulinus* have bipolar mating systems.

(ii) **Tetrapolar.** *Coprinus lagopus, Polyporus abietinus, Schizophyllum commune, Crucibulum vulgare,* and *Cyathus striatus* have tetrapolar mating systems.

LIFE CYCLE PATTERN

The general life cycle pattern of the Basidiomycotina has resemblance with that of the Ascomycotina. The basidiospore on germination by germ tube gives rise to the haplophasic somatic body represented by primary mycelium. The primary mycelium often produces oidia. The oidia may behave as spores and give rise to primary mycelia, or by spermatization with compatible primary mycelia take part in the formation of dikaryotic secondary mycelia. The secondary mycelium may also be formed by somatogamy between two compatible primary mycelia. The dikaryotic apical cell of the hypha of the secondary mycelium develops into basidium. It may so happen that resting spores may be formed from the dikaryotic hyphal cells. Each resting spore on germination

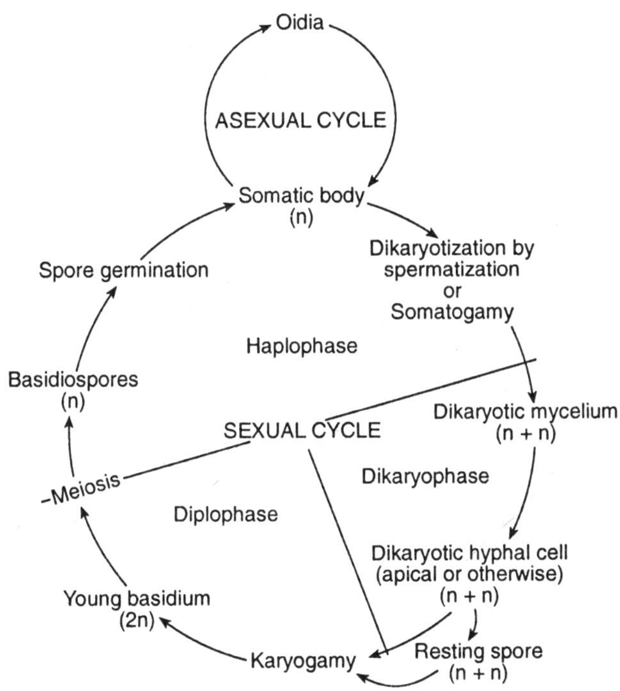

Fig. 14.9: Life cycle pattern in the Basidiomycotina

gives rise to a basidium. Karyogamy takes place in the basidium producing diplophasic condition which is followed immediately by meiosis, and ultimately haploid basidiospores are produced. The basidiospores germinate to produce haplophasic somatic body. Thus the life cycle is complete. The life cycle comprises of the three phases: the haplophase, dikaryophase and diplophase; of which the dikaryophase is very much prolonged and is nutritionally independent of the haplophase. The haplophase is also well represented. Just as in almost all the fungi the diplophase is very short and is followed by meiosis. The general life cycle pattern in the Basidiomycotina is presented in Fig. 14.9.

CLASSIFICATION

The earlier classifications of the Basidiomycotina were based mainly on gross characters of the basidiocarps. The main features used by Fries (1874) for separating taxa were the development of the basidiocarp (gymnocarpous or hemiangiocarpous or angiocarpous), the position of the hymenium in relation to the substratum, and the convolution of the hymenophore. Subsequently, Patouillard (1900) emphasized the importance of the nature of basidium for classifying the Basidiomycotina and recognized the classes Heterobasidiomycetes and Homobasidiomycetes. These classifications are summarized below:

Class: **Heterobasidiomycetes**
- Orders: **Uredinales** and **Ustilaginales**—plant parasites having teleutospores
- Orders: **Tremellales**—mostly saprobic with well-developed, often gelatinous basidiocarps

Class: **Homobasidiomycetes**
- Series: **Hymenomycetes**—developing gymnocarpously or hemiangiocarpously
  - Order: **Aphyllophorales**—with smooth, denate, or tubular hymenophore
    - Family: **Clavariaceae**—hymenium amphigenous, basidiocarps erect, club-like or branched
    - Family: **Thelephoraceae**—hymenium inferior, smooth; besidiocarps effused, effuse-reflexed or flabellate
    - Family: **Cyphellaceae**—hymenium inferior, smooth; basidiocarps cupular or of discrete tubules
    - Family: **Hydnaceae**—hymenium inferior and spread over dentate processes
    - Family: **Polyporaceae**—hymenium inferior lining the inside of tubes not easily separable from the context
  - Order: **Agaricales**—hymenium spread over lamellae or lining tubes, inferior (if tubular, associated with fleshy basidiocarps and the tubes easily separable from the context tissue)
    - Family: **Agaricaceae**—hymenium spread over lamellae
    - Family: **Boletaceae**—hymenium lining tubes
- Series: **Gasteromycetes**—developing angiocarpously
  - Order: **Lycoperdales**—epigean, gleba powdery
  - Order: **Phallales**—epigean, gleba gelatinous
  - Order: **Nidulariales**—epigean, gleba divided into lenticular parts
  - Order: **Hymenogastrales**—hypogean

The Friesian system of classification has obvious advantages in case of comprehension and application in the field, but suffers from the disadvantages of being artificial. It is gradually being replaced by more sophisticated classifications used by specialists.

Pegler (1966) classified Basidiomycotina in the following manner:

Class: **Basidiomycetes**

Subclass: **Heterobasidiomycetidae**

Order: **Uredinales**

Order: **Ustilaginales**

Order: **Tremellales**

Subclass: **Homobasidiomycetidae**

Order: **Agaricales**

Order: **Aphyllophorales**

Talbot (1969) designated Basidiomycetes as Subdivision Basidiomycotina and classified as follows:

Basidiocarp represented by a sorus of teleutospores, mostly parasitic on vascular plants
Class: **Teliomycetes**

Teleutospores terminal, basidiospores on sterigmata
Order: **Uredinales**

Teleutospores mostly intercalary, basidiospores not on sterigmata
Order: **Ustilaginales**

Basidiocarp usually well-developed, mostly saprobic

Mostly gymnocarpous or hemiangiocarpous, basidiospores ballistospores
Series: **Hymenomycetes**

With phragmobasidia Class: **Phragmobasidiomycetes**

Metabasidia broad, septation longitudinal or oblique
Order: **Tremellales**

Metabasidia cylindrical, septation transverse,

Basidiocarp not associated with scale insects
Order: **Auriculariales**

Basidiocarp associated with scale insects
Order: **Septobasidiales**

With holobasidia Class: **Holobasidiomycetes**

Basidia cylindric-clavate, always with two sterigmata
Order: **Dacrymycetales**

Basidia subspherical to broad, with four sterigmata, sterigmata inflated cut off by basal adventitious septa
Order: **Tulasnellales**

Basidia with small uninflated sterigmata or with long filamentous sterigmata

Internal parasites of leaves of higher plants

Basidia emerging in tufts from stomata, basidia cylindrical
Order: **Brachybasidiales**

Basidia emerging through the epidermis to form a hymenium, basidia clavate
Order: **Exobasidiales**

Mostly saprobic, if internal parasites then form external basidiocarp

Basidiocarp developing gymnocarpously; hymenium unilateral or amphigenous, smooth or covering dentate processes or lining cups or tubes, woody, corky or membranous

Order: **Aphyllophorales**

Basidiocarp developing hemiangiocarpously or gymnocarpously, hymenium covering lamellae on the lower surface of the pileus, or if lining tubes then these are separable from the fleshy pileus; basidiocarp fleshy

Order: **Agaricales**

Mostly angiocarpous, basidiospores not ballistospores

Series: **Gasteromycetes**

Gleba fleshy indehiscent, hypogean bisidiocarp

Order: **Hymenogastrales**

Gleba mucilaginous Order: **Phallales**

Gleba powdery lacking capillitium Order: **Sclerodermatales**

Gleba powdery with capillitium Order: **Lycoperdales**

Glebal chambers forming peridioles Order: **Nidulariales**

Webster (1970) followed Ainsworth (1966) and designated Basidiomycetes as Subdivision Basidiomycotina and proposed the following classification:

A. Basidia arising from a thick-walled cell (teliospore or chlamydospore)

Class: **Hemibasidiomycetes**

B. Basidia bearing two to four basidiospores

Order: **Uredinales** (rusts)

BB. Basidia bearing numerous basidiospores

Order: **Ustilaginales** (smuts)

AA. Basidia not arising from a thick-walled cell

B. Hymenium freely exposed at maturity, basidiospores function as ballistospores

Class: **Hymenomycetes**

C. Basidia not segmented or forked

D. Fruit bodies fleshy Order: **Agaricales**

DD. Fruit body usually leathery, woody or corky

Order: **Aphyllophorales**

CC. Basidia segmented or forked Order: **Tulasnellales**

BB. Hymenium enclosed, basidiospores not violently projected

Class: **Gasteromycetes**

C. Fruit bodies not opening at maturity, mostly subterranean

Order: **Hymenogastrales**

CC. Fruit bodies opening at maturity, epigean

D. Basidiospores contained in peridioles

Order: **Nidulariales**

DD. Basidiospores not enclosed in periodioles

E. Basidiospores presented in a sticky mass dispersed by insects

Order: **Phallales**

EE. Basidiospores dispersed dry

F. Fruit body with a central columella and capillitial threads

Order: **Lycoperdales**

FF. Fruit body lacking a columella and capillitial threads

Order: **Sclerodermatales**

Elizabeth-Moore-Landecker (1972) classified Basidiomycotina based on morphology of the basidia. The Basidiomycotina are also conveniently subdivided into two classes, the Heterobasidiomycetes and the Homobasidiomycetes. The chief characteristic of the Heterobasidiomycetes is their basidium which may be septate or deeply lobed. The basidium may also be represented by a teleutospore germinating into a promycelium. Septation of the basidium may be transverse or longitudinal or oblique to the long axis. In the class Homobasidiomycetes, the basidium is typically club-shaped aseptate. Fungi with aseptate basidia developed in a basidiocarp are separated into two series, the Hymenomycetes and the Gasteromycetes. The taxonomic category 'series' is being used for the convenience of classification of the Homobasidiomycetes. In the series Hymenomycetes, the hymenium remains exposed before the spores are mature, as against the hymenium of the series Gasteromycetes which remains closed within the basidiocarp, at least until the spores are mature. The rest of the fungi possessing aseptate basidia but lacking basidiocarp are placed in a small but distinct order Exobasidiales.

A simplified form of key to the classes, series and orders is presented below:

AA. Basidia septa or deeply lobed or developed from resting spores (telestospores) on germination into a promycelium

Class: **Heterobasidiomycetes**
(Phragmobasidiomycetes)

B. Basidiocarp lacking or poorly developed

C. Teleutospores formed only by the terminal cells of dikaryotic hyphae, sterigmata present, number of basidiospores limited

Order: **Ustilaginales**

BB. Basidiocarp usually well-developed

Order: **Tremellales**

AA. Basidia without any septation

Class: **Homobasidiomycetes**
(Holobasidiomycetes)

B. Basidiocarp lacking

Order: **Exobasidiales**

BB. Basidiocarp present

C. Basidiocarp gymnocarpous or hemiangiocarpous

Series: **Hymenomycetes**

CC. Basidiocarp angiocarpous

Series: **Gasteromycetes**

PHYLOGENETIC RELATIONSHIPS OF THE BASIDIOMYCOTINA

It is almost agreed among the mycologists that the Basidiomycotina have evolved from the Ascomycotina. Both these groups of fungi have similarities in their hyphal structure, in having an elaborate dikaryophase and in the development of asci and basidia. It is also accepted that a basidium has evolved from an ascus, during which the number of spores has been reduced from eight to four and the spores have taken external position on the sterigmata of a basidium.

The Ustilaginales and the Uredinales are considered to have been derived from some Ascomycotina possibly before the Pezizales, Sphaeriales, and Dothideales. Linder (1940) believed that the Uredinales have evolved from the fungi close to the Dothideales and the Ustilaginales again from the Uredinales by reducing life cycles. He traced the origin of the teleutospore and promycelium from the ascus with a two-layered wall. In such ascus, the endoascus containing ascospores passes out through the softened apex of the ascus and projects some distance beyond the outer wall of the ascus and eventually ruptures and the spores are set free. Linder also assumed that such an ascus becomes more thick-walled to become a resting spore and by further evolution has given rise to resting spores of the Uredinales and the Ustilaginales. According to Linder, with the gradual loss of obligate parasitism and simplification of the life cycle the Uredinales led to the origin of the Auriculariales. Linder also suggested that all other Basidiomycetous fungi arose from the Uredinales.

According to another school of thought it is held that the Heterobasidiomycetes other than the Uredinales and the Ustilaginales may have arisen from some Ascomycotina somewhat like *Ascocorticium.*

Rogers (1932, 1934) suggested that the genus *Tulasnella* has been derived from an *Ascocorticium*-like ancestor. He further stated that two lines have been derived from *Tulasnella, Tremella* and Auriculariales; one being the Uredinales and the other the Homobasidiomycetes.

A tenable scheme has been proposed by Saville (1955). According to him, the phylogenetic age of the host plant may be used as an index of the approximate age of the parasite. The most primitive existent Basidiomycete genus is assumed to be *Uredinopsis,* a rust parasitic on the phylogenetically oldest rust-parasitized genus, the fern *Osmunda.* Presumably, the hypothetical ancestral Proto-basidiomycete was a rust that shared some resemblance to *Uredinospsis.* This rust has no pycnia, aecia, nor uredia, and it had simple teliospores which may have merely been rounded up hyphal cells. The mycelium inhabited fern fronds and was self-sterile, clampless, and dikaryotic. The hyphal cells rounded up to form teliospores, which produced a basidium bearing four subspherical asidiospores upon sterigmata. Presumably the spores germinated upon the fern frond, and the hyphae from two compatible spores fused either immediately upon or after infection. This hypothetical rust ancestor is remarkably similar to the Ascomycotina genus *Taphrina,* which also forms a dikaryotic mycelium in its host and rounded up cells from which the ascus forms. Under abnormal conditions, *Taphrina* may bud externally, producing subspherical spores upon sterigma-like projections. Both the rusts and *Taphrina* parasitize plants of ancient linkage, the ferns and woody dicots, indicating that they existed during the same geological age span; of greatest importance, a fungus similar to *Taphrina* would be unspecialized and plastic enough to be ancestral to phylogenetic lines which must adapt to different environments. A *Taphrina*-like fungus, or a Prototaphrina ancestor, is assumed to have given rise to both present-day *Taphrina* and the Basidiomycotina.

Both the Prototaphrina and Protobasidiomycete ancestors are extinct. The Protobasidiomycetes gave rise to the Uredinales, beginning with the *Ureinopsis*-like forms. Life cycles of the rusts lengthened with the introduction of more spore forms. The first heteroecious rusts originated when part of the life cycle transferred from the telial to the aecial host.

The Protobasidiomycete ancestor also gave rise to a primitive parasitic fungus in the Heterobasidiomycetidae. The unstable and highly variable heterobasidium is assumed to be primitive and not likely to be derived from the undivided basidium which has become a stable feature in the Homobasdiomycetidae. This primitive fungus was the type found in the family Auriculariaceae in the Tremellales, which has a transversely septate basidium. This basidium type is considered to be primitive because it resembles a mycelium and because the fungi in which it appears are morphologically simple parasitic forms on phylogenetically primitive hosts and, most importantly, because clamps appear sporadically in this group. This primitive, *Auricularia*-type ancestor gave rise to (i) the Septobasidiaceae (parasitic and with no clamps), (ii) the Ustilaginales (parasitic with clamps), (iii) the remainder of the Tremellales (saprobic with clamps), and (iv) the Homobasidiomycetidae.

The early members of the Homobasidiomycetidae line were parasitic fungi like some Polyporales (the family Thelephoraceae) which form a flat hymenium on a resupinate basidiocarp, either with or without clamps. Through evolutionary pressure, the mymenium developed a greater spore-producing surface through the formation of ridges, teeth, gills, and pores. In addition, there was a tendency for the spore-producing surface to project downward as this increased the efficiency of spore discharge. This evolutionary line gave rise to the more advanced Hymenomycetes. The Gasteromycetes represent a polyphyletic group in which convergent evolution under the pressures of the same habitat produced superficial similarities. Both the Gasteromycetes and the Clavariaceae represent polyphyletic groups originating from the Hymenomycetes.

Teixeira (1962), Gaumann (1964), and Olive (1965) have suggested that the clamp connection of Basidiomycotina is homologous to the crozier at the tip of ascogenous hyphae in *Ascomycotina,* and have argued that the Basidiomycotina have evolved from Ascomycotina ancestors.

COMPARISON BETWEEN THE ASCOMYCOTINA AND THE BASIDIOMYCOTINA

1. The Ascomycotina are characterized by the development of asci and ascospores. But the Basidiomycotina produce basidia and basidiospores.
2. Both the ascospores and basidiospores are formed by karyogamy and meiosis. But the ascospores are borne endogenously in an ascus, whereas the basidiospores are produced exogenously on a basidium.
3. Majority of the Ascomycotina and the Basidiomycotina produce fruit bodies.
4. In both these classes of fungi the somatic structure is represented by well-developed septate mycelium.
5. A great number of Ascomycotina produce spores by which they reproduce asexually. But in the Basidiomycotina the asexual reproduction is rather insignificant.
6. The general life cycle of the Ascomycotina and the Basidiomycotina follows almost the same pattern comprising of haplophase, dikaryophase, and diplophase.

7. The dikaryophase of the Ascomycotina is dependent on the haplophase for nutrition. But in the Basidiomycotina the dikaryophase is completely independent.
8. The cells (apical or otherwise) of the dikaryotic hyphae are responsible for the development of asci and ascospores, whereas the Basidiomycotina produce dikaryotic secondary and tertiary mycelia. The former are responsible for the development of basidia and basidiospores, and the latter give rise to the sterile tissue of the fruit body.
9. The dikaryotic condition in the Ascomycotina is established by the development of ametangia, or by spermatization, or by somatogamy. No gametangia are produced in the Basidiomycetes. Dikaryotization is either by spermatization or by somatogamy.
10. There is partial degeneration of visible sexuality in the Ascomycotina. But in the Basidiomycotina, there is complete degeneration of visible sexuality.
11. The Ascomycotina are mostly characterized by crozier formation, whereas the Basidiomycotina possess clamp connections in thier dikaryotic mycelia.

Order: **Uredinales**

The Uredinales, or rust fungi comprise a large group of obligate parasites. They are called rusts, because of the rusty, reddish-brown colour of their spores, which are formed chiefly upon surfaces of host leaves and stems in pustular outgrowths (Figs. 14.10A, D and E). Individual species of rusts may have a very wide or a very limited range of hosts. Parasitism and specialization are so highly developed

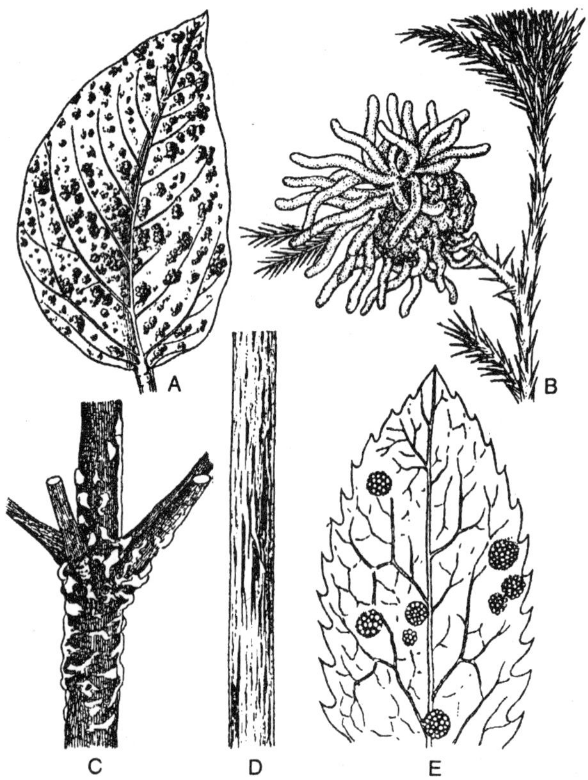

Fig. 14.10: Rusts on different hosts. A. Bean rust (*Uromyces appendiculatus*) on leaf. B. Cedar-apple rust (*Gymnosporangium juniperi-virginianae*) on red cedar. C. White pine blister rust (*Cronartium ribicola*) on white pine twig. D and E. Black stem rust of wheat (*Puccinia graminis tritici*). D. On stem of wheat. E. On barberry leaf

among them that thus far, except a few cases, almost all attempts to grow them on artificial media have failed. Rusts are parasitic on seed plants or ferns. They cause great loss to crop, forest, and ornamental plants. No other group of fungi is so injurious to plants useful to man. They are considered as one of the virulent plant parasites.

Some rust fungi complete their life cycle on a single kind of host, they are known as **autoecious rusts.** But others require two unrelated hosts for the completion of life cycle, they are called **heteroecious rusts.** Most autoecious rusts are of relatively little economic importance, but the heteroecious ones cause some of the most destructive of all plant diseases. Among the important heteroecious rusts are: white pine blister rust (Fig. 14.10C), *Cronartium ribicola,* which has currents and gooseberries (the genus *Ribes*), especially *Ribes nigrum,* as alternate hosts with white pine (*Pinus strobus*); and black stem rust of wheat, *Puccinia graminis tritici* has wheat (*Triticum aestivum*) and barberry (*Berberis vulgaris*) as alternate hosts. Among the more common autoecious rusts are: hollyhock rust, *Puccinia malvacearum* on species of *Althaea, Levatera, Malva,* all genera of the Malvaceae; and flax rust, *Melampsora lini* on species of *Linum.*

Rust fungi do not produce fruiting bodies. Life cycle of most of them is extremely complicated. It usually involves as many as five kinds of spore forms which appear in a definite succession. The life cycle is thus **polymorphic.** The five spore forms are: **pycniospores (spermatia), aeciospores, uredospores, teleutospores,** and **sporidia (basidiospores).** These spore forms are also expressed as follows:

Stage 0. Spermogonial stage producing spermogonia bearing spermatia and receptive hyphae(n).

Stage I. Aecial stage producing aecium bearing aeciospors (n + n).

Stage II. Uredial stage producing uredia bearing uredospores (n + n).

Stage III. Telial stage producing telia bearing teleutospores (n + n), 2n.

Stage IV. Sporidial stage producing promycelium bearing sporidia (n).

The rust fungus producing either all the five spore forms or at least one type of dikaryotic spore in additon to the teleutospore, is referred to as a **long-cycled** or **macrocyclic rust.** Macrocyclic rusts again may be heteroecious (e.g., *Puccinia graminis*), which produces all spore forms on two unrelated hosts; or autoecious (e.g., *Puccinia asparagi*) whose all spore forms are produced on asparagus. The prefix 'eu' is sometimes attached to the generic name as a descriptive epithet, i.e., *Puccinia graminis* is an *eu-Puccinia,* whereas in a **short-cycled** or **microcyclic rust,** the teleutospores are the only dikaryotic spores produced in the life cycle. The life cycle of rusts is said to be **demicyclic** in which uredospores are lacking; as in heteroecious forms in *Gymnosporangium,* and amongst autoecious forms in *Xenodochus carbonarius.* It is believed that the heteroecious macrocyclic forms represent the primitive condition, and that the autoecious conditions arose later in evolution. It is also believed that forms with shorter life cycles arose during the course of evolution from macrocyclic ancestor.

The mycelium of rust fungi usually ramifies intercellularly in the host tissue sending haustoria in the adjacent cells and inducing malformations in the infected parts (Fig. 14.10B). The infection may be localized or systemic. The mycelium may be short-lived or perennial consisting of well-marked phases that alternate: a monokaryotic phase and a dikaryotic phase. Clamp connections have been found in the dikaryotic mycelium of very few rusts. The monokaryotic mycelium develops the Stage 0 (Spermogonial stage) and the dikaryotic mycelium whose activity is rather more predominating than the monokaryotic mycelium produces Stage I (Aecial stage), Stage II (Uredial stage), and Stage III (Telial stage). In the late telial stage karyogamy takes place. The Stage IV (Sporidial stage) is the one where meiosis takes place and monokaryotic condition is established. In the Uredinales the sporidia fall into two compatible groups and are determined genetically by a pair of allelomorphs (A, a). The dikaryotic condition is initiated in the late Stage 0. Craigie (1927) working on *Puccinia graminis* showed that the dikaryotic condition in rusts is achieved by the fusion of spermatia with receptive hyphae of opposite strains. Brown in 1932 found that in *Puccinia helianthi,* an autoecious rust, a dikaryotic mycelium of uredosorus can induce aecial formation in a monokaryotic mycelium produced from a sporidial infection. Such a process of diploidization of a monokaryotic mycelium by a dikaryotic mycelium is known as the **Buller phenomenon.** Late on Brown (1935) found evidence that in *P. helianthi,* the dikaryotic condition is also established by the fusion of monokaryotic hyphae of spermogonia of two opposite strains. In many rust fungi, only the aecial stage alone is known. Again, in many others, the uredial stage has been worked out. It has become necessary at present to name and classify the uredial, aecial and other non-telial forms of which the telial stage is not yet known. For convenience of non-telial forms of which the telial stage is not yet known. For convenience of reference they are classified under an artificial group 'Uredinales Imperfecti' in which are included the form-genera: *Caeoma Aecidium, Peridermium, Roestelia,* and *Uredo.*

LIFE CYCLE PATTERN

In the usual course of the life cycle, as a result of infection of the host by sporidia, monokaryotic mycelia are produced which give rise to **spermogonia** (sing. **spermogonium**). The spermogonia are also known as **pycnia** (sing. pycnium). The spermogonia are subepidermal or subcuticular structures which may be acervular and pycnidial in form (Fig. 14.11). The pycnidial type is rather common. The whole structure in gross appearance is much like the pycnidium being globose to flask-shaped with an ostiole and is composed of pseudoparenchymatous tissue. The spermogonia are reddish or orange in colour. The inner wall of the spermogonium is lined by elongated uninucleate hyphal structures, called **spermatiophores** from the apices of which are abstricted in series single-celled uninucleate **spermatia,** also known as **pycniospores** (Figs. 14.11A and D). Slender thread-like periphyses protrude from the ostiole of the spermogonium. Just below or among the periphyses are developed slender and branched hyphae known as **receptive hyphae** (Fig. 14.18A) or **flexuous hyphae.** The receptive hyphae, along with the periphyses grow out through the ostiole and extend beyond the host surface. The spermatia developed in the spermogonia ooze out through the ostiole along with a sticky sugary exudate which sometimes gives off a fragrant odour. Insects attracted by the sticky exudate mixed with spermatia visit the spermogonia and serve as agencies through which the spermatia are transferred from one spermogonium to the other. The function of spermatia was questionable for quite some time. Craigie (1927) observed that spermatia are carried from one spermogonium to the other possibly by insects which suck the 'nector drops' (exuded by spermogonia), in which the spermatia are held. Again, along with the development of a spermogonium in its vicinity, an **aecial primordium** or a **protoaecium** (Fig. 14.18A) which has hyphal connection with it, develops. But further growth of the aecial primordium into aecium bearing aeciospores never takes place unless dikaryotic condition is established by spermatization between a spermatium and a receptive hypha. Craigie also demonstrated that the 4 sporidia that are developed from a promycelium during the germi-

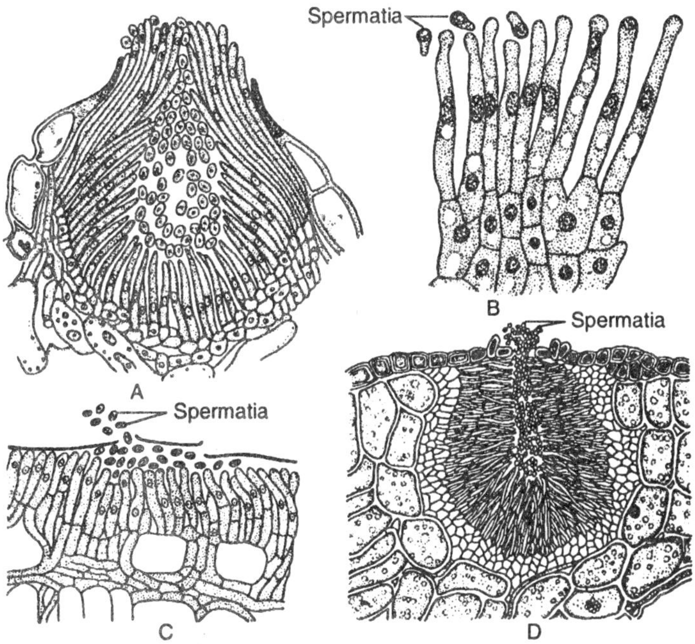

Fig. 14.11: Acervular and pycnidial types of spermogonia. A. Spermogonium of *Puccinia graminis* on barberry leaf. B. Subcuticular spermogonium of *Phragmidium violaceum* on leaf of *Rubus*. C. Spermogonium of *Cronartium ribicola* on white pine twig. D. Subepidermal spermogonium of *Milesia marginalis* on needle of *Abies balsamea*.

nation of a teleutospore belong to two groups: two of them possess (+) nuclei and the other two (–) nuclei, one in each. The sporidium bearing (+) nucleus produces a (+) spermogonium and the one with (–) nucleus gives rise to a (–) spermogonium and dikaryotic condition is established only between a (+) spermatium and a (–) receptive hypha and *vice versa*. It has been demonstrated by Brown (1935) that the dikaryotic condition may be established between hyphae of (+) and (–) spermogonia. Immediately after the establishment of dikaryotic condition, it is perpetuated into the aecial primordium through the hyphal connection that is already present between the spermogonium and the aecial primordium. The aecial primordium then develops into an **aecium** (pl. **aecia**) (Fig. 14.18).

The aecia are usually deeply sunken in the host tissue. They are cup-shaped structures and are commonly known as **cluster cups**. The whole structure is usually red or yellow. The outer layer of the cup is composed of sterile tissue and is termed the **peridium** (Fig. 14.19A). At the base of the cup a hyphal plexus is developed from which vertically elongated cells arise. From these cells **aeciospores** (Fig. 14.19A) are produced in chains. In long chain the spores are separated from each other by intercalary sterile cells known as **disjunctor cells** (Fig. 14.19B). The whole interior of the young aecium is filled up by numerous parallel chains of spores (Fig. 14.19A). The peridium at first completely encloses the aecium. But as the aeciospores within the aecium increase in number, the cup is thrown open by bursting an area of the peridium. Thus a cup with a toothed margin with its edges torn-apart is developed. Peridium may also be lacking in an aecium. The aecia of the Melampsoraceae lack peridia so that they are diffuse instead of cup-shaped. Such aecia are termed **caeomata** (sing. **caeoma**).

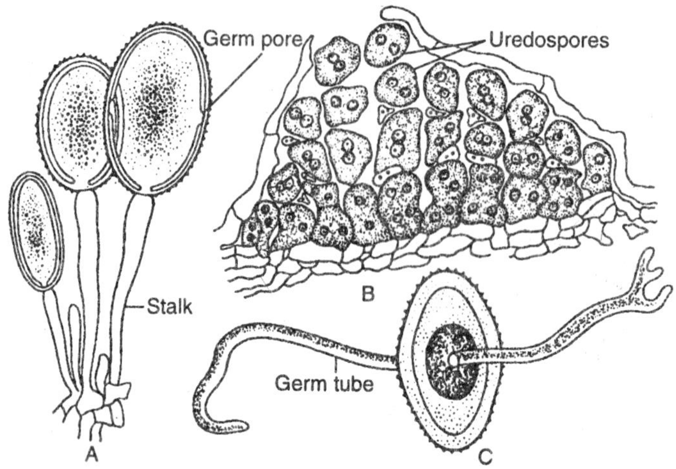

Fig. 14.12: Uredosorus and uredospores of various rusts. A. Stalked uredospores of *Puccinia graminis*. B. Vertical section of uredial pustule of *Coleosporium solidaginis*. Intercalary cells separate the uredospores that are produced in chains. C. Germination of uredospore of *Puccinia graminis*

Rusts which produce uredinoid aecia are **brachycyclic,** e.g., *Puccinia punctiformis* is a *brachy-Puccinia*. The aeciospores are unicellular usually globular, or angular by compression. They are light orange, yellow, smooth-walled with thin places, and are dikaryotic (Fig. 14.12B). The aeciospores become separated by breaking down of the disjunctor cells between them. They are capable of germination at once and on germination give rise to germ tubes which emerge through the thin places in the spore wall, and may infect susceptible hosts, leading to the development of mycelium. The aeciospores may infect the same kind of host on which the aeciospores were produced (autoecious), or entirely unrelated hosts (heteroecious). The dikaryotic mycelium produced on germination of aeciospore develops within the host producing localized infection. Ultimately a subepidermal hyphal plexus is formed, known as **uredium** (pl. **uredia**) or **uredosorus** (pl. **uredosori**) (Figs. 14.12B and 14.15), from which arises a bed of spores, the **uredospores** (Fig. 14.12A). The uredospores are yellow to orange-red, stalked, unicellular dikaryotic oval or globose with echinulate or verrucose wall bearing 2 to 15 variously placed germ pores (Fig. 14.12A). They may be produced in chains with intercalary cells which separate them from each other (Fig. 14.12B). The uredospores with maturity, are exposed through ruptured epidermis of the host surface. These spores falling on susceptible hosts may germinate at once by producing a germ tube (Fig. 14.12C) which develops into mycelium. The infection is usually stomatal and production of uredia may be continued for some time. The uredospores are usually short-lived and function to spread summer infection. They are also sometimes known as a **summer spores** or **repeating spores.** In a few species there are also resting forms of thick-walled uredospores which are also known as **amphispores.** These spores are more persistent than those of the usual uredospores.

Towards the latter part of the growing season another kind of spore appears, often in the same sorus with the uredospore and from the same mycelium producing what is called **mixed sorus.** This kind of spore is known as **teleutospore** or **reliospore.** Ultimately the dikaryotic mycelium produces only the **teleutosori** (sing. **teleutosorus**) or **telia** (sing. **telium**) (Fig. 14.16) bearing teleutospores. The telia are generally darker than uredia, and when mature may

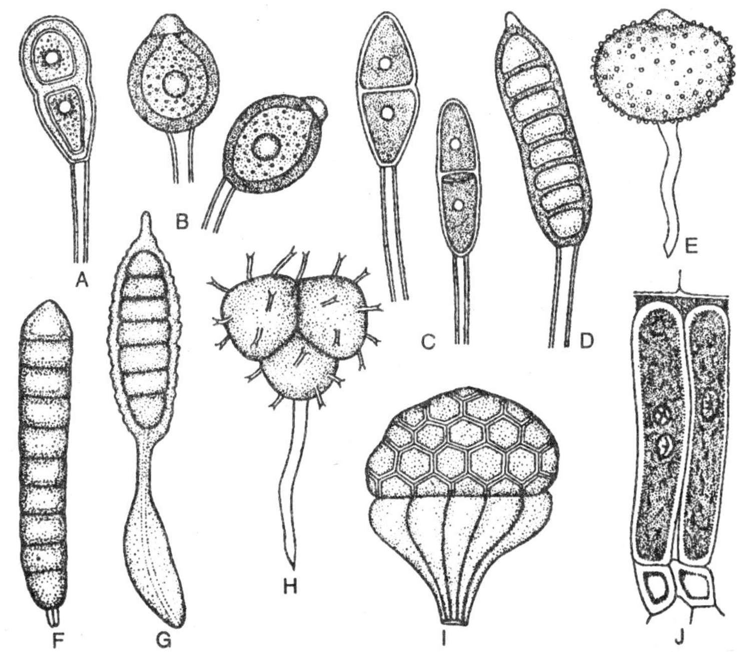

Fig. 14.13: Various types of teleutospores in different genera of rusts. A. *Puccinia*. B. *Uromyces*. C. *Gymnosporangium*. D and G. *Phragmidium*. E. *Pileolaria*. F. *Xenodochus*. H. *Nyssopsora*. I. *Ravenelia*. J. *Melampsora*

be exposed like the uredia or may still be covered by the host epidermis. They may also form crusts or cushions or columnar masses. The teleutospores are of various forms in different genera, one or more-celled, and are extremely variable in shape, thickness of wall, surface marking and colour (Fig. 14.13). They may be sessile (Fig. 14.13J) or stalked (Figs. 14.13A-I), may be completely free from one another (Figs. 14.13A-I), embedded in a gelatinous matrix, united laterally forming small groups, layers (Fig. 14.13J), or columns (Fig. 14.14B). The teleutospores are almost colourless to a dark reddish-brown. They vary in size and in external characteristics of their walls which may be smooth, spiny, or variously sculptured (Fig. 14.13). The teleutospores when combined together to form a telia column are not readily detachable (Fig. 14.14B). But they are dikryotic and are uniform in the character of the germination which is very different from any other rust spores. The teleutospores are resting spores and most rust fungi perennate in the telial stage. All rust fungi except those included under the Uredinales Imperfecti produce teleutospores. The teleutospores are usually considered as the perfect stage of the rust fungi, since karyogamy and meiosis take place during their germination. The morphology and behaviour of the teleutospores furnish the principal characteristics for the differentiation of families of the order Uredinales.

In teleutospore germination of most rust fungi, each cell of the teleutospore produces a hypha-like structure of limited growth known as **promycelium** (Fig. 14.14A). The promycelium may be borne externally, **external promycelium** of internally, **internal promycelium** from a teleutospore. In certain rust fungi the teleutospore on germination becomes septate without the formation of an external promycelium. Karyogamy takes place during

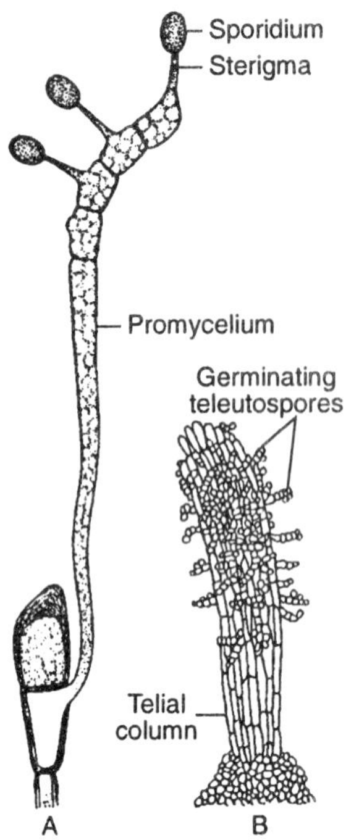

Fig. 14.14: Germination of teleutospores. A. Germinating teleutospore of *Puccinia graminis* showing promycelium and sporidia. B. Telial column of *Cronartium* sp. showing germinating teleutospores

teleutospore germination and the diploid nucleus migrates in the promycelium. Where four haploid nuclei of two groups (+) and (–) are produced as a result of meiosis. Septa are laid down in the promycelium separating the four haploid nuclei into four cells. Each cell rating the four haploid nuclei into four cells. Each cell then sends out a sterigma on which one-celled thin-walled oval to rounded spore is developed. The four spores of two groups (+) and (–) so formed are known as **sporidia** (sing. **sporidium**) or **basidiospores.** The sporidia at maturity are dispersed by wind. Morphologically the promycelium is a basidium bearing its four sterigmata and basidiospores. The sporidia germinate immediately by germ tubes which on suitable hosts give rise to spermogonia of two groups (+) and (–) completing the life cycle.

NUCLEAR CYCLE

In all the rust fungi the dikaryotic condition (n + n) is rather more well represented than the monokaryotic stage (n). It is established by spermatization between a (+) spermatium and a (–) receptive hypha and *vice versa.* Immediately after dikaryotization, the dikaryotic condition is perpetuated to the protoaecium through the hyphal connection that already existed between the spermogonium and protoaecium. The protoaecium which has become dikaryotic, develops into an aecium bearing dikaryotic hyphal cells. The aeciospores borne in the aecium are also dikaryotic. The dikaryotic condition is further continued through dikaryotic mycelia that are formed from the aeciospores. The dikaryotic mycelia induce infection in the host and produce uredia and urediospores which are also dikaryotic. The dikaryotic condition is further perpetuated by the development of telia and teleutospores. End of the dikaryotic condition comes with the development of teleutospores. The teleutospores when young are dikaryotic, and at maturity karyogamy takes place when they become diplophasic (2n). Meiosis of the diploid nucleus takes place during the teleutospore germination when the monokaryotic condition (n) is established. The monokaryotic condition is represented by sporidia and continues up to the development of spermogonia, spermatia, receptive hyphae and finally the protoaecia with which the life cycle is completed. Of the stages 0 to IV, only the stages 0 an IV are monokaryotic while the rest of the stages I to III of the life cycle are dikaryotic with a very short diplophase. All rust fungi perennate in the dikaryotic condition either as mycelium or as uredospores or as teleutospores. But the teleutospores serving as perennating organs may also be in the diplophasic condition. Besides this, it is during the dikaryotic condition when maximum damage is caused to the host.

BIOLOGICAL SPECIALIZATION

Specialization of parasitism is one of the outstanding characteristics of the rust fungi. Certain rusts may be confined to a single host species even variety, while others are capable of infecting many different hosts which may be related or unrelated. Again, there are others, each of which is found on a large number of hosts. Upon its numerous hosts the fungus may show no morphological variation, yet attempts to inoculate from one host to another may uniformly produce negative results. Such a biological specialization of parasitism has necessitated to introduce the term **biological species.** Hence a biological species be regarded as a collection of morphologically similar organisms or strains that differ in ability to produce infection in different hosts. The existence of biological species has been amply demonstrated among cereal rusts for example, *Puccinia graminis.*

CLASSIFICATION

The Uredinales are classified into three families on the basis of their teleutospore characteristics and behaviour. The three families are: **Coleosporiaceae, Melampsoraceae,** and **Pucciniaceae.** Key to the families is presented below:

AA. Teleutospore germination by the formation of an external promycelium

- B. Teleutospores usually stalked, less commonly sessile, free or variously united but never form layers or crust **Pucciniaceae**
- BB. Teleutospores sessile laterally united forming crusts or cushions or columnar masses **Melampsoraceae**

AA. Teleutospore germination by the formation of an internal promycelium **Coleosporiaceae**

Family: *Pucciniaceae*

This is the largest and most important family of the order Uredinales. Rust fungi included in this family infect numerous valuable agricultural plants and cause enormous losses to crops. Members of the family are manifold and the complexities, owing to polymorphism, heteroecism and biologic specialization are very great.

Some of the economically important rusts are: *Uromyces appendiculatus* is rust of beans (*Phaseolus mungo* and *P. radiatus*); *Puccinia graminis* f. sp. *tritici* is black stem rust of wheat (*Triticum aestivum*); *Puccinia malvacearum* is hollyhock rust, attacks malvaceous hosts; *Gymnosporangium juniperi-virginianae* is juniper apple, and cedar apple rust (Fig. 14.10B); and *Gymnosporangium sabinae* is juniper, and pear rust.

Teleutospores in the family are stalked, stalk sometimes short or evanescent. They are one to several-celled being borne singly on a pedicel or in groups on either simple or several united to form a parasol-like head on a compound stalk. The teleutospores are subepidermal and covered or free, and erumpent or embedded in a gelatinous matrix. Sporidia are formed from promycelia produced as a result of the germination of the teleutospores. Uredospores are solitary.

Genus: *Puccinia*

This genus possesses spherical spermogonia with minute ostioles and aecia with peridia. It has uredospores which are unicellular, spherical ellipsoid to variously-shaped having rough wall with many germ pores. The teleutospores are separate, and produced in flat sori. They are pedicellate consisting of two superimposed cells each of which is provided with a germ pore. The germ pore of the upper cell is at the apex; whereas the lower cell has its germ pore placed immediately below the septum present between the two cells.

Some Indian Species: *Puccinia graminis* Pers.; *P. graminis* f. sp. *poae* (Pers.) Erikss.; *P. graminis* Pers. f. sp. *tritici* Erikss. and Henn.; *P. malvacearum* Mont.; *P. recondita* Rob. ex. Desm.; *P. striiformis* West.

PUCCINIA GRAMINIS Pers.

Puccinia graminis, commonly known as rust fungus, produces one of the most important plant diseases designated as rust disease. It was a wide host range of germinaceous plants including wheat, oats, barley, rye, and numerous wild and cultivated grasses. It is most important economically on wheat, on which crop it has been a subject of most extensive study for centuries. Special mention may be made of classical researches of Anton de Bary and the Tulasne brothers who established the parasitism of the rust fungus. The researches of de Bary have repeatedly served as the basis of fundamental investigations in parasitism, cytology and biological specialization. The fungus is heteroecious, parasitizes two unrelated species of hosts alternating from cereal or grass to barberry. The life cycle of the fungus is extremely complicated involving five distinct stages, viz., 0, spermogonia bearing spermatia and receptive hyphae; I, aecia bearing aeciospores; II, uredia bearing uredospores; III, telia bearing teleutospores; IV, promycelia bearing basidiospores. This is known as long-cycled or macrocyclic rust.

The somatic structure of the fungus consists of septate hyphae which penetrate the host tissues and cause destruction but do not become organized into structures of definite form. The hyphae branch intercellularly and produce small haustoria which penetrate the host cells. They occur in two well-marked alternating phases, the monokaryon (n) and the dikaryon (n + n). The fungal hyphae live in the stem and leaf tissues of graminaceous hosts but do not produce any visible change until late in the season, particularly in the wheat plants, mostly when the wheat is in spikelet stage.

Elongated rusty red streaks or pustules appear on the stem, leaf sheaths and leaves, particularly on the stem. They very soon burst breaking through host epidermis due to pressure of the developing spores from inside the host tissue. Groups of minute yellowish spores formed just beneath the host epidermis are exposed. They are known as **uredosori** (sing. **uredosorus**) or **uredia** (Fig. 14.15). The spores are known as **uredospores**. These spores give external evidence of the presence of mycelium growing among the cells of the host tissue. The uredosori are rusty red from which the name 'rust' is given to the fungus and the disease it

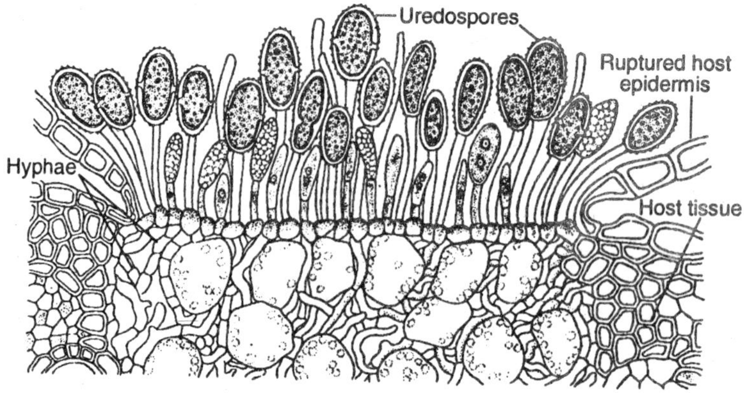

Fig. 14.15: Section through uredosorus of *Puccinia graminis* on wheat stem, showing uredospores in various stages of development and ruptured host epidermis

causes. The production of uredospores is spoken of as the 'red rust' stage of the disease. The uredospores are dikaryotic (n + n), single-celled, generally oval in form, and stalked. They are rather thick-walled. The wall bears numerous short, spiny projections and four equatorially arranged germ pores. They are yellow to orange-red. The uredospores are disseminated by wind. They directly infect the graminaceous host specially the wheat plants and are chiefly responsible for the rapid spread of the disease. These spores are capable of germination within four or five days after they are shed, if sufficient moisture is present. The uredospores germinate rapidly in water or moist air, sending out a germ tube (Fig. 14.12C). The germ tube does not enter in the host tissue directly, but the tip swells up into an elongated **appressorium** (pl. **appressoria**). From this, a narrow branch passes through stomata and immediately swells up to form a substomatal vesicle. Soon a dikaryotic hypha (n + n) arises from one end and grows towards a neighbouring cell, into which it sends a haustorium. Infection is now complete and the dikaryotic mycelium (n + n) continues its growth within the host. After 10 days or 2 weeks a new crop of uredospores is formed. Thus it is possible to have a considerable number of successive crops of uredospores in a single season, and from one or a few infected plants the disease may be spread over a large area. The uredospores are capable of infecting only the graminaceous host and are responsible for repeating infection of the susceptible host.

Within a short time, further formation of uredospores ceases, and from the mycelium that has been producing uredospores, or from mycelia originating from uredospores, gradually black pustules or black sori appear on the same stem and leaves. By the rupture of the host epidermis black sori consisitng of black masses of spores are exposed. These black sori consist of dark-coloured spores known as **teleutospores** or **teliospores**. The sori are called **teleutosori** (sing. **teleutosorus**) or **telia** (Fig. 14.16). During the transitional time betweeen the development of uredosori and teleutosori, both uredospores and teleutospores may be formed in the same sorus. Such a sorus is often known as **mixed sorus.** The teleutospores are stalked, thick-walled smooth, two-celled with

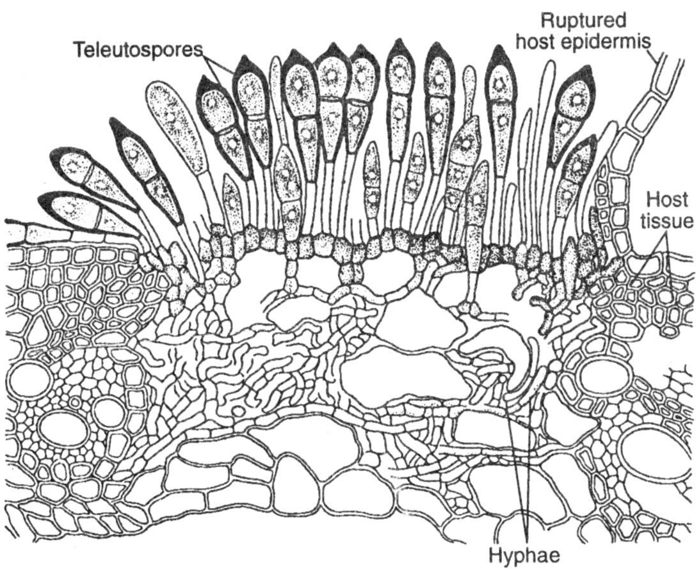

Fig. 14.16: Section through teleutosorus of *Puccinia graminis* on wheat stem, showing teleutospores in various stages of development and ruptured host epidermis

superimposed cells, the top cell being rounded or blunted or pointed and thickened at the apex. They are ellipsoid to oblong clavate being slightly constricted at the septum, dark-brown (black is in the mass). Each constituent cell of the spore is furnished with a germ pore, situated apically in the top cell, and just below the septum in the lower cell. When young, each cell is dikaryotic (n + n), but karyogamy takes place as the spore matures. Unlike uredospores, the teleutospores are not capable of germination immediately they are formed, but must go through a period of rest of several months. Hence they are not responsible for the rapid spread, but are rather resting spores to carry the fungus over to the following season. After resting in the open on dead host tissues or on the soil they germinate freely. But their survival in such a condition in the plains of Indian is rather difficult since the temperature during March to June is considerably high. The teleutospores perennate in the uninucleate diplophasic condition (2n) after karyogamy has occurred.

On germination, each cell of the teleutospore gives rise to an elongated germ tube-like structure called a **promycelium** (Fig. 14.14A), also known as **basidium.** Each cell of the teleutospore is also known as a **probasidium.** Those who hold this designation, according to them the probasidium germinates to form an **epibasidium** which is nothing but the promycelium, and the teleutospore cell from which the epibasidium is produced is the **hypobasidium.** The diploid nucleus of the teleutospore cell then migrates in the promycelium where it undergoes meiosis producing four haploid nuclei (n). The promycelium is then transversely partitioned to develop into a four-celled structure, cells having a single haploid nucleus one in each. Each cell of the promycelium gives rise to a sterigma on which a single ellipsoid **sporidium** (pl. **sporidia**) (Fig. 14.14A) or **basidiospore** is formed. The haploid nuclei ultimately squeeze through the sterigmata into the sporidia (n). The sporidia are obliquely set on the sterigmata. Two of the sporidia are of one strain and two are of the other, may be designated as **plus strain** and **minus strain** respectively. The sporidia are forcibly discharged at maturity by water droplet method. They are caught up by air currents, and if they alight on the leaves of barberry they will germinate. The sporidia do not germinate on wheat or other graminaceous hosts.

The sporidia germinate on the upper surface of the barberry leaf in presence of moisture. Each sporidium produces a germ tube which gains entrance into the host tissue directly through the epidermis (Fig. 14.17A), not by way of stoma. The germ tube then produces an extensive intercellular (Fig. 14.17B) monokaryotic mycelium (n) which depending on the sporidium, carries **plus** or **minus factor** in its nuclei. After a short period of growth, the mycelium submerged below the surface on the upper side of the leaves develops flask-shaped structures called **spermogonia** (sing. **spermogonium**) (Fig. 14.17A), also known as **pycnia** (sing. **pycnium**). The spermogonia are monokaryotic (n) in nuclear condition. The small yellow or reddish spots that are visible on the upper surface of the infected leaves are the areas where spermogonia appear. The spermogonia open to the surface of the leaves. The cavity of each spermogonium is lined with numerous spermatophores which cut off minute colourless rod-shaped or oval monokaryotic (n) **spermatia** (sing. **spermatium**) (Fig. 14.17A), also known as **pycniospores.** Besides these structures, **receptive hyphae** (monokaryotic) growing from the side walls and several periphyses project outward through the opening of each spermogonium. Depending on the nature of sporidia carrying **plus** or **minus factor,** the corresponding spermogonia produced by them are also **plus** or **minus,** which may develop on the same leaf or on

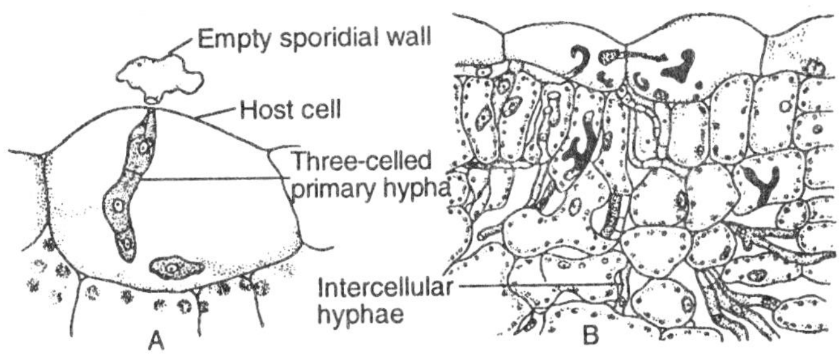

Fig. 14.17: Stages of barberry leaf infection. A. Sporidial germination and host penetration. B. Section through infected tissue showing intercellular hyphae

different leaves. Similarly, the spermatia produced by the spermogonia will carry **plus** or **minus factor.** The mature spermogonia produce spermatia and a kind of sweet sticky liquid. This liquid along with spermatia exuded out from the spermogonia attracts insects. The insects unwittingly transfer the spermatia in moving from one spermogonium to other and thereby the spermatia come in contact with the receptive hyphae, which extend out through the opening of each spermogonium. By the time the spermogonia are being developed just beneath the upper epidermis; the monokaryotic mycelium (n) also produces near the lower epidermis, opposite the spermogonia, a number of **aecial primordia** (Fig. 14.18A), also known as **immature aecia.** Each one may also be termed an **aecium initial.** Buller (1938) suggested for this structure the term **protoaecidium.** The aecial primordium consists of an upper closely packed weft of hyphae, the basal cells, and a lower group of larger parenchyma-like cells. Unless spermatization takes place by the contact of a **plus spermatium** with a **minus receptive hypha,** or *vice versa,* the aecial primordium develops no further, and is functionless. During spermatization spermatial contents pass into the receptive hypha by a pore developed in the walls at the point of contact of the spermatium and the receptive hypha. The spermatial nucleus passes down through the hypha from one cell to another establishing dikaryotic condition (n + n) and ultimately the hyphal cells of the aecial primordium become dikaryotized (n + n). Within a few days after initiation of the

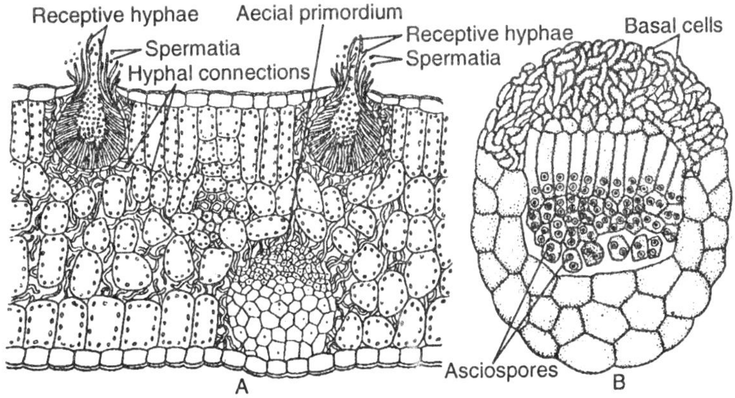

Fig. 14.18: Section through barberry leaf. A. Two spermogonia at the upper epidermis and an aecial primordium near the lower epidermis and hyphal connections in between. B. Structural details of young aecium with basal cells and aeciospores

dikaryotic condition, the aecial primordium is stimulated to further growth, and after about two days later, on the lower surface of the leaf, **aecium** (pl. **aecia**) (Fig. 14.19A) or **cluster cup** (n + n) is formed. That spermatization between a **plus spermatium** and a **minus receptive hypha** or *vice versa* is a prerequisite to the development of aecium has been confirmatively established by Craigie (1927) and Buller (1950). The young aecium is closed (Fig. 14.18B) and buried in the tissue of the leaf, but as growth proceeds it pushes through the leaf surface and opens, exposing the spores within. The aecium is cup-shaped, composed of whitish hyphae, and the uppermost cells forming the rim of the cup, readily slough away, leaving a ragged margin. It has distinct peridium. Within the cup are the dikaryotic (n + n) **aeciospores** (Fig. 14.19A), yellow to orange in colour, extending in chains from specialized stalk cells at the base of the cup. The cells of the chain producing aeciospores are alternately large and small, the former maturing into an aeciospore but the latter remains sterile and is called intercalary cell or **disjunctor cell** (Fig. 14.19B). When mature the aeciospores fall from the cup and are carried away by currents of air. The aeciospores are infectious only to graminaceous hosts. If they fall upon graminaceous hosts, and if suitable conditions of temperature and moisture prevail, they germinate by sending germ which pass through stomata into the tissues of the stem and leaves and produce dikaryotic hyphae (n + n). The dikaryotic hyphae in one to two weeks produce rusty red uredosori on the stem, leaf sheaths

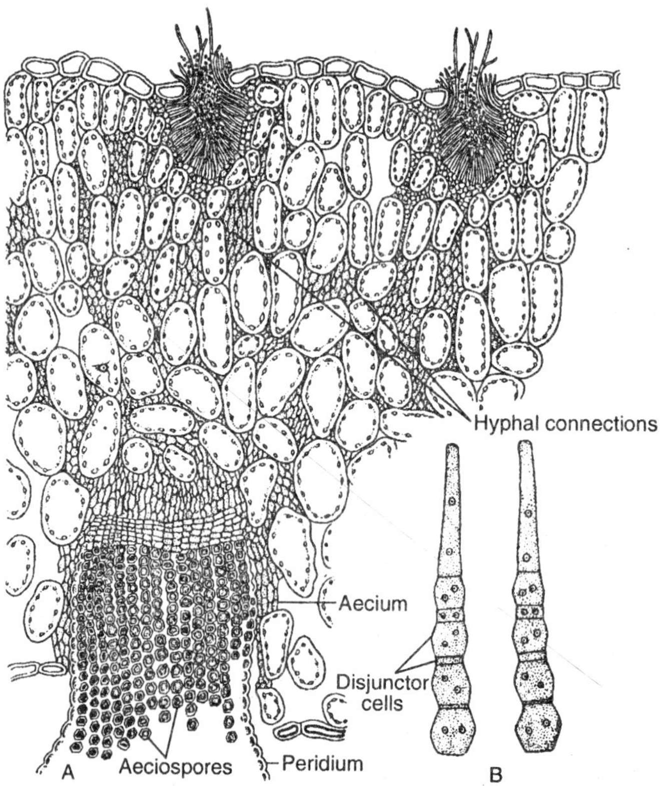

Fig. 14.19: Section through barberry leaf. A. Two spermogonia at the upper epidermis and a mature aecium at the lower epidermis, and intercellular hyphal connections between these organs. The aecium is with bell-shaped peridium and aeciospores in chains. B. Two aeciospore chains showing young aeciospores alternating with disjunctor cells

and basal part of the leaf blades. Thus the cycle is continued. *Puccinia graminis* seldom does much damage to the barberry but is often very destructive to wheat and other graminaceous hosts. Aeciospores may infect graminaceous hosts hundreds of miles away from the barberry bushes where they are produced. The principle of heteroecism in *Puccinia graminis* was demonstrated by Anton de Bary. Although the sporidia cannot infect the graminaceous hosts, the uredospores may frequently survive a mild winter and infect the wheat during the following spring. In some parts of India, the uredospores survive throughout the whole year by infecting various graminaceous hosts which grow at different altitude in the hills and in the plains. As such it is evident that *Puccinia graminis* may be propagated without the aecial stage. Life cycle of *Puccinia graminis* is presented in Fig. 14.20.

Biological specialization: That the aeciospores can bring about infection of graminaceous hosts has been demonstrated as early as 1816. Inoculations in the reverse order between one host and another were made in 1865. This study has produced very interesting results. *Puccinia graminis* is a good morphological species within which there are **biological forms** or **biological strains** or **subspecies.**

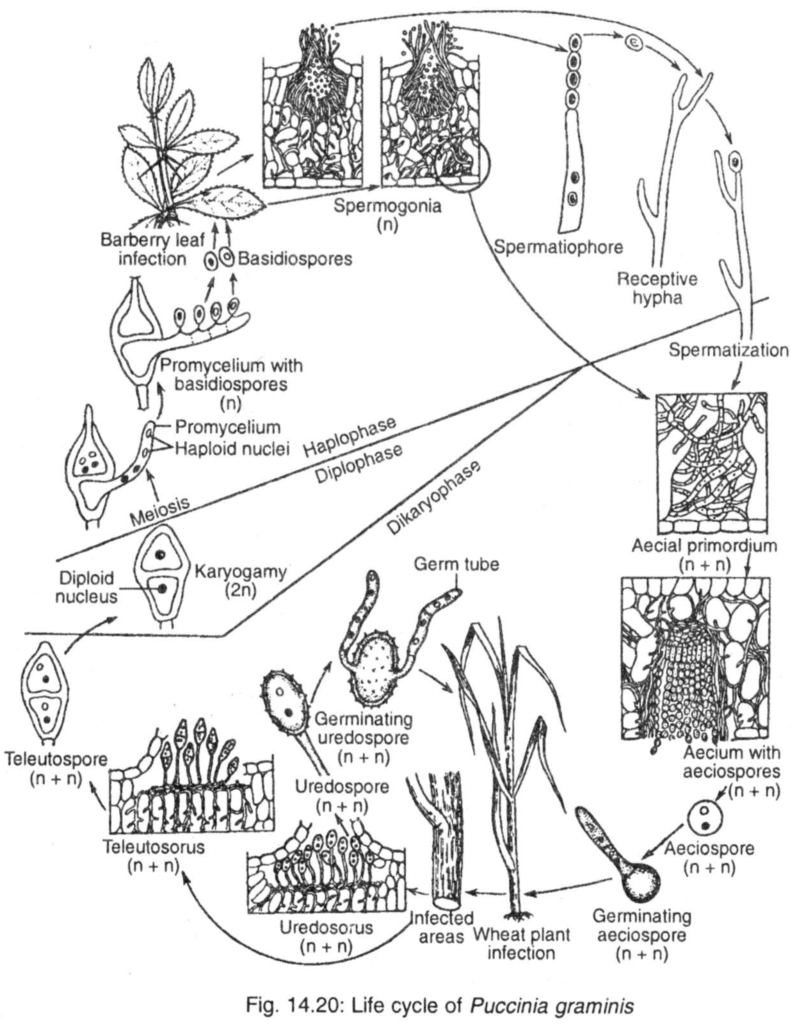

Fig. 14.20: Life cycle of *Puccinia graminis*

These biological forms are morphologically identical but differ in physiological, biochemical and pathological characters. They are specialized in their nature of parasitism and are specific to particular host or group of hosts. Such a behaviour is known as **biological specialization.** Stakman and many others have added much to our knowledge of biological specialization in *Puccinia graminis.* The biological forms or subspecies of *Puccinia graminis* designated **formae speciales** abbreviated as **f. sp.**, are named by **trinomials**, i.e., by adding to its binomial a third name after the chief host parasitizing. These are genetically different strains with differing pathogenic qualities.

Eriksson and Henning have divided *Puccinia graminis* into following subspecies:

Puccinia graminis f. sp. *tritici* parasitizes *Triticum aestivum, Hordeum vulgare* and many wild grasses.

Puccinia graminis f. sp. *avenae* parasitizes *Avena sativa* and wild grasses.

Puccinia graminis f. sp. *secalis* parasitizes *Secale cereale* and wild grasses.

Puccinia graminis f. sp. *agrostidis* parasitizes *Agrostis caninis* and wild grasses.

Puccinia graminis f. sp. *airae* parasitizes *Aira caespitosa.*

Puccinia graminis f. sp. *bromi* parasitizes species of *Bromus* and other wild grasses.

Puccinia graminis f. sp. *phleipratensis* parasitizes *Phleum pratense* and wild grasses.

Puccinia graminis f. sp. *poae* parasitizes *Poa compressa* and related species.

Within each of these biological forms or subspecies there are numerous **physiologic races** which differ in their nature of parasitism on various agronomic varieties of the host. But all these biological forms and physiologic races have the common barberry plant as their alternate host, and whatever differences they exhibit are on the graminaceous hosts. They are believed to originate during growth upon the barberry plant.

Order: **Ustilaginales**

The Ustilaginales or smut fungi, as they are commonly called, have received the name from the black soot-like masses of **teleutospores** (also known as chlamydospores) they produce in sori or **smut balls.** Exception being *Entyloma,* where the teleutospores are hyaline to yellowish or reddish-brown. The spore masses in silvery membrane either break up into fine dust-like powder, readily scattered by the wind (loose smuts) (Figs. 14.21C and D); or remain firm and are covered with a membrane which ultimately ruptures to expose the spores (covered smuts) (Fig. 14.21E). Smuts lack definite fruiting bodies. The fungi occur on a wide range of hosts, usually of flowering plants, except a few which have been recorded on *Selaginella,* a peridophyte. The smuts, in general, are parasitic fungi, but all of them are not obligate parasites. Smuts never have alternate hosts as do many rusts. Besides the parasitic smuts there are also smut fungi which are facultative saprophytes. Many of the smuts have been induced to complete their life cycle on artificial media. The economically important smuts mostly parasitize wheat, oats, rye, barley, and maize. They most frequently destroy either the grains or the inflorescence, thus lower the quality and yield of commercial product (Figs. 14.21A and B). The smut fungi seldom kill the hosts outright, but they lower their vitality and often the host plants become stunted due to the attack of these fungi. During the course of infection the developing grains are usually transformed into sooty, in some cases ill-smelling, masses of smut spores.

Some of the common smuts are: *Ustilago maydis* (corn smut), *Ustilago nuda* var. *tritici* (loose smut of wheat), *Ustilago hordei* (covered smut of barley), *Ustilago avenae* (loose smut of oats), *Ustilago scitaminea* (whip smut of

Fig. 14.21: Various types of smuts. A. Corn smut (*Ustilago maydis*). B. Grain smut of jowar (*Sphacelotheca sorghi*). C and D. Loose smut of wheat (*Ustilago nuda* var. *tritici*). E. Covered smut of barley (*Ustilago* hordei)

sugar cane), *Tilletia foetida* (bunt or stinking smut of wheat—smooth spored), *Tilletia caries* (bunt or stinking smut of wheat—rough spored), *Sphacelotheca sorghi* (grain smut of jowar), *Neovossia indica* (kernel bunt of wheat), *Neovossia horrida* (kernel bunt of rice), and *Urocystis brassicae* (root-gall smut of mustard).

The teleutospores are dikaryotic; single-celled globose; subglobose to ellipsoidal in shape being hyaline, yellow, brown to dark; and with smooth, echinulate to reticulate wall. The spore wall consists of two layers. An outer layer—the **epispore**—which is thick, pigmented smooth or variously sculptured, and an inner layer—the **endospore** which is smooth and delicate. The spores may be free from each other (Figs. 14.22D-H) or having a tendency to cling together in groups (Fig. 14.22C), but they are easily separable. In some smuts, the spores are cemented together and many of these spores cling together to form balls. These compound spores, present in *Urocystis* are often differentiated into sterile and fertile regions, only the cells of the fertile region are capable of germination (Figs. 14.22A and B).

The teleutospores may germinate immediately after they are developed or after a more or less protracted rest interval. In germination in water or nutrient solution, manure water, etc., a short hyaline mycelial structure of limited growth is protruded, the promycelium, which differs in structure being septate or non-septate. Karyogamy and meiosis take place during the germination of the teleutospores. Usually karyogamy takes place in the spore and meiosis in the promycelium. From the promycelium of most of the smut fungi sporidia are developed. The sporidia are not differentiated morphologically, but they are of two kinds. Some bear (+) nuclei and others have (–) nuclei, one in each. In smuts producing septate promycelium, the promycelium is usually divided into four uninucleate cells, each of which produces large number of ovate, continuous, hyaline sporidia and the promycelial cells are not normally emptied (Figs. 14.23A and B). The sporidia are borne both terminally and laterally on a promycelium, but not on sterigmata. They are not violently discharged. The sporidia often, in suitable nutrient solution, bud profusely in yeast-like

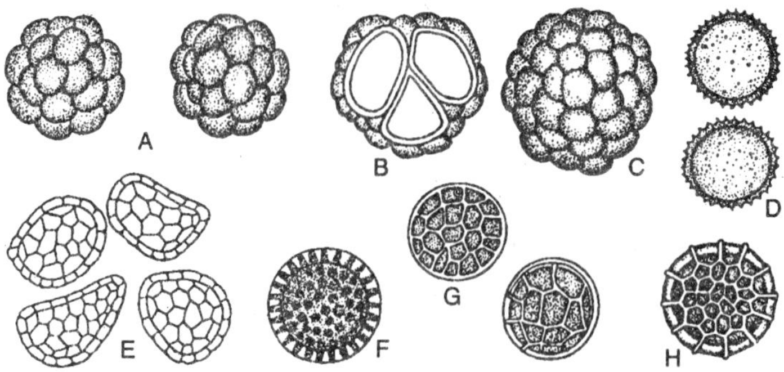

Fig. 14.22: Various types of smut spores (teleutospores). A and B. *Urocystis agropyri*. A. Surface view of smut balls. B. Section through a smut ball. C. Surface view of smut ball of *Sorosporium everhartii*. D. *Ustilago maydis*. E. *Ustilago reticulata*. F. *Neovossia indica*. G. *Tilletia* sp. H. *Tilletia eleusines*

fashion (Fig. 14.23C), whereas the non-septate promycelium bears eight or more sporidia terminally. Here the sporidia are long and sickle-shaped (Figs. 14.23G-K). Again, a four-celled promycelium may not produce sporidia at all. The promycelial cells besides producing sporidia produce conjugation tubes, and fusion takes place between the compatible conjugation tubes, and thereby dikaryotic condition is established, as in *Ustilago nuda* (Figs. 14.23D-F). Hyphae arising from the fused conjugation tubes are predominantly binucleate. Again, cells of the promycelium may develop into infection hyphae, as in *Ustilago nuda* var. *tritici*.

The behaviour of sporidia in the smut fungi is extremely variable. The sporidia may germinate to produce monokarytic infection thread or hypha which may infect the host, or these sporidia may bud indefinitely in nutrient solutions, and the buds give rise to infection hyphae when the conditions are favourable. The infection caused by monokaryotic hypha remains invisible and becomes visible only when it is induced by a dikaryotic hypha. Even if the infection is effected by a monokaryotic hypha, dikaryotic condition is subsequently established in the host tissue. As a general rule infection is established in the host only by the dikaryotic hypha. Hence dikaryotization is a prerequisite to normal infection. Dikaryotization may take place outside or inside the host tissue and may be mainly in any of the processes: (i) fusion of two compatible sporidia, (ii) union between a compatible sporidium and a compatible monokaryotic hyphae, (iii) fusion of two monokaryotic compatible hyphae, (iv) fusion between compatible promycelial cells by conjugation tubes.

The mycelium is mostly inconspicuous and is often distributed very widely in the host tissue without giving external evidence of its presence until the time of formation of teleutospores. The mycelium is both intracellular and intercelular, passing through and between the cells and producing haustoria which enter the host cells. Depending on the nature of infection, the growth of mycelium may be widespread in the host tissue or may remain localized at a particular area. Again, the mycelium may be monokaryotic or dikaryotic. But before the infection becomes visible the mycelium must be dikaryotic. The dikaryotic mycelium ultimately develops in great abundance at certain special places in the host, either the ovary or in some other part of the host, leading to the development of a large mycelial structure which subsequently develops into a mass of teleutospores. During the development of the teleutospores numerous transverse septa are formed in the mycelial cells resulting in the formation of small cells. These small cells round off and become coated with a

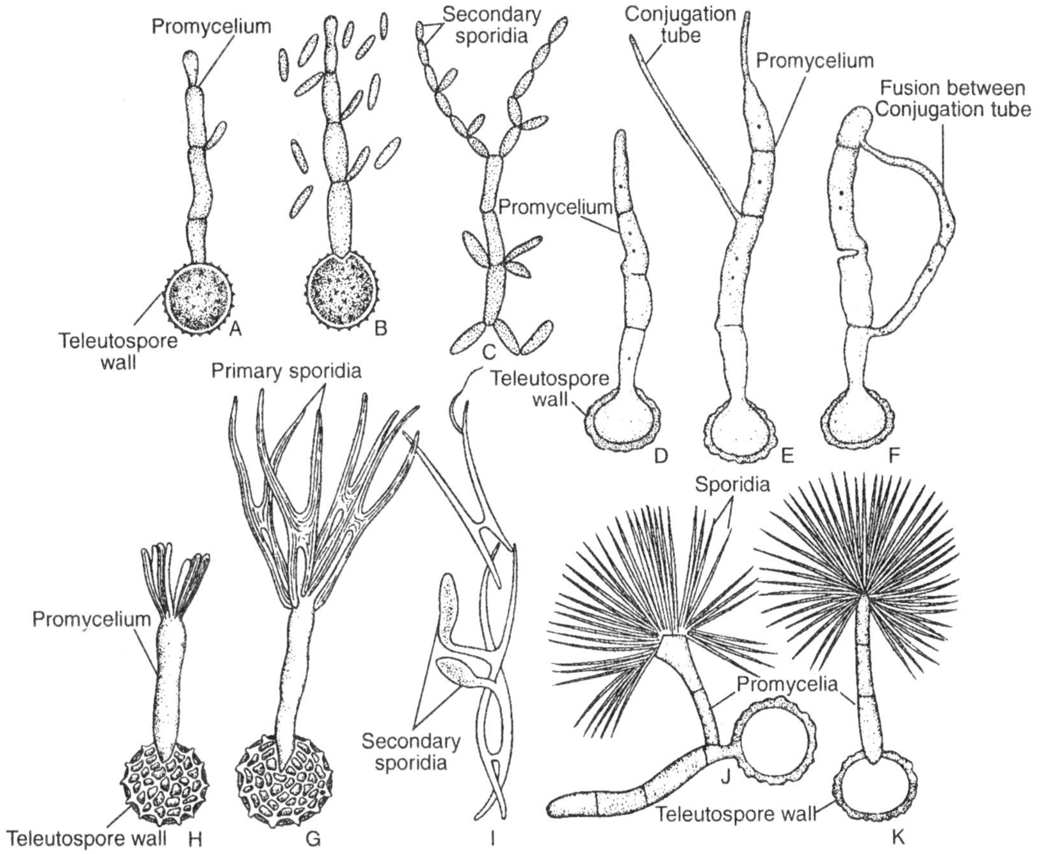

Fig. 14.23. Germination of teleutospores of smuts. A-C. *Ustilago avenae.* A and B. Germination of teleutospores and formation of sporidia. C. Formation of secondary sporidia by budding. D-F. *Ustilago nuda.* D. Formation of promycelium. E. Conjugation tube development. F. Fusion between conjugation tubes. G-I. *Tilletia caries.* G. Non-septate promycelium with young sporidia. H. Fusion of primary sporidia. I. Formation of secondary sporidia. J and K. *Neovossia horrida.* Promycelia with innumerable sporidia

gelatinous envelope which later disappears and the spores develop a new, thick double wall. The teleutospore mass so formed is covered by a definite whitish membrane. Where the membrane is delicate and ruptures readily, the spores are very easily dispersed. Again, where the membrane remains firm, the spores are exposed late by the rupture of the membrane.

The points at which host infection can occur are very diverse. Depending upon the life history of the smut fungus, especially as to the time, place and method of infection, the types of infection may be:

1. **Seedling infection:** In this type, infection occurs only during the seedling stage from the spores that were either carried on the seed or were present in the soil in close proximity to the germinating seed. Along with the germination of the seed the spores also germinate producing promycelia and sporidia. The sporidia produce infection threads which enter the young seedling, reach the growing point and grow simultaneously along with the growth of the seedling keeping pace with the growth of the host tissue remaining completely invisible from outside. With the appearance of the inflorescence the infection becomes visible by forming characteristic masses of spores which replace the grains. Seedling infection from seed-borne spores is encountered in covered smut of barley caused by *Ustilago*

hordei, bunt or stinking smut of wheat caused by *Tilletia caries*, and others. Seedling infection may also be from internally seed-borne dormant mycelium in the seed coat, found in loose smut of oats caused by *Ustilago avenae*. Besides these, the spores may be present within the floral bracts from which the infection may start, as in covered smut of oats caused by *Ustilago kolleri*. Whatever may be the source of inoculum, the seedling infection is systemic in nature.

2. **Blossom infection:** In this type, infection occurs in the flowers at blossoming time. The smutted ears are mature at blossoming time and the spores are disseminated by the wind and lodge within the flower and infect young ovaries of healthy inflorescence producing therein an internal mycelium which is developed within the seed without causing any injury to it. During seed germination the mycelium also keeps growing and ultimately reaches the inflorescence of the host and give rise to masses of smut spores there in place of grains. Here the fungus exists in the seed as a vegetative mycelium which resumes its activity along wth the germination of the seed. This is also a case of systemic infection. This type of infection is found in loose smut of wheat and in loose smut of barley caused by *Ustilago nuda* var. *tritici* and *U. nuda* respectively.

3. **Localized infection** or **shoot infection** or **general infection:** Infection may occur on any of the tender growing parts of the plant. This is common smut of corn (maize) caused by *Ustilago maydis*. Leaves, stems, and flower parts may become infected locally from spores which lodge upon these parts while they are very tender and are yet in a growing and susceptible condition. The spores germinate producing infection threads which enter the host tissue and get established by producing mycelium. The mycelium remains localized in and around the particular area of infection without spreading much within the host tissue from the original centre of infection. In response to the invasion of the fungus the host cells become enlarged (hypertrophy) and increase in number (hyperplasia) giving rise to gall formation. Depending upon the location of the gall, various morphological and physiological effects follow. Galls on young seedlings may result in extreme dwarfing or death of the plant. At first the galls are covered with a white membrane and the central part consists of fungus mycelium. At maturity the central portion appears as a black powdery mass consisting largely of smut spores. The covering membrane gradually becomes dry, breaks and the spores are disseminated. This is a case of localized infection.

CLASSIFICATION

The order Ustilaginales is subdivided into three families: **Ustilaginanceae, Tilletiacea,** and **Graphiolaceae.** Of these, the families Ustilaginaceae and Tilletiaceae are well known, and are widely distributed, whereas the family Graphiolaceae is comparatively less known and very much confined to particular regions of the world. All the three families may be distinguished in accordance with the characters employed in the following key:

A. Basidiocarp lacking

B. Promycelium septate, sporidia produced laterally and terminally from each cell of the promycelium **Ustilaginaceae**

BB. Promycelium non-septate, sporidia produced terminally in cluster **Tilletiaceae**

AA. Basidiocarp present **Graphiolaceae**

Family: *Ustilaginaceae*

In the Ustilaginaceae the infection may be localized or systemic. All the three types of infection as already indicated are encountered in this family. The teleutospore masses or sori may be developed in leaves, stems, or inflorescences. They are covered by a thin membrane which may be delicate or firm. The spores may be buried within the host tissue or sometimes a central columella may be present along with the spores. The teleutospores are single or united into spore balls. They germinate by means of septate promycelia bearing terminal or lateral sporidia, rarely a single sporidium, conjugation tubes, or infection hyphae.

Genus: *Ustilago*

Sori are produced in various parts of the host forming at maturity dusty, dark-coloured spore masses which may be naked or covered. Teleutospores are single. Spores on germination produce septate promycelia. Cells of the promycelia may produce sporidia, infection hyphae, or conjugation tubes formed terminally or laterally near the septa. The sporidia may germinate directly into infection hyphae or go on budding indefinitely.

Some Indian Species: *Ustilago avenae* (Pers.) Rostr.; *U. hordei* (Pers.) Lagerh.; *U. maydis* (DC.) Cda.; *U. nuda* (Jens.) Rostr., *U. nuda* var. *tritici* Schaf.

USTILAGO AVENAE (Pers.) Rostr.

It is the causative fungus of loose smut of oats which is sometimes known as naked smut or black head. This fungus does not attack any other cereal crop. It infects only the panicle, where the individual flowers are replaced by sori of black powdery masses of teleutospores (smut spores). The sorus is covered temporarily with a delicate membrane which ruptures to expose to spores. The teleutospores are dark-brown to black with fine echinulations. These spores produced on diseased heads are dispersed by the wind. Many of them lodge on healthy heads and are carried beneath the lemma and palea of the young kernels. Some of the spores germinate immediately there and remain inactive until the seeds are sown the following year. Other spores do not germinate immediately but remain on the seeds, in the grooves or between the kernels and the glumes, until the seeds are sown. On germination the spores produce septate promycelia from the cells of which oblong to elongate, hyaline, uninucleate sporidia are developed (Figs. 14.23A and B). Karyogamy and meiosis take place during spore germination and each cell of the promycelium contains one haploid nucleus. The sporidia increase in number by budding producing secondaiy sporidia (Fig. 14.23C). The sporidia of compatible haploid lines fuse in pairs and the nucleus of one passes into the other to initiate dikaryotic condition. The fungus infects the oat seedlings and after penetration it grows intracellularly at first in mesocotyl, coleoptile, and first leaf. It then becomes intercellular, advancing with the growing tip, ultimately entering the flower primordia and by heading time replaces the oat spikelets with black masses of spores. The spores, when shed, lodge on the grain. They germinae as the grain germinates and infect the seedling. Usually all the heads on an infected plant are smutted, but sometimes one or two are normal. Generally, too, all the spikelets of a panicle are diseased, but occasionally the upper ones appear healthy. Diseased plants cannot be told from healthy ones before heading out, they are usualy shorter than healthy plants.

USTILAGO NUDA (Jens.) Rostr.

This fungus causes loose smut of wheat and barley. It occurs on wheat and barley and a few wild grasses destroying heads. But it does not attack any of the other grain crops. The characteristic dusty-black appearance of the diseased heads is produced by the teleutospores (chlamydospores) that are formed in sori principally in the position of the grains. Each sorus is enclosed in a delicate, silvery membrane which ruptures early to expose the finely echinulate and dark-coloured teleutospores. The spores are scattered mostly by air current during the flowering time or during harvest, to healthy heads. Those that alight on the heads lie dormant on the surface or become lodged under the glumes. Spores may live over in the soil under dry conditions. Again, if temperature and moisture are favourable to the fungus, the spores germinate at once by the formation of a promycelium of one to four cells (Fig. 19.23D) during which karyogamy and meiosis take place. No sporidia are produced. The promycelial cells are haploid, and fusion between compatible cells takes place by conjugation tubes (Figs. 14.23E and F). The conjugated cells then produce slender dikaryotic infection threads (hyphae) which enter in the host (young ovary) through the young tissue at the base of the ovary. The infection threads become established in the pericarp or in the embryonic tissues of the grains. Usually the infection threads remain dormant in the scutellum, until the grains are sown in the following growing season. In cases, where the spores remain on the surface of the grains, the spores germinate along with the seeds. The fungus grows through the first leaf into the growing point of each young seedling. The fungus perennates as dormant infection threads (hyphae) in seeds rather than as teleutospores on the surface of the seeds. In any case, the fungus grows with the growing point of the wheat and barley plants and ultimately the grains are replaced by sori of spores that are developed from the dikaryotic mycelium. The chlamydospores of *U. nuda* are short-lived, rarely survive more than a few days under normal conditions.

USTILAGO MAYDIS (DC.) CDA.

It is the causal organism of corn (maize) smut disease. It produces sori on any part of the host forming large, irregular galls. These galls are at first protected by a white membrane composed of host tissue and semi-gelatinized fungus mycelium. The white membrane remains intact until a late stage, when it breaks and exposes the powdery black mass of teleutospores. The spores are unicellular, ellipsoidal to subglobose, black with medium-thick epispore being heavily echinulate. The teleutospores germinate and produce septate promycelia, when karyogamy and meiosis take place. Each cell of the promycelium produces continuously ovate sporidia. The sporidia are released directly in air. *Ustilago maydis* is heterothallic. The sporidia produced from a teleutospore belong to two groups: (+) and (–). The sporidia bud profusely in a yeast-like fashion under suitable organic medium. The sporidia are readily carried by the wind, and when they fall upon any young tissue of maize plant under suitable conditions of moisture, they germinate sending out germ tubes which penetrate the epidermal cells and thus infect the plant. Each germ tube devlops into a mycelium which is either (+) or (–). The dikaryotic condition may be initiated early or late, and according to the circumstances, may arise from fusion of spores (+) and (–), fusion of hyphae from different mycelia (+) and (–), which come in contact with each other. The dikaryotic mycelium stimulates enlargement (hypertrophy) and increase in number (hyperplasia) of host cells resulting in the formation of a gall. The mycelium grows through or between the cells of the host tissue, and branches profusely. Most of the hyphal cells of the gall are transformed into teleutospores. The teleutospores are capable of germination as soon as mature, if conditions are favourable. They may remain viable for two or more years. Life cycle of *Ustilago maydis* is presented in Fig. 14.24.

Family: *Tilletiaceae*

Teleutospores are single or combined in sori of dusty, erumpent spore masses or permanently embedded in the host tissues. During spore germination, karyogamy, meiosis and a short promycelium development take place. The promycelium is simple, usually non-septate up to the time of formation of sporidia. Large number of sporidia even up to 128 or more are produced in clusters at the apex of the promycelium (Figs. 14.23J and K). The sporidia are elongate with or without tapering

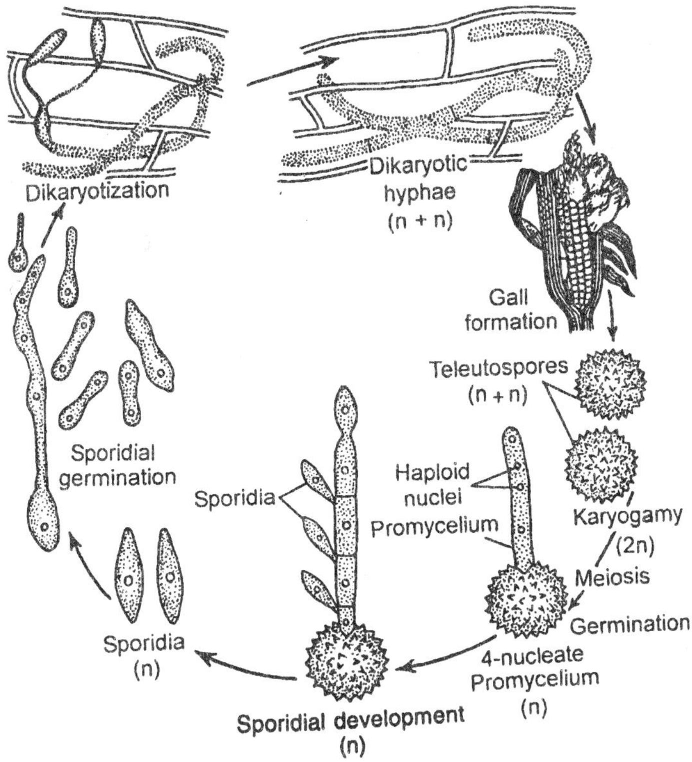

Fig. 14.24: Life cycle of *Ustilago maydis*

appendages and each one bears a single haploid nucleus. They may germinate while still attached to the promycelium with or without fusing in pairs by conjugation tube and produce simialr or dissimilar secondary sporidia (Fig. 14.23I) or germinate directly into infection threads.

Genus: *Tilletia*

Sori develop in various parts of the host, but are usually in the ovaries, forming dusty spore masses. The teleutospores are formed from the dikaryotic phyphae which later disappear more or less completely through gelatinization. Germination of teleutospore is by a short non-septate promycelium which bears a terminal cluster of elongate sporidia (Figs. 14.23G and H). The sporidia fusing in pairs give rise to conidia with or without the formation of mycelium.

Some Indian Species: *Tilletia ajrekari* Mundk.; *T. caries* (DC.) Tul.; *T. eleusines* Syd.; *T. foetida* (Wallr.) Liro; *T. koeteriae* Mundk.

TILLETIA CARIES (DC.) TUL.

This is the causal organism of the disease bunt, or stinking smut of wheat. A characteristic fishy odour is given off by the teleutospores of the fungus formed in the infected heads. Black powdery teleutospores occupy the entire kernel within the pericarp being transformed into a spore ball. The spore balls of infected heads are shattered during threshing, and teleutospores thus liberated lodge on healthy kernels of wheat.

The teleutospore wall has reticulations ranging from minute shallow meshes to deep indentations. The teleutospores measure 15 to 23 μ in diameter. The young teleutospore is binucleate, karyogamy is in mature teleutospore. The diploid nucleus by meiosis and mitosis gives rise to haploid nuclei which pass into sporidia borne on promycelium, each sporidium having one haploid nucleus. Planted in the soil with the wheat kernel itself, the teleutospores germinate when soil conditions become favourable. By the time a wheat sprout has emerged, the spores germinate by the production of promycelia at the apex of which filiform hyaline sporidia are produced 8 to 16 in number. The sporidia fuse in pairs while already attached with the promycelium, to form characteristic H-shaped structures. The fused sporidia germinate producing dikaryotic mycelium on which sickle-shaped hyaline dikaryotic conidia are borne. These conidia on germination produce dikaryotic mycelia which cause infection of the young wheat plants. After gaining entrance in the host, the mycelium continues to grow as an internal parasite. From the infection point, the mycelium approaches the growing point and follows the development of its host, sending its branches into each spikelet and finally into the growing ovules. Eventually, the wheat kernels transform into spore balls.

Soil can become infested with teleutospores when spore balls shatter off before or during harvest or when spore balls are planted with wheat sowing. Life cycle of *Tilletia caries* is presented in Fig. 14.25.

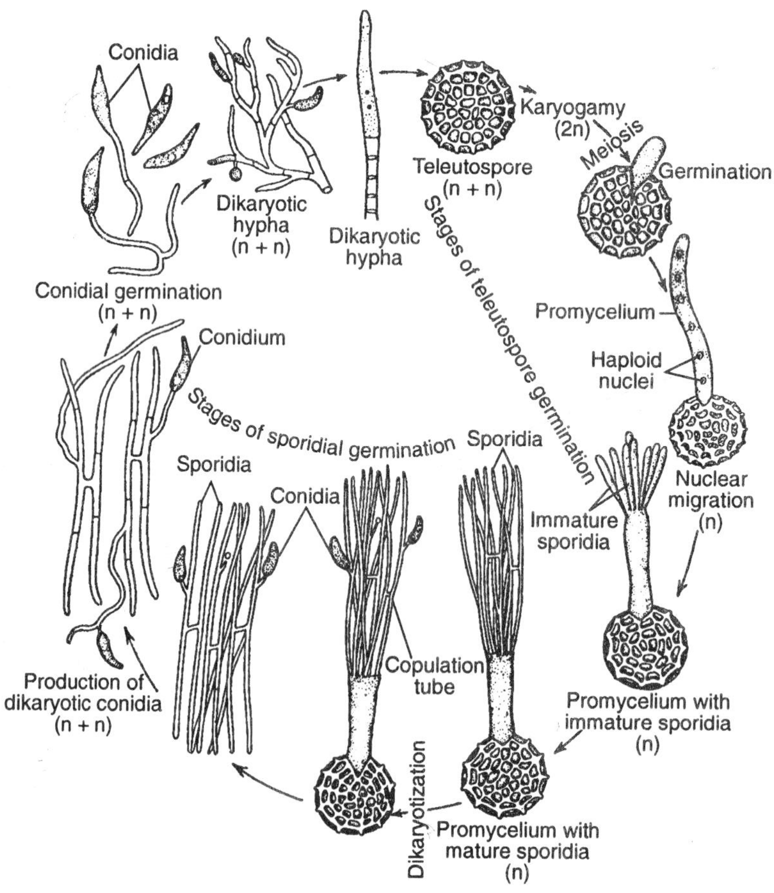

Fig. 14.25: Life cycle of *Tilletia caries*

Genus: *Neovossia*

Sori in the ovaries form cavities in the tissues and even sometimes destroy the entire grain. The sori rupture and the dusty spore masses are exposed. The teleutospores are produced from the dikaryotic hyphae. Germination of teleutospores is by the formation of a promycelium producing numerous sporidia, number being even up to 128 or more. The sporidia cluster terminally on a promycelium (Figs. 14.23J and K). They germinate without conjugation giving rise to mycelia which produce conidia.

Some Indian Species: *Neovossia barclayana* Bref.; *N. brachypodii* (Mundk.) Mundk. and Thirum.; *N. horrida* (Takahashi) Padwick and Khan; *N. indica* (Mitra) Mundk.

Order: **Tremellales**

The basidiocarps in members of this order are usually well-developed and frequently, but not always, gelatinous in texture. They may be whitish, brightly coloured to blackish. Mostly, the basidia are formed in well-defined hymenium. They are variously septate to deeply lobed—tuning fork type with large variable sterigmata. The basidiospores are mostly forcibly discharged and may frequently become septate and bud off conidia at maturity. Members of this order may be saprobic or parasitic.

Only Alexopoulos has kept the traditional order Tremellales as such and has subdivided into families: Tremellaceae, Auriculariaceae, Septobasidiaceae, Tulasnellaceae, and Dacrymycetaceae. But Talbot placed the orders: Tremellales, Auriculariales, and Septobasidiales under the class Phragmobasidiomycetes and the orders: Dacrymycetales, Tulasnellales, Brachybasidiales, Exobasidiales, Aphyllophorales and Agaricales under the class Holobasidiomycetes, whereas Webster put the families: Auriculariaceae, Tremellaceae, and Dacrymycetaceae under the order Tulasnellales.

Family: *Dacrymycetaceae*

All members of this family are saprobes being almost confined to dead wood, with or without bark. The basidiocarps are mostly gelatinous or waxy, drying down to a thin sheet or horny mass. They vary from thin, broadly effused sheets to cushion-like, cupulate or pileate structure, or clavarioid with smooth or wrinkled surface, coralloid or spathulate upright forms. The basidiocarps are often bright yellow or orange. The major pigment responsible for the yellow or orange colour characteristic of these fungi is beta-carotene. Pigment formation is dependent on light, but once the pigment is formed remains stable even after the organism is subjected to continuous darkness. The basidium consists of a cylindrical base—hypobasidium terminally on septate hypha, is elongated and somewhat thicker than the parent hypha. The hypobasidium forks into two long arms, the epibasidia, each of which terminates into a pointed sterigma on which a sausage-shaped basidiospore is obliquely perched. This is the so-called tuning fork type of basidium.

Genus: *Dacryopinax (= Guepinia)*

This genus is widely distributed throughout the tropical zones of the world. Most common species is *D. spathularia.* Species of this genus are commonly found growing on the decaying stumps of wood, in shady and moist environments. The mature specimens can easily be identified by their stipitate, spathulate and petaloid form and the characteristic unilateral inferior hymenium.

Basidiocarps are at first discoid, becoming stipitate; pileus obliquely cupulate, spathulate, petaloid, occasionally lobed or somewhat morchelloid (Figs. 14.26A, B). Stipe is covered with a layer of cylindrical, thick-walled hyphae. Hymenium is unilateral, inferior, smooth forming gill-like folds or occasionally lobed (Fig. 14.26C). Cortex is stiffy gelatinous to

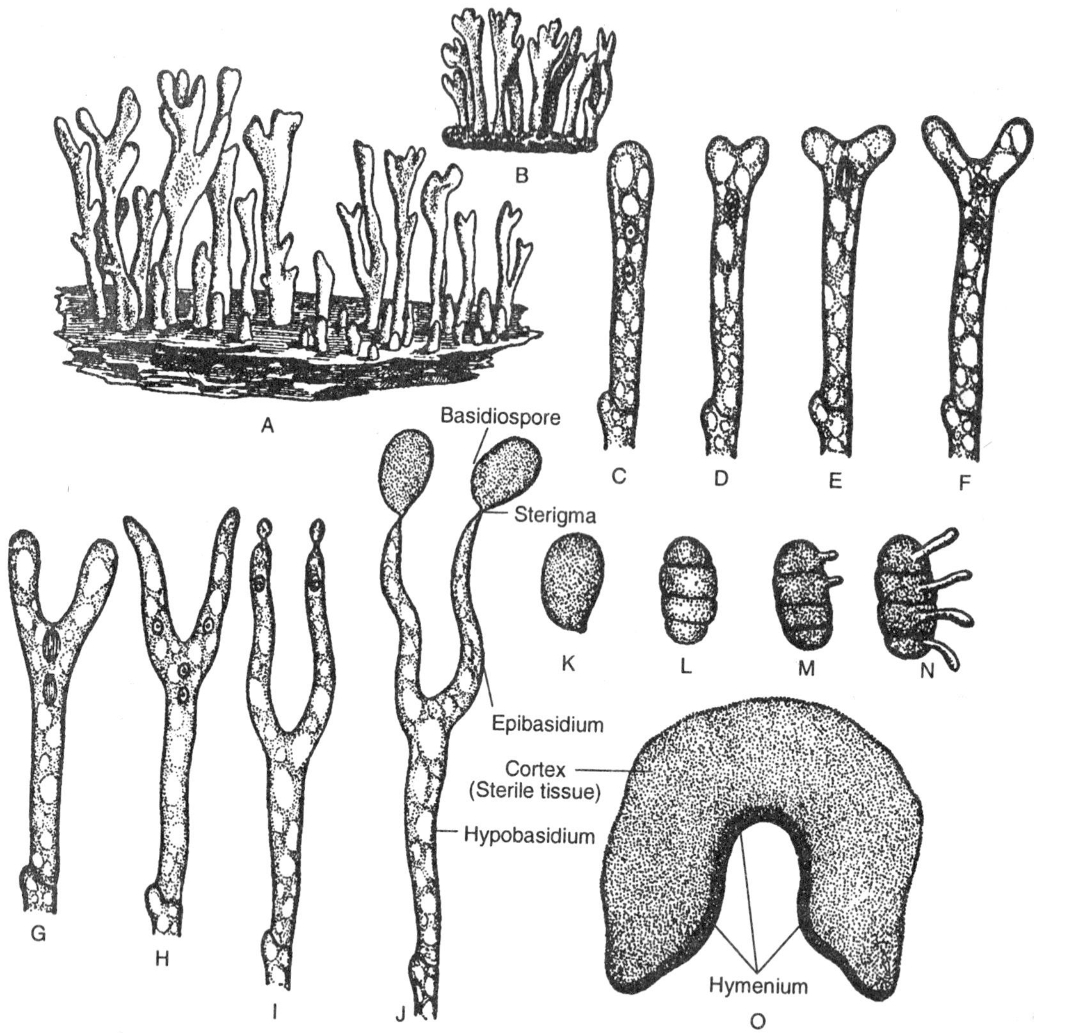

Fig. 14.26: *Dacryopinax* sp. (*Guepinia* sp.) A and B. Basidiocarps at different stages of maturity attached to substratum. C to I. Successive stages in the development of a basidium and basidiospores. C. Dikaryotic terminal cell. D. Karyogamy. E-I. Stages of nuclear division, formation of epibasidium and hypobasidium, sterigmata and basidiospores. J. A mature basidium bearing basidiospores. K-N. Stages of basidiospore germination. O. Sectional view of a basidiocarp showing cortex (sterile tissue) and hymenium

cartilaginous homogeneous composed of fibrous thick-walled hyphae with bulbous septa. The basidiocarps are erumpent, usually gelatinous when wet, drying to horny consistency but soon reviving when moistened. Pigmentation is variable, yellow and orange predominating, but on drying the colour changes to apricot yellow.

The basidia are intermingled with sterile hyphae in the hymenium (Figs. 14.26O). They start as binucleate terminal cells, soon thicken than the rest of the hyphae (Fig. 14.26C). They are first long cylindrical or somewhat clavate. The two nuclei unite and then divide once stichobasidially, and then a second time in the same direction (Figs. 14.26D-G). The branched apex of the basidium becomes lobed to produce two epibasidia of the same diameter and often of almost the same length as the hypobasidium. Each epibasidium is terminated by a sterigma. The upper two nuclei migrate into the two epibasidia and then through the sterigma at the tip of each into the basidiospores (Fig. 14.26H). The two nuclei remaining in the hypobasidium degenerate (Fig. 14.26I). The mature basidium takes a tuning fork appearance (Fig. 14.26J).

Basidiospores are very pale yellow under the hand lens, elliptical to kidney-shaped, uni- to tri-septate (Figs. 14.26K-N), germinating by conidia. Spore print is light ochraceous buff.

Note: *Dacryopinax,* a name substituted for the more familiar but pre-occupied name *Guepinia* by G. W. Martin (1948).

Some Indian Species: *Dacryopinax spathularia* (Sch.) Martin; *D. ramosa* Currey.

Series: HYMENOMYCETES

The fungi included in the Hymenomycetes are with basidiocarp gymnocarpous or hemiangiocarpous. The basidia are often arranged in a palisade-like fashion to form a hymenium which is fully exposed at maturity, in contrast with the Gasteromycetes where the basidiocarp is angiocarpous. Interspersed among the basidia are sterile structures, usually the paraphyses and sometimes other specialized sterile structures of taxonomic importance. Basidia of all ages may be found in the hymenium. The basidia vary greatly in shape in the Hymenomycetes. They may be cylindrical, tapering at the very base and rounded at the apex, or clavate or almost globose. The basidia are aseptate and may be bi-, or tetra-sterigmate. The basidiospores are placed asymmetrically on the sterigmata and are discharged violently by the fluid drop mechanism.

The basidiocarps show a very great variability of size and complexity of structure. The simplest form of basidiocarp is a thin crust-like, smooth layer spread over the surface of the substratum with the hymenium fully exposed—**resupinate** (Fig. 14.27A). The basidiocarps developed on the surface of the substratum may have the tendency to have the edges pulled away from the substratum to form—**effused-reflexed** or **effuso-reflexed** forms (Fig. 14.27B). This tendency persists until shelf-like—**reflexed** or **dimidiate** (Figs. 14.27C and D) basidiocarps with little or no resupinate portion are produced. These reflexed structures vary widely in form (Figs. 14.27E and F) and may be narrowed toward the point of attachment often producing a distinct stalk-like structure, known as **stipe** which bears the rest of the tissue containing hymenium—the **pileus.** The stipe may be attached eccentrically at the edge or centrally to the under surface of the pileus. All these forms of basidiocarp are known as **stipitate** (Fig. 4.27G). The stipe may be missing and the basidiocarp may then be shell-like (**conchate**)

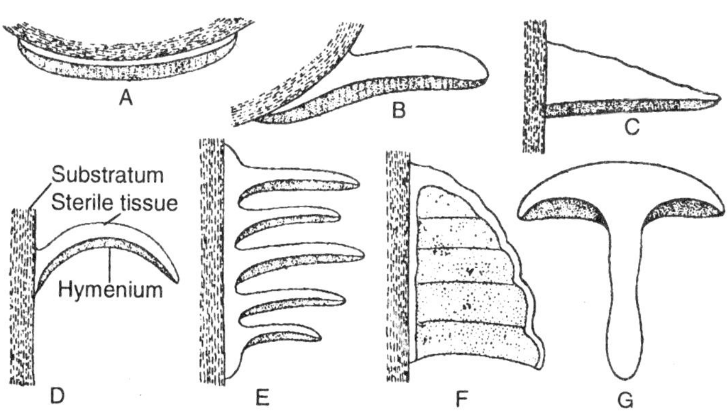

Fig. 14.27: Various types of basidiocarps. A. Resupinate. B. Effused-reflexed. C and D. Reflexed. E and F. Variations in the reflexed type. G. Stipitate

or fan-shaped (**flabellate**). Other forms are: cupular (**cyphelloid**) and club-like or coralloid (**clavarioid**). Associated with the development of the basidiocarps, the orientation of the hymenium also changes making growth all-round or on two sides—**amphigenous**; or one-sided—**unilateral** with a tendency that the hymenium faces toward **inferior**, especially in the effused-reflexed, reflexed and stipitate forms. Along with the change in form of the basidiocarps and the orientation of the hymenium, the hymenial surface also undergoes changes to become smooth, wrinkled or toothed, porous, or lamellate. The consistency of the basidiocarp may be fleshy, woody, corky, leathery or papery. So also in colour, the basidiocarp may be white or shades of grey or bright-coloured or almost black. Basidiocarp may be short-lived or may persist for many years.

In the taxonomic studies of the Hymenomycetes importance is given to the hyphal analysis of the basidiocarps. In some of the Hymenomycetes, particularly the Aphyllophorales (Polyporales), the hyphae of the tertiary mycelium may be differentiated into three different **hyphal types** combined in three different **hyphal systems.** Hyphal types are: **generative hyphae** which are branched, thin-walled, usually septate with or without clamp-connections, they give rise to other types of hyphae and the hymenial layer; **skeletal hyphae** which are branched or unbranched, thick-walled, aseptate, straight or slightly flexuous with thin-walled apices; and **binding hyphae** which are much branched, thick-walled, aseptate, interwoven, narrow, often coralloid, they bind the generative and skeletal hyphae together (Fig. 14.28). The skeletal and binding hyphae build a framework. The generative hyphae produce **mediate hyphae** which are sparingly branched, thick-walled, aseptate. The mediate hyphae ultimately give rise to skeletal and binding hyphae, and to the hymenium.

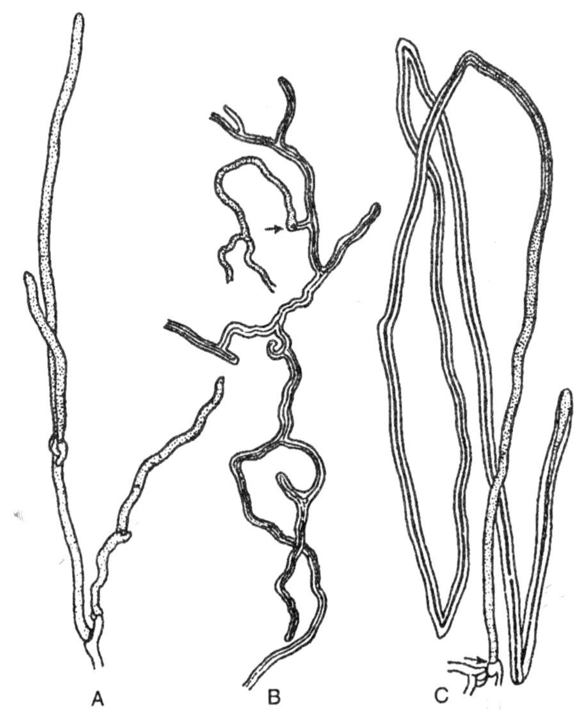

Fig. 14.28: Hyphal system in the Hymenomycetes. A. Generative hyphae. B. Binding hyphae, the arrow shows the origin from a generative hypha. C. Skeletal hyphae, the arrow shows the origin from a generative hypha

These hyphal types are combined into **monomitic, dimitic** or **trimitic** hyphal systems. The hyphal system of a basidiocarp is **monomitic** when it is made of only one kind of hyphae—generative hyphae. This is generally encountered in most soft and fleshy basidiocarps. Again, the hyphal system is known as **dimitic** when a basidiocarp is composed of two kinds of hyphae—generative and skeletal hyphae, whereas it is **trimitic** having hyphae of three kinds—generative, skeletal and binding hyphae. Hard and tough basidiocarps may be monomitic with the generative hyphae becoming thick-walled, dimitic with thick-walled skeletals, or trimitic. Basidiocarps of every species of fungus have a well-defined constant construction, which is maintained regardless of changes in the external morphology of the basidiocarps due to environmental conditions. Hence the importance is given to the hyphal system. An important distinction between the Agaricales and Aphyllophorales is that the basidiocarps of Agaricales are fleshy being usually composed of thin-walled hyphae which inflate. In contrast, the basidiocarps of many Aphyllphorales may be monomitic, dimitic or trimitic.

In many Agaricales the developing basidiocarp is surrounded by one or more veils, but these are not present in the Aphyllophorales. A further distinction is that in the Agaricales the nuclear spindles in the basidia are transverse (chiastobasidial) whilst in the Aphyllophorales forms longitudinally oriented spindles (stichobasidial) occur.

CLASSIFICATION

While trying to workout a suitable classification system of the Hymenomycetes, one should go through the works of Elias Fries who devoted sixty years and made extensive studies on the Hymenomycetes. The original classification of Fries was, however, based upon the studies of Persoon. Mycologists working on the Hymenomycetes have follwed the Friesian system in one way or the other.

In place of the traditional Friesian system of classification based on gross macroscopic features of the basidiocarp and details of hymenial arrangement, modern systems of classification including that of Donk's (1964) lay stress on microscopic features, especially the details of anatomy, colour and staining reactions of hyphae and spores. Critical examination of Basidiomycete basidiocarps in this way demonstrates close relationships between fungi hitherto classified in quite separate groups under the Friesian system.

All of the Hymenomycetes were so long included in one order, the Agaricales, but the modern tendency is toward further division, even to the recognition of two or more orders.

Two orders are recognised in accordance with the characters employ'd in the following key:

A. Hymenium unilateral or amphigenous, the surface smooth or ridged to warty, spinose or porose or rarely lamellate **Aphyllophorales (Polyporales)**

AA. Hymenium inferior, borne on lamellae, or lining the interior of tubes **Agaricales**

Order: **Aphyllophorales (Polyporales)**

The Polyporales include fungi having basidiocarps with hymenia borne in various ways and remaining exposed to the air since their inception. The hymenium may be unilateral being developed on one side, or amphigenous making growth all-round or on two sides of the basidiocarp. Growth of the hyenium is often without definite morphological limits. The basidiocarp may be simple being developed without much differentiation of morphological and anatomical characteristics or may be a complex structure of highly differentiated tissue with a tendency to develop upright or shelf-like portions. The basidiocarps may be of clavate or dendroid or pileate forms being sessile or laterally, eccentrically or even centrally stipitate. They are usually papery, leathery, corky, woody, or fleshy in texture. The hymenium may be smooth, ridged, toothed, or spiny, or may be lining shallow or deep pits or pores.

Donk (1964) modified Friesian classification and divided the Aphyllophorales into twenty-one families: Ganodermataceae, Fistulinaceae, Schizophyllaceae, Punctulariaceae, Coniophoraceae, Gomphaceae, Hymenochaetaceae, Thelephoraceae, Bondarzewiaceae, Auriscalpiaceae, Hericiaceae, Echinodontiaceae, Hydnaceae, Bankeraceae, Cantharellaceae, Clavulinaceae, Clavariaceae, Sparassidaceae, Corticiaceae, Polyporaceae, and Stereaceae. This great change was widely appreciated because both microscopic and macroscopic characters were used to define the families.

Following is a simplified key to the families of the order Aphyllophorales based on field characters:

A. Hymenium smooth, roughened, or ridged

  B. Basidiocarp thin papery, leathery, or slightly woody **Thelephoraceae**

  BB. Basidiocarp clavate, or coralloid, hymenium smooth or ridged **Clavariaceae**

  BBB. Basidiocarp stipitate, funnel-shaped agaricoid, hymenial surface inferior **Cantharellaceae**

AA. Hymenium otherwise

  C. Hymenium spread on pegs or teeth **Hydnaceae**

  CC. Hymenium lining shallow or deep pits or pore tubes

    D. Shallow anastomosing pits **Meruliaceae**

    DD. Pore tubes shallow or deep, may form lamellae **Polyporaceae**

Family: ***Polyporaceae***

The Polyporaceae, or pore fungi grow on both living and dead trees (**lignicolous**), or less commonly on organic matter in the soil, a few are parasitic on other fungi. They are of enormous importance as they are largely responsible for causing disease of standing timber and decay of logs and timbers. The tremendous destruction of wood by these fungi is probably greater than by any other agencies including fire and man. Some members of the Polyporaceae are saprophytes after the tree is dead. Again, others grow on living trees typically attacking only the dead cells of which the wood is composed, but the enzymes they produce may kill some of the living cells and ultimately contribute to the death of the tree. They very often do serious damage to mature standing timber, and also to lumber which is kept moist or in contact with the ground. The visible portion of the fungus is the fruiting body (basidiocarp), produced by vegetative hyphae located in the bark and wood (Figs. 14.29A and C). Removal of such a basidiocarp from a tree in which the hyphae are living parasitically does not destroy the fungus or constitute a method to control its incidence. The vegetative hyphae develop extensively in the substrate before a basidiocarp is produced, and appearance of it usually indicates that considerable decay has already occurred. Even after removing the basidiocarp the hyphae will continue to feed upon the wood and will subsequently form new basidiocarps. The fungus gains entrance through wounds caused by various agencies like fire, lumbering, prunning, insects, and storms. The hyphae spread into adjoining tissues, which they digest by means of cellulase and other enzymes. The enzymes are the chief weapons with which the hyphae invade the plant cells and bring about damage. The enzymatic activity is exhibited at different stages of growth of hyphae; for example, young vegetative hyphae, hyphae about to bear fruiting bodies, and hyphae with fruiting bodies.

The basidiocarps are usually large bearing pores which are often so small as to be barely visible to the naked eye. In texture they may be fleshy, corky, woody, or gelatinous to cartilaginous. In shape also the basidiocarps are extremely variable. They may be: (i) a flat crust-like thin structure developed on the substrate with the pores on the outer side is known as resupinate; (ii) a stretched out, partly crust-like and partly turned up at the edge forming a small bracket-like projecting margin is effused-reflexed; (iii) the entire fruiting body growing erect with broad or narrow base forming an angle with the substrate producing a bracket- or shelf-like structure is termed reflexed; and (iv) the one possessing a stalk or stipe of sterile tissue is stipitate, and depending on the position of the stipe lateral or centrally placed, a stipitate fruiting body may be **laterally stipitate** or **centrally stipitate** respectively. The main bulk of the basidiocarp—**hymenophore** is composed of vertically arranged tubes, the ends of which usually appear as pores on the lower surface (Fig. 14.29B). The internal surface of the tubes of these pores is lined by the hymenium composed of basidia bearing basidiospores (Fig. 14.29E). Setae or cystidia are also sometimes present in the hymenium.

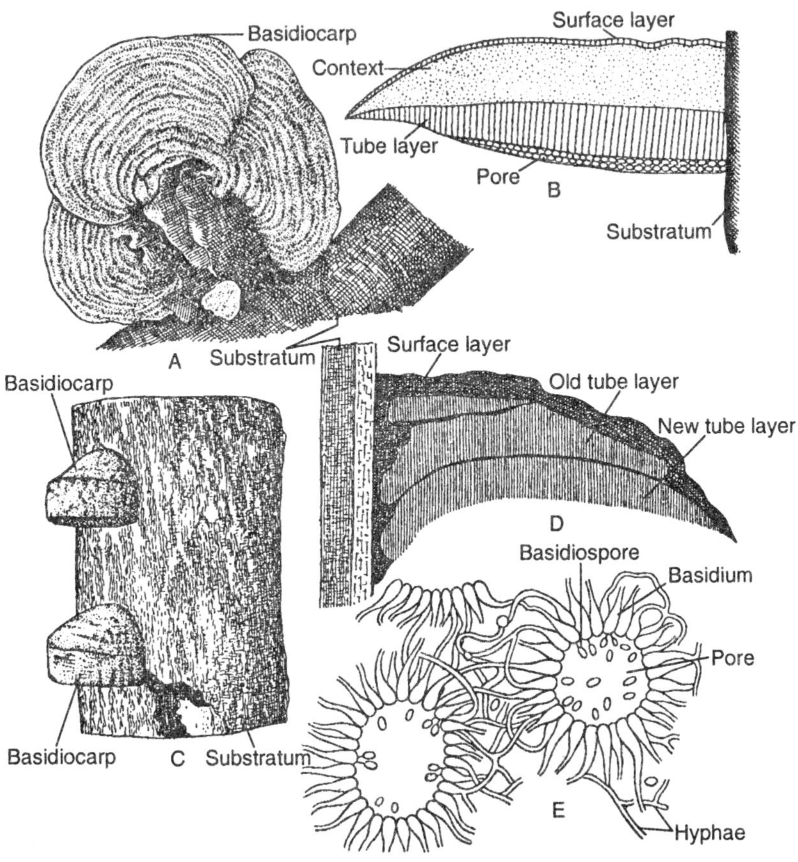

Fig. 14.29: Fruiting bodies (basidiocarps) of the Polyporaceae. A. A basidiocarp with zonate upper surface attached to substratum. B. Diagrammatic representation of a basidiocarp in vertical section. C. Thick basidiocarps attached to substratum with unilateral inferior hymenium. D. Diagrammatic representation of the stratifications of tubes in a basidiocarp. E. Section through hymenium of a basidiocarp showing pores lined with basidia

Mature basidiospores escape through the pores. The pore tubes develop from an interwoven mass of hyphae forming a fibrous tissue which is known as the **context** (Fig. 14.29B). The context may be thick or thin. The pore tubes again may be shallow corrugations, uniformly deep-seated, sunken to unequal depths in the context, or the walls of the tubes form into toothed plates. The shape of the pores may be: (i) circular, called **poroid** (Fig. 14.30D); (ii) angular, known as **hexagonal** (Fig. 14.30A); (iii) intermediate between rudimentary lamellae (gills) and pores, termed **daedaloid** (Fig. 14.30C); or (iv) almost resembling gills, called **lamellate** (Fig. 14.30B). Pores vary from round to angular in *Poria, Polyporus, Trametes, Gonoderma,* and *Fomes;* hexagonal in *Hexagonia*; and labyrinthiform in *Daedalea* and *Lenzites.* While the number of pores per mm is an important character for a species, this is governed by the thickness of wall, the number of pores decreasing with increasing width of wall; thus both characters are important. The pore surface often changes in colour on drying. Therefore, its colour when fresh and also on drying should be recorded. The basidiocarps are also extremely variable in colour ranging from deep-red, white, brown, rusty-brown, yellowish-brown, black, or shades of different other colours. The upper surface of a basidiocarp may be smooth, laccate, hairy, zonate, or velvety. The basidiocarps are annual or perennial. In perennial forms, stratification of horizontal layer of tubes corresponding to each year's growth takes place (Fig. 14.29D). It is often possible to count the layers and determine the age of the basidiocarp, encountered in *Fomes applanatus.* In some cases, the cavities of the new tubes are continuous with those of the preceding season. In others, the new layer is separated from the old by a thin compact layer of hyphae. The spores in perennial forms are discharged continuously, day and night, over a period of several months. Stratifications of tubes do not always indicate perennial growth of the basidiocarp. In *Polyporus gilvus* where pore tubes are stratified, the new growth is produced from mycelium within the host tissue at the base of the basidiocarp and not from any live mycelium of the previous year's basidiocarp.

The hymenial configuration may exhibit toothed condition as in *Irpex,* but such forms are not included in the Hydnaceae, since the growing region is distinctly poroid. Again, the genera *Daedalea* and *Lenzites* exhibit lamelloid or gilled hymenial surface like the Agaricaceae but are not included in the latter since the hymenial configuration in these two genera can be distinctly poroid or at any rate in the younger tubes at the margin. Further, the consistency of the fruit body in *Daedalea* and *Lenzites,* their persistent nature, lignicolous habit and ability to decay wood place these genera in the Polyporaceae and not in the Agaricaceae where the fruit body is usually fleshy and deliquescent and members do not commonly grow on wood and even then seldom decay.

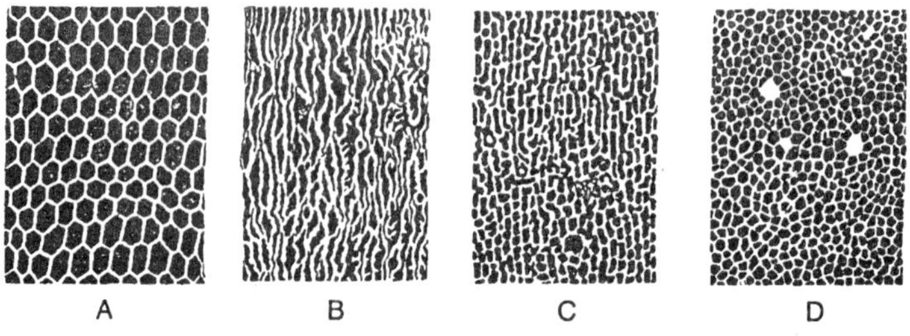

Fig. 14.30: Shape of pores of the pore tubes in the basidiocarps of the Polyporaceae. A. Hexagonal. B. Lamellate. C. Daedaloid. D. Poroid

The conventional genera under Polyporaceae are only treated. The recognition of *Polystictus* for species under *Polyporus* having very thin context is arbitrary and is thus no longer valid, and therefore, the species usually recognized under *Polystictus* are placed under *Polyporus.*

A simplified key to the genera of the family Polyporaceae is presented below:

Basidiocarp always entirely resupinate, completely appressed to the substratum, without any pileus *Poria*

Basidiocarp pileate, rarely resupinate pore tubes not separate

Basidiocarp perennial, each year a tube layer added to previous layer so that pore tubes appear stratified, distinctly or indistinctly

Spores smooth on inner wall *Fomes*

Spores rough on inner-coloured wall, outer wall hyaline

Spores ovate to cylindric-ovate with the wall not uniformly thick, apex always thicker, truncate or rounded *Ganoderma*

Spores globose to subglobose, occasionally oblong with the wall uniformly thick *Amauroderma*

Basidiocarp annual, pore tubes in a single layer

Hymenium with folds that are joined into shallow pits or pores, over which hymenium is continuous or pores alveolate

Pileus waxy, hymenium with shallow net-like pores *Merulius*

Pileus gelatinous, pores alveolate *Laschia*

Pores daedaloid or aproaching lamellate *Daedalea*

Pore surface lamellate

Lamellae concentric *Cyclomyces*

Lamellae not concentric *Lenzites*

Pores hexagonal

Basidiocarp sessile *Hexagonia*

Basidiocarp stipitate or substipitate, pores arranged in radiating rows *Favolus*

Pore surface with teeth, spine or warts in mature areas, usually poroid near growing region *Irpex*

Pores circular to angular

Pore tubes sunk into an even depth into context forming a distinct stratum, their bases forming an unbroken straight line *Polyporus*

Pore tubes sunk to unequal depths into context not forming a distinct stratum, their bases do not form a straight line *Trametes*

Basidiocarp pileate or sessile, pore tubes isolated, free from each other, spores coloured *Fistulina*

Genus: *Polyporus*

The genus *Polyporus* includes both saprophytic as well as parsitic species, some of which are homothallic and others heterothallic. Basidiocarps occur on lumber, logs, living trees, and on ground. They are usually annual, but occasionally surviving for two to three years. The basidiocarps may be effused-reflexed, reflexed, or stipitate being fleshy to tough, leathery or woody in texture. Pore tubes simple or stratified usually sunken into thick context to equal depths, so their bases form a straight line. They are not separable from the surface of the context. Pores are circular to angular but rarely daedaloid. Hymenium forms a distinct layer along the pore tubes. Clavate tetrasterigmate basidia bear hyaline or coloured basidiospores. Cystidia, setae or setal hyphae may or may not be present.

The species *Polyporus gilvus* (Schw.) Fr. is a parasite on *Shorea robusta* (sal), *Acacia arabica* (babul), *Dalbergia sissoo* (shisham), and many other hardwood trees. It also grows saprophytically on a large number of hardwood trees, of which mention may be made of *Quercus* sp. The fungus causes a white rot. It usually gains entrance into the living trees through wounds. It also behaves as a root-parasite, and basidiocarps are common on exposed roots. Sometimes it causes severe damage to hardwood trees in conjunction with *Ganoderma lucidum* or with *G. applanatum.*

The basidiocarp of *Polyporus gilvus* is annual usually reflexed and hard to brittle in texture. The upper surface is with shades of yellow, brown or red and are either coarsely hairy with faintly zoned or smooth with concentric zonations. Context is thick yellowish-brown. It is composed of thick-walled yellowish-brown hyphae. Hymenial surface is yellowish-brown with long uniform or stratified pore tubes. Pores are round and regular. Thick-walled dark-brown setae project into the pore tubes which are lined by basidia. The basidia are hyaline, clavate and tetrasterigmate. Each sterigma bears single hyaline oval basidiospore.

The basidiospores on germination produce monokaryotic primary mycelia. The monokaryotic compatible mycelia by somatogamy, produce dikaryotic secondary mycelium from which the basidiocarp is developed. Life cycle of a species of *Polyporus* is presented in Figure 14.31.

Some Indian Species: *Polyporus agariceus* Berk.; *P. betulinus* Fr.; *P. circinatus* Fr.; *P. cubensis* Mont.; *P. gilvus* (Schw.) Fr.

Order: **Agaricales**

This order comprises of a large number of saprophytic fungi which play an important role in the decay of woodland and grassland litter wood, dung and composts. Many agarics form mycorrhiza with forest trees. The somatic body of the Agaricales consists of a loose and extensive growth of septate hyphae which penetrate the substratum and after sufficiently ramifying form basidiocarps. The basidiocarps grow upward or outward into the air. They are largely fleshy, some leathery or even corky or woody at

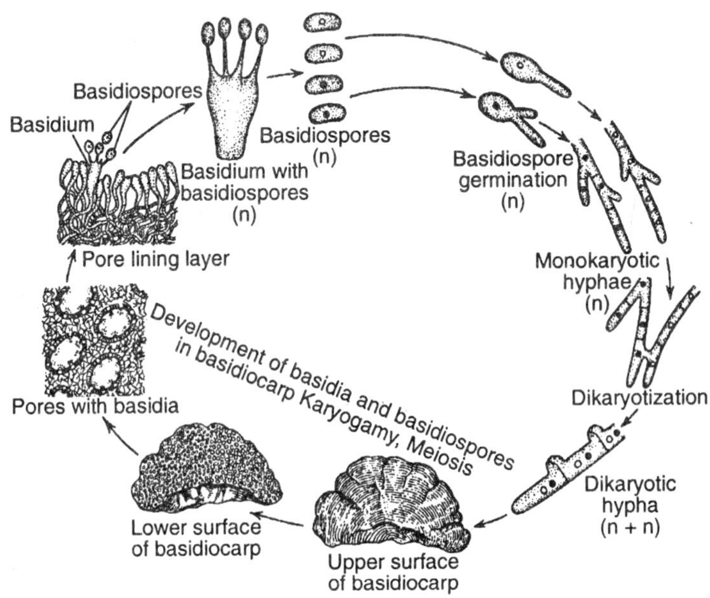

Fig. 14.31: Life cycle of *Polyporus* sp.

maturity. Hymenium may remain uncovered while the spores mature, **gymnocarpic**; or closed being covered by **veil** but opens before the spores are mature, **hemiangiocarpic**. Basidiocarps are mostly stipitate (Figs. 14.34H and 14.35A-E), but sometimes **stipe** or **stalk** may be absent (Figs. 14.35F and G). In stipitate basidiocarp the umbrella-like head is the **pileus.** It has very much dome-like upper portion composed of plectenchymatous tissue of elongated cells, the **trauma** also known as **context**, and down below facing the substratum are the gills (lamellae) (Fig. 14.34H) or pores borne on hymenophore. Besides the elongated cells, the trama may also contain groups of large, globose or oval cells—the **sphaerocysts.** A trama is **homoiomerous** when it is composed of hyphae only; and **heteromerous** when composed of hyphae and either sphaerocysts, or laticiferous hyphae. Gills or pores, whatever may be the case, are lined with hymenium consisting of basidia interspersed with paraphyses or cystidia, or both. Each basidium usually bears four spores borne obliquely on sterigmata.

Singer (1962) classified Agaricales on the basis of combinations of microscopic and macroscopic characters and recognized sixteen families of Agaricales. Among the microscopic characters used by Singer for this purpose are: spore ornamentation and amyloidity; distribution of clamp connections; the outermost layer of the pileus surface; cystidial types; and especially the tissue arrangements of the trama.

Following is a simplified key to the families of the order Agaricales:

A. Basidiocarp soft an putrescent, hymenophore usually separating readily from context
  B. Hymenophore poroid or less commonly pitted **Boletaceae**
  BB. Hymenophore lamellate, the lamellae often united by conspicuous veins **Paxillaceae**

AA. Basidiocarp fleshy to tough or membranous, hymenophore not separating readily from context
  C. Context containing sphaerocysts **Russulaceae**
  CC. Context without sphaerocysts
    D. Lamellae waxy, distant, broadly triangular in section **Hygrophoraceae**
    DD. Lamellae not waxy, distant to close, narrow in section **Agaricaceae**

Family: *Agaricaceae*

The family Agaricaceae includes fungi, commonly known as agarics or gill-fungi, mushrooms or toadstools, produce conspicuous basidiocarps. Mushroom is a general term applied to the fleshy agarics. A mushroom may be edible, poisonous, unpalatable, or otherwise. But popular usage applies the term 'mushroom' to edible ones, calling the others 'toadstools'. What is ordinarily called the **mushroom** is simply the basidiocarp. This structure is not the whole plant but merely a spore-producing part. The vegetative part of the plant body is the mycelium composed of septate branched hyphae which grow saprophytically in soil rich in organic matter, in or on dead wood, in or on rotting logs or stumps. The hyphae spread in every direction of the substratum and absorb food from it. The individual hyphae are usually not isolated, but woven together into strands which may develop into root-like structures often termed rhizomorphs. These fungi are mostly saprophytic growing commonly in lawns, pasture and gardens. Some of them are *Agaricus campestris, Lepiota morgani*. But some are also facultative parasites on the roots of trees and elsewhere. Very common one is *Armillariella mellea* which is known as 'honey agaric'. It causes root rot of

Larch. Again, some are involved in mycorrhizal associations like *Amanita muscaria* and *Lactarius rufus*. The vegetative mycelium buried in the substratum sends up basidiocarps at irregular intervals of months or years, usually after soaking rains. A mycelium established at a single point tends to ramify radially in all directions and to die out at the centre with the depletion of the available food supply. Successive crops of basidiocarps may thus be produced in successively larger rings, often called **fairy rings** (Fig. 14.32) based on the ancient belief that these circular rings marked the path of dancing fairies. The fairy rings or fungus rings are very frequent in grassland, and not uncommon in woods. They are of three chief types: (i) those in which the development of the basidiocarps has no effect on the vegetation, e.g., *Lepiota morgani;* (ii) those in which there is increased growth of the vegetation, e.g., *Tricholoma personatum;* and (iii) those in which the vegetation is damaged very badly, e.g., *Agaricus praerimosus*. Rings of the third type are frequently made up of outer and inner zones in which the growth of the vegetation is strong with a zone of dead or badly damaged vegetation in between where the fungus ring is developed.

Many of the agarics are edible and are considered great delicacies but others extremely poisonous. The edible species are popularly known as mushrooms, whereas the poisonous ones are the toadstools. Some of the edible agarics are: *Agaricus campestris, Agaricus bisporus, Coprinus comatus,* species of *Lepiota, Marasmius, Tricholoma;* and the poisonous ones are: *Amanita muscaria, A. phalloides, A. verna.*

Spores that are produced in the basidiocarps on germination produce germ tubes which develop into monokaryotic mycelia. The monokaryotic mycelia, in many species, produce at the surface of the substratum single-celled uninucleate structures usually called **oidia** (sing. **oidium**). The oidia may be distributed by wind or insects. An oidium which comes into contact with a cell of compatible hypha, fuses with that cell initiating the dikaryotic stage by spermatization. In some members of the Agaricaceae the oidium germinates to produce a new mycelium. The oidia may thus behave as gametes or spores. Besides spermatization the dikaryotic condition may be initiated by somatogamy, i.e., by the fusion of cells of compatible hyphae. Two compatible mycelia or mycelium and oidium thus are involved in the establishment of the dikaryotic condition. Once the initial dikaryotic cell has been formed, a new dikaryotic mycelium may then be derived directly and solely by growth from this cell, or the existing mycelium may be converted to the dikaryotic condition by migration and mitotic divisions of the proper nuclei. The newly introduced nucleus in the dikaryon initially divides mitotically, and one of the daughter nuclei passes through the pore in the septum into the adjoining cell, where it again undergoes mitosis and the daughter nucleus passes in the adjoining cell and the process con-

Fig. 14.32: Fairy ring of basidiocarps

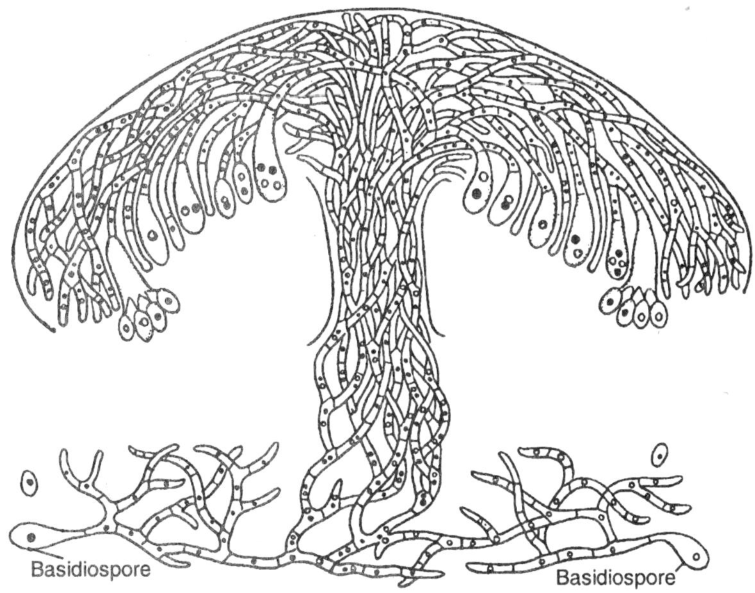

Fig. 14.33: Diagram of basidiocarp development by dikaryotic mycelia

tinues when finally all the cells become dikaryotic. The basidiocarps usually arise from a dikaryotic mycelium which may be short-lived or perennial (Fig. 14.33). Their development is controlled by the available supply of food, moisture, temperature, and other factors. Under favourable conditions little knots or swellings (Fig. 14.34A) appear at one or many places on the dikaryotic mycelial threads. These swellings increase in size until they project outside the substratum in which they started to grow. Each one of these swellings is destined to develop into a full-grown basidiocarp. The swellings gradually develop as small balls, which constitute the **button stage** (Fig. 14.34B). A longitudinal section through a swelling from top to bottom shows an undifferentiated mass of fungal tissue enclosed in a

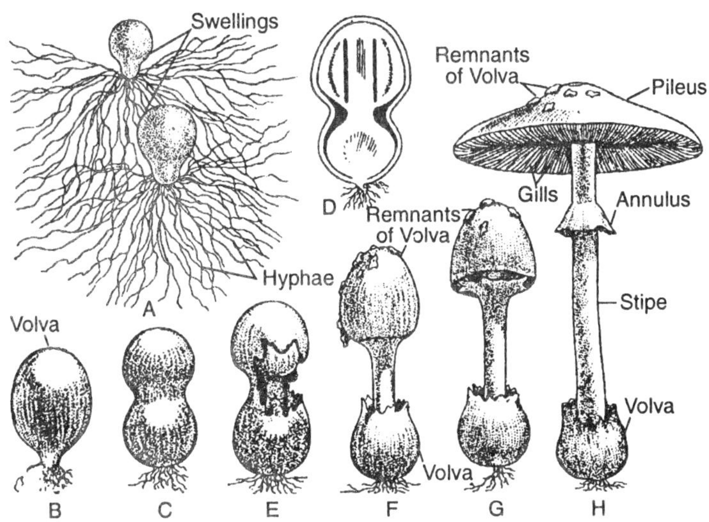

Fig. 14.34: Successive stages in the development of basidiocarp in the Agaricaceae

membrane or **veil** which is known as **universal veil** or **volva.** Gradually the swelling grows in the vertical direction and a depression appears all-round more or less near the central region (Figs. 14.34C and D). A vertical section of it indicates the presence of a **stalk** or **stipe** and a small very immature umbrella-like cap or the **pileus** together with a thin membrane, the **partial** or **inner veil**, which connects the margins of the pileus to the upper part of the stipe. Growth of the pileus and stipe ruptures the volva part of which, in some species, forms a cup at the base of the stipe and the rest breaks up into small pieces which remain on the roof of the pileus (Figs. 14.34E-H). The partial veil, also ruptures, in most cases, disappears entirely, whereas in other parts of it persists as a collar or little ring around the stipe and is known as an **annulus** (Fig. 14.34H). In still others, fragments of the partial veil may be found hanging from the margin of the pileus. The presence (Fig. 14.35A) or absence (Figs. 14.35D and E) of volva and annulus or any one (Figs. 14.35B and C) of them are features used to identify the members of the Agaricaceae. In some cases along with the development of basidiocarp, the individual hyphae may be aggregated into mycelial strands or rhizomorphs which pass deep into the soil.

Some shade of tan or brown is the commonest colour of the basidiocarps. But there is scarcely any colour that does not occur among them. A mature basidiocarp usually consists of a stalk or stipe and an umbrella-like cap called the pileus. The tissue of the pileus is like that of the stipe, but rather denser. On the under side of the pileus, there are immense number of plates, called **gills** or **lamellae** (Figs. 14.34H and 14.36A and D), which radiate from the stipe toward the margin of pileus and are

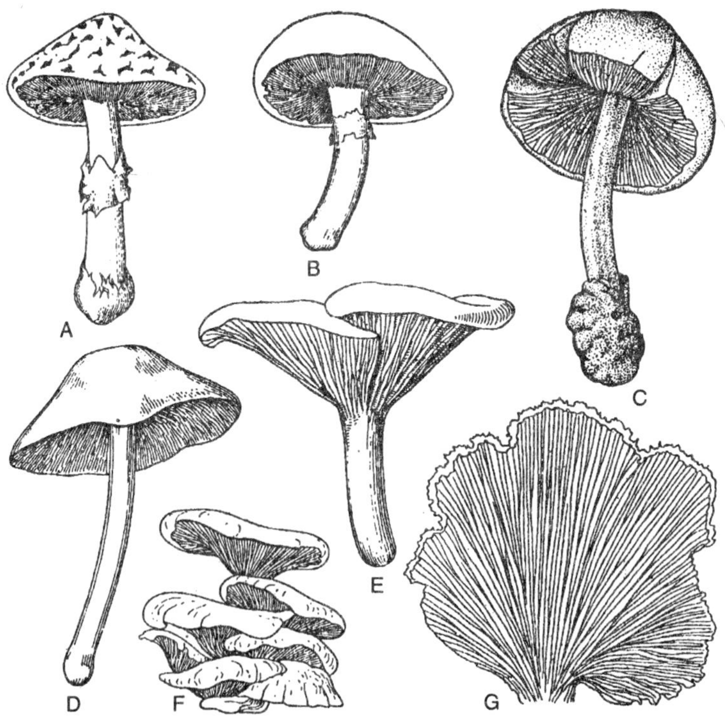

Fig. 14.35: Basidiocarps of the common genera of the Agaricaceae. A. *Amanita* with volva and annulus. B. *Agaricus* with annulus only. C. *Volvaria* with volva only. D. *Collybia* without volva and annulus. E. *Clitocybe* without volva and annulus. F. *Pleurotus* attached by margin of basidiocarp. G. *Schizophyllum* attached by margin of basidiocarp

covered by the hymenium. The tissue of the basidiocarp is pseudoparenchymatous, consisting of an aggregation of hyphae. The hyphae are usually thin-walled and dikaryotic (generative hyphae), and may or may not bear clamp connections. The basidiocarp expands due to inflation of the cells. Although no differentiation into skeletal or binding hyphae occurs, specialized tissues or cells may arise. In a study of the fine structure of the basidiocarp of *Agaricus campestris*, Manccha (1965) has shown that the stipe contains two kinds of cells: wide inflated cells and narrower thread-like cells. A similar differentiation is found in *Coprinus cinereus.*

Two main types of gill structure in a basidiocarp of the Agaricaceae may be distinguished.

(i) **Aequi-hymeniiferous (aequihymenial).** Most genera have gills of this type. In longitudinal section the gills are wedge-shaped. The aequi-hymeniiferous refers to the fact that the hymenium develops in an equal manner all over the surface of the gill, i.e., basidial development is not localized at any one point on the gill. The wedge-shaped section may be an adaptation to minimize wastage of spores should the basidiocarp be tilted from the vertical.

(ii) **Inaequi-hymeniiferous (inaequi-hymenial).** The gills are not wedge-shaped in section, but parallel-sided. The term inaequi-hymeniiferous refers to the fact that the hymenium develops in an unequal manner, with basidia ripening in zones.

The hymenium consist of a layer of very many basidia intermingled with paraphyses (Figs. 14.36B and C). In some cases, certain specialized structures of taxonomic importance are also present in the hymenial layer. The basidia are aseptate, commonly longer than thick, with four terminal ellipsoid basidispores obliquely set on the sterigmata (Fig. 14.36E). Mature basidiospores are forcibly discharged from the basidia. Under favourable conditions the basidiospores germinate to produce monokaryotic mycelia. The basidiocarps die and decay soon after the liberation of basidiospores. The monokaryotic mycelia subsequently becomes dikaryotic from which successive crops of basidiocarps are developed. In general, the basidiocarp is **stipitate,** i.e., with a stipe. The stipe is usually attached to the under

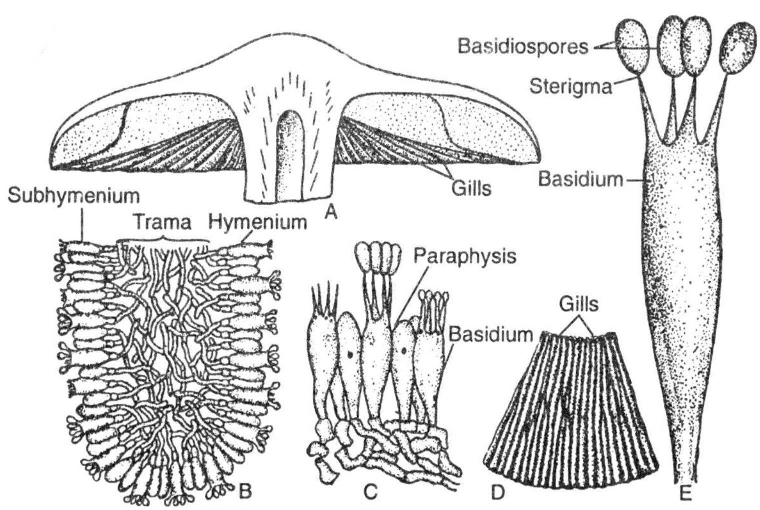

Fig. 14.36: Structural features of the pileus of *Agaricus* sp. A. Sectional view of the gills. B. Structural details of gill. C. Portion of hymenium. D. Gills in surface view. E. Basidium with sterigmata and basidiospores

surface of the pileus at its centre, and the basidiocarp is known as **centrally stipitate.** When the stipe is attached to the margin, it is **marginally stipitate,** and when laterally, **eccentrically stipitate.** The stipe may be slender or thick, long or short, hollow, pithy or solid. The stipe may also be absent, when the basidiocarp is attached to the substratum directly by its margin (Figs. 14.35F and G). The pileus may be leathery (**coriaceous**), hairy (**tomentose**), with or without scales. When dry, the pileus may change colour **hygrophanous.** The pileus may be bell-shaped (**campanulate**); cone-shaped (**conic**), inverted cone-shaped (**obconic**), with edge turned up (**revolute**), with edge turned downward and inward (**involute**); funnel-shaped (**infundibuliform**) with a pit or small depression at the centre (**umbilicate**), with a knob protruding from the centre (**umbonate**); with radiating marks or furrows near the edge (**stristulate** or **striate**); and split at the edge, with weavy edge, smooth, scaly (**squamose** or **squamulose**). The pileus may be large having four or more inches in diameter, medium of one to four inches in diameter and small with less than one inch in diameter. In form, gills may be broad or narrow, varying with different species. An important distinction is made between gills that extend to the stem, known as **adnexed,** those that are attached broadly to it **adnate** and others that do not reach it but are said to be **free.** Gills that run down the stem are termed **decurrent,** those that are notched in their edge near the stem are known as **sinuate** or **emarginate.**

Basidiocarps and mycelium of some of the genera of the Agaricaceae, such as species of *Panus* and *Pleurotus,* and *Clitocybe illudens* glow in the dark which is known as **phosphorescence** and the basidiocarps are **phosphorescent.** In *Armillariella mellea,* mycelium in wood, when exposed to air, gives out light. The fungi that give out light sometimes cause the attacked wood or leaves to become luminous, they are also known as **luminescent fungi.** Again, some species of *Lactarius* possess milky juice in their basidiocarps, whereas some genera of the Agaricaceae are commonly known as inky-caps for having typical pileus. In certain genera, particularly in *Coprinus* when the basidiocarp is mature, a process of **autodigestion** or **autodeliquescence** begins. The pileus becomes liquid resulting in the formation of inky-drops, until only the stipe remains (Fig. 14.37). The spores around the margin of the pileus of the basidiocarp of the Agaricaceae are the first to be discharged and are carried away by air current, after which this area is digested. The spores lying higher on the gills are released next, after which the zone where they were formed liquefies and so on, until the entire pileus has disappeared.

The spores vary in size and shape in different species. The characteristics used to distin-

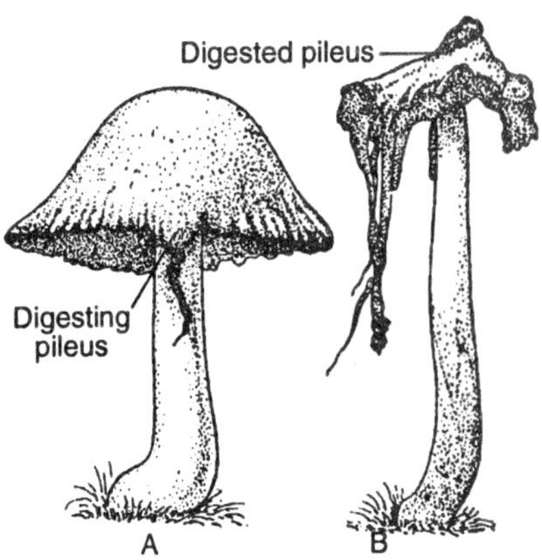

Fig. 14.37: Stages in autodigestion of *Coprinus comatus*

Fig. 14.38: Spore print of *Agaricus* sp.

guish one species of agaric from another include, not only external features but also the colour of the spores. The spore colour is not necessarily correlated with the colour of the gills, which often changes with age, darkening as the pileus matures. Individual spores are too small to be seen with the naked eye. Ordinarily, spore colour is determined by bringing the fungi in the laboratory and making a **spore print** (Fig. 14.38). The stipe is removed from the basidiocarp close to the pileus. The pileus is placed, gill side down on a sheet of white paper. Since some spores are white or very light-coloured, a duplicate pileus is often placed upon black paper as well. The pileus is placed under a bell jar to prevent the spores from being distributed by air current. After several hours, the pileus is carefully lifted and the spores which have fallen are seen in a radiating pattern which duplicates the arrangement of the gills. The colour of the spores may be observed very easily. The spore colours of the various genera of the Agaricaceae are white, yellowish, brown, pink, purple, purplish-brown, black, etc.

Genus: *Agaricus*

The genus *Agaricus* includes many species that live on the decaying matter in the soil, some of which are edible. Again, many of the species belonging to the genus *Agaricus* are the causes of parasitic diseases and saprophytic rots. There are both homothallic as well as heterothallic species of the genus *Agaricus*. This genus includes fungi whose basidiocarps are soft. They are stipitate. Stipe is readily separating from the pileus. The basidiocarp is characterized by the production of radiating

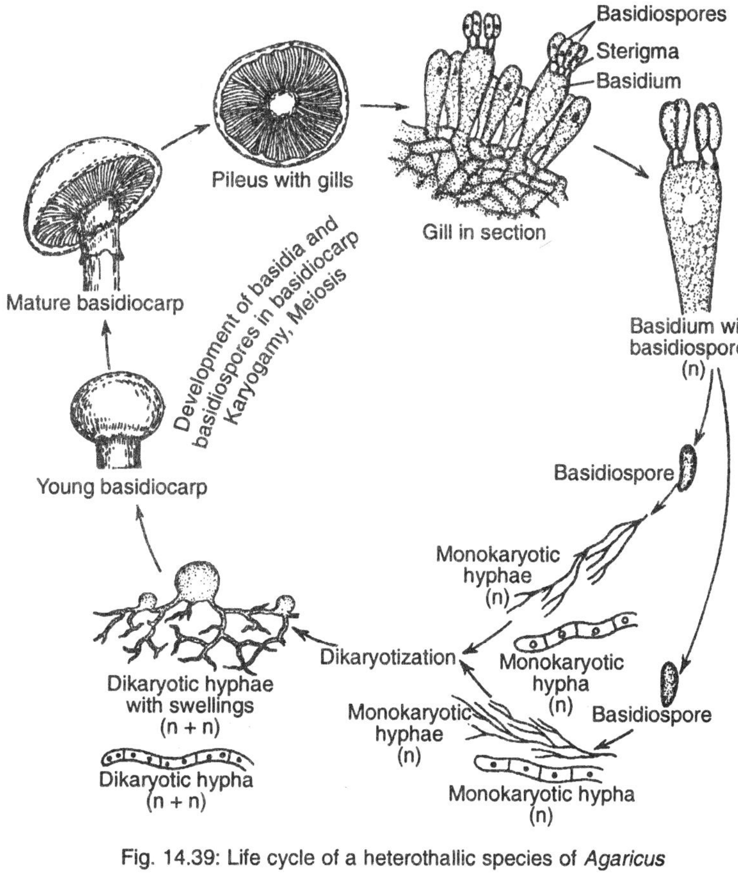

Fig. 14.39: Life cycle of a heterothallic species of *Agaricus*

lamellae (gills). Lamellae are not deliquescing. When young, the lamellae are covered by a veil that extends from the stipe to the margin of pileus. During growth of the basidiocarp the veil breaks away from the edge of the pileus as this expands or spreads open. Part of it remains as a collar or little ring around the stipe and forms the annulus. Volva is absent in the basidiocarp. The lamellae are pink in colour, but turning brown with age. They are free from the stipe. Club-shaped basidium usually produces four elliptic basidiospores, whose number may also be two. The spores are purple-brown in mass.

Some Indian Species: *Agaricus arvensis* Schacff.; *A. campestris* L. ex. Fr.; *A. exaltatus* Berk.; *A. latipes* Berk.; *A. pratensis* Schacff.; *A. squalidus* Massee.

AGARICUS CAMPESTRIS L. EX FR.

Agaricus campestris, commonly known as meadow or field mushroom grows on ground in grassy places, in pastures, on lawn and manured ground, never in thick forests. Basidiocarp producing hyphae are dikaryotic which usually escape notice. The reproductive phase is initiated by the formation of small knob-like swellings at different points of interwoven mycelial strands. These swellings increase in size and break through the surface of the substratum as small balls constituting the button stage. These bill-like swellings increase in size and each one of them is destined to develop into a full-grown basidiocarp. Since the growth of the mycelium follows more or less a regular pattern of centrifugal extension starting from a particular point and accompanied by a death and decay of old hyphae in the central portion, this may ultimately under favourable conditions, result in the development of a circular ring of basidiocarps which is known as fairy ring. The basidiocarps develop rhizomorphs. A matured basidiocarp is white or whitish in colour and consists of a thick short stipe with an annulus rather more than half-way up. The stipe supports the pileus which appears as a hat-like expansion. The pileus is broadly expanded or nearly flat, white or with a tendency to become pinkish when cut. On the underside of the pileus are immense number of radiating gills or lamellae, pink when young, but purple-brown when mature. The pink colouration of the young gills is due to cytoplasmic pigment in the spores. Later the gills turn purple-brown due to the deposition of dark pigments in the spore wall. The gills are free from the stipe. The gills are formed by an extension of the hyphae of the pileus. A cross-section of a gill shows that the bulk of the tissue is composed of compact hyphae. The middle part of each gill is formed of hyphae coming down from the pileus, and following on the whole a longitudinal course, their lateral branches, however, diverging towards the two surfaces. This central tissue of the gill is called the **trama.** Towards the free surfaces, the cells of the diverging hyphae are shorter and more closely packed, forming the **subhymenial layer,** and beyond this again is the **hymenial layer** (Fig. 14.36B), which is composed of the terminal cells of the same hyphae constituting the trama and subhymenial layer. In this layer the hyphae have diverged from their original direction to such an extent that they now stand at right angles to the surface of the gill. The hymenial layer is composed of a palisade-like club-shaped cells rich in protoplasm. A few of these are somewhat more slender than the rest, and remain sterile, they are known as **paraphyses** (sing. **paraphysis**) (Fig. 14.36C). The others are stouter at maturity, and are the elements which produce spores, these are the **basidia** (Fig. 14.36C). Each basidium bears four basidiospores on **sterimata** (sing. **sterigma**) (Fig. 14.36E). The young basidiospores are pink, but when ripe are purple-brown in mass. They contain oil, and have each two or four nuclei. The basidia do not develop simultaneously on all parts of the gills. In the pink zones the basidia are nearly ready to shed their spores, whereas in the paler zones the basidia are only just beginning to form their spores. At maturity the spores are violently shot from the sterigmata. After spore discharge the basidia collapse, and this part of the gill again turns pale. Sub-sequently, the same area once more becomes pink owing to the development of another crop of basidia. At a later stage the whole of the gills of the basidiocarp become brownish or purplish black on account of darkening of the cells in the matrix of the gills. Ultimately the basi-diocarp falls to the ground and decays. The spores give rise to fresh mycelium from which new basidiocarps are developed.

Life cycle of a heterothallic species of *Agaricus* is presented in Figure 14.39.

The Scientific name. E. M. Fries (1794-1878) placed *Agaricus campestris* under the genus *Agaricus* but within a tribe which he called *Psalliota.* The tribe *Psalliota* was later elevated to the rank of genus. Consequently, books not directly influenced by modern taxonomy and not concerned with the international rules of nomenclature, named *Agaricus campestris* as *Psalliota campestris.* The name *Psalliota campestris* is taxonomically and nomenclaturally wrong, and should disappear from the literature. The genus *Agaricus* which has been split into two or more genera, for one of them, according to the Internatinal Rules of Botanical Nomenclature, Article 51, the original generic name *Agaricus* should have been retained. But very few have followed this rule. According to this rule, *Agaricus* is the valid name for the genus of *Agaricus campestris.* Instead of naming the genus correctly it is often called *Psalliota* which actually is a synonym of *Agaricus.* In spite of this, a small group of mycologists wished to retain the name *Psalliota* against the rules of priority creating a confusing situation. This was rejected at the International Congress in Paris in the year 1954.

Cultivated and meadow mushrooms. *Agaricus campestris* L. ex. Fr. as described by Linnaeus and accepted by Fries, is a very common fungus of the open pastures and grasslands. Its basidia are four-spored, whereas the commonly cultivated white mushroom whose correct name is *Agaricus bisporus* (Lange) Sing, (also called *Agaricus campestris bisporus*) has basidia with two spores. For some time the situation was very confusing, when both *Agaricus campestris* and *Agaricus bisporus* were thought to be same considering the possibility that the four-spored *Agaricus campestris,* once taken into cultivation, changes the type of basidia under the new conditions, and becomes two-spored. No body has ever experimentally shown that this can happen. The other possibility is, a change from four-spored to two-spored basidium takes place as a mutation occurring under the influence of environmental conditions, not as a consequence of adaptation. But all recent attempts to grow *Agaricus campestris* from wild strains under the same conditions as those prevailing in the cultivation of the cultivated white mushroom *Agaricus bisporus* have failed. Besides the difference in the number of spores, this also indicates the specific difference between *Agaricus campestris* and the cultivated white mushroom *Agaricus bisporus.*

MUSHROOM CULTIVATION

Mushroom cultivation was initiated for the first time in France during 1638-1715. But special attention was paid in different countries for large-scale mushroom cultivation only during the first World War (1914). In India, the prospect of mushroom cultivation was pointed out by S. R. Bose (1921), but here actually K. M. Thomas and his associates (1943) first attempted to grow mushrooms. Of late, mushroom cultivation has become very popular in India because of easy method of cultivation, short growing period, high productivity, and nutritive value and prospect of marketing of mushrooms. Mushrooms in general contain besides water, protein, carbohydrate, fat, Calcium, Phosphorus and Iron. They are rich in vitamins (Thiamin, Riboflavin, Niacin, and Pantothenic acid) and also contain enzymes and a few amino acids.

The conventional nutrient substrate used for mushroom growing all over the world has been horse-dung manure. But raw materials like straw of all kinds of cereal varieties, wood shavings or other cellulose-containing raw materials are also used after being subjected to biological composting by microorganisms like, certain cellulose-digesting bacteria and members of the Actinomycetes, etc.

Mushrooms can be grown the year round, if the temperature suitable for their development is maintained. Some common edible mushrooms that have already been cultivated in India belong to the genera: *Agaricus, Calocybe, Pleurotus* and *Volvariella. Agaricus bisporus* grows well in cold climate (15-18°C), it can be grown at a low cost; *Calocybe indica* (on a bed of fine garden soil, clean sand and wheat bran or maize meal put in proper proportion) at a temperature range of 30 to 31°C; species of *Pleurotus* (on saw dust) at a tempe-rture range of 21 to 35°C; and *Volvariella volvaceae* (on paddy straw bed) at a temperature range of 35 to 40°C. The duration of each crop of mushroom usually ranges between 4 and 6 weeks.

Cultivation Procedure

A. **On paddy straw.** Paddy straw mushroom—*Volvariella volvacea* (Bull. ex. Fr.) Sing. cultivation comprises of the following stages: preparation of spawn; preparation of mushroom bed; spawning and watering of mushroom bed; mushroom crop production; and harvesting of mushrooms.

i. **Preparation of spawn.** Mushroom spawn can be simply defined as a medium impregnated with mushroom mycelium and serves as the 'seed' or 'inoculum' for mushroom cultivation. It is actually a fibrous mass of mycelium growing on dead organic matter. Spawn may be obtained either by tissue culture method or by pure culture from germinating spores obtained from desired mushroom fruit body. The entire process of preparation of spawn should be aseptic from the beginning to the end. Spawn may be: grain spawn and straw spawn. Grain spawn is prepared with grains of rye, sorghum or wheat cooking in water and sterilizing in autoclave for 30 minutes, inoculating with pure culture of desired mushroom and incubating at proper temperature. When the mycelium has run over the whole surface of the grain, the spawn is read for use. Straw spawn is prepared with paddy straw or wheat straw cutting into suitable pieces, soaking in water for 5 to 10 minutes, mixing in 2 per cent lime and proceeding further as done in case of grain spawn. Preparation of spawn is usually done in glass bottles.

ii. **Preparation of mushroom bed.** At first prepare a bamboo or wooden frame of suitable size, say about 1 sq. metre. Then prepare paddy straw bundles from uncrumpled paddy straw of not more than one year old from which leafy material has been removed. Each bundle of straw weighing around 1 kg should be tied at both ends by strings. Clip off unequal parts of straw at both ends of each budle. Dip 8 such bundles in water in a tank for 18 to 24 hours. Take them out from the tank and spread on the frame prepared earlier and this constitutes the first layer of the mushroom bed.

iii. **Spawning of mushroom bed.** Spawning is the process of planting spawn on the mushroom bed materials. Remove the spawn from the glass bottles in which spawn has been prepared. Break it into small pieces by crushing and crumbling with the fingers. Spawning should be done on the first layer of straw bundles by broadcasting keeping about 4 to 6 inches margin all around the bed unspawned. Gram or pigeon pea powder should now be dusted over the spawned area. Treatment of the first layer of straw is now complete. During the spawning period proper temperature and humidity should be maintained. Never let the bed surfaces dry out. As the spawn grows, it produces heat which contributes to the water loss. Water mushroom bed lightly with water sprinklet. Prepare a second layer of 8 straw bundles as above and place on the first layer arranging the straw bundles crosswise with the first layer to facilitate proper aeration of spawn. The second layer should be spawned and treated with gram or pigeon pea powder in the similar manner as done in the first. In all, mushroom bed of 4 layers of straw are to be arranged in one frame following exactly the same procedure so that the straw bundles are arranged crosswise. All the 3 layers of straw whould be spawned and treated with gram or pigeon pea powder in the same manner. But the 4th layer should not be spawned or treated with gram or pigeon pea powder. It should be gently pressed and tied with a rope with the 3rd layer. Then place a polythene or plastic sheet on the mushroom bed in such a manner that it does not touch the straw bundles.

iv. **Watering of mushroom bed.** The mushroom bed should be watered once or twice daily to keep the temperature range between 35 and 40 °C and moisture suitable for the development of mushroom fruit bodies.

v. **Mushroom crop production.** Small mushroom buttons (primordia of fruit bodies) will start appearing within 10 to 15 days after spawning. But it will take 4 to 6 weeks to get a crop of mushroom. Mushrooms appear in rhythmic cycles which are called "flushes" or "breaks".

vi. **Picking of mushrooms.** During picking, mushrooms should be lightly twisted so that neighbouring young ones are not disturbed.

vii. **Precautions of mushroom cultivation.** Precautionary measures of mushroom cultivation are: during cropping period suitable temperature, humidity and ventilation of mushroom bed should be sprinkled to check insect incidence; growth of any other fungus in the mushroom bed should be checked by spraying Captan and Zineb (0.2 per cent); and during picking of mushrooms immature fruit bodies should not be touched. Poor cropping of mushrooms and production of abormal fruit bodies may be at times due to virus infection. Symptoms like watery stipe and die-back are also produced as a result of virus infection.

B. On wood log. Cultivation of mushroom—*Pleurotus sajor-caju* (Fr.) Singer on wood logs of *Mangifera indica* may be done in the following manner:

Prepare logs 60 cm long and 2.5 to 3.75 cm thick. Dry the logs in the Sun for 15 days and then keep them immersed in water for 7 days. Make holes (2 cm diameter each) on the logs at 7.5 cm apart from each other. Fill in the holes with wheat grain spawn of *Pleurotus sajor-caju*. Keep the spawn inoculated logs vertically in a large earthen pot and cover them with a polythene bag and incubate at room temperature of 20 to 30 °C. Watering of logs should be done at intervals of 5 days. Remove the polythene bag as soon as the mushroom buds appear on logs and continue to water the logs daily. Two flushes of mushroom fruit bodies appear in course of 80 to 90 days, the yield being higher in the first flush.

C. Cultivation of Oyster mushroom. Species of *Pleurotus* are wood-destroying. Saprophyte fungi with broad adaptability spread over all the world. The best known species is *Pleurotus ostreatus*, widely known as oyster mushroom. Traditionally the oyster mushroom is cultivated in tree trunks and logs of deciduous trees. The leading countries producing oyster mushroom are: Japan, Italy and France.

Cultivation procedure. About 1.5 kg of good quality paddy straw or wheat straw is to be taken and chopped into 2–3″ pieces, soaked in water for 4-8 hours or overnight. The excess of water is drained off. This straw is filled in a small gunny bag and dipped in boiling water for 30 minutes (pasteurization) or is autoclaved at 12°C for half an hour. Excess water is drained off and straw is allowed to cool. The straw is filled in plastic bags of 35 × 35 cm, which are disinfected with 4% formaldehyde. The straw is spawned heavily layer by layer. Make holes especially in the lower and upper regions of the plastic bag contained spawned straws for aeration. Arrange the spawned straw pieces in cylindrical position. An average humidity of 80 to 85% and temperature of 25 to 28°C should be maintained. The white mycelium covers the substrate (spawned straw pieces) within about 15 to 20 days after spawning. They are cut out from the plastic bag by a clean knife. The crop can be harvested from 10 to 15 days. At this stage tear out the plastic bags and water lightly twice a day from the gunny bag. The fruit bodies start developing in about 15 to 20 days after spawning. They are cut out using a clean knife. The crop can be harvested for 10 to 12 days. About 1 to 1½ kg of mushroom can be harvested from one bed.

Wheat straw gives more yield, but the fruit bodies formed on paddy are larger in shape.

Large scale cultivation. Outdoor cultivation of mushrooms can be done in a large shed or in a permanent building where bed space can be increased by erecting tiers of wood, allowing 75 cm to 90 cm space in between each.

For a profitable mushroom growing, most important feature is careful preparation of bed with nutrient substrate. The commonly used crude nutrient substrate is horse-dung manure and a composing mixture of chopped wheat, paddy straw and some fertilizers like superphosphate, urea and ammonium sulphate. The beds should be filled from top tier downwards and made up to the same depth with nutrient substrate. Sterilized silt clay loam soil should be added to the beds followed by careful labelling of bed surface. All doors and ventilators should be left open for several hours.

Small pieces of spawn about the size of a walnut should be inserted pressing firmly into the compost. A fairly humid atmosphere and depending on the genus of mushroom used, a temperature range of 10 to 15°C encourage profuse growth of mycelium after three weeks. The beds should be regularly dusted with gammexane powder to take precaution against insect attack.

Small buttons of mushrooms will appear after 5 to 8 weeks from the date of spawning. Mushroom beds should now be carefully sprinkled with water. In about 7 to 10 days mushroom buttons develop into mature mushrooms. Mature mushrooms should be plucked taking care so as not to loosen the base of the stalks. Depending upon the food value of the compost, 3 to 4 flushes of mushrooms may be plucked at the end of which the compost may be used as garden or field fertilizer.

Prospects of mushroom cultivation. Above all, mushroom lends itself particularly well to food stuff production independent of soil, and can produce more than six generations per year with low cost of cultivation. Besides these, the raw materials for culture medium are easily available with low cost. Even the spent manure may be reused in the field or garden as fertilizer and paddy straw as cattle feed.

Keeping in mind the above aspects of mushroom cultivation every effort should be made to popularize: (a) mushroom eating habit, (b) mushroom cultivation, and (c) mushroom marketing which will by and large release pressure from the consumption of conventional food stuff and will go a long way to solve the problem of food crisis to a great extent.

The use of mushrooms as food is probably as old as civilization itself. These were earlier preferred for their flavour and taste while their nutritive value was recognised later. Mushrooms provide a rich addition to the diet in the form of proteins about 3.7 per cent; Carbohydrates about 4.4 per cent; good source of minerals P, K, Fe and Copper; Vitamins (B_1, B_2, Niacin, B_{12}, Pantathenic acid and Vitamin C) and without Cholesterol. As such they constitute a valuable source of nutritive and protective food, unlike vegetables the entire mushroom can be used as food. With rich source of protein, mushroom can be a good substitute for protein deficiency and also useful food for diabetics.

N.B.: It may be noted that all mushrooms what are cultivated may be done basically by the same method or described above with a little variations. Besides this, all mushrooms which are cultivated have the same nutritional value.

Genus: *Lentinus*

Members of the genus *Lentinus* grow profusely on dead and decaying wood during rainy season. Some members of this genus, especially *L. lepideus* Fr., are important in the decay of railway line joining timbers (railway sleepers).

Basidiocarps are soft corky when fresh; shrivel and turn horny on drying, revive when moistened. They are large, eccentrically stipitate, funnel-shaped. The pileus is white bearing brown hairs on the upper surface. The margin of the pileus is weavy, edge turning towards the lower suface. On the lower surface of the pileus are the snuff-coloured gills which run from pileus margin down to the stipe, edges of the gills are serrate-form. Volva and annulus are lacking.

Structural peculiarities of the gills are very much similar to those of the genus **Agaricus** (Fig. 14.38). Special feature in this genus is, besides paraphyses and basidia, the hymenium has elongated subcylindrical globose to subglobose, hyaline to uniformly coloured cystidia which project beyond basidial layer (Fig. 14.35A), very similar to *Armillariella mellea.*

Basidia are aseptate, club-shaped bearing sterigmate basidiospores, similar to what is present in the genus **Agaricus** (Figs. 14.38C and E). Basidiospores are white in mass.

Some Indian Species. *Lentinus cladopus* Lev.; L. *glabratus* Mont.; L. *lepidus* Fr.; L. *nepalensis* Berk; L. *subnudus* Berk.

All members of the Gasteromycetes are saprobic, and grow on such substrate as soil, rotting wood and other vegetation, and on dung. *Rhizopogon,* which forms subterranean basidiocarps, and *Scleroderma* can form mycorrhiza with forest trees.

The fungi included in the Gasteromycetes produce large, often very large basidiocarps which either remain closed, or open only after the spores are mature. Some of them have hypogeous basidiocarps that emerge as they approach maturity; others possess hypogeous; while a third type develops entirely epigeous basidiocarps. The surrounding wall of the basidiocarp composed of sterile tissue is the **peridium,** which may be composed of one to three layers known as the **endoperidium** (on the inside), the **mesoperidium,** and the **exoperidium** (on the outside). The peridium may be firm and hard or soft and papery or may be so delicate that it may disappear during the development of the basidiocarp. Enclosed within the peridium is the fertile tissue known as **gleba.** The gleba consists of more or less fleshy mycelial growth anastomosing to form numerous chambers or cavities lined by a layer of basidia. In some cases instead of glebal chambers with a definite layer of basidia being formed, clusters of basidia appear here and there in the gleba. Again, there may develop in the gleba a network of veins or interglebal tissue performing functions of support and food conduction, whereas in others, a sterile axial tissue or columella may be developed which behaves like a stipe. Both veins and columella may also be developed in the same basidiocarp. In many of the Gasteromycetes, the gleba undergoes autodigestion after the basidiospores have been formed. Again, in some others, after autodigestion of the gleba, only the spore mass remains which at time is interspersed with stiff, thick-walled thread-like structures, the **capillitia** (sing. **capillitium**), they help spore dissemination. In the Gasteromycetes, the gleba may be fleshy to cartilaginous, slimy, powdery, or waxy. The nature of gleba has great taxonomic importance.

Once basidiospores have developed, the gleba undergoes autodigestion or conversion into dried-up, thread-like strands. The tissue between the glebal chambers constitutes the trama and serves, in different species, for conduction, mechanical suport or even dehiscence in some cases. There are variations in the origin of glebal tissue in the basidiocarps of the Gasteromycetes. These variations are as follows:

(a) **The lacunar type** where, in the extreme case, a single cavity or more usually, cavities lined with hymenium, arise within the basidiocarp, encountered in the Nidulariaceae (Fig. 14.40A).

(b) **The coralloid type,** in which a single cavity arises in the upper half of the basidiocarp. This is enclosed above by tramal tissue which may unite with the inner side of the peridium but, at the base of the cavity, branching masses of hyphae grow upwards and outwards to form 'coral-like' branches covered by a continuous hymenium (Fig. 14.40B).

(c) **The multipileate type** is one in which a number of tramal branches grow rapidly, reach the inner surface of the peridium and there spread out over it. Coral-like branches then grow from these main branches into the spaces into which the basidiocarp has been divided, exhibited by the genus *Clathrus* (Fig. 14.40C).

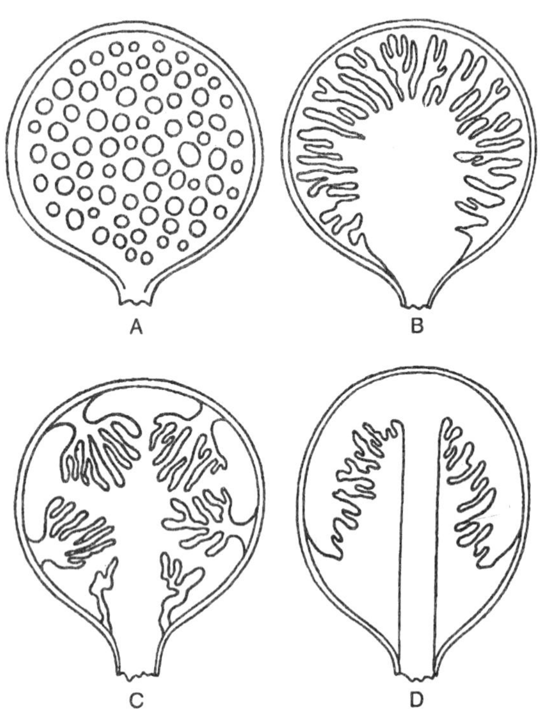

Fig. 14.40: Origin of glebal tissue in the basidiocarps of the Gasteromycetes. A-C. Cavities lined with hymenium. A. Small cavities, B. An irregular cavity arising in the upper half. C. Several irregular cavities, D. Hymenium on the underside of the pileus-like structure

(d) **The unipileate type** resembles the multipileate but here a single tramal branch develops as a column of tissue from base to apex and spreads out in a pileus-like manner, on the inner surface of the peridium. Hymenium develops on coralloid branches on the underside of the pileus-like structure, encountered in the genus *Phallus* (Fig. 14.40D).

In the Gasteromycetes the basidia may be clavate or cylindrical, but in majority of cases the basidia are ovoid or globose with short or long sterigmata. The sterigmata may also be lacking (Fig. 14.46B). The basidia are mostly four, occasionally 2 to 3 or 6 to 8-spored. Very rarely a basidium may produce up to 12 basidiospores. The basidiospores are placed symmetrically at the tips of the sterigmata and are not forcibly discharged as encountered among the Hymenomycetes. In case of basidium with

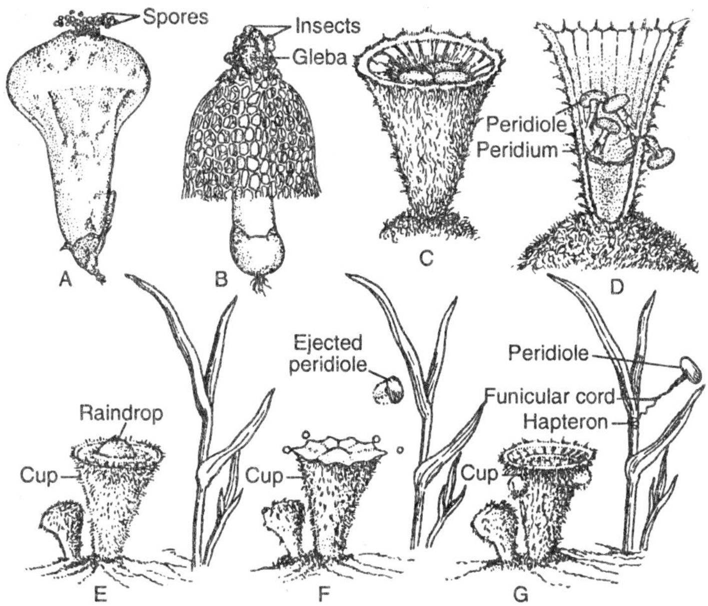

Fig. 14.41: Dissemination of spores among the Gasteromycetes. A. By wind in *Lycoperdon*. B. By insects in *Dictyophora*. C-G. By rain splash in *Cyathus*

long sterigma, either the entire sterigma or a portion of it remains attached to the spore. In the Gasteromycetes there is tetrapolar heterothallic condition. Sexual compatibility is governed by two pairs of factors Aa Bb situated on different chromosomes and segregating independently. Four types of basidiospres (AB, Ab, aB, ab) are produced which give rise to four types of primary monokaryotic mycelia with different dikarotizing possibilties. Basidiocarps are produced from dikaryotic mycelium.

Dissemination of spores among the Gasteromycetes usually takes place only upon the distingration of the peridium and through the action of external agents such as wind, flies, raindrops, etc. The spore mass of a large number of Gasteromycetes is dry and powdery and is adapted for dissemination by air currents, through a natural opening (**stoma**) of the peridium. The most common example is the genus *Lycoperdon* (Fig. 14.41A). The strong and penetrating odour in the glebal portion of the members of the order Phallales, commonly known as stinkhorns, attracts insects, which help dissemination of spores (Fig. 14.41B). In the genus *Sphaerobolus,* the entire gleba is discharged by a special hurling mechanism through the activities of the different layers of the peridium (Figs. 16.2A to C), whereas in the genus *Scleroderma,* the spores are dispersed by irregularly rupturing of the thick and firm peridium. The masses of glebal tissues (**peridioles**) (Fig. 14.41D) of the Nidulariales, known as the 'bird's nest fungi' are disseminated by rain splash. The peridioles on being displaced by raindrops adhere to nearby objects by the sticky bases of the funicular cords, **haptera** (sing. **hapteron**). Simultaneously with this, the funicular cords warp closely round the objects against which the peridioles strike (Figs. 14.41E to G).

CLASSIFICATION

The Gasteromycetes are subdivided into five orders as follows:

A. Gleba, more or less fleshy to cartilaginous, or gelatinous; original glebal structure maintained until maturity; fructifications hypogenous or epigeous

Order: **Hymenogastrales**

AA. Gleba powdery, slimy, or waxy at maturity; original glebal structure disorganized or fallen apart at maturity; peridium usually opening at maturity

B. Gleba not broken up into waxy bodies

C. Gleba slimy and odoriferous; exposed on a receptacle

Order: **Phallales**

D. Receptacle stalk-like in form — Family: **Phallaceae**

Genus: *Phallus*

D. Receptacle net-like — Family: **Clathraceae**

CC. Gleba powdery; not exposed on a receptacle

D. Hymenium well-developed

Order: **Lycoperdales**

E. Peridium two-layered; outer opening variously, inner remaining intact

Family: **Lycoperdaceae**

Genus: *Lycoperdon*

DD. Hymenium poorly-developed

Order: **Sclerodermatales**

BB. Gleba broken up into a number of waxy 'egg-like' bodies or peridioles

Order: **Nidulariales**

Family: **Nidulariaceae**

Genus: *Cyathus*

Order: **Phallales**

The Phallales are commonly called stinkhorns because of the strong foetid odour of decaying meat they possess in their exposed glebal tissue. These are some of the most beautiful fungi. They grow where there is ample organic matter such as humus, rotted tree stumps, or decaying sawdust and often in damp soil of bamboo grooves. The young basidiocarps arise as terminal swellings of rhizomorphs. They are spherical to ovoid masses known as **eggs** surrounded by a firm, somewhat leathery peridium. At first the young basidiocarps are completely or partially subterranean. They contain the fertile tissue or gleba and a specialized structure known as **receptacle** or **receptaculum** with maturity of the basidiocarp, the receptaculum elongates. The pressure caused by the elongation of the receptacle finally ruptures the peridium elevating the gleba above ground. The gleba at maturity deliquesces to a slimy evil-smelling mass with spores borne on a specialized receptacle. The receptacle may be a long, spongy pileate stipe, or a wide meshed lattice-like structure with or without a stipe. The evil-smelling slimy gleba filled with spores attracts flies which serve to carry the spores far and wide. Two families are recognized: the Phalaceae and Clathraceae. They are distinguished on the nature of receptacle.

Family: ***Phallaceae***

The members of the Phallaceae are distinguished by a stout, hollow stipe. At the top of the stipe is the pileus with viscid mass of spores. The pileus may or may not be adherent to the stipe. The stipe may also develop from below the pileus, a white skirt-like network, known as

indusium (pl. **indusia**). The Phallaceae include 10 genera, of which *Phallus, Mutinus* and *Dictyophora* are best known. In *Phallus*, the stipe elongates rupturing through the peridium. The apex of the stipe is covered with bell-shaped pileus bearing gleba (Fig. 14.42E). *Mutinus* is very similar in structure to *Phallus*, but the basidiocarp is smaller, and the pileus is adherent to the stipe (Fig. 14.42D), whereas *Dictyophora* has in addition to stipe and pileus, a beautiful skirt-like indusium originating between the pileus and the stipe and encircling the stipe (Fig. 14.42B). Both *Phallus* and *Dictyophora* have a bell-shaped pileus free from the stipe except at the top.

Genus: *Phallus*

Phallus, because of its strong odoriferous slimy glebal mass, is known as **stinkhorn** or **dead man's finger**. It grows on damp soil rich in humus. Most widely known species is *Phallus impudicus*. Besides the typical odour of the slimy glebal mass, *Phallus* is well-known for its attractive basidiocarp and the mechanism of dispersal of spores by insects.

The first sign of appearance of basidiocarps is the development of ellipsoid swellings on the mycelial strands which ultimately develop into rhizomorphs of the mature basidiocarps. Each of these swellings is of uniform compact weft of hyphae which ultimately differentiate into a dome-shaped central column rising vertically from the point of insertion. Gradually a bell-shaped layer of felted gelatinous tissue (gelatinous layer) enveloping the column, and a wite membrane (outer wall of the peridium) develop. While the whole developing basidiocarp increases in size becoming more narrowly ovoid in form, the outer wall of the peridium

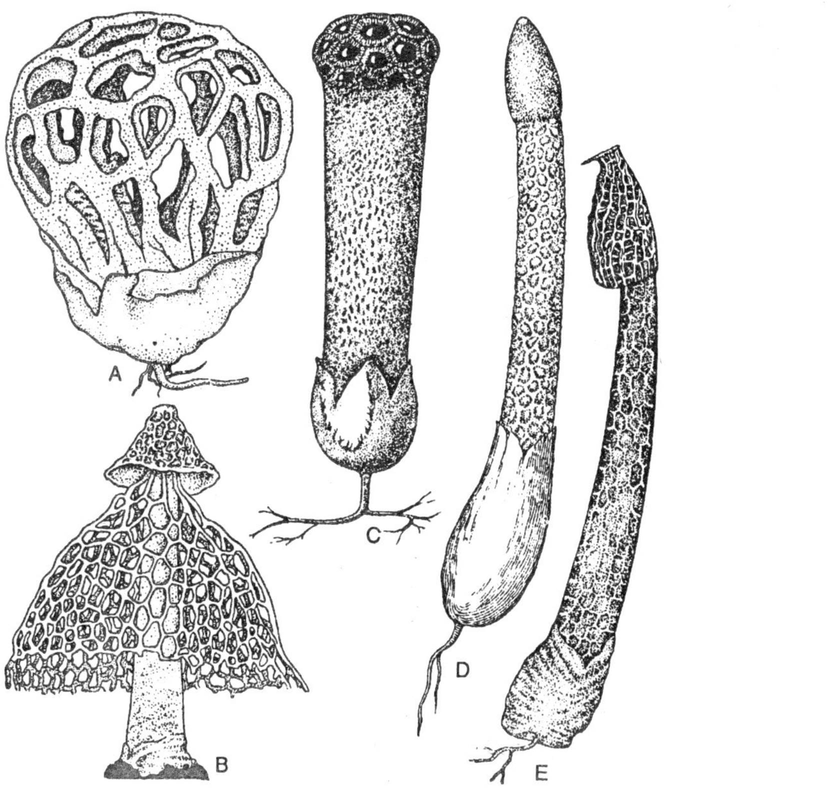

Fig. 14.42: Basidiocarps of the Phallales. A. *Clathrus*. B. *Dictyophora*. C. *Simblum*. D. *Mutinus*. E. *Phallus*

and the gelatinous layer grow in circumference and thickness; the central column acquires the form of a globular head supported on a cylindrical stipe. Ultimately gleba, pileus and the inner layer of the peridium are differentiated. The stipe is at first a very narrow, afterwards a more broadly fusiform body which runs through the longitudinal axis of the entire central column from its apex to its base. The pileus carrying the gleba becomes free. The young basidiocarp now looks like an egg (Fig. 14.43A) which may be semi-subterranean spherical to ovoid in shape surrounded by a tough peridium composed of two layers—the inner layer is the **endoperidium** and the outer **exoperidium.** The exoperidium is continuous at the base with the central column which forms the receptacle of hollow stipe filled with mucilage (Fig. 14.43B). The stipe and exoperidium are again continuous with the rhizomorphs produced from the mycelium embedded in the substratum. The gleba arises from the peripheral layer just within the endoperidium. The central part of the young basidiocarp is differentiated into a cylindrical hollow stipe and a folded honeycomb-like receptacle which bears the fertile part of the gleba (Fig. 14.43B). Within the young gleba are cavities lined by basidia bearing up to nine spores. As the glebal mass ripens the basidia disintegrate. When the spores are mature, the gleba breaks down forming a viscid, olive-green slimy mass. A sudden increase in length of the stipe thrusts the pileus and gleba against the peridium and bursting it raises the gleba into the air. The cup-like ruptured peridium remaining at the base of the stipe is the **volva** (Figs. 14.43C and D). Expansion of the stipe is accompanied by breakdown of glycogen and its conversion to sugar. The gleba when exposed to the air undergoes autodigestion, the characteristic foetid smell and sweet taste of the disintegrated gleba attract insects which aid in dissemiantion of the spores (Figs. 14.43E and F).

Some Indian Species: *Phallus impudicus* L. ex. Pers.; *P. rubicundus* (Bosc.) Fr.

Family: *Clathraceae*

The Clathraceae, very much similar to Phallaceae, attract attention from a distance by the strong odour emitted by the slimy glebal mass which lines the inner surface of the

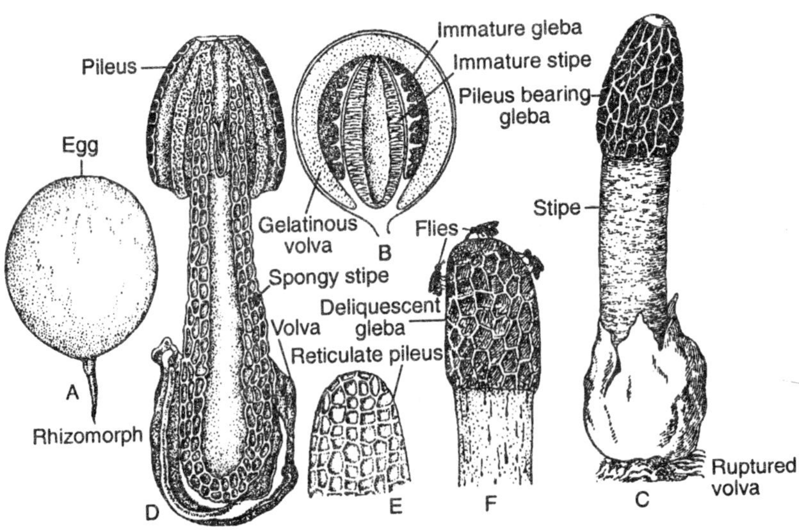

Fig. 14.43: *Phallus* sp. A. Egg bearing rhizomorph. B. Sectional view of an egg. C. Stipe with pileus exposed in the air, ruptured peridium (volva) at the base. D. Vertical section of exposed stipe and pileus, and volva. E. Portion of pileus. F. Deliquescent gleba with visiting flies

hollow receptacle. The family is characterized by the presence of an ovoid, hollow, lattice-like receptacle. The gleba occupies the entire interior surface of the interlocking network. The receptacle may or may not be borne on a stalk or stipe. In *Simblum* the lattice-like receptacle is held on a long stipe (Fig. 14.42C), whereas in *Clathrus* the receptacle arises directly from the ruptured volva without the developmet of any stipe (Fig. 14.42A).

Order: **Lycoperdales**

The Lycoperdales include fungi which are commonly called **puff-balls** and **earth-stars.** They are chiefly saprophytes, obtaining their food from decomposing plant parts, animal remains, and other organic materials in soil. The vegetative mycelium is buried in the substratum and sends up basidiocarps on its surface. The basidiocarps are usually more or less globose to depressed-globose, often tapering to a broad stalk-like base. They vary greatly in size, colour, form and texture. In long diameter they may range between 2 to 3 inches and even 3 to 4 feet. Many species of puff-balls are highly edible when young, and none is known to be poisonous. A mature basidiocarp consists of a compact peridium, enclosing a mass of loose, often more or less spongy mycelial tissue, the gleba. The gleba consists of very numerous basidia and intermingled sterile hyphae, the latter forming the **capillitium** (pl. **capillitia**). The peridium and capillitium are superficially similar to the peridium and capillitium of the Myxomycetes, being quite different in structure and origin. The basidiospores are symmetrically attached to the sterigmata and are merely released from the sterigmata at maturity, rather than being forcibly discharged. The glebal chambers in mature basidiocarp become transformed into a powdery mass. At maturity the peridium tends to rupture, leaving one or more openings into the gleba, especially from the top. Wind or other disturbance causes the basidiospores to be discharged into the air, but the capillitium tends to restrict the number of spores which are discharged at any one time.

Smith (1951) subdivided the order into six families, whereas Martin (1961) recognized only three families: Lycoperdaceae, Geastraceae and Arachniaceae.

Family: ***Lycoperdaceae***

The members of the Lycoperdaceae are the common puff-balls. They are characterized by the peridium of two layers, the outer opening variously and the inner remaining intact, and by the presence of capillitium in a powdery mass of spores.

Genus: *Lycoperdon*

Lycoperdon, the puff-ball, is a very cosmopolitan genus. It grows saprophytically on decaying organic matters in soil, often in grassland areas in forests, and even on the surface of the soil of flower pots. The basidiocarps arise either laterally or terminally on mycelial threads in which there are no clamp connections. They may be external on the soil surface from the very beginning of development or may be shallowly subterranean when young, becoming external at maturity. The basidiocarp initially arises on rhizomorph. It is at first homogeneous throughout. Gradually the peridium is differentiated being composed of radially oriented hyphae becoming pseudoparenchymatous. The whole interior of the basidiocarp constructed of a thick tissue gradually becomes differentiated into upper glebal area and lower sterile base (Figs. 14.44A and B). As the basidiocarp increases in size, cavities bounded by parallel hyphae appear in the glebal portion. These cavities ultimately develop into glebal chambers (Figs. 14.44C and D). The glebal chambers may be long and narrow labyrinthine, radiating and often with twisted slits which are lined by hymenium. The sterile base may project as a columella into the glebal area. Two kinds of hyphae are formed in the glebal area when still young. The slender delicate hyphae and their branches take part of the development of hymenium, whereas the stout ones make up the chief mass of glebal area. When spores begin to ripen, the elements of the hymenium become dissolved with copious effusion of water and ultimately large quantities of dry powdery spores are produced. The powdery spores are intermingled with capillitium of irregularly branched threads arising from stout hyphae. A mature basidiocarp has a well-marked sterile base. The basidiocarps may be globose to pyriform and with the elongation of the sterile basal portion each one may develop into a cylindrical to capitate structure possessing a definite stalk (Figs. 14.44A and B). The two layers of peridium are quite distinct. The outer peridial layer, the **exoperidium,** is spiny or verrucose, and being fragile disappears entirely, especially in wet weather when the basidiocarp is mature. It ruptures in various ways, scaling off in granules or larger pieces. The inner peridial layer, the **endoperidium,** opens apically by one or more small pores (ostioles) through which the spores escape (Fig. 14.44E). The endoperidium, may also break up in pieces, but more often forms apical pore. As wind or firmer objects strike the basidiocarp the spores are puffed out through one or more pores, suggesting the name puffball. The basidia are short, swollen and ovoid each bearing four long sterigmata on which four basidiospores are developed. The basidiospores may also be less than four (Fig. 14.44D). They are globose with variously ornamented spore wall. The basidiospores fall off the basidium, each taking a portion of its sterigma with it. The basidia eventually disintegrate.

Some Indian Species: *Lycoperdon alveolatum* Lev.; L. *elongatum* Berk.; L. *microspermum* Berk.; L. *mundkuri* Ahmad.; L. *perlatum* Pers.

Order: **Nidulariales**

The Nidulariales are commonly known as 'bird's nest fungi', also known as fungi with 'Worchestershire Corn bell' or 'fairy purse'. They are so-called because in them the basidiocarp is cup-shaped and contains several tiny lenticular bodies resembling a bird's nest with eggs. These lenticular bodies are the peridioles which are formed by the breaking of gleba. The cup is the peridium. In this order the large basidiocarps are external, not subterranean. The Nidulariales comprise two families: the

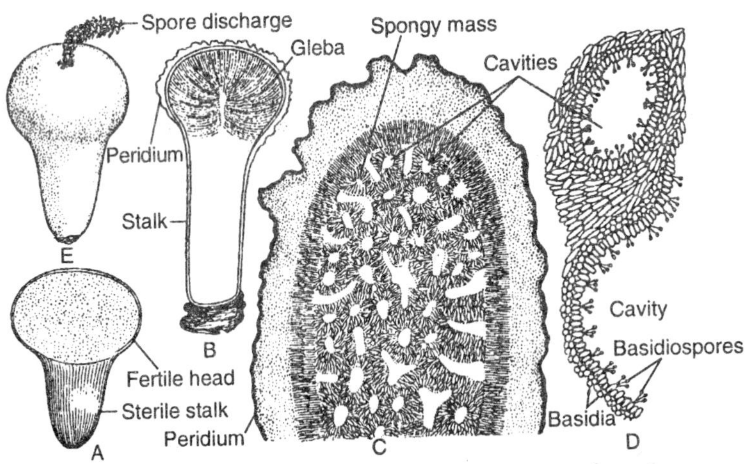

Fig. 14.44: *Lycoperdon* sp. A. Basidiocarp. B. Sectional view of a basidiocarp. C. Closer view of gleba in section. D. Portion of glebal cavities showing basidia and basidiospores. E. Spore discharge from a basidiocarp

Nidulariaceae and Sphaerobolaceae. In the family Nidulariaceae several peridioles are formed in each cup-shaped basidiocarp, whereas in the Sphaerobolaceae the number of peridioles has been reduced to one. Here the entire gleba is transformed into a single peridiole (Figs. 16.2A-C).

Family: *Nidulariaceae*

The members of this family are characterized by the presence of several peridioles. The epigeous basidiocarps arising from mycelial strands become differentiated into a cup-shaped peridium and numerous peridioles borne inside the cup (Figs. 14.45A and B). Each peridiole is a segment of the gleba. It becomes separated from other peridioles by hyphal gelatinization but remains attached to the inner surface of the peridium by a long stalk known as **funiculus.** The lenticular body of the peridiole has a thick-wall—**tunica** which encloses basidia with basidiospores and mass of coiled and twisted hyphae. The number of basidiospores may be four or more and may or may not be borne on sterigmata. The peridioles are completely enclosed by the peridium and a thin membranous layer, the **epiphragm** which at maturity breaks exposing the peridioles. The peridioles are disseminated by splashing of raindrops. Brodie (1948-1951) working on a number of species of the Nidulariacea has shown that all species he worked out were tetrapolar. The principal genera are: *Nidula, Nidularia, Cyathus,* and *Crucibulum.*

Cyathus is similar to *Crucibulum* in many respects. But the basidiocarps of *Cyathus* are larger in height than those of *Crucibulum.* The peridium is differentiated into three layers in *Cyathus,* but in *Crucibulum* it is two-layered. The peridioles of *Cyathus* are dark-grey to black, whereas in *Crucibulum* the peridioles are white. Both these two genera have basidiocarps with epiphragm and each peridiole is attached to the peridium by a funiculus.

In *Nidula* epiphragm is present, but the peridiole is not provided with a funiculus, whereas in *Nidularia* both epiphragm and funiculus are lacking.

Genus: *Cyathus*

Both *Cyathus* and *Crucibulum* are cosmopolitan. They exhibit tetrapolar heterothallism. The genus *Cyathus* is rather common in bamboo grove and similar other damp shady places. The basidiocarps often arise epigeously on rotton wood as small knobs of tissue.

The basidiocarp of *Cyathus* is deep inverted bell-shaped with a flattened top (Fig. 14.45A) and is provided with a large solid mass of hyphae around is base—the **emplacement** (Fig. 14.45B). Its cup-shaped peridium is differentiated into three layers: a middle pseudoparenchymatous layer lies in between the exo- and endoperidium. The exoperidium is composed of loosely woven hyphae. It bears on its surface long hairs, the **setae.** But the endoperidium is of more or less gelatinized hyphae. The surface of the peridium is with **striae** (Fig. 14.45B). The base of the cup is composed of solid mass of hyphae—the **stipe** (Fig. 14.45B). The first sign of development of the fruit body is the appearance of brown mycelial strands at the soil surface, on which knots of hyphae differentiate. Soon develop at the periphery stout hairs with tooth-like branches and brown membrane, which cover the surface of the knots as a brown felt. Next step is the separation of the peridium from the rest of hyphae which are again separated by gelatinization into rounded portions. These are the primordia of the chambers of the gleba, or primordia of peridioles. During further growth gelatinization goes on. The apex of the young basidiocarp is at first clothed with brown hairs; as it increases in size with the general growth the hairs are pushed aside. The summit of the basidiocarp is then covered by a thin white membrane, the **epiphragm** which eventually becomes gelatinized and is torn apart when the basidiocarp is mature. The originally rounded

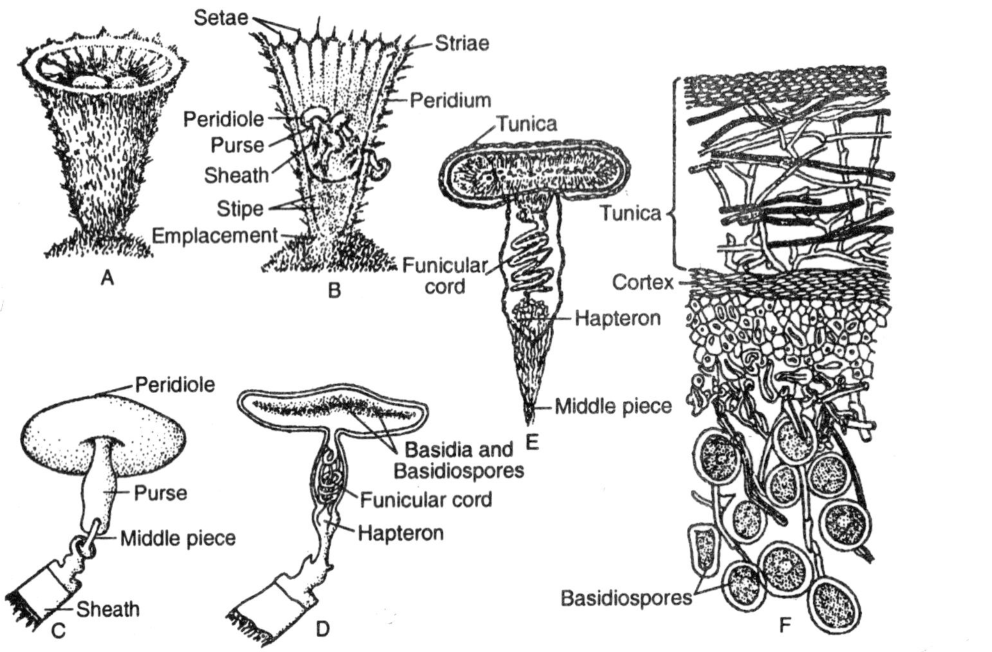

Fig. 14.45: *Cyathus* sp. A. An entire basidiocarp. B. Sectional view of a basidiocarp showing structural details. C. An entire peridiole with purse, middle piece and sheath. D. A peridiole in section showing basidia, basidiospores, funicular cord and hapteron. E. Sectional view of a peridiole showing details of tunica, funicular cord, hapteron and middle piece. F. Sectional view of a portion of a peridiole showing details of tunica, cortex, basidiospores mixed with nurse hyphae

portions (the primordia of peridioles) become lenticular, whereby thick-walled, **tunica** is formed. While at the flattened side of the peridiole there is a median depression (Fig. 14.45C). The centre of this depression is occupied by tuft of hyphae. These hyphae continue their course closely united and parallel to one another as a smooth somewhat sinuous strand, the funiculus or **purse** (Fig. 14.45C). The funiculus expands greatly in presence of moisture producing **funicular cord** and the base of the funicular cord is the **hapteron** (pl. **haptera**) (Figs. 14.45D and E) which is quite sticky and adheres easily to solid surface. Before expansion, the funicular cord and hapteron are enclosed in bag-like structure. Some of the hyphae of the bag-like structure form a structure, the **middle piece** (Fig. 14.45C). The middle piece again is attached to a platform of hyphae, the **sheath** (Fig. 14.45C) that is directly produced on the inner surface of the peridium (Fig. 14.45B).

A mature periodiole is hollow. Each peridiole is surrounded by a **tunica** made up of loosely intertwined hyphae and a dark thick-walled **cortex**, in turn lined by a mass of very thick-walled hyaline cells (Fig. 14.45F). The inner part of the periodiole is made up of thin-walled hyphae between which basidia develop. The basidia project inwards from the wall composed of thickened layer of the trama (Fig. 14.46A). Basidium produces 4 to 8 basidiospores. The basidiospores are not sterigmate (Fig. 14.46B). The basidia collapse while the spores are not yet mature or fully grown. The spores become covered by **nurse hyphae** arising from the inner wall of the periodiole (Fig. 14.45F). They are then nourished by the nurse hyphae and increase in volume and wall thickens. Nurse hyphae gradually disappear.

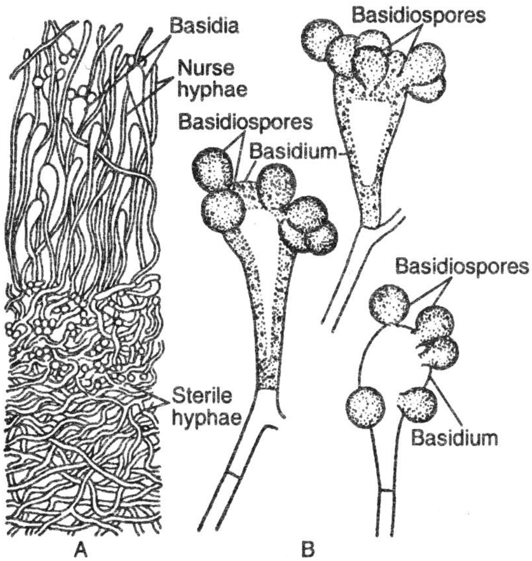

Fig. 14.46: *Cyathus* sp. A. Portion of a section through peridiole. B. Basidia with basidiospores, sterigmata absent and basidiospores more than four

Simultaneously with this, gelatinization of the tissues lining the walls of the periodiole takes place. The thick-walled mature spores remain immersed in a gelatinous matrix. The spores are freed by the decay of the peridiole wall, germinate to form mycelia. The peridioles retain viability for many years.

Both *Cyathus* and *Crucibulum* show tetrapolar heterothallism with relatively few alleles (generally not more than 15) at each locus. In *Cyathus striatus* sexual compatibility is governed by 2 types of sex factors Aa Bb based on different types of chromosomes and segregation. Four types of basidiospores are developed which on germination give rise to four types of primary mycelia (Ab, AB, aB, ab). Diploidization is effected by two compatible mycelia (AB × ab and Ab × aB), where no factor is common.

The dissemination of basidiospores in *Cyathus* was not understood for many years. Martin (1927) was the first to suggest that falling raindrops can spatter the peridioles from the cup of the basidiocarp and thus eventually effect the dissemination of spores. Shea (1941) noted the hanging of peridioles from the lower surface of the leaves of some angiospermous plants. He pointed out that the peridioles are shot into the air and become attached to whatever objects are present near the cups. But he could not explain the exact mechanism how the periodioles are shot from the cups. Dodge (1941) reported the presence of peridioles on the upper surfaces of leaves 10 to 15 ft above the ground and concluded that the peridioles must be discharged with considerable violence. Brodie (1951) clearly explained the mechanism of peridiole dissemination in *Cyathus striatus.* According to him, raindrop is responsible for the dispersal of the peridioles and Brodie described the phenomenon as 'splash-cup' mechanism. Brodie observed that a large raindrop having a diameter of about 4 mm with a velocity of about 6 metres per second falls into the basidiocarp (Fig. 14.41E). The raindrop creates a strong thrust upward along the inclined side of the cup and the peridioles are forcibly ejected (Fig. 14.41F). When a peridiole is being splashed from a basidiocarp, the force jerks the funiculus, and the purse is torn open at its lower end where it is attached to the middle piece. The rupture of the purse causes almost instantaneous expansion of the special mass of hyphae that had been coiled up within the lower part of the purse—the funicular cord with an attachment organ—hapteron. The sticky hapteron at the base of the funicular cord helps to attach the peridiole to surrounding vegetation and the momentum of the periodiole may cause the funicular cord to wrap around objects (Figs. 14.41G). The peridiole can fly to the height of 13 to 15 ft in the air. If the flying peridiole with funicular cord and hapteron strikes an object, the hapteron will adhere to it (Fig. 14.41G) and the peridiole will dangle down and the funicular cord will tie around the object with the momentum gathered, or the peridiole may glide along the surface of a leaf. The solid mass of hyphae around the base of the basidiocarp acts as an emplacement and prevents the cup from being knocked over by rain.

The peridioles of *C. stercoreus* are eaten by herbivorous animals and the basidiospores on release from the peridiole are stimulated to germinate by incubation at about body temperature. The germinated basidiospores pass out along with faeces on which mycelium is produced from which basidiocarp develops. But whether animals play a significant role in the dispersal of peridioles of other bird's nest fungi is uncertain.

Some Indian Species: *Cyathus coleusoi* Berk.; *C. hookeri* Berk.; *C. intermedius* (Mont.) Tul.; *C. limbatus* Tul.; *C. microsporus* Tul.

Chapter 15

Subdivision of Deuteromycotina

INTRODUCTION

Present day mycologists prefer not to use the prefix form.

This is a heterogenous group of many types of fungi whose normal sexual stage (also called perfect stage) is absent or unknown. But in many of them heterokaryosis and parasexuality have been reported. For detail of these aspects see Chapter 16. Due to lack of information about the normal sexual stage these fungi cannot be assigned to the three main subdivisions of fungi which have already been described. Since these fungi apparently lack sexual stage, they are called Fungi Imperfecti or Deuteromycetes and have been taxonomically categorized as **Subdivision Deuteromycotina** adding the prefix **form** to the taxonomic category subdivision and categories below the rank. The use of the prefix form indicates an entirely artificial and temporary grouping of these fungi. This taxonomic category is entirely artificial and has beeen created for the temporary grouping of these fungi pending discovery of perfect stage that they belong either to the Ascomycotina or the Basidiomycotina. Many of these fungi may be reduced forms and might have lost the power to reproduce sexually. But, in general, it has been seen that a large number of fungi whose sexual stage was not worked out and hence are grouped for a long time under the Subdivision Deuteromycotina. In course of time and investigation perfect stages of some of them have been worked out. When the perfect stage of a fungus previously referred to the Deuteromycetes is discovered, depending on the nature of spore produced by karyogamy and meiosis, the organism is transferred to the appropriate subdivision of fungi. According to the rules of nomenclature, a name based on the perfect stage always takes precedence over one based on an imperfect stage, regardless of which name is the older. In the phylogenetic classification scheme based upon the oospore, zygospore, ascospore, and basidiospore, there is no space for the fungi solely reproducing asexually. They have been given the name Fungi Imperfecti subsequently as Deuteromycotina. By this was formerly meant that these forms were 'incomplete' in their life cycles, lacking the sexual or perfect stage. It also meant that they were incompletely known but that the future would bring us more complete data regarding the sexual stage.

It is very much established that these imperfect stages are not **autonomic** (i.e., independent) species, but are connected with a sexual stage, whether this stage is known or not. It must not be overlooked, however, that some of them may lack the perfect stage, i.e., the fungus always lives in its asexual stage, the reason being that they never produced spores by sexual method or that this power has been lost. In the broadest sense, the Fungi Imperfecti include all the existing fungi which reproduce solely by asexual means and even those of the fungi which produce no spores at all. These organisms are placed in so-called form-genera with form-species, which are based on certain morphological characters. But their true relationships being not known, hence they do not represent natural groups of fungi. The form-genera are included under form families which are again placed under form-orders of which the highest taxonomic category is form-class.

A large proportion of the fungi which have been removed from the Deuteromycotina on discovery of the perfect stage prove to be Ascomycotina but some, on the contrary, are Basidiomycotina.

Even though the zygospore or oospore stages of certain lower fungi have never been observed, these species are commonly placed among the Basidiomycotina rather than among the Deuteromycetes. This is because of the aseptate nature of hyphae. Besides these, 'many rusts are known only in the aecial or uredial stage. They are classified under an artificial group 'Uredinales Imperfecti' in whch are included the form-genera: *Aecidium, Peridermium, Caeoma, Roestelia,* and *Uredo.*

The Deuteromycetes occur as saprophytes and parasites. The parasitic Deuteromycetes cause diseases of plants, animals and human beings.

ECONOMIC IMPORTANCE

The Deuteromycotina are of great economic importance since, in addition to causing plant diseases, they cause several diseases of human beings. Again some are useful in controlling some of the soil-borne plant diseases.

1. **Plant pathogens:** A number of species is parasitic on higher plants, causing serious diseases of leaves and fruits. Special mention may be made of, *Alternaria solani* causes early blight of potatoes, *Helminthosporium oryzae* causes brown spot of rice, *Pyricularia oryzae* causes blast of rice, *Septoria tritici* causes leaf blotch of wheat, *Colletotrichum gloeosporioides* causes anthracnose of mango, *Fusarium oxysporum* var. *cubense* causes panama wilt of banana, *Fusarium udum* causes wilt of pigeon pea, and *Cercospora personata* causes leaf spot or tikka disease of ground-nut.

2. **Human pathogenus:** Most of the mycoses are induced by the Deuteromycetes. Fungi causes diseases of the skin of human beings, animals, or both are known as the **dermatophytes** or **ringworm** fungi and the diseases are **dermatomycoses** (sing. **dermatomycosis**). These fungi cause ringworm of the scalp (found chiefly in children) or other hairy parts of the smooth skin, of the nails and of the hands and feet. Ringworm of the feet is known popularly as athlete's foot. *Trichophyton gypseum* and *T. purpurium* cause athlete's foot. Another disease, called **moniliasis,** chiefly of the fingers is caused by *Candida albicans.* Species of *Candida* also cause a throat and mouth disease called thrush, pulmonary infections, infection of the mucous membranes of the genital organs, and various other diseases.

3. **Biological control of soil-borne diseases:** Several of the Deuteromycotina are able to trap nematodes (microscopic roundworms that often infest the roots of crop plants) by forming hyphal rings which constrict about the nematode when stimulated by contact. Special

mention may be made of *Arthrobotrys oligospora, Dactylella cionopaga* whose mycelia form sticky processes, loops or branches to which nematodes adhere, other hyphae then invade and digest them. Thereby these Deuteromycotina reduce the disease incidence due to nematode of the crop plants.

SOMATIC STRUCTURES

The thallus of the Deuteromycotina is composed of well-developed mycelium of septate branced hyphae, exception being false yeasts (*Cryptococcus, Torulopsis,* and *Brettanomyces*) which possess yeast-like somatic body. The hyphae may be simple and short, or long and much branched. The hyphal cells are usually multinucleate. They possess perforated septa. Through these perforations cytoplasmic streaming and nuclear migration take place from cell to cell. The nature of septa is very much similar to that of the Ascomycotina which suggests affinities between the Deuteromycotina and the Ascomycotina.

REPRODUCTION

Deuteromycotina are conidial forms, i.e., produce conidia. Besides conidia, some of the Deuteromycetes may produce different other spores. But there are also numerous forms which are known to produce merely sterile mycelia and no conidia at all. The sterile mycelia may or may not make sclerotia. These sterile forms are often worst of plant pathogens.

Conidial forms. The conidial forms produce conidia on conidiophores arising directly from the soamtic hyphae which may be hyaline or bright-coloured. The somatic hyphae may be loose (Figs. 15.1A-D); separate; innate or not; or closely aggregated to form sporogenous structures, such as, **acervulus, sporodochium**

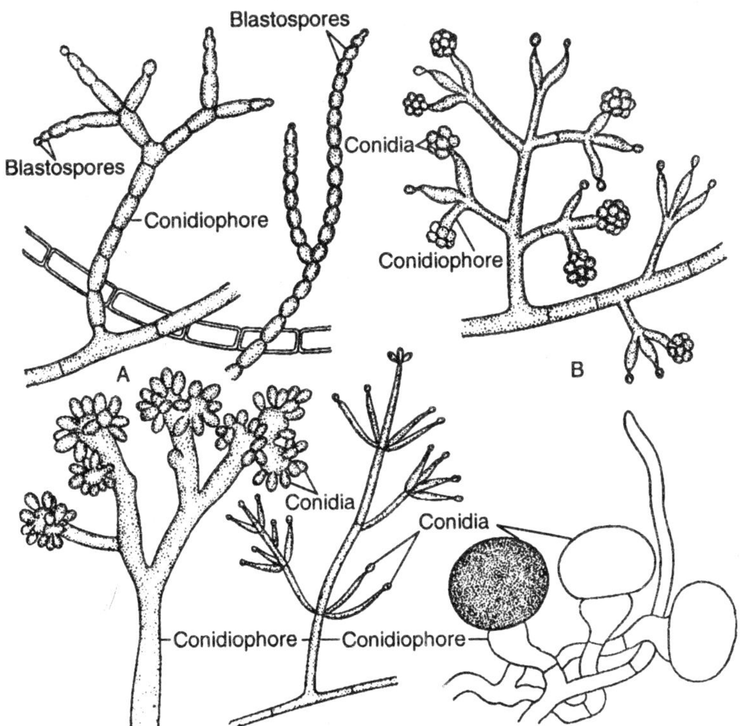

Fig. 15.1: Different types of conidia and conidiophores in the genera of the Deuteromycetea: A. *Monilia*. B. *Trichoderma*. C. *Botrytis*. D. *Verticillium*. E. *Nigrospora*

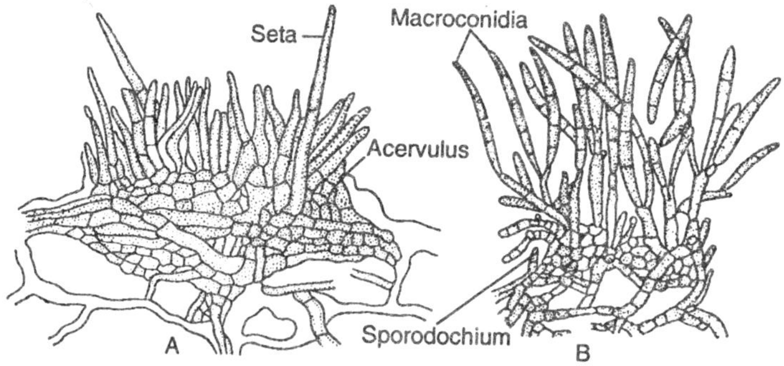

Fig. 15.2: Sporogenous structures of the Deuteromycetes. A. Acervulus of *Colletotrichum* sp. B. Sporodochium of *Fusarium* sp.

and **synnema.** Besides these, conidia may be produced on more or less loose cotton hyphae. They are often designated as **Hyphomycetes.**

(i) **Loose hyphae producing forms:** In the loose hyphal forms, conidiophores grow for some distance above their supporting substratum, so that the terminal parts at least stand out at separate threads. In majority of cases the conidia are borne externally on the conodiophores which may (Fig. 15.4A) or may not be clustered (Fig. 15.4D-F). But there are also forms where the conidia are produced in the centre of the conidiophore and pushed out successively from an opening at the apex. These are known as endoconidia.

(ii) **Acervulus and sporodochium producing forms:** These Deuteromycetes produce small or large cushion-like masses of hyphae with a bed of short conidiophores from which conidia are produced. These are the **acervuli** (Fig. 15.2A). The acervuli may be subepidermal, subcortical or superficial and may or may not be provided with hairs (**setae**) of various kinds. The subepidermal and subcortical acervuli are erumpent. Often an acervulus may develop a compact mass of conidiophores on a

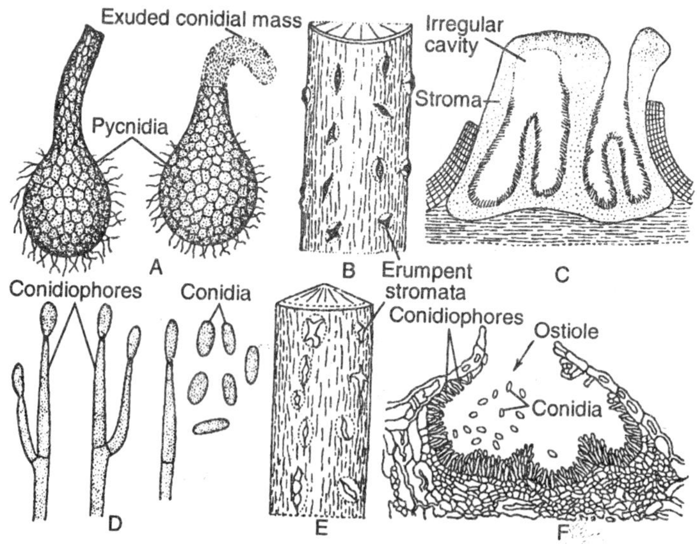

Fig. 15.3: Various types of pycnidia. A. Beaked. B and E. Stromatic pycnidia in erumpent stromata. C. Labyrinthiform. D. Conidiophores producing conidia. F. Unilocular pycnidium with ostiole

well-marked basal stroma or mass of hyphae producing a structure known as **sporodochium** (Fig. 15.2B). It is sometimes difficult to distinguish between an acervulus and a sporodochium.

(iii) Pycnidia producing forms: These are the Deuteromycetes which produce pycnidia. The pycnidia are more or less globose or flask-shaped, hollow sporocarps in which conidia are borne at the tips of the conidiophores arising from the cells lining the inner wall of the cavity. They may be of various colours though most commonly black, and may be superficial or embedded in the substratum being formed directly by the loose mycelium or may be definitely stromatic (Figs. 15.3B and E). The pycnidium is usually provided with an opening (**ostiole**) through which conidia escape (Fig. 15.3F). The ostiole may be narrow or wanting, or it may be very large, round, irregular. The pycnidium is often with (Fig. 15.3A) or without a beak (**rostrum**). The pycnidial wall may be extremely delicate to very thick pseudoparenchymatous. It may be smooth or provided with hairs, spines, etc. The pycnidium may be uniloculate, **simple** (Fig. 15.3F) or **labyrinthiform** (Fig. 15.3C). The conidiophores borne in the pycnidia are generally very short, in some cases almost absent, whereas in others the conidiophores are quite long and branched (Fig. 15.3D).

NATURE OF CONIDIA AND DIFFERENT OTHER SPORES

The Deuteromycotina possess several different spore forms of which mention may be made of conidia, blastospores and chlamydospores.

Conidia: Conidia vary greatly in size, shape, number of septations, wall thickness, colour, etc., as do the cells of the conidiophores

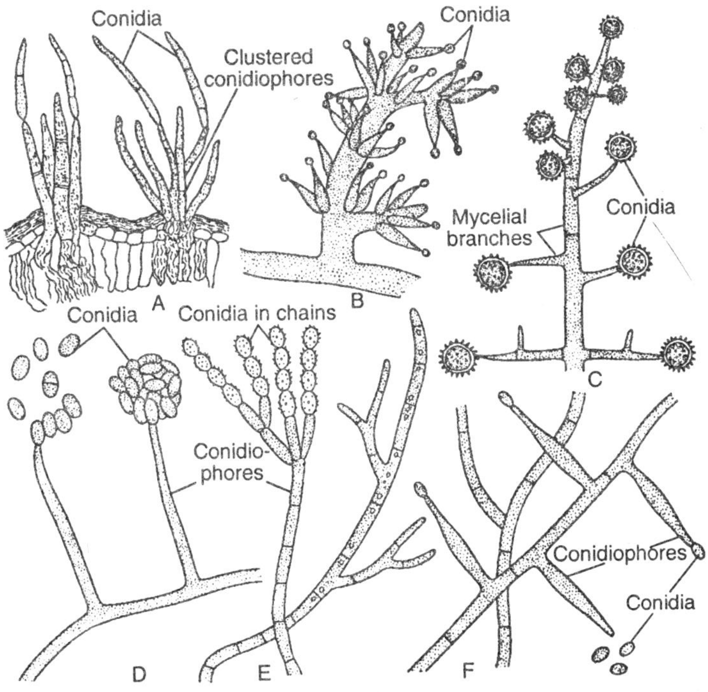

Fig. 15.4: Different types of conidia and conidiophores. A. Septate conidia produced on erect clustered conidiophores that protrude from the leaf tissue. B-C. Conidia produced on branches of the mycelium. D. Conidia produced successively at the tip of the conidiophore are collected in a slime drop. E. Conidia developed in chains. F. Non-septate conidium produced one on each conidiophore

which bear them. Such differences serve to separate the numerous genera and species among the Deuteromycotina. In general, conidia are produced in chains and are then said to be **catenulate.** The youngest conidium is always the latest one formed and the oldest conidium is always on the far, or distal end of the chain (Fig. 15.4E). But cases of development of solitary conidia are also not infrequent (Figs. 15.4B, C and F). A superficially similar but quite different type of chain formation is exhibited when conidia are not cut off from the conidiophores, but are developed from them by a process of budding. The conidia may form globular clusters at the ends of the conidiophores, or the conidia become surrounded by a mass of slime, which binds the conidia into a rounded mass giving the fungi a very characteristic appearance (Fig. 15.4D).

Some form-genera develop two- or more-celled conidia. The mode of septation may be only transverse or both transverse and longitudinal. Correspondingly the conidia may be thick- or thin-walled. Again some of the Deuteromycetes produce more than one type of conidia in the same thallus. Conidia which are small spherical, oval, slightly elongated are usually single-celled or with one-septate, are called **microconidia,** whereas conidia which are large, usually multi-celled are called **macroconidia.** The microconidia may be spherical, elliptical pyriform or clavate. The macroconidia may be divided into two or more cells by transverse septations and may appear fusiform or long crescent-shaped. Conidia may be thick- or thin-walled having surface smooth or warty and may be extremely variable in shape from globose, oval, cylindrical, clavate, filamentous thread-like, filamentous spirally coiled, to stellate or irregular in shape. They may be hyaline to bright- or dark-coloured. Conidia are also named in accordance with their shape or structure, nature of septation and number of cells in a conidium. There may be: one-celled conidia, **amerospores** (Fig. 15.5A); two-celled conidia **didymospores** (Fig. 15.5D); conidia having two or more transverse septa, **phragmospores** (Fig. 15.5B); conidia possessing septa both lengthways and crossways, **dictyospores** (Fig. 15.5C); long, thread- or worm-like conidia are **scolecospores** (Fig. 15.5G); conidia being rolled-up or corkscrew-like cylindrical generally septate are **helicospores** (Fig. 15.5E); and **staurospores** (Fig. 15.5F), when conidia are star-like having three or more arms.

Other spores: Besides conidia other types of spores are frequently encountered in the Deuteromycetes. These spores are often referred to as **thallospores.** Thallospores are asexual spores formed by transformation of existing cells of the thallus, and are set free by decay or disarticulation of the parent hyphae, not by a process of active cutting-off (abstriction). This is clearly a type of natural fragmentation of the mycelium. Thallospores may be terminal or intercalary. There are three princi-

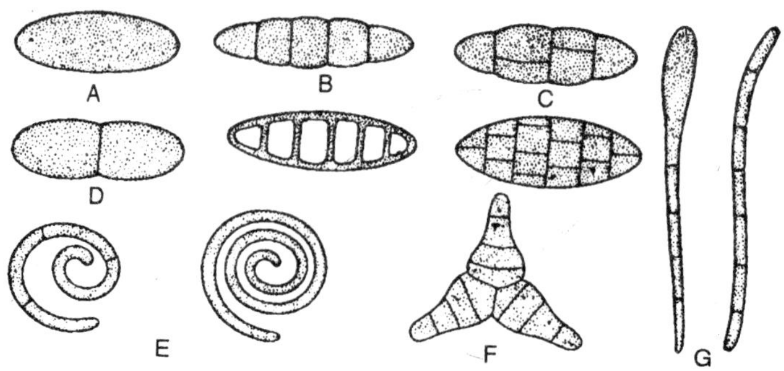

Fig. 15.5: Grouping of conidia on the basis of colour, shape and septation

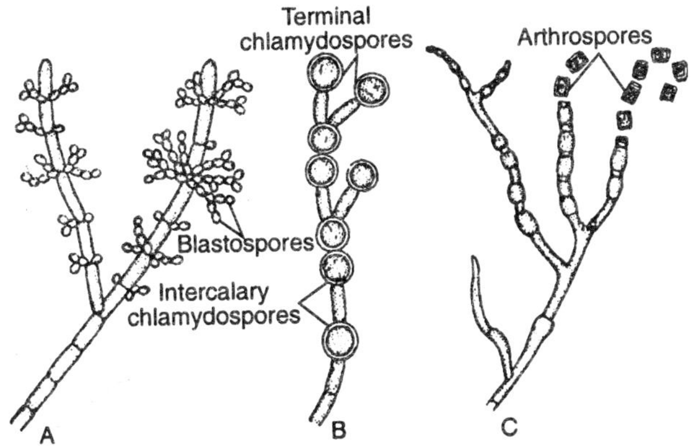

Fig. 15.6: Various types of spores. A. Blastospores. B. Chlamydospores. C. Arthrospores

pal types of thallspores: **blastospores, chlamydospores** and **arthrospores.**

Blastospores: These are round or oval cells that are budded off from the parent cells, enlarge and reproduce themselves, likewise by budding (Fig. 15.6A). They are rather common in unicellular form-genera. There are also mycelial forms which produce blastospores.

Chlamydospores: These are thick-walled, non-deciduous, intercalary (**intercalary chlamydospores**) or terminal (**terminal chlamydospores**) spores made by the rounding up of cells of the somatic hyphae (Fig. 15.6B). In some cases they may also be formed on the side of the hyphae and are then said to be **lateral chlamydospores.** These are the resting spores that remain viable after the remainder of the somatic hyphae has died and disintegrated.

Arthrospores: These are the spores that are formed by segmentation of the somatic hyphae, resulting in the cutting off of rectangular, somewhat thick-walled separate cells (Fig. 15.6C).

CLASSIFICATION

This subdivision is often designated as Deuteromycotina and also as Form-Division, and was formerly called the Imperfect Fungi. The taxon has been as a Form-Subdivision because of its artificial composition. These large number of fungi have been assembled together for the sake of convenience. This group lacks any phylogenetic origin or relationship. These fungi are combined in a single group because they apparently lack a sexual stage and therefore have no convenient place in the remainder of the fungus classification system which is principally based on the mode of sexual reproduction.

The Deuteromycotina are classified mainly by the types of spores produced, they present many difficulties from the point of view of classification. A pure culture of a single fungus may contain more than one spore type, which would justify its placement in at least two different genera. Since the conception of species is here most loose the arrangement of form-species is purely tentative, the names are to serve merely as handles for convenience in treating these imperfect fungi.

Several systems of classification of the Deuteromycetes have been proposed, of which the oldest one is that of Saccardo (1886) published in **Sylloge Fungorum.** Saccardo's classification has been modified by Ainsworth (1966) and outlined in the following manner:

A. Reproduction by thallospores or conidia
  B. Thalospores or conidia not borne in any form of pycnidium or acervulus **Hymenomycetes**
  BB. Thallospores or conidia not borne in a pycnidium or acervulus **Coelomycetes**
AA. No reproductive structures known **Agonomycetes** or **Mycelia Sterilia**

The **Hyphomycetes** include only one form-order: Moniliales (or Hyphales) which may be subdivided into the following form-families:

A. Conidiophores discrete
  B. Mycelium and conidia hyaline or light-coloured **Moniliaceae**
  BB. Mycelium and/or conidia dark-coloured **Dematiaceae**
AA. Conidiophores synnemata **Stilbellaceae**
AAA. Conidiophores in sporodochia **Tuberculariaceae**

The **Coelomycetes** may be subdivided in the following orders:

A. Pycnidial fungi **Sphaeropsidales**
AA. Acervular fungi **Melanconiales**

The form-order Sphaeropsidales includes the following form-families:

Pycnidia dark, leathery or membranous **Sphaeropsidaceae**
Pycnidia light or brightly coloured, waxy or flexy **Zythiaceae**
Pycnidia flattened, shield-shaped or elongated **Leptostromataceae**
Pycnidia deeply cupulate when mature **Excipulaceae**

The order Melanconiales has one form-family **Melanconiaceae.**

The **Agonomycetes** or **Mycelia Sterilia** include only one order: Agonomycetales having form-genera: *Rhizoctonia, Sclerotium, Racedium, Ozenium,* and *Rhizomorpha.*

But Saccardo in his original work divided the families into a number of sections taking colour, shape, and septation of conidia as the basis for such groupings. Following are the sections:

Amerosporae: conidia one-celled, globose, oval or short-cylindrical
  Allantosporae: conidia cylindrical, curved (allantoid) hyaline to pale.
  Hyalosporae: conidia hyaline or bright.
  Phaeosporae: conidia dark-coloured.
Didymosporae: conidia two-celled, ovate or elongate.
  Hyalodidymae: conidia hyaline or bright.
  Phaeodidymae: conidia dark-coloured.
Phragmosporae: conidia oblong to fusoid, transversely septate into three or more cells.
  Hyalophragmiae: conidia hyaline or bright.
  Phaeophragmiae: conidia dark-coloured.
Dictyosporae: conidia ovate to elongate, divided by transverse and longitudinal septa.
  Hyalodictyae: conidia hyaline or bright.
  Phaeodictyae: conidia dark-coloured.
Scolecosporae: conidia filamentous, thread-like, one- to several-celled, hyaline or dark.

Helicosporae: conidia cylindric, spirally coiled, one- to several-celled, hyaline or dark.

Staurosporae: conidia stellate or irregular, one- to several-celled, hyaline or dark.

The Deuteromycetes are also subdivided in the following manner into four form-orders based on the presence or absence of spores and structures from which or in which the spores are produced.

A. Conidia and other spores present.

B. Reproduction by conidia formed within pycnidia which open by a pore or a long sit, or finally become disk-like.

Order: **Sphaeropsidales**

BB. Conidia, when formed, not in pycnidia; conidia in acervuli which are immersed in substratum or erumpent, black or light-coloured, accompanied by setae or not.

Order: **Melanconiales**

BBB. Conidia not borne within a pycnidium or acervulus; conidia although produced sometimes as buds are usually found on conidiophores; besides conidia, other spores developed from the cells of the somatic hyphae or yeast-like forms.

Order: **Moniliales**

AA. Conidia and others spores lacking, sterile mycelium may or may not form sclerotia or masses of fungal cells present.

Order: **Mycelia Sterilia**

Order: **Sphaeropsidales**

The Sphaeropsidales are predominantly leaf-spotting fungi though many of them grow on fruit or stems causing blight, rot, cankers, etc. A great number are saprophytes or parasitic on tissues of weak vitality, but not a few are active parasites. The pycnidia chracteristic of this group are with or without ostioles or other openings and are separate or joined by stromatic tissue. The conidia produced in the pycnidia are slime spores. A simplified key to the families and genus is outlined below.

A. Pycnidia globose, conic or lenticular; membranous, carbonous or coriaceous; stromatic or non-stromatic; black.

Family: **Sphaeropsidaceae**

B. Pycnidia non-stromatic without setae; conidia filiform, one- to several-celled, several times longer than wide, hyaline.

Genus: *Septoria*

AA. Pycnidia very similar to the Sphaeropsidaceae but light-coloured, soft or waxy.

Family: **Zythiaceae**

AAA. Pycnidia more or less irregular or shield-shaped to elongate, flattened, black.

Family: **Leptostromataceae**

AAAA. Pycnidia more or less deeply cup-shaped, black.

Family: **Excipulaceae**

Order: **Melanconiales**

The Melanconiales include a fairly large number of forms which are plant pathogens. The common name 'anthracnose' is applied to any disease caused by a member of this group. The mycelium is internal remaining within the host or substratum. The conidiophores form a stratum. The strata typically bear conidia in acervuli which are immersed or erumpent like sporodochia or may be like pycnidia, but true pycnidia are never developed. They are corneous or even submembranous, may or may not be accompanied by setae. The conidia are variable.

This order has a single family **Melanconiaceae** having following genera.

A. Conidia one-celled, short, hyaline without appendages; produced apically on conidiophores; acervulus with dark setae.

Genus: *Colletotrichum* (Fig. 15.8G)

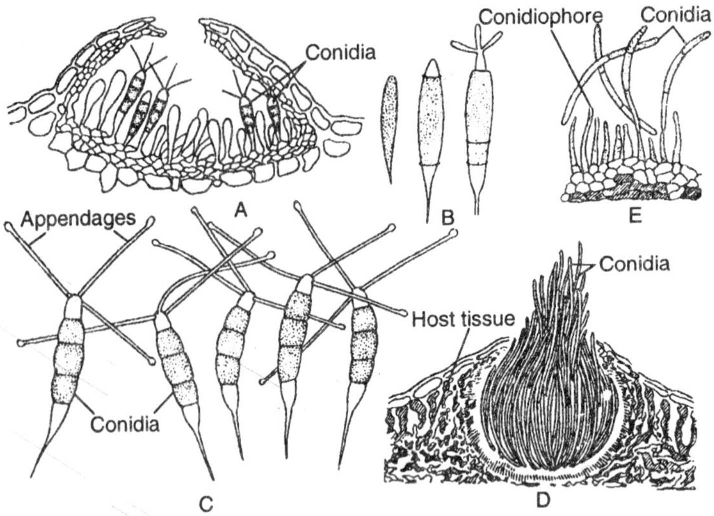

Fig. 15.7: A-C. *Pestalotia* sp. D-E. *Septoria* sp. A. Subepidermal acervulus bearing conidia. B-C. Conidia at different stages of maturity. D. Non-stromatic pycnidium in the host tissue bearing conidia. E. Several-celled filiform conidia borne on conidiophores

AA. Conidia three- to several-celled with transverse septa only, not filiform; dark, but end cells hyaline, with appendages; two to three appendages at the apex of conidia.

Genus: *Pestalotia* (Figs. 15.7A-C)

Order: **Moniliales**

This is the largest order under the Deuteromycetes. The vast majority are of immense importance: (i) for being human pathogens, (ii) in industry, (iii) as plant pathogens, and (iv) as laboratory contaminents. This group is one of very great diversity and contains a multitude of forms. Many are saprophytes while some are aggressive parasites. The Moniliales differ from the Sphaeropsidales in he absence of the pycnidium and from the Melanconiales in their somewhat loose, separate hyphae, not innate and closely aggregated as in the Melanconiales. As such they are designated as conidial fungi without pycnidia or acervuli. The conidia, though produced sometimes as buds or by a hyphae becoming broken up, are commonly on conidiophores, which have a wide distribution over the hyphae, or are united in groups, or massed in other ways.

A simplified key to the families and genera is presented below:

A. Conidia and conidiophores hyaline or brightly coloured; conidiophores not united into sporodochia or synnemata; conidia seldom as oidia, usually on conidiophores which may or may not be differentiated from somatic hyphae; conidia one- to three- or more-celled, globose to cylindrical or elongated to filiform, sometimes irregular in shape hyaline or bright-coloured.

Family: **Moniliaceae**

B. Conidia pyriform to nearly ellipsoid, two- to three-celled, hyaline; conidiophores long slender, simple or rarely brached, septate, single or in tufts; conidia lateral as well as apical; presence of conidial scars on the conidiophores.

Genus: *Pyricularia* (Fig. 15.8B)

AA. Conidia or conidiophores (or both) dark-coloured (brown or black); conidiophores not united into sporodochia or synnemata; conidiophores may or may not be sharply differentiated from the somatic hyphae, erect, simple or much-branched; conidia one- to several-celled, septation either only in a transverse plane or in longitudinal, transverse, and diagnonal planes; conidia extremely variable in shape.

Family: **Dematiaceae**

B. Conidia one-celled globose to subspherical, smooth with distinct dark pigment, borne singly at the apex of short conidiophore; conidiophores inflated below the tip exhibiting jar-shaped appearance.

Genus: *Nigrospora* (Fig. 15.1E)

BB. Conidia long, slender to filiform, tapering upward, one- to several-celled, hyaline to dark; somatic hyphae produce stroma from which conidiophores arise in tufts of a few or many; conidiophores simple or branched, continuous or septate, denticulate or geniculate, usually at the points of denticulation or geniculation scars are formed which bear conidia acrogenously.

Genus: *Cercospora* (Fig. 15.8F)

BBB. Conidia three- to several-celled with transverse septa only, ellipsoid

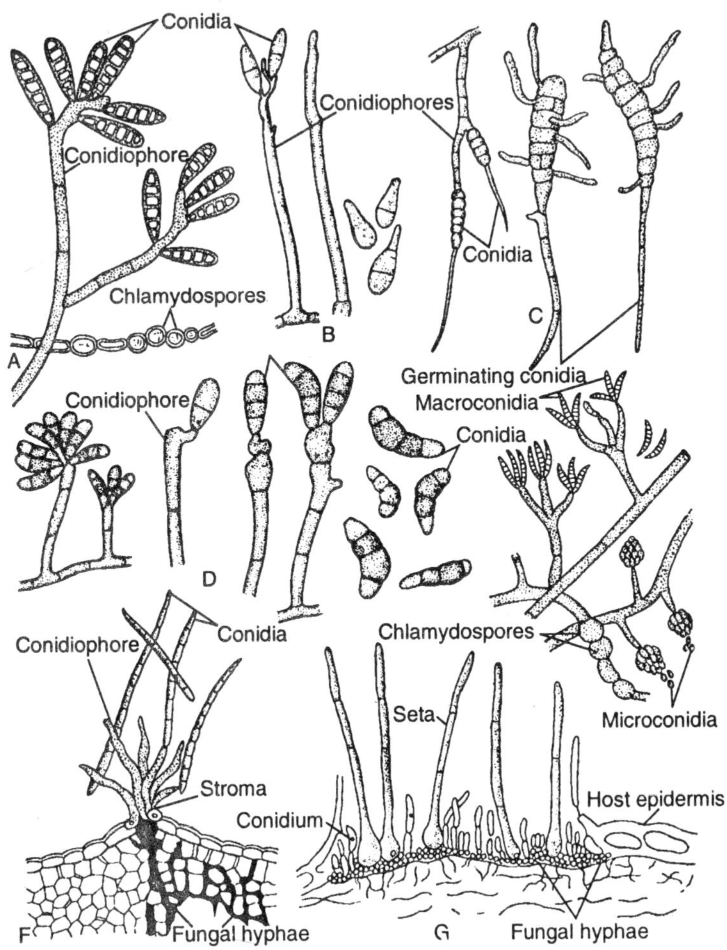

Fig. 15.8: Conidia and conidiophores of different genera of the Deuteromycetes. A. *Helminthosporium*. B. *Pyricularia*. C. *Alternaria*. D. *Curvularia*. E. *Fusarium*. F. *Cercospora*. G. *Colletotrichum*

or cylindrical, straight, or slightly curved or bent, ends rounded, smooth-walled, dark-coloured, exogenous, not catenulate; conidiophores stout, erect, short or long, septate, branched or unbranched, more or less irregular or bent, bearing conidia successively on new growing tips, often at the knee-like bents.

Genus: *Helminthosporium* (Fig. 15.8A)

BBBB. Conidia three- to five-celled with transverse septa only; dark, more or less fusiform, typically bent or curved with one or two of the central cells distinctly larger and darker than the terminal cells; conidiophores brown, unbranched sometimes branched, septate, bear conidia at the apex.

Genus: *Curvularia* (Fig. 15.8D)

BBBBB. Conidia dark, typically with both transverse and longitudinal septa, variously shaped, obclavate to elliptical or ovoid having a simple or branched appendage; conidiophores dark, simple, septate, short or elongated, conidia borne on conidiophores acropetally in long chains, less often borne singly.

Genus: *Alternaria* (Fig. 15.8C)

AAA. Conidia produced on sporodochia, the external form of the sporodochium is usually definite, occasionally extended as an unlimited crust; conidiophores usually thickly crowded and often forming a closed hymenium, simple or branched, rod-like; conidia various, usually terminal, single, but also in chains and lateral.

Family: **Tuberculariaceae**

B. Conidia hyaline, variable, principally of two kinds—macroconidia several-celled slightly curved or bent at the pointed ends, spindle- or sickle-shaped, and microconidia usually one-celled, may be two-celled, ovoid or oblong borne singly or in chains; conidia often held in a mass of gelatinous material; conidiospores variable, slender, stout, short, simple or branched irregularly or bearing a whorl of phialides, developed singly or grouped into sporodochia; sporodochia without setae; chlamydospores may or may not be developed.

Genus: *Fusarium* (Fig. 15.8E)

AAAA. Conidia produced on conidiophores which are pressed into a synnema.

Family: **Stilbellaceae**

AAAAA. Plant body consisting of more or less less yeast-like forms, genera separated according to their ability to form pseudomycelium and true mycelium and on the basis of the kinds of spores.

Family: **Cryptococcaceae**

Order: **Mycelia Sterilia**

Orders included under this group do not produce conidia and other spores. The sterile mycelium may or may not develop sclerotia. These sterile forms are among the worst of plant pathogens. Following are the genera under this group:

A. Sclerotia variable in form, horny-fleshy, frequently small, formed of loose hyphae brown with long cells.

Genus: *Rhizoctonia* (Fig. 15.9)

AA. Sclerotia large, globose or irregular, cartilaginous-fleshy, formed of compact hyphae, mycelium usually light.

Genus: *Sclerotium*

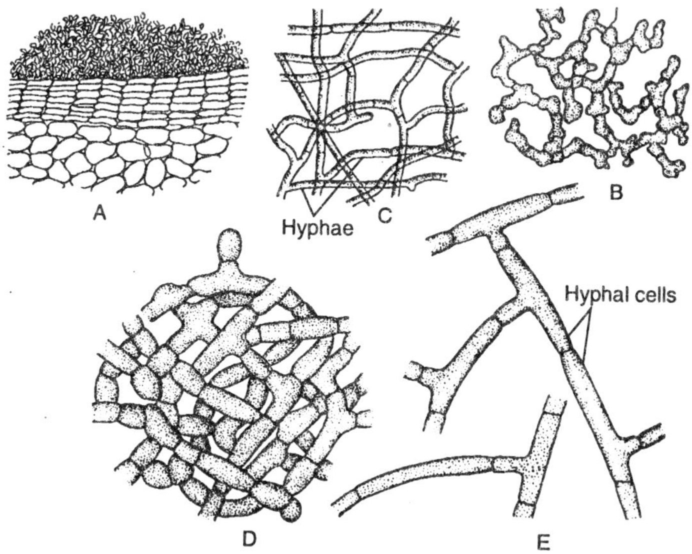

Fig. 15.9: *Rhizoctonia* sp. A. Sclerotial tissue on host surface. B. Closer view of sclerotial tissue. C-E. Nature of hyphae at different stages of growth

AAA. Sclerotium-like compact clusters of small cells, mycelium light.

Genus *Papulaspora*

Genus: *Fusarium*

Fusarium is soil inhabitant. Its septate mycelium is colourless at first, but with age it becomes cream coloured, pale yellow, pale pink or somewhat purplish. The fungus spreads over short distance by means of water and contaminated farm equipment, and over long distances through infected transplants. *Fusarium* produces three kinds of asexual spores. They are: microconidia, which have one or two cells and are most frequently and abundantly produced under all conditions, even inside the vessels of infected host plants; macroconidia are the typical **Fusarium** spores, they are three- to five-celled having gradually pointed and curved ends and appear in sporodochia-like groups on the surface of plants killed by the pathogen, and thick-walled, round chlamydospores produced within or terminally on older mycelium or in macroconidia. All the three types of spares are produced in cultures of the fungus and probably in soil, although only chlamydospores can survive in the soil for long (Fig. 15.9).

Fusarium causes vascular wilts of vegetables and flowers, herbaceous perennial ornamentals, plantation crops. Different host plants are attacked by special forms or races of the fungus. Some of them are: *Fusarium oxysporum f. lycopersici* attacks tomato causing tomato wilt; *F. oxysporum f. vasinfectum* attacks cotton. Several nematode *Fusarium* disease complexes are also known. *Fusarium* causes post-harvest pink or yellow molds on root crops.

Fusarium solani generally produces only asexual spores, but under certain conditions it produces its perithecial stage—*Nectria haematococca.*

Three groups of toxins: zearalenones, trichothecenes, and femonisins are produced by several species of *Fusarium.*

Chapter 16

General Considerations of the Fungi

A. DISCHARGE AND DISSEMINATION OF SPORES IN FUNGI

The actual process of spore discharge or spore liberation in fungi is the detachment of spores from the spore-bearing structure, and spore dissemination is the subsequent movement of the spores before coming to rest on a substrate.

Spore discharge may be an active or a passive process. Actively discharged spores are projected a greater or lesser distance by a variety of devices generally based upon the hydrostatic pressure of the spore-bearing structures or upon devices triggered off through changes in turgidity or the development of negative pressures in cells adjacent to the spores or in the spores themselves, whereas amongst the passive spore discharge are number of different methods depending on whether the spores are dry or slimy, enclosed by a membrane or associated with material which attracts other organisms to it.

Spore discharge exhibits a surprising number of periodic patterns. Many of these are related to the diurnal rhythm of alternating day and night. Others are related to the water relations of the reproductive structures, and yet others reflect changes in climatic cycle. Temperature and rainfall, the two external factors which greatly effect the development of reproductive structures, themselves appear to be the prime determinants in long term periodic fluctuations of spore discharge. Temperature, humidity and light are often determinants of diurnal rhythms. It has been observed that the day-maximum periodicity of *Erysiphe* spores is related to light, that of *Fomes annosus* basidiospores to temperature.

Spore dissemination again is controlled by various physical agencies of which most important is wind. The velocity of wind is again an important factor for distributing the spores in distant areas from the source of spore development (Fig. 16.1).

I. **Active Spore Discharge.** Spores may be violently discharged by the mechanisms like: (i) the bursting of spore-producing structures; (ii) sudden changes in shape of turgid spores or of turgid structures asso-

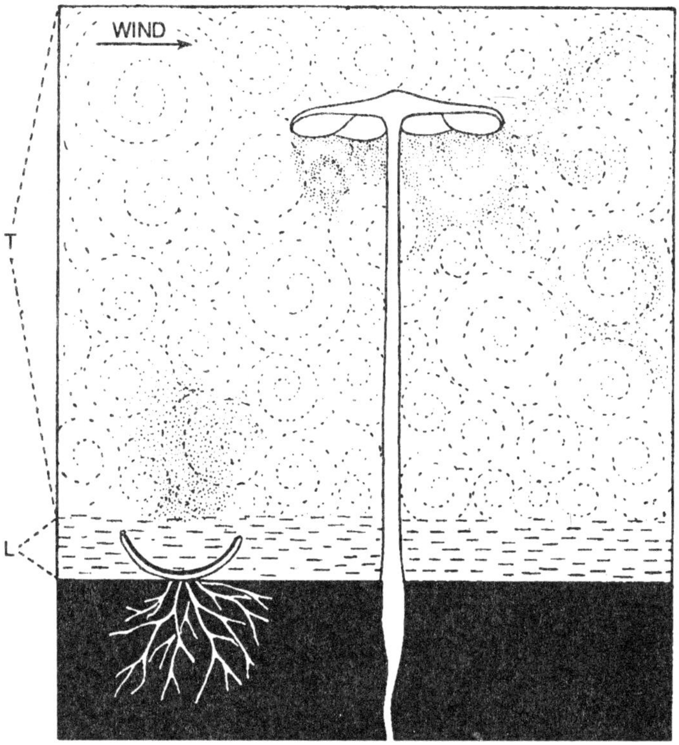

Fig. 16.1: Discharge of spores from the apothecium of *Peziza* sp. and pileus of *Collybia* sp. The ground is shown black. Above ground is thin layer of laminar air (L) and on top of this turbulent air (T). Discharged spores are disseminated by wind

ciated with the spores; (iii) rapid twisting movements produced as a result of drying in filamentous sporangiophores or by hygroscopic movement; (iv) sudden breaking of tensile water in conidia or conidiophores, distorted on drying, which are thereby permitted to return to their original form; and (v) by impaction. The distance of projection of spore discharge depends on the initial velocity of the projective and on its size, shape and density. In violently discharged fungal spores and spore masses there is a general positive correlation between size and distance of shooting. The principal external conditions effecting violently spore discharge are temperature, light, water supply and humidity of air.

(i) **Bursting of spore-producing structures.** At maturity or soon thereafter, the spores of many fungi are forcibly discharged. Some of the commonly encountered devices are

(a) ***Trigger devices.*** The observations of Meredith (1961-63) have shown that in some cases, dry spores are liberated by an active discharge mechanism. The cell concerned loses water by evaporation so that it develops a negative pressure and tends to be distorted. As the process continues, water molecules in the cells tend to cohere both amongst themselves and the walls. Thus the water is exposed to an increasing tension until a point is reached where the tension is released by the water expanding into a gas phase and so the distorted walls recover. This rapid process imparts a jerk, at recovery from tension, discharges one or more conidia into the air; in others, however, gas bubbles develop in one or more of the conidia immediately prior to discharge, e.g., *Curvularia, Alternaria.* In all cases the jerk of recovery is the affective agency of liberation but in some cases is, perhaps, assisted by hygroscopic movements of the conidiophores.

The most striking discharge mechanism due to increased turgor pressure of cells is known by the Gasteromycete, *Sphaerobolus.* Here the basidiocarp is a minute sphere with peridium enclosing a spherical glebal mass—peridiole, embedded in a gelatinous matrix. The peridium comprises of six histologically distinct layers. When mature, the peridium splits to give six to eight teeth and, at the same time, a split develops between the outer three and the inner layers except at the apices of the teeth. The innermost layer gelatinizes so that the peridiole rests in a drop of fluid held in an inner and an outer cup. The innermost layer of the inner cup is composed of large parenchyma-like cells which, initially, are full of glycogen. This becomes hydrolysed to glucose, hence its suction pressure increases and it takes up water. The cells tend to expand but the outer layer of the inner cup is composed of rigid, tangentially running inter-woven fibres which resist expansion. The tension so set up is finally released by a sudden inversion of the inner cup, accompanied by a minute detonation, which thereby projects the peridiole (Figs. 16.2A-C).

(b) *Syringe devices.* The essential process involved in this type of spore liberation is the development of a high hydrostatic pressure within a cell. This type of spore liberation is found particularly in *Pilobolus.* In *Pilobolus* the walls of the apices of the sporangiophores are

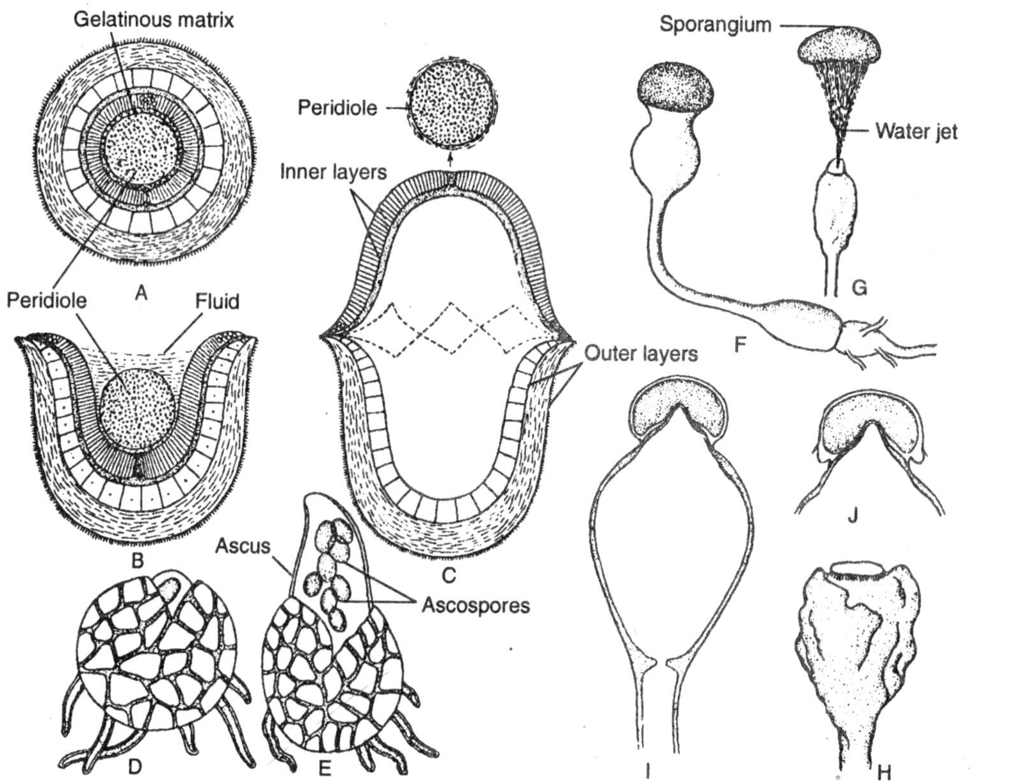

Fig. 16.2: Spore discharge mechanisms. A-C. Trigger device in *Sphaerobolus* sp. D-H. Syringe devices. D-E. In *Sphaerotheca* sp. F-J. In *Pilobolus* sp.

extremely elastic and have a circular transverse line of weakness around their apices. The hydrostatic pressure of the sporangiophore increases until it ruptures, then it contracts violently, and squirts out a jet of water and so projects the sporangium. Particular interest applies to *Pilobolus* because of its phototropic response which results in the tip of the sporangium being directed towards the maximum light intensity (Figs. 16.2F to J).

The mechanism of syringe discharge is better understood in the Ascomycetes. In the Erysiphales, the asci within the cleistothecium take up water and swell. The cleistothecium is ruptured and the ascus or asci within protrude or may even squeeze out and finally be violently ejected. Continued swelling of the contained or ejected asci results in their bursting and scattering the enclosed ascospores in all directions (Figs. 16.2D and E).

In the Discomycetes the exposed hymenium of the apothecium permits the simultaneous discharge of ascospores and this gives rise to the phenomenon of 'puffing', i.e., synchronized discharge of ascospores from mature asci (Fig. 16.3A). As ascus becomes turgid, a positive hydrostatic pressure often associated with the development of a vacuole forces much of the cytoplasm and the ascospores to the apex of the ascus. The ascus frequently enlarges and may protrude from the rest of the hymenium in a state of high tension as in *Ascobolus.* This tension is released when the operculum gives way and the elastic walls of the ascus contract, driving out the upper half of the contents as a cylindrical jet which dissipates into a cloud of ascospores carrying droplets. The distance of discharge may be even up to 17 cm.

In inoperculate Discomycetes asci dehisce by an irregular apical split, e.g., *Geoglossum, Trichoglossum.* In these genera the ascospores are filiform and septate and after dehiscence an ascospore is pushed through the apical pore, slowly at first but with increasing velocity until the last spore has beeen ejected; the ascus then collapses. Not all inoperculate Discomycetes behave like this, however, and some are capable of effective 'puffing', e.g., *Helotium, Sclerotinia* and *Rhytisma acerinum.*

Pyrenomycetes never exhibit 'puffing' for the ostiole of the perithecium, enclosing the asci, which is usually so narrow that only one ascus can pass through it at a time. There are three types of active discharge of spores in the Pyrenomycetes. The first is illustrated by *Sordaria fimicola.* The asci remain attached to the base of the perithecium. They elongate up the neck canal, one at a time, until the tip is forced through the ostiole (Fig. 16.3B). Increasing internal hydrostatic pressure leads to the rupture of the ascus and the spores are forced out singly or in groups (Figs. 16.3B and C). A second type is seen in *Sporomia intermedia.* The ascus wall is double layered, the outer **ectoascus** being thick and rigid, the inner **endoascus** thinner and extensible. As the ascus imbibes water it tends to swell but this is resisted by the ectoascus until it is finally ruptured by the internal pressures. The endoascus then elongates suddenly to two to three times its original length and penetrates the ostiole. The ectoascus is ruptured, its upper part separating as a thimble-shaped lid (Figs. 16.3F and G), the endoascus elongates pushing through the ostiole and the spores are discharged. Finally the empty ascus contracts and disintegrates (Figs. 16.3D to G). The third type is found in *Ceratocystis.* The asci become detached and lie free within the perithecium. They pack the perithecial cavity and are forced into the neck. Finally, one protrudes, it swells but its lower half is firmly held by the ostiole so that it bursts apically, liberating the ascospores, and is itself extruded when empty, its place being taken at the ostiole by next ascus (Figs. 16.3H and I). In some other Pyrenomycetes the asci collect in a mucoid drop and are disseminated by water.

The active discharge or basidiospores of

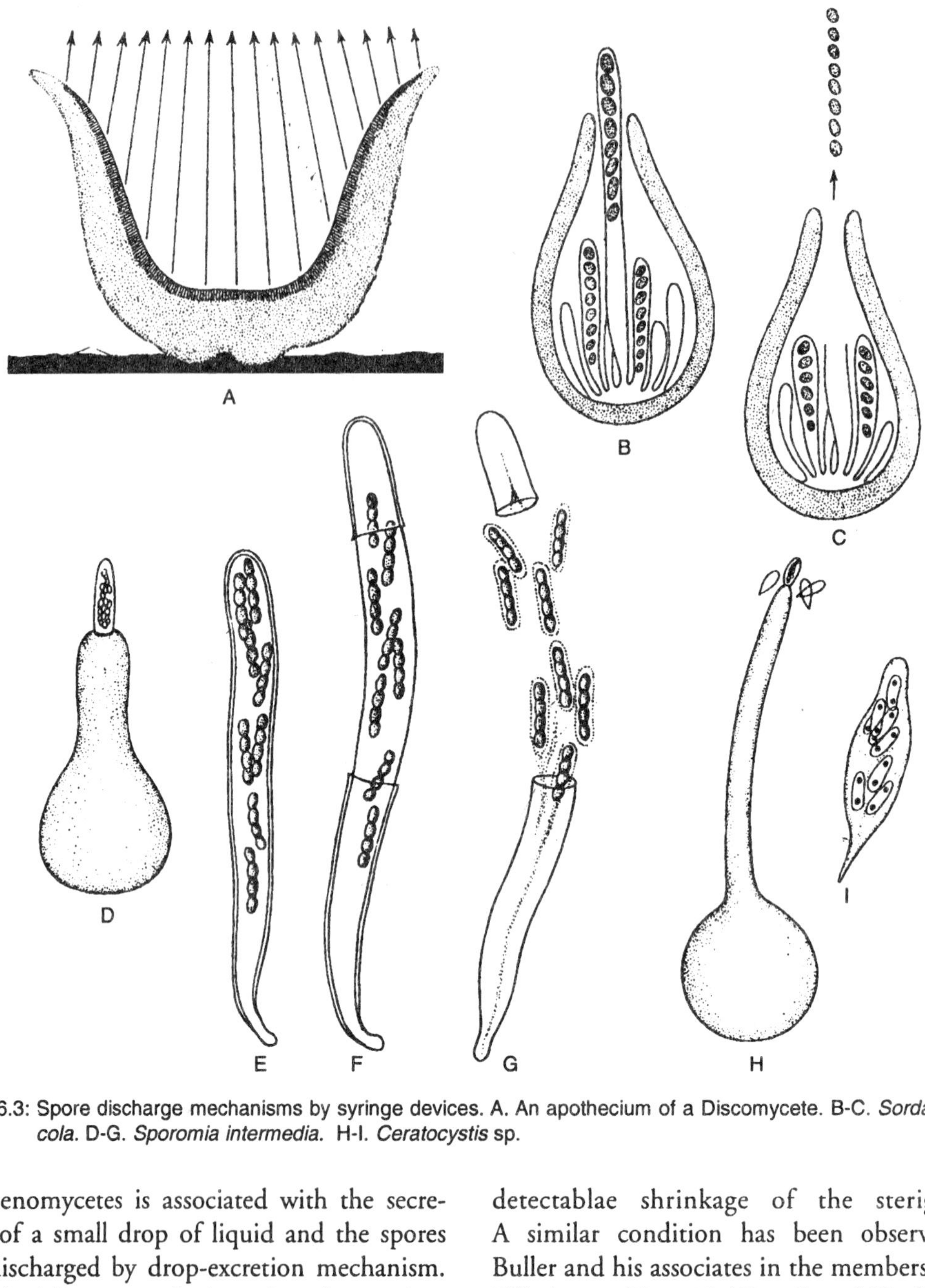

Fig. 16.3: Spore discharge mechanisms by syringe devices. A. An apothecium of a Discomycete. B-C. *Sordaria fimicola*. D-G. *Sporomia intermedia*. H-I. *Ceratocystis* sp.

Hymenomycetes is associated with the secretion of a small drop of liquid and the spores are discharged by drop-excretion mechanism. One basidiospore only is projected at a time but all four follow each other successively, usually in opposite pairs (Figs. 16.4A-F). Once the basidiospores have been launched into the air they are free to be dispersed by air current (Fig. 16.1). After discharge there is no detectablae shrinkage of the sterigmata. A similar condition has been observed by Buller and his associates in the members of the Tilletiaceae (Figs. 16.4G to I) and Uredinales.

(ii) **Sudden changes in shape of turgid spores or of turgid structures associated with spores.** Spore discharge is often associated with sudden rounding off of turgid cells of fungi. In *Entomophthora coronata* a single and

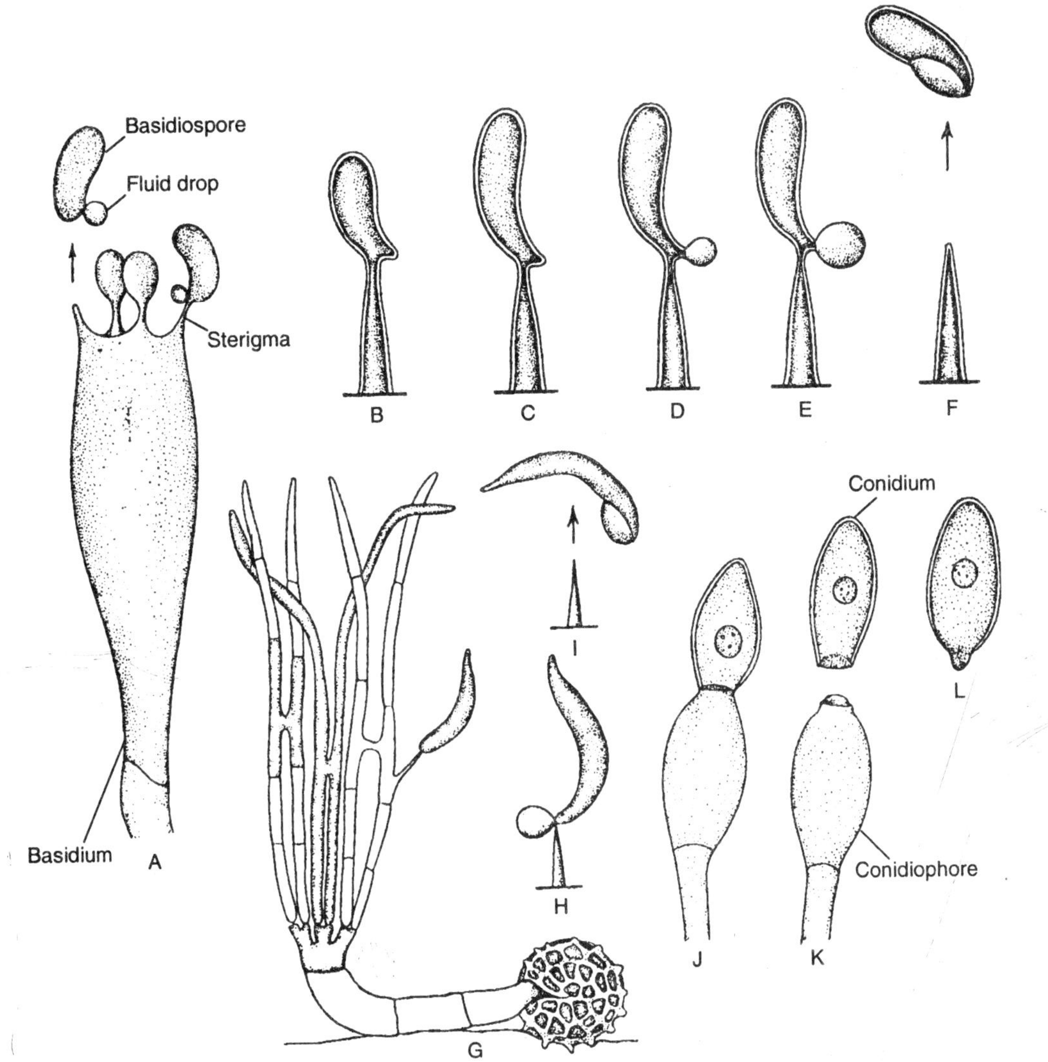

Fig. 16.4: Spore discharge mechanisms. A-I. Drop-excretion mechanism. A-F. In the Hymenomycetes. G-I. In the Tilletiaceae. J-L. Change in condition of turgid structures in *Entomophthora* sp.

relatively large conidium is borne at the end of a straight conidiophore the tip of which projects into the conidium as a distinct columella. As the conidium tends to round off, the re-entrant part bulges outward and, as a result, the conidium bounces off to a distance (Figs. 16.4J to L). A similar mechanism is encountered in the genus *Sclerospora*. Here the area of contact between the sporangium and its sterigma is relatively small and the tip of the latter does not project into the sporangium; there is merely a flat contact, rounding off in this region brings about discharge of sporangium (Figs. 16.5A and B). In most rusts (Uredinales) the aeciospores are violently discharged within the aecium. The spores are polyhedral because of mutual pressure. There is a tendency for them to round off when conditions are sufficiently damp. This happens suddenly and spores bounce out either singly or in groups to a distance of several millimetres. In the Erysiphales, conidia are actively discharged by rounding off.

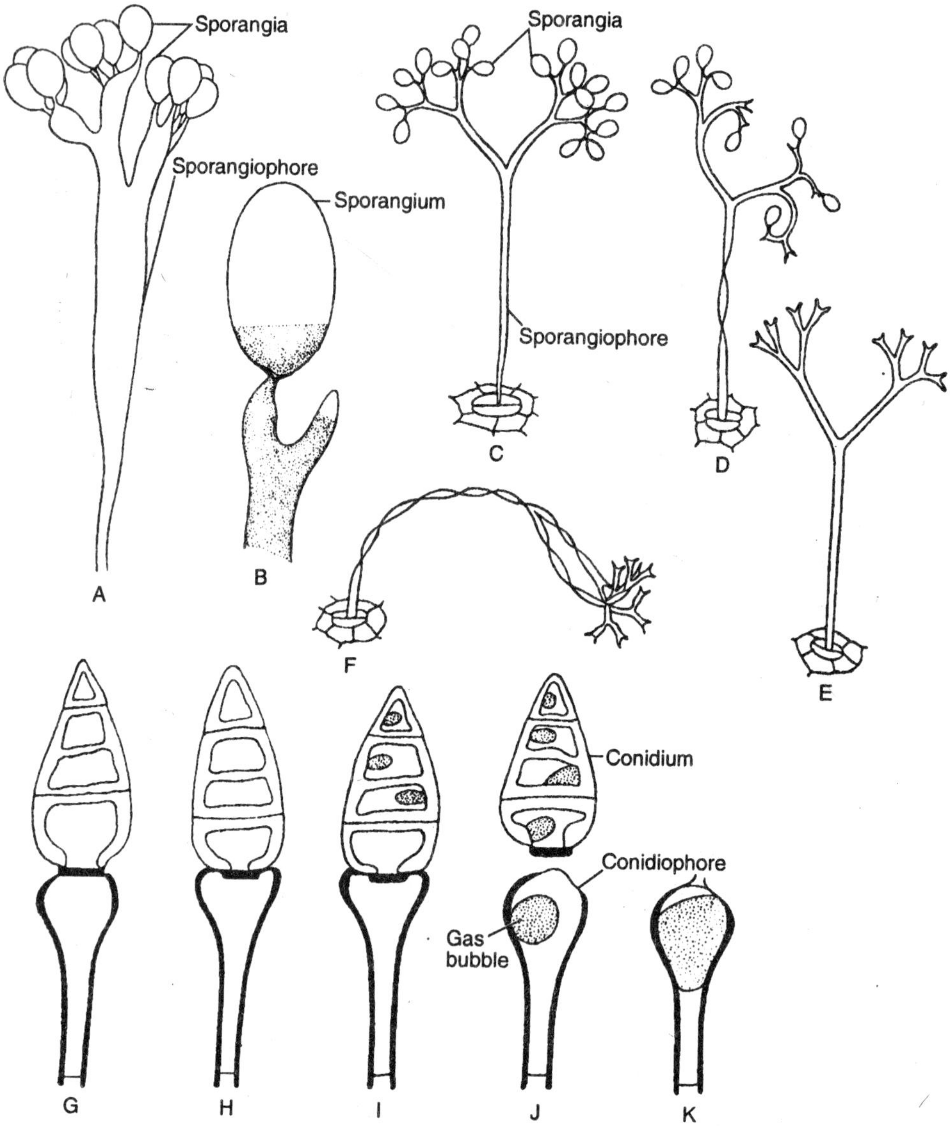

Fig. 16.5: Spore discharge mechanisms. A-B. Change in condition of turgid structures in *Sclerospora* sp. C-F. Hygroscopic movements of sporangiophores in *Peronospora tabacina*. G-K. Gas bubble mechanism in *Deightoniella torulosa*

(iii) **Spore liberation by hygroscopic movements.** This mode of liberation occurs in dry spored fungi. In *Peronospora tabacina,* the dichotomously branched sporangiophore twists dislodging the sporangia through hygroscopic movements consequent upon drying out (Figs. 16.5C to F). In some Gasteromycetes, spore liberation is effected by hygroscopic movement of the capillitial threads.

(iv) **Sudden breaking of tensile water in conidia or conidiophores.** It has been shown recently that, in a number of dematiaceous fungi, conidia are set free by the breaking of tensile water and release strains. Meredith in 1961 observed that *Deightoniella torulosa* the end cell of the conidiophore, which bears a single conidium, has a thickened wall. Because the thickening is uneven, evaporation leads to a temporary distortion so that the upper region of the end cell is drawn downward. However, this cell, because of the rigidity of its wall, tends to return to its original form, and the fluid contents are under increasing tension as evaporation proceeds. Finally, a break occurs either because the cohesion between the water molecules or their adhesion to the cell wall is overcome. When this happens, a minute **gas bubble** makes its appearance and instantaneously enlarges as the distorted cell returns to its original shape with a sudden jerk which throws off the attached conidium to a distance of several centimetres (Figs. 16.5G to K).

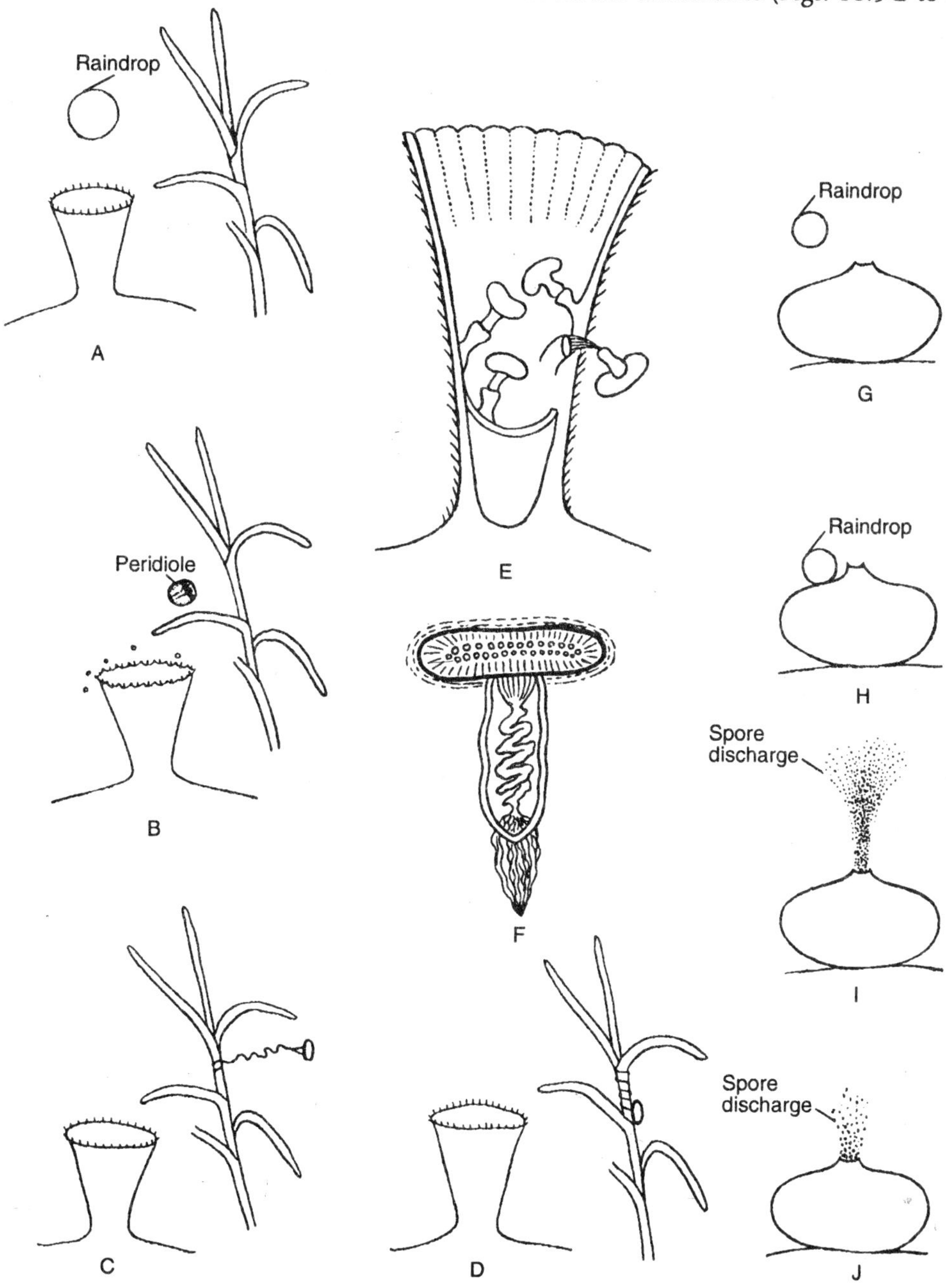

Fig. 16.6: Spore liberation mechanism by impact of raindrop. A-F. In the Nidulariales. G-J. *Lycoperdon* sp.

(v) **Spore liberation by impaction.** The best known examples of this type of spore liberation are the puff-balls and their alies (Lycoperdales), and the various types of 'splash-cups' such as those of the bird's nest fungi (Nidulariales). In the Lycoperdales the contents of the basidiocarp are a mass of dry spores interspersed with fine dried-out hyphae, the capillitial threads. The basidiocarp opens by a simple (*Lycoperodon*) or complex (*Geastrum*) ostiole. Spores may be blown out by the wind across the ostiole. They may also be forced out when the basidiocarp is struck by raindrops. When air is expelled with spores, the peridium then springs back to shape (Figs. 16.6G to J). Bordie (1951) has studied splash-cup dispersal mechanism in the Nidulariales (Figs. 16.6A to D). For details see general account of the series Gasteromycetes. The impact of large drops of water from a height on a twig covered in conidial stromata of *Nectria* causes the drops to break into several droplets containing conidia. These droplets are scattered at a great distance.

II. **Passive Spore Liberation.** In a large number of fungi the spores are not actively discharged. In them, spore liberation depends on wind rain splash, or the activity of insects and other organisms. Many fungi have dry spores which, although not violently discharged, are fairly easily set free by sufficiently strong air currents. As to the fungi producing slime spores, these spores cannot be blown off. Spore liberation by rain splash or insect dispersal of these fungi seems to be the rule. The mechanism of passive spore liberation in principle, however, is not very different from the process of spore dissemination.

(i) **Liberation of dry spores.** Some of the most important epidemic producing pathogenic fungi such as the rust fungi (Uredinales), smut fungi (Ustilaginales), and the downy and powdery mildews (Peronosporales and Erysiphales) have dry spores. These spores are detached and dispersed by the wind. This is also true of many of the common molds (*Penicillium*) and (*Aspergillus*). Dry-spored types are most effectively liberated at low relative humidities and high wind-speeds. Separation devices include various modifications like gelatinization of the wall between the spore and hypha, intercalary cells or disjunctors as in *Albugo* and aeciospores of many rusts. The pressure exerted by the intercalation of disjunctor leads to a rupture of the membrane in the plane of the septum and the sporangia or spores as the case may be are separated by the disjunctor. In some cases, small hygroscopic movement of the conidiophore or sporangiophore is sufficient to liberate conidia or spores from sporangia. Subsequently, splashing of raindrops may also help dispersal of conidia or spores.

(ii) **Liberation of slime spores.** In many of these, the spores are liberated as a consequence of the gelatinization of thier walls or through the autolysis of adjacent structures or cells. Good examples are found in the Mucoraceae where the sporangium wall dissolves, leaving the spores held together in a drop of mucilage containing water, e.g., *Mucor* sp. A similar kind of behaviour is exhibited by many conidia producing fungi where the conidia accumulate to form slimy drops at the ends of the conidiophores, e.g., *Verticillium* sp. These drops are not blown away by the wind and, in a dry atmosphere, the spores or conidia involved are left sticking firmly to each other. While sticky they may be liberated through contact with a passing animal, or by spreading out like oil over a film of water. When dry they may become detached as a consequence of some physical shock.

In many Pyrenomycetes spore formation is associated with the production of mycilaginous material produced by autolysis of sterile cells of the perithecium as in *Chaetomium* or pycnidium in *Diatrype*. In all such cases, the spores, embedded in slime, are exuded through the ostiole.

A rather different form of spore liberation is achieved in those cases where the spore-drop or associated slime is attractive to animals, especially insects. The spores become detached in the attractive material which is then removed by the animal. Examples are the sugar-containing droplets exuded by the spermogonia of many rusts, the oidia of agarics and the *Sphacelia* stage of *Claviceps purpurea.* The best known examples of these are the stinkhorns of various types (*Phallus, Dictyophora, Mutinus*) and the hollow, spherical lattice-wall of *Clathrus.* Spore liberation is here dependent upon a complex series of processes. The basidia in all cases are embedded in a dark-green sugary gluten which encloses the hymenial surfaces and is itself encased in a gelatinous material and a peridium, this being spherical or egg-shaped. The 'eggs' are subterranean and evidently take up water so that the volva swells. When this happens the characteristic sticky-sweet odour which is attractive to flies develops. The stalk or network, as the case may be, now enlarges, the peridium splits and the hymenium, with its covering of gluten, is carried above the ground where the material, including the spores, is rapidly removed by flies (Fig. 16.6).

III. Spore Dissemination. Spore dissemination is essentially a passive process determined by physical agencies such as (i) wind, (ii) water, (iii) rain splash, (iv) living organisms, and (v) other means. It is not often clearly distinguishable from some of the processes like spore liberation by wind, rain splash or insects.

(i) **Dissemination by wind.** Spores once discharged are either below or above the boundary layer of turbulent air. In the former case, the air is still and the spores fall to the ground. In the latter, however, they are air borne in the turbulent air of the atmosphere and become dispersed by air currents (Fig. 16.1). Spores which are projected violently through the air follow a quasi-parabolic curve like most other projectiles.

(ii) **Dissemination by water.** Ingold and others have studied the dissemination of spores of aquatic fungi. According to them, the dissemination is dependent upon concentrations of spores and water speeds and stimuli of chemical substance that may remain in water.

(iii) **Dissemination by rain splash.** A classical example of this process is exhibited by the members of the Nidulariales, details of which have been observed by Brodie. The process has been described in the general account of the series Gasteromycetes.

(iv) **Dissemination by living organisms.** Dissemination by living organisms is very common. The living organisms responsible for spore dispersal are: insects and other animals.

(a) ***Dissemination by insects.*** The Gasteromycetes, particularly the members of the Phallales are a group with particular adaptations to this mode of dissemination. Another common kind of attractor is the production of nectar-like secretation as in *Claviceps purpurea* (conidial stage), the pycnial stage of many rust fungi and, in some cases, their aecial stage. The insects visit the fungus primarily to feed on the attractive material such as nectar. During feeding the insects' body gets smeared with conidia or spores as the case may be. When the feeding is over, the insects naturally carry conidia or spores over to whatever substrate they touch, thereby carry with them conidia or spores and indirectly help dissemination of conidia or spores. The spores of Dutch elm disease pathogen, *Ceratocystis ulmi* are dispersed by bark beetles, *Scolytus* spp.

(b) ***Dissemination by other animals.*** Certain animals, such as dogs and pigs are attracted by the odour of fruit body of truffles (*Tuber* spp.) which they dig out from the ground and feed on and the spores pass through their alimentary canal and pass out through fecal matter at distant places and thus the spores are disseminated. The spores of a number of coprophilous fungi, notably species of *Pilbolus, Ascobolus, Sordaria,* and *Coprinus* are dispersed by horses, cattle, sheep, goats, etc. These animals swallow the spores and herbage together, the spores with chewed herbage pass through alimentary tract and are ultimately voided in the feces. By the time the feces are discharged from the alimentary tract, the animals might have travelled quite a distance thereby the spores are disseminated in distant places.

(v) **Dissemination by other means.** Fungal spores are often disseminated through spore-infected seeds, soil, manure, agricultural implements and often human hands and feet. Besides these, human beings quite often unknowingly disperse fungal spores through nursery stocks and similar other media.

B. HETEROKARYOSIS

Blakeslee (1906) first reported the germination of zygotes of the Mucoraceae obtained under controlled conditions. He obtained a number of 'neutral' mycelia from the germination of zygotes of *Phycomyces blakesleeanus.* These mycelia grew abnormally, tended to produce zygophore-like branches (**pseudophores**) and were bright yellow as a consequence of the abundant carotene in them. It was supposed that such mycelia carried nuclei of different mating types, + and –, in their hyphae. Burgeff (1912, 1914) found the same phenomenon in the same species. He removed the tip of a hypha of + mating type and inserted it into the cut of a hypha of – mating type, so that cytoplasm and nuclei intermingled. The resultant mycelium, derived from this 'myxochimera', grew in a manner, similar to those of Blakeslee's 'neutral' cultures. Burgeff described such mycelia as **heterokaryotic.** The phenomenon of the co-existence of genetically different nuclei **heterokaryons** in the same cytoplasm of the same individual is known as **heterokaryosis.** Heterokaryosis, as a natural phenomenon, is virtually confined to the fungi. Its significance for the biology of the fungi was not appreciated until pathologists like Brierley (1929), realized that it could account for some of the natural variation shown by phytopathogenic fungi. Hansen and Smith (1932) showed that if single spores of *Botrytis cinerea* were isolated and cultured, they gave rise to colonies with various attributes which could be grouped into three main types, a, b and x. Types a and b were different but constant in their cultural and morphological characteristics. Type x, however, was inconsistent. It gave rise to sectors to types a and b in addition to areas sharing its characteristic, irregular morphology; this type of segregation was also shown by conidia isolated from type x cultures. They obtained strains of types a and b that were constant for several subcultures and then grew them together. As a consequence of hyphal fusions which they saw and in which they detected nuclei, presumably migrating, they reconstituted a mycelium of type x. In 1935, Kohfer carried out a similar analysis of *Mucor mucedo* and obtained by selection **homokaryotic mycelia** from the original **heterokaryotic strain.**

In 1938, Hansen reported similar studies on 35 genera of Fungi Imperfecti, in which 32 showed this phenomenon; it, therefore, seemed to be widely distributed. The assumption was made that two distinct types of nuclei can coexist in *B. cinerea* hyphae and that when spores which are multinucleate are abstricted, differing proportions of these nuclei may be incorporated in them. Sometimes by chance all nuclei will be the same (types a and b), but more usually they will be different (type x). The products of germination will be correspondingly stable or unstable. Jinks (1959) offered another explanation for it. Since the origin of the nuclei which entered the spores was not determined; this is the somatic segregation of different cytoplasmic determinants (**plasmons**) from a **heteroplasmon.** Such segregations have been demonstrated by Jinks and his co-workers and by others.

Heterokaryosis may arise in any of the following ways:

(i) By the germination of a heterokaryotic spore producing a heterokaryotic somatic condition.

(ii) By the fusion of vegetative cells of genetically distinct nuclei and cytoplasm which intermingle in the fusion cells, produce hyphae bearing heterokaryons resulting from multiplication of cells and repeated division of heterokaryotic nuclei.

(iii) By mutation.

(iv) By fusion of nuclei of same genetic constitution and the subsequent multiplication of the diploid nuclei so formed resulting in a mixture of haploid and diploid nuclei. No new genetic combination would be formed in such heterokaryosis.

Developmental details of heterokaryosis. The development of heterokaryons or the phenomenon of heterokaryosis proceeds along following lines:

1. **Hyphal fusion.** The frequency with which hyphal fusion occurs is subject both to genetics and to environmental control. The ability of a heterokaryon to retain a well-mixed nuclear population, and thus phenotype uniformity, depends in part on fusion among the hyphae near the frontier and the number of nuclei per cell. Any statical tendency toward segregation of nuclear types during branching of hyphae will be reversed to a greater or less extent by fusions.

However, hyphal fusion is only effective when it is followed by nuclear migration at least into the fusion cell.

2. **Nuclear migration.** Buller (1931) first described this phenomenon in *Coprinus* spp. using nuclei of different mating types. He showed that, in compatible matings, nuclei can apparently migrate from a small mycelium into a large one. The presence of both nuclei is newly formed cells, could readily be detected by the presence of clamp connections. Since then the phenomenon has been studied in several other Homobasidiomycetes and in *Ascobolus stercorarius.* Migration proceeds at remarkably rapid rates. It may or may not be affected by light. The determination of which mycelium is to be the donar and, which the acceptor may be genetic or environmental.

The actual mechanism of nuclear migration is not known. It is often assumed that the nuclei are carried passively by cytoplasmic streaming.

3. **Cytoplasmic streaming.** The growth of filamentous fungi is correlated with cytoplasmic streaming; this will tend to randomize the nuclear population if the nuclei move freely with the stream. Even if the nuclei are not carried easily with the cytoplasm, the movement of cytoplasm will facilitate phenotypic, homogeneity over large regions.

4. **Nuclear division.** Nuclear division is, of course, the ultimate focus of selection, and aside from differences among nuclei in division rate, changes in nuclear ratios must reflect differences in division rates among regions of different nuclear constitution. In many fungi, particuarly in the Ascomycetes, nuclear division takes place rapidly in hyphal tips and for a greater or lesser distance behind them. In fact, division is reported to be simultaneous. The timing of nuclear division is determined by the common cytoplasmic condition.

Genetic control of heterokaryosis. In *Neurospora* heterokaryosis is determined by two pairs of alleles, C/c and D/d. Only mycelia carrying like alleles will form heterokaryon, i.e., CD and CD, Cd and Cd, cD and cD or cd and cd but not CD with Cd or cd and so on. When hyphae of unlike alleles are brought together, hyphal fusions are not prevented but the fusion cell is rapidly sealed off. In a short time, bubble-like structures appear and the cell dies of autolysis. It has been called 'Heterokaryon incompatibility', and is found in *Aspergillus nidulans* and other species of *Aspergillus.*

Significance of heterokaryosis. While the demonstration of heterokaryosis may be credited to Burgeff (1912, 1914) the significance of heterokaryosis in nature was first fully appreciated by Hansen and Smith (1932), working with isolates of *Botrytis cinerea.*

Stable heterokaryons are with a selective advantage, usually in rare of growth over their components. These may arise in case of heterokaryons containing nuclei with nutritional defects as a result of complementation, i.e., the deficiency of one nucleus is made good by the ability of the other and *vice versa.* In general, this effect must operate in the cytoplasm. By heterokaryosis heterotrophic organisms can survive in nature by creating balanced nutritional conditions among two or more nutritionally deficient strains. Such condition is brought about by the fusion of cells without any nuclear intermixing.

It has been demonstrated that strain of *Aspergillus niger* incapable of producing conidia although the mycelium grows almost as well as the wild type and is apparently normal when the nuclei are combined experimentally in a diploid nucleus, a mycelium containing such nuclei grows and produces conidia normally. In this case, therefore, complementation, as a physiological process, can only occur at the intra-nuclear level, not in the cytoplasm. Similar effects may also be seen in relation to dominance and the quantitative expression of dominant genes can be studied by changing heterokaryons.

Many of these interactions are manifested in what Dodge (1942) called 'heterokaryotic vigour' and this has been compared with heterosis in diploid organisms. Since this heterokaryotic vigour to a greater or less degree may be expected to be the rule in nature it will not be surprising to find heterokaryosis in most natural populations of fungi. It is clear that the heterokaryotic situation is usually explicable in terms of complementation. Nevertheless, complex situations do arise involving non-alletic genes and their suppressors, and little study has yet been made of such complications. Attention should also be drawn to the fact that recessive genes, including lethals, can arise in heterokaryosis and persist until they segregate in spores.

The major initial work of complementation was that of Dodge (1942) with *Neurospora tetrasperma,* followed by that of Beadle and Coonradt (1944) with *N. crassa.* Dodge found that two slow-growing mutants were able to complement one another in heterokaryotic association to yield a rapidly growing mycelium.

In fact, heterokaryons have been used widely in tests of allelism, as well as in the study of complex loci, alleles of which may complement to some extent through interaction of mutant proteins, since a wild type (normal) nucleus may mark the deficiencies of a mutant nucleus and as such a heterokaryon will resemble the wild type. If two nuclei carrying non-allelic mutations are combined in a heterokaryon, the resulting phenotype will approach the wild-type condition. The latter behaviour is designated as "complementation".

Heterokaryosis is likely to be of some importance in nature for it represents an ideal system ensuring rapid and successive adaptations of a fungus to a continually hanging environment. This is a situation that most fungi must constantly maintained their own activities, both catabolic and secretory. Moreover, experimental studies with heterokaryons have shown that it is exceedingly difficult to eradicate entirely any particular type. Such nuclei, which may be of little or no adaptive significance in one situation, may be of enormous significance in another. It is for this reason that more experiments on selection with heterokaryons are desirable so that both the possibilities or, and rates of change of nuclear ratios can be fully assessed. That selection can be efficacious and that new properties can be shown by heterokaryons compared with those of their parental strains, has been in seminatural situations. For example, *Puccinia graminis tritici* produces a race virulent to a wheat, hitherto only resistant to one of the parental strains of the heterokaryons. Buxton (1954, 1956) also showed the virulent parental strains which could be combined to produce heterokaryons which were quite a virulent as naturally occurring pathogenetic strains.

C. THE PARASEXUAL CYCLE

Up to the middle of the present century the normal sexual cycle as encountered in the lower and higher fungi, was considered the only mechanism of genetic combination in fungi. Lack of knowledge of recombination processes was a serious drawback in certain areas of study in the Deuteromycetes (Fungi Imperfecti) in which normal sexual cycle has not been observed. They are: control of virulence, breeding of improved industrial strains, the study of genetic diversity within related groups, etc. There was no suitable answer to the question "how did they meet the challenge of evolutionary forces?" With the discovery of processes other than normal sexual cycle, often known as "alternatives to sex" has opened up a new line of genetic approach of recombination mechanisms in fungi and related organisms. One of these novel processes of recombination, was discovered and elucidated in *Aspergillus nidulans*, the imperfect stage of *Emericella nidulans* and designated by Pontecorvo and Roper (1952)—the **parasexual cycle.** The parasexual cycle, as defined by Pontecorvo, is the spontaneous occurrence in hyphae heterokaryosis, fusion of genetically dissimilar nuclei, and recombination and segregation of the diploid nuclei so formed to the development of haploid ones.

An obvious prerequisite for recombination is fusion of haploid nuclei. This normally requires, as a first step, inclusion of the genetically different nuclei in the same cytoplasm. This is achieved through heterokaryosis. There are several ways in which heterokaryosis is established. The most common way perhaps is by the fusion between cells of genetically different somatic hyphae. Due to nuclear migration, cytoplasmic streaming and division of newly migrated nuclei the hyphae become heterokaryotic. The heterokaryotic condition may be established by mutation in one or more nuclei and another way is by the fusion of some of the haploid nuclei of the same hypha and their subsequent division and spread among the haploid nuclei. Since the last method of heterokaryosis involves no new genetic combination, this type of heterokaryosis is not of much significance.

In a heterokaryotic mycelium, nuclear fusion takes place between haploid nuclei of different genotypes as well as of the same genetic constitution resulting in heterozygous and homozygous diploid nuclei respectively. As such a heterokaryotic mycelium may contain as many as five types of nuclei. They are: two types of haploid, two types of homozygous diploid, and heterozygous diploid nuclei; all of them multiply at about the same rate. But initially the diploid nuclei are much smaller in numbers that the haploid ones. The way in which nuclear fusion occurs is a mystery. It might be possible that in a crowded hyphal apex with 20 to 30 nuclei, nuclei divide synchronously. The overlapping divisions become intermingled and a new nuclear membrane encloses two haploid sets of chromosomes. They have been actually demonstrated and designated as 'streaming-mitoses'. The frequency of such nuclear fusion is usually low.

The diploid nuclei divide mitotically and increase in number. During mitotic division crossing over occasionally takes place resulting in new combinations known as **mitotic recombination.** This is probably the most important phase in the parasexual cycle. The recombinations so formed give the fungus some of the advantages of sexuality without possessing it at all. But overall, the recombinations are entirely dependent on the existence of heterokaryosis. Indeed the recombinations obtained by parasexual mechanism are by far much smaller in number than those achieved through sexual mechanism. This mechanism is, however, very significant in the Deuteromycetes where sexual mechanism is not known.

The next stage is the sorting out of haploid and diploid nuclei. This takes place by the development of haploid and diploid conidia which produce haploid and diploid mycelia respectively. This may take place in nature and also can be isolated by artificial means. J. A. Roper (1952) first isolated diploid strains of *Aspergillus nidulans.* It has also been observed that a diploid colony of mycelium may have areas where haploid conidia are produced. This indicates that the diploid nuclei undergo **haploidization** in the mycelium and are sorted out in course of mycelial growth. The frequency of spontaneous development of somatic diploids is usually low and has been traced in *A. nidulans, A. niger, Penicillium chrysogenum,* and in *Coprinus lagopus.* In *Ustilago maydis,* however, diploids have been obtained as a consequence of failure of normal meiosis. Mitotic recombination is also a rare process. There is also suggestive evidence for a parasexual cycle in *Puccinia graminis tritici.* The significance of the two modes of recombination, somatic and meiotic, in the same fungus is problematical.

The sequence of events that follow in a parasexual cycle may be outlined as: heterokaryosis, fusion between nuclei (like or unlike), multiplication of both haploid and diploid nuclei with occasional mitotic crossing over of diploid nuclei, sorting out of diploid nuclei with occasional haplodization of diploid nuclei, and sorting out of haploid nuclei.

It is now quite certain that parasexuality occurs in several Deuteromycetes which lack a normal sexual cycle. In them, parasexuality is responsible for generating variability in the mycelium though its mechanism is far from being adequately understood. Besides the fungi already mentioned, parasexuality has also been detected in the Ascomycetes and Basidiomycetes with a normal sexual cycle as well. Special mention may be made of *Cochliobolus sativus, Ustilago maydis,* and *Schizophyllum commune.* It was Bradley who in 1962 for the first time demonstrated the presence of parasexual cycle in fungi possessing normal sexual cycle.

D. MYCORRHIZA

The name 'mycorrhiza' in its broadest sense is the non-pathogenic or feebly pathogenic association of fungi and the roots or rhizomes of flowering plants, thalli or other parts of non-flowering plants. Both the organisms may derive benefit from the association, or the fungus becomes pathogenic when the other component is weak. The components are associated in a nutritional, as designated by **mycotrophic** relationship with each other. But strictly speaking, the term mycorrhiza is used only where the organ concerned is a root. Roots provide an important habitat for these special ecological groups of fungi. The presence of fungus may be essential for good growth of the host plant, assisting in mineral uptake by its roots. The host plant produces carbonaceous compounds, vitamins, etc., which it supplies to the fungus and the fungus absorbs the nutrients from the soil and provides the host plant with these. In this way both components develop. Associations of fungi with the thalli, rhizomes, or other parts of plants should be reasonably termed 'mycothalli' and 'mycorrhizomata', though these are not widely accepted. One of the main distinguishing features of mycorrhizal organs is that the plants which possess them do not suffer acute disease due to the presence of fungal hyphae in them. Indeed, mycorrhizal organs are very often the main absorbing organs of many plants which form a few uninfected roots under natural conditions.

The knowledge of the existence of mycorrhizae dates back to the fourth century B.C. The mycorrhizal associations appear to have occurred in fossil club mosses (*Lepidodendron* spp.), of the early Devonian period. They are characterized by several infection points and the development of some coils of intracellular hyphae. But definite information with regard to their true structural nature started with the classical work of Frank in 1885. Frank introduced the name mycorrhiza to the composite fungus-root organs of the Cupuliferae. He showed that seeds of certain Orchids germinate only when in association with right type of fungi. Frank also pointed out that in some forest trees, mycorrhizae are formed with a number of different fungi, but in others the association is specific. But Burges, A (1936) established that the mycorrhizal association is an example of limited parasitic attack having no symbiotic value.

The physiological and biological significance of mycorrhiza was not thoroughly understood for a long time and therefore, caused some disagreement among botanists in interpreting the behaviour of its components. Some botanists regarded mycorrhizal fungi as parasites upon the roots, whereas others believed the relationship between the fungi and roots in a mycorrhizal association is more complex than fungal parasitism and that both fungi and roots benefit from their close association in mycorrhiza. According to this latter view, the fungi receive food from root tissues, and in turn facilitate certain physiological activities of roots. Through experimental trials it has been established that mycorrhizal roots have greater absorption rate and disease immunity than nonmycorrhizal ones. It is also true, the fungal component behaves as a parasite as soon as the associated nonfungal component becomes weak.

Occurrence. Mycorrhizal infection is not universal. Trees on soils rich in mineral nutrients may be quite uninfected. Aerable land, well manured soil or disturbed soil are less favourable for mycorrhizal infection. The occurrence of mycorrhiza is very common in poor soil, particularly in the forest trees.

Mycorrhizal components. The components of mycorrhiza are as follows:

(i) **Nonfungal.** Mycorrhizas are common among the angiosperms (members of the families—Cupuliferae, Orchidaceae, Ericaceae, Monotropaceae, Leguminosae, Genteniaceae, Gramineae, and Palmae); among the gymnosperms (members of the families—Pinaceae and Taxaceae); among the pteridophytes (species of the genera *Lycopodium* and *Ophioglossum*); and among the bryophytes (species of the genera—*Marchantia, Anthoceros, Pellia, Anuria, Lobelia* and *Lolium*).

(ii) **Fungal.** The fungal components of mycorrhiza are: among the basidiomycetes (species of the genera *Russula, Lactarius, Clitocybe, Tricholoma, Amanita, Armillaria, Marasmius, Boletus, Rhizopogon* and *Scleroderma*); among the ascomycetes (species of the genera *Tuber* and *Elaphomyces*); among the phycomycetes (species of the genera *Pythium* and *Endogone*); and among the deutoromycetes (species of the form-genera *Phizoctonia* and *Coenococcum*). There are wide variations in the behaviour and distribution of the mycorrhizal fungi. Some (*Amanita muscaria* and *Boletus edulis*) exhibit wide host range; again, *Boletus elegans* forms mycorrhiza with *Larix* species only; again others may behave differently in different situations, e.g., *Armellaria mellea* in Japan grows as a pathogen on conifers, whilst at the same time associated with the mycorrhizal tubers of the Orchid *Gastroidea elata*. Even a single fungus species (*Acaulospora laevis*) has been reported in Australia, Brazil, England, Scotland, New Zealand, South Africa, the U.S.A. and Pakistan. It forms mycorrhiza in almost all families of flowering plants including gymnosperms (except the family Pinaceae), as well as in many ferns and liverworts.

Mycorrhizal development. Seedlings remain uninfected until active photosynthesis begins. In the early stages of growth when the tap root and long roots are uninfected, the short roots become infected from the soil. The fungal hyphae enter into rootlets either through root hairs or rhizoids or through outer layer by direct penetration. They may grow intercellularly or intracellularly remaining restricted in the cortical cells, and do not affect stelar tissues and chlorophyllous tissues. Mycorrhizal development is dependent upon the growth cycle of the host, season, and habitat variations, arising from human activity.

Fungal behaviour. The hyphae may form external sheath on the root surface or loose weft of hyphae may remain externally or connect with soil merely by a few fungal strands. Intercellular hyphae produce specialized organs: **arbuscules** (haustoria) or **vesicles** or both arbuscules and vesicles and many other specialized structures—**sporangioles, peloton,** etc. in the host cells.

Nature of Mycorrhizal root. Due to mycorrhizal infection, the pattern of development of the parent root is changed. Those rootlets which become infected remain active and branch repeatedly growing slowly in length by dichotomous branch system. Hence there is increase in short root production and thereby absorption surface of root system is increased. This is commonly encountered in human soil. The mycorrhizal roots may also become coralloid and ball-like in shape. The best known mycorrhizae are those of forest trees and Orchids: these represent the two main types of fungus-root association—the **ectotrophic ectomycorrhiza** and the **endotrophic endomycorrhiza.**

I. **Ectotrophic mycorrhiza (Ectomycorrhiza).** In the ectotrophic mycorrhiza many of the fine lateral rootlets are enveloped in a mantle or sheath of light pseudoparenchymatous fungal tissue which prevents growth in length of the root causing it to assume a swollen or coralloid form. The light pseudoparenchymatous fungal tissue sends branches inward between the cortical cells of the root and outward in the soil and isolates the root from the soil and comprises a considerable part of the total weight and volume of the whole organ. The hyphae which penetrate the cortex form a network, often called the **Hartignet,** in the middle lamella of the cell walls and the intercellular spaces of the external cortex. Hyphae pass outward into the soil and ramify among the soil particles. These may be scarce or abundant. In some species an internal Hartignet only is present in the cortex, in others only an external sheath is found. The main mother roots by which the root system spreads are not usually completely infected by the mycorrhizal fungus. As new lateral roots are produced from the mother roots, they become mycorrhizal during their passage through the cortex or sheath so that a permanent state of mycorrhizal infection persists after initial infection in the seedling stage. Forest trees produce a large number of fine roots as their main function is absorption. Shortest roots become mycorrhizal. In them, hairs are absent and caps are poorly developed. They develop many branches. Clowes (1949) showed that the difference in anatomy between mycorrhizal and nonmycorrhizal roots of species of *Fagus* is in a cell shape. In infected roots some cortical cells show reduced longitudinal extension and increased transverse expansion and, in addition, meristematic activity might be reduced. These changes account adequately for the swollen shape of the mycorrhizal roots. Slankis (1958) showed that *Boletus variegatus* induces increase in the frequency of short-root laterals having dichotomous forking in species of *Pinus*. This suggests the morphogenetic effect of the fungus on the roots. Ectotrophic mycorrhiza is common in many of the forest trees belonging to the genera: *Pinus, Picea, Cedrus, Abies, Fagus, Quercus*. Among the basidiomycetes, species of *Russula, Lactarius, Clitocybe, Tricholoma, Amanita, Boletus, Scleroderma* and *Rhizopogon;* and among the ascomycetes species of *Tuber* and *Elaphomyces* are involved in ectotrophic mycorrhiza formation. *Coenococcum graniforme*, a deuteromycetous fungus is also involved with mycorrhizal association.

Some of these fungi exhibit wide host range, e.g., *Amanita muscaria, Boletus edulis;* others show host specificity, e.g., the genus *Pinus* has its selected group of "pine" mycorrhizal fungi and *Boletus elegans* forms mycorrhiza only with larch.

The basidiomycetous fungi of ectotrophic mycorrhiza are physiologically specialized in their dependence upon simple carbohydrates for growth. Most of them utilize glucose and fructose, some also mannose and sucrose. Others can hydrolyze starch, but few are able to utilize cellulose.

Mycorrhizal trees have greater vigour. The soil mycelium attached to the mycorrhiza can tap a greater volume of soil than the host root. All nutrients entering the host root first pass through the fungal hyphae. The interdependence of host and fungus is well understood in cases where the fungus produces extrahyphal enzymes which break down lignin and hydrolyze cellulose of humus and plant debris making suitable for the utilization of the forest plants, e.g., *Boletus subtomentosus* with *Pinus montana* and *Lactarius deliciosus* with *Pinus sylvestris*.

The ectotrophic mycorrhizae are basically **coralloid** and **ball types**. According to Melin and others, on the basis of external forms, the mycorrhizae may be differentiated into: (i) short dichotomously branched roots invested with mantles of various colours, the colour being determined by the species of fungus involved in its production, it is very common in nature; (ii) knots or tuber-like growths developed by the fusion of mantles produced by merged clusters of forked roots; and (iii) long, thin unbranched structures which occur upon the roots.

II. **Endotrophic Mycorrhiza (Endomycorrhiza)**. This is the most common and most important mycorrhiza virtually ubiquitous being present in tropical, temperate and arctic regions. It is found in almost all families of flowering plants including gymnosperms (absent in the Pinaceae), as well as in many ferns and liverworts. Endotrophic mycorrhiza is also characterized by the presence in plants of economic importance such as grasses, cereals and legumes. Here the external hyphal mantle or sheath is lacking, but present in *Monotropa* and among the members of the family Ericaceae. Again among the members of the Orchidaceae the external mycelium is scanty. In general, the fungal component forms externally on the root surface a loose weft of hyphae without ensheathing it. The branches of these hyphae enter the root cortex and penetrate cortical cells.

Fungal hyphae which may be septate or nonseptate, on gaining entrance in the root through root epidermis, very rarely through root hairs, grow inter- or intra-cellularly. They do not exploit green tissues, nor do they attack storage organs such as tubers. The endotrophic mycorrhizal fungi are not destructively parasitic as they occur at certain times as part of the normal process of root development. Roots become dichotomously branched when they become mycorrhizal increasing surface area of absorption. It has also been found that endotrophic mycorrhizas are associated in the nutrient absorption of saprophytic flowering plant *Monotropa.* The fungi also influence the germination of seeds and spores, and the early development of the young and adult plant by providing food material and enzymes essential for metabolism. Besides this, they also break down cellulose and lignin in the soil and supply carbon compound to their nonfungal components.

The main types of endotrophic mycorrhiza are distinguished as follows: (i) That caused by fungi with septate hyphae, usually basidiomycetous (*Armillaria mellea*) or deuteromycetous fungi with basiodiomycetous affinity (species of *Rhizoctonia* which have produced fruit bodies of *Corticium, Tulasnella, Ceratobasidium*). The nonfungal component belongs mainly to the Ericales, Orchidaceae, Gentianaceae; and a few pteridophytes and bryophytes. (ii) That caused by aseptate hyphae of phycomycetous fungi (species of the genera *Pythium* and *Endogone,* and *Rhizophagus-Endogone* complex); and the nonfungal component belongs to the families Gramineae, Palmae, Leguminosae, Compositae; in gymnosperms (other than Pinaceae); pteridophytes and bryophytes.

In some forest trees, mycorrhizae are formed with a number of different fungi, but in others the association is specific, e.g., strains of *Rhizoctonia*-like fungus with certain orchids; and *Armillaria mellea* with a Japanese orchid *Gastroidea elata.*

Depending on the nature of fungal and nonfungal components, fungal hyphae behave in the following manner. (a) The hyphae of larger diameter grow intercellularly producing intercalary or apical large and thick-walled swellings—the **vesicles.** The vesicles contain oil and are believed to function as storage organs or resting spores when the roots decay. Such mycorrhiza is known as **vesicular mycorrhiza (VM).** (b) Branches of main hyphal system usually of smaller diameter pass into cells and produce **arbuscules** (haustoria) which are often dichotomously branched. The apices of the arbuscules become swollen to form "sporangioles". The sporangioles may become free within the host cells, but disintegrate there. This is known as **arbuscule-sporangiole complex.** The whole arbuscule-sporangiole complex may disintegrate in the host cells. This is common in the Leguminosae (Papilionaceae) and Gramineae. (c) The hyphae may also produce combination of vesicle and arbuscule and thus known as—**vesicular-arbuscular mycorrhiza (VAM).** It is also known as phycomycetous mycorrhiza. It is found widespread in the plant kingdom. (d) Again form-species of *Rhizoctonia* produce intracellular hyphae in the roots of Orchids which become coiled as **pelotons** to be digested in the root cell.

Fossilized vesicular-arbuscular mycorrhizae have been found in the Carboniferous and Devonian strata. Their modern counterparts have been investigated, especially in the Gramineae. Two types have been identified, although both were formerly included in the genus *Rhizophagus*: they are species of *Pythium* and certain other fungi which are possibly small forms of *Endogone.*

Significance of mycorrhiza. Convincing proof of the functional nature of mycorrhiza, however, is lacking. Among bryophytes internal mycelium has been found of common occurrence in certain genera. But whether this relation is mutualistic still lacks positive evidence. Similar is the condition with the pteridophytes. Numerous theories have been put forward as to the true nature of mycorrhizae. But in general, there are two lines of thought: (a) mycorrhizae are pathological structures induced by the parasitic action of the fungus upon the root tissues; and (b) that the mycorrhizae are symbiotic structures.

Mycorrhizal infection induces increase in dichotomous branching of roots followed by profuse growth of rootlets increasing surface area of absorption of roots. Thereby mycorrhizal plants receive better nutrition than non-mycorrhizal ones.

Again majority of the feeding roots of ectotrophic mycorrhizal host are covered by a layer of fungal sheath which helps higher absorption rate. Besides this, the soil mycelium attached to the mycorrhiza can tap a greater volume of soil than the host root. All nutrients entering the host root first pass through the fungal hyphae. As such the mycorrhizal fungal hyphae play a very important role in nutrient absorption. Also this shows that the host is dependent upon the fungal partner for nutrients absorbed from the soil.

Mycorrhizae are especially important for plant nutrition in deficient soils where they may greatly increase the availability of phosphorus and several other nutrients not readily accessible to the plant. Many reports show that mycorrhizae enhance the uptake of Nitrogen, Phosphorus, Calcium, Na, Fe, Cu, Bo, Zn, Al, and Strontium, but reduce the uptake of K, and Mn. Since adequate phosphorus nutrition is required for root nodule formation, mycorrhiza may be a precondition for nodulation under deficient soil condition (Brown 1968, 1973; Sanders and others 1975). Besides this, mycorrhizae remain active on old roots for relatively long period of time increasing absorption rate.

It has been widely established that mycorrhizal plants when growing on nutrient deficint soil produce greater dry weight and absorb disproportionately larger amounts of nutrients from the soil than uninfected plants. Hence mycorrhizae are especially important for plant nutrition in deficient soil due to increase in the total and relative absorbing area of the root system and increased ability to accumulate nutrients per unit area for which mycorrhizal trees have greater vigour.

Slankis (1973) showed that auxin production enhanced during the dichotomous branching of roots induced by ectotrophic mycorrhizae helps in the elongation of radial walls of their cortical cells.

Mycorrhizal association releases inhabitory compounds to establish barriers against infection by pathogenic fungi. Thereby the host plant docs not suffer from any acute disease.

The fungal partner gets suitable habitat for growth on the roots surface where essential requirements are provided by the root itself during mycorrhizal association. Roots also excrete various organic substances which stimulate fungal spore germination and encourage root penetration. Again the mycorrhizal fungus influences the germination of seeds, the early development of young plant, or may stimulate the growth and activity of the adult. It provides enzymes essential for metabolism of the partner. Mycorrhizal association is almost universal in Orchids as because the mycorrhizal fungi provide germinating Orchid seeds with surgars and sometimes Vitamins (thiamine and biotin) and amino acids (Smith 1974). Form-species of *Rhizoctonia* form mycorrhizal association with terrestrial Orchids (Talbot 1966-67). This is also true with regard to the germination of *Cinchona* seeds.

The interdependence of host and fungus is well understood in cases where the fungus produces extrahyphal enzymes which break down lignin and hydrolyze cellulose of humus and plant debris making suitable for the utilization of the forest plants, e.g., *Boletus subtomentosus* with *Pinus montana* and *Lactarius deliciosus* with *Pinus sylvestris.* The interdependence of host and fungus in mycorrhizal relationship is, therefore, seen to involve the supply of major and minor nutrients as well as morphogenic factors to both partners and demonstrates how closely integrated are the two components, fungus and host, which form nutrient absorbing organs. This behaviour of the host and fungus is better understood in cases where an exchange of enzymes and Vitamins (thiamine and biotin) and amino acids takes place. Hence, there is little doubt that mycorrhizal infection has a morphogenetic effect leading to prolonged life, growth, and branching of the rootlets. Larger number of basidiomycetous fungi are mycorrhizal, they are dependent on simple carbon compounds which have been provided by their hosts.

Species of *Endogone* are totally dependent on association with plant roots for growth. In the initial stages of infection they require the stimulation of exudation from the host for the determination of the direction of growth of the hyphae, for the production of appressoria and for penetration of the host tissues.

An interesting phenomenon is encountered in the saprophytic flowering plant *Monotropa* which is associated with endotrophic mycorrhiza in which fungal component helps in nutrient absorption. Hence carbonaceous food passes from neighbouring spruce and pine trees through the mycorrhizal fungal hyphae to the cells of *Monotropa.* Such a relationship is also found in some saprophytic Orchids, and is termed **epiparasitism.**

Mycorrhizae play important role in afforestation. It has been demonstrated that pine and spruce seedlings exhibit nitrogen starvation and eventually die unless they are invaded with mycorrhizal fungi. The seedlings of *Pinus strobus* do not grow well without mycorrhizal fungi. Of the mycorrhizal fungi experimented, *Boletus viscidus* produces best results. By field trial it has also been established that some members of the Boletaceae form mycorrhizae with forest trees and may be important in the establishment of pine plantation. Besides this, it has been further established that the seedlings with mycorrhizal fungi are immune to various diseases and pests.

Mycorrhizae play a great role in agriculture for crop production and forestry for forest development plant.

1. **Role of mycorrhizae in agricultural crop production.** Ectotrophic micorrhizae helps the mycorrhizae infected roots to absorb mineral ions, N_2, Ca, Zn, Fe, Na and others, also in the formation of dichotomous branching and profuse root growth and hence absorption area helping improved plant of growth.

2. **Role of mycorrhizae in Forestry.** Trees with facultative endotrophic mycorrhiza act as first invader in wastelands as a pioneer in plant succession. The application of mycorrhizal fungi in forest bed enhances the growth of mycorrhiza and prevents the entry of fungal root pathogens. This method is very effective against the attack of *Phytophthora cinnamoni* in the roots of *Pinus clausa.* Mixed nitrogenous compounds—nitrate, ammonia, etc., are available to the plants. Hence it helps plant growth, especially in a acid soil.

Orchids produce a large number of seeds per capsule. The embryo of seeds has no storage food. The majority of seeds are unable to germinate without the exogenous supply of carbohydrates. Hence mycorrhizal association is obligatory to the seeds to germinate. Initially the mycorrhizal fungus enters the embryo and colonizes being restricted to the cortical cells receiving C-nutrition from the invaded mycorrhizal fungus. Species of *Rhizoctonia, Corticium, Ceratobasium* are associated in this process.

Mycorrhizal fungal infection does not cause disease, but its absence in certain fields results in plant stunting and poor growth, which can be avoided if the appropriate mycorrhizal fungi are added to the plants. Also soil fumigation results in the evadication of mycorrhizal fungi disturbing normal plant growth. The application of systemic fungicides metalaxyl and fecetyl. At increase mycorrhizal infection rate and yield in some crop plants.

Pseudomycorrhizae. The term pseudomycorrhiza was first introduced by Melin to short roots attacked by certain root parasites and other fungi in the absence of mycorrhizal fungi, so that the cells were penetrated, growth was slow, and branching was rare or absent. The fungus particularly involved was *Rhizoctonia sylvestris.* This weak pathogen does not compete successfully with mycorrhizal fungi for the colonization of the surface tissues of the roots in normal conditions. Usually the root becomes invested to some extent with a kind of sheath of fungal tissue and the cortex is permeated by hyphae which penetrate the cells to form structures called haustorial. When seedlings of pine and spruce are infected, some decrease in vigour of growth which varies with the species, occurs. *R. sylvestris* does not cause damage to the hosts. Actually pseudomycorrhiza formation is stimulated in conditions where mycorrhizae fail to form.

E. WHAT ARE FUNGI?

The fungi are regarded as members of the Plant Kingdom and certainly, in general aspects, the majority of them bear a superficial resemblance to plants.

That the fungi do not really belong to either of the Plant and Animal Kingdoms, but constitute a third co-equal Kingdom, is often observed by mycologists. Following are some of the aspects of fungi that deserve special consideration to finally decide what the fungi are.

One fundamental difference from the plants is that none of the fungi bear chlorophyll and are not capable of effecting photosynthesis. Plants, in general, utilize simple substances and build them up into substances of greater complexity by various processes of which photosynthesis is of primary importance. True, fungi are incapable of synthesizing carbohydrate in a process similar to that of green plants. But it has been experimentally established that a large number of fungi whose metabolic activities have been worked out, can synthesize complex organic substances in special processes of their own which can seldom be compared with those of plants and animals. Fungi are, however, similar to green plants in being fixed and possessing cell walls though they differ fundamentally in their lack of chlorophyll.

The fungi show great differences in size, structure, and metabolic activities. Some of the fungi are unicellular, while others, such as the mushrooms and other higher fungi form large fruit bodies of complicated structure with elaborate mechanism for propagation. The faculty of fruit body development, an outstanding feature of a large number of fungi, has virtually provided an ample scope to consider fungi to be different from plants and animals.

Again, according to Langeron (1945), a further fundamental difference between plants and animals on the one hand and fungi on the other is that the latter never form tissue. All structures, including the highly organized fruit bodies of the higher fungi, consist entirely of a system of tubes. Even when the tubes are apparently divided into individual cells by cross walls, these invariably have a central pore, through which both cytoplasm and nuclei can freely pass. Langeron maintains that fungi are thus essentially unicellular.

The cell wall material of the fungi is made up of chitin, which, though has resemblance with that of animals is chemically quite different has already been established by Weitstein as early as 1910. The animal chitin is much more complex than the fungus chitin, the similarity being very superficial.

With gradual evolutionary progress in fungi the interval between plasmogamy and karyogamy increases and there is interpolation of a third phase, the dikaryophase, in between the haplophase and diplophase. This can very well be traced in the different groups of fungi. In the lower fungi, the life cycle comprises only of two phases, the haplophase and the diplophase, the latter being very short and the former is predominant, whereas in the Ascomycotina, there is clear interpolation of dikaryophase in between the haplophase and diplophase, though the dikaryophase is entirely dependent on the haplophase in every respect. Finally, in the Basidiomycotina there is clear occurrence of the three phases (haplophase, dikaryophase, and diplophase) which are entirely independent from each other. The dikaryophase is rather predominant in the entire life cycle. The dikaryophase, an unique feature of fungi, is not encountered in any other living organism.

Cytological and genetical studies indicate that the life cycle pattern, and divisional pattern of mitosis and meiosis in fungi are distinctly different from other Eucaryotes, i.e., plants and animals. The dicaryotic stage in the life cycle is unique and provides immense potentiality of sexual recombination. Consistence of a nuclear membrane during the divisional stages, presence of spindle-pole bodies (SPB), climbing-pole type mitosis, equational separation of chromosomes in first division of meiosis and parasexual recombination to form a 2n-zygote and haplodization through aneuploidy are very unique features of fungi.

That the fungi have not evolved from the algae by the loss of chlorophyll is now an accepted fact, whereas the fungi had a common ancestry with the protozoa but split off at a very early stage of organic evolution. If this line of evolution is found to be correct, the fungi do not belong to any of the two Kingdoms: Plant and Animal.

To a precise definition for Fungi (Sing. Fungus) is very difficult.

Alexopoulos and Mims (1979) defined fungi as eukaryotic spore leading a chlorophyllous organism that may reproduce sexually and asexualy, having vegetative body filamentous, branched whole cell wall is composed of chitin, cellulose or both of these substances and many complex carbohydrates.

Fungi vary informs, behaviour and life cycle pattern. Their vegetative and reproductive structures are so unique that they cannot be assigned to positions like Algae and Higher green plants. They exclude themselves from bacteria by their cell structure. Fungi are also different from the Myxomycetes (Slime Molds) have animal-like nutrition.

According to Martin and others (1932), fungi are neither plants nor animals and hence belong to a separate **Kingdom.** They recognized three **Kingdoms: Fungi, Plants** and **Animals** having a common origin and undergone a parallel development.

Fungi are different from Plants and Animals for having features.

1. Vegetative body mostly hyphal growing in and along the substratum. Hyphae are nonseptate to septate. Septa are porous, pores are often plugged by cellulin or some other substances. The hyphal cell wall is laminated, composed of several layers of microfibrils arranged in various ways. The chemical heterogenecity of hyphal wall has been recognized for several decades, since De Bary (1866, 1887) distinguished 'true cellulose' from 'fungal celulose'. Three classes of compound: **cellulose, chitin** and insoluble **glucan.** The fungal cell wall is generally composed of a complex nitrogenous substance known a **chitin** which is often impregnated with some salts or similar other substances. The chitin present in the fungal cell wall is not strinctly identical with animal chitin. This has been established by Weistein (1910).

2. Hyphae often produce pseudoparenchytous (false parenchymatous) tissue cannot prepare own food.

3. Lack of photosynthesis cannot prepare own food. Dependent on other external source.

4. Nutrition absorptive from substratum, as saprophytes, as parasites, and as symbionts and mycorrhizal.

5. Unique life cycle pattern.

6. Reproduction both by asexual and sexual methods.

7. Motile stages are present in vegetative stage in some lower group of fungi in asexual spores (zoospores) and sometimes in gametes.

8. According to Langeron (1945) fungi possess unique process of fruit body development along with pseudoparechymatous tissue development.

9. Fungal evolution of sex is by the degeneration of sex from lower to higher groups of fungi.

10. The interval betweeen plasmogamy and karyogamy, in lower to higher fungi, is occupied by three phases—haplophage (n), dikalycphase (n + n) and diplophase (2n) followed by meiosis.

According to above facts, it is logical to accept that fungi belong to a **Kingdom** of their own which is different from **Plant Kingdom** and **Animal Kingdom**.

F. ROLE OF FUNGI AS DECOMPOSER IN FOREST ECOSYSTEM

All kinds of leaves—dead and alive that follow the forest floor are collectively known as **litter.** A large amount of litter that accumulates on the forest floor is decomposed by forest floor-microflora and fauna (fungi, bacteria, protozoa, etc.). These microorganisms inhabiting on or in forest floor soil are designated as **decomposers** as their function is to decomposes litters on the forest floor under suitable moisture and temperature conditions. Finally, these decomposers convert forest litter into a dark organic matter known as **humus.** By the time along with this process the microorganisms grow and particularly fungi produce sexual spores which are disseminated by wind causing air pollution. Fungi taking part in this process have been identified as species of **Cladosporium** and **Aureobasidium.** These fungi are designated as 'resident inhabitors' in contrast with casual inhabitors' based on their behaviour. All inhabitors like yeasts and filametous fungi are **Chemo-organotrophs.** During their growth free sugars, amino acids and inorganic ions are produced. Fungi taking part in this process are species of genera *Alternaria, Curvularia,* and *Nigrospora.* They grow propusely on dead and decomposed tissues of leaves. Species of *Mucor, Mortierella, Penicillum* and *Trichoderma* together with litter decomposing fungi like species of *Collybia* and *Mycena* grow in association with cellulolytic Agaricales as commensals by taking a share of hydrolytic products of cellulose and thus acting as secondary saprophytes. Again decomposition of cellulose, leganin and other substances is done by *Rhizopus* sp. and *Fomes pomaceus.*

Three types of wood decay of forest trees have been recognized—white rot, brown rot and soft rot. *Lenzites betulina, Conolus hirrutus* and *C. vervicolor* are common white rot fungi. Species of *Chaetomium, Ceratocystis,* and *Xylaria* cause soft rots. Members of Agaricales and Aphyllophorales cause white rots. Wide variety of invertebrates and bacteria also play an important role in wood decomposition. Armillaria *mellia* known as *'shacisting fungus'* cause root rot of forest trees. In living forest trees, *Fomes* spp. cause root and butt rot; *Inonotus* spp. cause white heart rot and *Ganderm* spp. induce root and basal rot. Vast quantities of liquid are incorporated into forest soil. These may be introduced into forest soil in the form of organic residues from wood decay.

The natural litter comprises of complex organic residues of plant and animal remains which are chiefly different forms of carbohydrates, proteins, fats, waxes, etc. They are attacked by soil inhabiting microorganisms which serve as biogeochemical agents for this conversion into simple inorganic components or into their constituent elements. This overall process is called **mineralization** performed by a large number of bacteria, some of them are: *Proteus* spp., *Pseudomonas* spp., and *Bacillus* spp; and many fungi and actinomycetes.

With the help of decomposition process, the energy source bound in plants are released and recycled and become available to the forest plants. The recycling process is very helpful for good health of the forest plants as well as microfauna of forest soil. Besides this, the fungi also become a member of mycorrhizal system and supply nutrients to the forest plants.

G. MYCOTOXINS

Toxins are highly specific poisonous substances produced by some microorganisms. The toxins produced by fungi are called **mycotoxins.** Again, the disease resulting from the consumption of mycotoxin contaminated food is **mycosis.**

Many fungi produce toxins that are responsible for food poisoning and cause different types of effects on animals including human beings. Mycotoxins are broadly classified into three groups: (a) Food toxins, (b) Ergot toxins, and (c) Mushroom toxins.

(a) **Food toxins.** Four groups of Food toxins are associated with human diseases.

These are: (i) **Aflatoxins**, (ii) **Ochratoxin**, (iii) **Zearalenone**, and (iv) **Trichothecenes**.

(i) *Aflatoxins.* By far the most commonly known mycotoxins that are produced by fungi—*Aspergill flavus, A. fumigatus, A. parasiticus* and *Penicillum islandicum* of them, aflatoxins produced by *Aspergillus flavus* and *A. parasitica* grow on stored food products largely in ground nuts. Aflatoxins, like most mycotoxins, are low-molecular weight compounds. They are heat stable so that they will survive most cooking conditions.

(ii) *Ochratoxins.* They are produced by *Penicillium viridicatum.*

(iii) *Zearalenone.* Produced by *Fusarium graminearum* and *F. moniliforme* prominantly found in maize.

(iv) *Trichothecenes.* They are produced by the species of the genera—*Fusarium, Cephalosporium.*

Species of *Aspergillus* are most important producers of mycotoxins. In the year 1960 turkey pullets died in England being the cause of feeding them with groundnut meal. Detailed work indicated that a toxin produced by *Aspergillus flavus* might be the cause of trouble. This led to the discovery of the toxin—**aflatoxin.** Aflatoxin TS_1 is the one of the most carcinogens. In addition to suspected carcenogenic nature, due to aflatoxin poisoning are jaundice, portal hypertension, etc. some more commonly occuring mycotoxins are as under:

(b) **Ergometrene** isolated from sclerotia of *Claviceps purpurea.* Due to consumption of ergot sclerotia as whole or as in power along with rye flower, the domestic animals show symptoms like—abortion in cows, diarrhoea and abdominal pain and vomiting, etc.

(c) **Mushroom toxins.** Several mushrooms—*Amanita phalloides* (mycotoxin causes the death of cub), *Helvella infula, Clificybe* sp. (mycotoxins causes various abnormalities).

H. INDUSTRIAL USES OF FUNGI

Common industrial uses are listed below:

1. Fungi have two major uses in industry, as reviewed by Civay (1959), Christensen (1965) and Emerson (1973): (a) As a source of chemicals and (b) as food or food processing agents.

(a) **Source of chemicals.** Fungi produce a wide range of useful compounds like organic acids, alcohol, antibiotics, etc. Of these, antibiotics are the most familiar. The most popular useful antibiotics produced by fungi are indicated in the following table:

Useful antibiotics of fungal origin	Produced by	Active against
Penicillin	*Penicillium chrygenum*	Gram-positive and Gram-negative bacteria
Griseofulvin	*Penicillium griseofulvum*	Most fungi
Mycophenolic acid	*Penicillium stoloniferum*	Antitumour
Hadacidin	*Fusarium coccineum*	Gram-positive
Fusidia acid	*Mucor ramanarianus*	bacteria

Apart from antibiotics, five major activities of fungi are exploided by industry: (i) production of alcohol and/or carbon dioxide; (ii) production of organic acids; (iii) production of glycerol, lipids, (iv) modification of steroid compounds and (v) production of extracellular enzymes.

(b) **Uses of fungi as food and in food processing.** A wide range of foods are obtained from fungi which are relatively more popular. To increase their protein content and also to produce 'single cell protein' (SCP), fermenting activities of fungi is of much use. Most common fungus lasts (mainly *Saccharomyces cereviciae*) and *Aspergillus niger* used in food industry for fermenting activity. Yeast is mixed with some cereal flours to increase their protein content.

(c) **Penicillum roquefortii** is used in *Cheese-making industry.*

(d) **Used for Biological assays**—*Nerospora crossa* used for biological assay of B-complex and *Aspergillus niger* used for trace elements as Zn, Pb, Cu, Mn, Mb.

The vital role played by fungi together with the other microbes is the decomposition of organic matter, thus releasing the nutrients locked up in the dead bodies of animals and plants and bringing about the recycling of nutrients in nature. In the soil, microorganisms convert the nitrogen of proteins to ammonia and nitrate, they oxidize organic carbon to carbon dioxide, and liberate organically bound materials. These products of microbial activity constitute the nutrients of higher plants. From these chemical compounds and elements, plants synthesize their tissues. Were there not be microbes to decompose waste organic materials, the earth in a very short time would be covered to an astonishing depth with an accumulation of vegetable debris and other waste products that had their origin either directly or indirectly in plant material.

The decomposition process actually starts while the plants are still intact standing in the fields in nature, that is at the senescent stage and is continued till the plant parts are totally decomposed. Colonization of the living plants as well as their dead remains by microorganisms has been studied by numerous workers. The constituents of the substratum

Table 16.1: General trend of fungal succession on a corpius of plant tissue lying within or upon the soil as proposed by Garrett (1963)

Senescent tissue	*Dead tissue*		
Stage 1a	Stage 1	Stage 2	Stage 3
Weak parasites	Primary saprophytic sugar fungi, living on sugars and carbon compounds simpler than cellulose	Cellulose decomposers and associate secondary saprophytic sugar-fungi, sharing products of cellulose decomposition	Lignin decomposers and associated fungi

Table 16.2: Scheme of fungal succession proposed by Hudson (1968)

Living	*Senescent*	*Dead*
PARASITES Ascomycetes and Fungi Imperfecti. May be host specific or host restricted, e.g., *Leptosphaeria sacchari* on sugarcane, *Gomonia errabunda* on beech leaves, *mirabilis* on Eucalyptus leaves	COMMON PRIMARY SAPROPHYTES Ascomycetes and Fungi Imperfecti *Cladosporium herbarum, Alternaria tenuis, Epicoccum nigrum, Aureobasidium pullulans* and *Botrytis cinerea* (+ in Tropics *Nigrospora* and *Curvularia* sp.) Readeriella RESTRICTED PRIMARY SAPROPHYTES Ascomycetes and Fungi Imperfecti, e.g., *Leptosphaeria microscopica* on grasses, *Fusicoccum bacillare* on Pinus, *Piggotia subtellata* on Eucalyptus	SECONDARY SAPROPHYTES (i) Ascomycetes and Fungi Imperfecti (ii) Basidiomycetes (iii) 'Soil Inhabitants', e.g., Mucorales, Penicillia, etc.

are attacked at different rates and any species of microorganisms is capable of decomposing only certain components in the plant or animal residue, and a number of species of microorganisms are required to bring about the complete decomposition of the substratum. Thus there is a succession of microflora on a substrate at various stages of decomposition (Table 16.1) and (Table 16.2). The time required for the decomposition of different plant materials ranges from a year for leaves and stem to about 8-9 years for a pine needle.

The organic fraction of the soil, often termed humus, is a product of the synthetic and decomposing activities of the microflora. According to Waksman (1938), humus is important in soil processes in three ways: (i) physical, in that it modifies soil colour, texture, structure, moisture holding capacity and aeration; (ii) chemical, in that it influences the solubility of certain soil minerals, forms compounds with certain elements (such as iron) which makes them available to plants and increases the buffering properties of the soil; and (iii) biological, in that humus is a source of energy for microorganisms and makes the soil a better medium for the growth of higher plants. Microorganisms, such as fungi, bacteria and actinomycetes play a vital role in the formation of brown or black organic complexes which remain in a dynamic state, through the decomposition of organic materials and through the synthesis of substances that also become part of the humus. Humus is sparingly soluble in water but is rendered into solution by alkali. It contains amino acids, purines (adenine, guanine), pyrimidines (cytosine, uracil, thymine), aromatic molecules, uronic acids, amino sugars, pentose and hexose sugars, sugar alcohols, methyl sugars, aliphatic acids and probably other substances of unknown nature. Humus can be fractioned into humins, humic acids, fulvic acids and hymetomelanic acids. Although resistant to microbial attack, along with bacteria and actinomycetes (*Bacillus, Pseudomonas* and *Streptomyces*) *Aspergillus* and *Penicillium* are the fungal genera involved in the degradation of humic acid.

The role of fungi as active agents in the degradation of organic debris—whether in connection with the removal of organic waste, the return of carbon dioxide to the atmosphere, or the formation of humus—involves the biological degradation of plant materials. These plant materials or residues comprise carbohydrates, proteins, fats, oils, alcohols, aldehydes, ketones, organic acids, lignin, phenols, tannins, hydrocarbons, alkaloids, pyrimidines, purines, enzymes and pigments. The bulk of the plant residue is made up of lignocelluloses. The decomposition of cellulose, hemicellulose, pectin and lignin is discussed below.

A diverse group of fungi utilizes cellulose for its carbon and energy sources. The cellulose decomposing fungi includes members of Deuteromycotina, Ascomycotina and Basidiomycotina and a few members of the Lower Fungi. Strongly cellulolytic fungi are represented by the species of *Aspergillus, Penicillium, Chaetomium, Trichoderma, Fusarium, Stachybotrys, Memnoniella, Cladosporium, Alternaria, Acremonium, Ceratocystis, Coniothyrium, Myrothecium, Humicola, Phoma, Thielavia,* etc. In the destruction of forest litter, wood and woody tissues, cellulolytic basidiomycetes (i.e., Aphyllophorales) are specially prominent. It has been proposed that fungi are the main agents of cellulose degradation in humid soils while bacteria are of greater significance in semi-arid localities. The enzymes involved in the decomposition of cellulose are collectively known as cellulases.

Decomposition of cellulose is brought about by cellulases which form a group of enzymes hydrolysing β-1, 4 glycosidic bonds in cellulose or its derivatives. The enzyme cellulase (correctly called β 1-4 glucan, 4 gulcan hydrolase) which enbales microorganisms to attack this complex system is not fully understood. However, two stages seem to be involved, first the crystalline micelle is broken down, and then the amorphous cellulose is hydrolysed either end-wise (exo-) or in a random manner (endo-) to cellobiose.

The nature of the initial attack on natural cellulose by cellulolytic enzymes is still not clear, but subsequent cleavage by β-1, 4 glucanases (exo-β-1, 4 glucanases end endo β-1, 4 glucanases) and β-glucosidases follows a pattern analogous to the hydrolysis of pectic sub-

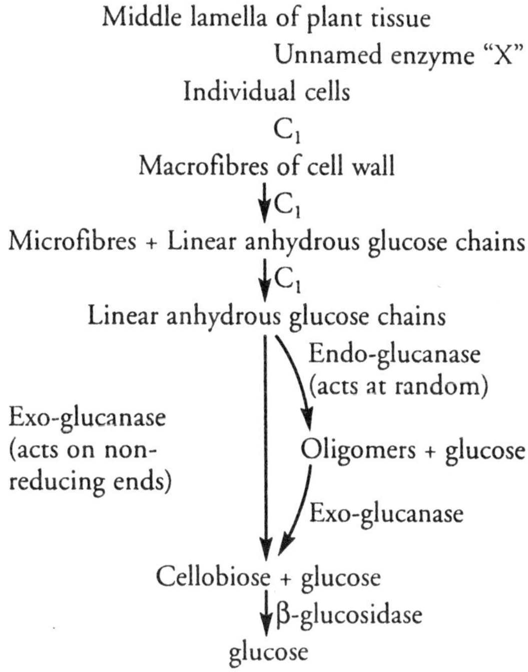

Fig. 16.7: Scheme of breakdown of cellulose in lignocellulosic materials (based on Chahal and Overend, 1982)

stances. C_1 enzyme (still unclassified) acting on highly oriented solid cellulose and exposing it to attack by other cellulases (Reese, Siu and Levinson, 1950; King and Vessal, 1969; Reese, 1975; Chahal *et al.*, 1976 and Chahal and Overend, 1982). Scheme of breakdown of cellulose in nature is given in Fig. 16.7.

The hemicelluloses occur abundantly in plant walls interspersed through cellulose and lignin. The term hemicellulose is an unfortunate choice because these molecules bear no structural relation to cellulose. These are low-molecular weight polysaccharides, most of which are water-insoluble but they are easily solubilized by alkali. They are more readily acid hydrolyzed than cellulose. In these polymers the simple sugars (or monosaccharides)—or uronic acids which are the derivatives of simple sugars are bound into large molecule. A polysaccharide may have 50-200 sugar units and exists as a linear chain as in cellulose, but the hemicelluloses are usually branched.

Decomposition of the hemicelluloses involves hydrolysis of the polymer to smaller units by specific enzymes. The actual enzymes involved and the biochemical steps depend on the particular hemicellulose which is being degraded. Three types of enzymes may be involved: (i) endo-enzymes, which randomly cleave the bonds, between building blocks in the polymer; (ii) exo-enzymes, which cleave either a dimer or monomer from the end of the polysaccharide chain; and (iii) enzymes collectively known as glycosidase which hydrolyzes the oligomers or disaccharides produced from the hemicellulose by the polysaccharide-cleaving enzymes and produces the simple sugar or uronic acid. The hemicellulose-decomposing fungi belong to all major groups, and the total number of species that can decompose hemicellulose is far greater than those capable of degrading cellulose. The most important fungi utilizing hemicelluloses are species of *Alternaria, Aspergillus, Chaetomium, Fusarium, Glomerella, Penicillium* and *Trichoderma.* The enzyme in some organisms is constitutive but in others it is inducible. Hemicelluloses tend to disapper rather rapidly during the early stages of decomposition.

Pectic substances are the primary constituents of the middle lamella and are the structural elements of the primary wall. The pectic carbohydrates are complex polysaccharides composed mainly of galacturonic acid units bound to one another in a long chain linked by α-1, 4 glycosidic links in linear chains. The carboxyl group may be esterified with methanol. There are three types of pectic substances: (i) pectic acid, which is the unmethylated polygalacturonic acid polymer; (ii) pectinic acid, which is the substantially methylated colloid; and (iii) protopectin, which is the insoluble pectic material as it occurs in the plant tissue. Pectins are water-soluble, pectinic acids of varying methyl ester content which may form gels with sugar and

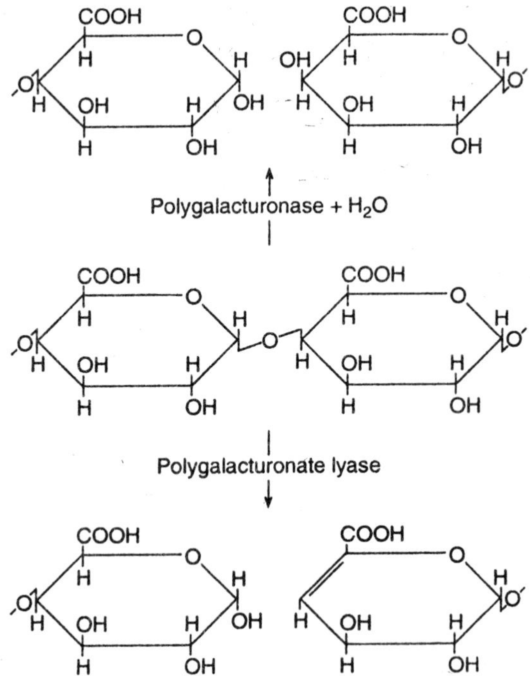

Fig. 16.8: Depolymerizing pectinolytic enzymes

acid. Decomposition of pectin is brought about by microbial enzymes, which can be divided into three major categories: (i) hydrolytic that carry out a hydrolytic cleavage of the polymer; (ii) transeliminative that brings about transeliminative cleavage of glycosidic links by the action of a lyase (Albersheim *et al.*, 1960) (Fig. 16.8) and (iii) pectinesterases which hydrolyse off the methoxyl groups from the pectins, thereby converting pectin or pectinic acids to pectic acid (Fig. 16.9).

$$(RCOOCH_3)_n + nH_2O \rightarrow (RCOOH)_n + nCH_3OH$$

Fig. 16.9: Action of pectinesterase

The pectic enzymes are usually extracellular and inducible, but intracellular and constitutive ones are also known, however, only the extracellular enzymes are probably important in the break-down of pectic substances in plant residues reaching the soil. A number of fungal genera are known which have the capacity of producing, pectolytic enzymes (Aneja, 1978). A single organism can and often does make several of the extracellular enzymes. Endo- or exo-polygalacturonases (PG) are known to be produced by various species of *Rhizopus, Aspergillus, Penicillium, Monilia, Fusarium* and *Rhizoctonia* while Endo-polymethyl-galacturonate lyase (Endo-PMGL) is produced by *Fusarium, Geotrichum* and *Rhizoctonia;* Endo-polymethyl-galacturonase (Endo-PMG) by *Aspergillus, Botrytis, Fusarium* and *Rhizoctonia,* Endo-polygalacturonate lyase (Endo-PGL) by species of *Aspergillus, Penicillium, Fusarium* and *Rhizoctonia* and Pectin esterases by *Alternaria* and *Fusarium.*

Legnin is a major cell wall constituent characteristic of woody tissues. The lignin contents of the plant increases both as the plant-ages and with the increase in abundance of woody tissues.

Lignin or lignin-derived molecules have long been considered to be of significance in the formation of humus which is important in maintaining soil fertility. The decomposition of lignin is brought about almost entirely by fungi and especially Basidiomycetes, such as members of Aphyllophorales and Agaricales. A single lignin-degrading enzyme, a so-called "lignase", has not been described so far. However, the most probable enzymes involved in its decomposition are phenol oxidases, laccases and peroxidases. The phenol oxidases oxidize aromatic compounds containing one (phenols) or two hydroxyls, while the laccases oxidize only those aromatic compounds with more than one hydroxyl group, peroxidases can oxidize aromatic molecules also, but they do so in the presence of hydrogen peroxide. These enzymes most probably cleave ether linkages, remove side chains, and cause the loss of methoxyl groups. Schubert's (1968) scheme for the microbial decomposition of lignin is given in Fig. 16.10.

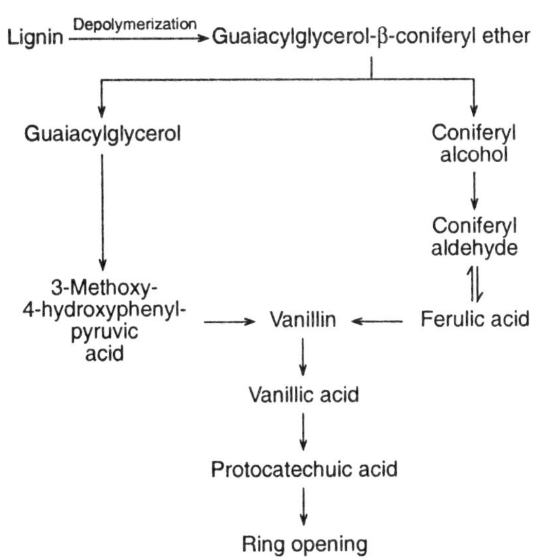

Fig. 16.10: Scheme of microbial decomposition of lignin (based on Schubert, 1968)

BIODETERIORATION OF CELLULOSIC AND NON-CELLULOSIC MATERIALS OF ECONOMIC IMPORTANCE

Fungi play an important role in biodeterioration of cellulosic materials, such as wood, used as timber or as wood products, such as paper and board, and other plant fibres such as cotton, flax, jute, hemp and sisal used as rope, cordage, textile and packing and filling materials and non-cellulosic materials, such as plastic, glass, electrical equipment, fuel, paints, paint films, leather and glue, etc.

Deterioration of wood: Among the most valuable of all our resources are our forests. Many of the materials used by man to make its products are of plant or animal origin. Although we have entered into the age of plastics but wood, an important product of forests, will continue to be used for one purpose or the other. Wood is the xylem conducting tissue of the trees and contains fibres or vessels or tracheids which are usually dead and in growing trees a small amount of live parenchyma also. The cell walls contain lignin and cellulose together with a selection of polysaccharides including mannans, xylans and other hemicelluloses. Wood from Angiosperms is generally referred to as a hard wood and from Gymnosperms (mainly conifers) as soft wood. Hard and soft woods differ in their cellular composition and decay fungi are restricted to one or other of them. But for a few ascomycetes (e.g., *Ustulina* and *Xylaria*) these are mainly the basidiomycetous fungi which are responsible for the wod decay. The important genera of the Basidiomycetes belonging to the orders Agaricales and Aphyllphorales are: *Armillariella, Collybia, Lentinus, Pholiota, Schizophyllum, Echinodontium, Hydnum, Daedalea, Fistulina, Fomes, Ganoderma, Lenzites, Merulius, Polyporus, Polystictus, Poria. Serpula, Trametes, Coniophora, Hymenochaete, Peniophora* and *Stereum.* In Britain decay of sawn timber by one fungus alone, i.e., *Merulius lacrymans* causes a loss of several million pound sterling per year.

There are two types of wood decay: (i) in which there is destruction of lignin (or infrequently cellulose) when the rotted wood becomes white and sponge-like, e.g., by *Trametes pini,* red ring rot of conifers; and *Ganoderma applanatum* white, mottled rot of hard woods, and (ii) 'dry rot' in which there is destruction of cellulose when the rotten wood is brown, soft and easily powdered, e.g., by *Phaeolus schweinitzii* responsible for red-brown butt rot of conifers; *Lentinus lepideus* responsible for brown cubical rot of conifers, especially pines used in buildings and *Serpula lacrymans* and *Poria incrassata.*

Spoilage of Paper: Raw materials, i.e., pulpwood (90% of the raw material for paper) and paper are vulnerable to fungus attack at every stage of paper manufacture. Species of *Alternaria, Aspergillus, Cladosporium, Penicillium. Trichoderma, Mucor* and a basidiomycetous fungus *Polyporus* are commonly encountered in bulk storage vats and are responsible for loss of cellulose and in association with bacteria form a coloured slime in a watery suspension of pulp which is foul smelling. The slime effects the quality of the paper and increases the cost of the maintenance of the machines. Once the paper is finished, it provides suitable substratum for cellulolytic fungi that may cause black, brown or yellow discolouration and spotting through "mildewing" or destruction of paper mainly by members of Deuteromycotina (Aspergilli and Penicillia) and species of *Chaetomium* belonging to Ascomycotina.

Deterioration of textiles: Textiles and cord made of linen, cotton, manilla, or jute composed entirely of cellulose are attacked by several fungi, especially cellulolytic. In the Second World War this sort of decay was of major importance in military operations and especially in the humid tropics where clothing, tents and other fabrics were moist much of the time. Principal damage to textiles include the production of musty odours, spotting and discolouration, loss in water repellency, loss in strength, and decrease in flexibility either by the production of pigments, enzymes or presence of fungus structures or both. Some of the common fungi attacking textiles are species of *Aspergillus, Penicillium, Trichoderma, Humicola, Chaetomium globosum* and *Cladosporium herbarum.* A hyphomycetous marine fungus, *Zalerion,* does serious damage on twines and ropes used in sea-water.

Fungi are also responsible for causing damage to wollens but the damage is insignificant in comparison to fungus damage on cottons. The most common genera responsible for the decay of woolens are *Microsporum* and *Trichophyton.*

Deterioration of leather: Hide which is a proteinaceous material is prone to considerable attack during its manufacture into leather, especially during the very early stages. In use decay of leather and leather goods is often due to fungi which may damage and mark leather. This is serious especially where storage conditions are poor. If leather is well tanned there is less likelihood of growth of fungi on it. Chrome-tanned leather is almost free of fungal attack but this type of leather may not be suitable for many other purposes. Much of the damage in leather materials are often due to the rotting of the stiching threads as a result of fungal growth. The common leather destroying fungi are species of *Aspergillus* and *Penicillium.*

Deterioration of optical equipment: Although glass as such is not utilized by fungi as a nutrient source. However, considerable damage may be caused by fungi of optical equipments such as binoculars, range finders cameras, gunsights, microscopes, telescopes and surveying instruments. Damage is due to the growth of fungi, living on accumulations of dirt on the lens surface or nearby. This growth either obscures vision by its presence or more seriously by etching the lens surface by its corrosive secretions. Species of *Aspergillus, Penicillium* and *Scopulariopsis* are the common molds associated with the optical equipments. Regular inspection, cleaning and good storage conditions are essential for saving the high-cost materials from its damage by molds.

Deterioration of electrical equipment: Electrical equipment are also subject to attack by molds especially species of *Aspergillus, Penicillium* and *Scopulariopsis,* particularly in tropical climates. The mold growth results in losses in insulating propertis of components and also direct short circuits via fungal hyphae. The use of non-nutrient components, fungicides, and the total sealing of assemblies after manufacture are done to prevent biodeterioration of electrical equipment.

Deterioration of plastics: The clean, smooth surface of many plastics is spoiled by even slight amounts of mold growth which would not be noticed or thought to be of objectionable on other materials. Such surface growth often draws on surface debris for nutrients, the plastic itself being unaffected. The most common molds often seen on plastic materials, like refrigerators, are the species of *Aspergillus, Penicillium, Mucor* and *Rhizopus,* especially when it remains closed due to power failure for 3 or 4 days continuously during the summer days. More serious problem occurs where a mold is able to utilize components of the plastic material. There is a considerable amount of literature on the effects of the mycelial fungi on plastics, but the consensus of the opinion is that the polymers are mainly resistant, with the exception of the polyurethanes (Kaplan *et al.*, 1968), while the component most likely to be attacked is the plasticizer, such attack resulting in the material becoming discoloured, brittle and cracked.

Deterioration of fuels: Hydrocarbon fuels such as kerosene and lubricants such as cutting oil emulsion are often attacked by fungi. Growth may occur throughout an emulsion, or in the case of fuels, as a fuel/water interface. Whether in bulk store, or in use, for example in air craft wing tanks, a water layer is often formed. The mold growth indirectly affects by forming matted mycelium which clog fuel lines and filters, and the corrosion of several types which may be initiated by fungal growth may weaken the tanks or surrounding structures. *Amorphotheca resinae* (conidial state *Hormoconis resinae = Cladosporium resinae*), commonly called 'kerosene fungus' or 'cresote fungus' is the most common deteriogen which is found on cresoted wood and corroding air craft fuel tanks.

The growth of molds and bacteria in cutting oil emulsions results in the change in the properties of the oil, leading to shortened tool life and poorer finish and creation of an odour problem and also increases skin infection risks in the operators.

Biodeterioration of paint and paint films: Paints are also attacked by molds either in can or as an applied film. Mold growth in can may lead to gassing, discolouration and a change in the properties of the paint. Films may be rendered brittle and cracked by fungal growth and may be marked by dark sporing fungi such as species of *Cladosporium, Phoma* and *Pullularia.*

Food spoilage: Every type of food is susceptible to fungi of one kind or another. It may be simply made unpalatable by a discolouration, by production of off-flavours, or by the simple distaste of eating food with an obvious fungus growth. Contamination of the food by the microbes not only decreases the food value, especially of fats, carbohydrates and proteins, is made unfit for consumption due to the production of mycotoxins.

Most familiar example of spoilage of food by fungi is that of *Rhizopus stolonifer,* the 'bread mold' on bread. Quite a good number of other fungi grow in bread, i.e., *Aspergillus niger,* Penicillium species, members of Mucorales which are a great nuisance in bakery industry and to the housewives at home. Calcium propionate or other mold inhibitors are added in bread to keep if free from molds for a few days. Better house keeping and good bakeries maintaining a degree of cleanliness is the answer to avoid this nuisance of molds on the baked materials.

Pickels and sauerkraut are commonly spoiled by film yeasts, pink yeasts, i.e., *Rhodotorula* sp. Spoilage of cured meat is commonly caused by species of *Aspergillus, Penicillum* and *Rhizopus.*

In addition to bacteria, fungi also occur on the dairy products and cause deterioration by producing off-flavour. The common members are the yeasts, *Geotrichum candidum, Penicillium, Alternaria* and *Cladosporium* species.

Fruits and vegetables also suffer from fungus infestation during storage, transport and marketing. The fungi attacking these may be field parasites in addition to those which occur predominantly in storage. Common examples of fungi occurring in storage are *Penicillium digitatum* on citrus fruits and *Rhizopus stolonifer* on sweet potatoes and other vegetables.

Grains are commonly infested during storage by a variety of fungi and are responsible for causing heavy losses and are made unfit for human consumption if not properly stored. The fungi associated with the grains include the species of *Aspergillus, Penicillum, Phoma, Alternaria, Cladosporium, Fusarium, Curvularia, Drechslera, Mucor* and several others. These fungi produce several types of mycotoxins in food and feed stuffs which are responsible for causing different mycotoxicoses, i.e., poisoning of the host which follow entrance into the body of toxic substances of fungal origin in man and animals. Some of the important mycotoxins, their producers and the diseases caused by them are given in Table 16.3 and their structures in Fig. 16.11.

Fungi as parasites of plants: A parasitic fungus is one which obtains its nutrients from its host, i.e., another living organism. Some parasites may cause virtually no disease to the organism on which they are feeding. The majority of the fungi, however, are pathogens. A pathogen is an organism which is able to cause disease in a particular host or range of hosts. It may or may not be parasitic on the host. Thus there is a subtle difference between a parasite and a pathogen. An organism may or may not be parasitic but may be pathogenic. The toxic metabolite of a fungus in soil may cause disease on a host plant. *Periconia circinata* which does not grow in the roots of sorghum plants, may produce a powerful toxin in the soil which comes in contact with the roots of plant and cause a disease without parasitizing the roots of sorghum plants.

Many parasitic fungi are highly phytopathogenic and induce severe and frequently economically important diseases. They are the most important agents of plant diseases and more than 60 per cent of literature on plant

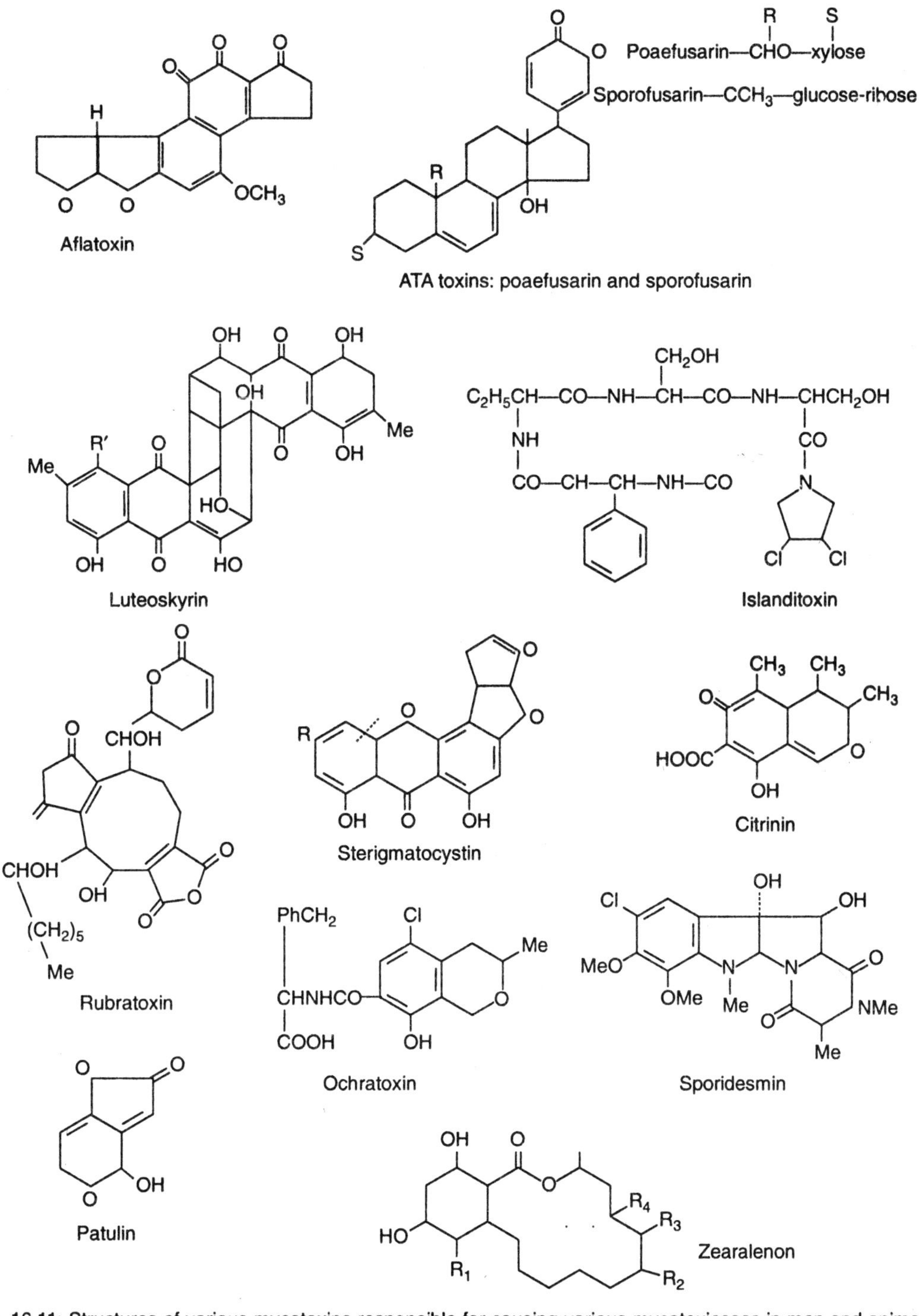

Fig. 16.11: Structures of various mycotoxins responsible for causing various mycotoxicoses in men and animals

diseases is devoted to fungal infections. Plant pathogens are represented in all the major groups of fungi and some groups, e.g., Peronosporaceae (downy mildews), Albuginaceae

Table 16.3: Mycotoxins of food and feed stuffs and the organisms producing them

Food/Feed stuff	*Mycotoxins*	*Mycotoxicoses*	*Fungi*
Peanut (groundnut) meal, cotton seed; copra	Aflatoxins	Aflatoxicoses	*Aspergillus flavus, A. parasiticus*
Millets and other grains (Wheat, rye, oats)	ATA toxins poae-fusarin, sporofusarin	Alimentary toxic aleukia (ATA)	*Fusarium sporotrichoides, F. poae*
Rice	Luteoskyrin, Islanditoxin	Yellow rice disease (hepatitis)	*Penicillium citreoviride, P. citrinum, P. islandicum*
Corn	Rubratoxin A and B	Mouldy corn toxicosis	*Penicillium rubrum, Aspergillus flavus*
	Zearalenone	Esterogenic (vulvovaginitis and infertility)	*Fusarium graminearum* (Gibberella zeae)
Cereal and legume crop	Sterigmatocystin	Hepatic carcinoma nephropathy	*Aspergillus vericolor, A. nidulans*
Cereals	Citrinin		*Penicillium citrinum, P. viridicatum*
Barley	Ochratoxin A and B	Nephrotoxicosis	*Aspergillus ochraceus, Penicillium viridicatum*
Pasture	Sporidesmins	Facial eczema	*Pithomyces chartarum (Sporidesmium bakerii)*
Apple fruits and juices bread and other bakery products	Patulin	Neurotoxicosis	*Penicillium patulum, P. claviforine, P. expansum, Aspergillus clavatus*
Cattle feeds	Fumitremorgin, Verruculotoxin	Neurotoxicosis (tremor)	*Aspergillus fumigatus, Penicillium verruculosum*

(white russ), Erysiphaceae (powdery mildews), Ustilaginales (smuts), Uredinales (rusts), are known by the common names of the diseases for which they are responsible. Other groups important as plant pathogens include hymenomycetes (frequently wood-attacking), certain ascomycetes, and particularly deuteromycetes. The symptoms produced by fungi in plants are diverse and may be roughly classified as necrosis (death of tissues) which includes the diseases like anthracnose, blight, canker, damping-off, scab, shot hole; wilts (loss of turgidity and collapse of leaves), hyperplasia-overgrowth (galls, witche's broom) and hypoplasia-dwarfing and chlorosis. The details will be given in plant pathology.

Fungal diseases of man and animals: Fungal infection is caused by molds or yeasts that happen to find the right set of conditions for growth in the human body and virtually no animal is free from the possibility of incurring an invasion most frequently by fungi that are found in soil. Many of the fungi initiating diseases in man and animals are

Table 16.4: Specialization of parasitism among fungi

Number of primitive characters	*Characters*	*Examples*	
4	Wide host range	Many soil-inhabiting species of *Pythium* and *Fusarium*	Primitive parasites
3	Capable of saprophytic growth under natural conditions	*Ophiobolus graminis, Thanatephorus cucumeris* (formerly known as *Corticium solani*)	↓
2	Cause necrosis of host cells	*Venturia inaequalis, Diplocarpon rosae*	
1	Limited to immature or senescent tissues	*Phytophthora infestans* *Taphrina deformans*	
0	A fungus with none of the above characters	Most rusts, smuts, powdery mildews, *Plasmodiophora brassicae*	Specialized parasites

closely related and in a few cases some species can infect both. In such cases animals act as reservoir of the inoculum and increase the chances of exposure of humans to infections. Some fungal diseases are confined to specific geographical areas, others are worldwide in distribution while some are seen only in certain populations, relating to age, sex, race, occupation, or general physical conditions. Fungal infection of men and animals are usually categorized into four categories: (i) superficial, where only the outer surface of the skin is invaded; (ii) cutaneous, where the keratinized tissues of the body (the outer epidermal cells of the skin, hair and nails) are invaded; (iii) subcutaneous, where invasion occurs in the deeper layers of the skin, extending into muscles, sometimes into the bone, and may involve the lymphatic system; and (iv) systemic, where the infection of lungs takes place by direct inhalation of fungal spores and from the lungs the infection may be disseminated to all organs of the body and may cause death in men and animals. The fungi that are able to invade human tissues are divided into two groups: (a) the primary pathogens which are capable of producing special kinds of fungal infection in healthy individuals and have specific clinical characteristics; and (b) the opportunists which will only invade the tissue of a compromised host.

The importance of the study of fungi inciting human mycoses is becoming recognized in recent years. Medical mycology which was considered to be the domain of the medical man is now taken up by the mycologists also. The subject is gaining importance rapidly. At present there are over 80 human pathogens among the approximately 1,00,000 species of fungi. Some of the important books dealing with the fungal diseases of man and animals are: 'Medical Mycology Manual' (Beneke and Rogers, 1970), Manual of Clinical Mycology (Conat *et al.*, 1971). Medical Mycology (Rippon, 1974; Emmons *et al.*, 1977). The Fungus Diseases of Man (Wilson, 1974). Laboratory Handbook of Medical Mycology (McGinnis, 1980), and The Medical Mycology Handbook (Campbell and Stewart, 1980). Thirumalachar (1968) has discussed the saprophytic habitats of causative agents of human mycoses in India and the gamut/galaxy of the fungal infections in Tamil Nadu state by Kamalam (1984). Ecology of human pathogenic fungi has been discussed by

Table 16.5: List of the important mycoses according to disease classification

Classification	*Disease*	*Affected area*	*Organism*
Superficial	Black piedra	Scalp and beard	*Piedraia hortae*
	White piedra	Beard, scalp, pubic hair	*Trichosporon beigelii*
	Tinea versicolor	Smooth body skin	*Pityrosporum orbiculare*
	Tinea nigra	Thick stratum corneum, palms and feet	*Cladosporium werneckii, C. mansonii*
Cutaneous mycoses	Dermatomycoses	Keratinized layers of body; skin, hair and nails	*Epidermophyton, Microsporum,* and *Trichophyton* spp.
	Tinea capitis	Scalp	*Microsporum canis, M. audouinii, Trichophyton tonsurans*
	Tinea favosa	Scalp, skin and nails	*T. schoenleinii*
	Tinea unguium	Nails	*T. rubrum, T. mentagrophytes, Epidermophyton floccosum*
	Tinea barbae	Beard and coarse body hair	*T. verrucosum, T. mentagrophytes*
	Tinea corporis	Smooth body skin	*T. rubrum, T. mentagrophytes*
	Tinea imbricata	Smooth body skin	*T. concentricum*
	Tinea pedis	Feet, interdigital spaces and soles	*T. rubrum, T. mentagrophytes, E. floccosum*
	Tinea manuum	Palms and fingers	*T. rubrum, T. mentagrophytes, T. floccosum*
	Tinea cruris	Groin	*T. rubrum, T. mentagrophytes*
	Tinea facei	Nonbeard area of adult males, face of adult females and children	*T. violaceum, T. rubrum, T. mentagrophytes, T. tonsurans, M. gypseum*
	Tinea axillaris	Groin and genital areas	*T. mentagrophytes, T. tonsurans, E. flosccosum*
	Intertriginous candidiasis	Moist skin areas; groin, glans penis scrotum, folds of buttocks, under the breast, axilla, interdigital spaces	*Candida albicans*
	Candidal granuloma	Hands, feet, face and scalp	*C. albicans*
	Onychomycosis	Nails	*C. albicans, Candida* sp., *Scopulariopsis* sp., *Aspergillus* sp.
	Mucocutaneous candidiasis	Mucocutaneous areas	*C. albicans*
	Thrush	Mouth and tongue	*C. albicans*
	Perleche	Corners of mouth	*C. albicans*
	Vaginal candidiasis	Vagina	*C. albicans*

Contd.

Contd. Table 16.5

Classification	*Disease*	*Affected area*	*Organism*
	Perianal candidiasis	Anal area	*C. albicans*
	Esophageal candidiasis	Esophagus	*C. albicans*
	Otomycosis	Ear	*C. albicans*, other *Candida* sp., *Aspergillus* sp. and other common yeasts and molds
Subcutaneous mycoses	Chromomycosis	Skin surface, mostly lower extremities	*Fonsecaea pedrosoi, F. compactum, Phialophora verrucosa, Wangiella dermatitidis, Cladosporium carrionii*
	Mycetoma (maduramycosis, madura foot)	Skin surface, mostly lower extremities of foot and is of various types; white or yellow or black	*Allescheria boydii, Aspergillus* spp., *Madurella mycetomii M. grisea* and actinomycetes.
	Sporotrichosis	Skin, primarily hands, arms and legs	*Sporothrix schenckii*
	Rhinosporidiosis	Nasal mucosa	*Rhinosporidium seeberi*
	Lobomycosis (Lobo's disease)	Smooth skin	*Loboa loboi*
Systemic mycoses	Histoplasmosis	Lung, reticuloendothelial system, bone, kidney and other organs	*Histoplasma capsulatum* (perfect state *Ajellomyces* (*Emmonsiella*) *capsulata*
	Coccidioidomycosis	Lungs	*Coccidioides immitis*
	Blastomycosis	Lung, may spread to other organs, skin lesions are common	*Blastomyces dermatitidis, Ajellomyces dermatitidis*)
	Cryptococcosis	Lungs, central nervous system, skin	*Cryptococcus neoformans* (*Filobasidiella neoformans*)
	Paracoccidioidomycosis	Subclinical infection in lung, mucous membranes and skin	*Paracoccidioides brasiliensis*
	Aspergillosis	Lung, skin, mucocutaneous tissue, other body organs	*Aspergillus fumigatus* and other *Aspergillus* spp.
	Mucormycosis	Face, sinuses, gastro-intestinal tract, lungs	*Absidia corymbifera* and other members of Mucorales, e.g., *Mucor, Rhizopus, Mortierella, Cunninghamella* and *Syncephalastrum* sp.
	Cerebral chromomycosis	Brain or central nervous system	*Cladosporium trichoides Wangiella (Phialophora) dermatitidis, Fosecaea pedrosoi.*

Srivastava (1984). Some of the important mycoses and their causative agents are given in the Table 16.5.

Entomogenous fungi. Fungi growing on insects are known as entomogenous (Gr. *entomo* = insects + *genous* = living in or on insects, especially as pathogens) fungi. These range from commensals or symbionts, through ectoparasites which do not seriously affect their insect hosts (e.g., members of Trichomycetes and Laboulbeniales) to pathogens which are severe enough almost to eliminate a population of insects. For example the chytrid *Coelomomyces* and the oomycete *Lageidium gigantium* which mainly occur on mosquito larvae are important insect parasites. Fungus diseases of insects are both common and widespread and are caused by the members of fungi which belong to all the major groups, i.e., Lower fungi, Ascomycotina, Basidiomycotina and Deuteromycotina. Table 16.6 lists the major taxa containing insect parasitic fungi.

Table 16.6: List of major taxa which contain insect-parasitizing fungi

A. Lower Fungi

1. ***Chytridiomycetes***
   Coelomomyces
   Myiophagus
2. ***Oomycetes***
   Lagenidium
3. ***Zygomycetes***
   Culicicola
   Entomophaga
   Entomophthora
   Erynia
   Massospora
   Neozygites
   Strongwellsea
   Tarichium
   Triplosporium
   Zoophthora
   Zygaenobia

B. Ascomycotina

Ascosphaera (*Pericystis*)
Caloncectria
Cordyceps
Hypocrella
Massaria
Myriangium
Nectria
Ophiocordyceps
Podonectria
Sphaerostilbe
Torrubiella
Laboulbeniales (approx. 125 genera)

C. Basidiomycotina

Septobasidium
Stereum
Uredinella

D. Deuteromycotina

Acrostalagmus
Aegerita
Akanthomyces
Aschersonia
Aspergillus
Beauveria
Cephalosporium
Culicinomyces
Derexia
Fusarium
Gibellula
Hirsutella
Hymenostilbe
Isaria
Metarhizium
Microcera
Nomuraea
Paecilomyces
Penicillium
Spicaria
Synglioсladium (*Sorosporella*)
Synnematium
Tetracrium
Verticillium

Fungi as allergens: Some hypersensitive people develop asthma as a result of an allergy to mold spores in house dust. Agricultural and industrial workers may develop severe asthma when exposed to repeated inhalation of mold spores. The asthma can be followed by a chronic respiratory condition, which may develop into a fatal disease, as in bronchopulmonary aspergillosis. The mold spores responsible to induce allergy, asthma, hay fever and vasomotorhinitis in man chiefly belong to the saprophytic fungi. They are mostly non-pathogenic and cause symptoms of allergy in patients without growing on the tissues. The genera *Alternaria, Aspergillus, Cladosporium, Penicillium, Fusarium, Cephalosporium, Trichoderma, Chaetomium, Phoma, Botrytis, Mucor, Rhizopus, Claviceps,* the yeasts, the rusts and smuts are responsible for causing various types of allergies.

The various allergic diseases known to medical persons are cheese-washer's lung due to *Penicillium casei,* maltster's lung due to *Aspergillus clavatus,* maple-bark stripper's lung due to *Cryptostroma corticale,* mushroom worker's lung due to *Doratomyces* sp., suberosis due to *Penicillium frequentans,* lycoperdonosis due to *Lycoperdon* sp. and some other puff balls, and byssinosis (dusty cotton disease) due to species of *Aspergillus, Penicillium, Scopulariopsis, Paecilomyces, Alternaria, Cladosporium* and *Syncephalstrum.*

Dermatophytes may cause secondary allergic effects on skin manifested as eczema, urticaria, dermatophytid, etc. which is an allergic reaction to spores or toxin of a dermatophyte. The site of fungal invasion, therefore, may be far away from the inflammatory reactions. Spores of *Arthrinium* (imperfect state of *Apiospora montagnei*) cause alergic skin defects in those persons who cut *Arundo donax* canes in France.

Poisonous fungi: Quite a good number of fungi are poisonous. Some people are allergic or intolerant to mushrooms and illness may result from eating decayed or mouldy specimens. A number of larger fungi contain toxins and quite a few species are deadly poisonous. Lincoff and Mitchel (1977) divided the major types of mushroom poisoning into four categories based on the physical effects of the toxin and the time elapsing from consumption of the mushrooms to the appearance of symptoms. The four categories were further divided into a total of six or seven groups according to the nature of the various toxins (Alexopoulos and Mims, 1979; Hawksworth *et al.*, 1983): (i) cyclopeptide poisoning (gastroenteric symptoms) due to *Amanita phalloides, A. virosa* and *A. verna;* (ii) haemolytic poisoning (anaemia) due to *A. rubescens* and *A. vaginata;* (iii) muscarine poisoning (increased perspiration, salivation, dehydration, nausea) due to *A. muscaria* and *A. pantherina;* (iv) coprine poisoning (ingestion) due to *Coprinus atramentarius;* (v) psychotrophic poisoning (hallucinations and delirium) due to *A. muscaria, A. pantherina, Psilocybe* and *Panaeolus;* and (vi) gastroenteric irritants due to *Entyloma sinuatum, Paxillus involutus, Agaricus xanthodermus, Boletus satanus, Hebeloma crustuliniforme, Tricholoma* spp., *Hypholoma fascicicülare, Lactarius* and *Russula* spp. and the ascomycete *Gyromitra esculenta.*

Fungi growing on food and feed products produce toxins. A toxin is a non-enzymic metabolite produced by one organism and is injurious to another. The toxins produced by fungi and injurious to man or animals are called mycotoxins and the poisoning of the host resulting from the ingestion of mycotoxins is called mycotoxicoses. But nowadays the term mycotoxicoses is used for the poisoning of animals and men by feed and food products contaminated by toxin producing fungi. Symptoms produced by the mycotoxins can be distinguished into four categories: (i) mycotoxins producing incordination and tremor (Tremorgen); (ii) mycotoxins producing exces-

Table 16.7: Mycotoxins, their producers and the diseases caused by the mycotoxins

Mycotoxin	*Organism*	*Disease*
Aflatoxins	*Aspergillus flavus, A. parasiticus*	Aflatoxicoses
Citreovirdin	*Penicillium citreoviride*	Cardiac beri-beri in man
Citrinin	*Penicillium citrinum* and *P. viridicatum*	Nephrotoxicosis in pigs
Leuteoskyrin, Islandicin and isolanditoxin	*P. islandicum*	Hepatitis (Yellow rice disease) in man
Lysergic acid	*Claviceps* spp.	Paspalum staggers
Maltoryzine	*Aspergillus oryzae* var. *microsporus*	Toxic for cattle
Ocratoxins—a, b	*Aspergillus ochraceus* and *Penicillium virdicatum*	Nephrotoxicosis in sheep, cattle and pigs
Patulin	*Aspergillus clavatus, Penicillium expansum, P. claviforme, P. patulum*	Neurotoxicosis in cattle and carcinogenic to mice
Roridins	*Myrothecium roridum* and *M. verrucaria*	Dendrochiotoxicosis (ill thriff in sheep, pig and man)
Rubratoxin—B	*Penicillium rubrum*	Hepatitis in cattle and pigs.
Slaframine	*Rhizoctonia leguminicola*	Slobber syndrome in livestock
Sporidesmin and Sporodesmolides	*Pithomyces chartarum* (syn. *Sporidesmium bakeri*)	Facial eczema in sheep and cattle in New Zealand
Sterigmatocystin	*Aspergillus versicolor*	Carcinogenic hepatotoxin (precursor of aflatoxin B_1)

sive salivation and diarrhoea; (iii) mycotoxins producing excessive salivation and diarrhoea; (iii) mycotoxin producing emitic symptoms; and (iv) mycotoxins producing carcinogenic symptoms (Prasad, 1982). In addition to the toxins of food/feed stuffs discussed earlier there are a number of other mycotoxins as given in Table 16.7.

Fungi in medicine: Quite a good number of fungi are used for medicinal purposes, i.e., in the production of antibiotics, several enzyme preparations such as polyzyme and takadiastase, microbiological transformations of steroids, synthesis of vitamins, therapeutic uses of yeasts, and alkaloids of *Claviceps* for obsteric purposes to induce uterine contraction at the time of childbirth and to prevent postpartum haemorrhage during childbirth, in migraine and as hallucinogens.

The sclerotia of *Claviceps purpurea* (ergot of rye) contains a number of alkaloids, of which the most important ones, used medicinally are the lysergic acid derivatives, the ergometrine, ergotamine and ergotoxine. In addition to these, ergosterol, the commonest sterol of fungi was first isolated from sclerotia of *Claviceps purpurea.* Yeast ergosterol is converted to vitamin D_2 by ultraviolet radiation.

In the ancient days a number of other fungi were used for medical purposes. For example, *Fomes officinalis* (Agarcum, the female, white or purging agaric) was a noted "universal remedy". *Lycoperon* spores and capillitium were used for stopping blood from wounds.

The antibiotics are the substances produced by microorganisms and are active against other microorganisms, some of the important fungal antibiotics are given in Table 16.8.

Antibiotics are not only obtained from fungi, bacteria and actinomycetes but they are also known to be produced by lichens. In

Table 16.8: Some Antibiotics obtained from fungi

Name of the antibiotic	*Microorganisms*	*Active against*
Penicillin	*Penicillium notatum* and *P. chrysogenum*	Active against gram +ve bacteria
Griseofulvin	*P. griseofulvum* and *P. nigricans*	Antifungal used as systemic fungicides agaisnt plant pathogens and orally against dermatophytes infections in man and animals.
Cephalosporin	*Emericellopsis minimum*	Antibacterial
Trichothecin	*Trichothecium roseum*	Antifungal
Antiamoebin	*Emericellopsis poonensis, E. synnematicola* and *Cephalosporium pimprinum*	Anti-protozoa and helminths
Viridin	*Gliocladium virens*	Antifungal
Calvacin	*Langermannia gigantea*	Active against mouse, rat and hamster tumours.
Dendrochin	*Dendrodochium toxicum*	Antifungal
Flammulin	*Flammulina velutipes*	Anti-tumour antibiotic
Fumagillin	*Aspergillus fumigatus*	An amoebocite
Famigatin	*A. fumigatus*	Antibacterial
Fusidic acid	*Fusidium coccineum*	Antibacterial
Gliotoxin	*Glocladium virens, A. fumigatus, Penicillium cinerascens*	Antibacterial and antifungal
Patulin	*Aspergillus clavatus, Penicillium patulum, P. claviforme, P. expansum*	Antibacterial and anti-fungal
Penicillic acid	*Penicillium puberulum, P. cyclopium*	Antibacterial
Helenin(e)	*Penicillium funiculosum*	Antiviral
Lepiochorin	*Lepiota* sp.	Antibacterial
Poricin	*Poria corticola*	Antitumour antibiotic
Proliferin	*Aspergillus proliferans*	Antitubercle bacillus antibiotic
Sparassol	*Sparassis ramosa*	Antifungal
Statolon	*Penicillium stoloniferum*	Antiviral
Ustilagic acid	*Ustilago maydis*	Antifungal and antibacterial
Variecolin	*Aspergillus variecolor* (Emericella variecolor)	Antitubercle bacillus antibiotic
Wortmannin	*Talaromyces wortmannii*	Antifungal especially against *Botrytis, Cladosporium* and *Rhizopus*

general they are most effective against gram-positive bacteria. One such antibiotic is usnic acid which is a yellow dibenzofuran derivative occurring in lichens (e.g., some species of *Usnea, Cladonia*) and is antibacterial (especially Gram + bacteria) and antifungal in action. Usnic acid is used commercially ('Usno', 'Binan', 'Usniplant') and strongly inhibits *Mycobacterium.* Usnic acid salt, i.e., sodium usnate is effective agaisnt *Corynebacterium michiganense* responsible for the tomato canker. Usnic acid inhibits *Neurospora crassa.* Usnic acid and other lichen extracts inhibit wood-rotting rungi. Several lichen acids are active against *Trichosporum.*

Transformation of steroids: Steroids are organic compounds which have 4-membered ring and one of extreme importance in the economy of animal body. These are secreted by the adrenal cortex, ovaries, testes and placenta and are biologically acitve as hormones and regulated the carbohydrate and mineral metabolism as well as the sexual activity. Steroid compounds have found great therapeutic uses as anti-inflammtory agents, antifertility agents, anaesthetics and for the treatment of sterility. The steroids differ in the nature of the side groups or side chains and these differences in structure confer different biological properties on the steroids. These may be obtained directly from a natural source or they may be synthesized. The steroid nucleus can be produced chemically and many of the transformations or additions of side chains may be brought about chemically. However, some changes are impossible to accomplish by chemical means and some may be very time consuming and expensive.

Vitamins from fungi: One of the best sources of vitamin B-complex are the yeasts. The increasing importance of adequate supply of these vitamins have led the food manufacturer to put on the market a number of preparations of high potency made from dried yeasts, yeast extracts or autolyzed yeasts. One of the B-group vitamin, riboflavin is now manufactured in pure form by fermentation. The fungi used are *Ashbya gossypii* and *Eremothecium ashbyi.* Ergosterol, the precursor of vitamin-D was synthesized by a number of filamentous fungi as well as yeasts by fermentation. At present most of the vitamin is obtained as a byproduct in the manufacture of streptomycin and aureomycin.

Industrial enzyme production from fungi. In the present day world highly purified enzyme preparations are being used in industrial processing, clinical medicine and laboratory practice. Of the six major classes into which enzymes are divided according to the proposals of the Enzyme Commission of the International Union of Biochemistry in 1965, those in industrial use belong mostly to group III, the Hydrolases. Producers of industrial use belong mostly to *Trichoderma viride, Aspergillus oryzae* and *A. niger,* although species of *Mucor, Rhizopus* and *Penicillium* and others have also been cited (Table 16.9).

Table 16.9: Some fungal sources of industrial enzymes

Fungus	*Type of Enzyme*
Aspergillus oryzae	Amylase, pectinase, protease, adenylic acid deaminase
A. niger	Amylase, amyloglucosidase, pectic enzymes, protease, glucose oxidae, maringinase
Rhizopus sp.	Amyloglucosidase, lipase, rennet protease, pectic enzymes
Penicillium chrysogenum	Glucose oxidase
P. citrinum	5 prime, phosphodiesterase
Mucor sp.	Rennet protease
Trichoderma viridae	Cellulase

α-amylase, an endo-enzyme, which hydrolyses the internal α 1, 4-glycosidic links of both amylose and amylopectin constituents of starch but not the α 1-6 linkage of amylopectin, with degradation of the starch to dextrins and oligosaccharides. It is mainly obtained from *Aspergillus niger* and *A. oryzae,* and also from *A. awamori.* This enzyme is utilized in the production of sugars and oligosaccharides from starches in various industrial contexts such as bread-making, production of beer and spirit.

Pectolytic enzymes, i.e., which split or hydrolyse the α-1, 4 glycosidic bonds of the pectic substances as well as bring about transeliminative splitting of glycosidic links. These are of various types, i.e., polygalacturonase (PG) (exo- and endo-), polymethylgalacturonase (PMG) (exo- and endo-), polygalacturonate lyase (PGL) (endo- and exo-) and polymethylgalacturonase lyase (PMGL) (endo- and exo-). Industrial uses of pectolytic enzymes lie in conversions in fruit pulp and fruit juices as well as clarification of fruit juices and wines. The organisms which have been used for the production of these enzymes include the species of *Botrytis, Mucor, Penicillium* and *Aspergillus.*

Cellulases form a group of enzymes hydrolyzing β-1, 4 glycosidic links in cellulose or its derivatives. These are made up of at least three components; (i) C_1 enzyme (unclassified) acting on highly oriented solid cellulose and exposing it to attack by other cellulases; (ii) β-1, -4 glucanases (exo- and endo-) which will hydrolyse soluble cellulose derivatives and act synergistically with C_1 component on native cellulose; and (iii) β-glucosidases acting on cellobiose and other β-dimers of glucose. *Aspergillus niger* cellulase has been used in a paper making process and in preparation of coffee concentrates. Tayama (1969) and Blain (1975) have discussed their uses in the feed supplement for cattle, digestive aids, extraction of green tea components, isolation of soya protein, modification of foods and production of single-cell preparations from vegetables. The use of a mixture of cellulase, hemicellulase and pectinase for the conversion of food by-products to edible materials.

The proteases or peptide hydrolases are enzymes hydrolysing the peptide links of proteins or peptides and are divided into endo-enzymes promoting attack within the chain of the macromolecule and exo-enzymes hydrolysing off terminal amino acids. The organisms which have been most studied and used sources of proteases are *Aspergillus oryzae, A. saitoi, A. niger, Trametes sanguineara, Mucor pusillus* and *Rhizopus* sp. The fungal proteases have found application mainly in food industry, i.e., meat-tenderizing, in the production of protein hydrolysates to control bitter flavours, in cheese making as a substitute for natural rennet and in bread making.

Production of organic acids: The commercial production of organic acids by mycelial fungi, in most instances, limited to members of the genus *Aspergillus,* indeed, for the larger volume fermentations, to one species, *Aspergillus niger.* The organic acids produced from fungi are gluconic acid (*Aspergillus niger),* citric acid (*A. niger*), itaconic acid (*A. terreus*), fumaric acid (*Rhizopus arrhizus*), lactic acid (*Rhizopus oryzae*), gallic acid (*A. niger*), L-malic acid (*Schizophyllum commune*) and trans-epoxysuccinic acid (*Asperillus fumigatus* and *Paecilomyces varioti*). These organic acids are used in various preparations, such as itaconic acid in the manufacture of plastics; gluconic acid in the manufacture of toothpaste and medicinally important since it increases the solubility of calcium when combined as a salt (given to lactating mothers as calcium gluconate tablets); citric acid in the manufacture of citrates, flavouring ingredients, effervescent salts, in ink, in dyeing, in silvering mirrors as well as medically.

Flow diagram for the manufacture of gluconic, itaconic and citric acids are given in Fig. 16.13.

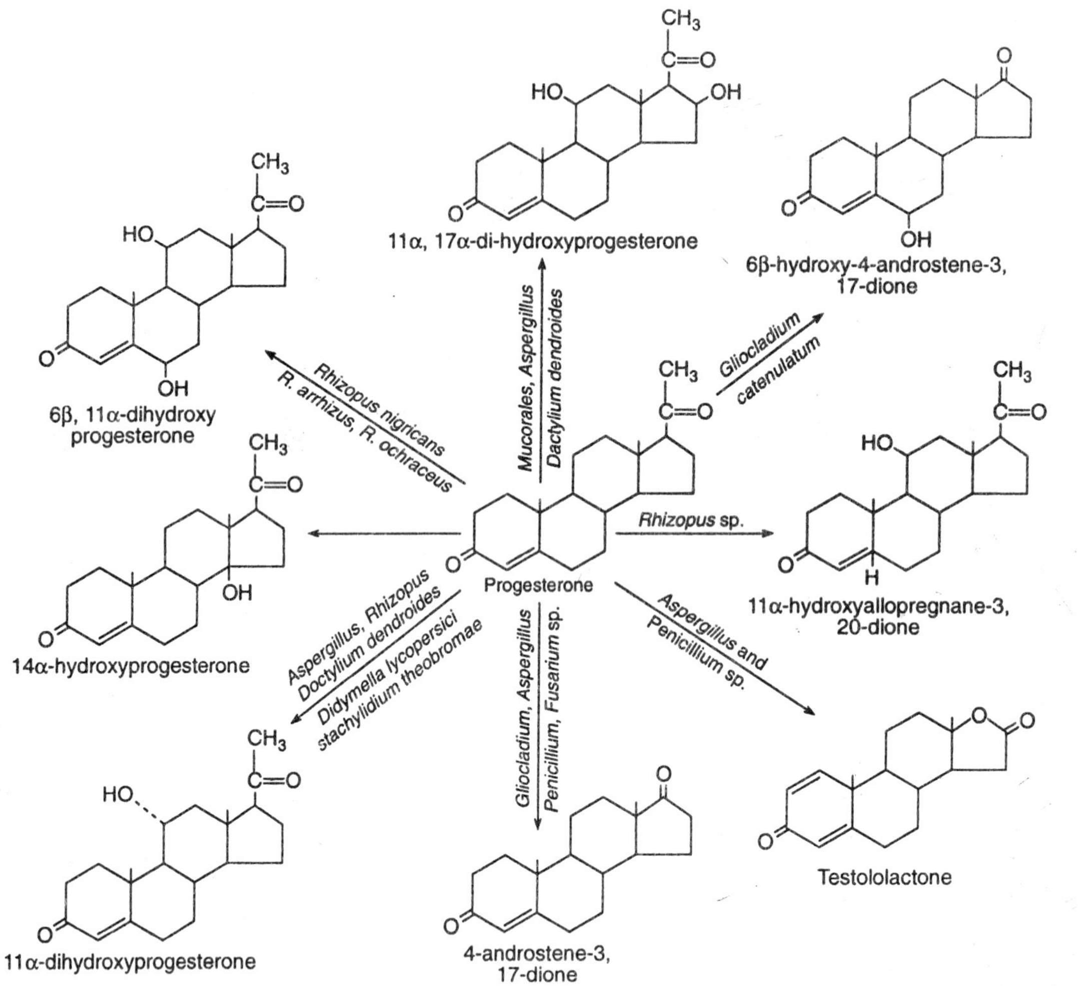

Fig. 16.12: Some steroids produced from progesterone by transformation by fungi

Fungi in fermentation and food processing: Fungi are very important in fermentation, in the production of alcohol, beverages, organic acids and various fermented food products. Yeasts are the most important in this capacity to ferment sugars to produce alcohol. To appreciate their importance to man's welfare one need ony to smell and taste newly baked bread or quaich an enjoyable alcoholic drink. Man has been interested in alcohol from the remote past. The yeasts ferment glucose to yield ethyl alcohol and carbon dioxide:

$$C_6H_{12}O_6 \xrightarrow[\text{Enzymes}]{\text{Yeast}} 2C_2H_5OH + 2CO_2 + \text{Energy}$$

Both the products are used in baking and brewery industries, respectively. Ethyl alcohol is used in the industrial production of beverages, i.e., beer, wine, whisky, rum and gin. Carbon dioxide is necessary for the baking industry because it causes the bread dough to rise creating "light" bread of commerce. Although alcohol is produced during this fermentation it is lost because it gets vaporized during the baking process. Most common substrates employed throughout the world are corn, molasses, sugar beet, potatoes, and grapes. The strains or yeasts used for brewing are usually called top and bottom yeasts. The

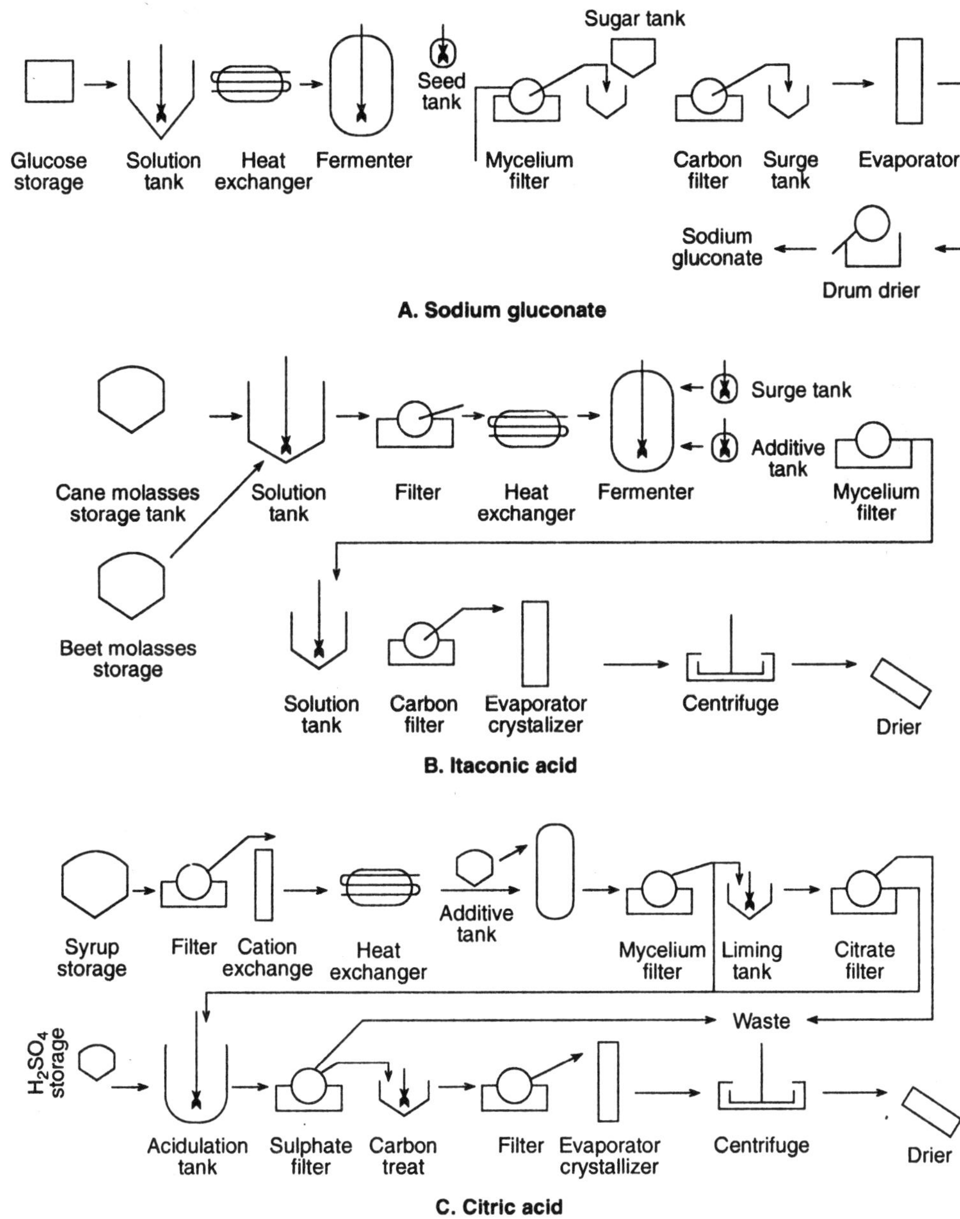

Fig. 16.13: Flow diagrams for the manufacture of A. Sodium gluconate (produced by *Aspergillus niger*), B. Itaconic acid (*A. terreus*) and C. Citric acid (*A. niger*)

top yeasts carry out a vigorous fermentation in which the yeasts cells are carried to the top of the liquor and form a surface pellicle. These are commonly used in the manufacture of typical English beers. The bottom yeasts ferment more slowly and the cells stay at the bottom of the liquor. These are used in the manufacture of so-called "lager" beer such as the fermentation brought about by *Saccharomyces carlsbergensis.*

The yeasts are one-celled fungi and are responsible for industrial alcoholic fermentation. They derive the energy for their growth and reproduction from the anaerobic breakdown of sugars. It may be mentioned that from the point of view of yeast, alcohol is a waste product since it is the energy released in respiration that is vital to them. The carbon dioxide produced in alcoholic fermentation is given off in the form of bubbles which escapes from the liquid or other medium in which the yeasts live. CO_2 bubbles liberated are used in the baking industry for rising of the dough.

Alcoholic beverages are prepared either by fermentation or by distillation. Beer is brewed from barley malt or from a mixture of malt or malt substitute and is a product of fermentation. The ethyl alcohol in this beverage is derived from malt sugars through the fermenting action of *Saccharomyces cerevisiae* or *S. carlsbergensis.* The bitter flavour is imparted by the addition of hops. Beer is about 90% water and the ethyl alcohol ranges from 3-8 per cent. The beverages also contain proteins, minerals and CO_2.

Wines are produced from grapes, logan berries or other fruits and is also a product of fermentation brought about by various strains of *Saccharomyces ellipsoideus.* The alcoholic content in wines is generally 14% or less. In certain wines the alcoholic content may be raised to 20% or so by the addition of alcohol.

The alcoholic beverages grouped under the general name of spirituous liquors include whisky, brandy, rum, gin, vodka and absinthe. In these, the alcoholic contents may be up to 50 per cent by volume.

It is wrong to speak of alcohol as a stimulant. The so-called stimulation due to its effect represents the first stage in depression that it brings about. Alcohol causes a "warm" feeling but lowers blood temperature. Alcohol is high energy food which does not have to be digested. It stimulates the flow of saliva and gastric juices and increases the excretion of urine. It dilates the blood vessels and brings about an increase in the rate of heart beat and blood pressure. When alcoholic beverages are taken it is directly absorbed into the blood without digestion, it gets converted into acetaldehyde in the liver which is a poisonous substance and is responsible for cirrhosis of liver if habitually taken. The acetaldehyde does not stay in the liver and is carried to different parts of the body where it is converted into acetic acid which is ultimately converted into carbon dioxide and water.

Large quantities of fermented products like idli, 'jalebies', 'kanji', 'murcha', 'Punjabi warries', 'papadams' and toddy are consumed daily in Asian countries like India, Nepal, Pakistan, Tibet, etc. The fungi responsible for these fermentatioins are *Torulopsis candida* and *Trichosporon pullulans* for idli; *Saccharomyces bayanus* for jalebies; *Hansenula anomala* var. *schneggii, Mucor rouxianus* and *Rhizopus arrhizus* for 'murcha'; *Candida* sp. and *Saccharomycs cerevisiae* for 'Punjabi waries' and papadams; *Saccharomyces cerevisiae* for toddy.

Asian and Oriental foods and drinks based on fungal fermentation include ang-kak, hamanatto, lao-chao, ontjom, sufu, tofu, tempeh, Japanese sake (rice wine), shoyu (soya sauce), moromi, scocho (a distilled spirit) and mizaume (a sugar syrup from rice) which have been discussed in detail by Hesseltine (1965), Hesseltine and Wang (1967), Gray (1970), Batra and Milner (1974), Wood and Yong (1975). Production of some of the important oriental foods will be considered in more detail.

Tempeh. It is a food which is produced in great quantities in Indonesia, New Guinea and Surinam. In tempeh preparation dehulled soyabean grits (not flour, presumably to allow maximum penetration of the fungal hyphae) are soaked in tap water for 2 hours at 25 °C, cooked for 30 minutes, drained and cooled. The beans are now inoculated with spores of *Rhizopus oligosporus* and lightly packed into shallow trays and incubated at 31 °C for 20-24 hours. In Indonesia the inoculation of beans is accomplished by adding pieces of previously made tempeh or from the wrappers in which a former batch was held. At the end of the incubation, the soyabean grits are totally bound together by fungal mycelium into an apparently solid cake; so solid in fact that the cooled slice is almost like bread. These slices are either roasted or dipped in salt water and then fried in a vegetable oil. The slices will become golden brown in colour and are delicious to eat when hot. Salt may be added to suit one's taste and some people like to add soya sauce for getting additional flavour. The newly prepared tempeh is consumed after this 24 hour period and is prepared fresh daily. Percentage of both soluble solids and soluble nitrogen increase during the period of fungal growth and thus the soyabeans become more digestible upon consumption. In the treated beans an increase in riboflavin (B_2), nicotinic acid (nyasin), and cobalmine (B_{12}) conents occurs while a decrease of thiamine (B_1) has been reported (Hesseltine, 1965). It is believed that the fungus secretes very active extracellular lipases and proteinases. According to Wang *et al.* (1969) and Wang and Hesseltine (1970), *R. oligosporus* produces an antibiotic active against many gram-positive bacteria, including some typical intestinal clostridia thus contributing to the beneficial nutritional effects of tempeh.

Shoyu or soya sauce. Among Oriental food stuffs made from he soyabean, Shoyu is the one best known in the Occident (west). The starting material for Shoyu production is beans + wheat. In terms of tonnes of soyabeans consumed, it ranks second to tofu in Japan (Shibasaki and Hesseltine, 1962). It is a dark, brownish liquid with a distinct pleasant aroma. It is very widely used for adding flavour to many foods, such as meat, poultry, fish and for barbecue and other sauces, for flavouring cooked vegetables and for seasoning in general. Yokotsuka (1960) outlined the procedure for preparing Shoyu on industrial scale in Japan. Flow sheet for the production of Shoyu according to Yong (1971) for a laboratory scale fermentation is outlined in Fig. 16.14. At the end, the mash is compressed as a cake, and the liquid is extracted as Shoyu, which is pasteurized to destroy any microorganisms and enzymes present.

Miso or soy cheese. Miso production ranks third in the annual tonnage of soyabeans consumed in Japan. This, like Shoyu, is perhaps more important for adding and intensifying flavour than for its nutritional contribution. As with soya sauce, Miso is also prepared both on a factory scale and in the home. The starting material for Miso production can be either beans + rice, beans + barley and beans alone.

Miso is a food prepared by a two-step fermentation process. The first involves the fermentation of rice to produce enzymes followed by a second fermentation of the moldy rice or Koji along with soyabeans, salt and suitable inoculum. Miso is principally used to make a soup which takes place of hot cereal for breakfast. It is also used as a flavouring agent with various foods. In Miso production fermentation is brought about by *Aspergillus oryzae* and *Saccharomyces rouxii.*

Soy milk, tofu and sufu (Chinese cheese): A milk made by soaking soyabeans in water, grinding, filtering and pasteurizing or boiling the filtrate to destroy the trypsin inhibitor has long been a staple article of diet in the East. It is made both on a factory scale and at home. It has an attractive creamy colour and is very palatable.

Tofu is an unfermented curd made from soy milk by calcium or magnesium sulphate precipitation (Wang and Hesseline, 1970). It can be flavoured with shoyu or miso, or cooked in admixtue with other foods such as meat, fish and vegetables.

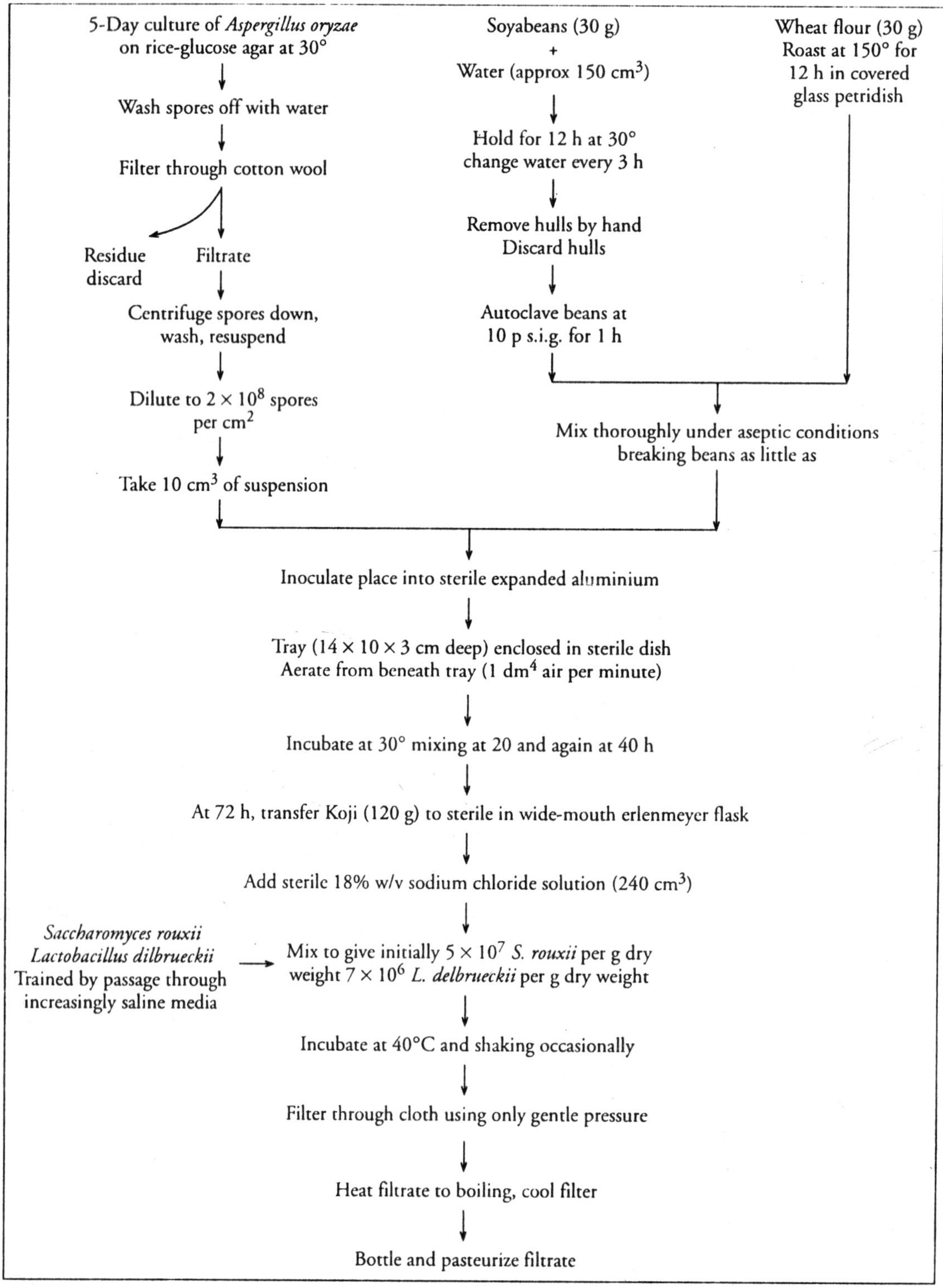

Fig. 16.14: Flow diagram for the production of soya sauce in the laboratory (based on Yong, 1971)

Sufu commonly called Chinese cheese, is another oriental fermentation, typical of China, Formosa and similar countries. It is prepared from the Sufu after fermenting the latter with the Mucorales, e.g., *Actinomucor elegans* or *Mucor hiemalis, M. subtilissimus.*

For the preparation of Sufu, the Tofu is sliced into piece about 2.5 cm cube. The cubes are then soaked for an hour in a solution of sodium chloride (6%) and citric acid (2.5%). The cubes are then dried in an oven at 100°C for 10 minutes, then individually suspended in a cabinet maintained at 20°C and inoculated with a suitable mold and incubated for 3-7 days until the cubes are completely covered with a firm layer of the mold mycelium. The cubes are finally aged for one to two months in a 5-10% brine to which 10% ethyl alcohol may be added in the form of rice wine or distilled liquor.

Ragi: It is another interesting Indonesian fermentation in which Mucorales are used. Ragi is sold in Indonesia as white, somewhat, flattened balls about 3 cm in diameter and is made with rice flour and contains yeast, bacteria and fungi. It is not consumed as such but is utilized to saccharify starch in the manufacture of arak, the latter prepared by the fermentation of liquids containing sugar, such as molasses and rice.

Ang-kak (or red rice) is a product prepared by permitting the mold *Monascus purpureus* to grow on polished rice which has been soaked overnight, drained, autoclaved, cooled and inoculated with a suspension of mold spores. Ang-kak is used for colouring foods which include fish and Chinese cheese, and also used in the manufacture of red wine in he Orient.

Lao-Chao is another food stuff that is consumed primarily for its own nutritional value, rather than as an accompaniment to other things. It is made from glutinous (waxy) rice probably the rice used for rice puddings in British kitchens. A mixture of two organisms, *Rhizopus chinensis* and a species of the yeast *Endomycopsis* are used in the commercial fermentation (Wang and Hesseltine, 1970). For the preparation of Lao-Chao steamed, cooked, rice is inoculated with the molds, lightly packed into a bowl, covered and allowed to ferment for 2-3 days at room temperature. The finished product is soft, juicy, sweet and mildly alcoholic (1-2% ethanol and 20-30% reducing sugars mainly glucose) and possessing a fruity aroma. It is eaten as such, or cooked with other things, i.e., sweet dishes.

Hamanatoo: It is an edible oriental product obtained by the fermentation of soyabeans with *Aspergillus oryzae.* It is used as snacks, and for adding flavour to foods such as beef, fish and Cantonese-style lobster.

Ontjom: It is an Indonesian fermentation product where the substrate used is the residue left after the oil has been pressed from peanuts, i.e., peanut press cake whch would otherwise be a waste product. *Neurospora sitophila* is utilized in the preparation of the food. For the Ontjom preparation, the fresh cake is broken up and mixed with water for 24 hours. The peanut solids are then washed several times, steamed, pressed into shallow trays and inoculated with material from a previous fermentation. The mold mycelium invades the entire cake and presumably binds it together as does the mold mycelium in tempeh and the sporulation of the mold imparts a pink colour to the finished product. The finished product can be cooked in several ways, for example, frying, roasting and serving with ginger sauce, or roasting and covering with boiling water and adding salt and sugar. The flavour of the fried material is comparable with that of minced-meat.

Proteins from Fungi

Keeping in view the ever-increasing human population and lack of a corresponding increase in food production, there is a growing need of more and more protein for human nutrition. One of the ways to augment protein production is by microbial fermentation. The earlier work on fungal protein production was done by Pringsheim and Lichtensten in 1920 when they fermented straw with *Aspergillus fumigatus* for animal feed (Litchfield, 1968). It is only during the last decade that a large number of workers started work on this line, i.e., on fungal protein production from cellulosic materials by the use of *A. fumigatus, Trichoderma viride, Sporotrichum thermophile, S. pulverulentum, Aspergillus terreus* and *Chaetominum cellulolyticum.*

Microbial fermentation results in quick growth of selected microorganisms whch are rich in protein and the dried biomass of the end product is referred to as single cell protein (SCP) (Roth, 1982). In addition to the use of bacteria (*Hydrogenomonas, Methanomonas, Methylomonas*), green algae (*Chlorella, Scenedesmus*) and blue green alga (*Spirulina*), the use of yeasts particularly *Rhodotorula* and *Saccharomyces* has long been established, recently *Candida* has also been used. The biomass of the microorganisms can be freed of unwanted cell contents and used as single cell protein for feeding animals. Because of the high nucleic acid content, microorganisms are not directly suitable for human nutrition but can be utilized as feed in animal nutrition. According to Roth (1982), the extent to which microbial proteins can be competitive with the conventional protein sources cannot be determined at this point, since data on market prices are not available. The prices for yeasts and bacteria, however, would probably fall in the same range as fish meal. Certainly, the price would be cheaper than that of skim milk powder. Large scale production and utilization of microorganisms as SCP in future will depend not only on price relationship but also on agricultural development and commercial trends.

Fungi in Cheese Production

Records of the production of Roquefort cheese, made in caves in the south of France from sheep's milk, go back about a thousand years. According to an ancient legend the art of cheese making was discovered accidently. An Arab merchant as the story goes set out on a journey and put his supply of milk in a pouch—made of sheep's stomach. At the end of the journey the milk had separated into a liquid whey and solid curd. Whether there is any truth in the story or not it is a fact that from early times the curdling of milk to form curds was promoted by a substance called rennet, obtained from the stomach of ruminants. We now know that the curdling action of rennet was due to an enzyme known as rennin. Cheese is a food product made from milk and contains butter, fat, lactose, the protein albumin, caesin and the watery whey. Some cheeses may be processed with a characteristic bacterial or fungal flora which will impart distinctive flavour during the ripening period. *Penicillium roqueforti* and *P. camembert* are utilized for the production of Roquefort and Camembert cheeses. Spores of the mold *P. roqueforti* are added either before curdling or later, when the curd is placed in hoops. During ripening, lipases produced by the mold hydrolyze the fat, releasing butyric, caproic, capsylic, capric, and higher fatty acids, some of which are oxidized, forming compounds, contributing to flavour. Considerable proteolysis also occurs during ripening.

Fungi in Biological Assay

For quantitative analysis of certain complex compounds and elements a number of microbes including fungi have been used as test organisms. For example, for the detection of arsenic, species of *Scopularopsis* are utilized, and for the estimation of copper (Cu) in soil, *Aspergillus niger* is used. The estimation of thiamine (vitamin B_1) is done by *Phycomyces blakesleeanus* and *Nematospora gossypi.*

Table 16.10: Successful examples of entomogenous fungi utilized for the biological control of insect pests

Entomogenous fungus	*Insect hosts*		*Country*
	Common name	*Zoological name*	
Aschersonia aleyrodis	Citrus whitefly	*Dialeurodes citri*	Japan
A. placenta	Citrus whitefly	*Dialeurodes citri*	Soviet Union
Aschersonia sp.	Greenhouse whitefly	*Trialeurodes vaporariolum*	" "
Beauveria bassiana	Colorado potato beetle	*Lepinotarsa decemlineata*	" "
B. bassiana	Codling moth	*Carpocapsa pomonella*	" "
B. tenella (*B. bronghiartii*)	Soyabean beetle	*Anomala rufocuprea*	Japan
B. tenella	Larger striated chafer	*Anomala costata*	Japan
Hirsutella thompsoni	Citrus rust mite	*Phyllocoptruta oleivora*	USA
Metarhizium anisopliae	Rhinoceros beetle	*Oryctes rhinoceros*	Tonga
M. anisopliae	Black rice bug	*Scotinophara lurida*	Japan
Nomuraea rileyi	Tobacco budworm	*Heliothis virescens*	USA
Paecilomyces farinosus	Citrus ground mealy bug	*Rhizoecus kondonis*	Japan
P. fumosoroseus	Peach fruit moth	*Carposina niponensis*	"
P. lilacinus	Potato tuberworm	*Phthorimae opeculella*	"
Verticillium lecanii	Brown soft scale	*Coceus hesperidum*	Czechoslovakia

Fungi in Biological Control

Biological control has been variously defined by different workers (See Baker and Cook, 1974). According to Kerr (1982), biological control may be defined as "any method of control that involves biological rather than physical or chemical methods." This would include disease resistance and the control of disease by cultural practices such as crop rotation. In the more restricted sense, biological control is defined as the use of one specific organism to control another specific organism. A large number of fungi have been utilized in the control of plant pathogens, insects, nematodes, and weeds as discussed below.

FUNGAL BIOTECHNOLOGY

It has been stated by John Naisbitt that recent history has taken idustrialized civilization through a series of technology based eras, from chemical change (plastics) to an atomic age (nuclear energy) and microelectronics (Computers), and presently we are at the beginning of an era, based on biotechnology.

In 1978 under the auspices of the British Mycological Society and the Society for Chemical Industry, academic and industrial workers, representatives of most of the main areas of fungal biotechnology met at Glasgow and discussed various aspects of fungal biotechnology met at Glasgow and discussed various aspects of fungal biotechnology, the reader is referred the publication of Smith *et al.* (1980) for greater details so as to know the importance and various other aspects of fungal biotechnology. Gene manipulations in fungi has been discussed in an edited volume *Gene Manipulations in Fungi* by Bennett and Lasure (1985).

Applications of Fungal Biotechnology

Fungi are sources of food, organic acids, alcohol, antibiotics, growth promoting substances, enzymes, steroid transformation, amino acids, etc. as discussed earlier. The recent techniques of protoplast fusion and the novel method of recombinant DNA technology has given a boost in the area of fungal biotechnology.

While drug makers have been using animal cells to produce pharmaceutical products, industrialists are turning towards the genetic engineering of fungi. Biotechnologists have discovered that foreign genes can be put into filamentous fungi and then used to secrete the desired products. Some filamentous fungi like the mould *Penicillium* and *Aspergillus* are already important work horses of biotechnology. *Aspergillus* spp. are used to make enzymes, organic acids, antibiotics and food additives. One of the more rewarding areas of research in biotechnology has been concerned with strain improvement. Genetic manipulation of fungi and other microbes producing antibiotics, amino acids, organic acids and other secondary metabolites have led to substantial increases in the yields of these secondary metabolites.

Genetic manipulation of fungi has led to the insertion of the genes for making cellulolytic enzymes which are of major potential use in hydrolysis of renewable cellulosic biomass. It is important to note that the major bottleneck to development is due to the recalcitrant nature of lignocelluloses. True ligninases have been demonstrated (Katsuki and Gold, 1982; Tien and Kirk, 1983) and cloning systems for the white rot fungi *Schizophyllum commune* (Seligy *et al.*, 1983) and *Phanerochaete chrysoprium* (Paterson *et al.*, 1984; Rao and Reddy, 1984) have recently been developed. These advances, taken in conjunction with the ability to clone cellulases, should permit consideration of efficient lignocellulose utilization, either by conversion to degradative products or by partial degradation to yield lignin and/or cellulosic polymeric compounds. Commercial fungal carbohydrases, the organisms producing them and their potential applications are given in Table 16.11.

Many countries are using microorganisms to produce protein, to close a worldwide gap. The organisms are mainly yeasts and bacteria, which can be grown in fermenters on cheap carbon sources such as alkanes or methanol. The yeasts used for single cell protein belong mostly to the *Candida, Saccharomyces* and *Rhodotorula*. *Candida* yeasts produce neither sexual nor asexual spores, which is a handicap for any kind of strain improvement. Wickerham *et al.* (1969, 1970) and Gaillardtin *et al.* (1973), found that under certain conditions wild strain of *Candida lipolytica* can fuse and meiosis occurs to produce the sexual ascospores. Since *Candida lipolytica* is now recognized as a perfect fungus, it can be handled for concerted breeding like any other perfect fungus. Microbial biomass as a protein source has been ably discussed by Roth (1982) and Olsen and Alterman (1987). Several other microorganisms have been used for SCP such as algae (*Chlorella, Scendesmus* and *Spirulina*), several other yeasts such as *Saccharomyces* and *Torulopsis* in addition to *Torula.* Efforts have also been made with several other filamentous fungi. Toxic compounds (mycotoxins) are produced by many fungi and any strain which is proposed as food or feed must be examined very thoroughly for possible mycotoxin production. Several substrates are being utilized for growing these fungi for SCP, such as fruit hydrocarbons, gaseous hydrocarbons, methanol, ethanol, renewable carbon sources such as carbon monoxide, molasses, whey, poly-saccharide hydrolysates, effluents from breweries, distilleries, confectionary industries, potato and canning industries and sulphite liquor from wood plup mills.

Table 16.11: Commercial Carbohydrases obtained from fungi

Enzyme	*Application*	*Source*
Glucoamylase	Saccharification in the bread making, soft drink, and brewing industries	*Aspergillus awamori* *A. oryzae* *A. niger* *Endomyces* sp. *Rhizopus niveus* *Saccharomyces distaticus*
α-Amylase	Saccharification in the bread making, soft drink and brewing industries	*Aspergillus oryzae* *A. niger*
Cellulase	To facilitate extraction in the food industry (green tea, soy protein)	*A. niger* *Trichoderma reesei* *T. vinide*
β-Glucanase	To enhance filtration in brewing	*Aspergillus niger*
Dextranase	Hydrolysis of dental plaque; removal of Leuconostoc slimes in sugarcane/beet processing	*Penicillium* sp. *Trichoderma* sp.
α-Galactosidase	In sugar beet processing to convert raffinose to sucrose to enhance crystalization	*Aspergillus* sp. *Mortierella vinca* *Saccharomyces uvarum*
Invertase	Sucrose conversion in the confectionery industry (soft centred candy)	*S. cerevisiae* *S. uvarum*
Lactase (β-galactosidase)	Utilization of whey and improved quality of milk products in dairy industry	*Aspergillus niger* *Kluyveromyces fragilis* *Kluyveromyces lactis*
Naringinase	Debittering of citrus products	*Aspergillus niger*
Pectinase	To and in the filtration of fruit pulp and to clarify fruit juices and wine	*Aspergillus* spp.
Yeast lytic enzyme	Research studies	*Basidiomyces* sp. *Trichoderma harzianum*

Compared with other fungi, the higher basidiomycetes have been the subject to very little biotechnological exploitations. Esser (1977) has found that a single gene is responsible for the initiation of the fruit bodies of *Polyporus ciliatus, Agrocybe aegerita* and *Schizophyllum commune.* The application of these findings will not only improve the chances of concerted breeding of edible mushrooms like *Agaricus brunnescens* (syn. *A. bisporus*) but also makes it possible to domesticate other delicious fungi such as *Agrocybe aegerita, Pleurotus ostreatus,* or even *Cantharellus cibarius.*

Furthermore, the knowledge of a 'threshold gene' could also be exploited in fermentation industry. It is known that some secondary metabolites like the aroma substances of *Agaricus,* morels and the toxids of the toadstools are produced only in the fruit bodies. Naturally one cannot expect to get fruit bodies of higher basidiomycetes in a fermenter but it might be possible by concerted breeding, to create strains which switch over to fruit body formation in a fermenter and, therefore, produce many substances typical of the fruit bodies. This is not mere speculation, but is based on the fact that in the ergot fungus (*Claviceps purpurea*) it is already possible to stimulate the mycelium to switch over to the so called sclerotial growth, a peculiar hyphal morphology in which the alkaloids only are produced.

Last but not the least, higher basidiomycetes are very well known decomposers, not only disintegrating cellulose but in many cases, lignin also. It would clearly be desirable to exploit these fungi for recycling procedure, for example, straw, paper and cotton waste.

Another area where biotechnology has been utilized is the area of prevention of senescence in organisms and strains (fungi) which are useful industrially. Esser and Keller (1976) had shown that the occurrence of senescence in *Podospora* may be prevented by the synergistic action of at least two linked genes. Furthermore, the ageing may be impeded when the mycelia are grown on media containing inhibitors of mitochondrial functions like ethidium-bromide on the antibiotics streptomycin or liamulin. *Podospora* work opens up the possibility of preventing senescence in useful strains of fungi and is full of great practical applications.

One great advantage of filamentous fungi is that they secrete the products of their genes to the outside of the cells rapidly thus simplifying the business of purifying the products. The use of genetically engineered fungi over bacteria for this purpose also helps in keeping the costs down.

Part Four

Lichens

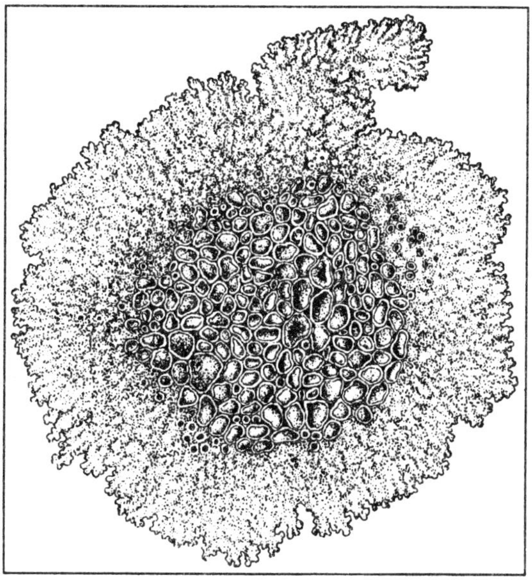

Part Four

Lichens

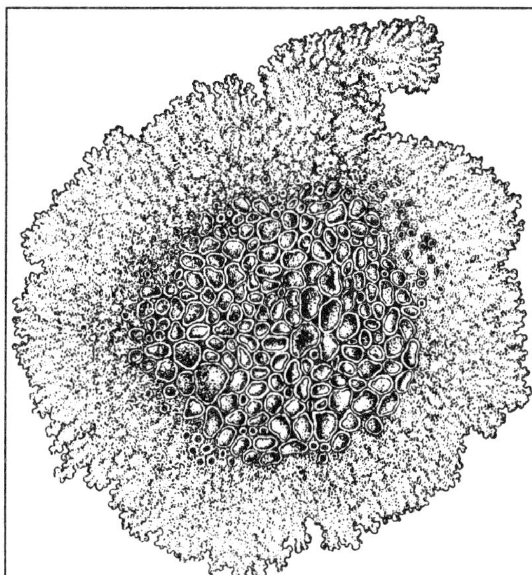

Chapter 17

Lichens

The **Lichen** can be defined as a structurally-organized entity consisting of the permanent association of a fungus and an alga. The lichens are commonly taken as a distinct group, though each one is a self-supporting intimate association of a fungus and an alga.

HISTORY OF LICHENOLOGY

Examination of lichen literature takes us back to Theophrastus, the disciple of Plato and Aristotle, who lived from 371 to 284 B.C., and who in his **History of Plants** used the term 'lichen', a word of Greek origin, to signify a superficial growth on the bark of olive trees. Among the plants described by Theophrastus, there are evidently two lichens, one of which is either an *Usnea* or an *Alectoria,* and the other certainly *Roccella tinctoria,* the last-named economically important lichen is likely to be well-known for is valuable dyeing properties. The same or somewhat similar lichens are also referred to by the Greek physician Dioscorides, in his work on **Materia Medica,** A.D. 68. During the centuries that followed, there was little study of lichens. In the sixteenth century there was a great awakening of scientific interest all over Europe. After the printing-press had come into general use, a number of books on Botany were published. The study of plants was at first entirely from a medicinal standpoint. One of the herbs recommended for various ailments at that time is 'Muscus arborum', the tree-moss (*Usnea*). In addition to the tree-moss, several species of reputed value in the art of healing now known as *Sticta* (*Lobaria*) *pulmonaria, Lobaria laetevirens, Caldonia pyxidata, Evernia prunastri* and *Cetraria islandica* were also described. Lichen literature started increasing in volume in subsequent centuries as more and more people got interested in this line. Mention should be made of the name of Wallroth (1825), who first described and delimited the elements composing the lichen thallus.

Modern lichenology begins with the enunciation of Schwendener's theory (1867) of the composite nature of the lichen. Schwendener's announcement of dual hypothesis, that is, two organisms (an alga and a fungus) of independent origin were combined in the lichen, was not universally accepted. It was indeed bitterly and scornfully rejected by some of the prominent lichenologists of the time. Schwendener held that the lichen was a fungus parasitic on an alga. His opponents considered such a view as wholly inadequate to explain the biology of lichens. It was not till a later date that the truer conception of the 'symbiosis' was proposed. Systematic literature on lichens has been enriched in course of time by a series of important monographs, they are too numerous to mention here. The publication of the volume dealing with Lichens in Engler and Prantl's **Pflanzenfamilien** has been a boon to those interested in lichens.

RELATIONSHIP BETWEEN COMPONENTS OF LICHEN THALLUS

A lichen is an intimate association of a fungus and an alga in which both the organisms intertwine to form a single thallus. The fungal component is called the **mycobiont**, and the algal component is known as the **phycobiont**.

Many species of Ascomycotina and a few species of Basidiomycotina are found growing in close association with one of several genera of green or blue-green algae constitute lichen thalli. The fungus always envelops the algal component of the association and the resulting thallus from the combined growth of the two are more or less constant in form and in internal structure. Every individual of a given lichen species contains the same alga and the same fungus.

The relationship between alga and fungus in a lichen is a matter of controversy. Some are of opinion that the fungus lives during all or a part of its life in parasitic relation with the algal host and also sustains a relation with an organic or an inorganic substratum (Fink, 1913 and others). Others regard the relationship between an alga and a fungus a lichen thallus as purely of parasitism, the fungus being parasitic on the alga, as evidenced in some cases by the presence of fungus haustoria in the algal cells (Geitler, 1937). The fungus not only parasitizes the algal cells but also lives saprobically on the dead algal cells. But in healthy lichens, dead or digested algal cells are observed only rarely and the true extent of haustorial penetration is not at all clear. Again it has been demonstrated experimentally that the alga of lichen thallus can be grown without the fungus, whereas the lichen associated fungus cannot survive without the alga which indicates the dependence of the fungus on the alga. Experimental evidences also indicate that the alga in a lichen thallus maintains a high level of soluble carbohydrates in the fungal hyphae—this being a major factor in the ability of lichens to withstand environmental extremes. The movement of photosynthate of the alga to the fungal hyphae is predominantly as a simple carbohydrate—glycose from blue-green algae such as *Nostoc* and from green algae either erythritol as in *Trentepohlia,* ribitol as in *Trebouxia,* or sorbitol as in *Hyalococcus.* But the flow of carbohydrate out of the algal cells ceases rapidly after isolation from the thallus resulting in a substantial increase in polysaccharide formation within them. All the blue-green algae of lichen thalli so far investigated are very active in nitrogen fixation. Over 90 per cent of the fixed nitrogen passes rapidly to the fungal hyphae probably as ammonia. Apparently during the association of a fungus and an alga in a lichen thallus the fungus stimulates the flow of simple carbohydrate and ammonia from algal cells to the fungal hyphae. But the basic cause of this phenomenon is yet to be established. Although the algae only occupy a small volume of the lichen thalli they are able to work a remarkable morphogenetic response from the associated fungi. It is evident that in the evolutionally sense lichens originated through **helotism** (master and slave), a partnership between an alga and a fungus in which the association is decidedly at the expense of the alga. Thus, it appears that the benefits of the association are heavily in favour of the fungi. For this reason, and because the form and structure of the lichen thallus are usually governed largely by the fungus rather than the alga, the International Code of Botanical Nomenclature provides that the name of a lichen is that of its fungal component.

Lichens have long been takes as a standard example of true symbiosis, in which both symbionts benefit from the association (Scott, 1960). The lichen thallus was considered to represent an idyllic marriage, a **consortium** between the fungus and alga. It is clear that the fungus benefits, since it is wholly dependent on the alga for food, but it is not always equally clear that the alga benefits. Some lichens grow in dry places where algae alone could not survive, but some others occur side by side with the algal component. The algal cells may receive some benefit from the fungus by way of absorption and retention of water. Experimental work suggests that some lichen fungi produce substances which stimulate or are necessary to the growth of the included algae. The alga survives because of its association with the fungal pseudoparenchyma formed in such a way as to give protection to the algal cells against unfavourable conditions. A very delicate balance thus exists in the relationship which if disturbed in favour of either may result in the destruction of the less favoured component.

The concept of **polysymbiosis** has been put forward by some workers who detected a combination of an alga, a fungus and nitrogen fixing bacteria. *Azotobacter* in a lichen thallus where all the three partners act in union as **polysymbionts.** But in the same lichen thallus exists the phycobiont *Nostoc* which fixes appreciable amount of nitrogen and there is no evidence whether nitrogen fixation be attributable to *Azotobacter* or *Nostoc.*

Of late, a number of lichen fingi were grown in pure culture away from algal component. Attempts to synthesize lichens artificially in the laboratory have produced some positive results, but more to be done before drawing any conclusion.

Nearly 40 genera of algae and cyanobacteria have been reported as photobionts/phycobionts in lichens. Three genera **Trebouxia, Trentepohlia** and **Nostoc,** are the most frequent phycobiants. The genera **Trebouxia** and **Trentepohlia** are of eukaryotic nature and belong to the green algae and the genus **Nostoc** belongs to prokaryotic algae. On the contrary nearly 98% of lichen-forming fungi are Ascomycotina, 0.3% Basidiomycotina, 1.5% Anamorphic fungi and 0.75% species are sterile in nature. Among the lichenized as Ascomycotina 42% belongs to subphylum Pezizomycotina.

DISTRIBUTION

The distribution of all lichens over the surface of the earth is controlled by climate and substratum. Though so numerous and so widely distributed, lichens have not evolved that great variety of families and genera characteristic to the allied fungi and algae. Study of the distribution of lichens reveals that certain families are more abundant in some regions than others, but, in general, nearly all are represented. Certain species are universal, where similar conditions prevail. Although they do not usually form a conspicuous part of the vegetation, except in high mountain elevations, in the arctic tundras, in desert regions, and in many other environments where conditions are unfavourable for the growth of other types of plants, lichens play important roles in nature and in human life. In arctic regions they develop as cushion-like masses on the ground, and because of the lack of competition of other plants they become an important part of the flora. In India, lichens are very common in the Eastern Himalayas with an annual rainfall of approximately 83 cm, than in the Western Himalayas where the rainfall is much less.

HABIT AND HABITAT

Lichens are widely distributed on every convenient and in varied habitats. Many of them are able to grow in situations where no other plants could survive. Consequently, they are usually the pioneers in a succession of plants beginning in a habitat of bare rocks or a severely burnt-over area. Lichens are found growing on the leaves, the bark of trees (Fig. 17.1), especially on the shaded sides; on decaying wood, on rocks and on the soil. They are adapted to survival under great extremes of heat, cold, and drought. In damp forests of warm regions, they often hang from the sides of trees often superficially resembling mosses. A very few species of lichens develop in aquatic habitats. Lichens are resistant to extremes of temperatures, and thier very slow growth rate (less than one millimetre per year for most mature cruts on boulders) enables them to tolerate habitats where the supply of mineral nutrients is low. They also withstand long periods of desiccation. They tolerate frequent cycles of wetting and drying which is a positive requirement for healthy growth.

ECONOMIC IMPORTANCE OF LICHENS

Lichens are useful to nature and human life.

1. **Pioneer initiators of rock vegetation.** Lichens are of considerable ecological importance as pioneers in colonization of rocky habitat by plants. They excrete organic acids which disintegrate rocks, thus forming soil and preparing substrata in which other kinds of plants can subsequently become established. Thus, the growth of lichens on bare rocks initiates the weathering away of such rocks. These activities are collectively called 'biological weathering'. The kind of lichen to appear first depends on the flora existing in the neighbourhood. Crustaceous species are most subject to this struggle for existence and are the first to lay foundation of vegetation. First lichen to appear on rocks, for instance, is *Lichen candelarius*. Other species of *Lichen* follow soon after.

2. **Soil-formers.** Lichens perform the work of breaking down the hard rock surfaces and preparing a soil on which more highly developed plants can grow. This process of new soil formation is known as 'pedogenesis'. Usually, crustaceous lichens begin to cover the

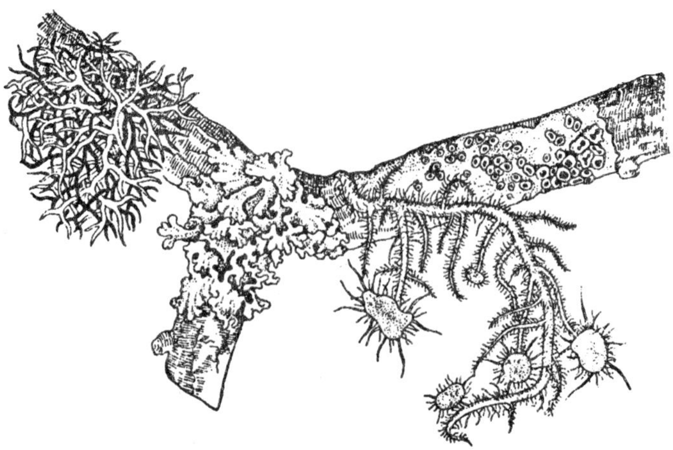

Fig. 17.1: Lichens growing on bark of tree twig

area. When they die, their decaying remains, together with rock particles, form a soil in which other plants may grow. The first successors are generally mosses, but sooner or later vascular plants begin to grow in the soil. A succession of *Lecanora saxicola,* then the moss *Grimmia pulvinata,* which forms compact cushion on which later grow *Poa compressa,* small crucifers, etc., may be cited as an example.

3. **Food for insects and similar other organisms.** Some species of lichens, such as *Aspicilia calcarea* and *Lecanora saxicola* are included in the diets of mites, caterpillars, termites, and snails.

4. **Food for the higher animals.** Many lichens, if deprived of the bitter principle they contain, by soaking in water, or with the addition of sodium or potassium carbonate, might be used with advantage as fodder for animals. Such lichens are: *Lobaria pulmonaria, Evernia prunastri, Ramalina fraxinea* and *Ramalina fastigiata,* all of which owe their nutritive quality to the presence of **lichenin,** a carbohydrate allied to starch. *Cladonia rangiferina,* the well-known 'reindeer moss', is the lichen most favourite as food for reindeer, cattle, etc. *Alectoria jubata, Stereocaulon paschale,* and *Certraria islandica* are good for horses, oxen, cows and swine.

5. **Food for human being.** Lichens contain, lichenin, a carbohydrate very much allied to starch. But no true starch nor cellulose is present in the lichen thallus. *Cetraria islandica,* the commercial 'Iceland Moss' is supplied from Sweden, Norway or Iceland is used as food by man. After removing the bitter principle by soaking in a weak solution of sodium or potassium carbonate it is dried and reduced to powder. The powder when boiled in water yields a jelly which forms the basis of various soups or other dishes prepared by boiling in milk, which are useful to dyspeptics or sufferers from chest diseases. The Egyptians have used *Evernia prunastri* in baking when yeast as fermentative agent was not known to them. In India, a species of *Parmelia* has been used as food, generally prepared as a curry by the natives. It has also been utilized as a source of medicine. In Japan, *Endocarpon miniatum* which they name in English as 'stone mushroom' is sold in the market like vegetables.

6. **Source of medicine.** Lichens owe their repute as curative herb due to the presence of lichenin in the thallus and of some bitter or astringent substances, which, in various ailments, proved of real service to the patient, though they have now been discarded in favour of more effective drugs. Various medicinal benefits of lichens have been ascribed since Pre-Christian times. Lichens have been used in the treatment of jaundice, diarrhoea, fevers, epilepsy, hydrophobia, and skin diseases. In Icelnad, lichen is used as a laxative. Lichen is also used as an ingredient in culture media for bacteria. A preparation of *Peltigera canina* has been used to cure hydrophobia. *Parmelia saxatilis,* the 'skull lichen' has a medicinal property which can cure epilepsy. Several species of *Pertusaria* and of *Cladonia* as well as *Cetraria islandica* were recommended in cases of intermittent fever; species of *Usnea* and *Evernia furfuracea* were used as astrigents in haemorrhages; and *Cladonia pyxidata* was found specially valuable in whooping cough.

7. **Useful in perfumery.** Lichens are used in perfumery in different ways. French perfumers extract an excellent perfume from *Evernia prunastri.* A still finer perfume is extracted from *Lobaria pulmonaria.* The virtue of the lichens also lay in their capacity to absorb and retain perfume. Different species of lichens, are ground to powders to be used for cosmetic purposes. The thalli of species of *Usnea* possess the power of retaining scent, and are profitably utilized in perfumery. Powdered thallus of *Ramalina calicaris* is often used as a substitute for starch that is used in perfumery.

A product obtained from European species of *Evernia* is widey used as a stabilizer in perfumes. In France, certain aromatic substances are extracted from lichens and are used in soaps.

8. **Preparation of dyes.** Some lichens produce dyes which have been used, since Pre-Christian times, for colouring fabrics and paints; among them are **orchil**, a beautiful blue dye, and **cudbear**, another blue dye. The value of *Roccella* as a dye-yilding lichen has been recognized from the time of Theophrastus. The product extracted from its thallus was called **Orseille** for which the English name is **orchil** or **archil** and **orcein** is a purified product of orchil. Lichens serve as a source of the litmus commonly used in the chemical laboratory. Litmus solution is made by grinding the lichen, *Roccella tinctoria,* and extracting the colouring matter, after which paper is soaked in the neutralized solution and is then known as litmus paper.

9. **Useful in tanning, brewing and distilling.** In parts of Russia and Siberia, lichens are used in the brewing of beer. Some lichens contain tannins and are used for tanning animal hides in France and other European countries. The astringent substances extracted from the thalli of *Cetratia islandica* and *Lobaria pulmonaria* have been made use of in tanning leather. *Lobaria pulmonaria* has also been use in the brewing of beer. *Cladonia rangiferina, Usnea florida* and *Ramalina fraxinea* have been used in the preparation of alcohol.

10. **Radioactive fall-out accumulation.** The 'reindeer moss' *Cladonia rangiferina,* and 'Iceland moss', *Cetraria islandica* accumulate large quantities of radioactive strontium and caesium from atomic fall-outs of the tundra.

11. **Sensivitity to sulphur dioxide.** Lichens are extremely sensitive to atmospheric pollutants, such as sulphur dioxide. They are absent from towns and cities and around industrial areas.

Harmful effects of lichens. A large number of lichens have been found causing harmful effects in various way.

1. Lichens of the genera *Cladonia* and *Amphiloma* cause total destruction to the colonies of mosses by direct parasitic attack. The hold-fast hyphae of *Usnea* often pierce through the cortex to the inner tissue of the host, destroy the middle lamella of the cell wall, and ultimately enter in the cells causing total destruction.

2. Lichens often cause serious damage to the window glasses in old buildings. They attach themselves to the glass and gradually corrode it with the help of excretion of acids.

3. Some species of *Usnea,* popularly known as oldman's record, are responsible for fire in forests during both seasons. It catches fire due to inflammable nature.

COMPONENTS OF LICHEN THALLUS

The algal components of the lichen thalli belong to the blue-green algae and green algae. Twenty-seven different genera of algae, both blue-green and green have been reported so far from lichen thalli. Most of the algae are land forms, and, in a free condition they inhabit moist or shady situations, tree-trunks, walls, etc. They multiply by division or by sporulation within the lichen thallus. Usually *Nostoc, Gloeocapsa,* and *Rivularia* are common blue-green algae involved in the development of lichen thalli; and *Pleurococcus* (*Protococcus*), *Cystococcus, Trentepohlia,* and *Cladophora* are among the green algae found in this relationship (Fig. 17.2).

The fungal components of the lichen thalli belong to the Ascomycotina and Basidiomycotina. The ascomycetous fungi produce either apothecia or perithecia bearing asci

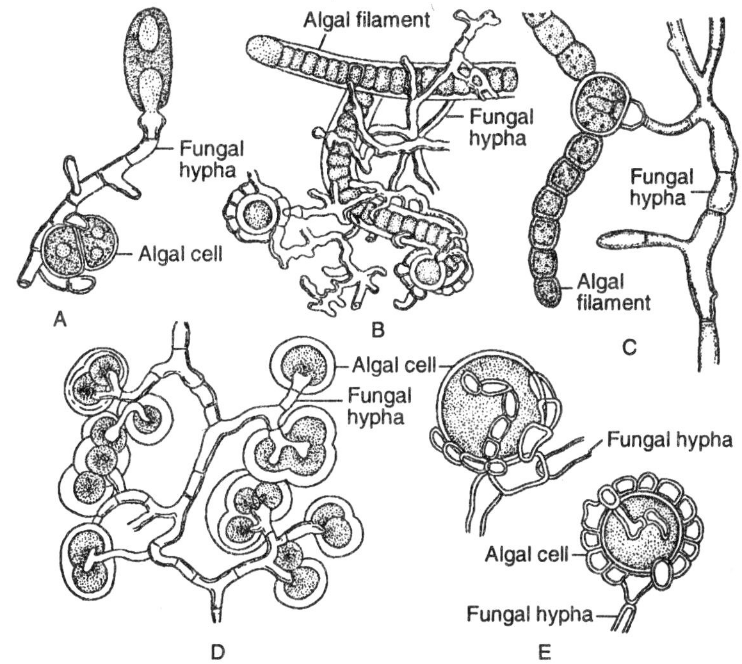

Fig. 17.2: Lichen-forming Algae. A. *Pleurococcus viridis.* B. *Scytonema* sp. C. *Nostoc* sp. D. *Gloeocapsa* sp. E. *Pleurococcus* sp.

with ascospores and paraphyses in mature lichen thalli. The development of ascocarps in a lichen thallus is accomplished exclusively by the fungal component, the alga does not take any direct part in the process. Lichen fungi combine with the same species of alga, and the continuity of genera and species is usually maintained. There are some exceptions where certain lichen fungi are able to associate separately with different algae and adopt strikingly different morphology with the different components. For example, the lichen *Sticta filix* contains the green alga *Coccomyxa,* but a lichen formed by the same fungus with the blue-green alga *Nostoc* is so different in appearance that it is assigned to a different genus, *Dendriscocaulon.*

The fungal hyphae are thin-walled and are closely applied to the algal cells. The algal cell wall may be penetrated by peg-like haustoria. The extent of haustorial penetration is not at all clear (Figs. 17.2 and 17.4C, D).

NATURE OF LICHEN THALLUS

1. **External features.** Lichens are of many different forms and colours. They range in size from minute types to large and conspicuous forms, and some attaining a length of several feet. Lichens vary greatly in colour; some are greyish-green, others are white, orange, yellow, yellowish-green, brown, or black. They commonly form thin thalli. It is convenient and customary to divide lichens into three groups according to the external appearance of the thallus, although there is no sharp distinction between these groups, and this classification bears no relations to the taxonomic position of the fungi and algae involved. Lichens which form a crust closely adpressed to the substrate may be partly or wholly embedded in it are called **crustose lichens** (Fig. 17.3A); lichens with a more leaf-like thallus often resembling dried-up thallus of the liverwort with lobed or

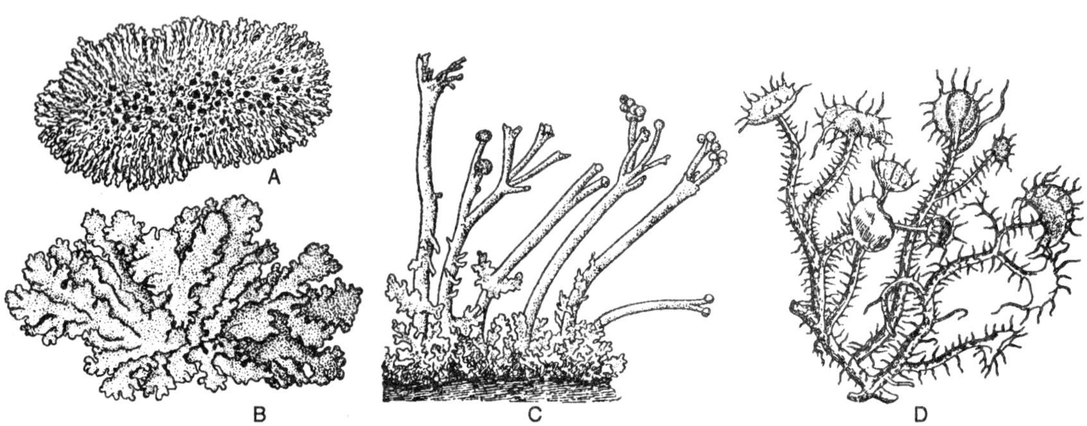

Fig. 17.3: Different forms of lichens. A. Crustose in *Graphis* sp. B. Foliose in *Parmelia* sp. C. Fruticose with a much-branched ribbon-like thallus bearing erect or pendant fruiting portion in *Cladonia* sp. D. Fruticose in *Usnea* sp.

irregular margins, usually attached to the substrate by a relatively smaller portion are called **foliose lichens** (Fig. 17.3B); and lichens which are more or less bushy-branched to upright in habit are called **fruticose lichens** (Fig. 17.3D). Special forms of fruticose lichens often designated as **pendant forms** possess long and slender branches which frequently remain hanging from the twigs or branches of trees being attached only at localized spots.

It should be noted that the distinctions of lichen thalli are not absolute for there are gradations between these groups. Most species of *Cladonia* are at first crustose to foliose but later form upright secondary branches called **podetia** upon which the apothecia are developed by that time the prostrate foliar growth disappears (Fig. 17.3C). A crustose lichen thallus may also develop into a discrete lobed structure partially or wholly free of the substrate producing scaly appearance and is designated as a **squamulose** thallus.

2. **Internal structure**. The bulk of the plant body is composed of closely inter-woven hyphae of the fungus in association with the algal component. The specialized hyphae, the rhizoids, extend downward and serve to attach the thallus to the substratum. The thallus is sponge-like and absorbs water. Mineral salts are obtained in part through the rhizoids also

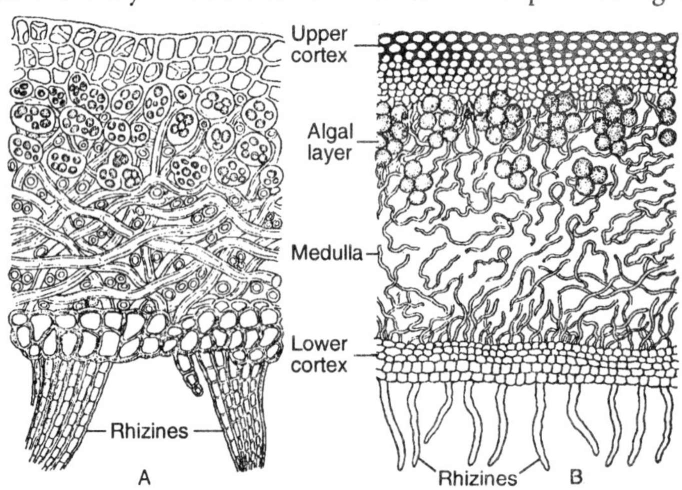

Fig. 17.4: Structural details of lichen thalli in section. A. Upper cortex with intercellular spaces. B. Upper cortex without intercellular spaces

known as **rhizines**, in part from rain water and windblown dust particles.

(a) **Foliose lichen:** A crosssection of a foliose lichen thallus shows internal differentiation of four zones. The uppermost zone consists of more or less vertical hyphae that are with (Fig. 17.4A) or without (Fig. 17.4B) intercellular spaces, which when present, are filled with gelatinous material. This is known as **upper cortex** which may or may not have an epidermis-like layer of hyphae. Beneath the upper cortex is the **algal layer,** also known as **gonidial layer** (Fig. 17.4B). The algal layer consists of rather loosely interwoven hyphae intermingled with algae. The distribution of algal cells and the nature of algal layer in the lichen thalli may be variable. Thalli having the algal cells scattered uniformly among the enveloping fungal hyphae are known as **homoiomerous.** Whereas, in others the algal cells are restricted to a single layer, they are called **heteromerous.** The heteromerous thalli are rather common than the homoiomerous ones. The layer beneath the algal layer is composed of very loosely interwoven hyphae and is designated as a **medulla.** Beneath the medulla is the **lower cortex** consisting of compact hyphae (Fig. 17.4B). The thallus is attached to the substratum by certain hyphal structures known as **rhizines** which arise from the underside of the lower cortex (Figs. 17.4A and B). Rhizines may be simple or branched. In some genera (*Anzia* and *Pannoparmelia*) an unusual development of hyphae from the lower cortex occurs which on anastomosing form a loose sponge-like layer known as **hypothallus** (Fig. 17.5).

(b) **Crustose lichen:** The crustose lichens, which have very poorly defined thallus, in general, have a similar structure but the development of the different layers is not so complete like that of foliose lichen. The tissues forming the thallus are arranged more or less in strata one above the other. The upper cortex is

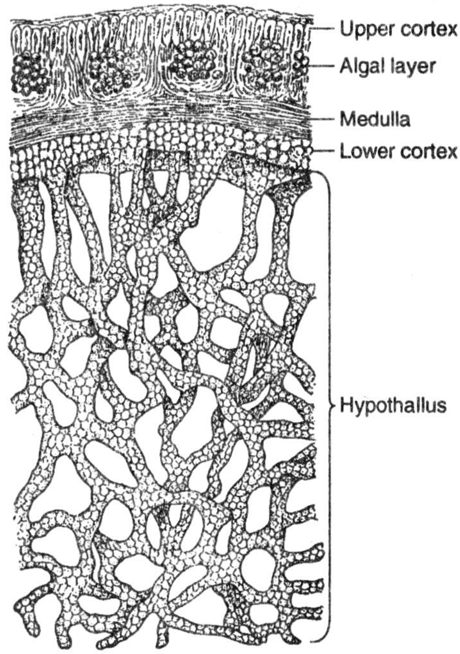

Fig. 17.5: *Pannoparmelia anzioides.* Vertical section of thallus and hypothallus

of hyphal layer, either rudimentary or highly elaborated, beneath which is the algal layer composed of algae and fungal hyphae in close association. Deeper down the algal layer is the medulla, generally a loose tissue of branching hyphae. The lower cortex which lies next to medulla may be as fully developed as the upper or it may be absent. The hyphae on the edge remain meristematic and provide for horizontal as well as vertical extension of the thallus, along with which there is also continual increase of the algal cells.

(c) **Fruticose lichen:** The fruticose lichens which are usually much branched, are either busy and erect or pendant. Their thalli exhibit no differentiation into an upper and lower surface, but are attached to the substrate by a definite basal portion composed of strands of densely-packed hyphae. The conditions of strain and tension in the upright thallus are entirely different from those in the decumbent thallus, and to meet the new requirements, new adaptations of structure are provided either in the cortex or in the medulla. The cortex consisting of parallel hyphae with thickened walls forms the strengthening element. It is further strengthened by the compact arrangement of the medullary hyphae which run parallel with the surface.

SPECIALIZED STRUCTURE OF LICHEN THALLUS

1. **Aeration structures.** Though lichens are composed of two actively growing organisms, the lichen thallus increases very slowly. The absorption of water and mineral salts and corresponding formation of carbohydrates by the deep-seated chlorophyll cells of algae are in small amount. Hence active aeration is not very essential for the lichen thallus. There are many indirect channels by which air can penetrate to the deeper tissues. In spite of this, gaseous exchange in the lichen thallus may take place through structures specially developed for the purpose which are as follows:

(i) **Breathing pores:** These are localized openings developed in the upper cortex where the hyphae are loosely interwoven. There may not be cone-like elevation on the surface of the thallus accompanied with the breathing pores.

(ii) **Cryphellae** (sing. **cyphella**): These are also organs through which aeration in the lichen thallus takes place. They are concave circular depressions confined to the lower cortex of a few foliose lichens (Fig. 17.6B). The hyphae which grow out from the medulla line the cup and from the terminal cells of which short roundish cells with comparatively thin walls are abstricted in a spore-like manner. These roundish cells give characteristic white powdery appearance to the cup.

(iii) **Pseudocyphellae:** These resemble cyphellae in every respect except in the fact that in them no margin is formed, the cortex is simply burst by the protruding filaments.

2. **Cephalodia.** These are external or internal gall-like swellings of the lichen thallus. They are dark-coloured consisting of the same fungal hyphae as in the lichen thallus but the algal component is always different. The cephalodia do not have any organic connection with the lichen thalli bearing them.

3. **Isidia.** These are coral-like outgrowths that are developed on the surface of the lichen thallus (Fig. 17.6A). They are composed of external cortical layer of fungal hyphae with an internal algal layer. They are distinguishable from the cephalodia by the fact that here the algal component is the same as that in the thallus, whereas the cephaloidal alga is different from that of the thallus. They are primarily meant for increasing the photosynthetic surface of the lichen thallus. At times they on being detached from the lichen thallus behave as reproductive bodies.

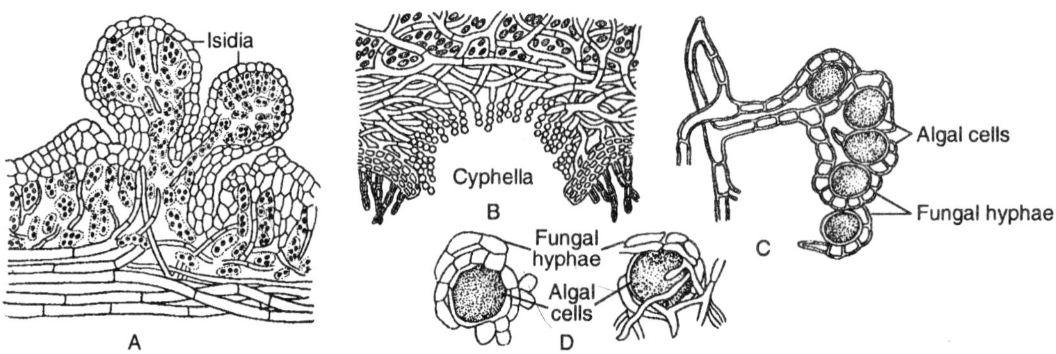

Fig. 17.6: Specialized structures of lichen thallus. A. Isidia of *Peltigera* sp. B. Cyphella of *Sticta* sp. C. Portion of algal layer showing association between fungal hyphae and algal cells. D. Soredia of *Physcia pulverulenta* and *Ramalina farinacea*

4. **Soredia.** The soredia are minute, rounded separable outgrowths of the lichen thallus. They are composed of algal cells clasped and surrounded by fungal hyphae (Fig. 17.6D). Both the fungus and the algal components are the same as in the parent thallus. Soredia are formed in fairly large number on the surface or margins of the lichen thallus. They may occur in patches or may arise so abundantly as to spread up like a thin greyish layer of dust on the surface of the lichen thallus. The soredia are widely disseminated by wind or rain and on being deposited on a suitable substrate and then germinate for giving rise to new thalli with all the characteristics of the parent.

5. **Soralia.** In foliose and fruticose, and in a few crustose forms of lichens, the soredia are massed together into compact bodies known as soralia.

6. **Cillia.** Cillia are hair-like thalline appendages, decolourized or carbonized strands of hyphae that originate along the lobed margins or on apothecia.

REPRODUCTION

Lichens reproduce both by asexual and sexual methods. Many lichens reproduce by simple process of progressive growth and decay of the thallus. Older portions die off leaving the tips to continue growth as separate individuals.

1. **Asexual method.** The asexual method of reproduction is rather common. Under this category the vegetative method of reproductions is also included.

(i) **Fragmentation:** The fragmentation of the lichen thallus may be accomplished by (a) death and decay of the older portions, and (b) accidental breaking. The broken-off pieces develop into new thalli provided they contain both the components.

(ii) **Development of special reproductive structures:** The most common ones are, soredia. Like soredia, isidia also may give rise to new lichen thalli.

(iii) **Asexual spore formation:** Many lichens produce conidia in pycnidia immersed in the thallus. The conidia on being dispersed germinate under favourable conditions by sending out hyphal branches in all directions. Any of the hyphal branches on coming in contact with the appropriate algal cell, branches further ultimately producing a lichen thallus.

2. **Sexual method.** With a few exception of tropical lichens whose fungal components are of Basidiomycotina, all known lichens possess fungal elements belonging to the Ascomycotina and these lichens reproduce by ascospores developed in asci. Synthesis of lichen thallus occurs through union of germinating fungal spore and algal cells.

KINDS OF LICHEN

Lichens are of two kinds: **Ascolichen** and **Basidiolichen.** Ascolichens possess fungal elements belonging to the Ascomycotina and those of Basidiolichens are of the Basidiomycotina.

Ascolichen: Ascolichens (Fig. 17.7) may be grouped into: **ascohymenial lichens** and **ascolocular lichens.** Ascohymenial lichens are those in which asci are borne in apothecia or perithecia, whereas in the ascolocular lichens asci are scattered in the locules of the stroma and a true hymenium is lacking.

(i) **Ascohymenial lichens.** In these lichens there is formation apothecia (Fig. 17.7A) and perithecia which may be embedded in the thallus, stand somewhat above it or be subtended by long stalks. The production of ascocarps in a lichen thallus is accompanished exclusively by the fungal component, the alga taking no direct part in the process. The development of ascogonium is many-celled with the upper end differentiated into a trichogyne. The tip of the trichogyne projects beyond the

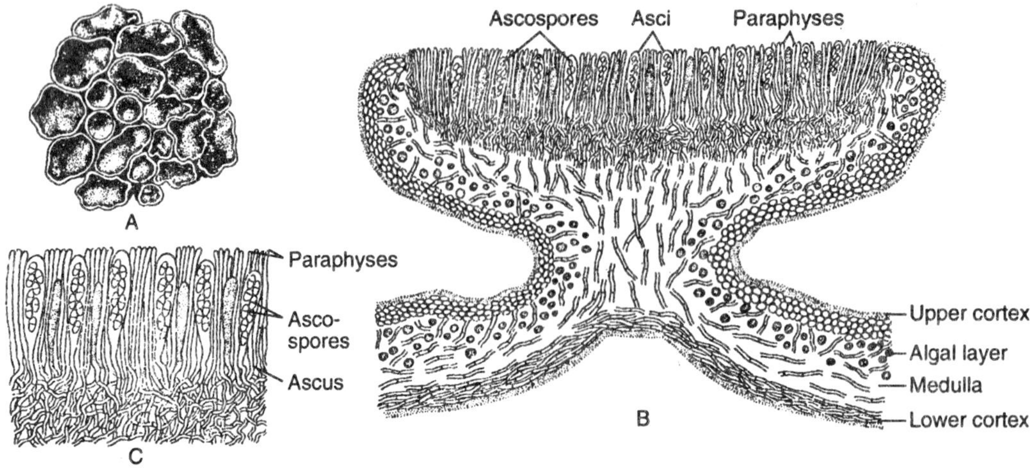

Fig. 17.7: Structural features of Ascolichen thallus and apothecium. A. Apthecia of Ascolichen thallus. B. Sectional view of apothecium and Ascolichen thallus. C. Portion of hymenium of apothecium showing asci bearing ascospores and paraphyses mixed with asci

thallus and receives spermatia produced in spermogonia developed near the ascogonium. The asci and ascospores are produced by the formation of ascogenous hyphae and croziers. The ascocarp is lined with a palisade-like sterile hyphae growing vertically are the paraphyses. The asci grow mixed with the paraphyses (Fig. 17.7B). The development of asci and paraphyses may continue for several years. The asci usually bear eight ascospores which may be one-celled or two- (Fig. 17.7C) to many-celled.

(a) **Apothecia.** The apothecia generally share the characters of the parent thallus in colour and consistency. They may be: **immarginate**, without a well-marked margin; **marginate**, having a well-marked edge; **lecanorine**, possessing a thalline margin; **biatorine**, apothecia being soft or wax-like, and frequently brightly coloured; **lecideine,** apothecia dark-coloured or carbonaceous and generally having no thalline margin; or **lirelliform,** a long, narrow apothecium.

Apothecia are typically open disc- or cup-shaped structures. The hymenium is a thin layer of asci and paraphyses lining the inner surface of the cup. Spores are ejected from the asci forcibly by pressure of swollen moistened ascus walls. There is a small group of lichens in which asci disintegrate at maturity and spores are liberated free in a mucilaginous or dough-like layer over the hymenial surface, made up of tips of the paraphyses and their secretions. This is so-called **mazaedium.**

(b) **Perithecia.** The perithecia wherever formed are also identical with those of the Ascomycetes. A perithecium is described as **entire** when it is partially or entirely immersed in the lichen thallus or in the substratum on which the lichen grows, and is either globose or conical in shape and wholly surrounded by a hyphal wall, whereas it is designated as **dimidiate** when the perithecium is somewhat hemispherical in form and the outer wall is developed only on the upper exposed part of the lichen thallus. As the perithecial wall gives sufficient protection to the asci, the paraphyses are of less importance and are frequently very sparingly produced.

(ii) **Ascolocular lichens.** The ascolocular lichens differ from ascohymenial lichens, as indicated above, in lacking a true hymenium and in having bitunicate asci interspersed with pseudoparaphyses, remaining scattered in locules of stroma. Organized, though often weakly delimited, such fruiting bodies are designated as **pseudothecia** or **ascostromata.**

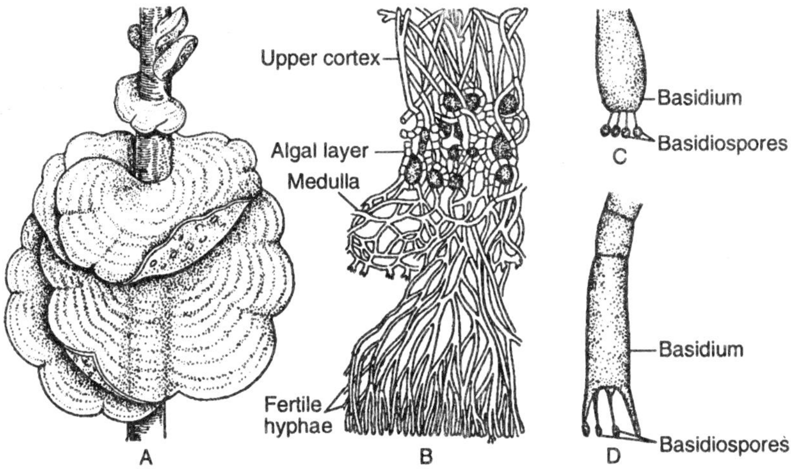

Fig. 17.8: Morphology and structure of Basidiolichen *Cora pavonia*. A. Thin bracket-like thalli. B. Vertical section of thallus. C-D. Basidia with basidiospores

Basidiolichen: Fungal elements of Basidiolichens (Fig. 17.8) are mostly related to the Thelephoraceae, along with which blue-green algae are associated. Their thalli stand out from the trees like small thin bracket fungi (Fig. 17.8A). There is no cortex in the thallus, except in cases where a rather compact cover is formed on the upper surface by hyphal anastomosis. The algal layer which lies at the base of the upward branches is surrounded with thin-walled short-celled hyphae closely interwoven into a kind of cellular tissue (Fig. 17.8B). The medulla of loose hyphae passes over to the lower cortex producing an uneven lower surface. A lichen thallus when spread over soil produces hypothallus without rhizines. Again thalli growing on trees develop extensive hold-fast tissue. The spores are exogenous and are borne at the tips of basidia which are produced on the undersurface of the thallus. In some cases the fertile filaments may form a continuous series of basidia over the surface of the thallus, but generally they grow out in separate, though crowded, tufts. With the broadening of these tuffs a continuous hymenial layer is formed. Each basidium bears four sterigmata and basidiospores (Figs. 17.8C and D); paraphyses exactly similar to the basidia are abundant in the hymenium.

DEVELOPMENT OF THE LICHEN THALLUS

Lichen-fungal spore on coming in contact with suitable substratum and under favourable conditions rapidly puts out germ tube; it may in some cases form numerous branches, which always perish after a certain time if they do not encounter suitable algae, even when the germination has taken place on a substratum favourable to the nutrition of the lichen thallus. In cases, where a certain number of algal cells adhere with the fungal spores that are ejected from the lichen thallus, they are ready at once to be seized upon by the germ tubes.

The germ tube puts out branches as soon as it comes in contact with the right alga. The branches embrace the alga and enclose it in fresh ramifications. At the same time the germ tube puts out other branches whch penetrate into the nutrient substratum and evidently obtain the necessary mineral matter from it. If there is no substratum of this kind, the above processes do not go beyond the very first stage. The union of the two components (alga and fungus) stimulates both, but more especialy the fungus, to new developments of vegetative form, in which the fungus, as the predominant partner, provides the framework of the lichen thallus. Varied structures evolve in order to secure life conditions favourable to both constituents. As the close association of the assimilating and growing tissues is maintained, the thallus thus formed is capable of indefinite increase. The morphological characters of the alga when it is associated by the fungus experience a more or less profound change from its condition in the free state. Size, structure and arrangement of cells are affected in a manner which varies in the same species of alga according to the species of fungus to which it is associated. The alga departing from its usual behaviour takes part in the development of the lichen thallus. The algal component multiples by division or by sporulation within the thallus. It absorbs water and nutriments from the fungal hyphae and assimilates carbohydrate food. The nature of alga and the extent of algal layer are extremely variable in different lichen thalli. The ground tissue of the lichen thallus, apart from the algal layer is composed of hyphal elements. The fungal component is dependent on the alga for the supply of carbohydrate food. It also takes advantage of the presence of humus, whether in the substratum or in the aerial dust. Since the rate of growth of a lichen thallus is very slow, there is not any large demand for nourishment.

RESYNTHESIS OF LICHEN THALLUS

The lichen mycobionts differ from the free-living fungi, in that they are (i) extremely slow growing, (ii) nearly always require biotin and thiamine, (iii) grow leather around 18 °C to 25 °C, and (iv) hardly ever forms conidia or other reproductive structures in culture. These fungi may be isolated either from lichen ascospores or from hyphae within the complete thallus. Similar to the mycobionts, the physiology of the photobionts also seem to differ from the free-living forms. Drew (1966) discovered that he could obtain *Nostoc* cells from lichen thallus of *Peltigera polylactyla* free from significant amounts of fungal filaments, by centrifuging a thallus macerate at various speeds. Such technique was successfully used by Richardson in 1971.

PSEUDOLICHENS

These are forms which have great resemblance with the true lichens. Most of them live on bark of trees. But they do not have the mode of life of true lichens. Their thalli do not possess algae. The fungi out of which their thalli are composed are of uncertain position since they form clustered spore-groups. Some of the Pseudolichens are saprophytes, whereas others are adapted to parasitic mode of life. These forms deserve further investigation.

FOSSIL LICHENS

Among fossil plants there are very few records of lichens due to their indistinct preservation, a large number of them being of doubtful determination. The earliest record of lichens from mesozoic are *Ramalinites lacerus* and *Opegrapha,* of which the former has been questioned. *Ramalina tertiaria, Lichen dichotomus, Opegrapha thomasiana, Cladonia rosea, Parmelia lacunosa* and many others have been reported from coenozoic.

ORIGIN OF LICHENS

Though lichens are very old members of the plant kingdom, their origin is from a time subsequent to the evolution of their components. The algal components which belong to the Myxophyceae and Chlorophyceae had become aerial before their association with fungi to form lichens.

It is unnecessary to look for the algae as their ancestors as algae have persisted through ages in the same form in the lichen thalli. In the condition of lichen thalli they may be considerably modified, but they revert to their normal form, and assume their normal life history of spore production, etc., under suitable and free culture. The fungus being the dominant partner, the principal line of development must be traced through it, as it provides the reproductive organs of the lichen thallus. Though lichens have no common origin, the manner of life is common to them all and has influenced them all in certain directions. They are fitted for a much longer existence than that of the fungi from which they started. In any case, great changes have taken place after relationship with algae became established. In respect of their fungal constituents lichens are also polyphyletic, and more especially in the Ascolichens which can be traced back to several starting points. Whether there has been a series of origins within the different groups or a development from one starting point in each, it would be difficult to determine. The formation of lichen-acids which are excreted by the fungus is rather a noteworthy feature of the lichens. These substances are peculiar lichens and go far to mark their autonomy. The production of the acids and the many changes evolved in the vegetative thallus suggest the great antiquity of lichens.

The Basidiolichens are of comparatively recent origin, since the fungi belonging to the Basidiomycetes had, in the course of time, become less liable and less capable of originating a new method of existence. They lag immeasurably behind the Ascomycetes in the formation of lichens.

DISEASES OF LICHENS

Lichens are subject to various diseases. Following are some of the causes of diseases of lichens:

1. **Caused by Parasitism:** Diseases of lichens due to parasitism by fungi or by other lichens are very frequent. But only in rare cases there is total destruction or any permanent harm to the host resulted by such parasitisim. The thallus of *Parmelia encausta* is dwarfed and deformed by the parasitic attack of *Physcia villosa.*

2. **Caused by Crowding:** Lichens very often suffer from being overgrown by other lichens. They may also be crowded out by other plants.

3. **Caused by Adverse Conditions:** Certain adverse conditions frequently induce disease symptoms. There may be cracking of the lichen thallus due to certain adverse conditions prevailing in and around the substratum of the lichen thallus. Besides this, the adverse conditions also induce the undue thickening of the cortex in the thallus of *Pertusaria communis,* whereby the formation of the fruiting bodies is inhibited and even development of the thallus is hampered. Unusual swellings are produced in the thallus of *Usnea articulata* due to abnormal air currents.

ECOLOGY OF LICHENS

The climatic factors most favourable to the development of lichens are: direct light, a moderate or cold temperature, constant moisture and a clear pure atmosphere. The development and distribution of lichens are so intimately associated with their habitat that it often surpasses all other ecological factors. Hence maximum importance is given to the habitat.

On the basis of habitat, lichen communities fall into the following categories:

1. **Arboreal,** those that grow on leaves, bark or wood.

2. **Terricolous,** ground-lichens.

3. **Saxicolous,** rock-lichens.

4. **Omnicolous,** lichens that can exist on the most varied substrata, such as bones, leather, iron, etc.

5. **Localized communities,** those responding very readily to special environments, and

associations arise either of species met with elsewhere or of species restricted to one type of surroundings.

CLASSIFICATION

Since the time when lichens were first recognized as a separate class, many schemes of classification have been outlined and attempts were made to understand their puzzling structure.

Linnaeus (1753) classified them among algae under the general name **Lichen,** dividing them into eight orders based on the character of their thalli.

Mention may be made about the classification of Acharius (1803), who along with the form of the thalli also introduced the fundamental differences in fruiting structures of the lichen thalli.

Reinke in 1896 stated that 'lichens are a special class different from fungi'. He published a classification system which may be briefly summarized as follows:

Class LICHENES

1. Subclass: **Coniocarpi** (ascocarp a mazaedium)
2. Subclass: **Discocarpi** (ascocarp an apothecium)
   (a) Series: **Grammophori** (apothecia lirelliform)
   (b) Series: **Lecideales** (apothecia lecideine)
   (c) Series: **Parmeliales** (apothecia lecanorine)
   (d) Series: **Cyanophili** (lichens with blue-green algae, including the **Peltigeraceae** and **Stictaceae**).
3. Subclass: **Pyrenocarpi** (ascocarp a perithecium).

Special mention is made of the work of A. Zahlbruckner (1907), who is responsible for the systematic account of the lichens in Engler and Prantl's **Naturlichen Pflanzenfamilien.** He included lichens under a class **Lichenes** and divided into two subclasses on the basis of fungal component.

Subclass I. **Ascolichens**—fungal component Ascomycetes.

Subclass II. **Hymenolichens**—fungal component Basidiomycetes.

Ascolichens were subdivided into two series:

Series 1. **Pyrenocarpeae,** with perithecial fruiting body.

Series 2. **Gymnocarpeae,** with apothecial fruiting body.

These are further subdivided into orders and families.

The development of lichen thallus is the result of two organisms mutually affecting each other, but as the fungus provides the reproductive system, it is the dominant partner. The main lines of classification should be primarily determined by the nature of fungal component and its fruiting structure. The algal component occupies a subsidiary position, but it is also of importance in shaping the form and structure of the lichen thallus. On the basis of above considerations Zahlbruckner's system of classification is widely acceptable and is generally followed by present-day lichenologists with minor modifications.

In the light of Zahlbruckner's system of classification, lichens are subdivided into two subclasses:

1. **Ascolichenes,** thallus with fungal component an ascomycete.
   (i) **Gymnocarpeae,** ascocarp is an apothecum (Discolichens).
   (ii) **Pyrenocarpeae,** ascocarp is a perithecium (Pyrenolichens).
2. **Basidiolichenes,** thallus with fungal component a basidiomycete.

Later it was realized that lichens can no longer be maintained as a separate group of plants since they consist of two discordent elemnts, fungi and algae. The thalli and fruiting bodies are predominantly fungal in structure and it is only logical that fungi be classified by the same system whether they are symbiotic or not. Nannfeldt (1932) laid the foundation for classifying lichens with fungi by separating ascohymenial and ascolocular types, a scheme first put to practical use by Santesson (1952). Nannfeldt system was slightly modified by Luttrell (1955) and the same was used by Dennis (1960) for the British Ascomycetes. Luttrell's system is as follows:

Class: ASCOMYCETES

1. Subclass: Ascomycetidae
   - Order: Lecanorales
   - Order: Sphaeriales
   - Order: Caliciales
2. Subclass: Loculoascomycetidae
   - Order: Pleosporales
   - Order: Hysteriales

M. E. Hale of the Smithsonian Institution, Washington in 1967 presented a detailed classification of lichens as given below:

Class: Ascomycetes (fruiting bodies containing asci)

Subclass: Ascomycetidae (asci with one wall (unitunicate), regularly arranged in a hymenium with free unbranched paraphyses)

Order: Lecanorales (ascocarps apothecia)—23 families

Order: Sphaeriales (ascocarps perithecia)—3 families

Order: Caliciales (ascocarps mazaedia)—2 families

Subclass: Loculoascomycetidae (pseudoperithecia poorly differentiated, asci irregularly distributed)

Order: Myriangiales

Order: Pleosporales

Order: Hysteriales

Class: Basidiomycetes (fruiting bodies containing basidia)

Family: Herpothallaceae
Coraceae
Dictyonemataceae
Clavariaceae
Tricholomataceae

Class: Fungi Imperfecti (fruiting bodies unknown; thallus crustose to squamulose, poorly differentiated).

Alexopoulos and Mims (1979) who have followed Poelt (1973) and Henssen and Jahns (1974), have classified lichens into 3 main groups on the basis of the component fungal partners, i.e., Ascolichens, Basidiolichens and Deuterolichens (Lichenes Imperfecti).

Ascolichens

1. Hymenoascolichens (untunicate asci and paraphyses in apothecia)
   - Order: Caliciales
   - Order: Graphidales
2. Loculoascolichens (bitunicate asci in apothecia or hysterothecia)
   - Order: Arthoniales
3. Loculoascolichens (bitunicate asci in pseudothecia)
   - Order: Dothideales
   - Order: Verrucariales
   - Order: Pyrenulales

Basidiolichens

- Order: Aphyllophorales
- Order: Agaricales

Deuterolichens

Bold *et al.* (1980) stated that it is difficult to assign lichens a natural position in any plan of classification of the plant kingdom. However, because of the dual nature of these organisms, i.e., consisting of a mycobiont belonging to the Ascomycotina, Basidiomycotina and Deuteromycotina and a photobiont belonging to the Cyanochloronta and the Chlorophycophyta, they proposed that the lichens may be classified in a separate division of the plant kingdom—the Mycophycophyta (Gr. *mykes* = fungus + *pycos* = alga + *phyton* = plant) or Phycomycophyta.

Recently, Hawksworth *et al.* (1983) in the 7th edition of the Dictionary of the Fungi have not even recognized lichens as a separate group but have integrated both lichenized and non-lichenized forms with fungi. According to them, the Ascolichens belongs to the following 16 orders of the sub-division Ascomycotina:

1. Arthoniales	9. Opegraphales
2. Caliciales	10. Ostropales
3. Dothideales	11. Peltigerales
4. Graphidales	12. Pertusariales
5. Gyalectales	13. Pyrenulales
6. Helotiales	14. Sphaeriales
7. Lecanidiales	15. Teloschistales
8. Lecanorales	16. Verrucariales

The lichens are not classified as a separate division but usually grouped with the fungi since in a majority of lichens fungal morphology appears to determine the morphology of the lichens as a whole.

PRESENT TREND OF LICHEN CLASSIFICATION

Since the lichen thalli and fruiting bodies are predominantly fungal in structure, it is only logical to classify lichens on the nature of various structures of the fungal component of the lichen thalli. There is no point to give taxonomic importance to the algal component of the lichen thallus. The old lichen-oriented systems of classification do not work well to key the lichens down to orders, families, genera, and species. Naturally a rational integration of lichens with fungi for a workable classification system, is highly desirable. But the complexity of both the groups makes it very difficult.

Nannfeldt (1932) for the first time initiated the classification of lichens with fungi. Bessey (1950) in his Morphology and Taxonomy of Fungi included Ascolichens under the Ascomycetes. Comparatively new and revised systems proposed by Korf (1958) and Tomaselli (1962) have received little support from mycologists or lichenologists.

Though it is very difficult to present a suitable classification system of lichens which will be acceptable to both mycologists and lichenologists, yet it will not be unreasonable to adopt a relatively conservative system outlined by Dennis (1960) for the British Ascomycetes which is again based on Nannfeldt's original classification of the Ascomycetes and subsequently modified by Luttrell (1955). A simplified key is presented below:

A. Fruiting bodies of the thallus containing asci

Class: **Ascomycetes**

B. Asci unitunicate, regularly arranged in a hymenium

Subclass: **Euascomycetes**

C. Fruiting bodies apothecia

Order: **Sphaeriales**

CC. Fruiting bodies perithecia

Order: **Sphaeriales**

CCC. Fruiting bodies mazaedia

Order: **Caliciales**

BB. Asci bitunicate, regularly or irregularly arranged in an ascostroma (pseudothecium) with branched pseudoparaphyses.

Subclass: **Loculoascomycetes**

C. Pseudothecia poorly differentiated, asci irregularly distributed

Order: **Myrangiales**

CC. Pseudothecia well-delimited, resembling perithecia, asci more or less regularly arranged in the stromatic layer

Order: **Pleosporales**

CCC. Pseudothecia well-delimited, round and resembling apothecia

Order: **Hysteriales**

AA. Fruiting bodies containing basidia

Class: BASIDIOMYCETES

AAA. Fruiting bodies unknown; thallus crustose to squamulose, poorly differentiated

Class: FUNGI IMPERFECTI

Some of the Indian Lichens:

I. **Crustose lichens**
1. *Graphis duplicata* Arch.
2. *G. scripta* Arch.
3. *Haematomma punicerum* (Sw.) Mass.
4. *Lecanora chlarona* (Ach.) Nyl.
5. *L. distans* (Pers.) Nyl.

II. **Foliose lichens**
1. *Gyrophora cylindrica* (Linn.) Ach.
2. *Peltigera scutata* (Dicks.) Duby.
3. *P. rufescens* (Weis.) Humb.
4. *Physcia aegiliata* (Ach.) Nyl.
5. *P. aspera* H. Magn. and Zahlbr.
6. *Parmelia cirrhata* Fr.
7. *P. furfuracea* (Linn.) Ach.

III. **Fruticose lichens**
1. *Cladonia aggregata* (Sw.) Ach.
2. *C. pityrea* (Flk.) Fr.
3. *Usnea aspera* (Eschw.) Wain.
4. *U. dichotoma* Fr.
5. *Ramalina calicaris* Rohl.
6. *R. himaayensis* Ras.

IMPORTANCE OF LICHENS IN MONITORING ENVIRONMENTAL QUALITY

Lichen denudation due to anthropogenic activities are well known globally and can easily be detected by means of remote sensing. As for example, in the case of the disappearance of epilithic lichens over a very large area of the candian shield as a direct consequence of atmospheric pollution formerly emanating from the smelting operations at sudbury, which has fortunately subsided due to pollution abatement measures all along the barren rock surfaces to be recolonized by lichens, thereby restoring their light absorbing ability.

LICHEN SENSITIVITY TO AIR POLLUTION

Lichens have been recognized as being very sensitive to air pollution for many years. In the 1800s independent observations in England, Munich, and Paris documented that lichens were already disappearing from urban areas. By the early 1900s this "city" effect was a widely recognized phenomenon in Europe and was first attributed to coal dust, which was emitted by most homes as well as many industries. Only later did the colourless gas, sulphur dioxide, become recognized as a principal phytotoxic agent. Today the list of air pollutants is much longer and includes oxidants, hydrogen fluoride, some metals, acid rain, and organics. Certainly the list of potentially toxic substances is not yet fully circumscribed.

The high sensitivity of lichens is related to their biology. Most species live for decades or hundreds of years and a few longer; thus, as perennials, they are subject to the cumulative effect of pollutants. Lichens have no vascular system for conducting water on nutrients; as a consequence, they have developed efficient mechanisms for taking up water and nutrients from atmospheric sources. Fog and dew, major water sources for lichens, often have much higher pollutant concentrations than precipitation, and the lichens' nutrient concentration mechanisms also will concentrate pollutants. Unlike many vascular plants, lichens have no deciduous parts, and hence cannot avoid pollutant exposure by shedding such parts. Furthermore, the lack of stomata and cuticle in lichens means that aerosols may be absorbed over the entire thallus surface; thus, lichens have little biological control over gas exchange, and air pollutant gases are assumed to readily diffuse down to the photobiont layer. Although dehydration allows lichens to survive dry periods, it also concentrates solutions to the point that toxic concentrations may occur. Finally alteration of the symbiotic balance between the photobiont and mycobiont may readily lead to a breakdown of the lichen association.

That lichens are sensitive to air pollution is, of course, a generalization that requires cautious interpretation and limited extrapolation. The distribution pattern of a lichen species may well reflect varying levels of an air pollutant, but variation in its distribution may also be caused by a variety of abiotic or biotic factors as well. Furthermore, all lichens are not equally sensitive to all air pollutants; rather, different lichen species may become locally extirpated when a pollutant is present, but at least some tolerant species are likely to persist. This differential sensitivity is, however, very useful when interpreting air pollution effects. Just as canaries are useful in detecting toxic gases in coal mines, lichens can be used to monitor ecosystem health were air pollutants are present (Rosentreter and Eldridge, 2002; Will-Wolf, 2002). Similarly, lichens can be used to monitor potential human health problems (Cislaghi and Nimis, 1997).

Lichens in relation to sulphur dioxide

The strongest case for using lichens as bioindicators of air pollution involves sulphur dioxide. Some forms of coal (and other fuel products) have particularly high levels of sulphur, and its oxidation leads to the formation of sulphur dioxide, one of the major gases associated with acid rain. In fact, sulphur dioxide has only an average atmospheric residency time of about 12 hours, because its high solubility in water leads to its trapping in water vapour aerosols and rapid conversion to sulphuric acid, one of the stronger acids.

Evidence that lichens have been responding to sulphur dioxide (SO_2) has now been obtained from throughout the world. For example, Skye (1968) elegantly demonstrated how bark of trees became acidified and lost its buffering capacity with proximity to the city centre (in his case a detailed investigation of Stockholm). This caused a shift in lichen epiphytic species composition to those that were more acidophilic.

Probably the most convincing field data came from England where a grid of sulphur dioxide monitoring stations was established throughout the country in the mid 1900s. Because SO_2 measurements across the grid covered a wide range of atmospheric SO_2 levels, it was possible to examine patterns in epiphyte community composition on adjacent trees and infer relative SO_2 sensitivities based on those patterns across sites. On the basis of differential changes in the epiphytic lichen communities, they established two 10-point scales (one for basic bark and the other for acidic bark) relating long-term SO_2 measurements to changes in the epiphytic lichen communities. Their scales essentially provided elaborate hypotheses about putative differential sensitivity to SO_2 among different lichen species. Even though initial experimental investigations involving SO_2 fumigations did not strongly support the scale, subsequent SO_2 experiments have, in fact, strongly supported the scale. Although some of the changes noted in the scale may well have been related to factors other than sulphur dioxide (e.g., Kostner and Lange, 1986), the general trends observed by Hawksworth and Rose have proved to the quite robust. The most dramatic evidence in support of this assertion has been the widespread reinvasion of lichens as tropho-spheric SO_2 levels have recently decreased throughout England. Recognition of the detri-mental effects of SO_2 had led to legislation limiting SO_2 emissions and implementation of that legislation. The result was a dramatic decrease in measured SO_2 levels, starting in the early 1970s.

The use of lichens as bioindicators of sul-phur dioxide has gained much validity from a plethora of experimental investigations invol-ving exposure to controlled levels of SO_2. Because experimental conditions have varied considerably, one needs to evaluate critically the degree to which different investigations are equivalent. For example, initial investigations used closed containers, where only the initial concentration was known, as the subsequent uptake of SO_2 by the lichens and/or the walls of the cotainer was unknown, and thus the exposure treatment remained undefined. A second approach involved exposing the lichens to aqueous solutions of SO_2, which are pre-pared by bubbling gaseous SO_2 through water and weighing the resultant solution. The method is intuitively attractive insofar as physiologically active lichens have aqueous solutions in their protoplasts. Nieboer *et al.* (1977) have proposed a method for calibrating aqueous solutions with gas phase concentra-tions, and this may be appropriate for defining initial conditions; however, SO_2 in aqueous solutions may decline with time.

About 10% of lichen species contain cyanobacteria, either as the main photobiont or secondarily in specialized structures called cephalodia, and many are very sensitive to air pollution. Many cyanolichens, due to the pre-sence of nitrogenase, are capable of nitrogen fixation. Kytövita and Crittenden (1994) demonstrated that nitrogenase activity was inhibited by acidification. Nitrogen fixation rate, as measured indirectly by acetylene reduc-tion activity (ARA), is often used as a response variable as well, for example, under aqueous conditions.

Oxidants and lichens

Oxidants are a group of strongly oxidizing air pollutants formed photochemically by series of chemical reactions involving mixtures of nitrogen oxides and hydrocarbons. Ozone, which was initially identified in the Los Angeles area, is probably the most infamous oxidant, and it is the most important air pollu-tant causing injury to agricultural crops. In addition, a number of other oxidants, such as peroxyacetyl-nitrate (PAN), are known to cause injury to plants.

Lichen studies have been conducted in the Los Angeles area both in coastal mountain areas and in higher mountain forests. Because of negligible home heating requirements in this region, SO_2 emissions are extremely low, and most vascular plant injury patterns are ascribed to oxidants. On the basis of historical data and the extensive, recent sampling of lichens, species exhibiting differential sensitivity to t'.e Los Angeles evirons are identified. Very sensi-tive species, such as *Evernia prunastri, Peltigera* spp., *Pseudocyphellaria* spp., and *Ramalina* spp., are no longer found, although herbarium records document that many of them were abundant in the late 1800s and early 1900s. The sensitive species, such as *Collema nigrescens* and *Usnea* spp., are found in vestigial amounts in protected habitats; moderately tolerant ones, such as *Hypogymnia enteromorpha* (= *H. imshaugii* in this area but called the former in the literature), common, but in reduced abun-dance compared with control areas, and they often exhibited marked morphological ano-malies (Sigal and Nash, 1983). In contrast, the few tolerant species, such as *Letharia vulpina*, were relatively unaffected. These observations were supported by sampling along oxidant gradients within the highly polluted San Bernardino Mountains of southern California. The gradient reflected not only differences in oxidant dose, but also a range of oxidant injury patterns exhibited by ponderosa pine, as demonstrated by fumigations with ozone (Miller *et al.*, 1983). Along this oxidant gradient there was a marked reduction in cover of the dominant two lichens (*Letharia vulpina* and *Hypogymnia imshaugii*) occurring on conifers and also by species occurring on black oak. Furthermore, experiments demonstrated photosynthetic reduction in *Hypogymnia imshaugii, Flavoparmelia caperata* and *Parmelia sulcata* with ozone exposure (Nash and Sigal, 1979) and similarly, in *Collema nigrescens, H. enteromorpha* and *P. sulcata* when treated with PAN. In contrast, *Ramalina menziesii*, a putatively sensitive species, was unaffected by very short-term ozone fumigations. However, overall these results together with the field work provided the first, relatively strong case for asserting that oxidants were having an adverse impact on lichens.

Hydrogen fluoride and lichens

In contrast to the ubiquitous occurrence of ozone in terrestrial environments, fluoride distribution is primarily limited to point sources, such as aluminum smelters, fertilizer plants, glassworks, and brick and ceramic manufacturing. Volatile hydrogen fluoride, which is very difficult to retain, is the primary form of fluoride that causes injury, which may occur at extremely low ambient concentrations (5 ppb or less). Fluoride accumulates in tissues and causes necrosis, literally death of cells that often become strongly discoloured. For lichens, critical accumulation apparently occurs at 20-25 ppb, above which necrotic patterns develop. Ascospore germination is also inhibited by fluorides (Belandria *et al.*, 1989). In addition, ultrastructure is markedly affected by fluorides by granulation and then breakdown of the thylakoids, the appearance of platogobuli among the thylakoids and the presence of pseudocrystalline structures in the chloroplast and cyctoplasm. All of these effects are different from SO_2 effects on ultrastructure. Fluoride accumulation in lichens has also been documented around volcanoes. Areas around aluminium smelters have been extensively investigated, but other sources have been researched to a lesser extent.

Organopollutants and lichens

Organopollutants exist in a wide variety of forms related to many aspects of oil exploitation and applications of the modern chemical industry, such as the development of pesticides. To a limited extent lichens have been used to monitor deposition patterns. Coastal lichens show little ability to recover after being exposed to crude oil spills. For a marine lichen, *Lichina pygmaea*, Brown (1972) found that the surfactant in the emulsifier used to disperse the soil was more detrimental to the lichen than the oil itself. Once the oil has been refined and combusted in vehicles, a wide range of organics are released to the atmosphere, and at least in the case of aldehydes some lichen sensitivity is known. Although agricultural impacts on lichens have primarily emphasized ammonia deposition, negative effects of herbicides and pesticides are beginning to be investigated. In addition, pesticide usage is known to adversely affect lichen species richness although the effects up to the food chain may be of greater concern.

Part Five

The Myxomycota, Microbiology, Bacteria, Viruses

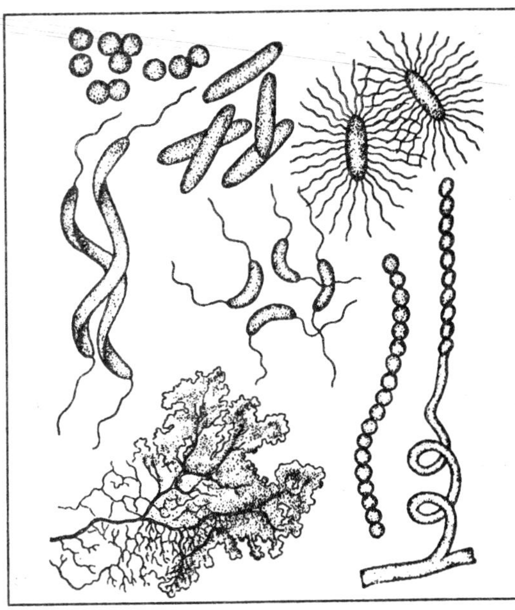

Chapter 18

The Myxomycota

The Myxomycota or slime molds, are fungus-like organisms. They are characterized by the absence of cell wall from their amoeboid, animal-like vegetative or assimilatory phase. The amoeboid assimilatory phase may consist of a free-living multinucleate mass of protoplasm called a **plasmodium** (Fig. 18.1) or simple uninucleate amoeboid cells, the **myxamoebae** (sing. **myxamoeba**) often aggregating into a **pseudoplasmodium** or amoeboid cells interconnected by slime filaments giving rise to a structure known as **net plasmodium** or **filoplasmodium**. Plasmodial slime molds are the **true slime molds,** pseudoplasmodial slime molds the **cellular slime molds** and the net plasmodial slime molds are known as the **net slime molds.**

The true slime molds are mostly saprophytic with a few parasitic ones. In the former the entire plamodium is consumed in the formation of fructifications, whereas in the latter there is absence of development of fructifications, the plasmodium is converted into masses of spores. All the true slime molds produce flagellated cells, the **swarm cells.** But the formation swarm cells is absent in the cellular slime molds. In the cellular slime molds fructifications are developed from the pseudoplasmodia. But in the net slime molds each cell is converted into a mucilaginous **sporocyte** in which spores are developed. The spores escape as biflagellate zoospores which after swimming for some time, lose their flagella, become surrounded by delicate envelope and give rise to new net plasmodium.

Following is a simplified key to the classes of the Myxomycota:

A. Free-living plasmodium bearing haploid or diploid nuclei

B. Plasmodium non-parasitic, fructification present.

Myxomycetes

BB. Plasmodium parasitic, fructification lacking

Plasmodiophoromycetes

AA. Amoeboid cells aggregating into a pseudoplasmodium

Acrasiomycetes

AAA. Amoeboid cells spindle-shaped to oval interconnected by slime filaments forming a net commonly known as 'net plasmodium' or 'filoplasmodium'

Hydromyxomycetes

PHYLOGENY

Although the members of the Myxomycota have some characteristics in common, certain differences are of major magnitude, with the result that a single evolutionary series seems quite unlikely. Macbride (1899) believed that the slime molds are fungi. But Karling (1944) suggested that the slime molds are closely related to Protozoa than fungi. Martin (1932, 1960-61) has pointed out that if the algal origin of fungi is accepted, then the slime molds are not fungi; but if the fungi have been descended from colourless flagellates, the slime molds can well be regarded as fungi, derived from the same general ancestral types as other fungi, but independent and highly specialized to constitute a group of their own—the Myxomycota. Sparrow (1958) proposed that the parasitic slime molds be regarded as a separate Class, the Plasmodiophoromycetes. The Plasmodiophoromycetes are again grouped with zoosporic fungi by some taxonomists though the Plasmodiophoromycetes have controversial affinities. Both the Myxomycetes and the Plasmodiophoromycetes are apparently of flagellate origin. However, the cytological details are so different that the two groups are related only to the extent of a distant flagellate ancestry. It is possible that the Plasmodiophoromycetes arose very early from a primitive fungus of the Phycomycetes, most likely the Chytridiales. Talbot (1971), Webster (1983), and others grouped all the four classes of slime molds under the Division Myxomycota close to fungi. Bonner considers the Acrasiomycetes to have had their origin from the free-living amoebae of the soil. This opinion is based largely on the absence of a flagellated stage. It is generally assumed that the free-living amoebae themselves came from flagellates. This assumption deserves careful consideration. The points of divergence were apparently quite early with the result that all member of the Myxomycota show separate and independent origins along parallel lines of descent.

CLASS: MYXOMYCETES

INTRODUCTION

The Myxomycetes, or true slime molds often designated as slime molds, or slime fungi, or Mycetozoa as they are called, are a unique group of fungus-like organisms concerning whose origin and relationships there is no common agreement. They exhibit characteristic of both animals and plants. Their somatic phase of multinucleate mass of protoplasm without cell wall exhibiting creeping movement, known as plasmodum, is definitely animal-like, resembling a giant amoeba in its structure and in its physiology. The reproductive process of the Myxomycetes, however, is plant-like producing spores with definite cell walls.

T. Panckow in 1654 was the first to discover the Myxomycetes, but they were first described in detail by Micheli in 1729. The Myxomycetes were first recognized as a natural group in 1829 by E. M. Fries. Later in 1833 Wallroth introduced the name Myxomycetes for this group of organisms. De Bary (1887) designated them as Mycetozoa or Fungus animals. The greatest contribution to the morphology and taxonomy of the group was made by J. Rostafinski (1875-1928), a student of De Bary.

ECONOMIC IMPORTANCE

The Myxomycetes are of relatively little economic importance, but they have been the subject of intensive laboratory studies. They contribute to the carbon and nitrogen cycles by using various organic matter including bacteria as food. They provide a large amount of protoplasm free from cell walls which has been used as an ideal medium to solve variety of fundamental problems of biochemists, biophysicists, mycologists and even the geneticists. Some of the areas of studies are: the structure and chemical composition of protoplasm, the velocity of the protoplasmic movement, the chemical changes governing the production of sporangia and spores, the behaviour of nuclei and chromosomal changes during plasmodial growth, various aspects of plasmodial compatibility, etc.

OCCURRENCE

The Myxomycetes are common but inconspicuous inhabitants of moist dead wood, rotting logs, damp soil, leaf mold, moist sawdust, bark of trees, decaying fleshy fungi, or other organic matter. They often spend most of their lives within the substrate and emerge only when about to produce sporangia. Following periods of rainy weather they may occur on leaves of grasses or other plants on lawns.

SOMATIC PHASE

The somatic phase consists of coenocytic mass of highly granular protoplasm containing hundreds or thousands of nuclei and vacuoles without cell walls, but delimited only by a thin plasma membrane—a **plasmodium** (pl. **plamodia**) (Fig. 18.1A). The plasmodium is a flat amoeboid sheet-like structure that usually spreads over a large area in the form of a very thin network with irregular and slowly shifting contours. It commonly assumes the form of a fan-shaped network, with vein-like branches continuing along the outer margin. The granular protoplasm streams rapidly through network of veins which ultimately join the larger veins through which also protoplasm flows, comes to rest for a short while and then immediately reverses its course. The flow is then continued from the larger to the smaller veins which branch again and again till they are lost in the broad margin of the advancing plasmodium. A rhythmic flow is thus maintained backwards and forwards at nearly equal intervals, but always of somewhat longer duration in the direction in which the plasmodium is creeping. The everflowing and everchanging of plasmodial protoplasm result the plasmodium to creep over the surface of the substratum in an amoeboid fashion and on its way feed like amoeba by engulfing living or dead microscopic organisms or other small bits of organic matter. This food is digested within food vacuoles, and waste products or undiges-

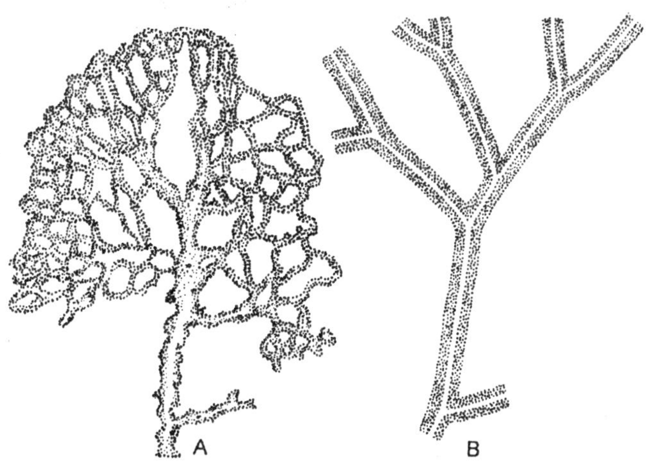

Fig. 18.1: Plasmodium and trial of a myxomycete. A. Phaneroplasmodium, B. Trail

ted particles are left behind in the form of a **trail** (Fig. 18.1B), as the plasmodium moves ahead. In the laboratory, plasmodia have been maintained by feeding them bacteria or yeast cells or other organic matter in small bits. They have also been successfully grown in a chemically defined liquid medium. The absence of cell walls in the plasmodium, together with the method of obtaining food (**holozoic**) are features characteristic of animals, rather than of plants. The colour of the plasmodium is usually either white, yellow, or pink; in some cases, it is purple or green and often brightly coloured; it is generally constant in each species.

The advancing margin is consistently succeeded by a great many strands which have all the properties of the former region; further away, the strands become more and more separated. At first each strand is profusely branched and its branches are connected by anastomosis into a frequently delicate net; the thickness of the strands increases with reduction in branching. The advancing region of the plasmodium, particularly the margin, is homogeneous, viscous, and in steady motion. The general appearance of the plasmodial protoplasm is comparable with that of *Amoeba* and similar forms, as is the manner of movement, although the form of the strand is different and its rate of movement may be slower or faster, depending on the age of the plasmodium and the temperature of the surrounding medium. The plasmodium consists of two interpenetrating substances, a border layer of homogeneous ground substance devoid of granules known as **hyaloplasm** and an interior substance of granular protoplasm containing numerous nuclei and vacuoles. This interior mass of granular material flows with regularly-timed pauses and reversals. The plasmodial strands possess a wide circumscribing layer of hyaloplasm and highly granular flowing fluid of **central core.** The thickness of the hyaloplasm gradually diminishes in the network region and finally both the layers merge near the margin of the advancing plasmodium. The hyaloplasm of the advancing margin exhibits amoeboid movement, projecting and withdrawing pseudopodia.

The nuclei are the same in the plasmodium with a diameter of 3 to 4.5 μm. Each nucleus is surrounded by a membrane and contains a nucleolus. As a plasmodium grows, the nuclei within it increase in number keeping pace with its growth by repeated division. The nuclei of the plasmodium are diploid. The plasmodium thus represents the **diplophase** in the life history of a myxomycete.

Three different types of plasmodia are recognised:

I. **Protoplasmodium:** This is a type of plasmodium which is microscopic in size. It is composed of homogeneous mass of protoplasm which exhibits very slow, irregular streaming movement, e.g., *Echinostelium minutum.* During fruiting the entire plasmodium gives rise to a single sporangium.

II. **Aphanoplasmodium:** This type of plasmodium has great resemblance with the protoplasmodium at its initial stages, but with maturity elongates, becomes branched producing a network of very fine transparent strands and is often not even visible. Such a plasmodium does not possess very granular protoplasm which is not conspicuously differentiated into gelified and fluid portions even in the veins though it exhibits rapid and rhythmically reversible streaming movement, e.g., *Stemonitis fusca.*

III. **Phaneroplasmodium:** Such a plasmodium is large, massive and well-spread out. The protoplasm is very granular, the gelified and fluid regions are easily distinguishable particularly in the veins. It exhibits very conspicuous rhythmic and reversible streaming movement, e.g., *Physarum polycephalum.*

THE SCLEROTIUM

The plasmodium withstands unfavourable conditions like dry or cold weather, by passing into a hardened mass, the **sclerotium** or resting stage. The change of plasmodium into sclerotium may also occur even when water and food material are present. The sclerotium can remain dormant for a long time and can resume the plasmodium state again with the return of favourable conditions.

THE FRUCTIFCATION

In many cases, as long as the supply of food is sufficient, the plasmodium continues to feed and grow. When the nourishment is exhausted, it prepares to pass into the fruiting stage. But cases of fructification development in presence of abudant supply of food are not infrequent.

Preparatory to the development of fructifications the plasmodium usually leaves the moist surroundings where it has been feeding and creeps to some drier place. Eventually the plasmodium heaps itself up somewhat into fructification initials which develop into a number of fructifications. The plasmodial protoplasm may also concentrate around some of the main veins of the plasmodium and develop into a fructification whose shape is the same as that of the plasmodial vein.

As the fructification development continues, a thin wall, called the **peridium,** is secreted by the protoplasm at the surface. The peridium is often calcified and may also contain cellulose. A network of slender vacuoles is often formed within the developing fructification and this network is usually augmented by slender invaginations of the peridium. The vacuolar network of fibrous, often calcified strand, called the **capillitium** (pl. **capillitia**) which is formed by deposition of capillitial material from the surrounding protoplasm within or on the surface of the vacuole network

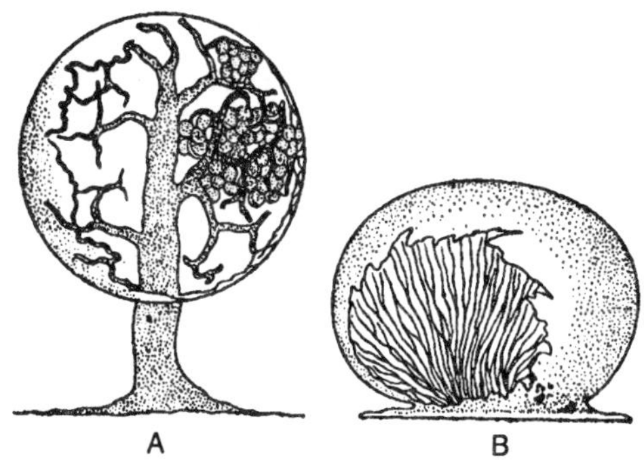

Fig. 18.2: Diagrammatic representation of fructifications of the Myxomycetes. A. Sporangium with stalk extended (columella) in the sporangium containing capilitia and spores. B. Aethalium

(Figs. 18.2A and B). The capillitium may also be absent in certain fructifications. In many Myxomycetes, sometimes irregular threads, plates or other structures resembling capillitium known as **pseudocapillitium** (pl. **pseudocapillitia**) may be present among the spores within the fructifications. A central column, the **columella** is often developed in the fructification (Fig. 18.2A).

The formation of spores is preceded by the division of the nuclei. The nuclei within the developing fructification undergo reduction division, and a very large number of spores, each with a haploid nucleus, is formed by cleavage of the protoplasm. The process of cleavage is not quite simultaneous throughout the whole fructification but begins at the periphery and passes inwards. Each spore is surrounded by a definite wall, which in various species has been reported to contain cellulose, chitin, or neither of these substances.

There are three types of fructifications among the Myxomycetes, the **sporangium, aethalium** and **plasmodiocarp.**

The erect fructifcations of definite form usually stalked are the **sporangia** (Figs. 18.3B and C). They are either scattered or clustered in groups. At the base of the sporangia there may remain a membrane or membranous strand not

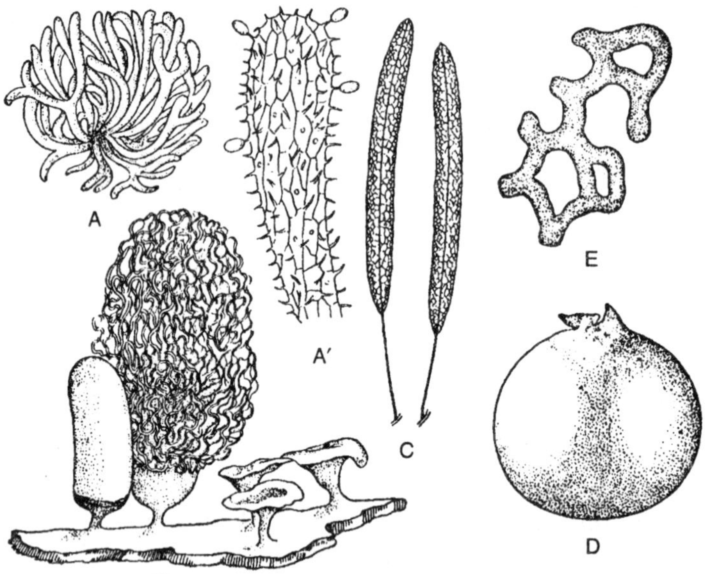

Fig. 18.3: Different types of fructification in the Myxomycetes. A-A′. *Ceratiomyxa fruticulosa*. A. Columnar structures composing fructification. A′. Details of single column showing attachment of sporangia. B. Sporangia in different stages of maturity in *Arcyria* sp. C. Sporangia bearing hypothalli in *Stemonitis*. D. Aethalium of *Lycogala* sp. E. Plasmodiocarp of *Hemitrichia* sp.

used up in sporangium development, this is termed the **hypothallus** (pl. **hypothalli**). This sporangia are of many patterns, some being marvels of delicacy and intricacy. The capillitium and columella are usually present in the sporangia. A sporangium represents only a small part of the plasmodium and is of definite form and structure for a particular species.

An **aethalium** (pl. **aethalia**) is a cushion-like structure resembling a sessile sporangium (Figs. 18.2B and 18.3D). Here all or considerable part of a given plasmodium is involved. Numerous aethalia may be closely combined without much clear delimitation of peridium of the individual aethalium.

The **plasmodiocarp** is a sessile and vein-like fructification that is spread on the substratum with an irregular outline (Fig. 18.3E). The plasmodiocarp is developed by the concentration of the plasmodial protoplasm along the veins, as such its outline is like the veins of the plasmodium. But it has the interior structure of a sporangium.

There are a few Myxomycetes where the fructification has no peridium. The spores are developed externally being borne on slender stalks arising from the surface of the fructification (Fig. 18.3A and A′). Some consider these structures as sporangia. The mature spore contains four nuclei.

THE SPORE

The spores are generally globose with smooth, spiny, warted, recticulate or ridged-reticulate wall (Fig. 18.4). Spore wall is of cellulose and not of chitin. The colour of spores are usually yellow, rosy, deep-violet, brown to black, or often shades of various other colours. The spores are uninucleate. They are exceptionally resistant to unfavourable

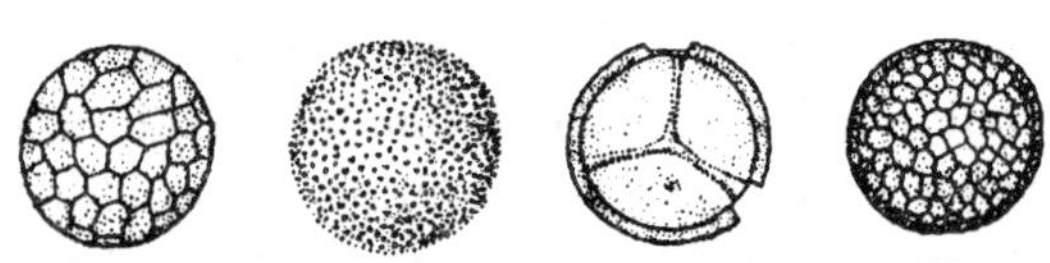

Fig. 18.4: Types of myxomycete spores

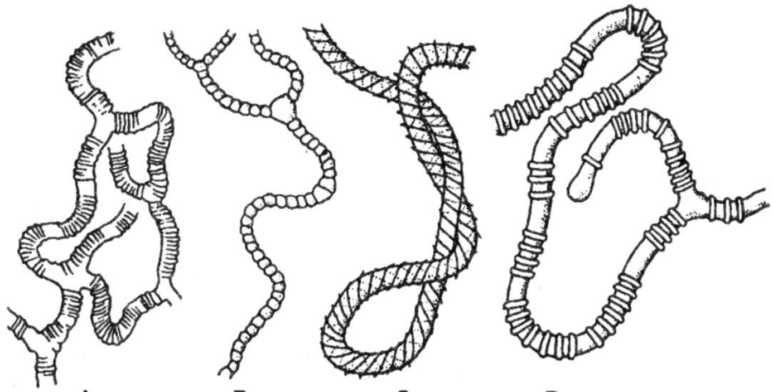

Fig. 18.5: Types of myxomycete capillitia

conditions, particularly to prolonged periods of desiccation.

The capillitia when developed are of different types (Fig. 18.5). They regulate the release of the spores, preventing all of them from being discharged at once. In certain species granules of carbonate of lime in plenty remaining incorporated in various parts of the fructification either in the peridium, in the capillitium or in expansions of capillitial threads as **lime-knots**, or in all these parts.

SWARM CELLS AND MYXAMOEBAE

The different genera and species of the Myxomycetes are mainly distinguished by the form and colour of the fructification and capillitium; by the colour, size and markings of the spores; and presence or absence of granules of lime.

The spores while remaining in a dry state retain their vitality for several years. The length of time elapses before the germination of the spore after it has been placed in water may be from a few hours to several days. The time required for germination of spores, and the percentage of germination vary not only with conditions and with the age of the spores but also with the species. Spores germinate either by cracking of the spore wall or by means of a timy pore through which the protoplast of the spore emerges. The protoplast then either starts exhibiting amoeboid movement and develops into a **myxamoeba** (pl. **myxamoebae**) or after remaining quiescent for a few minutes develops two unequal flagella giving rise to a **swarm cell** (Fig. 18.6). The swarm cell development may also take place within the spore before the emergence of the protoplast. Uniflagellate swarm cells have also been reported in some species of the Myxomycetes. But irrespective of bi- or uniflagellate swarm cells, the flagella are of whiplash type. Division of swarm cell begins a few hours after it leaves the spore wall. Myxamoeba also divides repeatedly producing a large number of myxamoebae.

Although the myxamoebae and the plasmodia of many Myxomycetes have been grown in pure culture, none has been induced to complete its entire life cycle in media of defined

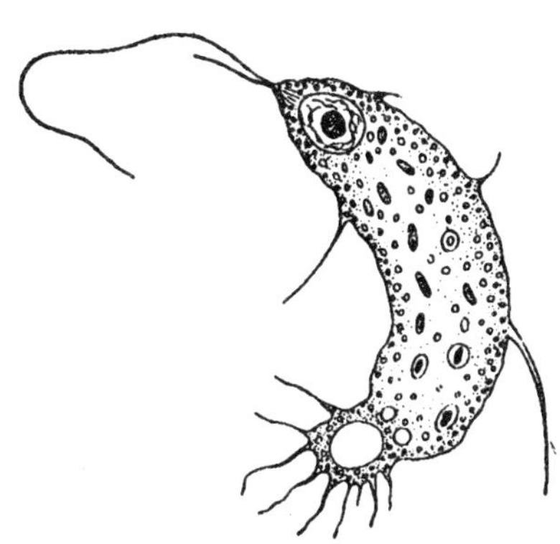

Fig. 18.6: Myxomycete swarm cell

chemical composition in the absence of living or killed bacterial cells.

SEXUAL REPRODUCTION

The myxamoebae or swarm cells as the case may be, behave as gametes and as such take part in the sexual reproduction. Fusion may take place between the two swarm cells or two myxamoebae or between a swarm cell and a myxamoeba resulting in the formation of a zygote, when plasmogamy is followed by karyogamy. Fusion between the two swarm cells or between a swarm cell and a myxamoeba leads to the development of a flagellated zygote which sheds its flagella and becomes amoeboid. The zygote formed in any one of the processes grows, along with which the diploid nucleus also divides mitotically giving rise to a multinucleate amoeboid structure, the plasmodium.

According to some, karyogamy takes place during zygote formation. To them, the plasmodium represents the diplophase in the life history of a myxomycete and meiosis takes place during the formation of spores in the fructification (fruit body), whereas others are of opinion that karyogamy takes place just before meiosis during the development of spores in the fructification and as such the plasmodium is in the dikaryophasic condition.

LIFE CYCLE PATTERN

The general life cycle pattern of the Myxomycetes is more or less clear. The spores on germination produce one to four swarm cells or myxamoebae. The swarm cells or myxamoebae behave as gametes. They fuse in pairs, plasmogamy is followed by karyogamy. The zygote so formed grows accompanied with repeated mitotic division of the diploid zygotic nucleus resulting in the development of a plasmodium. Many zygotes may also coalesce to produce a plasmodium. A mature plasmodium

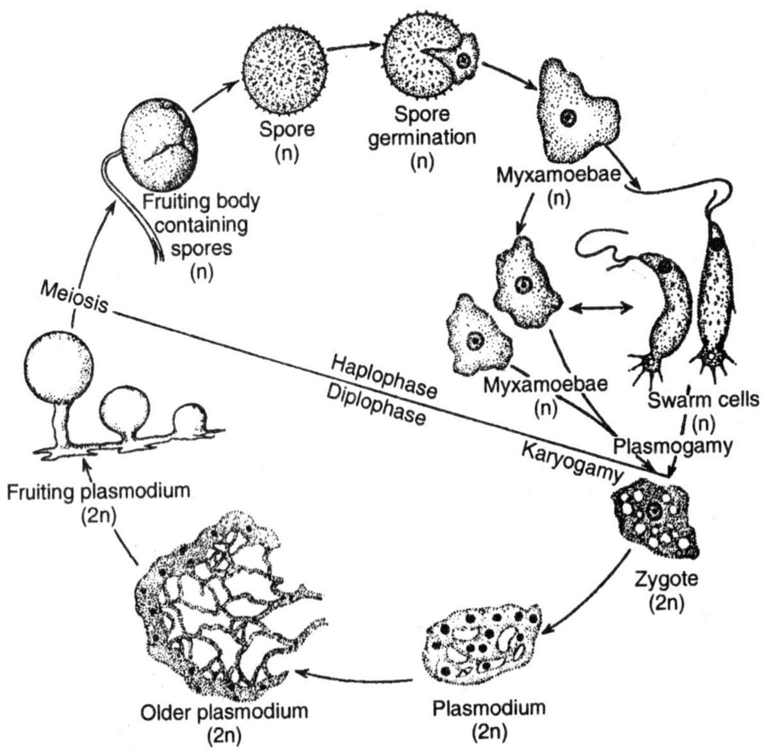

Fig. 18.7: Life cycle of a myxomycete

thickens and gives rise to the fructification of varied nature. During the development of the fructification (fruiting body) and spore formation the diploid nuclei divide reductionally. Each haploid nucleus with a portion of cytoplasm enveloped by a thick wall develops into a spore. The spores vary greatly in colour and characteristics of the wall which are of taxonomic importance. Life cycle pattern of a myxomycete is indicated in Figure 18.7.

CLASSIFICATION

The Myxomycetes may be classified into the following orders depending on the nature of sporangia, presence or absence of lime, and its location; presence or absence of columella and capillitium; and colour of spores in a mass.

A. Sporangia monosporous, borne on delicate structure.

Ceratiomyxales

AA. Sporangia multisporous, simple or compound.

B. Spores rusty to deep violet in mass.

C. Without lime, except in the sporangiophore.

Stemonitales

CC. With lime in the peridium or capillitium.

Physarales

BB. Spores pale or bright coloured, not rusty or violaceous.

C. Columella and capillitium absent.

Liceales

CC. Columella and capillitium present.

Trichiales

CCC. Columella usually present, capillitium present, sporangia very small.

Echinosteliales

AFFINITY

Since naming of the Myxomycetes as Mycetozoa or Fungus animals by De Bary (1887), slime molds were grouped by the Zoologists in the animal kingdom under the name Mycetozoa. True, these organisms possess some animal characters like somatic body resembling a giant amoeba, absence of cell wall, exhibit amoeboid movement, and have holozoic mode of nutrition. But in course of time, Macbride (1899), Martin (1932, 1960-61), Alexopoulos (1962), and others have demonstrated that the Myxomycetes have characters very much like plants and as such they should be placed in the plant kingdom and that too close to the fungi with whom they have closer affinity. The major plant-like characters they possess are: can be grown spore-to-spore in artificial cultures; and life cycle pattern, habit and habitat; existence of sclerotial stage; nature and behaviour of warm cells; development of fructification and spore development by meiosis in the fructification.

Order: **Stemonitales**

This order is characterized by the absence of lime from the peridium and from the capillitium. But lime may be present on the stalk of the fructification. The capillitium is usually abundant and dark-coloured. The spores are also dark-coloured. The order Stemonitales includes three families: Collodermaceae, Stemonitaceae, and Amaurochaetaceae.

Family: ***Stemonitaceae***

The sporangia of the Stemonitaceae are stalked. The upper part of the stalk extends within the sporangium as a columella from which the branching threads of the capillitium originate. The sporangial wall is composed of a delicate membrane which often may be evanescent.

Genus: *Stemonitis*

This is one of the most common Myxomycetes growing on dead wood and dead leaves

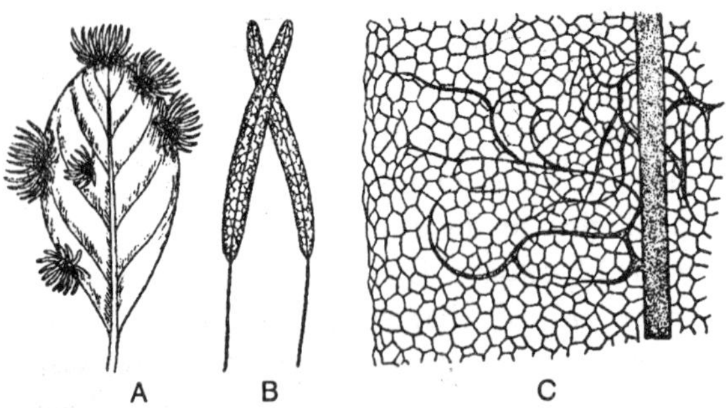

Fig. 18.8: *Stemonitis* sp. A. Group of sporangia on a dead leaf, B. Sporangia. C. Columella with capillitia

(Fig. 18.8A). The plasmodium is of aphanoplasmodium type. The colour of the plasmodium is usually watery to creamy-white, often may be pale-yellow. The plasmodium does not possess the typical reticulated fans but is composed of network of fine strands. The young plasmodium consists of long main veins with branches here and there which bear secondary branches with some anastomosis. In the young, actively growing stage of the plasmodium, the veins are almost completely devoid of an outer layer. Streaming of protoplasm is almost imperceptible in the smaller vein. Protoplasmic streaming, however, is very rapid and rhythmically reversible in the larger veins. The protoplasm is not very granular. As the plasmodium grows, it often spreads in thin, sheet-like, highly vacuolated areas. The plasmodium eventually becomes concentrated into thick, blob-like masses which are transformed into sporangia.

The sporangia are attached in small clusters on a common hypothallus. They are cylindrical, stalked, and fasciculate (Fig. 18.8B). The solid stalk extends within the sporangium usually near the apex as a columella. The capillitium is formed of numerous threads radiating from all parts of the columella and is combined into a loose network, the ultimate branches of which are united into a surface net

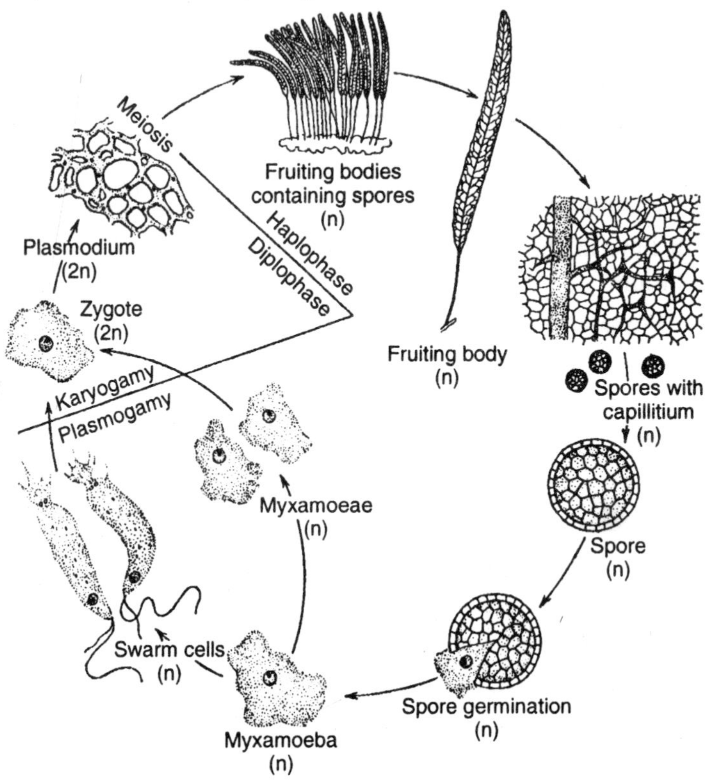

Fig. 18.9: Life cycle of *Stemonitis* sp.

(Fig. 18.8C). The surface net of capillitium may be complete or incomplete above. But the surface net may also be absent. Spores remain embedded in the capillitium.

The spores are abundant in a sporangium, and vary in colour, wall marking and size. They are grey, violet-grey, rufous-violet to pale ferruginous. The spore wall may be smooth, warty or reticulated. The dimension of spores usually ranges from 4 to 12 μm in diameter. The mature spore on germination produces one or two swarm cells or myxamoebae. The fusion between the swarm cells or myxamoebae as the case may be, results in the formation of zygote. The zygote grows and undergoes a series of mitotic nuclear divisions developing into a plasmodium. Eventually the plasmodium heaps itself up somewhat developing into sporangia. The nuclei within the developing sporangium undergo meiosis and a very large number of spores, each with a nucleus bearing haploid number of chromosomes is formed by cleavage of the protoplast. Simultaneously with the development of spores, the capillitium and columella also develop. Life cycle of *Stemonitis* sp. is presented in Figure 18.9.

Some Indian Species: *Stemonitis allida* Wingate; *S. axifera* (Bull.) Macbr.; *S. fusca* Roth; *S. herbatica* Peck; *S. splendens* Rostaf.

RECENT CLASSIFICATION

Myxomycetes are classified into three subclasses: Ceratiomyxomycetidae, Myxogastromycetidae and Stemonitomycetidae (Alexopoulos, 1973). Hawksworth *et al.* (1983), however, have directly classified the class Myxomycetes into six orders; Echinosteliales, Echinosteliopsidales, Liceales, Trichiales, Stemonitales and Physarales, on the basis of the type of sporocarp, the structure of the peridium and capillitium of the sporangium, the presence or absence of calcium carbonate (lime), and the size, colour and ornamentation of spores.

Key to the subclasses of Myxomycetes

1	Spores borne externally on individual stalks	Ceratiomyxomycetidae
1′	Spores borne internally in various types of sporophores	2
2	Sporophore development subhypothallic (myxogastroid); assimilative stage of various types, but never a true aphanoplasmodium ...	Myxogastromycetidae
2′	Sporophore development epihypothallic (stemonitoid); assimilative stage always an aphanoplasmodium	Stemonitomycetidae

Chapter 19

Microbiology

WHAT ARE MICROORGANISMS?

Microorganisms are microscopic forms of life, include bacteria, fungi, algae, protozoa, and the infectious agents at the border-line of life that are called viruses which are not cellular organisms. They are also known as **microbes.** The term microbe is taken from the French and means a microscopic organism or microorganism, being usually applied to the pathogenic forms. Again the term **germ,** in popular usage, refers to any microorganism but especially to one of the pathogenic or disease-producing bacteria. Hence the terms microbe and germ are probably synonymous with bacterium (pl. bacteria). A microorganism that harms its host is also called a **pathogen.** Microorganisms differ widely in form and life cycle. Some are single-celled, but others are multicellular. Some of them have no well-organized nucleus, but others do. Some of the microorganisms are predominantly plant-like, others are animal-like, again others share characteristics common to both plants and animals. They grow rapidly and reproduce at an unusually high rate, within 24 hours some of them complete almost 100 generations. Microorganisms have pattern of metabolic processes like that of higher plants and animals. Regardless of the complexity of structure of a microorganisms, the cell is the basic structural unit of life. All living cells are fundamentally similar. Microorganisms may or may not contain chlorophyll and can use inorganic or organic carbon as the source of carbon. Some have the unique ability of utilizing either radiant or chemical energy and may be autophytes or heterophytes. Again some are able to utilize atmospheric nitrogen for the synthesis of proteins and other complex organic nitrogenous compounds. Some microorganisms synthesize all their vitamins, while others need to be furnished with vitamins. Microorganisms play an important and often dominant role in all fields of human endeavour like industry, agriculture, problems connected with food, shelter and clothing, and in the conservation of human health and combating disease.

ORIGIN, DISTRIBUTION AND NUTRITION OF MICROORGANISMS

Microorganisms originate from the earth's surface and are dispersed in nature through various agencies like air, water, rain, insects and various other media. They occur nearly everywhere in nature where they find food, moisture, and a temperature suitable for their growth and multiplication. Some are found in the bottom of the ocean and others miles away on the mountain heights, thereby can stand wide range of temperature conditions. Microorganisms occur in water (fresh and marine), stagnant and/or free flowing. Since they thrive under conditions suitable for human survival, it is inevitable that we live among a multitude of microorganisms. They infest the air we breathe and the food we eat. Microorganisms inhabit on our body surfaces, in our alimentary tract, in our mouth, nose and all other body orifices. Fortunately most microorganisms are harmless to us; and we have means of resisting invasion by those that are potentially harmful.

The microbial population in our environment is both large and complex. For example, a single sneeze may disperse approximately from 10,000 to 100,000 bacteria. One gram of faeces may contain million of bacteria. Our environment—air, soil, water—likewise consists of a menagerie of bacteria and other microbes. Soil microorganisms include many species of bacteria, fungi, algae, protozoa and viruses. A study of the microorganisms in these habitats requires specialized knowledge of the specific microbes present.

The bacterial population of the soil exceeds the population of all other groups of microorganisms in both number and variety. Several billions of bacteria per gram of soil have been reported by direct count method. They are autotrophs and heterotrophs; mesophiles, thermophiles, and psychrophiles; aerobes and anaerobes; cellulose digesters and sulphur oxidizers; nitrogen fixers and protein digesters; and other kind of bacteria in soil which are yet to be discovered.

Large number of actinomycetes, as many as millions per gram, are present in dry warm soils. The most predominant genera of this group are *Nocardia, Streptomyces,* and *Micromonospora* which are responsible for the characteristic misery or earthy odour of a freshly ploughed field. They are capable of degrading many complex organic substances and consequently play an important role in building soil fertility.

Fungi are most abundant near the soil surface where an aerobic condition is likely to prevail. They exist in both the mycelial and spore stage. It is difficult to estimate their numbers. Fungi are active in decomposing the major constituents of plant tissues, namely, cellulose, lignin, and pectin. The physical structure of soil is improved by the accumulation of fungal mycelium within it.

The population of algae in soil is generally smaller than that of either bacteria or fungi. The major types present are the green algae and diatoms. Their photosynthetic nature accounts for their predominance on the surface and just below the surface layer of soil. On barren and eroded lands algae may initiate the accumulation of organic matter because of their ability to carry out photosynthesis and other metabolic activities. The blue-green algae are known to grow on the surfaces of freshly exposed rocks where the accumulation of their cells results in simultaneous deposition of organic matter. This establishes a nutrient base that will support bacterial species. The growth and activities of the initial algae and bacteria pave the way for the growth of other bacteria and fungi. The mineral nutrients of the rock are slowly dissolved by acids resulting from microbial metabolism. This process continues with a gradual accumulation of organic matter and dissolved minerals unit a condition results that supports growth of lichens, then mosses, then higher plants. The blue-green algae play a key role in the transformation of rock to soil, a first step in rock-plant succession.

Most soil protozoa are flagellates or amoebas. Their number per gram of soil ranges from a few hundred to several hundred thousand in moist soils rich in organic matter. Their mode of nutrition involves ingestion of bacteria with some selectively. Hence protozoa is one of the factors in maintaining equilibrium of microorganisms in soil.

Plant and animal viruses, as well as bacterial viruses (bacteriophages) gain entry into soil microbial population through plant and animal wastes.

In general, some microorganisms survive utilizing solar energy, others use various chemical substances as their fuel, again others have special mechanism of livelihood. As such, microorganisms may behave either as autophytes, as saprophytes, as parasites, or as symbionts.

Microorganisms bring about in nature many changes, some desirable (beneficial) and others undesirable (harmful) to us. The diversity of their activities ranges from the enhancement of soil fertility by decomposing dead organic matter, transformation and deposition of minerals in the soil, formation of coal, production of various useful substances, causing diseases of humans and animals and plants to reducing soil fertility and various other activities.

WHAT IS MICROBIOLOGY?

Since both the problems of small size and the methods of study and since unrelated microorganisms frequently occupy the same habitat and thus influence each other, it is convenient to study them within the same discipline, i.e., **Microbiology.** The science of microbiology is the study of microorganisms and their activities. It is concerned with the form, structure, metabolism, growth, reproduction and identification of microorganisms. It also includes the study of their distribution in nature, their relationship to each other and to other living organisms. For the most part, microbiology deals with microscopic organisms. **Microbiologists** are those who specialize to work with the microorganisms. They have been remarkably successful in exploiting the useful microorganisms and combating the harmful ones and have also successfully solved intricate problems of biochemistry an genetics using microorganisms as tool for their study.

Microbiologists may specialize in the study of different groups of microorganisms. For example, **Bacteriology** is the study of bacteria often broadly designated as **Microbiology; Mycology** is the study of fungi; **Phycology** is the study of algae; **Protozoology** is the study of protozoa; and **Virology** is the study of viruses. Although viruses are not cellular organisms, they are included under microbiology for two reasons: (i) the techniques that are to study viruses are very similar to those that are applied in the study of microorganisms; and (ii) viruses are causal agents of diseases, hence their diagnostic procedures and identification are also similar to employed in the clinical microbiological laboratory as well as the plant pathology laboratory. There may be further specialization in some aspects of above groups of microorganisms; for example, bacterial genetics, bacterial cytology, algal physiology, medical mycology, etc.

HISTORY OF MICROBIOLOGY

Microbiology or study of microorganisms has an interesting past history. During the thirteenth century Roger Bacon suggested that disease is induced by invisible living organisms. Similar suggestion was made by Fracastoro of Verona (1483-1553) and von Plenciz (1762) without any evidence. But in the mean time in 1658 Kircher designated the disease inducing living organisms as 'worms', which according to him are invisible to the naked eye. It was Kircher who first recognized the importance of microorganisms in disease development. But Anton van Leeuwenhòek (1632-1723) was the first person to report descriptions of microorganisms in detail. His discovery brought inspiration to many workers to take interest in the origin of living things. The concept of origin of animals spontaneously from the soil, plants, or other unlike animals sponsored by Aristotle (384-322 B.C.) was still accepted in the seventeenth century. In course of time John Needham (1713-1781), Lazaro Splallanzani (1729-1799), Franz Sculze (1815-1873) and Zheodor Schwann (1810-1882), Pouchet (1859) spoke for and against the theory that living things can originate spontaneously. Finally Louis Pasteur (1822-1895) through his series of experiments proved his germ theory by establishing the fact that different germs came into the world from their parent germs and germs induce various diseases which was suggested by earlier workers. He also founded that fermentation of fruits and grains, resulting in alcohol, was brought about by microbes. Today the pasteurization process, widely used in fermentation industries, is the contribution of Pasteur. Pasteur also tackled the problem of anthrax—a disease of cattle, sheep, and sometimes human beings. In the mean time Robert Koch (1843-1910) was busy with the anthrax problem in Germany. It was he who discovered the typical bacilli responsible for the anthrax disease of cattle and this was the first time a bacterium had been proved to be the cause of an animal disease. His series of observations led to the establishment of **Koch's postulates:** (i) A specific organism can always be found in association with a given disease. (ii) The organism can be isolated and shown in pure culture in the laboratory. (iii) The pure culture will produce the disease when inoculated into a susceptible animal. (iv) It is possible to recover the organism in pure culture from the experimentally infected animal. Again the name of Joseph Lister (1878) is associated with the concept on pure culture technique. Lister obtained pure cultures of bacteria and emphasized the importance of growing an organism (bacterium, fungus, alga, protozoan, or higher forms) in an environment free of any other living organism, i.e., **pure culture.** The period from 1880–1900 was actually a golden time for microbiology when classical contributions were made leading to the establishment of microbiology as a science.

Besides the classical works of Pasteur and Koch, some of land marks of the history of microbiology are: (i) opening of the field of soil microbiology in the late 1880 by the Russian Serge Winogradsky; (ii) application of pure culture technique in industrial fermentation by Emil Christian Hansen (1842-1909) of Denmark, Adametz (1889) of Austria, and H. W. Conn of the USA and H. Weigmann of Germany; (iii) in 1888 the symbiotic relationship between bacteria and leguminous plants was demonstrated by Hellriegel and Wilfarth; (iv) a famous Dutch microbiologist, Beijerinck (1851-1931) described the usefulness of free-living nitrogen-fixing bacteria (*Azotobacter*) in promoting soil fertility; and (v) in the late nineteenth century Burrill, an American scientist working on fire blight of pears established that bacteria cause plant diseases.

Of late, microorganisms have been used as ideal tools to carry out the study of intricate life processes. This has inspired physicists, chemists, and biologists to join with microbiologists in what is known as **molecular biology.** Mention may be made of some of the contributors and their contributions in microbiology and molecular biology during the period from 1944 to 1975: Avery and associates (1944)—DNA carries genetic information in pneumococcus; Fritzhipman and Hanskrebs (1953)—physiology and metabolism of living cells; Joshua Lederberg, George Beadle, and Edward Tatum (1958)—genera in bacteria and their recombination. Ochoa and Kornberg (1959) isolated and synthesized RNA and DNA; Robert W. Holley, Har Govind Khorana and Marshall W. Nirenberg (1968)—study of genetic code and its function in protein synthesis; Salvador E. Luria (1969)—functions of organisms in terms of molecular structure including elucidation of enzyme structure and mode of action; Gerald M. Edelman and Rodney R. Porter (1972)—chemical structures of antibodies; Renato Dulbecco; Howward M. Temin, and David Baltimore (1975)—enzymes in RNA tumour virus particles. Microbiologists are also engaged in making valuable contributions in medical science, industry, agriculture and in science in general for the welfare of the human society.

The basic knowledge of molecular biology and genetics accumulated during the past few decades is rapidly being translated into practical objectives and is revolutionizing industrial microbiology.

The most outstanding current development in applied microbiology is the ability to alter an organism's genetic makeup which is commonly referred to as **genetic engineering** which area of scientific contribution holds great potential for production of drugs and vaccines, for improvement of agricultural crops, and various other areas.

Robert L. Folk, a geologist at the University of Texas observed lumps and knots in electron microscope pictures of soil and rocks representing bacteria with diametres as small as 10 nanometres, usually ranging from 5 to 200 nm which he designated as **Nanobacteria.** His findings were subsequently confirmed by E. Olavi Kajander, University of Kuopio, Finland. One other microbiologist, James W. Coulton of McGill University also confirmed the truth of nanobacteria. So far nanobacteria escaped research attention due to the fact that their effective dimensionlessness makes them quite impossible to be caught by standard microscopes, many of the chemicals used to stain cell walls or their elements of bacteria fail to cling to nanobacteria, they do not thrive on agar medium that is used to culture bacteria, and finally their lethargic growth of about double number every three days makes it tedious to study their metabolism.

E. O. Kajander and Neva Ciftcioglu pointed out that nanobacteria live in urine and by precipitating calcium and other minerals around themselves induce the formation of kidney stone building.

CLASSIFICATION OF MICRO-ORGANISMS

The subject of microbial classification has a long history. One of the first systems was proposed in 1773, and many more appeared since then. Until the eighteenth century, the living organisms were classified into two kingdoms: plant and animal. Since there are organisms which are typically neither plants nor animals, naturally they do not fall into either plant or the animal kingdom. Hence it was proposed that new kingdom be established to include these organisms. In 1866, E. H. Haeckel, a German zoologist, suggested that a third kingdom, **Protista,** be formed to include those unicellular microorganisms that are typically neither plants nor animals. These organisms, the **protists,** include bacteria, algae, fungi and protozoa. Since viruses are not cellular organisms, hence they were not included with the protists. He referred bacteria to lower protists, and called algae, fungi and protozoa higher protists. But Haeckel's concept of Kingdom Protista failed to put forward criteria that could be used to distinguish a bacterium from a yeast or certain microscopic algae. Late in the 1940s with the help of powerful magnification provided by electron microscopy it was possible to make more definitive observation of internal cell structure. It was discovered that in some cell, particularly in bacteria, the nuclear substance was not enclosed by a nuclear membrane, whereas in other cells, for example typical algae and fungi, the nucleus was enclosed in a membrane. This discovery enabled to separate one group of protists (bacteria) without membrane bound nucleus from all the others (algae, fungi, protozoa) which possess membrane bound nucleus. These two cell types have been designated as: **procaryotic** and **eucaryotic** and organisms of each cell type are called **procaryotes** and **eucaryotes** respectively. Hence bacteria are procaryotic organisms, and eucaryotic organisms include algae, fungi and protozoa.

A comprehensive system of classification, designated as five-kingdom system, was proposed by R. H. Whittaker (1969). This system of classification is based on three levels of cellular organization which evolved to accommodate three principal modes of nutrition: photosynthesis, absorption, and ingestion. The procaryotes are included in the kingdom **Monera,** they lack the ingestive mode of nutrition. Unicellular eucaryotic microorganisms are placed in the kingdom **Protista,** all nutritional types are represented here. It includes microalgae whose mode of nutrition is photosynthetic and protozoa with ingestive mode of nutrition. The multicellular and multinucleate eucaryotic organisms are found in the kingdom **Plantae** (multicellular green plants and higher algae), **Animalia** (multicellular animals), and **Fungi** (higher fungi). The diversified modes of nutrition lead to a more diversified cellular organization of the three kingdoms: **Monera** (bacteria and cyanobacteria), **Protista** (microalgae and protozoa) and **Fungi** (yeasts and molds). The scheme of classification is presented below:

A. Unicellular organisms
- B. Procaryotes — **Monera**
- BB. Eucaryotes — **Protista**

AA. Multicellular organisms
- B. Higher algae and green plants — **Plantae**
- BB. Animals — **Animalia**
- BBB. Higher fungi — **Fungi**

Bergey's Manual of Determinative Bacteriology in 8th edition (1974) has recognized the kingdom Monera, but has called it kingdom **Procaryotae** because of the procaryotic nature of the cells. This kingdom is divided into two divisions: the blue-green algae or cyanobacteria (some microbiologists consider the blue-green algae as bacteria) and bacteria (procaryotic organisms other than blue-green algae).

In 1984 the scope of Bergey's Manual was greatly broadened by bringing together various other information concerning bacterial classification and identification. It was published under a new name—Bergey's Manual of Systematic Bacteriology in which all bacteria were placed in the kingdom Procaryotae which has been divided into 4 divisions as follows:

Division 1. Gracilicutes

Procaryotes with a complex cell wall structure—Gram-negative bacteria

Division 2. Firmicutes

Procaryotes with a cell wall structure characteristic of Gram-positive bacteria.

Division 3. Tenericutes

Procaryotes that lack a cell wall.

Division 4. Mendosicutes

Procaryotes with characteristics of an earlier phylogenetic origin than above Divisions.

The study of cell organization by the electron microscope and by advanced biochemical techniques has revealed fundamental resemblances and differences between various microorganisms and offers a sound basis for their division into major groups (i.e., procaryotes and eucaryotes, and viruses) and for their further subdivision.

L. E. Hawker and A. H. Linton (1978) included under the kingdom Protista the so-called microorganisms (both procaryotic and eucaryotic) having relatively slowly differentiated body and subdivided it in the following manner:

A. Organization subcellular **Viruses**

AA. Thallus unicellular, multicellular, or plasmodial
- B. Nucleoplasm not bounded by a membrane **Prokaryota**
  - C. Chlorophyll absent, or if present, of type different from that of plants **Bacteria**
  - CC. Chlorophyll present, together with characteristic blue-green pigment, pigments not located in discrete plastics **Cyanobacteria** (Blue-green algae)
- BB. Cells or plasmodia containing one or more discrete membrane-bounded nuclei **Eukaryota**
  - D. Cell(s) of vegetative thallus (with a few exceptions possessing a cell wall(s)
    - E. Chlorophyll present and located in discrete chloroplasts **Algae** (excluding the blue-green algae)
    - EE. Chlorophyll absent **Fungi**
  - DD. Cell(s) of vegetative thallus lacking true cell wall
    - F. Thallus unicellular, remaining so **Protozoa**
    - FF. Thallus unicellular at first, becoming a plasmodium or pseudoplasmodium and eventually forming a fructification **Slime molds**

In Bergey's Manual of Systematic Bacteriology (1984) the Cyanophyta (Blue-green algae) have been placed in volume 3—Bacteria with unusual properties in the Section: Oxygenic Phototrophic Bacteria. They were designated as the Cyanobacteria as they contain chlorophyll, can use light as an energy source, and evolve O_2 in a manner similar to that of green plants. This arrangement has not been accepted by the algologists. According to them, the Cyanophyte (Blue-green algae) have closer affinities with algae than bacteria and as such they should be placed under the group Algae and not under Bacteria.

PRESENT DAY APPROACH TO MICROBIAL CLASSIFICATION

Approaches to Microbial Classification

With changing history of biology microbial classifications have been attempted with different approaches. These are as follows:

1. **Artificial approach.** The earliest systems of biological classification were artificial. Aristotle, during fourth century B.C. classified the living organisms into two kingdoms—the animals and plants. This division was simply based upon (i) ability of active movement and (ii) ability to photosynthesise. The microbes were accommodated in both these kingdoms.

2. **Natural approach.** In the middle of 18th century, Linnaeus (1758) devised a system, that was more useful. This system expressed the biological nature of the objects that it classified. There was scanty knowledge of microbes in this system also.

3. **Phylogenetic approach**. The post-Darwinian biologists thought in terms of evolutionary affinities between organisms. In the 19th century, the concept of a **natural** system accordingly changed to phylogenetic ones. However, this approach needs a sound knowledge of fossil records of organisms. We would like to refer here again to **Protista.** A German,

Ernst H. Haeckel separated the unicellular organisms from plants and animals, placing them under a third kingdom, Protista. The fungi, protozoa, bacteria and algae were included in this kingdom.

A modern system of this type is devised by R. H. Whittaker (1969). We have already referred to this five-kingdom system of the living world. The microorganisms have been included in the kingdoms **Monera, Protista** and **Fungi** in this system.

4. **Adansonian approach (Numerical taxonomy)**. This was first suggested by a French biologist, Michael Adanson, a contemporary of Linnaeus, in the 18th century. Here the taxonomic arrangement is based upon quantification of the similarities and differences among organisms. There are used as many characters as possible, each of which is of an equal weight. It is assumed that if each phenotypic character is given equal weighting, it should express numerically the taxonomic distances between organisms, in terms of the number of characters they share, relative to the total number of characters examined. The results are fed into a computer so as to define the similarities and differences between microorganisms and thus to indicate possible natural groupings.

5. **New macromolecular approach.** The growth of molecular biology has now opened up a number of new approaches to microbial taxonomy. For identification and classification of bacteria a range of criteria are now used. The shape and size of organism, its oxygen, pH and temperature requirements, and laboratory culture characteristics are considered. Photosynthetic ability, capsule production, staining reaction, spore-forming ability and type of movement are other characteristics. Biochemical digestion of certain carbohydrates, proteins and fats are also considered. There are used certain techniques that give insights into the genetic characters of organisms. Two kinds of analysis performed upon isolated nucleic acids furnish information about genotype.

REPRODUCTION OF MICROORGANISMS

Microorganisms reproduce by the simple method of binary fission. Besides this, there are other methods by which they reproduce which will be treated separately in Chapters to follow.

IMPORTANCE OF THE STUDY OF MICROORGANISMS

Microorganisms though microscopic in size are most interesting life-forms which attracted attention of workers through ages. They possess immense potentialities which are continuously being explored for the welfare of human society. Not only that they are most fascinating group of organisms, but also have a very simple life-form, easy to handle, and are pliable. With increase in knowledge of the microbial world, the task of befriending the microbial regime—particularly bacterial has now become an urgent agendum in the struggle for existence of the human beings together with the necessary organisms. Following are some of the essential aspects of human society that have opened up based on the study of microorganisms and have emphasized the importance of their study.

INDUSTRIAL MICROBIOLOGY

Microorganisms were used in industrial processes even before their existence was known. The production of fermented beverages and vinegar, and the baking of bread are all traditional processes which have come down to use from time immemorial. From an industrial view point the substrate in which microorganism is grown is regarded as a raw material and the microorganism as the "chemical factory" since it converts the raw material into new products. The microorganisms possess three outstanding qualities: (i) rapid growth in large quantities in suitable organic substrate, (ii) maintain physiological constancy of growth to produce approximate amounts of products, and (iii) can convert cheap raw materials into useful products. This is why they have been exploited on commercial scale to obtain valuable products like alcoholic beverages, food supplements, medicine, organic acids, enzymes, etc. Microorganisms that are chiefly employed in various industrial processes include fungi (yeasts and molds) and bacteria. Again, production of bacteria and viruses on a commercial scale for the preparation of vaccines is also one of the industrial processes.

Industrial microbiology may be grouped in the following manner:

I. **Alcoholic beverage Industry.** One of the oldest and largest microbial industries is the production of ethyl alcohol from carbohydrate materials by fermentation process. This process is used by brewers of malt, beverages, distillers, wine makers, and many others. Starches and other complex carbohydrates are used as raw materials which should be hydrolyzed to simple fermentable sugar. The source of raw material may be molasses, sugar beets, grapes, etc. The hydrolysis can be accomplished with enzymes from molds or by heat treatment of acidified material.

Strains of *Saccharomyces cerevisiae* are used for fermentation. They must be vigorously growing with high tolerance for alcohol having a capacity for producing a large yield of alcohol. Next to the production of ethyl alcohol is the production of alcoholic beverages—beer, wine, rum, whisky, and gin. Beer is prepared from barley malt by using strains of *S. cerevisiae* or *S. carlsbergensis,* rum and whisky are prepared from molasses and maize starch respectively by using strains of *S. cerevisiae,* and wine from grape juice by using strains *S. ellipeoideus.*

II. **Bakery Industry.** Pure cultures of selected strains of *Saccharomyces cerevisiae* (backeries' yeast) are mixed with bread dough to bring about desired changes in texture and flavour in the bread. Bakers yeast has the ability to ferment sugar in the dough vigorously and to grow rapidly. The carbon dioxide produced during fermentation is responsible for the rising of the dough.

III. **Food supplements Industry.** In recent years various microbial proteins (single-cell protein) are produced by using waste hydrocarbon from industrial wastes. These proteins are useful for animal feed. Yeasts, molds, bacteria and algae are used for this purpose. Of all these organisms, yeasts are better suited for single-cell protein extraction. Mass cultivation of these organisms offers a food supplement for animals bridging 'the protein gap' *Methophilus methylatrophus* acted upon methyl alcohol derived from oil and natural gas treated with various chemical substances produces large quantities of protein from the substrate. Such proteins are used to produce a milk substitute for calves, thus releasing more cow's milk for human consumption.

IV. **Pharmaceutical Industry.** Most important aspect of this category is the production of antibiotic of this category is the production of antibiotic substances used for the therapeutic purpose by utilizing the source of microorganisms belonging to the groups of Fungi (molds) and Bacteria. Penicillin was the first antibiotic to be produced commercially. The remarkable chemotherapeutic effectiveness of penicillin was demonstrated by Florey and Chain during 1939 and 1941 and subsequently due to pressure of war yield of penicillin was increased many times. Initially *Penicillium notatum* was used for penicillin production, but in course of time a better penicillium-producing species *P. chrysogenum* was isolated. Besides penicillin, tetracyclines are produced by using *Streptomyces rimosus* and *S. aureofacieus.* In addition to these, various other antibiotic substances are prepared commercially details of which are incorporated in Chapters 19, 20, 21 and 22.

V. **Industrial production of Acids and Enzymes**

(a) Citric acid is an important chemical used in medicines, flavouring extracts, food, manufacture of ink, and engraving. It is manufactured industrially by converting sugar (molasses) to citric acid by *Aspergillus niger* incorporated in a medium containing inorganic nitrogen compound and inorganic salts under aerobic condition.

(b) Perhaps the most favour of all industrial fermentations is that of acetone-butanol production by *Clostridium acetobutylicum* needed for the manufacture of cordite.

(c) Enzymes (amylases, invertases, proteases, and pectinases) synthesized and excreted in large quantities by species of *Apergillus, Penicillium, Mucor* and *Rhizopus* are industrially purified and used in the processing or refining of variety of materials, for example, amylases; and pectinases in clearifying fruit juices; proteases used for treatment of hides to provide finer texture and grain.

(d) Lactic acid essential for food and pharmaceutical industries is prepared from hydrolysed starch with barley, and molasses with nitrogenous material by the use of *Lactobacillus delberueckii.*

(e) Vinegar, Cider vinegar, apple vinegar are the various types of vinegar that are prepared by fermenting fruit juices, sugar containing syrup and starchy materials with *Acetobacter.*

(f) In paper industry, paper manufacture involves: the physical or chemical treatment of cellulosic material (e.g., wood, cotton, etc.) for separating and purifying the cellulose, fibres; and fabrication of the resulting fibrous pulp, after further refinement for redeposition of the fibres in the form of sheet. Microbes play a vital role in the paper industry for the preparation of paper-pulp slime.

VI. **Industrial production of Vaccines.** Certain groups of microorganisms (bacteria and viruses) are put to large scale cultivation for the commercial production of vaccines.

Study of Microbes and their likelihood and utilization of their various aspects for industrial purposes comes within the perview of **Biotechnology.** This aspect is treated in the following outlines.

INDUSTRIAL USES OF MICROBES AND THEIR METABOLIES

Enzymatic activities of microbes are utilized in various processes in industries. Some of them are described below:

I. **Enzymes used in food and food additive industries**

1. **In Bakery.** The Bakeries yeasts (*Sachharomyces cerevisiae*) secretes enzymes—**amylase, maltase** and **zymase.** Those enzymes breakdown substrate to make the bread spongy and soft.

2. **In Vinegar production.** The production takes place in two steps. The first step involves the initial anaerobic fermentation of the carbohydrates of food stuffs (fruits, vegetables, malted cereals, honey, molasses, etc.) into alcohol by *Saccharomyces cerevisiae,* followed by second step, the secondary transformation of alcohol to acetic acid by species of **Acetobacter** and **Glucobacter.**

3. **In Single cell protein (SCP) production.** The microbial proteins are also known as **single cell proteins** as because a single kind of microorganism can be artificially cultivated to get such proteins. Different types of microorganisms, like bacteria, algae, yeasts and other fungi can serve as a source of such proteins. Because microorganisms, specially bacteria, have the ability to utilize a wide range of organic compounds for growth and because they have a remarkably higher growth rate than animals and plants. There is considerable scope

of obtaining a large amount of proteins with a short time and at a low cost. Moreover, microorganisms can be artificially grown under controlled conditions without depending on vagaries of nature.

4. **Microbial food spoilage.** Most foods are generally excellent mediator for microbial growth. Microbial activity very often makes food unacceptable and sometimes harmful due to production of toxins by the microbes. Some microorganisms are also used for making different types of foods and beverages. Thus, microorganisms have both useful and harmful effects on food. Some of food spoilage by microbes are:

(i) **Post-harvest microbial spoilage of food-grains and oil-seeds** is mainly due to molds belonging to the genera, *Aspergillus, Penicillium, Fusarin* and *Rhizopus* which are carried from soil during harvest. Microorganisms, specially fungi, mainly attack carbohydrates and fats of the stored grains. Some of the fungi can elaborate toxins in the stored grains. **Aflatoxins** produced by *Aspergillus flavus* in oil-seeds is a well-known example. (ii) **Spoilage of Bread and Bakery products** like cakes, pastries, etc. are also subject to microbial spoilage. Although baking removes many microorganisms present in rare material, bacterial endospores and many fungal spores can survive and proliferate later in the moderate substrate attacking the carbohydrates. Particularly, species of *Bacillus* produce of slimy consistency. Species of *Aspergillus, Penicillium, Mucor, Neurospora,* etc. grow in bread causing spoilage of the fatty materials like cream used in preparation of cakes and pastries, degraded by microorganisms, to fatty acids which impart sources or rancidity. (iii) **Milk and dairy products spoiled** by bacteria of different species like *Lactobacillus, Lactococcus* and *Salmonella.* (iv) **Fish, meat, eggs and poultry products are spoiled** by different species of bacteria—*Pseudomonas, Alteromonas, Actinobacter, Vibrio, Salmonella,* etc. Bacteria-like species of *Pseudomonas, Serratia* and fungi-like species of *Cladosporium* and *Penicillium* attack the egg proteins producing foul smell of H_2S.

5. **In production of Cheese, Butter and Butter-milk.** During cheese preparation, milk is acidified inoculating lactic acid bacteria to yield casein (*curd*) separated from the liquid portion which is called whey. The raw casein is known as **Cottage cheese.** Usually, the casein is ripened by inoculating it different strains of bacteria and some fungi to give characteristic texture, taste and aroma. Certain yeasts and species of *Brevibacterium* are employed for a softer consistency. Now, butter is the milk fat. During the process, the butter is separated from milk mechanically by churning, butter being lighter floats on the surface and is separated. The residual milk is *butter-milk.* The flavour of butter is imported by diacetyl which is an oxidation product of acetone. For this, butter is fermented by lactic acid bacteria like *Lactococcus cremoris, Lactobacillus plantarum, Lauconostoc cremoris* and *L. citrovorum.*

II. Industrial production of Antibiotics and their use in Medicines

1. **Industrial production of penicillin.** Penicillins are active against Gram-positive bacteria. Because its efficiency and because of non-toxic property penicillin was of wide choice for quite a long time. However, the pharmacokinetic property of penicillin C is not good, since it is only partially absorbed when only administered. Penicillin C is most widely used B lactam antibiotic active against Gram-positive bacteria.

The years 1940 onward have been considered as a historical landmark in industrial microbiology. The technical knowledge of large-scale cultivation of microorganism in submerged aerated condition proved useful for other antibiotic producing organisms, mostly species of *Streptomyces* which are also mycelial in nature. Strain-improvement through constant efforts has become an essential part of any industrially useful organism. The industrial preparation of **penicillin** produced culturing high-yielding strain of *Penicillium chrysogenum* is higher than mostly produced species of *streptomyces.* Medium composition is of critical importance for maximum production of penicillin. A high concentration of glucose produces a large biomass, but is not suitable for maximal use of penicillin. At the same time a limited supply of nitrogen source stimulates penicillin production. The penicillin formation takes about 6 to 7 days to be completed during which penicillin concentration is highest. The fermentation medium is continuously agitated mechanically with simultaneous aeration. This general operational pattern is followed also in case of production of other antibiotics. Penicillin, like all other antibiotics, is a secondary metaboline of the producing organism.

PRODUCTION OF STREPTOMYCIN

This antibiotic is named after the actinomycete genus *Streptomyces* which produces. This antibiotic was discovered by Waksman in 1943. Streptomycin is effective against Gram-negative organisms like *Escheria coli, Aerobacter aerogenes, Pseudomonas aeroginosa, Shigella gallinarium* and *Serratic marcescens.* The production of worth of streptomycin is a pure culture of *Streptomyces grisens* grown of aerobic fermentation. All operations are carried out under aseptic conditions. For fermentation high yielding strains of *Streptomyces grisens* are selected from natural various and from mutant strains produced by controlled exposure to UV light irradiation. Such selected mutants are used as starter cultures for production of streptomycin. Streptomycin fermentation is subject to severe phage attack and for preventing this, phage resistant strains are necessary. Spores from starter culture (agar slant) is suspended in steriled distilled water and the suspension is divided into three different parts containing 300 ml of nutrient solution.

The flasks are then placed on a shaker at 28 °C in a temperature controlled room. This is the first stage of submerged fermentation.

To build up the size of inoculum for a large fermenter, quantity of the laboratory flask cultures should be transferred through a series of increasingly large fermenter. Each vessel in this series is designed so that it will provide the required quantity of growing organisms or inoculum for efficient utilization of the next larger fermenter in the series.

Fermentation conditions are generally same throughout the series of fermenters. The temperature is controlled around 25 to 28 °C by circulating temperature controlled water through jackets or coils built into the fermenter. Aeration is accomplished by passing sterile air through the broth. Agitation in the fermenter helps to promote disperson of air. Sterile air is introduced at the bottom of the fermenter and this air is distributed evenly in fermentation broth by the action of agitators. For maximum of antibiotic, contact between the liquid and air is essential, as because the oxygen demand of *Streptomyces grisens* is quite high.

The pH of the medium which is neutral in the beginning rises to 8.0 to 8.5 as fermentation progresses. Sterilization of the medium is accomplished by heating the medium to 120°C in the fermenter and maintaining the temperature for 1½ hours and then cool it to 25 to 28 °C. Continuous sterilization of the medium and sterility of medium and equipment is essential for satisfactory production of streptomycin.

One of the difficulties encountered during formation is the tendency of medium organ. This is controlled by balancing of aeretion and agitation. The addition of antifoaming agents also reduces foam production. Such substance are sterilized added to the medium.

The fermentation cycle of the antibiotic varies from one to several days depending upon the culture, the medium and the condition employed.

A commercial streptomycin fermentation passes through three phases. The first phase lasts approximately for 24 hours and consists of rapid growth of mycelium for fermentation. During the second phase streptomycin is produced rapidly and lasts from 24 hours to 168 hours of incubation. There is no mycelial growth during this phage and the weight of the mycelium remains constant.

Before filtration the broth is acidified with HCl or H_2SO_4 to pH 2.0. It is then stirred with charcoal and mixture is filtered. Significant quantities of impurities adsorbed on charcoal. It is further purified by different methods of chromatography. The preparation of crystalline streptomycin suitable for therapeutic use is of considerable importances, since crystalization is generally a means of achieving high degree of purity and generally ensures constancy in chemical composition.

Table 19.1: Major Fermentation Products

Fungal species	Product
ANTIBIOTICS	
Penicillium chrysogenum	Penicillins
Cephalosporium acremonium	Cephalosporins
Emericellopsis sp.	Penicillin N
Fusidium coccineum	Fusidic acid
Paecelomyces varioti	Variotin
Penicillium patulum	Griseoflvin
Tolypocladium inflatum	Cyclosporin
METABOLITES OTHER THAN ANTIBIOTICS	
Claviceps purpurea	Alkaloids
Ashbya gossypii	Riboflavin
Fusarium moniliforme	Gibberellins
Phycomyces blakesleeanus	β carotene
	Organic acids
Aspergillus niger	Citric acid
A. terreus	Itaconic acid
A. itaconicus	Itaconic acid
A. oryzae	Kojic acid
Penicillium stolonifer	Mycophenolic acid
Rhizopus sp.	Fumaric acid
ENZYMES	
Trichoderma reesei	Cellulases
Aspergillus sp.	Cellulases
Aspergillus oryzae	α amylase
Aspergillus sp.	β amylase
A. niger	β amylase
A. niger	Lactase
Aspergillus sp.	Lipase
Aspergillus sp.	Pectinase
Fusarium sp.	Penicillin acylase
Mucor sp.	Rennet
OTHER COMPOUNDS	
Aureobasidium pullulans	Pullulan
Saccharomyces sp.	Ethanol
Trignopsis sp.	Glycerol
Neurospora crassa	Insulin
Saccharomyces cereviseiae	Interferons

USE AS BIOFERTILIZER AND BIO-PESTICIDES

I. Biofertilizer

Some bacteria like, spp. of *Rhizobium* and *Bradyrhizobium* have the capacity to fix symbiotically atmospheric nitrogen in collaboration with pea, bean, soyabean, medicago and other roots of leguminous plants and others like, spp. of *Azotobacter* and *Azomonas* as free living in soil. *Frankia*, an actimycete, forms root nodules in different non-leguminous plants, forms root nodules in some forest trees like spp. of *Casuarina*, *Coriaria*, and *Alnus glutinosa*. Nodules are also formed by bacteria on leaves of *Paveta indica*. These absorb free nitrogen from the air and convert into nitrogenous compounds. A portion of the nitrogen passes into the fixed surrounding soil and the rest remains in the nodules. On the other hand, bacteria receive some amount of carbohydrates from the host plant. Rhizospheric bacteria increase in number and also the amount of fixed nitrogen. In course of time, nodules decay in the soil and thereby produce fertility of the soil as an **organic biofertilizer.**

II. Biopesticides

Pesticides are the chemicals used in plant protection work. Depending upon the plant to be protected, they are known as fungicides and insecticides. The synthetic pesticides are highly resistant. But the biopesticides are highly resistant to natural biodegradation leading to their accumulation in soil and water causing pollution. Thus, people get interested in the use of alternative process—use of biopesticides which are harmless to plants, human and other animals.

These biopesticides are the microorganisms or their products. Several bacterial pathogens like *Bacillus thuriugenesis* is widely used to control pest of crucifers, Different entomogenous fungi have been used to protect from insect damages. *Vertcillium lacanii* is very effective against whitefly and aphids.

MICROBIOLOGY OF SOIL

Early work in soil, microbiology was almost entirely confined to the study of soil bacteria. The pioneer workers of the late nineteenth century in this line are: Beijerinck and Winogradsky. But it was not until about thirty years later that comparable interest was shown in the soil fungal flora and its activities. The soil microfauna was neglected for even longer and only recently attention has been drawn in the soil algae.

Soil, water and air which contain living organisms are collectively known as **biosphere.** It is produced due to interaction of parental material (rock itself), climate, age, and growth

Table 19.2: Organic acids and other Metabolites

Enzymes	Source	Mode of Action
1. **Amylases**	*Aspergillus niger* (β amylase) *A. oryzae* (α amylase), *Bacillus subtilis*	Break down starch to glucose
2. **Proteases**	*Bacillus licheniformis,* spp. of *Bacillus* and *Aspergillus*	Used in leather industry
3. **Amino acids**	From genetically engineered strains amylases of microbes	Used in animal feeds
4. **Lysine**	From *Bacillus subtilis* and *B. diastaticus* *Corynebacterium glutamicum*	Bread and other food staffs are supplemented with lysine
5. **Organic acids**		
Citric acid	*Aspergillus niger*	Used to lower the pH content to maintain low level. After clearing the mycelial mat amount, citric acid is precipitated as a calcium salt
6. **Organic solvents**		
Ethanol	Through fermentation of yeasts	Used for beverages
7. **Alcohol, Bear etc.**	By the fermentation of different substrates with the help of yeasts	Used for alcoholic drinks, Beer etc.
8. **Dextrose**	Is produced by a wide range of microorganisms **Leuconostoc** spp.	Used in physical fraction of different molecular weight
9. **Glutamic acid**	It is the biosynthate product of Amino acid	Used as other Amino acids.

and interaction of microorganisms. Soil is an excellent medium for the growth of microorganisms which include bacteria, fungi, algae, protozoa and various insects whose number and kinds in the soil depend mainly on the nature and depth of soil, seasonal condition, state of cultivation, temperature, amount of organic matter, moisture content and aeration.

A fertile soil has various components—mineral materials, organic materials and organisms. The mineral materials are derived from the parent rock by weathering to which organic materials are added from plant and animal sources. The organic materials are always associated with soil organisms. The amount of the above components in the soil varies greatly from one situation to another. The organic materials are incorporated into the soil by the action of soil organisms and if it is not first oxidized, it is converted to humus—the dark amorphous colloidal material.

Assessment of microbial population. The complexity of soil environment and the diversity of its population make sampling and assessment of the microbial population very difficult. There are three main methods: **cultural studies, direct examination,** and **activity measurement.** No information on the species of organisms is obtained by these methods. Cultural studies however, permit identification of the isolates though not very satisfactory for fungi. Direct examination method is time consuming and it is difficult to locate and count organisms in opaque soils. Activity measurement study is even more difficult and inconclusive. Although bacteria are more numerous than any other group of organisms, the biomass of the fungi is larger than that of bacteria and also the combined biomass of protozoa, nematodes and other soil animals are quite different.

Interaction and distributional pattern of microorganisms. The behaviour and distributional pattern of microorganisms depend on the nutrients available, temperature, moisture, gas content, pH, and depth of soil. Generally there is decrease in numbers with depth.

(a) **Interaction among microorganisms.** The microorganisms that inhabit the soil exhibit different types of associations or interactions which depend upon the biotic and abiotic components of soil. An association is **neutral** when two different species of microorganisms occupy the same environment without affecting each other and without producing metabolic end products that are inhibitory. When an association involves symbiotic relationship between two species, it is **mutualism.** The benefits derived by the associated organisms are extremely variable. An association involving the exchange of nutrients between two species is **syntrophism.** Again, a relationship between organisms in which one partner receives benefit, the other is not affected. Such a relationship is designated as **commensalism** encountered in association between fungi and bacteria, where fungi break down cellulose to glucose. Many bacteria are unable to utilize cellulose, but they utilize the fungal breakdown products of cellulose. The association is said to be **antagonism** when one species affects the environment for another species by producing antibiotics or other inhibitory substances. There may be **competition** among species for essential nutrients. The best adapted microbial species will predominate eliminating other species. A relation between microorganisms may also be **parasitic** when one organism lives in or on another organisms encountered in case of Gram-negative bacteria *Bdellovibrio bacteriovorus* which is widespread in soil and sewage. Again, viruses which attack bacteria, fungi, and algae are intracellular parasites.

(b) **The rhizosphere.** Plant roots cause an increase in the population of microorganisms in the soil adjacent to them. The region of soil under the influence of roots is the **rhizosphere,** where many types of reactions occur. The influence of the root effects on the soil population and is expressed as the R/S ratio, that is, the number of organisms in rhizosphere soil as compared with the number of the same in soil beyond the influence of the roots. R/S values of over 100 are sometimes known for some bacteria. In general, bacterial response to the rhizosphere condition is greater than that of other groups of microorganisms. Again, with a high bacterial response the R/S ratio for protozoa is also high. But fungal population increases only slightly in the rhizosphere. As a whole the microflora benefits from the presence of plant roots in the soil. Again, microflora also influences plant root branching and root hair production.

(c) **Distributional pattern.** A general account of the distribution pattern of the microorganisms is given below:

(i) As to the bacterial types, members of the genera *Pseudomonas, Achromobacter,* and *Bacillus* are found in most aerobic soils; where conditions are anaerobic, *Clostridium* spp. occur. But when suitable substrates are added to the soil, a multiplication of chemoautotrophic organisms such as the nitrifying bacteria—*Nitrosomonas* and *Nitrobacter* and the sulphur oxidizers of the genus *Thiobacillus* take place. Organisms whose numbers increase in this way as a result of special soil conditions may be regarded as showing a **zymogenous** (fermentative) response.

(ii) Most actinomycetes are soil inhabitants. Their numbers increase with the warmth of the climate and decrease with depth of the soil. They break down the residual nutrients remaining in the soil.

(iii) Isolation experiments suggest that, in general, species of *Mucor, Penicillum, Trichoderma,* and *Aspergillus* predominate and those of *Rhigopus, Zygorhynchus, Fusarium, Cephalosporium, Cladosporium,* and *Verticillium* are also often present in the soil. Addition of organic matter to a soil stimulates the growth of fungal flora in soil in the same way as it does the zymogenous bacterial population. The fungal population of arid environments is greater than that of bacteria. Majority of filamentous fungi are aerobes and their mycelium is interwoven among the soil particles and it binds these together, thus improving the texture of soil. Again, the major part of the fungal flora occurs in the upper soil horizons where there is most organic matter.

(iv) Algal species living in soil include flagellate, coccoid, or filamentous ones. Some of the common algae are: *Oscillatoria, Nostoc, Anabaena, Cylindrospermum, Chlorococcum, Chlorella* and certain diatoms (e.g., *Hantzschia* and *Navicola*). The growth of soil algae affects the surface soil by depletion of some nutrients. The algal biomass on surface soil is great.

(v) The soil fauna contains numerous protozoa and representative of metazoa. The smallest species, protozoa and nematodes are widely distributed in the soil. The former generally occur in the water film surrounding soil particles and the latter usually in the soil of the upper horizons among plant roots.

The more microorganisms there are in the soil, the more productive it is. These soil inhabiting microorganisms play important roles in the soil. The former generally occur in the water film surrounding soil particles and the latter usually in the soil of the upper horizons among plant roots.

The more microorganisms there are in the soil, the more productive it is. These soil inhabiting microorganisms play important roles in the soil. Some of them are: (a) decomposition of organic matter, transformations in the soil and maintenance of soil fertility balance; (b) improvement of soil fertility, reclamation of barren soil and check soil erosion; and (c) biological control of diseases and afforestation.

(d) **Decomposition of organic matter and transformations in soil.** The soil population is responsible for the removal of natural litter from the surface of the earth and also for transformations which are important for continued soil fertility. The natural litter comprises of complex organic residues of plant and animal remains which are chiefly different forms of carbohydrates, proteins, fats, waxes, etc. They are attacked by soil microorganisms which serve as biogeochemical agents for this conversion into simple inorganic compounds or into their constituent elements. The overall process is called **mineralization.** It provides for the continuity of elements (or their compounds) as nutrients for plants and animals including humans. Since the constituent elements of carbohydrates, proteins, fats, etc. cannot be utilized by plants or animals in the elemental form, microorganisms through various cycles of transformations (**Nitrogen Cycle, Carbon Cycle, Sulphur Cycle, Oxygen Cycle** etc.) convert them into complex forms. Nitrogen Cycle is the one in which free molecular Nitrogen of the atmosphere passes through a cycle of transformations mediated by microorganisms to fixed inorganic Nitrogen, to simple organic compounds, to complex organic compounds in the tissues of plants, animals, and microorganisms. Eventually this Nitrogen is released back to the atmosphere through microbial action. Microorganisms thus perform an essential role in maintaining a cyclic process for the reutilization of elements under natural conditions. The predominant microbial types involved and the rate of attack depend on the chemical nature of the residues, the environmental conditions and the nature of the underlying soil. For example, nitrogenous materials are generally degraded more rapidly than residues with high C : N ratios. Of the environmental conditions, moisture is particularly important as in dry conditions the attack will be slow. Temperature also affects the rate of degradation and the composition of the attacking flora. Again the underlying soil determines to a large extent the microbial population available for degradation.

Proteins, nucleic acids, purine and pyrimidine bases, and amino sugars represent the complex organic nitrogenous substances which are deposited in soil in the form of animal and plant wastes or their tissues. Synthetic processes of microorganisms also contribute some amount of complex organic nitrogen compounds. The nitrogen that is locked in proteins is not available to plants as a nutrient. Hence to set free this organically bound nitrogen, enzymatic hydrolysis of protein—**proteolysis** must take place. This is accomplished by microorganisms capable of converting protein to smaller units—peptides with the help of enzymes—proteinases. The peptides are attacked by enzymes peptidases resulting in the release of amino acids. These proteolytic enzymes are produced by *Clostridium histolyticum* and *C. sporogenes* and *C. sporogenes* and species of the genera *Proteus, Pseudomonas,* and *Bacillus;* and many fungi and actinomycetes. The amino acids are degraded by microbial attack to produce ammonia, a process known as **ammonification.** Ammonia being volatile, some portion leaves the soil and the rest is solubilized to form NH_4^+ which is utilized by plants and microorganisms and, under favourable conditions is oxidized to nitrates by nitrifying bacteria by a process referred to as **nitrification.** This happens in two steps: (i) oxidation of ammonia to nitrite by ammonia-oxidizing bacteria and (ii) oxidation of nitrite to nitrate by nitrite-oxidizing bacteria. Ammonia oxidizers are: *Nitrosomonas europaea, Nitrosovibrio tenuis, Nitrosococcus nitrosus,* and *Nitrosococcus oceanus;* and nitrite-oxidizing bacteria are: *Nitrobacter winogradskyi* and *Nitrospina gracilis.*

Again, under anaerobic conditions, in waterlogged soil heterotrophic bacteria convert nitrates into nitrites or ammonia, the process known as **denitrification**, an undesirable process which results in loss of nitrogen from the soil declining nutrient level for plant growth.

Besides these, the microorganisms also convert atmospheric molecular nitrogen into ammonia by a process known as **nitrogen fixation**. They belong to two groups: (i) **symbiotic** microorganisms, those living in roots of plants; and (ii) **nonsymbiotic** microorganisms, those living freely and independently in the soil.

During bacterial nitrogen fixation the nitrogen fixing enzyme—nitrogenase enzyme complex (characterizes by two components) which react together along with a strong reducing agent—ferredoxin or flavodoxin and ATP.

Symbiotic nitrogen fixation is accomplished by bacteria of the genus *Rhizobium* in association with leguminous plants with some degree of specificity between the bacteria and legumens. Again, nonsymbiotic nitrogen fixation is done by *Clostridium pasteurianum* and species of the genus *Azotobacter.*

Although green plants are the most important agents of assimilation of carbon dioxide present in the atmosphere or dissolved in water, bacteria are also capable of synthesizing organic matter from inorganic carbon through Carbon Cycle. The organic carbon compounds so formed are deposited in the soil due to annual litter fall and death of different groups of plants and animals. They are degraded by microbial activity. The end product, carbon dioxide, is released into the air and soil. The microflora is responsible for the evolution of 95 per cent of carbon dioxide, and animals are only 5 per cent. Most of the microorganisms involved in the carbon cycle in terrestrial environments are heterotrophic. Some portion of carbon contained organic residues is incorporated in microbial tissues which themselves will eventually be decomposed. The remainder is incorporated into humus—the dark-coloured amorphous organic material which is important for soil fertility. The soil fauna is also associated with humus formation. The humus is not a single chemical substance, it is essentially polyphenolic in nature and contains some amino acids and amino sugars. Bacteria and fungi are the principal microorganisms that degrade organic carbon compounds. The most abundant organic material in plants is cellulose which is readily attacked by bacteria and fungi. During cellulose decomposition fungal hyphae rapidly colonize the surface of the organic residues and hydrolyse cellulose. Some of the genera of fungi taking part in the process are: *Botryotrichum, Chaetomium, Humicola, Stachybotrys,* and *Stysanus.* After a few weeks the mycelium dies off and there appear bacteria and animals to cause decomposition and degradation of these organic materials. The initial enzymatic attack by enzyme cellulose splits cellulose to cellobiose. Again, cellobiose is split to glucose by the enzyme β-glucosidase; glucose is metabolized readily by many microorganisms. Complete oxidation yields carbon dioxide and water. Similar degradation paths occur for the other plant tissue substances such as hemicellulose, lignin and pectin. Carbon dioxide may also originate from the decarboxylation of amino acids, as well as from the dissimilation of fatty acids. All of these transformations may occur in the soil.

Sulphur in its elemental form cannot be utilized by plants or animals. It passes through a cycle of transformations like nitrogen and carbon mediated by microorganisms. Many of the microorganisms taking part in the Sulphur Cycle are not active in the soil, e.g., the photosynthetic bacteria in the oxidation of sulphides. In most agricultural soils sulphur is present in sufficient concentrations to meet the requirements of the soil population and crop plants. This supply is derived from the parent rock, from organic residues and from volatile compound in the industrial regions. Microorganisms are, as a group, more versatile and can utilize most of the sulphur compounds and are instrumental in making sulphur available to higher plants in a utilizable form. Again, some microorganisms oxidize and others reduce various sulphur compounds. *Thiobacillus thiooxidans*, an autotroph can oxidize sulphur to sulphate which is assimilated by plants and is incorporated into sulphur-containing amino acids and then into proteins, whereas members of the genus *Desulfovibrio* are responsible for reducing sulphate in the soil. In water-logged paddy field soil where anaerobic conditions prevail, they are responsible for the generation of hydrogen sulphide which may damage the roots of rice plants growing there. They are also responsible for causing extensive erosion to underground iron pipes. Again, degradation of proteins liberates amino acids, some of which contain sulphur. This sulphur is released from amino acids by enzymatic activity of many heterotrophic bacteria. Examples of bacteria involved in the sulphur cycle in various ways are: species of the genus *Desulfotomaculum;* purple and green sulphur bacteria *Chromatium* and *Chlorobium;* nonsulphur purple bacteria *Rhodospirillum, Rhodopseudomonas* and *Rhodomicrobium.*

Trace elements (e.g., Zn, Cu, Co, etc.) are released by microorganisms during decay and their uptake by higher plants can be affected by bacteria growing on the root surface. Some of the bacteria taking part in the above process are: *Pseudomonas, Corynebacterium,* and *Flavobacterium.*

The metabolic activity of the microorganisms also solubilizes phosphate from insoluble Calcium, Iron, and Aluminium phosphates. Phosphates are released from organic compounds such as nucleic acids by microbial degradation. Bacteria change insoluble oxides of Iron and Manganese to soluble ferrous and manganous salts. The reverse is also possible.

Thus it is apparent that microorganisms perform numerous and essential functions that contribute to the productivity of soil.

The wide-scale application of herbicides and pesticides for improving crop yield produces certain side effects which need careful consideration and fruitful solution. These substances when deposited in soil cause destruction of soil microorganisms, they run off with rainwater and thereby also cause pollution to streams and rivers and as such affect aquatic flora and fauna.

(e) **Improvement of soil fertility and other aspects of soil.** Blue-green algae—*Oscillatoria princeps; O. formosa;* species of *Anabaena; Spirulina; Nostoc; Cylindrospermum* increase soil fertility by fixing atmospheric nitrogen. Besides this, blue-green and green algae which grow on soil surface check soil erosion preventing the upper fertile layer of soil from getting washed during rainy season and thereby playing an essential role in agricultural soil improvement.

Biological control of diseases and afforestation. Soil microorganisms like *Arthrobotrys oligospora, Dactylella cinopaga* are very effectively utilized to catch and destroy nematodes in the soil as a programme of biological control of diseases of agricultural crops. Besides these, mycorrhizal fungal microorganisms associated with the roots of forest trees, e.g., *Boletus subtomentosus* with *Pinus montana* and *Lactarius deliciosus* with *Pinus sylvestris* increase the absorption surface and absorption rate of various minerals and also make them immune to the attack of diseases. This group of microorganisms is profitably utilized in afforestation programme.

MICROBIOLOGY OF AIR

The atmosphere of the earth contains many particles of solid matter, a large proportion of which being of biological origin. Most viable air borne particles are spore of different organisms which are to some extent suited for survival in such an environment. Not only spores of fungi, myxomycetes, bryophytes and pteridophytes, but also pollen grains, moss gemmae, propagules of lichens, cells of algae, vegetable cells and cells and spores of bacteria, cysts of protozoa, and virus particles may occur in the air and constitute the **air spora** which pollute the air we breathe. Spores remain suspended in the air for as long as their fall speeds are less than the speeds of frequently recurring upward air currents.

Air borne microorganisms constitute mainly viruses (influenza, measles, etc.); bacteria (species of *Streptococcus, Mycobacterium, Corynobacterium,* etc.) and spores of *Thermoactinomyces vulgaris* and *Micropolyspora faeni;* and spores of fungi (*Aspergillus fumigatus, Cladosporium herbarum,* etc.). They are responsible for various diseases (pulmonary or otherwise) of humans. It is essential to study the nature, origin, and behaviour of the aerial microorganisms to prevent the incidence of the disease induced by them.

Source and distribution of air spora. Most of the air spora derives from the surface of vegetation or vegetable debris above ground level or dust. True, numerous organisms expel their spores forcibly for distances ranging from a few to many millimetres by active mechanisms, but most air borne microorganisms have no special structures to facilitate their dispersal into the atmosphere. They are dependent on physical disturbance for their take-off and dust serves as vehicle of air borne contamination.

Nature of air spora. The nature of air spora is extremely variable as it is governed by various physical and physiological factors.

(i) **Air spora in external air.** Air spora constituting fungal spores near ground level display a diurnal periodicity. Spores of yeasts are most abundant in the air spora before dawn, spores of *Phytophthora infestans* late in the morning and spores of *Cladosporium, Alternaria,* and *Ustilago* in the afternoon. Often after rain near the ground level—**wet-air-spora** is rich in basidiospores and ascospores which is replaced by **dry air spora** which consists of pollen grains and dry spores of powdery mildews, *Alternaria,* smuts and rusts. Air borne fungal spores are responsible for pulmonary disease. These spores are derived from moldy grains, straw, damp vegetation and compost heaps. Aspergillosis in pigs develops through straw contaminated with *Aspergillus fumigatus* which can also produce many human diseases. Spores of the fungus *Cladosporium herbarum,* thermophilic actinomycetes, e.g., *Thermoactinomyces vulgaris,* and *Micropolyspora faeni* and pollen grains penetrate deeply into the respiratory tract producing allergic symptoms in man. Besides these, there are many more bacteria whose spores contribute to form air spora in the external air and when inhaled cause serious diseases.

(ii) **Air spora inside buildings.** The microbial content of the air inside buildings may include viruses, pathogenic and nonpathogenic bacteria and fungi. Many viruses including those of influenza, the common cold, measles, smallpox, etc. are dispersed by air and may produce infection when inhaled. Endospores of the genera *Bacillus* and *Cladosporium,* especially *Cl. perfringens,* are commonly found in occupied rooms, hospital wards. *Rhodotorulla* and other yeasts, and spores of the species of *Apergillus, Penicillium, Mucor* and other molds are commonly present. These may contaminate food and moist perishable organic materials, such as leather, and inhalation may cause respiratory infections of men and animals and allergic reactions, such as asthma. In buildings housing animals, the air pollution is particularly high owing to the presence of hay, straw or other fodder, bedding and dried excreta and contamination from animal coats. The spores of *Aspergillus fumigatus* have been found in 80 per cent or more of samples of dust examined from city houses. When inhaled there is development of aspergillosis of lung. Some of the air borne bacterial pathogens include the organisms causing tonsillitis (*Streptococcus pyogenes*); tuberculosis (*Mycobacterium tuberculosis*); diphtheria (*Corynebacterium diphtheriae*); and Q-fever (*Rickettsia burnetti*). The denser particles settle rapidly but of lighter ones remain suspended permanently. The dry sweepings of floors, the dusting of objects, shaking clothes, making beds, movement of people and droughts can break up the original substrate finely divided particles or disturb settled dust and cause it to become air borne.

Survival of air spora. The survival of air borne microorganisms work as air purifiers forming one of the active components of ecosystems in nature. For example, photosynthetic bacteria (*Chromatium* sp., *Rhodospirillum molischianum,* etc.) with bacteriochlorophyll; and blue-green algae and green algae with chlorophyll convert atmospheric carbon dioxide to various forms of carbohydrate and thereby purify air, maintain gas equilibrium taking part in the ecosystem in the nature.

MICROBIOLOGY OF WATER

Microbiology of water comprises the study of nature, distribution and activities of microorganisms in fresh, estuarine, and marine waters. The earth's moisture is in continuous circulation. This process of circulation of moisture is known as **water cycle or hydrologic cycle** which has different stages: **atmospheric, surface**, and **ground water.** The volume of moisture and the nature of microorganisms in each stage are extremely variable. The microbial flora of atmospheric moisture is contributed by the air along with particles of dust. It is removed from the earth's surface in large numbers by rain and hail. The surface water includes waters of lakes, streams, rivers, and oceans. These waters are susceptible to periodic contamination with microorganisms from atmospheric water by precipitation, surface washings of soil, and any wastes that are dumped into them. The composition of microbial flora in the surface water is dependent upon the microbial nutrients that are present in the water, geographical, biological, and climatic conditions. Ground water is subterranean water that occurs in the soil or rock-containing materials. Depending upon the permeability of the soil and the depth to which the water penetrates, the microbes are removed by filtration and as such the ground water may be free from any microbes.

The microbial population in a body of surface water is largely determined by the physical and chemical conditions which prevail in that habitat. Some of these conditions are: temperature, hydrostatic pressure, light, salinity, turbidity, pH, and inorganic and organic constituents. Again optimum growth of microbes depends upon the interaction between these conditions and the nature of surface water (fresh or marine). Besides these, the distribution of microorganisms varies according to the depth of water. In deep waters, the top layers and the bottom sediments harbour the higher concentration of microorganisms than the rest. The aggregation of floating and drafting microbial group in the surface of water is known as **plankton** which when comprising primarily of algae is **phytoplankton,** when protozoa and other minute animals predominate, it is **zooplankton,** and when there is a mixture of both plant and animals lives, it is **zoophytoplankton.** Macrobial inhabitants of the bottom region (**benthic zone**) of a body of water are the **benthic organisms,** richest in an estuarine-marine system. In the benthos two distinct life forms are found, one attached and non-motile, the other unattached and capable of horizontal and vertical movements. The attached species growing on rock or stone surfaces are **epilithic,** on plants are **epiphytic,** and on animals are **epizooic.** Owing to the intensive concentration of biomass of benthic organisms, it forms one of the most important grazing ground for protozoa, vegetarian fish, etc. In between the benthic zone and the zone of producers (photosynthetic algae) is the **profundal zone** of open water where photosynthetic activity decreases progressively. In freshwater, the profundal zone and benthic zone are largely populated by heterotrophic organisms, whereas in marine water, the area between the upper strata and the area just above the sea floor is relatively barren, a vast **microbiological oceanic desert region**.

Nature of microorganisms. All microorganisms of the aquatic environment have patterns of seasonal growth which again are controlled by the nature of water—fresh or marine and stagnant or free flowing. As such it is extremely difficult to enlist the nature of aquatic microorganisms. Still a very broad idea is given in the following manner.

(i) Freshwater planktonic microorganisms include primarily the producer organisms (photosynthetic algae)—green and blue-green algae, diatoms, pigmented bacteria, etc., whereas marine plaktonic ones are: diatoms, blue-green algae, dinoflagellates, chlamydomonads; species of the genera *Pseudomonas, Vibrio, Flavobacterium,* and *Achromobacter* which give protection against lethal portion of solar radiation; protozoa—species of *Foraminifera* and *Radiolaria,* and many flagellated and ciliated species. The characteristic colour of the Red Sea is associated with heavy blooms of a blue-green alga, *Oscillatoria erythraea.* The chytrid *Rhizophidium,* small animals and viruses parasitize on planktonic algae causing profound effect on the algal population.

(ii) In the benthos epilithic forms include many diatoms (e.g., species of the genera *Synedra, Meridion, Licmorphora*); species of *Chamaesiphon, Rivularia, Chaetophora, Cladophora, Sphacelaria* and many microscopic Rhodophytes. Some produce mucilage in which filaments are embedded whilst others precipitate Calcium carbonate around themselves and build up nodular growths. Again, some small algal cells are attached amongst the coating of bacteria on both freshwater and marine sand grains—**epiplasmic flora**—the majority belong to the Bacillariophyta, though some of Cyanophyta and Chlorophyta. Besides this, small Crustacea bear epizoic algal populations.

(iii) Both pigmented and nonpigmented bacteria including species of *Pseudomonas, Chromobacterium, Achromobacter, Flavobacterium* and *Micrococcus* are very common in less polluted water. Bacteria washed into water are mainly species of *Bacillus* (aerobic) and *Clostridium* (anaerobic). In water polluted with animal or human excreta some of the common bacteria are: *Escherichia coli, Streptococcus faecalis, Proteus vulgaris, Clostridium perfringens, Salmonella typhi* and *Vibrio cholerae.* Where oxygen is available, a wide range of aquatic fungi is available of which chytrids are most important. Some more common ones are the species of *Mycosphaerella, Ceriosporopsis, Saprolegnia, Monoblepharis.* Where oxygen is scarce, species of *Clathrosphaerina* and *Helicodendron* may be present. Aquatic protozoa are common in both fresh and salt waters. Planktonic protozoa (ciliates, flagellates and the Heliozoa) and grazing species of other animals fluctuate with the nature of phytoplankton.

Pollution of natural waters. The increasing use of fertilizers and disposal of wastes in streams are resulting in rapidly increasing in nutrient content in many natural water bodies. Enormous algal crops result, followed by their decay giving massive pollution of water. Similar problems of excessive algal growth arise in industrial plants, e.g., in power station cooling towers, and ponds. The degree of pollution of both fresh and marine waters can be determined by studying the algal growth. Benthic algae are very sensitive indicators. In the extremely polluted zones only algae such as: *Oscillatoria chlorina,* and *Spirulina jenneri* occur and in the slightly less polluted zones the diatoms *Nitzschia palea* and *Gamphonema parvulum* appear.

The recent growth in number of nuclear power stations, from which large amount of excess heat has to be dissipated, is giving cause for concern over 'heat pollution' of rivers and estuaries. It causes direct danger to fish and microorganisms, thereby may disrupt the food chain.

MICROBIOLOGY OF ANIMALS

Certain microorganisms remain regularly associated with animals (including insects) constituting the 'normal flora' exhibiting a dynamic equilibrium. Such equilibrium may be disbalanced when the host resistance is lowered and the normal flora assumes a pathogenic role. The types of microorganisms of normal flora are influenced by the kind of animal body and environmental conditions.

(i) **Microorganisms associated with human body.** Normal flora of human skin includes Gram-positive cocci mainly *Staphylococcus albus* and *Sarcina* spp. In the secretions of skin *Pityrosporum ovalis* and *P. orbiculare* are present. In the secretions of ears saprophytic acid-fast *Mycobacterium smegmatis* is common.

The tears secreted on the conjunctiva of the eyes contain the enzyme **lysozyme**, which attacks bacterial cell walls causing lysis, and this keeps the number of organisms in this site at a low level.

Human nostrils are always heavily colonized with *Staphylococcus albus* and *S. aurens.* Again, the normal flora of the upper respiratory tract is represented predominantly by *Streptococcus salivatius, S pyogenes, Neisseria pharyngis,* and *Haemophilus influenzae.* There is no normal flora of the trachea and bronchi.

Flora of human alimentary tract includes: nonhaemolytic streptococci, nonpathogenic species of *Neisseria,* anaerobic spirochaetes, *Fusobacterium* and lactobacilli. Besides these, *Candida albicans* is also present. The small intesting flora includes: *Escherichia coli, Clostridium perfringens,* species of the genera *Bacillus* and *Bacteroides.* But in general, diet influences the intestinal flora.

(ii) **Microorganisms associated with other animals.** Besides human body, one of the major habitats for microorganisms is other animals. As such there exist many different types of interaction between them. These interactions are traditionally classified into: **symbiotic, parasitic** and **commensal.**

Approximately 10 per cent of all insects exist in a symbiotic association with one or frequently more kinds of microorganism. Some of the common examples are: fungi (*Termitomycetes* spp.) and termites; flagellate (*Hypermastigina* and *Polymastigina*) as well as bacteria and rickettsias with termites; mycoflora (*Saccharomyces* spp. *Aspergillus* spp.) of beehives; wood-inhabiting insects (bark-boring beetles) and ambrosia fungus (*Ceratocystis ulmi*).

An interesting aspect is the rument (first chamber of stomach of herbivorous animals) symbiosis of microorganisms. Microorganisms like *Bacteroides succinogenes; B. ruminicola, Streptococcus bovis, Selenomonas ruminantium* grow in the first chamber (rumen) of multi-chambered stomachs of cows along with stored forage. They digest cellulose, synthesize proteins and vitamins and alter fats of the forage in the rumen and make them suitable for nutritional requirements of the cows. In the rumen, microorganisms also synthesize amino acids, proteins, and vitamins used by the host as well as by some species of the rumen microflora. Again, rumen has an atmosphere suitable for development of continuous crop of microorganisms. This is how both the components of the symbiotic association are benefited. Study of the nature and activities of the microorganisms and their quality improvement is useful for the benefit of dairy animals. Pasteur did not believe that animals could live in the absence of microorganisms.

The parasitic relationship between microorganisms and animals is exhibited by different pathogenic microorganisms. Again, the relationship between aerobic bacteria (species of the genera *Pseudomonas, Achromobacter, Vibrio,* etc.) and marine fish may be cited as one of the best examples of commensal interactions of microorganisms.

MEDICAL MICROBIOLOGY

This is the branch of Microbiology that deals with disease-producing microorganisms in the human beings, other animals, and plants. Microorganisms are best known to the average person by the disease they cause. Medical Microbiology is concerned with the prevention and control of diseases. The importance of the various types of microorganisms in causing diseases will be clear only with the study of the concerned microorganisms. This branch of Microbiology is treated separately as: (a) Study of disease of plants (Plant Pathology) and (b) Study of diseases of animals including humans.

SEWAGE MICROBIOLOGY

It is essential to treat biological wastes by chemicals or by microorganisms to remove disease-producing microorganisms which cause offensive odour. Proper treatment which varies with the type and source of wastes, can eliminate health hazards and remove the compounds harmful to fish and other aquatic life. Knowledge of these organisms is essential to maintain balance of life forms. Some of the microorganisms associated with sewage are: species of *Enterobacter, Pseudomonas, Escherichia, Methanobacterium, Methanococcus.*

SPACE MICROBIOLOGY (EXOBIOLOGY)

It is the study of possible occurrence of microorganisms in the outer space and on planets (extra-terrestrial life), or the establishment of earth types of planets through space vehicle. It also includes the study of the potential use of microorganisms for food and energy and for maintenance of a suitable oxygen-carbon dioxide balance in the space vehicle, e.g., use of species of *Chlorelia.*

MICROBIOLOGY AND ORIGIN OF LIFE

Many explanations have been offered for the origin of life on planet earth. One of the more acceptable of these proposals suggests that life originated in the sea following million years of chemical evolutionary process. The wide range of biochemical capabilities found among the microorganisms, the great simplicity of some types make microorganisms attractive "experimental tools" for the study of the transition between chemical synthesis and living forms.

GEOCHEMICAL MICROBIOLOGY

Study of microorganisms has answered many vital questions like the role of microorganisms in the formation of coal, petroleum, and utilization of raw materials (hydrocarbons) for transformation into valuable chemicals. Organic matters derived from microorganisms accumulated in mud deposits of the ocean floor were buried in course of time by sedimentary action and were gradually converted into oil and gas. Continued study in this area of microorganisms will help us to tap our natural resources.

MICROBIAL FERTILIZERS

Microorganisms are used in various ways for the preparation of fertilizers for application in soil to improve soil fertility. Microbial fertilizers contain primarily active strains of microorganisms (blue-green algae, bacteria), mainly bacteria in sufficient numbers. They are used either to fix atmospheric nitrogen or to solubilize plant nutrients like phosphates or to otherwise stimulate plant growth through synthesis of growth promoting substances.

Principle of microbial interaction. In a system of blue-green algae, *Azotobacter* and photosynthetic bacteria *Rhodopseudomonas,* the photosynthetic evolution of oxygen by the blue-green algae can provide an aerobic environment to *Azotobacter.* The utilization of oxygen by *Azotobacter* may result in a partial or complete anaerobiosis in the immediate vicinity, where the anaerobic *Rhodopseudomonas* can thrive well. *Rhodopseudomonas* can use lower fatty acids and carbohydrates. In this course the bacteria can fix nitrogen and assimilate carbon dioxide. In the heterotrophic course, the heterotrophic bacteria *Azotobacter* can use the carbohydrates and excrete lower fatty acids. The carbon dioxide released in the heterotrophic dark course can be assimilated by the photosynthetic bacteria and blue-green algae. The beneficial associative effects of blue-green algae, photosynthetic bacteria and *Azotobacter* for increasing nitrogen fixation can very well be utilized for increasing nitrogen content of soil.

It has also been demonstrated that total nitrogen fixed in the mixed culture of *Azotobacter chroococcum, Rhodopseudomonas capsulatus,* and *Cylindrospermum muscicola* is greater than in pure cultures of each organisms as because the mutual association of *Azotobacter* and the photosynthetic bacteria *Rhodopseudomonas capsulatus* results in providing the energy source.

As long as there is decay and microbial cell synthesis, both mineralization and immobilization take place. For example, solubilization of insoluble inorganic compounds is brought about by some of the species of *Bacillus, Pseudomonas, Mycobacterium, Micrococcus, Flavobacterium, Penicillium, Sclerotium* and *Aspergillus.* These bacteria fungi bring about solubilization by the production of organic acids. The organic acids play the role for liberating phosphate improving phosphorous content of soil which controls the uptake or release of the nutrient by the roots.

Some of the common microbial fertilizers and their uses are given below.

Nitragin. Among the bacterial fertilizers, the most important are those containing the legume root nodule bacteria *Rhizobium* which functions in association with the host plant and helps in fixing atmospheric nitrogen. Nitragin is a peat or soil-based preparation of *Rhizobium* specific for different leguminous crops. Soil-based cultures are suspended in 12.5 per cent sugar or gur solution in water. The sugar or gur solution is boiled and cooled before the culture is sprinkled on the seeds and the seeds are thoroughly mixed with bacterial suspension so as to have uniform coating of bacteria. About 1000 viable bacterial cells per seed is to be attained at the time of treating the seed. Usually 400g of peat-based culture or 900g of soil-based culture would be sufficient for the quantity of seed required per hectare. Finely ground lime (calcium carbonate) is added so that the seeds may have coating of lime before sowing. The pellets so formed should appear dry without loose lime left on the surface, and should be firm. Pelleting with finely-divided lime provides a fair protection against acidity in soil of the fertilizer at the time of sowing. Seed-coat toxicity which often affects the survival of *Rhizobium* on the seed can be removed by soaking seeds in water for a few hours before treating them with peat-based *Rhizobium* culture.

Azotobacterin. It is a preparation containing cells of *Azotobacter chroococcum* grown on agar. The bacterial growth on agar surface is to be scraped by adding water and the solution so obtained is to be sprinkled on the seeds spread in thin layer. The seeds are mixed thoroughly with bacterial solution to ensure uniform distribution of bacterial solution on seed surface. They are then allowed to dry in the shade. The seeds are now ready to be sown. *Azotobacter* should be sown in soils rich in organic matter and moisture. Since *Azotobacter* is not effective in acid soils, these should be limed.

Phosphobacterin. It is also a bacterial fertilizer, a Kaolin based preparation containing cells of *Bacillus megaterium* var. *phosphaticum.* The duration of the use of phosphobacterin is longer than that of azotobacterin, for example, a packet of phosphobacterin can be used for 12 months while azotobacterin must be used within two months. Phosphobacterin when used with farmyard manure and ammonium sulphate enhances the utilization of added superphosphate by crops like wheat, cowpea, etc. Five grammes of phosphobacterin suspended in one litre of water should be allowed to stand for two hours so that the bacterial spores may germinate. The phosphobacterin solution should be then sprinkled on seeds before sowing. The quantity of phosphobacterin to be used, however, varies from the nature of seed to seed.

Blue-green algal fertilizer. Blue-green algae in mass scale for the use as fertilizers can be cultured by (i) tank culture method, and (ii) dry sand culture method. By the dry sand culture method dried algae are obtained which remain viable more than two years. This method is economical and facilitates easy handling of bulk material for distribution. Again, by tank culture method though large quantity of algal material can be obtained with minimum labour, but the algae obtained should always be kept wet at a certain percentage of humidity. During application of blue-green algal fertilizer, algal inoculum may be inoculated in small algal beds in the field sprinkled with lime to promote the growth of inoculum. The beds may then be waterlogged. The inoculated algae will grow to form a scum, which may be scooped out, dried, and broadcast over the field together with lime. Lime will suppress growth of other algae and also lower the acidity of the water to a favourable level.

MICROBIAL GENETIC ENGINEERING

Recent advances have been made in the techniques for the isolation and joining of unlike pieces of DNA molecules so that biologically active recombinant DNA molecules can be made in vitro. These molecules can then be introduced into bacteria where they can replicate. The recombinant DNA method might be used for various purposes. For example, by the application of the recombinant DNA method new strains of bacteria may be developed which are capable of synthesizing a wide variety of biological and chemical substances not yet producible on a industrial scale. One can also think of the use of the knowledge of Micibial Genetic Engineering in agriculture, for example, the genes for nitrogen fixation (*nif*) are transferable. There are possibilities for preparing new nitrogen-fixing bacteria by genetic engineering. In the field of Agriculture, efforts are being made to produce genetically engineered microorganisms that can fix nitrogen in cereal crops and thereby improve soil fertility. This technique of genetic engineering may further be applied in different areas of Microbiology and the results so obtained may be utilized for the welfare of human society.

By the application of genetic engineering technique the pharmaceutical industry has already produced several products for human therapy, such as insulin, interferon, urokinase, and somatostatin, and new techniques for vaccine development have emerged. Microorganisms produced through this technique can be profitably applied to decompose oil in spills and also in mining and oil recovery industries.

GENETICS OF MICROOGANISMS

The organization of genetic material is basically different in the prokaryotic and eukaryotic cells through both the groups of organisms possess the double-stranded DNA.

In contrast, cellular viruses possess wider variety of genetic materials. These may be DNA and RNA which may be single-stranded or double-stranded and DNA may be linear or circular. But in all viruses, the size of genome is smaller than that of prokaryotic organisms. Again, the genome size of prokaryotic organisms is smaller than the dikaryotic ones.

1. In bacteria

The DNA of bacteria, e.g., *E. coli,* is a covalently closed circular molecule. It forms the bacterial chromosome, though this chromosome is much simpler in structure and in level of organization than the eukaryotic chromosome of plants and animals. Also, each bacterial cell normally has single chromosome containing a single circular DNA molecule.

2. In Eukaryotes

In Eukaryotic organisms—plants, animals and fungi, the major portion of DNA is present in chromosomes which are well organized structures and quite different from the prokaryotic counterparts. Besides the chromosomes, mitochondria of both plants and animals and the chloroplasts of green plants also contains DNA. The organization and the nature of the DNA of these cell-organelles are similar to those of bacterial DNA. In both these organelles, the DNA is a covalently circular molecule. In eukaryotic cell, the chromosomes are present in distinct, double-membrane bond structure, the nucleus.

3. In Viruses

In virus, an infection particle contains of a core of nucleic acid (DNA or RNA) enclosed in a protein shell.

In bacteriophage which attacks *E. coli,* has a *single-stranded* **circular DNA genome,** whereas the T-series bacteriophages which also infect *E. coli* causing lysis have much larger genomes consisting of linear ds-DNA molecules. The lamda (λ) phage of *E. coli* is a DNA-virus has a linear ds-DNA. The ends of the lamda (λ) phage are complementary to each other and during circularization, they form base pairs. The free ends are ligated to form **covalently closed cohesive site (cos).**

MICROBIAL ECONOMIC LOSS

Though years of study it has been possible to trace how microorganisms are responsible for: reducing soil fertility by denitrifying bacteria; spoilage of foodstuffs causing food poisoning; animal diseases (tuberculosis of cattle, anthrax of sheep, chicken cholera, etc.); various human diseases (tuberculosis, leprosy, tetanus, etc.); plant diseases (blights, rots, galls, canker, etc.); and cause various other economic loss.

There are a large number of microorganisms that cause tremendous economic loss in various ways. Some of them are incorporated below, others are described under the Chapters 11 and 12.

(i) **Damage of cellulose products.** Economically the most important industrial materials, other than foodstuffs, affected by microorganisms are cellulose and wood products (including wood itself), wood pulp and paper, and textiles made from jute. These materials are attacked by fungi, and to a less extent by bacteria, causing economic loss. Conditions affecting the rate of decay are those governing the growth of the organisms, e.g., pH, temperature, nutrient available, oxygen supply, and moisture. The balance of these factors determines the particular species concerned, for example, bacteria are more common than fungi when the oxygen concentration is low. Some of the organisms largely associated with decay are: fungi—*Chaetomium globosum,* and *Stachybotrys atra;* bacteria—*Cellvibrio cellulomonas, Cytophaga, Clostridium.* Bacterial decay may precede fungal attack in the wood. The bacteria destroy pit membranes and reduce the C/N ratio making fungal invasion more rapid.

(ii) **Degradation of petroleum.** Decay depends on the presence of water and minerals. Since crude oil is contaminated with wine from the oil beds and water is invariably present in the storage tanks from condensation, in transport and storage crude oil may be attacked by *Actinomyces* sp., *Mycobacterium* sp., *Pseudomonas aeruginosa,* and *Desulfovibrio* sp. These are originally introduced from soil contamination, but become establizhed as a resident population in tanks and pipelines. Damages to the crude oil caused by these microorganisms include: (a) decrease in viscosity, (b) loss of oil by oxidation, and (c) changes in the relative amounts of the aliphatic and aromatic fractions of oil. Degradation of petroleum products such as: gasolene, kerosene, paraffin and lubricating oil is more serious, though it occurs only in the presence of water. Attack on aircraft fuel has become important and is particularly associated with the fungus *Cladosporium resinae* and bacteria (*Pseudokonas*). The growth of the microorganisms causes blocking of fuel lines. The organic acids which are produced by microorganisms damage fuel tanks and pipes.

(iii) **Degradation of other materials.** Large number of industrial products are degraded by microorganisms including substances like glass (lens), paints, stones, plastics, rubber and electrical insulations. Some of the organisms responsible are: species of *Aspergillus, Pullularia, Cladosporium,* and *Phoma*—among fungi; and species of *Pseudomonas, Flavobacterium,* and *Thiobacillus thiooxidans*—among bacteria. Natural rubber is subject to the attack by bacteria. The hydrocarbon content may be lowered by the attack of species of *Actinomyces, Serratia,* and *Pseudomonas;* and *Penicillium* and *Aspergillus.* Sulphur oxidizer *Thiobacillus thiooxidans* may attack sulphur in vulcanized rubber causing the production of sulphuric acid which destroys textile reinforcing and metal joints and hoses.

ENVIRONMENTAL MICROBIOLOGY

Of the three major components of environment—water, air and soil contain a great variety of microorganisms. Microorganisms in the air have no permanent microflora. When all the components of environment are in a state of harmonious balance, only then the living organisms can live and flourish. This is called **ecological balance.** If the balance is upset, then the existence of life becomes endangered.

Humans are unique in the living world by virtue of their ability to manipulate nature and master the natural resources so their advantage. Often humans indiscriminately use natural resources to the extent that it endangers the ecological balance.

Microorganisms have always played a vital role in maintaining the ecological balance. Nature is provided with an unique self-purification mechanism operating in her own way.

1. Biogeochemical Cycles

Cyclic transformation of elements and compounds is essential for sustenance of life. Without such recycling process, biologically useful elements—carbon, nitrogen, sulphur or oxygen, would have been fixed in organic materials and had been non-available as nutrients of living organisms.

2. Carbon Cycle

Carbon is the most important element for all living beings, because it forms the skeleton of all organic substances that constitute their body including those of viruses. Carbon alone accounts for about 50% of the dry matter of cells. The ultimate source of this important element in all biological organisms is CO_2 present in air and the water bodies. Two fundamental biological processes help to maintain the CO_2 balance in nature. They are photosynthesis and respiration. Oxygen released in photosynthesis is used for respiration. Green plants including algae and photosynthetic protozoa as well as cyanobacteria carryout oxygenic photosynthesis using water as reductant CO_2 producing organic matter and oxygen. They are the primary producers of organic compounds, which are used by all heterotrophic compounds, which are used by all heterotrophic consumers as the source of carbon and energy. Oxygen released in photosynthesis is used for respiration resulting inhibition of energy from organic compounds.

Iron-bacteria, like *Gallinalla ferruginea* and *Leptothrix ochracea* can also oxidise ferrous ions to ferric ions. The magnetostatic bacteria are able to form magnetic particles in the intracellular magnetosomes. Certain animals, like birds, fishes and dolphins also have magnetic particles in their head which help them in navigation.

3. Microbiology of Air

Air has no resident microflora, as water and soil have. Nevertheless, air generally has different microorganisms, viruses, bacterial and fungal spores, pollens of angiosperms and gymnosperms, as also various spores of lower plants. Fine dust particles floating in air are used as vehicles by microorganisms and viruses. Microbial content of air depends on the location. Most of the microorganisms present in air come from soil, but pathogenic organisms and viruses may originate from diseased persons, particularly those suffering from droplet infections. An important aspect of air-borne pollens and other biological agents is that they may act as **albergens** (antigens causing hypersensitive reactions) in certain susceptible persons causing allergic manifestations.

4. Microbiology of Soil

Soil is a very congenital home for not only a great variety of microorganisms, but also of many other living organisms, including nematodes, insects, worms or even larger animals like rodants, serpent, etc. The contents of resident microorganisms in soil may, however, widely vary depending on physical and physicochemical nature and conditions, such as nitrogen and organic contents, pH, temperature, moisture-level, availability of oxygen, etc.

For Nitrogen Fixing Bacteria (See Chapter Bacteria)

5. Microbiology of Water

Like soil, water in the rivers, springs, lakes, ponds and the ocean harbour, a great many types of living organisms, including microorganisms.

6. Microorganisms and Waste Disposals

Wastes are materials that are thrown off as by-products of mainly human activities. Wastes are highly diverse in nature. Wastes can be divided into two categories, **Collected wastes** and **Disposed wastes.** Disposed wastes include the plant and animal residues left in soil. They are various organic and inorganic compounds like hydrocarbons, herbicides, insecticides, chemical fertilizers, etc. The collected wastes can be subjected to a well-planned treatment and recycled to produce useful products as we see in paper mills. But it is extremely difficult to handle dispersed wastes. Soil houses millions of diverse organisms which coexist and carry out complex interaction. These interactions are named according to their nature. When two or more organisms coexist without interfering one another, is **neutralism.** It is **syntrophism** when an organism produces some product which is utilized by others for growth, whereas when an organism produces some product which inhibits growth of other organism is **antagonism.** When an organism enters into another for its own nutrition is **parasitism.** In addition, there also exist **predator-pray** relationship among different organisms, for example, many protozoans feed on microbial cell, and some fungi trap nematodes and digest them for their nutrition. Viruses can attack bacteria, cyanobacteria and other eukaryotic organisms and kill them.

Soil is also the substratum supporting growth of all kinds of terrestrial plants including algae, lichens, bryophyte, pteridophytes, gymnosperms and angiosperms, as also of cyanobacteria, fungi, slime moulds, etc. All these organisms drew nourishment from soil partly or fully and excrete different metabolic products which are used by soil inhabiting organisms for their growth. For examples, plants excrete several amino acids, vitamins and other organic compounds through their roots into the surrounding soil. This region of soil is known as **rhizosphere.** The microfloral composition of the rhizosphere differs from quantitatively and qaualitatively from the soil outside zone. Plants belonging to the family Fabaciae, commonly known as leguminous plants can enter into symbiotic association with rhizobia leading to formation of root nodules. These bacteria can fix atmospheric nitrogen and supply it in organic form to the leguminous plants. The bacteria are in turn benefitted by supply of photosynthetic products from the host plant. Another genus of bacteria *Frankia* can also form nitrogen-fixing root nodules in a number of non-leguminous plants belonging to different families. Many cyanobacteria can also fix molecular nitrogen either growing freely in soil or in association with other plants.

Green plants (algae) also form symbiotic association with fungi producing endo- and actomycorrhizae. Such association enables the green plants to absorb nutrients from soil more efficiently from poor soils. Mycorrhizal play an important role in forest ecology. Lichens represent another group of organisms, composed of a mycosymbiont and a phycosymbiont which are permanently associated with each other to form a composite organism.

Microorganisms permanently residing in soil take part a great variety of processes responsible for cyclic transformation of different organic and inorganic substances. From biological point of view the most important among such processes is **mineralization.** This process involves transformation of the organic wastes originating from plants, animals and microbes into simpler compounds and ultimately to inorganic compounds, like CO_2, H_2O, N_2, NH_3, H_2S, HNO_3, etc. These compounds are then used by microorganisms and plants as nutrients. Without such recycling process, the sustenance of life should have been impossible.

Some microorganisms present in nature have developed the faculty of degrading pollutants like organic solid, dead wastes (biomass) of plants and animals by a process known as **biodegradation** is a very commendable process of the microorganisms to save the pollution of nature.

Recently microorganisms are utilized for the recovery of nucleic acid directly from the environment, like isolating ribosomal DNAs (r-DNA).

Chapter 20

Bacteria

THE SCHIZOPHYTA

INTRODUCTION

It is Anton van Leeuwenhòek of Holland, who is credited for the discovery of bacteria in 1676. During his work van Leeuwenhòek the tiny microbes which he called **animalcules.** All kinds of unicellular microbes that we know today—protozoa, algae, yeasts and bacteria were first described by him.

Since Leeuwenhòek's discovery of bacteria, these organisms did not receive much attention for about 20 years or so. It was Louis Pasteur in France and Robert Koch in Germany, who established that bacteria and other microbes which could bring about chemical changes as fermentation were the cause of diseases. Koch is also credited for his contributions in the development of microbiology. Some of these are: Establishment of Koch's Postulates or Rules of Proof, Development of pure culture techniques, and Koch also established the final proof that bacteria could be isolated and shown to cause **Anthrax** disease of animals. He thus presented his germ theory of diseases.

Koch's experimental methods made possible the formulation of rules or postulates, which are followed whenever possible, before a suspected organism could be finally accepted as the cause of a specific disease. These are known as **Koch's Postulates or Rules of Proof**. The Koch's postulates are as follows:

1. The organism must be consistently associated with the disease in question.
2. The organism must be isolated from the diseased plant or animal and grown in pure culture and accurately described.
3. The organism of the pure culture, when inoculated back into healthy plant or animal, must be capable of reproducing the characteristic symptoms of the disease.
4. The organism must be re-isolated from the diseased plant or animal tissue, grown in pure culture and must be identical with the original inoculated organism.

All these postulates have been accepted as working guides by workers of diseases of human beings and other animals, and also of plants. But they are not applicable in case of viruses, insects, and fungi which cannot be grown on artificial media.

Bacteria may be defined with a sophisticated four-word definition: **Prokaryons without Chlorophylla.** Bacteria lack, generally, a sexual mechanism as found in higher organisms and therefore the definition of bacterial species is not so rigid.

Bacteria differ from other living organisms those are not containing chlorophyll a (though some species of bacteria contain bacteriochlorophylla).

Microbiologist in turn leaves considerable amount of freedom to the taxonomy to create "new" species on the basis of even one or four differences in characteristics. The higher taxonomic ranks, like genus, family, order suffer from the same uncertainties due to the inherent laxity in their definitions. Besides scientific names, bacteria are often known by their **trivial** or **informal** names. For example of Mycobacteria tuberculosis of *Streptococcus* pneumococcus. Koch's bacillus is the common name.

Earlier classification of Bacteria was based mainly on morphological characteristics. Subsequently more information came from researches at mainly from biochemical and molecular level.

In the year 2001, Bergey's Manual of Systematic Bacteriology, Prokaryotes are separated into two Domains **Archaea** and **Bacteria.** Only the domain bacteria is treated in the chapter.

GENERAL CHARACTERISTICS

Bacteria are among the smallest of living organisms. They are unicellular (some multicellular) or thalloid living organisms, sometimes with some differentiation of cells, but without differentiated tissues. Bacterial cells are procaryotic possessing relatively simple nucleus without any nucleous and nuclear membrane. Some species of bacteria are parasites. They attack the living cells of other plants or of animals and secure their food from that source. But most bacteria grow as saprophytes on dead remains or the products of plant and animal life without a direct relationship with living cells. Parasitic bacteria are responsible for some of the diseases of plants and animals, whereas the saprophytic kinds may be beneficial in one way or the other. While most of the bacteria are heterotrophic, a few are photosynthetic and chemosynthetic autotrophs and are capable of synthesizing carbohydrates out of carbon dioxide and water. Photosynthetic autotrophs contain photosynthetic pigment known as **bacteriochlorophyll** and can carry on a type of photosynthesis—**photoautotrophic** while others lack such pigment, but can manufacture carbohydrates by chemosynthesis—**chemoautotrophic.**

OCCURRENCE, DISTRIBUTION & MAIN GROUPS OF BACTERIA

Bacteria are ubiquitous in their distribution. They are found in all the habitats—soil, water and air. In soils, physical and chemical characteristics are determined to some extent of their bacterial counts. The **rhizosphere** is the zone of soil around roots of plants where microbial activities are quite prominent. Rain water is generally free of bacteria. Importable (drinking) water, more prominent bacteria as taste and odour products and filter clogging are species of *Caulobacter, Clonothrix, Cronotothrix, Leptochrix, Callionella.* Several bacteria also occur in sewerage. Bacteria also occur in a variety of foods and food products including milk, butter, cheese and milk beverages. They occur in perishable (meat, fish) semiperishable (potatoes) and stable (cereals, sugar, flour) foods. They grow on various types of fruits and vegetables. A number of industrial wastes are also utilized by bacteria and produce some useful acids.

Some bacteria grow below 0 °C; others require temperatures above 45°C and may even grow at 80 °C. Some require atmospheric oxygen; others are indifferent to or inhibited by it.

Almost all naturally occurring organic compounds can be used as food by one or another kind of bacteria. On or in a substratum suitable for their growth and development they may become extremely numerous. Decaying vegetable and animal materials and solutions rich in organic matter are usually excellent places for the growth of saprophytic species. Bacteria are useful criteria of water and air pollution and cause many serious diseases of human beings, domesticated animals, and cultivated plants. These are known as **pathogenic bacteria.** Bacteria constitute a normal part of: (1) the soil flora, and (2) the intestines of animals promoting digestive and possibly physiological process and are particularly abundant in the tropics. Some can withstand extreme heat (40-75 °C)—**thermophilic** and others extreme cold (0-30 °C) **psychrophilic,** whereas the majority of bacteria are intermediate in their temperature requirements (20-46 °C)—**mesophilic.**

Most bacteria are not only harmless but absolutely necessary for the existence of living things—**non-pathogenic bacteria.** Life could not exist in the complete absence of bacteria. Plants and animals owe their existence to the fertility of the soil, and this in turn depends upon the activity of microorganisms which inhabit the soil. Bacteria are involved in nitrogen transformations in the soil and help maintenance of soil fertility.

Bacterial physical environments range from hot springs at 80 °C to refrigerated foods, and from distilled water with trace contaminants to the Dead Sea.

REPRODUCTION OF BACTERIA

Bacteria reproduce by binary fission. Some forms produce spores. Sexuality in bacteria is a very intriguing problem which is being thoroughly investigated. Transfer of genetic material from one cell to the other has, however, been reported in some bacteria. It has also been reported that bacteria maintain certain specific, inheritable attributes through countless series of cell generations. They also possess some nuclear mechanism of inheritance. Both DNA and RNA are present in bacterial cells.

ECONOMIC IMPORTANCE OF BACTERIA

Bacteria are of major importance to mankind for they include some of his most destructive enemies and also some of his greater benefactors. Injurious species include those responsible for disease of human beings and other animals and those that destroy food and reduce crop production by inducing various diseases of crop plants. Beneficial species liberate fertilizer elements for growing crops, destroy sewage and other wastes. Activities of many bacteria have been harnessed for various industrial purposes to produce valuable chemicals, medicines and various other products essential for human society.

A. **Beneficial activities.** Some of the beneficial activities of bacteria are:

1. **Bacteria and soil fertility:** Of all the living organisms found in soils, bacteria are among the most active. They are especially abundant in the surface layers of soil, decreasing in numbers with depth of soil. These bacteria, along with other soil organisms, play a dominant role in soil fertility. In general, they succeed in converting insoluble or unavailable materials into forms that can be used by higher plants. Of the essential elements secured by the green plants from the soil, nitrogen is, next to hydrogen and oxygen, the one which is needed in the largest quantities. This element makes up a larger part of the molecule of the different proteins of which protoplasm largely consists. Large quantities of the nitrogen from

the soil which has been built into plant proteins of crop plants are removed from the field in the grain or other useful parts. Although the waste products from domestic animals containing much of the nitrogen of the food which they have eaten and utilized in their metabolism, are in part returned to the soil as fertilizer, much of the nitrogen of the crops removed from the soil is never returned to it. Maintenance of the supply of soil nitrogen is thus one of the principal problems of soil fertility.

The element nitrogen passes repeatedly through a cycle which is known as **nitrogen cycle,** in which it is first converted from simple to complex compounds and then returned to the simple form in a cyclic order by the activity of bacteria and fungi in collaboration with higher plants and animals. Most plants absorb nitrogen from the soil principally in the form of nitrate ions, although they can also absorb ammonium ions. Since nitrate ions are continually being removed from the soil by green plants, the supply of nitrates would eventually be exhausted if there were no way of replenishing it, neither the elaborated nitrogenous wastes excreted by animals are directly available for reuse by most green plants. The conversion of nitrogen from proteins and other nitrogenous compounds into the form of nitrates involves several steps and is due in large part to the activities of bacteria, although some of the earlier stages are also carried on by fungi. Certain groups of bacteria are involved in the decomposition of nitrogenous organic compounds of both plant and animal bodies and the nitrogenous waste excreted by animals and nitrogen transformation in the soil and thereby play an important role in the maintenance of soil fertility. They are: **ammonifying bacteria, nitrifying bacteria,** and **nitrogen fixing bacteria.** The bacterial decomposition of nitrogenous organic compounds in the absence of abudant oxygen usually results in formation of materials of offensive odour chiefly of sulphur compounds. Such anaerobic decomposition is termed **putrefaction.** The decomposition of organic compounds in the presence of oxygen and without the development of odoriferous substance is called **decay.** Actually there is no sharp line of differentiation between putrefaction and decay. Both the processes are carried on by bacteria in the soil. The decomposition of organic compounds is important to human society from at least two standpoints. It prevents the accumulation of organic matter, both plant and animal, on the earth, and it results in the formation of simple compounds or sets free elements that are returned to the soil to be used again by plants.

The ammonifying bacteria transform various proteinaceous substances into ammonia in the soil. The process is known as **ammonification.** In the first step of the process, the proteins of the soil organic matter are broken down to amino acids. This occurs during ordinary decay of organic matter and is brought about chiefly by bacteria and fungi. In the next step, the ammonifying bacteria, together with certain fungi, convert the amino-nitrogen to ammonia. Some of the ammonia escapes into the air and is lost. But most of it usually reacts with water to form ammonium hydroxide without the intervention of bacteria. Ammonium ions can be absorbed directly by most plants and are used as a source of nitrogen. The ammonia of the ammonium salts is oxidized to stable nitrates by the nitrifying bacteria and the process is known as **nitrification.** In the first step, the ammonia is oxidized to nitrite by bacteria known as *Nitrosomonas* group. Finally, the oxidation of nitrite to nitrate is due chiefly to *Nitrobacter.* Nitrate is most readily utilized by green plants. Ammonifying and nitrifying bacteria are directly concerned in the transformation of protein compounds of plant and animal dead bodies and the animal wastes into nitrates, and thus are essential in maintaining a supply of nitrates in the soil.

Nitrogen-fixing bacteria assimilate atmospheric nitrogen (which green plants cannot utilize) by converting into organic nitrogen compounds, which are subsequently decomposed by other bacteria and transformed through a series of stages into nitrates. The process by which they assimilate free nitrogen is called **nitrogen-fixation.** There are two types of nitrogen-fixing bacteria. (i) **Non-symbiotic** nitrogen-fixing bacteria live independently in the soil. Two prominent genera of this group are *Azotobacter* and *Clostridium.* These free-living bacteria fix independently atmospheric nitrogen in their body and convert to organic nitrogen compounds. When they die, the organic nitrogen compounds of their body and convert to organic nitrogen compounds. When they die, the organic nitrogen compounds of their body are made available to higher plants through the activities of other bacteria. (ii) **Symbiotic** nitrogen-fixing bacteria, consisting of species of *Rhizobium,* live symbiotically in small swellings or **nodules** on the roots of various seed plants, chiefly leguminous plants. These bacteria fix atmospheric nitrogen to synthesize organic nitrogen compounds. When the leguminous plants are harvested, the roots with nodules containing bacteria are left in the soil. When they decay, the organic nitrogen compounds made by bacteria become available as nitrates through the process of nitrification. A leguminous crop in a rotation thus possesses an advantage besides that of the crop itself, it increases the amount of nitrogen in the soil.

Each type of leguminous plant requires a particular strain of nodule-forming bacteria in its roots. If this strain is not present in the soil, it can be added artificially by inoculating the seeds before they are sown. The relationship between nodule-forming bacteria and their leguminous plants is one of symbiosis. The bacteria secure food from the tissues of the legumious plant, and the leguminous plant obtains nitrogen which is fixed by the bacteria. The nodule-forming bacteria are thus also known as symbiotic nitrogen-fixing bacteria. These bacteria are widely distributed in soils, so that even in the absence of seed inoculation with these bacteria, leguminous seedling roots sooner or later become infected with symbiotic nitrogen-fixing bacteria.

2. **Bacterial metabolism—its commercial importance:** The metabolic processes and their products of bacteria have been utilized in many industries. In their metabolic activities bacteria excrete waste products some of which have exceedingly important commercial uses. The large-scale production of valuable substances from bacterial metabolism is a relatively new field in the economic utilization of bacteria, and as such, is one in which new discoveries are frequent and often very startling.

Bacteria are grouped: **Gram-positive bacteria** and **Gram-negative bactria** based on the following characters. In Gram-positive bacteria cell wall is rigid, single layerd, more homogeneous and amorphous having peptidoglicogen up to 80% and rest are of protein, polysaccharide and teichoic acid. Principal surface antigen is of teichoic acid, whereas in Gram-negative bacterial cell wall is single layered, elastic, less homogenous, multilayered less homogeneous, and more complex having peptidoglycon is 5 to 15%, and rest are protein, phospholipid and lipopolysaccharide. Principal surface antigen is lipopolysaccharide.

Use of Bacteria for the Society

Model for biological research. Secrets of discoveries in molecular biology, genetics, physiology and biochemistry have been revealed by researches made in microbes.

(i) **Antibiotics.** Bacteria are useful source of these important medicines.

Streptomycin	*Streptomyces griseus*
Chloromycetin	*S. venezuelae*
Neomycin	*S. fradiae*
Tervamycin	*S. numosus*
Bacitracin	*Bacillus subtilis*
Polymyxin	*B. polymyxa*

Soil is perhaps the most important source of microorganisms which produce antibiotic substances. These include filamentous bacteria (actinomycetes). Some of the antibiotic substances are secreted outside the cells and into the environment; others are retained largely within cells and must be separated by extraction. The modern period of antibiotics began in 1939 with the finding of **tyrothricin,** produced by a soil bacterium. Prominent among the antibiotics from actinomycetes are: **streptomycin,** discovered in 1944 and obtained from *Streptomyces griseus,* **streptothricin** from *S. lavendulae;* **chloromycein,** in 1947 obtained from *S. venezuelae;* **aureomycin,** in 1948 obtained from *S. aureofaciens;* **neomycin,** in 1949 obtained from *S. fradiae* and **terramycin,** in 1950 obtained from *S. rimosus.* Besides these, **bacitracin** and **polymyxin** are antibiotics produced by *Bacillus svbtilis* and *Bacillus* sp. resepectively were reported long ago.

(ii) **Fermentation—its industrial application:** The process of bacterial fermentation and its products have been utilized in various industries. Some of them are: *Clostridium acetobutylicum* ferments carbohydrates producing acetone, methyl alcohol, and *n*-butyl alcohol, which have important industrial uses. Very recently, vitamin B_2, a commercially important product, has been discovered as a product of fermentation of carbohydrates by this species of *Clostridium.* This vitamin has been one of the more expensive vitamins to obtain in quantity, and is valuable as a preventive of certain nervous disorders and other diseases.

The manufacture of vinegar is one of the oldest processes in human history which involves bacterial metabolism. Vinegar production begins with the fermentation of sugars in apple juice to alcohol by yeasts. In the presence of oxygen, the vinegar bacterium, *Acetobacter aceti,* oxidizes alcohol to acetic acid, which is responsible for the characteristic odour and flavour of vinegar.

Lactic acid as one of the products resulting from the souring of milk, has many uses in the processing of foods, in pharmaceuticals and in the chemical industry. It is named after the milk constituent lactose, or milk sugar. It is probably the oldest known acid, having been discovered by Scheele in 1780. A group of bacteria designated as 'lactic acid bacteria', ferment lactose of milk to lactic acid. All these bacteria are classified in the family Lactobacteriaceae which includes the genera *Lactobacillus, Leuconostoc,* and *Streptococcus.*

The dairy industry finds bacteria an essential aid in a number of processes. Butter is sometimes made from cream which has been allowed to undergo 'ripening'—that is, a lactic acid fermentation, causing it to become sour. The cream is pasteurized, incubated and then churned. This fermentation eventually results in the formation of substances responsible for the characteristic odour of butter. The process of **pasteurization** was named after Louis Pasteur, the French scientist, who developed it to prevent the spoilage of wines. It is chiefly associated with the treatment of milk to eliminate pathogenic bacteria. Two methods are used to pasteurize milk. In the older of these, the **holding process;** the milk is heated to 140 °F, and held at the temperature for 30 minutes. Another **short time, high temperature method**, sometimes termed **flash pasteurization**, the milk, in thin layers, is exposed to a temperature of 160 °F for a minimum of 15 seconds. The milk is then cooled as rapidly as possible to a temperature which retards bacterial growth. Pasteurization destroys pathgenic bacteria without appreciably affecting the physical and chemical properties of the milk. The pasteurization of milk is widely recognized as one of the most important public health measures. It is not a process of sterilization. Spores of bacteria are not killed, and since the temperature falls far short of boiling, neither are all the vegetative cells. This is notably true of the lactic acid bacteria although the number of these is greatly reduced. Pasteurized milk will, therefore, be normally sour, as will raw milk, although more slowy. Besides all these processes, a number of complex processes involving lactic acid or other kinds of bacteria occur during the ripening of cheese.

(iii) **Retting of fibres:** Bacteria play an important role in the retting of jute, flax, and hemp fibres. They hydrolyze the pectic substancs which act as cement-like materials that bind the fibres together. Retting is carried out by immersing the stalks of jute, flax or hemp in water and weighing them down. Water is absorbed by the tissues, causing swelling and the extraction of water soluble substances. The water becomes highly coloured containing substances that have been extracted from the submerged stalks. This highly coloured water now becomes a good culture medium for the growth of many kinds of organisms. The aerobic organisms reduce the concentration of dissolved oxygen and create an environment suitable for the growth of anaerobes. The pectic substances are slowly fermented and dissolved by the anaerobes, leaving the fibres intact. During fermentation, various, organic acids and gases are produced. The submerged stalks should be removed from water when the reaction has gone to completion; otherwise over-retting will result. They are then throroughly washed to remove the organic acids, odours, and other undesirable substances and the fibres removed from them are then spread out in the sun or air to dry. The dried material is then ready for dressing. Bacteria responsible for retting of fibres are: *Bacillus subtilis, B. polymyxa, Clostridium tertium,* and *C. felsimium.*

(iv) **Other fermentation processes:** Certain streptococci and lactobacilli are used for the preparation of silage for the consumption of cows.

Tobacco is cured and fermented to improve its colour, texture and aroma. Bacteria of the *Bacillus megatherium* group are used for the curing process. A satisfactory fermentation is associated with a rapid increase in numbers of the *Micrococcus candicans* and *Bacillus subtilis* types.

(v) **Tanning of leather.** Tanning as a result of fermentation broken down to insoluble compounds with gelatins, alkaloids and some metallic salts which are used then in producing hides from purification during tanning process.

(vi) **Natural gas production.** Bacteria are used for producing methane, source of power, formed due to decay of organic matter under anaerobic conditions. *Methanobacillus* occurs in dung, pond, marshes, etc. where organic matter is fermented.

(vii) **Production of industrial chemicals.** Sorbose-Sorbitol in fruits of *Sorlus* sp. is converted to sorbose by *Acetobacter* suboxylon. Acetone and butanol by *Clostridium acetobutylicum,* Vinegar by *Acetobacter aceti,* butter, cheese and other dairy products over their flavour to fermenters like *Lactobacillus acidophilus, L. casoi* and others. Amino acids and vitamins are also obtained from such bacteria.

(viii) **Enzymes.** Several enzymes of industrial value are produced by bacteria. Pectinase, a product of *Clostridium* sp. is used in retting of flax. Proteases are produced by *Bacillus subtilis.* These are used in liquid glues, meat tenderisers, spot removers and bear clarifiers.

3. **Bacteria as human symbionts:** Bacteria promote digestive and possibly other physiological processes in the intestinal tracts of animals. The digestion of cellulose by such herbivorous animals as horses and cattle results in part from cellulose-digesting enzymes excreted by bacteria inhabiting the intestines of these animals. Bacteria also dwell in large numbers in the human intestinal tract, particularly in the lower part of the small intestine and in the large intestine.

Escherichia coli, formerly called *Bacterium coli* and *Bacillus coli,* is a regular inhabitant of the lower intestinal tract of man and other vertebrates. About 40 per cent of human feces consisting bacteria mainly *E. coli* in milk, water, or food is used as an index of fecal contamination. *E. coli* has commonly been considered chiefly a commensal, a partner of reciprocal parasitism (**commensalism**). It is also occasionally known as a pathogen. It is now known, however, that *E. coli* and other intestinal bacteria synthesize considerable quantities of some of the B vitamins and release these into the intestine. These bacteria may, therefore, be properly regarded as normally being symbionts rather than mere commensals. The destruction of the bacterial flora of the colon by vigorous dosage with certain antibiotic drugs sometimes causes functional disturbances which suggest that the usefulness of these organisms in human physiology is not limited to the production of known vitamins.

The activity of these bacteria in the human gut and the significance of this activity are not known positively; some physiologists believe that these bacteria carry on certain digestive activities of value of the human body, while others believe that lactic acid and other metabolic products of these normally occurring bacteria inhibit the growth of putrefactive and possibly certain pathogenic bacteria. Whatever may be their specific physiological significance, it is certain that the maintenance of a normal bacterial flora in the human intestinal tract is essential to the health of the human organism and that any major disturbance in this intestinal bacterial flora results in derangements of health.

B. **Harmful activities.** Among the harmful or wasteful results of bacterial activity are these:

1. **Reduce soil fertility:** Certain soil bacteria reduce soil fertility by depleting the nitrogen content of soil. They are known as **denitrifying bacteria.** These bacteria are especially abundant and active under anaerobic conditions in wet soil and soil with high organic matter content. They break down nitrates through intermediate compounds to free nitrogen gas, which escapes into the air and thereby the soil fertility is reduced to a great extent. This process of breaking up nitrates to free nitrogen is designated as **denitrification.** Denitrification is favoured by poor aeration of the soil, and nitrates may eventually disappear from water-logged soils.

2. **Spoilage of foodstuffs:** Of all agents involved in the spoilage of foodstuffs, the activities of living organisms are undoubtedly the most important. Spoilage is caused principally by bacteria, yeasts and molds. The organisms may be pathogenic or non-pathogenic. Different kinds of organisms produce different types of changes in food. Bacteria are more exacting in their requirements than either the yeasts or the molds. They require relatively large amounts of moisture, hydrogen-ion concentration usually near the neutral point, and relatively low osmotic pressures. Bacteria cause spoilage of human food in the form of rotting meat and fish, spoiling butter, all kinds of vegetables and fruits. Certain bacteria cause severe types of food poisoning in persons who eat bacteria-contaminated food.

The bacteria found on meat surfaces are usually of *Achromobacter* and *Pseudomonas.* These organisms, growing on meat, are probably not dangerous to health. Some of the bacteria concerned with putrefaction of meat are: *Bacillus subtilis, B. cereus, Eschrichia coli, Proteus vulgaris,* and *Aerobacter cloacae.* Spoilage of fish is caused largely by bacteria. They belong to the genera *Achromobacter, Flavobacterium, Micrococcus, Microbacterium* and *Pseudomonas.* Bacteria chiefly responsible for food poisoning are: *Micrococcus pyogenes* var. *aureus, Clostridium botulinum, Salmonella enteritidis* and *Solmonella typhimurium.* The most common kind of food poisoning is **Staphylococcus food poisoning** which follows the consumption of food contaminated by staphylococci. A second and less common food-borne disease is caused by bacteria of the genus *Salmonella,* and is commonly known as **Salmonella food infection.** Although infection has been traced to several kinds of foods, including milk, meat is usually involved. *Salmonella* also infects domestic animals and infections in human beings may result from the consumption of such meat when it is insufficiently cooked. Another type of food poisoning is **Botulism food poisoning** which follows the consumption of bacteria contaminated food. It is caused by *Clostridium botulinum.* This organism thrives on meat and other high-protein foods like peas and bean in the absence of air, and the spores often survive the temperatures used in home canning. *C. botulinum* multiplies in food liberating a potent toxin which causes botulism.

3. **Cause animal diseases:** That bacteria cause diseases of animals was detected as early as 1850, but the final proof of the bacterial cause of animal disease was obtained in 1876 by Robert Koch.

Bacteria cause tuberculosis of cattle, anthrax of sheep, chicken cholera, pneumonia, glanders in horses, sheep and goats. One of the oldest diseases of animals is anthrax which is caused by *Bacillus anthracis.*

Species of the genus *Rickettsia* cause diseases like Rocky Mountain spotted fever, classical typhus fever of man and other animals. They are transmitted to humans via arthropod vector. Strains of *Chlamydia trachomatis* cause keratoconjunctivitis, trachoma, that often results in blindness. A number of microorganisms isolated from animal mucous membrane and sewage were termed pleuropneumonia-like organisms (PPLO). They were subsequently designated as mycoplasmas. *Mycoplasma pneumoniae* is the causative agent of primary atypical pneumonia.

4. **Cause human diseases:** Many of the serious human diseases are caused by bacteria. Some of them are: tuberculosis (caused by *Mycobacterium tuberculosis* var. *hominis*), diphtheria (caused by *Corynebacterium diphtheriae*), leprosy (caused by *Mycobacterium leprae*) and tetanus (caused by *Clostridium tetani*).

The medical importance of nanobacteria was highlighted by E. O. Kajander and his associate Neva Ciftcioglu. They pointed out that nanobacteria live in urine and by precipitating Calcium and other minerals around themselves induce the formation of kidney stones building a mineralized shell, "castles" as described by Kajander. As such, nanobacteria acted as nucleation factors for kidney stones. They are also responsible for dementia and pathological extraskeletal calcification.

5. **Cause plant diseases:** Many of the plant diseases are induced by bacteria. Bacteria are usually selective in attacking only particular hosts. It has been seen that bacteria that attack plants do not attack animals. Comparatively bacterial diseases are more common in animals than plants. Plant pathogenic bacteria are distributed among 8 genera: *Pseudomonas, Xanthomons, Rhizobium* (under certain environmental conditions), *Agrobacterium, Erwinia, Corynebacterium, Streptomyces,* and *Mycoplasma.* Most bacteria enter the host through wounds, stomata, hydathodes, lenticels, and nectaries. Species of *Streptomyces* are known to penetrate the cuticle directly and the legume nodule bacteria (*Rhizobium*) penetrate the non-cuticularized root hairs. Since the discovery of Professor T. J. Burrill of the University of Illinois, U.S.A. in 1879, that bacteria are capable of causing plant diseases, many bacterial diseases of plants have been discovered. Bacteria cause various plant diseases. Some of them are:

(i) **Leaf spots:** Invasion and multiplication within the substomatal chamber and the intercellular spaces lead to necrosis of the invaded plant tissue causing spots. The organisms responsible for leaf spot disease produce a vigorous attack on the plant, with the result that the cells become heavily infected and strongly discoloured. The discoloured areas dry up and frequently fall out, leaving holes in the leaves.

Some of the organisms causing leaf spot disease are *Pseudomonas angulata,* the causal agent of angular leaf spot of tobacco; *P. maculicola,* causes cauliflower spot; *P. mellea,* the etiological agent of leaf spot of tobacco; *Xanthomonas cucurbitae,* causes leaf spot of squash; *X. mavacearum,* the agent of angular leaf spot of cotton; *X. ricinicola,* the causal agent of leaf of castor bean.

(ii) **Extensive blights:** Here the progress of the bacteria is more rapid than leaf spots causing a more extensive and rapid necrosis. Organisms producing blight diseases are capable of penetrating considerable distances between cells, leaving the neighbouring tissue intact. The bacteria grow in the plant juices without producing any digestion of the tissues. Usually a discolouration of the leaves and branches is produced. Death is due probably to an interference with the flow of the plant sap. Some of the organisms producing blights include *Erwinia amylovora,* the agent of fire blight or pear blight; *E. lathyri,* the cause of the streak disease of sweet peas and clover.

(iii) **Soft rots:** The major effect is a slimy softening of the tissue by the secretion of an enzyme which diffuses in advance and dissolves the middle lamella of the cells, plasmolysis and death of the cells follow, and the bacteria grow upon the dead plant tissue rather than upon the living cells. Organisms responsible for soft rots reduce the plant tissue to a soft, very moist, pulpy mass. The organisms producing soft rots differ from the other forms found in the soil, in that they have the power to attack healthy plant tissue by the secretion of an extracellular enzyme. The enzyme dissolves the pectin or cement-like material that binds the plant cells. The action is probably hydrolytic, resulting in the liberation of soluble sugars, which are utilized by the bacteria for food. The results is that the plant tissue is reduced to a mass of separate cells, which become converted later into a slimy, pulpy material. The important species causing soft rots include the following: *Erwinia aroideae,* the cause of soft rot of potato, egg plant, cauliflower, radish, cucumber, cabbage, parsnip, turnip, and tomato; *E. atroseptica,* responsible for black rot of stem and tuber of potato and other vegetables; *E. betivora,* the etiological agent of sugar beet rot; *E. carotovora,* the cause of soft rot in carrot and cabbage.

(iv) **Vascular diseases:** In some cases the bacterial invasion in the vascular system becomes systemic, in others, bacteria multiply in the vascular system resulting acute wilt. The bacterial wilts constitute a group of very important and destructive plant diseases. The infecting organisms multiply and accumulate in large numbers in the vascular system, causing an interruption in the flow of sap in the plant. A complete interruption in the flow of sap results in a rapid wilting of the plant. Some important organisms causing wilt diseases are *Bacterium stewartii,* the cause of wilt disease of maize; *Corynebacterium insidiosum,* the agent of vascular disease of alfalfa; *Erwinia tracheiphila,* the etiological agent of wilt of cucumber, pumpkin, and squash.

(v) **Bacterial galls:** Here the primary effect of the baceria on the host is the stimulation of the cell division—hyperplasia leading ultimately to hypertrophy, the combined effect of which results in the gall formation. These abnormal growths are produced by the action of organisms on the meristematic tissue of the plants. In some infections the galls remain small; in others they may assume large proportions. The important organisms producing intumescence diseases include: *Agrobacterium rhizogenes,* the agent of hairy root of apple; *Bacterium pseudotsugae,* the agent of galls of Douglas fir.

(vi) **Canker:** During early part of disease development bacteria are contained primarily in the phloem, rather than the xylem and are more active in the living host tissue at the outer edge of the canker. Example, *Cornebacterium michiganense,* the cause of canker of tomato.

(vii) **Mycoplasmal disorders:** Species of the genus *Mycoplasma* induce various plant diseases. Some of them are: Little leaf of Brinjal, Greening disease of *Citrus,* and Grassy shoot disease of sugarcane (GSD).

GENERAL MORPHOLOGY AND STRUCTURE OF BACTERIA

Most bacteria are unicellular, there are some mycelioid forms too. Cross-walls are absent except in the mycelioid ones (actinomycetes), where septation does take place previous to sporulation and as prerequisite to it. Bacterial cells may be motile bearing flagella or nonmotile without any flagella.

Shape and Arrangement (Fig. 20.1): According to the shape of cells, bacteria may be of four types: (i) **Coccus** (pl. **Cocci**)—cells spherical, (ii) **Bacillus** (pl. **Bacilli**)—cells rod-shaped, (iii) **Spirillum** (pl. **Spirilla**)—cells, Cork-screw-like, and (iv) **Vibrio** or **Comma** (pl. **Commas**)—cells 'comma'-like. Cocci may be of different forms: (a) **Micrococcus,** cells are small and occur singly; (b) **Diplococci,** cells divide in one plane and remain attached predominantly in pairs; (c) **Streptococci,** cells divide is one plane and remain attached to form chains; (d) **Tetracocci,** cells divide in two planes and characteristically form groups of four cells; (e) **Staphylococci,** cells divide in three planes in an irregular pattern producing bunches of cocci; (f) **Sarcinae,** cells divide in three planes in a regular pattern producing a cuboidal arrangement of cells. Bacilli are not arranged in patterns as complex as those of cocci, and most occur singly or in pairs—**diplobacilli.** Some species form chains—**streptobacilli,** others form **trichomes** which are similar to chains but have a much larger area of contact between the adjacent cells, and again in others the cells are lined side by side like matchsticks—**palisade arrangement.**

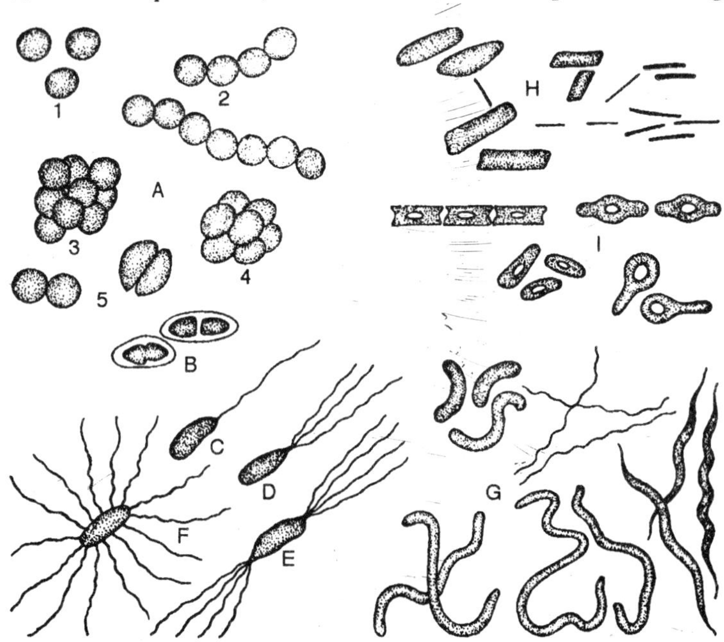

Fig. 20.1: Shape and arrangement of bacteria. A. (1-5) Spherical forms. 1. Coccus, 2. *Streptococcus,* 3. *Staphylococcus,* 4. *Sarcina,* 5. *Diplococcus,* B. Encapsulated diplococci. C-F. Flagellated bacteria. G. Types of spirilla. H. Types, shapes, and groupings of bacilli. I. Bacilli showing various positions of endospores

There are other kinds of bacteria that exhibit a considerably different morphology. For example, some species of bacteria possess appendages, bacteria of the genus *Saprosphira* form helical filaments, individual cells of the genus *Caulobacter* are rod-shaped or fusiform with a stalk sometimes protruding from one pole, species of *Streptomyces* produce a well-developed branched mold-like mycelium through much narrower than fungal mycelium. Although most bacterial species have cells that are of a fairly constant characteristic shape, some have cells that are **pleomorphic**, i.e., that can exhibit a variety of shapes. A bacterial species is generally associated with a definite cell form when grown on standard media under certain specified conditions, such as temperature of incubation of the medium. Bacteria usually exhibit their characteristic morphology in young cultures and on media possessing favourable conditions of growth. They are of three forms: (1) the embryonic, (2) the mature, and (3) the senescent. The embryonic forms correspond to the growth phase and are of long slender uniform cells. The mature form corresponds to the resting phase and is characterized by short cells of small size but more variable in form. The senescent form corresponds to the death phase and shows great variation in both form and size. Some morphologists have considered them as definite stages in an orderly life cycle of an organism. Young cells are, in general, larger than old ones of the same species. The changes in age are only temporary; the original forms appear when the old cells are transferred to fresh medium.

Size: Bacteria vary greatly in size according to the species. Regardless of their size, none can be clearly seen without the aid of a microscope. The bacteria most frequently studied in the laboratory measure approximately 0.5 to 1.0 μm in diameter. Staphylococci and streptococci may have diameters ranging from 0.75 to 1.25 μm. Rod (bacillus) forms, such as typhoid and dysentry bacteria often have a width between 0.5 and 1 μm and a length of 2 to 3 μm. Some filamentous forms may exceed 100 μm in length and diameter between 0.5 and 1.0 μm.

Structure: A bacterial cell has certain definite structures inside (internal) and outside (external) the cell wall. Some cellular parts, such as the cell wall and cytoplasm are common to almost all cells, again some structures are present in only certain species; still others are more characteristic of certain species than of others (Fig. 20.2).

(a) **Cell wall:** Except mycoplasmas, all other bacterial cells are invested in a rigid cell wall. Bacterial cytoplasm is protected by three layered cell wall, the outermost **slime** or **capsule**, the middle **cell wall** and **inner cell membrane.**

These external to the cytoplasmic membrane is the **cell wall** which provides a rigid framework to the bacterial cell and also supports and protects the more labile protoplasmic entities from osmotic damage. Thickness of cell wall ranges from 10 to 24 μm. Bacterial cell wall is essential for bacterial cell growth and division. It is a mucocomplex structure and is composed mainly of **diaminopimelic acid** (DPA), **muramic acid,** and **teichoic acid,** in addition to which are present amino acids, aminosugars, carbohydrates, and lipids which form a polymeric substance known as **peptidoglycan** (mucopeptide murein). The mucopeptides form the rigid part of the bacterial cell wall. Two different groups of bacteria are commonly differentiated on the basis of their response to a differential stain (Gram stain) named after Danish Bacteriologist Christian Gram. Those species of bacteria which retain the crystal violet stain are designated as **Gram-positive,** while others which do not are **Gram-negative.** The cell wall of Gram-positive bacteria is thicker, more rigid and chemically less complex than that of Gram-negative ones. It is composed of homogeneous layer of 10-50 nm wide having chemical composition—peptidoglycan, teichoic acids, and polysaccharides. Whereas in Gram-negative bacteria the cell wall is more complex than that of Gram-positive ones. It is thin being composed of inner layer 2-3 nm wide and outer layer 7-8 nm wide with chemical composition—peptidoglycan, phospholipids, proteins and lipopolysaccharides. The lipopolysaccharide which is also known as **endotoxin** material determines antigenicity, toxigenicity, and sensitivity to phase infection.

(b) **Cytoplasmic materials:** The cytoplasmic contents are **nucleoid, plasmid, ribosome, mesosomes,** etc. The nuclear material, like prokaryotes consists of single, long, supercoiled circular double-stranded DNA molecule which is associated with RNA and some proteins (histones). The nuclear cell is devoid of endoplasmic reticulum, mitochondria, centrosomes and golgi bodies. Some cytoplasmic inclusions of bacteria cell are—**Volutin granules, Sulfur granules** are found in many photosynthetic bacteria. Some *Bacillus* like *B. theringiensis* contain a **Crystalline proteins,** called a **parasporal body;** which possesses insectical property. Immediately beneath the cell wall is a thin membrane or covering called the **cytoplasmic membrane** (Fig. 20.2A), also called **bacterial cell membrane** or simply the **plasma membrane.** It is semipermeable, selective membrane that controls the passage of nutrients and waste products into and out of the cell. The bacterial cell membrane is an important centre of metabolic activity, where all enzymes for the replication of DNA and other metabolic reactions are located. It also contains many different kinds of proteins, each of which probably has a specific catalytic function responsible for the transport of many organic and inorganic nutrients into the cell and biosynthesis of membrane lipids and various classes of macromolecules that compose the bacterial cell wall (peptidoglycans, teichoic acid, lipopolysaccharides and simple polysaccharides). It often contains important components of the machinery of ATP generation. In some bacteria, the membrane appears to have a simple contour, which closely follows that of the enclosing cell wall. In others, it is infolded, at one or more points into cytoplasmic region resulting in the increase of surface area of the membrane. The infoldings of the bacterial cell membrane are known as **mesosomes.** The mesosome performs the functions of DNA replication and septum formation in a bacterial cell. It is a link between cytoplasmic membrane and nuclear material. Mesosome may be lamellae type, as in the species of *Lactobacillus* or vesicular type structure found in the species of *Bacillus.* Bacterial cells are characterized by the absence of mitochondria and Golgi bodies. Hence a large number of enzymes involved in the generation of energy in a bacterial cell are attached to the cytoplasmic membrane (membrane-bounded enzymes). Mesosomes, like mitochondria, also control the respiratory activities of the bacterial cells, they are also known as **chondrioids.**

Cell material contained within the cytoplasmic membrane may be divided into the **cytoplasmic area** and **chromatinic** or **nuclear area (nuclear material)** (Fig. 20.2A) which is less dense than the surrounding cytoplasm. The bacterial nuclear body is devoid of nuclear membrane, nucleolus and nuclear sap, known as **genophore** or **nucleoid.** Bacterial cell cytoplasm may contain material excepting the genophore, called **plasmid** or **episomes,** also designated as plasmid **extragenophoral material.** The cytoplasmic area is granular and is rich in RNA. The RNA in combination with protein forms macromolecular bodies. These RNA-protein particles are called **ribosomes** which contain enzymes useful in protein biosynthesis. Ribosomes are small granular structures rang-

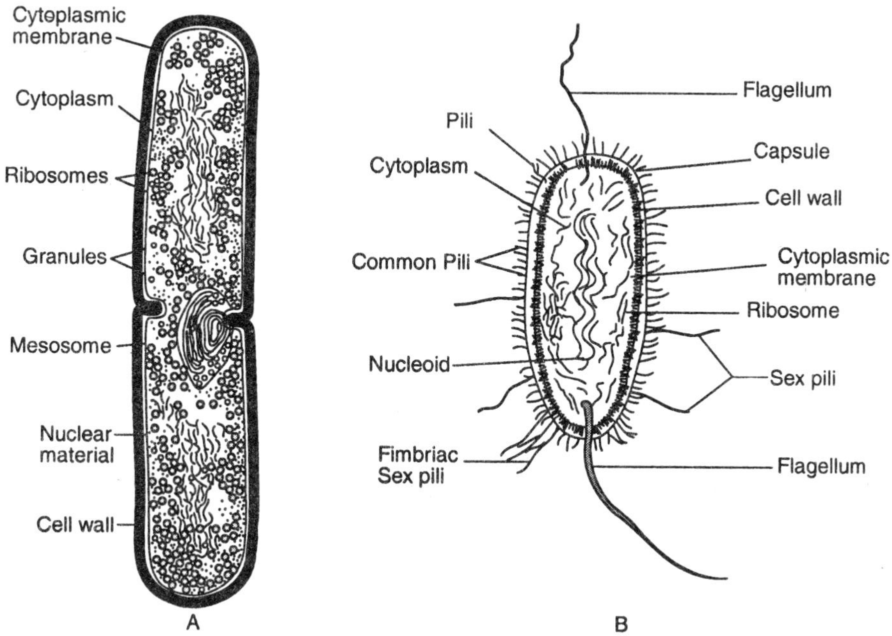

Fig. 20.2: Diagrammatic representation of structural organization of bacterial cells. A. Internal and external structure details. B. External structural details

ing from 10 μm to 13 μm in size and are about 10,000 to 15,000 in number in a cell. Electron microscope study reveals that each ribosome is composed of two unequal and roughly spherical halves having nearly equal amount of RNA and protein. The greater the rate of protein synthesis greater is the number of ribosomes. Ribosomes active in protein synthesis occurring groups are the **polysomes** (polyribosomes). They are held together by messenger RNA. All rapidly growing bacteria contain numerous ribosomes. The cytoplasm of a bactrial cell also contains **volutin granules,** also known as **metachromatic granules,** various kinds of polysaccharides as food reserves; and one or more vacuoles. Globules of fat also accumulate in bacterial cell cytoplasm. In anoxygenic photosynthetic bacteria there occur extensive intracular membrane systems which are the site of the photosynthetic apparatus, the infoldings of which accommodate bacteriochlorophyll.

The chromatinic or nuclear area is rich in DNA which can be made visible under the light microscope by Feulgen staining. Bacterial cell does not contain nucleus charactereistic as in eukaryotic cell. There is no evidence of nuclear membrane separating the nuclear area from the cytoplsm. Since the DNA of the bacterial cell does not form a discrete nucleus and is not capable of mitotic and meiotic divisions, it has been suggested that this structure be designated as **chromatin body, nucleoid, nuclear equivalent,** and even **bacterial chromosome.** Besides giving rise to enzymes and other proteins of specific nature in biological systems, DNA molecules are responsible for passing on the necessary information to make a new cell from one generation to the next. The chromosome is double stranded having succession of genes (about 10,000 in all) arranged either singly or in groups (**operons**). The individual genes and operons are controlled by regulatory mechanisms of variable complexity.

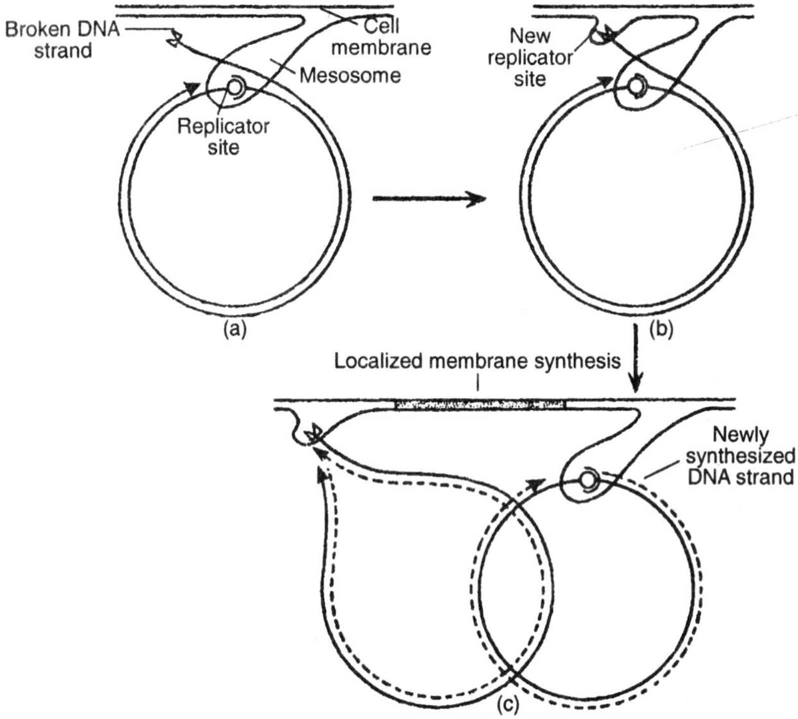

Fig. 20.3: Diagrammatic representation of stages of bacterial chromosome replication. (a) Chromosome attached to replicator site, one DNA strand broken. (b) End of broken DNA strand attached to new replicator site. (c) Repli-cation of chromosome completed with synthesis of new DNA strands and separation of old and new replicator sites by localized membrane synthesis

Replication of the circular DNA of the bacterial chromosome is unidirectional and autonomous. When replication begins, the chromosome is attached to a specific point on the bacterial cell membrane, this is known as **replicator site** with which the enzyme machinery responsible for DNA replication is associated. The replication fork is located at the replicator site, and its passage along the chromosome involves movement of the chromosome past the attachment site. Prior to the initiation of replication, a new replicator site adjacent to the old one is formed on the membrane, and the free end of the broken DNA strand is attached to it. Separation of daughter chromosome is effected by the localized synthesis of the membrane in a region situated between the old and new attachment sites. Membrane growth consequently spreads the attachment sites farther and farther apart (Fig. 20.3).

Most bacteria carry pieces of DNA, either as **extrachromosomal elements** called **plasmids** or in the form of bacterial viruses (bacteriophages) carried in a quiscent form by the cells. Plasmids are closed circular autonomously replicating double-stranded DNA molecules like the chromosome having molecular weight ranging from 10^6 and 10^8 and are 1 per cent of the chromosome in size. They usually contain three to four genes. Plasmids are attached somewhere on the inner face of their hosts cell membrane. Some of the plasmids are non-transferable and are lost, others have the ability to transfer themselves from one cell to another. Plasmids transfer genetic material between bacteria to spread resistance to antibiotics and genes for the degeneration of complex organic compounds in geochemical cycle. Like bacteriophages, plasmids are employed as cloning vehicles or vectors. Both chromosome and plasmids carry genetic information to daughter cells. But plasmids are independently replicating DNA pieces which may sometimes fail to replicate or be distributed to daughter cells leading to loss from the bacterial cell. Besides plasmids, bacterial cells also have attached to the cell membrane pieces of DNA called **episomes** which are one-hundredth times the size of the bacterial chromosome. Episomes sometimes replicate autonomously in the cytoplasm and at other times become integrated into the chromosome DNA and replicate with it. This behaviour distinguishes an episome from a plasmid because the latter does not integrate into the chromosome.

(c) **Structures external to the cell wall:** Structures that may be present external to the bacterial cell wall are: **flagella, pili, fimbriae, prosthecae** and **capsule.**

Flagella: Most motile bacteria possess long, helical appendages, called **flagella.** Electron microscopy shows that the flagella consist of three distinct regions—**filament, hook** and **basal body.** The filament is attached to the bacterial one end through the cell wall the cell membrane by the hook, which, is attached to the basal body. The rings of basal body remain attached to the cell membrane inside bacteria cell wall. The filament is made up of identical subunits (3 or more) arranged hilically along the axis to give a hollow tube. These subunits are made up of protein molecules, called **flagellin.** The flagellae filament ends with a capping protein. Some bactria have sheath surrounding the flagella. Structurally, the hook and the basal body are quite different from the filament. The hook is slightly wider than the filament and is made up of protein subunits. The structure of the basal body is quite different in (Gram-negative) and Gram-positive) bacteria.

Flagella are extremely thin hair-like helical appendages that protrude through the cell wall and are responsible for swimming motility. They are 0.01 to 0.02 μm in diameter and up to 70 μm in length. Each flagellum has three parts: a basal body associated with the cytoplasmic membrane and cell wall, a short hook, and a helical filament usually several times as long as the cell. A bacterial flagellum usually consists of a single proteinaceous fibril—**monofibrillar,** which in some species has been demonstrated to be composed of a sheath and a core. Flagella are composed of protein sub-

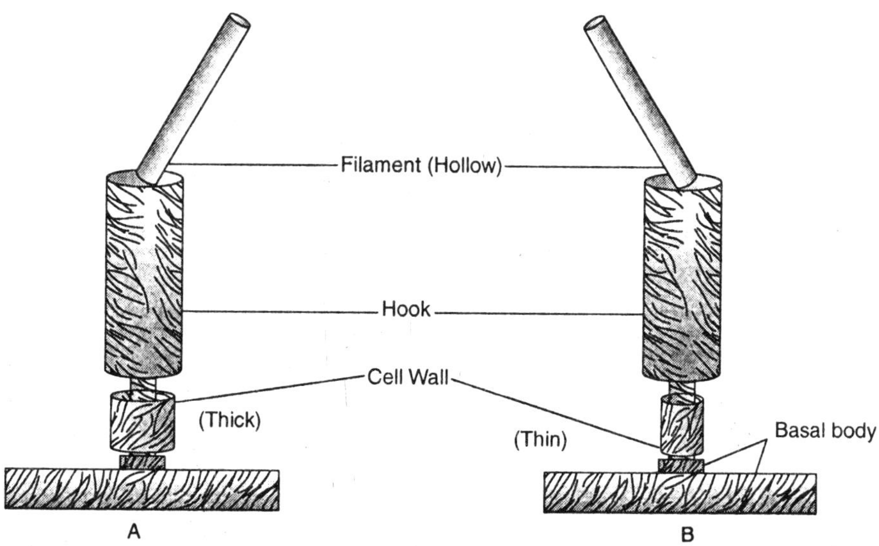

Fig. 20.4: Diagrammatic presentation of fine structure of the bacterial flagella. A. Gram-positive, and B. Gram-negative

units called **fiagellin** which has the properties of fibrous proteins as kerasin and myosin. The number and arrangement of flagella vary with different bacteria, but they are generally constant for each species. Motile bacteria exhibit their motion with the help of flagella which commonly have a rotary motion. Certain helical bacteria exhibit swimming motility, particularly in highly viscous media, yet they lack flagella. They have flagella-like structures located within the cell, just beneath the outer cell envelope. These are called **periplasmic flagella** or **axial flagella** or **endoflagella**. Some other bactria are motile only when they are in contact with a solid surface. As they glide they exhibit a sinuous, flexing motion, known as **gliding motility**.

Bacteria without flagella are known as **atrichous** (as in the genus *Streptococcus*), and those with a single polar flagellum—**monotrichous** (Fig. 20.5A) (as in the genera *Vibrio, Spirillum*), with a tuft of flagella at one pole—**lophotrichous** (Figs. 20.5B and E) (as in the genus *Pseudomonas*), a tuft of flagella at each pole—**amphitrichous** (Fig. 20.5C) (as in *Alkaligenes faecalis*), or several flagella originating at various points over the surface of cell—**peritrichous** (Fig. 20.5D) (as in *Escherichia coli* and genus *Salmonella*). Majority of the spherical bacteria are non-motile flagella being rare; among bacilli, there are many motile as well as non-motile species; and most spiral forms are motile.

Bacteria may also possess very fine, hollow, nonhelical, filamentous appendages much smaller, shorter and more numerous than flagella which do not form regular waves as flagella do. These appendages are called **pili** (sing. **pilus**) or **fimbriae** (Fig. 20.1B). They measure less than 10 nm in diameter and usually less than one μm long and can be seen only by electron microscopy. Pili have no function for motility. There are many morphological types of pili, at least ten have been recognized. They are named according to their function—**Sex pili**—pili where stands for fertility factor and **Common pili.** Pili have several functions. One kind of pili known as **F-pilus** (or sex pilus) serves as the port of entry of genetic material (DNA) between donor and recipient cells during bacterial mating. The sex pili are little hollow protein tubules which serve as conjugation tubes in which there is just enough room for the DNA to transfer itself. Some pili

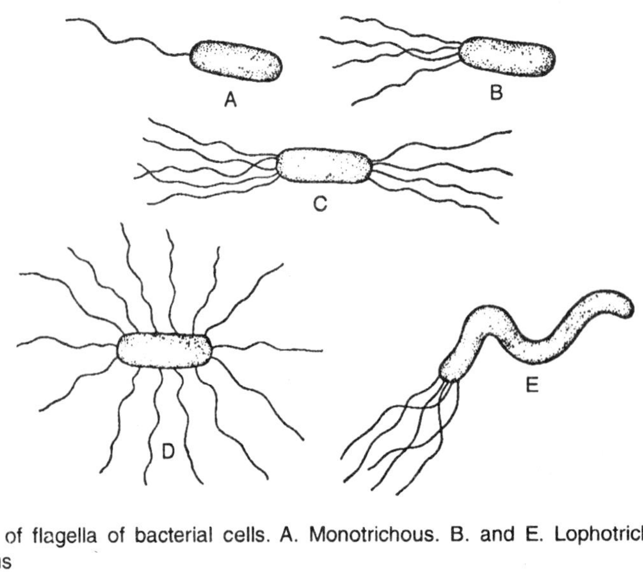

Fig. 20.5: Arrangement of flagella of bacterial cells. A. Monotrichous. B. and E. Lophotrichous. C. Amphitrichous. D. Peritrichous

are concerned with attaching two bacterial cells together prior to the transfer of DNA from one cell to the other. Other types of pili cause bacteria to adhere to one another and to foreign cells, such as red blood cells, epithelial cells, etc. Whereas others serve as attachment sites for bacterial viruses and pathogenic bacteria. Again others may also serve to keep bacteria near the surface of liquid or where oxygen is most available.

Many pathogenic bacteria use fimbriae for attachment to the surface of tissues for initiation of infection. The fimbriae of *E. coli* contain a special protein, called **adherin** at the distal end which helps bacteria to attach to the intestinial wall. *E. coli* is a normal intestinal flora. Both pili and fimbriae are made up of protein known as **pilin.** Pili and fimbriae play no part in the locomotion of bacteria.

Some bacteria produce semirigid extensions of the cell wall and cytoplasmic membrane known as **prosthecae** (sing. **prostheca**). Prosthecae increase surface area of cells for nutrient absorption and may also serve as aids in attachment to surfaces when they have adhesive substance at the end. Certain nonliving ribbon-like or tubular appendage excreted by the cell is termed **stalk.** Fimbriae are observed mostly in Gram-negative roots (*Salmonella typhae*). Gram-positive bacilli (*Corynebacteria renale*), also have pili or fimbriae.

Many bacteria produce a nonliving secretion of viscid material around the external surface of their cell wall. According to the amount of material produced and the degree of its association with the cell, one of the three terms may be applied: **capsules, slime layers** and **sheaths.** If the material is closely adherent to the cell and detectable only with difficulty, it is known as a **microcapsule;** if equally sharply defined but extensive and readily visible, it is called a **macrocapsule;** if copious in quantity and only relatively loosely associated with the cell and parts freely from it, it is **slime layer** or **free slime** or **gum** and it is designated as **sheath** when the amount of material is very insignificant. The above viscous substance is composed of polysaccharides of several types—dextan, levan, and cellulose. It is believed that capsule and free slime are morphologically and biochemically distinct—the capsule is a part of the cell, whereas the free slime is a secretion. Capsule formation depends upon the composition of the culture medium, but slime layer may be modified layer of the cell wall formed as an extracellular material of slimy or gelatinous nature. Capsules are not essential structures because: (i) they are not synthesized under all environmental conditions, (ii) mutants exist which have lost the ability to produce them, and (iii) cells from which they have been removed by enzymic digestion remain viable. They may be 10μ thick and are stainable by copper salt. The presence of a capsule is often associated with the virulence of pathogenic bacteria as they are able to resist phagocytosis by the white cells of the blood. Capsules may act as ion exchangers. Capsule formation may be responsible for considerable economic loss in dairy and other food industries. Encapsuled bacteria are better protected against unfavourable conditions. They are pathogenic to humans. A number of bacteria often adhere together with their gelatinous material forming the **zoogloea stage.** The capsule may often function in the storage of food substances or waste materials. In some aerobic bacteria the capsules form a raft in which are found the actively growing cells of other organisms. Bacterial capsules are species-specific and can, therefore, be used for immunological distinction of closely related species, for example. *Acetobacter xylinum* and *Pseudomonas aeruginosa.*

(d) **Bacterial Endospores and Cyst Formation:** Formation of spores and cysts is still rare in bacteria. The spore-forming genera may form spores within the cell usually singly. These spores are called **endospores.** They are formed typically by all members of the genera *Bacillus, Clostridium* and *Azotobacter.* A few other bacteria belonging to *Sporolactobacillus, Desulfotomaculum, Thermoactinomyces* and *Sporosarcina* are also known to form endospores (Fig. 20.6). Spores are also formed outside the cells by some bacteria particularly the mycelium forming actinomycetes. These bacteria, like fungi, produce spores externally either singly or in chains. Such spores are known as **Conidia.** In some gene of actinomycetes, conidia are formed within sporangia. Such spores are generally known as **sporangiospores.**

A group of bacteria, known as **myxobacteria,** are characterized by the formation of microfruit bodies and **myxospores.** Myxobacteria may also form myxospores without fruit-bodies. Myxospores are also known as **microcysts.**

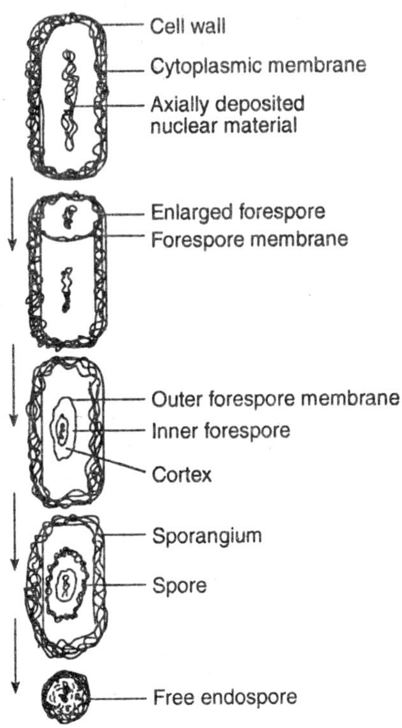

Fig. 20.6: Stages of Bacterial endospore formation

BACTERIAL CELL DIVISION

Bacterial cells are haploid. In rapidly growing bacterial cells nuclear division proceeds ahead of cell division. The bacterial cell division is completed by the doubling of all the cell constituents followed by partitioning of the cell to produce two daughter cells. The duplicated DNA separates into two strands and replicates individually. Each of the DNA strands are distributed in two daughter cells. During cross-wall formation, transverse plasma membrane is laid down which is followed by the centripetal growth of cell wall and the membrane is splitted into two halves. Sometimes chain of cells are formed by continued division.

STAINING OF BACTERIA

Due to smallness, bacteria are invisible to the naked eye. This necessiciates one to study the size, shape and structural characteristics of bacteria with the help of microscope which permits a wide range of magnifications. Many techniques have been developed by which specimens of bacteria can be prepared for examination in detail under the microscope. Depending upon the principle of magnification, microscopes are of two categories: **light** (or **optical**) and **electron.** Optical microscopy includes: bright-field, dark-field, ultraviolet, fluorescence and phase-contrast. Bright-field microscope is the most widely used instrument to study bacteria. For this, two general techniques are employed to make microscopic preparations: (i) hanging-drop of suspension of organisms in a liquid; and (ii) dried, fixed, and stained films or smears. Hanging-drop preparations are made by placing a drop of the bacterial suspension on a cover slip and inverting it over the concave area of a hollow-ground slide. Fixed, stained preparations are more widely used for the observation of the morphological characteristics of bacteria. The essential steps in the preparation of a fixed, stained smear are: preparation of the film or smear, fixation and application of one or more staining solutions. A large number of coloured organic compounds (dyes) are available for staining bacteria. These dyes may be acidic, basic, or neutral. Acid dyes generally stain basic cell components, and basic dyes generally stain acidic cell components. A neutral dye is a complex salt of a dye acid with a dye base.

The staining process is dependent upon the nature of the bacterium. Some forms strain easily, others do not. Spores and flagella are difficult to stain.

Simple Staining Method

A dried fixed smear of bacteria is flooded with a dye solution (say, methylene blue) for a specified period of time, after which this solution is washed off with water and the slide blotted dry. The cells of bacteria usually stain uniformly. This may not be true with some bacteria where the interior part of the cell may be deeply stained than the rest of the cell.

Other Staining Methods

Several bacterial staining methods are used for studying bacteria. Some of the commonly used staining methods are given below in brief.

(i) *Gram staining*. In the year 1884, a Danish physician, Hans Christian Gram developed a technique (**Gram Staining Method**) which is very useful in taxonomic grouping of Bacteria. The basic distinction between Gram (+ve) and Gram (–ve) bacteria is the difference of cell wall structure along with other characteristics. In this technique crystal violet iodine, ethyl alcohol (95%) and safranin are used. In this technique at first bacterial cells are stained with crystal violet, then with Gram's iodine and finally stained by Saffranin after washed in 95% ethanol which differentiates two groups of bacteria: one which retains stain and becomes dark-violet or purple coloured is **Gram-positive** bacteria and the other is which loses crystal violet stain is **Gram-negative** bacteria. Thus Gram-positive bacteria remain dark-violet or purple coloured, whereas Gram-negative bacteria become red (pink). In this process, the fixed bacterial smear is subjected to the solutions of: Crystal Violet, Iodine solution, 95 per cent alcohol (decolourizing agent) and Safranin or some other suitable counter stain in a sequence. Bacteria stained by the Gram staining method fall into two groups: Gram-positive bacteria retain crystal violet stain and Gram-negative bacteria lose the crystal violet but appear red as they are stained by the Safranin. The Gram stain has its greatest use in characterizing bacteria.

In addition to Gram-positive and Gram-negative bacteria, some bacteria are Gram-variable being Gram-positive at sometimes, and Gram-negative at others. The over all differences of gram-positive and gram-negative bacteria is shown in Table 20.1.

(ii) *Acid-fast staining*. In this technique the fixed bacterial smear is subjected to stained with the solutions—Carbol-fuchsin (heated), acid-alcohol, and methylene blue in regular sequence. Bacteria retaining the Carbol-fuchsin appear red are classified as *acid-fast bacteria*, and those decolourized by the acid-alcohol and counterstained by the methylene blue are designated as non-acid-fast bacteria.

(iii) *Ziehl's Carbol-Fuchsin*. This is a good spore stain. It is prepared by adding 10 cc saturated alcoholic solution of basic fuchsin to 100 cc of 5 per cent aqueous carbolic acid. Flame fixed bacterial smear on a coverslip is to be treated for 5 to 10 seconds with 5 per cent aqueous chromic acid, washed in water, then stained in Ziehl's Carbol-Fuchsin for one minute heating the solution until is steams, destained in 5 per cent aqueous sulphuric acid, washed in water, counterstained with methylene blue, to be washed dried in the air, and finally mounted in Canada balsam. Staining combination of 5 per cent acqueous malachite green counterstained with 0.5 per cent aqueous Safranin also gives good results.

Table 20.1: Difference between Gram-positive Bacteria and Gram-negative Bacteria

Characters	Gram-positive Bacteria	Gram-negative Bacteria
1. Cell shape	Spiral, rod-shaped or filamentous banded in some membranes.	Oval, spherical, straight curved rods, helical or filamentous, some have capsule or sheath.
2. Cell wall	Walls are homogeneous, rigid, single layered mainly composed of peptidoglycogen, protein not present.	Walls thick, elastic, flex, with lipopolysaccharide of complex lipid having antigenic specificity.
3. Appendages	Commonly non-motile, motile members have peritrichous flagella.	Develop various types of appendages like pili, fimbriae, etc.
4. Metabolism	Usually chromoorganotrophic.	Phototrophic, chemolithotrophic or chemoorganotrophic.
5. Endospore	Endospore formed in some groups of bacteria.	Endospore does not form.
6. Mode of Reproduction	By binary fission.	Commonly by binary fission, sometimes by budding.

Ziehl's Carbol-Fuchsin stain with some procedural modifications is also used for staining bacterial flagella.

To demonstrate bacteria in the host tissue, stain combination which give excellent result are: Carbol-Fuchsin and light green, Giemsa and Safranin.

(iv) *Endospore staining.* The endospore smear is allowed to remain in contact with crystal violet or methylene blue stain for a long time by gentle heating and then steaming for 30 to 60 seconds. Heating results in loosening of the vegetative cells from the smear and ultimately only the endospores remain attached to the slide. Endospores are coloured green, after washing they may be counterstained with safranin. After rinsing and drying the preparation is ready for observation.

(v) *Nuclear staining.* Bacterial smear is fixed in osmic acid vapour, hydrolysed with NHCl for 8 minutes at 58°C, then stained in Fuchsin, washed, dried and mounted in Canada balsam.

(vi) *Negative staining.* By this technique bacterial cells are made readily visible without staining in a dark-field preparation. Bacterial suspension is mixed with India-ink or nigrosin, which is then spread in a thin film across a slide and allowed to dry. Under the microscope the bacterial cells appear transparent and outlined to dry. Under the microscope the bacterial cells appear transparent and outlined by the dark background. Bacterial cells do not receive any physical or chemical treatment. This technique is advantageous for studying capsuled bacteria.

NUTRITION OF BACTERIA

Depending on their mode of nutrition, bacteria may be **autotrophic (autotrophs)** or **heterotrophic (heterotrophs)**.

Heterotrophic bacteria. Most bacteria are heterotrophic. They are unable to synthesize organic substances from carbon dioxide and live at the expense of organic compounds manufactured by other organisms. They are either parasites or saprophytes. Among the parasitic bacteria there are also gradation of parasitism. For example, **obligate parasites,**

they can survive only at the cost of living protoplasm and **facultative parasites** usually live upon the living tissues of a host, but may, under certain circumstances, derive nourishment from non-living organic matter. Similarly saprophytic bacteria may be **obligate saprophytes** and **facultative saprophytes.**

The strict or obligate heterotrophic bacteria cannot synthesize their complex protoplasm from simple inorganic salts but must have organic compounds, such as proteins, peptones, amino acids, and vitamins for growth, whereas the **facultative heterotrophic** ones show characteristics intermediate between the two autotrophic and heterotrophic, being able to utilize both inorganic and organic compounds. This group comprises the great majority of bacteria that have been studied and classified.

Some species of bacteria are very selective in the types of organic compounds which they utilize, deriving their nourishment from certain specific substances, whereas other species can use a great variety of organic substances. Many bacteria live largely upon carbohydrates, while others are primarily feeders upon proteins, fats, amino acids, and other organic substances. In general, the common types of saprophytic bacteria which are responsible for processes of decomposition in nature are rather cosmopolitan in their nutritional requirements, obtaining nourishment from a variety of organic substances. Many heterotrophic bacteria take in simple organic molecules and treat them in two ways: some molecules are used as building blocks to make the necessary macromolecules for cell growth, while the others are degraded to provide the necessary energy. A few molecules have a fate which depends on the metabolic state of the organism at the time and the external environment in whch it finds itself. Many parasitic, disease-producing bacteria are very exacting in their food demands and thus can utilize only a few specific types of organic compounds. Thus, certain parasitic bacteria can live only in blood, different species requiring different types of blood, or on certain specific tissues of host plants or animals.

Like other kinds of plants, bacteria manufacture enzymes, with the aid of which they digest foods, converting complex or water-insoluble foods into simpler or water-soluble foods. A number of enzymes have been isolated and prepared from bacteria. The enzymes are classified on the basis of what they do rather than what they are. For example, lactase hydrolyzes lactose to glucose and galactose. There are also general names for groups of enzymes, such as esterases, carbohydrases, proteases, dehydrogenases, and oxidases. It is customary to name the enzyme after the substance acted upon and to add the suffix -ase. Enzymes of bacteria are of two groups, those secreted into the surrounding medium are the **extracellular enzymes** or **exoenzymes** and others remaining confined within the cell, the **intracellular enzymes** or **endoenzymes.** The extracellular enzymes usually dissolve or destroy the orgaic compounds into simpler substances, whereas the intracellular enzymes are concerned with the respiratory activities of bacteria. The enzymes produced by bacteria act upon a tremendous diversity of organic compounds. Some of the enzymes of bacteria include amylase, which converts starch to malt sugar; lipases, which digest fats into fatty acids and glycerol; invertase, which tranforms cane sugar into glucose and fructose; maltase, which digests maltose into glucose; and proteolytic enzymes which digest proteins and protein-like compounds into simpler substances such as amino acids.

Autotrophic bacteria. A minority of bacterial species is autotrophic, that is, capable of synthesizing organic compounds from carbon dioxide and other simple, inorganic substances. Such autotrophic species thus resembles green plants which are also autotrophic. Autotrophic bacteria may be separated into two groups: **chemosynthetic** or **chemoautotrophic species,** which obtain the energy required for their synthesis of organic compounds from the oxidation of certain chemicals; and **photosynthetic** or **photoautotrophic species,** which utilize light energy in promoting their food-making processes. They are commonly found in sulphur springs where hydrogen is commonly available as electron donor.

Chemosynthetic bacteria. Among the common chemosynthetic bacteria are these:

1. **Sulphur bacteria,** which inhabit sulphurous waters, soils, and sewage and which oxidize hydrogen sulphide to free sulphur and then to sulphuric acid thereby obtain energy which they utilize in synthesizing organic compounds from carbon dioxide and other inorganic substances.

2. **Iron bacteria,** which live in iron-bearing waters and oxidize ferrous compounds, thus securing energy for the synthesis of organic compounds.

3. **Hydrogen bacteria,** which oxidize molecular hydrogen in soils to form water, with the release of energy.

4. **Nitrifying bacteria,** of which there are two types, those oxidizing ammonia to nitrites, and those oxidizing nitrites to nitrates; both types occur widely and abudantly in soils and secure energy from their oxidative activities.

Photosynthetic bacteria. There are two principal groups of photosynthetic bacteria. They are known as **purple bacteria** and **green bacteria,** which can be sharply distinguished from one another by the nature of their photosynthetic pigments. In purple bacteria, there is always one kind of chlorophyll, **bacteriochlorophyll,** accompanied by a variety of carotinoids, all of which are **aliphatic** (open chain compounds). The greeen bacteria contain either one or two chlorophylls, known as **chlorobium chlorophylls,** accompanied by **alicyclic** carotinoids (that is, molecules which contain a closed ring). As a consequence of the chemical differences between their respective pigment systems, purple bacteria and green bacteria which perform photosynthesis have completely different absorption spectra.

Bacterial photosynthesis is always an anaerobic process, since it results neither in the production nor in the consumption of oxygen. Most photosynthetic bacteria are, in fact, strict anaerobes. Consequently, the presence of light and the absence of oxygen are both essential factors for their development in nature. The photosynthetic bacteria are aquatic organisms and can develop only in the deeper layers where they can find anaerobic conditions. It is very common for them to receive light which has passed through a **biological filter** composed of some overlying algae. However, since the wavelength of light that are particularly effective for bacterial photosynthesis are ones that are not absorbed by any of the overlying algae, the bacteria can still photosynthesize effectively such conditions. The unique pigment systems of the photosynthetic bacteria are thus essential for their survival in competition with algae in natural environments. Photosynthetic bacteria like green plants use light as a source of energy. As a consequence all the polymerized carbon in the bacteria can be built up from CO_2 by a reaction which is basically:

$$CO_2 + 2H_2X \xrightarrow{\text{Light}} (CH_2O) + H_2O + 2X$$

where (CH_2O) denotes cell material and a source of reducing power (H_2X) reduces the CO_2 to a form that can be ultimately used for biosynthesis.

In green plants the source of reducing power is water, and the general equation is in the form:

$$CO_2 + 2H_2O \xrightarrow{\text{Light}} (CH_2O) + H_2O + O_2$$

oxygen gas is evolved. Distinction between bacterial photosynthetic systems and those of green plants is that bacteria produce no oxygen while green plants do.

Photosynthetic bacteria use light energy for the formation of ATP and the accumulation of reducing power by trapping the light in the pigments. The ATP and the reducing power are then redused to reduce CO_2 which do not involve light.

The photosynthetic bacteria can be conveniently grouped in the following manner:

1. **Purple sulphur bacteria,** containing a purplish pigment which apparently functions in a manner comparable with the activity of chlorophyll in green plants in the synthesis of organic compounds.

2. **Purple non-sulphur bacteria,** which utilize molecular hydrogen in the presence of light, reducing CO_2 in the process and syntesizing organic compounds.

These chemosynthetic and photosynthetic bacteria comprise only a small number of species and thus form a distinct minority of all bacterial species. They are widely distributed and their importance in nature is tremendous. Sulphur bacteria, for example, are responsible for the conversion of hydrogen sulphide, a common product of protein decay, to sulphur and then to sulphuric acid, which undergoes chemical reactions in soils to form sulphates, the principal source of sulphur for green plants. The sulphur bacteria (*Beggiatoa*) thus play a very important role in nature, since without their activity, most of the sulphur in nature might be locked up as *n*-hydrogen sulphide, and the supply of sulphates, necessary for the nutrition of higher plants, might disappear. This important activity of sulphur bacteria proceeds not only on land, but also in the freshwater and salt-water bodies, with the result that hydrogen sulphide is constantly being transformed into sulphur compounds which both land and sea plants can utilize. Similarly, the nitrifying bacteria maintain in soils a continuing supply of nitrates, the most readily available nitrogen compounds for most green plants. Many biologists believe that the world's iron-ore deposits are products of iron-oxidizing bacteria. Hence iron bacteria (*Leptothrix*) of past ages have been important in determining the chemical nature of portions of the earth's surface.

Nitrogen-fixing bacteria. These bacteria fix (utilize) atmospheric free nitrogen by converting into organic nitrogen compounds which are again decomposed and transformed into nitrates through a series of stages. Thereby soil fertility is increased. The nitrogen-fixing bacteria are of two types: (i) **Non-symbiotic** and (ii) **Symbiotic.**

The non-symbiotic nitrogen-fixing bacteria live freely in the soil and fix atmospheric free nitrogen and convert into organic nitrogen compounds. There are many non-symbiotic nitrogen-fixing bacteria of which most common are: *Azotobacter* and *Clostridium. Azotobacter* is an extremely active nitrogen fixer *in vitro*, to fix 5-20 pounds of nitrogen per acre it would utilize 454 kg of carbohydrate.

There are several genera of symbiotic nitrogen-fixing bacteria of which *Rhizobium* is widely known. *Rhizobium* is a Gram-negative rod which forms symbiotic association with the roots of angiospermic plants mainly belonging to the family Leguminosae. Strains of *Rhizobium* are widespread as free-living bacteria in the soil. In this condition they cannot fix atmospheric nitrogen. When they enter into the roots of leguminous plants, they induce the roots to produce nodules where they live securing food from the tissues of the nodules. They use food as a source of energy for the fixation of atmospheric nitrogen. During infection of a root, free-living *Rhizobia* aggregate around root hairs in response to some unidentified attractant exuded by the root and gradually work their way inward by means of 'infection thread'. Coincident with the passage of the thread, cortical cell division is stimulated to form the bulk of the eventual nodule tissue of polyploid cells. The infection thread usually terminates in a polyploid host cell. It may branch and spread through other tetraploid cells as they are formed. Many small groups of bacteroids are also formed in the nodule cells. The bacteria utilize the carbohydrates and other food of the leguminous plant and undergo a change in form, considered a type of degeneraion, resulting V- and Y-shaped organisms. In this condition they utilize atmospheric nitrogen to synthesize organic nitrogen compounds. Some of these nitrogen compounds become available to the leguminous plant in which the bacteria are growing.

A pigment **leghemoglobin** is formed by host tissue and colours the nodule pink. Only under this condition there is fixation of nitrogen in the nodules. At the same time bacteria produce nitrogen fixing enzyme **nitrogenase** which is very sensitive to oxygen. It is believed that the pigment leghemoglobin combines with molecular oxygen in such a manner that the function of the enzyme nitrogenase is not inhibited, but the combined oxygen can become available for the oxidative metabolism in the host cytoplasm near the bacteroids. The original source of energy for fixation of nitrogen is host photosynthesis, which also provides the carbon skeletons necessary to combine the ammonia which is the first stable product of nitrogen fixation. In the entire process the bacteria provide the keyenzyme (nitrogenase) and the host plant root provides the optimum environment in which there is an abundant supply of energy coupled with protection of enzyme against molecular oxygen. Structural organization of the nodule also promotes rapid removal of the products of nitrogen fixation (the accumulation of which inhibits fixation).

Certain *Rhizobium* strains from effective association with certain legume species, whereas others can infect more than one group of legumes. It was previously believed that *Rhizobia* could form nodules only on the roots of leguminous plants, but there is recent report of their occurrence in nodules on he rootless of *Trema aspera,* a member of the family Ulmaceae. A number of non-leguminous plants (e.g., *Alnus, Ceanothus,* and others) in association with the Actinomycetes also form root nodules capable of nitrogen fixation. Non-leguminous root nodules do not contain any haemoglobin-like pigments, though in many other aspects the process of nodule formation resembles that of leguminous one. The leaves of *Pavetta,* a member of the family Rubiaceae produce nodules which contain *Mycobacterium rubiacearum.*

The Chemical diversity of bacterial metabolism shows how extensively bacteria have evolved. In some cases the need for metabolic enzymes has been sacrificed at the expense of having taken up molecules performed from the environment, and in some pathogenic bacteria this specialization has reached the point that they cannot grow outside the organism that they are infecting. At the other extreme a totipotent synthetic apparatus within the cell allows bacteria to grow in chemical environments that are far too spartan for organisms which do not derive their energy requirements from light or simple inorganic reactions.

HOST-BACTERIA INTERACTIONS

Entire human life is spent with bacteria. Many kinds of bacteria colonize human body surfaces, alimentary tract, and orifices. Human health and well-being are influenced by the presence or absence of bacteria in the surrounding environment. Depending on their nature and activities, bacteria may be: **nonpathogenic** and **pathogenic.** Nonpathogenic bacteria are harmless. They often inhabit in various parts of human body giving protection to it against the attack of various pathogens. The activities of pathogenic bacteria are varied. The virulence, or the disease-producing capacity of pathogenic bacteria is determined by genetic properties which may be fully expressed only under certain environmental conditions including physical conditions and host susceptibility. To initiate infection, pathogenic bacteria must first survive the competition with other microorganisms on the surface of the host and then penetrate into the tissues. Once within tissues, the invading bacteria grow and multiply to produce disesease symptoms either by increasing in a local lesion or by spreading throughout the host via the lymphatics and blood.

Some bacteria produce poisonous substances of high molecular weight known as **toxins.** The capacity of bacteria to produce toxin and the potency of the toxin are important factors in the ability of bacteria to cause disease. Some bacteria do not produce toxins that can be demonstrated *in vitro.* Bacterial toxins when excreted into the surrounding medium are known as **exotoxins** or retained within the cell are **endotoxins.**

Exotoxins are proteins and are readily separable from the living producer cells. They produced mainly by Gram-positive bacteria, highly toxic to the host, heat sensitive, and are unstable to chemicals. Exotoxins produce symptoms of food poisoning within 6 hours when ingested in foods. Since these toxins are absorbed by the gut wall they are called **enterotoxins.** Common examples are tetanus and diphteria toxins. Exotoxin' when heated or treated wth acid lose thier toxicity and the resulting substances are called **toxoids.** Toxins and toxoids have the ability to stimulate the production of **antitoxins,** which neutralize toxins in the body of the host. This is important in the protection of susceptible hosts from disease caused by bacteria toxins.

Endotoxins are produced mainly by Gram-negative bacteria. They are liberated when the bacterial cells disintegrate, relatively heat-stable, do not form toxoids, and are less toxic than the exotoxins.

Plant pathogenic bacteria can be grown with varying degrees of ease on fairly simple culture media such as nutrient agar. Many appear to be rather unspecialized parasites, some seem to have a rather wide host range, whereas others, often considered as parasitically more highly specialized, may attack only a few closely related species of plants. For example, *Xanthomonas malvacearum* is mainly restricted to *Gossypium.* Bacteria with a wide host range probably comprise a number of strains each specialized to a limited range of plants. The problem is complicated by the difficulty of defining a bacterial species satisfactorily. In the past, some so-called species have been based on the host plant attacked. Some bacteria at present described as distinct species are perhaps better regarded as **formae speciales** of a larger species comprising a number of parasitically specialized strains. The application of computer techniques which assess the overall simplicity of organisms as based on numerous characteristics may shed light on this problem.

There are a few bacteria which attack fungi. These include *Erwinia carotovora* on *Agaricus, Xanthomonas uredovorus* on uredospores of rusts.

Bacteria, including those which attack plants, are very variable organisms, new biotypes arising through mutation, recombination and in other ways. Acquisition and loss of pathogenicity are not uncommon. Some bacterial plant pathogens tend to beocme non-pathogenic in culture but can often be rejuvenated by passage through their host plants. Pathogenicity can often be increased by repeated passage through resistant plants and decreased by passage through susceptible ones, presumably due partly to selective survival of races best adapted to the plant used. There is also evidence that some bacteria can be 'trained' to attack new hosts. For example, *Xanthomonas malvacearum* which infects

Gossypium can be made pathogenic to bean plants after four successive passages through beans. The pathogenicity of bacteria tends to be rather unstable and is perhaps affected by the plant with which the bacterium is associated; possibly the associating plant donates genetical material to the bacterium, thereby modifying its pathogenicity.

BACTERIAL CULTURE

The mode of occurrence of bacteria in our environment is very complex. Hence study of bacteria requires techniques for unravelling the complex mixed population or **mixed culture** of bacteria and other microorganisms into separate, distinct species as pure cultures. A pure culture is thus a population of cells all derived from a single parent cell. It actually represents an artificial condition for the growth of bacteria imposed on them by laboratory manipulation. To determine the characteristics, it is imperative that bacteria be isolated and grown in the laboratory as a pure culture. There are a variety of methods whereby different bacteria can be isolated and grown as pure cultures: the **streak-plate** and **spreading techniques**, the **pour-plate technique, enrichment-culture technique,** the **serial-dilution technique,** the **single-cell isolation technique** by the use of micromanipulator.

In most bacteriology laboratories a large collection of identified pure cultures of bacteria, referred to as the **stock-culture collection** is maintained for screening of new, potentially effective bacteria.

One of the major features of bacteria is their growth characteristics on various media. Such characteristics are: colour, abundance of growth, and even the odour of the culture provide useful clues for identification. To determine the growth characteristics of a pure culture it is customary to observe growth of bacteria on agar media (colonies on plate cultures), agar slants, in nutrient broth and in gelatin stabs. After inoculation on the medium and after inocubation, the cultural characteristics of the organism are determined. The features of growth on each of the above media to be studied are: on agar plate cultures—size, margin or edge, elevation, pigmentation, and optical features; on agar slant—amount, margin or edge of growth, distribution of growth througout broth, and odour; and in gelatin stabs—growth (no liquefaction) along line of inoculation and liquefaction of gelatin.

BACTERIAL GROWTH

Bacteria are capable of growing over a wide range of physical conditions and are capable of utilizing many different nutrients, but optimal growth requires certain specific conditions for a given species. Bacterial cells inoculated into a fresh medium under optimum conditions, selectively take up nutrients from their environment. A major sequence of biochemical synthesis follows. The nutrients that enter the cell from the medium are converted into cell structures—RNA, DNA, protein, enzymes, and other macromolecules. Cell mass and cell size increase and new cell wall substance is synthesized. Subsequently, the process of binary fission is initiated, ultimately resulting in two new cells. The increase in population is by geometric progression. The time interval required for the cell to divide or for the population to double is known as the **generation time.** Not all bacteria have the same generation time which is strongly dependent upon the nutrients in the medium and on prevailing physical conditions. During this generation time bacteria pass through different phases of growth:

(i) **The Lag Phase.** It is called the phase of adjustment. The individual cells increase in size beyond their normal dimensions. Physiologically they are very active and are synthesizing new protoplasm. The bacteria cells are metabolizing, but there is a **lag** in cell division. As such the population remains temporarily unchanged. At the end of the lag phase the cells divide with gradual increase in population.

(ii) **The Logarithmic,** or **Exponential** or **Log Phase.** During this phase the cells divide steadily at a constant rate and the log of the number of cells plotted against time results in a straight line. Growth rate is maximum during this phase.

(iii) **The Stationary Phase.** The population remains constant for a time, perhaps as a result of complete cessation of division or the balancing of reproduction rate by an equivalent death rate. This may be due to exhaustion of some nutrients or production of toxic products during growth.

(iv) **The Phase of decline** or **Death.** A variety of conditions contributes to bacterial death, the most important are the depletion of essential nutrients and the accumulation of inhibitory products, e.g., acids. The number of viable cells decreases essentially the inverse of growth during the log phase. Bacteria die at different rates, just as they grow at different rates. Some species of Gram-negative cocci die very rapidly, so that there may be very few viable cells left in a culture after 72 hours or less. Other species die so slowly that viable cells may persist for months or even years.

(v) **The transitional period.** Between each of the above phases there is a transitional period required, before all bacterial cells enter the new phase.

(vi) **Death phase.** In this phase the cells begin to die and as a result there is an increase in divergence between the total and viable court. However, when death occurs, autolysis takes place, i.e., cells begin to digest themselves, in other words, **cannibalism** occurs.

Physical conditions affecting bacterial growth. Principal factors that affect the physical conditions of bacterial growth are: temperature, moisture, light, and hydrogen-ion concentration.

Temperature. Bacteria can survive at a temperature range of 0°C to 85°C or even higher. Again, bacteria die slowly when exposed to a lower temperature range of 0°C to 5°C, but death rate is very high when the temperature of the medium in which bacteria are growing is quickly lowered. Spore forming bacteria survive even the boiling temperature. The greater the percentage of water in a medium, the lower will be the temperature required to kill bacteria. Moist heat is a more reliable agent of sterilization than dry heat. Dry bacterial cells are more tolerant to heat than young actively growing cells having more moisture.

Moisture. Bacteria are more aquatic than terrestrial. They thrive in water of high temperature range and are destroyed in water-logged soils under anaerobic condition. Bacterial cells when dried may temporarily cease activity, but regain normal active condition when moist condition is restored. Again, prolonged drying may be lethal to bacterial cells.

Light. The visible light rays are not injurious to the bacterial cells, whereas invisible ones like ultra-violet and infrared rays affect bacterial cells adversely.

Hydrogen-ion concentration—Range of hydrogen-ion concentration suitable for bacterial growth is 5.0 to 9.0.

Since bacterial growth involves the synthesis of complex molecules from more simple precursors, a positive input of energy is needed for bacteria to grow and consequently the bacteria themselves will have to degrade a sufficient number of energy-yielding bonds to provide the necessary biosynthetic potential, and thus anaerobic activity of heterotrophic bacteria must be balanced by an equivalent amount of catabolism. The source of this energy is the medium in which bacteria grow and break down molecules of the substrate to receive necessary energy. In many bacterial species these molecules are commonly monosaccharides—notably glucose, but with others amino acids and fatty acids are used for this purpose. The molecular changes that either require or liberate energy are numerous in bacteria. Bacteria use primarily ATP for this purpose.

ATP generated by metabolic interconversions in bacterial cells is known as **substrate-level phosphorylation** reactions. But by far the majority of the ATP generated comes from the conversion of hydrogen atom to H_2O. This process is known as **oxidative phosphorylation,** and approximately three ATPs are generated for every pair of hydrogens oxidized to form water. The conversion of these hydrogen atoms to water is achieved by the cytochrome system of the bacteria working in connection with other hydrogen carriers.

The most efficient alternative source is light energy—photosynthesis, but other bacteria use the chemical interconversion of small inorganic molecules—the process known as chemoautotrophy.

BACTERIAL RESPIRATION

Like all living organisms, bacteria carry on respiration, release energy which is necessary for growth, reproduction and other cellular processes. The respiratory process is controlled by enzymes which are synthesized by bacteria. The activities of these enzymes in relation to respiration are very complex. Depending upon the conditions under which the respiratory process is carried out, the bacteria may be **aerobic, anaerobic** and **facultative anaerobic.** Most aerobic species of bacteria, which live only in an environment containing free oxygen, complete the process of respiration through the aerobic phases, liberating carbon dioxide, water, and energy, whereas anaerobic bacteria, which thrive in the absence of free, gaseous oxygen, carry on anaerobic respiration or fermentation, in which under the control of certain oxidizing enzymes, foods are broken down, with the formation of carbon dioxide and intermediate organic compounds, such as alcohols and organic acids, and with the release of energy. Again, some facultative anaerobes, when they grow under anaerobic conditions, form carbon dioxide and intermediate organic compounds, given sufficient free oxygen, these organisms may complete the oxidation of these intermediate compounds to water and carbon dioxide.

As a result of their various metabolic activities, chiefly respiration, different species of bacteria form differnt types of products. Some of them are: lactic, butyric, and acetic acids; gases including carbon dioxide, methane, hydrogen sulphide, hydrogen, and others; and red, yellow, orange, and blue pigments; and other substance. These products are frequently important in distinguishing among different species of bacteria, and are of great importance in industry.

BACTERIAL REPRODUCTION

As a rule bacteria reproduce very rapidly. The common method of reproduction in bacteria is by the process of cell division, commonly known as **fission,** binary fission, or simply fission, in which a bacterium simply divides into two new one-celled bacteria. The individual cells may divide at times repeatedly and multiply in number in one plane or in two or three planes. Cell division is by ingrowth and constriction, and may result in semi-permanent cross-walls, dividing the organism into one to many cells. The daughter cells may separate immediately, or they may remain attached and produce chains or colonies of cells. The fission may be slowed down or often entirely stopped due to shortage of food supply, by the accumulation of poisonous waste products of metabolism, presence or absence of oxygen, or variation of optimum temperature. Nuclear division in a bacterial cell, as already indicated, is quite different from that of eucaryotic cell.

In the genera, *Bacillus* and *Clostridium* due to dehydration of the cell protoplast and accumulation of calcium and dipicolinic acid (DPA), the vegetative cell is transformed into a thick-walled structure known as **endospore** (Fig. 20.2C). The endospore is resistant to high temperatures, disinfectants, and other unfavourable factors of the environment. The spore germinates in a suitable environment by one of several methods, depending upon the species. Only one bacterium comes from each spore. Opinions differ with regard to the functions of spores and as to the factors involved in the process of sporulation. Spore formation in bacteria is not to be considered a method of reproduction, since the germination of a spore results merely in the reappearance of the protoplasm of the original bacterial cell. It is not a process to increase bacterial numbers, because a cell rarely produces more than one spore. It may be regarded as a stage for the survival of unfavourable conditions. Bacterial spores thus differ from the reproductive spores of the algae, fungi, and other spore-bearing plants. Bacteria having mold-like mycelial body do not produce endospores, but produce chain of spores, or **conidia** with which they reproduce asexually.

Reproduction by **budding** is not uncommon. But reproduction by more complex methods than budding or fission is more common in bacteria. Many bacteria produce tiny motile **gonidia,** and almost all have a **resting** or **distributive stage** which may be more or less **spore-like.** Again, some aquatic bacteria show **alternation of generations** between **motile** and **non-motile stages.**

Typical sexual processes involving fusion of gametes and their nuclei to form a zygote, followed eventually by meiosis, have not yet been clearly demonstrated in bacteria. It is, however, clear that in some bacteria transfer of genetic material takes place from one cell to the other.

BACTERIAL GENOME, PLASMIDS AND GENETIC ENGINEERING

Gene from a bacterial genome carrying DNA is transferred through a bacterial body constituent designated as **plasmid.** Plasmids are defined as extrachromosomal genetic elements, occurring chiefly in bacteria and rarely in eukaryotic organisms. In bacteria, plasmids are circular-double strained DNA controlling wide range of genetic functions. In yeast (**Saccharomyces cerevisiae**) an RNA plasmid has been found. Plasmids are self-replicating elements, yet they are largely dependent on the host bacterial cell for their replication process.

The term **plasmid** for extrachromosomal genetic elements occurring chiefly in bacteria was termed by different authors as **sex factors, conjugans, extrachromosomal replicons, transfer factors.** The first plasmid was discovered and several **sex-factor** or F-plasmid (F stands for fertility) of *E. coli* k12. This plasmid confers the ability to an *E. coli* cell (F^+) to conjugate with another cell lacking plasmid (F-cell). The F-plasmid can exist in two alternative states, viz., it can either remain free in the bacterial cell or it can be integrated into the *E. coli* chromosome. Plasmids with such property are known as **episomes.**

Genes responsible for transfer resistance to some antibiotics are called **resistance transfer factors.** The plasmids normally contain only about 1 per cent of the total genetic information and multiply independently of the chromosome. They pass quickly from cell to cell as does the F-factor (fertility factor) and are found to be a critical factor in transferable drug resistance (TDR). Thus by genetic recombination, a pathogenic bacterium may acquire the genes for resistance to a particular drug from harmless to a harmless non-pathogenic organism.

Genetic recombination in bacteria takes place through interchange of genetic material. This is accomplished through three processes:

(1) Conjugation, (2) Transformation, and (3) Transduction.

1. **Conjugation.** (i) In this process, the exchange of genetic material takes place through a conjugation tube between two cells of bacteria, one is male type as F^+ or donor cell, which donates some DNA. The other is female type, F^- or recipient cell, which receives DNA through sex pilus. After receiving DNA, the recipient cell may behave as donor cell, i.e., F^+ type. The F-factor is the fertility factor or F-plasmid present in the cell of F^+, i.e., donor cell or male type. The plasmid takes part in conjugation. (ii) In another process of conjugation is known as **High frequency of recombination.** In this process the recipient cell does not acquire new genes other than those on the F-factor. However, there exists in bacteria a type of conjugation that accounts for the passage of chromosomal material. Strains of bacteria that display this ability are called **high frequency of recombination,** or Hfr strains. Such strains are developed when a male (F^+) mutated to a **super-male** which showed a high frequency of recombination with a female (F^-), hence the mutant is called Hfr. In Hfr strains, the F-factor attaches to the chromosome. This integration is a rare event and requires an insertion sequence to be present on the chromosome to recognize the F-factor. At the point of attachment, the chromosome opens and a copy of one strand is made by the rolling circle mechanism. A portion of single-stranded DNA then passes via sex pilus into the recipient cell (Fig. 20.7).

Many microbiologists refused to believe that bacteria had stable hereditary systems which could undergo permanent changes. But in 1943 Max Delbruck and Salvador Luria through a classical experiment demonstrated that bacteria have stable hereditary mechanisms. The establishment of the similarity between the genetic systems of bacteria and higher forms of life resulted in the experimental use of *Escherichia coli.* The first demonstration of genetic recombination in bacteria was achieved by Lederberg and Tatum in 1946. They combined two different auxotrophic strains (strains that use only inorganic materials as a source of nutrients) of *E. coli* and allowed them to mate. They found prototrophic

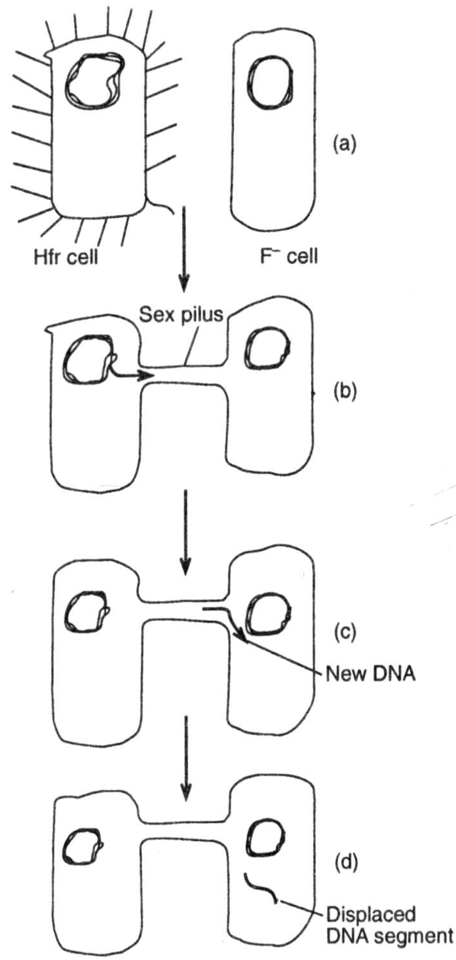

Fig. 20.7: Conjugation between Hfr and F-cell. (a) Hfr-cell contains the F-factor on chromosome, (b) An Hfr contacts as F-cell and a pilus attaches them. The Hfr chromosome opens, and a copy of one strand is made by rolling circle mechanism. A portion of the new strand passes into the recipient cell, (c) the new strand arrives in the recipient cell and align along side the chromosome, (d) the new strand displaces a segment of the recipient chromosome and the displaced segment dissolves

(nutritionally independent) colonies growing there, these must have been the result of transfer of genetic material (DNA) from cell to cell and a recombination between the auxotrophs by conjugation and is indeed a true sexual fusion. But genetic recombination in bacteria is basically different from that of higher plants and animals. In genetic recombination, there is the occurrence of progeny whose combinations of genes are different from those that are present in the parents. In higher plants and animals, genetic recombination is the fusion of two haploid gametic nuclei resulting in the development of a diploid nucleus. The complete genomes of both gametes are involved and recombination comes about from the independent assortment of chrosomes or by the process of crossing-over between homologous chromosomes, whereas bacteria are free-living organisms and are not differentiated into gametic and somatic cells. Each cell is a potential gamete. Bacteria displaying sexuality have cells which behave as gametes and are differentiated into two functional types: **donors** (or males) and **recipients** (or females). As a rule, in bacteria only a fraction of the genetic constitution of the donor cell is transferred to recipient cell which contributes its entire genome and cytoplasm resulting in the development of a **merozygote**, a zygote that is incompletely diploid. Genetic recombination is limited to the diploid portion of the genome and occurs when a new, recombinant chromosome is formed. The recombinant chromosome is formed from DNA contributed by two different organisms.

The transfer of DNA from donor cells to recipient cells and recombinant forms can be accomplished by three distinct mechanisms: cell **conjugation** (mating), **transformation** (uptake of naked soluble DNA from the environment), or **transduction** (bacteriophage infection).

In conjugation there is an actual cell-to-cell contact betweeen the mating bacteria (Fig. 20.8). The double-stranded closed DNA opens and part of it is transferred through pili called sex pili which behave as bridges through which DNA is transferred from donors to recipient cells. The amount of DNA transferred is directly related with the duration of conjugation between the mating bacterial cells. In 1952 Hayes first presented evidence for sexual differentiation in *E. coli.* It is now known that mating types exist and all pairs are not fertile. Donors (male cells) contain a small circular piece of DNA called the fertility, sex, or F-factor and labelled as F^+. Recipient (female cells) are labelled as F^- and are devoid of F-factor. Crosses between two F^- strains are sterile. Only $F^+ \times F^-$ crossed yield recombinants. Donor strains are of two kinds: F^+ and Hfr (high frequency of recombination). Donor F^+ strain donates only a small portion of its genome, whereas donor Hfr strain donates large amount of genome to the recipients. Sex-factor-DNA is an episome which sometimes replicates autonomously in the bacterial cytoplasm and at other times integrated into the bacterial chromosome DNA and replicates with it. In addition to *Escherichia,* conjugation has also been reported to occur in the genera: *Salmonella, Pseudomonas,* and *Serratia.*

2. **Transformation.** In transformation, the DNA passes between the bacteria as a naked molecule, and nucleic acid can be isolated in active form from the medium separating the bacteria (Fig. 20.8). This phenomenon was discovered by Fred Griffith (1928) working on *Diplococcus pneumoniae.* During this process, lysis (disruption) of donor cell takes place and its DNA becomes fragmented and the fragments are thrown out in the culture medium. The fragments of DNA are taken up by the recipient bacterial cells. The introduced genetic fragment, called an **exogenote,** undergoes recombination with the recipient bacterial chromosome replacing a homologous segment.

This takes place during late log phase when the recipient bacteria are said to be **competent** to take up and incorporate donor DNA and its characteristics. The process of recombination which basically involves the replacement of one region in the molecule by different pieces of DNA has given rise to the exciting prospects of 'Genetic Engineering'. In 1944 Avery, Macleod, and McCarthy implicated DNA as the transforming principle.

3. **Transduction.** Transduction involves the transfer of DNA from a donor bacterial cell into a recipient bacterial cell through the agency of a virus (bacteriophage) which acts as a vector (Fig. 20.8). This phenomenon was discovered by N. D. Zinder and J. Lederberg in 1952, when they searched for sexual conjugation among *Salmonella* species. During transduction bacteriophage undergoes a rapid lytic growth cycle in its host bacterial cell in which it injects its DNA. The DNA replicates rapidly and directs the synthesis of new phage proteins. The new DNA combines with new proteins to make phage particles which are released by the destruction of cell wall and lysis of the host bacterial cell. The new phage particles carry the genes of bacterium in which they are produced. There are some phages known as **temperate phages** which ordinarily do not lyse the host bacterial cell, carry DNA that can behave as a kind of episome in bacteria. These viral genomes can become integrated into the host bacterial genome. They are then known as **prophages.** There are several types of transduction: (i) when a temperate phage transfers any gene on a bacterial chromosome it is called **generalized transduction,** and (ii) when the phages transduce only those bacterial genes adjacent to the prophage in the bacterial chromosome it is known as **specialized** or **restricted transduction.** Transduction has been demonstrated in several bacterial species. This technique is used for making new bacterial strains.

Again, in **lysogenization,** the agency is a temperate phage whose DNA on entering into the bacterial cell remains attached to the DNA of the host bacterial cell, reproduces along with it, and is transferred to the progeny without causing lysis of the host bacterial cell. The phage DNA is called the **prophage.** The phage

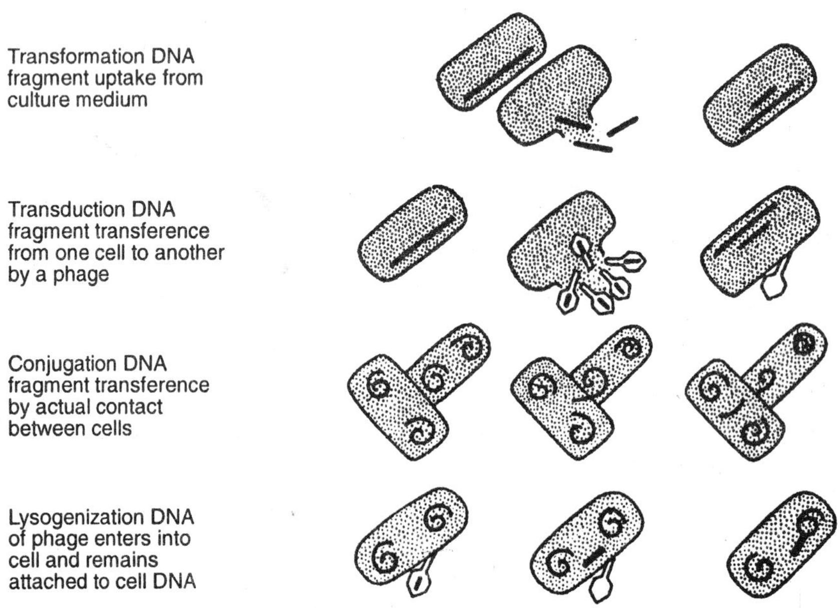

Fig. 20.8: Mechanisms of DNA transfer in bacteria

DNA may be carried along several generations of the host bacteria. Sometimes, when the balance of the host bacterial cell is disturbed, the phage DNA detaches from the bacterial DNA and goes into a virulent state. Because of this latent potential for lysis, cultures containing prophage are called **lysogenic** and the phenomenon is **lysogeny.**

VARIABILITY IN BACTERIA

As a living organism, any bacterium is subject to a temporary or permanent changes. Temporary changes (variations) are physiological or morphological and are induced by such things as changes in either the physical or chemical environment of the bacterium. Reversion to the normal form occurs quickly upon its return to usual conditions. On the other hand, permanent changes (variations) are brought about by changes in the DNA of the bacterium and are accomplished in any number of ways. In such cases reversion to the normal form occurs very rarely.

The offsprings resulting from binary fission in bacteria are duplicates in morphology and physiology of the parent cell. In a low percentage a part of the DNA of the nucleoid of the parent cell may change or **mutate.** If this occurs, a sister cell is produced that may give rise to a culture with characteristics that differ from the parent cell. Mutations can occur spontaneously in nature—**spontaneous mutation,** and can be induced experimentally—**induced mutation.** Mutations express themselves as changes in colony morphology, antigenicity, colour, cell morphology, virulence, phage resistance, bactericide resistance, etc.

Mutations can occur in any part of the bacterial DNA and in some areas of the molecules. As long as the base sequence of the DNA does not change, bacteria will 'breed true' from generation to generation. Any change in base sequence, however, will alter the informational content of the DNA and this is likely to produce heritable changes. Such change in base sequence is **mutation** in DNA. When the change in the nature of a single base occurs it is **point mutation,** removal of a section of the DNA is **deletion,** and removal of a piece of DNA from one position on a replicon to another position on the same replicon, or to a position on another replicon in the same cell is **translocation.** Again, when the sequence of the DNA is altered either by adding or removing a single base pair it is known as a **frameshift mutation.**

Point mutation may be caused by a variety of agents known as **mutagens** which may be chemical compounds. Point mutation can revert by a further change in base sequence when DNA returns to its original structure, this is **back mutation** to **parental type.** In deletion a piece of DNA is removed from a replicon and the gap is closed by joining together the ends generated by the process. The effect of deletions is to remove certain genes completely from the genome and to damage the two genes that span the gap. Deletions are caused by chemical agents and by irradiation. Changes due to translocation are commonly noticed in bacteria where antibiotic resistance markers are present. Such changes are lethal as they destroy essential genes in the cell.

BACTERIOPHAGE

Bacteriophages or bacteria-eating agents or bacterial viruses or just 'phages', as they are commonly called, are widely distributed in nature. They are dependent upon living bacteria for their existence, as such, they occur in substrata where bacteria are found. Bacteriophages were independently worked out by F. W. Twort (1915) in England and by F. d'Herelle (1917) in Paris. Structurally, they resemble other viruses being composed of a nucleic acid (DNA) core surrounded by a protein coat. Morphologically, bacteriophage

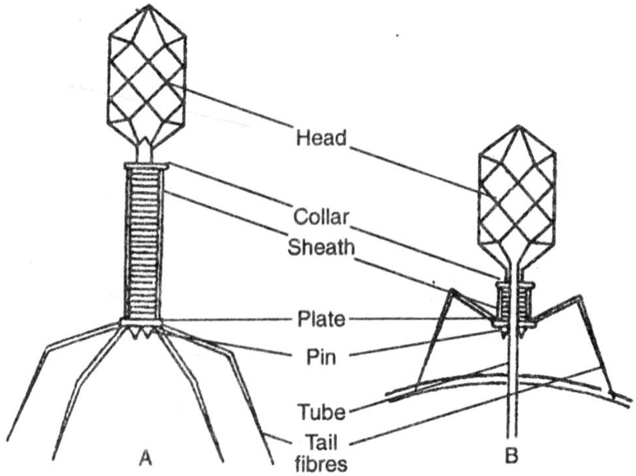

Fig. 20.9: Diagrammatic representation of a coliphage. A. Structural details, B. Bacterial cell infection by tail-like tube

resembles a tadpole in having a head and a tail-like structue (Fig. 20.9A) through which it injects the host bacterial cell with nucleic acid through a puncture hole in the cell wall (Fig. 20.9B). The nucleic acid then takes conrol of the cell metabolism and directs the bacterium in the synthesis of more bacteriophages. The newly formed bacteriophages are released by a sudden rupture of the bacterial wall-**lysis** (Fig. 20.10). These bacteriophages are free to infect other susceptible bacteria. All these stages of phage infection in a bacterial cell comprise what is designated as **lytic cycle** (Fig. 20.10). There are two types of phage: **virulent** and **temperate.** The virulent bacteriophage causes the lysis of the bacterial cell very quickly, whereas the temperate bacteriophage replicates in step with its host bacterium being transmitted through cell divisions without necessarily causing lysis. Further details about the bacteriophages are incorporated in the Chapter on Viruses.

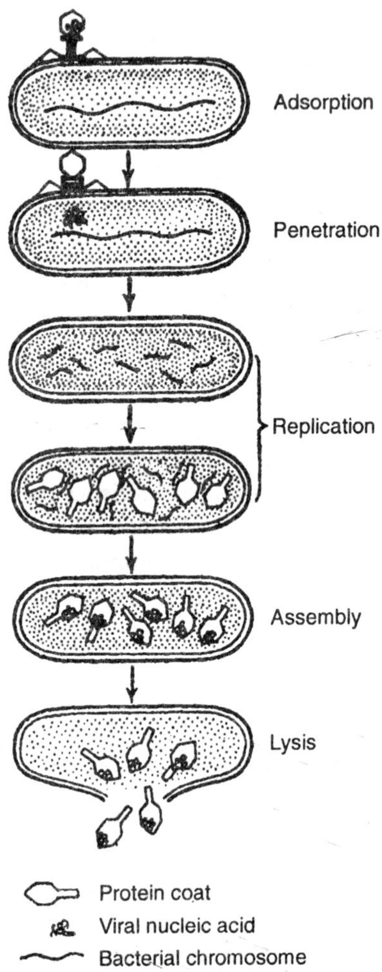

Fig. 20.10: Stages of phage infection in a bacterial cell

ORIGIN AND RELATIONSHIPS

Little is known positively concerning the origin of bacteria upon the earth's surface. It is generally agreed that bacteria were among the living organisms that appeared on the surface of the earth, constitute a group of great anti-

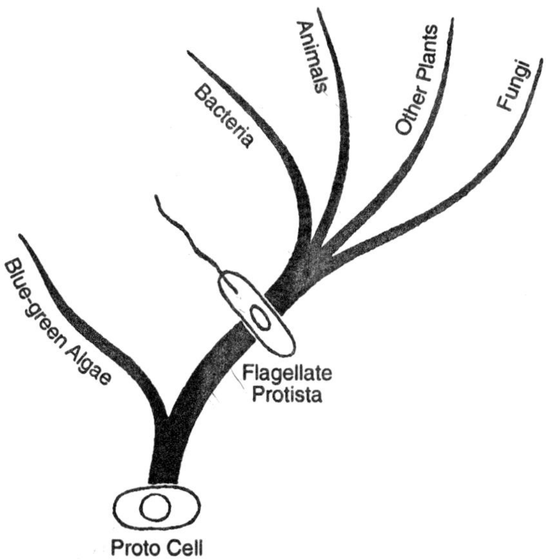

Fig. 20.11: Suggested origin and relationship of bacteria

quity. A large group of Biologists are of opinion that chemosynthetic organisms probably appeared earlier than the green plants and iron bacteria are responsible for the earth's tremendous ore deposits. The simplicity of bacterial cell and mode of reproduction indicate close relationship with blue-green algae, and some fungi. Some Biologists regarded bacteria as a primitive group of plants from which both chlorophyllous and non-chlorophyllous plants may have evolved, whereas others prefer to accept bacteria as degenerate organisms and have developed by the loss of chlorophyll.

Since the bacteria have flagellate representatives, there is a possibility that all the bacteria have flagellate ancestry. These ancestors formed a group which had flagellate representatives capable of a photosynthetic, chemosynthetic, and saprophytic mode of existence, and thus had much in common with the smaller **flagellate protista.** Accordingly, it is not unlikely that the bacteria arose independently from the flagellate protista just as all other living organisms except the blue-green algae which have branched off much earlier than the flagellate protista having a common source of origin which may be designated as 'proto cell' (Fig. 20.11).

As to the relationships, many authorities hold that the bacteria are closely related to the blue-green algae in various aspects. But the important and fundamental points which distinguish the bacteria from the blue-green algae are that the latter are unique among major groups of living organisms in showing no sign of even possessing any flagellated structure or sexuality, whereas bacteria, in common with other plants and animals, have representatives which possess flagella or sexuality.

CLASSIFICATION

The classification and identification of bacteria involve certain difficulties not found in other groups of plants. These difficulties arise in part from the very small size of bacteria and consequent difficulty of recognizing clearly their structural characteristics in part from the relatively restricted range of morphological differences among different species. In most other groups of plants, morphological differences among species are rather apparent. In bacteria, however, there is relatively little variation in body structure, and thus different species of bacteria frequently appear similar or identical in structure when they are examined microscopically. There are, of course, some morphological differences among bacteria, such as in numbers, lengths, and positions of flagella, in body form, and size. These differences are used, of course, in distinguishing among certain groups of bacteria, but they are often inadequate taxonomically, since several species of bacteria may resemble each other in thier flagellum characteristics, their forms, and their body sizes. Thus, bacteriologists must often depend upon physiological differences to identify and classify different species of bacteria and must, therefore, investigate the types of organic compounds which bacteria utilize, the kinds of organic acids, gases, pigments, and other compounds produced in their metabolism, their growth in relation to oxygen supply, the shapes and structure of the colonies which they form and their reactions to various biological stains. The complete classification and identification of bacteria thus are intricate procedures, founded upon a knowledge of both their morphological and physiological characteristics.

INITIAL APPROACH TO CLASSIFICATION

CLASSICAL TAXONOMY

Classification of any group of uses selected stable characteristics which vary among the **taxa.** These are known as **taxonomic characteristics.** Classically the bacteria have been classified as the basis of similarity in phenotypic characteristics, like morphological features, response to Gram stain, cultural characteristics, physiological and biochemical properties, pathogenicity, antibiotic sensitivity, serological relationships, etc.

No one system for the classification of bacteria is generally accepted. The three most widely used classifications are those of Lehmann and Neumann, Migula, and Bergey.

A comprehensive classification was published by Lehmann and Neumann in 1896. Almost at the same time Erwin F. Smith proposed a classification of bacteria, more or less a modification of Lehmann and Neumann. A fresh system of classification was suggested by Migula in 1900. Later Bergey in 1923 in Manual of Determinative Bacteriology introduced a new system of classification of bacteria. Bergey's system of classification attracted attention of a large number of workers. Subsequently, in 1948 an abridged classification was prepared by a committee appointed by the Society of American Bacteriologists. This committee, known as the Board of Editor-Trustees, has the co-operation of a group of approximately 65 bacteriologists interested in developing the systematic relationship of the various groups of bacteria. The work of the committee was published under the title of 'Bergey's Manual of Determinative Bacteriology' in honour of Dr. D. H. Bergey who was responsible for developing the first edition of the manual.

In Bergey's manual, bacteria were separated into five orders which may be briefly characterized as follows.

1. **Eubacteriales:** Simple, undifferentiated forms; spherical; rods may be short or long, straight, curved, or spiral; motile or non-motile; endospores sometimes produced. This is the largest order of bacteria. Most of the bacterial species which are economically important and which have been most intensively studied are members of this order.

2. **Actinomycetales:** Organisms forming elongated cells which have a definite tendency to branch. Thier vegetative structures resemble fungal hyphae. These hyphae do not exceed 1.5 µm and are about 1 µm or less in diameter. Some are soil saprophytes, others are pathogens of animals (**actinomycosis** of man, 'lumphy-jaw disease' of cattle) and of plants (potato scab). Many of these bacteria are very important in processes of decay in soils.

3. **Chlamydobacteriales:** Filamentous water forms which may show false branching. May or may not be ensheathed. In the ensheathed forms, the sheath is composed of, or is impregnated with ferric hydroxide; in others the sheath may contain sulphur particles.

4. **Myxobacteriales:** The relatively long, slender, flexible, non-flagellate cells produce a thin spreading colony. They frequently exhibit a creeping type of movement, and they are embedded in large masses of slime which they excrete. They are widely distributed in soil.

5. **Spirochaetales:** Slender, flexuous cell; body in the form of a spiral with at least one complete turn, 6 to 500 μm in length. These organisms resemble both bacteria and protozoa. They are free-living, saprophytes and parasites of animals causing syphilis, chicken septicemia, and many other diseases.

But in course of time bacteria were further grouped into ten orders based on the morphology of vegetative cells and nature of flagellation.

1. **Eubacteriales:** Rods and Cocci not in groups; motile forms with pertrichous flagella.

2. **Pseudomonadales:** Rods and Cocci in some groups; motile form with polar flagella.

3. **Caryophanale:** Cells in groups, with peritrichous flagella.

4. **Actinomycetales:** Fungus-like myceloid body, but with very narrow cells.

5. **Chlamydobacteriales:** Motile forms in groups.

6. **Myxobacteriales:** Nonflagellate slime producing cells (rods, coccoid) with gliding movement.

7. **Beggiatoales:** Cells without flagella forming groups with gliding movement.

8. **Hyphomicrobiales:** Cells not in groups, with polar flagella.

9. **Spirochaetales:** Spiral forms without flagella but with rotary motions.

10. **Mycoplasmatales:** Cells with filterable stages.

In the 8th edition of Bergey's Manual of Determinative Bacteriology published in 1974, there has been a radical departure from all previous editions for the classification of bacteria. Instead of allocating to orders and families, bacteria were separated into 19 major Groups (Parts) based on a few readily determinable criteria.

Part 1. Phototrophic Bacteria (18 genera)—predominantly aquatic organisms, photosynthesize by a method different from that of blue-green algae and green plants, most of them fix atmospheric nitrogen; morphology spiral, spheres to ovoid; Gram-negative; flagellate to nonflagellate; absence of spore development. Some genera: *Rhodospirillum, Chromatium, Chlorobium.*

Part 2. Gliding Bacteria (27 genera)—**Myxobacteriales**—mostly soil organisms; characterized by the ability to glide on a solid or liquid surface and by the tough slime layer that surrounds the cells; cells aggregate to behave as a single colonial organism. Some genera: *Beggiatoa, Stigmatella, Leucothrix.*

Part 3. Sheathed Bacteria (7 genera)—found in both clean and polluted waters and in activated sewage sludge; cells grouped together in gelatinous sheaths. Some genera: *Leptothrix, Sphaerotilus, Streptothrix.*

Part 4. Budding and/or **Appendaged Bacteria** (17 genera)—mainly soils or aquatic organisms; the products of binary fission are usually unequal and may have specialized functions, e.g., stalks, holdfasts, swarming cells, or may resemble fungal hyphae. Some genera: *Hyphomicrobium, Caulobacter, Gallionella.*

Part 5. Spirochaetes (5 genera)—flexuous helical organisms; largely aquatic; medically important genera: *Treponema, Leptospira,* and *Spirochaeta.*

Part 6. Spiral and **Curved Bacteria** (6 genera)—rigidly helical bacteria having single or grouped flagella which impel the cells with characteristic corkscrew motility; free-living aquatic. Some genera: *Spirillum,* and *Campylobacter.*

Part 7. Gram-negative Aerobic Rods and **Cocci** (20 genera)—soil and fresh and marine water organisms; cause decomposition of organic matter. Some genera: *Pseudomonas, Azotobacter, Rhizobium.*

Part 8. Gram-negative Facultative Anaerobic Rods (26 genera)—organisms of medical importance. Some genera: *Salmonella, Shigella, Yersinia, Vibrio, Escherichia, Erwinia.*

Part 9. Gram-negative Anaerobic Bacteria. (9 genera)—irregularly shaped cells non-sporing rods found among the normal flora of the alimentary tract of man and animals; some constitute sulphur-reducing ones found in brackish waters. Some genera: *Bacteroides, Desulphovibrio.*

Part 10. Gram-negative Cocci and Coccobacilli (6 genera)—most of these organisms are parasitic on man and many are pathogenic. Some genera: *Neisseria, Branhamella, Moraxella, Acinetobacter.*

Part 11. Gram-negative Anaerobic Cocci (3 genera)—these are parasitic bacteria found in the alimentary tract of warm-blooded animals. Genus: *Veillonella.*

Part 12. Gram-negative Chemoautotrophic Bacteria (17 genera)—organisms important in the nitrogen and sulphur cycles in soils and water. Some genera: *Nitrobacter, Nitrosomonas, Thiobacillus.*

Part 13. Methane producing Bacteria (3 genera)—found in anaerobic sediments of muddy waters and in anaerobic sewage digestors as well as in the gastrointestinal tract of animals. Some genera: *Methanobacterium, Methanosarcina, Methanococcus.*

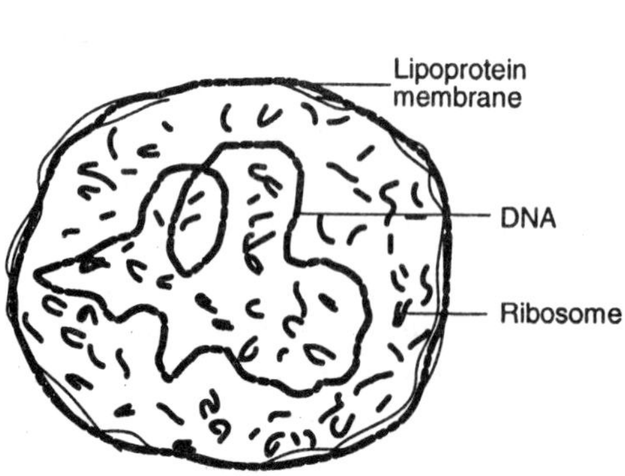

Fig. 20.12: Diagrammatic presentation of Mycoplasma

Part 14. Gram-positive Cocci (12 genera) —aerobic bacteria common in soil and freshwater (*Micrococcu*), facultative anerobes commensal or pathogenic or warm-blooded animals (*Staphylococcus*), and anaerobic cocci include normal flora of man and animals.

Part 15. Endospore forming Rods and **Cocci** (6 genera)—these include aerobic genus *Bacillus* and the anaerobic genus *Clostridium,* both of which contain organisms essential to soil fertility and well-known pathoges such as *B. anthracis, Cl. tetani,* and *Cl. botulinum.*

Part 16. Gram-positive Non-sporing Rods (4 genera)—bacteria belonging to this part under Lactobacillaceae are associated mainly with milk and milk products.

Part 17. Actinomycetes and Related Organisms (39 genera)—(Actinomycetales and Coryneforms)—a tendency to form mycelial growth; most soil organisms, some of which are used in the production of therapeutic antibiotics.

Part 18. The Rickettsiae (18 genera) (Rickettsiales and Chlamydiales)—intracellular parasites that can be cultured in the laboratory only by virological methods. Many cause disease in man or other vertebrate or invertebrate hosts and are often transmitted by arthropod vectors, e.g., *Rickettsia prowazekii* causing house-borne typhus fever, *Chlamydia trachomatis* causing trachoma. They multiply only inside host cells.

Part 19. The Mycoplasmas (4 genera)—these prokaryotic organisms lack a cell wall and have probably been derived from a number of different origins. Some resemble wall-less forms of known bacteria, but others seem to be less closely related and their relation to bacteria is unclear. Cells are bounded by a triple-layered membrane enclosing cytoplasm, randomly distributed ribosomes and strands of nuclear material and can pass through a bacteriological filter. Some genera: *Mycoplasma, Spiroplasma.*

Part 20. The Mollicutes are single-celled prokaryotic microorganisms whose genetic material (DNA) is bound by lipoprotein membrane only. In 1967, Doi and his colleagues in Japan observed mollicutes in the phloem of plants exhibiting yellow and witches' broom symptoms.

Mycoplasmas are commonly found in soil. They cause plant diseases: little leaf disease of brinjal, *Bunchytopol* papaya, big bud of tomato, witches' broom of legumes, yellow dwarf of tobacco and strip disease of sugar cane; human diseases: Primary atypical pneumonia (PAP), pleuropneumonia, prostatitis, and inflammation of genitals; animal diseases: agalactia of goat and sheep, inflammation of genitals.

Virologists include *Maoplasmas* under viruses.

In 1984, major change occurred. The scope of the Manual was greatly broadened by incorporating information dealing with ecology, enrichment, isolation, preservation, and characteristics of bacteria. All these aspects were concerned with bacterial classification and identification. A new name—Bergey's Manual of Systematic Bacteriology was given to the Manual.

The most widely used reference for bacterial classification is Bergey's Manual of Systematic Bacteriology. It is now published in four volumes which are briefly outlined below. Each volume is divided into a number of major sections. Within each section the bacteria are divided into formal taxa at various levels, most attention being given to families, genera, and species.

Volume 1. It includes the familiar Gram-negative chemoheterotrophic eubacteria having a relatively simple morphology. Some are saprophytes, others are parasites. They do not form prosthecae, sheaths, endospores and cysts. The mode of reproduction is by transverse binary fission. The major sections of this Volume are: The Sprochetes; Aerobic/Microaerophilic, Motile, Helical/Vibrioid, Gram-negative bacteria; Nonmotile (or Rarely Motile), Gram-negative curved bacteria; Aerobic Gram-negative Rods and Cocci; Facultatively anaerobic Gram-negative Rods; Anaerobic Gram-negative straight, curved, and Helical Rods; Dissimilatory Sulphate- or Sulphur-Reducing bacteria; Anaerobic Gram-negative Cocci; The Rickettsias and Chlamydias; The Mycoplasmas; and Endosymbionts.

Volume 2. It includes Gram-positive chemoheterotrophic eubacteria having a simple morphology without forming prosthecae, sheaths or cysts, some from endospores. Some are saprophytes, and others are parasites. Reproduction is by transverse binary fission. The major sections are: Gram-positive Cocci; Endospore-Forming Gram-positive bacteria; Nonspore forming Gram-positive Rods of regular shape. Nonspore forming Gram-positive Rods of irregular shape; Mycrobacteria; and Nocadioforms.

Volume 3. Most of the organisms are Gram-negative eubacteria, but some produce methane gas and may be Gram-negative or Gram-positive belonging to the group Archaeobacteria. Some of these organisms are phototrophic, an others are chemolithotrophic. Gliding motility is more common than motility by flagella. They may form sheaths, prosthecae and stalks. Reproduction is by budding. The major sections are: Anoxygenic Phototrophic bacteria; Oxygenic Phtotrophic bacteria; Gliding fruiting bacteria; Gliding non-fruiting bacteria; the sheathed bacteria; Budding and/or Appendaged bacteria; Cemolithotrophic bacteria; and Archaeobacteria.

Volume 4. It includes Gram-positive bacteria possessing branched filamentous hyphae (approximately 1 μm in diameter) resembling fungal hyphae. They are saprophytic and chemoorganotrophic soil organisms which degrade plant or animal residues and some produce a wide range of antibiotics. One genus (*Frankia*) fixes N_2 symbiotically in woody plants. These organisms reproduce by asexual spores which are known as conidia if they are naked, or sporangiospores if they are enclosed in a sporangium. The major sections are: Filamentous bacteria that divide in more than one plane; Filamentous bacteria that form true sporangia; *Streptomyces* and similar genera; and additional filamentous bacteria having uncertain taxonomic placement.

PRESENT DAY APPROACH TO CLASSIFICATION

CLASSICAL TAXONOMY

In classification of any group of organism selected stable characters are used. These characters are known as **taxonomic characteristics.** Classically, the bacteria have been classified on the basis of similarities in phenotypic characters, like morphological features, response to Gıam stain, cultural characteristcs, physiological and biochemical properties, pathogenicity, antibiotic sensitivity, serological relationships.

Before considering various types of approaches to classification of bacteria and other microbes, consider the characters used by bacteriologists for this purpose. The following are the types of characters used in classification:

1. **Morphological characters.** These concern cell shape and size, staining reactions, presence or absence of spores or reproductive forms, type of motility, etc.

2. **Cultural characters.** These include the cultural requirements for multiplication (e.g., nutrients, oxygen, temperature, etc.) and the way growth occurs in liquid media, and particularly on solid media (e.g., colony form).

3. **Biochemical characters.** These include the more specific biochemical characteristics such as metabolic end-products and the presence or absence of a particular enzyme or pathway.

4. **Serological characters.** These concern the nature of the surface antigens as revealed by suitable specific antibodies.

5. **Molecular characters.** These include the sequences of bases in the DNA, GC ratios and nucleic acid-hybridisation.

Subsequently, Bergey's Manual of Systematic Bacteriology (2001), prokaryotes have been placed in two Domains—**Archaea** and **Bacteria.**

1. The **Domain Archaea** is divided into Phylum 1. **Crenarchaeota** and Phylum 2 **Euryachaeota.**

Phylum Crenarchaeota include mainly the extreme thermophiles. Whereas, the phylum Euryachaeota have extreme halophiles and some thermophiles.

2. The Domain Bacteria has been divided into 23 phyta.

3. All **Cyanobacteria** have been placed in a single phylum, including a single class having 56 form-genera. Similalrly, the green photosynthetic bacteria (except the antenus *Chlorloflexus*) have been classified in another pluglum: **Chlorobi.**

4. Most of the Gram-negative bacteria have been brought under a single phylum.

5. Most of the Gram-positive bacteria are classified in the Phylum, **Firmicutes** which includes three classes—**Clostridia, Mollicutes** and **Bacilli.**

6. The Phylum **Actinobacteria** comprises of Gram-positive bacteria which includes, among others, the actinomycetes and related bacteria.

7. The rest of the phyla of **Eubacteria** (nine in number) are a variety of organisms most of them are pathogens. The bacteria in these nine phyla are Gram-negative, but they

are phytogenetically distinct from the **Proteobacteria.**

Proteobacteria has been divided into 5 classes—**Alphaproteobacteria, Betaproteobacteria, Gammaproteobacteria, Deltaproteobacteria** and **Epsilonproteobacteria.**

(i) **Alphaproteobaceria.** The bacteria incuded in this class are morphologically as well as physiologically widely diverse. Some are pathogenic including even intracellular obligate parasites. Some of the common genera are **Rhizobin, Rickettsia** and **Nitrococcus.**

(ii) **Betaproteobacteria.** The class betaproteobacteria have genera which may be chemoorganotrophic (*Aquespirillum*), chemolithotrophic (*Nitrosomonas*), photoheterotrophic (*Rhodocyclus*) or methylotrophic (*Methylobacillus*).

(iii) **Gammaproteobacteria.** It constitutes the largest class of the phylum proteobacteria. Many organisms of this class are physiologically and pathologically important. The largest and most important of this group is the order is **Enterobacterials** which has a single family, Enterobacterioceae having 41 genera. Some of these are *Escherichia, Salmonella, Citrobacter, Proteus* and others.

(iv) **Deltaproteobacteria.** This class of Gram-negative bacteria consists of 7 orders and 17 families. More than half of these families include anaerobic sulfate reducing bacteria.

(v) **Epsilonproteobacteria.** It is the smallest class of proteobacteria comprising of single order **Campylobacteriales.** There are two well-known genera *Campylobacter* and *Helicobacter,* both are of medical importance.

MOLECULAR TAXONOMY

Approach to bacterial taxonomy has undergone drastic changes since the development of molecular biology in the second half of century. The concept of twentieth macromolecules, like proteins and nucleic acids could be used as indicator of evolution of living organisms, was first suggested by Zucker Kanal and Pauling in 1965. They described these macromolecules as "molecular chronometers" because the sequence of monomers in them have changed slowly and randomly and the number changes in a particular macromolecule has increased linearly with geological time scale. A comparison of the sequence of monomers of a particular macromolecule from two organisms should, therefore, give a measure of their phylogenic relationship. If the sequences differ considerably, it indicates that the two organisms are phylogenetically distant. This new approach has given rise to **molecular taxonomy.**

NUMERICAL TAXONOMY (TAXOMETRICS)

An entirely different approach in classification of organisms is numerical taxonomy or taxometrics, also known as Adansonian analysis after Adanson (1727-1800). It differs from both classical taxonomy and molecular taxonomy. Though numerical taxonomy also uses phenotypic (phenotic) characters, it aims at devising an objective approach in classifying organisms into similarities in phenolic characters or **traits.** A large number of such traits are examined for each taxon without assuming that any one trait is more important than any other trait, i.e., all traits are given equal weightage. This eliminates the experimenters' bias and makes classification more objective, rather than subjective. This is the main difference of numerical taxonomy from the classical taxonomy where the choice of phenotypic characters and the tendency of an experiments to give more importance to certain selected traits make taxonomy often a subjective matter. Because in numerical taxonomy a large number of characters of numerous taxa have to be considered, a heavy load of routine calculations becomes essential. Successful application of numerical taxonomy has become possible only through the use of computers. Numerical taxonomy has, therefore, come to be known as computer taxonomy or taxomatrics.

The different steps are followed in numerical taxonomy for classifying groups or clusters of organisms which were clearly defined by Sneath (1972). In the language of numerical taxonomy, the taxa intended to be classified have been given as non-committed term, **operational taxonomic uses** (OTUs). In case of bacteria, an OTU may be genus, a species or a strain. Therefore, a generalized term, OTU, is considered more practical. The steps in numerical taxonomy are given below.

Step 1: **Collection of Data.**

Step 2: **Codig of the Data.**

PRESENT DAY APPROACH TO BACTERIAL WORLD

In Bergey's Mammal of Systematic Bacteriology (2001), all prokaryotic organisms are recognized in two **Domains—Archaea (Archaebacteria)** and **Bacteria**. Again the domain Archaea has been divided into two Phyla—**Crenarchaeta** and **Euryarchaeota.** The Crenarchaeta are extremely thermophiles, acidophiles and sulfur metabolizing archaebacteria. They are mostly anaerobic and occur generally in geothermally heated environment, whereas Euryachaeota are more diverse and include anaerobic methanogens, extremely halophiles and extreme thermophiles.

Domain Archaea

Archaebacteria may be Gram-positive or Gram-negative cells and are generally invested with a wall. Archaebacteria may be spherical rod-shaped, spira, irregularly lobbed or filamentous cells multiplied by binary fission, budding, fragmentation. The organisms may be aerobic, anaerobic, chemolithotrophic or chemoorganotrophic. They mostly occur in extreme environments, though some are mesophilic. Gram-positive archaebacteria have a thick homogeneous cell wall mainly containing complex polysaccharide, whereas Gram-negative forms have a thinner wall mainly of protein or glycoprotein. An outer membrane characteristically present in Gram-negative true bacteria is absent in archaebacteria. Murine is absent in the cell wall of both Gram-positive and Gram-negative archaebacteria. Due to the absence of murine, the archaebacteria are insensitive to β-lactam antibiotic like penicillin, cephalosporin etc. Like all other prokaryotes, archaebacteria have a covalently-linked colored circular genome, but its size is smaller than that of other prokaryotes. Three members of this group have been reported to be able to fix atmospheric nitrogen. Archaebacteria do not possess photosynthetic ability. Their metabolic activity may range from typical organotrophy to strict chemolithotrophy.

BACTERIAL NOMENCLATURE

The 1947 International Microbiological Congress adopted a tentative International Bacteriological Code of Nomenclature which is, in general, similar to the botanical code. It has the following points:

1. The starting point for the nomenclature of bacteria is 1753.

2. Latin diagnosis is never required.

3. Genetic names which are later homonyms of genera of bacteria, plants, or protozoa are illegitimate.

4. Avoid the introduction of genetic names that are in use in Zoology.

Chapter 21

The Actinomycetes

GENERAL CHARACTERISTICS

The Actinomycetes or Streptomycetes or Actinomycetales as they are called are a group or Gram-positive bacteria which form branched filamentous hyphae having resemblance with fungal hyphae. But their hyphal diameter is approximately 1 μm, whereas in fungi it is 5 to 10 μm. These organisms reproduce by asexual spores which are termed conidia when they are naked or sporangiospores when enclosed in a sporangium. Although these spores are not heat-resistant, they are resistant to desiccation and survived during periods of drought. These filamentous bacteria are mainly harmless soil organisms, although a few are pathogenic for humans (*Streptomyces somaliensis* causes actinomycetoma of human), other animals (*Actinomyces bovis* causes humpy-jaw disease of cattle), or plants (*Streptomyces scabies* causes common scab in potatoes and sugar beets). In soil they are saprophytic and chemoorganotrophic, and they have the important function of degrading plant or animal residues. Again, some are best known for their ability to produce a wide range of antibiotics useful in treating human disease. These organisms excrete extracellular enzymes which are decomposers of dead organic material. These enzymes lyse bacteria and thereby keep the bacterial population in check and thus help to maintain the microbial equilibrium of the soil.

The Actinomycetes superficially resemble fungi for having subterranean and aerial hyphae and chains of spores. But their hyphal diameter, cytology and chemical composition of cell walls are quite decidedly bacterial in pattern.

BRIEF HISTORICAL REVIEW

The early exploratory studies by McCormack (1935) and Alexopoulos and Herrick (1938-1942) were followed by the intensive studies by Professor S. A. Waksman and his students (1943-1951) which culminated in the discovery of streptomycin and other new and potentially useful chemotherapeutic agents. Nearly 100 antibiotic substances have been reported in the literature as metabolites of the Actinomycetes. A few of these have been isolated in pure form and their chemistry is studied in detail, while others have been described only as concentrates or in a preliminary way.

ECONOMIC IMPORTANCE

The Actinomycetes, forming soil microflora have gained the greatest importance in recent years as producers of therapeutic substances. Many of the Actinomycetes have the ability to synthesize metabolites which hinder the growth of bacteria; these are called **antibiotics,** and, although harmful to bacteria are more or less harmless when introduced into the human or animal body. Antibiotics have in modern times great therapeutical and industrial value.

The past decade has seen considerable interest in the Actinomycetes as producers of antibiotic substances. The successful use in chemotherapy of streptomycin, chlorompheni-col (chloromycetin is the trade name of this substance), aureomycin and tetramycin—all metabolites of the Actinomycetes, has stimulated the search for new Actinomycetes and new antibiotics among the Actinomycetes. The genus *Streptomyces* is the largest and the most important one, antibiotically speaking. Details of this aspect have already been discussed in Chapter 20.

DISTRIBUTION AND MODE OF NUTRITION

The Actinomycetes are essentially mesophilic and aerobic in their requirements for growth and they resemble both bacteria and fungi. They along with other microorganisms, form the soil microflora and produce powerful enzymes by means of which they are able to decompose organic matter. The majority of these are soil organisms and are associated with blotting material. The characteristic odour of soil after it is ploughed or wetted by rains largely due to the presence of the Actinomycetes. Some are pathogens. The Actinomycetes grow slowly and on artificial media produce hard and chalky colonies which smell decaying leaves or musty earth. They are particularly abundant in forest soil because of the abundance of organic matter. They occur mainly in soils of neutral pH, although some prefer acidic or alkaline soil. The Actinomycetes can grow in soils having less water content than that needed for most other bacteria.

The Actinomycetes are capable of utilizing a large number of carbohydrates as energy sources when the carbohydrates are present in the media as sole sources of metabolizable carbon.

Most of the Actinomycetes are quire proteolytic and attack proteins and polypeptides, and are also able to utilize nitrates and ammonia as sources of nitrogen. Nearly all synthesize vitamin B12 when grown on media containing cobalt salts, and many are able to synthesize rather complex organic molecules which have antibiotic properties. The mechanism of synthesis of these substances is not understood.

SOMATIC STRUCTURES

Most of the Actinomycetes are mycelioid. They begin their development as unicellular organisms but grow into branched filaments or **hyphae** which grow profusely by producing further branches constituting the **mycelium.** The width of the hyphae is usually 1 μm. The delicate mycelia often grow in all directions from a central point and produce an appearance that has been compared with the rays of the sun or of a star. Therefore, the Actinomycetes are also called 'ray fungi'. They often produce complicated design and resemble some of the drawings in modern art exhibitions. They are Gram-positive. The protoplasm of the young hyphae appears to be undifferentiated, but the older parts of the mycelium show definite granules, vacuoles and nuclei. Many Actinomycetes at first produce a very delicate, widely branched, mycelium that may embed itself into the soil, or, if grown in culture, into the solid medium.

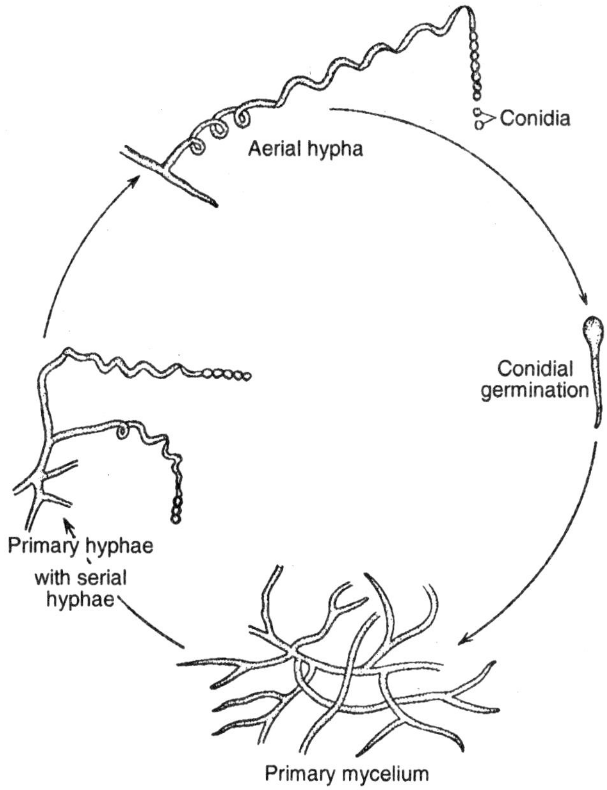

Fig. 21.1: Life cycle of an actinomycete

This kind of mycelium is, therefore, called the **substratum** or **primary mycelium** (Fig. 21.1). After a period of growth, hyphae of a different kind develop, which raise themselves up from the substratum mycelium and grow into the air. These are called **aerial hyphae**, and the corresponding mycelium is the **aerial** or **secondary mycelium.** The aerial mycelium may be white yellow, violet, red, blue, green, or grey and many form pigments that are excreted into the medium. The aerial mycellium is usually slightly wider than the substratum mycelium. The aerial hyphae possess an extra cell wall layer (sheath). The hyphal tip undergoes septation within this sheath to form a chain of conidia. Conidial cell contains a plump, deeply staining, oval or rod-shaped nuclear body.

REPRODUCTION

Most species reproduce by conidia which are developed in chains from the aerial hyphae. The chains may be straight, flexous (wavy) or coiled to various degrees. The conidia bearing filaments are often spirally twisted. Sometimes the whole length of the aerial hypha, sometimes only its upper part is transformed into conidia. Each conidium has a roundish nucleus and is surrounded by a firm outer wall. The conidial wall may be smooth, warty, spiny, or hairy. The conidia can persist in the dry state for many years. Even the vegetative forms of the Actinomycetes are quite hardy and are able to adapt themselves to the changing soil conditions. The conidia appear as a fine powdery coat on the surface of cultures. When the conidia have been scattered on the ground and conditions are favourable they germinate producing 1-3 or even occasionally four little germ tubes which give rise to mycelioid condition (Fig. 21.1).

The primary mycelium in some species commonly breaks up into small fragments called **arthrospores,** which often look like bacterial cells and which might easily be mistaken for the latter.

Chapter 22

Antibiotics

Organic compounds produced by fungi and other organisms, especially bacteria which inhibit the life processes of microorganisms are known as **antibiotics.** But Waksman (1947) defined antibiotics as organic compounds produced by microorganisms, which in dilute concentration have the capacity to inhibit the growth and even destroy other microorganisms. Other workers in this field have suggested that antimicrobial substances from higher forms of life, such as plants and animals should be included among the antibiotics. General usage of the term antibiotics actually applies to those organic compounds of fairly simple structure produced by organisms which inhibit growth of microorganisms.

In 1896, while investigating the cause of pellagra, Gosio isolated an acid crystalline substance from moldy rice, which showed slight antibacterial activity. It appeared to be produced by the fungus *Penicillium brevicompactum.* Later it was also isolated from a number of other *Penicillium* species and was named **mycophenolic acid.** It showed high activity against a large number of fungi, many of which are pathogenic to plants and humans, but it was much less active against bacteria. Mycophenolic acid was actually the first antibiotic to be isolated in a pure form.

During early twentieth century antibiotic activity of microorganisms was observed accidentally by various investigators. Most famous of which is the one that led to the discovery of the antibacterial antibiotic **penicillin,** details of which have been discussed in Chapter 11. But during and shortly after the World War II it was realized that antibiotics could be of tremendous importance for human life. During that time screening of microorganisms producing antibiotics was started on a large scale by University laboratories and by the pharmaceutical industry.

The principal use of antibiotics is to control disease in man and animals. In addition to killing or inhibiting human and animal pathogenic organisms, an antibiotic may also be useful for controlling plant diseases. Besides these, in recent times antibiotics are used in commercial scale for preservation of fruits, vegetables, poultry products, fish, etc. This aspect has been partly treated in Chapter 11, and will also be discussed in the pages to follow.

Antibiotics are referred to specially as **antibacterial, antifungal,** and **antiviral substances.** The antibacterial antibiotics are much more widely known than the antifungal and antiviral antibiotics. Some of the most common antibacterial antibiotics are: **streptomycin, chloromycetin, aureomycin, tetramycin,** and **neomycin.** Their details have been incorporated in Chapters 11, 19, 20 and 21. Some of the antibiotics extracted from algae have been discussed in Chaper 10.

Antibiotics are very specific in action. Penicillin, for example, is active against many Gram-positive bacteria and only a relatively few Gram-negative organisms. The occurrence of antibiotics is probably far more widespread than suspected at present. The largest source of antibiotics is, however, the Actinomycetes. The percentage of *Streptomyces* strains producing antibiotics is high. To a comparatively much lesser extent, antibiotics are produced by fungi. Certain fungi produce substances which are capable of inactivating some of the plant viruses. The Basidiomycetes are especially noteworthy in this respect. Again, extracts of *Trichothecium roseum* reduce infectivity of tobacco mosaic and tobacco necrosis viruses.

It is possible that one microorganism produces more than one antibiotic. *Streptomyces griseus,* for example, forms the antibacterial streptomycin as well as the antifungal antibiotics **cycloheximide** and **candicidin,** and *S. rumosus* forms both the antibacterial **oxytetracycline** and the antifungal **rimocidin.** On the other hand, the same antibiotic may be formed by various microorganisms, such as different species of one genus, or even species of several different genera. For example, the antibiotic **patulin,** which is formed by *Penicillium patulum, P. expansum, P. claviforme, Aspergillus clavatus, A. terreus, A. giganteus,* and by *Gymnoascus* spp.

Besides being widely used for controlling bacterial disease in men and animals, penicillin has been used successfully, to a limited extent, in controlling necrosis of giant cactus, caused by *Erwinia carnegieana.* Injections of penicillin into the necrotic tissue apparently diffuse through the plant tissues for some distance, kill the bacteria. This is one of the new cases in which an antibiotic has been used successfully in therapeutic treatment of plant disease. This and similar other uses of antibiotics have opened up new approach of controlling plant diseases.

Up to the present, antifungal antibiotics have received considerably less attention than the antibiotic active against bacteria. In recent years, however, the interest in antifungal antibiotics has increased for various reasons.

The antifungal antibiotic **cycloheximide,** more commonly known under its brand name **Actidione,** was discovered 20 years ago. It is produced by *S. griseus.* Cycloheximide has also been isolated from *S. noursei.* It is toxic to a wide range of organisms, including yeasts, filamentous fungi, algae, protozoa, higher plants, and animals; but is inactive against bacteria. It is a powerful antifungal agent, most fungi being affected at very low concentrations. Because of its higher activity against fungal plant pathogens and its systemic properties, numerous investigators have tested cycloheximide against a wide variety of fungal diseases in plants.

Griseofulvin in another antifungal antibiotic isolated from *P. griseofulvin.* Synonyms for griseofulvin are **fulvicin, fulcin,** and **grisovia.** It shows action against Basidiomycetes, Ascomycetes, Fungi Imperfecti, and certain Phycomycetes, causing stunting, excessive branching, abnormal swelling and distortion of hyphae. But its antifungal activity is rapidly lost in sunlight.

A highly antifungal group of antibiotic comprizes of **polyene macrolides** produced by *Streptomyces* species. All known polyene macrolides possess a pronounced activity against yeasts and filamentous fungi. Polyene macrolides show high activity *in vitro* against a wide range of fungal plant pathogens. They are, moreover, remarkably non-toxic to host plants. The storage of dressed chicken can be prolonged by the use of this group of antibiotic. Several polyene macrolides have been used against mycoses in man.

The use of antifungal antibiotics in therapy of human mycoses has also received considerable attention. Of special interest in this respect is the antibiotic griseofulvin, which first was developed as a plant disease control agent, but which proved later of greater value as a chemotherapeutant of human fungal diseases of the skin, hair, and nails by oral application. Griseofulvin is translocated in the human body without being toxic. Some of the polyene macrolides, such as: **nystatin, pimaricin, trichomycin, amphotericin** B, and **hamycin** are of clinical use against diseases like: **moniliasis, vaginal candidiasis, coccidiomycosis, trichophytosis,** and **systemic mycosis.**

True, most important antibiotics have been obtained from: (i) the bacteria, and (ii) the fungi. But at least some have been reported from algae, lichens and seed plants.

Details of some of the common antibiotics are given below:

Antibiotic	Source	Biological activity
Actidione	*Streptomyces giseus*	Not antibacterial, antifungal
Alternaric acid	*Alternaria solani*	Not antibacterial, antifungal
Amphotericin	*Streptomyces nodosus*	Not antibacterial, antifungal
Aspergillic acid	*Aspergillus flavus*	Antibacterial and antifungal
Aureomycin	*Streptomyces aureofaciens*	Antibacterial, not antifungal
Bacitracin	*Bacillus subtilis*	Antibacterial
Cephalosporin	*Caphalosporium acremonium*	Antibacterial
Chloromycetin	*Streptomyces venezuelae*	Antibacterial, not antifungal
Chlorotetracycline	*Streptomyces aureofaciens*	Antibacterial
Citrinin	*Aspergillus flavus*	Antibacterial and antifungal
Enniatin B	*Fusarium orthoceras*	Antibacterial, not antifungal
Erythromycin	*Streptomyces erythraeus*	Antibacterial
Fusaric acid	*Gibberella, Nectria, Fusarium* spp.	Antibacterial and antifungal
Griseofulvin	*Penicillium griseofulvin*	Not antibacterial, antifungal
Grisein	*Streptomyces griseus*	Antibacterial, not antifungal
Illudin	*Clitocybe illudens*	Antibacterial and antifungal
Mycophenolic acid	*Penicillium brevicompactum*	Antibacterial and antifungal
Neomycin	*Streptomyces fradiae*	Antibacterial, not antifungal
Nystatin	*Streptomyces noursei*	Not antibacterial, antifungal
Oxytetracycline	*Streptomyces rimosus*	Antibacterial
Patulin	*Penicillium patulum*	Antibacterial and antifungal
Penicillin	*Penicillium chrysogenum, P. notatum*	Antibacterial, not antifungal
Streptomycin	*Streptomyces griseus*	Antibacterial, not antifungal
Streptothricin	*Streptomyces lavendulae*	Antibacterial and antifungal
Terramycin	*Streptomyces rimosus*	Antibacterial, not antifungal
Ustalagic acid	*Ustilago zeae*	Not antibacterial, antifungal
Vancomycin	*Streptomyces orientalis*	Antibacterial

Chapter 23

Viruses

GENERAL CHARACTERISTICS OF VIRUSES

The word **virus** is from Latin **virus,** a poison. As a preliminary working definition, viruses may be characterized as ultra-microscopic disease-producing entities, capable of being introduced into living cells of particular kinds of organisms, and capable of reproducing or being reproduced only within such cells. They cannot be made to multiply on artificial media once, the three properties of virus were thought to be unique and absolutely distinctive features of virus. These are:

Ultramicroscopic nature (i.e., invisible with optical microscope), **filterable** (as they could pass through unglazed porcelain and other bacteria-retaining filters) and **noncultivable** on in animate media. They are now easily retained on specially prepared **molecular filters** made of very fine pare collodion on plastic material and are thus **nonfilterable.** They can now be easily propagated in living culture media at least.

In the light of the above developments, the viruses can be characterized as follows.

Plant viruses have not been definitely observed in plants other than flowering plants and bacteria, but this may be due to lack of study rather than a real absence. Again, animal viruses inhabit vertebrates, arthropods, and many other animals.

Particles of both plant and animal viruses vary from spherical to slenderly rod-shaped, according to the kind of virus. Some animal viruses are brick-shaped. Some of the smallest viruses are only about 0.01 micron in length, while some of the largest ones approach 0.5 micron.

The viruses are responsible for a large number of important plant and animal diseases. In many cases the virus is more or less **latent** (i.e., it exists and reproduces but causes no detectable harm) in a particular host and causes a recognizable disease only when introduced into some other kind of host. In general, the plant viruses are transmissible by sap, by grafting, or by insects. The virus diseases and infections in plants are recognized and described on the basis of symptoms and transmissibility. Viruses possess some of the qualities of living organisms: they are able to reproduce, they occur in distinct strains or varieties, and they undergo changes similar to mutations. Unlike living organisms, they do not respire, nor do they possess cellular structures. Many Biologists regard viruses as intermediate between non-living matter and living organisms.

HISTORICAL REVIEW OF VIRUSES

The first known record of the existence and behaviour of virus is a variegation in the colour of tulips reported by Carolus Clusius in 1576. That the variegation might be due to a disease was suggested only in 1670. In 1715, an account of an infectious chlorosis of *Fasminum* was published. About fifty years later the so-called 'curl' disease or potatoes came into prominence. But there was a great controversy over its cause.

About 1886 Adolf Mayer described a disease of tobacco plant which occurred in the tobacco-growing regions of Holland, as **Mosaikkrankheit** which means mottling type of virus disease. He described the disease, and from the mosaic pattern common on leaves of affected plants, Mayer suggested the name 'mosaic'. Mayer showed that this mosaic disease of tobacco could be communicated to a healthy tobacco plant by inoculation with the sap of the infected plant. But Mayer did not suggest that the diseased condition was due to virus. Two years later, Erwin F. Smith showed that the 'peach yellows' was also communicable and could be transmitted by transplanting a bud from a diseased tree to a healthy tree.

The first scientific proof of the existence of a virus was given by the Russian Botanist Dmitri Iwanowski in 1892. Iwanowski working with the mosaic disease of tobacco, described by Mayer, proved that sap from such a diseased plant was capable of inducing the mosaic disease in healthy tobacco plants. He passed the sap through a bacteria-proof Chamberland filter candle and found the filtrate to retain infectivity. It is the first record of the passage of either a plant or an animal virus through a bacteria-proof filter. Six years later in 1898, Loeffler and Frosch showed that the foot and mouth disease of cattle is caused by an agent which could pass through bacteriological filters. In 1892, a Dutch Bacteriologist Martinus Willem Beijerinck took up the study of tobacco mosaic. He found the sap of infected plant, when filtered through bacteria-proof filter, to be sterile but still infectious, which he designated as **contagium vivum fluidum** and subsequently, referred it as a **virus.** Beijerinck confirmed the findings of Mayer and Iwanowski and claimed more emphatically than either of them that the causal agent was not a bacterium or any conceivable corpuscular material.

The relationship between an insect and a plant virus has been experimentally established by a Japanese farmer, Hashimoto, who worked in 1894 with the dwarf disease of rice and the leafhopper *Nephotettix apicalis* var. *cincticeps.* In 1895, Takata in Japan transmitted virus by means of the leafhopper *Deltocephalus dorsalis.* During 1906-07, Ball, Adams, and Shaw working on curly top of sugar-beet established the leafhopper transmission of virus. Further evidences of leafhopper transmission of virus were put forward by Bancquet and Hartung in 1915. That aphids are also responsible for the transmission of virus was demonstrated by Allard in 1914. Iwanowski continuing his study of tobacco mosaic virus described in 1903 certain intracellular bodies in the tissue or diseased plants. One type was amoeboid, the other was in the form of crystalline plates. Holmes in 1929 described the primary infection lesion of tobacco mosaic virus and thereby he indicated the usefulness of the symptomatology.

In the area of strain differences in viruses, McKinney (1926) is the pioneer worker. He suggested that strains arise by mutation. Takahashi and Rawlins in 1933 exhibited the physical phenomenon of tobacco mosaic virus. By 1935 it was evident that the virus was a particle distinct from any known living entity and comparable in many ways to larger molecules or colloidal particles. In 1935, Stanley for the first time isolated a crystalline protein in more or less purified condition possessing the properties of the tobacco mosaic virus. In his opinion, tobacco mosaic virus is an autocatalytic protein which may be assumed to require the presence of living cells for multiplication. In the succeeding five years, protein crystals were isolated from preparations of several plant viruses. These preparations were infectious and capable of producing the disease concerned upon inoculation into the respective host. Other pioneer workers in this line are Bawden and Pirie. They have shown that all virus proteins crystallized so far, are nucleoproteins.

HOW VIRUSES DIFFER FROM BACTERIA?

The viruses differ from Bacteria in the following characters:

(i) Viruses do not have cellular organization.

(ii) They are smaller than bacteria.

(iii) They are not free-living in nature, and occur only as obligate intracellular parasites.

(iv) Viruses have only single kind of nucleic acid either DNA or RNA, but never both.

(v) Virus nucleic acid is a single molecule, i.e., one replicon, which may be single or double-stranded.

(vi) The outer shell-capsid is mostly of protein, except a few animal virus, where additional polysaccharides are present.

(vii) Viruses lock machinery for protein synthesis.

(viii) Viruses cannot grow or multiply in nutrient medium. In some cases some viruses can grow in a medium supplemented with extract of host cell in which they grow, but can grow only in living cell, they thus are **obligate parasites.** Recently plant virus, TMV has been cultivated in plant protoplast cultures. Wound-tumour virus can also be grown in cell cultures of insect vectors.

(ix) Viruses do not contain metabolic enzymes or protein synthetic machinery of their own. They use host machinery for synthesis of their proteins.

(x) The replication of the virus involves the independent synthesis of its nucleic acid and protein, which are assembled into organized structures after the completion of their synthesis, the infectious particles, **virons.** The virons do not arise from pre-existing virons. The viron contains only one kind of nucleic acid—DNA or RNA.

(xi) Viruses cannot divide by binary fission.

(xii) Viruses are resistant to antibiotics.

(xiii) Viruses are among the smallest biological agents. The largest ones, like smallpox-viruses, all almost of the size of the smallest bacteria, like rickettsiae.

(xiv) Viruses are composed of protein that covers a central nucleic acid molecules called **genome,** either RNA or DNA. The amount of protein varies from 60 to 95% and the rest consist of **nucleoid** whose pairs range from 1,000-250,000. The number of genes per pair ranges from 4-8 for small viruses and 100-200 for the large viruses-nucleic acid. A simple virus particle is often called a **virion.** They grow and multiply within living cells.

(xv) They cause diseases of animals, including man, and plants of different groups, except bryophytes and gymnosperms.

VIRUS INDUCED SYMPTOMS

Effects of Viruses on Plants. Viruses are similar to obligate parasites, in that they cannot be grown on non-living media. They are intimately associated with the host cell and a few kill the infected plant although some cause severe distortion and dwarfing. The changes brought about by viruses are treated as symptoms which may be: (i) morphological changes or external symptoms, (ii) histological and cytological or internal symptoms, and (iii) metabolic changes; but these are all correlated.

The symptoms of the majority of plant virus diseases are most conspicuous on plants making rapid growth. Plants that are almost mature at the time of infection usually do not develop symptoms on any part except on new growths.

Viruses being infectious induce a variety of symptoms covering a wide range of host reactions. Some of the symptoms frequently encountered are mentioned below:

I. **External Symptoms.** The most common symptom in the green tissues of higher plants is the alteration in the normal development of chlorophyll—**chlorosis.** This may be accompanied with various other malformation. Besides this, necrosis of tissue and dwarfing, distortion of a particular organ or the entire plant are also common symptoms. The external symptoms may be **primary** or **localized** and **systemic.** The primary or initial symptom is a local reaction at the actual site of inoculation consisting of spots or rings of various types. They are usually necrotic but are occasionally chlorotic and known as **local lesions.** There may also occur a second type of primary symptom known as **clearing of the veins,** a condition where the veins of the youngest leaves become yellow, whereas the systemic symptoms

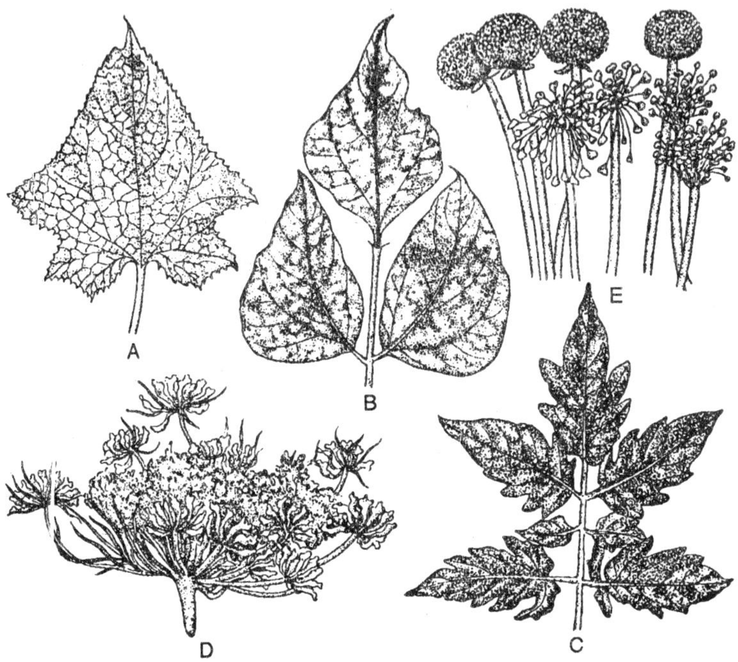

Fig. 23.1: Symptoms of virus infection. A-C. Common mosaic of leaf. A. Cucumber, B. Bean, C. Tomato, D-E. Flower distortion and sterility, D. Carrot, E. Onion.

are of widespread occurrence in the host tissue. Some of the external symptoms are described below:

Chlorosis: The disbalance of normal development of chlorophyll leading to yellowing or formation of different shades of green without pattern is the chlorosis.

Mosaic: The interspersion of various degrees of chlorosis with the normal green colour of the leaf resulting in a mosaic pattern of yellow and green, forms the mosaic symptom (Figs. 23.1A to C).

Mosaic mottling: When the leaves show a mottling of light or dark-green, yellow or even white, it is known as the mosaic mottling.

Necrosis: Death of the host cells, or necrosis, is a symptom of many virus diseases, and may consist of small areas on the leaves, streaks on the stem or large areas of dead tissue which ultimately cause the death of the whole plant. The necrosis may spread causing various patterns as it develops. A relatively rapid killing of a bud, branch, or the entire top of the plant is the **top necrosis.** The necrosis of phloem elements is known as **phloem necrosis.**

Ring spots: These spots consist of various types of chlorosis and necrosis. In each ring spot numerous concentric rings develop on the leaves with a central spot. Spots of circular chlorotic areas are known as **chlorotic ring spots,** whereas in cases where necrosis appears in rings alternating with normal green are the **necrotic ring spots.**

Veinclearing and veinbanding: It infected leaves, when a clearing or chlorosis of the tissue in or immediately adjacent to the veins takes place, is referred to as veinclearing. Again, the symptom consisting of a broader band of chlorotic tissue along the veins or bands of green tissue in that position, set off by chlorosis or necrosis in the interveinal parenchyma is called veinbanding.

Yellowing: Leaves and other green parts instead of developing green colour turn yellow, this is known as yellowing.

Distortion and overgrowths: Distortion of the leaves is a common symptom of virus disease and may take the form of **crinkling** and **curling** or upward **rolling** of margins. Some viruses induce the formation of outgrowth known as **enations,** masses of hypertrophied tissue developed on the surface of leaf or stem. Virus attack may also induce **proliferation** of stem buds, or cause distinct **tumours** or **galls** on roots and stems. Often virus infection results in **stimulation** of dormant buds and hyperplastic tissue leading to unusual differentiation. Virus infection may also cause distortion and sterility of flowers (Figs. 23.1D and E).

Stunting: This is shown by reduction in size of leaves or other organs or of the entire plant. Stunting is frequently accompanied with **rosetting.** This is shown by shorter internodes, smaller leaves and fruits. Stunting with more or less rosetting is characteristic of **bunchy top** of bananas. Reduction in size also results in **dwarfing.**

Premature defoliation and death: Premature dropping of leaves is also a symptom of virus disease. This may also lead to premature death of the affected plant.

Masked symptoms: Under certain environmental conditions often no distinguishable symptoms appear in spite of the presence of virus in the infected plant. Such an infected plant is referred to as having masked symptoms. Again, when plants show symptoms for a short time when first infected, but eventually become and remain symptomless, the infected plants of this type are known as **symptomless carriers.**

II. **Internal Symptoms:** This will include intracellular inclusions. Certain abnormal intracellular inclusions are characteristic of virus infections, they do not occur in diseases caused by other infectious agents. There are

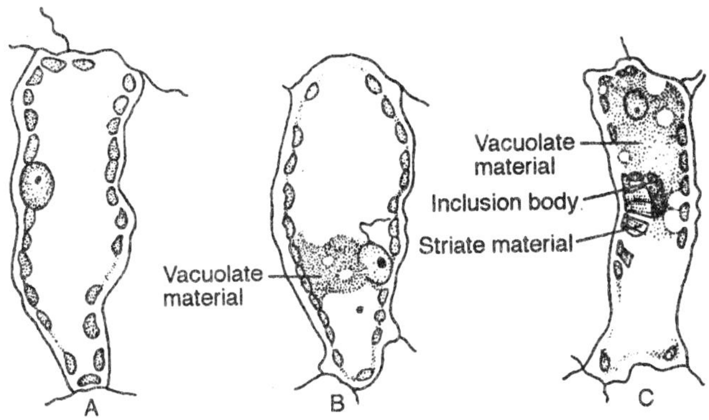

Fig. 23.2: Intracellular inclusions. A. Virus-free palisade cell of potato. B. Mild mosaic virus-infected palisade cell of potato showing vacuolate material close to the nucleus. C. Common tobacco mosaic virus-infected palisade cell of potato showing inclusion body (X-body)

several different kinds of intracellular inclusions: **crystalline** or **fibrous** or **striate material; amorphous** known as **X-bodies; vacuolate material** (Figs. 23.2B and C); **intranuclear inclusions;** and **other types of inclusions.**

Crystalline inclusions or **Striate materials:** These occur mainly in the cells of plants infected with tobacco mosaic virus and are usually in the form of plates of varying size (Fig. 23.2C).

Amorphous inclusion bodies or **X-bodies:** The amorphous bodies are protoplasmic (Fig. 23.2C), more or less about 10 μm in length, may be several in one cell. They are relatively stable and are preserved by ordinary cytological fixatives. They often resemble the nucleus of the cell. In addition to the tobacco mosaic virus, three inclusions occur in all tissues of plants infected with *Hyosyamus* mosaic virus.

Intranuclear inclusions: These inclusions may consist of thin rectangular plates and usually several in each nucleus, or of isometric crystals. They are common in the nuclei of leguminous plants affected with the viruses of pea mosaic and yellow bean mosaic.

Other types of inclusions: A number of miscellaneous inclusions occur in the cell cytoplasm of virus-infected plants. Some of them are: a **spherical hyaline** and **homogeneous body** called a **spherule,** present in the wound tumour virus-infected root tumour cells of *Rumex acetosa;* **spindle-shaped bodies** in the virus-infected cells of *Epiphyllum;* **fusiform** and **variably shaped protein bodies** in the cytoplasm of the virus infected epidermal cells of the leaves of *Opuntia brasiliensis.*

VIRUS ENTRANCE, TRANSLOCATION AND DISTRIBUTION IN PLANT TISSUE

Viruses enter plants either through damaged cells including leaf hairs, or are introduced into the plant by vectors. Slight damage is necessary for successful establishment of the virus within the cell. This involves removal of the proteinaceous coat of the particle and liberation of nucleic acid within host cell. After infection there is usually replication of viruses in the host cell. There is some evidence that virus protein and nucleic acid are both formed within the nucleus, synthesis of the virus particles being initiated in the nucleolus and perhaps completed in the cytoplasm after extrusion from the nucleus. Some viruses are probably incomplete viruses which cannot replicate without the help of second virus (the 'activator'), but others appear to be complete.

When a virus particle infects a host cell, the particle multiplies in that cell and leads to the expression of disease symptom involving a multitude of cells either in a localized area around the point of entry of the virus particle—localized infection, or throughout the host plant and in areas far removed from the point of infection—systemic infection. Translocation of viruses is thus of two types: short-distance translocation which involves cell-to-cell movement causing localized symptoms and long-distance translocation which involves phloem and xylem elements and incites systemic symptoms. During long-distance translocation, viruses multiply *en route* and do not cause any infection. Cell-to-cell movement is through plasmodesmata and by cytoplasmic streaming. But in meristematic tissue viruses are distributed during cell division. Phloem and xylem are the only tissues that act as channels for long-distance spreading of viruses. Viruses reach phloem in two different ways: by cell-to-cell movement through plasmodesmata or by direct deposition by an insect vector. Viruses are passively carried in phloem in the liquid stream carrying photosynthates. The same virus may travel over long distances in phloem, and can also travel in xylem. Viruses fall in four categories: parenchyma-restricted, phloem-restricted, xylem-restricted, and not restricted to any particular tissues but are widely present in tissues of all types. In general, transportation of viruses is more rapid in the direction of nutrient utilization or storage. But movement in the reverse direction can occur, although virus movement from shoot to root is generally more rapid than from root to shoot. The rate of spread of viruses is greater in young than in old tissues and the movement is faster in higher than at low temperatures because of the fast streaming of cytoplasm at higher temperatures. Nutrition which promotes rapid vegetative growth of the host plant favours the virus multiplication.

PLANT CELL CHANGES CAUSED BY VIRUSES

The changes of plant cells caused by viruses may be indicated in the following manner:

1. **Histological and Cytological Changes.** The chlorotic tissue of mottled leaves is generally thinner than the normal green tissue and has shorter palisade cells containing fewer and smaller chloroplasts. Some viruses, those which cause leaf curling and yellowing, bring about necrosis of the phloem tissue, whereas in others necrosis is preceded by tissue proliferation and is perhaps caused by crushing than directly by the virus.

Leaf enations caused by the cotton leaf curl virus originate in part from increased cambial activity producing extravascular tissue which may be abnormal in nature. Xylem is generally less affected by virus than is parenchyma or phloem, but tyloses and formation of gum have been described in the xylem of plants infected with viruses.

Nuclear division may fail or be abnormal in cells infected with certain viruses. Amitotic nuclear division takes place in *Petunia* stem enations caused by virus infection. The nuclei after elongating, constricting and cleaving produce daughter nuclei. Amitosis may be associated with a shortage of RNA. During division of virus infected cells there is competition between the virus particles and the nuclear DNA for the RNA present in the nucleolus. Since the formation of the nuclear spindle in mitosis is dependent on RNA, any shortage of the latter could lead to amitotic nuclear division. Other abnormalities shown by virus infected cells include lobing and distortion of the nucleolus, and spindle breakdown resulting in either scattering of the chromosomes or their failure to separate, the latter leading to the formation of 'giant' nuclei. The sterility of plants infected by certain viruses may be associated with such nuclear abnormalities.

2. **Metabolic changes:** Viruses affect the nucleic acid metabolism of the infected cell, thereby bringing about changes in protein synthesis which in some way result in the formation of more virus. Reduced efficiency of chloroplasts in virus infected plants and the photochemical activity of chloroplasts lead to mosaic, mottling and yellowing of leaves. Changes in the amounts and activities of several enzymes take place due to virus infection. Increased oxidase activity is a general feature of many virus infections in plants. The abnormal growth pattern of plants infected by some viruses suggests an interference with the growth regulating processes of the plant. Virus infection causes increased respiration rate in plants. The combination of decreased photosynthesis and increased respiration results in a fall in the carbohydrate content of virus infected leaves.

HOST ENTRY OF VIRUSES

This topic is treated separately in two separate aspects as follows:

Transmission. Plant virus as, in general, do not have any specific mechanism for attachment to host plant cell. They are passively transmitted by several means. Tobacco mosaic virus (TMV) and some other plant viruses are carried by wind, raindrops, gardening equipments and grazing animals to host plants. They enter into leaves damaged by mechanical injuries, like lesion. Some other plant viruses are transmitted to healthy hosts through seeds, tubers and other propagules. The tobacco ring spot virus is carried by nematodes infecting roots. Tobacco necrosis virus is transmitted through parasitic fungi. The most important agent of transmission of plant viruses is the insects feeding on plant cell sap. Among these insect **vectors**, the most important ones are the aphides, and arthopods e.g., grasshoppers.

Translocation of virus within the plant

Viruses move from hot plant cell to cell through plasmodesmata. During movement they multiply in most cells. Initially they infect parenchyma cells, where they multiply and then pass in the neighbouring cells. In leaf parenchyma cells, the virus moves both upward and downward directions through phloem. Thereby, virus can reach easily in the apical meristem, as well as the storage organ cells, organ like rhizome, tuber etc.

METHODS OF TRANSMISSION OF VIRUSES

The various methods of transmission of plant viruses are as follows:

1. **Seed Transmission.** Viruses may be externally seed borne as in tomato, cucumber, etc.; or internally seed borne in testa, endosperm and/or embryo as in barley, cowpea, bean (bean mosaic), etc. The internally seed borne viruses are more effective than the externally seed borne ones.

2. **Transmission by Grafting.** Since viruses are intimately associated with the living cells of the host, it is rather easy for their transmission through grafting between living cells of virus infected and virus-free plants. In fruit and ornamental trees where grafting is the normal method of propagation. Transmission of virus by grafting becomes a means of natural transmissive.

3. **Transmission by Vegetation Propagation.** Viruses are very commonly perpetuated in the vegetative organs of perennial plants (fruit trees). When such plants are virus-infected all the vegetative parts used for their propagation also become virus-infected. As such, viruses are readily transferred from locality to locality in virus-infected nursery stock, bulbs, tubers (leaf roll of potato) and roots. Hence the infected perennials are the common reservoirs for perennating of many viruses.

4. **Transmission by Parasitic Phanerogams.** Species of *Cuscuta* when parasitizing virus-infected host plants sends haustoria into

the host tissue and thereby receives virus infection. The same virus-infected species of *Cuscuta* when extends its stem to parasitize other plants, the virus may be transmitted to such plants through the newly formed penetrating haustoria. *Cuscuta* thus functions as the transmitting agent.

5. **Transmission by Insects.** Most viruses are transmitted by insects. The insects responsible for the transmission of viruses either possess mouth parts adapted for biting or stylets for piercing and sucking. The sucking insects are the usual insect vectors. But the principal insect vectors are: thrips, plant bugs, leafhoppers, white flies, aphids and coccids.

There is specificity of certain insects for particular viruses. Some viruses may be carried mechanically on the mouth parts of the insects and the latter remain viruliferous for a period of only a few minutes to a few hours. These are known as **nonpersistent viruses.** They are rapidly lost by the vector, usually after a short period of feeding. The non-persistent viruses are carried to the first plant and rarely to the second if the feeding periods are of some hours' duration. Again, some vectors may not transmit the viruses to a healthy plant until some time has elapsed after they have fed upon the diseased plant. These are **persistent viruses.** The persistent viruses retain infectivity for a long period of time and there is delay in the development of infective power. In such case it is possible that the virus is ingested by the insect and is later transmitted through the body into the saliva, by which channel it eventually reaches the next host plant. The delay in the development of infective power—the **latent period** also known as the **period of incubation,** varies greatly with the different viruses. These viruses may also multiply within the body of the vectors. The persistent viruses are not transmitted to the first two or three plants, but to all the others for a considerable period. The longest latent period, so far discovered, is that of a strawberry virus known as 'Virus 3' transmitted by the aphid *Capitophorus fragariae* Theob., which takes 10 to 19 days. Persistent viruses may be: **semipersistent viruses** where viruses persist in the vector for a few days; **circulative persistent viruses** where the insect vectors accumulate the viruses internally, and after passage of the virus through the insect tissues is introduced in the plants through the mouth parts of the insect vector; and again some circulative viruses may multiply in their respective vectors, they are then called **propagative viruses.**

6. **Transmission by Mechanical Means.** Transmission by this means consists of transference of sap from a virus-infected plant to a healthy plant by artificial or natural means. Since infection through natural openings, like stomata is rather rare, mechanical transmission often involves wounding of the host tissue for the easy entrance of the virus from host to host. Viruses transmitted by mechanical means are usually in high concentration in the plant. Some viruses can spread from a diseased plant to a healthy one by contact of the leaves brought about by the wind. Cultivation procedures and the movement of animals may play some part in the spread of viruses. The tobacco mosaic virus is transmitted very rapidly by rubbing the extracted juice of a diseased plant over the leaf of a healthy tobacco plant. By this process hairs or epidermal cells are sufficiently wounded to bring about infection.

Some viruses may spread below ground by mechanical contact between the roots of infected and healthy plants.

Usually viruses of mosaic group are most readily transmitted by mechanical means.

7. **Soil Transmission.** The soil-borne viruses infect host through root system. These viruses do not usually persist in the soil more than a few months at the most. The viability of the soil-borne viruses, however, depends largely on the soil texture. Roots of infected perennial

hosts serve as permanent reservoirs of soil-borne viruses.

8. **Transmission by Mites.** Eriophyid mites transmit several viruses. The big-bud mite, *Phytopus ribis* transmits virus that causes disease of *Ribes.* Mites cannot fly and presumably spread viruses by crawling from plant to plant or more likely, by being dispersed by wind.

9. **Transmission by Nematodes.** Nematodes belonging to the genera *Xiphinema, Longidorus* and *Trichodorus* transmit a number of viruses. Spread might result from systemic root infection of plants with extensive root systems; the roots could thus be made available to nematodes feeding at some distance from the original site of infection. The nematodes feed on the epidermal cells near the root tip and acquire virus.

10. **Transmission by Fungi.** Several viruses including those causing big-vein disease of lettuce and tobacco necrosis are transmitted by *Olpidium* and *Synchytrium* which infect plants. The virus is borne internally by the zoospores of the fungus when they are developed in the virus infected host.

11. **Pollen Transmission.** Cases of dissemination of viruses through pollen grains are few in comparison with other means. Common example is bean mosaic virus.

12. **Transmission through weeds.** Weeds serve as collateral hosts for transmission of sugarcane mosaic virus.

IMPORTANCE OF VIRUSES

Viruses have both harmful activities and useful activities.

1. **Harmful activities.** Viruses are responsible for various diseases of both plants (tobacco mosaic virus, yellow vein mosaic of lady's finger, leaf roll of potato, leaf evil of papaya, etc. and (Smallpox, Meningitis, Pneumonia, Mumps, Bronchitis, etc.). The plant viruses cause damage to different parts like root, leaf, fruit, seeds, etc. and cause economic loss by reducing the quality and quantity of the plant products.

2. **Useful activities.**

(i) In space research, lysogenic phage cultures are used as **radiation detector** in the space ship.

(ii) Avirulent a temperate phages help in genetic recombination (transduction) and are used widely in **genetic research.**

(iii) Phages are used as **scavengers** to eradicate the bacteria present in the polluted water.

(iv) To a limited extent phages are used in **therapy and prophylaxis** of some bacterial diseases.

(v) Viruses have both the living and non-living characters. Hence they have the importance of determining the **origin of life.**

(vi) Viruses are utilized in the **production of vaccines** useful in developing immunity against viral infection.

PROPERTIES OF VIRUSES

The properties of the plant viruses are conveniently divided into the following categories:

1. **Host Range.** Some viruses like beet curly top, cucumber mosaic, and tobacco mosaic virus have wide host range and host plants may fall within widely different families. Others have extremely restricted host range, for example, corn mosaic virus. Host specificity is a genetic character of the virus and is determined by its nucleic acid. In some cases, virus is actively present in the host plant without causing obvious effect. The absence of distinctive symptoms is due to masking effects of unfavourable environmental conditions. Symptoms appear when conditions become favourable. A virus showing the phenomenon of masking is a **masked virus** and the host plant is the **masked carrier.** Again the presence of a

virus in the host with the total absence of visible symptoms over the entire range of environment (favourable and unfavourable) to which the host is exposed is designated as a latency of a virus and such a virus is a **latent virus.** A latent virus never induces symptoms or makes its presence known in a host over the entire range of environmental conditions. Host plants which harbour the virus but remain symptomless throughout the entire range of environmental conditions are the **symptomless carriers.**

2. **Physical Properties.** The viruses could be distinguished in part by the symptoms they produce on the infected hosts. But the symptoms very often overlap and also change appearance with variation in environment. Viruses were found to differ markedly in the point at which they were inactivated by various agents. Those most extensively used are explained below: (i) the **thermal inactivation point** is the constant temperature at which a virus extract is completely inactivated when exposed for 10 minutes; (ii) the **longevity in vitro** is the number of hours or days at which a virus remains infectious at room temperature either as a virus extract or in the host tissue; (iii) the **dilution inactivation point** is the degree of dilution of the virus extract with water or buffer solution at which infection no longer occurs. These are commonly designated as physical properties.

3. **Structure.** Each virus is called a **virion.** It is composed of a single type of nucleic acid (DNA or RNA), but never both, which gives the virion infective capability. The nucleic acid which may be either linear or circular is surrounded by a protein coat called a **capsid** to form a **nucleocapsid** (Fig. 23.3C). A capsid is again made up of protein subunits called **capsomeres** which are in turn composed of a number of protein molecules. The virus particles or units of virus—the virions possess a true symmetry of structure. Virions may either be naked (naked virions) or surrounded by an envelope of carbonhydrates and lipids or lipoproteins (Figs. 23.3A and B). The enveloped virions are sensitive to lipid solvents such as ether and chloroform. The nucleic acid of a virus may occur as either double-stranded DNA, single-stranded DNA, double-stranded RNA or single-stranded RNA. Plant viruses have been found to contain only single- or double-stranded RNA. Bacterial viruses contain single- or double-stranded DNA or single-stranded RNA. Animal viruses have all types of nucleic acids except single-stranded DNA.

The amount of nucleic acid in a virus particle varies enormously: RNA 1 to 5 per cent and DNA 5 to 50 per cent. The viral protein usually forms the largest part—50 to 90 per cent. Viruses lack components necessary for energy generation and protein synthesis (for example, ribosomes). Generally, the enzymatic capabilities of viruses are extremely limited and confined to enzymes involved in the viruses' entry into cells and replication of their own nucleic acid. Because of their metabolic limitations viruses are unable to replicate indepen-

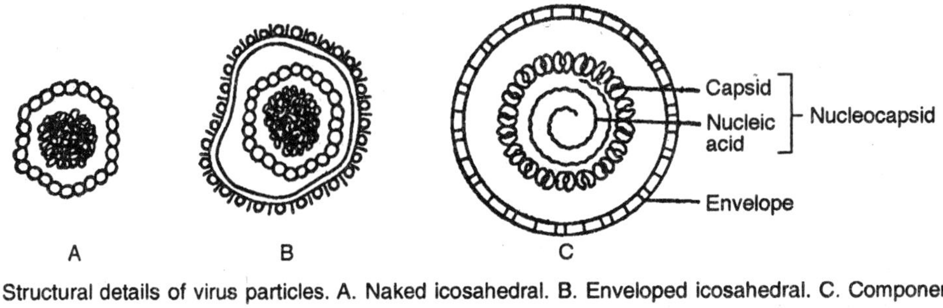

Fig. 23.3: Structural details of virus particles. A. Naked icosahedral. B. Enveloped icosahedral. C. Components of a virus particle.

dently, and they must invade a living cell and utilize the cellular ribosomes, energy sources and certain other components of this cell in order to produce new viruses.

Viruses differ considerably in size. The smallest viruses are similar in size to large protein molecules or ribosomes, and their nucleic acid codes for only a few genes. The more complex virions may be larger than some of the most minute bacteria. Largest virus is Smallpox virus 300 nm in diameter and the smallest ones are 20 nm in diameter. Some simple viruses can be prepared as crystals.

The virions conform to one of the following shapes.

(i) ICOSAHEDRAL—a regular polyhedron with 20 triangular faces and 12 corners. This shape is determined by the capsid.

(ii) HELICAL—these virions resemble rods. Their capsid is a hollow cylinder with a helical structure.

(iii) ENVELOPED—the internal nucleocapsid of these viruses, which may be either icosahedral or helical, is surrounded by a membranous envelope. Enveloped virions are **pleomorphic** (have varying shapes) since the envelope is not rigid, although they generally appear somewhat spherical.

(iv) COMPLEX—some virus particles have a very complicated structure. They have several coats around the nucleic acid.

(v) BACTEROID—the virus particles are with a more or less spherical 'head' and a slender 'tail' of equal or greater length.

Some virus—like infectious agents, known as **viroids** which cause a variety of plant disease, e.g., potato spindle tuber chrysanthemum stunt, etc., have no protein capsid and consist only of a small single-stranded RNA of low molecular weight—and are subviral in size. They were described by Diener in 1967.

4. **Virus Replication.** Viruses replicate only in living cells. The term 'replicate' actually means multiplication of virus particles which does not take place by division of existing virus particles but are formed directly by aggregation and organization or molecules within the protoplasm of the host cell. The basic steps in virus replication are similar for all viruses, whether they infect the plant, bacterial, or animal cells.

In case of plant viruses, after entry into the host through natural openings, wounds (mechanical or insect), or through pollen grains, the virus comes in contact with host cytoplasm by pinocytosis. The infective part (RNA) is freed from its protein coat soon after inoculation. This is the **eclipse stage.** This stage is followed by an extensive virus synthesis. The site of viral RNA synthesis is nucleus, more correctly nucleolus; that of protein synthesis is cytoplasm in the vicinity of nucleus and/or nucleolus and that assembly of virus particles is the endoplasmic reticulum and/or nucleus. Complete virus particles gradually diffuse throughout cytoplasm and may ultimately lead to the formation of large virus aggregates or virus crystals. The details of plant virus synthesis are given below.

The degraded protein coat of the virus probably remains in the host cell and becomes a part of the host cell protein. The naked RNA then induces the host cell to form enzymes: RNA-polymerases, RNA-synthetases, or RNA-replicases. These enzymes in the presence of viral RNA and its nucleotides produce additional viral RNA. The new viral RNA induces the host cell to produce the specific protein molecules required for its coat. During this process the inhibition of cell protein and RNA synthesis makes nucleotides, amino acids, and free ribosomes available for synthesis of viral components. The viral protein and nucleic acid are constructed from the same 20 amino acids and four nucleotides which occur in normal cells. Energy needed for energising various chemical reactions for synthesis of viruses comes from ATP of the cells. Thus viruses are completely dependent upon host cells for this replication.

In case of bacterial and animal viruses, the virion (virus particle) adsorbs specifically to receptors on the host cell and it is after this adsorption that the viral nucleic acid penetrates into the cell. Either free nucleic acid enters the cell (in the case of most bacteriophages) or whole virions (for all other viruses) enter and then release their nucleic acid. Replication of viral nucleic acid and synthesis of other viral constituents follow. These viral nucleic acid and protein constituents are made separately within the host cell and are then assembled into complete virions (virus particles) during the stage of virus maturation. Finally, the newly formed mature virions are released from the host cell. The method of release varies depending on the virus in question. Cells may be lysed releasing many virions, or virus particles may be gradually extruded from the living host cell. The details of replication in T-even phage which offer an excellent model for the general process of infection by a phage and destruction of the host cell by lysis, are incorporated under Bacteriophage.

Virus and host cells may often reach some sort of equilibrium with a minimum damage to the host allowing both to survive. The virus is then a **commensal.**

5. **Serological Tests.** Viruses respond to serological tests. Antiserum could be secured by injecting a partially purified virus extract to a suitable animal. It has been demonstrated that when juice extracted from a tobacco plant infected with the tobacco mosaic virus is injected into the blood of the rabbit in proper doses, the serum taken later from the animal causes flocculation when added in proper proportion to freshly extracted juice. Besides this, when strains of the virus differing in various disease symptom pattern are tested against the same serum, they react similarly. When a group of viruses give similar reactions to a given antiserum, they are regarded usually as strains of the same virus, whereas, when a virus gives no reaction, it is regarded as distinct from the virus used to produce the antiserum. This response to the serological test by the viruses may be utilized as a useful tool in determining virus similarities and dissimilarities.

6. **Cross-protection Tests.** If plants affected with virus, such as white mosaic, were inoculated with common mosaic virus, the symptoms of the latter did not become evident. It is generally assumed that when one virus, preceding another in the tissue, prevents or markedly impairs increases of virus subsequently introduced, it is an indication that the two viruses are closely related and are probably strains of a single virus. Such a protective phenomenon is known as the **cross-protection test.** The cross-protection test in conjunction with others may be used in the characterization of a virus.

7. **Synergistic Effect.** Two unrelated viruses when present separately incite relatively mild effects on the hosts. But the combined effect of both of the two viruses produces very severe effect on a particular host. The combined action of the two viruses is known as a **synergistic effect.**

8. **Special Features.** In tulips where a virus transmitted through the bulbs, it can cause a desirable colour variegation of the flowers. Bulbs are purposely maintained in a virus-infected state to obtain plants bearing flowers with variegation to fetch high market value. Again virus-infected aphids or aphids fed on virus-infected plants (for example, *Aphis fabae* and *Macrosiphum granarium*) have longer life span, greater egg laying capacity, earlier attainment of adulthood and more rapid breeding.

METHODS OF STUDY OF VIRUSES

Since viruses cannot be cultivated on nonliving media, they can be studied by cultivating in specific host cells, either bacterial, plant, or animal cells, depending on the nature of virus. From this, host range and symptoms produced by viruses on different hosts may be studied. Besides these, certain physical and chemical properties and other features of viruses like: thermal inactivation point, longevity in vitro, dilution of end-point, transmission characteristics, cross-protection reactions, serological reactivity may also be studied by using virus infected host tissue extracts. Viruses can also be critically studied by isolating them from virus infected plant tissue extracts in a suitable environment of pH, temperature, and ionic strength by treatment with organic solvents, filtration and centrifugation. Various techniques including gel filtration, electrophoresis, and density gradient centrifugation can be employed to further purify a virus. Again, for critical studies of viruses, application of methods including electron microscopic counting, plaque assay, quantal assay and hemagglutination produces good results.

Some of the tools that are used for the study of the structure and properties of viruses are: centrifuges (low-speed and high-speed), electron microscope, X-ray diffraction, equipments for serological studies and tissue culture techniques, and indicator plants.

For identification and grouping of viruses; characteristic features like: size and shape and relative percentage of nucleic acid and protein, host range and virus-symptoms, chemical and physical properties of viruses, and structural peculiarities of the virus particles are taken into consideration.

VIRUS STRAINS AND VIRUS MUTATION

Viruses that resemble one another in host range, symptomatology, physical properties, chemical composition, serological reactions, and particle morphology, but still differ in some small way, are called **strains.** They can differ from one another in the arrangement of amino acids in the subunits, amino acid composition, ability to be transmitted by different species of an insect, or in virulence or severity of symptoms produced on different host plants. Virus strains can be naturally occurring or can be produced by the use of mutagenic agents, all of which in some manner alter the RNA of the virus.

Again, due to mutation, a virus responsible for a particular disease, may exist in several or many slightly different strains. In viruses, mutation is spontaneous and the frequency of mutation can be increased by exposure to X-rays or other known mutagens. It is probable that mutations in viruses are similar in nature to the gene mutations.

New virus strains can also be obtained by hybridization between two strains when inoculated into the same host plant. The new strains recovered possess properties different from either of the two strains originally used for inoculation. These new strains are developed by recombination of the genetic material (RNA or DNA).

BACTERIOPHAGE

The **bacteriophages** or **bacterial viruses** or simply **phages** are widely distributed in nature. They are ultramicroscopic, but some (*Vaccinia*) are larger than small bacteria. A bacteriophage has a hexagonal to polyhedral head and a rigid tail which are almost same in length. The tail has a central core surrounded by a contractile sheath. The tail serves as an adsorption organ. The portion closer to the head has a projected structure called collar and the tail is terminated by six plates each of which again has contractile fibres. Bacteriophages occur in six morphological types (Fig. 23.4).

(i) A head with a rigid tail having contractile sheath and tail fibres (Fig. 23.4A).

(ii) A head with a flexible tail without contractile sheath. It may or may not have terminal appendages (Fig. 23.4B).

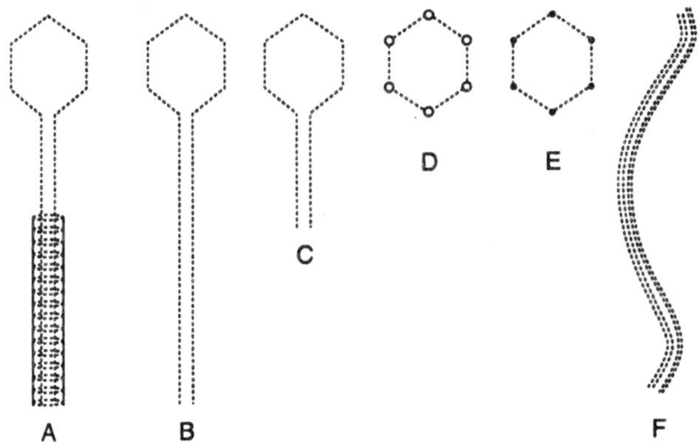

Fig. 23.4: Morphological types of bacteriophages. A. Head with rigid tail having contractile sheath. B. Head with flexible tail without contractile sheath. C. Head with short tail without contractile sheath. D. Head without tail, but with capsomeres. E. Head without tail and capsomeres. F. A long flexible tail without a head.

(iii) A head with a short tail, the tail is without contractile apparatus and may or may not have appendages (Fig. 23.4C).

(iv) A head with large capsomeres (individual protein subunits of capsids) at each apex of the hexagon; it has no tail (Fig. 23.4D).

(v) A simple head without capsomeres and a tail (Fig. 23.4E).

(vi) A long flexible tail without a head (Fig. 23.4F).

A bacteriophage has nucleic acid (DNA or RNA) and a protein coat. But a few phages contain both DNA and RNA. Again, some phages contain lipid and ribosome. In most of the DNA phages the nucleic acid of the virion is double-stranded, although in a few cases it is single-stranded. The RNA phages have single-

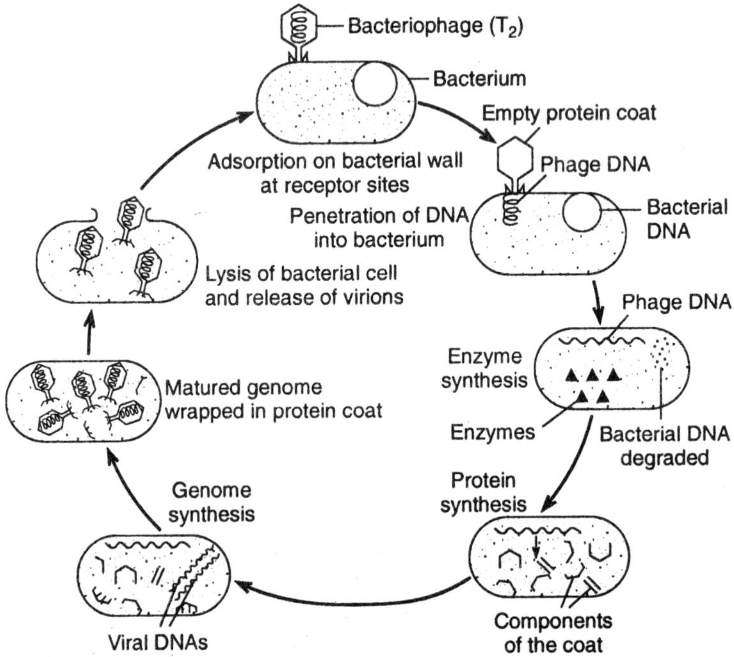

Fig. 23.5A: Lytic life cycle of bacteriophage (T_2)

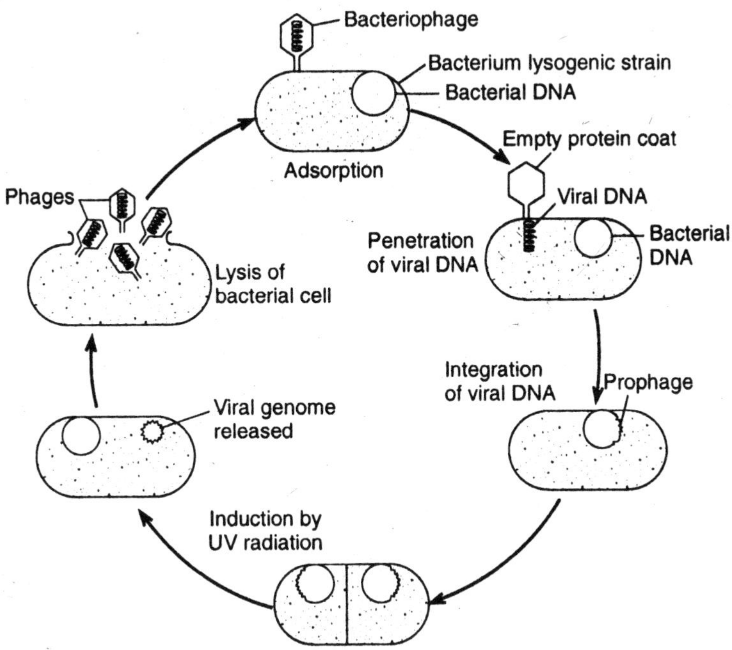

Fig. 23.5B: Lysogenic life cycle of bacteriophage (λ-phage)

stranded RNA in their virions. In all kinds of phages the nucleic acid is contained within the head. Bacteriophages are easily isolated and cultivated on young, actively growing cultures of bacteria in broth or on agar plates. The best and the most usual source of bacteriophage is coliphage—phage pathogenic for *Escherichia coli* cultures.

There are several steps of bacteriophage infection which ultimately leads to bacteriophage multiplication (Fig. 23.5A-B). These are: adsorption, penetration, replication of DNA, maturation, and release. When a phage particle comes in contact with a susceptible strain of *E. coli*, fibres at the end of the phage tail are the adsorption sites of the phage that bind to specific receptors on the bacterial cell wall (Fig. 23.7). Following adsorption, an enzyme (phage lysozyme) located in the phage tail degrades a small portion of the bacterial cell wall. The tail sheath of the phage then contracts driving the core of the tube into the cell injecting the DNA much as a syringe injects a vaccine (Fig. 23.7C). The protein coat of the phage remains outside the cell. Within minutes after penetration of phage DNA into the host cell, all transcription of RNA from the host chromosome ceases. The host DNA is degraded. All RNA subsequently synthesized is in RNA transcribed from the phage DNA. By this mechanism the phage subverts all metabolism of the bacterial cell to its own purpose—the synthesis of more phage. The host enzymes supply energy for phage replication through the breakdown of glucose, synthesize the subunits of protein and nucleic acid of replicating phage; and even participate in the synthesis of phage nucleic acid and phage coat protein. For the active replication of nucleic acid and the synthesis of viral proteins viruses require cellular ATP, ribosomes, transfer RNA, enzymes, and certain biosynthetic processes. In addition, the phage DNA codes for enzymes concerned with the assembly of the phage protein capsid. During the replication of any virus, the viral protein and nucleic acid components develop separately from each other. During maturation of phage, assembly of phage protein and phage

DNA takes place independently; so also the head and tail by stepwise processes. Once the head is formed, it is packed with phage DNA, after which the tail is attached, a new phage is assembled. The foregoing processes are repeated resulting in the formation of a large number of new phages. During the latter stages of infection period, another phage-induced enzyme, coded for the phage DNA, makes its appearance. This is the phage **lysozyme** that digests the host cell wall from within, resulting in the host cell lysis and release of the new phages (Fig. 23.5), which infect new bacteria host cells.

From the standpoint of infectivity and host relationships two types of bacteriophages are recognized: **lytic** or virulent and **temperate** (lysogenic) or avirulent. When lytic phages infect bacterial cells a large number of new phages are produced. They are released bursting bacterial cells. This is called a **lytic cycle.** As already indicated above in detail and also in Figs. 23.5A and B, the phage gets attached to bacterial cell wall by its tail fibres. The phage enzyme (lysozyme) at the tip of the tail-like tube dissolves the bacterial cell wall to form a pore through which the phage DNA is injected in the bacterial cell by contracting the tube. The phage DNA synthesises an enzyme known as nuclease which degrades the bacterial cell DNA and finally destroys it. The replication of phage DNA takes place along with multiplication of genes at the cost of bacterial ribosomes. Synthesis of new bacteriophage particles takes place within the phage infected bacterial cell from the components of the new phage particles. The newly formed phage particles are released by bursting the host bacterial cell wall. Thereby lytic cycle is completed. But in the temperate type of bacteriophage, the phage infection of bacterial cell is not apparent. There exists a state of **lysogemy**—a process in which the viral nucleic acid does not resume the functions of the same bacterial synthetic processes but becomes an integral part of the bacterial chromosome. As the bacterium reproduces, viral nucleic acid is transmitted on the daughter cells at each cell division. In the lysogenic state viral nucleic acid becomes inactive because of synthesis of repressor protein which causes repression of all the phage-genes and becomes integrated as one of the bacterial genes and is known as **prophage** or **provirus.** The cycle thus involved is **lysogenic cycle.** Under altered environmental conditions inhibition of repressor protein synthesis takes place resulting in the expression of lytic genes and prophage becomes a lytic phage and resumes lytic cycle.

There are three possible ways of inheritance of bacteriophage DNA: (i) the phage chromosome inserts itself into the host chromosome, after which it can be passively replicated and distributed at cell division as part of the host DNA; (ii) the phage chromosome establishes itself independently, it replicates and is distributed in daughter host cells like the host chromosome keeping its separate entity; and (iii) the phage chromosome replicates separately from the host chromosome, but the distribution of phage chromosome takes place in the daughter cells only when the number of newly replicated chromosome is large.

Actinophages. Particular viruses attack actinomycetes and have been found in *Actinomyces bovis* and *Nocardia farcinica.* Such viruses—actinophages—are abundant in the soil. Some possess tail and in others there is no tail at all.

Bacteriocins. Both Gram-positive and Gram-negative bacteria produce bacteriocidal substances known as **bacteriocins.** The formation of bacteriocins is due to genetic determinants, the **bacteriocinogens** which are not integrated into the bacterial chromosome and are thus plasmids. Bacteriocins have relationship with phages and are considered as produces of defective phage genomes. The intimation of bacteriocin synthesis kills the producer cell. Bacteriocins take their specific names from the organisms producing them, e.g., **colicins** from *Escherichia coli*, **pyocins** from *Pseudomonas aeruginosa*, **megacins** from *Bacillus megaterium*.

Cyanophages. These are viral agents that attack a wide range of blue-green algae. They were first discovered by Safferman and Morris in the year 1963. They are very similar to bacteriophages both in structure and infection cycle. Cyanophages are named according to their known hosts, for example LPP-1 is specific on hosts: *Lyngbya, Phormidium, Plectonema.* The nucleic acid of cyanophages are double-stranded DNAs.

Mycophages (Mycoviruses). These are viruses attacking fungi. Mycophages were first discovered in *Agaricus bisporus* in 1957 by Sinden. Subsequently in 1960 mycophages were discovered in *Penicillium chrysogenum, P. cyaneofulvum, P. funiculosum,* and *P. stoloniferum.* The viral particles of the mycophages are polyhedral or spherical with diameters of 33 to 41 nm. They all contain double-stranded RNA. Young apical regions of hyphae are generally free of virus particles; older regions contain many particles enclosed in vesicles.

NATURE OF VIRUSES

There has been some argument as to whether viruses should be considered to be living or non-living. True, viruses are individual organic compounds whose chemical composition resembles protoplasmic constituents. They behave as microorganisms only when in association with the complex mechanisms of living cells. Viruses reproduce in a host cell and are capable of mutation. They contain the true essence of life by the possession of an extremely potent complement of genes. In their ability to reproduce themselves in living tissues they also resemble microorganisms. With the messages contained in the single strand of the nucleic acid, the virus is able to divert the enzyme systems of the host cells into new pathways and synthesize more virus particles instead of host substance. In this respect viruses seem to resemble self-duplicating genes and chromosomes, but they differ in that, they are able to penetrate some unicellular of multicellular hosts from the outside.

But on the other hand, viruses do not possess cellular structures. They by themselves do not carry on respiration, lack capacity for independent metabolism, and do not multiply by classical growth and fission methods from pre-existing virus particle. Those viruses which have been prepared in pure crystalline form and found to be nucleoproteins are not living organisms in the ordinary sense. Again, the virus crystals are chemically inert and can apparently be kept indefinitely without significant change.

Again, viruses are often accepted as molecules having ability to duplicate themselves. But Stanley and others suggested the significant difference between viruses and other molecules pointing out that a virus comes to life the moment it infects a cell. Besides this, the term 'molecule' implies a precise knowledge of the structure of a compound.

Viruses are thus ultra-microscopic disease-producing particles of organic matter which can multiply only in living plants and animals and are responsible for a large number of important plant and animal diseases.

ORIGIN OF VIRUSES

Broadly speaking, only three general hypotheses of the origin of viruses are taken into consideration.

(i) The ancestors of viruses were at one time cellular organisms. As a result of parasitic existence in other cells, they gradually lost more and more of their own cellular machinery until they eventually became reduced to their present form.

(ii) The ancestors of viruses were once free-living pre-cellular forms of life, which managed to survive after the evolutionary emergence of cellular organisms only by becoming parasitic on them.

(iii) The viruses have not evolved from organisms, either pre-cellular or cellular, but have arisen from detached fragments of the genetic material of cellular organisms. These genetic fragments, as a result of detachment from the rest of the genetic system, acquired the ability to multiply more rapidly than the other constituents of the cell, and their unregulated growth caused disease and death of the cell. Liberated after cell death, the genetic fragments were able to ensure their own perpetuation by entering adjacent healthy cells and again multiply there. Originally passed from cell to cell in the form of nucleic acid, they eventually acquired the capacity to direct the simultaneous synthesis of the infected cell of a special protein, which served to enclose the nucleic acid fragments, and thus made their transfer from cell to cell a much less hazardous operation.

The above hypotheses have not yet been supported by factual information.

1. General characteristics of viruses.
2. Historical review of viruses.
3. Nature of viruses—how viruses differ from Bacteria.
4. Occurrence and methods of transmission of viruses.
5. Symptoms inducted by virus infection.
6. Properties of viruses.
7. Plant viruses.
8. Virus life cycle.
9. Subvirus pathogenic erotities.
10. Lamda (λ) virus and circulation of its name.
11. Signification of lysogeny.
12. Virus multiplication cycle. One step multiplication of curve of T_2 phage.
13. Mechanism of molecular lysogeny and induction of lambda (λ) phage.
14. Tobacco mosaic virus.
15. Classification of viruses.

PLANT VIRUSES

Plant viruses, in general, do not have any specific mechanism for attachment of host cells. They are transmitted passively by several means. Tobacco Mosaic Virus (TMV) and some other plant viruses are carried by wind, rain drops, gardening equipments and grazing animals to host plants. They enter into leaves damaged by mechanical injuries, like abrasion. Some other plant viruses are transmitted to healthy hosts through seeds, tubers and other propagules. The tobacco mosaic ring spot virus is carried by nematodes infecting roots. Tobacco necrosis virus is transmitted through parsitic fungi. Potato X-virus enters through haustonia. However, the most important agent of transmission of plant viruses is the insects feeding on plant cell sap. Among these insect **vectors,** the most important ones are the aphids and grasshoppers. Some beetles also can propagate plant viruses.

Host Entry

The most common mode of entry of plant viruses is that the virus particles (virions) adhering to the feeding organs of insect. Vectors are introduced passively when they feed on leaves or stems of healthy plants.

As already stated in the viruses life cycle, the replication of viruses takes place by two types of cycle—**Lytic cycle** and **Lysogenic cycle.** Based on interaction of phages with bacterial cell, they have been distinguished into two types—**Virulent** or **Lytic** and **Avirulent** or **temperate phages.** The virulent phages perform the lytic cycle and the avirulent or temperate phages perform the lysogenic cycle. Many bacteriophages show a different form of parasitism. **Virulent** types include the phages which infect the hosts leading to a lytic cycle resulting in immediate lysis of host cell, whereas **temperate** phages do not cause any immediate host cell lysis of the infected host cell. But enter into a symbiotic relationship, called **lysogeny.** The injected DNA persists in the host cell as a **prophage** for generations, and under suitable conditions the viral DNA in the **lysogenic** cells multiply like virulent phages leading to lysis of most cell and are liberated to attack new host cells. The temperate phages when they infect appropriate bacterial strains may show the alternative responses. One of the options is to carry out a lytic cycle like a virulent phage, and the other is to enter into a lysogenic state. When the lysogenic host cells are disrupted artificially, no infectious phages are recovered which shows that the phage must be present in a non-infectious state. A characteristic feature of the temperate phage is that they produce **turbid plaque** on a lawn of the lysogenic bacterial host. In contrast, the lytic virulent phages produce **clear plaques.** A turbid plaque is produced because only some cells of the host are lysed at any given time producing progeny phages, while the majority of the population enters into lysogenic state.

Another important feature of lysogenic bacteria is that they are immune to infection by the same or related bacteriophages. Sometimes, lysogenic hosts acquire new properties which are absent in uninfected bacteria. This is known as **lysogenic conversion.**

The most widely known temperate phage is the lambda (λ) phage infecting *E. coli* K12. It is a doubt stranded DNA virus with an icosahedral head. It is a double-stranded virion is linear with 12 nucleotide long single-stranded cohesive ends. The cohesive ends are complementary and with the help of these single-stranded segments. The linear molecule can form a circular DNA molecule with two staggered nicks which are sealed by DNA ligase. The virus has long tail without a contractile sheath. The linear virion DNA produces a covalently closed circular DNA soon after entering the host cell. The structural feature of lambda (λ) virus is presented in Fig. 23.6.

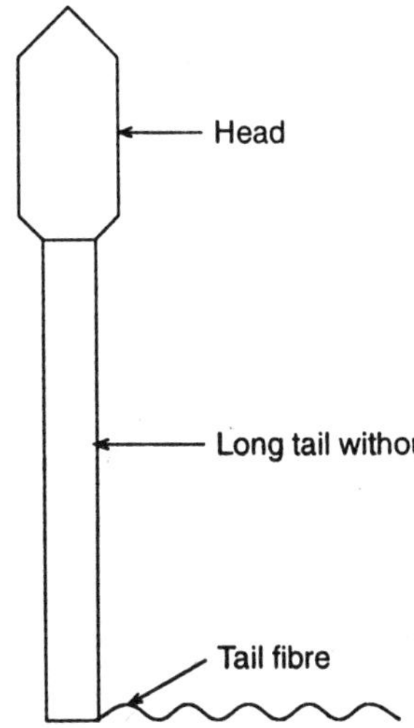

Fig. 23.6: Characteristic features of lambda (λ) phage

VIRUS LIFE CYCLE

Being non-cellular the viral particle has no power of independent metabolism and reproduction like other organisms. The replication of virus takes place by two types of cycles: (i) **Lytic cycle** and (ii) **Lysogenic cycle.**

Both the cycles are based on interaction of phages with bacterial cell they have been distinguished as **virulent** or lytic and **avirulent** or **temperate phages.** The virulent phages perform the lysogenic cycle. A portion of lysogenic cycle is like lytic cycle.

1. **Lytic cycle.** This type of cycle is seen in T-even phages (T_2, T_4, etc.) which attack *E. coli.* The lytic cycle consists of five steps: (i) Adsorption, (ii) Infection, (iii) Synthesis of phage components in host cell, (iv) Formation of new phage particle, and (v) Liberation of new phages from the host cell.

(i) **Adsorption.** The interaction between the phages organella—the tail and the receptor site of the host cell is called adsorption. The adsorption is facilitated by the negative charged carboxyl groups on the host surface and the positively charged amino-group of protein present at the tip of phage tail. In T-even phages, the tip of the tail fibre first attaches to the cell surface. The tail fibre then allows attachment on the host surface by an irreversible attachment.

(ii) **Infection.** After adsorption, the phage particle secretes enzymes while hydrolyzes murin complex of the host cell and forms pore on the host cell.

The sheath of the tail then contracts and pushes the central tubular part, i.e., **core** (tube) of the tail through which nucleic acid is released in the host cell. The empty protein shell of the phage is called **ghost** with receptor remaining attached even after release of nucleic acid. Once the host cell receives the nucleic acid of a phage, it becomes resistant to the other phages.

(iii) **Synthesis of phage components in host cell.** Once the phage nucleic acid forces the entry inside the bacterial cell, it supresses the synthesis of bacterial protein and directs to synthesise the proteins of the phage particle.

The empty protein shell of the phage is called **ghost.** The DNA of phage replicates by **semi-conservative** process. Majoriy of the DNA acts as template for its own synthesis and the rest is used as template for the synthesis of the viral specific m-RNA by utilizing, the RNA-polymerase of the host. The newly formed m-RNA direct the host cell to synthesise the proteins which are used to build up the protein coat of the phage particle. Amount at the end of replication of phage nucleic acid, a

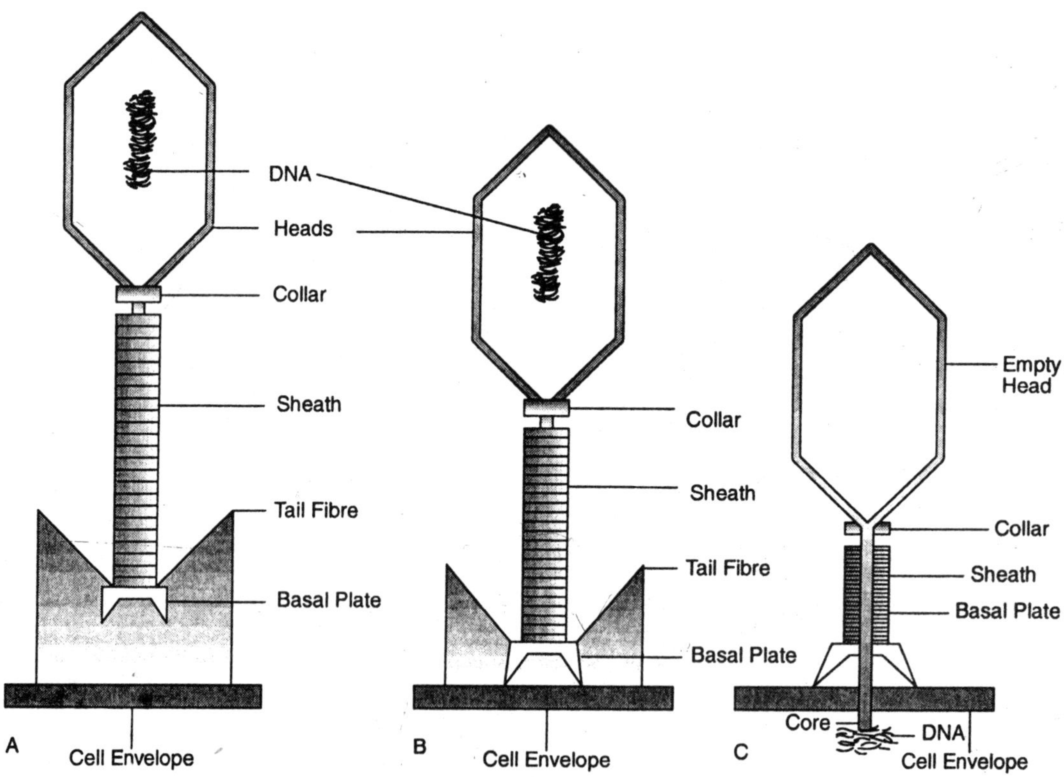

Fig. 23.7: Diagrammatic presentation of mechanism of attachment and introduction of DNA into host cell in a T-even bacteriophage; A. The tail fibre is affixed to the surface of host cell. B. The basal plate is brought into contact with the cell surface. C. The sheath contracts driving the tubular core through the cell wall and the DNA from the head enters the host cell through the core

protein, the **phage lysozyme** is synthesised (Fig. 23.7).

Infection of host cell

The first step in infection is a highly specific interaction of the phage and adsorption organelle, such as the tail, with **receptors** on the surface of the host cell, which leads to attachment of the phage to the **cell-adsorption.** Then the DNA is released from the capsid and enters the cell.

1. **Adsorption.** The initially reversible attachment of the phage to the receptors rapidly becomes irreversible (i.e., the phage cannot be washed away). All virions have a specialized structure for adsorption. In the T-even coliphases it is a complex structure—the **tail fibre.** The tips of the fibres attach first and reversably to the cell surface and are followed by the **tail pins** which attach irreversibly. The adsorbed phage acquires its head perpendicularly to the cell wall. The host cell **receptors** are often complex polysaccharides with phage-binding and antigenic speciality. With capsulated host cells, enzyme in the tip of the phage tail hydrolyses specific capsular polysaccharides, digging a tunnel through which the phage reaches the cell wall. The acceptor for phage is a mature transport protein.

2. **Separation of nucleic acid from coat.** After the nucleic acid takes entry in the bacterial cell, it suppresses the synthesis of bacterial protein and directs to synthesise the proteins of the phage particle. Almost at the end of the replication of phage nucleic acid, a protein, the **phage lysozyme** is synthesised. The phage particles are formed by the assembly of nucleic aid and protein.

Subviral Pathogenic Entities

Although viruses vary widely in size and in the nature and size of their genomes, but biological entities are much smaller than the viruses that have been found to cause various diseases of plants, animals and humans. Among these agents are **viroids, virusoids** and **prions.** Viroids and virusoids cause disease of plants, and may also cause diseases of animals or humans. Prions, on the other hand, cause neurological disorders in animals and humans.

1. **Viroids.** They differ in having no coat enclosing the nucleic acid which is small **single stranded circular** RNA. Viroids are among the smallest pathogenic agents which can replicate within infected hosts with the help of host enzymes, because the viroid RNA does not code for any proteins. Replication takes place mainly in the nucleus of infected cells. The naked RNA can infect many plants causing various diseases. Some of these diseases result in heavy economic loss. Although, the viroid RNA is single-stranded and generally circular, it possesses considerable amount of secondary structure due to intrastrand lease-pairing with intermittent looks produced by unpaired bases.

The mechanism of pathogenicity of viroids is not well-understood. Among the important diseases caused by viroids are—**the potato spindle tuber, Cucumber pale fruit and Chrysanthemum stunt.**

2. **Virusoids.** Like viroids, virusoids are also small circular single-stranded RNA. But they differ from viroids in their inability to replicate by themselves and in thier requirement of a **helper** virus. Hence, viroids occur independently in the infected cell, virusoids are always associated with a helper virus. The helper virus associated with virusoids are—**velvet tobacco mettle virus, Solanum nodiflorum virus, Lucerine transient steak virus** and others. The virusoids may occur in infected host cells either as free circular molecules or within the capsid of the helper virus. The length of virusoid RNA is more or less of the size of RNA containing nucleotides.

3. **Prions.** Prions were di.covered in 1982 by Stanly Prusiner as the causative agents

of infectious disease of sheep as **scrapie.** The most significant feature of prions is that they have been claimed to consist of only protein. Hence the name **prior** having proteinacious infectious particles. Prions can stand heat, radiation and chemical treatments. They are susceptible to some protein hydrolyzing enzymes, but are resistant to nucleases which show that they do not contain nucleic acid, either DNA or RNA. This raises an important question, how can infectious agent containing proteins direct its own replication, because an universal characteristic of all cellular organisms as well as viruses or even viroid, store **genetic** information in nucleic acids, DNA or RNA.

4. **Viroids.** While working with potato spindle tuber disease (PSTV), T.O. Dietrer in 1971 was the first to isolate and characterize the causal organism, **Potato spindle tubur viroid** (PSTV). These are smaller than viruses and he coined the name the term **viroid** to those subviral partices. Since then large number of plant diseases have been detected by viroids. Only two animal diseases such a **Scrapie** disease and **Kuru** disease are continued to be caused by viroids.

Viroids are small, low molecular weight, circular, single stranded RNA molecule. The viroids do not have capsids (i.e., protein coat) around the RNA molecule. The viroid is not yet clearly known.

The virus RNA is enclosed in protein coat. While viroids lack any protein coat they remain free RNA. **Virusoid** is the synonym of **viroid.**

5. **Mycoplasma.** The Mycoplasmas are the smallest among the known aerobic prokaryotes. They were first discovered by L. Pasteur in 1843, during his work on the causal organism of pleuropneumonia of cattle. Mycoplasm as are commonly found in soil, hot spring, savage water and also in plants, animals, including humans. Based on nutritional requirement, mycoplasmas are **Mycoplasma, Acholeplasma** and **Thermoplasma.**

Some Scientist included Mycoplasma in the Group—Bacteria. Mycoplasmas lack cell wall. The mycoplasma cell is covered by cytoplasmic lipoprotein membrane—which covers the cytoplasm containing nucleoplasm-like structure and ribosome. The genetic material is composed of DNA and RNA. Mycoplasmas are non-motile, but some members show gliding movements. Since mycoplasmas lack cell wall, so their cells are readily deformable showing irregular and variable shapes. Mycoplasmas are sensitive to antibiotics, chloramphenical, streptomycin, erythromycin, but are insensitive to penicillin, ampicillin due to the absence of cell wall.

SATELLITE VIRUSES

Satellites are sub-viral agents, composed of nucleic acid molecules that depend for their productive multiplication on co-infection of a host cell with a helper virus. When a virus is dependent on another virus for its proper replication in a host cell, it is named **satellite viruses** and such a solution is **satellism.** Here the nucleic acid is not encapsulated and the virus depends for encapsulation on a helper virus. Satellite viruses (SV) are single stranded RNAs and single stranded DNAs.

Structure of lambda (λ) phage and circulation of its genome

A lambda (λ) phage is a double stranded DNA virus with an icosahedral head and a long tail, but without any counter-active sheath. The tail has a thin **tail fibre** at the distal end. The double stranded DNA of the virus is a linear thread-like structure with nucleoides long, single stranded cohesive ends. The cohesive ends are complementary to each other. The single stranded complementary regions of both ends come to each other and form a circular double stranded DNA. The ends of both the threads join by DNA ligase. Circulation of DNA takes place after its entry inside the host bacterium.

Activity of the Circular (λ) DNA

After circulation of lambda (λ) DNA inside the bacterial cell (*E. coli* K12), it functions in either of two alternative ways:

(i) Through lytic cycle the DNA of (λ) phage undergoes transcription, assembly of progeny and disease by lysis of the short bacterium.

(ii) Through lysogenic cycle, the DNA of (λ) phage is integrated with bacterial DNA at a specific site to become a **prophage** and thus the infected bacterium becomes lysogenic, thus the prophage replicates along with bacterial DNA. During this process the genes of phage DNA controlling lytic cycle become inhibited by a repressor—the (λ) repressor. During cycle, the (λ) phage initially gets attached to the bacterium with the help of tail fibre. It then injects its DNA thread into the host bacterium (*E. coli*). After entry, the ds-DNA thread is converted into a circular DNA. Replication is initiated near the origin. The daughter DNA molecules synthesised and developed from the parental DNA undergoes replication by rolling-circular model and develop long thread-like **Concatemeters** which contain several (λ) genomes. The proper lengths with single-stranded cohesive ends are cut off by enzymes and are packed into the heats of (λ) phage.

Transcription of the phage DNA develops coding messages that help to form the capsid and their proteins of phage. It is initatiated by host polymerase which binds with the two promoters on (λ) DNA that transcribes two different strands in opposite directions.

After the synthesis of sufficient number of virion components they assemble and release by lysis of the host bacterium. The released (λ) phages infect new bacterium and continue another lytic cycle or may enter into lysogenic cycle.

In lysogenic cycle, the (λ) phage after attachment with the host bacteria (*E. coli* K12) pushes to its DNA inside the cytoplasm of host cell. The phage DNA is then circularized in the usual process. The Cl-gene of (λ) phage then produces the (λ) repressor which inhibits the action of gene controlling phage, phage multiplication and lysis. Thus the operation of lytic cycle stops and ensures clearance to lysogenic state.

The circular (λ) phage becomes integrated with bacterial DNA. In course of time, the lysogenic bacteria are able to produce phage particles through the process of induction. Due to induction, the inserted DNA of (λ) phage—called **prophage** and thus the infected host bacterium becomes lysogenic. During this process the genes of phage DNA controlling, lytic cycle becomes inhibited by a repressor—the (λ)-repressor.

During lytic cycle initially the (λ) phage gets attached to the bacterium with the help of tail fibre. The (λ) phage then injects its DNA thread into the host bacterium (*E. coli* K12). After entry, the **ds-DNA** thread is converted into a circular DNA. The circular DNA of phage then gets attached to the membrane at the specific sit and starts replication. Replication is initiated near the origin and progress in both directions symmetrically and then terminates when the two replicating meet each other. The daughter DNA molecules synthesized and developed from the parental DNA, undergo replication by '**rolling circular medel**' and develop long thread-like **concatemeters.** Those concatemeters contain several (λ)-genomes.

Transcription of the phage DNA develops coding messages that help to form the capsid and other proteins of phage. After the synthesis of sufficient number of virion components, they assemble and release by lysis of host bacterium.

Generally, the virus continues lytic cycle within a number of infected cells, but major portion enters into lysogenic relationship and continues the lysogenic cycle.

During lysogenic cycle, after attachment of (λ) phage with the host bacterium (*E. coli* K12), the phage pushes its DNA inside the cytoplasm of the host cell. The phage DNA is then circulated in usual process. The Cl-gene of (λ) phage then produces the (λ) suppressor and binds with two different operators of its own genome which are invoved in the initiation of transcription of phage controlling the phage multiplication. Thus the operation of lytic cycle stops and ensures clearnce to establish the lysogenic state.

The circular (λ) phage then becomes interested with this bacterial DNA. The gene of (λ) phage produces an enzyme **integrase** which helps in the integration. During this process the circular DNA of (λ) phage is inserted as a linear DNA into the bacterial DNA. The inserted DNA of (λ) phage is called **prophage.** The (λ) phage DNA remains with bacteria DNA for a long time and replicates along with bacterial DNA. Due to induction, the prophage is released from bacterial chromosome and again becomes circular. The circular (λ) phage then proceeds through lytic cycle and develops new phage particles.

Significance of Lysogeny

The lysogeny has an important role in the transfer of genetic material from one bacterium to the other. The temperate phage acts as agent for gene transfer through the process known as **specialized transduction.** In this process, when a temperate phage comes out from a host bacterium as prophage, it may include a part of bacterial DNA along with its DNA by mistake. After lysis, the newly developed phage can infect a new host bacterium transferring a portion of previous bacterial DNA to the newly infected bacterium, thereby recombination takes place.

One-Step Growth Curve Viral Multiplication Cycle

The process of multiplication, initiated by the penetration of the nucleic acid, involves many sequential steps, which end in the release of the newly synthesized progeny virions. This sequence is called the **multiplication cycle. One-step multiplication of curve of T_2 phage (one-step growth curve)** is as follows.

Susceptible bacterium (*E. coli*) is mixed with bacteriophage (T_2) and is allowed for period to get attached with the host cell. The culture is othen diluted much, thereby any virus particle—if released by lysis of host cell will not be able to infect new cells. The number of phage particle released from bacteria is subsequently determined at different time intervals by plaque counts.

The result is plotted as bacteriophages released from bacteria versus time, indicating several distinct phages. Growth curve of this type consists of two periods.

1. **Latent period.** This period immediately follows phage addition and thus there is no release of virions. This is the shortest period required for virus reproduction and release.

2. **Eclipse period.** This is the early part, the bacteria do not contain any complete infective virions. This can be shown by lysing them with chloroform. The initial stage of the latent period is called **Eclipse period.** At this stage the virions are not detectable even by disrupting the bacteria by treating with chloroform. At the end of the eclipse period, the phage virions increase in number within the host bacterium and the host bacterium is prepared for lysis.

3. **Intracellular accumulation period (Rise period).** After the latent period, the host cells rapidly lyse and relase all infective phages, known as **rise period or burst,** host cells lyse and release infective phages. At the end of this period, no more viruses are liberated.

The term burst is the number of viruses released from single infected bacterium. The total number of bacteriophages released can be used to calculate the **bust size.**

The experiment is carried out in enriched medium at 37°C and growth is reached in approximately 30 minutes.

Majority of the DNA acts as a template for its own synthesis and the rest is used as template for the synthesis of viral specific m-RNA utilizing the RNA-polymerase of the host. The newly formed m-RNA directs the proteins which are used to build up the protein coat of phage particle. Almost at the end of replication of phage nucleic acid, a protein, the **phage lysozyme,** is synthesized.

4. **Formation of new phage particle.** The new phage particles are formed by the assemblage of nucleic acid and protein. This process is called **maturation,** which is controlled by viral genome.

5. **Liberation of phages from the host cell.** Large number of new phages developed during the Lytic cycle. The new phage DNA secretes lysozyme (an enzyme) that cause lysis of host cell walls as a result of which the new phages are liberated.

6. **Lysogenic cycle.** During this cycle, (λ) phages attack *E. coli.* The phage involved in this cycle is called **temperate phage,** and the bacterium involved is the **lysogenic strain.** The entire process is **lysogeny.**

At first the phage is adsorbed on the wall of the host bacterium and its DNA is injected into the host cell. Here phage DNA, like the lytic cycle, does not take over the protein synthesis mechanism of the host cell, instead, it becomes integrated with the nucleoid of the host genome. This integrated phage DNA is called a **prophage.** Thus the new composite genome replicates as one unit. This composite genome then multiplies for indefinite number of daughter lysogenic bacteria. After a number of generations, the viral genome gets detached from the composite genome and releases in the cytoplasm. This dissociation is called **induction.** The viral genome then enters the lytic cycle and forms temperate phages. They are released by lysis of wall of the host bacterium.

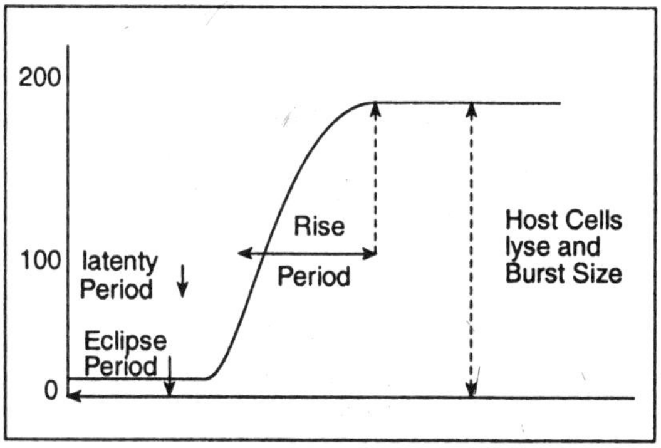

Fig. 23.8: Diagrammatic presentation of one-step multiplication curve of T-even phage

Mechanism of molecular lysogeny and induction of lambda (λ) phage

Virus particles have no independent metabolism and reproduction like other organisms. Hence they need help of others for this multiplication. The process of host infection, multiplication and lysis of host bacterium cell is designated as **lytic cycle.** But in others the mechanism is different. They show different mode of parasitism—infection, integration with host genome, and get separated from the host genome causing multiplication along with host genome. Later it gets separated from the host genome, causing multiplication and comes out through lysosis of host cell, known as **lysogenic cycle.** The phage involved in the process is **temperate phage.** The bacterium is lysogenic strain. The injected phage DNA integrated with the host genome is **prophage** and the entire process is known as **lysogeny.**

TOBACCO MOSAIC VIRUS (TMV)

Structure. The TMV is rod-shaped. Each rod consists of two parts: the protein coat (capsid) and nucleic acid which is a single-stranded RNA molecule.

Host Entry. The TMV infects a tobacco plant, tomato and many other plants mostly belonging to the family solanaceae. It enters the host plants through (i) mechanical wounds caused by insects or (ii) through infected pollen grains. It is introduced into the host plant (cells) by insect vector like *Myzus pseudosolani, M. circumflexus*, and grasshoppers. It may also be transmitted through rubbing, transplanting and during tobacco cultivation.

Host Infection. During host infection, usually the entire virus particle (iron) enters into the host cell. Just after entry, the protein coat (capsid) of the virus particle degenerates, and thus the nucleic acid (RNA) gets free in the host cell. The RNA then induces the formation of RNA-polymerase (also called RNA-replicates or RNA-synthetases) by the host cell. The viral RNA then acts as a template and with the help of RNA-polymerase, it synthesizes additional RNA, which is not the copy of the viral RNA but a **mirror image,** which remains temporarily connected to the viral strand. Thus, the two RNA threads form a double-stranded RNA that soon separates into the original virus RNA and then usually forms strand, i.e., mirror image strand which behaves as a template for the synthesis of more viral RNA in the insect tissue and are carried to the salivary gland before being introduced into the host plant.

Replication of plant viruses. One of the best known studied plant viruses is TMV. It has been cultivated in plant protoplast cultures. Some viruses, like the wound-tumour virus can also be grown on cell cultures of insect vector:

Assembly of TMV occurs in the cytoplasm of host cells. When a large number of capsomer proteins and ss-RNA have accumulated in the infected cell, the virions are assembled spontaneously in a highly organized manner. The capsomeres are assembled to form disks made up of two layer of helically arranged protomers. The assembled virions of TNV can produce microscopically visible intracellular accumulation of viral aggregates often in crystalline form.

Synthesis of viral protein

The new nucleic acid (RNA) produced induces the host cell to synthesize the protein molecules that will be the protein subunits required to develop their own capsids. Only a portion of viral RNA strand is required to form the viral protein. The RNA of TMV consists a large number of nucleotides to code the sequence of amino acids in protein subunits.

During synthesis of viral protein, the portion of viral RNA utilizes the amino acids, t-RNAs and ribosomes of host. Some enzymes may activate or initiate the chemical reaction and thereby affect the process.

Formation of new particle

The newly synthesized protein subunits then arrange in an organized way on the viral RNA strand and then both are assembled together and form a complete virus particle.

Site of synthesis of viral RNA and Proteins.

After the host infection the free RNA moves into the nucleus and perhaps the nucleolus, where it replicates itself. Proteins are synthesized in the cytoplasm and assembly of viral RNA and protein also take place in the cytoplasm to form viral particle. It takes about 10 hours from the time of host infection for the formation of new virus particle. The virus particle so formed, may exist either singly or in groups. They may form crystallines or amorphous inclusion bodies within different cells (nucleus, nucleolus and cytoplasm).

Viruses are separated as: 1. Geminiviruses, 2. Bacteriophage.

1. **Geminiviruses** are plant viruses having single stranded circular DNA. The particles occur in pairs, and usually found in the nucleus. One molecule of DNA is present per pair of particles, e.g., Banana bunchy tap virus (BBTV).

2. **Bacteriophage.** Details are discussed earlier.

CLASSIFICATION OF VIRUSES

Depending on the host viruses are divided into **animal viruses, plant viruses, bacterial viruses (Bacteriophage)**, etc.

According to International Committee on Taxonomy (ICTV) viruses have been generally agreed that the viral families will have the ending **viridae**, and viral genus will have the ending—**Virus.** For the sake of convenience, the classification of viruses infecting three hosts, viz., animals, plants and bacteria has been treated separately.

1. **Animal viruses.** Animal viruses can be classified into DNA viruses and RNA viruses.

2. **Plant viruses.** Classification of plant viruses, in general, is not well-organized as Animal viruses or Bacterial viruses. They are mainly classified into several groups than families. Some plant viruses have been placed in the same families as those of animal viruses. For example, the families **Rhabdoviridae** and **Reoviridae** and **Bunyaviridae**, including both animal and plant viruses have been assorted into groups without assigning to any family.

A brief outline classification of the plant viruses are: 1. DNA Plant viruses, 2. RNA Plant viruses.

3. **Bacterial viruses.** The characteristics used for classification of bacterial viruses, attacking bacterial cells are similar to those used for higher organisms. These include the nature of nucleic acid, strandedness of the nucleic acid, the symmetry of the nucleocapsids, the presence of an envelope and the nature of the host. Many bacterial viuses have additional feature not present in animal and plant viruses. It is the presence of **tail** in some members, the tail may be covered by a sheath which may be contractile. Many bacterial viruses have more complex structure than other viruses. The tail has a **binary symmetry.** The main body of the virus consists of an ico nucleocapsid, called the **head** and the **tail** having a helical symmetry. Presence of an envelope is rare among the bacterial viruses. There are also some **filamentous** bacterial viruses. An outline classification of bacteriophages is as follows:

I. Bacterial phages with DNA genome.

II. Bacterial phages with RNA genome.

Practical classification

Viral taxonomy is still in progressive stage and many are still to be undivided.

For convenience, viruses are separated on the basis of hosts infected.

(1) Vertebrates and other hosts, (2) Vertebrates only, (3) Invertebrates, (4) Plants, (5) Fungi, and (6) Bacteria.

Some viruses cross such boundaries.

Fenner (1976) distributed different families of viruses among each of the above said host groups of viruses.

A comprehensive account of classification and nomenclature of viruses is given by Matthews (1982) and the following account is based on this paper.

1. **Viruses multiplying in vertbrates and other hosts**

Families: Iridoviridae; African swine fever virus; Parvoviridae; Reoviridae; Birnaviridae; Picornaviridae; Togaviridae; Flaviviridae; Rhabdoviridae; Bunyaviridae.

2. **Viruses multiplying only in vertebrates**

Families: Herpesviridae; Adenoviridae; Papoviridae; Hepadnaviridae; Coronaviridae; Calicivirus; Arenaviridae; Paramyxoviridae; Orthomyxoviridae; Filoviridae; Retroviridae.

3. **Viruses multiplying only in invertebrates**

Families: Baculoviridae: Polydnaviridae: Nudaurelia β virus group; Nodaviridae.

4. **Viruses multiplying only in plants**

Families: These include Caulimovirus, Geminivirus, Tomato spotted wilt virus group, Machlovirus, Necrovirus, Sobemovirus, Tombusvirus, Tymovirus, Bromovirus, Cucumovirus, Ilavirus, Alfalfa mosaic virus group, Comovirus, Dianthovirus, Nepovirus, Pea enation mosaic virus group, Tobamovirus, Tobravirus, Hordeivirus, Protexvirus, Potyvirus, Carlavirus, Closterovirus.

5. **Viruses multiplying only in fungi**

Families: Totiviridae; Partitiviridae.

6. **Viruses multiplying only in bacteria**

Families: Myoviridae; Styloviridae; Podoviridae; Tectivitidae; Plasmaviridae; Corticoviridae; Microviridae; Inoviridae; Cystoviridae; Leviviridae.

7. **Unclassified, virus-like and suspected viruses**

(a) Viruses such as plant satellite viruses and free RNAs, such as viroids.

(b) Scrapie of sheep, Creutzfeldt—Jakob and Kuru of man causing disease agents.

(c) Agents of diseases of central nervous system such as schizophrenia, presenile dementia, multiple sclerosis, Alzheimer's disease.

CLASSIFICATION AND IDENTIFICATION OF VIRUSES

In spite of continuous efforts made by Johnson (1927), Smith (1937), Fawcett (1940), Holmes (1939, 1948), Valleau (1940), Lwoff, Horne and Tournier (1962), Pereira (1966), Tourinier (1966), Hansen (1956, 1968), Thornberry (1968), Gibbs (1969), Martyn (1968, 1970), Harrison (1971) and many others, there is no final agreement about nomenclature and classification of viruses as because viruses are genetically variable and new strains differ in host range, virulence and other characteristics arising from different arrangements of the nucleotides in the nucleic acid molecule.

A specific virus exhibits fixed characteristic properties. These include the size, structure, and chemical composition of the virus particles, the host range, the tissue specificity, and the nature of the infection caused. When the properties of a large number of different viruses are examined, it is found that they fall into groups, each characterized by the possession of a number of properties in common. The major groups of viruses may broadly be separated on the basis of characters like: type of nucleic acid present in the virus particle, nature of host and disease induced, properties of virus particles (shape, size, etc.), and other related characters. Although nothing is known about the origin and relationships of the viruses, it is tempting to imagine that these groups are **natural** ones, each of which unites a series of viruses that are genetically related to one another. Some have gone so far to create families, genera, and species for viruses, conferring on individual viruses Latin binomial designations, just as if they were cellular organisms. Since a homology between viruses and cellular organisms is still questionable, it is too early for such an approach to the viruses.

The situation often becomes all the more difficult as viruses are genetically variable and new strains differing in host range, virulence and other characteristics arise, these perhaps representing different arrangements of the four nucleotides in the nucleic acid molecule. Such aberrations no doubt occur during virus replication; many are probably harmful and disappear in a uniform environment but some survive and prosper under changed environmental conditions or in a different species and variety of plant, thus extending the host range of the virus. There are probably other mechanisms, perhaps including some form of genetical recombination, which bring about variation in viruses. The problem is complicated by the difficulty of distinguishing between viruses and virus strains, and by the lack of any satisfactory system of nomenclature and classification.

Hence the classification of viruses has been subject to change over the years. For one thing, as more is learned about the properties of different viruses, their classifications change. For this reason only the principles of virus classification are considered here.

The most widey used taxonomic criteria for viruses depend upon the structure of a virus itself. Four major criteria are used:

(i) the nature of a nucleic acid—DNA or RNA, single-stranded or double stranded;

(ii) particle structure—helical, icosahedral, or complex;

(iii) presence or absence of viral envelope; and

(iv) dimensions of the viral particle.

Beyond these physical characteristics, other criteria (immunologic, cytopathologic, or epidemiologic) are used to subdivide the groups. Such a classification provides great convenience and utility, although it is not necessarily based upon the evolutionary origin of individual viruses.

Following are methods that are helpful to identify plant viruses.

1. Viruses are inoculated into indicator plants which develop typical symptoms when infected by specific viruses and in virus assay.

2. Serological tests are carried out using antisera of known viruses.

3. Transmission aspects of the virus are considered; whether by sap inoculation, and the vectors, if any, involved; whether the virus is persistent or non-persistent in the vector; whether stylet borne, circulative or propagative, and other aspects of its transmission.

4. Such properties as the thermal inactivation point, the dilution end point, and survival outside the plant can be used to characterize viruses.

5. Interaction with other viruses is considered, notably cross-protection.

6. Host range and symptoms are studied.

7. Study of morphology and chemical constitution of the virus particle.

CONTROL OF VIRUS DISEASES

Some of the general principles of controlling virus diseases are incorporated below.

I. Cultural Methods

1. **Use of Virus-free Planting Material.** The simplest way to ensure this is securing planting material from uninfected plants. Tuber indexing is one of the methods for selecting virus-free potato tubers. It consists of selecting tubers from vigorous plants appearing healthy. From these tubers eyes or buds are taken out and planted under suitable conditions. If the plants grown from these buds are healthy, the tubers produced by them are suitable for cultivation as virus-free planting material. If not, the parent tuber should be rejected. Virus-free planting materials can also be obtained by culturing of excised stem tips or by heat treatment.

2. **Isolation.** Cultivation of susceptible crops at a distance from each other delays or reduces the severity of virus diseases. In cases where insects spread the virus from old to new crops, appreciable control can be achieved by breaking the continuous pattern of cultivation by omitting a crop from the cycle.

3. **Roguing and Field Sanitation.** Both these practices eliminate virus sources. Roguing of diseased plants from seed beds and fields cuts down the source of virus. Some viruses are

carried in weeds which when detected should be immediately removed. Perennial plants should be dug out and destroyed just as ratooning in sugarcane should be discouraged. Elimination of alternate host plants also produces effective results.

4. **Cover crops and other barriers.** The non-susceptible plants towering over the undersown economically important crops break the flight of insects as a result of which a few vectors can reach the crop. Barriers erected at intervals in the field may also have the same effect. The barriers may be screens of cloth or some other suitable material, but more often rows of plants including sunflowers, maize, oats, and barley which are neither susceptible to the virus nor colonized by vectors.

5. **Density of Crop.** Dense growth of a crop reduces disease incidence because: (i) dense foliage generally provides an unfavourable microclimate for the insects to develop, and (ii) viruliferous insects infect a smaller percentage of plants.

6. **Area of Field.** Insects infect mostly plants at the edges of the field resulting greater concentration of diseased plants in the periphery of a field than in its middle, as such many plants escape infection in large fields as against smaller fields.

7. **Planting and Harvesting Time.** Alteration of the dates of sowing or harvesting of a crop enables it to escape disease. Early sowing leads to early maturity of plants at the time when a virus normally strikes. This causes reduced incidence of disease severity because host plants possess greater resistance to infection with age.

II. Heat Therapy

Heat therapy has proved most effective for controlling virus diseases of vegetatively propagated crops. Plant material treated up to one hour with hot air at a temperature range of 35 to 40°C produces good result. Sugarcane mosaic has been controlled by treating the cuttings at 53 to 54°C. Immersion of sets in water at 52°C for 20 minutes kills the virus of chlorotic streak of sugarcane.

III. Chemotherapy

Chemotherapy for the control of plant virus diseases has two aspects:

(i) protection of plants with chemicals decreases virus multiplication or insect vectors fail to acquire viruses from infected host tissue,

(ii) use of insecticides for killing vectors.

IV. Apical or Meristem Tissue Culture

Many viruses cannot attack apical meristems. Localized viruses cannot move far away from their points of entry into host tissue; though systemic viruses can do so but majority of them fail to reach the apical region where apical meristem is located. Apical meristems and a few inches of adjoining system portions of a rapidly growing host and/or axillary buds of systematically infected plants remain free from viral infection. These when grown in tissue culture medium produce virus-free plantlets which when transferred to soil produce virus-free adult plants. This apical or meristem tissue culture technique has proved extremely valuable in plants what are propagated by cutting vegetative parts.

V. Immunization

Cross-protection (vaccination) method in which infection of the plant with a mild strain of the virus protects it against subsequent attack by a virulent strain, may produce good result.

VI. Resistant and Tolerant Varieties

Cultivation of virus disease resistant varieties obtained by breeding and/or selection, is the best way of cotrolling virus diseases.

VII. Indexing and Certification Programme

Many countries of the world follow certification programme by certifying stocks and seeds of various plants to be healthy and free from virus infection. This is associated with indexing which is testing of plants or plant parts for virus presence or absence. This is done by: (i) the use of indicator plant, (ii) by the application of polyvalent antiserum technique, (iii) observing the presence or absence of virus disease symptoms, (iv) by the use of immunodiffusion procedures, and (v) by this layer chromatography.

VIII. Other Methods

Plant Protection. Thin film of oil spray on host surface inhibits both acquisition and transmission of stylet borne viruses by aphids.

Elimination of insect vectors. The easiest way to control spread of the large number of insect-transmitted viruses would be through elimination of vectors by spraying both healthy and infected plants with insecticides. Some of the effective insecticides are: Parathion, Menazon, Malathion, Pyrethrum, DDT, Demeton.

Part Six

Plant Pathology

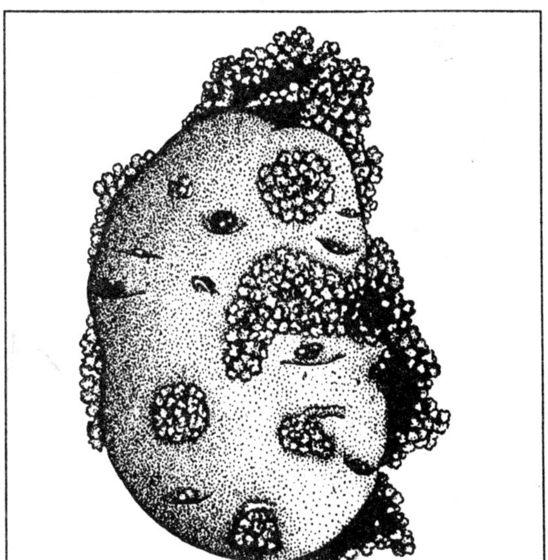

Chapter 24

Plant Pathology

INTRODUCTION

A plant is in health when all its organs or parts are doing their proper work, and the processes of growth and reproduction are going forward in the natural and regular manner. It is also equally true that when any of the organs and parts of a plant are not doing their proper work, and when either the growth or reproduction is not going forward in the natural and rgular manner, the plant is diseased, regardless of the cause or causes of its abnormal condition. Disease, therefore, involves a malforming of various plant processes and may occur while the plant is growing or while it is in a dormant condition. Hence the term plant disease is applied to any deviation from normal growth or structure of plants that is sufficiently pronounced and permanent to produce visible changes or to impair quality and economic value. Since disease is a process that occurs over a period of time, there must be an active cause that keeps the disease process going. It, therefore, becomes necessary to distinguish between the cause of a disease of a plant and the diseased condition of the plant. Fungi, bacteria, slime mold, viruses, flowering plants, and insects growing upon or within the tissues of a plant frequently cause some of the most serious diseases. Likewise the character of the soil, the amount and distribution of moisture and nutrients in the soil; and the climatic conditions may be the causes of severe and destructive diseases. Insects which eat parts of plants or suck the juices may also interfere with the performance of the normal functions of the plant and cause a weakened or unhealthy condition.

BRIEF HISTORICAL REVIEW OF PLANT PATHOLOGY

I. Abroad

Plant diseases have been known since the dawn of history. References to blights, mildews and rusts occur in the Bible and in writings of the ancient Greeks and Romans. These early records show that, while the ancient people recognized that plant diseases were taking toll of their crops, their ideas concerning them were largely of a superstitious nature. Plant diseases were mentioned by Aristotle in his writings about 350 B.C., and have been referred to by various writers since that time, but did not receive much attention until about the middle of the last century; and greatest development has been within the last quarter of a century. For many years workers were divided into two groups: the botanists who were interested in the study of the organisms which caused the diseases and the practical growers who were interested in finding remedies.

The origin of disease in plants was a great mystery for quite some time. Many groundless theories were advanced to explain the occurrence of the disease in plants. The general tendency was to attribute the troubles to God or gods, as the case might be, or to the influence of the stars and planets. The nearest approach to truth was in attributing certain diseases to environmental factors such as soil and climatic or weather conditions. Besides this, most widely accepted was the spontaneous generation and autogenetic theory of dis ases of plants. The living organisms generated *de novo* from dead matter and organisms are generated from the sap of the diseased plants.

Italian botanist Micheli (1679-1737) was the first to collect spores of fungi, sow them on pieces of melon, and watch the development of mycelium together with sporangia and spores characteristic of the species. He was convinced that fungi arose from their own spores.

In 1748, Needham upheld strongly the theory of spontaneous generation. Spallanzani (1755), an Italian investigator, challenged Needham. He showed that decay of meat by bacteria and other lower forms of life could be prevented by heating the material in a sealed flask to exclude contamination from air.

Tillet in 1755 working on Bunt (Stinking Smut) demonstrated that the disease may be increased by dusting diseased to healthy plants. He thus showed that the disease is due to transmissible agent.

In 1767, Fontana observed rusts by lens. Although Persons gave the name *Puccinia graminis* to the fungus associated with the black stem rust of wheat in 1794, but he did not regard the fungus as the incitant of the disease. In 1818, he expressed the opinion that some fungi grew from spores, but that others arose by spontaneous generation.

In 1807, Prévost working on wheat bunt first demonstrated the germ theory of disease—life arises from pre-existing life only. He observed germination of the spores of bunt of wheat and followed the penetration of the organism in the young wheat plant, and was convinced about the actual cause of the bunt of wheat. His revolutionary idea was refuted and did not receive favourable attention until 40 years later.

During 1815 and 1885 the Tulasne brothers confirmed the observations of Prévost with regard to spore germination of wheat-bunt fungus and relation of the fungus to the host plant. The work of the Tulasne brothers was followed by Anton De Bary. De Bary described the parasitic nature of smut of maize and rust of wheat. During 1861-1876 he followed the development of germ tube and penetration of host. De Bary also followed the life cycle of rust of wheat and barberry and finally established the germ theory of disease.

Pasteur (1860) confirmed the germ theory of Prévost. He furnished evidence that microorganisms arise from pre-existing living entities and that fermentation is a biological phenomenon rather than a purely chemical one.

Robert Koch (1876-1882) working on bacterial diseases postulated certain fundamental laws known as Koch's postulates.

That fungi, environmental conditions and insects are all causes of disease of plants was emphasized by Kuhn (1825-1910). Kuhn had a great influence on the development and application of control measures.

Robert Hartig (1839-1901), a contemporary of De Bary and Kuhn contributed greatly in forest pathology and was the author of important textbooks of diseases of forest trees.

The Irish potato famine of 1845 had a great impact on the establishment of a strong footing of plant pathology as an independent branch of study. Besides this, the observations of M. J. Berkeley on the fungus *Botrytis* (now *Phytophthora*) *infestans* was the cause, not the result, of the potato blight disease that caused Irish potato famine in 1845 established the view that fungi were able to cause disease in plants.

Very little was made in the treatment of diseases previous to 1882, when the value of lime and copper sulphate as a fungicide was accidentally discovered by Professor Millardet, and very quickly resulted in the development of Bordeaux mixture, which is now used extensively throughout the world, especially in fruit-growing areas. The discovery of Bordeaux mixture made Professor Millardet most widely known.

Oskar Brefeld (1839-1925) pioneered the development of pure culture technique for the fungi. His methods were gradually taken over by others and became a basis for present-day research in plant pathology, along with improvements introduced by Koch, Petri, and others.

Names of workers on bacterial and virus diseases of plants have already been discussed under Chapters on Bacteria and Viruses.

The major trends of work that had been established in the Nineteenth century were continued and expanded at the beginning of the Twentieth century with new or increased emphasis in widely different areas. Special mention may be made of:

1. Genetics of host plants in relation to the inheritance and the nature of disease resistance, the development of disease resistant varieties, and the nature of host-parasite interactions.

2. Environment in relation to plant disease development; influences on host, parasite, and host-parasite interactions.

3. Genetics of pathogens in relation to the inheritance and the nature of pathogenecity and to the nature of host-parasite interactions.

4. Physiology of parasitism.

5. Improvement in materials and methods of chemical control of plant diseases.

Advances in molecular plant pathology have also provided a new set of diagnostic tools and techniques that are used to detect and identify pathogens even when they are present in very small numbers or in mixtures with other closely related pathogens. Such tools include detection with monoclonal antibodies, analysis of isozymes or of fatty acid profiles of pathogens, analysis of fragments of their nucleic acids produced by specific enzymes, calculation of percentages of hybridization of their nucleic acids, and determination of nucleotide sequences of the nucleic acids of the pathogens. Since the mid-1980s, segments of DNA (probes), complementary to specific segments of the nucleic acid of the microorganisms, have been labelled with radioactive isotopes or with colour-producing compounds and are used extensively for the detection and identification of plant pathogens. Numerous techniques, often referred to by their acronyms, have been developed and are used; some of them are better suited for diagnosing one or more types of pathogens. For at least some pathogens, PCR, with selected differential random sequences of different species, can be effective for the detection and identification of each of these species. At other tests, PCR of sequence segments of rDNA internal transcribed spacer (ITS) regions are used or PCR of other genes or spacers of the fungal DNA is carried out. The product is then differentiated by digestion with restriction enzymes and gel electrophoresis and detection of differential random fragment length polymorphisms (RFLP) or use of PCR together with DNA hybridization in a reverse dot blot hybridization (RDBH) assay using PCR of selected RAPD markers. Reverse transcription PCR (RT-PCR) or immunocapture RT-PCR (IC/RT-PCR), direct binding PCR (DB-PCR), and a combination of PCR and enzyme-linked immunosorbent assay (ELISA) tests are often used successfully, especially for viruses.

An area of molecular plant pathology that is going to pay multiple dividends in the future is that of genomics, i.e., sequencing of the entire genomes of plants and their pathogens. Already, the genomes of the experimental plant *Arabidopsis thalliana,* of several plant viruses and viroids, and of the plant pathogenic bacteria *Ralstonia solanacearum* and *Xylella fastidiosa,* the white rot fungus *Phanerochaete chrysosporium,* and the model nematode *Caenorhabditis elegans* have been sequenced in entirely. Significant progress has already been made in sequencing the entire genomes of the very destructive plant pathogenic fungi *Magnaporthe grisea,* cause of rice blast; *Ustilago maydis,* cause of corn smut; *Cochliobolus heterostrophus,* another pathogen of corn; *Botrytis cinerea,* the gray mold of many fruits and vegetables; *Fusarium graminearum,* cause of head scab of wheat; and *Phytophthora infestans,* cause of the crops. Once the genomes have been sequenced, it will be easier to locate, identify, compare, isolate, and manipulate the genes for pathogenicity in the pathogens and of resistance in their host plants, as well as manipulate the introduction of them into specific locations of the plant genome where they would be most effective.

The molecular phase of plant pathology is expected to develop a great deal more and to make contributions in ways that we can hardly imagine at present. One area in which molecular plant pathology is expected to contribute greatly and to provide tremendous benefits is the area of detection, identification, isolation, modification, transfer, and expression of genes for disease resistance from one plant to another. Several such resistance genes have already been identified, isolated, transferred into susceptible plants, and, when expressed, made the plants resistant. The possibility that molecular plant pathology can modify and combine resistance genes makes likely the future utilization of resistance genes from unrelated plants or from other organisms, and perhaps even the synthesis of artificial genes for resistance for incorporation into crop plants. The practical implications of such developments cannot be overestimated, as they are likely to revolutionize the control of plant diseases by providing us with cultivars that can resist disease in the presence of the pathogen, without the need to use any pesticides.

II. In India

As in most countries, in India also the development of plant pathology was preceded by Mycology and the progress has necessarily been linked up with the availability of technical man-power resources of the country. The science of Plant Pathology is still young in India and it is, therefore, hard to define 'past' and 'present'. A convenient classification would be to treat the period prior to 1930 as the 'past' and thereafter up-to-date as the 'present'. The mycological work in India started practically from the middle of the Nineteenth century. The mycological and plant pathological researches on an organized and sustained basis became possible only after the establishment of the Imperial (now Indian) Agricultural Research Institute at Pusa in the first decade of this century. E. J. Butler, the first Imperial Mycologist of the Institute, can rightly be regarded as the founder of Mycology and Plant Pathology in India. During his stay in India for about twenty years, he continued to investigate and publish profusely. It was indeed a remarkable achievement, not only because of quantity, but also because of very high quality of the work. He, in course of his work trained huge number of workers who carried on research work in different parts of India and thereby spread knowledge and enthusiasm throughout the entire country. The culmination of Butler's brilliant career in India was the publication of the book entitled Fungi and Disease in Plants. It has been widely used both in research and teaching institutions with great interest. The area of investigations of Butler and his colleagues was indeed very wide, ranging from taxonomic work either alone or in collaboration with H. and P. Sydow, to the description of symptoms, causal organism and control measures of a wide variety of crop diseases, notable among them being wilt of pigeon-pea, cotton and sesamum; diseases of palms, betelvine, sugar-cane, paddy and potato; downy mildew of maize; tikka disease of ground-nut; spike disease of sandal; and rusts of wheat.

Some contemporaries of Butler also need a special mention. They are: Leslie Coleman for his researches on 'Koleroga' of arecanut; McRae on bud rot of palm, and secondary leaf fall disease of rubber; J. F. Dastur on *Phytophthora parasitica* and various aspects of late blight of potato disease; J. F. Shaw for his investigations on diseases caused by sclerotial fungi; G. S. Kulkarni for investigations on downy mildew of bajra and jowar and smuts of jowar; S. L. Ajrekar on wilt of cotton, smut of sugarcane and sugary disease of sorghum; and A. C. Tunstall on diseases of tea. In fact, up to 1920, knowledge not only of fungi but also of various crop diseases (symptomatology, life history, causal organism and control measures) had been substantially enlarged.

The above mentioned earlier workers devoted a great deal of their time to the most meticulous field observations of the symptoms of the disease and description of the fungi. Although by 1920 phenomenal advances had been made in plant diseases of economic importance, much of the work was of descriptive nature with definite mycological bias. During the period 1920 to 1930, the pattern of investigations continued to be on more or less similar lines through researches undertaken by a much larger number of workers, some being former colleagues of E. J. Butler. Subsequently due to political reforms, agricultural research and development became a provincial subject and naturally mycology and plant pathology sections were established in many provinces under the Departmnt of Agriculture. From Madras S. Sundararaman contributed to the knowledge of newer diseases of millets, rice, turmeric, palms and ginger. J. F. Dastur from the former Central Provinces worked on diseases of chillies, cotton and betelvine; G. S. Kulkarni from former Bombay Province contributed on smuts of millets. He and B. B. Mundkur initiated studies on wilt of cotton with a view to producing wilt-resistant varieties of cotton. K. M. Thomas of Mysore State worked on diseases of coffiee. At the Imperial Agricultural Research Institute, new diseases of various crops were investigated by M. Mitra, L. S. Subramanian and others.

The establishment of the Indian Council of Agricultural Research was the beginning of a new era in Plant Pathology in India, firstly because sufficient funds were provided by it for promoting researches and secondly, a large body of trained and qualified scientists, many of whom had worked under distinguished scientists of U.K. and U.S.A. became available both to the Agriculture Departments and the Universities. Academic and applied aspects of Mycology and Plant Pathology, therefore, started receiving adequate attention. Many Universities started teaching and created facilities for the research work in Mycology and Plant Pathology.

The period 1930 to to-date is marked by greater emphasis on experimental work as opposed to descriptive. In short, the fundamental and applied aspects of Plant Pathology received due recognition. These include physiology of fungi, physiology of parasitism, cytology, genetics of pathogenic fungi, associative aspects of fungi, physiologic races, breeding of resistant varieties, ecology of fungi, etc.

Till 1930, investigations were mainly confined to diseases caused by fungi, although there were occasional references to those caused by bacteria, viruses and nematodes. Notable among them are brown rot of potato, wilt of tobacco, angular leaf spot of cotton, spike disease of sandal, mosaic of sugarcane, 'Ufra' disease of paddy and earcockle of wheat. Much of the work was only of descriptive nature. After 1930, researches on bacterial and virus diseases were greatly intensified and undertaken by persons with special training in the field. The science of Nematology is still in its infancy in India, but a body of trained workers is now engaged in researches in this line. By and large, the physiological diseases have not received the attention which they deserve.

While the Indian Council of Agricultural Research continued to promote researches in various fields, the emphasis also gradually shifted to some particular crops by the establishment of Commodity Committees. The first to be established was the Indian Central Cotton Committee (1921), followed by Jute Committee (1936), Sugarcane Committee (1941), Tobacco Committee (1945), Cocoanut Committee (1945), Oilseeds Committee (1947) and the Areca nut Committee (1947). These Committees started financing researches, among others, on plant diseases of the concerned crops and many of these have now their own Research Stations of Institutes. Besides the Research Stations and Institutes of the Committees; two Central Institute, one for rice and the other for potato, were also established in 1946 and 1949, respectively. To further strengthen research on some crops and effect greater co-ordination, Regional Co-ordinated Research Stations for Cotton, Oilseeds and Millets were set up by the Indian Council of Agricultural Research and all these have been provided with senior Plant Pathologists. There thus exists today a large body of Plant Pathologists, who instead of being general practitioners as before 1930, are now working in specialized fields on diseases of particular crops.

The period from 1930 to approximately the time of partition of India was one of notable advances on researches such as, on rusts of wheat by late K. C. Mehta; on virus diseases by Uppal and others; on diseases of temperate fruits by late U. B. Singh; on diseases of citrus by late H. Chaudhury; on black tip (physiological disorder) of mango and saltation in fungi by S. N. Das Gupta; on diseases of sugarcane by L. S. Subramanian, V. S. Desai, B. L. Chona and others; on root rot of cotton and virus diseases of potato by R. S. Vasudeva; on gram (*Cicer arietinum*) blight and wilt by J. C. Luthra and co-workers; on smuts of cereals and downy mildew of millets and maize by B. N. Uppal; on bacterial diseases of plants by M. K. Patel, V. P. Bhide and subsequently by G. Rangaswami; on *Helminthosporium* diseases of cereals by late M. Mitra; on diseases caused by *Fusarium* sp. by Padwick and co-workers; on diseases of betelvine by R. P. Asthana; on genetics of sorghum smuts by S. Vaheduddin; on rusts of linseed and some cultivated leguminous crops by R. Prosad and co-workers; on tea by Agnihothrudu and co-workers; on cocoanut by K. P. V. Menon; and on rice by S. Y. Padmanabhan and S. B. Chattopadhya. Luthra's discovery of solar heat treatment to control the loose smut of wheat, which has stood the test of time, could be regarded as an outstanding contribution. During the same period, M. J. Thirumalachar greatly added to our knowledge of mycology and cytology of some rusts in particular. Thirumalachar and

Mundkur, and T. S. Ramakrishnan worked on the genera of rusts and smuts. R. K. Saxena and co-workers investigated on the cytology of lower fungi. Again, K. D. Bagchee and B. K. Bakshi made contributions on pathology of forest trees. This list is by no means exhaustive, but only to illustrate fields in which sustained and extensive researches were undertaken.

It was the great famine in Bengal in 1943, together with the impact of the Second World War, which focussed attention on the need to protect crops from pests and diseases in the fields and in the storage. From 1948 onward many Plant Pathologists directed all their energies towards organizing control programmes and undertaking what could be called 'operational research' leading to the establishment of Plant Protection work resulting in the creation of industries on a fairly large scale for the manufacture of plant protection equipment and pesticides and fungicides. The operational research has emphasized the needs to study of epidemiology and control of plant diseases by chemotherapy, antibiosis and eradicant fungicides, and the like.

R. Prosad and co-workers enhanced our knowledge of rust disease of crop plants. T. S. Sadasivan established school of advanced study in Botany at the University of Madras and made significant contribution on the ecology of soil-borne pathogens. Plant pathology was popularized with the establishment of Agricultural universities all over the country. Besides this, All India Co-ordinated Projects operating on different crops have opened up unique opportunities for the plant pathologists to take up serious and in-depth studies of various plant diseases and their management. Researches on virus diseases of various crop plants have been actively pursued by R. S. Vasudeva, S. P. Kapoor, P. M. Varma, S. P. Raychaudhuri, T. K. Nariania and others, whereas R. N. Tandon, K. S. Thind, R. S. Mehrotra, S. B. Saksena and others made significant contributions on the taxonomy and physiology of pathogenic fungi.

DEFINITIONS AND TERMS

The word pathology is derived from the Greek words—*pathos* meaning "suffering" and *logos* meaning "speech". Thus plant pathology is concerned with the study of the suffering of plants. The plant pathologist is concerned with the science of plant pathology such as understanding the nature and cause of a plant disease, and the diagnosis, treatment or control of plant diseases. **Diseased plants** are those plants which, due to certain physiological or structural or both changes brought about by unfavourable environmental conditions or by one or another organism, have become distinguishable from the normal or healthy plants as their normal growth has been interrupted. A **disease** may thus be defined as sustained physiological or structural imbalance in the plant as a whole or any part of it, caused by certain external agencies may be designated as **causal agencies.** Again, any structural imbalance in the plant induced by one or more external agencies is known as **lesion.** The agencies may be: unfavourable environmental conditions and similar other situations, also designated as **non-infectious** or **non-parasitic agencies**, or organisms, the **infectious parasitic agencies** such as: Bacteria, Myxomycetes, Fungi, Viruses, Insects, Angiospermic plants, etc. The agency that incites the disease is known as the **incitent.** Similarly, an agency which generates a disease is also referred to as a **pathogen.** The word pathogen was most likely borrowed from medical science to signify animate pathogens like, Fungi, Bacteria, etc. But the idea is gaining ground to consider pathogens as being both animate and inanimate. In the opinion of others there are animate pathogens, inanimate pathogens, and viruses. According to them, viruses belong to a separate category of their own. But till now the term pathogen is practically always used to denote a living entity. The ability of a pathogen to cause incite disease is **pathogenicity** and the process or chain of events leading to the development of disease is **pathogenesis.** Successful expression of pathogenicity is **virulence.** A pathogen is **avirulent** when it fails to cause disease. With a few exceptions, all pathogens are parasites although all parasites are not necessarily pathogens. Any parasite, however, may have pathogenicity under the right conditions and should be regarded as a **potential pathogen.** Again, the investigation of the causes of a disease is **etiology.** Before a suspected organism could be finally accepted as the causal agency of a particular disease, it must satisfy certain conditions which are known as **Koch's postulates** or **Rules of Proof** or **Pathogenicity tests. Pathogenicity tests** or **Koch's postulates** or **Rules of Proof** are thus the experimental evidences of any agency to be designated as an incitent or pathogen of a disease. The Koch's postulates are: (i) the organism must be consistently associated with the disease in question; (ii) the organism must be isolated from the diseased plant or animal and grown in pure culture and accurately described; (iii) the organism must be isolated from the diseased plant or animal and grown in pure culture and accurately described; (iii) the organism of the pure culture, when inoculated back into healthy plant or animal, must be capable of reproducing the characteristic symptoms of the disease; and (iv) the organism must be re-isolated with the original inoculated organism. All these postulates have been accepted as working guides by workers of diseases of human beings and other animals, and also of plants. But these postulates cannot be suitably applied in plant diseases where the causal agency is an obligate parasitic fungus, virus or a nematode as none of them can be brought under artificial culture—**isolated** from natural habitat. As such, special method is applied to test pathogenicity in each case. In non-culturable fungal pathogens, spores can be removed from the infected host. In such cases spore morphology is a valuable aid to the identification of the non-culturable fungal pathogens. The problem of isolating viruses from their host plants is overcome by utilizing **indicator plants.** These are **alternative hosts** which develop symptoms specific for a particular virus. The original host plant may be reinoculated with the infective juice from the alternative host. The original and induced symptoms may be compared to prove pathogenicity of that particular virus. Nematodes also cannot be cultured in artificial media. But they are visible in or on the affected parts of the host. These organisms can be separated and tested for their parasitism, and symptom development on healthy plants under controlled conditions. Cultures of nematodes can also be maintained on living plants.

The establishment of the pathogen in the host tissue is known as **infection.** The infection is induced by the infective propagules, collectively known as **inoculum** (pl. **inocula**), produced by the pathogen and insertion of inoculum in a body (plant or animal) or in any other suitable place is **inoculation.** The number of propagules and their capacity to cause infection, known as **inoculum potential** which is determined by the type of pathogen, its genetic capacity to multiply rapidly and its nutrition which is again dependent upon host susceptibility. With proper inoculum potential the pathogen will complete the process of infection by: (i) **penetration** resulting from breaking down of physical and chemical barriers, and (ii) establishment of physiological relationship with the host, the infection proper followed by **colonization** in which the pathogen advances through the host tissue in varying extents. The success of host infection again depends on: inoculum potential, aggressiveness of the pathogen, favourable environmental conditions and host susceptibility. The environmental effects that make plants more amenable to

Koch's postulates

Robert Koch (1843-1910) was a physician and a bacteriologist. He was the first to show, in 1876, that anthrax, a disease of sheep and other animals, including humans, was caused by a bacterium that he called *Bacillus anthracis*. He subsequently discovered, in 1882, that tuberculosis and, in 1883, that cholera are each caused by a different bacterium, which led to the general conclusion that each disease is caused by a specific microbe. These experiments confirmed for the first time the germ theory of disease proposed earlier by Louis Pasteur.

Before Koch's experiments, and while Koch himself was carrying out the work on the diseases mentioned earlier, there was confusion and uncertainty about the occurrence and the cause of each disease. Much of the time when bacteria or fungi were isolated from diseased or dead human, animal, or plant tissues, the isolated bacteria or fungi were subsequently shown to be saprophytes, i.e., they coexisted with the microorganism that caused the disease but could not by themselves cause the disease for which they were being considered. Based on his experiences, in 1887, Koch set out the four steps or criteria that must be satisfied before a microorganism isolated from a diseased human, animal, or plant can be considered as the cause of the disease. These four steps, rules, or criteria are known as "Koch's postulates."

1. The suspected causal agent (bacterium or other microorganism) must be present in every diseased organism (e.g., a plant) examined.
2. The suspected causal agent (bacterium, etc.) must be isolated from the diseased host organism (plant) and grown in pure culture.
3. When a pure culture of the suspected causal agent is inoculated into a healthy susceptible host (plant), the host must reproduce the specific disease.
4. The same causal agent must be recovered again from the experimentally inoculated and infected host, i.e., the recovered agent must have the same characteristics as the organism in step 2.

Koch's rules are possible to implement, although not always easy to carry out, with such pathogens as fungi, bacteria, parasitic higher plants, nematodes, most viruses and viroids, and the spiroplasmas. These organisms can be isolated and cultured, or can be purified, and they can then be introduced into the plant to see if they cause the disease. With the other pathogens, however, such as some viruses, phytoplasmas, fastidious phloem-inhabiting bacteria, protozoa, and even some plant pathogenic fungi that are obligate parasites of plants (such as the powdery mildew, downy mildew, and rust fungi), culture or purification of the pathogen is not yet possible and the pathogen often cannot be reintroduced into the plant to reproduce the disease. Thus, with these pathogens, Koch's rules cannot be carried out, and their acceptance as the actual pathogens of the diseases with which they are associated is more or less tentative. In most cases, however, the circumstantial evidence is overwhelming, and it is assumed that further improvement of techniques of isolation, culture, and inoculating of pathogens will some day prove that today's assumptions are justified. However, in the absence of the proof demanded by Koch's rules and as a result of insufficient information, all plant diseases caused by phytoplasmas (e.g., aster yellows) and fastidious vascular bacteria (e.g., Pierce's diseae of grape) were for years thought to be caused by viruses.

Despite the difficulties of carrying out Koch's postulates with some causal agents, they have been and continue to be applied, sometimes with certain modifications, in all cases of disease. They have had and continue to have a tremendous effect in deciding and in convincing others that a particular microorganism is the cause of a specific disease. By attempting to carry out Koch's postulates in all newly discovered diseases, a great deal of work with potential saprophytes has been avoided, while, at the same time, doubt and criticism are reduced to a minimum while confidence in and use of the identification increase greatly and quickly.

infection are called **predisposition.** A host may show good growth despite good development of the pathogen on it. Such a host is designated as a **tolerant host.** The point at which the inoculum establishes itself inducing infection is the **infection court** or **infection locus.** Infection loci are local concentrations of disease. They need not be conspicuous. A large number of small infection loci might appear simultaneously, with multiplication they become conspicuous. An infection locus extends its limits contiuously and grows bigger; but it also extends discontinuously to start new secondary loci beyond its own limits. The two processes, growing bigger and starting new loci may involve different methods. When disease multiplies, it multiplies in loci. As loci enlarge, infection spreads largely from the perimeter. The rate of growth of a locus and the behaviour of the inoculum at the infection court are very much controlled by certain conditions mainly: humidity, temperature, suitability of the host tissue for penetration, and aggressiveness of the pathogen. The climate prevailing immediately surrounding the infection court, or in and around the source of inoculum is known as **microclimate** and that around the microclimate is the **macroclimate.** The infection when visible externally on the host surface is known as **visible infection** which is commonly known as the appearance of disease, also as **symptom** which is nothing but the visible effects of the pathogen on the host plant. Again, when the infection is established in the host tissue but is not visible from outside the host, it is the **invisible infection.** The infection may again be **localized** and **systemic.** When the effect of infection inducing disease development remains restricted in or around the area surrounding the infection court, it is known as **localized infection,** whereas in **systemic infection** symptoms (**visible symptoms**) appear at a place other than the infection court after an interval of time during which the pathogen develops in the host body remaining externally invisible (**invisible infection**) without causing any apparent injury. The pathogen is usually latent until some nutritional and developmental changes occur in the host. Again, there is another type of infection where the infection is strictly specific to particular organ of the host induced by a particular pathogen. Besides this, a host may be infected by a pathogen but does not show any symptoms. This is known as **latent infection.**

There may be a long period between infection and appearance of disease but ordinarily one follows the other rather closely. The interval between infection and appearance of disease is known as the **incubation period.** Again, the external manifestation of a disease is the **disease symptom.** The term **causal organism** is also equally applicable to the organism that induces disease. Though most causal organisms are parasites the terms pathogen and parasite are not synonymous as in some cases organisms may be parasitic without becoming causal factors in certain diseases. At times two or more agencies may be combined to cause a disease. It may also be necessary for a plant first to be weakened by unfavourable environmental conditions before it can be attacked by a pathogen. The entire plant or only a portion of it may be affected, and disease may or may not necessarily be significantly injurious. Usually a combination known as **causal complex** of several factors both external and internal becomes the cause of a disease when it alters the physiological processes and morphological development of a particular part or the plant as a whole. Unfavourable environmental conditions or an organism inciting disease is a part of the causal complex. Any plant susceptible to any disease is commonly known as a **host.** A newer term **suscept** is applied to the plant which is susceptible to any disease brought about by any causal complex. Usually a pathogen completes its life cycle on one host only. But where it needs two hosts to complete its life cycle, one host designated as **principal host** and the other as **sub-**

sidiary or **alternate host.** Again, a pathogen may need to have a host for **perennation,** i.e., to tide over unfavourable conditions. Such a host plant is known as **collateral host.** In general, plants make attempts to resist themselves against any disease incidence. Some are successful, others are not. The ability of a plant to resist itself from the effect of disease incidence is known as **resistance.** Again, on the other hand, the condition of a plant that makes it suitable for the attack by a pathogen is **susceptibility.** Resistance and susceptibility are measures of the extent to which the host plant is damaged by the pathogen or, more often, the extent to which the host plant prevents the entry or subsequent growth of the pathogen within its tissues. Resistance has been classified as **horizontal** (uniform) when it is evenly spread against all races of pathogen, or as **vertical** (differential) when it is effective against some races of the pathogen but not against others. It may be: **polygenic**—determined by many genes, **oligogenic**—determined by several genes or **monogenic**—determined by one gene. Again, the genes involved may be **major** or **minor** with respect to their effectiveness in conferring resistance, whereas **tolerant** plants offer little resistance to the pathogen but, although infected, are only slightly affected. The visible reactions of the host plant to the pathogen attack, are relatively mild. The extreme of tolerance is shown by **symptomless carriers** in which the disease inciting agent (usually a virus) present in the host causes no symptoms. Again, **hypersensitive** host plants develop minute island of dead cells resulting from rapid death of cells in the vicinity of the point of infection—the **infection court** of the pathogen so that the progress of the pathogen is halted although it does not necessarily die immediately. This phenomenon is termed **hypersensitivity.** Hypersensitivity thus confers high resistance and is sometimes used in the sense of immunity. The hypersensitive response is the effect of plant defence responses initiated by the host plant resulting from response of specific pathogen-produced signal molecules, known as **elicitors.** This happens, perhaps, because of the presence in the host plant of a resistance gene (R) which recognizes and triggers into action by the elicitor molecule released by the pathogen. The pathogen produced elicitor is the product of a pathogen gene which triggers the development of resistance in the host that makes this pathogen avirulent. This is called **avirulence gene.** The host plant may be called resistant to the disease if it shows hypersensitive reactions to infection by the pathogen. Such reactions are often encountered while pathogens infest resistant host plants. Soon after the establishment of infection in the host, the pathogen grows absorbing food from the host cells by excreting certain enzymes or other diffusible material. This causes certain reactions on the part of the host cells which produce some specific chemicals, which may kill the pathogen. Again, in some hosts, as a reaction to the entry of the pathogen, the host cells around the infection court may be killed and if the pathogen is an obligate parasite, it is unable to obtain nutrition from the dead cells surrounding the infection court. Thus the spread of the pathogen is checked. This type of hypersensitivity has been reported in uredospore infection of *Puccinia graminis* on wheat plant. Many authors do not consider the symptom of hypersensitive reactions in the host as indicative of high resistance in host plant.

The host plants are said to be **immune** to a disease when the pathogen is incapable to establish itself on them even under the most optimum conditions. Immunity may thus be considered as cent per cent resistance. If the host plant is susceptible to the pathogen when the host plant is young and develops resistance with maturity and age, then it is said to possess **field resistance** or **adult plant resistance.** In some cases the host plant is capable of carrying most of the normal metabolic processes, in spite of the diseased condition. Such a phenomenon is known as **disease endurance**, whereas the tendency of a plant to escape from disease incidence is **klendusity**. In certain cases, a variety of crop may escape from disease incidence not because of any inherent or specific resistance but because of maturity ahead of the advent of period (e.g., weather conditions) suitable for infection of pathogen. This is known as **disease escape** and the variety of crop concerned is the disease-escaping variety. Soils in which certain diseases are suppressed because of the presence in the soil of microorganisms antagonistic to the pathogen are **suppressive soils.**

Some pathogens cannot complete their life cycles except in a parasitic relation to their respective hosts, they are known as **obligate parasites.** They are found in various groups of fungi. Viruses also have properties in common with obligate parasites. Again, there are organisms which are ordinarily saprophytes but under certain conditions may become parasitic, they are called **facultative parasites.** Such organisms have weakly pathogenic properties which are expressed only under particular conditions. There are, however, many other organisms known as **facultative saprophytes** which are capable of subsisting as saprophytes or of infecting living organisms according to circumstances. The facultative parasites and facultative saprophytes represent the extremes in the range of microorganisms and there are no well-defined limits between them. The facultative saprophytes have much higher potentiality as a pathogen than does the facultative parasites. On the basis of mode of nutrition, parasites are often grouped as: (i) **biotrophs**, parasites deriving nutrients from living host cells; (ii) **necrotrophs**, those derive their nutrients from cells that they have already killed. Most biotrophs cause minimal damage to the host tissue and their host range is narrow, e.g., fungi causing rust and powdery mildew diseases, whereas necrotrophs rapidly kill the host cells and then live saprophytically on the dead tissues. They have wide host range, e.g., species of *Botrytis* and *Pythium*. The cereal smut fungi can change their biotrophic to necrotrophic habit in course of their life cycle. These fungi although present in the host throughout their development induce no visible symptoms until the host teaches its reproductive stage, when they attack the floral parts and become necrotrophic. There are cases where two living organisms may live non-pathogenically in association with one another. In such cases, both the partners are mutually benefited through their life processes. The participating organisms are known as **symbionts** and the phenomenon is **symbiosis.** More often it is used to describe associations in which there is a considerable degree of tolerance between the organisms involved—**mutualism.** But in **antagonism** (**antibiosis**) one organism is injured by another, especially when the antagonist exerts its harmful effects by secreting toxic substances (**antibiotics**). Again, **synergism** occurs when the combined effect of two organisms acting together is greater than the sum of the two separate effects. There are microorganisms which grow upon other plants without being parasitic or symbiotic. They are known as **epiphytes.** Some of the epiphytes which do not cause any direct injury or inconvenience to the plants on which they grow but cause hindrance to metabolic activities by producing shedding effect on the green parts and producing unsightly appearance of the different plant parts, whereas organisms living within plant parts either parasitically or symbiotically are the **endophytes.**

Diseases incited by pathogens are much larger in number than those produced due to unfavourable environmental conditions like: disbalance in soil-moisture relations, sudden change in temperature—high or low, deficiencies in mineral nutrition, effect of poisonous gas, etc. These diseases will be treated separately.

What is the starting point of the incidence of a disease when a pathogen is a part of the causal complex? Naturally the pathogen is expected to be present in some form or the other, either anywhere in soil, debris of infected plants of the previous season, harbouring in weeds, or in the propagating stocks and seeds. This is how a pathogen perennates year after year and induces diseases. The organs through which pathogens perennate and withstand unfavourable conditions are also known as **overwintering** or **oversummering** stages since it is through these forms they tide over unfavourable conditions during winter or summer as the case may be.

With the advent of favourable season the perennating organs become the source of production of infective organ or organs of the pathogen. Such an organ of infection of a pathogen is commonly known as **inoculum** (pl. **inocula**). The perennating organs serve as the **source of inoculum**. The inoculum is either disseminated through some agencies like: wind, rain, insects, etc., or directly comes in contact with host tissue and establishes itself with it ultimately inducing infection. The inoculum produced from the perennating organs is the **primary inoculum**. The infection caused by the primary inoculum in the host tissue is the **primary infection**. The primary infection induces disease and also produces infective bodies in the host tissue which are collectively known as the **secondary inoculum**. The primary inoculum as well as the secondary inoculum may again cause infection on the same host—the **secondary infection**. The production of secondary inoculum and inducing secondary infection may be repeated several times during the current season. During disease development chain of events occurs betwen the time of infection and the complete expression of disease. This chain of events follows in regular succession in cyclic order and is known as **disease cycle**. The disease cycle is intimately associated with the **life** known as **disease cycle.** The disease cycle is intimately associated with the **life cycle** of the pathogen since both are interdependent on each other. In case of a pathogen being an obligate parasite, disease cycle and life cycle are interdependent. But where a pathogen has its life cycle comprising partly parasitic phase and partly saprophytic phase, in such case the disease cycle is dependent primarily on that part of the life cycle during which the pathogen passes parasitically. The disease cycle usually comprises of two cycles—a **primary disease cycle** which is pro-

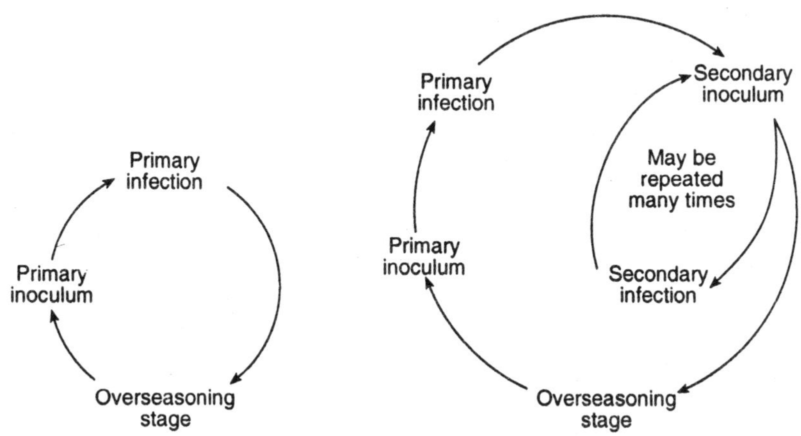

Fig. 24.1: Pattern of (left) monocyclic and (right) polycyclic disease cycle

duced by the primary infection, and a **secondary disease cycle** resulting from the secondary infection.

VARIOUS ASPECTS OF STUDY OF PLANT DISEASES

The development of a plant disease is dependent upon three aspects: (i) a susceptible host, (ii) a virulent pathogen, and (iii) a favourable environment of sufficient duration to allow infection by pathogen and subsequent development of the disease. As such, the study of plant diseases can be divided into four interrelated aspects: (i) the causes of disease—**etiology**, (ii) the interaction of host and pathogen (including pathogenesis, the development of disease in the plant); (iii) the interaction of populations of plants, pathogens and environment—**epidemiology**; and (iv) the prevention and control of plant diseases. The practical grower is especially interested in the prevention and control of plant diseases only. But it is almost impossible to treat a plant successfully for a disease without a thorough knowledge of the cause. However, in some instances, methods have been devised for keeping diseases under control without a complete knowledge of the organism causing them.

A disease may be serious on one species or variety of plant, and of little importance or not present at all one one closely related; or it may be serious under certain climatic conditions, and harmless under other conditions. For these reasons, a disease-producing organism, when introduced from one part of the world to another, may become more or less virulent than in its original home. It may find new host plants which may be more or less susceptible than its previous hosts, or it may find the climatic conditions more or less favourable to its growth. It is also true that when such an organism is introduced into another part of the world, it is liable to meet enemies which will endanger its existence.

If the disease is due to a specific organism, the organism may bring about the unhealthy condition in its host: (a) by feeding upon its tissues or sucking its juices or otherwise for food; (b) by covering the surface of the leaves or stem and thus interfering with the action of the light and proper transpiration; (c) by clogging up the passage of the fibrovascular bundles and interfering with the circulation of water and other fluids; (d) by attacking the growing regions, especially the root tips, and interfering with or preventing their development; (e) by attacking the cambium; (f) by living inside the host and feeding upon it and secreting toxic substances; (g) by attacking the fruits but doing little or no damage to the plant itself; (h) by injuries to the leaves, interfering with their activity; (i) by malformations of various parts of the host.

Although the symptoms and effects of certain diseases caused by agencies other than specific organisms are well-known, the factor or combination of factors producing them is still questionable. Details of etiology, interaction of host and pathogen, epidemiology, and prevention and control of plant diseases are incorporated in pages to follow.

CLASSIFICATION OF PLANT DISEASES

Plant diseases have been classified by different workers in various ways.

In general, plant diseases may be classified in various ways on the basis of:

1. **Host plants affected,** such as potatoes, wheat, cereals, vegetables, fruit, etc.

2. **Part of host plant affected,** such as roots, leaf, stem, flower, fruits, etc.

3. **Symptoms produced** in the host plants such as wilt slight, soft rot, anthracnose rust milder, damping off, etc.

4. **The mode of spread and severity of infection,** such as endemic, epidemic, sporadic and pandemic.

Table 24.1: Examples of Severe Losses Caused by Plant Diseases

Disease	Location	Comments
Fungal		
1. Cereal rusts	Worldwide	Frequent severe epidemics; huge annual losses
2. Cereal smuts	Worldwide	Continuous, although lesser, losses on all grains
3. Ergot of rye and wheat	Worldwide	Infrequent, poisonous to humans and animals
4. Late blight of potato	Cool, humid climates	Annual epidemics, e.g., Irish famine (1845-1846)
5. Brown spot of rice	Asia	Epidemics, e.g., the great Bengal famine (1943)
6. Southern corn leaf blight	U.S.	Historical interest, epidemic 1970, $1 billion lost
7. Powdery mildew of grapes	Worldwide	European epidemics (1840s-1850s)
8. Downy mildew of grapes	U.S., Europe	European epidemic (1870s-1880s)
9. Downy mildew of tobacco	U.S., Europe	European epidemic (1950s-1960s); epidemic in North America (1979)
10. Chestnut blight	U.S.	Destroyed almost all American chestnut trees (1904-1940)
11. Dutch elm disease	U.S., Europe	Destroying American elm trees (1918 to present)
12. Pine stem rusts	Worldwide	Causing severe losses in many areas
13. Dwarf mistletoes	Worldwide	Serious losses in many areas
14. Coffee rust	Asia, S. America	Destroyed all coffee in southeast Asia (1870s-1880s) since 1970 present in South and Central America
15. Banana leaf spot or Sigatoka disease	Worldwide	Great annual losses
16. Rubber leaf blight	South America	Destroys rubber tree plantations
17. Fusarium scab of wheat	North America	Severe losses in wet years
Viral		
18. Sugar cane mosaic	Worldwide	Great losses on sugar cane and corn
19. Sugar beet yellow	Worldwide	Great losses every year
20. Citrus tristeza (quick decline)	Africa, Americas	Millions of trees being killed
21. Swollen shoot of cacao	Africa	Continuous heavy losses
22. Plum pox or sharka	Europe, North America	Spreading severe epidemic on plums, peaches, apricots
23. Barley yellow dwarf	Worldwide	Important on small grains worldwide
24. Tomato yellow leaf curl	Mediterranean countries, Caribbean Basin, U.S.	Severe losses of tomatoes, beans, etc.
25. Tomato spotted wilt virus	Worldwide	On tomato, tobacco, peanuts, ornamentals, etc.
Bacterial		
26. Citrus canker	Asia, Africa, Brazil, U.S.	Caused eradication of millions of trees in Florida in 1910s and again in the 1980s and 1990s
27. Fire blight of pome fruits	North America, Europe	Kills numerous trees annually
28. Soft rot of vegetables	Worldwide	Huge losses of fleshy vegetables

Contd.

Contd. Table 24.1

Disease	Location	Comments
Phytoplasmal		
29. Peach yellows	Eastern U.S., Russia	Historical, 10 million peach trees killed
30. Pear decline	Pacific coast states and Canada (1960s), Europe	Millions of pear trees killed
Nematode disease		
31. Root knot	Worldwide	Continuous losses on vegetables and most other plants
32. Sugar beet cyst nematode	Northern Europe,	Continuous severe annual losses on sugar beets
33. Soybean cyst nematode	Asia, North and South America	Continuous serious losses on soybean

5. **Causal organism or factors** responsible for disease development. **Classification of plant diseases is also done on the basis of spread and severity of infection.**

(a) **Endemic Disease**. The disease is known as endemic where it is constantly present in a moderate or severe form and is confined to a particular country or district; for example, the wart disease of potato caused by *Synchitrium endobioticum.*

Synchitrium endobioticum is endemic the Darjeeling. In plant pathology, this term is generally applied to simple interest diseases which either indigenous of ancient introduction.

(b) **Epidemic or Epiphytotic Diseases.** An epidemic or epiphytotic disease usually occurs widely, but periodically, in a destructive form. The pathogen, may be present as in epidemic diseases, but the environments factors responsible for the development of true disease occur only periodically for example, social rust, powdery mildews.

(c) **Sporadic Diseases.** Sporadic diseases in reality, belong to epidemic group of disease. But in cases where disease occurs at a very irregular intervals and locations, and in reality in a fewer cases are **Sporadic Diseases,** as in Angular leaf spot and blotch disease of cucumber.

(d) **Pandemic Diseases.** In pandemic disease occurs in vast areas causing severe loss of crop very much like epidemic disease, for example, later blight of potato.

Nalker, in 1969, has classified plant diseases into three main types:

1. Non-parasitic disease.
2. Parasitic disease.
3. Mycoplasmal and viral diseases based on the major causal factors. There are further subdivided on the basis of casal agency organism involved. The topic is further elaborated.

(i) **Diseases classified in relation to their occurrence.** Study of plant diseases in relation to their occurrence (interaction of populations of plants, pathogens and environment) is known as **epidemiology.** There may be **infectious disease** and **contagious disease.** A disease which spreads slowly and is incited by a transmissable pathogen is referred as **infectious disease,** and that which spreads rapidly is a **contagious disease.** Considering from the point of view of occurrence the diseases are often classified as: **endemic, epidemic** or **epiphytotic,** and **sporadic.** A disease which constantly occurs from year to year in moderate to severe form is an **endemic disease.** Greeneria fruit rot of grape (*Vitis vinifera*) is an endemic disease in India caused by *Greeneria uvicola.* But an **epi-**

demic disease or epiphytotic disease is the one whose incidence is periodical in wide areas spreading very fast. The term epiphytotic is rather used in case of plant disease. The epiphytotic disease is usually very responsive to variations of environmental conditions which actually control its incidence. Like endemic disease, the pathogen may be more or less constantly present but to ensure disease development environmental conditions must be favourable.

The pandemic outbreak of an epidemic disease which occurs over vast and extensive areas causing severe losses are the **pandemic diseases.** These occur in very wide areas causing tremendous loss of crop plants; for example, late blight of potatoes.

The epiphytotic disease incidence is controlled by: (a) availability of susceptible host, (b) rate of production of inoculum, and (e) micro- and macro-climatic conditions. According to their occurrence in crops, the epiphytotic disease may be of two kinds; the diseases ordinarily endemic but assume epidemic proportions from time to time, and those completely newly introduced in the locality—the **exotic diseases.** Wart disease of potato caused by *Synchytrium endobioticum* is an exotic disease in India. In the former kind, the disease incidence depends mainly on the variations of the climatic conditions, whereas the latter is resulted either from a new parasite having been introduced from another country, or from the passage of a parasite of some pre-existing plant of a place to a newly introduced host.

Some of the important epiphytotic outbreaks are: the Irish famine of 1945 and 1946 due to the destruction of the potato crop by the late blight of potato disase; the destruction of vineyards in England in 1845 and in France in 1848, by downy mildew disease; the devastation of the flourishing coffee plantation of Ceylon caused by coffee rust disease; and the destruction of sugarcane industry of the West Indies in 1895 by the rind disease. Similar situations arose in the United States where almost the entire chestnut trees were completely killed by chestnut blight disease, so also the fast spreading of the Dutch elm disease in the wiping out of elm trees from the country.

The **sporadic disease** is rather an epiphytotic disease except that its occurrence is at very irregular intervals and in lesser areas. In a sporadic disease, pathogen affects only a few plants in a large population of host plants and others remain unaffected.

(ii) **Diseases classified based on the nature of dispersal medium.** Diseases are often classified as: soil-borne disease, seed-borne disease, wind-borne disease, etc. Here importance is given only on the dispersal medium of the pathogen and nothing else.

(iii) **Diseases classified according to parts of host affected.** The classification of plant diseases based on parts of the host affected may be outlined as: (a) Foliage disease, (b) Stem disease, (c) Root disease, and (d) Vascular disease. Such a classification neither does clearly indicate the nature of the causal agency nor the effect produced on the host tissues and does not have much practical value.

(iv) **Diseases classified on the basis of hosts affected.** The classification of plant diseases on the basis of hosts affected is simply a matter of convenience. There is no fundamental principle involved in this system of classification. Some of these diseases are: Vegetable diseases or Diseases of vegetable, Fruit diseases, Cereal diseases, Timber diseases, Diseases of ornamental plants, Diseases of shade trees, etc.

(v) **Diseases classified on the basis of symptoms.** On the basis of symptoms, plant diseases are classified into three categories: (a) diseases resulting in the death of the affected tissues—**necrotic** (b) diseases resulting in the slowing down or stunting of the growth of the plant or plant parts—**atrophic** and (c) diseases resulting in the overgrowth or over development of the plants or organs affected—**hypertrophic.** According to symptoms diseases are often designated as: blight disease, mildew diseases, rust disease, smut disease, rots, cankers, etc. This system of classification gives an idea about the nature of damge done to the host and as such has some practical value.

(vi) **Diseases classified on the nature of causal agencies.** The diseases of plants may be attributed to one or more of the following causal agencies:

FIRST GROUP:

(a) Fungi.

(b) Bacteria.

(c) Slime molds.

(d) Viruses.

(e) Parasitic angiosperms, e.g., *Viscum album, Dendrophthae* sp., *Phoradendron* sp., *Orobanche* sp., *Cuscuta* spp., and *Striga* spp.

(f) Algae (*Cephaleuros* spp.)

(g) Insects.

(h) Mites.

(i) Nematodes.

SECOND GROUP:

(a) Soil-moisture disbalance.

(b) Nutritional disorders.

(c) Optimal temperature disbalance.

(d) Light intensity disbalance.

(e) Gas, smoke and other air pollutants.

(f) Careless spraying of chemicals.

The first group comprises of **parasitic causal agencies** (with the exception of Virus, may be) and the disease produced is the **parasitic disease**, whereas the second group consists of **non-parasitic causal agencies** and the corresponding disease produced is the **non-parasitic disease.** The first group includes a great host of living organisms which live within or are attached to the host plants, affecting their health and frequently causing death. The extent of their parasitism is extremely variable, which again is dependent upon the nature of host and parasite and their interaction. Some parasites destroy their hosts in a very short periods of time; while others live upon their hosts for a long period, even for many years, before death results. Still other parasites never cause the death of the host plant, but do reduce its vitality growth, or fruit production. Some parasites do not require a great amount of food from the host plant but grow in such a manner as to interfere with the performance of its ordinary functions. Again, some of which are of little importance in themselves, open the way for the attacks of organisms which cause other and more destructive diseases. A parasitic causal agency may also produce conditions suitable for the growth of causal organisms which are in no way the cause of the disease in question. The presence of two or more organisms in or on a diseased plant thus frequently prevents a satisfactory diagnosis and treatment.

Diseases resulting from the second or non-parasitic group of causal agencies are frequently very difficult to diagnose. Often they are of little importance in themselves, but make a plant so weak as to make it possible for one of the specific organisms of the first group to attack it.

Plant diseases are classified with a view to: (i) get a clear idea about the cause or causes of the disease, (ii) follow the pattern of disease development, and (iii) determine the extent of damage done to the host. These facts are needed to diagnose a disease and to recommend control measures of the disease concerned. Of the various ways of classification of plant diseases stated above, only the one classified on the nature of causal agencies gives adequate information about the cause of the disease development and related facts. Hence such a classfication has the most practical value and is often considered as modern system of classification of plant diseases. Next best is the one based on the nature of symptoms which is also very useful to diagnose a disease.

A. Non-parasitic diseases

Diseases caused by non-parasitic causal agencies are known as **non-parasitic diseases** or **physiological diseases,** also known as **deficiency diseases.** They are diseases having specific svmptoms. Although the symptoms and effects of certain diseases are well-known, the factor or combination of factors producing them is still questionable. Until early part of 1900 the field of non-parasitic diseases did not receive much attention. It is now well-recognized that much loss is also brought about through the effect of non-parasitic causal agencies, they are precisely: soil-moisture disbalance; nutritional disorders; high or low temperature; change in intensity and quality of light; atmospheric impurities like poisonous gases; certain disagreeable chemicals (fungicides and insecticides) of unusual doses, etc.

1. **Soil-moisture disbalance.** The general condition of water imbalance in diseased plants is associated with a derangement in its absorption transport, and/or transpiration and is often accompanied by a disturbance to the various other basic processes, such as carbohydrate and nitrogen metabolism, respiration, and mineral uptake. It is reasonable to presume that the primary effect of a disease is on the water status of the cells, which reflects in the various derangements in metabolism functions, leading to a progressive degeneration of vital activities, culminating in death. It will not be exaggeration to say that the fire of plant life burns in water. Different plants react differently with regard to soil-moisture variations. The most apparent effect of water deficiency is loss of form in herbaceous plant tissues due to loss of turgour exhibited by wilting, die-back, etc. Continued shortage of water eventually results in cessation of growth and decrease in the production of auxins. The physical structure of the soil not only affects its water retaining capacity and aeration but, if sufficiently hard and compact, can retard the growth of roots and so result in stunted and weak plants. Some of the symptoms due to soil-moisture disbalance are:

(i) **Wilting.** This is the most common symptom caused by deficiency of water supply. Wilting is recognizable from drooping of leaves and stem tips. There may be physiological and pathological wilting. The former condition is seen due to interference of water balance caused by irregularity of any of the physiological processes connected with water circulation in the body, such as absorption, conduction and transpiration. This deficiency can be easily corrected by rectifying the hindrance of absorption mechanism existing in the soil, whereas pathological wilting is caused by pathogenic agencies. This symptom would naturally arise not only from the interference with water uptake consequent to the damage to subterranean portions of the plant, but as a result of the presence of metabolites of certain infecting parasitic fungi which act as **wilt toxins.** The disease also arises out of the dysfunctions of root system resulting from injury and foot rot incited by

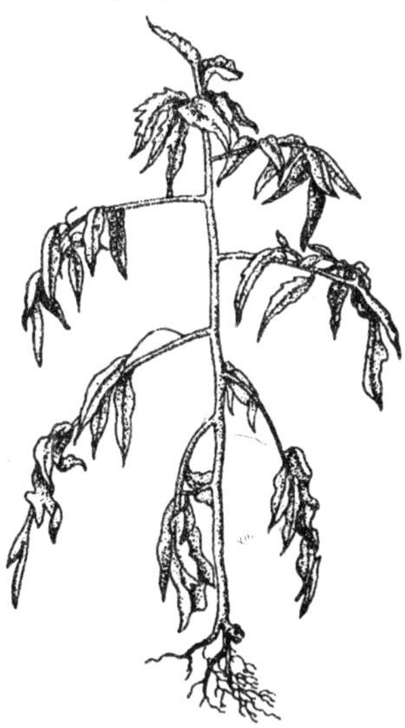

Fig. 24.2: *Verticillium wilt* of tomato

some fungi. Water shortage may also be caused by the dysfunctioning of conductive elements (rupturing of xylem elements), brought about by the growth of hyphae and masses of bacteria which also obstruct water flow through the xylem (Fig. 24.2).

(ii) **Die-back.** The dying backward from the tip of twigs or branches of trees and shrubs during the growing season is the symptom of plants and their branches caused by definite interference with water supply, either by tissue disintegration or by inducing gum formation in the xylem region.

(iii) **Scorch.** This is a symptom that is exhibited by the sudden death and browning of large indefinite areas of leaves and fruits (a) due to inadequate supply of water, and (b) due to a sudden and rapid loss of water from the leaves which cannot be replaced by the roots, since the soil moisture is low. The affected leaves show dead, brownish areas on thier margins or similar areas between the veins, but the leaves remain alive and do not drop so that the tree is little injured (Figs. 24.3B and C). The symptom is most pronounced on the side of the tree facing the wind.

(iv) **Blast.** Blast is the sudden death of young buds, inflorescences, or young fruits due to water shortage.

(v) **Firming.** The drying and dying of leaves specially of grasses, maize and tobacco for want of water is **firming.**

(vi) **Blossom-end rot.** Water-soaked spots appear at the junction of the floral parts increasing rapidly ultimately forming depressions turning dark sunken areas which are very suitable for the infection of parasites. This is the **blossom-end rot.** This happens due to excessive loss of water from the plant body.

(vii) **Dropsy.** An excess of water is unfavourable to trees, since it stimulates the excessive development of tender, succulent tissue, which is more readily invaded by parasitic fungi and more susceptible to extremes of heat and cold. Roots must get oxygen necessary for their development from the soil air, and, since most trees cannot obtain sufficient oxygen from soil saturated with water which results in the death of the roots by asphyxiation. Bark becomes very soft and spongy and several times thicker than normal, such a condition is commonly known as **dropsy.**

2. **Nutritional disorders.** Abnormalities of one kind or the other appear in plants due to either pronounced shortage of supply or in excess of certain essential elements which really upsets the normal nutritive processes. These abnormalities are known as mineral-deficiency or mineral-excess diseases. These diseases are expressed by some characteristic symptoms like **chlorosis, yellowing, defoliation, rosetting, dark spots,** etc. Here only the most common and widely prevalent and comparatively easily recognizable ones are considered. Unsuitable

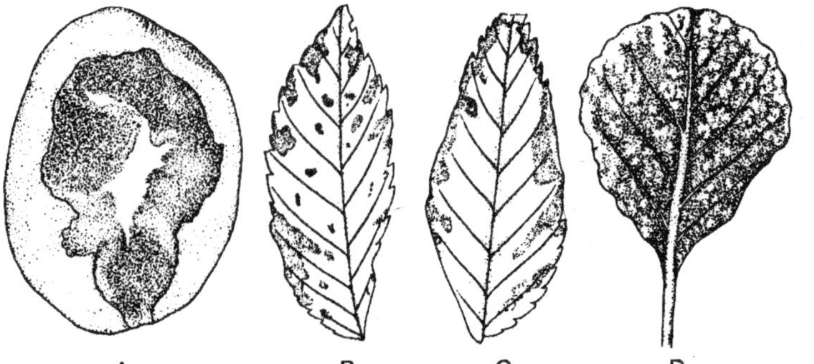

Fig. 24.3: Symptoms of non-parasitic diseases. A. Blackheart of potato. B-C. Scorch of leaves. D. Iron-deficiency disease of cabbage

chemical composition or pH of the soil can be damaging to plants. Some plants can withstand considerable alkalinity or acidity, but most grow best around neutrality or slight acidity. An unfvourable pH results in poor growth and sometimes death. It also has indirect effects on the availability of mineral elements in the soil to plant roots, as in calcareous soils where it reduces the availability of iron and manganese (lime induced chlorosis). It is sometimes possible to correct high acidity by addition of lime and, less easily, high alkalinity by addition of sulphur, although this may have phytotoxic effects. The salts responsible for the unfavourable pH can sometimes be washed out by irrigation.

Faulty mineral nutrition can predispose plants to attack by some pathogens. Potato blight caused by *Phytophthora infestans* is said to be more severe on plants growing in magnesium deficient soil, and excess of nitrogen results in 'soft' plants which tend to be more susceptible to pathogens and pests.

(i) **Iron-deficiency diseases. Chlorosis** is a most common iron-deficiency disease symptom. It is exhibited by a very acute yellowing (Fig. 24.3D) of green parts and usually happens when the soil becomes alkaline due to conversion of available ferrous salt to unavailable ferric salts. It is not always due to an absolute deficiency of iron in the soil. The chief causes of chlorosis are usually due to presence of excess of lime—**lime-induced chlorosis** or of manganese—**manganese-induced chlorosis** in the soil. The lime-induced chlorosis is rather common in vineyards, orchards and in clovers, whereas the manganese-induced chlorosis, comparatively less common, is best known in pineapple cultivation as 'yellows'. This is very well-known pineapple disease.

(ii) **Boron-deficiency diseases.** Boron-deficiency, in general, affects storage organs of plants such as roots, tubers, fruits, and similar other organs. The diseases are recognizable by various symptoms of which more common ones are, **heart** and **dry rot** caused by distortion and necrosis of fleshy tissue. The chief anatomical response to boron-deficiency is visible in the meristematic regions of certain plants leading to hyperplasia, hypertrophy and abnormal differentiation of parts. Most common unfolded leaves fail to develop normally and ultimately become necrotic and fall off. Similar is the condition of the dormant buds in the axils of older leaves. Disturbance of normal functioning and development of vascular tissue takes place. In apples due to boron-deficiency light brown spots of dead tissue appear in the fruits, and in trees **rosettee** and **die back.** In case of cruciferous plants, particularly in cauliflower, leaves become **deformed, brown-coloured,** sometimes **discolouration** takes place. In comparison with this, cabbage is less susceptible to boron-deficiency. In cabbage, **discolouration of head** and **break down in the core** take place.

(iii) **Potassium-deficiency diseases.** Symptoms of potassium-deficiency diseases are very clearly exhibited by the leaves and can be recognised from **mottle** and **chlorosis** of leaves particularly near the margins. In some cases the entire leaf turns brown, dies, and becomes brittle. Necrotic spots may also occur in the inner portion of the leaf blade. Often the first symptom of potassium-deficiency is exhibited by dark-brown colour of the leaflets which is followed by yellowing or bronzing.

(iv) **Phosphorus-deficiency disease.** A deficiency of phosphorus in many plants results in abnormally high quantities of anthocyanins producing purpling of leaves.

3. **High or low temperature effects.** Every growing plant has its own temperature range of normal growth. Any variation either above or below of which results in the irregularities, since temperature affects almost every function of life. These irregularities in growth are expressed by the various symptoms of the

diseases induced by high or low temperature. The nature of the symptoms is also dependent on the nature of plants, their respective parts affected and the extent of variations of temperature. Here only a few are discussed.

(i) **High-temperature effects.** Growing plants when subjected to temperatures much higher the optimum are liable to sustain certain injury of cell degeneration which is considered as symptom of the disease induced by high-temperature. The symptom may be visible both in aerial as well as subterranean parts. Most common high soil temperature disease of potato tuber is the **blackheart of potato** expressed by blackened centres (Fig. 24.3A). High temperature above 90°F in transit may also cause blackheart. Similar symptom may also develop when potatoes are put in poorly ventilated storage places. Specially the tubers at the bottom of deep piles are very commonly liable to face shortage of oxygen developing blackheart. Another common high-temperature injury is **sunscald of vegetables,** particularly of tomato. To start with, a water-soaked appearance of tissue is visible which is followed by rapid desiccation leading to the formation of sunken areas which ultimately turn grey in green fruits and yellow in ripe ones. Premature casting of leaves may occur in trees when subjected to high-temperature. Leaves of the inner crown are first affected, because they are more sensitive to heat, this is known as **heat of defoliation.** Again bark when suddenly exposed to the summer sun by removal of neighbouring foliages becomes overheated and dries out. Such heating and drying out of the bark, resulting in open wounds, occurs on certain young trees with smooth, thin bark, known as **sunscald** also as **bark scorch.** Sometimes leaves and twigs of Douglas fir are covered with white sugar. Sugar is exuded in solution directly from the leaf tips during the hot, dry, cloudless days of summer—**sugar exudation.** The water then evaporates leaving the sugar encrusted on the leaves. The effect of sugar exudation is rather negligible.

(ii) **Low-temperature effects.** Potato tubers when stored in cold storage are often liable to freezing injury of temperature range of 0°C to 5°C when stored for a long period of time. The freezing injury is seldom externally visible. The freezing injury results in **necrosis** of cells. Necrosis may be **ring necrosis; net necrosis** and **blotch type of necrosis.** The ring necrosis of discoloured tissue occurs in the peripheral region, net necrosis of blackened small areas are rather scattered throughout, and blotch type usually comprises irregular areas of various sizes and colour—opaque grey to black in the tubers.

4. **Effects of change in intensity and quality of light.** Both reduced or increased intensities and any appreciable change in the quality of light result in the development of various symptoms of the disease. Reduced light intensity promotes succulent growth along with developing internodes longer than usual and turning weak in physical structure and retards chlorophyll formation as a result of which normal green colour does not develop resulting in the **etiolation** which is rather a very common symptom. The change in quality of light, for example, when the atmosphere is clear short waves reach the earth with greater intensity inducing **sunscald** of leaves and fruits of vegetables.

5. **Effects of atmospheric impurities.** Gas, smoke and other air pollutants of the atmosphere cause serious damage to the plant parts. Orchards and shade trees are commonly injured by unfavourable conditions of the air or soil arising from industrial processes. Forest trees escape most of these adverse factors. Death or injury is often by the escape of artificially illuminating gas into the soil or by the escape of steam over-heating the soil, encountered in young succulent plants which are sensitive to some of the constituents of illu-

minating gas. Forest trees are often affected by smoke. The **smoke injury** is caused by the gases arising from the incomplete combustion of coal. From the smelting of ores containing sulphur, sulphur dioxide frequently causes severe damage. The symptoms consist of a variety of forms from **yellowing** and **reddening** of tissues of leaves to **defoliation** and **stunting.** Three stages of smoke injury are reognized: (i) the **acute stage** due to high concentration of smoke in the air causes rapid discolouration of the foliage followed by defoliation and in extreme cases by the death of the plant; (ii) the **chronic stage** due to small quantities of gas present in the atmosphere produces unhealthy condition of the plant expressed in stunted growth; and (iii) the **invisible stage,** due to the action of gas in a very dilute quantity, is characterized by reduction in growth. The gas confines its effect to the chlorophyll-bearing or green parts of plants, woody tissues are not directly affected. In addition to direct injury to foliage, indirect injury is claimed because of soil changes, such as the acidity of the smoke depleting the soil of calcium carbonate and thus reducing the number and hampering the activity of nitrifying bacteria. Sometimes gas injuries occur in certain fruits and vegetables when they are fumigated in storage and transit, since fumigation becomes necessary to protect them against certain diseases during storage and as a quarantine requirement. The symptoms are usually discolouration of parts, appearance of sunken brown spots, etc. Again, the effects of air pollutants on plants can be seen around the major population and industrial centres. Plants affected by air pollutants generally respond according to the principal toxicant present. Burning, bronzing, silvering, and growth abnormalities and necrosis are common symptoms. Sulphur-dioxide causes black tip or mango necrosis.

6. **Effects of fungicides and insecticides.** Certain chemicals—fungicides and insecticides, when sprayed on plants as a part of plant protection measure, if sufficient care is not taken during use, cause injuries particularly to the leaves. The injuries are in the nature of discolouration of the affected parts, abnormal growth leading to unusually rapid development of certain parts of the leaf induced by the careless spraying of pesticide 2, 4-D. Russeting in apples may be caused by parathion; burning, spotting discolouration by improper application of Bordeaux mixture sprays.

B. Parasitic diseases

Diseases induced by parasitic causal agencies are the **parasitic diseases.** The development of a parasitic disease induced by parasitic causal agencies particularly by fungi, bacteria, slime molds and viruses has a sequence of events following in cyclic order and as such comprises of disease cycle. The sequence of events of the disease cycle is greatly conditionable. But the disease cycle basically consists of two phases—the **dormant phase** and the **active phase.** During dormant phase the causal agency (pathogen) remains in dormant condition and overcomes the unfavourable conditions in the perennating organs. These perennating organs may be: spores of varied forms, resting vegetative cells, resting mycelium, virus particles, etc. The perennating organs may remain in or on propagating organs of the host, in diseased plant debris, in soil, in collateral hosts, in insect body, etc. With the advent of favourable environmental conditions the pathogen becomes active and there starts the active phase and the perennating organs produce primary inoculum. When disease occurs, its quality, its severity and its epidemic potential or how fast it will advance is very much dependent on the interactions among host, pathogen and physical environment. The process starts with pre-inoculation (prepenetration) period when the inoculum is produced which is again controlled by pre-inoculation environment, such as condition of the source of inoculum (perennating organs) and its microclimate. After successful development of inoculum, will follow the dispersal of inoculum. The inoculum arrives the host tissue after either being dispersed by some agency or directly coming in contact without the help of any medium. The density, viability, and potential infectivity of the inoculum are important for successful infection. It is meaningless unless the inoculum has found susceptible hosts and can penetrate the host and establish infection. This is the most vital link in the entire process where environmental dependencies are quite strong since most pathogens have an obligate free moisture requirement. This process is also temperature dependent and so the interaction of time span and temperature of moisture on host surfaces become important. The situation may be further complicated by light effects, by the charge that can take place in the infection court and by the age and physiology of hosts. Precisely speaking, it is the microclimate of the infection court that plays a very important role in the establishment of host infection. The details of development of the disease and the duration of the disease cycle are often extremely variable. In some case, the inoculum (secondary inoculum) produced by the primary infection is dispersed by various agencies and on coming in contact with the same host, it produces secondary

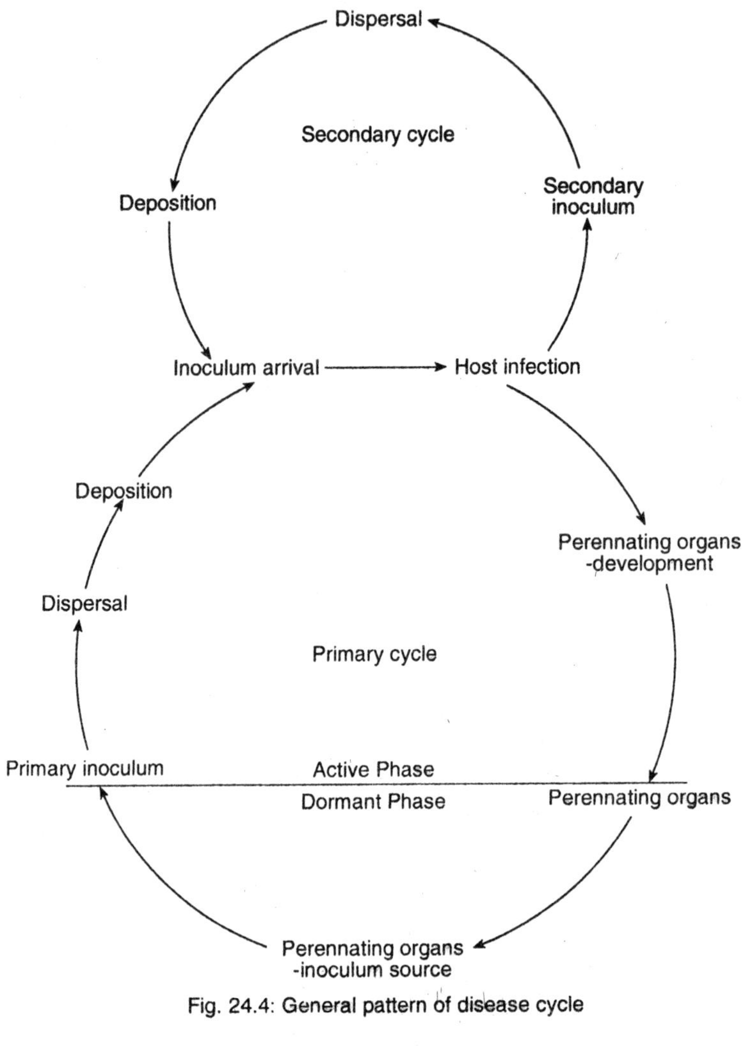

Fig. 24.4: General pattern of disease cycle

infection and secondary disease cycle as against the primary disease cycle produced by the primary inoculum developed by the perennating organs of the pathogen, whereas in others, the secondary disease cycle may be absent.

The general pattern of disease cycle is presented in Figure 24.4.

Some of the parasitic diseases are incorporated in the chapter to follow:

Some parasitic angiosperms have a wide and apparently unselective host range but in others there seems to be some degree of host specialization. *Viscum album* is a semiparasite on trees. It can adversely affect growth and yield as well as bring about premature death of severely infected trees. *Dendrophthae* sp. and *Phoradendron* sp. also can cause extensive damage to various trees. *Orobanche* sp. damages tobacco crop. *Cuscuta* spp. can cause severe damage to leguminous plants. *Striga* spp. are partial root parasites of sugarcane, cereals, maize and millets.

There are large number of parasitic algae of which special mention may be made of the genus *Cephaleuros*. Different species of *Cephaleuros* grow as parasites on various angiosperms causing severe damage to leaves and shoots. *C. virescens* forms the red rust of tea plants causing economic loss. *C. mycoidea* attacks tea, citrus, and other plants causing red rust disease. This alga is sometimes epiphytic and sometimes parasitic on the leaves of the tea plant in India, but is usually parasitic on the stems. In Sudan *C. mycoidea* has been found to attack coffee, citrus and guava plants. *Cephaleuros* causes leaf spotting and shedding and the photosynthetic activity of the foliage is reduced. In case of severe infection, branches and twigs may be girdled inducing die back symptom. This disease is spread by air-borne sporangia which produce zoospores which invade the leaf or stem through stomata.

A large number of insects, mites and nematodes feed on various plants, thereby causing injury and symptoms which can resemble those caused by other pathogens. The damaged tissue may subsequently be colonized by fungi and bacteria, so that it is sometimes difficult to find out the primary cause of symptoms. Many wound parasites enter the injured plants through the injuries caused by insects, mites and nematodes. Insects damage plants by chewing, sucking, injecting phytotoxic saliva besides behaving as vectors and inoculators of pathogens, especially viruses. The injected phytotoxic saliva may diffuse into the tissues beyond the actual point of feeding and cause symptoms similar to those caused by fungi, bacteria and viruses. Such insects are known as **toxicogenic insects.** Insects also play a great role as disseminators of inoculum. Mites are frequently present on diseased plant tissue and it is sometimes difficult to assess their significance. Nematodes attack and sometimes kill host plants. More often their attack results in weakening, reduced growth and diminished yield, and it may also enable other more destructive pathogens to enter the host plant. When feeding on roots, nematodes can cause several types of injury: necrotic lesions induce formation of hypertrophic cells, and suppression of cell division of apical meristem of roots. Nematodes which attack the aerial parts of plants may cause discolouration, necrosis, blotches, spots, distortion, and galls on leaf, stem and seed. Buds growing points or flower primordia are attacked by some nematodes, resulting in abnormal growth of the affected plant.

SYMPTOMS AND SIGNS OF PLANT DISEASES

Diseases make themselves manifest by certain peculiar characters or **symptoms** which readily distinguish a sick plant from a normal or healthy one. The most common characters of indications of disease are: (a) discolouration of the foliage and of new growths; (b) wilting;

(c) dropping of foliage; (d) spotting of foliage; (e) perforation of foliage (shot-hole); (f) variegation of foliage (mosaic); (g) wilting or 'damping-off' of seedlings; (h) death of leaves, twigs, stems, etc. (necrosis); (i) reduction in size of parts (dwarfing or atrophy); (j) increase in size of parts (hypertrophy); (k) formation of excrescences namely galls, pustules, corky outgrowths, etc.; (l) cankers; (m) malformation of fruits; (n) witches' brooms; (o) leaf curls; (p) leaf rosettes; (q) hairy-root; (r) exudations (gums, resins); (s) sun-burns; and (t) rots.

A symptom may appear as the result of any one of a number of causes, for example, wilting may be due to attacks from insects, fungi, nematodes, or unsuitable soil conditions; hypertrophies may be due to the attacks of insects, nematodes, fungi, or bacteria; witches' brooms may be due to the attack of insects of fungi; etc. Again, expression of disease symptoms is influenced by several factors (both internal and external) that play a great role on the causal agency, the host, and the interaction between causal agency and host.

However, it must be remembered that the part in which the disease becomes manifest to the eye is not necessarily the seat of the disease. The cause of the disease may be in a part of the plant very remote from the part in which some one or more of the preceding symptoms appear, e.g., a disease of the roots may cause a discolouration, wilting, or dropping of the leaves. It is also true that some diseases may be present in the plant for a long time before the plant shows any external evidence of the disease. This fact increases the difficulties of detection of certain diseases by inspection, and thereby increases the dangers in the introduction of foreign plants.

Diagnosis is one of the important steps in the study of diseases in plants. It is a process of determining what causes a specific plant disease. It also consists in discrimination between diseased and healthy plants and in distinguishing one disease from another by means of their characteristic symptoms and signs. Plant pathologists must be good diagnosticians. One of the tools of diagnosis is the careful use of plant **symptoms** which are the external expressions of the plant disease exhibited by the host plant itself. Symptoms are diseased conditions, evidences of sickness or injury which the plant presents and are usually visible and may also be detected by taste, smell or touch. Every disease in plants is expressed in a variety of symptoms and signs which constitute the clinical picture or **syndrome** of the disease. **Signs** are again evidences of disease other than those expressed by the plant itself. Usually signs consist of structures (mycelium, sclerotia, bacterial ooze, etc.) of the pathogen inducing disease. They may also be the products of the interaction of parasitic agency and suscept, such as cell inclusions, and exudates. Sign is actually the clinical manifestation of a disease. The symptoms may be expressed in the form of distinct localized structural changes known as **lesions.** Such symptoms are **lesional symptoms,** and **habitual symptoms** when they are expressed in the form of an unnatural or abnormal habit of the general appearance of the plant. When a variety of symptoms are exhibited in the same plant, these symptoms are combined to form a **symptom complex.** The symptom complexes are more common than single symptoms. Symptoms are again **primary** or **secondary.** The primary symptoms are exhibited in the host tissues immediately invaded by the pathogen, whereas the secondary symptoms are expressed in distant and uninvaded parts at a distance from the invasion. Depending upon whether one is observing the manifestation of disease in cells, tissues or entire organs or plants, the symptoms may be **cytological, histological** or **morphological.**

Pathogenesis, pathological anatomy and physiology, symptomatology and diagnosis may be investigaed in principle on individual plants. In contrast to this, in **epidemiology** the plant population as such moves into centre of consideration. Epidemiology nearly always deals with a great number of plants, e.g., with a plant formation in a field or in a plantation. The sickness of an individual plant usually is unimportant only with its epidemic occurrence the disease gains interest. The investigation of epidemiological events is carried out mainly on the basis of mathematical-statistical models.

Symptoms fall into three categoreis: (1) **necrotic,** (2) **atrophic,** (3) **hypertrophic.** Sometimes disease caused by one organism may have more than one type of symptoms.

1. **Necrotic symptoms.** These symptoms are evidences of death (**necrosis**) of the affected tissues caused by both parasitic and non-parasitic causal agencies. The best known of these in death of green tissue, which is shown by a marked change in colour, usually first yellowing followed by browning, reddening or occasionally greying when the tissue finally dies. The entire organ may be killed—'leaf blight', 'bud blight', or only a portion of it may be affected.

2. **Atrophic symptoms.** A slowing down in development of the affected plant or parts thereof, resulting from subnormal cell division (**hypoplasia**) or from degeneration of the cells. **Dwarfing** and **nanism** are the best known symptoms in this category, among its causes are unfavourable environmental conditions, such as repeated frosts, unfavourable soil conditions; and infection of viruses, and pathogenic organisms. Chlorosis expresses underdevelopment of the chlorophyll apparatus, and is caused by various agencies, such as viruses, bacteria, fungi and excess or deficiency of certain mineral elements in the soil. Leaves that are both chlorotic and dwarfed are symptoms of certain virus diseases and of diseases caused by parasitic organisms.

3. **Hypertrophic symptoms.** Overgrowths of all kinds resulting from an abnormal increase in the number of cells are due to excessive cell division (numerical **hyperplasia**) or from an abnormal increase in the size of cells (simple **hypertrophy**) or from both. They are striking symptoms of disease caused by both parasitic and non-parasitic causal agencies, expressed by various prominent malformations and excrescences, such as **galls, burls, intumescences, hypertrophied lenticels, fasciations, witches' brooms, hairy root, leaf curls, leaf blisters** and **deformations of fruit and flowers.**

NECROTIC SYMPTOMS

Yellows (Yellowing). The yellowing of parts containing chlorophyll; it may be confined to spots, rings or streaks in leaves, stems or fruits or it may be generalized.

Browning. Various shades of browning commonly accompany dying of plant tissues.

Silvering. A silvery green or ash grey luster is developed on the surface of leaves or other organs. It is sometimes due to air-spaces below the cuticle or epidermis.

Hydrosis (Water-soaking). A water soaked translucent condition of the tissues is due to the passage of water from the cells into the intercellular spaces. This is an early symptom of many necrotic diseases.

Wilt (Wilting). A flaccid or drooping condition of leaves, shoots, or other parts is usually due to a diseased vascular system (Fig. 24.2). The wilting may be temporary, the plants recovering at night, or it may be permanent and progress to death of the parts involved. Again, wilting may be physiological or pathological, details of which have been incorporated under non-parasitic diseases. Vascular pathogens particularly *Verticillium* and *Fusarium,* also known as **wilt fungi** often cause discolouration of xylem due to accumulation of **melanins** or related materials which are forerunners of cell death and necrosis.

Scorch (Scorching) or **Burn (Burning).** A term used to describe the rapid death and browning of large indefinite areas in leaves

(Figs. 24.3B, C and 24.5D) or fruits. Drought, sun burn, spray injury, freezing are the more common causes of scorch, but some pathogens also cause such symptom.

Scald (Scalding). A pale or white appearance of fruits or leaves resembling that produced by scalding water is scalding.

Blotch (Blotching). A superficial discolouration of leaves or fruits in which a slight necrotic injury of the epidermal tissues is usually emphasized by the presence of dark mycelium of the fungus pathogen (Fig. 24.5E). This is the symptom from whch the sooty blotch disease of apples derives its name.

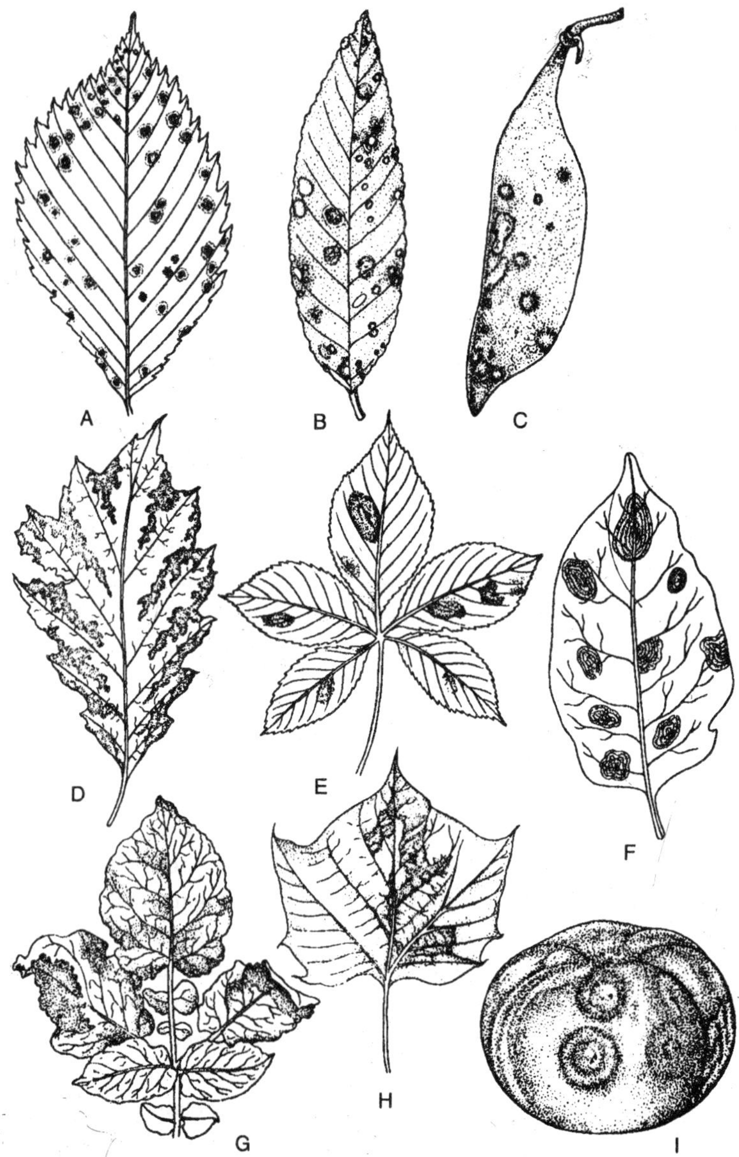

Fig. 24.5: Various disease symptoms. A. Leaf-spots—scattered, circular to oval-brown dead areas in the leaf. B. Leaf-spots and shot-holes in the leaf. C. Pod-spots of pea. D. Scorch—browning and yellowing between the veins and along the margins in the leaf. E. Blotch of leaf. F. Early blight on potato leaf showing target board effect. G. Late blight on potato leaf. H. Anthracnose of leaf. I. Anthracnose of tomato

Blight (Blighting). A condition developed due to the rapid killing of foliage, blossoms, etc., by pathogens (Figs. 24.5F and G). The killed tissue is often transformed into a slimy mass which emits pungent odour.

Blast (Blasting). This term is most commonly aplied to the sudden death of young buds, inflorescences or young fruits. This is an ancient term but it is still in use.

Die-back. A symptom resulting from the dying of twigs or branches from the tip towards the base, some pathogens may also cause such symptom. **Stag-headedness** is applied to the dying back of large trees.

Streak (Streaking) or **Stripe (Striping).** An elongated narrow lesion appears on stems or leaf-veins. It may be hydrotic, yellow, necrotic, etc.

Spot (Spotting). A very common symptom in many diseases in which more or less circular diseased areas develop. While spotting is very common on leaves it also occurs on petals, stems, fruits and other organs (Figs. 24.5A-C). Often the affected tissue dies, turns brown and dries out. Sometimes the central dead area is surrounded by one or more zones of reddish or yellowish tissue. There may be marked zonations or rings as in early blight of potato (Fig. 24.5F) or in certain virus ring-spot diseases. While usually circular, spots may be characteristically angular as in angular leaf spot of cotton or abutilon mosaic. Frequently the name of the affected organ is coupled with the term 'spot' as **leaf-spot, fruit-spot, stem-spot,** or the colour of the lesion is made the qualifying word as in **brown-spot, black-spot** and so on.

Shot-hole. The necrotic spotting of the leaves of certain plants, especially stone fruits, very commonly results in the diseased tissue dropping out. Such lesions are well described by the designation/shot-hole (Fig. 24.5B).

Anthracnose. Elongate, more or less angular spots appear on the veins on the lower side of the leaf (Fig. 24.5H) spreading into surrounding tissue and eventually appearing on the opposite side. Petioles, stem, and often fruits (Fig. 24.5I) may be affected.

Pitting (Pit). The dying of the fleshy tissues just beneath the epidermis of fruits, tubers and similar organs, often results in definite depressions or pits giving the surface a pocked appearance.

Mildews. In this type of symptom, growth of pathogen remains restricted on the surface of the host tissue. Mildews may be **downy mildews** and **powdery mildews.** In downy mildews the pathogen produces a cottony mycelial growth affecting host epidermis, whereas in the powdery mildews mycelium produces enormous numbers of spores producing a powdery appearance of the host surface.

Rusts. Rusty spores resembling crusts of iron appear in a crater formed by breaking of host epidermis swollen as pustules due to growth of spores from host tissue below the epidermis. Colour of rusts may be orange red, brown, yellow or black.

White rusts. White blister-like pustules appearing on infected host tissue break open exposing powdery masses of spores. White rusts resemble rusts by pustule formation and powdery masses of spores, but differ in white colour of pustules.

Smuts. Sooty or charcoal-black powdery masses of spores are developed by replacing ovaries. When the spore mass is covered by a reasonably tough covering it is known as **covered smut,** when covered by a brittle covering—**loose smut,** and when the spore masses have fishy smell it is **stinking smut.**

Scab. Here the pathogen produces a roughened or crust-like lesion.

Rot. Rot is a condition in which dead tissues are in a more or less advanced stage of disintegration under the influence of various enzymes which decompose cellulose, lignin and other associated substances of the plant tissue. The great majority of rots are of parasitic origin and are induced by either bacteria or fungi. Rotting may, however, result from the operation of non-parasitic causal agencies, as in the blossom-end rot of tomato and various kinds of rots in vegetables or fruits while in storage or during transportation to market. The character of rot depends on the structures involved, the causal agencies and external conditions. Basically, two types of rots are recognized: **white rot** and **brown rot**. There is another form of rot known as **soft rot** (Fig. 24.6A). For details see Weed Decay. Rots may also be grouped according to the special structures or organs involved and are often designated by such descriptive adjectives as: (a) **root-rots,** illustrated by decay of root crops and destruction of woody roots of forest trees, in the root-rot the taproot and secondary roots may rot owing to fungal attack resulting sudden death of affected plant, the entire crown shows symptoms of wilting; (b) **foot-rot,** here the stem at the level of the soil rots, though the parts above and below are more or less intact when the foot-rot finally girdles the entire plant, the translocation of sap is interfered with, and the plant ultimately succumbs to the disease; (c) **leaf-** or **stem-rots,** illustrated by the decay of succulent or hervaceous leaves or stems as in the late blight of potatoes, by the decay or modified stems such as tubers, rhizomes, etc., and by the decomposition of the woody stems including living standing timber; (d) **bud-rots** or decay of fleshy buds of which bud-rot of cocoanut is the example; (e) **fruit-rots,** which are soft or dry, varying with the nature of the fruit, the causal agency, and the environmental factors which are operative; (f) **heart-rots,** caused by wood-destroying fungi which make their best development in the heartwood, or in the portion of the tree trunk containing only dead tissue and are unable to advance any further (Fig. 24.6B); and (g) **sap-rots,** also caused by wood-destroying fungi who are able to establish themselves at once in living sap wood without first growing for a time in dead tissue, thus bringing about disintegration.

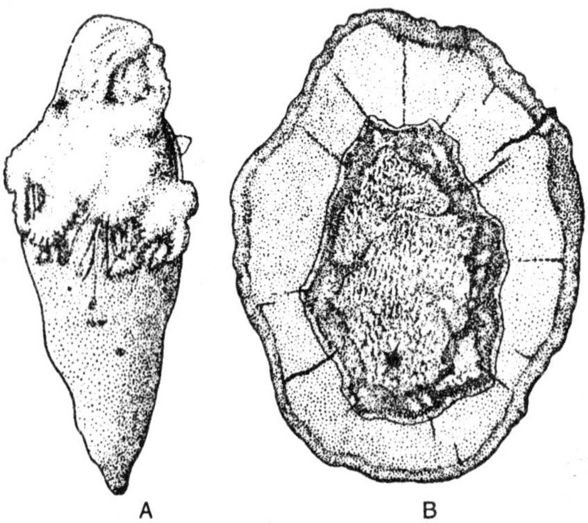

Fig. 24.6: Common rot. A. *Rhizopus* soft-rot of sweet potato. B. Heart-rot of Plum caused by *Fomes pomaceus*

Leak. A term applied to a stymptom of certain soft-rot diseases in which the freed juices leak out of diseased organs, as in leak of potatoes.

Mummification (Mummies). This is a symptom exhibited in the final stages of certain fruit rots where the rotted fruit dries out to form a dark, wrinkled, hard mass, called a 'mummy'.

Canker. This is a sunken necrotic lesion developed in the cortical tissues of stems, fruits or leaves. The term is most commonly applied to such lesions in the bark of trees, or less commonly to such lesions in the stems of annuals. A sharp demarcation of the affected area sometimes by a crack separating diseased and healthy tissue often marks a stop in the development of a canker. There may be callus formation at the healthy margin (Figs. 24.7A-C).

Canker is actually a disease that causes the death of definite and relatively localized areas of bark on branches or trunks of trees. Repeated callusing is necessary before a lesion can be

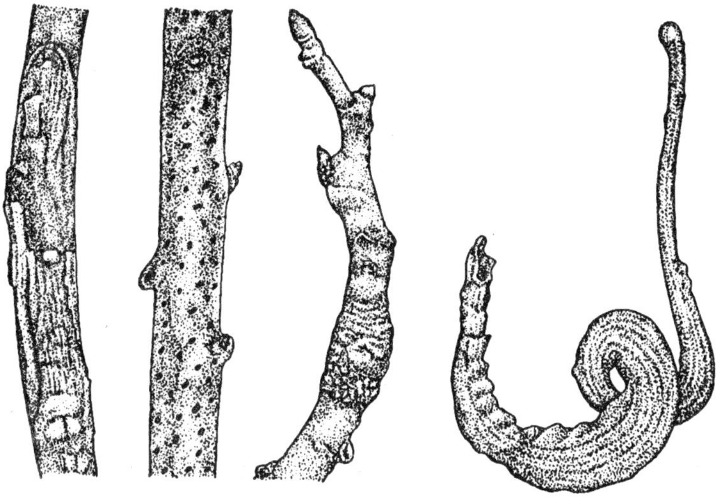

Fig. 24.7: Disease symptoms. A-C. Different stages of canker of twigs. D. Fasciation of apple twig

classes as a canker. Although most cankers are of fungus origin, they can be caused by non-infectious agents such as frost, sunscald or illuminating gas. Death of the bark and cambium is followed by death of the underlying wood, although the causal organism may or may not penetrae the wood. Canker disease may cause superficial lesion only and may result only in roughening the bark. Those cankers that penetrate to the cambium, sooner or later the bark sloughs off exposing the wood. Canker may be **annual** or **perennial**, the latter usually being most destructive and conspicuous. With annual cankers the agency causing the disease is operative for one season only, the injured tissues then being sloughed off and grown over by a single callus in the same way that a mechanical injury is healed, whereas the causal agency of perennial canker functions year after year. If caused by a fungus that is able to penetrate vigorously through living tissues during the growing season of the host, little or no callus is formed and, therefore, the stem, no matter how large, is finally girdled. The canker may be either **sunken** or **slightly swollen,** but ridges of callus will not be in evidence. Occasionally advance of mycelium may be checked by the formation of impenetrable cork layers in the bark. Then the stem recovers and the lesion is healed over. A perennial canker may persist indefinitely on a stem without ever girdling it.

Repeatedly the new callus tissue that the host develops around the advancing border of the lesion is killed, until the canker consists of more or less regular concentric ridges around the first infection, is the so-called 'concentric' and 'target' canker. Sometimes these callus ridges are irregular or broken in outline, resulting in a mishappen canker, swollen in some places and sunken in others.

Fungi causing long-standing perennial cankers have certain aggregate characteristics that distinguish them from other pathogens. The killing of living tissue is either by direct hyphal penetration or by secreting some substance toxic to cells in their vicinity so that the hyphae can penetrate the dead cells. The mycelium grows primarily in the inner bark and cambium, but, if the hyphae penetrate the wood at all, they are unable to spread extensively and reinfect the inner bark, otherwise the hyphae would spread through the stem and girdle it immediately. The mycelium is perennial and spreads slowly fro:..

cell to cell, with its rate of circumferential extension approximately equal to the rate of circumferential growth of the stem, for, if the mycelium grows more rapidly, it would soon girdle the stem. Its radial growth is faster than that of stem, otherwise it would never reach the cambium.

Canker causes malformation of the stem and discolouration of the wood, and trees break down along the canker. Canker may infect a stem when it is very young.

Damping-off. The rapid rotting of the stems of seedlings at the ground level so that tops fall over. The individual lesions are usually of the wet-rot type.

Bleeding. It is the chronic flow of sap from wounds or lesions. This is usually accompanied by an odour of fermentation. Elms and other trees with narrow crotches between limbs are liable to have these repeatedly split and closed by the wind. Organisms develop in the escaping sap and the products of fermentation may damage the living tissues producing a condition known as **slime-flux** of trees.

Ooze. Viscid masses composed largely of bacteria or the spores of fungi combined with the juices from the host plant form drops of various colour on the surface of the lesion.

Odour. Many diseased tissues give off characteristic odour of diagnostic value, e.g., Phloem necrosis of elm.

Gummosis. The oozing of gum from wounds or other openings in the bark of diseased trees is common in citrus- and stone-fruit trees.

Resinosis. The abnormal exudation of resin from wounds or lesions, as is seen in coniferous trees.

ATROPHIC OR HYPOPLASTC SYMPTOMS

Variegation. Patterned white areas in leaves or other organs are produced due to the failure of chlorophyll to develop in certain cells. Usually this symptom results from some defect in the hereditary system of the plant.

Chlorosis. Here a lighter green colour than normal is produced due to partial failure of chlorophyll to develop. It may be systemic throughout the plan as in aster-yellows or it may be patterned (as in mosaic which is a type of chlorosis). Aster-yellows, peach-yellows, etc. are really chlorosis rather than yellows although these names are firmly established.

Veinclearing. This is the most common symptom of virus infection. The green colour fails to develop along the veins and they may appear translucent.

Mosaic. Chlorosis in which the light green areas alternate with darker green areas in various patterns. The symptom is most common on leaves but also appears on fruits, stems and other green organs. When the chlorotic areas are white or nearly so and large, the mosaic may be termed **calico.** If the areas of the pattern are very small the mosaic may be termed a **mottle.**

Achromatosis. Results from the failure of cell pigments other than chlorophyll to appear where they normally should, e.g., the failure of red colour to develop on apples beneath the sooty-blotch fungus.

Abortion. A halting in the development of an organ after partial differentiation, e.g., the ovary of rye in ergot.

Suppression. Complete prevention of the development of certain organs, e.g., seeds in aster-yellows.

Etiolation. A syndrome involving lack of chlorophyll development, and abnormally elongated stem growth usually results from growth in the dark, though certain pathogens or malnutrition may also produce it.

Puckering. A symptom exhibited by leaves where the veins do not grow commensurately with the interveinal areas so that the leaf appears as it might if the veins were threads that had been pulled out through the petiole. This effect may be produced by a failure of the veins to grow properly as in certain virus diseases or it may result from an overgrowth of the interveinal tissue as in peach leaf curl.

Wrinkling (Savoying, Blistering, Crinkling). A very common symptom accompanying mosaic due to the failure of the chlorotic areas to develop normally and sometimes also to an overgrowth of the darker green areas.

Dwarfing (Dwarf) or Stunting (Stunt). This is the development of sub-normal size of the entire plant or some of its organs.

Rosetting (Rosette). Failure of the internodes of shoot or branch to elongate normally so that the foliage is crowded into a rosette.

HYPERTROPHIC AND HYPERPLASTIC SYMPTOMS

Restoration. It is the more or less complete development of tissues or organs normally rudimentary. The rudimentary stamens in the pistillate flowers of *Lychnis* become fully developed when attacked by a smut fungus; in certain other smut infections ovaries are developed in staminate flowers.

Prolepsis. Prolepsis is the premature development of a shoot from a bud.

Alteration. This is a symptom exhibited either by the entire floral axis or by the flowers only. Most common example is the sugar-cane smut, where the floral axis is transformed into a black, dusty whip induced by *Ustilago scitminea.* Large sclerotia replace the ovaries in rye inflorescence affected by *Claviceps purpurea.* Inflorescence of *Brassica compestrics* var. *sorson,* when attacked with *Albugo candida,* the petals turn into sepals and stamens into carpels.

Abscission. Here premature separation of leaves, flowers, or fruits takes place from the plant due to the formation of a special cell layer, the abscission layer.

Russeting (Scurfing). A brownish superficial roughening of the skin of fruits, tubers or other organs, e.g., russeting of apples takes place following spray injury.

Heterotopy. It is the development of more or less normal tissues or organs in an abnormal location. The development of potato tubers on the stems above ground is an example. Another is the production of leaf-like structures in place of sepals, petals, stamens or carpels. This latter symptom is called **phyllody, sepalody, staminody,** etc. may occur.

Bronzing. A peculiar coppery tinge is sometimes exhibited by the epidermis of green leaves.

Virescence (Greening). The development of green colour in tissues or organs in which it is not normally present.

Anthocyanescence. An abnormal red or purple colour is developed in leaves or other organs.

Intumescence (Edema). Intumescence is a small epidermal swelling or eruption. Commonly this is due to the marked hypertrophy of groups of epidermal or sub-epidermal cells under conditions of excess moisture.

Sarcody. It is an abnormal development of tissue above a wound of a stem or trunk.

Tumefaction (Tumora). A local abnormal overgrowth that takes place on any part of the plant. A variety of names are applied indiscriminately to such growths, such as **gall, wart, knot, tubercle, boil, blister, club,** etc. (Fig. 24.8).

Callus. Tissue overgrowth that is produced in response to damage and proceeding towards the healing of the damaged part.

Curl (Curling, Rolling). Abnormal bending or curling of shoots or leaves, it is due to localized overgrowth on one side or in certain tissues. Sometimes the same symptom may be due to subnormal growth on the concave surface.

Scab. Definite more or less circular, usually slightly raised and roughened, ulcer-like lesions that are developed on fruits, tubers, leaves,

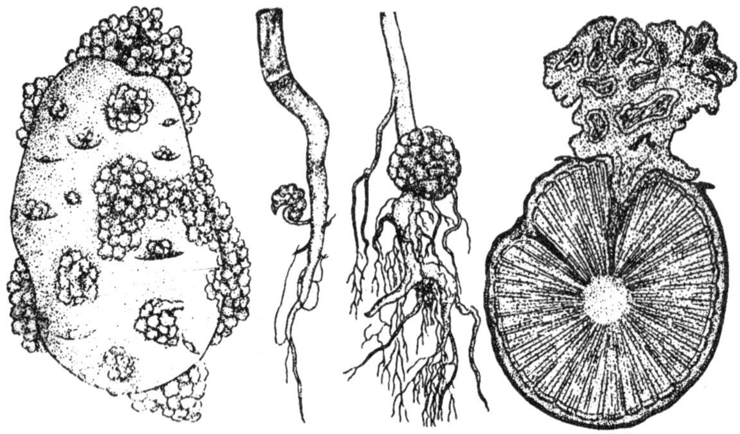

Fig. 24.8: Disease symptom—tumefaction. A. Wart of potato tuber. B. Gall formation on roots. C. Crown gall of raspberry in cross-section

stems, etc. Scab resulsts from the overgrowth of epidermal and cortical tissues usually modified by rupturing and suberization of cell walls and often aggravated by the presence of dark coloured mycelium or stromata, e.g., apple scab, potato scab, etc.

Fasciation. This symptom is exhibited by a cohesive growth of organs normally separate. Usually the resulting structure presents a flattened, much branched appearance (Fig. 24.7D), stems, roots, flowers or fruits may be affected.

Fasciculation. Here clustering of organs such as twigs, roots, flowers and fruits about a common focus takes place, due usually to the adventitious development of such organs.

Witches' brooms. These are closely grouped clusters of fine slender branches, generally arranged more or less parallel to each other, and frequently originating from an

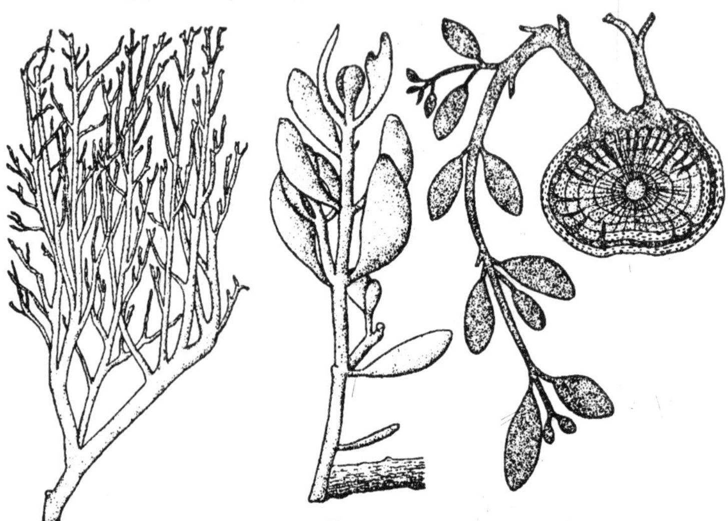

Fig. 24.9: Disease symptom and causal agency. A. Witches' brooms. B. Viscum album (Mistletoe). C. Sectional view of a wig infested with *Viscum album*

enlarged axis (Fig. 24.9A). They are induced by parasitic causal agencies.

Hairy root. This is the development of an abnormal number of fine fibrous roots in compact clusters.

Proliferation. A condition resulting from the continued development of a part after it has reachd the stage at which it normally ceases to grow.

DEVELOPMENT OF PLANT DISEASE

A plant becomes diseased very commonly when it is attacked by a pathogen or, very rarely, when it is affected by an abiotic agent. Hence for the development of a plant disease, at least two components (plant and pathogen) must come in contact and must interact. But if at the time of contact and for sometime later conditions are too cold, too dry, or some other extreme, the pathogen may be unable to attack or the plant (host) may be able to resist the pathogen attack, and in spite of the two (host and pathogen) being in contact, no disease develops. Naturally, then, a third component, namely, a set of environmental conditions within a favourable range, must also occur for the development of a disease. Each of the three components, however, can display considerable variability and as one component changes, it affects the degree of disease severity within an individual plant and within a plant population. For example, the host plant may be more or less resistant to the pathogen or the pathogen may be in small number, or it may be in a dormant state, or it may require a film of water or a specific vector. The environment may affect the rate of growth or multiplication and degree of virulence of the pathogen.

The interactions of the three components of disease development have been visualised as a triangle (Fig. 24.10) designated as the "desease triangle". Each side of the triangle represents as one of the three components. The length of each side is proportional to the sum total of the characteristics of each compoent that favour disease development. For example, if the plants are resistant the amount of disease would be small or zero, whereas, if the plants are susceptible, the host side would be long and that potential amount of disease could be great. Similarly, the more virulent, abundant, and active the pathogen, the longer, the pathogen side would be and the greater the potential amount of disease. Also, the more favourable the environmental conditions (e.g., temperature and moisture) that help pathogen or that reduce host resistance, the longer the environmental side would be and the greater the potential amount of disease. If the three components of the disease triangle could be measured, the area of the triangle would represent the amount of disease in a plant or in a plant population. If any of the three components is zero, there can be no disease.

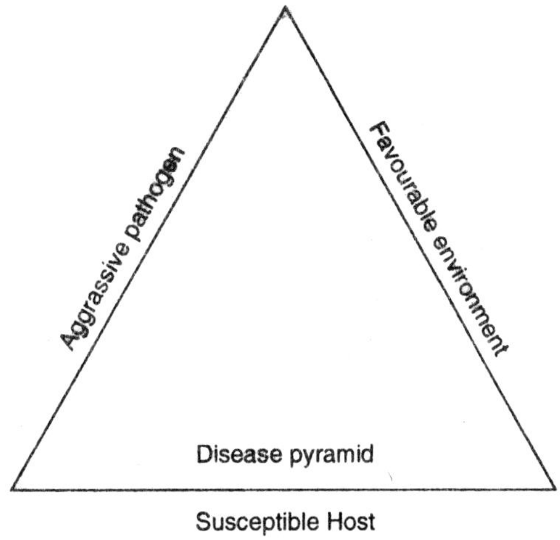

Fig. 24.10: Pathological process

During development of plant disease a series of more or less distinct events occur in succession leading to perpetuation of the disease and the pathogen or pathological process. This chain of events is called a **disease cycle** which corresponds fairly closely to the life cycle of the pathogen. But it refers primarily to the appearance, development, and perpetuation of the disease as a function of the pathogen rather

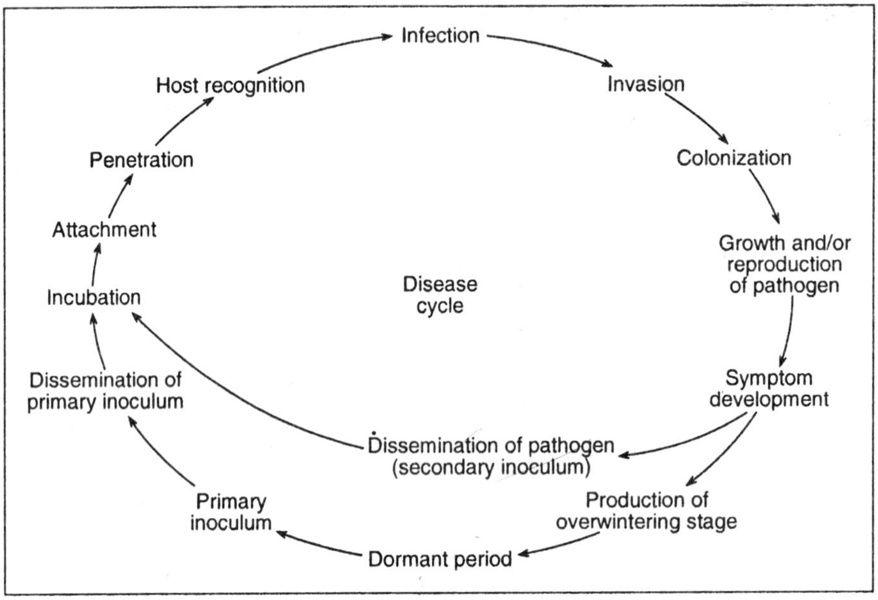

Fig. 24.11: Stages in development of a disease cycle

than to the pathogen itself. The disease cycle involves the changes in the host plant and its symptoms as well as those in the pathogen within a growing season. The primary events in a disease cycle are: inoculation, penetration, establishment of infection, colonization (invasion), growth and reproduction of the pathogen, dissemination of the pathogen, and survival of the pathogen in the absence of the host (overwintering or oversummering) of the pathogen (Fig. 24.11).

STUDY OF PLANT DISEASES

One of the important aspects of study of plant diseases is how to either cure the disease or prevent its recurrence. Before any cure can be recommended, the cause or in other words, the causal agency of the disease should be known—whether parasitic or non-parasitic.

Plant diseases can be studied in extremely variable methods which depend on the nature of disease, the causal agency, and the host involved. But the initial study of symptoms, periodicity, incidence, and extent of damage done to the host—all these aspects are to be studied irrespective of the nature of disease, causal agency, and host infected. This initial study often produces information to identify the disease under study either as a parasitic or a non-parasitic disease. If the disease is a parasitic one, the next step will be to determine the nature of causal agency. This area of study extends: (i) a closer examination of disease infected host tissue in the laboratory in naked eye and under microscope, (ii) isolation of pathogen from the disease infected tissue, and (iii) growing of pathogen in pure culture.

The presence of a pathogen can frequently be determined by naked-eye examinations of lesions. But in many cases the pathogen is too small to be seen without magnification, or may be internal within the tissues or organs. Microscopic examination of surface scrapped mounts or dissected mounts of lesions without staining in water or staining in lactophenol cotton blue or potassium hydroxide and phloxine may also reveal much information about the pathogen. Often microscopic examination of thin free hand or serial microtome sections stained in suitable dye combinations is necessary to work out the details of the pathogen. The microscopic examination is expected to reveal the nature of pathogen, at least, the group to which it belongs and the details of mode of growth in the host tissue, nutrition, and damage done to the host cells by the pathogen.

Isolation is a technique by which a pathogen is transferred from its natural habitat (infected host tissue) to a suitable artificial medium. Isolation of pathogen should be done from vigorously infected areas. Small beats of infected tissue should be placed on sterilized nutritive agar medium in Petri dishes or culture tubes. Earlier to this the selected infected host tissue bits should be cleaned from dirt with water and surface sterilized either by applying mercuric chloride or silver nitrate and sodium chloride technique. For isolation work two per cent potato dextrose agar medium has been found very suitable. The entire isolation process should be done asceptically in an isolation room or chamber especially built for the purpose. The Petri dishes or culture tubes, whatever used, should then be put in an incubator having incubation temperature range of 20°C to 25°C. In a few days time the pathogen grows out on the agar surface from the infected host tissue.

The pathogen is then transferred to fresh nutritive agar medium and grown in pure culture under the same cultural condition. The pathogen should then be accurately described. The procedures described above is, however, very suitable for the study of disease caused by fungi.

But just because a fungus is present in a disease infected tissue does not really prove that this particular fungus is responsible for inducing this particular disease. The pathogenicity or causal relationship of the fungus concerned has to be established. The pathogenicity tests of a pathogen are also designated as Koch's postulates or Koch's Rules of Proof. They comprise of certain postulates or conditions that are to be fulfilled, only when the causal relationship of a pathogen is established. They are as follows:

1. The causal organism must be constantly associated with the disease in question.

2. The causal organism must be isolated from the diseased plant and grown in pure culture and accurately described.

3. Inoculations with inocula from pure culture of the causal organism must reproduce the disease in question in the same species or variety of plant from which the causal organism was isolated.

4. The causal organism must be re-isolated from the plant in which disease has been produced by inoculating with inocula from pure culture.

These postulates cannot, however, be fulfilled with organisms which cannot be grown on artificial media. In that case constant accompaniment of the organism with the diseased plant may also be taken as proof of causal relationship.

Procedures for the first and second postulates have been discussed above. As to the third and fourth postulates, the inoculum should be a suspension of spores or mycelium of the fungus isolated from the infected host tissue. Suspension of spores or mycelium should be prepared by following standard technique. The suspension so prepared should be sprayed with an atomizer on the surface of the host plant which has been grown in a pot in the greenhouse. The variety of host plant must be the same as the one on which the disease had first appeared. The sprayed plant should then be covered with a bell-jar or polythene covering. The soil of the pot in which the host plant is grown should be kept moist. This enable to maintain humid condition around the plant. The sprayed potted plant with covering is then placed for twenty-four to forty-eight hours in a place where the temperature and humidity both are favourable for the appearance of the disease. The sprayed plant should not be exposed to direct sunlight, as the spore germination and mycelial growth may be hampered. Usually lesions start appearing after twenty-four to forty-eight hours. Ultimately typical disease symptoms that have been recorded earlier will be visible. The fungus is then re-isolated from the lesions produced by artificial inoculation. Its character is then studied and compared with what has been recorded earlier. When all the stages produce expected results, then only the pathogenicity of a particular fungus is established.

The procedure described above is, however, suitable for diseases where the infection is not soil borne. But in soil borne plant diseases the isolation of pathogen should be done from soil around the disased plant. These isolations will form the inocula and they are aseptically mixed with sterilized potted soil in which the seeds of the exactly same variety of plant should be sown after surface sterilization by suitable fungicides. Under suitable temperature and humidity the seeds develop into seedlings and receive infection through roots from the pathogen that has been growing in the soil and ultimately the disease appears as expected. In all the above experiments good number of uninoculated controls should be kept side by side to make a comparative study.

While growing in pure culture, the fungus concerned often produces spores. The study of sporulation and spores in details is also essential for identifying the fungus. The mode of formation of spores and spores in details is also essential for identifying the fungus. The mode of formation of spores and their nature and function, all these aspects besides being useful in the identification of the fungus also yield clue about the mode of dissemination of the disease and perpetuation of its inoculum. All these data may also be utilized in connection with recommending control measures of the disease.

DISSEMINATION OF PLANT DISEASES

Dissemination of plant diseases is basically dependent upon the dissemination of inoculum whose origin and behaviour are again controlled by various factors. To understand this aspect one has to consider the various aspects of inoculum: (i) source, (ii) development, (iii) dissemination, and (iv) behaviour. It has already been indicated that these aspects follow in regular sequence in a cyclic order and are interdependent for disease dissemination followed by disease development. Here only mechanism of dissemination of plant diseases with special reference to dissemination of inoculum is discussed in detail.

The spread of parasitic diseases from host to host and from locality to locality is primarily dependent upon the dissemination of pathogens (inocula). The dissemination may be: (i) **direct** when it takes place without the help of any external agency, e.g., explosive spores of *Taphrina deformans* that cause peach leaf curl; (ii) **indirect** when dissemination of pathogens (inocula) takes place through agencies or media like, seeds, propagating stocks, plant debris, manure, soil, field operations and implements, insects, air currents, water, etc. For effective dissemination the inoculum should be: (i) sufficiently productive, (ii) suitable for liberation, (iii) suitable for dissemination, (iv) able to stand influence of the agency, and (v) virulent. Following are some of the common mechanisms of dissemination.

Dissemination by seed. Pathogens may be carried inside or outside the seeds or fruits functioning as seeds. The infective materials may be mycelium, spores or any other part of the pathogens. One of the examples of externally seed-borne pathogen may be cited from *Tilletia caries* causes bunt of wheat; that of internally seed-borne is *Ustilago nuda* var. *tritici*, and of pathogen borne between the glumes and kernels in *Ustilago hordei* and *U. avenae* causing smuts of wheat, barley and oats respectively.

Dissemination by propagating stock. This is rather common in cases where plants are multiplied by using propagating stocks and is partly common with the virus diseases, as in Mosaic disease of sugarcane. Here the propagating stocks behave as carriers of the pathogens.

Dissemination with plant debris. During clearning of soil prior to sowing of seeds, previous year's infected plant debris should be thoroughly removed and burnt. But very often the infected plant debris instead of being burnt off is carelessly transported from place to place resulting in the dissemination of pathogens and spread of the diseases. One of the common examples is the late blight of potato where the pathogen perennates as oospore in the plant debris.

Dissemination with manure. Pathogens are disseminated when manures like, compost and cowdung are used particularly in seed bed soil contaminated with various pathogens. This is rather common with the damping-off of seedling disease.

Dissemination with soil. Transference of soil from one area to the other is a very common practice. When such soil is contaminated with pathogen knowingly or unknowingly contamination through soil is spread. Clubroot of crucifers is an example.

Dissemination by field operations and implements. Pathogens may be disseminated by field operations, such as transplanting of seedling, watering, weeding, harvesting operations and through implements. Example may be cited from the dissemination of the tobacco mosaic virus diseases.

Dissemination by insects. Insects play a very important role in the dissemination of plant diseases. The mechanism of dissemination, however, is extremely variable which depends on the nature of insects and the nature of inocula of the diseases. Again, the insects may play direct or indirect role. As to direct role, the inoculum may be carried: (i) on the body of the insects, (ii) in the body passing through the alimentary tract when they feed on the inoculum along with other exudates, or, (iii) as they feed on diseased host tissue they carry the inoculum and spread infection. Examples of direct role of insects for plant disease dissemination are as follows.

(i) The inoculum may get adhered to the body of the insects as they come and visit the diseased host, and when the same insects visit disease-free plants the inoculum carried by them comes in contact with the healthy host plant. Best example is Dutch elm disease caused by *Ceratocystis ulmi*, disseminated by bark beetles.

(ii) The spread of Ergot of rye disease is an example where the insects feed on the conidia along with honey during honey-dew stage, and the conidia pass through the alimentary tract of the insects and are brought in contact with the disease-free plants when they visit them.

(iii) The transmission of virus diseases is a classical example of dissemination of plant diseases throuh insects. Insects like leaf-hoppers and aphids feed on virus infected leaves and they go and feed on leaves of disease-free hosts and thereby spread infection behaving as **vectors** of virus diseases.

Insects indirectly help dissemination of plant diseases by causing wounds on the host surface through which the pathogen gains entrance in the host and induces disease. This may be illustrated from the dissemination of the Citrus canker. Another example of insects indirectly helping dissemination of plant diseases is the dissemination of the rust disease. Where insects transfer spermatia to the compatible receptive hyphae, dikaryotic condition is established and ultimately physiologic races are developed which are disseminated through aeciospores from the barberry host.

Dissemination by animals other than insects. Nematodes, snails, birds, and wild and domestic animals often help dissemination of plant diseases. Yellow ear rot of wheat disease caused by *Corynebacterium tritici* is dispersed by nematode *Anguina tritici,* seeds of *Dendrophthae* sp. are disseminated by birds, and cattle feeding with contaminated fooder often pass out viable fungal propagules (conidia, and sclerotia) in their dung are disseminated at different places.

Dissemination by air currents. The spores of many parasitic fungi are disseminated by air currents from diseased to diseae-free host. This may be illustrated from the spread of Powdery mildew disease. Besides this, the air currents may indirectly help dissemination of plant diseases. For example, diseased plant parts may be caught up by air currents and brought to distant places to spread disease.

Dissemination by water. The dissemination of pathogens by water may be in two ways: (i) flagellate spores of pathogenic fungi are entirely dependent on the supply of water for their dissemination; example, Wart disease of potato; (ii) pathogens like bacteria are often disseminated by splashing of raindrops, as in case of Citrus canker disease; and (iii) dissemination of conidia of red rot fungus (*Colletotrichum falcatum*) takes place through flowing irrigation water or rain water.

Dissemination by exporting and importing of commodities. That diseases are disseminated during exporting and importing various commodities is very well known. This may happen in different ways: (i) with contaminated commodities, (ii) with contaminated containers of the commodities, or (iii) with both contaminated commodities and containers. Some of the examples are: the introduction of Dutch elm disease in the United States, the introduction of Wart disease of potato in India, etc. As a control measure against this quarantine laws, inspection, and health certificate system have been introduced.

Dissemination by natural root grafting. This type of dissemination is encountered in the Oak wilt disease caused by *Endoconidiophora fagacearum.* Where natural grafting between the roots of diseased and disease-free plants takes place resulting in the spread of the disease. The oak wilt fungus living in the sapwood passes through the grafts from a diseased to a neighbouring disease-free tree.

Dissemination by shooting out of spores. In *Taphrina deformans,* the causal organism of Peach leaf curl disease, the ascospores are shot out from the asci, come in contact with the host, and cause infection and thereby the disease is disseminated.

Dissemination through angiospermic plant. Pea mottle virus is transmitted by *Cuscuta campestris.*

Dissemination through pollen grains. Tobacco ring spot virus and Potato spindle tuber virus are transmitted through pollen grains.

Dissemination through other media. Sclerotia of *Claviceps purpurea* unknowingly remain mixed with healthy rye seeds to help dissemination of disease being the source of inoculum. Besides this, mummified apples and pears also serve as source of inoculum for seasonal carry over and incidence of disease in orchards.

HOST PATHOGEN RELATIONSHIP

When a pathogen (inoculum) comes in contact with a potential host plant, a complex of interacting factors including temperature, moisture, host susceptibility, effects of other microorganisms present, aggressiveness of the pathogen and others determine the course of action of the pathogen.

The entire host-pathogen relationship actually depends largely on inoculum potential, successful host penetration followed by the capacity of the plant pathogens to derive nutrition from their hosts which is accomplished particularly with the help of various enzymes. During this process the pathogen is gainer, and the host besides being always on the defensive, is loser. As because the synthesis of carbohydrate and other metabolites is impaired, increase in respiration rate inhibits the synthesis of carbohydrates, and cell permeability is disturbed due to which distribution pattern of metabolites is affected. Besides these, transport of metabolites is impeded on account of diseased conducting vessels. Again, rot infection disturbs mineral nutrition of the host plant.

The behaviour of the pathogen with the host may thus be conveniently grouped into three stages: **pre-penetration, penetration**, and **post-penetration,** although these stages are not completely distinct.

A. Pre-penetration stage

During pre-penetration stage the pathogen (inoculum) on arrival on the host surface interacts sharply with the surrounding environment and host itself. The environment which is an aggregate of all external conditions inlcuding temperature, moisture (relative humidity), light and the competing microoranisms, affects the life and form of the pathogen of the inoculum. The optimum environmental condition for ideal growth of a pathogen again varies on the nature of the pathogen and the host surface. For example, the development and abstriction of conidia are favoured by high air temperature and humidity in downy mildew, whereas in powdery mildew both the number of spores and their germination are greater in bright sunlight. In cereal rusts uredospores germinate at low temperature, but the infection process is delayed at this temperature. Again, soil pH plays a very vital role in the growth of bacterial plant pathogen in the rhizosphere (area of soil immediately surrounding the roots).

Whether a pathogen will survive and grow on the host surface also depends on its behaviour with the exudates of the host surface and the microbial population present on it. The exudates of the host surface may encourage or inhibt the growth of the pathogen. The root exudates mainly sugars and amino acids are nutrients for the growth of fungi and bacteria. But root exudates like hydrocyanic acid, various organic acids and antibiotics are antifungal and antibacterial. For example, spores of *Rhizopus arrhizus* germinate only in presence of proline (amino acid) present in the rhizosphere region whereas exudates of root of onion varieties inhibit spore germination of *Colletotrichum circinaus.* Leaves also exude substances which may go in favour or against the growth of the pathogen. The glands of leaf hairs of gram contain malic acid which is antifungal and arrests the growth of *Uromyces ciceris arietini.* Protocatechuic acid, an exudate of onion skin is also antifungal. The pathogen has to neutralize these exudates or has to be resistant to them for survival. Besides these, the rhizosphere region contains microbial population which is antagonistic to the growth of the pathogen. As such, the pathogen has to overcome the above barriers during pre-penetration stage before it can survive for host penetration.

Once a favourable relationship is established with the host surface, multiplication or growth of the pathogen begins. Rapid proliferation of cells of bacterial pathogen results in a relatively short time. The bacterial cells so formed being delicate structures may be easily killed by unfavourable conditions. Hence they survive under layers of slime. Again, development of fungal phytopathogen usuallly includes spore germiantion by germ tube which grows producing infection hypha as a result of hyphal tip growth or may give rise to an appressorium which anchors the fungus to the host surface. Penetration peg is produced from the appressorium with which the pathogen causes host penetration. But multiplication of viruses takes place only in the living host cells.

B. Penetration stage

The success of host penetration leading to disease development is a very complicated process which is a combined effect of various factors like: (i) the nature and behaviour of the pathogen including its multiplication capacity, (ii) favourable physical conditions, and (iii) host susceptibility. Of all these factors, the factors nature and behaviour of the pathogen are the most important ones which control the overall disease development. The nature and behviour of pathogen encompass the **inoculum potential** of the pathogen. The inoculum potential is again a measure of the biological energy available for the colonization of a host. It is a function of: (i) inoculum density which refers to the number of viable propagules per unit area of leaf or stem or per unit volume of soil; (ii) the nutrients available to the infectious units that allow them to germinate or grow; (iii) the environment (temperature range of 15 to 25°C, moisture content 70 per cent, and relative humidity 90 to 95 per cent); (iv) the virulence (aggressiveness) or genetic capacity of the pathogen to cause disease; and (v) the susceptibility of the host.

Besides these, the physiological state of the host may have an effect on the ability of the pathogen to attack it or on the extent to which a pathogen may harm it. The concept that encompasses this phenomenon is termed **predisposition.** Factors involved in predisposition are: (i) age of the host; (ii) environmental conditions to which the pathogen has been exposed, i.e., light, humidity, soil environment, and temperature; (iii) infection by other pathogens; and (iv) presence of chemicals, i.e., pesticides, herbicides. Generally adverse conditions predispose a plant to greater susceptibilty to attack by a pathogen. Temperature predisposition phenomenon in nature is extremely variable depending on the nature of host and pathogen. In most bacterial and fungal diseases, free moisture is necessary for pathogen development. Most known cases of predisposition usually result in increased susceptibility of plants that were genetically resistant. A few cases are known in which a predisposing treatment has greatly increased resistance of plants that were genetically susceptible.

Certain fungal pathogens exhibit specificity in the part of host tissue infection. For example, some of them may remain restricted in the cortical tissue only throughout the entire period of attack and cause damage to it. Again, others remain restricted in the vascular tissue only, whereas still others do not attack plant unless its heart wood is developed and the pathogen remains confined there causing damage.

With favourable inoculum potential associated with aggressiveness under favourable predisposition, the pathogen will now enter the host tissue which is designated of infection. The host penetration and establishment of host-pathogen relationship and the subsequent development of disease symptoms involve a chemical interaction between host and pathogen. The pathogen employes offensive chemical weapons (molecules) to breach host barriers. The molecules that are produced by the pathogen induce defence response in the host are the **elicitors.** This happens, perhaps, because of the presence in the host plant of a resistance gene (R) which reorganizes and trigger into action by the elicitor molecule released by the pathogen. The pathogen-produced elicitor is the product of a pathogen gene which triggers the development of resistance in the host that makes this pathogen a virulent, which is called **avirulence gene.** Besides this, some phenolic compounds, like chlorogenic acid, Caffeic acid, and Ferulic acid, that are toxic to pathogens accumulate at a faster rate after infection, especially in a resistant variety of plant. Parasitism and host resistance run parallelly and are inseparable. A kind of biochemically warfare continues between the host and the pathogen in which one tries to outwit the other. Enzymes, toxins, growth regulators, polysaccharides, antibiotics are the important biochemical offensive weapons secreted by the pathogens (fungi and bacteria). Fungi produce great variety of enzymes, no doubt related to the nature of the materials colonized. Those normally produced have been called **constitutive enzymes** as distinct from **adaptive** or **inductive enzymes** whose formation is stimulated by the presence of the appropriate substrate. Enzymes catalyze many of the biochemical processes of living organisms and thus play a fundamental role in the host-pathogen interaction. They are concerned not only in the initial entry of the pathogen and its spread within the host but also the degradation of host tissue into metabolites which the pathogen can utilize. Enzymes break down cell walls and connective cell layers. They also seriously affect the respiratory processes, growth, ionic exchange of the host cell causing interference with the cell permeability and other normal functions. Viruses, however, on account of their basic nature cannot produce any of these substances. They multiply by directing the host cell to manufacture viral nucleic acid and viral protein.

Host penetration takes place: (i) through natural openings, (ii) through wounds, (iii) by direct penetration of surface cells causing tissue disintegration, and (iv) through specific parts or organs. Both bacteria and viruses enter the host tissue mainly through wounds, whereas the fungal pathogens gain entrance in the host through natural openings, wounds and by direct penetration through cuticle and outer wall of the surface cells, or root hairs, or through specific parts or organs of the host. The direct penetration process is more complicated than the entry through natural openings and wounds. It is performed by breaking the host structural barriers through wide range of chemical actions.

(i) **Entry through natural openings.** Both bacteria and fungi gain entrance into the host through natural openings such as stomata, lenticels, hydathodes, nectaries, leaf scars, stigma, etc. This is a process in which the pathogens have an easy access to the host, except in cases where substomatal hairs may cause resistance against the host entry.

(ii) **Entry through wounds.** Wounds caused due to natural calamities (storm, fire, etc.); during field operations; by insects, by accidental breaking of parts or otherwise; offer easy passages of pathogens in the host. But so far as viruses are concerned, the host entry is only through wounds.

(iii) **Entry by direct penetration of surface cells.** The entry of pathogen by direct penetratin of the outer wall of the host surface cells is rather a difficult process for which the pathogen usually requires high moisture or free water supply. It is even more difficult in leaves with waxy covering on their surface which allows water to run off freely. Fungal pathogens penetrate into host either by boring through the outer wall of the surface cell or penetration is effected by pressure and sometimes due to chemical softening or solubilizing of the barrier caused by the solvent action of enzymes secreted by the infecting organ. After the hypha made contact with a suitable host surface, some growth in close contact with it takes place. This is followed by the development of an appressorium or increase in diameter of hypha serving as adherent area from which develops penetration tube. The penetration tube penetrates through the cuticle at a point softened by enzymatic action and followed by mechanical pressure. In some cases, germ tube produced by spore germination passes down between the radial wall of the adjoining cells without actually entering the cells. Other fungi send only haustoria in the epidermal cell through the outer cutinized wall. In certain others, the penetration hypha penetrating the cuticle induces the formation of a local internal swelling of the underlying cellulose wall, so that a small papilla projects from the under side of the wall into the cell. This papilla is penetrated by the penetration hypha which enters the host cell and ultiamtely develops into a haustorium (Fig. 24.12).

Bacteria are mostly weak parasites which cannot employ force to effect penetration. Their penetration is effected by chemical action. The plant parasitic nematodes pierce the host surface with spears or stylet.

The entire process of direct penetration, however, depends on: (a) the nature of cell wall layers, (b) the potentialities of the enzyme system of the pathogen, and (c) the potential force the pathogen can exert. The main components of the host cell wall are: pectin, cellulose, hemicellulose, lignin and small quantity of protein. The three main softening or solubilizing enzymes found in fungi are: pectolytic, cellulolytic, and lignolytic. The outer layer of host plant cell may have a layer of wax which is followed by cuticle impregnated with wax. The cutin gradually decreases with the depth of the epidermis and is replaced by pectin which occurs as a homogeneous layer in some host plant. Subsequently in the secondary wall of the cell, pectin is replaced by cellulose. Cellulose layers contain appreciable amounts of protein.

No pathogen secretes enzymes that would degrade wax. Only mechanical entry can enable a pathogen to breach this layer. Cutin layer is penetrated either by pressure (mechanically exerting force) or by the action of degrading enzymes: cutinase, cutin esterase, and carboxyl cutin esterase. Cutinase breaks cutin into fatty acid and hydroxyfatty acid.

Pectic substances which form basic material of middle lamella, primary and secondary walls are degraded by pectinolytic enzymes. The pectinolytic enzymes act against pectin and pectic substances in many steps: (a) the enzymes pectin methyl esterases (PME) hydrolyse the pectic substances into methanol and pectic acid; (b) the enzymes pectin glycosidases, polygalacturonases (PG) and polymethyl galacturonases (PMG) degrade the pectic acid and methylated chains of pectin. Besides these, wall modifying enzyme (WME) modifies the pectic material for subsequent degradation. The degradation of pectic substances provides nutrients for many fungal pathogens and due to weakening of the cell wall facilitates inter- and intra-cellular invasion by hyphae. Pectic enzymes are produced both constitutively and adaptively. High C/N ratio of the substrate favours increased mycelial growth and low enzyme synthesis, whereas lower ratios give poorer growth but increased enzyme synthesis. Pectic enzymes have no effect on protoplast and thus cannot do much killing of host cells. They may be secreted by some of the pathogens involved in spotting and blotching of leaves, although the macerating effect may be masked by necrosis caused by toxins.

Cellulose forms the structural framework of cell walls. Cellulolytic enzymes act upon the cellulose and break it to simple compounds and make the way for the easy penetration of the pathogen into the host cell. There are two theories to explain the mechanism of cellulose degradation and maceration by cellulolytic enzymes which are adaptive enzymes: the unienzyme theory and the multienzyme theory. According to unienzyme theory, complete degradation of cellulose into glucose units is by a single enzyme: Cellulose $\xrightarrow{\textit{Cellulase}}$ Cellobiose $\xrightarrow{\textit{Cellobiase}}$ Glucose, whereas the multienzyme theory explains cellulose degradation in a series of steps through two groups of enzymes. One group of enzymes loosens the cellulose fibrils of the crystalline area by hydrolytic mechanism. The other group of enzymes penetrates the cellulose lattice and causes hydrolytic cleavage of hydrogen bondage.

In the spaces between the cellulose microfibrils of the cell wall, hemicelluloses are present which are degraded by hydrolytic cleavage by enzymes hemicellulases. The degradation of hemicelluloses by the enzymes hemicellulases produces simpler sugars. In vascular wilt diseases the molecules released by cellulose degradation can cause plugging of the vessels.

The most complex chemical compound in the plant cell wall is lignin which occurs chiefly in the matrix surrounding the cellulose fibrils. It is one of the most structurally complex biopolymers whose degradation is caused by polyphenol oxidases produced by wood-rotting fungi. But fungi cannot utilize the degradation products.

The attack of nematodes is both internal as well as external. Both endophytic and ectophytic nematodes use force as well as enzymes to dissolve or break the cell wall to come in contact with cell protoplasm through their stylet. But endophytic nematodes enter bodily in the host cell, whereas the ectophytic ones remain outside the host.

(iv) **Entry through specific parts or organs.** In such cases the pathogens exhibit specificity of host parts or organs (stem, leaf, root hairs, floral parts, coleoptile, buds, etc.) during their entry in the host. Some examples are: *Synchytrium endobiuticum* infects and remains confined in the epidermal cells of potato tuber, *Erysiphe graminis* infects the epidermal cells of leaves of cereals and grasses, *Ustilago nuda* causes infection only in the flowers of wheat, *Claviceps purpurea* infects the ovaries of rye flower, certain wood rotting fungi are—*Merulius lacrimans* and *Fomes annosus;* soil inhabiting pathogens enter through root hairs—*Plasmodiophora brassicae.* Besides these, there are highly specialized insect parasitizing fungi—members of the order Entomophthorales and Laboul beniales.

C. Post-penetration stage

Usually with the success of the penetration process, post-penetration is successful. But the entry of the pathogen in the host tissue may not always ensure immediate infection leading to disease development. The process may be delayed or there may be failure for various reasons. The delay may be in cases where the pathogen has incubation period and infection is established only after the expiry of the incubation period. The success of post-penetration process depends largely upon competition of pathogen for nutrition, and production of enzymes and toxic substances and their effects on host metabolic activities. Again, due to toxic effect of host cytoplasm, the pathogen may fail to establish biological relationship with the host. Host-pathogen interaction may also result **hypersensitivity** of the host tissue, whereby rapid death of the affected cells prevents the further spreading of the pathogen due to shortage of nutrition.

But in most of the plant diseases, host infectioin is followed by **invasion,** a condition when a pathoen grows rapidly in the host tissue. For example, bacteria invade host tissues intercellularly and destroy them, whereas fungal mycelia invade inter- or intra-cellularly but may or may not cause destruction immediately after invasion. Viruses always invade host tissues intracellularly. They multiply in the living host cells by directing them to manufacture viral nucleic acid and viral protein, their movement from cell to cell is through plasmodesmata. Again, fungal hyphae and spores, and bacterial cells may move through vascular tissues once they gain entrance in them. Successful host invasion of the pathogen is invariably associated with disease syndromes of various types of varying degrees.

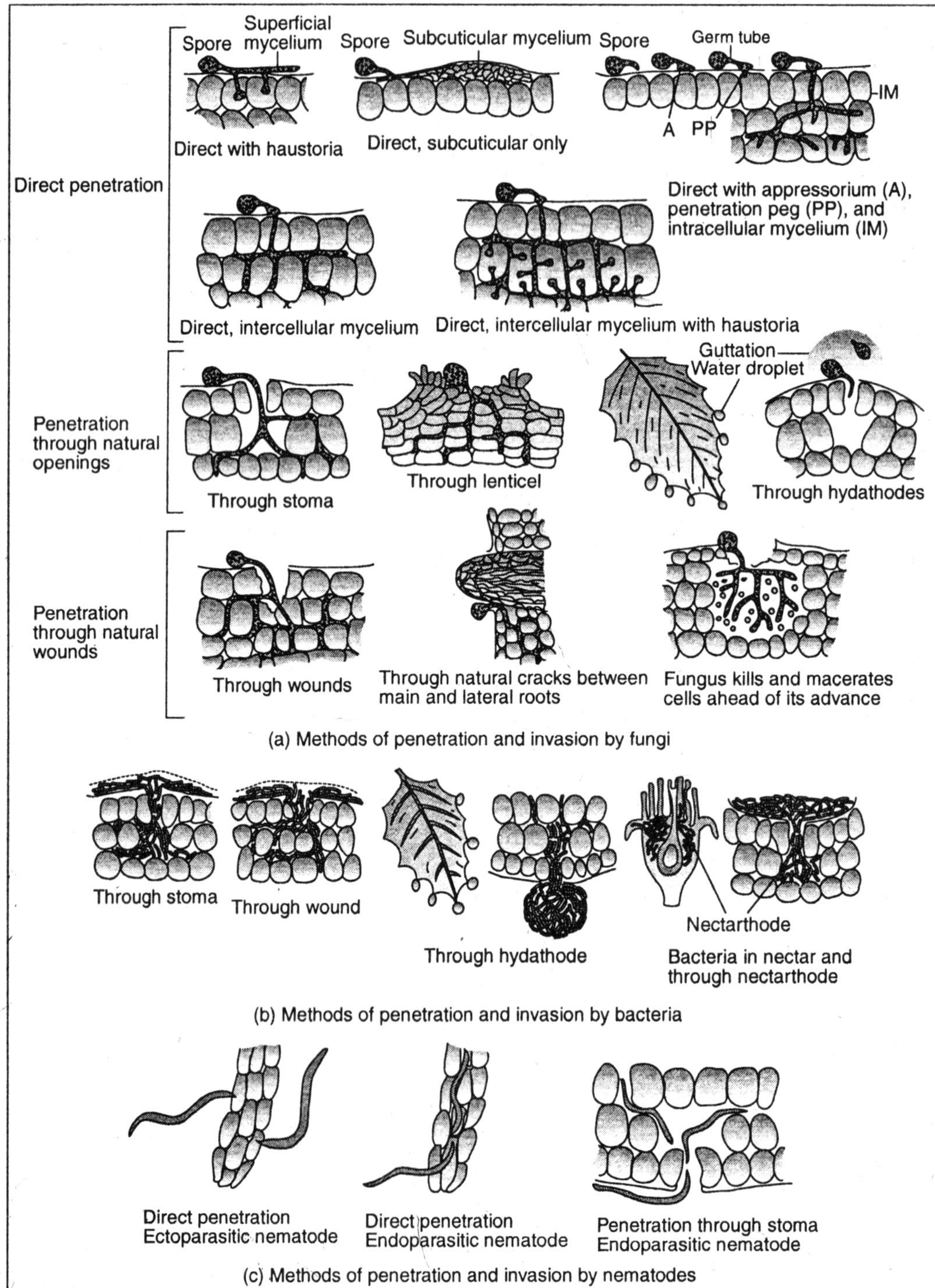

Fig. 24.12: Overall penetration and invasion by various pathogen is shown above

After penetrating the host cell walls, the pathogen comes in contact with the host cytoplasm from which it gets its required nutrition. In response to the activities of the pathogen, the host metabolic processes (osmoregulation, respiration photosynthesis, etc.) get upset. Depending on the nature of host and virulence of pathogen and their relationship, the host interacts in the following manner.

(i) **Increase respiratory rate.** Lavel of many enzymes associated with respiration increass resulting in accumulation and oxidation of phenols. Break down of ATP is accelerated causing in increase in respiratory rate. A host-pathogen relationship is known as **incompatible relationship** when increase in respiratory rate takes place in the early stage of infection. Again, it is designated as **compatible relationship** where the pathogen is an obligate parasite and there is little change in respiratory rate of the infected tissue.

(ii) **Disbalance of photosynthetic process.** The photosynthetic process is impaired by the breakdown of chlorophyll through the enzyme chlorophyllase causing **chlorosis.** There may also be increase in the amount of chlorophyll exhibited by **green islands** on the affected leaf surface.

(iii) **Dislocation of normal growth.** The normal synthesis of growth promoting substances (auxins, gibberellins, and cytokinins) is disbalanced resulting in overgrowth, subnormal growth and correlated symptoms. There may be rapid development of growth inhibitors (dormin and other substances) leading to premature ripening of fruits and causing formation of abscission layers resulting in premature fall of leaves, flowers, and fruits.

(iv) **Influence on reproductive structures.** There may be halting of development of reproductive structures (abortion) or due to localized or systemic infection there may be destruction of ovaries.

(v) **Effect on transporting organs.** The transporting organs are affected in various ways: (a) plugging and rupturing of xylem vessels due to growth of pathogen in their lumen; (b) presence of pathogen in the vascular tissue causes formation of toxins and development of tyloses in the xylem vessels from the adjacent parenchyma causing obstruction to the passage of water; (c) vascular infection produces pectic enzymes in the xylem vessels whose middle lamella is degraded releasing pectic acid and other substances which form gels and gums to plug the lumen of xylem vessels; (d) polysaccharides secreted by vascular pathogens have more often been regarded as "wilt toxins", dissolved or suspended in the xylem sap exert their action at a distance from the cells of the pathogen. Bacterial and fungal wilts are often associated with the movement of the pathogen in and colonization of vascular elements of the host. Again phloem necrosis caused by viruses obstructs the movement of carbohydrates synthesized in the leaves causing many deformities of the leaf surface.

(vi) **Dislocation of tissues.** Secretion of lytic enzymes causes loss of coherence of the tissues, breakdown of cell walls, and death of protoplasts. Death of aerial parts may result from the transport of toxic substances from the diseased tissues or from ionic imbalance caused by the failure of selective absorption.

(vii) **Result abnormal growth.** Increase or decrease in host cell number and size are attributed to the metabolism imbalance regardless of whether growth disturbances are caused by fungal, bacterial, viral or nematode infection. Metabolism imbalance disturbs normal hormonal balance. The hormone **kinetin** works together with indoleacetic acid (IAA) for cell expansion and cell division causing tissue proliferation.

(viii) **Cause tissue destruction.** As the pathogen grows in the host tissue, cell wall degrading enzymes, like **cellulases, hemicellulases, pectinases**, etc., induce maceration processes. Tissue-disintegration of the host continues further accompanied by the activity of several oxidative enzymes such as **peroxidase, polyphenol oxidase** and **ascorbic acid oxidase.** Perhaps the most refined type of pathogenesis may be observed with the viruses. With the aid of their nucleic acids they give the host cells incorrect information which carry out damage. As the pathogen grows, it also interferes with protein metabolism of the host tissue causing reduction in the protein and nitrogen level in the affected tissue.

During post-penetration stage offensive biochemical weapons—enzymes, toxins, growth regulators, polysaccharides, and antibiotics are employed simultaneously by the pathogen for degrading and macerating of host cell wall and ultimately killing and/or interacting with the host cells. This behaviour of the pathogen, besides causing disbalance of the host metabolic processes, also triggers several metabolic systems in the host cells resulting in formation of defence chemicals—**phenolic substances** with which the host tries to repel the attack of the pathogen. These two processes continue simultaneously. These aspects are treated separately under separate headings.

It has already been discussed that cell wall destruction and maceration are done by various enzymes. These processes are closely followed by the action of toxins and other offensive biochemical weapons.

Toxins are biochemical substances produced by pathogens during host infection. They are responsible for disease symptoms in the host plant. That toxins are responsible for the death of host cells was first pointed out by De Bary (1996). The term **toxin** can have different meaning for different workers. Owens (1969) defined toxins as non-enzymatic substances which injure plant cells or disrupt their metabolism. Toxins also inactivate the enzymatic reactions of host tissue. They are produced by microorganisms (fungi and bacteria) in a variety of environments and are injurious or lethal in low concentration to macroorganisms. Hence it is the suspect, a macroorganism, instead of a microorganism that distinguises toxins from antibiotics. Again, substances that are toxic in excessive doses, although they are excreted by microorganisms are not considered as toxins. Toxin may thus be defined as a chemical substance produced by microorganisms during their interaction with host. It affects the host metabolism in a subtle manner. Toxin also acts directly on living host protoplasm to influence either the course of disease development or symptom expression. A toxin has a low molecular weight, extremely mobile and easily reaches the subcellular levels of the hosts. Toxins are different from enzymes due to the fact that they do not attack the structural integrity of the host tissue but affect the metabolic activities of the host cells. They can be antimetabolites in action and can reduce the amount of specific growth factor. They also disturb the osmotic relations of host cells and most of them affect respiration directly or indirectly.

In contrast with the enzymes, the effect of toxins is direct on the protoplast. Scheffer and Briggs (1981) defined toxin (for plant pathological purposes) as a microbial product other than enzyme, which causes obvious damage to plant tissues, and which is known with reasonable confidence to be involved in disease development. Most toxins are not specific in their action. They harm many species belonging to different unrelated genera or family. A few toxins are species-specific. They only affect plants susceptible to the pathogen; resistant plants are not damaged. Host specific toxins are also designated by some workers as **pathotoxins.** These toxins are able to produce in susceptible host plants all characteristic disease symptoms, but have little or no effect on resistant species, unless used in massive concentrations. The pathogen and pathotoxins exhibit similar host specificity. The virulence of the pathogenic strain varies directly with their ability to produce toxin. Host-specific toxins are complex chemicals. It is relatively difficult to isolate any of these toxins in pure form and still retain the property of specific toxicity. Some of them are low molecular weight peptides. They rapidly act upon both cell and plasma membrane causing changes in permeability and respiration of diseased tissues. Some toxins affect various kinds of plant tissue, whereas others damage a single type of tissue. They may induce a particular syndrome when applied to plant in suitable concentrations. Wilting, for example, is a common effect of many toxins which cause loss of cell turgor and leaf flaccidity, and eventually a general wilting of the whole shoot. Such toxins are kown as **marasmins** or **wilting toxins. Lycomarasmin** and **fusaric acid** are also of this category of toxins. Again, toxins causing necrosis are the **necrotoxins** and are produced by species of *Alternaria, Phytophthora* and *Pseudomonas.*

Toxin concentration may not be correlated with virulence. A single toxin may induce all symptoms. The ability of a microorganism to produce a toxin or an enzyme is, of course, fixed genetically, but certain chemicals may be able to repress activation of genetic information or to prevent it from functioning the cell metabolism.

Toxins are classified in three categories:

(i) animal toxins or **zootoxins,** produced by animals;

(ii) plant toxins or **phytotoxins,** produced by fungi or higher plants; and

(iii) **microbial toxins,** produced by microorganisms.

Phytotoxins may be (a) **pathotoxins** which are capable of producing all of the disease symptoms, and (b) **vivotoxins** produce portion of disease symptom *in vivo.* According to Diamond and Waggoner (1963) vivotoxin is a substance produced in the infected host tissue by the pathogen and/or its host, which functions in the production of disease, but it is not itself the initial inciting agent of diseae. The phytotoxins have the properties as follows: (a) It can induce the same plant response but may have different mechanisms of action; (b) different pathogens can synthesize the same phytotoxin; (c) most phytotoxins are composed of simple carbohydrate and/or amino acid residues; (d) the pathogen and the toxin may not exhibit the same host range; (e) toxin concentration may not be correlated with virulence; (f) toxigenicity is not always a requisite for pathogenicity; (g) the rates of synthesis and degradation may vary depending upon the nature of microorganism, environment and host; (h) may not act alone but in concert with other pathogen- and host-produced substances in a dynamic, interrelated system; (i) have low molecular weight and are mobile in plant tissues; (j) may be intrinsically unstable; (k) may be irreversibly reacting with cell constituents; and (l) may act in combination with other components found in diseased tissue.

Other types of toxins are: **endotoxins,** intracellular toxins formed in bacterial cells and are not liberated until the latter die; and **exotoxins,** extracellular toxins which diffuse from the living bacterial cell. Most of the toxins produced by plant pathogens seem to be **pleiotropic,** that is, they have multiple effects on the host cells, but the 'wildfire' toxin produced by *Pseudomonas tabaci* is **monotropic,** as are many of the toxins produced by bacteria which attack animals. This is an unstable toxin.

The varied mechanisms by which microbial toxins damge a host are not clearly understood. However, toxins must eventually influence adversely the membrane system, whether these be the limiting cytoplasmic membranes, the internal membranous cytoskeleton or the diverse membranes of organelles, including vacuoles. Toxins damage permeability of plant membranes.

Bacterial phytotoxins respond to plants in

only a few days, mainly by the development of chlorosis, growth abnormalities, necrosis, water soaking or wilting. The response will depend upon phytotoxin concentration, environmental conditions and/or physiological state of the host. These responses by the plant are not only limited in kind but are non-specific in that different agents acting in different ways can induce the same response.

A summary of some of the toxins and their associated organisms as given in a table below:

A number of storage fungi as well as some other fungi are known to produce organic substances which have toxic effects on animals and humans. These toxic substances are known as **mycotoxins** and the diseases they cause in animals are the **mycotoxicoses.** Interest in mycotoxins resulted from the discovery of **aflatoxin,** a metabolite of *Aspergillus flavus* Link ex Fries produced in kernels of groundnut, reported for the first time by K. Sargeant and others (1961). 'Aflaroot' disease induced by *A. flavus* which is a seed and soil borne disease of groundnut causes post-harvest aflatoxin problem in groundnut by developing in poorly dried pod shells and kernels. Aflatoxins are thus produced in kernels of groundnut, particularly in improperly dried pods. Natural occurrence of aflatoxins has been reported in a wide range of commodities, particularly in improperly dried pods and seeds both in field and storage conditions. In course of time aflatoxin contamination of food and feedstuff has posed a storage problem as because aflatoxin contaminated food consumed produces deleterious effects. Intake of aflatoxin contaminated food causes carcinoma, specially liver cancer. Among the various derivatives, aflatoxin B_1 is the most toxic and generally produced in higher amounts. It is most potent carcinogen. Mycotoxins in general, and aflatoxins in particular, in food and feedstuff have carcinogenic and toxinogenic properties. Some of the mycotoxins produced by storage fungi are: sterigmatocystin, ochratoxin, citrinin, and patulin.

There are similarities between bacterial and fungal infections and the resulting symptoms. In general, virus infection reduces the capacity of the host for starch synthesis accompanied by reduced degeneration of starch, resulting in accumulation of starch in the foliage and other plant parts. Virus infection also enhances respiratory activity and alters auxin levels in the host cells.

The host-pathogen interaction ultimately may produce symptoms like: (a) **necrotic,** (b) **atrophic,** (c) **hypertrophic,** (d) **hyperplastic.** The pathogen may also establish a relationship or association with the host in such a manner that none of the above conditions are

Table 24.1: Toxin and their associated organisms

Organism	Toxin	Action
Pseudomonas tabaci (Wolf et Forst.) Stev.	Wildfire toxin	not host-specific
Rhizobium japonicum (Kirch.) Buch.	Rhizobitoxine	" " "
Alternaria tenuis Nees	Alternatic acid	" " "
Fusarium oxysporum Schl. f. sp. *lycopersici* (Sacc.) Syn et Hans.	Lycomarasmin	" " "
Fusarium spp.	Fusaric acid	" " "
Pyricularia oryzae Cav.	Pirucularin	" " "
Helminthosporium carbonum Ullstr.	HC-toxin	host-specific
H. victoriae Meehan et Murphy	HV-toxin or Victorin	" "
Periconia circinata (Mang.) Sacc.	PC-toxin	" "

Mycotoxins and mycotoxicoses

Many grains and sometimes other seeds and also plant products such as bread, hay, purees, and rotting fruit are often infected or contaminated with one or more fungi that produce toxic compounds known as mycotoxins. Animals or humans consuming such products may develop severe diseases of internal organs, the nervous system, and the circulatory system and may die. Also, many pasture grasses are infected with certian endophytic fungi that grow internally in the plant and, although they do no seem to seriously damage the grass plants, they produce toxic compounds that cause severe diseases in the wild and domestic animals that eat the plants. Similarly, toxic and sometimes lethal to animals are some grasses whose seeds are infected with bacteria carried there by a nematode; these bacteria are often themselves infected with a virus (bacteriophage) that induces the bacteria to produce compounds very toxic to animals.

Ergotism is an example of a mycotoxicosis caused by food and feed made extremely unhealthy by mycotoxins produced by the fungus *Claviceps purpurea.* Ergotism causes very direct and dramatic symtoms and has been known for many centuries, if not millennia. There have been, however, innumerable other cases in which people or animals became chronically or acutely ill by eating food or feed that contained unsuspected toxic substances. The existence and identity of the toxic substances had remained unknown, the sources of such unsafe food and feed had been little noticed, and the ailments affecting humans and animals remained unexplained. It was not until the 1960s that a severe disease of young turkey birds was shown to be caused by moldy feed and called attention to the importance of mycotoxins in the health of people and animals.

Mycotoxins are toxic fungal metabolites that are released by relatively few but universally present fungi growing on grains, legumes, and nuts. Such produce, especially when harvested while still containing a high percentage of moisture or if it is damaged and stored at relatively high humidity, becomes moldy, i.e., it supports the growth of mycotoxin-producing fungi. Such moldy produce is likely to carry high concentraitons of mycotoxins. Several of the mycotoxins are proven carcinogens, may disrupt the immune system, and may retard the growth of animals or humans that consume them. Even very small amounts of mycotoxins bring about the detrimental effect of mycotoxins on the immune system and metabolism of humans and animals, thereby posing a continuous health hazard. At higher concentration, which occurs often on moldy produce, mycotoxins cause visible clinical symptoms (mycotoxicoses) in both humans and animals in the form of nervous agitation, dermal and subcutaneous lesions, impaired growth, damage to kidneys and liver, cancer, and other symptoms.

Although the last recorded outbreak of gangrenous ergotism occurred in Ethiopia in 1978, it was not until 1960 that the first general interest in mycotoxicoses was shown when the so-called "turkey X disease" appeared in farm animals in England. It was eventually shown that the disease was caused by feed contaminated with aflatoxins, and when these were shown to cause cancer in the liver of humans and animals, interest in mycotoxins skyrocketed. Aflatoxins are extremely toxic, appear in the milk of animals consuming contaminated feed, attack primarily the liver, and are mutagenic, teratogenic, and carcinogenic. In the last several decades, several outbreaks of aflatoxicosis have occurred in tropical countries where many adults in rural populations often consume moldy corn. Blood examiantions in adults and children living in some tropical areas and showing various symptoms of varying intensity have revealed the presence of aflatoxins in them, with significant seasonal variations.

In addition to aflatoxins produced by the two aforementioned species of *Aspergillus,* several other equally toxic mycotoxins, e.g., ochratoxins, are produced by these and by other species of *Aspergillus,* by *Penicillium,* and by other fungi. Ochratoxins occur in cereals, coffee, bread, and in many preserved foods of animal origin. About 20,000 people in the northern Balkans seem to be suffering from diseases caused by chronic exposure to ochratoxin. Poisoning from moldy sugar cane is caused by a mycotoxin produced by species of Arthrinium, and in one rural area in China it affected more than 800 persons who had ingested moldy sugarcane. *Aspergillus* and *Penicillium* are extremely common in nature and are almost always present to some extent in any feed and in most foods. Aflatoxins are the most common mycotoxins, but even more potent mycotoxins, e.g., patulin, roquefortin C, and others, are also produced by species and strains of *Penicillium.*

A number of potent mycotoxins, the trichothecins, are produced by several species of *Fusarium* and, to a lesser extent, by species of *Trichoderma, Trichothecium, Myrothecium,* and *Stachybotrys.* The most common trichothecin is deoxynivalenol, also known as vomitoxin. Another type of mycotoxin, zearalenone, is produced by somewhat different species of *Fusarium* (*F. graminearum*). Vomitoxin and zearalenone often occur together, especially in scabby wheat and in corn infected with *Gibberella* ear rot, but they have also been found in moldy rice, cottonseed, flour, barley, malt, beer, and other foods. In addition to humans, vomitoxin and zearalenone affect cattle, swine, chickens and other birds, cats, dogs, and fish. Individuals, fed contaminated food or feed over a period respond by vomiting, refusal to eat, suppression of their immune system, diarrhea, loss of weight, and low milk production in the case of cows. A still different group of mycotoxins, called fumonisins, are produced by *Fusarium verticillioides* (*F. moniliforme, F. proliferatum*) and related species, primarily in corn and corn-based products. Fumonisins affect all or most of the animals affected by the other *Fusarium* toxins but they also affect and are particularly toxic to horses. In horses, low concentrations of fumonisins cause liquefaction of the brain, resulting in the "blind staggers" and "crazy horse disease" in which horses display blindness, head butting and pressing, constant circling and being agitated, and finally die. In swine, fumonisin attacks the heart and the respiratory system, in which it causes swellings, and it also causes lesions in the liver and pancreas. In humans, fumonisins have been linked to cancer. In the last 10 years, outbreaks of fumonisins in feed or food have been reported in several states from Arizona to Virginia and from South Carolina to the upper Midwest and in some Canadian provinces.

In most of the cases just mentioned, most of the damage is caused by the mycotoxins in food or feed consumed by humans and animals. However, for people and animals spending considerable time surrounded by moldy food or feed, there is the added danger of directly breathing spores of these fungi. It is not clear how detrimental to their health this is, but humans and animals, especially horses, exposed to spores of *Stachybotrys chartarum* develop irritation of the mouth, throat and nose, shock, skin necrosis, decrease in leukocytes, haemorrhage, nervous disorder, and death *Stachybotrys* grows on straw and feed and on moist surfaces on walls and in air-conditioning ducts and is considered one of the most important causes of the "sick building syndrome".

produced. Either after some of its growth or at least when it reaches a particular organ of the host or both, then only the pathogen produces symptoms that will be visible from outside.

In infected green tissues chlorophyll may be reduced or destroyed turning brown. Or in certain areas due to preservative action of certain parasities on the chlorophyll apparatus of leaf tissues there is marked formation of green islands of tissues after the rest of the green tissue withers, this is known as **green island effect.** There is evidence that not only is the chlorophyll preserved but also the chloroplasts are stimulated to increased activity by the pathogen. There may be a development of red or purple pigment due to the formation of **anthrocyanin** resulting from infection.

There are a group of cellulose dissolving fungi which are responsible for rots of wood. This is a large group comprising both parasites and saprophytes. Though their activities are more or less restricted to cellulose and substances similar to cellulose of the wood they also attack all the major components of the cell walls of the wood. There is no uniformity of the wood simultaneously, whereas others cause delignification before the cellulose is depleted, others again show a delayed delignification.

The variations are due to differences in the nature and time of enzyme becoming active and its course of action.

Wood decay by fungi. Wood which is a rich organic substrate provides a suitable food base for fungi. Fungi draw their nourishment from the cell wall material, which they breakdown by enzymatic activity. The hyphae exude various enzymes into their substrate. These enzymes convert the chemically complex wall material into simpler water-soluble products, which are further acted upon by endoenzymes in course of fungus metabolism. The enzymes are selective in action. The ability of fungi to attack various components of wood is determined by the nature of enzymes they secrete.

There are various stages of decay as wood is changed from sound to completely decayed. In the early or incipient stage, also known as **incipient decay**, the wood appears like normal, except in a few cases by a change in colour. This stage can be easily overlooked and as such incipient decay is dangerous. From this stage the wood becomes noticeably affected until it is finally changed completely in structure and appearance. This is the **advanced stage of decay.** Change in colour of wood is not always a criterion for decay, since it may be due to other causes, either parasitic, e.g., fungi causing sap stain in wood, or due to chemical action. Some decays are associated with formation of dark-brown to black zone lines, and also possess mushroomy odour. In incipient decay the wood may not always weaken. But in advanced stage of decay all mechanical properties of wood are seriously affected and such wood is useless for any purpose where strength is required. Decay results in decrease of specific gravity of wood, so that the amount of wood substance per unit volume becomes less. Decayed wood, based on volume, has thus less calorific value and yields low grade fuel. The microscopic characters of decay include presence of hyphae in the cells. The hyphae pass from cell to cell consuming food materials during decay; the penetration occurring probably by chemical action, resulting in bore holes in the wall. In utilizing the cell wall material as food, the hyphae cause corrosion marks, uneven thickness of secondary walls leading to their disintegration, enlargement of pits and dissolution of bordered pits. The damage in the cell wall mainly accounts for reduced strength of decayed wood.

Besides enzymes, for causing decay, fungi require an optimum concentration of air (oxygen) and moisture for their growth. Insufficient air may act as a limiting factor and if air is removed by saturation of the wood with water, the fungi die. Fungi cannot grow in the wood if the moisture is below 20 per cent based on oven-dry weight of wood. If, however, the infected wood is air-dried, the fungus does not die but its active growth ceases and may remain dormant for several years in this condition. If such wood is moistened, the fungus may revive and continue its growth in the wood. With respect to temperature, wood-decaying fungi make optimum development over a range of 60°F to 90°F temperature. Rainfall also greatly influences the occurrence of decay fungi. A certain amount of moisture in the substrate and in the atmosphere is necessary for the fungi to develop sporophore. In dry localities, therefore, as in deserts, decay fungi are few. In tropical conditions with high temperature and rainfall, decay fungi are diverse and also large in number. They remain active for greater part of the year. In temperate regions, fungi are fewer. They remain active for a shorter period, i.e., during summer.

Some wood decaying fungi which can tolerate wide range of temperature are: *Fomes lividus, Polyporus gilvus, P. hirsutus.* Others, like *F. badius, F. pachyphloeus, Ganoderma lucidum, Polyporus palustris, P. tulipiferae,* thrive at high temperatures. Fungi which thrive at low temperatures are: *Fomes annosus, F. robustus,*

F. roseus, F. pini, Lenzites sepiaria, Merulius lacrymans, Polyporus abietinus, P. pargamenus, P. versicolor, Poria monticola, P. xantha.

For practical purpose two kinds of wood decay are generally recognized: **white rot** and **brown rot.** Both kinds of decay are caused by a large number of Ascomycetes and Basidiomycetes. In both these kinds of wood decay, hyphae penetrate deep into the tissues of the wood and inhabit the lumen of vessels of host cells. They pass cell to cell through bore holes. The white rot causing fungi breakdown lignin with oxidising exoenzymes. They also remove a small amount of the structural carbohydrates leading to a lightening of the wood. But the brown rot inducing fungi remove the structural carbohydrates like cellulose and hemicellulose, leaving the lignin relatively unchanged. Hence the wood, after decay, appears more brown than normal wood. In both the cases the decay in wood is generally associated with a change of colour. In case of white rot the colour is bleached and in case of brown rot the colour turns dark. The initial darkening caused by the white rot fungi may be due to oxidation of phenols. The decayed wood shrinks. This is more in case of brown rot than in white rot. In case of decay caused by brown rot fungi, at the initial stage there is no thinning of the cell wall. The white rot can be fibrous or spongy (white fibrous or spongy rot) or pockets may be developed in the wood in which the tissues are white and fibrous and such pockets are delimited by thin brown areas, less decayed (white pocket rot).

The decayed wood also cracks in all directions into cuboidal blocks, which easily become powder under pressure. When such decay occurs in structural timbers, as in buildings it is termed **dry rot.** The term dry rot is a misnomer, since such rot, as well as others occur only under damp condition. There is another form of decay in timber, termed **soft rot,** which is recognized by its brittle fracture when broken across the grain. The fungi attack mainly cellulose and the hyphae are mostly confined to secondary wall in which dissolution occurs in the form of pits or cavities. The fungi causing soft rot belong to the Ascomycetes and Deuteromycetes. Soft-rot fungi are nearly restricted to the surface of the secondary wall and they thrive producing large cavities by enzymes which lie in parallel to the microfibrils.

The diagnosis of decay can be done when the decay is in the advanced stage, by gross characters. Incipient decay can be determined by microscopic characters of decayed wood and by isolation of the fungus in culture.

Fungi may attack sapwood or heartwood in trees or timbers. The decay of heartwood is—**heart rot.** Fungi attacking trees may or may not decay timber after conversion. When such rot occurs in converted wood, the term 'inactive rot' may be applied to it.

Sap stain fungi form another group causing defect in trees and timber. They do not attack the cell wall but live on cell contents stored in the sapwood only; thus the heartwood where no such food material is present remains uneffected. Since cell wall is not attacked, most of the properties of wood, except toughness, remain uneffected. However, stain depreciates value of wood due to unsightly appearance. Sap stain is caused by some members of the Ascomycetes and Deuteromycetes.

Cell to cell colonization of mesophyll cells may cause leaf spot symptoms, wilting may develop if vascular cells are invaded, or a soft rot may be caused by breakdown of the living parenchymatous cells of cortex or pith.

Changes in vascular system. The vascular system of a disease-free plant serves as a pathway for the movement of water carrying nutrients. But in some diseased plants the vascular system may serve as a pathway for the movement of the pathogens. Again, the pathway of water and nutrients may be clogged by metabolites produced by the pathogen, itself by its growth, or by the debries of the host tissue produced by infection and pathogen. This will ultimately result wilting of plant. The wilt diseases have a multiplicity of causes rather than a single cause; but are largely due to deficiency of water supply. Water becomes deficient in wilt-diseased plants as a result of a number of events. (i) Vessels become completely plugged by tyloses and gums. (ii) Resistance to flow of water through vessels increases as a result of development of vascular gels, the presence of mycelium, and the lodging of propagules in the lumen of vessels. (iii) Resistance to transfer of water from vessel to vessel and to redistribution of water in the plant also increases in diseased plants as a result of the plugging of pit membranes by compounds of high molecular weight like polysaccharides excreted by the pathogen, the resinous melanoid pigments associated with vascular browning, and the partially hydrolysed cleavage products of cell walls by hydrolytic enzymes of the pathogen. (iv) Flow of water and nutrients may be impaired in cases where vascular tissues are crushed locally due to growth of the pathogen. (v) Wilt diseae may be induced by wilt toxin produced in the xylem vessels.

In addition to above factors, glycopeptides produced by some pathogens accumulate at termini of the vascular system in leaf margins, where they accumulate in the membranes of parenchyma cells and damage differential permeability. The nonsymptomatic leaves on infected plants may have transpirational loss about one-third of those of healthy plants receiving a normal supply of water. Stomata remain closed. Tissues are subjected to an ever-increasing shortage of water. Infected plants may have a lower photosynthetic rate than healthy plants. Depressed rates of photosynthesis and transpiration are attributed to limited supply of water. Both terminal and cambial growth are reduced in infected plants. Again, the cells in which the parasite penetrates get killed prior to which there may be change of pigmentation of cells and destruction of protoplasm. In others, the parasite may produce certain substances which produce streaking or staining of certain parts of the host tissue.

Some woody tissues react to parasitic attack by the secretion of gum and by the development of tyloses in the vessels of the wood. In this respect the conifers are important because several fungi and bacteria cause an outflow of resin from them. Again, there are certain fruit trees from which gum is exuded due to both parasitic and nonparasitic causes whch is known as **gummosis.** When the exudate is resin it is known as **resinosis.** Again, there may be **slime flux** when the exudation product is a slimy substance.

DEFENCE MECHANISMS OF HOST

Interaction between host and pathogen is a complicated process during which disbalance of host metabolic processes and formation of defence chemical—phenolic substances take place. With the help of defence chemicals the host tries to repel the attack of the pathogen. These chemicals are not only active against the inducer organism but are also against any other pathogenic organisms leading to delay in symptom appearance and even total protection. Besides this, whether a plant is resistant to the attack of a pathogen or susceptible, and the degree of these qualities, was determined by the genetic characters of the host plant. The defence mechanisms in the host plant may be considered broadly under two heads.

Defence through Defence structures. The external and internal barriers present may be differentiated into: **existing defence structures,** some of them are—waxes and cuticle, thick and tough cell wall, lignified or silicified cell wall, and shape and size of stomata and lenticels; and **post-infection defence structures,** these may be—histological defence (formation of cork layers, development of abscission layers, formation of tyloses and deposition of gums), cellular defence structures (callus-like swellings of the cell wall and ensheathing of penetrating hyphae), cytoplasmic defence reaction, necrotic or hypersensitive reactions. In some resistant host plants infection stimulates the development of Calcium oxalate crystals which protect them against disease incidence. There may be antagonistic effect during host infection in which pectic enzymes are inactivated by toxins. High sugar content in the host tissue often suppresses production of pectic enzymes by pathogens. Also high Calcium content confers disease resistance, possibly due to enhanced resistance to attack by certain polygalacturonases resulting from increased formation of Calcium pectate.

Defence through induced Biochemical and Structural factor. This may be designated as post-infection biochemical defence reaction (dynamic defence) or physiological or biochemical defence mechanism. As the pathogen grows rapidly, accumulation of **phenolic substances** takes place in the host tissue particularly in case of incompatible host-pathogen combination than in compatible combination. Higher plants contain large number and variety of aromatic substances which are precursors for the synthesis of phenolic substances. These phenolic substances pre-existing in the host plant tissue whose synthesis is accelerated by host infection are called **common phenolics,** whereas those which do not exist in the host tissue but are formed as a result of host-pathogen interaction are known as **phytoalexins** (*alexin* means to *ward off*). Müller and Borger first introduced in 1940 the phytoalexin concept in plant pathology. Müller (1956) defined phytoalexins as "antibiotics, which are produced as a result of the interaction of two different metabolic systems—the host and parasite, and which inhibit the growth of the microorganisms pathgoenic to plants." Hence phytoalexins are phenolic substances produced in the host tissue due to pathogen infection have antimicrobial properties. The synthesis of phytoalexins is common in the hypersensitive reactions of pathogens and is more rapid in resistant than in susceptible host plants. Phytoalexins are nonspecific in their toxicity towards fungal pathogens although the latter may be differentially sensitive to the former. The defence reaction is confined to the living host tissue colonized by the pathogen, and the speed with which phytoalexins are produced depends upon the degree of resistance of the host. It is determined by the concentration of phytoalexins. The sensitivity or capacity of the host to produce phytoalexins is governed by genetic factors, and a gene-for-gene relationship in host-pathogen reactions appears to be present. Some examples of phytoalexins are: **pisatin** in *Pisum sativum* following infection by *Ascochyta pisi,* **ipomeamatone** in *Ipomoea batatas* following infection by *Thielavia basicola.* These substances may also be produced in other host plants in response to infection by other fungal pathogens.

It is also speculated that a pathogen metabolite designated as "elicitor" stimulates phytoalexin biosynthesis in a host tissue. The degree of stimulation may depend largely upon the quantity and speed of release of elicitors, presence or absence of receptors in the host cell membrane and the environmental conditions. A new term "metabolic elicitor" has also been proposed for the chemical substance released by any living system for the elicitation of phytoalexin production.

Again both "elicitor" and "suppressor" were detected during biosynthesis in a pea pathogen *Mycosphaerella pinoides.* The elicitor and suppressor were high and low molecular weight substances respectively. The suppressor was found to counteract the activity of elicitor.

Phytoalexin production is not only induced by living organisms but also by certain microbial components (elicitors) in the absence of living organisms. Some of them are: glucan, glycoprotein and lipoglycoprotein isolated from culture filtrates of several fungi. Like biotic elicitor, abiotic ones (like Cupric chloride, Sulphur dioxide, Gibberellic acid, etc.) may also induce phytoalexin production.

Cellulase and pectinase are obtained from *Trichoderma viride* and *Aspergillus niger* respectively elicited capsidin in *Capsicum annum*, strongly suggest that enzymes are associated with elicitation of phytoalexins.

Meyer and Müller and their associates using various techniques presented strong evidences that resistance of potato to *Phytophthora infestans* is caused by the production of fungitoxic compounds by the host. Müller defined phytoalexins as compounds produced during host infection where components of fungi, bacteria and viruses, or microbial metabolites act as elicitors. Hence it is reasonable to consider phytoalexins as possible plant-defence compounds against diseases caused by fungi, bacteria and viruses.

More than 350 phytoalexins have been chemically characterized from approximately 30 plant families. Most phytoalexions have been isolated from dicotyledons, but they have also been isolated from monocotyledons including rice, maize, wheat, barley, onions and lilies. Phytoalexins have been isolated from stems, roots, leaves, and fruits though they have not always been reported produced, or the same compounds produced, by all organs of a plant. In general, phytoalexins are lipophilic localized at and immediately around sites of infection and have great structural diversity. Diverse phytoalexins are: phenyl propanoid, isoflavonoid, and sesquiterpenoid.

Defence in various ways. In general, plants defend themselves against pathogen attack by a combination of weapons basically of two areas: (1) Structural characteristics act as physical barriers from gaining entrance in the host plant and (2) Biochemical reactions produced in the host tissue due to pathogen attack.

I. Pre-existing Defence Structures—cuticle, wax, hairs.

II. Induced Structural Defence—necrotic or hypersensitive defence reaction.

1. **Inhibitors related by host plant.** *Fungitoxic exudates* released on leaves of tomato and sugar beet inhibit the germination of spores of species of *Botrytis* and *Carcospora.*

2. **Inhibitors present in host plant cell before infection.** Several **phenolic compounds, tannins**, and some fatty acid-like compounds such as **dienes** cause resistance against infection of *Botrytis* sp. or young fruits, leaves, or seeds. Several preformed saponins (glycosylated steroidal or triterpenoid compounds) **tomatine** in tomato and **avenacin** in oats repel the attack of fungi.

3. **Histological Defence Structures.** Formation of cork layers, abscission layers, tyloses, deposition of gums.

4. **Induced Biochemical Defence.** Hypersensitive response is responsible for limiting the growth of the pathogen, i.e., visible when nonhost plants are injected artificially with plant pathogenic bacteria resulting in necrotic and collapsed condition of the host cells. The hypersensitive response may also occur when virulent strains of plant pathogenic bacteria are injected into resistant varieties of host plants.

Each plant species is affected by different kinds of fungi, bacteria, viruses, and nematodes. Plants attacked by various pathogens may suffer damage to a lesser or greater extent, many survive all these attacks and manage to grow well and produce appreciable yields.

In general, plants defend themselves against pathogens by combination of (1) structural characteristics that act as physical barriers and (2) biochemical reactions that take place in the cells and tissues of host plant and produce substances which either are toxic to the pathogen or create conditions that inhibit growth of the pathogen in the host plant.

Defence through Lack of Essential Factors

1. Lack of Recognition between Host and Pathogen—Plants of a species or variety may not become infected by a pathogen if their surface cells lack specific **recognition factors** (specific molecules or structures) that can be recognized by the pathogen. Some of these recognition factors are: **aligosaccharides** and **polysaccharides** and **proteins** or **glycoproteins.**

2. Lack of Host Receptors and Sensitive Sites for Toxins—In host pathogen combinations, usually a fungus produces a host-specific toxin which is responsible for disease symptoms only where it reacts with specific receptors or sensitive sites in the plant cell.

3. Lack of Essential Substances for the Pathogen—When plants that do not produce substances essential for the survival of an obligate parasite, for example, *Rhizoctonia* needs a substance necessary for formation of a **hyphal cushion** from which the fungus sends penetration hyphae for inducing infection. Again, certain mutants of *Venturia inaequalis* fail to cause apple scale in plants which fail to synthesize certain growth factor. Bacterial soft rot of potatoes, caused by *Erevinia carotovora* **var. atrosetica,** is less severe on potatoes with low reducing sugar content.

Besides phytoalexins, phenols like **chlorogenic acid, catechol, rufianic acid** and **cytokinins (kinetin, benzyladenine)** act in defence reactions. Phenolic compounds may play an important role as natural inhibitors of pectolytic and cellulolytic enzymes. In general, increasd phenol level in the host tissue gives resistance qualities in the host. Several **amino acid analogues** like **ethionine** and **phenylserine** may also suppress the formation of enzymes. Various sugars and their derivatives may similarly suppress enzyme synthesis.

Many plants contain substances which can inhibit attack of fungi, bacteria and viruses, and disease resistance has frequently been attributed to such substances. These substances are: **furocoumarin, isopimpinelli, oxazolinones, chlorogenic acid, caffeic acid, catechol** and **phloridzin.**

There are reports that some phenolic compounds are converted into more highly fungitoxic substances, such as **quinones,** as a response to infection. There is some evidence that sterols may be involved in defence reaction. **Tomatine,** a steroid of solanaceous plants may protect tomato leaves against some pathogens.

Fungal pathogen produces IAA-oxidase to inactivate the host auxin development while the host in turn inhibits the IAA-oxidase through formation of polyphenols like chlorogenic acid and caffeic acid. The enzyme peroxidase provides resistance in resistant host. Again, dehydrogenases and amino acid-oxidases are also part of defence reaction against fungal pathogen. Supply of calcium and potassium also increases disease resistance of host plants. Again, as a result of defence reaction, fusaric acid is detoxified by the host plant to a nontoxic compound—N-methyl fusaric acid amide.

Besides above defence mechanisms, host plant may offer resistance to pathogen attack through: (i) **hypersensitive reaction** and (ii) **inherited genetic character.** In case of hypersensitive reaction, usually a small part of the plant tissue quickly dies after infection (necrosis) resulting in minor local lesion around the infection court. Inactivation or localization of the pathogen takes place pre-

Phytoalexins

Phytoalexins are toxic antimicrobial substances produced in appreciable amounts in plants only after stimulation by various types of phytopathogenic microorganisms or by chemical and mechanical injury. Phytoalexins are produced by healthy cells adjacent to localized damaged and necrotic cells in response to materials diffusing from the damaged cells. Phytoalexins are not produced during compatible biotrophic infections. Phytoalexins accumulate around both resistant and susceptible necrotic tissues. Resistance occurs when one or more phytoalexins reach a concentration sufficient to restrict pathogen development. Most known phytoalexins are toxic to and inhibit the growth of fungi pathogenic to plants, but some are also toxic to bacteria, nematodes, and other organisms. More than 300 chemicals with phytoalexin-like properties have been isolated from plants belonging to more than 30 families. The chemical structures of phytoalexins produced by plants of a family are usually quite similar; e.g., in most legumes, phytoalexins are isoflavonoids, and in the Solanaceae they are terpenoids. Most of the phytoalexins are produced in plants in response to infection by fungi, but a few bacteria, viruses, and nematodes have also been shown to induce the production of phytoalexins. Some of the better studied phytoalexins include phaseollin in bean; pisatin in pea; glyceollin in soybean, alfalfa, and clover; rishitin in potato; gossypol in cotton; and capsidiol in pepper.

Phytoalexin production and accumulation occur in heathy plant cells surrounding wounded or infected cells and are stimulated by alarm substances produced and released by the damaged cells and diffusing into the adjacent healthy cells. Most phytoalexin elicitors are generally high molecular weight substances that are constituents of the fungal cell wall, such as glucans, chitosan, glycoproteins, and polysaccharides. The elicitor molecules are released from the fungal cell wall by host plant enzymes. Most such elicitors are nonspecific, i.e., they are present in both compatible and incompatible races of the pathogen and induce phytoalexin accumulation irrespective of the plant cultivar. A few phytoalexin elicitors, however, are specific, as the accumulation of phytoalexin they cause on certain compatible and incompatible cultivars parallels the phytoalexin accumulation caused by the pathogen races themselves. Although most phytoalexin elicitors are thought to be of pathogen origin, some elicitors, e.g., oligomers of galacturonic acid, are produced by plant cells in response to infection or are released from plant cell walls after their partial breakdown by cell wall degrading enzymes of the pathogen.

The formation of phytoalexins in a susceptible (compatible) host following infection by a pathogen seems, in some cases, to be prevented by suppressor molecules produced by the pathogen. The suppressors also seem to be glucans or glycoproteins, or one of the toxins produced by the pathogen.

The mechanisms by which phytoalexin elicitors, phytoalexin production, phytoalexin suppressors, genes for resistance or susceptibility, and the expression of resistance or susceptibility are connected are still not well understood. Several hypotheses have been proposed to explain the interconnection of these factors, but much more work is needed before a satisfactory explanation can be obtained.

Species or races of fungi pathogenic to a particular plant species seem to stimulate the production of generally lower concentrations of phytoalexins than nonpathogens. For example, in the case of pisatin production by pea pods inoculatd with the pathogen *Ascochyta pisi,* pea varieties produce concentrations of pisatin that are approximately proportional to the resistance of the variety to the pathogen. When the same pea variety is inoculated with different strains of the fungus, the concentration of pisatin produced is approximately inversely proportional to the virulence of each particular fungal strain on the pea plant.

venting its further spread in the host tissue exhibiting defence reaction. This may result in the death of the pathogen by virtue of isolating in dead tissue only by its inactivation. Several aspects of hypersensitive reaction are: (a) only living microorganisms and viruses are capable of inducing this reaction, (b) it will occur only incompatible host-pathogen combination,

(c) at the initial stage of infection the pathogen multiplies at the same rate in either susceptible or resistant host, and (d) the reaction appears earlier in resistant plants than the typical symptoms in the susceptible plants. Such a defence reaction is often considered as an indication of a disease resident variety.

The ability of a host plant to resist pathogen attack is an inherited genetic character. There are two types of genes for resistance—major and minor genes. Host plants having major gene resistance are usually highly resistive to specific races of a pathogen, resistance being conditioned by a single dominant gene and is stable over a wide range of environmental conditions. Resistance can also be governed by minor genes with lesser efficiency, but not stable under varying environmental conditions. Major gene resistance is also designated as single dominant gene resistance or monogenic resistance. While resistance governed by minor genes is known as polygenic resistance. This is due to the fact that resistance is usually inherited by a dominant character and is governed by one, two or more genes and the resistance is thus monogenic, bigenic, oligogenic or polygenic.

Defence through lack of essential factors. This happens when the pathogen does not produce infection substances, such as enzymes polysaccharides and proteins or glycoproteins or structures like appressoria, penetration pegs and haustoria.

Defence through Genetically Engineering Disease Resistant Plants. The first gene for resistance (R) isolated was Hml gene of maize in 1992, which codes for an enzyme that inactivates the HC-toxin produced by the leaf-spot fungus *Cochliobolus carbonum.* In 1993, the PTO gene of tomato was isolated. This gene confers resistance to strains of the bacterium *Pseudomonas syringae* pv. *tomato.* Besides this, the area of inducing plant resistance to pathogens through genetic transformation with pathogen derived genes as done with coat protein gene of tobacco mosaic virus (TMV) integrated in the tobacco genome carried through seeds will provide excellent additional tool for plant disease control.

Defence through Immunization of Plants against Pathogens. Defence through plant bodies. Plants do not have an immune system like that of humans and animals, that is, they do not produce antibodies. In the early 1990s, however, transgenic plants have been produced that have been genetically engineered to incorporate in their genome, and to express, foreign genes, such as mouse genes that produce antibodies against certain plant pathogens. Such antibodies, encoded by animal genes but produced in and by the plant, are called **plantibodies.** It has already been shown that transgenic plants producing plantibodies against coat proteins of viruses, for example, artichoke mottle crinkle virus, to which they are susceptible, can defend themselves and show some resistance to infection by these viruses. In the future, this type of plant immunization work will produce fruitful results.

Defence through Genetically Engineering Diseasse—Resistant Plants

1. **With Plant-derived Genes**—The number of plant genes for resistance (R genes) that have been isolated is increasing rapidly. The first plant gene for resistance to be isolated was the Hml gene of corn in 1992, which codes for an enzyme that inactivates the HC-toxin produced by the leaf-spot fungus *Cochliobolus carbonum.*

2. **With Pathogen-Derived Genes**—Tobacco plants transformed (genetically engineered) to express the coat protein gene of tobacco mosaic virus (TMV) showed various degrees of resistance to subsequent inoculation with the same virus. Once the TMV coat protein gene was integrated in the tobacco genome, it was carried through the seed and behaved like any other tobacco gene.

Transgenic plants show high levels of resistance to the virus from which the coat protein gene was derived and in many cases, to other more or less related viruses.

With Plant-derived Genes

The first plant gene for resistance (R genes) isolated was the Hml gene of maize in 1992, which codes for an enzyme that inactivates the HC-toxin produced by the leaf-spot inducing fungus *Cochliobolus carbonum*. In 1993, the PTO gene of tomato was isolated. This gene confers resistance to strains of the bacterium *Pseudomonas syringae* pv. *tomato*. In course of time several more plant-derived genes have been isolated.

With Pathogen-derived Genes

In 1986, the coat protein gene of tobacco mosaic virus (TMV) was integrated in the tobacco genome and it was carried through the seed, and behaved like any other tobacco gene. Since then, numerous other plants such as tomato, legumes, barley, maize, oats, and rice have been transformed with the coat protein gene of one or more of the viruses that infect them. In the vast majority of cases, transgenic plants shows quite high levels of resistance to the virus from which the coat protein gene was derived. Again, transgenic plants transformed with viral genes other than the coat protein gene often exhibit even higher levels of resistance to the virus providing the gene(s) and to perhaps, additional viruses. It is expected that the area of inducing plant resistance to pathogens through genetic transformation with pathogen-derived genes will provide an excellent additional tod for plant disease control.

Pathogens may thus be permanently present through the year in and around a population of plants which are perenneal, although not necessarily active all the time. But so far as the population of annual plants is concerned some means of perennation are necessary for the pathogens to carry over intercrop period when the crop is absent. This is very essential for the continued survival of pathogens serving as potential source of inoculum of the cropping season.

Organs of perennation of pathogens. Principal organs of perennation of the pathogens are: (i) for fungi—mycelium, sclerotia, spores of different kinds with or without being associated with fruiting structures; (ii) for bacteria—vegetative cells with or without remaining along with host plant parts; and (iii) for viruses—symptomless carriers or tolerant plants, different infected host plant parts, insect vectors, eggs of viruliferous leaf hoppers.

Media of pathogen perennation. Media in which organs of perennation of pathogens survive during intercrop period vary on the nature of pathogen and climatic conditions. But in general, they may be: soil, organic manure, infected crop residues of harvest, seeds and other propagating stocks, collateral hosts, subsidiary hosts, plants which are missed during harvesting—commonly known as grondkeepers, mummified fruits, etc.

Perennating organs and climatic conditions. Climatic conditions can directly affect the longivity of perennating organs of the pathogens which again depend on the nature of perennating organs and the media in which they perennate. For soil-borne pathogens, water-holding capacity and pH of soil are very important for the survival of pathogen and development of inoculum. Similar is the situation with the organic manure. As to the crop residues of harvest, pathogens usually persist longer in woody residues than in leaves which disintegrate much more rapidly. The pathogens which perennate in infected crop residues are usually relatively unspecialized parasites that are able to survive in dead tissue, but obligate parasites may also survive in the form of thick-walled resting structures. In the seeds, propagating stocks and various plant parts the pathogens remain in a dormant condition during the intercrop period and become active with the return of suitable climatic conditions causing infection to crop plants during cropping season by producing inoculum. Subsidiary, collateral and similar other hosts in which pathogens survive during intercrop period normally do not cause any economic loss but on the contrary produce inoculum during cropping season to induce diseases.

Crop infection by pathogens. With the return of cropping season and favourable climatic conditions—both macro- and microclimatic, inoculum development, spread and host infection start simultaneously. The virulence of disease development depends on the virulence of pathogens, favourable medium of inoculum dissemination, susceptibility of host plants and the surrounding climatic conditions (temperature and moisture). When the incidence of disease is quick and widespread it will lead to epiphytotic condition. As such, the development of an epiphytotic condition is a chain reaction of survival of pathogens during intercrop period—development of inoculum under suitable climatic conditions during cropping season—dissemination of inoculum through suitable agency—inoculum arrival on host surface—host infection leading to disease incidence and further development of inoculum which in turn will be disseminated resulting quick spread of disease and destruction of the entire crop plant population.

FORECASTING OF PLANT DISEASES

Forecasting of plant diseases is predicting the occurrence of disease in an epiphytotic form in a particular area. Plant diseases vary in incidence from season to season due to differences in the mature and amount of inoculum, environmental conditions, numbers and activity of vectors, and other factors whch affect the development and spread of pathogens. Reliable forecasting of the likely incidence of plant diseases can save a great deal of money provided it can be done sufficiently early to organize effective control measures. This will also avoid wasting time and money on unnecessary control measures. Early forecasting gives crop growers sufficient time to rearrange their crop schedules and to avoid susceptible crop in a season when disease is likely to be severe. The aim of disease forecasting should also be to arrange control measures before the inoculum is likely to infect the crop. It is thus very important that forecasts sould be correct so that an effective disease warning system is reliable to the crop growers.

In most diseases, however, the pathogen goes through more than one generation per growth season. Such pathogens are called polycyclic, or multicyclic pathogens which can complete from 2 to 30 disease cycles per year and with cycle the amount of inoculum is multiplied manyfold. Polycyclic pathogens are disseminated primarily by air or air-borne vectors (insects) and are responsible for the kinds of diseases that cause most of the explosive epidemics on most crops. In polycyclic fungal pathogens, the primarily inoculum generally consists of the sexual spores, sclerotia, or mycelium in infected tissue. Although the polycyclic pathogens do not increase greatly within a year, but at the start of each year, they have almost as much inoculum as they had at the end of the previous year. Therefore, the inoculum may increase steadily from year to year and may cause severe epidemics when considered over several years.

Forecasting methods. A rational method of disease forecasting methods should be based on: (i) factors (microclimatic) which influence the initial appearance and subsequent spread of inoculum; (ii) thorough knowledge of the life cycle of the pathogen; (iii) the ways in which the pathogen perennates; (iv) rough estimate of the quantities of inoculum expected to be disseminated through propagating stock, soil, air, vectors, etc.; (v) mechanism of host infection; (vi) knowledge of the susceptibility of the host plant at different stages of growth; and (vii) meteorological data (macroclimatic conditions) of the area.

Methods of disease forecasting may be (i) **empirical,** in which correlation between the resulsts of disease surveys and the corresponding weather conditions in a particular area has to be related to the biology of host plant and pathogen, or (ii) **fundamental,** in which the effects of different weather conditions (moisture, temperature, etc.) on plant and pathogen, separately and together, are investigated in the laboratory and conclusions tested under field conditions. Besides these, many methods are based on a survey of the viable inoculum available at the beginning of the crop season. These methods provide an early warning but have the disadvantage that many other factors may subsequently modify the course of expected outbreak, e.g., unsuitable weather may prevent disease development despite the presence of ample inoculum. Again, other forecasting systems are based on weather conditions during the crop season, but these assume the presence of sufficient inoculum and will be irrelevant if this is not available. Forecasting systems should, therefore, take account of inoculum sources as well as the weather conditions which affect the development and dissemination of inoculum. Hence forecasting methods should be basically based on: (i) weather conditions during the intercrop period, particularly as affects survival of inoculum; (ii) weather conditions during the crop season; (iii) the amount of disease in the young crop; and (iv) the nature of propagating organs of the pathogen in the air, soil, or planting material. Weather conditions during the intercrop period are usually related to the survival of the pathogen or its vectors between crops. Intercrop weather which reduces pathogen or its vectors is likely to minimize inoculum sources for the following crop. Weather conditions during the crop season are important in the development and spread of inoculum. The effects of weather conditions on plant diseases are complex, influencing not only the events of the infection cycle, but also the resistance of the plant, its ability to throw off or survive attacks, and the numbers and activity of any vectors of the pathogen. Hence the results of laboratory experiments on the pathogens are not always a completely reliable indication as to what happens under more varied conditions of the field. Detailed observations over a number of years may be necessary before forecasting systems based on weather conditions can be prepared. Some diseases originate from inoculum blown in from distant sources and information on the incidence of the disease in such areas, if known can be useful in forecasting the date and severity of the disease in the area expected to receive inoculum. This may be supplemented by spore trapping to determine when the inoculum will be arriving in large quantities. Such study should be supported by the knowledge of the circulation of the air currents which is responsible for the transport of inoculum. The distribution and concentration of soil borne pathogens can be estimated by suitable laboratory experiments and the data collected may be utilized for both forecasting and controlling of diseases. Some pahogens are partly or exclusively seed-borne and the degree of seed-borne infection can be estimated in the laboratory by germinating seeds under conditions suitable for disease development. According to virulence of infection or contamination the propagating materials may be rejected or treated to destroy contaminating pathogen to render them safe for sowing. In such a case both forecasting and control are combined.

PREVENTION AND CONTROL OF PLANT DISEASES

Various plant diseases affect the lives and well-being of so many plants; and in the interest of plants themselves as well as the other living organisms on earth every effort should be made to control thier incidence, dispersal, and perpetuation. Plants are subjected to diseases which reduce their vitality, check development, or cause abnormal growths, and frequently result in the death of the affected individuals. When these diseases attack the cultivated plants on which mankind depends for food, clothing, building materials, fuel, etc., they always cause loss. Such losses may be comparatively insignificant or they may be sufficiently great to cause financial disturbances and suffering in the community or nation. At times these losses have been so great as to cause local famine, and sometimes they have been so great as to increase the cost of the agricultural products. When a disease is not prevalent it does not attract the attention of the causal observer, but when it becomes epidemic it becomes an economic factor in the lives of the grower, the tradesman, the carrier, the consumer, and frequently the manufacturer. When the loss is very great it involves extensive studies by experts to determine the cause and devise methods of control or eradication.

Any attempt to check diseases of plants should be based on the intimate knowledge of the true nature of the diseases concerned which includes causes of the diseases, the nature of the agencies responsible in detail, and the circumstances which influence the establishment of the diseases.

Control of non-parasitic diseases. The control of non-parasitic diseases depends upon accurate diagnosis of the cause of the disease. This is difficult in comparison with the diagnosis of parasitic diseases. But once the diagnosis of the cause of the non-parasitic disease is done correctly, the control measures are generally a matter of common sense. This often involves: correction of high soil acidity by liming, improvement of the drainage of water logged soil, improvement of soil moisture supply in irrigated crops, correction of deficiency of the supply of mineral elements, and correction of similar other deficiencies.

Control of parasitic diseases. Intelligent control of a parasitic disease depends on a thorough knowledge of life cycle of the pathogen, the host plant, the surrounding climatic conditions and their interaction. This also applies to any vectors of the pathogen. Disease control measures should be economical—value of the crop should exceed the cost of control, sufficiently effective to reduce the disease to an acceptable level, and easy to apply with no host and mammalian toxicity.

The prevention and control of the parasitic diseases may be considered along three major lines: the prevention of plants from being exposed to the agencies inciting various plant diseases, *prophylaxis*; curing of diseased plants, therapy; and improvement of the power of disease resistance of the plants to protect themselves against the development of the diseases, **immunization.**

PROPHYLAXIS

A wide variety of control measures are used as a prophylactic measure. They are directed against the pathogen to reduce population or inoculum to a very low level by preventive, cultural, biological, and chemical methods. These measures are the following:

A. Exclusion. Certain measures are taken to keep the pathogen away from entering the sphere in which the host is growing or at least to reduce to minimum the extent of pathogen to come in contact with the host. These measures are:

I. Quarantine and Prohibitions. There is sufficient evidence to indicate that many of the plant diseases have been introduced from one country to the other unknowingly through trade of diseased plants. At that time the diseases of imported plants were not detected. Gradually the matter received attention and against this menace plant quarantines were established. With the establishment of plant quarantine, regulations were imposed on the imported plant materials like, seeds, other planting or propagating material, ursery stock and other plant and plant products to regulate, restrict, or prohibit the entry in the importing country owing to occurrence in them certain diseases desired to be excluded.

II. Certification and Inspection. The system of compulsory inspection and certification of all living plants that are intended for planting is also introduced for the purpose of controlling the entrance of diseases in the country. Unless accompanied by an official certificate from the phytopathological service of the country in which they are grown their ingress in another country is prohibited. The phytosanitary certificate issued by the country of origin must be within 20 days before shipment. Even such procedure sometimes fails to prevent the entry of diseases in the country owing to the masking of symptoms of certain viruses under certain weather conditions prevailing. To ensure thorough check up and perfect controlling of the diseases the majority of countries, by means of acts and ordinances, have reserved to themselves the right to a phytosanitary inspection of incoming plants and plant parts. The service should have the right to take samples for closer inspection. The inspectors entrusted for the work will have very close co-operation with customs authorities in the ports of entry of the country. The efficiency of work depends on the knowledge and skill of the inspectors. Arriving articles which are liable to certification and inspection at the port of entry are to be put in isolation in an isolated room until all fear of disease transmission was thought to be excluded. The inspectors will inspect the articles and according to their nature will disinfect them by applying fumigation and heat treatments based on foreign requirements or agreements with services of importing countries and then allowed entry in the country.

B. Eradication. There are various methods by which eradication of pathogen is carried out. The methods are extremely variable since their suitability depends on the nature of pathogen, host and the conditions prevailing. Some of them are: crop rotation, sanitation, elimination of wild and alternate hosts and perennating organs, soil treatment, selection of disease-free seeds and propagating stocks, application of substances to eradicate incidence and spread of infection, etc. Some of the methods of eradication of pathogen are as follows:

I. Crop rotation. This is a very old practice that is being carried on in agriculture for better crop production since continuous cropping in a particular land by the same crop may lead to exhausting of certain mineral elements which are very much in need of high rate of crop production. At the same time it is also true that continuous cropping helps the perpetuation and intensification of pathogenic organisms. This is particularly true for those soil inhabitors which survive year after year causing injury to the crop that is grown in that soil. Crop rotation accompanied with change in the type of manuring has been found to be successful to eradicate many of the soil-inhabiting pathogenic organisms. There are pathogenic soil-inhabiting organisms which are very specific to particular crop and can thrive at a particular hydrogen-ion concentration of the soil, for example, *Plasmodiophora brassicae* causes disease Club Root of crucifers at a soil pH above 7.0. When in a crop rotation programme that particular type of cruciferous crop is not grown and is replaced by some other crop plant for some time and at the same time different types of manures are used as a result of which the hydrogen-ion concentration of the soil is changed to such a degree that the organism cannot survive any more. The organism which is specific to that particular crop, if at all survives, will fail to infect the newly introduced crop. The elimination of pathogen through crop rotation may be a result of starvation of the pathogen or due to toxic effects of root exudates of the intervening crop. Again rotation with a leguminous plant will have an additional effect of soil enrichment with nitrogenous substances.

Besides crop rotation, **mixed cropping** by growing a susceptible crop mixed with a non-susceptible crop (pigeon pea mixed with sorghum) reduces the incidence of wilt disease of pigeon pea by creating root barriers and through toxic root exudates of sorghum.

II. Sanitation: Sanitation has been advocated repeatedly as an essential means of reducing the amount of inoculum to which the host population is exposed. This operation comprises of the removal of the body or reproductive parts of the pathogen. During the preparation of soil as a part of pre-sowing operation, care should be taken to remove the remnants of the previous year's diseased crop very thoroughly. Best procedure will be to collect them together and burn completely as otherwise if they are stacked in the outscarts of the plot there is every possibility that the source of primary inoculum will still be within the reach of the crop. Along with the growth of the crop, the primary inoculum will also start developing from the last year's plant debris and will start spreading infection as a result of which the new crop will be infected. Sometimes these left-over plant parts besides being the source of primary inoculum give shelters to some insects and serve as their suitable breeding ground. These insects behave as the carriers of pathogens. Again, inocula may spread through contaminated field equipment and carelessly throwing, chewing and smoking tobacco leaves here and there in the cultivating plots. To avoid such happenings, whatever equipment be used in the field operation should be free from any contamination and careless throwing of tobacco leaves here and there should be avoided. As a part of the sanitary measure the sickly and diseased plants should be rogued out as soon as detected in the field since the sickly plants are very susceptible to infection and the diseased plants if not removed will produce the inoculum for spreading infection to the healthy plants. Citrus canker disease can be controlled by prunning out infected branches of host plant. Whether a part or an entire infected plant is removed, it should be burnt immediately to completely destroy the pathogen and whatever instrument is used for this operation should also be disinfected very thoroughly.

Rogueing, so-called, or the systematic removal of diseased individuals from a host population is an obvious and often practiced control procedure because it is characteristically a hand operation, it involves, on any extensive scale, undeniably high labour costs. Though frequently recommended, it is doubtful whether rogueing ever achieves a degree of control which would be fully satisfactory. This is because, by the time disease symptoms are so conspicuous as to indicate removal of the diseased plant, inoculum will have spread to nearby healthy individuals.

Akin to rogueing is the removal of volunteer plants, usually hold overs from the crop of the preceding season. These often serve as a source of inoculum threatening the new host population, whereby pathogenic organisms survive the unfavourable environmental conditions. They also serve between successive cropping periods as a site for the development of viruliferous vector populations. Weeds (*Cyanodon dactylon* and *Setaria italica*) often help the pathogen to perennate, when detected, should be uprooted.

As sanitation is the simplest of all methods of disease control it should be first attempted. In many cases sanitation is of unquestioned utility as an aid to diseae control by spraying. The amount of inoculum can thus be markedly reduced by relatively inexpensive sanitation procedure.

III. Biological control. Biological control of plant diseases can be done by utilizing mutual antagonism or one or more organisms against others in nature by the following methods:

(i) **Lysis method.** By this method soil-borne pathogens are inactivated along two lines: (a) **heterolysis,** is the secretion of extra-cellular enzymes by other organisms which when grown with the pathogens act on them and check their growth, and (b) **autolysis,** is the production of intracellular enzymes in the host which increases power of resistance of the host plant and thereby rate of infection is reduced.

(ii) **Antibiosis method:** Soil microflora can affect the growth of other soil-borne pathogenic organisms by producing some diffusible substances. For example, kernel smut of maize can be controlled by *Clostridium globosum.*

(iii) **Through antagonistic plants—trap crops.** *Solanum nigrum* produces exudates that stimulate the hatching of eggs of nematodes. The larvae enter this plant but they die before developing into adult. Crop plants having this quality are the **trap crops.** These plants are used to destroy soil inhabiting pathogen.

(iv) **Predaceous fungi mehod.** There has been, and continues to be considerable interest in the biological control of plant diseases especially by the use of **predaceous fungi**—fungi which obtain their livelihood by capturing animals. There are various groups of predaceous fungi of which those belonging to the Zoopagales trap numatodes very effectively. These fungi are obligate predators, requiring a living animal in order to satisfy their nutritional requirements. Often they are host specific and are restricted to a certain kind of animal. These fungi may either develop within their host (endogenously) or exist primarily outside their host (exogenously). The predaceous members of the Moniliaceae are also predominantly captors of nematodes. These fungi have a comparatively broad or unrestricted host range and are not obligate predators, meaning that they may live saprobically in the absence of prey.

Soil inhabiting living organisms, particularly nematodes are responsible for the various root diseases of plants of economic importance, for example, Root knot disease of vegetables (tomato, potato) caused by *Meloidogyne javonica* and *M. incognita.* The rate of nematodes infection can be appreciably reduced by the use predaceous fungi which are common in soil.

The problem of biological control of the plant diseases by the use of predaceous fungi can be tackled in different ways: by the use of soil inhabiting predaceous fungi, and by the use of terrestrial predaceous fungi.

By the use of soil inhabiting predaceous fungi. The predaceous fungi may be suitably inoculated to soil where there has been report of the incidence of soil-borne root diseases of plants by nematodes. The mechanism of attack of predaceous fungi on nematodes may be along the following lines:

(a) Infection is initiated by a spore of nematophagous hyphomycete *Monacrosporium salinum.* At the initial stage of infection the fungus does not cause any hindrance to the normal growth of the nematode. But as the fungus continues to grow, it feeds on the contents of the nematode.

(b) The mechanism for the capture of nematodes is the formation of short lateral hyphal branches that curl and anastomose with similar branches forming simple loops or a network of loops. Nematodes are captured in the sticky secretion of these loops or by special enlargements known as **snares.** This type of mechanism may be found in *Arthrobotrys oligospora.* This fungus may be cultured in horse manure. After a nematode is trapped by the loop, the dying nematode's integument is perforated by the fungus and with the further growth of the fungus the trapped nematode is killed. Death of the nematode occurs within about 2 hours after capture.

By the use of terrestrial predaceous fungi. The use of terrestrial predaceous fungi to control infection by insect that causes economic loss has produced some success and needs further experimentation and trial. The technique basically involves the introduction of the fungus into a population of insects under environmental conditions which will allow the fungus to establish itself. Spraying or dusting insect infested plants with a spore suspension prepared from an agar culture or from insects on which the fungus is sporulating. Before using a given fungus for a control agent, it must be determined whether the fungus has a wide host range and, if so, whether it will attack any beneficial insects or other organisms. There is an advantage of using an insect-parasitic fungus of restricted host range as it usually infects and kills only the particular species of insects unlike a chemical insecticides which would kill all insects, including beneficial species. Of the insect-parasitic fungi that have produced positive results are *Beaveria bassiania* and other *Beaveria* species. They have shown some promise in controlling pests such as the common border. In orchards, Citrus mealybug infection may also be effectively controlled by the use of *Entomophthora fumosa.*

IV. Elimination of wild and alternate hosts and perennating organs. Certain pathogens overwinter or oversummer in collateral hosts like weeds, grasses, etc. In order to break the life cycles of the pathogens the overwintering or oversummering hosts are to be destroyed. Again, there are pathogens which need two hosts to complete their life cycles. Of the two hosts principal host is usually of economic importance and the subsidiary or the alternate host is practically of no importance or of much less importance. In such cases the best way to destroy the pathogen is to eliminate the alternate host as a result of which the pathogen will be unable to complete its life cycle and will be ultimately eliminated. Again, instead of destroying, if the alternate host is planted at a distance far from the principal host the link between the two hosts will be lost.

V. Soil treatment. Soil inhabiting pathogens can be removed by various soil treatments. The treatments are rather more successful under greenhouse conditions than in large scale in the field since there are factors which are beyond control. Some of the soil treatments are: (a) exposing of soil to summer heat, (b) burning of straw of combustible material on the soil surface to produce heating effect, (c) heating the soil electrically, (d) application of steam either free-flowing or under pressure, (e) application of various chemicals, and (f) deep ploughing.

In soil treatment by heat dry or wet complete sterilization of soil is not usually necessary, duration and temperature range vary according to the nature of pathogens. Streaming of soil to the point of complete sterilization usually results in the accumulation of toxic materials, chiefly ammonia, which injure sensitive seeds and plants if they are placed in the soil immediately after treatment. To guard against this toxicity, a period of several days is allowed to elapse after which the trated soil is used. In general, heating soil with stream at a temperature of 60°C for 30 minutes is adequate to destroy most soil inhabiting plant pathogens as well as weed seeds and insects.

Soil treatment to remove soil-borne pathogens is done by the application of various chemicals in a number of ways. It is possible to eradicate the pathogens by the use of either volatile chemicals like: formaldehyde; methyl bromide or chloropicrin; dichloropropene-dichloropropane (D-D); and ethylene dichloride. Most common chemical treatment is drenching the soil with formaldehyde about 40 per cent in water. Soil treatment with formaldehyde should be applied several days before sowing seeds to allow the formaldehyde to evaporate from the soil, failing which the formaldehyde will damage the seeds.

It is, however, possible to apply chemicals which are not phytotoxic but effectively fungicidal or fungistatic and will protect the host plants against the attack of the pathogen. These chemicals are: chlorinated nitrobenzenes, various compounds of copper, mercury or thiocarbamate. Seeds are often treated with fungicides for giving protection against soil-borne pathogens. Details of this aspect are incorporated separately.

Some soil-borne pathogens are found only in the upper regions on the soil, the top 20-30 cm. If this upper infected area could be inverted with a lower noninfested area by deep ploughing, disease might be controlled especially in case of shallow rooted host.

VI. Selection of disease-free seeds and propagating stocks. Many of the diseases are carried from year to year by means of diseased seeds or propagating stocks since many of the pathogens harbour in them and unless they are not destroyed or to start with clean and healthy seeds or propagatng stocks are not selected out, there is every possibility that the pathogens will invade the plant body of the host and cause damage to it. Easiest way is to obtain seeds from areas where a particular disease is unknown. Besides this, special care should be taken in the selection of disease-free seeds and propagating stocks to prevent the spread of diseases. Sometimes this can be done by simple process of careful selection of healthy seeds and other propagating stocks and rejecting the undesirable ones. In most cases this method does not work well. Naturally to get them free from pathogens certain other more effective processes are to be applied. To destroy the pathogens certain other more effective processes are to be applied. To destroy the pathogens that may remain in or on the seeds and propagating stocks themselves and to prevent their invasion in the plant body, various treatments are to be applied on them before they are ready for use. These treatments are collectively designated as **seed treatment.** The seed treatments broadly comprise the application of heat is sometimes useful to destroy the pathogens that may remain externally as spores or internally as mycelium. But this treatment is not widely used for not being very suitable, whereas treatment by chemicals is rather more common and since it has been found more effective it is widely used in different varieties of seeds. Usually formaldehyde or compounds of mercury, copper, or sulphur are of common use. The selection of the nature of chemicals, whether they will be applied as dusts or liquids, and their doses depend on the nature of the seeds and the type of pathogen to be controlled. In general, both heat and chemical treatments are very effective in seeds only and the treatments for other propagating stocks are different.

Depending on the nature of seeds, type of seed treatment may be: **seed disinfection, seed disinfestation,** and **seed protection.** But a particular treatment may serve more than one purposes. Again, for other propagating stocks and plant parts entirely different treatments are suitable.

(i) **Seed disinfection:** This is applicable in cases where the treatment is needed to eradicate the pathogen which has already established in the seed. In such case, hot-water treatment has been found effective. But special care must be taken to see that the heat is high enough to kill the pathogen and not the embryo. Chmicals like 1 to 5 per cent ethyl mercury phosphate, tetramethylthiuram disulphide, have produced good results.

(ii) **Seed disinfestation:** This treatment is effective against those pathogens which are present as spores on the surface of the seeds. Copper sulphate and copper carbonate dusts are widely used. In recent days organo-mercuric compounds and Arasan (tetramethylthiuram disulphide) are also commonly used.

(iii) **Seed protection:** This treatment is applied with a view to prevent seeds and seedlings against infection and damage by soil inhabiting pathogens. Cuprous oxide and Zinc oxide have been used for a long time with success. Organo-mercuric compounds, Spergon, Arasan, and Phygon are also good protectants. Some of the other seed-treating chemicals are: Captan, Ferbam, Fermate, and Zineb. In certain cases liquid formaldehyde produces desirable results. Any successful treatment needs long applicational experience of different treatment at various combinations or separately applied with different timing and stages of growth.

(iv) **Treatment of other propagating stocks and plant parts:** Types of this treatment are entirely different from those applied on seed. They are extremely variable since objects to be treated are of various types. For various kinds of tubers, dilute formaldehyde solution and mercuric chloride solution of low concentration are useful provided they have not sprouted. Treatment at a low concentration and dipping for a short time in organic mercury liquid, produces comparatively good result without causing any injury to the propagating stocks. Plant materials including seeds, vegetatively propagating plant parts such as sugarcane setts and onion bulbs are fumigated with formalin vapour.

(a) **Bulbs:** Bulbs of flowering plants are treated with a combination of hot water and formaldehyde producing good result. Satisfactory results have been obtained by dipping bulbs in mercuric chloride solution or dusting mercuric chloride powder. Any successful treatment needs long experience of different treatments at different combinations.

(b) **Other plant parts:** Soaking in mercuric chloride or borax solution of propagative plant parts produces good results. Hot-water treatment and immersing in Bordeaux mixture have also yielded good results.

VII. Control by cultural practices. Cultural practices of crop planting should be such as to reduce attack by pathogens and pests. Cultural practices which discourage incidence of pathogens are most satisfactory methods of plant disease control. Some of the cultural practices which can be manipulated to control plant disease development are: (i) only pathogen-free; (ii) overcrowding of seedlings should be avoided with judicious seed sowing; (iii) since pathogens are able to infect susceptible plants only under certain environmental conditions, and their requirements in this respect are sometimes exacting, hence sowing date of seeds should be manipulated so that the crop passes through its susceptible phase at a time when the pathogen is either absent or unable to infect because of unsuitable environmental conditions; (iv) modification of the concentration of available nutrients in the soil by appropriate fertilizer treatments can be done to reduce disease incidence; (v) by the introduction of crop rotation practice pathogen content of the soil may be reduced to a fairly low level at which it causes only minor damage or is insufficient for infection; (vi) rate of infection may be reduced by reducing host injury; (vii) planting trap crop induces germination of soil borne propagative organs of pathogens after which the trap crop is to be ploughed or otherwise destroyed before the crop itself is sown; and (viii) accumulation of water in a field when there is a crop standing, helps in accumulation of toxins around the roots and base of stems making them liable to easy pathogen infection. Drainage is, therefore, essential to keep the field free from pathogens.

VIII. Application of substances to eradicate incidence and spread of infection. This operation is carried out with the help of certain chemical substances whose nature and method of application are extremely variable. The selection of chemical substances and their method of application depend on the nature of disease, pathogen, and host involved. Application of chemical substances to host plants may be **direct** or **indirect.** When applied indirectly through soil it may be in the absence or presence of the standing crop. Direct application may be in the form of liquid, solid or gas, i.e., as spray, dust of fumigant. The spray material has to be put into a solution, suspension or emulsion. Depending on the nature of the chemical, water or other solvent is used. Mixing of chemical and solvent must be thorough and spraying should be uniform. In the case of dust preparations, the fineness of the particles and the force with which they are ejected from the blower and the atmospheric conditions under which dusting is done, are very important factors in successful control of plants against the incidence of diseases and to eradicate the existing inoculum in and around the standing crop. Equipment required for this work are: seed dressing machines, sprayers (power or otherwise) dusters and soil injectors. While using fumigants on seeds, vegetatively propagating plant parts, etc. should be put in a special airtight container for a specified time depending on the materials fumigated and the nature of fumigants used. After the fumigation the materials should be transferred in a sterilized container for future use. Seed treatment is done in seed dressing machines in which seeds and necessary chemicals are put in a rotary seed drum which is rotated manually or by power for a specified time (5 to 10 minutes) so that the seeds are thoroughly coated with a film of chemical, after which the seeds are ready for use.

The chemical substances that cause destruction of fungi are known as **fungicides.** A fungicide is thus any substance which kills fungi, but its modern connotation implies a substance which, when applied to higher living plants in any stage of growth, will kill parasitic fungi and prevent the development of fungus diseases without causing damage to the host plants; similarly **bactericides** in respect of bacteria; and **nematicides** are the chemicals which kill nematodes. But those fungicides which cause inhibition of fungal growth are the **fungistatic agents (fungistats).** The same agent is commonly capable of producing both actions. But the fungistatic activity of an agent is broader than its fungicidal activity. Whether an agent is fungicidal or fungistatic is primarily a matter of degree of intensity and duration of exposure. If the action is less severe and may be reversed, the result is fungistasis, while if it is irreversible, the action is fungicidal. Fungicides also act as protective agents against the attack of a pathogen.

In addition to heat, light, and electricity, which have fungicidal properties, a great number of chemical substances are also fungicidal. Fungicides stop the fungus from forming viable spores, or they stop germ tubes or other mycelium from infecting the host plants; in all these ways they reduce inoculum development. Hence fungicides are applied mainly: to protect the host against the attack of pathogens, and to check the spread of inoculum that is already present in the infected host. The effectiveness of a given fungicide treatment is again dependent upon the environment during application of the treatment, inoculum level or age of the infection, or susceptibility of the host to infection. As a matter of fact, there is no useful universal fungicide. The choice of a fungicide depends upon: the species of fungus to be controlled and the nature of the material to be protected. The solubility of the fungicide is of great importance. For most efficient preservation of wood or protection as a spray, a fungicide must have a low solubility in order that the protection may extend over a long period of time. When a fungicide is to be used on a living plant, the relative sensitivity of the host and of the fungus to the fungicide must be considered. Host sensitivity limits the use of many potent fungicides. A useful fungicide must be more toxic to the fungus than host.

Most of the fungicides act directly on the fungal cell or spores after entering them. Many are rather insoluble in water but may be sufficiently soluble to inhibit the germination of sensitive spores. Several possible factors bring about solubilization of fungicides on plant surfaces with the help of carbon dioxide and ammonia in rain water and dew, guttation fluids and other exudates from the plant. The cell wall of many fungi probably offers little resistance to the entry of fungicides of low molecular weight but in others it may prevent certain substances from reaching the protoplast. The cell wall may contain fatty or other substances which reduce its permeability to water and fungicides. To what extent the cytoplasmic membrane regulates the entry of fungicides into cells is uncertain. The effect of fungicides on fungal cell is a complex process which may be: (i) harmful by damaging cell walls or by interfering with their extension or initiation; (ii) interfering with the cytoplasmic membrane releasing soluble substances from the cells; (iii) act as rather nonspecific inhibitors of enzymes; (iv) cause antimetabolic effect and interfere with the utilization of metabolites; and (iv) cause antimetabolic effect and interfere with the utilization of metabolites; and (v) proteins may be hydrolysed by acids or bases and other direct interactions between fungicide molecules, and vital constituents of cells may lead to toxic reactions.

One of the major aspects of the operation is the selection of a suitable fungicide. In doing so one has to bear in mind certain aspects of the fungicide like relative cheapness; easy to handle; tenacity or adhesiveness to have a lasting effect; property to cover the expected surface well; free from phytotoxicity of the host but should be toxic to the pathogen or in other words, it should be **pathocidal** and not **phytocidal**; harmless to the host as well as human beings and normal growth of the host must not be hampered. In recent years there has been considerable interest in the control of disease—pest complexes by the application of fungicides in combination with insecticides (insect killing substances).

On the basis of their application to plants, fungicides are classified in three groups; **protective, eradiative** and **systemic.** The fungicides may be (i) **protective fungicides**—those used to protect host agianst infection by a pathogen forming a protective layer which kills the pathogen or inhibits its development, (ii) **eradicative (therapeutic) fungicides**—those applied to a substratum in which the fungus is already present, and (iii) **systemic fungicides**—those which are fungicidal after absorption and subsequent translocation and are able to control the growth of pathogen in the host tissue.

(i) and (ii) Protective and Eradicative fungicides. A protective fungicide is applied to the plant or other material before the inoculum arrives at the infection court. It acts against infections of the future and against disease still to come. While doing so the operation may have a direct effect upon pathogens which have already invaded the host, whereas an eradicative fungicide acts against infections of the past. These fungicides are applied in the form of dusts, sprays or fumes.

When the fungicides are used as dusts, the active ingredient is diluted with a finely divided powder, called a **carrier.** Dusting machinery usually consists of a fan or blower, which may be manually operated or power driven with the help of engines. Dusting is usually more convenient, the dusts being easy to handle, and is not dependent on the supply of water. Machinery for the application of dust is usually lighter than that used for sprays. Dusts, however, should not be sticky or form lumps or balls. With sprays, on the contrary, the active ingredient is added to the requisite quantity of water and the fluid is forced under pressure through a nozzle, which results in the formation of very small droplets. Usually, compressed air is forced into the spray tank for the purpose. Efficiency in spraying depends on the size of the droplets, which is dependent on the size and design of nozzles to ensure uniform deposition. The active ingredient is made water-dispersible, and the stability of the suspension is usually maintained by the addition of complementary materials or the use of mechanical agitators. Spraying is more laborious than dusting because the machinery is heavier and not so easily portable, whilst the supply of water often poses a difficult problem. Nevertheless, spraying usually gives more effective and lasting protection than dusting. The spraying fluid may be applied in the form of very fine droplets by **mistblowing** or **atomization.** In both spraying and dusting the size of the particles is important regardless of the chemical nature of the fungicide. The smaller the size of the particles, the better the coverage, but too small a size of the particles causes quick decomposition. Sometimes, soaps, caesin, resinates, various mineral oils, petroleum, sulphonic acids, etc. are added to the spray fluid to reduce surface tension and improve contact between the fluid and the sprayed surface so that the fluid does not collect into large drops and run off easily. These substances are known as **spreaders.** They have also wetting and penetrating properties which contribute to the efficiency of the spray. Fungicides to be effective for a long period of time should have good adhesiveness to improve their tenacity and retention. The adhesiveness of the fungicides can often be increased by the addition of gummy or resinous substances, e.g., gum arabic, resins, etc. At times, oil, e.g., linseed oil is also used for the same purpose. These substances which increase adhesiveness of the fungicides are known as **stickers.** They can be used both in dusts and sprays.

The suitability of a fungicide and whether it should be applied as a spray or dust and the time and frequency of application to be effective, all these aspects are, however, dependent on the nature of the host, its parts to be sprayed or dusted with chemical compounds or mixtures, and the pathogen against which the operation is being applied. Quite often spraying is more effective than dusting since in the former process the chemicals will remain in closer contact with the host body for a longer period and thereby be more effective. Spraying or dusting operation, according to necessity, may be applied before the emergence of leaves—**post-emergency spray** or **dust** to ensure better result. Again, post-emergence spray or dust may be repeated more than once according to necessity. Special care should be taken to make selection of sprays or dusts and the time of application. Simultaneously attention should be paid for the selection of equipment for the operation. Most handy and useful should be the choice. Most of the fungicides belong to any one of the groups: (a) **the inorganic sulphur group,** (b) **the inorganic copper group,** (c) **the organic group with or without various heavy metals,** and (d) **antibiotics used as fungicides.**

(a) **The Inorganic Sulphur Group:** Sulphur is one of the oldest and most widely used fungicides. The sulphurs are used in various forms. **Sulphur dust** has been applied extensively and with success against may pathogens. A fine dust is of high fungicidal efficacy. **Lime-sulphur** has a very good fungicidal action. But burning of foliage and fruits may occur in hot weather following lime-sulphur spray. Leaf and fruit drop may also occur. Since it has certain drawbacks, proper care should be taken before selecting lime-sulphur as spray. Sublimed or ground sulphurs are cheaper and said to be more adhesive and less liable to cause injury. **Self-boiled lime-sulphur** is less phytotoxic than lime-sulphur. It is prepared by adding water to lime and then starring with sulphur. **Wettable sulphur** is even better than self-boiled lime-sulphur and is more extensively used. In general, sulphur in some form or the other is used as spray in the fruit trees. Even there are sulphur sensitive varieties of fruit trees where sulphur spray should be avoided.

(b) **The Inorganic Copper Group:** The use of copper as **copper sulphate** as a fungicide had been recognized and successfuly used in much earlier days which was improved upon in later years in 1885-87 by the use of a mixture of copper sulphate and lime under the name, **Bouillie bordelaise** commonly known as **Bordeaux mixture.** The story behind the discovery of the fungicide Bordeaux mixture has already been related. The Bordeaux mixture became very famous and widely popular in use due to its efficacy against the attack of diseaes of crops of economic importance. It is the most widely used and most generally serviceable fungicide to the present day. A common mixture is the '4-4-50': copper sulphate 4 lb., quicklime 4 lb. (or hydrated lime 6 lb.), water 50 gal. When making, copper sulphate is put in some of the water, the lime in the rest, and the two solutions are then mixed thoroughly. Any time Bordeaux mixture is used, it should be freshly prepared. Of the several advantages some are: relatively less expensiveness; easy to prepare; easy and safe to handle; harmless to almost all hosts, does not cause injury to the foliage but very efficient in fungicidal action; adhesive quality; economical in use; and effective as a fungicide. But it has certain disadvantages too, they are: it is injurious to certain fruit trees of which most common are apples, has the tendency to delay ripening of fruits, in sprayed leaves photosynthesis is temporarily retarded, not suitabe for ornamental plants, and causes injury to the spraying machines made of iron or zinc. Besides, Bordeaux mixture there is another copper spray fluid, best known as **Burgundy mixture.** It is prepared very much like Bordeux mixture except that lime is replaced by sodium carbonate crystals. For a long time it was also named as **Soda Bordeaux.** Burgundy mixture is equally well effective as Bordeaux mixture but may be due to more risk of spray injury in use it was never widely used. It is generally applied on the blighted leaf or other related disease and is employed particularly in the industrial areas where the acid fumes interfere with the action of Bordeaux mixture. Various combinations of copper carbonate and ammonium hydroxide and ammonium carbonate are also largely used. **Copper-lime Dusts** are also in use which in presence of moisture form Bordeaux mixture on the foliage. It is more effective when applied on foliage with moist surface due to presence of dew as otherwise the adhesiveness is poor. The value of **insoluble copper** like **Cuprous oxide** as a seed protectant has been well appreciated and it is also being used as a foliage spray. Insoluble copper spray is less phytotoxic and has been found more useful than Bordeaux mixture especially on sensitive foliage.

(c) **The Organic group with or without heavy metals:** Bordeaux mixture and lime sulphur which are to be prepared fresh every time often cause injury to leaves, premature leaf-fall and russeting of fruits, if improperly used. Since 1930, they have been replaced by certain ready-to-use copper organic combinations. The fungicidal properties of **elementary mercury** is well known, but because of its extreme toxicity to animal it is used only to a limited extent as a plant protectant. There are two groups of mercurials available for the purpose: inorganic and organic mercurials. Of the inorganic or metallic mercurials, mercuric chloride and mercurious chloride are used for seed treatment. Seed treatment with organo-mercury compounds, e.g., **Agrosan GN, Ceresan,** and **Tillex,** has given perfect control of a number of seed-borne diseases. The shortage of copper and mercury in the market led to the development of dithiocarbamate fungicides, such as **Dithane Z-78, Dithane M-22, Dithane S-31,** and **Dithane M-45.** These are very broad spectrum fungicides which can control downy mildews, rusts and *Alternaria* blights of cereals, vegetables and fruit crops. **Thiram** as a dust is extensively used in seed treatment against seed-rot, damping-off, seedling blights, and root-rots of many crops. **Captan (Flit 406)** has been found very useful for vegetable-seed treatment against pre-emergence and post-emergence injuries. A number of soil and seed-borne diseases caused by fungi and nematodes are controlled by fungicide such as **Brassicol (PCNB)** and nematicide like **Nemagon** and **Vapam.** The dinitrophenolic compounds (**Elgetol,** and others) are useful as eradicant sprays for destroying the inoculum on overwintering or oversummered trash of diseased plants of the previous crops, or for treatment of dormant fruit trees. Bayer's **Antracol** is as efficatious as nickel chloride or copper fungicides in the control of blister blight of tea. **Karathane** wettable powder and liquid are most effective in controlling powdery mildew and mites. The use of the **carbamates** as insecticides and fungicides is well established. The compound **ferric dimethyl-dithiocarbamate,** known as **Ferbam** and marketed under the name **Fermate** has been successfully introduced as a general foliage spray being effective against a large number of fungus diseases affecting foliage, some of which have previously been difficult to control. Besides being highly effective it has tenacity and low phytotoxicity. Another compound **zinc dimethyldithiocarbamate,** known as **Ziram** widely introduced under the trade name **Zerlate** is equally well-effective fungicide. Again, another important fungicide is **disodium ethylene bisdithiocarbamate,** commonly known as **Nabam** is marketed under the trade name **Dithane.** It is effective against many diseases of fruits and vegetables. **Zineb** which is marketed under the trade names **Parzate** and **Dithane Z-78** is **zinc ethylenebis-dithiocarbamate.** It is also commonly used with success. Another group of materials which possesses fungicidal value is **Quaternary-Ammonium Derivatives.** Some of the compounds which have already been proved successful as foliage fungicides are: **Zephirin,** a complex of **benzyllauryldimethyl-ammonium chloride; Puratized NSA and NSE phenyl mercuri-triethanol ammonium lactate; Isothan Q15, laurylisoquinolinium bromide; Cationic nicotinium salts. Sodium dinitro-orthocresolate** is commonly known as **Elgetol** and **Krenite,** used as spray to eradicate or suppress sporulation of pathogenic fungi overwintering on plant debris in the orchards. **Phygon** is nothing but **2, 3-dichloro-1, 4-naphthoquinone** is also of use in the orchards as fungicide spray.

Spoilage of fruits and vegetables during transit and storage can be prevented by previously dipping them for about two minutes in a tank containing a concentration of 1 to 1.25 per cent sodium orthophenylphenate. Bleaching powder in a weak watery solution is equally beneficial in cuts, bruises and spoilage of fruits and vegetables.

(d) **Antibiotics used as Fungicides:** In addition to the above fungicides, certain antibiotics, in widely variable doses and forms, may be used to control fungi and bacteria which cause plant diseases. The antibiotics have high specificity of action against plant pathogens, relatively low phytotoxicity, absorption through foliage and systemic translocation, and activity at low concentrations. They may be used as spray or dust behaving like protective and eradicative fungicides or as systemic fungicides by allowing them to be absorbed as solutions through roots, and translocated in all parts of the diseased plant. Dip application is, however, effective in the treatment of seeds or other propagating parts. Antibiotics may also be applied in the form of a paste with a lanolin base. Some of the antibiotics used as fungicides are: **Penicillin, Streptomycin, Citrinin, Cycloheximide (Actidione), Polyene macrolides, Griseofulvin, Patulin, Gliotoxin** and **Viridin.** It may often be desirable to replace certain commercial fungicides, which are toxic to humans and animals, by the antibiotics which are comparatively less toxic. Application of antibiotic might facilitate control of internally seed or other part borne diseases, by internal therapeutic action, since many are taken up by plants and are readily transported in the plant system. Other advantages of these fungicides might be resistance to washing off by rain and, upon translocation in the plant, protection of untreated parts or new growth.

The use of antibiotics for the control of plant disease has made possible the internal protection of the host, due to the systemic diffusion of streptomycin, tetracyclines, cycloheximides and griseofulvin in the host plants. Streptomycin has been observed to be very effective against bacterial diseases (both prophylaxis and therapy) many of which were uncontrollable for a long time. Again, cycloheximide (actidione) and griseofulvin have been proved to be most potent antifungal antibiotics. Actidione and its derivatives have been successfully used in controlling rust diseases, the powdery mildew of roses and ornamental plants, cherry leaf spot, and the monilia disease of fruit trees. The antibiotic blasticidin-S is reported to be used in Japan on a large scale for control of *Pyricularia oryzae* which causes blast diseases of rice. Species of *Xanthomonas* which cause various plant diseases can be controlled using streptomycin or streptomycin-terramycin mixture.

Antibiotics may act directly on the pathogen in the plant tissue or they may also be transformed into substances of greater potency. They may act directly by either nullifying the toxin or inducing host resistance.

The application of antibiotics against plant diseases, however, has met serious difficulties. Preferably the antibiotic fungicide should be toxic to the pathogen and harmless to the host. A number of powerful antibiotic fungicides do not meet the latter requirement and have been found to be phytotoxic. For example, the use of cycloheximide, which is active against plant diseases, is limited due to its phytotoxicity. Remarkably non-phytotoxic and reasonably well-translocated in plants are the polyene macrolides. A drawback to their use, however, seems to be the fact that their fungicidal activity may be marked after introduction into plant tissue. Again, when applied on the plant surface some polyene macrolides appeared unstable. The most pronounced example of a non-phytotoxic systemic antibiotic fungicide seems at the present, to be griseofulvin. This antibiotic was reported to be more active against powdery, mildew disease of various plants when administered to the roots in solution than when applied to plant surfaces. It is fungistatic rather than fungicidal. But this feature does not necessarily hamper effective disease control. Besides being expensive in comparison with commonly used fungicides, antibiotics, particularly streptomycin used as fungicide may cause destruction of chlorophyll resulting in chlorosis. The antibiotics in many cases cannot be effectively applied to the soil as they may be chemically or biologically or physically inactivated in the soil.

(iii) Systemic fungicides. In the past, the emphasis in fungicidal treatment has been on external application. This has somewhat altered since success in providing the host plant with internal protection against disease has lately been obtained. The systemic fungicides are allowed to absorb as solutions through roots and translocated in all parts of the host plants. They affect host plants along three lines:

(a) Change the host plant's metabolism.

(b) Reorientate the physiological relationship between the host and pathogen.

(c) Introduce deactivating tissue-destroying enzymes.

Some of the effective fungicides are: **Benlate, Bavistin** and **Vitavax.** Bavistin and benlate have been successfully used to control pineapple disease. Systemic fungicides known as **Oxathins** distributed under the trade name **Vitavax** penetrate into the developing seedling and selectively kill or inhibit the growth of fungus in the host tissue, particularly smut fungus. Many of the synthetic hormones can also act as systemic fungicides. Systemic fungicides are also used in combination with systemic insecticides to control both fungi and insects (Table 24.2).

Namaticides and their uses. Some of the nematicides are: carbon disulphide, chloropricrin, methyl bromide, dichloropropene-dichloropropane (D-D) and ethylene dibromide (EDB). The nematicides are applied in the soil as fumigants a few weeks before the crop is sown or planted. Since nematodes are confined mostly to the top 12 inch-layer of the soil, the nematicides must be brought into contact with the nematodes in the top soil layer. The fumes of the nematicides kill the nematodes, but the diffusion may not usually cover a radius of over ten inches. The nematicide treatment of the soil is effective for one reason only and can kill only about 90 per cent of the nematode population. Soil can be ploughed up after harvest and the nematicide applied in case of annual plants. But in case of perennial plants nematicide of low phytotoxicity should be used.

IX. Protection against post-harvest disease losses. The major overall cause of post-harvest disease is invasion by the pathogen after harvest through wounds or natural openings in the surface of the host. A reduction of initial infection can be accomplished by: lessening of the number of harvesting injuries and by a reduction in he amount of available inoculum.

Reduction in the amount of available inoculum may be possible by the following techniques:

(i) **Use of Fungicides to reduce the Inoculum Level.** Major sources of pathogen inoculum are the harvesting equipment and containers, and the atmosphere of the houses where the harvested articles are kept. Post-harvest fungicides are used either as aqueous solutions or as fumes. They are more often applied in the form of aqueous solutions than by any other method. The popularity of solution treatment rests upon the fact that solutions are easy to prepare and apply uniformly, possess penetrating powers approaching or exceeding those of fumigants. Solution treatments are conveniently applied to commodities that are wetted for other purposes. Treatments of harvesting equipment and container with solutions of sodium 0-phenylphenate and 8-hydroxyquinoline, and harvested articles with captan, copper 8-hydroxyquinolinate, copper naphthenate, sodium 0-phenylphenate, or tributyltin succinate are recommended.

But fumigation is the most efficient means of applying a fungicide to a commodity after harvest. A fumigant can be applied to fruits and vegetables in the field containers immediately after harvest, thereby minimizing handling of a delicate commodity. A fumigant applied soon after harvest would inactivate or inhibit the development of a pathogen, thereby increasing the effectiveness of a solution treatment applied subsequently. The ideal fumigant is one which is barred from penetrating the intact surface of the commodity. Impregnating

Table 24.2: Major categories of fungicides used in plant disease control

Categories	Name of Fungicides	Composition	Mode of application (in diseases)
A. Inorganic Copper compounds	1. Bordeaux mixture	• Copper sulphate (2.268 kg) Hydrated Lime (2.268 kg) Water (50 gallons)	Spray (in variety of diseases)
	2. Burgundy mixture	• Copper sulphate (4.536 kg) Sodium carbamate (5.666 kg) Water (50 gallons)	Spray (in variety of diseases)
B. Inorganic Mercury compounds	3. Mercuric chloride ($HgCl_2$), or Mercurous chloride (Hg_2Cl_2) solution	1 : 1000 dilution	Seed/root stock treatment (in variety of fungal and bacterial diseases)
C. Inorganic Sulphur compounds	4. Sulphur powder	• Sulphur (3.56 kg) Water (50 gallons)	Spray
D. Organic Sulphur compounds	5. Carbamate (Sulphur containing carbonic acid)	• Thiram (Tetramethyl thiuran disulphide) 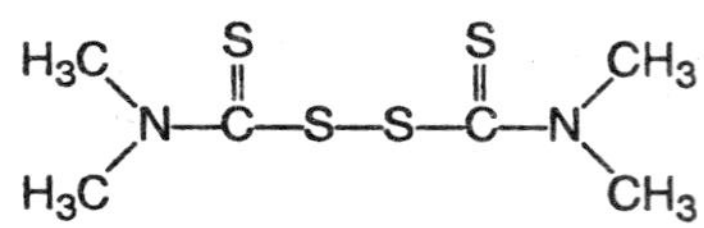 (Trade names: Arasan, Tundas, Terson, Tulisan)	Spray on foliage or seed treatment.
		• Ferbam (Ferric dimethyl carbamate) 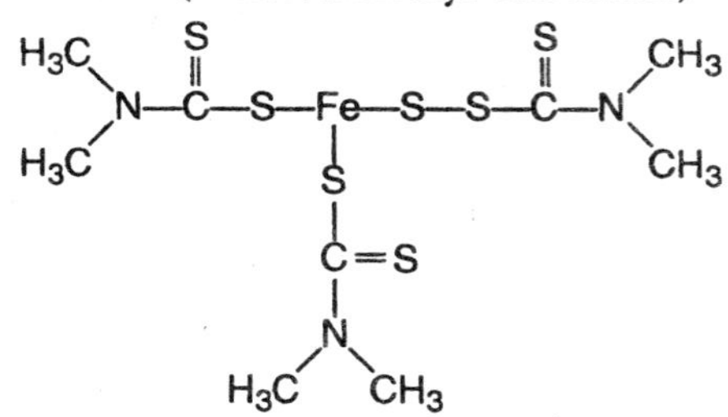 (Trade names: Fermate, Ferbam, Karbam or Coromet)	Spray on foliage (as fungicide)

Contd.

Contd. Table 24.2

Categories	Name of Fungicides	Composition	Mode of application (in diseases)
		• Ziram (Zinc dimethyl dithiocarbamate) 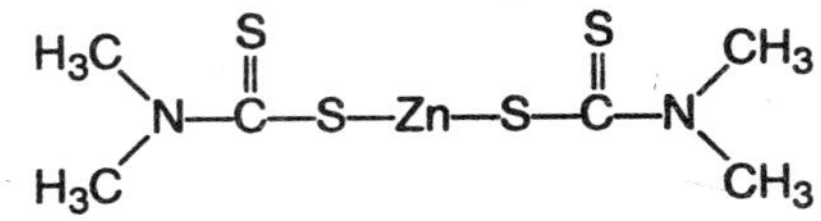 (Trade names: Cuman, Zerlate, Karbam)	Foliac spray
		• Zineb (Zinc ethylene bisdithiocarbamate) 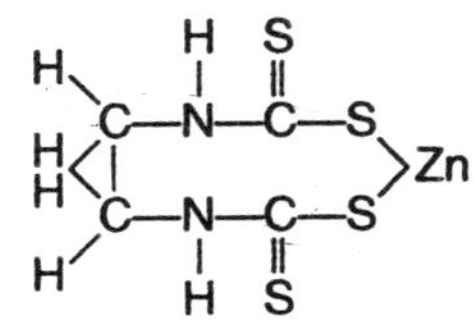 (Trade names—Dithane Z-78, Lonacol)	Foliac spray (Blight of potato and tomato, blast of rice, blight of rice, downy mildew)
		• Nabam (Disodium ethylene bisdithiocarbamate) 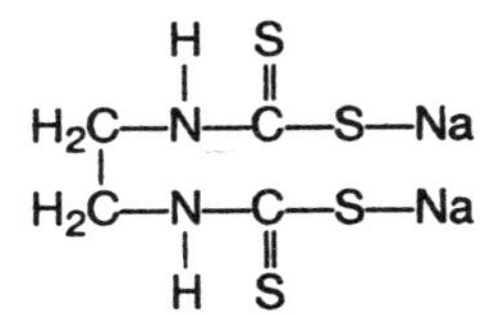 (Trade names: Dithane D-14, Dithane A-40)	Foliac spray (Blight disease)
		• Maneb (Manganese ethylene bisthiocarbamate) 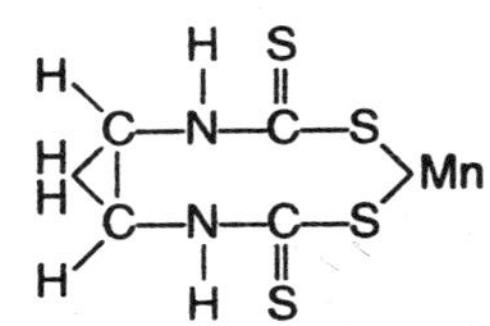 (Trade names: Dithane M-45)	Foliac spray

Contd.

Contd. Table 24.2

Categories	Name of Fungicides / Composition	Mode of application (in diseases)
E. Quinone and phenolic fungicides	6. Chloranil (Tetrachloro-p-benzoquinone) 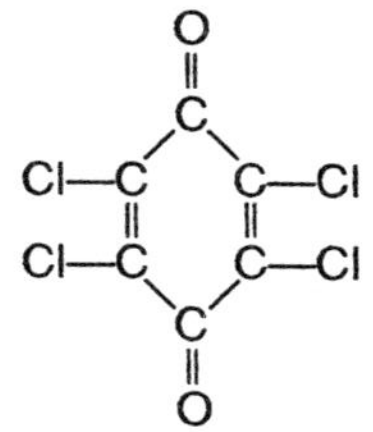	Seed treatment
	7. Dichlone (2, 3 dichloro-1, 4 naphthaquinone) 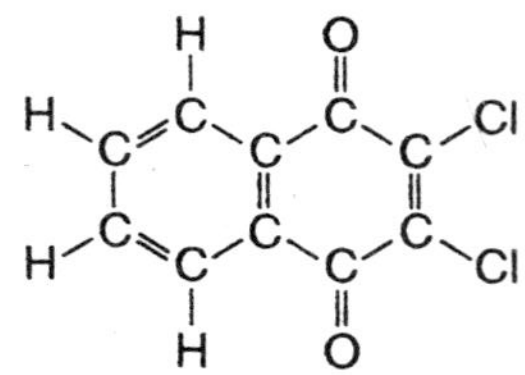	Foliac spray
	8. Captan (N-trichloromethyl thio-4-cyclohexane 0-1.2 dicarboximide) 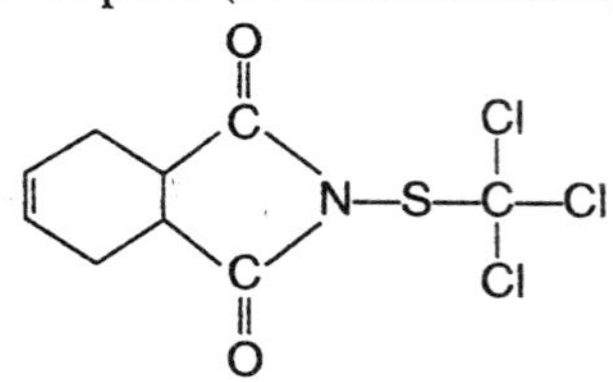 (Trade names: Captan 50w, Esso 406)	Foliar spray or Seed treatment
	9. Chloroneb (1, 4-dichloro-2, 5 dimethoxy benzene) 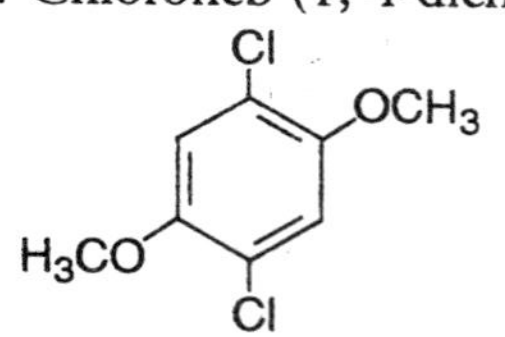	Seed treatment

Contd.

Contd. Table 24.2

Categories	Name of Fungicides	Composition	Mode of application (in diseases)
	10. Dexon (p-dimethylamino-benzene-diazosodium-sulfonate) 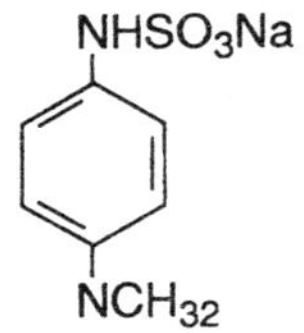		Soil treatment
F. Organomercurals	11. Ceresan-M (ethyl mercury-p-toluene sulphanamide)		Seed treatment
	12. Agrosan-GN (phenyl mercury acetate and ethyl mercury chloride mixture)		
G. Systemic fungicide	13. Benomyl (Methyl N-butylcarbomoyl-2 benzimidazole carbamate) 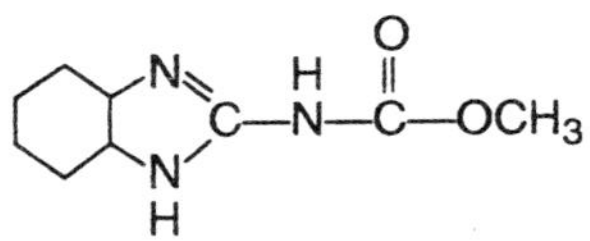 (Trade name: Bavistin)		Foliac spray or Soil treatment
	14. Tetrazole (Tetrazole-5-ethoxy-3-trichloromethyl-1-2-4 thiacliazole) 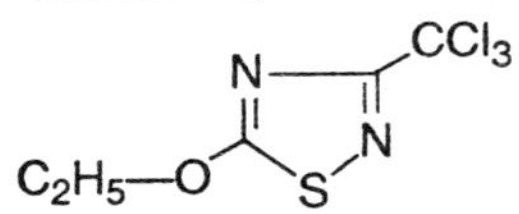		Soil treatment
	15. Oxathins (2, 3 dihydro-5-carboxy anilido-6-methyl 1-4 oxathin) 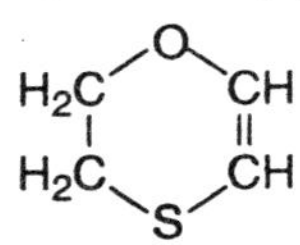 (Trade name—Vitavax)		Foliac spray
	16. Organophosphate fungicides • Kitazin (S-benzyl diethyl phosphorothiorate)		Foliac spray (Blast of rice)

Contd.

Contd. Table 24.2

Categories	None of Fungicides (in diseases)	Composition	Mode of application
	• Triazoles		
	• Morpholines		
	• Phosphites (Aluminium ethyl phosphite)		
	17. Pyrimide derivatives		Foliac spray
	• Triarimol		
H. Antibiotics	18. Streptomycin		
	19. Cycloheximide		Foliac spray
	20. Griseofulvin		or
	21. Aureofungin		Seed treatment
	22. Nystatin		
I. Soil Fumigants	23. Formalin		Soil treatment
	24. Methyl bromide		
	25. Carbon disulphide		
	26. Chloropicrin		
J. Oils	27. Petroleum derivatives		Foliac spray

harvesting containers with paraffin wax or fumigating with formaldehyde or sulphur dioxide often reduces contamination.

(ii) **Interrupting the Infection Process.** After inoculation has taken place it is possible to prevent disease by interrupting the infection process with certain fungicide treatments, provided that development of the pathogen is still quite limited. Some of the recommended post-harvest fungicides are: Borax; Sodium metaborate; Sodium pentaborate; Sodium carbonate, Sodium bicarbonate, Sodium hydroxide (1-3 per cent), Tributyltin oxide; Copper 8-hydroxyquinolinate; Sulphur dust; Liquid lime sulphur compounds; Chloropicrin; Methyl bromide; Dibromotetrachloroethane; Sorbic acid; Formaldehyde; Captan; 2, 6-Dichloro-4-Nitroaniline.

(iii) **Heat Therapy.** Heat treatment appears to be the best and most convenient means to control post-harvest diseases. The range of temperature and the duration of treatment are extremely variable. Most of the recommendations for effective disease control have fallen in the range of 120°F-130°F for 2 to 5 minutes.

(iv) **Radiation treatment.** The great penetrating powers of high-energy gamma radiation is recommended for control of post-harvest disease.

(v) **Attenuation of Established Infections by Refrigeration.** Refrigeration alone does not provide adequate protection against post-harvest diseases of many fruits and vegetables. Low temperatures are most effective in retarding the development of diseases caused by microorganisms which grow well only at moderately high temperatures. Refrigeration is of great value in prolonging the marketable life of perishable commodities after harvest, but it is not the ultimate solution to the problem of microbial decay.

X. **Protection of fruits and vegetables during storage.** Occasionally huge loss is encountered due to decay of fruits and vegetables during storage. This is mainly due to not taking proper care during storage against pathogens which cause severe damage to the stored articles. Protection against the loss can, however, be taken either by the use of fumigants or by the application of fungicides to the stored fruits and vegetables. Of late, irradiation is considered as suitable alternative to fungicides and fumigants.

(i) **Fumigants:** These are nothing but gaseous materials which are sometimes very useful to check the decay of fruits and vegetables during storage. Of many fumigants **Sublimated sulphur, Paradichlorobenzene** are used with success. Both these fumigants are also useful to control disease incidence under greenhouse condition. **Sulphur dioxide** gas released in storage rooms controls certain fungal diseases.

(ii) **Fungicides:** A variety of fungicides are in use to perishable fruits and vegetables after harvest. Very common in use is **Copper sulphate** in different ways like: applied on the surface of fruits, wrapped by wrappers impregnated in copper sulphate solution, etc. Besides this, **sodium ethyl mercurithiosalicylate, borax, sodium orthophenylphenol, sodium metaborate** have also been found effective when used in solutions and dipping the fruits in the solutions.

(iii) **Irradiation:** The fungicides and fumigants that are used to check the decay of fruits and vegetables during storage may often be very harmful to humans if they are not carefully chosen and applied properly. Irradiation involves sterilization of fruits and vegetables with ionizing radiations such as gamma rays or X-rays or accelerated electrons to destroy insects, molds or microorganisms. Irradiation has been successfully used in storage of onions, potatoes and spices to ward off pests and microbial contamination.

THERAPY

Therapy is the cure of a diseased plant, the relief of its symptoms, or the repair of the damage caused by the pathogen. It is the treatment of the diseased plant as an individual, not the plant as a member of population. Therapy removes the cause of disease so that the normal mechanism of living can run smoothly. Two courses are open for the therapy of plant diseases—**physical therapy** and **chemotherapy.**

(i) **Physical therapy.** By this treatment the disease is fought with physical means, such as **surgery, temperature modification, radiation,** and **moisture modification.**

(a) **Surgery.** Surgery is the removal of infected tissues to prevent additional damage.

(b) **Temperature modification.** This is a process by which the pathogen is inactivated in or on the host by differential heat treatment. The curing of the diseased plants by growing at above-normal temperatures has often been reported. Hot-water immersion, with or without antiseptics in solution, has also been quite effective against cotrolling the disease development.

(c) **Radiation.** Ultraviolet radiation is often found useful as a plant disease therapy.

(d) **Moisture modification.** Intercellular humidity is a factor concerned with disease development in a large number of plants. In such cases the diseased plants may be cured by controlling intercellular humidity. The modification of the moisture-oxygen content of woody stems acts in therapy of certain trees by way of controlling symptoms. The modern large-scale vacuum cooling of food crops results to retard the development of bacterial soft rot and other diseases. The plants are exposed to a partial vacuum. This converts the inferior liquid water films to intercellular water vapour which is then withdrawn from the plant structure. This vaporisation cools and dries the intercellular spaces.

(ii) **Chemotherapy.** Chemotherapy is the control of plant diseases by chemical treatment that reduces or nullifies the effect of pathogen after infection. The chemicals that initiate a curative or mitigating effect, either directly or indirectly are known as **chemotherapeutants.** A chemical therapeutant must be effective inside the plant where it may (a) kill the pathogen as it enters the host; (b) inactivate the toxins, enzymes and host pathogen. The action of a chemical therapeutant may also include transformation of a product after it has been introduced in the host tissue, into a substance capable of a greater or different activities than the original material, whereas a fungicide should possess fungicidal activity against the pathogen as well as capable of systemic translocation into the host plant so that it can kill the pathogen on entrance in the host tissue. A chemical therapeutant should be selective in its action but not harmful to the host plant. It may alter the host protoplasm resulting in increase in host resistance.

A pathogen may injure only surface cells or a localized area of tissue. On the other hand, it may systemically injure the entire plant or any major part thereof. Treatment of the former is **topical chemotherapy**, whereas treatment of the latter is **systemic chemotherapy.** Some of the chemotherapeutants are: phenyl mercuritriethanol ammonium lactate (Puratize), sodium dinitrocresylate, cycloheximide, griseofulvin, thiolutin and phenyl mercury acetate. Depending on the nature of the disease and the conditions of the symptoms, the chemotherapeutants are used either directly on the surface lesions or as solutions to be absorbed and translocated through the roots and pass in all parts of the diseased plant.

The chemotherapeutants are different from systemic fungicides. A systemic fungicide is one that distributes itself systemically through the plant and kills the pathogen by direct toxic action, whereas a chemotherapeutant acts to make the host resistant rather than to kill out the pathogen. Some of the chemotherapeutants may be directly toxic to the pathogen, such as nitrosopyrazoles and pyridine-2-thial-N oxide. Some others 8-qunolinol benzoate, Cerolate are useful to control fungal diseases; and thiouracil and 8-azaguanine prevent multiplication of viruses.

IMMUNIZATION

This is one of the most effective control measures of the plant diseases by which hosts are made immune or resistant to the infection of the parasitic causal agencies. The efforts to produce resistant varieties must be intelligent, efficiently systematic, and with wide scope and great promise for the future. To be successful, one should have a thorough knowledge of the variability and genetic characters of the pathogen as well as the inheritance and expression of resistance, or susceptibility in the host is essential in a breeding programme for disease resistance. Resistant varieties are obtained through: (i) plant introduction, (ii) selection of diseases resistant varieties, (iii) breeding disease resistant varieties, (iv) integrated control of plant diseases, (v) induced mutation, and (vi) chemical immunization.

(i) Plant introduction. Plant introduction is one of the oldest methods of crop improvement. It has been a common practice to introduce new varieties of plants (from a particular area) where they have shown promise of disease resistance to an area where various disease incidences have been prevalent. This practice of plan introduction has often been between distant localities of variable climatic conditions, even between different countries of extremely variable climatic conditions having natural barriers of oceans and mountains. For some time the newly introdued plants showed to some extent desired disease resistance. But the disease resistant quality did not last long. The newly introduced plants quite often failed to retain their disease resistant quality under new agro-climatic conditions. Besides this, unknowingly new pests and pathogens also migrate from place to place along with plant introduction. Ultimately it was realized that this practice does more harm than expected and as such practice of large scale plant introduction has to be abandoned.

(ii) Selection of diseases resistant varieties. It has often been found that within a good variety under cultivation some plants show better degree of disease resistance than others. Careful selection of such plants and their large-scale cultivation often produce varieties of high disease resistant quality.

Selection of disease resistant varieties in this manner has been successful in some cases and not in others. In order to produce a desirable variety at least four conditions must be fulfilled. They are:

(a) there must be resistant lines or biotypes within the plant populations,

(b) there must be reliable methods for identifying the resistant individuals,

(c) there must be suitable methods for propagating resistant lines, and

(d) the lines selected must have the other characters of a good variety.

But as a prerequisite in selection there should be an abundant development of the disease in question in order to make it possible to distinguish between resistance and disease escape. With a view to take protection against the development of physiologic races of the pathogen concerned and thereby loss of resistance in some varieties, it is essential to test varietal populations against an adequate sample of races of the pathogen under conditions of maximum disease development. Generally speaking, plant populations, although often relatively homogeneous for most visually observable characters, are quite heterogeneous with respect to disease resistance. The early successes with selection led many to believe that this method might be universally applicable. But this was proved not to be true; consequently it was necessary to resort to hybridization.

(iii) Breeding disease resistant varieties. The ideal way of controlling all plant diseases would be by means of permanently and universally resistant varieties.

To secure disease resistant varieties, common practice is, breeding of high yielding varieties with some of the most important pathogens present in the area. The varieties that are resistant to these pathogens are related to growers for crop production. This is an attempt to incorporate into the varieties, genes that would make them resistant to the pathogens without changing any of their desirable characteristics, these resistant varieties possess genes for resistance (R genes). Disease resistance may be genetically controlled by the presence of one gene—**monogenic** or by many genes—**polygenic** secure **true resistance.** There are two kinds of true resistance: **horizontal** and **vertical.**

Horizontal resistance is controlled by many genes, and hence known as **polygenic** or **multigene** resistance. Each of these genes alone may be effecive against the pathogen and may play minor role and is thus known as **minor gene resistance.** Many genes involved in horizontal resistance exert their influence by controlling the numerous steps of the physiological processes in the plant to make up the defence mechanisms of the plant. Generally, horizontal resistance does not protect plants from becoming infected, but it shows down the development of indvidual infection lesion a plant and thereby shows down the spread of the disease. But **vertical resistance** is always controlled by one or a few genes and thereby named as **monogenic** or **oligogenic resistance.** These genes, referred to as R genes, control a major step in the recognition of the pathogen by the host plant and thus play a major role in the expression of resistance. Such resistance is strong, major, specific and qualitative. Vertical resistance is effective against specific races of pathogens and ineffective against others. This type of resistance inhibits the development of epidemics by limiting the initial inoculum development or by limiting the further production of inoculum resulting complete resistance.

Both horizontal and vertical have their advantages and limitations. A plant with horizontal resistance is never completely resistant, whereas a plant with vertical resistance may be completely resistant to a pathogen, hence it is easy to manipulate in breeding programme.

In nature host plants and their pathogens coexist side by side and changes of their resistance and virulence also take place a dynamic equilibrium. This change is often explained by the **gene-for-gene concept,** according to which each gene that confers resistance to the host, has a corresponding gene in the pathogen that confers virulence. Whenever a new gene for virulence that attacks the existing gene for resistance, often by mutation of an avirulence gene breaks down. Plant breeders then introduce in the plant another gene for resistance to extend the resistance of the host plant.

Resistance through hypersensitive response during host infection by a pathogen has been shown to be the result of gene-for-gene systems in which avirulence gene (**avr**) in the pathogen corresponds to a resistance gene (**R**) in the host plant. Such gene-for-gene systems occur in the **response to diseases caused** by obligate intracellular and facultative pathogens. It is believed that whatever may be the type of pathogen, the hypersensitive response is the result of recognition by the plant of special signal molecules, the **elicitors**, produced by the avirulence genes of the pathogen and recognized by R gene-coded by specific receptor molecules in the host plant. It also leads to activation of numerous other defence-related genes that result in other types of resistance, including horizontal and system acquired resistance.

Resistance controlled by specific single gene has high resistance to a particular strain or race of the pathogen but is susceptible to other races. Monogenic characters are stable over a wide range of conditions. The breakdown of resistance to the disease in plants with single gene may be sudden and complete. But the resistance controlled by many genes is not so highly resistant but does not break down into new races so easily. Polygenic resistance is highly variable and is influenced by environmental conditions and by host nutrition. The reaction between pathogen and a single gene type of resistant host accelerates the metabolism of infected host cell, followed by abnormal metabolism and death of the cell concerned, whereas in plants with polygenic resistance any change in the host plant metabolism due to infection of a pathogen alters the degree of resistance. Besides this, there are several plant diseases in which resistance is controlled by genetic material contained in the cytoplasm. Such resistance is known as **Cytoplasmic resistance**, encountered in maize yellow leaf blight induced by *Phyllosticta maydis.*

Repeated selections through breeding for resistance in a crop result in accumulation of more genes for resistance in the newer varieties. There may be a gene which may inhibit completely or partially the expression of resistance to a disease. Similarly an inhibitor gene may suppress the expression of susceptibility in some varieties. Some crop varieties may be immune to a pathogen at certain temperatures, but may completely break down and become susceptible at a different temperature. If the pathogen is capable of producing many races, the monogenic resistance in the host may result in its becoming susceptible to one or more of the newer races of the pathogen. If a large number of genes for resistance are accumulated in one variety by repeated breeding and selection, then the possibilities of its continued resistance to most races of the pathogen are better. Host resistance may be inherited as a qualitative or quantitative character. Actually the three factors that are necessary for disease development in plants are: genetically susceptible host, virulent pathogen, and favourable environmental conditions. If they fail to coincide and interact at the proper time or for sufficient duration, the plants do not become infected. Such a condition is designated as **disease escape.** Disease escape may also happen due to physical barriers by susceptible host plant. Certain anatomical structures such as thick cuticle, high silica content of the wall of the epidermal cells give resistance to pathogen against host penetration. The size and number of stomata are often directly related to the rate of invasion of a pathogen in the host tissue. Resistant plants may aso show defence reaction following infection by the production of: (a) toxic substances in the host cell sap which ultimately check the growth of pathogen in the host tissue, and (b) callus-like swellings, an abscission layer, cork barrier, or gummy deposits.

In general, host plant resistance against infection may be:

(i) structural, (ii) functional, and (iii) protoplasmic.

Structural peculiarities of the host like epidermal wall may be cuticularized, suberized or silicified or there may be layer of cork cells which form natural barriers against infection by any pathogen.

Functional resistance actually means certain physiological functions which are related with the opening and closing of stomata which is a natural passage of pathogens, can control the entrance of pathogens in the host.

Protoplasmic resistance is offered by some host plants whose protoplasm contains some toxic substances which prevent the growth of the pathogen. Besides this, during host-pathogen interaction, due to disbalance of host metabolic processes certain defence chemicals are produced with which the host repels the attack of the pathogen.

Since several external factors affect the degree of resistance in host plants, by careful manipulation of agronomic factors the pattern of host resistance may be changed. For exmaple, application of fertilizers in the soil in a well-planned manner helps the host plants to resist infection. In general, application of high doses of nitrogenous fertilizers to the soil makes host plants more susceptible to infection. Potassium promotes formation of thicker cell walls to resist infection by the pathogen. With the change in pattern of the soil, resistance qualities of the host plants can be improved. The enzymatic degradation of pectic substances is affected strongly by calcium. Supplying host plants with increasing amounts of calcium enhances resistance against diseases in which pectolytic enzymes contribute to pathogenesis.

In breeding for resistance, knowledge is thus required about the physiological and anatomical characters of the host that are useful to give resistance against the attack of the pathogen. Again, detailed knowledge of the pathogen, and about the disease process will also be needed to draw up any programme of breeding work. The genetics, the ecology, and the behaviouristic physiology of both the host and the pathogen should be known. Moreover, it is of the utmost importance to know the interactions between the two. The disease process must be studied not only under one set of environmental conditions but under many. A number of diseases must be taken into consideration in breeding for resistance, and the testing of parental varieties and breeding lines against any adequate sample of the pathogens under an adequate sample of environmental conditions is a prerequisite to success in many cases.

It has been observed that the resistant varieties are poor in yield and the susceptible ones possess high yielding quality. Naturally such undersirable linkages have to be broken by crossing a parent having disease resistant and poor yielding quality with a susceptible possessing high yielding quality and a variety should be produced having a combination of disease resistance and high yield. Ultimately the desired character combination of disease resistant and high yielding quality is to be incorporated in the progenies. This is possible by following a programme of study of characters of the progenies and by selection and if necessary, by further crossing between the progenies. The disease resistant varieties so obtained should be tested at a number of places in different agro-climatic zones under conditions most favourable for disease development. They are to be exposed to as many physiologic races of the pathogen are known in the areas for which the varieties are to be developed. It is only then a variety resistant to the particular disease can be obtained for large-scale cultivation.

The greatest single handicap to the development of permanently resistant varieties has been the numerous physiologic races within species of most pathogens. All races of certain pathogens probably can never be catalogued at any given time for two principal reasons: the practical difficulty obtaining an adequate world-wide sample from the thousands of cultivated and wild biotypes of the host plants, and the difficulty of being sure that the methods of identification are adequate at any given time. Even if there could be assurance that all races and their effects were known at one time, the information might soon be inadequate because of changes in populations of host varieties and of pathogens. There is a continuous and changing stream of varieties of crop plants and of races of pathogens. Hence, there must be continuous effort in studying them. Likewise, the concept of the number of biotypes within species of crop plants and within species of pathogens must be expanded.

One of the simplest ways of trying to assure adequate testing is to maintain so-called disease gardens in which parental materials and hybrid lines are tested against physiologic races of pathogens. Since it is not possible to control environmental factors completely in the field, it is desirable to test critical materials in plant houses where temperature, light, and humidity can be cotrolled. As it is necessary to wait several seasons before conditions in the field permit maximum development of certain disease, the probable seasonal effect must be predicted by testing materials under various combinations of conditions in plant houses. It is a common practice to test parental varieties and hybrid lines in several places in the hope of providing adequate samples of races, of pathogens, of weather, and of soil conditions. For special purposes it is highly desirable to make tests in areas where the weather is always favourable for the development of the disease in question. The testing varieties should not be restricted by national boundaries; international cooperation is of paramount importance in the development of resistant varieties.

There are many factors for virulence in species of pathogens that comprise many physiologic races. The final outcome of breeding for disease resistance will be determined by the ability of breeders to find and combine genes for resistance that counteract the genes for virulence in pathogens. As different genes for resistance are often dispersed among many varieties of crop plants and those for virulence often are widely dispersed among many physiologic races of pathogens, it is not an easy task to prognosticate the outcome of many breeding programmes. Information is continuously being accumulated about the genes for resistance, and same has been accumulated about the genes for virulence, but much more is needed.

Different genes for resistance and for virulence may exist in different distant places, only a broad-based programme will be needed to assemble them. Varieties with different genes for resistance must be assembled and then tested extensively in different geographical areas and intensively under optimum conditions for disease development. Most promising plant materials may be tested under necessary combinations of environmental conditions. It is possible that a succession of varieties may have to be produced to control some diseases, as there can be no assurance that new pathogens or new races of old ones may not appear periodically to disrupt the *status quo*. The important thing is to anticipate the changes and prepare to meet them. Even if permanent immunity is not attainable, practical field resistance or tolerance may reduce the destructiveness of many diseases. Quality may sometimes have to be sacrificed in order to attain greater resistance and productivity, especially when the standard of quality is not based on real values.

Naturally, there must be a large number of different genes and combinations of genes to produce so many different characters and combinations of characters. There may be—and sometimes are—a number of known genes for resistance to a single pathogen and to numerous physiologic races of each. In some cases this may be relatively easy, but in other cases it is not. For the interaction between genes, linked or otherwise, increase in the complications is still more. The possibility of finding resistance in wild varieties has been explored to utilize it in hybridization technique for being incorporated in the cultivated varieties. If satisfactory genes or combinations of genes for resistance cannot be found, can they be created by radiations or other physiochemical method? Radiation or other tools may be useful in breaking undesirable linkages or possibly in deleting inhibitors. Present studies in 'radiation genetics', and similar fields should be continued and amplified to determine the possibilities, but, in the present state of knowledge, reliance should also be placed on the principles of conventional genetics as a guide to breeding procedures.

As already explained breeding for resistance.

(iv) Integrated control of Plant diseases. An integrated control of plant diseases is usually aimed against all diseases affecting a crop. However, if a specific disease constitutes the major or only threat to a crop, an integrated control programme is directed against that threat. The main goals of an integrated plant disease control programme are to (i) eradicate the initial inoculum source, (ii) reduce the effectiveness of initial inoculum, (iii) delay the onset the disease, (iv) slow of and eradicate the secondary inoculum development, and (v) increase the resistance of the host.

In an integrated control of orchard crop, like apple citrus certified nursery stock to be planted in a carefully inspected land to ensure proper drainage. Soil should be treated with fumigants before planting. Planted stocks should be inspected regularly. Once the trees are placed and until they bear fruits, they should be fertilized, irrigated, pruned and sprayed with suitable fungicides and insecticides. Any tree that develops symptoms of a disease should be immediately removed. Fungicide—Bordeaux mixture and insecticide—Superior oil should be sprayed before bud-break. Spraying operation should be continued for 3-5 days a week up to young fruit appearance. Fruit must be sprayed every 10 to 14 days with materials that will control fruit-spotting or fruit rotting fungi. Injured fruits should be removed and discarded before packing. Fruit-store house should be fumigated with formaldehyde, sulphur dioxide, or some other fumigant.

In an integrated control programme of an annual crop, like potato, one must start with healthy stock (potato tuber) to be planted in a suitable field as indicated earlier. Certified potato seed produced under strict quarantine and inspection rules should be obtained and treated with a fungicide or biocontrol agent. Only then they are planted in a field absolutely clean from previous year's plant debris.

The soil has to be treated with suitable fumigant to protect against the attack of soil borne pathogens. The field must have good drainage to reduce damping-off, seed rot, and root rots. To reduce pathogen infection cut seeds and knives should be disinfected with fungicides and insecticides. Seeds and cut seeds are to be planted at a suitable date and distance. A few weeks after young plants emerge and under high moisture conditions they become susceptible to attack by early blight or late blight. Under such condition spraying operation with suitable fungicide should be started. Spraying must continue throughout the growing season as needed and as weather dictates. Before harvest the infected vines must be burnt. The tubers must be stored at about 15°C and the damaged ones are discarded. Storage rooms must be cleaned and disinfested before the tubers are brought in.

In an integrated control programme, several control methods are applied including regular inspection for healthy seeds or nursery stocks cultural practices (crop rotation, sanitation, prunning this affected parts); biological control (use of resistant varieties, use of suitable fungicides and insecticides); physical control (harvest products to be stored at storage temperature) and chemical controls (soil fumigation, seed or nursery treatment, sprays, disinfection of cutting tools and store houses. Routine use of all these measures produces desired results. The Indian Council of Agricultural Research (ICAR) established a National Centre of Integrated Pest Management (NCIPM) in 1988 to cater to the emerging protection needs of agriculture in the country.

(v) **Induced mutation.** Gene mutation occurs in all living organisms and can often be accelerated by treatment with chemical or physical mutagens. The effectiveness of host plant's resistance can be changed in this way.

(vi) **Chemical immunization.** In recent years attempts are being made in artificially immunize host plants by the use of certain chemicals. This is known as **chemical immunization.** By this process certain organic chemicals and antibiotics are allowed to be absorbd through the roots of the host plant so that these substances are distributed in the entire plant system giving immunity to the cells of the actidione has given some positive result. *Pyricularia oryzae,* causal organism of the rice blast disease, produces two toxins, x-picolinic acid which is closely related to fusaric acid; and priciularin, a nitrogen-containing compound. Piricularin which is the main component in pathogenesis, can be detoxified by chlorogenic acid and ferulic acid. Application of ferulic acid to rice plants increases resistance to infection by the fungal pathogen.

Importance of Plant Biotechnology in Plant Disease Management

On the basis of recent researches it appear that plant biotechnology impringes or helps plant pathology in various ways:

1. To obtain pathogen-free mother plants through rapid clonal propagation and meristem culture.
2. To control plant diseases by incorporating resistance genes into sensitive plants by genetic engineering techniques (r-DNA technology).
3. To develop disease resistant plant through protoplast fusion techniques.

Several virus infected crop types are thus developed uncontaminated by meristem culture and clonal propagation methods (e.g., cauliflower from infection of CuMv, TuMv; Tobacco from alfalfa mosaic; potato from potato virus, etc.). Downey mildew resistant plant of letuce, Club root resistant Brassica and Bacterial soft rot resistant Potato, etc., were development through protoplast fusion technique. Identically, Blight resistant potato, smut resistant sugarcane and virus resistant tobacco, etc. are developed through r-DNA technology. In addition, a number of biocontrol products are available for control of various plant diseases. The details are given in Table 24.3.

Table 24.3: Biocontrol products produced by Bacteria or Fungi as Available Commercially

Name	Source	Target pathogen(s)	Crop(s)	Application
Bacterial				
Galltrol	*Agrobacterium radiobacter* strain 84	*A. tumefaciens* crown gall	Fruit and ornamental nursery stock, grapes, brambles	Slurry to seeds, seedling, drench
Nogall	*A. radiobacter* strain K1026	*A. tumefaciens* crown gall	Fruit, nut and ornamental nursery stock	Suspension, drench
Companion	*Bacillus subtilis* str. GB03, other	*Pytium, Phytophthora, Fusarium, Rhizoctonia*	Many in greenhouse and nursery	Drench at planting time
HiStick N/T	*B. subtilis* str. MB1600	*Fusarium, Rhizoctonia, Aspergillus*	Legumes	Slurry to seeds
Kodiak	*B. subtilis* GB03	*Rhizoctonia solani, Fusarium, Alternaria, Aspergillus*	Cotton, legumes	Slurry to seeds
Deny	*Burkholderia cepacia,* Misc.	*Pythium, Rhizoctonia, Fusarium,* several nematodes	Legumes, cotton, grain crops	Seed treatment
Intercept	*B. cepacia*	*R. solani, Fusarium, Pythium*	Maize, vegetables, cotton	Seed treatment, drench
BioJect Spot-less	*Pseudomonas aureofaciens*	Dollar spot, anthracnose, *Pythium*, pink snow mold	Turf, other	Overhead irrigation
Bio-save 10LP, 110	*P. syringae*	Postharvest *Botrytis, Mucor, Penicillium, Geotrichum*	Pome fruit, citrus, cherries, potatoes	Drench, dip, spray
BlightBan A506	*P. fluorescence* A506	Frost damage, *Erwinia amylovora*, russeting bacteria	Pome and stone fruits, potatoes, tomatoes, strawberries	Spray
Degree G	*P. fluorescens*	*Rhizoctonia, Pythium*	Field crops, vegetables	Seed treatment
Cedomon	*P. chlororaphis*	Barley, oat leaf spots, *Fusarium*	Grain cereals	Seed treatment
Fungal				
AQ10 Biofungicide	*Ampelomyces* quisqualis M-10	Powdery mildews	Apples, grapes, ornamentals, cucurbits strawberries, tomatoes	Spray
Aspire	*Candida oleophila* 1-182	*Botrytis, Penicillium*	Citrus, pome fruit	Drench, drip, spray

Contd.

Contd. Table 24.3

Name	Source	Target pathogen(s)	Crop(s)	Application
Biotox C Fusaclean	Nonpathogenic *F. oxysporum*	*F. oxysporum*	Basil, carnation, tomatoes, cyclamen	Drench
Contans WG, Intercept WG	*Coniothyrium minitans*	*Sclerotinia sclerotiorum, S. minor*	Many crops. All soils	Spray
DiTera Biocontrol	*Myrothecium verrucaria*	Parasitic nematodes	Cole crops, grape, ornamentals, turf, trees	Soil application
Polygandron	*Pythium oligsndrum*	*Pythium ultimum*	Sugar beet	
Primastop	*Gliocladium catenulatum*	Soil-borne pathogens causing rots and wilts	Ornamentals, vegetables, tree crops	Drench, spray, irrigated water
Root Shield, Plant Shield, T-22 Planter box	*Trichoderma harzianum*, Rifai strain—KRL–AG2(T-22)	*Pythium, Rhizoctonia, Fusarium*	Tree, shrub, ornamental, transplants, cabbage, tomato, cucumber	Mixed w/soil, soil drench
F-Stop	*T. harzianum*	*Rhizoctonia, Pythium*	Ornamental and food crops	Seed treatment
SoilGard (GlioGard)	*Gliocladium* (*Trichoderma*) *virens* GL-21	*Rhizoctonia solani, Pythium*	Ornamental and food crops, greenhouses, nurseries	Slurry, seed treatment
BINAB T	*T. harzianum/T. polysporum*	Wood decay fungi	Trees	Spray, wound
Promote	*T. harzianum* and *T. viride*	*Pythium, Rhizoctonia, Fusarium*	Transplants, trees	
Rotstop	*Phlebia gigantea*	*Heterobasidion annosum*	Trees	
Trichodex	*T. harzianum*	*Colletotrichum, Monilin, Plasmopara Rhizop. Sclerotinia*	Various	
Trichopel,	*T. harzianum* and *T. viride*	*Armillaria, Botryosphaerim, Fusarium*	Various	
Trichoject		*Nectria, Phytpphthora, Pythium, Rhizoctonia*		

Chapter 25

Plant Diseases

WART DISEASE OF POTATO

Wart disease of potato was for the first time discovered and described in Hungary in the year 1895. In England, the disease was first reported in 1902. It invaded the U.S.A. about 1909. In India, the disease was reported to occur in 1953 in the Darjeeling district in the variety of potato 'Furore' which was imported from Denmark. The disease is essentially specific to the tubers and hardly any part above soil-level is affected. It appears on all underground parts like buds on stems, stolons (Fig. 25.1A), and the tubers which are the centres of infection, but never on roots.

Symptoms. Wart disease of potato is characterized by warty, tuberous, and dirty cauliflower-like proliferations of abnormal tissue of variable size and slape on the tubers (Fig. 25.1B) and lower parts of stems. The warty growths and contamination are spread for succeeding potato crops. At the infection courts there appear numerous pustules consisting of protuberances in which brownish depression centres appear. The entry of the parasite into a particular cell causes hypertrophy and hyperplasia of the epidermal cells which are in immediate contact with that cell. Ultimately a raised 'rosette' overarching of host cells is formed, at the centre of which therefore the infected cell remains sunken as it itself takes no part in this outward growth (Fig. 25.1C).

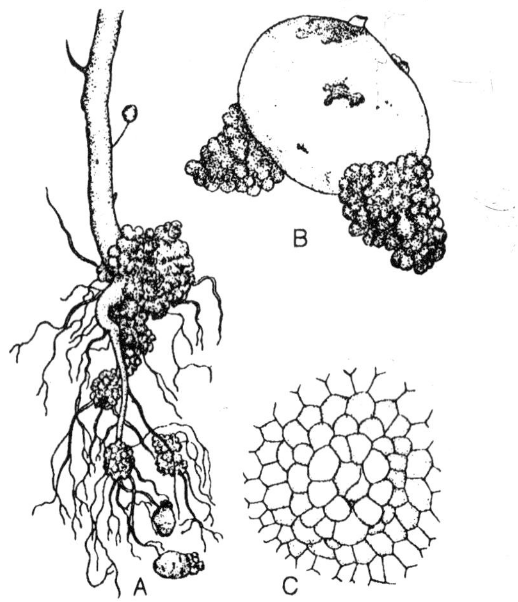

Fig. 25.1: Wart disease of potato. A. Heavily warted plant. B. Warted potato tuber. C. Rosette of host cells

The parasite enlarges considerably filling almost the entire cell space simultaneously resulting sporangia around which hyperplasia of the host tissue has been initiated. The warts are protuberances of variable size. A typical wart is roughly spherical, but it is usually not a solid structure. It is soft pulpy and can be cross-sectioned more readily than a tuber. Morphologically it consists of distorted, proliferated hypertrophic tissue grown together, may be partly underground and partly exposed above the soil-level. As the warts become older, they darken somewhat and decay and the infection is spread. When affected tubers are dug, warts may disintegrate in part causing spread of infection.

The Causal Organism. This disease is induced by *Synchytrium endobioticum* (Schilb.) Perc., a fungus belonging to the most primitive group of fungi, the Archimycetes. It is an obligate parasite, it starts its parasitic life as a minute uninucleate naked, uniflagellate, structure which is holocarpic, can behave as a zoospore and cause infection to the host cell resulting in the formation of wart around the infected cell. On entering the host cell, the cytoplasm of the parasite enlarges and is surrounded by a golden brown thick wall and is known as a prosorus. The prosorus after undergoing various changes develops into a sorus of sporangia or gametangia or both. The sporangia liberate zoospores whch cause fresh infection in the host cells, whereas the gametes that are liberated out from the gametangia after undergoing planogametic isogamy develop into zygotes which infect host cells in a process similar to that of zoospores. A thick wall is laid down surrounding the zygotic nucleus and cytoplasm to form a resting sporangium. With maturity of the resting sporangium, the diploid nucleus undergoes meiosis and with continued mitotic division of the daughter nuclei innumerable nuclei are formed around which zoospores are organized. The zoospores on liberation from the sporangium spread infection of the host cells. In the life history of the causal organism there are two resting stages through which it tides over unfavourable conditions and thus perennates from year to year. The development of the two resting stages—prosori and resting sporangia of the causal organism may or may not follow each other in strict sequence. *S. endobioticum,* besides infecting potato, also attacks *Lycopersicum esculentum, Solanum alatum* and *S. nigrum.*

Disease Cycle. The perennating organs—prosori and resting sporangia of the causal organism which are very resistant to extremes of temperature, high and low, may remain viable in soil even in the absence of a potato crop for many years, periods of 9 to 12 years being recorded in many instances. They may be carried along with manure from one place to another by any means that ensures the transport of contaminated soil and manure, adherent to seed tubers or on the underground parts like stolons being the source of infection in different localities. Even the tubers of very resistant varieties may not be free from infection. The moderately resistant varieties in which warts are readily overlooked are most effective in spreading the pathogen. Prosori and resting sporangia are commonly produced in small warts which cause little or no commercial loss to the potato crop but are responsible for the spread of the disease.

The disease is more prevalent during wet season than the dry since a high degree of soil moisture is necessary to bring about infection. A periodic flooding followed by drainage at a relative soil humidity of 45 per cent and aeration are very favourable for the disease incidence. The entrance of the parasite takes place through a pore formed on the cell wall of the epidermal cell of the host tissue. Infection is limited to a temperature range of 12°C to 24°C and over a range of soil reaction from pH 3.9 to 8.5 but from neutral to slightly acid soil

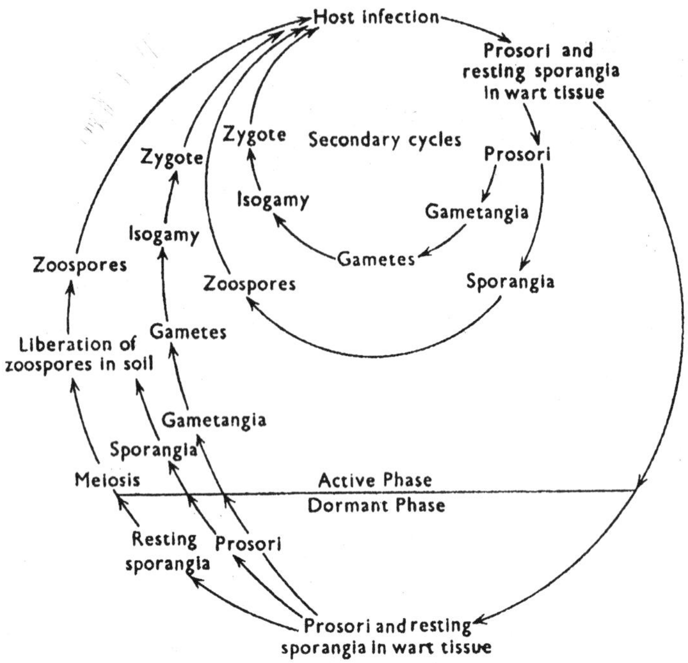

Fig. 25.2: Disease cycle of Wart disease of potato

is most favourable. The organism is sensitive to high degrees of alkalinity. Certain amount of moisture is essential for the germination of prosori and resting sporangia, liberation of swarmers and ultimately causing infection on the host cells. With the favourable moisture conditions, temperature and soil reaction the rate of infection may be very high and widespread. The protuberances that are formed due to primary infection are the centres of secondary inoculum. Both primary as well as secondary inocula may spread infection at the same time with favourbale environmental conditions inducing the disease and causing destruction of potato crop.

Disease cycle of Wart disease of potato is presented in Figure 25.2.

Control. Various degrees of quarantine measures may be practized in prohibiting the export of potatoes from those localities where wart is known to occur. The disease has been successfully controlled by the introduction of potato varieties which possess complete resistance to or immunity from it. In resistant varieties, if invasion of pathogen occurs, it dies without producing prosori and resting sporangia. A resistant variety may deteriorate or breakdown its quality when it is grown at a place which is different from where it is raised, still such a variety is more suitable than the susceptible one. Since the resting sporangia are viable in soil even up to periods of 9 to 12 years, no system of crop rotation can starve them out of the soil. Various experimental results have indicated that application of soil treatment to eradicate the disease is too expensive and impracticable to apply. The application of a sulphur has often given good results by the formation of thiosulphuric acid in the soil, so also the application of ammonium thiocyanate to the soil, but it has a retarding effect on the growth of the potato crop. But the application of lime to raise pH of the soil and the use of soil fungicides before planting often reduce infection rate. Again, none of the ordinarily used seed disinfectants is capable of destroying the resting sporangia in soil adherent to the tubers.

Thoroughly cleaning the cultivation plots and burning the previous year's plant debris also help to reduce the rate of infection.

DAMPING-OFF

Damping-off is a term applied to a very common trouble in which seedlings are attacked by soil-inhabiting fungi usually just at the soil level on to hypocotyl or upper taproot causing partial or complete rot and the seedlings suddenly topple over in a characteristic manner. Of the various species of soil-inhabiting fungi responsible for damping-off of seedlings, *Pythium debaryanum* ranks first in importance, in relation to this disease as a damping-off parasite.

The damping-off of seedling disease has been known for a long while both in Europe and America. It was first studied in Germany by Hesse in 1874 and De Bary in 1881. In the United States this type of disease attracted attention by the end of the nineteenth century and the beginning of the twentieth century. As early as 1901 to 1905, however, the damping-off disease began to receive attention as a trouble in forest nursery seed beds.

This disease is encountered in seed beds where plants are propagated to be transplanted later to the field. But green houses and nursery seed beds are ideal places for outbreaks of damping-off if moisture and temperature conditions are favourable.

Almost any kind of plant may be attacked by damping-off fungi while in young, tender, succulent stage of development. But the disease incidence is dependent more upon the conditions under which the seedlings are grown than upon the particular species of plant concerned.

Symptoms. The most striking symptom is the sudden toppling over of the seedlings. The fungus attacks the seedlings at or near the surface of the ground, according to the moisture level and depth of planting, on the hypocotyl or taproot (Figs. 25.3A and B). The cell wall of the rapidly growing seedlings is generally thin, and as such the tissue is particularly vulnerable. In a very short time the pathogen causes upon invasion, rapid killing and rot of cells (Fig. 25.3C), and collapse of the tissue to such an extent that the seedlings suddenly fall over (Fig. 25.3B). This occurs so quickly that the seedlings are found in a prostrate condition even before the top has had time to wilt. Often the damage may be complete before emergence of the seedlings above ground. After the seedlings have fallen over they may continue to decay, if the ground is wet and the air humid, until entirely rotted away, or if drier conditions prevail the top may wilt and dry up after toppling over.

Damping-off of seedlings actually occurs in two stages: (i) Pre-emergence stage and (ii) Post-emergence stage.

(i) **Pre-emergence stage.** This is the stage when the young seedlings are killed before they

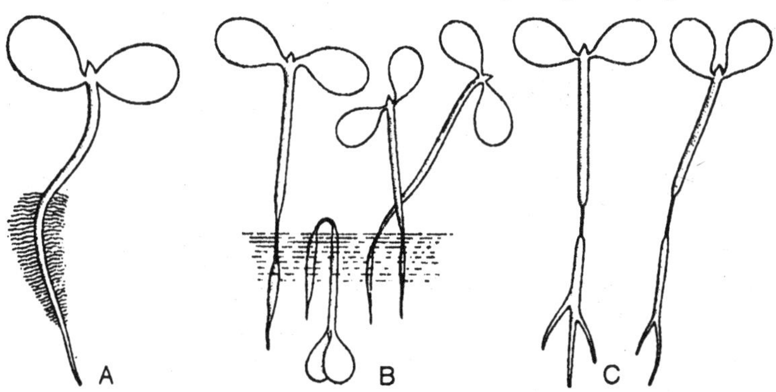

Fig. 25.3: Damping-off. A. Disease-free seedling. B. Diseased seedlings under natural condition. C. Closer view of the diseased seedlings

reach the surface of the soil when the hypocotyl has just emerged. The radicle and the plumule are destroyed as soon as they come out of the seed coat. Since this happens below the surface of the soil, it remains unnoticed and as such the cause of failure of seedling emergence is given to the poor quality of the seed.

(ii) **Post-emergence stage.** During this stage the seedlings are out on the surface of the soil and on being infected by the pathogen they topple over and lie on the surface of the soil. This toppling is due to the killing of host tissue by the pathogen at or below the ground level. Initially the infected areas appear soft and water-soaked. With the further growth of the pathogen the stem or the hypocotyl, as the case may be, becomes constricted and the seedlings collapse. The seedlings remaining erect on the surface of the soil may thus collapse any time.

The disease usually radiates from initial infection points causing areas of dead seedlings that have been toppled. The disease is most severe when the moisture content of the soil is medium to high. Under favourable conditions, the rapidity of onset of the disease may be to such an extent that within a very short time the seedlings of the entire seed bed may collapse.

The losses due to damping-off are confined largely to seed beds, wherever they may be.

The Causal Organism. *Pythium debaryanum* Hesse, is the common cause of damping-off. It is eminently a soil-borne organism and is of common occurrence in field and garden soils, in which it survives in the form of oospores. It lives in soil saprophytically on dead organic matter or parasitically on the young seedlings of a great many susceptible hosts. The non-septate somatic hyphae of the organism may survive both intra- and intercellularly without any haustoria. The organism reproduces asexually by the development of sporangia which are either terminal or intercalary on the somatic hyphae. Depending on the environmental conditions, the sporangia produce either biflagellate secondary zoospores or germ tubes. The contents of the sporangium are discharged into a protruding vesicle within which a large number of zoospores are developed. After release, the zoospores swarm for some time, come to rest, round off, and germinate to produce somatic hyphae. The sex organs, oogonia and antheridia usually appear at a later stage than sporangia and may arise within or outside the substratum. Paragynous antheridia, of which one or more may invest an oogonium, may arise either from the same hyphae bearing the oogonium or from a neighbouring branch. Following fertilization by a fertilization tube, each oogonium produces a single spherical oospore which becomes thick-walled. On the decay of the host tissue, the oospores pass into the soil where they germinate after a period of rest. At high temperatures the oospore germinates directly by germ tube which develops into somatic hyphae. At low temperatures, however, a vesicle is developed at the tip of the germ tube in which zoospores develop. Meiosis occurs in the first division of the diploid oospore nucleus.

Disease Cycle. The survival of the fungus in the soil suggests that it lives largely as a saprophyte on plant debris or humus in the soil, and becomes parasitic when the proper host and proper growing conditions are present. The temperature and moisture conditions play a very large part in determining the severity of infection. Abundant moisture and a fairly warm temperature are conducive to invasion of the seedlings. Soil temperatures of 20°C to 28°C are ideal for the host infection by direct penetration. Abundance of moisture in the surface layer of the soil due to over-watering, thick sowing in the seed bed causing heavy stands, lack of aeration of the surface soil, and too much shade are conditions which favour damping-off.

Dissemination of inoculum takes place through oospores in the plant debris and myce-

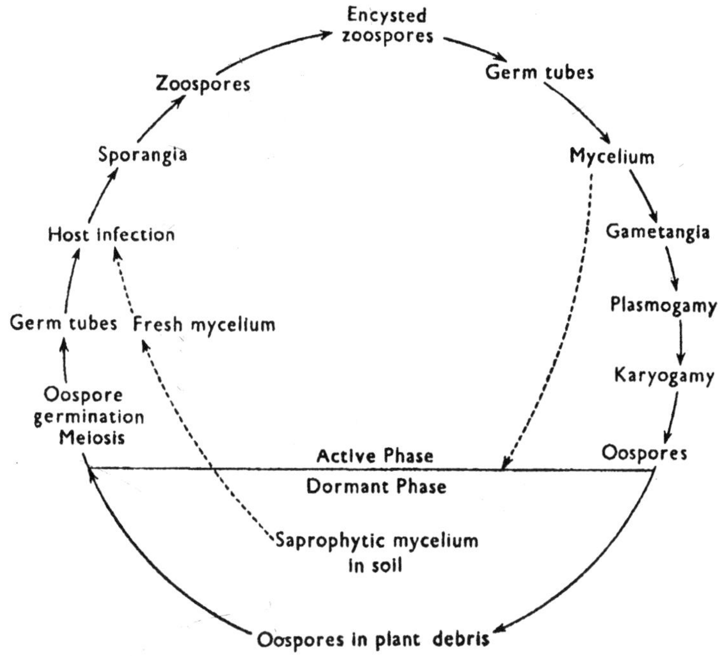

Fig. 25.4: Disease cycle of Damping-off

lium remaining in the soil. A very common source of contamination in seed beds is in the soil or compost used in making up the seed beds. Sometimes these materials are obtained from an infested source and of course the organism is introduced along with the soil or compost.

Disease cycle of Damping-off is presented in Figure 25.4.

Control. Since the fungus responsible for causing damping-off trouble is soil-borne and since the temperature and moisture factors of the environment play important role in determining the severity of infection, control measures are largely based on: (i) proper seed bed soil preparation, (ii) soil treatment, (iii) seed treatment, (iv) the manipulation of watering system, and (v) application of improved cultural practices.

(i) **Seed bed soil preparation.** Every effort should be made to secure seed bed soil free from damping-off fungus. The seed bed soil should be light, having good proportion of sand. Pure fine sand or saw dust is also a suitable medium for growing seedlings. An alkaline reaction in the soil favours the growth of the fungus causing damping-off. As such every effort should be made to reduce alkaline reaction of the soil. Manure used in the seed bed should be well decomosed. The seed bed should be at a raised level to maintain good water drainage.

(ii) **Soil treatment.** The most perfect control of damping-off is by soil treatment. The seed bed soil should be sterilized by suitable methods to make it free from the fungus inducing the disease. Some of the common methods of soil sterilization are by using steam, dry heat, and chemicals. Steaming of soil should be done continuously for thirty minutes. Sterilization by dry heat is usually accomplished by burning wood upon seed bed soil. Seed bed soil can be sterilized by using formaldehyde diluted in the proportion of 1 part to 50 parts of water to be sprinkled over the loose soil in sufficient amount to soak it to a depth of at least 4 inches, which will mean one gallon for 2-3 square feet of soil. Soil treatment with formaldehyde should be applied several days before sowing of seeds and the soil should be covered with paper, tarpaulin or other covering until ready for seed sowing. In using sulphuric acid in seed beds, three-sixteenths fluid ounce dissolved in 1 to 2 pints of water is allowed for each square foot of seed bed if the soil is heavy; a little less is used if the soil is sandy. This acid treatment can be applied without danger immediately after the seed is sown. The formaldehyde and steam treatment are most commonly used in green-houses and in seed beds other than in forest nurseries; the sulphuric acid is in common use in forest nurseries. Drenching of seed beds with 0.1 per cent. Captan is also recommended for soil sterilization. Two sprayings with Bordeaux mixture at the rate of one-half gallon per square foot, one before seed germination, often produce good results. Post-emergence damping-off can be controlled by drenching seedlings sprayed with 0.3 per cent Blitox-50 or 0.2 per cent Ziram. Spraying at frequent intervals gives satisfactory results. The sprays should begin as soon as the first leaves are formed and should be repeated every week as long as damping-off is likely to affect the seedlings. Disinfection of the seed bed soil and control of post-emergence damping-off can also be achieved by drenching the seed bed soil with Bordeaux mixture, 0.2 per cent Fytolan, 0.5 per cent Perenox, and 0.2 per cent Flit 406.

(iii) **Seed treatment.** One of the most effective control measures of the disease is the use of seed-protectants and is a very effective measure against pre-emergence damping-off. The chemicals used as seed-protectants are applied to the seed in dry or in wet form producing a protective layer around the seed coat to enable the seedlings to grow uninterrupted. Some of the chemicals that have produced effective results are: Arasan, Agrosan GN, NI Ceresan, Captan, Blitox-50, Copper sulphate, Mercuric chloride, and Semesan.

(iv) **Manipulation of watering system.** A system of watering should be maintained to ensure that the seedlings and the surface of the seed bed soil remain dry without permitting the seedlings to suffer from lack of water. A thin layer of sand should be sprinkled over the surface of the soil. This aids in keeping the surface of the seed bed soil dry. Light irrigation at frequent intervals should be given, water-logging should be avoided.

(v) **Improved Cultural practices.** Avoid heavy seedling so that there is no overcrowding of seedlings in the seed bed and thereby no shedding effect is produced. Seedlings are liable to severe infection in poorly aerated soil and growing in shed. To avoid water stagnation, seed bed soil should have proper drainage facilities.

LATE BLIGHT OF POTATO

This is one of the most serious of all the diseases of potato when conditions are favourable for its development. Since control measures are now better understood, it is possible to ward off such a dangerous disease which at one time caused a severe epidemic on the European continent.

It is difficult to trace with certainty the history of late blight of potato disease. But it is probable that it originated in South America, the native home of the potato plant. There is evidence that a disease of potatoes which might have been late blight, existed in the Andes Mountains of South America as early as 1571. It was introduced into Europe and North America some time between 1830 and 1840. In succeeding years it increased in extent and severity, culminating in a widespread epidemic on the European continent in 1845. The notable Irish famine of 1845 and 1846 was caused due to the destruction of the potato crop by the late blight of potato disease. It is probable that the disease was introduced into the United States and Canada at about the same time as the European introduction. This diseases is now known in all potato-growing sections of the world. In India, late blight of potato disease was first introduced in the Nilgiri hills during 1870 and 1880 and subsequently detected in Darjeeling and Hooghly.

Symptoms. The disease attacks the tops at about blooming time causing a blight and may also invade the tubers and cause a dry or wet rot. It damages potato crops by killing the foliage and so reducing yield, and by rotting tubers.

The symptoms on leaves, stems, and pedicels are similar in character. The first indication of the disease on leaflets, petiole, or on stem consists of brownish- to purplish-black lesions (Fig. 25.5A). The lesions are not deliminted in size and under favourable conditions enlarge rapidly. If weather conditions are favourable, these infected areas enlarge rapidly so as often to involve the whole surface. Extension to the stem quickly occurs in bad cases, and the entire crown may fall over in a rotten pulp in a few days. The influence of the weather is most marked. In dry, clear weather successful infections are limited in number and the resulting spots remain small, brown and dry, while the stems may escape altogether. In warm, humid weather the colour rapidly changes to black, the lesions are wet, the stems are quickly attacked, and a pronounced smell of decaying vegetable matter is given off. This characteristic odour is a helpful diagnostic character in identifying the disease. The fungus forms a whitish layer on the infected host tissue consisting of the sporangiophores bearing sporangia in great number. This growth is scanty or even absent in dry weather and dense but often evanescent on moist, cloudy days. The diseased area turns darker and may become dry, blackened and shrivelled if the weather becomes dry. With the return of favourable weather conditions, the fungus beocmes active, and corresponding symptoms and progress reappear. When the blight is advancing rapidly in a potato field, a mildly pungent characteristic odour is given off.

The underground parts, especially the tubers, are also affected. A brown to purple discolouration of the skin followed by a brownish dry rot extends to about ½ inch below the surface of the affected tubers. The dry rot does not soften the tissues but causes rusty-brown markings just below the skin and extends inwards for a variable distance in an irregular fashion (Fig. 25.5F). A wet rot is usually set up by the action of secondary organisms which follow late blight development. In moist atmosphere, white tufts of mycelium and sporangiophores of the fungus appear on the surface of the infected tubers. Many of the tubers may decay before harvest. This wet rot may continue to spread in storage by contact, but the blight infection by itself does not pass from tuber to tuber. Tuber infection is much heavier in clay soil than in sandy soil.

The losses due to late blight are of two kinds: actual reduction in yield due to attacks on foliage and stem, and damage to tubers. If the leaves and stem are blighted down before the crop is mature, the tubers are prevented from reaching normal size and the yield is reduced proportionately. Sometimes great loss is sustained on account of direct attack on the tubers themselves.

The Causal Organism. *Phytophthora infestans* (Mont.) D By.

The fungus was described by Montagne in 1845 as *Botrytis infestans.* De Bary in 1876 gave its name, *Phytophthora infestans.* De Bary worked on the parasitism of the organism and its causal relation to the late blight disease.

The aseptate mycelium of the pathogen ramifies in the host tissue both intra- and intercellularly. It often produces rudimetary haustoria in foliage cells (Fig. 25.5E). But in potato tuber cells the haustoria are more elaborate, being **club-shaped**, hooked, or spirally twisted.

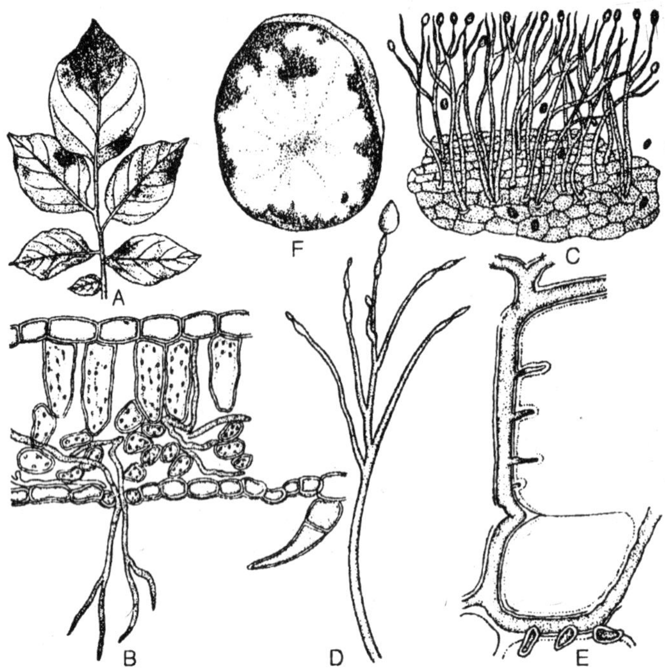

Fig. 25.5: Late blight of potato. A. Blighted leaf. B. Section of diseased leaf showing sporangiophores passing out through stomata. C. A general view of the sporangiophores passing out through stomata of diseased tissue and bearing sporangia. D. A sporangiophore. E. Hyphae producing haustoria penetrating in host cells. F. Diseased potato tuber

The sporangiophores arise directly from the internal mycelium, emerging from the leaf through the stomata in groups (Figs. 25.5B-D). Emergence through or between the cells of the epidermis is also sometimes found. On the tubers they mainly arise from lenticels or abrasions in the rind. The sporangia are colourless papillate, lemon-shaped. When the fungus is active and environment is favourable, the sporangia germinate by the development of biflagellate secondary zoospores. Otherwise germ tubes are given out from the sporangia and thereby the sporangia behave as conidia. The zoospores when formed, after a swarming period come to rest and germinate by germ tubes which penetrate through the stomata or directly through the epidermal cells of the host tissue.

Sexual reproduction is oogamous. Oospores were first seen by Clinton in 1911 in artificial cultures of the fungus but later were obtained on diseased tubers under natural condition by Murphy in 1927 from Ireland. But Barrett (1948) has indicated the heterothallic nature of the fungus, when he was able to produce oospores by mating compatible strains of *Phytophthora infestans.* Oospores are very rarely found in Europe, and have recently been discovered in quantity in Mexico, where compatible strains of the fungus coexist.

Disease Cycle. How the fungus perennates has long been one of the classic problems of plant pathology. Various theories have been advanced to explain the annual occurrence of the disease. The explanation most generally accepted has been that proposed by De Bary that the fungus could perennate in diseased tubers and then invade developing sprouts to form lesion and sporangia above ground. The mycelium from infected seed potatoes may grow under suitable conditions to produce sporangiophores and sporangia. The sporangia so produced on coming in contact with new foliage through the agencies like wind and rain cause infection and thereby the disease is spread. Since oospores are formed in nature, there is a possibility that the pathogen perennates in that stage also.

Primary inoculum of the disease in the field usually comes from the planting of infected tubers and from oospores in the previous year's plant debris. It has been conclusively established that the mycelium developed during previous year's crop infection, frequently survives in the tubers without causing any great damage. On planting the diseased tubers, the enclosed mycelium renews active growth passing into the tissus of the young sprouts chiefly between the cells of the cortex and producing sporangiophores and sporangia on the sprouts above ground if the weather conditions are favourable. The fungus establishes itself more quickly on young than on old foliage. This tendency for the infection to start at the base of the plant is due largely to the more favourable conditions of the microclimate, such as abundance of humidity provided by the dense growth in that situation.

It has long been known that the temperature and humidity of the environment play important part in the occurrence of late blight epiphytotics. Aerial emergence of the hyphae takes place only at an atmospheric humidity exceeding 85 per cent, while sporangia apparently form only above 95 per cent and do not become abundant under about 97 per cent. The most favourable temperature for the production of sporangia is about 18°C to 22°C and for development of zoospores about 6°C to 15°C, but direct germination by germ tube occurs at an optimum temperature of 24°C. Conditions favourable for the germination of sporangia are also optimum for host infection. Darkness often inhibits production and germination of the sporangia. Germ tubes developed from zoospores or directly from sporangia may directly cause infection or may develop into infection hyphae. The infection hyphae may enter the leaves or stems through the stomata or even penetrate the unbroken epidermis. In tuber infection, the entrance is probably through the lenticels. This is how the primary infection is completed. The inocula (sporangia producing germ tubes and zoospores) produced during primary infection constitute the secondary inocula which induce host infection and secondary cycles. The sporangia are developed on the sporangiophores protruding out through the stomata of the infected host tissue. Slightly higher temperature favours growth after infection, but protracted cold weather is unfavourable. The hyphae of the intercellular mycelium send small filamentous haustoria into the host cells. The rate of mycelial growth in the host tissues is directly proportional to the water content of the host tissues or perhaps to the ratio of water to nitrogen. There is evidence that the plants are more susceptible to the pathogen when nitrogen is withheld and growth is most rapid at 20°C to 23°C.

The susceptibility of the host is governed by factors like: (a) the water nitrogen ratio—increased nitrogen supply affords resistance, (b) water content of the leaves—succulence favours infection, (c) soil-moisture—dry soils increase resistance, and (d) supply of potash—a shortage reduces infection rate.

Disease cycle of Late blight of potato is presented in Fig. 25.6.

Control. Late blight of potato disease can be effectively controlled by two principal methods: (a) direct attack on the pathogen by spraying and dusting of fungicides, and (b) cultivation of resistant varieties. Following are some of the important recommended control measures of the disease.

(i) *Sanitary measures:* Previous year's plant debris should be thoroughly cleared to cut the source of primary inoculum.

(ii) *Use of Disease-free seeds:* Since the pathogen perennates as mycelium in the tubers, seed tubers should be used raising from disease-free fields.

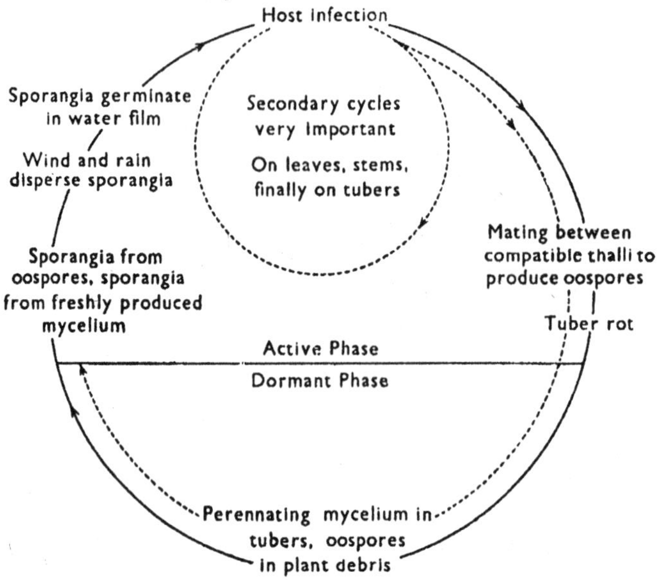

Fig. 25.6: Disease cycle of Late blight of potato

(iii) *Improvement of storage of Seed potato:* Seed tubers should be disinfected with 0.1 per cent mercuric chloride immediately after harvest. They should be stored at 40°F or below to check rot of tubers. The tubers should be stored in a dry and well-ventilated place.

(iv) *Soil management:* (a) Frequent earthing-up of growing crop at a four to six inches high riding diminishes the risk of tuber infection. (b) Spraying of soil with 10 to 20 per cent sulphuric acid or 5 per cent copper sulphate, or 12 lbs of copper sulphate with ½ lb of caustic soda in 40 gallons of water at the rate of 100 gallons per acre cuts down infection rate.

(v) *Use of Fungicides:* The best method of control of the disease incidence, is foliage spraying with suitable fungicides. In areas where late blight is common, spraying of the crop with suitable fungicides should be started well ahead of usual time of appearance of the disease. However, in order to control this disease, it is better to start spraying the crop when it is about six weeks old or when the plants are 6-8 inches high. Spraying should be repeated every ten to fifteen days, depending on the weather. If there is continuous rain, the spray may be washed off. In such cases spraying may have to be repeated immediately and also done more frequently. More sprayings may be needed if the disease is in severe form. The spray must be delivered with considerable force in the form of a fine mist so that both surfaces of the leaves and other above ground parts are completely covered by the fungicide. Bordeaux mixture has been the most and effective spray material to control the disease incidence. The earlier spraying should be done with 4 : 4 : 50 formula of the mixture while later sprayings should be done with 6 : 6 : 50 strength. Commercial copper fungicides such as, Cupravit, Fycol 8E, Perenox, Blitox-50 (strength 1 lb in 20 gallons of water), Fytolan (Strength 4-8 oz in 10 gallons of water) are also used for foliar spray material. The copper fungicides are being replaced by more effective fungicides like Dithane D-14, Dithane Z-78 (strength 1.5 lb in 100 gallons of water) and Dithane M-22. Depending upon the type of spraying equipment, the quantity of liquid required for one acre will vary from 70 gallons to 300 gallons.

(vi) *Improvement of Harvesting Practice:* Due **precautions** must be observed at harvest time to prevent tuber infection. If the foliage has been blighted, digging should be delayed until the foliage has died and dried out. Care should be taken to prevent the tubers from coming in contact with the diseased foliage. Visibly infected tubers should be sorted out from the healthy ones before storage.

(vii) *Use of Resistant Varieties*: The use of resistant varieties is a very important control measure. Potato varieties in which *demissum* resistance has been fixed produce promising result.

RESISTANCE AGAINST POTATO BLIGHT

When blight first struck the potato fields of Europe in the 1840's, *Phytophthora infestans* met a host that had existed for centuries in a blight-free world. The potato varieties were very susceptible, and great havoc was brought. Since then conscious and unconscious selection has raised the level of resistance considerably. Most of the old varieties mature late and intensely susceptible to blight. Large-scale selection and breeding work were started to get blight reistant varieties. It has been found that in Central Mexico *Solanum demissum* is prevalent at high altitudes as an uncultivated weed. Many clones of *Solanum demissum* exist which differ considerably in resistance to *Phytophthora infestans.* Potato breeders started using widely *Solanum demissum* as a source of R-genes for commercial varieties to be incorporated in them. The interaction of R-genes with races of *P. infestans* has been proposed by W. Black and others in 1953 is now internationally accepted. The interrelationships of genes and races of *P. infestans* are, for example, the gene R_1 makes a potato variety resistant to all races of *P. infestans* that do not have the number 1 in them: races (0), (2), (3), (4), (2, 3), (2, 4), (3, 4), and (2, 3, 4). It leaves the variety susceptible to races with the number 1 in them: (1), (1, 2), (1, 3), (1, 4), (1, 2, 3), (1, 2, 4), (1, 3, 4), (1, 2, 3, 4). Similarly, a potato variety with the three genes R_1, R_3, R_4, is susceptible to races with the three nunmbers 1, 3, and 4 in them: races (1, 3, 4) and (1, 2, 3, 4), but resistant to all races without these three numbers.

This system was proposed at a time when only the four genes, R_1, R_2, R_3, R_4 had been much studied. At least two more genes, R_5 and R_6, are now known. The system is readily extended to include them, e.g., $R_1R_2R_3R_4R_5R_6$ types are susceptible only to races (1, 2, 3, 4, 5, 6). The system refers only to resistance in the normal foliage. Tuber resistance does not follow the pattern in the foliage exactly, and tubers may be susceptible to races that cannot attack the foliage.

The Mutability of Races of Infestans. When genes for resistance from *S. demissum* were first incorporated in the common potato, they conferred resistance amounting almost to immunity from *P. infestans* for a while. In the breeder's nurseries the new varieties with *S. demissum* genes behaved excellently. But as soon as the varieties has been increased to a few acres, blight began to appear conspicuously. From the blight lesions new races were isolated. Since then it has been realized that, in conditions that favour blight, new derivatives from *S. demissum* invariably succumb before long to appropriate races of the blight fungus. Quite possible that the blight population intercrossed freely in nature, with sexual recombination new races are formed.

WHITE RUST OF CRUCIFERS

White rust is common on cruciferous plants like *Brassica napus, B. alba, B. juncea, B. campestris;* different varieties of *B. oleraceae, Raphanus sativus;* and cruciferous weeds *Nasturtium indicum* and *Capsella bursa pastoris.* This being a common and conspicuous disease, was observed and studied by early mycologists including Persoon, Berkeley, Tulasne, and De Bary. The disease is of relatively less economic importance in proportion to its widespread occurrence. Earlier workers thought that white rust of crucifers belongs to the true rusts, hence the name.

Symptoms. The disease attacks all parts of the host plant except the roots. Infection is both localized and systemic. Due to localized infection prominent white postules of variable shape and size appear on the leaves (Fig. 25.7A), stem and inflorescence. Pustules may arise in close proximity and coalesce to form large patches. Ruptured pustules become powdery in appearance, since sporangia enclosed within the pustules are exposed (Fig. 25.7D). Depending on the severity of infection, the pustules may appear on one leaf surface only or on both surfaces. When the young stems and floral organs are infected, the infection becomes systemic in the tissue and stimulates hypertrophy and hyperplasia resulting the enlargement and varied distortion of infected organs. Invaded stems often exhibit extended swellings or enlargements which may result in sharp bends, turns or even complete spirals. A proliferation from lateral buds that are normally dormant may result in a bushy growth. The inflorescence axes before distorted, development of flowers may be suppressed, or discolouration and malformation of floral parts (Figs. 25.7B and C) are often resulted from systemic infection. Pustules may develop on hypertrophied organs, but they may be relatively rare on such tissue. When the white rust causes infection alone, it induces very little injury. But the attack is severe when it is associated with downy mildew.

The Causal Organism. *Albugo candida* (Lev.) Kunze. It is an obligate parasite. Infection of host tissue results in the development of a copious growth of internal, intercellular, aseptate mycelium which sends in globular haustoria within the adjacent cells of the host tissue. Hyphae from the internal mycelium collect below the epidermis of the host tissue into which numerous haustoria are sent. From these hyphae numerous short sporangiophores arise in a closely compacted palisade layer beneath the epidermis and at right angles with it. The sporangiophores developed beneath the epidermis, raise it, to make whitish pustules or extended blister-like areas, the sori. The sporangiophores are broadly clavate from which sporangia are abstricted in basipetal succession and adhere in chains. Pads of gelatinous material formed between successive sporangia function as disjunctors. Pressure exerted on the superimposed epidermis by the accumulation of sporangia finally causes the host epidermis

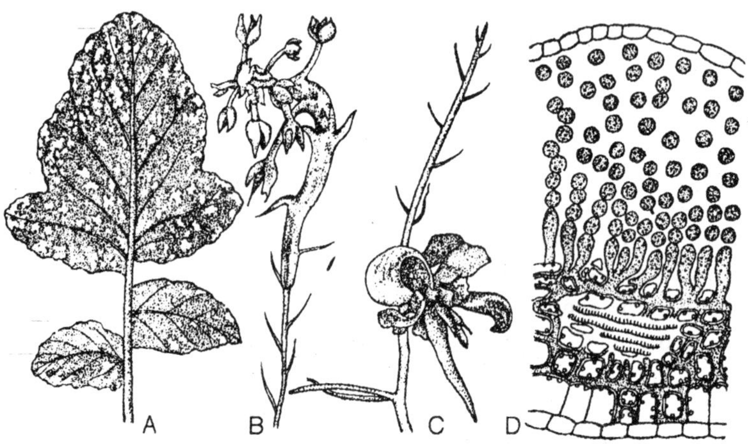

Fig. 25.7: White rust of crucifers. A. Diseased leaf of *Brassica* sp. showing white pustules. B-C. Diseased inflorescence and flowers of *Brassica* sp. D. Section through a pustule on leaf of *Brassica* sp. showing hyphae producing sporangiophores and sporangia, and ruptured host epidermis

to rupture along the softened middle lamellae exposing the sporangial bed. The sporangia are disseminated by wind. The sporangia are hyaline, nearly spherical and are 14 to 16μ by 16 to 20μ in size. They usually germinate by the formation of zoospores, rarely by germ tube. The zoospores are reinform, biflagellate with laterally inserted flagella. After a period swarming, they encyst, form a germ tube which is capable of penetrating a suitable host through the stomata. Towards the end of the growing season antheridia and oogonia begin to appear when the production of sporangia has almost ceased. They arise within the host tissues. The oogonia are globular, multinucleate. One or more small globular antheridia arise close to an oogonium. Gradually the oogonial protoplast becomes differentiated into a peripheral or external multinucleate layer, the periplasm, surrounding a central uninucleate egg cell, or **ooplasm.** The antheridium produces a short fertilization tube, which penetrates the periplasm and comes in contact with the egg. The antheridial contents are discharged through the fertilization tube into the egg cell. A single antheridial nucleus fuses with the egg nucleus. Immediately after fertilization the periplasm is absorbed, and the fertilized egg cell develops a dark wall, becomes filled with oily and fatty food and develops to form an oospore. The oospore wall becomes thick, turns brown, and is provided with low, blunt ridges. The mature oospores measure from 40 to 55 μ in diameter. They survive in the soil and germinate after a period of rest of several months. During the germination of oospore the diploid nucleus divides into a large number of nuclei, the first division of which is reductional. Each oospore produces a large number of secondary zoospores, enclosed in a vesicle which becomes extruded during germination of the oospore. The zoospores cause infection to the susceptible host.

Disease cycle. The causal organism perennates as oospore left either in soil or in previous year's plant debris. Thus the oospore is the source of primary inoculum, and primary infections are caused by zoospores that are developed during the germination of the oospore with the return of favourable conditions. In perennial hosts, the causal organism can, however, perennate as mycelium and may be source of primary inoculum. In such case the primary inocula are the zoospores that are developed during the germination of the sporangi produced from the sporangiophores borne on perennating mycelium. But in general, the secondary infection is caused by the zoospores produced during the germination of sporangia that are produced during primary infection. These zoospores thus constitute the secondary inocula. Temperature conditions affect not only the germination of the both oospores and sporangia but also the susceptibility of the hosts. The optimum temperature being 10°C, moisture on the host surface is essential for the production of inocula and infection. Host penetration is through stomata. The greatest development of the disease is during the cool periods of the year. The existence of physiologic races has been demonstrated by many workers, but the extent of specialization and the number of races have not been thoroughly investigated.

Disease cycle of White rust of crucifers is presented in Figure 25.8.

Control. One of the most effective methods of controlling the disease is the destruction of infected crop refuse of the previous year by burning to prevent the development of primary inoculum from the oospores. Next is the destruction of susceptible cruciferous weeds from in and around the crop growing areas. Crop rotation to avoid the ill-effect of the previous year's diseased crop is also recommended to control the disease incidence. In cases of severe attacks, destruction of diseased plants accompanied with spraying with fungicides such as, 0.8 per cent freshly prepared Bordeaux mixture or 0.3 per cent Blitox-50 often becomes very effective.

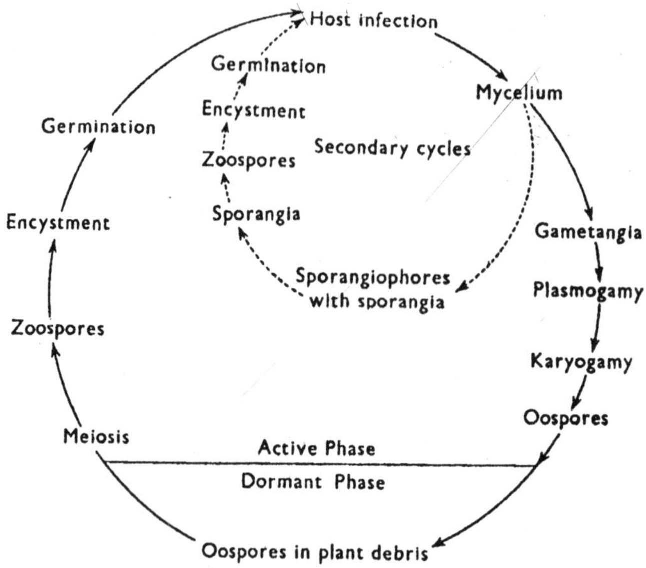

Fig. 25.8: Disease cycle of White rust of crucifers

DOWNY MILDEW OF PEA

The pea (*Pisum sativum* L.) is subject to various types of diseases—downy mildew, blights, root rots, and wilts. Of these diseases, the downy mildew is of minor economic importance. This disease is rarely very injurious to pea.

Symptoms. Yellow to brown scattered patches of discoloured areas appear on the upper surface of the leaflets and stipules. Simultaneously with this, downy growth in patches of variable size are visible on the under surface corresponding with the lesions of the upper surface. The downy coating is whitish when young, but the colour gradually turns greyish-violet with age. This growth is composed of branches sporangiophores, arising from the mycelium of the fungus which has entered within the leaf tissues. Sporangia are borne at the tips of the ultimate branches. The infected leaflets and stipules become reduced in size with their margins curled downwards. The infected areas gradually develop into elongated blotches, and often to irregular spots.

With the spread of infection, blotches also appear on the pods. When young, the blotches on the pods are pale-green, more or less elliptical to irregular, but gradually the blotches turn dark to bright-brown, mottled with light-green islands. Blotch followed by the green island effect is a typical symptom of the disease. Seeds lying corresponding to the infected tissues of the pods abort and become very much reduced in size. Conidia are very seldom produced on the pods.

The Causal Organism. Downy mildew of pea is caused by *Peronospora pisi* Sydow, whose mycelium consists of aseptate branched hyaline hyphae, confined to the intercellular spaces of the host tissues. The hyphae produce branched, finger-shaped haustoria which penetrate into the adjoining cells of the intercellular spaces (Fig. 25.9B). The sporangiophores arise directly from the internal hyphae and emerge in clusters from the stomata of the under surface of the leaflets (Fig. 25.9A). They are dichotomously 2 to 10 times branched at acute angles (Fig. 25.9C), ultimate branchlets acute and more or less reflexed. The sporangia are nearly oval (Fig. 25.9D) but narrowed a little below, greenish-yellow to pale-violet in mass and 22 to 27μ by 15 to 19μ in diameter. They are very delicate structures and germinate by a germ tube in presence of moisture. The sporangia are

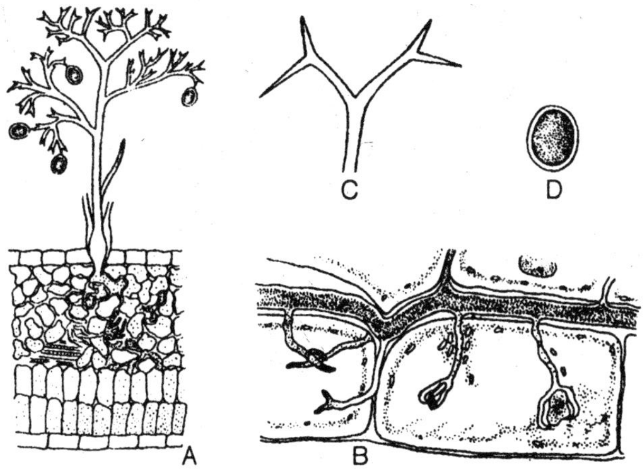

Fig. 25.9: Downy mildew of pea. A. Section through an infected leaf showing hyphal ramification in the host tissue, sporangiophores with sporangia protruding out through a stoma, B. Hyphae producing haustoria in the host cells, C. Branching of sporangiophore, D. Sporangium

blown about freely by the wind and spread the disease rapidly from plant to plant. They germinate directly by germ tube behaving like conidia. In the old withered infected leaves oospores are found in the host tissues singly within a thin-walled oogonium, which soon disappears. The oospores are roundish, light-brown having a thick epispore marked by a large, raised reticulation. They measure 26 to 43μ in diameter. The oospores germinate by a germ tube in the season following that in which they were formed. They can stand the favourable conditions, like desiccation.

Disease Cycle. Cool humid condition is very favourable for the development of the disease, while the disease incidence is slowed down with warm condition. This is due to the fact that the sporangia which behave as conidia spread the disease, but cannot stand desiccation. The disease may be described as: (i) the primary cycle, when infection starts in the leaflets and stipules from the inoculum pro-

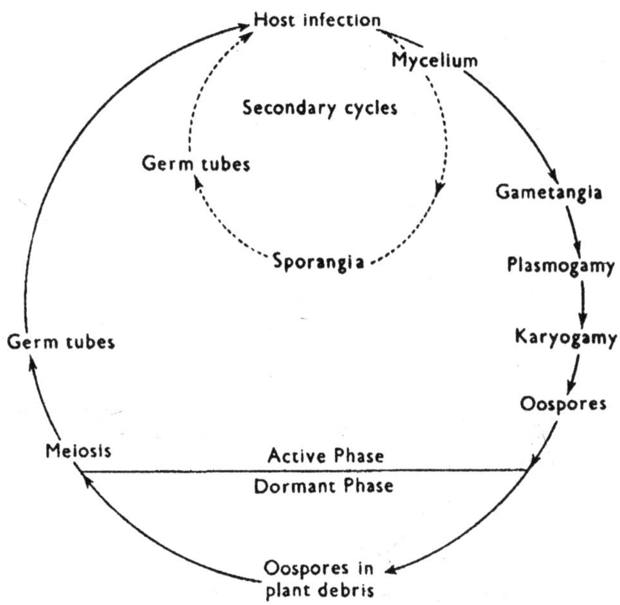

Fig. 25.10: Disease cycle of Downy mildew of pea

duced by the germination of oospores of previous year; and (ii) the secondary cycle, when the pods become infected from the sporangia produced by the primary cycle. The discoloured patches on the leaflets and stipules denote successful infection. Host penetration is through stomata by the germ tube produced by sporangial germination. The sporangia are light and are developed profusely. They spread the disease very rapidly, whereas the oospores are responsible for carrying the infection from year to year by tiding-over unfavourable conditions.

Disease cycle of Downy mildew of pea is presented in Figure 25.10.

Control. Since the causal organism survives from season to season in the form of oospores in the plant debris, the destruction of previous year's plant debris and following crop rotation of two or three years are very effective control measures. Spraying and dusting pea plants with fungicides are effective in limiting the spread of the disease.

DOWNY MILDEW OF GRAPE

This is one of the best known of the many diseases of plants of economic importance. It has an interesting historical background associated with the accidental discovery of Bordeaux mixture as a fungicide by the French plant pathologist, Millardet who perfected the Bordeaux mixture as a spray for this disease. The best information available indicates that the downy mildew of grape is endemic in North America, where it was for the first time reported by Schweinitz in 1837. This disease was introduced in France in 1874 and subsequently in other parts of Europe by 1917, where it became a serious disease because of frequent epiphytotics. It is also well established in North and South Africa, Australia and New Zealand.

The disease is confined largely to species and varieties of grapes (*Vitis*) although it has been reported also on the five-leaved ivy (*Parthenocissus quinquefolia* Planch), and on English ivy (*P. tricuspidata* Planch).

Symptoms. The disease attacks all green parts of the plant: leaf blades, petioles, tendrils, green shoots, and fruits at different stages of development.

Early symptoms of the disease on the leaves consist of round light-green spots of an oily appearance on the upper surface, which enlarge even 1/2 cm or more in diameter. On the corresponding underside white downy mildew consisting of the tufts of sporangiophores soon appear, bearing sporangia in great numbers (Fig. 25.11B). In moist weather it persists; in extremely dry weather it may disappear. Later the spots become yellow, or variegated with tints of yellow and yellowish-brown forming patches of irregular shape, especially between the larger veins of the leaf (Fig. 25.11A). Old spots turn brown owing to the killing of the leaf tissue. The leaf spots may be a few in number or so numerous as to coalesce and involve nearly the entire leaf area. At this time the leaves are in a morbid condition, and the fungus within enters upon the sexual phase, producing oogonia, anteridia, and finally oospores which survive the winter in the fallen leaves.

On the stems, the lesions are brown and sunken, and along with the death of the affected parts, portions of the vine become brittle and break off easily. In extreme cases of infection the whole shoot may be dwarfed, the leaves remaining very small and densely covered with the mildew.

Flowers may be completely blighted by early attacks of the disease. Fruits may be attacked when young or when approaching maturity. The young fruits show brownish spots and later become covered with downy mildew, their growth is checked. The young fruits then darken and finally dry up. They may also assume a reddish-brown colour and failing to ripen, develop a soft, rot. On full-

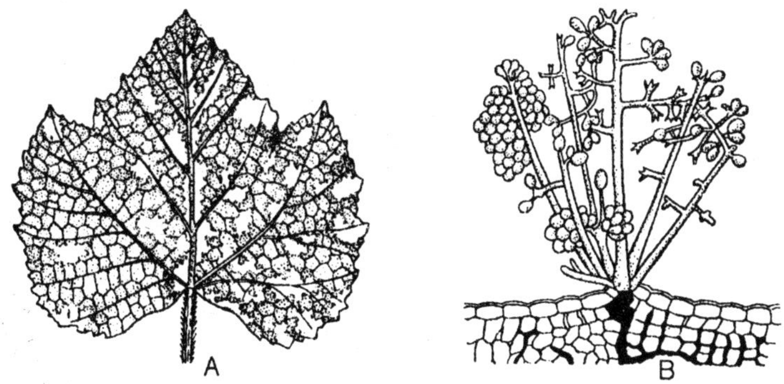

Fig. 25.11: Downy mildew of grape A. A diseased leaf, B. Section of a diseased leaf with sporangiophores emerging from a stoma

grown fruits, brownish patches appear and the fruits harden ultimately becoming mummified.

The disease may often strip plants of their leaves and tendrils, flowers may fail to set, fruit may be destroyed in the early stages of growth, causing severe loss.

The Causal Organism. *Plasmopara viticola* (Berk. and Curt.) Berl. The characteristic aseptate, intercellular mycelium produces knob-like haustoria to absorb food from the host cells. Fasciculate sporangiophores arise from the hyphae in the intercellular spaces just beneath the lower epidermis and emerge through the stomata. They are branched monopodially along the main trunk and dichotomously at the extremities. The branches in turn give off other branches more or less at right angles so that the system of branches shows many small branches with arms projecting out in cross-like fashion. The tips of the branches are provided with short finger-like projections on which the sporangia are borne. The sporangia are hyaline, ovate, and attached to the sporangiophore branches at the small end, in which the wall is thinnest. The profuse growth of sporangiophores and sporangia on the surface of the host accouts for the downy appearance which characterizes the downy mildew. The sporangia germinate by secondary zoospores which after an active period shed their flagella and each one produces a germ tube from which aseptate mycelium is developed.

The oospores are produced from the mycelium embedded within the leaf tissue. The oospore consists of a thick endospore wall surrounded by a thinner rough exposure wall. On germination, an oospore sends up a short unbranched stalk, at the apex of which a single sporangium is produced. The sporangium so formed germinates by zoospores exactly in the same manner as the sporangia borne on the sporangiophores.

Disease Cycle. In general, the sporangia are not adopted for long survival but serve to spread the disease in localities where the leaves may remain on the vines all the year round. In such cases infection by sporangia may be continuous from season to season. On the other hand, the oospores are capable of survival in the soil for at least a year. Oospores which remain in the fallen leaves, vine debris, or shrivelled fruits on the spring. Moreover, the fungus may, in some localities, survive in the form of mycelium perennating in the winter buds.

The oospores germinate at a temperature range of 13°C to 33°C. Primary infections are believed to occur when the zoospores formed during oospore germination, are conveyed by splashing raindrops from the soils on to the lowermost leaves of the vines. It may so happen that the sporangia produced during oospore germination may themselves be carried on the leaves near the ground and then zoospores are produced there. In any case, the earliest symptoms usually appear on the leaves near the ground.

Host penetration of the pathogen takes place through the stomata. The mycelium invades the intercellular spaces, ramifies in the host tissue deriving its nutrients in a large measure through haustoria which penetrate the cell lumen. In almost all cases, infection of the leaves takes place through the lower surface. The optimum temperature for the development of the disease is 18°C to 24°C. Prolonged warm wet weather is conducive to an epidemic of the disease. Dry weather, however, checks the growth of the fungus. After the mycelium has established itself in the host tissue, sporangiophores are sent to the host surface through the stomata (Fig. 25.11B). Sporangia produced on the sporangiophores serve for rapid dissemination of the disease. They also serve as the source of secondary inocula of the disease. Through the secondary inocula, secondary infections are induced resulting in secondary cycles. The sporangia are transported from host to host and to different parts of the host by both wind and rain.

Disease cycle of Downy mildew of grapes very similar to that of Downy mildew of pea, except that the sporangia germinate by secondary zoospores,

Control. The chief methods of control of the disease are as follows:

(i) *Sanitation:* Destruction of old leaves is recommended to eradicate the source of inoculum since the causal organism of the disease usually hibernates as oospores in the fallen leaves.

(ii) *Spraying of Bordeaux mixture:* Bordeaux mixture, 5-5-50 strength is usually recommended as a protective spray to the susceptible parts; but even up to 8-8-50 is often recommended for epiphytotic conditions. Sporangia or zoospores absorb copper until a toxic limit is reached, the more epiphytotic the disease, the stronger must be the spray to be effective. But emphasis is laid on the value of timely and early spraying, and of the thinning of the foliage during the growing season in order to gain better access for sprays, and the removal of the shoot tips in order to increase resistance in the remaining parts. The first application of spray should be made soon after the buds open, a second before the flowers open, a third after the fall of the petals, and a final application should be given 14 days later. The time of application, however, varies with prevailing environment and the relative susceptibility of the host concerned.

(iii) *Spraying of other Fungicides:* The spread of the disease can be effectively controlled by spraying the vines with 0.3 per cent. Blitox-50, Dithane Z-78, Ferbam or Captan first when the shoots are 6 to 8 inches high, again when they are about to flower, and finally when the fruits are just about to change their colour. Depending on the local weather conditions and the nature of virulence of the disease, number of spray applications may be increased to 6 to 8 times.

(iv) *Improved cultivation practices:* Certain improved methods of cultivation are helpful to control the disease incidence. Vineyards should be sufficiently open to allow the foliage to dry quickly after rains. This method not only ensures free access of air but facilitates spraying as well. The risk of infection from resting oospores is reduced. All green shoots or suckers that develop at the base of the stocks should be removed, as they are very susceptible to primary infections.

(v) *Use of Resistant Varieties:* The use of resistent varieties is of course a very useful practice. But even with resistant vines, spraying with Bordeaux mixture is advisable.

PEACH LEAF CURL

Peach leaf curl disease is distributed throughout the world wherever peach is grown. The disease received its name on account of its characteristic effect on leaf—curl. There is no definite knowledge of the place of origin of the disease. But it is believed to have been introduced in the European countries far back in 1841 from China, the original home of the peach. The disease is now widely distributed in Europe and in the United States, in parts of Asia, China and Japan, and also in Africa, Australia, and New Zealand.

Symptoms. The disease affects leaves, tender growing shoots and very rarely flower parts and fruits. It is most conspicuous on the leaves. The disease first appears, although relatively inconspicuous in symptoms, in the early spring when the leaves begin to unfold. Soon after the leaves are well out of the bud some of them appear to be distorted, puckered along the midrib and curled up (Fig. 25.12A) assuming a red tint. Leaf blade and petiole may both be involved in the curling. While in more heavily infected leaves the curling is so severe that the whole lamina with the exception of the tip looks like a partially inflated paper bag. The blistered portions of the leaf are softer than the disease-free parts of the leaf blade. Only a few leaves of a tree may be affected or the infections may be so numerous as to involve almost the entire foliage. The individual lesions may include a small portion of a leaf or its entire surface. As the disease progresses, the leaf tissue becomes yellowish and shows some reddish colour. The reddish-velvety surface of the lamina soon becomes whitish-grey. The leaves gradually turn brown, wither and fall off. In heavy infections the trees may suffer severely from premature defoliation. The loss of leaves is recovered by a fresh crop of leaves produced from dormat buds. Twigs may also be affected. Infected twigs become pale-green to yellow, swollen, stunted, sometimes exude gummy material and ultimately die. Flowers and fruits when infected drop promptly.

The Causal Organism. *Taphrina deformans* (Berk.) Tul. That the disease is caused by an ascomycete, at that time called *Exoascus deformans* was first described by the Rev. M. J. Barkeley. The old generic name *Exoascus* is still in use by many authors. The fungus does not produce any ascocarp. An intercellular, septate mycelium is quite common in the infected host tissue. The intercellular mycelium travels across the mesophyll and proceeds to spread out chiefly between the palisade layer and the upper epidermis (Fig. 25.12D), where it develops in greater abundance than in any other parts of the leaf. In preparation for the reproductive stage, the branches of the mycelium

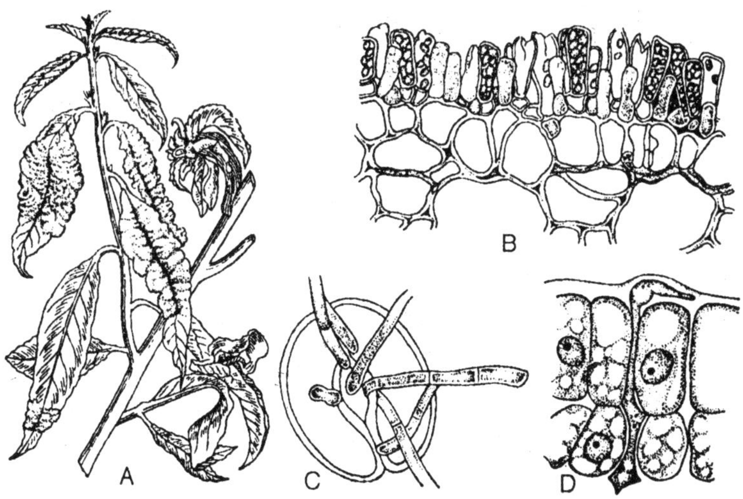

Fig. 25.12: Peach leaf curl. A. Twig with diseased leaves. B. Section through diseased leaf showing arrangement of asci. C. Host penetration. D. Hyphal ramification in the host tissue

which become established between the epidermis and mesophyll now make their way between the epidermal cells to form a network of short cells beneath the cuticle. The cells of the hyphae are with dense cytoplasm and two nuclei. The subcutaneous and intercellular hyphal cells become uninucleate resulting from karyogamy. These are the ascogenous cells from which asci are developed. The upward growth of the developing asci raise the cuticle, and this is either pierced or torn and disappears, leaving the asci exposed to the surface as a more or less continuous hymenium. In section the asci appear as a palisade-like layer (Fig. 25.12B). The asci are usually flattened or somewhat truncate at the free end, broader above than below usually eight unicellular ascopores are formed in each ascus, but occasionally the ascospore number may be only two. The ascospores are forcibly expelled through an apical slit or rupture in the ascus and may accumulate on the surface of the leaf, giving a white or greyish powdery condition. They commonly germinate by budding to form a series of uninucleate thin-walled secondary spores which are also known as conidia. The ascospores may also germinate directly by the formation of germ tubes. The young hyphae become binucleate in the host tissue.

Disease Cycle. The incidence of the disease is associated with comparatively low temperatures such as occurring in the spring. The optimum temperature for the growth of the fungus is about 20°C. Opinions differ with regard to the method by which the fungus survives from season to season. For a long time it was believed that the fungus existed as a perennial mycelium in the twigs, the fungus attacking the leaves again at the opening of the buds. The general opinion is that the primary infections which occur at the opening of the buds in spring are effected by spores, probably by conidia, which have been sheltered on some part of the host or may also persist in the soil or in the soil cover. Germ tubes from ascospores or conidia penetrate the young leaf directly into the cuticle or pass through the stoma (Fig. 25.12C). The fungus makes its initial growth within the cuticular layer and then sends hyphae between the epidermal cells. Ultimately the hyphae become established as intercellular mycelium. The first effect of the mycelium is to irritate the host cells and stimulate them to an abnormal activity. The infected host cells increase in size and change in form and structure. The affected cells suffer almost a complete loss of chlorophyll accompanied with puckering with the midrib behaving like a puckering string.

Peach being a host of economic importance, economic losses at one time may be very heavy. But this is one of the diseases which is rather easily controlled with the adoption of proper protective measures.

Disease cycle of Peach leaf curl is presented in Figure 25.13.

Control. A single application of Bordeaux mixture, at the right time—which is just before the buds begin to expand in the spring—can almost completely suppress the disease for the year. The disease can also be effectively controlled by a single application of lime-sulphur (5 : 50) and Burgundy mixture (10 oz copper sulphate: 11 oz soda: 3 gallons of water). A dormant spray of Bordeaux mixture of 6 : 10 : 100 with lubricating oil as an adhesive in the proportion of 3 gallons of oil to each 100 gallons of mixture is also very effective. Spraying during December or before mid-January with Bordeaux mixture (6 : 6 : 50) before the buds burst, or with limesulphur (1 : 8) before the winter buds swell also produces good result. The application of a very alkaline Bordeaux mixture with casein as an adhesive at the beginning of winter is also recommended as a control measure of the disease. The spray application with Blitox-50 after harvest or before

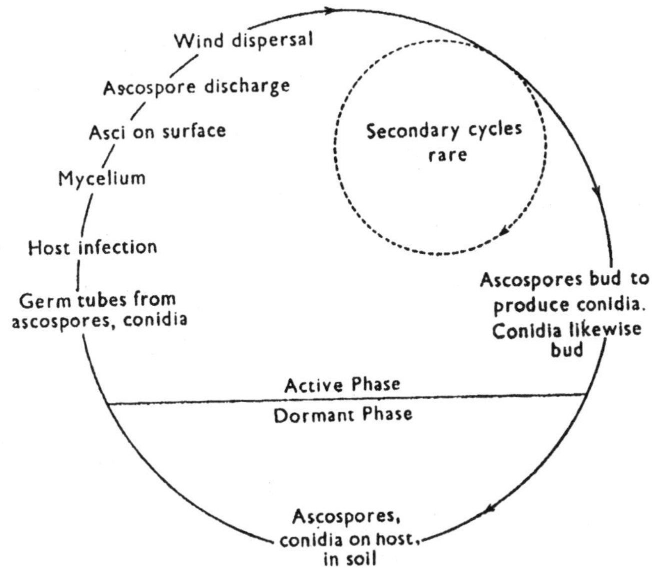

Fig. 25.13: Disease cycle of Peach leaf curl

the leaf-buds swell also produces effective results. But in any case it is useless to spray chemicals when the leaves have already opening. Besides these spray applications, cutting out infected shoots and removing and burning of all infected leaves adequately control the disease incidence.

POWDERY MILDEW OF CEREALS AND GRASSES

Powdery mildew disease caused by various members of the family Erysiphaceae is exceedingly common on a great variety of plants. It produces white powdery effect on leaves, stems, and sometimes flowers of the host. The causal organisms of the powdery mildew disease are highly specialized in their choice of hosts. All of them are obligate parasites.

Powdery mildew of cereals and grasses is rather common in the regions where humid conditions prevail during the growing season and in parts of the fields where the growth is dense. The disease affects a wide range of cereals and grasses including wheat, oats, barley, rye and the grasses like *Agropyron, Bromus, Dactylis, Elymus* and *Poa.* The damage done to the hosts is difficult to assess, since the hosts are not usually destroyed.

Symptoms. The early symptom of the disease is the appearance of white mildew areas on aerial pars of the host infected (Fig. 25.14A). Gradually these areas enlarge and coalesce producing a dusty effect on the infected plant parts. The diseae develops most conspicuously on the leaves, usually on the upper surface but sometimes also underneath, and also on the leaf sheaths and stems. The effect on the inflorescences is extemely variable. The young inflorescences may often escape infection, or sometimes so severely affected that their development may be checked or they may wither completely. Infection results in temporary stimulation of respiratory activity of the tissue and in chlorosis in and around the infected areas. The mycelium of the causal organism is entirely superficial, forming a flocculent matted growth, at first white. Simultaneously with the host infection conidia are cut out profusely in chains from the conidiophores that are developed from the spreading mycelium (Fig. 25.14B). The conidia are dispersed by the wind and on reaching susceptible host germinate to produce infec-

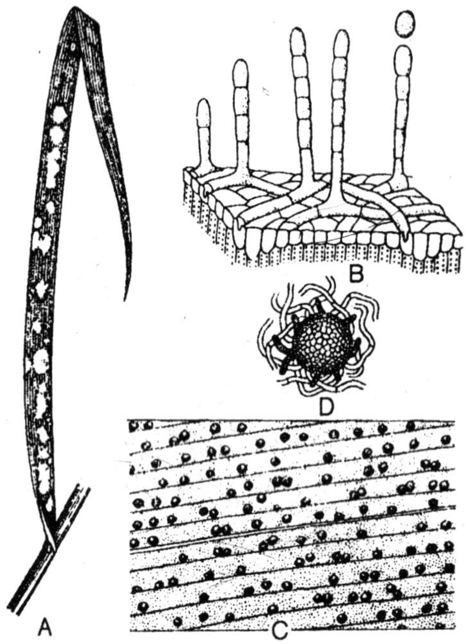

Fig. 25.14: Powdery mildew of cereals and grasses. A. Grass leaf with powdery mildew patches. B. Hyphae producing conidiophores and conidia on the host surface. C. Cleistothecia on the host surface. D. Cleistothecium

tions. This is how the disease is spread. The activity of mycelium gradually declines with the rise in the summer heat. Conspicuous green spots appear in the infected host surface. The mycelium turns grey and even macroscopically visible dark spherical bodies appear which remain scattered over the mycelial weft without being attached to the host (Fig. 25.14C). The symptoms are much alike on all the common cereal plants. From such discolouration and coverage of the host epidermis by the causal organism, photosynthesis is feeble and there is much chlorosis, with consequent weakening of the plant. In extreme cases the leaves are crinkled, twisted, or variously deformed; the top of the shoot may droop and wither, and ear development may be wholly or partially checked according to the severity of attack. At times the young seedling when attacked very early may be killed outright. Usually the disease is in greater vigour on well-established plants. The disease fails to infect plants that are suffering from nitrogen deficiency.

The Causal Organism. The disease is caused by *Erysiphe graminis* DC. The mycelium is septate branched. It produces conidiophores which are hemispherical swellings developed at right angles with the prostrate hyphae from whch they arise (Fig. 25.14B). Each conidiophore forms a chain of 10 to 20 conidia, the oldest terminating the conidiophore. The conidia are elliptical, hyaline, mostly uninucleate, occasionally binucleate measuring 8 to 10μ by 25 to 40μ. They readily fall off and are dispersed by wind. The conidia germinate on susceptible hosts and produce infections. As the development of conidia slows down and ultimately ceases, the superficial mycelial weft develops the ascigerous stage when small, dark, spherical cleistothecia are formed (Fig. 25.14C). The cleistothecia are said to appear in greater number on matering plants or on older leaves of younger plants. They are provided with short mycelioid appendages (Fig. 25.14D). The cleistothecium measuring above 200μ in diameter contains several asci which usually contain eight, sometimes four ascospores.

In its parasitism, *Erysiphe graminis* breaks up into distinctive physiologic races that are specialized to genera, species, and varieties of host plants. *E. graminis* has several physiologic races which are separated on the basis of the principal host genera affected by them. These are *E. graminis avenae, E. graminis hordei, E. graminis secalis, E. graminis poae* and *E. graminis agropyri;* cause powdery mildew of *Avena sativa* (oats), *Hordeum vulgare* (barley), *Secale cereale* (rye), species of *Poa* and species of *Agropyron* respectively.

Disease Cycle. In general, the fungus perennates in the form of dense, brown mycelial wefts on the leaves of grasses and winter wheat, these being capable of producing conidia in the following spring forming the source of primary inoculum. It is also true, the fungus may perennate as cleistothecia on previous year's plant debris and under favourable conditions causes infection by the germination of the ascospores forming the source of primary inoculum. It is well known that powdery mildew, in general, appears to be worse during dry seasons, although a certain amount of moisture is necessary to establish infection. The causal organism is, however, capable of adaptation to conditions of moderate drought and has been reported to have thrived at various degrees of atmospheric humidity with little change. The young barley plants are infected without the presence of liquid water on leaves, whereas oats are infected under very humid conditions. The disease develops abundantly during early summer. The superficial mycelium is anchored to the host only by the lobed haustoria which enter the epidermis, and penetration is rarely deeper into the mesophyll of the leaf.

But in general, the spreading of cereal mildew disease takes place through conidia forming the secondary inoculum, and in such a case the primary infection is caused by the ascospores developed in the cleistothecia at which stage the fungus perennates. Since the rate of production of conidia is very high under favourable conditions, the disease dissemination is very effective through them though they are unable to withstand extremes of temperature. But they are capable of surviving over long periods under mild climates.

The damage the disease does to the host is due to the fact that the fungus robs the leaves of food manufactured for the use of the host as well as to the fact that attack by the fungus reduces the amount of leaf area available for food manufacture.

Disease cycle of Powdery mildew of cereals and grasses is presented in Figure 25.15.

Control. Because of its superficial habit, the disease could be very easily controlled by dusting with sulphur, which acts both as an eradicant and as a protectant. To prevent plant injury, the use of sulphur should be avoided in hot sun. The use of Thiovit or Karathane also produces good result. Heavy manurial treatment, especially nitrogenous, should be avoided, and phosphatic compounds should be

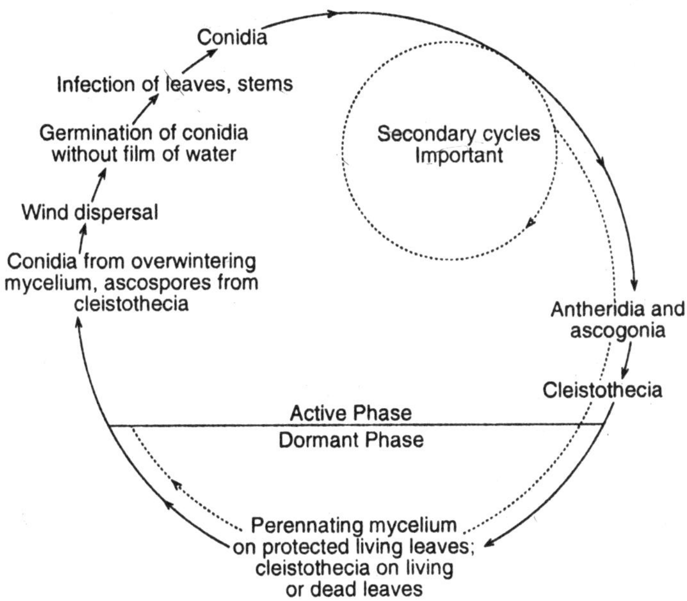

Fig. 25.15: Disease cycle of Powdery mildew of cereals and grasses

applied. Considerable control can, no doubt, be obtained by ascertaining the needs of particular soils for culture of cereals, and a balanced fertilizing schedule should be observed. Avoiding varieties which are known to be extremely susceptible in the locality is an effective control. The disease can be satisfactorily controlled by using resistant varieties. Crop rotation often tends to minimize the chances of heavy infection.

ERGOT OF GRAINS AND GRASSES

Ergot is a disease which commonly affects cereals and a large number of wild and cultivated grasses. Among economic cereals it is most prevalent on rye. It confines its attack only on the flowering parts. The ovaries are replaced by dark-coloured trilateral horn-shaped sclerotia known by the name of ergots. The name ergot is derived from old French **argot** or **ergot,** meaning 'spur' refers to the resemblance of the sclerotium, to a cock's spur. Ergot, when ingested by human beings and various animals, has long been known to be poisonous, causing a disease known as **ergotism** or ergo poisoning.

The ergot disease has an interesting past history. The earliest experiences with ergot are connected with its relation to epidemics among people and animals, and it was used as a drug long before its true nature was known. There is likelihood that the ancient Greeks and Romans knew this disease. It is highly probable that an outbreak of ergotism was first reported in the **Annales Xanthensis** in A.D. 857 as 'a great plague of swollen blisters consumed the people by a loathsome rot, so that their limbs were loosened and fell off before death'. Doubts are also there with regard to the cause of the epidemic named as 'holy fire' (**ignis sacer**) that appeared later. The gangrenous condition of limbs, resulting in death or the loss of hands and feet, undoubtedly was ergotism, although some other diseases may have accounted for a portion of the mortality. The ergot disease has also a direct bearing on ergotism or ergot poisoning of cattle during the Middle Ages, recognized as St. Anthony's Fire.

The modern history of ergotism begins with the epidermic in the Sologne district of France in 1676. From various parts of the world like New Zealand, Germany, Poland, Russia, Siberia, United States and Argentina have come reports of the poisonous effect of ergotized pasture grasses and of ergotized hay when consumed by domestic animals. The loss to stockmen from abortion of cows and mares is heavy when ergot is abundant. Gastric disturbances and varying degrees of paralysis are other symptoms of ergot poisoning in horses and cattle. Ergotism is still present at times in parts of Central Europe and Russia where rye bread is a food. As recently as 1951, many people in a small village in France were poisoned from eating bread made from ergot-infected rye, five died, thirty were driven temporary insance with hallucinations of fire, demons, and snakes.

Because of the conspicuous nature of the disease, it was observed and recorded in the earliest literature and is one of the oldest known plant diseases. The disease is worldwide in occurrence. There are two economic aspects of the disease. The loss of yield of rye grains not only due to the replacement of grains by sclerotia, but also the infection of the flowers and the sclerotial development are usually followed by sterility of neighbouring flowers. The loss in yield is, ordinarily not more than 20 per cent. The other aspect is, the injuries are caused to the livestock and to people when sclerotia are consumed by them. The injury is caused by alkaloidal materials contained in the sclerotia. The alkaloids of ergot also contain valuable therapeutic properties. These alkaloids extracted from sclerotia are purified as drugs that are of use in gynaecology and obstetrics. Of the number of alkaloids,

ergotamine and **ergotoline** are the most important physiologically. The alkaloid content of ergot varies with the locality and various other conditions. When stored at high humidities, ergot deteriorates, but keeps for long periods when stored dry. Most ergot of commerce comes from naturally diseased rye, but lately ergot has been produced by aritificial infection of rye inflorescence with the fungus *Claviceps purpurea.*

Symptoms. The disease is not generally noticed in the field until the formation of sclerotia has progressed considerably. But usually the first visible symptom of the disease is in the inflorescence in the form of a sticky yellow fluid secreted from infected flowers. This sticky fluid secretion or 'honey-due' as it is called, is sweet to the taste. It oozes out through the glumes. An immense number of minute conidia are embedded in the sticky fluid. This is the conidial stage and is known as the 'Sphaecelia' condition from which the casual organism was originally named as *Sphacelia segetum* before the connection between it and the ascigerous stage was not established. This comprizes the primary infection. The sticky fluid is attractive to flies and other insects which feed on it and aid in disseminating the conidia to cause secondary infections. Dew and rain also serve to convey conidia to other flowers. The infected overy is invested with a superficial weft of mycelium and the ovarian tissues are penetrated and become completely replaced by mycelium. The mycelium consists of intricately woven hyphae which emerge from the surface of the ovary in contorted slimy masses. The hyphae give rise to a more or less close palisade of short conidiophores which profusely abstrict conidia. The conidia liberate into the sticky exudate which accumulates on the surface of the inflorescence. A decline in the production of conidia is followed by a progressive swelling and hardening of the mycelial mass which is gradually converted into a sclerotium (Figs. 25.16A and B). During the development of the sclerotium the young ovarian tissues are completely destroyed. The conversion of the young ovary into a fungoid sclerotium takes place gradually from below upwards. The sclerotia are generally longer and larger than the normal grains. They usually protrude somewhat from the glumes (Fig. 25.16D). The sclerota are varying in number ranging from one to many in a head.

The Causal Organism. *Claviceps purpurea* (Fr.) Tul. The fungi involved in producing ergotized inflorescences of grasses and rye are mainly *Claviceps purpurea* and *C. paspali. Claviceps purpurea* can infect *Secale cereale*

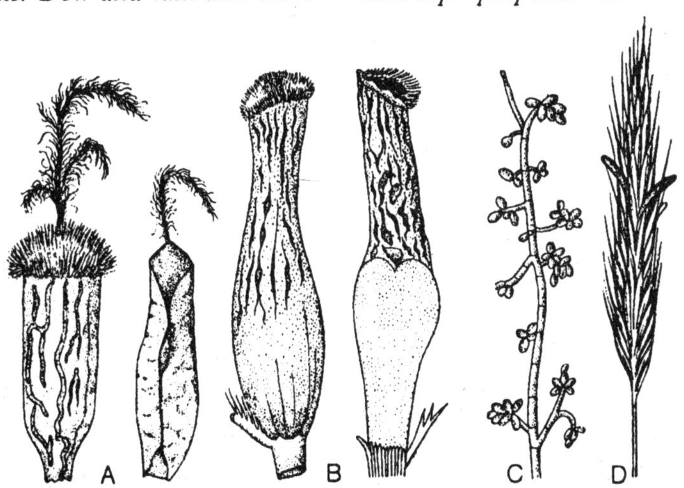

Fig. 25.16: Ergot of grains and grasses. A-B. Stages of sclerotial development. C. Hyphae and conidia. D. Inflorescence of rye bearing sclerotia

(rye), *Lolium perenne,* a few species of *Poa, Agrostis alba, Agropyron occidentale, Anthoxanthum odoratum,* and *Elymus canadensis,* whereas *Claviceps paspali* occurs on several species of *Paspalum.* Besides these two species of *Claviceps, C. wilsoni* produces infection on *Glyceria fluitans* and *C. sesleriae* on *Sesleria coerulea.* Again, *Claviceps purpurea* itself has several physiologic races. For example, *Claviceps purpurea* of rye although morphologically indistinguishable from the *Claviceps* that produces sclerotia on *Anthoxanthum odoratum,* almost never produces sclerotia on *Anthoxanthum odoratum.*

Fries in 1822 named the fungus-producing ergot as *Spermoedia clavus,* but later in 1849 he changed the name to *Claviceps purpurea.* The conidial or honey-dew stage of the casual organism was discovered by Leveille in 1827 and named as *Sphacelia segetum.* But earlier than Leveille, the sclerotia or ergots had been named by De Candolle in 1815 as *Sclerotium clavus.* The complete life history of the ergot fungus was worked out by Tulasne in 1853 and the fungus was named by him *Claviceps purpurea* (Fr.).

The sclerotia fall to the ground, and after a period of dormancy germinate in the following spring. Germination is first evident as bulges in the sclerotial cortex. These bulges burst through the rind and develop into clavate stromata, which consist of slender stalks surmounted by spherical heads (**sphaeridia**). The entire stroma is yellowish white at first but changes to greyish violet. The stromata arising from each sclerotium vary in number from a few to more than twenty. At maturity the sphaeridium is covered with wart-like projections, each of which marks the position of the ostiole leading to a perithecium. The perithecia embedded in the sphaeridium are flask-shaped. They are filled with curved hyaline asci, narrow below and above and broader in the middle. The asci are interspersed with slender paraphyses. Each ascus contains 8 very slender filiform ascospores measuring from 50 to 76 μ by 0.6 to 0.7 μ in diameter. The ascospores are extruded under conditions of high humidity and may collect at the ostioles in viscid masses for dispersal in splashing raindrops, or by insects, or they may be swept away by the wind. The ascospores may also be forcibly ejected from the asci.

The discharge of ascospores often synchronizes with the opening of the flowers of the host plant. The ascospores germinate on coming in contact with the ovaries and the resulting hyphae soon pervade the tissues of the ovary. The hyphal mass is densely covered with elongated conidiophores from whose apices conidia are abstricted (Fig. 25.16C) constituting the conidial, or sphacelial stage, or honey-dew stage. The conidia are dispersed by insects and are thus carried to other flowers spreading infection during the flowering time. Gradually the hyphal mass is transformed into a sclerotium.

Disease Cycle. The fungus perennates as sclerotia. The sclerotia when mature either fall to the ground or are mingled with the grain and may be returned to the field with the seed. They can withstand very low temperatures. Low temperature is rather conducive to their germination. Sclerotia remain in dormant condition for a period from a few weeks to months, after which germination takes place. Even portions of broken sclerotia which may pass undetected in threshed grain and seed samples may grow quite well as whole ones. Burial in moist sand constitutes a favourable substratum for germination. But in general, sclerotia do not remain viable on the ground for more than a year. During the following spring from the year of development of perithecia, asci, ascospores and finally the dispersal of ascospores having coincidence with the opening of flowers of the susceptible hosts facilitate the spread of primary infection through ascospores and secondary infection through conidia. The dissemination of ascospores is usually by wind and that of conidia by insects. The infection is localized and the actual number of ergotized kernels varies from one to several in an inflorescence. The sum total effect of infection may be: (i) the replacement of grains by sclerotia; (ii) the sterility of florets adjacent to infected ones, may be due to incipient infection or exhaustion of food supply; and (iii) reduction in size of the heads. The disease lowers yield and quality in case of grain crop, and in hay crop quality is lowered. Even though the conidia are known to preserve their infective power for at least a year if kept dry, they are not believed to survive the winter under natural conditions.

The actual infection may be before pollination, or even after fertilization has taken place, but only in the young condition of the ovary. The principal infection court is not the stigma, or the ovarian wall, but the point of insertion of the ovary. Ascospores or conidia may often germinate on the stigma and produce long germ tubes which grow down, encircling the ovary, and subsequently enter at its base. Flowers infected from ascospores produce conidial stage and conidia from the conidial stage are carried by insects to other flowers spreading infection, this may continue as long as there are flowers in the susceptible host accompanied with favourable conditions. A few primary infections with ascospores may, therefore, be sufficient to provide inoculum to iduce infection in a large area, provided conditions are favourable for the development of conidial stage and the dissemination of the conidia. In general, the severity of the disease incidence largely depends upon the atmospheric conditions during flowering period. Usually lower temperatures prolong the flowering period and keep the glumes open and thereby prolong the period of susceptibility to infection. It is generally agreed that grasses in low, damp areas are more susceptible than those growing in drier places. Those species and varieties in which glumes tend to remain closed during flowering largely escape infection. The degree of uniformity of flowering and the duration of flowering period are also important, since in rye, increased tillering and consequent prolongation of the flowering period result in more infection.

Disease cycle of Ergot of grains and grasses is presented in Figure 25.17.

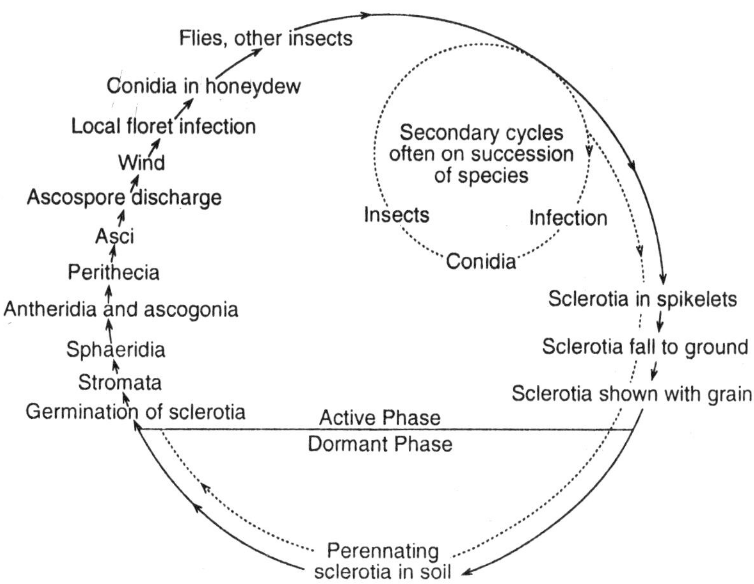

Fig. 25.17: Disease cycle of Ergot of grains and grasses

Control. The most effective control measures are: (i) *Use of Ergot-free seeds.* Seeds may be made ergot-free by dipping them in 20 to 32 per cent solution of sodium chloride or 32 to 37 per cent solution of potassium chloride. The ergots floating on the surface are easily separated from the seeds that sink at the bottom of the solution. The ergot-free seeds are then washed with water immediately and dried. They are now ready for sowing. (ii) *Crop rotation.* Practicing rotation of susceptible grain crops with legumes or other non-susceptible hosts reduces infection. Since the sclerotia lose their viability after one year, sowing susceptible host at an interval of one year produces effective result. The previous year's sclerotia left in the field will germinate as usual, but for want of susceptible host will fail to produce any infection. (iii) *Cultural method.* Deep ploughing buries the sclerotia and thus prevents sclerotial germination. Sowing seeds 2 to 3 years old, even if contaminated, is reported to be sufficient precaution against recurrence of the disease since sclerotia lose viability by then. Close planting should be avoided. (iv) *Sanitation.* Clean out susceptible grasses from the margins of cereal fields before blossoming to prevent from the dissemination of inoculum from sclerotia and diseased grasses in the borders to the cereal crop. (v) *Harvesting method.* Harvest the outside of the field separately, and avoid mixing with the rest of the harvest. It is a precautionary measure to reduce ergotized seed rate. As the disease affects inflorescence only, hay should be cut where still green, before ergots develop.

RUSTS OF WHEAT

Wheat is attacked by three different rusts: (i) black rust or stem rust or black stem rust *Puccinia graminis* Pers. f. sp. *tritici* Eriks. and Henn.; (ii) yellow or stripe rust, *P. striiformis* West; and (iii) brown or orange or leaf rust, *P. recondita* Rob. ex Desm.

Rusts of various crops are mentioned in ancient history. The writings of the Romans bear evidence that their cereal crops were attacked with rusts. During subsequent years rusts were reported from different countries of the world, but without realizing the actual reason behind the incidence. It was Persoon in 1797, who recognized that the stem rust was due to the attack of a fungus *Puccinia graminis.* But the true life history of *P. graminis* was not known until 1865 when De Bary, by inoculation experiment, was successful to demonstrate that *P. graminis* needs two hosts—wheat and barberry to complete its life history.

Black Rust or Stem Rust or Black Stem Rust. (*Puccinia graminis* Pers. f. sp. *tritici* Eriks. and Henn.). This disease is present in greater or less amount practically in every country throughout the world, where wheat is grown, and epiphytotics have been recorded in many countries. The disease is most damaging in moderately moist areas and in moist season in areas with low average rainfall.

Symptoms. Black stem rust is characterized by the development of the 'red rust' or summer stage represented by the uredial stage appearing on leaves and culms, followed a little later by the 'black rust' or winter stage represented by the telial stage.

With the onset of the disease, elongated brown or reddish-brown pustules or **sori** burst through the epidermis of the host tissue (Fig. 25.18C). These pustules or uredia or uredosori split open in an irregular manner having an elongated crater-like opening with ruptured host epidermis clinging to one or both sides of the opening. The size, shape and number of sori vary with environmental conditions and particularly with the inherent resistance of the host. The uredosori may develop without any surrounding chlorotic or dead cells or they may be seated in chlorotic areas which soon become dead. They may be a few in number, or they may be very numerous and may coalesce to form more or less elongated brown or rusty powdery streaks. Within these sori are masses of reddish or rust-coloured powder consisting of thousands of minute dust-like uredospores. Later in the season the black rust stage or winter stage or telial stage appears. This consists of elongated pustules or telia or teleutosori similar in shape to the uredosori but black in colour. The black colour is due to the presence of masses of dark teleutospores they bear. The teleutosori are dark-brown to black and are smooth rather than powdery, although the epidermis of the host tissue is broken and the teleutospores are exposed. The dark teleutospores are more firmly attached in their beds than the uredospores, and their stalks are more rigid and thicker. They are essentially resting spores adapted for perennating on straw or stubble. The teleutospores on germination produce promycelia which again produce sporidia (basidiospores). The sporidia cannot infect the cereal host, and infect only the alternate host, the barberry (*Berberis vulgaris* L.).

On the barberry, small circular yellowish spots appear on the upper surface of the leaf. These spots increase in size, become margined with a brighter colour or reddish-purple and show a central cluster of minute honey-coloured pustules—spermogonia from which minute droplets of nectar ooze out. At more or less opposite places towards the lower surface of the leaf, yellowish or orange-coloured spots appear. Close examination of these spots on the lower surface of the leaf shows a cluster of small cups with saw-toothed edges—the aecia or cluster cups. Often the host tissue developing aecia may become hypertrophied. Similar hypertrophied cluster-cup lesins may appear on the fruits and petioles.

The Causal Organism. Black stem rust is caused by *Puccinia graminis* Pers. f. sp. *tritici* Eriks. and Henn. This is a heteroecious fungus. Its principal hosts are graminaceous

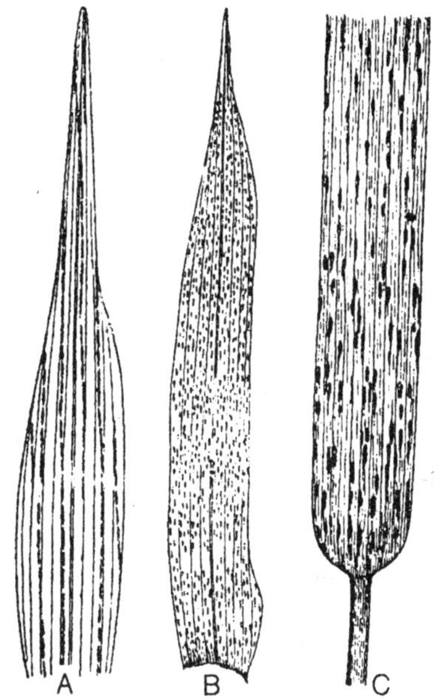

Fig. 25.18: Symptoms of Rusts of wheat on wheat leaves. A. Uredosori of yellow rust. B. Uredosori of brown rust. C. Uredosori of black stem rust

plants including wheat, oats, barley, rye, a large number of grasses and the alternating host is barberry.

A transverse section of an uredosorus shows that the mycelium is intercellular, the hyphae produce small-rounded or branched haustoria which penetrate in the host cells. The uredospores are stalked single-celled, dikaryotic and golden-brown. The cell of the uredospore is oblong or ellipsoid with four equatorially arranged germ pores, and strongly echinulate. These spores must have a film of water in the host surface before they can germinate. The uredospores can infect graminaceous hosts repeatedly.

The teleutospores follow the uredospores on the same or similar dikaryotic mycelium. Two kinds of spores may frequently be found together in the same sorus. A teleutospore is stalked consisting of two thick-walled, smooth, superimposed cells, the top cell being rounded or blunted and thickened at the apex. Each spore is dark-brown, but black in mass. Each constituent cell is furnished with a germ pore, situated apically in the top cell, and just below the septum in the lower cell. Each cell during development possesses a pair of nuclei contributed by the dikaryotic mycelium. When the teleutospore turns brown and matures in the teleutosorus, the paired nuclei in each cell fuse. The mature teleutospore thus represents the diploid phase in the life history of *P. graminis tritici.* The teleutospores do not germinate forthwith but undergo a period of rest and remain dormant on stubble or straw for several months.

Like the uredospores, the teleutospores also germinate in presence of a film of water. One or both the cells of a teleutospore may give rise to a promycelium, during the formation of which the diploid nucleus undergoes a reduction division and four sporidia (basidiospores)—two of positive strain and two of negative strain, are formed. The sporidia germinate on the barberry leaf in the presence of moisture. Two types of spermogonia positive and negative are developed near the upper epidermis of the barberry leaf. The spermogonia are flask-shaped structures embedded in the host tissue and opening out into a small ostiole at the epidermis. They bear spermatia and flexuous (receptive) hyphae. The spermatia are exuded at the ostiole in a copious nectar or honey-dew and are carried to respective (+) to (–) to (+) flexuous hyphae through insects establishing dikaryotic condition. The dikaryotic condition so established perpetuates through hyphal connection to the protoaecium which develops close to the lower epidermis of the barberry leaf along with the development of spermogonia. Ultimately the protoaecium develops into an aecium. The aecium when young is covered by a peridium which breaks with the maturity of the aecium. The aecium bears aeciospores in chains. The aeciospores are dikaryotic, single-celled and hexagonal in shape. The aecispores are readily dispersed by wind. They have about six germ pores in the wall. These spores serve to return the rust to the cereal or other graminaceous host. The aeciospores germinate on the graminaceous host in the same manner as the uredospores. The spores being dikaryotic are responsible for the establishment of a dikaryotic mycelium which ultimately gives rise to uredosori on the graminaceous host.

Physiologic Races. True, *P. graminis* can attack any of the graminaceous hosts indicated, but it is also true that much fewer than that number of graminaceous hosts are found to be susceptible in any particular locality. This is due to the fact that *P. graminis* is composed of many different physiologic forms, races or varieties. Any particular form, race or variety can attack certain species or varieties of host plants but not all of the susceptible graminaceous hosts.

The mighty Rust investigation began in Australia following the epidemic in 1889. The classical researches of Anton de Bary and the Tulasne brothers had long since established the parasitism of the rust fungi.

Between 1892 and 1896 Carleton grew various cereals in pots in a greenhouse and tried to infect the wheat plants with uredospores from the rust on the oats. The attempt failed—the spores from the rust on the oats would not infect wheat, barley or rye. And with the exception of the rust on the wheat, which did infect the barley, he found that the rust on one cereal would not infect the others. There was not one *Puccinia graminis* but several, to be distinguished from one another only by the kinds of wild or cultivated on which they would grow.

Meanwhile, Professor J. Eriksson (1894-96) of Sweden had already sorted the single *Puccinia graminis* of his botanical forefathers into six special forms: *Puccinia graminis* special form *avenae,* to be found on oats and some half-a-dozen wild grasses; *Puccinia graminis* special form *tritici,* to be found on wheat and barley and a few others of the grasses; *Puccinia graminis* special form *secalis,* on rye, and so forth. These special forms of the fungus which differed so widely in their ability to parasitize the different cereals could perhaps be regarded as species in the making. Forms in which adaptation and natural selection had gone far to make profound physiological differences, but not far enough to alter the external appearance of the parts of the organisms in so far as those alterations could be seen by man with the aids to vision available. Were these special forms of *Puccinia graminis* really species in an early stage of differentiation? Subspecies? Adapted species? Physiologic forms? Biologic races? Biologic forms? Physiologic races?

H. M. Ward (1905) encountered a similar type of specialization toward species of *Bromus*. He was of opinion that the host substrate influenced the pathogenicity of a given race.

E. C. Stakman and his associates made intensive study of the physiologic races of *Puccinia graminis* and indicated that the physiologic races are best defined by the effect upon different hosts. The physiologic races may or may not be stable. New physiologic races may arise in nature through gene mutations and also by genetic, recombinations of characters in the process of sexual reproduction. Physiologic races in so far as barberry is concerned are unknown. All physiologic races of graminaceous hosts, in so far as is known, infect barberry. When this host is infected, the pathogenic difference between races no longer exists. As such, it appears that the monokaryotic mycelium of rusts has as a rule pathogenic properties distinct from those of the dikaryotic mycelium.

For details of biological specialization and physiologic races of *P. graminis* please see in earlier Chapter.

Disease Cycle. The pathogen perennates as teleutospores which remain dormant on stubble or straw of graminaceous hosts for several months. They are known to be viable for at least 18 months. Teleutospores serve, indirectly, to convey the disease to the alternate host, the barberry. They must be wetted before they germinate. The most favourable temperature for germination is 19°C to 21°C. *P. graminis* being heterothallic, of the four sporidia formed by the teleutospore germination two belong to (+) and two of (–). The sporidia are dispersed by air currents and fall on the barberry host and infect young stems, petioles and leaves by direct penetration of the cuticle. Infection may take place during the day or night but is favourable mostly by daylight. Two types (+) and (–) of spermogonia are produced by the respective (+) and (–) sporidia. The function of spermogonia was established by Craigie in 1927. Following transfer of (+) spermatia to (–) flexous hyphae and *vice versa*—spermatization by flies or raindrops, dikaryotic condition is established resulting in the formation of aecia and aeciospores. The aeciospores dropping from the aecia are caught up by air currents and are carried to the graminaceous hosts. The aeciospores germinate in presence of a film of water, form appressoria, penetration always through the stomata, and each forms a vesicle in the substomatal chamber from which dikaryotic hyphae ramify intercellularly into the parenchymatous tissue resulting in the formation of uredosori. The uredospores formed in uredosori germinate on graminaceous hosts to spread infection. These spores germinate by a germ tube in a film of water. The tip of the germ tube swells to produce an appressorium from which a narrow hypha arises which passes through stomata and invades the substomatal space where it soon expands into a vesicle. From the latter an infection hypha emerges, which having penetrated a mesophyll cell establishes within it. The hypha proceeds sending haustoria into the cells with which it comes into contact. From such an initial infection an uredosorus with mature spores becomes established in eight to fourteen days. With the dissemination of the uredospores throughout the crops, fresh uredosori continue to be formed, but they usually cease to develop

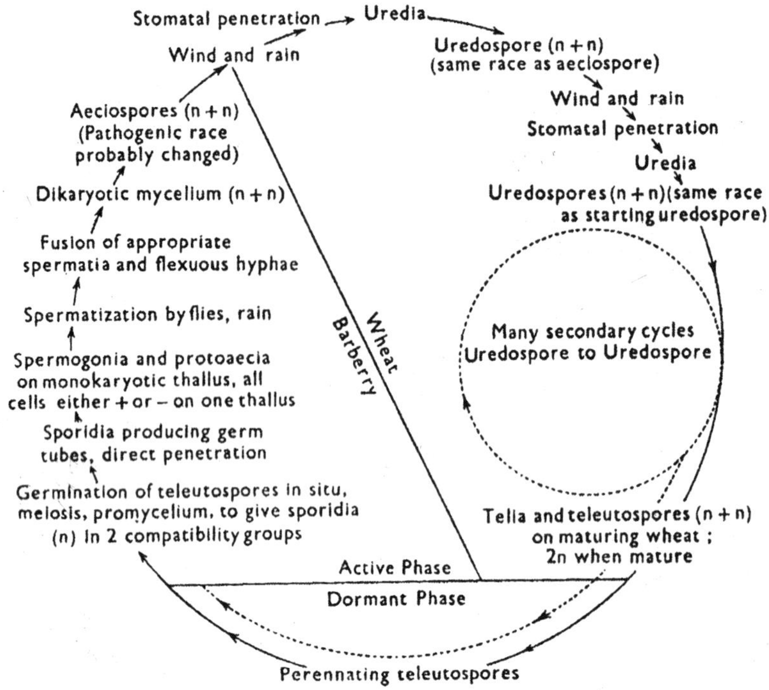

Fig. 25.19: Disease cycle of Black stem rust of wheat

before the wheat begins to change colour and ripen. The development of teleutosori follows along with the slow-down of the formation uredosori as the host plant approaches maturity.

Disease cycle of Black stem rust of wheat is presented in Figure 25.19.

Annual Recurrence. Since sporidia cannot infect the graminaceous hosts, and the source of inoculum of wheat infection is the aeciospores, it is logical to conclude that both wheat and barberry hosts are essential for the annual recurrence of the disease. The pathogen perennates as teleutospores in the stubbles and straw of graminaceous host.

Eriksson in 1894 in his 'Mycoplasm theory' suggested that during the active wheat growing season, the fungal hyphae attack the wheat host and thereby an intimate mixture of the fungus protoplasm and the host protoplasm is established resulting in the formation of what he designated **mycoplasm.** When the seeds from rust infected wheat plants bearing **mycoplasm** are sown, the mycoplasm splits up into the host protoplasm and fungus protoplasm. The fungus protoplasm then becomes organized into fungal hyphae and thus the infection is re-established. But Eriksson's theory did not receive wide support.

In areas where both wheat and barberry hosts are present, the infection of the former may be initiated by aeciospores. But this ideal condition does not prevail all over India. For example, in the plains of India where barberry plants do not grow for hundreds of miles but the wheat rust appears year after year. What is the explanation for it? Since the barberries are absent, aecial stage does not come into picture at all. The summer heat in the plains is so intense that the survival of rust on weeds or any other substrate is not possible. It is obvious then that the rust does not spend summer in plains. Therefore, the first appearance of rust on wheat crop during next winter is due to the dissemination of inoculum which is most likely from uredospores from some source other than plains. The possibility of air-borne dissemination from other countries is rather nil due to natural barriers like Indian Ocean and the Himalayas.

K. C. Mehta (1931) working in the Nilgiri Hills conclusively proved that the uredospores and teleutospores cannot survive in the plains during summer, but the uredospores can do so in the hills at an elevation of 3,000 to 5,000 ft and above where the summer temperature is congenial and where the rust survives in the uredial stage on self-sown wheat plants or other graminaceous hosts particularly grasses, of which *Briza minor* has been found most suitable. When regular wheat crop is sown in hills it is infected by the inoculum (uredospores) from *Briza minor.* From this infected wheat crop in the hills the disease spreads to the wheat plants growing near the foot of the hills—the Tarai regions, by wind-borne uredospores. Again, from these infected wheat plants of the Tarai regions, the disease gradually spreads to the plains. This is how Mehta explained the infection of wheat crop annually in the plains in spite of the absence of barberry plants in the plains. Mehta devised an **aeroscope,** a small weather cock fitted with a greased slide, on which he collected uredospores flying with air currents at different areas from hills, the Tarai regions and the plains.

Control. Since black stem rust infection takes place during the growing season from spores produced on the barberry or on wheat plants, prevention of infection is a logical method of control. Toward this end measures applicable are: the growing of wheat varieties resistant to attack; interruption of the life cycle of the rust so that an important infective spore type cannot be produced; eradication of grasses which serve as alternate hosts; treatment of growing wheat so that spores alighting on it do not produce infection; and improved cultural practices.

Wheat varieties entirely **immune** to black stem rust do not exist. However, varieties resistant in various degrees are numerous. Since a large number of physiologic races greatly complicates the problem of developing varieties resistant to black stem rust, continuous processes of breeding, improvement and replacements will be useful to evolve new resistant varieties. Varieties of wheat that are not resistant to the fungus but that mature early may escape serious rust damage.

Control by interference with the life history of the rust fungus is another means to reduce the incidence of the disease. Since the perennating stage of the rust is able to infect only the barberry, eradication of the barberry should remove the link necessary for carrying infection over from the crop of one year to that of the next. Eradication of barberry not only serves to remove infection of wheat by aeciospores, but also, it eliminates the possibility of recombination of genes in the fungus and the production of new races. Eradication of barberry, however, does not prevent infection from uredospores.

In India, one of the effective control measures is to discontinue wheat cultivation in the hills and eradication of grasses like *Briza minor* and *Bromus japonicus* which serve as hosts on which fungus harbours when wheat season is over.

Experimental work has shown that reduction in stem rust infection accompanied by increases in yields can be obtained if growing wheat is dusted with sulphur at intervals of 4 days before heading time. Some positive results have also been obtained by chemical control of the disease. Immediately with the appearance of rust symptoms, the crop should be sprayed 3 to 4 times with dithiocarbamate fungicides (Dithane Z-78, Dithane M-22, and Dithane S-31) at the rate of 1 kg per acre. Dusting of crop with sulphur for 2 to 3 times at the rate of 6 kg per acre is also recommended.

Four to five applications of Nabam and zinc sulphate starting from initial stage of attack of the disease have proved to be effective in controlling the disease. Besides these, the use of organic nickel salts; some of the antibiotics, e.g., actidione and related products; sulpha-compounds such as sulphadiazine, sulpha-pyrazine, and sulphapyridine have given satisfactory results. Soil or seed treatment with systemic fungicides—plantvax gives effective control of the disease.

Certain items in the cultural practices of wheat-growing may contribute to a reduction in losses due to rust infection. Drainage facilities, date of sowing and maturing of crop, and the use of fertilizers all play a part. A better opportunity is afforded for rust infection and spread on low-lying wheat fields. Hence wheat cultivation should be avoided in low-lying fields. Early maturing varieties of wheat are preferable as they have a greater chance of escaping rust infection than late-maturing varieties. Excessive use of nitrogenous fertilizers promotes excessive growth of the rust fungus. A judicious application of phosphates, however, is necessary to counteract the harmful effects of excessive nitrogenous fertilizers.

Much work has been done on the breeding of wheat varieties resistant to various categories of rusts and other diseases. Some important Mexican draw varieties introduced in India and developed later for rust resistance. These are as follows: Sonara 64, Lerma Rojo 64A, Sharbati Sonara, Kalyan sona, Sonalika, Chotilarma, safed Lerma, Hira, Up 262, Np-809, Np-770 and JNk (4-w-184).

Yellow or Stripe Rust. (*Puccinia striiformis* West). Yellow or stripe rust (yellow rust) rarely ruins the wheat crop in India to the same extent as black stem rust. It is particularly limited to Northern and Eastern parts of India. The disease is often accountable for very serious losses in the field, due to destruction of the foliage, followed in some cases by sterility of spikelets or in the production of badly shrivelled grain. But in general, the infection is chiefly confined to the leaves, only in case of severe attack it spreads over to the leaf sheath, the stalk, and even the glumes.

Due to infection, the green colour of the leaves first of all fades in long streaks, and later, as small pustules of uredosori arise along them, a distinct yellow striped effect is produced in the leaves (Fig. 25.18A). The uredosori appear as bright-yellow pustules. They are subepidermal and remain covered for a much longer time than in the other rusts, but when they finally break through, the yellow uredospores are shed and dispersed by wind. The uredospores are spherical to ovate in shape, dikaryotic, and very variable in size. The spore wall is colourless, minutely echinulate and may have six to sixteen germ pores which are scattered, whereas in the black stem rust the germ pores are equatorial in position and are four in number.

As the host matures the teleutosori appear on the leaf sheaths, glumes, but rarely on the leaf blade. The teleutosori always develop abundantly on the leaf sheaths. It is very infrequent that the teleutospores and uredospores occur together in the same sorus, as is often the case in the black stem rust. The teleutosori are compact, appear as dull black spots, arranged in rows very similar to the uredosori. They do not, however, as in the black stem rust, break through the epidermis at all, but remain as flat black crusts. The teleutospores are dark-brown and flattened at the top in contact with the host epidermis (Fig. 25.20A). They are two-celled inserted with brown unicellular paraphyses. The teleutospores on being released can germinate immediately without any rest. It is a heteroecious rust but no alternate host has so far been found.

The uredospores are capable of withstanding very low temperatures and serve as the source of inoculum to spread infection through grass hosts from the hills, as is encountered in case of black stem rust incidence in India. There is also evidence that this rust can

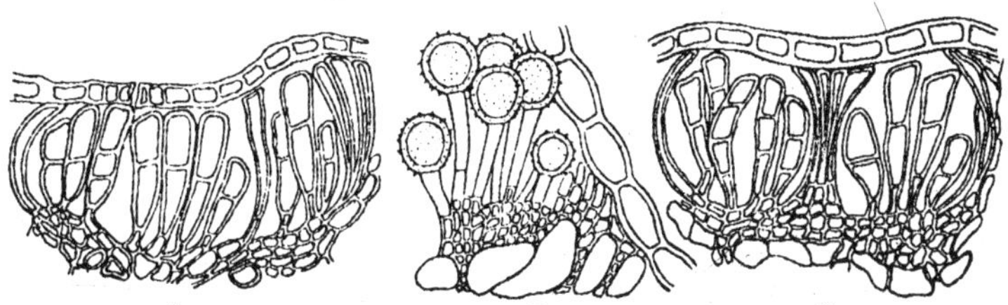

Fig. 25.20: Uredosorus and teleutosori of yellow and brown rusts of wheat in section. A. Teleutosorus of yellow rust showing teleutospores flattened at the top in contact with intact host epidermis. B. Uredosorus of brown rust with ruptured host epidermis. C. Teleutosorus of brown rust divided into small groups by the interpolation of paraphyses

perennate as mycelium in the wheat leaf or in various grasses. With the advent of favourable season the perennating mycelium produces uredosori.

The effective control of yellow rust disease is through the development of resistant varieties of wheat by hybridization of favourable types of wheat. Satisfactory results have, however, been obtained by using calcium cyanamide as an effective fungicide.

Brown or Orange or Leaf Rust. (*Puccinia recondita* Rob. ex Desm.). This is known as brown rust or orange rust or leaf rust of wheat. In general, it is rather common in the Northern and Eastern parts of India and occurs along with the yellow rust causing damage to the wheat crop in some parts of Bihar, Punjab and the United Provinces.

This disease is favoured by the same climatic conditions as those which favour black stem rust, but brown rust can develop over a wide range of temperature, particularly at temperatures below 15°C.

Rust infection is mainly confined to the leaves, rarely occurring on the stems and ears. In cases of heavy infection photosynthesis is retarded, the affected leaves turn yellow and wither early. Consequently the ears become retarded reducing in the grade, yield, and quality of grain.

The uredosori appear only on the leaves, very rarely on the leaf sheaths and stalks. They do not arise in rows or stripes as in yellow rust but are grouped in small clusters or irregularly scattered (Fig. 25.18B). The uredosori break out from under the epidermis as orange or brown specks. They are round to slightly oblong in shape. The uredospores are brown and spherical with minutely echinulate wall (Fig. 25.20B) furnished with seven to ten germs pores.

The teleutosori are very rarely produced. When formed, they are scattered chiefly on the underside of the leaf and on the leaf sheaths having similarity with those of yellow rust and do not break through the epidermis. The teleutospores resemble those of yellow rust in size and shape but the teleutosorus is divided into small groups by the interpolation of paraphyses (Fig. 25.20C). The teleutospores are oblong with slightly constricted at the septum and possessing rounded apex with prominent thickening. *Thalictrum flavum* is the alternate host of this rust. Its role in the annual recurrence of this rust has not been established in India. This rust, however, perennates as uredospores on self-sown wheat plants in the hills at an altitude of 5,000 ft and above. It is probable that brown rust can perennate as mycelium in the host leaf, and if the infected host can survive, the mycelium may produce fresh uredosori. Though the fungus has numerous physiologic races, yet raising resistant varieties to these physiologic races would be the effective control of the disease.

HOLLYHOCK RUST

Hollyhock rust disease has been studied with great interest since 1852. The original home of the disease is either Chile, where it was found to occur in 1852, or Australia, where its presence was recorded in 1857. It was found in Spain in 1867, in France in 1872, in England and Germany in 1873, in Belgium, Holland, Denmark and Italy in 1874; and in South Africa in 1875. The first appearance of the disease in the United States on the hollyhocks was reported in 1886. This disease was introduced in the United States with imported seeds of *Malope.* Since then the disease spread all over the world. The fungus which produces this disease is too well known in general appearance and by the effects it produces. The sudden and widespread appearance year after year of the disease particularly on the Malvaceous plants has grown the keenest interest for the study of the disease. In 1852 Bertero first observed in Chile the appearance of the Hollyhock rust on different kinds of Malvaceous plants on which it was very common. He sent his collections to Montagne, who diagnosed, described and named the causal organism of the disease *Puccinia malvacearum* Bertero, which is the cause of a widespread disease of *Althaea, Malva* and related genera of the family Malvaceae. The fungus is one of the most noticeable of the Uredinales being truly **plurivorous,** i.e., attacks a number of hosts; so far from being confined to a species it is not even confined to a genus. It can spread on all green parts of all Malvaceous plants if they are grown side by side. In all the plants the appearance of the disease is identical and by artificial inoculations it was proved that it can be transferred from genus to genus.

Symptoms. It is an autoecious rust which produces only one form of spore, the teleutospore, and has therefore been classified in the group **Lepto-peccinia.** The *Uredo* and *Aecium* forms are wanting. The teleutospores are formed in sori. The sori are again formed scattered simultaneously on the leaf blade, petiole and stem. They are very prominent and stand up almost like beads from the surface of the host and are more abundant upon the underside than the upper side of the leaf, but arise anywhere on the leaf even upon the long petioles. The sori are about 1 to 2 mm in diameter. The infection is very much localized, that is, confined to a small area immediately in connection with the sorus. This is especially true of the leaves but on the petioles and stems the sori may be much longer than broad and running some distance up and down the periole and stem from the point of infection. The sori on the petioles and stems are covered by the epidermal layer of the host for a longer time than is the case with the sori on the leaves. Upon the petiole the mycelium works more readily and more extensively, giving the sorus more strikingly the appearance of an eruption.

The Causal Organism. *Puccinia malvacearum* Bert. The teleutospores are oblong to oblong-fusiform, two-celled, rarely somewhat elliptical, smooth light-yellow slightly constricted at the septum and acute to slightly rounded but not thickened apex. The two cells are more or less of the same shape and size, narrowed each way from the septum, hyaline and persistent. Pedicels are firmly attached to the tissues of the host plant which bear them by well-developed mycelium and hence they remain attached to their host tissue until the mycelium causes the dissolution of all vegetable substances. The teleutospores germinate in a damp atmosphere upon the host plant as soon as they are mature without any resting period by throwing a four-celled promycelium. The sporidia are formed in one of the two ways: (i) the four cells of the promycelium may separate into four free and independent cells, each of these cells, later sends out a little protrusion which elongates and swells up as the content of the mother cell is gradually passed into it, and thus a mature sporidium is formed, which readily breaks away from the basal cell and germinates; (ii) in the other type, the four cells of the promycelium do not break apart but each sends out little sterigma, which gradually swells up until the sporidium is formed into which the content of the promycelial cell is passed. The sporidium is then abstricted and germinates. The sporidia are oval or spherical or subreniform. The sporidia are carried by wind readily and are lodged on the leaf, where infection takes place. They are the source of secondary infection and can spread the disease very rapidly. The sporidia produce germ tubes in 24 to 48 hours, which immediately bore through the cuticular cells of the host. Plowright observed two kinds of teleutospores (i) roundish masses of teleutospores are clear brown and germinate at once, (ii) the other form occurs more upon the veins of the leaves, but especially upon the stem, in more elongated sori. These spores are darker in colour and do not germinate immediately. The entire sorus becomes detached from the stem and falls upon the ground where it probably remains until the next year.

Disease Cycle. The fungus becomes active as soon as the winter is over. Epidemics are regulated by the frequency of rains and dampness in the summer season. Usually it is not active in the spring and early summer as it is in late summer or autumn. The host plants suffer severely in September and October. The teleutospores germinate very rapidly at 15°C with suitable moisture. The first visible sign of infection is tiny white spot or pustule on the under surface of the leaf, indicating the grouping of hyphae. The rate of development of the hyphae is independent of the host, but varies appreciably with the temperature.

Germ tube formation starts within 24 hours after infection, penetration occurring at the point of the epidermal wall to which the sporidium is adherent. A minute pore is formed at the point of contact. The germ tube then slowly works itself in. Infection never occurs through a stoma. The germ tube then elongates and in three days it becomes branched with many transverse septa. The branches (infection hyphae) then pass into adjacent cells and in the intercellular spaces. Profuse branching of mycelium takes place in all directions. This vegetative mycelium becomes both inter- and intracellular and ramifies very rapidly in all directions of the host tissue.

The change in colour of the pustule from white to pale-yellow marks the initiation of the dikaryophase. This somewhat suggests the homothallic nature of the fungus. The colour becomes deeper and deeper until finally the epidermis is ruptured, and the rusty brown teleutospores are exposed.

The teleutospores germinate as soon as they are ripe. At first, fusion between the two nuclei takes place. The diploid nucleus migrates in the promycelium whereas a result of reduction division four haploid nuclei are formed. Ultimately from the four-celled promycelium four sporidia are formed. Sporidia are discharged at binucleate stage. Under cultural condition sporidia produce germ tubes within 10 to 24 hours after infection. The fungus is readily transmitted from *Althaea rosea* to *Malva rotundifolia* or *Malva crispa*, and *vice versa* by artificial cross-inoculation.

Both intra- and intercellular hyphae cause disorganisation of the host tissue. The hyphae extend inwards in the intercellular spaces of the cortical parenchyma, making a very definite attack on the starch sheath. Strands of mycelium pass into the vascular bundle by way of the less resistant phloem rays, and then ramify among the tissues of the phloem. Branches of mycelium extend to the parenchyma within the circle of vascular bundles. The mycelial branches attack the tissues of the stem in the similar manner. In the petiole there is a definite attack on the starch sheath, medullary rays and the phloem elements of the vascular bundle. The attack of the pathogen causes a marked diminution of starch in the cells of the endodermis and the phloem-rays. The nature of invasion is practically restricted in certain areas. Due to the fungus attack the cellulose thickenings on the cell walls of the collenchyma tissue disappear gradually as the mycelium advances between the cells. During the early stage the cells elongate. The pericycle fibres of the vascular bundle of both stem and petiole turn out to be almost devoid of thickening.

Due to the attack of the pathogen transpiration rate of the host tissue is increased. This may be due to the fact that the infected host cells have lost their power to retain water. At the same time the enhancement of the rate of transpiration is further helped by the ruptured epidermis due to fungal attack.

The infected leaves wither and dry up very rapidly. Very often the leaves take a scrorched appearance long before the appearance of flowers. Such an abnormality may be due to rapid growth of mycelium in the host tissue. In many cases even flowers fail to appear.

In some extreme cases entire annihilation of the hollyhocks has been reported, as a result of which many growers have discontinued to cultivate them.

Disease cycle of Hollyhock rust is very simialar to the secondary cycle of Black stem rust of wheat as indicated in Figure 25.19.

Control. Various control measures have been suggested by many workers which may be summarized as follows:

(i) *Eradication:*

Destroy all old infected stems and leaves immediately after flowering to reduce the source of primary inoculum.

(ii) *Protection:*

(a) Planting in crowded plots should be categorically avoided, as closer planting gives more chance for the spread of infection. So best way is to plant the seedlings in isolated rows.

(b) Application of copper sulphate fungicide to the soil in weak solution (1 to 3 per cent) has some effect in reducing the virulence of the disease.

(c) Dusting once a week with a fine grade of sulphur before flowering is useful.

(d) Infection may also be satisfactorily controlled by frequent spraying with lime-sulphur (1 in 60 to 1 in 100) from spring to summer, or with Bordeaux mixture (4-4-40) spraying several times during the season. The application of sulphur spray should be completed before blooming, otherwise it may cause burning of the flowers.

(e) Application in the infected areas with a sponge the following mixture is a cheap and effective method of control of the disease:

Permanganate of Potash—2 to 5 table spoonfuls.

Water —¼ gallon.

(f) Application of lime-sulphur in the seed-bed at a concentration of 1 in 40, yields good control of the disease and at the same time causes no injury to the host plants.

(g) Application of summer spray with Bentonite sulphur prepared by fusing fluid sulphur is very effective. This fungicide does not produce any caustic effect to the host tissue and is more adhesive than any other sulphur. Moreover it is cheaper and is not washed away readily.

LOOSE SMUT OF WHEAT

Loose smut of wheat is the most early recognized of all wheat diseases because of the characteristic dusty appearance of diseased heads, an appearance that has given it such common name as **smut, black head, blasted head, snuffy ear** and **black smut.** An important difference between this and other smuts of wheat is that the infection is carried over from season to season within the seed and not as spores on the surface of the seed. The incidence of the disease is common in almost all the wheat-growing regions of the world. Prior to 1888 not much differences were indicated between the loose smuts of wheat, barley and oats. It was Maddox in 1895 who first showed the nature of loose smut of wheat. Maddox's work was later confirmed by Brefeld in 1903.

Symptoms. Loose smut is recognizable as soon as the affected head emerges from the leaf sheath, usually about the same time as, or very often in advance of, the heads of healthy plants. Loose smut of wheat is easily distinguishable from stinking smut and flag smut of wheat being characterized by the complete destruction of spikelets, whereas stinking smut is confined to the kernels, leaving the glumes intact, and flag smut attacks the leaves and culms. The smutted head consists of deformed spikelet filled with black, dry, powdery masses of spores, known as **brand spores** or **chlamydospores;** at first covered by a delicate membrane, which soon bursts and exposes the powdery spores. The glumes and kernels are completely disintegrated and soon after emergence the wind blows the powdery mass of spores leaving only the bare rachis which is not attacked by the fungus. Usually all spikelets in a head are destroyed but occasionally only a part of a head is affected. Ordinarily all the heads on a plant are affected, and while the smut is confined mostly to the ears, sometimes dark streaks of spore formation may occur also on the leaves and less often on the stem. In dry weather the spores are blown in clouds throughout the crop when the healthy heads are in bloom. During intermittent brief intervals when the spikelets are open for pollination, infection of the flowers takes place.

The Causal Organism. Loose smut of wheat is caused by the fungus *Ustilago nuda* var. *tritici* Schaf. The chlamydospores are pale-olive, more or less spherical in shape or occasionally oval, minutely echinulate, and measure 5 to 9 μ in diameter. They germinate by germ tube which develops to become a promycelium of four uninucleate cells. No sporidia are produced. The spores retain their vitality for 5 or 6 months, a much shorter period than in the other cereal smuts. The fungus is not capable of a saprophytic life in the soil.

Note: The name for the fungus causing loose smut of wheat is mostly a matter of taxonomic opinion rather than of nomenclature. For quite some time the fungus was named as *U. tritici* (Pers.) Rostr.

Disease Cycle. The spore masses on smutty heads are broken up and scattered by wind, rain, and other agencies just at the time when the healthy heads of adjacent plants are in flowering stage. Innumerable spores lodge between the glumes and reach the feathery stigmas. Spores caught up on the stigmas of the flowers, germinate on the moist stigmatic fluid, put forth promycelia. The promycelium becomes septate, but the dikaryophase is established by fusion of germ tubes derived from the individual unincleate cells of the promycelium and ultimately infection hyphae are produced. Spores are short-lived, rarely survive more than a few days under normal conditions. Spore germination occurs in moisture and only on the feathery stigmas of normal wheat flowers. It was earlier thought that the point of entry of the infection hyphae was the stigma of healthy flowers, but it has been shown that the normal entry point is the young tissue at the base of the ovary. After entering in the ovary the infection hyphae pass through intercellular spaces and through the integuments and ultimately reach the ovule. The infection hyphae ultimately pass into the space between the endosperm and the nucellus and move round the bottom of the endosperm to reach the scutellum and the embryo. Growth of the infection hyphae is exclusively intercellular, there are no haustoria, and the host cells are not affected in the slightest degree by the presence of the parasite. Simultaneously with this process, pollination, fertilization and embryo formation take place in the heads. The fungal mycelium continues to grow along with the development of seeds. Fungal infection in no way hampers the grain formation. The fungal mycelium becomes somewhat thick-walled, remains dormant in the seed, and from superficial examination at harvest time, the infected grain cannot be distinguished from the perfectly normal grain. In the matured seed there is copious mycelium in the scutellum. The mycelium remains dormant in the seed, until the following season. When an infected wheat seed is sown, it begins to sprout, the dormant mycelium resumes activity, grows into the young shoots up to the growing points and keeps pace with the development of the plant. As each head forms, there is a great accumulation of hyphae which ultimately replace the spikelets with masses of dusty spores. The spores are ready to infect the flowers of healthy plants. The spore masses are covered by a delicate silvery membrane which bursts usually before the head emerges from the sheath, forming very dark olive-brown powdery masses in place of the spikelets, of which only the ends of the awns ordinarily escape transformation. The spores separate easily and in dry weather may be almost all blown off, leaving a bare stalk (the rachis) behind. In dry weather, clouds of spores are thrown into the air. In dry weather again the glumes of healthy flowers are rather wide open and the stigmas exposed on which the spores lodge and behave in the manner already described. A diseased seed or plant cannot be distinguished from a healthy one until the plant begins to develop heads. As such, it is a case of seed-borne systemic infection and the fungus perennates as dormant mycelium in the seed. The severity of infection is influenced by moisture and temperature prevailing during the dissemination and germination of spores, and the duration of flowering stage. In many regions the loose smut of wheat is of little economic importance because it occurs in small amounts. Again, there are areas where the damage is considerable since the infection is very severe.

Disease cycle of Loose Smut of wheat is presented in Figure 25.21.

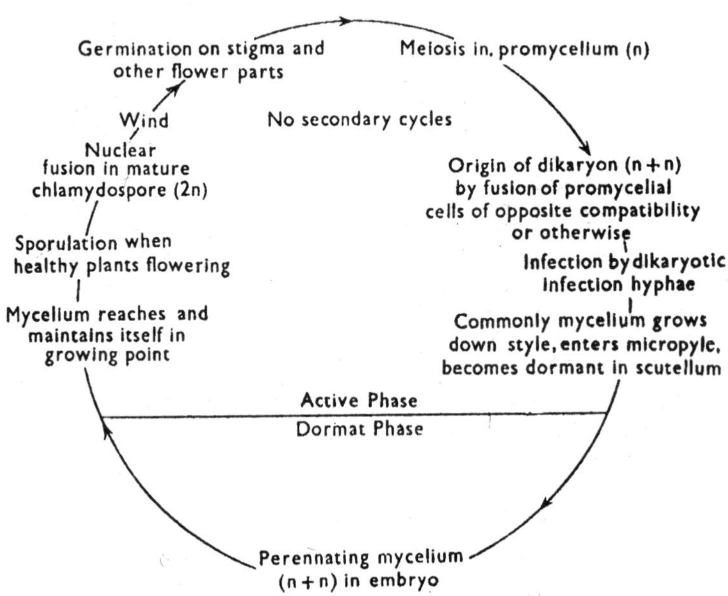

Fig. 25.21: Disease cycle of **Loose Smut** of wheat

Control. Seed treatment with fungicides in the form of dips or dusts as employed with other diseases of cereals is of no value, since the infection is internally see-borne as a dormant mycelium. An effective method must kill the internal mycelium. The only effective method thus far developed for controlling loose smut of wheat is the **hot water seed treatment.** This consists essentially of soaking the seeds in hot water for a sufficient time, the object being to destroy the infection within the seeds without harming the embryos. The exact method of providing the hot water and handling the seeds can be variable according to convenience. The hot-water seed treatment method requires great care since by the same treatment mycelium will be destroyed but the embryo should be saved from the heating effect. The embryo being a very delicate structure special care should be taken at every step. The recommended procedure will be: (i) carry out the treatment in small lots of seeds, (ii) soak the seeds in cold water for 4 to 6 hours to ensure complete wetting of all the seeds, (iii) immerse the seeds in warm water at 120°F for one minute for a brief dip, (iv) finally immerse the seeds in water held at 129°F for 10 minutes or within a range of 124°F to 130°F, (v) remove the seeds from warm water and dip in cold water to stop the action of heat, (vi) spread out seeds in a thin layer to dry quickly to avoid moldiness or deterioration, and (vii) sow the seeds or dry thoroughly and store for later use. Since even good sound seed has from 6 to 10 per cent of the grains injured by the heat, it is necessary to increase the sowing rate. Bad seed may suffer more heavily. Besides hot water seed treatment the disease may, however, be controlled by having special seed-raising areas being protected from the prevailing winds at flowering time to reduce the risk from blown spores, and no wheat fields should be closer than 500 yards. Field of other crops and surrounding trees acting as windbreakers often will help to reduce external infection. The seed-plot should be large enough to allow for loss in cleaning and selecting the seed and still provide enough for the field sowings. Growing a crop in a dry area for a season, where the conditions of humidity would be so low as to inhibit spore germiantion at the flowering time may yield infection-free crop without seed-treatment. It is always recommended that seeds for sowing should be procured from certified disease-free supplying concern.

Some claims have been made from Punjab that loose smut of wheat can be controlled by a **sun heating method.** In Punjab, where summer temperatures are usually very high, sun heating treatment is in practice with great amount of success. The seeds are spread in shallow flatbottom open containers and covered with water having water-level about 2 inches above the layer of seeds. The containers are then placed in the sun for about five hours from early in the morning. The water is then poured off from the containers and the soaked wheat seeds are spread in sun to dry. The direct action of hot summer sunrays produces two effects: (i) the newly grown young hyphae developed from dormant mycelium get killed, and (ii) the seeds get dried at the same time.

The most effective control is, however, to produce and introduce resistant varieties. Great attention is being paid along this line with a considerable success. The whole programme of breeding resistant varieties often becomes complicated, since along with the new resistant varieties there develop fresh physiologic races of the causal organism almost simultaneously to defeat the whole purpose.

Loose smut of wheat can be effectively controlled by soaking wheat seeds in fungicides like Chlorex, Purex, Ceresan, and Vitavax.

COVERED SMUT OF BARLEY

As the name suggests, the spore masses are covered with membranes for a long time, often until harvest. This disease commonly occurs in the United Provinces, Punjab and some part of Behar where it causes considerable losses annually. The covered smut was first recognized as distinct from the oat smut in 1591 by Lobelius. In 1888 the causal organism was separated from the other smut on barley.

Symptoms. The diseased plants head at about the same time as healthy plants. They are recognizable by their blackened ears that emerge from the leaf sheaths. All the ears in a diseased plant and all the grains in a diseased ear are attacked by the disease. All the grains in a diseased ear are transformed into masses of spores which are held in place by persistent, tough, greyish-white membrane. The covered masses of spores are not released from their enclosing membranes until threshing time, unless the membrane is broken accidentally. Because the membrane is often prematurely ruptured, covered smut is difficult to distinguish from loose smut.

The Causal Organism. This disease is caused by *Ustilago hordei* (Pers.) Lagerh. The chlamydospores are dark-brown to charcoal-black in mass usually subspherical or spherical smooth-walled, and lighter coloured on one side. When young, the chlamydospores are dikaryotic and with maturity karyogamy takes place. Germination takes place readily in water or damp soil by the formation of a septate promycelium during which meiosis of the diploid nucleus takes place. A large number of sporidia are produced which multiply by budding, germinate by germ tubes, or fuse with each other producing dikaryotic condition.

Disease Cycle. When the masses of spores are broken open as the barley is threshed after harvesting, innumerable spores are released. Many of the spores lodge on healthy kernels and remain dormant until the seed is sown. When the barley seed begins to germinate the spores also germinate and infect the seedling along the epicotyl by dikaryotic infection hyphae. After the fungus has entered the seedling, its hyphae continue to grow with the shoot and eventually replace the grains by masses of spores. A warm, moist, acid soil favours seedling infection. The greatest number of seedlings are infected at a soil temperature range of 50°F to 70°F during the germination period.

Disease cycle of Covered Smut of barley is presented in Figure 25.22.

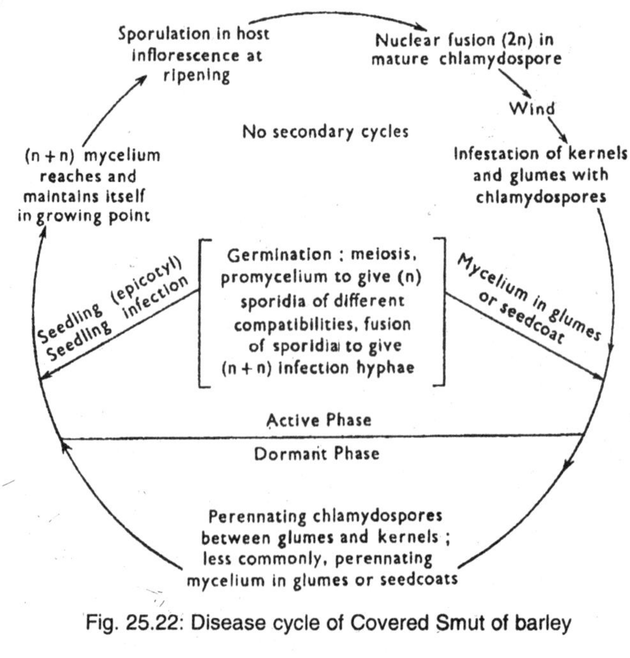

Fig. 25.22: Disease cycle of Covered Smut of barley

Control. Since covered smut of barley is an externally seed-borne disease, most effective control measure will be seed treatment with fungicides. Treatment of barley seeds with fungicides like Ceresan, Spergon, Agrosan successfully controls the disease. Use of disease resistant varieties also helps to reduce losses.

Sowing seeds in a moderately dry soil will aid in checking the disease, since moist soils are favourable for infection. Rate of infection is reduced when barley is raised in a neutral or alkaline soil than in an acid soil.

BUNT OF WHEAT

Bunt of wheat, also called stinking smut of wheat gets its name from the characteristic odour of decaying fish given off by infected ears. This is one of the most severe diseases of Middle Ages to attract attention. In 1755 Tillet was the first to establish distinction between the bunt or stinking smut and the loose smut. Prevost, in 1807, described the germination of the bunt spores and the production of sporidia. The true nature of the fungus was determined by Tulasne (1854), Kiilu (1874) and Brefeld (1883).

There are three principal bunts of wheat. They are: rough-spored bunt, caused by *Tilletia caries* (DC). Tul.; smooth-spored bunt, caused by *T. foetida* (Wallr.) Liro; and Karnal bunt, caused by *T. barclayana* Sacc. and Sydow. Only the rough-spored bunt will be discussed in detail.

This disease occurs wherever wheat is grown all over the world. Its severity varies greatly, however, in different localities. The damage done by bunt fluctuates from season to season and in various localities. Losses are attributable to two separate and distinct items, namely, actual percentage of decrease in yield and the quality of the threshed grain. Besides these, losses may be due to explosion of harvesting and threshing equipment instigated by the concentration of spores. Again, the species may produce allergic effect on man during handling of infected grains.

Symptoms. The disease is not evident until the wheat plant is in the heading stage. The diseased heads are usually slim compared with healthy one, and they retain their greenish colour longer. They stand more nearly erect than healthy heads, because of their lighter weight. Stunting of growth of diseased plants is shorter and plumper but lighter in weight than a normal wheat grain. The spore balls are recognizable by their colour which is darker than that of the healthy grains.

Diseased grains retain more or less the shape and size of normal kernels. When broken open the smut balls are found to be filled with black or dark-brown powdery substrate composed entirely of spores. The offensive odour announces the presence of heavy infections in the field.

The Causal Organism. *Tilletia caries* (DC) Tul. The chlamydospores are of various shades of brown and are black in mass. The spore wall has reticulations ranging from minute shallow meshes to deep indentations. The young chlamydospores are dikaryotic, the nuclei fuse in the mature chlamydospores. On germinating the chlamydospore sends out a non-septate promycelium during which meiosis takes place. The promycelium produces at the apex a cluster of filiform, hyaline 8 to 16 sporidia—the primary sporidia which frequently fuse in pairs to form characteristic H-shaped structures. These H-shaped structures germinate to produce mycelium on which sickle-shaped, hyaline, secondary sporidia are borne. The secondary sporidia develop infection threads which penetrate the young seedling.

Disease Cycle. Many of the spore balls are shattered during threshing and spores thus liberated lodge on the healthy kernels or are left in the soil and thereby carry the pathogen over from one crop to the next. This is how the pathogen perennates. With the return of favourable season the chlamydospores germinate to produce promycelium bearing primary sporidia where dikaryotic condition is established by the formation of H-shaped structures between the compatible sporidia. The H-shaped structures then germinate to produce mycelium from which secondary sporidia are produced. The secondary sporidia give rise to dikaryotic infection hyphae which penetrate the wheat seedlings. Infection takes place on the young seedlings before the first leaf emereges. Conditions suitable for infection are: (i) temperature of 25°C, and (ii) moderate amount of soil moisture. Following upon penetration of the coleoptile, the hyphae pass to the young shoot keeping pace with the growth of the host and cause little interference to it until the formation and development of the ears. Eventually hyphae accumulate in the ovaries, the cells of which are crushed and replaced by hyphal masses. A rapid transformation of the hyphal mass into spores takes place filling the grain internal to the pericarp to constitute a spore ball.

Disease cycle of Bunt of wheat is presented in Figure 25.23.

Control. In controlling the disease it should be carefully considered that the spores are either externally seed-borne or are in the soil where wheat seeds are sown. Some of the effective control measures are as follows:

(i) *Use of Clean Seeds:* Visibly spore infested seeds should be rejected. Seeds from fields where there was not bunt infection are always preferable.

(ii) *Seed Treatment:* Seed treatment with appropriate fungicides is very effective not only in destroying seed-borne spores but also in checking infection from soil-borne spores. Some of the commonly used fungicides are:

(a) Copper sulphate solution, 2 per cent, seeds to be dipped for 5 to 10 minutes.

(b) Fomaldehyde solution, 1 pint of formaldehyde in 40 gallons of water, seeds to be dipped for 5 minutes.

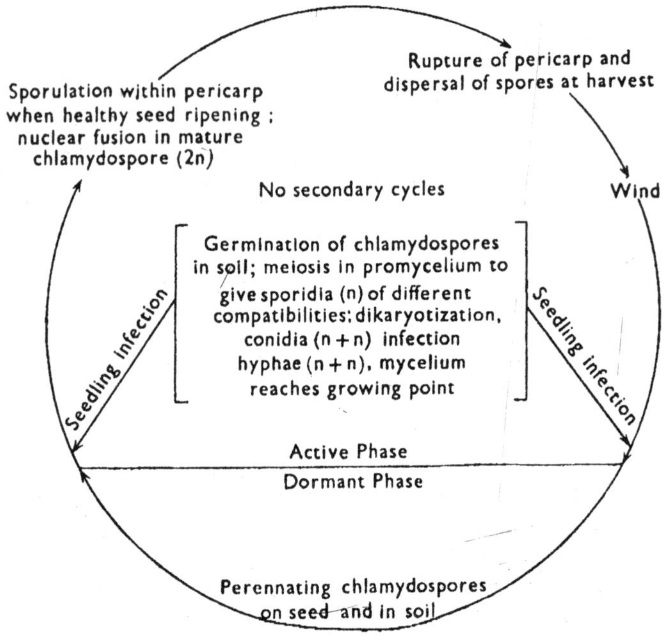

Fig. 25.23: Disease cycle of Bunt of wheat

(c) Dusting of seeds with copper carbonate.

(d) Dipping seeds in Bordeaux mixture of 4-4-50 to 8-8-50 strength for 10 to 15 minutes.

(e) Seed treatment with Ceresan M, Agrosan GN, Spergon, and Pangon has been found very effective.

(iii) *Crop Rotation:* Rotation of wheat cultivation with some other crop enables to make the spores in the soil ineffective.

(iv) *Cultural Practices:* Cultural practices that may reduce the rate of infection are:

(a) seed sowing in relatively dry rather than moderately moist soil,

(b) shallow rather than deep sowing in soil, and

(c) early sowing.

(v) *Use of Resistant Varieties:* Many satisfactory disease resistant varieties are now in use which has enabled to raise disease-free wheat. Continued effort along this line should be made to combat with the physiologic races that may upset the entire programme of breeding of disease resistant varieties. Event if resistant varieties are used, the seeds should be treated with a fungicide.

RED ROT OF SUGARCANE

This is one of the most severe of the known diseases of sugarcane. It was first described from Java by Went in 1893. It is widely distributed throughout the sugarcane-growing countries of the world, and in fact, it is extremely doubtful if there are any sugarcane-growing areas where it does not exist, although it may be much more destructive in some places than others. The disease was very widespread and virulent in North Behar and Eastern part of the United Provinces during 1939 and 1942. It was so destructive that it almost wipped out the sugarcane plantations in those areas.

Symptoms. The first external evidences of disease are the drooping, withering, and finally yellowing of the upper leaves. This is followed by a similar wilting of the entire crown, and finally the entire plant shows indications of disease and dies. When not severe, the eyes frequently die and blacken and the dead areas extend out from the nodes.

Infection in the stem being internal, the presence of the disease is not visible externally. Upon splitting a diseased cane during the early stages of the disease, it will be found that the fibrovascular bundles near the base are reddish in colour. The host tissue reacts vigorously to the presence of the fungus and some kind of reaction or change sets in the host cells in advance of the hyphal invasion. The protoplasm changes colour and a gummy dark-red material oozes out of the cells filling the intercellular spaces. The soluble pigment present in this ooze, is absorbed by the cell wall producing the characteristic red rot appearance. However, the presence of a red colour in the fibrovascular bundles is not necessarily an indication of this disease, since the colour may be due to any one of many other causes. As the disease advances the red colour spreads to the surrounding tissues extending through many internodes, and irregular discoloured blotches are formed, which may be reddish or yellowish or white with red margins (Fig. 25.24A). These white areas with red margins are a positive proof of the disease. When the stem is completely rotted inside, the natural bright colour of the rind disappears and turns dull as it shrivels. Black specks appear on shrivelled rind. The stem shrinks at the nodes (Fig. 25.24C). Split cane gives sour smell and shows red tissue with white cross-bands. About this time the upper leaves of the stem turn pale and gradually droop down. These leaves then wither at the tips and along the margins. Ultimately the entire plant withers and droop down. In areas where the disease appears in a severe epidemic form, the entire crop withers and droops resulting in a complete loss of crop.

Though the fungus attacks all parts of the host above ground, stems and midribs of leaves are more susceptible to fungal attack. Infection in the leaves is visible along the midribs as dark-reddish zones having tendency to elongate rapidly turning blood-red enclosed by dark margins (Fig. 25.24B). When the infection becomes old the central blood-red colour changes to straw colour. The hyphae after ramifying in the infected host tissue collect beneath the epidermis and form a stroma of densely packed cells and ultimately an acervulus is developed resulting in the rupture of host epidermis. The acervulus bears long septate setae along with short conidiophores on which falcate (sickle-shaped) conidia are borne (Fig. 25.24D to F). After growing for a period within the host tissue, the hyphae produce a large number of chlamydospores in the pith parenchyma. The clamydospores persist in the soil for a long time.

An examination of the diseased tissues with a microscope will reveal more or less mycelial threads of the fungus, or if the diseased canes are split and put in a moist chamber the fungus will develop readily and be easily recognized.

The Causal Organism. Red rot of sugarcane disease is caused by *Colletotrichum falcatum* Went, the perfect stage of which is *Glomerella tucumanensis* (Speg.) Arx and Muller. There has been considerable difference in opinion as to the nature of the fungus that causes this disease. Some insisted that this fungus is more strictly saprophytic than parasitic, and that it cannot attack healthy canes. Others said that it cannot attack mature canes except through wounds, but that it can attack young plants. However, the young canes are usually protected by the leaf sheaths. In some places the fungus has been reported to grow on the dead canes only and the disease is not known. The mycelium of the fungus grows both inter- and intracellularly in the parenchymatous cells of the host tissue. The hyphae are colourless, slender, freely branched and septate. Acervuli appear just above or below the nodes along the depressions or ridges. They are black velvety bodies, develop in clusters. Acervuli are cuspidate with irregularly arranged setae (Fig. 25.24D). Aseptate conidiophores 20 μ long and 8 μ wide, on which one-celled falcate conidia are borne. Conidia are 16 to 48 μ long and 4 to 8 μ broad. They bear large oil globule in the centre. Chlamydospores are terminal or intercalary.

The perfect stage was reported from India under cultural condition in 1952 and under natural conditions on sugarcane leaves in 1953.

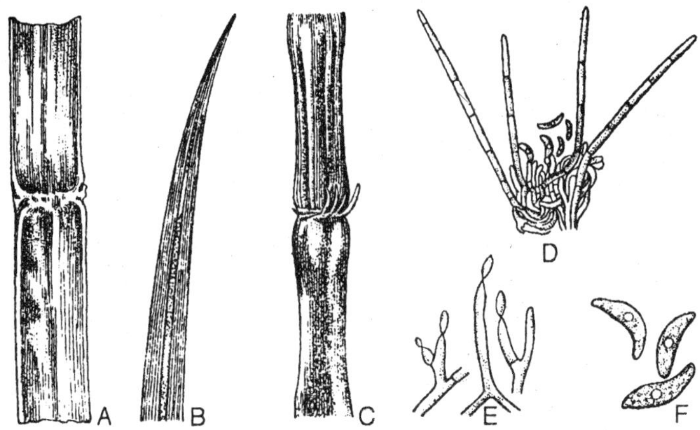

Fig. 25.24: Red Rot of sugarcane. A-C. Disease symptoms. A and C. On stem. B. On leaf, D. An acervulus. E. Conidiophores producing conidia. F. Conidia

It comprizes of perithecia which are globose superficial with bottom embedded in the host tissue. Asci are numerous, clavate and paraphysate bearing 8 ascospores which are aseptate, hyaline and elliptical.

Disease Cycle. The sources of primary inoculum are the old fragmented stalks and leaves and other rubbish on which the fungus grows saprophytically; and unknowingly planted diseased stock during cultivation. Ratoon crops also serve as a source of primary inoculum. Opinions differ whether the fungus is strictly saprophytic or parasitic.

The conidia that are produced in the acervuli along the midribs of the diseased leaves during primary infection, form the secondary inoculum. They are disseminated by wind, rain splashes, irrigation water and also by insects. The conidia germinate readily by germ tube which on coming in contact with any hard surface, e.g., soil particles or plant parts forms appressorium from which infection hypha is produced. The pathogen may gain entrance through the nodes at the leaf scars, through any kind of wound, through root primordia and seed-cuttings. The diseased canes are frequently found to be injured by insects, especially borers, and no doubt these wounds facilitate the entrance of the fungus, which in turn does much more damage than the insects. Red rot is not a root disease, though roots are often infected by the fungus. High humidity due to water-logging, weak growth of host plant for want of proper cultural operations, continuous cultivation of the same variety of sugarcane in a particular locality, and cultivation of susceptible cane variety in the neighbouring areas are some of the aspects that help disease incidence and often to epiphytotics.

Disease cycle of Red Rot of sugarcane is presented in Figure 25.25.

Control. Red rot of sugarcane is hard to control because the stalk from which seeds are prepared has been largely affected from the time of planting, and fungicides cannot reach the infected tissues inside a diseased seed sett. Therefore, careful selection of red rot-free seed setts is recommended for planting. Seed should always be taken from disease-free nurseries examined regularly by the cane protection staff. Before planting, each seed sett should be carefully examined and those setts which show reddening should be discarded.

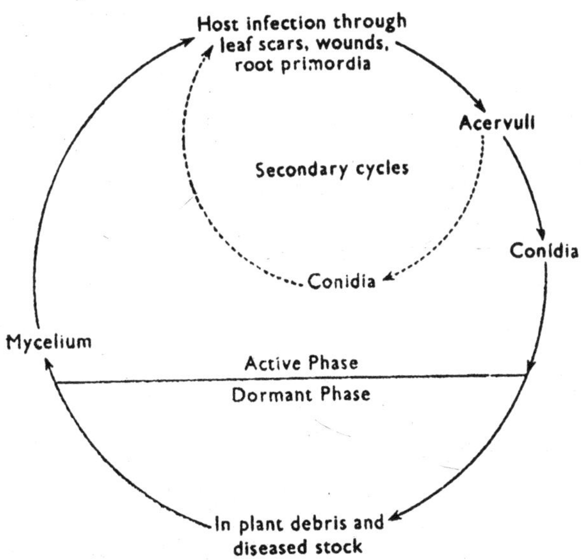

Fig. 25.25: Disease cycle of Red Rot of sugarcane

The spread of the red rot can be prevented during the growing season by timely rogueing and burning of the affected clumps with utilization of the green leaves for cattle fodder. In no case ratoons of sugarcane should be kept in the red rot affected fields. Attention should always be given to sanitation by digging out stubbles of diseased canes and burning them with other trash in the field.

Where facilities are available for hot water treatment of seeds, they can be utilized for controlling red rot of seed (treat in water at 50°C, for two hours). Treating seed wth fungicides like Arasan (0.25 per cent) is often effective.

The use of sugarcane varieties resistant to red rot is also recommended. Some of the resistant varieties are: Co. 975, 1148, 1158, 1336 and 6611; Co. S 561, 574; B.O. 3, 10, 47.

The possibilities of an epidemic is very much minimized with the practice of long crop rotations (2 to 3 years) where planting is done in plots.

One of the best ways to reduce the incidence of the disease is to raise healthy stock for planting in plots especially fertilized, cultivated, and kept disease-free by constant care.

NIGROSPORA DISEASE OF RICE

This disease of rice has widespread occurrence in the rice-growing tracts of the world. The occurrence of the disease in India has been reported by Thomas (1937), Padwick (1938), Padmanabhan (1949), Chattopadhyay and Sen Gupta (1952) and Kar (1966). A similar condition of rice plants has been recorded from other countries; Thomson from Malaya, Kimura from Japan and Martin from the United States.

Symptoms. Dark streaks and powdery spots of variable dimensions are visible on the surface of leaf blades, sheathing leaf petioles, and the glumes of grains (Figs. 25.26A-C). A similar diseased condition is also seen on young leaves which remain partly enclosed within the sheathing petioles, and on immature grains of rice. Closer examination reveals that the streaks are minute dark pustules developed in linear series and coalescent pustules form the powdery spots (Fig. 25.26B). The pustules are more prevalent along furrows than the ridges of the host surface. The dark pustules are powdery masses of conidia developed by the mycelium of the pathogen that has already gained entrance in the host tissue.

The Causal Organism. The disease is caused by the fungus *Nigrospora oryzae* (Berk. and Br.) Petch, 1924. The fungus possesses characteristic septate mycelium which is hyaline when young, becoming dark with age. The ultimate branches of the mycelium bear conidiophores either laterally or terminally. The condiophores are short, simple, and inflated below the tip; exhibiting a jar-shaped appearance (Fig. 25.26F). Each conidiophore produces apically a solitary conidium. The conidia are dark, one-celled, globose to subspherical, and smooth-walled with distinct dark pigment. Dimensions of conidia range from 10.5 to 14 μ by 11.5 to 15 μ.

Behaviour of pathogen within the host. The infection is very superficial, specific, and localized. The pathogen gains entrance through stomata. The activity of the pathogen is restricted particularly along the outer layers of the parenchyma with very little tendency to penetrate deeper (Fig. 25.26D). The mycelium ramifies along the parenchymatous tissue and produces sporodochium-like bodies after rupturing the host tissue (Fig. 25.26E). The sclerenchyma adjoining the parenchyma remains unaffected. The pathogen induces mild infection resulting little damage to the host tissue. The conidia, conidiophores, and the mycelium produce an unsightly appearance of the host surface.

Control. Since the disease causes very little damage to the rice crop one can get by without

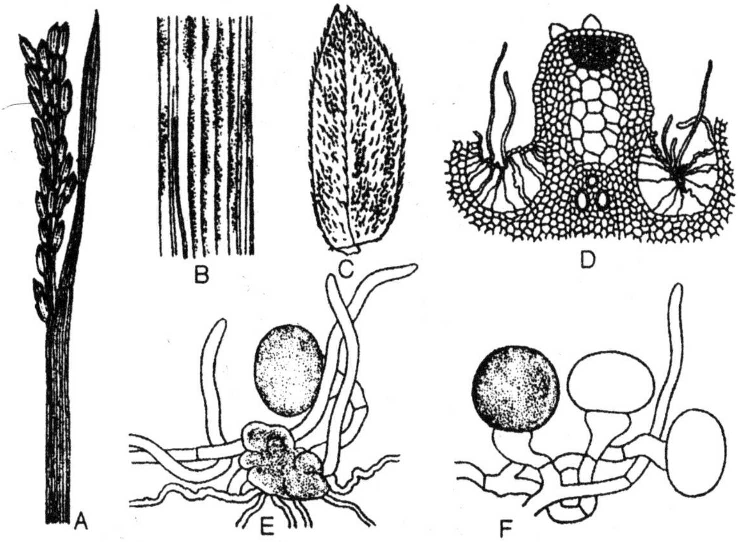

Fig. 25.26: Nigrospora disease of rice. A-C. Disease symptoms. A. On grains, leaf sheath and leaf. B. On leaf sheath, C. On grain, D. Host infection, E. Sporodochium-like structure, F. Conidia borne on conidiophores

taking any notice of it. Still it is always advisable to use disease-free seeds.

BROWN SPOT OF RICE

Brown spot of rice (*Oryza sativa* L.), also known as Helminthosporium disease of rice is distributed in almost all the rice-growing tracts of the world. But it occurs most severely in South-East Asia, Japan, the Philippines, and the Southern parts of the United States. In India, the disease was first reported in 1922 and since then it has been recorded to cause damage to rice crop in eastern and southern rice-growing regions of the country. It occurs in an epiphytotic form in States of Andhra Pradesh, Assam, parts of Kerala and Mysore, and West Bengal.

Besides causing heavy leaf spotting, the severity of the disease may reduce yield when ears may be sterile and grain formation may be stopped. The infected seeds have very poor germination rate.

Symptoms. The disease attacks all parts of the host, usually excepting the roots, in all stages of development. There may be complete failure of seed germination or the seedlings may succumb to the disease before they emerge out of the soil. In emerged disease affected seedlings, necrotic lesions appear in the coleoptile when its tip turns brown to dark-brown and the infection spreads to the mesocotyl and ultimately the seedlings are completely destroyed. It has been reported that in infected rice seedlings, the pathogen produces a toxin (Cochliobolin) which inhibits the growth of the roots and affects respiration of leaves.

The most conspicuous symptom of the disease is the appearance of discrete, dark-brown, ellipsoidal to eye-shaped spots on the upper surface of the leaves (Fig. 25.27A). The spots often vary in size and shape. They appear in the form of very small brown areas surrounded by a yellowish halo and gradually increase in size turning dark-brown. Along with the increase in size, the spots also increase in number very rapidly. A fully developed spot has a greyish-brown central portion surrounded by a deep, reddish-brown margin and may be as big as 4-6 mm in dimension. Some of the spots coalesce to form irregular patches of brown discolouration. When the infection is severe, the affected leaves become greyish-brown and show considerable drying and wilting causing severe damage to rice crop. The

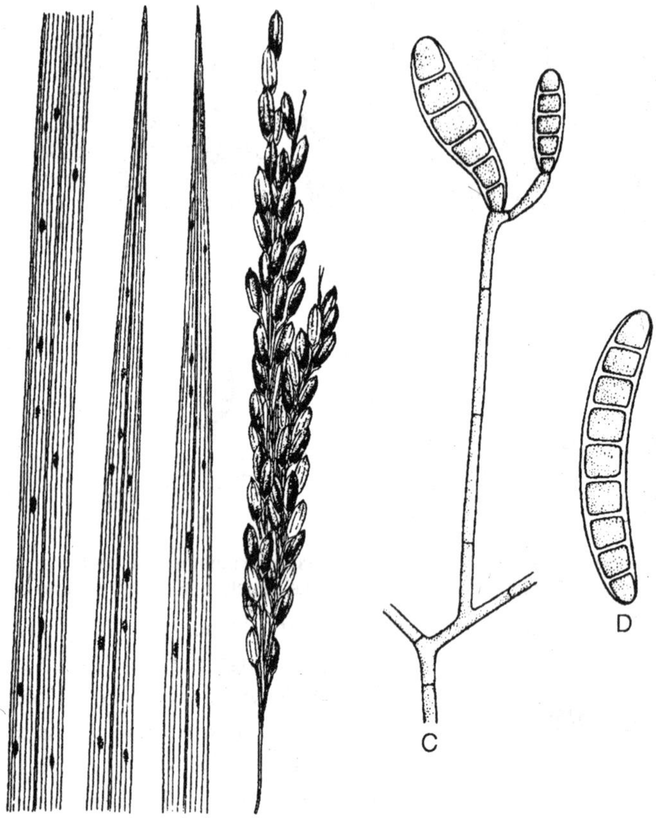

Fig. 25.27: Brown spot of rice. A-B. Disease symptoms. A. On leaves, B. On grains, C. Conidiophore bearing conidia, D. A conidium

symptoms also develop on leaf sheath and culm which turn dark-brown with velvety surface for being covered with the conidiophores of the pathogen. In case of severe attack the ears fail to emerge from the sheaths and perish without further development. If the ears have already emerged from the sheaths, they become distorted. The disease may also cause damage to the grains. Lesions on the glumes first appear close to margins of the outer and inner glumes and gradually dark-brown spotting spreads over the entire surface (Fig. 25.27B) causing brown discolouration of the kernels inside. The quality of grain is also affected. Depending on the stage of development of the grains, there may be complete sterility or partial filling up of the grains. In case of severe infection no grain formation takes place and the ears become sterile.

The Causal Organism. The disease is caused by *Helminthosporium oryzae* Breda de Haan, the conidial stage of *Cochliobolus miyabeanus* (Ito and Kuribay) Drechsler ex Dastur.

The fungus produces septate mycelium which ramifies inter- and intracellularly in the host tissue. It produces erect rigid conidiophores which arise in tufts through the stomata and at times through the ruptured epidermis or wounds. The conidiophores are short or long, septate, simple or branched, dark-brown to olivaceous at the base and somewhat paler at the growing tip. Under suitable atmospheric condition, mycelial mat may be developed externally over the surface of the spots, and conidiophores may also arise from such mycelial mat. The conidiophores produce conidia successively on new growing points (Fig.

25.27C). The points of attachment of successive conidia are marked by scars at regular intervals on the conidiophores. Conidia are sligtly curved tapering towards the rounded ends and are only transversely septate. They are olivaceous brown and are extremely variable in size and number of septation (Figs. 25.27C and D).

The perfect stage of the causal organism has been obtained under cultural conditions in Japan and in the disease stubbles in Mexico, but has not yet been found in India.

Disease Cycle. The pathogen perennates both externally and internally in the seeds (as conidia externally, and internally as mycelium), as well as in straw and stubble. It has also been reported that two graminaceous plants *Leersia hexandra* and *Echinochloa colona* are the collateral hosts. Besides these, some of the rice field weeds (*Cyanodon dactylon* and *Setaria italica*) often help in the perennation of the pathogen. The soure of primary inoculum is conidia on the surface of the seeds and mycelium inside the seed. Along with the germination of the diseased seeds the perennating organs (conidia and mycelium) of the causal organism become active and cause infection on the seedlings. This is the starting point of the disease. Continuous rainy days accompanied with cloudy weather and temperature range of 28-30°C are ideal for the development and spread of the disease. The disease is disseminated by wind and rain. Entrance of the pathogen in the host tissue may be by direct penetration of host epidermis or through stomata. Conidia on germination produce germ tube which is surrounded by a thick mucilaginous substance which enables the germinating conidia to remain adhered to the host surface. The germ tube then swells to produce an appressorium from which infection hyphae are developed. The infection hyphae then enter the host tissue either through stomata or by rupturing through epidermis. Immediately after the entrance in the host tissue, the infection hyphae produce more hyphae which ramify along the intercellular spaces of the mesophyll. Disease symptoms usually appear within 24 hours of infection. Maximum development of sports is in shade and with relative humidity of 92 per cent or more. Sunlight retards the growth of the spots. Under favourable condition, hyphae produce conidiophores which emerge out through stomata of the host tissue. The emerging conidiophores produce a succession of conidia which are disseminated by wind and rain and on coming in contact with susceptible host induce infection. The air-borne conidia arising from primary infection produce secondary infection of the leaves and grains. Host plants become more susceptible to infection.

Deep sowing resulting in slow emergence of seedlings favours high pre-emergence loss and good seedling infection. Plants growing under conditions of deficiency of nitrogen are extremely susceptible to infection. Susceptibility of infection of the plants also increases when grown in potash deficient soil. Rice plants are most susceptible to the disease at the flowering stage.

Disease cycle of Brown Spot of rice is presented in Figure 25.28.

Control. Since the pathogen is both internally and externally seed-borne, it is very difficult to make the seeds absolutely disease-free. The rate of disease incidence can, however, be reduced by the following control measures:

1. Destruction of diseased stubbles and other plant parts by burning.
2. Eradication of collateral graminaceous and rice field weed hosts.
3. Seeds should be obtained from an area where no disease incidence has been reported.
4. Seed treatment with fungicides like Agrosan GN (3 gm in 1 kg seeds) or Ceresan or with Aureofung in followed by foliar sprays with copper sulphate.

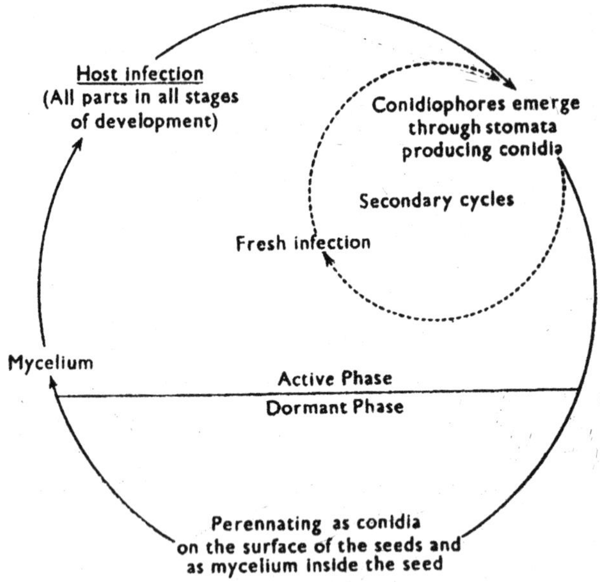

Fig. 25.28: Disease cycle of Brown Spot of rice

5. Soaking seeds for 48 hours in 1 : 1000 dilution of Uspulan reduces disease incidence.
6. As the pathogen is sometimes internally seed-borne, hot water treatment at 55°C for ten minutes and presoaking in cold water for 8 hrs gives better control than chemical seed disinfectants.
7. Seeds germinated in antibiotics Nystatin and Griseofulvin give some resistance to the seedlings though these antibiotics are often toxic to the seedlings.
8. Cultural practices like the application of optimum nitrogen fertilizer and sowing and transplanting at optimum dates reduce the intensity of the disease.
9. Dissemination of secondary inoculum can be controlled by spraying fungicides like Bordeaux mixture (5 : 5 : 50) or 0.2 per cent. Dithane Z-78 at regular intervals. Dusting with organo-mercuric dusts containing 0.1 per cent mercury often produces effective results. Spraying or dusting operation has to be done 3 to 4 times between tillering and emergence of inflorescence.
10. The use of resistant varieties is by far the best way to control the disease incidence. As resistance is controlled by a larger number of genes, it is difficult to obtain completely resistant varieties.

Rice varieties like CH 12, CH 45, T-141, T-298-2A, Co-20, BAM-10, T-998, T-2118 and T-960 were resistant to brown spot of rice.

TIKKA DISEASE OF GROUNDNUT

The **Cercospora leaf spot** of groundnut (*Arachis hypogaea* L.), commonly designated as **Tikka disease,** is one of the serious diseases of groundnut. It has been reported from almost all over India wherever groundnut is cultivated and is often responsible for the reduction in yield to the extent of 20 to 50 per cent, the disease incidence causes defoliation and early maturity, and thereby size and quality of nuts are affected.

Symptoms. All parts of the host plant above soil level are attacked by the disease. The first visible symptoms appear on the leaflets of lower leaves as dark spots which at a later stage, are surrounded by yellow rings. The spots are circular. They appear in a large number on the leaves. Mature spots are dark-brown to almost black, particularly on the upper surface of the leaflets (Fig. 25.29A), whereas on the lower surface they are lighter in colour. The spots are few on the leaf petioles and stem. Sometimes spots coalesce resulting in the defoliation. The shedding of leaves is a characteristic feature of the disease. Due to excessive spotting and consequent leaf fall, smaller and fewer nuts are formed. In cases where young plants are attacked by the disease, nuts fail to develop in them. But the mature plants when attacked by the disease produce immature nuts which are shrivelled and become loose in the shell. The total effect is the loss in yield.

The Causal Organism. The spotting is due to the attack of *Cercospora personata* (Berk. and Curt.) Ell. and Ever., the conidial stage of *Mycosphaerella berkeleyii* Jenkins; and *Cercospora arachidicola* Hori, the conidial stage of *Mycosphaerella arachidicola* Jenkins.

Cercospora personata possesses mycelium which is entirely internal and ramifies intercellularly by developing haustoria in the palisade and spongy mesophyll cells of the host. The mycelium forms dense stroma which produces long septate to non-septate geniculate hypophyllous conidiophores (Fig. 25.29B). The conidiophores emerge in tufts by rupturing host epidermis. Conidia are pale-brown, obclavate or cylindrical, septate, measuring 30-50 μ in length and 5-6 μ in breadth (Figs. 25.29D and E).

Cerospora arachidicola has both internal and external, intra- and intercellular mycelium without haustoria. The mycelium produces scanty stroma (Fig. 25.29C). The conidiophores are usually amphigenous, but on the younger spots they are developed exclusively on the upper surface. They are geniculate, non-septate to septate and produce hyaline to slightly olivaceous, obclavate 4- to 13-septate, often curved, conidia measuring 38-108 μ in length and 2-5 μ in breadth (Fig. 25.29D).

A ready and reliable means of distinguishing the spots produced by these two fungi are: spots due to *C. personata* are more circular and smaller than those produced by *C. arachidicola*.

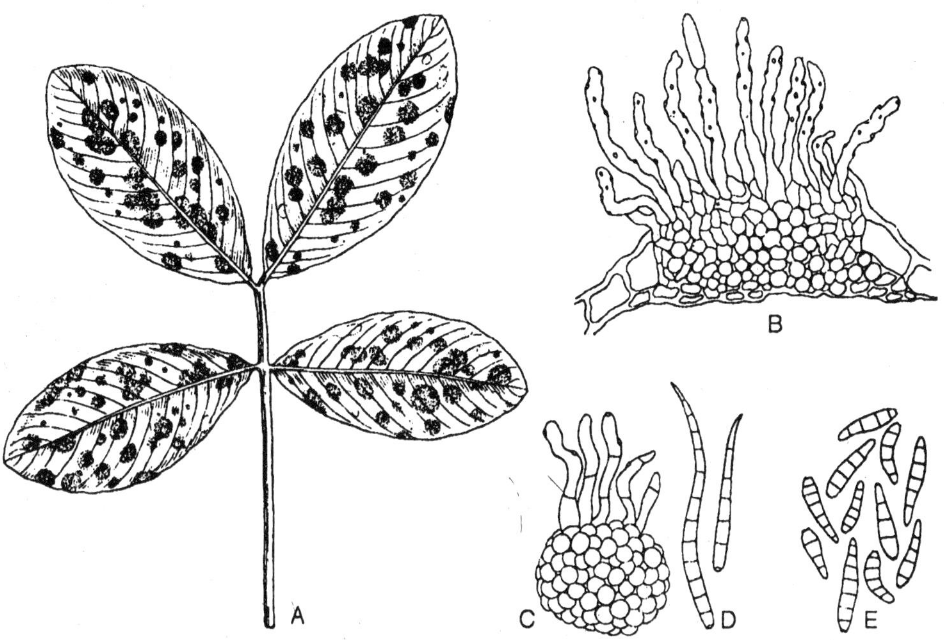

Fig. 25.29: Tikka disease of groundnut. A. Disease symtoms, B. Section through infected host tissue showing geniculate conidiophores, C. Stroma bearing conidiophores, D. and E. Conidia

These are the typical 'Tikka' disease spots which almost cover the entire leaf surface in an epiphytotic condition of the disease. Again in the latter a yellow halo is present around spots from the beginning while in the former halo develops at maturity of the spots. Besides this, the colour of the spots on the lower surface of the leaflets produced by *C. personata* is carbon black and those produced by *C. arachidicola* are light-brown. Although the spots produced by *C. personata* appear late, yet they are more dangerous than those of *C. arachidicola* being rapid in their development resulting in severe epiphytotic.

The perfect stage of both these species of *Cercospora* has been reported from the USA, but not yet from India.

Disease Cycle. The pathogen perennates through conidia on diseased plant debris lying in the soil. The conidia may also remain adhered to shell. They have also been found to remain associated with the seeds and are responsible for primary infection. A temperature range of 26°C to 31°C with high atmospheric humidity is favourable for disease development. Prolonged low temperature and dew also favour infection. The entrance of the pathogen in the host tissue takes place either by direct penetration through the epidermal cells or by way of stomata.

The leaf infection is largely through the upper surface of the leaflets. The fungus mycelium ramifies the host tissue in and around the infection court and aggregates underneath the epidermis and forms stroma. During the development of stroma the epidermis is ruptured by the pressure developed in the host tissue and conidiophores developed from the stroma emerge out, ultimately conidia are produced on them. These conidia form the secondary inoculum through which secondary infection is induced.

The disease is disseminated by wind which blows the conidia from leaf to leaf. Insects and splashes of rain have also been reported to play role in the dissemination of the disease.

Application of nitrogen and phosphatic fertilizers often makes host plants susceptible to infection.

Disease cycle of Tikka disease of groundnut is presented in Fig. 25.30.

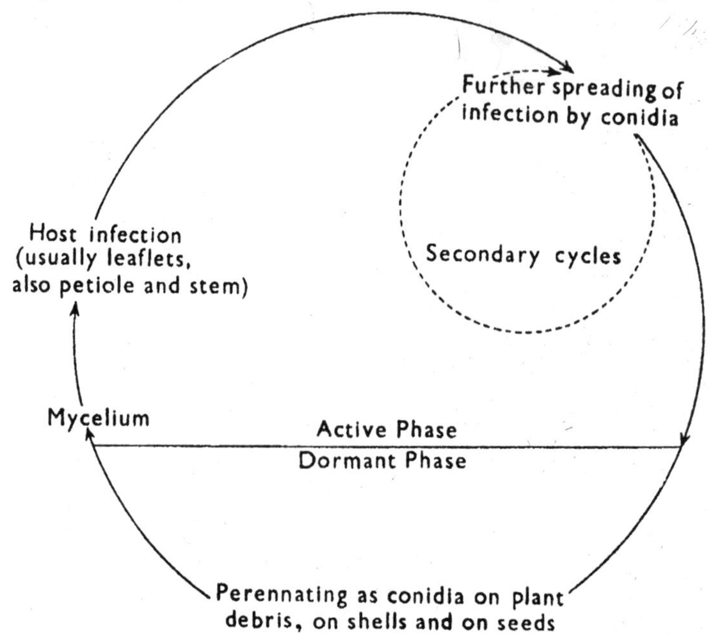

Fig. 25.30: Disease cycle of Tikka disease of groundnut

Control. Following are some of the suggested control measures of the disease:

(i) Burning of previous year's diseasd plant debris will, to a great extent, reduce the source of primary infection.

(ii) Two to four years' crop rotation often cuts down the rate of infection.

(iii) Avoid late sowing to reduce infection rate.

(iv) Seed disinfection checks the disease incidence. Care should be taken to remove the shells thoroughly and the adhering soil before seed disinfection. Treatment for half an hour in 0.5 per cent copper sulphate solution is recommended for seed disinfection. Agrosan GN also is an effective disinfectant. Seed dressing with Thiram (1 : 350) or Flit 406 (1 : 500) before sowing prevents *Aspergillus* seed rot and pre-emergence losses.

(v) The spread of secondary inoculum can be controlled by spraying the crop with different fungicides. Three sprayings of Bordeaux mixture (4 : 4 : 50) with linseed oil as sticker at an interval of 15 days are effective in checking the disease. A similar treatment with Dithane Z-78 has also produced good results. Spraying with 0.15 per cent Cupravit or 0.15 per cent Perenox also produces effective results. To obtain best results, timely and thorough spraying of the crop should be done. Care should be taken that under-surface of the leaves is thoroughly sprayed.

(vi) Dusting with sulphur (six applications at 10 days interval) or with sulphur containing 3.5 per cent metallic copper has been found to be very useful for controlling the disease.

(vii) The use of resistant varieties is the most effective means to combat the disease.

Varieties resistant to leaf spot are G-0120, G-1032 and G-0607.

EARLY BLIGHT OF POTATO

This is one of the most common foliage diseases of potato. The disease is designated as 'early blight' to distinguish it from the late blight caused by *Phytophthora infestans.* But it is also known as 'leaf spot' and 'Alternaria blight'.

This disease was first recorded in the United States in 1891 and has since been reported from Mexico, Canada, from all the other continents and from Java, Bermuda, Australia and New Zealand. It probably occurs in all potato-growing countries of the world.

Symptoms. The disease is characterized by the appearance of spots on the leaflets. The spots are dark-brown to almost black and are ovate or angular in shape usually 3 to 4 mm in diameter. They are circular to angular dead areas with concentric rings or ridges producing a characteristic target-board effect (Figs. 25.31B and C). There is usually a narrow chlorotic zone around the spot, which fades into the normal green. The spots may be a few in number and small or they may be numerous and involve a large part of the leaf area, owing to the coalescing of adjacent spots. When spots are numerous, the general effect on the leaflet is such as to make it go into a senescent condition. Lowest leaves are affected first, and the disease progresses upward. Affected leaves may show ragged margins or irregular margins or irregular breaks, or, sometimes, perforations due to the breaking away of the dead tissue and sometimes more or less curling or rolling. Under certain conditions, the spots may remain small without the target-board effect, being limited by some of the smaller veins. In severe cases of infection a complete blighting of the affected leaves may be resulted, preceded by the yellowing of the tissue between the spots.

On potato tubers, the lesions are slightly sunken, circular to irregular in shape (Fig. 25.31A).

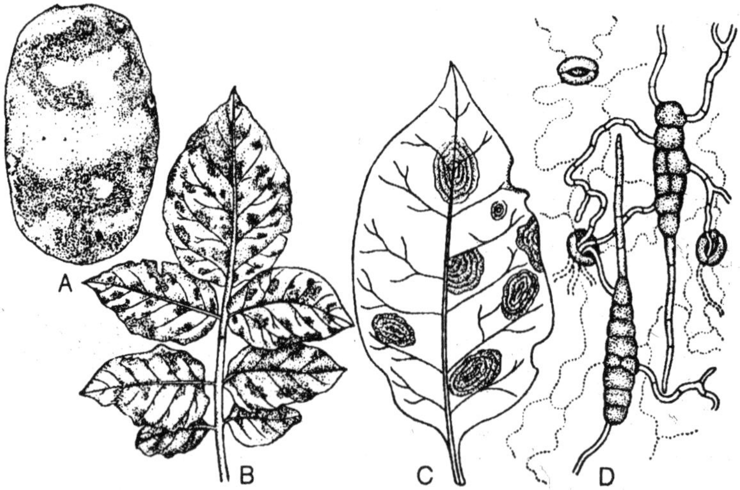

Fig. 25.31: Early Blight of potato. A-C. Disease symptoms. A. On tuber. B. On leaflets, C. On a single leaflet, D. Host penetration

Spots may develop on stem and affected plants turn yellow and dry up. The disease may cause heavy loss to the potato crop, often even heavier than late blight. With severe foliage injury, the tuber may remain small, immature, soft-skinned, deficient in starch having poor storage quality.

The Causal Organism. Early blight is caused by *Alternaria solani* (Ell. and G. Martin) L. R. Jones and Grout, a fungus belonging to the Deuteromycetes.

The fungus possesses septate branched mycelium which becomes dark-coloured with age. The mycelium usually, ramifies in the intercellular spaces and also penetrates cells of the host tissue. After the invaded tissue has been filled with mycelium, relatively short dark-coloured septate and erect conidiophores emerge from the stomata or arise from the older diseased host tissue. Conidia are beaked, muriform, dark-coloured, borne singly or in chains (Fig. 25.31D).

Disease Cycle. The source of primary inoculum for the primary infection in an early crop is either the conidia which remain viable for as long as seventeen months or new conidia developed from mycelium that has persisted in the soil or in the infected leaves of the previous year's crop. Infected potato tubers also should not be overlooked as a source of inoculum. Conidia germinate at an optimum temperature of 28°C to 30°C. Host penetration of the fungus as a rule is through the stomata (Fig. 25.31D), but direct penetration through the epidermis is not infrequent. Under conditions of favourable temperature and moisture, the period of incubation is relatively short, spots become visible within forty-eight to seventy-two hours after inoculation. But conidia are produced within the next three or four days. Conidia produced during primary infection from secondary inoculum for secondary infection, and are readily detached and carried to later maturing plants by air currents. Often insects facilitate infection by producing wounds and at the same time they may also carry conidia on their bodies.

The foliage is usually most susceptible after the plants have passed their period of greatest vigour and are being weakened by external conditions or by tuber formation.

Disease cycle of Early Blight of potato is presented in Fig. 25.32.

Control. Infected plant debris of the previous year should be thoroughly cleaned and

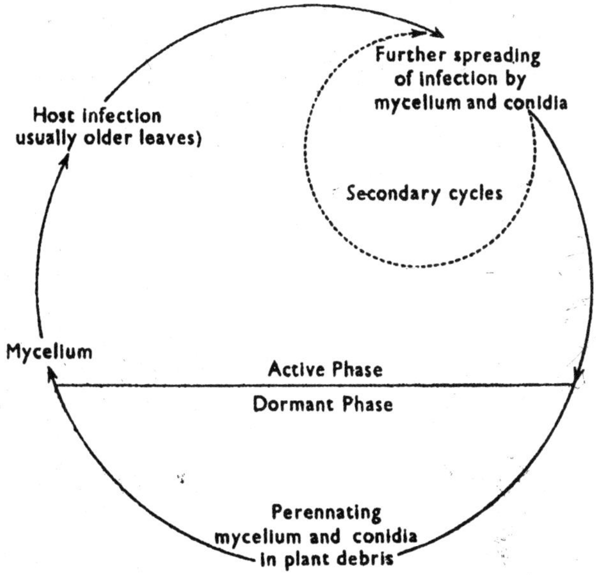

Fig. 25.32: Disease cycle of Early Blight of potato

burnt to cut down the source of priamry inoculum. Disinfection of seed is essential. It is accomplished by treatment of seeds with new improved Ceresan, applied either in 1 : 2000 concentration or as a dust at the rate of 0.5 per cent of the weight of the seed.

The use of Bordeaux mixture to some extent reduces the disease incidence. But greater success is attained with fungicides like zinc carbamites and calcium arsenate (one pound in fifty gallons of water).

Crop rotation is an important control measure. But because of the wind-borne conidia the incidence of the disease may be delayed than completely controlled by crop rotation.

There are several disease resistant varieties now cultivated, viz., Kufri Jeevan, Kufri Alanker, Kufri Khasi Garo and Kufri Chandramukhi.

WILT OF PIGEON PEA

Wilt of pigeon pea (*Cajanus cajan* (L.) Millsp.) is common throughout India. This disease is very destructive particularly in parts of Bihar, the United Provinces, Madhya Pradesh and Maharastra, and is rather less prevalent in Tamilnadu. The loss from the disease is often very high if the crop is grown on the same soil for two or three years in succession.

Symptoms. The main symptom of the disease is wilting of seedlings and adult plants as if they were suffering from drought even though there may be plenty of water in the soil. Examination of the main roots and of the base of the stem reveals that the tissues of the wilted plants are blackened, either uniformly or in dark streaks, especially in early stages. Streaking or blackening whatever may be the condition, the colour change takes place in the wood where it becomes pronounced. The blackening is sometimes visible through the bark, but is very clearly seen on removing the same. The streaks can be traced down to the roots and are found to arise from main or lateral roots. The streaking can often be detected up the stem even to a height of several feet. At times plants are visible with one side only withered with the stem on this side stained black exhibiting partial wilting. The earliest branches to wither are those which arise from the blackened parts of the stem.

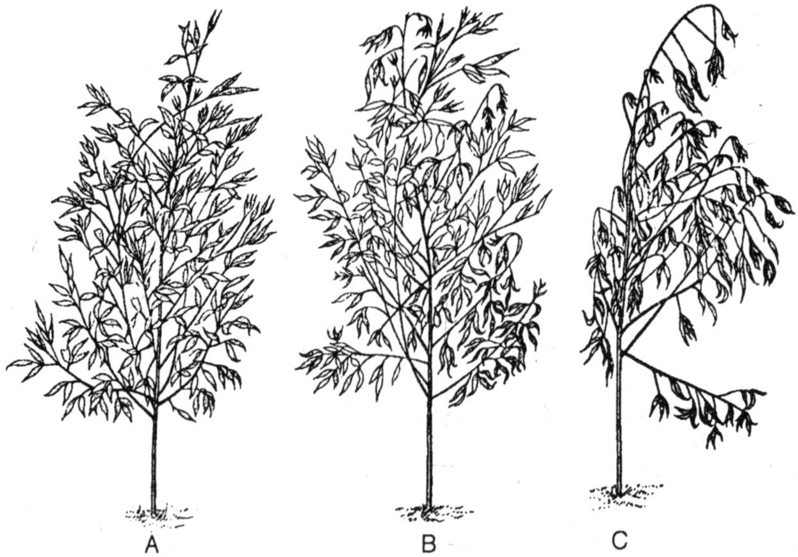

Fig. 25.33: Wilt of Pigeon pea. A-C. Different stages of disease symptoms

Most of the affected plants die, but in cases of partial wilting only the parts affected dry up. Diseased plants may show prolonged symptoms of attack (Fig. 25.33) before they are dead, but with no chance of recovery. In five to six weeks old plants when attacked by the disease, the leaves turn yellow and wither, ultimately the plants dry up. Patches of diseased plants are often scattered throughout the field indicating locations where colonies of the pathogen were present and infection started. These patches develop around the earlier attacked plant in a centrifugal manner. In cases of mild infection affected plants fail to produce pods, dropping of their flowers takes place immaturely and often fail to develop flowers even.

The Causal Organism. Wilt of pigeon pea disease was first attributed by Butler to a *Nectria* which he found on the stems, and it has also been attributed to *Neocosmospora vasinfecta*, but as a result of later investigations, Butler has determined it to be due to a *Fusarium* which he described under the name of *Fusarium udum* Butler which was ultimately named as *Fusarium oxysporum* f. *udum* (Butler) Synder and Hansen. This is a soil-inhibiting fungus which is capable of a saprophytic life in the soil for long periods, in the absence of a suitable host plant. When, however, it encounters the **roots** of the host plant, it is capable of entering **and** passing through a period of parasitic life (Figs. 25.34A and B). The causal organism is thus a facultative parasite. The fungal mycelium that grows within the lumen of the xylem vessels, plugs up the tracheary tissues, thus interfering with the free circulation of water causing the plants to wild and ultimately results in death. The mycelium is hyaline and produces microconidia, macroconidia and chlamydospores within the host tissues (Figs. 25.34C to E). The microconidia are borne terminally on short hyphae within the large vessels. They are small, elliptical or curved, and are usually unicellular but may have one or more septa. The microconidia measure from 5 μ to 15 μ by 2 μ to 4 μ in diameter and are held together in a mass of gelatinous material. The macroconidia are formed on the surface of the bark on small cushions of stromatic mycelium, which burst out through the epidermis. These conidia are many-celled, long slightly curved or bent at the pointed ends. They are divided by **from** three to five transverse septa

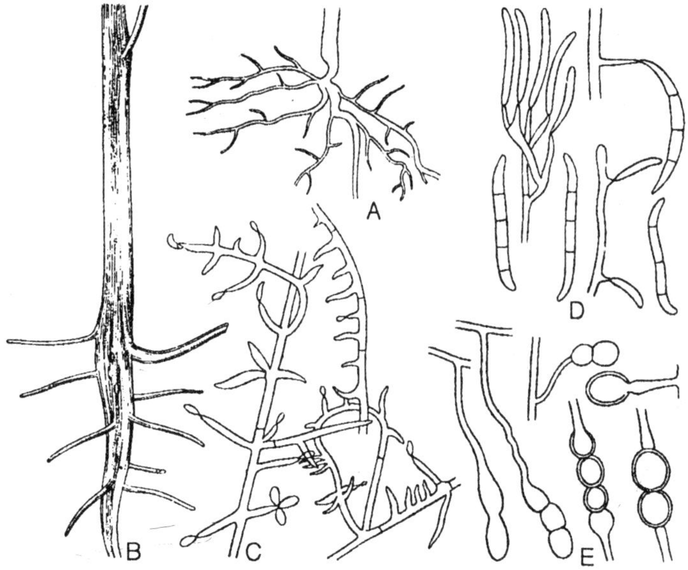

Fig. 25.34: Wilt of Pigeon pea. A-B. Stages of root infection, C. Conidiophores bearing microconidia, D. Conidiophores bearing macroconidia, E. Chlamydospores

and measure 15 μ to 50 μ in length and 3 μ to 5 μ in breadth. The macroconidia are formed at the tips of short conidiophores and fall individually without being held together. The chlamydospores are developed within the vessels of plants which have been suffering from the disease for long time. They are thick-walled spherical or oval in shape, formed from segmented hyphal cells, developed singly being apical or intercalary in position, or united in a short chain. Both micro- and macroconidia remain viable for a considerable period of time. But being thick-walled the chlamydospores are more durable than micro- and macroconidia, and are capable of germinating after a long interval of time. All three—microconidia, macroconidia and chlamydospores may be found together. Usually the microconidia develop first, then the macroconidia, while the chlamydospores are formed profusely only when the supplies of food and moisture get reduced.

Disease Cycle. The prevalence of the disease and the extent of damage done by it are largely due to soil infection from the preceding crop at a temperature range of 17 to 29 °C. The fungus passes from decaying roots into the soil and continues to grow and produces spores under saprophytic conditions until the next crop is planted. It is not, however, capable of indefinite life in cultivated soils, as otherwise it would be practically impossible to grow pigeon pea crop on a large scale in the infected areas. Infection being soil-borne, the rate of infection increases very rapidly in fields where plants are grown very much crowded, actually the fungus spreads more readily from root to root than through intervening soil. The fungus remains alive in the soil, even in the absence of the host plant for a considerable period, even more than a year. It can tolerate pH range of 4.6 to 9.0. The micro- and macroconidia are capable of germination even after one year, wheras the chlamydospores retain their viability for a still longer period of time.

Infection occurs through the fine lateral roots, which are penetrated by the hyphae. The infected lateral roots become rapidly blackened and shrivelled. The fungus then gains entrance in the wood and passes along this to the larger roots, which are in turn killed. But in case of large roots or the base of the stem the fungus enters through wounds. But no evidence of infection has been recorded in parts above soil level.

The hyphae run through the cells, growing with great rapidity along the inside of the walls of the large vessels, up into the stem. They accumulate chiefly in the wood, plugging the xylem vessels. Off-shoots from the plugging hyphal masses extend into the neighbouring cells. This general tendency to rapid longitudinal spread up the xylem explains why the discolouration, which marks the presence of the fungus, may be found at a comparatively early stage in the disease, extending in long streaks up the stem from the point of origin of the infected roots. The preference shown by the fungus for the xylem vessels, and the manner in which these are plugged by matted coils of hyphae, are sufficient to account for the symptoms of drooping and wilting of the leaves. Though the water supply to the roots may be ample, its passage to the green parts is impeded by the choking of the xylem vessels. Such a condition is sufficient to produce the same effect as severe drought. But it is also possible that toxins secreted by the fungus may also play an important role.

Various theories have been advanced to explain wilting: (i) plugging of the xylem vessels by fungal hyphae; (ii) the fungus secretes toxins which cause wilting of plants; (iii) diseased plants transpire more water than healthy plants, indicating that wilt, rather than being the result of plugging, is on the contrary due to excessive water loss; (iv) the major effects of vascular invasion rupturing xylem vessels; and (v) distribution of metabolic products of the fungus by the xylem stream may cause change in permeability of the host cells leading to more rapid water loss and wilting.

The total or partial stoppage of the water supply from the roots to the aerial parts of the plant causes withering, drying and ultimately death of the green parts. The disease first appears in the young seedlings a few inches high. At this stage the disease is usually overlooked unless the number of deaths is large, which is rather uncommon. If adequate precaution is not taken at this stage, the neighbouring plants, generally those nearest to the early cases, dry up. By the time the crop has reached its full height, deaths will be found to have occurred chiefly in small groups, each group corresponding to an early case. These small groups again spread infection centrifugally up to the harvesting time. Along with such an organized spread of disease, new isolated cases of infection may also occur. These isolated cases of infection again produce fresh wilted patches in the field. These new patches of infection may arise as a result of infection by conidia blown from the stems of wilted plants on the soil. Such air-borne infection, however, usually occurs late in the season, as superficial conidia are only produced in large quantities to be freely exposed to air currents when the wilted plants are of good size. There is no evidence that the disease is seed-borne. But it is quite possible that the conidia or chlamydospores may adhere to the seeds during harvesting and reach the soil when the seeds are sown.

That the wilt has been caused due to fungal attack and the plants are not suffering from drought can be confirmatively said from the fact that there is no uniformity in the spread of the wilt in the field and the wilting has occurred even though there is enough of water in the soil.

Disease cycle of Wilt of pigeon pea is presented in Figure 25.35.

Control. Just as any other soil-borne root-infecting disease, the control of wilt of pigeon pea is rather difficult. Before applying any treatment care should be taken to kill the fungus without injuring the roots and that too should be done early, as once the fungus enters the host plant it is out of reach of any kind of treatment. The only effective control measure

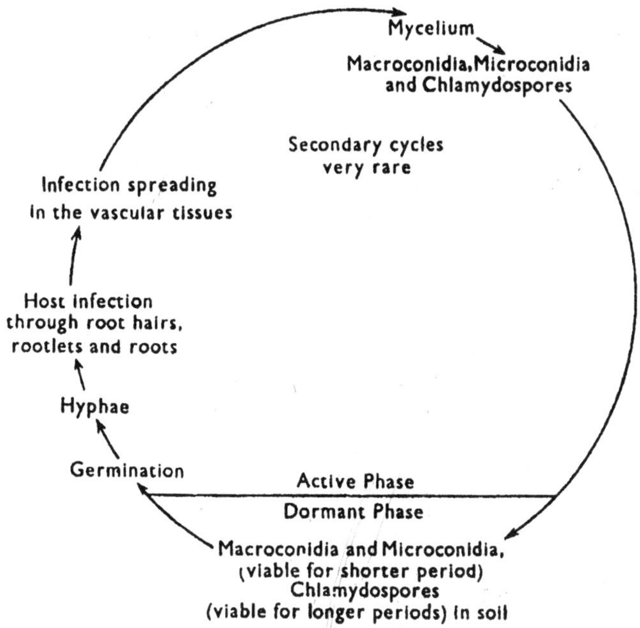

Fig. 25.35: Disease cycle of Wit of Pigeon pea

is to introduce improved cultural practices. The use of resistant variety is also equally important. Crop rotation is undoubtedly the best way of eliminating soil-borne infection. A 4 to 5 year rotation has been found to free the field completely of the pathogen. The length of rotation can be reduced by ensuring removal of affected roots. Rotation with tobacco produces good result as because tobacco exudates cause adverse effects on the pathogen. Mixed cropping with sorghum (jowar) reduces disease incidence by creating root barriers and through toxic root exudates of sorghum. Since the fungus survives on deep-seated roots of the host, below the depth of ordinary cultivation, the success of rotation will depend on field sanitation (removal of affected plants with roots). Green manuring produces adverse effect on disease development as because green manuring increases microorganisms in the soil. These microorganisms produce antagonistic effect on wilt pathogen ultimately reducing wilt incidence. Some of these antagonistic microorganisms are: *Aspergillus niger, Rhizopus nigricans, Bacillus subtilis, Streotpmyces* spp.

Along with these measures, the wilted plants should be systematically removed and burnt, as they are the source of danger since infection inocula are produced on them.

STEM-ROT OF JUTE

This disease of Jute, initially designated as die-back, was reported for the first time from Formosa in 1916 by K. Sawada of Japan. In course of time it was reported from India where it is known as Stem-Rot of Jute disease.

Stem-Rot of Jute may attack both *Corchorus capsularis* L. and *C. olitorius* L. and is common in all the jute tracts in India (Assam, Bihar, West Bengal, Orissa and the eastern districts of Uttar Pradesh.) No other disease causes more damage both in out-turn and quality of fibre than the stem-rot. It is a destructive disease of seedling jute and is usually sporadic in occurrence; but occurs in an epidemic form only under favourable weather conditions. The disease attacking the jute plants both in the seedling and in the adult stages leads to the formation of gaps in the jute field and reduction in yield. When the disease incidence is mild and the deaths of the jute plants are few in the crop, there is not much cause of anxiety. But if the death is severe, large gaps may occur and the plants develop stems of uneven thickness which take a longer time to ret and produce corser fibres. Again, with wider spacings, the plants are induced to branch rendering them useless for fibre production. In the adult stage the damage done is of a more serious nature. In plants that have not succumbed to the disease, the stem usually splits and shreds. Such stems take more time to ret and the fibre itself instead of being glossy, fine, and white is coarse, brownish, weak lacking in lusture. Again, fibres from dead stem are uniformly dull, coarse and brittle.

Symptoms. The disease may attack jute plants at any stage of growth. Very young seedlings, about 2 in high, die with the symptoms of "damping-off", the hypocotyl becoming soft and completely rotted. In these diseased seedlings an abundant mycelium is found growing in the tissues of the young stem. The earliest symptoms appear either on the hypocotyl in the form of lesions near the collar or at the node as thin blackish-brown to black streaks. Under wet conditions the younger seedlings damp-off. But when conditions are dry and the seedlings have reached a three-leaf stage, they develop blight resulting in the shedding of the leaves and the death of the plants. Blight also occurs when the tender roots get infected. On the cotyledonary leaves the necrotic lesions are orange-citrine and roundish. As rotting of the leaves and hypocotyl proceeds pycnidia are formed in abundance. Sclerotia are, however, formed in all the parts of the seedlings.

Again, in adult plants, buffy citrine necrotic lesions occur along the apex and margins of the leaves ultimately covering the entire leaf. The leaf infection spreads to the node through the petioles; the infected leaves are shed. Numerous pycnidia cover the entire lamina and the petiole with pycnidial ostiole on the upper surface of the leaf. Sclerotia are not developed on the affected leaves. Lesions on the stem usually occur at the nodes as small, blackish-brown depressions which may coalesce and gradually girdle the stem causing rot. The root of the stem leads to either shredding, canker or wilting. In the advanced cases, the plants wilt leaving bare dead stalks in the field with abundant pycnidia all over. In plants where lesions appear as streaks along the length of the stem without girdling it, the cortex get shredded exposing the fibrous tissue. They break down easily when strong wind blows. In some vigorously growing plants, the lesions may become localized due to cork formation producing cankers. A furrow may be formed by coalescing several cankers. Pycnidia are usually embedded in the epidermis of the stem being visible as small black dots. Sclerotia are found in the soft tissues of the infected stem. In cankers, pycnidia and sclerotia occur sporadically.

In plants whose aerial parts are attacked, the roots very often are healthy. It is only after all parts die, the infection spreads in the roots. Sometimes, however, direct infection of roots occurs in which case the plants wilt and the infected roots are invariably full of black sclerotia.

Infected capsules either drop off or split. In the latter case, the seeds get infected. In some of the unsplit diseased capsules seeds may also be diseased. The infected seeds are brown without any lusture and light. Sclerotia are developed within the capsule and on the seeds. Pycnidia occur on the outer surface of the capsule and sporadically on the seeds.

The Causal Organism. The stem rot of Jute disease is caused by the fungus: *Macrophomina phaseolina* (Tassi) Goid. [= *M. phaseoli* (Maubl.) Ashby] whose sclorotia phase is known as *Rhizoctonia bataticola* (Taub) Butter. It belongs to the form-family Sphaeropsidaceae under the form-order Sphaeropsidales of the form—class Deuteromycetes. The fungus is characterized by colourless, septate, branched hyphae measuring 3-8 μ across which pass freely through the host cells. It produces subepidermal, punctiform, black pycnidia with erumpent ostiole. The pycnidia measuring 98-225 × 89-275 μ have multicellular wall. Numerous conidiophores are borne in each pycnidium. The conidiophores are hyaline, simple, straight to slightly curved with tapering apex, 10-14 × 2.5-35 μ, terminated by a conidium. The conidia are unicellular, hyaline, smooth-walled, ovoid to oblong, straight or slightly curved, 16-32 × 10 μ. Along with the development of pycnidia, frequent dark sclerotia are produced by the hyphae. The sclerotia are 41 to 86 μ in diameter.

Disease Cycle. The pathogen perennates on stubbles and in soil in the jute fields in its sclerotial form. It also harbours on the jute seeds under the seed coat and within the cotyledons. Where rotations are practised, other crops like linseed, sesamum or cowpea may form the collateral hosts. The pathogen has also been found to survive on a common weed *Cyperus distans*. But primary infection is chiefly due to the seed borne inoculum than any other source.

There are two phases of disease development. During the first phase the pathogen kills a large number of seedlings and young plants in the first two months. During the second phase, which starts after the plants are four to five months old, there are no deaths, but quality of the fibre is ruined owing to the formation of cankers and shredding.

Due to primary infection, symptoms appear in the seedlings on the cotyledonary leaves and the hypocotyl within a fortnight after sowing leading to the development of pycnidia. Conidia released from the pycnidia of the diseased seedlings cause secondary infection on the leaves in the vicinity of the seedlings where primary infection has been established. Both rain and air play a great role for dispersal of conidia. The conidia readily infect the leaves and the leaf infection spreads to the node through the petiole. In an infected stem, the pathogen grows in the epidermis and cortex, often invading the phloem parenchyma and the wide rays between the phloem wedges. It is also found in the central woody portion, the hyphae penetrating even the wood fibres. Ultimately girdling of the stem occurs causing rot of the stem which leads to either shredding, canker or wilting. Usually the secondary infection is four times the primary infection.

Prolonged warm and humid conditions favour disease development leading to the severe outbreak of the stem-rot. Initially abundant leaf infection is visible during and after heavy rains which gradually leads to the spread of the disease. The most critical period for the crop is first eight weeks of its growth, for it is during this period that the disease is serious. Again if the weather during this period is comparatively dry, the incidence of the diseaese is less. The interval between the appearance of the stem-rot and the death of the plant ranges between one and twenty weeks. Where the top of the plant is attacked the death is quicker. Attacked plants often survive by means of callus formation.

Root-rot usually occurs sporadically late in the season, but also appears early in the season when high temperature prevails.

Disease cycle of Stem-rot of jute is presented in Figure 25.36.

Control. Since the pathogen can perennate in soil as well as on stubbles, jute field sanitation is an important measure to cut down the source of primary inoculum. Besides this, some of the suggested control measures are given below

(i) Use clean seeds from localities where disease incidence has not been reported.

(ii) Seeds should be thoroughly dried and before storage should be treated with

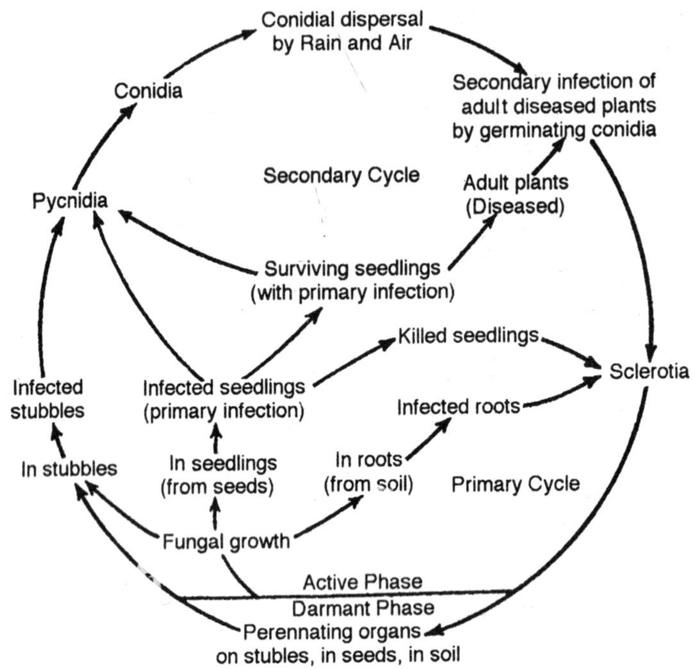

Fig. 25.36: Disease cycle of Stem-rot of jute

Agrosan GN or ceresan or any other organomercuric compounds.

(iii) Betterment of cultural practice in the following manner:

(a) Improvement of soil pH with lime;

(b) Application of potash between 50 to 100 kg/ha in soil;

(c) Good drainage and improvement of porosity (in clay soil) with organic matter;

(d) Proper spacing 30 cm × 7 cm, weeding, and thinning;

(e) Crop rotation jute with paddy.

(iv) Spraying copper oxychloride (50% Cu) at 0.75 concentration at first leaf infection produces fruitful results. Leaf infection can be reduced spraying 0.2% Lime Sulphur, 0.5% Perenox, and Bordeaux mixture (5 : 5 : 40) under field condition.

(v) Early harvesting reduces loss of fibre.

(vi) Since the fungus can survive on linseed, sesamum and cowpea, these plants should not be grown in jute fields.

(vii) Jute field weed *Cyperus distans* should be eradicated to reduce the source of inoculum.

(viii) Root-rot can be effectively controlled by the use of Brassicol.

(ix) The standard varieties like D-154 and JRC-918, a selection from the Brazilian material are sources of resistance for stem-rot. Further report of varieties like JRC-8694, JRC-212, and JRC 5854 are shown to be highly resistant to stem-rot infection.

CLUB ROOT OF CRUCIFERS

This disease attacks a wide range of cultivated and garden plants, all of which belong to the family Cruciferae. Among the cruciferous weeds that are infected by this disease, mention may be made of shepherd's purse (*Capsella bursa-pastoris*). The disease is widely spread througout both tropical and temperature regions of the world. Some of the common hosts are: cabbage, cauliflower, raddish, turnip, etc.

The first occurrence of the disease was reported from Scotland in 1780 and again in 1855. But the first scientific investigation of the disease on cabbage and the nature of its causal organism was done in Russia by Woronin in 1878. He established the fact that the disease is caused by a slime mold (*Plasmodiophora brassicae*) and worked out its general life history. The disease is known under a great many common names: in Germany it is 'Kelch' or 'Kropf des Kohles'; in France it is 'Maladie Digitoire'; in Belgium it is 'Uingergiekt'; in Russia it is 'Kapoustnaya Kila'; in England it is 'Anbury', 'Hamburg', and 'Finger and Toe'; and in thè United States 'Club Foot', 'Club Root', and 'Clump Foot'.

Symptoms. Club root, as the name suggests, is essentially a disease of the rooting system. The presence of the disease within the host tissue leads to a remarkable increase in growth in the parts affected, causing the roots, for the greater part, to become abnormally swollen. But the symptoms vary somewhat according to the nature of the host infected. In the majority of cases the disease is manifested by a hyperplasia and hypertrophy in the root system. The different types of symptoms are: (i) complete clubbing of both the main and lateral roots, the whole root system often resembles a mass of fingers and toes; (ii) the main root becomes so

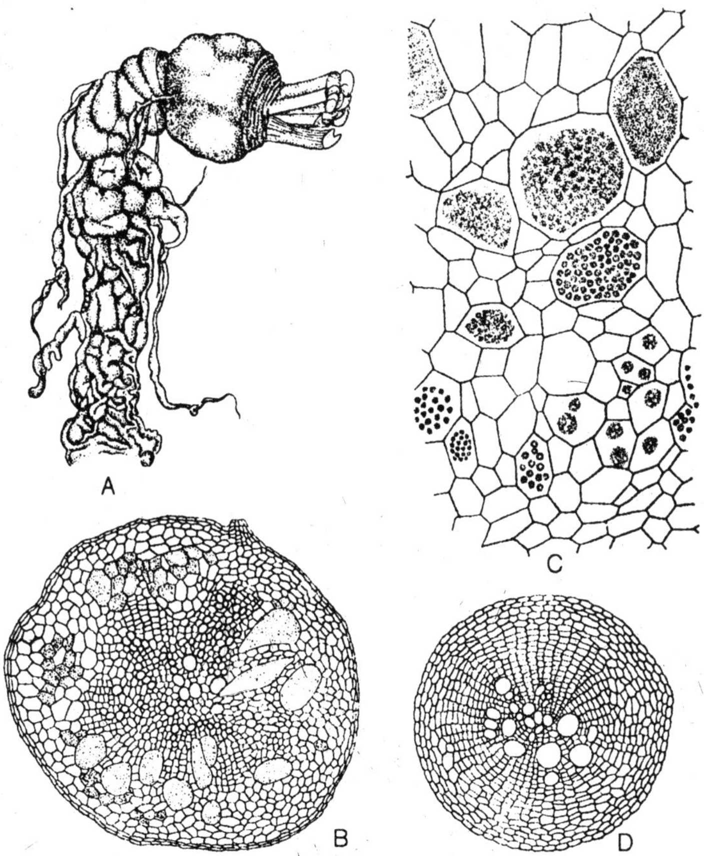

Fig. 25.37: Club Root of crucifers. A. Diseased roots of turnip. B. Sectional view of a diseased root showing infected cells. C. Infected cells magnified showing the causal organism in different stages. D. Sectional view of a healthy root for comparison with the diseased one

much enlarged and clubbed as to assume the form of very large, irregular knots or tumours, the laterals remain healthy; (iii) only the lateral roots are transformed into clubs, while the main root remains free from clubbing; (iv) both main and lateral roots are affected, but healthy rootlets are developed from the diseased parts; (v) tumours or lobulate swellings of the tap root (Fig. 25.37A); and (vi) characteristic cracks of fissures and darkened areas develop in the affected parts of the root, but true hypertrophy is lacking.

A transverse section of hypertrophied root shows the infected areas to consist of several scattered groups of cells, chiefly in the cortex and medullary rays, having denser contents than normal. These cells are filled with plasmodia or with a multitude of minute spores (Figs. 25.37B-C). In advanced stages of disease, due to increase in the size of the galls, disruption of the outer tissues of the host takes place and a general decay of the galled tissue sets in and the spores are released in the soil.

The disease may progress to a considerable extent before above ground symptoms become noticeable. But seriously diseased plants look sickly. Symptoms of wilting or a yellowing of the foliage are visible in them. Gradually there is loss of foliage. Plants of this kind rarely form a head. If plants are older, when infected, they may head out but the heads are usually small and of inferior quality.

The Causal Organism. *Plasmodiophora brassicae* Wor. It is a parasitic slime mold. The vegetative phase consists of a plasmodium. There is no fruit body development in the life history of this slime mold. The plasmodium lives within the host cell and consumes the host protoplast and occupies the entire lumen of the infected cell. The plasmodium at maturity cleaves down into fragments in such a manner that each nucleus with its surrounding portion of cytoplasm is enclosed by a wall and thus becomes a spore. This is how a mass of resting spores is developed. The spores are entirely free from one another, and they are held together by the host cell wall, until it is decomposed in the soil.

The spores are very small and spherical. A single diseased host cell contains a large number of these spores. The spores lie dormant in the diseased root cells or in the soil until the next season or until conditions are favourable for germination. The spores are liberated from the diseased root cells when the tissues decay or when the roots are eaten by animals or broken into pieces by some physical means. When conditions are favourable for germination, the spores germinate in soil by the production of a swarm cell. Host penetration is effected by the swarm cell through the root hair. The flagella are left behind the root hair. Only the body of the swarm cell enters the root hair. The body of the swarm cell without flagella exhibits amoeboid movement and is thus nothing but a myxamoeba. The myxamoeba migrates to other host cells by direct penetration of the cell wall and there develops into plasmodium. If the plasmodium develops in this manner, it is probably haploid. But mycologists believe that there may also be a diploid plasmodium. They are of opinion that the myxamoeba produces haploid plasmodium in the root hair cells. This haploid plasmodium gives rise to gametangia, again each gametangium produces a single gamete which morphologically resembles the swarm cell. The gametes are liberated out by the decay of root hair cells and gametangia. Gametic union takes place in the soil. The zygote so formed enters in the host tissue and gives rise to a diploid plasmodium from which resting spores are developed by meiosis.

Life cycle of *Plasmodiophora brassicae* is presented in Figure 25.38.

Disease Cycle. Various factors such as soil acidity, temperature and moistre content of the soil control the disease incidence. The pH of the soil above 7.0 and a moisture content

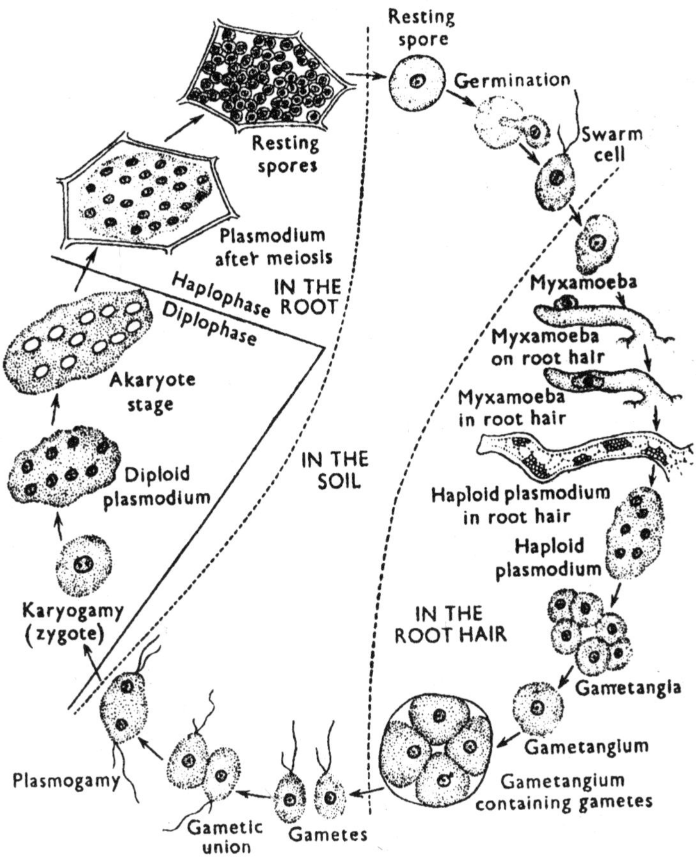

Fig. 25.38: Life cycle of *Plasmodiophora brassicae*

between 45 and 60 per cent are very suitable for the development of this disease. Loam and clay soils are generally less prone to harbour club root than light soil. Usually low-lying, poorly drained soil favours club root, and well-drained soils inhibit it. The effect of temperature is rather indirect and its influence is related to the host development. Most pronounced development of club root is at a temperature range of 20°C to 25°C, which is the temperature at which normal root growth occurs vigorously.

The causal organism is one of the most persistent soil-invaders. It remains viable in soil for 10 years or longer without the presence of host plants. Infected subterranean plant parts disintegrate in the soil, releasing the free spores. While the spores germinate with little or no resting period, a large percentage of them remain dormant for many years.

In the early stages of root infection, the invaded host cells are naturally those of the primary cortex, the parasite, later seeks out the primary cambium. The occupation of the cambium, however, initiates a much more serious phase of the disease than the occupation of the primary cortex. The presence of a plasmodium in a cell stimulates not only the occupied cell to meristematic activity but also the cells contiguous to it. This is effected in two ways: (a) by the one or more divisions of the infected cell being accompanied by a sharing of the plasmodium into the daughter and subsequent cells, before the divisional walls are laid down; and (b) by bits of the plasmo-

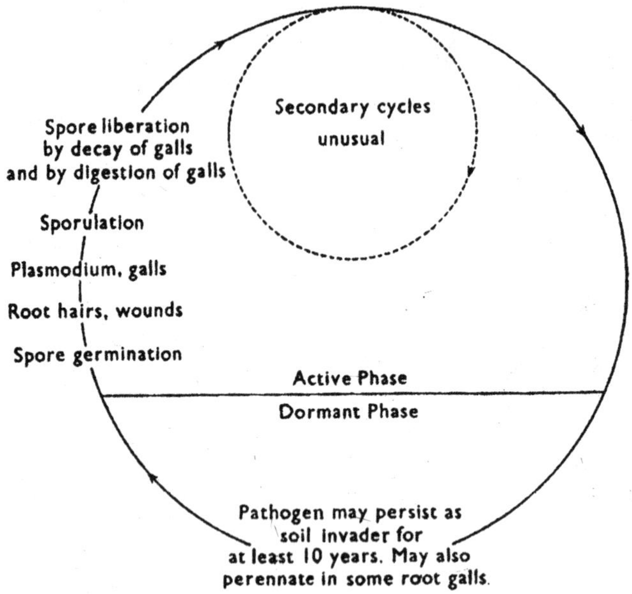

Fig. 25.39: Disease cycle of Club Root of crucifers

dium, cut off within an infected cell, migrating through pores in the wall into contiguous cells, with a repetition of the same process within prescribed areas. Both of these methods appear to be operative, though several authors deny the capacity of plasmodia to penetrate cell walls. Plasmodial portions within dividing cambial cells are supplied to practically every living cell added by the cambium. The parasitized cells, wherever they may occur, have no plan of division, and their irregular multiplication at advanced stages of infection takes place on so wide a scale that the proliferating tissues intrude upon the vascular continuity. Such abnormal meristematic activity, at numerous centres of infection within a root, causes considerable displacement of host tissues which are, therefore, forced outwards and finally appear at the broken surface as clubs or galls of hypertrophied tissues.

Disease cycle of Club Root of crucifers is presented in Figure 25.39.

Control. Any effective control measure needs careful consideration of the facts like: (i) The causal organism is perpetuated by means of resting spores which remain either in soil or in infected host tissue. (ii) The spores may remain dormant for several years. (iii) The disease is disseminated by agencies which transport diseased plants or parts of plants, or contaminated oil. (iv) The virulence of infection is favoured by acid soils and high moisture content of the soil. Some of the suggested control measures are:

(i) *Sanitation:* At harvesting time roots of diseased plants should not be left to decay in the soil. To guard against the spread of the disease, the diseased roots if fed to cattle should be thoroughly boiled before feeding. Soil should not be transported from infected field to new areas. Long periods of rest from crucifers are advisable in order to starve the parasite out of the soil, and every effort should be made to eradicate susceptible weeds from the vicinity of the crops.

(ii) *Crop rotation:* Since the pathogen remains viable in soil for several years, a comparatively long rotation with non-cruciferous crops should be adopted to control the disease incidence.

(iii) *Soil treatment:* Since acid soil is favourable for disease incidence, neutralizing the acid soil with lime in the form of raw-ground lime, caustic lime, and hydrated lime often produces better result. The use of Calcium cyanamide as a substitute for lime is also effective. Often good results follow the applications of disinfectants such as the highly poisonous chlorides of mercury, corrosive sublimate and calomel.

(iv) *Soil drainage:* The use of too much of free water should be avoided. Necessary soil drainage facilities are to be made to avoid water logging.

(v) *Use of Resistant varieties:* Grow resistant varieties.

CITRUS CANKER

This disease though a native of the Orient, is now widespread in the citrus-growing countries of the world. It causes consternation in the citrus fruit-growing sections in India, China, Java and Japan.

Grape fruit and sweet orange (*Citrus aurantium* L.) are the most susceptible species of citrus crop.

Symptoms. The disease affects all above-ground parts of the tree, but most susceptible are the leaves, twigs and fruits. It particularly produces scabby lesions on the surface of the fruits (Fig. 25.40D) and thereby reduces their market value. The lesions first appear on host surface as small watery translucent spots which increase in size, turn dark-green with age. The host tissues surrounding the spots become raised resulting the spots to develop convex surface. The central region of the spots gradually turns light-brown and spongy and breaks down producing a crater-like appearance. With age, the spots become corky and brown, sometimes pinkish, assuming a cankerous appearance.

On the leaves, the lesions appear first on the lower surface (Figs. 25.40A and B). The

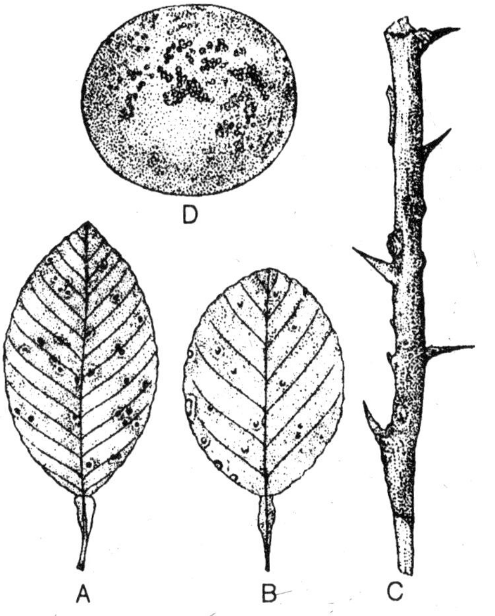

Fig. 25.40: Citrus Canker. A-D. Disease symptoms. A-B. On leaves. C. On twig. D. On fruit

lesions on the twigs (Fig. 25.40C) and fruits (Fig. 25.40D) are very similar to those of leaves, only difference is somewhat dull yellowish colour of the lesions on the leaves is absent in those of twigs and fruits. On fruits the lesions often coalesce forming very conspicuous, irregular patches of rough, scabby raised areas.

The Causal Organism. The disease is induced by the bacterial pathogen, *Xanthomonas citri* (Hasse) Dows. The organism is rod-shaped monotrichous with yellow, water-soluble pigment. It is an aerobic bacterium.

Disease Cycle. Bacteria gain entrance to the host through stomata, lenticels, and wound. They perennate in the affected host tissue. The inoculum is disseminated by wind, rains and sometimes by insects. But the disease is commonly disseminated through infected nursery stock. Mild temperature accompanied with humid condition is very favourable for the disease. The most suitable range of temperature for the disease incidence is between 20°C and 35°C.

Disease cycle of Citrus Canker is presented in Figure 25.41.

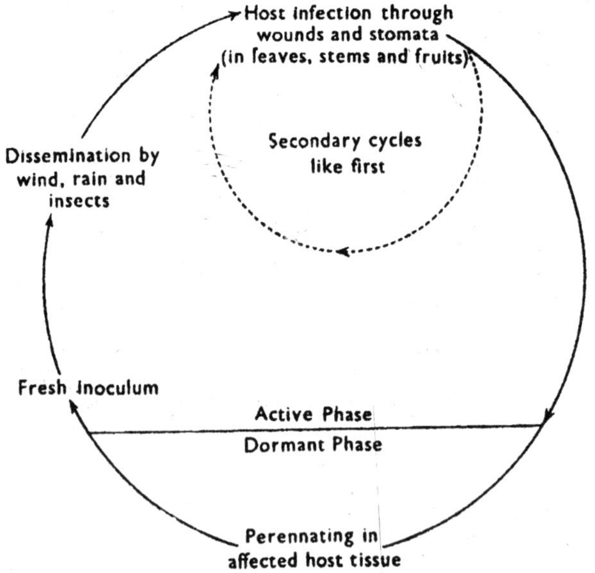

Fig. 25.41: Disease cycle of Citrus Canker

Control. It is rather very difficult to control the disease. Following are some of the effective control measures:

(i) *Sanitation:* Destruction of all affected trees by burning. Pruning of infected parts, particularly during dry season reduces source of inoculum.

(ii) *Spraying of Fungicides:* Spraying of fungicides like Bordeaux mixture and lime-sulphur is often very effective to protect the fruits against infection. It should be done during the first three months of the development of fruits.

(iii) *Use of Disease Resistant Varieties:* There is a possibility that cultivation of disease resistant citrus varieties may produce good results.

BACTERIAL BLIGHT OF RICE

Bacterial blight of rice disease is widespread in Asia. But it has not been reported from any rice growing areas outside Asia. In India, bacterial blight of rice was first reported from Maharastra in 1951. In Japan, farmers observed the occurrence of bacterial blight of rice in 1884, but its incidence was reported in widespread areas during 1908-1910. There this disease is known as "white with ring disease". But at that time, the causal organism was that identified. The causal organism was identified only in 1959. Subsequent studies have revealed the occurrence of bacterial blight of rice disease in most rice growing states of India.

Symptoms. The first appearance of disease symptoms is in the seedbed as tiny, water soaked spots at the margin of fully developed lower leaves of the seedlings. As the spots enlarge, the leaves turn yellow dry rapidly and wither.

The disease then spreads on other leaves affecting leaf blade. The lesions enlarge both in length and width, having a wavy marginal turning yellow within a few days. The adjoining healthy areas shows water-soaked appearance. Lesions may start growing at one or both the surfaces of leaf blade. As the disease advances, the lesions cover the entire leaf blade, turning white and later become greyish. In susceptible varieties of rice the lesions extend to the leaf sheath. The infected leaf blades of susceptible varieties wilt and roll. The entire leaf blade may soon become infected by the disease and gets dried up. Symptoms become visible one week after infection. The disease is characterized by the drying of the leaf tip and inwards rolling and twisting of the leaf blade. Lincar, yellow strips may develop along with midrib of the leaf blade the disease extends to the leaf sheath killing/destroying tillers of the entire plant. In severe cases, the infection may be spread in the grains. The glumes get discoloured and water soaked spots appear on them.

Disease Cycle and Predisposing Factors

The bacterial leaf blight is principally a vascular disease. The pathogen enters the host leaf through wounds caused by Chaffing and string of leaves against one another in the wind, through broken roots, at transplanting or through hydathodes or stomata. Once inside the leaf, the pathogen multiplies and moves along both directions. With the progress of pathogen (bacteriological) development, bacterial exudates are produced on the surface of leaves. Bacterial exudates along with bacteria are dispersed by air irrigation water and by some such processes during rice cultivation. In India, the pathogen propagates from season to season **through infected seeds.** Several workers from Japan (Mizu Kani, 1961), and from India (Chattopadhyay and Mukherjee, 1961). Rao and Kauff, 1971 have worked on this disease. Their work indicated that some seeds like *Leersia oryzoides, L. hexandra* and *Cyperus rotandus* are the hosts which are sources of *primary inoculum.* According to Devadath, the Chaffs dumped near the threshing paces and rice field may play an important role in being the source of primary inoculum. Devadath reported that bacterial cells on coming in contact with the rice seedling are activated and multiply and invade through wounds. In site of the extensive work in Japan and elsewhere on the perpetuation of the bacteria from one season to the next, there is yet no confirmative information on the sources that contribute to the primary inoculum source.

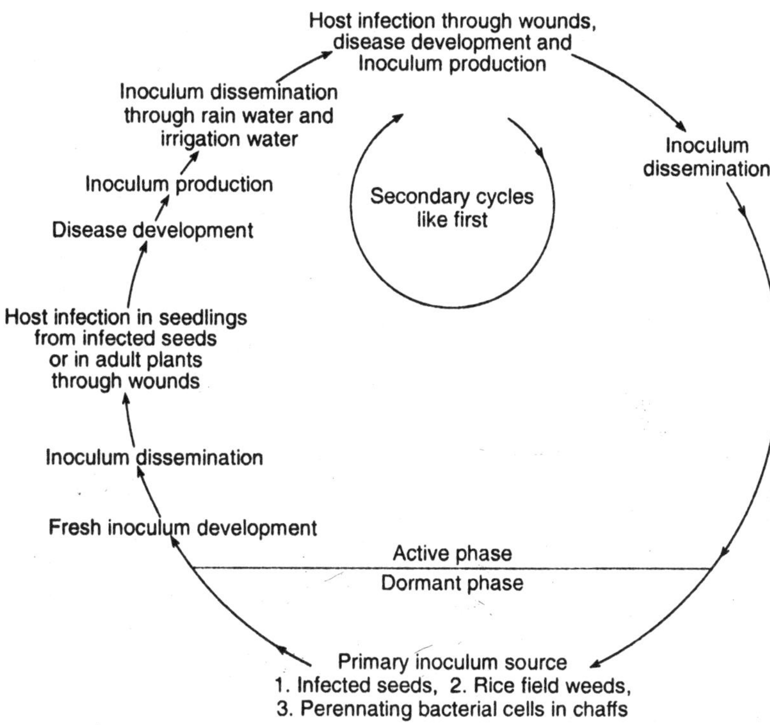

Fig. 25.42: Disease cycle of Bacterial Blight of rice

The Causal Organism. The Bacterial Blight disease of rice is caused by *Xanthomonas compestris pv. oryzae* (Ishiyama) Dye. It is a rod-shaped bacterium with rounded ends measuring 1-2 × 0.8-1 μm and possesses a monotrichous flagellum. It is Gram-negative and non-spore forming bacterium. Bacterial cells are surrounded by mucous capsules which are joined to form an aggregated mass, relatively stable inmates. The strains of bacterial leaf blight causing organism vary in virulence, they also differ in their susceptibility to bacteriophages.

Disease Management/Control. In India, *Xanthomonas compestris pv. oryzae* is suspected to be a self-borne pathogen. Various seed treatment procedures have been applied of which soaking seeds for 12 hours in ceresan plus Agrimycin 100 followed by hot diety treatment for 30 minutes at 52°-54°C has been recommended. Spraying in the field is also recommended with a mixture of 100 ppm each of Duthranone and Cellucidin. Rice plants should be sprayed as soon as the sign of disease appears. Antibiotic spray preparations were tried by different workers (Thirumalachar, 1968, Srivastava, 1972). The results of these experiments showed that the antibiotic is not at all effective against this disease.

It has been established that application nitrogenous fertilizer influences disease development.

Extent of Damage caused by the disease attack. Since the introduction and cultivation of new high-yielding, but susceptible varieties which are used in recent years, in many of them, the disease has become one of the most serious problems of rice cultivation in India. According to estimates of 1966, in India the losses had been estimated to vary from 6 to 60 per cent.

The disease resistant varieties Tainan 3 and Chinung-292 have been found useful. It has been suggested that the sowing of nursery beds in disease-free, isolated areas prevents the inflow of primary inoculum into the fields might reduce the disease incidence.

TOBACCO MOSAIC VIRUS

This is the best known of all virus diseases. The tobacco mosaic virus affects all dicotyledonous plants of which most important are tobacco and tomato and many others, mostly solanaceous plants. But it does not affect any monocotyledonous plants. Tobacco mosaic virus causes serious losses in their hosts by damaging the leaves, flowers, and fruits and by causing stunting of the host plant. The losses are greatest when the plants are infected young.

Although Adolph Mayer in 1886 first pointed out the mosaic pattern on leaves of affected tobacco plants, it was not until 1898 the first scientific proof of the existence of a virus was given by Beijerinck. Earlier than this, in 1892 Iwanowski demonstrated that tobacco mosaic virus would pass through a bacteria-proof filter. He was able to demonstrate that a diseased tobacco plant juice was able to induce mosaic disease in healthy tobacco plants. But Iwanowski could not find out the true significance of this. Holmes in 1929 described the primary infection lesions of tobacco mosaic virus and in 1935 Stanley first isolated crystals of tobacco mosaic virus and indicated their paracrystalline nature. Again, Takahashi and Rawlins in 1933 demostrated the physical phenomenon of tobacco mosaic virus, whereas C. A. Knight showed that the tobacco mosaic virus is made up of sixteen amino acids.

The tobacco mosaic virus affects photosynthetic tissue of the host leading to distortion, blistering and necrosis. It also causes dwarfing of affected plants. It is one of the most damaging viruses of plants, causes enormous loss of tobacco crop by reducing yield and quality.

Symptoms. The symptom is systemic mosaic type. The primary symptom on young leaves is faint circular chlorotic lesions appear with gradual vein clearing. This is followed by

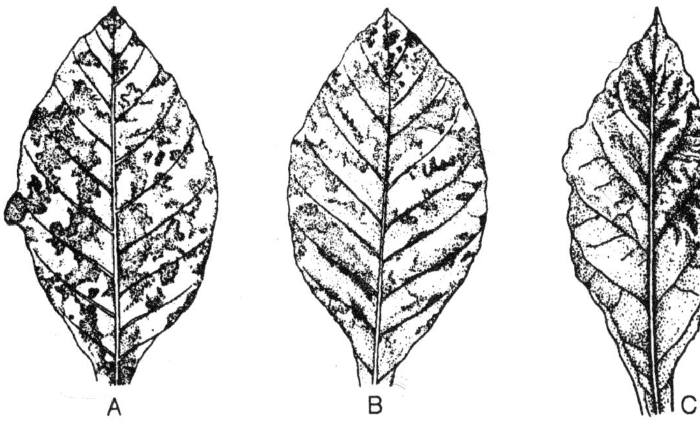

Fig. 25.43: Tobacco Mosaic Virus. Disease symptoms on tobacco leaves induced by ordinary or field strain

the development of characteristic systemic mosaic. With the maturity of the leaves, abnormally dark-green spots appear which develop into irregular crumpled blister-like areas while the rest of the tissue becoming more or less chlorotic (Fig. 25.43). Various degrees of leaf malformation like enations follow and some leaves exhibit only a mild diffuse mottle. The development of symptoms is governed by many variable factors of which the most important is the difference in virulence of the virus strains. For example, one strain of tobacco mosaic virus may cause yellow mottling on the leaves, a second may cause necrosis only, whilst a third induces a gross malformation. Another variable factor is the variety of plant affected. In flowers, petals show mosaic symptoms. Severe strains cause streaking of stem. The disease is seldom fatal to the host.

The Virus. The typical tobacco mosaic virus is Tobacco mosaic virus 1, *Marmor tabaci* Holmes.

The virus remains active in extracted host plant juice even up to 25 years. It is a very resistant virus, can stand desiccation for 25 years or more. It occurs in very high concentration in plant and its dilution end point is 10^{-6}. The thermal inactivation point of the virus is 90°C. The virus particles are rod-shaped (Fig. 25.44) measuring about 280 × 15-18 μm with a central hollow tube about 20Å in diameter. It has a molecular weight of 40 millions. The RNA of the virus is single stranded and molecular weight is 2.4 millions. It consists of 6500 nucleotides. The centrally placed RNA strand is covered with a protein coat containing 2130 indentical protein subunits, each with a molecular weight of 18,000. The subunit is com-

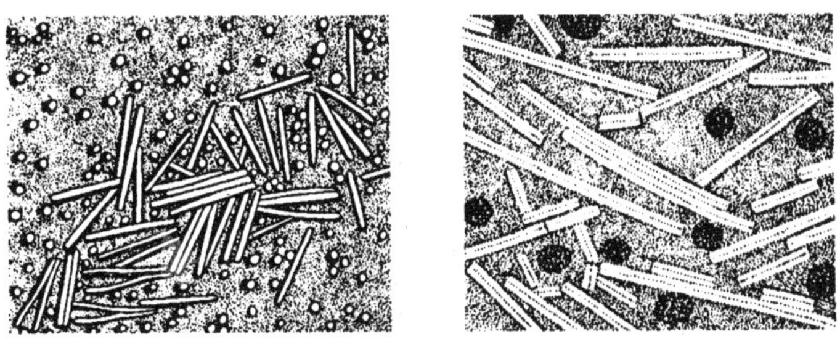

Fig. 25.44: Tobacco Mosaic Virus. A-B. Virus particles. A. Shadowed with palladium-gold. B. Stained with phosphotungstic acid

posed of a single chain of 158 amino acids of known sequence. The X-ray studies reveal that the virus particle consists of a number of protein subunits set in helical array with 49 subunits to one turn of the helix and 2,130 subunits in one rod. The ribonucleic acid thread intertwines more or less centrally between the protein subunits.

The cells of tobacco plants infected with tobacco mosaic virus are characterized by the presence of certain cell inclusions. They are: (i) two types of intracellular inclusions, and (ii) intranuclear inclusion. The intracellular inclusions are: (a) X-bodies are amorphous, protoplasmic more or less vacuolate inclusions, whereas striate material of crystalline plates gives protein reaction. These crystals resemble the purified virus-protein crystals. The intranuclear fibrous and crystalline inclusions are produced by a yellow-mottling strain of tobacco mosaic virus.

Disease Cycle. The virus perennates in infected tobacco plant debris, tobacco refuse from warehouses, cigarettes, cigars, pipe and chewing tobacco and in perennating hosts which form the source of primary inoculum.

This is one of the most infected of the plant viruses. The virus is disseminated from plant to plant by mechanical transmission through wounds by handling tobacco plants during transplanting; through other field operations; and contact by man and cultivation implements. The virus enters in the host tissue, multiplies very rapidly moving from cell to cell through plasmadesmata producing disease symptoms.

Disease cycle of Tobacco Mosaic Virus is presented in Figure 25.45.

Control. Following are some of the suggested control measures:

(i) Seed beds should be located at a great distance from the tobacco warehouses.

(ii) Seed beds should be free from any tobacco refuse.

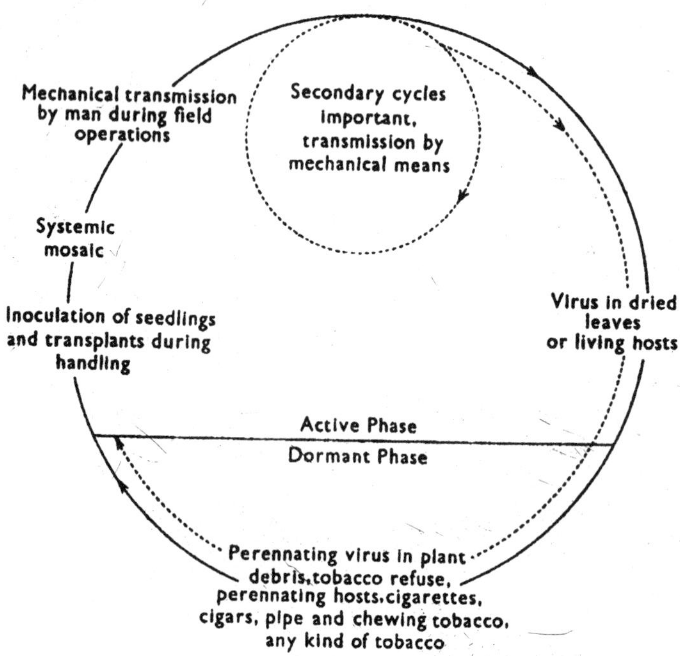

Fig. 25.45: Disease cycle of Tobacco Mosaic Virus

(iii) Seed bed soil should be sterilized by steam.

(iv) Care should be taken to avoid contamination through hands and cultivation implements.

(v) Since pipe tobacco, cigarettes and chewing tobacco are all sources of primary inoculum, smoking or chewing of any kind of tobacco should be avoided.

(vi) Susceptible hosts, weed or otherwise in which virus may harbour, should be destroyed.

(vii) Previous year's plant debris should be destroyed by burning.

(viii) Diseased plants should be removed and burnt to stop further spread of the disease.

(ix) Growing resistant varieties produces good results.

(x) In recent past promising experimental control has been obtained by genetic engineering tobacco and tomato plants with the gene coding for the TMV coat protein.

TUNGRO VIRUS DISEASE OF RICE

Rice tungro virus (RTV) disease is out of the most important virus diseases of rice prevalent all over South and South-East-Asian countries. It has been reported under different names, such as yellow gauge leaf in Thailand, and Leaf yellow in India. It is a composite disease caused by *Rice tungro bacilliform virus* (RTBV) and *Rice tungro spherical virus* (RTSV).

Predisposing Factors

The vector *Nephotetlix virescens* is monophagus to rice and its density can reach high levels depending on the environment. In most irrigated areas in the Asian tropics, rice planting areas are favourable for Tungro virus and its vectors.

The use of high yielding varieties also favours the development of population of vectors.

RTBV. It is a plant pararetrovirus with bacilliform particles, the structure of which is based on T-3 icosahedral symmetry cell across the threefold axis. The particle is a circular double stranded DNA like *Badna virus.* Two major strains of RTBV have been recognized, one from the Indian subcontinent and the other from South-east Asia.

RTSV. Its particles are Isometric containing single-stranded RNA, as in *Waika virus.*

Virus Transmission. RTBV and RTSV are transmitted in a semipersistent manner by the vectors *Nephotetlix virescens, N. migropictus, Recilia dorsalis.* Immediately after feeding on host plants infected with RTBV and RTSV. The vector rice leafhopper (*N. virescens*) transmits the viruses, either both viruses RTBV and RTSV together or RTBV alone. The vector retains RTBV for 4-5 days and RTSV for 2-4 days. The RTBV transmission by leafhopper is possible only when RTSV is also present in the donor plant. In the infected plants RTBV is localized in the vascular bundles and RTVS phloem tissues alone. In infected cells both RTBV and RTSV particles are scattered or aggregated in the cytoflem.

Mode of Virus Transmission

RTSV and RTSV viruses are transmitted in a simple manner through insect vectors by *Nephotetlix virescens, N. migropictus* and *Racilia dorsalis.* Immediately after feeding on virus infected rice plant *N. viruceous* transmits either both the viruses, together RTBV and RTSV or alone. The insect vectors retain RTBV for 4 days and RTSV for 2 to 4 days. The RTVS virus transmission by leafhopped is possible only when the RTSV is also present in the donor plant. The use of higher yielding rice varieties favours the development of vector population.

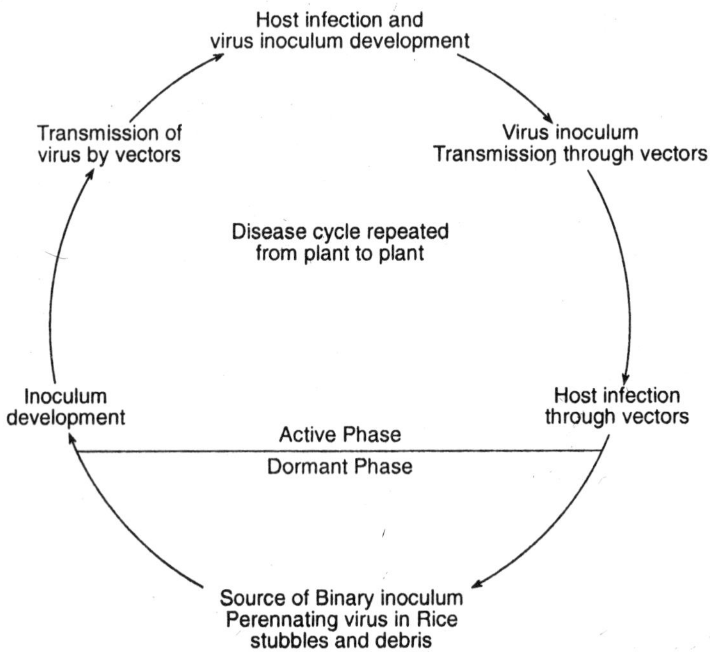

Fig. 25.46: Disease cycle of Tungro Virus

The virus usually overwinters in rice stubbles.

Symptoms. Rice plants infected with RTBV and RTSV together show tungro symptoms, including stunting, yellow or yellow-orange discolouration, and reducing tillering. Discoloured leaves may show irregular shapd dark-brown blotches. The leaves specially the younger ones, may show stripping or mottling and intervenal chlorosis. Plants infected with RTBV alone show similar milder tungro symptoms. But plants infected with RTSV alone show no obvious symptoms except very mild stunting.

The Tungro virus infected rice plants are studied and show mottling and yellow-discolouration of the leaves. Typical tungro symptoms can appeared caused by RTBV. But they are intensified by the infection of RTSV. RTSV often occurs alone causing only mild symptoms. The disease caused by RTSV alone was earlier known as **Rice Tungro Spherical Waika Virus (RTSV).** It is double-stranded DNA virus. The Waika viruses infect only certain species of grain crops and weeds. The virus particles occur in granular inclusions in the cytoplasm of phloem cells and occasionally in masophyll cells. They are trasmitted by specific leafhoppers in the semipersistent manner.

Control. Suggested control measures are:

1. The best way to control is by keeping the Tungro virus disease of rice from any rice growing area by following the systems of quarantine, inspection and certification.

2. Eradication of diseased plants by hand picking to eliminate inoculum vectors from the fields.

3. Rice plant is the major host of Tungro virus disease. Perhaps some rice field weeds can be infected by the Tungro virus. This virus is also found infecting rice stubbles and volunteer rice plants in the rice field. Best way is to go for hand picking or burning to cut down the source of Tungro virus inoculum.

4. In some Asian tropical countries, rice cultivators use *Nephotetlix virescens* resistant rice varieties with success.

5. Rice cultivators also apply various fungicides as Ribavirin (viricides) is applied as a spray, or injected into rice plant to reduce Tungro virus infected symptoms.

6. Spraying of 0.1 p.c. insecticides—Carbafururam and Creptomethrin give significant reduction of vector population.

7. Control of waiker viruses depends on the use of virus or vector-resistant rice varieties.

Chapter 26

Principal Diseases of Common Crop Plants

Disease		Causal agency
		I. Rice (*Oryza sativa* L.)
		(i) Fungal disease
Downy mildew	...	*Sclerospora oryzae* Brizi
False smut or green smut	...	*Ustilaginoidea virens (Cke.)* Tak.
Arkanas foot rot or brown sheath	...	*Ophiobolus oryzinus* Sacc.
Bakanae disease and foot rot	...	*Gibberella fujikuroi* (Saw.) Wr.
Scab	...	*Gibberella zeae* (Schw.) Petch
Seedling blight and foot rot	...	*Corticium rolfsii* Curzi, its imperfect stage is *Sclerotium rolfsii* Sacc.
Oriental sheath and leaf spot	...	*Corticium sasakii* (Shirai) Matsumoto
Bunt	...	*Tilletia barclayana* Sacc. & Sydow
Leaf smut	...	*Entyloma oryzae* H. & P. Sydow
Kernel bunt	...	*Neovossia horrida* (Tak.) Padwick & Azmatullah Khan
Helminthosporiose or Brown spot or Sesame leaf spot	...	*Helminthosporium orzyae* Breda de Haan., its perfect stage is *Cochliobolus miyabeanus* (Ito & Kuribay.) Dreschler ex Dastur
Irregular stem rot or Sclerotial disease	...	*Helminthosporium sigmoideum* Cav. var. *irregulare* Cralley & Tullis
Minute lead and grain spot or Nigropora disease	...	*Nigrospora oryzae* (Berk. and Br.) Petch
Blast	...	*Pyricularia oryzae* Cav.

Disease		Causal agency
Stackburn disease	...	*Trichoconis padwickii* Ganguly
Narrow brown leaf spot	...	*Cercospora oryzae* Miyake
Stem rot	...	*Sclerotium oryzae* Catt., its conidial stage is *Helminthosporium sigmoideum* Cav., and perfect stage is *Leptosphaeria salvinii* Catt.
Foot rot	...	*Fusarium moniliforme* Sheld.
Rhizoctonia sheath spot	...	*Rhizoctonia oryzae* Ryler & Gooch
Udbatta disease	...	*Ephelis oryzae* Syd., its perfect stage is *Balansia oryzae* (Syd.) Nar. & Thirum.
	(ii)	Bacterial disease
Bacterial blight	...	*Xanthomonas oryzae* (Uyeda & Ishiyama) Dowson
Black rot of rice grains	...	*X. itoana* Tochinai
	(iii)	Virus disease
Dwarf or stunt disease	...	Rice dwarf virus
Stripe disease	...	Rice stripe virus
	(iv)	Nematode disease
Ufra or Dakpora disease	...	*Ditylenchus angustus* (Butler) Filipjev
Akhet-pet disease	...	*Anguillulina* sp.
	(v)	Deficiency disease
Straight head	...	Soil flooded and poorly aerated
Khaira disease	...	Due to non-availability of Zinc
Pan-sukha or Dry leaf disease	...	Due to accumulation of sulphuretted hydrogen; to oxygen deficiency; to rapid decomposition of organic matter in flooded water under high temperature; to a condition of the soil causing the reduction of nitrate to nitrite, sulphate to sulphite, ferric to ferrous salts, which may be toxic to the plants; to the production of methane.

II. Wheat (*Triticum aestivum* L.)

Disease		Causal agency
	(i)	Fungal disease
Powdery mildew	...	*Erysiphe graminis tritici* (DC.) E. P. Marchal
Ergot disease	...	Race of *Claviceps purpurea* (Fr.) Tul.
Black or Stem rust or Black stem rust	...	*Puccinia graminis* Pers. f. sp. *tritici* Eriks. & Henn.
Yellow or Stripe rust	...	*Puccinia striiformis* West.
Brown or Orange or Leaf rust	...	*Puccinia recondita* Rob. & Desm.

Disease		Causal agency
Loose smut	...	*Ustilago nuda* (Hens.) Rostr.
Flag smut or Leaf smut	...	*Urocystis tritici* Koern.
Bunt or Stinking smut	...	*Tilletia caries* (DC.) Tul. (rough-spored) and *Tilletia foetida* (Wallr.) Liro. (smooth-spored)
Dwarf bunt	...	*Tilletia controversa* Kuhn
Karnal bunt	...	*Tilletia barclayana* Sacc. & Sydow
Leaf blight	...	*Alternia tenuis* Auct. *Alternaria triticola* Vasant Rao *Alternaria triticina* Prasada & Prabhu
Leaf blotch	...	*Septoria tritici* Rob. and Desm., its perfect stage is *Leptosphaeria tritici* Weber.
Foot rot	...	*Pythium graminicolum* Subr. and *P. arrhenomanes* Dresch.
Helminthosporium leaf spot	...	*Helmintosporium sativum* Pammel, King and Bakke.
		(ii) Bacterial disease
Bacterial rot or Bacterial spike blight or Yellow slime disease	...	*Corynebacterium tritici* (Hutch.) Burkholder
Yellowing rot	...	*Corynebacterium rathayi* (Smith) Dowson
		(iii) Virus disease
Soil-borne mosaic	...	*Marmor tritici*
Streak mosaic	...	*Marmor virgatum*
		(iv) Nematode disease
Ear cockle of wheat	...	*Anguina tritici* (Steinb.) Filipjev
Root knot disease	...	*Meloidogyne arenaria*

III. Potato (*Solanum tuberosum* L.)

(i) Fungal disease

Disease		Causal agency
Wart of potato	...	*Synchytrium endobioticum* (Schilb.) Perc.
Late blight	...	*Phytophthora infestans* (Mont.) D By.
Early blight	...	*Alternaria solani* (Ell. and Martin) Jones and Grout.
Black scurf	...	*Rhizoctonia solani* Kuhn.
Leaf blotch	...	*Cercospora concors* (Casp.) Sacc.
Charcoal rot	...	*Macrophomina phaseoli* (Maubl.) Ashby
Fusarium wilt	...	(a) *Fusarium oxysporum* f. *tuberosii* Snyder and Hansen (b) *Fusarium solani* var. *eumartu* Wr. emend Synder and Hansen

Disease	Causal agency
Collar rot	... *Sclerotium rolfsii* Sacc.
	(ii) Myxomycete disease
Powdery scab	... *Spongospora subterranea* (Wallr.) Lagerheim
	(iii) Bacterial disease
Bacterial brown rot and Wilt	... *Xanthomonas solanacearum* (E. F. Smith) Dowson
Black leg and soft rot	... (a) *Erwinia carotovora* (Jones) Holland (b) *Erwinia atroseptica* (van Hall) Jennison (c) *Erwinia aroideae* (Townsend) Holland (d) *Bacillus mesentricus* (Flugge) Migula
Ring rot	... *Corynebacterium sepedonicum* (Spieck and Kotlh) Skapt and Burkh.
Brown rot	... *Pseudomonas solanacearum* E. F. Smith.
	(iv) Virus disease
Leaf roll	... Leaf roll *virus* also known as Potato *Virus* 1 or *Solanum Virus* 14
Mild mosaic	... *Potato Latent Virus* also known as *Potato Mottle Virus, Potato Virus X, Solanum Virus* 1
Rugose mosaic	... *Potato Virus* X, it is also known as Vein banding virus or *Solanum Virus* 2
Crinkle	... *Potato Virus* X and *Potato Virus* A (*Solanum Virus* 3)
	(v) Nematode disease
Root knot disease	... (a) *Meloidogyne incognita* (Kofoid and White) Chitwood (b) *Meloidogyne incognita acrita* Chitwood (c) *Meloidogyne javanica* (Treub.) Chitwood
	(vi) Deficiency disease
Black heart	... Storage, transit and market disease of potatoes; caused by unfavourable oxygen relations. It occurs in tubers stored improperly ventilated room in packed condition, in the field when the temperature of the soil goes above 90°F.
Spots on leaves	... Due to Manganese deficiency (light-green spots on leaves)

Disease		Causal agency
IV. Sugarcane (*Saccharum officinarum* L.)		
		(i) Fungal disease
Red rot	...	*Colletotrichum flacatum* Went, its perfect stage is *Glomerella tucumanensis* (Speg.) Arx and Muller
Smut	...	*Ustilago scitaminea* Syd.
Wilt	...	*Cephalosporium sacchari* Butler
Ring spot	...	*Leptosphaeria sacchari* Breda de Haan
Sett rot or Pineapple disease	...	*Ceratocystis pardoxa* (Dade) Moreau
Downy mildew	...	*Sclerospora sacchari* Miyake
		(ii) Bacterial disease
Red stripe	...	*Xanthomonas rubrilineans* (Lee *et al.*) Starr and Barkholder
Gummosis or Gumming disease	...	*X. vasculorum* (Cobb.) Dowson
		(iii) Virus disease
Mosaic	...	(a) Sugarcane Mosaic Virus (b) Saccharum Virus 1 (c) Sugarcane Virus 1
		(iv) Nematode disease
Root knot	...	*Meloidogyne javanica* (Treub.) Chitwood

Part Seven

Bryophyta

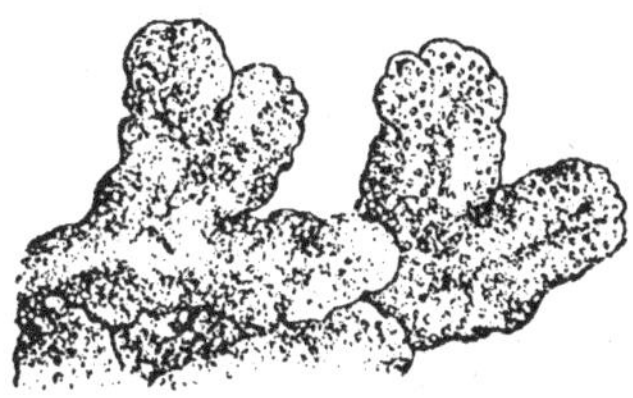

Chapter 27

Introduction to the Bryophyta

It was realised in the later eighteenth century that it was not quite scientific to divide the Plant Kingdom into the two subkingdoms *Cryptogamia* and *Phanerogamia* as was done in the classical systems of classification. Four main divisions of the Plant Kingdom—*Thallophyta, Bryophyta, Pteridophyta* and *Spermophyta* (seed plants) seemed to be more appropriate. A. P. de Candolle (1819) divided the Plant Kingdom into the *Cellulares* (Thallophyta and Bryophyta) and the *Vasculares* (Vascular plants). Endlicher (1826-40) divided the Plant Kingdom into *Thallophyta* and *Cormophyta.* In the 1880's, Engler also suggested that the Plant Kingdom be divided into two groups—the *Thallophyta* and the *Embryophyta,* the latter to include all the plants excepting the thallophytes as in all these the zygote develops a multicellular embryo. The *Embryophyta* has been called *Cormophyta* by Endlicher, Eichler and other authors because of the definite solid *Cormus* type plant body. Another term, *Telemophyta,* in place of Embryophyta, has been suggested by Takhtajan (1953) on the basis of the Telome Theory (fully explained in the concluding chapter of Pteridophyta). In his *Syllabus der Pflanzenfamilien,* Engler divided the *Embryophyta* into two divisions (abteilungen)—(i) *Embryophyta asiphonogama* or the *Archegoniatae* which includes the Bryophyta and the Pteridophyta and the (ii) *Embryophyta siphonogama* or the *Spermaphyta.* In the latest edition of the *Syllabus* (1954) the Thallophytes have been divided into 13 divisions and the Embryophyta into 4 (Bryophyta, Pteridophyta, Gymnosperm, and Angiosperm). By the early twentieth century it became clear that the vascular plants (i.e., the Pteridophytes and the Spermaphytes) form a rather distinct group within the *Embryophyta* and some authorities suggested that these should form a distinct group named *Trachaeophyta.*

According to the International Code of Botanical Nomenclature the Plant Kingdom has to be divided into a number of *divisions* each ending in the suffix—*phyta.* There is no rule about intermediate groupings (subkingdom, etc.). Tippo (1942) divided the Plant Kingdom into the subkingdoms, Thallophyta and Embryophyta, as proposed by Engler and then subdivided each of them into *phyta,* a term not regular for Botany and should be changed to *divisions* according to the International Code. The *Embryophyta* are divided by him into *Bryophyta* and *Trachaeophyta*. Takhtajan (1953) and others, however, do not think Trachaeophyta as natural and puts Bryophyta within other Trachaeophytes considering it as a reduced form from Trachaeopytes. In the latest classification by Cronquist, Takhtajan and Zimmermann (1966) the Plant Kingdom has been divided as follows:

Subkingdom: Thallobionta (= Thallophytes)

Subkingdom: Embryobyonta (= Embryophytes or Cormophytes or Telemophytes)

Division: Rhyniophyta (= Psilophytales)

Division: Bryophyta

Division: Psilotophyta (= Psilotales) ... and other Trachaeophyte divisions.

It may thus be seen that while the terms *Cryptogamia, Phanerogamia* have become obsolete and the term *Pteriodophyta* still open to dispute, *Bryophyta* still stands as a definite entity. But, still there are some doubts as to whether the *Bryophyta* should be considered as a *division* or as a *subkingdom.* Bold (1957) has suggested that *Bryophyta* should be considered as a *subkingdom* comprizing of two separate divisions—*Hepaticophyta* and *Bryophyta.* In that case, the subkingdom should be called *Bryophytae,* along with the two other subkingdoms—the *Thallophytae* and the *Trachaeophytae.* But, it has already been pointed out that Cronquist, Takhtajan and Zimmermann has not accepted the idea of separation into Bryophyta and Trachaeophyta. Whether the *Thallophytae* should be broken up into a number of subkingdoms is not considered in this Part of the book. In this book the *Bryophyta* is considered as a *division.*

GENERAL CHARACTERS OF THE BRYOPHYTES

The Bryophytes, generally, form a natural group where the common plant body is a *gamatophyte,* whereas in the Trachaeophytes the common plant is a *sporophyte.* They represent a phase in evolution when plants migrated from water to land. Water is so much necessary for the completion of their life history that we may describe them as the *amphoebians* among plants. The bryophytic gametophyte is either a thallose structure or becomes foliose in the foliose hepatics and the mosses. In mosses there is also a temporary filamentous form of the gametophyte, called the *protonema.* The protonemal stage is also present in the Hepatics but the protonema in this group is merely a continuous first stage of gametophyte formation and not a separate stage as in the mosses. The gametophytic body is green with chlorophyll and possesses *rhizoids* and *scales* for absorption although it may be able to absorb through its whole surface. In the foliose forms, specially, in the mosses, the terms 'stem' and 'leaf' are often applied to the organs. But, this is not quite correct as here the plant body is a gametophyte and not a sporophyte as in the higher plants (Trachaeophytes) where the terms 'stem' and 'leaf' correctly apply. It is perhaps more appropriate to use the terms *phyllid* (for leaf-like appendages of gametophytes) and *caulid* (for the apparent stems) as suggested by Koch (1958). The Bryophytic gametophytes are independent autophytes. The sex organs, *antheridia* and *archegonia,* are multicellular structures within which the *gametes* (eggs and biciliate sperms) are developed. The embryo develops within the venter of the archegonium and ultimately develops the *sporophyte.* Unlike the Trachaeophytes, this sporophyte is a very rudimentary structure completely parasitic on the gametophyte and devoid of such common organs as leaf or root or even regular vascular tissues. At most it is differentiated into a haustorial *foot,* a *seta*—the forerunner of the stem, and a *capsule* which carries the sporangium developing spores by reduction division.

CLASSIFICATION OF THE BRYOPHYTA

The *Bryophyta* is such a distinctive group of plants that it may easily be considered as a definite *division* of the Plant Kingdom. It is, again, phylogenetically so different from the Thallophytes and the Trachaeophytes that it is considered by some as a definite *subkingdom* parallel to the two others and named *Bryophytae* (the ending-*phytae* distinguishes it from the division ending-*phyta*). The *Bryophyta* then becomes the single *division* of the subkingdom *Bryophytae*. As against this, we have the view of Cronquist, Takhtajan, Zimmermann and others that the Bryophytes being possibly reduced from some higher form, should not be separated from the Trachaeophytes. It may at least be said that, representing a line of evolution of the Land Plants like the Pteridophytes, they deserve a parallel position.

Within the *Bryophyta*, again, two types of plants, *Hepatics* or *Liverworts* (from their imaginary similarity with the liver and the consequent supposed property of curing liver diseases in medieval medicine; Green *hepatos*=liver) with dorsiventral differentiation and *Musci* or the mosses which usually show radial symmetry. In classical taxonomy *Hepaticae* and *Musci* are considered as two *classes* under the Bryophytes. This position has been retained even in the 1954 edition of Engler's *Syllabus der Pflanzenfamilien*. Cavers (1910-11) suggested that the Bryophytes be divided into ten groups of equal rank:

I. *Sphaerocarpales*
II. *Marchantiales*
III. *Fungermanniales*
IV. *Anthocerotales*
V. *Sphagnales*
VI. *Andreaeales*
VII. *Tetraphidales*
VIII. *Polytrichales*
IX. *Buxbaumiales*
X. *Eubryales*

But, this ignores the difference between the *Hepaticae* and the *Musci* which attitude does not seem to be justified. The differences between the two groups are so great that some modern taxonomists want to raise the two groups to two *subdivisions* or even *divisions*. Bold (1956) suggested *Hepaticophyta* and *Bryophyta* (*Musci*) as two *divisions*. Leaving aside this question, most modern Bryologists, beginning from Howe (1899) and Campbell (1918), are almost unanimus that the *Anthocerotae*, formerly included within the class *Hepaticae*, are sufficiently distinctive to be raised to the status of a class so that three classes, *Hepaticae*, *Anthocerotae* and *Musci*, are recognized within the division *Bryophyta*. These *class* names are changed to *Hepaticopsida*, *Anthocerotopsida* and *Bryopsida* (proposed by Rothmaler 1951 and modified by Prostauer 1957) in accordance with the International Code of Nomenclature which recommends that *class* names should end in the suffix-*psida*.

Thus,

Division: **Bryophyta**

Class: **Hepaticopsida** (*Hepaticae*). Dorsiventrally differentiated (sometimes radial and erect) gametophytes which are either simply thallose or with leaf-like appendages (foliose) and are always nerveless. Internal tissues are homogeneous or differentiated. Sex organs borne terminally or developed from superficial dorsal cells. Sporophyte simple or differentiated into foot, seta and capsule—always of limited growth. Sporogenous cells of endothecial origin and without any columella.

Class: **Anthocerotopsida** (*Anthocerotae*). Gametophyte dorsiventrally differentiated, simple and not foliose, internal tissue homogenous. Sex organs embedded in gametophyte—female formed from superficial and male from hypodermal dorsal cells. Sporophyte differentiated into capsule and foot. Lower part of capsule remains meristematic so that the growth of the sporophyte is not limited. Sporogenous cells of amphithecial origin (with exceptions in *Notothylas*), all internal cells forming a sterile columella.

Class: **Bryopsida** (*Musci*). Gametophyte, after a distinct though usually short protonema phase, develops erect shoots, usually of radial symmetry, with leafy appendages (phyllids) on stem-like caulids. Sexual organs develop from superficial cells at the top of apical or axillary meristems. Sporophyte of limited growth, usually differentiated into foot, seta and capsule. Sporogenous cells of endothecial or amphithecial origin but always (with rare exceptions) surrounding a columella.

The question as to what is the order of evolution of these classes and what is the trend of evolution within each class is discussed in chapter 32 as such discussions need a more thorough knowledge of the groups. In the following chapters, the classes are discussed in the order shown above.

Chapter 28

Hepaticopsida

The Class *Hepaticopsida* (*Hepaticae*) includes all the Hepatics excepting those which are taken out to form the class *Anthocerotopsida* (*Anthocerotae*). The class, formerly, used to be divided into two orders *Marchantiales* and *Jungermanniales*. Cavers (1910) suggested the taking out of the *Sphaerocarpales* to form a separate order. Similarly, Campbell (1936) suggested the taking out of the *Calobryales* out of the *Jungermanniales*. So the current idea is to distinguish four orders under the *Hepaticopsida:*

1. *Sphaerocarpales*
2. *Marchantiales*
3. *Fungermanniales*
4. *Calobryales.*

The recently discovered genus *Takakia* presents a problem as to its systematic position. It is an erect plant like the Calobryales but apparently does not belong to that order. Its antheridium and sporophyte are not yet known, so that it cannot be definitely said whether it is a representative of a taxon parallel to Hepaticopsida and Anthocerotopsida. It is provisionally placed here as the fifth order of the Hepaticopsida:

5. *Takakiales*

Reimers (Engler, 1954), however, still retains Anthocerotales within the class *Hepaticae* and also retains Calobryales within Jungermanniales.

Recent years, biologist consider class Hepaticopsida as Marchatiopsida under phyllum Marchantiophyta. An outline classification is thus added.

CLASSIFICATION

PHYLUM (DIVISION): MARCHANTIOPHYTA

CLASS: MARCHANTIOPSIDA Stotler & Stotl.-Crand., Bryologist 80 : 426.1977.

SUBCLASS: SPHAEROCARPIDAE Stotler & Stotl.-Crand., Bryologist 80 : 426.1977

Order: Sphaerocarpales Cavers, New Phytol. 9 : 81.1910

Suborder: Sphaerocarpineae R. M. Schust. ex Stotler & Stotl.-Crand.

Sphaerocarpaceae (Dumort.) Heeg, Verh. K. K. Zool.-Bot. Ges. Wien 41: 573. 1891. *Sphaerocarpos* Boehm., *Geothallus* Campb.

Suborder: Riellineae R. M. Schust. ex Stotler & Stotl.-Crand.

Riellaceae Engl., Syll. Pflanzenfam. Grosse Ausgabe. p. 45. 1892. *Riella* Mont.

SUBCLASS: MARCHANTIIDAE Engl. [Unterklasse "Marchantiales"] in Engler & Prantl, Naturl. Pflanzenfam. I, 3. I: 1.1893.

Order: Monocleales R. M. Schust., J. Hattori Bot. Lab. 26: 296.1963.

Monocleaceae (Nees) A. B. Frank in Leunis, Syn. Pflanzenk. (ed. 2)p. 1556.1877. *Monoclea* Hook.

Order: Marchantiales Limpr. in Cohn, Krypt.—Fl. Schlesien 1 : 239, 336. 1876 [1877].

Suborder: Marchantiineae Buch ex Schljakov, Bot. Zhurn. 57: 507.1972.

Aytoniaceae Cavers, New Phytol. 10 : 42. 1911. *Asterella* P. Beauv., *Cryptomitrium* Austin ex Underw., *Mannia* Opiz nom. cons., *Plagiochasma* Lehm. & Lindenb. nom. cons., *Reboulia* Raddi nom. cons.

Wiesnerellaceae Inoue, Ill. Jap. Hep. 2 : 192. 1976. *Wiesnerella* Schiffn.

Conocephalaceae Müll. Frib. ex Grolle, J. Bryol. 7 : 207. 1972. *Conocephalum* Hill nom. cons.

Lunulariaceae Klinggr., Höh. Crypt. Preuss. p. 9. 1858. *Lunularia* Adans.

Marchantiaceae (Bisch.) Lindl., Nat. Syst. Bot. (ed. 2) p. 412. 1836. *Bucegia* Radian, *Dumortiera* Nees, *Marchantia* L., *Neohodgsonia* Perss., *Preissia* Corda

Monosoleniaceae Inoue, Bull. Natl Sci. Mus. Ser. 2, 9 : 117.1966. *Monosolenium* Griff., *Peltolepis* Lindb.

Cleveaceae Cavers. New Phytol. 10: 42. 1911. *Athalamia* Falconer, *Sauteria* Nees

Exormothecaceae Müll. Frib. ex Grolle, J. Bryol. 7 : 208. 1972. *Aitchisoniella* Kashyap, *Exormotheca* Mitt., *Stephensoniella* Kashyap

Suborder: Corsiniineae R. M. Schust. ex Schljakov, Bot. Zhurn. 57 : 507. 1972.

Cyathodiaceae (Grolle) Stotler & Stotl.-Crand., stat. et fam. nov. *Cyathodium* Kunze

Corsiniaceae Engl., Syll. Pflanzenfam. Grosse Ausgabe. p. 44. 1892. *Corsinia* Raddi, *Cronisia* Berk.

Suborder: Monocarpineae R. M. Schust. "Carrpinae," J. Hattori Bot. Lab. 26 : 299. 1963.

Monocarpaceae D. J. Carr ex Schelpe, J. South African Bot. 35 : 110. 1969. *Monocarpus* D. J. Carr

Suborder: Targioniineae R. M. Schust. ex Schljakov, Bot. Zhurn. 57 : 507. 1972.

Targioniaceae Dumort., Anal. Fam. Pl. p. 68, 70. **1829**. *Targionia* L.

Order: Ricciales Schljakov emend. Stotler & Stotl.-Crand.

Oxymitraceae Müll. Frib. ex Grolle, J. Bryol. 7 : 215. 1972. *Oxymitra* Bisch. ex Lindenb.

Ricciaceae Reichenb., Bot. Damen. p. 255. 1828. *Riccia* L., *Ricciocarpos* Corda.

CLASS: JUNGERMANNIOPSIDA Stotler & Stotl.-Crand., Bryologist 80 : 425. 1977.

SUBCLASS: METZGERIIDAE Barthol.-Began, Phytologia 69 : 465. 1990.

Order: Haplomitriales H. Buch ex Schljakov, Bot. Zhurn. 57 : 499. 1972.

Haplomitriaceae Dedecek, Arch. Prir. Proskoumání Cech 5 : 71. 1884. *Haplomitrium* Nees nom. cons.

Order: Blasiales (R. M. Schust.) Stotler & Stotl.-Crand., stat. et ordo nov.

Blasiaceae H. Klinggr., Höh. Crypt. Preuss. p. 14. 1858. *Blasia* L., *Cavicularia* Steph.

Order: Treubiales Schljakov emend. Stotler & Stotl.-Crand.

Suborder: Treubiineae R. M. Schust. ex Stotler & Stotl.-Crand.

Treubiaceae Verd., Man. Bryol. p. 427. 1932. *Apotreubia* S. Hatt. & Mizut., *Treubia* K. I. Goebel nom. cons.

Suborder: Phyllothalliineae R. M. Schust., Trans. Brit. Bryol. Soc. 5 : 283. 1967.

Phyllothalliaceae E. A. Hodgs., Trans. Roy. Soc. New Zealand, Bot. 2 : 247. 1964. *Phyllothallia* E. A. Hodgs.

Order: Fossombroniales Schljakov emend. Stotler & Stotl.-Crand.

Suborder: Fossombroniineae R. M. Schust. ex Stotler & Stotl.-Crand.

Fossombroniaceae Hazsl. nom. cons., Magyar Birodalom Moh-Flórája. p. 20. 1885. *Austrofossombronia* R. M. Schust., *Fossombronia* Raddi. *Petalophyllum* Nees & Gottsche ex Lehm., *Sewardiella* Kashyap

Allisoniaceae (R. M. Schust. ex Grolle) Schljakov, [Liverworts, morphology, phylogeny, classification] p. 119. 1975. *Allisonia* Herzog, *Calycularia* Mitt.

Suborder: Pelliineae R. M. Schust. ex Schljakov emend Stotler & Stotl.-Crand.

Pelliaceae H. Klinggr., Höh. Crypt. Preuss. p. 13. 1858. *Noteroclada* Taylor ex Hook. & Wils., *Pellia* Raddi nom. cons.

Sandeothallaceae R. M. Schust., New Man. Bryol. p. 951. 1984. *Sandeothallus* R. M. Schust.

Order: Metzgeriales Schljakov emend. Stotler & Stotl.-Crand.

Suborder: Pallaviciniineae R. M. Schust. emend. Stotler & Stotl.-Crand.

Pallaviciniaceae Mig., Krypt.-Fl. Deutschl., Moose. p. 423. 1904.

Greeneothallus Hässel, *Hattorianthus* R. M. Schust. & Inoue, *Jensenia* Lindb., *Moerckia* Gottsche, *Pallavicinia* Gray nom. cons., *Podomitrium* Mitt., *Seppeltia* Grolle, *Symphyogyna* Nees & Mont., *Symphyogynopsis* Grolle, *Xenothallus* R. M. Schust.

Makinoaceae Nakai, [Ord. Fam. Trib. Nov.] p. 201. 1943. *Makinoa* Miyake, *Verdoornia* R. M. Schust.

Hymenophytaceae R. M. Schust., J. Hattori Bot. Lab. 26 : 296. 1963. *Hymenophyton* Dumort.

Suborder: Metzgeriineae R. M. Schust. ex Schljakov, Bot. Zhurn. 57 : 501. 1972.

Aneuraceae H. Klinggr., Höh. Crypt. Preuss. p. 11. 1858. *Aneura* Dumort., *Cryptothallus* Malmb., *Lobatoriccardia* (Mizut. & S. Hatt.) Furuki, *Riccardia* Gray nom. cons.

Mizuaniaceae Furuki & Z. Iwats., J. Hattori Bot. Lab. 67: 291. 1989. *Mizutania* Furuki & Z. Iwats.

Vandiemeniaceae Hewson, J. Hattori Bot. Lab. 52 : 163. 1982. *Vandiemenia* Hewson

Metzgeriaceae H. Klinggr., Höh. Crypt. Preuss. p. 10. 1858. *Apometzgeria* Kuwah., *Austrometzgeria* Kuwah., *Metzgeria* Raddi, *Steereella* Kuwah.

SUBCLASS: JUNGERMANNIIDAE Engl. emend. Stotler & Stotl.-Crand.

Order: Lepicoleales Stotler & Crand.-Stotl., ordo nov.

Suborder: Ptilidiineae R. M. Schust., J. Hattori Bot. Lab. 26 : 227. 1963.

Ptilidiaceae H. Klinggr., Höh. Crypt. Preuss. p. 37. 1858. *Ptilidium* Nees

Mastigophoraceae R. M. Schust., J. Hattori Bot. Lab. 36 : 345. 1972. *Dendromastigophora* R. M. Schust., *Mastigophora* Nees nom. cons.

Chaetophyllopsidaceae R. M. Schust., J. Hattori Bot. Lab. 23 : 68. 1960 [1961]. *Chaetophyllopsis* R. M. Schust., *Herzogianthus* R. M. Schust.

Suborder: Lepicoleineae R. M. Schust., J. Hattori Bot. Lab. 36 : 336, 338. 1972 [1973].

Vetaformataceae Fulford & J. Taylor, Mem. New York Bot. Gard. 11 : 27. 1963. *Vetaforma* Fulford & J. Taylor

Lepicoleaceae R. M. Schust., Nova Hedwigia 5 : 27. 1963. *Lepicolea* Dumort.

Suborder: Perssoniellineae R. M. Schust., J. Hattori Bot. Lab. 26 : 229-230. 1963.

Schistochilaceae H. Buch, Commentat. Biol. 3 : 9. 1928. *Gottschea* Nees ex Mont., *Pachyschistochila* R. M. Schust. & J. J. Engel, *Paraschistochila* R. M. Schust., *Pleurocladopsis* R. M. Schust., *Schistochila* Dumort.

Perssoniellaceae R. M. Schust. ex Grolle, J. Bryol. 7 : 216. 1972. *Perssoniella* Herzog

Suborder: Lepidolaenineae R. M. Schust., J. Hattori Bot. Lab. 36 : 345. 1972[1973].

Trichocoleaceae Nakai, [Ord. Fam. Trib. Nov.] p. 201. 1943. *Eotrichocolea* R. M. Schust., *Trichocolea* Dumort. nom. cons. [*Leiomitra* Lindb. = *Trichocolea*]

Lepidolaenaceae Nakai, [Ord. Fam. Trib. Nov.] p. 200. 1943. *Gackstroemia* Trevis., *Lepidogyna* R. M. Schust., *Lepidolaena* Dumort., *Trichocoleopsis* S. Okamura

Neotrichocoleaceae Inoue, Ill. Jap. Hepat. 1 : 176. 1974. *Neotrichocolea* S. Hatt.

Jubulopsidaceae (Hamlin) R. M. Schust., Phytologia 56 : 68. 1984.

***Jubulopsis* R. M. Schust.**

Order: Jungermanniales H. Klinggr. emend. Stotler & Stotl.-Crand.

Suborder: Herbertineae R. M. Schust., J. Hattori Bot. Lab. 26 : 232. 1963.

Herbertaceae Müll. Frib. ex Fulford & Hatcher, Bryologist 61 : 284. 1958. *Herbertus* Gray, *Triandrophyllum* Fulford & Hatcher

Pseudolepicoleaceae Fulford & J. Taylor, Nova Hedwigia 1 : 411. 1960. *Archeophylla* R. M. Schust., *Blepharostoma* (Dumort.) Dumort., *Chaetocolea* Spruce, *Herzogiaria* Fulford ex Hässel, *Isphyllaria* E. A. Hodgs. & Allison, *Pseudolepicolea* Fulford & J. Taylor [*Archeochaete* R. M. Schust. and *Lophochaete* R. M. Schust. = *Pseudolepicolea*], *Temnoma* Mitt.

Trichotemnomataceae R. M. Schust., J. Hattori Bot. Lab. 36 : 340. 1973. *Trichotemnoma* R. M. Schust.

Grolleaceae Solari ex R. M. Schust., Phytologia 56 : 66. 1984. *Grollea* R. M. Schust.

Suborder: Balantiopsidineae R. M. Schust., J. Hattori Bot. Lab. 36 : 353. 1972[1973].

Balantiopsidaceae H. Buch, Thüring. Bot. Ges. 1 : 23. 1955. *Anisotachis* R. M. Schust., *Austroscyphus* R. M. Schust., *Balantiopsis* Mitt., *Eoisotachis* R. M. Schust., *Hypoisotachis* (R. M. Schust.) J. J. Engel & Merr., *Isotachis* Mit., *Neesioscyphus* Grolle, *Ruizanthus* R. M. Schust.

Suborder: Lophocoleineae Schljakov, Bot. Zhurn. 57 : 504. 1972 (= Geocalycineae R. M. Schust., 1972[1973]).

Geocalycaceae H. Klinggr., Höh. Crypt. Preuss. p. 34. 1858 [incl. Lophocoleaceae (Jörg.) Vanden Berghen in Robyns, Fl. Gén. Belgique, Bryophytes 1 : 208. 1956]. *Campanocolea* R. M. Schust., *Chiloscyphus* Corda nom. cons., *Clasmatocolea* Spruce, *Conoscyphus* Mitt., *Evansianthus* R. M. Schust. & J. J. Engel, *Geocalyx* Nees, *Harpanthus* Nees, *Hepatostolonophora* J. J. Engel & R. M. Schust., *Heteroscyphus* Schiffn. nom. cons., *Lamellocolea* J. J. Engel, *Leptophyllopsis* R. M. Schust., *Leptoscyphopsis* R. M. Schust., *Leptoscyphus* Mitt., *Lophocolea* (Dumort.) Dumort., *Pachyglossa* Herzog & Grolle, *Pedinophllopsis* R. M. Schust. & Inoue, *Perdusenia* Hässel, *Pigafettoa* C. Massal., *Platycaulis* R. M. Schust., *Pseudolophocolea* R. M. Schust. & J. J. Engel, *Saccogyna* Dumort. nom. cons., *Saccogynidium* Grolle, *Stolonivector* J. J. Engel, *Tetracymbaliella* Grolle, *Xenocephalozia* R. M. Schust.

Gyrothyraceae R. M. Schust., Trans. Brit. Bryol. Soc. 6 : 87. 1970. *Gyrothyra* M. Howe.

Plagiochilaceae (Jörg.) Müll. Frib. & Herzog in Müller, Leberm. Eur. p. 877. 1956. *Acrochila* R. M. Schust., *Pedinophyllum* (Lindb.) Lindb., *Plagiochila* (Dumort.) Dumort. nom. cons., *Plagiochilidium* Herzog, *Plagiochilion* S. Hatt., *Steereochila* Inoue, *Szweykowskia* Gradst. & Reiner-Drehw., *Xenochila* R. M. Schust.

Acrobolbaceae E. A. Hodgs., Rec. Domin. Mus. 4 : 177. 1962. *Acrobolbus* Nees, *Austrolophozia* R. M. Schust., *Enigmella* G. A. M. Scott & K. G. Beckm., *Goebelobryum* Grolle, *Lethocolea* Mitt. nom. cons., *Marsupidium* Mitt., *Tylimanthus* Mitt.

Arnelliaceae Nakai, [Ord. Fam. Trib. Nov.] p. 200. 1943. *Arnellia* Lindb., *Gongylanthus* Nees, *Southbya* Spruce

Suborder: Lepidoziineae R. M. Schust., J. Hattori Bot. Lab. 26 : 228. 1963.

Phycolepidoziaceae R. M. Schust., Bull. Torrey Bot. Club 93 : 442. 1967.

Phycolepidozia R. M. Schust.

Calypogeiaceae (Müll. Frib.) Arnell in Holmberg, Skand. Fl. 2a. p. 189. 1928. *Calypogeia* Raddi nom. cons., *Eocalypogeia* (R. M. Schust.) R. M. Schust., *Metacalypogeia* (S. Hatt.) Inoue

Lepidoziaceae Limpr. in Cohn, Krypt.-Fl. Schlesien 1 : 310. 1876[1877].

Acromastigum A. Evans, *Arachniopsis* Spruce, *Austrolembidium* Hässel, *Bazzania* Gray nom. cons., *Chloranthelia R. M. Schust., Dendrobazzania* R. M. Schust. & W. B. Schofield, *Drucella* E. A. Hodgs., *Hyalolepidozia* S. W. Arnell ex Grole, *Hygrolembidium* R. M. Schust., *Isolembidium* R. M. Schust., *Kurzia* G. Martins, *Lembidium* Mitt. nom. cons., *Lepidozia* (Dumort.) Dumort. nom. cons., *Mastigopelma* Mitt., *Megalembidium* R. M. Schust., *Micropterygium* Lindenb., Nees & Gottsche, *Mytilopsis* Spruce, *Neogrollea* E. A. Hodgs., *Odontoscrics* Fulford, *Paracromastigum* Fulford & J. Taylor, *Protocephalozia* (Spruce) K. I. Goebel, *Pseudocephalozia* R. M. Schust., *Psiloclada* Mitt., *Pteropsiella* Spruce, *Telaranea* Spruce ex Schiffn., *Zoopsidella* R. M. Schust., *Zoopsis* Hook. f. ex Gottsche, Lindenb. & Nees

Suborder: Cephaloziineae Schljakov, Bot. Zhurn. 57 : 503. 1972 (= Cephaloziineae R. M. Schust., 1972 [1973]).

Cephaloziaceae Mig., Krypt.-Fl. Deutschl., Moose. p. 465. 1904.

Alobiella (Spruce) Schiffn., *Alobiellopsis* R. M. Schust., *Anomoclada* Spruce, *Cephalozia* (Dumort.) Dumort., *Cladopodiella* H. Buch, *Haesselia* Grolle & Gradst., *Hygrobiella* Spruce, *Iwatsukia* N. Kitag., *Metahygrobiella* R. M. Schust., *Nowellia* Mitt., *Odontoschisma* (Dumort.) Dumort., *Pleurocladula* Grolle, *Schiffneria* Steph., *Schofieldia* J. D. Godfrey, *Trabacellula* Fulford

Cephaloziellaceae Douin, Bull. Soc. Bot. France, Mém. 29 : 1, 5, 13. 1920.

Allisoniella E. A. Hodgs., *Amphicephalozia* R. M. Schust., *Cephalojonesia* Grolle, *Cephalomitrion* R. M. Schust., *Cephaloziella* (Spruce) Schiffn. nom. cons., *Cephaloziopsis* (Spruce) Schiffn., *Cylindrocolea* R. M. Schust., *Kymatocalyx* Herzog, *Stenorrhipis* Herzog

Jackiellaceae R. M. Schust., J. Hattori Bot. Lab. 36 : 395. 1973. *Jackiella* Schiffn.

Adelanthaceae (Jörg.) Grolle, J. Hattori Bot. Lab. 35 : 327. 1972.

Adelanthus Mitt. nom. cons., *Wettsteinia* Schiffn.

Suborder: Antheliineae R. M. Schust. ex Stotler & Stotl.-Crand.

Antheliaceae R. M. Schust., J. Hattori Bot. Lab. 26 : 236. 1963. *Anthelia* (Dumort.) Dumort.

Suborder: Brevianthineae J. J. Engel & R. M. Schust., Bryologist 85 : 382. 1982 [1983].

Brevianthaceae J. J. Engel & R. M. Schust., Phytologia 47 : 317. 1981.

Brevianthus J. J. Engel & R. M. Schust.

Chonecoleaceae R. M. Schust. ex Grolle, J. Bryol. 7 : 206. 1972. *Chonecolea* Grolle.

Suborder: Jungermanniineae R. M. Schust. ex Stotler & Stotl.-Crand.

Jungermanniaceae Rchb., Bot. Damen. p. 256. 1828 [incl. Lophoziaceae Cavers, New Phytol. 9: 293. 1910]. *Anastrepta* (Lindb.) Schiffn., *Anastrophyllum* (Spruce) Steph., *Andrewsianthus* R. M. Schust., *Anomacaulis* (R. M. Schust.) R. M. Schust. ex Grolle, *Barbilophozia* Loeske, *Bragginsella* R. M. Schust., *Cephalolobus* R. M. Schust., *Chandonanthus* Mitt., *Cryptochila* R. M. Schust., *Cryptocolea* R. M. Schust., *Cryptocoleopsis* Amak., *Denotarisia* Grolle, *Diplocolea* Amak., *Gerhildiella* Grolle, *Gottschelia* Grolle, *Gymnocolea* (Dumort.) Dumort., *Gymnocoleopsis* (R. M. Schust.) R. M. Schust., *Hattoria* R. M. Schust., *Horikawaella* S. Hatt. & Amakawa, *Jamesoniella* (Spruce) F. Lees, *Jungermannia* L., *Lophonardia* R. M. Schust., *Lophozia* (Dumort.) Dumort., *Mylia* Gray nom. cons., *Nardia* Gray nom. cons., *Nothostrepta* R. M. Schust., *Notoscyphus* Mitt., *Pisanoa* Hässel, *Protosyzygiella* (Inoue) R. M. Schust., *Pseudocephaloziella* R. M. Schust., *Rhodoplagiochila* R. M. Schust., *Roivainenia* Perss., *Scaphophyllum* Inoue, *Sphenolobopsis* R. M. Schust. & N. Kitag., *Syzygiella* Spruce, *Tetralophozia* (R. M. Schust.) Schljakov, *Tritomaria* Schiffn. ex Loeske, *Vanaea* (Inoue & Gradst.) Inoue & Gradst.

Mesoptychiaceae Inoue & Steere, Bull. Natl Sci. Mus., Tokyo, B 1 : 62. 1975. *Mesoptychia* (Lindb.) A. Evans.

Gymnomitriaceae H. Klinggr., Höh. Crypt, Preuss. p. 16, 1858. *Acrolophozia* R. M. Schust., *Apomarsupella* R. M. Schust., *Eremonotus* Lindb. & Kaal. ex Pearson, *Gymnomitrion* Corda nom. cons., *Herzogobryum* Grolle, *Marsupella* Dumort., *Nanomarsupella* (R. M. Schust.) R. M. Schust., *Nothogymnomitrion* R. M. Schust., *Paramomitrion* R. M. Schust., *Poeltia* Grolle, *Prasanthus* Lindb., *Stephaniella* J. B. Jack, *Sephaniellidium* S. Winkl. ex Grolle.

Scapaniaceae Mig., Krypt.-Fl. Deutschl., Moose. p. 479. 1904 [incl. Blepharidophyllaceae (R. M. Schust.) R. M. Schust., nom. inval. Syst. Assoc. Spec. Vol. 14: 74. "1979" 1980]. *Blepharidophyllum* Ångstr. [*Clandarium* (Grolle) R. M. Schust. = *Blepharidophyllum*], *Delavayella* Steph., *Diplophyllum* (Dumort.) Dumort. nom. cons., *Douinia* (C. N. Jensen) H. Buch, *Krunodiplophyllum* Grolle, *Macrodiplophyllum* (H. Buch) Perss., *Scapania* (Dumort.) Dumort. nom. cons., *Scapaniella* H. Buch.

Order: Porellales (R. M. Schust.) Schljakov emend. Stotler & Stotl.-Crand.

Suborder: Porellineae R. M. Schust., J. Hattori Bot. Lab. 26 : 229. 1963.

Porellaceae Cavers nom. cons., New Phytol. 9 : 292. 1910. *Ascidiota* C. Massal., *Macvicaria* W. E. Nicholson, *Porella* L.

Suborder: Jubulineae (Spruce) Müll. Frib., Lebermoose 1 : 403. 1909.

Goebeliellaceae Verd., Man. Bryol. p. 425. 1932. *Goebeliella* Steph.

Jubulaceae H. Klinggr., Höh. Crypt. Preuss. p. 40. 1858. *Frullania* Raddi, *Jubula* Dumort. nom. cons., [*Amphijubula* R. M. Schust., *Neohattoria* Kamim., *Schusterella* S. Hatt., Sharp & Mizut., and *Steerea* S. Hatt. & Kamim. = *Frullania*]

Bryopteridaceae Stotler, Bryophyt. Biblioth. 3: 57-8, 1974. *Bryopteris* (Nees) Lindenb.

Lejeuneaceae Casares-Gil nom. cons., Fl. Ibér. Brióf. 1: 703. 1919. *Acanthocoleus* R. M. Schust., *Acantholejeunea* (R. M. Schust.) R. M. Schust., *Acrolejeunea* (Spruce) Schiffn. nom. cons., *Amphilejeunea* R. M. Schust., *Anoplolejeunea* (Spruce) Schiffn., *Aphanolejeunea* A. Evans, *Aphanotropis* Herzog, *Archilejeunea* (Spruce) Schiffn., *Aureolejeunea* R. M. Schust., *Austrolejeunea* (R. M. Schust.) R. M. Schust., *Blepharolejeunea* S. W. Arnell, *Brachiolejeunea* (Spruce) Schiffn., *Bromeliophila* R. M. Schust., *Calatholejeunea* K. I. Goebel, *Capillolejeunea* S. W. Arnell, *Caudalejeunea* (Spruce) Schiffn., *Cephalanthoejeunea* R. M. Schust., *Cephalolejeunea* Mizut., *Ceratolejeunea* (Spruce) Schiffn., *Cheilolejeunea* (Spruce) Schiffn., *Cladolejeunea* Zwick., *Cololejeunea* (Spruce) Schiffn., *Colura* (Dumort.) Dumort., *Crossotolejeunea* (Spruce) Schiffn., *Cyclolejeunea* A. Evans, *Cyrtolejeunea* A. Evans, *Cystolejeunea* A. Evans, *Dactylolejeunea* R. M. Schust., *Dactylophorella* R. M. Schust., *Dendrolejeunea* (Spruce) Lacout., *Dicladolejeunea* R. M. Schust., *Dicranolejeunea* (Spruce) Schiffn., *Diplasiolejeunea* (Spruce) Schiffn., *Drepanolejeunea* (Spruce) Schiffn., *Echinocolea* R. M. Schust., *Echinolejeunea* R. M. Schust., *Evansiolejeunea* Vanden Berghen, *Frullanoides* Raddi, *Fulfordianthus* Gradst., *Haplolejeunea* Grolle, *Harpalejeunea* (Spruce) Schiffn., *Hattoriolejeunea* Mizut., *Kymatolejeunea* Grolle, *Leiolejeunea* A. Evans, *Lejeunea* Lib. nom. cons., *Lepidolejeunea* R. M. Schust., *Leptolejeunea* (Spruce) Schiffn., *Leucolejeunea* A. Evans, *Lindigianthus* Kruijt & Gradst., *Lopholejeunea* (Spruce) Schiffn. nom. cons., *Luteolejeunea* Piippo, *Macrocolura* R. M. Schust., *Macrolejeunea* (Spruce) Schiffn., *Marchesinia* Gray nom. cons., *Mastigolejeunea* (Spruce) Schiffn., *Metalejeunea* Grolle, *Metzgeriopsis* K. I. Goebel, *Microlejeunea* Steph., *Myriocolea* Spruce, *Myriocoleopsis* Schiffn., *Nephelolejeunea* Grolle, *Neurolejeunea* (Spruce) Schiffn., *Nipponolejeunea* S. Hatt., *Odontolejeunea* (Spruce) Schiffn., *Omphalanthus* Lindenb. & Nees, *Oryzolejeunea* (R. M. Schust.) R. M. Schust., *Otolejeunea* Grolle & Tixier, *Phacolejeunea* Mizut, *Physantholejeunea* R. M. Schust., *Pictolejeunea* Grolle, *Pluvianthus* R. M. Schust. & Schäfer-Verwimp, *Potamolejeunea* (Spruce) Lacout., *Prionolejeunea* (Spruce) Schiffn., *Ptychanthus* Nees, *Pycnolejeunea* (Spruce) Schiffn., *Rectolejeunea* A. Evans, *Rhaphidolejeunea* Herzog, *Schiffneriolejeunea* Verd., *Schusterolejeunea* Grolle, *Siphonolejeunea* Herzog, *Sphaerolejeunea* Herzog, *Spruceanthus* Verd., *Stenolejeunca* R. M. Schust., *Stictolejeunea* (Spruce) Schiffn., *Symbiezidium* Trevis., *Taxilejeunea* (Spruce) Schiffn. nom. cons., *Thysananthus* Lindenb., *Trachylejeunea* (Spruce) Schiffn. nom. cons., *Trocholejeunea* Schiffn., *Tuyamaella* S. Hatt., *Tuzibeanthus* S. Hatt., *Verdoornianthus* Gradst., *Vitalianthus* R. M. Schust. & Giancotti.

Order: Radulales (R. M. Schust.) Stotler & Stotl.-Crand., stat. et ordo nov.

Radulaceae (Dumort.) Müll. Frib., Lebermoose 1: 404. 1909. *Radula* Dumort. nom. cons.

Order: Pleuroziales (R. M. Schust.) Schljakov, Bot. Zhurn. 57: 505. 1972.

Pleuroziaceae (Schiffn.) Müll. Frib., Lebermoose 1: 404. 1909. *Pleurozia* Dumort. [*Eopleurozia* R. M. Schust. = *Pleurozia*]

Order 1. SPHAEROCARPALES

Three genera formerly included within the anacrogynous *Fungermanniales* have been taken out to form a separate order, the *Sphaerocorpales,* comprizing of two families (i) *Sphaerocarpaceae* with the genera *Sphaerocarpos* and *Geothallus* and (ii) *Riellaceae* with the genus *Riella.* These have simple gametophytic bodies-resembling the anacrogynous *Fungermantiales* but simpler sex organs and sporophytes resembling the *Marchantiales. Riella indica* St. in Kashyap was described from Lahore. A second species of this genus *R. vishwanathai* Pande, Misra & Srivastava (near *R. affinis* fid. Udar) has been described from near Varanasi.

Family: SPHAEROCARPACEAE

The type genus *Sphaerocarpos* (Fig. 28.1) is abundant in the Pacific coast of the USA and has a small, simple, green gametophyte with a midrib several cells in thickness which may branch dichotomously. There are lateral wing-like, one-cell thick expansions on the two sides of this midrib and, depending on environmental conditions, these wings may be entire or lobed into leaf-like expansions. There are no scales, but smooth rhizoids occur on the ventral side of the thallus. There is no internal differentiation of the gametophytic chlorophyllose tissue. The gametophytes are dioecious and are usually formed in groups of four—two male and two female. The antheridia and archegonia (Figs. 28.1C & D) are developed in the dorsal surface of the thallus, each enclosed in its own involucre, ovoid for the archegonia and flask-shaped for the antheridia. These involucres (Figs. 28.1A & B) are typical of the order. Male plants are smaller than the female. The mature sporophyte (Fig. 28.1F) shows a globular capsule and a small foot to which it is attached by a short neck or seta. The spore tetrads usually remain adhering together causing the four gametophytes (2 male and 2 female because of the presence of sex chromosomes) to grow in a cluster.

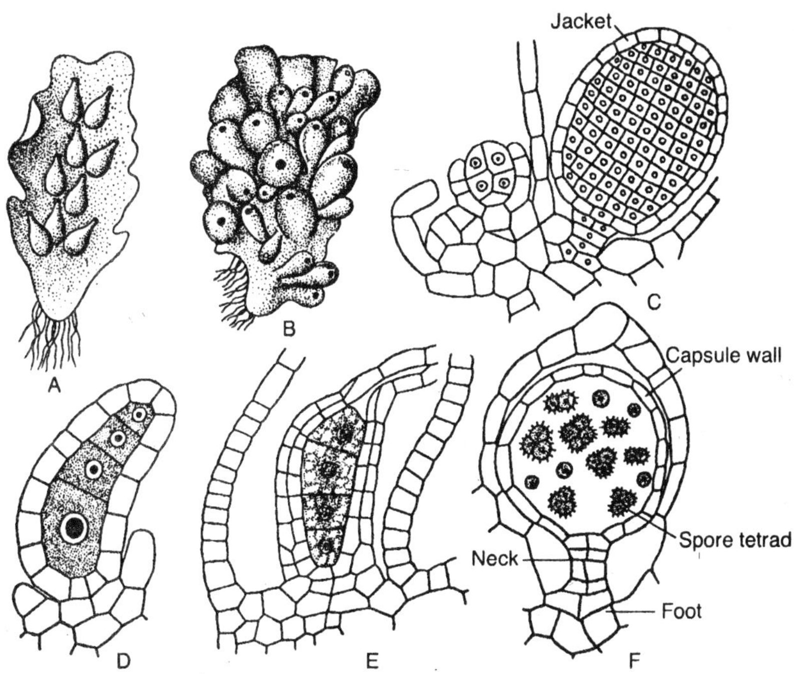

Fig. 28.1: *Sphaerocarpos.* A. Male plant with antheridia within involucres. B. Female plant with archegonia within involucres. C. L.s. of young and mature antheridia. D. Young archegonium. E. Four-celled embryo within venter. F. Mature sporophyte showing foot, neck (seta) and capsule

Family: RIELLACEAE

This family with only one genus *Riella* differs from the former in its gametophyte being erect or ascending, aquatic, held to the mud by rhizoids. There is a thick midrib with a dorsal undulate wing (usually interrupted) and lateral leafy appendages.

The involucres of sex organs are as in *Sphaerocarpos* (Fig. 28.2C). This family occurs in India with two species of *Riella* as already stated.

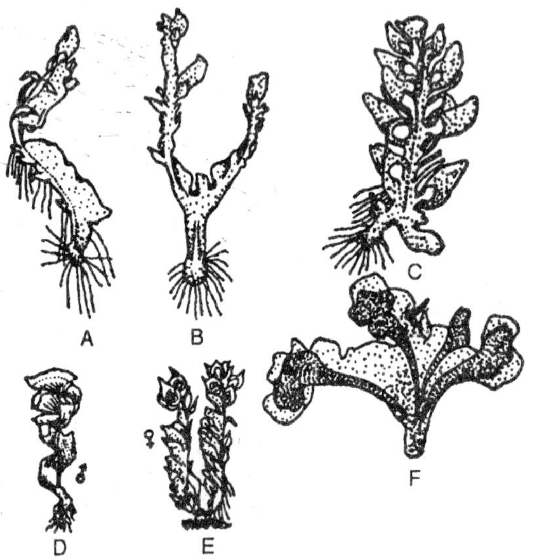

Fig. 28.2: A to C. *Riella indica* St. in Kashyap. Plants showing habit (after Kashyap). D & E. *Riella americana*. F. *Riella affinis* (= *R. vishwanathaii*) var.

Order 2. MARCHANTIALES

The *Marchantiales* gametophyte shows a prostrate, ribbon-shaped, dichotomously branched, dorsiventrally differentiated thallus. The **dorsal** tissue is green and internally differentiated into air chambers except in the genera *Monoclea, Monoselanium* and *Dumortiera* where the air chambers seem to have become obliterated by degeneration. The tissue on the ventral side is of the parenchymatous storage type and may sometimes be differentiated into two or three types of cells, in some cases even showing elongated conducting cells in the axial region. Most genera also show scales and rhizoids of two types on the ventral side. The sporophytes are very simple in *Riccia,* showing a spore case only, while they are more elaborate in *Marchantia,* showing differentiation into foot, seta and capsule. The question whether the simple forms or the complex forms are more primitive is discussed in chapter 32. In all cases the capsule wall is only one cell in thickness and there is no elaterophore or columella.

The order includes about 32 genera with some 400 species which are divided into five families—(i) **Ricciaceae**, (ii) **Corsiniaceae**, (iii) **Targioniaceae**, (iv) **Monocleaceae**, and (v) **Marchantiaceae** (this last may be further subdivided into three families as discussed later) by Cavers (1910), Campbell (1939) as well as Smith (1955). Some later Bryologists, however, prefer more families. Thus, Reimers (1954) suggests 12 families, dividing *Ricciaceae* into *Ricciaceae* and *Oxymitraceae, Targioniaceae* into *Targioniaceae* and *Cyathodiaceae,* and *Marchantiaceae* into six families enumerated later on.

The five original families of Marchantiales show a progressive differentiation of the gametophyte and a remarkable progressive sterilization with elaboration of the sporophyte beginning from the Ricciaceae to the Marchantiaceae. The leading features of the families are as follows:

1. **Ricciaceae**—Simple gametophyte and the simplest sporophyte (Fig. 28.9) represented only by a spherical bag-like structure on the gametophyte axial surface. The family is described in details later.

2. **Corsiniaceae**—Gametophyte with air-chambers (with or without assimilatory filaments) in one tier (Fig. 28.3A). Sporophyte with a foot (Fig. 28.3B). At the centre of archegonial chamber there is a swelling denoting the future evolution of the archigoniphore (Fig. 28.3B). Capsule shows sterile cells in

place of elaters (Fig. 28.3C). The family is not found in Asia.

3. **Targioniaceae**—Gametophyte with air chambers in one tier usually with assimilatory filaments. Antheridia dorsal or lateral adventitious shoots (Fig. 28.3D). Archegonia ventral near top (Fig. 28.3E). Sporophyte involucre split bivalved (Fig. 28.3F). Capsule with foot, short seta and elaters having two-three spiral thickenings.

4. **Monocleaceae**—The family shows certain anomalous features so that its systematic position is sometimes disputed and it is sometimes placed with the Jungermanniales. The genus does not occur in Asia or Europe. Air chambers are not present in the gametophytes. Scales are not found but there are mucilage hairs. Both antheridia and archegonia are dorsal (developing anacrogynously near tip) and plants are dioicous. The sporophyte has a tubular involucre, a long seta and an elongated capsule (Fig. 28.3H). Elaters are formed but in a peculiar manner. The dehiscence is not jungermannioid.

5. **Marchantiaceae**—This is certainly the most differentiated both in the gametophyte and the sporophyte and is being described in details later.

The *Marchantiales* are represented in India by the following genera (The reported number of Indian species is shown within brackets):

Family: *Ricciaceae*

Riccia (18 spp.), *Ricciocarpus* (1 sp.)

Family: *Targioniaceae*

Targionia (1 sp.), *Aitchinsoniella* (1 sp.), *Cyathodium* (1 sp.).

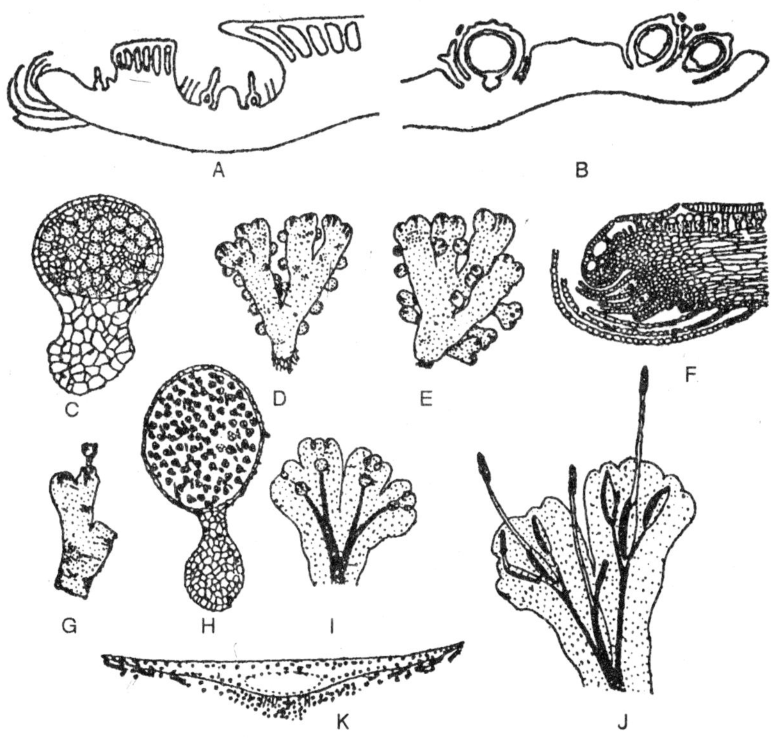

Fig. 28.3: *Corsinia:* A. Vertical section of female gametophyte before development of involucres. B. After development of involucres. C. Sporophyte.
Targionia: D. Male plant. E. Female plant. F. Thallus & archegonic. G. Sporophyte with two valves of involucres. H. L. S. of sporophyte.
Monoclea: I. Male plant. J. Female plant. K. T. S. of thallus (after Cavers & Smith)

Family: *Marchantiaceae*

(a) *Astroporae* (= *Sauteriaceae*)

Sauteria (1 sp.), *Sauchia* (1 sp.), *Athalamia* (2 spp.), *Clevea* (1 sp.), *Gellaniella* (1 sp.).

(b) *Opereulatae* (= *Rebouliaceae*)

Reboulia (1 sp.), *Plagiochasma* (10 spp.), *Asterella* (= *Fimbriaria*) (24 spp.), *Grimaldia* (2 spp.), *Mannia* (1 sp.), *Cryptomitrium* (1 sp.), *Massalonga* (1 sp.).

(c) *Compositae* (= *Marchantiaceae*)

Marchantia (11 spp.), *Preissia* (1 sp.), *Lunularia* (1 sp.), *Wiesnerella* (1 sp.), *Conocephalum* (1 sp.), *Exormotheca* (1 sp.), *Stephensoniella* (1 sp.), *Dumortiera* (1 sp.), *Monoselenium* (1 sp.).

Of the above, only the first and the last families (representing the simplest and the most complex) are discussed in details below as they are so very common.

Family: RICCIACEAE

Plants belonging to the Ricciaceae have the simplest sporophyte showing only a bag-like capsule while the gametophyte is a dichotomously branched thallus with the sex organs along the entire length of the median strip. There are three genera: *Riccia*, *Ricciocarpus* (= *Ricciocarpos*) and *Oxymitra* (= *Tesselina*) comprizing about 140 species. Of these, *Riccia* alone accounts for 130 species while *Ricciocarpus* is monotypic with the single species *Ricciocarpus natans* (= *Riccia natans*) found floating on water in marshes. Reimers (1954) suggests raising *Oxymitra* into a separate family *Oxymitraceae*.

Genus: *RICCIA*

Riccia is a cosmopolitan genus quite common in moist places, specially during and after rains. All the species are terrestrial excepting one, *Riccia fluitans* which is aquatic floating on water like *Ricciocarpus natans* (Fig. 28.4B). About 18 species of *Riccia* are known in India

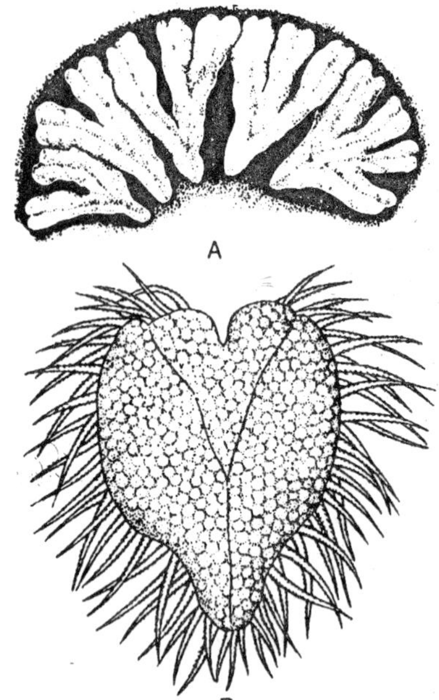

Fig. 28.4: Gametophytic thalli of (A) *Riccia diecolor* and (B) *Ricciocarpus natans* (floating form)

of which *Riccia discolor* (= *R. himalayensis*) (Fig. 28.4A) seems to be the commonest.

The gametophytic plant body is a dorsiventral prostrate thallus branching dichotomously. As a number of such dichotomous plants begin to grow together from one place, the plants usually occur in a rosette form (Fig. 28.4A). Later, however, the basal parts of the dichotomies may die and single dichotomies may grow separately. Each dichotomy (Fig. 28.5A) is linear to wedge-shaped (because of further dichtomy) in which the median region is thickened and there is a conspicuous longitudinal furrow on the dorsal side ending in a notch at the tip where the growing point is located. On the under (ventral) surface there is a transverse row of scales, one cell in thickness which are more crowded near the apex and overlap the growing point. The scales are often pigmented violet and are further apart from the apex. In the mature part, each scale is split into two so that there seem to be two rows of scales along the two margins of the thallus (Fig. 28.5C). In addition to the scales, there are two

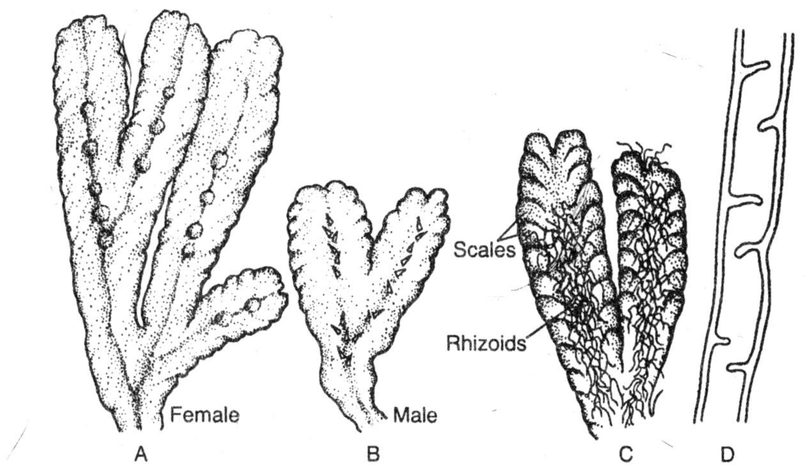

Fig. 28.5: Female (A) and male (B) plants of *Riccia discolor*. C. Ventral surface of same showing scales and rhizoids. D. Longitudinal sectional view of a tuberculate rhizoid

types of rhizoids: some with smooth walls and other *tuberculate* with internal peg-like or plate-like projections of the wall (Fig. 28.5D). Rhizoids are devoid of protoplasm when mature and are altogether absent in aquatic forms.

Anatomically (Fig. 28.6A), the ventral region of the thallus is formed of a compact colourless parenchymatous tissue which serves as the storage region and often contains starch. The scales (one-cell in thickness) and unicellular rhizoids (smooth or tuberculate) develop from the base of this tisssue. On the dorsal side this compact tissue passes on into a green assimilatory region. This region is formed of vertical rows of chlorophyll-bearing cells separated by narrow vertical air canals so that the top (dorsal side) of the thallus is porose. Usually each air canal is very narrow being surrounded by four vertical rows of chlorophyllose cells but in some species the canals may be wider, being surrounded by 8 rows of cells. The canals are open on the dorsal surface. (Fig. 28.6B) though there are no organized air pores as in *Marchantia*. The outermost cell of each

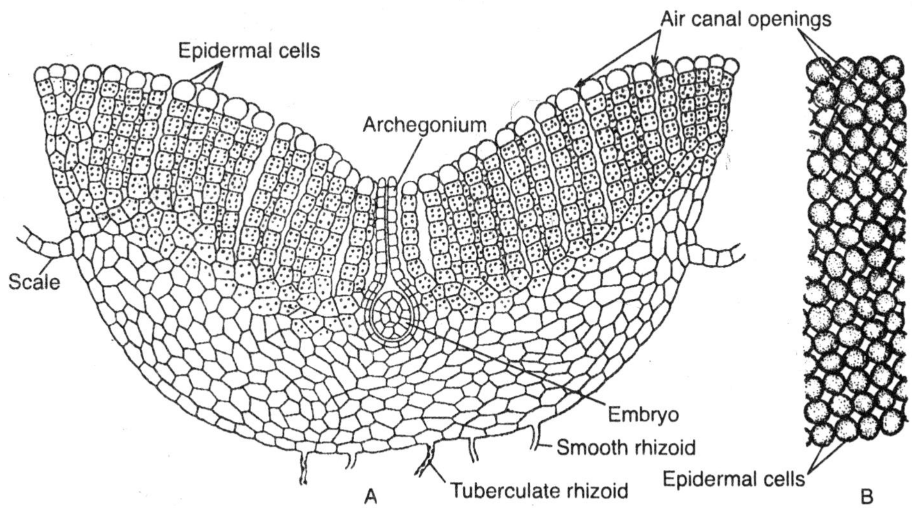

Fig. 28.6: A. Vertical transverse section through a *Riccia* gametophyte showing a sporophyte embryo within the venter of an archegonium. B. Top surface view of a bit of thallus showing the epidermis interspersed by air canal openings

vertical row of chlorophyllose cells is larger and colourless and these form an interrupted, one-cell deep *epidermis*.

The growing tip is formed of 3 to 5 apical cells more or less triangular in outline. Now and then, some median cell of these fails to divide or splits vertically so that two separate growing regions are formed and thus the thallus becomes dichotomous.

Vegetative propagation

Vegetative propagation is usually by the continuous growth from the growing point. The older basal parts of the dichotomies die and rot away while the tip goes on growing dichotomously giving rise to separate plants. This growing tip may survive the dry period one season and continue growing next year. Vegetative growth is also known to take place by the formation of adventitious branches from the ventral side of the thallus, and also by the development of tubers on the thallus which easily survive a period of drought and germinate when the season is favourable. New gametophytes have also been found to develop from the tips of rhizoids in the same manner as moss gametophytes develop on the protonema.

Sex organs

Most *Riccia* species are monoecious or homothallic but some including the common species *R. discolor* are dioecious or heterothallic. Sex organs develop singly and acropetally in a linear row on the dorsal median furrow. They are at first superficial in development, developing on the floor of the furrow depression, but are soon enveloped by the outgrowth of tissues on all sides and are at last embedded, lying single in the bottom of a cavity formed by the overarching tissues (Fig. 28.6A). These cavities are termed *antheridial* and *archegonial chambers* which are open on the dorsal surface by narrow cylindrical canals.

In the development of the **antheridium,** a single superficial dorsal cell only 2 or 3 cells behind the apical cell and placed in the median furrow protrudes out and becomes the *antheridial initial* (Fig. 28.7A). This now divides transversely forming an upper *outer cell* which later develops the antheridium and a lower *basal cell* which develops the stalk (Fig. 28.7B & C). Meanwhile the tissue surrounding the antheridial initial begins to overgrow and gradually envelopes the antheridium leaving only a pore above. The mature antheridium (Fig. 28.7D) is pear-shaped and its single-layred jacket (wall) encloses a number of *sperm* or *androcyte mother cells.* These undergo divisions by diagonal walls, each forming two triangular *androcyte* cells. Each androcyte forms a biflagellate *spermatozoid* or *antherozoid* (Fig. 28.7E). When the antherozoids are mature all

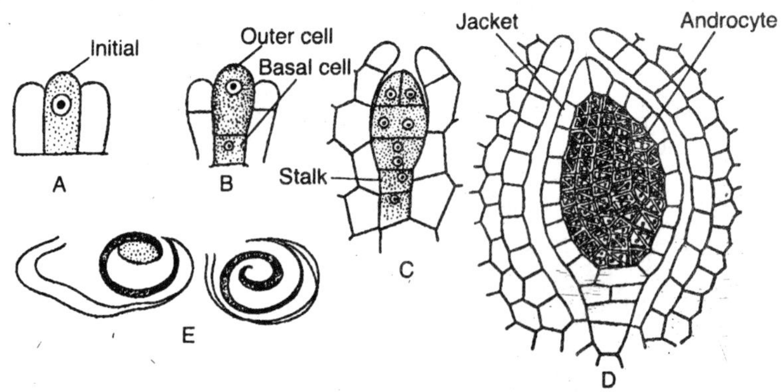

Fig. 28.7: Stages of development of *Riccia* antheridium. A. Antheridia initial. B. First division of the initial. C. A later stage of the development of the stalk and the antheridium. D. Mature antheridium with the androcytes formed. E. Two spermatozoids

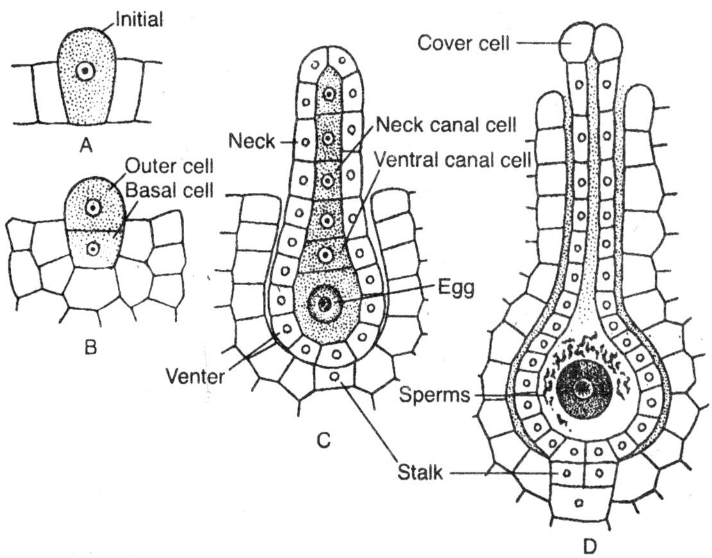

Fig. 28.8: Stages of development of *Riccia* archegonium. A. Archegonial initial. B. First division of the initial. C. A mature archegonium. D. Mature archegonium at the stage of fertilization

the cell walls inside the antheridium disintegrate leaving the antherozoids free inside the antheridial jacket in a gelatinous fluid.

The **archegonium** also develops from an exactly similar *archegonial initial* (Fig. 28.8A) in the furrow and close to the apical cell. This divides to form an *outer cell* and a *basal cell*, the former of which forms the archegonium proper while the latter forms the stalk (Figs. 28.8B and C). The mature archegonium (Fig. 28.8C) is a flask-shaped body which shows a short *stalk*, a swollen basal *venter* containing the large *egg* or *oosphere* or *ovum*, a *ventral canal cell* and an elongated *neck* containing a row of four *neck canal cells*. The neck has a wall formed by six vertical rows of cells and its tip is covered by four specialized *cover cells* whch are larger than the other cells surrounding the neck. Although the archegonium is embedded by outgrowing adjacent cells, the embedding is not as complete as in the case of the antheridium so that the tip of the neck always protrudes out.

Fertilization

When the archegonium becomes mature, the neck canal cells and the ventral canal cell disintegrate. The disintegrated product forms a mucilaginous mass which comes out by forcing open the cover cells. *Fertilization* now takes place. This needs the presence of water but the gametophyte need not be submerged under water as the sex organs are placed in the groove where water can accumulate. The antheridia rupture and the liberated spermatozoids swim in the channel water. As they come near the mature archegonia they are chemotactically attracted by the proteins and other chemicals coming out of the contents. A number of sperms pass into the neck, reach the venter and ultimately one fertilizes the egg (Fig. 28.9A). The fertilized egg invests itself with a thin cellulose wall and becomes the *oospore* or the *zygote* which is the first *diploid* cell (since two sets of chromosomes from the two gametes have united) and, thus, the *sporophytic generation* begins.

Sporophytic generation

Immediately on investing itself with a wall, the zygote increases greatly in size and almost completely fills up the venter (Fig. 28.9B). Simultaneously, the wall of the venter of the archegonium (part of the former gametophyte) undergoes periclinal divisions developing a 2- or 3-layered casing for the growing sporophyte.

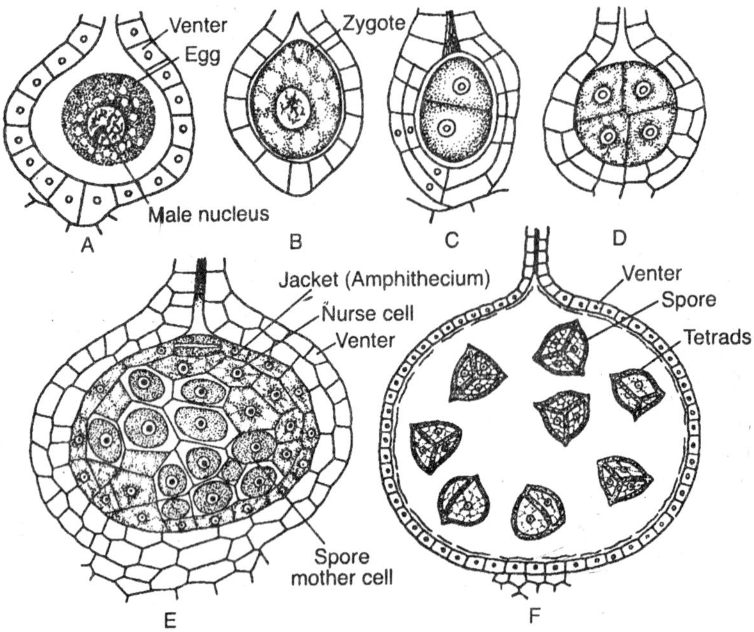

Fig. 28.9: Stages of development of *Riccia* sporophyte. A. Fertilization. B. Zygote. C. 2-celled stage. D. Quadrant stage. E. The sporophyte at maturity (spore mother cell stage). F. Spore tetrads within the venter, sporophyte cells having disintegrated

The zygote now divides. In most *Riccia* species the first division is transverse (Fig. 28.9C). Then, another wall at right angles causes a *quadrant* stage of 4 cells (Fig. 28.9D) which is followed by another division in a vertical plane at right angles to the former so that an *octant* stage of 8 cells results. Subsequent divisions are in all directions resulting in a cell mass of 20 to 36 cells which periclinal divisions cut off an outer layer or *amphithecium* and an inner mass of *endothecium.* The amphithecium forms the *jacket* of the sporophyte while the endothecium forms the sporogenous tissue and is the *archesporium.* The archesporial cells divide further forming two types of cells, the *spore mother cells* or *sporocytes* and the *nurse cells* with a watery vacuolate cytoplasm (Fig. 28.9E). This is the maximum development of the **sporophyte** after which it begins to disintegrate. The nurse cells and the amphithecial jacket layer disintegrate and this is followed by disintegration of even the inner cell layers of the venter wall and all these form a nutritive viscous fluid within which the *spore-mother cells* remain suspended.

The spore mother cells now undergo meiosis (reduction division) resulting in haploid *spore-tetrads* (Fig. 28.9F). The spore-tetrads usually remain attached to one another for a long time within a common spherical sheath and in some species, may remain attached like that even when they are liberated and germinated. The mature *spores* show three layers of walls—*exosporium* (outermost), *mesoporium* (middle) and *endosporium* (innermost).

Usually the structure containing the mature spores, as in Fig. 28.9F, which lies embedded within the thallus, is termed the *sporophyte.* But, this is by no means correct as the structure does not contain a single diploid sporophytic cell. Here we have got only some haploid spores surrounded by the outermost layer of the venter wall (which is also a gametophytic haploid tissue, but of the previous generation).

Formation of the new gametophyte

The spores (Fig. 28.10A) are liberated only by the complete decay of the venter wall and the surrounding tissue of the gametophytic

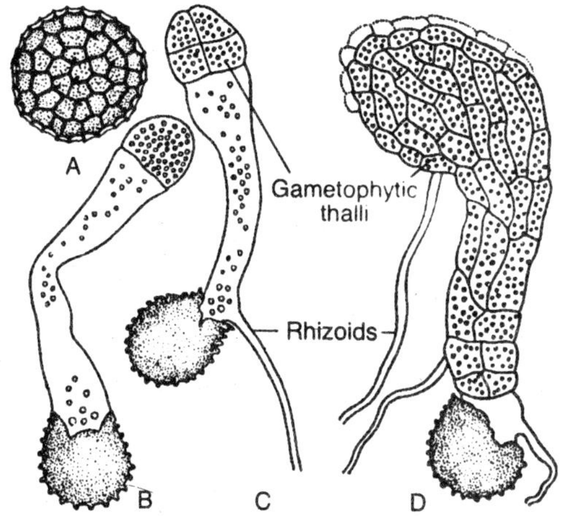

Fig. 28.10: Stages of germination of *Riccia billardieri* spore (after Udar). A. A spore. B. Germ tube after cutting off the first tip cell. C. 4-celled tip and first rhizoid. D. An advanced stage of the germling thallus

thallus. The spores germinate in favourable environment when the two outer walls burst and the endosporium comes out as a germ tube. Most of the protoplasm passes to the tip of the germ tube which is now cut off by a transverse wall (Fig. 28.10B). The first rhizoid develops from the base of the germ tube (Fig. 28.10C) while the tip cell develops the gametophytic thallus (Fig. 28.10D).

Figure 28.11 illustrates the life cycle of *Riccia.*

Family: Marchantiaceae

The *Marchantiaceae* differs from the *Ricciaceae* in its more differentiated *gametophyte* and in having the antheridia and the archegonia localized on special branches called the *antheridiophore* and the *archegonophore* respectively. (In some genera special *conceptacles* take the place of antheridiophores). The *sporophyte* also is more elaborate, being differentiated into *foot, seta* and *capsule* and always showing sterile cells called *elaters* in the capsule. The family is cosmopolitan in distribution and may be subdivided into three subfamilies after Leitgeb (1880):

(i) *Astroporae* Leitg: Genera—*Sauteria, Sauchia, Clevea,* etc. (Family *Astroporae* Verdoorn or *Sauteriaceae* Evans or *Cleveaceae* Reimers).

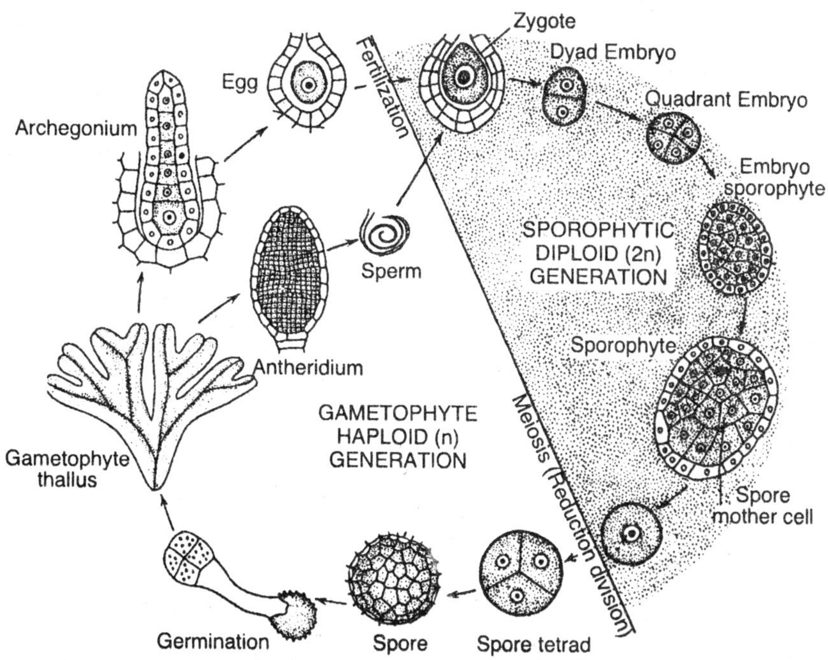

Fig. 28.11: Life cycle of *Riccia* showing alternation of generations

Air pores simple with cells in a single ring. Radial walls of these cells are strongly thickened, giving the pores a star-like appearance in surface view. Air chambers in one tier, deep but without assimilating filaments. Ventral scales in several series but without apical appendages. Capsule splitting by longitudinal slits into 4 irregular valves. Archegoniophore present but antheridia dorsal on surface or sunk in the thallus.

(ii) *Operculatae* Leitg: Genera—*Reboulia, Grimaldia, Asterella (= Fimbriaria), Plagiochasma,* etc. (Family *Operculatae* Verdoorn or *Rebouliaceae* Evans or *Grimaldiaceae* Reimers).

Capsule opening not by valves but by the irregular detachment of an apical portion which may be somewhat like an operculum. Pores small, neither star-shaped nor barrel-shaped. Air chambers well-developed, spongy by the development of secondary horizontal walls so that the chambers are in two or more tiers. In some cases, however, chambers may be in a single tier even without assimilating filaments. Archegoniophore developed. Antheridiophore, if developed, sterile or dorsal on thallus. Appendages may be present on scales (e.g., *Reboulia*).

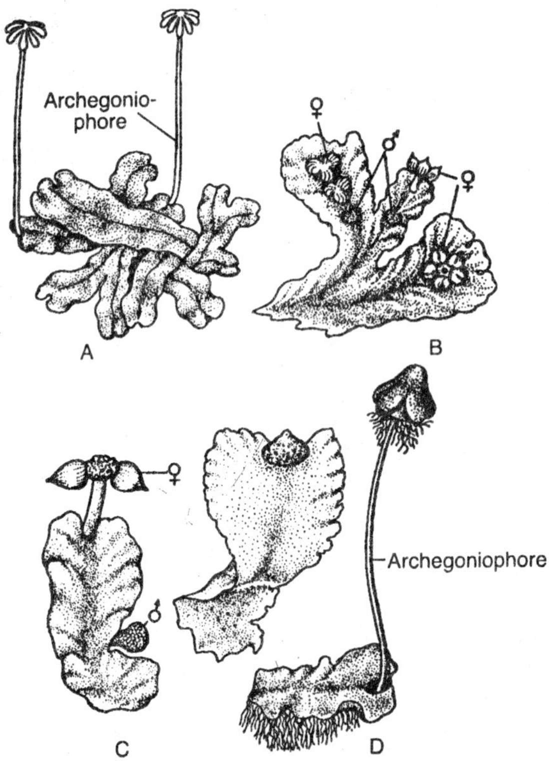

Fig. 28.12: Some common Indian representatives of Marchantiaceae. A. *Marchanti* female plant. B. *Plagiechasma* showing male receptacles and female archegoniophores. C. *Asterella* (= *Fimbriaria*) with an archegoniophore and a male receptacle on a lobe D. *Reboulia* showing male and female lobes separately

(iii) *Compositae* Leitg: Genera—*Marchantia, Conocephalum, Lunularia, Dumortiera, Exormotheca, Stephensoniella, Monoselenium,* etc. (Family *Marchantiaceae* prop.).

Capsule dehiscing by valves. Pores composite, barrel-shaped or simple. Air chambers normally well-developed with assimilatory filaments but in one tier. Thallus tissue is well-differentiated. Ventral scales with appendages. Archegoniophore highly differentiated Artheridia stalked and usually on well-developed antheridiophores but are in serrile conceptacles in the simpler genera.

Verdoorn (1932), Evans (1939) and some others consider these three subfamilies as separate families—*Sauteriaceae, Rebouliaceae* and *Marchantiaceae* respectively. Some other Bryologists, again, prefer to break up the old *Marchantiaceae* into more than three families. This, Reimers (1954) suggests six families—*Cleveaceae, Grimaldiaceae, Conocephalaceae, Lunulariaceae, Marchantiaceae* and *Exormothecaceae* (the last 4 families out of *Marchantiaceae* prop.). However, the orthodox classification is being used in this book.

There are about 23 genera with some 250 species within this family.

The Indian genera have previously been enumerated. The type genus *Marchantia,* which is very common on the hills, is being described in the next page.

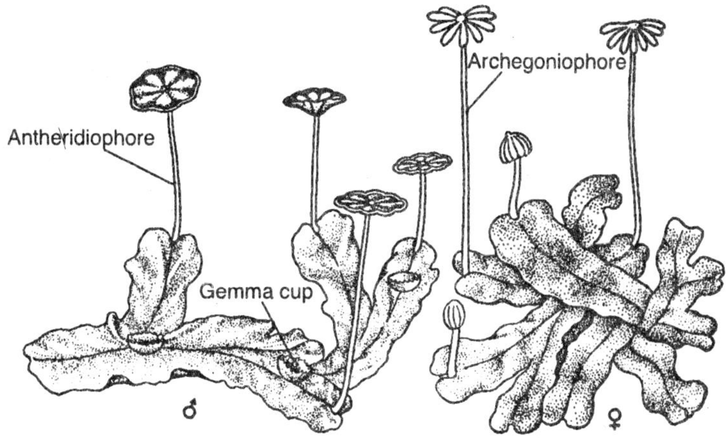

Fig. 28.13: Male and female plants of *Marchantia polymorpha*

Genus: *MARCHANTIA*

Marchantia is a genus of cosmopolitan distribution with about 65 species. About 11 species are knwon in India occurring on the Himalaya and all the hills in India. It grows on moist rocks and soil. The commonest species is *Marchantia polymorpha* L., which is found all over the world.

The *Marchantia* plant body (Fig. 28.13) is a dichotomously branched prostrate thallus. Close observation of the dorsal surface of the thallus shows it to be divided into regular

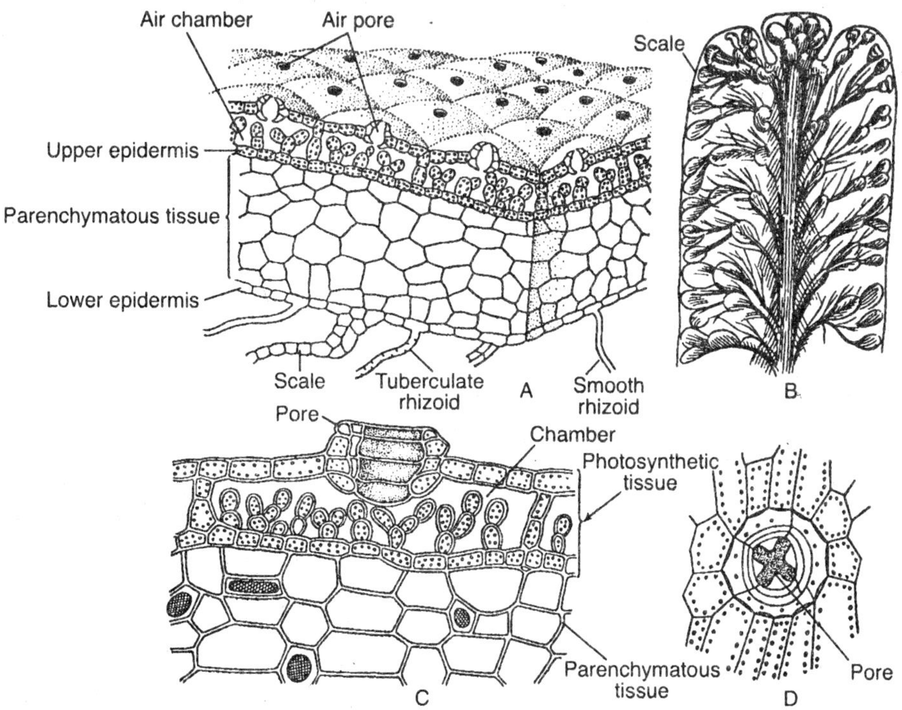

Fig. 28.14: *Marchantia*. A. Three-dimensional figure of thallus showing dorsal surface and sectional views. B. Ventral surface of thallus showing median ridge, scales with appendages and rhizoids (after Goebel). C. T.s. of thallus showing photosynthetic tissue with an air pore and lower parenchyma tissue. D. Surface view of an air pore (after Strasburger)

rhomboidal or polygonal areas (Fig. 28.14A), each with a spot at the centre which locates a pore. The branches grow indefnitely, apparently by a transverse row of apical cells in a terminal notch, and usually become cut off from the mother plant by the decay of older parts. There is a prominent median groove (midrib) on the upper dorsal surface of the branches and a corresponding ridge on the ventral surface. The ventral surface shows 3 to 4 rows of scales on both sides of the midrib (Fig. 28.14B)—the scales are larger and more elaborate near the midrib and smaller towards the margin. There are also two types of rhizoids between the scales—tuberculate and smooth, as in *Riccia.*

A section of the thallus (Fig. 24.14A) shows different tissues in it. The uppermost dorsal layer of cells is the *epidermis* formed of quadrate cells containing a new chloroplastids. Each polygonal area of the epidermis has just below it an *air chamber* of schizogenous origin which is connected with the outside atmosphere by an *air pore* at the centre of the polygon (Figs. 28.15A & C). The air pore is a barrel-shaped structure formed by 4 to 8 superimposed tiers of concentric rings, each ring again composed of 4 or 5 cells (Fig. 28.14C). The lowermost ring has cells projecting inwards giving a star-shaped appearance to the pore when viewed from above (Fig. 28.14D). It has been reported that these wall cells swell and contract somewhat like the guard cells of stomata. Each air chamber is demarcated from others by single layered partitions of cells full of chloroplastids. From the floor of each air chamber arise simple or branched filaments formed of cells full of chloroplastids. These form the main photosynthetic tissue. The ventral tissue of the thallus below the air chambers is compact and formed of thin-walled parenchymatous cells which contain few or no chloroplastid although they contain starch grains and some isolated cells among them contain large oily bodies or mucilage. The middle (midrib) portion of the thallus is formed of cells elongated lengthwise and with reticulate thickenings. The lower epidermis bears the different rows of scales and the two types of rhizoids.

Vegetative propagation

Vegetative propagation by the unlimited growth of the branches and decay from the base is quite common. There may also be propagation by the development of adventitious

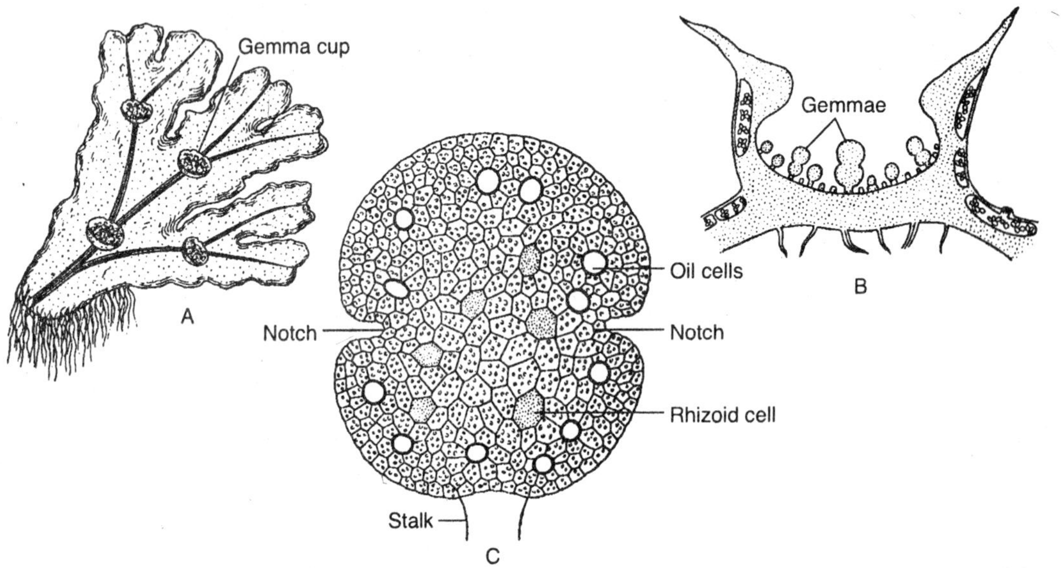

Fig. 28.15: *Marchantia*. A. Thallus showing gemma cups. B. Vertical section of a gemma cup showing gemmae. C. A single gemma highly magnified.

branches from the ventral side of the thallus or even from reproductive buds. But, the most specialized method of reproduction is by special asexual bodies known as *gemmae* which are produced in large numbers in small receptacles called *gemma cups* (Fig. 28.15) or *cupales* borne on the dorsal surface of he thallus along grooves. The gemmae are multicellular discoid bodies constricted at the middle. The two notches in the constriction contain two growing points. Most of the cells contain chloroplastids but some may be colourless, containing oil bodies. Each gemma is vertically attached to the base of the cup-shaped cupule (gemma cup) by a one-celled stalk. On the floor of the cupule there are also some club-shaped hairs. The gemmae get detached from the cupule and germinate when falling on a suitable substratum. Some colourless cells on the lower face of the gemmae then develop rhizoids and the two growing points in the notches grow to develop the thalli of two plants in the usual way.

Sex organs

Sex organs are developed on *Marchantia* plants only under special environmental conditions (day length, humidity, nutrition, etc.) The sex organs are borne on special branches called *antheridiophore* and *archegoniophore* which are developed on separate plants so that *Marchantia* is *dioecious.* Although the antheridiophore and the archegoniophore are erect branches, they are actually direct prolongations of the horizontal thallus as may be seen from the dorsiventral nature of the erect shoots with scales within grooves on the outer (ventral) side.

The **antheridiophore** (Figs. 28.13 and 28.16A) shows a prismatic *stalk* bearing at its apex a *diso* which is usually 8-lobed. Each lobe reprsents a branch along whose upper (dorsal) median line the antheridia are borne in a row, the oldest one being at the centre and the younger and youngest ones towards extremity (acropetal). The antheridia develop in the same way as in *Riccia.* Each *antheridium* (Fig. 18.16B)

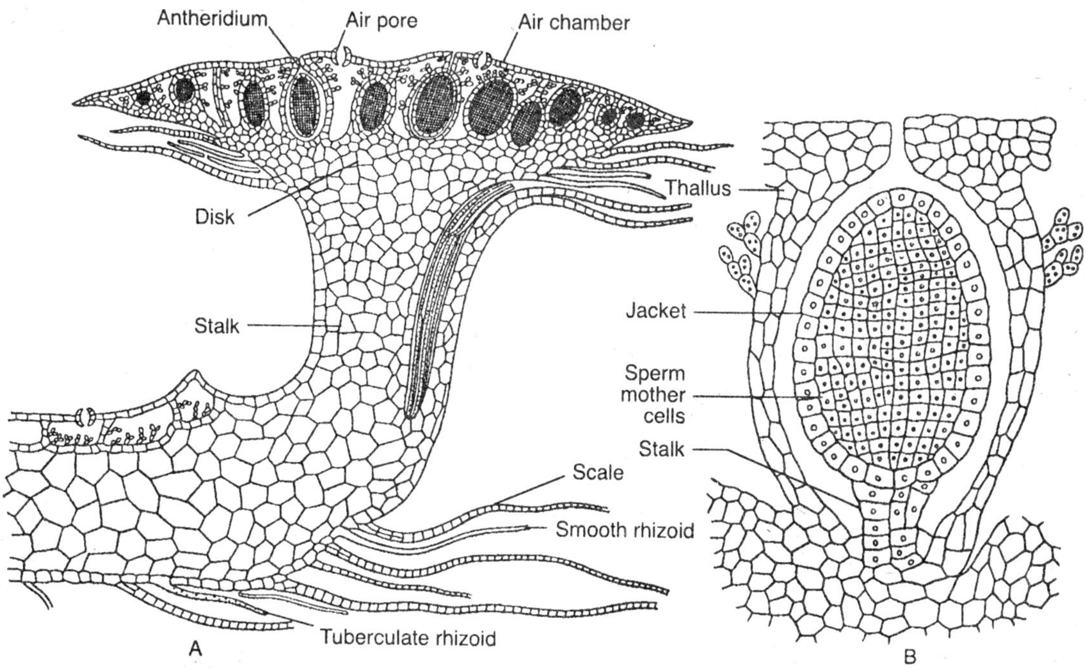

Fig. 28.16: *Marchantia polymorpha*. A. Vertical section through antheridiophore. B. Vertical section through a single antheridium inside an antheridial chamber

is borne in a cavity and is separated from one another by air chambers with air pores as on the dorsal side of a normal thallus. A mature antheridium shows a short stalk and a globular body which later, again, has a jacket of one layer only and encloses a number of *sperm mother cells* each of which forms two triangular *androcytes* as in *Riccia*. Each androcyte develops a biciliated *sperm* or *antherozoid*.

The **archegoniophore** or the *carpocephalum* (Figs. 28.13, 28.17 and 28.18A) also contains a *stalk* and a *disc* and develops in the same way as the antheridiophore. This disc bears 9 *rays* (Fig. 28.17D) instead of lobes. In the early stage, the archegonia develop as in *Riccia* on the upper (dorsal) face of the disc in the same manner as the antheridia (Fig. 28.17A, A′). Soon after, the cental sterile part of the disc begins to increase enormously so that the growing rows of archegonia are pushed down (Fig. 28.17B, B′) to the lower side of the disc, hang upside down and the youngest one almost reaches the stalk (Fig. 28.17C, C′), the older ones being nearer the periphery. The tissue between the rows of archegonia develops afterwards and hangs down as rays (Fig. 28.17D, D′). But, although the number of lobes is originally 8, usually the rays number 9 by the splitting of one. After this curvature, a one-layered plate of tissue develops on the either side of each row or archegonium and forms a *perichaetium* enclosing all the 12 to 15 archegonia in a row. A mature archegonium (Fig. 28.18B) shows a swollen *venter* with an *egg cell*, a *ventral canal cell* and a row of *neck canal cells* inside the elongated whose wall is similar to that of *Riccia*.

Fertilization

The manner of fertilization is essentially the same as in *Riccia*. Usually the antherozoids from the top of the antheridiophore discs are splashed by rain-water on to the archegoniophores which are very short at this stage. The antherozoids then swim in the moisture and ultimately reach the eggs. The development of the archegoniophore in size takes place after fertilization.

Sporophyte

The fertilized egg enlarges and fills up the cavity of the venter, invests itself with a thin cellulose wall and becomes the *oospore* or *zygote* initiating the sporophyte generation. The stimulus of fertilization causes development of the venter wall to form a 2- to 3-layered *calyptra* which envelops the growing sporangium. A ring of cells at the base of the calyptra also develops to form a second 1-layered covering round the growing sporangium and is called the *perigynium* or the *pseudoperianth*. Thus, the sporogonium develops within three gametophytic coverings—calyptra, perigynium and perichaetium.

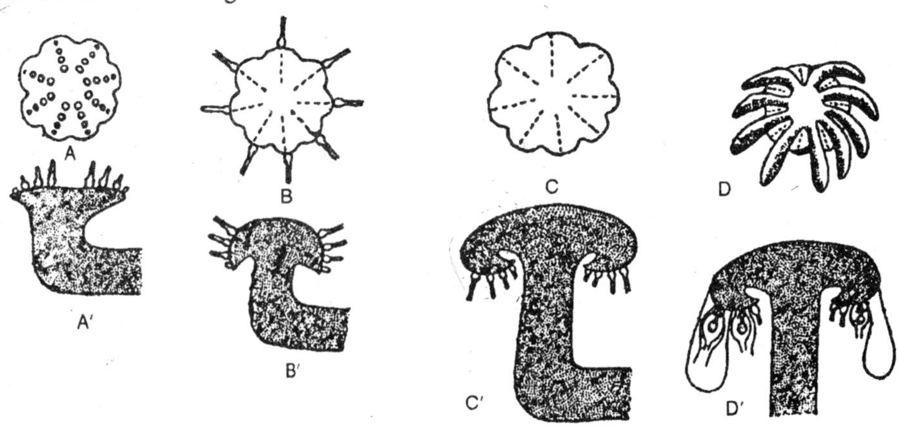

Fig. 28.17: *Marchantia polymorpha*. Stages of development of the *Archegoniophore*. Upper row—top views. Lower row—vertical l.s. (Modified from Smith)

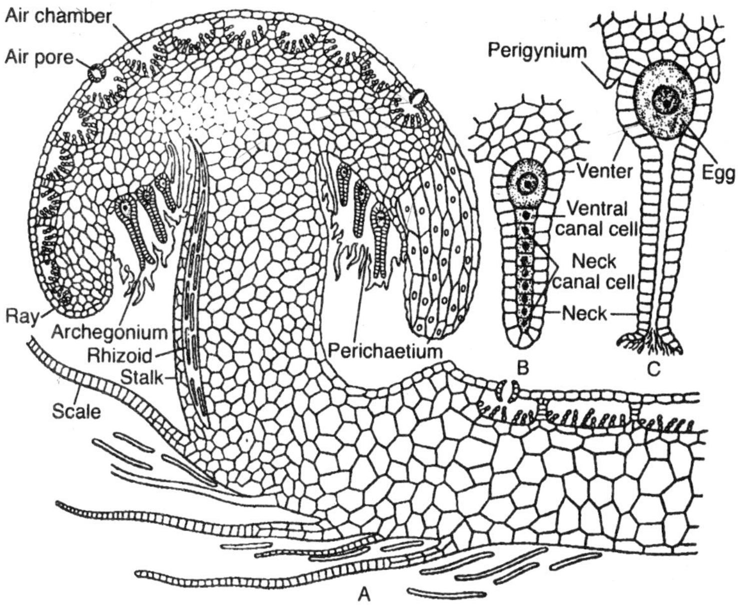

Fig. 28.18: *Marchantia polymorpha*. A. Vertical section through archegoniophore showing two radial rows of archegonia. B. A mature archegonium. C. An archegonium after disintegration of neck and ventral canal cells showing rudiments of the perigynium. (Modified from Smith)

The zygote cell inside the venter develops first by a transverse (Fig. 28.19A) and then by a vertical wall forming a *quadrant* of cells. These cells now divide and redivide. Ultimately the two upper *epibasal cells* of the quadrant form the *capsule* and the upper part of the *seta* while the two lower *hypobasal* cells form the lower part of the *seta* and the *foot*. In the early stage all these organs contain chloroplasts. The mature sporophyte (Figs. 28.19C and 28.20A) consists of three parts: (i) the *foot* is an expanded bulbous mass of cells at the base working as an absorbing and anchoring organ for the sporophyte obtaining nutrients from the gametophyte; (ii) the *seta* is a slightly elongated stalk formed of cells elongated lengthwise; (iii) the upper spherical *capsule* contains the *spores* and the *elaters*. The whole sporogonium hangs downwards. The cells within the capsule are early differentiated into the *amphithecium* which forms a single layered outer *jacket* and the inner mass of *endothecium*. Most of the endothecium (barring some cells near the seta) form a massive *sporogenous tissue*. Ultimately, half the sporogenous cells develop a large number of *spore mother cells* while the other half of cells become elongated, endowed with two spiral thickenings (Fig. 28.20D) and become the *elaters*. The spore mother cells and the elaters are distributed within the capsule. Each spore mother cell now undergoes reduction division and the *spores* are formed in tetrads.

On the maturity of the spores, the seta elongates quickly. The development of the seta pushes the sporogonium out of the calyptra, the perigynium and the perichaetium (Figs. 28.20B and C). In the mature sporogonium the jacket cells (amphithecial in origin) form vertical ring-like thickened bands. The jackets of the hanging mature sporophytes then split longitudinally into an irregular number of lobes. The long, narrow, spindle-shaped elaters are hygroscopic because of their peculiar thickening and their movement loosens the spore masses which are discharged and dispersed by air. The spores are very small (12 to 30μ in diameter) and are each endowed with a thin *intine* and a slightly thicker rough-*exine*.

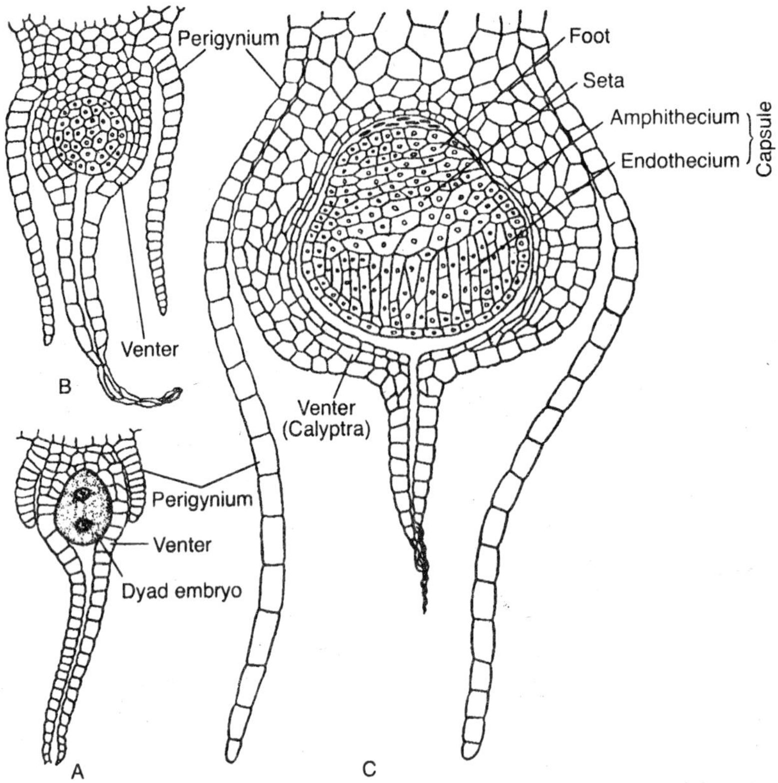

Fig. 28.19: Stages of development of *Marchantia* sporophyte. A. Dyad stage of embryo. B. Further development of sporophyte. C. Sporophyte sufficiently formed to show differentiation. (Redrawn after Smith)

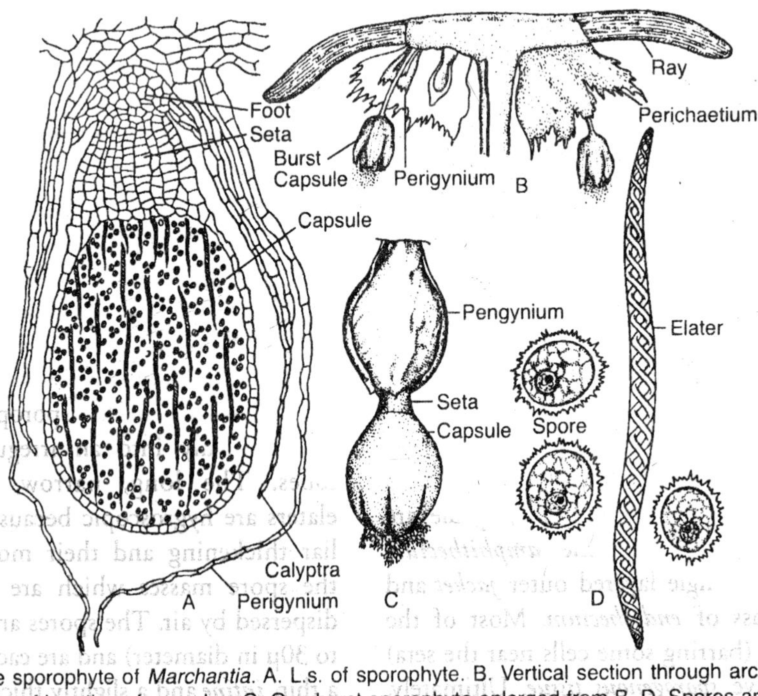

Fig. 28.20: Mature sporophyte of *Marchantia*. A. L.s. of sporophyte. B. Vertical section through archegonial head at the time of bursting of capsules. C. One burst sporophyte enlarged from B. D. Spores and an elater

The new gametophyte

Under conditions favourable for germination the spore first increases in size (Fig. 28.21B), gradually forms an irregular thallus (Fig. 28.21E) which ultimately forms the normal *Marchantia* thallus. Out of each spore tetrad two male and two female plants are formed.

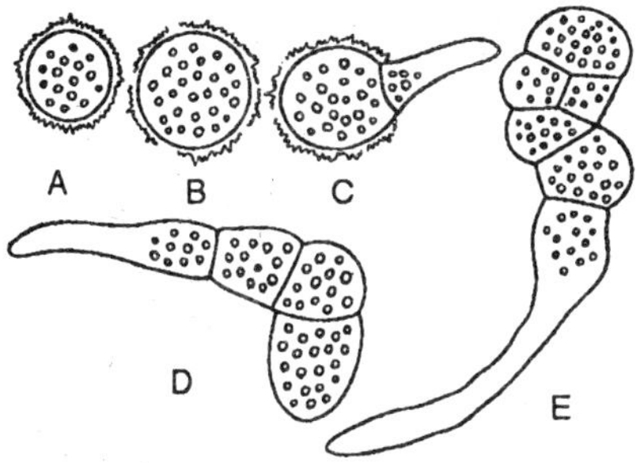

Fig. 28.21: A to E. Stages of germination of *Marchantia* spore up to the beginning of differentiation of the new gametophytic thallus

Figure 28.22 represents the life cycle of *Marchantia*.

Order 3. JUNGERMANNIALES

The *Jungermanniales* are usually known as the leafy liverworts as the gametophytes of a majority of these are of a foliose nature. However, there are some genera of *Jungermanniales* where the gametophyte is thallose like those of *Marchantiales*. But, in no case does the gametophytic tissue show any air-space or any internal differentiation whatsoever as is usual in the *Marchantiales*. The scales are usually absent and the rhizoids are all smooth-walled. Growth is by a single apical cell. The antheridia are usually globose and borne on long stalks. The development of the antheridium differs from that of the *Marchantiales*. The archegonial neck is formed by five vertical rows of cells; the venter is not much wider than the neck. The sporophyte is differentiated into a capsule, a foot and a seta which is much elongated. The archesporium develops spore mother cells and elaters. The capsule wall is multilayered and dehisces at maturity by splitting into four valves.

The *Jungermanniales* seem to have been specially successful in evolution as is shown by

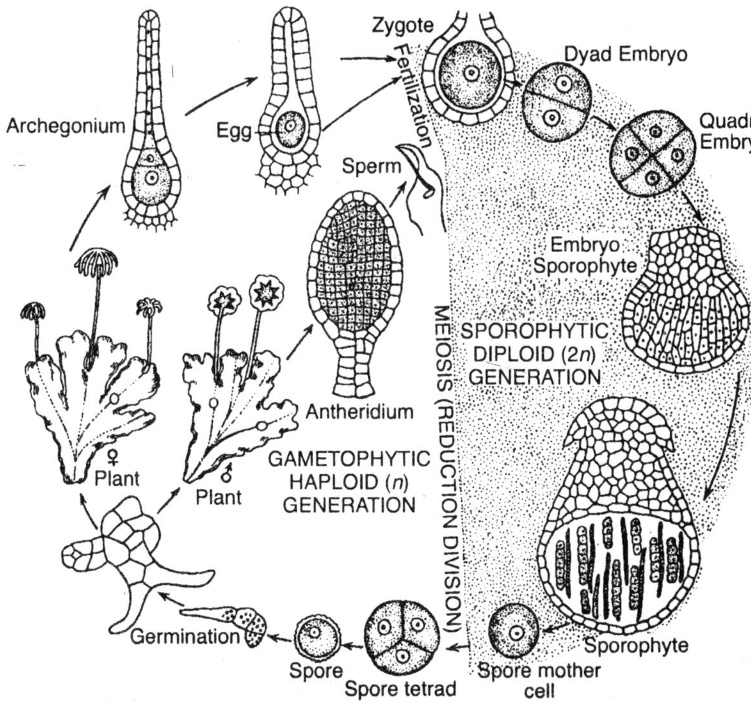

Fig. 28.22: Life cycle of *Marchantia* showing alternation of generations

the fact that out of some 8500 species of *Hepaticopsida* today, less than 500 belong to the four other orders (*Sphaerocarpales, Marchantiales, Calobryales* and *Takakiales*). The approximately 8000 *Jungermanniales* species are grouped in about 200 genera. As the *Jungermanniales* are readily distinguished into two types according to the position of the archegonia. Leitgeb (1872-77) distinguished two series, (1) *Anacrogynae* and (2) *Acrygynae*. Underwood (1894) proposed the names *Metzgeriaceae* and *Fungermanniaceae* respectively for these. These are now recognized as two suborders of this order. In the *Anacrogynae* (now called *Metzgerineae*), the archegonia are borne on the dorsal surface of the prostrate shoot so that apical cells are not involved in their formation. In the *Acrogynae (= Fungermannineae)*, the archegonia are borne on the apex of the shoot, being developed out of the apical cell. Accordingly, in a modern classification, the *Jungermanniales* are divided into two suborders:

Suborder 1. **Metzgerineae**
(= Jungermanniales Anacrogynae)

Gametophyte usually thallose, rarely foliose but always dorsiventral and prostrate. Apical growth by a single cell with two cutting faces except in *Petalophyllum*. Archegonia on dorsal surface of thallus. There are about 21 genera and 550 species. The families are named differently by different authors. While Cavers (1911) recognized only three families (*Codoniaceae, Aneuraceae* and *Blyttiaceae*), Verdoorn (1932) recognized six (*Treubiaceae, Codoniaceae, Haplolaenaceae, Dilaenaceae, Metzgeriaceae* and *Aneuraceae*) and Evans (1939) recognized seven, viz., *Treubiaceae, Fossombroniaceae, Pelliaceae, Blasiaceae, Pallaviciniaceae, Metzgeriaceae* and *Riccardiaceae*. In the recent Reimers (1954) classification of ten families have been recognized. Two families, *Monocleaceae* and *Calobryaceae*, formerly included within this suborder have been removed to *Marchantiales* and *Calobryales* respectively.

Suborder 2. **Jungermannineae**
(= Jungermanniales Acrogynae)

They include the predominating foliose *Jungermaniales* comprizing some 7500 species in about 180 genera. If not the whole gametophytic body, at least the reproductive branch is foliose and the archegonia are borne on the tips being formed by the apical cells. Apical growth (except in *Pleurozia*) is by a single cell with three cutting faces. The numerous genera, again, are placed differently in different families by different authors. While Cavers (1911) recognized only eight families (*Lophoziaceae, Cephaloziaceae, Ptilidiaceae, Scapaniaceae, Radulaceae, Pleuroziaceae, Porellaceae* and *Lejeuneaceae*), Verdoorn (1932) recognized thirteen, Evans (1939) recognized seventeen and Reimers (1954) twenty.

Such a large group of plants as the *Jungermanniales* is well represented in India by the following genera arranged according to Cavers classification. The approximate number of Indian species is shown within brackets. Chopra (1943) had listed 496 species of *Jungermanniales* but the present number on record is more than 550.

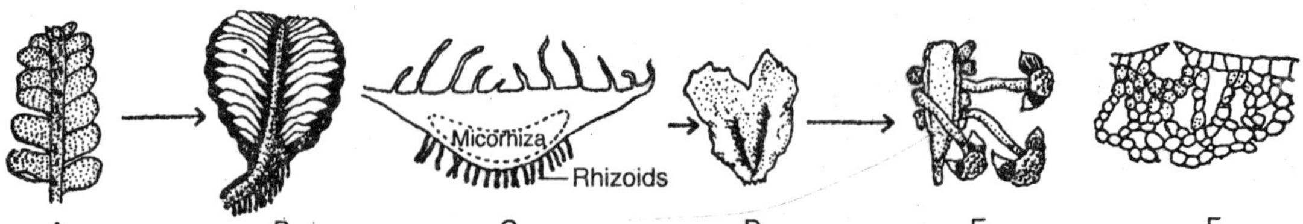

Fig. 28.23: Some Metzgerineae: A. *Forsombronia*, B. *Petalophyllum*, C. T. S. through middle of leaf bearing part of *Petalophillum* (after Cavers), D. *Stephensoniella* (after Kashyap), E. *Asterella* (= *Fimbriaria*) After Kashyap, F. Vertical section through a pore of *Asterella*

Family Genera

Metzgerineae

1. Aneuraceae: *Aneura* (= *Riccardia*) (8 spp.), *Metzgeria* (8 spp.).

Antheridia and archegonia are restricted to short special branches which arise from the margin of the thallus in *Aneura* and are ventral in origin in *Metzgeria* and the other genera. Elaterophores are apical on the valves. Accordingly Verdoorn, Evans and Reimers have divided the family into: 1. *Riccardiaceae* Evans (= *Aneuraceae* prop. Verdoorn and Reimers) and 2. *Metzgeriaceae*.

2. Blyttiaceae: *Blyttia* (*Pallavicinea*) (7 spp.).

In *Blyttiaceae* the thallus has a strong midrib which only is present in the cylindrical basal portion. Upwards, it has two lateral wings and is repeatedly divided dichotomously. The sex organs are borne on the upper side of the ordinary thallus and not on short special branches. Evans calls this *Pallaviniaceae* while Campbell calls it *Dilaenaceae* and Reimers treats it up into three families of which *Dilaenaceae* (including *Blyttia* = *Pallavicineae*) is one.

3. Codoniaceae: *Pellia* (3 spp.), *Blasia* (1 sp.), *Calycularia* (4 spp.), *Petalophyllum* (1 sp.), *Fossombronia* (3 spp.), *Sewardiella* (1 sp.).

Codoniaceae as understood by Cavers is a large family comprising of both thallose and foliose forms. The family is distinguished by spherical to ovoid capsules which dehisce in most genera by four valves extending from apex to base. But in *Fossombronia* and *Petalophyllum* the wall breaks up in or regular plates. A basal tuft of fixed elaters is present. Verdoorn divides the family into 3 (*Treubiaceae, Codoniaceae* and *Haplolaenaceae*—the last including *Pellia*), Evans into 4 (*Treubiaceae, Pelliaceae, Blasiaceae* and *Fossombroniaceae*) and Reimers recognises 5 families here (*Treubiaceae, Pelliaceae, Blasiaceae, Noterocladaceae* and *Codoniaceae* which last includes the foliose forms.)

Jungermannineae

A. Leaves various in form and uncertain. Perianth rarely with longitudinal wings. Capsule usually long-stalked. Elaters numerous but never fixed above or extending from root to floor. Each elater with two or more spiral fibres.

(a) Leaves entire or toothed or with lobed margins or deeply lobed but with lobes of equal size and form.

1. Lophoziaceae: *Acolea* (1 sp.), *Acrobolus* (1 sp.), *Aticularia* (1 sp.), *Anastrepta* (2 spp.), *Anastrophyllum* (5 spp.), *Aplozia* (10 spp.), *Chiloscyphus* (12 spp.), *Cuspidatula* (1 sp.), *Eucalyx* (1 sp.), *Gymnomitrium* (1 sp.), *Jamesomella* (1 sp.), *Jungermannia* (21 spp.), *Leptoscyphus* (1 sp.), *Lephocolea* (11 spp.), *Lophozia* (5 spp.), *Notoscyphus* (2 spp.), *Plagiochila* (100 spp.), *Solenostoma* (1 sp.), *Southbya* (1 sp.), *Sphenolobus* (1 sp.), *Syzygiella* (1 sp.), *Tylimanthus* (1 sp.).

Amphigastria absent or small. Leaves succubous or transversely oriented, entire or two-lobed. Perianth usually epigonanthous (triangular in cross section, with a dorsal angle and 2 ventral angles).

2. Cephaloziaceae: *Calypogeia* (11 spp.), *Cephalozia* (3 spp.), *Jackiella* (1 sp.), *Lepidozia* (16 spp.), *Telaranea* (= *Microlepidozia*) (1 sp.), *Mastigobryum* (31 spp.), *Nowellia* (2 spp.), *Odontoschisma* (1 sp.), *Pleurocarda* (1 sp.), *Saccogyna* (1 sp.), *Schiffneria* (1 sp.).

Amphigastria usually present but smaller than leaves. Leaves usually incubous, leaf-margin usually lobed or toothed. Perianth usually hypogonanthous (triangular in cross section with two dorsal and one ventral angle).

3. Ptilidiaceae: *Anthelia* (1 sp.), *Blepharostoma* (1 sp.), *Chandeenthus* (1 sp.), *Herberta* (12 spp.), *Herpocladium* (1 sp.), *Isotachis* (1 sp.), *Mastigophora* (3 spp.), *Ptilidium* (1 sp.), *Trichocolea* (1 sp.).

Amphigastria always present and resembling leaves in form and size. Leaves generally with several long teeth or lobes. Perianth with longitudinal folds and often concrescent with involucre.

(b) Leaves typically with two lobes folded upon each other and differing in size and also in form.

4. Scapaniaceae: *Diplophyllum* (4 spp.), *Scapania* (12 spp.), *Schistochila* (4 spp.).

Amphigastria small or absent. Upper lobe of leaf smaller than lower. Perianth cylindrical or dorsiventrally compressed.

5. Radulaceae: *Radula* (17 spp.).

Amphigastria absent. Upper leaf lobe much larger than lower. Rhizoids in tufts on lower leaf-lobes instead of stem. Perianth dorsiventrally compressed. Monogeneric.

6. Pleuroziaceae: *Pleurozia* (1 sp.).

Amphigastria absent. Upper leaf-lobe larger than lower; lower saccate often with trap-door mechanism. Perianth elongated and folded. Monogeneric.

7. Porellaceae: *Porella* (= *Madotheca*) (35 spp.).

Amphigastria present. Lower leaf-lobe small, flat or concave, never saccate or bearing rhizoids. Perianth dorsiventrally compressed. Capsule valves not extending below base of capsule. Monogeneric.

B. Amphigastria usually present. Leaves clearly divided into a larger upper lobe and a small lower lobe, the latter usually rolled up or saccate. Perianth usually with wing-like ridges. Capsule stalk short, ruptured irregularly. Elaters few, with single spiral fibres all fixed by their upper ends to the inside of the capsule wall and extending to the floor of the capsule-cavity.

8. Lejeuneaceae: *Frullania* (39 spp.), *Jubula* (1 sp.), *Lejeunea* (15 spp.), *Eulejeunea* (4 spp.), *Taeniolejeunea* (2 spp.), *Drepanolejeunea* (2 spp.), *Rectolejeunea* (1 sp.), *Cololejeunea (Physocolea)* (15 spp.), *Ceratolejeunea* (1 sp.), *Aphanolejeunea* (2 spp.), *Diplasiolejeunea* (1 sp.), *Pettolejeunea* (1 sp.), *Leptolejeunea* (7 spp.), *Stictolejeunea* (1 sp.), *Archilejeunea* (6 spp.), *Leucolejeunea* (1 sp.), *Ptychanthus* (2 spp.), *Lopholejeunea* (2 spp.), *Brachiolejeunea* (2 spp.), *Marchesinia* (2 spp.), *Trocholejeunea* (1 sp.), *Dicranolejeunea* (2 spp.), *Spruceanthus* (2 spp.), *Mastigolejeunea* (4 spp.), *Thysanthus* (2 sp.), *Bryopteris* (2 spp.), *Ptychocoleus* (5 spp.), *Prinolejeunea* (2 spp.), *Harpolejeunea* (1 sp.), *Strepsilojeunea* (3 spp.), *Taxilejeunea* (3 spp.), *Otigonolejeunea* (1 sp.), *Hygrolejeunea* (3 spp.), *Enosmolejeunea* (2 spp.), *Pycnolejeunea* (6 spp.), *Cheilolejeunea* (4 spp.), *Microlejeunea* (7 spp.), *Leptocolea* (4 spp.), *Colura* (1 sp.).

Suborder 1. **Metzgerineae**

The *Metzgerineae* include ***anacrogynous*** forms most of which are thallose, but there are a few foliose forms.

Family: **PELLIACEAE** Evans and Goebel (Codoniaceae Cavers)

Genus: *PELLIA*

This anacrogynous genus is a thallose form belonging to the *Metzgerineae.* Cavers (1911) included the genus within the family *Codoniaceae,* a larger family including such simple thallose forms as *Pellia* as well as foliose forms like *Fossombronia.* Verdoorn as well as Evans broke up the family and Verdoorn (1932) placed *Pellia* within the family *Haplolaenaceae* while Evans (1939) has named the new family *Pelliaceae.*

There are four species within the genus *Pellia* of which three are known in India: *P. epiphylla, P. fabbroniana* (= *P. calycina*) and *P. neesii.* The first two species occur all over the Himalaya but *P. epiphyla* is more abundant in the Eastern Himalaya and *P. fabbroniana* in the Western. The usual habitat is on moist soil or on rocks by the side of ditches or springs, growing as green patches among mosses or grasses.

The gametophytic plant body (Fig. 28.24) is a thin, dorsiventral, prostrate, dichotomously branched thallus with a somewhat wavy margin. The dorsal surface is almost smooth but a median midrib (where the thallus is thicker below) may be recognized. Each branch terminates in a median notch where the growing point is located. On the ventral surface numerous smooth, unicellular rhizoids are borne below the midrib. Scales and tuberculate rhizoids are absent. Mucilage hairs are present at the apex protecting the growing tip. The appearance of the thallus is known to vary according to the amount of moisture available.

The anatomy of the thallus is usually very simple. The median midrib is many layered (8 to 16 layers—Fig. 28.25) in depth. The layers decrease on the two sides and the wings are only one row in thickness. The upper cells contain abundant chloroplasts, the lower cells containing less or none. All the cells contain starch and some contain oil. In *P. epiphylla* (Fig. 28.25B) and *P. neesii,* some midrib cells are elongated longitudinally and there are some fibre-like cells. Such cells are, however, absent in *P. fabbroniana* (Fig. 28.25C), where all cells are more or less uniform.

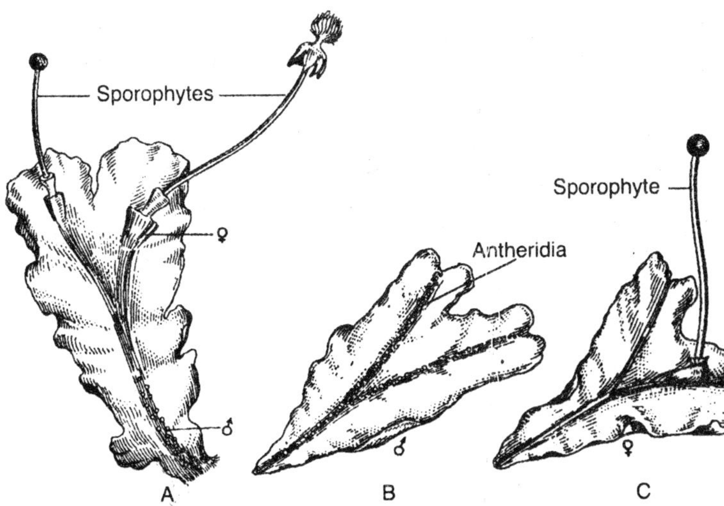

Fig. 28.24: A. *Pellia epiphylla* showing male antheridial pits and sporophytes developed from female archegonial pits (modified from Engler). B. (Male) and C (Female). *Pellia fobbroniana* plants (after Kashyap)

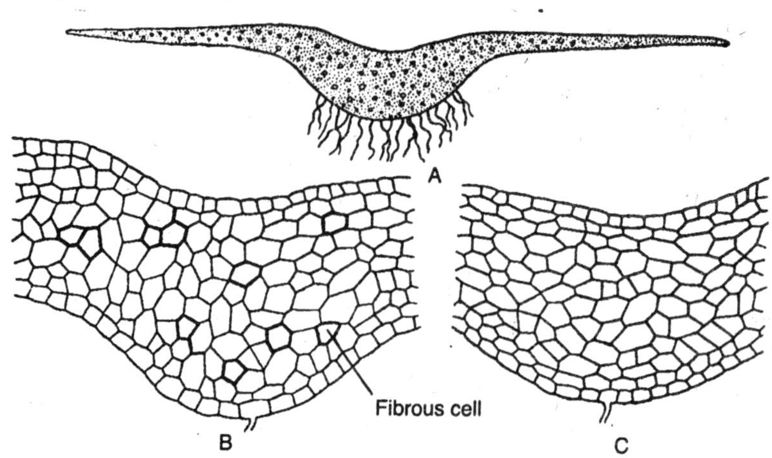

Fig. 28.25: A. Diagrammatic t.s. of gametophytic thallus of *Pellia epiphyla*. B. T.s. of thallus of *P. epiphylla* showing thick-walled fibrous cells in cross-section. C. T.s. of thallus of *P. fabbroniana* showing uniform thin-walled cells only

Vegetative propagation

As in other hepatics, vegetative propagation by unlimited growth of the branches and decay of older parts is quite common. Adventitious shoots also develop here and there on the plant body.

Sex organs

While *Pellia fabbroniana* and *P. neesii* are dioecious, *P. epiphylla* is monoecious. In the dioecious forms, the male plant (Fig. 28.24B) bears antheridia all along the dorsal face of the midrib while the archegonia develop in

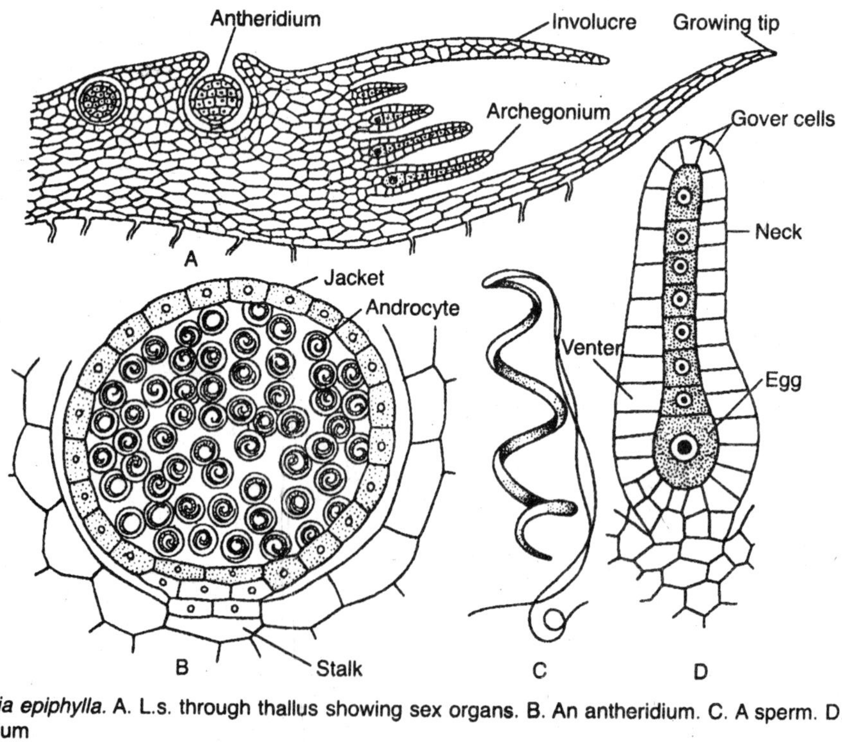

Fig. 28.26: *Pellia epiphylla*. A. L.s. through thallus showing sex organs. B. An antheridium. C. A sperm. D. An archegonium

clusters (also on the midrib) just behind the growing tip of the female plant (Fig. 28.24C). In monoecious forms (Figs. 28.24A and 26A), the antheridia are behind the archegonia.

The positions of the *antheridia* on the thallus (Fig. 28.24) are shown by numerous wart-like projections. Each projection marks an antheridial cavity where a single antheridium develops from a dorsal superficial cell behind the apical growing point. The mature antheridium (Fig. 28.26B) is globose with a jacket one cell in thickness and a multicellular short stalk. It lies at the bottom of the antheridial chamber which is open on the dorsal surface by a pore.

The *archegonia* develop in clusters of 4–12 out of a group of cells very close to the apical cell. The group of archegonia (Fig. 28.26A) is protected by an *involucre* (an outgrowth of the thallus tissue on the dorsal surface developing from the base of the archegonia) which is cylindrical in *P. neoesii* and *P. fabbroniana* (Fig. 28.24C) but flap-like in *P. epiphylla* (Figs. 28.24A and 28.26A). Older and younger archegonia remain mixed together and they are more or less horizontal. The mature archegonium (Fig. 28.26D) shows a *venter* containing the *egg*, a *ventral canal cell* and 4 to 6 *neck canal cells*. The wall of the venter and the lower part of the neck become 2-layered before fertilization while the neck wall is single layered with five vertical tiers of cells. The top of the archegonium is closed by 4 *cover cells*.

Fertilization

The mature antheridium bursts in contact with water and the mucilaginous mass containing a large number of *biciliate sperms* (Fig. 28.26C) is violently extruded out, reaching the archegonia (which are by now mature and have also become open) within minutes. Fertilization takes place as in *Riccia*.

Sporophyte

Within each involucre only one **sporophyte** develops to maturity. The other archegonia disintegrate early. The zygote invests itself with a cell wall and enlarges to fill up the venter. Its first division is by a transverse wall forming a *hypobasal* and an *epibasal* cell (Fig. 28.27A). The lower *hypobasal* cell becomes a *suspensor* which acts as a haustorium (Fig. 28.27B). The epibasal cell now divides and redivides to form a group of 8 cells the upper 4 cells of which form the *capsule* and the lower 4 cells form the *seta* and the *foot* (Fig. 28.27C). The ventral wall of the archegonium also grows and keeps pace with the growing sporophyte forming the *calyptra*. This forms a sheath covering the growing sporophyte (Fig. 28.27D).

The upper cells of the embryo, which are destined to form the capsule, demarcate an outer layer, the *amphithecium*, by periclinal division while the inner mass of cells is the *endothecium* (Fig. 28.27B). The endothecium (Fig. 28.27C) becomes the archesporium within which the basal cells are differentiated early forming a bouquet of elongated, spirally thickened sterile cells to which many *claters* are attached. This bunch is the *elaterophore* (Fig. 28.27D). The upper archesporial cells form a mass of sporogenous cells (*spore-mother cells*) intermixed with sterile elaters. The *elaters* are elongated, narrow cells with double spiral thickenings (Fig. 28.27E) as in *Marchantia*. The mature capsule (Fig. 28.27D) is globose with an outer jacket, two layers thick. Of these layers, the outer layer is of polygonal cells with brown, radial thickening bands. The inner layer is of somewhat flattened cells, with semi-annular thickenings in most species. At the base of the capsule is the elaterophore which is a bundle of 20 to 100 elaters. Above, are the *spores* (formed by reduction division of the spore mother cells) intermixed with free elaters.

The *seta* is short at first (Fig. 28.27D), formed of regular longitudinal rows of cells, when the spores become mature, the seta cells begin to elongate suddenly and the seta reaches

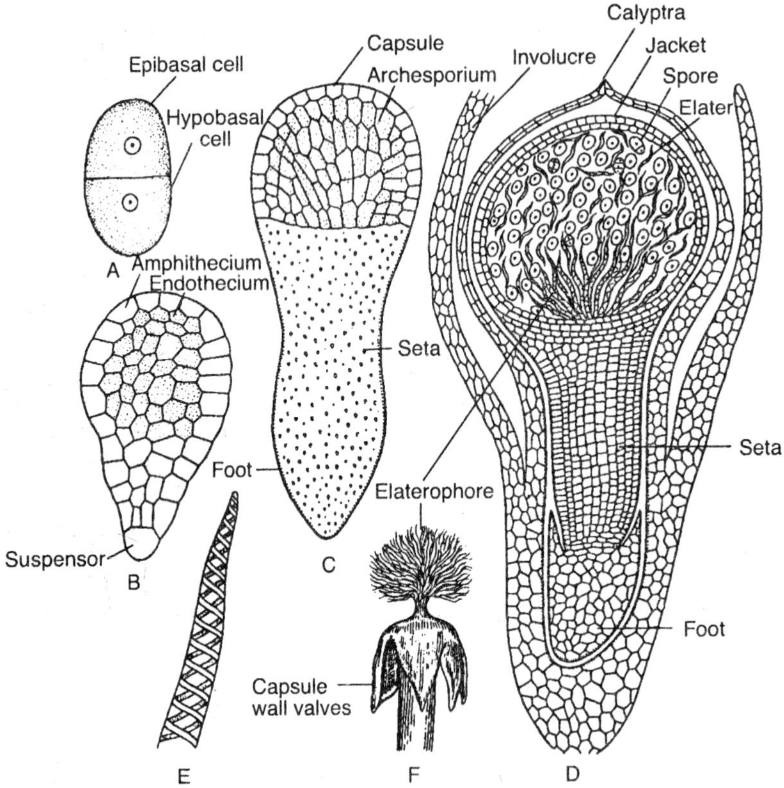

Fig. 28.27: A to D. Stages of development of the sporophyte of *Pellia epiphylla* (after Hofmeister). A. Dyad stage. B. and C. Subsequent stages. D. The mature sporophyte. E. Part of an elater. F. A ruptured capsule of *P. fabbroniana*

a length of up to 8 cm. in course of 2 or 3 days (Fig. 28.24). The elongation of the seta bursts the calyptra and brings the capsule up in the air. The *foot* (Figs. 28.27C and D) is a conical mass of parenchymatous cells and its upper edge surrounds the base of the seta like a collar.

The sporophyte takes about a year to mature. When fully mature it has a long seta and the split calyptra remains as a sheath at the base of the seta (Fig. 28.24A). The capsule wall shows four vertical bands of thin-walled cells along which the wall suddenly bursts into four valves and the valves become reflexed and hang downwards (Fig. 28.27F). The elaters are hygroscopic and the movements of those in the elaterophore serve for an efficient dispersal of the spores.

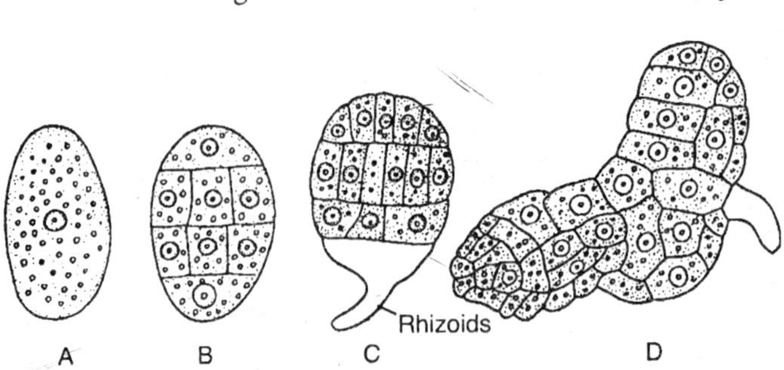

Fig. 28.28: A to D. Stages of germination of the *Pellia* spore beginning the development of the new gametophytic thallus

The new gametophyte

The spore (Fig. 28.28A) begins to germinate while already encased within the capsule. Soon, it becomes a 6-9 celled, ellipsoid structure (Fig. 28.28B) though still inside. All the cells contain chlorophyll but the basal one may be lighter. The capsule bursts at this stage and the germinated spores at this multicellular stage fall on moist soil. The rhizoid develops from the lower lighter cell (Fig. 28.28C). The cell mass soon develops the new gametophyte directly (Fig. 28.28D).

Figure 28.29 depicts the life cycle of *Pellia.*

Family: FOSSOMBRONIACEAE Evans and Goebel (Codoniaceae Cavers)

Genus: *FOSSOMBRONIA*

Fossombronia, a genus of about 50 species, is described briefly to give an idea of the *foliose* anacrogynous *Jungermanniales* (i.e., *Metzgerineae*). This genus, included within the family *Codoniaceae* by Cavers, has now been separately placed in the family *Fossombroniaceae* because of its foliose nature. Three species, *Fossombronia himalayensis* (Western and Eastern Himalaya and South Indian mountains) (Fig. 28.30A), *F. crispata* (W. Himalaya) and *F. indica* (South Indian hills) are known from India.

The gametophytic thallus is prostrate and branched but each branch shows a distinct stem with two lateral rows of leaf-like appendages (Fig. 28.30). The stem is thick but shows no tissue differentiation. There are smooth-walled rhizoids on its lower face while multicellular mucilage hairs protect the tip. Plants often propagate by tubers. Spherical antheridia and flask-shaped archegonia develop on the dorsal surface of the stem in the anacrogynous manner as in *Pellia.* The archegonia are enclosed in a large companulate *involucre.* After fertilization, the zygote divides into an *epibasal* and a *hypobasal* cell, the former of which is larger. But, unlike *Pellia,* there is no suspensor, the hypobasal cell forming the *foot*

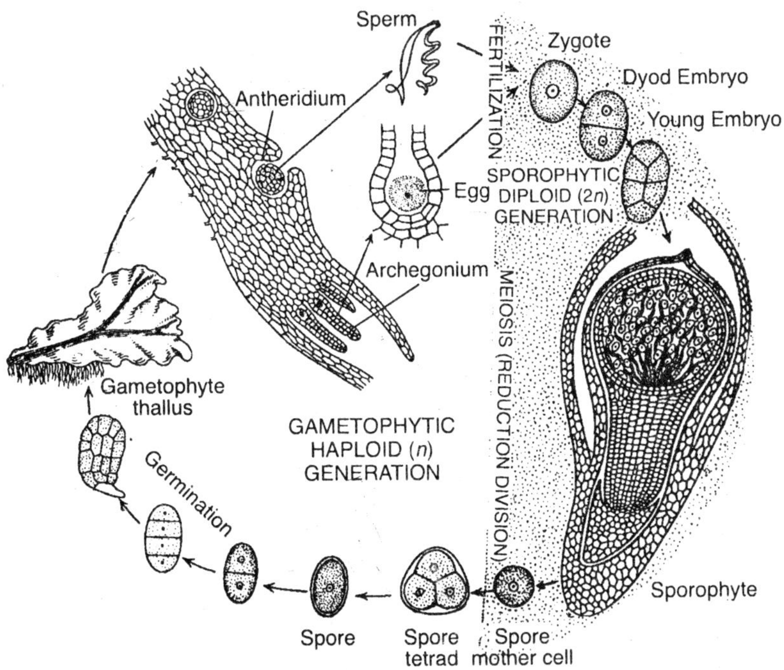

Fig. 28.29: Life cycle of *Pellia* showing alternation of generations

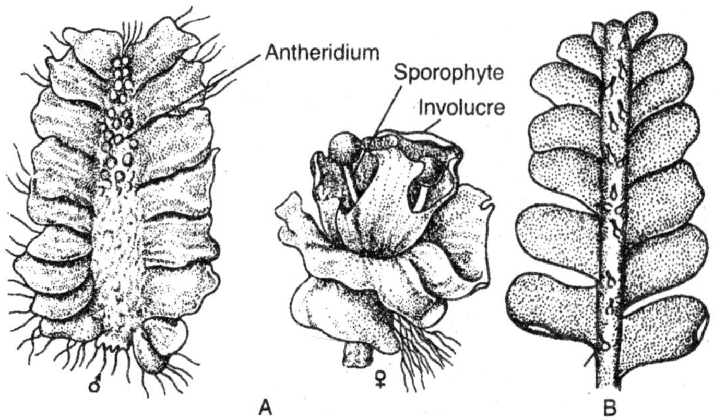

Fig. 28.30: A. Male and female gametophytes of *Fossombronia himalayensis,* the latter showing mature sporophytes (after Kashyap). B. Female gametophyte of *Fossombronia intestinalis* (aftr Smith and Stephani)

and the epibasal cell the *capsule* and the *seta.* The mature sporophyte is like *Pellia* but the *elaterophore* is not so prominent. The *elaters* are peculiar in having 5 to 9 spiral bands of thickening instead of the usual 2. The spore, on germination, forms a short filament of 2 to 12 cells from the apex of which a new gametophyte develops.

Petalophyllum also belongs to this family.

Suborder 2. **Jungermannineae**

The *Jungermannineae* consist of *acrogynous* forms all of which are leafy as at least the shoots bearing sexual organs bear leaves. The three cutting faces of the apical pyramidal growing cell cuts off periclinal cells in a spiral order (Fig. 28.31) each of which forms a bit of a stem (in the inner face) and a leaf in the outer face.

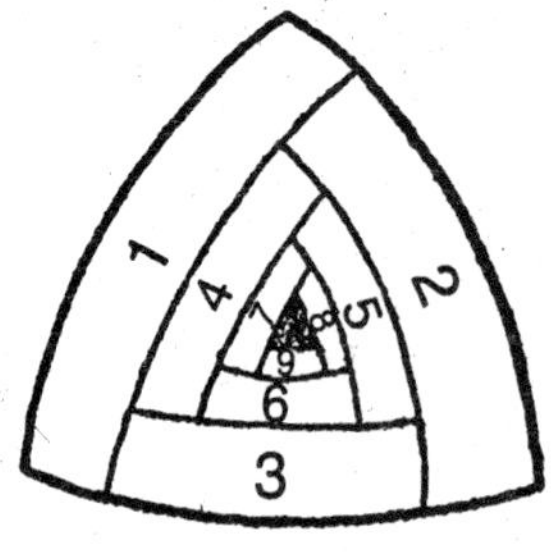

Fig. 28.31: The apical pyramidal cell (x) successively cuts of cells (1, 2, 3, 4, 5, 6) in a spiral upward order. Each such cell is a *merophyte* forming a bit of a stem in the inside and a leaf on the outside

The bit of leaf and stem, developed from a single derivative out of the apical cell is called a *merophyte.* Leaves are, thus, borne spirally and there are two *lateral* and one *ventral* row of them. The ventral leaves are usually smaller and called *amphigastria* or *underleaves.* The archegonia are borne on the apices of stems and branches and, as the apical cells are involved in their formation, apical growth terminates with them. Antheridia are borne singly or in groups in the axils of special *perigonial bracts.*

Family: LOPHOZIACEAE (PLAGIOCHILACEAE)

Genus: *PLAGIOCHILA*

The genus *Plagiochila,* with about 100 species, is the most numerous Jungermanniale in India. Of these about 72 species are present in the Eastern Himalaya alone. The genus is readily identified by the very large plants with toothed, unlobed, decurrent leaves. *Amphigastria* are absent or extremely rudimentary. Fig. 28.32 shows a Darjeeling species.

Family: MADOTHECACEAE
(= **Porellaceae** of Cavers)

Genus: *MADOTHECA* (= **Porella**)

Madotheca is a widespread genus of about 180 species occurring mainly in the subtropics but also in temperature regions. Even now the genus is often called **Porella** though the more **correct name** seems to be *Madotheca.*

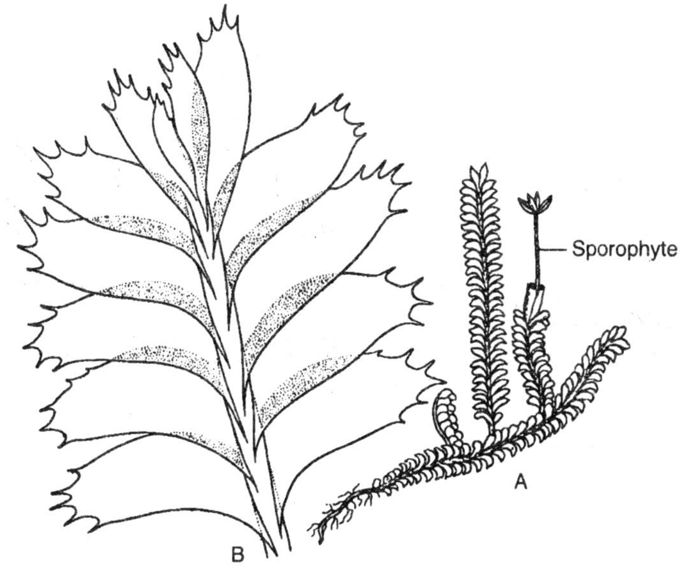

Fig. 28.32: *Plagiochila thomsonii* from Darjeeling. A. Entire gametophytic plant with a burst sporophyte. B. A part of the gametophyte shoot (magnified)

Madotheca and the family name Madothecaceae have always been recognized as a distinct family consisting of the single genus *Madotheca* (i.e., *Porella*). About 35 species are reported from India—from the Himalaya and from South India. Most of the species are recorded from the eastern Himalaya. *Madotheca platyphylla* from the Western Himalaya has a wide distribution covering Europe, North America and Asia. *M. campylophylla* and *M. plumosa* occur all over India (Western Himalaya, Eastern Himalaya and South India). *Madotheca platyphylla* is being described below.

The gametophyte grows as a green patch on shaded moist rocks, trees and even on soil. The plant is flat, dorsiventral with a bi- or tripinnately branched leafy axis (Fig. 28.33A). There are three rows of leaves—two *lateral* on the dorsal side and one smaller, formed of the *amphigastria,* on the ventral side. Each leaf is very simple in structure, being formed of one layer of polygonal cells full of chloroplastids.

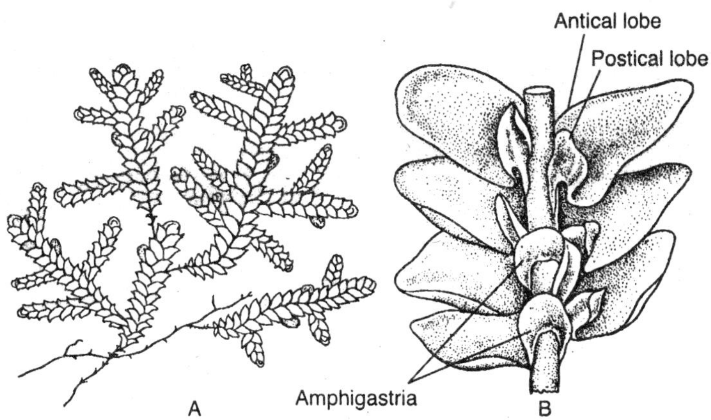

Fig. 28.33: A. *Madotheca platyphylla* gametophyte plant bearing several sporophytes (dorsal view). B. A part of gametophyte shoot (magnified ventral view, top amphigastrium removed)

The lateral leaves are bilobed: the upper anterior (*antical*) lobe is larger, usually ovate with rounded apex while the lower posterior (*postical*) lobe is smaller, narrower with acute apex and is sometimes called the *lobule*. The lobes can be seen clearly from the ventral side of the plant (Fig. 28.33B). When viewed from the top (dorsal side), the lower edge of each leaf is found covered by the upper edge of the leaf next below (the arrangement is seen in a reversed manner from the ventral side in Fig. 28.33B). This type of arrangement is called *incubous*. The *amphigastria* on the ventral surface (Fig. 28.33B) are small leaves somewhat resembling the postical lobes but are broader with decurrent bases. Rhizoids develop from the ventral surface of the stem and anchor the plant to the substratum. The stem shows little differentiation of tissue, only the cortical cells are smaller with slightly thicker walls while the central cells are larger with thinner walls. Branched originate from cells which are to develop the postical lobes of the lateral leaves.

Vegetative propagation is usually by the growth of the vegetative thallus with the progressive decay of the older parts or by the formation of adventitious buds.

Sexual reproduction

Porella is dioecious.

The male plant (Fig. 28.34A) is smaller with the antheridial branches projecting out at right angles from the axis. The leaves on the antheridial branch are closely imbricate and a single antheridium is borne in the axil of each leaf. The mature antheridium (Fig. 28.34B) is globular with a long stalk. The jacket is one-layered at the top and 2- to 3-layered below. The spermatozoid (Fig. 28.34C) is biciliate. There is not much difference in the development of the antheridia (and also of archegonia) with that of Marchantiales.

The female plant is more robust than the male and the archegonial branches are smaller and not so conspicuous. Such a branch bears a number of specialized, enlarged perichaetial bracts (Fig. 28.34D). The lower bracts form an *involucre* while the upper two coalesce to form a *perianth* within which the *archegonia* develop. A number of archegonia develop from the apical cell, the last one of them being terminal. A mature archegonium shows a *venter* with the *egg*, the *ventral canal cell* and a *neck* which is scarcely narower than the venter. The neck is

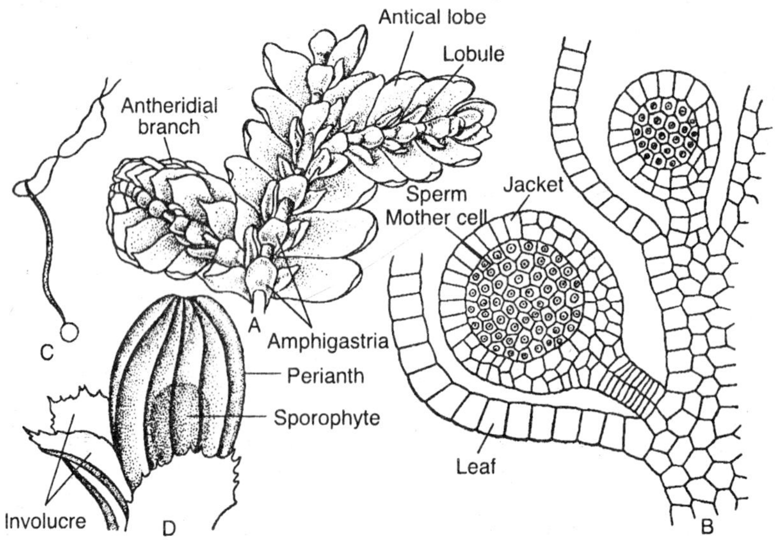

28.34: A. Part of male plant of *Madothecae platyphylla* (ventral view) showing anthecidial branches and details of leaves. B. L.s. of antheridial shoot showing antheridia in leaf axils. C. A sperm. D. An archegonial branch from a female plant showing involucre and perianth with the young sporophyte inside (after Kashyap)

formed of five vertical rows of cells and contains 6 to 8 *neck canal cells.* The wall of the venter is 2-layered.

Fertilization takes place in the usual way through the medium of water.

Sporophyte

After fertilization, the zygote invests itself with a wall and enlarges to fill up the venter. It then divides by a transverse wall to form an *epibasal* and a *hypobasal cell* (Fig. 28.35A). The hypobasal cell forms a *suspensor* which acts as a haustorium while the epibasal cell, by repeated divisions, gradually forms the sporogonium (Fig. 28.35B).

The mature sporophyte (Fig. 28.35C) consists of a spherical *capsule,* a short *seta* and a bulbous *foot.* The capsule wall is 2 to 5 layers thick. The wall cells are generally thickened but there are four vertical layers of thin-walled cells which are the lines of dehiscence. The *Endothecium* forms the *archesporium* which again develops *spore mother cells* and *elaters* intermingled with one another. The spore mother cells, on reduction division, form the spores with which the *magetophytic generation* begins. The elaters show two or three spiral thickenings. With the development of the sporophyte, the archegonial wall becomes multilayered and this forms the *calyptra* enclosing the sporophyte. The calyptra, again, is more or less covered by the *perianth.*

When the spores become mature, the seta elongates suddenly so that the capsule comes out bursting the calyptra and the perianth (Fig. 28.35D). The capsule now splits into four valves along the lines of dehiscence and the spores are discharged being helped by the hygroscopic movement of the elaters.

The new gametophyte

The spore possesses an *exine* (*exospore*) and an *intine* (*endospore*). The exposspore is tougher and often papillose. When the spore falls on a suitable substratum, it germinates to form a multicellular thalloid structure (Fig. 28.36).

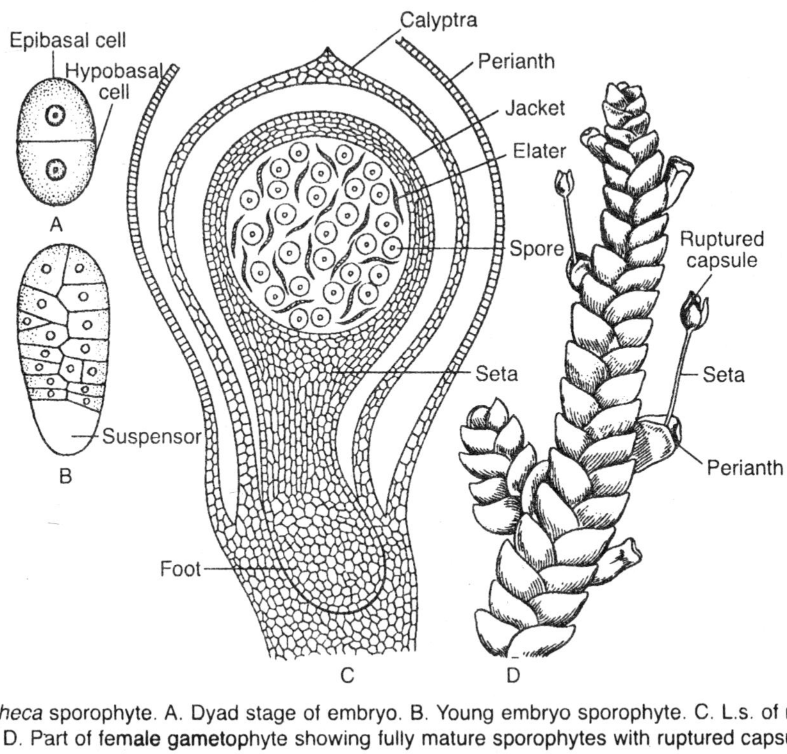

Fig. 28.35: *Madotheca* sporophyte. A. Dyad stage of embryo. B. Young embryo sporophyte. C. L.s. of mature sporophyte. D. Part of female gametophyte showing fully mature sporophytes with ruptured capsules

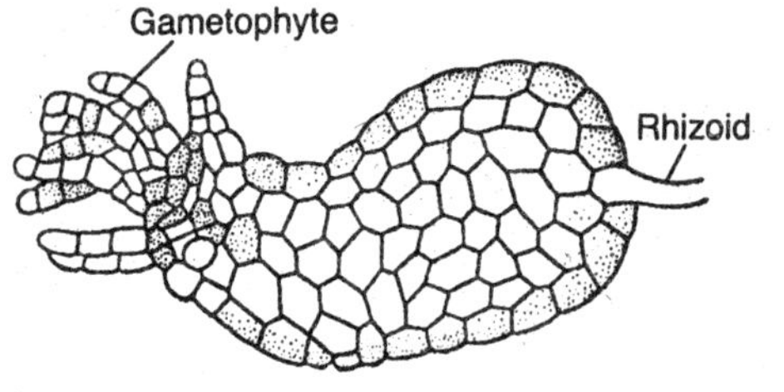

Fig. 28.36: Multicellular thalloid structure formed by spore of *Madotheca* which develops into new gametophyte

This germination may begin while the spore is still within the capsule. One or more rhizoids develop from this thallus and the latter becomes green. Shortly afterwards, an apical cell is differentiated and a new gametophytic shoot develops.

Figure 28.37 shows the life cycle of *Madotheca* (= *Porella*).

Family: LEJEUNEACEAE

Genus: *LEJEUNEA*

The family *Lejeuneaceae* (divided into two families, *Frullaniaceae* and *Lejeuneaceae* by the later authors) consists of the Jungermanniales with the most complicated characters and these are placed in a larger family group called *Jubuleae* by Verdoorn (1932) and *Jubulineales* by Reimers (1954). It is the largest family of the *Jungermanniales* with about 2000 species. The original genus *Lejeunea* itself shows so much variation that it has now been divided into a very large number (about 60) of genera. Chopra (1943) has listed 15 species of *Lejeunea* proper from India but the number of allied species (formerly included within *Lejeunea*) as shown in the list on page 760, is indeed very large. Even the 15 species placed under *Lejeunea* in Chopra's list, a few are doubtfully placed here. Most of these species (e.g., *Lejeunea cavifolia*) are from South India, only *Lejeunea flexuosa*, *L. khasiana* and *L. neplaensis* are definitely reported from the North (Nepal and Eastern Himalaya). Two more species, *L. subacuta* and *L. appendiculata*, originally described by Mitten, are also from this region. The species *chinensis* and *perrottetii* described by

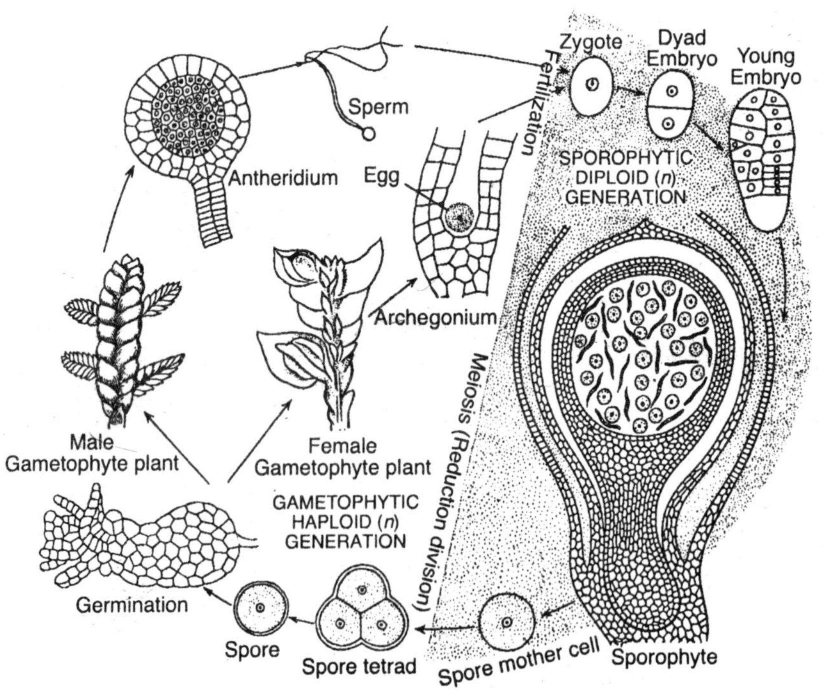

Fig. 28.37: Life cycle of *Madotheca* showing alternation of generations

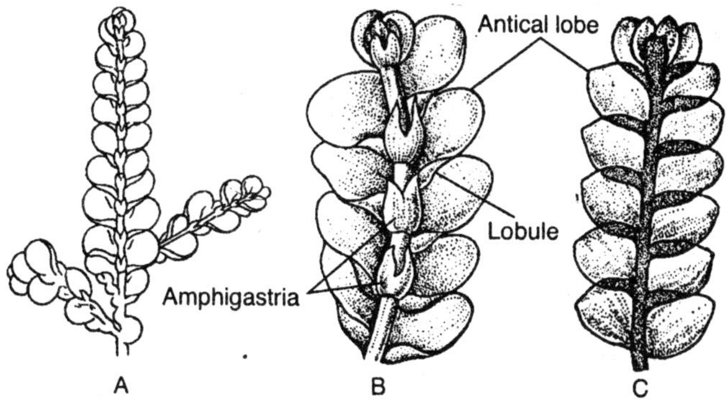

Fig. 28.38: *Lejeunea*. A. Ventral view of a part of gametophyte plant. B. A part of A magnified. C. Dorsal view of shoot showing incubous arrangement of leaves

Mitten, are also from this region. The species *chinensis* and *perrottetii* described by Kashyap (1932) from South India are placed in the allied new genus *Ptychanthus*. An allied genus *Lopholejeunea* grows round about Calcutta but it is not found here in the fruiting stage. The difficulty in correctly assigning some species to their respective positions is because most of them are not known in the fruiting stage.

Lejeunea is widespread in the tropical and temperature regions growing mainly on tree barks and also on moist rocks, walls and even on leaves of other plants (*epiphyllous*). The gametophytes are usually minute and delicate, creeping along the moist surface of the substratum and branching pinnately forming green patches (Fig. 28.38A). The branches in *Lejeunea* do not develop in place of the postical lobes as in the case in *Porella* and *Frulleria*. The shoots are usually not more than 2 cm long but are endowed with stems and leaves in three rows. In the mature, flattened, dorsiventral shoot there are two *lateral* rows of alternating leaves on the dorsal side and one row of *amphigastrial* leaves on the ventral side (Fig. 28.38B). The stem anatomy is very simple, showing no differentiation and smooth-walled rhizoids develop in clusters from the bases of the amphigastria on the ventral side. The leaves are also very simple, being formed of one layer of polygonal cells full of chloroplasts and oil globules. The arrangement of the leaves is *incubous* (lower edge of each leaf covered by the upper edge of the leaf next below—when viewed from the top) (Fig. 28.38C). Each leaf has two lobes, the upper anterior (*antical*) lobe is much larger and is usually ovate-rounded; the lower posterior (*postical*) lobe, also called the *lobule* is much smaller and somewhat swollen (*saccate*). In some species of *Lejeunea*, this lobule may be missing. The ventral row *amphigastria* are rounded leaves, each cleft or notched in the top to about the middle. The amphigastria are smaller than the antical lobes of the lateral leaves but much larger than the lobules.

The usual method of *vegetative propagation* is by death and decay of the older parts, the separate branches forming individual plants.

Sexual reproduction and sporophyte

Lejeunea is monoecious or dioecious. The *antheridia* are usually borne singly or in groups of two to three in the axils of the *antheridial* or *perigonial bracts* (Fig. 28.39A) on short lateral branches or, sometimes even on the main stem. A number of such antheridial bracts may occur in the antheridial shoot and they are usually subequal. Each *antheridium* (Fig. 28.39B) is globular with a short stalk enclosing a number of bi-flagellate antherozoids. During fertilization the top of the antheridium bursts due to hygroscopic action.

The *archegonia* are borne on the tips of *archegonial branches*. Each such shoot ends in a

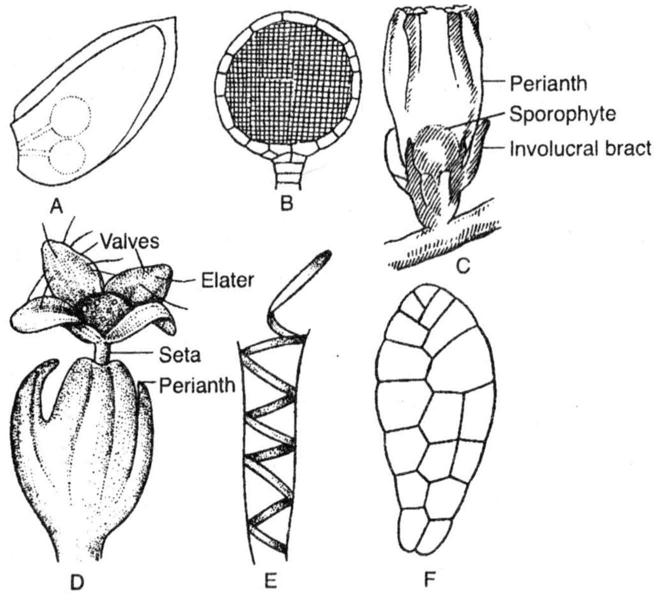

Fig. 28.39: *Lejeunea*. A. A perigonial bract subtending two antheridia, B. An antheridium, C. An archegonial branch, D. Dehiscene of mature sporophyte, E. Part of an elater showing single spiral thickening. F. Thalloid stage of gametophyte (protonema)

single archegonium formed out of the apical cell. The leaves in this shoot are specialized *perichaetial bracts*. The lower ones are the *involucral bracts* or *bracteoles* while the top ones fuse together forming a *perianth* (Fig. 28.39C). The perianth in this case is not cylindrical but frontally compressed and is 5-keeled. There are also sub-involucral innovations. The mature sporophyte (Fig. 28.39D) has a foot, a seta and a capsule as in other jungermanniales. The capsule is globose, with a two-layered jacket, the inner layer of which is spongy. The young sporophyte is covered by a *calyptra* which splits at maturity. When mature, the jacket splits into four valves up to ⅔ the length of the capsule, the lower one-third of it being solid. The *elaters* differ from other Hepaties in having a single spiral thickening (Fig. 28.39E) and these are attached to the inner apex of the valves (Fig. 28.39D), their lengths spreading from the top to the bottom of the capsule. The elaters are differentiated before the *spore-mother cells* which latter give rise to the *spore-tetrads* by reduction division. The origin of the different layers is as in *Porella* or *Marchantia*.

The new gametophyte

The spore germinates as in other *Jungermanniales*, first forming a thalloid *protonema* (Fig. 28.39F) from which the new gametophyte develops.

Figure 28.40 represents the life cycle of *Lejeunea*.

Order 4. CALOBRYALES

The family *Haplomitriaceae* (= *Calobryaceae*) has been taken out of the *Jungermanniales* to form a separate order because of its many peculiarities. Reimers (1954), however, still prefers to retain it within the *Jungermanniales* as a sub-order with two families, *Haplomitriaceae* (anacrogynous) and *Calobryaceae* (acrogynous) in between the *Anacrogynae* and the *Acrogynae*. The order involves two genera: *Haplomitrium* (anacrogynous) from Northern Europe, America and India; and *Calobryum* (acrogynous) from India, Malaya, Japan, Phillipines, New Zealand and tropical America. Some of the important peculiarities may be seen in the best known species, *Calobryum blumii* from Malaya. Udar and his associates have now

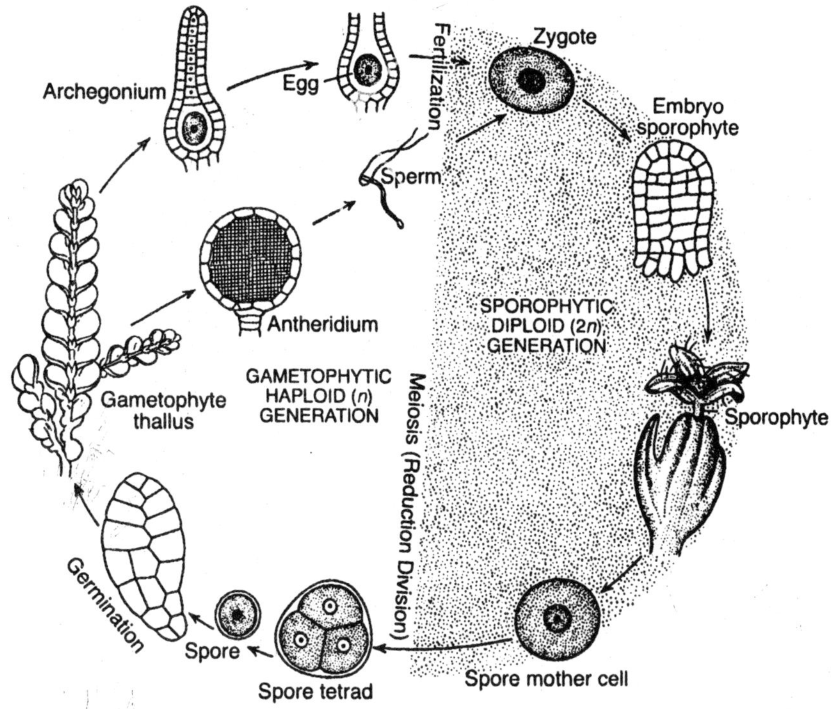

Fig. 28.40: Life cycle of *Lejeunea* showing alternation of generations

established the presence of three species of *Calobryum* (*blumii* Nees, *indicum* Udar & Chandra, *dentatum* Kumar & Udar) and three species of *Haplomitrium* (*hookeri* Nees, *grollei* Kumar & Udar and *kashyapii* Udar & Kumar). All these species occur only in the Eastern Himalaya except *H. hookeri,* which was known from Europe and Northern America, has been found both in the Western and the Eastern Himalayas.

The gametophytic plant shows a freely branching creeping rhizome devoid of rhizoids giving rise to upright shoots bearing three vertical rows of leaves (Fig. 28.41A). The rhizomatous base bears mucilage hairs and has been compared with a root. This is similar to that of *Takakia* (loc. cit.).

The leaves in the three rows may be of the same size or those on one row may be slightly smaller recalling the amphigastrial row of the *Jungermanniales.* Growth is by a single pyramidal cell with three cutting faces. The anatomy of the stem shows thicker cortical cells and thinner central cells resembling some mosses. The plant is dioecious and the uppermost leaves are clustered together. The antheridia are borne in clusters (Fig. 28.41B) in the tip but the archegonium is borne singly—all developing from the apical cell. They strongly remind mosses like *Mnium* or *Funaria.* The archegonial neck shows only four vertical rows of cells instead of the five or six shown by other Hepatics. The neck is usually very long and twisted containing 16 to 20 neck canal cells. The ventral wall is two-layered. The developed sporophyte (Fig. 28.41C) has a very short acuminate foot, a long seta and a cylindrical capsule with a wall of a single layer of cells (resembling *Sphaerocarpales* and *Marchantiales*) excepting at the top. There is no perianth but a massive calyptia. At maturity the capsule splits by a single longitudinal cleft, possibly because the four usual valves do not separate. There is

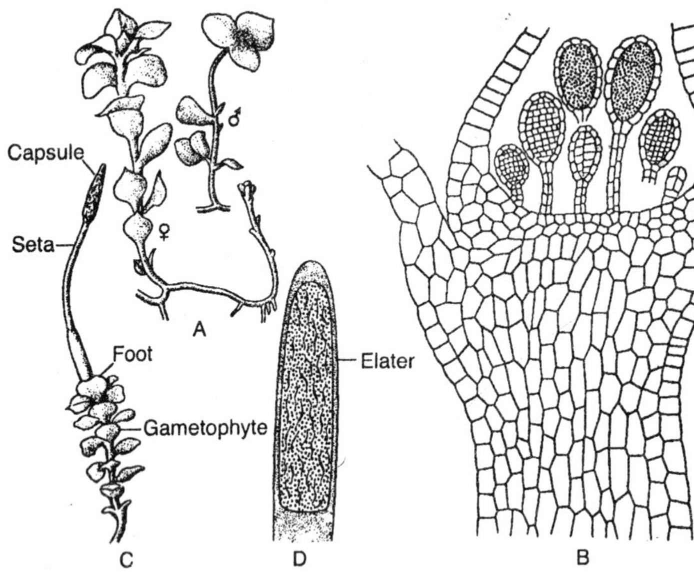

Fig. 28.41: *Calabryum blumii.* A. Male and female gametophytes (after Campbell). B. L.s. of fertile apex of male gametophyte showing antheridia. C. Female gametophyte with sporophyte on top. D. L.s. of capsule with spores and elaters (B & D after Smith)

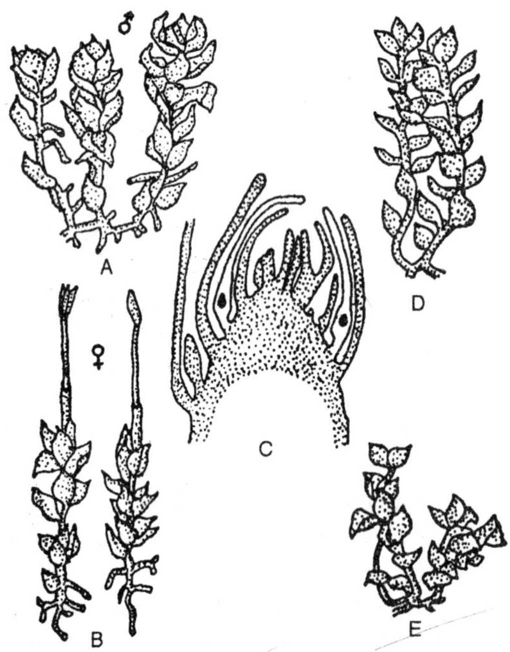

Fig. 28.42: Haplomitrium species. A & B. Male and female plants of *Haplomitrium hookeri* (after Reimers). C. L.s. of shoot apex of *H. hookeri* showing anacrogynous conditions (after Uder). D & E. Vegetative and female plants of *Haplomitrium kashyopi* (after Udar)

no elaterophore but normal elaters (Fig. 28.41D) with double spiral thickening.

Haplomitrium (Fig. 28.42) shows almost the same habit as *Calobryum* but the striking difference is in the development of the archegonia. While in *Calobryum,* the apical cell is used up in its development (acrocarpic), in *Haplomitrium* the archegonia always develop from some lateral segment of the apical cell (anacrogynous Fig. 28.42C), the apical cell always remaining free for further development. Of the 13 species of Calobryales (*Calobryum & Haplomitrium*) recognized by Udar as many as 6 are from Eastern India. Santra *et. al.* also reported *Calobryum blumii* from northern Sikkim.

Order 5. TAKAKIALES

Recently, a very peculiar plant, *Takakia* (named after Takaki who first discovered it in 1951) has been discovered first in Japan and then in Alaska, Canada, N. Borneo, Eastern Himalaya and the Aleutian Islands. The gametophyte (Figs. 28.43A & 28.44A) is erect and

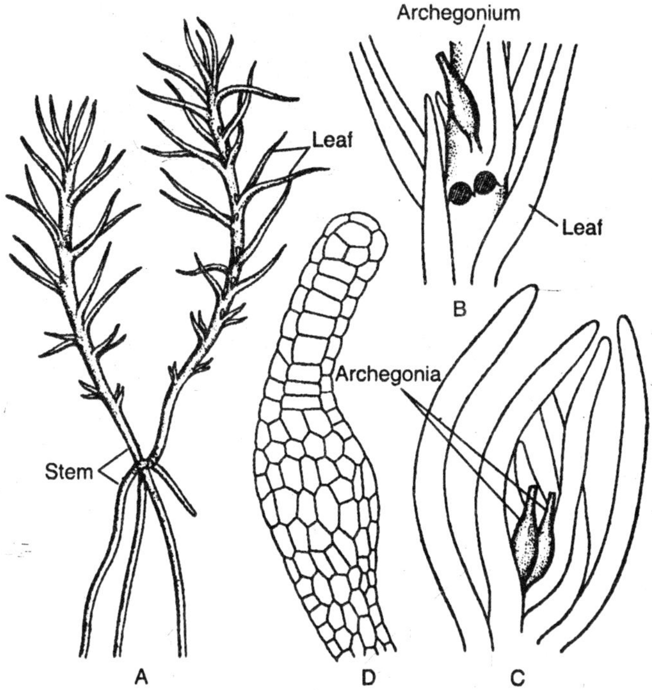

Fig. 28.43: *Takakia lepidozoides* Hatt. & Inoue A. The gametophyte, B (cauline) and C (apical). Archegonia on shoot. D. An archegonium (after Hattori, Mizutani and Inoue)

radial as in the Calobryales. The 'leaves' (better called *phyllids*) are of one kind (isophyllous) and are linear-cylindrical. They are irregularly arranged but they are found on all sides of the stem and are often found in pairs or even in threes. These are deep green, cylindrical and multicellular in thickness (Figs. 28.44B and 28.44C). Filamentous (Fig. 28.44F on the stem) and branched (Fig. 28.44E both on stem and rhizome) mucilage hairs are present. Growth of the stem is by an apical cell with three cutting faces. The stem is laxly erect, branching is exogenous and it shows a central zone of smaller parenchymatous cells surrounded by several layers of larger cortical cells (Fig. 28.44D). There is an underground rhizomatous part but no rhizoid is present on it though scale leaves and mucilage hairs are present (Fig. 28.44A). There is an endophytic fungus in the stem spidermis as well as in the rhizome. The phyllid resembles the stem anatomically showing 1 to 5 axial cells surrounded by an epidermis of one layer. The chromosome number is $n = 4$ which is the lowest among Bryophytes.

Antheridia and sporophytes are so far unknown so that our knowledge of the group is still incomplete. Archegonia, however, have been found. They are in groups of 1 to 3 on the stem apices (Fig. 28.43C) and are also found singly on the stem (Fig. 28.43B). In the apex the apical growing cell does not take part in its development. Each archegonium (Fig. 28.43D) is more massive than in the Hepatics and is as large as in the mosses. It is flask-shaped and stalked, with a multilayered ventral jacket and a neck about 10 cells long with 4 vertical rows of wall cells. The stalk is uncommonly massive. Sometimes it forms a long pedical. Peculiarly, the archegonia are green.

Although not fully known, it may at least be placed in a separate order of the Hepaticopsida as the Takakiales. Proskauer (1962),

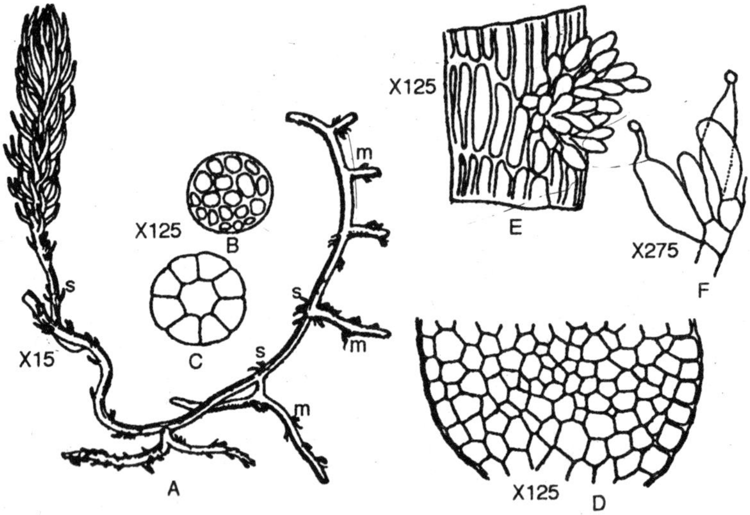

Fig. 28.44: *Takakia ceratophylla* (Mitt.) Grolle. A = plant showing aerial and underground rhizomatous portion (s = scale leaf and m = mucilage hair on rhizomatous portion). B = t.s. of leaf (phyllid). C = t.s. of leaf (phyllid). D = t.s. of stem. E = branched mucilage hair. *T. lepidozoides* Hatt. & Inoue. F = filamentous mucilage hair in stem (Figures after Hattori *et al.*)

Schuster (1966) and Grubb (1970), however, prefer to place the genus within Calobryales. This is supported by the presence of the rhizomatous base devoid of rhizoids and also by the presence of mucilage hairs on this rhizomatous base. Grubb has even called this rhizomatous base a root. But, still the position is not certain.

The first discovered species in Japan is *T. lepidozoides* (Fig. 28.43A). *Takakia ceratophylla* (Mitt.) Grolle (Fig. 28.44A), collected from Sikkin (Lachen) by Heoker was described more than a century ago by Mitten as a Jungermanniale which he had named *Lepidozia ceratophylla* Mitt. Recently (1973) both *Takakia ceratophylla* and *T. lepidozoids* have been discovered by Iwatsuki from Arun Valley (400 m) in East Nepal. Outside India *T. ceratophylla* is known only from the Aleutian Islands. The presence of both the species in Eastern Nepal may point out to the origin.

Chapter 29

Anthocerotopsida: Anthocerotales

The class *Anthocerotopsida* hornwarts (also called *Anthocerotae*) is a small but very distinct (c.f. chapter 28) group of Hepatics differing from the others, specially, in its sporophyte which is of indefinite growth because of a characteristic intercalary meristem at the base of the capsule. The sporogenous cells are amphithecial in origin. The cells of the gametophytes, usually, have large single chloroplasts with a conspicuous pyrenoid each. The sex organs are embedded in the gametophyte.

The Class has a single order *Anthocerotales* in which four generan, *Anthoceros, Megaceros, Dendroceros* and *Notothylas,* are universally recognized. But, the largest genus *Anthoceros* has now been divided into three: Stephani (1916) first took out 55 species to form the genus *Aspiromitus* and, lately, Proskauer (1951) has proposed the creation of a third genus *Phaeoceros* out of *Anthoceros.* Thus, the total number of genera are now six and, all these used to be included within the single family *Anthocerotaceae.* But, Müller (1940) proposed that the genus *Notothylas* should be separated to form a second family *Notothylaceae.* He is supported by Proskauer (1951) but Reimers (1954), while accepting the two families, has retained only the four original genera. There are about 300 species within *Anthocerotales* and these are distributed throughout the world, specially in tropical to temperate regions.

The *Anthocerotales* are represented in India by the following:

Family: *Anthocerotaceae*

Aspiromitus (4 spp.), *Antheroceros* (including *Phaeoceros*—25 spp.), *Megaceros* (1 sp.) and *Dendroceros* (1 sp.).

Family: *Notothylaceae*

Notothylas (3 spp.)

Interrelationships among genera and delineation of generic boundaries remain problematic in hornworts. Cladistic analyses of morphological and ultrastructural data have been conducted and some of these phylogenetic hypotheses have been used to propose alternate classifications for anthocerotes (Hässel de Menéndez 1988, Hyvonen and Piippo 1993, Hasegawa 1994). Different concepts of classification are demonstrated in Table 29.1 wherein the four taxonomic schemes recognize from five to nine genera. In total, 11 genera of hornworts have been named, of which only six have gained wide recognition; *Anthoceros, Phaeoceros, Folioceros, Notothylas, Megaceros,* and *Dendroceros.* The remaining five genera are defined by anomalies of the spores and pseudoelaters or by specific features of the antheridia or thallus.

An over view of four alternative classifications are shown in Table 29.1.

Table 29.1: Four alternative classification of the hornworts

Hässel de Menéndez (1944)
Anthocerotales Limpricht in Cohn
  Anthocerotaceae Dum.
    Anthoceros (Mich.) L.
    Sphaerosporceros Hässel
  Notothyladaceae (Milde) Müll.
    Subfam. Notothyladoideae Grolle
      Notothylas Sull.
    Subfam. Phaeocerotoideae Hässel
      Phaeoceros Prosk.
Foliocerotales Hässel de Menéndez
  Foliocerotaceae Hässel
    Folioceros Bharadwaj
Leiosporocerotales Hässel de Menéndez
  Lciosporocerotaceae Hässel de Menéndez
    Lciosporoceros Hässed de Men.
Dendrocerotales Hässel de Menéndez
  Dendrocerotaceae (Milde) Hässel de Men.
    Dendroceros Nees
    Megaceros Campbell

Schuster (1992)
Anthocerotales Limpricht in Cohn
  Anthocerotaceae Dum.
    Subfam. Anthocerotoideae Dumort
      "*Aspiromitus* Steph. emend. Rink
        Subgen. *Aspiromitus*
        Subgen. *Folioceros*
      "*Anthoceros* (Mich.) L
        Subgen. Anthoceros
        Subgen. Leiosporoceros
    Subfam. Dendrocerotoideae Schuster
      Megaceros Campbell
        Subgen. *Megaceros*
        Subgen. *Notoceros* Schuster
      Dendroceros Nees
        Subgen. *Dendroceros*
        Subgen. *Apoceros* Schuster
    Subfam Notothyladoideae Grolle
      Notothylas Sull.

Hyvönen and Piippo (1993)
Anthocerotales Limpricht in Cohn
  Anthocerotaceae Dum.
    Anthoceros (Mich.) L
    Folioceros Bharadwaj
    Leiosporoceros Hässel de Men.
    Mesoceros Piippo
    Phacoceros Prosk
    Sphaerosporoceros Hässel de Men.
  Dendrocerotaceae (Milde) Hassel de Men.
    Dendroceros Nees
    Megaceros Campbell
Notothylales
  Notothyladaceae (Milde) Müll.
    Notothylas Sull

Hasegawa (1994)
Anthocerotales Limpricht in Cohn
  Notothyladaceae (Milde) Müll
    Notothylas Sull
  Anthocerotaceae Dum.
    Subfam Anthocerotoideae
      Anthoceros (Mich.) L.
        Subgen. *Anthoceros*
        Subgen. *Folioceros*
      Leiosproceros Hässel de Men.
      Hattorioceros Hasegawa
      Phaeoceros Prosk.
    Subfam. Dendrocerotoideae
      Dendroceros Nees
        Subgen. *Dendroceros*
        Subgen. *Apoceros* Schuster
      Megaceros Campbell
      Notoceros (Schuster) Hasegawa

Note: This author recognizes *Aspiromitus* in place of *Anthoceros* and *Anthoceros* in place of *Phacoceros*

Family: **ANTHOCEROTACEAE**

Genus: *ANTHOCEROS*

Anthoceros is the largest genus of the *Anthocerotopsida* comprizing about 200 of the 300 species included within the Class. About 25 species of these are reported from the tropical, subtropical and temperate hilly areas all over India—the Himalayas, Chhotanagpur, the Ghats, Nilgiri, etc. Of these, mention may be made of *Anthoceros erectus* Kashyap (Figs. 29.1C and D) and *A. himalayensiss* Kashyap (Figs. 29.1E and F), found all over the Himalayas, in South India and even in the Gangetic plains. They grow in very moist, shady places by clay banks and ditches or in the crevices of rocks. Campbell (1939) noted that the species of *Anthoceros* fall into two groups, with black spores and with yellow spores, the two groups also showing other morphological differences. Proskauer (1951) has placed the yellow spored ones into the new genus *Phaeoceros.*

The Gametophyte

The gametophyte is a very simple, small, prostrate, dorsiventral thallus of a greasy dark-green colour. The thallus is scarcely branched though usually dichotomously lobed (Fig. 29.1B), often suborbicular (Fig. 29.1A). There is no definite midrib although the middle part of the thallus is always many layers thick. Sometimes, as in *A. erectus* (Figs. 29.1C and D), the thallus is raised on a thick ascending stalk. Again, in *A. hallii*, the thallus is long and pinnately lobed. The dorsal surface of the

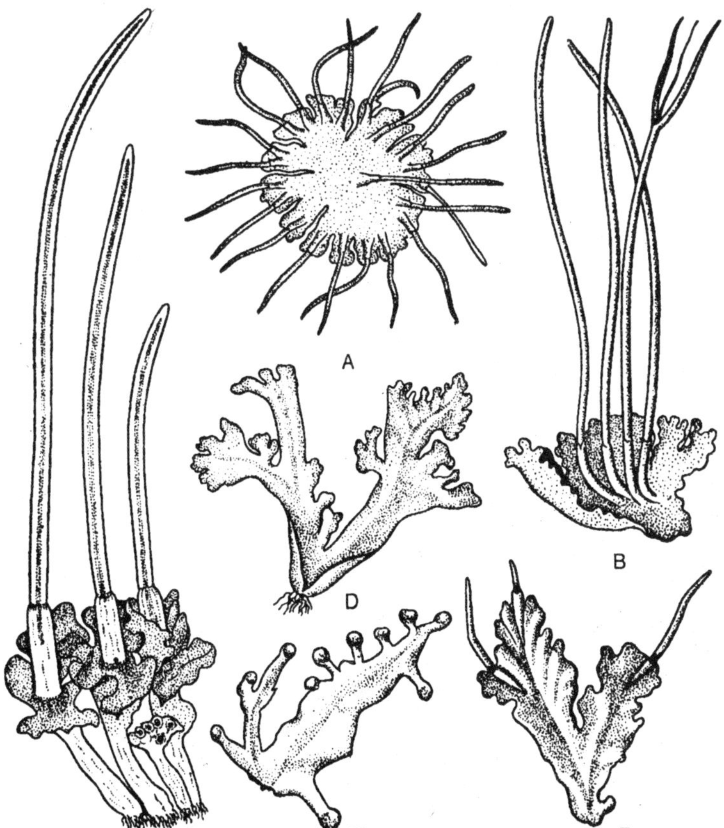

Fig. 29.1: A. *Anthoceros crispulus* (after Bharadwaj), B. *Anthoceros fusiformis* (after Smith). C (after Bharadwaj) and D (after Kashyap). Two forms of *Anthoceros erectus.* E (Sterile, bearing tubers) and F (Fertile). Forms of *Anthoceros himalayensis* (after Kashyap)

thallus may be smooth or rough with ridges or even with some lamellar growths. The lower ventral surface always bears numerous rhizoids which are only of the smooth-walled type. Scales, tuberculate rhizoids or mucilage hairs are absent in *Anthoceros*.

The thallus tissue (Fig. 29.2A) shows little or no differentiation. The thallus is composed of uniform, thin-walled, parenchymatous cells which are many cells deep in the middle (from 6 to 8 cells in *A. laevis* to 30-40 cells deep in *A. crispulus*). In some species, the upper epidermal layer may differ in being formed of smaller, compact cells with larger chloroplasts. Each cell of the thallus (Fig. 29.2B) shows a *single* oval and fattened chloroplast with a central *pyrenoid* resembling some green alga like *Coleochaete*. The nucleus remains very near the chloroplast. Such a chloroplast with pyrenoid is unknown in the whole group of *Embryophyta* except in *Selaginella*. But, this 'pyrenoid' is now known to be different from those in the algae as it is actually an aggregate of 25 to 300 disc- or spindle-shaped bodies each of which is capable of being transformed into a rudimentary starch grain. In several species of *Anthoceros*, the deeper layers of the thallus show two or more chloroplasts instead of the normal one. There are no air chambers or pores in the tissue of the thallus but the lower ventral surface of the thallus shows small, rounded, opaque bluish green spots which may be detected with a hand lens. These are intercellular cavities filled with mucilage and opening on the ventral side by means of narrow slits (Figs. 29.2A and C). These mucilage cavities always contain colonies of endophytic *Nostoc*. It was previously assumed that this *Nostoc* was of no use to *Anthoceros* but now it has been claimed that nitrogen fixed by this *Nostoc* may be utilized by the latter. In certain other species of *Anthoceros*, in addition to these mucilage ducts, there are schizogenous tubular cavities behind the growing point and running longitudinally through the thallus and also, commonly, filled with mucilage. In older parts this mucilage may be replaced by gaseous contents. In a few species, again, there may be some enlarged cells filled with mucilage.

The thallus usually grows by a single apical cell with two cutting faces, one on the dorsal and the other on the ventral side. But, *A. erectus* and *A. himalayensis* have been shown to grow by a *group* of apical cells (Mehra and Handoo, 1953). The apical cell may split into two giving rise to the dichotomous lobes.

Vegetative reproduction

Vegetative reproduction may be by the decay of older parts and growth. But, this method is not so common in *Anthoceros* as in the other Hapatics.

Under favourable conditions, many species of *Anthoceros* are known to produce *tubers* by thickening of marginal patches of the thallus tissue. Such a tuber is protected by a layer of

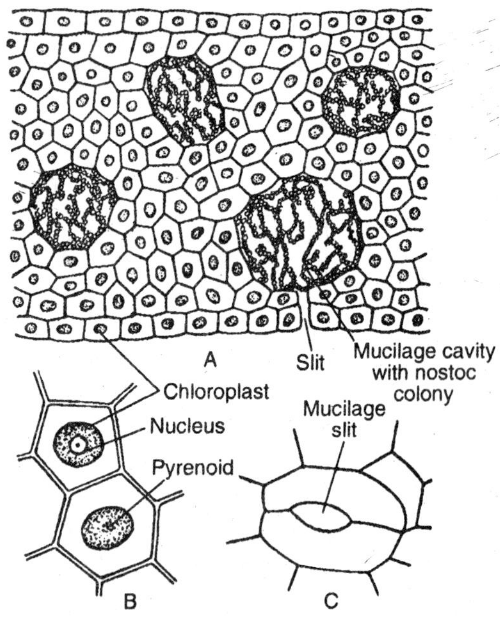

Fig. 29.2: *Anthoceros*. A. Vertical section through gametophyte thallus showing mucilage cavities with Nostoc colonies and opening slit in ventral surface. B. Gametophyte thallus cell showing single chloroplastid, pyrenoid and nucleus. C. Mucilage slit or cleft opening to mucilage cavity from ventral side

cork cells and may resist desiccation. This germinates into a thallus on the return of favourable conditions. Such stalked tubers are known in *A. himalayensis* (Fig. 29.1E).

Gemmae are known to develop on the margin of the thallus of some species of *Anthoceros.* Sometimes they are stalked and develop mucilage pores even before they are detached from the parent plant. These gemmae germinate into new gametophytes as in *Marchantia.*

Sexual organs

Anthoceros is dioecious or monoecious. *A. erectus* is a dioecious species while *A. himalayensis* is monoecious. Monoecious species are usually protandrous. Both antheridia and archegonia are embedded in the dorsal surface of the thallus. They are initiated just behind the growing points where they develop in continuous rows.

As opposed to the *Hepaticopsida*, the *antheridium* develops from a *hypodermal cell.* A superficial dorsal cell usually divides by a periclinal wall (Fig. 29.3A). The upper daughter cell becomes the *roof initial* which divides and redivides forming a multicelled *roof* (one or more layers thick) over the antheridium below. The lower cell is the *antheridial initial* which may develop a single antheridium or may divide to give rise to several antheridia. Sometimes younger antheridia may also develop from the base of the older antheridium so that there may be antheridia of different ages in a cluster and the number, which is originally 1, 2 or 4, may go up to 25 or even more within a single *antheridial chamber* which is gradually formed below the roof with the accumulation of mucilage in the intervening space (Fig. 29.3B). The antheridial initial divides transversely to form a *primary stalk cell* below and a *primary antheridial cell* above. Further development of the antheridium and spermatogenesis is as in the *Hepaticopsida.* A mature antheridium (Fig. 29.3B) shows a more or less slender *stalk* (four to more rows of cells) bearing a club-shaped or sperical antheridium which has a jacket one or more layers thick containing the androcytes inside. Each jacket cell contains a plastid which develops chlorophyll and becomes green as the antheridium matures. In the ripe antheridium these plastids remain green or become orange. Each androcyto forms a biflagellate *antherozoid* (Fig. 29.3C).

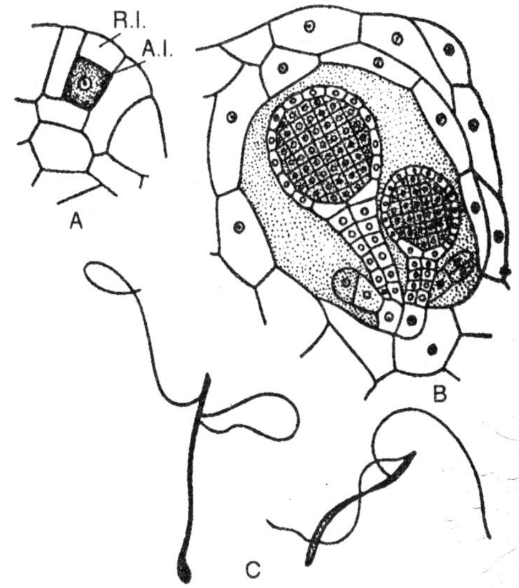

Fig. 29.3: *Anthoceros*. A. Differentiation of roof initial (*R.I.*) and antheridial initial (*A.I.*). B. An antheridial chamber showing a cluster of antheridia. C. Two antherozoids

The *archegonia* are produced acropetally from superficial dorsal cells close to the apical growing point. In case of monoecious species, the archegonia develop later in the same thallus which had developed the antheridia. The archegonia develop singly and are completely embedded in the gametophyte in direct contact with the surrounding vegetative cells. The *archegonial ivitial* (Fig. 29.4A) functions directly as the primary archegonial cell, there being no stalk. Three vertical walls (Figs. 29.4B and C) cut off three outer *jacket initial cells* and a *primary axial cell.* This axial cell now divides transversely (Fig. 29.4D), the lower one becoming the *primary ventral cell* and the upper divides again (Fig. 29.4E) forming a top *cover initial* and a lower *primary neck canal cell*. The development of the jacket is difficult to follow as it is indistinguishable from the surrounding vegetative cells. The cover initial forms a rosette of four cover cells, the primary neck canal cell forms a vertical row of 4 to 6 or more *neck canal cells* and the primary ventral cell forms the *ventral canal cell* and the egg (Fig. 29.4F). The neck wall is formed by six vertical rows of cells. There is usually a *mucilage mound* (Fig. 29.4G) covering the growing archegonium.

Fertilization

At maturity, the roof of the antheridial chamber breaks down exposing the antheridia which, on absorbing water, rupture by apical apertures. The androcytes are now extruded out into the covering film of water where the antherozoids soon get liberated. The mature archegonium, only whose cover cells are protruding (Fig. 29.4G), also bursts. The neck canal cells and the ventral cell disintegrate and are extruded into the mucilage mound while the cover cells are thrown off. The egg now becomes directly exposed. Fertilization then takes place as in the *Marchantiales*.

The Sporophyte

The zygote (fertilized egg) increases in size and completely fills up the venter. Then it secretes a cellulose wall round itself and becomes four-celled (Fig. 29.5A) by two successive walls at right angles to one another. Another series of vertical walls brings about the 8-celled *octant* stage composed of two tiers of 4 cells each. The lower tier of four cells forms the sterile *foot* after repeated divisions. The lowermost cells of the foot grow short rhizoid-like projections increasing the absorptive surface for sucking food from the gametophyte. The upper tier of cells divides and its lower cells usually form an *intermediate part* which, later on, develops and intercalary meristematic tissue. Very soon, periclinal divisions in the upper cells separate out the inner *endothecium* and the outer *aphithecium* (Fig. 29.5B). The endothecium gives rise to the sterile *columella* (Fig. 29.5C) which is composed to 16 vertical rows (the number of rows is even more in some species) of cells as is seen in the t.s. of a mature

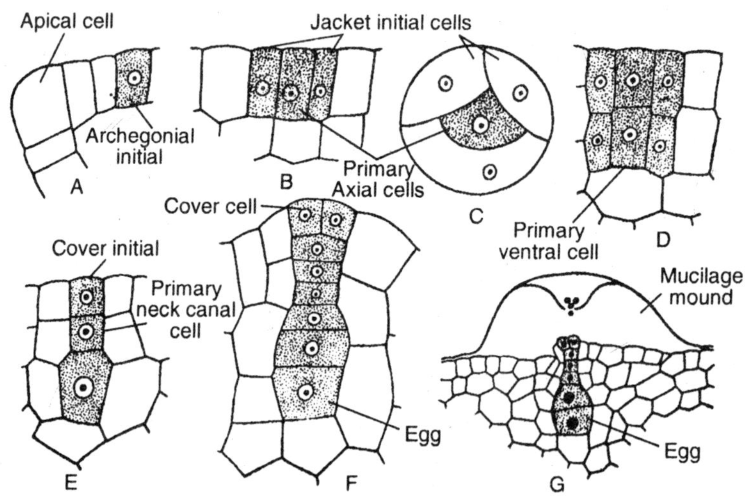

Fig. 29.4: Stages of development of *Anthoceros* archegonium. A. Archegonial initial. B. Differentiation of primary axial and jacket initial cells. C. T.s. of B. D. Division of axial cell to form primary ventral cell. E. Further division of upper axial cell forming cover initial and primary neck canal cell. F. Nearly **mature archegonium**. G. A mucilage mound covering a fully mature archegonium

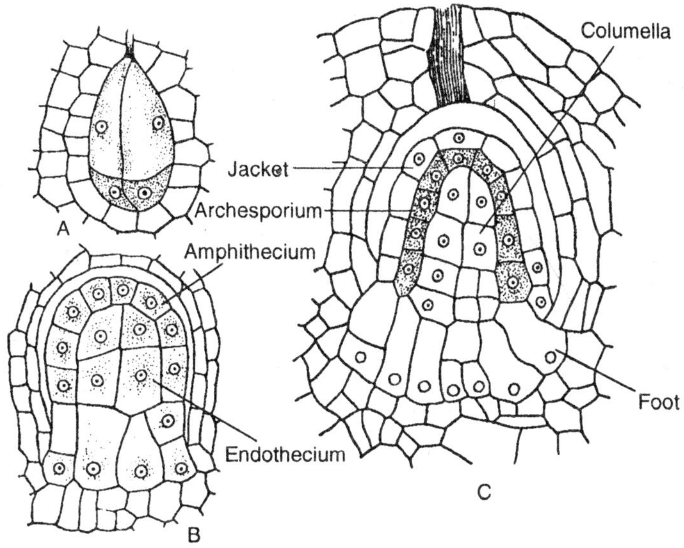

Fig. 29.5: *Anthoceros* sporophyte embryo. A. 4-celled stage. B. Differentiation of amphithecium and endothecium. C. Further differentiation showing jacket, archesporium, columella and foot

sporophyte (Fig. 29.6B). The amphithecial cells divide periclinally and its outer layers form the *jacket* which is 4 to 6 cells in thickness in the mature sporophyte (Figs. 29.6A and B). The outermost layer of the jacket becomes an epidermis with cutinized outer walls. Regular stomata with guard cells (Fig. 29.6C) occur on this epidermis. The inner cells of the jacket are parenchymatous and, usually, contain two chloroplastid each in place of the single chloroplastid of the gametophytic cells—but this number may vary.

The inner cells of the amphithecium become the *archesporium* (Fig. 29.5C) which overarches the rounded apex of the columella and may or may not extend to the base of the columella. This tissue may become up to 4 layers in thickness. Alternate transverse tiers of the archesporium become up to 4 layers in thickness. Alternate transverse tiers of the archesporium become *spore mother cells* and *sterile cells*. The spore mother cells form the *spore tetrads* in the usual way, while the sterile cells join end to end forming simple or branched *pseudoelaters* of 3 to 5 cells (Figs. 29.6A and B) which lose their protoplasmic contents when mature and remain thin-walled or are variously thickened.

The **mature sporophyte** (Fig. 29.6A) shows a bulbous *foot*, the lowermost cells of which are haustorial, though often sharply delimited by the gametophytic tissue. Above the foot is the slender, erect, smooth, cylindrical *capsule* standing out like a bristle (or a 'horn'—hence the English name 'hornwort'), which projects out 2 to 15 cm in some cases. The base of the capsule is ensheathed by an *involucre* to a certain height. This involucre is formed by the development of the gametophytic tissue immediately surrounding the base of the growing capsule. Between the foot and the capsule there is no seta but a short *intermediate zone* which is meristematic and is regenerating the capsule from the base. Thus, the capsule is always in different stages of growth: it is in the embryonic stage at the base and very mature at the tip. The centre of the capsule is taken up by the *columella* and round it is the *jacket* wih a regular *epidermis* and *stomata* (Fig. 29.6C) below which is the green parenchyma-

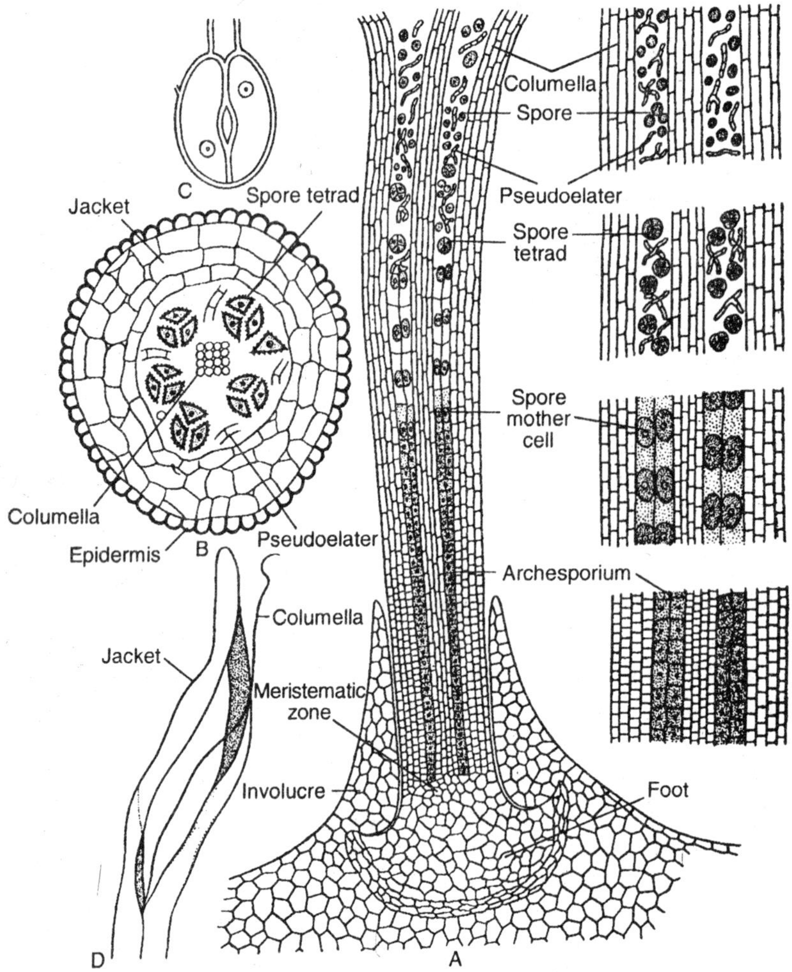

Fig. 29.6: Mature sporophyte of *Anthoceros*. A. L.s. (semidiagrammatic). Different portions are shown magnified on the right-hand side. B. T.s. at the middle of the sporophyte. C. A stomata in the jacket epidermis. D. Ruptured tip of a mature sporophyte capsule

tous photosynthetic tissue. The columella is not a regular conducting tissue although it may represent the first tendency towards it. In between the jacket and the columella is the cylindrical *archesporial zone* which is young and one-layered at the base, gradually showing differentiation into spore *mother cells* and *pseudoelaters* upwards. At the top there are regular spores and *pseudoelaters*. When mature, the top of the capsule becomes black or brown. A split appears below the tip and gradually widens upwards. Hygroscopic movement of the pseudoelaters releases the mature spores at the top and the split into two halves progresses downwards. The tip of the columella projects out like a flagellum. The top of the capsule looks twisted at this stage (Fig. 29.6D).

The new gametophyte

The spore (Fig. 29.7A) germinates on a suitable substratum. The exospore bursts and the endospore comes out as a germinal tube (Fig. 29.7B) at the tip of which the new gametophyte is formed by the development of a thallus.

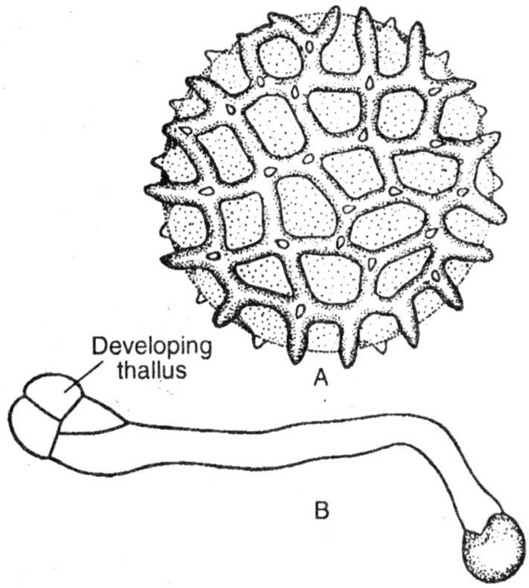

Fig. 29.7: *Anthoceros.* A. A spore (after Bharadwaj), B. Germination of the spore (after Campbell)

Figure 29.8 shows the life cycle of *Anthoceros.*

Family: NOTOTHYLACEAE

Genus: *NOTOTHYLAS*

Notothylas is a common genus in India growing on shady moist soil or rock in tropical and subtropical regions. Three Indian species were reported previously: *N. indica, N. levieri* and *N. chaudhurii.* To these Udar and Singh have added two more species: *N anaporata* and *N. pandei.* Because of its peculiarities, Muller (1940), Proskauer (1951) and Reimers (1954) are of opinion that this genus should be taken out of the *Anthocerotaceae* and placed to form a new family *Notothylaceae.*

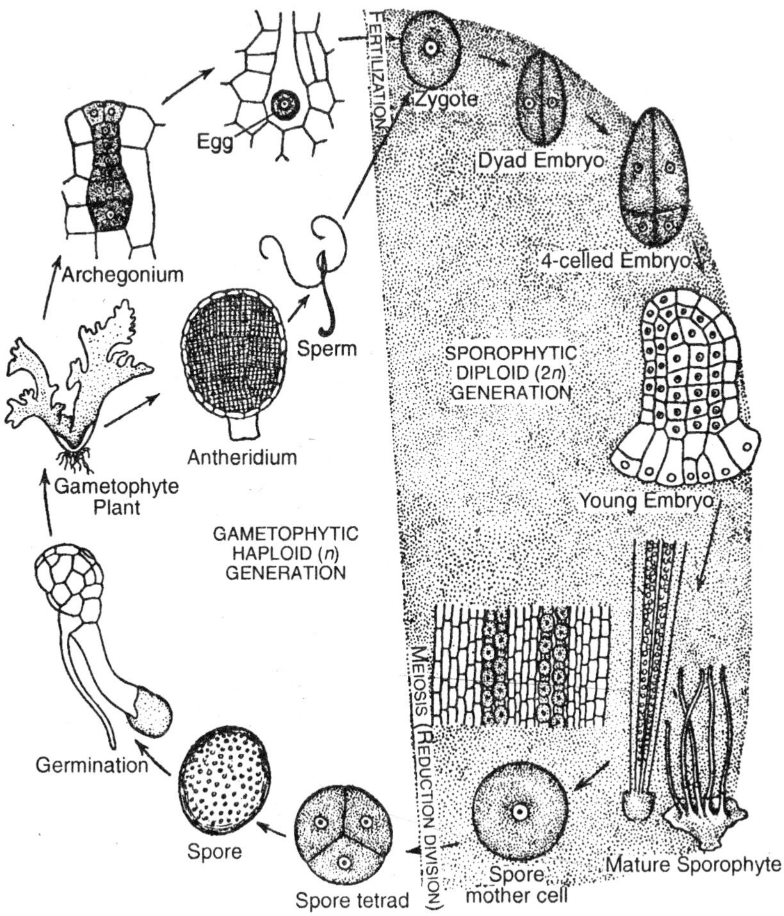

Fig. 29.8: Life cycle of Anthoceros showing alternation of generations

The *gametophytic thallus* (Fig. 29.9A) is yellow-green forming an orbicular or suborbicular rosette which has got a characteristic pleated and not flat appearance. Anatomically, the tissue generally resembles that of *Anthoceres*, not showing any differentiation. It is 6 to 8 cells deep in the middle, thinning out to 1 to 3 cells in the edges.

Notothylas may be *dioecious* but all the Indian species are *monoecious* and *protandrous.* The *antheridia* (Fig. 29.9B) and the *archegonia* (Fig. 29.9C) generally resemble those of *Anthoceros*, only the neck canal cells are wider.

The **sporophyte** differs in different species. In *N. indica* (Fig. 29.10A), the *endothecium* forms the *columella* and the *amphithecium* forms the *archesporium* on the inside and the jacket on the outside. In some species, the *archesporium* is developed partly from the *amphithecium* and partly from the outer *endothecium.* In *N. levieri* (Figs. 29.10B and C) and *N. choudhurii*, there is no columella, the entire *endothecium* forming the *archesporium* while the *amphithecium* forms the *jacket.* The archesporium develops from the endothecium in the Hepaticopsida and from the inner amphithecium in *Anthoceros.* Thus, the position of *Notothylas* is clearly between the two. In the cellwall thickening of the capsule wall also some *Notothylas* species approaches the Jungermanniales of Hepaticopsida.

The mature sporophyte is only 2 to 3 mm long, tapering at both ends and lies flat (horizontal) on the thallus, scarcely projecting beyond the thallus margin (Figs. 29.9A and 29.9C). It is usually completely ensheathed by the thin, membranous *involucre.* The *foot* is much smaller than in *Anthoceros* although the haustorial outgrowths are well developed. The *intermediate meristematic zone* also is much less developed. The meristematic development being negligible so that the *capsule* does not grow much. The *pseudoelaters* are unicellular of irregular form, and have thickenings on the walls. The dehiscence is not as prominent as in *Anthoceros. N. indica* and *N. levieri* dehisce along one suture like a follicle. Spores are often liberated by the decay of the capsule wall.

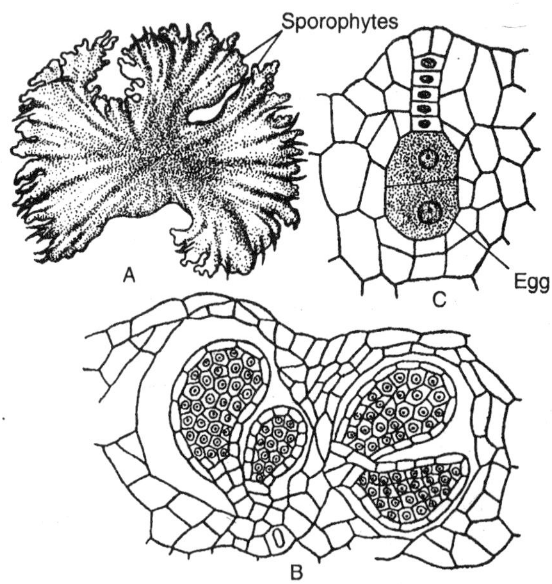

Fig. 29.9: A. *Notothylas levieri* with sporophytes (after Kashyap). B. Antheridia and C. Archegonium of *Notothylas*

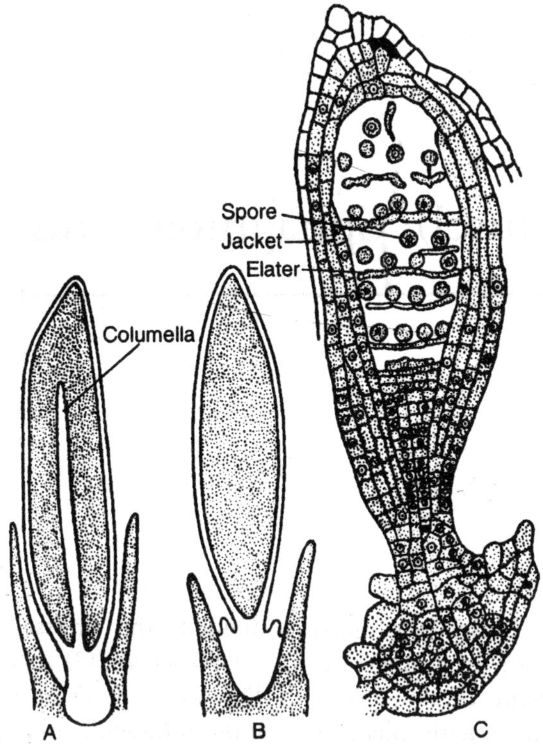

Fig. 29.10: Diagrammatic figures of *Notothylas* sporophyte in l.s. A. *N. indica* showing columella. B. *N. levieri* showing absence of columella (after Kashyap). C. L.s. of *N. levieri* sporophyte (after Pande)

Chapter **30**

Bryopsida (Musci): Sphagnidae and Andreaeidae

The class **Bryopsida** or **Musci** (also called *Musci Veri*) is the largest among the Bryophytes comprizing about 14,500 species in 660 genera. Their gametophytic phase is characterized by a prostrate, branched, filamentous or simple thallose *protonema* which later develops the characteriztic erect, leafy gametophytes. Unlike the Hepatics where the protonemal phase is a continuous process in the development of the gametophyte, the protonemal phase in the mosses is a distinct and separate phase prior to the formation of the proper gametophytic plant body. The rhizoids on the gametophyte are multicellular, branched, and provided with diagonal cross-walls. Sex organs on it develop from superficial cells and their early development, unlike the Hepatics, is by single apical cells with two cutting faces. The sporophyte is of limited growth and usually shows foot, seta and capsule, the last of which has a greater proportion of sterile tissues. The archesporium develops spores only, elater cells being absent.

Naturally, such a large group shows great diversity. All the earlier systematists (viz., Engler, Wettstein, Bower, Campbell) have divided the mosses into three groups: (1) *Sphagnales*, (2) *Andreaeales*, and (3) *Bryales*. The groups have been called *orders* by some authorities while others preferred to call them *subclasses*. Cavers (1911), although not accepting the subdivision between the *Hepaticae* and the *Musci*, recognized six groups of mosses—all of equal rank; (1) *Sphagnales*, (2) *Andreaeales*, (3) *Tetraphidales*, (4) *Polytrichales*, (5) *Buxbaumiales*, and (6) *Eubryales*. Smith (1955) prefers the names (1) *Sphagnobrya*, (2) *Andreaeobrya*, and (3) *Eubrya* for the three classically recognized groups and calls them subclasses. But, according to the International Code of Botanical Nomenclature, while they should be called *subclasses* each subclass name is to end in the suffix -idae. Hence, the correct classification of the *Class* **Bryopsida** is:

(1) *Subclass* **Sphagnidae**

(2) *Subclass* **Andreaeidae**

(3) *Subclass* **Bryidae**.

The last subclass *Bryidae* (Chapter 31) is very diverse and really needs further subdivision, something like that suggested by Cavers. Some of these are so striking that Reimers in the last edition of Engler's *Syllabus der Pflanzenfamilien* (1954) suggests that the Class *Bryopsida* (he still calls it *Musci*) be divided into *five subclasses* instead of the three: (1) *Sphagnidae*, (2) *Andreaidae*, (3) *Bryidae*, (4) *Buxbaumidae*, and (5) *Polytrichidae*. In this book, however, the classical system of three subsclasses (i.e., including the *Buxbaumiidae* and the *Polytrichidae* within the *Bryidae*) is being followed.

The three subclasses, as named above, are being described separately in what follows

Subsclass 1. SPHAGNIDAE

Genus: *SPHAGNUM*

Sphagnidae comprise of a single order *Sphagnales* of a single family *Sphagnaceae* with a single and very distinct genus *Sphagnum*. They differ from the other mosses in showing a broad, thallose protonema (Fig. 456) which develops only one normal gametophytic plant. Leaves on the gametophyte are without midrib and are formed of two different types of cells. The gametophytic body on the whole shows remarkable adaptations for the absorption of water. Antheridia are formed between leaves on special shoots while archegonia are formed acrogynously. In the sporophyte a gametophytic structure, the *pseudopodium* takes the place of the seta and the endothecium forms only a sterile columella while the archesporium develops from the inner layers of the amphithecium and arches over the tip of the columella. *Sphagnum* presents features which are considered more primitive than in the other *Musci*. But there is no fossil evidence to show that *Sphagnum* existed earlier than the *Bryidae*.

Sphagnum is commonly called *bog moss*, *peat moss* or *turf moss* because of its *ecological importance* in temperate to subarctic regions in the development of *bogs* and *peat*. Sometimes they are known even at sea-level in some tropical countries like Guinea and Trinidad. The plants are perennial and of aquatic or subaquatic habit growing in extremely moist situations where water accumulates or where water drips along rocky slopes. When established on the shores of lakes it gradually encroaches more and more of the water as *creeping bogs* and may completely cover up the lake transforming it into a *bog* in course of a century. Angiospermic plants then encroach the bog and the topographical feature gets changed. With this development, the water of the bogs becomes acid (the pH value becoming as low as 3.7), a condition when sphgnum thrives. In this acid condition there is little decay of the old basal parts of the gametophyte and accumulation of these, year after year, followed by compression from plants on top leads to the developments of dark-coloured *peat* which, when dried, is a very handy fuel in countries like Iceland and, of course of age, may form a type of coal. Because of its water absorbing capacity, *Sphagnum* plants are favoured as a packing material for transporting living plants and as a water retaining constituent for preparing seed beds. After propoer treatment, *Sphagnum* has even been used to replace absorbent cotton for dressing wounds, specially during wars, Thus, *Sphagnum* has some *economic importance* as well.

342 species of *Sphagnum* were recognized from all over the world by Warnstorf (1911). *Index Muscorum* (1967) has brought this number down to 207. They are distributed all over the high, cool and very wet regions of the earth (cosmopolitan). 22 species are kown to occur in India. Of these, only 3 species (*S. fimbritum*, *teres* and *squarrosum*) occur in the drier Western Himalaya, only 1 species (*S. ceylonicum*) occurs in warmer South India but as many as 20 species, e.g., *S. nemoreum* Scop. (= *S acutifolium* Schrad), *S. Junghuhnianum* D. and M., *S. teres* (Schimper) Aongstr, *S. squarrosum* Crom. In Hopp., *S. cuspidatulum* C.M., *S. khasianum* Mitt., *S. subsecundum* Nees, *S. palustre* L. (= *S. cymbifolium* (Ehrh.) *Hedw.*), *S. ovatum* Hamp, are from the moist Eastern Himalaya (including Khasia Hills). *S. fimbrietum* and *ceylonicum* are not known in the Eastern region.

Basic chromosome number $n = 19$.

The gametophyte

The shoots of the mature gametophytic plant are of a whitish or brownish green colour. These do not bear rhizoids and are of perennial, unlimited growth. Very young gametophytes, however, show delicate rhizoids with oblique cross-walls. It is differentiated into stem and leaves (Fig. 30.1). Growth is by means of an apical cell with three cutting faces. Each segmental cell cut off by this cell gives rise to a single leaf and the subtending portion of stem. At the growing apex leaves are in three rows but, later on, his arrangement is disturbed because of the growth of the stem. At intervals, on the stem, a leaf bears an axillary cluster of lateral branches some of which are pendent and others upwardly divergent (Fig. 30.1B). Sometimes, one of the divergent branches becomes as strong as the main stem and ultimately gives rise to a new plant when it becomes separated from the mother plant. At the apex of the main stem the branches are short and densely crowded forming a compact head (Fig. 30.1A). Although these shoots are weak in a bog, the stems grow erect being supported by one another and may grow considerably long (up to 7 feet).

The small leaves (Fig. 30.2A) are of a typical construction. Each dividing leaf cell gives rise to two asymmetrical cells, one small towards the margin and one large. The large cell again cuts off another small cell towards the apex (Fig. 30.2B). Eventually, the small cells mature into elongated green cells and remain living while the large cell enlarges in all directions, becomes spirally thickened, develops pit on the walls and becomes hyaline by losing its protoplast. The ***hyaline*** and ***chlorophyllose*** cells form a regular reticulate pattern (Fig. 30.2C). The leaf remains one-cell shick (Fig. 30.2D). The hyaline cells perform an important role in the absorption and retention of water—a typical characteristic of *Sphagnum.* That is why rhizoids are not necessary in the

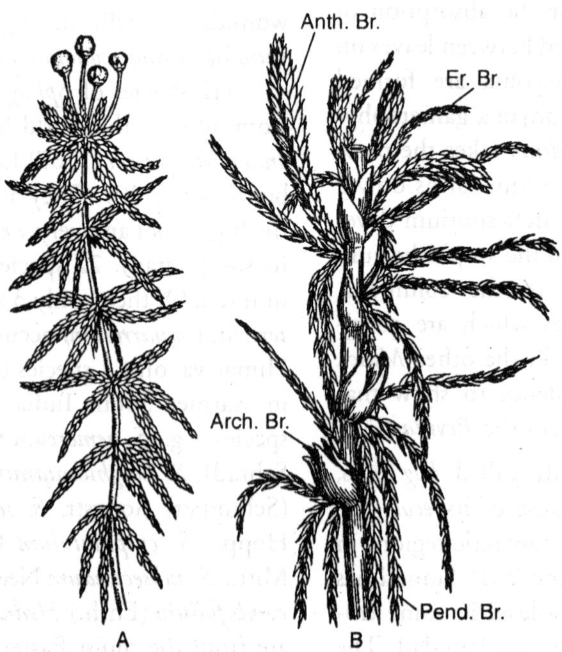

Fig. 30.1: A. Gametophyte plant of *Sphagnum palustre*, bearing terminal cluster of sporogonia. B. Part of *Sphagnum nemoreum* gametophyte showing antheridial branches (Anth. Br.), archegonial branch (Arch. Br.), erect branches (Er. Br.) and pendent branches (Pand. Br.) (after Schimper)

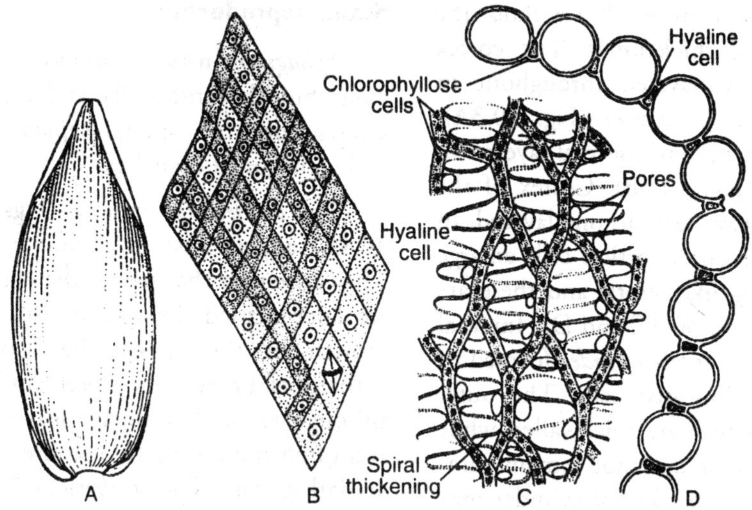

Fig. 30.2: A. Leaf of *Sphagnum nemoreum*. B. Surface view of young *Sphagnum* leaf cells showing cutting off of hyaline and chlorophyllose cells (after Smith). C. Surface view of mature *Sphagnum* leaf showing the two types of cells. D. T.s. of *S. nemoreus* leaf

mature plant. The characteristic areolation (cell pattern) of *Sphagnum* leaf *alone* is a sufficient diagnostic feature to identify the genus. No other plant has such a leaf.

The stem in *t.s.* shows an ensheathing cortex of large hyaline cells which are also greatly capable of water absorption. The genus *Sphagnum* is divided into two subgenera: (1) Subgenus *Sphagnum* (= *Inophloea*) with cortical hyaline cells fibrose and porose as in the leaves and (2) Subgenus *Lithophloea* (meaning 'smooth-skinned') where these cells are similar

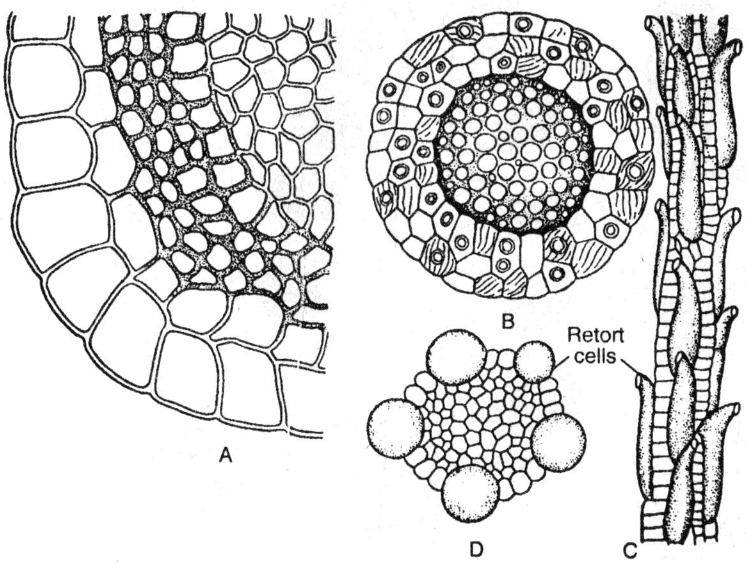

Fig. 30.3: A. T.s. of *Sphagnum ovatum* (subgenus *Lithophloea*) stem. B. T.s. of stem belonging to a species of subgenus *Sphagnum* showing 3 cortical layers. C-D. L.s. and t.s. of *S. nemoreum* stem showing retort cells (after Schimper)

to the hyaline cells of the leaf but without the pores or the spiral thickenings. The cortex remains one or two-layered throughout in some species (e.g., *S. ovatum*—Fig. 30.3A), but, more often, it becomes several layered in the mature stem (Fig. 30.3B). Cortex of the lateral branches always remains one-layered. In a few species, some outer cortical cells enlarge peculiarly appearing like retorts and these are called *retort cells* (Figs. 30.3C and D). Internal to the cortex is the *central cylinder* formed of smaller, vertically elongated cells. The outer cells of central cylinder are thickwalled while the inner ones may or may not remain thin-walled. Some cells of the central cylinder may be devoid of cell contents.

Vegetative propagation

As the shoots are perennial, the common mode of propagation is vegetative. Branch shoots get detached from the mother plant by decay of the lower parts and thereby form independent plants.

Sexual reproduction

Sphagnum may be monoecious or dioecious but the antheridia and archegonia are always borne on special separate antheridial and archegonial branches.

The *antheridial branches* (Figs. 30.1B and 30.4A) appear first in a monoecious shoot. They arise near the apex of the main shoot but are later carried downwards by the eventual growth of the apex. The leaves on the antheridial branch are closely imbricate and, although resembling foliage leaves, are shorter and often more brightly coloured (yellow, red or dark-green). The *antheridia* develop acropetally, apparently in the axils of the leaves (it has now been shown that they are below the leaves and not in the axis) (Fig. 30.4B), the top leaves usually do not develop antheridia and the apex is free to grow further after the maturity of the antheridia. Each *antheridial* initial is a superficial cell of the stem. This first develops a short filament (Fig. 30.4C), the apical cell

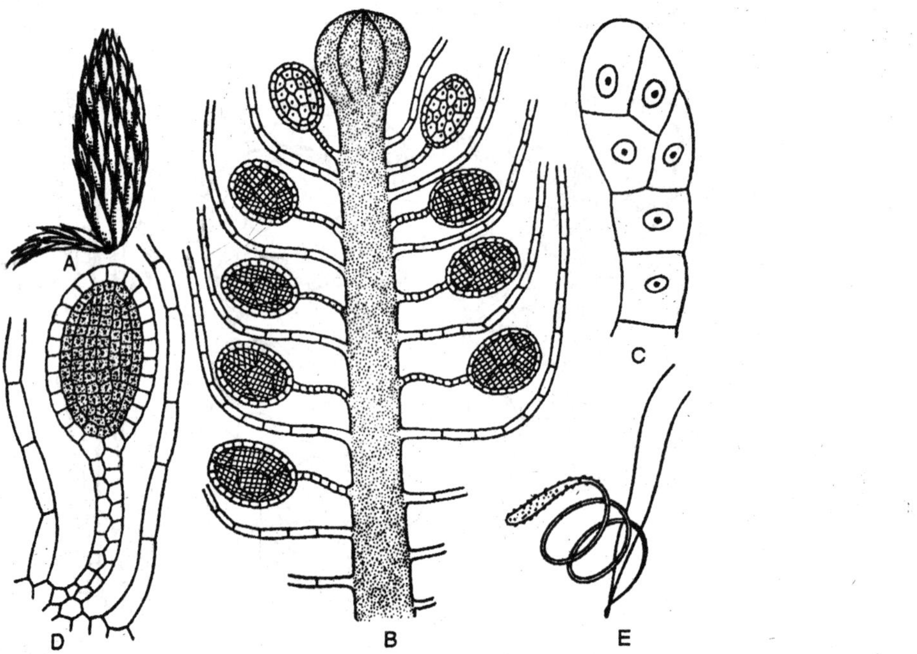

Fig. 30.4: *Sphagnum nemoreum*. A. Antheridial branch. B. L.s. of A showing leaves and antheridia. C. Filamental stage in the development of an antheridium. D. A mature antheridium with two leaves. E. An antherozoid

growing by two cutting faces. The top cells, later, form the antheridium in a manner resembling the *Jungermanniales*. A mature antheridium (Fig. 30.4D) shows a long *stalk* of two to four rows of cells and a spherical or ovoid *antheridium* with a jacket of one layer of cells enclosing a large number of *androcytes* formed from the *sperm mother cells*. The androcytes contain the sperms of *antherozoids* which are coiled structures with two long cilia (Fig. 30.4E) of equal lengths each, resembling the Hepatics.

The *archegonia* develop on the tips of special *archegonial branches* which develop at the apex or laterally (Fig. 30.1). The leaves on these branches are larger than the foliage leaves and the hyaline cells on them are loss fibrose. These constitute the *perichaetium* enclosing the archegonia. The spical cell of this branch forms the *primary archegonium*. Two to four *secondary archegonia* develop from 2 to 4 segments cut off by the same initial cell. Usually, only two secondary archegonia are formed so that the mature archegonial branch shows three archegonia at the tip (Fig. 30.5C). The initial of the primary archegonium divided either by two cutting faces or transversely, forming a short filament four to six cells high at first. Then, the terminal cell cuts off three *jacket initial cells* and an *axial cell* (Fig. 30.5A). Further division is similar to the acrogynous *Jungermanniales* and the archegonium (Fig. 30.5B) is developed. A *mature archegonium* (Figs. 30.5B and C) shows a *stalk*, a twisted neck (wall formed by five or six vertical rows of cells with the cover cells not so distinct) with 8 to 9 *neck canal cells*, a *central canal cell* and the *egg* in the *venter* whose wall has by now become multi-layered.

Fertilization

The mature antheridium opens at the top, the androcyte mass comes out and the antherozoids are liberated immediately. The archegonium also opens at the tip, the neck canal cells are disorganized and fertilization takes place as in other Bryophytes. Bryan (1920) found in *S. secundum* the persistence of the ventral canal cell and the ultimate fusion of the egg with the ventral canal cell to form a zygote.

The sporophyte

Usually, the zygote of only one archegonium in an archegonial branch develops an

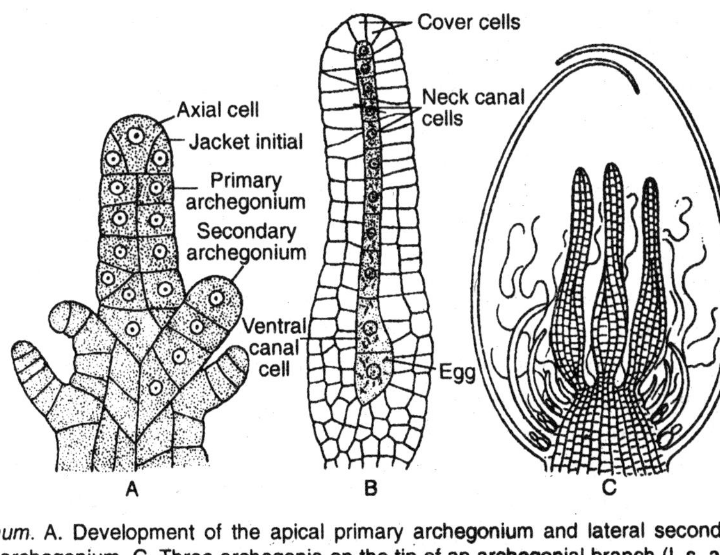

Fig. 30.5: *Sphagnum*. A. Development of the apical primary archegonium and lateral secondary archegonia. B. A mature archegonium. C. Three archegonia on the tip of an archegonial branch (L.s. after Schimper)

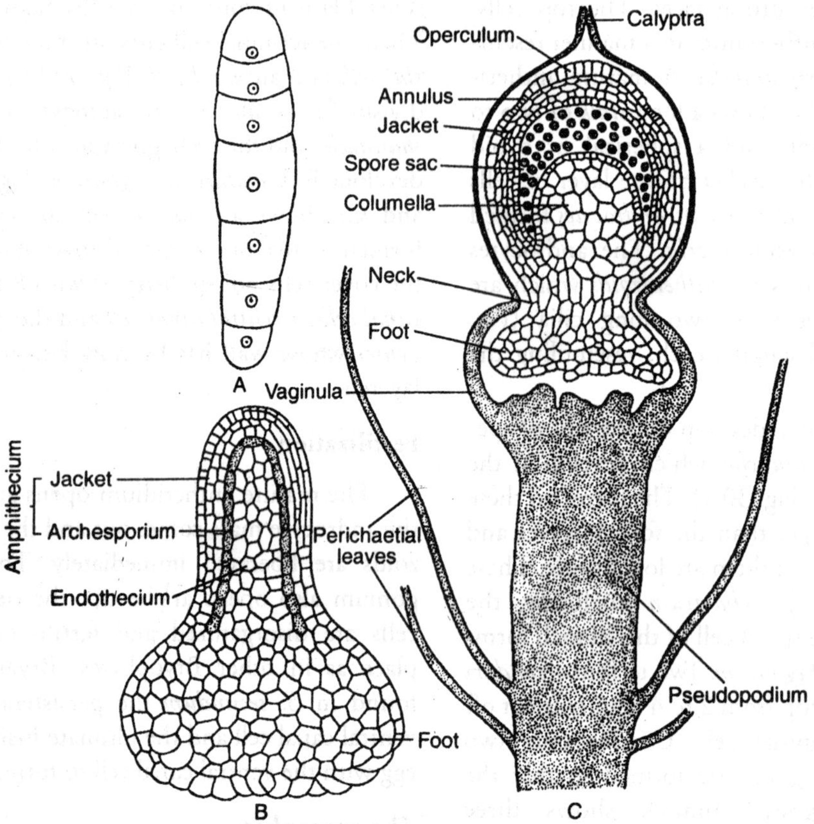

Fig. 30.6: Stages of development of the *Sphagnum* sporophyte. A. Early filamentous sporophyte. B. Later stage showing differentiation of foot, amphithecium and endothecium. C. Mature sporophyte on pseudopodium in l.s.

embryo—the other archegonia wither early. The zygote invests itself with a cell wall and then divides transversely. Transverse divisions continue until a filament of 6 or 7 cells (Fig. 30.6A) is formed. The development of the embryo resembles the *Jungermanniales* in the earlier stages and the *Anthocerotales* in the later stages. The lower half of the filament divides irregularly forming a bulbous *foot* (Fig. 30.6B), the lowest cells of which are haustorial. The haustorial part helps absorption of food material from the gametophyte and is obliterated when the sporophyte matures. The upper cells of the filament divide periclinally. Soon an inner *endothecium* and an outer *amphithecium* are differentiated (Fig. 30.6B). The endothecium remains sterile and forms the *columella* with a dome-like top over which the amphithecium overarches. The inner layers of the amphithecium form the 2 to 4 layers thick *archesporium* while the outer layer forms a jacket 4 to 6 layers thick. The jacket cells remain green till the sporophyte matures. There is only a short neck and no proper seta connecting the upper *capsule* and the lower foot. The function of the seta is performed by the tissue of the gametophyte just below the foot which develops into a long stalk rising the sporophyte and is called the *pseudopodium.* The archesporial cells form the *spore mother cells* which divide meiotically developing the spores.

The *mature sporophyte* (Fig. 30.6C) shows an almost spherical, black to dark-brown capsule and a bulbous *foot* connected by a very shor *neck*. The base of the foot is seen disintegrating. The whole sporophyte is covered by the *calyptra* which is the archegonial wall having developed with the sporophyte. The lowest part of the calyptra surrounding the foot is called the *vaginula*. Below the sporophyte is the *pseudopodium* which is a stalk increasing in length and pushing out the sporophyte as it develops but is not a part of the sporophyte as it originates out of gametophytic tissue. The capsule shows a *jacket* of 4 to 6 layers, and the *spore-sac*, full of *spores*, overarches the dome-shaped *columella* (representing the endothecium). The outermost layer of the jacket becomes thickened, loses chlorophyll and develops some *non-functional stomata* in which there are guard cells but the apertures are missing. The convex top of the jacket becomes differentiated into an *operculum* after being delimited from the rest of it by a ring of cells with thinner walls forming the *annulus*. The *pseudopodium* carries up this sporophyte (Fig. 30.7) whose capsule has now developed a dark colour. In sunny days the space has increased by the drying up of the columella. This air, expanded by heat, makes the capsules burst so that the operculum breaks at the annulus and these, along with most spores, are thrown off inches high.

The new gametophyte

The spore (Fig. 30.8A) falling on a moist substratum germinates, developing a small thalloid *protonema* (Fig. 30.8B). This develops further, forming a prostrate, green, irregularly lobed thalloid structure (Fig. 30.8C). Is is one cell in thickness and is fastened to the substratum by multicellular rhizoids. On this protonema, a single bud is developed from a marginal cell and this develops the *new leafy gametophyte*. It has been found (Noguchi, 1958) that the protonema may develop a filament from any marginal cell and on this filament another *secondary protonema* may develop.

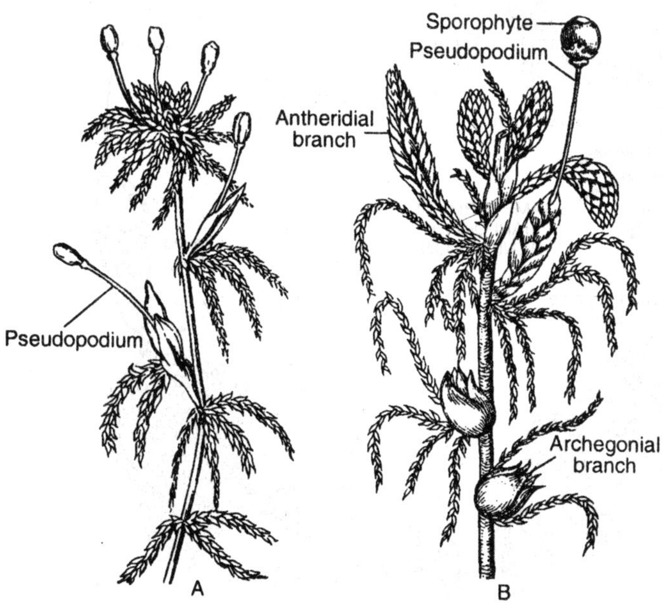

Fig. 30.7: A. Mature plant of *Sphagnum squarrosum* showing lateral and apical sporophytes (after Campbell). B. Mature plant of *S. nemoreum* showing a sporophyte (after Schimper)

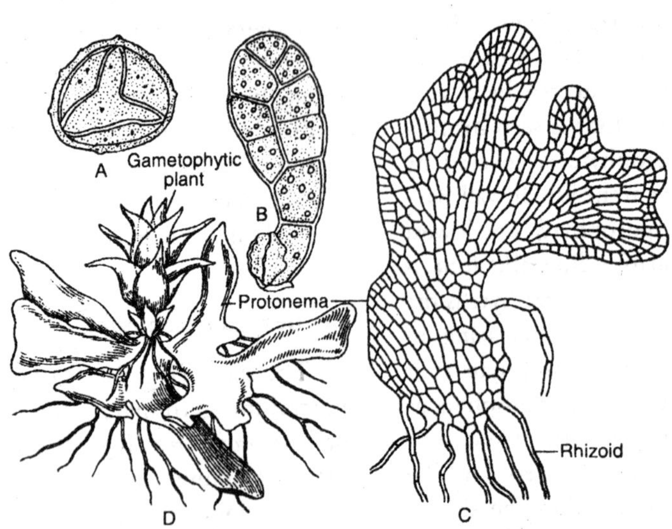

Fig. 30.8: *Sphagnum palustre.* A. Spore. B. Germination of spore. C. The thalloid, lobed protonema. D. Development of a gametophyte plant on a lobed protonema of *S. nemoreum* (after Schimper)

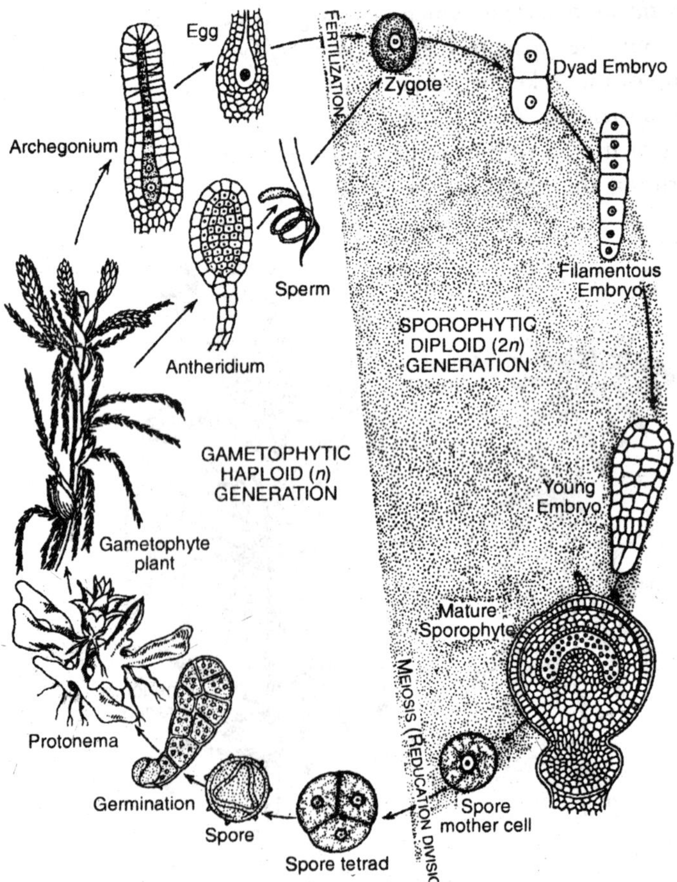

Fig. 30.9: Life cycle of *Sphagnum* showing alternation of generations

Subclass 2. ANDREAEIDAE

Like the *Sphagnidae*, the *Andreaeidae* is represented by a single order Andreaeales with a single family *Andreaeaceae*. While previous authors recognized only a single genus *Andreaea*, Brotherus (1924) recognized two—*Andreaea* (109 spp.) and *Neuroloma* (1 sp. from the Sub-Antarctic region). But, Reimers (1954) favours three genera: *Andreaea, Acroschisma* (1 sp., also from the sub-antarctic region, taken out of *Andreaea*) and *Neuroloma*. These plants are distributed all over the earth in very cool climates, in the Arctic and Antarctic regions and in the alpine heights of the high mountains.

The *Andreaeaceae* has characters intermediate between the *Sphagnidae* and the *Bryidae*. In the sporophyte, both the *archesporium* and the *columella* develop from the *endothecium* but the archesporium arches over the columella like a dome. The seta is almost as short as in *Sphagnum* and the sporophyte is pushed up by the development of a similar *pseudopodium*. The capsule opens by means of a number of longitudinal slits as in some *Jungermanniales*, there being no operculum of peristome. The *protonema* is thallose.

Genus: *ANDREAEA*

As opposed to *Sphagnum*, *Andreaea* grows on dry and exposed arctic situations on siliceous rocks, specially granite (Hence the name 'granite moss'). The 109 species are distributed only in the very cool alpine and subalpine regions, i.e., the subarctic regions and the lofty peaks throughout the world. They grow in dark-brown to blackish dense tufts which become very brittle when dry. Four species: *Andreaea indica, A. densifolia, A. commutata* and *A. rigida* were discovered by J. D. Hooker from above 15,000 ft in Sikkim. A fifth species (cosmopolitan *A. rupestris*) was later discovered in Sikkim and has now also been found at 13,000 ft in Garhwal and also in Western Nepal, i.e., from the Western Himalaya.

Basic chromosome number $n = 10$.

The gametophyte

The general appearance of the gametophyte is like that of a common moss (Bryidae). The stem is prostrate on the rock surface and branching is dichotomous sympodial with one branch growing more strongly than the other (Fig. 30.10A). There are many rhizoids on the lower part of the stem but these are different from those in other Bryophytes. Some of them are cylindrical while others are flattened, plate-like—the latter making the stem stick to the rock surface. The stem grows by an apical cell with three cutting faces, leaves developing in three rows. Leaves are small, one layer of cells in thickness and without any midrib in some species, while in some other species the median longitudinal axis becomes more than one layer

Fig. 30.10: A. *Andreaea rupestris* gametophyte with sporophytes (after Smith). B. T.s. of *Andreaea* stem

thick. The stem anatomy shows a uniform mass of parenchyma cells, there being no differentiation into a cortex and a central cylinder. But, the superficial cells are often somewhat thick-walled (Fig. 30.10B) and darker in colour. Large oil drops are present in the cells.

Sexual reproduction

Most *Andreaea* species are monoecious but there are a few dioecious species. In the monoecious species, the antheridia and archegonia are formed in groups terminally on separate branches. The development of the sex organs resembles the *Bryidae.*

The sporophyte

The zygote divides by a a transverse wall. The lower (hypobasal) cell divides irregularly and forms a haustorial tissue which later organizes into the *foot.* The upper (epibasal) cell is destined to form the capsule. It first functions as an apical cell with two cutting faces and soon, after periclinal divisions, develops an inner *endothecium* and an outer *amphithecium.* The amphithecium forms the *jacket,* 3 to 8 cells in thickness. The inner layers of the *endothecium* become the sterile *columella* while its outer cells become the *archesporium* which develops a dome-like *sporogenous tissue,* two layers thick, and arches over the top of the columella. Large chloroplasta are present in the sterile cells showing that the sporophyte is not wholly parasitic.

The mature sporophyte (Fig. 30.12A) shows a swollen *foot* which is haustorial in function and an elliptical *capsule.* The two are connected by a short *neck* which is the seta. The *jacket* of the capsule is 3 to 8 layers in thickness, the superficial cells of which are thick-walled except along four vertical lines (reaching neither the tip nor the base of the capsule) where the cells remain small and thin-walled. The club-shaped *columella* is in the centre and the dome-shaped *spore-sac* is arching over it. The whole is raised by a stalk

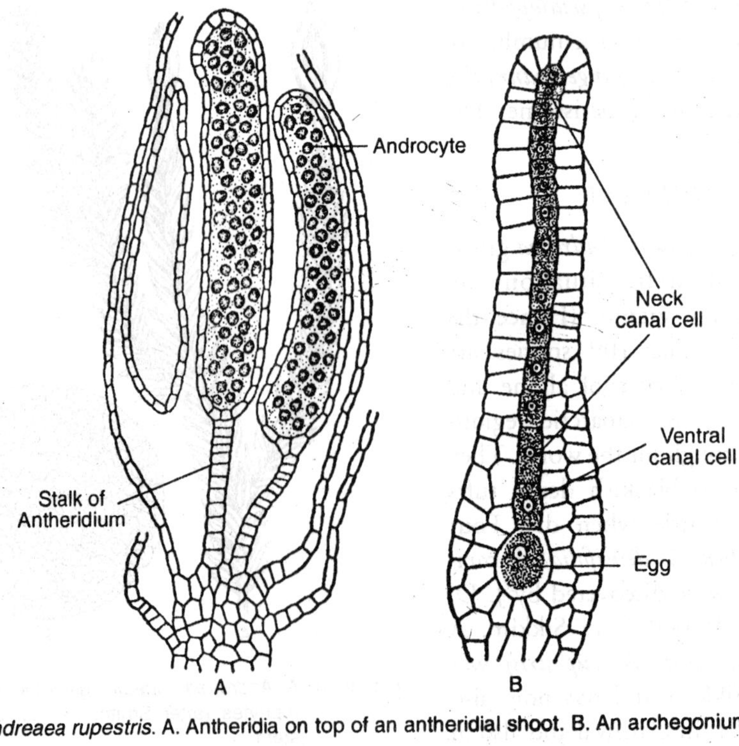

Fig. 30.11: *Andreaea rupestris.* A. Antheridia on top of an antheridial shoot. B. An archegonium (After Smith)

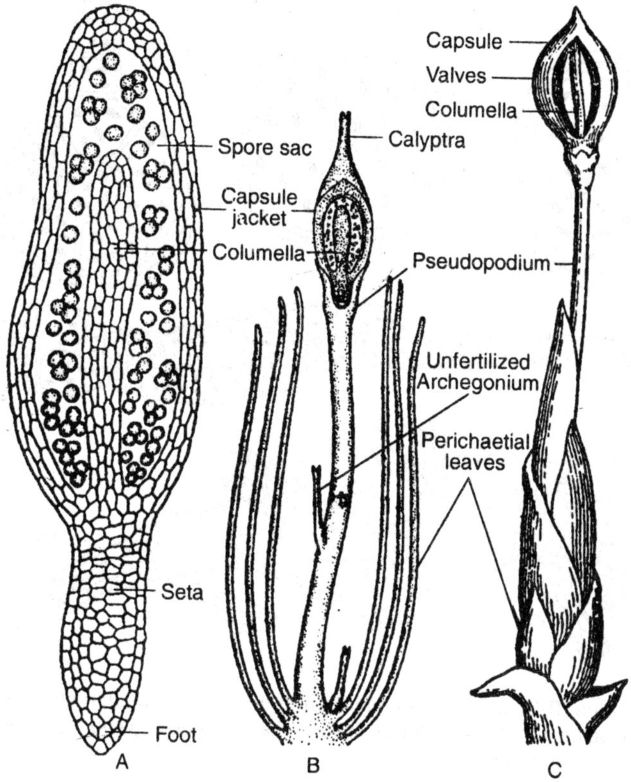

Fig. 30.12: *Andreaea rupestrics*: A. A sporophyte. B. Diagrammatic section through a sporogonial shoot showing an elongated pseudopodium. C. A sporogonial shoot bearing a sporophyte dehiscing by valves

which is the *pseudopodium*, a growth of gametophytic tissue as in *Sphagnum* (Figs. 30.12B and C). The pseudopodium pushes up the main sporophyte on its tip. Just below the tip, the pseudopodium becomes swollen (the foot of the sporophyte being embedded within it). Lower down, some of the lateral unfertilized archegonia may also be seen on the pseudopodium (Fig. 30.12B). The calyptra, remnant of the archegonial wall, is visible on the sporophyte as in other Bryophytes.

At maturity, the capsule splits along the four vertical lines of weakness so that, usually, there are four values, although in a few species there may be more. But, the valves are not completely separate as the slits do not reach the apex or the base (Fig. 30.12C). These valves are hygroscopic and the spores are scattered out with their movements.

The new gametophyte

The spore (Fig. 30.13A) has two coats as usual and contains chloroplasts and oil globules. As soon as the spore is liberated, it begins to divide while still within the exospore and forms a globular mass of cells (Fig. 30.13B). The exospore, then, bursts and the cell mass begins to develop one or more filaments from different points (Fig. 30.13C). The ultimate appearance of the gametophytic *protonema* varies. It may be a branched ribbon-like structure (Fig. 30.13D) or it may be a thalloid leaf-like structure resembling that of *Sphagnum*. Unlike most other protonema, the *Andreaea* protonema may even lie in a dormant stage if the environment becomes too rigorous. Ultimately, buds develop on any part of the protonema and these develop into new leafy gametophytic plants.

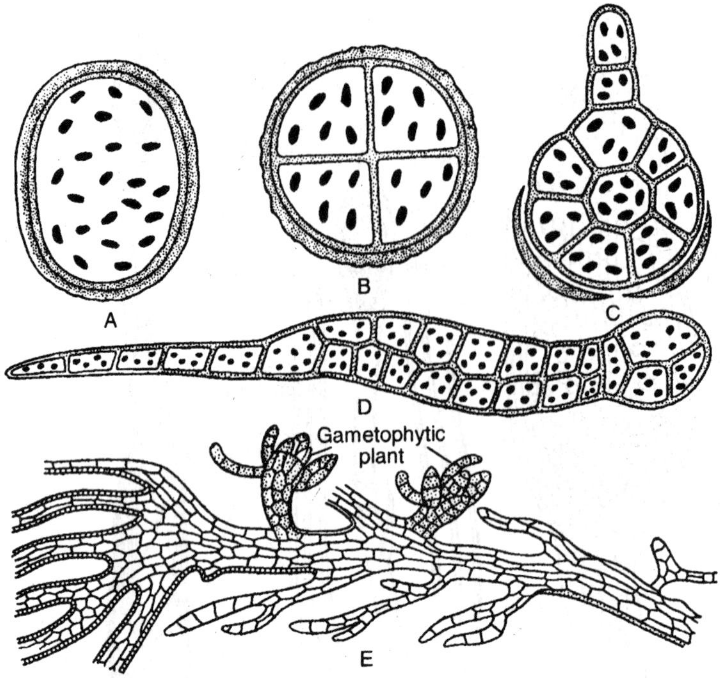

Fig. 30.13: *Andreaea*. A. A spore. B-C. Stages of germination of the spore, the latter showing the development of a filament. D. Development of the filament. E. Ribbon-like protonema giving rise to new gametophytic plants from buds (after Kuhn and Berggren)

Chapter 31

Bryopsida (Musci): Bryidae

The third and the last subclass *Bryidae* (*Bryales* of Campbell and Brotherus, *Eubrya* of Smith) is the largest group of the Bryophytes comprizing more than 14,000 species in more than 650 genera. They are the most highly developed of *Bryophyta* and are distinguished from other *Bryopsida* in their usually filamentous (thallose in Tetraphidales and *Diphyscium* of Buxbaumiales) protonemata, absence of pseudopodium and the usual good development of the seta. The gametophytic plants of most genera grow by an apical cell with three cutting faces but there are a few showing distichous arrangement of leaves where the shoots grow by apical cells with two cutting faces. The sporophyte embryo grows by an apical cell with two cutting faces. The sporogenous tissue is always from the outer layer of the endothecium and the jacket shows more differentiation with several types of tissues. Most species show characteristic development of peristome and operculum.

CLASSIFICATION

In such a vast group, naturally, there are many variations so that classification is difficult. Structures of both gametophytes and sporophytes have to be considered and in mosses these two phases have attained specialization in different degrees in the different species. Thus, it is not possible to arrive at a phylogenetic system based on a strict parallel evolution of both gametophyte and sporophyte. But, just as in the Angiosperms the flower characters are considered more reliable in classification, in Bryidae the sex organs and the sporophyte are considered more important in any general classification. On the other hand, a system based only on sporophytes ignoring the gametophytes altogether also becomes artificial and not properly phylogenetic in uderstanding the modern trend of moss taxonomy.

The following systems of classification may be considered as important in understanding the modern trend of moss toxanomy:

(1) *Hedwig's system* (1801)

The really first attempt at a natural system of classification of mosses was by J. Hedwig in his *Species Muscorum* (1801). He arranged the genera (only 35 were known at that time) mainly on the peristome and a few other sporophytic characters. The system is no longer followed as it is clumsy and has also become artificial as the gametophytic characters have been ignored altogether. Thus, such widely differing genera as *Funaria* and *Hypnum* have been placed close together. Yet, his genera are so well defined that most of them stand even today. His work is, therefore, considered as the starting point of modern Muscology.

According to International Rules, the nomenclature of mosses (with the exception of *Sphagnum* which alone starts with Linnaeus) starts with Hedwig's *Species Muscorum.* All previous names are to be considered invalid.

(2) *Bridel* (1826) and *Bryologia Europaea* (1836-1855)

S. E. Bridel (1826) suggested division of the Bryidae into **Acrocarpi** (erect plants with apical sporophytes) and **Pleurocarpi** (prostrate, branching plants with numerous sporophytes on short lateral branch tips). P. Bruch, W. P. Schimper and T. Guembal adopted this division in their *Bryologia Europaea* (1836-1855) and elaborated the system of classification rendering it more natural than in Hedwig.

(3) *Mueller* (1849)

C. Mueller, in his *Synopsis Muscorum Frondosorum* (1849), divided the true mosses into **Cleistocarpi** (no organized annulus, peristome or operculum; dehiscence of capsules by rupture or decay of jacket) and **Stegocarpi** (rugular annulus and operculum, spore dispersal usually by peristome).

(4) *Mitten* (1859)

W. Mitten, who wrote the only Bryological Flora of India, divided the Bryidae into the **Arthrodonti** (articulate thin peristome teeth) and the **Nematodonti** (wormlike solid peristome teeth) in his *Musci indiae orientalis* (1859). The spellings got changed later on. This was the first real attempt to classify mosses according to the nature of the peristomes.

(5) *Lindberg* (1879)

Classification of the common mosses into the *Cleistocarpi* and the *Stegocarpi* was soon found to be artificial. Excepting in *Archidium,* all cleistocarpic genera were found to be connected to some stegocarpic genera by their manner of development. S. O. Lindberg distributed these cleistocarpic genera among the various stegocarpic families in his *Musci scandinavici* (1879).

(6) *Philibert* (1884-1902)

Although the importance of the peristomes in the classification of the common mosses was recognized by Mitten, it was the detailed study of the peristomes published in a series of articles by H. Philibert in the *Revue bryologique* during 1884 to 1902, which clearly showed that these mosses could be fully classified better by laying a greater stress on the peristome structure. He further classified the *Arthrodonteae* into (1) *Haplolepideae* and (2) *Diplolepideae* as explained later on. He also considered the distinctiveness of Encalyptaceae which contains the characters of the Haplolepideae and the Diplolepideae and even some characters of the Nematodonteae. He, therefore, suggested that the Encalyptaceae may be the starting point of the Haplolepideae and the Diplolepideae.

(7) *Fleischer* (1906-1908)

The first really phylogenetic classification of mosses laying importance on sporophytic characters but not ignoring the gametophytic characters was adapted by M. Fleischer in the *Musci der Flora von Buitenzorg* Vol. II (1906-1908). This is the system which was later adapted by V. F. Brotherus in Engler's *Pflanzenfamilien* (described below).

(8) *Cavers* (1910-1911)

As already explained (page 928), F. Cavers directly divided the Bryophytes into ten groups or 'orders' ("The Inter-relationships of the Bryophyta" published in *New Phytologist,*

1910-1911) the last four of which (*Tetraphidales, Polytrichales, Buxbaumiales* and *Eubryales*) comprize the Bryidae as we understand it today. Accepting the findings of Philibert, Cavers classified them as follows:

A. NEMATODONTEAE

I. **Tetraphidales**

II. **Polytrichales**

III. **Buxbaumiales**

B. ARTHRODONTEAE

IV. **Eubryales**

Series 1. **Haplolepideae** (Peristome teeth in a single ring. The teeth are formed by membranes belonging to two concentric cell layers. Each tooth formed by a single row of external plates and two rows of internal plates. *Lepos* = scale).

Cohort (i) *Archidioideae* (Cleistocarpic) —*Archidiaceae.*

Cohort (ii) *Dicranoideae* (Teeth cleft into two or three) —*Dicranaceae, Fissidentaceae, Leucobryaceae.*

Cohort (iii) *Monocranoideae* (Teeth undivided) —*Leucophanaceae, Syrrhopodontaceae, Calymperaceae.*

Cohort (iv) *Ditrichocranoideae* (Teeth deeply divided into two, three or more filaments) —*Trematodontaceae, Angstroemiaceae, Ditrichaceae, Trichostomaceae.*

Cohort (v) *Platycranoideae* (Teeth broad and usually undivided, may be perforated by slits, rarely split into two or three) —*Seligeriaceae, Grimmiaceae.*

Series 2. **Heterolepideae** (Peristome teeth variable, haplolepideous or diplolepideous) —*Encalyptaceae.*

Series 3. **Diplolepideae** (Peristome teeth normally in two rings. Outer ring (*exostome*) teeth with two rows of external plates. Inner ring (*endostome*) teeth formed of thin plates, their outer layer corresponding to the inner layer of the exostome).

Subseries (i) *Epicranoideae* (Exostome teeth superposed on endostome processes) —*Funarineae, Splachnineae.*

Subseries (ii) *Metacranoideae* (Exostome teeth alternating with endostome processes) —*Bryinae, Isobryinae, Hookerinae, Hypnobryinae.*

This system was followed by Campbell and forms the basis of the classification proposed by Dixon. It is certainly a high scientific system and has been adapted in this book after some modifications to bring it into line with the families as they are known today. Figure 31.1 shows the peristome structures of the main groups of Bryidae.

(9) *Fleischer's system as adapted by V. F. Brotherus in Engler's Pflanzenfamilien* (2nd Edn., 1924).

Subsclass **Bryales** (= **Bryidae**)

Reihengruppe (Chort) I. *Eubryinales* (= *Eubryiidae*)

Reihe (Orders): 1. *Fissidentales*, 2. *Dicranales*, 3. *Pottiales*, 4. *Grimmiales*, 5. *Funariales*, 6. *Schistostegales*, 7. *Tetraphidales*, 8. *Eubryales*, 9. *Isobryales*, 10. *Hookerlales*, 11. *Hypnobryales.*

Reihengruppe (Cohort) II. *Buxbaumiinales* (= *Buxbaumiidae*)

Reihe (Order) 12. *Buxbaumiales.*

Reihengruppe (Cohort) III. *Polytrichinales* (= *Polytrichiidae*)

Reihe (Orders): 13. *Polytrichales*, 14. *Dawsoniales.*

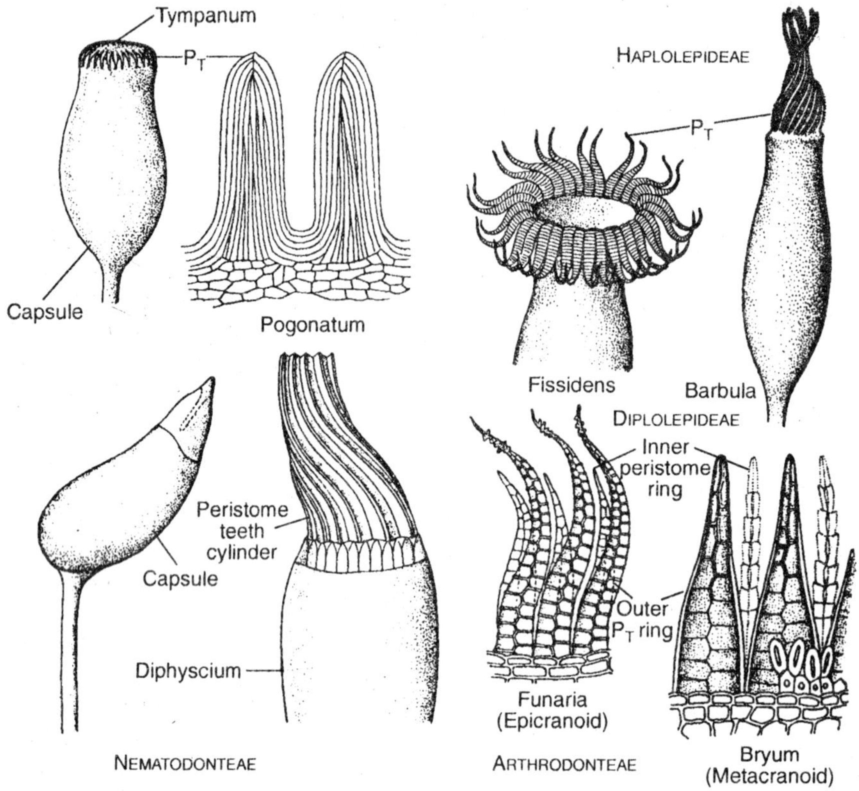

Fig. 31.1: Peristome structure of the main groups of Bryidae: *Pogonatum* and *Diphyscium* of Nematodonteae; Arthrodonteae represened by *Fissidens* and *Barbula* of Haplolepideae; *Funaria* (epicranoid) and *Bryum* (metacranoid) of Diplolepideae

The names in brackets are modifications according to the International Code of Nomenclature.

(10) *Dixon's* (1932) *System*

Although following Fleischer's System for the orders, Dixon laid greater stress on the structure of the peristome in the broader groupings following the system developed by Mitten, Philibert and Cavers. He also supported Cavers by bringing the *Nematodonteae* to the first position instead of the last (this aspect is discussed in Chapter 32). This classification, is, no doubt very natural although his use of the term 'clan' is not happy. The general classification is as follows:

Subclass: **BRYALES**

Clan. I. *Nematodonteae* (Peristome teeth solid, formed of longitudinal fibres derived from concentric layers of cells of the capsule).

Orders: 1. *Tetraphidales*, 2. *Calomniales*, 3. *Schistostegales*, 4. *Buxbaumiales*, 5. *Polytrichales*.

Clan. II. *Arthrodonteae* (Peristome teeth thin, membranous, articulated).

Subclan (i) *Heplolepideae*

Orders: 6. *Fissidentales*, 7. *Grimmiales*, 8. *Dicranales*, 9. *Syrrkopodontales*, 10. *Pottiales*.

Subclan (ii) *Heterolepideae*

Order: 11. *Encalyptales*

Subclan (iii) *Diplolepideae*

Orders: 12. *Orthotrichales*, 13. *Funariales*, 14. *Eubryales*, 15. *Isobryales*, 16. *Hookeriales*, 17. *Hypnobryales*.

While the orders up to Heterolepideae are acrocarpous, the Diplolepideae begins in acrocarpous and ends in pleurocarpous forms. The acrocarpous teeth show vertical striations while the pleurocarpous stritation is horizontal.

(11) *Reimer's* (1954) *System in Engler's Syllabus der Pflanzenfamilien* (12*th Edition*)

Reimers has suggested that the three Cohorts in Brotherus' System should be raised to subclasses parallel with *Sphagnidae* and the *Andreaeideae.* He names these subclasses (1) *Bryidae* (= *Eubryinales*), (2) *Buxbaumiidae* (= *Buxbaumiinales*) and (3) *Polytrichidae* (= *Polytrichinales*). He also increases the number of Orders to 15 by raising the family *Archidiaceae* to the first order *Archidiales* in conformity with Cavers.

(12) Cronquist, Takhtajan and Zimmermann (1966) divided the Plant Kingdom into the Subkingdoms Thallobionta and Embryobionta. Within the Embryobionta are the Divisions Bryophyta (without tracheaea) and the Trachaeophytic Divisions. The Division Bryophyta is divided into Subdivisions of which Muscophytina (= *Bryopsida*) is one. This Subdivision is divided into four classes (1) *Sphagnopsida* (= *Sphagnidae*), (2) *Andreaeopsida* (= *Andreaeidae*), (3) *Archidiopsida* (= *Archidiales* here) and (4) *Peristomiopsida* (Bryidae without *Archidiales*).

Due to recent cladistics studies coupled with molecular data, the phylum Bryopsida is classified in much more elaborate fashion. The details are shown below:

PHYLUM: BRYOPHYTA

CLASS: TAKAKIOPSIDA Leaves divided into terete filaments; capsules dehiscent by a single longitudinal spiral slit.

ORDER: TAKAKIALES

Takakiaceae S. Hatt. and Inoue. *Takakia* S. Hatt and Inoue

CLASS: SPHAGNOPSIDA Branches usually fascicles; leaves composed of a network of chlorophyllose and hyaline cells; setae lacking; capsules elevated on a pseudopodium; stomata lacking.

ORDER: SPHAGNALES Plants mostly branched, with branches in fascicles; stems with wood cylinder; leaves unistratose; antheridia subglobose; archegonia terminal on branches; capsules ovoid.

Sphagnaceae Dum. *Sphagnum* L.

ORDER: AMBUCHANANIALES Plants sparsely branched, with branches not in fascicles; stems without wood cylinder; leaves partially bistratose; antheridia oblong-cylindric; archegonia terminal on stems; capsules cylindrical.

Ambuchananiaceae Sepelt and H.A. Crum. *Ambuchanania* Seppelt and H.A. Crum

CLASS: ANDREAEOPSIDA Plants on acidic rocks, generally autoicous; cauline central strand absent; calyptrae small; capsules valvate, with four valves attached at apex; seta absent, pseudpodium present.

ORDER: ANDREAEALES

Andreaeaceae Dum. *Acroschisma* Lindl., *Andreaea* Hedw.

CLASS: ANDREAEOBRYOPSIDA Plants on calcareous rocks, dioicous; cauline central strand lacking; calyptrae large and covering whole capsule; capsules valvate, apex eroding and valves free when old; seta present.

ORDER: ANDREAEOBRYALES

Andreaeobryaceae Steere and B. M. Murray. *Andreaeobryum* Steere and B. M. Murray

CLASS: POLYTRICHOPSIDA Plants typically robust, dioicous; cauline central strand present; leaves costate, often with ventral lamellae on costa; capsules operculate; seta present; peristome nematodontous.

ORDER: TETRAPHIDALES Gametophore relatively small: stems not rhizomatous; costa slender, elamellate; capsules long-exserted.

Tetraphidaceae Schimp. Leaves unicostate; calypatrae small conic; capsule symmetric and erect, neck short; peristome single, of four erect teeth; *Tetraphis* Hedw., *Tetrodontium* Schwägr.

Oedipodiaceae Schimp. Leaves unicostate; calyptrae cucullate; capsule symmetric and erect, neck very long; capsules gymnostomous; *Oedipodium* Schwägr.

Buxbaumiaceae Schimp. Leaves ecostate; calyptarae cucullate or mitrate; capsule strongly asymmetric and horizontal, neck short; peristome double; *Buxbaumia* Hedw.

ORDER: POLYTRICHALES Gametophore typically large; stems typically rhizomatous; costa broad, with adaxial chlorophyllose lamellae; peristome mostly of 16 or 32-64 teeth.

Polytrichaceae Schwägr. *Alophosia* Card., *Atrichopsis* Card., *Atrichum* P. Beauv., *Bartramiopsis* Kindb., *Dawsonia* R. Br., *Dendroligotrichum* (Müll. Hal.) Broth., *Hebantia* G. L. S. Merr., *Itatiella* G. L. Sm., *Lyellia* R. Br., *Meiotrichum* (G. L. Sm.) G. L. S. Merr., *Notoligotrichum* G. L. Sm., *Oligotrichum* Lam. and DC., *Plagioracelopus* G. L. S. Merr., *Pogonatum.* P. Beauv., *Polytrichadelphus* (Müll. Hal.) Mitt., *Polytrichastrum* G. L. Sm., *Polytrichum* Hedw. (Fig. 31.10), *Pseudatrichum* Reimers, *Pseudoracelopus* Broth., *Psilopilum* Brid., *Racelopodopsis* Thér., *Racelopus* Dozy & Molk., *Steereobryon* G. L. Sm.

CLASS: BRYOPSIDA Plants small to robust; leaves costate, typically lacking lamellae; capsules operculate; peristome arthrodontous.

SUBCLASS: DIPHYSCIIDAE Gametophore small, perennial; leaves costate; capsules immersed among long perichaetial leaves; peristome double.

ORDER: DIPHYSCIALES Plants perennial; leaves costate; capsules immersed among long perichaetial leaves.

Diphysciaceae M. Fleisch. *Diphyscium* D. Mohr, *Muscoflorschuetzia* Crosby, *Theriotia* Cardot

SUBCLASS: FUNARIIDAE Plants terricolous, acrocarpous; stem typically with central strand; annulus often well developed.

ORDER: TIMMIALES Plants robust; laminal cells mammillose; calyptrae cuculate, often persistent; capsules bent to pendent; cilia present.

Timmiaceae Schimp. *Timmia* Hedw.

ORDER: ENCALYPTALES Plants small to medium-size; laminal cells thick-walled, isodiametric above, rectangular hyaline or reddish below; annulus not differentiated; calyptrae very large, enclosing the entire erect capsule.

Encalyptaceae Schimp. *Bryobrittonia* R. S. Williams, *Encalypta* Hedw.

ORDER: FUNARIALES Peristome lacking cilia, opposite, divisions in the IPL symmetric.

Funariaceae Schwägr. Protonema short-lived; costs well developed; laminal cells smooth and thin-walled; paraphyses with swollen apical cell; calyptrae smooth and naked; stomata with single guard cell; peristomes opposite, following a 4 : 2 : 4 pattern, or lacking; *Aphanorrhegma* Sull., *Brachymeniopsis* Broth., *Bryobeckettia* Fife, *Corynotheca* Ochyra, *Cygnicollum*

Fife & Magill, *Entosthodon* Schwägr., *Funaria* Hedw., *Funariella* Sérgio, × *Funariophyscomitrella* Wettst., *Loiseaubryum* Bizot, *Nanomitriella* E. B. Bartram, *Physcomitrella* Bruch and Schimp., *Physcomitrellopsis* Broth. and Wager, *Physcomitrium* (Brid.) Brid., *Pyramidula* Brid.

Disceliaceae Schimp. Protonemata persistent; costa weak to absent; calyptrae persistent below the urn; stomata lacking; peristomes reduced and opposite; *Discelium* Brid.

Gigaspermaceae Lindb. Gametophore united by subterranean rhizomatous aphyllous steam, cladocarpous; annulus not differentiated; capsules often immersed, gymnostomous; *Chamaebryum* Thér. and Dixon, *Costesia* Thér., and Dixon, *Costesia* Thér., *Gigaspermum* Lindb., *Lorentziella* Müll. Hal., *Neosharpiella* H. Rob. and Delgad., *Oedipodiella* Dixon.

SUBCLASS: DICRANIDAE Peristome haplolepideous, with a formula of 4 : 2 : 3; exostome typically absent; late state division in the IPL asymmetric.

ORDER: GRIMMIALES Plants slender to robust; laminal cells with thick and often wavy walls; peristome of 16 entire or divided teeth.

Grimmiaceae Arn. Laminal cells with sinuose walls; leaves often terminated by hair-point; *Aligrimmia* R. S. Williams, *Coscinodon* Sprengh., *Coscinodontella* R. S. Williams, *Dryptodon* Brid., *Grimmia* Hedw., *Indusiella* Broth. and Müll. Hal., *Jaffueliobryum* Thér., *Leucoperichaetium* Magill, *Racomitrium* Brid., *Schistidium* Bruch and Schimp.

Ptychomitriaceae Schimp. Laminal cells with straight walls; leaves lacking hair-points; *Campylostelium* Bruch and Schimp., *Glyphomitrium* Brid., *Ptychomitriopsis* Dixon, *Ptychomitrium* Fürnr.

Scouleriaceae S. P. Churchill. Plants blackish, acro- or cladocarpous, saxicolous in riparian habitats; calyptrae mitrate, smooth; annulus not differentiated; capsules urceolate to globose; *Scouleria* Hook., *Tridontium* Hook. f.

Drummondiaceae (Vitt) Goffinet, comb. nov. (Orthotrichaceae subfam. Drummondioideae Vitt, Canad. J. Bot. 50 : 1194-1972). Stem with central strand, cladocarpous; costa with differentiated adaxial stereids; laminal cells thick-walled; peristome reduced; *Drummondia* Hook.

ORDER: ARCHIDIALES Plants small, ephemeral; seta lacking; capsules cleistocarpous, with less than 200 large spores (often 4-8); columella lacking.

Archidiaceae Schimp. *Archidium* Brid.

ORDER: SELIGERIALES Plants generally saxicolus; alar cells differentiated; capsules often systylious.

Seligeriaceae Schimp. Plants small; leaves constate; costa homogenous in transverse section; peristome mostly deeply inserted, relatively well developed; spores small; *Blindia* Bruch and Schimp., *Brachydontium* Fürnr., *Hymenolomopsis* Thér., Seligeria Bruch and Schimp., *Trochobryum* Breidl. and Beck.

Wardiaceae W. H. Welch. Plants robust, rheophilic; leaves ecostate; peristome attached at mouth, reduced; spores large; *Wardia* Harv. and Hook.

ORDER: DICRANALES Plants small to large; laminal cells generally smooth; alar cells often differentiated; peristome single, lacking basal membrane, segments trabeculate and striate.

Bryoxiphiaceae Besch. Leaves distichous with small dorsal extension (?) along costa; capsules gymnostomous; *Bryoxiphium* Mitt.

Fissidentaceae Schimp. Leaves distichous and complanate, with vaginant lamellae; apical cell two-sided; *Fissidens* Hedw.

Dicranaceae Schimp. Plants generally robust; cauline central strand present; leaves with well differentiated alar cells; peristome of 16 flat teeth divided in upper two-thirds, typically with vertically pitted outer surface; *Anisothecium* Mitt., *Aongstroemia* Bruch and Schimp., *Aongstroemiopsis* M. Fleisch., Arctoa Bruch and Schimp., *Atractylocarpus* Mitt., *Braunfcelsia* Paris, Brothera Müll. Hal., *Brotherobryum* M. Fleisch., *Bryohumbertia* P. De la Varde and Thér., Bryotestua Thér. and P. de la Varde, *Camptodontium* Dusén, *Campylopodiella* Cardot, *Campylopodium* (Müll. Hal.) Besch., *Campylopus* Brid., *Chorisodontium* (Mitt.) Broth., *Cnesrum* I. Hagen, *Cryptodicranum* E. B. Bartram, *Dicranella* (Müll. Hal.) Schimp., *Dicranodontium* Bruch and Schimp., *Dicranoloma* (Renauld) Renauld, *Dicranum* Hedw., *Eucamptodontopsis* Broth., *Holomitriopsis* H. Rob., *Holomitrium* Brid., *Hygrodicranum* Cardot., *Kiaeria* I. Hagen, *Kingiobryum* H. Rob., *Leptotrichella* (Müll. Hal.) Lindb., *Leucoluma* Brid., *Macrodictyum* (Broth.) E. H. Hegew; *Mesotus* Mitt., *Microcampylopus* (Müll. Hal.) Fleisch., *Mitrobryum* H. Rob., *Muscoherzogia* Ochyra, *Orthodicranum* (Bruch and Schimp.) Loeske, *Paraleucobryum* (Limpr.) Loeske, *Parisia* Broth., *Pilopogon* Brid., *Platyneuron* (Cardot) Broth., *Pocsiella* Bizot, *Polymerodon* Herzog, *Pseudephemerum* (Lindb.) I. Hagen, *Pseudochorisodontium* (Broth.) C. H. Gao, Vitt, X. H. Fu and T. Cao, *Schliephackea* Müll. Hal., *Sclerodontium* Schwägr., *Sphaerothecium* Hampe, *Steyecrmarkiella* H. Rob., *Symblepharis* Mont., *Werneriobryum* Herzog.

Bruchiaceae Schimp. Alar cells not differentiated; capsules with elongate necks; spores mostly with trilete markings, usually strongly ornamented; *Bruchia* Schwägr., *Cladophascum* Sim, *Eobruchia* W. R. Buck, *Pringleella* Cardot, *Trematodon* Michx.

Dicnemonaceae Broth. Plants robust, cladocarpous; central strand generally absent; laminal cells elongate, thick-walled and porose; perichaetial leaves differentiated and sheathing; calyptrae mitrate; germination endosporous; *Dicnemon* Schwägr, *Eucamptodon* Mont.

Leucobryaceae Schimp. Plants robust, glaucous; cauline central strand lacking; costa broad, occupying most of the leaves, with median chlorophyllose cells and adaxial and abaxial layers of hyaline cells; *Cladopodanthus* Dozy and Molk., *Leucobryum* Hampe, *Ochrobryum* Mitt., *Schistomitrium* Dozy and Molk.

Ditrichaceae Limpr. Plants slender; alar cells not differentiated; peristome of 16 completely divided, terete teeth; *Astomiopsis* Müll. Hal., *Austrophilibertiella* Ochyra, *Bryomanginia* Thér., *Ceratodon* Brid., *Cheilothela* Broth, *Chrysoblastella* R. S. Williams, *Cladastomum* Müll. Hal., *Cleistocarpidium* Ochyra and Bendnarek-Ochyra, *Crumuscus* W. R.

Buck and Snider, *Cygniella* H. A. Crum, *Distichium* Bruch and Schimp., *Ditrichopsis* Broth., *Ditrichum* Hampe, *Eccremidium* Hook. f. and Wilson, *Garckea* Müll. Hal, *Kleioweisiopsis* Dixon, × *Pleuriditrichum* A. L. Andrews and F. J. Herm., *Pleuridium* Rabenh., *Rhamphidium* Mitt., *Saelania* Lindb., *Skottsbergia* Cardot, *Strombulidens* W. R. Buck, *Trichodon* Schimp., *Tristichium* Mülll, Hal, *Wilsoniella* Müll. Hal.

Rhabdoweisiaceae Limpr. Plants small to medium size; stem lacking contral strand; capsules ribbed, widest at mouth; *Amphidium* Schimp., *Cynodontium* Schimp., *Dichodontium* Schimp., *Dicranoweisia* Milde, *Holodontium* (Mitt.) Broth., *Oncophorus* (Brid.) Brid., *Oreas* Brid., *Oreoweisia* (Bruch and Schimp.) De Not., *Pseudohyophila* Hilp., *Rhabdoweisia* Bruch and Schimp., *Verrucidens* Cardot

Rhachitheciaceae H. Rob. Laminal cells rectangular in lower half, short to isodiametric above; alar cells not differentiated; perichaetial leaves differentiated; capsules ribbed, rarely smooth; endostome teeth fused or not; IPL of 8 or 16 cells only (peristome formula: (4) : 2 : 2 or (4) : 2 : 1); *Hypnodontopsis* Z. Iwats. and Nog., *Jonesiobryum* B. H. Allen and Pursell, *Rhachitheciopsis* P. de la Varde, *Rhachithecium* Jolis, *Tisserantiella* P. de la Varde, *Uleastrum* W. R. Buck, *Zanderia* Goffinet.

Erpodiaceae Broth. Plants cladocarpous, costa lacking; laminal cells often papillose; calyptrae mitrate; *Aulacopilum* Wilson, *Bryowijkia* Nog., *Erpodium* (Brid.) Brid., *Solmsiella* Müll. Hal, *Venturiella* Müll. Hal., *Wildia* Müll. Hal. and Broth.

Schistostegaceae Schimp. Gametophores dimorphic, small, annual arising from persistent luminescent protonemata; leaves ecostate, distichous or in five rows; capsules globose, gymnostomous, lacking stomata and annulus; *Schistostega* D. Mohr.

Eustichiaceae Broth. Leaves distichous; laminal cells quadrate and thick-walled; capsules ribbed; peristome of 16 teeth; *Eustichia* (Brid.) Brid.

Viridivelleraceae I. G. Stone. Protonemata persistent; stems producing gametangia and associated leaves only; capsules gymnostomous; *Viridivellus* I. G. Stone.

ORDER: POTTIALES Plants minute to robust, generally orthotropic; upper laminal cells usually isodiametric and papillose; alar cells not differentiated; perichaetial leaves typically not differentiated; capsules erect; peristome typically papillose, not trabeculate.

Pottiaceae Schimp. Plants small to robust, primarily terrestrial; cauline central strand often present; leaves narrowly lanceolate to ligulate; laminal cells typically papillose; calyptrae cucullate, naked, smooth; peristome of 16 of 32 segments; *Acaulon* Müll. Hal., *Aloinella* Cardot, *Aloinia* Kindb., *Anoectangium* Schwägr., *Aschisma* Lindb., *Barbula* Hedw., *Bellibarbula* P. C. Chen, *Bryoceuthospora* H. A. Crum and L. E. Anderson, *Bryocrythrophyllum* P. C. Chen, *Calymperastrum* I. G. Stone, *Calyptopogon* (Mitt.) Broth., *Chenia* R. H. Zander, *Chionoloma* Dixon, *Crossidium* Jur., *Crumia* W. B. Schofield, *Dialytrichia* (Schimp.) Limpr., *Didymodon* Hedw., *Dolotortula* R. H. Zander, *Erythrophyllastrum* R. H. Zander, *Erythro-*

phyllopsis Broth, Eucladium Bruch and Schimp., Ganguleea R. H. Zander, *Gertrudiella* Broth, *Globulinella* Steere, *Goniomitrium* Hook. f. and Wilson, *Gymnostomiella* M. Fleisch., *Gymnostomum* Nees and Hornsch., *Gyroweisia* Schimp., *Hennediella* Paris, *Hilpertia* R. H. Zander, *Hymenostyliella* E. B. Bartram, *Hymenostylium* Brid., *Hyophila* Brid., *Hyophiladelphus* (Müll. Hal.) R. H. Zander, *Hypodontium* Müll. Hal., *Leptobarbula* Schimp., *Leptodontiella* R. H. Zander and E. H. Hegew., *Leptodontium* (Müll. Hal.) Lindb., *Luisierella* Thér. and P. De la Varde, *Microbryum* Schimp., *Microcrossidium* J. Guerra and M. J. Cano, *Mironia* R. H. Zander, *Molendoa* Lindb., *Neophoenix* R. H. Zander and During, *Pachyneuropsis* H. Mill, *Phascopsis* I. G. Stone, *Plaubelia* Brid., *Pleurochaete* Lindb., *Pottiopsis* Blockeel and A. J. E. Sm., *Pseudocrossidium* R. S. Williams, *Pseudosymblepharis* Broth., *Pterygoneurum* Jur., *Quaesticula* R. H. Zander, *Reimersia* P. C. Chen, *Rhexophyllum* Herozog, *Sagenotortula* R. H. Zander, *Saitobryum* R. H. Zander, *Sarconeurum* Bryhn, *Scopelophila* (Mitt.) Lindb., *Splachnobryum* Müll. Hal., *Stegonia* Venturi, *Stonea* R. H. Zander, *Streptocalypta* Müll. Hal., *Streptopogon* Mitt., *Streptotrichum* Herzog, *Syntrichia* Brid., *Teniolophora* W. D. Reese, *Tetracoscinodon* R. Br., *Tetrapterum* A. Jaeger, *Timmiella* (De Not.) Schimp., *Tortella* (Lindb.) Limpr., *Tortula* Hedw., *Trachycarpidium* Broth., *Trachyodontium* Steere, *Trichostomum* Bruch, *Triquetrella* Müll. Hal., *Tuerckheimia* Broth., *Uleobryum* Broth., *Weisiopsis* Broth., *Weissia* Hedw., *Weissiodicranum* W. D. Reese, *Willia* Müll. Hal.

Ephemeraceae Schimp. Plants minute, developing from persistent protonemata; capsules immersed, cleistocarpous or with poorly differentiated operculum; columella disintegrating; *Ephemerum* Schimp., *Micromitrium* Austin, *Nanomitriopsis* Cardot.

Serpotortellaceae W. D. Reese and R. H. Zander. Plants robust, cladocarpous, epiphytic; cauline central strand present; leaf margins entire and unistratose; perichaetial leaves differentiated; peristome well developed, reflexed when dry; *Serpotortella* Dixon.

Calymperaceae Kindb. Plants epiphytic; stem lacking central strand; leaves narrowly to broadly lanceolate; laminal cells papillose or smooth; often with hyaline cancellinae on either side of costa at leaf base; calyptrae persistent or not; peristome of 16 (rarely fused into 8) segments, smooth, papillose or vertically striate; *Arthrocormus* Dozy and Molk., *Calymperes* Sw., *Exodictyon* Cardot, *Exostratum* L. T. Ellis, *Leucophanes* Brid., *Mitthyridium* H. Rob., *Octoblepharum* Hedw., *Syrrhopodon* Schwägr.

Cinclidotaceae K. Saito. Plants blackish and robust, generally cladocarpous riparian to aquatic; cauline central strand lacking; leaves with thickened margins; capsules ribbed when dry; *Cinclidotus* P. Beauv.

Bryobartramiaceae Sainsb. Plants very small, acrocarpous; calyptrae remaining attached to vaginula, persisting as an epigonium; capsules cleistocarpous; *Bryobartramia* Sainsb.

SUBCLASS: *BRYIDAE* Peristome double, of alternating teeth and segments; endostome ciliate; late stage division in the IPL asymmetric.

SUPERORDER: *BRYANAE* Plants acrocarpous or cladocarpous; pseudoparaphyllia generally lacking; leaves erect to spreading, lanceolate to ovate, mostly costate, costal anatomy mostly heterogeneous; laminal cells generally short.

ORDER: SPLACHNALES Laminal cells rhombic to elongate, typically smooth; capsules erect with differentiated neck; peristome single or double; cilia rudimentary or lacking.

Splachnaceae Grev. and Arn. Plants mostly coprophilous; laminal cells thin-walled, rhomboidal; annulus not differentiated; capsules erect, neck often differentiated into broad hypophysis; endostome fused to exostome or lacking; *Aplodon* R. Br., *Brachymitrion* Taylor, *Moseniella* Broth., *Splachnum* Hedw., *Tayloria* Hook., *Tetraplodon* Bruch and Schimp., *Voitia* Hornsch.

Meesiaceae Schimp. Capsules more or less horizontal, with elongated neck; annulus differentiated and revoluble; peristome double; exostome typically shorter than endostome; *Amblyodon* P. Beauv., *Leptobryum* (Bruch and Schimp.) Wilson, Meesia Hedw., *Neomeesia* Deguchi, *Paludella* Brid.

Catoscopiaceae Broth. Plants small, slender; leaves in three ranks; laminal cells quadrate and smooth; capsules black, asymmetric, horizontal; peristome double and reduced; *Catoscopium* Brid.

ORDER: ORTHOTRICHALES Plants medium size, epiphytic or saxicolous; cauline central strand lacking; laminal cells typically papillose; capsules erect; peristome double or reduced; exostome recurved.

Orthotrichaceae Arn. Plants acrocarpous or cladocarpous; laminnal cells mostly isodiametric, thick-walled; calyptrae typically plicate and hairy; capsules erect, rarely immersed, often ribbed; OPL thick and teeth recurved when dry; cilia lacking; *Bryomaltaea* Goffinet, *Cardotiella* Vitt, *Ceuthotheca* Lewinsky, *Codonoblepharon* Schwägr., *Desmotheca* Lindb., *Florschuetziella* Vitt, *Groutiella* Steere, *Leiomitrium* Mitt., *Leptodontiopsis* Broth, *Leratia* Broth. and Paris, *Macrocoma* (Müll. Hal.) Grout, *Macromitrium* Brid., *Matteria* Goffinet, *Muelleriella* Dusén, *Orthotrichum* Hedw., *Pentastichella* Müll. Hal., *Pleurorthotrichum* Broth., *Schlotheimia* Brid., *Sehnemobryum* Lewinsky-Haapasaari and Hedenäs, *Stoneobryum* D. H. Norris and H. Rob., *Ulota* D. Mohr, *Zygodon* Hook and Taylor.

ORDER: HEDWIGIALES Plants medium to robust, plagiotropic, acrocarpous or cladocarpous; laminal cells thick-walled, papillose or smooth; capsules gymnostomous and immersed.

Hedwigiaceae Schimp. Protonemata globular; laminal cells pluripapillose; calyptrae smooth, and naked; *Braunia* Bruch and Schimp, *Headwigia* P. Beauv., *Headwigidium* Bruch and Schimp., *Pseudobraunia* (Lesq. and James) Broth.

Helicophyllaceae Broth. Leaves dimorphic, with lateral leaves strongly inrolled when dry, dorsal and ventral leaves reduced and appressed; laminal cells smooth; *Helicophyllum* Brid.

Rhacocarpaceae Kindb. Leaves bordered by narrow cells; laminal cells roughened; alar cells inflated; *Pararhacocarpus* Frahm, *Rhacocarpus* Lindb.

ORDER: BRYALES Plants primarily terricolous; cauline central strand present;

laminal cells rhombic to elongate, smooth; annulus differentiated; capsules pendent, neck differentiated; peristome double, typically well developed and ciliate; exostome incurved.

Aulacomniaceae Schimp. Plants medium to robust; laminal cells isodiametric, incrassate; capsules ribbed when dry; annulus differentiated; peristome double, bryoid; *Aulacomnium* Schwägr.

Bartramiaceae Schwägr. Plants often roburst; laminal cells isodiametric, quadrate or rectangular, smooth or prorulose; annulus typically undifferentiated; capsules subglose, erect or slightly curved, typically ribbed; neck undifferentiated; *Anacolia* Schimp., *Bartramia* Hedw., *Breutelia* (Bruch and Schimp.) Schimp., *Conostomum* Sw., *Fleischerobryum* Loeske, *Flowersia* D. G. Griffin and W. R. Buck, *Leiomela* (Mitt.) Broth., *Philonotis* Brid., *Plagiopus* Brid, *Quathlamba* Magill.

Orthodontiaceae (Broth.) Goffinet, comb. nov. (Bryaceae subfam. Orthodontoideae Broth. in Engler, Die natürlichen Pflanzenfamilien, ed. 2, 11 : 347, 1924). Plants small to robust; laminal cells elongate, lax; capsules often ribbed; annulus lacking; endostomial membrane reduced or lacking; *Orthodontium* Wilson, *Orthodontopsis* Ignatov and B. C. Tan.

Bryaceae Schwägr. Plants erect, mostly unbranched, acrocarpous; laminal cells mostly rhomboidal, smooth, thin-walled; costa single, strong; capsules inclined to pendulous, smooth, with differentiated neck; *Acidodontium* Schwägr., *Anomobryum* Schimp., *Brachymenium* Schwagr., *Bryum* Hedw., *Mielichhobryum* J. P. Srivast, *Mniobryoides* Hörmann, *Perssonia* Bizot, *Plagiobryum* Lindb., *Pleurophascum* Lindb., *Rhodobryum* (Schimp.) Limpr., *Roellia* Kindb., *Rosulabryum* J. R. Spence.

Phyllodrepaniaceae Crosby. Plants small; leaves complanate, in four rows; peristome single, of 16 segments; *Mniomalia* Müll. Hal, *Phyllodrepanium* Crosby.

Pseudoditrichaceae Steere and Z. Iwats., Plants very small; leaves ovate lanceolate; laminal cells thick-walled; capsules erect; peristome double, cilia lacking; *Pseudoditrichum* Steere and Z. Iwats.

Mniaceae Schwägr. Plants acro- or cladocarpous; leaves often bordered and often toothed; laminal cells thin-walled, rhomboidal to elongate; *Cinclidium* Sw., *Cyrtomnium* Holmen, *Epipterygium* Lindb., *Leucolepis* Lindb., *Mielichhoferia* Nees and Hornsch., *Mnium* Hedw., *Orthomnion* Wilson, *Plagiomnium* T. J. Kop., *Pohlia* Hedw., *Pseudobryum* (Kindb.) T. J. Kop., *Pseudopohlia* R. S. Williams, *Rhizomnium* (Broth.) T. J. Kop., *Schizymenium* Harv., *Synthetodontium* Cardot, *Trachcystis* T. J. Kop.

Leptostomataceae Schwägr. Plants forming dense mats; steam heavily tomentose; leaf margins entire, unbordered; annulus poorly differentiated to lacking; stomata cryptoporous; peristome strongly reduced; *Leptostomum* R. Br.

SUPERORDER: *RHIZOGONIANAE* Plants erect, generally from rhizomatous stem, acro-, clado-, or pleurocarpous; leaves oblong to lanceolate, costate, costal anatomy heterogeneous; laminal cells generally short and thick-walled.

ORDER: RHIZOGONIALES

Hypnodendraceae Broth. Plants robust, rhizomatous and stipitate; secondary stem erect, frondose to dendroid; laminal marginal cells differentiated or not, unistratose, often toothed; capsules often ribbed when dry; annulus differentiated; peristome well developed, ciliate; *Hypnodendron* (Müll. Hal.) Mitt.

Rhizogoniaceae Broth. Plants medium to large; cauline central strand present; marginal laminal cells often differentiated, bi- or multistratose, often toothed; costa typically toothed above; annulus differentiated; capsules generally smooth; peristome typically well developed, ciliate, or reduced to endostome or exostome; *Cryptopodium* Brid., *Goniobryum* Lindb., *Hymenodon* Hook. f. and Wilson, *Hymenodontopsis* Herzog, *Leptotheca* Schwägr., *Mesochaete* Lindb., *Pyrrhobryum* Mitt., *Rhizogonium* Brid.

Calomniaceae Kindb. Plants small, arising from persistent protonema; leaves tristrichous and dimorphic, with ventral ones much smaller; perichaetial leaves highly differentiated; capsules gymnostomous; *Colomnion* Hook. f and Wilson.

Mitteniaceae Broth. Plants small, with luminescent protonemata; cauline central strand lacking; leaves complanate, decurrent; perichaetia polysetous; peristome double, outer row homologous to bryoid endostome; *Mittenia* Lindb.

Cyrtopodaceae M. Fleisch. Plants robust and large, with rhizomatous stem; secondary stems erect, frondose to dendroid; lamina bistratose above; margin unistratose, toothed or entire; costa toothed above; capsules smooth, peristome double or reduced, eciliate; *Bescherellia* Duby, *Cyrtopodendron* M. Fleisch., *Cyrtopus* (Brid.) Hook. f.

Spiridentaceae Kindb. Plants robust and large, with rhizomatous stem; secondary stems erect, sparsely branched; lamina bi- or pluristratose; margin typically toothed; costa toothed above; capsules smooth, immersed or exserted; peristome double, teeth and segments long, eciliate; *Franciella* Thér., *Spiridens* Nees.

Pterobryellaceae (Broth.) W. R. Buck and Vitt. Plants robust and large, stipitate from rhizomatous stem, frondose to dendroid; cauline central strand lacking; capsules short oval; annulus differentiated; peristome double, with long teeth and segments but reduced cilia; *Pterobryella* (Müll. Hal.) A. Jaeger.

Racopilaceae Kindb. Stems plagiotropic; leaves dimorphic with dorsal ones reduced; costa excurrent; capsules long-exserted; peristome double, well developed; *Powellia* Mitt., *Racopilum* P. Beauv.

SUPERORDER: *HYPNANAE* Plants pleurocarpous, typically freely branching; pseudoparaphyllia usually present; leaves mostly ovate, costate or not, costal anatomy homogeneous; laminal cells generally elongate.

ORDER: HOOKERIALES Alar cells mostly not differentiated; opercula mostly rostrate; exostome teeth often furrowed.

SUBORDER: Hookeriineae Plants mostly medium-sized to slender, never (?) phyllodioicous; stems monopodially branched; leaves seldom plicate; laminal cells often large and lax, thin-walled; costae often strong and double, sometimes single or short and double; capsules mostly smooth; exothecial cells often collenchymatous; opercula often rostrate; exostome teeth often furrowed; endostomial segments with baffle-like cross-walls; calyptrae often mitrate.

Hookeriaceae Schimp. Stems with central strand; pseudoparaphyllia filamentous or absent; laminal cells large and lax; costa usually short and double; calyptrae multistratose at middle, naked; *Achrophyllum* Vitt and Crosby, *Cyathophorella* (Broth.) M. Fleisch., *Cyathophorum* P. Beauv., *Dendrocyathophorum* Dixon, Hookeria J. E. Sm., *Schimperobryum* Margad.

Leucomiaceae Broth. Stems lacking central strand; pseudoparaphyllia absent; laminal cells linear, lax; costa lacking; calyptrae cucullate; *Leucomium* Mitt., *Philophyllum* Müll. Hal., *Rhynchostegiopsis* Müll. Hal.

Daltoniaceae Schimp. Stems lacking central strand; pseudoparaphyllia absent; laminal cells oval to long-hexagonal, differentiated at leaf margins; costa single, calyptrae unistratose at middle, fringed at base, usually naked; *Calyptrochaeta* Desv., *Crosbya* Vitt, *Daltonia* Hook. and Taylor, *Distichophyllidium* M. Fleisch., *Distichophyllum* Dozy and Molk., *Ephemeropsis* K. I. Goebel, *Leskeodon* Broth., *Leskeodontopsis* Zanten, *Metadistichophyllum* Nog. and Z. Iwats.

Adelotheciaceae W. R. Buck. Stems lacking central strand; pseudoparaphyllia filamentous; laminal cells isodiametric, not differentiated at leaf margins; costa single; calyptrae unistatose at middle, not fringed at base, densely hairy; *Adelothecium* Mitt., *Bryobrothera* Thér.

Pilotrichaceae Kindb. Stems lacking central strand; pseudoparaphyllia none or foliose; laminal cells various, costa strong and double, or short and double; calyptrae unistratose at middle, usually hairy; *Actinodontium* Schwägr., *Amblytropis* (Mitt.) Broth., *Beeveria* Fife, *Brymela* Crosby and B. H. Allen, *Callicostella* (Müll. Hal.) Mitt., *Callicostellopsis* Broth., *Crossomitrium* Müll. Hal., *Cyclodictyon* Mitt., *Diploneuron* E. B. Bartram, *Helicoblepharum* (Mitt.) Broth., *Hemiragis* (Brid.) Besch., *Hookeriopsis* (Besch.) A. Jaeger, *Hypnella* (Müll. Hal.) A. Jaeger, *Lepidopilidium* (Müll. Hal.) Broth., *Lepidopilum* (Brid.) Brid., *Pilotrichidium* Besch., *Pilotrichum* P. Beauv., *Sauloma* (Hook. f. and Wilson) Mitt., *Stenodesmus* (Mitt.) A. Jaeger, *Stenodictyon* (Mitt.) A. Jaeger, *Tetrastichium* (Mitt.) Cardot, *Thamniopsis* (Mitt.) M. Fleisch., *Trachyxiphium* W. R. Buck, *Vesiculariopsis* Broth.

Hypopterygiaceae Mitt. Plants dendroid; amphigastria differentiated; leaves often limbate; costa single; laminal cells short, mostly smooth; alar cells nor differentiated; exostome teeth not furrowed; endostomial segments lacking baffle-like cross-walls; *Canalohypopterygium* W. Frey and Schaepe, *Catharomnion* Hook. f. and Wilson, *Hypopterygium* Brid., *Lopidium* Hook. f. and Wilson.

SUBORDER: Ptychomniieae Plants usually robust and turgid, often phyllodioicous; stems sympodially branched, usually lacking a central strand; leaves usually plicate, often strongly toothed; costae short and double; laminal cells elongate, often thick-walled and porose; alar cells often colored; capsules mostly ribbed; endostomial segments lacking baffle-like cross-walls; calyptrae often cucullate.

Ptychomniaceae M. Fleisch. Alar cells mostly little differentiated, except for color; capsules long-exserted, strongly eight-ribbed; calyptrae cucullate; spores isosporous; *Cladomnion* Hook. f. and Wilson, *Cladomniopsis* M. Fleisch., *Glyphothecium* Hampe, *Hampeella* Müll. Hal., *Ptychomnium* (Hook. f. and Wilson) Mitt., *Tetraphidopsis* Broth. and Dixon.

Garovagliaceae (M. Fleisch.) W. R. Buck and Vitt. Alar cells typically well differentiated in shape and color; capsules often immersed, smooth or somewhat sulcate; calyptrae mitrate or cucullate; spores anisosporous; *Endotrichellopsis* During, *Euptychium* Schimp., *Garovaglia* Endl.

ORDER: HYPNALES Stems monopodially or sympodially branched; alar cells often differentiated; opercula various, mostly not rostrate; exostome seldom furrowed; calyptrae mostly cucullate, naked.

Rutenbergiaceae (Broth.) M. Fleisch. Stems sympodially branched, lacking a central strand; secondary stems little branched; costa single; laminal cells prorulose; alar cells well differentiated; capsules immersed; calyptrae mitrate, hairy; *Neorutenbergia* Bizot and Pócs, *Pseudocryphaea* Broth., *Rutenbergia* Besch.

Trachylomataceae (M. Fleisch.) W. R. Buck and Vitt. Stems sympodially branched; secondary stems stipitate frondose, complanate-foliate; alar cells weakly differentiated; asexual propagula of stem-borne, filamentous gemmae; exostome teeth pale, densely papillose; *Braithwaitea* Lindb., *Trachyloma* Brid.

Amblystegiaceae G. Roth. Plants typically growing in moist areas; stems monopodially branched; costa often variable; laminal cells mostly short; alar cells not or weakly differentiated; setae relatively long in comparison to size of plants; capsules strongly curved and asymmetric; exostome teeth yellow-brown, cross-striolate; *Amblystegium* Schimp., *Donrichardsia* H. A. Crum and L.E. Anderson, *Gradsteinia* Ochyra, *Hygroamblystegium* Loeske, *Hypnobartlettia* Ochyra, *Koponenia* Ochyra, *Leptodictyum* (Schimp.) Warnst., *Limbella* (Müll. Hal.) Müll. Hal., *Ochyraea* Váňa, *Orthotheciella* (Müll. Hal.) Ochyra, *Platylomella* A. L. Andrews, *Sciaromiella* Ochyra, *Sciaromiopsis* Broth., *Sinocalliergon* Sakurai, *Vittia* Ochyra.

Cratoneuraceae Mönk. Stems monopodially branched, often wiry, often densely tomentose; paraphyllia present, foliose; stem and branch leaves differentiated; costa single; laminal cells relatively short, smooth or prorulose; alar cells well differentiated; exostome teeth yellow-brown, cross-striolate; *Cratoneuron* (Sull.) Spruce, *Sasaokaea* Broth.

Helodiaceae (M. Fleisch.) Ochyra. Stems monopodially branched; paraphyllia present, filamentous to narrowly foliose, the cells elongate, not papillose; costa single; laminal cells mostly prorulose; alar cells often well differentiated; exostome teeth cross-striolate; *Actinothuidium* (Besch.) Broth., *Bryochenea* C. H. Gao and K. C. Chang, *Helodium* Warnst., *Palustriella* Ochyra.

Hylocomiaceae (Broth.) M. Fleisch. Plants mostly robust; stems monopodially or sympodially branched; paraphyllia often present; leaves often strongly toothed; costae often strong and double; laminal cells elongate, smooth or prorulose; alar cells weakly differentiated; setae very elongate; capsules typically curved and asymmetric; exostome teeth yellow- to red-brown, often with reticulate pattern; *Hylocomiastrum* Broth., *Hylocomium* Bruch and Schimp., *Leptocladiella* M. Fleisch., *Leptohymenium* Schwägr., *Loeskeobryum* Broth., *Macrothamnium* M. Fleisch., *Meteoriella* S. Okamura, *Neodolichomitra* Nog., *Orontobryum* M. Fleisch., *Pleurozium* Mitt., *Rhytidiadelphus* (Limpr.) Warnst., *Rhytidiopsis* Broth., *Schofieldiella* W. R. Buck.

Rhytidiaceae Broth. Plants robust; stems monopodially branched; paraphyllia none; leaves plicate, rugose; costa single; laminal cells linear, strongly porose, prorulose; alar cells well differentiated; exostome teeth yellow-brown, cross-striolate; *Rhytidium* (Sull.) Kindb.

Leskeaceae Schimp. Plants terrestrial or epiphytic; stems monopodially branched, often terete-foliate; paraphyllia nonpapillose; leaves mostly short-acuminate; costa mostly single; laminal cells short, usually unipapillose; alar cells weakly differentiated; capsules curved and asymmetric when plants terrestrial but in epiphytes often erect; exostome striate in terrestrial taxa but in epiphytes often pale, weakly ornamented; endostome often reduced; *Claopodium* (Lesq. and James) Renauld and Cardot, *Dolichomitriopsis* S. Okamura, *Fabronidium* Müll. Hal., *Haplocladium* (Müll. Hal.) Müll. Hal., *Hylocomiopsis* Cardot, *Leptocladium* Broth., *Leptopterigynandrum* Müll. Hal., *Lescuraea* Bruch and Schimp., *Leskea* Hedw., *Leskeadelphus* Herzog, *Leskeella* (Limpr.) Loeske, *Lindbergia* Kindb., *Mamillariella* Laz., *Miyabea* Broth., *Okamuraea* Broth., *Orthoamblystegium* Dixon and Sakurai, *Pseudoleskea* Bruch and Schimp., *Pseudoleskeella* Kinbd., *Pseudoleskeopsis* Broth., *Ptychodium* Schimp., *Rigodiadelphus* Dixon, *Schwetschkea* Müll. Hal.

Regmatodontaceae Broth. Plants epiphytic; stems monopodially branched; paraphyllia absent; costa single; laminal cells short, smooth; alar cells not or weakly, differentiated; capsules erect; exostome teeth much shorter than endostome segments; *Regmatodon* Brid.

Pterigynandraceae Schimp. Plants terrestrial or epiphytic, mostly relatively small; stems monopodially branched, mostly terete-foliate; paraphyllia absent; costa short and double; laminal cells short, prorulose; alar cells weakly differentiated; gemmae stem-borne; setae smooth; capsules often erect; peristome often reduced; *Habrodon* Schimp., *Heterocladium* Bruch and Schimp., *Iwatsukiella* W. R. Buck and H. A. Crum, *Myurella* Bruch and Schimp., *Pterigynandrum* Hedw., *Trachphyllum* A. Gepp.

Rigodiaceae H. A. Crum. Plants terrestrial or weakly epiphytic, more or less stipitate; stems monopodially branched; paraphyllia absent; stipe, stem and branch leaves differentiated; costa single; laminal cells short, smooth; alar cells not or weakly differentiated; setae smooth; capsules curved and asymmetric; exostome teeth densely cross-striolate; *Rigodium* Schwägr.

Thuidiaceae Schimp. Plants terrestrial; stems monopodially branched; paraphyllia present, the cells papillose; stem and branch leaves differentiated; costa single; laminal cells short, papillose; alar cells not or weakly differentiated; setae often roughened; capsules typically curved and asymmetric; exostome teeth densely cross-striolate; calyptrae naked or sparsely hairy.; *Abietinella* Müll. Hal., *Boulaya* Cardot, *Cyrtohypnum* (Hampe) Hampe and Lorentz, *Fauriella* Besch., *Pelekium* Mitt., *Rauiella* Reimers, *Thuidiopsis* (Broth.) M. Fleisch., *Thuidium* Bruch and Schimp.

Campyliaceae (Kanda) W. R. Buck, comb. nov. [Amblystegiaceae subfam. Campylioideae Kanda, J. Sci. Hiroshima Univ., Ser. B., Div. 2, Bot. 15:250. 1975 (1976)]. Plants typically growing in moist areas, relatively robust; stems monopodially branched; costa mostly single; laminal cells elongate; alar cells often differentiated; setae not especially long; capsules asymmetric; exostome teeth yellow-brown; *Anacamptodon* Brid., *Callialaria* Ochyra, *Calliergon* (Sull). Kindb., *Campyliadelphus* (Kindb.) R. S. Chopra, *Campylium* (Sull.) Mitt. *Conardia* H. Rob., *Cratoneuropsis* (Broth.) M. Fleisch., *Drepanocladus* (Müll. Hal.) G. Roth, *Hamatocaulis* Headenäs, *Hygrohypnum* Lindb., *Loeskypnum* H. K. G. Paul, *Pictus* C. C. Towns., *Pseudocalliergon* (Limpr.) Loeske, *Sanionia* Loeske, *Scorpidium* (Schimp.) Limpr., *Straminergon* Hedenäs, *Tomentypnum* Loeske, *Warnstorfia* Loeske.

Brachytheciaceae G. Roth. Plants mostly growing in mesic woodlands, terrestrial; stems monopodially branched; leaves often plicate; costa single, often projecting as a small spine; laminal cells elongate; alar cells mostly weakly differentiated; setae sometimes roughened; capsules often relatively short, curved, asymetric; opercula conic to rostrate; exostome teeth mostly red-brown; calyptrae mostly naked; *Aerobryum* Dozy and Molk., *Aerolindigia* M. Menzel, *Bestia* Broth., *Brachythecium* Schimp., *Bryhnia* Kaurin, *Bryoandersonia* H. Rob., *Bryostreimannia* Ochyra, *Cirriphyllum* Grout, *Clasmatodon* Hook. f. and Wilson, *Cratoneurella* H. Rob., *Eriodon* Mont., *Eurhynchiella* M. Fleisch., *Eurhynchium* Bruch and Schimp., *Flabellidium* Herzog, *Homalotheciella* (Cardot) Broth., *Homalothecium* Schimp., *Isothecium* Brid., *Juratzkaeella* W. R. Buck, *Lindigia* Hampe, *Mandoniella* Herzog, *Meteoridium* (Müll. Hal.) Manuel, *Myuroclada* Besch., *Nobregaea* Hedenäs, *Palamocladium* Müll. Hal., *Plasteurhynchium* Broth., *Platyhypnidium* M. Fleisch., *Pseudopleuropus* Takaki, *Pseudoscleropodium* (Limpr.) M. Fleisch., *Puiggariopsis* M. Menzel, *Rhynchostegiella* (Schimp.) Limpr., *Rhynchostegium* Bruch and Schimp., *Rozea* Besch., *Schimperella* Thér., *Scleropodiopsis* Ignatov, *Scleropodium* Bruch and Schimp., *Scorpiurium* Schimp., *Squamidium* (Müll. Hal.) Broth., *Steerecleus* H. Rob., *Stenocarpidiopsis* M. Fleisch., *Trachybryum* (Broth.) W. B. Schofield, *Zelometeorium* Manuel.

Stereophyllaceae (M. Fleisch.) W. R. Buck and Ireland. Plants terrestrial or epiphytic; stems monopodially branched; costa typically single; laminal cells elongate, mostly smooth, rarely unipapillose; alar cells differentiated, collenchymatous, extending across base of costs; setae smooth; capsules inclined to erect; exostome teeth cross-striolate to papillose; *Catagoniopsis* Broth., *Entodontopsis* Broth., *Eulacophyllum* W. R. Buck and Ireland, *Juratzkaea* Lorentz, *Pilosium* (Müll. Hal.) M. Fleisch., *Sciuroleskea* Broth., *Stenocarpidium* Müll. Hal., *Stereophyllum* Mitt.

Myriniaceae Schimp. Plants often epiphytic, small; stems monopodially branched; costa single, often slender; laminal cells elongate, smooth; alar cells weakly differentiated; capsules often erect; peristomes mostly variously reduced; calyptrae rarely hairy; *Austinia* Müll. Hal., *Helicodontium* Schwägr., *Macgregorella* E. B. Bartram, *Merrilliobryum* Broth., *Myrinia* Schimp., *Nematocladia* W. R. Buck.

Fabroniaceae Schimp. Plants epiphytic., often small; stems monopodially branched, sometimes fragile; leaves mostly acuminate; costa single, slender; laminal cells short, smooth; alar cells mostly weakly differentiated; capsules typically erect; peristome often reduced; exostome teeth often paired; *Dimerodontium* Mitt., *Fabronia* Raddi, *Ischyrodon* Müll. Hal., *Levierella* Müll. Hal., *Rhizofabronia* (Broth.) M. Fleisch.

Meteoriaceae Kindb. Plants epiphytic, often pendent; stems monopodially branched, often very elongate, costa short and double or single; laminal cells mostly elongate, sometimes short, often variously papillose; alar cells not or weakly differentiated; setae often short, roughened; capsules often immersed, erect, symmetric; exostome teeth cross-striolate to papillose; endostome often reduced; calyptrae mitrate or cucullate, often hairy; *Aerobryidium* M. Fleisch, *Aerobryopsis* M. Fleisch., *Ancistrodes* Hampe, *Barbella* M. Fleisch., *Barbellopsis* Broth., *Chrysocladium* M. Fleisch., *Cryphaeophilum* M. Fleisch., *Cryptopapillaria* M. Menzel, *Diaphanodon* Renauld and Cardot, *Duthiella* Renauld, *Floribundaria* M. Fleisch., *Lepyrodontopsis* Broth., *Meteoriopsis* Broth., *Meteorium* (Brid.) Dozy and Molk., *Neodicladiella* W. R. Buck, *Neonoguchia* S. H. Lin, *Pseudospiridentopsis* (Broth.) M. Fleisch., *Pseudotrachypus* P. de la Varde and Thér., *Sinskea* W. R. Buck, *Toloxis* W. R. Buck, *Trachycladiella* (M. Fleisch.) M. Menzel and W. Schultze-Motel, *Trachypodopsis* M. Fleisch., *Trachypus* Reinw. and Hornsch.

Plagiotheciaceae (Broth.) M. Fleisch. Plants terrestrial; stems monopodially branched, mostly complanate-foliate; leaves decurrent; costa short and double or absent; laminal cells elongate, often strongly chlorophyllose; alar cells differentiated into the decurrencies; setae smooth; capsules often curved and asymmetric; peristome teeth mostly pale yellow; exostome typically cross-striolate below; endostome well developed; *Plagiothecium* Bruch and Schimp.

Fontinalaceae Schimp. Plants aquatic; stems sympodially branched; costa single or short and double (and then the leaves concave to carinate); capsules immersed or short-exserted; endostome forming a trellis; calyptrae mitrate or cucullate; *Brachelyma* Cardot, *Dichelyma* Myrin, *Fontinalis* Hedw.

Entodontaceae Kindb. Plants often

epiphytic; stems monopodially branched; costa short and double or absent; laminal cells linear, smooth; alar cells subquadrate, numerous; capsules erect and symmetric, long-exserted; columella often exserted; peristome inserted below mouth of capsule; endostome mostly strongly reduced; *Entodon* Müll. Hal., *Erythrodontium* Hampe, *Mesonodon* Hampe, *Pylaisiobryum* Broth.

Climaciaceae Kindb. Plants dendroid; stems sympodially branched, with paraphyllia on stipe; leaves not decurrent; costa single; laminal cells relatively short, smooth; capsules erect and symmetric; *Climacium* F. Weber and D. Mohr.

Pleuroziopsidaceae Ireland. Plants dendroid; stems sympodially branched, with longitudinal lamellae on stipe; leaves decurrent; costa single; laminal cells relatively short, smooth; capsules inclined and asymmetric; *Pleuroziopsis* E. Britton.

Hypnaceae Schimp. Stems monopodially branched; pseudoparaphyllia foliose or rarely filamentous; paraphyllia none; leaves often falcate or homomallous; costa short and double (or absent); laminal cells mostly linear; capsules mostly inclined and asymmetric; exothecial cells usually not collenchymatous; opercula apiculate to short-rostrate; exostome teeth mostly cross-striolate; calyptrae mostly naked; *Andoa* Ochyra, *Bardunovia* Ignatov and Ochyra, *Breidleria* Loeske, *Bryocrumia* L. E. Anderson, *Bryosedgwickia* Cardot and Dixon, *Callicladium* H. A. Crum, *Calliergonella* Loeske, *Campylophyllum* (Schimp.) M. Fleisch, *Caribaeohypnum* Ando and Higuchi, *Chrysohypnum* (Hampe) Hampe, *Crepidophyllum* Herzog, *Ctenidiadelphus* M. Fleisch., *Ctenidium* (Schimp.) Mitt., *Cyathothecium* Dixon, *Ectropotheciella* M. Fleisch., *Ectropotheciopsis* (Broth.) M. Fleisch., *Ectropothecium* Mitt., *Elharveya* H. A. Crum, *Elmeriobryum* Broth., *Entodontella* M. Fleisch., *Eurohypnum* Ando, *Fallaciella* H. A. Crum, *Giraldiella* Müll. Hal., *Gollania* Broth., *Herzogiella* Broth., *Homomallium* (Schimp.) Loeske, *Hondaella* Dixon and Sakurai, *Horridohypnum* W. R. Buck., *Hycomium* Bruch and Schimp, *Hypnum* Hedw., *Irelandia* W. R. Buck, *Isopterygiopsis* Z. Iwats., *Isopterygium* Mitt., *Leiodontium* Broth., *Macrothamniella* M. Fleisch., *Mahua* W. R. Buck, *Microctenidium* M. Fleisch., *Mittenothamnium* Henn., *Nanothecium* Dixon and P. de la Varde, *Orthothecium* Bruch and Schimp., *Phyllodon* Bruch and Schimp., *Plagiotheciopsis* Broth., *Platydictya* Berk., *Platygyriella* Cardot, *Platygyrium* Bruch and Schimp., *Podperaea* Z. Iwats. and Glime, *Pseudotaxiphyllum* Z. Iwats., *Ptilium* De Not., *Pylaisiella* Grout, *Rhacopilopsis* Renauld and Cardot, *Rhizohypnella* M. Fleisch., *Sclerohypnum* Dixon, *Serpoleskea* (Limpr.) Loeske, *Stenotheciopsis* Broth., *Stereodontopsis* R. S. Williams, *Syringothecium* Mitt., *Taxiphyllopsis* Higuchi and Deguchi, *Taxiphyllum* M. Fleisch., *Trachythecium* M. Flesch, *Tripterocladium* (Müll. Hal.) A. Jaeger, *Vesicularia* (Müll. Hal.) Müll. Hal, *Wijkiella* Bizot and Lewinsky.

Catagoniaceae W. R. Buck and Ireland. Stems monopodially branched; pseudoparaphyllia filamentous; leaves conduplicate; costa short and double or absent; laminal cells linear, smooth;

alar cells not differentiated; exostome teeth cross-striolate; *Catagonium* Broth.

Symphyodontaceae M. Fleisch. Stems monopodially branched; laminal cells mostly prorulose; alar cells not or weakly differentiated; setae mostly roughened; capsules symmetric, typically spinose; exostome teeth papillose to cross-striolate; calyptrae cucullate or mitrate; *Chaetomitriopsis* M. Fleisch., *Chaetomitrium* Dozy and Molk., *Dimorphocladon* Dixon, *Symphyodon* Mont.

Sematophyllaceae Broth. Stems monopodially branched; leaves often golden green; costa short and double or none; laminal cells mostly linear, smooth or papillose; alar cells well differentiated; exothecial cells mostly collenchymatous; opercula mostly obliquely rostrate; exostome teeth often furrowed, cross-striolate; *Acanthorrhynchium* M. Fleisch., *Acritodon* H. Rob., *Acroporium* Mitt., *Allioniellopsis* Ochyra, *Aptychella* (Broth.) Herzog, *Aptychopsis* (Broth.) M. Fleisch., *Brotherella* M. Fleisch., *Chinostomum* Müll. Hal., *Clastobryella* M. Fleisch., *Clastobryophilum* M. Fleisch., *Clastobryopsis* M. Fleisch., *Clastobryum* Dozy and Molk., *Colobodontium* Herzog, *Donnellia* Austin, *Foreauella* Dixon and P. de la Varde, *Gammiella* Broth., *Hageniella* Broth., *Heterophyllium* (Schimp.) Kindb., *Isocladiella* Dixon, *Leptoischyrodon* Dixon, *Macrohymenium* Müll. Hal., *Mastopoma* Cardot, *Meiotheciella* B. C. Tan, W. B. Schofield and H. P. Ramsay, *Meiothecium* Mitt., *Papillidiopsis* (Broth.) W. R. Buck and B. C. Tan, *Paranapiacabaea* W. R. Buck and Vital, *Potamium* Mitt., *Pseudohypnella* (M. Fleisch.) Broth., *Pterogonidium* Broth., *Pterogoniopsis* Müll. Hal., *Pylaisiadelpha* Cardot, *Pylaisiopsis* (Broth.) Broth., *Radulina* W. R. Buck and B. C. Tan, *Rhaphidostrichum* M. Fleisch., *Schraderella* Müll. Hal, *Schroeterella* Herzog, *Sematophyllum* Mitt., *Struckia* Müll. Hal., *Taxitheliella* Dixon, *Taxithelium* Mitt., *Timotimius* W. R. Buck, *Trichosteleum* Mitt., *Trismegistia* (Müll. Hal.) Müll. Hal. *Trolliella* Herzog, *Warburgiella* Müll. Hal, *Wijkia* H. A. Crum.

Hydropogonaceae W. H. Welch. Plants aquatic; stems monopodially branched; costa short and double or none; laminal cells linear, smooth; alar cells weakly differentiated; capsules cladocarpous, immersed; endostome none; calyptrae mitrate; *Hydropogon* Brid., *Hydropogonella* Cardot.

Myuriaceae M. Fleisch. Stems sympodially branched; secondary stems little or not branched; leaves mostly long-acuminate; costa short and double or none; laminal cells linear, smooth; alar cells well differentiated, mostly colored; capsules long-exserted, erect; exostome teeth reduced, smooth, often perforate; endostome rudimentary; *Eumyurium* Nog., *Myurium* Schimp., *Oedicladium* Mitt., *Palisadula* Toyama, *Piloecium* (Müll. Hal.) Broth.

Cryphaeaceae Schimp. Stems sympodially branched; secondary stems little or not branched; costa single; laminal cells short, smooth or sometimes prorulose; alar cells numerous; capsules immersed or seldom emergent; exostome teeth pale, papillose; endostome rudimentary to absent; calyptrae mitrate; *Cryphaea* D. Mohr, *Cryphidium* (Mitt.) A. Jaeger, *Cyptodon*

(Broth.) M. Fleisch., *Cyptodontopsis* Dixon, *Dendroalsia* E. Britton, *Dendrocryphaea* Broth., *Dendropogonella* E. Britton, *Pilotrichopsis* Besch., *Schoenobryum* Dozy and Molk., *Sphaerotheciella* M. Fleisch.

Prionodontaceae Broth. Plants epiphytic; stems sympodially branched; axillary hairs as in *Breutelia* (Bartramiaceae); leaves usually plicate and with strongly toothed margins; costa single; laminal cells short, papillose; alar cells differentiated in large areas; capsules immersed to emergent; annulus revoluble; exostome teeth papillose; endostome segments united into a reticulum; *Prionodon* Müll. Hal.

Leucodontaceae Schimp. Plants mostly epiphytic; stems sympodially branched; secondary stems often not or little branched, mostly curled when dry; leaves rapidly spreading when moist, mostly plicate; costa short and double or none; laminal cells oval to linear, mostly smooth, rarely prorulose; alar cells numerous, capsules usually exserted; annulus not differentiated; exostome teeth pale, papillose; endostome mostly rudimentary; spores often large and anisosporous; *Antitrichia* Brid., *Dozya* Sande Lac., *Eoleucodon* H. A. Mill. and H. Whittier, *Felipponea* Broth., *Leucodon* Schwägr., *Pterogonium* Sw., *Scabridens* E. B. Bartram.

Pterobryaceae Kindb. Plants mostly epiphytic; stems sympodially branched; secondary stems often well branched, and thus stipitate; stem and branch leaves often differentiated, branch leaves sometimes five-seriate; costa mostly single, sometimes short and double or absent; laminal cells mostly linear, mostly smooth, sometimes prorulose; alar ells usually differentiated, often thick-walled and colored; capsules mostly immersed; exostome teeth pale, often smooth; endostome most rudimentary; calyptrae cucullate or mitrate, often hairy; *Calyptothecium* Mitt., *Cryptogonium* (Müll. Hal.) Hampe, *Henicodium* (Müll. Hal.) Kindb., *Hildebrandtiella* Müll. Hal., *Horikawaea* Nog., *Jaegerina* Müll, Hal., *Micralsopsis* W. R. Buck, *Muellerobryum* M. Fleisch., *Neolindbergia* M. Fleisch., *Orthorrhynchidium* Renauld and Cardot, *Orthostichidium* Dusén, *Orthostichopsis* Broth., *Osterwaldiella* Broth., *Penzigiella* M. Fleisch., *Pireella* Cardot, *Pseudopterobryum* Broth., *Pterobryidium* Broth. and Watts, *Pterobryon* Hornsch., *Pterobryopsis* M. Fleisch., *Pulchrinodus* B. H. Allen, *Renauldia* Müll. Hal., *Rhabdodontium* Broth., *Spiridentopsis* Broth., *Symphysodon* Dozy and Molk., *Symphysodontella* M. Fleisch.

Phyllogoniaceae Kindb. Plants epiphytic; stems sympodially branched; secondary stems irregularly branched, strongly complanate—foliate; leaves conduplicate, cucullate, auriculate; costa short and double or absent; laminal cells linear, smooth; alar cells differentiated in small groups; capsules immersed or shortly exserted; exostome teeth place, not or scarcely ornamented; endostome rudimentary or absent; calyptrae cucullate or mitrate, naked or hairy, *Phyllogonium* Brid.

Orthorrhynchiaceae S. H. Lin. Plants terrestrial; stems monopodially branched; leaves conduplicate, cucullate; costs short and double or absent, laminal cells linear, smooth; alar cells undifferentiated; capsules short exserted, erect; exostome teeth pale, unornamented; endostome none; calyptrae mitrate, hairy; *Orthorrhynchium* Reichardt.

Lepyrodontaceae Broth. Plants terrestrial or epiphytic; stems sympodially branched; secondary stems not or little branched; leaves sometimes plicate; costa single and weak or short and double to absent; laminal cells linear, smooth, thick-walled and porose; alar cells few or scarcely differentiated; capsules long-exserted; peristome usually only endostomial; *Dichelodontium* Broth., *Lepyrodon* Hampe.

Neckeraceae Schimp. Plants terrestrial or epiphytic; stems mostly sympodially branched, sometimes monopodial; stipes sometimes differentiated and plants then frondose; leaves mostly complanately arranged; costa typically single, sometimes short and double; laminal cells fusiform to linear, rarely shorter, mostly smooth, rarely prorulose or papillose; alar cells mostly few or weakly differentiated; capsules immersed (mostly in epiphytes) to long-exserted (mostly in terrestrial taxa); exostome teeth often pale, usually cross-striolate at least at extreme base, papillose above; endostome often reduced; calyptrae mostly cucullate; *Baldwiniella* M. Fleisch., *Bissetia* Broth., *Bryolawtonia* D. H. Norris and Enroth, *Caduciella* Enroth, *Crassiphyllum* Ochyra, *Cryptoleptodon* Renauld and Cardot, *Curvicladium* Enroth, *Dixonia* Horik. and Ando, *Dolichomitra* Broth., *Handeliobryum* Broth., *Himantocladium* (Mitt.) M. Fleisch., *Homalia* (Brid.) Bruch and Schimp., *Homaliadelphus* Dixon and P. de la Varde, *Homaliodendron* M. Fleisch., *Hydrocryphaea* Dixon, *Isodrepanium* (Mitt.) E. Britton, *Metaneckera* Steere, Neckera Hedw., *Neckeropsis* Reichardt, *Neomacounia* Ireland, *Noguchiodendron* Ninh and Pócs, *Pendulothecium* Enroth and S. He, *Pinnatella* M. Fleisch., *Porothamnium* M. Fleisch, *Porotrichodendron* M. Fleisch., *Porotrichopsis* Broth. and Herzog, *Porotrichum* (Brid.) Hampe, *Thamnobryum* Nieuwl., *Touwia* Ochyra.

Echinodiaceae Broth. Plants epipetric or less often on soil or bases of trees; stems sympodially branched, wiry; secondary stems irregularly branched; leaves mostly subulate; costa single, mostly excurrent; laminal cells short, smooth; alar cells weakly differentiated; capsules long-exserted, inclined to horizontal; exostome teeth reddish, cross-striolate; endostome well developed; *Echinodium* Jur.

Leptodontaceae Schimp. Plants mostly epiphytic; stems sympodially branched, often curled when dry; secondary stems irregularly branched to bipinnate; costa typically single; laminal cells isodiametric to long-hexagonal, smooth, unipapillose or prorulose; alar cells numerous; capsules immersed to short-exserted; exostome teeth pale, unornamented to spiculose; endostome rudimentary; calyptrae hairy; *Alsia* Sull., *Forsstroemia* Lindb., *Leptodon* D. Mohr, *Taiwanobryum* Nog.

Lembophyllaceae Broth. Plants often turgid; stems monopodially branched; leaves mostly strongly concave; costa mostly short and double (rarely single); laminal cells elongate, smooth; alar cells often somewhat differentiated; capsules mostly erect and immersed to short-exserted; endostome mostly reduced; calyptrae rarely mitrate, naked or hairy; *Acrocladium* Mitt., *Camptochaetre* Reichardt, *Fifea* H. A. Crum, *Lembophyllum* Lindb., *Neobarbella* Nog., *Orthostichella* Müll. Hal., *Pilotrichella* (Müll. Hal.) Besch., *Weymouthia* Broth.

Anomodontaceae Kindb. Plants mostly epiphytic; stems sympodially or monopodially branched, secondary stems and/or branches often curled when dry, not complanate-foliate; paraphyllia none; leaves often acute to obtuse; costa single or short and double; laminal cells mostly short, papillose or prorulose; alar cells mostly poorly differentiated; capsules exserted, erect; exostome teeth pale to white, cross-striolate sometimes with overlying papillae to papillose; endostome often reduced; *Anomodon* Hook. and Taylor, *Bryonorrisia* L. R. Stark and W. R. Buck, *Chileobryon* Enroth, *Curviramea* H. A. Crum, *Haplohymenium* Dozy and Molk., *Herpetineuron* (Müll. Hal.) Cardot, *Schwetschkeopsis* Broth.

Theliaceae (Broth.) M. Fleisch. Plants terrestrial or on bases of trees; stem monopodially branched; paraphyllia present; leaves imbricate, little altered when moist, deltoid-ovate; costa single; laminal cells short, stoutly unipapillose; alar cells differentiated; capsules exserted, erect; exostome teeth white, smooth to papillose; endostome strongly reduced; *Thelia* Süll.

Microtheciellaceae H. A. Mill. and A. J. Harr. Plants epiphytic; stems monopodially branched; costa single; laminal cells short, smooth; alar cells weakly differentiated; capsules short-exserted, erect; exostome teeth truncate, reduced, weakly ornamented; endostome rudimentary; *Microtheciella* Dixon.

Sorapillaceae M. Fleisch. Leaves distichous and complanate; capsules cladocarpous, immersed; peristome double, of 16 slender segments and 32 stout exostome knobs, cilia absent; *Sorapilla* Spruce and Mitt.

However in the present text, the following account of the Indian mosses (arranged according to a new modification of Cavers, Brotherus, Dixon and Reimers systems) the number of species occurring in the Indian subscontinent, Burma and Ceylon is given within brackets after the name of each genus. Families and genera not represented in this area have been omitted.

Subclass: **BRYIDAE**

Section I. *NEMATODONTEAE*

The Nematodonteae peristome is formed of several concentric layers of amphithecial cells. The individual teeth are usually formed of bundles of dead fibrous cells with unthickened portions resorbed (Polytrichales) but may also be formed of cellwalls only (Buxbaumiales). They look solid without articulations but with striations or plicae. The final structure differs in the different orders.

Dixon has included five orders under the Nematodonteae. Three of these orders (Tetraphidales, Buxbaumiales and Polytrichales with *Dawsonia*) are well known and it is justified to raise them to the rank of cohorts. The two other orders—*Calomniales* and *Schistostegales* do not show any peristome and are not found in India. They are not discussed here.

Cohort I. TETRAPHIDAE

Order 1. **TETRAPHIDALES**

In the Tetraphidales, consisting of the single family *Tetraphidaceae* (*Georgiaceae*), the peristome, as seen in *Tetraphis* (= *Georgia*), is formed of the entire tissue lying within the single layered operculum and remains adherent to the upper part of the columella. This mass of tissue splits into four thick triangular segments (Fig. 31.2) almost looking like the four valves of the Jungermanniale capsule. The peristome is certainly more analogous to that of *Polytrichum* than to those with the articulated teeth of the Arthrodonteae. The protonema is partly thallose in this order.

Tetraphidales are not represented in India.

Cohort II. BUXBAUMIIDAE (BUXBAUMIINALES)

These are perculiar mosses with a peristome differing much in its construction as described below.

Order 2. BUXBAUMIALES

The Buxbaumiales peristome is double, formed of 3 to 6 concentric amphithecial layers. The outer peristome (*exostome*) is formed of 1 to 4 series of slender teeth while the inner peristome is a thin tabular membrane resembling a truncated cone showing 16 or 32 folds or plicae. Both peristome are formed of thickened cellwalls only and not of entire cell groups which is the case in Tetraphidales and Polytrichales.

Basic chromosome number $n = 8$ or 9.

Family 1. Buxbaumiaceae

The protonema is filamentous, persistent and the gametophyte is extremely reduced (with scarcely and leaf). The exostome and endostome are as described above but in *B. aphylla*, the outer peristome is missing.

Genus 1. *Buxbaumia* (1 species—*B. himalayana* recently discovered by Udar *et al* from Deoban in U.P. Himalaya).

Family 2. Diphysciaceae

The protonema is thallose shield-shaped and the gametophytic plant is normal. In the peristome the exostome is missing and the endostome is a pleated (16 plicae) asymmetrical tube which shows a spiral twist (Fig. 31.1)

Genus 1. *Diphyscium* (3 spp.)

Genus 2. *Theriotia*—found only in Kashmir (W. Himalaya).

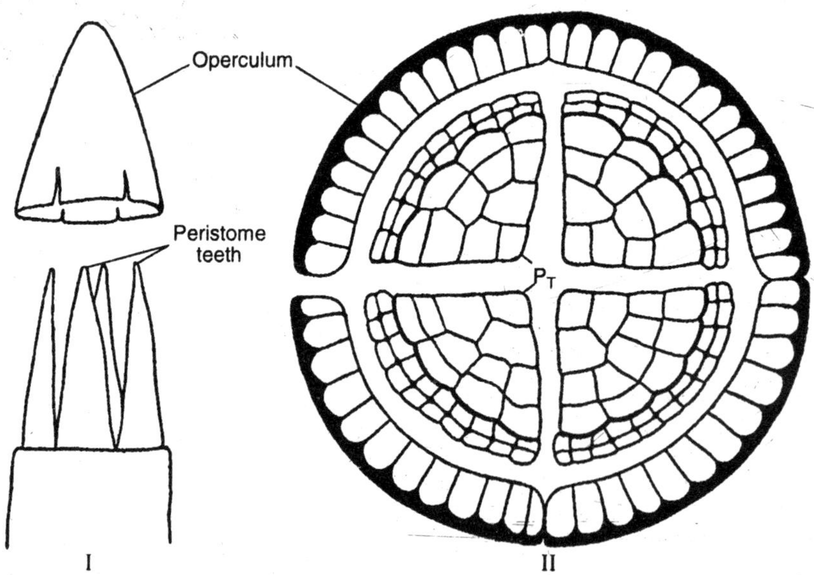

Fig. 31.2: *Tetraphis pellucida* (after Cavers). I. Mature peristome and operculum. II. T.s. of top of capsule showing splitting of the mass of tissue inside single layered operculum to form four thick triangular segments developing the four peristome teeth

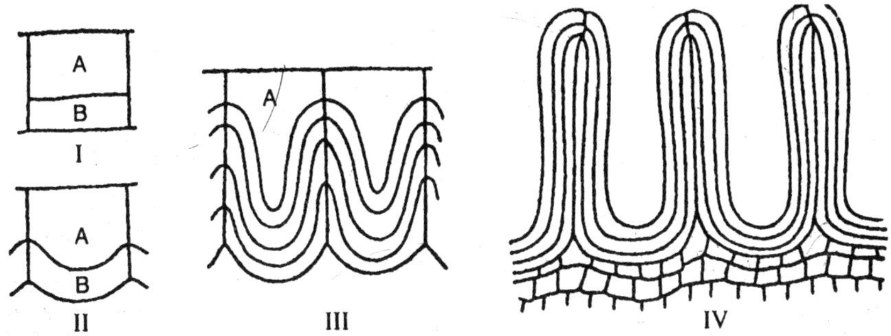

Fig. 31.3: Stages of development of the Polytrichaceae peristome teeth (after Goebel and Cavers). I. Tangential section to show the cutting off of a thin-walled upper cell A from the peristome teeth forming cell B. II. Curving out of the peristome forming cell. III. Formation of bundles of peristome cells by repeated divisions. IV. Three peristome teeth after dissolution of the thin-walled upper A cells

Cohort III. POLYTRICHIDAE (POLYTRICHINALES)

Mosses of this group differ from the common *Eubryinales* in the solid peristome teeth derived from several concentric layers of cells. There are two orders:

Order 3. POLYTRICHALES

Usually sturdy mosses of the hills. In Polytrichales the peristome is formed out of a single annular series of cells. In this layer curved dividing lines in the lower cells form continuous bands of many cells (Fig. 31.3-I, II, III). The uppermost cells remain thin-walled and are ultimately resorbed (Fig. 31.3-IV) giving rise to 32 or 64 (rarely 16) separate solid teeth in a ring. The tips of these teeth remain attached to the drum-top like *tympanum* formed out of the top cells of the columella. Leaves show longitudinal lamellae on the upper face of entire leaf or on vein only.

Basic chromosome number n = strictly 7.

Family: *Polytrichaceae*

Genera: 1. *Atrichum* (= *Catharinaea*) (6 spp.), 2. *Oligotrichum* (1 sp.), 3. *Lyellia* (1 sp.—Fig. 31.4), 4. *Pogonatum* (34 spp.), 5. *Polytrichum* (5 spp.).

Genus: POGONATUM

Pogonatum and *Polytrichum* are the two dominating genera of the family *Polytrichaceae.* Dixon did not recognize *Pogonatum* as a separate genus and most British books include *Pogonatum* within *Polytrichum* although *Pogonatum* is internationally accepted as a separate genus. *Pogonatum* is the largest genus with about 199 species all over the world of which 34 are known from the temperate hills of India and Ceylon. Of these, 28 species are known from Eastern India, 10 species from the

Fig. 31.4: Female and male plants of *Lyellia crispa* from Darjeeling

Western Hamalaya and Kashmir and 10 species from South India and Ceylon. *Pogonatum* is very common on rock soil in the hillstations of India where 3 speies—*P. aloides, P. microstomum* and *P. junghuhnianum* are common. *Pogonatum microstomum*, the common sturdy moss of the Himalayas and the Ghats, are referred to in the following descriptions.

The gametophytic plant

The base of the erect *Pogonatum* stem (Fig. 31.5A) is usually described as *rhizomatous*. It is not a rhizome in the usual sense but is slightly stouter and stiffer (with more thick-walled tissue) than the stem above. Numerous *rhizoids* develop from this region. The individual rhizoid cell is thick-walled, elongated and with oblique walls. The multicellular rhizoids occur as strands and a number of these occur twisted in a rope-like manner forming strong cable-like strings as shown for *Polytrichum* (Fig. 31.10D). On the upright stem the lower leaves are very small, scale-like and much paler in colour. The leaves grow in length by a two-sided apical cell. The upper leaves are crowded together spirally spreading out from the stem and are rather stiff. The leaves are comparatively larger than in other mosses. The base of each leaf is broad, somewhat sheathing and paler in colour. The upper part is deep green to brown or reddish in colour, gradually tapering with a serrated margin. Most of the leaf is occupied by the thicker midrib (Fig. 31.5C), there being only a very narrow wing-like thinner lamina on the two sides. The upper surface of this 'midrib' (i.e., most of the leaf) is completely covered by parallel longitudinal vertical plate-like structures called *lamellae* (Fig. 31.5C and E) somewhat like the gills of Agaric fungi. The top row of cells of these lamellae differ from species to species. In *Pogonatum microstomum* each tip cell is split into 2-flask-shaped cells. A *t.s.* of leaf (Fig. 31.5E) shows the midrib with patches of thick-walled cells and sectional view of the lamellae—each 4 or 5 cells high.

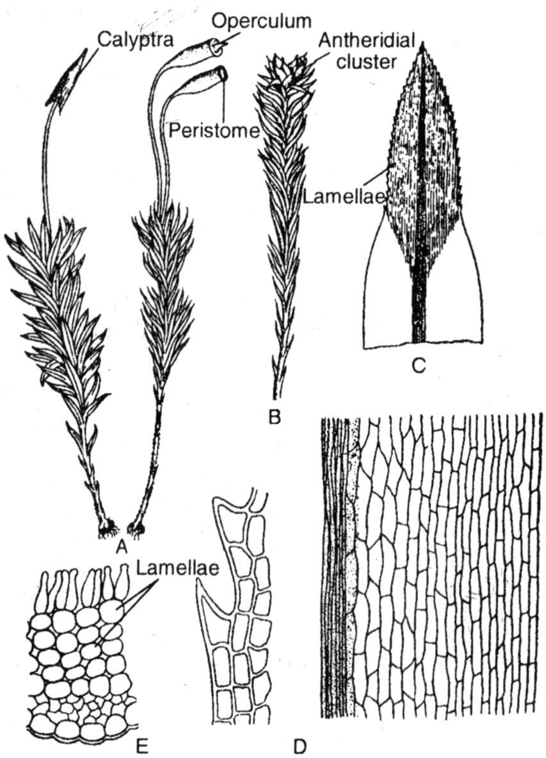

Fig. 31.5: *Pogonatum microstomum*. A. Female and B. Male Plants. C. A leaf showing longitudinal lamellae. D. Leaf cells at upper margin and at base. E. Part of t.s. of leaf showing lamellae

The growth of the stem is by an apical pyramidal cell with 3 cutting faces. The anatomy of the mature stem of both *Pogonatum* and *Polytrichum* is rather complicated. *Pogonatum* is somewhat simpler. The stem (Fig. 31.6) is surrounded by an *epidermis* which gets broken up in the upper part where the stem outline becomes irregular due to the presence of a number of leaves. Inside this is the zone of the *cortex* which is formed of living cells full of starch. The outer thick-walled, more deeply coloured, elongated cortical cells pass on to lighter, compact, parenchymatous cells of the cortex. The innermost cells of the cortex contain more starch and have been likened to an irregular, rudimentary pericycle. The cortex shows here and there groups of cells representing the *leaf traces*. The leaf traces join the

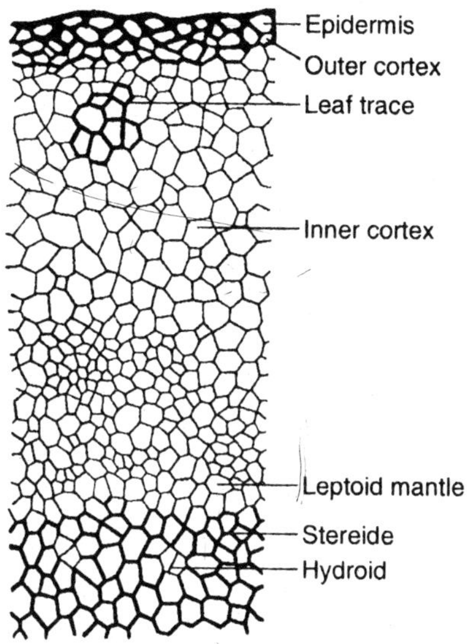

Fig. 31.6: Part of t.s. of *Pogonatum* stem

midribs of leaves to the central hydrom cylinder of the stem. Inside the cortex there is a zone of elongated cells with protoplasm but no starch which is likened to the sieve tubes and are called *leptoids*. It has very recently been demonstrated (Stevenson, 1974) that the leptoid cells are demarcated by oblique intervening cells which are nothing but porous sieve plates almost as in higher plants. The leptoid cells are interspersed with cells containing starch and this zone is called the *leptoid mantle*. The centre of the stem is the central *hydrom cylinder* where there are two types of cells; elongated thick-walled cells with living contents called *stereides*, and similar cells devoid of cell contents called *hydroids*. The hydroids are considered as the conducting strands by some but others doubt if there is any conduction of water at all. The leptoids and hydroids are much more clearly seen in *Polytrichum* (Fig. 31.11).

Vegetative reproduction

Pogonatum gametophytic plants are usually perennial. The vegetative shoots grow indefinitely, specially by proliferation of the male plant, and they may break apart to form independent plants. Another way of vegetative reproduction is by the development of bud-like *gemmae* on the rhizoids as seen in *Polytrichum* (Fig. 31.10D). All gametophytic parts are also known to give rise to secondary *protonemata* behaving like the original one. This is called *regeneration* a very common phenomenon in most mosses. The original protonema is also known to divide into a number of protonemata or to give rise to gemmae on it. Protonemata are also known to develop out of sporophytic parts. This is *apospory* as the protonema is being developed on the sporophyte without the intervention of spores. A gametophytic plant developed on an aposporous protonema is diploid and not haploid like the normal gametophyte.

Sexual reproduction

Most *Pogonatum* species are dioecious, the antheridia and archegonia developing on separate plants.

In the male plant (Figs. 31.5B and 31.7), the top leaves are specialized, coloured red or orange and form something like an involucre which is called the *perichaetium*. This forms a flower-like head (Fig. 31.7A). Inside this structure there is a cluster of *antheridia* and filamentous *paraphyses* at the base of each *perichaetial leaf* (Fig. 31.7B) so that the structure is compound and not simple as in the cases of *Semibarbula* and *Funaria* described later. This compound nature is not clear when mature. The tip of the stem in the centre of this structure is left free for further growth (*proliferation*) into a new shoot. Each *antheridium* is a club-shaped structure with a short *stalk*. In its development, the *antheridium initial* divides transversely into two cells the lower of which is the *stalk initial* and the upper the *primary antheridial cell*. The stalk does not develop much but the primary antheridial cell elongates by a two-sided apical cell. The latter first forms a filament, then by periclinal divisions it forms

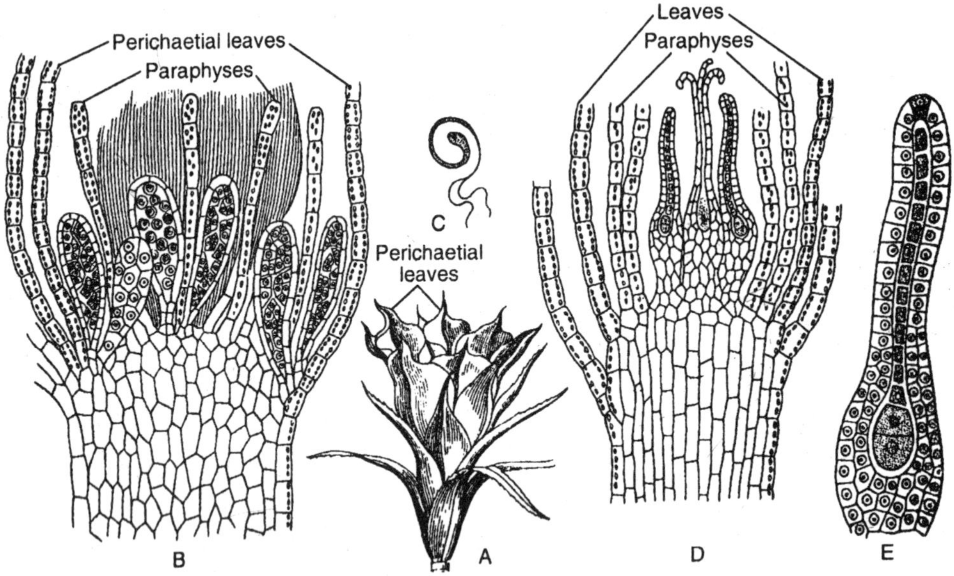

Fig. 31.7: *Pogonatum*. A. Male flower. B. L.s. of male head. C. A sperm. D. L.s. of archegonial cluster. E. An archegonium

a *jacket* of one layer enclosing *androcyte mother cells*. Each androcyte mother cell divides further forming *androcytes* or *sperm cells* within each of which develops a biflagellate, coiled *antherozoid* or *spermatozoid* (Fig. 31.7C). The androcyte also contains lot of mucilage. The mature antheridium (Fig. 31.7B) is a club-shaped structure with a short stalk.

The *archegonial cluster* (Fig. 31.7D) develops at the tip of the female plant. There are usually three archegonia in a cluster which are surrounded by filamentous *paraphyses* and then the *perichaetial leaves*. The apical cell of the stem itself is consumed in the development of the archegonia so that there may be no further proliferation. The *archegonial initial*, which is the tip growing cell or a cell quite near it, divides by a transverse wall forming an upper *primary archegonial cell* and a lower *primary stalk cell*. The primary stalk cell forms the short *stalk* while the upper archegonial cell lengthens by an apical cell with three cutting faces. This upper part ultimately organizes the *archegonium* proper (Fig. 31.7E) with the *venter* and the *neck* containing the *egg*, the *ventral canal cell* and the *neck canal cells* inside. The top growing cell remains as the *cover cell*. Archegonia of mosses are longer than those of the Hepatics and that of *Pogonatum* shows a specially large (about 12) and variable number of neck canal cells.

Fertilization

When mature, and in presence of a film of water covering the plants at this time, the antheridia burst by the disintegration of the top cells. The neck and the ventral canal cells in the archegonia together with the top cover cell also disorganize leaving a free passage to the egg. The decomposed substance coming out possibly contains some chemotactic substance like sugar attracting the swimming sperms which now get access to the egg. Ultimateley, single sperm fertilizes the large spherical egg forming the *diploid zygote*.

The sporophyte

The *zygote* invests itself with a cell wall, becomes brownish, enlarges and divides by a transverse wall into an *epibasal* and a *hypobasal cell* which divide repeatedly organizing a young *embryo* (Fig. 31.8A) with two terminal growing cells at the two ends each with two cutting faces. The lower end forms the *foot* and the upper end the *seta* and the *capsule*. The archegonial wall also enlarges simultaneously and becomes fibrous forming the *calyptra* (Fig. 31.5A) enveloping the young sporophyte. As the sporophyte lengthens, the calyptra breaks, only its upper part remaining attached to the tip like a cap. A long, slender sporophyte is gradually formed (Figs. 31.5A and 31.8C). The lower part of this (formed from the hypobasal cell) burrows down into the tip of the gametophyte forming the *foot*. The upper part elongates as a narrow slender structure bearing the calyptra at the tip. The *capsule* differentiates last in the region covered by the calyptra. In this capsule part an *amphithecium* surrounding an *endothecium* is organized. The amphithecium forms the multilayered *wall* or *jacket* of the capsule while the endothecium forms the *archegonium* from its outer layers and the *columella* from the axial layers.

In the mature *sporophyte* (Figs. 31.5A and 31.8C), there is the *foot*, a sucking organ penetrating the tip of the gametophyte, the *seta* which is equivalent to the stem of the higher plants and the *capsule* completely covered by the fibrous *calyptra*.

The *seta* is green when young, reddish when mature and then turns deep brown. A t.s. of the seta (Fig. 31.8B) shows a strong *epidermis* of one layer, *hypodermis* of several layers of very thick-walled cells, a *cortex* of somewhat loose parenchymatous cells and a central strand of very small thin-walled cells.

An l.s. of *Pogonatum* capsule (Fig. 31.8E) shows a *wall* several layers thick of which the one-layered *epidermis* with thick outer walls form the outermost layer. The inner cells of the wall are chlorophyllose and there are several layers of these. After that is a large cylindrical

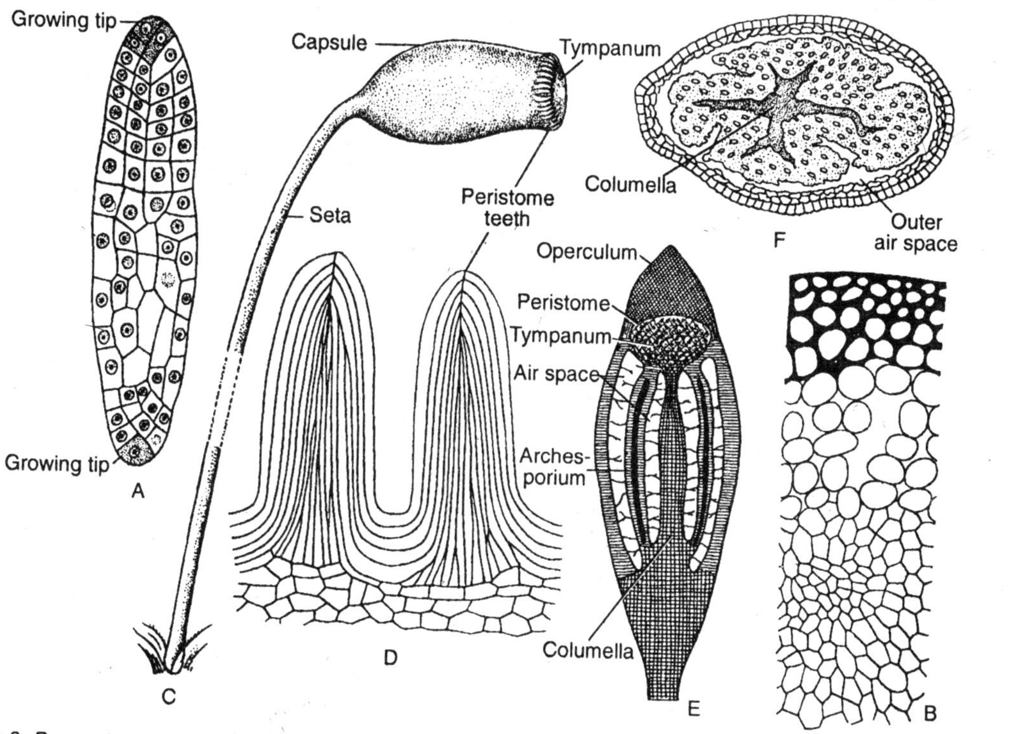

Fig. 31.8: *Pogonatum*. A. L.s. of young embryo. B. T.s. of seta. C. A sporophyte. D. Two peristome teeth. E. L.s. of capsule. F. T.s. of capsule

air space (*outer lacuna*) connected by string-like filaments of green cells with the *outer wall* of the internal *spore sac* cylinder. The spore sac has an outer wall and an *inner wall* of 2 thin-walled cell layers each. The spore sac contains the *archesporium* (usually two layers thick) when young and *spores* (developed as in other Bryophytes) when mature. After the inner wall of the spore sac there is another *air space* (*inner lacuna*) also traversed by green filaments. Thus, the spore sac actually hangs free but bound down by the green filaments on both sides. The inner green filaments connect with the *columella* which is formed of colourless elongated cells (Fig. 31.8E). In *Pogonatum*, usually the columella is not cylindrical in cross section (Fig. 31.8F) but there are four wing-like expansions meeting the inner spore sac walls. Thus, a longitudinal section of the capsule may be illusive if it passes through a wing.

The columella extends and swells upwards and forms a drum-like roof of the capsule called the *epiphragm* (also called *tympanum*) so that the spore sacs are not exposed (Fig. 31.8C and E). Above the tympanum there is a conical *operculum* which is usually beaked and is connected with the capsule by a ring-like *diaphragm* but there is no organized annulus. Just above there is the *peristome* with a ring of 32 short *teeth* (Figs. 31.8C and E). These teeth (Fig. 31.8D) are solid, formed of bundles of fibrous cells (Fig. 31.3) and are different from the peristome teeth of the *Eubryinales*. Hence the name *Nematodonteae*.

The *Pogonatum* capsule shows no neck (apophysis) with stomata which is the case in *Polytrichum*. Stomata are not present on the capsule of *Pogonatum*.

When the capsule dries up, the operculum is broken loose by the pressure of the columella and the cells between the epiphragm and the teeth also break. The spores then come out of the cylindrical slit, the coming out being partially controlled by the hygroscopic teeth.

The new gametophyte

The spore has a thick *exins* or *exospore* and a thin *intine* or *endospore*. The exine bursts at one point and the intine comes out like a green algal filament (Fig. 31.9B). This filament grows like a tube, cuts off cells and then becomes highly branched. The filament branches are green being rich in chloroplasts and form the *chloronema*. Other filaments also develop from these which are thin and colourless and tend to go down and become the rhizoids. This entanglement of filaments is the *protonema* (Fig. 31.9C). Later, multicellular buds develop on the chloronema each one of which develops a new erect leafy *gametophyte* by a normal apical cell with three cutting faces.

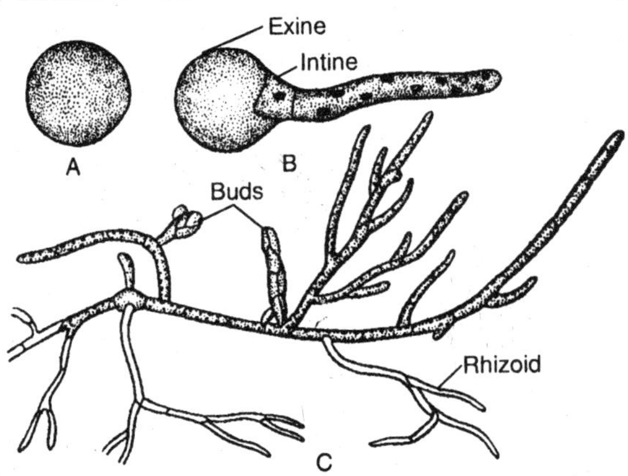

Fig. 31.9: *Pogonatum*. A. Spore. B. Germination of spore. C. Protonema showing buds of new gametophytes on the chloronema and also colourless rhizoids

Genus: *POLYTRICHUM*

The genus *Polytrichum* is better known in the Western countries as it is so common and prominent there. It is, however, very rare and uncommon in India being found only in the comparatively inaccessible alpine regions of the Himalayas. All similar plants seen by the ordinary student on the hills belong to the genus *Pogonatum*. Of the 111 species distributed over the frigid temperate regions of the word only five species (*alpinum, densifolium, xanthopilum, formosum* and *juniperinum*) are known from above 3600 metres in the Eastern Himalaya and only two species (*alpinum* and *juniperinum*) are known from the Western Himalaya. The genus is not known in the warmer south of India or in the Khasia Hills or the extreme north-eastern India where alpine height is never attained.

Polytrichum generally resembles *Pogonatum* but differs in certain important characters pointed out below so that it is distinctly a different genus.

The gametophytic plant

The gametophyte plant (Fig. 31.10) generally resembles *Pogonatum* but is of a hardier nature. The stem anatomy is more developed and differentiated (Fig. 31.11). The *leptoid* and *hydroid* cells are more clearly developed (Fig. 31.12B). The *leptom mantle* is demarcated on the inside by one or two layers of suberised cells full of starch and this zone is called the *amylom layer* or the *hydrom sheath*. Hydroids sometimes occur in groups of two and three when the cell walls between them are found to be very thin and delicate. On the whole the central strand simulates some characters of a real vascular

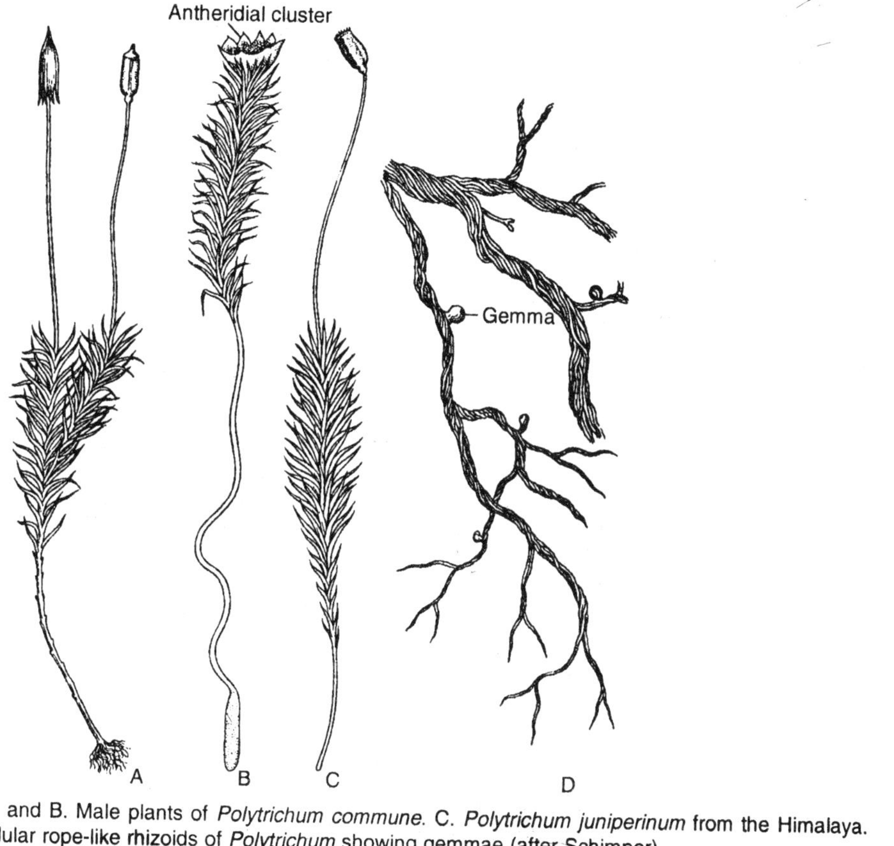

Fig. 31.10: A. Female and B. Male plants of *Polytrichum commune*. C. *Polytrichum juniperinum* from the Himalaya. D. Multicellular rope-like rhizoids of *Polytrichum* showing gemmae (after Schimper)

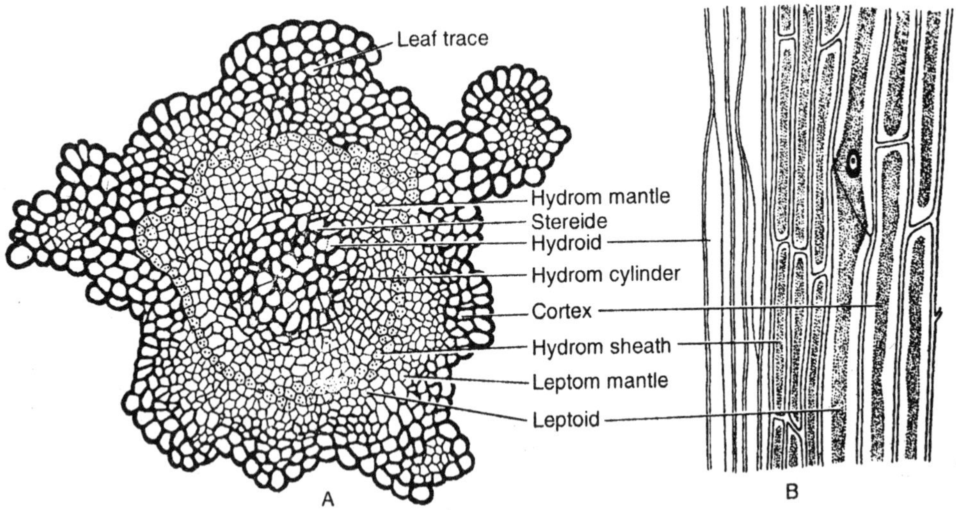

Fig. 31.11: *Polytrichum*. A. T.s. of gametophyte stem (after Tansley and Chick), B. L.s. of central strand (after Ruhland)

strand as seen in an l.s. of the central of *Polytrichum formosum* depicted by Ruhland (Fig. 31.11B).

The rhizoids of the *Polytrichum* gametophyte show remarkable twisting and *gemmae* for vegetative propagation are very common on these (Fig. 31.10D).

The sporophyte

It is in the sporophyte capsule that *Polytrichum* and *Pogonatum* differ very distinctly.

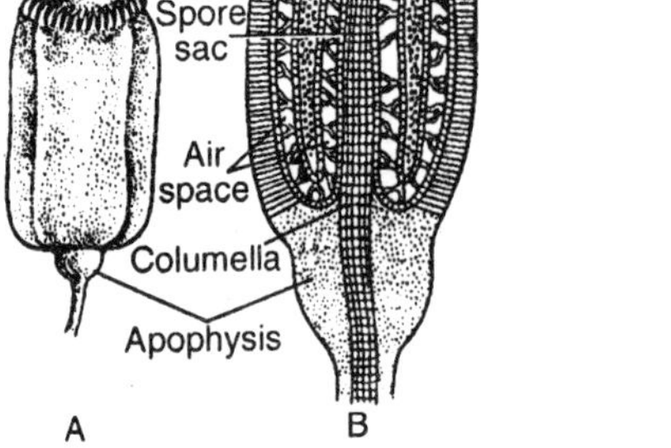

Fig. 31.12: *Polytrichum commune*. A. The capsule, B. L.s. of a capsule with operculum

The capsule in *Polytrichum* (except in *Polytrichum alpinum*) is usually angular (Fig. 31.12A) and not round. Unlike *Pogonatum* where there is no neck or *apophysis* with stomata, *Polytrichum* neck is generally marked off by a constriction from the main capsule (Fig. 31.12A and B). This apophysis, internally, shows a central strand which is always (even in species where the apophysis is not prominently developed) surrounded by a mass of lax chlorophyllose cells and is surrounded by an epidermis with many stomata. The peristome shows solid teeth as in *Pogonatum* but the number of these teeth is 64 and not 32 as in the case of *Pogonatum*.

Order 4. Dawsoniales

The fourth order *Dawsoniales* of the Cohort Polytrichiidae comprizes of the single family *Dawsoniaceae* with bristle-like (like a broomstick) peristome teeth. Here the peristome forming amphithecial layer is broad and formed of a large number of cell layers. In this layer interspersed vertical rows of cells become thick-walled (Fig. 31.13-II), the intervening, thin-walled vertical rows are resorbed so that thickened bands of cells stick out like, bristles in several rings giving rise to a broomstick-like

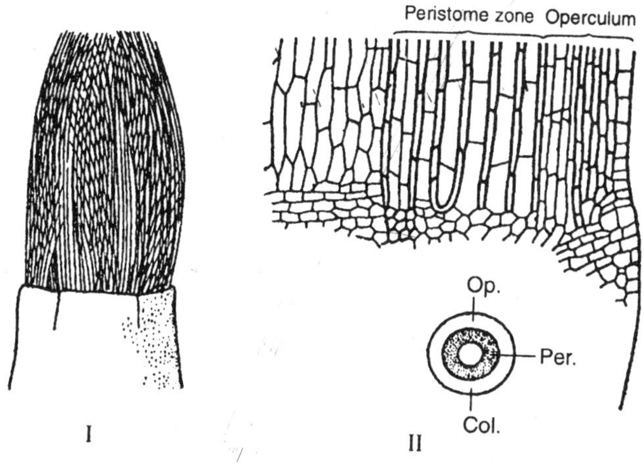

Fig. 31.13: *Dawsonia* (after Cavers). I. The mature peristome. II. Showing the operculum, peristome and columella zones. The thinner walls in the peristome zone break down to form the bristles

peristome (Fig. 31.13-I). The order is not represented in India.

Basic chromosome number n = strictly 7.

Section II. *ARTHRODONTEAE*

Cohort IV. Eubryiidae (Eubryinales)

The Arthrodonteae peristome is formed of only 2 or 3 layers of amphithecial cells. The teeth are membranous or scaly being actually formed of the thickened walls of the cells which are like plates (called *lepos* in latin), the thinner parts being resorbed. They are articulated or transversely barred showing the joined up walls of cells of which they are formed.

The Arthrodonteae are divided into two series: *Haplolepideae* and *Diplolepideae.*

Series I. **Haplolepideae**

In Haplolepideae the teeth are placed in a single row. Each tooth shows a single row of external plates and, usually, a double row of internal plates as may be seen by looking at the outer and inner faces of the teeth. The two sets of plates or scales (*lepos*) come from two concentric layers of cells. But, these teeth may be differently split longitudinally in the different orders.

Order 5. Archidiales

The single family *Archidiaceae* with the single genus *Archidium* was formerly included within the next order Dicranales. But the sporophyte structure is so simple and different

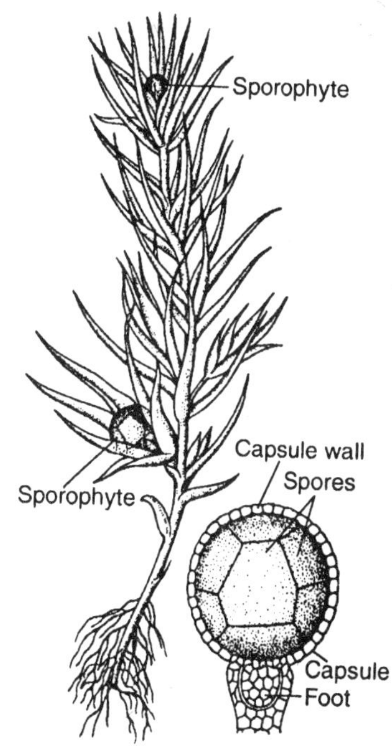

Fig. 31.14: *Archidium birmannicum.* Gametophyte with two sporophytes from near Calcutta. B. L.s. of sporophyte of A. (After Gangulee)

that Cavers raised it to a cohort and Reimers considered it as an order and placed it at the top of the Eubryiidae. The sporophyte is cleistocarpic without any columella or peristome. At least in these plants it is not reasonable to consider this cleistocarpy to have arisen by reduction. It is, therefore, better to consider it as a separate and problematic order. The capsule is a mere bag with a one-layer thick jacket enclosing a few large spores (Fig. 31.14B) and a short foot. There is no columella and the endothecium is entirely sporogenous. 3 spp. of *Archidium* (*indicum, microthecium* and *octosporum*) are known from South India, 1 sp. (*birmannicum*) from near Calcutta (Fig. 31.14A) and from N.E. Assam and 2 spp. from Burma.

The position of the Archidiales is somewhat problematic. Archidium has a gametophyte which quite fits within the Dicranales. The chr. no. (n = 13) also resembles the Dicranales. But, its sporophyte is unique, being as simple as the simplest Hepatics. There is litle logic in supposing that it is a degenerate from the peristomic. Dicranales as is the case with many other cleistocarpic species. Its peristome differs so much from the other Bryidae that several Bryologists think that it deserves a rank even higher than an order to a cohort. Hence the name Class Archidiopsida by Cronquist, Takhtajan and Zimmermann. Although the arguments have weight, the taxon is kept only as a separate order here, pending farther ellucidation by farther work.

Order 6. DICRANALES

The order Dicranales is acrocarpous, haplolepideous and usually with 16 dicranoid peristome teeth but in Ditrichaceae and some genera of Dicranales the teeth are more finely or irregularly split (*ditrichocranoid*) while in *Pleuridium* the peristome is missing.

Basic chromosome numbers n = 11, 12, 13.

Family: *Ditrichaceae*

Genera: 1. *Pleuridium* (3 spp.), 2. *Pleuridiella* (1 sp.), 3. *Garckea* (2 spp.), 4. *Ditrichum* (12 spp.), 5. *Ditrichopsis* (1 sp.), 6. *Ceratodon* (2 spp.), 7. *Distichium* (5 spp.).

Family: *Dicranaceae*

Genera: 1. *Bruchia* (1 sp.), 2. *Trematodon* (13 spp.), 3. *Wilsoniella* (1 sp.), 4. *Anisothecium* (3 spp.), 5. *Aongstroemia* (2 spp.), 6. *Aongstroemiopsis* (1 sp.), 7. *Microdus* (7 spp.), 8. *Dicranella* (11 spp.), 9. *Campylopodium* (3 spp.), 10. *Microcambylopus* (1 sp.), 11. *Campylopodiella* (2 spp.), 12. *Campylopus* (including *Thysanomitrium*—46 spp.), 13. *Dicranodontium* (16 spp.), 14. *Brothera* (2 spp.), 15. *Paraleucobryum* (2 spp.), 16. *Amphidium* (1 sp.), 17. *Atractylocarpus* (2 spp.), 18. *Rhabdoweisia* (1 sp.), 19. *Dicranoweisia* (4 spp.), 20. *Symblepharis* (3 spp.), 21. *Dicranum* (14 spp.), 22. *Oncophorus* (4 spp.), 23. *Holomitrium* (1 sp.), 24. Oreas (1 sp.), 25. Oreoweisia (2 spp.), 26. *Leucoloma* (11 spp.), 27. *Dicranoloma* (7 spp.), 28. Cynodontium (2 spp.) 29. *Dicnemoloma* (1 sp. from Ceylon), 30. *Braunfelsia* (1 sp. from Ceylon).

Aongstroemia julacea (Hook.) Mitt. discovered at an elevation of 19,800 ft. in the Rongbuk Valley off Mt. Everest perhaps marks the highest limit of plant survival.

Family: *Leucobryaceae*

Genera: 1. *Leucophanes* (4 spp.), 2. *Leucobryum* (17 spp.), 3. *Ochrobryum* (4 spp.), 4. *Octoblepharum* (1 sp.), 5. *Arthrocormus* (1 sp. from Ceylon), 6. *Exodictyon* (2 spp.).

Order 7. FISSIDENTALES

The *Fissidentals* consist of a single family *Fissidentaceae* which includes four genera among which the genus *Fissidens* (Fig. 31.15) is

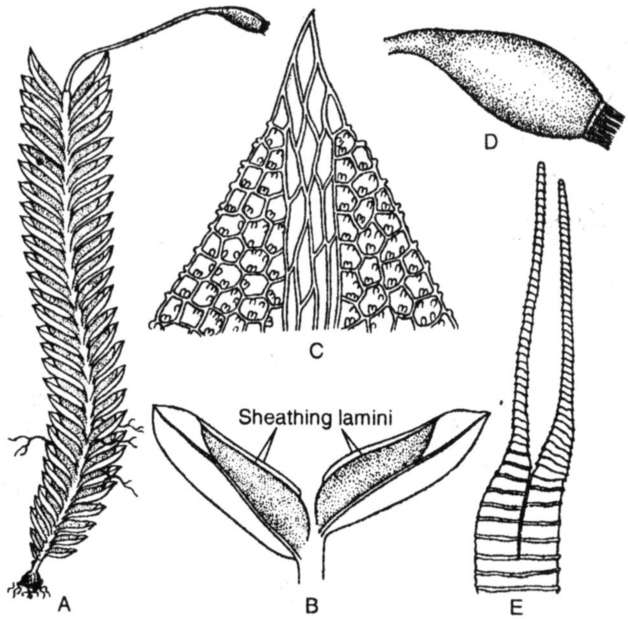

Fig. 31.15: *Fissidens ceylonensis* (after Gangulee). A. Plant. B. A pair of distichous leaves showing sheathing lamini. C. Leaf spex showing papillose cells. D. A capsule. E. Dicranoid haplolepideous peristome tooth

predominating and only this is found in India. This is mainly a tropical genus although some species are found on the temperate hills. Most species grow on soil and as many as 83 species are recorded from India.

Basic chromosome number anomalous (n = 5 to 24 known).

The clear diagnostic feature of the genus is the pair of sheathing lamini of the distichous leaves (Fig. 31.15B). Plants are *acrocarpous.* There is one ring of 16 peristome teeth (*haplolepideous*), each of the teeth being split into two (*dicranoid*) almost to the base (Fig. 31.15E).

Order 8. SYRRHOPODONTALES

Acrocarpous, haplolepideous, usually epixylic and often gemmiferous mosses. Peristome not split (16) or absent altogether (*Monocranoid*).

Family: *Calymperaceae*

Genera: 1. *Syrrhopodon* (16 spp.), 2. *Thyridium* (9 spp.), 3. *Calymperes* (37 spp.), 4. *Calymperopsis* (1 sp.).

Order 9. POTTIALES

Acrocarpous, haplolepideous mosses. Peristome teeth not split, split into filaments (*ditrichocranoid*) or missing altogether. A few genera (e.g., *Phascum*) are cleistocarpic (no operculum or peristome, capsule sunk within leaves).

Family: *Pottiaceae*

Genera: 1. *Anoectangium* (9 spp.), 2. *Molendoa* (3 spp.), 3. *Merceya* (including *Merceyopsis*—6 spp.), 4. *Astomum* (2 spp.), 5. *Hymenostylium* (7 spp.), 6. *Hymenostomum* (4 spp.), 7. *Weisia* (4 spp.), 8. *Gymnostomum* (2 spp.), 9. *Reimersia* (1 sp.), 10. *Rhamphidium* (1 sp.), 11. *Oxystegus* (6 spp.), 12. *Trichostomum* (8 spp.), 13. *Pseudosymblepharis* (3 spp.), 14. *Timmiella* (1 sp.), 15. *Tortella* (7 spp.), 16. *Pleurochaete* (1 sp.), 17. *Chionoloma* (1 sp.), 18. *Leptodontium* (2 spp.), 19. *Bryoerythrophyllum* (8 spp.), 20. *Hyophila* (14 spp.), 21. *Didymodon* (14 spp.), 22. *Barbula* (22 sp.), 23. *Geheebia*

(1 sp.), 24. *Bellibarbula* (1 sp.), 25. *Semibarbula* (2 spp.), 26. *Hydrogonium* (11 spp.), 27. *Prionidium* (1 sp.), 28. *Weisiopsis* (1 sp.), 29. *Beddomiella* (1 sp.), 30. *Pottia* (1 sp.), 31. *Hyophilopsis* (1 sp.), 32. *Desmatodon* (8 spp.), 33. *Syntrichia* (7 spp.), 34. *Tortula* (1 sp.), 35. *Cinclidotus* (1 sp.).

Genera: *Barbula, Semibarbula and Hydrogonium*

The original genus *Barbula* Hedwig shows so many variations that it has been thought proper by Hilpert and Chen to divide it into seven independent genera, viz., *Barbula* Hedw., ***Streblotrichum*** P. Beauv., ***Bellibarbula*** Chen, ***Prionidium*** Hilp., ***Semibarbula*** Herz. ex Hilp., ***Hydrogonium*** (C. Muell.) Jaeg., et Sauerb. and ***Bryoerythrophyllum*** Chen. An eighth genus ***Geheebia*** Schimp. should be added to these. All these genera, excepting *Streblotrichum* and *Prionidium*, are represented in this area but the three genera *Barbula, Semibarbula* and *Hydrogonium* are prominent.

Barbula proper is mainly represented on the Himalayas. 22 species are known here of which *Barbula constricta* (Fig. 31.16A). Mitten is the commonest. *Barbula* is of a very sturdy habit, the leaf is rough because of papillose development on the leaf and midrib cells. The peristome teeth are like long papillose threads which are twisted together and would like a cork-screw.

Hydrogonium is found in the Indo-gangetic plains and also on the Himalayas and other high mountains in this area. *Hydrogonium arcuatum* (Griff.) Wijk. et. Marg. (Fig. 31.16B—Syn. ***Barbula comosa*** and *Barbula gangetica*) is a common species in the Indogangetic plains. It grows on soil. The peristome teeth are as in *Barbula* but the leaf-base shows elongated pollucid cells distinguishing it from *Barbula* and the plants are not so stiff.

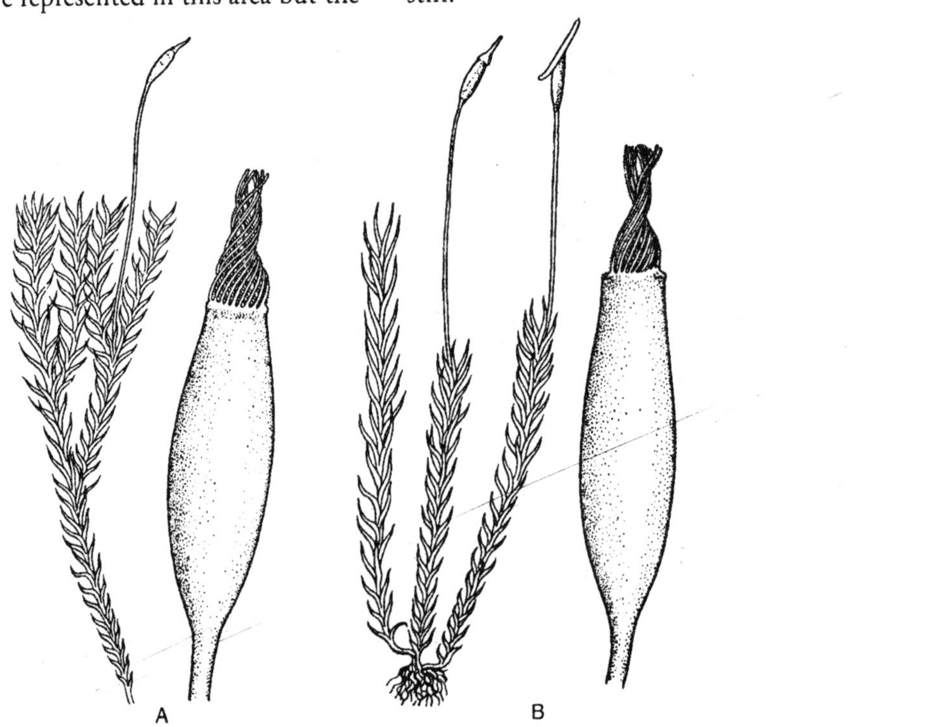

Fig. 31.16: Gametophytic plant with sporophyte and a magnified capsule of A. *Barbula constricta* and B. *Hydrogonium arcuatum* (after Gangulee)

Semibarbula differs from these two other genera in having a reduced peristome represented by a number of short papillose threads, too short to be spirally twisted (Fig. 31.21F). The leaves resemble *Barbula* but the plants are generally shorter. The genus is mainly represented by *Semibarbula orientalis* (Web.) Wiji et Marg. (Syn. *Barbula indica* Brid.) which grows all over India and is perhaps the commonest species of this group of mosses in the Indogangetic plains, growing on old walls, bricks and mortar. Its complete life history has been worked out by Banerji and Sen (under the old name *Barbula indica*) and is being described here.

The gametophytic plant

Semibarbula orientalis is a *calcicolous* (growing on substratum containing lime) moss growing during the rains and maturing in the cooler months all over the tropics in the Eastern hemisphere forming dense, green to yellow-green gregarious patches on old brick walls, mortar, etc., where it may get sufficient moisture and partial shade. The erect shoots are usually unbranched (Fig. 31.17A). The axis bears small, spirally arranges leaves (Fig. 31.17B) and at the base there are smooth-walled, multicellular rhizoids with oblique walls. The rhizoids and the lower part of the stem becomes brown with age. The growth of the stem is by an apical cell with three cutting faces so that the original leaf arrangement is tristichous which arrangement loses clarity in some portions. The leaf shows elongated thin-walled cells at base (Fig. 31.17D) and smaller, coarsely papillose cells (Fig. 31.17C) above. The vein back also is coarsely papillose and, therefore, rough. A transverse section of the stem (Fig. 31.17E) shows a thin-walled *central strand* of vertically elongated cells (*hydroid*) and a *cortex* surrounding it. Outer cortical cells contain chloroplasts when young but become thicker walled and brown when mature. A *t.s.* of leaf (Fig. 31.17F) shows a prominent midrib with a row of larger cells (*deuter cells*) in the middle flanked by thick walled (*stereid*) cells.

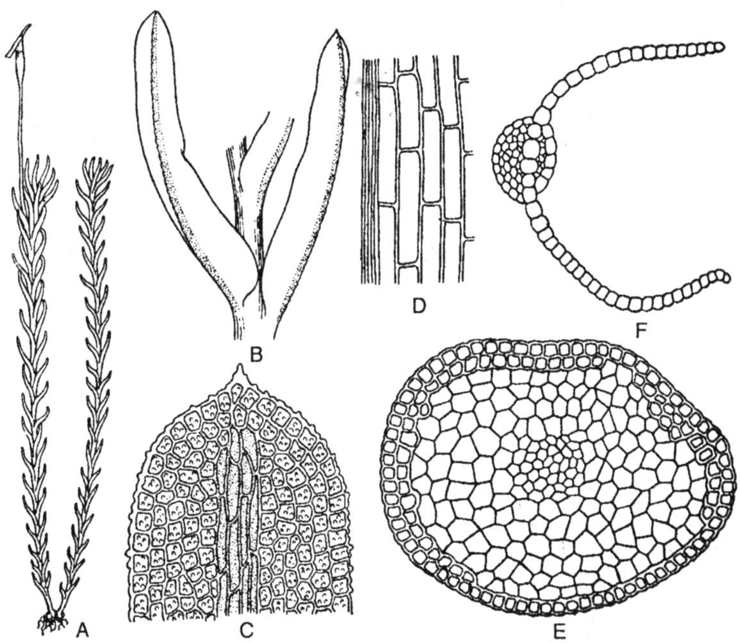

Fig. 31.17: *Semibarbula orientalis* (Syn. *Barbula indica*). A. Female plant. B. Leaves. C. Leaf apex showing papillose cells. D. Leaf base cells. E. T.s. of stem. F. T.s. of leaves (after Gangulee)

Reproduction

A. Vegetative reproduction

Vegetative reproduction is common in *Semibarbula*. It may take place in a number of ways.

(a) New branches develop from the apices (*proliferation* by *innovations*) of older decaying plants and may develop new independent plants.

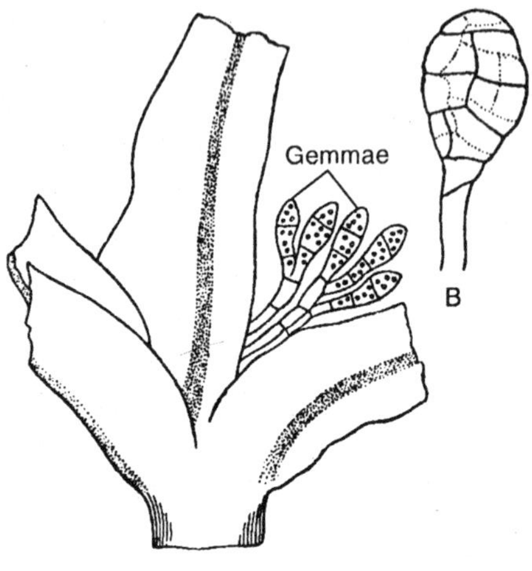

Fig. 31.18: *Semibarbula*. A. Cluster of gemmae in the axil of a leaf. B. A gemma. (After Gangulee)

(b) Protonemal threads are known to develop from different gametophytic parts. These develop buds and new plants in the usual way.

(c) Multicellular *gemmae* or *brood-bodies* develop in clusters in the axils of leaves (Figs. 31.18A and B) as well as tips of shoots. These gemmae get detached and germinate, developing filamentous protonemata which give rise to new plants in the usual way.

B. Sexual reproduction

Semibarbula is dioecious. Male (Fig. 31.19A) and female (Fig. 31.17A) plants usually grow intermixed.

The *antheridia* grow in *antheridial clusters* (Fig. 31.19B), intermixed with long, multicellular, filamentous *paraphyses*, in apparently axillary positions. The cluster is covered by a few short bract leaves and in its growth the apical cell with two cutting faces intermittenly cut off segmental cells some of which act as the antheridial initial (Fig. 31.19C) while other develop the paraphyses. Thus, a short *head* is formed in which the antheridia and paraphyses lie in a bud-like structure surrounded by small leaves. The original apical cell usually grows

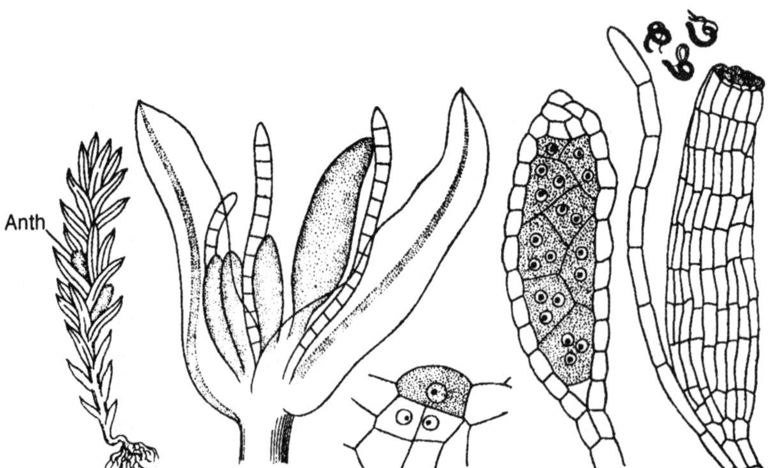

Fig. 31.19: *Semibarbula orientalis*. A. Male plant with antheridial clusters (Anth.). B. An antheridial cluster showing antheridia and paraphyses. C. Antheridial initial cell (shaded). D. A young antheridium. E. A mature antheridium with liberated sperms and a paraphysis. (A-B after Gangulee, C-E after Banerji and Sen)

developing the shoot further so that the antheridial bunch takes up a secondary lateral postion. The antheridial initial cell with two cutting faces develops a short filament which, later, forms a *jacket layer*, the *androcyte mother cells* and few cells at the tip forming the *cover*. The androcyte mother cells at first do not form all the partition walls so that they look multinucleate (Fig. 31.19D), but, ultimately, each androcyte mother cell is partitioned off by a wall and soon divides into two *androcytes* or *sperm cells* each of which develops a single biflagellate, coiled *antherozoid* or *spermatozoid* inside it. The androcytes remain surrounded by a lot of mucilage. The mature antheridium (Fig. 31.19E) is a club-shaped structure with a short stalk.

The *archegonial cluster* (Fig. 31.20A) develops at the tip of the main stem or a branch. Here the archegonia are surrounded by the paraphyses which again are surrounded by leaves (*perichaetial bracts*). The *archegonia* develop as the antheridia but each *archegonial initial* grows by three cutting faces. The original apical cell, usually, does not grow further so that the archegonia are always at the tip. The mature archegonium (Fig. 31.20B) shows a swollen *venter* with multilayered wall containing the large *egg* and a *ventral canal cell*, a long neck containing up to 12 *neck canal cells* in a row surrounded by a *jacket* of one layer and four *cover cells* at the tip.

The sporophyte

Fertilization takes place as in *Pogonatum*. The development of the sporophyte also is, on the whole, like *Pogonatum*. The sporophyte shows the *foot*, *seta* and *capsule* (Fig. 31.21D) as usual. The *calyptra* (Fig. 31.21D) is quite different from that of *Pogonatum*. It is hood-shaped, split and struck to one side to the capsule like a cape.

The *capsule* (Figs. 31.21C and E) has on its top a conical lid-like structure with a long beak called the *operculum*. This operculum is exposed when the calyptra falls off as a result of mechanical jerkings. The operculum is attached to the main capsule by a ring of thin-walled cells called the *annulus*. This is a ring of weakness and the operculum breaks loose at maturity along this line. When the operculum is shed a ring of 32 thread-like papillose teeth

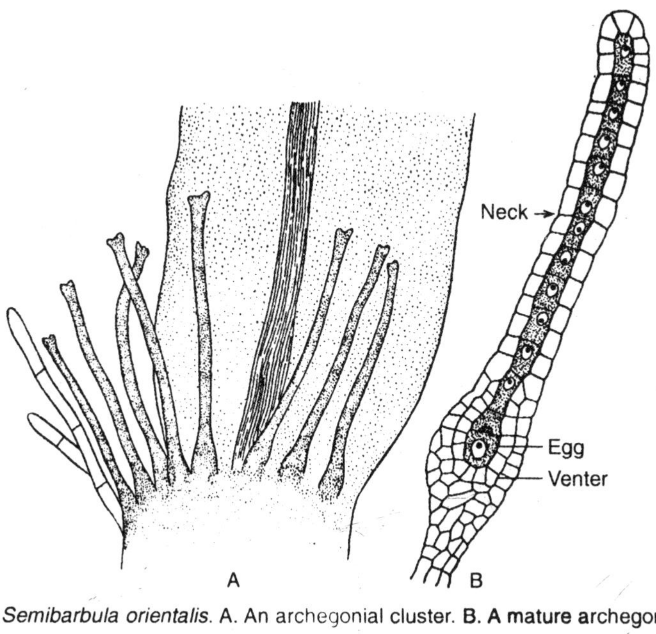

Fig. 31.20: *Semibarbula orientalis*. A. An archegonial cluster. B. A mature archegonium (After Gangulee)

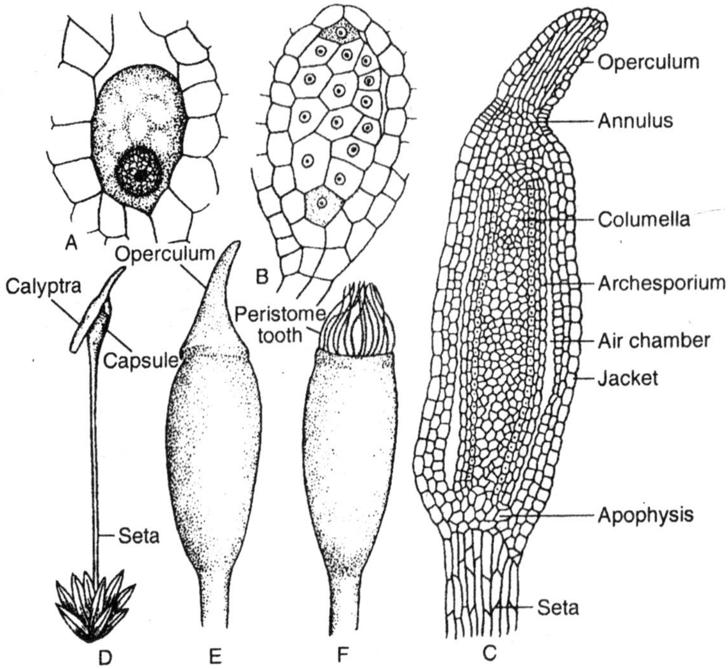

Fig. 31.21: *Semibarbula orientalis*. A. Zygote (shaded). B. Young embryo showing two terminal growing cells (shaded). C. L.s. of mature capsule. D. A sporophyte on the tip of a gametophyte. E. Capsule with operculum exposed. F. Capsule with peristome exposed after shedding of operculum. (A-C after Banerji and Sen)

(Fig. 31.21F) shorter than in the two previous genera, and too short for any coiling, are seen to cover the opening at the tip.

The base of the capsule is called the *apophysis* or the *neck*. The cells of the apophysis are green and show stomata on the surface so that the capsule is not completely helpless from the point of its nutrition. The main body of the capsule shows a central axial *columella* of larger cells, spreading from the apophysis to the operculum and cylindrically surrounded by a single layer of *archesporium* which again is surrounded by two layers of sterile cells. All these inner tissues of the capsule develop out of the *endothecium*. The single layer of *archesporium* then develops 6 to 8 layers of *sporogenous cells*. These cells round up and from the *spore mother cells* which again give rise to the *spores* by reduction division. These endothecial layers are separated from the amphithecial layers by a cylindrical *air chamber*. Outside, the amphithecium forms the *jacket* of the capsule several layers in thickness.

At maturity, the sporogenous region widens forming a cylindrical *spore sac* in which the loose spores lie. The capsule is red-brown at this stage. With the shedding of the operculum, the spore sac becomes exposed at the tip and the spores are dispersed mechanically, their release being controlled by the peristome teeth which open and close the opening by hygroscopic movement.

The new gametophyte

Germination of the spore is generally as in *Pogonatum*. The spores are small, light brown, with a thin *exine* or *exospore* and an *initine* or *endospore* (Fig. 31.9A). It germinates when it falls on a suitable substratum. The endospore comes out through an opening in the exospore in the form of a filament (Fig. 31.9B). Cases are known where several filaments come out of the same spore. The filament forms a multicellular thread and becomes branched. The cells contain numerous discoid chloroplasts and these filaments from the *chloronema.* Later, other thinner pale-green filaments also develop and the *rhizoids* develop on them. This entanglement of filaments is known as the *protonema.* On the chloronema multicellular *buds* develop (Fig. 31.9C). Such a bud ultimately develops a new erect, leafy *gametophytic plant* by a normal apical cell with three cutting faces. The *protonema* is formed of branched filaments.

The life cycle is shown in figure 31.22.

Order 10. GRIMMIALES

Acrocarpous mosses growing on rocks with leaf cells showing sinuose walls. Haplolepideous. Peristome teeth irregularly split (*platycranoid*).

Family: *Grimmiaceae*

Genera: 1. *Coscinodon* (1 sp.), 2. *Schistidium* (2 spp.), 3. *Grimmia* (17 spp.), 4. ***Racomitrium*** (9 spp.) (***Pleurocarpous***).

Series II. **Heterolepideae**

Order 11. ENCALYPTALES

Family: *Encalyptaceae*

Genus: 1. *Encalypta* (5 spp.)

The family with the single genus ***Encalypta*** has proved to be of great interest. The calyptra of this genus is strongly developed, persistent and falls off carrying the operculum. While the genus is uniform in all other characters, the peristome is highly variable in the different species—two rowed, one rowed or missing. There are diplolepideous as well as haplolepideous species. Some species may also be related to the Arthrodonteae. So, Philibert considered this as a synthetic group related to the Arthrodonteae and from which the Haplolepideae and the Diplolepideae had arisen. Cavers placed the family in a separate

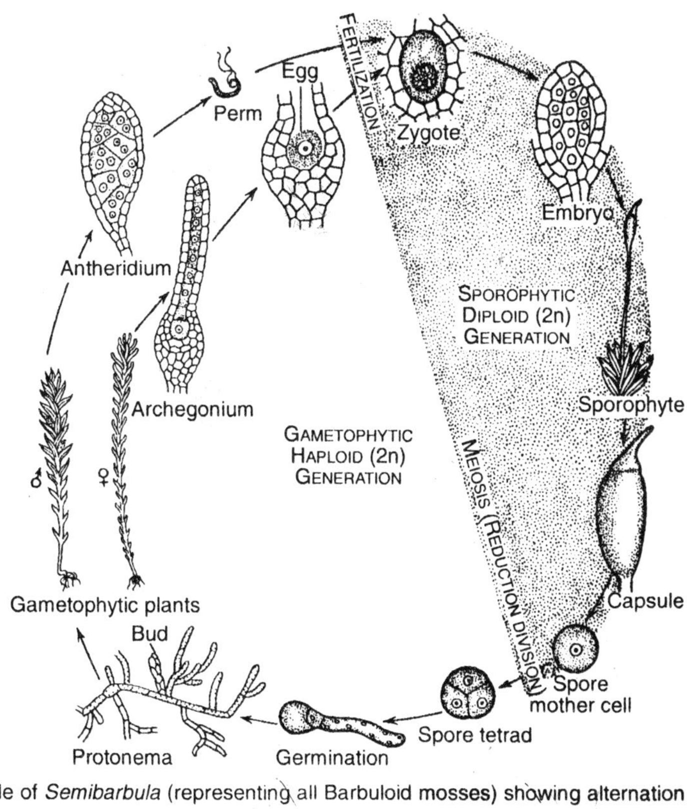

Fig. 31.22: Life cycle of *Semibarbula* (representing all Barbuloid mosses) showing alternation of generation

series in between Haplolepideae and Diplolepideae. Dixon also followed this plan. *Encalypta* is not known from Eastern India but as many as five species are known from the Western Himalaya and Kashmir.

Series III. **Diplolepideae**

In Diplolepideae the teeth are usually in two rows although, in some cases, one row (or even both rows) may be suppressed. The peristome is formed of three concentric layers of cells. The outer peristome (*exostome*) teeth (usually 16 in number) look similar to the haplolepideous teeth but the outer face is formed of 2 rows of plates (hence, diplolepideous) while the inner face is formed of a single row of plates. The inner peristome ring (*endostome*) is formed of very indistinct plates of more or less hyaline cells and is usually united at base (*basal membrane*) and then split into segments above. The segments may sometimes be uniform and completely free so that the endostome is formed of 16 free teeth (c.f. Funariaceae). Those endostome segments may have between them very thin or filamentous segments called *cilia* which latter again may be appendiculate or simple.

The Diplolepideae is divided into two subseries: *Epicranoideae* (exostome teeth overlapping endostome teeth or segments c.f. *Funaria*) and *Metacranoideae* (exostome teeth alternating with endostome segments c.f. *Bryum*).

Subseries I. EPICRANOIDEAE

Order 12. FUNARIALES

Acrocarpous, diplolepideous mosses. Capsule not cylindrical and the operculum without beak. Peristome teeth usually in two epicranoid rings but may be in one ring or absent by reduction.

Family: *Ephemeraceae*

Genus: *Micromitrium* (= *Nanomitrium*) (1 sp.)

Family: *Funariaceae*

Genera: 1. *Physcomitrium* (15 spp.), 2. *Entosthedon* (16 spp.), 3. *Funaria* (7 spp.).

Family: *Splachnaceae*

Genera: 1. *Gymnostomiella* (1 sp.), 2. *Splachnobryum* (5 spp.), 3. *Voitia* (1 sp.), 4. *Tayloria* (8 spp.), 5. *Tetraplodon* (2 spp.)

FUNARIA HYGROMETRICA

Funaria hygrometrica Hedwig is a cosmopolitan moss known from all over the world and from all the major hills of India. It is a very common Himalayan moss thriving best in the temperate heights. *Funaria leptopoda* Griffith (with the capsules more erect) found in the Himalayas, though reported as a separate species, should be considered as a variety of *Funaria hygrometrica.* It is commonly found on semiexposed rocky ground, walls, etc. It is often found as the first vegetation on burnt up land in these areas, specially where ash is present.

The gametophytic plant

The erect gametophytic plant (Fig. 31.23A) is about an inch in height with an erect radial stem (growing by an apical cell with three cutting faces) and spirally arranged simple leaves. The leaves are more crowded near the apex where they appear like a rosette though actually arranged in three rows corresponding to the three cutting faces when young. This arrangement is lost in the mature parts. Branches arise on the main stem from extra-axillary positions. Leaves (Fig. 31.23C) are sessile, attached to the stem by a broad base, ovate elongate, pointed at the apex and with a smooth margin. All normal mature leaves show midribs but very young leaves and the sparsely arranged small leaves at the base of the stem are devoid of midribs. The gametophytic plant bears at its base a strong, much branched rhizoid system which becomes brown and cable-like when mature.

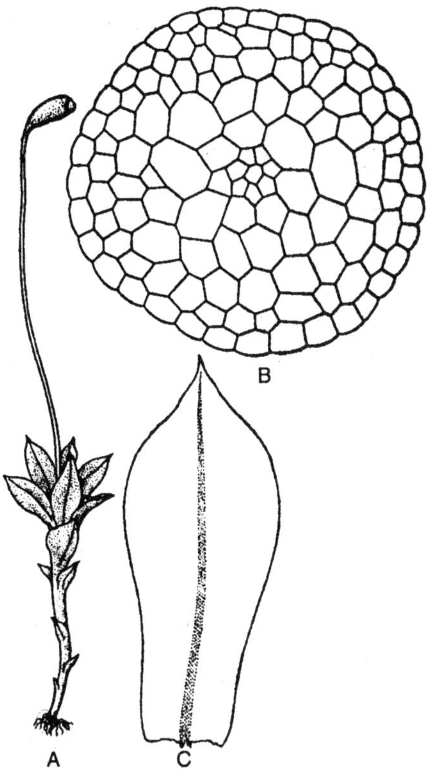

Fig. 31.23: *Funaria hygrometrica.* A. Gametophyte plant with a mature sporophyte. B. T.s. of gametophyte stem. C. A leaf

A t.s. of the stem (Fig. 31.23B) shows a small *central strand* of long, narrow, colourless, thin-walled cells devoid of protoplasm. This is surrounded by a *cortex* of larger cells which contain chloroplasts when young. In a mature plant this chlorophyll is lost and the outer cells become somewhat thick-walled and reddish-brown. The cortex also shows, here and there, some smaller cells or cell groups which are termed *leaf traces* although they end blindly and do not reach the central cylinder. The outermost layer of the stem is the *epidermis* and this also contains chloroplastids. There is no stomata. The leaf, in cross-section, shows a well-defined *midrib* while the *lamina* is only one cell in thickness. The midrib has a small central strand of narrow, thin-walled cells surrounded by a sheath of narrower cells with thicker walls. Lamina cells are elongated, lax, thin-walled; rectangular or rhomboidal and contain chloroplasts which divide even in mature cells.

Reproduction

The plants may propagate *vegetatively* by branching and being cut off from the mother plant by decay and also by the development of secondary protonemata as in *Semibarbula.* The secondary protonema develop from anywhere and are diploid or haploid according to the tissue from which they are developing. Asexual regeneration by *gemmae* (Fig. 31.24) is also common. These commonly develop on the rhizoidal branches of the protonema and are multicellular, brown, bud-like structures. The gemmae are useful for propagation during unfavourable periods. The normal method of reproduction, however, is sexual. *Funaria* is *monoecious* but *autoicous,* i.e., the antheridia and archegonia develop on separate schoots of the same plant. The main shoot of the plant bears the antheridia. The female branch develops as a side shoot but, it is more vigorous and soon becomes higher than the main shoot.

In the *antheridial shoot* (Fig. 31.29D, E), a number of bract-like leaves from a rosette called the *perichaetium.* This perichaetium surrounds a number of club-shaped, stalked *antheridia* and multicellular hairs known as *paraphyses.* The paraphyses in *Funaria* have swollen tips. The *archegonia* also are borne in clusters on the *archegonial shoot* (Fig. 31.28C) but no Perichaetium is differentiated here. The archegonia and the manner of development of both archegonia and antheridia generally resemble that of *Semibarbula. Fertilization* takes place in the usual way.

The sporophyte

The development of the sporophyte is as in *Semibarbula* but the mature capsule has a different appearance. The mature sporophyte shows a poorly developed concial *foot* embedded in the apex of the archegonial branch; a long *seta*

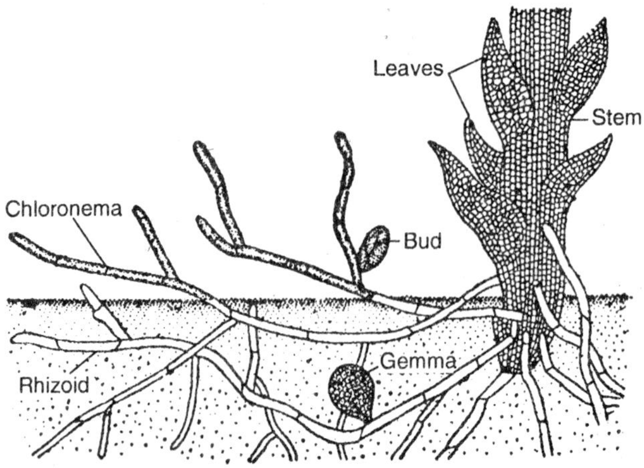

Fig. 31.24: Gemma on gametophyte base

which is reddish-brown in colour and often twisted; and a pear-shaped, asymmetrical, bent, often bright orange-coloured *capsule* at the tip. The tissues of the seta (Fig. 31.25) show more thick-walled cells than in the gametophytic stem.

The mature pear-shaped, nodding capsule (Figs. 31.26A and l.s. in 31.26B) is asymmetrical. The lowermost part of it is the *apophysis* or the neck which connects it with the seta below. The axis of the apophysis shows in the lower part a central strand of thin-walled elongated cells connected with the similar tissue of the seta below. Above, the axial strand disappears into the columella. This axis develops out of the *endothecium* and it is surrounded by a green spongy tissue formed out of the *amphithecium.* This *spongy tissue* contains abundant chloroplasts and air gaps like the photosynthetic tissue of leaves and serves the same function. The spongy tissue is surrounded by one compact layer of *epidermis* regularly provided with *stomata.* The stomata (Figs. 31.26C and D) develop as in the higher plants but, when mature, often the two guard cells combine to form a ring-like single annular guard cell (Fig. 31.26D). The stomata are regularly connected with air spaces.

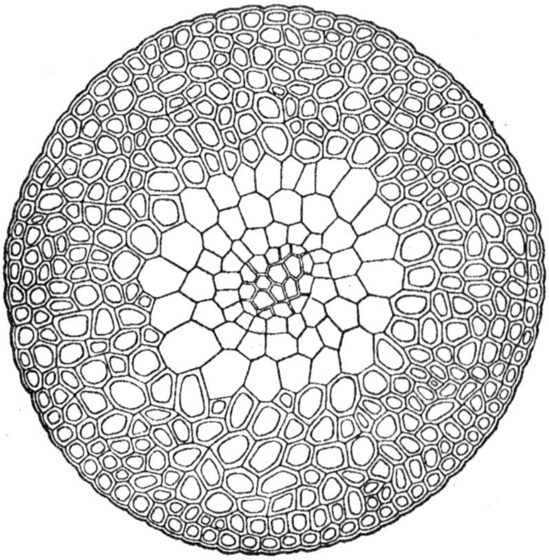

Fig. 31.25: T.s of *Funaria hygrometrica* seta

Above the apophysis, the main part of the capsule is a slightly bent, cylindrical structure with the central *columella* of thin-walled parenchymatous cells which is constricted at the base just above the apophysis. The columella is surrounded by the *archesporium* (one-layered) which latter is the *sporogenous layer,* becoming a *spore sac* when *spores* are formed by reduction division of the *spore mother cells* organized out of the archesporium. The sporogenous layer has an *inner wall* of one layer of small cells and an *outer wall* of 3 to 4 layers of such cells. The columella, the inner wall and the spore sac are developed out of the *endothecium* while the outer wall of the spore sac and the tissues

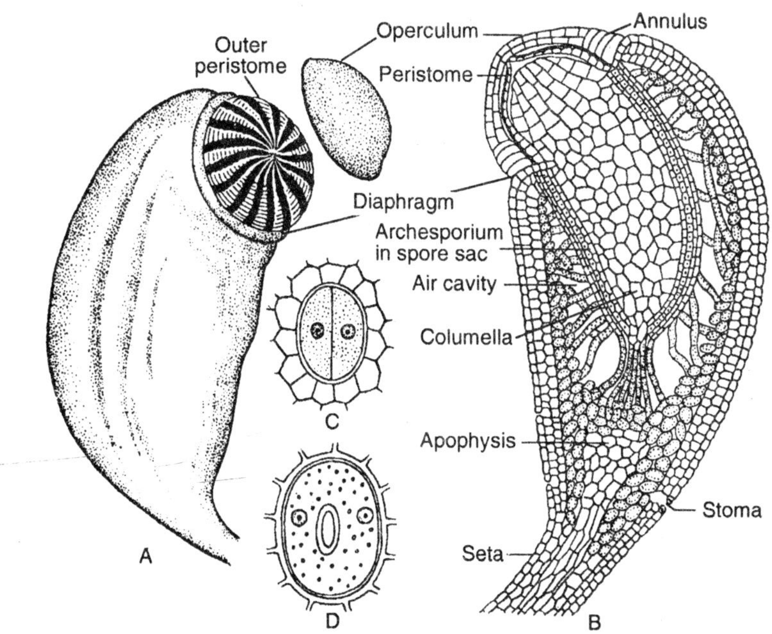

Fig. 31.26: *Funaria hygrometrica.* A. Mature capsule with operculum removed (shown on side). B. L.s. of capsule. C. Young stomata. D. Mature stomata

surrounding it are formed out of the *amphithecium.* The outer wall of the spore sac is surrounded by a big cylindrical *air cavity* traversed by strings of filaments of elongated green cells which form the *trabeculae* bridging the air space. Outside this air space is the *wall of the capsule* formed of 2 or 3 layers of compact, colourless, parenchymatous cells bounded down by the outermost compact *epidermis* of one layer, which is devoid of stomata, may be green when young but becomes bright-brown or orange when mature.

The tip of the capsule has a complicated structure. It is not exactly on the apex but on one side as the capsule is asymmetric (Fig. 31.26A and B). The columella extends into this part but the cells are a bit larger here. This is the only tissue out of the *endothecium* in this region—all the outer tissues being formed out of the *amphithecium.* There is a notch where this top part joins the capsule proper. At this notch (Fig. 31.26A) there is an annular *rim* (or *diaphragm*) of small cells edging the mouth of the capsule.

Above the rim is another ring of slightly extended cells called the *annulus.* The three innermost layers of the amphithecium forms two sets of *peristome teeth*—in the *outer circle* is one ring of 16 larger brown peristome teeth ornamented with transverse bars—and in the *inner circle* are another 16 peristome teeth which are thinner and of a paler colour (Figs. 31.27A and 31.27B). The two rings of peristome teeth are the outer and inner rings called the *exostome* and the *endostome* respectively. The whole structure is the *peristome.* The outer peristome teeth are superposed on the inner (*epicranoid*). The outermost cells of this capsule apex form a dome-shaped *operculum* (Figs. 31.26A and B) which keeps the teeth covered.

At maturity, the distended cell walls of the annulus break and the loosened operculum is shed leaving the peristome teeth exposed (Fig. 31.26A). The peristome teeth are twisted spirally appearing somewhat like an iris diaphragm and the tips of the outer peristome teeth are attached to a small circle of a few small cells. This attachment soon disintegrates

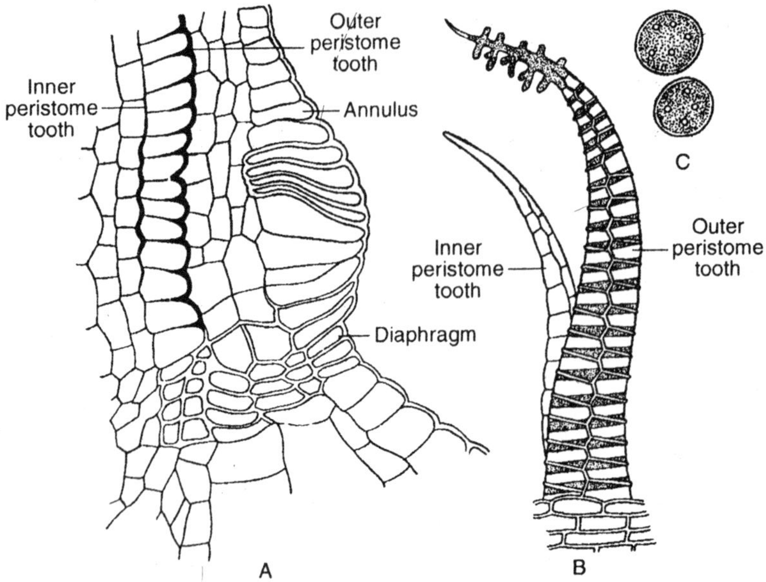

Fig. 31.27: *Funaria hygrometrica*. A. L.s. of capsule at diaphragm-annulus region (after Sacs). B. Diplolepideous epicranoid peristome teeth (an inner and an outer peristome tooth shown). C. Spores

and the outer peristome teeth, which are hygroscopic, show movements according with the presence or absence of moisture—thus helping the dissemination of spores.

The *calyptra* develops out of the old archegonial ventral wall as in other Bryophytes. It is found attached to the top of the young capsule like a hood.

The new gametophyte

The small round spores (Fig. 31.27C) containing chloroplasts germinate and develop

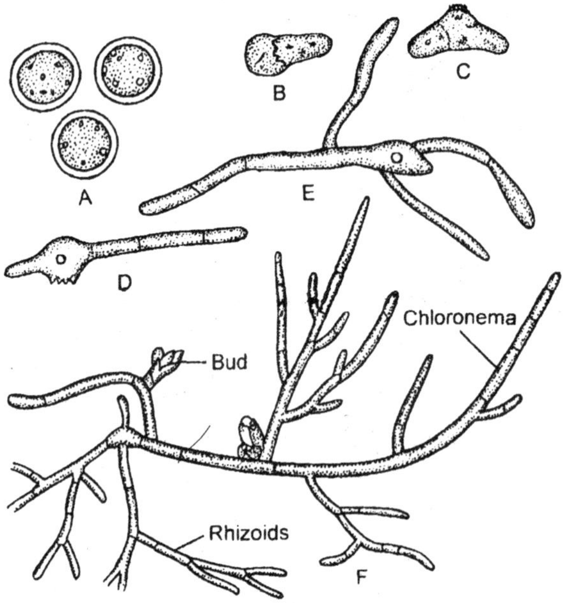

Fig. 31.28: *Funaria*. Germination of the spore and development of the Protonema

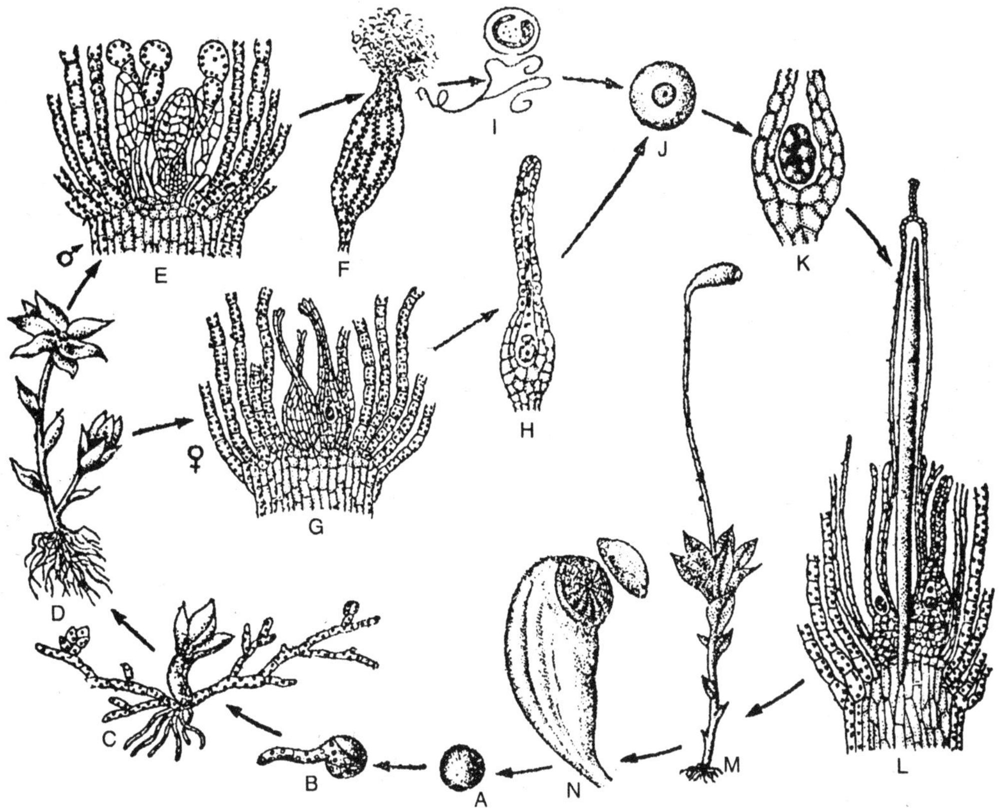

Fig. 31.29: The life cycle of *Funaria hygrometrica*

a *protonema* (Fig. 31.28) composed of a *chloronema* and *rhizoidal branches* as in other mosses. Detailed observation of this protonema shows that the first formed *chloronema* soon organizes a more mature type of protonemal threads which have been named *caulonema* by Sironval (1947). Buds giving rise to new gametophytic plants develop on this *caulonema.* Others (viz. Alsopp and Mitra, 1958) did not note any distinct *caulonema*, the rhizoidal branches developing from the base of the chloronemal threads. The partition walls are at right angles in the green chloronema threads which are full of chloroplastids and are lateral in the thinner, brownish rhizoidal branches.

Subseries 2. METACRANOIDEAE

Order 13. EUBRYALES

Both acrocarpic and pleurocarpic mosses are included in this large order but the acrocarpic forms predominate and only these are represented in India. The peristome is diplolepideous, but it differs from the Funariales in the peristome rings by being *metacranoid.*

Family: *Bryaceae*

Genera: 1. *Orthodontium* (2 spp. from Ceylon), 2. *Mielichhoferia* (9 spp.), 3. *Pohlia* (includes *Webera*) (21 spp.), 4. *Mniobryum* (3 spp.), 5. *Epipterygium* (2 spp.), 6. *Brachymenium* (29 spp.), 7. *Anomobryum* (13 spp.), 8. *Leptobryum* (1 sp.), 9. *Plagiobryum* (1 sp.), 10. *Bryum* (54 spp.—Fig. 31.30), 11. *Rhodobryum* (4 spp.).

Genus: *BRYUM*

Bryum is a very large genus with about 700 species distributed throughout the world. This genus has given the name to the order and the higher taxa. The species vary much from one

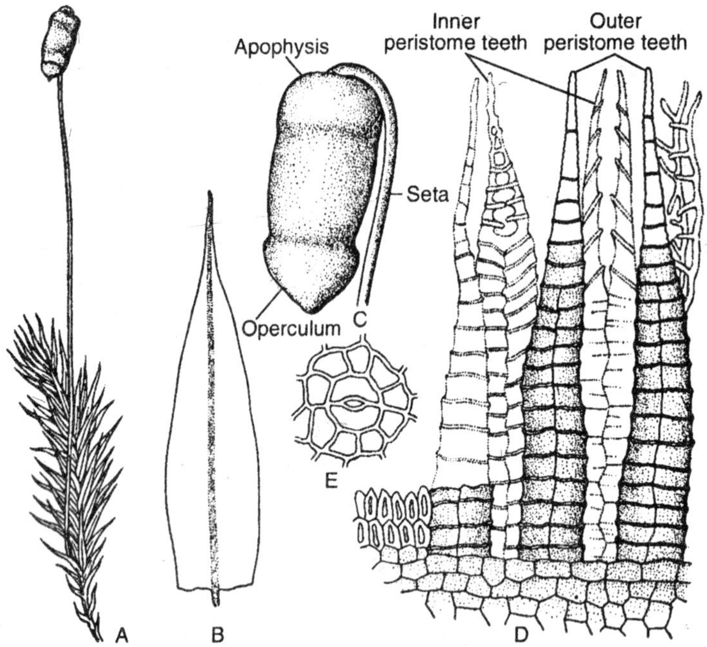

Fig. 31.30: *Bryum coronatum*. A. Gametophytic plant with sporophyte. B. A Leaf. C. A capsule. D. Peristome showing metacranoid teeth (note cilia). E. Stomata on apophysis

another but **are constant** in leaf areolation and specially in the *bryoid* peristome.

More than 50 species of the genus are known in India. The commonest species in the plains is *Bryum coronatum* shown in Figure 31.30 which clearly shows the plant habit.

The sporophytes of *Bryum coronatum* are orange in colour and quite beautiful to look as the plants grow on soil or walls in large patches. The capsule has a swollen apophysis (Fig. 31.30C) and the peristome (Fig. 31.30D) is metacranoid with two rings of teeth of equal size. The outer ring is deeply coloured while the inner ring is hyaline and cilia are present.

The life history is an in other mosses.

Family: *Leptostomaceae*

Genus: 1. *Leptostomum* (1 sp. from Ceylon)

Family: *Mniaceae*

Genera: 1. *Orthomnium* (1 sp.), 2. *Orthomniopsis* (1 sp.), 3. *Mnium* (19 spp.)

Fig. 31.31: *Bartramidula* gametophytic plant with sporophyte from Darjeeling

Family: *Drepanophyllaceae*

Genus: 1. *Mniomalia* (1 sp. from Ceylon).

Family: *Rhizogoniaceae*

Genus: 1. *Rhizogonium* (1 sp.)

Family: *Hypnodendraceae*

Genera: 1. *Hypnodendron* (1 sp. from Ceylon), 2. *Mniodendron* (1 sp. from Ceylon)

Family: *Aulacomniaceae*

Genus: *Aulacomnium* (2 spp.)

Family: *Meeseaceae*

Genus: 1. *Meesea* (2 spp.)

Family: *Bartramiaceae*

Genera: 1. *Plagiopus* (1 sp.), 2. *Anacolia* (3 spp.), 3. *Bartramia* (8 spp.), 4. *Bartramidula* (3 spp.), 5. *Philonotis* (18 spp.), 6. *Breutelia* (2 spp.), 7. *Fleischerobryum* (2 spp.).

Family: *Timmiaceae*

Genus: *Timmia* (3 spp.)

Most of the mosses of the pleurocarpic families that follow are long plants creeping along tree barks, rocks, etc., and are often seen hanging down from trees in tropical, subtropical and temperate rain forests.

Order 14. ISOBRYALES

Pleurocarpic (rarely acrocarpic), diplolepideous, metacranoid mosses but the inner peristome ring (*endostome*) usually not developed.

The pleurocarpous mosses are the most modern Bryophytes. They form a special feature of the tropical-subtropical rain forests. They form a thick-mat, creeping along (often covering metres) bark of tall trees and often hang down from the boughs (Fig. 31.32) giving a sombre aspect to the semidark interior of the thick forests.

Family: *Erpodiaceae*

Genera: 1. *Erpodium* (1 sp.), 2. *Aulacopilum* (3 spp.), 3. *Solmsiella* (1 sp. from Ceylon).

Family: *Ptychomitriaceae*

Genera: 1. *Ptychomitrium* (3 spp.), *Glyphomitrium* (1 sp. from Ceylon).

Family: *Orthotrichaceae*

Genera: 1. *Amphidium* (2 spp.), 2. *Zygodon* (11 spp.), 3. *Rhachithecium* (1 sp.), 4. *Orthotrichum* (18 spp.), 5. *Stroemia* (1 sp.), 6. *Ulota* (2 spp.), 7. *Drummondia* (3 spp.), 8. *Macromitrium* (39 spp.), 9. *Groutiella* (1 sp.), 10. *Schlotheimia* (2 sp.), 11. *Desmotheca* (1 sp.), 12. *Trigonodictyon* (1 sp.).

Family: *Racopilaceae*

Genus: 1. *Racopilum* (3 spp.)

Family: *Climaciaceae*

Genus: 1. *Climacium* (1 sp.)

Family: *Hedwigiaceae*

Genera: 1. *Hedwigia* (1 sp.), 2. *Hedwigidium* (1 sp.), 3. *Braunia* (4 spp.), 4. *Bryowijkia* (1 sp.).

Family: *Cryphaeaceae*

Genera: 1. *Schonobryum* (2 spp.), 2. *Sphaerotheciella* (1 sp.), 3. *Pilotrichopsis* (2 spp.), 4. *Forsstroemia* (4 spp.).

Family: *Leucondontaceae*

Genus: 1. *Leucodon* (3 spp.)

Family: *Ptychomoniaceae*

Genus: 1. *Glyptothecium* (1 sp. from Ceylon).

Family: *Trachypodaceae*

Genera: 1. *Diaphanodon* (2 spp.), 2. *Trachypus* (3 spp.), 3. *Pseudospiridontopsis* (1 sp.), 4. *Trachypodopsis* (2 spp.), 5. *Duthiella* (4 spp.)

Family: *Myuriaceae*

Genus: 1. *Myarium* (1 sp.).

Family: *Pterobryaceae*

Genera: 1. *Trachyloma* (2 spp.), 2. *Penzigiella* (2 spp.), 3. *Osterwaldiella* (1 sp.),

4. *Endotrichella* (2 spp.), 5. *Garovaglia* (6 spp.), 6. *Jaegerina* (1 sp.), 7. *Pterobryopsis* (21 spp.), 8. *Symphysodontella* (5 spp.), 9. *Horikawaea* (1 sp.)

Family: *Meteoriaceae* (Fig. 31.32)

Genera: 1. *Papillaria* (4 spp.), 2. *Meteorium* (4 spp.), 3. *Meteoriella* (1 sp.), 4. *Aerobryopsis* (3 spp.), 5. *Aerobryidium* (1 sp.), 6. *Barbella* (13 spp.), 7. *Floribundaria* (10 spp.), 8. *Pseudobarbella* (4 spp.), 9. *Chrysocladium* (6 spp.), 10. *Meteoriella* (1 sp.), 11. *Meteoriopsis* (4 spp.), 12. *Aerobryum* (1 sp.).

Family: *Phyllogoniaceae*

Genus: 1. *Orthorrhynchium* (1 sp. from Ceylon).

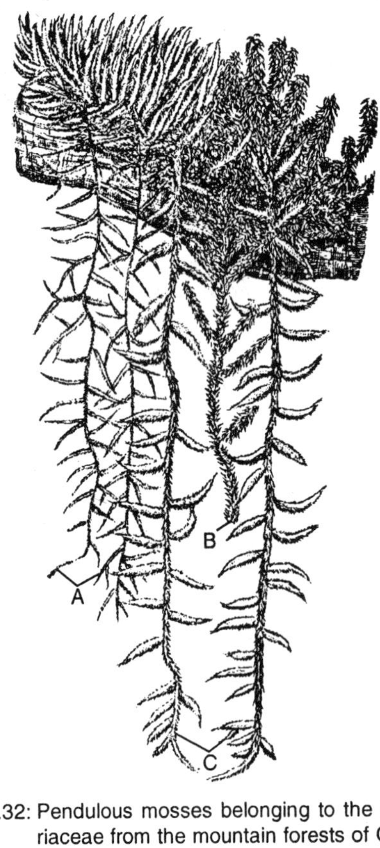

Fig. 31.32: Pendulous mosses belonging to the Meteoriaceae from the mountain forests of Ceylon: A. *Papillaria semitorta*, B. *Meteoriopsis reclinata*, C. *Meteorium miquelianum*. (After Geheeb and Herzog)

Family: *Neckeraceae*

Genera: 1. *Cryptoleptodon* (3 spp.), 2. *Calyptothecium* (13 spp.), 3. *Neckera* (10 spp.), 4. *Neckeropsis* (8 spp.), 5. *Himantocladium* (5 spp.), 6. *Homaliodendron* (13 spp.), 7. *Homaliadelphus* (1 sp.), 8. *Handeliobryum* (1 sp.), 9. *Pinnatella* (10 spp.), 10. *Porothamnium* (1 sp. from Ceylon), 11. *Thamnobryum* (7 spp.).

Family: *Lembophyllaceae*

Genera: 1. *Isothecium* (3 spp.). 2. *Isotheciopsis* (2 spp.), 3. *Dixonia* (1 sp.).

Order 15. HOOKERIALES

Soft, pleurocarpous, diplolepideous, metacranoid mosses from tropical, subtropical and temperate rain forests.

Family: *Hookeriaceae*

Genera: 1. *Daltonia* (15 spp.), 2. *Distichophyllum* (15 spp.), 3. *Eriopus* (4 spp.), 4. *Hookeria* (1 sp.), 5. *Callicostela* (2 spp.), 6. *Hookeriopsis* (3 spp.), 7. *Lepidopilidium* (1 sp.), 8. *Actinodontium* (2 spp.) 9. *Pseudohypnella* (1 sp. from Ceylon), 10. *Chaetomitrium* (4 spp.), 11. *Chaetomitriopsis* (1 sp.), 12. *Orontobryum* (2 spp.).

Family: *Symphyodontaceae*

Genera: 1. *Symphyodon* (11 spp.).

Family: *Leucomiaceae*

Genus: 1. *Leucomium* (2 spp.).

Family: *Hypopterygiaceae*

Genera: 1. *Lopidium* (2 spp. from Ceylon), 2. *Hypopterygium* (5 spp.), 3. *Cyathophorella* (7 spp.), 4. *Dendrocyathophorum* (1 sp.).

The Hyprobryales form a climax of moss evolution. The mats of pleurocarpic mosses specially abound in South-Eastern Asia (Indian and Pacific Ocean regions) pointing out the final craddle of their evolution.

Order 16. HYPNOBRYALES

Pleucocarpous, diplolepideous mosses with metacranoid Bryoid type peristome.

Family: *Theliaceae*

Genus: 1. *Myurella* (2 spp.).

Family: *Fabroniaceae*

Genera: 1. *Fabronia* (12 spp.), 2. *Anacamptodon* (1 sp.), 3. *Juratzkea* (1 sp.), 4. *Schwetschkea* (2 spp.), 5. *Schwetschkeopsis* (2 spp.).

Family: *Leskeaceae*

Genera: 1. *Regmatodon* (4 spp.), 2. *Lindbergia* (3 spp.), 3. *Leskea* (3 spp.), 4. *Leskeella* (1 sp.), 5. *Lescuraea* (5 spp.), 6. *Pseudoleskeopsis* (1 sp.), 7. *Okamuraea* (1 sp.).

Family: *Thuidiaceae*

Genera: 1. *Leptopterygynandrum* (3 spp.), 2. *Haplohymenium* (2 spp.), 3. *Anomodon* (4 spp.), 4. *Herpetineuron* (1 sp.), 5. *Claopodium* (3 spp.), 6. *Haplocladium* (5 spp.), 7. *Thuidium* (20 spp.), 8. *Abietinellaa* (2 spp.), 9. *Actinothuidium* (2 spp.), 10. *Pelekium* (2 spp.).

Family: *Amblystegiaceae*

Genera: 1. *Cratoneuron* (3 spp.), 2. *Sciaromium* (1 sp.), 3. *Campylium* (8 spp.), 4. *Leptodictyum* (1 sp.), 5. *Hygroamblystegium* (3 spp.), 6. *Amblystegium* (4 spp.), 7. *Amblystegiella* (1 sp.), 8. *Drepanocladus* (5 spp.), 9. *Hygrohypnum* (3 spp.), 10. *Platyhypnidrium* (2 spp.), 11. *Calliergon* (3 spp.), 12. *Ortholimnobium* (1 sp.), 13. *Calliergonella* (1 sp.).

Family: *Brachytheciaceae*

Genera: 1. *Homalothecium* (7 spp.), 2. *Brachythecium* (37 spp.), 3. *Bryhnia* (3 spp.), 4. *Rhynchostegium* (11 spp.), 5. *Rhynchostegiella* (10 spp.), 6. *Eurrhynchium* (10 spp.).

Family: *Entodontaceae*

Genera: 1. *Erythrodontium* (1 sp.), 2. *Pterigynandrum* (2 spp.), 3. *Trachyphyllum* (4 spp.), 4. *Campylodontium* (2 spp.), 5. *Orthothecium* (1 sp.), 6. *Rozea* (3 spp.), 7. *Entodon* (19 spp.), 8. *Levierella* (1 sp.), 9. *Nanothecium* (1 sp.), 10. *Pleurozium* (1 sp.).

Family: *Plagiotheciaceae*

Genera: 1. *Stereophyllum* (13 spp.), 2. *Plagiothecium* (11 spp.).

Family: *Sematophyllaceae*

Genera: 1. *Aptychella* (5 spp.), 2. *Clastobryum* (5 spp.), 3. *Clastobryopsis* (2 spp.), 4. *Chionostomum* (1 sp.), 5. *Hageniella* (3 spp.), 6. *Struckia* (1 sp.), 7. *Gammiella* (1 sp.), 8. *Pylaisiopsis* (1 sp.), 9. *Heterophyllium* (2 spp.), 10. *Wijkia* (8 spp.), 11. *Trismegeistia* (1 sp.), 12. *Meiothecium* (4 spp.), 13. *Pylaisiadelpha* (1 sp.), 14. *Brotherella* (12 spp.), 15. *Foreauella* (1 sp.), 16. *Rhaphidorrhynchum* (2 spp.), 17. *Warburgiella* (5 spp.), 18. *Sematophyllum* (13 spp.), 19. *Acroporium* (10 spp.), 20. *Trichosteleum* (8 spp.), 21. *Acanthorrhynchium* (1 sp.), 22. *Taxithelium* (7 spp.– Fig. 31.33), 23. *Clossadelphus* (7 spp.), 24. *Macrohymenium* (2 spp. from Ceylon), 25. *Trolliella* (1 sp.)

Family: *Hypnaceae*

Genera: 1. *Bryosedgwickia* (2 spp.), 2. *Platygyrium* (3 spp.), 3. *Pylaisia* (3 spp.), 4. *Hypnum* (16 spp.), 5. *Ectropothecium* (19 spp.), 6. *Trachythecium* (1 sp. from Ceylon), 7. *Isopterygium* (17 spp.), 8. *Isopterygiopsis* (1 sp.), 9. *Taxiphyllum* (3 spp.), 10. *Vesicularia* (13 spp.), 11. *Dolichotheca* (1 sp.), 12. *Ctenidium* (3 spp.), 13. *Ptilium* (1 sp.), 14. *Neodolichomitra* (1 sp.), 15. *Leiodontium* (2 spp.).

Family: *Rhytidiaceae*

Genera: 1. *Rhytidium* (1 sp.), 2. *Rhytidiadelphus* (1 sp.), 3. *Gollania* (3 spp.).

Family: *Hylocomiaceae*

Genera: 1. *Hylocomium* (3 spp.), 2. *Macrothamnium* (4 spp.), 3. *Leptohymenium* (3 spp.), 4. *Stenotheciopsis* (1 sp.), 5. *Macrothamniella* (1 sp.).

Taxithelium Nepalense

Taxithelium is a tropical and subtropical genus belonging to the family *Sematophyllaceae* and occurring on tree barks in wet forests. It occurs extensively in India, Burma, Ceylon and the islands spreading to Australia. It is also known in Africa and Central America. 9 species are known in India of which *Taxithelium nepalense* (Schwae-grichen) Brotherus is found extensively on tree trunks in tropical and subtropical regions. It is briefly described here as an example of hypnoid (i.e., belonging to the *Hypnobryales*) pleurocarpous moss.

The *gametophytic plant* grows during the rains on tree trunks. The stems are creeping, branching profusely and pinnately (Fig. 31.33A), forming a dense yellowish-green mat on the tree. The small leaves are deeply concave, broadly ovate and bluntly acute, uniformly and densely surrounding the branches (Fig. 31.33B) so that the branches do not appear flat. The leaves do not show any vein and are one cell thick. The leaf cells (Fig. 31.33C) are papillose, elongate rhomboidal except at base on two sides where there are two patches called *alars* (Fig. 31.33D) each composed of a few quadrate cells.

The plants are monoecious but the *antheridia* and *archegonia* are borne on separate small side shoots. Such a plant is called *autoicous.* The *sporophytes* develop during the winter and occur in large numbers on each plant. There is a small *foot*, a long (up to 1 inch) *seta* and an oval, asymmetrically bent (almost horizontal) *capsule.* The operculum (Fig. 31.33E) is short conical. When the operculum is shed, the *peristome* is exposed and *two metacranoid rings* of *peristome teeth* are seen. The peristome teeth curl out exposing the spores when dry (Fig. 31.33F). The spores are soon discharged by the hygroscopic movement of these peristome teeth. Each spore germinates forming a filamentous *protonema* in which the new gametophytic plants develop in the usual way.

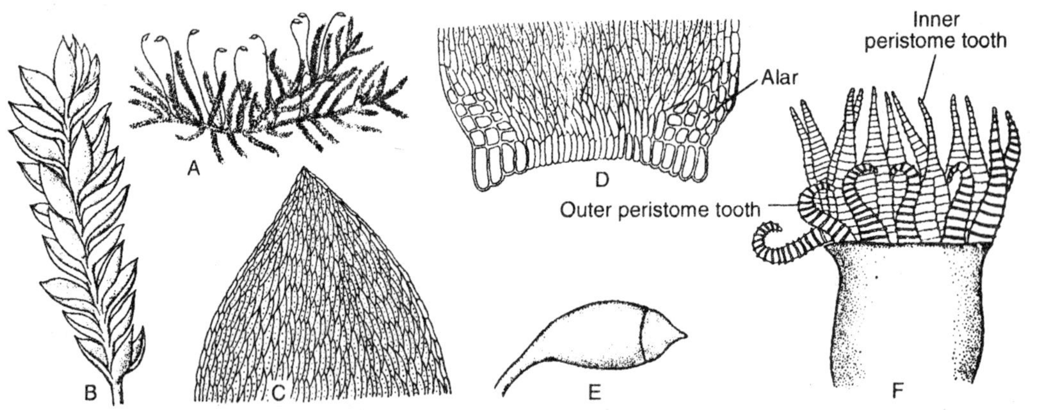

Fig. 31.33: *Taxithelium nepalense*—Common epiphyte in Eastern India plains. A. Gametophyte plant with sporophytes (Pleurocarpous). B. Bit of twig. C. Leaf apex. D. Leaf base. E. Capsule with operculum. F. Diplolepideous peristome showing inner and outer rings of teeth (After Ganguilee)

Chapter 32

Evolution and Affinities of the Bryophytes

The problem of the origin of the Bryophytes, i.e., their evolution out of other groups has not yet been solved satisfactorily, mainly due to the lack of fossil evidences. Possibly, because of the delicate nature of Bryophytic plant bodies which are devoid of lignified or cutinized tissues, the earlier Bryophytes are not known in the fossil stage. The few fossil Bryophytes that are known are discussed below.

FOSSIL BRYOPHYTES

Hepaticopsida

Fossils of the *Hepaticopsida* were previously not known before the Carboniferous while the lower vascular plants (the Pteridophytes) were known from a much earlier age. This fact, naturally, gave rise to speculations. Recently however, Hueber (1961) has discovered *Hepaticites devonicus* (Fig. 32.1A) from Devonian rocks in New York State. This is a thallose Jungermanniale resembling *Pallavicinia* and definitely belongs to the Metzgerineae. *Upper Carboniferous* rocks show the presence of the genus *Hepaticites* side by side with the moss genus *Muscites*. Five species of *Hepaticites* (foliose and thallose Jungermanniales) are described by Walton (1925, 1928) from this age, viz., *H. willsi, H. langi, H. lobatus, H. kidstoni* (Fig. 32.1B) and *H. metzgerioides*. These are known

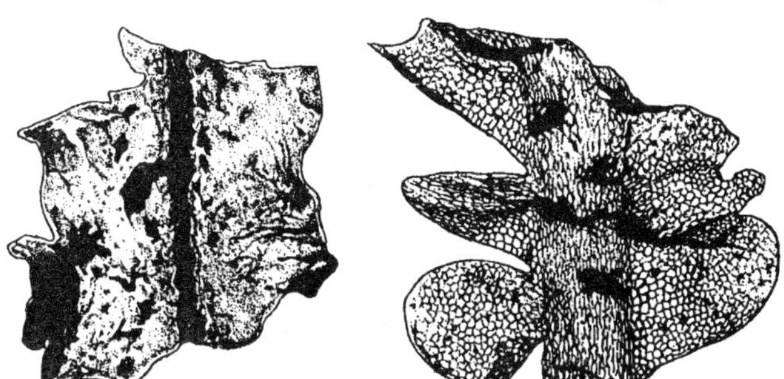

Fig. 32.1: Fossil Hepaticopsida. A. *Hepaticites devonicus* from Devonian (after Hueber). B. *Hepaticites kidstoni* from Upper Carboniferous (after Walton)

only in the vegetative stage. Of these, *H. kidstoni* is a foliose Jungermanniale with a central axis and four rows of leaves of which two rows are larger and two smaller. The other species resemble the anacrogynous Jungermanniale genera, *Treubia, Aneura* (*Riccardia*) and *Metzgeria*. Some *Hepaticites* are also known from the *Permian*.

During *Mesozoic*, more hepatics are known. Nine more species of *Hepaticites* (*amaures, selenotus, laevis, rosenkrantzii, glebosus, arcuatus, haiburensis, hymenoptera* and *warncotti*) are named by Harris from the *Triassic* and *Jurassic* of Greenland and Britain and these also seem allied to the anacrogynous *Jungermanniales*. These rocks also show the first definite presence of the *Marchantiales* in *Ricciopsis florinii, Ricciopsis scanica* and *Marchantiolites porosus*. Of these, *Ricciopsis* resembles *Riccia* while *Marchantiolites* shows chambered thallus with air pores. The best known Bryophyte fossil, however, is *Naiadetia lanceolata* (Fig. 32.2) from the *Upper Triassic* (*Rhaetic*) of England which has been described in detail by Harris (1938). This is the only hepatic whose reproductive parts (gemmae and archegonia) and sporophyte are known. This seems to be allied to the *Riellaceae* of the *Sphaerocarpales* but has an upright axis with spirally arranged (3/8 phyllotaxy) leafy appendages reminding the *Calobryales*. The plants are 1 to 3 cm high and there are unicellular rhizoids from the base of the stem. The stem tissue is parenchymatous and is not differentiated. The leaves are linear at the base of the stem, lanceolate in the middle and somewhat rounded on the top. They are one cell thick and without midrib. There are multicellular gemmae in terminal *gemmae cups* (Figs. 32.2B and C). The *archegonia* (Fig. 32.2D) are lateral at extra-axillary positions. They are naked at first but then become surrounded by a perianth. The *sporophyte* shows a spherical capsule and a minute foot but no seta (Fig. 32.2E). Columella or elaters are absent. The capsule opened irregularly.

The Devonian fossil *Sporogonites*, sometimes supposed to be allied to the Psilophytales

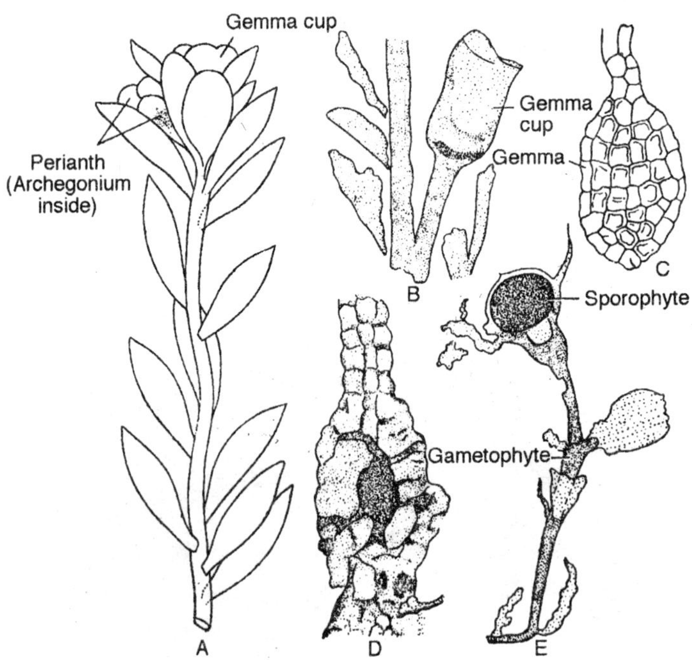

Fig. 32.2: *Naiadetia lanceolata* (after Harris), A. Gametophyte plant, B. Gemma cup, C. A gemma, D. An archegonium. E. Plant with mature sporophyte

(c.f. Chapter 37) has been considered to be a hapatic by Andrews (1959).

The still younger *Mesozoic* rocks, i.e., the *Cretaceous*, shows the presence of more *Marchantiales* in four species of the genus *Marchantites* (*blairmorensis, sewardi, hallei* and *yukononsis*) from N. and S. America. Jungermanniales are represented by *Jungermannites cretaceus*. *Hepaticopsida* sub-fossils identical with the present-day genera begin to appear in the modern *Cainozoic* age beginning from *Miocene* onwards.

Bryopsida

Bryopsida fossils are found beginning from the Carboniferous side by side with the *Hepaticopsida*. The specimens are fragmentary and incomplete like most *Hepaticopsida* fossils. Two species of *Muscites* (*bertrandi* and *polytrichaceous*) occur in the Upper Carboniferous (France). These rather large specimens show stems with anatomy definitely like that of moss, simple leaves with single veins and rhizoids with oblique walls. But their affinity is uncertain and, although it is suspected, it can not be definitely said that *M. polytrichaceous* is related with the modern *Polytrichaceae*. It can only be said that they are types of *acropcarpic* mosses. A suspected moss capsule (*Capsulites gondwanensis*) has been reported from the Indian Permo-Triassic Gondwana by Saksena. Neuberg (1956) has reported 14 species in 9 genera from the Lower and Upper Permian of the Russia. Three of these genera (*Protosphagnum, Vorentannularia* and *Jungajia*) have leaves which are somewhat like *Sphagnum*. But, distinct veins are present so that their relationship with *Sphagnum* is doubtful and they should be placed in a new order *Protosphagnales* as suggested. They may as well be related to the Polytrichales or the Dicranales (Gams 1962). In this assemblage Neuberg has also described *Intia* which seems near *Msium* or *Bryum*—but no sporophyte was found. The other genera named by Neuberg are *Salairia, Uskatia, Polyssaiuria, Bajdaieira* and *Buchtia*. Mesozoic moss fossils are rarer than the *Hepaticopsida* fossils of that time. There is one species of *Muscites* (*M. lesquereuxi*—still of uncertain affinity) and specimens of *Sphagnum* sp. leaves from Greenland—both from *Cretaceous*. More moss species appear during *Cainozoic*. *Jungermannites* occur during the Tertiary. Some *Muscites* fossils and that of *Palaeohypnum* (possibly a *pleurocarpic* moss) occur during *Eocene*. *Oligocene* shows *Muscites yallournensis* from Australia—the first definite moss fossil with a capsule. During *Miocene* there are fossils of *Muscites, Polytrichites* (possibly belonging to *Polytrichales*), *Plagiopodopsis* (possibly *acrocarpous*) and *Palaeohypnum*. In the newer *Pliocene* and *Pleistocene* rocks a large number of *subfossils* (i.e., not always fully fossilized) of modern moss genera are known. In India, *Sphagnum* peats of recent origin are known from Kashmir and the Nilgiri Hills.

The above record of the fossils throws little light on the evolution of the *Bryophyta*. It is only known that there is evidence of the thallose Jungermanniales (anacrogynous) in the Devonian and during the Upper Carboniferous both the *Hepaticopsida* (possibly) the *anacrogynous Jungermanniales*) and the *Bryopsida* (*acrocarpic* mosses) are found. The Marchantiales, *Sphagnum* and pleurocarpic mosses are known later. We are still in the dark about the *Anthocerotopsida* except that some Tertiary spores have been assigned to this order. In this connection, a note should be kept of the contention by Naumova (1949) that fossil spores found in the Cambrian rocks represent both the Bryophytes and the Pteridophytes. If further verified, this would take down the origin of the Bryophytes to a much earlier age.

ORIGIN OF BRYOPHYTES

The ecological history of Bryophytes does not throw any light on the origin of Bryophytes because of the fact that bryophytes possess very soft and fragile nature of plant body. Only the comparative study of different groups of plants through some light on the topic—Origin of Bryophyta suggests:

I. Algal origin of Bryophytes

II. Pteridophytic origin of Bryophytes

I. Algal origin of Bryophytes

Many scientists believed that the bryophytes have originated from the algae, though they were not unanimous about whether the origin of bryophytes was monophyletic or polyphyletic. Even, there is controversy regarding the type of ancestoral alga that has given rise to bryophytes.

The concept of the algal origin of bryophytes is based on the similarity in algae (especially green algae) and bryophytes, like: (i) Amphibion nature, (ii) Presence of thalloid gametophyte, (iii) Cell wall made of cellulose, (iv) Presence of filamentous protonema, (v) Starch as reserve food, (vi) Similar photosynthetic pigments (Chlorophyll a, b and Carotinoids α, β), (vii) Presence of thalloid gametophyte, and (viii) Presence of flagellated sperms.

Many scientists suggested various algal ancestors that have given rise to the bryophytes. In 1903, Lignier proposed the hypothetical terrestrial plant, *Protohepatics* which evolved from the aquatic algal ancestor. In course of time, the *Protohepatics* has given rise to the bryophytes (**Hepatics**) on the one hand to the pteridophytes on the other hand.

In 1908, Bower supported the algal origin of land plants—bryophytes and pteridophytes, probably from freshwater green algal chlorophyllan ancestor.

In 1916, 1945, Fritsch has suggested that the **Chaetophoraceous ancestor** has given rise to both bryophytes and pteridophytes. The green algal members belonging to Chaetophorals which possessing heterotrichous habit, a prostrate and upright systems have rise to erect land plants by elaborating the erect (upright) portion and by diminishing the prostrate portion. An apical growing point is established as has been observed in some members of Chalphorals e.g., *Trentopholia.* The heterotrichous branching habit as in mosses where filamentous protonema represents the prostrate system, while foliase gametophore represents the erect system. Ultimately, the gametophytic plant body becomes much more elaborated and the sporophytic plant becomes parasitic on the gametophytic portion of the moss plant.

In 1919, Church proposed the marine algal origin of land plants—both bryophytes and pteridophytes. He postulated that marine algae were **planktonic** in nature and subsequently became **benthic** in the shallow seas. Further transmigration of these benthic forms to the subaerial condition led to the establishment of a large group of extinct sea weeds which are much more than the present day thallophytes. Church named the hypothetical group as "**Phallophyta**".

According to Church, the transmigrant algae had (a) metabolic efficiency of Chlorophyceae for the production of starch, (b) somatic structure of the Phaephyceae, like multiseptate organization differentiation into root-like holdfast, stem-like stipe, leaf-like blade, conducting tissue, etc., (c) reproductive system of life history is more advanced than that of *Dictyota,* such as isogamy, advanced fertilization, and alternation of haploid sexual and diploid asexual phages. Both bryophytes and pteridophytes of today originated from this hypothecal "**Thallophyta**".

In 1955, Smith strongly supported the algal origin of bryophytes. He believes that bryophytes have originated from the green algal group, such as Uldrichales of Chlorophyceae.

In the earlier classification system, bryophytes were placed in the single phyllum

Bryophyta intermediate in position between the algae and the pteridophytes. Modern studies of cell ultrastructure and molecular biology, however, confirm that the bryophytes comprized of these separate evolutionary lines, which are today recognized as liverworts (Phyllum Marchantiophyta or Hepatophyta), hornworts (Phyllum Anthocerotophyta and mosses Phyllum Bryophyta). According to Palmer 1999, Bryophytes are not considered to have given rise to the vascular plants, but they probably were the earliest land plants. Like the rest of the land plants, they probably evolved from green algal ancestor *Coleochacte,* closely related to Charophyta.

II. Pteridophytic Origin

This theory (concept) suggests that the bryophytes are the decendent of pteridophytes and evolved by the progressive degeneration of the pteridophytes. This theory is designated as **Regressive Evolution.** The simplest forms—liverworts and the more complex ones—mosses. This theory was supported by Wettstein (1908), Scott (1911), Christensen (1954), Zimmermann (1960) and Udar (1970) and others.

According to this theory, the earliest vascular plants, Rhynopsida of the Upper Solurian to Lower Devonian age, and the sporophytes of certain mosses like *Andreaea, Funaria, Sphagnum* and others and the members of hornworts like *Anthoceros, Dendroceros* and others. The characteristics of Rhynopsida are comparable with the sporophytes of bryophytes. As found in (i) the members of Rhynopsida are rootless (Paverhizoids), leafless and dichotomously branched shoots bearing sporangia, (ii) the sporangia of *Horneophyton* and *Sporogonites* have columella showing bryophytic affinities, (iii) Capsules of *Funaria, Sphagnum* and *Anthoceros* are composed of photosynthetic tissue and are provided with stomata. These characters are common features of pteridophytes, (iv) cells of the outer layer of the columella and the lining layer of the jacket in *Dendroceros crispers* of Anthocerotophyta form incomplete spiral thickening. The lining layer of the jacket is found to be identical with taputum (we celled thick layer lining the jacket).

ORIGIN OF ALTERNATION OF GENERATIONS

A plant arising either from a zygote or from a spore may give rise to an independent plant exactly like itself. Each member of such a series in a natural use of the term **generation.** Hence, the generation extending from a zygote to a spore or from a spore to a gamete may consists of a series of **generations** of subordinate order. Furthermore, each generation, may include hundreds or thousands of **cell generations.** The two alternating arcs of the life cycle of sexually reproducing organisms, one **diploid** and closing with meiosis; the other **haploid** ending in the production of gametes and their union. In what follows, each of these arcs will be called a **generation** however many successive independent individuals it may include. The advantages resulting from alteration of generations are: (i) it produces many spores, increasing the possibilities of reproduction, and (ii) meiosis of a zygotic nucleus can given rise in consequence of a new groupings of chromosomes and chromosome parts 60 at most four genetically different lines of new individuals.

Sexually, reproducing organisms, in general, pass through two phases of development to complete their life cycle. These are: (i) the spore producing phase, the **sporophyte** or **sporophytic generation** or **diploid generation** (2n), whose nucleus bears two chromosme compliments (2n); and the **gamete producing phase,** the **gametophytic** or **gametophytic generation** or **haploid generation** (n) having only one chromosome complement (n) in each nucleus. These two phages or generations which occur in regular succession, one alternating with the other is known as **alteration of generations.** The process of alternation of generations was discovered by Hofmeister in 1851.

As early as 1894, Strasburger moles the generalization that the plants showing alternation of generations, the generations differ in nuclear constitution.

Members of sexually reproducing algae exhibit various life cycle patterns like:

(i) An alternation of predominantly gametophytic generation or haploid phase with a very short insignificant sporophytic generations or diploid phase which is the diploid zygote. This type of sporangium of *Hornophyton* (Rhynopsida).

The similarities of sporangia of bryophytes and pteridophytes have led many botanists to interpret that the Anthocerophyte or briophytes life cycle pattern is known as **haplontic or monogenetic life cycle.**

(ii) In the second type of life cycle the haploid generation is insignificant and is represented by gametes only. It occurs alternating with a very predominating diploid generation. Life cycle of such pattern is caled **diplontic life cycle.**

(iii) In the third kind of life cycle both the sporophytic and gametophytic generations are equally predominant, enumerated as **haplo-diplontic** or **digenetic life cycle.** In this kind of life cycle both the haploid and diploid phases may be morphologically similar or dissimilar, that is, **isomorphic** or **heteromorphic** respectively.

(iv) In the fourth kind of life cycle both the genetophytic and sporophytic generations are elaborately developed. The difference being, the diploid phase is long drawn, part of which is dependent on the gametophytic generation and later on gives rise to an independent diploid phase. Such a life cycle pattern is enumerated as **diplohaplontic** with **trigenetic life cycle** or **triphasic alternation of generations.**

The life cycle of bryophytes shows a **heteromorphic alternation of generations** where two phases follow one another in a regular alternate succession. The gametophytic phase is independent, autotrophic, haploid and bears gametes. This phase initiates from the haploid spaces (n) produced by the sporophyte and ultimately ends with the production of gametes within the multicellular sex organs. The sporophyte of bryophytes is known as **sporangium.** The sporophyte is completely dependent on gametophyte. The sporophytic phase begins with the diploid zygote (2n) produced from the fertilization, and the spore mother cell represents the last stage of sporophytic generation.

There are two theories in connection with the alternation of generations and the origin of sporophyte: (i) **antithetic** or **interpolation**, and (ii) **homologous** or **transformation** theory.

(i) Antithetic or Interpolation theory

This theory was proposed by Celakovsky in 1874 and subsequently developed by Bower in 1890, Strasburburger in 1894, Cavers in 1910, Chamberlain in 1935 and Campell in 1940. According to the antithetic theory the gametophytic generation is the original generation, while the sporophytic generation is a new phase which has been gradually evolved afterwards by the progressive elaboration of the sporophyte (zygote) of the green algal ancestor. Thus, a structurally different sporophyte is interpolated into the life cycle between the fertilization and meiosis in course of evolution.

The stages of progressive evolution of sporophyte from the zygote of green algal are as follows:

(a) **Stage one.** In this stage, unicellular diploid zygote (2n) which passes through a resting period and subsequently develops four haploid spores (n) by **meiosis.**

(b) **Stage two**: In this stage, the zygotic nucleus first divides by mitosis and subsequently a multicellular structure is developed in which archegonium is developed. In a separate another zygote by a similar method an antheridium is developed.

(c) **Stage three**: In this stage, a zygote is developed in the separate multicellular structures by the development of female gamete (ovum) in the archegonium and male gamete (sperm) in the antheridium. Ultimately fertilization of male and female gametes takes place producing a zygote (2n).

(ii) **Homologous or Transformation Theory**

This theory was put forwarded by Ringsheim in 1876 and was supported by Scott (1896), Church (1919), Zimmermann (1930), Evans (1939), Fritsch (1945) and Bold (1948).

According to this theory, the gametophytic and sporophytic generations are fundamentally similar in nature. The sporophyte is not a new phase, but a direct modification of the gametophyte. Thus, during the migration to land, the two generations evolved differently and one because the independent gametophyte, while the other specialized as a sporophyte that became a parasite on the gametophye.

The various evidences in support of homologous theory are as follows:

(a) **Isomorphic alternation of generations.** The Algae belonging to Cladophorales, Ulvacea, Ectocarpales, Dictiotales show two alterating homomorphic forms, one a haploid gametophyte, reproducing sexually by gametes and the other a diploid sporophyte, reproducing asexually by asexual spores.

(b) **Nutrition of sporophyte.** The of photosynthetic tissue in the sporophytes of Liverworts, Hornworts and Mosses suggests the fundamental similarity between the gametophytic and sporophytic phases with respect to self-nutrition.

(c) **Apogamy and apospory.** Apogamy is a phenomenon where a haploid sporophyte is produced from a gametophyte directly, without syngamy (sexual union). While in apospory a diploid gametophyte is produced from a sporophyte without the formation of spaces. These two phenomena have been observed in a large number of bryophytes and pteridophytes, which support the homologous theory.

(d) **Similarity between primitive gametophytes and sporophytes of pteridophytes.** The gametophytes of *Psilotum, Tmesipteris, Ophioglossum* are cyndrical, diochromously branched with continuous apical growth. These gametophytes, showing structural similarity with the sporophytes of primitive pteridophytes like *Psilotum* and members of Rhyniopsida, strongly support the homologous theory.

UP-GRADE OR PROGRESSIVE EVOLUTION THEORY

Most classifications of Bryophytes begin with the simpler forms and end with the complex forms although there is a large number of Bryologists now holding that evolution took place the other way (regressively).

According to the Up-grade view, the first evolved Bryophyte was of the hypothetical *Sphaero-Riccia* type (Lotsy, 1909) which combined the very simple thalloid gametophyte of *Sphaerocarpos* with the simple bag-like sporophyte of *Riccia.* Cavers (1910-11) developed this view and the phylogenetic tree proposed by him is shown in Figure 32.3. Campbell (1940, etc.) also supported this view but Campbell added that the primitive form could also be like *Riccardia* or *Metzgeria* of the anacrogynous Jungermanniales. Smith (1955) also favours this Up-grade view. This primitive plant in one direction developed the *Marchantiales* with differentiation of the tissues of the gameophytic thallus as well as that of the sporophyte. In another direction the same primitive plant evolved into the *Jungerman-*

niales by retention of the simplicity of the gametophytic thallus (lack of air pores, etc.) but by developing wings and then lateral leafy expansions. The *Sphaerocarpales* is a side line from this branch. Finally, in this Jungermanniales direction, when a third row of mucilage hairs on the ventral surface became modified into leaves, the erect *Calobryales* were evolved. Such a plant might have easily given rise to the *Sphagnales* and the other mosses. The development of three rows of leaves and the erect habit is also to be associated with the change in the apical meristematic cells from two cutting faces to three cutting faces which first occurs in the acrogynous Jungermanniales. The *Anthocerotopsida* developed as another side line of Jungermanniales by elaboration of the sporophyte and retention of the primitive simple gametophyte. According to this view, the primitive Bryophyte developed out of the Algae had a *haplobiontic life cycle*, i.e., the plant was a gametophyte within which a transient sporophyte involved only meiosis directly after the sextual fusion. The sporophytic plant was gradually evolved afterwards by a lengthening of the sporophytic phase so that a regular sporophyte was intercalated between sexual fusion and meiosis. This is strictly in accordance with the *antithetic theory* first proposed by Celakovsky (1874) and then developed by Bower, Cavers, Campbell and Chamberlain. The origin is to be sought in the green algae. During the migration to land, in the *first phase*, the life history is pictured as in *Oedogonium*, a haploid plant wih the diploid stage represented by a unicellular zygote which after a period of rest develops four haploid spores by reduction division. In the *second phase* of evolution the form was probably similar to *Coleochaete* where the zygote develops four haploid cells, which, by further division, develops 16 to 32 zoospores. In the *next phase*, one may suppose that the zygote did not undergo reduction division all at once but first became multicellular by mitosis, thus initiating the first *embryo* and the first *sporophyte* where the ultimate cells become spore mother cells. In the *fourth* or the *final phase*, the outer cells of this spherical sporophyte form a sterile jacket while the inner cells become sporogenous—a stage which is attained in *Riccia*.

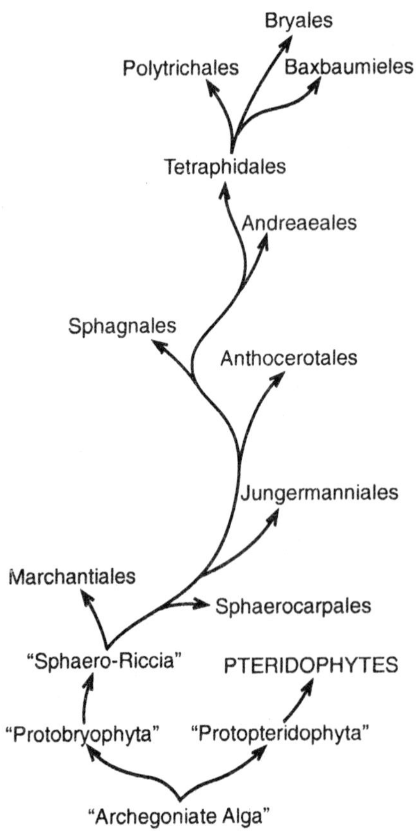

Fig. 32.3: Phylogeny of Bryophytes according to Cavers (1911)

DOWN-GRADE OR REGRESSIVE EVOLUTION THEORY

Origin of Bryophytes out of the Pteridophytes is a regressive view already discussed. Cronquist, Takhtajan and Zimmermann (1966) still hold this view retaining the Bryophytes in between Pteriodophytean groups, being a derivative. Proskauer (1960) has pointed out some vascular affinities between Anthocerotales and Psilophytales. Mehra

(1953) explained these by a very close and common origin of Anthocerotales and Psilophytales through *Anthorhyniaceae* (Fig. 32.4 on page 1054). But, whatever the origin, according to this view the most primitive Bryophyta had the most complicated appearance within the group.

Wettstein (1903-1908) first suggested that the primitive Bryophyte had an erect gametophyte like that of the mosses. Later, change to dorsiventrality gave rise to the acrogynous *Jungermanniales*, reduction of leaf development to the anacrogynous *Jungermanniales* and then to the *Marchantiales. Anthocerotales* also arose similarly from a side line. Church (1919) and Evans (1939) also think that the primitive Bryophyte was an erect plant with three rows of leaves and a pyramidal apical cell with three cutting faces.

The nearest approach to such a plant is to be found in the *Calobryales* and *Naiadatia* among. Hepatics and in some acrocarpic mosses. This primitive plant, later, developed dorsiventrality because of general weakness of structure and gradually evolved the different types of Hepatics by regressive evolution. In most Hepatics, although the gametophyte is generally dorsiventral, the reproductive shoots are radial and erect. This is because in the plant world the reproductive structures are the most conservative. The reduction view found support in Goebel, Kashyap (1919), Harris (1938), Mehra (1953 onwards), Fulford (1965) and others.

Kashyap (1919), from this extensive studies on Indian Hepatics, was convinced of reduction in the evolution of the Hepatics and some of his arguments are noted in connection with the discussion of the *Marchantiales* that follow. He suggested that the elaterophore of the Jungermanniales is derived by the reduction of the columella of the more elaborate Anthocerotales. This is further reduced to the elaters and then vanish.

According to the most adherents of the regressive view, the primitive ancestor arose from the algae and it had a *homologous* type of alternation of generations. The homologous view of alternation of generations in Bryophytes was first advocated by Pringsheim (1876), and then affirmed by Scott (1896), Church (1919), Zimmermann (1930), Bold (1938), Evans (1939), Fritsch (1945), etc. There are many genera among the advanced algae showing two alternating isomorphic forms, one a gametophyte and the other a sporophyte. According to this view, these two generations evolved differently when plants migrated to land and one became the gametophyte plant body of the Bryophytes while the other specialized as a sporophyte. In the Bryophytes, the latter became a parasite on the former. This view has a further confirmation in the presence of *apospory* and *apogamy* within the Bryophytes. Fritsch explained that this tendency towards developing an erect plant is already seen in the *Chaetophorales* (*Chlorophyta*) in their heterotrichous habit where some filaments are prostrate while others are erect. Some *Chaetophorales* (e.g., *Fritschiella* and *Trentepohlia*) show the organization of an apical growing point in the erect branches. If these tendencies are combined with the tissue development of the *Phaeophyta*, it is conceivable to have a plant, still an alga, but having the characters of *Bryophytes*. According to Fritsch (1945), the *Bryopsida* (moss) shows the prostrate habit in the protonema and the erect habit in the gametophytic plant. Recent studies of Allsopp and Mitra (1958) show that even the moss protonema is heterotrichous, resembling the *Chaetophorales* so that the resemblance is more than what was supposed by Fritsch. It has already been explained that Church's 'Thalassiophyta' hypothesis also pictures a similar origin of the first Bryophytic land plant. The sex organs of the *Chlorophyta* are generally unicellular. Davis (1903) explained that both the antheridia and the archegonia of the Bryophytes might have evolved from some multicellular algal gametangium by progressive sterilization.

Zimmermann (1930-49) on the development of his *Telome Theory* (c.f. last Chapter on Pteridophyta), postulates that both the *Bryophyta* and the *Psilophyta* (the most primitive vascular plant according to him) developed out of the algae, perhaps independently. The *Psilophyta* evolved first, by a change of the isomorphic life cycle into a heteromorphic one, the sporophyte forming an erect body of several *telomes* and *mesomes.* The *Bryophyta* arose at a later time probably directly out of the Algae. Here also the life cycle became heteromorphic, the sporophyte became parasitic on the gametophyte and was reduced to a single *telome.*

The latest view of Cronquist, Takhtajan and Zimmermann (1966, c.f. page 1st page of Bryophytes) is that the Bryophytes, arising as reduction of some higher plant forms, should be placed within the Embryophytes as a division in between the Psilophytales and the Psilotales. But such a position is disliked by many as it suggests an origin of the Bryophytes out of the Pteridophytes with vascular strands containing lignin.

Thus, according to the regressive view, the primitive Bryophyte had an erect gametophyte with three rows of leaves and with a sporophyte like that of *Anthoceros.* Reversion to prostrate form gave rise to the Hepatics, of which the *Anthocerotopsida* represents the most primitive type. Elaboration of the sporophyte with development of the peristome structure gave rise to the mosses. Mehra (1968), however, suggests a different origin for the Anthocerotales (Fig. 32.4) discussed later.

EVOLUTION AND AFFINITIES OF THE BRYOPHYTES

To properly understand the interrelationships of the different bryophytic taxa, it is necessary to have a clear comparative idea of both the gametophytic and the sporophytic forms. In the gametophytes, increase of complexity is apparent in the series Anthocerotopsida, Hepaticopsida [*Riccia-Marchantia*-Jungermanniales (Anacrogynous-Acrogynous)–Calobryales and Takakiales], Bryopsida. When sporophytes are considered, Bower has pointed out the **progressive sterilization** in the series. *Riccia-Sphaerocarpos-Marchantia*-Jungermanniales (Anacrogynous-Acrogynous)—Anthocerotales and Bryopsida. In *Riccia* the sporophyte is a simple bag of spores—almost wholly fertile. In *Sphaerocarpos* there is a short sterile stalk. In *Marchantia* there is differentiation into capsule, seta and foot. Moreover, some sterile elaters develop inside the capsule. In the Jungermanniales the sterile wall increases in thickness. Maximum differentiation is in *Anthoceros* and the mosses where most of the sporophyte tissue the columella and the thick, photosynthetic wall of the capsule, even a meristematic region in *Anthoceros,* the peristome in mosses is sterile.

HEPATICOPSIDA

It has been stated above that according to the Regressive Theory the *Hepaticopsida* evolved out of the more complex *Anthocerotopsida* or some erect complicated primitive Bryophyta while, according to the Progressive Theory, the simplest of the *Hepaticopsida* was the first Bryophyte evolved.

The Sphaerocarpales and the Marchantiales

The *Sphaerocarpales* and the *Marchantiales* show the simplest plant bodies among the Bryophytes. The gametophyte of *Sphaerocarpos* is of the simplest type. *Sphaerocarpos, Riccia* and *Marchantia* grow by a transverse row of apical cell each with two cutting faces. The sporophytes of the *Sphaerocarpales* and the *Marchantiales* are also extremely simple, their capsule walls being only one cell in thickness. The bag-like sporophyte of *Riccia* is no doubt the simplest so that Lotsy pictured a hypothetical *Sphaero-Riccia* embodying the gametophyte of *Sphaerocarpos* with the sporophyte of *Riccia* as the ancestral Bryophyte. The spermatozoids of *Sphaerocarpales*, again, closely resemble those of *Marchantiales* as against the *Jungermanniales*, bringing them closer to the former.

Considering all points, it may be said that the *Sphaerocarpales* and the *Marchantiales* represent the simplest of all the Bryophytes. Whether they are the most primitive (Progressive Theory) or the most reduced (Regressive Theory) is a matter of dispute.

Within the *Marchantiales*, again, the *Ricciaceae* is certainly the simplest family while the *Marchantiaceae* is the most complex considering both the gametophyte and the sporophyte. The air pores and assimilatory filaments are well-developed in *Marchantiaceae*, the sex organs are borne on special antheridiophores and archegoniophores, the sporopytes are more elaborate in having foot, seta, elaters and have special protective structures surrounding the immature sporophytes. All these structures are much simpler in the *Ricciaceae*. While the proponents of the Regressive Theory like Kashyap hold that this happened by reduction, adherents of the Progressive Theory, like Bower and Smith, hold that the *Marchantiales* developed these characters as advance in course of evolution. Bower held that the most primitive Bryophyte had a sporophyte like that of *Riccia* (*Ricciaceae*) where the sporophyte is wholly formed of sporogenous cells. Evolution then took the line of *progressive sterilization* of potential sporogenous tissues. The next stage is shown by the family *Corsiniaceae* where the sporophyte resembles that of *Sphaerocarpos*, sterile foot and seta are differentiated and some of the sporogenous cells remain sterile thought their walls do not become thickened. Next, in the *Targioniaceae*, the sterile sporogenous cells form elaters. The final stage is represented by the *Marchantiaceae* where the foot, seta and the elaters are very conspicuous although the wall of the capsule is still only one layer thick. As against the views of Bower, Campbell, etc., Kashyap (1919), after a close study of the Indian Hepatics, became convinced of the Regressive Theory. He concluded that the more elaborate *Marchantiales* like *Marchantia* and *Preissia* are more primitive and were gradually reduced to the simple form of *Riccia* by progressive reduction as becomes apparent on the comparative study of the gametophytes and sporophytes of different genera of Marchantiales, viz.,

(a) *Loss of assimilatory filaments in the air chamber:* The air chambers (several tiers in Rebouliaceae but one tier in Marchantiaceae) are full of assimilatory filaments which are very well developed in *Reboulia*, *Preissia* and *Marchantia* but are progressively more reduced in hydrophilous *Conocephalum* and *Wiesnerella*. It ultimately disappears in aquatic *Dumortiera*.

(b) *Simplification of pores:* The pores are complex barrel-shaped in *Marchantia* and *Preissia*. In *Conocephalum* and *Reboulia*, they are barrel-shaped in the discs (archegoniphore, and the antheridiophore) but simple on the thallus. In *Exormotheca* and *Stephensoniella*, they are simple both on the thallus and the discs. In *Riccia*, the pore status is lost, they are simply the branches of the pits.

(c) *The gradual shifting of the stalkes of the antheridiophore and the archegoniophore from the terminal to a dorsal position and then to the complete elimination of the stalk. The archegonia and the young sporophytes also gradually lose their protective coverings.*

Mehra (1957 to 1969), a pupil of Kashyap, supports the above hypothesis and supposes that the *Marchantiales* were derived from the *Jungermanniales* by reduction. He is of the view that although the Marchantiales have evolved by reduction of the Jungermanniales stock, this does not mean that the order Marchantiales represents a single line of regressive evolution from Marchantiaceae to Ricciaceae. He thinks (Fig. 32.4) that after the regressive derivation of the Marchantiales, there was a line of progressive evolution ending in the Marchantiaceae proper (i.e., Compositae). Sauteriaceae, Rebouliaceae and Exormothecaceae are the earlier branches of this line. The exormothecaceae (*Exormotheca* and *Stephensoniella* taken out of the Compositae, c.f. 2nd page of Marchantiaceae—classification) line developed the families Corsiniaceae and Targioniaceae by reduction. Ricciaceae being developed as an extreme reduction of this line. Monocleaceae was developed by reduction directly out of the Marchantiaceae stock. Thus, according to Mehra, both *Marchantia* and *Riccia* are the latest forms in evolution in two different lines. The reduction of foliose forms to thallose forms (i.e., from Jungermanniales to Marchantiales) is explained by Mehra (1957) by the **Condensation Theory** originally suggested by Kashyap. The leaf expansions of foliose forms fuse to form thallose expansions. Within the Metzgerineae foliose *Fossombronia* may be experimentally induced to take up the shape of thallose *Petalophyllum* by the fusion of *succubous* lower edges of the foliose expansions into the thallus so that the upper foliose expansions look like the vertical lamellae of *Petalophyllum*. Mehra suggests that from this structure, by farther condensation (Fig. 32.4C), completely flat *Stephensoniella*—(Marchantiaceae—Fig. 32.4D) and *Asterella* (*Fimbriaria*—Rebouliaceae Fig. 32.4C)—hyper thalli develop which first gives off the Sphaerocarpales line and then branches into the Sauteriaceae, Marchantiaceae, Rebouliaceae and the Exormothecaceae (which further develops Targioniceae, Corsiniaceae, Ricciaceae etc.) lines. Within the Sphaerocarpales *Geothallus* is somewhat leafy while *Sphaerocarpos* is completely thallose.

Harris (1938, 1939) while discussing the fossil (*Naiadatia*) suggested that the ancestral Hepatics had an erect leafy shoot with spiral leaves as in mosses. Burgeff (1943) after twenty years study of mutations of *Marchantia* concluded that the mutants clearly show a phyletic reduction series which proves that *Marchantia* gave rise to the other genera by reduction.

Cavers, we have alaready seen (Fig. 32.3), considered the Marchantiales as a blind line of evolution from the hypothetical *Sphaero-Riccia*.

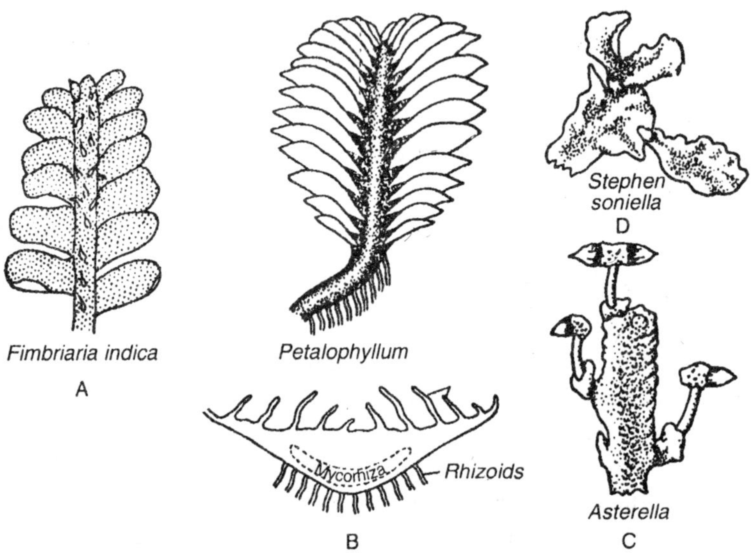

Fig. 32.4: Explaining *Condensation Theory*

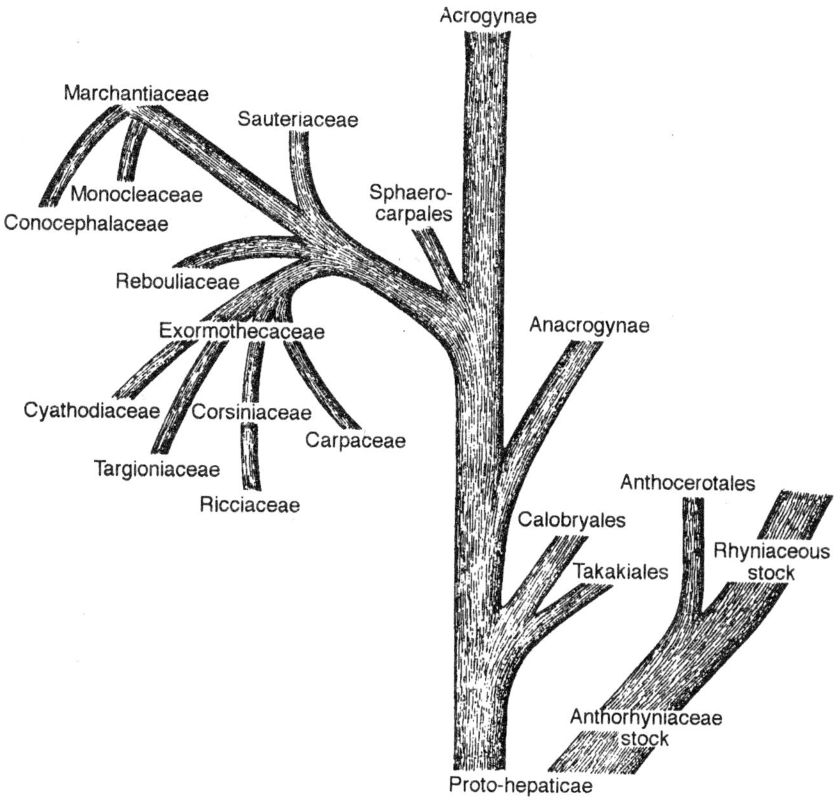

Fig. 32.5: Evolution of Hepaticopsida and Anthocerotopsida according to Mehra (1968)

Jungermanniales

The *Jungermanniales* have the plant body more complex than that of the *Marchantiales.* The sporophytes have multilayered capsule walls and a general structure which is also more elaborate than that of the *Marchantiales.* Here again, the *Metzgerineae* (anacrogynous *Jungermanniales*) are comparatively simpler with thallose or simple foliose, prostrate gametophytes and apical growth by a cell with two cutting faces while the archegonia develop on the dorsal surface of the thallus. The *Jungermannineae* (acrogynous *Jungermanniales*) are more complicated in having foliose gametophytes growing by apical cells having three cutting faces and with archegonia and sporophyte growing on tips of special foliose shoots which have erect tendencies.

As in other cases, the Regressive Theory holds that the simpler *Metzgerineae* (anacrogynous) came out of the more complicated *Jungermannineae* (acrogynous) by gradual reduction of the leafy appendages and the sexual shoots. According to this view, the most primitive Hepatic was an erect plant with radial symmetry like the *Calobryales* or like *Naiadatia.* This then obtained dorsiventral symmetry by reduction of the ventral row of leaves which became the *amphigastria* in the acrogynous *Jungermanniales* (*Lejeunea* and *Lepidozia*), then became further reduced into the *slime papillae* (*Plagiochila* and *Cephalozia*) and got completely lost in *Radula.* Simultaneously with this change, the axis beame flattened and then the *Metzgerineae* (anacrogynous forms) was evolved. The reproductive shoots, being more conservative in all plants, tried to retain the radial habit till the last. The *Marchantiales*, the *Sphaerocarpales* and the *Anthocerotopsida* gametophyte developed from the *Metzgerineae* by reduction. It should be remembered in this connection that thallose Jungermanniales are known from the Devonian while the foliose forms are known from the Carboniferous—earlier than the Marchantiales. Mehra (1957) has explained the origin of thallose forms from foliose forms in Metzgerineae by *compaction* (as in *Metzgeria* by flattening of the midrib and loss of leaves) and *condensation* (as in *Petalophyllum*).

The Progressive Theory argues just the other way. Bower's view of sterilization of the sporophyte is further extended giving rise to the elaterophores of *Pellia* and *Riccardia*. The sporophyte also becomes more elaborate with a multilayered capsule wall and a longer seta. There is also tendency of independence shown by the presence of chloroplastids and stomata on the sporophyte. Cavers (Fig. 32.3) held a similar view.

Some of the Metzgerineae like *Metzgeria* and *Aneura* (*Riccardia*) have so simple gametophytes that they have been associated with the most primitive Bryophyte (like *Sphaero-Riccia*) by Campbell.

Calobryales

The *Calobryales* is a peculiar group of Hepatics with erect plants showing almost radial symmetry. They strongly resemble *Naiadatia* and simple acrocarpous mosses so that they have been held to be very near the most primitive Bryophyte by the proponents of the Regressive Theory who hold that this was a plant with radial symmetry. The sporophyte, again, is simpler than that of the *Jungermanniales* and is similar to those of the *Sphaerocarpales* and the *Marchantiales*. The order is partly acrogynous and partly anacrogynous so that some Bryologists place it between the acrogynous and anacrogynous *Jungermanniales*. But, because of its special features, it is considered better to remove it out of the *Jungermanniales* as a separate order.

The growth of the antheridia and archegonia in Calobryales and Mosses by 2-sided or 3-sided apical cells and the peculiar way of archegonial development of *Anthoceros* makes Mehra (1957, 1958) believe that these sex organs represent reduced branches.

The rhizomatous base with mucilage hairs and without rhizoids is similar to that of the Takakiales so that Pyoskauer (1962) and others think that the latter actually should be within the Calobryales.

The proponents of the Progressive Theory, Smith for instance, hold that the *Calobryales* developed in the same line as the *Jungermanniales*, but they separated earlier so that they have retained the simplest type of sex organs among the Hepatics and a simple sporophyte but, have developed the most elaborate type of gametophyte.

ANTHOCEROTOPSIDA

The *Anthocerotopsida* is a group of plants with such distinctive features that it has been thought necessary to take it out of the general Hepatics and to place it in a separate class parallel to the *Hepaticopsida* and the *Bryopsida*. Four genera within this class are universally recognized: *Anthoceros, Megaceros, Dendroceros* and *Notothylas*. Later systematists have proposed to break down *Anthoceros* into three genera : *Aspiromitus, Phaeoceros* and *Anthoceros*. *Notothylas* is now placed in a separate family *Notothylaceae*.

The *Anthocerotopsida* has been called a **synthetic group** as, apparently, it shows characters linking it to widely differing plant groups, viz., the *Algae*, the *Hepaticopsida*, the *Bryopsida* and the lower vascular plants (*Pteridophytes*).

The most striking similarity with the algae is the chloroplast. A single alga-like **chloroplast** is the rule within the group. The chloroplast has a tendency to increase in number in deeper layers and in *Megaceros* there are multiple chloroplasts so that this genus is considered nearer to the higher embryophytes. The simplicity of the gametophytic structure brings it at par with some of the green thalloid algae.

The general appearance of the gametophyte and those of the mature sex organs resemble the *Hepaticopsida* and that is why they were included within the *Hepaticae.* The sporophyte may be derived by extending Bower's view of greater sterilization and more chlorophyllose tissue tending to independence. This is a stage just in advance of the *Hepaticopsida.* But, this sporophyte differs in certain points. While the primary sporogenous tissue of the *Hepaticopsida* is derived from the *endothecium,* that of the *Anthocerotopsida* is derived from the *amphithecium.* In this respect, *Notothylas* shows the nearest approach to the *Hepaticopsida*—specially to the simpler members like *Sphaerocarpos* and *Cyathodium.* It has a very reduced sporophyte and in some *Notothylas* species the endothecium also is sporogenous, there being no sterile columella. In still other species of *Notothylas* the amphithecium forms only the jacket while the endothecium is entirely sporogenous—just as in the simpler *Hepaticopsida.* It should be noted that *Cyathodium* also approaches the *Anthocerotopsida* in having fewer chloroplasts in the cells, approaching *Megaceros.* It is possible that the *Anthocerotopsida* developed out of the *Hepaticopsida* by way of plants like *Notothylas.* But, it has also been argued that *Notothylas* was derived from *Anthoceros* by reduction and further reduction might have given rise to the *Hepaticopsida.* Cavers (Fig. 32.3) derived the *Anthocerotopsida* as a side line from the *Jungermanniales* branch of the *Hepaticopsida* just as is the case of the *Sphaerocarpales* but with more evolution of the sporophyte.

Anthocerotopsida is linked with the *Bryopsida* by the structure of the sporophyte. The amphithecial development of the sporogenous tissue is common with the *Sphagnideae,* and the embryos are also similar in other details. The sporophyte of many mosses have a long continued growth, an elaborate photosynthetic tissue with regular stomata and a regular sterile columella—all in common with *Anthocerotopsida.* Cavers derived the Bryopsida through the Anthocerotopsida (Fig. 32.3)

The green, cylindrical, upright sporophyte of unlimited growth as found in *Anthoceros* and its tendency towards an independent life is a near approach to the lower vascular plants (Pteridophytes). The establishment of a central sterile columella (a forerunner of vascular cylinder), the elaborate photosynthetic tissue with stomata and an intercalary mertistematic tissue is an advance in the direction of the independent, green, vascular sporophytes of unlimited growth as found in the most primitive land plants. The sporophyte of *Anthoceros* greatly resembles that of the Devonian *Rhyniaceae* and is short of it only by the absence of a vascular system, of completely independent life and of the branching habit. Haskell (1949) gave emphasis on the similarity between Pteriodophytic plant structure as shown by *Horneophyton* and the *Anthoceros* sporophyte. Proskauer (1960) noted spiral cellulose thickening lining the archesporium both outside and inside in *Dendroceros.* According to him this indicates the tapetal layers in *Rhynia* and *Horneophyton.* Even within *Anthoceros,* in certain species (e.g., *A. fusiformis*) it has been noted that the sporophyte may live independently for sometime after the gametophyte, on which it had developed, decays. The bivalved dehiscence of the capsule of *Anthoceros* may be the beginning of dichotomous branching. Naturally, there are theories on the evolution of the Pteriodophytes out of *Anthoceros* (discussed later in connection with the origin of Pteriodophyta. Mehra (1968) is of opinion that the Anthocerotopsida did not originate from the same stock out of which the Hepaticopsida and the Bryopsida arose but out of a separate stock called *Anthorhyniaceae.* This stock gave rise to the *Antherocerotopsida* on one hand and the *Rhyniaceous stock* (which gave rise to the Psilophytales and the Pteridophytes other than Lycopsida) in the other hand. The Anthocerotopsida is, thus, closely related to the Psilophytopsida.

In conclusion, it may be said that the *Anthocerotopsida* resembles the type of plants which existed before the actual evolution of the vascular plants. If not the actual ancestor of the modern land plants, they seem to be related to this ancestor. The *Hepaticopsida* and the *Bryopsida* must also be related to this ancestor. But, the primitive *Anthocerotopsida* was closer to the *Hepaticopsida* than to the other groups.

In assigning any primitiveness to this class, it should be remembered that it is unknown as fossil and the earliest indirect evidence of its presence is in some tertiary spores.

BRYOPSIDA

The Bryopsida represents a very specialized group of *Bryophyta* in the life history of which the *protonema* has an important place in the gametophytic phase. The filamentous protonema resembles algae giving rise to speculations (e.g., that of Goebel) that the mosses might have arisen directly out of the algae. But, these filaments are associated with highly developed sex organs and the gametophyte as well as the sporophyte of the *Bryopsida* are much more elaborate than those of the *Hepaticopsida* and of the *Anthocerotopsida.* Moreover, the protonemata of the simpler Bryopsida, those of the *Sphagnidae* and the *Andreaeidae,* are thallose. So, there seems to be little evidence that the *Bryopsida* arose directly out of the algae. Campbell supposes that the thalloid protonema of the primitive mosses developed out of the thalloid Hepatics and the thallose form gave rise to the filamentous protonema as it was more convenient for giving rise to a large number of gametophytic plants.

Fossil evidence does not take us anywhere. It has already been seen that although the first fossil Hepaticopsida is known from the Devonian, Bryophytes appear abundantly only in the Upper Carboniferous and then, simultaneously, both the *Hepaticopsida* and the *Bryopsida* are equally developed. Of the Bryopsida, the *acrocarpic* ones with erect, radial gametophytes are found first and the *pleurocarpic* ones with prostrate, branching gametophytes followed them. The ancestor of the *Bryopsida* is likely to be an erect, radial plant as found in the *Calobryales* of *Hepaticopsida.* But, the gametophytes of all mosses are much more complicated structures than that found in any known *Hepaticopsida* or *Anthocerotopsida.* The gametophyte of mosses is adapted well to its amphibious condition of life. The erect sporophyte of mosses has the nearest approach in the *Anthocerotopsida.* But, here again, the sporophyte of the mosses with its intricate peristome is such a complicated structure that it needs a very specialized origin. Among mosses, *cleistocarpic* forms with no apparent peristome, seem nearer to the *Hepaticopsida* than to the *stegocarpic* forms with elaborate peristome structures. But, it is not the general opinion of the Bryologists that the different cleistocarpic genera developed separately and independently by the reduction of different stegocarpic genera. However, it may be that some of these cleistocarpic forms, specially *Archidium,* are more closely related to the ancestral moss. Among stegocarpic mosses, *Tetraphis* has a capsule similar to that of *Andreaea* with apparent valvular dehiscence. The Cavers concept of the course of evolution of the Bryopsida is seen in Figure 32.4.

Considering the manner of water conduction, mosses are divided into *endohydric* and *ectohydric.* The great majority of mosses do not conduct water through the central stand, water being absorbed by all the gametophytic surface. These are called *ectohydric.* In many of these no central strand is differentiated. This group is much nearer to the *Hepaticopsida* and the *Anthocerotopsida* in this respect. The *endohydric* mosses (e.g., *Polytrichum loc., cit., Bryum*), on the other hand, have a better organized central strand through which there is some possibility of water being actually conducted. But, as yet, it is not known how far this distinction is due to heredity and evolution and how far merely due to ecological response.

Sphagnidae and Andreaeidae

Of the *Bryopsida*, the *Sphagnidae*, with its simpler gametophyte, and a simple sporophyte in which the sporogenous tissue originates entirely from the amphithecium, is the nearest approach to the *Anthocerotopsida.* Even its protonema is thallose like the Hepatics gametophyte plant body. Next to the *Sphagnidae* is the *Andreaeidae* which may be considered intermediate between the *Sphagnidae* and the *Bryidae.* The sporophyte and the protonema of the *Andreaeidae* resembles the *Sphagnidae*, the sporogenous tissues arches over the columella as in the *Sphagnidae* and the *Anthocerotopsida* but it is not developed out of the amphithecium as in those cases; it develops out of the endothecium as in the *Bryidae.* The pseudopodium is similar to the *Sphagnidae* but the capsule does not show any peristome or operculum and is valvular in dehiscence as in the *Jungermanniales.*

Bryidae

Within the *Bryidae*, Cavers and Campbell recognize four groups: (1) *Tetraphidales*, (2) *Polytrichales*, (3) *Buxbaumiales*, and (4) *Eubryales.* Of these, the *Tetraphidales* is considered by them to have the simplest peristome but there is no evidence that this group, in spite of the apparent valvular dehiscence of its capsule simulating the *Andreaeidae*, forms a link between the *Andreaeidae* and the *Bryidae*, specially because there is no peristome in the former. Cavers, however, considers that the *Tetraphidales* arose out of the *Andreaeales* as the ancestor of the *Bryopsida* (Fig. 32.3). According to him, there were two lines of evolution out of the *Tetraphidales:* the *Polytrichales* arose in one line while the *Buxbaumiales* and the *Eubryales* arose in the other line. The *Polytrichales* have very advanced characters but their relatively small number leads Campbell to suppose that they are more primitive than the general *Eubryales.* The *Buxbaumiales* have some anomalous characters but they may be related to the *Polytrichales* through *Dawsonia.* The very stable low basic chromosome number ($n = 7$) of the Polytrichiidae also points out to their primitiveness. The *Eubryales* is the predominant group but they are probably the most recent in evolution like leptosporangiate ferns. Cavers and Dixon consider the *Nematodonteae* to be the more primitive. The common mosses (i.e., the *Eubryales*) comprise the *Arthrodonteae*, where again, probably the *Haplolepideae* (one row of peristome teeth) came first and the *Diplolepideae* (two rows of peristome teeth) came last. Philibert, however, considered the *Heterolepideae* (*Encalyptales*) as the possible common origin of both the *Haplolepideae* and the *Diplolepideae.* Within the *Diplolepideae*, again, the pleurocarpous forms are the latest in evolution, all the previous forms being acrocarpous. Fossil evidence also shows that the more primitive mosses (*Nematodonteae* or *Arthrodonteae*) were acrocarpous.

Fleischer, Brotherus and Reimers (Engler's System), on the other hand, consider that the cohort *Eubryinales* (including *Tetraphidales* within the *Eubryales*; Reimers has called this subclass *Bryidae*) is more primitive and the other two cohorts *Buxbauminales* and *Polytrichinales* (called subclass *Buxbaumiidae* and *Polytrichiidae* by Reimers) are later developments.

So, here again the problem is whether the more advanced *Polytrichales* are the ancestors of the simpler *Eubryales* (or *Bryidae*) i.e., whether evolution was progressive or regressive. Most Bryologists agree that the *Polytrichales* might be very primitive but whether the *Eubryales* came out of them or arose from a common ancestor is still open to question.

Present concept about bryophytes lineage

The fossil record offers valuable insights into the taxonomic diversity of the early land plant flora, yet fails to offer unambiguous evidence for the presence of bryophytes among early plant lineages. The overall lack of complexity and their haplo-diplobiontic life cycle suggest, however, that bryophytes compose a rather basal lineage of land plants. Except for these features, the liverworts, hornworts, and mosses seen to have little in common, raising doubts about close affinities among them (Steere 1969, Crandall-Stotler 1980, 1984, 1986). Due to the lack of relevant fossil taxa, reconstruction of the relationships among bryophytes thus relies primarily on the analyses of characters of extant taxa.

To address the phylogenetic relationships of bryophytes we need first to examine how many lineages should be considered. The hornworts are generally regarded as a well-defined natural group (Mishler and Churchill 1985, Renzaglia and Vaughn, Chapter 1 in this volume). Although mosses are generally considered to compose a monophyletic group, the affinities of *Takakia* have puzzled botanists ever since its discovery. Crandall-Stotler (1986), for example, accommodated the isolated position of *Takakia* by erecting the Takakiophyta, based on morphogenetic, anatomical, and cytological features (see Crandall-Stotler and Bozzola 1988). Based on morphological characters of the gametophyte, Murrary (1988) interpreted *Takakia* as representing a primitive group of mosses, a hypothesis congruent with morphological features of the sporophyte (Smith and Davison 1993). Schuster (1997) recognized that *Takakia* was unique, but only by the combination of characters it exhibited, arguing that most individual characters could be found in either mosses or liverworts. Schuster (1997) chose to retain *Takakia* with the liverworts, albeit in a basal position, serving as the "glue that holds the bryophytes together." Recently, several studies (Ligrone *et al.* 1993, Hedderson *et al.* 1996, Renzaglia *et al.* 1997) converged toward a hypothesis wherein *Takakia* belongs to the mosses.

Another alternative to the "three-lineage" concept of bryophytes was raised by Capesius (1995), based on analyses of nucleotide sequences of the nuclear gene coding for the RNA composing the small subunit of the ribosomes. In this and subsequent analyses (Kranz *et al.* 1995, Bopp and Capesius 1996, Capesius and Bopp 1997) the thalloid and leafy liverworts (i.e., the marchantioid and jungermannioid liverworts) are resolved as a paraphyletic group, with the leafy liverworts sister to the mosses, a hypothesis proposed earlier by Hori *et al.* (1985). Lewis *et al.* (1997) tested this hypothesis using sequences of the chloroplast gene *rbc*L. Although inferences under certain assumptions yielded topologies congruent with Capesius's hypothesis, analyses accommodating rate heterogeneity among codon positions resulted in a monophyletic lineage of liverworts. Analyses by Garbary *et al.* (1993) resolved liverworts as a paraphyletic group when morphological characters of the gametophyte were analyzed, or a monophyletic clade when sporophytic features were included. More recently Nishiyama and Kato (1999) coumbined sequences of the nuclear 18S and various chloroplast genes for addressing the phylogeny of land plants. Although their toxon sampling was extremely limited, including only a single exemplar for each liverwort clade, they concluded that liverworts form a monophyletic group, corroborating the results obtained by earlier workers (e.g., Mishler and Churchill 1985, Garbary *et al.* 1993, Samigullin *et al.* 1998, Backert *et al.* 1999, Renzaglia *et al.* 2000). Morphological, ontogenetic, and ultrastructural characters also point to a single liverwort lineage (see Crandall-Stotler and Stotler, Chapter 2 in this volume). Further study may be warranted, but in the light of the growing consensus in favour of the monophyly of hepatics, the following discussion will rely on this hypothesis.

Within a monophyletic Embryophyta, mosses, liverworts, and hornworts compose either a monophyletic lineage, defined by a common ancestor whose descendants do not include the Polysporangiophyta, or a paraphyletic group, with one or two lineages of bryophytes originating from an ancestor shared with the polysporangiophytes (Fig. 32.6). For a group of four taxa, 12 different relationships are possible. Most if not all of these have received some support from one or the other set of characters (see for example results from different analyses performed by Mishler *et al.* 1994). It is impossible to evaluate every phylogenetic hypothesis in terms of its implications for character evolution. Instead I will briefly present the major trends suggested by these alternative hypothesis and discuss the implications of the currently preferred topology in terms of the evolution of morphological characters focusing on those traditionally used to infer the phylogeny of land plants.

A monophyletic concept for bryophytes has gained little support in recent years (Fig. 32.6). Except for the analyses of 5S nucleotide sequences (Hori *et al.* 1985) and data from sperm cells (Garbary *et al.* 1993, Renzaglia *et al.* 2000), this hypothesis has generally been rejected. Mishler and Churchill (1984) proposed the first formal phylogenetic hypothesis for the origin of bryophyte lineages. Based on morphological, physiological, and biochemical characters, they proposed that the bryophytes compose a paraphyletic group; among the descendants of their common ancestor were the vascular plants (Fig. 32.6E). Their phylogenetic scenario showed the liverworts as the most basal lineage among extant land plants, followed by the hornworts, and mosses forming a sister group to the tracheophytes. Support for this topology has been gained from analyses of data sets restricted to general morphology (Bremer 1985, Kenrick and Crane 1997*b*) or in combination with sequence data of the nuclear genes, 18S and 26S (Mishler *et al.* 1994). A paraphyletic bryophyte assemblage with liverworts composing the most basal group (Fig. 32.6C) was also resolved by analyzing *rbc*L sequence data (Lewis et al. 1997). Unlike morphological characters (Fig. 32.6E), these analyses resulted in the hornworts being sister to the tracheophytes (Fig. 32.6). Except for small data sets analyzed by Mishler *et al.* (1994), this result has found little additional congruent data.

Most other studies resolve bryophytes as paraphyletic, too, but differ in the position of hornworts, which compose the most basal lineage of land plants (Fig. 32.6B, D, F). The relationships within the sister goup to the hornworts differ, however, among the studies. Reconstructions based on sporophytic characters suggest that liverworts share a common ancestor with polysporangiophytes (Garbary and Renzaglia 1998), a hypothesis congruent with topologies obtained based on 18S and *cox*3 data (Mishler *et al.* 1994, Malek *et al.* 1996). Combination of 18S data with morphological characters for a reduced set of taxa results is a sister group relationship between mosses and polysporangiophytes (Mishler *et al.* 1994). The remaining studies converge toward a phylogenetic hypothesis wherein mosses and liverworts compose a monophyletic clade sister to the polysporangiophytes (Fig. 32.6F). This hypothesis was first proposed by Headderson *et al.* (1996) on the basis of 18S sequence data analyses, and has subsequently been corroborated by various other data sets (Fig. 32.6F), including the latest morphological data set (Renzaglia *et al.* 2000), and a combination of sequences derived from all three genomes (Nickrent *et al.* 2000).

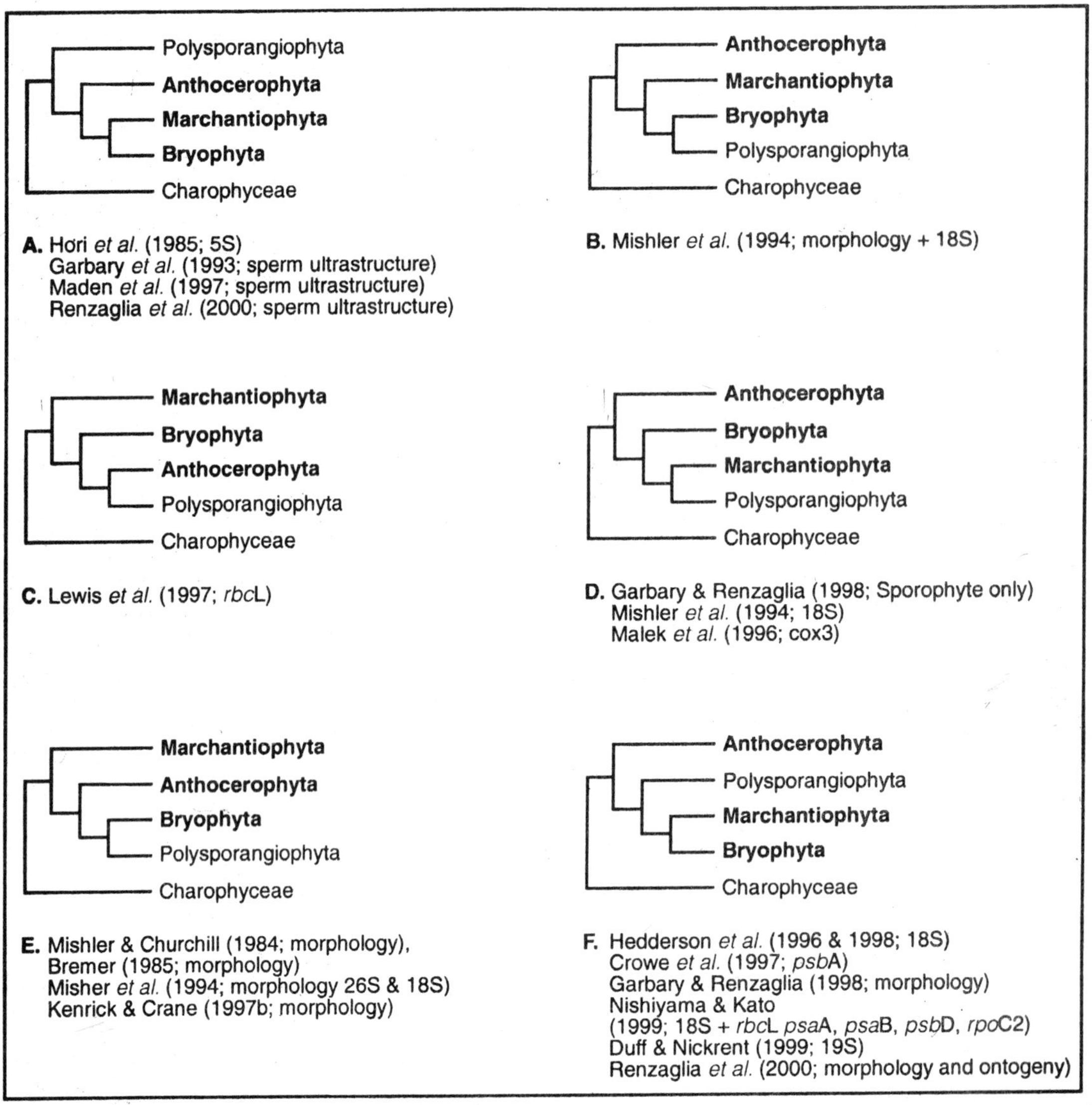

Fig. 32.6: Summary of alternative phylogenetic relationships among extant lineages of land plants, based on the assumption that land plants are composed of four monophyletic lineages, i.e., hornworts, liverworts, mosses, and vascular plants. Only studies including exemplars of all four lineages are considered here

Chapter 33

Some Aspects of Bryophytes

Perennation of the Bryophyta

Besides sexual reproductivity species, most bryophytes have to modes perennial vegetative reproduction. Species growing in very most situations seem never to die, its activity being checked only during, severe winter or the plant may curl up when the season is very dry. Most of the plants are perennial mosses drying up in the dry season leaving traces which regenerate or develop protonemal threads from any part of the dried up body as soon as the first showers fall. Thus in the tropics, bare rocks or walls are seen suddenly to turn green as soon as the dry season is broken up by rains. The hepatics form regular tubers which hibernate through unfavourable periods but burst up into new shoots with the first shower. Gammae are a common mode of perennation in Hepaticopsida as well as Bryopsida. Many species in the plains (e.g., *Calymperes*) do not form sporophytes but perennate by gemmae alc.ie. Even in the hills many species of Marchantiaceae rely more in the gemmae than the spores for perennation.

Water absorption and conduction in Bryophyta

As has been stated, Bryophytes thrive only in a saturated atmosphere. Most Bryophyte absorb water through their whole surface. Thus, *Sphagnum* absorb water by their whole body, many times their own weight. Many creeping Jungermanniales or mosses also grow in only a wet situation. Marchantiales absorb specially by their vertical scales and rhizoids. The question of any water conduction comes in only in the case of very erect bryophytes like the Calobryala or Takakiales, Polytrichales.

BRYOPHYTE ECOLOGY

Bryophytes are affected by the same climatic and edaphic factors as the other plants, but sometimes more so. Species vary in their adaptability to the different factors. Some, e.g., *Hypnum cupressiforme* is said to withstand pH values 4.6 to 8.6 in the substrate and can grow equally on wood, stone or soil. But, others grow on narrow ecological limits so that they may be termed as *indicator plants* of certain conditions. Among these may be mentioned the very common *Simibarbula orientalis, Hyophila involuta, Bryum coronatum, Tortella* or *Neckera* which are *calcicoles*. They grow only on walls, bricks or stone containing some amount of lime. These calcicolous plants hate acid soil. On the other hand *Polytrichum, Pogonatum* and *Sphagnum* are accedophilous as they prefer acidic soil. *Pogonatum aloides, Funaria hygrometrica, Brachymenium acuminatum, Bryum ghatense* grow of laterite and show xerophytic adaptation here because of. excessive iron. *Merceya* also grows on laterite here. The genus *Merceya,* quite common in the Himalayas, which is termed *copper moss* because it indicates the presence of copper (sometimes iron) in the substrate. It should, however, be noted that bryophytic rhizoid system penetrates only a few mm of the substrate—a very different pH or chemical condition may exist only a few inches below. Because of this, some such plants are called **Indicator plants** as their presence indicate the chemical nature of the substratum. Some epiphyte mosses are known as highly sensitive to smoke so that they thrive only at a non-smoky height on the trees. Such mosses are indicators of pollution.

The following ecological factors need special mention:

1. **Water** is of primary importance. It has already been stated that Bryophytes are amphoebean in nature as a part of their life cycle (the sexual phase) cannot be completed unless they are fully saturated in water. Nevertheless, bryophytes can withstand great desiccation. Such highly desiccated bryophytes form a common sight in the bare exposed rocks during the dry phases. They are not really xerophytic. A few hours rain revive these plants to their normal moist shape. It is only a phase of hibernation. Bryophytes can absorb moisture by their whole bodies of which the rhizoids may bear insignificant part. Bryophytes can perennate in this manner through years. They are really totipotent. It has been found that almost every cell of the gametophyte or the sporophyte is capable of giving rise to a secondary protonema thread which buds into newer gametophytic plants (haploid or polyploid). Dried bryophyte specimens (not chemically killed) are known to give rise to new plants on moistening after decades of hibernation in herbaria.

Most mosses are helophytes. Some of the mosses, however, need a saturated atmosphere for ideal growth and are abundant near waterfalls. Aquatic bryophytes are not very common in India. Among the few are *Riccia fluitans* and *Ricciocarpous natans. Fissidens diversefolius* grows on river-beds and withstands a submerged condition for sometime. *Fissidens grandifrons* grows floating on the edges of Kashmir streams. *Sphagnum* is aquatic on peat bogs in other countries, but such bogs are rare in present India (they are known more extensively from the Quaternary Himalayas) and grow only in or near running streamlets where *Actinothuidium hookeri* also grows.

In its water selection it is not merely the amount of water that it receives. A heavy downpour in an open tropical rain forests may be less significant than the heavy saturation in a comparatively cool shady location. Epiphytes and epiphyllous bryophytes thrive only in the latter situation.

2. **Temperature** in the microclimate (i.e., just round about of the bryophytic vegetation) is more important. These may be cooler shades in tropical situations and warmer situations in temperate or arctic lands which are favoured. While bryophytes may withstand both very high and very low temperatures for some time, luxuriant growth is possible only in the optimum temperature of the particular species.

3. **Light** is certainly a factor affecting bryophytic growth but its actual effect is to be studied species by species. Generally, it is found that most species prefer a situation of shade with partial exposure of light. In forest canopies, it is usual to find that while gametophytic growth is heavy in shade, sporophytes are abundant only in situations where sunlight is available for sometime. Some shade loving

species like *Fissidens nobilis* which grow only in the shade of East Himalayan caves fruit only on the edges which are somewhat lighted. On the other hand, many species of *Plagiochasma, Targionia, Anoectangium, Ditrichum, Microdus* etc. grow quite well on exposed rock faces as they can withstand desiccation.

It is apparent that the above three factors, along with wind have a combined effect on bryophytic life and should be considered in its totality and not individually.

4. **Edaphic factors** are mainly the conditions of the substrate for which most species have their individual likings. While pH value, chemical and physical constitution of the substrate must be a guiding principle, it is found that species are adapted to the particular substrate—terrestrial (on humus, argillaceous or arenaceous soil), rupestrine (it may differ in being silicic, calciferous, granitic, etc.) or epixylic. The pH value of *Sphagnum* bogs has been greatly studied. Extreme edaphic situations may be determined by the point of toleration and not necessarily any particular liking.

BRYO-CENOLOGY (BRYO-SOCIOLOGY)

Cenological study is usually taken up from three different angles: (1) *Synecological* studies involving the studies of the different communities and societies in relation to the ecological conditions; (2) studies on *succession* involves the study of different species which dominate successively from the beginning of bare or virgin tracts; and (3) *autecological* study involving the same species occurring in different situations.

Not considering the latin names we may consider the following synecological societies.

1. Floating, not fixed by rhizoids—there are very few species in India as have already been mentioned (*Riccia fluitans, Ricciocarpus*). Some mosses found on firm ground may also grow floating.

2. Fixed on firm ground with rhizoids

(1) Periodically submerged: *Fissidens diversifolius* and *F. grandifrons* have already been mentioned. *Sphagnums* also come in here.

(2) Growing on rocks: Numerous mosses on the hills are fixed on rocks (rupestrine). Species may have likenings for different types of rocks (siliceous, granitic, calcareous on shales, copper mosses etc.) of different pH values.

(3) Growing on loose soil: These are characteristic on the vast planes of India. The soil may be of different physical and chemical constitution. Many soils on the Central India dry hills are actually on disintegrated rocks. In the Gangetic plains many species like *Hydrogonium* could grow only on clay soil while *Hyophila* or *Semibarbula* would only grow on calcarious substrates (bricks and mortar).

(4) Epiphytic bryophyta: These are the very luxuriant forms of mosses which grow on the trees in the wet tropical, subtropical and temperature forests. In a dense forest they form a very lovely sight. Most pleurocarpous mosses are *epixylic* (growing on tree barks, and there also some acrocarpic forms like *Calymperes*,

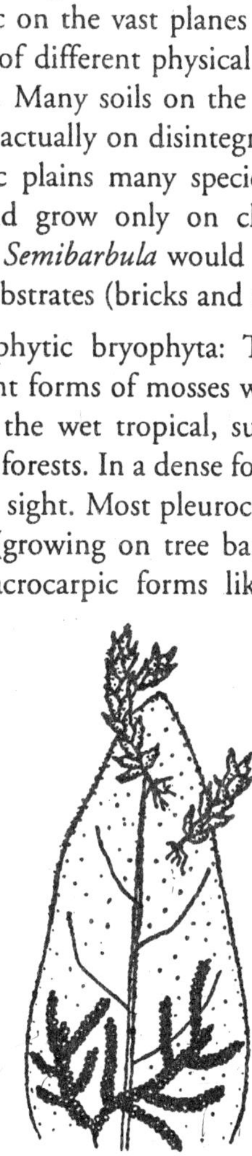

Fig. 33.1: Epiphyllous Bryophytes. A leaf in a semidark wet underbush in Mungpoo. A *Lejeunea* is growing at the base of the leaf and two small morses on the apical leaf

Stereophyllum, Syrrhopodon etc.) Epiphytic forms often grow down into rotten logs and soil—usually wet humus. In many species the growth is so vigorous that after creeping on the barks of the branches the plants hang down in pendulous form (Fig. 31.32).

In very wet forests bryophyta develop on leaves (*epiphyllous*). In the Eastern Himalaya epiphyllous Lejeunoid hepatics are very common (Fig. 33.1).

(5) Ephemeral and Xerogeophytic forms: It has already been stated that many bryophytes dry up in a turfaceous or dry wooly form on bare rock surface during drought period. They mostly revive after a few hours of rain. They actually perennate by different devices. Some xerophytic characters (among which may be mentioned formation of tubers) may develop. Among them are many Marchantiaceae, Targioniaceae, Dicranaceae, Syrrhopodontataceae, Pottiaceae, Grimmiaceae, Bryaceae, Sematophyllaceae etc. There are also some *ephemeral* forms which last for one season only (*Ephemerum*) and apparently perish but often perennate by the persistent protonema.

(6) Peatforming on very wet or inundated soil: *Sphagnum* dominates in this group but they rarely form peats in India today.

(7) There may be special groupings like *Chasmiophyta* (growing in rock crevices c.f. *Fissidens nobilis*) and *Psammophyta* (growing on sand viz. *Fissidens splachnobryoides*).

Bryophyta societies are studied by noting different plants (of all groups) growing in the same community and the occurrence, of the Bryophytes in different heights of trees etc. A community is named by the dominating plant, e.g., *Sphagnata* where *Sphagnum* dominates. Studies of Bryophytic societies have been done very thoroughly in Europe and in Japan.

Some erect and strong Eubryales: The question of a conducting tissue arises only in these later cases. It has already been shown that Ruhland and others pictured a hydrom core and a leptom mantle in the tuberous part of *Polytrichum* stems. A core of elongated, dead cells have been found in the rhizomatous (rhizoid-less) base of the Calobryales and the Takakiales and the stems of mosses like *Pogonatum, Funaria, Mnium* etc. Some have detected a slow transpiration current in this central strand. But, these cells are not lignified and it is difficult to consider it as a xylem core. It is quite possible that these are vestigial organs if we believe that there had been a regressive evolution—the present day Bryophytes arose out of more developed plant types. But, then, the complete absence of lignin has to be explained.

BRYOGEOGRAPHY AND CONSERVATION OF BRYOPHYTES

Introduction

Except for the marine environment, bryophytes, as a group, are nearly cosmopolitan in distribution. Because of the small size of spores and the frequent occurrence of vegetative propagules, bryophytes are easily dispersed across the landscape. Consequently, many show much wider distribution patterns than the seed plants (Watson 1974). This is especially true at the family and generic levels. The purported greater age of bryophytes may have contributed further to the wider range of bryophytes (Schofield 1992).

The ecological and biological factors shaping the distribution patterns of seed plants have also affected the bryophytes. Indeed, like the seed plants, many bryophyte species found in the arctic and boreal zones are the same in North America, Asia, and Europe. In the tropics, there is a significant number of moss taxa that are pantropical, but the number is far lower when compared to the circumboreal taxa.

Although a great number of bryophyte genera and species are cosmopolitan and distributed throughout different climatic regions, many do exhibit a disrupted or narrow range. The latter represent the uncommon, rare and endemic bryophytes. Their ultimate survival is critically dependent on the preservation of their natural habitats.

Factors affecting the dispersal of bryophyte diaspores

Diaspores are defined as any propagative parts of bryophytes, be they spores or gemmae, capable of giving rise to a new individual. Among the bryophytes, few members, like the moss family Splachnaceae, produce sticky spores that are dependent on flies for their dispersal. Instead, many bryophytes are dispersed over great distances by air currents. A few bryophytes that display a bipolar distribution in arctic and subantarctic regions have been alleged to be the outcome of transequatorial dispersal brought by migratory birds. Others are thought to travel over long distances via the jet stream in the upper atmosphere. However, the majority of bryophytes probably disperse across the land by slow expansion through forest and by island or mountain hopping. This hypothesis is supported by the fact that spores of many bryophytes are ill-equipped against desiccation, extreme temperature, and strong UV radiation (van Zanten 1976, 1978). Furthermore, many epiphytic mosses and epiphyllous liverworts produce green spores that are rather short-lived and will not survive long-distance travel.

According to van Zanten and Pócs (1981), Hepaticae, in general, is less effectively dispersed by air currents than Musci. One of the reasons is the small spores of many tropical rain forest hepatics do not seem to tolerate desiccation. Accordingly, a spore size less than 25 μm is best suited for air transport over long distances within a climatological belt.

Imura (1994) reviewed the vegetative diaspores of 186 out of 1183 species of Japanese mosses. He concluded that specialized vegetative diaspores are found more frequently in dioicous than monoicous mosses.

After dispersal, the next critical factor is the establishment of a viable population at the new site. This is where the ecological requirement and adaptation of the bryophyte come into play. A few on-site studies showed that diaspores of many more species of bryophytes had landed but failed to germinate and mature (Miller and Ambrose 1976).

Interestingly, more than half of the bryophytes in the world are dioicous. This poses a problem in sexual reproduction when the gametophyte of only one sex arrives at a distant locality. The subsequent survival and expansion of the population must depend on asexual propagation. Such is situation of the present-day widely disjunctive female populations of *Takakia lepidozioides* found in Alaska, British Columbia, Japan, Himalayas, and on Mount Kinabalu of Borneo.

The changing geographical position of land masses brought about by plate tectonics and the climatic vicissitudes through Tertiary and Quaternary periods are two other factors shaping the modern distribution of bryophytes. Fossil records, such as corticolous mosses and liverworts preserved in ambers, have revealed the existence of subtropical to tropical floras in many parts of Europe during the Tertiary period (Grolle 1988, Frahm 1996).

In modern times, humans have become the unwitting vector of bryophyte diaspores. A number of bryophytes have been introduced well beyond their natural range and have become weedy in foreign lands. Well-known examples of alien bryophytes are the European *Pseudoscleropodium purum* in the southern hemisphere and North America, and the southern hemisphere *Orthodontium lineare* and *Campylopus introflexus* in Europe, especially in the UK (Söderström 1992). Furthermore, several bryophytes have become adapted to man-made habitats and are found more commonly around human settlements than in their natural habitats. Examples are *Barbula unguiculata, Bryum argenteum, Hyophila propagulifera, Leptobryum pyriforme, Tortula pagorum,* and *T. muralis.*

Major patterns of bryophyte distributions

In many parts of the world the diversity and number of taxa are greater in mosses than in liverworts and hornworts. The exceptions are the extreme wet or dry environments, e.g., the exposed forest canopy, alpine/tundra habitats, super-humid montane forest, and wet coastal vegetation, where the increasing trend of diversity is in favour of liverworts.

Modern distribution of plants is affected primarily by the prevailing climatic pattern and, secondarily, by the available habitat. Herzog, in his monumental work *Geographie der Moose* (1926), outlined several distribution patterns of bryophytes that are still recognized today. To understand the bryogeography of a region, it is imperative to dissect the flora into its floristic components by grouping together the species exhibiting similar distribution patterns. From such a floristic analysis, the origin and evolution of the flora can be deduced.

The worldwide distributions of bryophytes can be grouped into five major patterns following closely the prevailing climatic zones of the world. Minor patterns of distribution such as bipolar and other disjunctions are discussed separately below. Within each major category, the range of a species can be continuous or discontinuous depending on other factors such as substrate and moisture. Because of the presence of microhabitats that are often different from the macroenvironment in the region, a large number of taxa, especially those with broad ecological amplitude, can be observed to have a total distribution combining two to three of the five patterns outlined below.

1. Arctic-alpine distribution

There is a good number of bryophytes that are widespread across the Arctic Circle. Many have spilled south towards the Equator along high mountain summits that support alpine vegetation. Examples are species of *Andreaeobryum, Anthelia, Arnellia, Arctoa, Conostomum, Kiaeria, Oedipodium, Oreas, Stegonia, Tetralophozia, Voitia,* and *Plagiothecium berggrenianum.* Because of the proximity of lands in the Arctic and the availability of ice-free areas serving as refugia during the ice ages, the arctic flora today has an impressive number of bryophytes close to 600 (Longton 1988).

2. Boreal distribution

These are the species found abundantly across the northern coniferous forest belts in all major continents. Many have extended their ranges northward into the colder arctic zone or southward into the warmer temperate zone, especially in the mountains. Most have continuous ranges across the three continents, namely Europe, Asia, and North America; hence, the term circumboreal taxa. Examples are species of *Abietinella, Aneura, Blasia, Catascopium, Cinclidium, Pellia, Pleurozium, Ptilidium, Ptilium, Tortella, Schistostega, Splachnum,* and *Warnstorfia.*

3. North and south temperate distributions

These are the warmth-living bryophytes found mainly at low latitudes in northern and southern hemispheres. They are important plant components of semi-evergreen, broad-leaved, and mixed deciduous temperate forests. They may show a continuous distribution confined to one mountain range or a disruptive range between two continents.

3(a). North temperate distribution

Examples of bryophytes showing a north temperate distribution are species of *Acrobolbus, Anastrophyllum, Brotherella, Entodon, Gollania,*

Hookeria, Leucodon, Neckera, Schwetschkeopsis, and *Tritomaria.*

In historical bryogeographical parlance, bryophytes that exhibit boreal and temperate ranges in the northern hemisphere are considered Laurasian elements. This is because many have their alleged origin from the ancient Mesozoic landmass called Laurasia. Gradstein and Vána (1987) mapped the world distributions of several Laurasian hepatics with broad ranges in the northern hemisphere reaching the tropics at high altitudes. Examples are *Blepharostoma trichophyllum, Lepidozia reptans,* and *Nowellia curvifolia*

3(b). South temperate distribution

Because of the land configuration, the distributions of bryophytes in the southern hemisphere are discontinuous. Owing to their alleged origin from another Mesozoic supercontinent south of the equator, Gondwana, the bryophytes of the southern hemisphere are called Gondwanic taxa. Examples are species of *Balantiopsis, Calomnium, Catagonium, Carrpos, Echinodium, Gigaspermum, Herzogiaria, Hydropogon, Hygrolembidium, Lepyrodon, Marsupidium, Monoclea, Neomeesia, Pleurophascum, Rhaphidorrhynchium, Verdoornia, Vetaforma* and *Wardia.* Like their counterparts in the northern hemisphere, many Gondwanic taxa also reach the tropics in their distributions. Some even successfully penetrated into the northern hemisphere. Examples of the latter group are species of *Bescherellia, Dawsonia, Leptostomum, Plagiotheciopsis,* and members of Phyllogoniaceae.

4. Tropical distribution

There are as many bryophytes found in the tropics as in the temperate regions of the world. The magnitude of diversity of temperate bryophytes rivals well the acclaimed high diversity of tropical bryophytes. This evidence has been taken to support a temperate origin of bryophytes.

About 80% of bryophytes of all the tropical rainforests belong to tropical families such as *Calymperaceae, Fissidentaceae, Meteoriaceae, Neckeraceae, Pterobryaceae,* and *Sematophyllaceae* among *mosses,* and *Frullaniaceae, Lejeuneaceae, Lepidoziaceae, Plagiochilaceae,* and *Radulaceae* among *liverworts* (Gradstein and Pócs 1989). Many of them are best represented in the tropics but are also fairly widespread in temperate zones.

Less than 20% of tropical bryophytes are pantropical like *Erpodium biseriatum.* The majority are either confined to the Old World tropics (paletropical) or the New World tropics (neotropical). Examples of paleotropical bryophytes are *Cephaloziella kiaerii* and genera such as *Cladopodanthus, Conoscyphus, Cuspidatula, Desmotheca, Macvicaria, Notoscyphus, Papillidiopsis, Ptychanthus,* and *Trichocoleopsis,* whereas examples of genera of neotropical bryopytes are *Aureolejeunea, Caribaeohypnum, Cyclolejeunea, Curviramea, Hildebrandtiella, Puiggariella, Ruizanthus, Vanaea,* and *Vitalianthus.*

In the Old World, equatorial Africa has both a lower diversity and a smaller number of bryophytes than equatorial Asia. Recent investigation shows that a handful of paleotropical mosses, especially species of taxonomically difficult genera, such as *Acroporium* and *Trichosteleum,* has been treated under different names in tropical Africa and tropical Asia. There is thus a need for critical taxonomic comparisons of moss floras of the regions under study before an accurate plant geographical analysis can be made.

5. Subantarctic and antarctic distributions

To data, some 160 genera and 380 species of bryophytes have been reported from the subantarctic and antarctic regions. The bryoflora of the Antarctic souch of 60th latitude in the southern hemisphere has around 100 mosses and 25-30 hepatics (Seppelt *et al.* 1998).

The rather limited diversity of subantarctic and antarctic mosses includes several taxa of acrocarpous mosses such as species of *Andreaea, Bartramia, Bryum, Campylopus, Ceratodon, Distichium, Ditrichum, Grimmia, Racomitrium, Polytrichum, Pottia, Sphagnum,* and *Tortula,* in addition to a few pleurocarpous mosses. The latter belong mostly to the widespread genera *Amblystegium, Brachythe-cium, Calliergon,* and *Drepanocladus.* The liverworts with restricted subantarctic and antarctic ranges are mostly members of Gymnomitriaceae, Jungermanniaceae, Lepidoziaceae, Lophoziaceae, Lophocoleaceae, and Cephaloziaceae. A good example is the genus *Herzogobryum* of Gymnomitriaceae.

Many species of bryophytes found in the Subantarctic and Antarctic are locally endemic. In his discussion on the possible origin of the antarctic moss flora, Robinson (1972) doubted the antiquity and validity of the many local endemics. True enough, later revisions have shown that a number of the local endemics are but incorrectly determined morphotypes of much more widespread taxa (Seppelt *et al.* 1998).

Surprisingly, there are two endemic families of liverworts described from the subantarctic region. They are Vetaformaceae (monotypic) and Phyllothalliaceae (with one genus and two species). The latter reaches as far as Patagonia in South America.

Disjunctive distributions

This is the most fascinating and much-studied pattern of plant distribution. The causes of disjunction in plant distributions are many and are difficult to ascertain because of the incomplete pictures of past and present distributions of plants around the world. Most likely, the causal events leading to range disruption very from species to species. Usually, plate tectonics and climatic changes in the region are the two principal causes. Allen (1996) suggested that some disjunctions were better explained by long-distance dispersal followed by short land and air dispersals across the continent.

Plant disjunction can be local between two nearby mountains or thousands of kilometers apart across an ocean. Some bryophytes even show tri- to quadricentric disjunctions across the continents or oceans.

The well-known transcontinental or transoceanic disjunctions are the bipolar amphi-Atlantic Mediterranean-Californian-Australian, East Asiatic-Eastern North American, circum-Northern Pacific, and South American-Australasian disjuncts. These disjunctive patterns have been reviewed repeatedly by many, including Iwatsuki and Sharp (1967), Schofield and Crum (1972), and Schofield (1974, 1984, 1985, 1988). Other less dramatic disjunctions, such as the Eastern-Western North American (Schofield 1980, Belland 1987), North-South American, and intra-Malesian disjunctions (Touw 1992, Tan 1998), are of regional interest.

At present, fewer than 60 bryophytes show an amphi-Pacific disjunct between the paleotropics and neotropics. In contrast, there are more bryophytes, such as species of *Arachniopsis, Bryopteris, Renauldia,* and *Symbiezidium,* that have a disjunctive range between tropical Africa and tropical America. Frahm (1982), Gradstein *et al.* (1983), Buck and Griffin (1984), Reese (1985), Ochyra *et al.* (1992), and Allen (1996) contributed much to the understanding of African and South American disjunctions. According to Delgadillo (1993a), the number of mosses exhibiting a broad neotropical-African disjunction pattern total more than 334 species. This is a reflection of the long history of juxtaposition of the two continents before the breaking-up of Gondwana in early Cretaceous time.

Pócs (1976, 1992) has made significant contributions to the knowledge on tropical African and tropical Asiatic disjunction of bryophytes. A total of 70 liverworts and 108 mosses were identified to have Afro-Asian bicontinental disjunctions.

Recently, Schofield (1994) identified the Pacific coast of North America as the more likely place of origin for xerophytic bryophytes now inhabiting disjunctively the Mediterranean climatic regions in the world. His interpretation is based on the wider ecological tolerance and the morphological variability of a high proportion of so-called Mediterranean taxa in Pacific North America than in Eurasia and Africa. His viewpoint on the origin of Mediterranean flora differs from that of Frey and Kürschner (1988) who suggested an ancient Tethyan Sea origin. This underscores the fact that, with the accumulation of distribution data, there will be new insight into and reinterpretation of existing hypotheses and principles governing the study of plant geography.

Vicariance and vicariads

In plant geography, two closely related forms or species found in widely separated places are called vicarious species, or simply, vicariads. Like plant disjunction, the causes of vicariance in plants are many. Theoretically, vicariads are formed by the fragmentation of a single parental taxon. The process involves: (1) formation of geographical, climatic, or ecological barriers leading to the fragmentation of a parental population, and (2) the speciation of fragmented populations in isolation.

Vicariance has been extensively investigated in seed plants but not among the bryophyte taxa. Schofield (1980) and Iwatsuki (1992) provided 12 examples of moss vicariads in Eastern Asia and Eastern North America, e.g. *Bruchia microspora* and *B. sullivantii*, *Climacium japonicum* and *C. americanum*, *Drummondia sinensis* and *D. prorepens*, *Diphyscium involutum* and *D. cumberlandianum*, and *Syrrohpodon kiiensis* and *S. texanus*, Gradstein *et al.* (1983) listed 16 pairs of vicariad taxa of liverworts between tropical America and Africa. Some of their examples are *Calypogeia peruviana* and *C. afrocoerulea*, *Frullania riojaneirensis* and *F. africana*, *Plagiochila columbica* and *P. integerrima*, and *Symbiezidium barbiflorum* and *S. madagascariense*.

Plate tectonics and bryophyte distribution

The theory of continental drift published by Alfred Wegener is 1915 which influenced the study of sea-floor spreading and plate tectonics is perhaps the most important geological concept to have revolutionized the study of modern biogeography. Indeed, the reconstruction of the history of continental drift through geological times has provided a solid basis for plant geographers to understand and explain the taxic similarity and floristic affinity between regions that are widely separated today.

Raven and Axelrod (1974) presented a useful review of the continental drift events and their significance in the phytogeography. According to them, the breaking-up of the supercontinent called Pangaea into two large landmasses, Laurasia and Gondwanaland, some 180 million years ago, and the final drifting of the continents to their present positions some 5 million years ago, constitute a critical sequence of events in the long history of the earth. The intermittent collision of continental and oceanic plates not only generated massive mountain ranges serving either as geographical barriers or corridors to plant migration, but also brought together disparate floras from various geographical sources to mix freely and co-exist.

Recently, Frahm (1994) compared the *Campylopus* floras of Sri Lanka and Madagascar. Although the two islands are widely separated today by the Indian Ocean, both had a common origin as part of the former Gondwana. It is no wonder that two-thirds of the *Campylopus* taxa of these two islands were found to be similar or very closely related to the African flora.

On a worldwide basis, Schuster (1982, 1983a) reviewed the family and generic distributions of bryophytes in relation to the plate tectonics. Earlier, Schuster (1972) presented a good discussion of the historical migration of Gondwanic elements into the rainforests of south-east Asia via the northward drifting of the Indian and Australian plates. His studies revealed further that many antipodal liverworts possess "primitive" characters and widely disjunctive ranges in the southern hemisphere.

However, not all of the currently observed patterns of plant distributions can be attributed to the influences of plate tectonics. In a number of cases, long-distance dispersal of diaspores may be the primary causal mechanism.

Bryophyte endemism

Bryophytes having small or narrow ranges, be they continuous or discontinuous, are recognized as endemics. Distinction should be made between two kinds of endemics: neo-endemic and paleo-endemic. The former refers to recently evolved species that have not had time to achieve a wider distribution, whereas the latter refers to ancient, surviving species with a shrinking range. In practice, it is difficult to establish the true nature of an endemic unless the fossil records can provide unequivocal support. As a general statement, the endemics described from young geological formations, such as the Andes, Himalayas, or Mount Kinabalu, can be referred to as neo-endemics.

Schofield (1985) stated that endemism in the bryoflora is positively correlated to the following three conditions: (1) the length of time during which the locality was available for colonization, (2) the environmental diversity, especially in the availability of moisture, and (3) the length of time in isolation.

Interestingly, the percentages of generic and species endemism of bryophytes are lower than those calculated for the seed plants in many parts of the world. Furthermore, the estimated percentages of bryophyte endemism for the temperate and tropical rainforests appear to be similar.

Areas in the world with reportedly high bryophyte endemism and diversity. Several of them are large islands with a long history of isolation. Others are topographically diverse, mountainous areas. Considering the alarming rate of destruction of natural habitats worldwide, efforts should be organized to protect as many endemics in these areas as possible.

Although more than 75% of the families of bryophytes are widespread in both northern and southern hemispheres, there are a few endemic families with rather narrow ranges. These endemic families are either stenotypic or monotypic. They are: Pseudoditrichaceae, restricted to arctic Canada; Gyrothyraceae, restricted to the Pacific cost of North America; Rutenbergiaceae, restricted to coastal Tanzania, Madagascar, and adjacent islands, and Hydropogonaceae, restricted to tropical and subtropical South America.

In Western Melanesia, Piippo (1992) calculated a high endemism of 38% for the entire hepatic flora. Percentages of endemism, however, vary from family to family, such that 55% in Plagiochilaceae, 61% in Frullaniaceae, 27% in Lepidoziaceae, and none in Herbertaceae and Marchantiaceae.

Island biogeography and bryophytes

Many islands are known to produce endemics in spite of the fact that some of the recognized endemics may simply be the result of insufficient survey of the floras of adjacent continents. Two islands well known for their high endemism of seed plants and bryophytes are Madagascar and New Caledonia. In the case of New Caledonian *Fissidens,* Iwatsuki (1990) reduced significantly the percentage of species endemism from 80.6% to 10.7% after a careful revision. However, the percentage of species endemism in New Caledonian *Frullania* remained high even after a modern taxonomic revision (Iwatsuki 1990).

There are two kinds of islands from the viewpoint of plant geography, namely, continental and oceanic islands. The former became part of a continent when the sea level lowered during the ice ages. The past land connections permitted movement of plants from the continent to the island in small numbers. Predictably, the flora of a continental island is like a reduced version of the large continental flora. Such is the case of the moss flora of Sri Lanka vis-a-vis the flora of continental southern India. For similar reason, the bryoflora of Taiwan is closely related to the much larger flora of continental China.

Oceanic islands, on the other hand, lack any land connection with a continent. Their origin from the ocean floor precluded any direct landward migration of plants into the island. Most of their diaspores arrived via long-distance dispersal. The flora of oceanic islands therefore is a mixture of chancy and adventive plant taxa from various distant sources. A good example is the moss flora of the Hawaiian Islands.

The theory of island biogeography made famous by MacArthur and Wilson (1967) has not been tested with bryophytes. There has been no experimental study on rate of migration and rate of extinction of bryophytes under the island conditions. Based on limited data of bryophyte distributions, the alleged positive correlation between island size and species diversity appears to be valid (cf. Pócs 1988). However, Frahm *et al.* (1996) found that the topography and elevational variation, more than the island size, affect the number of bryophyte taxa present on an island. As to the relationship between the distance of the island from the continent and the degree of diversity of its bryofloras, the few recent studies (Menzel and Passow-Schindhelm 1990, Dirkse *et al.* 1993, Frahm *et al.* 1996) unfortunately do not show any clear correlation. Perhaps the barrier created by the distance of the island from the continent has been overcome by the efficacy of dispersability of air-borne diaspores of many bryophytes over the years.

Piippo (1992) and Piippo and Koponen (1997) reported that the floristic affinities of mosses of the island groups in Western Melanesia and Oceania are more closely related to Asia than to Australia and South America. Equally interestingly, Tan and Engel (1990) observed some differences in the patterns of distributions of mosses and liverworts in Borneo, New Guinea, and the Philippine archipelago. For large moss genera, they reported many species shared by the three large island groups. However, the same pattern is not seen among large liverwort genera. Instead, each of the large liverwort genera has its own set of endemics present on each of the island groups. According to them, these differences in distribution patterns between mosses and liverworts probably resulted from the differences in their ecological adaptation and reproductive biology evolved under island conditions.

DISTRIBUTION OF BRYOPHYTES IN INDIA

From what has been stated before, it is easy to understand that for a luxuriant bryophytic community, there must be a good amount of moisture (even to the point of saturation), a good amount of shade (occasional sunlight is necessary for fruiting, *i.e.*, development of sporophytes) and comfortable coolness. The distribution of species varies according to the adaptation of the species to different variations of these conditions.

Considering the geography and the topography of India we can divide the subcontinent into the following regions :

(1) *The Himalayas* spreading from Kashmir to the farthest east of India reaching Burma. Climatic conditions vary greatly along the Himalayas. Western Himalaya is drier and cooler. As we move to the east the climate is wetter and wetter and gradually somewhat warmer. So that the perpetual snow line goes higher and higher. Again, rainfall is heavier in the outer Himalaya bounding the Terai gradually decreasing upwards and there is very scanty rain behind the Inner Himalaya. In the Terai there is a tropical forest which is dense and of the monsoon type in the east abounding in Bryophytes.

(2) *The Indus-Ganga Plains.* This vast area has been greatly deforested and heavily populated for more than one thousand years and is generally warm. Long ago it was a region of monsoon forests (specially in the east) which have disappeared. Tropical Bryophytes are formed here but they are quickly disappearing now. The Gangetic delta mangroves are practically devoid of Bryophytes as they avoid the saltish water.

(3) Central India is generally dry and Bryophytes which can withstand some amount of drought are only to the expected here. Actually this area is the northern fringe of the Deccan Plateau and extends to the Chhotanagpur Hills and all northern Orissa Hills. Rajputana is a very arid zone. Even in the generally dry area there are a few high peaks (viz., Parashnath Hills, Koraput hills, etc.) with a sizeable Bryophytic flora.

(4) South India or the Deccan spreading from Western to Eastern Ghats. Central India is only the northern fringe of it. It is generally dry but there are some very moist suitable spots (viz., the Nilgiris and the Palni Hills) where there is a big range of tropical and subtropical Bryophytes.

HEPATICS

The Hepatics are even more moisture-loving and follow the topographical conditions stated above. The species that can stand drought (viz., *Plagiochasma, Asterella, Targionia*) are found on the bare rocks with limited rainfall viz., some exposed parts on the Western Himalaya, the Ghats etc. Beyond the W. Himalaya, Kashyap did not find a single species in Western Tibet. Most Marchantiales species are found in the Eastern Himalaya and only the moister regions of the Western Himalaya and South India. *Marchantia polymorpha,* however, is cosmopolitan reaching the highest ranges all over India. The long trailing Jungermanniales are very rare in the Western Himalaya but are abundant in the Eastern Himalaya. Epiphyllous Jungermanniales (mostly of the Lejeunoid group) are often to be found in the perpetually wet regions viz., Mungpoo area of Darjeeling; Moosmai, Mawflong and similar regions in Khasia Hills, and similar spots in N.E. India, the beautiful Falls and similar wet spots in South India. Even *Calobryum* and *Haplomittrium* were found by Udar and his associates from the Eastern Himalaya. Completely aquatic hepatics are rare but *Ricciocarpus natans* has been reported from different areas spreading from the Dal Lake in Kashmir to water channels in the suburbs of Calcutta. *Riccia fluitans* has been found in water in the W. and E. Himalayas and South India. *Riella viswanathai* was found in shallow water in Varanasi while *Riella indica* has been illustrated from Shalimar garden water channels in Lahore.

Maheswari *et al.* analyses the hepatic species (Gazette of India, 1965) as follows: Total species in India—about 672 (Parihar listed earlier 689 species in 116 genera from 29 of a total of 55 families recognized by Evans). Of the 672 species 330 are in the Eastern Himalaya, 225 in South (including Central) India and 170 species in the Western

Himalaya. Of the Eastern Himalayan species a great number are Indomalesian and a few common to China (12 spp.), Japan (26 spp.) and Australia (12 spp.) Some Darjeeling-Sikkim species are also found in Yunnan which was always connected with Eastern Himalaya. The Western Himalayan species show a great number of European and a few common to China (12 spp.) and Japan (13 spp.). South Indian species are Indomalesian with a few common to China (7 spp.), Japan (23 spp.) and East Africa (13 spp.). The Western Ghats species show African and the Eastern Ghats species Indomalesian relationships.

MOSS

Regional distribution of the Mosses follow the same pattern as that of the Hepatics. While Mitten in 1969 listed only 777 species from India, Tibet, Ceylon, Burma, Malay, Singapore and Hongkong (of these 500 are from Eastern India) Gangulee has now described and illustrated 990 species (in 274 genera of 52 families) from the Eastern India region alone. In his analysis of the floristic distribution of these species (1980, pp. 2041-2082) the following features come out :

1. **Sphagnideae**

Sphagnum: They need maximum moisture (trickling water), and a reasonable amount of coolness and shade. Naturally their distribution is rather limited. There are 25 species in India (out of a total of 207 spp. all over the world) of which 20 spp. are in Eastern India, only 3 spp. in Western Himalaya and only 1 sp. in South India. Of the 20 Eastern Indian species, one is endemic, 12 are cosmopolitan (some of these limited to the North Hemisphere), 6 common to Yunnan and 1 common to Africa.

2. **Andreaeideae**

Andreaca: These are found in Alpine region. In India they are only found in the granitic core of Inner Himalaya which is perpetually covered by snow. Of the 113 world species, only 5 are found in India, all of which occur in the Eastern Himalaya (Northern Sikkim and Bhutan) while only 1 of them is known from the Western Himalaya. Of the 5 Indian species 4 are endemic to Eastern Himalaya while 1 is cosmopolitan.

3. **Bryideae**

A. ***Nematodonteae***

Order 1. *Tetraphidales.* Not found in India.

Order 2. *Buxbaumiales:* Of the three genera found in India, 2 (*Theriotia* of Diphysciaceae and *Buxbaumia* of Buxbaumiaceae) are found only in the Western Himalaya while 1 (*Diphyscium* of Diphysciaceae) is found in Eastern Himalaya (2 spp. in Khasia Hills which are also found in Ceylon and S. E. Asia).

Order 3. *Polytrichales:* Of the five genera of Polytrichales found in India (*Atrichum*, *Oligotrichem, Pogonatum, Polytrichum* and *Lyellia*). *Lyellia* is found only in the Eastern Himalaya (outside, it is found only in Yunnan and North America). Of the 4 other genera, 39 spp. are in the Eastern Himalaya, 16 spp. are in the Western Himalaya and 11 spp. (*Atrichum* and *Pogonatum* only) are in South India.

Order 4. *Dawsoniales:* No found here. Found only in Australia.

B. ***Arthrodonteae***

(i) Haplolepideae

Order 5. *Archidiales:* Of the single peculiar genus *Archidium* there are 34 tropical-subtropical species throughout the world. Of these only 1 species is found in Eastern India and this is endemic. Three other species of this genus are found in Western India. There are 8 other species of this genus is Central Africa and 5 other species in North America.

Order 6. *Dicranales:* Of the 127 species under 38 genera of 3 families, 54 species are endemic in the sub-continent, 32 spp. are cosmopolitan (mainly North Hemisphere), 26 species are Indonesian-Pacific (a few species crossing over to American coasts), 13 species cross over to Yunnan and 2 are pantropical. *Aongstroemia julacea* (Hook.) Mitt. discovered during 1921 Everest expedition at 19,800 ft. in the Rongbok Valley by Somervell is probably the highest record of plant survival. There are 66 species of these genera in the E. Himalaya and 108 spp. in South India.

Order 7. *Fissidentales:* Of the 84 spp. of the genus *Fissidens* which are found in India, 51 spp. are found in Eastern India, 22 spp. in N. Western India and 55 spp. are found in S. India and Ceylon. Of the 51 Eastern Indian species, 29 spp. are endemic, 12 spp. are Indomalesian Pacific, 5 spp. spread to Yunnan, 1 spp. is Pantropical and the remaining 5 are cosmopolitan (mostly predominant in N. Hemisphere).

Order 8. *Syrrhopodontales:* 4 genera of the single epiphytic family *Calymperaceae* are found in India of which the genus *Calymperopsis* is found only from S. India. There are 300 spp. of the other 3 genera (*Syrrhopodon, Thyridium* and *Calymperes*) in Eastern India and there are 33 spp. in S. India. There are only 2 spp. of *Syrrhopodon* and 1 spp. of *Calymperes* in W. India. The majority of the species are from Indomalaysia and the Pacific Islands but there is also a very large concentration in Central Africa. *Calymperes sundarbanensis* gangulee is the only moss found in the Sundarbans where it grows on tree trunks but only in the position above tidal level.

Order 9. *Pottiales:* 74 species in 24 genera of the single family Pottiaceae are present in Eastern India while there are 97 species in N. W. India and 44 species in South India. Of the 74 E. Indian species, 31 are endemic, 10 cosmopolitan, 8 N. Hemishere, 10 India-Yunnan, 10 Indomalesian-Pacific, 2 Palaeotropic, 1 Pantropic, 1 Indotibetan and 1 Indian–C. Asia species. It seems the order has more European species so that they are more in the Western Himalaya than in the Eastern.

Order 10 *Grimmiales:* This is another Euro-American order but well represented in Eastern India where there are 23 spp., while there are 14 spp. in W. Himalaya and only 1 sp. in S. India. Of the 23 E. Indian spcies, 11 are endemic, 8 cosmopolitan, 3 N. American and 1 common with Yunnan.

(ii) Heterolepideae

Order 11. *Eucalyptates:* Represented by the genus *Encalypta*, it is completely absent from Eastern India and Indomalesia. But, as many as 5 spp. are known from Western Himalaya, while there are 12 spp. in North Aaia, 12 spp. in Europe and 19 spp. in the two Americas. There are 3 spp. of the genus in Oceania and 2 in Australia.

(iii) Diplolepideae

Order 12. *Funariales* (Acrocarpic): There are 24 species in 9 genera of 3 families in Eastern India, while there are 27 spp. of these genera in the N. W. India and 25 spp. in South India. Of the 24 spp. in E. India, 9 spp. are endemic, 1 cosmopolitan, 4 from N. Hemisphere, 7 Indomalesian-Pacific, 1 Indotibetan, 1 Indoafrican and 1 Palaeotropical.

It may be noted that of the two dominating families Splachnaceae and Funariaceae, the family Splachnaceae is better represented in North Asia and the family Funariaceae is cosmopolitan.

Order 13. *Eubryales* (Acrocarpic): There are 109 spp. in 24 genera of 7 families while there are 122 spp. in 20 genera in Eastern India and 74 spp. in 11 genera in South India. Of the 122 Eastern India species, some spread to Western and North-western India, 22 are endemic, 26 are cosmopolitan, 29 from N. Hemispheres, 12 are Indomalesian-Pacific, 12 are connected with Yunnan, 3 are Afro-

Asiatic, 1 Indoamerican and 1 spreads from Madagascar to Japan. It seems the affinity with western forms is comparatively stronger.

Order 14. *Isobryales* (mostly pleurocarpic, a few accrocarpic genera): There are 162 spp. in 58 genera of 12 families in Eastern India while there are 72 spp. in only 29 genera in N. W. India and 154 spp. in 51 genera in South India. Of the 162 spp. in Eastern India, 66 are endemic, 51 are Indomalesian-Pacific, 27 spread to Yunnan (may be beyond) while there are only 4 cosmopolitan and 5 N. Hemispheric species. There are 3 Afro-Asian, 2 Asia-American, 2 Palaeotropical, 2 pantropical and 1 Indotibetan species. Clearly the link is with the East and not the West.

Order 15. *Hookeriales* (pleurocarpic): There are 47 spp. in 15 genera of 4 families in Eastern India while there are only 7 spp. in only 3 genera in N. W. India and 29 spp. in 12 genera in South India. Of the 47 Eastern Indian species as many 31 are endemic, 13 Indomalesian-Pacific, 1 spreading from Madagasar to Tahiti over the Pacific, 1 is pantropical, 1 palaeotropical and some from the northern or western part of the globe. Distinctly this is a South-east Asiatic tropical order.

Order 16. *Hypnobryales* (pleurocarpic): This is the largest and the most modern order of the mosses. There are as many as 277 spp. in 89 genera of 12 families in Eastern India while there are 189 spp. in 58 genera in N. W. India and 140 spp. in 51 genera in South India. Of the 277 Eastern Indian species, 116 are endemic, 65 are Indomalesian-Pacific, 34 spread over Yunnan and beyond, 19 are cosmopolitan, 30 are found in the N. Hemisphere, 1 is Indotibetan, 5 Asia-American, 2 Asia-African, 3 spread from Madagascar to the Pacific and 2 are palaeotropical. So, here also there is greater link with the East and not the West.

We may now analyse the Eastern Himalayan mosses in general terms. The Sphagmdae and the Andreaeidae have already been enumenated. Of the Nematodonteae and the acrocarpic Arthrodonteae, there are 493 spp. of which 192 are endemic, 58 are connected with Yunnan, 99 are Indomalesian-Pacific, 5 are connected with Africa and 128 are cosmopolitan and North Hemispheric. Of the Pleurocarpic Diplolepideae, there are 472 spp. of which 206 are endemic, 62 are connected with Yunnan, 133 are Indomalesia-Pacific, 8 are connected with Africa, 10 only to America, and 54 are cosmopolitan and North Hemispheric. So, we see the properties of Western mosses decrease as we come down to the pleurocarpic mosses. This is natural as the Pleurocarpic mosses, like the larger Jungermanniales require a condition of humidity which is only available in regions from the Eastern Himalaya to the Pacific. The large pleurocarpic species *Macrothamnium* is found only from India to Oceania. It may, thus, be concluded that the mosses (which must be of very ancient origin as pointed out by a large number of cosmopolitan species) are still in the evolving phase with the Hypnobryales in the Eastern India-Pacific region. Of a total of 990 Eastern India species 403 are endemic (40%), 127 are connected with Yunnan, 227 are Indomalesian-Pacific, 14 are African, 19 are scattered and 20 are cosmopolitan or north hemispheric. The lower Yunnan group is significant showing an ancient connection with the region.

As for the Hepatics it is found that the Western Himalayan ones are closer to the European and Siberian ones while the South Indian taxa have a large Indomalesian element and a few Africans species.

BRYOPHYTE CYTOLOGY

Bryophytes from ideal material for cytological, karyological or genetic studies as the dominant plants are gametophytes or haploids (n) with one set of chromosomes only. So all plants are monosomic, i.e. they are pure with only one of each gene in a plant not involving the complication of heterozygosity. This is the reason why sex chromosomes were first clearly demonstrated in this group as in these cases the plants are either male (showing the male chromosome) or female (showing the female chromosome).

Another advantages of this group is its totepotency. It is possible to regenerate a new plant from any part of the old plant. Thus, it is possible to generate a protonema and then new gametophytes from a bit of the seta (sporophyte—hence $2n$) and these new gametophytes could be $2n$ (diploid) and develop $4n$ (teraploid) sporophytes. Repetition of the same process with the $4n$ seta would ultimately induce an $8n$ sporophyte. Such experiments were performed in the beginning of the twentieth century by the Marchals (1906-1912) who obtained polyploid gametophytes from decapitated bits of sporophyte which gave rise to callus and the new $2n$ on the $4n$ gametophyte mosses. Wettstein (1923-29), again crossed these different ploid plants with one another and obtained triploids etc.

Sex chromosomes

X (female)—Y (male) types of sex chromosomes have been clearly demonstrated in a large number of Bryophytes. These are often associated with heterochromaticity. Allen (1919, 1926) showed the sex-chrosomes in *Sphaerocarpus* (Fig. 33.2) which were further studied by Lorbeer (1927). Heitz (1928) studied similar sex chromosomes in *Pellia*. Since the X-Y types of sex chromosomes have been demonstrated in numerous Bryophyta genera, sometimes involving whole families. Polytrichaceae, Bryaceae (Fig. 33.2 *Rhodobrym* of Bryaceae), and Mniaceae abound in such species. A point of interest is noted in some of the species. In many species of Polytrichaceae and Mniaceae haploid as well as diploid gemetophytes, (cytological forms) are known. The haploid forms are dioicous while the diploid forms are monoicous. The latter characteristic is easily found to be because of the presence of both X and Y in the diploid sets of the gametophytes. Seganda (1955-1971) noted "structural sex-chromosomes" (where X and Y chromosomes are of the same size and cannot be structurally distinguished) in various genera and species of Jungermanniales and even in a *Pogonatum* (*P. inflexum*).

Karyotypes

Karyotypes have been analysed in numerous mosses (a large number by Yano, [illegible] 1962). A. Chatterjee and Gangulee (1970) noted the karyotypes (*Kn*) of both *Atrichum* and *Pogonatum* to be V(H) + 3V + 2J + m(h) which is the same as suggested by Yano for the Polytrichaceae; *Fisidens*—V(H) + 2V + 2v + 4j + m(h); *Pohlia flexuosa*—V(H) + 2V + 5v + 2j + m(h), *Mniobryum ludwigii*—V(H) + 2J + 2v + 4j + (h), *Bryum plumosun*—V(h) + V + 2J + 2v + 3j + (h), *Rhodobrym roseum*—V(H) + 5V + 3J + (h), *Mnium rostratum*—6V + 6J (diploid), 3V + 3J (haploid), *Fleischerobryum longicolle*—V(H) + 2J + 3v, *Cratoneuron filicinum*—V(H) + 2V + 4v + 2J + (h), *Amblystegium serpens*—V(H) + 3V + 2J + 3v + j + m(h), *Brachythecium buchanani*—V(H) + 2V + 4v +

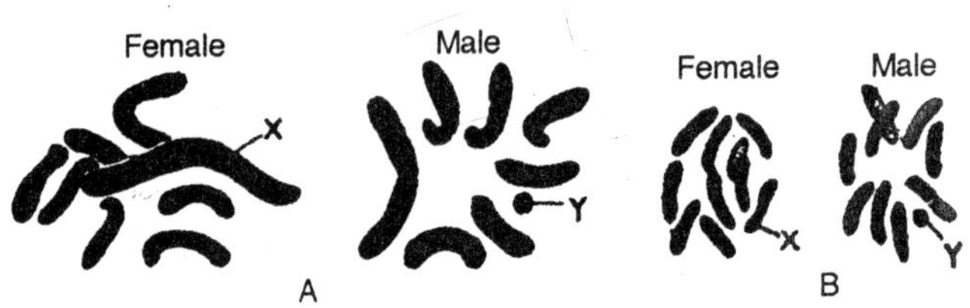

Fig. 33.2: Sex chromosomes in dioicous gametophytes of Bryophytes. A. The liverwort *Sphaerocarpos* (after Allen). B. The moss *Rhodobryum* (after N. Chopra)

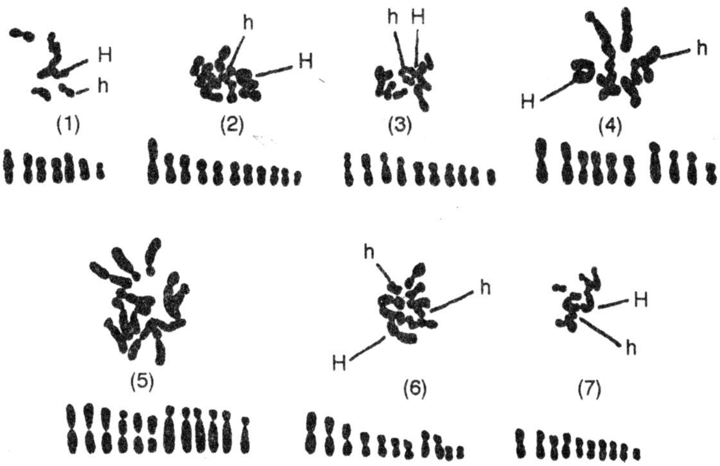

Fig. 33.3: Karyotypes of some well known mosses as observed by Chatterjee and Gangulee: (In each the figure is the somatic mitotic plate and the lower figure is the Ideogram). (1) *Pogonatum gymnophyllum*, (2) *Pohlia flexuosa*, (3) *Bryum plumosum*, (4) *Rhodobryum roseum*, (5) *Mnium rostratum*, (6) *Brachythecium buchananii*, (7) *Taxithelium nepalense*

2j + 2m(h), *Taxithelium nepalense*—V(H) + 2V + J + 2v + 3j + m(h), *Bryosedgwickia aurea*—V(H) + 2J + 4v + 3j + (h) and *Vericularia montagnei*—V(H) + 2V + 2J + 4v + 2j + (h).

A feature in the mosses is the presence of some very minute chromosomes called *m*-chromosomes whose number sometimes vary within the same species. Usually they do not seem to carry any gene but they pair like ordinary chromosomes during meiosis and are often heterochromatic. Their function seems to be problematic. It has been suggested that they may play some role in the formation of new chromosomes.

Chromosome numbers of Bryophytes and their trends in cytological evolution

Although by no means complete, a large number of studies of the Bryophytes has lead to the speculation on the basic chromosome (*n*) numbers of different groups. They may be summarized as follows:

Hepaticopsida

Sphaerocarpales	—	The derivation number is 8.
Marchantiales	—	The dominant numbers are 8 and 9.
Jungermanniales	—	The dominant numbers are 8 and 9
Calobryales	—	The dominant numbers are 8 and 9.
Takakiales	—	Ch. No. 4—V(H) + V + J + J(h). This is the lowest number among the Bryophyta and, as already stated, it is a very interesting and problematic group.

Thus, barring the Takakiales, there is a uniformity in the chromosome numbers of the whole group of Hepaticopsida.

Anthocerotopsida

The dominant numbers in the whole group are 5 and 6.

Bryopsida

Sphagnidae	—	The basic *n* number is clearly 19 + 2 m.
Andreacidiae	—	The basic number is 10.
Bryidae Polytrichales	—	A consistent *n* No. of 7 point out to the stability and the antiquity of the order.
Dawsoniales	—	The n no is again 7.
Buxbaumiales	—	The *n* nos. are 8 in Buxbaumiaceae and mostly 9 in Diphysiciaceae.
Tetraphidales	—	The two species of *Tetraphis* have the reported numbers $n = 7$ and 8. But S. Inoue (1981) finds 8 in all specimens examined by him. He suggests that the original number may even be 4—V (H) + V + J + J(h) (the same as that of Takakiales) which was subsequently doubled.
Archidiales	—	The basic chromosome number is 13.
Dicranales	—	In the numerous studies *n* numbers of 10, 11, 12, 13, 14, 15, 16 etc. have been met with. But Mehra and Khanna (1961) and Fritsch put the basic number as 12—14. Gangulee (1972) is undecided.
Fissidentales	—	Various chromosome nos. (5, 6, 8, 12, 14 etc.) have been reported. Chatterjee and Gangulee (1970) got V(H) + 2V + 2u + 4J + m(h) as the Karyotype of *F. bryoides.* Both Mehra and Khanna (1961) and Gangulee (1971) are undecided about the basic no but Fritsch (1972) suggests 12.
Pottiales	—	Here also various numbers like 11, 12, 13, 14 etc. are obtained. 13 (or 14) seems to be the basic number.
Grimmiales	—	The most suggested numbers are 12, 13 and 14. Acrocarpic *Grimmia* and pleurocarpic *Rhacomitrium* do not differ. The basic no. may be 13 (or 14).
Encalyptales	—	The nos. 13 and 14 have been obtained and either of them might have been the original number.
Funariales	—	There are anomalous numbers in the order. In Ephemeraceae it is 11 and 27. In Funariaceae a basic number of 4 may be suggested although there are a few anomalous numbers. In Splachnaceae again 8, 9, 10, 11, 12, 16, 19, 20 numbers are found. The basic number is difficult to arrive at.
Eubryales	—	Here again the numbers are very variable. Bryaceae shows an abundance of $n = 10, 11, 12$ so the basic no. may be near that. The families like Meesiaceae and Timmiaceae also show similar nos. But there is another group with families like Mniaceae, Bartramiceae, Rhizogoniaceae where the chr. nos. are lower so that a basic no. near 6 is suggested. Various karyoptypes have been noted by Yano and Gangulee *et al.* Most interesting is the X–Y sex chromosomes observed by N. Chopra in *Rhedobryum roseum* (Fig. 33.2).

Isobryales — Here again we come across various *n* numbers in the various families but 10, 11 and 12 may be suggested as the basic numbers out of which the others are derived.

Hookeriales — It is a small order but the *n* numbers reported vary. 9, 10 and 11 seen near the basic number.

Hypnobryales — This is a vast order with a large number of worked out species showing various chromosome numbers. The commonest numbers being 11, 10 and 12, the basic number is likely to be nearer these. A. Chatterjee and Gangulee (1970) noted the following karyotypes: *Taxithelium nepalense* (Sematophyllaeae) K(*n*) = 10 = V(H) + 2V + J + 2v + 3j + m(h); *Bryosedgwickia aurea* (Hypnaceae) K(n) = 11 = V(H) + 2J + 4v + 3j + m(h) and *Vesicularia montagnei* (Hypnaceae) K(n) = 12 = V(H) + 2V + 2J + 4v + 2j + m(h).

Mehra and Khanna (1961) suggests a hypothesis: The original karyotypes of Bryideae was 6 = V(H) + 2V + 2J = m(h). All other karyotypes are derived in course of evolution of polyploidy (euploidy or aneuploidy). Of course, polyploidy is very common in mosses—sometimes the same species may show *n* or 2*n* number cytoforms without any morphological variation.

EFFECT OF AIR POLLUTION ON BRYOPHYTES

Researches on the response of bryophytes to air pollution has been carried out in recent years by a number of workers in different countries particularly with regard to mosses (an hepatics to some extent) and it has been suggested that the knowledge gained from such studies could be used successfully as a tool by the foresters or prospectors in determining the character and condition of land. There is now ample evidence that quite a number of bryophytes can be used as reliably as that of lichens in measuring pollution in industrial and urban areas. Panada (1978) even suggested the use of Bryophytes as Bryometers—instruments for measuring phytotoxic air pollutants.

Ecologists in several countries have noted the *impoverishment of Bryophytic communities in and around cities and industrial areas. Many, bryophytes have become extinct in urban industrial environment* because of their sensivity to polluted air. According to Barkman (1967) within the last century the *Dutch flora* has lost 15% of its terrestrial Bryophytes and 30% of its epiphytic Bryophytes. Twentythree bryophyte species which occurred in *Amsterdam* in 1900 are now extinct there, and comparable changes have been observed in *England* due to man's activities (Rose and Wallace 1974) Delvosalle *et al.* (1969) observed that among some 600 species of bryophytes which were in *Belgium* in 1950, nearly 114 disappeared and 34 are now very rare. They attributed this change to the direct and indirect effects of human activities, specially to air pollution. LeBlanc and Sloover (1971) reported the some bryophytes on *Mount Royle* in the city of Montreal (in Canada), of about 50 years ago are now extremely rare and some species have disappeared altogether. Similarly most of the common Bryophytes in *Northumberland* in England are totally absent from a large part of the lower. *Tyne Valley* where scattered colliery town burning pit heaps are located and are the source of large scale air-pollution (Gilbert 1968). Here, in Calcutta most of the mosses in Greater Calcutta have disappeared or become extremely within the ast fifty years. Mosses named by Prain are now extinct within area, more duly to deforestation.

Bryophytes occupying certain substrates appeared to be more sensitive to air pollution than others. Species diversity in an air polluted area varies not only with the distance from the source of pollution but also with the type of substrate. The order of increasing sensitivity appear to be from terricolous (on trees) to corticolous (on bark) species. Species growing on trees are generally far more sensitive than those growing on other substrates. Buck (1974) showed that on the whole a delicate but perceptable cuticle is present on the leaves of those mosses, in which the warm water supply comes from below. In these *endohydric mosses*, of which the genera *Bryum* and *Mnium* are good examples, the rhizoids are used to some extent to extract water from the substratum and it is then conducted up the stems to leaves. On the other hand, there are mosses called *ectobydric mosses* in which cuticle can be demonstrated and here normal water supply is from above. Under this category falls *Orthotrichum, Tortella, Trichostomum,* etc. A further group, combining, in some measure, the characteristics, of these two groups, is said to be *mixohydric.* The same division can be made among liverworts. Most Marchantiales are endohydric, while leafy liverworts are mainly ectohydric. It is clear that we have two biologically contrasted groups that are likely to differ from one another in other anatomical characters. They might be expected to have very different development of tissues specialized for the conduction of water.

Pollutants reach terrestrial bryophytes in the form of dry-deposited gases and particles, and in wet deposition. In the developed world the major pollutants affecting plants since the beginning of the Industrial Revolution have been smoke and sulfur dioxide (SO_2) from combustion of traditional fuels, together with various metals released into the atmosphere by smelting and other heavy industries. More recently, with socioeconomic changes, the introduction of desulfurization technology, and the switch to low-sulfur fuels, other gaseous pollutants have assumed importance. Nitrogen oxides (NO_x) have increased greatly in cities with the rise in the use of motor cars. SO_2 and NO_x may affect poikilohydric bryophytes as their solution products, but both gases are readily converted to strong acids by oxidation and solution in cloud droplets to form "acid rain." This may be deposited in precipitation and affect vegetation many miles from the point of production of the original pollutants. Ozone (O_3) is a secondary pollutant formed by a series of reactions involving hydrocarbons and NO_x ultimately derived from car exhaust gases. Some of the reactions are photochemical, and ozone is most frequently produced by day in sunny climates, although it is now known to be important over much of Europe in the summer. As ozone is also destroyed by NO_x under certain conditions it tends to reach largest concentrations in rural areas. Ammonia (NH_3) is of increasing importance as a rural pollutant being released principally by intensive animal rearing activities. Other pollutants that are of importance to bryophytes include agricultural pesticides and fertilizers (see Brown 1992), heavy metals, radionuclides, and the various forms of aquatic pollution.

Bryophytes are extremely sensitive to some of the above pollutants but less so to others. Heavy metals and radionuclides are accumulated to levels in the plant which far exceed those in the surrounding environment. These properties appear to result from: (1) the lack of a waterproof cuticle in ectohydric species and, therefore, a moist surface for deposition and free access to the cells; (2) the absence of stomata in the gametophore which would otherwise allow the atmospheric environment to be excluded at night or when the plant is stressed; and (3) their high surface areas which render them highly efficient at trapping deposited molecules, particles, and droplets. Some of the most pollution-sensitive bryophytes are epiphytes, yet these have received much less attention than lichens in pollutions research although of similarly high sensitivity.

Sulfur dioxide: SO_2 is among the most strongly phytotoxic of the atmospheric pollutants and the one that has caused most damage to bryophytes. Earlier studies (reviews : Rao 1982, Winner 1988) aimed principally at relating bryophyte distributions and performance to zones of pollution intensity using mapping and transect surveys. Following careful study of the distribution of bryophytes and lichens around the heavily polluted conurbation of Newcastle upon Tyne in north-east England. Gilbert (1968, 1970) set out the main features of bryophyte response to SO_2-polluted environments: (1) they show differential sensitivity, thus the number of species present shows a strong negative correlation with average SO_2 concentrations; (2) historically, losses of luxuriance and sporophyte production always precede extinctions; (3) bryophytes in sheltered habitats like valleys are less affected than those in exposed situations; and (4) a high pH and buffer capacity of the substratum can ameliorate the effects of SO_2 on bryophytes. Certain bryophytes posses a high tolerance to SO_2 pollution and may actually benefit from components of urban pollution (e.g., *Funaria hygrometrica, Ceratodon purpureus, Leptobryum pyriforme, Bryum argenteum*), whereas the apparent tolerance of others (e.g., *Tortula muralis*) probably results from the buffering properties of their substratum (e.g., alkaline mortar). In temperate climates few bryophytes occur where the annual mean SO_2 concentration exceeds 50 μgm^{-3} (0.17 ppm). Many cryptogams are so sensitive to SO_2 that recently a *critical level* of 10 μgm^{-3} (0.034 ppm) has been adopted to protect them by air pollution regulation in Europe (UKCLAG 1996).

Nitrogen oxides: Little work has been done on effects of realistic concentrations of nitrogen oxides on bryophytes (Farmer *et al.* 1992). Bell *et al.* (1992) subjected *Polytrichum formosum* to NO_2 at a concentration (122.4 μgm^{-3}) frequently experienced on rural roadsides. Initially growth of existing shoots and a 36% reduction of new shoot formation. The sensitivity of the nitrate reductase enzyme (NR) to fumigations wiith NO and NO_2 (both at 35 ppb) was investigated in four mosses by Morgan *et al.* (1992). NO_2 stimulated NR activity probably because upon dissolution in the extracellular water it forms predominantly nitrate ions. NO fumigations caused an immediate reduction of NR activity presumably because the main solution product is the toxic nitrite ion. Continued exposure to either gas eventually led to a loss of NR activity, indicating that these pollutants may be harmful by disrupting nitrogen metabolism, but no reduction of photosynthesis was recorded.

Ammonia: Potentially ammonia (NH_3) is highly phytotoxic and in preliminary fumigation experiments Greven (1992) found that most of the species investigated were injured at concentrations above 30 μgm^{-3} (42 ppb). Leaf tip chlorosis was followed by necrosis within 3 days in *Racomitrium lanuginosum* and *Dicranum spurium*, but only after 12 days in *Campylopus flexuosus. Pleurozium schreberi* showed no visible injury except where NH_3 was accompanied by SO_2. These observations must be treated wth caution as Lee *et al.* (1998) have pointed out that atmospheric concentrations are usually $<$10 ppb and that, because of its high solubility, it is difficult to maitain steady, realistic concentrations in fumigation system.

From growth and distribution studies of bryophytes in Christchurch, New Zealand. Daly (1970) observed that species on stonewall can tolerate higher levels of pollution that those on tree trunks. In case of epiphytes, species growing at tree bases can cope with the polluted condition better than those on the trunks of trees. It has been suggested that shelter, reduces the level of pollution and a high pH of the substrate, increases in the degree of ionization and the rate of oxygenation of acidic pollutants provide favourable conditions for survival of species (Gilbert). Perhaps, this explains why in a city ecosystem the shady and alkaline niches are more preferred by bryophytes.

Besides habitat, the life forms of bryophytes can to certain extent modify the pollution effect.

According to Gilbert (1970), the general trend of increasing resistance to pollution among life forms would be in the order of (i) rough mats, wide turfs, large cushions and leafy liverworts; (ii) smooth mats and small cushions; (iii) short turfs and and thalloid forms. This sequence, however, only suggests a general tendency. The moss protonema, precursors of the leafy gametophores appeared to be more sensitive than the mature gametophytes. It is now generally accepted that air pollution inhibits sexual reproduction in Bryophytes. De Sloover and Le Blanc (1970) stated that fertility is related to vitality. However, this situation may vary from species to species, depending on the prevailing climatic condition in the area (though a decrease in fertility under pollution stress seems to be common in all bryophytes). In *Laskea polycarpa* there is an increase in fertility from the most polluted to the least polluted area.

Most gametophytes are known to accumulate minerals from 5 to 10 times more readily than do vascular plans, Mayer and Gorham (1951) analyzed the content of moss species from the Lake District of England and found that these plants accumulate between 0.005 and 0.4% Mn by dry weight. Shacklette (1965) determined the element content of 27 species of bryophytes and noted that the concentration of Al, Ba, Cr, Cu, Fe, Ni, Pb, Ag, Ti, V, Zn and Zr (Zirconium) were higher in Bryophytes than those of in Angiosperms. Also the bryophyte species were able to concentrate rarer elements even when they were below the detection level in the substrate. Centain element very rarely found in plants were also detected; Bi in the thallose liverworts *Conocephalum conicum* and *Marchantia polymorpha*; Strontium (Sn) in saxicolus mosses, *Grimmia laevigata* and *Hedwigia ciliata*; Ag in *Atrichum angustum* and *Polytrichum commune, Thelia asperella* growing on *Juniperus virginiana* contains 8 elements not found in the bark of the tree on which it was growing. It is likely that these elements were obtained from wind blown material lodged in the moss mats or from the solutes or rain water.

Among mosses, the profusely branched and ramifying pleurocarpous and densely packed acrocarpous forms are generally more efficient absorbers and entrappers of metal particles than the unbranched and erect acrocarps.

CHEMICAL CONSTITUENTS AND BIOCHEMISTRY OF BRYOPHYTES

Only 30 years ago the chemistry of bryophytes was virtually unknown. Recent research on the biology of bryophytes and progess in analytical techniques has resulted in a deeper knowledge about the chemical constituents of bryophytes, although our understanding of their biochemical processes, especially biosynthetic pathways, compared to vascular plants, is still rather poor. In the first part of this chapter the present state of the art regarding the chemistry of bryophytes will be presented, the second part deals with some aspects of chemosystematics, and in the third part knowledge of the biochemistry of bryophytes is summarized. A recent thorough and comprehensive review on the topic has been published by Asakawa (1995). A further review, in 1997, on "heterocyclic compounds in bryophytes" was published by the same author. He deals mainly with secondary metabolites and does not mention inorganic compounds. Primary metabolites are partly discussed, e.g., lipids, carbohydrates, and phaeophytins.

Biochemistry was not a topic of his review; this field of research was recently surveyed in Chopra and Bhatla (1990) and Rudolph (1990).

Inorganic compounds

Among bryophytes there does not appear to be a general accumulation of special inorganic compounds as known for example in lycopods with aluminium oxide, or in horsetails with silica (Hegnauer 1962). Inorganic ions found in most bryophytes do not differ qualitatively and quantitatively from those found in vascular plants (Huneck 1983). A few recent papers deal with this topic and in an increased number of studies bryophytes are used to monitor dispersion patterns of atmospheric pollutants (Muhle 1984, Steninnes 1995). It is frequently observed that certain bryophytes accumulate potentially toxic levels of metal ions without apparent damage. Some species show specific tolerance to metals present in their natural environment (Wells and Brown 1995). There is one liverwort species with an unusually high dry-weight content of 1.3% mercury: *Jungermannia vulcanicola* Steph. The mercury is deposited in the cell walls in a mercury-sulfur component (Satake and Miyasaka 1984). Satake *et al.* (1984) found in some water mosses in New Caledonia rather unusual high contents of inorganics. In the moss *Ectropothecium subobscurum* Thér. they detected high contents of nickel (690 μgg^{-1} dry weight), manganese (470 μgg^{-1}), chromium (254 μgg^{-1}), and bromine (210 μgg^{-1}) and in the liverwort *Lopholejeunea* sp. an incredibly high amount of manganese with 15300 μgg^{-1}. In general, according to those authors, the amounts of nickel, chromium, and bromine in plants do not exceed 100 μgg^{-1} dry weight.

Primary metabolites, their structural analogues, and other ubiquitous compounds

Primary metabolites are regarded as essential for almost all living organisms and are derived from reactions of primary metabolism. Among them should be noted carbohydrates, nucleotides, proteins, intermediates of the tricarboxylic acid cycle, lipids, common pigments for photosynthetic processes, and lignin. For a detailed introduction to 'primary" and "secondary" plant products see Mothes (1980). As bryophytes definitely belong to the "green line" of land plants developing from early chlorophytes (Graham *et al.* 1991, Hedderson *et al.* 1998), their primary metabolism is very similar to that of vascular plants (Hegnauer 1962, 1986). This holds true for such essential compounds as cellulose in the cell walls, chlorophylls a and b, main carotenoids, starch, nucleic acids, sugars, and certain lipds. So far no real **lignin** has been found in the cell walls of bryophytes, only "lignin-like" aromatic compounds (Wilson *et al.* 1989, Edelmann *et al.* 1998). Some of the few recent papers on **carbohydrates** in bryophytes are discussed in Asakawa (1995). Konno *et al.* (1987) analyzed cell cultures of the liverwort *Marchantia polymorpha* L. for its pectic polysaccharides in cell walls and classified them as three. types. According to these authors, carbohydrates correspond in general to those of vascular plants, even in their absolute configuration. The protein **amino acids** of bryophytes do not seem to be significantly different from those of higher plants (Huneck 1983). In a survey of the occurrence of amino acids, Dutt (1996) found common protein amino acids in 23 species of mosses and liverworts, and in the liverwort *Pellia epiphylla* (L) Corda also the two non-protein amino acids; γ-aminobutyric acid (GABA) and ornithine. Salm *et al.* (1998) isolated the amino acid N-(4-hydroxy-3, 5-domethoxy-benzoyl) aspartic acid, called fontinalin (**1**), from *Fontinalis squamosa* Hedw.

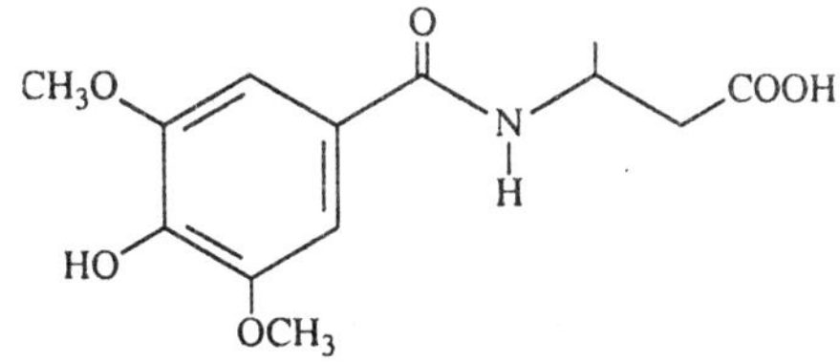

(1)

Trennheuser *et al.* (1994) detected six glutamic acid amides in *Anthoceros agrestis* Paton. Hegewald and Kneifel (1987) investigated three liverwort and nine moss species for their **polyamine** content. They found nine polyamines, spermidine being the dominant one. Kraut *et al.* (1997a) isolated and synthesized the oligopeptide rufulamide (**2**) from *Metzgeria rufula* Spruce. Adam and Becker (1993) screened 13 liverwort, one hornwort and seven moss species for the occurrence of lectins. From *Marchantia polymorpha* they purified the first bryophyte lectin and showed it to be a monomeric protein of around 16 kDa. Among the **lipids** one may distinguish between *n*-alkanes and their derivatives, and fatty acids, either free or bound as triglycerides.

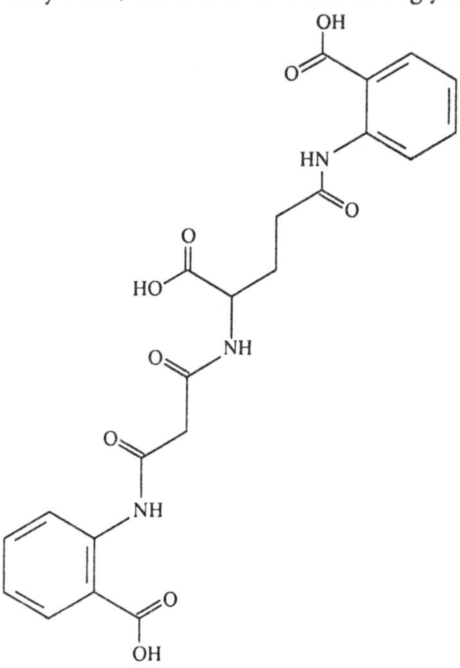

(**2**)

The occurrence of aliphatic compounds in liverworts and mosses was summarized by Asakawa (1995), whereas no report of compounds of this class seems to have been published for hornworts. A few additional aliphatic compounds have since been reported. According to Asakawa *et al.* (1996) the liverwort *Chiloscyphus pallidus* Mitt. in Hook. f. emits a "very strong stink bug smell" from crushed fresh material. The authors found as compounds responsible for the odor simple aliphatic aldehydes, of which (E)-dec-2-enal (**3**) is the major one. Nissinen and Sewón (1994) studied hydrocarbon patterns in different tissues of *Polytrichum commune* Hedw. Neinhuis and Jetter (1995) found *n*-alkanes, *n*-alkyl esters, *n*-alkanals, primary *n*-alkanols, and *n*-alkanoic acids, closely resembling those of moss gametophytes and higher plants, in waxes of the sporophytes of *Pognatum belangeri* (C. Muell.) Jaeg. and *P. urnigerum* (Hedw.) P. Beauv. The dominant compound was nonaco-san-10-01. In addition to the common fatty acids ubiquitous in green land plants, mosses and liverworts contain typical unsaturated ones, most of them mentioned by Asakawa (1995). Examples for other recent papers on lipids in bryophytes are those of Kohn *et al.* (1994), Dembitsky and Rezanka (1995), and Toyota *et al.* (1997b). Further primary compounds, including enzymes, will be discussed under *Biochemistry.*

(**3**)

Secondary products

Table 33.1 shows the classes of secondary compounds and their distribution in bryophytes.

Table 33.1: Secondary compounds from bryophytes

Compound class	Liverworts	Hornworts	Mosses
Terpenoids	++	(+)	+
Aromatic compounds	++	+	++
Nitrogen-containing compounds	+	(+)	(+)
Sulfur-containing compounds	(+)	–	–
Chlorine-containing compounds	+	–	–

Notes: ++ >100 compounds; + 10–100 compounds; (+) <10 compounds

Terpenoids

The distribution of terpenoids in bryophytes and the approximate numbers of single compounds of the different terpenoid classes known from bryophytes are shown in Table 33.2.

Table 33.2: Terpenoids from bryophytes

Terpenoid class	Liverworts	Hornworts	Mosses
Monoterpenes	60	–	3
Sesquiterpenes	~600	5	9
Diterpenes	~300	–	4
Triterpenes	18	2	25
Steroids	17	4	10
Tetraterpenes	26	?	28
Polyterpenes	+	?	+

Monoterpenes and related compounds: Besides aliphatic compounds (see above) and certain sesquiterpenes (see below) monoterpenes are mainly responsible for the characteristic odour of many liverworts. They are typically stored in the oil bodies and are therefore almost limited to liverworts.

Sesquiterpenes and related compounds: Sesquiterpenes are by far the largest class of secondary metabolites in bryophytes, although they mainly occur in liverworts (Fig. 33.4). Liverworts synthesize both the same sesquiterpenes as higher plants and others with characteristic liverwort skeletons. In some cases the compounds correspond to the same enantiomers as in higher plants, in others they represent the optical antipodes.

Diterpenes: Diterpenes detected in bryophytes are the second largest terpenoid class in bryophytes. As with sesquiterpenes, almost all diterpenes have been detected in liverworts. They are unknown for hornworts and except for phytol, which is a ubiquitous part of chlorophylls a and b in all green bryophytes, only four are known from mosses (Fig. 33.5).

Triterpenes: Triterpenes are found in species of all three bryophyte classes, but only two are known from hornworts. So far no bryophyte-specific triterpenoid skeleton has been detected. About 40 different triterpenes are known from bryophytes, with around 20 from mosses only.

Nog: This was the first record of this triterpene type for bryophytes. From the hornwort *Phaeoceros laevis* (L.) Prosk. Trennheuser (1992) isolated the triterpenoid acids betulic and ursolic acid, both common in higher plants. Ursolic acid is also known from mosses (Asakawa 1995) and liverworts (Schmidt 1996).

Steroids: The main phytosteroids, campesterol, sitosterol, and stigmasterol, are found in almost all bryophytes investigated. In some mosses and liverworts, Patterson *et al.* (1990) found only the 24α-ethyl epimer (sitosterol), whereas in other moss and liverwort species

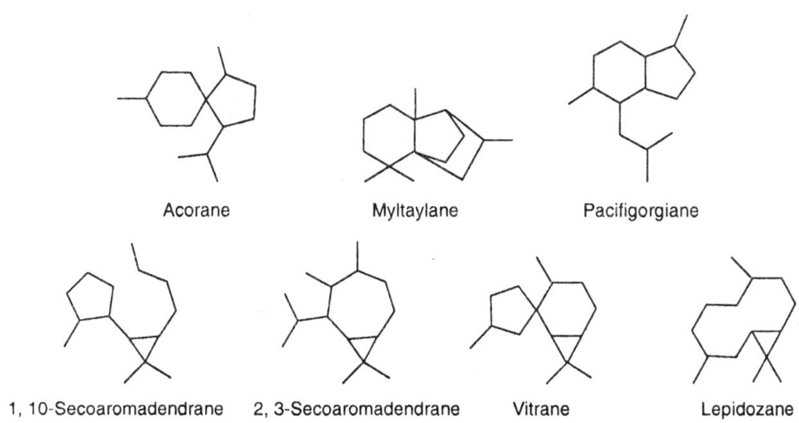

Fig. 33.4: Sesquiterpene skeletons unique to liverworts

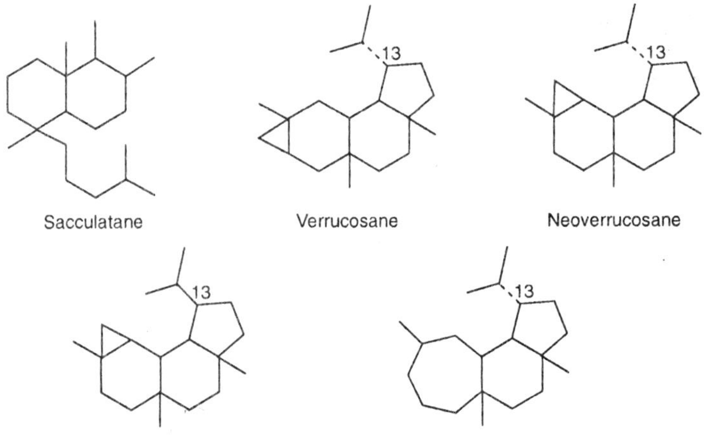

Fig. 33.5: Diterpene skeletons unique to liverworts

mixtures with the 24β epimer (clionasterol) occur.

Tetraterpenes (carotenoids): For the distribution of carotenoids in bryophytes see Table 33.3. Besides the 12 carotenoids mentioned by Asakawa (1995) for mosses and liverworts, Czeczuga (1980, 1985*a, b*) and Czeczuga *et al.* (1982) found an additional 22 carotenoids in 10 liverworts and 32 moss species. Among them seven are found only in liverworts and nine only in mosses. There seem to be no reports of carotenoids in hornworts.

Polyterpenes: Such polyterpenes as rubber and gutta-percha from certain higher plants are unknown from bryophytes.

Aromatic compounds

Besides the terpenoids, aromatic compounds constitute the most important secondary products of bryophytes. They occur as a wide range of different types, but rather unequally distributed among the bryophyte classes (Table 33.3). Only one type has been found in species of each class; most types are

Table 33.3: Aromatic compounds from bryophytes

Compound class	Liverworts	Hornworts	Mosses
Benzoic and cinnamic acid derivatives	+	(+)	+
Phenolethers, alkylphenols, phenylglycosides	+	–	–
Bibenzyls, bisbibenzyls, bisbibenzyl dimers, stilbenes, and related compounds	++	–	(+)
Phenanthrenes	+	–	–
Naphthalenes	(+)	–	–
Acetophenones	(+)	–	–
Lignans	+	(+)	–
Flavonoids	++	–	++
Coumarins, isocoumarins, coumestans	(+)	–	+
Benzonaphthoxanthenones	–	–	(+)

Notes: + + >100 compounds; + 10-100 compounds; (+) <10 compounds

known only from liverworts, others only from mosses. No type is found exclusively in hornworts. Most of these types are also known from vascular plants. Only the sphagnorubins from peat mosses, the benzonaphthoxanthenones from *Polytrichum*, and with very few exceptions also the bisbibenzyls and bisbibenzyl dimers of liverworts are regarded as bryophyte-specific.

Benzoic and cinnamic acid derivatives: Asakawa (1995) recorded about 30 compounds of this class. Only a few more papers have reported such compounds in bryophytes. Perry *et al.* (1996) found the new cytotoxic monoterpene phenyl ether (4) in *Trichocolea mollissima*

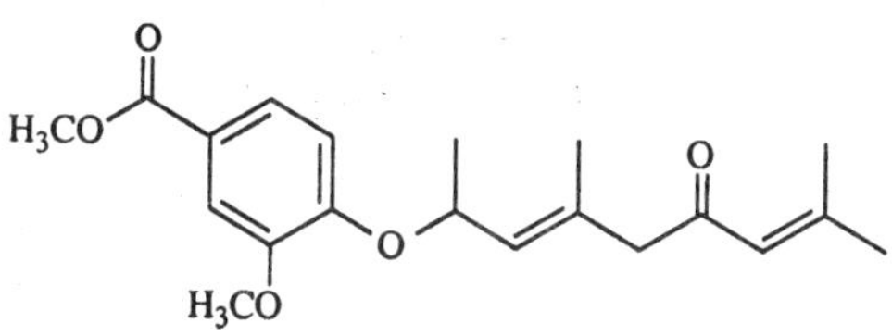

(4)

(Hook. f. and Tayl.) Gott. They also corrected the structures of benzoates earlier reported from *Trichocolea tomentella* (Ehrh.) Dum. Kraut *et al.* (1993) isolated from two *Frullania* species (one of them at that time still known as *F. muscicola* Steph., now confirmed to be a new species and renamed *F. azorica* Sim-Sim, Sérgio, Mues and Kraut [Sim-Sim *et al.* 1995] three glycerol glucosides acylated with caffeic acid, the main compound being (5).

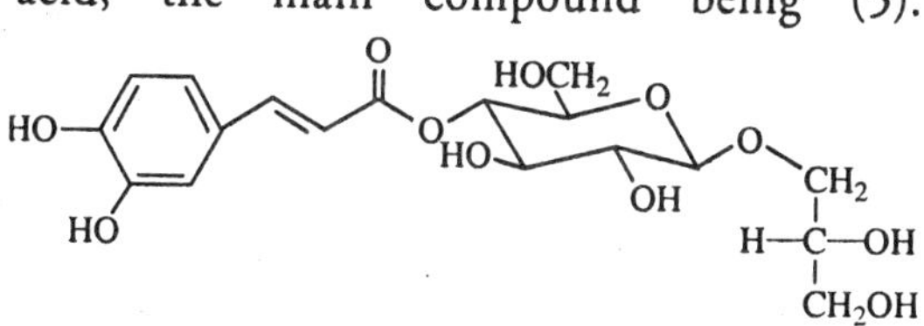

(5)

Schoeneborn (1996) detected the isocitric acid esters of caffeic (6) and ferulic acid, respectively, in *Marchesinia bongardiana* (Lehm. and Lindenb.) Trev. Further benzoic and cinnamic acid derivatives have been reported by Trennheuser (1992), Brinkmeier (1996), and Kraut and Mues (1999).

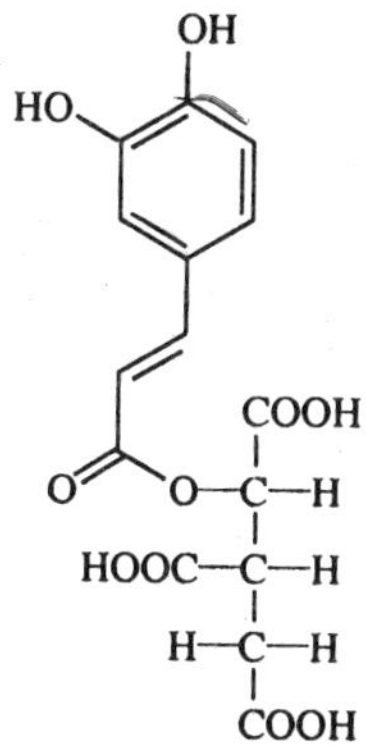

(6)

Phenylethers, alkylphenols, phenylglycosides: Asakawa (1995) mentioned a few examples of this type of aromatic compounds in liverworts. According to Toyota *et al.* (1997*c*), the intensely fragrant odour of crushed *Leptole-jeunea elliptica* (Lehm. and Lindenb.) Schiffn. is due to 1-ethyl-4-hydroxy-, 4-methoxy- and -4-acetoxybenzenee. The first allopyranosyl aromatic glycoside for bryophytes was isolated by Toyota *et al.* (1996b) from *Conocephalum conicum:* 2-(3, 4-dihydroxyphenyl)-ethyl-O-β-allo-pyranoside.

Bibenzyls, bisbibenzyls, bisbibenzyl dimers, stilbenes, and related compounds: After flavonoids this class of aromatic compounds is the second most important one in liverworts, still unknown from hornworts and mosses with the exception of pallidisetins A [7, 2—(E) and B [7, 2—(Z)], two novel cinnamoyl bibenzyls

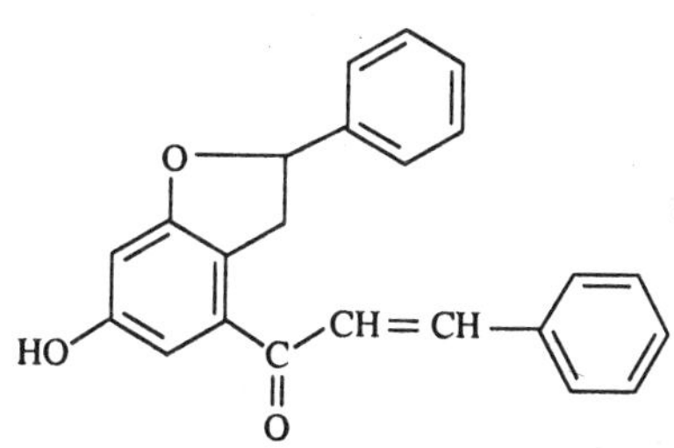

(7)

isolated by Zheng *et al.* (1994) from *Polytrichum pallidisctum* Funck. Both have cytotoxic activities against the human tumor cell lines RPMI-7951 melanoma and U-251 glioblastoma multiforme. Bibenzyls also occur regularly in species of certain angiosperm families, e.g., Cannabaceae, Dioscoreaceae, and Orchidaceae. Until recently bisbibenzyls and dimeric bisbibenzyls had been regarded as liverwort-specific, but Bai *et al.* (1996) isolated a bibenzyl-dihydrophenanthrene dimer, biogenetically related to the bisbibenzyls, from tubers of the orchid *Pleione bulbocodioides* (Franch.) Rolfe. Asakawa (1995, 1997) recorded more than 90 bibenzyls and their derivatives. including about 50 bisbibenzyls and seven bisbibenzyl dimers. Since then about 30 papers describing this class of compounds from liverworts have appeared. Kunz and Becker (1994) isolated the new 2, 5, 4′-trihydroxybibenzyl from *Ricciocarpos natans* (L.) Corda and Cullmann *et al.* (1997) found the new 7-hydroxypellepiphyllin (8) in *Pellia epiphylla.*

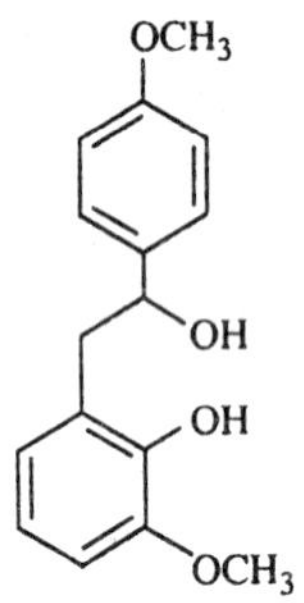

(8)

Other new bibenzyls have been reported by Kraut *et al.*(1997*b*), Rycroft *et al.* (1998*a*), and Schoeneborn (1996). A new bisbibenzyl was detected by Kraut *et al.* (1997*b*) in *Lethocolea glossophylla* (Spruce) Grolle, the bisprenylated glossophyllin (9) with a chromene and a chromane moiety. Anton *et al.* (1997) found the two new bisbibenzyls isoplagiochins E (10) and F(11). Nabeta *et al.* (1998) isolated from cell

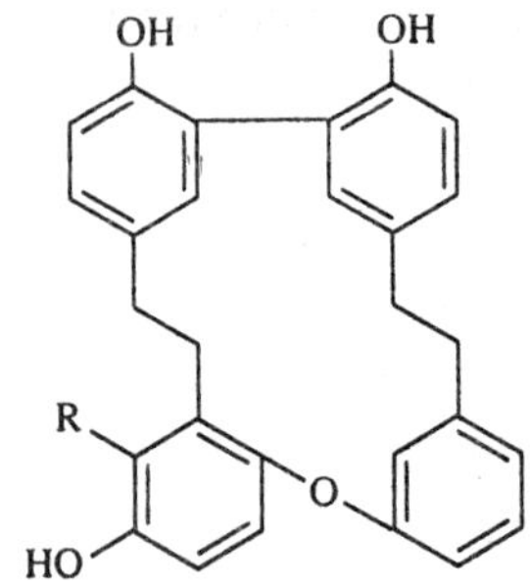

R = H : 16
R = OH : 17

(10/11)

cultures of *Heteroscyphus planus* (Mitt.) Schiffin. two bisbibenzyls, the known isopla-

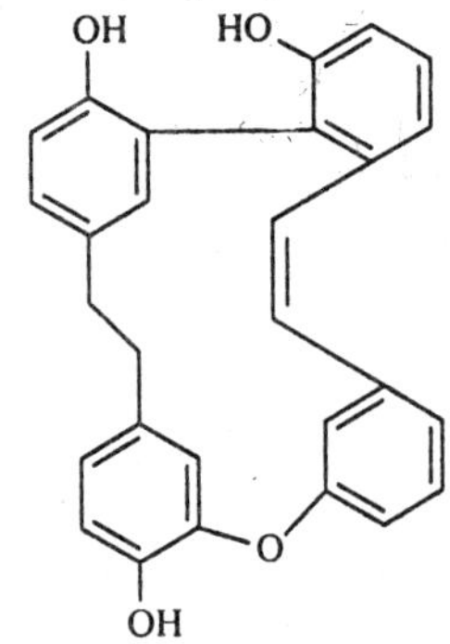

(12)

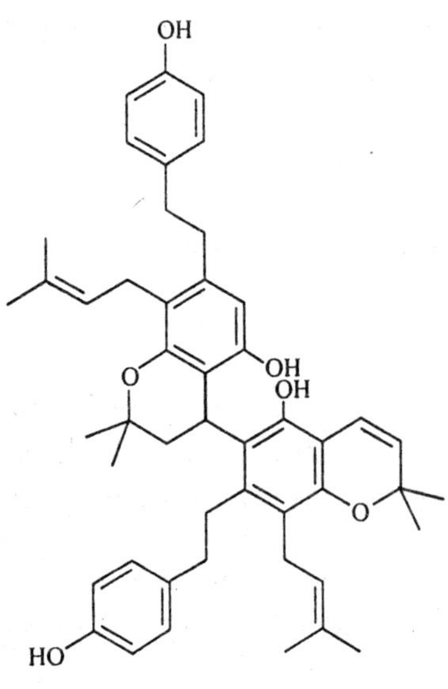

(9)

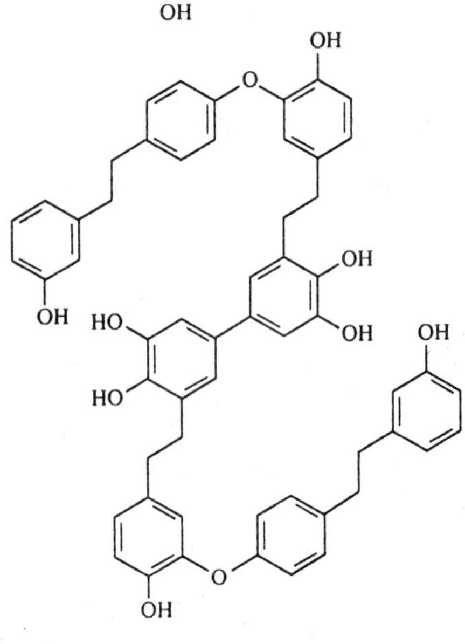

(13)

giochin A and the structurally related planusin A (12). Cullmann *et al.* (1997) revised the structures of some perrottetin E-type bisbibenzyls from *Pellia epiphylla* and isolated the new bisperrottetin, 13′, 13‴-bis-(10′-hydroxyperrottetin E) (13). The first and still the only monomeric stilbene from bryophytes, 3, 4-dihydroxy-3′-methoxystilbene, was isolated by Speicher and Schoeneborn (1997) from *Marchesinia bongardiana*, its structure being confirmed by independent synthesis.

Phenanthrenes: Although biogenetically related to the bibenzyls via the dihydro derivatives, phenanthrenes are treated separately here. Asakawa (1995) recorded only three dihydrophenanthrenes and three phenanthrenes. Subsequently, a few papers have reported this compound type: Adam and Becker (1994), Anton *et al.* (1997), and Connolly (1997).

Naphthalenes: Besides naphthalene itself, only 2, 4, 7-trimethoxy-naphthalene and the methylene-dioxy substituted naphthalenes wettstein A—C were discussed in Asakawa (1995, 1997). Later, Rycroft *et al.* (1998b) isolated the new naphthalenes 1, 2, 4-trimethoxy- and 1, 2, 3, 4-tetramethoxynaphthalene from *Adelanthus decipiens* (Hook.) Mitt.

Acetophenones: Acetophenones are reported by Asakawa (1995) only from *Trocholejeunea sandvicensis* (Gott.) Mizut. Lorimer and Perry (1994) isolated two antifungal acetophenones (14, 15) from *Plagiochila fasciculata.*

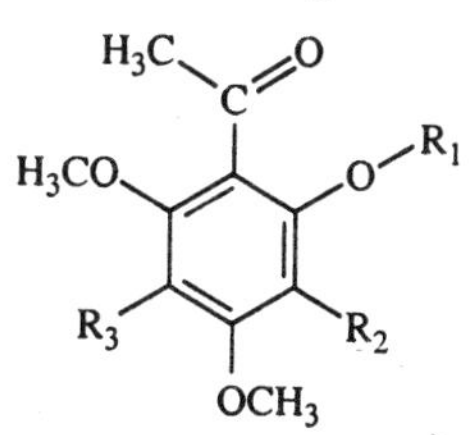

14 $R_1 = R_3 = H; R_2 = OCH_3$
15 $R_1 = R_2 = R_3 = H$
16 $R_1 = H; R_2 = R_3 = OCH_3$
17 $R_1 = CH_3; R_2 = OCH_3; R_3 = H$
18 $R_1 = CH_3; R_2 = R_3 = OCH_3$ **(14-18)**

Lindenb: These were also detected by Rycroft *et al.* (1998b) in *Adelanthus decipiens*, which also contained additionally the new acetophenones 16, 17, 18. Cullmann and Becker (1999) detected the first acetophenone C-glycoside from bryophytes, the 2, 4, 6-trihydroxyacetophenone 3, 5-di-C-β–D-glucopyranoside, in *Lepicolea ochroleuca* (Spreng.) Spruce.

Lignans: The lignans are a class of phenolics of which at the time of Asakawa's reviews (1995, 1997) only a few compounds were known from bryophytes. Among them the cyclolignan 2, 3-dicarboxy-6, 7-dihydroxy-1-(3, 4-dihydroxy-phenyl-1, 2-dihydronaphthalene (19) first isolated by Cullmann *et al.* (1993) from *Pellia epiphylla*, seems to be a common basic element of this class. This compound was later detected in other liverwort species.

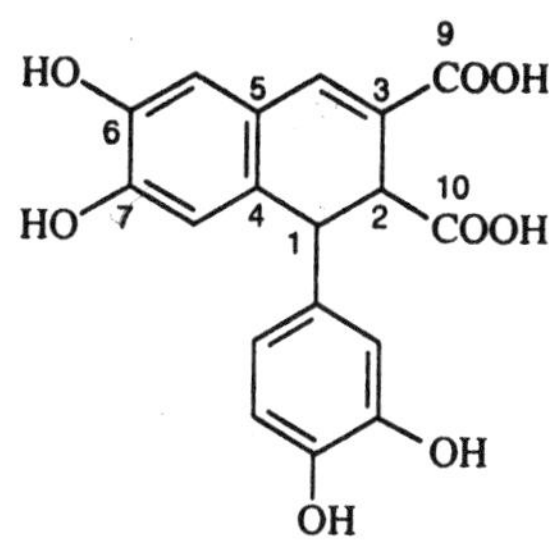

(19)

Flavonoids: Flavonoids are the most widespread phenolics in bryophytes, although still unknown from hornworts (Table 33.4). Since Asakawa's review (1995), about 30 papers on the occurrence of flavonoids in mosses and liverworts have appeared, among them some reports of other flavonoid monomeric aglycones, e.g., Morais and Becker (1991), Cullmann and Becker (1998b), and Hashimoto *et al.* (1998b). Sievers *et al.* (1994) fround traces of kaempferol in *Hypnum cupressiforme*, the first time for a moss as a free aglycone. They detected the new dihydroflavonol derivatives, hypnum acid (20) and its methyl ester. About 15 new flavone glycosides have since been isolated, e.g., by Anhut *et al.* (1992), Kraut *et al.* (1993, 1995, 1996), and Brink-meier *et al.* (1998). From *in vitro* cultures Kunz *et al.* (1994) isolated the red cell-wall pigments of *Ricciocarpos natans.*

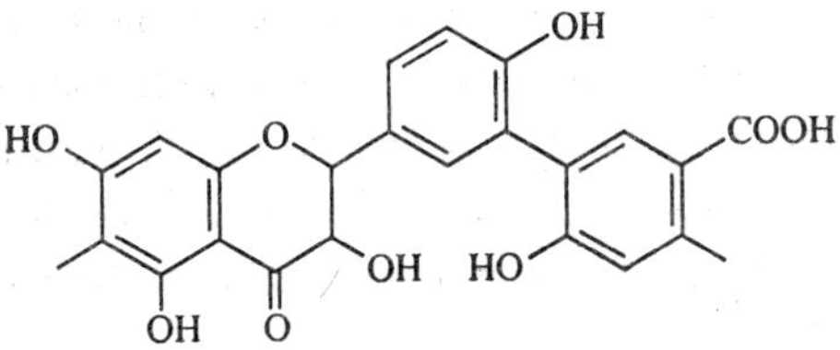

(20)

Table 33.4: Flavonoids from bryophytes

Flavonoid type	Liver-worts	Mosses
Flavone aglycones and glycosides	218	77
Flavonol aglycones and glycosides	10	9
Anthocyanins and derivatives	2	5
Aurones	1	3
Biflavonoids	1	36
Flavanones	4	–
Dihydrochalcones	2	–
Dihydroflavonols	–	3
Isoflavones	–	7
Triflavones	–	5

Coumarians, isocoumarins, coumestans: The first coumarins isolated and identified from bryophytes were reported by Jung *et al.* (1994, 1995) from *Atrichum undulatum* (Hedw.) P. Beauv., *Polytrichum formosum* Hedw., and *Tetraphis pellucida* Hedw.

Benzonaphthoxanthenones: The only known compounds of this type from bryophytes mentioned by Asakawa (1995, 1997) are the ohioensins A–E from *Polytrichum ohioense* Ren. and Card., based on the novel benzonaphtho-xanthenone skeleton.

Nitrogen-containing compounds

Bryophytes do not regularly produce nitrogen-containing secondary products, e.g. alkaloids. Thus, Asakawa (1995, 1997) recorded only two alkylindols, skatol and indoleacetic acid (IAA) from liverworts, the new alkaloid anthocerodiazonin from *Anthoceros agrestis*, four maytansinoids from *Isothecium subdiversiforme* Broth. and *Thamnobryum sandei* (Besch.) Iwats., an alkylamide from *Thuidium kanedae* Sak., and indole and pyrrolidine from *Splachnum rubrum.* Only one other paper on nitrogen-containing compounds has since been reported for bryophytes. Besides the known harmane alkaloid, 7-hydroxyharmane (= harmol), Salm *et al.* (1998) found for the first time in a bryophyte, the new harmol propionic acid ester (21) in *Fontinalis squamosa.*

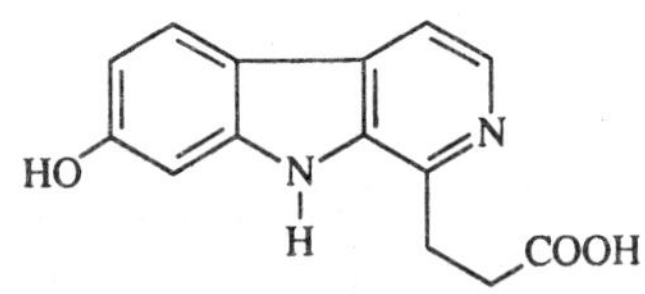

(21)

Sulfur-containing compounds

The only sulfur-containing compounds reported from bryophytes are dimethyl sulfide and the thioacrylates isotachins A—C from *Isotachis* and *Balantiopsis* species (Asakawa 1995).

Chlorine-containing compounds

Asakawa (1995, 1997) recorded as the only chlorine-containing compound from bryophytes the drimane-type sesquiterpene 7α-chloro-6β-hydroxyconfertifolin detected in *Makinoa crispata* (Steph.).

Part Eight

Palaeobotany

Chapter 34

Elements of Palaeobotany

INTRODUCTION

Palaeobotany or *Paleobotany* (American spelling) is a study of the plant life of the geological past. It is a branch of the wider geological science of **Palaeontology** (Paleontology) which deals with both animal and plant remains. In Palaeontology animal remains get predominance over plant remains as the former are more numerous.

The Earth cooled down from a hot mass and there have been gradual as well as catastrophic changes before the Earth came to its present condition. The first surface of land masses was formed by the igneous rocks. It was mostly covered by water and first life arose in this water. Then the Earth took its shape, the land masses became conspicuous and life migrated from water to land. The original rocks gradually got worn out and probably does not exist anywhere on the land surface today. The torrential rain of those days brought down the debris of the worn-out rocks to the seas, lagoons and river mouths of those days. The debris settled down and consolidated to form *sedimentary rocks.* The sedimentary rocks that have accumulated so far could have been as much as 70 miles deep. But since the face of the Earth changed from time to time, there is perhaps no place where uniform sedimentation went on for all these years. The nature of sedimentation also changed from time to time with changes in the constitution of the sediments, in the topography of the area, in the water-flow etc. This caused *stratification* of the sedimentary rocks. The sequel of the strata may be irregular due to similar reasons. At a place water might have receded and come back again so that a number of strata may be missing or some strata might have got eroded by a river valley. Finally, catastrophies like earthquakes, landslides, etc., may cause such an upheaval that an upper strata may go down below an older one or the whole sequence may be reversed (Fig. 34.1).

It is possible to ascertain the ages of the rocks, sedimentary or igneous, by scientific methods. Study of radio-isotopes has enabled more or less correct estimation of the ages of rocks. For rocks not more than 50000 years old it is possible to make accurate estimates of their ages by studying the radiocarbon (C^{14}) in their

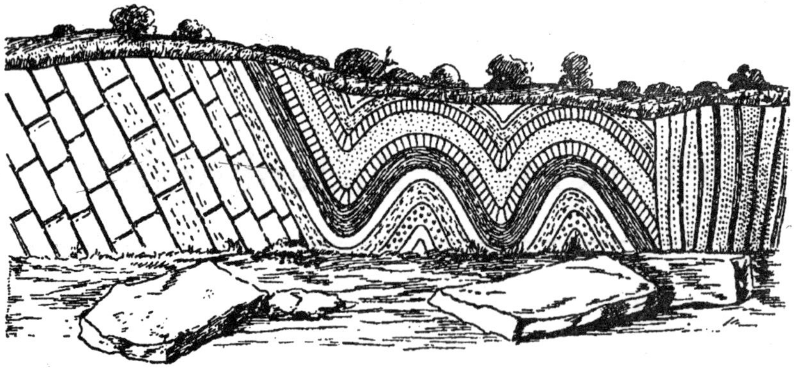

Fig. 34.1: Vertical section of surface rocks showing stratification in sedimentary rocks disrupted by later events

Table 34.1: Changes in Organic Life Through the Ages

Era	System or Period	Series or Epoch	Time scale in million years	Climatic changes	Evolutionary changes
CAINOZOIC (CENOZOIC)	*Quaternary*	Holocene (Recent) Pleistocene		Modern climatic zones with frigid poles preceded by alternating periods of glaciation and warmer spells.	Age of primates and dominance of Man. Dominance of herbaceous plants.
			1		
	Tertiary	Pliocene Miocene Oligocene Eocene		Modern mountains arise. Primeval forests spreading to Poles. Warmer climate.	Rise of Angiosperms, Mammals and modern birds.
			60		
MESOZOIC	*Cretaceous*	Upper Lower		Varying	Dwindling of Gymnosperms and increase of Angiosperms including Monocots. Decline of reptiles and increase of Mammals.
			125	from	
	Jurassic	Upper Middle Lower		temperature	Leptosporangiate ferus increase. Dominance of Gymnosperms and reptiles. First Angiosperms and first birds.
			150	to	
	Triassic	Upper Middle Lower		warm	Age of Cycadophyte Gymnosperms. Plant types fewer due to Permian glaciation. Origin of mammals and huge dinosaur reptiles.
			180		

Contd.

contd. table 34.1

Era	System or Period	Series or Epoch	Time scale in million years	Climatic changes	Evolutionary changes
PALAEOZOIC (PALEOZOIC)	*Permian*	Upper Middle Lower		Rigorous Climate and glaciation	Decline and often extinction of older groups (older gymnosperms and arborescent Pteridophytes) due to rigorous climate. Reptiles and other vertebrates. Most Indian coal.
			205		
	Carboniferous	Upper Middle Lower		Warm and humid	Age of giant Lycopods. Equisetums and Cycadofilicales forming great coal forests and swamps. Liverworts and Mosses. Amphibians and early reptiles.
			255		
	Devonian	Upper Middle Lower		Warm and humid	First established land flora of vascular plants (early Pteridophytes—Psilophytes and Lycopods). Liverwort known. Early Amphibians. Fish dominant.
			315		
	Silurian	Upper Middle Lower		Warm with	Primitive land plants. Higher invertebrates.
			350		
	Ordovician	Upper Middle Lower		local drought	Marine algae and higher invertebrates.
			430		
	Cambrian	Upper Middle Lower		Probably warm	Traces of tracheides and spores point out to the existence of land plants which probably arose in the Precambrian.
			510		
PROTEROZOIC	PRECAMBRIAN			Warm	Early sedimentary rocks with traces of Algae (including Cyanophyceae), Bacteria and Protozoans.
			1500		
ARCHAEOZOIC (ARCHEOZOIC)				Aquatic	Algae, fungi and bacteria have been reported from rocks up to 3,400 million years old. Life might have

Contd. table 34.1

Era	System or Period	Series or Epoch	Time scale in million years	Climatic changes	Evolutionary changes
ARCHAEOZOIC (ARCHEOZOIC)				condition	begun towards the close of this Era. Igneous and metamorphosed rocks without fossils. Indian gneiss is known to be 3,900 million years old. Australian igneous rocks 4,100 to 4,200 million years old are known.
			4,000 to 5,000		

organic constituents. The age of the Earth thus estimated has been divided as shown in Table 34.1. With the advancement of age, Life evolved and the main evolutionary changes are also noted on the Table. The time scale, in millions of years, is according to what is accepted by the United States Geological Survey. But it should be clarified that the time scale given for the different periods is only approximate. Thus, the total age up to the lowest Precambrian rocks, which has been assumed to be 3000 million years here, has been extended to 4500 million or even 5000 million years by others.

There are local names for the different strata in different parts of the world and some Indian names (*Gondwana Group*) are cited later.

The plant remains that are found in these rocks are known as **fossils.** The word *fossil* came from the Latin verb *fodere* which means 'to dig'. Thus, in the original sense, it means anything dug out. But now the word is exclusively used in connection with organic (plant or animal) remains. In its correct sense *fossils* include not only the remains of organisms or of parts of it but also anything connected with an organism proving its existence. Thus, a chemical which could not exist without an organism or a prehistoric trail mark of a worm on a stone is also a fossil. The actual nature of fossilization depends on the environmental condition in which it takes place. Dead plant remains are liable to get disintegrated and it is only rarely that they get fossilized. Naturally, the softer parts disintegrate early so that chances of fossilization are better for harder parts and specially for those with stiff skeletons. It cannot be expected that a large plant or a tree would get fossilized in its entirety. Usually, fossils of only bits of plants are found and those bits are then placed in some **form genus** without actually indicating to what plant it belongs. *Stigmaria* (rhizophores of Lepidodendrids), *Lepidostrobus* (strobili of Lepidodendrids), etc., are such from genera. In naming such form genera usually suffixes, signifying which part of the plant it came from, are applied. Thus *-phyllum* means a leaf; *-pteris* is a fern-like stem or found; *-dendron* is a tree trunk; *-xylon* is a woody part; *-spermum, -carpon, -carpus, -stoma* are seeds; *-theca* is microsporangium while *-strobilus* or *-strobus* signifies a cone. In such cases it is the work of the palaeobotanist to collect bits of such fossils, (i.e., form genera) and to *reconstruct* the whole plant as it existed. Such work is certainly very difficult and sometimes may even lead to errors.

THE FOSSILIZATION PROCESS

Ancient plant remains are being turned into *fossils* or are being *fossilized* ever since the *sedimentary rocks* (c.f. Vol. I : Evolution) began to deposit. The process of fossilization is going on in nature even now. Plants parts may be deposited on the site where they grow (*in situ*), e.g., in the case of the European coal forests. Usually such fossilization *in situ* took place in swamps and small inland lakes where the low oxygen content and the presence of toxic substances in the water containing decayed plant parts stopped growth of microbes and helped preservation of the plant remains until they were covered by layers of sediments. The other way is when plant parts are carried down by rivers, etc., and deposition takes place in estuarine areas. This was the case in most of the Indian Gondwana coal deposits. When plant materials sink to the bottom of a lake or estuarine water they are more immune to decay by fungi and microbes. If sedimentation over the organic deposits be too quick the fossils may get too much compressed and destroyed before they are hardened enough to resist the enormous pressure.

During fossilization the protoplasmic contents disappear first. Then come the softer parenchymatous cells while the harder wood and other sclerenchymatous or cutinized tissues resist to the last. The first hazard in fossilization is the decay caused by organisms and then comes the growing pressure of the heavy sedimentary rocks above. The pressure first reduces the vacant spaces inside the cells and forces the watery substances out. Some organic substances may also escape as marsh gas or humic acids. Naturally, all fossils get highly compressed and the final result depends on how far the conditions were favourable for good fossilization. In spite of all hazards sometimes fossils are found which retain their cellular structure beautifully and sometimes even some of the cell contents.

According to the nature of fossilization, fossils may be of the following types:

(1) Petrification

Petrification is the best but perhaps the rarest type of fossilization. This literally means transformation of the organic tissues into stone. Although the actual process of petrification is not very well understood, it is clear that no 'molecule by molecule' replacement of the organic molecules by mineral molecules takes place. The buried plant material absorbs mineral solutions like silicates, carbonates, sulphates, phosphates, etc., and infiltration followed by precipitation takes place so that silica, calcium carbonate, magnesium carbonate, iron sulphide, etc., get impregnated within the tissues. Most of the organic material may get destroyed but at least some original cell wall compounds often remain. The whole structure becomes stone-like so that fine sections may be obtained by stone sectioning methods and exact tissue structures (Fig. 34.2) may be observed under the microscope. Anatomical structures of ancient plants are beautifully obtained from such petrifications.

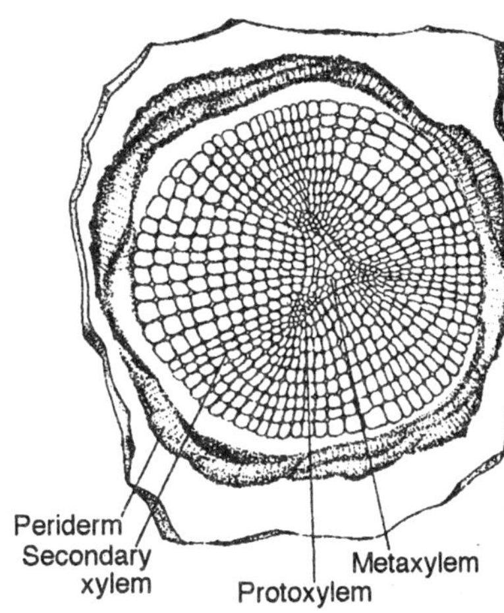

Fig. 34.2: Section of a coal ball showing t.s. of a petrified *Sphenophyllum* stem

Petrifications are usually bits of stems, twigs, seeds, sporangia, etc. Silicified bits of

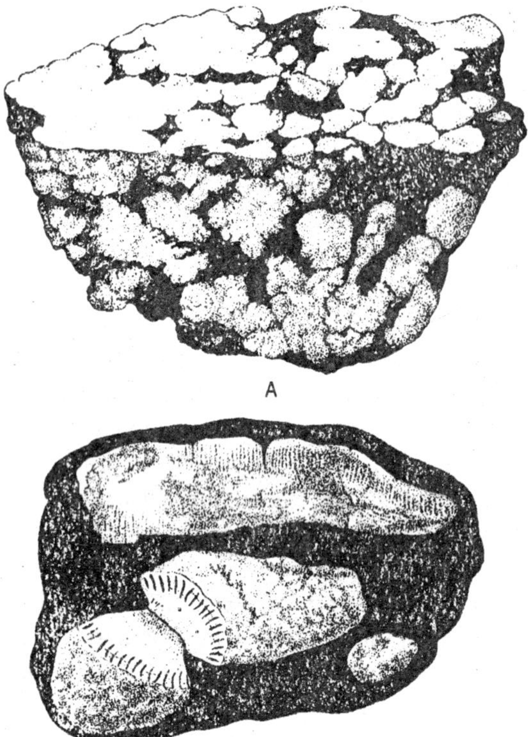

Fig. 34.3: A. A mass of coal showing coal balls. B. Section of A showing petrified stems within coal balls

wood are often found. Calcified fossils are also known. The best examples are, however, the *coal balls*. Coal balls (Figs. 34.2 and 34.3) are irregularly rounded masses ranging in diameter from a few millimetres to a metre. These occur often in great numbers within chunks of coal. Each ball is a mass of calcium and magnesium carbonate with, sometimes, iron sulphide. These show petrified remains of a great number of plant fragments representing the debris of those days. Even delicate parts remain in tact in the coal balls so that the anatomy as well as the morphology is clear.

(2) Cast or Incrustation

One of the commonest types of fossils is the *cast* or *incrustation*. In the formation of this, the plant part gets covered up by sand or mud. In course of time the plant material inside rots away leaving a hollow. This cavity, again, gets filled up by some rock forming material. In course of time the inside as well as the outside solidifies into stone from which the external part may be peeled off leaving an exact cast of the plant material showing all its surface features. The casts are as correct as one may obtain from clay or plaster of paris moulds today. Figure 34.4 shows a cast fossil of an

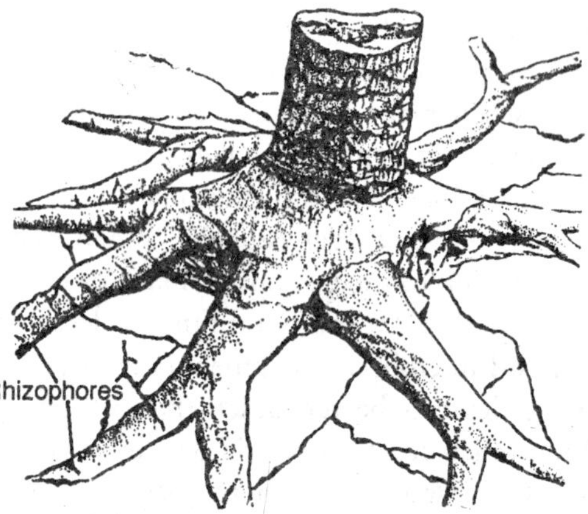

Fig. 34.4: Cast of *Stigmaria* (stump of a lepidodendrid showing typical stigmarian system)

ancient Lycopod stump. Internal casts of pith cavities may also be preserved in the same way (Fig. 34.5). A cast fossil does not actually contain any part of the original plant but it is of great use as it correctly shows the plant part as it was although it cannot show the internal anatomy.

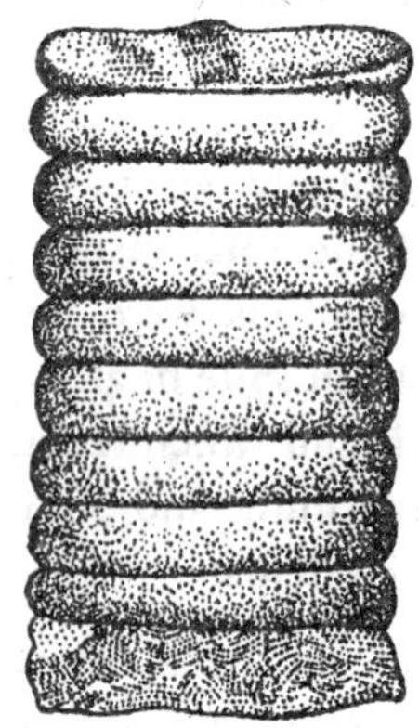

Fig. 34.5: *Sternbergia* (internal path case of *Cordaites*)

(3) Impression

A leaf or any organic part falling on semistiff clay easily leaves an *impression* on its surface. In course of time this impression becomes permanent when the clay turns into stone. Such impressions are often very clear showing full details of venation, etc. (Fig. 34.6). The impression is often of a darker colour than the surface of the rock below because it very often retains some of the organic material. Some specimens are extremely beautiful looking like paintings or base-relief of the actual twigs. In some well preserved material at least the skin or the epidermis remains intact so that structures like stomata are clearly seen in good preparations.

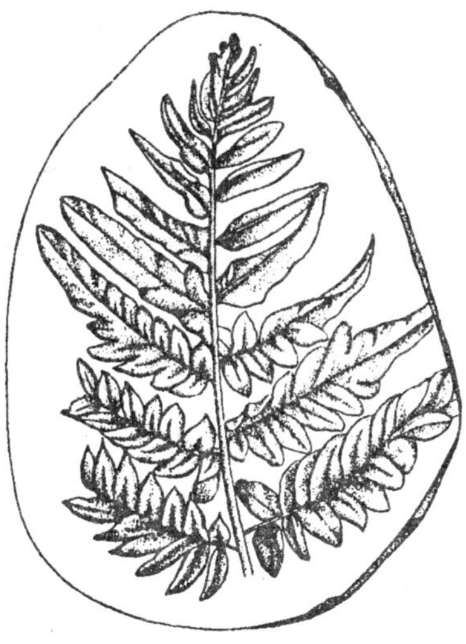

Fig. 34.6: Impression of *Neuropteris* leaf

(4) Compression

Compression is only a degree of *impression* when the organic remain of the plant part actually remains in the fossil but is highly compressed. The great pressure under which fossilization takes place flattens out all round or solid organs so that what remains in the fossil is usually a carbonaceous film. But, in good compressions it has been possible to swell out the organ by some chemical treatments so that some details become visible. A good type of compression fossil is the *clay nodule*. In this the plant material gets encased in a ball of clay, gets compressed and the clay ball turns into stone. On splitting open this nodule the organic remains is found very much intact (Fig. 34.7) although not as perfectly as in a petrified fossil.

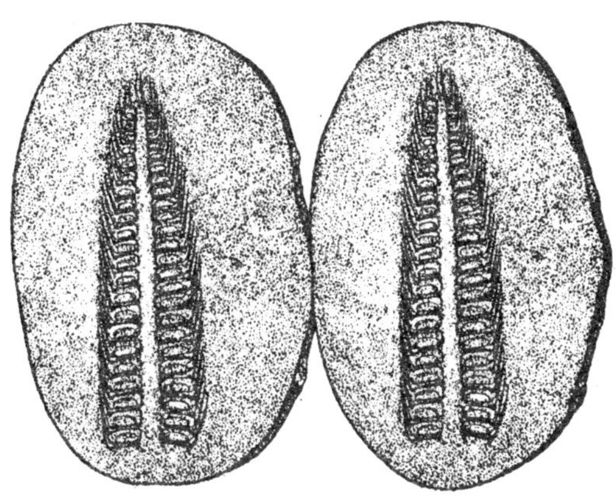

Fig. 34.7: A clay nodule split open showing a *Lepidostrobus* cone compression inside

(5) Cellular preservation

With a few exceptions, none of the preceding types of fossil preservation provide the opportunity to examine the internal structure of a plant or plant part. In the case of permineralizations and petrifactions, however, it is possible to study the internal anatomy of ancient plants (Schopf, 1975). In these fossils, one can examine cells and tissue systems within a plant, as well as produce a series of serial sections that can be used to reconstruct the three-dimensional organization of a structure. This type of fossil is called a permineralization or a petrifaction.

In both types, the process begins when a plant part becomes immersed in water containing a high concentration of dissolved minerals, the most common being silica (silicon dioxide, SiO_2), which is often readily available in areas of volcanic activity. The plant part thoroughly waterlogged with water and dissolved minerals permeating all the cells and tissue systems. Then some thing triggers precipitation of the dissolved mineral, so that it hardens around and within the plant fragment. The cell walls of the plant itself may serve as nucleation sites for the growth of the mineral crystals. When the mineral is completely solidified, the plant fragment, in effect, is entombed within solid rock. The fossil can now be sectioned by various means and examined under the microscope to see internal details of the plant. In a permineralization, minerals fill the cell lumina and the intercellular spaces, but do not completely replace the cell walls. The cell walls still consist of organic matter, although they may be chemically altered to various degrees.

Coal ball represent permineralized peat deposits and are composed almost entirely the plant parts preserved in calcium carbonate. Coal balls are variously shaped nodules which occur in bituminous coal seams. Scott *et al* (1996) examined the origin of carboniferous and permian coal balls from Euramerica and China and concluded that several different mechanisms were involved, depending on the region and the location of the coal balls within the coal seam.

(6) Rocks, minerals, etc. of organic origin

Any object which might have connection with ancient organisms is considered as a fossil. A stone showing a footprint of an animal or a trail of a worm is a fossil. Gums of ancient resinous, coniferous trees are found in fossilized forms as *amber* which is of great commercial value. Amber sometimes encloses beautiful fossils of flowers or insects. *Coal* is nothing but a highly compressed fossil derived from primeval forests. *Diatomaceous Earth* is formed of skeletons of billions of diatoms depositing on the sea bed. Even *petroleum* may be considered as a fossil as organisms are responsible for its formation. *Graphite* used in lead pencils is a fossil in the same sense as this type of carbon is supposed to be of organic origin. There are different types of *algal limestones* (Fig. 34.8) in the formation of which specific algae took part. All such minerals are to be considered as fossils.

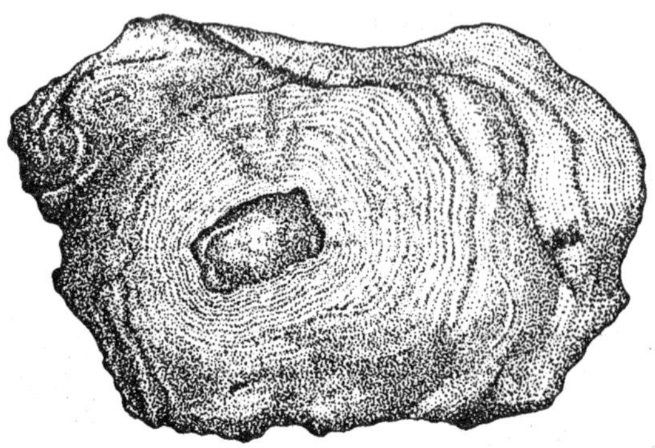

Fig. 34.8: *Cryptozoon*, an algal limestone

On the other hand, it should also be noted that sometimes certain minerals, etc., during their formation or crystallization resembles some plants (e.g., algae) or animals. There are instances of these being considered wrongly as fossils. Such structures have been termed as *pseudofossils*.

PALAEOPALYNOLOGY

Palynology is the science of studying pollens of plants. A branch of this is **Palaeopalynology** which is the study of *fossil pollens* only. These pollens are found abundantly in peats. They are also found in alluvial soil and even in strata below. They are found in stratified layers, so that their age may be calculated. Study of them is useful in tracing the history of plant communities, showing the climate at the time they were shed, studying its influence on the environment and a future find of the same combination of fossil pollens is used in dating of the deposits.

A very important use of palaeopalynology in modern days is that the presence of oil bearing strata is indicated by certain deposits. During boring for oil pits it has been useful to examine the fossil pollens in every depth which may serve as a surer indication of the presence of oil or even other minerals.

Chapter 35

Fossil Flora

A study of the fossil plants of the different Ages is the most direct evidence of the evolution of the Plant Kingdom. This evolution and the flora of the different ages have been discussed in Fig. 35.1.

The history of the different Plant Groups as evidenced by Palaeobotanic studies may be summarized as follows:

THALLOPHYTA

Schizomycetes (Bacteria)	Archaeozoic onwards
Cyanophyceae (Blue-green Algae)	Archaeozoic onwards
Chlorophyceae	Proterozoic onwards (Flagellates)
Charophyceae	Devonian onwards
Phaeophyceae	Silurian onwards
Rhodophyceae	Proterozoic onwards
Mycophyta (Fungi)	Archaeozoic onwards

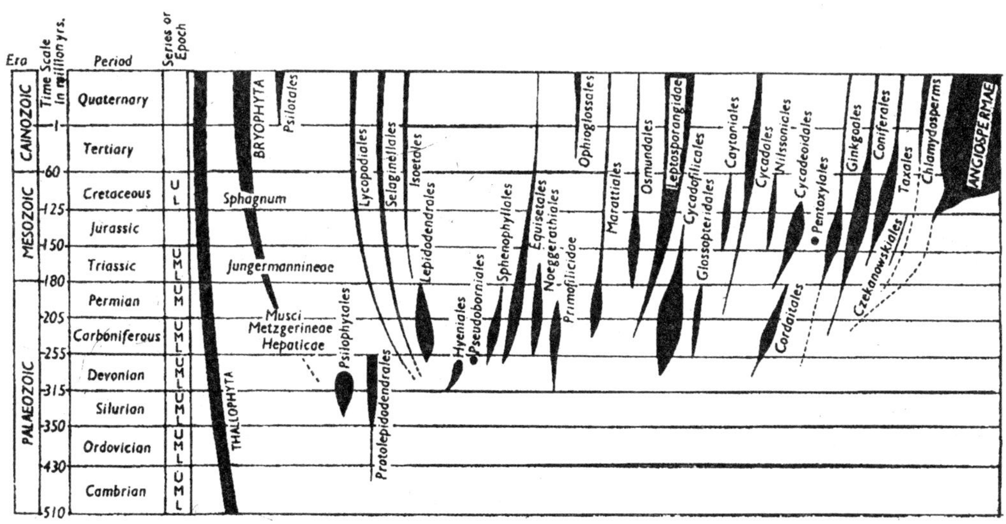

Fig. 35.1: Distribution of plant groups in the ages

BRYOPHYTA	
Hepaticopsida	Devonian onwards
Bryopsida (Moss)	Carboniferous onwards
PTERIDOPHYTA	
Psilophytopsida	Silurian to Devonian
Psilotopsida	Recent
Lycopsida	Silurian onwards
Sphenopsida	Devonian onwards
Filicopsida	Devonian onwards
GYMNOSPERMOPHYTA	Devonian onwards
ANTHOPHYTA (Angiosperms)	Jurassic onwards

The distribution of the different groups in the different ages is shown more elaborately in Figure 35.1. The major groups forming the flora of the different ages can also be ascertained from the same figure. Figure 35.2 is an imaginary picture of a Carboniferous forest.

The groups which are of great interest from the Palaeobotanical point of view are given below. These are discussed in the two parts (Pteridophyta and Gymnosperms) that follow. An asterisk (*) marks extinct groups which are represented only by fossils.

Pteridophyta

*Psilophytopsida	Psilophytales
Lycopsida	*Protolepidodendrales
	Lycopodiales—**Lycopodites*
	Selaginellales—**Selaginellites*, **Midesmia*
	*Lepidodendrales Isoetales —**Pleuromeia*
Sphenopsida	*Hyeniales
	*Pseudoborniales
	*Sphenophyllales
	Equisetales—**Calamitaceae*
*Noeggerathiopsida	*Noeggerathiales
Filicopsida	*Protopeteridales
	*Coenopteridales
	*Archaeopteridales
	Marattiales—**Psaronius*
	Filicales—*Fossil forms

GYMNOSPERMOPHYTA

*Pteridospermopsida	Cycadofilicales
	Glossopteridales
	Caytoniales

Fig. 35.2: An imaginary picture of a Carboniferous forest depicting the flora of that Period

Cycadopsida	*Nilssoniales
	Cycadales—*Fossil forms
	*Cycadeoidales
	*Pentoxylales
Coniferopsida	*Cordaitales
	Coniferales—*Fossil forms
	Taxales—*Fossil forms
	Ginkgoales—*Fossil forms
	*Czekanowskiales
Chlamydospermopsida	Gnetales—*Fossil forms
ANTHOPHYTA (Angiosperms)	*Fossil forms

THE INDIAN GONDWANA FLORA

The most important fossil flora of India is what is known as the *Gondwana Flora* which covers the period Upper Carboniferous to Jurassic. Indian plant fossils are known before (a few) and after (quite a large number) this period but no flora is as interesting as the Gondwana Flora. Towards the end of the Palaeozoic were the great Hercynian mountain building waves. This was followed in the Upper. Carboniferous and the Permian by severe glaciation and the Permian in India is the greatest coalforming age. The geography of the Earth at that time was quite different from what it is today (Fig. 35.3). There were three continents—the Euro-American Continent (modern Europe and North America) to the North-West, Angaraland (Siberia and North China) to the North-East and a vast Gondwana land on the South which combined the land masses of India, Africa, Australia, South America and Antarctica. In between the continents was the great Sea of Tethys. The Gondwana land has been named after the Gond tribe of Madhya Pradesh ruled by the famous Rani Durgabati during the reign of Akbar. The name was coined by H. B. Medlicott in 1872 but actually published by O. Feistmantel in 1876. The Gondwana land began to split in the Jurassic as shown by the intensive lava injections at that age. There was drifting apart of the continents in the Cretaceous, gradually bringing them to their present positions. Another series of great mountain building waves built the present dominant Mountains (Himalayas, Alps, etc.) on the location of the Sea if Tethys.

Fig. 35.3: How the world geography has changed. The continents during Carboniferous Jurassic is shown by think lines while the thin lines depict the world map today

The whole Gondwana land (now distributed over five continents) shows a uniformity of the Flora (and also the Fauna). In Indian Geology the Gondwana rocks are considered to form a group which is almost equivalent to an Era like Palaeozoic. Feistmantel (1882) divided the Gondwana into **Lower Gondwana** (Permo-Carboniferous with *Glossopteris Flora*), **Middle Gondwana** (Triassic with *Glossopteris-Thinnfeldia* (*Dicroidium Flora*) and **Upper Gondwana** (Jurassic with *Ptilophyllum* Flora). This three-fold division has been followed by many authorities like Wadia and Lele. But many other authorities prefer a two-fold division into **Lower Gondwana** (Upper Carboniferous to Lower Triassic) and **Upper Gondwana** (Lower Triassic to Jurassic, probably reaching Cretaceous). The two-fold division is supported by the two (*Glossopteris Flora* and *Thinnfeldia* (*Dicroidium*)—*Ptilophyllum Flora*) distinct floras (and faunas) in the two divisions. There is a sudden break between the two floras as during the Upper Permian and Triassic intensive glaciation and drought killed most of the previous early Gymnosperms and arborescent Pteridophytes giving place to the more modern Gymnosperms and herbaceous Pteridophytes (mainly ferns).

Table 35.1: The Gondwana Group

		Time scale (million years)	International System	International Series	Indian System	Indian Series	Indian Stage
UPPER	UPPER	125	Jurassic	Upper		Jabalpur	Jabalpur—also Kathiawar, Kutch (Bhuj, Ukra, Umia) Chaugan—also Godavari Valley, Kutch, East coast
				Middle			Kota, Chikiala—also Godavari Rajmahal (mixed with volcanic)—also
				Lower		Rajmahal	Orissa, Ongole, Madras-Chinglpur
	MIDDLE	150	Triassic	Upper		//////////// Mahadeva	Maleri, Denwa, Bagra (plant remain obscure) Pachmarhi (no fossil)
				Middle		////////////	Parsora (Rewa)
				Lower		Panchet	
LOWER			Permian			////////////	Panchet
	LOWER	180		Upper	Damuda (Damodar)	Raniganj	(Coal)—also Kamthi (Nagpur), Satpura, Wardha, Rewa, Mahanadi, Brahmani, Pachmarhi, Godavari
				Middle		Middle (Barren Measure)	Motur
				Lower		Barakar Karharbari	(Coal)—also Nepal, Sikkim, Darjeeling, Bhutan, NEFA (Coal)—also Kashmir (*Gangamopteris* beds)
		205	Carboniferous	Upper	Hercynian Mountain Building	Talchir Boulder bed	Umaria Rikba Talchir—also Deoghar, Punjab Salt Range, Simla Hills, Kashir-Hazara, Tehri-Garhwal
				Middle		////////////	
				Lower		////////////	
		255					

//// Non-fossiliferous rocks. Coal is found also in other stages but the main position are located as (Coal)

Indian Gondwana beds are named as shown in Table 35.1. The two-fold system (i.e., only *Upper* and *Lower Gondwana*) has been followed in this book but the Middle Gondwana beds and fossils have also been pointed out.

As seen from the Table, places in Indian showing Gondwana rocks occur mainly in the Damodar, Sone, Narbada, Godavari and Mahanadi Valleys. There are also some exposure along the Himalayan foothills of Nepal, Bhutan and Arunachal and also in the Punjab, Himachal, Kashmir and Baluchistan. Upper Gondwana rocks are more detached occurring in patches along Rajmahal-Cuttack to Kanya Kumari, in Madhya Pradesh, Rewa, Saurashtra, Kutch and in Ceylon. The Map in Figure 35.4 shows the principal outcrops of Gondwana rocks. It should be noted against that the Triassic beds (Maleri, Mahadeva, Pachmarhi, Parşora, Panchet) spreading from Sone to Godavari Valleys are placed by some in *Middle Gondwana.*

THE LOWER GONDWANA FLORA (GLOSSOPTERIS FLORA)

The Lower Gondwana Flora is called the **Glossopteris Flora** because of the predominance of the Pteridosperm *Glossopteris.* The constituents of this Flora in India, according to the strata where they occur, are:

Talchir series

Equisetales—*Schizoneura* sp.

**Unclassified Ferns—*Alethopteris* sp.

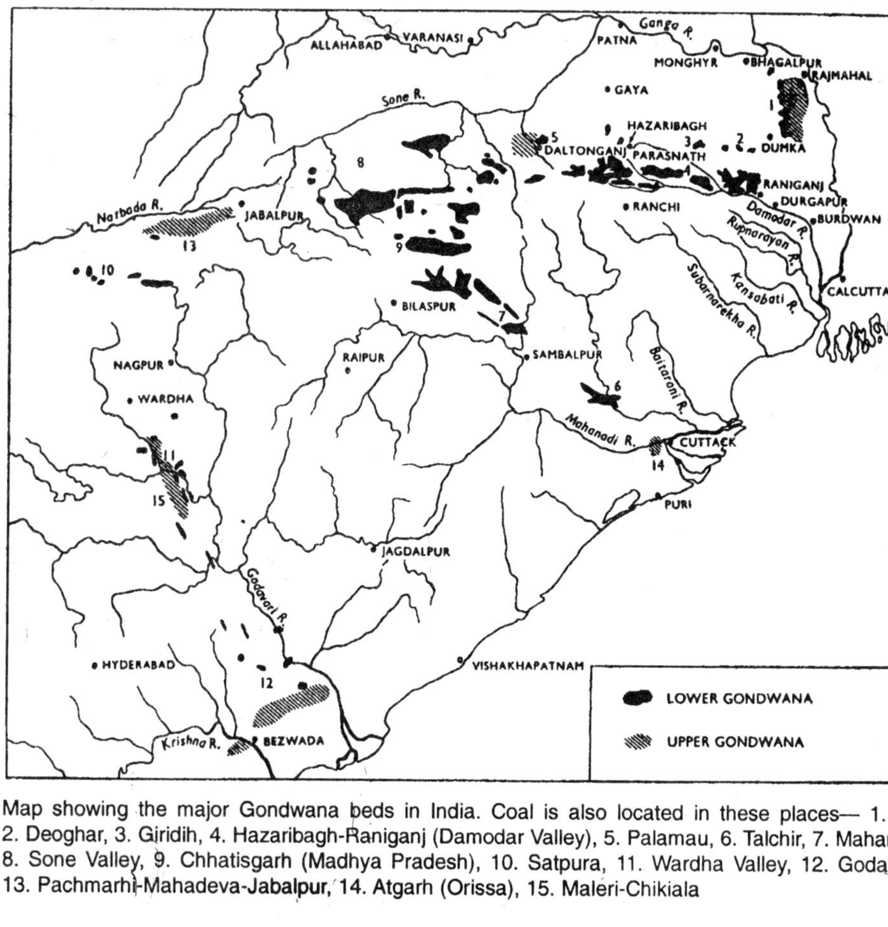

Fig. 35.4: Map showing the major Gondwana beds in India. Coal is also located in these places— 1. Rajmahal, 2. Deoghar, 3. Giridih, 4. Hazaribagh-Raniganj (Damodar Valley), 5. Palamau, 6. Talchir, 7. Mahanadi Valley, 8. Sone Valley, 9. Chhatisgarh (Madhya Pradesh), 10. Satpura, 11. Wardha Valley, 12. Godavari Valley, 13. Pachmarhi-Mahadeva-Jabalpur, 14. Atgarh (Orissa), 15. Maleri-Chikiala

Glossopteridales—*Glossopteris indica, G. communis, G. angustifolia; Gangamopteris cyclopteroides, G. buriadica; Vertebraria indiac; Ottokaria kathwaiensis.*

Cordaitales—*Noeggerathiopsis hisolpii, Samaropsis* sp., *Samaropsis emarginata* (?), *Cordiacarpus sahnii.*

DAMODAR SYSTEM

Karharbari series

Equisetales—*Schizoneura gondwanensis, S. wardii; Phyllotheca sahnil.*

Lepidodendrales—*Cyclodendron* sp.

**Unclassified Ferns—*Gondwanidium vallidum.*

Cycadofilieales—*Callipteridium* sp.

Glossopteridales—*Glossopteris indica, G. decipiens, G. longicaulis, G angustifolia; Gangamopteris cyclopteroides, G. buriadica; Vertebraria indica; Ottokaria bengalensis.*

Cordaitales—*Noeggerathiopsis hisolpii, N. stoliczkana, N. whittiana; Samaropsis milleri, S. raniganjensis; Cordaicarpus indicus.*

Coniferales—*Buriadia sewardiana; Moranocladus oldhamii.*

ADDITIONAL FROM KASHMIR GANGAMOPTERIS BEDS:

Glossopteridales—*Gangamopteris kashmirense.*

Ginkgoales—Psygmophyllum haydenii.

Barakar series

Equisetales—*Schizoneura gondwanensis, S. wardii; Phyllotheca griesbachii; Stellotneca robusta.*

Sphenophyllales—*Sphenophyllum speciosum*

**Unclassified Ferns—*Sphenopteris polymorpha.*

Glossopteridales—*Glossopteris indica, G. communis, G. ampla, G. retifera; Gangamopteris cyclopteroides* (only lower beds).

Cycadopsida—*Taeniopteris danaeoides, T. feddeni; Pseudoctenis ballii.*

Cordaitales—*Noeggerathiopsis hisolpii, N. whittiana; Dadoxylon indicum; Cordoicarpus* sp.

Ginkgoales—*Rhipidiopsis ginkgoides.*

Barren measures

Lepidodendrales—*Cyclodendron leslii* (*Bothrodendron* sp.)

Glossopteridales—*Glossopteris indica, G. ampla; Gangamopteris cyclopteroides.*

Cordaitales—*Noeggerathiopsis hisolpii.*

Raniganj series

Equisetales—*Schizoneura gondwanensis; Phyllotheca indica, P. griesbachii; Raniganjia bengalensis, R. indica.*

Sphenophyllales—*Sphenophyllum speciosum.*

Filicopsida—*Pteuchocarpus srivastavae* (Marattiales); *Alethopteris lindleyii* (uncertain).

**Unclassified Ferns—*Sphenopteris hughesii, S. polymorpha, S. lobifolia, S. Merianopteris major; Belemnopteris wood-masonii; Pecopteris phegopteroides.*

Cycadofilicales—*Palaeovittaria kurzii*

Glossopteridales—*Glossopteris indica, G. communis, G. browniana, G. retifera; G. angustifolia, G. stricta, G. tortuosa, G. formosa, G. divergens, G. conspicua; Gangamopteris whittiana; Vertebraria indica.*

Cycadopsida—*Taeniopteris danaeoides, T. feddeni,*

Cordaitales—*Noeggerathiopsis hisolpii; Samatopsis raniganjensis.*

** Most of these specimens being leaf types, it is not certain if a particular specimen belongs to the Filicopsida or the Pteridospermopsida

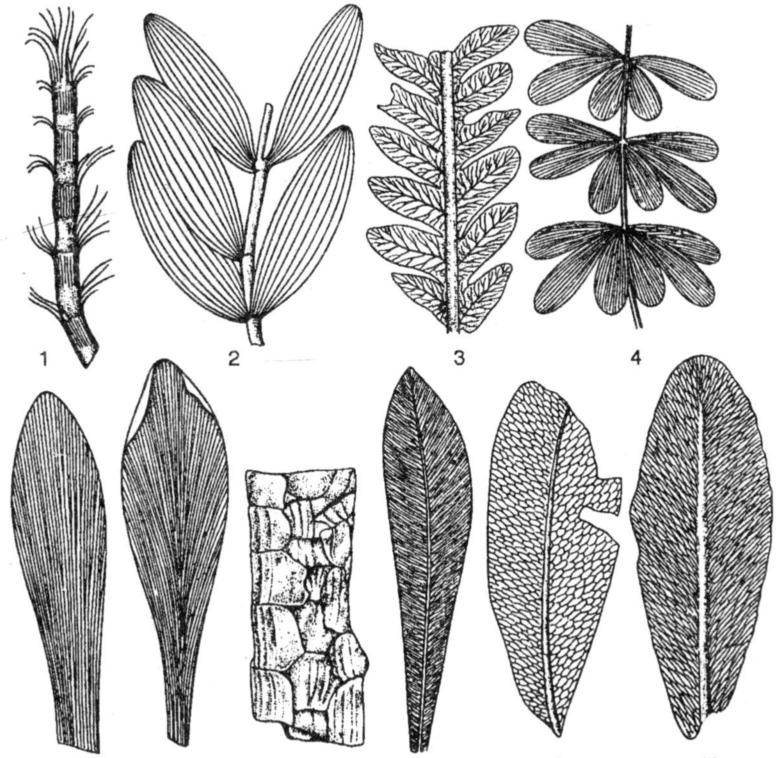

Fig. 35.5: Some representatives of the *Glossopteris Flora* (Lower Gondwana): 1. *Phyllotheca indica*, 2. *Schizoneura gondwanensis*, 3. *Alethopteris lindleyii*, 4. *Sphenophyllum speciosum*, 5. *Noeggerathiopsis hisolpii*, 6. *Gangamopteris cyclopteroides*, 7. *Vertebraria indica*, 8. *Glossopteris indica*, 9. *Glossopteris retifera*, 10. *Glossopteris browniana*

Ginkgoales—*Rhipidiopsis densinervis.*

Coniferales—*Buriadia sewardiana.*

Panchet series

(*Middle Gondwana* c.f., pages 1105, 1110)

PANCHET STAGE

Equisetales—*Schizoneura* sp.

**Unclassified Ferns—*Pecopteris concinna; Cyclopteris pachyrachis; Sphenopteris* sp.

Glossopteridales—*Glossopteris* sp.

THE UPPER GONDWANA FLORA

[THINNFELDIA (DICROIDIUM)—PTILOPHYLLUM FLORA]

The Upper Gondwana Flora is called the **Thinnfeldia (Dicroidium)—Ptilophyllum** Flora because of the preponderance of these two genera. In India this flora seems to have started in the later Triassic. In the Panchet Series, while the Panchet Stage shows the *Glossopteris Flora*, the Upper Parsora Stage in Rewa, also supposed to belong to the Lower Triassic, shows the Upper Gondwana type of Flora with *Thinnfeldia* (*Dicroidium*) and without *Glossopteris* but with *Noeggerathiopsis* still persisting.

Those who prefer a three-fold division of the *Gondwana Group* include all the Triassic beds (which may also spread to the Upper Permian) form Panchet to Mahadeva into the **Middle Gondwana**. This *Middle Gondwana* shows the **Thinnfeldia (Dicroidium) Flora** while the rest, the *Upper Gondwana*, shows the **Ptilophyllum Flora**, (e.f. pages 1105).

Panchet series

PARSORA STAGE (*Middle Gondwana*)

Equisetales—*Neocalamites foxii.*

Lycopodiales—*Lycopodites gracilis.*

Marattiales—*Marattiales* sp.; *Caladophlebis* sp.; *Danaeopsis gracilis.*

**Unclassified Ferns—*Sphenopteris* sp.

Cycadofilicales—*Thinnfeldia* (*Dicroidium*) *hughesii, T. odontopteroides, T. sahnii.*

Cordaitales—*Noeggerathiopsis hisolpii.*

The moss Capsulites gondwanensis, has been reported by Saksena from the Rewa Pali beds.

Mahadeva series (*Middle Gondwana*)

Most of this series is non-fossiliferous. Some animal fossils are known. Plant fossils are obscure.

Rajmahal series

RAJMAHAL STAGE

Equisetales—*Equisetites rajmahalensis.*

Lycopodiales—*Lycoxylon indicum, Lycopodites gracilis.*

Filicopsida—Marattiles—*Marattiopsis macrocarpa.*

Osmundales—*Cladophlebis denticulata, C. indica, C. srivastavae, C. sahnii; Osmundites sahnii.*

Gleicheniaceae—*Gleichenites gleichenoides.*

Protocyatheaceae—*Protocyathea rajmahalense.*

Dicksoniaceae—*Coniopteris hymenophylloides; Tinpaharia sinuosa.*

Diptheridaceae—*Hausmannia indica.*

Matoniaceae—*Phlebopteris* sp.

Uncertain—*Alethopteris medlicottii; Rhizomopteris chakshu, R. sahnii, R. rajmahalensis; Solenostelopteris nipanica, S. sahnii; Dictyostelopteris rajmahalensis, D. jacobii, D. fasciosteloides.*

**Unclassified Ferns—*Sphenopteris hisolpii, S. rajmahalensis, S. khairbanensis, S. arguata; Pecopteris lobata.*

Caytoniales—Sagenopteris bambanii.

Cycadopsida—*Ptilophyllum acutifolium, P. cutchense; Otozamites bengalensis; Dictyozanites falcatus, D. indicus; Taeniopteris lata, T. spatulata, T. musaefolia, T. morrisii, T. ovata, T. crassinervis; Nilssonia princeps. N. rajmahalensis, N. morrisiana, N. meddlicottiana, N. bindrabunensis, N. fissa; Williamsonia sewardiana.*

Pentoxylales—*Pentoxylon sahnii; Nipanioxylon guptai; Nipaniostrobus sahnii; Nipeniophyllum raoi, N. spathulatum; Carnoconites compactum, C. laxum; Sahnia nipaniensis.*

Coniferales—*Palissya* (*Elatocladus*) *conferta, P.* (*Retinosporites*) *indica; Pagiophyllum peregrinum; Brachyophyllum rajmahalensis, B. expansum.*

KOTA STAGE

Filicopsida—

Osmundaceae—*Cladophlebis reversa.*

Protocyatheaceae—*Protocyathea trichinopoliensis.*

Uncertain—*Alethopteris medlicottii.*

**Unclassified Ferns—*Sphenopteris* sp.

Coniferales—*Palissya* (*Elatocladus*) *jabalpurensis, P. conferta, P. indica; Araucarites cutchensis.*

Taxales—*Taxites* (*Elatocladus*) *tenerrima.*

Rocks of Rajmahal Series age are also known from Orissa (Atgarh—the bed supplying Konark building Stones, and also Ongole), Madras-Chinglput and and also from Godavari Valley. They show the following additional fossils:

Filicopsida—

Osmundaceae—*Cladophlebis indica, C. whittbyensis, C. reversa.*

Dicksoniaceae—*Dicksonia* sp.

Uncertain—*Rhizomopteris ballii.*

**Unclassified Ferns—*Sphenopteris* sp.

Cyadofilicales—*Pachypteris ellorensis.*

Cycadopsida—*Taeniopteris ensis, T. macclelandi; Williamsonia* sp.; *Otozamites abbreviatus, O. bunburyanus; Zamites proximus.*

Coniferales—*Palissya (Elatocladus) jabalpurensis, P. indica; Elatocladus plana; Araucarites macropterus, A. cutchensis, A. latifolia; Brachyphyllum rhombica.*

Taxales—*Taxites (Elatocladus) tenerrima*

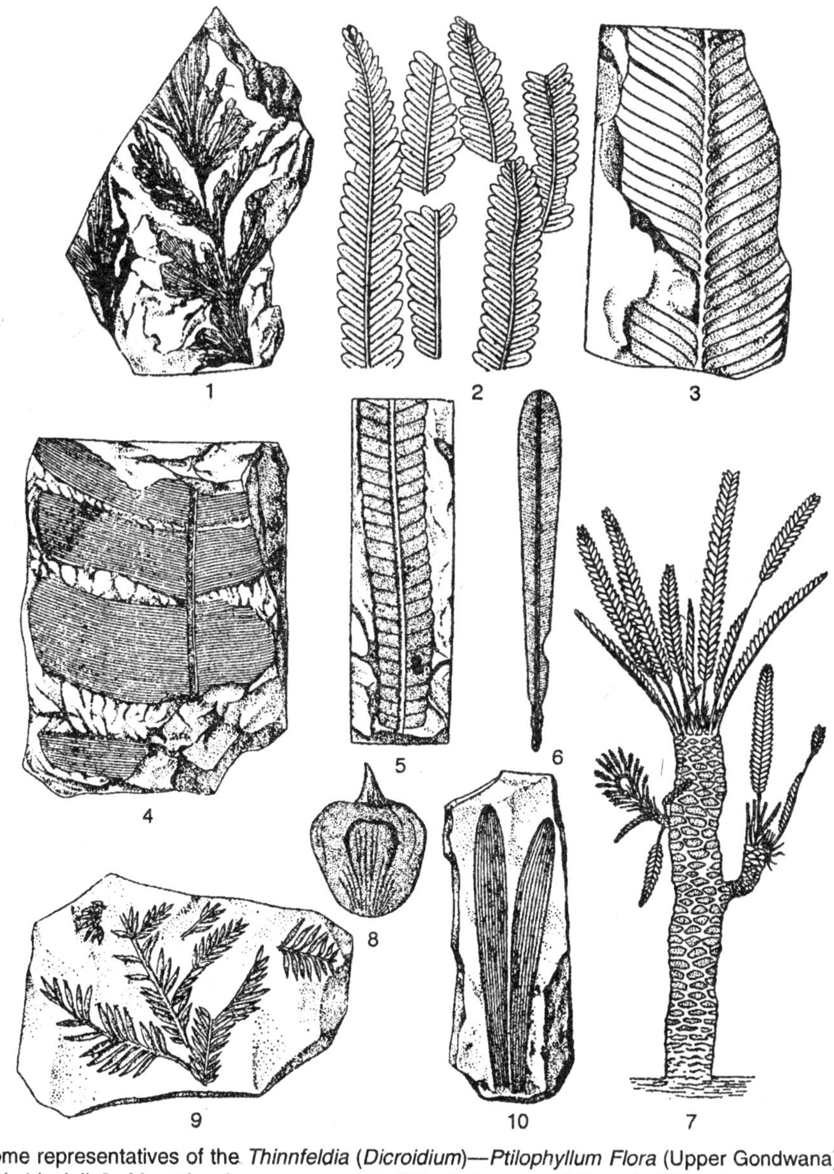

Fig. 35.6: Some representatives of the *Thinnfeldia (Dicroidium)—Ptilophyllum Flora* (Upper Gondwana): 1. *Sphenopteris hisolpii*, 2. *Marattiopsis macrocarpa*, 3. *Ptilophyllum acutifolium*, 4. *Nilssonia princeps*, 5. *Otozamites bengalensis*, 6. *Toeniopteris spatulata*, 7. *Williamsonia sewardiana*, 8. *Araucarites cutchensis*, 9. *Palissya (Elatocladus) jabalpurensis*, 10. *Podozarnites lanceolatus*

Ginkgoales—Ginkgo crassipes.

Jabalpur series

CHAUGAN STAGE

Cycadofilicales—*Thinnfeldia* sp.

Cycadopsida—*Dictyozamites indicus; Taeniopteris spatulata, T. crassinervis; Anomozamites* sp; *Nilssonia princeps; N. orientalis, N. rajmahalensis.*

Coniferales—*Pagiophyllum divaricatum.*

JABALPUR STAGE

Filicopsida—

Osmundaceae—*Cladophlebis* sp.

Gleicheniaceae—*Gleichenites gleichenoides, G. rewahensis.*

Dicksoniaceae—*Coniopteris hymenophylloides.*

Dipteridaceae—*Hausmannia buchii, H. dichotoma.*

Uncertain—*Alethopteris medlicottiana, A. lobifolia, A. whittbyensis; Onychiopsis paradoxus, O. psilotoides; Weichselia* sp.

**Unclassified Ferns—*Sphenopteris* sp.

Cycadopsida—*Ptilophyllum acutifolium; Otozamites hisolpii; Williamsonia sewardiana.*

Coniferales—*Palissya (Elatocladus) jabalpurensis, P. (Retinosporites) indica; Elatecladus plana; Brachyophyllum expansum; Pagiophyllum* sp.; *Podozamites lanceolatus; Araucarites cutchensis, A. macropterus; A. latifolius.*

Ginkgoales—*Ginkgoites lobata; Pheoniocopsis* sp.

The Jabalpur Stage is represented in Kathiawar and Kutch (Bhuj, Ukra and Umia). These contain the following plant fossils:

UMIA STAGE

Isoetales—*Isoetites serratifolius, I. indicus.*

Filicopsida—

Matoniaceae—*Matonidium indicum.*

Uncertain—*Alethopteris whittbyensis; Weichselia reticulata.*

**Uncertain Ferns—*Pecopteris phagopteroides, Actinopteris* sp.

Cycadofilicales—*Pachypteris specifica.*

Cycadopsida—*Ptilophyllum cutchense; Taeniopteris vittata; Williamsonia blanfordii.*

The Kutch Bhuj belt shows the silicified palm, *Palmoxylon mathuri*, but this bed is probably of the Lower Cretaceous age and outside the Gondwana Flora.

Part Nine

Pteridophyta

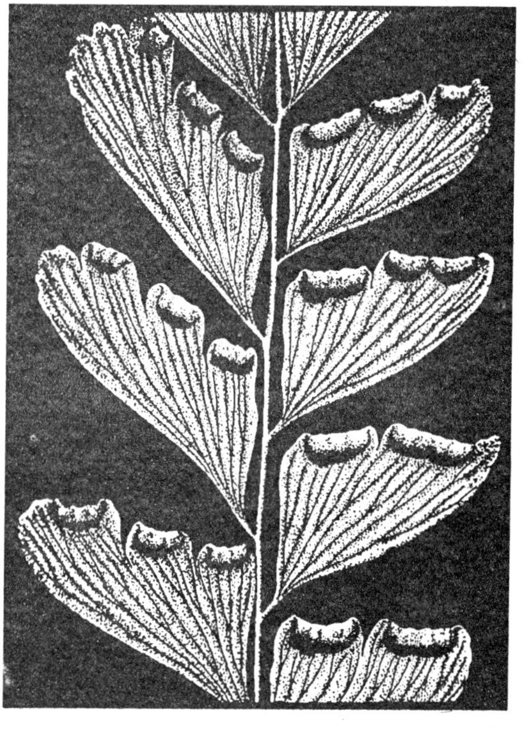

Chapter 36

Introduction to the Pteridophyta or the Lower Vascular Plants

The Pteriodophytes comprise of the lower vascular plants (Trachaeophyta) which show distinctly vascular, independent sporophytic plants arising from embryos (Embryophyta or Cormophyta) which are not enclosed within any 'seed' structure. These alternate with independent gametophytes which are almost bryophytic in appearance and bear antheridia and archegonia (Archegoniatae).

These lower vascular plants are called the Pteridophytes or the *Pteridophyta.* Although numerically much below the Bryophytes (there are 23,000 living species in about 900 genera of Bryophytes against about 10,500 species in some 400 genera of living and fossil Pteridophytes) the Pteridophytes claim a special position as the first land plants forming forests. But the idea of the relative position of this group Pteridophyta within the Plant Kingdom has greatly changed in recent days. The artificial concept of dividing the Plant Kingdom into the Cryptogams and the Phanerogams persisted throughout the 19th century but was practically abandoned in the twentieth. Even during the 19th century we find abandonment of the idea of considering the Cryptogams as a group. Auguste Pyramos de Candolle, who coined the term 'Taxonomy' (1813) divided the Plant Kingdom into the Cellulares (Thallophyta and Bryophyta) and the Vasculares (Vascular Plants) in 1819. Thus, de Candolle was the first to recognize that the Pteridophytes should rather be linked with the higher plants than with the artificial Linnaen group of Cryptogams. Stephen Endlicher divided the Plant Kingdom into Thallophyta and Cormophyta raising the Bryophytes also to the higher Cormophytes. In the latter part of the 19th century, A. Engler also supported Endlicher and stated that it was much better to divide the Plant Kingdom into two groups: Thallophyta and Embryophyta. Our concept of both the Thallophyta and the Embryophyta has undergone much change during the 20th Century. One idea had been to consider the Pteridophyta as a division of the Plant Kingdom parallel to the three other divisions—Thallophyta, Bryophyta and Spermophyta. Jeffrey (1902) first proposed that the original units be abandoned and the vascular plants be divided into two stocks: the Lycopsida (Lycopods and Equisetums) and the Pteropsida

(Ferns, Gymnosperms and Angiosperms). Bessey (1911) divided the whole Plant Kingdom directly into 14 phyla, the 9th, 10th and 11th of which comprise of the Pteridophytes. Engler, in the 1936 edition of the Syllabus similarly divides the Plant Kingdom directly into 14 Abteilungen (Divisions) the first 12 of which are Thallophytes, the 13th is Archegoniatae (or Embryophyta Asiphonogama) and the 14th Embryophyta Siphonogama (Gymnosperms and Angiosperms). The Archegoniatae are again divided into two subdivisions (Unterabteilungen): Bryophyta and Pteriodophyta. In 1935, Sinnott proposed the term 'Trachaeophyta' to include all the Vascular Plants. Tippo (1942) suggested that the Plant Kingdom be divided into two subkingdoms: Thallophyta and Embryophyta and the latter be divided into two phyla—Bryophyta and Trachaeophyta. Takhtajan (1950) coined the term 'Telemophyta' to rename the Embryophyta in view of the Telome Theory. Bold (1957) suggested dividing the Plant Kingdom into three subkingdoms: Thallophyta, Bryophyta and Trachaeophyta. Consideration of the Trachaeophyta as a distinct group emphasises the most important character acquired by the land plants—the vascular bundles. The Trachaeophytes, no doubt, show certain characters embodying the main trend of evolution since the land habit was acquired. But, this is too vast a group to be considered as a single division as a vast majority of the modern plants are included here. Moreover, Takhtajan (1953) and others do not think that the Bryophyta should be separated from the Trachaeophyta, the former being a possible derivative of the latter. This point has now been emphasised by Cronquist, Takhtajan and Zimmermann (1966) in their latest classification detailed later in this chapter.

The living Pteridophytes distinctly show four groups of plants representing four lines of evolution. The first three of these lines—the Psilotums, the Lycopods and the Equisetums—ended blindly. The last—the Ferns—evolved further into the Gymnosperms and the Angiosperms. Because of this, Bessey (1911) proposed the abolition of the old taxon Pteridophyta replacing it by three divisions ('phyla')—Pteridophyta proper (ferns), Calamophyta and Lepidophyta. Since then other divisions have been added, most important of which is the division Psilophyta. Some systematists have retained Pteridophyta as a division calling the main taxa 'classes' (some have called the main taxa 'phyla', a term used in Zoology but cannot be retained in Botany according to International Rules of Botanical Nomenclature) while others have deleted the Pteridophyta as a taxon raising the lower taxa themselves into divisions. The main classifications proposed by different authorities during the twentieth century are as follows:

1. E. C. Jeffrey (1902)

   Vascular Plants

   Stock I. Lycopsida (Lycopods and Equisetums)

   II. Pteropsida (Ferns, Gymnosperms and Angiosperms)

2. Ch. E. Bessey (1911 and 1915)

   Phyla: I-VII. Thallophytes

   Phylum: VIII. Bryophyta

   IX. Pteridophyta (Ferns only. Classes—Eusporangiatae and Leptosporangiatae)

   X. Calamophyta (Classes—Sphenophyllineae, Equisetineae and Calamarineae)

   XI. Lepidophyta (Classes—Eligulatae and Ligulatae)

XII. Cycadophyta
XIII. Strobilophyta (Class—Pinoideae)
XIV. Antophyta (Classes—Monocotyledoneae and Dicotyledoneae)

3. Adolph Engler (1936 edition of Syllabus)

Division:(Abteilung):

13. Archegoniatae

Subdivision: 1. Bryophyta

2. Pteriodophyta

Class 1. Psilophytinae
2. Articulatae
3. Lycopodiineae
4. Psilotinae
5. Isoetinae
6. Filicinae

14. Embryophyta Siphonogama (Spermaphytes)

4. A. J. Eames (1936)

Trachaeophyta

Group 1. Psilopsida (Psilophytales and Psilotales)
2. Lycopsida (Lyeopodiales, Selaginellales, Lepidodendrales, Pleuromeiales and Isoetales)
3. Sphenopsida (Hyeniales, Sphenophyllales and Equisetales)
4. Pteropsida (Filicineae, Gymnospermae and Angiospermae)

5. Oswald Tippo (1942)

Subkingdom Embryophyta

Phylum: Bryophyta

Phylum: Trachaeophyta

Subphylum: Psilopsida

Class: Psilophytinae (Psilophytales and Psilotales)

Subphylum: Lycopsida

Class: Lycopodineae (Lycopodiales, Selaginellales, Lepidodendrales, Pleuromeiales and Isoetales

Subphylum: Sphenopsida

Class: Equisetineae (Hyeniales, Sphenophyllales and Equisetales)

Subphylum: Pteropsida

Class: Filicineae (Coenopteridales, Ophioglossales, Marattiales and Filicales)

Class: Gymnospermae

Class: Angiospermae

6. A. L. Takhtajan (1950 translated 1953)

   Division: Telomophyta

   Phylum: 1. Psilopsida
   2. Bryopsida
   3. Lycopsida
   4. Tmesopsida
   5. Sphenopsida
   6. Pteropsida (Classes—Filicineae, Gymnospermae and Angiospermae)

7. A. Engler (Melchior, Werdermann and Reimers in 1954 edition of Syllabus)

   Division: (Abteilung):

   XIV. Bryophyta

   XV. Pteridophyta

   Class: 1. Psilophytopsida (Psilophytales)
   2. Lycopsida (Ptotolepidodendrales, Lycopodiales, Selaginellales, Lepidodendrales and Isoetales) Lepidospermales?
   3. Psilotopsida (Psilotales)
   4. Articulatae (Hyeniales, Pseudoborneales, Sphenophyllales, Calamitales and Equisetales)
   5. Filices (Subclasses—Primofilices, Eusporangiatae, Osmundidae and Leptosporangiatae)

   XVI. Gymnospermae

   XVII. Angiospermae

8. G. M. Smith (1955)

   Division: Bryophyta

   Division: Psilophyta

   Class: Psilophytinae (Psilophytales, Psilotales)

   Division: Lepidophyta

   Class: Lycopodinae (Lycopodiales, Selaginellales, Lepidodendrales and Isoetales)

   Division: Calamophyta

   Class: Equisetinae (Hyeniales, Sphenophyllales and Equisetales)

   Division: Pterophyta

   Class: Filicinae (Subclasses—Primofilices, Eusporangiatae and Leptosporangiatae)

9. R. E. G. Pichi-Sermolli (1958)

   Higher taxa of Pteridophyta:

   1. Lycopsida
   2. Sphenopsida
   3. Noeggerathiopsida

4. Psilotopsida
5. Psilophytopsida
6. Filicopsida

10. A. Cronquist, A. Takatajan and W. Zimmermann (1966)

 Subkingdom: Thallobionta

 Subkingdom: Embryobionta

 Division: Rhyniophyta (= Psilophytales)

 Division: Bryophyta

 Division: Psilotophyta (= Psilotales)

 Division: Lycopodiophyta (= Lycopsida)

 Division: Equisetophyta (= Sphenopsida)

 Division: Polypodiophyta (= Filicopsida)

 Division: Pinophyta (= Gymnospermae)

 Division: Magnoliophyta (=Angiospermae)

11. D. W. Bierhorst (1971)

 Division: Trachaeophyta

 Class: Rhyniopsida

 Class: Zosterophyllopsida

 Class: Lycopodiopsida (Lycopsida includes Asteroxyllales)

 Class: Cladoxylopsida

 Class: Equisetopsida (Sphenopsida)

 Class: Coenopteridopsida

 Class: Filicopsida

 Class: Ophioglossopsida

 Class: Marattiopsida

 Class: Aneurophytopsida

 Class: Cycadopsida

 Class: Coniferopsida

 Class: Gnetopsida

 Class: Angiospermopsida

The above discussion shows that the 19th century classification of the Plant Kingdom into Cryptogams and Phanerogams has been discarded. The Embryophyta (= Cormophyta or Telemophyta) is being considered as a taxon. A modern tendency is to consider the Vascular Plants ('Vasculares' of de Candolle) or the Trachaeophyta as a subkingdom or division separate from the Bryophyta although, even recently, Takhtajan has retained the Bryophyta within his taxon Telemophyta. It is now commonly agreed that the Pteridophyta is not a phylogenetic taxon but a combination of at least four independent taxa. Because of this, the question whether the term Pteriodophyta should still be retained as a divisional name or whether each of the higher taxa of the Pteridophytes should be considered as a division is still open to dispute. There is also some anomaly about the naming of these higher taxa. According to International Rules of Botanical Nomenclature there is no taxon called 'phylum' and the endings for division is -phyta, for subdivision -phytina, for class -opsida and for subclass -idae. There is no rule for subkingdom ending but certain authorities suggest -tae. Against the deletion of the term Pteridophyta although it may be said that, even if not monophyletic, the Pteridophytes, as a group, possess certain common peculiarities. They are all embryophytes, archegoniate and show a life history in which both the sporophytic and the gametophytic generations are independent but the sporophyte predominates. They have not yet attained the seed habit. They, no doubt, comprise the lower strata of the Trachaeophyters. But since, as already stated, there is some objection against considering the Trachaeophytae as a subkingdom after excluding the Bryophyta, it is better not to take that for granted but simply to consider the Pteridophyta as a division of the Embryophytae and the higher Pteridophytic taxa as classes (ending in -psida). This system is being followed in the next chapters.

 Subkingdom: Embryophytae

 Division: Pteridophyta

 Class: 1. Psilophytopsida

 2. Psilotopsida

 3. Lycopsida

4. Sphenopsida
5. Noeggerathiopsida
6. Filicopsida

EARLY LAND PLANT AND THE SALIENT FEATURES OF VASCULAR PLANTS

One of the most significant events in the long evolutionary development of the plant kingdom was the origin of the first land plants. These organisms, which in all probability arose from some ancient group of green algae, were extremely simple in organography. For example, the ancient plant *Rhynia gwynne-vaughanii* was a rootless, leafless plant that was low in stature, with a simple and primitive vascular system and simple reproductive structures located at the tips of the aerial branches.

A land flora was possible only after the detrimental effects of solar radiation, the ultraviolet (UV) radiation, became shielded by an ozone layer in the atmosphere. During the exceedingly long Precambrian and Cambrian Eras little oxygen was evolved from photosynthesis, and hence little ozone was produced. Aquatic algae would have been shielded from the UV effects by immersion. With an increasing density of algae, more oxygen would have been formed and an ozone layer created.

In addition to formation of the protective ozone layer, there was the evolution of biochemical conversion, by plants living in the littoral zone, of certain primary metabolites (amino acids and sugars) to secondary products (alkaloids, anthocyanins, tannin, lignin, and other phenolic substances) which could provide some protection from UV, parasitic fungi, protozoa, and other predators.

For vascular plant evolution the formation of the polymer lignin is of paramount importance. This substance provides the strength to keep a plant upright so that it can present a maximum photosynthetic surface. It made possible the dominance of vascular plants on earth, thereby permitting the evolution of animals. Lignin is a tough, hard substance that fills the microcapillaries between cellulose microfibrils in the cell walls of water-conducting cells (tracheary elements) and fibers; it is not present in bryophytes (Niklas and Pratt, 1980).

Whether the bryophytes (liverworts and mosses) and vascular plants were derived from an aquatic multicellular ancestral algal group(s) which then migrated to land, or originated from algae already present in terrestrial habitats, may never be resolved. This subject will be discussed more fully later in this chapter in the section on Alternation of Generations. However, there are certain features that land plants developed during millions of years for successful existence on land. Some of these modifications are shared by bryophytes and vascular plants, whereas others are exclusive to vascular plants and have made possible their eminent success. The following features pertain mainly to vascular plants:

1. Development of an anchorage and water- (and dissolved minerals) absorbing system, such as underground stems (rhizomes) with rhizoids (extensions of epidermal cells), performing the function of roots. True roots apparently developed later.

2. Development of a water-and-mineral—conducting system (xylem), and a system (phloem) for the conduction of photosynthates. The xylem would also provide support by the incorporation of lignin in walls of cells.

3. Prevention of desiccation by the formation of a waxy layer (cuticle) on aerial branch systems.

4. Development of structures for aerial gas exchange (stomata).

5. Development of specialized photosynthetic tissue in aerial stems; leaves, as well know them, were undoubtedly formed later in evolution.

6. Production of spores with the cell wall impregnated with sporopollenin, a substance that prevents desiccation and is virtually indestructible by microorganisms.

Except for the previously existent photosynthetic system in algae, all of the above adaptive modifications developed over an exceedingly long time. (For expanded discussions of this subject, the reader is referred to Niklas, 1976, 1981; Lowry, Lee, and Hébant, 1980; Stebbins and Hill, 1980; Niklas and Pratt, 1980.)

Sporophyte Generation

The normal origin of this generation is from the zygote, a diploid cell which results from the fertilization of the egg by a male gamete (a motile sperm in ferns, for example) (Fig. 36.1). There is considerable variation in vascular plants with respect to the length of the period of attachment and physiological dependency of the young multicellular sporophyte upon the gametophyte generation. Ultimately, the sporophyte gains physiological independence (except for certain parasitic plants) and

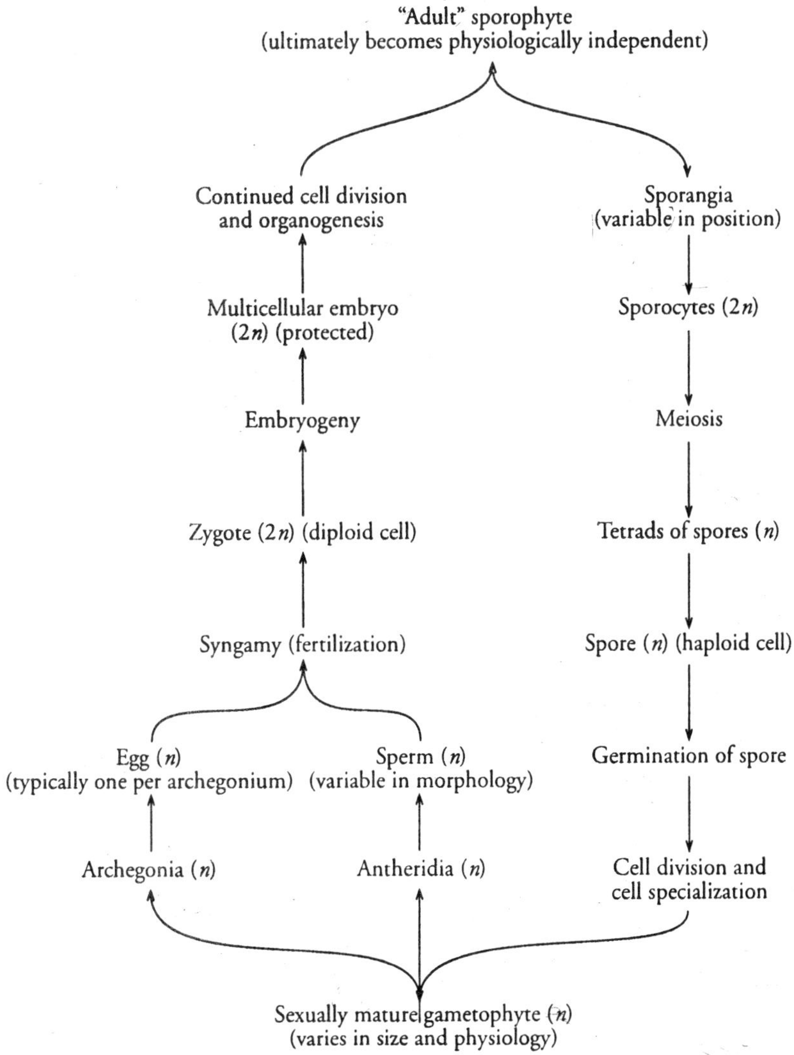

Fig. 36.1: Alternation of sporophyte and gametophyte generations in the complete life cycle of lower vascular plants. The processes and structures represented are discussed in detail in the text

develops into the dominant, typically photosynthetic, phase of the life cycle. *This physiological independence and dominance of the sporophyte constitutes one of the most definitive characters of vascular plants.*

From an organographic viewpoint, the sporophyte typically consists of a shoot system (usually aerial) made up of stems, various types of foliar organs, and roots. The latter may vary widely in their origin during embryogeny and later phases of growth. Shoots and roots are theoretically capable of unlimited apical growth, branching, and organ formation because of the maintenance at their tips of apical meristems. This capacity for continued indefinite apical growth is not found in the small dependent sporophytes of mosses and liverworts and hence represents a fundamentally important distinction between these organisms and vascular plants.

From an anatomical viewpoint, one of the most definitive features of the sporophyte is the presence of a vascular system. The form and arrangement of the vascular system vary not only among different groups of vascular plants but among different organs of the same plant as well. Because of its remarkably well-preserved condition in fossils, the vascular system often provides important clues to relationships and to trends of phylogenetic development. Except for the vein endings, which in leaves or floral organs consist largely of tracheary cells, the vascular system consists typically of two distinctive tissues; *phloem*, with conducting sieve elements, and *xylem*, with tracheary elements. Vascular tissues are essential to the conduction of water, mineral solutes, and soluble organic compounds—upon their coordinated functioning depends the normal physiology and growth of the sporophyte. Evidently, in the early occupation of the land by plants, the rapid evolution of an effective vascular system must have been of primary importance.

The characteristic reproductive structure of the sporophyte is the sporangium. Sporangia differ significantly in their position, methods of origin and development, and mature structure including sporangium wall and spore type. These differences provide reliable criteria, in many instances, for classifying vascular plants.

Gametophyte Generation

The normal origin of the gemetophyte generation is form a spore, which, in turn, is a product of the meiotic divisions of sporocytes in the sporangium (see Fig. 36.1). The gametophyte may be a free-living photosynthetic plant or subterranean and apparently dependent upon the presence of an endophytic fungus for its existence. In marked contrast, the gametophytes of angiosperms are much smaller and are physiologically dependent upon the sporophyte generation. Aside from these morphological and physiological variations, the chief importance of the gametophyte is the production of male gametes (or sperms) and female gametes (or eggs). These gametes are developed in distinctive multicellular gametangia in all lower vascular plants, the sperms arising in antheridia and the solitary egg developing within the archegonium. In angiosperms, the male and female gametes are produced directly by the greatly reduced and modified gametophytes, and morphologically definable sex organs are not developed. The position, ontogeny, and structure of gametangia and gametes furnish important characters used in the comparison and classification of the groups of lower vascular plants.

Although typical vascular tissue is usually restricted to the sporophyte generation, tracheary tissue has been observed in the gametophytes of certain plants. The best known example is the subterranean gametophyte of *Psilotum*, a vascular plant of uncertain phylogenetic origin and systematic relationships. Gametophytes, 1 millimeter or more in diameter, may have a central conducting strand, demarcated by an endodermis. A well-developed strand consists (in transection) of one to three tracheids with annular or scalariform thickenings, surrounded by elongated cells appearing to be the exact counterparts of the sporophytic sieve elements (Hébant, 1976). The vascular strand arises at the apical meristem of the gametophyte but commonly is discontinuous, fading out and reappearing usually several times in the length of a few millimeters.

Several "explanations" for the occurrence of a vascular strand in *Psilotum* gametophytes were made by Holloway (1939) and deserve brief examination especially in the light of additional studies made since the publication of his paper.

First of all, it might be argued that the vascularized gametophytes of *Psilotum* are "abnormal" in chromosome number which "conditions" the development of a conducting strand. Manton (1950) discovered that *all sizes* of gametophytes were diploid (2*n*) and that the sporophytes, to which they were related, were tetraploid (4*n*), and Bierhorst (1968) only found vascular tissue in the larger diploid gametophytes of *P. nudum*. Haploid gametophytes of the same species were entirely devoid of a vascular system.

A second possible interpretation of vascularized gemetophytes in *Psilotum* is that some type of physiological change occurs after a given gametophyte has reached a certain size in its growth. Holloway (1939) correlated "the presence of the conducting strand in *Psilotum*... with special robustness of growth." Bierhorst (1953) reached a similar conclusion and stated that "all the gametophytes are potential producers of vascular strands, and that a given gametophyte apex can, if the bulk of it meristem reaches a given threshold, produce a strand."

A third interpretation of the vascular strand in the gametophytes is that it represents the persistence of a type of vasculature that was characteristic of the gametophytes of at least *some* of the ancient and primitive vascular plants of the Devonian, such as *Rhynia.* (For a more complete description of the morphology of rhyniophytes. During the past 20 years suggestions have been made that the axes of *Rhymia gwynne-vaughanii,* which had previously been interpreted by Kidston and Lang (1921) as portions of the sporophyte, might actually represent vascularized gametophytes, comparable with those of *Psilotum.* The axes of *R. gwynne-vaughanii* have hemispherical projections on their surfaces, and Pant (1962) suggested that these might be young attached sporopytes. Lemoigne (1968, 1969, 1970) described what appear to be archegonia on certain vascularized axes of *R. gwynne-vaughanii,* although no antheridia were observed as is characteristic of *Psilotum.* However, Pant (1962) cautiously suggested that the presumptive gametophytes of *Rhynia gwynne-vaughanii* might eventually prove to be the sexual generation of *Rhynia major.* Recently this idea has been questioned by Edwards (1980, 1986) who was able to establish that sporangia were attached to some axes of *R. gwynne-vaughanii*—hereby identifying it as a sporophyte. Also, Edwards (1986) believes that *R. major* is not a vascular plant because the conducting strand in the stems resembles that of certain nonvascular plants such as mosses. It was renamed *Aglaophyton major.*

There is, however, evidence of gametophytic plants from the Devonian Rhynie Chert in Scotland (Remy and Remy, 1980; Remy, 1982). These gametophytes consist of radially symmetrical axes that terminate in concave, disklike pads (gametophores) that bear either antheridia or archegonia, or both, that resemble those of certain bryophytes. The internal anatomy of the axes and the presence of stomata, however, resemble that of *Rhynia.* The investigators are careful to point out that it is difficult to decide where the taxonomic affinities of these fossils lie. For the present, they suggest that one should perhaps regard them as gametophytes of plants "representing an evolutionary stage previous to the separation of land plants into bryophytes and higher land plants." However, the results of their investigations offer the hope that future paleobotanical discoveries may reveal whether there is a relationship between these gametophytic plants and the well-known sporophytes (such as *Rhynia* and *Horneophyton*) of the Rhynie Chert.

Alternation of Generations

Alternation of generations, or phases in the life cycle is a consistent feature of all groups of vascular plants and hence represents the basic pattern of reproduction in these dominant plants of the modern world. Figure 36.1 represents schematically the structures and processes common to the complete life cycle of *lower vascular plants.* The reader must understand this generalized life cycle clearly before becoming involved in the infinite variations of detail which occur in the cycle of reproduction of specific genera or groups of plants.

Because the sporophyte generation is the obvious and dominant phase of the life cycle, we may properly begin our analysis of Fig. 36.1 with the zygote. This diploid cell results from the union of male gamete with the egg. The next event is the process of embryogeny, which involves the production from the zygote of a multicellular embryo, the early development, form, and organography of which are often specific in a given group. By means of further growth and differentiation from the shoot and root apices of the embryo, the adult and independent sporophyte is developed. Ultimately the sporophyte plant forms sporangia in which spore-mother cells, or sporocytes, are produced. These cells, like all normal cells of the vegetative sporophyte, are diploid. However, each sporocyte can by meiotic division given rise to a tetrad or group of four haploid spores. When circumstances are favourable a spore germinates and by cell division and cell specialization produces the gametophyte generation. The salient function of the phase is the production of male gametes and eggs. The union—called fertilization or syngamy—of a male gamete and an egg restores the diploid chromosome number and produces a zygote from which a new sporophytic plant may develop.

It should be clear from this description that the alternation of sporophyte and gametophyte phases in the life cycle is normally coordinated with a periodic doubling followed by a halving of the chromosome number. The diploid zygote ($2n$), from which the sporophyte arises contains twice the number of chromosomes typical of the spore (n) that produces the gametophyte. Is this difference in chromosome number of spore and zygote a clue to the remarkable morphological and functional differences between the generations which arise from these cells? Or, to ask the question in another way: Are syngamy and meiosis *always essential processes* in the production, respectively, of sporophyte and gametophyte? These questions apparently deserve an answer in the negative because of certain deviations from a "normal" reproductive cycle, namely, *parthenogenesis, apospory,* and *apogamy.*

In certain angiosperms and a few ferns, the embryo arises from an infertilized egg, a phenomenon designated as parthenogenesis (i.e., "virgin birth").

The phenomenon termed *apospory* is the development of gametophytes, without a haploid spore stage, from vegetative cells of the sporophyte. Very commonly, aposporous gametophytes bear functional anatheridia and archegonia, and chromosome studies have shown that their gametes are diploid and that a sporophyte resulting from their union is tetraploid (see Steil, 1939, 1951 for further information on apospory).

A third deviation from the usual reproductive cycle of a vascular plant is the phenomenon of *apogamy*, which is the fomation of a sporophyte, without the act of fertilization, from vegetative cells of the gametophyte. Depending upon the chromosome number of the gametophyte (i.e., whether normal *n* or aposporous 2*n*), the apogamous sporophyte may be either haploid or diploid in chromosome number.

Quite apart from the interesting questions raised by the phenomena of parthenogenesis, apospory, and apogamy, there remains the broader and still unsolved problem of the phylogenetic origin of the type of alternation of generations which is "normal" and prevalent throughout all living vascular plants. If, as is commonly believed, the remote ancestors of terrestrial plants were green algae, the central question at issue is the evolutionary origin of the diploid, vascularized, independent sporophyte typical of the vast majority of vascular plants.

A century ago, Celakovsky (1874) recognized two principal form of alternation which he distinguished by the terms "antithetic" and "homologous". In his opinion, antithetic" alternation, which is characteristic of archegoniate plants (i.e., bryophytes and vascular plants with archegonia), arose during evolution by the interpolation of a new phase, i.e., the sporophyte, between successive gametophytes. In contrast, he limited his concept of homologous alternation to the succession of morphologically similar phases which occurs in the reproductive cycles of certain fungi and algae.

Today the terms "antithetic" and "homologous" are usually applied to two divergent theories which both seek to explain phylogenetically the characteristic type of life cycle in vascular plants. Bower (1935) has urged that in place of "antithetic" and "homologous" theories, it would be more appropriate to speak, respectively, of the "interpolation theory" and the "transformation theory". In his view, these substitute terms are explicit in that they convey the alternative methods of phylogenetic origin of the sporophyte generation. This point may be clarified by the following brief contrast between the two theories. For a more detailed analysis, the reader is referred to Bower (1935), Fritsch (1945), and Wardlaw (1952).

According to the *interpolation theory*, which was strongly championed by Bower, the origin of the sporophyte generation was fundamentally the result of the postponement of meiosis and the development from the zygote of a "new" diploid vegetative phase. During subsequent evolution, this vegetative phase or sporophyte became increasingly complex, both organographically and anatomically but "meiosis and spore formation, though delayed, would still be the final result" (Fig. 36.2). Bower also maintained that the gradual specialization of an independent spore-producing diploid phase was closely connected, in a biological sense, with the transition from aquatic to terrestrial life by the algal-like "progenitors" of the first land plants.

The *transformation theory* of alternation, in contrast, postulates that the spore-producing and gamete-producing *individuals* of the original parental green algae were morphologically similar and hence that the life cycle was *isomorphic* in type (Fig. 36.3). This idea was developed in great detail by Fritsch (1945) and some of his interesting ideas deserve brief attention at this point. According to Fritsch, the first step in the long evolutionary development of archegoniate plants from aquatic green algae was the attainment of a morphological distinction between *erect* branches and a creeping prostrate system. This resulted in what is termed a *heterotrichous habit* of growth. Further elaboration of the upright system by means of longitudinal divisions in some of the cells then produced a more-or-less "parenchymatous" type of organization. Fritsch regarded this second step a "most significant evolutionary advance,

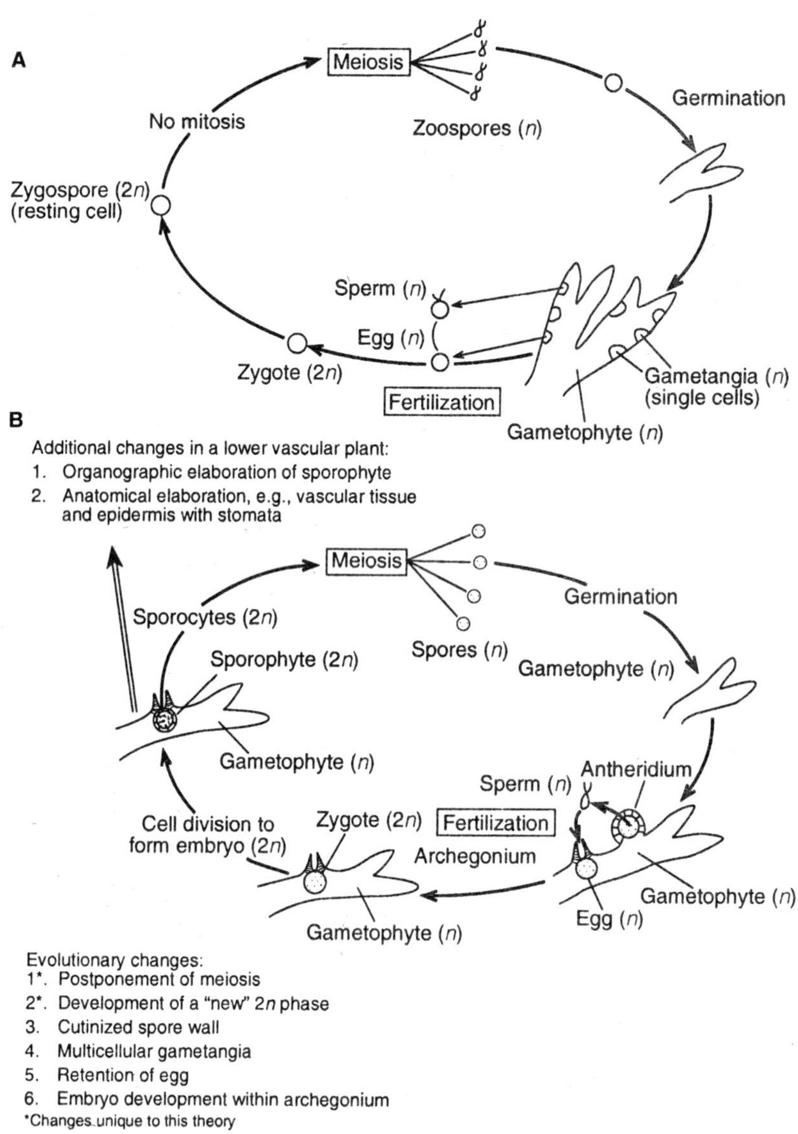

Fig. 36.2: Schematic representation of the origin of alternation of generations or phases in the life cycle according to the interpolation (or antithetic) theory. A. Hypothetical haploid green alga with oogamy. B. Development of a simple multicellular sporophyte; note additional features required for the evolution of a simple vascular plant (Modified from class handout by Dr. R. Schmid.)

since in it lay the germ for the development of a plant body of almost unlimited size." Evolutionary modifications then occurred that would have led to morphological divergence between sporophytic and gametophytic individuals and the adoption of a heteromorphic type of life cycle (Fig. 36.3). In Fritsch's view, the gametophyte, at this stage in evolution, corresponded in form and dichotomous branching to the primary underground system of the sporophyte and had "lost the capacity for emergence over-ground."

This conclusion is particularly interesting in light of discoveries of possible "rhizomelike" gametophytes (Lemoigne, 1968, 1969; Remy, 1982) from the Devonian.

For many years it has been assumed that some group in the class Chlorophyceae of the green algae (Chlorophyta) was the progenitor of land plants, from which bryophytes and vascular plants later evolved. Indeed, an argument in support of this assumption is that vascular plants, bryophytes, and the Chlorophyceae share certain basic features: chlorophylls *a* and *b*, a cellulose wall, and starch as a reserve product. Some of the living Chlorophyceae have an isomorphic alternation of generations, so essential to Fritsch's concept. While the idea of shared attributes is appealing, recent research has cast doubt on the concept that the Chlorophyceae is the ancestral group from which bryophytes and vascular plants originated. It is probably impossible to select a specific living algal progenitor, but recent studies have pointed toward the green-algal class Charophyceae (stoneworts), including *Coleochaete* (Stewart and Mattox, 1975; Pickett-Heaps, 1976), as being more likely candidates. These

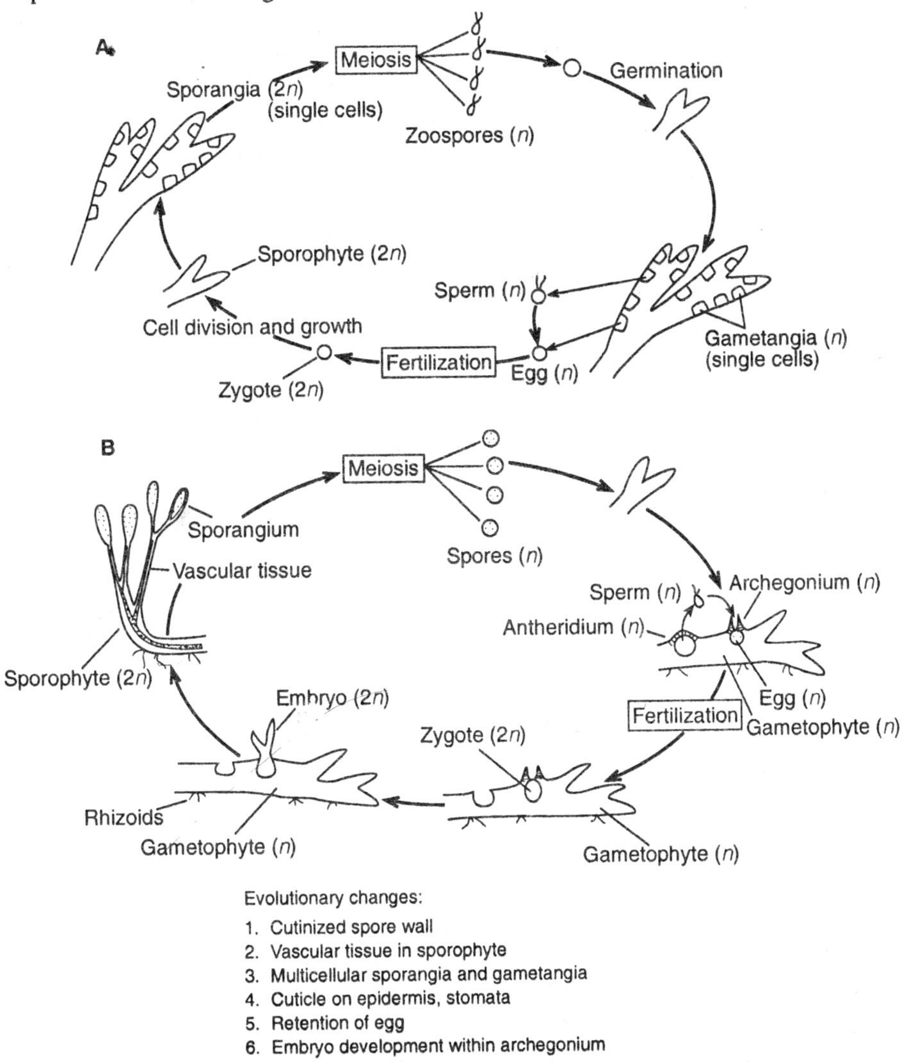

Fig. 36.3: Schematic representation of the origin of alternation of generations or phases in the life cycle according to the transformation (or homologous) theory. A. Hypothetical green alga with isomorphic alternation of generations and oogamy. B. Homosporous lower vascular plant (generalized), (Modified from class handout by Dr. R. Schmid.)

algae possess the attributes enumerated previously for the Chlorophyceae, but ultrastructural and biochemical studies have revealed basic differences between the two classes. Charophytes, bryophytes, and vascular plants display a persistent mitotic spindle, followed by formation of the new cell wall. In the Chlorophyceae, however, the spindle collapses at telophase, the daughter nuclei often approach each other, and microtubules are formed at right angles to the original spindle. The new cell wall is then formed parallel to the new microtubules by cell plate formation or cleavage. In addition, the Charophyceae posses glycolate oxidase similar to that in higher plants but which has not been identified in Chlorophyceae. Furthermore, planes of cell division and branching patterns in *Coleochaete* have been cited as examples of how land-plant meristems and parenchyma may have evolved from branched, filamentous charophycean green algae (Graham, 1982; Hagemann, 1978). Also, details of sexual reproduction in *Coleochaete* provide support for the interpolation theory of the origin of alternation of generations (Graham, 1985).

Classification of Vascular Plants

Certainly from a broad evolutionary viewpoint, the origin of a vascular system must have been highly significant in the development of the sporophyte generation and in its continued adaptation to life on land. In the second edition of this book the term Tracheophyta (a division) was used to designate all vascular plants. Twelve classes were recognized as constituting the Tracheophyta. These classes consisted of groups of plants that are generally recognizable based on consistent morphological features. Acceptance of the Tracheophyta as a major taxon does imply a monophyletic origin of vascular plants (Stewart, 1983). In recent years there has been a tendency to recognize several divisions of vascular plants, instead of only one. In these systems, what were formerly classes, for example, have been raised to divisional status. Embodied in these systems is the concept of polyphyletic origin of vascular plants, whether stated or implied. Paleobotanical research in recent years has resulted in the recognition of what appear to be separate evolutionary lines of vascular plants as far back as the Lower Devonian and even into the mid-Silurian period. Thus far no megafossils of vascular plants from earlier periods have been discovered, although microfossils (e.g., isolated tracheidlike cells, spores) that could be from vascular plants or bryophytes have been described. Thus there are many uncertainties as to whether all vascular plants can be traced back to one origin from an algal group, or whether the acquisition of vascular tissue is the expression of an adaptive tendency appearing in seemingly unrelated groups, as is the case with the development of leaves, vessels, and the seed habit. Only with additional paleobotanical evidence may we ultimately be able to answer this question.

All phylogenetic schemes of classification are, at best, tentative. No one scheme satisifies contemporary morphologists and paleobotanists with respect to the number of divisions, orders, and families. One of the difficulties, just mentioned, is the absence of convincing phyletic interconnections between the various groups of vascular plants that have been proposed. Moreover, there is considerable confusion at the level of formal schemes of classification because of the shifting of lower taxa (e.g., classes) to higher categories (e.g., divisions) or *vice versa*, merely by changing the suffix of certain names.

In view of these difficulties and nomenclature problems, no attempt will be made to contrast the various schemes of classifying vascular plants. However, instead of using "Tracheophyta" as the one division for vascular plants, in this edition we recognize fifteen divisions. Each of these formerly represented a class or even an order. This system is more in line with those of some contemporary botanists. The plants in each division have characters that are recognizable and can be used to differentiate them from those in other divisions, as well as to provide the bases for making comparisons and for establishing tentative phylogenetic relationships.

The conspectus shown in Table 36.1 lists the divisions of vascular plants recognized in the present text and relates them to the categories used in the previous edition. We introduce the conspectus at this point simply to orient the reader to the names and representative examples of the divisions of vascular plants. More detailed taxonomic treatment of each of the divisions will be found in subsequent chapters.

Table 36.1: Generalized Classification of vascular plants

Classification from comparative Morphology of Vascular Plants, 2nd edition	Classification used in present edition (Names in parentheses are used in some other systems for the same group)
Tracheophyta (division)	Divisions
Rhyniopsida*	1. Rhyniophyta (extinct plants, e.g., *Rhynia, Cooksonia*)
Zosterophyllopsida	2. Zosterophyllophyta (extinct plants, e.g., *Zosterophyllum*)
Trimerophytopsida	3. Trimerophytophyta (extinct plants, e.g., *Psilophyton, Trimerophyton*)
Psilopsida	4. Psilophyta (Psilotophyta) (two living genera, *Psilotum* and *Tmesipteris*)
Lycopsida	5. Lycophyta (Microphyllophyta) (extinct forms such as *Protolepidodendron* and *Lepidodendron*, and the living genera *Lycopodium, Selaginella, Phylloglossum, Isoetes*, and *Stylites*)
Sphenopsida	6. Sphenophyta (Arthrophyta) (mostly extinct plants, e.g., *Calamites, Sphenophyllum*; the one living genus is *Equisetum*)
Filicopsida	7. Filicophyta (Pteridophyta) (living and extinct ferns)
Progymnospermopsida	8. Progymnospermophyta (extinct plants, e.g., *Archaeopteris*)
Cycadopsida	
Pteridospermales (order)	9. Pteridospermophyta (extinct seed ferns, e.g., *Lyginopteris, Medullosa*)
Cycadales (order)	10. Cycadophyta (extinct and living cycads, e.g., *Cycas, Zamia*)
Cycadeoidales (order)	11. Cycadeoidophyta (extinct cycadeoids, e.g., *Cycadeoidea*)
Coniferopsida	
Ginkgoales (order)	12. Ginkgophyta (extinct members and living *Ginkgo biloba*)
Coniferales (order)	13. Coniferophyta (extinct and living conifers, e.g., pine, fir)
Gnetopsida	14. Gnetophyta (*Ephedra, Gnetum, Welwitschia*)
Angiospermopsida	15. Magnoliophyta (Anthophyta) (angiosperms or flowering plants)

*The taxonomic level in the left column is class unless otherwise indicated.

Chapter 37

Psilophytopsida

The Psilophytopsida comprise of the fossilized ancient Silurian-Devonian order Psilophytales. The discovery of the different genera from such wide places as Canada, Scotland, Europe and Australia shows that these plants were of world-wide distribution in those days. The Psilophytopsida and the living Psilotopsida have been included within a single class Psilopsida (or a division Psilophyta) by several authors. But the recent trend is to consider them as separate classes for reasons explained at the end of this Chapter and that of the next Chapter (Psilotopsida).

Previously all the known genera within the class used to be placed within a single order Psilophytales. The better known genera of this order used to be placed in five families: 1. Rhyniaceae, 2. Zosterophyllaceae, 3. Psilophytaceae, 4. Asteroxylaceae, and 5. Trimerophytaceae. The two metres high, tree-like Middle Devonian genus *Pseudosporochnus*, formerly supposed to be a member of the Psilophytales is now considered to be a member of the ancient fossil fern taxon *Cladoxylales*. Moreover, recent investigations have completely changed our idea of the Psilophytales. Of the principal genera, our original idea of the important genera *Psilophyton* and *Asteroxylon* has been shown to be completely wrong. As explained below. *Psilophyton* as constructed by Dawson has been shown to be a mixture of three plants so that some even doubt whether the genus can be retained at all and *Asteroxylon* proper has been proved to be a Lycopod. Thus only four families may now be recognized under Psilophytopsida: 1. *Rhyniaceae*, 2. *Psilophytaceae* (in a restricted and revised sense), 3. *Trimerophytaceae* and 4. *Zosterophyllaceae*. The so-called *Asteroxylon elberfeldense* should also belong to the Psilophytales but it needs a new name since the original *Asteroxylon* (*A. mackiei*) has now been found to be altogether different and has now been transferred to be Lycopsida.

The confusion about the genus *Psilophyton* has led some eminent Pteriodologists to discard the class name Psilophytopsida. Again because of the fact that this taxon represents different tendencies of evolution leading to all the higher plant groups, some of them want to raise this taxon into one or more taxa of a status higher than a class, i.e., division or subdivision. Thus,

Banks (1967) proposed three subdivisions: (1) *Rhyniaphytina*, (2) *Zosterophyllophytina* and (3) *Trimerophytina* directly under the Division *Trachaeophyta*. But, Bierhost (1971) has suggested two classes: (1) *Rhyniopsida* (including Trimerophytaceae) and (2) *Zosterophyllopsida* also directly under the division Trachaeophyta.

Whatever may be the nomenclature of the taxon status of the four families recognized here, they are not necessarily closely related and probably represent an artificial association of three parallel lines of evolution evolved in the earliest history of the land plants as suggested by Delevoryas (1969). Of the four families (1) Rhyniaceae and (2) Psilophytaceae represent a plexus with terminal sporangia while (3) Trimerophytaceae is also with terminal sporangia but associated with more bushy, radial branching. The two may represent two different trends of evolution as suggested by Delevoryas but may also be two tendencies of the same terminal sporangiate plexus as considered by Mehra (c.f. last Chapter of Pteriodophyta) and apparently, also by Bierhorst (1971). The fourth family (4) Zosterophyllaceae, no doubt, represents a lateral sporangiate plexus out of which the Lycopsids arose (c.f. Mehra *loc. cit.*).

Considering all these, this class is divided here into two orders:

I. **Psilophytales** (with terminal sporangia—including Trimerophytaceae) and

II. **Zosterophyllales** (with lateral sporangia)

Order I. PSILOPHYTALES

The Psilophytales represent the plexus of primitive plants with apical sporangia and probably gave rise to the great majority of modern vascular plants. Three families (*Rhyniaceae, Psilophytaceae* and *Trimerophytaceae*) are clearly recognizable here with a forth one (represented by the so-called *Asteroxylon elberfeldense*) not yet defined.

Family: RHYNIACEAE

The Rhyniaceae are the simplest of the Psilophytales, often compared with the sporophyte *Anthoceros*. The most important genera are *Rhynia* and *Horneophyton* from the Middle Devonian in Scotland.

Genus: *RHYNIA*

The genus *Rhynia* from the Rhynie chert beds (Middle Devonian) in Scotland was discovered by Mackie in 1913 and fully described by Kidston and Lang in 1917. This discovery established the Psilophytales as a separate and distinct taxon. Three species are known of which *Rhynia major* and *R. gwynne-vaughani* are the better known.

The plants, apparently, grew in swampy marshes near volcanoes where the atmosphere contained sulphurous vapour and the soil was acid. The reconstructions are from silicified petrifications.

The sporophyte

The plant body was a herbaceous sporophyte with dichotomously branching horizontal rhizomes bearing rhizoids on the underside and some of the branches growing up abruptly forming aerial shoots. The aerial shoots of *R. major* (Fig. 37.1A) were up to 50 cm long and 6 mm in diameter while those of *R. gwynne-vaughani* (Fig. 37.1B) were shorter and more slender. The aerial branches were cylindrical and sparsely dichotomously branched. These were naked being devoid of any appendage of leaf and usually tapering upwards ending in a point or in an erect sporangium.

R. gwynne-vaughani shows hemispherical, oval or lenticular protuberances arising from the lower parts (more mature) of the aerial shoots or from the rhizomatous parts (Fig. 37.1B). These are constricted at the points of attachment and, in mature ones, have their own vascular bundles not connected with those of the main stems. They are found to be readily

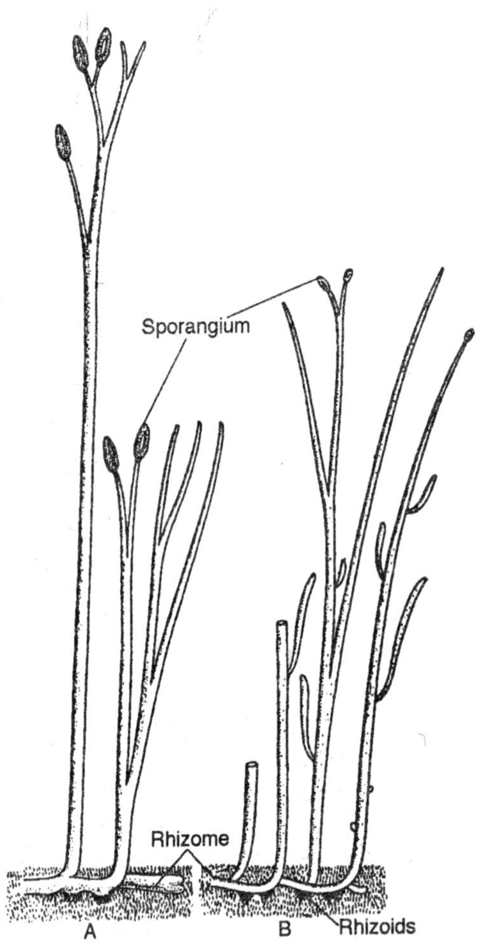

Fig. 37.1: A. *Rhynia major*, B. *Rhynia gwynne-vaughani* (After Kidston and Lang)

detachable and, possibly; was a means of vegetative propagation germinating into new shoots.

The *anatomy* of the stem is very simple (Fig. 37.2A). In the centre is a slender, hadrocentric protostele with a small central xylem formed of simple annular tracheides which, in some larger specimens, become smaller towards the centre. This is surrounded by four or five layers of elongated cells with oblique ends which represent the phloém although sieve plates have not been observed. There is no endodermis or pericycle. All round this vascular bundle is a massive inner cortex of loose, rounded, parenchymatous cells with lots of air spaces. This fills most of the stem and was probably the main photosynthetic tissue. The outer cortex is formed by one or two layers of somewhat, larger, comparatively angular, compact (except below stomata) cells at the hypodermal region. The epidermis is a compact layer of cells broken here and there only by the stomata with pairs of guard cells as in other vascular plants. This is externally covered by a heavy cuticle. Often the smallest branches show no vascular supply.

The *sporangia* are oval or cylindrical structures with pointed ends (Fig. 37.2B) at the apices of the dichotomies. They may be slightly constricted at the bases though continuous with the stem and are always wider. Those of *R. major* were rather big (about 12 mm long and 4 mm in diameter). The sporangium wall is thick and multilayered with the outer cells thick-walled and no method of dehiscence is observed. The thinner, inner cells probably represent the tapetum. The whole interior is filled with spore tetrads or free spores. The spores are spherical, large (up to 65μ in diameter in *R. major*) and covered with a thick cuticle.

The gametophyte

As in all the Psilophytopsida, the gametophytes are not known. Lyon (1957) found some germinating spores within the same Rhynie chert beds which show multicellular structures developing at the ends of germ tubes. These may represent the gametophytic germination. Merker (1959 and 1961) has suggested that it is not possible that the gametophytes of such a big group were not fossilized while the large algae had been preserved. He argued that the underground creeping parts of *Rhynia* and *Horneophylon* are the gametophytes and not rhizomes. But no sex organ has been found on these underground parts and the strong vascular bundle is not normal in a

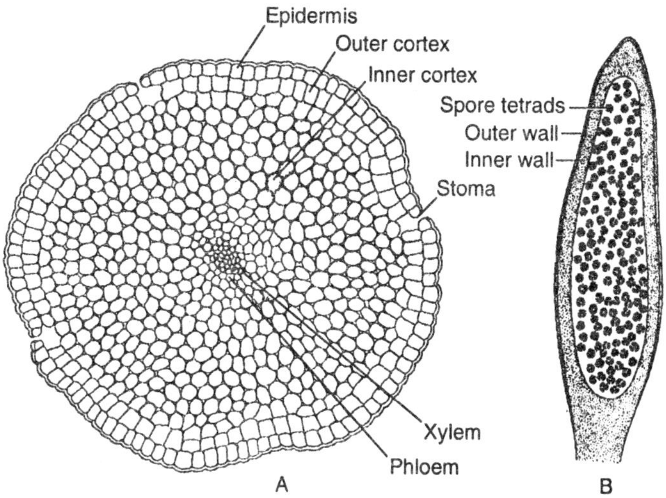

Fig. 37.2: A. T.s. of stem and B. L.s. of sporangium of *Rhynia gwynne-vaughani* (after Kidston and Lang)

gametophyte. His view is, till now, mere speculation. Puri (1961, supported by Pant, 1962) is of opinion that *Rhynia* had a homologous gametophyte and, according to him, some of the plants described as sporophytes may be gametophytes, e.g., some smaller *R. Gwynne-vaughani* may be gametophytes of the larger *R. major*. These two views suggest a vascular gametophyte which is uncommon, though not completely absent (c.f. polyploid *Psilotum*) among the Pteridophytes. Moreover all *R. gwynne-vaughani* show stomata, something unusual in gametophytes.

OTHER GENERA OF RHYNIACEAE

Horneophyton (originally named *Hornea* but name changed due to nomenclatural defect) was similar to *Rhynia* and was discovered from the same place but was smaller and the rhizomes were short and tuberous presenting a jointed appearance (Fig. 37.3A). The rhizome is devoid of any vascular supply. The vascular supply enters the rhizome tuber from the stem but fades out after expanding like a bell. There are some mycorrhazal fungi inside the rhizome parenchyma. The sporangium (Fig. 37.3B) is slightly wider than the stem apex and in all respects show that it is simply a modification of its apex. It has a sterile columella which is a projection of the stem phloem below with the spore sac overarching on it as in *Sphognum* or *Andreaea*. This columella sometimes shows a tendency to bifurcate like the stem. The tapetal cells form an extension of the columella lining the inside of the spore sac. The presence of this columella is of phylogenetic importance.

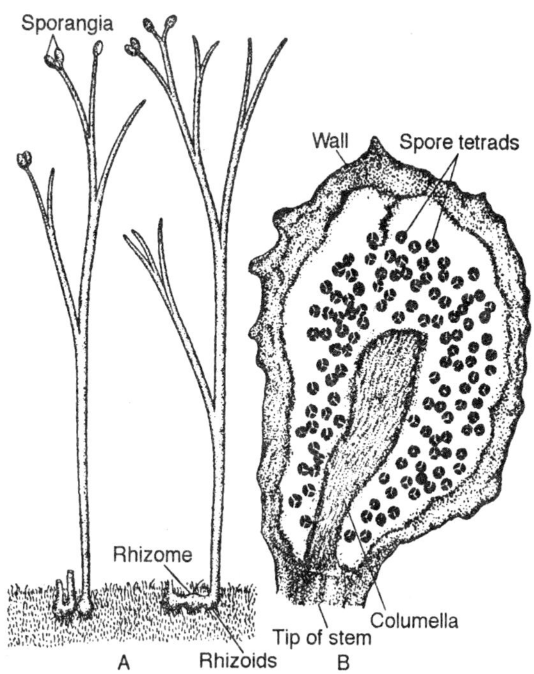

Fig. 37.3: *Horneophyton lignieri* (after Kidston and Lang). A. Sporophyte plant. B. L.s. of sporangium

Sporogonites specimens, discovered from Upper Silurian of Australia and Lower devonian of Norway, are rather imperfectly preserved. The sporangium shows a sterile basal half, there being a dome-shaped spore sac only in the upper half. No vascular strand is noted in the stem so that a relationship with the Bryophyta is strongly-suggested. This lack of vascular supply was explained to be due to imperfect preservations. But, Andrews (1959) found the stalked sporangia borne parallelly on a crust-like film which he supposed to be the thallus of a Hepatic.

Besides *Rhynia, Horneophyton* and the doubtful genus *Sporogonites*, a number of other genera assignable to this family have been discovered. *Cooksonia* (Fig. 37.4A) from Upper Silurian and Lower Devonian of Wales show slender, naked, dichotomously branched stem fragments with apical sporangia similar to *Rhynia*. *Hedeia* (Fig. 37.4B) and *Yarravia* (Fig. 37.4C) were discovered by Lang and Cookson from the Upper Silurian of Australia as fructifications only. The former has a corymbose bunch of apical sporangia while in the latter there is an apical sporangial structure which is a synangium of radial symmetry formed by the lateral fusion of 5 to 6 lineal-oval sporangia. These structures were, apparently, more complex than the later *Hornea* or *Horneophyton*. *Hicklingia* (Fig. 37.4D) from Lower to Middle Devonian of Scotland had a bushy appearance formed by the repeated dichotomy with apical oval but bell-shaped sporangia on the dichotomies. *Nothia aphylla* (Fig. 37.4E) is the name given by Lyon (1964) to the Lower Devonian unattached,

Fig. 37.4: Rhyniaceae. A. *Cooksonia pertoni* (after Croft and Lang). B. *Hedeia corymbosa* (after Cookson). C. *Yarravia oblonga* (after Andrews). D. *Hicklingia* (after Banks). E. *Nothia aphylla* (after Lyon). F. *Taeniocrada* (after Krausel and Weyland)

dichotomously branched, leafless fructification which was wrongly associated by Kidston and Lang with *Asteroxylon mackiei. Taeniocrada* (Fig. 37.4F) from Upper Silurian and Lower Devonian with a flattered, dichotomously branched, thallus-like stem, was probably of a submerged, aquatic habit, and showed apical sporangia on branched fertile clusters.

Family: PSILOPHYTACEAE

The concept of the Class Psilophytopsida begin with the discovery by Sir William Dawson in 1858 from the Lower Devonian Gaspe sandstone bed in Canada. This very primitive pteridophyte was quite unthought of in those days and was named as *Psilophyton princeps* by Dawson in 1859 and was fully described by him later (1871). Later on this was also found in Scotland and still later the generic name was being applied very loosely to numerous fragmentary and imperfect specimens from all over the world, sometimes on very incomplete evidence. Now even Dawson's reconstruction has been proved to be incorrect. Dawson's original reconstruction is described belows:

Psilophyton Princeps Dawson

The *sporophytic plant* (Fig. 37.5A) grew in dense champs and, possibly, in wet places along streams or in marshes. They were probably not more than a metre high with the stem not exceeding 1 cm in diameter. The general habit is the same as that of *Rhynia* with a creeping rhizome bearing rhizoids and erect, aerial, dichotomously branched shoots. The aerial shoots are much branched and the tips of stems circinately coiled. The base of the stem bears short, rigid, curved, spinous structures (Fig. 37.5B) 0.3 to 2 mm long. These were described as leaves by Dawson but seem to be mere superficial emergences flattened in a vertical plane and devoid of any vascular supply or stomata. The presence of a dark substance inside some of them suggests that they were of a glandular nature.

The *anatomy* of the stem is not very clear but, on the whole, resembles *Rhynia*. The central xylem contains spiral as well as annular (Fig. 37.5C) and, perhaps, some scalariform tracheides. The epidermis with stomata and the inner cortex, possibly, resembled *Rhynia* and the latter probably served as the photosynthetic tissue. The outer cortex is collenchymatous and the stele is mesarch.

The *sporangia* (Fig. 37.5A) are usually borne in pairs (because of the bifurcation of the stem tip) suspended from stem tips. They are broader than the stem, obovoid with acute pointed ends, somewhat curved, up to 6.5 mm long and 1.5 mm in diameter. The sporangial wall is multilayered. The spores are 60 to 100µ in diameter and, in spite of the irregularity in size, do not suggest heterospory.

The *gametophyte* is completely unknown.

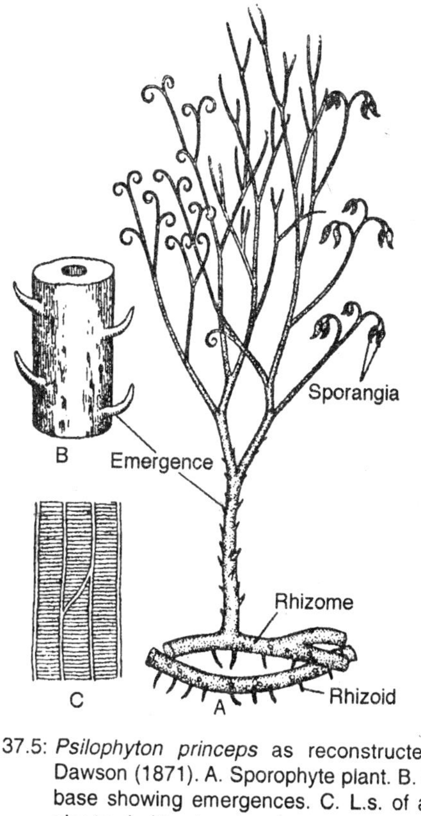

Fig. 37.5: *Psilophyton princeps* as reconstructed by Dawson (1871). A. Sporophyte plant. B. Stem base showing emergences. C. L.s. of annualar tracheides in xylem

The above is a description of the later reconstruction made by Dawson in 1871 out of fragments which were never found together. Recent intensive investigations by Hueber (1964) and Hueber and Banks (1967) in the same Gaspe bed from where Dawson obtained his material and also of the very material used by Dawson as available has shown that Dawson's reconstruction actually involved three different plants. The rhizome structure shown by Dawson did not belong to this plant but probably belonged to *Taeniocrada.* Apart from this, Hueber and Banks found in the Gaspe bed three different plants all assignable to the genus *Psilophyton:* (1) A plant corresponding to the plant described by Dawson but without any basal rhizome. The t.s. of the stem (Fig. 37.6B) is similar to *Rhynia* but with scalariform (sometimes pitted) tracheides. This (Fig. 37.6A) is now *Psilophyton princeps* as amended by Hueber and Banks (1967). (2) Another similar plant with terminal sporangia with smooth stem (not showing any enation or spine) first named *Dawsonites* (Fig. 37.6C) but is now considered not as a separate genus but as a species or variant of *Psilophyton* as amended by Hueber and Banks. (3) A third plant which was noted by Dawson and called var, *ornatum* has got lateral and not apical sporangia very unlike *Psilophyton* although the stem is spinous. Provisionally named *Psilophyton ornatum,* it has now been given a new generic name *Sawdonia* by Hueber (1971) so that the name now becomes *Sawdonia ornata* (Dawson) Hueber (Fig. 37.9B) and belongs to the Zosterophyllales.

"Asteroxylon elberfeldense"

The family Asteroxylaceae has so long been considered as a family within the order Psilophytales. It has now been clearly demonstrated that *A. mackiei* on which the family is

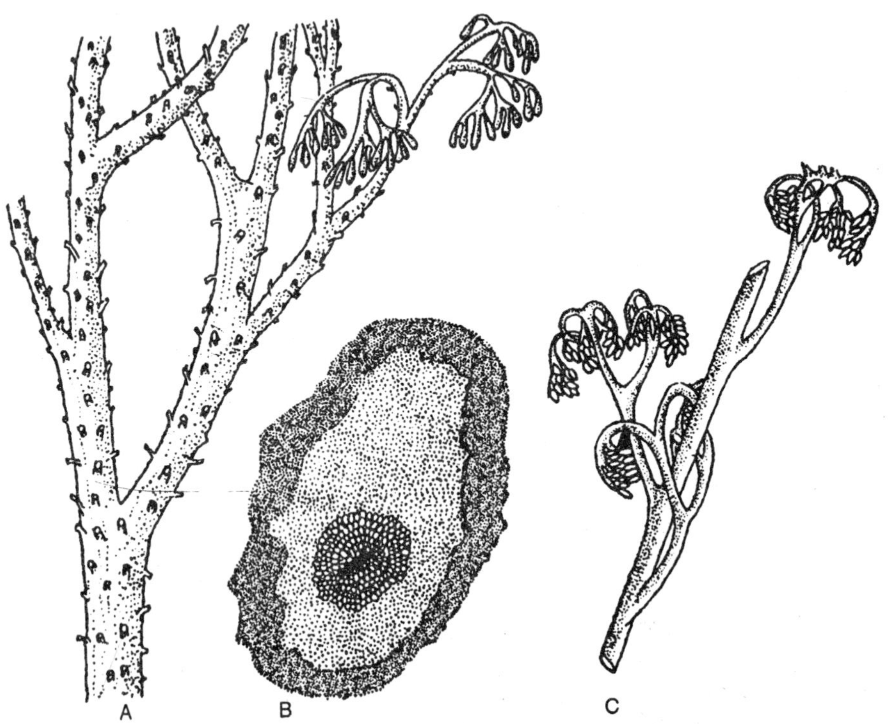

Fig. 37.6: *Psilophyton.* A. *Psilophyton princeps* Dawson as modified by Hueber. B. T.s. of *P. princeps* stem (both smooth and spiny forms) (after Hueber and Banks). C. Smooth form of *P. princeps* once called *Dawsonites* (after Hueber)

based did not have terminal sporangia and is a Lycopod so that it has been transferred to the Lycopsida (*loc. cit.*). But, there is a problematic plant named *Asteroxylon elberfeldense* by Krausel and Weyland (Fig. 37.7) from German Middle Devonian rocks. This has got the vegetative stem closely covered with enations and underground branches as in *Asteroxylon mackiei* (for description see Lycopsida) but in organic association with a sporangial shoot showing apical sporangia as in the Psilophytales. This plant is, therefore, a psilophyte and not an *Asteroxylon.* It needs a new generic name and also a new family status. The stem of *elberfeldense* is even more developed in being siphonostelic with a central pith and showing the presence of scalariform and round-pitted tracheides.

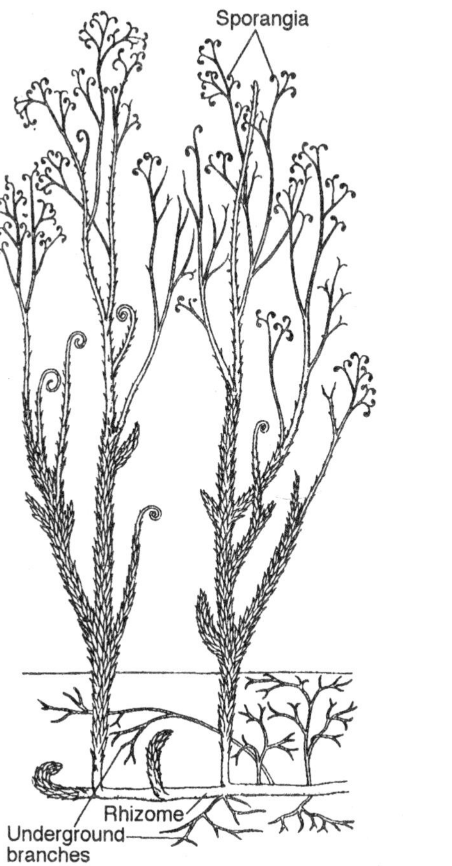

Fig. 37.7: *Asteroxylon elberfeldense* Kräusel and Weyland *nom. invalid*

Family: TRIMEROPHYTACEAE

The genus *Trimerophyton* from the Dovonian deserves placing in a separate family as it probably represents a separate line of evolution leading to the Progymnosperm *Archaeopteris* and then the seed ferns (c.f. Hueber and Banks, 1967). It is included within the Psilophytopsida as it belongs to the terminal sporangial plexus. But, as explained before, some Pteridologists consider it as a separate line altogether, parallel to Psilophytopsida and Zosterophyllopsida.

Trimerophyton robustius (Fig. 37.8) has a stout erect axis, about 1 cm in diameter bearing lateral branches, each one of which in its turn branch trichotomously and then dichotomously ultimately ending in sporangia. There is no leafy appendage but still it presents a bushy appearance. The radial appearance of the plants gives them a distinct position.

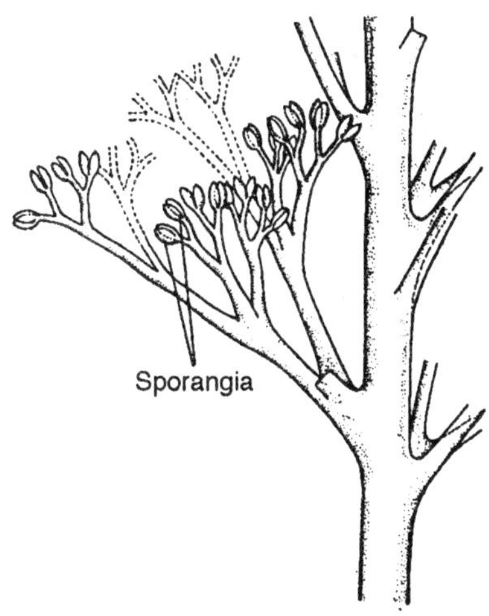

Fig. 37.8: *Trimerophyton robustius*

Bierhorst (1971) would like to include the bushy plants of Psilophytaceae (c.f. *Psilophyton* itself) within this family. In that case *Asteroxylon elberfeldense* also should come here.

Order II. ZOSTEROPHYLLALES

The *Zosterophyllales* represent the lateral sporangiate plexus of this primitive group which prevailed during the Lower Devonian. As has been explained, some Pteridologists consider this as a taxon large than an order, *viz.*, Class Zosterophyllopsida evolving parallel to the Psilophytopsida. There is good reason to believe that these plants gave rise the Lycopsida. The genera may be placed within a single family *Zosterophyllaceae.*

Family: ZOSTEROPHYLLACEAE

Of the genera included within this family *Zosterophyllum* (Fig. 37.9A) is the best known. *Zosterophyllum* fossils have been obtained from the Upper Silurian to Lower Devonian Old Red Sandstone in Scotland. The plants were probably partially submerged. The most important peculiarity is that the sporangia are borne on very short lateral stalks spirally arranged on the leafless aerial branches—giving the appearance of spikes. The sporangia are rather wide and

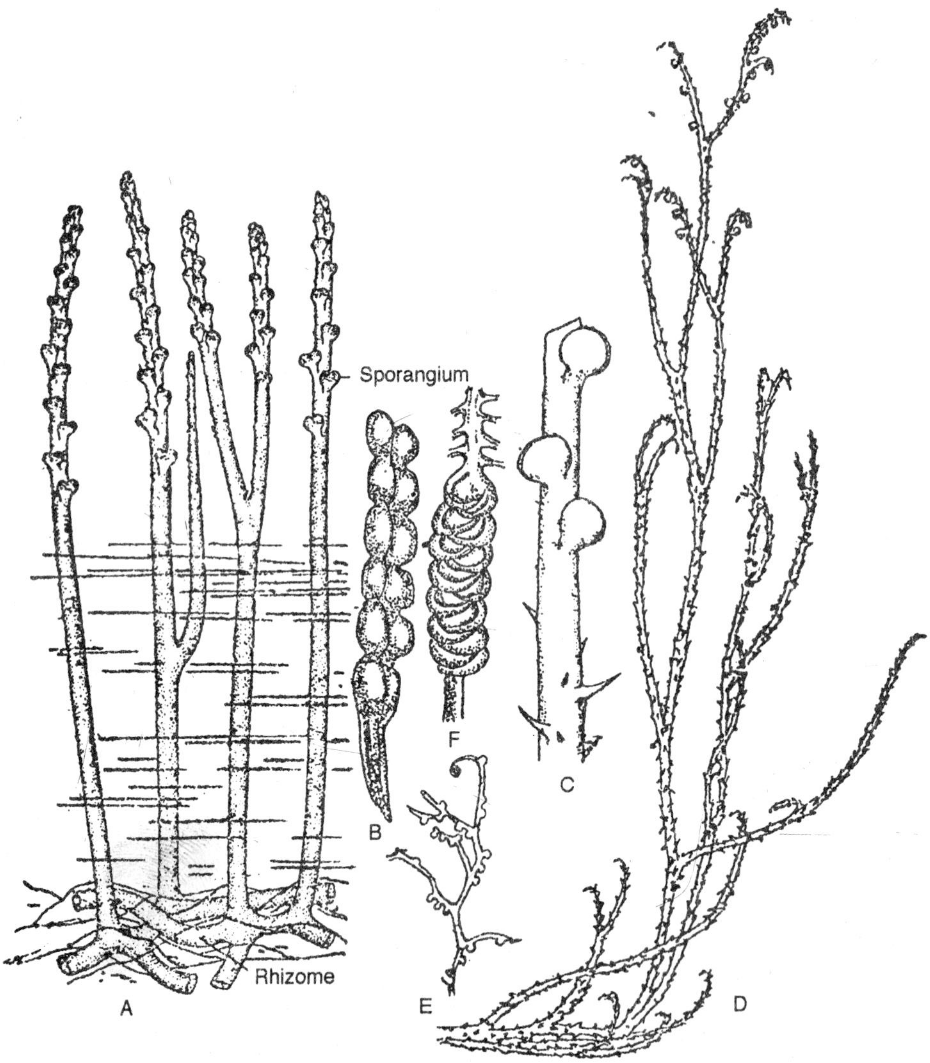

Fig. 37.9: A. *Zosterophyllum myretonianum* (modified from Lang). B. *Bucheria ovata* (after Dorf). C. Part of *Sawdonia ornata* (Dawson). Hueber = *Psilophyton princeps* var. *ornatum* Dawson (after Hueber). D. Restoration of entire *S. ornata* plant after Aneniev and Setepanov. E. *Gossilinga* (after Andrews). F. *Pectinophyton bipectinatum*

each provided with an apical split for spore dispersal. The spores are of variable size but the size difference is not connected with heterospory. The stem anatomy is as in Rhyniaceae.

Of the other genera within this Lower Devonian family, mention may be made of (1) *Bucheria* (Fig. 37.9B) of which only the shoots bearing lateral sporangia are known, (2) *Gossilinga* (Fig. 37.9E) with lateral sporangia on a dichotomously branched, delicate plant and (3) *Pectinophyton* (Fig. 37.9F) bearing peculiar lateral, flat sporangia on incurred stalks. The most interesting however is (4) the so-called *Psilophyton ornatum* (Fig. 37.9C, D) with lateral and not apical sporangia which has already been discussed (c.f. Psilophytaceae). This has got lateral and not apical sporangia although it was formerly mixed up with *Psilophyton* Dawson. The taxon now needs a new generic name which has now been supplied by Hueber as *Sawdonia* so that the name now becomes *Sawdonia ornata* (Dawson) Hueber.

PHYLOGENY OF THE PSILOPHYTOPSIDA

The discovery of *Psilophyton* in Canada by Dawson was at first disbelieved as a vascular plant was not expected at such an early age. Later discoveries from Rhynie chert beds (*Rhynia, Horneophyton, Asteroxlon*) brought this group into great prominence and its simplicity gave rise to the belief that these are the ancestors of all the Pteridophytes. Later, however, the positions are: *Psilophyton* and *Asteroxylon* had to be modified.

It has been suggested by several authors (Zimmermann, Eames, Smith, Takhtajan, etc.) that the Psilophytopsida is the ancestral stock of vascular plants from which all the other groups have evolved. The living Psilotopsida, also showing a rootless, dichotomously branched sporophyte, has often been considered as a direct descendent of the Psilophytopsida. Several authors have combined Psilophytopsida and Psilotopsida into a common group Psilopsida (c.f. discussion under Psilotopsida). The Lycopsida may be connected with the lateral sporangiate plexus of the Psilophytopsida (i.e., the Zosterophyllales) through the simplest of the former—the fossil Asteroxylaceae and Protolepidodendraceae. The Sphenopsida may be connected to the terminal sporangiate plexus of the Psilophytopsida through the Hyeniales and the Filicopsida through the fossil Protopteridaceae. In the latter direction, probably the bushy Trimerophytace played a special part and, ultimately, this line lead to the Spermaphyters.

It should also be mentioned that, of late, strong arguments have been put against the theory of the origin of the other taxa of Vascular Plants from the Psilophytopsida as they are known today. It is now established that a well-developed vascular flora was already present in the Cambrian so that the origin of this group of plants must have taken place in the Pre-Cambrian ages—*Baragwanathia*, a Lycopsida, with well-developed leaves, flourished during the Silurian—much earlier than the leafless Rhyniaceae. Moreover, the Psilophytopsida itself seems to be a complex group of yet uncertain phylogeny. The Psilophytopsida grew in a very abnormal atmosphere full of sulphurous vapour and in a marshy acidic soil. One may even suspect that the leaflessness of the Rhyniaceae may be rather due to reduction. Even the course of evolution within the Psilophytopsida is not yet clear. It is reasonable to presume that this Silurian-Devonian group had a long earlier history before the vascular plants actually evolved and the predominating Lower to Middle Devonian forms probably represent a very specialized group which became adapted to the peculiar environment. Nevertheless, the environment alone cannot explain the general simplicity in all its characters. Later discoveries (c.f. *Zosterophyllum* by Lang 1937, *Yarravia* with synangia from Australia by Lang and Cookson) from still earlier (Lower Devonian and Silurian) beds have brought to light very complicated forms, which existed earlier than the simpler *Rhyniaceae*, so that their interrelations are still not well understood. It is now the opinion of many authorities (*viz.*, Andrews, Leclercq, Delevoryas, Axelrod) that during the Pre-devonian period their was a complex plexus of Psilophyton-like plants, not necessarily related to one another, which we are now placing in an artificial group—the Psilophytopsida. These were probably of polyphyletic origin. Further discovery and research only may clarify the position. Delevoryas (1969) thinks that there are at least three independent and powerful evolutionary lines (classes *Rhyniophytina, Zosterophyllophytina* and *Trimerophytina*) within the Psilophytopsida. *Trimerophytina* probably evolved the Pterophytes, *Zosterophyllum* represented a lateral plexus giving rise to the Lycopsida (also c.f. *Asteroxylon mackiei*) and the *Rhyniaceae* gave rise to the other Pteridophyta. Mehra (1968) has also adopted the idea that an apical sporangiate plexus and a lateral sporangiate plexus gave rise to all groups of plants.

The Psilophytopsida no doubt represents the simplest group of vascular plants. But whether this simplicity is due to reduction or whether this group represents the actual ancestors of the present-day vascular plants is open to dispute. They certainly form a very artificial group. However, they throw some light on the origin of the vascular plants from the algae or some *Anthoceros*-like pre-Bryophyte. The simplicity of this group lies in the very simple rootless, dichotomously branched sporophytic plants lacking true leaves, the very simple protostelar vascular strand and in the sporangia which are evidently just the specialized branch tips.

The vegetative sporophytes of the three families Rhyniaceae, Psilophytaceae and Asteroxylaceae (which is now included within the Lycopsida) represent the trend of evolution of leaves through enations although the exact relationships between the three families are not yet definite. The Rhyniaceae is the simplest with no leaf or such appendage, with the simplest stelar constitution (annular tracheides only) and simple sporangia which show the columella (a prolongation of the stem vascular system) in *Horneophyton* demonstrating the stem nature of the sporangium. The extreme simplicity of the Rhyniaceae suggests an elaboration of the *Anthoceros* type sporophyte. *Sporogonites* apparently shows no vascular supply and this is now assumed to be a Bryophyte. In Psilophytaceae the simplicity is retained but the xylem shows spiral tracheides and there are some enations at the base of the stems which are certainly not true leaves. The sporangium is as simple. In Asteroxylaceae the structures become more complex. The vascular system is actinostelic and even siphonostelic with a pith; spiral, scalariform and even pitted tracheides are found to have evolved. The enations also become more leaf-like showing stomata and leaf traces although reaching only up to the bases of the leaves.

Trimerophyton of Trimerophytaceae is another very interesting plant which probably connects Psilophytales with Archaeopteris and ultimately the Gymnosperms, according to Delevoryas.

It may be concluded that the Psilophytopsida represents a primitive and problematic group of plants, the understanding of which throw much light on the origin of all the vascular plant taxa. It at least shows two lines of evolution: (1) a terminal sporangiate plexus of plants (including the bushy *Trimerophyton*) leading to the Filicopsida and the Spermaphyta and (2) a smaller lateral sporangiate plexus leading to the Lycopsida and, according to some, may have some connection with the living Psilotopsida.

Chapter 38

Psilotopsida

The psilotopsida comprise of present-day living plants which, like the Psilophytopsida, show footless, dichotomously branched sporophytes differentiated into a rhizomatous horizontal part and aerial shoots. The aerial shoots are with rudimentary 'leaves'. The vascular supply is simple. Gametophytes are cylindrical and resemble the sporophyte rhizomes. Because of the similarities of the sporophytes they are often combined with the Psilophytopsida to form the taxon Psilopsida (also called division Psilophyta). But the absence of any fossil history of connection with the Silurian Devonian Psilophytopsida and the complicated sporangial structure leads one to treat this taxon as a separate class. They had been included within the Lycopsida by some earlier authors but they certainly differ greatly from that taxon.

There is a single order Psilotales.

PSILOTALES : PSILOTACEAE

The Psilotopsida is represent by only two genera, *Psilotum* and *Tmesipteris* placed within in same order Psilotales. The two genera are usually placed within the same family Psilotaceae but Reimers (in Engler's Syllabus, 1954) as well as Pichi-Sermolli (1959) favour placing them in two separate families, Psilotaceae and Tmesipteridaceae. The differences between the two genera would be apparent from the following descriptions:

Genus *PSILOTUM*

Although several species of the genus *Psilotum* are reported, these are considered to be the variations or only two species—*Psilotum nudum* (L) Griseb. (= *P. triquetrum* Sw.) and *P. flaccidum* Wall $2n = 104$ (= *P. complanatum* Sw.). Of these, *Psilotum nudum* is widespread in the tropical and subtropical regions of both the hemispheres including India. This species is often grown in nurseries. It is usually terrestrial growing on humus but may also be epiphytic. *P. flaccidum* is usually epiphytic and pendulous growing specially on tree ferns and palms but may also grow in humus pockets. It is more restricted in distribution being found in tropical regions from Malaysia to Mexico spreading over Indian and Pacific Ocean islands. *Psilotum nudum* ($2n = 96$) is described below. Basic Chr. no. $n = 13$.

The sporophyte

The sporophytic plant body (Fig. 38.1) has a basal dichotomously branched rhizome system usually hidden under humus. The degree of branching of the rhizome is dependent on the obstruction received by it in the substratum. Any branch of the rhizome may give rise to a green aerial shoot which is usually erect but may be pendant in epiphytes. The slender aerial shoots are perennial and become shrubby by repeated dichotomies and may be 8 cm to a metre high according to the environment. The branching is at different planes at the bese but rather complanate at the tops. The stem is almost cylindrical at base, furrowed in the upper parts but triangular (may be somewhat flattened) at top bearing minute, scaly appendages and sporangia. The green stems are photosynthetic and the general appearance is xerophytic although the plants can grow in very moist situations. No vestige of root structure is even noticed and the rhizome system, provided with rhizoids, serve both for absorption and anchorage. A mycorrhizal fungus enters through its rhizoids and becomes established in the rhizome cortex.

Multicellular gemmae (Fig. 38.1C) are produced in large numbers on the surface of the rhizome. They germinate into new rhizomes which ultimately develop new shoots. This is a rapid method of vegetative multiplicaion.

The minute, leaf-like appendages (Fig. 38.1B) are irregularly arranged on the stems and are scale-like (awl-shaped). They are mere expansions of the stem epidermis and upper cortex. The surface layer, although outwardly cutinised, is devoid of stomata. The inner tissue is parenchymatous and contains some **chloroplastids**. There is no vein although in

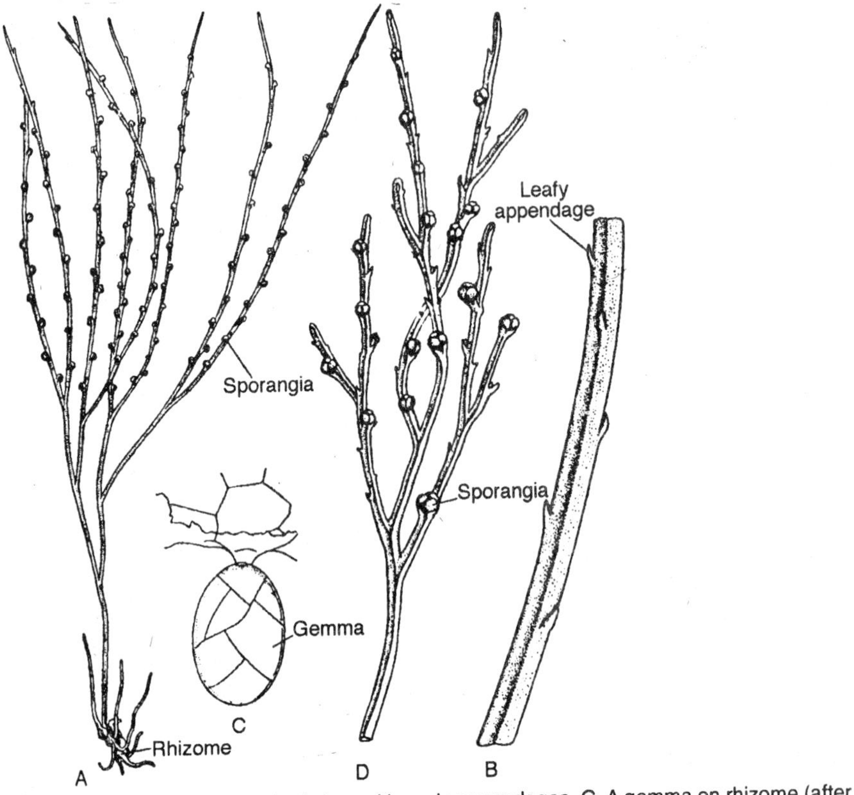

Fig. 38.1: *Psilotum nudum*. A. Sporophyte plant. B. A bit of stem with scaly appendages. C. A gemma on rhizome (after Solms-Laubach). D. A fertile twig

P. flaccidum a 'leaf' trace reaches the base of each appendage. These appendages become lobed (bifurcate) bearing axillary sporangial structures at the apices of the stems.

The growth of both the aerial shoot and the rhizome is by a single large apical cell with three cutting faces. This gives rise to further meristematic cells which bifurcate, giving rise to the dichotomies.

The *anatomy* of the stem (Figs. 38.2 and 38.3A) shows a siphonostelic vascular system which becomes actinostelic in the younger branches. Outermost is a single layer of an epidermis with heavily cutinised outer walls which is broken regularly by longitudinal rows of stomata in the furrows. Below this, the cortex is massive and may be resolved into three zones. The outer cortex (hypodermal) is formed of 2 to 5 layers of chloren-chymatous, vertically elongate, thin-walled cells with intercellular spaces and full of chloroplastids and starch grains. Below this there is a zone of 4 to 5 layers of almost compact vertically elongate sclerenchymatous cells with little or no intercellular space (Middle cortex). Farther inwards there is a still broader zone of thin-walled parenchymatous cells without intercellular spaces but containing more starch grains (Inner cortex). The cortical tissue is bounded inwardly by the endodermis showing casparian strips on the radial walls. The xylem is generally 6-rayed (but actually varying with 2 to 10 rays according to the thickness of the stem) and exarch with the protoxylem elements on the tip of each ray. The protoxylem is composed of spiral and scalariform tracheides while the metaxylem is formed of scalariform and pitted tracheides. In the centre of the xylem is a patch of elongated sclerenchyma cells. As this is considered as a sclerotic pith, the xylem is a fluted siphonostelic cylinder. In the thinner branches this pith is absent so that the structure is actinostelic or haplostelic. At the extreme base also the stem is protostelic. At this region, sometimes xylem elements are present between the

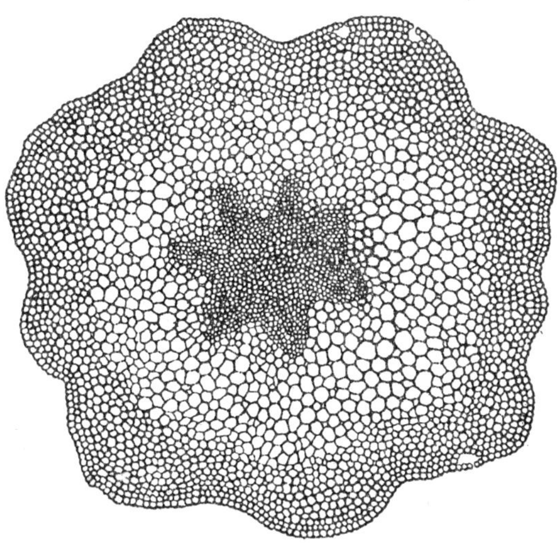

Fig. 38.2: *Psilotum nudum*. T.s. of stem

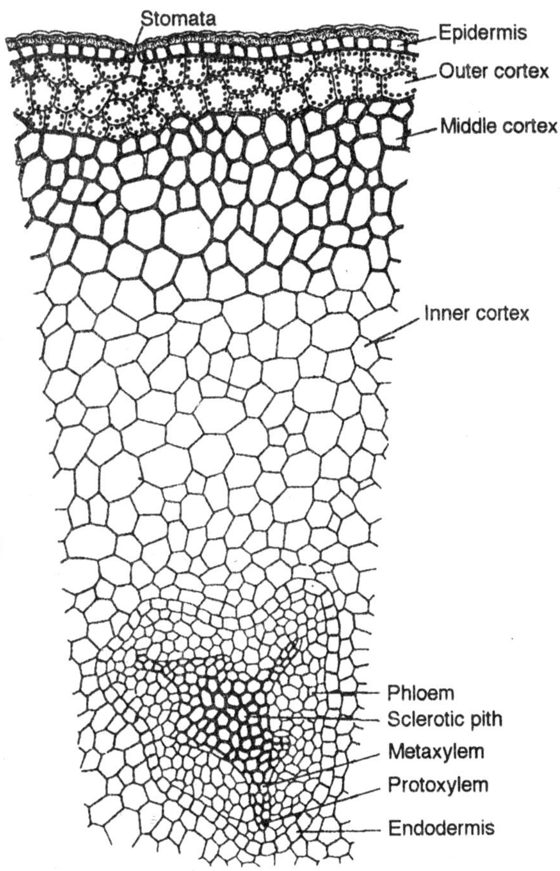

Fig. 38.3: *Psilotum nudum*. A. A segment of t.s. of stem

protoxylem and the phloem. As these elements do not seem to be of secondary origin, the xylem may be considered to be mesarch in such cases. The tissue between the xylem and the endodermis is composed of elongated, tubular, thin-walled cells which, presumably, function as the pholem though sieves have not been demonstrated.

In the rhizome (Fig. 38.3B) the epidermis is indistinct and the outer cells give rise to 2-celled rhizoids. Mycorrhizal fungus is present in the cortical cells. The central vascular strand is protostelic (haplostelic or actionostelic).

The *sporangial structures* of *Psilotum* are mostly located at the distal ends of the shoots (Fig. 38.1). But, it cannot be said that the reproductive shoot is isolated from the vegetative as isolated vegetative appendages are found on the distal end and some isolated sporangia are also found lower down on the stem where most of the appendages are vegetative. Each mature sporangial structure (Fig. 38.4A) is 3-lobed, 2 to 3 mm wide and closely clasped by a comparatively minute, forked, foliar appendage. Each lobe contains a spore chamber (Figs. 38.4C and D) with numerous spores within a multilayered wall and dehisces by a vertical split (Fig. 38.4B) at maturity.

The morphology of the sporangial structure has been greatly disputed. The views are:

(1) The subtending appendage is a sporophyll on which the trilocular fructification is borne on the adaxial face. This was the view specially supported by Bower and Solms-Laubach.

(2) The whole structure is a modified shoot on which the appendage is vegatative and the trilocular fructification is borne on the tip of a very much reduced sporangiophore. This is the view of Strasburger, Goebel and Bierhorst. This view seems to be very logical because of the presence of a vascular strand connecting the bases of the sporangial lobes (Fig. 38.4C) through the short sporangiophore. Then the sporangiophore is similar to that of *Equisetum* and the sporangial lobes are apical as in Rhyniaceae.

The second dispute is as to whether the sporangial lobes represent a single sporangium with three loculi or a synangium or three sepa-

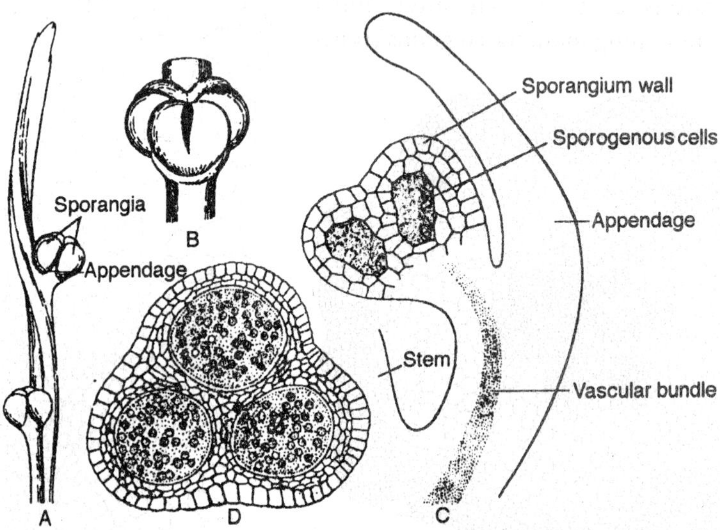

Fig. 38.4: *Psilotum nudum*. A. Bit of stem showing sporangia and forked appendages. B. Sporangia showing dehiscence. C. L.s. passing through two sporangial lobes showing vascular connection D. T.s through the sporangial complex

rate sporangia. Goebel as well as Bierhorst found that the three sporangia develop from separate initial cells. Each initial cell divides into an upper primary wall cell and a lower primary sporogenous cell. But, while Goebel considered this structure to be a synangium, Bierhorst (1956) noticed three separate trades for the three lobes so that he considers the structure to be an apical vertical of three sporangia. This has been clearly demonstrated in some specimens of *P. nudum* where the stalk of the sporangial complex becomes greatly elongated and, sometimes, even the three terminal sporangia become completely separate.

The wall of each sporangium (or chamber) is multilayered and out of the sporogenous cells, some disintegrate forming a nutritive fluid while others become the spore mother cells undergoing reduction division. The spores are bilaterally symmetrical while the walls are thick and reticulate. Each sporangial lobe dehisces at maturity by splitting along a vertical row of thinner cells on the wall.

The gametophyte

The spore germinates to form the small, saprophytic, subterranean gametophyte. The structure of the gametophyte came to be known only since 1917. The peculiarity of the mature gametophyte is that it does not outwardly differ from a piece of the sporophytic rhizome. It is a brown, cylindrical structure up to 18 mm long and 1.5 mm in diameter, branching dichotomously but irregularly like the sporophyte rhizome and growing by single apical cells at the lighter tips (Fig. 38.5A). The surface is covered by brownish rhizoids. The gametophyte is homothallic with no differentiation of vegetative and reproductive tissues so that the surface is strewn with superficial antheridia and archegonia (Fig. 38.5B). A t.s. of the structure (Fig. 38.5C) shows that the peripheral cells are brownish but all the internal cells are thin-walled parenchymatous and many of these are filled with the hyphi of a symbiotic fungus. It has been found that in certain very mature and well-developed gameto-

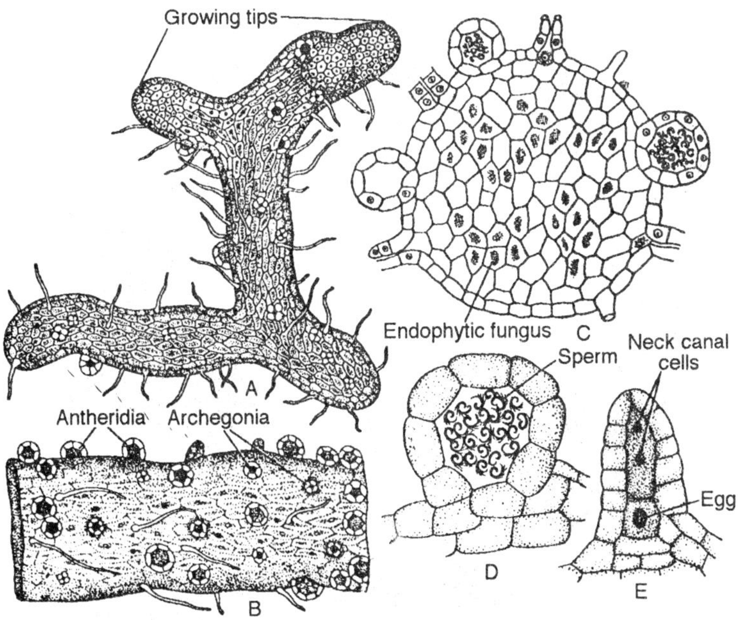

Fig. 38.5: *Psilotum nudum* gametophyte. A. The gametophyte. B. A bit of the gametophyte showing sex organs and rhizoids. C. T.s. of gametophyte. D. Mature antheridium and sperms. E. Mature archegonium

phytes the centre is occupied by a xylem with annular, scalariform and reticulate tracheides surrounded by a phloem and even an endodermis. This is a very striking feature which cannot be ascribed to physiological reasons and, combined with the general resemblance of the gametophyte with the sporophyte rhizome, has been supposed to lend support to the Homologous Theory of the origin of alternation of generations (i.e., the view that originally the gametophyte and the sporophyte were essentially similar) and also to the great antiquity of this group of plants. But, it is now observed that such abnormal gametophytes are diploid and not haploid, so that the vascular elements probably arise as an effect of polyploidy. This feature at least, then, has nothing to do with homologous generations. Gemmae similar to those on the sporophytic rhizome are also found on the gametophyte.

Both the antheridia and the archegonia arise from superficial cells but the antheridia usually develop earlier. In the antheridium the first division gives rise to an outer cell which later becomes the one-layered wall and an inner cell which later forms a mass of sperms. The mature antheridium (Fig. 38.5D) is spherical and protruding like that of a fern and the sperms are spiral and multiciliate. The archegonium develops likewise and in this the outer cell forms the neck and the inner develops the axial row. There are probably two neck canal cells and the egg. In the mature archegonium (Fig. 38.5E) only the neck protrudes out. Usually the tip of the neck disintegrates in this stage so that neck appears very short.

The new sporophyte

The zygote is formed after fertilisation as usual. The first division of the zygote is more or less transverse and the upper cell gives rise to the axis of the sporophyte while the lower cell develops the foot. The foot enlarges downwards sending out finger-like haustorial processes and the axis breaks upwards beginning to branch dichotomously, some lower dichotomy forming the rhizome system. At this stage the sporophyte axis breaks off from the foot which latter

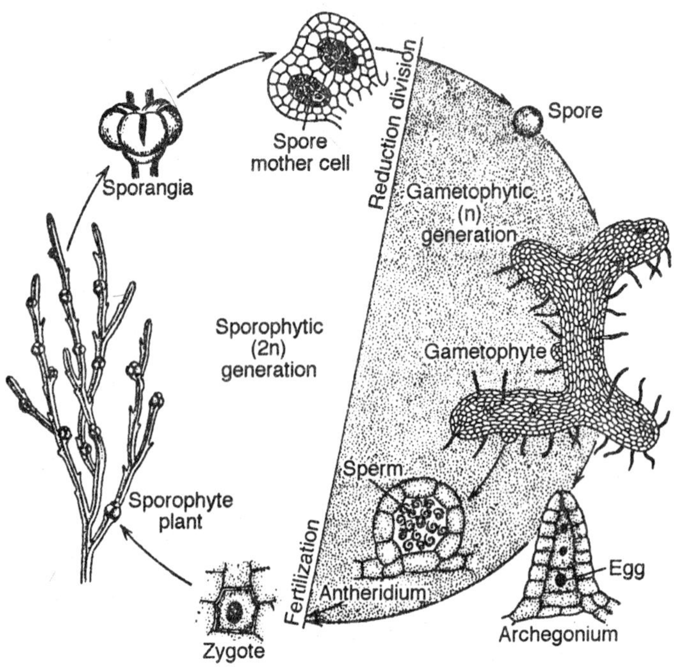

Fig. 38.6: Life cycle of *Psilotum*

remains with the gametophyte. The developing sporophyte gets an infection of the mycorrhizal fungus before it becomes independent. The mode of development of the Psilotales embryo is the simplest among the Pteridophytes. The absence of any root structure even in the assumption that the root was never present in this group and was not lost by reduction.

Fig. 38.6 depicts the life cycle of *Psilotum.*

Genus: *TMESIPTERIS*

The number of species of *Tmesipteris* may also prabably be reduced to only one, *Tmesipteris tannensis* Bernh., although some other species (e.g., *T. viellardi*, an erect species investigated by Sahni) have been reported. These plants occur in the Philippines, Australia, New Zealand and some nearby Pacific islands as pendent epiphytes on tree ferns, etc. It has already been stated that the genus has been placed in a separate family Tmesipteridaceae both in Engler's Syllabus (1954) and by Pichi-Sermolli.

The sporophyte

The general habit of the sporophyte, with rhizome and aerial systems, is the same as that of *Psilotum* but it is usually a pendent epiphyte which may be as small as 5 cm and as big as 4 ft according to Holloway. The aerial shoot is usually unbranched (Fig. 38.7A) but may show a single dichotomy. The aerial shoot is bare above the rhizome, then shows minute, scale-like veinless leaves which become progressively larger, broadly lanceolate with mucronate tips and median veins in the upper parts. Most of the leaves at the distal end are bifurcate and bear a vertical pair of sporangia at the axils (Fig. 38.7B). The stem is cylindrical and lobed.

The 'leaves' are more 'leaf'-like than is *Psilotum* but they are still rather like cladodes. The 'leaves' develop in a vertical plane like phyllodes and the leaf bases are strongly decurrent so that these are often indistinguishable from the stem. The stem tip often ends in a leaf. Some authorities doubt whether such flattened stem-like structures should be considered as true leaves. Each 'leaf' has a median vein which is continuous with a leaf trace from the stem. Each leaf bundle has 4 to 5 scalariform tracheides surrounding 1 or 2 protoxylem elements. This is surrounded by a phloem finally ensheathed in a poorly developed endodermal layer. The mesophyll is formed of angular chlorophyllose cells with air spaces. The epidermis shows sunken stomata on one or both the faces and the leaf surface is cutinised.

The *anatomy* of the stem (Fig. 38.8) shows a siphonostele with a central pith of thin-walled parenchyma. The xylem cylinder shows about five groups of mesarch protoxylem but the cylinder is often completely broken up by leaf traces at the apex of the stem. Both the protoxylem and the metaxylem are composed of scalariform tracheides. The phloem surrounding the xylem consists of sieve tubes with lignified walls and sieve plates on the lateral

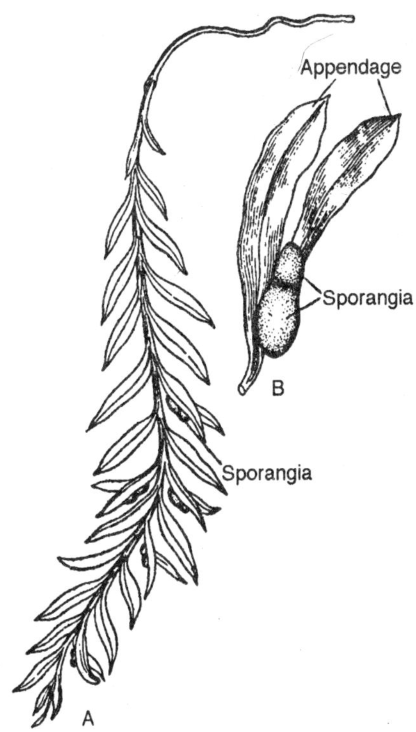

Fig. 38.7: *Tmesipteris tannensis*. A. Sporophyte plant, B. Sporangial structure with appendage

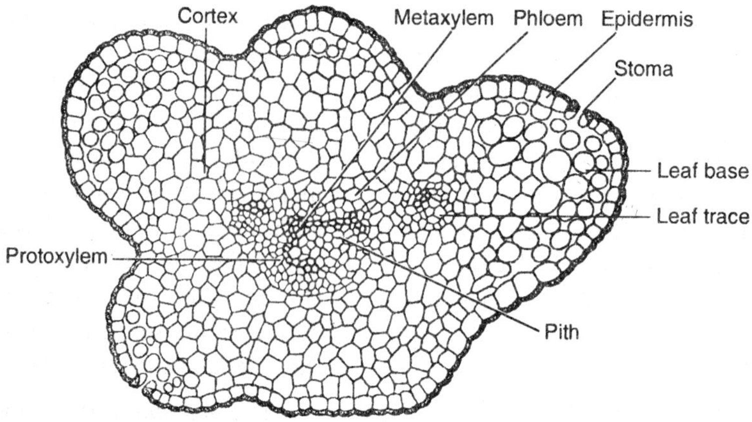

Fig. 38.8: *Tmesipteris tannensis*. T.s. of stem (after Smith)

walls. The massive cortex is formed by angular cells without intercellular spaces except below the leaf bases where only it seems to be photosynthetic. The one-layered epidermis is heavily cutinised on the outside and does not show stomata except at the leaf base region. The rhizome is protostelic as in *Psilotum*. The rhizome shows an endodermis but no such layer is distinguishable in the aerial shoot.

The sporangia have the same morphological interpretation as in the case of *Psilotum*. The sporangial structure (Fig. 38.7B) is a modified shoot with a bifurcate vegetative appendage (the sabtending leaf) and a very short sporangiophore bearing two sporangia on the tip. The vascular bundle from the stem divided into three, two of them passing to the two segments of the subtending leaf and the third to the sporangiophore. The latter trace again trifurcates—two traces passing to the margins of the two sporangia and the central one ending medianly in the septum.

The gametophyte

The gametophyte of *Tmesipteris* (Fig. 38.9A) does not differ from that of *Psilotum*. The sporophyte develops on the gametophyte in the same way but, there again, the similarity

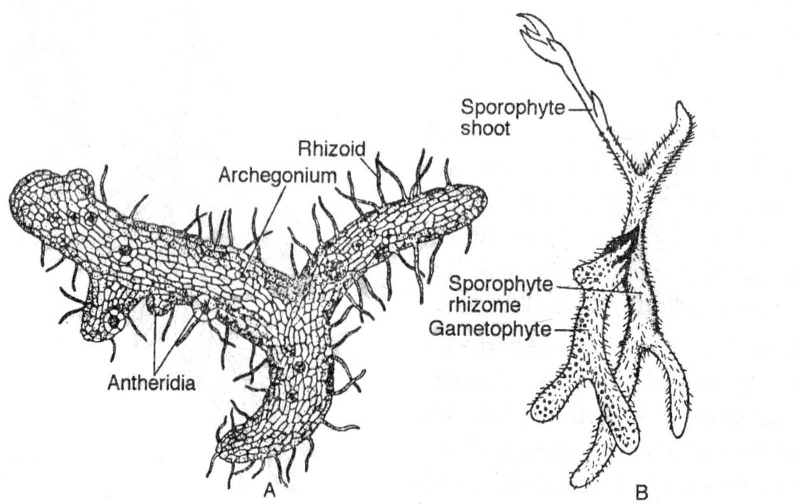

Fig. 38.9: *Tmesipteris tannensis*. A. The gametophyte (after Lawson). B. A new sporophyte on a gametophyte (after Holloway). Note the similarity between the gametophyte and the sporophyte rhizome

between the gametophyte and the sporophytic rhizome (Fig. 38.9B) is striking.

AFFINITIES OF THE PSILOTOPSIDA

The Psilotopsida definitely shows certain very primitive features. It is rootless and there is no true leaf in *Psilotum*. The 'leaves' of *Tmesipteris* are actually lateral flattened lobes of the axis so that they cannot be considered as true leaves. There is no vestige of any root or leaf even in the embryo so that these features cannot be ascribed to reduction. The fruiting structures are now considered as simply cauline and terminal. The steler structure, though more advanced than that in Psilophytopsida, is still rather primitive. The gametophytes strikingly resemble sporophytic rhizomes supporting the Homologous Theory—another point in favour of primitiveness. Moreover, sterile and fertile regions are not segregated in the gametophyte—a fact also poining to primitiveness. The embryo, without root or leaf, is the simplest among the living Pteridophytes and almost approaches the Bryophyte *Anthoceros.* Before the recognition of the Psilophytopsida, the Psilotales had been considered related to the Lycopsida and the Sphenopsida and had even been included within one or the other of them. The relation with the Lycopsida was considered valid for a long time on the ground that the appendages subtending the sporangia are sporophylls bearing the sporangia on the upper surface and also the subterranean, mycorrhizal, radial gametophyte being considered Lycopodaceous. But now this hypothesis is not tenable as the subtending appendages are no longer considered as sporophylls and the absence of real leaves, superficial antheridia, multiciliate sperms, the different nature of the gametophyte, etc. do not support Lycopodaceous affinity. Because of the multiciliate sperms, Seward considered the Psilotales as a taxon *sui generis* (i.e., not connected with others). The sporangiophoric nature of the spore-bearing structures, the anatomy of the stem and the ribbed nature of the stem were advanced in favour of a Sphenopsid relationship but there are too many points of dissimilarities which go against this hypothesis as well. After the establishment of the Psilophytopsida and considering all the points there is no other group excepting the Psilophytopsida with which it may be compared. It is certainly the simplest of all the living Pteridophytes. It is mainly the long gap of time (Devonian and modern) combined with a few advanced characters (e.g., the sporangia and the stelar structure) that the Psilotopsida is being kept separate from the Psilophytopsida. Actually several authorities prefer to combine the two together into the taxon Psilopsida. Cronquist, Takhtajan and Zimmermann (1966) are of opinion that although a distinct relation between the Psilotopsida and the Psilophytopsida is evident, differences are too strong to place them within the same group. "The synangial, axillary or foliar sporangia and the apparently telomic leaves of *Psilotum* and especially *Tmesipteris*" do not fit in with the Psilophytan structure.

Recently, Bierhorst (1968 and 1971) has concluded from anatomical studies that the development of the sporangial structure is *Psilotum* resembles that in the fern *Straumatopteris* and *Psilotum* belongs to the Filicopsida unconnected with the Psilophytopsida. All its simplicities are due to reduction. This is indeed a very spectacular conclusion but still many Pteridologists do not yet seem to be fully satisfied with this hypothesis.

Considering the two genera *Psilotum* and *Tmesipteris,* the former shows greater adaptation to xerophytism (photosynthesis mainly in stem, leaves scaly) and the latter shows differences in the 'leaf' (photosynthesis in the leaf and the leaf base), the sporangia, etc. But, there is so much resemblance in other points that it seems proper to include them within the same order and even the same family.

Chapter 39

Lycopsida

The Class Lycopsida (= Division Lepidophyta or Lycophyta or Microphyllophyta also called Lycopodineae) comprise of plants which are rather small in size today although in ancient days they were the prevalent trees. They are plants with roots and microphyllous leaves having single veins. The steler structure is protostelic, siphonostelic or polystelic but never showing leaf gaps. *The most important characteristic is the sporophyll with a single sporangium apparently on the adaxial face near its base.*

The Lycopsids are definitely the oldest known vascular plants though not the simplest. *Baragwanathia* occurred during the Middle Silurian, i.e., much earlier than the Psilophytopsida. But their features are more advanced than those of the Psilophytopsida. This emphasises the fact that the early history of the first evolution of the land plants is yet unknown. Although, theoretically, the Lycopsids may be derived from the Psilophytopsida (actually occurring later), the actual source of evolution is highly problematic. An origin for the Lycopsids different from that of the other Pteridophytes has been suggested. As has been explained elsewhere, the Lycopsida probably arose from a plexus other than that of the other Pteridophytes. This is a *lateral sporangia plexus* retaining the old biciliate sperms (c.f. Mehra, Fig. 43.1) represented within the Psilophytopsida by the Zosterophyllales (*loc. cit.*).

From the earliest times the Lycopsids seem to have evolved along two different lines. In one group each leaf shows a tongue-like ligule (*Ligulatae*) developing from the base of each leaf on the adaxial face. The base of each ligule shows a peculiar *glossopodium* (Fig. 39.14B). It is the ligulate *Lycopsida* which shows great evolution in size as well as in the evolution of heterospory. As against these, there are other Lycopsids not showing this ligule (*Eligulatae*) which line never developed into trees or never become heterosporous. According to this character some Pteridologists have divided the class Lycopsida into the *Ligulatae* (= class Glossopsida of Bold = class Ligulopsida) and the *Eligulatae* (= class Aglossopsida of Bold = class Eligulopsida). But, Pichi-Sermolli corrently points out that even within the Ligulatae the Lepidodendrons and the Selaginellas were independent from the start and both of them probably independently arose out of the eligulate Protolepidodendrons which have a much earlier history. He, therefore, recognizes four subclasses within the class Lycopsida:

Subclass 1. Protolepidodendridae (eligulate, earliest, fossil Lycopsids)—Order Protolepidodendrales.

Subclass 2. Lycopodiidae (eligulate fossil and living Lycopsids derived from Protolepidodendridae)—Order Lycopodiales.

Subclass 3. Selaginellidae (ligulate, fossil and living Lycopsids independently derived from Protolepidodendridae)—Order Selaginellales.

Subclass 4. Lepidodendridae (ligulate, fossil and living Lycopods independently derived from Protolepidodendridae)—Order Lepidodendrales and Isoetales.

Among Indian fossils, Lepidodendrales are represented by *Cyclodendron* (Permian Lower Gondwana), Lycopodiales by *Lycopodites* and *Lycoxylon* (Upper Gondwana) and Isoetales by *Isoetites* (Upper Gondwana).

The above classification by Pichi-Sermolli now needs the addition of the taxon *Asteroxylales*. The subclass of this taxon has not yet been decided but it may find its place within the Protolepidodendridae or it may be a link between the Zosteropyllales and the Protolepidodendrales. It should however be remembered that the placing of *Asteroxylon* within the Lycopsida is still provisional as some authorities like Andrews (1974) think that this taxon does not fit with any modern established taxon but is a taxon which stands quite apart from all others.

ASTEROXYLALES: ASTEROXYLACEAE: *ASTEROXYLON*

Kidston and Lang (1917) established the genus *Asteroxylon* along with *Rhynia* and *Horneophyton* basing on the material discovered by Mackie in the Rhynie chert beds (Middle Devonian of Scotland) in 1913. As originally described, it had long been considered as the most advanced genus of the Psilophytales. The material out of which the type plant *Asteroxylon mackiei* has been reconstructed is extremely fragmentary as in the case of *Psilophyton* and this has made the modification of the original description and its taxonomic position necessary.

The *sporophytic plant body* of *A. mackiei* (Fig. 39.1A) is quite large and shows a prostrate, dichotomously branched, underground, leafless rhizome giving rise to two types of branches. The tips of some of its dichotomies grew up and gave rise to the leafy, aerial, branch shoots very much resembling a modern *Lycopodium* while some other dichotomies grew down giving rise to comparatively slender, underground, leafless, dichotomous branch systems also resembling the rhizophores of *Lycopodium*. The latter might have functioned as roots. No rhizoid is seen. The aerial shoot is more or less branched and its surface is closely beset with irregularly arranged leaf-like appendages. The base of each aerial stem is naked, up to 1 cm in diameter and was transitional between the aerial stem and the underground rhizome. The leaf-like appendages are up to 5 mm long, becoming shorter and scaly at base. They are awl-shaped or conical—more like spines than like leaves. They are connected by 'leaf' traces with the stem vascular system but the traces end at the base of each 'leaf', the 'leaves' themselves being veinless. The leaf-like structures are provided with stomata.

The *anatomy* of the stem is more complex than that of *Rhynia* although the t.s. of the rhizome is simples, resembling *Rhynia*. A t.s. of the aerial shoot (Fig. 39.1C) shows a mesarch actinostele. The central xylem shows 4 to 10 rays and the smaller protoxylem tracheides are situated a little below the tip of each ray. 'Leaf' traces depart from the tips of the rays and end at the bases of the 'leaves' after passing obliquely through the cortex. The xylem is formed of spiral tracheides (without conjunctive parenchyma) and annular tracheides in the protoxylem. The xylem is surrounded by the phloem which again is surrounded by a massive cortex presenting an inner and an outer region. The inner cortex shows three zones. Innermost is a cylinder of compact parenchymatous cells merging with the stellar tissue between which the endodermis can only be immagined than

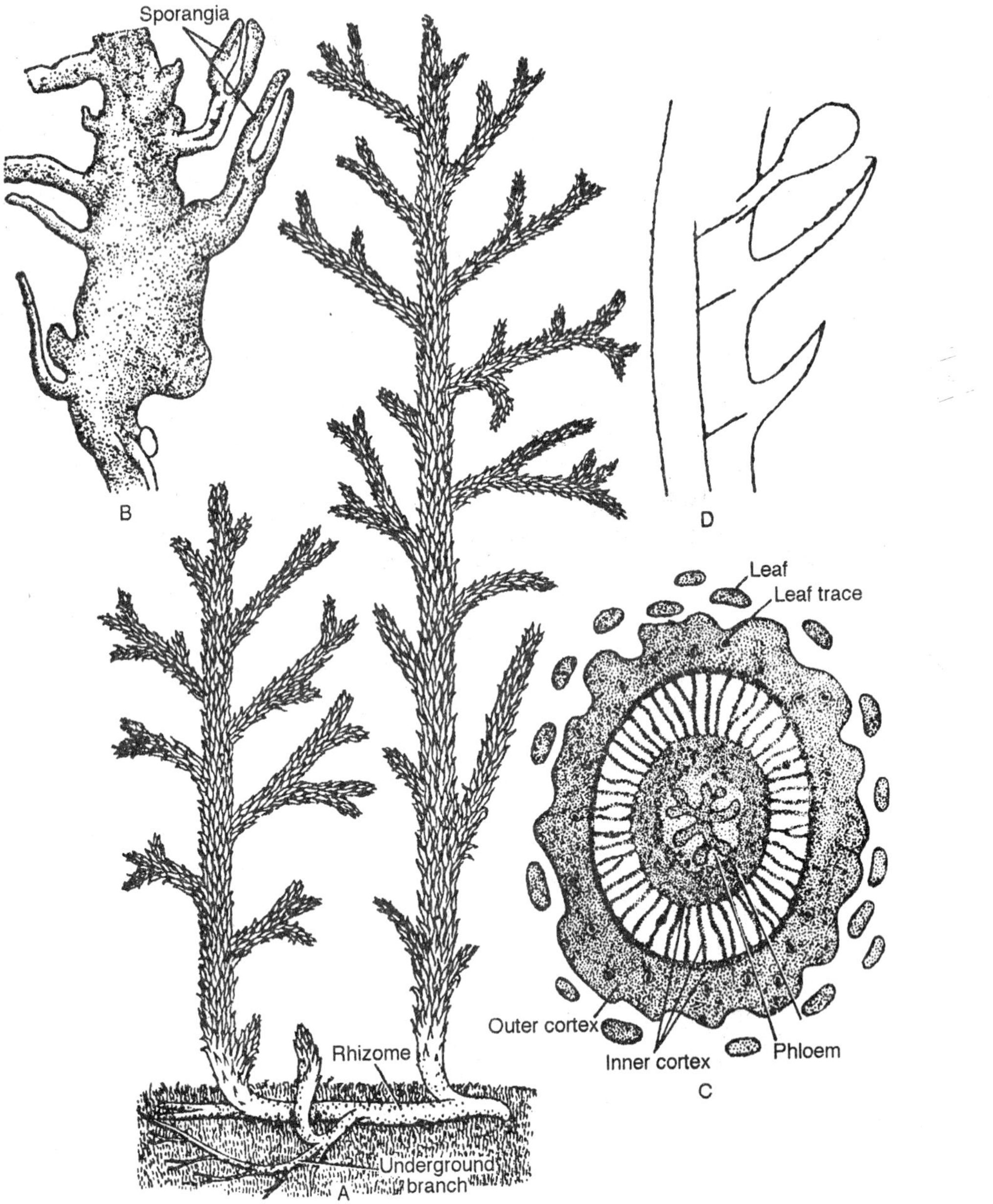

Fig. 39.1: *Asteroxylon mackiei.* A. Sporophyte plant (after Kidston and Lang). B. Sporangia of A (after Lyon). C. T.s. of stem. D. Stem leaves and sporangia in l.s (after Banks)

clearly defined. This is surrounded by the middle zone of the inner cortex which shows vertically radiating plates of cells (like trabeculae) separated by large intercellular spaces. The zone is again surrounded by a cylinder of compact tissue which is the outer zone of the inner cortex. The outer cortex surrounds the inner cortex, equally massive and is formed of tangential cells. The cortex contains hyphae and spores of a fungus *Palaeomyces.* One layer of epidermis with stomata and cuticle on the outer surface is the outermost layer of the stem.

No sporangial shoot was found associated with the sporophytic plant. But, a sporangial shoot found in the same beds with apical sporangia (now named *Nothia aphylla*) was used in the reconstruction, giving it a complete psilophytean appearance. In spite of this apparent psilophytean reconstruction some authorities like Zimmermann (1959) considered *Asteroxylon* to be a Lycopod because of its advances sporophyte. The anomaly was solved by the discovery of Lyon (1964) of the real sporangia of *A. mackiei* (Fig. 39.1B and 1.s. by Banks 39.1D) which are lateral, borne on vasculated stalks hidden within the leaves. Modern Pteridologists (c.f. Bierhorst 1971), therefore, place *Asteroxylon* within the primitive Lycopods although it is not as old as the Silurian *Baragwanathia.* There is now good reason to believe that Zosterophyllales, Asteroxylales, Protolepidodendrales are offshoots of the same ancestral lateral sporangiate plexus which was probably extent in the Cambrian.

It has already been mentioned that *A. elberfeldense*, the later described 'species' of *Asteroxylon* is an altogether different plant. Its apical sporangia retains it as a psilophyte quite different from *A. mackiei.* It has to be given a new generic name belonging to a new family of the Psilophytales.

Subclass 1. PROTOLEPIDODENDRIDAE: PROTOLEPIDODENDRALES: PROTOLEPIDODENDRACEAE

The Protolepidodendridae comprizing of the single order Protolepidodendrales consist of Palaeozoic fossils extending from the Silurian

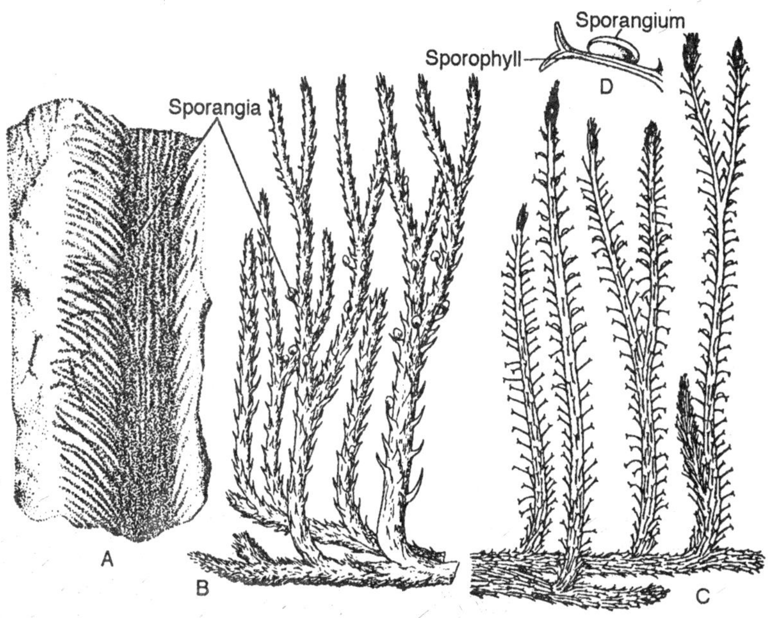

Fig. 39.2: A. *Baragwanathia longifolia*. B. *Drepanophycus spinaeformis*. C. *Protolepidodendron scheryanum*. D. Sporophyll of C. (As after Lang and Cookson; B–D, after Kräusel and Weyland)

to the Devonian. They are usually placed within a single family Protolepidodendraceae which may be rather artificial as the details of most of the genera are not known. The sporophytes were herbaceous with the branches densely covered by microphyllous, eligulate leaves. Sporophylls were dispersed and not localized on definite strobili. The sporangia, where known, were homosporous. Three genera (*Baragwanathia, Drepanophycus,* and *Protolepidodendron*) are better known:

Baragwanathia longifolia Lang and Cookson (Fig. 39.2A) from the Silurian of Australia is better known. Even this is older than the Psilophytes. The plant was probably larger than the present-day *Lycopodium.* Microphyllous, narrow, leaves with single median veins up to 4 cm long and 1 mm wide are laxly borne on the stout, dichotomous, stems 1 to 5 cm in diameter. Reniform sporangia, about 2 mm in diameter, are present in certain areas of the stem in axillary position of leaves intermixed with vegetative leaves. Although it is not definite whether the sporangia were adaxial on the leaves, the general pattern seems to be Lycopodiaceous. The stele, though not well preserved, seems to be actinostelic with annular tracheides in the 12 or more xylem rays.

Drepanophycus (= *Arthrostigma* of Dawson) has been reported from the Lower and Middle Devonian of Canada, Germany, Norway, Great Britain, Belgium and China. *D. spinaeformis* (Fig. 39.2B) shows a creeping stem giving rise to erect, often dichotomously branched shoots up to 45 cm high. The stems with leaves are 2 to 5 cm in diameter and are clothed with stout, falcate, spiny appendages which are leaf-like in having median veins. While the basal leaves are simple, the upper ones are often forked. Some of these appendages bore single sporangia on the upper surface at about the middle of the leaf.

Protolepidodendron is best known from the Middle Devonian of Germany although it has also been found in the Lower and Middle Devonian rocks of Bohemia, Scotland and Eastern United States. *P. scheryanum* (Fig. 39.2C) from the Middle Devonian of Bohemia shows a branched prostrate system from which aerial shoots (simple or with a single dichotomy) arise upwards to a height of 20 to 30 cm. The tips of the aerial shoots and the prostrate system are closely covered with small, spinous, eligulate leaves which bifurcate at the tips. On the middle part of the stem sporophylls are laxly borne intermixed with vegetative leaves. The sporophylls (Fig. 39.2D) resemble the vegetative leaves with bifurcate tips and bear signle sporangia on the adaxial face is little above the bases. The stems are protostelic with triradiate exarch xylems. The metaxylem contains scalariform tracheides. Leaf traces depart from the angles of the xylem and median veins are present in the leaves but it is not clear if the two were continuous. Leaf scars were not formed on the stem when the leaves were shed.

It is a point of interest, to be discussed again, that none of the Protolepidodendrids show any vascular connection leading to the sporangia.

Mention should be made of a few other fossil genera. *Archasosigillaria* (first described by Kidston from Upper Devonian of New York) is an imperfectly preserved type often indistinguishable from *Protolepidodendron.* Its upper leaves show hexagonal bases resembling those of *Sigillaria* (described later). *Colpodexylon,* also from New York State Middle and Upper Devonian, is also included within this family. This is a herbaceous plant with crowded 3-forked leaves, adaxial sporangia and a solid, lobed xylen.

The discovered of *Aldanophyton antiquissimum* by Kryshtofovich from the Middle Cambrian of Eastern Siberia in 1953 caused some speculation on the evolution of vascular plants. Although Pteridophytic spores and tracheides with bordered pits had already been reported from the Cambrian rocks of India and elsewhere, *Aldanophyton* was at first accepted as a vascular plant from that age. The specimen is, however, poor and reported by only shoots about 8.5 cm long and 13 mm wide covered by microphyllous leaves attaining a length of 9 mm. Internal anatomy and reproductive structures are unknown. The specimen was placed within the Lycopsids on not very complete evidence. Later considerations however have brought forth the conclusion that *Aldanophyton* is quite possibly a marine alga and evidences are too inadequate for considering it as a vascular plant. This reminds the case of another such fossil—*Nematothallus* from Upper Silurian discovered by Lang. Its 7 feet long and 3 feet diameter 'trunk' was discovered earlier and its anatomy formed of tubes led Dawson to name it *Prototaxites*, a definite vascular plant. Further researches have shown that this is actually some marine alga and no vascular plant at all.

Subclass 2. LYCOPODIIDAE: LYCOPODIALES: LYCOPODIACEAE

This subclass is extensively represented in the present-day flora by the modern herbaceous Lycopods. The microphylls are eligulate. They probably arose in the Upper Devonian from the herbaceous Protolepidodendrids. The subclass includes only one order Lycopodiales with the single family Lycopodiaceae.

The genus *Lycopodites* is known usually from the Carboniferous but extends down to the Upper Devonian. This was a homosporous genus not differing from the modern *Lycopodium*. In some forms the sporophylls are clustered together to form strobili while in others they are mixed with sterile leaves not forming distinct strobili. *Lycopodites gracilis* is known from the Indian Upper Gondwana (Rajmahal series—Lower Jurassic). *Lycopodites sahnii* is known from the Middle Gondwana Parsora stage (Triassic). Lycoxylon from Rajmahal series also belongs to this family.

There are two living genera: *Lycopodium* and *Phylloglossum*.

Genus: *LYCOPODIUM*

Lycopodium is a large genus with about 180 species having world-wide distribution in tropical to temperate regions. They are herbaceous, terrestrial plants or erect to pondent epiphytes. The stems are densely covered with microphylls and are protostelic. The genus is homosporous with the sporangia on the adaxial faces of the sporophylla.

This large genus *Lycopodium* shows a great diversity in forms. Because of this, it is customary to distinguish within this genus two subgenera: *Urostachya* and *Rhopalostachya*.

Urostachya shows erect (e.g., *L. selago*) or pendent (e.g., *L. phlegmaria*) plants which are never creeping. They may be unbranched or dichotomously branched. The adventitious roots come out only through the base of the stem and are not to be found along the surface of the stem. Even if any root develops on the upper stem, it traverses through the cortex and emerges only at the base. The sporophylls are green and usually of the same size as the vegetative leaves. In *L. selago* (Fig. 39.3) there are sterile (vegetative leaves) and fertile (sporophylls) patches on the stem. In most species of this subgenus there are no organized strobili. In some species (e.g., *L. phlegmaria*), however, the sporophylls, though green, are shorter and localized at the tips forming distinct dichotomously branched strobili. In **Rhopalostachya**, on the other hand, the stem is prostrate with adventitious roots developing on the under surface of the prostrate stem. The branching, though dichotomous at first, becomes monopodial later. The sporophylls are smaller, paler in colour, usually with dentate margins

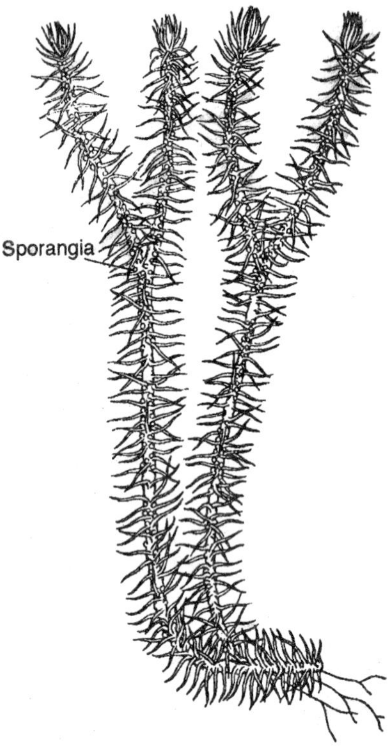

Fig. 39.3: *Lycopodium selago* (after Bower)

although the vegetative leaves are smooth-margined and form distinct strobili. The chromosome numbers of the *Urostachya* are much higher ($2n$ = up to 528 as against up to 68 in the *Rhopalostachya*).

Because of this variation, some modern Pteridologists are not satisfied by merely breaking the genus into two subgenera but they suggest several genera—even families. Pichi-Sermolli (1959) supports breaking up into four genera: *Huperzia, Lycopodium, Lepidotis* and *Diphasium*—the first one of *Urostachya* and the other three out of *Rhopalostachya.* The four genera are also supported by Löve and Löve (1958) from the cytological point of view.

On the basis of chromosomes, modern Pteriodologists have divided the living lycopods as follows:

LYCOPHYTON

1. Huperziaceae (Basic chr. no. 11)
   Huperzia selago ($2n = 264$)
   H. serrata ($2n = 528$)
2. Phlegmariurus ($n = 17$)
   Phlegmarius hamiltonii ($2n = 272$)
   P. phlegmaria ($2n = 272$)
   P. squarrosus ($2n = 272$)
3. Lycopodiaceae
   Phyllostomum ($n = 17$)
      P. drummondii ($2n = 502$)
   Lycopodium ($n = 17$)
      clavatum ($2n = 68, 136$)
      complanatum
   Pahlinhaea ($n = 13$)
      cernua ($2n = 312, 208, 260, 338, 416$)
4. Selaginella (7, 8, 9, 10)
   ciliaris ($2n = 18$)
   delicatula ($2n = 20$)
   krauriana ($2n = 20$)
   rupestris ($2n = 18$)
5. Equisetum ($n = 9$)
   arvense ($2n = 216$)
   debile
   diffusum ($2n = 216$)
   maximum ($2n = 216$)
   palustre ($2n = 216$)
6. Hippchaete ($n = 9$)
   raminissirum ($2n = 216$)
7. Psilotum nudum ($2n = 104, 156, 208, 96, 100$)
8. Tmesipteris ($n = 13$)
9. Helminthostrachys ($n = 17$)
   H. zeylanica ($2n = 188$)
10. Botrychium ($n = 15, 45$)
11. Ophioglossum ($n = 15, 45$)
    O. graminium ($2n = 240$)
    O. nudicaule ($2n = 240$)
    O. reticulatum ($2n = 870 - 1260$)
    O. polyphyllum (= aitchinsoni) ($2n = 210$)
12. Ophioderma (Ophyoglossum) pendulum ($n = 15, 24$) ($2n = 960$)

But the general opinion is Ptesidophytic classification is yet far from satisfactory.

Lycopodium clavatum is the commonest species in India. This is taken as a type in the description that follows. Choudhury (1937) recorded 33 species found in India. Under *Urostachya* are: *L. setago, L. hamiltonii, L. aloifolium, L. petiolatum, L. vernicosum, L. ceylanicum, L. sikkimense, L. serratum, L. setaceum, L. squarrosum, L. phlegmaria*, etc. Under *Rhopalostachya* are: *L. clavatum, L. cernuum, L. annotinum, L. casuarinoides, L. alpinum, L. wightianum, L. complanatum*, etc.

Lycopodium has some economic value. It is used in medicine and the dried spores, '*Lycopodium powder*', is a commercial product used in medicine as well as in pyrotechnique. Spores of *L. clavatum* and *L. cernuum* are collected for this purpose in India.

The sporophyte

Lycopodium clavatum (Fig. 39.4A) is a ternperate to sub-tropical, terrestrial species very common on the Indian hills, specially the Himalayas. The sporophyte has a weak, prostrate stem trailing along the surface and rooted down with adventitious roots growing anywhere on the lower surface. The branching is dichotomous becoming monopodial by the strong development of aerial branches here and there. The stem is closely covered spirally by small, simple (microphyllous), sessile, lanceolate leaves (Fig. 39.4B) with mildly serrate margins and single median veins.

Lycopodium plants may grow almost perennially by the dying out of older parts and the growth of the branches. Gemma-like reproductive buds are also known in several species.

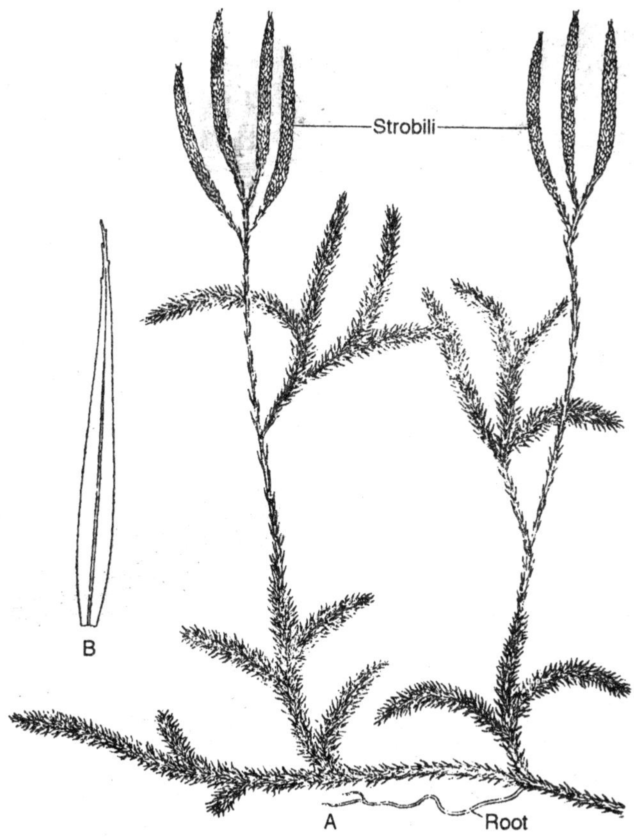

Fig. 39.4: *Lycopodium clavatum*. A. Habit of sporophyte with strobili. B. A. vegetative leaf

The stem of *Lycopodium* is protostelic without any cambium. It grows at the tip by several apical cells. A. t.s. of the stem of *L. clavatum* (Fig. 39.5A), which may be cylindrical or somewhat fluted, shows as epidermis of one layer of thick-walled cells broken here and there by stomata. Below it there is a thick cortex traversed by leaf trace bundles. The outer layers of the cortex are sclerenchymatous while the cortex below is parenchymatous. In mature specimens the innermost region of the cortex also has thickened cell walls. The cortex is bounded on the inside by a layer of endodermis with the usual thickened radial walls (casparian strips). Internal to the endodermis is a pericycle of one or more layers of parenchymatous cells. The stele in this case is protostelic of the **plectostele** type. The xylem elements are arranged in more or less parallel plates with the phloem between these patches. The xylem is formed only of tracheides and is exarch with the protoxylem (smaller spiral and annular tracheides) at the ends of the plates and the metaxylem (large scalariform tracheides) forming the general mass. The phloem shows sieve cells and parenchyma. The sieve cells are elongated with tapering ends and with sieve plates on the lateral walls. Leaf traces do not cause leaf gaps in the stele.

In other species of *Lycopodium* the stelar structure shows great variation. In *L. serratum* (Fig. 39.5a) the stele is **actinostelic** (star-shaped xylem). In *L. annotimum* (Fig. 39.5b) the stelar structure is broken up. The breaking up is the maximum in *L. cernuum* (Fig. 39.5c) where the stele is a meshwork of innumerable xylem patches with surrounding phloem tissue. This type is termed as a **mixed protostele**. These

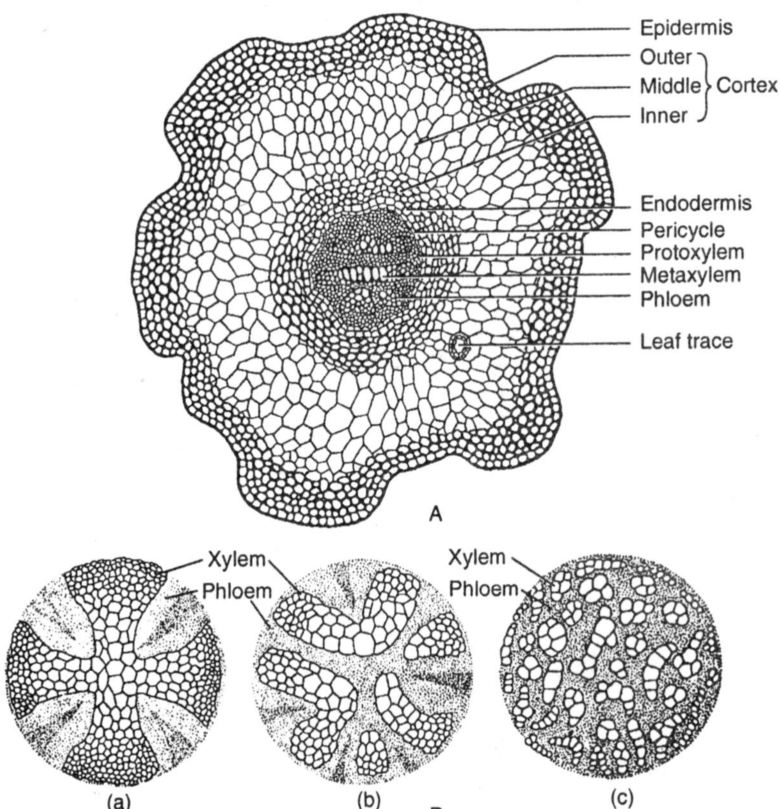

Fig. 39.5: A. T.s. of *Lycopodium clavatum* stem. B. T.s. of stelar regions of stems of (a) *L. serratum*, (b) *L. annotinum* and (c) *L. cernuum*

types of stelar structure may be considered as showing the course of stelar evolution—the *actinostele* being the most primitive and the *mixed protostele* the most modern. The *plectostele* is placed before the *mixed protostele.* The stelar structures are so characteristic that *Lycopodium* species may be identified from anatomy alone.

The *anatomy* of the leaf is very simple with a surrounding one-layer epidermis broken by stomata, a uniform parenchymatous mesophyll with numerous air spaces and chloroplastids, and a median concentric vascular bundle.

The slender roots show simpler structures with usually monarch steles.

The reproductive shoots arise as erect branches from the horizontal stem late in the season (Fig. 39.4A). The lower part of the reproductive shoot is comparatively sparsely leaved and the tip branches dichotomously into two or more spike-like strobili (*sporangiferous spikes*) compactly covered by sporophylls (Fig. 39.6A). The sporophylls are of one type only (*homosporous*). The sporophylls (Fig. 39.6B) are differentiated from the vegetative leaves by the wider bases and more serrations in the margins. The sporangia are comparatively large, reniform of subglobose, orange to light-yellow in colour and with short stalks when mature. These develop on the adaxial (ventral) face of the sporophyll a little above the axil. In other species the sporangium is known to develop from the axil or even from the stem just above the axil. The sporangium is a massive structure developing from a group of initial cells (Fig. 39.6C). This is known as the **eusporangiate** mode of sporangium development. In the nearly mature sporangium (Figs. 39.6D and E) there is a *stalk*, a *jacket* (2 or more layers thick), a massive **sporogenous tissue** and a nutritive *tapetum* formed partly by the inner layer of the jacket and partly by some outer sporogenous cells. The sporogenous cells soon become *spore mother cells* (Fig. 39.6E) which become rounded, separate from one another and undergo reduction division to form the *spore tetrads.* The mature sporangium

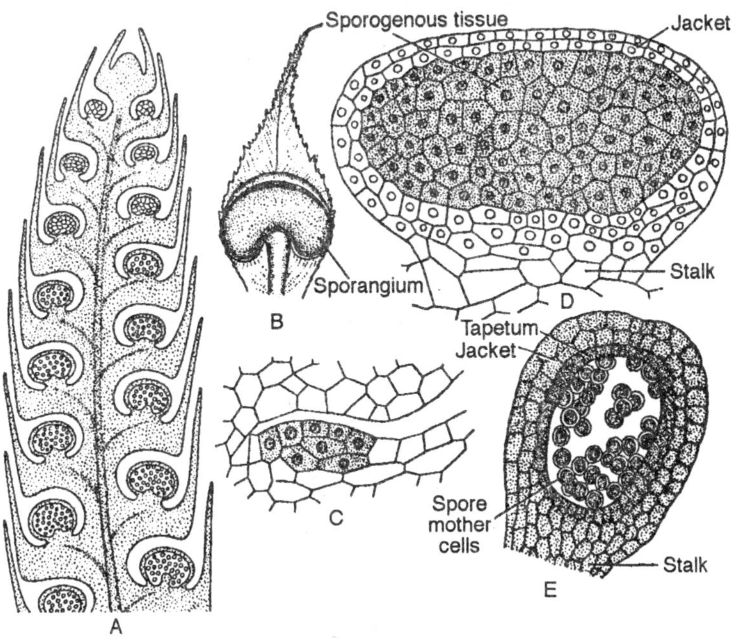

Fig. 39.6: *Lycopodium clavatum*. A. Median l.s. of strobilus. B. A sporophyll. C. Leaf base showing sporangial initials (shaded). D. Nearly mature sporangium before spore-mother cells are formed. E. Mature sporangium

splits along a vertical line or weakness (*stomium*) and the spores are released.

The gametophyte

The spore is tetrahedral with the usual intine and a sculptured exine showing a triradiate ridge (Fig. 39.7A). A few chloroplastids are usually present.

The spore of *L. clavatum* remains viable for a long time and may germinate only after a year or more. The exine splits at the triradiate fissures and a tissue, developing very slowly (taking another year or more), comes out forming a top-shaped (Fig. 39.7B), underground, a non-green, tuberous gametophyte or *prothallus*. With age this gametophyte loses its shape and becomes much convoluted (Fig. 39.7C). The gametophyte ceases to grow if it does not become infected by a fungus at an early stage of development. A vertical l.s. of the mature gametophyte (Fig. 39.7B) shows an outer epidermis with some rhizoids; an outer cortex filled by mycorrhizal fungi; an inner cortex with an outer parenchymatous and an inner palisade zone; and a central parenchymatous storage tissue where the outermost cells are elongated. The top of this top-shaped gametophyte is lobed and the antheridia, the archegonia and the growing embryos are located on these lobes. This type of gametophyte is noted in many creeping species.

A second type of gametophyte is represented by *L. cernuum* (Fig. 39.7D). In these the spore germinates without passing through any resting stage. The gametophyte is usually smaller, annual, partly aerial and partly underground. The lobed crown with antheridia and archegonia becomes green.

A still third type of gametophyte is found among the epiphytic species like *L. phlegmaria* (Fig. 39.7E). The prothalli are saprophytic, growing on trunks below a coating of humus. Here a central, small, tuberous body develops a number of colourless, slender, cylindrical arms

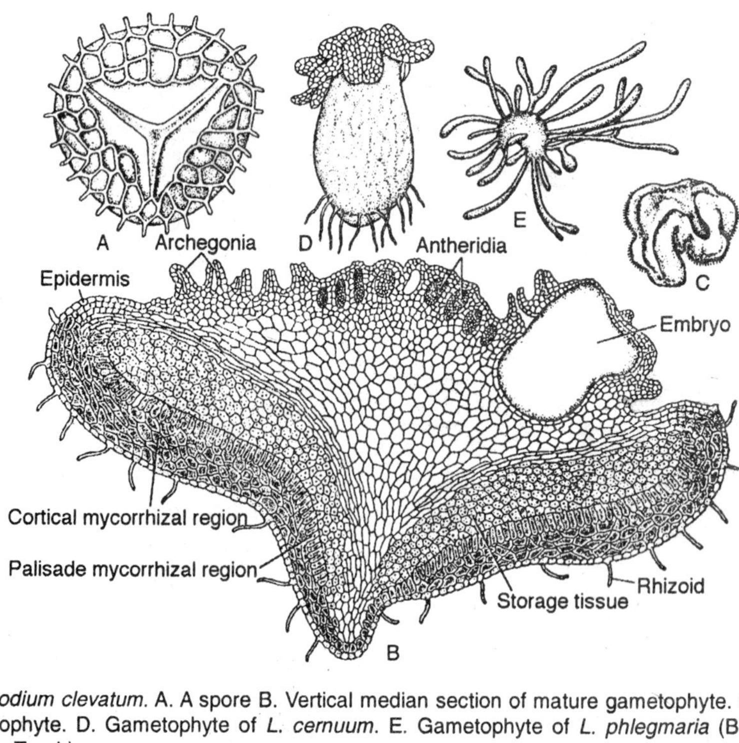

Fig. 39.7: *Lycopodium clevatum*. A. A spore B. Vertical median section of mature gametophyte. C. An old convoluted gametophyte. D. Gametophyte of *L. cernuum*. E. Gametophyte of *L. phlegmaria* (B-D after Bruchmann; E. after Treub)

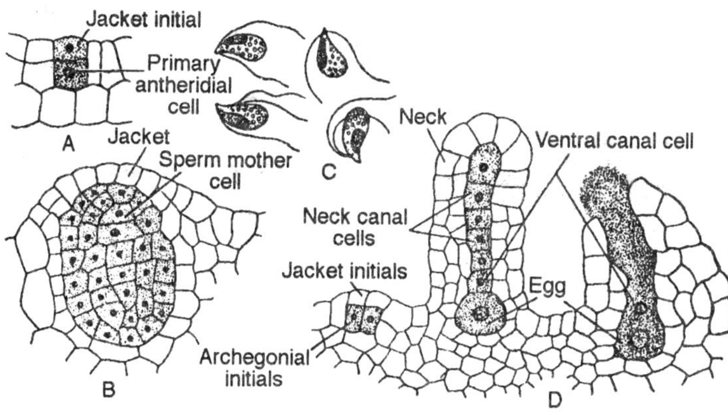

Fig. 39.8: *Lycopodium clavatum*. A. Early stage of antheridium formation. B. Mature antheridium. C. Sperms. D. Part of gametophyte showing stages of development of archegonia

on whose surfaces the antheridia and the archegonia develop.

L. selago shows gametophytes of both the first and the second types according as it grows on the surface or below the soil level.

All gametophytes are monoecious. In the development of an *antheridium*, an epidermal cell divides transversely forming an upper *jacket initial* and a lower *primary antheridial initial* (Fig. 39.8A). The jacket initial ultimately forms a *jacket* layer one cell in thickness with a triangular cell at the top centre. The lower cell forms a mass of tissue which ultimately become very small cubical *sperm mother cells* (Fig. 39.8B). Each sperm mother cell gives rise to a biflagellate (rarely triflagellate) *sperm* (Fig. 39.8C) resembling rather the Bryophytes. The antheridia are almost wholly sunken in the gametophytic tissue. The sperms are liberated by the breaking down of the triangular cell at top.

The *archegonium* (Fig. 39.8D) also develops similarly from a superficial *archegonial initial cell*. The first division gives rise to an upper *primary cover cell* and a lower *cenral cell*. The central cell divides to form a lower *primary ventral cell* and an upper *primary canal cell*. The primary canal cell divides transversely to form usually four (1 to 3 in *L. cernuum*, 7 in *L. selago*, up to 16 in *L. complenatum*) *canal cells* while the ventral cell forms the *egg* often after cutting off a *ventral canal cell*. The primary cover cell develops the *neck* 3 to 4 cells high. In the mature archegonium the neck portion protrudes out while the venter remains sunken.

Fertilization takes place in the usual way. The neck canal cells and the ventral canal cell (if any) disintegrate and come out exuding citric acid and citrates which probably chemically attract the sperms one of which fertilizes the egg developing the zygote.

The new sporophyte

The zygote divides transversely to form an upper *suspensor* and a lower embryonic cell (Fig. 39.9A). The embryonic cell divides into eight cells in two tiers of which the upper 4 near the suspensor gives rise to the *foot* (for absorption of food material from the gametophyte) and the lower 4 to the *stem* on one side and a *cotyledonary leaf* (Figs. 39.9B and C) on the other side. As the embryo grows it rises erect above the gametophyte, the first root developing from the point where the cotyledon and the foot joins. The first leaves are scaly. The new sporophyte soon established as an independent plant (Fig. 39.9D).

In *L. cernuun*, the 8-called embryo develops a massive globose structure called the **pro-**

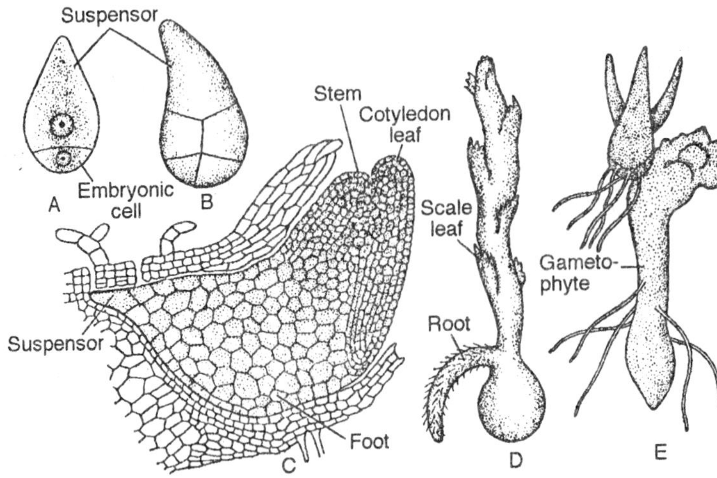

Fig. 39.9: Stages of development of *Lycopodium* embryo. A. 2-celled stage. B-C. Subsequent stages. D. New sporophyte. E. Protocorm of *L. cernuum*

tocorm which becomes green and develops rhizoids below. The structure develops a few erect, conical outgrowths functioning as leaves (Fig. 39.9E). The protocorm remains in this condition for some time and then the apical meristem bursts into a normal shoot. A mycorrhizal fungus grows inside this structure. This intermediate protocorm structure has been considered to be of some evolutionary significance by some (*Protocorm Theory*) while others consider it as a mere passing phase. Treub considered this as an undifferentiated primitive stage

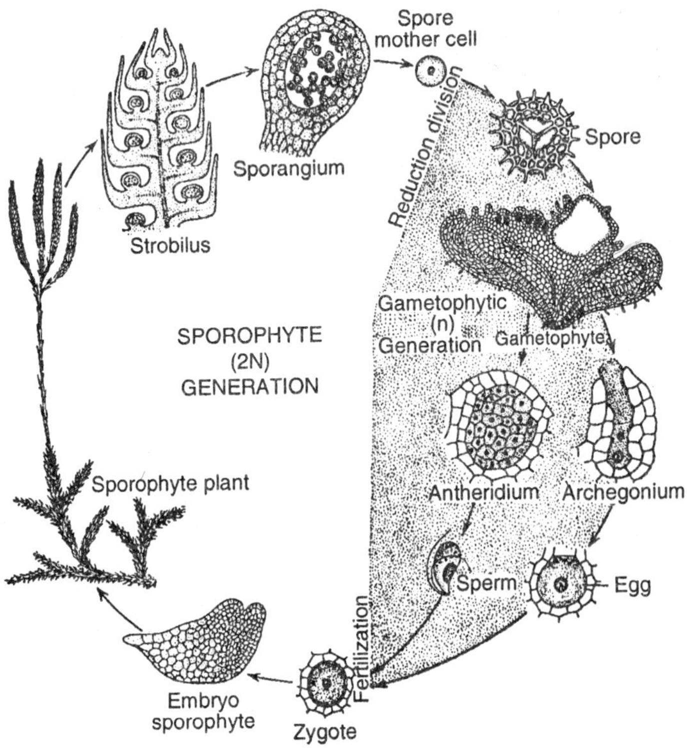

Fig. 39.10: Life cycle of *Lycopodium*

of sporophyte which was present in all Pteridophytes but has been lost in most of them. *Phylloglossum* (described below) shows a permanent protocorm while this occurs occasionally in *Ophioglossum.* Goebel, Bower, etc., however, consider the protocorm merely as an occasional adaptation to, meet the strain of sporophytic development under special conditions. The Protocorm Theory is now only of historical importance.

Figure 39.10 shows diagrammatically the life cycle of *Lycopodium.*

Genus: *PHYLLOGLOSSUM*

This is a peculiar plant represented by a single species *Phylloglossum drummondii* found only in Australia, Tasmania and New Zealand. It is a small plant with a fleshy, tuberous, perennial stem and a few stiff, awl-shaped leaves (Fig. 39.11) reminding the protocorm stage of *Lycopodium cernuum.* A compact sporangiferous spike is developed on a stalk at the tip of the stem.

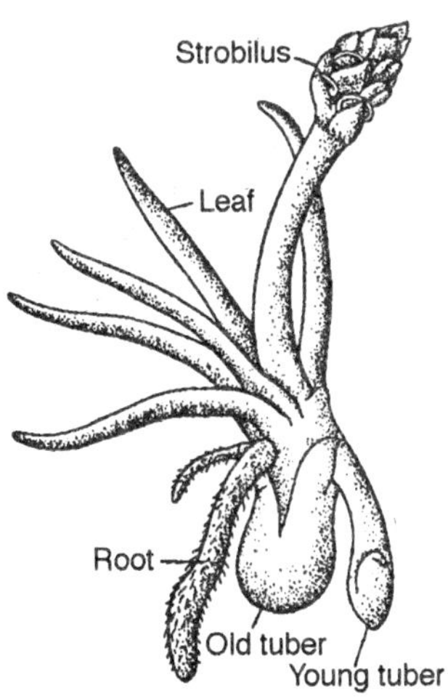

Fig. 39.11: *Phylloglossum drummondii* (after Bertrand)

Subclass 3. **SELAGINELLIDAE: SELAGINELLALES**

This subclass is represented by the single order Selaginellales. The plants are herbaceous as in the Lycopodiales and usually there is no secondary thickening. The microphylls are provided with *ligules.* The subclass probably arose independently out of the Protolepidodendrideae between the Upper Devonian and the Carboniferous. Roots are borne not directly on the stem but on the distal ends of special branches called *rhizophores.* They are *heterosporous.* Two families are recognized—Selaginellaceae and Miadesmiaceae. Miadesmiaceae is extinct and showed a very advanced type of heterospory for which it is put by some authors with Lepidocarpaceae in a group called Lepidospermales.

Family 1. SELAGINELLACEAE

Selaginellaceae comprizes of two genera—(1) *Selaginellites,* a fossil heterosporous (showing 4 to 24 megaspores in the megasporangium) genus from the Upper Carboniferous of which about half a dozen species are known and (2) *Selaginella* which is very common in the present-day flora.

Genus: *SELAGINELLA*

Selaginella is a very large genus of about 700 tropical to temperate species growing almost everywhere on this Earth. Although so large, the genus shows a uniformity lacking in the genus *Lycopodium.* The genus is more abundant in the tropical rain forests in weak light and is also common in the warmer plains of India where *Lycopodium* does not survive. Some species are also xerophytic. Most of them grow perennially by vegetative propagation while some are annual. Some of the species are rather small (a few centimetres) resembling mosses while others may grow even to a length of 20 metres. 58 species are recorded by Alston (1945) from India which include *Selaginella rupestris, S. pentagona, S. megaphylla, S. proniflora, S. ciliaris, S. delicatula, S. chrysocaulos, S. pallidissima, S. sanguinolenta,* etc. *S. kraussiana* from Africa is commonly grown in greenhouses as an ornamental plant.

Selaginella shows considerable variation in the symmetry of the plant but most species are prostrate and creeping, a few are even climbing. A few xerophytic species (e.g., *S. lepidophylla* and *S. pilifera*) typically roll up the leaves and become ball-like but open up into normal green plants when put in water even when dead. These are often sold inside bottles as curiosities.

Selaginella shows great uniformity even cytologically. Most species show the basic number $n = 9$. A few species are polyploids. However, the genus *Selaginella* has been divided into two subgenera according to the sporophytic form. In the subgenus **Homoeophyllum** (*Selaginella rupestris*, etc.), the plant body has a tendency to be upright with radially arranged, more or less uniform (isophyllous) leaves. In the subgenus **Heterophyllum** the plant body is more or less dorsiventral and prostrate with a distinct heterophylly (anisophylly) of the leaves. There are only about 50 species in the subgenus *Homoeophyllum* while all the rest are in the subgenus *Heterophyllum.* As the latter is the commoner and more typical form this is being described in what follows. The most typical species is *Selaginella kraussiana.*

The sporophyte

The common *Selaginella* (viz., *S. kraussiana*—Figs. 39.12A and B) has a horizontal creeping stem branching along one plane giving rise to a more or less dorsiventral plant body. The branching is dichotomous but often appears to be monopodial because of stronger growth of the main stem. Here and there *rhizophores* grow vertically down from the forkings of the dichotomies. Adventitious roots develop where the rhizophores meet the soil. Plants are often 'floating' being kept propped up by the rhizophores.

In the *Heterophyllum* subgenus of *Selaginella* the leaves are of two types (*dimorphic*). There are four rows of leaves—two rows of smaller leaves on the upper side and two rows of larger leaves on the lower side. Actually the leaves are arranged in an *anisophyllous opposite* order (Fig. 39.12C) though in some species the arrangement may look spiral. At each node there is one small leaf on the upper side and one large leaf on the lower side. Flattening (dorsiventrality) of the stem may disturb this arrangement.

The leaves (Fig. 39.12C) are microphyllous with a single median vein and lanceolate to ovate in shape becoming narrower at stem attachment. In some species the smaller leaves become filiform. Each leaf bears on its upper surface near its base a small, membranous, tongue-like outgrowth called the *ligule.* The ligule shrinks and becomes inconspicuous in mature leaves.

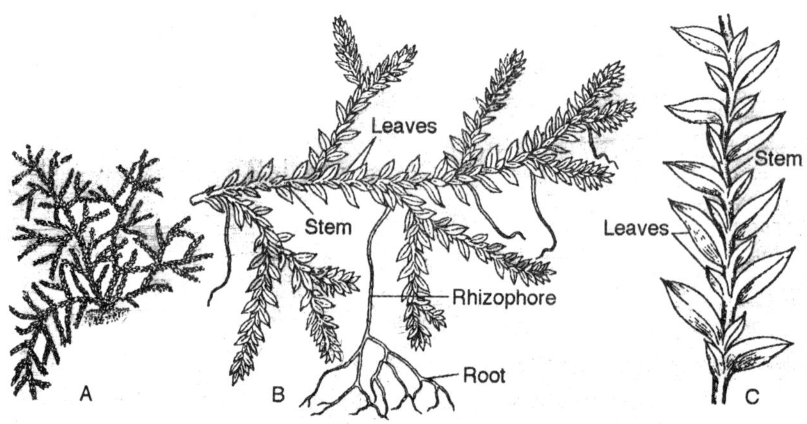

Fig. 39.12: *Selaginella kraussiana.* A. **General habit.** B. A bit of sporophyte. C. A bit of stem showing leaf arrangement

The original embryonic root is short-living. All other roots are adventitious from the tips of the rhizophores.

The apical growth of most species is by a single pyramidal cell with three cutting faces. But, in a few species a group of meristem cells has been observed.

The *anatomy* of the mature stem is very distinct. A t.s. of *S. kraussiana* (Fig. 39.13A) shows a single layer of epidermal prosenchymatous fibres without any stomatal opening. Below it is the *cortex* of angular parenchymatous cells usually wihout intercellular spaces. The outer cortex cells are thickened in some, specially, xerophytic species. The vascular cylinder in *Selaginella* species varies from protostelic to siphonostelic and monostelic to polystelic (with 2 to 16 separate steles). But, on the whole the stelar arrangement is simpler than in *Lycopodium. S. kraussiana* is *distelic.* The course of the steles is shown in Fig. 39.13B. Each stele is surrounded by an *endodermis* which is a normal compact ring is very young stems but each cell becomes greatly elongated in the axis radial to the stele and also laterally separated from one another by big air gaps in the mature stem. Thus, the endodermal cells in the mature stem are the *trabeculae* of long radiating cells bridging the wide space between the steles and the inner face of the cortex. In spite of this radial elongation, the endodermal cells still retain the transverse girdles of *casparian strips.* Inside each endodermis is one layer of *pericycle* enclosing a simple protostele. In the stele there is an exarch *xylem* at the centre, the *metaxylem* showing scalariform tracheides and a *protoxylem* lies towards circumference. The xylem is surrounded by one or two layers of parenchyma which again is surrounded by a single layer of sieve tubes with sieves on lateral walls forming the *phloem* outside which is the pericycle.

In some other species of *Selaginella* some complexities are noted in the stele. While secondary growth is not normal in *S. selaginelloides* a few secondary xylems do develop. In certain species (e.g., *S. rupestris, S. densa, S. arizonica, S. rupicola*) the metaxylem is almost wholly formed of *true vessels* formed by the dissolution of end walls with only a few tracheides intermixed. These are usually species showing spiral arrangement of vegetative leaves.

The adventitious roots (all roots in a mature plant are adventitious) are delicate,

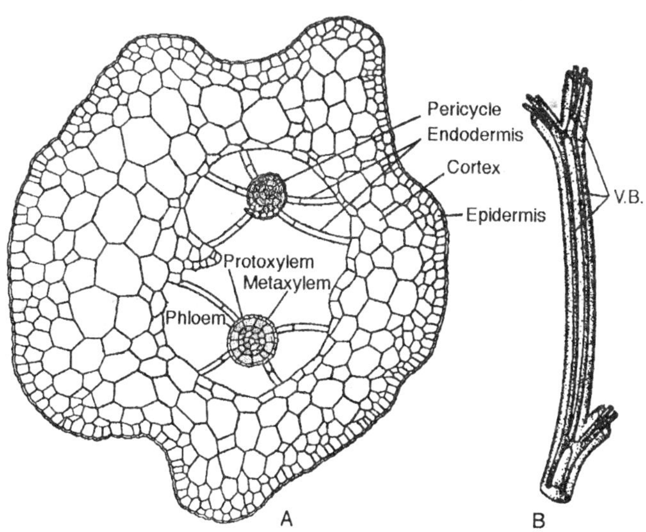

Fig. 39.13: *Selaginella kraussiana.* A. T.s. of stem. B. L.s. of stem showing course of steles (marked V.B.)

sparingly branched structures. A t.s. of a root shows a very simple arrangement with a small monarch stele.

A t.s. of the *rhizophore* which shows a monarch arrangement is also as simple as that of a root. This fact, combined with its positively geotropic nature, has led some to believe that it is a root. But, its exogenous origin, lack of a cap and the experimental production of leafy shoot from the apices of decapitated rhizophores seem to prove that it is actually only a modified stem. Bower (1908) and Goebel (1905), however, concluded that it is an organ *sui generis* which is neither a root nor a stem.

A t.s. of the leaf (Fig. 39.14A) shows a simple organization. There is a simple vascular bundle at the centre. The *epidermis* is one layer thick and the stomata are usually present on the lower epidermis only. The *mesophyll* is uniformly formed either of all spongy or all palisade-like elongated cells full of air species. Mesophyll cells have single cupshaped or many chloroplastids according to species. In all cases there are several spindle-shaped, pyrenoid-like bodies at the centre of each chloroplast. The bodies may be transformed into rudimentary starch grains. The only other Embryophyte with similar chloroplasts is the Anthocerotopsida.

The ligule (Fig. 39.14B) is differentiated rather early. It matures and withers away also early. It has a cup-shaped *sheath* at base. The *sheath* cells in *S. kraussiana* have casparian strips as in an endodermis. Immediately above the sheath is the hemispherical *glossopodium* formed of large cells full of a greatly vacuolated protoplasm. The structure of the ligule has suggested that it is a water-absorbing or exuding structure serving for the protection of the growing leaf or the sporangium.

When the plants become mature the branches terminate in the *sporangiferous spike* or *strobili*. The *sporophylls* are all alike resembling the larger vegetative leaves but the *sporangia* are of two types—*microsporangia* and *megasporangia* (or *macrosporangia*) so that the sporophylls also are called *microsporophylls* and *megasporophylls*. The sporophylls are spirally arranged on the stem roughly forming four vertical rows. In most species microsporophylls and megasporophylls are borne on the same strobilus but in a few cases they may be in different strobili. The position of the micro- and megasporophylls vary in the different species. In some the microsporophylls are on top and the megasporophylls below, in others the microsporophylls are on one side of the strobili while the megasporophylls are on the other vertical half. In *S. kraussiana* (Fig. 39.15A) the megasporophylls occupy a part of the base on one half.

In the development of the *sporangium*, each sporangium develops from a group of *initials* (it has been disputed that in some cases there is a single initial) either on the stem immediately on the sporophyll or at the axil of the sporophyll (Figs. 39.15B and C). When mature, the sporangium is always located at the

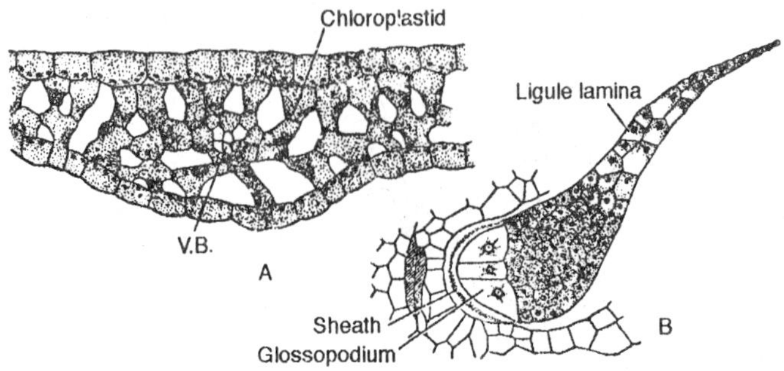

Fig. 39.14: *Selaginella kraussiana*. A. T.s. of leaf. B. Vertical l.s. through ligule at base of leaf

axil of the sporophyll below the ligule and is reniform or obcvoid in appearance with a *stalk*. The initial cells divide into the upper *jacket initials* and the lower *archesporial cells*. The former develops a *jacket* two cells in thickness and the latter differentiates into an outer *tapetum* and an inner group of *spore mother cells*. In the microsporangium (Fig. 39.15D) most of the spore mother cells are functional and give rise to the *microspore tetrads* on *reduction division*. In the megasporangium only one spore mother cell remains functional. On reduction division this gives rise to a tetrad of *megaspores* arranged tetrahedrally. The megaspores increase greatly in size with the megasporangium so that a megaspore is much larger than a microspore and a megasporangium is larger than a microsporangium. Enlargement of the megaspores cause the megasporangium (Fig. 39.15E) to have a four-lobed appearance, each lobe enclosing a megaspore. This pressure and the drying up of the sporangial wall cause a splitting of the megasporangium into two valves along an unthickened vertical line on the wall. In some cases some of the four megaspores may even fail to enlarge giving rise to megasporangia with 1, 2 or 3 megaspores.

The male and the female gametophytes

The *microspores* are very small and with a 2-layered wall (*exine* and *intine*). The exine is usually ornamented with spines. Its shape is more or less tetrahedral with one side rounded. This soon germinates to form the *male gametophyte*. A small lens-shaped *prothallial cell* is cut off on one side and the other large cell becomes the *antheridial initial* (Fig. 39.16A). The antheridial initial develops an *antheridium* which shows four central *primary spermatogenous cells* surrounded by a *jacket* of 8 cells

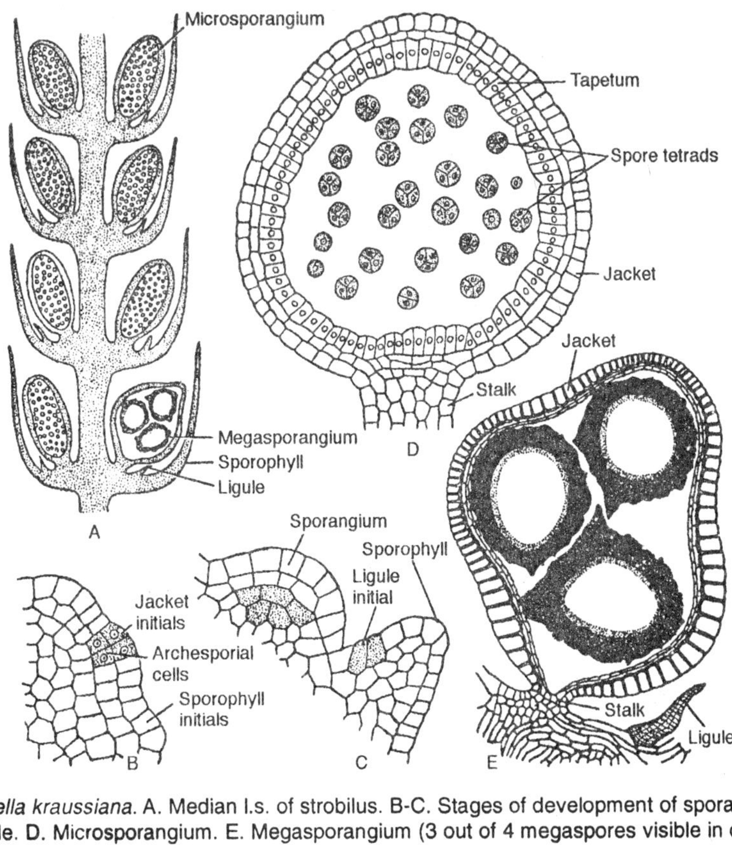

Fig. 39.15: *Selaginella kraussiana*. A. Median l.s. of strobilus. B-C. Stages of development of sporangium, sporophyll and ligule. D. Microsporangium. E. Megasporangium (3 out of 4 megaspores visible in one plane)

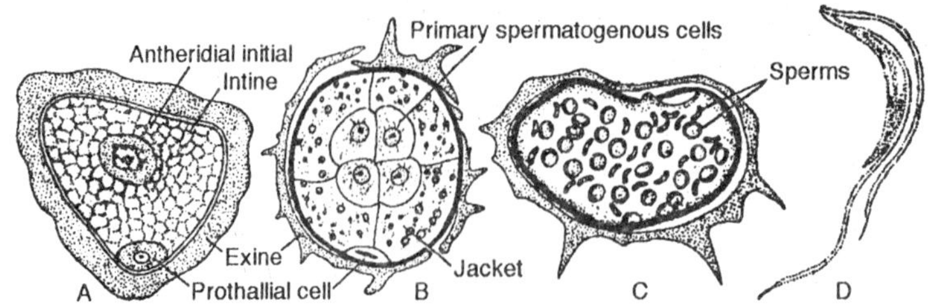

Fig. 39.16: Development of male gametophyte of *Selaginella kraussiana*. A-B. Stages of development, C. Mature stage, D. A sperm

(Fig. 39.16B). The microscope may be shed from the plant during any of these stages. After shedding, the primary spermatogenous cells undergo repeated divisions forming up to 256 sperm cells. Meanwhile, the exine cracks and the jacket cells with the prothallial cell disintegrate so that the biciliated sperms (Fig. 39.16D) swim freely inside the mature gametophyte (Fig. 39.16C).

The *megaspore* also shows a 2-layered wall with a very thick and sculptured *exine* and a thin *intine*. A triradiate ridge is prominent on the top of the tetrahedral structure (Fig. 39.17A). The stage at which the megaspore is

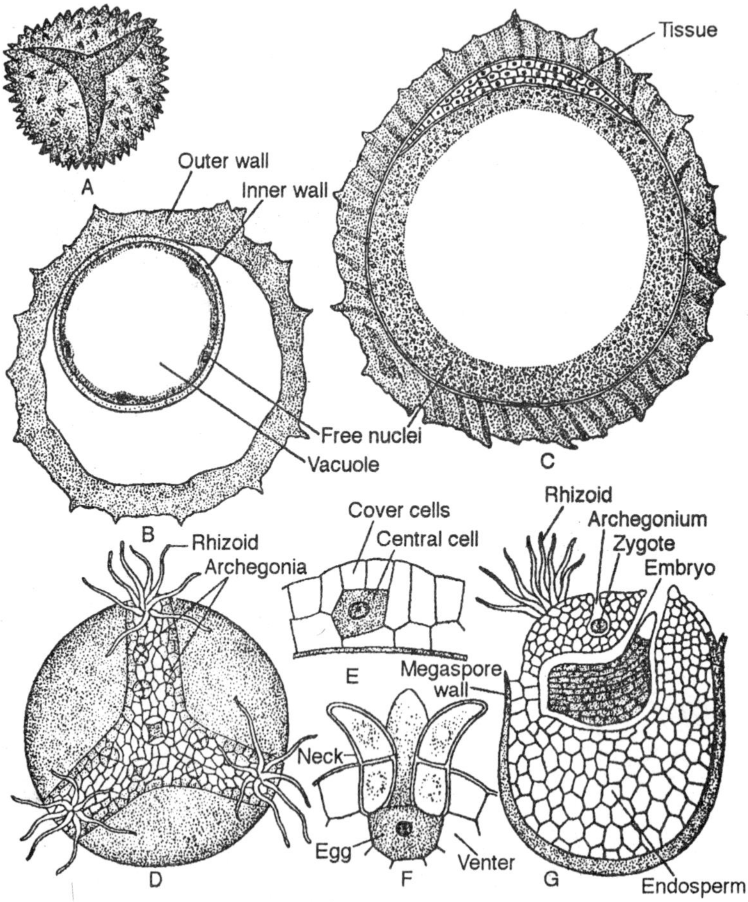

Fig. 39.17: *Selaginella kraussiana*. A. A megaspore. B-D. Stages of development of female gametophyte. E-F. Stages of development of archegonium. G. A fully mature female gametophyte

shed is extremely variable. It may be shed from the sporangium before any trace of cellular organization, shortly after the first archegonia are formed or it may be retained within the sporangium until after fertilization and a considerable development of the embryo. Development of the *female gametophyte* begins with a conspicuous enlargement of the megaspore. The megaspore nucleus divides repeatedly without immediate formation of cell walls (free cell formation). A conspicuous central vacuole with many free nuclei around it in a gradually thickening peripheral cytoplasm is organized (Fig. 39.17B). Very oftern, the outer thickened part of the megasporangial wall grows much faster than the thinner inner wall leaving a gap between the two as seen in Figure 39.17B. In the next stage, a cellular tissue 2 or 3 cells in thickness is formed by wall formation at the pyramidal end of the female gametophyte. This tissue is separated from the rest of the gametophyte by a conspicuous diaphragm below which the free nuclei remain round the central vacuole (Fig. 39.17C). Eventually the central vacuole is obliterated by the increasing cytoplasm but this lower part remains multinucleate until the embryo begins to develop. The apical tissue now becomes exposed by the cracking of the spore wall above it along the triradiate fissure. The exposed tissue may become green, and, on falling to ground, rhizoids may develop on it in patches (Fig. 39.17D). However, this photosynthetic tissue and rhizoids are of minor importance as far as the nutrition of the gametophytes is concerned as the reserved food in the lower part is of much greater importance for the nutrition of the gametophyte and the growing embryo. All superficial cells on the apical tissue are potential *archegonial initials*. An archegonial initial divides periclinally to form a *primary cover cell* and a *central cell* (Fig. 39.17E). The central cell again divides periclinally to form a *primary canal cell* and a *primary ventral cell*. The primary ventral cell forms a *ventral canal cell* and an *egg* by another periclinal division. The cover cell forms a 2-tiered *neck* which protrudes out and spreads apart to open the inner passage (Fig. 39.17F). The primary canal cell does not divide so that there is a single *canal cell*. While the archegonia are being formed, the free nuclei below form walls and organize into large cells (Fig. 39.17G). This tissue below is nutritive to the archegonia and the future embryo above so that this is the *endosperm* of *Selaginella* which is a gametophytic tissue as distinguished from the triploid endosperm of the Angiosperms.

Fertilization may take place while the female gametophyte is still within the unshed megasporangium on the plant or after it is shed. *Apogamy* and *parthenogenesis* are known in several species. Embryos are known to arise out of archegonial initials, eggs in unopened archegonia and also out of sporophytic cells in species not forming microspores.

The new sporophyte

The first division of the zygote is usually transverse (Fig. 39.18A). The upper cell gives rise to the one or more celled *suspensor* and the

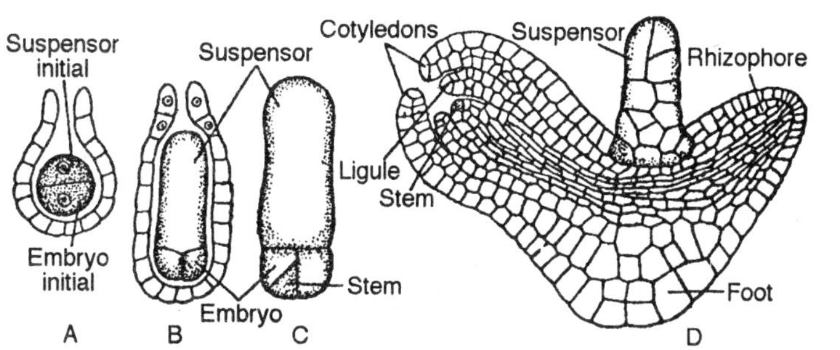

Fig. 39.18: A-D. Stages of development of *Selaginella* embryo (after Bruchmann)

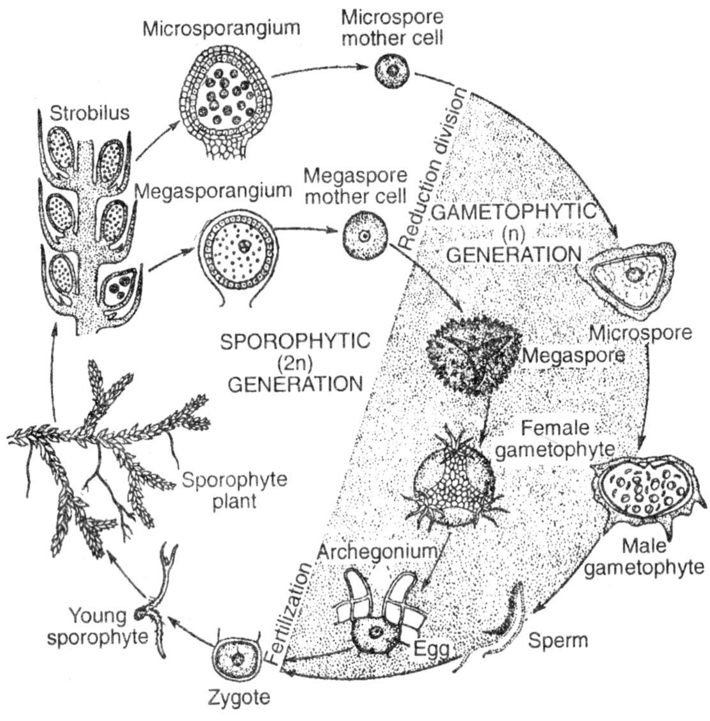

Fig. 39.19: Life cycle of *Selaginella*

lower cell to the *embryo*. The *embryo initial* divides vertically (Fig. 39.18B). The two resultant cells again divide vertically to form four cells one of which cuts off the *stem initial* from its base (Fig. 39.18C). Gradually, the embryo organizes into sectors on one side of which is the *stem* between two *cotyledons*, on the other side is the foot (for sucking food material from the endosperm) occupying a major part of the base and the *rhiziphore initial* on the top of it (Fig. 39.18D).

The new sporophyte eventually pierces through the surrounding gametophyte and establishes itself as an independent plant after growing adventitious roots from the rhizophore. It is to be noted that, unlike most Pteridophytes, it shows two cotyledons and even a *hypocotyl*.

Fig. 39.19 shows diagrammatically the life cycle of *Selaginella*.

Family 2. MIADESMIACEAE : *MIADESMIA*

The family Miadesmiaceae comprizing of the single genus *Miadesmia* is represented entirely by Carboniferous fossils. It shows certain characters reaching very near the seed habit so that, along with the Lepidoc paceae, it is often included in a common group *Lepidospermales* or *Lepidospermae*. But the *Miadesmia* sporophyte is herbaceous (though not much branched) and ligulate like *Selaginella* which it also resembles in the stem anatomy, viz., the endodermis is represented by trabeculae. It is, therefore, included within the order Selaginellales of the Selaginellidae while the Lepidocarpaceae is placed within the Lepidodendrales of the Lepidodendridae which the latter resembles in the arboreal habit of the sporophyte.

It is the megasporophyll and the female gametophyte of *Miadesmia* which deserve the greatest attention. The *strobili* of *Miadesmia* are lax and it is not certain whether the micro- and the megasporophylls were borne on the same strobilus. Both types of sporophylls were shed with the sporangia at maturity. The megasporangium was never detached from the megasporaphyll but was completely enveloped by an envelope growing from the surface of the megasporophyll which envelope bore on its surface a number of long outgrowths called 'tentacles' (Fig. 39.20). A single megaspore matured within the megasporangium and developed the single *female gametophyte* completely filling it up and remaining exposed at the ruptured tip. The envelope also was open at the tip leaving an opening like a 'micropyle'. This envelope, therefore, looked like an 'integument' but as it is something quite different from a true integument it has been named *velum* as in *Isoetes.* The female gametophyte never came out of the megasporangium but there is no reason to suppose that fertilization was siphonogamous. Fertilization probably took place by free-floating sperms as in other Pteriodophytes. This structure has been compared to a 'seed' because of the non-shedding of the female gametophyte. But, the

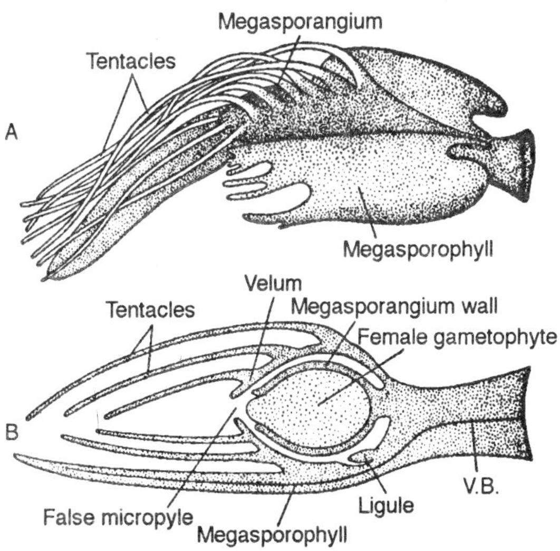

Fig. 39.20: *Miadesmia membranacea.* A. Megasporophyll with megasporangium inside velum. B. Median l.s. of A

absence of a true integument, absence of any siphonogamy and the apparent absence of a 'dormant seed stage' do not justify its being considered as a real 'seed' of a Spermaphyte. It may, at best, be called a *Pseudospermaphyte.*

Subclass IV. LEPIDODENDRIDAE

The Lepidodendrids gained a predominance in the Carboniferous when they had attained the arboreal habit with secondary growth. They probably evolved in the Devonian from the Protolepidodendrids. These ligulate, heterosporous Lycopods are divided into two orders: *Lepidodendrales* showing the Stigmarian rhizophores and the *Isoetales* showing condensed rhizomorphs.

Order: LEPIDODENDRALES

The Lepidodendrales are known from the Upper Devonian, became the predominant trees in the Carboniferous forests and disappeared in the Permian. The sporophytes are heterosporous and arboreal with secondary growth in both stem and root. The rhizophores bearing roots are of the characteristic Stigmarian type (Fig. 39.7) which are usually indistinguishable in the different genera so that all such fossils have been placed in the form genus *Stigmaria.*

In spite of the secondary growth, the stelar structure was rudimentary and weak showing an exarch protostele or siphonostele. The leaves, however large, were microphyllous with single median veins. The genera are distinguished mostly by the stem surface and branching habit. The order may be divided into four families—*Lepidodendraceae, Lepidocarpaceae, Bothrodendraceae* and *Sigillariaceae.*

Family 1. LEPIDODENDRACEAE

Lepidodendron, the type genus of Lepidodendraceae is the predominant tree in the Carboniferous coal fossils. They are known since the Upper Devonian and gradually disappeared during the Permian. The root system (mainly rhizophores) are included within the form genus *Stigmaria* (Fig. 39.22) and the strobili within the form genus *Lepidostrobus.* Over 100 species of *Lepidodendron* are known all over the world.

Most of the species were tall trees (Fig. 39.21A) with a trunk up to about half a metre in diameter and 40 metres high. The main trunk was unbranched but repeated dichotomies at top gave rise to a large crown. Leaves were simple, ligulate, acicular to linear and 1 to 80 cm long. Some of them are included in the invalid form genus *Lepidophyllum.* These were arranged spirally on the shorter branches and were probably deciduous. There was a single vein traversing the length of a leaf. Leaves were borne on the summit of pyramidal cushion-like leaf bases of rhombic outline which persisted on the trunk after abscission. Species and genera are distinguished by the shape and arrangement of these cushions. In *Lepidodendron,* the vertical diagonal of the rhombus is longer than the transverse and the scars are arranged on oblique rows on the stem (Fig. 39.21B). The actual *leaf scar* is above the middle line of the rhombus. In the *leaf scar* the bundle scar mark is flanked by two scars and at a lower level by two more larger scars. These

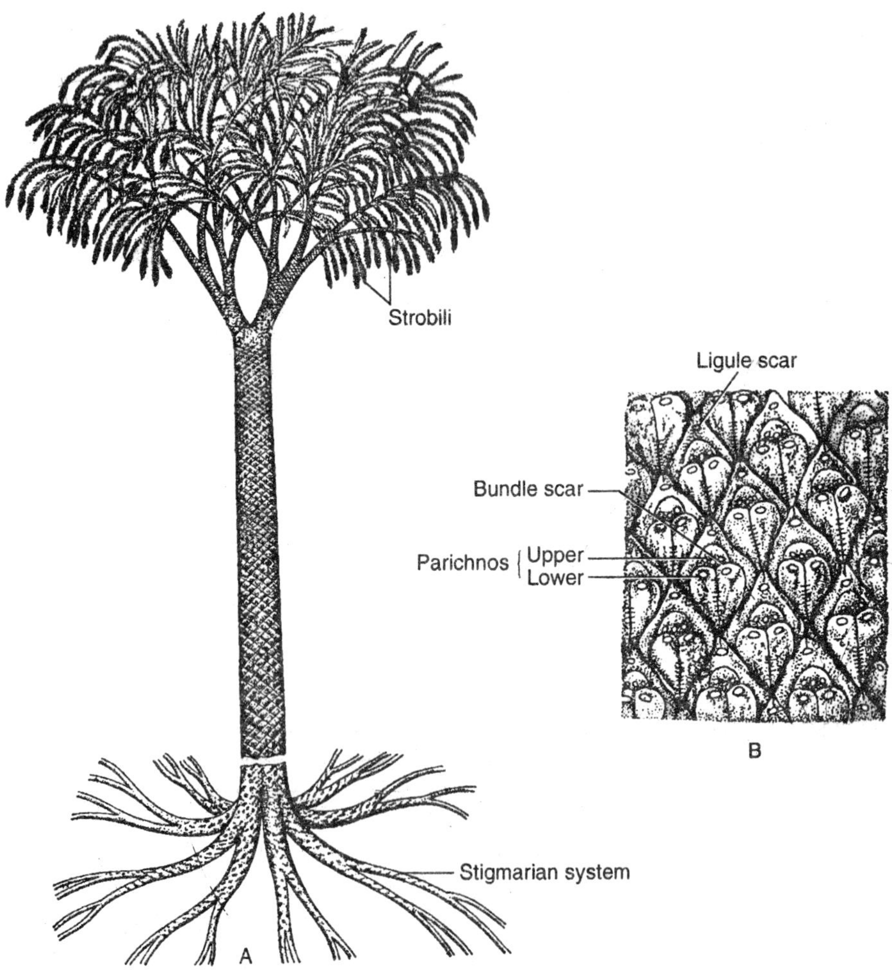

Fig. 39.21: *Lepidodendron* (after Dillon, Hirmer and Dahlgren). A Reconstruction of a tree (the base is reconstructed from a Stigmarian cast). B. Trunk surface scars showing leaf cushions in *L. obovatum*

four are the *parichnos* scars. The *parichnos* were longitudinal channels of thin-walled, loose tissues traversing the length of the leaf parallel to the vein. These might have served the purpose of aeration.

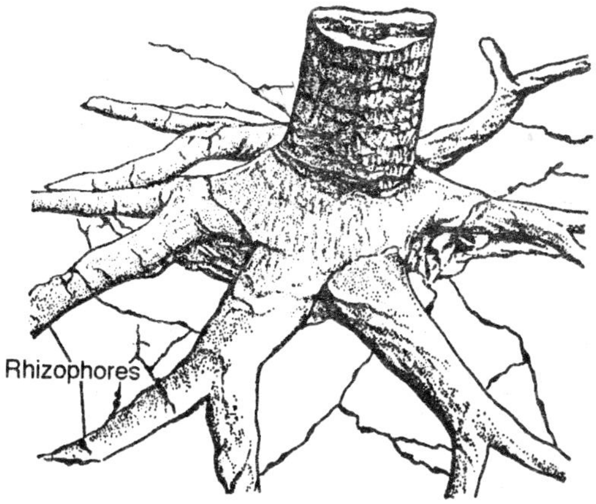

Fig. 39.22: Stump cast of *Stigmaria*

The trunk passed on at base to a Stigmarian system. *Stigmaria* is the name applied to the form genus of trunk bases of all Lepidodendrids. These are found mainly as stump casts but also as a few silicified fossils. The base of the trunk divided dichotomously into four large massive rhizophores (up to 80 cm. in diameter) which again divided dichotomously within a few feet of the trunk (Fig. 39.22). The structures are rather like *rhizophores* than like 'roots' as they are sometimes called. These were monostelic with a large amount of secondary wood. The final branches of these bore on their surfaces lateral adventitious roots (also monostelic) which are sometimes called 'Stigmarian rootlets'.

A t.s. of the stem at the trunk (Fig. 39.23) shows a protostele which is exarch and polyarch. The stele is slender in comparison to the

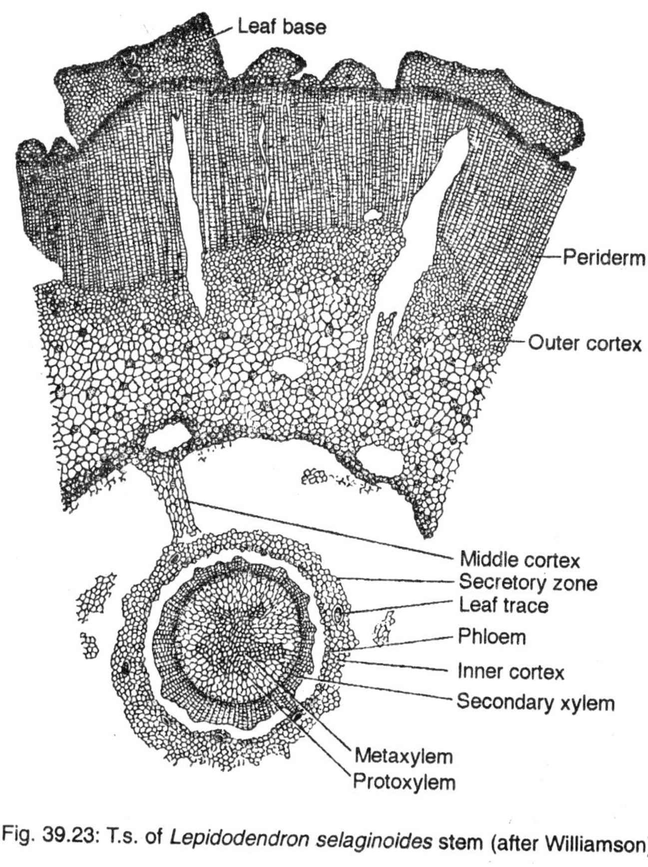

Fig. 39.23: T.s. of *Lepidodendron selaginoides* stem (after Williamson)

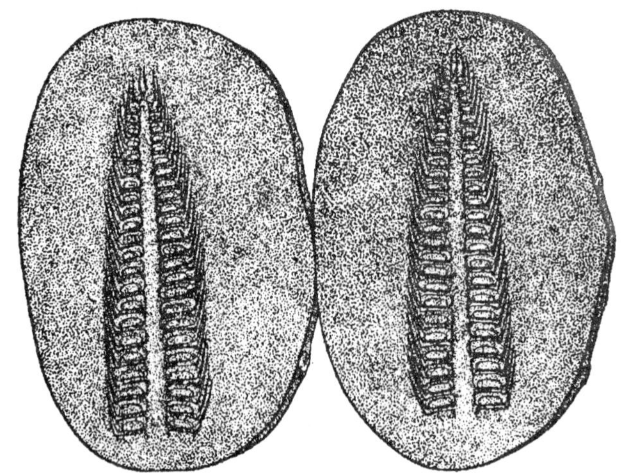

Fig. 39.24: A clay nodule fossil of *Lepidostrobus variabilis* (after Stopes)

massive cortex. The *metaxylem* is at the centre with a number of *protoxylem* points on the periphery. This is usually surrounded by a *cambium* round which is a narrow belt of *secondary xylem.* Both the xylems show scalariform and spiral tracheides. The cortex surrounding the stele is large and shows three zones. Innermost is a small patch of the *inner cortex* of parenchymatous cells surrounded by a *secretory zone.* Outside this is a large *middle cortex* formed of even more delicate cells which is often destroyed and filled up with the cast mineral. *Leaf traces* occur in the middle and the inner cortex. Outside this is the massive *outer cortex* which, with the *periderm,* is the main mechanical tissue being formed of alternating radial patches of thin-walled and thick-walled tissues or of uniform thick-walled tissue. Outermost is a hard *periderm* formed early in the life of a plant by a *phellogen* in the outer cortex. The periderm was another main supporting tissue of the trunk. Probably Evolution did not support this dependence on cortex and periderm instead of the stele for the support of the erect trunk and this fact caused the extinction of this whole group of land plants. The surface of the trunk was covered by the persistent *leaf bases* or leaf cushions.

The leaf was quite simple with a ligule at the base. The stomata were arranged alone two longitudinal patches on the lower *epidermis* on the two sides of the vein. In some species there is a thick-walled *hypodermis* all round the leaf. In the centre is a spongy *mesophyll* in which there are four longitudinal bands of *parichnos* ending before tip. The centre is occupied by a prominent *vascular bundle.*

Numerous *strobili* of the Lepidodendraceae have been discovered mainly from the Carboniferous and have been placed in the form genus *Lepidostrobus* (Fig. 39.24). The strobili are elliptical, 1 to 7 cm in diameter and 2.5 to 30 cm or more long. It is heterosporous and the *sporophylls* are arranged as in *Selaginella,* but the sporangia are much larger and the sporophylls are almost peltate with the upper terminal lobe clasping the sporophyll above (39.25A). The *microsporangia* and the *megasporangia* are of the same size. The microsporangium encloses a large number of *microspores* while the megasporangium shows a few much larger *megaspores* although the number is more than four. Some sporangia are incompletely septate (Fig. 39.25B) but this trabecular tissue differs from that to *Isoetes* and may be only of a haustorial nature. The megaspores show gametophytic development while still within the sporangium as in *Selaginella.*

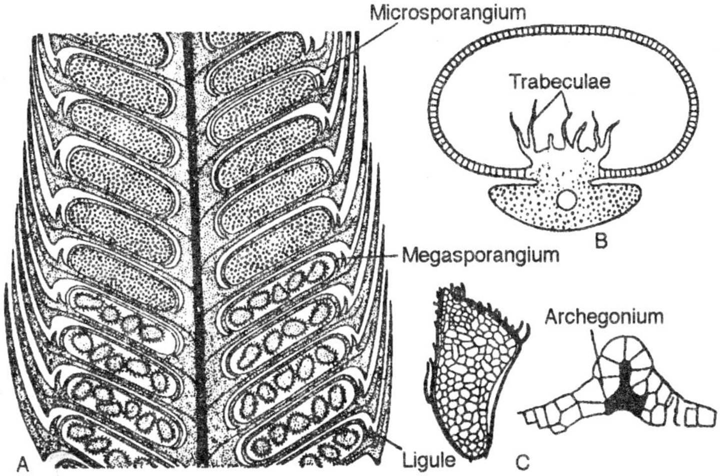

Fig. 39.25: *Lepidostrobus*. A. Median l.s. of part of strobilus. B. A sporangium showing trabecular growth. C. A female gametophyte and an archegonium (showing separately)

The remnant of a female gametophyte (Fig. 39.25C) shows the similarity with *Selaginella*.

Lepidophlois is another contemporaneous genus of Lepidodendraceae showing similar habit. But, here the transverse diameter of the leaf base is longer than the vertical and the leaf scar is below the middle line.

Family 2. LEPIDOCARPACEAE

Lepidocarpon from the Carboniferous is a very distinctive genus of the Lepidocarpaceae which approaches the seed habit as in *Miadesmia* so that this has also been called a *Lepidospermale* although it should not be included in the same order as the latter. It is also a *Pseudospermaphyte*. The tree on which it grew was most probably *Lepidophlois* or some tree resembling that. Its interest lies in the megasporangium.

The strobili, which were heterosporous and outwardly resembled *Lepidostrobus*, bore only one kind of sporophyll—*microsporophyll* or *megasporophyll*. Microsporophyll, microsporangium and microspore were as in *Lepidostrobus*. But, the megasporophyll was very different. The *megasporangium* was elongated and its wall formed an elongated beak, the tip of which split open at maturity. Inside, a single *megaspore* mother cell developed four *megaspores* of which three disintegrated and only one enlarged and germinated into a female gametophyte completely filling up the sporangial cavity while the megasporangium

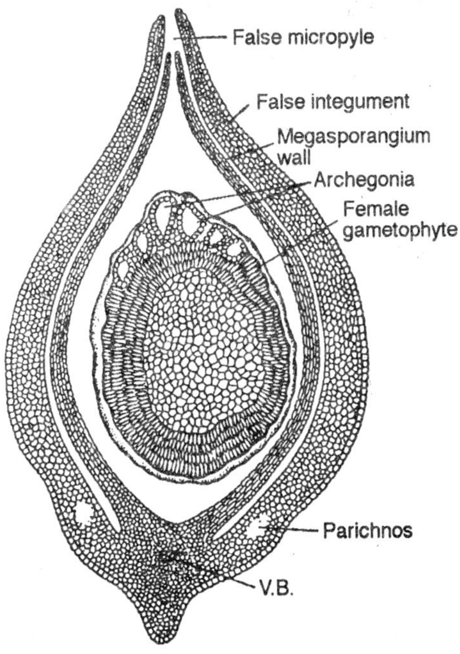

Fig. 39.26: *Lepidocarpon magnificum*. Median l.s. of false seed structure (modified from Andrews)

remained permanently attached to the megasporophyll. Moreover, the megasporophyll tissue below the sporangium grew up and developed a *false integument* which was open at the tip by a long slit forming a false *microphyle.* On the whole the structure looks like a 'seed' (Fig. 39.26) although it is not really so because of the same reasons as explained for *Miadesmia.* Most probably fertilization took place by motile sperms without any siphonogamy after the megasporophyll with the sporangium was shed.

Family 3. BOTHRODENDRACEAE

Bothrodendraceae comprizes of the large tree genus *Bothrodendron* (Fig. 39.27A) and its strobilus *Bothrostrobus.* It occurred from the Upper Devonian to the Permian and its represented in the Indian Lower Gondwana Flora. The heterosporous *strobili* were borne in two rows on the bases of the major branches. *Microsporophylls* and *megasporophylls* were borne on the same strobili. *Female gametophytes* are preserved and resemble *Selaginella.* Bothrodendraceae is represented in India by *Cyclodendron* from the Lower Gondwana Karharbari series (Lower Permian). This is the only fossir representative (with its probable spore *Gondisporites* from Raniganj series—Upper Permian) of the Lepidodendridae from the Indian Gondwana.

Family 4. SIGILLARIACEAE

Sigillariaceae is a Lepidodendrid family extending from the Carboniferous to the Permian. It was arboreal but the trunk was practically unbranched with a cluster of very long leaves at the apex (Fig. 39.27B). There is great variation in the strobili discovered but all the sporophytes are placed under the single genus *Sigillaria* showing more than one hundred species. Some of the trunks are more than

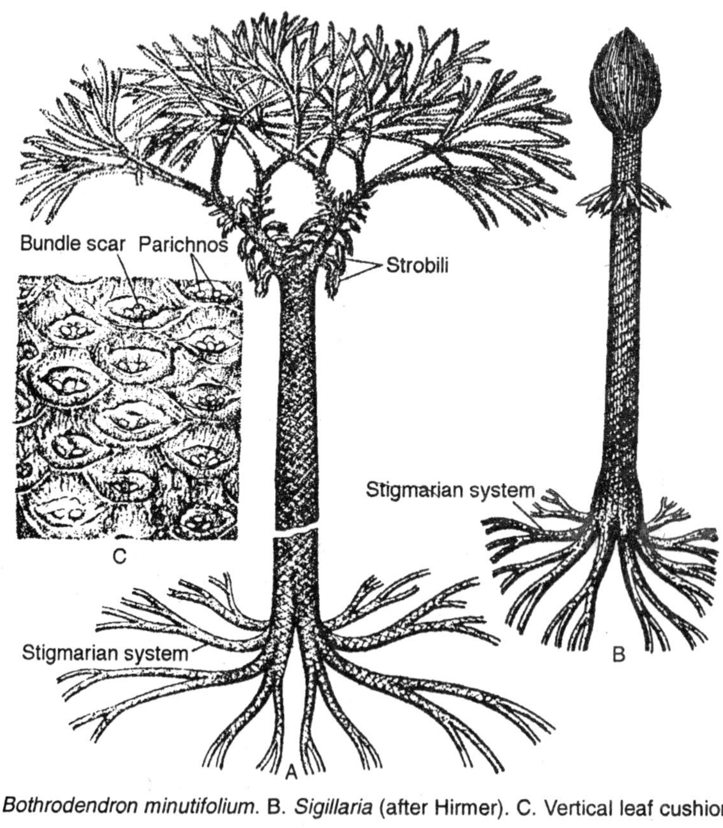

Fig. 39.27: A. *Bothrodendron minutifolium.* B. *Sigillaria* (after Hirmer). C. Vertical leaf cushions of *Sigillaria*

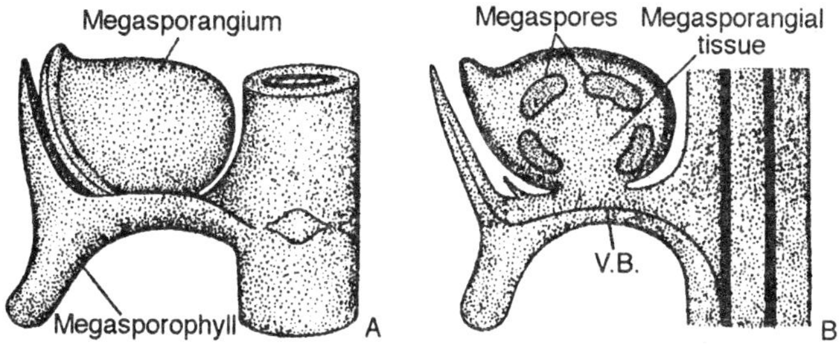

Fig. 39.28: *Mazocarpon oedipernum*. A. Megasporophyll. B. Vertical median section of A

22 metres high and two metres in diameter. The trunk is ribbed in some species (subgenus *Eusigillaria*). The leaf cushions are arranged in vertical rows (Fig. 39.27C). The tufted leaves (*Sigillariophyllum*) at apex are sometimes more than a metre long and in some species there are two parallel veins instead of one. The root system is Stigmarian. The stem is siphonostelic, exarch and polyarch showing secondary growth and generally resembles *Lepidodendron.*

Two types of strobili of *Sigillaria* are known. Both are heterosporous and 16 to 20 cm long. *Sigillariostrobus* resembles *Lepidostrobus* generally. But, the other type, *Mazocarpon,* is peculiar. Here the *microsporophylls* are borne above the *megasporophylls* and the two types of sporangia are externally similar. But the *megasporangium* (Fig. 39.28) shows eight large (up to 2 mm in diameter) crescent-shaped megaspores arranged in two tiers *embedded within a tissue* formed inside the megasporangium. The megaspores developed into female gametophytes in this position, each showing a single archegonium, but eventually the megasporophyll got detached from the plant. *Mazocarpon* also is considered as a *Pseudospermaphyte* as the megaspores, in all stages, show some tissue round them. The tissue is something like *nucellus* but still condition is far from the 'seed habit'.

Order: ISOETALES

The order Isoetales differs from the Lycopodiales in a modification of the Stigmarian trunk base into a condensed *rhizomorph.* The stem shows secondary thickening, leaves are microphyllous and ligulate and it is heterosporous although the sporophylls may not be organized into strobili. Two families are distinct—*Pleuromeiaceae* and *Isoetaceae.*

Family 1. PLEUROMEIACEAE

The Pleuromeiaceae from the Middle Triassic and Cretaceous represent a stage intermediate between the modern *Isoetes* and the Permo-Carboniferous *Sigillaria.* It is nearer to *Isoetes* than to the lepidodendraceae so that Seward suggested placing it nearer to the former. Two genera, *Pleuromeia* (Triassic) and *Nathorstiana* (Cretaceous) seem to point out the gradual evolution.

The original specimen of *Pleuromeia* was discovered in a stone in the Magdeburg Cathedral. *Pleuromeia* (Fig. 39.29A) shows an unbranched stem like *Sigillaria* attaining a height of 2 metres and maximum diameter of 9 cm. The lower stem shows spiral leaf scars and ligulate, long, narrow, awl-shaped leaves (Fig. 39.29B) are clustered at the top. Leaves show two parallel median veins, are up to 11 cm long and more than 2 cm wide at the clasping base. The base is a condensation of the Stigmarian system. It divides dichotomously into four (rarely more) very short and unbranched lobes turning upwards, each of which bears numerous rootlets on the lower surface as in modern *Isoetes.*

The *anatomy* of the stem, though not well preserved, seem to be exarch actinostelic. Secondary thickening is noted only from the extrastellar *periderm* which, as in the Lepidodendraceae, was the main mechanical system.

The stem terminated in a massive *strobilus* of very short, closely overlapping, shield-shaped sporophylls. The plant was heterosporous and dioecious. The sporangia were apparently borne on the *lower* surface of the leaf base but it has been doubted whether this abnormal position is not due to the bulging out of the adaxial sporangia to the abaxial position. There is also a hypothesis that *Pleuromeia* might have an ancestor with peltate sporophylls from which the upper sporangia had been lost in evolution and the lower persisted. The sporangia are large and show *trabeculae* linking them to *Lepidostrobus.* The megasporangia (about 2 cm in diameter) were larger than the microsporangia and the megaspores were about twenty times as large as the microspores.

Nathorstiana (Fig. 39.29C) from the Createaceous is a much shorter plant (about 20 cm tall) coming nearer to *Isoetes.* The sporophylls of this plant are not known. The rhizomorph shows a number of ridges and spirally arranged rootlets.

Family 2. ISOETACEAE

The family Isoetaceae comprise of three genera: (1) The lower Createaceous *Isoetites*

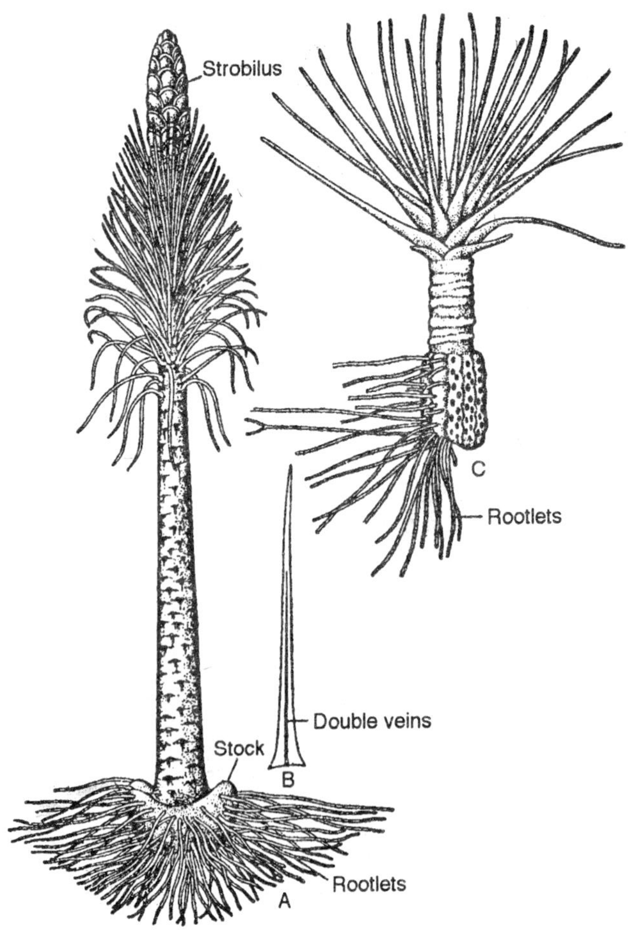

Fig. 39.29: A. *Pleuromeia sternbergi* (after Hirmer). B. A leaf of *Pleuromeia*. C. *Nathorstiana arborea* (after Mägdefrau)

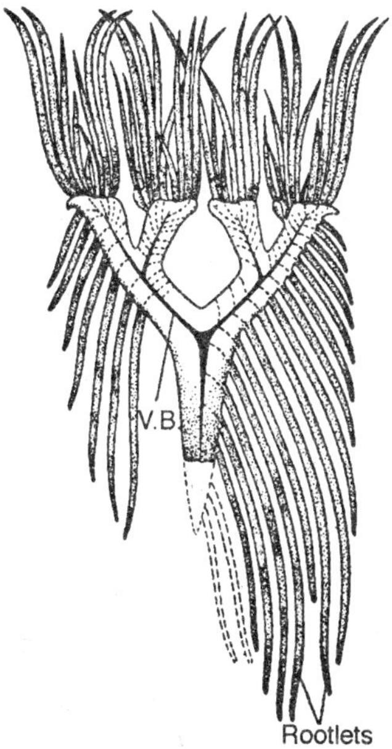

Fig. 39.30: *Stylites gemmifera* (after Rauh and Falk)

resembling and Isoëtes today. *Isoetites* is known from the Upper Gondwana Umia stage in India (Upper Jurassic). (2) The rare and recently discovered *Stylites.* (3) The modern and well-spread *Isoetes* known since Upper Cretaceous.

Stylites (Fig. 39.30) has been discovered by Amstutz (1957) and Rauh (1959) only recently from the Peruvian high Andes mountains. The stem branches dichotomously at least three times and bears a rosette of *Isoetes*-like leaves while the lower parts of it bear rootlets on the outer surface only. The sporangia are borne on the upper surface of sporophylls but far above the base. Its habit emphasises the hypothesis of the *Pleuromeia-Nathorstiana-Stylites-Isoetes* course of evolution.

Genus: *ISOETES*

Fossils of the genus *Isoetes* are known since the Upper Cretaceous. More than 65 species are known in the present-day flora, a majority of which are hydrophytes growing submerged in water or on swampy land. It is known almost from all over the Indian coasts and also from interior places. The commonest species seems to be *Isoetes coromandeliana* known from the eastern and the western coasts of South India, from lower Bengal, Uttar Pradesh and Madhya Pradesh. Other Indian species are *I. sahyadri, I. indica, I. sampatkumarini, I. panchanani.* Morphologically, the genus seems to be very uniform. Cytologically, great variations in chromosome numbers have been reported, the *n* number being reported as 10, 11, 22, 33, 54-56. This is probably due to a high degree of polyploidy.

The sporophyte

The sporophytic plant is often mistaken for ordinary tufted grass. The short, corm-like stem is not externally differentiated into an axis and rhizomorph. This *stock* is 2- or 3-lobed and on its upper surface bears long, stiff, awl-shaped leaves. The whole plant is 5 to 50 cm high (Fig. 39.31). The leaves (50 cm long to very small) are borne in acropetal order over-

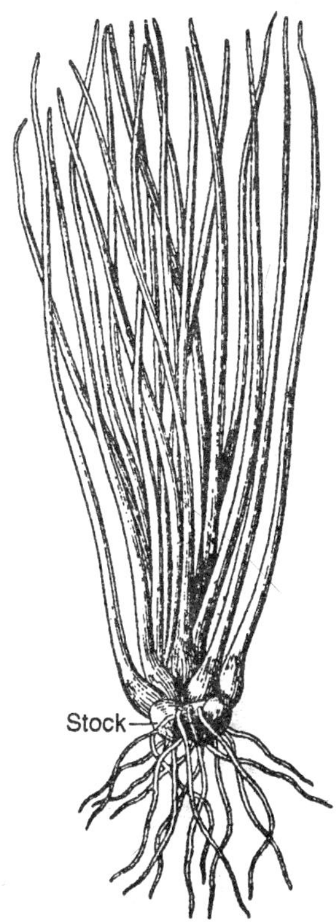

Fig. 39.31: *Isoetes coromandeliana*

lapping one another and taper abruptly from a broad base to an elongated acicular limb. The *ligule* is as in *Selaginella* and is on the adaxial surface at the juncture of the acicular and basal portions. A single median vein traverses the whole length. The lowest (outmost) leaves on the stock are sterile. Successively within them are the *megasporophylls*, the *microsporophylls* and leaves with immature sporangia. The growing tip is at the apex at the bottom of a depression and is formed of a group of meristematic cells. The lower surface of the stock bears many dichotomously branched rootlets which appear to cover the entire lower surface and spread up its sides though in actuality, they are in two- or three group according as the stock is two- or three-lobbed.

The corm-like *stock* is such a condensed structure that its real nature is confusing. There is little doubt about the morphological nature of the *Isoetes* corm but to realise that it has to be examined from all angles anatomically. Figure 39.32A shows a vertical section of the corm which shows a median groove. Figure 39.32B shows a vertical section of the same through the plane of the basal groove. The inner shaded portion is the vascular system. The upper erect portion is the stem axis while the lower horizontal part is formed of telescoped rhizophores. Figure 39.32C is a similar

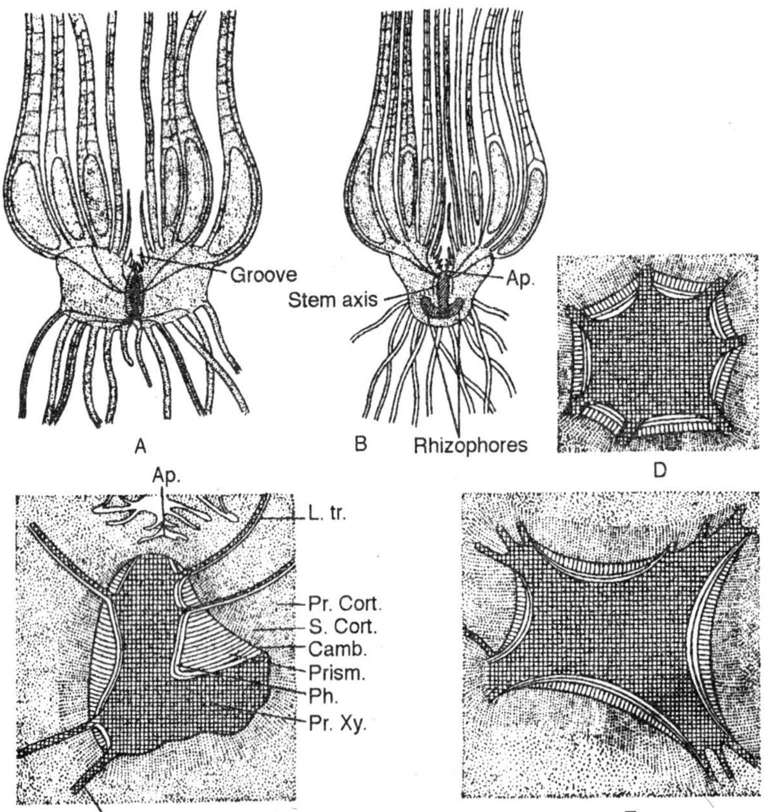

Fig. 39.32: The *Isoetes* stock (diagrammatic figures). A. Vertical median section showing groove. B. Vertical section through plane of groove. Shaded portion is the vascular system. C. Similar section through another plane showing one rhizophore only. D. T.s. of stem axis. E. T.s. through rhizophore part. *Ap.* = apical meristem. *L. tr.* = leaf trace. *Pr. Cort.*= primary cortex. *S. Cort.* = Secondary cortex. *Camb.* = cambium. *Prism.* = Prismatic layer. *Ph.* = phloem. *Pr. Xy.* = primary xylem. *R. tr.* = root trace

section through another vertical plane showing one arm of the rhizophore. The stem axis given rise to the leaf traces and the rhizophores to the root traces. Figure 39.32D is a t.s. of the stem while Figure 39.32E is a t.s. through the basal part showing the four rhizophore limbs. A study of these show that the stock is essentially a stem on the top and four rhizophores result from two dichotomies at the base. The rootlets originate from the meristematic surface of the rhizophores.

A t.s. of the mature part of the stem (Figs. 39.32C and D) shows a *protostele* in which the *xylem* is formed of reticulate and spiral tracheides with xylem parenchyma at the centre. This is encircled by a *phloem* of sieve tubes with sieve plates on the walls. *Leaf traces* are present but they do not give rise to leaf gaps. In the mature stem and rhizophore, a *cambium* is developed outside the phloem. The cambium is often active on the inner side only where it gives rise to a secondary xylem of reticulate tracheides or to a secondary phloem or to a mixed tissue composed of tracheides and sieve tubes. Because of this variable nature of the secondary internal tissue, this has been named the *prismatic layer.* The cambium develops on the outer surface, at least in some cases, a *secondary cortex.* External to this stele, there is a massive parenchymatous *cortex* (primary) which forms the bulk of the stock. This is homogeneous, full of starch and has big intercellular spaces. There is no endodermis delimiting the stelar and cortical tissues. The top of the rhizophore part (Fig. 39.32E) does not differ much from the stem. The apical meristems of the rhizophore branches give rise to the rootlets which are renewed every year.

A root shows a *monarch protostele* on one side of a central air space (Fig. 39.33A). The *xylem* and *phloem* are almost collateral. There is a clear *endodermis.*

A t.s. of the leaf of *Isoetes* (Fig. 39.33B) shows a single collateral *vascular bundle* continuous with the leaf trace. There are four longitudinal *air chambers* running along the whole length of the acicular part. These may be transversely barred. In sporophylls the vascular bundles may be flanked on two sides in the basal part by two mucilage canals. These remind the *parichnos* of the lepidodendrales. The mesophyll may be wholly parenchymatous or may contain four longitudinal strands of sclerenchyma alternating with the air chambers. The leaf *epidermis* shows stomata on parts not submerged under water.

All leaves of *Isoetes* are potentially sporophylls as they bear normal or aborted sporangia on the upper face. A *sporophyll* bears a single

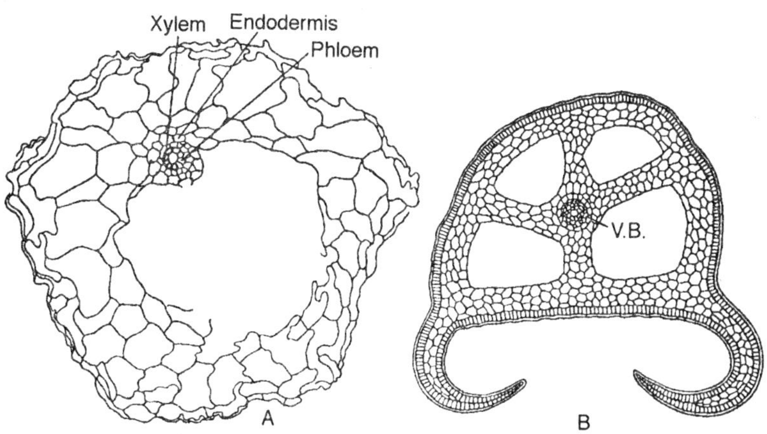

Fig. 39.33: *Isoetes*. A. T.s of root. B. T.s. of leaf

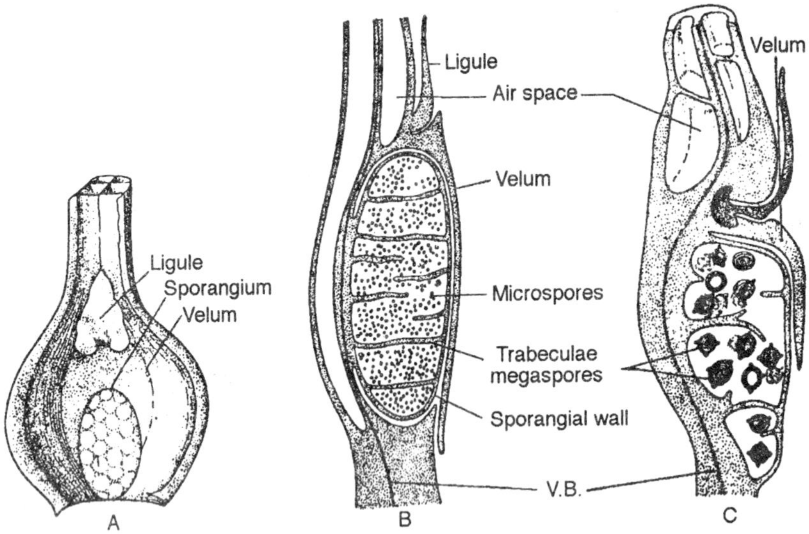

Fig. 39.34: *Isoetes*. A. Sporophyll base, B. Median l.s. of microsporophyll base, C. Median l.s. of megasporophyll base

flattened, oval or rounded *sporangium* between the ligule and the leaf base on the adaxial face (Fig. 39.34A). The sporangium lies within a depression in the leaf base and is completely or incompletely roofed by a membranous outgrowth arising from the base of the ligule which is called the *velum*. The sporangium originates from a group of cells below the ligule which divided periclinally to give rise to a *jacket layer* on the outside and an *archesporium* on the inside. A massive *sporogenous tissue* is formed from the latter which becomes broken up into blocks of fertile cells by transverse plates or incomplete columns of sterile cells. These sterile cells eventually mature into *trabeculae* which incompletely divide the cavity within the mature sporangium. The sporogenous cells adjacent to the trabeculae, as well as those adjacent to the sporangial wall form a 2-layered *tapetum*. The sporangial wall is 1-layered for a considerable time but becomes 3- or 4-layered later on. Differentiation between *micro-* and *megasporangia* become apparent after the *spore mother stage*. In the *microsporangia* (Fig. 39.34B) all the spore mother cells are functional forming fifteen to thirty thousand (even a million has been reported) *microspores* while in the *megasporangia* (Fig. 39.34C) 40 to 80 enlarged spore mother cells become functional forming 160-320 *megaspores*. The megaspores (up to 0.9 mm in diameter) are 10 to 20 times as large as the microspores. Sporangia of most species are indehiscent and the spores are liberated after the decay of the sporophylls.

The gametophytes

The *microspores* are usually bilateral or lenticular while the *megaspores* are tetrahedral. Both of them are variously sculptured externally (Fig. 39.35A). The spores, freed from the decayed sporophylls, usually remain close to the mother plant or may be distributed by wind and worms.

The *microspore* (Fig. 39.35B) soon germinates to develop the *male gametophyte*. The nucleus migrates to one end, divides and a wall partitions off the single *prothallial cell*, the remaining larger cell functioning as the *antheridial initial* (Fig. 39.35C). The antheridial initial divides and redivides to develop an *antheridium* with a single layered *jacket* and four *sperm mother cells* (*androcytes*) (Fig.

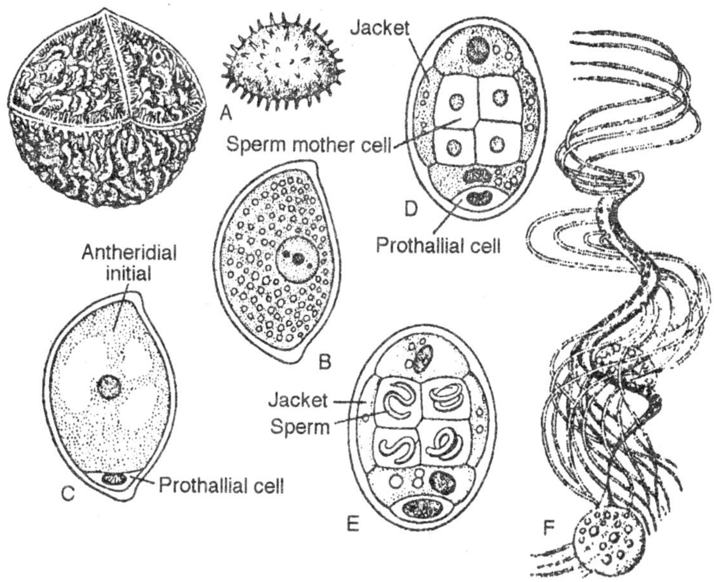

Fig. 39.35: *Isoetes*. A. Megaspore and microspore (magnifications of microspore is higher). B-D. Stages of development of the male gametophyte. E. Mature male gametophyte. F. A sperm

39.35D). Each of the androcytes develops a single *spermatozoid* in the mature male gametophyte (Fig. 39.35E). The four spermatozoids are spirally coiled and have about 15 cilia each (Fig. 39.35F). The multiciliate spermatozoid of *Isoetes* is unlike the other Lycopsids but is common with the Sphenopsida and the Filicopsida. All the inner cell walls now disintegrate and the sperms are finally liberated by dehiscence of the microspore wall.

The *megaspore* germinates into the *female gametophyte* somewhat in the manner of *Selaginella*. The nucleus divides by a series of free nuclear divisions, but unlike *Selaginella*, there is no conspicuous central vacuole at any stage. More nuclei gradually accumulate near the apex and wall formation starts at this end forming an apical tissue (Fig. 39.36A). But, this apical tissue, though exposed, does not protrude out of the triradiate fissure on top and never turns green although a few rhizoids may develop. As in *Selaginella*, the lower *endosperm* remains in the free nuclear condition for a long time and may not become cellular till the func-

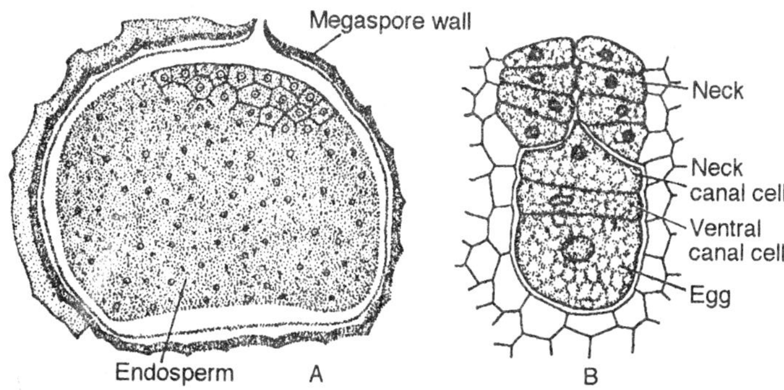

Fig. 39.36: *Isoetes*. A. Young female gametophye. B. An archegonium

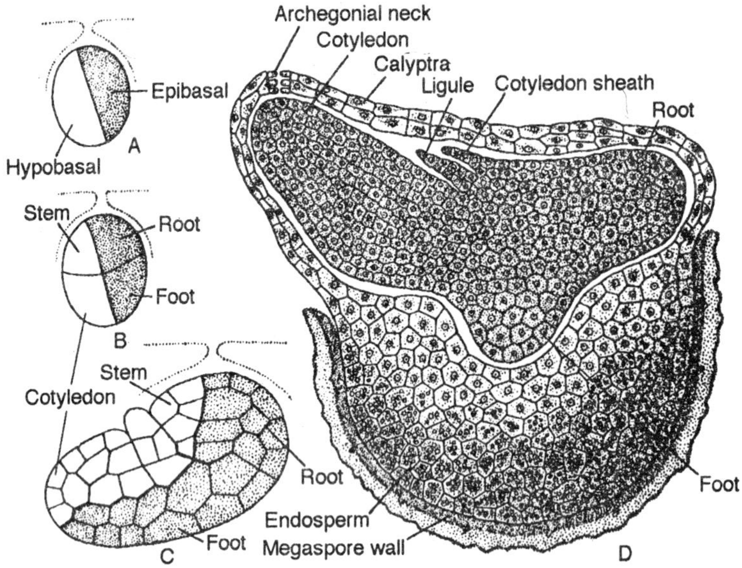

Fig. 39.37: *Isoetes*. A-C. Stages of development of embryo (The epibasal parts are shaded). V. A female gametophyte with a mature embryo

tional embryo has developed considerably (Fig. 39.37D). Also as in *Selaginella*, any of the top central cells may function as an *archegonium initial*. This cell develops an *archegonium* (Fig. 39.36B) in the same manner as in *Selaginella*. Several archegonia may be formed but only one remains functional.

Fertilization takes place in the usual way and the zygote is formed.

The new sporophyte

The *embryo* differs from all other Lycopsids in having no suspensor. The zygote divides in a diagonal plane (Fig. 39.37A). There is again another division in a transverse plane giving rise to four cells (Fig. 39.37B). In some cases a *quadrant stage* has been noticed. Subsequent divisions and interpretations are somewhat confused but it seems that the upper (*epibasal*) half of the embryo gives rise to the *foot* below and the *root* above (shown shaded) and the lower (*hypobasal*) half gives rise to the *cotyledon* and the *stem* (Fig. 39.37C). The stem, even though recognizable early, is rather slow in development. The foot grows down and sucks nutrition from the *endosperm*. The root and the crotyledon (with a *ligule*) lie in one plane and the whole embryo is covered by the archegonial venter which is now the *calyptra* (Fig. 39.37D). The first leaf soon breaks through the calyptra and the root bends downwards and reaches the substrate making the young sporophyte independent. The real development of the stem takes place only then.

The terminal position of the single cotyledon (opposite the primary root) and the late development of the stem combined with the grass-like habit of the plant reminds one of the monocotyledons. But, a real phylogenetic relationship between *Isoetes* and monocotyledons seems to be a rather remote idea.

Figure 39.38 is a diagrammatic representation of the life cycle of *Isoetes.*

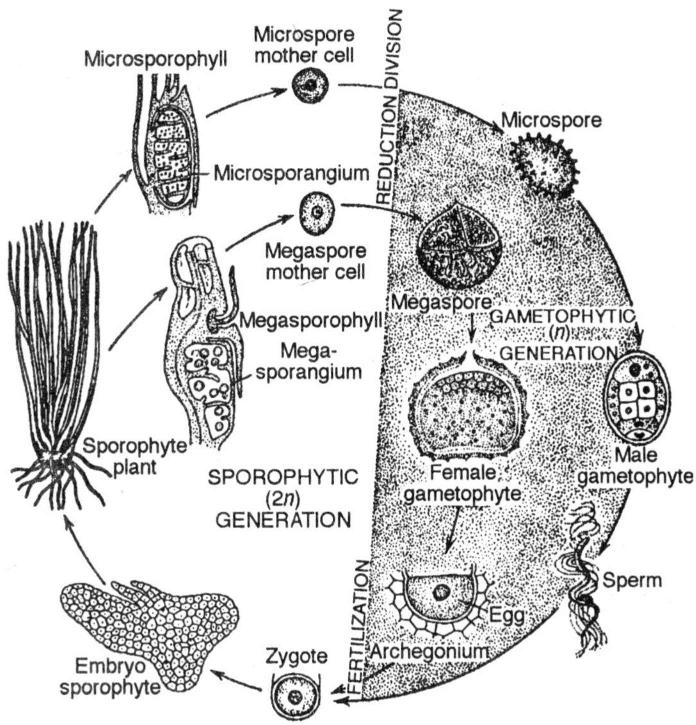

Fig. 39.38: Life cycle of *Isoetes*

Chapter 40

Sphenopsida

The class Sphenopsida (= Division *Calamophyta* or *Arthrophyta* or *Sphenophyta,* also called *Articulatae* or *Equisetinae*) is represented today only by the herbaceous and insignificant *Equisetum* but included magnificent trees in the geological times. The evolution is almost parallel to the Lycopsids. They arose in the Lower Devonian, attained the climax of their evolution during the Carboniferous when they grew into trees 30 metres high, began to dwindle in the Permian and since Jurassic only *Equisetum* are some of its immediate associated are known. Besides those included definitely within the Sphenopsida, there are two fossils specimens, *Climaciophyton* and *Sphondylophyton* from the Lower Devonian which are not fully known but are suspected to be Sphenopsids.

The class is well defined in certain morphological features. The stem is articulated with nodes and internodes and sometimes with a longitudinally ribbed surface. Equisetales fossils are identified by the ribs on stem casts. The leaves are borne in whorls at the nodes and the branch buds are not axillary but alternate with the leaves. The sporangia are borne in unique structures called *sporangiophores* arranged in strobili. Heterospory possibly developed the least in this group.

The class may be divided into four orders:

1. Hyeniales: Lower and Middle Devonian
2. Pseudoborniales: Middle Devonian
3. Sphenophyllales: Upper Devonian to Lower Triassic
4. Equisetales: Upper Devonian to Lower Triassic excepting the Equisetaceae which is still extant.

Of these, the Hyeniales are the most primitive as well as the simplest. From these perhaps there where two lines of evolution—one represented by the Sphenophyllales and the other by the Equisetales. The Sphenophyllales are protostelic and the leaves are megaphyllous with dichotomous venation. In the equisetales the stem is siphonostelic and the leaves are microphyllous with single unbranched veins. The leaf in the Sphenopsida, therefore, is not as simple as that in the Lycopsida where it is uniformly microphyllous. It is not certain whether the leaf of the Equisetals evolved as the microphyllous leaf of Lycopsida (c.f. Chapter 39) or by reduction from macrophyllous types.

Order 1. HYENIALES

The order Hyeniales is also called *Protoarticulatae* because it is supposed to be the starting point of the Sphenopsida. It is represented by three genera—*Protohyenia* from Lower Devonian; *Hyenia* and *Calamophyton* from the Middle Devonian. Of these, the latter two are the better known. All the three genera were comparatively small shrubby plants. Although apparently simple, the taxonomic position of the Hyeniales is now being supposed to be very complicated. The taxon may be a very artificial assemblage of unrelated plants.

Protohyenia (Fig. 40.1) from the Lower Devonian of Western Siberia shows horizontal rhizome, aerial shoots with terminal sporangia as in the Psilophytales but with a tendency towards the Sphenopsid verticillate arrangement of the sporangia. It was similar but smaller than *Hyenia* (see below). The leafy appendages looked more like forked branches than like leaves. Recently, it has been suggested that *Protohyenia* shows close affinities with *Pseudosporochnus* of the Cladoxylales.

Hyenia (Fig. 40.2A) from the Middle Devonian of Germany and Norway had a stout, horizontal rhizome bearing numerous erect shoots which were usually unbranched but sometimes digitately branched. Articulation was indistinct. The leaf-like appendages (Fig. 40.2B) were forked several times and these were borne in close irregular whorls of three or more. The fertile appendages (Fig. 40.2C) were on separate shoots (lax strobili). Each of them was forked and reflexed towards the axis, each arm bearing two or three hanging sporangia. The anatomy is not known.

Calamophyton (Fig. 40.2D) from the Middle Devonian of Germany and Belgium was a larger plant attaining up to 2 ft. in height. Articulation was very clear and the general branching was digitate. The sterile appendages were whorled but while they appear notched and wedge-shaped (Fig. 40.2E) in some specimens they have been found to be 3-dimensional with forkings in different planes in other specimens. The fertile appendages (Fig. 40.2F) were in lax strobili, were forked only once and each fork bore only one hanging sporangium. The anatomy is not clear but seems to be siphonostelic and possibly with secondary growth. Reticulate and scalariform tracheides have been discovered in the compressed stem fossils. Recently (Leclercq and

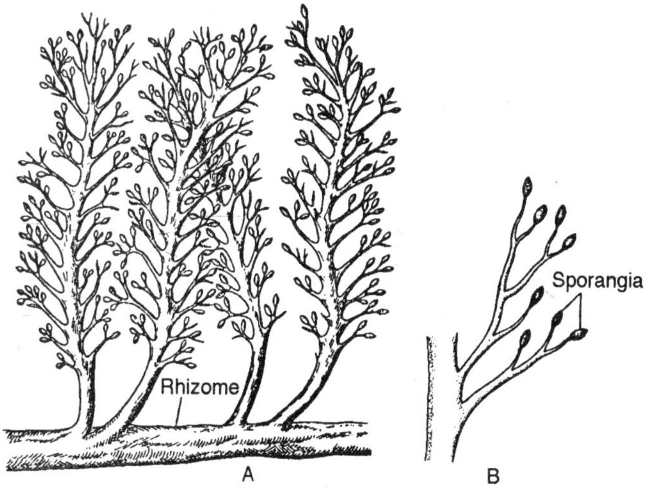

Fig. 40.1: *Protohyenia janovii*. A Sporophyte. B. Sporangia (After Mehra)

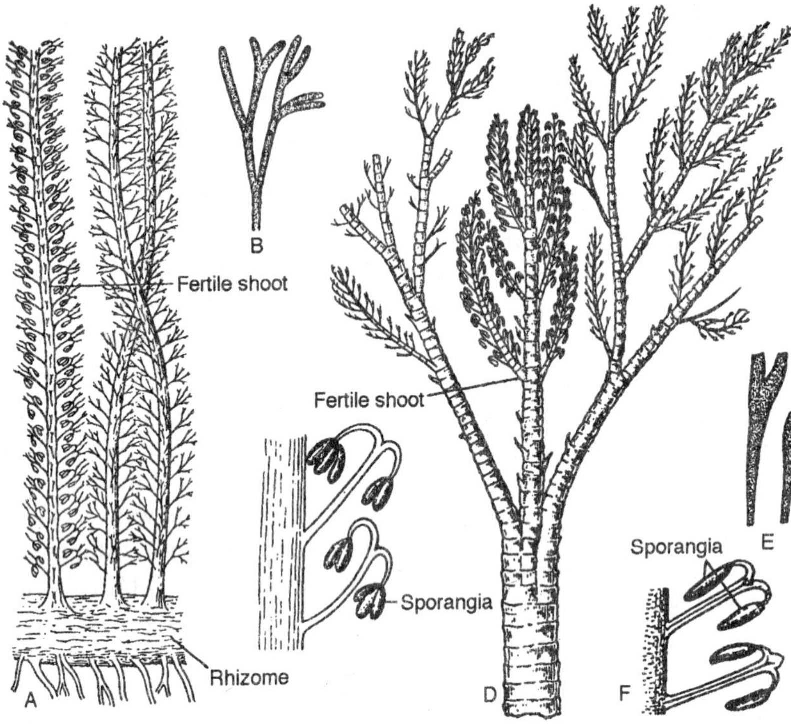

Fig. 40.2: *Hyenia elegans*. A. Sporophyte plant. B. Sterile appendage. C. Fertile appendages. *Calamophyton primaevum*. D. Sporophyte plant. E. Sterile appendages. F. Fertile appendages. (A after Leclercq; B-F after Kräusel and Weyland)

Schweitzer 1965) *C. bicephalum* has been shown to possess a Cladoxylalian stem. This puts the status of the Hyeniales into further doubt.

Hyeniales approach the Psilophytopsida in outward appearance and should be nearer to a common origin if not dircetly evolved from the latter.

Hyeniale fossils are not known from India.

Order II. PSEUDOBORNIALES

The single species *Pseudobornia ursina* described by Nathorst as a compression from the Upper Devonian of the Arctic Bear Island is not fully known but is interesting. A horizontal stem up to 10 cm in diameter bore aerial shoots which were distinctly articulated and bore leaves in whorls of four. Each leaf forked dichotomously, the forkings again being pinnnately dissected into fine segments presenting a feather-like appearance. The strobilus was a lax spike up to 32 cm long bearing whorled sporangiophores each which looked like a reduced leaf which had a sterile bract-like portion and sporangia on the under surface. The spores were large and might be megaspores.

Nathorst considered the Pseudoborniales as a distinct group showing affinities both with the Sphenophyllales and with *Asterocalamites* of the Equisetales.

Order III. SPHENOPHYLLALES: *SPHENOPHYLLUM*

The order Sphenophyllales does not seem to be in the direct line of evolution of modern *Equisetum* but seems to be on a side line. Its habit suggested that it was probably not an erect plant although showing secondary growth but was probably a trailer producing a dense mat on the muddy floor vegetation of the coal swamps. It was probably partly aquatic. It first appeared in the Upper Devonian, had its climax in the Carboniferous and the Lower Permian, persisted to the Lower Triassic and then disappeared. *Sphenophyllum speciosum* is known from the Barakar and Raniganj Series (Lower and Upper Permian) of Indian Lower Gondwana.

The stems and leaves are assigned to the type genus *Sphenophyllum* (Fig. 40.3A) of which there are a number of species. Most of them were discovered from Germany, France, Scotland and the Arctic in the coal bearing shales or coal balls. They were herbaceous creeping plants and shoots up to 85 cm long are known. The slender stems (only up to 5 mm in diameter pointing to the trailing habit) were articulated, longitudinally ribbed and bore whorls of wedge-shaped leaves at the nodes. The ribs did not alternate at the nodes. 6 to 18 leaves (multiples of 3) are known at each whorl. The leaves were sessile, less than 2 cm in length, dichotomously veined and entire or variously dichotomously lobed according to the species (Fig. 40.3B). Some plants had whorls of both entire and lobed leaves—a case of heterophylly—but presumably not due to aquatic nature as dissected and entire whorls were mixed up. The dichotomous venation points out to the megaphyllous nature of the leaves which is not found in the Equisetums. The isolated leaves were formerly mistaken for fern leaves. The strong mechanical construction of the leaf does not favour the idea that it was a floating plant. The roots were adventitious and small, arising from the nodes.

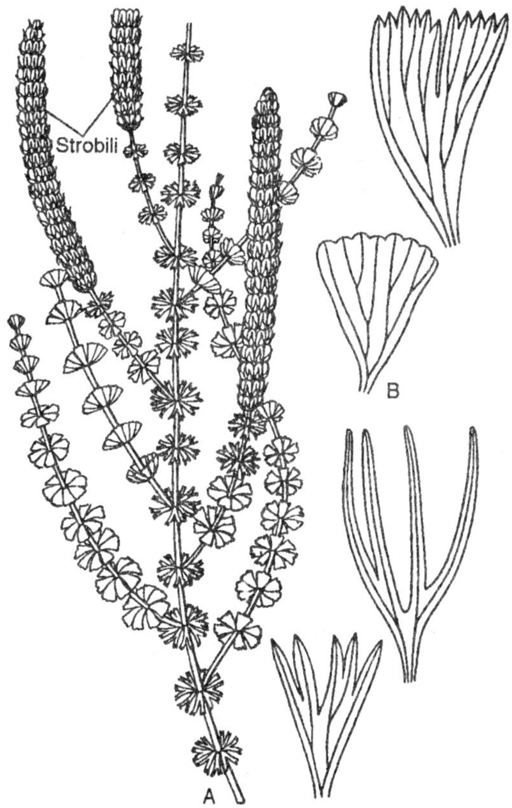

Fig. 40.3: A. Shoot of *Sphenophyllum cuneifolium* (redrawn after Hirmer). B. Leaf forms of *Sphenophyllum*

In the middle of the stem there is a solid strand of *protostele* (Fig. 40.4A). The *primary xylem* is triangular, exarch, actinostelic, triarch or hexarch with *protoxylems* at the tips of the arms. The *metaxylem* had multiseriate bordered pits. There was a *primary phloem* round it but it is not well preserved. The *cortex* round the stele is formed in thin-walled cells in the interior and thick-walled cells on the exterior. In the secondary structure (Fig. 40.4B) a *cambium* surrounds the primary stele. This given rise to a *secondary xylem* of multiseriate bordered pitted tracheides arranged in rays with parenchymatous rays in between so that the stele becomes rounded. A *secondary phloem* developed outside the cambium. A *cork cambium* was formed in the cortex rather early. Successive internal phellogens gave rise to more *periderm* and ultimately all the cortex was used up.

The *anatomy* of the small roots resembles stems except that they were diarch. The pholem is better preserved.

Although the vegetative structure of *Sphenophyllum* is fairly uniform, the strobili

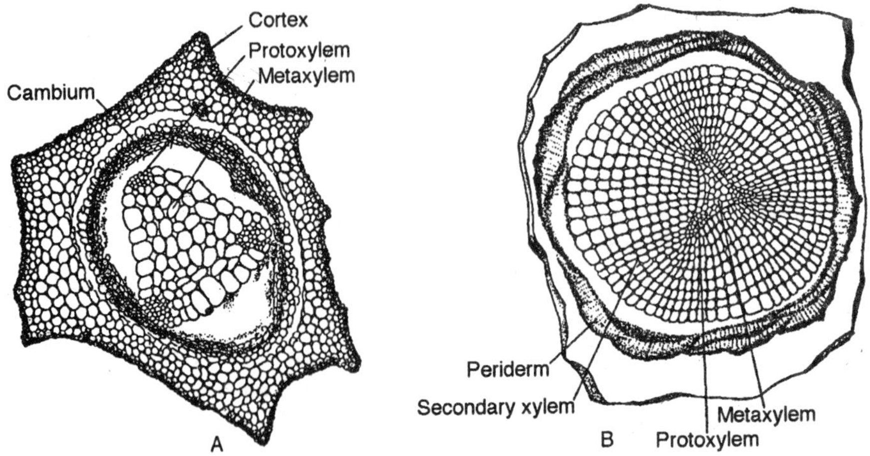

Fig. 40.4: A. T.s. of primary stem of *Sphenophyllum*. B. T.s. of *Sphenophyllum plurifoliatum* stem showing secondary growth

discovered are very diverse and most of them have not been found in actual association with vegetative structures. These have been placed in a number of *form genera* which have even been grouped into families as follows:

Family: Sphenophyllaceae

- Genus: *Sphenophyllostachys* (= *Bowmanites*)
- Genus: *Peltrastrobus*
- Genus: *Sphenostrobus*
- Genus: *Litostrobus*

Family: Cheirostrobaceae

- Genus: *Cheirostrobus*

Family: Tristachyaceae

- Genus: *Tristachya*

Family: Eviostachyaceae

- Genus: *Eviostachys*

The strobili were either apical or they formed side shoots. Most strobili are placed in the form genus *Sphenophyllostachys* (= *Bowmanites*) of which 31 species are on record but only 3 or 4 have been found in actual association with vegetative *Sphenophyllum*. The best known specimen is *Sphenophyllostachys dawsonii* (Fig. 40.5). In it the cone was about 2 cm in diameter with an articulated axis of very short nodes. The nodes carried whorls of sporangia supported by whorls of sterile bracts. The sterile bracts were joined laterally at the base but were free at the apex. The apex was curved forming a cup-shaped structure which often covered the next higher whorl of bracts. Attached to the axis between these verticils of 14 to 20 coalescent bracts were the sporangiophores which differed markedly in length and branching. They might or might not be partially fused with the bracts. These was apparently, only one whorl of partially fused sporangiophores (Fig. 40.5) which branched at the tips so that several whorls were simulated. Branching was normally of two types, some had a short median arm and two long distal ones while others had two short median arms and one long distal arm. *Sphenophyllostachys aequensis* (Fig. 40.6A) and *S. majus* (Fig. 40.6B) are simpler, the former showing single nodal whorls of free bracts and a single sporangiophore with a single sporangium at the axil of each bract while the latter (*majus*) shows a single whorl of dichotomised bracts at each node with one whorl of sporangiophores coalescent to bracts and each sporangiophore bearing four sporangia at the tip. *S. fertilis* (Fig. 40.6C),

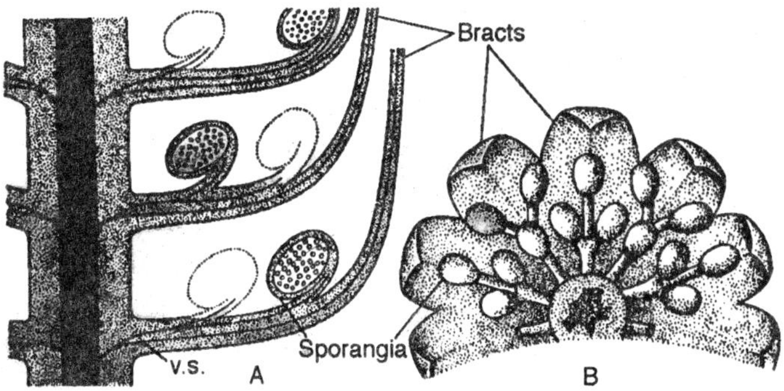

Fig. 40.5: Strobilus of *Sphenophyllostachys (Bowmanites) dawsonii.* A. L.s. B. Top view of one whorl

however, was very much complicated. The cone was large (about 6 cm long and 2.5 cm in diameter). At each node the whorl consisted of six groups of three appendages each. Within each group the two lower appendages were actually formed by the splitting of a bract from the base while the uppermost one was a compound sporangiophore having a short axis bearing a cluster of some 16 branches each ending in a pair of pendant sporangia. *Cheirostrobus* also had a very large and complicated strobilus in which each appendage or bract bore fertile branches on top and sterile ones below. *Eviostachys* (Figs. 40.6D and E)

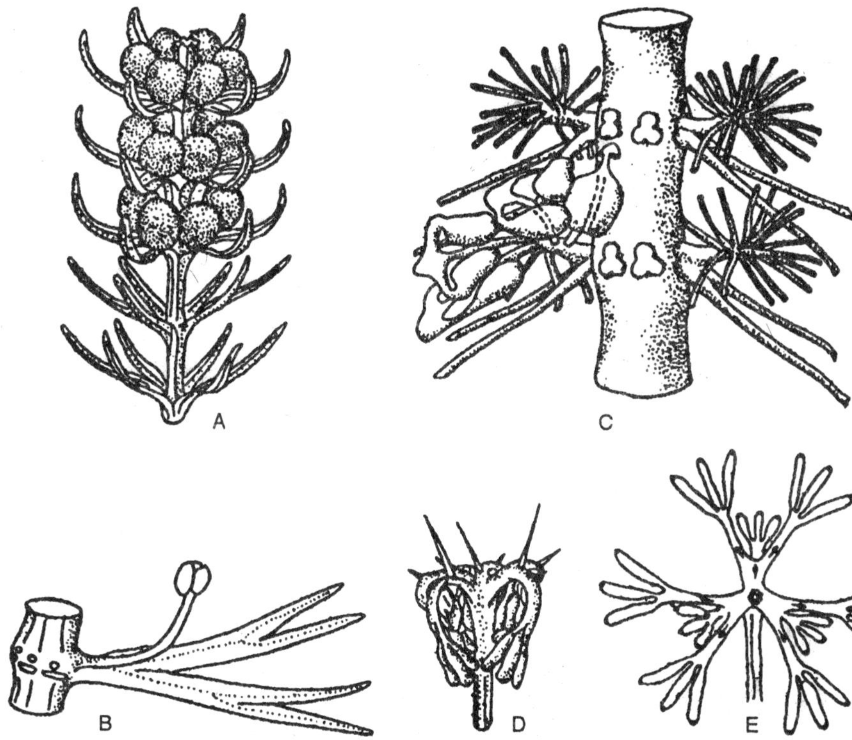

Fig. 40.6: Sphenophyllales. A. *Sphenophyllostachys aequensis* (after Remy and Remy). B. *S. majus* (after Walton). C. *S. fertilis* (sporangia shown only on one sporangiophore : after Leclercq). D. and E. *Eviostachya hoegi* sporangiophore as found. (D) and artificially extended (E) (after Leclercq)

showed the same triangular stele and sporangiophore organization as in others but the strobilus was devoid of bracts. Each sporangiophore was trichotomously divided into three and again into nine (with each middle branch shorter in the second stage) and each branch terminated into a cluster of three sporangia so that there were 27 sporangia on each sporangiophore. There were spinous projections from the sporangiophore branches.

Homospory seems to be the rule is *Sphenophyllum*. But, certain strobili specimens are now known in which there is a great variation in the size of spores. In some the spore size in one sporangium is 20 per cent larger than in the adjacent one. This, possibly, signifies some sort of incipient heterospory.

Order IV. EQUISETALES

The Equisetales comprise of three families—*Asterocalamitaceae, Calamitaceae* and *Equisetaceae.* Asterocalamitaceae and Calamitaceae are sometimes considered to form a separate order *Calamitales* but it is so close to Equisetaceae in anatomy and morphology that all the three should better be placed in the same order.

The sporophytic plant body of the Equisetales is generally branched and the branches arise in whorls from the nodes alternating with the leaves. Internodes are ribbed and the ribs of successive internodes alternate in the Calamitaceae and Equisetaceae but not in the Asterocalamitacae. The leaves are comparatively larger and green in Asterocalamitaceae and Calamitaceae while they are smaller and scaly in Equisetaceae. The stele is endarch siphonostelic and entire at the nodes but longitudinally perforated or split through the internodes. While Asterocalamitaceae and Calamitaceae show secondary growth and are arboreal, Equisetaceae is herbaceous without any secondary growth. Sporangiophores are in whorls in the strobili but while the strobili of Calamitaceae are of various types and with sterile bracts between sporangiophore whorls such sterile bracts are absent in the Equisetaceae and in some of the Asterocalamitaceae. There is heterospory in Calamitaceae but Equisetaceae is homosporous although some incipient heterospory is indicated.

Family 1. ASTEROCALAMITACEAE (= ARCHAEOCALAMITACEAE)

The family Asterocalamitaceae with the single genus *Asterocalamites* (= *Archaeocalamites*) is sometimes merged within Calamitaceae. Reported from the Upper Devonian and Lower Carboniferous of Europe, it is one of the oldest Calamites. The stem was probably arboreal and resembled *Calamites* but the ribs on the internodes did not alternate at the nodes. The whorled leaves were different from those of *Calamites.* They were large (5 to 40 cm long), dichotomously forked several times and reminds one of *Pseudobornia.* The strobili in some cases, e.g., *Pothocites,* resembled *Equisetum* not showing any sterile bract between the whorls of sporophylls. In some other specimens there were only a few sterile bracts among the sporangiophores. It is likely that both the Calamitaceae with sterile bracts among sporangiophores and the Equisetaceae without such sterile bracts, arose out of this taxon.

Family 2. CALAMITACEAE

The Calamitaceae with giant Carboniferous trees occupies a position in Sphenopsida comparable to that of the Lepidodendrales in the Lycopsida. They are even more important in this class as the climax of evolution shown by the Calamitaceae is very poorly represented today by the straggling *Equisetum,* the sole survivor of the Sphenopsids. The Calamitaceae appeared in the Upper Devonian, attained its zenith in the Carboniferous and disappeared in Lower Triassic. Parts of these plants are known from the Carboniferous all over the world.

Calamites is considered as the only genus by many Pteridologists although numerous form genera of stems, leaves and strobili are recognized. The name *Calamites* is due to Sukow (1784) who used it for pith casts. The sporophytic plant, known mainly from casts, were mostly trees attaining heights of 20 or 30 metres although a few of them might have been smaller shrubs. Although so large, it had a strong horizontally growing rhizome from which the aerial shoots arise as in the *Equisetum* today. The rhizome differentiated into nodes and internodes with a whorl of adventitious roots at each node. The rhizomes also had secondary growth. According to the habit three subgenera are recognized (Fig. 40.7). (1) *Stylocalamites*—a single main stem branched at the base just above the rhizome giving rise to a clump of parallel, erect branches on the surface which did not branch any more. The leaf verticils were at the top (Fig. 40.7A). (2) *Calamitina*—there was a single strong stem which showed many smaller branches in whorls at every node. Leaf verticils were on the secondary branches. Here and there, very irregularly, there might be a node bearing comparatively stronger side branches (Fig. 40.7B). (3) *Eucalamites*—these represent the dominant type. There was a main trunk unbranched for a short height then bearing verticils of two or more strong branches which again branched repeatedly giving rise to a bushy tree. The leaf verticils were on the final branchlets (Fig. 40.7C). The manner of origin of the branches was peculiar. Each branch suddenly narrowed down at the base meeting the rhizome or the stronger branch below at an articulated narrow point. The stele also narrowed down likewise (Fig. 40.8). Thus, erect branches which grew from the upper side of

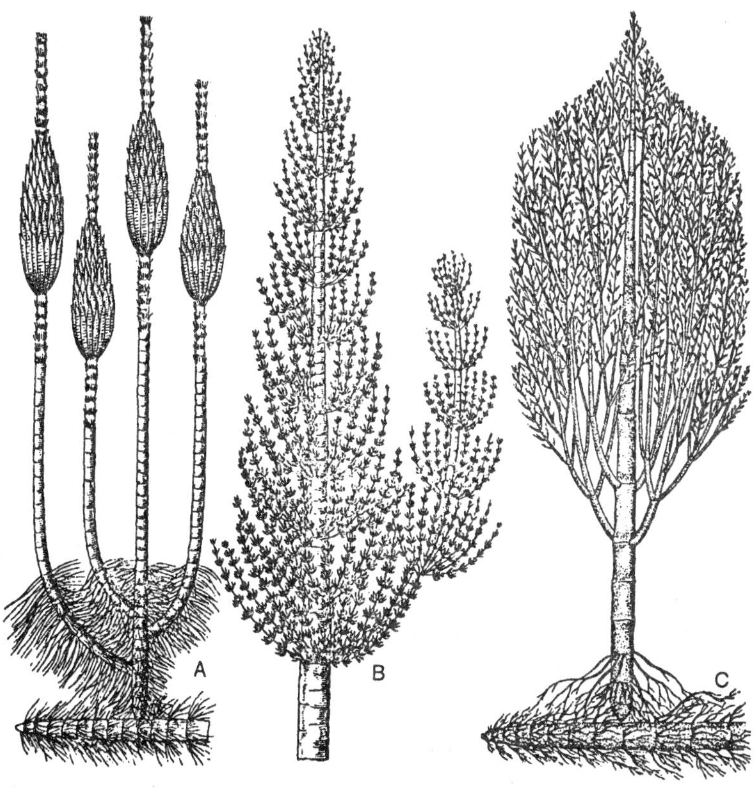

Fig. 40.7: Subgenera of *Calamites* tree forms. A. *Stylocalamites*. B. *Calamitina*. C. *Eucalamites*

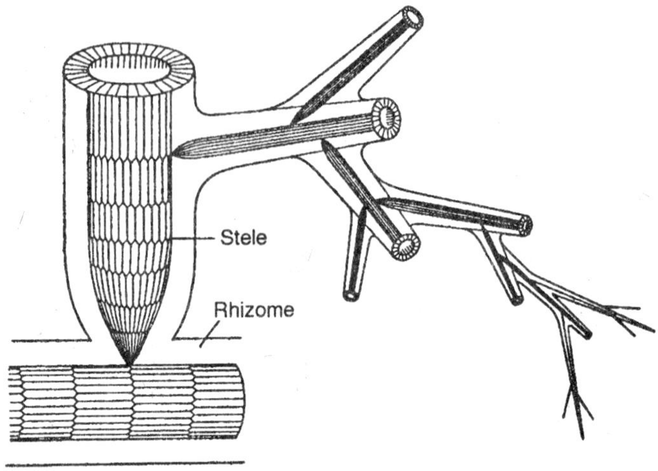

Fig. 40.8: Manner of basal branching of *Calamites*

the rhizome had a very constricted base. This certainly caused some weakness in the tall trees. The stem surface was ribbed and the ribs of alternate nodes were alternating.

The stem anatomy (Fig. 40.9) shows a prominent delicate *pith* as in Equisetum which broke down in mature shoots leaving central cavities extending from node to node. This was surrounded by an *endarch siphonostele*. The *protoxylem* strands (annular and spiral tracheides) disintegrated forming *carinal canals* as in *Equisetum*. This was surrounded by a *metaxylem* (scalariform and pitted tracheides) which merged into the pitted tracheides (some of them bordered, resembling Gymnosperms) of a continuous ring of secondary wood. A *cambium* and the consequent *secondary wood* are the principal differences with *Equisetum*. It was seriated into tracheides and rays (the nature of rays distinguishing some species) but there were no annual rings, thus signifying uniformity of climate. A young stem shows about a dozen or more collateral bundles. The cortex had an outer thick-walled and an inner thin-walled zone. The former was sometimes in longitudinal strands as in Equisetum. The vallecular ring of cavities was absent in *Calamites*. A *periderm* was formed in the cortex during

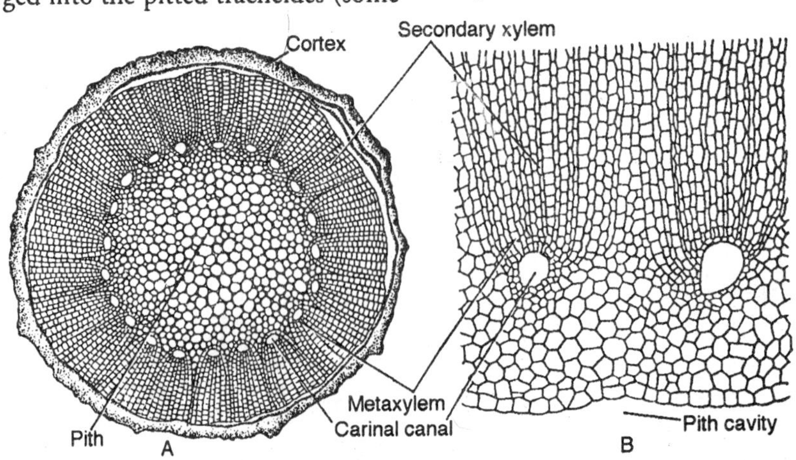

Fig. 40.9: A. T.s. of young ***Calamites*** stem. B. Part of t.s. of mature *Calamites* (*Arthropitys*) stem

secondary growth. *Calamites* stems are known which are about 50 cm in diameter with the secondary wood about 6 cm thick and the periderm also a few centimetres.

A few other types of stems are also found within this genus. Thus, *Protocalamites*, a stem genus from the Lower Carboniferous, shows the primary xylem exarch instead of endarch.

Roots of *Calamites* are usually placed in the form genus *Asteromyelon*. These adventitious roots do not show nodes and internodes. The *primary xylem* is *exarch* and carinal canals are absent. It is di- to polyarch, *phloem* masses alternating with *protoxylem* patches. Cambial growth was limited. The *endodermis* is 2-layered as in *Equisetum*. The pitch is small and not hollow. The cortex, usually, had large air spaces as in aquatic plants.

Leaves of *Calamites* are placed in a number of form genera, e.g., *Annularia* and *Asterophyllites*. *Annularia* (Fig. 40.10) shows many species. There are 8 to 32 linear to spathulate leaves, about 5 mm to a few centimetres in length, in a whorl. There is a single median vein in each leaf and the leaves are fused to each other at the base. *Asterophyllites* has been found in petrified condition so that the anatomy of these narrow leaves are known (Fig. 40.11). The t.s. is triangular or semicircular. A small *vascular strand* in the centre is surrounded by a starch-bearing *bundle sheath*. This is encircled by the *mesophyll* tissue of long, radiating palisade cells. Finally, there is a continuous 1-layered epidermis all round with numerous stomata. The structure has some similarity with Pine leaves. A doubtful specimen of *Asterophyllites* has been recorded from the Po series of Indian Lower Gondwana.

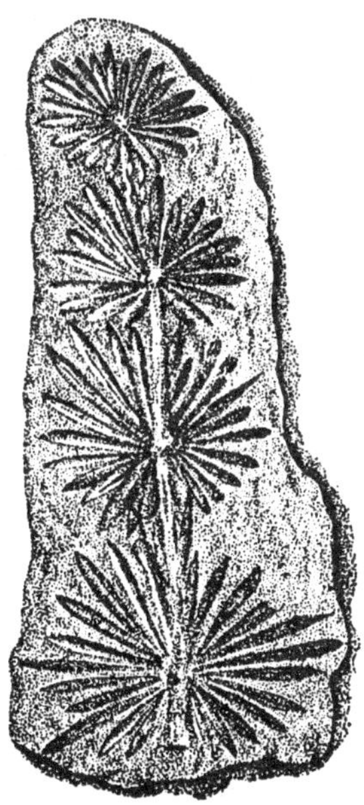

Fig. 40.10: *Annularia radiata* (*Calamites* leafy shoot) (redrawn after Dahlgren)

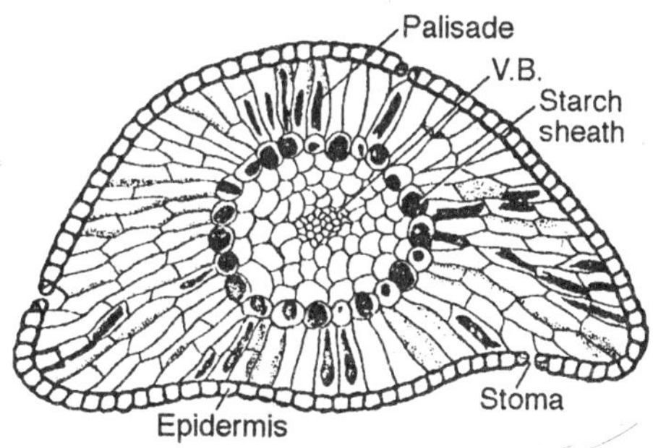

Fig. 40.11: T.s. of *Calamites* (*Asterophyllites*) leaf

Several types of strobili of *Calamites* are known and these have been placed in different form genera. The strobili were borne on special fertile shoots at the nodes in verticils of two or more in the axils of bracts. Each strobilus shows an articulated axis as in all Sphenopsids and the sporangiophores are borne in whorls at each node as in modern *Equisetum*. But, in *Calamites*, the sporangiophore whorls always have whorls of sterile bracts in between. The best known forms are *Palaeostachya* and *Calamostachys*. In *Palaeostachya* (Fig. 40.12A and B) a sporangiophore whorl alternates with a sterile bract whorl and the former whorl is axillary to the latter. But, the sporangiophores were not appendages to the bracts as they had independent vascular supplies coming from the axis. In *Calamostachys* (Fig. 40.12C and D)

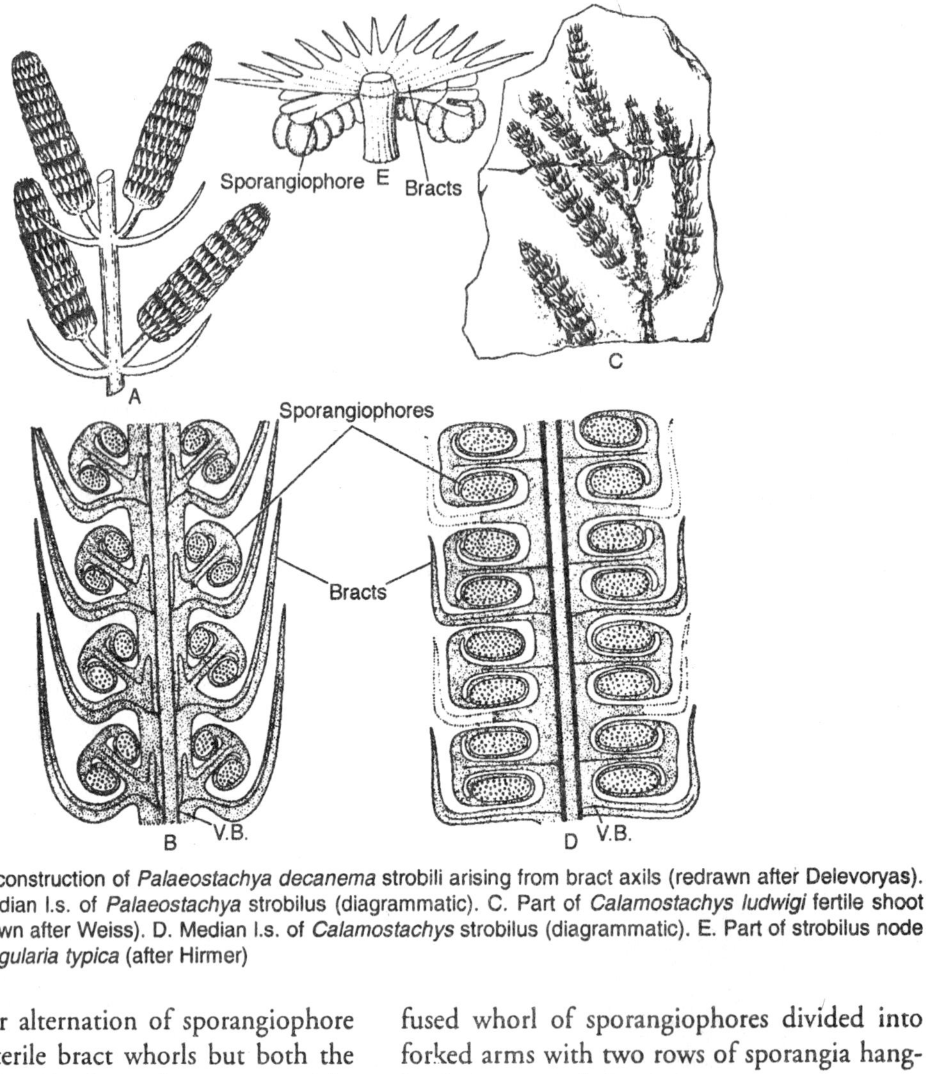

Fig. 40.12: A. Reconstruction of *Palaeostachya decanema* strobili arising from bract axils (redrawn after Delevoryas). B. Median l.s. of *Palaeostachya* strobilus (diagrammatic). C. Part of *Calamostachys ludwigi* fertile shoot (redrawn after Weiss). D. Median l.s. of *Calamostachys* strobilus (diagrammatic). E. Part of strobilus node of *Cingularia typica* (after Hirmer)

there is similar alternation of sporangiophore whorls with sterile bract whorls but both the whorls arise directly from the stem at right angles to it so that they look quite independent. The sterile bracts are usually laterally fused at the base and the numbers of sporangiophores and bracts vary. The number of sporangia peltately borne by each sporangiophore also varies but the usual number is four. Many variations of this structure are known. In some forms several whorls of sporangiophores occur between two whorls of sterile bracts. A very peculiar form in *Cingularia* (Fig. 40.12E), in which, it seems, a horizontally developed fused whorl of bracts has got fused with a similarly fused whorl of sporangiophores divided into forked arms with two rows of sporangia hanging below each fork.

Calamites shows homosporous as well as heterosporous strobili. In *Calamostachys binneyana* the spores are of irregular size but that does not necessarily singnify heterospory. But, in *Calamostachys casheana* some of the sporangia bear fewer spores which are four times as large as in others. This seems to be definite heterospory. *Macrostachya* (another form genus of strobili) shows very large spores (up to 0.4 mm in diameter) and one species, *M. infundibuliformis*, seems to be heterosporous. *Palaeostachya andrewsii* also shows very large spores (up to 0.32 mm) which may be megaspores. Recently, Dr. Baxter (reported by Holt *et al.* 1963) has found separate male and female strobili resembling *Calamostachys* in Kansas Carboniferous in which each megasporangium contains only one functional megaspore which seems to be retained within the megasporangium. In that case this comes at par with the *Pseudospermaphytes* like *Lepidocarpon.*

Family 3. EQUISETACEAE

The Equisetaceae differs from the Calamitaceae in that the plants are entirely herbaceous being devoid of any secondary tissue and in that the strobili do not contain sterile bracts between sporangiophores. They became more frequent after the extinction of the Calamitaceae but fossils members of Equisetaceae are known since the Upper Carboniferous. The fossils members are as follows:

Phyllotheca (Figs. 40.13), known from the Upper Carboniferous, is abundant during the Permian and extends to the Lower Cretaceous. The articulated stem was branched and the slender leaf whorls were fused at the base. Peltate sporangiophores with numerous sporangia similar to *Equisetum* were borne in clusters between leaf whorls. *Phyllotheca griesbacki* and *P. indica* are known from the Barakar and Raniganj series (Lower and Upper Permian) of Indian Lower Gondwana. *P. sahnii* (Fig. 40.13C) is known from the Karharbari series.

Schizoneura also had an articulated branch stem up to 2 metres high bearing catkin-like strobili terminally on some of the slender branches. These occurred during Upper Carboniferous to Jurassic. The jointed leaf whorl at each node is split into two large, flat, many-veined structures which are again split into 6 or 8 segments, each with one vein, *Schizoneura* is represented throughout the Indian Lower Gondwana (Upper Carboniferous to Lower Triassic). *Schizoneura gondwanensis* and *S. wardi* are known in the Karharbari,

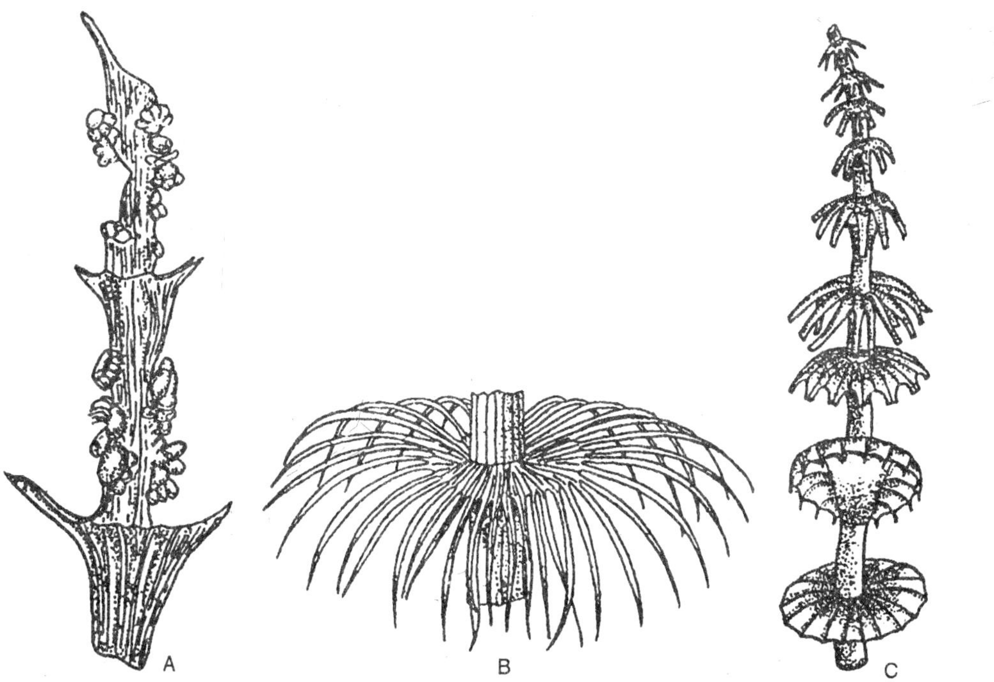

Fig. 40.13: *Phyllotheca*. A. *Phyllotheca delequescens* (after Schmalhausen) B. *P. equisetoides* (after Rasskazova). C. *P. sahnii* (after Saksena)

Barakar and Raniganj series (Lower and Upper Permian).

Stellotheca robusta (Barakar), *Raniganjia indica* and *bengalensis* (Raniganj) are similar specimens from Indian Lower Gondwana.

The Triassic *Neocalamites* has features combining Equisetaceae and the earlier Calamitaceae. The leaves were like *Annularia* but free to the base. *Neocalamites foxii* (Parsora—Middle Gondwana) is known from India.

Most clearly related to the modern *Equisetum* is *Equisetites* which spreads from the Upper Carboniferous to the Cainozoic. It resembled *Equisetum* in the scale leaves and the strobili but nothing is known about the anatomy. A Triassic form had stems 20 cm in diameter with 120 leaves in a whorl. *Equisetites rajmahalensis* is known from the Indian Upper Gondwana (Rajmahal series—Lower Jurassic).

Spores with elaters (*Elaterites, Equisetosporites*), a diagnostic feature of *Equisetum*, are also known as Mesozoic fossils.

Genus: *EQUISETUM*

All modern species of Equisetaceae are placed within the genus *Equisetum* (c.f. note below). Fossils specimens of *Equisetum* itself are known as far back as Upper Cretaceous.

Equisetum (horsetails) is a very natural and uniform genus but it is divided into two sections or subgenera: (1) *Equisetum* (*Eu-equisetum*) having stomata with accessory cells and at the same level with the other epidermal cells, leaves blunt; and (2) *Hippochaete* having stomata sunk below the level of the epidermis and structurally different, leaves pointed. Rothmaler (1944) has, accordingly, divided the genus into two genera—(1) *Equisetum* and (2) *Hippochaete*. But, otherwise, the species are so uniform that the creation of a new genus *Hippochaete* is not generally supported. Moreover, there is no cytological support. All the species have a uniform and stable chromosome number $n = 108$. Only the subgenus *Hippochaete* has got somewhat larger chromosomes.

The horsetails, despite the range of variation in branching and shoot dimorphism, are usually not mistaken for some other group, and most botanists recognize the one genus *Equisetum*. However, some specialists have proposed that at least two genera be recognized. Consequently, the genera *Equisetum* and *Hippochaete*, or even three genera, have been described. Hauke (1969a) has pointed out that there are some real and consistent differences between the two major groups of species, but that they are minor, and the number of similarities outweigh the differences. Therefore, he could see no gain in establishing two genera instead of one. Two subgenera, however, recognize the differences. Table 40.1 summarizes the main differences between the subgenera.

Table 40.1: Subgenera of *Equisetum*

Equisetum subgenus *Equisetum*

1. Aerial stems generally branched, annual or short lived.
2. Strobili borne on chlorophyllous vegetative stems or on nonchlorophyllous stems, lacking apiculum (not pointed).
3. Stomata on stems scattered in the furrows between the ridges, or in bands one to three stomata wide.
4. Stomata flush with the epidermis.
5. Upright, platelike lamellae on gametophytes.
6. Projecting antheridia, generally with more than two opercular cells.

   Examples: *E. arvense, telmateia, sylvaticum.*

Equisetum subgenus *Hippochaete*

1. Aerial stems generally unbranched, and may function for several seasons.
2. Strobili borne on chlorophyllous vegetative shoots, apiculate.

3. Stomata on stems in one or two longitudinal rows in the furrows between the ridges.
4. Stomata sunken below surface of the epidermis.
5. Upright, columnlike lamellae on gametophytes.
6. Antheridia sunken, with two opercular cells.

Examples: *E. hyemale, scirpoides, laevigatum*

32 species are recognized in Engler's Syllabus (1954). They are found all over the world except in Australia and New Zealand. Most species occur in the North Temperate zone but they are also known in the Arctic, the South Temperate and the Tropics of both the hemispheres. They are found in all habitats. Some species grow in or near ponds and marshes while some others are xerophytic. *Equisetum arvense* is a cosmopolitan species growing on the Himalayas. *Equisetum debile* with sparse branching grows along banks of rivers in sandy or swampy soil and is to be found even in the Gangetic plains. Other Indian species are *E. diffusum, E. ramossisimum, E. maximum, E. dubium* and *E. elongatum.*

The sporophyte

The sporophyte (Fig. 40.14) has an aerial part and an underground rhizome part. The rhizome is perennial, horizontal, much branched and often penetrates to more than a metre below the soil surface. It has distinct nodes and internodes and there are nodal whorls of small scale leaves which unite to form sheaths round the node. The branch primordia alternate with the leaves at the whorls. Such branch primordia on rhizomes develop into aerial shoots or subterranean branch rhizomes or into small, rounded *tubers* for vegetative propagation. The rhizome nodes also bear whorls of slender, much branched adventitious roots which usually function for one growing season only. Aerial branches arise from the upper surface of the rhizome and are usually annual although in some cases they may be perennial. Aerial shoots are usually less than a metre in height (may be even as small as a few centimetres, e.g., *E. scirpoides*) but in *E. giganteum* from South America it has been reported to be as long as 12 metres. But, even then, the stem is only 2 cm in diameter and trails on the underbushes. *E. debile* sometimes grows to two or three metres in moist conditions but the stem does not exceed a centimetre in thickness. *E. schaffneri* is reported to have stems up to 10 cm in diameter and 2 metres in length. In no case, however, has secondary growth been reported. The branches may be all sterile, some of the later ones forming cones on the apices. The sterile branches are green and have whorls of lateral branches at each node which may branch again. Many species have the strobili only on special and separate fertile branches.

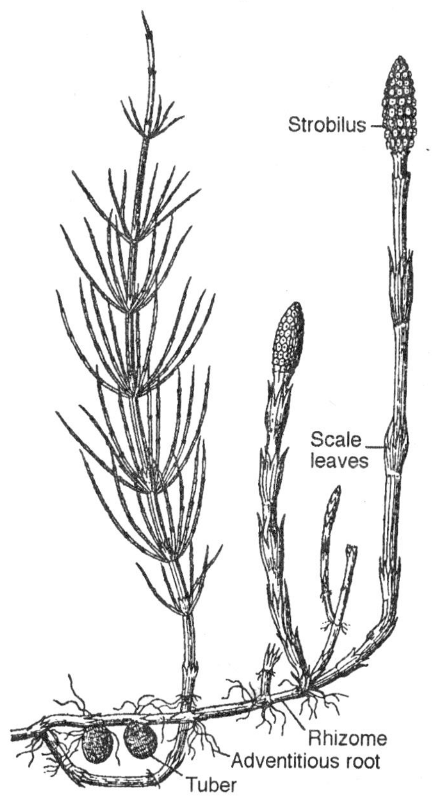

Fig. 40.14: *Equisetum arvense* sporophyte showing subterranean rhizome fertile and sterile shoots

Such fertile branches are usually brown, unbranched and have single strobili at their apices. These branches generally develop, mature and shed their spores before the vegetative branches appear. Some other species have green, branched, fertile shoots with a strobilus at the apex of each lateral branch. Fertile shoots of this type persist after the spores are shed and function as vegetative shoots. The stems of both the aerial and the subterranean parts are ribbed and the ribs alternate with leaves of the same number on the upper node. The leaves and the ribs on the upper internode, again, alternate with those below. The leaves are minute and scale-like and brown when mature. They form a whorl at each node and are fused laterally at the bases forming a sheath encircling the stem with only the tips free. These leaves are of a very simple construction with a single median vein and, although some chloroplastids and stomata are present, are of no photosynthetic use. The number of leaves vary according to the species and, of course, corresponds with the number of ribs. Each leaf, again, corresponds to a main vascular bundle the upper part of which passes to the leaf as the leaf trace. The branch primordia alternate with the leaves. Some of these branch primordia mature, developing the whorl of branches.

The terminal growth of rhizomes and branches is initiated by a pyramidal apical cell with three cutting faces. A t.s. of the stem (Figs. 40.15A and B) shows a clear *epidermis* of which the outer walls are cuticularized and impregnated with silica to give the characteristic rough feeling. Below each ridge there is a patch of *hypodermal sclerenchyma* which in *E. giganteum* extends to the stele. There is also another smaller patch of sclerenchyma below each furrow. The *outer cortex* is *chlorenchymatous* and is placed in a semicircle below each ridge sclerenchyma patch. The *stomata* are sunken in the *Hippochaete* subgenus and on level with the epidermis in the *Equisetum* subgenus. These are arranged in two vertical rows on the two sides of each ridge, just where the chlorenchyma meets the epidermis. A stomatal initial divides twice, then an inner pair forms the guard cells and an outer pair the accessory cells (Fig. 40.16A). The chlorenchyma is an efficient photosynthetic tissue. The *inner cortex* below the chlorenchyma is formed of thin-walled parenchymatous cells. There is one large longitudinal air cavity in it below each furrow forming a ring of *valecular canals*. The inner cortex is bounded on the inside by a continuous wavy 1-layered *endodermis*. In a few species (c.f., *E. giganteum*), the endodermis is broken up and develops so as to surround each vascular bundle separately. In a few others (e.g., *E. sylvaticum*) there is one continuous endodermis on the outside and another similar continuous one delimiting the pith.

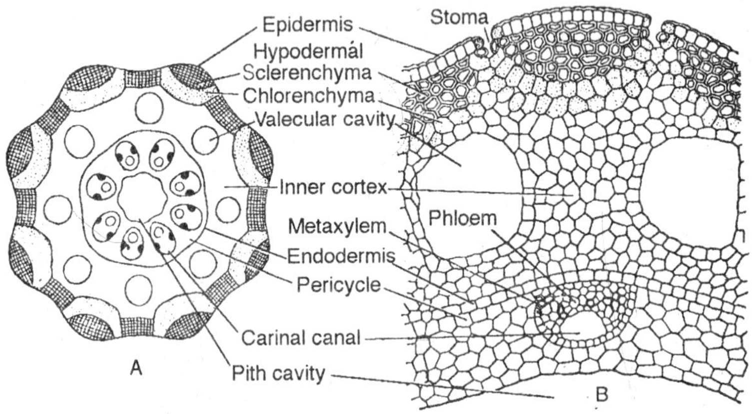

Fig. 40.15: *Equisetum* stem. A. T.s. (diagrammatic); B. A sector of the t.s.

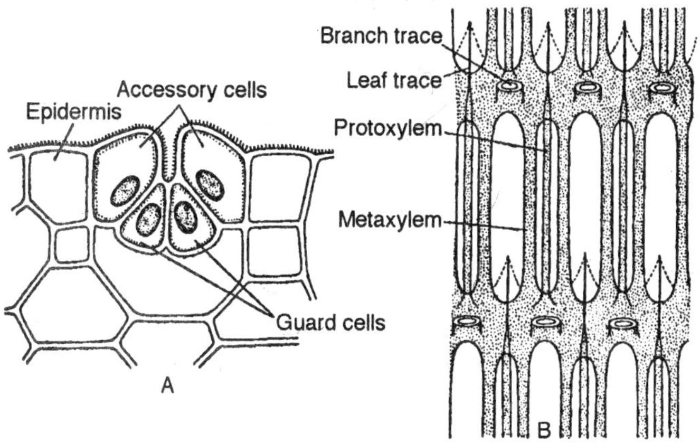

Fig. 40.16: *Equisetum*. A. A stoma in cross-section. B. The vascular skeleton (shown shaded)

Below this endodermis showing casparian strips there is a *pericycle* of one layer. The stele is remarkably reduced and feebly developed. There are as many vascular bundles as the number of leaves and they form a ring, there being one under each ridge. The vascular system (Fig. 40.16B) is actually siphonostelic as seen in the nodes. In the internodal space there is a longitudinal gap below each branch but these are not real branch gaps. A *branch trace* is given out from above the furrow at each node while the protoxylem of each bundle passes out as a *leaf trace* above the ridge at each node. The bundles are endarch, closed and collateral with the *xylem* feebly developed as in an aquatic monocot. The *protoxylem*, formed of annular and spiral tracheides, is mostly disorganized to form a *carinal canal*. The carinal canals, therefore, form an inner ring of intercellular spaces. The *phloem* with sieve tubes, having sieve plates on diagonal walls, is placed on top of the carinal canals. On both sides of the phloem are two small patches of *metaxylem* (scalariform tracheides). There are *medullary rays* between the vascular bundles. The *pith* mostly disorganizes and forms a hollow *central canal* at the internodal regions.

The slender, adventitious endogenous roots are tri- to hexarch with 3 to 6 *protoxylem* points surrounding a single axial *metaxylem* element (*exarch*) at the centre. The angles between protoxylems are completely filled up with *phloem*. There is no pericycle but there is a 2-layered *endodermis* the inner layer of which may behave as a pericycle in giving rise to root branches. The cortex is 3 to 4 layered, the outer layers of which are thickened and form the *exodermis* in maturer regions.

The nature of the sporangiophores has been very much disputed as has been that of the sterile bracts of the Calamitaceae. Some hold that the bracts are sterile sporangiophores or sporangiophores are fertile leaves, i.e., sporophylls. But the similarity between *Hyenia* and *Psilophyton* sporangiophores should be noted. It may also be argued that the sterile bracts of the Calamitaceae invaded the strobilus from below and *Equisetum* has either lost them in course of evolution or never had them. Thus, the sporangium bearing organ of *Equisetum* is similar to that of the primitive Psilophytales and should be called a *sporangiophore* and not a sporophyll (c.f. further discussion in Chapter 42).

In the *strobilus* or *sporangiferous spike* or *cone* of *Equisetum* (Figs. 40.17A, B and C) there are several whorls of these densely crowded, peltate appendages bearing sporangia or *sporangiophores* on a stem axis. Sporangiophores have been found abnormally to become leaf-like. Immediately below the whorls of sporangiophores the axis bears a small or large ring-like outgrowth, the *annulus*. This structure may be explained as a vestigeal reminder of the

Calamitean bracts or as an aborted ring of sporangiophores since the annulus sometimes bears small sporangia. The *sporangiophore* (Fig. 40.17D) has a slender stalk bearing a peltate, hexagonal disc on the under surface of which there are 5 to 10 elongate, cylindrical, hanging *sporangia* with rounded apices. Mature sporangia open along a single longitudinal line of dehiscence. The manner of development of the sporangium is eusporangiate as it is not entirely formed from a single initial. A superficial cell of the young sporangiophore divides periclinally and the inner cell forms only *sporogenous tissue* while the outer cell contributes partly to the sporangenous tissue and partly to a portion of the *sporangial jacket* radial to the sporogenous tissue. The rest of the jacket is derived from the cells adjacent to the original initial. The jacket is several layers in thickness in which the innermost layer becomes the nutritive *tapetum.* In the mature sporangium the tapetum as well as the inner layers of the jacket are disintegrated and only two jacket layers remain. The outer layer of these two has spirally thickened walls. The sporangenous tissue is transformed into *spore mother cells* but one-third of these also disintegrate before the actual reduction division. The remaining spore mother cells separate from one another, become rounded, float in the plasmodial liquid and then develop the spores in tetrads.

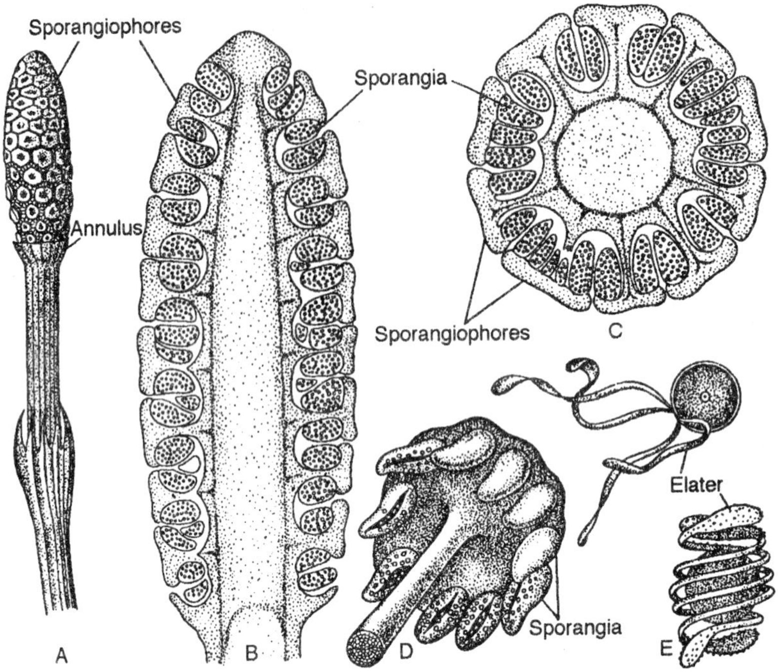

Fig. 40.17: *Equisetum*. A. Strobilus. B. Median l.s. of strobilus. C. T.s. of strobilus. D. A sporangiophore showing dehiscence of some sporangia. E. Dry and wet spores showing movement of elators

The gametophytic generation

The spore secretes a wall which shows four concentric layers. Innermost is the delicate cellulose *intine*, outside this is the firmer *exine.* This is surrounded by a delicate *cuticular layer* and outermost is the *epispore* or the *perispore.* The epispore splits spirally into four ribbon-like strips with spoon-like tips which split away from the remainder of the wall except from a common point of attachment and become the *elaters* (Fig. 40.17E). The elaters are hygroscopic. They coil when moist and uncoil when dry. When the sporangium dehisces the spores fall out in lumps being intertwined by the elaters. The spores are fairly large and contain chloroplasts besides the nuclei.

The gametophytes of *Equisetum* generally develop in muddy soil along streams or water courses. The spores are viable only for 10 to 12 hours. During germination the exine splits and is cast off (Fig. 40.18A). The first division of the cell divides it into a small and a large cell (Fig. 40.18B). The small cell loses its chlorophyll and becomes the first rhizoid while the large cell gradually develops into the new gametophytic thallus. Soon a cushion-shaped tissue is formed with rhizoids on the under surface. The top cells of the cushion are meristematic, green and give rise to irregularly shaped upright plates. Cells of the lower portion are colourless. Gametophytes of *Equisetum* are long-lived although the first sex organs develop within a month or 40 days.

Equisetum spores are all alike and gametophytes of most species are normally monoecious (Fig. 40.18C). It is, therefore, homosporous and homothallic. But, it has been found that certain species under certain circumstances do produce heterothallic male and female gametophytes (Fig. 40.18D). Schratz (1928) found that in *E. arvense*, when the prothalli grow crowded together and conditions of growth are unfavourable, all the prothalli are small and male as shown in Figure 40.18D. But when conditions are favourable 50% of these become larger and develop archegonia as shown in that figure. These latter (i.e., those with archegonia), however, may later develop antheridia if fertilization fails. The other 50% always remain male; however favourable the cultural condition is made. He, therefore, concluded that there is some genetic sex-controlling factor so that 50% are definitely male and the other 50% are female although the latter do not lose their male potentiality. This may be termed as **incipient heterospory** as, even though all the spores are apparently alike, they are latently sex-determined. But, others have concluded that *Equisetum* is normally mono-

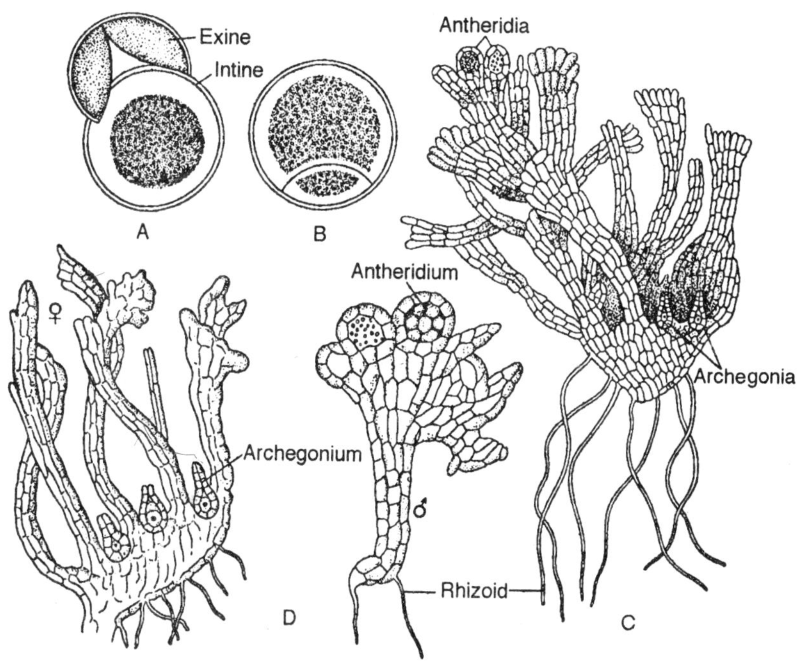

Fig. 40.18: *Equisetum*. A-B. Stages of germination of spore. C. Monoecious gametophyte. D. Heterothallic gametophytes of *Equisetum arvense*

ecious. When starved, growing in unsuitable condition they remain small and produce antheridia only. When vigorously growing they produce archegonia first and antheridia later, i.e., they are monoecious. This question of latent heterothallism is, however, still argued and it has been suggested that 50% of the spores show different physiological activity (showing less power of oxidation, possessing a reserve fat, etc.) and are female while the others are male.

As described, the prothalus has a lobed structure. The *archegonia* and the *antheridia* develop from the superficial cells of the thallus meristem when lobation begins. The mature archegonia are formed at the bases of the lobes or between them while the antheridia are usually formed on the upper portions of the lobes though, sometimes, they may also be found lower down. In the development of the antheridium, the *antheridium initial* divides periclinally into an outer *jacket initial* and an inner *androgonial initial.* The jacket initial develops into a *jacket* layer, one cell thick, with a triangular *opercular cell* at top as in the Eusporangiates. The androgonial initial forms a *spermatogenous tissue* which develops a large number of *sperm mother cells* or *androcytes* (Fig. 40.19A). Within the sperm mother cell two *blepharoplasts* are visible and when the sperm mother nucleus divides each of the two developing sperms receive one of them. Each sperm mother cell form two *spermatozoids.* There are about 256 sperms in an antheridium each of which is rather large and shows the blepharoplast coiled round with numerous cilia at the end (Fig. 40.19B). The sperms are liberated by a rupture of the top of the antheridium through the opercular cell.

The superficial *archegonial initial* also similarly divides periclinally into a *primary cover cell* and a *central cell.* The cover cells forms a protruding neck of 3 or 4 tiers of 4 cells each. The central cell divides transversely to form a *primary canal cell* and a *primary ventral cell.* The primary canal cell may or may not divide to form two *canal cells.* The ventral cell divides to form a small *ventral canal cell* and a large *egg cell* (Fig. 40.19C).

Finally, all axial cells except the egg disintegrate and the tip neck cells gape out to make a passage for the sperms (Fig. 40.19D).

Fertilization takes place in the usual way and a *zygote* is formed.

The new sporophyte

Unlike most other Pteridophytes, several sporophytes usually grow on the same prothallus. Each zygote divides transversely and then at right angles giving rise to quadrants (Fig. 40.20A) as in the Leptosporangiates but these quadrants do not give rise to the stem, root, cotyledon and foot as there. There is no suspensor. The lower part of the embryo is the *foot* from the side of which the *root* comes out. The whole upper part is the *shoot* (Fig. 40.20B).

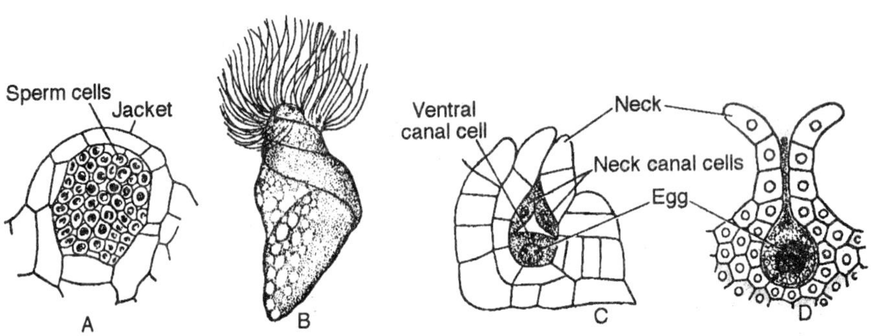

Fig. 40.19: *Equisetum*. A. Mature antheridium. B. A sperm (drawn after Sharp). C. An archegonium (drawn after Jeffrey). D. A mature archegonium ready for fertilization

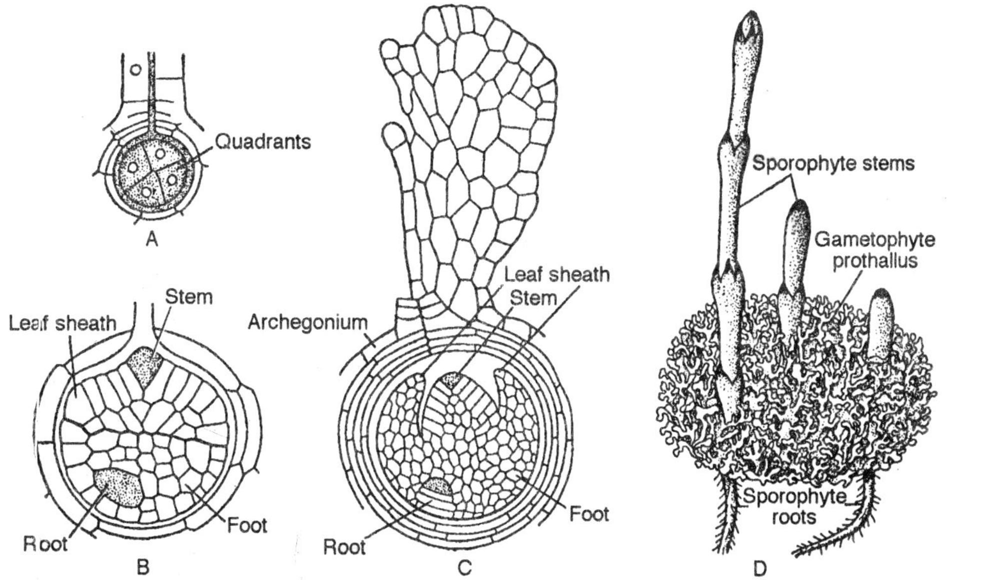

Fig. 40.20: *Equisetum:* A-C. Stages of development of embryo within venter. D. Young sporophytes developing from a gametophyte

The first *leaf sheath* of about three scale leaves is differentiated round the stem apex (Fig. 40.20C). The stem soon pushes through the calyptra and the root grows out through the base. A number of such sporophytes may be seen developing on a single prothallus (Fig. 40.20D).

Fig. 40.21 shows the life cycle of Equisetum (dioecious cycle).

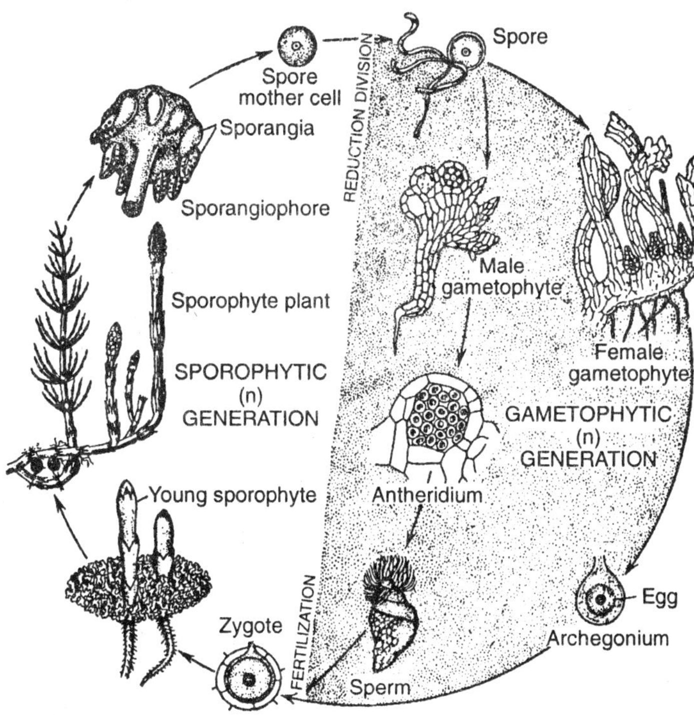

Fig. 40.21: Life cycle of *Equisetum arvense* (dioecious cycle)

Chapter 41

Noeggerathiopsida

The class Noeggerathiopsida or the *Noeggerathiales* form a group of plants extremely peculiar and not yet fully known. Their affinities are rather uncertain. They have been placed within the Sphenopsida, within the Filicopsida and also between the Lycopsida and the Sphenopsida. But, the more general opinion is to consider them as an isolated group of *pteridophyta incertae sedis* (i.e., Pteridophyta of uncertain affinities). It is, therefore, reasonable to consider them as a separate *class* of uncertain affinities as proposed by Pichi-Sermolli.

Two genera, *Tingia* with its strobilus *Tingiostachya* and *Noeggerathia* with its probable strobilus *Noeggerathiostrobus* are recognized. There are also some other fossils of fertile shoots which have been ascribed to this group. They are known from Carboniferous to Triassic in continental Europe. Leaves are cycad-like with repeated dichotomous venation and with two rows of leaves on an axis resembling compound leaves of ferns. The strobili were heterosporous. One strobilus (Fig. 41.1) bears two compact rows of semicircular sporophylls attached to an axis. The margins of the megasporophylls are serrate and

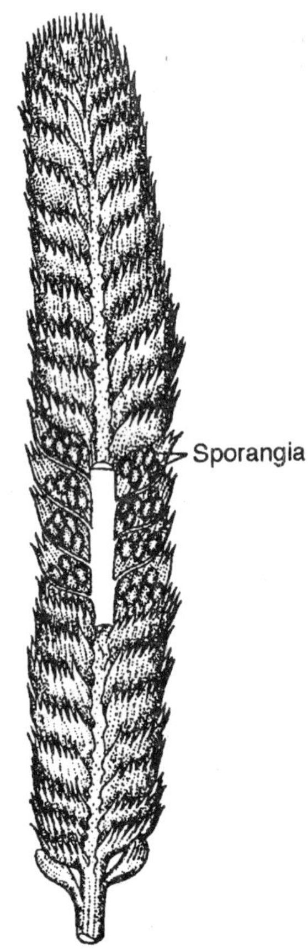

Fig. 41.1: Restoration of *Noeggerathiostrobus bohemicus* (after Halle)

fringed with numerous sporangia, each of which latter contains 16 megaspores. Microspores and also megasporangium containing only a single megaspore have been found in other specimens. The megaspores were eight times the size of microspores and both of them were almost smooth as in the Sphenopsida. In *Tingia* the leaves were in four rows. The strobili occurred in pairs and had four rows of free sporophylls with upturned tips, on the upper surfaces of which quadrilocular sporangia were borne possibly on stout stalks.

Chapter 42

Filicopsida

The class Filicopsida, commonly called the Ferns, is the most predominant group of the Pteriodophytes today with 250 to 300 living genera and more than 9,000 widely distributed species according to various estimates. The group has also been called *Filicinae, Filices, Pteropsida* or *Pterophyta.* As Pteropsida or Pterophyta they signify the main line of evolution of terrestrial plants, i.e., the megaphyllous line leading to the seed habit. As such, the term Pteropsida and Pterophyta have been used by several authors to include within them the Spermaphyta, i.e., the Gymnosperms and the Angiosperms. The term Filicopsida with the proper class ending suffix *-psida* seems better to signify the ferns only. The term *Filicophyta* (as a division) has also been used by those who no longer want to retain in the division Pteridophyta.

The Filicopsida probably arose in the Lower Devonian almost at the same time as the other groups of Pteridophyta but did not attain the prominence of the Lycopsida and the Sphenopsida in the Carboniferous. They became numerous only during the Triassic and the Jurassic when the other major Pteridophytes were getting extinct. Most of the modern families of Filicopsida evolved very late. Polypodiaceae, by far the most predominant group today, began in the Jurassic. But, almost not numerous, their rate of evolution was perhaps more rapid since the seed plants evolved from this megaphyllous line possibly before the Carboniferous. The Palaeozoic had once been called the 'age of ferns' because of a preponderance or fern-like leaves. But most of those 'fern' leaves have been assigned now to the 'seed ferns' or the *Pteridosperms* (Gymnosperm). These 'seed ferns' possibly arose from the same stock from which the Filicopsida arose or might have even evolved from some members of the Filicopsida itself.

It is usual to divide the Filicopsida into three groups: (1) The Palaeozoic fossil forms are so distinctive that they are placed under one group—the *Primofilices.* The other fossil and living forms are placed under (2) the *Eusporangiatae* (the ancient ferns) where the sporangium is massive growing from a group of cells and (3) the *Leptosporangiatae* (the modern ferns) where the sporangium is delicate arising from a single initial cells. But, the family

Osmundaceae, placed within the Leptosporangiatae, shows characters intermediate between the Eusporangiatae and the Leptosporangiatae so that Hirmer (1927) proposed a separate group *Protoleptosporangiatae* for it. In the last edition of Engler's Syllabus (1954) the 'class' Filices has been divided into four 'subclasses'.

1. Primofilices
2. Eusporangiatae
3. Osmundidae
4. Leptosporangiatae.

Pichi-Sermolli (1959), however, argues that Goebel's concept of the division into Eusporangiatae and Leptosporangiatae is artificial. The two orders Ophioglossales and Marattiales placed together under Eusporangiatae are not necessarily closely related. He, therefore, proposes to divide the class Filicopsida into seven subclasses:

1. Primofilicidae
2. Ophioglossidae
3. Marattiidae
4. Osmundidae
5. Filicidae
6. Marsileidae
7. Salviniidae.

But, considering that the affinities of some of the genera included within the Primofilicidae have so far baffled all Pteridologists and Palaeobotanists, even this classification is not complete. In this book, considering the modern trends, the following classification has been adapted:

Class: Filicopsida

Subclass: 1. Primofilicidae
2. Ophioglossidae
3. Marattiidae
4. Protoleptosporangidae
5. Leptosporangidae.

However, most recent classification of Filicophyta /Filicopsida are presented below:

FILICOPHYTA: Living and extinct ferns; plants showing a conspicuous alternation of generations (in modern representatives at least); sporophyte most conspicuous and often elaborately developed; megaphylls present; stems protostelic or siphonostelic, often with complex vascular cylinders; sporangia terminal on ultimate axes, terminal on veins, marginal, or on abaxial surface of fronds; eusoporangiate, or more commonly leptosporangiate in living species; homosporous, a few heterosporous; gametophytes: (1) majority exosporic and green, (2) others exosporic, nonchlorophyllous, and subterranean, and (3) endosporic (restricted to heterosporous types); multiflagellate free-swimming sperm; embryo exoscopic, endoscopic, or intermediate.

OPHIOGLOSSALES: Living ferns, meager fossil record; sporophyte axis usually short and fleshy; stem siphonostelic; fronds simple or pinnately compound; vernation noncircinate to weakly circinate (*Botrychium*); each fertile frond consists of a fertile segment or spike, bearing sporangia, and a sterile or vegetative segment; eusporangiate, producing great quantities of spores per sporangium; homosporous; gametophytes subterranean, bisexual, nonchlorophyllous, tuberous or wormlike with endophytic fungus; embryo exoscopic or endoscopic.

OPHIOGLOSSACEAE: Characteristics as in Ophioglossales.

Ophioglossum, Botrychium, Helminthostachys.

MARATTIALES: Living and extinct ferns; sporophyte stem in most erect and short, or may be dorsiventral; stem with complex dictyostelic vascular cylinder; fronds commonly large, simple pinnate to tripinnate, and circinate in vernation; paired, clasping stipules at base of each leaf; eusporangia free and grouped into elongate sori or united into synangia on abaxial surface of fronds; many spores formed per sporangium; homosporous; gametophyte terrestrial, green, cordate to ribbon shaped, bisexual, and with endophytic fungus; endoscopic embryo.

MARATTIACEAE: Characteristics as in Marattiales.

Representative genera: *Angiopteris, Marattia, Psaronius* (extinct), *Eoangiopteris, Scolecopteris* (from genera for sori).

FILICALES: Living and extinct plants, of diverse growth habits and habitats; stems vary from protostelic to intricately dictyostelic; fronds simple to compound pinnate; sporangia scattered or grouped into sori; sori marginal or on abaxial side of fronds; sori with or without a protective structure, the indusium; leptosporangiate, most with a definite dehiscence mechanism, the annulus; spores numerous to few per sporangium, tetrahedral or bilateral; homosporous, gametophytes primarily green, exosporic, commonly thalloid, some filamentous; embryo "prone."

MARSILEALES: Living and extinct ferns; grow on damp soil or submerged in water; solenostelic; lamina consisting of four pinnae, bipinnate, or without pinnae; circinate vernation; sori enclosed in sporocarps; leptosporangiate; heterosporous; microsporangia and megasporangia in same sporocarp, endosporic gametophytes.

MARSILEACEAE: Characteristics as in Marsileales.

Marsilea, Regnellidium, Pilularia

SALVINIALES: Living and extinct ferns; small plants that float on surface of water, sori enclosed in specialized indusia ("sporocarps"); leptosporangiate; heterosporous; microsporangia and megasporangia in separate sporocarps; endosporic gametophytes.

SALVINIACEAE: Characteristics as in Salvinials.

Azolla, Salvinia.

Subclass 1. PRIMOFILICIDAE

The affinities of the fern-like Devonian and Carboniferous fossils included under the Primofilices are very perplexing. Their affinities with the modern ferns seem to be very uncertain. Sufficient data are not available to classify them into Eusporangiates and Leptosporangiates. These Palaeozoic ferns may be divided into four orders:

1. Protopteridales
2. Coenopteridales
3. Cladoxylales
4. Archaeopteridales.

The affinities of some of the genera included within the above are so complex that it is even doubted whether they should be considered as ferns or as a separate group of precursors of the Gymnosperms—the *Progymnosperms*. They are, however, still to be retained within the Pteridophytes as no 'seed' structure has so far been demonstrated in them.

A number of leaf and stem form genera of uncertain affinities (c.f. Palaeobotany) are known from the Indian Lower and Upper Gondwanas, viz., *Gondwanidium, Sphenopteris, Pecopteris, Alethopteris, Rhizomopteris, Onychiopsis, Weichsiella, Solenostelopteris, Dictyostelopteris, Cyclopteris.* Some of them may belong to the Primofilicidae.

Order 1. PROTOPTERIDALES

Protopteridium (Figs. 42.1A and B) from the Middle and Upper Devonian of Eastern Canada, Scotland, Belgium, Bohemia and China almost approaches the Psilophytales in simplicity with fern-like vegetative characters. The stem is advanced in having a sympodially branched axis with the branches divided dichotomously. Some of the ultimate divisions are flattened into leaf-like appendages. There are six to eight species under the genus. In *P. hostimense* (Fig. 42.1B) the ultimate branches are pinnate and bear fern-like lamellae near the tips with tissue formed between pinnate segments (c.f., 'webbing' in Chapter 42). These ultimate branches may even be circinately coiled. Oval sporangia are borne at the tips of some of the smallest subdivisions and on one side of such a sporangium there is a specialized band of cells along which, probably, it dehisced.

The steles of the Protopteridales are of the actinostelic type. The tips of the arms show patches of protoxylem mixed with parenchyma. There are a number of other genera, viz., *Arachnoxylon, Reimannia, Iridopteris, Aneurophyton, Eospermatopteris* (all from Middle and Upper Devonian) all of which show complicated, lobed xylems (actinostele). *Iridopteris* stem (Fig. 42.1C) is very unlike the Psilophytean type. It had a five-lobed xylem, gave out leaf traces and had two parenchymatous patches on each arm. *Eospermatopteris* developed a tree some six metres high, a metre in diameter at base and bore sporangia which looked like seeds. It is often thought to be the same as *Aneurophyton. Rhacophyton,* formerly supposed to be a *Coenopteric* with phyllophores (c.f., below) is now supposed to belong to this group. It is interesting that *Rhacophyton* shows the habit of *Trimerophyton.*

There is no evidence of heterospory in any of the Protopteridales. The stem character may be linked with the Coenopteridales.

Order 2. Coenopteridales

Coenopteridales is a large assemblage of fern-like plants mainly from the late Palaeozoic (Permocarboniferous) sometimes extending to Devonian. It is an artificial assemblage which may be divided into the families *Stauropteridaceae, Zygopteridaceae, Botryopteridaceae* and *Anachoropteridaceae.* Some authorities favour dividing the group into two orders: the *Phyllophorales* and the *Inversicatenales.* The former (Stauropteridaceae and Zygopteridaceae) includes those in which the main rachis of the

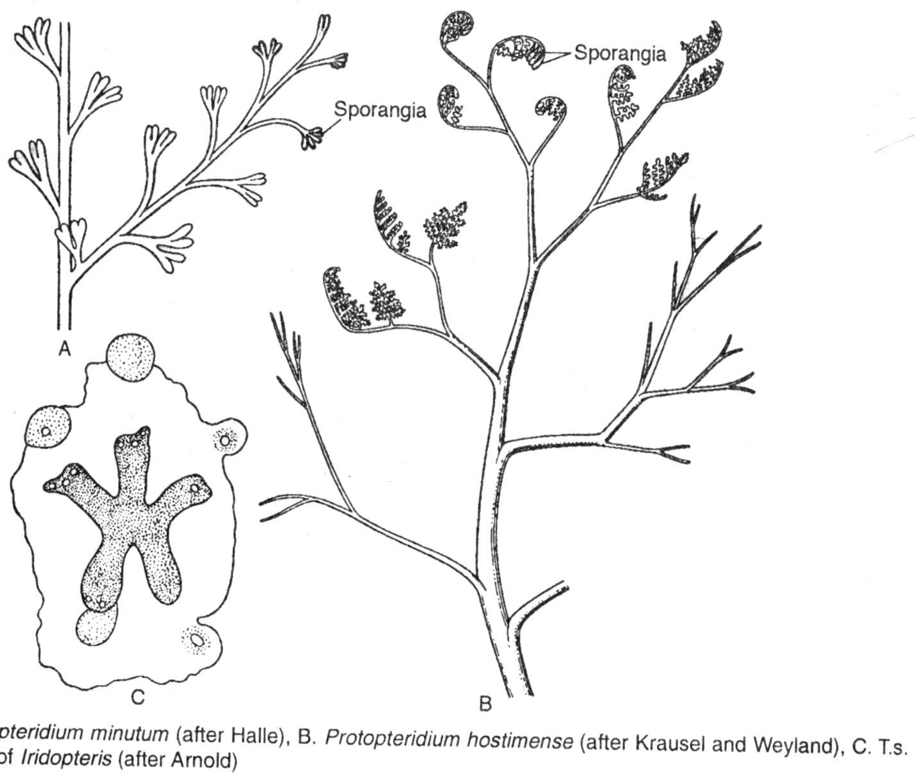

Fig. 42.1: A. *Protopteridium minutum* (after Halle), B. *Protopteridium hostimense* (after Krausel and Weyland), C. T.s. of stem of *Iridopteris* (after Arnold)

large compound leaves is a *phyllophore* which is somewhat intermediate between a stem and a rachis. The branch raches are borne on the *phyllophore* not in one plane but on different planes like branches on a stem axis. In the Inversicatenales (Botryopteridaceae and Anachoropteridaceae) the main leaf rachis is normal with branches in one plane only.

Phyllophorales

Family 1. STAUROPTERIDACEAE

Stauropteris (Fig. 42.2A) is abundantly represented in the Lower and the Upper Carboniferous. The plant is bush-like with highly branches phyllophores which bear alternate branch pairs in different planes and each branches again branches similarly. At each branching there are two scale-like structures called *aphlebiae.* The ultimate ramifications of *S. oldhami* do not bear lamellae but each end in a spherical *sporangium* with multilayered wall. The sporangium (Fig. 42.2B) shows a *stomium* on top through whch the spores escaped. This appears to be a Psilophytean character and this species seems to be homosporous. But *S. burntislandica* (Lower Carboniferous) has definitely been found to be heterosporus. There is no annulus but the *stomium* for spore escapement is present. Its *megasporangium,* called *Bensonites,* is flask-shaped and is about 1.3 mm long. Its lower part is sterile and there are two *megaspores* near the tip. The stem of *Stauropteris* shows a four-lobed *xylem* with *phloem* between the lobes. The lobes sometimes separated from one another forming four separate xylem patches. *Stauropteris,* with its heterospory, is one of the most advanced among the Primofilices and may be on the main line of the evolution of the Pteridosperms.

Family 2. ZYGOPTERIDACEAE

Some 17 form genera with prominent phyllophores and occurring from the Middle Devonian to the Permian are included within the Zygopteridaceae. The type genus is *Zygopteris* while an important leaf genus is

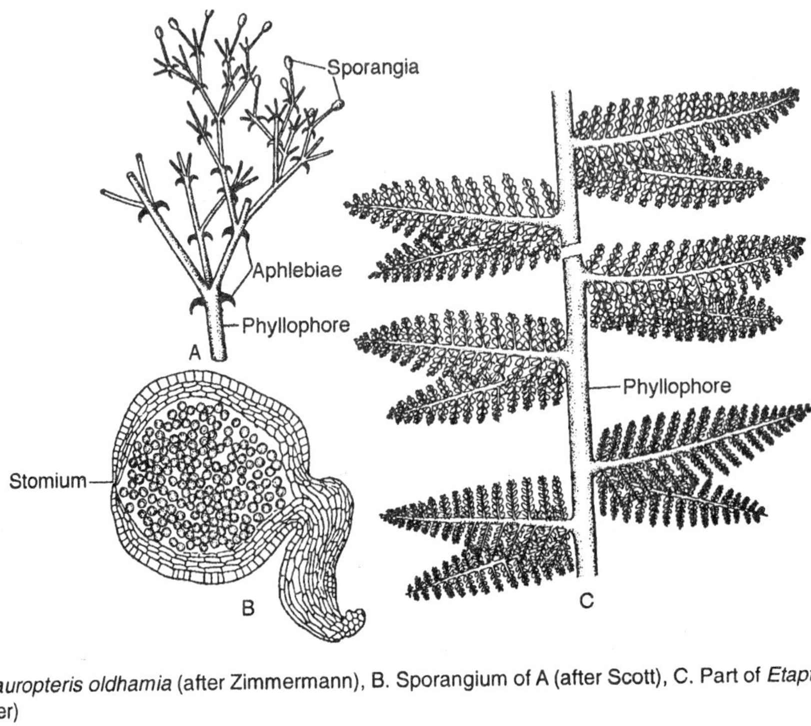

Fig. 42.2: A. *Stauropteris oldhamia* (after Zimmermann), B. Sporangium of A (after Scott), C. Part of *Etapteris* leaf (after Hirmer)

Etapteris (Fig. 42.2C) whose stem is probably *Botrychioxylon.* Other important related form genera are *Asteropteris, Ankyropteris,* etc.

The type genus *Zygopteris* and some other genera probably had stems which were erect or prostrate or trailing with a more or less complicated central cylinder which may be termed as actinostelic although the central core sometimes contained parenchyma mixed with tracheides. The protoxylem groups were at the tips of the lobes. The stellate primary xylem was encircled by radially arranged secondary wood elements.

The most important characteristic of the group was the elaborately branched leaf and the distinctive vascular organization in its petiole. The main axis of the frond is the *phyllophore* which was sometimes 2 cm in diameter and on which the secondary raches were in two to four rows in a vertical plane at right angles to the phyllophore axis. The pinnae of the second order were again turned at right angles to the axis on which they were borne. In *Etapteris* (Fig. 42.2C) the fronds were very large and the secondary and tertiary pinnules were on the same plane. The phyllophore had an H-shaped bundle (Fig. 42.3A), the uprights of the H being sometimes curved inwards. There were protoxylem patches on the tips of each of the four extremities of the H each of which, again, gave rise to a leaf trace. The traces from the extremities of each vertical bar joined to form a strap-like plate (shown left on the figure) which again divided higher up to form two C-shaped strands one of which passed to each branch. The appearance is so characteristic that the Zygopterids may be identified only if the phyllophore bundles are known.

Fertile pinnules were mixed with sterile ones. The sporangia of *Etapteris lacattei* (Fig. 42.3B) were borne in pedicellate tufts along branch raches which lacked expanded blades. The sporangia were ellipsoid, about 2.5 mm long and borne in tufts of three to eight. The wall was multilayered. There was an annulus of thickened cells extending from base to apex on both sides constituting a vertical annulus of unique type. The spores were homosporous and resembled those of ordinary ferns.

Other types of fructifications have also been attributed to the Zygopteridaceae. One is *Corynepteris* in which the sporangia were sessile, large, ovate and five or six of them were grouped together round a central point to form something like a *synangium* (Fig. 42.3D). These synangia were arranged on a stalk (Fig. 42.3C). The annulus extended over the point of contact of the sporangia and then over the apices so that dehiscence was probably around the edge of the contact. Suggestions have been made that this shows a connection between Zygopteridaceae and Marattiaceae.

Inversicatenales

Family 3. BOTRYOPTERIDACEAE

The Botryopteridaceae, known principally from the Permocarboniferous, has the simpler

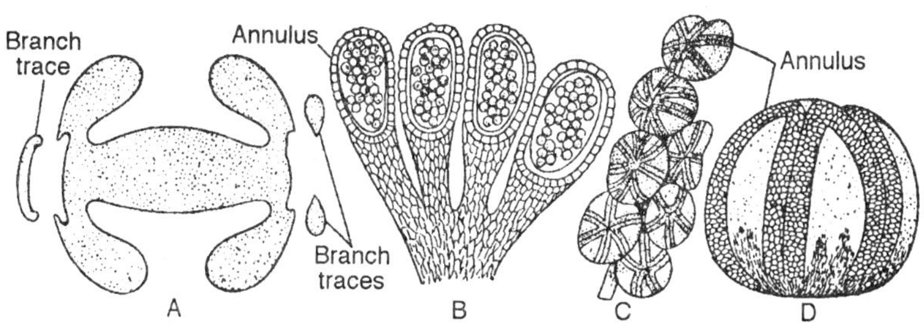

Fig. 42.3: A. Stele (shaded) *Zygopteris* phyllophore with traces of branches (after Bertrand), B. Sporangia of *Etapteris lacattei* (after Renault), C–D. *Corynepteris corallioides* sporangia on rachis and a single sporangium

construction of the Inversicatenales although the more complex Zygopteridaceae is actually more primitive in fossil records. There was no phyllophore and the stem anatomy was much less complex.

Botryopteris (Fig. 42.4A) is the principal genus with some 20 species spread over European and American Permocarboniferous, specially in coal balls. The stem was slender, only a few mm in diameter, was much branched and bore spirally arranged fronds sometimes showing 2/5 phyllotaxy. Fronds forked at least once and the pinnae were borne on the final branches. In the stem a broad homogeneous cortex surrounded the stele without any endodermis. The stele was protostelic with a layer of phloem surrounding a cylindrical mass of xylem. The xylem was mesarch with 2 or more protoxylem points some distance in from the surface. In the rachis the branch traces were curved with three arms resembling a curved E (shown in Fig. 42.4A). The sporangia were oval or pyriform, short-stalked and borne in clusters (sometimes even a thousand sporangia in a cluster) directly on the branches of the fertile rachis (Fig. 42.4B). The annulus was plate-like on one side of sporangium somewhat like *Osmunda*. The clusters look like sori but probably they were more akin to the pedicellate sporangia of the Zygopterids. There was a large number of spores of the same size (homosporous).

Grammatopteris from the lower Permian of France may be placed within the Botryopteridaceae because of the simple stele and the absence of phyllophores. It had a slender protostelic stem inside a cortex showing leaf traces and a thick armour of leaf bases and adventitious roots. Although unlike Osmundaceae, a relationship of the Zygopteridaceae with the Osmundaceae through *Grammatopteris* has been suggested by Sahni on a close study of the anatomy of the latter.

Family 4. ANACHOROPTERIDACEAE

Anachoropteridaceae, also from the Permocarboniferous, is another Inversicatenale family. Several genera, viz., *Anachoropteris*, *Tubicaulis*,

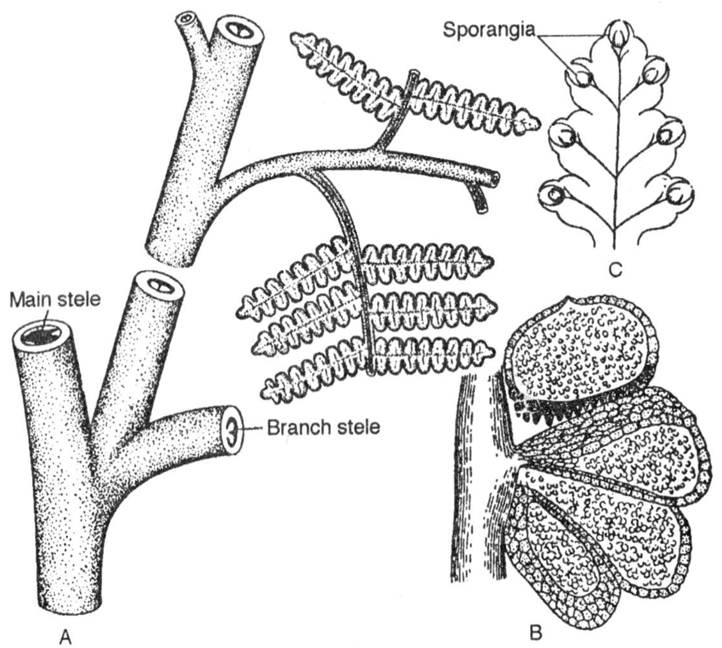

Fig. 42.4: A. *Botryopteris* (stelar structure also shown) (after Delevoryas and Morgan), B. Sporangia of *Botryopteris forensis* (after Renault), C. Fertile pinnule of *Anachoropteris pulchre* (after Hirmer)

Gyropteris, are included here. *Tubicaulis* shows a stem about 1 cm in diameter. The central xylem was roughly terete and mixed with parenchyma. The cortex of the stem and the petiole contained air spaces. The petiole and rachis of both *Tubicaulis* and *Anachoropteris* had conspicuous C-shaped bundles, the convex side of which was turned towards the stem (reverse of modern ferns) and the two or more protoxylem groups were on the convex side.

Sporangia were borne in tetrads on the apices and along the margins of flattened leaf-like pinnules at the ends of veins (Fig. 42.4C). Each tetrad was surrounded by a cup forming something like a synangium. The sporangial wall seems to be 1-layered and there was no individual annulus although the whole synangium opened by spreading apart of the sporangia and apical ruptures. There was a large number of small spores of the same size (homosporous).

The fertile pinnule of *Anachoropteris* seems to be the nearest approach to living ferns, specially Marattiales.

Order 3. Cladoxylales

Cladoxylales is represented by the type genus *Cladoxylon* with about a dozen species from Middle Devonian to Lower Carboniferous. The position of the group has been greatly disputed. Scott placed it within the Pteridosperms for its polystelic secondary wood. Pichi-Sermolli is of opinion that *Protopteridium, Stauropteris, Archaeopteris* and *Cladoxylon* do not belong to the Filicopsida but to a separate group (*Progymnosperms*) which is on the line of the evolution of the Pteridosperms. On the other hand, its primary xylem and leaf traces show a link to the Zygopterids while the fertile leaves are Psilophytean. It is better leaf as another problematic group.

The stem of *Cladoxylon* (Fig. 42.5A), up to

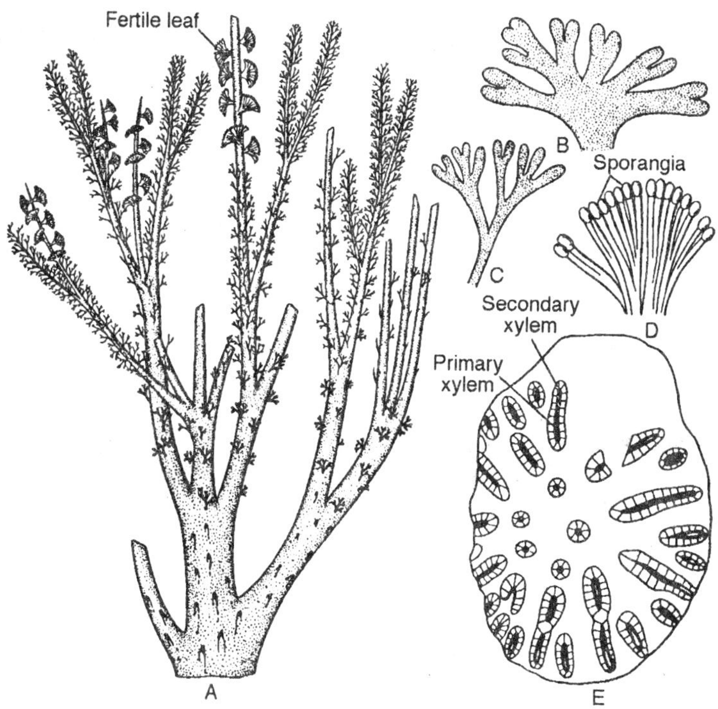

Fig. 42.5: *Cladoxylon scoparium* (after Kräusel and Weyland). A. Reconstruction of plant, B. Lower sterile leaf, C. Upper sterile leaf, D. Fertile leaf, E. T.s. of stem of *Cladoxylon taeniatum* (after Bertrand)

5 cm in diameter, was dichotomously branched and spirally bore numerous small (rarely over 2 cm) leaves which were dichotomously dissected in one plane and showed heterophylly (Figs. 42.5B and C). Fertile leaves (Fig. 42.5D) were borne on some upper branches. These were dichotomously dissected and bore at the tips of the dissections small sporangia. They were homosporous.

A t.s. of the stem (Fig. 42.5E) shows a broad homogeneous cortex, the central area of which was polystelic. The steles were many and narrow, often sinuous, and their number increased towards the base. They contained scalariform and pitted tracheides and each stele was surrounded by a secondary xylem.

Pseudosporochnus, another Middle Devonian fossil discovered in a bad state of preservation from New York, Scotland, Norway and Australia, was considered to be a member of the Psilophytales. It has a tree-like habit with a bulbous base, a 2-3 metres high trunk and a bushy crown of branches dichotomously branching once or twice. It is not known if there was any secondary growth as the anatomy has not been preserved. Some of the ultimate tips of the branches probably served as leaves while other tips swelled into sporangia containing minute spores 4 to 7μ in diameter. Leclerq and Blanks (1962), after studying its anatomy, consider that this specimen does not belong to the Psilophytales at all but to the Cladoxylales.

Order 4. ARCHAEOPTERIDALES

Archaeopteridales, with the type genus *Archaeopteris* represented by more than 15 species from the Upper Devonian is another problematic group which vacillates between Ferns and Pteridosperms.

Archaeopteris (Fig. 42.6) was so long known only from sterile and fertile fronds. The fronds were rather large, up to 1 metre in length and bore pinnae on two sides. Some of he fronds were sterile, others bore fertile pinnae either at the middle (Figs. 42.6A and D) or at the end of at both ends. The sterile pinnae were wedge-shaped and dichotomously veined. Carluccio, Hueber and Banks (1966) have shown that the fronds had a radial anatomy. The fertile pinnae bore two rows of club-shaped sporangia (about 2 mm long) in place of pinnules. Arnold (1939) has shown that the sporangia in *A. latifolia* were of two types (Fig. 42.6B). A few stouter ones contained only 8 to 16 megaspores about 0.3 mm in diameter. Others were narrower and contained some 100 microspores 1/10 the size. Heterospory was, therefore, definite.

Beck (1960, 1962, 1966) has subsequently made a spectacular discovery. He has found *Archaeopteris* fronds attached to a stem formerly identified as *Callixylon*. *Callixylon* had so long been known as a very abundant Devonian stem genus belonging to the Pityaceae of Cordiatales of Gymnosperms (Chapter 46) which is contemporary with *Archaeopteris*. It does not mean that all *Callixylon* stem specimens have become synonymous with *Archaeopteris* but simply that this Devonian heterosporous fern genus *Archaeopteris* had a stem like *Callixylon* with a distinct pith and abundant secondary wood having bordered pitted tracheides. It may now be said after Beck that *Archaeopteris* was a tall tree (Figs. 42.6C and D). The longest piece of wood discovered so far is 28 ft. long and 5 ft. in diameter at base.

It has, therefore, been argued that the Pteridosperms and other Gymnosperms did not arise from the general group of ferns but from a group of plants combining characters of ferns and Psilophytales, having secondary wood and showing heterospory. A name **Progymnosperms** has been suggested for this group. *Stauropteris, Cladoxylon* and *Archaeopteris* may be members of this group. Beck has proposed the class name *Progymnospermopsida* for this taxon but the members come from assorted orders and have not yet been shown to form a coherent taxon. Delevoryas (1969) suggests this group might have arisen out of the Psilophytale group *Trimerophyton* (*Trimerophytina* or even *Trimerophytopsida*) and it may even be possible that the seed habit developed even before the evolution of leaves!

EUSPORANGIDAE

Living ferns (with their fossil allies) in which the *sporangium* initial is not single cell but a group of cells resulting in a sporangium with a massive multicellular wall containing a large number of spores used to be placed within a subclass *Eusporangidae*. The antheridia in these are sunk and also contain a large number of sperms. The first division of the zygote is transverse. Two orders: (1) *Ophioglossales* and (2) *Marattiales* were included within this subclass. But, it is now generally realised that these two orders are not related and should not be placed within the same subclass. Some modern authors (c.f. Bierhorst classification) are not even in favour of using the term *Filicopsida* as done here but are in favour of independent paralled subclasses *Primofilicopsida*, *Ophioglossopsida*, *Marattiopsida* and *Filicopsida* and even more subclasses. It is, therefore, at least reasonable to break up the subclass into two subclasses, *Ophioglossidae* and *Marattidae*.

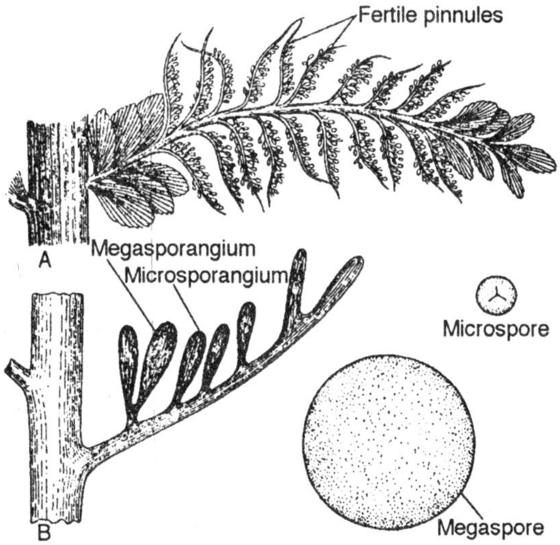

Subclass 2. OPHIOGLOSSIDAE

Ophioglossidae is considered as a separate subclass parallel to the Marattidae and the following subclasses as it shows some important differences to keep it sufficiently separate.

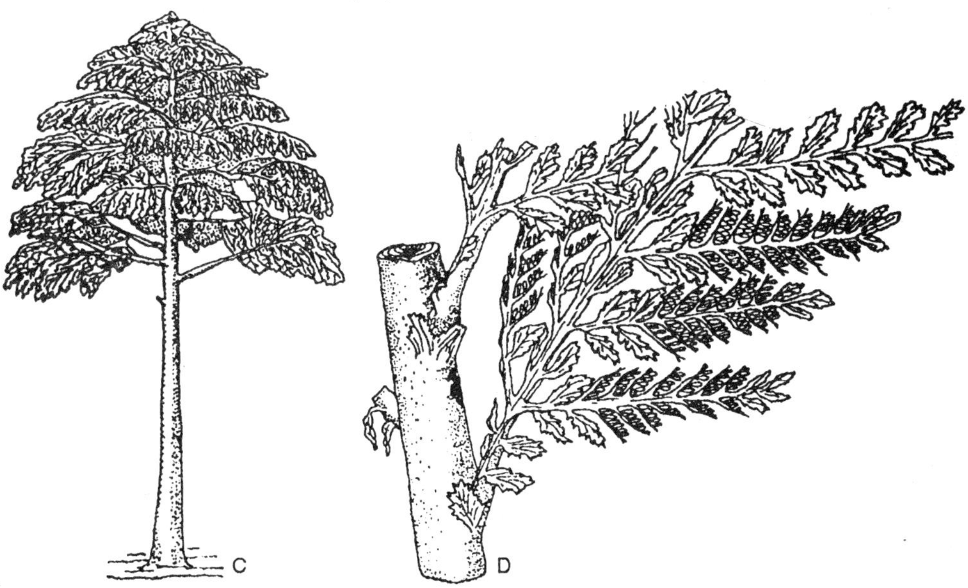

Fig. 42.6: A. *Archaeopteris hibernica* leaf with fertile pinnules in the middle (after Schimper), B. *Archaeopteris latifolia* fertile pinna with a megasporangium and several microsporangia. A megaspore and a microspore (both in the same scale) are shown separately (after Arnold), C. Reconstruction of *Archaeopteris* tree (after Beck), D. Twig with fronds (after Beck)

Among the most important features in which it differs from the other subclasses (c.f. U. Sen 1968) are: (1) peculiar fertile spike in which each sporangium is provided with a vascular supply at its base, (2) absence of circinate vernation even in the embryonic stage, (3) absence of sclerenchymatous cells in any part of the plant body, (4) presence of oval or circular bordered pits in the metaxylem tracheides, (5) hairless mycorrhizal roots, (6) presence of subterranean, cylindrical, mycorrhizal gametophytes with indefinite apical growth and scattered sex organs.

A single order with a single family is recognized:

Order: **OPHIOGLOSSALES**

Family: OPHIOGLOSSACEAE

Ophioglossaceae is characterized by the sporangia being placed on a *fertile spike* growing from the adaxial face of the leaf at the juncture of the lamina and the petiole. It is surprising that there is no fossil record of this group. But, such a characteristic group could not have sprung up all at once and there must be some link with the fossil ferns that is hitherto unknown. Bower (1908) tried to link Ophioglossales with the Lycopsida on the assumption that the sporangia are of adaxial origin. But the better established hypothesis that the sporangiferous spikes are modified lateral lobes explodes this hypothesis. Scott (1920) suggested that the Ophioglossales might have come out of the Zygopterids but had no evidence in support.

The family is represented by 3 genera—*Ophioglossum, Botrychium* and *Helminthostachys.* Besides these, three other suggested genera—*Rhizoglossum, Cheiroglossa* and *Ophioderma,* all taken out of *Ophioglossum,* do not need serious consideration here.

Of the three genera, *Helminthostachys* seems to be the most primitive and *Ophioglossum* the most modern, possibly due to reduction.

Genus: *OPHIOGLOSSUM*

Ophioglossum (sometimes called 'Adder's Tongue') is taken up first as it is the commonest, although it is possibly more advanced (mainly by reduction) then the other two genera. Clausen (1938) recognized only 28 species but Reimers in Engler's Syllabus (1954) recognizes 45. The species may be divided into the sections *Euophioglossum, Rhizoglossum* (*O. bergianum*), *Cheiroglossa* (*O. palmatum*) and *Ophioderma* (*O. pendulum* and two other species). The proposal to raise the four sections into four separate genera has already been mentioned). The species are of world-wide distribution in temperate as well as tropical zones. Most of them grow on humus soil and are perennial although the aerial shoots may be annual. A few tropical species are epiphytes. There are about a dozen Indian species of which *Ophioglossum vulgatum* (Fig. 42.7A), *O. reticulatum* and *O. pedunculosum* are very common. Other Indian species are—*O. costatum* (= *fibrosum*), *O. gramineum* (a very small species from Maharashtra), *O. nudicaule, O. capense, O. aitchinsoni* and *O. lusitanicum. O. pendulum,* a large, pendulous, epiphytic species (Fig. 42.7C) often grown in greenhouses, is known from South India and Ceylon.

The sporophyte

The sporophyte (Fig. 42.7A) has a short, vertical, subterranean rhizome. Several leaves develop spirally from the top (in Euramerican species only one leaf grows in a year because of slow growth in the cold climate). Numerous fairly stout adventitious roots grow from the base of the rhizome and these are increased by one root growing from the base of every leaf. Some of the roots grow horizontally like runners and young plants grow from buds on their upper surface. Each leaf has a thin *stipule* at its base and is differentiated into a petiole and a

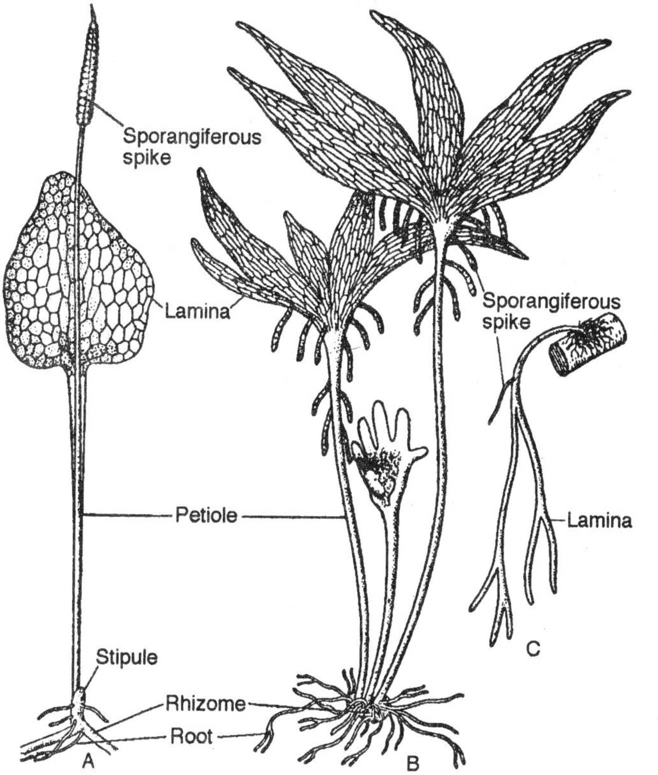

Fig. 42.7: A. *Ophioglossum vulgatum*, B. *O. palmatum*, C. *O. pendulum*

lamina which does not show circinate ptyxis. The venation is closed reticulate with some finer veins ending blindly within the meshes. This is an advanced character. The lamina in most species is entire and linear to ovate. But, in *O. palmatum* (Fig. 42.7B), growing on rotten wood, it is palmately dissected. In *O. pendulum* (Fig. 42.7C), an epiphyte with dorsiventral rhizome, the lamina is pendulous, narrow and dichotomously dissected truly like Adder's tongue.

The sporangiferous spike develops later at the junction of the petiole and the lamina and is, again, unbranched and single in most species including *O. pendulum*. But, in *O. palmatum*, several of these grow from the same region and there may even be several from the same spot.

The rhizome apex shows a single inconspicuous pyramidal apical growing cell with three cutting faces. The part of the rhizome may show a protostele surrounded by an endodermis but the stem is generally siphonostelic. A t.s. of the mature rhizomatous stem (Fig. 42.8) shows a broad homogeneous *cortex* which may sometimes show an outer *periderm*. The endodermis is not defined in the upper part of the rhizome. The siphonostele inside may be solenostelic due to one leaf gap but more often it is *dictyostelic* by several leaf gaps showing small or big crescent-shaped *meristeles*. The meristeles are endarch and collateral. The *protoxylem* lines the entire inner face of the xylem mass. The *metaxylem* has irregularly shaped reticulate tracheides. The *phloem* lies on the outer side. In *O. vulgatum* a small amount of secondary thickening has been observed. The central *pith* is homogenous but in *O. pendulum* small strands of xylem are found mixed with it.

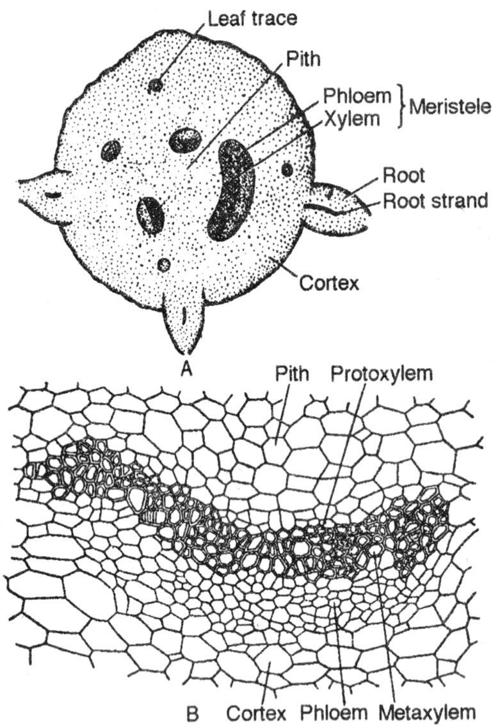

Fig. 42.8: A. T.s. (diagrammatic) of mature rhizome of *O. vulgatum*. B. A part of shown magnified

The roots are endogenous. They (Fig. 42.9A) have no root hair. There is a mycorrhizal phycomycetous fungus in the cortex. The endodermis is distinct. The stele is monarch to tetrarch and the phloem patches alternate with protoxylem patches. The xylem is exarch.

The leaf trace is not usually branched but in *O. palmatum* and *O. pendulum* two traces depart at each gap. In all cases the traces branch further before they run into the leaf base. The petiole (Fig. 42.9B), thus, shows a number of vascular bundles arranged in a semicircle. The leaf t.s. (Fig. 42.9C) shows stomata on both the epidermises, a uniform spongy mesophyll with air spaces and a number of vascular bundles.

The fertile spike (Fig. 42.10A) meristem grows with an apical cell having three or four cutting faces originating from a superficial cell of a young leaf. Bower stated that, as the spike begins to project, two continuous *sporangiogenic bands* (Fig. 42.10B) are noticed on the two sides. Later, the band is differentiated into

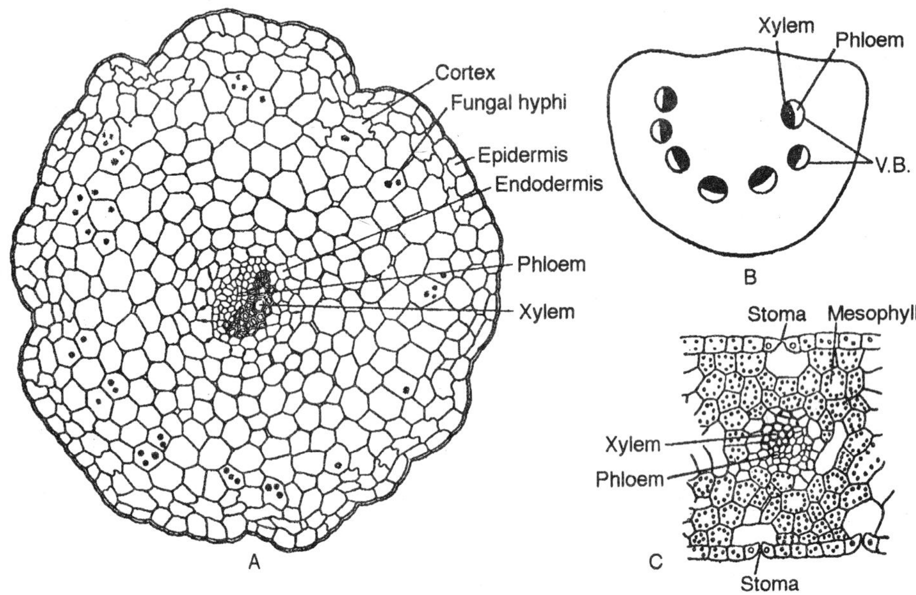

Fig. 42.9: *Ophioglossum vulgatum*. A. T.s. of root, B. T.s. of leaf petiole (diagrammatic), C. Vertical section of part of leaf

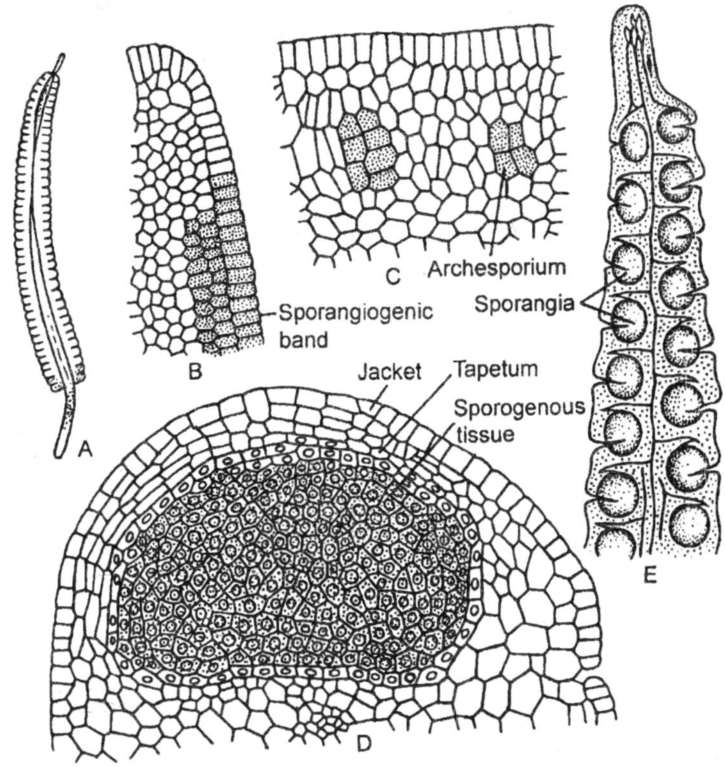

Fig. 42.10: *Ophioglossum*. A. Sporangiferous spike, B-C. Stages of development of sporangium (after Bower), D. A nearly mature sporangium, E. A ripe spike showing dehiscence

alternate blocks of *archesporial* and sterile cells (Fig. 42.10C) but U. Sen (1968) has noticed that in all cases examined by him the sporangiogenic band is not continuous but the sporogenous patches are separated by sterile patches from the beginning (Fig. 42.11A). Each archesporium supported by contiguous cells forms a *sporangium* (Fig. 42.10D) with a multilayered *jacket*, a weak *tapetum* and the *sporogenous* tissue inside. The tapetum ultimately breaks down into a plasmodial mass with persistent nuclei which invades the space between the isolated *spore mother cells*. The sporangia become somewhat dissociated from one another when mature and ultimately open by transverse clefts after the spores are mature (Fig. 42.10E). There is no annulus.

The nature of the petiolar and the foliar part of the leaf and, more specially, that of the sporangiferous spike has proved to be a problem. Sometimes, to avoid the problem, the whole structure is called an *aerial complex*. As shown by U. Sen (1968), a single leaf trace runs for the base of the stalk of each *aerial complex* but this trace splits into three before actually entering the base of the stalk (Fig. 42.11B). The 3 traces, again, form 12 traces or bundles by repeated divisions at the level where the stalk divides into fertile and sterile lobes. Of these 12 bundles, 9 pass on to the sterile lobe (leaf) where they do not anastomose while 3 pass to the fertile lobe (sporangiferous spike) (Fig. 42.11C). The 3 traces in the base of the sporangiferous spike divide and anastomose here and there in its axial region. Small vascular strands extend horizontally from this axial system into the sterile tissue in between the sporangia where their tips are sharply recurved towards the respective sporangia below and frequently branch into 2 or more strands

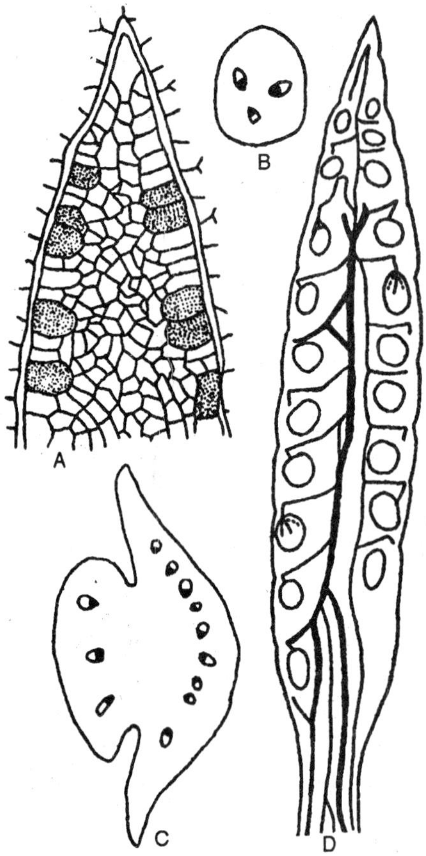

Fig. 42.11: *Ophioglossum vulgatum*. A. L.s. of young spike with sporangiogenic cells (dotted), B. T.s. of the base of aerial complex stalk showing, 3. Vascular bundles, C. T.s. at junction of fertile and sterile lobes showing 3 and 9 bundles, D. L.s. of sporangiferous spike showing courses of vascular bundles (after U. Sen)

(Figs. 42.10E and 42.11D). In each case one of these ultimate strands supplies the sporangial wall while others end blindly.

The morphological nature of the fertile spike of the Ophioglossales has been a matter of much discussion. Instead of going into the historical details of these arguments, the salient features may be stated. Roeper (1826) interpreted it as a pair of fushed leaves but, later, changed his opinion. Bower (1896) suggested that it is a septate sporangium growing on the adaxial face of the leaf bringing it near to Lycopsida but himself discarded this idea as untenable in 1911. Campbell (1890) also considered it as a homologue of the Lycopsid sporangium but, later (1911) considered it as similar to the sporangium of *Anthoceros*. The idea that the fertile spike is formed by the fusion of two fertile pinnae remained popular for a long time (Roeper 1859, Holle 1875, Chrysler 1911) but now has given way to a more acceptable view. Chrysler also has changed his opinion. Zimmermann (1930) explained the spike of the basis of his *telome theory* as being derived from a dichotomous shoot (as in the Psilophytales or the Coenopteridales) in which the fertile spike and the vegetative leaf represent different arms of the dichotomy. Ontogenetic studies show that the spike often develops from the mesophyll cells on the adaxial side whereas a pair of modified pinnae could arise only from the margins (U. Sen 1968, Bierhorst 1971). The pinna-like nature is, therefore, not supported. The present view is that the *aerial complex* represents a condensed dichotomous branch system (Zimmermann 1930, Bower 1935, Chrysler 1945, U. Sen 1968). Nozu (1950) and Nishida (1957) consider the stalk of the leaf (or the *aerial complex*) to be a stem-like organ which they name as *phyllomophore*. Thus, it is reasonable to conclude that the stalk or petiole of the complex (phyllomophore) is a structure intermediate between a stem and a leaf. It dichotomises to form the fertile spike by one arm and the sterile leaf by the other. It, therefore, differs from the Lycopsida and the Filicopsida in general.

The chromosome number of *Ophioglossum* is very high. The *n* number has been found to be from 125-130 in *O. lusitanicum* to more than 630 in *O. reticulatum*. In *O. vulgatum* it has been found to be 250-260.

The gametophytic generation

The spores occur in large numbers (1500 to 15,000) and each of them is endowed with an exine and an intine. The exine is pitted, moderately thick, often sculptured but colour-

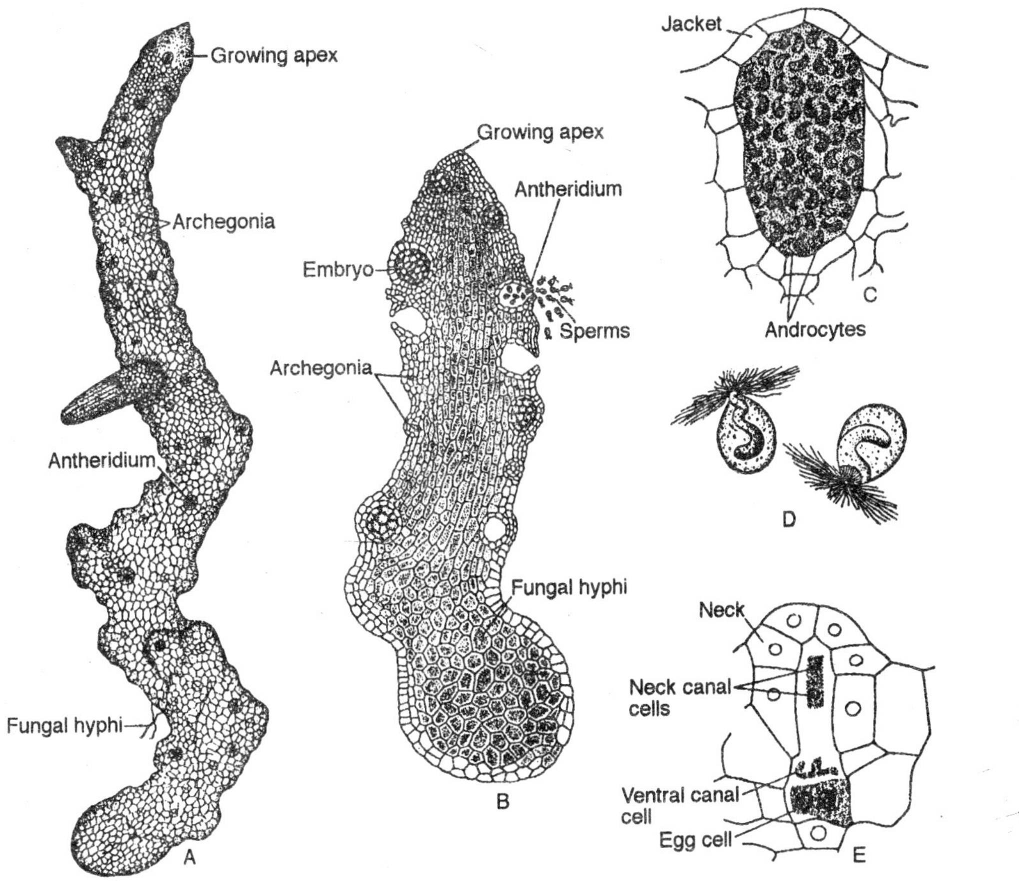

Fig. 42.12: *Ophioglossum vulgatum*. A. Gametophyte, B. L.s. of A (after Bruchmann from Bower), C. Mature antheridium, D. Sperms, E. A mature archegonium (the ventral canal cell disintegrating)

less. The spore, on germination, produces a subterranean gametophyte. Ophioglossales is the only group of known Filicopsida with a subterranean gametophyte and this again reminds the Lycopsida. The gametophyte does not grow beyond the rudimentary stage if it does not get infected with a mycorrhizal fungus at an early few-celled stage. The mature gametophyte (Figs. 42.12A and B) is irregularly cylindrical to conical and branched or unbranched. *O. vulgatum* prothallus is not branched at all looking like a piece of root while *O. pendulum* freely branches in different directions. The growing apex grows by a single apical cell with 3 or 4 cutting faces and this growth is indefinite. The prothallus may be long living (often 20 years) or annual. It is brown with little differentiation of tissue except that the axial cells may be a little elongated. The endophytic fungus is located in the inner tissue of the lower part only. There is no rhizoid and chlorophyll may develop only if some part gets exposed to light.

Antheridia and archegonia are formed on the surface indiscriminately intermingled and scattered. They arise from superficial cells. The antheridium (Fig. 42.12C) develops as in other Eusporangiates with a *jacket initial* and a *spermatogenous initial* and the final triangular *opercular cell* in the single-layered jacked disintegrates to form an opening. The *androcytes* or *sperm mother cells* form two sperms each, each of which receives a *blepharoplast* as in *Isoetes.* Ultimately, a coiled multiciliate *sperm* (Fig.

42.12D) is formed and there are thousands of these in an antheridium. All antheridia are embedded. The *archegonia* (Fig. 42.12E) also develop as in other cases. The *primary canal cell* rarely divides to form two *neck canal cells*—usually there is only one of them. The *ventral canal cell* disintegrates as soon as it is formed. The neck barely protrudes out of the thallus.

The new sporophyte

The *zygote* divides by a transverse wall giving rise to an *epibasal* and a *hypobasal* cell. The growth of the embryo is very slow taking a year or several years to develop. There is no suspensor in the embryo and it takes a long time for the different organs to differentiate. An embryo of *O. vulgatum* a few months old (Fig. 42.13A) shows the *foot* and the *root* in the hypobasal region. The epibasal tissue is still undifferentiated. The root is the first and the prominent organ to pierce the calyptra and at this stage the embryo seems all root with a short foot and an undifferentiated epibasal region (Fig. 42.13B). The epibasal region then gives rise to the *stem apex* and the cotyledon when the new sporophyte becomes established. In some other species, the growth of the epibasal region may be at the same rate as the hypobasal. The first root gets infected by the endophytic fungus at a very early stage.

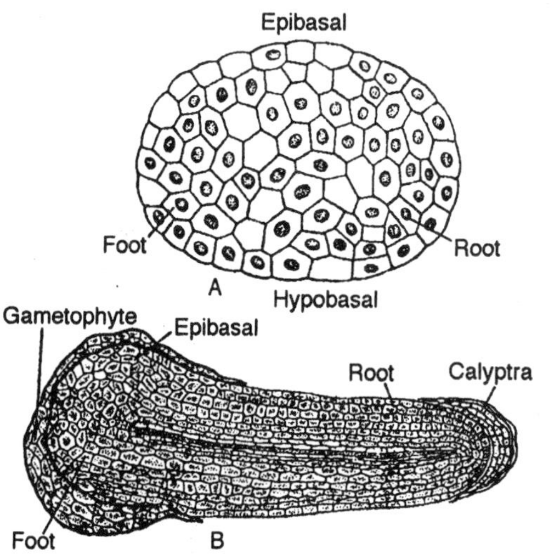

Fig. 42.13: *Ophioglossum vulgatum*. A. An advanced embryo, B. Embryo after it has pierced the calyptra (after Bruchmann from Bower)

Genus: *BOTRYCHIUM*

Botrychium ('Moonwart' or 'Grape fern'), with 35 cosmopolitan species from temperate regions, has pinnately branched leaves and the fertile spike also is similarly branched. Secondary growth is clear in the stem and a cambial cylinder is present. The endarch xylem shows bordered pitted tracheides. The leaf traces in *Botrychium* are complex.

The genus is well represented on the Himalayas and the South Indian as well as the Ceylonese hills. The commonest species is

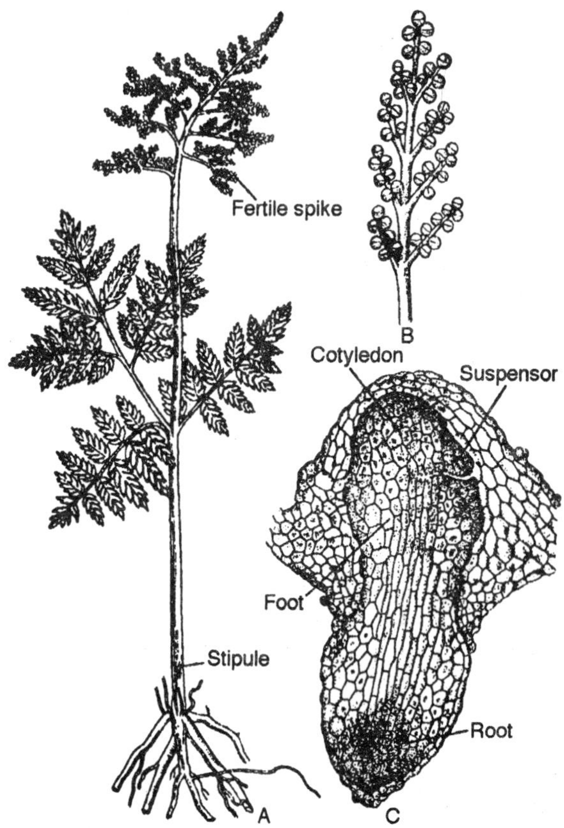

Fig. 42.14: A. *Botrychium virginiacum*, B. Bit of fertile spike of A, C. Embryo of *Botrychium* (*Sceptridium*) *obliquum* showing suspensor (after Lyon from Bower)

B. virginiacum (Figs. 42.14A and B). Other Indian Species are *B. lunaria* and *B. daucifolium.*

The genus is divided into two sections—*Eubotrychium* and *Sceptridium.* The first section shows an embryo as in *Ophioglossum.* The second section (*B. obliquum, B. dissectum,* etc.) shows an embryo in which the epibasal cell forms a *suspensor* (Fig. 42.14C) and the hypobasal cell forms an embryo in which the foot is not apparent. There is a suggestion that *Sceptridium* should be placed in a separate genus because of the nature of its embryogeny.

Genus: *HELMINTHOSTACHYS*

Helminthostachys is a monotypic genus found only in Indomalesia and the Polynesian Islands including New Zealand. The single species is *Helminthostachys zeylanica* (Fig. 42.15A). It is well-represented in the Lower Bengal plains and all over such places in India. The rhizome is dorsiventral and the leaves are very large and palmately compound showing pronounced dichotomy. The fertile spike is compound with clusters of sporangia on the rachis (Fig. 42.15B). The xylem is mesarch in this genus, bordered pitted tracheides are present but no secondary growth has been detected.

The embryo (Fig. 42.15C) shows a suspensor and the normal foot. The genus is considered the most primitive in the family.

Subclass 3. **MARATTIDAE**

Order: MARATTIALES

Marattiales are eusporangiate ferns showing circinate vernation and with the sori on the abaxial face of the leaf. The sporangia are adherent to form circular or linear synangia on the veins. Leaves have stipules and the stem structure is typical.

Marattiales is a very ancient order, fossil-specimens of which are found since the Middle Carboniferous. Some of the fossil Marattiales were larger trees. They had *Pecopteris* type of leaf and synangia on its lower sarface. Circular synangia were present in *Ptychocarpus* and *Asterotheca* while linear synangia are seen in *Danaeites* (all Middle Carboniferous to Permian). Indian Upper Gondwana shows *Marattiopsis* (Rajmahal stage) and *Danaeopsis* (Rajmahal and Parsora stages) belonging to this order. *Ptychocarpus* from the Raniganj Lower Gondwana, is also suspected to be of Marattiaceous affinity.

Psaronius is a group of tree ferns which were abundant during Permocarboniferus.

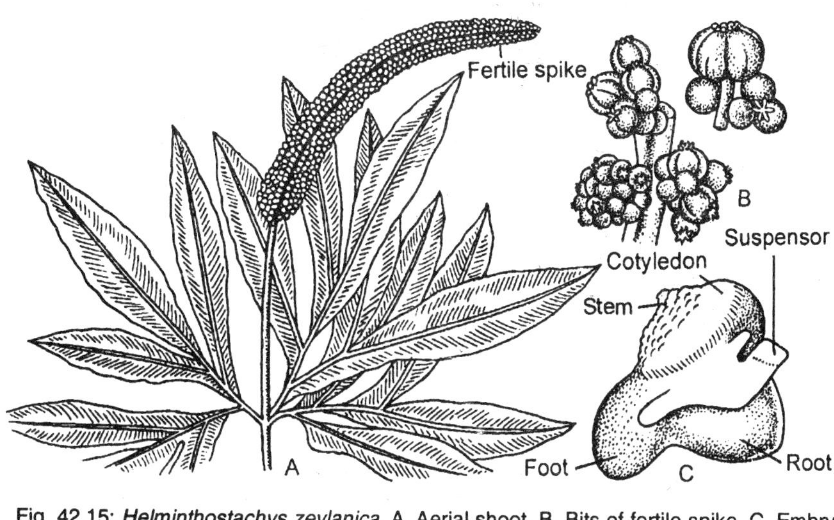

Fig. 42.15: *Helminthostachys zeylanica*. A. Aerial shoot, B. Bits of fertile spike, C. Embryo

They were up to 10 metres high, bore *Pecopteris* type leaves and are known by peculiar trunk fossils bearing thick root mantles. Their affinities are not clear as fructifications are not known. They are often associated with the Marattiales but it is not certain.

The origin of the Marattiales has been sought through *Corynepteris* of the Zygopteridaceae as already discussed in page 1213. The eusporangiate origin of sporangia, the associate manner of development of antheridia and archegonia, the transverse division of the zygote and the late differentiation of the embryo links the order to the Ophioglossales. On the other hand, the more fern-like structure of the sporophyte, the sori, the gametophyte and the very complicated stelar structure shows that the relation must be extremely remote.

7 genera from the tropical rain forests are recognized within the single family *Marattiaceae.* But, modern Pteridologists favour splitting the original Marattiaceae into a number of families. Reimers in Engler's Syllabus (1954) as modified by Pichi-Sermolli (1959) makes the classification as follows:

Family 1. Angiopteridaceae: Sporangia almost free, arranged in two rows on two sides of vein forming linear sori near margin.

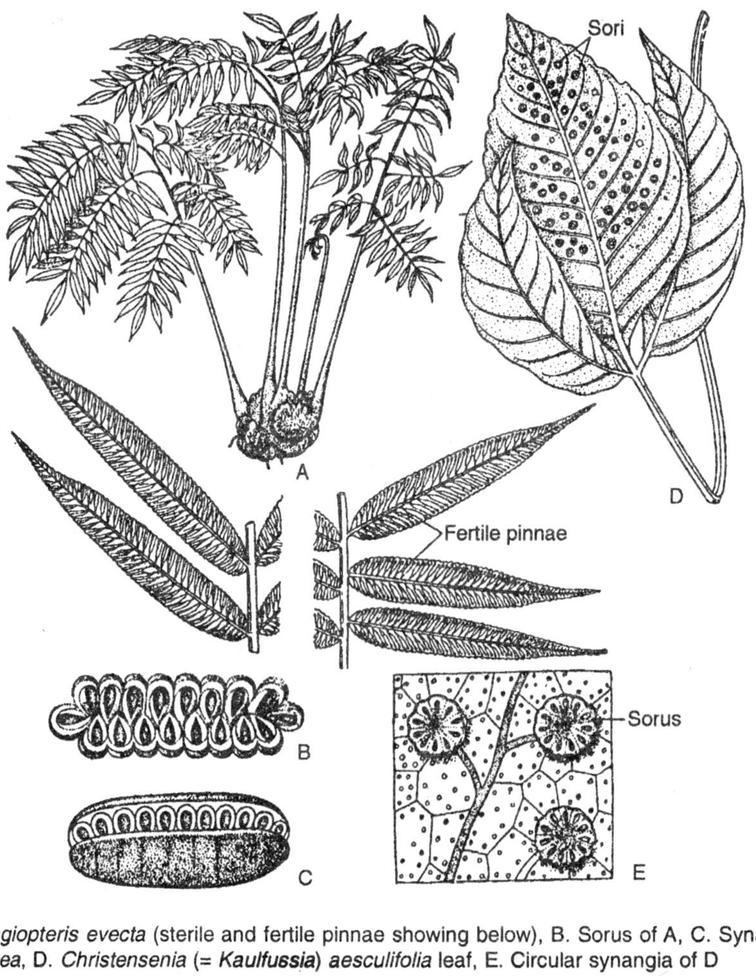

Fig. 42.16: A. *Angiopteris evecta* (sterile and fertile pinnae showing below), B. Sorus of A, C. Synangium of *Marattia fraxinea*, D. *Christensenia* (= *Kaulfussia*) *aesculifolia* leaf, E. Circular synangia of D

Genera: *Angiopteris* (100 spp. from Polynesia to Madagascar and the Himalayas), *Archangiopteris* (4 spp. from S.E. Asia) and *Macroaglossum* (2 spp. from Melanasia).

Family 2. Marattiaceae: Sporangia forming linear oval *synangia*.

Genera: *Marattia* (60 spp.) and *Protomarattia* (1 sp.) Pantropical.

Family 3. Christenseniaceae (= Kaulfussiaceae); sporangia radially arranged to form circular *synangia*.

Genus: *Christensenia* (= *Kaulfussia*) (1 sp. from Indomalesia).

Family 4. Danaeaceae: Linear long *synangia* (leaf middle to margin) very close together, almost covering lower outer half of leaf.

Genus: *Danaea* (32 spp. from tropics of New World).

Angiopteris evecta (Figs. 42.16A and B) is found all over India and specially in the Eastern Himalaya (up to 2500 metres), Assam and Chittagong. *Marattia fraxinea* (Fig. 42.16C) is found in South India (4-6000 ft.). *Christensenia* (= *Kaulfussia*) *aesculifolia* (Figs. 42.16D and E) is found in North as well as East India. *Angiopteris*, which is more primitive, shows a chromosome number $n = 40$ and 80. *Marattia*, which is more advanced, shown $n = 78$.

The sporophyte

The sporophyte stem is short (although it may be half a metre or more in diameter), radial (sometimes dorsiventral) and covered by persistent leaf bases. There is a crown of compound leaves (pinnate leaves of *Angiopteris evecta* are reported to reach a length of 5 metres) at the apex. Leaves show circinate ptyxis and each leaf has a pair of stipules at its base and the stem surface below is covered by a mantle of leaf bases or leaf scars each with two persistent stipules. The stem has numerous long adventitious roots, one or more inserted below each leaf or leaf scar.

The growing apex has at first a single cell which is later replaced by a group of meristematic cells. In the stem, the *cortex* is uniformly parenchymatous and contains a number of mucilage canals and tannin-filled cells (Fig. 42.17). Young stems and lower portions of old stems have *endodermis* but late mature portions show no delimitation between cortex and stele. The vascular cylinder is protostelic at base but it abruptly becomes siphonostelic a little higher up. In the lower regions there is a single trace departing to each leaf but higher up two strands depart for each leaf breaking up the siphonostele into meristeles. The vascular strands further subdivide when mature, breaking up into strands arranged in concentric circles looking like the disintegration of a *polycyclic* system. Complications further increase by peculiar branches called *commissural strands* which run diagonally across the pith and often form an internal ring. *Root traces* come out only from meristeles of the inner rings. The *pith* is homogeneous like the cortex but it is full of branch or commissural strands in mature parts.

The root cortex also is full of mucilage canals and, in the young stage, contains an oomycetous endophytic fungus.

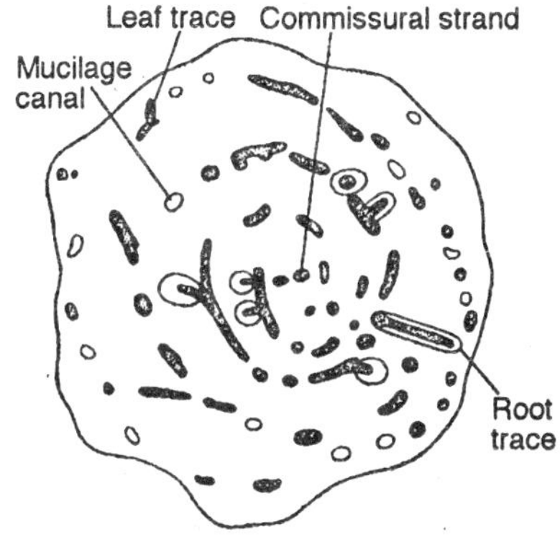

Fig. 42.17: T.s. of *Angiopteris evecta* stem

The sporangia grow in linear sori or linear and circular *synangia* (formed by the coalescence of sporangia) formed by 12 to 36 sporangia (Figs. 42.16B, C and E). These develop as massive structures from epidermal patches of tissues and may even contain stomata on the multilayered walls. The innermost layer of the jacket acts as the *tapetum*. There is no annulus, the sporangia rupturing though the middle of the jacket after splitting apart of the individual sporangia in case of a synangium. A huge number of spores is formed in each sporangium.

The gametophyte

Gametophytes or prothalli of Marattiales are bigger than those of modern ferns. They are dark-green, fleshy (multilayered) and externally resemble liverworts like *Pellia*. They may be ribbon-shaped, spathulate or heart-shaped. Rhizoids are on the under surface. There is an endophytic fungus in the thallus but it is supposed to be more parasitic than symbiotic in nature. There is a strong midrib. The gametophytes are monoecious. Antheridia are formed early on the ventral side of the midrib but may later grow on the dorsal side as well. Archegonia grow only on the dorsal side of the midrib. Antheridial and archegonial development resembles *Ophioglossum* generally. The sperms are multiciliate and they grow in large numbers.

The new sporophyte

The first division of the *zygote* is transverse. Differentiation of the embryo is late as in the Eusporangiates. It seems that the *epibasal* part gives rise to *foot* and the *hypobasal* part to the *root*, the *cotyledon* and the *stem*. The new sporophyte comes out by the cotyledon piercing the calyptra. The stem develops after the root. The *suspensor* has been reported in the Marattiales inconstantly. It has been reported in *Angiopteris* only occasionally. It has also been reported in *Danaea* and *Macroglossum* but not in most species of *Angiopteris* or any species of *Marattia*. It is possibly a vestigeal organ reciting the primitive nature.

Subclass 4. **PROTOLEPTOSPORANGIDAE**

Order: OSMUNDALES

Family: OSMUNDACEAE

The Protoleptosporangidae or the *Osmundidae* comprizing of the single order Osmundales with the single family Osmundaceae is taken out of the Leptosporangiates as proposed by Hirmer (1927) as it is not truly leptosporangiate.

The Osmundales is a very ancient order represented by *Thamnopteris* and *Zalesskya* from late Permian and by *Osmundites* itself from Jurassic onwards. The climax of the group was probably in the Jurassic after which they dwindled. Sahni has suggested its derivation from *Grammatopteris* (Lower Permian) of Botryopteridaceae after a detailed study of the anatomy (c.f. page 1214). Several species of *Caladophlebis* fronds and petrified stems and roots of *Osmundites sahnii* are found in Rajmahal stage Upper Gondwana beds.

Osmundaceae is divided into three genera: *Osmunda* (14 spp. from temperate and tropical of both hemispheres), *Todea* (1 sp. from South Africa) and *Leptopteris* (61 spp. from Australasia and Melanasia).

The Chromosome number of Osmundaceae is very constant (n = 22) proving their primitiveness.

Osmunda claytoniana is known in the Himalayas from Kashmir to Khasia hills (1500 to 3000 metres). *O. regalis* is also known all over the Himalayas as well as in South India (1200 to 1900 metres).

The sporophyte of *Osmunda* has a short, upright or creeping, usually subterranean stem covered by persistent leaf bases and adventitious roots from the bases of the leaves. The other two genera have erect, 1 to 3m high trunks. Stems of *Osmunda* may bifurcate once

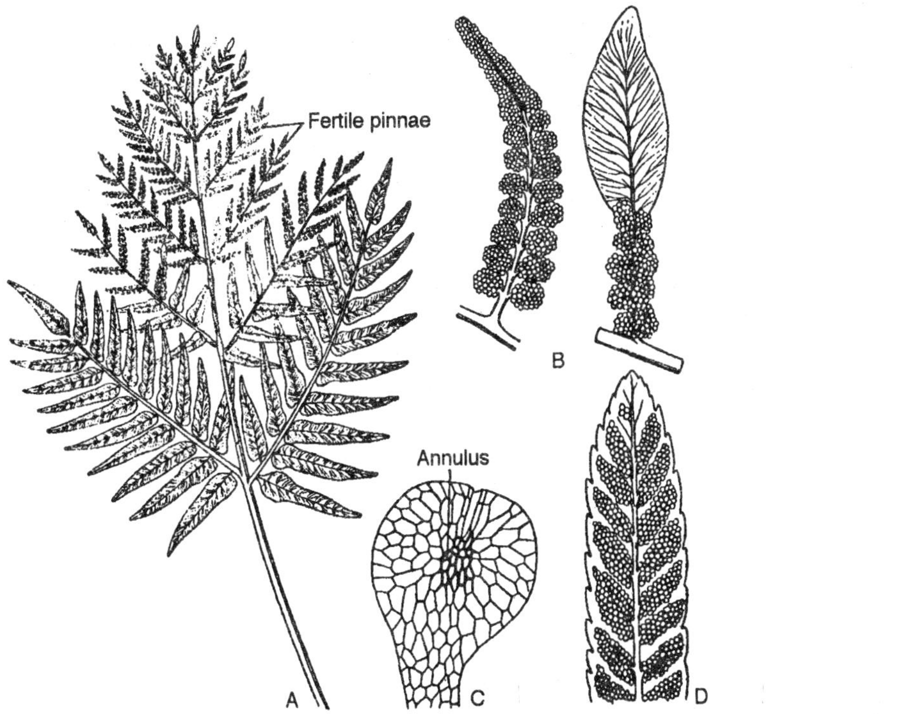

Fig. 42.18: *Osmunda regalis*. A. Leaf, B. Fertile pinnae, C. Sporangium, D. Fertile pinna of *Leptopteris*

or twice. The apex of each stem bears a crown of pinnate or each stem bears a crown of pinnate or bipinnate leaves (Fig. 42.18A) which may be up to 3 metres long and show circinate ptyxis. The leaves, when young, have a covering of pale-brown, uniseriate hairs some of which may be glandular and the leaves also develop stipule-like wings at base on which, again, there may be glandular hairs. Some species bear two kinds of leaves, sterile and fertile, while in other species fertile pinnae develop at the middle or at the end of sterile pinnae (Figs. 42.18A and B). In *Todea* and *Leptopteris* all leaves are of one kind (Fig. 42.18D). The leaves of *Osmunda* and *Todea* are as in other ferns but those of *Leptopteris* are filmy and only one layer thick as in *Hymenophyllum.*

The first developed portion of a stem may be protostelic. Later developed portions are dictyostelic with numerous *leaf gaps* (Fig. 42.19). Outer portion of *cortex* is dark-brown with highly sclerified cells and covered by numerous persistent sclerenchymatous leaf bases. Inner cortex is colourless and parenchymatous. There is a distinct *endodermis* which is continuous in the numerous spirally arranged *leaf traces.* The leaf traces are C-shaped. There is a 2 to 4-layered *pericycle.* The *phloem* is outside the xylem patches. The xylem patches are oval to U-shaped, mesarch to endarch. *Metaxylem* is composed of large scalariform-pitted tracheides. *Pith* is usually homogeneous but endogenous.

The sporangia are marginal or terminal on stalks in Osmunda and superficial on the lower surface is the other two genera. They do not form sori and there is no covering (Figs. 42.18B and D). The sporangia (Fig. 42.18C) are large, pyriform and with stout, short stalks. In the

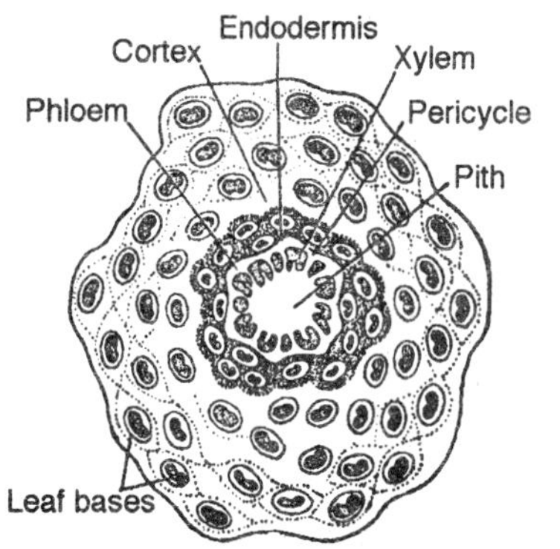

Fig. 42.19: T.s. of stem of *Osmunda regalis*

development of the sporangium there is a conspicuous *initial cell* which develops the sporangium capsule with a single layered jacket as in the Leptosporangiates. But the massive stalk and in some cases even a part of capsule wall is formed by cells outside the initial so that the mode of development is intermediate between eusporangiate and leptosporangiate. A group of cells on one side of the jacket thickens at a late stage and forms a shield-shaped annulus. A vertical zone of dehiscence is also formed on the top by some elongate cells. All the sporangia in a pinna mature together.

Spores are formed in large numbers, contain chloroplasts when ripe and germinate rapidly. The gametophyte is more or less heart-shaped, dark-green, multilayered and with a thick midrib. It, thus, resembles *Marattia* and not the leptosporangiate ferns. Adventitious lobes are cut off from the prothallus and reproduced vegetatively. The gametophyte is monoecious but the *antheridia* develop first. The antheridia develop marginally, terminally or on the ventral surface and the development reminds the eusporangiate type though the mature antheridia project about the general surface. The sperms are produced in large numbers and are coiled with numerous cilia at tips. Their escape is by breaking the apical opercular cell. The *archegonia* develop on the ventral midrib or along its edges and have long necks about 8 cells high which project out of the prothallus.

The first division of the *zygote* is vertical as in the leptosporangiates but the differentiation of the embryo is slow as in the eusporangiates. Usually the epibasal half gives rise to the cotyledon, stem, and root while the hypobasal half forms the foot.

Apogamy and apospory are known in the Osmundales. The former takes place naturally and the latter from wounds. Aposporous gametophytes are diploid.

Thus the Osmundales, although resembling the leptosporangiates in some characters, resemble the eusporangiates in (1) the pseudo-eusporangiate development of the sporangium which has a shield-shaped lateral annulus, (2) the stipular structure, (3) the morphology of the gametophyte, (4) the large number of sperms and their escape by breaking an opercular cell, and (5) the relatively late differentiation of the embryo.

Subclass 5. **LEPTOSPORANGIDAE**

The Leptosporangidae comprise of the modern ferns which are truly leptosporangiate. True leptosporangiate characters are: the sporangium develops from a single initial, is delicate with a single layered jacket and a slender stalk, the tapetal layer comes from the archesporial part and the number of spores is usually limited; the gametophyte is more delicate; the antheridia are small and projecting with a limited number of sperms; the first division of the zygote is vertical and the embryo quickly forms quadrants which differentiate into foot, root, cotyledon and stem.

Bower has stressed two trends of evolution among the Filicopsida generally and among the Leptosporangiates in particular. These concern the position and the order of maturity of the sporangia or the sori. First, all ferns may be divided into *Marginales* (sporangia on sporophyll margin) and *Superficiales* (sporangia on the lower surface of the sporophyll). The Marginale condition is more primitive considering that the sporangia were first located at the tips of branches and megaphyllous leaves developed by flattening and webbing of branches (discussed in Chapter 42). The Superficiale condition evolved later by the pushing out and, then, down of the sporangia from the margins to the under surfaces of the sporophylls. In many modern fern families, both Marginale and Superficiale conditions are present side by side so that it is not possible to conceive parallel evolution of the Marginales and Superficiales as two independent lines. The other trend, *Simplices* (all sporangia on a sporophyll or in a sorus mature *simultaneously*) → *Gradatae* (the sporangia in a sorus mature in a *basipetal* order, the topmost one being the oldest) → *Mixtae* (sporangia do not mature simultaneously and there is no order in maturation, the oldest may be next to the youngest) is even more striking. The Eusporangiates, the Osmundales and the primitive familes of the Leptosporangiates are all of the Simplices type. Next come the Gradate families. Most modern are the Mixtae families.

After taking the Osmundales out of the Leptosporangiates, Pteridologists are now almost unanimous that the true Leptosporangiates are of the following groups which are considered as orders by most authorities although Pichi-Sermolli considers them as independent subclasses:

Order 1. Filicales—homosporous, cosmopolitan (Simplices → Gradatae → Mixtae)

Order 2. Marsileales—heterosporous, aquatic (Gradatae)

Order 3. Salviniales—heterosporous, aquatic (Gradatae)

Some older Pteridologists combined together the Marsileales and the Salviniales into a single order *Hydropteridineae* or *Salviniales* (including *Marsilea*) or *Rhizocarpae* because both the orders are aquatic, gradate, heterosporous and sporangia below. But, it is unnatural to combine the two groups as the sporocarps of the two orders are entirely of separate origin and no close relationship between the two is tenable. The last name Rhizocarpae also leads to the wrong idea that the sporocarps are on roots.

Order 1. FILICALES

With various estimates bringing the number of genera between 234 and 298 and that of the species to about 9000 the Filicales comprise more than 95% of all the Pteridophytes today. Some of the families are modern and still at the climax of evolution which is not the position in any other group of Pteridophytes. The classification of the Filicales is a highly controversial issue and all the controversies are not within the scope of this book. The controversy is so high because some of the families, specially the most predominant family *Polypodiaceae* is definitely polyphyletic (different genera coming from different evolutionary lines) so that in a phylogenetic system these families must be broken into a number of families. But, there is no unanimity as to what the broken up families should be. In this book, therefore, the families adapted by Reimers in the last edition of Engler's Syllabus (1954) are being enumerated. Bower's and Christensen's systems form the basis of the classification of the Filicales by most authors today, including Reimers. The taking out of the Osmundales, Marsileales and Salvineales has already been explained. The families my be divided into Simplices, Gradatae and Mixtae groups.

SIMPLICES

Family: 1. Schizaeaceae
2. Gleicheniaceae
3. Matoniaceae
4. Dipteridaceae

GRADATAE

Family: 5. Hymenophyllaceae
6. Hymenophyllopsidiaceae
7. Loxsomaceae
8. Plagiogyriaceae
9. Dicksoniaceae
10. Protocyatheaceae
11. Cyatheaceae

MIXTAE

Family: 12. Polypodiaceae
13. Parkeriaceae

Family 1. SCHIZAEACEAE

Schizaeaceae is a family of terrestrial ferns with diverse habits—some of them are grass-like, other are long leaf-climbers. The stem is a creeping or ascending rhizome clothed with simple hairs or scales as in *Mohria.* On these are spirally borne leaves which are fundamentally dichotomous. Leaves of *Lygodium* are long climbers of indefinite growth which may easily reach 3 or 4 metres. The vascular system ranges from protostelic in *Lygodium,* siphonostelic in *Schizaea* to dictyostelic in *Mohria.* Roots are adventitious. Sporangia are borne singly along the leaf margins often covered by a flap-like

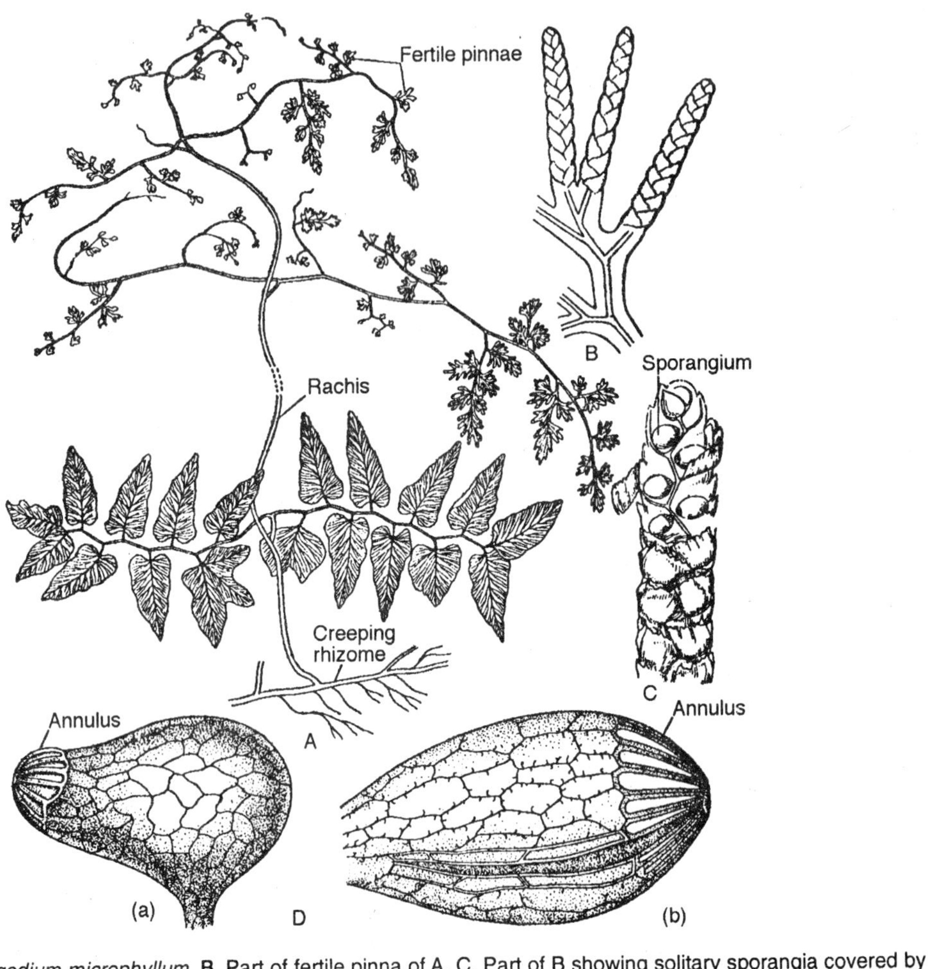

Fig. 42.20: A. *Lygodium microphyllum,* B. Part of fertile pinna of A, C. Part of B showing solitary sporangia covered by false indusia, D. (a) Sporangium of *Lygodium japonicum*; (b) Sporangium of *Schizaea* showing apical annulus and vertical dehiscence

growth of the leaf forming a false indusium. In *Anemia* the sporangia are naked developed along the margins of stalk-like, branched pinnae not showing any lamina. The most distinctive feature is the sporangium with a ring-shaped annulus on the top and causing vertical dehiscence (Fig. 42.20D). A large number of spores are contained in the sporangia. In a few cases the sporangia are apparently shifted to the back of the leaf from the marginal position. The gametophyte is more or less heart-shaped with a median midrib in all genera except *Schizaea* where it is filamentous and protonema-like. This last gametophyte is considered abnormal, Embryo development in the family is typically leptosporangiate.

Schizaeaceae has a very early history being represented by *Senftenbergia* in Upper Carboniferous, *Klukia* and *Norimbergia* in the Triassic and Jurassic.

The family comprizes four genera: *Schizaea* (30 spp. mostly tropical and temperate, specially from the Southern Hemisphere), *Lygodium* (39 tropical and subtropical species) with climbing leaves which are reported to reach up to 30 metres in length, *Anemia* (90 spp. mostly from tropical New World, one species reaching India) and *Mohria* (1 sp. from South and East Africa).

Lygodium (Fig. 42.20), often grown as an ornamental plant, is very common in India and some five species (*L. japonicum, L. pinnatifidum, L. incinatum, L. microphyllum, L. flexuosum*) are known. Other species and genera occurring in India are *Schizaea digitata, S. dichotema* and *Anemia tomentosa.*

All characters point out that the family is a very primitive Marginales-Simplices one.

Family 2. GLEICHENIACEAE

Gleicheniaceae is a family of terrestrial ferns with long, creeping, dichotomously branched rhizomes clothed with scales or hairs. Leaves are restricted to the upper side of the rhizome usually in a single row with adventitious roots on the underside of the rhizome. Leaves pinnate, in certain species showing pseudodichotomy by abortion of terminal bud and with forkings on two sides may be very large by indefinite growth, clambering on other plants forming a thicket; ultimate venation open. Stem in usually protostelic but may be incipiently actinostelic; siphonostelic in two species.

Sporangia in sori (with usually only few sporangia in each) without indusia, on the under surface of leaf, often on two sides of the midrib of the pinnule. All sporangia develop simultaneously. Each sporangium has a large number of spores and has an obliquely transverse complete annulus which causes vertical dehiscence. Prothallus green, flat, with a median midrib.

The family claims antiquity as *Oligocarpia* from Upper Carboniferous is assigned by some to this family. Other definite fossils are known from the Triassic. The zenith of development of the family was probably in the Cretacious. *Gleichenites* fossil fronds are known from the Indian Upper Gondwana (Rajmahal and Jabalpur stages).

Previously all the species in the family were referred to the single genus *Gleichenia* which was divided into a number of subgenera. Six separate genera are, however, advocated now. These are: *Gleichenia* proper (10 spp. from Australia and pantropical), *Dicranopteris* (10 pantropical species), *Sticherus* (100 pantropical species), *Hicriopteris* (10 spp. from tropical Old World up to Japan), *Platyzoma* (1 sp. from Australia) and *Stromatopteris* (1 sp. from New Caledonia).

Dicranopteris dichotoma (= *Gleichenia linearis*) is the species very common in subtropical and temperate India (Fig. 42.21). This species occurs all over the World in the tropical to

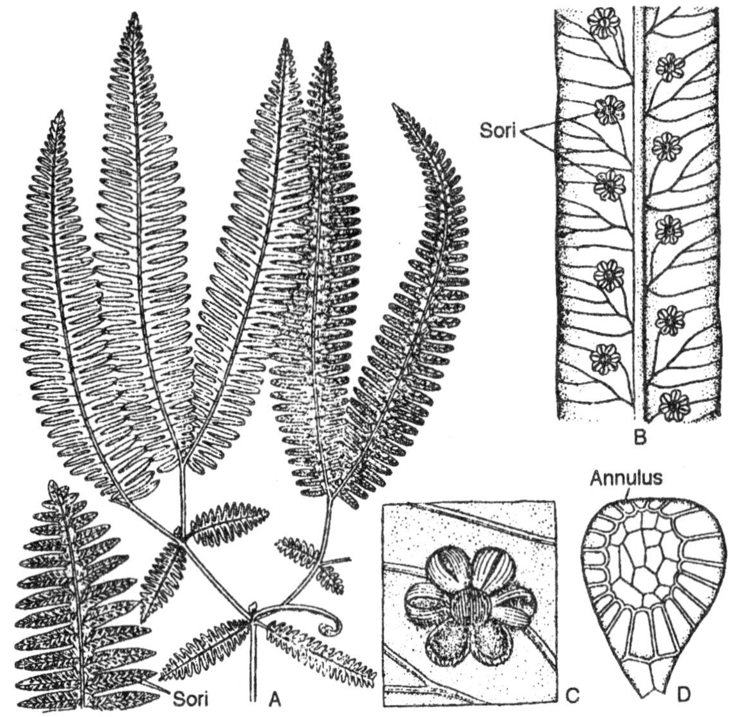

Fig. 42.21: *Dicranopteris dichotoma* (= *Gleichenia linearis*). A. Part of a leaf, B. Part of fertile pinna back, C. A sorus, D. A sporangium

temperature regions. *Dicranopteris glauca* also is known from the Eastern Himalaya.

Gleicheniaceae also is an ancient family of the Superficiales-Simplices group. Gleicheniaceae is structurally related to Cyatheaceae according to Bower and Holttum and the former has been suggested as the ancestor of the latter. Mehra also supported this view on cytological grounds. But, Sen is of opinion that the two families are derived from the same stock and not one from the other.

Family 3. MATONIACEAE

Matoniaceae is the only family in the Simplices group in which the sori are covered by umbrella-shaped indusia. The sori (Fig. 42.22A) are round and on the under surface of leaf. Sporangia (Fig. 42.22B) are sessile or

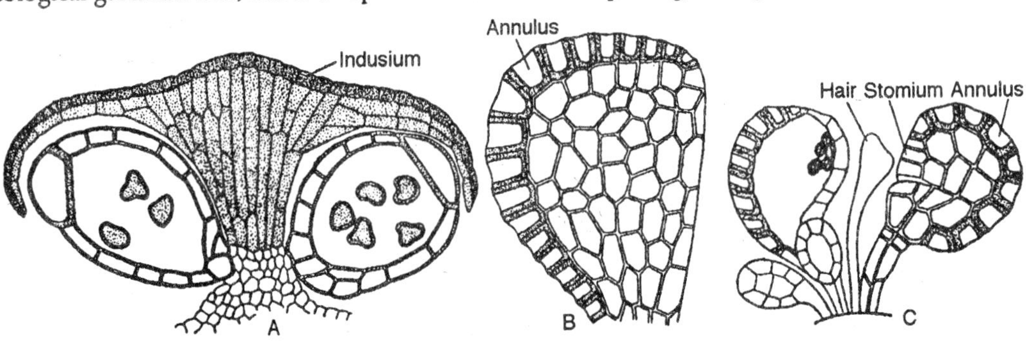

Fig. 42.22: A. Vertical section of sorus of *Matonia pectinata* (after Smith), B. A sporangium from A, C. Sporangia of *Dipteris* with hair

nearly so with an obliquely vertical, incomplete annulus with undifferentiated stomium. The dehiscence is transverse.

There are two genera—*Matonia* (1 sp. from East Indies) and *Phanerosorus* (2 spp. from Borneo and New Guinea); Two fossil genera, *Matonidium* (Umia) and *Phlebopteris* (Rajmahal), are found in the Indian Upper Gondwana Jurassic beds.

This is an advanced family of the Superficiales-Simplices group.

Family 4. DIPTERIDACEAE

The Dipteridaceae has naked sori on the under surface of the leaf like the Gleicheniaceae all maturing together (Simplices). But, at least one species with the 'mixed' type of sorus is known. Although there is no indusium, there are numerous capitate hairs within the sporangia and each sporangium has an obliquely vertical annulus (Fig. 42.22C) with stomium causing transverse dehiscence.

Dipteridaceae is known since the Triassic as fossils. Several species of *Hausmannia* from Rajmahal and Jabalpur, represent the family in the Indian Upper Gondwana Jurassic.

There is one genus *Dipteris* with 8 species spreading from East Asia to Polynesia. *Dipteris wallichii* is known from the Khasia hills.

This is another advanced family in the Superficiales-Simplices group and merges into the Polypodiaceae (Mixtae).

Family 5. HYMENOPHYLLACEAE

Hymenophyllaceae is the first of the Gradate families with minute to medium sized terrestrial or epiphytic plants growing in very wet habitats with prostrate, creeping rhizomes which are not scaly. The stem is protostelic. The leaves are only one cell thick and beautifully transparent, hence the common name 'filmy ferns'. These leaves are produced without any order (young and old mixed) on the rhizome. The sporangia are borne in basipetal (i.e. gradate) sori on the terminal projection of veins (marginal) covered by an indusium growing from the base as a flap. The indusium is two-lipped in *Hymenophyllum* and tubular in *Trichomanes.* The axis of the *Trichomanes* sorus often extends out as a sterile projection. The sporangium has an obliquely vertical sorus and transverse dehiscence. The gametophyte is filamentous or ribbon-shaped due to reduction.

The geological history of the family is not clear but *Hymenophyllites* from Upper Carboniferous has been supposed to be a relative by some although Seward doubts any such relationship.

Hymenophyllum (about 300 species from Palaeotropics to temperate) and *Trichomanes* (about 350 species from the same region) were the only two recognized genera (Fig. 42.23). But now Pteridologists are prone to divide these into a large number of genera and Copeland, in his monograph on the family has recognized as many as 33.

Some 8 species of *Hymenophyllum* and 17 species of *Trichomanes* are recorded from India. *Hymenophyllum exsertum* is the commonest species here.

Hymenophyllaceae is a Marginales-Gradate family of some antiquity. Apogamy is very common in *Trichomanes.*

Family 6. HYMENOPHYLLOPSIDIACEAE

These are terrestrial ferns with *Hymenophyllum*-like habit but the rhizome is solenostelic and with scales. Sori are terminal at the ends of veins with leafy green, lobed indusia from base. Sporangia gradate, with broad, complete, obliquely vertical annulus.

Single genus *Hymenophyllopsis* with 2 species from South America.

Family 7. LOXSOMACEAE

These are terrestrial ferns with creeping, solenostelic, hairy rhizomes. Sori marginal at the ends of veins with cup-like indusium as in

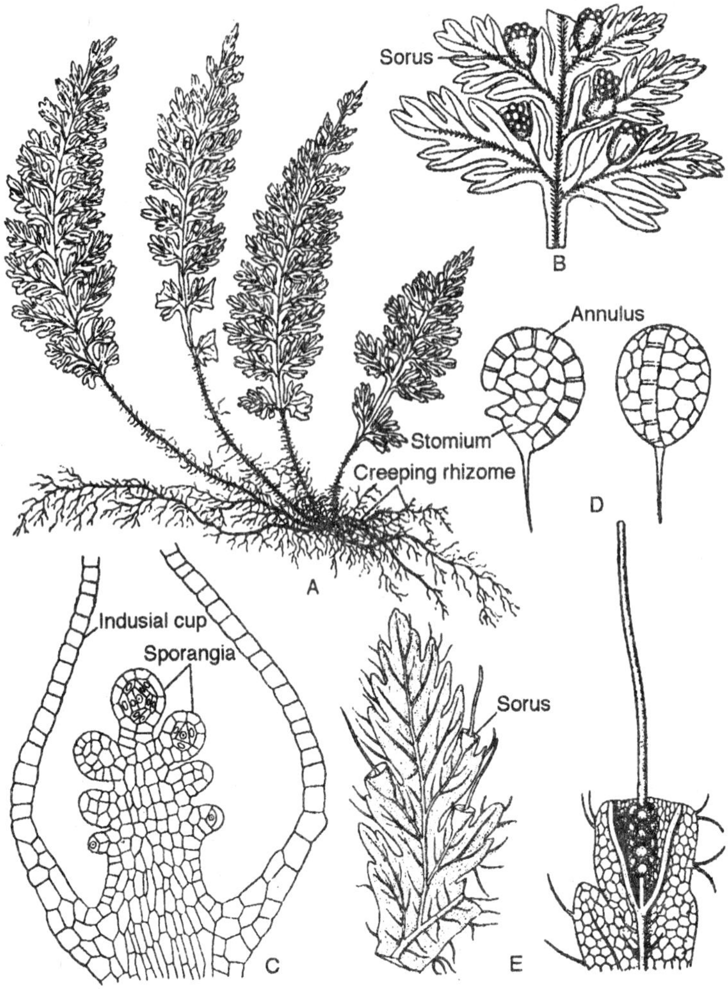

Fig. 42.23: A. *Hymenophyllum exsertum*, B. Fertile pinnae, C. Vertical section through sorus, D. Sporangia of A, E. Fertile pinnae of *Trichomanes* and vertical section through a sorus

Trichomanes. Sporangia with complete obliquely vertical annulus, gradate. Gametophyte heart-shaped with bristles below. This was formerly placed within Hymenophyllaceae.

To genera: *Loxsoma* (1 sp. from New Zealand) and *Loxsomopsis* (3 species from South America).

Family 8. PLAGIOGYRIACEAE

Terrestrial ferns with hard, erect, dictyostelic rhizome devoid of scales or hairs, with solenostelic runners. Pneumatophores develop from stipes. Sori are superficial on vein branches without indusia but protected by revolute leaf margin. Sporangia on long stalks with obliquely vertical annulus and stomium. Gametophyte normal.

Only one genus *Plagiogyria* with 35 species from East Asia and a few from Mexico to South America. Bower placed the genus within Polypodiaceae. 3 species of *Plagiogyria* are known from India, mainly in the Himalayas.

Family 9. DICKSONIACEAE

A family mostly of tree ferns with lofty, upright, radial, complicated dictyostelic stems covered by hairs but devoid of scales. Leaves pinnate, very large, forming a crown with their

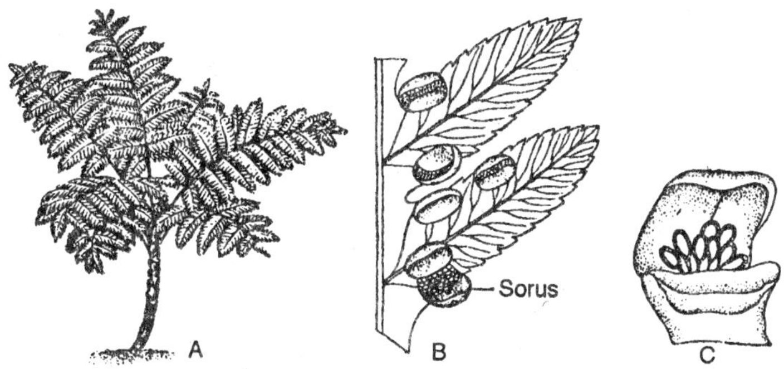

Fig. 42.24: A. *Cibotium* tree, B. Fertile pinnae of A, C. Bivalved sorus of *Cibotium barometz*. (B—C after Hooker)

bases hidden in a dense mass of hairs. Sori marginal and terminal on veins with cup-shaped (*Thyrsopteroideae*) or two-valved (*Dicksonioideae*) indusia. Sporangia basipetal (Gradate) with obliquely vertical annulus, stomium and transverse dehiscence. Gametophyte heart-shaped, thicker at median line.

The family is divided into two subfamilies:

(1) *Thyrsopteroideae* with the monotypic genus *Thyrsopteris* (1 sp. from Juan Fernandez) in which the sori with cup-shaped indusia arise from stalks (apparently without lamina) in the fertile part of the frond. *Thyrsopteris* is a possible relation of *Coniopteris*, a Jurassic fossil fern.

(2) *Dicksonioideae* with marginal, gradate sori having bivalved indusia. 3 genera are included here: *Dicksonia* (25 spp. from temperate Southern Hemisphere), *Cibotium* (13 spp. from East Asia, Hawaii and Central America) and *Culcita* (9 spp. from tropical and subtropical regions).

Cibotium barometz (Fig. 42.24) is known from Upper Assam.

Family 10. PROTOCYATHEACEAE

Protocyatheaceae is often included within Cyatheaceae but there are good reasons for considering it as a separate family. Bowers keeps this family separate as of Gleicheniaceous affinities. Short, erect stem with hairs but without scales, solenostelic. Sori are without indusium. The sporangia are of simultaneous origin, superficial on the under side of the leaf, mixed with hairs, large and short-stalked with vertical annulus.

2 genera: *Lophosoria* (1 sp.) and *Amphidesmium* (= *Metaxya*) (1 sp.)—both from South America. The family is known in India in the fossil stage from the specimen *Protocyathea trichinopoliensis* found in Tiruchirapalli Upper Cretaceous beds.

Family 11. CYATHEACEAE

A family of lofty tree ferns with dictyostelic stele covered by hairs or scales at top. Lower trunk shows leaf scars and is often covered by adventitious roots. Very large, pinnate compound leaves form a spiral crown. The stipes are often spiny and densely scaly, sometimes with *aphlebia* (abortive pinnae). Venation usually free, simple or dichotomous. Sori superficial on lower surface on veins, gradate, indusium absent or cup-shaped indusium present. Annulus obliquely vertical with stomium, dehiscing by transverse slit. Sporangia often mixed with hairs. Gametophyte normal but shorter and broader then in Polypodiaceae and several cells in thickness below notch. Multicellular hairs are present on prothallus. The family is known since Jurassic.

3 pantropical genera are recognized within the family:

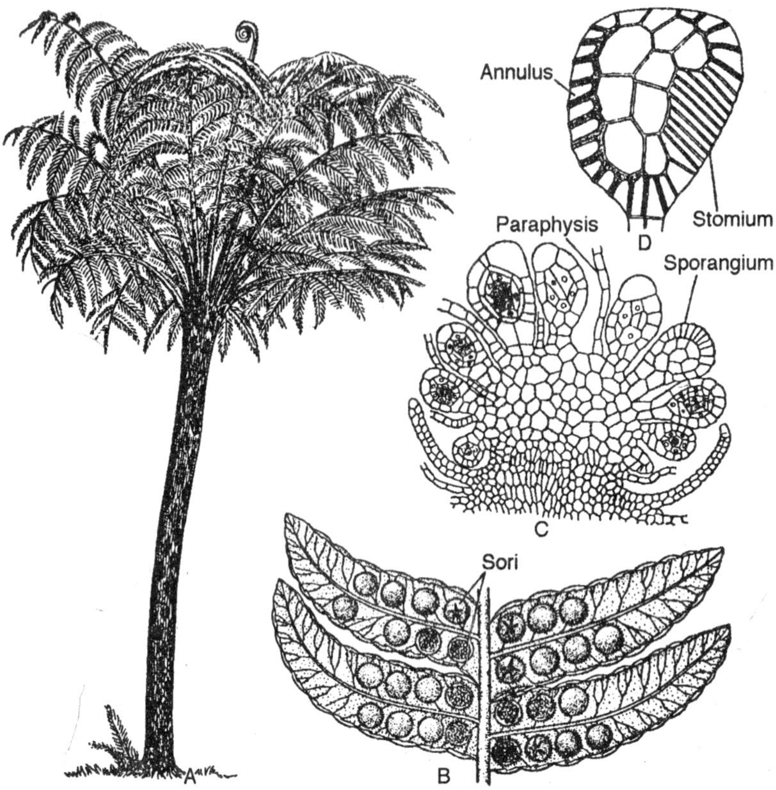

Fig. 42.25: *Cyathea*. A. Habit of the tree, B. Part of lower surface of sporophyll (some indusia have fallen off), C. Vertical section of gradate sorus, D. A Sporangium

Cyathea (about 300 spp.) with cup-shaped indusium completely covering the sorus when young (Fig. 42.25). Trees up to 20 metres high are known in the Darjeeling region and elsewhere. Five species are recorded by Beddome from India: *C. brunoniana, C. spinulosa, C. hookeri, C. sinuata* and *C. walkeri*. The genus *Amphicosmia* has been included within *Cyathea* by Christensen.

Alsophila (about 300 spp.)—without indusium. A predominant genus of Indian tree ferns with 7 species (c.f. Beddome)—*A. latebrosa, A. glauca, A. ornata, A. andersoni, A. oldhamia, A. glabra* and *A. crinita.*

Hemitalia (about 100 spp.) with a basal, scaly indusium. Some American species shown an anastomosis of veins.

Some Pteridologists (viz., Copeland) favour merging the three genera into one—*Cyathea.*

The relationships between Cyatheaceae and Dicksoniaceae has not yet been finally decided and still, the latter is considered to be only a part of Cyatheaceae by many pteridologists.

Family 12. POLYPODIACEAE

Polypodiaceae includes all (excluding Parkeriaceae) ferns of the Superficiales-Mixtae (i.e., sori on under surface with irregular development of sporangia) type. With about 170 genera and 7000 species this is by far the biggest family and the most preponderant of all the Pteridophytes. The oldest members of the family are from Jurassic but most of the species

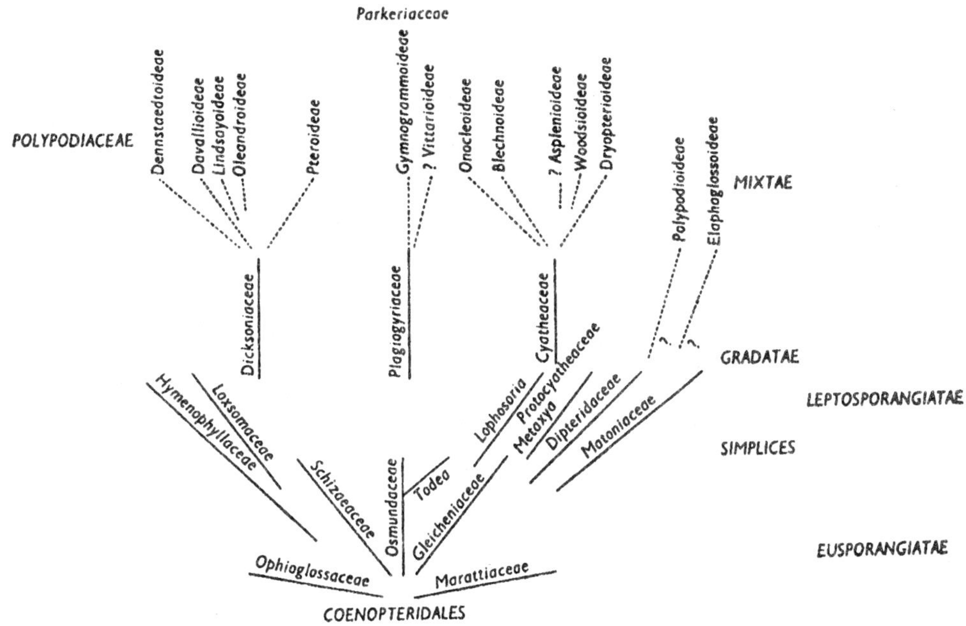

Fig. 42.26: Chart showing evolution of living ferns and polyphyletic origin of Polypodiaceae (after Bower but modified after Christensen)

are from Cainozoic so that this is a modern family now in its climax of evolution.

There is no doubt that Polypodiaceae is polyphyletic in origin. Because of this, many Pteridologists are of opinion that the Polypodiaceae should be divided into a number of families according to the phylogenetic lines. Thus, Ching (1940) has proposed 33 families out of the present Polypodiaceae. Others (viz., Pichi-Sermolli, 1958) prefer to break up the

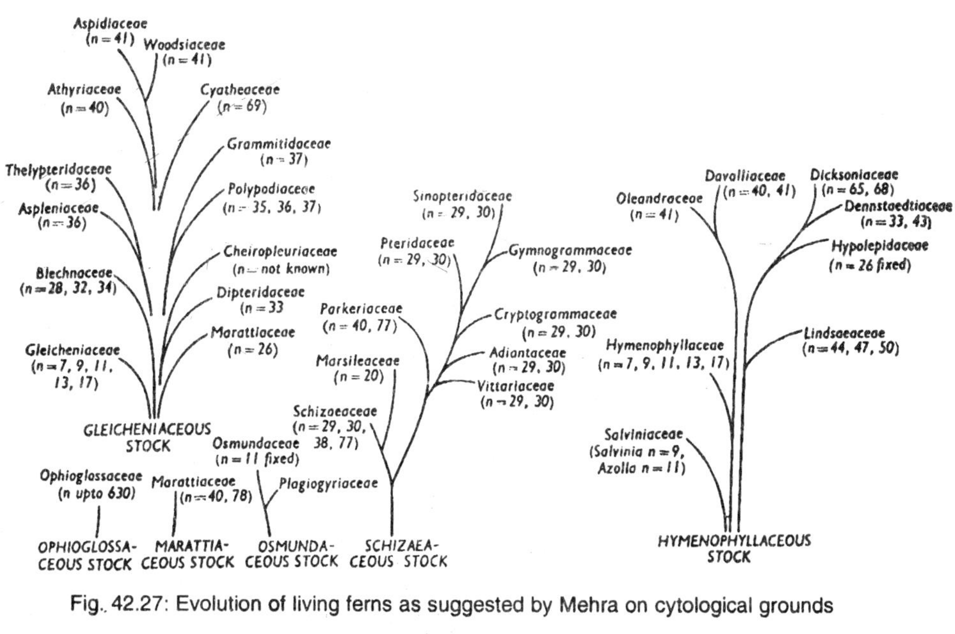

Fig. 42.27: Evolution of living ferns as suggested by Mehra on cytological grounds

Filicidae into a number of orders each of which includes some families. The families in each order may include some Polypodiaceous families along with some other families in the same phyletic line.

Bower as well as Christensen have divided the family Polypodiaceae into a number of subfamilies according to their phylogenetic tendencies. Figure 42.26 shows Bower's concept of the origin of the subfamilies. Figure 42.27 shows another view of the origin of the 'families' within Polypodiaceae as given by Mehra (1961) based on cytological evidences.

The subfamilies are shortly described below as enumerated by Christensen (1938) and embodied by Reimers in Engler's Syllabus (1954). The total number of species under each genus as well as the number of Indian species (Continental India only) as reported by Beddome (1883), after necessary nomenclatural changes, are also noted. While describing the subfamilies as such, the author apologises that this is only done to simplify the Polypodiaceae tangle for the beginner. Most Pteridologists now consider the subfamilies as families with varying affinities.

Subfamily 1. *Dennstaedtoideae*

Terrestrial ferns with creeping solenostelic rhizomes covered by hairs (rarely erect, dictyostelic and with scales). Sori round, terminal on veins, in punctiform or transversely oblong receptacles which are never fused, marginal (slightly shifted in *Microlepia*), basipetal (gradate), or mixed. Indusium double and indistinctly two-lipped or single. Stalk long, thin, often mixed with paraphyses.

Genera: *Dennstaedtia* (700 spp.—3 spp. in India), *Microlepia* (45 spp.—7 spp. in India), *Hypolepis* (45 spp.) and 5 other smaller genera.

Ching considered this subfamily as a separate family Dennstaedtiaceae. This family is now considered to be closely related with Cyatheaceae (Dicksoniaceae).

Subfamily 2. *Lindsayoideae*

Terrestrial and epiphytic ferns with proto- to solenostelic rhizomes covered by bristles or narrow scales. Sori single and round or coenosori on fused vein tips placed in a line, marginal or submarginal and terminal, mixed and with indusium (usually double, one side formed by upturned leaf margin). Sporangium stalk short.

Genera: *Lindsaya* (200 spp.—4 spp. in India, Fig. 42.28A), *Schizoloma* (6 spp.—3 spp. from South India but most of these species are not recognized by Christensen), *Stenoloma* (18 spp.—1 sp. from South India), *Taenites* (3 spp.—1 sp. from Assam, sporangia on two narrow bands parallel to midrib) and 5 other smaller genera.

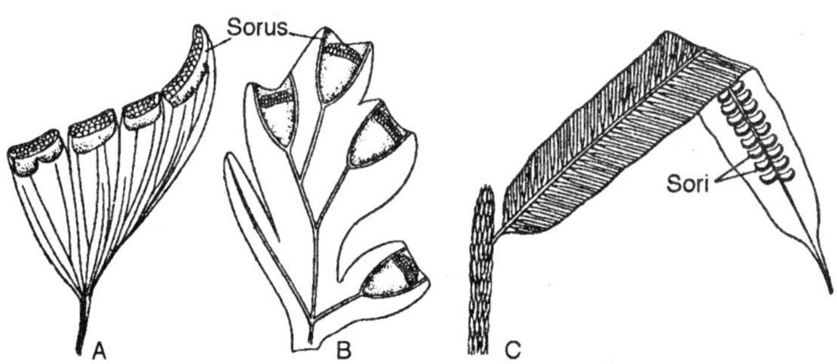

Fig. 42.28: A. fertile pinna of *Lindsaya cultrata*, B. Fertile pinna of *Davallia bullata*, C. A sporophyll of *Oleandra neriformis* showing sori (after Beddome)

Subfamily 3. *Davallioideae*

Mostly epiphytic ferns with creeping dictyostelic rhizomes covered by broad, peltate scales. Sori and indusia as former but never fused, indusium reniform or suborbicular, stalk of sporangium very thin (one row of cells).

Genera: *Davallia* (40 spp.—4 spp. in India, Fig. 42.28B), *Humata* (40 spp.—1 sp. from South India), *Leucostegia* (20 spp.—9 spp. from India), *Nephrolepis* (30 spp.—4 spp. from India), *Prosapita* (10 spp.—2 spp. from India, *Arthropteris* (15 spp.) and two other smaller genera.

Subfamily 4. *Oleandroideae*

Creeping or erect rhizome with imbricates scales. Sori dorsal superficial, round, in a wavy line usually near midrib with reniform indusia. Fertile frond may or may not be much narrower.

Genus: *Oleandra* (40 spp.—3 spp. from India, Fig. 42.28C).

Subfamily 5. *Pteridoideae*

Terrestrial ferns with erect or creeping solenostelic (often perforated) rhizomes covered by hairs (e.g., *Pteridium*) or scales (e.g. *Pteris*). Sori marginal or submarginal, usually continuous as coenosori borne on a vascular commissure connected with vein ends and protected by reflexed leaf margin and also in two genera (e.g., *Pteridium*) by a thin inner indusium.

Genera: *Pteridium* (single cosmopolitan species *Pteridium aquilinum*, very common in India, Fig. 42.29), *Pteris* (includes *Campteria* and *Litobrochia*—280 spp.—19 spp. from India), *Acrostichum* (3 spp.—one sp. *A. aureum*, Fig. 42.30A, is common in tidal

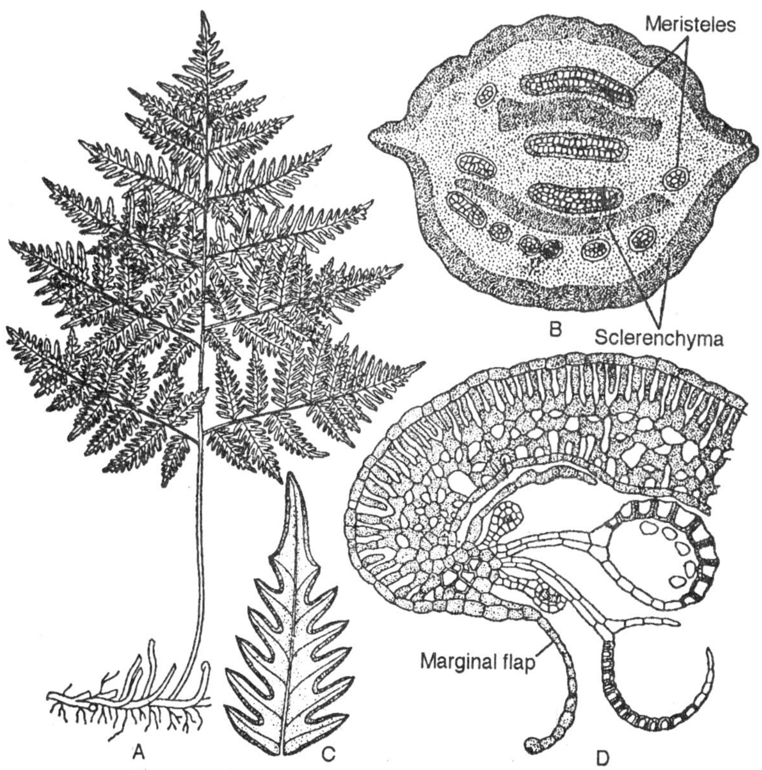

Fig. 42.29: *Pteridium aquilinum*. A. Habit, B. T.s of rhizome (should be compared with *Dryopteris*), C. A fertile pinna, D. T.s of sorus

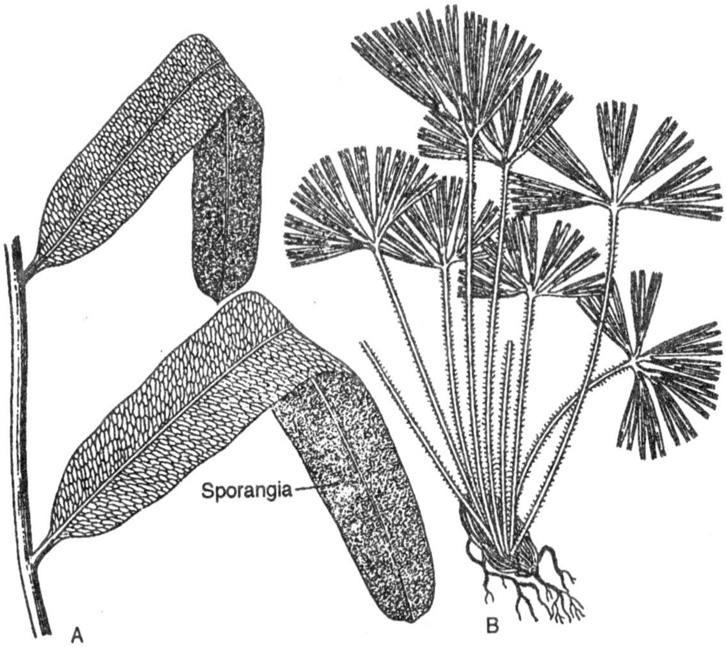

Fig. 42.30: A. Leaflets of *Acrostichum aureum* (sporangia on lower surface), B. *Actinopteris dichotoma*

coasts as a halophyte all over the tropics including India. The golden sporangia are spread all over the lower surface without any distinct sorus. Upper leaflets are fertile or fertile fronds are separate. The genus is included within the Pteridoideae for apparent Pteriodoid origin), *Stenochlaena* (4 spp. of long climbers—one species, *S. palustre* in India. Upper fertile leaflets are much narrower, sporangia on lower surface spreading from vein to margin. Copeland removes this genus to Blechnoideae), *Actinopteris* (single species, *A. dichotoma* founnd in India, Fig. 42.30B) and 7 other genera.

Subfamily 6. *Gymnogrammeoideae*

Ferns of diverse habits, mostly terrestrial with solenostelic (sometimes dictyostelic) rhizomes. Sori are superficial without real indusia but may be covered by marginal lobes or continuous reflexed margin, either confined to distal end regions of veins or forming a continuous marginal row where there is no commissural vein (except *Onychium*), sometimes following course of veins up to midrib. Sporangia mixed, large and short-stalked. Spores mostly globose tetrahedral. Bower derives this subfamily from Plagiogyriaceae but Christensen thinks it is of Pteridoid affinity. The subfamily is divided into four tribes:

A. *Cryptogrammeae:* Sori on apical parts of free veins protected by reflexed leaf margin. Rhizome scaly. Leaves usually dimorphous.

   Genera: *Onychium* ('golden fern'—6 spp.—2 spp. in India), *Cryptogramma* (4 spp.—1 sp. in India) and 2 other genera.

B. *Gymnogrammeae:* Sori linear along veins becoming reticulate when venation is reti-

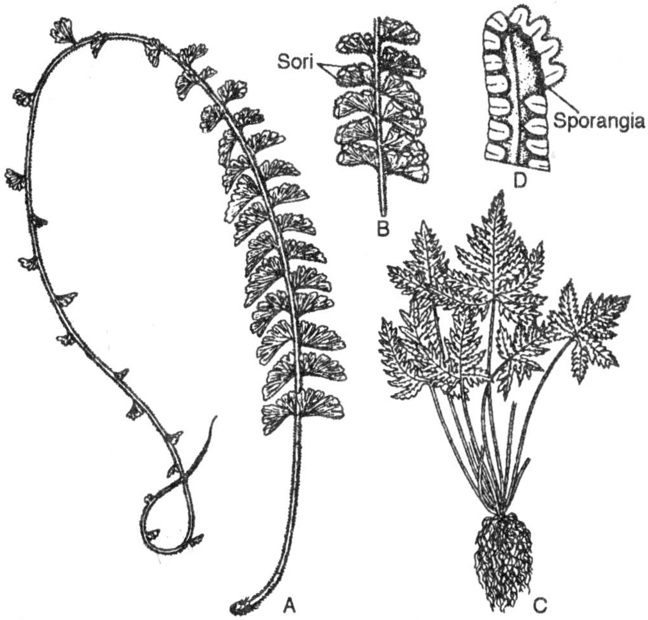

Fig. 42.31: A-B. *Adiantum caudatum* ('maidenhair fern') leaf the fertile pinnae, C-D. Same of *Cleilanthes argentea* ('silver fern')

culate. Sporangia scattered, large. Rhizome hairy or scaly.

Genera: *Gymnogramme* (60 spp.—3 spp. in India), *Syngramma* (20 spp.—2 spp. in India), *Gymnopteris* (about 10 spp.—mostly from India), *Hemionitis* (8 spp. in India), *Jurmesonia* (18 spp. from South America) and 11 other genera.

C. *Adianteae:* Sori under the sharply reflexed leaf margin, continuous or not. Rhizome scaly.

Genus: *Adiantum* ('maidenhair fern'—200 spp.—8 spp. in India, *A. caudatum* being very common, Figs. 42.31A and B. Also c.f. cover picture of this part. Popular in fern houses).

D. *Cheilantheae:* Sori at ends of veins but may extend up to midrib, protected by reflexed modified leaf margin.

Genera: *Cleilanthes* (130 spp.—10 spp. of India mostly 'silver ferns' because of a white powder on undersides of leaves, Figs. 42.31C and D), *Pellaea* (80 spp.—7 spp. in India), *Notholaeana* (60 spp.—2 spp. from Alpine Himalaya), *Doryopteris* (35 spp.—1 sp. in India) and 4 other genera.

Subfamily 7. *Vittarioideae*

Epiphytic ferns with creeping protostelic (viz., *Monogramma*) or dorsiventrally dictyostelic (viz., *Vittaria*) stems with characteristic scales.

Sori without indusium, coenosori immersed or superficial, usually mixed with paraphyses, spores bi- or triplanate.

Genera: *Vittaria* (80 spp.—2 spp. in India, Figs. 42.32A and B), *Monogramma*, (2 spp.—grass-like), *Antrophyum* (40 spp.—3 spp. in India—sori along reticulate veins) and 5 other genera.

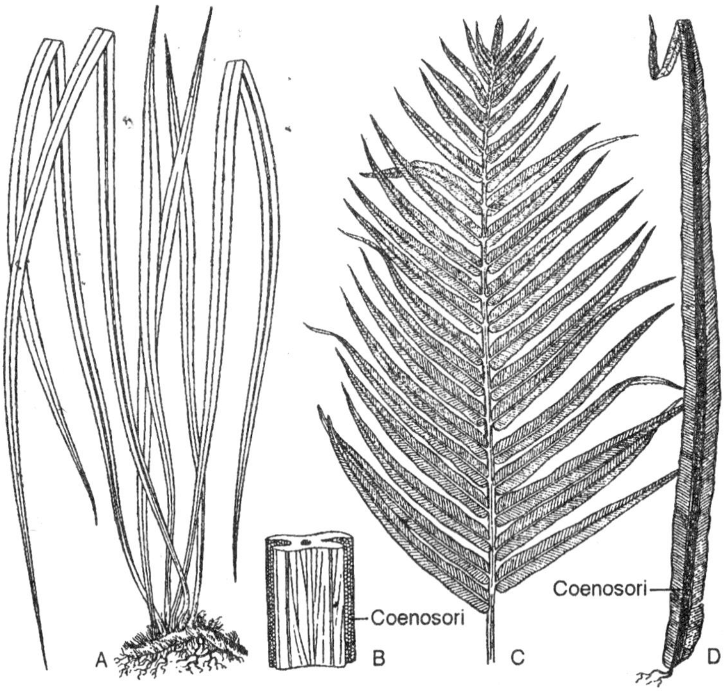

Fig. 42.32: A-B. *Vittaria elongata:* Habit and a bit of leaf showing immersed marginal coenosori, C-D. Frond and fertile pinna of *Blechnum orientale*

Subfamily 8. *Onocleoideae*

Terrestrial ferns with dictyostelic rhizomes covered by scales. Leaves dimorphous. Sori superficial, dorsal on veins, globose, in cylindrical receptacles with membranous, inferior, shell-shaped indusium, basipetal (Cyatheaceous origin). Spores bilateral.

Genera: *Matteuccia* (= *Struthiopteris* Willd.) (3 spp.—1 sp. in India), *Onoclea* (1 sp.) and another genus.

Subfamily 9. *Blechnoideae*

Terrestrial, sometimes subarborescent with erect, creeping, dictyostelic rhizome covered by scales. Sori as short or long coenosori borne on commissures parallel to midrib in a single line (1 to 3 rows in *Doodia*) protected by membraneous indusium. Spores large, bilateral.

Genera: *Blechnum* (200 spp.—4 spp. in India, Figs. 42.32C and D). *Woodwardia* (12 spp.—1 sp. in India), *Doodia* (11 spp.), *Brainia* (1 sp. found in India) and 2 other genera.

Subfamily 10. *Asplenioideae*

Erect, creeping or climbing with dictyostelic, scaly rhizomes. Sori superficial, linear, oblique to the midrib, with membranous single or double flap-like indusia. Spores, bilateral, with perisperm except in *Cystopteris.* Divided into two tribes:

A. *Asplenieae:* Sporangia direct on veins, usually on one side only. Rhizome scales hard.

Genera: *Asplenium* (includes *Phyllites, Thamnopteris* and part of *Hemidictyum*; about 700 spp.—about 42 sp. found in India, Figs. 42.33 A and B) and 5 other genera.

B. *Athyrieae:* Sporangia on receptacles on a vascular strand branched off vein.

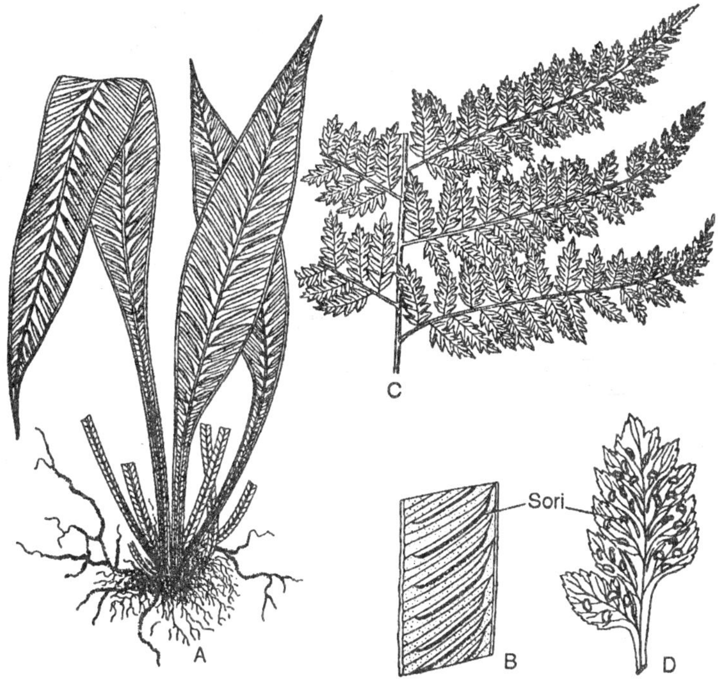

Fig. 42.33: A-B. Habit and sori of *Asplenium (Thamnopteris) gravillei*, C-D.Part of frond and fertile segment of *Athyrium fimbriatum*

Indusia variable. Rhizome scales large and soft.

Genera: *Diplazium* (375 spp.—20 spp. in India), *Athyrium* (includes *Anisogonium* and *Allantodia* (180 spp.—17 spp. in India, Figs. 42.33C and D), *Cystopteris* (18 spp.—1 sp. from Alpine Himalaya) and another genus.

Subfamily 11. *Woodsioideae*

Small terrestrial ferns with dictyostelic, erect rhizomes having scales. Sori dorsal superficial, near vein ends, circular. Indusium from base, often split like calyx, may wholly embrace sorus. Sporangia basipetal, few. Spores bilateral, with perispore.

Genera: *Woodsia* (40 spp.—3 spp. in India), *Diacalpe* (1 sp.—found in India), *Peranema* (1 sp.—found in India) and 3 other genera.

Subfamily 12. *Dryopteridoideae*

Usually terrestrial ferns with erect or creeping dictyostelic rhizomes having scales. Sori superficial, dorsal (sometimes terminal on veins), usually round, indusiate or not, may be acrostichoid in some evolved genera. Indusia basal in origin but later centrally fixed on an elevated receptacle, reniform or orbicular. Sporangia mixed, with 3-rowed stalks. Spores bilateral with perispore.

A large family of about 1500 species in 25 genera. Divided into two tribes representing two phyletic lines:

A. *Dryopterideae:* Rhizome and leaves show large and broad to hair-like scales. Rachis with 4 to 7 or more vascular bundles.

Genera: *Dryopteris* (includes *Nephrodium*; 150 spp.—18 spp. from India are described by Beddome under

Nephrodium. Life history of *Dryopteris* is given below), *Polystichum* (225 spp.—8 spp. from India. Merges into *Dryopteris*), *Cyrtomium* (1 sp. from India), *Polybotrya* (24 spp.—1 sp. from India), *Ctenites* (150 spp.), *Tectaria* (= *Aspidium*, includes *Pleocnemia*, 200 spp.—16 spp. in India), *Pteropsis* (= *Drymoglossum*, 6 spp.—2 spp. in India) and about 12 other genera.

B. *Thelypterideae:* Rhizome and leaves with few scales. Rachis with 1 to 2 vascular bundles.

Ching and others consider Thelypteridaceae as a separate family of independent origin while Holttum considered it to be derived from Cyatheaceae. But Sen, again, consider the two as sister families not derived from one another. Pichi-Sormolli suggests a derivation of the family from Dennstaedtiaceae.

Genera: *Lastrea* (= *Thelypteris*, including *Leptogramma*; about 500 spp.—44 spp. from India), *Cyclosorus* (including *Phegopteris* and *Stegnogramma*; 200 spp.—11 spp. from India under *Phegopteris*), *Meniscium* (American—3 spp. from India), *Bolbites* (85 spp.) and a few other genera.

The life history of *Dryopteris filix-mas* is detailed below to give an idea of the Polypodiceae in general.

DRYOPTERIS FILIX-MAS (L.) SCHOTT

Dryopteris filix-mas (the 'Male Shield Fern'), a cosmopolitan species, is chosen to typify the Polypodiaceous life history.

The sporophyte

The sporophyte (Fig. 42.34) has a partly horizontal and partly erect rhizome covered by scales and adventitious roots with the apex rising just above soil surface. Leaves are borne at the apex acropetally and the lower parts are covered by persistent leaf bases and adventitious roots growing from the leaf bases. The younger leaves at top are circinately coiled. Young leaves (or leaf buds) are edible as

Fig. 42.34: *Dryopteris filix-mas*. A. Habit of sporophyte (greatly reduced). B. A frond (fertile)

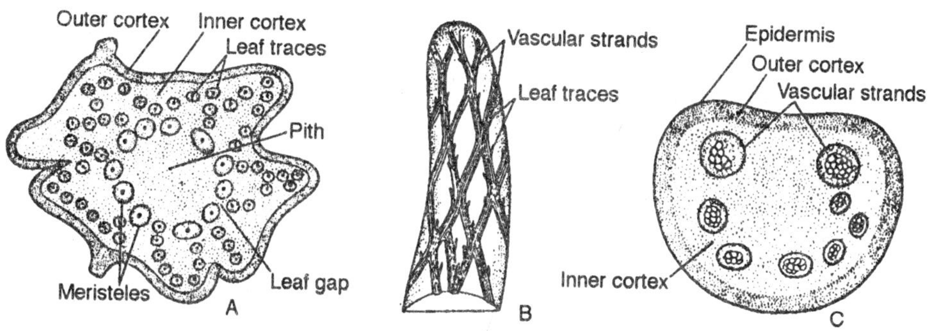

Fig. 42.35: *Dryopteris filix-mas.* A. T.s. of rhizome (diagrammatic), B. Vascular skeleton showing courses of the meristeles, C. T.s. of rachis

potherb (*dhenki shak* in Bengali). The *fronds* are large (may exceed a metre in length) and pinnately compound and there is usually an annual flush of a cluster of such fronds on the top of the rhizome, the leaf bases persisting when the older leaves perish. The lower portion of the rachis of each frond is *paleaceous,* i.e., covered by numerous scales which are brown and called *ramenta.*

A t.s. of the rhizome (Fig. 42.35A) shows below the *epidermis* a brown zone of *outer cortex* formed of lignified cells below which is the general mass of thin-walled parenchymatous *inner cortex.* The *vascular cylinder* is a simple *dictyostele* formed by a number of *meristeles* broken up by numerous *leaf gaps.* The resultant course of the vascular strands is shown in Figure 42.35B. There is no definite 'node' in the rhizome and the numerous *leaf traces* are seen in the inner cortex everywhere. The meristeles are rounded or elliptical and placed in a ring. Each is surrounded by an *endodermis* of two layers of which only the outer layer is distinct and the inner leayer merges with the pericycle of one or two layers. The bundle has a *phloem* of sieve tubes with sieve plates on lateral walls. This phloem surrounds a *mesarch xylem* in which there are one to three *protoxylem* patches of spiral tracheides surrounded by *metaxylem* of scalariform tracheides. Like *Selaginella,* vessels formed by the dissolution of the end walls of tracheides are known in *Pteridium.* Such vessels have also been reported in *Dryopteris.*

The rachis (Fig. 42.35C) has a similar arrangement of tissues but the large number of vascular strands are arranged in the shape of a horse-shoe and not in a ring.

The adventitious roots are provided with root hairs and the single central stele is exarch and diarch. The roots are dark in colour, branched and the branches are endogenous in origin rising form opposite the two protoxylem points.

The leaf is isobilateral with stomata on both the *epidermises* and a uniform spongy *mesophyll* (Fig. 42.36B).

When the leaves are mature the *sori* are borne on the under surface of the fertile pinnae. Usually whole fronds produced at this time are *sporophylls.* In *Dryopteris filix-mas* the sori are borne in two rows on two sides of the median vein of a pinnule (Fig. 42.36A) in between the margin and the midrib. Each sorus is covered by a kidney-shaped *indusium* which, although basal in origin, is raised on a cushion of tissues forming an umbrella-shaped structure covering the sorus. In the centre of the sorus (Fig. 42.36E) the vein ends in a *placental tissue* from which a number of *sporangia* arise without any order of development (i.e., *mixed*). A sporangium (including its stalk) develops from a single initial cell in the typical leptosporangiate way. The superficial *sporangial initial*

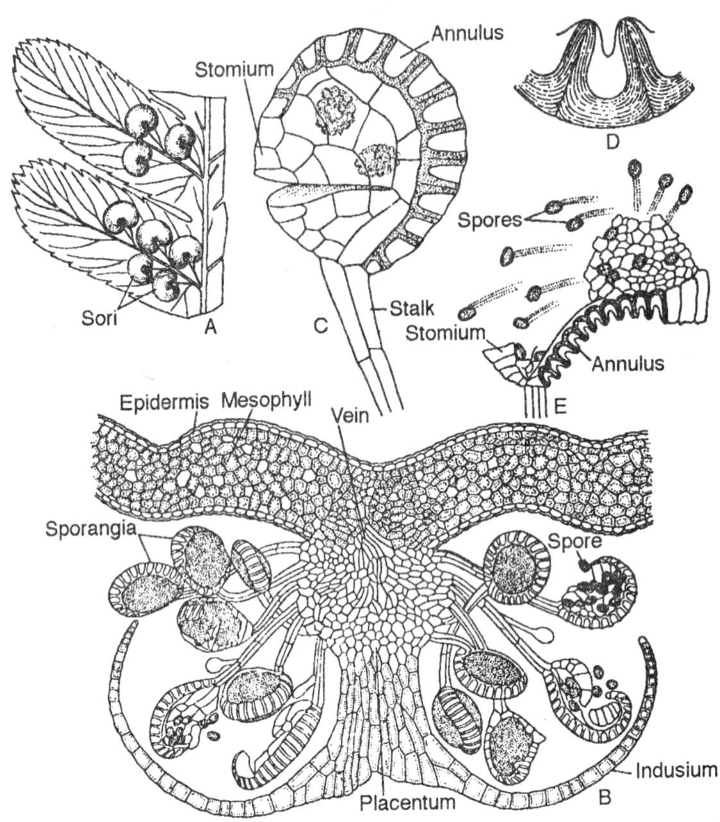

Fig. 42.36: *Dryopteris filix-mas*. A. Fertile pinnules with sori covered by indusia, B. Vertical section through a sorus (also showing leaf structure), C. A sporangium when moist, D. An annulus cell when dry causing contraction of the annulus and opening of the sporangium, E. The open dry sporangium with spores being catapulted out

(Fig. 42.37A) projects out of the surface of the receptacle and divides transversely. The outer cell gives rise to the sporangium. This outer cell forms a short filament with an *apical cell* with three lateral cutting faces (Fig. 42.37B). A 3-rowed *stalk* is developed from its base and the upper part ends in a flat outer *jacket initial cell* and an inner tetrahedral *archesporial cell* (Fig. 42.37C). The jacket initial forms the one layer thick *jacket*. The archesporial cell form a 2-layered nutritive *tapetum* on the outside and the *sporogenous cells* in the inside (Fig. 42.37D). The sporogenus cells develop 16 *spore mother cells* but some of them may abort so that the number of *spores* after reduction division may be 64 or less. Each mature sporangium (Fig. 42.37C) has a long slender *stalk* or 3 rows of cells and a biconvex *capsule* above. The wall of the mature capsule is one cell in thickness. While all the cell walls are thin, one vertical row of cells circumscribing the capsule has got three inner walls greatly thickened and is called the *annulus*. The annulus is incomplete at the base of one side where the cells are somewhat transverse and form the *stomium*. Because of the peculiar thickening of the annulus cells, when the capsule matures and is dry, the upper thin wall of each annulus cell is drawn in (Fig. 42.36D), the annular band contracts the stomium which ruptures and the spores are catapulted out (Fig. 42.36E). The hygroscopic movement may be repeated according to the moisture content of air and ultimately all spores are thrown out.

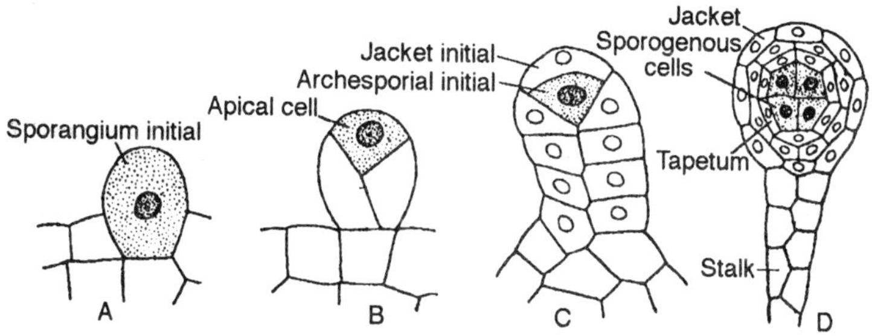

Fig. 42.37: *Dryopteris filix-mas.* A-D. Stages of development of the sporangium

The gametophyte

All spores are similar (*homosporous*) and a spore is more or less tetrahedral (Fig. 42.38A) with a thin *intine* and a hard, brown, irregularly thickened *exine.* The spore may remain viable for a long time. When suitable conditions for germination are attained, the exine bursts and the intine comes out with an apical cell of two cutting faces and forms a filament of 3 to 5 cells and one or two rhizoids at base (Fig. 42.38B). The apical cell is now replaced by a transverse row of apical initials which soon forms a one-cell thick, heart-shaped, green prothallus with an apical notch hiding the growing point. Later, the region posterior to the notch may become more than one layer thick. This *prothallus* may sometimes give rise to other daughter prothalli by *gametophytic budding.* The mature prothallus (Fig. 42.38C) shows on the ventral (lower) surface *archegonia* below the notch and numerous sturdy *rhizoids* with interspersed *antheridia* at the pointed end. There are a few short unicellular *hairs* on the margin of the prothallus. The fern gemetophyte is an independent plant like the Bryophyte thalloid gametophyte.

In the development of the *antheridium,* a superficial *antheridium initial* (Fig. 42.39A) divides transversely, the outer (i.e., lower, since the sex organs are pointing down on the lower

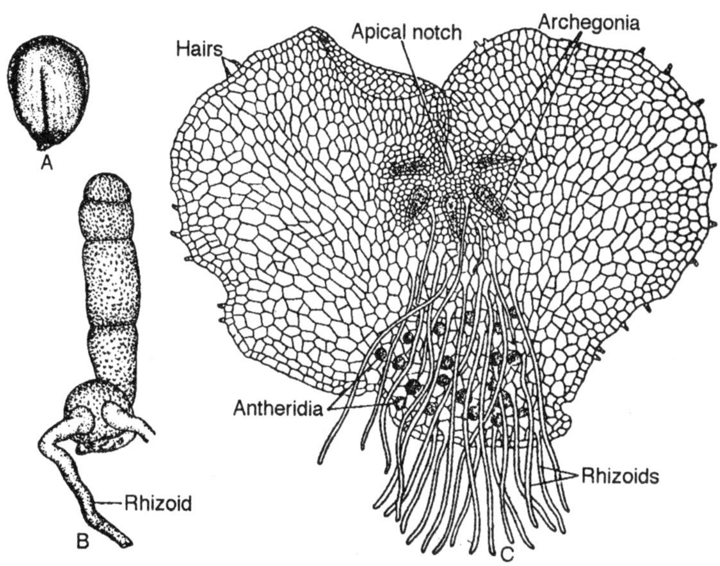

Fig. 42.38: *Dryopteris filix-mas.* A. A spore, B. Filamentous stage of prothallus, C. Mature prothallus

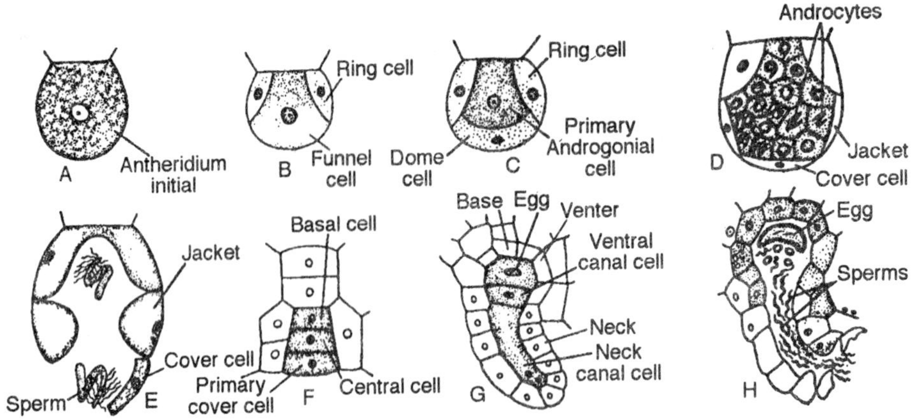

Fig. 42.39: *Dryopteris filix-mas.* A-E. Stages of development of antheridium up to its rupture, F-H Early and mature stages of archegonium

or ventral face) cell of which is pushed in by turgidity to form a *funnel cell* penetrating the inner one which then becomes a *ring cell* (Fig. 42.39B). The funnel cell cuts off a peripheral *dome cell* and an inner *primary androgonial cell* (Fig. 42.39C). The dome cell cuts out a top *cover cell* and another *ring cell* below. The two ring cells form the jacket and the primary androgonial cell forms 20 to 50 *androcytes* or *sperm mother cells* (Fig. 42.39D). Each androcyte forms a multiciliated *sperm.* The mature antheridium bulges out of the thallus and, at maturity, the cover cell is pushed off and the sperms are liberated (Fig. 42.39E).

The *archegonium* develops by the division of the *archegonial initial* into a *primary cover cell,* a *central cell* and a *basal cell* (Fig. 42.39F). The primary cover cell forms four *neck initials* which divide transversely forming a *neck* 5 to 7 cells high. The central cell gives rise to a binucleate *neck canal cell,* a *ventral canal cell* and an *egg.* The basal cell divides to form a few celled *base.* In the mature archegonium (Fig. 42.39G) only the neck protrudes out and remains bent away from the notch.

During fertilization the ventral canal cell and the neck canal cell along with the tip of the neck disintegrate and the passage becomes full of mucilage and chemicals like malic acid. These chemicals attract the sperms in large numbers (Fig. 32.39H) and fertilization takes place in the usual way.

The new sporophyte

Development of the *embryo* is rather uniform within the Polypodiaceae. The *zygote* divides into two by a vertical wall (Fig. 42.40A) and then by another vertical division at right angles to the first. A subsequent transverse division gives rise to 8 cells with 2 upper and

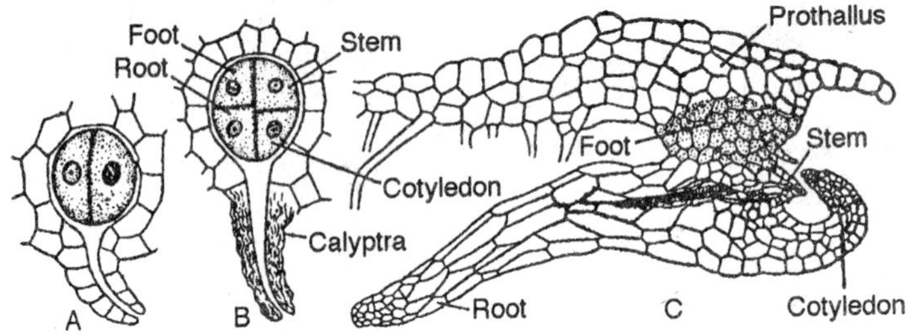

Fig. 42.40: A-C. Stages of development of embryo of *Dryopteris*

2 lower quadrants of 2 cell each (Fig. 42.40B showing 4 front cells of the quadrants only). These quadrants give rise to foot and stem from (*hypobasal* region), cotyledon and root (from *epibasal* region) respectively. The venter of the archegonium now forms the *calyptra*. The embryo (Fig. 42.40C) pierces out of the calyptra by the root coming out first. The root now establishes in the soil while the cotyledon and the stem bend and come out of the notch establishing the new independent sporophyte. Usually, one sporophyte is developed out of a gametophyte.

Figure 42.41 shows the life cycle of *Dryopteris* which may be considered as typical of the Filicales.

Subfamily 13. *Polypodiodeae*

A large subfamily of polyphyletic origin comprizing some 1200 species of various habits. Mostly tropical and subtropical ferns of small to medium size with many epiphytes. Sporangia on under surface of lamina, superficial or sunk, sometimes acrostichoid but usually in sori which are always exindusiate, globose or oblong, borne terminally or subterminally on free veins. Sori may be fused to form coenosori. Spores bilateral without perispore.

Most of the species are often included within the single genus *Polypodium* but this is not proper considering the polyphyletic origin of the species. Christensen considers the following two tribes and some subtribes of various derivations:

I. *Chaetopterides* (Dipteroid ferns): Rhizome protostelic, covered by hairs or bristles. Principal veins dichotomous, finer ones reticulation. Fronds dimorphous. No definite sorus. Filiform paraphyses.

Genera: *Cheiropleuria* (1 sp.), *Platycerium* (17 spp.) and *Christiopteris* (4 spp.; dictyostelic rhizome with scales, a link with the next tribe).

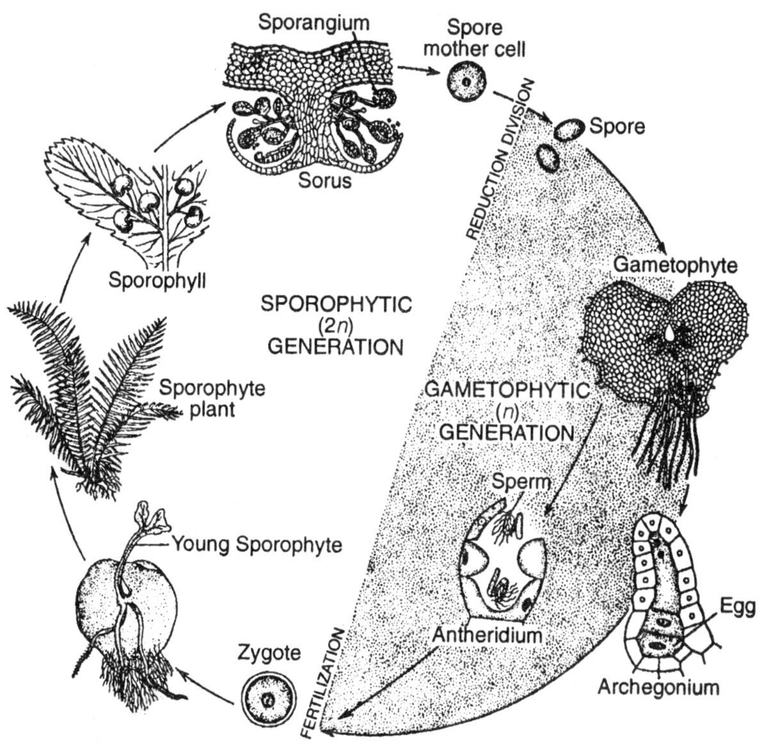

Fig. 42.41: Life cycle of *Dryopteris*

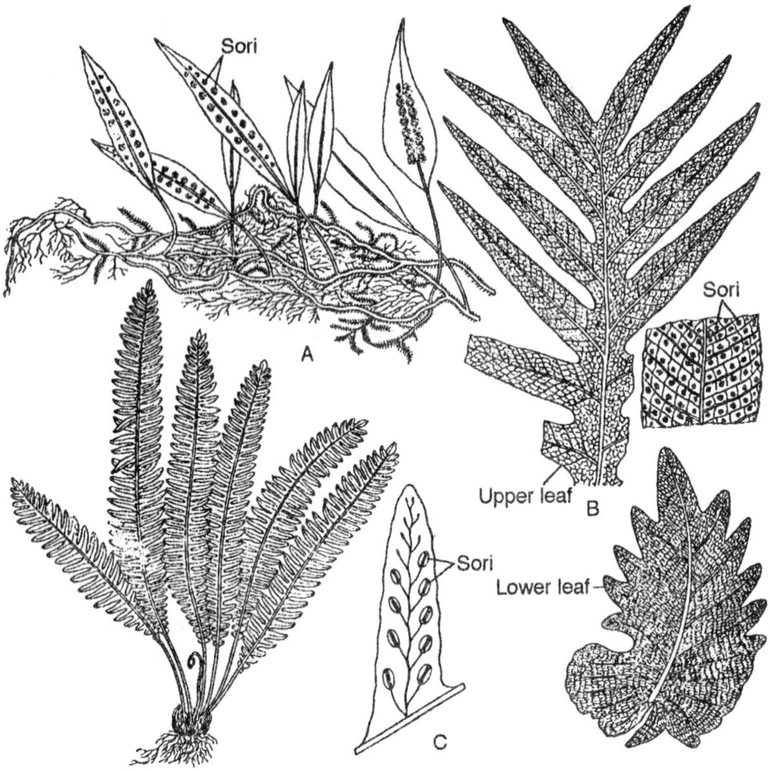

Fig. 42.42: A. *Pleopeltis rostrata*, B. *Drynaria quercifolia* showing two types of leaves, C. *Polypodium obliquatum* with a pinna showing sori

II. *Lepidopterides:* Rhizome with scales. Principal veins pinnate.

Subtribe A. *Pleopeltideae*. Venation reticulate. Sori on anastomosing veins or a plexus of veins, oblong, sometimes acrostichoid.

Genera: *Pleopeltis* (including *Lepisorus*; 40 spp.—27 spp. in India, Fig. 42.42A), *Hymenolepis* (15 spp.), *Loxogramme* (40 spp.—2 spp. in India, *Phymatodes* (100 spp.), *Selliguea* (2 spp. in India), *Microsorium* (40 spp.), *Cyclophorus* (including *Nipholobus*; 100 spp.—11 spp. in India), *Drynaria* (20 spp.—5 spp. in India; dimorphic leaves, epiphytic; Fig. 42.42B) and number of other genera.

Subtribe B. *Polypodieae*. Veins free or anastomosing. Sori usually oblong or circular on free veins.

Genera: *Polypodium* (includes *Goniophlebium*; 50 spp.–25 spp. in India, Fig. 42.42C), *Ctenopteris* (200 spp.), *Grammitis* (100 spp.) and a few other genera.

Subfamily 14. *Elaphoglossoideae*

Usually large terrestrial or epiphytic ferns with suberect or creeping woody rhizomes densely covered by scales. Leaves usually simple and articulated to petiole with free veins. Fertile fronds acrostichoid. Spores bilateral. A very advanced subfamily.

Genera: *Elaphoglossum* (includes *Acaniopteris*; about 500 spp.–7 spp. in India) and 3 other genera.

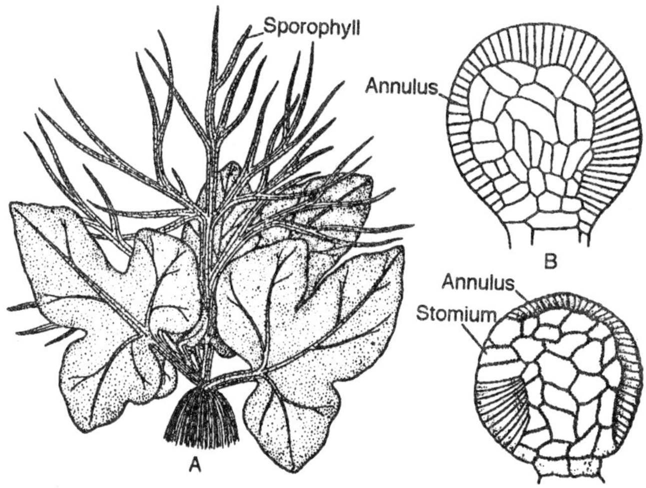

Fig. 42.43: *Ceratopteris thalictroides*. A. Habit, B. Sporangia

Family 13. PARKERIACEAE

The family Parkeriaceae has been made out of the genus *Ceratopteris* which Christensen and Bower had placed within the *Gymnogrammeoideae* (Family Polypodiaceae, page 1238) as a separate tribe *Ceratopterideae*. A single species *Ceratopteris thalictroides* is usually recognized although 2 to 4 species have been suggested. It is an aquatic fern, free-floating or rooted, growing in estuarine and other marshes through the tropics and sub-tropics. It is very common in Bengal and elsewhere up to an elevation of 1000 metres. It is cultivated in fish aquaria and also in rice patches to be used as a vegetable. Viviparous buds grow into a new plants from floating or submerged leaf margins. The stem is short, protostelic at first and then dictyostelic. Leaves are dimorphous (Fig. 42.43A). The first leaves are large, less dissected any may continue as long as the environment is not suitable for fertile fronds. Development of fertile fronds needs semi-dry condition. They do not develop under water. Fertile fronds are large and dissected like a stag-horn (hence the name *stag-horn fern*). Sporangia are solitary, submarginal and subsessile. The sporangial structure, specially the annulus, is very variable and the vertical annulus may be complete or incomplete with or without a stomium (Fig. 42.43B). Antheridia are on the margins of the gametophyte.

Because of the degeneration of the sporangial structure some Pteridologists insist on keeping it as a separate family although the degeneration may be merely due to its special environment.

Order 2. **MARSILEALES**: MARSILEACEAE

The Marsileales comprise of the single family Marsileaceae of small, subaquatic ferns growing in mud or on water covered mud. The rhizome is long, creeping, branched, solenostelic and hairy. Fronds are in two rows, circinate in bud and either simple and linear or with two or four opposite leaflets at the ends of long stipes. Veins are dichotomous, anastomosing at the tips. Sporangia are borne in hard, sessile or stalked structures called *sporocorps* one or more of which are borne ventrally on the lower part of the stipes. These are highly modified fertile pinnae. The heterosporous, gradate sori contain megasporangia and microsporangia.

The geological history of the group is not at all clear.

Three genera are included within the family: (1) *Pilularia* with linear grass-like leaves (Fig. 42.44A) with 6 species from temperate Europe, America and Australia; (2) *Regnellidium* with bifoliate leaves (Fig. 42.44B), monotypic with the single species *diphyllum* from Brazil; and (3) *Marsilea* with quadrifoliate leaves (Fig. 42.44C) with 70 spp. from the temperate and tropical regions all over the world.

Genus: *MARSILEA*

Of about 70 species of *Marsilea* spread all over the world, 9 species are recorded by Gupta (1962) from India. These are *M. minuta, aegyptiaca, brachycarpa, condensata, coromandelica, gracilenta, rajasthanensia* and *quadrifolia.* Of these, *M. minuta* is the commonest, being known from almost all the States while the temperate European species *M. quadrifolia* is definitely known only from Kashmir. They vary from hydrophytes (*M. minuta*) to xerophytes (*M. aegyptiaca*). In the following descriptions *M. minuta* is taken as the type.

The sporophyte

The stem of the sporophyte (Fig. 42.44C) is a thin creeping rhizome growing on the surface or slightly embedded in the mud. The internodes may be fairly long or short and there are two alternate rows or erect leaves along the upper side. Adventitious, branched roots develop in clusters from below the nodes. The rhizome branches freely and the branches are associated with leaf bases, axillary or lateral. A single plant may grow in all directions covering a large area. Leaves, when young, show circinate ptyxis. A mature leaf shows a long petiole on the tip of which there are four leaflets, apparently in a rosette but actually one pair is placed a little higher and alternates with the pair below. The venation of the triangular leaflets is dichotomous but the fine veins anastomose at the apices. The leaflets margin is entire, serrate or crenulate, more hydrophytic forms showing smoother margins.

Adventitious tubers on rhizomes meant for

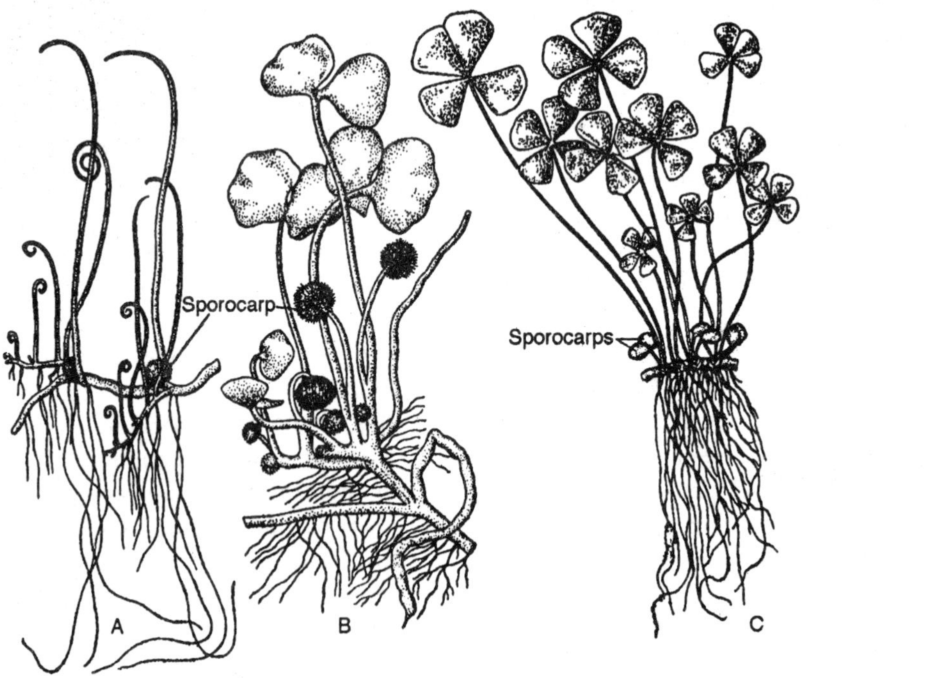

Fig. 42.44: A. *Pilularia globulifera* (after Meunier), B. *Regnellidium diphyllum* (after Lindman), C. *Marsilea minuta*

vegetative propagation have been reported in some forms of *M. minuta.*

The stem grows by an apical tetrahedral cell with three cutting faces. The vascular cylinder is solenostelic and amphiphloic with exarch xylem. A t.s. of the stem (Fig. 42.45) shows a single layer of *epidermis* and then a larger cortex divided into an *outer cortex* and an *inner cortex.* The outer cortex shows a ring of large *air spaces* interspersed by only single rows of cells forming an *aerenchymatous* zone. The outer part of the inner cortex is *sclerenchymatous.* All other tissue of the cortex is thin-walled parenchymatous. The cortex is lined on the inside by a layer of *outer endodermis,* then a single layered *outer pericycle* followed by the stelar ring and again by an *inner pericycle* and an *inner endodermis* on the inside. The stele is *amphiphloic* with a ring of *outer phloem* on the outside and another ring of *inner phloem* on

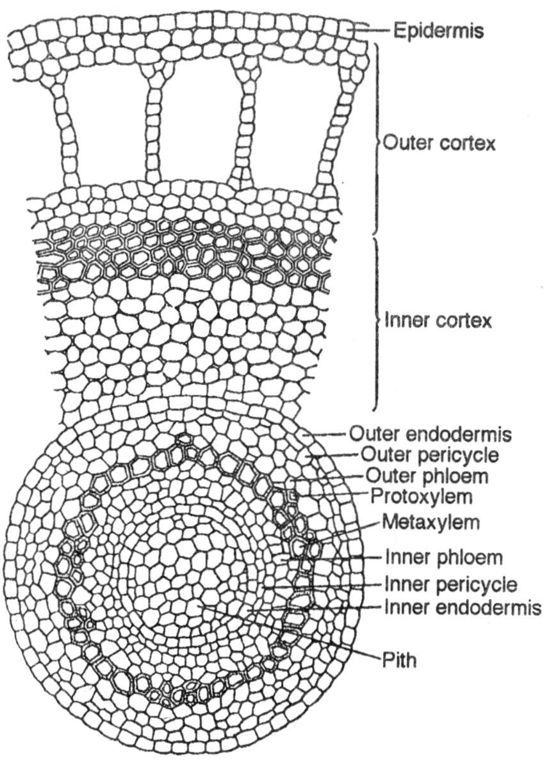

Fig. 42.45: T.s. of rhizome of *Marsilea minuta*

this inside. The xylem ring is formed of *metaxylem* while there are several patches of *protoxylem* on the outside (*exarch*). The central *pith* differs according to the environment where the plant grows. In aquatic forms it is all parenchymatous while in forms growing on mud or dry places the pith cells become sclerenchymatous (*sclerotic*).

The roots are monarch or diarch and exarch. Vessels, as in *Selaginella* and *Dryopteris,* formed by the dissolution of end walls, have been reported in the roots of several *Marsilea* species.

The leaf shows stomata on the upper side only in aquatic forms and the mesophyll is uniform. But in xerophytic forms sunken stomata have been reported on both the epidermises and the mesophyll is differentiated into spongy and palisade tissues.

The reproductive organs usually arise during the end of the winter when the ponds begin to dry up. The *sporocarps* develop on short or long pedicels from the petiole base (e.g., *M. minuta*) or higher up. There may be 1 to 20 sporocarps with their pedicels free or jointed. These are oval, bean-shaped, rounded or squarish structures. They begin as soft, green, hairy structures but become considerably hard and brown when mature. In *M. minuta* there are very few sporocarps (usually 2 to 3) on short pedicels arising from the base of the petiole (Figs. 42.44C and 42.46A). The pedicel joins the sporocarp at one end and is adherent to the shorter side of it forming something like a rimmed *raphe.* At the end of the raphe there are two short *horns* (Fig. 42.46B). The vascular strand enters the sporocarp from here and the upper longitudinal edge forms the *midrib* as if two pinnae have been folded and fused with the rachis. From this midrib the branch strands pass on to the two sides of the sporocarp. The sporocarp, therefore, has a bivalved structure. The outer side of the sporocarp is extremely hard, tough and even resistant to acids or any

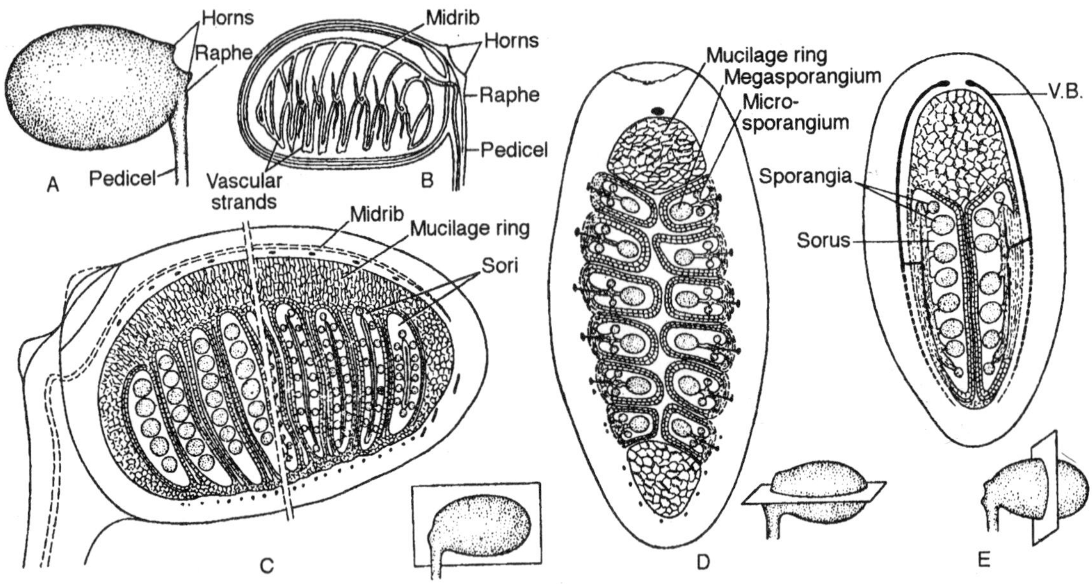

Fig. 42.46: *Marsilea*. A. Sporocarp, B. Diagrammatic longitudinal view of sporocarp showing vascular strands on one valve (after Puri and Garg), C. L.s. of sporocarp in vertical plane, D. L.s. of sporocarp in horizontal plane, E. T.s. of sporocarp (vertical plane)

form of weathering so that sporocarps of *Marsilea* have been reported to remain viable in herbarium sheets for even 50 years.

On each valve of the sporocarp, nearer to the margin than to the midrib, there is one row of elongated sori (Figs. 42.46C and E). A longitudinal horizontal section of the sporocarp (Fig. 42.46D) cuts each sorus transversely and it is seen that each sorus is elongate, covered by a delicate *indusium, gradate* basipetal in arrangement with a row of largest sporangia (megasporangia) on top and two rows of smaller sporangia (microsporangia) in two sides.

A scrutiny of the morphology of the sporocarp shows that it has been developed by the folding of two pinnae of a leaf (Fig. 42.47A) so that there is no doubt that the sporocarp is a fertile segment of a leaf of sporophyll nature. The sori are marginal and gradate. There is, however, some dispute whether the fertile segment represents a separate sporophyll or just a fertile basal leaflet of the vegetative leaf.

The sporangia develop in the leptosporangiate way with a one-layered jacket and a tapetum. Each sorus is heterosporous. The upper row of sporangia are larger and mature first into *megasporangia.* The lower two rows of sporangia are smaller and become *microsporangia.* Development of the two is similar and in each the *spore mother cells* develop 32 to 64 spores by reduction division. But, in the megasporangium, only one spore survives to become a large *megaspore* while in the microsporangium all the *microspores* mature. In the mature sorus all the sporangia fuse together with the megasporangia at top and the microsporangia below (Fig. 42.47C). Simultaneously with the development of the sori, the wall of the sporocarp becomes horny and the internal tissue round the rim, specially on the top, becomes mucilaginous. To this mucilaginous band the sori remain attached.

The opening of the sporocarp may be long delayed. It gets detached and remains in the mud. When the hard covering is sufficiently decayed it splits into two valves from the base and the sori come out slowly with the swelling

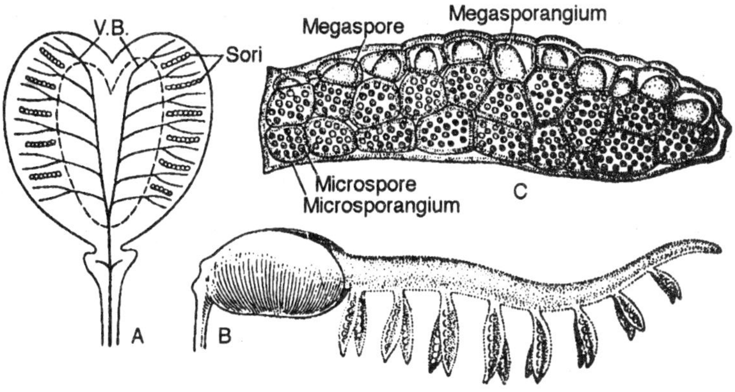

Fig. 42.47: A. Hypothetical fertile leaf segment from which the sporocarp of *Marsilea* has been derived, B. Extrusion of the sori on the mucilaginous cord, C. A mature sorus

and expansion of the long mucilaginous cord-like structure (Fig. 42.47B). The indusia of the sori also become gelatinous and absorb water.

The gametophytes

The microspores and the megaspores germinate on coming in contact with water.

The microspore germinates to form the *male gametophyte.* The microspore is spherical with a pyramidal apex and contains a large nucleus and numerous starch grains. It first cuts off a small lower cell, the *prothallial cell* and the remaining large cell is the *apical cell* (Fig. 42.48A). The apical cell divides into two to form two antheridial initials (Fig. 42.48B). Each of the antheridial initials now cuts off a few external *jacket cells* and one *spermatogenous cell* (Fig. 42.48C). Each spermatogenous cells forms 16 *androcytes* and in this stage the jacket and the prothallial cells disintegrate so that the two masses of androcytes representing the two *antheridia* remain free in a cytoplasmic mass formed from the disintegrated cells within the original spore wall (Fig. 42.48D). Each androcyte is now transformed into a multiciliated coiled *sperm* (Fig. 42.48E).

The macrospore germinates into the *female gametophyte.* The macrospore or megasore is

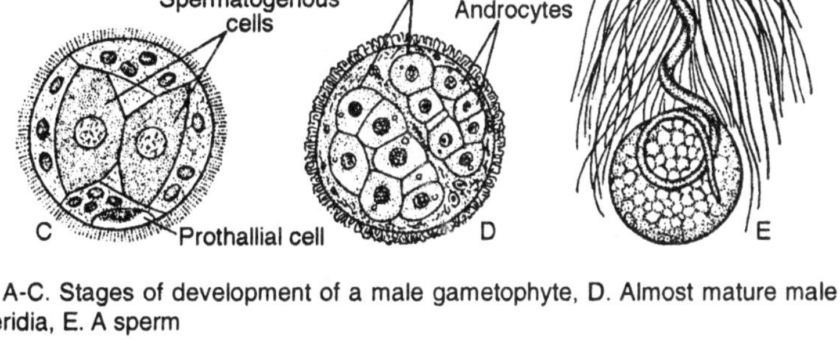

Fig. 42.48: *Marsilea.* A-C. Stages of development of a male gametophyte, D. Almost mature male gametophyte with two antheridia, E. A sperm

elliptical with a short protuberance at the tip where the wall is thinner and the nucleus is located in a dense part of cytoplasm. The rest of the spore is filled with a frothy cytoplasm full of starch grains (Fig. 42.49A). This apical nucleus divides into two and the upper smaller nucleus is cut off by a cell wall from the lower watery cytoplasm where the larger nucleus goes. The lower part forms a large cell with the nucleus which does not divide any further and this large cell acts as the *nutritive endosperm cell.* The upper cells soon develops an *archegonium* with a short *neck* of two tiers of four cells each, a *canal cell* (which may or may not divide), a *ventral canal cell* and the *egg* inside the *venter* (Fig. 42.49B). At this stage the tips of the megaspore splits by a triradiate fissure and the archegonium becomes exposed. The whole female gametophyte remains enveloped in a gelatinous mass which has a more watery, funnel-shaped entrance to the archegonium on top (Fig. 42.49C).

Fertilization takes place in the usual way with the sperms liberated from ruptured male gametophytes entering through the gelatinous matrix surrounding the female gametophyte (Fig. 42.49C)

The new sporophyte

The *zygote* resultant on fertilization develops the embryo in the typical leptosporangiate way. The first division is in a vertical plane and very soon four quadrants—cotyledon and root (*epibasal*), stem and foot (*hypobasal*) are clear in the embryo (Fig. 42.49D). The *calyptra* round the embryo is first pierced by the cotyledon and a few rhizoids are formed at first on the root (Fig. 42.49E). The embryo soon settles on soil to give rise to the independent sporophyte.

The life cycle of *Marsilea* is shown in Figure 42.50.

Campbell and Bower agreed that *Marsilea* had a Schizaeaceous origin and it was, therefore, placed next to the Schizaeaceae. The

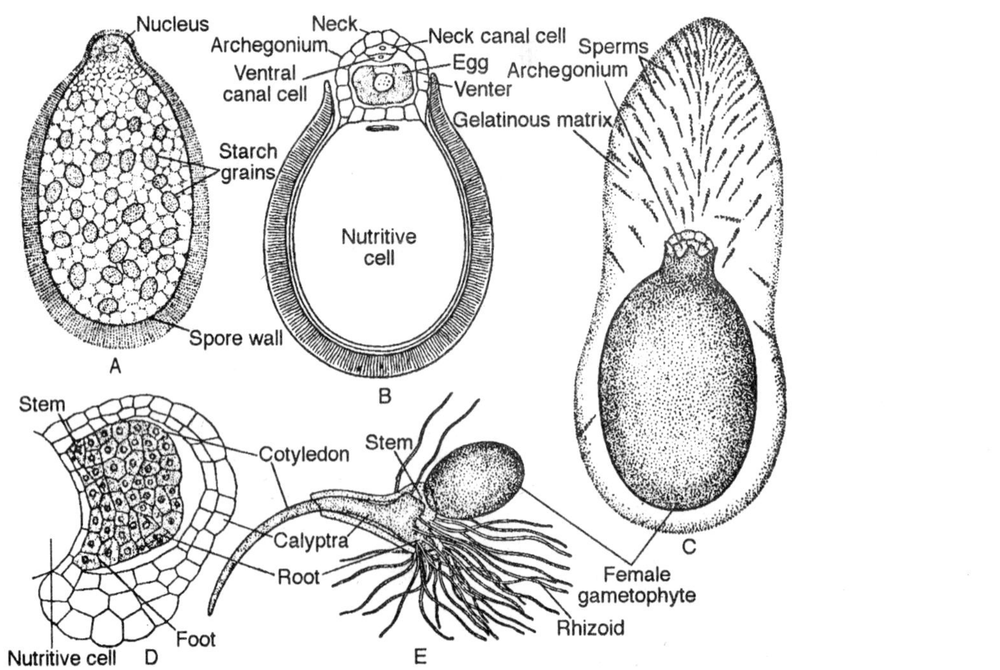

Fig. 42.49: *Marsilea.* A. Megaspore, B. Mature female gametophyte, C. Female gametophyte inside gelatinous matrix at the time of fertilization, D. Early stage of embryo, E. The new sporophyte on the gametophyte

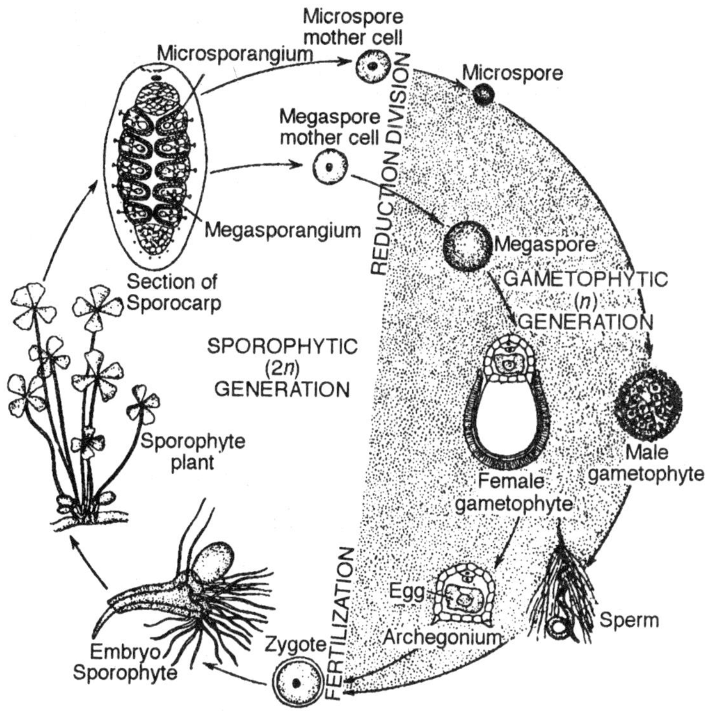

Fig. 42.50: Life cycle of *Marsilea*

similarities are morphological the anatomy, the back to back enfolding of pinnae in the sporocarp, the indusium and the marginal sori. *Pilularia* sporangium shows a vestigeal apical, ring-shaped annulus also linking with Schizaeaceae. But, this ancestral connection must be very remote since *Marsilea* shows a gradate sorus with a number of sporangia while in Schizaeaceae the sporangia are borne singly, the sori are simultaneous and homosporous. This means connection with gradate, leptosporangiate families. At most it may be said that the Marsiliaceae is a gradate, heterosporous derivative of the Schizaeaceous stock. Heterospory in Marsileales is less advanced since the same sorus contains megasporangia as well as microsporangia but, on the other hand, there is a single megaspore in the megasporangium and a single archegonium is developed in the female gametophyte.

It has already been pointed out that the Marsileales and the Salviniales cannot be grouped together and this point will become clearer on a comparison with the Salviniales.

Order 3. **SALVINIALES**

The Salviniales comprise of the two floating fern genera *Salvinia* and *Azolla* in which each sporocarp is a single sorus enclosing either megasporangia or microsporangia with the sporocarp walls formed by the indusia.

The above characters clearly show that there is no justification for including the Marsileales and the Salviniales within a single group (*Hydropteridineae* or *Rhizocarpae*) even though both of them are aquatic and heterosporous.

The terminal position of the sporocarps (sori), gradate development of the sporangia, the basal indusium enclosing a sorus and the elevated column on which the sporangia are borne suggest a Hymenophyllaceous origin for the order. Some species of *Azolla* show a vestigeal oblique, complete annulus-like structure resembling the Hymenophyllaceae. But, the whole sporocarp of the Salviniales is developed out of a single initial and this fact greatly prevents the idea of any direct connection with the Hymenophyllaceae.

The fossil history of the order is recent with fossils only from the Cainozoic.

Several Pteridologists have included *Salvinia* and *Azolla* within a single family *Salviniaceae*. But Christensen (1938), Reimers in Engler's Syllabus (1954) and Pichi-Sermolli (1959) are of opinion that the two genera are sufficiently distinct to be placed in two different families *Salviniaceae* and *Azollaceae*. This later classification is followed in this book.

Family 1. SALVINIACEAE

The single genus *Salvinia* forms this family. Basic $n = 9$. *S. cucullata* $2n = 45$. *S. natans* 2n = 18.

Genus: *SALVINIA*

About 12 species of *Salvinia* are reported from tropical Asia, Africa and America with one or two species in warm temperate Europe. Two species, *S. natans* (Europe and Asia—Fig. 42.51B) and *S. vucullata* (Fig. 42.51A) are known from India.

The sporophyte

The sporophyte of *Salvinia* (Fig. 42.51) is free floating (often covering the surfaces of whole ponds by copious vegetative multiplication), rootless, with a branched delicate, hairy rhizome up to 10 cm or more long, densely covered with leaves in alternate whorls of three thus giving rise to six rows of leaves. In each whorl the upper two leaves are green, floating, entire and ovate, cordate or oblong. Each of these two leaves has a midrib and anastomosing veins, the upper surface is papillose or grey-hairy or with stalked, branched hyaline hairs and the underside is often densely matted with brown hairs. The third lower leaf of each whorl is submerged, divided into numerous root-like filaments and covered by multicellular hairs to absorb nutriments from water.

The stem grows from a two-sided apical cell (c.f. Campbell). A t.s. of the stem shows extreme reduction as expected in such an aquatic plant (Fig. 42.52). Below the *epidermis* is an aerenchymatous *cortex* with large air

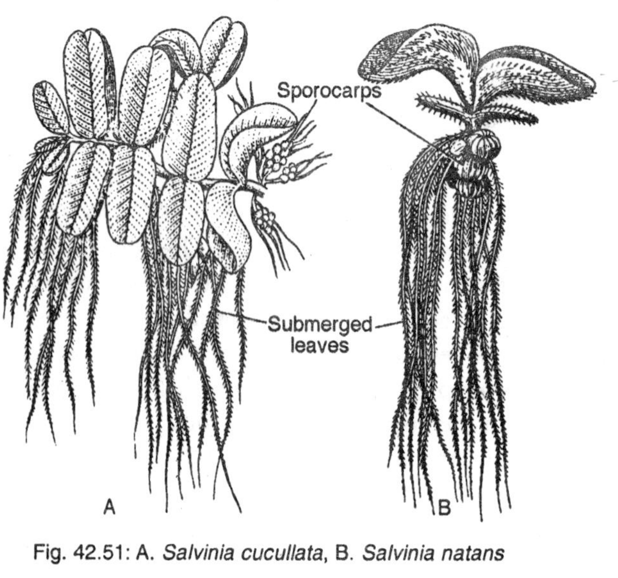

Fig. 42.51: A. *Salvinia cucullata*, B. *Salvinia natans*

spaces and the *stele* inside the *endodermis* is extremely reduced. It is siphonostelic with a delicate broken up *xylem* ring and *phloem* patches.

The floating leaves have no stomata in the *epidermis* and the *mesophyll* is spongy with two rows of large air spaces.

The sporocarps are borne in clusters near the base of the submerged leaf or in two-rows along some of the inner segments of it, there being 4 to 20 of them in a cluster. Actually, each sporocarp terminates a segment of the leaf being on the tip of the vein, the cluster being sympodial. The sporocarps are globose or

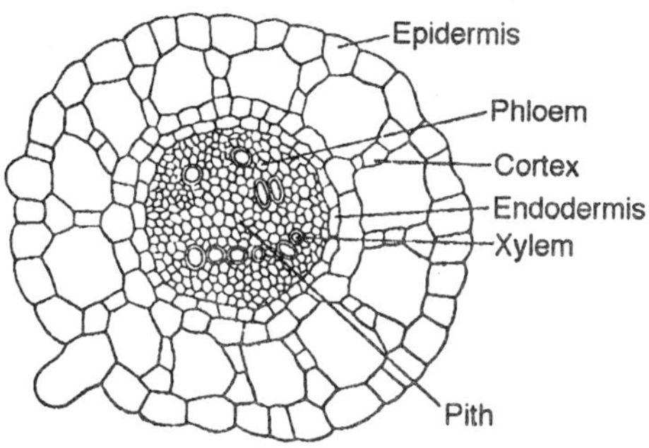

Fig. 42.52: T.s. of stem of *Salvinia natans*

ovoid, hairy when young and, although all of them are of the same size, the top ones contain only megasporangia and the lower ones only microsporangia.

In the development of a sporocarp, the whole structure begins from a single initial. Soon, a number of sporangia begin to develop basipetally (gradate) on a raised cushion and a cup-like *indusium* begins to grow from the base (Fig. 42.53). The development of the sporangium is leptosporangiate. Ultimately the indusium covers the whole sorus and inside it a number of stalked sporangia. with one cell thick *jackets* but no annulus, are seen on the raised cushion (Fig. 42.53B). A *tapetum* layer and *sporogenous* cells are organized inside each sporangium. Although development is the same in all sporocarps at the beginnings, later, in the *megasporocarps* a number of sporangia abort. In those surviving, although 32 megaspores are formed in each after reduction division of the *spore mother cells,* all except one large *megaspore* degenerate. In the *microsporocarps* (Fig. 42.53C) a large number of microsporangia are present and each develops 64

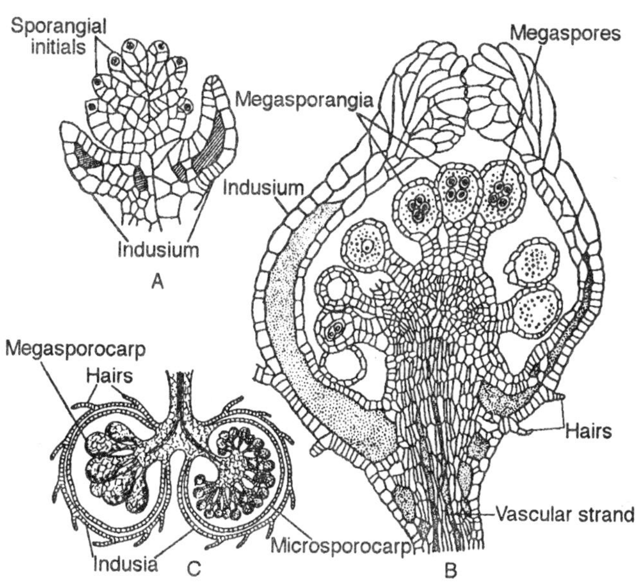

Fig. 42.53: *Salvinia natans.* A. Young sporocarp, B. Almost mature megasporocarp, C. A megasporocarp and a microsporocarp

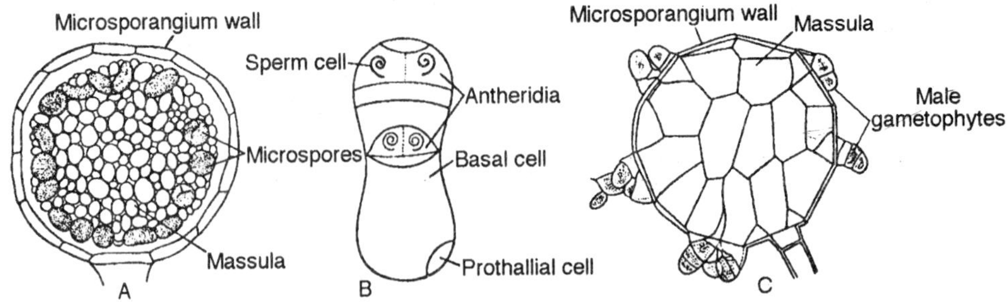

Fig. 42.54: *Salvinia natans*. A. Mature microsporangium, B. A male gametophyte (diagrammatic), C. Male gametophytes protruding out of the massula

spores on the reduction division of the spore mother cells. All these spores survive.

Indian species are annual and they are destroyed at the end of the winter when the sporocarps fall to the bottom and decay, liberating the sporangia which then rise to the surface.

The gametophytes

The microsporangium never opens. A copious tapetal cytoplasm formed on disintegration of the tapetal tissue forms inside it a frothy hardened mass called *massula* which encloses the microspores. The microspores show triradiate markings and they arrange themselves on the periphery of the massula (Fig. 42.54A).

Each microspore germinates into a *male gametophyte* at this position, i.e., within the massula. The mature gametophyte (Fig. 42.54B) shows a lower small *prothallial cell* and two *antheridia*, each with four *sperm cells* or *androcytes*. There is also a *basal cell* which pushes out the antheridia. Each androcyte gives rise to a coiled, multiciliate *sperm*. All the gametophytes developing from the peripheral microspores in a microsporangium burst out of the microsporangium wall so that the male gametophytes look like peeping out of the microsporangium although all of them are firmly embedded in the massula (Fig. 42.54C).

Similar changes go on inside the *megasporangium* itself when it matures. The single

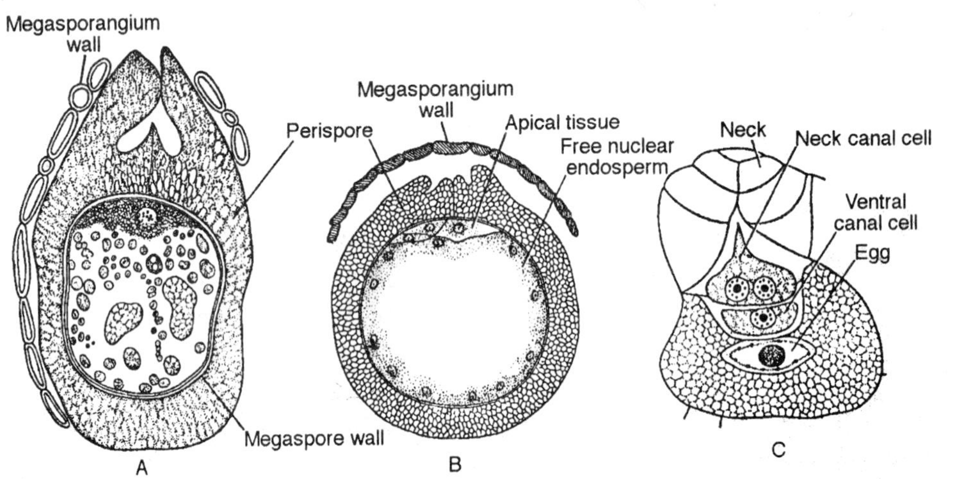

Fig. 42.55: *Salvinia*. A. Mature megasporangium, B. Top of young female gametophyte, C. An archegonium

surviving *megaspore* begins to enlarge and all the cells, i.e., the tapetum and the degenerated sister megaspores, harden into a vacuolated layer around the megaspore forming the *perispore* or the *epispore* which is lobed, triangular and pointed at the apex (Fig. 42.55A).

The megaspore now develops into the *female gametophyte.* The first division of the nucleus, which is at top, cuts off a small lenticular *apical cell* and the rest becomes a large lower *basal cell* filling up the spore cavity. The apical cell soon forms an apical tissue while the basal cell nucleus divides by free nuclear division. The free nuclei remain scattered in the copious cytoplasm forming a *nutritive endosperm* (Fig. 42.55B). The apical tissue projects out of the triradiate mound, often becomes green and an *archegonium* is formed in it. A mature archegonium (Fig. 42.55C) has a short *neck* protruding only slightly, one *neck canal cell* with two nuclei, a *ventral canal cell* and an *egg.* The first archegonium may be followed by the development of 2 to 4 archegonia on its sides at a later stage.

Fertilization takes place at this stage. The archegonium neck gapes open liberating the degenerated ventral and neck canal cells and sperms enter in large numbers until one fertilizes the egg.

The new sporophyte

The first division of the *zygote* is vertical and soon four quadrants give rise to *cotyledon, stem, root* and *foot* as in all Leptosporangiates. The root quadrant aborts at a very early stage so that a mature embryo has only cotyledon, stem and foot. The cotyledon comes out of the *calyptra* first and the stem next. The new sporophyte is then established.

Figure 42.56 depicts the life cycle of *Salvinia.*

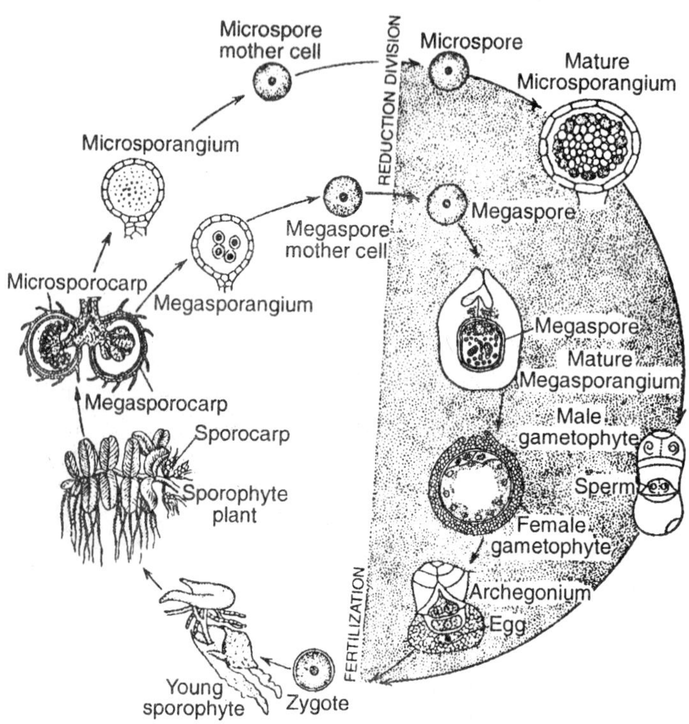

Fig. 42.56: Life cycle of *Salvinia*

Family 2. AZOLLACEAE

The family Azollaceae comprise of the single genus *Azolla.*

Basic $n = 11, 12$; *A. pinnata* $2n = 44$.

Genus: *AZOLLA*

Six species of *Azolla* are reported from tropical and temperate regions all over the world. The common Indian species is *Azolla pinnata.*

The sporophyte

The sporophyte of *Azolla* (Fig. 42.57A) is also free floating in ditches and ponds with a thin, delicate, pinnately branched rhizome. The plants are much smaller than *Salvinia.* Often the water surface is completely covered because of vegetative multiplication by decay of older parts and setting free of growing branches. The floating mass looks reddish when growing vigorously. The rhizomatous stem bears two rows of alternating leaves densely packed to overlap one another on the upper surface and long adventitious roots on the under surface here and there. Each leaf has two lobes (Fig. 42.57B) of almost equal size. The upper lobe (Fig. 42.57C) has photosynthetic tissue (palisade parenchyma) and stomata above and a large hollow below containing the alga *Anabaena azollae* which fixes free nitrogen and the association seems to be symbiotic. There is a single vein in the leaf. The leaf surface is papillate to prevent wetting. The lower submerged leaf is colourless and only one cell in thickness except for a few chloroplasts in the thicker base. This lobe seems to be an absorbing organ.

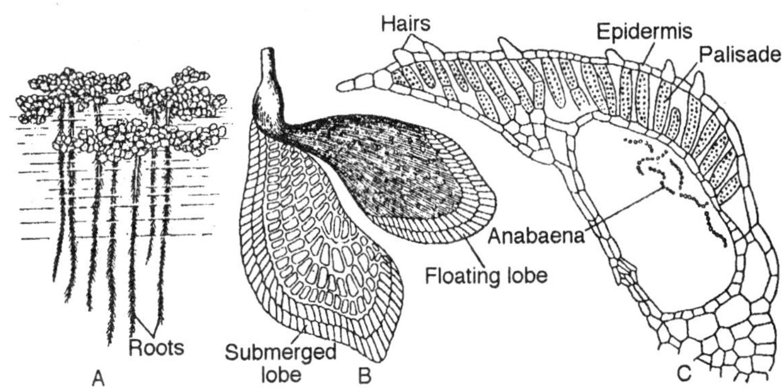

Fig. 42.57: *Azolla.* A. Habit, B. A leaf showing two lobes, C. Vertical l.s. of upper lobe of leaf

The stem grows by an apical cell of three cutting faces. The anatomy of the stem is similar to that of *Salvinia.* The *vascular cylinder* is slender and siphonostelic with *endodermis* and *pericycle* of one layer each. There are a few *xylem* elements in a ring broken up by *leaf traces.* The *phloem* is better developed on the lower surface of the stem. The *pith* is almost non-existent (Fig. 42.58).

The roots have root caps. It has a cortex of large thin-walled cells and a narrow, exarch, tetrarch xylem with a patch of phloem inside.

Sporocarps are borne only on the lowermost leaf of each branch at the end of the annual season. The submerged lobe of this leaf is split into two division (rarely four) and one sporo-

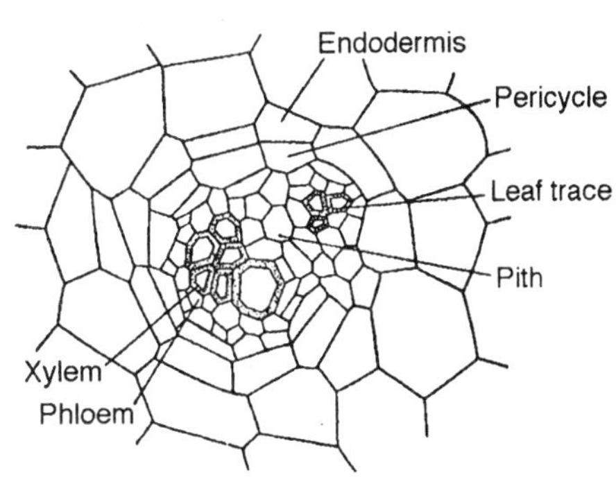

Fig. 42.58: T.s. of axial portion of *Azolla* stem (after Queva)

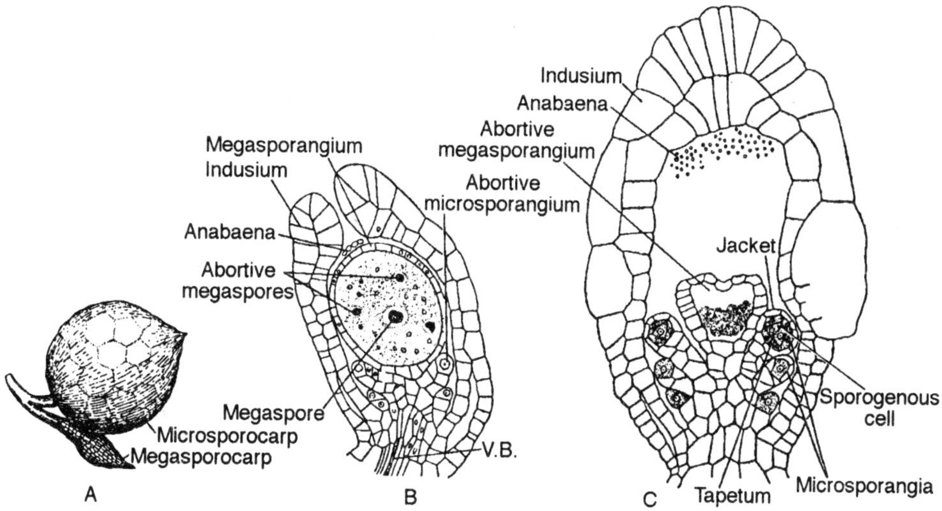

Fig. 42.59: *Azolla*. A. The fertile submerged lobe with one large microsporocarp and one small megasporocarp, B. L.s. of mature megasporocarp, C. L.s. of mature microsporocarp

carp is borne on the tip of each division. Of the two sporocarps one is larger and male (*microsporocarp*) and the other smaller and female (*megasporocarp*) (Fig. 42.59A). Each sporocarp is a *sorus* covered by its *indusium*. The development of the sporocarp is similar to that of *Salvinia*. The indusium develops from the base and is two-layered when mature. There is a central raised cushion on which the sporangia develop basipetally (*gradate*). Both the sporocarps seem to be potentially bisexual, the tip one developing as a *megasporangium* and the side ones as *microsporangia*. In the *megasporocarp* only the tip megasporangium develops and the lower microsporangia abort. In the megasporangium only one *megaspore mother cell* gives rise to four *megaspores* of which, again, only one megaspores enlarges and becomes functional while the other three degenerate (Fig. 42.59B). In the *microsporocarp* (Fig. 42.59C) the reverse takes place. The apical megasporangium aborts and all the lower microsporangia mature. Each microsporangium has a *tapetum* below the one-layered jacket and 16 *microspore mother cells* inside, which latter give rise to 64 *microspores*. Microsporangia of certain species have been reported to show vestigial oblique *annulus* suggesting a Hymenophyllaceous origin. But, usually there is no annulus.

In the *microsporangium* a multinucleate *periplasmodium* is formed out of the tapetal cells. This increases greatly with increase in the size of the sporangium and completely surrounds the microspores which soon take up a peripheral position (Fig. 42.60A). Subsequent to this, the periplasmodium organizes into four or more quadrately arranged *massulae* of alveolar, solidified structures each of which embeds some spores on the periphery (Fig. 42.60B). In some species the massulae develop on the surface tubular anchor-like structures called *glochidia*.

The single *megasporangium* within the megasporocarp also develops similarly. In it, the *periplasmodium* develops four *massulae* of which a large one at the base contains the functional megaspores while the three smaller ones hanging from top may each contain an aborted megaspore (Fig. 42.60C).

The sporocarp walls rot away and the sporangial walls burst at this stage liberating the massulae. The male massulae float away separately while four female massulae stick together causing better buoyancy. One or more male massulae stick to the female mass of massulae

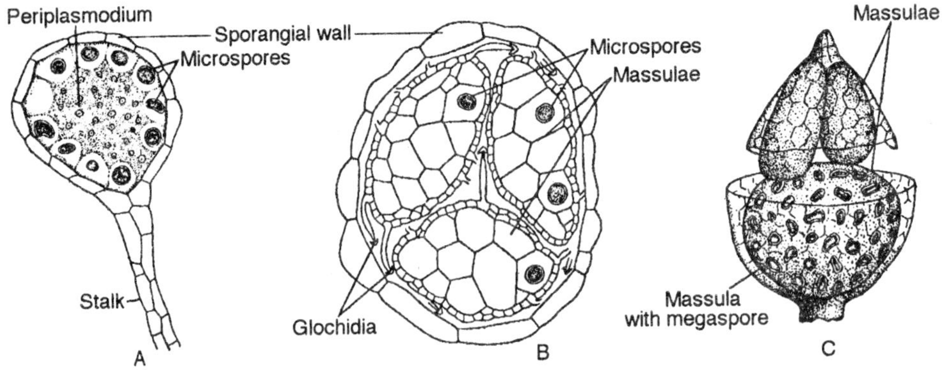

Fig. 42.60: *Azolla*. A. Nearly mature microsporangium, B. Massulae inside mature microsporangium, C. Massulae inside megasporangium

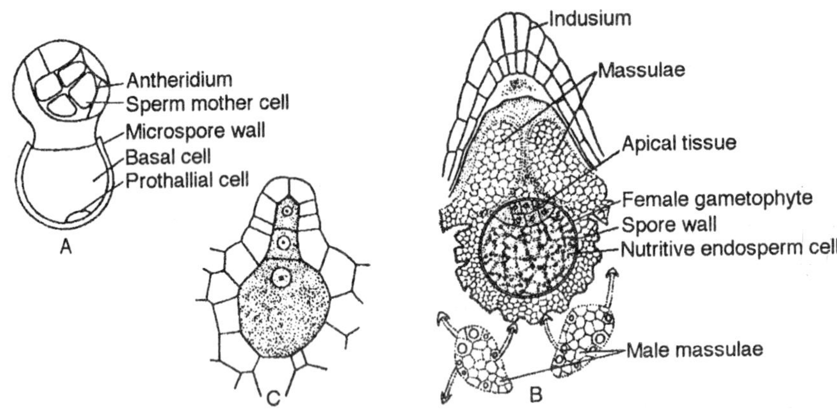

Fig. 42.61: *Azolla*. A. A male gametophyte, B. Early stage of the female gametophyte within massulae, C. A mature archegonium

which latter usually retains the top of the sporocarp wall above it like a cap (Fig. 42.61B).

The gametophytes

The microspores germinate into *male gametophytes* while still within a massula. It develops as in Salvinia but only a single *antheridium* with 8 *sperm mother cells* develops and this peeps out of the massula (Fig. 42.61A). Each sperm mother cell gives rise to a coiled multiciliate *sperm*.

The megaspore, still within the lower massula, develops a *female gametophyte* as in *Salvinia*. An *apical tissue* and a large, basal, free nucleate *nutritive endosperm* is formed (Fig. 42.61B). The apical tissue develops a number of archegonia and gets exposed bursting open the top of the megaspore wall but never turns green. A mature *archegonium* (Fig. 42.61C) shows a short, protruding *neck*, a *neck canal cell* whose nucleus may or may not divide and a *ventral canal cell* which sometimes may be absent and an *egg*.

Fertilization takes place as in *Salvinia*.

The new sporophyte

The first division of the *zygote* is transverse and not vertical. Subsequently, the four quadrants—*cotyledon, stem root* and *foot* are organized as in other leptosporangiates. The growing embryo comes out by pushing off the upper massulae, sporocarp wall top, etc. The cotyledon grows first and then the root and stem when the new sporophyte gets established.

Chapter 43

Discussions: Origin of the Pteridophyta and Origin of the Land Plants

The most important stage in the evolution of the Plant Kingdom is the migration of plants from water to land. Life, in the form of primitive algae, arose in the seas which then covered the earth in the very very early age. Algal, or bacterial remains as early as 3000 million (or at least 2000) years ago have been claimed to be discovered (Berghoorn and Schopf, 1966 and 1968—algae and bacteria 2000 to 3000 million years old from South Africa; Engel *et al*, 1968—algae 3000 million years old from South Africa). This means the very early archaeozoic days—the beginning of our Earth as we know it today. Even in India Gowda and Sreenivasa (1969) have found algal fossils in Chitaldrug Schist belt as old as 1400—2000 million years. Definite fossils of land plants (the Archegoniatae—vascular Pteridophytes and nonvascular Bryophytes) are known only from the Silurian. But, even in the Cambrian, spores of both Bryophytes and Pteridophytes and tracheides of vascular plants have been reported (Naumova, 1949 from Esthonia; A. K. Ghose, A. Bose, J. Sen, K. Jocob, etc. 1950-1953 from Indian Cambrian). This means the migration of plants from land to water took place during the Cambrian or pre-Cambrian days.

It would have been a simple matter if it could be proved that the algae gave rise to the Bryophytes and the Bryophytes to the Pteridophytes or the earliest vascular plants. But it has been explained in the concluding chapter on the Bryophytes that evolution was never as simple as that. The predominant idea is that from the algae arose a plexus of very primitive plants—monophyletically or polyphyletically. They may be called the *Thalassiophyta* or anything else. They gave rise to the Bryophytes and the Pteridophytes either separately or in a mixed manner. The views of Takhtajan, Zimmermann, Cronquist, Delevoryas and others have already been explained. The simplest, but not necessarily the earliest, vascular plants are the Psilophytales which, again, is another very problematic plexus of related or unrelated plants of monophyletic or polyphyletic origin. Takhtajan, Zimmermann and others believe that the Psilophytales gave rise to the Bryophytes as well by reduction in addition to giving rise to all the groups of the Pteridophytes as explained below. Mehra

(1969), on the other hand, has a very interesting theory as shown in Figure 43.1. He explains as follows: The pre-land plant ancestors (*Chaetophoraceous Ancestors*) gave rise to *Proto-Archegoniatae* which again gave rise to all the land plants. But this was in a bi-phyletic manner. One branch gave rise to the *Lycopsida* by evolution of a lateral sporangium ultimately on the adaxial surface of the leaf. Mehra supports this view on the ground that the primitive sporangia of this group do not show vascular connections and are, therefore, not telomes. The other branch, the true telomic line, had sporangia at the apices of the branches showing vascular connections in the vascular plants. This line split into two—one line giving rise to the Hepaticopsida and the Bryopsida and the other line is the *Anthorhyniaceae* branch. This branch gave rise to the Anthocerotales on the one hand and the *Rhyniaceous stock* (which later

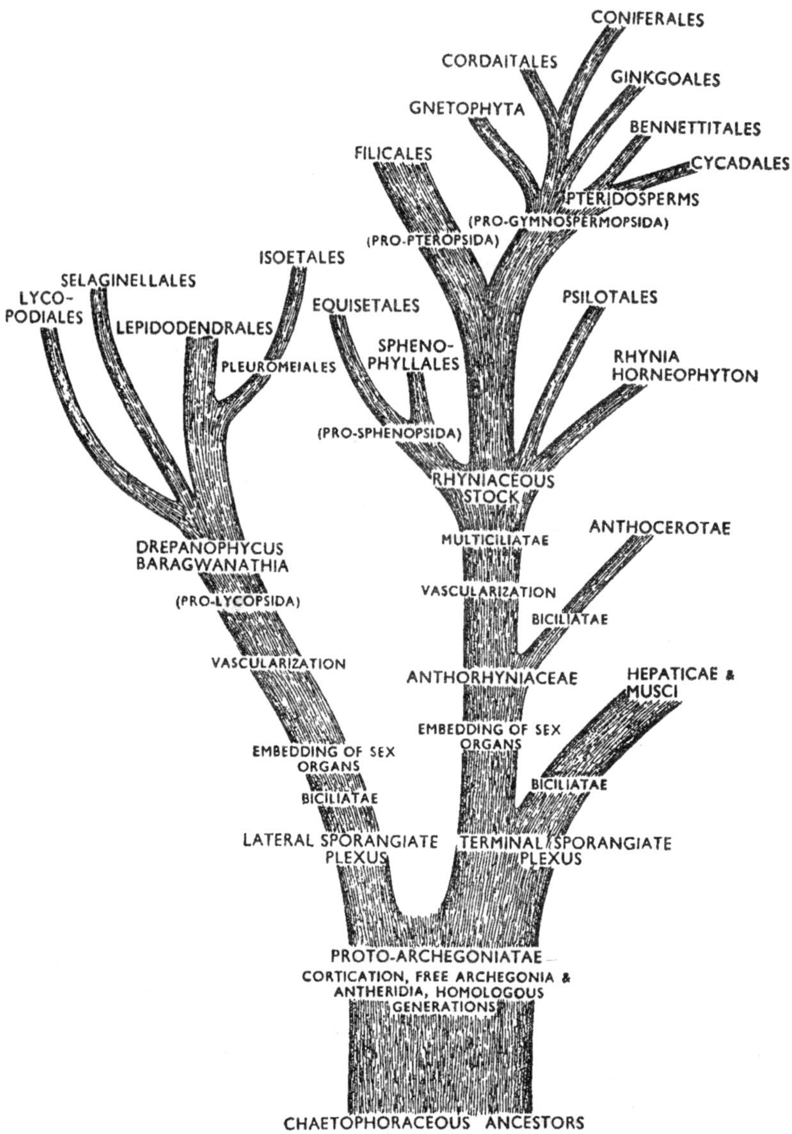

Fig. 43.1: Phylogenetic tree showing the origin of land plants as proposed by Mehra (1969)

evolved multiciliate sperms) on the other. The Rhyniaceous stock gave rise to most of the *Psilophytopsida* (some of them are problematic as stated before). The *Sphenopsida* and the *Filicopsida* (or *Pteropsida*). From the base of the Filicopsida stock rose the *Progymnospermopsida* which gave rise to the Gymnosperms.

To understand the views it is now necessary to discuss the different views on the origin of the Pteridophytes. The views may be discussed in two categories: (a) those of Algal origin and (b) those of Bryophytic origin.

(a) Theories of Algal origin of Pteridophyta

The dominant idea today is that the Pteridophytes arose out of the algae. Pteridophytic fossils are now known since the Silurian, decidedly before there is any known trace of the Bryophytes.

Lignier (1908) suggested the evolution of a hypothetical *Prohepatics* out of the algae which gave rise to the Hepatics on one hand and to the *Prolycopodinee* on the other out of which latter the modern Pteridophytes arose.

Church (1919) in his essay on the *Thalassiophyta* (discussed in the last chapter on Bryophyta) opined that after the algae became *benthic*, a large group of extinct seaweeds, much more advanced than the Thallophytes of today, arose. He has named this hypothetical group *Thalassiophyta*. He is not clear about the origin of the Thalassiophyta but it is supposed to be polyphyletic in origin and both the Bryophytes and the Pteridophytes of today originated out of this group.

Polyphyletic origin of the Pteridophytes out of different groups of algae has been suggested by Andrews (1947, 1956), Merker (1961) and others.

Monophyletic origin of the Pteriodophytes out of algae has been suggested by several authorities. Schenck (1908) suggested origin of the Pteridophytes from the Phaeophyceae. But more convincing is the origin of the Pteridophytes (as well as of the Bryophytes) from the heterotrichous Chaetophoraceous algae as suggested by Fritsch in various papers. Mehra (1969) does not support any polyphyletic origin of the Pteridophytes out of different algae because of the uniformity of the development of the sex organs in the former. The origin must have been from some *Protoarchegoniatae* where sex organs were already fixed.

Zimmermann (1930, 1938, 1949) has given a picture of the origin of the Pteridophytes based on his *Telome Theory*. To understand this it is necessary to have a clear picture of the Telome Theory.

The Telome Theory

The Telome Theory was preceded by similar ideas expressed by Bower (1884), Lignier (1903—1908) and Potonie (1913) who explained the uniformity of the plant body. Lignier explained the plant body of primitive land plants as a dichotomously branched system of cylindrical *cauloids* which bore leaf-like appendages (which he called *phylloids*) and the absorbing rhizoids.

Zimmermann defines the *telome* as the single-nerved extreme portion (at base or apex) of the plant body from the tip to the next point of branching (Fig. 43.2). The telome may be *fertile* bearing sporangia or *sterile* when it is called a *phylloid*. In Bryophytes the sporophytic plant is a single fertile telome (although not having a real nerve) terminating in a sporangium above and the foot below (Fig. 43.2A). In Pteridophytes (viz., Psilophytales) it is *syntelome* developed by growth of the telome. The telome divides dichotomously and as it grows the new segments becomes new telomes and the older segments below are termed *mesomes*. Thus the plant body is now formed of mesomes terminating in telomes above and below. The lower telomes bear rhizoids and the upper telomes

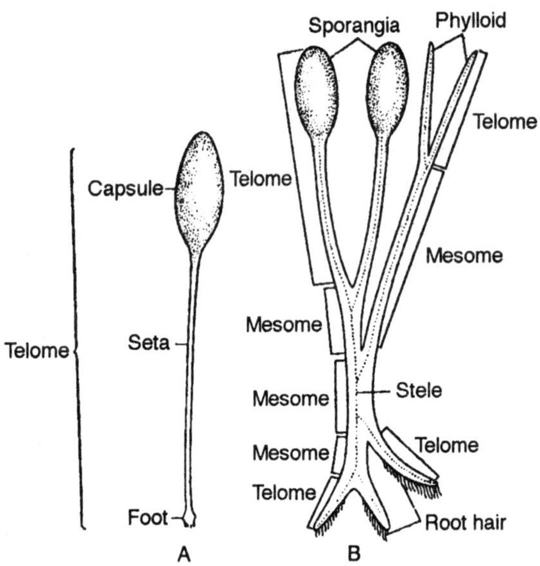

Fig. 43.2: A. The single telome Bryophyta sporophyte, B. The syntelome structure of Rhyniaceae (Redrawn after Zimmermann)

may be fertile bearing sporangia or be leaf-like *phylloids* (Fig. 43.2B).

From the primitive *syntelome* the pteridophytic plant body evolved by certain *elementary processes*:

Overtopping: Of the two usually equal dichotomies from the telome one became stronger and erect becoming the *axis* which grew further while the other remained overtopped as a short *lateral branch*. The axis, therefore, is sympodial (Figs. 43.3A and B).

Planation and syngenesis (*webbing*): The sympodial axis is liable to give rise to groups of telomes or *telome trusses* of short lateral branches. This branching is radial (Fig. 43.3B). Now, all such branches grow on the same plane by *planation* so that a flat structure with fan-like *telome trusses* is formed (Fig. 43.3C). Fusion of the telomes of such a telome truss by the development of connecting tissue as in the foot of a swan is called *syngenesis* or *webbing* (Fig. 43.3D). Webbing results in large megaphyllous leaves as found in ferns and higher plants.

Reduction: According to Zimmermann, the *microphyllous* leaves of Lycopods and pine needles were evolved by the reduction of telome trusses similar to those giving rise to the megaphyllous leaves. This view, however, is opposed by the *Enation Theory* which is discussed later.

The primitive dichotomy of the *telome* was derived from the algae. The structure of the Spermaphytes is a further development of the telome structure. Takhtajan (1953) has suggested renaming the *Embryophyta* (*Cormophyta*) as *Telemophyta* as their plant bodies are all formed by *telomes*.

Based on this *Telome Theory* Zimmermann pictures the origin and evolution of the land plants in this way:

1. The ancestors of land plants were green algae which lived in the tidal water of the

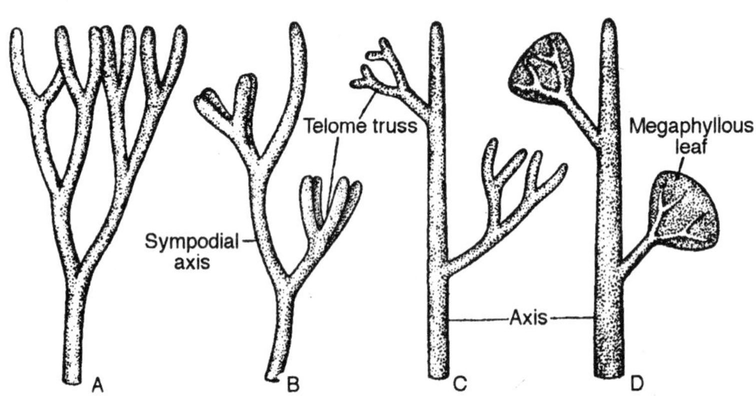

Fig. 43.3: A. A syntelome of radial symmetry, B. Sympodial axis formed by overtopping, C. Structure after plantation, D. Megaphyllous leaves after syngenesis (webbing) of telome truss

Cambrian and Silurian sea coasts. (N.B. Since vascular remains are now known even from the Cambrian the origin might have been earlier). They had *isomorphic* alternation of generations (*Homologous Theory*). It is very difficult to name the particular alga which was ancestral. (N.B. Monophyletic as well as polyphyletic origin have been suggested by various authors.)

2. The Psilophytan plant body was evolved out of the algae by the evolution of *heteromorphic* alternation of generations in which the two generations were independent of one another. The sporophytic plant body consisted of *telomes* (fertile and sterile, the upper sterile ones are *phylloids*) and *mesomes* (mature telomes). (N.B. Lignier termed the enations only as *phylloids*). This *syntelome* later differentiated into stem, leaf and root.

3. Bryophytes arose independently out of algae at a later time, probably Devonian. The sporophyte became parasitically dependent on the gametophyte and was reduced to a *single telome* (Fig. 43.2A). (N.B. other authorities have suggested that the Bryophytes and Pteridophytes might have arisen out of the same hypothetical plant.

(b) Theories of Bryophytic origin of Pteridophyta

Lignier (1908) had suggested that the ancestral Pteridophyte *Prolycopodinee* arose out of the *Prohepatic* which again arose out of the algae. Both these groups are, however, hypothetical.

When the Psilophytales were still unknown. Campbell (1895) suggested the origin of the Pteridophytes out of a Bryophyte of the type of *Anthoceros*. This idea has been reiterated by him (1940) and Smith supported this in his Cryptogamic Botany. The discovery of the Psilophytales, according to them, has strengthened this hypothesis.

This view of the origin of the Pteridophytes from *Anthoceros* gets further support from Bower (1908) in his **Theory of the Strobilus** by which he derived *Lycopodium selago* from *Anthoceros* (the Psilophytales were not known then). Bower suggested the following stages from *Anthoceros* to *Lycopodium selago* (in which there is no differentiated strobilus):

1. The continuous sporogenous tissue is just outside the central columella in *Anthoceros* (Fig. 43.4A).

2. The sporogenous tissue became more and more superficial (Fig. 43.4B).

3. The continuous sporogenous layer became broken into separate masses by intercalated sterile tracts (a condition sometimes found in the Anthocerotales) (Fig. 43.4C).

4. Separated sporogenous masses became more superficial, resulting in an alternation of green tissue and sporogenous tissue (Fig. 43.4D).

5. The intervening green tissue developed green expansions or *enations* (small sporophylls), each one having at its base a mass of sporogenous tissue (*sporangium*) (Fig. 43.4E). This is the condition is *Lycopodium selago* (Fig. 39.3).

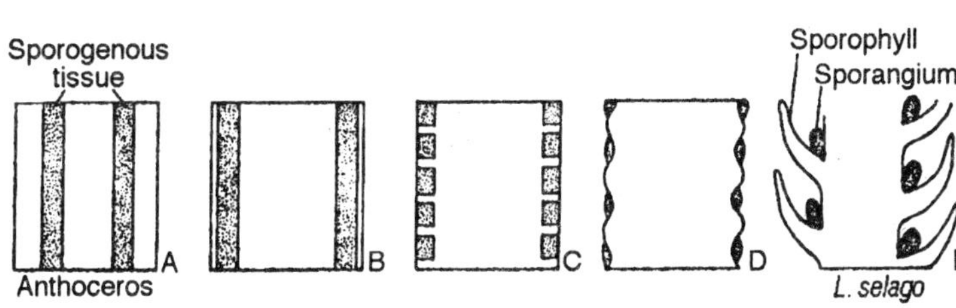

Fig. 43.4: A–E. Successive imaginary stages of evolution from *Anthoceros* (A) to *Lycopodium selago* (E) as suggested by Bower (Descriptions in text)

The discovery of the Psilophytales where the sporophyte is a rootless, leafless, dichotomously branched shoot has strengthened the theory of *Anthoceros* origin of Pteridophytes according to Smith (1955). It no longer needs the formation of the sporophylls as postulated by Bower for *Lycopodium*. The internal sporangial tissue is simply located at the tip leaving a long green sterile stem below (Fig. 43.5B), the tip bifurcates by dichotomy (Fig. 43.5C) and the base develops rhizoids so that the independent sporophyte of the Psilophytales (Fig. 43.5D) is obtained. Under favourable conditions *Anthoceros* sporophytes are known with the spore production restricted to the distal end and the columella differentiated into a conducting strand in the sterile lower part. This, together with the normal splitting of the top of the *Anthoceros* sporophyte simulating dichotomy, approaches Psilophytales. Moreover, the origin and development of both archegonia and antheridia in the Pteridophytes and in the Anthocerotopsida show similarity. Instances are known when the *Anthoceros* sporophyte lives as an independent plant for some time even after the loss of the subtending gametophyte.

Mehra (1969), as explained above, suggests a bi-phyletic origin of the Pteridophytes—*Lycopsida* directly from the Proto-Archegoniatae and the other groups from the *Anthorhymiaceae* stock (Fig. 43.1).

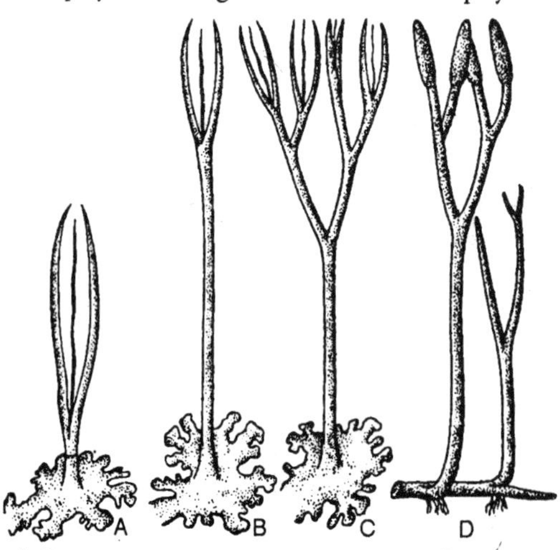

Fig. 43.5: A–D. Evolution of *Rhynia* (D) from *Anthoceros* (A) as suggested by Smith

Thus, the modern trend is a sort of a synthesis between the two types of hypotheses. Pteridophyta arose from the algae, but from an ancestral plexus which had also the characters of the Bryophytes or, where also the ancestors of the Bryophytes.

EVOLUTION OF THE PTERIDOPHYTIC TAXA

Since the discovery of the Psilophytales, a majority of the Pteridologists and Taxonomists (Zimmermann, Eames, Smith, Takhtajan, etc.) have held them as the most primitive and the starting point of the Pteridophytes. Zimmermann (1938) has explained the evolution of the Lycopsida, Sphenopsida and the Filicopsida in accordance with the Telome Theory as follows:

Lycopsida

The *syntelome* developed microphyllous leaves by *reduction* (explained before) and the *sporophyll* was derived by the fusion of a phylloid (sterile telome) and a fertile telome (Fig. 43.6A to C).

Mehra (1969), however, does not think that the Lycopsida arose out of the Psilophytales. The primitive Lycopsid sporangia did not have vascular connections, so they were not telomes. He, suggests that while the other groups arose out of the Rhyniaceous stock, the Lycopsida arose directly out of the ancestral Proto-Archegoniatae on a *Lateral Sporangiate Plexus* side shoot (Fig. 43.6). Lateral sporangiate plants are, however, represented within the Psilophytopsida (apparently, an artificial association) by the *Zosterophyllales.*

Sphenopsida

Leaf developed by webbing and by reduction. The sporangiophore was evolved by

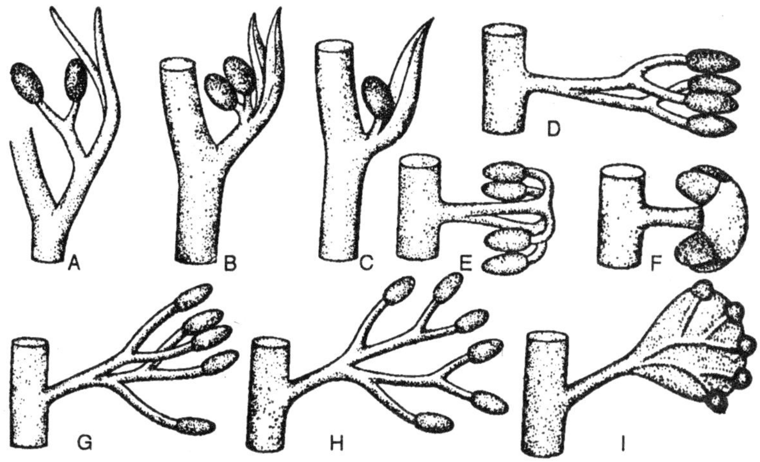

Fig. 43.6: A-C. Derivation of Lycopsida leaf and sporophyll; D-F. Derivation of Sphenopsida sporangiophora; G-I. Derivation of Filicopsida sporophyll (Modified from Zimmermann and Smith)

downward bending and fusion of several fertile telomes (Fig. 43.6D to F) which bent down by *incurvation*. The Hyeniales show this tendency.

Filicopsida

The megaphyllous leaves were clearly developed by planation and syngenesis of a number of sterile telomes and mesomes. The sporophyll was developed by similar webbing of a number of fertile telomes and mesomes (Fig. 43.6G to I). This explains the primitive *marginal* position of sporangia.

Takhtajan (1953), accepting the Psilophytales as the ancestors of all Telemophyta, does not agree that the leaves of the lycopsids are reduced megaphylls. They are *phylloids* or *enations* (explained below). The *Tmesopsida* leaf is syntelomous. *Sphenopsida* leaves are neither microphyllous nor megaphyllous, they are termed *meiophylls* after Benson. *Pteropsida* (Ferns and seed plants) leaves were formed by webbing as suggested by Zimmermann.

While previously the above arguments were held very good and the known psilophytales were generally accepted as the starting point of all Pteridophytes, gave doubts have arisen of late. Cambrian fossils of vascular elements and Silurian Lycopsid fossils have been discovered. This means that the Psilophytopsida is not the most primitive and this honour goes to the Lycopsida. *Archaeopteris* from Devonian and a few other Carboniferous fossils have been found to combine Psilophytan characters with heterospory and advanced wood characters so that they have been suspected to be the forerunners of the Pteridosperms and designated as **Progymnosperms**. Thus, the origin of the Pteridophytes has to be sought earlier than the Devonian, perhaps even earlier than the Cambrian, i.e., before the Psilophytales are known. Even in 1923, Scott doubted if the simplicity of the Rhyniaceae really signified primitiveness. Later discoveries of *Baragwanathia* with vascularized leaves, older than the Rhyniaceae have strengthened the idea that the simple aphyllous condition of the Rhyniaceae may be resultant of reduction rather than of simple primitiveness. Pichi-Sermolli (1958) is strongly of opinion that the Psilophytopsida is "one of the parallel lines of evolution in the Upper Devonian" whose simplicity may or may not be due to reduction but they are not the ancestors of the Pteridophyta. The modern trend is to call the ancestor of Pteridophytes not the Psilophytales but a *plexus* of ancient plants, probably representing several evolutionary trends (c.f. Delevoryas in the chapter Psilophytopsida) which *plexus*, of course, includes the Psilophytales.

ORIGIN OF THE LEAF STRUCTURES

Zimmermann's concept of the origin of the leaf structure on the basis of the telome theory (Fig. 43.3) has already been explained. He presumes that the microphyllous leaves were formed by flattening after reduction of the telome truss (Fig. 43.7A to C). But a still better explanation of the origin of the micropyllous leaf is by the **Enation theory** of Bower (1908). But this he explained how microphylls arose in the Lycopsida. His explanation of the origin of the sporophylls by enation has already been given on page 1248 (*Theory of the Strobilus*). According to this theory leaves first arose as superficial outgrowths (Fig. 43.7E) which flattened (Fig. 43.7F). A leaf trace now arose from the stele but ended at the leaf base (Fig. 43.7G). Finally, in the more established Lycopsida, the leaf trace extended into a median leaf vein (Fig. 43.7H). The conditions D, E, F and G are now clearly seen in *Rhynia* (leafless—D), *Psilophyton* (superficial enation—E), *Psilotum flaccidum* and *Asteroxylon* (leaf trace up to leaf base only—G).

The enation theory is now generally accepted as the most plausible explanation for the origin of the microphyllous leaves. The megaphyllous leaves, however, are well explained by flattening and webbing of telome truss as propounded by Zimmermann. In Sphernopsida, where both megaphylls and microphylls are present, the microphylls might have arisen by the reduction of megaphylls.

EVOLUTION OF THE STELE

Evolution of the stelar structure has been fully discussed in the Anatomy part of Volme I (Chapter 21) of this book and there is no need to repeat it. The same stelar evolution is clearly seen in the Pteridophytes. All primitive plants are *protostelic. Haplostele, actinostele* and *plectostele* conditions are progressively seen in all the groups of Pteridophytes. *Siphonostele, solenostele* and *dictyostele* are seen in the more advanced forms. *Polystelic* structures are also seen in a few cases. In some ferns the lower parts are *protostelic*, then *solenostelic* and finally *dictyostelic* and sometimes even *polystelic. Branch traces* and *leaf traces* with consequent *gaps* are seen in the advanced Pteridophytes.

Zimmermann's concept of the evolution of the stelar structure is shown in Fig. 43.8.

The vascular elements also show progessive evolution. Progressive evolution of annular, spiral, scalariform, reticulate and pitted tracheides is evident. There are some advanced

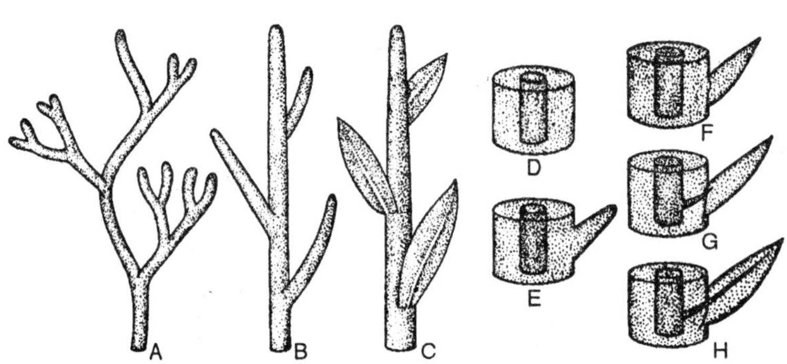

Fig. 43.7: Origin of the microphyll. A-C. According to Zimmermann: A. Sympodial axis with telome trusses after overtopping, B. Same after reduction of telome trusses, C. Microphylls formed by flattening of reduced telomes, D-H. Stages of formation of microphyll according to enation theory

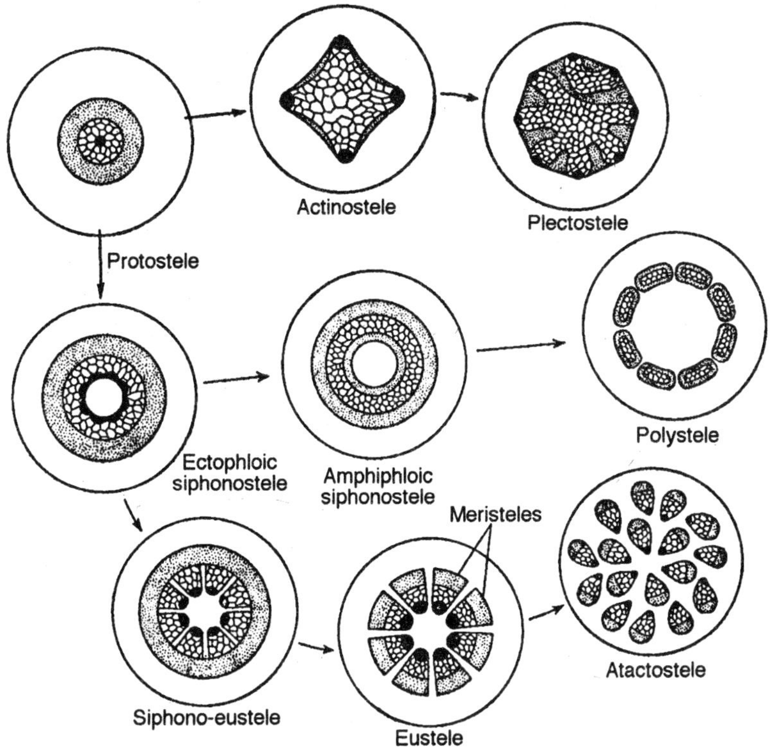

Fig. 43.8: Evolution of stelar types after Zimmermann

Pteridophytes showing bordered pitted tracheides and even vessels formed by the dissolution of end walls. Secondary growth also is noted in the advanced Pteridophytes.

HETEROSPORY IN THE PTERIDOPHYTA

Heterospory developed independently in Lycopsida, Sphenopsida and Filicopsida. In *Lycopsida* heterospory had developed so much that *Lepidocarpon, Miadesmia* and *Mazocarpon* are called 'Pseudospermaphytes'. Yet, they were not seed plants. The relationship between heterospory and seed habit is discussed in the next part on Gymnosperms. In *Sphenopsida* heterospory did not develop so much. In the living *Equisetum* an *incipient heterospory* may be perceived. Among some fossil Sphernopsida there is great size difference between the spores signifying some form of heterospory. Definite heterospory with a single megaspore in the megasporangium has only recently been claimed in *Calamites*. In the living *Filicopsida* heterospory is known only in the aquatic Marsileales and Salviniales. Heterospory is known in the very primitive fossil ferns. But, these heterosporous fossil ferns are associated with such complicated wood structures that doubts are now being expressed as to whether they should be retained within the Filicopsida or be placed in a new group 'Progymnosperms' (nevertheless Pteridophytic) as forerunners of the Gymnosperm group *Pteridosperms*, the leaves of which latter were once supposed to be those of ferns.

The gametophytes of all living heterosporous Pteridophytes are so greatly reduced that it is unfortunate that the gametophytes of the fossil forms are not known. Such fossils might throw some light on the stages of this reduction.

APOMIXIS IN PTERIDOPHYTA

Apomixis is very common only in the family Polypodiaceae. Outside the Polypodiaceae it has been reported only in the Selaginellaceae, Hymenophyllaceae, Osmundaceae and Marsileaceae. Within the Polypodiaceae it is widespread. It has been abundantly reported in the genera *Dryopteris, Pteris, Pellaea, Adiantum* and *Diplazium.* It is also reported in *Cyrtomium, Athyrium, Cheilanthes, Actinopteris, Onychium, Hemionites, Humata, Doodia, Asplenium* and *Lastrea.*

Aposporously produced gametophytes grow directly on the leaf of a sporophyte. Apospory may be artificially induced on such plants prone to it. Such aposporous gametophytes have $2n$ chromosomes and often show such sporophytic characters as development of tracheides. These gametophytes may produce $4n$ sporophytes normally or $3n$ sporophytes may be induced by crossing with normal n gametophytes. But, usually the aposporous gametophytes compensate by forming apogamous $2n$ sporophytes.

Apogamy has been found in aposporous gametophytes as well as in normal gametophytes. Such apogamous sporophytes may result from any cell of the gametophyte—a vegetative cell, any cell of the archegonium or even the egg itself (*parthenogamy*). *Selaginella* and *Marsilea* show parthenogenetic apogamy. An apogamous sporophyte has n chromosomes. In it, usually, there is a failure of mitosis before meiosis giving rise to $2n$ spore mother cells (the number of spore mother cells is halved) so that the spores and resultant gametophyte are normal n. Otherwise, apogamy may be compensated by apospory.

In *Lastrea* a sporophyte is known to have been formd on a gametophyte by a fusion of two of the latter's vegetative cells.

EVOLUTIONARY TENDENCIES WITHIN THE FILICOPSIDA

Generally speaking, the Filicopsida shows certain evolutionary tendencies which may be summarized as follows:

1. Habit and external morphology of sporophyte—upright radial to prostrate.

2. Vascular organization of stem—protostele to solenostele to dictyostele; exarch to mesarch to endarch.

3. Vascular supply of leaf—single trace to dissected trace.

4. Leaf blade—equal to unequal dichotomy; veins with free ends to closed veins forming reticulum.

5. Position and protection of sporangia—single sporangia at ends or veins or along margin to marginal sori to superficial sori pushed to lower side of leaf; unprotected sori to those covered by indusia.

6. Sequence of sporangial development—simultaneous to gradate to mixed.

7. Mode of sporangial development—eusporangiate to leptosporangiate.

8. Spore output—large to very limited.

9. Heterospory—only in very advanced groups.

10. Sporangial dehiscence—patch of annulus with vertical dehiscence to lateral annulus with transverse dehiscence to vertical annulus with stomium and transverse dehiscence.

11. Gametophyte—spore forming a thalloid sporeling to those germinating into filamentous structures later being heart-shaped or ribbon-like; thick prothallus with median thickening to thin prothallus with only a patch of thickening below the notch.

12. Antheridium—massive, embedded in gametophyte with a large number of sperms to small, emergent, containing a few sperms.

13. Embryo development—transverse division of zygote and massive embryo with late differentiation to vertical division of zygote and early differentiation of cotyledon, stem, root and foot quadrants of embryo; presence of suspensor to its absence.

Part Ten

Gymnosperms

Chapter 44

Introduction to the Gymnosperms

The **Gymnosperms** comprise of that group of plants in which a *naked,* (i.e., exposed and not encased in a closed ovary) *seed with a single integument* is present. *A seed is a mature integumented megasporangium.* The megasporangium is now called the *ovule* and is usually associated with a sporophyll or *carpel.* The ovule permanently retains within it the *megaspore* which develops the *female gametophyte* (*embryosac*) within a tissue (*nucellus*) formed inside the megasporangium. The female gametophyte is a complete parasite within the sporophyte and has no independent existence—thus completely reversing the condition in the Bryophytes. The integument develops from the base of the megasporangium. The *archegonium* structure is present but it disintegrates in the most advanced Gymnosperms, the Gnetales. The embryo develops within the seed. The *endosperm* is still a gametophytic tissue as in the Pteridophytes. The seed passes through a short or long period of *dormancy* before germination. Fertilization is *siphonogamic* through pollen tubes, by motile sperms in the primitive Gymnosperms and by non-motile male gametes in the more evolved ones. The *stigma* and *style* passage is not yet evolved in the Gymnosperms.

The Gymnosperms arose in the Palaeozoic, dominated the World during the Mesozoic—the age of the dinosaurs, and the earlier members of the group have become extinct today. Although the Angiosperms form the dominant vegetation today the Gymnosperms are not less important as they form a wide belt of forests in the North Temperate to Subarctic regions which is of the greatest economic importance as a source of timber, lumber and wood pulp. They are mostly arboreal, sometimes shrubby in habit.

The Gymnosperms had been considered as a group coordinate with the Angiosperms within the Phanerogams or Spermaphytes. There is now, however, a tendency to divide the Gymnosperms into a number of *divisions* as in the case of Pteridophytes. J. D. Hooker (about 1857) placed the Gymnospermae in between the Dicotyledons and the Monocotyledons. Eichler (1883) proposed Gymnospermae and Angiospermae as the two divisions of Phanerogamae. **Bessey** (1911) divides the Spermatophytes into three *phyla* (divisions) parallel to his three divisions of Pteridophyta. They are:

Phyllum XII. Cycadophyta

Class 27. Pteridospermeae

" 28. Cycadineae

" 29. Bennettitineae

" 30. Cordaitineae

Phyllum XIII. Strobilophyta

Class 31. Pinoideae

Phyllum XIV. Anthophyta

Class 32. Monocotyledoneae

" 33. Dicotyledoneae

Engler, in the 1936 edition of his Syllabus gives the following classification:

Abteil (Division) XIV. Embryophyta Siphonogama

Subdivision 1. Gymnospermae

Class 1. Cycadofilicales

" 2. Cycadales

" 3. Bennettitales

" 4. Ginkgoales

" 5. Cordaitales

" 6. Coniferae

" 7. Gnetales

Subdivision 2. Angiospermae

Class 1. Monocotyledoneae

" 2. Dicotyledoneae

Sahni (1920):

Gymnosperms

A. Phyllospermae

(a) Pteridospermae

(b) Cycadales

(c) Bennettitales (probable position)

B. Stachyospermae

(a) Cordaitales

(b) Ginkgoales

(c) Coniferales

(N.B. Sahni suggested the taking out of Taxales as a separate order).

(d) Gnetales

Chamberlain (1920):

Gymosperms

A. Cycadophytes

Cycadofilicales

Bennettitales

Cycadales

B. Coniferophytes

Cordaitales

Ginkgoales

Coniferales

Gnetales

Tippo (1942):

Subkingdom Embryophyta

Phylum Bryophyta

Phylum Trachaeophyta

Subphylum Psilopsida

Subphylum Lycopsida

Subphylum Sphenopsida

Subphylum Pteropsida

Class 1. Filicinae

" 2. Gymnospermae

Subclass Cycadophytae

Orders: Cycadofilicales

Bennettitales

Cycadales

Subclass Coniferophytae

Orders: Cordaitales

Ginkgoales

Coniferales

Gnetales

Class 3. Angiospermae

Subclass Dicotyledoneae

Subclass Monocotyledoneae

Takhtajan (1950):

Division Telemophyta

Phylum VI. Pteropsida

Class A. Filicinae

Class B. Gymnospermae

Subclass 1. Pteridosperminae

Order 1. Cycadofilicales (Lyginopteridaceae, Calamopityaceae, Medullosaceae, Peltaspermaceae, Corystospermaceae, Caytoniaceae, Glossopteridaceae)

Subclass 2. Phyllosperminae

Order 1. Cycadales (Cycadaceae)

Order 2. Bennettitales (Williamsoniaceae, Bennettitaceae)

Subclass 3. Stachyosperminae

Order 1. Cordaitales (Pityaceae, Poroxylaceae, Cordaitaceae)

Order 2. Ginkgoales (Ginkgoaceae)

Order 3. Coniferales (Voltziaceae, Araucariaceae, Podocarpaceae, Cephalotaxaceae, Taxaceae, Pinaceae, Sciadopityaceae, Taxodiaceae, Cupressaceae)

Subclass 4. Chlamydosperminae

Order 1. Ephedrales (Ephedraceae)

Order 2. Welwitschiales (Welwitschiaceae)

Order 3. Gnetales (Gnetaceae)

Class C. Angiospermae

Subclass 1. Dicotyledones

Subclass 2. Monocotyledones

Johansen (1951):

Phylum Pteridospermophyta

Order Cycadofilicales

Phylum Cycadophyta

Order Bennettitales

" Cycadales

" Nilssoniales

" Caytoniales

Phylum Ginkgophyta

Order Cordaitales

Order Ginkgoales

Phylum Coniferophyta

Order Voltaziales

Order Coniferales

Phylum Ephedrophyta

Order Gnetales

Engler's Syllabus (1954) (**Pilger** and **Melchior**):

Division XIV. Pteridophyta

Division XV. Gymnospermae

Class 1. Cycadopsida

Orders: Pteridospermae

Caytoniales

Cycadales

Nilssoniales

Bennettitales

Pentoxylales

Ginkgoales

Class 2. Coniferopsida

Orders: Cordailales

Coniferae

Class 3. Taxopsida

Order: Taxales

Class 4. Chlamydospermae

Order: Gnetales

Division XVI. Angiospermae

Pant (1957):

Division 1. Cycadophyta

Class 1. Pteridospermopsida

Orders: Lyginopteridales
Medullosales
Glossopteridales
Peltaspermales
Corystopermales
Caytoniales

Class 2. Cycadopsida

Order: Cycadales

Class 3. Pentoxylopsida

Order: Pentoxylales

Class 4. Cycadeoidiopsida

Order: Cycadeoidales

Division 2. Chlamydospermophyta

Class 1. Gnetopsida

Orders: Gnetales
Welwitschiales

Division 3. Coniferophyta

Class 1. Coniferopsida

Orders: Cordaitales
Coniferales
Ginkgoales

Class 2. Ephedropsida

Order: Ephedrales

Class 3. Czekanowskiopsida

Order: Czekanowskiales

Class 4. Taxopsida

Order: Taxales

Cronquist, **Takhtajan** and **Zimmermann** (1966):

Subkingdom Embryobionta

Division 7. Pinophyta (= Gymnospermae)

Subdivision A. Cycadicae (= Cycadophyta)

Class 1. Lyginopteridatae (= Cycadofilicales)

Class 2. Cycadatae (= Cycadales)

Class 3. Bennettitatae (= Bennettitales)

Subdivision B. Pinicae (= Coniferophyta)

Class 1. Ginkgoatae (= Ginkgoales)

Class 2. Pinatae

Subclass 1. Cordaitidae

Subclass 2. Pinidae (= Coniferales)

Subdivision 3. Gneticae (= Gnetales)

Subclass 1. Ephedridae

Subclass 2. Welwitschiidae

Subclass 3. Gnetidae

Division 8. Magnoliophyta (= Angiospermae)

It may be seen from some of the above that there is a tendency to consider the Gymnosperms into two groups, **Cycadophytes** and **Coniferophytes**. The former have comparatively shorter unbranched stems, pinnate leaves and large pith with scanty wood in a thick cortex. The latter, on the other hand, have comparatively large, profusely and excurrently branched stems with a small pith and abundant wood in a thin cortex. Sahni and Chamberlain have considered them as two independent lines of evolution. Arnold (1948) also held Cycadophyta and Coniferophyta as two parallel Gymnosperm line. But, modern concepts of phylogeny show the difficulty in considering only two separate Cycadophytic and Coniferophytic phyletic groups. There are probably more lines of evolution. Lam (1955) suggests three lines: (1) the Coniferopsida out of the Lycopsid stock, (2) the Cycadopsida (and Pteridospermopsida) out of the Psilopsida (Rhyniaceous) stock and (3) the Chylamydospermopsida and the Angiosperms are developments of a Proto-angiosperm stock which arose earlier from the Psilopsida stock.

The classification followed in this book is more conservative in retaining the Division status for Gymnospermae. With the state of our knowledge of phylogeny still so incomplete, it does not seem justifiable to dissolve the divisions Bryophyta, Pteridophyta and Gymnosperms each of which is so well established in its own peculiar type of life cycle. The taxon endings, however, should be changed according to International rules. The classification followed is as follows:

Division Gymnospermophyta

Class 1. Pteridospermopsida

Orders: Cycadofilicales (Families—Lyginopteridaceae, Medullosaceae, Calamopityaceae, Peltaspermaceae, Corystospermaceae)

Glossopteridales (Family—Glossopteridaceae)

Caytoniales (Family—Caytoniaceae)

Class 2. Cycadopsida

Orders: Cycadales (Cycadaceae)

Nilssoniales (Nilssoniaceae)

Cycadeoidales (Cycadeoidaceae)

Pentoxylales (Pentoxylaceae)

Class 3. Coniferopsida

Orders: Cordaites (Pityaceae, Cordaitaceae, Poroxylaceae)

Coniferales (Lobachiaceae, Voltziaceae, Cheirolepidaceae, Protopinaceae, Pinaceae, Taxodiaceae, Cupressaceae, Podocarpaceae, Cephalotaxaceae, Araucariaceae)

Taxales (Taxaceae)

Ginkgoales (Ginkgoaceae)

Czekanowskiales (Czekanowskiaceae)

Class 4. Chlamydospermopsida

Order: Ephedrales (Ephedraceae)

Welwitschiales (Welwitschiaceae)

Gnetales (Gnetaceae).

In recent years a new Division/Class is added as ***Progymnospermophyta*** with two orders ***Aneurophytales*** and ***Archeopteridales.***

Chapter 45

Pteridospermopsida

The Carboniferous was considered to be an 'age of ferns' because of a preponderance of fern-like leaves among fossils. It was later found that most of these fern-like leaves were associated with true seed structures so that they could no longer be placed within the Pteridophytes and had to be transferred to the Spermaphytes. These form the group of Pteridosperms where seed structures are borne on fern-like leaves. These leaves (sporophylls) are megaphyllous and never grouped into strobili. The group is known since the Devonian, had its climax in the Permo-Carboniferous and became extinct in the Jurassic.

Amongst the Pteridophytes there are a few fern-like primitive fossil plants which combine heterospory, megaphylly with some Psilophytan characters and secondary growth resulting in considerable wood formation (e.g., *Cladoxylon, Archaeopteris, Stauropteris,* etc). There are some authorities who consider that these form a group amongst themselves (the **Progymnosperms**) which, rather than the general run of megaphyllous ferns, are the real ancestors of the Pteridosperms.

Three orders, Cycadofilicales, Glossopteridales and Caytoniales may be considered to form the class Pteridospermopsida.

Order 1. CYCADOFILICALES

The Cycadofilicales or Cycad-like ferns may be taken to comprise of five families: Lyginopteridaceae, Medullosaceae, Calamopityaceae, Peltaspermaceae and Corystospermaceae. The first three of these families are Palaeozoic while the last two are Mesozoic.

PALAEOZOIC CYCADOFILICALES

Family 1. LYGINOPTERIDACEAE

Lyginopteridaceae plants have monostelic stems and petioles usually with single vascular strands. Seeds are small.

Genus *LYGINOPTERIS (CALYMMATOTHECA)*

The most well-known plant of this family is **Lyginopteris oldhamia**. Although known under this name given by Potonie in 1899, the more correct name is *Calymmatotheca hoeninghausi* as *Calymmatotheca* (Stur 1877) is the oldest name of a fragment of this plant out of

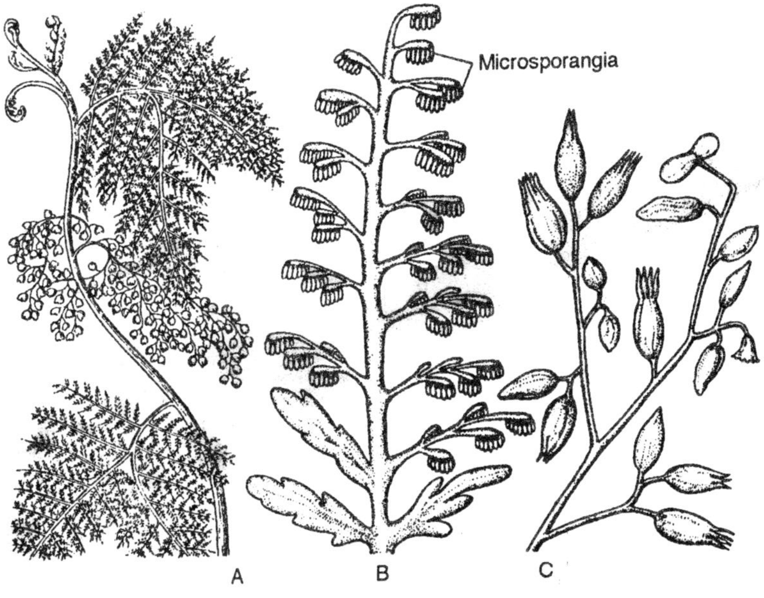

Fig. 45.1: *Lyginopteris oldhamia*. A. Reconstruction of part of shoot (*Sphenopteris hoeninghausi*), B. Microsporophyll (*Crossotheca* sp.), C. Megasporophyll (*Lagenostoma sinclairi*)

which segments the plant has been reconstructed. Different parts of this plant had been discovered at different times and given different names (stems—*Lyginodendron oldhamium* and *Lyginopteris oldhamium*; leaf—*Sphenopteris hoeninghausi*; petiole—*Rachiopteris aspera*; root—*Kaloxylon hookeri*; stamen—*Crossotheca* spp., *Telangium* spp.; seed cupule—*Calymmatotheca hoeninghausi*; seed—*Lagenostoma* spp.). It is an Upper Carboniferous plant found in England, Europe and North America.

The stem of the plant was long, slender (2 mm to 4 cm in thickness) and, evidently climbing. It bore large fronds of the *Sphenopteris* type which showed at least one forking and the ultimate pinnae were pinnately arranged (Fig. 45.1A). The leaf was covered by glands. These glands were characteristic features of *Lyginopteris* occurring on all organs except roots. Some of the fronds were sporophylls bearing microsporangia (Fig. 45.1B) and megasporangia (Fig. 45.1C) on the extremities of the ultimate segments of veined pedicels.

A t.s. of the stem (Fig. 45.2) shows an *outer cortex* with characteristic radial patches of sclerenchymatous cells, a parenchymatous *inner cortex* (which is very often crushed in the fossils) and a thick ring of *secondary wood* with rows or radially arranged multiseriate bordered-pitted tracheides. Inside the secondary wood there are about five separate *primary xylem* strands which are *mesarch* (*ptotoxylem* surrounded by *metaxylem*). The protoxylem tracheides are spiral while those in the metaxylem are scalariform or bordered-pitted. Some young stems show primary and secondary *phloems* and *cambium* but in older ones these are not preserved. Some cells have a tendency to get suberised. *Leaf traces*, originating from primary xylem strands and with mesarch xylem surrounded by pholem are seen outside the stele. The *pith* is large and parenchymatous with patches of sclerenchymatous cells here and there forming *sclerotic nests*.

Roots (*Kaloxylon*) found in climbers are adventitious, aerial and fairly long. They show radial, exarch stelar cylinders with alternating

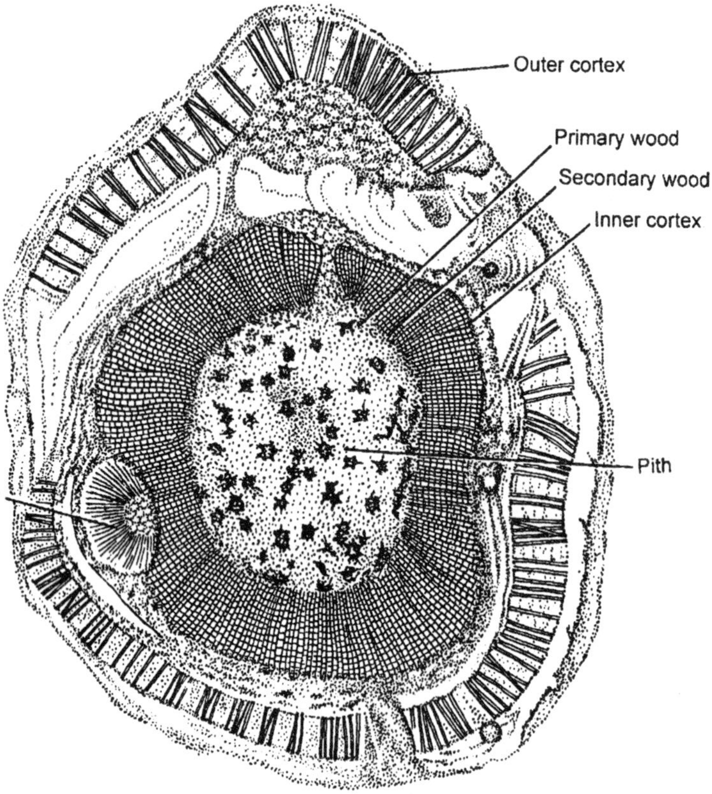

Fig. 45.2: T.s. of *Lyginopteris oldhamia* stem. Sclerotic cells are shown as dotted strands or patches

phloem and xylem rays. Larger roots show secondary growth.

The leaf anatomy is normal with a cutinized upper epidermis, palisade and spongy parenchyma.

Microsporophylls and *megasporophylls* have been identified from association with *Sphenopteris hoeninghausi* leaves and the characteristic glands. *Crossotheca* (Fig. 45.1B), found in such association, is considered to be its microsporophyll. The fertile prinnate are spathulate or hastate, each bearing 6 to 7 (more in some species—about 30 in *C. hastata*) pendent, bilocular microsporangia, the whole structure looking like a hair brush. There is some resemblance with the microsporophyll of *Araucaria.* A microsporangium is about 3 mm long and half as wide. Microspores showing multicellular gametophytes inside have also been found.

The *megasporophylls* named *Lagenostoma* (Fig. 45.1C) are better known. The seed structure *Lagenostoma lomaxii* (Fig. 45.3) was first described by Oliver and Scott in 1903. The seeds are about 5 mm long and borne terminally on the ultimate ramifications of the frond. Each looks like a tulip bud being covered by a *cupule* which splits longitudinally into about six coriaceous segments at maturity. The cupule is covered by capitate glands as found in the fronds (Fig. 45.3A). The *ovule* (*megasporangium*) is orthotropus and symmetrical (Fig. 45.3B). It is covered by an *integument* which is attached to the *nucellus* except at the apical *pollen chamber.* A vascular strand from base branches to the cupule and then, separately, to the integument. The outer part of the integument became stony in the seed as in *Cycas* leaving a fleshy inner part with the vascular supply. The nucellus is free and tapering at

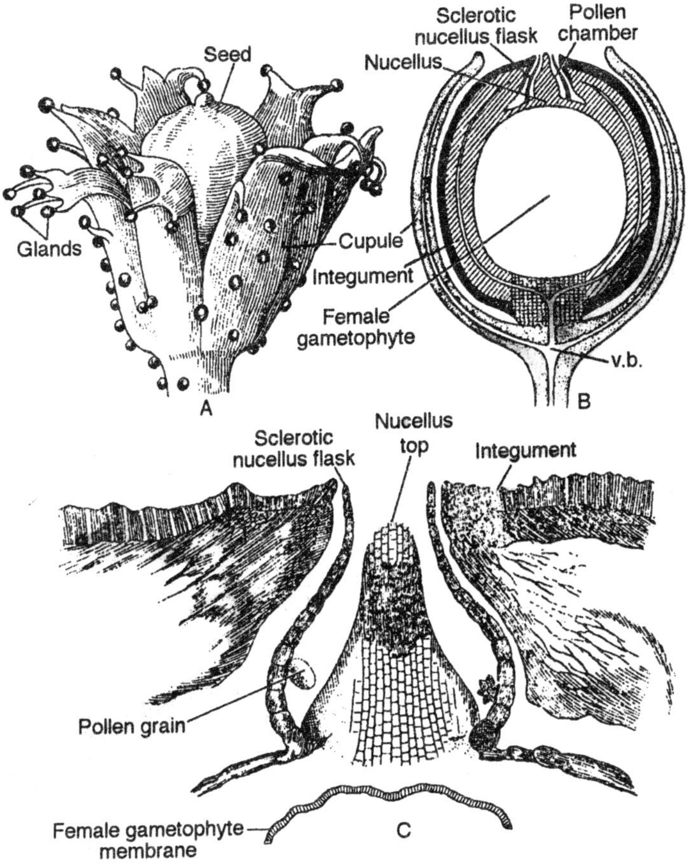

Fig. 45.3: *Lagenostoma lomaxii* (after Oliver and Scott). A. Seed surrounded by glandular cupule, B. Median l.s. of seed (outer stony layer of integument shown in black), C. Top of seed showing pollen chamber (*lagenostome*)

the tip leaving a cylindrical *pollen chamber*. The pollen chamber (Fig. 45.3C) became more complex by the development of a flask-like sclerotic collar from the nucellus leaving a chamber, sometimes called a *lagenostome*, where the pollens are found lodged. The tip of the nucellus is elongated as an axial column in the centre of this structure.

Pollen tubes are not known but from the other structures it may be assumed that fertilization was siphonogamic but the sperms were probably directly discharged as in the living *Cycas*.

Heterangium is another genus within this family which is also associated with *Sphenopteris* leaves and found from Lower Carboniferous in Scotland. *Sphaeroistoma* is supposed to be its seed. The stem (Fig. 45.4A) is protostelic and resembles *Gleichenia* with a central core of mesarch xylem broken up into groups by third-walled parenchyma. The secondary wood is thin but shows bordered pitted tracheides as in *Lyginopteris*.

Family 2. Medullosaceae

Medullosaceae is characterized by polystelic stems and large petioles containing a number of scattered vascular bundles. Seeds were often very large.

Medullosa was widespread in the Permocarboniferous. *Neuropteris*, *Sphenopteris* and *Pecopteris* belonging to this family are known from the Indian Gondwana Flora. The plants were probably erect with a crown of leaves on

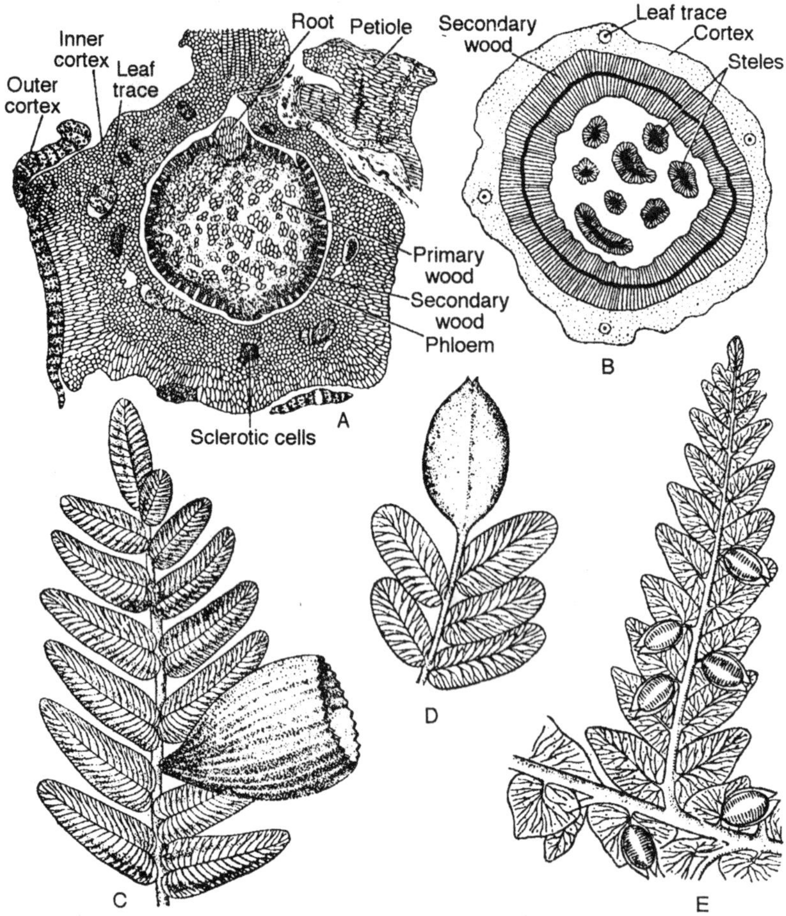

Fig. 45.4: A. T.s. of stem of *Heterangium grievii*, B. T.s. of stem of *Medullosa stellata*, C. *Neuropteris* leaf with *Whittleseya* synangium. D. Seed on *Neuropteris* frond, E. Seeds of *Emplectopteris triangularis* on frond (after Scott, Steward and Delevoryas and Andrews)

top. Stems named *Medullosa* are associated with leaf types *Neuropteris* (Fig. 45.4) and *Alethopteris.* The petiole is *Myeloxylon.* The stem is polystelic with three or more steles (Fig. 45.4B). Each primary stele is surrounded by its own secondary wood. Some species show a solenostelic complete or incomplete ring of wood outside, the smaller steles lying at the centre. There is also a periderm deep inside the cortex. Some of the microsporangia known (*Whittleseya, Codonotheca,* etc.) show peculiar synangial structures (Fig. 45.4C). Seeds associated with *Medullosa* are usually very large. Such seeds are apical or lateral on fronds (Figs. 45.4D and E). Various types of *Neuropteris* and *Sphenopteris* leaves and other Medullosaceae fossils named *Emplectopteris, Trigonocarpus, Spherostoma* and *Pecopteris* are known. *Neuropteris decipiens* has seeds 6 cm long. *Trigonocarpus* seeds are also large and show very long micropyles.

Family 3. CALAMOPITYACEAE

Calamopityaceae is an artificial association of Upper Devonian to Lower Carboniferous stem fossils only. Leaves and reproductive organs are not known. Quite a number of very incompletely known genera are included here. The stems are monostelic with compact secondary wood. Some associated petioles show a number of bundles.

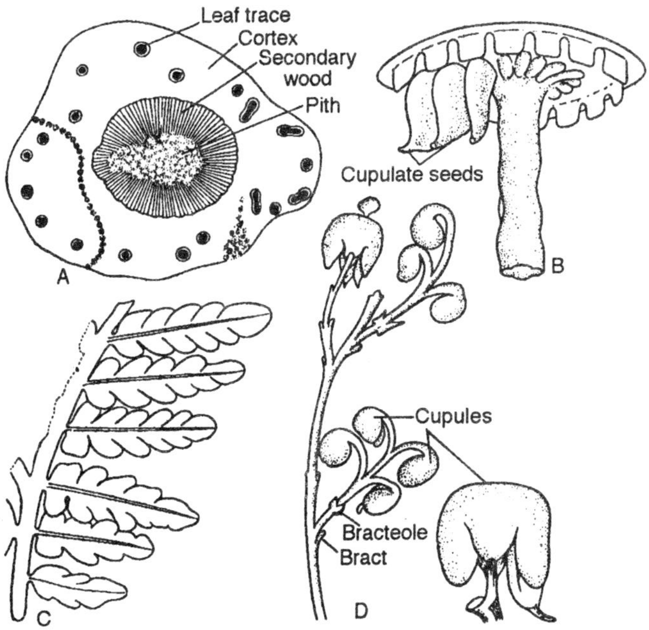

Fig. 45.5: A. T.s. of stem of *Calamopitys americana*, B. Capsulate disc of *Lepidopteris ottonis* from Greenland with three seeds intact (after Harris), C. *Thinnfeldia (Dicroidium)* leaf, D. Female inflorescence of *Umkomasia macleani* with one corystospermous cupule shown separately (after Thomas)

Calamopitys (Fig. 45.5A) is the commonest genus. The stem shows a large cortex with leaf traces enclosing a solenostele in which the secondary wood shows broad medullary rays causing a spongy stem. The pith is of a mixed nature containing tracheides and parenchyma.

MESOZOIC CYCADOFILICALES

Family 4. PELTASPERMACEAE

The Mesozoic Peltaspermaceae is represented by *Lepidopteris* from the Upper Triassic in Greenland and South Africa. The most interesting features is that the cupulate seeds are borne hanging under paltate circular disc (Fig. 45.5B).

Family 5. CORYSTOSPERMACEAE

Corystospermaceae from Triassic and Jurassic is another assemblage of Mesozoic Pteridosperms. The foliage is *Stenopteris* and *Thinnfeldia* (= *Dicroidium*). The latter (Fig. 45.5C) is a predominant constituent of the Indian Upper Gondwana Flora called the *Thinnfeldia—Dicroidium—Ptilophyllum Flora,* more specifically, the Middle Gondwana Flora which is called the *Thinnfeldia* (= *Dicroidium*) *Flora* (c.f. page 45.5C). At least three species, viz., *Thinnfeldia* (*Dicroidium*) *hughesi, T. odontopteroides* and *T. sahnii* are known from the Jurassic Rajmahal and Jabalpur (c.f. Palaeobotany). Micro- and megasporophylls are known in several genera, e.g., *Umkomasia, Pilophosperma, Petruchus.* The female inflorescence is in the axil of a bract and each branch of it is subtended by two bracteoles above which there are several cupulate seeds (Fig. 45.5D). The cupule is helmet-shaped (*Corystosperm* means helmet-shaped seed) and bivalved. The micropyle is curved and bifid. The microsporangia hang marginally from the flattened pinnae. The pollen grains have two lateral wings.

Order 2. GLOSSOPTERIDALES

Glossopteris, the dominant leaf form of the Lower Gondwana (Permo-Carboniferous) giving rise to the name *Glossopteris Flora* has been found associated with *Scutum* fructification and *Vertebraria* stem. *Gangamopteris*, a relative and also found in the Glossopteris Flora has been found associated with *Ottokaria* type of fructification. The complex fructifications have given rise to some doubt as to the real affinities of the plants but they are usually considered as Pteridosperms and are placed in the family Glossopteridaceae which is considered to form a separate order Glossopteridales.

Glossopteris leaves (Fig. 45.6C) are simple, entire, sessile or very shortly petiole, linear lanceolate to spathulate ovate and with a very strong midrib from which numerous fine veins pass out to the margin forming a network. As many as 14 species of such leaf fossils are known from the Indian Lower Gondwana, viz., *G. indica, communis, angustifolia, damudica, decipiens, longicaulis, ampla, retifera, browniana, stricta, tortuosa, formosa, divergens, conspicua* (c.f. Palaeobotany). The leaves are 3 to 40 cm long and in some species whorls of such leaves have been found in thin stems. A stem, *Vertebraria*, has been found associated with such leaves. *V. indica* (Fig. 45.6A) is common in Indian Lower Gondwana. It might have been a rhizome as it bears slender, branched roots. It is flattened and grooved. The stelar structure is unusual showing several plates of vascular tissue (xylem and pholem) radiating from a common centre with broad parenchymatous rays between the arms. Horizontal plates of vascular tissue connect the vascular arms. Pitting of the tracheides is gymnospermous.

The fructifications go under the name *Scutum*. *Scutum* has been found also in India. A number of species of *Scutum* (viz., *S. leslium, dutoitidis, rubidgeum, draperium, stowanum*) are known and *Lanceolatus lerouxides* is also known to be a fructification of *Glossopteris*. The reproductive structure is lanceolate to ovate and attached by a pedicel to a lower part of a leaf midrib. It is a cupulate structure with two bilaterally concave symmetrical (upper and lower) valves and a wing-like expansion along the line where the two valves join (Fig. 45.6B). It seems that the valves of the cupule closed after fertilization. Small sac-like (1 to 2 mm in diameter) structures occur in the concavity of both the halves (Fig. 45.6C). Mrs. Plumstead (1952, 1956), who examined the specimens closely, believes that the structure is bisexual, the lower sacs representing seeds while the upper ones are pollen sacs. The pollens are not winged and resemble *Magnolia*. The specimens, however, are not in so perfect condition as to be definite about their nature.

Although the **Glossopteria flora** has been described in the part Palaeobotany. The palaeo-

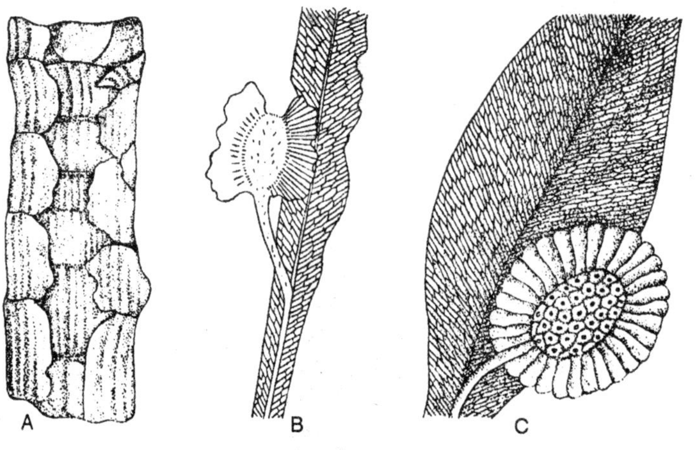

Fig. 45.6: *Glossopteris*. A. *Vertebraria indica*, B. *Scutum leslium* fructification, C. *Scutum leslium* fructification associated with *Glossopteris brunoniana* leaf (after Plumstead)

zoic southern hemisphere coal forming Lower Gondwana flora includes besides pteridophytic elements, conifers and cycadophytes a typical plant group known as Glossopteris flora named after its dominant member, genus *Glossopteris.* Glossopteris flora flourished in the Lower Gondwana during U. carboniferous to Permian age and contains in addition to the dominant leaf genus *Glossopteris* Brongn. Other allied leaf genera, viz., *Gengamopteris* McCoy, *Rijbidgea* tate, *Palaeovittaria* Feistm., *Macrotaeniopteris* Schimp., *Euryphyllum* Feistm., the *vertebraria* root, Gymnospermous type of wood and about 30 different types of reproductive structures a few of which occurs in organic connection with the leaf genera and species.

The significant contribution about the flora gathered during the last dacade have led to the serious consideration of the systematic position and affinity of the group of plants. However, yet much remains unexplored about this widely diverse flora for a better understanding of the origin and diversity of modern day plants.

The leaf genera are distinguished on the basis of morphographic features. Of all the leaf genera, genus *Glossopteris* has been studied most extensively and a large number of species is attributed to this genus.

Glossopteria Brongniart: Simple leaves pinnate, shape and size varies widely, mostly distinctly petiolate or with narrow base, distinct midrib present from base to apex of the lamina, base and apex of variable shape, margin entire excepting with apical notch in some species secondary veins fork and/or anastomose to form venation, reticulate venation, mesh formation distinct for a species.

More than hundred species are attributed to the genus uptil now among which about seventy species are reported from Indian Lower Gondwana. Amongst the earliest known species, *G. angustifolia, G. browniana, G. indica, G. communis, G. decipiens, G. damudica, G. formosa, G. divergens, G. retifera, G. conspicua, G. longicaulis, G. taeniodes, G. tortuosa* etc. were described by Brongniart (1822, 1828), Schimper (1869), Feistmantel (1876, 79, 81, 82), Zeiller (1902).

Gangamopteris McCoy: Simple leaves, shape and size variable, absence of distinct midrib, instead shows parallel anastomosing veins in the median portion, secondary veins form reticulate venation.

About 24 species are attributed to the genus, *Gangustifolia, G. obliqua, G. spathulata, G. cyclopteroides, G. major* etc. were described by McCoy (1847), Feistmantel (1879). Most of the species are known from Indian Lower Gondwana.

Palaeovitttaria Feistmantel: Simple leaves, midrib present at the basal part of the lamina, veins forked, parallel to subparallel to each other.

P. kurzii described by Feistmantel shows fronds arranged in the form as if attached to a common stalk.

Macrotaeniopteris Schimper: Taeniopteroid frond, lamina large in size, distinct midrib present, secondary vein parallel arising at right angle to the midrib.

Rubidgea Tate: *Rubidgea* has been described as obovate leaves without midrib but longitudinal veins in the median portion simulates a false midrib, veins dichotomised, anastomosis absent.

Euryphyllum Feistmantel: Leaf simple ± symmetrical, veins dichotomise, subparallel in the median region.

Lower Gondwana fossils other than leaf genera

Vertebraria Royle: The common lower Gondwana fossils, branched or unbranched axes with 2 or 3 series of rectangular blocks of alternates ridges and furrows has so far been reorganized mostly as rhizome or as stem.

However, the anatomy of the petrified specimen of *Vertebraria,* has strongly supported the root nature of the axes (Schopf, 1965).

Gymnospermous wood: A large number of petrified gymnospermous wood of significant length and secondary wood are recorded from Lower Gondwana. The woods with uniseriate to multiseriate bordered pits of "araucaria type" were recognized earlier as *Dadoxylon Endlicher.* In recent years all the *Dadoxylon* woods have been reconsidered and recognized under *Dadoxylon* and *Araucarioxylon* Kraus (Maheshwari, 1972) having multiseriate and uniseriate xylem rays respectively.

In addition, a number of other gymnospermous woods are recorded from lower Gondwana. All these woods commonly occur in the Lower Gondwana of India.

Reproductive organ of Glossopteris flora

About 30 genera of reproductive organs of various morphology have been assigned to *Glossopteris* flora which include leafy bract-like structures bearing micro- and megasporangia in variable organizational pattern viz., *Lidgettonia* Thomas, *Denkania* Surange and Chandra, *Eretmonia* DuToit, *Glossotheca* Surange and Maheshwari etc.; dorsiventrally flattened sporophyll bearing ovule or microsporangia on one side of the sporophyll, viz., *Ottokaria* Zeiller, *Dictyopteridium* Feistmantel, *Scutum* Plumstead, *Lanceolatus* Plumstead etc.

Ovules borne on two sides of an axis without sporophyll, viz., *Senotheca* Banerjee, fructufication with cone-like appearance, viz., *Kendostrobus, Bankolaea* Banerjee, most of the genera are reported from India.

Otktoaria, Scutum, Lanceolatus, Senotheca, etc. are found to be attached to different species of *Glossopteris.*

The reproductive organ of the glossopteris flora have been classified under Pteriodospermales and Glossopteridales considering their morphological appearance by Surange and Chandra (1975) or Pteridospermous and non-Pteridospermous groups (Rigby, 1978) or into 7 Orders under the Single Phyletic division Glossopteridopsida (Banerjee, 1979). Schopf (1979) also considers the Glossopteris group of plants to be a choherent phyletic group and classified the reproductive organs into 7 types.

The Glossopteris group of plants with varieties of leaf venation pattern including reticulate venation in a *Glossopteris,* lignin lamellae in the stomata of leaf cuticle, circular bordered pitted tracheids in the wood, disaccate pollen grains and occurrence of such pollen grains in the ovules recovered from the fertile organs has so far been confirmed as gymnosperms. The group occupies an unique position among gymnosperms having widely diffused characters of several phyla of gymnosperms (Delevoryas and Gould, 1977).

The affinity of the Glossopteris group of plants have been considered variously from time to time. They were speculated to be pteridophyte in the initial stage and sometime as angiosperm. However, the group shows strong evidences of gymnospermous affinity with pteridosperms Pectoxyleae, Cycad, and Cordaitalean group of plants. Schopf (1976) has suggested that Gnetaceous plants are more likely descendant from Glossopterid ancestor. Angiospermous affinity of the group has been suggested by Melville (1969, 1970, 1971, 1983) on the basis of venation pattern of Glossopteris species and Morphology of the reproduction structures described by him as gouophyll. However, the direct or indirect affinity of the Glossopteris group of plants to the angiosperms awaits further evidences.

Among the different genera of Glossopteris group of plants the genus *Glossopteris* is known more extensively than the other general.

A reconstruction of the plant bearing *Glossopteris* leaves is branched and supported by an aerenchymatous roots system *Vertebraria* with gymnospermous secondary wood, the fructification consists of an infolded megasporophyll bearing sessile seeds and microsporophyll is of *Eretmonia* type bearing *Arberiella* type of microsporangia with bisaccate pollen.

Gangamopteris is another genus found abundantly in the *Glossopteris Flora.* The leaves do not have the prominent midribs of *Glossopteris.* Four species (*G. cyclopteroides, buriadica, kashmirense* and *whittiana*) are known from Indian Lower Gondwana rocks. The fructification is *Ottokaria* of which two species (*O. bengalensis* and *kathwaiensis*) are known from Indian Lower Gondwana. The appearance of the reproductive structure is as complex as that of *Glossopteris.*

On the whole, the Glossopteridales seems to be a very problematic group. The fructification is on a leaf (*phyllospermous*), but the structure is much more specialized than in any Pteridosperm. Although placed within the Pteridosperms as present, its relationship with the Angiosperms is strongly suspected by Mrs. Plumstead as well as Melville (1962). The present position, therefore, may be tentative and subject to alteration in the future when more facts get known.

Order 3. CAYTONIALES

Caytoniales is another group of plants as problematic as the Glossopteridales. First reported by Thomas (1921) from English Jurassic rocks it has since been found to be widespread from the Upper Triassic to the Upper Cretaceous. It was first reported to be an Angiosperm because of the apparent closed nature of the seed but is now considered as a Pteridosperm of problematic affinities.

The fruit-bearing shoots are placed in the form genus *Caytonia* (includes *Gristhorpia*), the

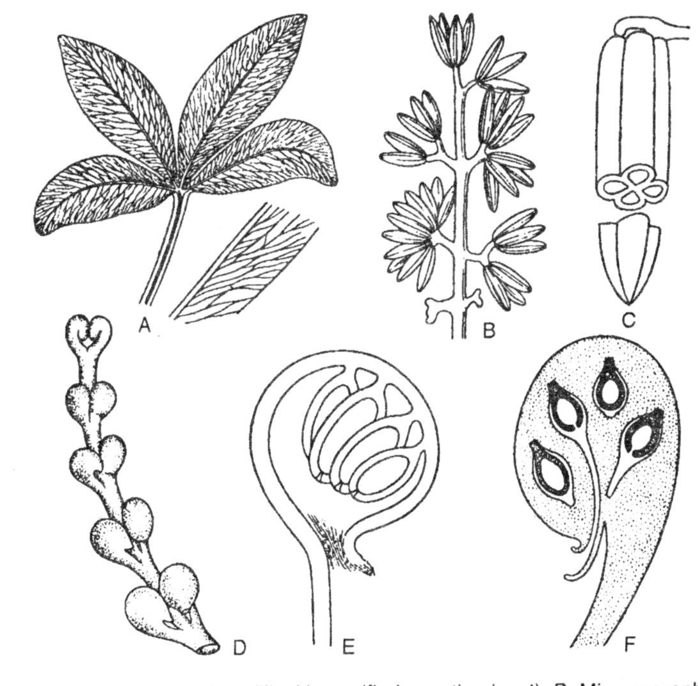

Fig. 45.7: *Caytonia*. A. Leaf, *Sagenopteris phillipsi* (magnified venation inset), B. Microsporophyll of *Caytonia sewardi* called *Caytonanthus (Antholithus) arberi*, C. A dissected microsporangium of *Caytonanthus arberi* showing the four pollen sacs, D. *Caytonia (Gristhorpia) nathorsti* (Megasporophyll with two rows of cupules, E. Vertical l.s. of *Caytonia sewardi* cupule showing seeds inside. F. Vertical l.s. of fleshy cupule fruit, *Caytonia thomasi*

microsporophylls in *Caytonanthus* (includes *Antholithus*) and the leaves in *Sagenopteris*.

Sagenopteris leaf (Fig. 45.7A) has a slender petiole and is palmately compound of 3 to 6 lanceolate leaflets each 2 to 6 cm long. The venation is reticulate.

The *microsporophyll* (*Caytonanthus*) is pinnately branched (Fig. 45.7B) and the ultimate divisions bear *synangia* each of which is formed of four sporangia or pollen sacs (Fig. 45.7C).

The *megasporophyll* or the fruit-bearing stalk (*Caytonia*) is a flat structure about 5 cm long with lateral short-stalked 'fruits' in two rows on the two margins (Fig. 45.7D). Each cupule or 'fruit' (Fig. 45.7E and F) is a rounded body with an opening or crack below, close to the pedicel. A row of 7 to 8 small orthotropous seeds are situated in a row on the curved inner face. Each seed has a short micropyle (not long as in the other Mesozoic Pteridosperms) pointing towards the opening below. The seeds and the wall on which they are borne are quite fleshy in some cases. But pollens are found inside the micropyles of the seeds and not at the opening of the cupule so that it is now presumed that the cupules was open at the time of pollination and the now apparent 'fruit' is formed only by the curving down of the rachis bearing the megasporangia. It was, therefore, never a true Angiosperm, the so-called closed carpel being only a cupule as found in most Pteridosperms. The seeds so far have been found to contain only pollinated ovules and no embryo has been seen. There is a single integument completely free from the nucellus. Pollination and fertilization seems to be as in the Pteridosperms and not as in the Angiosperms.

Although the Caytoniales cannot by considered as Angiosperms, the fruit-like structure in *Caytonia* may be a step towards the evolution of the Angiospermic carpal. Some palaeobotanists now consider the Pteridosperms as the possible ancestors of the Angiosperms and they think that Caytoniales should be quite near the ancestor.

Chapter 46

Cycadopsida

The *Cycadopsida* plants are often with stout trunks showing massive cortex but less developed wood, megaphyllous leaves and with at least the microsporophylls and, usually, even the megasporophylls grouped into strobili. The group is evident from the Permian, the climax was attained in the Middle and decline is clear in the Late Cretaceous. It is suggested that the group arose in the Carboniferous out of the Pteridosperms but Andrews (1961) does not support this idea and believes that the Cycadopsida had an origin different from that of the Pteridosperms. The fertilization of the group is by motile spermatozoids.

Four orders, *Cycadales, Nilssoniales, Cycadeoidales* and *Pentoxylales* may be considered to comprise this class.

Order 1. CYCADALES

The Cycadales are the only living representatives of this group. This order is treated first not because they are the most primitive of the Cycadopsida but because a complete idea is possible only by comparing the living Cycads.

It is suggested that the Cycadales probably arose out of the Cycadofilicales in the Carboniferous although some palaeobotanists do not approve the idea. *Bjuvia simplex* (Fig. 46.1) with *Palaeocycas* megasporophylls and *Taeniopteris* leaves from the Upper definite representative. The stout trunk of this (Fig. 46.1A)

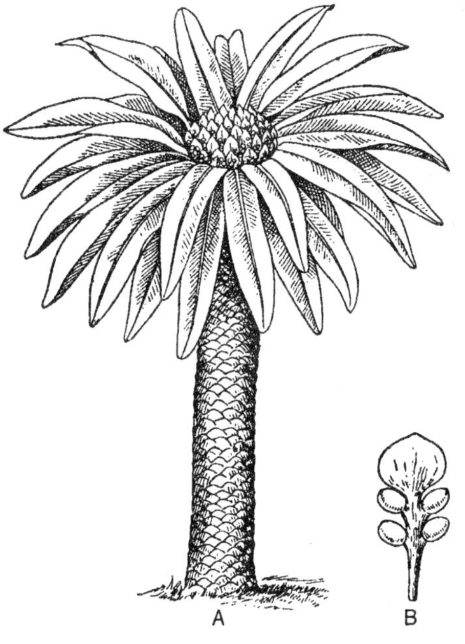

Fig. 46.1: *Bjuvia simplex* (after Florin). A. Restoration of the tree. B. A megasporophyll (*Palaeocycas integar*)

bore persistent lead bases and a crown of leaves on top. The *Palaeocycas* megasporophylls (Fig. 46.1B) are arranged in a loose terminal cluster and ovules are borne on two sides of the rachis resembling modern *Cycas*. Microsporophylls are not known. Some 5 or 6 species of *Taeniopteris* leaves are known from Indian Lower Gondwana (Permian) as well as Upper Gondwana (Jurassic) rocks. *Dioonites* leaves from Permian China are also suspected to belong to the Cycadales. *Cycadospadix* from Late Triassic resembles Cycas megasporophyll but the stomata resemble the Cycadeoidales.

Family: CYCADACEAE

Cycadaceae is the sole representative of the Cycadales today. They have been called the 'palm-ferns' because of the palm-like habit combined with the fern-like foliage of most of the genera. As has been mentioned above, the family seems to be represented as fossils since the Triassic, or even the Permian. Epidermal fossils of Cycadales and Cycadeoidales are quite common and the stomatal structures has been used to identify the two orders. The stomata of the Cycadales (and also Nilssoniales) do not show subsidiary cells formed from the guard cells initials with the guard cells heavily cutinized as found in the Cycadeoidales. This type of stomata with the subsidiary cell not form the same initial as the guard cells is called *haplocheilic*. Moreover, the epidermal cell walls of Cycadeoidales are characteristically wavy (Fig. 46.21A). These characteristics have made the inclusion of *Cycadospadix* and *Cycadites* (Also Triassic) within the Cycadales doubtful.

The Mesozoic distribution of the Cycadales seems to have been world-wide and they dominated the vegetation in those days. But, the present day distribution is rather limited. The Cycadaceae comprise of nine genera which are divided by Pilger and Melchior in Engler's latest Syllabus (1954) into five subfamilies as follows:

A. *Megasporophylls not forming any compact cone.* Ovules on megasporophyll margin. Leaves once pinnate. Leaflet with single median vein and no lateral vein.

Subfamily 1. **Cycadoideae**

Genus 1. *Cycas*—Arborescent up to 20 ft. tall. Polyxylic. Old World, spreading from Japan and South China over Indomalayasia to North Australia and to Madagascar on the west. 15 to 20 species.

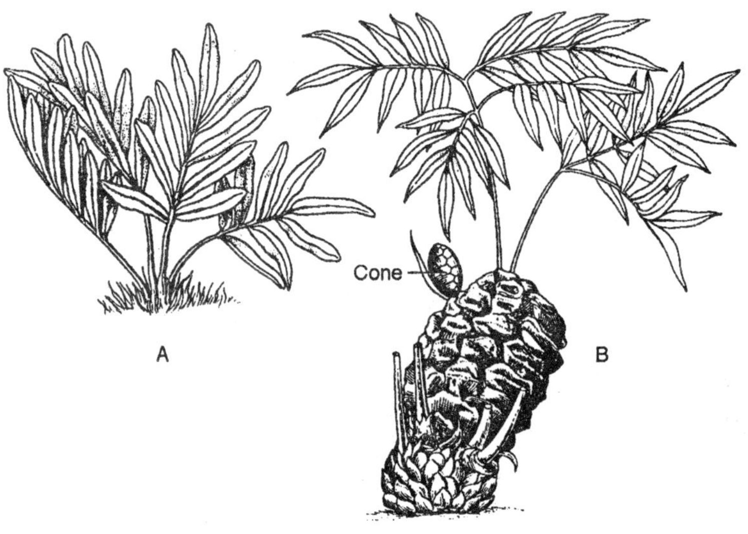

Fig. 46.2: A. *Stangeria paradoxa*, B. *Bowenia serrulata*

B. *Megasporophylls forming compact cones.* Each with 2 ovules.

Subfamily 2. **Stangeriodeae**

Genus 2. *Stangeria* (Fig. 46.2A)—Small fern-like plants with almost subterranean tuberous stems and once pinnate leaves. Leaflets with midribs and lateral veins. Once considered to be a fern. Found in South Africa only. Single species *S. paradoxa.*

Subfamily 3. **Bownenioideae**

Genus 3. *Bowenia* (Fig. 46.2B)—Shrubby plants with almost subterranean tuberous stems and bipinnate leaves. Found in Australia only. 2 species.

Subfamily 4. **Dioonoideae**

Genus 4. *Dioon*—Arborescent up to 40 ft. tall. Monoxylic. Leaves once pinnate. Leaflets without midrib and showing parallel venation. Female cone loose with two lateral stalked ovules at the base of each sporophylly (Fig. 46.18C). Central American. 4 to 6 species.

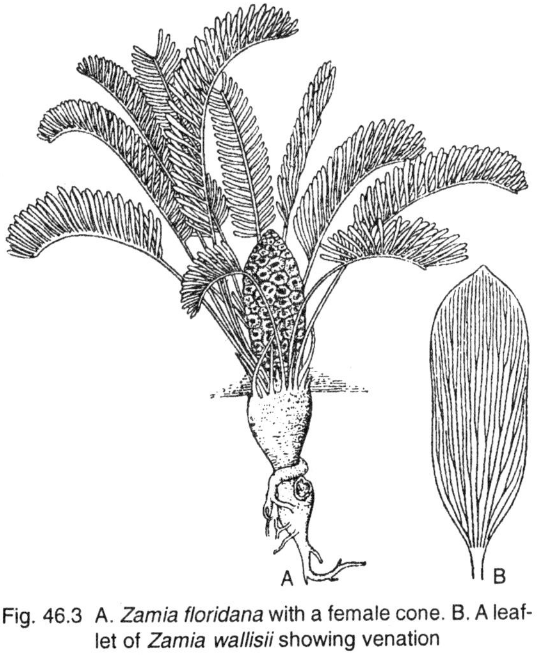

Fig. 46.3 A. *Zamia floridana* with a female cone. B. A leaflet of *Zamia wallisii* showing venation

Subfamily 5. **Zamioideae**—Leaves once pinnate. Leaflets without midrib and with parallel venation (Fig. 46.3B). Megasporophylls peltate or flat-topped. Ovules seesile.

(a) Leaflets not articulated. Stem usually polyxylic with medullary bundles.

Genus 5. *Macrozamia*—Arborescent up to 60 ft. tall as well as short tuberous species. Sporophylls terminating in long median spines. Australian. 12 to 16 species.

Genus 6. *Encephalartos*—Arborescent as well as short tuberous species. Sporophyll top flat or rounded. Mature female cones up to 45 kg in weight. Tropical and South African. 15 to 20 species.

(b) Leaflets articulated. Stem monoxylic.

Genus 7. *Ceratozamia*—Arborescent. Leaves with a pair of tooth-like stipules. Sporophylls with two horn-like projections. Central American. About 6 species.

Genus 8. *Microcycas*—Arborescent up to 30 ft. tall. Sporophyll top flat or shield-shaped. 1 species (*M. calocoma*) from Cuba.

Genus 9. *Zamia*—Usually with almost subterranean short tubers (Fig. 46.3A) but may be arborescent. Sporphylls peltate, in longitudinal rows, forming compact cones (Fig. 46.18G). From Florida to tropical Central and South America. About 30 species.

All the nine living genera included within the family are tropical. Five of these are in the Eastern Hemisphere. Of these, *Cycas* has the widest distribution while the other four,

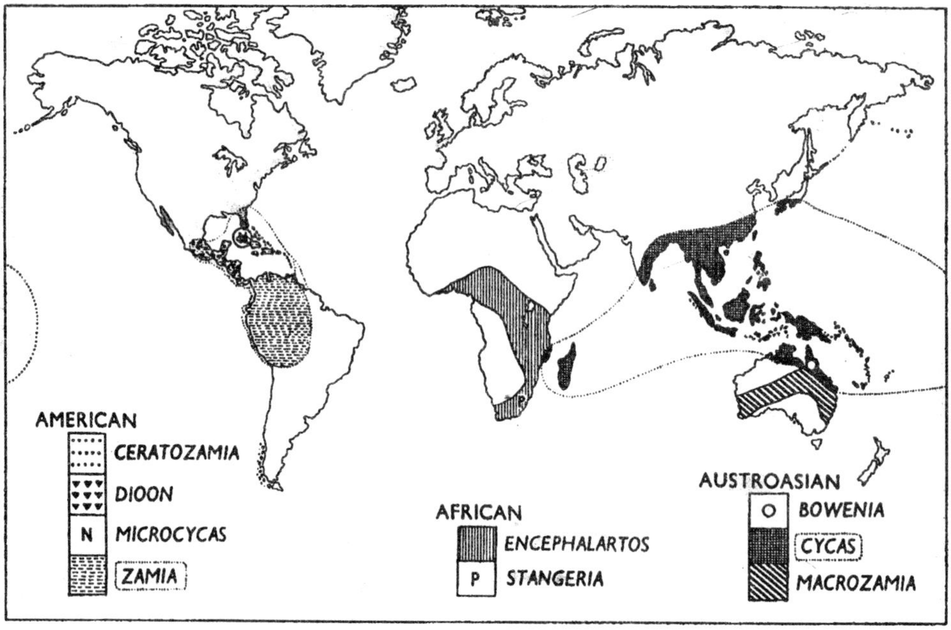

Fig. 46.4: Map showing distribution of living Cycad genera (after Schuster)

Macrozamia, Bowenia, Stangeria and *Encephalartos* are localized. Of the four genera in the Western Hemisphere, *Zamia* has the widest distribution while *Dioon, Microcycas* and *Cerotozamia* are more localized. The distribution of the genera is shown in the map on Figure 46.4.

Genus: *CYCAS*

Fossil remains which may be ascribed to the genus *Cycas* or, at least, to one closely related to it, are known since the Triassic (viz., *Bjuvia*). The genus represented by at least 15 (which number is raised by some to 20) species today. The species are localized in South East Asia with South Japan and South China on the north, Madagascar and East Africa on the west and the north tip of Australia with some Pacific Islands in the south and the east. *Cycas circinalis* occurs in Pacific Islands, China, India, Madagascar and East Africa. *C. revoluta* is Japanese although it is grown in India as an ornamental plant. *C. madagascariensis* is endemic in Madagascar. *C. media, C. kenneayana, C. normanbyana, C. carinsiana* are Australian species, the first being the commonest. *C. siamensis* is a dwarf tuberous species found in Burma, Yunnan, Siam and Indochina. This species is suspected to be a variation of *C. pectinata.* Species found in India are:

1. *Cycas circinalis* Linn. (Malabar coast, dry Orissa hills, Andhra, Madras to Ceylon 1500 to 3500 ft. besides places outside India as stated above). About 15 ft. tall. Leaves 5 to 9 ft. long.

2. *C. pectinata* Griff. (Nepal, Sikkim, Khasia, Chittagong, Arakan). 8 to 10 ft. high. Leaves 5 to 7 ft. long.

3. *C. rumphii* Miq. (Andaman, Nicobar, Ceylon, Lower Burma, Malaysia, New Guinea, North Australia). Up to 20 ft. high. Produces numerous and very large seeds. Leaves up to 7 ft. long.

4. *C. beddomei* Dyer. (Andhra, Madras, Malabar, Java). A low shrub with stems only up to 40 cm high. Leaves 3 ft. long.

KEY TO THE RELEVANT CYCAS SPECIES

Margins of leaflets revolute

Up to 10 ft. high small palm-like tree

Megasporophylls pectinate (large lobed) ... ... *C. revoluta*

Small shrubs up to 40 cm high. Megasporophylls dentate lobed ... ... *C. beddomei*

Margins of leaflets flat

Tuberous short trunks, very short leaflets

Megasporphylls broad and pectinate ... ... *C. siamensis*

10 to 20 ft. high small palm-like trees

Megasporophylls broad and pectinate

Leaflets less than 10 mm broad ... ... *C. pectinata*

Megasporophylls dentate lobed

Leaflets more than 12 mm broad

Megasporophylls densely hairy ... ... *C. rumphii*

Leaflets less than 12 mm broad

Megasporophyll teeth sharply spinose ... ... *C. circinalis*

The distribution of these species with the adjacent *C. siamensis* is shown in the map of Figure 46.5.

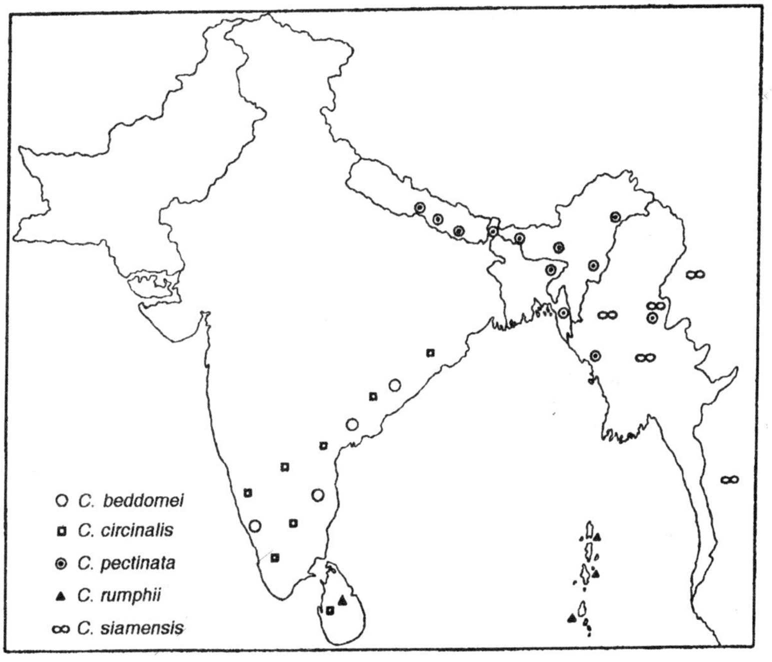

Fig. 46.5: Map showing distribution of Indian *Cycas* species and that of *C. siamensis* (modified from Pant)

The *Cycas* plant is arboreal, looking like a palm tree, hence the name 'palm-fern' although it is neither a palm nor a fern. The stem is columnar, normally unbranched, covered by an armour of persistent leaf bases and bearing a crown of large leaves above (Fig. 46.6A). The leaves are spirally arranged and dimorphic. The conspicuous type of leaves are very large (up to 9 ft.), simple pinnate and with tough, leathery leaflets which show only simple midribs without side veins. Alternating with these are clusters of scale leaves which are often roughly felted. The pinnate leaflets show typical circinate ptyxis (Fig. 46.6B). The rate of leaf development is slow but uniform and it is possible to determine the age of a plant by counting the persistent leaf bases and considering the number of leaves which are normally developed during a year. In the allied genus *Dioon* the age of some trees is determined to be more than one thousand years.

There is a normal tap root system at the beginning but it seems that, later on, this is replaced by a number of strong, branched adventitious roots. The adventitious roots develop a number of lateral ageotropic roots of the nature of pneumatothodes which grow up vertically or grow horizontally. There are lenticels on the surface of these roots. These roots get infected by bacteria. The bacterial growth is followed by incursion of the alga *Anabaena* which multiply rapidly forming a blue-green zone between the vascular cylinder and the outer cortex (Fig. 46.6D). The infection causes a distortion of the rootlets giving rise to a mass of exposed tubercles looking like coral which is known as the *coralloid root* (Fig. 46.6C). These are greenish or brownish and are more common in the seedling stage, especially in nurseries.

The stem (Fig. 46.7A) is covered by persistent leaf bases and shows a large *pith* and a large *cortex* of thin-walled cells with scanty rings of wood in between. The cortex as well as the pith show numerous *mucilage canals* which are interconnected through leaf gaps. The *primary*

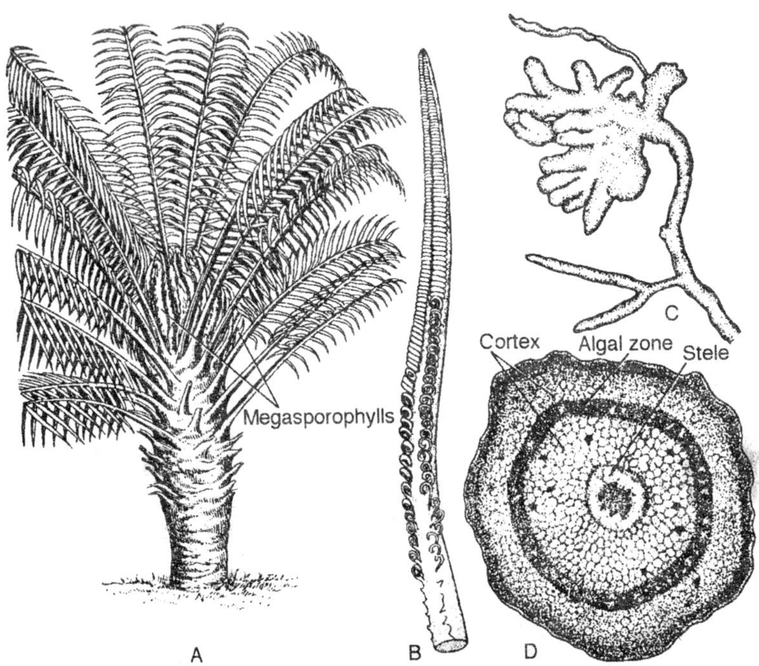

Fig. 46.6: A. Female plant of *Cycas circinalis*, B. A young leaf showing circinate ptyxis of leaflets, C. A bunch of coralloid root tubercles in *Cycas rumphii*, D. T.s. of a root tubercles showing algal zone

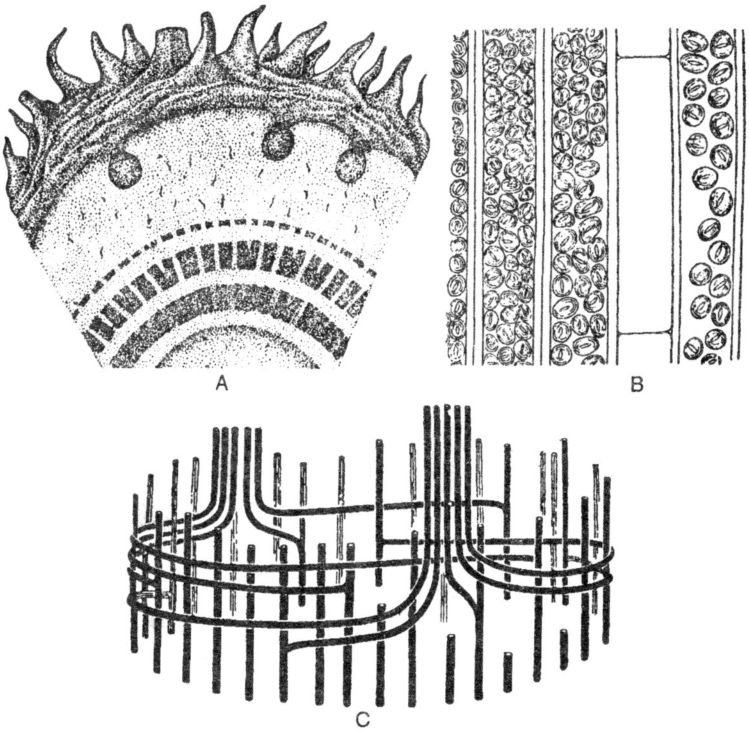

Fig. 46.7: *Cycas*, A. T.s. of stem showing three rings of wood, B. L.s. of mature wood showing multiseriate bordered pitted tracheides and also one row of thin-walled cells in between, C. Girdliwin of leaf traces

vascular cylinder is an *endarch siphonostele* in adult plants but *mesarch* in the seedling condition. The *bundles* are conjoint, collateral and open. The *primary cambium* is short-lived but *Cycas* is *polyxylic*. As soon as the primary cambium has stopped functioning, a second cambium is formed outside in the cortex and develops a second and stronger ring of wood. Such rings of wood are formed successively one outside another but the outer rings gradually become weaker and weaker. Usually only 3 or 4 rings of wood are found in a *Cycas* stem 3 metres high but a stem of *C. pectinata* showed as many as 14 rings. Annual rings are absent in *Cycas*. *Protoxylem* of seedlings shows spiral tracheides and those of older plants scalariform tracheides. *Secondary xylems* shows rows of multiseriate bordered pitted tracheides with rows of thin-walled cells in between (Fig. 46.7B). A peculiarity of *Cycas* and allied genera is the *girdling of leaf traces* (Fig. 46.7C). A leaf trace (formed of scalariform tracheides in the xylem), after arising from the stelar cylinder, usually does not pass into the nearest leaf directly but turns round a semicircle or 'girdles' the stem before passing to the next higher leaf. A number of leaf traces enter each rachis.

The petiole shows a large number of collateral open vascular bundles. The xylem is on the upper face and the phloem below. The leaf (Fig. 46.8A) has a xerophytic structure. The two epidermises are highly cutinized and there are sunken haplocheilic stomata (Fig. 46.8B) only in the lower epidermis. Below the upper epidermis is a hypodermis of one or two layers of sclerenchymatous cells below which is a row of palisade cells. Above the lower epidermis is a spongy parenchymatous tissue which may have palisade cells below. Between the palisade and the spongy layers are several layers of transversely elongated, thin-walled, colourless cells forming a *transfusion tissue* which serves for

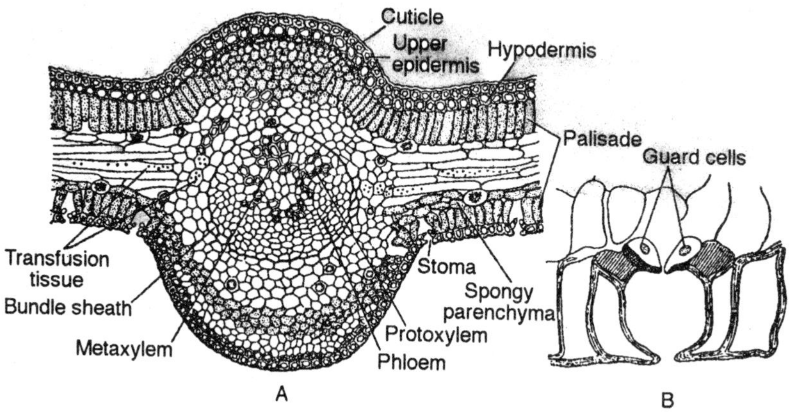

Fig. 46.8: *Cycas*, A. Vertical section of leaf, B. A sunken stoma

conduction. The *vascular bundle* of the median vein has a mesarch xylem on the top with a full metaxylem above and metaxylem patches below the protoxylem. The phloem is placed below the xylem. This bundle is encircled by a sclerenchymatous sheath.

The *Cycas* root is usually tetrarch. Secondary growth in it takes place early but secondary xylem formation is irregular.

Cycas is dioecious. The *male strobilus* of *Cycas* is a compact cone as in the other genera of Cycadaceae (Fig. 46.9A). Each strobilus or *cone* is quite large, and up to 50 cm in length. The strobilus develops at the apex but becomes lateral by the growth of a lateral bud so that the future stem becomes sympodial. In the strobilus the axis bears the compact *microsporophylls* arranged spirally and acropetally. A microsporophyll of *C. circinalis* (Fig. 46.9B) is 3 to 5 cm long and 12 to 23 mm wide. On its under surface more than a thousand *microsporangia* are borne in groups (*sori*) of 3 or 4 microsporangia each (Fig. 46.9C). Each microsporangium is unilocular and develops in the eusporangiate way. There is a multilayered *jacket* with a thickened *epidermis* and a *tapetum* enclosing numerous *microspore mother cells* which on reduction form numerous *microspores* or *pollen grains*. On maturity, each microsporangium splits by a slit at the top and the microsporophylls on a strobilus mature progressively from top to base so that it may take several days for all the pollens in a strobilus to shed.

Cycas is the only genus in the family which does not show any ovulate strobilus. The *megasporophylls* arise at the tip of the plant

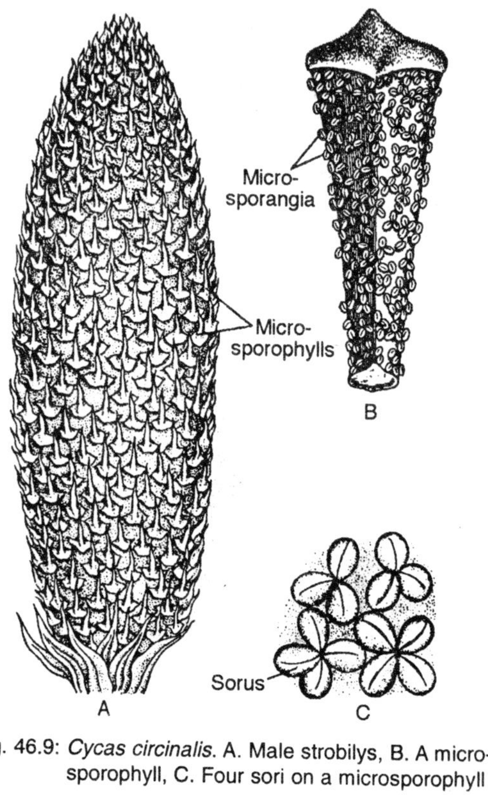

Fig. 46.9: *Cycas circinalis*. A. Male strobilys, B. A microsporophyll, C. Four sori on a microsporophyll

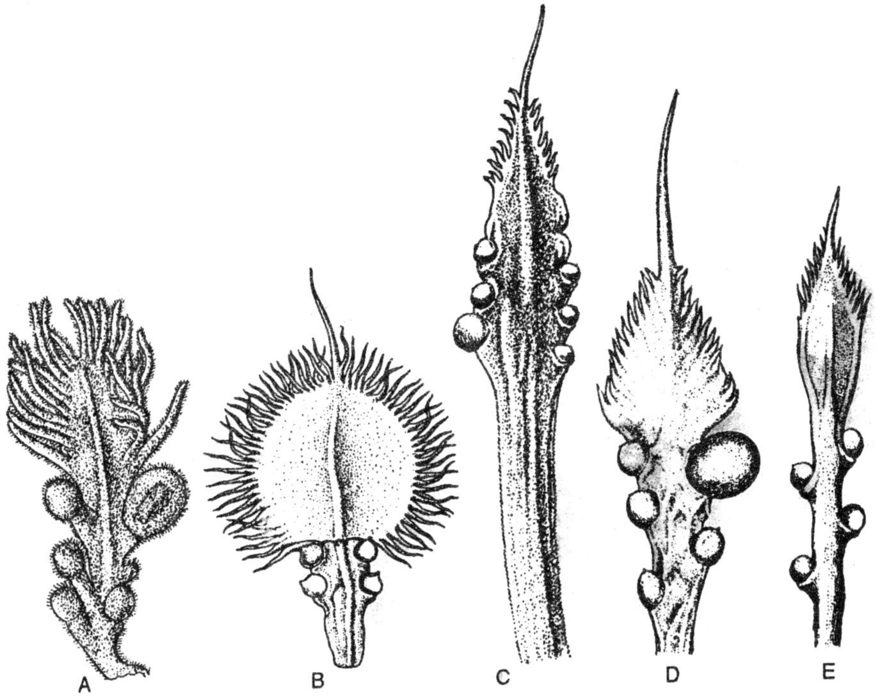

Fig. 46.10: Megasporophylls of A. *Cycas revoluta*, B. *C. pectinata*, C. *C circinalis*, D. *C. rumphii*, E. *C. beddomei*

acropetally forming a loose crown (Fig. 46.6A) leaving the apical growing point free for further growth. Each *megasporophylls* is a leaf-like structure 15 to 30 cm long which is pinnately dissected in the upper part into numerous segments which may be considered as pinnae or leaflets. Lower down, 1 to 5 pairs of ovules are borne on the two sides of the sporophyll. In abnormal cases some pinnae on the top have been found to be replaced by small ovules. Both sporophylls and ovules are covered by yellow hairs but as the ovules ripen the hairs are lost and the seeds develop a soft orange-red colour. In the various species of *Cycas* a gradual reduction in the sporophylls may be traced. In *C. revoluta* and *C. pectinata* (Figs. 46.10A and B) the lamina is large with elongated pinnae. In *C. circinalis, C. rumphii* and *C. beddomei* (Fig. 46.10C to E) the pamina becomes narrower and the pinnae are greatly reduced almost to mere serrations. In *C. normanbyana* (Fig. 46.18B) not only is the lamina reduced but the number of ovules also becomes only two.

The *megasporangium* or the *ovule* (Fig. 46.11) is orthotropous and becomes consider-

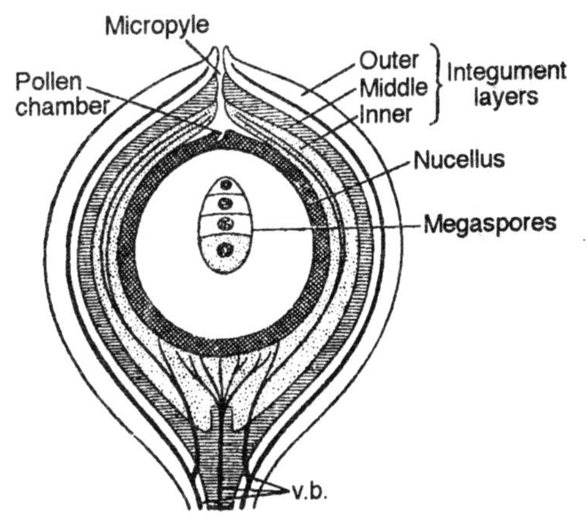

Fig. 46.11: Vertical median section of *Cycas* ovule (megasporangium)

ably large even before fertilization. A mature ovule (or seed) or *C. circinalis* may be up to 6 cm long and 4 cm in diameter. There is a single *integument* which is free only at the top from the *nucellus* (general tissue of the megasporangium). The integument is three-layered. There is a stony *middle* layer (shown dark) in the figure, a soft *outer* layer which is pulpy and variously coloured at maturity, and a soft *inner* layer which is used up and becomes papery before maturity. The *vascular supply* from the base of the ovule divides into three strands the middle one of which supplies the base of the sporangium and the two side ones pass to the integument. The latter again branch and the stronger branch supplies the outer pulpy layer of the integument while the thinner one ramifies into the inner soft layers of it. A single *megaspore mother cell* within the nucellus undergoes reduction division and forms a *linear tetrad* of four *megaspores.* There is an opening on the top of the integument forming the *micropyle* and a concavity in the nucellus top immediately below forming the *pollen-chamber.* The *n* number of chromosomes is 12.

The female gametophyte

Of the linear tetrad of megaspores (Fig. 46.11) usually the lowest one only becomes functional and the top three degenerate (Fig. 46.12A). The functional megaspore is never shed and germinates to form the *female* gametophyte entirely within the nucellus. Usually a spongy nourishing tissue is formed round the developing gametophyte. A very tough *membrane* is formed surrounding the megaspore and persists in the gametophyte. The megaspore enlarges considerably and its nucleus multiplies rapidly by free-nuclear division. A large central vacuole is formed at this stage and the free nuclei are pressed to the perphery in a layer of cytoplasm (Fig. 46.12B). Cell walls next begin to develop in the periphery and gradually proceed to the centre so that ulti-

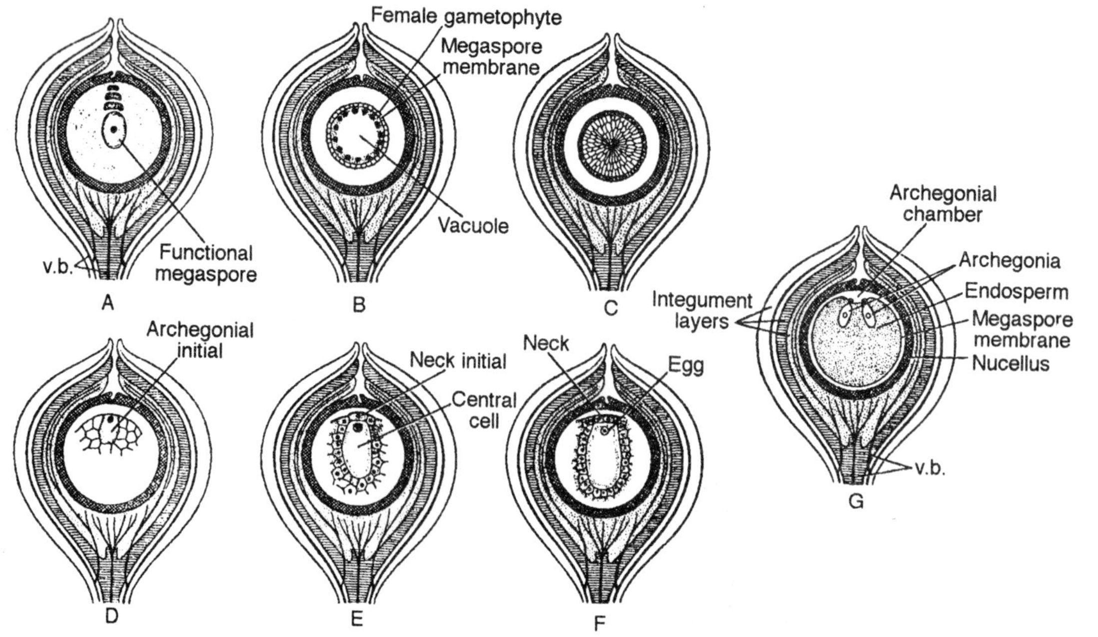

Fig. 46.12: Development of female gametophyte of *Cycas* within ovule. A. Functional megaspore stage (also showing disintegration of other megaspores), B. Free-nuclear stage, C. Cellular stage, D–F. Stages of development of an archegonium, G. Mature female gametophyte

mately the central vacuole becomes obliterated. The gametophyte is then completely filled up by a cellular tissue of which the upper cells are smaller (Fig. 46.12C). Several cells in the top periphery now become *archegonial initials* (Fig. 46.12D). Each initial divides transversely to form a *neck initial* and a *central cell* (Fig. 46.12E). The neck initial divides immediately to form two *neck cells* while the central cell increases in size. The nucleus of the central cell divides to form a *ventral canal nucleus* (which eventually disintegrates) and an *egg nucleus* (Fig. 46.12F). There is no wall at any stage between these two nuclei. In the mature female gametophyte (Fig. 46.12G) the *archegonia* are found below a depression called the *archegonial chamber* formed by the development of the surrounding tissue. The gametophyte membrane and the top nucellar tissue are destroyed at this region so that the archegonial chamber fuses with the pollen chamber. The gametophytic tissue below the archegonia is full of starch grains serving for the nutrition of the archegonia and then of the growing embryo. This is the *endosperm* (Fig. 46.12G) which is still a gametophytic tissue as in the Pteridophytes.

The male gametophyte

The *microspore* (*pollen*) has definite polarity as its *exine* is thicker at the bottom (Fig. 46.13A). The pollen is full of starch grains. Its nucleus undergoes mitosis and a small *prothallial cell* is cut off (Fig. 46.13B) which remains attached to the lower thicker exine and does not get disorganized. The other cell again divides forming two nuclei (Fig. 46.13C) one of which forms a small cell attached to the prothallial cell and becomes the *generative cell*. The free nucleus now is the *tube nucleus* (Fig. 46.13D) which is inside the large *tube cell*. Pollination takes place in this three-celled stage.

Pollination and fertilization

Cycas is wind pollinated. The air-borne pollens deposit on the top of the ovule and are drawn through the micropule by a drying speck of mucilage. They are then deposited within the pollen chamber where they are now sealed off by further drying and contraction of the top (Fig. 46.14A). The pollen germinates here and the tube cell of the male gametophyte (Figs. 46.14 and 46.15A) comes out through the thinner end of the pollen forming a *pollen tube*. Fertilization is, therefore, *siphonogamous* as in the Angiosperms. But, the nature of siphonogamy is very different. The pollen tube penetrates into the tissue of the nucellus destroying all the tissue between the pollen chamber and the female gametophyte. It thus acts rather like a haustorium than like a gamete carrier which latter is the case in Angiospermic siphonogamy. The haustorial pollen tube sometimes even branches like a root. It sucks in food material from the tissue it penetrates while its growth pushes own the base of the pollen to near the archegonium. In the base of the pollen tube (i.e., the cite of the original pollen) the *generative cell* divides into a *stalk cell* and a body cell (Fig. 46.14B). Months (four months in *Cycas revoluta*) elapse between pollination and fertilization in *Cycas*. The body cell

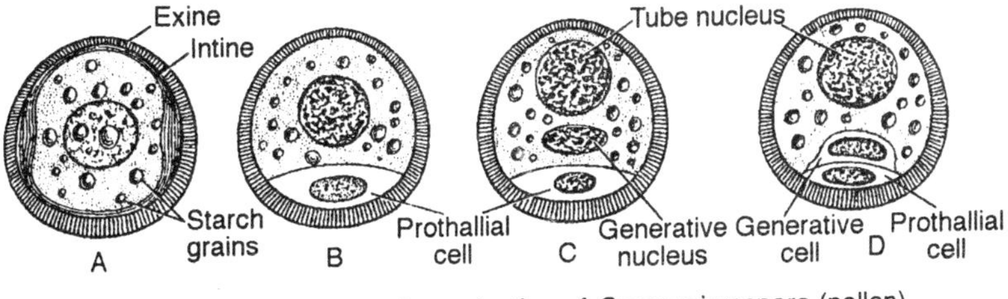

Fig. 46.13: Early stages of germination of *Cycas* microspore (pollen)

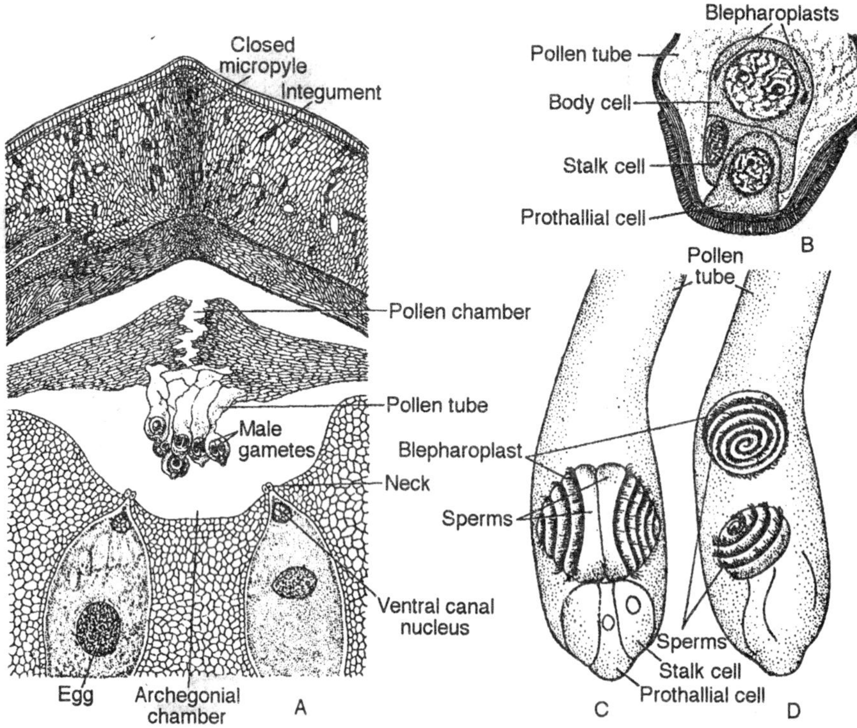

Fig. 46.14: *Cycas.* A. Closed top of ovule after pollination showing course of pollen tubes, B. Advancing base of pollen tube showing *body cell* with two *blepharoplasts*, C. Later stage with sperms formed, D. Same showing freed sperms at the time of fertilization

is the *spermatogenous cell.* Just before fertilization the body cell enlarges and two *blepharoplasts* appear at the two poles (Fig. 46.14B). Soon the body cell divides into two *sperms* to each of which a blepharoplast gets attached forming a spiral band with numerous cilia on it (Fig. 46.14C). The sperms of Cycads are remarkably large as are the eggs. They are even visible to the naked eye. The sperms are now liberated within the pollen tube (Fig. 46.14D) and then into the moist archegonial chamber by rupture of this basal end. The discharged of the pollen tube fluid causes the sperms to be forced into the egg cell. The cilia and cytoplasm are left behind on the top of the egg cell and the sperm nucleus fuses with the egg nucleus forming the *zygote.*

The embryo and the seed

After fertilization the zygote enlarges and undergoes a prolonged period of free cell formation. 64 to 256 nuclei have been reported in *Cycas.* A large central vacuole is reported by Treub in *C. circinalis* (Fig. 46.15A) where all the nuclei are crowded in a peripheral layer of cytoplasm. In other species the central vacuole has been found to be formed by the coalescence of a large number of smaller vacuoles. As the real embryo of *Cycas* is formed only from a part of this structure, the latter is not termed an embryo but a *proembryo.* Cell formation now begins at the base of this proembryo and gradually extends up the periphery until all the nuclei become encased in cell walls. The basal cells are smaller and full of a denser cytoplasm (Fig. 46.15B). It is from this basal region that the *embryo* is formed. The upper cells of this embryonic region elongate greatly forming a long *suspensor* which may be coiled and twisted while an *embryo* with two prominent *cotyledons* develop at the tip. The suspensors from the several egg cells in the different archegonia in

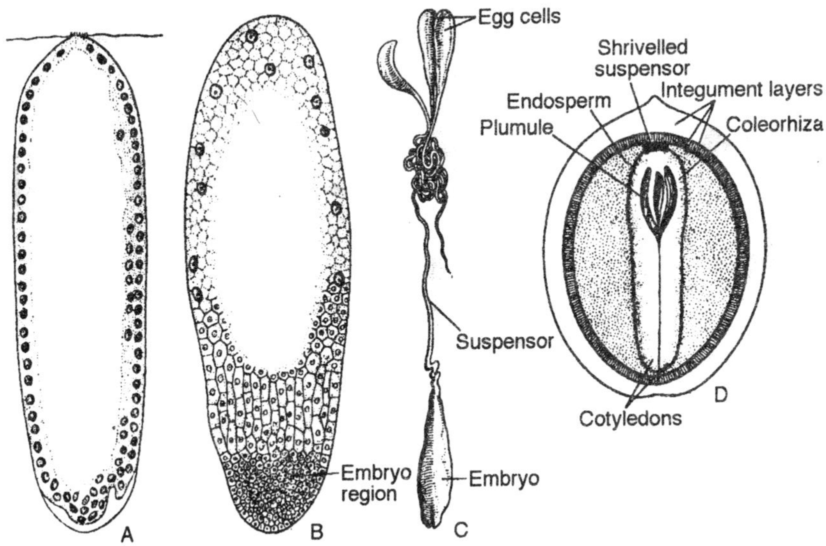

Fig. 46.15: A. Just before the beginning of wall formation in *Cycas circinalis* proembryo (after Treub), B. Later stage of proembryo showing organization of the embryo region,. C. Entangled suspensors arising from egg cells of several archegonia. One suspensor has a dicot embryo at the tip which has penetrated the endosperm, D. The seed of *Cycas*

the archegonial chamber of the same ovule become coiled up and push the embryos out of the egg membrane deep into the endosperm from which latter they obtain nutrition. Ultimately, however, only one embryo in an ovule attains maturity (Fig. 46.15C).

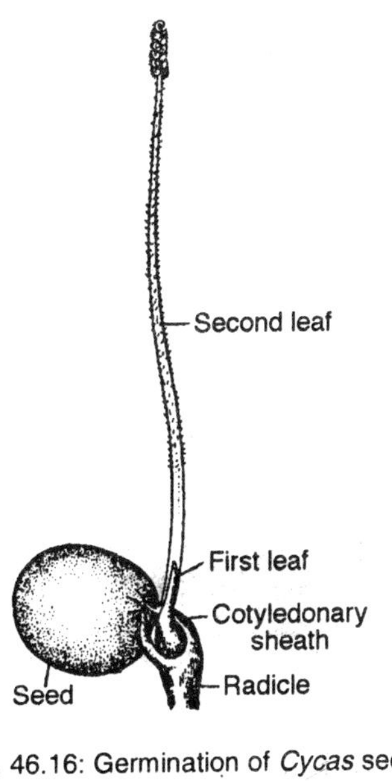

Fig. 46.16: Germination of *Cycas* seed

The single successful embryo in the ovule now enlarges greatly. Meanwhile, the nucellus is completely used up and the integument forms a covering of the *seed* or becomes the *seedcoat* showing three layers an outer fleshy, a middle stony and an inner fleshy (which later becomes papery) (Fig. 46.15D). The *plumule* develops late inside the base of the *cotyledons* and opposite this a hard structure, the *coleorhiza*, is formed. The coleorhiza hides the *radicle* inside. The suspensor dries up by now.

The seed is dispersed by birds, etc., which are attracted by the fleshy outer pulp. *Cycas* does not show any resting stage before germination. Germination is *epigeal* though the cotyledons do not rise high above the soil surface. The *radicle* forms a tap root after piercing the coleorhiza. The *first leaves* and the *plumule* come out last after piercing the *cotyledonary* sheath (Fig. 46.16).

A type of *sago* is obtained from *Cycas* stems. The seed and tender shoots are also sometimes eaten. The leaf is used in mat making. Figure 46.17 shows the life cycle of *Cycas*.

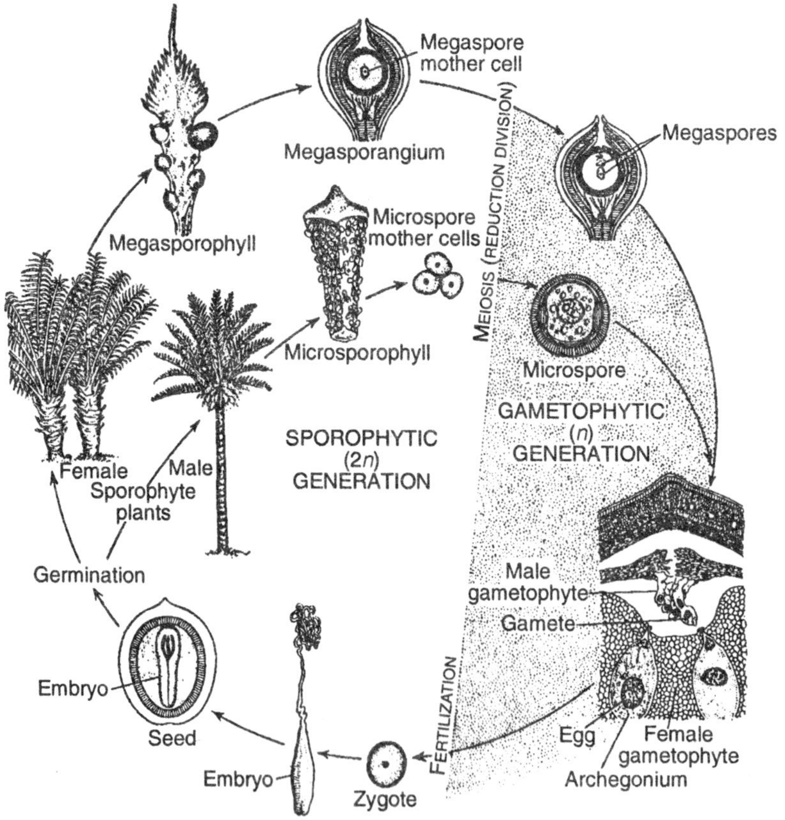

Fig. 46.17: Life cycle of *Cycas*

EVOLUTION OF THE CYCADALES

The fern-like characters of the Cycadales points out to an origin from the Cycado-filicales. Origin has been sought from the Medullosaceae (Worsdell, Stewart and Delevoryas) and from the Lyginopteridaceae (Scott, Arnold). Origin from the monostelic Lyginopteridaceae seems to be more plausible. It should also be considered whether the Cycadales might have arisen directly out of some ally of the Filicopsida. The relation between the Cycadales and the Cycadeoidales is discussed in page 1314.

Cycas, without any female strobilus and with a foliar megasporophyll, is clearly the most primitive of the living Cycadaceae while

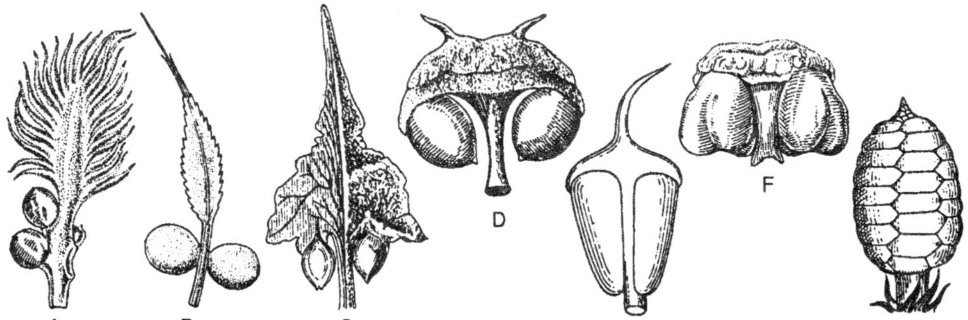

Fig. 46.18: Megasporophylls of A. *Cycas revoluta*, B. *Cycas normanbyana*, C. *Dioon edule*, D. *Ceratozamia*, E. *Macrozamia*, F. *Zamia*, G. Female cone of *Zamia floridana*

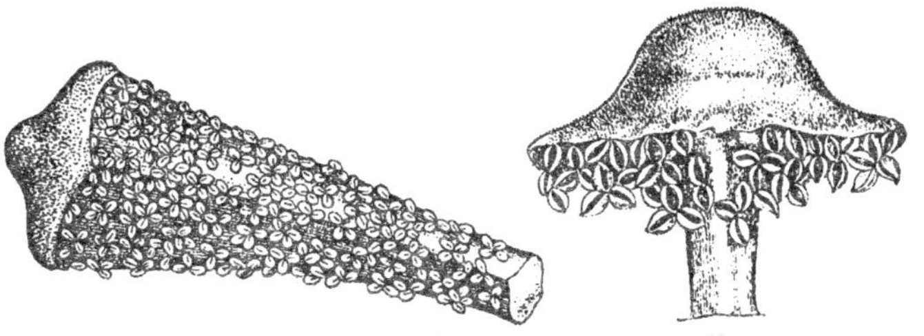

Fig. 46.19: Microsporophylls of A. *Cycas circinalis*. B. *Zamia integrifolia*

Zamia, is the most advanced. A comparative study of the sporophylls stresses this point. The megasporophyll shows progressive reduction from *Cycas* to *Zamia*. Within *Cycas* itself, the typically foliar megasporophyll of *C. revoluta* (Fig. 46.18A) is much reduced with only two ovules and a very simple leafy part in *C. normanbyana* (Fig. 46.18B). The megasporophyll gradually loses its foliar appearance through *Dioon* (Fig. 46.18C), *Cerotozamia* (Fig. 46.18D) and *Macrozamia* (Fig. 46.18E). In *Zamia* (Fig. 46.18F) it is simply a peltate structure with two hanging ovules. All the genera, except *Cycas*, from female cones, and this is most compact in *Zamia* (Fig. 46.18G). The microsporophylls also show similar reduction from the *Cycas* microsporophyll (Fig. 46.19A) with more than 1000 microsporangia to the peltate microsporophyll of *Zamia* (Fig. 46.19B) showing the maximum of 25 and the minimum of only one microsporangium. Genera in between show a gradual reduction in the number of microsporangia.

Order 2. NILSSONIALES

The Nilssoniales, occurring from the late Triassic (Rhaetic) to the Cretaceous but best known from the Jurassic, is often considered as a 'cycad' but Johansen and others have placed it in an order separate from the Cycadales. The epidermis and the stomata resemble the Cycadales (*haplocheilic*) and not the Cycadeoidales so that it should be placed nearer to the Cycadales.

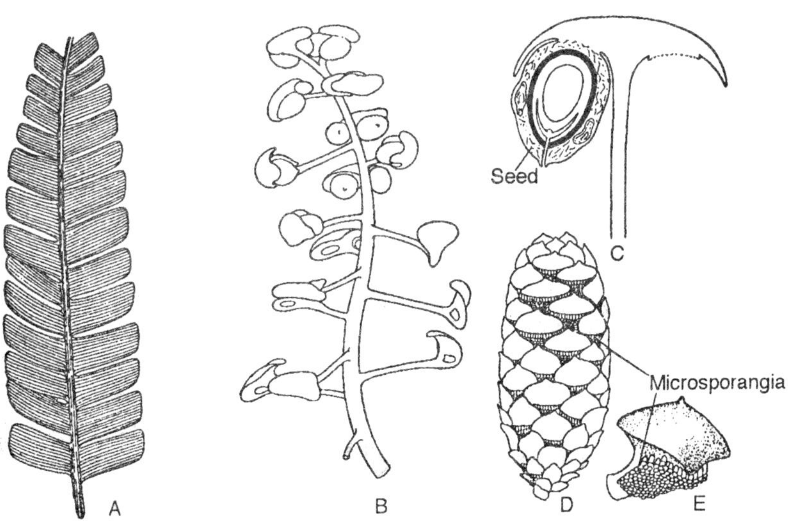

Fig. 46.20: Nilssoniales. A. Leaf (*Nilasonia compta*), B. Female cone and C. A single megasporophyll with one seed (*Beania gracilis*), D. Male cone and E. single microsporophyll showing microsporangia (*Androstrobus manii*) (After Harris and Andrews)

The primitive leaves with parallel venation, haplocheilic stomata and thin euticle, are placed in the form genus *Nilssonia* (Fig. 46.20A). These were abundant in the Jurassic and some seven-species (*N. princeps, rajmahalensis, morrisiana, medlicottiana, bindrabaunensis, orientalis* and *fissa*) are known from the Indian Upper Gondwana Rajmahal and Jabalpur-Chaugan stages (Lower and Upper Jurassic). The stem is not known with certainty. The male cones are placed in the form genus *Androstrobus* (Fig. 46.20D). The female cones (Fig. 46.20B), placed in the form genus *Beania* are so loose that it has been suggested that these were pendulous amentums on big trees. The megasporophylls (Fig. 46.20C) bore very few megasporangia hanging peltately just like *Zamia*.

Order 3. **CYCADEOIDALES** (= BENNETTITALES)

Although Arnold (1947) had explained that the valid name is Cycadeoidales and not Bennettitales, the latter name is still used by some authorities. A silicified trunk of *Cycadeoidea etrusca* was described by Capellini and Solms in 1867 from an Etruscan tomb where it was placed as an object of reverence 4000 years ago. Carruthers described *Bennettites gibsonianus* in 1868 but this latter was found to be another member of the genus *Cycadeoidea*. Hence, the prior name for the genus is *Cycadeoidea* and not *Bennettites* so that the names of the genus, family and order should be changed accordingly.

A distinctive character of the Cycadeoidales is the epidermis whose cells have sinuose walls and the guard cells and the subsidiary cells in the stomata are formed by division of the same initial (*syndetocheilic*—Fig. 46.21A). It may be noted that *syndetocheilic* stomata are round only in the Cycadeoidales, Pentoxylales and in *Gnetum* and *Welwitschia* while *haplocheilic* stomata are found in the Cycadales, Nilssoniales, Pteridospermopsida, Coniferopsida, Taxopsida and in *Ephedra*. Angiosperms show both types of stomata.

The Cycadeoidales are predominant Mesozoic fossils (Triassic to Cretaceous but predominant in Jurassic) and may be divided into three families: *Williamsoniaceae*, *Wielandiellaceae* and *Cycadeoidaceae* (= *Bennettitaceae*).

Family 1. WILLIAMSONIACEAE

Williamsoniaceae is characterised by columnar trunks in which the 'flowers' were borne on short side shoots which were not immersed within the leaf bases.

Williamsonia gigas was described in 1868 as *Zamia gigas* from the Yorkshire coast Jurassic by Williamson. Since then many species of this genus have been described from the Jurassic of various places. The leaf forms are *Ptilophyllum* and *Otozamites*. Several species of *Williamsonia* (viz., *W. sewardiana, indica, santalensis, blanfordia*), *Ptilophyllum* (viz., *acutifolium, cutchense*) and *Otozamites* (viz., *bengalensis, hisolpi, abbreviatus, bunburyanus*) are known from the Indian Upper Gondwana Rajmahal and Jabalpur series (Lower and Upper Jurassic—c.f. Gondwana Flora in *Palaeobotany*). *Bucklandia* stems and *Dictyozamites* leaves, several species of which are known from the same Indian rocks, are now known also to belong to the Williamsoniaceae.

The best known Indian specimen is *Williamsonia sewardiana* (Fig. 46.21B) described by Sahni from the Rajmahal and Jabalpur series (Lower and Upper Jurassic). The trunk was formerly described as *Bucklandia indica*. It is sturdy columnar, about 2 metres high and resembling modern *Cycas* in having an armour of persistent leaf bases. There is a crown of simple pinnate *Prilophyllum cutchense* leaves showing parallel venation and mixed with pointed scales. Fertile ('flower') and sterile lateral branches project out of thick layer of leaf base armour. These shoots are constric-

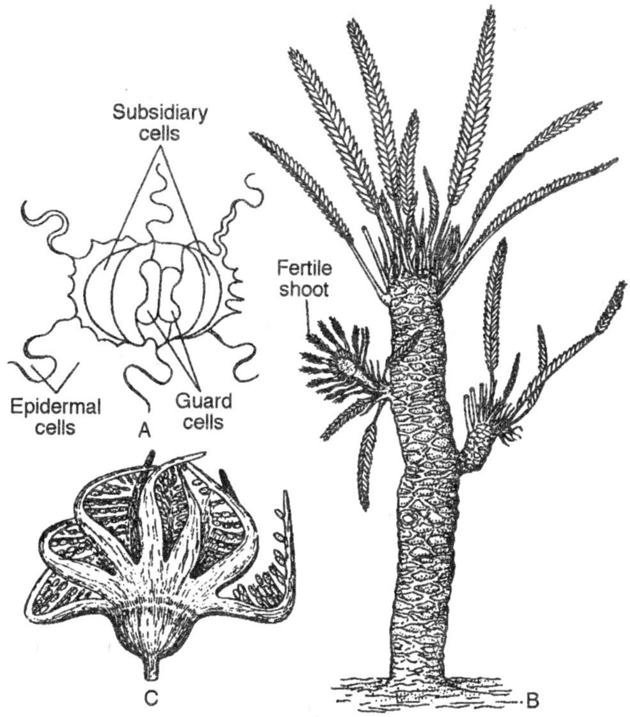

Fig. 46.21: A. Syndetocheilic stoma and sinuous epidermal cells of *Williamsonia pecten* (after Thomas and Bancroft), B. *Williamsonia sewardiana* Sahni (after Sahni), C. Male flower of *Williamsonia spectabilis* (after Thomas)

ted at base and usually bear many bracts and scales, the former being simple below but pinnate at top. The 'flower' is unisexual. Only the female is known. The female shoots somewhat resemble those of the Cycadeoidales (described below). There is a central hemispherical receptacle forming the axis on which there are long-stalked *ovules* surrounded by club-shaped *interseminal scales* which latter are swollen on the tops causing compactness. Seeds are few and in these the nucellus is greatly elongated engulfing the stalk and expanded in the top forming a funnel-shaped *macropyle*. The whole shoot is enclosed in dense hairy ramenta above and scaly bracts below.

The stem shows a central pith about 1 cm in diameter which is surrounded by a thin ring of xylem which latter is an endarch siphonostele. The xylem is traversed by numerous, usually uniseriate rays with compact xylem between rays imparting a coniferous appearance. Tracheides are usually scalariform but *Bucklandia indica* shows multiseriate-bordered pits. The pith and the cortex surrounding the stele contain secretory sacks.

Male 'flowers' are known in several species of *Williamsonia*, viz., *W. santalensis* (described by Sitholey and Bose from India) and *W. spectabilis* (Fig. 46.21C). This consists of a whorl of microsporophylls fused at the base but free upwards. The free segments bear lateral pinnae which in their turn bear synangial structures.

Some species of *Williamsonia* (viz., *W. gigas*) are reported to be with bisexual flowers but one cannot be definite because of the bad state of preservation.

Family 2. WIELANDIELLACEAE

Wielandiellaceae did not have their trunks covered by leaf armour. These were comparatively slender and dichotomously branched. There are two genera: (1) *Wielandiella* with

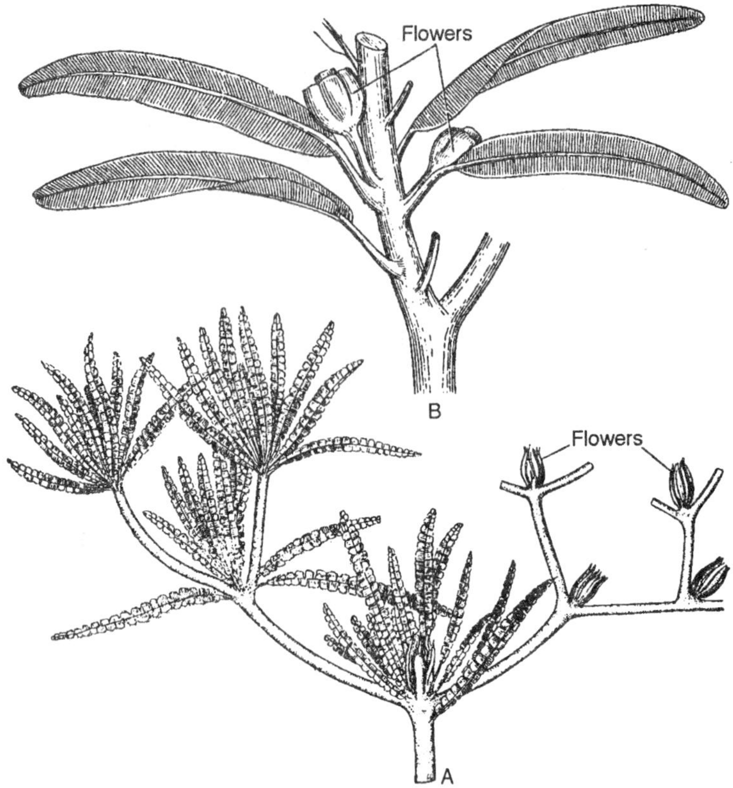

Fig. 46.22: A. *Wielandiella angustifolia* with *Anomozamites* leaves (after Nathorst), B. *Williamsoniella coronata* with *Taeniopteris vittata* leaves (after Zimmermann)

Anomozamites leaves (Fig. 46.22A). The stems are slender and repeatedly dichotomously forked. The bisexual flowers are borne on the forks and are surrounded by leaves. (2) *Williamsoniella* with leaf forms *Nilssoniopteris* and *Taeniopteris vittata* (Fig. 46.22B). The stems are as in the former whereas the bisexual flowers are on axils of spirally arranged leaves.

Family 3. CYCADEOIDACEAE

Genus: *CYCADEOIDEA* (Bennettites)

Cycadeoidaceae is characterized by short, stumpy, ovoid or columnar, stout trunks (once supposed to be fossils beehives) bearing flowers hidden within the armour of persistent leaf bases (Figs. 46.23A and B). The principal genus is *Cycadeoidea* which includes *Bennettites* and *Raumeria*—two generic names which should be discarded, *Cycadeoidea* being the oldest name. Specimens of these are abundant in the Mesozoic (most abundant in the Jurassic) rocks of Europe and North America.

Stems of *Cycadeoidea* (sometimes termed *Cycadella*) are up to 50 cm in diameter and covered by a thick mantle of persistent leaf bases with a ramentum of flattened scale-like hairs packed between them. Simple pinnate, parallel-veined, xerophyllous leaves up to 10 ft. long from a crown at the top. Some of the specimens (viz. *C. marshiana*) show branched stems. The stem (Fig. 46.24) shows a thin zone of *endarch siphonostele* surrounding a larger *pith*. There is a *cambium* ring with a thin zone of *secondary wood*. The *primary xylem* shows spiral tracheides but the *secondary xylem* is formed of scalariform tracheides and sometimes even pitted ones. Leaf trace bundles pass directly to the leaves through the *cortex*. The

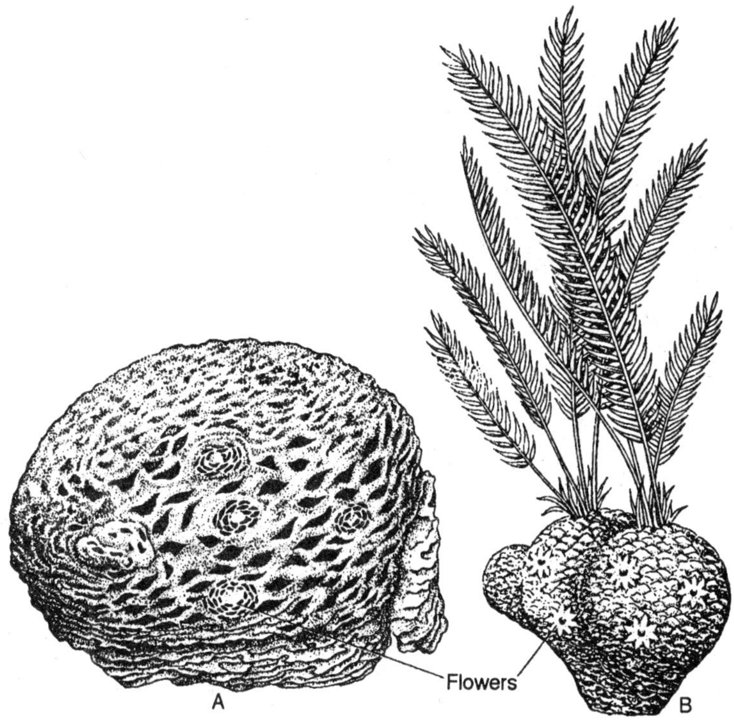

Fig. 46.23: A. *Cycadeoidea dacotensis* stem (after Wieland), B. Reconstruction of *Cycadeoidea marshiana* plant

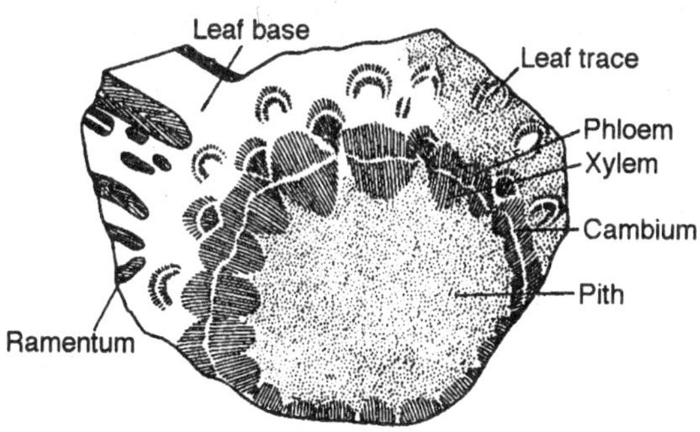

Fig. 46.24: T.s. of *Cycadeoidea* stem (= *Cycadella*) (after Wieland)

leaf shows xerophyllous structure with a heavily cutinized epidermis, epidermal cells with wavy walls and spongy parenchyma. The vascular bundles are mesarch as in Cycadales.

'Flowers' are always partially sunken in the leaf base armour. Each such reproductive shoot (Figs. 46.25A and B) is surrounded by a large number (sometimes as many as 100) spirally arranged pinnate bracts which closed over the fructification before maturity. These formed a sort of a *perianth* of the 'flower' surrounding an *androecium* and a *gynoecium* in the centre. The flower, therefore, is bisporangiate. The *microsporophylls* or *stamens* form a whorl united at the base into a sheath. They are pinnate, 8 to 18 in number, each 10 to 12 cm long and looking like small foliage leaves. There are 10 or more pairs of pinnae on the two sides of a sporophyll. Each pinna, again, bears two lateral rows of *synangi*. Each microsporophyll ends in a sterile point. Each synangium is multilocular with 20 to 30 loculi. The axis of the flower is expanded into the *ovuliferous receptacle* or *ovulate* cone. It is shaped like a cone or convex cushion. A ripe 'fruit' (Fig. 46.25C) with the stamens shed is closely covered by the bracts. On the receptacle there are a number of *megasporophylls*. Each megasporophyll has a long, slender stalks, representing the rachis of the sporophyll, bearing a terminal, orthotropous *ovule*. The ovules (Fig. 46.25D) are small with the *integument* tip expanding into a

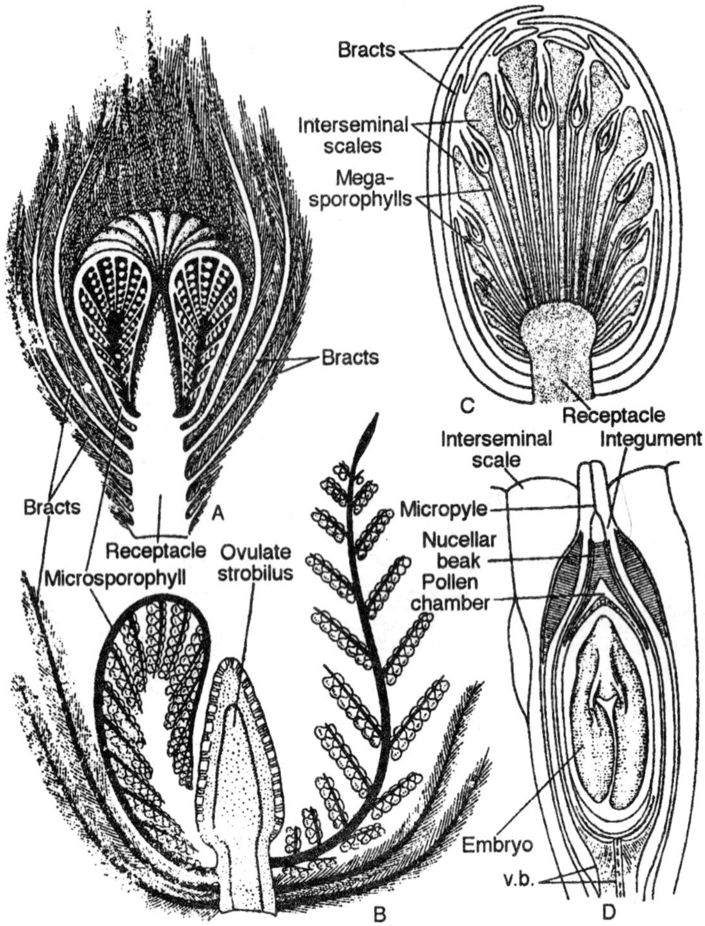

Fig. 46.25: *Cycadeoidea*. A. Unexpanded flower (after Wieland), B. Same with a microsporophyll expanded, C. A mature fruit of *C. gibsonianus*, each ovule showing a dicotyledonous embryo (after Scott), D. Vertical l.s. of *C. morrieri* seed (after Lignier and Solms-Laubach)

micropyle which projects out of the 'fruit' surface. The integument shows hard outer and inner layers with a pulpy middle. The nucellus expands into a *nucellar beak* above. *Interseminal scales*, about as numerous as the ovules, occur between the megasporophylls. The tips of these scales are enlarged like clubs which fit into the space between the seeds and later become fused in such a manner as to form a continuous 'pericarp'. Each seed has a hard integumental covering and a dicotyledonous *embryo* inside. The whole 'fruit' structure is an inch or two long. No pollen tube has been discovered.

AFFINITIES OF CYCADEOIDALES

Although Cycadales and Cycadeoidales show certain similarities in habit (stem, leaf, ovule, etc.), there are prominent characters (leaf trace, inflorescence, fruit, etc.) for which the Cycadales cannot be derived from the Cycadeoidales. Both the Cycadeoidales and the Cycadales probably arose out of the Cycadofilicales. Chamberlein has pictured a common origin for the two from which the two orders arose by parallel evolution (Fig. 46.26). It is supposed that the hypothetical ancestral Cycadofilicale had a foliar megasporophyll

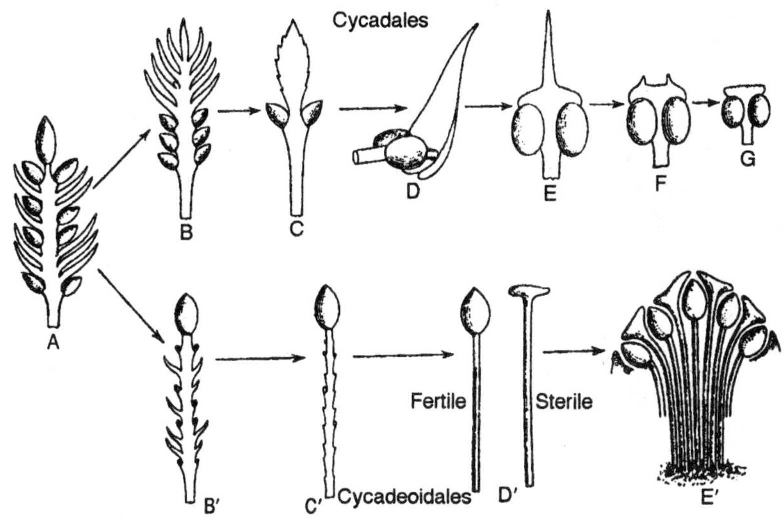

Fig. 46.26: Parallel origin of Cycadales and Cycadeoidales (modified from Chamberlein), A. Hypothetical ancestor, Cycadales line → B. *Cycas revoluta*, C. *Cycas normanbyana*, D. *Dioon edule*, E. *Macrozamia*, F. *Cerotozamia*, G. *Zamia*. Cycadeoidales line → B′, C′. Hypothetical stages of reduction of A by suppression of lateral ovules, D′. Condition in Cycadeoidales, fertile one with terminal ovule, sterile one (interseminal scale) without ovule, E′. *Cycadeoidea gibsonianus*

with ovules both at apex and on sides. From this, in the Cycadales line, there was complete suppression of the terminal ovule and gradual reduction of lateral ovules with ultimate transformation of the upper leaf-like part of sporophyll into a peltate structure. On the other hand, in the Cycadeoidales line, all the lateral ovules were suppressed and the sporophyll became a stalk-like structure with one terminal ovule. Even the interseminal scales are supposed to be evolved by complete reduction of the sporophylls into sterile structures.

The Cycadeoidales do show some fern-like characters (direct leaf traces, synangia, ramentum, leaf, etc.) because of which a direct origin from the Filicopsida may be suspected but there is not sufficient evidence to corroborate this origin.

The Cycadeoid inflorescence strikingly resembles the flower of *Magnolia*. But, closer examination does not show any real similarity. There is no carpel or angiospermic stamen. Such similarity seems to be merely due to similar evolution and not due to any phylogenetic connection.

Order 4. **PENTOXYLALES**

The Pentoxylales is a peculiar group of fossils discovered from the Rajmahal series (Lower Jurassic) of Indian Upper Gondwana and investigated by Sahni and his students. The phylogenetic position of the toxon is still uncertain. The fossils are placed in a single genus *Pentoxylon* comprizing the family Pentoxylaceae. They show characters both of Conifers and of Cycads which combination is not found in any other group. Further discoveries may necessitate placing this group in a higher taxon than an order, distinct from both Cycads and Conifers.

Different parts of *Pentoxylon* (Fig. 46.27) have been discovered under different names. The stem is *Pentoxylon* (Figs. 46.27A and C) *sahnii* and *Nipanioxylon guptai*; the leaf is *Nipaniophyllum* (Fig. 46.27B) *ravi* and *N. spathulatum*; the male flowers *Sahnia nipaniensis* (Fig. 46.27D) and the female cones were described as *Carnoconites* (Fig. 46.27E) *compactum* and *C. laxus*.

Nipanioxylon leaves have been found in clusters on dwarfshoots borne laterally by

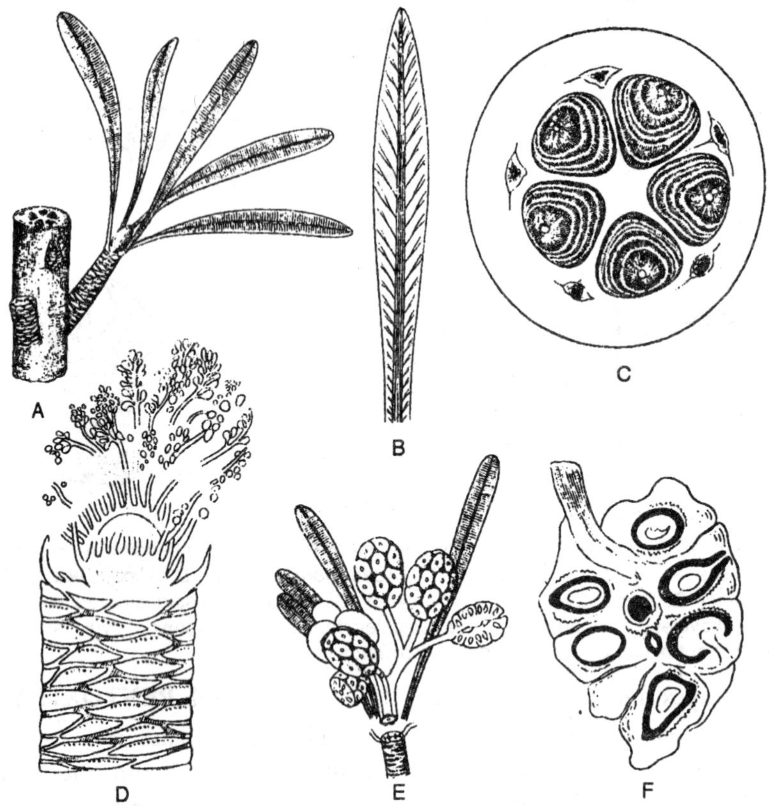

Fig. 46.27: *Pentoxylon*. A. Stem and leafy shoot (after Sahni), B. Lead (*Nipaniophyllum*), C. T.s. of stem showing endocentric steles (*Pentoxylon sahnii*), D. Male flower (*Sahnia nipaniensis* after Vishnu-Mittre), E. Female shoot (*Carnoconites*), F. L.s. of female cone

Pentoxylon stems (Fig. 46.27A). The leaf (Fig. 46.27B) is strap-shaped resembling *Taeniopteris* but the midrib is formed of several strong parallel veins which again have branches running parallelly to the margins and these branch veins may show forkings at the ends. Stomata on the epidermis are syndetocheilic. The parallel leaf vascular bundles are mesarch.

The stem of *Pentoxylon* (Fig. 46.27C) shows two rings of steles. The inner ring is formed of 5 or 6 larger steles in each of which there is conspicuous *endoventric secondary growth*, i.e., a greater part of the wood was formed towards the pith, on the inner side of the cambium. Alternating with these steles there is one outer ring of smaller steles, formed of secondary wood. Growth rings in the wood are well defined and are more coniferous than cycadophytic in structure. The tracheides are narrow with crowded bordered pits and there are uniseriate rays in between.

The male flower, *Saknia nipaniensis* (Fig. 46.27D) is borne terminally on shoots. Branched microsporophylls fuse at the base to form a tubular disc. The final branches bear unilocular microsporangia at the tips. Pollen grains are boat-shaped.

Carnoconites (Fig. 46.27E), the female flower, is borne apically on dwarf leafy shoots. The peduncle is branched, each branch bearing a female cone. Each cone has an axis bearing spirally arranged sessile ovules directly on it. No scale, bract or megasporophyll is associated with it. The seeds show thick, fleshy integument and protruding micropyles (Fig. 46.27F).

Chapter 47

Coniferopsida: Cordaitales

The Coniferopsida, as understood here, comprise of the *Coniferophytes* or *Stachyosperms* (Sahni, Chamberlain, Takhtajan, Pant) with the Gnetales taken out. As such, it includes four orders *Coedaitales, Coniferales, Taxales* (often included within Coniferales) and *Ginkgoales*. Of these, the Cordaitales and Coniferales seem to be closer to one another but no common ancestor at the gymnospermous level (i.e., among pteridosperms or cycads) can be envisaged. They probably originated in the Upper Devonian and dominated the Permocarboniferous and the Mesozoic. The Taxales, because of certain peculiarities, are now considered separate from the Coniferales. The position of the Ginkgoales, now a mere relic of a larger Mesozoic order, is even more problematic. They are placed within the Cycadopsida in Engler's system (1954) but the majority opinion is in favour of placing them among the Coniferophytes. But, again, no common ancestor is known. There is a possibility that all these groups had their origins in the 'Progymnosperm' Pteridophytes.

The general habit of the Coniferopsida is as described for the Coniferophyta (page 1281) which shows larger and more profusely branched trees than in the Cycadophyta, simple (usually microphyllous) leaves and better developed wood. As opposed to the Cycadopsida, which is generally tropic 1, the Coniferopsida predominates the temperate regions. They include the *Conifers* which form the most abundant forests today.

Order 1. CORDAITALES

The Cordaitales are known only as fossils today. Some Middle Devonian tree trunks have been ascribed to this order but they are widely known only from the Permocarboniferous when they dominated forming the world's first great forests. They became extinct in the Jurassic. Three families are recognized:

Family 1. PITYACEAE

Pityaceae is known only from trunks of wood and a few vegetative shoots. Occurring in the Devonian and the Lower Carboniferous, they comprise the oldest of the Cordaitales. Arnold (1947) is of opinion that placing the Pityaceae within the Cordaitales is only provisional pending future discoveries. The recent discovery of the *Callixylon-Archaeopteris* association (c.f. page 1216) has brought into doubt the systematic positions of even the other members of Pityaceae.

The trunks fossils representing the family were of considerable girth signifying tall trees. In a trunk a large central pith is surrounded by strands of mesarch primary xylem some of which are in contact with secondary xylem while others are free and embedded in the pith. The pith is parenchymatous but often mixed with scattered tracheides or even with complete mesarch strands. Leaf traces pass out as branches from the primary strands without causing leaf gaps. The secondary wood forms a compact cylinder without wood parenchyma and is composed of alternating narrow rays of bordered pitted tracheides and medullary rays as generally found in conifers. Some of the forms come very close to *Calamopitys.* The wood also shows certain definite Araucarian characters.

Of the three genera assigned to the Pityaceae, *Callixylon* and *Archaeopitys* are from the Upper Devonian while *Pitys* is from the Lower Carboniferous.

Callixylon from the United States Upper Devonian (Fig. 47.1), which was placed within the Pityaceae from its anatomy, has aroused great interest of late. Trunks up to 5 ft. in diameter have been discovered signifying tall trees. The tracheides show multiseriate bordered pits placed in intermittent groups leaving clear pitless surface in between (Fig. 47.1B). Beck (1960 c.f., page 1216) has discovered a stem of *Callixylon* in association with *Archaeopteris* fronds. It clearly assigns this to the pteridophytic Archaeopteridales. Although it cannot be said that all the stems placed under the form genus *Callixylon* are Pteridophytes, the discovery points out that many members of the Pityaceae may be actually pteridophytic. The taxon *Progymnosperm* has, therefore, been suggested as the heterosporous pteriodophytic group from which the gymnosperms might have arisen.

Pitys, from the Lower Carboniferous, is known not only by its wood (sometimes with feeble annual rings) but also by a leafy shoot, *Pitys dayi.* Trunks 50 ft. long and 10 ft. in diameter are known. The 'leaves' appear to be cylindrical fleshy cladodes 4 to 6 mm in diameter and up to 50 mm long. There is no lamina but there are hairs and sunken stomata in the epidermis. These leaves were regularly shed, only the tip cluster remaining attached. At the tips the cladodes were clustered to form buds. This type of leaf is not in conformity with that of Cordaitaceae and brings *Pitys* nearer to Araucariaceae.

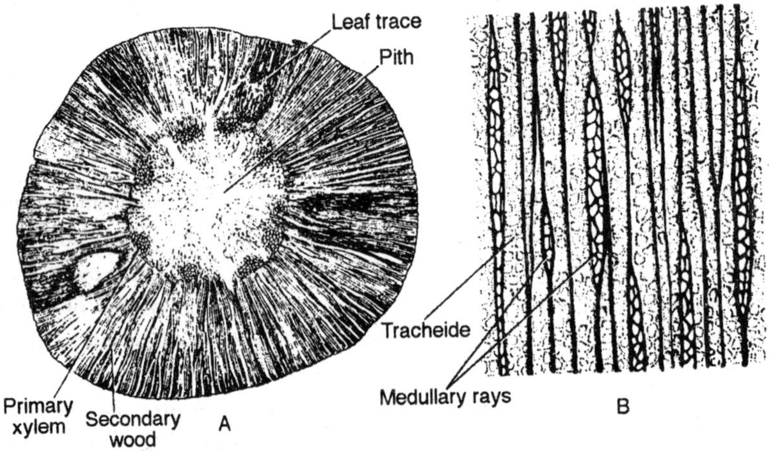

Fig. 47.1: *Callixylon.* A. T.s. of stem (after Arnold), B. Radial l.s. of secondary wood showing bordered pitted tracheides and medullary rays

Family 2. POROXYLACEAE

Poroxylaceae comprises of the single Permocarboniferous genus *Poroxylon* from France which, again, is known mainly as trunks and shoots. Although later in appearance, the stem structure of *Poroxylon* is more primitive. The primary xylem is exarch instead of mesarch and the secondary wood is made up of large cells and broad rays more like the Cycadofilicales than the Cordaitales. The leaves and leaf traces are, however, cordaitean. Some cordaitean seeds (*Rhabdocarpus*) have also been found in association with these stems and leaves. A gradation is suggested from *Poroxylon* to *Mesoxylon* and finally to *Cordaites.*

Family 3. CORDAITACEAE

Genus: *CORDAITES*

Cordaites is the main genus of Cordaitaceae and is often considered as the only genus, the other generic names being considered as form genera of its different parts. The name *Cordaites* was at first used only for the leaves but gradually it has come to be applied to whole plants. The wood of *Cordaites* has the named *Cordaixylon, Dadoxylon, Noeggerathiopsis* and *Mesoxylon.* Of these, *Mesoxylon* has a different anatomy (considered later) so that it is often considered as a genus separate from *Cordaites.* Mesozoic *Araucarioxylon* was sometimes included within Cordaitaceae but now its more prabable affinity seems to be with Araucariaceae. The root called *Amyelon* is usually associated with *Mesoxylon.* Inflorescences and strobili are named *Cordaianthus.* Defoliated leaf stalk compressions—*Cordaicladus.* Pith casts—*Artisia* (= *Sternberhia*). Seeds—*Cordaicarpus, Cardiocarpus, Mitrospermum* and *Samaropsis.* Prominent cordaitean Indian Lower Gondwana fossils are—*Noeggerathiopsis hisolpi, Samaropsis* spp. and *Gordaicarpus sahnii* from Upper Carboniferous: *Noeggerathiopsis hisolpi, N. stoliczkana, N. whittiana, Dadaxylon indicum, Samaropsis ranigan-jensis, S. milleri* and *Cordaicarpus indicus* from Permian. Some *Noeggerathiopsis hisolpi* is present in the Panchet-Parsora Stage which is in between Upper and Lower Gondwana and probably extends to the Lower Triassic.

Cordaites formed the dominant forest vegetation during Carboniferous and Permian when it was widely distributed. It became extinct after this age though some (*Noeggerathiopsis*) seem to have persisted in India up to the Lower Triassic. It formed tall trees with trunks often reaching 100 ft. in height and 3 ft. in diameter (Fig. 47.2A). The trunk was usually branched only at the top with the final branches bearing dense clusters of leaves suggesting that the trees were crowded together forming a deep shade.

The leaves are the best preserved. In contrast to the compound leaves of the Cycadophytes the leaves of *Cordaites* are simple. These are leathery with a highly xerophy structure, entire margins and dichotomous venation (Fig. 47.2B). But, there is little forking of the vein above the leaf base so that the fine veins look paralled and simulate monocot leaves. This, of course, cannot signify any affinity with the very distant monocots. The leaves are arranged spirally on the ultimate branches. A transverse section of the leaf (Fig. 47.2C) shows a cutinized thick-walled *epidermis* and a supporting *hypodermis* of thick-walled cells which latter expands near vascular bundles to reach them. In some leaves there is a distinction between palisade and spongy tissues while in others the *mesophyll* is uniform. *Haplocheilic* stomata occur in longitudinal rows or irregularly on the lower epidermis. When preserved, they show 4 to 6 subsidiary cells. The parallel leaf veins are mesarch or sometimes exarch. The veins are connected by transversely elongated cells somewhat like the modern Cycad leaves. Thus, the leaf anatomy, on the whole, resembles that of the modern Cycads.

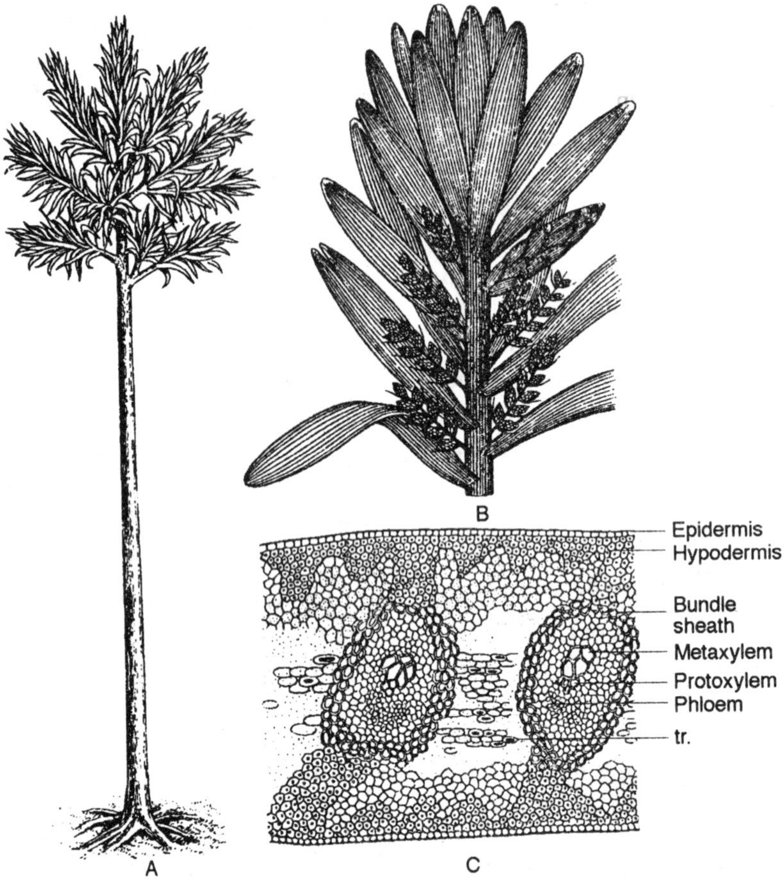

Fig. 47.2: *Cordaires*. A. *Cordaites* tree: *Dorycordaites* (after Scott). B. A shoot showing leaves and inflorescences (after Grand'Eury). C. T.s. of leaf (after Renault). *tr* = transversely elongated cells connecting bundles

Grand'Eury classified *Cordaites* according to leaf forms into *Eu-Cordaites* (broad leaf), *Dory-Cordaites* (lanceolate leaf) and *Poa-Cordaites* (linear leaf). The leaves were 2 cm to 1 metre long and up to 20 cm broad.

The stem of *Cordaites* should be considered under two heads—*Mesoxylon* and *Cordaites* proper (which latter includes *Dadoxylon*). *Mesoxylon* stems are most abundant among Carboniferous Cordaitales and are so different that they are often considered to form a separate genus. Its stem structure is more primitive than that of *Cordaites* and is intermediate (hence *meso-*) between *Poroxylon* and *Cordaites*. Its cortex is differentiated into two layers as in the Cycadofilicales (there is a periderm and the wood, with usually large discoid pith) as in other *Cordaites*. The leaf traces are mesarch as in *Poroxylon* and are double at first but divide further into eight on entering the leaf.

The *Cordaites* stem (Fig. 47.3) shows a large pith in most species (it may be smaller in some), a thin zone of primary wood and a large cortex resembling the Cycads. But, the secondary wood is well developed, thick and coniferous, rendering it distinct. The pith, 1-10 cm or more in diameter, is rather peculiar. It is entirely parenchymatous but in course of growth it cracked transversely (Fig. 47.4A) giving the appearance of a pile of concave discs. Pith casts named *Artisia* (Fig. 47.4B—former name *Sternbergia*) are obtained separately. The primary wood is a ring of endarch bundles. The mesarch primary bundles giving rise to leaf traces as found in *Mesoxylon* are absent here. The primary tracheides are annular, spiral, scalariform and finally pitted. Round this discontinuous ring is a thick layer of secondary wood. The tracheides in secondary wood are long and slender with multiseriate (in 1 to 3 or more rows) bordered pits on radial walls. The bordered pits are hexagonal or flattened oval resembling *Araucaria*. The rays are usually uniseriate (Fig. 47.4C). Dark resinous material may be present in the rays but resin canals and xylem parenchyma are conspicuously absent. Annual rings are not found in Carboniferous *Cordaites* but are frequent in Permian specimens signifying the change in climate. *Dadoxylon* shows both types. The cortex is parenchymatous with gum canals. A periderm develops during secondary growth.

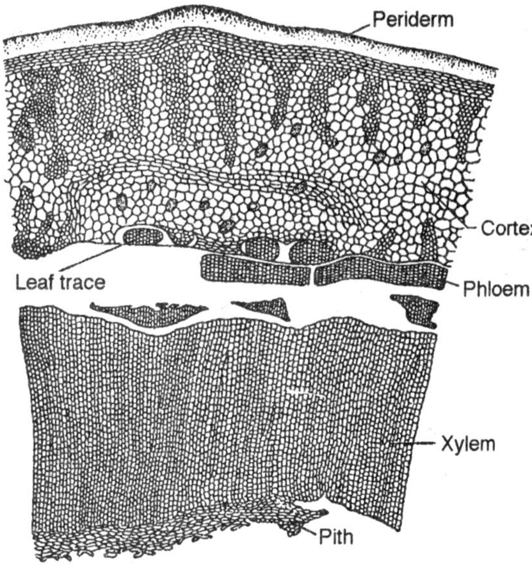

Fig. 47.3: T.s. of *Cordaites* stem

The wood, specially the pits are Araucarian and for this it merges into the Mesozoic fossil trunk *Araucarioxylon* which is now assigned to Araucariaceae.

The cordaitean root fossil (Fig. 47.5) is named *Amyelon*. It is suspected to belong to *Mesoxylon* although no actual association has been found. It shows a central triangular protostele of diarch to tetrarch xylem surrounded by a thick secondary wood. The cortex shows a periderm.

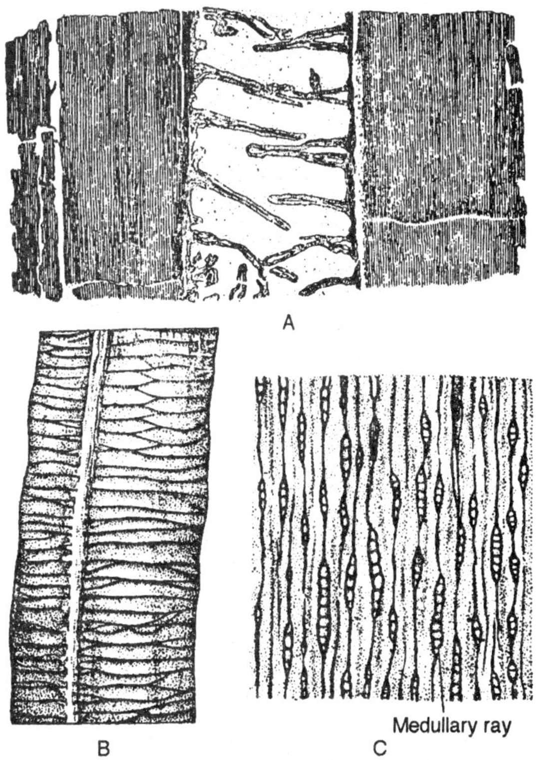

Fig. 47.4: A. L.s. of *Cordaites* stem showing wood and pith with septations in the latter (after Arnold). B. *Artisia* sp.—Cordaitean pith cast (after Arnold). C. Radial l.s. of Cordaitean secondary wood (after Andrews)

Cordaites is monoecious or dioecious but the fructifications are always unisexual. The fructifications are lax inflorescences 10 cm or more long, borne on the stem (not necessarily the axils of leaves) among the leaves (Fig. 47.2B). Each inflorescence axis is a slender stalk bearing stiff bracts each of which subtends a short axillary bud-like strobilus. The inflorescence and strobili fossils are placed in the form genus *Cordaianthus*. Sometimes the strobili may be arranged in four rows but usually they are arranged spirally in two rows (Fig. 47.6A).

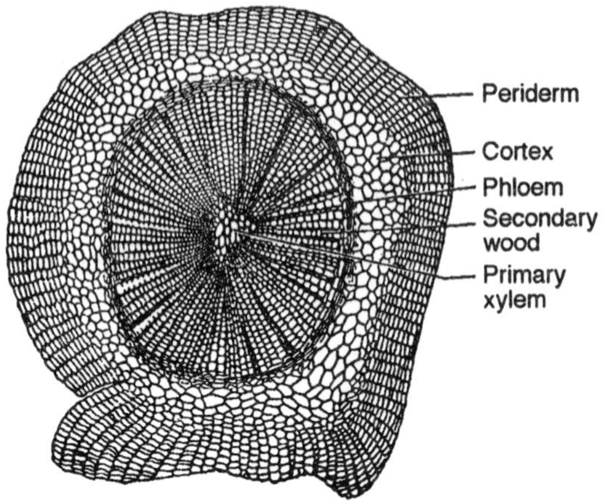

Fig. 47.5: T.s. of Cordaitean root—*Amyelon* (after Scott)

Each strobilus is a dwarf shoot consisting of a short woody axis bearing spirally arranged woody appendages some of which are sterile bracts and others sporophylls. A *male strobilus* (Fig. 47.6B) contains one to six or more *microsporophylls* or stamens each supporting one to six terminal *microsporangia* or unilocular anthers. The *female strobilus* (Fig. 47.6C) similarly contains one to four *megasporophylls* or carpels each bearing a terminal *megasporangium* or ovule. In some cases the megasporophylls may be forked bearing two terminal ovules. The ovule or megasporangium (Fig. 47.6D) shows an elongated oval nucellar mass com-

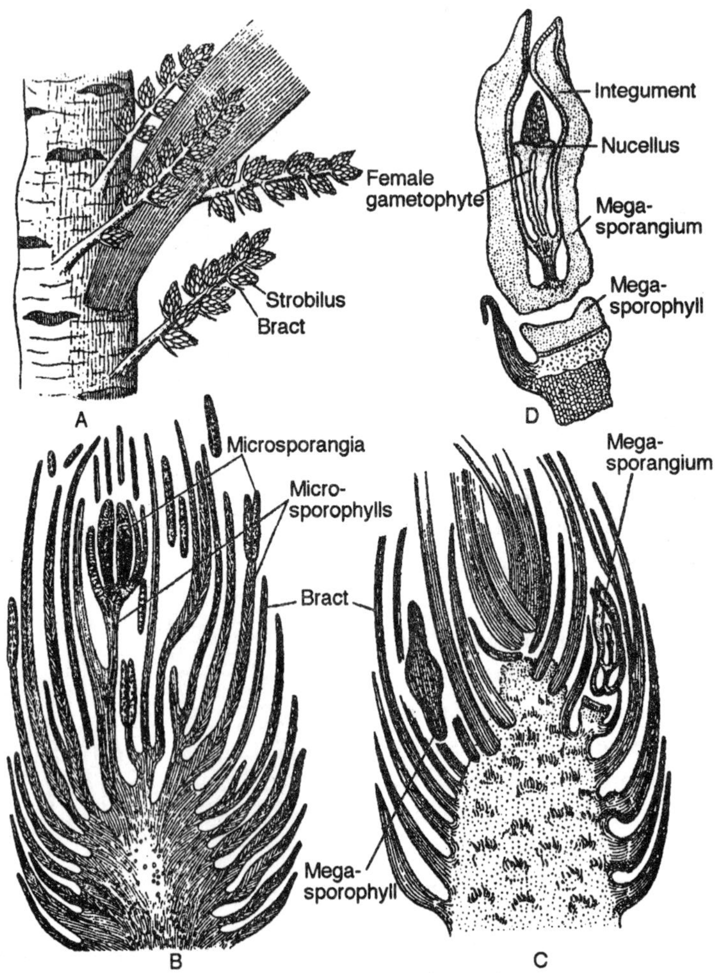

Fig. 47.6: A. *Cordaites* inflorescence. B-D. *Cordaianthus* strobilus (after Renault)—median l.s. of male strobilus (B), female strobilus (C) and a megasporangium or ovule (D)

pletely free from the single integument which is extended into a short micropylar canal at top.

A gap in the place of an elongated female gemetophyte is noticed in some specimens.

In some mature ovules the nucellar beak projects into the micropyle and there is a pollen cavity in this beak where pollens are found lodged (Fig. 47.7A). These pollens show thin intines and thick exines. The exine contains a big air sac which rendered the pollen buoyant. The pollen contains a multicellular gametophyte. There seems to be a lining of prothallial cells on the circumference and some spermatogenous cells in the centre. There is no evidence of any pollen tube formation. Even motile sperms are not known and may only be conjectured. If there was no pollen tube, fertilization in *Cordaites* was almost as in the Pteridophytes. If a pollen tube was formed, it was nearer to the Cycadophyta.

By the time the seeds were mature the stalk-like sporophylls greatly increased in length and the seeds were suspended in air far beyond the bracts (Figs. 47.7B and C). Such seed fossils go by the names *Cordaicarpus, Cardiocarpus, Samaropsis* or *Mitrospermum.* The seed (Fig. 47.7D) is flattened and heart-shaped, the nucellar portion of which is surrounded by an integument whose outer layer is expanded into a wing-like outer testa and the inner layer forms a hard, sclerotic inner testa.

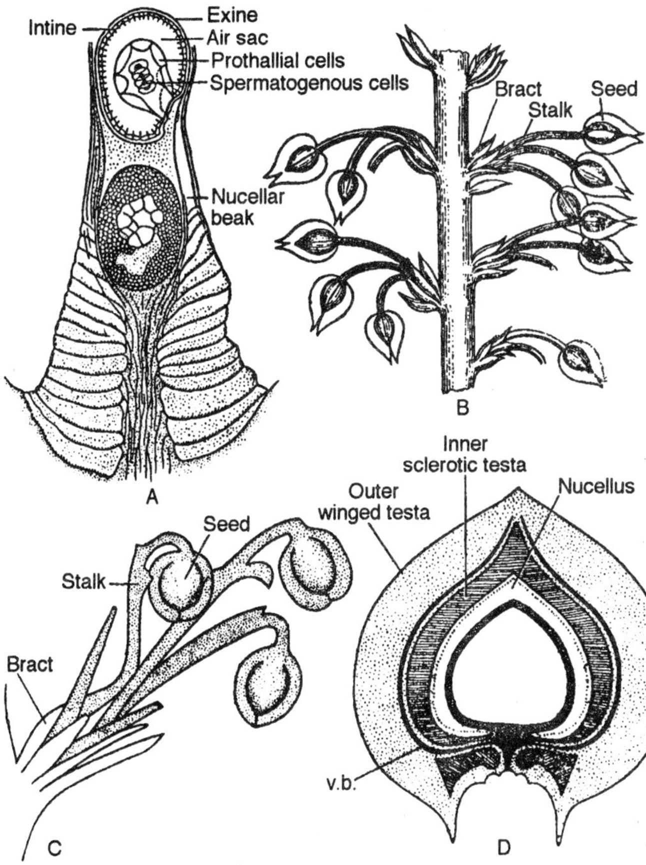

Fig. 47.7: A. L.s. of nucellar beak of *Cordaianthus* sp. showing pollen (ferminated into male gametophyte) logged therein (modified from Renault and Florin). B. *Samaropsis* sp. showing *Cordaites* seeds (after Carruthers). C. *Cordaianthus* sp. (*Cordaites* seed) (after Florin). D. L.s. of *Mitrospermum* (Probably referable to *Mesoxylon*) seed (after Arber)

A vein from the base divides into two to supply the integument. The structures clearly signify pollination and seed dispersal by wind.

PHYLOGENY OF THE CORDAITALES

The Cordaitales show affinities with the Cycadofilicales in the seed structure, the large pith, the double leaf trace of *Mesoxylon*, the haplocheilic stomata and the exarch primary xylem of *Poroxylon*. The Cycadales are approached in the large pith, the leaf anatomy including haplocheilic stomata, the large seeds, the secondary wood of *Poroxylon* and possible motile sperms. The Coniferales are reminded in the large tree habit, the strobili and the Araucarian wood anatomy. The Ginakgoales are also reminded in the double leaf trace, the leaf structure, the wood and the possible motile sperm.

Considering all points the Cordaitales may be assigned a position between the Cycadopsida and the Coniferales. But, in spite of such apparent affinities, it is difficult to consider the Cordaitales to have originated from any of these related groups. The now known relation between *Callixylon* and *Archaeopteris* seems to point out that the origin of the Cordaitales is not to be sought in any of the Gymnosperms hitherto known but at a level below the Gymnosperms—possibly in the 'Progymnosperm' group of Pteridophytes, sometimes in the Devonian.

Chapter 48

Coniferopsida: Coniferales and Taxales

The Coniferales and the Taxales are the common conifers. They are by far the most numerous of the modern Gymnosperms represented by some 50 genera. The Taxales are often included within the Coniferales but many are of opinion that they are sufficiently distinct to be placed in a separate order as is done here.

The conifers constitute the world's most extensive forest belts today. The trees are of typical coniferophyte type and are often gigantic in size. Some *Sequoia* trees are supposed to be 4000 years old, may be up to 300 ft. high with a trunk 30 ft. in diameter and may yield 5000 c. ft. of lumber. In the two orders there are only a few shrubs (e.g., some Junipers), all others being trees. The stem is excurrent, specially in *Pinus*.

Order II. CONIFERALES

The Coniferales, as understood here, do not include the Taxales for reasons explained in connection with the latter. The Coniferales are known since the Carboniferous, were predominant in the Mesozoid and form an important constituent of the flora today.

The Coniferales may be grouped under eleven families of which the first five are known only in the fossils state. The fossil families are—1. *Lebachiaceae*, 2. *Voltziaceae*, 3. *Palissyaceae*, 4. *Cheirolepidiaceae* and 5. *Protopinaceae*. The six other families are represented in the flora today and are known also in the fossil state. These are—6. *Pinaceae* (= *Abietaceae*), 7. *Taxodiaceae*, 8. *Cupressaceae*, 9. *Podocarpaceae*, 10. *Araucariaceae* and 11. *Cephalotaxaceae*.

Family 1. LEBACHIACEAE

The family Lebachiaceae, based on the genus *Lebachia* and formed by Florin out of Upper Carboniferous and Permian fossils preciously assigned to the next family Voltziaceae, is the best known of the fossil families and throws some light on the origin of the Coniferales. The different form genera which go in to form this plant are *Walchia, Walchiostrobus, Walchianthus, Gomphostrobus, Pollenites* and completely vegetative organs like *Palaeotaxites* and *Buriadia*. *Ernestiodendron* is a different plant also placed within this family.

Lebachia was a tree with pinnately arranged branchlets on which very small (a few mm long)

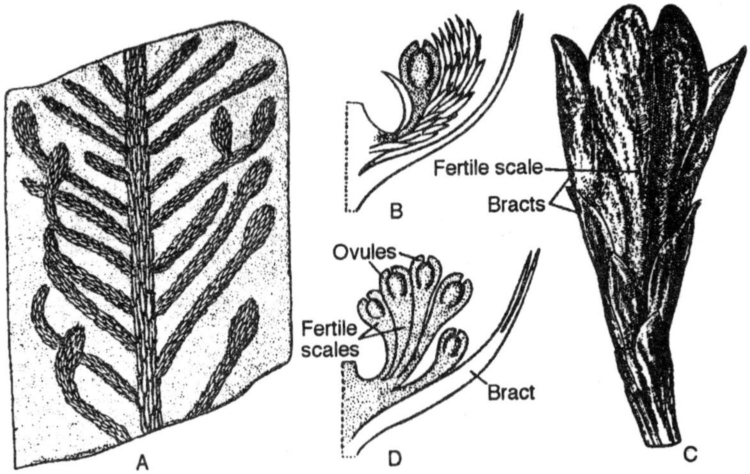

Fig. 48.1: A. *Lebachia piniformis* shoot showing cones, B. *Lebachia piniformis*—a dissected female dwarf shoot from a female cone, C. *Walchiostrobus* (*Lebachia*) *elongatus* female cone, D. *Ernestiodendron filiciforme* female shoot (After Florin)

needle-like leaves were borne spirally in an imbricate manner (Fig. 48.1A). Sometimes the needle leaves were bifurcate at the apex. The *cones* or *inflorescences* were cylindrical or ovoid at the tips of the branches. There were separate *male* and *female cones.* The cone axis bore spirally arranged bifurcate bracts (called *Gomphostrobus*) at the axils of which were the fertile dwarf shoots. A female fertile shoot (Figs. 48.1B and C) bore on its axis a few scaly appendages only one of which was fertile with a single terminal ovule. The ovule was notched at the apex and the two sides expanded into wings in the mature seed. *Ernestiodendron* was generally similar to *Lebachia* but the axillary female dwarf shoot was reduced to a fan-shaped 3- or 4-lobed structure each lobe of which bore a terminal ovule (Fig. 48.1D).

Lebachiaceae is the best known of the earliest conifers which, along with the Cordaitales, arose about the Carboniferous and became more established in the Permian when the earlier swamps were vanishing because of a spell of drought.

Buriadia sewardiana, which probably belongs to this family, is known from the Indian Lower Gondwana (Karharbari—Lower Permian and Raniganj—Upper Permian Stages).

Family 2. VOLTZIACEAE

The Voltziaceae (after taking out the Lebachiaceae) shows a group of fossils from the

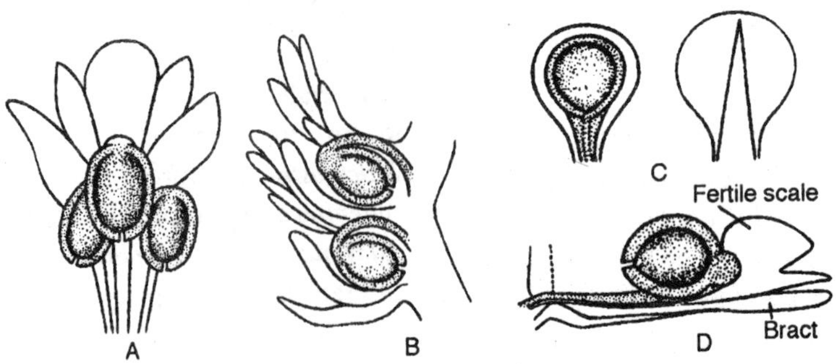

Fig. 48.2: A. Female dwarf shoot of *Pseudovoltzia*, B. L.s. of A, C. Upper and lower views of *Ullmannia* female dwarf shoots, D. L.s. of *Voltzia* female dwarf shoots. (After Florin)

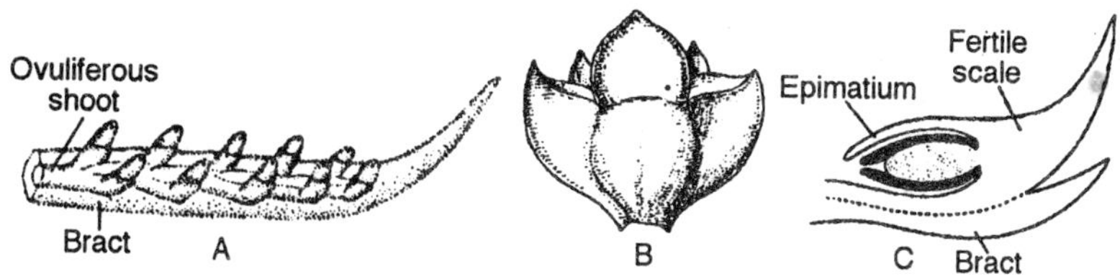

Fig. 48.3: A. Ovuliferous scale or shoot and subtending bract of *Palissya*, B. Female shoot or cone of *Cheirolepis*, C. Vertical section of part of B. (After Hirmer)

Permian to the Jurassic. The Upper Permian *Pseudovoltzia* shows dimorphic leaves, those on the lower stem are shorter while those on the upper parts are longer and more flattened. The *female dwarf shoot* is in the axil of a simple bract and is dorsiventrally flattened into five partially fused scales on the upper surface of which are three stalked, recurved ovules (Figs. 48.2A and B). In *Ullmannia* (also Upper Permian) the female dwarf shoot is further reduced. On the axil of a single simple bract there is a single round, flat fertile scale (probably formed by the union of several sterile appendages) on the upper face of which is a single recurved ovule (Fig. 48.2C). In the Lower Triassic *Voltzia* there are five flat scales as in *Pseudovoltzia* but three of them are fused with the three ovules. A sectional view shows the structure to be near the modern conifers (Fig. 48.2D).

Voltziaceae proper is represented in India by *Moranocladus oldhamii* (Karharbari—Lower Permian) in the Lower Gondwana and by *Pagiophyllum peregrinum*, *Brachyophyllum expansum* and *B. rajmahalensis* (Rajmahal—Lower Jurassic, Kota and Jabalpur—Upper (Jurassic) in the Upper Gondwana.

Family 3. PALISSYACEAE

The Upper Triassic and Jurassic genera *Palissya* and *Stachyotaxus* are placed in this family. The female branch of *Palissya* shows about ten distichously arranged ovuliferous axes each in the axil of a large bract. Each such ovuliferous shoot bears a fertile appendage (ovuliferous scale) bearing two rows of erect ovules on short side proliferations. This fertile appendage does not appear to be fused with the sterile bract below (Fig. 48.3A). *Stachyotaxus* ovuliferous shoots are similar but more reduced in showing two ovules only.

Palissya is well represented in the Indian Upper Gondwana by *Palissya conferta*, *P. indica* and *P. jabalpurensis* from the Rajmahal, Kota and Jabalpur Stages (Jurassic).

Family 4. CHEIROLEPIDACEAE

This is a family of Triassic-Jurassic fossils which is very close to the Podocarpaceae and also related to the Araucariaceae. Only the female cones are known. In a female cone of *Cheirolepis* (Fig. 48.3B) the bract scale is digitately five-lobed. On the upper surface of this, on the top of the lobes, two fertile scale are fused each of which extends down in an *epimatium* which forms a pocket enclosing the ovule keeping it open only at the micropyle (Fig. 48.3C).

Family 5. PROTOPINACEAE

Protopinaceae is an assorted group of coniferous wood fossils mainly from the Jurassic and the Cretaceous which shows more or less distinct affinities with Pinaceae, Taxodiaceae, Cupressaceae and Podocarpaceae. Among them may be named *Protopodocarpoxylon, Protophyllocladoxylon, Protocedroxylon, Pinoxylon, Protocupressinoxylon* and *Piceoxylon.*

The fossils families above have been discussed to show the evolution of the Coniferales especially as seen in the female cone. These began in the Carboniferous *inflorescences*, viz., those of Cordaitales like *Cordaianthus* where the dwarf shoots were arranged in two rows on the axis and then in Lebachiaceae where the arrangement was spiral. The dwarf shoots or individual female 'flowers' were radial and distinct in the Lower Permian. There were rapid changes by the Upper Permian evolving the dwarf shoot of *Ullmannia* where the sterile scales fuse and form a discoid scale not unlike the Pinaceae. This bract and the fertile scale then show different degrees of fusion. Thus, the sterile bract and the ovuliferous scale of the Pinaceac seem to have been evolved from what was originally a radial dwarf shoot. Finally, different variations seem to have evolved the different Coniferale families which are extent today.

Family 6. PINACEAE

Pinaceae (= *Abietaceae*) is the predominant conifer of the northern hemisphere in the temperate to cooler zones. The tall trees show the excurrent form with the leaves and sporophylls spirally arranged. The bracts and the ovuliferous scales are distinct in the female shoot. 9 genera with about 145 species are included in this family:

Subfamily: *ABIETOIDEAE*

Genera: 1. *Abies* (fir—40 spp.): Northern Hemisphere mountains. *A. balsamea* yields Canada Balsam. Genus known since Tertiary (Fig. 48.4A).

2. *Picea* (spruce—40 spp.): North Hemisphere mountains. Known since Cretaceous (Fig. 48.4B).

3. *Keeteleria* (3—4 spp.): China.

4. *Tsuga* (14 spp.): Himalaya to Japan, N. America. Genus known since Tertiary when they spread to Europe.

5. *Pseudotsuga* (7 spp.): East Asia, Pacific, N. America. Genus known since Cretaceous.

Subfamily: *LARICOIDEAE*

Genera: 6. *Larix* (larch—10 spp.): North Hemisphere, Genus known since Tertiary (Fig. 48.5A).

7. *Cedrus* (cedar—4 spp.): Mediterranean and Himalayan. Genus known since Createceous (Fig. 48.5B).

Fig. 48.4: A. A twig of *Abies delavayi* with female cone, B. Same of *Picea spinulosa*

Fig. 48.5: A. A twig of *Larix griffithiana* with female cone, B. *Cedrus deodara*—Part of female plant showing cones

8. *Pseudolarix* (1 sp.): China.

Subfamily *PINOIDEAE*

Genus 9. *Pinus* (pine—80-90 spp.): Dominant genus in Northern Hemisphere, also Sunda Islands. Genus known since Cretaceous.

Key to the Indian Genera of Pinnaceae

Leaves evergreen (pollen grains winged except in *Tsuga*)
- Cones pendulous, woody, persistent
  - Leaves acicular, in bundles — *Pinus*
  - Leaves not acicular, not in bundles
    - Cones globose and short (< 2·5 cm) (pollen grains not winged) — *Tsuga*
    - Cones spindle-shaped, long (6-15 cm) — *Picea*
    - Cones erect with scales shedding while on tree leaving the bare central axis
      - Leaves distichous — *Abies*
      - Leaves in tufts — *Cedrus*

Leaves deciduous, pollen grains not winged — *Larix*

Indian Species of Pinaceae

1. *Abies spectabilis* (D. Don.) Spach. (= *A. webbiana* Lindl.): (*Gobra Salla* in Nepalese): Sikkin, Bhutan to Western Himalaya 2750-4000 m.: 150 ft. tree, may be stunted.
2. *A. pindrow* Royle: Afghanistan to Nepal above 2200 m: Trees about 200 ft. high and trunk 19 ft. in girth known.
3. *A. densa* Griff.: (*Gobra Salla* in Nepalese): N. E. Burma to Nepal about 2750 m: up to 200 ft. high.
4. *A. delavayi* (Van Tieghem) Franchet: China and NEFA above 2750 m: about 60 ft. high (Fig. 48.4A).
5. *Picea smithiana* (Wall.) Boiss. (= *P. morinda* Link.): Afghanistan to Kumaon 2130-3350 m: up to 200 ft. high.
6. *P. spinulosa* (Griff.) Henry: (*Sehsing* in Nepalese): Sikkim, Bhutan 2400-3000 m: up to 200 ft. high. (Fig. 48.4B).
7. *Tsuga demosa* (D. Don.) Eichler (= *T. brunoniana* Carr.): (*Tangshing* in Bhutia): Kumaon to Nefa 1900-3050 m: about 100 ft. high.
8. *Larix griffithiana* Hert. ex Carr. (= *L. griffithii* Hook. f.): Eastern Nepal to N.E. Burma including Tibet Chumbi Valley, 2400-3650 m: 20 to 60 ft. high tree (Fig. 48.5A).
9. *Cedrus deodara* (Roxb. et Lambert) G. Don (= *C. libani* Banel, var. *deodara* Hook. f.): (*Deodar*): Afghanistan to Garhwal 1200-

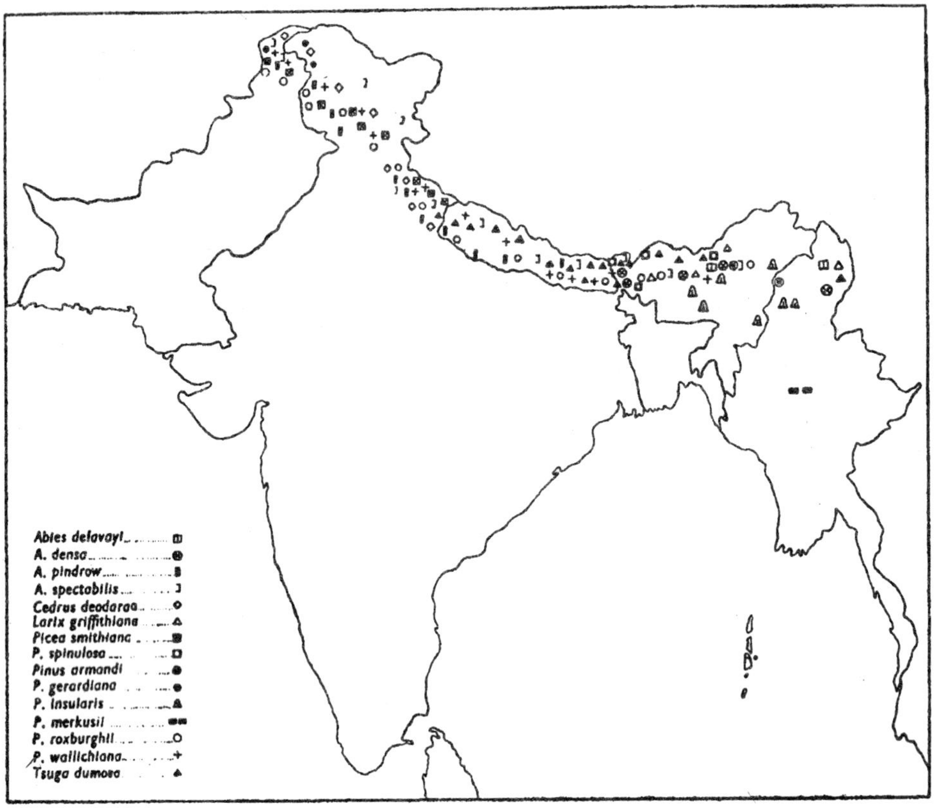

Fig. 48.6: Map showing the distribution of Pinaceae in India and adjacent areas

3300 m: Trees more than 700 years old known. Some are up to 250 ft. high with trunk 45 ft. in girth. Most valuable of Indian coniferous timber (Fig. 48.5B).

10-14. *Pinus* species are enumerated below under the genus.

Distribution of the Pinaceae in India is shown in the map on Figure 48.6.

Genus: *PINUS* LINN

Pinus is the common 'pine' with 80 to 90 species (number varying according to author) spread over the temperate to subalpine regions of the Northern Hemisphere and also in the Sunda Islands. The following five species of these occur in India:

1. *Pinus wallichiana* A. B. Jackson (= *P. griffithii* McClel. and *P. excelsa* Wall.: 'Blue Pine', *Kail* in Hindi): 1800-3000 m on the Himalayas except in Sikkim and most of Kumaon. Up to 150 ft. tree. A source of timber and resin (Fig. 48.7A).
2. *P. armandi* Franchet: *NEFA*, C. and W. China about 1500 m. About 55 ft. tall tree (Fig. 48.7B).
3. *P. gerardiana* Wall. ex. Lamb. (*Chilgoza* in Hindi): 1800-3500 m in the dry interior valley of N. W Himalaya. 50 to 70 ft. tree. The seed is edible and of commercial value (Fig. 48.7C).
4. *P. insularis* Endl. (= *P. khasya* Royle): (*saral* in Bengali, *ass* in Khasiya) : Chittagong, K. and J. Hills, Assam, Naga Hills, Manipur to Upper Burma and Philippines. 1000 to 2500 m. Up to 150 ft. tree or taller in Burma though shorter in Khasia Hills.

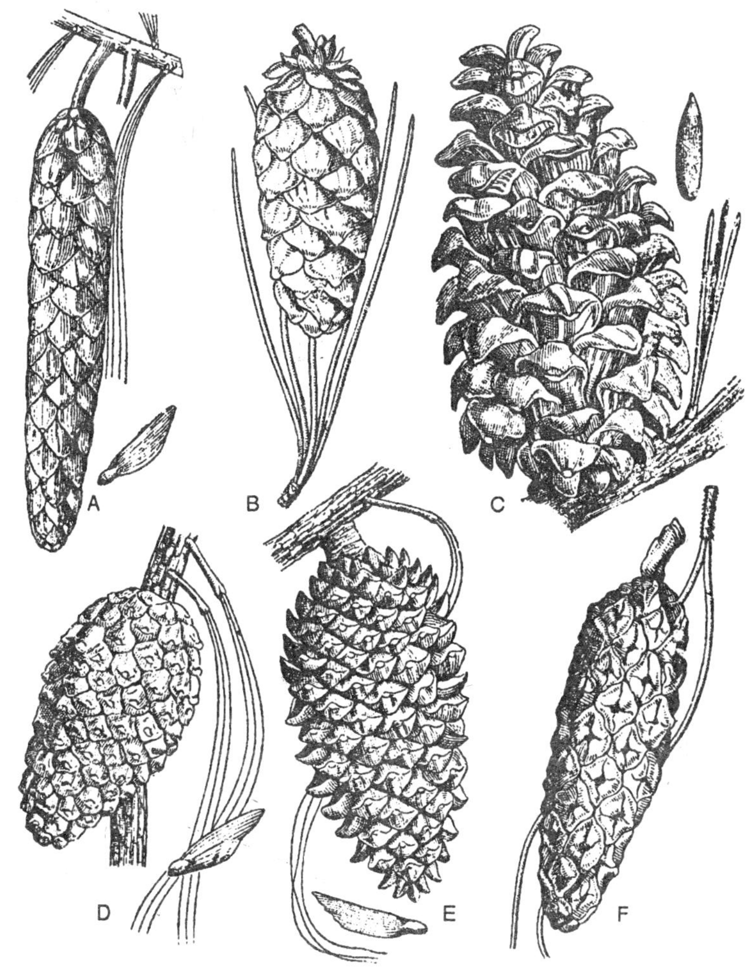

Fig. 48.7: Female cones, dwarf shoots and seeds of Indian *Pinus* species: A. *Pinus wallichiana*, B. *P. armandii*, C. *P. gerardiana*, D. *P. insularis*, E. *P. oxburghii*, F. *P. merkusii*

A source of timber and resin (Fig. 48.7D).

5. *P. roxburghii* Sargent (= *P. longifolia* Roxb.): (*chir* in Hindi): 460 to 2300 m in the outer monsoon Himalayas from Afghanistan to Bhutan. Trees up to 100 ft. high. Principal source of resin and turpentine and a source of lumber (Fig. 48.7E).

Mention may also be made of *P. merkusii* Jungh. et de Vr. (Fig. 48.7F) which is a small tree (about 50 ft.) common in Burma spreading to the Philippines and growing even in the plains. It is also known to grow in Bangladesh.

Key to the Relevant Species

Leaves in fascicles of five

Leaves 4-8" long, cones 6-12" long *P. wallichiana*

Leaves 4-6" long, sharply bent, cones 4-7" long *P. armandi*

Leaves in fasicles of three

Leaves stiff, less than 7" long; conex 5-9" long *P. gerardiana*

Leaves soft, more than 7" long

Leaves 7-10" long, rarely fascicle of 2; cones 2-3" long; wings 4 times as long as seed;

umbo of scales with prominent beak *P. insularis*

Leaves 8-14" long; cones 4-8" long. Wings slightly longer than seed; umbo of scales with prominent hooked beak *P. roxburghii*

Leaves in fascicles of two *P. merkusii*

Figure 48.6 shows the distribution of these species.

The young *Pinus* plant (Fig. 48.8A) has a strong tap root and short single needle leaves arranged spirally on it. The tap root may persist but it may become unimportant because of stronger adventitious roots. Very soon the nature of the leaves change. Small *scale leaves* appear on the stem. These are brown, small, thin and scaly, From the axils of some of these arise the *dwarf shoots* or *leaf spurs* or *brachyplasts* which are limited in their growth. A *dwarf shoot* is a small piece of stem bearing scale leaves and from its apex comes out a fascicle of two or more long, narrow, green *needle leaves* (Fig. 48.8B). The number and length of the needle leaves in a dwarf shoot vary according to the species as stated for the Indian species listed above. These needle leaves form the main photosynthetic organs of the plant. The seedling now begins to branch. From the base of the apical growing tip a number of branches appear at the axils of the scale leaves. The branches are, theoretically, almost in a whorl but only a few of them survive. At the end of every year, in the spring, new branches appear similarly from the branch of that year's growth.

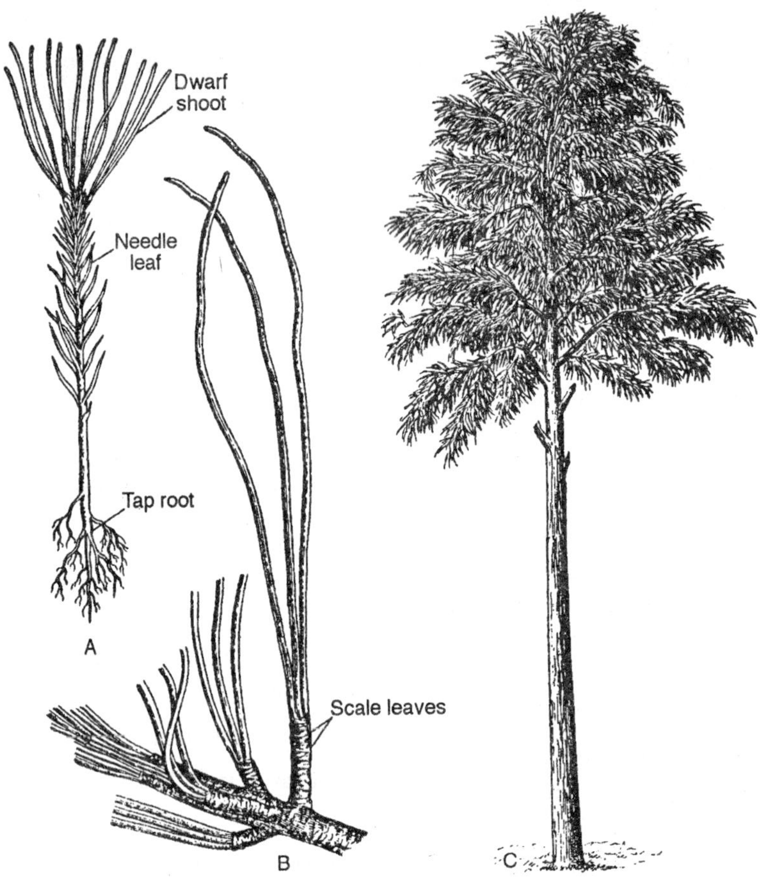

Fig. 48.8: A. Young seedling of *Pinus*, B. Dwarf shoots of *Pinus insularis* (one shown entire, others are dissected), C. A *Pinus* tree

This regular branching, if not disturbed, gives a beautiful excurrent shape to the growing tree. Although the trees are evergreen, the *dwarf shoots* are shed every two or three years leaving scars on the stem. The mature tree (Fig. 48.8C) shows a single stout, cylindrical stem covered by a scaly bark and regular lateral branches of unlimited growth which are called *long shoots.* Long shoots show leaf scars and scale leaves only. The younger branches show the *dwarf shoots* or definite limited growth. There are, therefore, two types of shoots. The youngest long shoots (branches) are of the current year's growth and show buds of dwarf shoots before they open up at the beginning of spring (Fig. 48.13). Leaves also are of two kinds, the *scale leaves* on the long and the dwarf shoots and the *needle leaves* on the dwarf shoots only.

The stem shows an endarch siphonostele broken up by leaf gaps and shows the same general arrangement as in the dicots. Described from the centre outwards a t.s. of a young stem (Fig. 48.9) shows (1) a *pith* of parenchymatous cells at the centre surrounded by (2) a ring of conjoint, collateral, open *bundles* showing *xylem, cambium* and *phloem.* (3) The individual bundles are separated by *primary medullary* rays. (4) An *endodermis* of one layer surrounds the stele but this is not distinct. (5) A cortex of parenchymatous cells occupies the rest of the stem. There is prominent *resin canals* or *ducts* in the cortex. Sometimes a few outer layers of the cortex form a sclerenchymatous *hypodermis.* (6) The *epidermis* is formed by a single layer of cells heavily cutinised on the outer face.

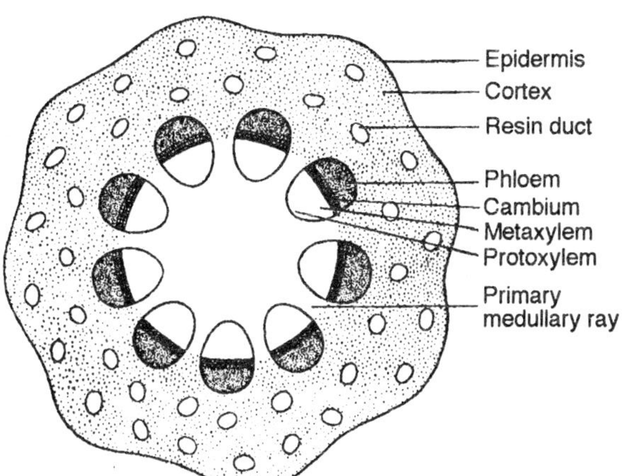

Fig. 48.9: T.s. of young *Pinus* stem (diagrammatic)

The elements of *xylem* and *phloem* are different from that of the Angiosperms. In the xylem true vessels are absent. The *protoxylem* near the pith shows annular and spiral tracheides while the *metaxylem* is formed of bordered pitted tracheides. Small resin ducts are also present in the primary wood. The phloem contains sieve tubes and phloem parenchyma but companion cells are absent. The sieve tubes have the sieve plates on the radial walls.

Secondary growth is as in dicots and is both *intrastelar* by the development of a *cambium ring* and *extrastelar* by a *phellogen* developing a *periderm* under the epidermis. The cambium gives rise to *secondary xylem* on the inside and *secondary phloem* on the outside. Distinct annual rings are seen (Fig. 48.10) in the secondary wood and resin ducts are interspersed. The wood structure shows some difference with that of the Angiosperms. Tracheides are more or less quadrilateral in cross-section and possess one or two rows of bordered pits along the radial walls only, such bordered pits being absent on the tangential faces (Figs. 48.11A and B). The summer tracheides are narrower and much thinner than the spring tracheides. The cellulose walls of young tracheides have thin, transverse, unthickened lines on the walls the borders of which are thickened and are known as *bars* or *rims of Sanio* (Fig. 48.11C— c.f. *crassulae.* There are *secondary medullary rays* as well as *resin ducts* and *canals* forming an anastomosing systems in the secondary wood. The *Pinus* secondary medullary rays are of rather peculiar construction (Figs. 48.11A and B). The rays are narrow (usually one cell wide) and a few cells in depth. Many of them extend right across the secondary tissue cutting

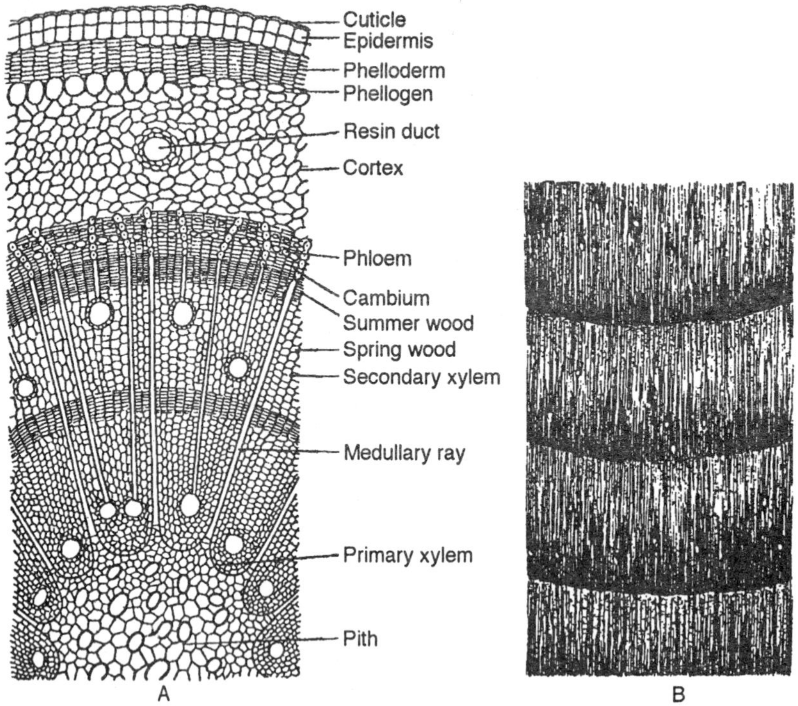

Fig. 48.10: *Pinus*. A. T.s. of two-year old stem, B. T.s. of wood

through the cambium and the phloem. The upper and lower layers of these rays are formed of bordered pitted ray tracheides elongated radially while the middle layers are elongated, pitted, parenchymatous cells containing starch. In the phloem regions the tracheides of these rays are substituted by albuminous cells often elongated vertically. In the extrastelar region the original *phellogen* is replaced by successive tanngential layers of phellogens deeper and deeper into the cortex giving rise to strips of *scaly bark*.

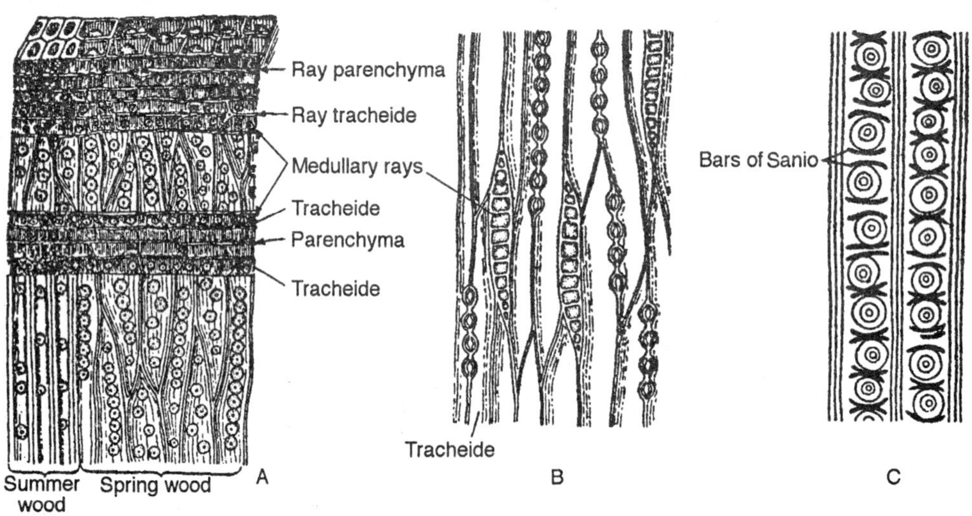

Fig. 48.11: *Pinus*. A. Longitudinal radial section of secondary wood, B. Tangential section of secondary wood, C. Tracheides showing bars of Sanio

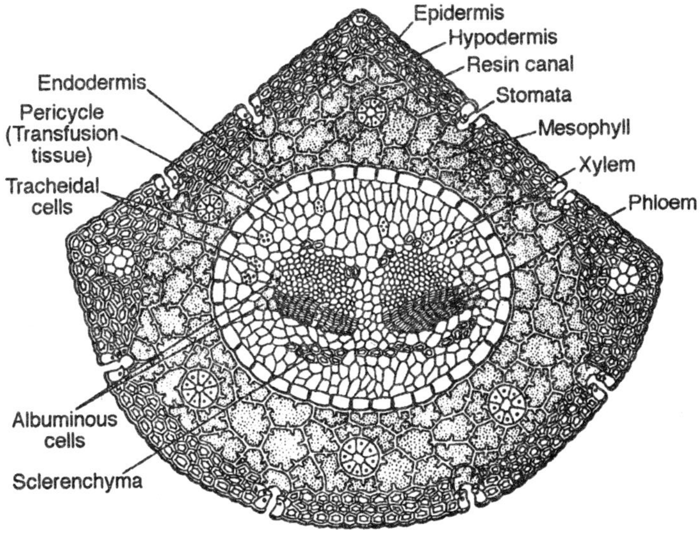

Fig. 48.12: T.s. of *Pinus roxburghii* leaf

The root structure also resembles dicots. There are 2, 3 or 4 (diarch, triarch or tetrarch) exarch bundles (more or less Y-shaped) with alternating xylem and phloem (radial). Secondary growth gives rise to some compact secondary wood and secondary phloem. Phellogen develops outside the pericycle giving rise to a periderm. Pith is generally absent but may be present in some cases. *Pinus* root is mycorrhizal and the root hairs disappear early. In the apical root meristem there is no dermatogen.

The leaf structure of *Pinus* is xerophytic of the centric type. The needles are more or less triangular in cross-section (Fig. 48.12). The *epidermis* is a layer of heavily cutinised thick-walled cells with very deeply sunk *haplocheilic stomata.* Below this is a *hypodermis* of 2 or 3 layers of very thick-walled cells. Internal to this is the *mesophyll* of large, thin-walled, crenately lobed chlorophyllose cells. *Resin canals* are present in the mesophyll and even in the hypodermis. The mesophyll is internally lined by an *endodermis* with prominent casparian strips. Internal to the endodermis there are one or two (according to species) *vascular bundles* surrounded by a broad *pericycle* which has the special name of *transfusion tissue* here. The transfusion tissue shows longitudinally elongated, bordered pitted, dead tracheidal cells near the xylem (serving for water conduction) and albuminous cells, with protoplasm full of starch and protein, translocating food near phloem. There is also a band of *sclerenchymatous fibres* below the phloems. The *xylem* is in the upper angular surface of the leaf and the *phloem* lies below.

Pinus leaf has one vascular bundle at the base which then divides into two parallel vascular bundles or strands within the common sheath. This is explained as double leaf trace which is the condition in the primitive Gymnosperms. The *Pinus* cotyledon, however, always shows a single leaf trace.

Although *Pinus* leaves with the simple veins appear to be microphyllous this is not the real situation as the leaf traces leave distinct leaf gaps in the stem vascular cylinder. It may be assumed that the apparently microphyllous leaf of *Pinus* had evolved by reduction from some megaphyllous ancestor.

Pinus is monoecious but male and female sporophylls are borne in separate *strobili.* The *male cones* appear spirally on the base of the current year's long shoot at early spring which at that time shows the dwarf shoots in the bud stage (Fig. 48.13A). At the end of the spring the male cones drop off, the dwarf shoot buds open up and at the tip of the long shoot the young *female cones* appear in a rosette round the growing tip (Figs. 48.13A and B). The female cones are slow in growth and persist for several years so that successive year's female cones may be seen below each year's growth of long shoot in the terminal branches.

The male or staminate strobilus

The male strobili or cones are seen clustered on the base of the young shoot early in the spring looking like fat worms about an inch long (Fig. 48.13A). The structure is rather simple. A number of *microsporophylls* are spirally arranged on an axis or thalamus (Fig. 48.14A). The male cone, therefore, is a 'flower' and not an inflorescence. On the under surface of each stalked, scale-like microsporophyll (Fig. 48.14B) there are two *microsporangia* containing numbers *microspores* or pollens. When mature, each microsporangium splits by a longitudinal slit. The development of the microsporangium and microsporogenesis are as in other eusporangiates. A semimature microsporangium (Fig. 48.14C) shows *sporangium wall, tapetum* and *microspore mother cells.* On reduction division the *pollens* or *microspores* are formed in large numbers. Each pollen is invested with an *exine* and an *intine.* Before separation of the tetrads the exine is inflated into two wings (Fig. 48.14D). This, naturally, points out to wind pollination. The pollens are full of starch grains and there is a definite apex and a base. The 2n

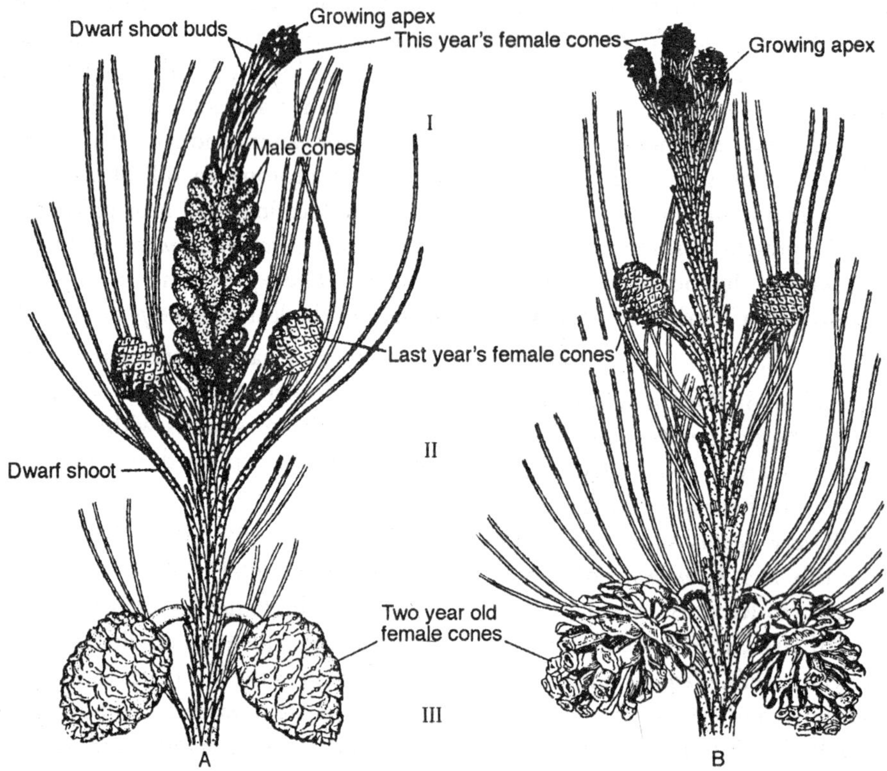

Fig. 48.13: Male and female cones of *Pinus insularis*. A. At early spring, B. At end of spring (rains). I. Long shoot development during current year, II. The same during last year, III. Part of long shoot which developed during year before the last

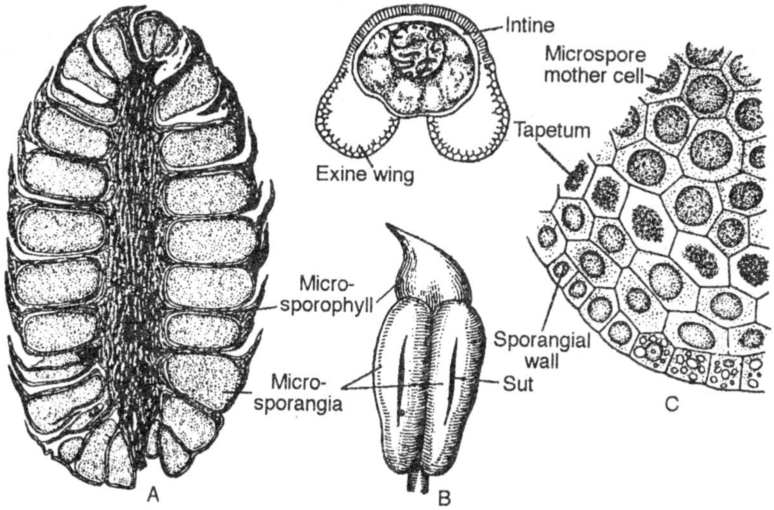

Fig. 48.14: *Pinus* A. Median l.s. of male cone, B. A. microsporophyll (lower surface), C. T.s. of a semimature microsporangium. D. A. microspore or pollen

number of chromosomes in *Pinus* is 24.

The female cone or ovulate strobilus

While the male cone is a simple structure which may be called a 'flower', the female cone is much more complicated. When first developed it is about a cm long (Fig. 48.13B) and red in colour. Gradually it turns green. In this (Fig. 48.15A) the axis bears on it spirally a number of *bract scales* on each of which, originating from the axil, is an *ovuliferous scale.* Each ovuliferous scale bears on its surface two anatropous *ovules* or *megasporangia* (Fig. 48.15C). There are separate vascular traces sup-

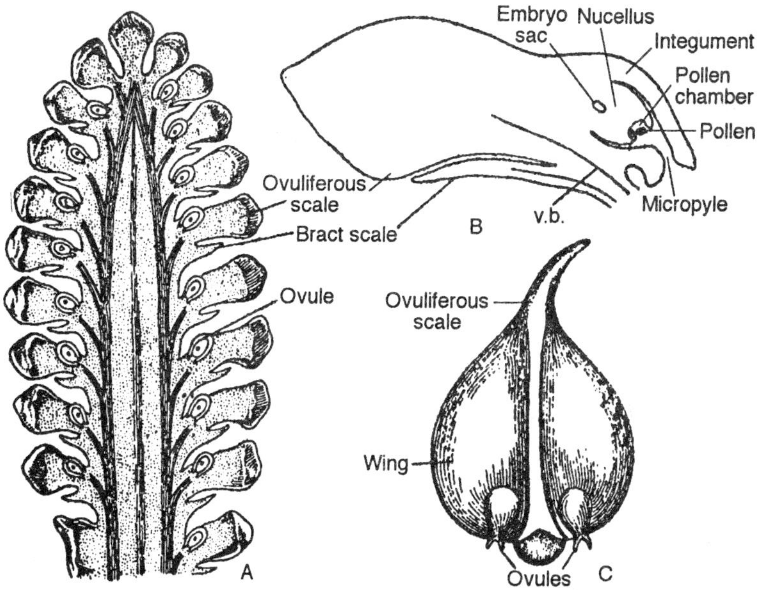

Fig. 48.15: *Pinus.* A. Median l.s. of female strobilus, B. Vertical l.s. of an ovule with scales, C. An ovuliferous scale with its two ovules

plying the bract scale and the ovuliferous scale. Each ovule (Fig. 48.15B) shows a mass of *nucellus* covered by an *integument* growing from its base. The integument is open at the top forming a *micropyle* and is free from the nucellus in the upper part of the ovule. The integument shows three layers—an outer fleshy, a middle stony and an inner fleshy (Fig. 48.18A). There is no separate vascular supply for the ovule or the integument. As the pair of ovules is associated with two scales, the two scales with the ovules are considered to form a separate shoot so that the female strobilus is not simple but compound, homologous with an 'inflorescence'.

The morphology of these ovuliferous structures has been much discussed. Considered chronologically, Robert Brown (1827) considered the ovuliferous scale as an open foliar 'carpel' bearing nacked ovules on it. This 'carpel' arises in the axil of the bract. But, nobody supported the idea that a leaf (carpel) could arise in the axil of another leaf (bract). So, Sehleiden (1829) was of the opinion that the ovuliferous scale was not a leafy carpel but a flattened 'axis' forming the placentum. A. Brown (1842) was of the opinion that the ovuliferous scale represented the first two leaves of an axillary shoot which had become fused. Dickson (1860) supported by Goebel and others concluded that the bract was the megasporophyll or carpel bearing at its axil a shoot which was the ovuliferous scale. Sachs (1868), later supported by Eichler, concluded that the bract was a carpel and the ovuliferous scale was a ligular outgrowth on its surface as found in *Isoetes* or *Selaginella*. Van Tieghem (1869), from an anatomical study, concluded that the ovuliferous scale was the first and the only leaf of a suppressed shoot in the axil of the bract. Celakovsky (1879) supposed that the ovuliferous scale represented the fused outer integuments of the ovules while Bessey (1892) supposed it to be a **chalazal outgrowth.** Chamberlain (1934) concluded that the bract is a megasporophyll bearing in its axil the ovuliferous scale which is a modified shoot with or without leaves and which bears the megasporangia or ovules. A comparative study of the fossil conifers known today, beginning from *Lebachia* (Figs. 48.1B and C) where the ovules are borne on radial dwarf shoots, ultimately leads to *Ulmannia* (Fig. 48.2C) and *Voltzia* (Fig. 48.2D) where the radial shoot becomes flattened and reduced into a sterile bract partially fused with a flattened fertile ovuliferous bract. From this evidence Florin (1951) has convincingly argued that in *Pinus* the inflorescence bears the bracts (bract scales) on the axils of which are the very much reduced dwarf shoots or 'flowers'. The ovuliferous scale represents a compound structure formed by what were originally two megasporophylls and a few sterile scales. We may, therefore, conclude that the reduced shoot nature of the ovuliferous scale is almost proved. The female strobilus is, then, an inflorescence. But the dispute even now is about the identity of the megasporophyll. Either it is fused in the ovuliferous scale (Florin and others) or the bract scale represents it (Chamberlain and others).

The gametophytes, pollination and fertilization

The pollen (microspore) begins to germinate while still within the pollen sac (microsporangium) a months before pollination actually takes place in March-April. The nucleus divides and a small *first prothallial cell* is cut off which remains pressed against the intine at top (Fig. 48.16A). A *second prothallial cell* is again cut off from the other cell and is similarly pressed against the first prothallial cell (Fig. 48.16B). The two prothallial cells are overgrown by the vigorous intine and slowly disorganize. The remaining cells is the *antheridial cell* which next divides to form a *generative cell*

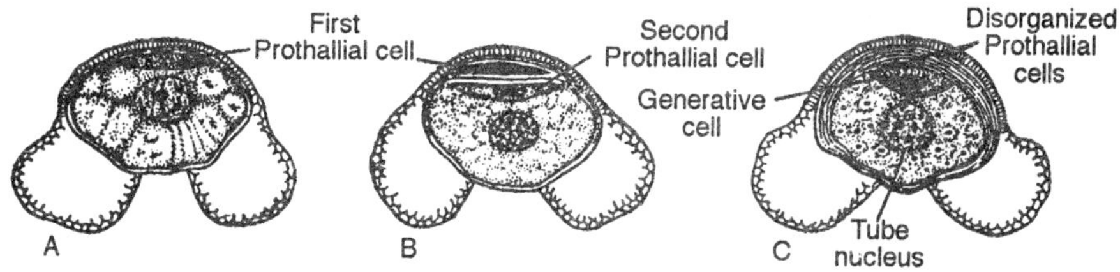

Fig. 48.16: A-C. Stages of development of the male gametophyte of *Pinus* before pollination

and a *tube nucleus* in the remaining cell (Fig. 48.16C). The smaller generative cell remains attached to the top intine below the degenerating prothallial cells. At this stage the pollen sac ruptures by a longitudinal fissure and the pollens are shed.

The female cones are very slow in growth. They take almost a year to be mature enough to receive the pollens. Sometimes after the female cones have emerged in the spring the *archesporium* of the *sporogenous tissue* organizes. It is deep seated in the nucellus of *Pinus.* A single *megaspore mother cell* differentiates and it is surrounded by a spongy tissue which is nutritive in function (Fig. 48.17A). The megaspore mother cell forms a *linear tetrad* of four *megaspores* by reduction division as in Angiosperms (Fig. 48.17B). The lowest megaspore only is functional. It enlarges to form the *female gametophyte* or the *embryosac* while the other three megaspores disorganise. The functional megaspore reapidly increases in size and passes through a period of free nuclear division. A large central vacuole filled with transparent fluid presses the protoplasm with the free nuclei on the wall so that the latter form a lining (Fig. 48.17C). The megaspore covering *membrane* is yet thin and there are two or three

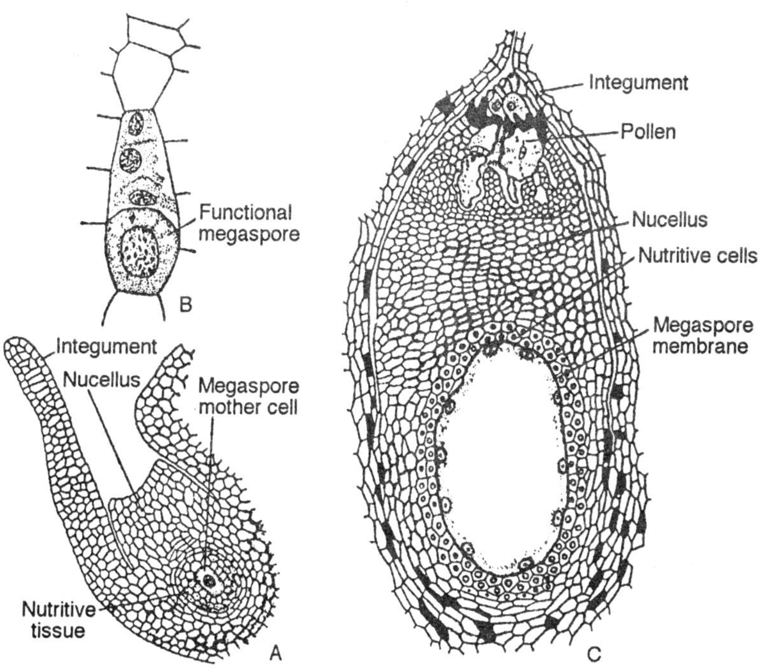

Fig. 48.17: *Pinus* ovule. A. Megaspore mother cell stage, B. Megaspore linear tetrad—the lowest one is functional, C. Free nuclear stage of female gametophyte

layers of spongy *nutritive cells* outside the growing gametophyte. This much development carries the female cones from one spring to another spring and they are now ready for pollination.

Pollination takes place about March-April in the Eastern Himalaya. It is carried on by wind (*anemophilous*). The amount of pollens liberated by the Pine forests at this time is prolific so that the air gets saturated with them and there is a yellow deposit of pollens on the forest floor. This phenomenon is known as 'sulphur shower'. Naturally, most of the pollens are wasted and only a small amount reach the female cones. These pollinate the female cones of the last year as the current year's cones are not yet ready. The scales of the last year's female cones slightly spread apart at this stage allowing the wind-carried pollens to reach the microphyles of the ovules. A colourless 'pollination drop' appears on the tip of the microphyle and the pollens coming in contact with this drop are drawn down to the tip of the nucellus where they germinate. After pollination the scales again close rendering the cone compact.

Post-pollination changes in the male and the female gametophytes are, again, slow, taking another year for them to complete *fertilization.*

In the *female gametophyte* the *megaspore membrane* now becomes thicker and more prominent. Next, cell wall formation begins inside. First, a parietal layer of cells is formed. Then, cell wall formation proceeds inwards ultimately completely obliterating the vacuole and filling up the whole of it with cells. At this stage a few *archegonia* develop round the apex of the female gametophyte and the cells below form a nutritive *endosperm* (Fig. 48.18A). The endosperm here, as in the Pteridophytes, is a gametophytic tissue (n chromosomes) and not a double fertilization product ($3n$ chromosomes) as in the Angiosperms. The *archegonial initial* is a superficial cell near the apex of the female gametophyte. This divides transversely into a top *neck initial* and a lower *central cell.* The neck initial divides to form 8 cells arranged in two tiers of four cells each. Neck canal cells are absent. The central cell enlarges immensely assuming a highly vacuolated appearance and an *archegonium jacket* forms around it (Fig. 48.18B). The nucleus of the central cell, situated on its upper part, divides just before fertilization showing a distinct cell

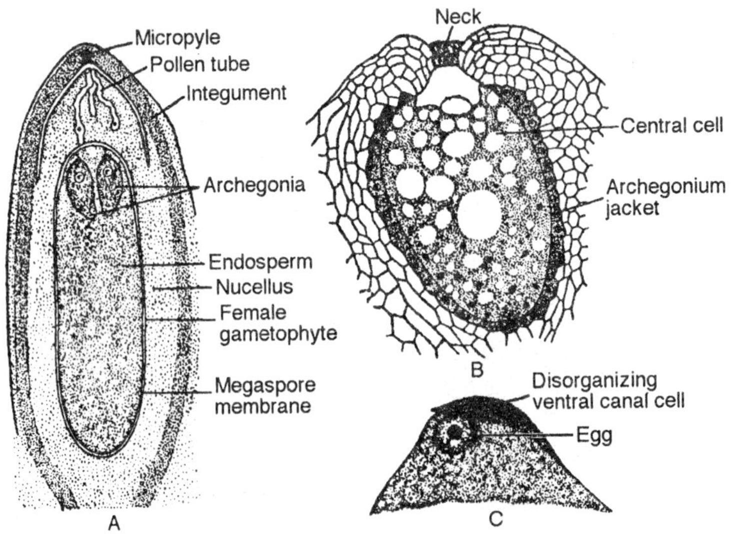

Fig. 48.18: *Pinus.* A. A mature ovule (the integument shows three layers), B. An archegonium, C. The division of the central cell into a transitory ventral canal cell and the egg cell

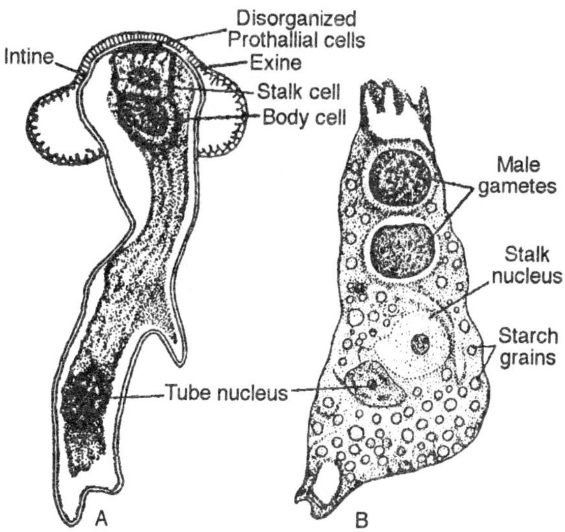

Fig. 48.19: *Pinus*. A. Pollen tube before development of gametes, B. Tip of pollen tube after development of male gametes

wall between the daughter nuclei. This gives rise to a small *ventral canal cell* above, which has a very short life and soon disorganizes (Fig. 48.18C). The lower cells is the *egg cell.*

Meanwhile, the *male gametophyte* also germinates further. After being lodged on top of the nucellus its intine develops a *pollen tube* as in the Angiosperms which corrodes and grows down the nucellus. The pollen tubes are full of starch grains, are not smooth and show a tendency to branching (Fig. 48.19A). At this stage the generative cell divides to form an upper *stalk cell* and lower *body cell.* The stalk cell soon breaks liberating its nucleus and loosening the body cell. Just before fertilization the body cell divides into two naked *male cells* or *gametes* each with a nucleus and some cytoplasm (though not very distinct) round it (Fig. 48.19B). In some cases the two male gametes have been reported to tbe unequal in size.

It should be pointed out that the male gametes in the Pinaceae are the simplest and the least organized among the Coniferales. They are, therefore, the most advanced as the tendency among the Spermaphytes is towards reduction. But, this is associated with a ventral canal cell with a distinct cell wall in the female gametophyte which is absent in the other families of the Coniferales. This character is more primitive. So the gametophytes of Pinaceae show both advanced and primitive characters.

During *fertilization* it is clear that the pollen tube is no longer simply haustorial but also the carrier of the non-motile gametes so that the method is truly *siphonogamous.* On reaching the archegonial neck (which takes place a year after pollination, i.e., two years after the female cones first emerged) the pollen tube destroys the neck cell, bursts at the tip and discharges its contents (often with some force) into the cytoplasm of the egg. The nucleus of one of the male gametes enters and fuses with the egg nucleus completing the act of fertilization. The other two nuclei from the pollen tube soon disorganize.

Development of the new sporophyte embryo

With fertilization the gametophytic generation ends and the next sporophytic generation begins. The fused nucleus resultant on fertilization is the *zygote* or *oospore* nucleus. This nucleus at once divides first into two and then into four nuclei by free nuclear division (Fig. 48.20A). At this stage *proteid vacuoles* and distinct in the egg cell and may be mistaken for nuclei. The four nuclei enlarge, pass to the base of the egg cell and arrange themselves in a single plane. The four nuclei now divide simultaneously forming two tiers of four nuclei each. Cell wall formation ensues at this stage (Fig. 48.20B). The two tiers again divide to form four tiers giving rise to a 16-celled *proembryo* (Fig. 48.20C). Of these four tiers the uppermost is open at top and is active for *nutrition.* The next tier is the *rosette* which is sometimes meristematic. The third is the *suspensor tier,* cells of which elongate immensely forming the *suspensor* or the *primary suspensor.* The lowest tier of cells divides further to form the *embryo* proper

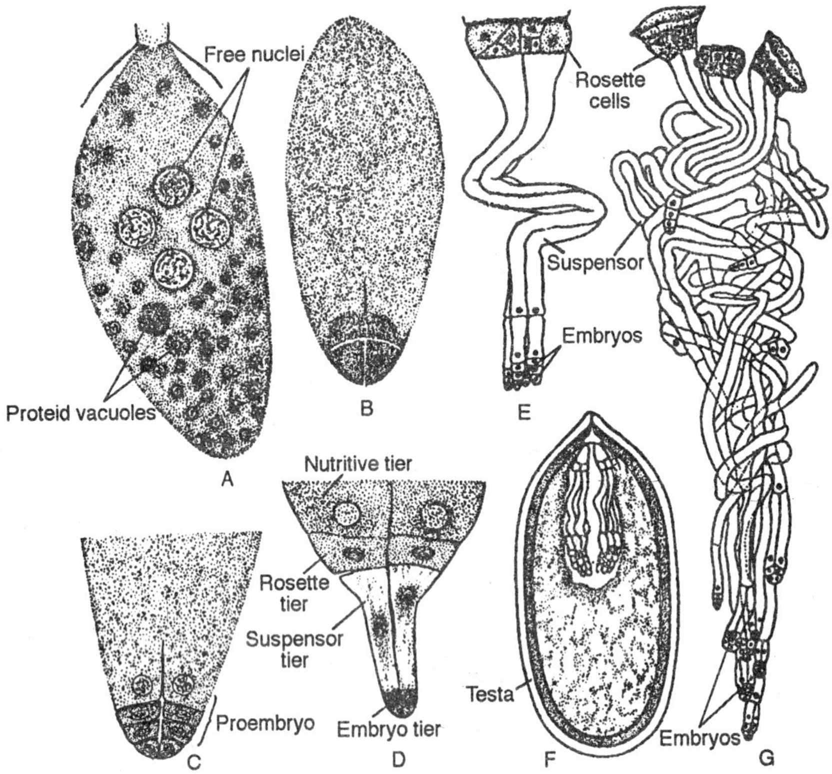

Fig. 48.20: *Pinus*. A-D. Stages of development of proembryo: A. Four free nuclei, B. 8-celled stage, C. Proembryo stage, D. Growth of the proembryo, E-G. Embryo development stages, E-F. A number of embryos developing in a young *Pinus* seed, G. About ten embryos with their suspensors resulting from three fertilized eggs in three archegonia. (after Buchholtz and Coulter)

(Fig. 48.20D). The *secondary suspensor cells* or *embryonal tubes* show a continuity with the primary suspensor cells. There are a number of archegonia in an ovule of *Pinus* giving rise to a number of such embryonal structures. *Simple polyembryony* (a number of embryos from a number of fertilized eggs) is common in some genera of Pinaceae, viz., *Larix* and *Picea*. But in *Pinus* (also *Cedrus* etc.) the proembryo tiers split from one another into four along the four cells of each tier giving rise to four separate embryos. This is known as *cleavage polyembryony* (Fig. 48.20E). Polyembryony of these and other types is common in the Coniferales.

The suspensors push the embryo down into the gametophyte tissue (endosperm). There is a large number of suspensors, from the fertilized eggs of the different archegonia and by cleavage, pushing down a large number of embryos (Figs. 48.20F and G) but usually only one of these embryos survive in the seed.

The seed

When the embryo matures into the seed it fills up the space inside the integument. The withered suspensor forms a dry cap on its top. An l.s. of the seed (Fig. 48.21A) shows the integument developed into the *seed coat*. The outer fleshy layer of the integument is completely dried up. The middle stony layer forms the *testa*. The inner fleshy layer dries up as a papery *tegmen*. At the micropylar end some remnant of the nucellus not used up even now remains as the *perisperm*. The testa, tegmen and perisperm may by easily seen in the *Pinus gerardiana* (*chilgoza*) seed which is sold in dry fruit

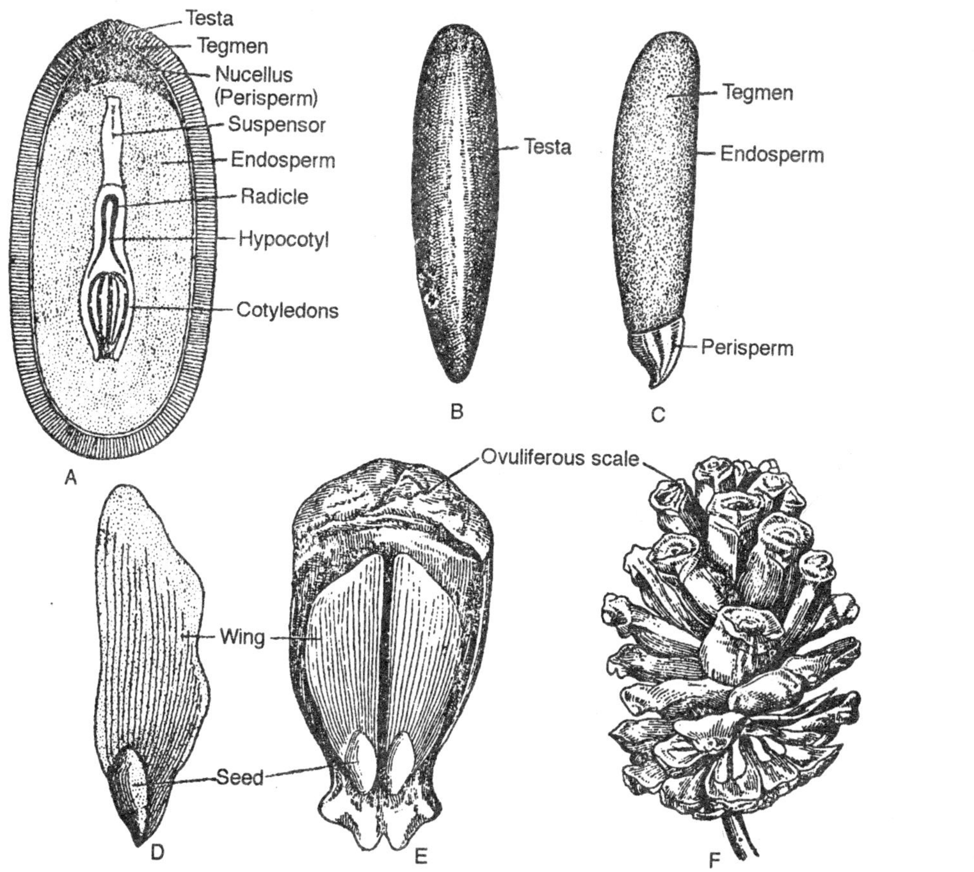

Fig. 48.21: A. L.s. of seed of *Pinus*, B-C. Seed of *Pinus gerardiana* (*chilgoza*) with the testa and after its removal, D. Seed of *Pinus* with the wing, E. The seeds on the ripe ovuliferous scale, F. A ripe pine cone with seeds ready for dispersal

stalls (Figs. 48.21B and C). Inside the seed is the *embryo* covered up by the fleshy *endosperm.* The embryo (Fig. 48.21A) shows the *shrivelled suspensor* at the micropylar end, then the *radicle* with a prominent *hypocotyl* ending in *several cotyledons* (ranging from 3 to 15) at the other extremity. In between the cotyledons is the *plumule* forming the other end of the embryo *axis.* The seed of Pine is *polycotyledonous.*

By the time the seeds are fully mature and dry the ovuliferous scales and the whole female cones are dry and hard woody structures. Each seed, when mature, acquires a large papery wing which is formed from the ovuliferous scale tissue but remains attached to the seed (Figs. 48.21D and E). The scales of the ripe cone now loosen and open wide apart (Fig. 48.21F). The loose seeds with the wings are buoyant and are easily carried away for dispersal by wind currents.

Germination of the seed

Pinus seeds usually germinate after a period of rest. Germination is epigeal. The radicle goes into the soil first. Then the hypocotyl pushes up the cotyledons which spread out after pushing off the testa. The stages of germination may be seen in Figures 48.22A, B, C and 48.8A.

Figure 48.23 shows the life cycle of *Pinus* in a diagram.

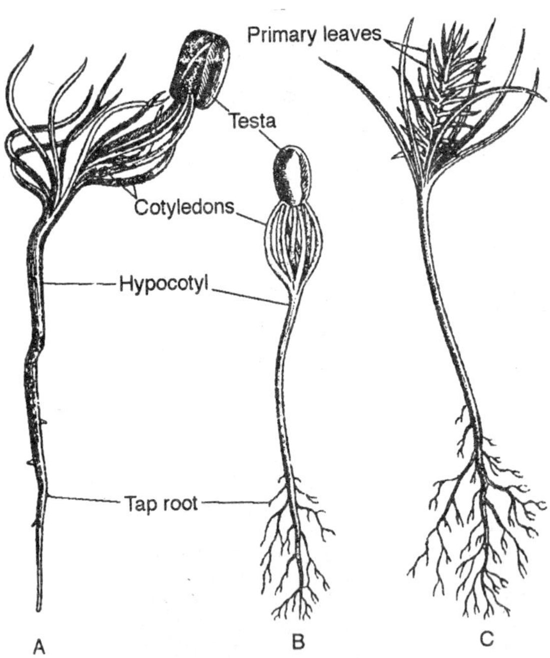

Fig. 48.22: A-C. Stages of germination of *Pinus* seed

ECONOMIC IMPORTANCE OF PINACEAE

Pinaceae wood is of great importance in commerce. *Pinus, Cedrus* and *Abies* yield a good quantity of timber used as building material, fuel, in furniture, poles, packing cases, match boxes and match sticks, pencils, etc. *Picea* wood pulp is specially preferred for paper manufacturing. Turpentine, resin, tannins (from *Tsuga*), wood alcohol and several other chemicals are extracted from several species of *Pinus, Abies* and *Larix. Abies balsama* (mainly from Canada) yields Canada balsam. In India, *Pinus roxburghii* is the principal source of turpentine and resin while *Cedrus deodara* is the most valuable timber of this type. *Pinus gerardiana* (*chilgoza*) seed is roasted and eaten. Fossilized resin of *Pinus succinifera* (now extinct) is known as *amber* which is of

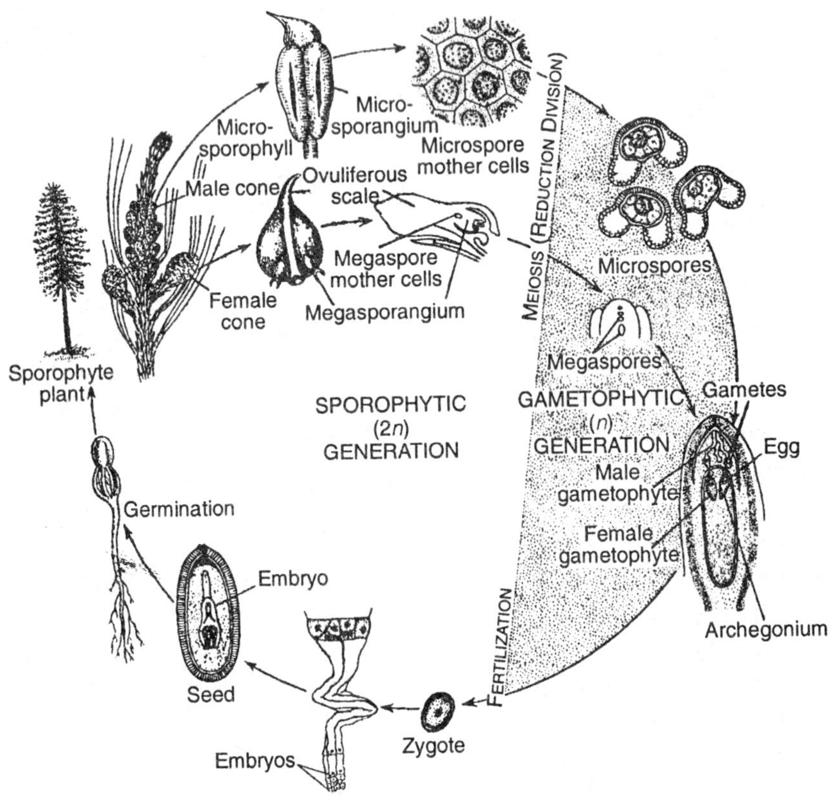

Fig. 48.23: Life cycle of *Pinus*

great commercial value and even used in jewellery.

The affinities of Pinaceae are discussed at the end of this chapter.

Family 7. TAXODIACEAE

Taxodiaceae also is a family of trees with small needle-like to falcate or scaly leaves borne spirally. Cones are small with spiral members. Bracts and ovuliferous scales are almost completely fused. There may be 2 to 9 ovules on an ovuliferous scale. Pollens are wingless. There are 10 genera with about 15 species:

Tribe: *Sequoiea*

Genera: 1. *Sequoia* (Redwood—1 sp.): from California (Fig. 48.24A). Remarkably gigantic ancient trees. Spread to Europe, Asia and other parts of North America in the Cretaceous and Tertiary.

2. *Sequoiadendron* (Bigtree—1 sp.): California to Sierra Nevada. Also a huge tree.

Tribe: *Metasequoiae*

Genus: 3. *Metasequoia* (1 sp.): previously known only as Cretaceous and Tertiary fossils, later dominant in the Eocena Arctic vegetation. Now known to exist in Central China and hence called a 'living fossil'.

Tribe: *Taxodieae*

Genera: 4. *Taxodium* (3 spp.): S. E. United States and Mexico. Grows in forests which often remain flooded with their swollen bases under water.

5. *Glyptostrobus* (1 sp.): S. E. China.

Tribe: *Cryptomerieae*

Genus: 6. *Cryptomeria* (1 sp.): Japan (Fig. 48.24B). It has been introduced into Darjeeling where it is now the commonest conifer.

Tribe: *Cunninghamieae*

Genus: 7. *Cunninghamia* (2 spp.): China and Taiwan.

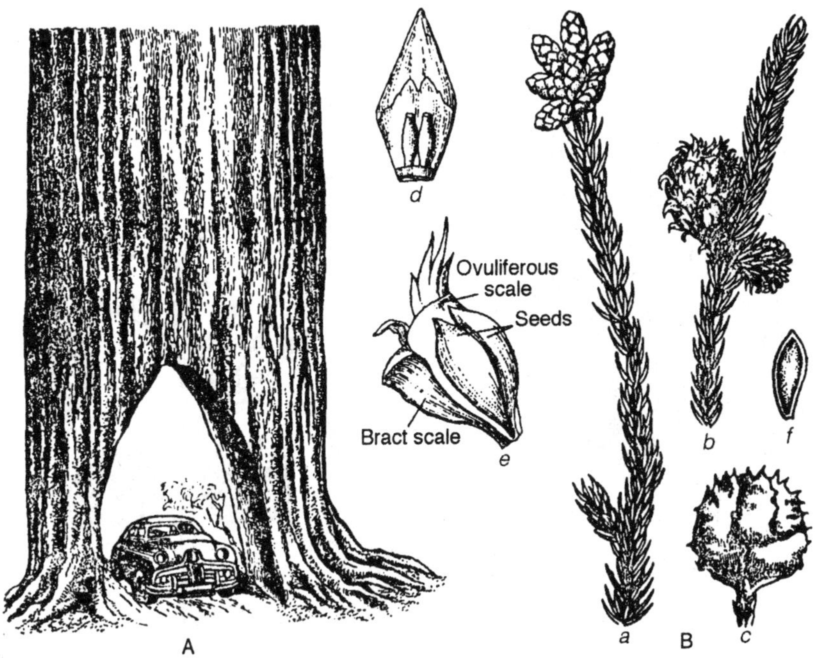

Fig. 48.24: A. *Sequoia*—base of a gigantic tree big enough for a car path, B. *Cryptomeria japonica*—a. Male shoot b. Female shoot, c. Female cone, d. Seeds on ripe scales, e. Ripe female flower, f. A seed

Tribe: *Sciadoptiyeae*

Genus: 8. *Sciadopitys* (2 sp.): Japan.

Tribe: *Arthrotaxeae*

Genera: 9. *Arthrotaxis* (3 spp.): Tasmania.

10. *Taiwania* (1 sp.): North Burma, Yunnan, Taiwan.

No species of this family is endemic in India. But, *Cryptomeria japonica* (Fig. 48.24B), introduced into India from Japan in the nineteenth century has become strongly naturalized in Darjeeling as a dominant forest vegetation in the temperate zone. Locally known as *dhupi*, it is a source of lumber and fuel.

Family 8. CUPRESSACEAE

Cupressaceae is another family of arborescent to shrubby conifers. The leaves are persistent, small, scale-like, opposite or inwhorls. Male strobili are small; microsporophylls are somewhat peltate bearing 2 to 6 microsporangia. Female cones are on short branches bearing few scales; ovuliferous scales are flattened or peltate bearing 1 to 12 ovules. Female cones are dry in *Thuja, Cupressus*, etc. but fused fleshy, berry-like in *Juniperus*, etc. There are 16 genera with about 133 species.

Subfamily: *Cupressoideae*

Genera: 1. *Cupressus* (Cypress—about 15 spp.): Eastern Mediterranean to Eastern Asia including China, Western North America, Central America. Widely cultivated (Fig. 48.25B).

2. *Chamaecyparis* (about 6 spp.): Japan, Taiwan, North America.

Subfamily: *Thujoideae*

Genera: 3. *Thuja* (6 spp.) East Asian, N. America. Widely cultivated as ornamental plant. Common species in Indian gardens—*T. orientalis* (Fig. 48.25A) and *T. plicata.* Known as fossil since Upper Cretaceous and Tertiary.

4. *Thujopsis* (1 sp.): Japan, Fossil from Tertiary Greenland.

5. *Libocedrus* (9 spp.): Burma, Molucca, China, Taiwan, New

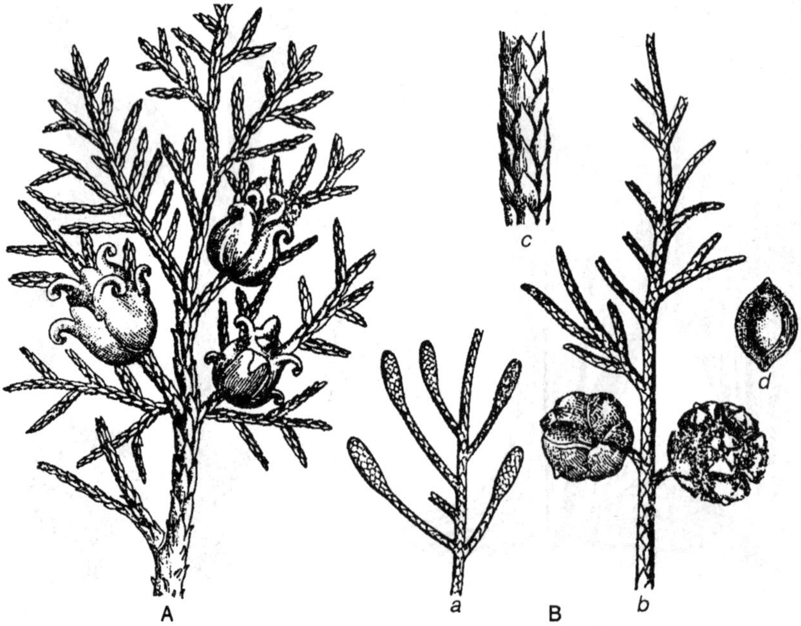

Fig. 48.25: A. *Thuja orientalis*—fertile shoot, B. *Cupressus torulosa*—a. Male shoot, b. Female shoot, c. Enlarged leafy branch, d. Seed on scale

Guinea, Australia, New Zealand, New Caledonia, California, South America. Fossil in Tertiary Europe and N. America.

6. *Pilgeodendron* (1 sp.): Chili.

7. *Callitris* (20 spp.): Australia, Tasmania, New Caledonia. Fossil in Upper Cretaceous of Europe. Cultivated in Nilgiri Hills for essential oil and fuel. Introduced in 1886.

8. *Neocallitropsis* (1 sp.): New Caledonia.

9. *Tetraclinis* (1 sp.): Spain.

10. *Actinostrobus* (2 spp.): W. Australia.

11. *Widdringtonia* (5 spp.): S. Africa.

12. *Fitzroya* (1 sp.): Chili and Patagonia.

13. *Diselma* (1 sp.): Tasmania.

14. *Fokienia* (3 spp.): China.

Subfamily: *Juniperoideae*

Genera: 15. *Arceuthos* (1 sp.): Greece, Asia Minor, Andes.

16. *Juniperus* (Juniper—60 spp.): Asia, Europe, N. and C. Africa, Alaska to C. America, West Indies. Widely cultivated (Fig. 48.26). Tertiary fossil *Juniperoxylon* common in Europe and N. America.

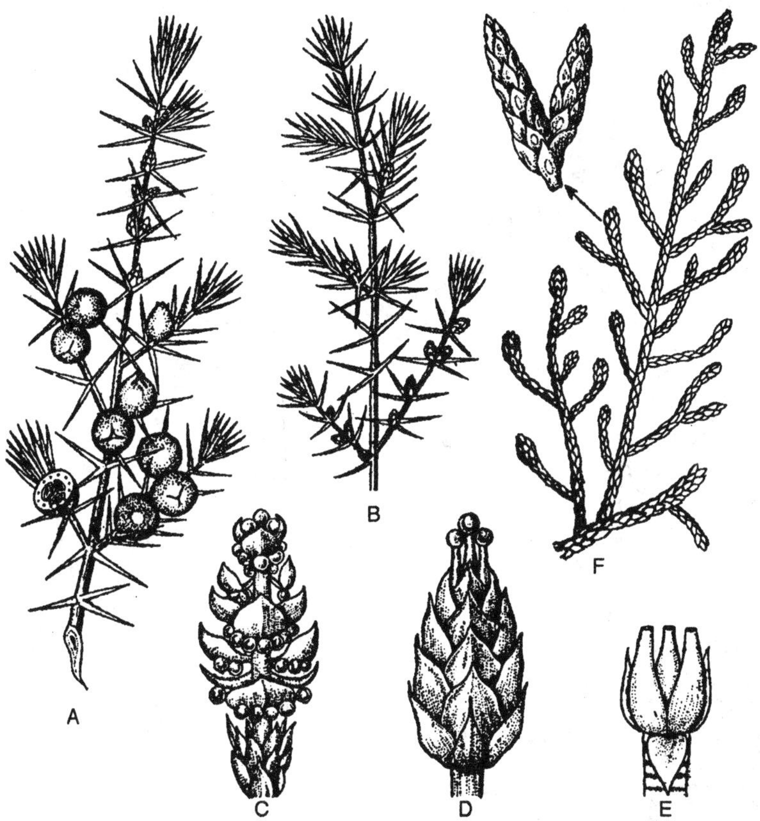

Fig. 48.26: *Juniperus communis*. A. Female shoot with berries, B. Male shoot, C. Enlarged male flower showing microsporophylls, D. Enlarged female cone with pollination drops on the tips of the three ovules, E. Top of the female cone (dissected out) showing the three ovules and bracts, F. Shoots of *Juniperus macropoda* (enlarged leafy shoot inset)

INDIAN SPECIES OF CUPRESSACEAE

1. *Cupressus torulosa* D. Don.: Outer ranges of Himalaya, Chamba to Nepal, Arunachal, Yunnan, 1800-2800 m. Up to 150 ft. tree. Timber very durable and easily workable, even better than *deodar.* Also used for pencil making (Fig. 48.25B). Several species of *Cupressus,* viz., *C. funebris, C. cashmeriana* (unknown in wild state—may be Tibetan in origin), *C. sempervirens*—are cultivated as ornamental plants.
2. *Juniperus communis L.* (= *J. sibirica* Burgod): Western Himalaya up to Kumaon 1700-4300 m, spreading to Afghanistan, Persia, Siberia. Bushy to 12 m high tree. Fruits used in gin manufacture in Europe. Burnt as incense in the Punjab (Fig. 48.26A to E).
3. *J. macropoda* Boiss: Inner ranges of Western Himalaya up to Kumaon, Western Tibet, Balauchistan, 1500-4300 m 1000 year old specimens with trunk 10 m in circumference are known from Lahul. 12-15 m high. Very suitable for pencil making (called 'Himalayan pencil cedar') but supply is inaccessible at present (Fig. 48.26F).
4. *J. recurva* Buch-Ham.: Temperate-alpine Himalaya 1700-4600 m in the Inner ranges from Afghanistan to Bhutan, Upper Burma, China. May be a tree up to 10 m high but bushy in higher elevations.
5. *J. squamata* Buch.-Ham.: 2400-4300 m in the Himalaya with *J. communis,* Upper Burma, Afghanistan, China, Taiwan, Japan. A dense thicket.

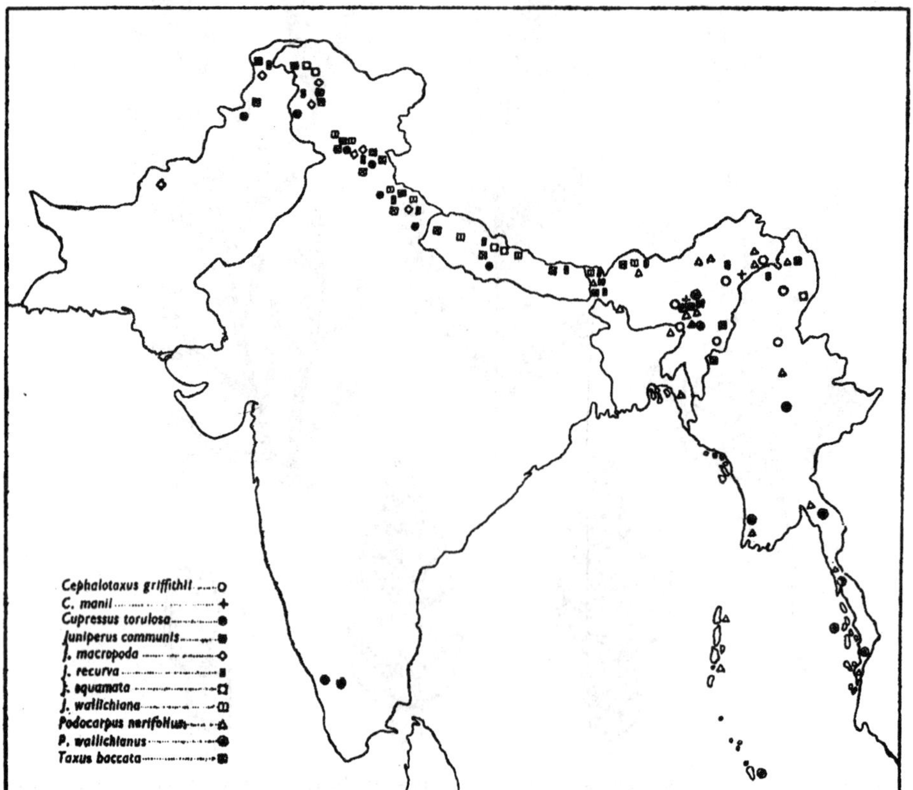

Fig. 48.27: Map showing distribution of Cupressaceae. Podocarpaceae, Cephalotaxaceae and Taxaceae in India and adjacent countries

6. *J. wallichiana* Hook f. ex Parl. (= *J. pseudosabina* Hook. f.): Temperate-alpine Himalaya from Kashmir to Bhutan 2800-4600 m. Shrub in Western Himalaya, a tree up to 18 m high in Sikkim.

The distribution of the Indian species is shown in the map in Figure 48.27.

Family 9. ARAUCARIACEAE

A family of beautiful trees with branches in whorls. Leaves needle-like with single vein or broad with parallel veins, spiral or distichous. Male cones large, catkin-like. Microsporangia linear, hanging from peltate microsporophylls. Female cones woody, very large, with spiral scales in which the bract and the ovuliferous scale are completely fused in *Araucaria* but are free in *Agathis*. The morphology of the scale is disputed by some who think it is a simple sporophyll with a ligule. Each scale has a single ovule. The male gemetophyte of *Araucaria* shows a prothallus tissue sometimes formed of about 40 cells (Fig. 48.28C) so that it is morphologically more primitive than even the living heterosporous Pteridophytes.

Predominantly a Southern Hemisphere family. The family is apparently very primitive. Upper Triassic *Araucarioxylon* wood is associated with the family. Widespread *Araucarites* from Jurassic also belong here. Three species of *Araucarites* fossils are known from Indian Upper Gondwana Jurassic (Rajmahal and Jabalpur stages) viz., *A. cutchensis, A. macropterus, A. latifolius.* Fossil *Agathis* is known from the Northern Hemisphere during the Cretaceous.

There are two genera with about 35 species:

Tribe: *Agatheae*

Genus: 1. *Agathis* (*Kauri* pine—20 spp.): Cochin-China, Malay, East Indies, Philippines, New Zealand, Australia, Fiji, New Caledonia, New Hebrides. A valuable timber tree. Leaves broad with parallel veins.

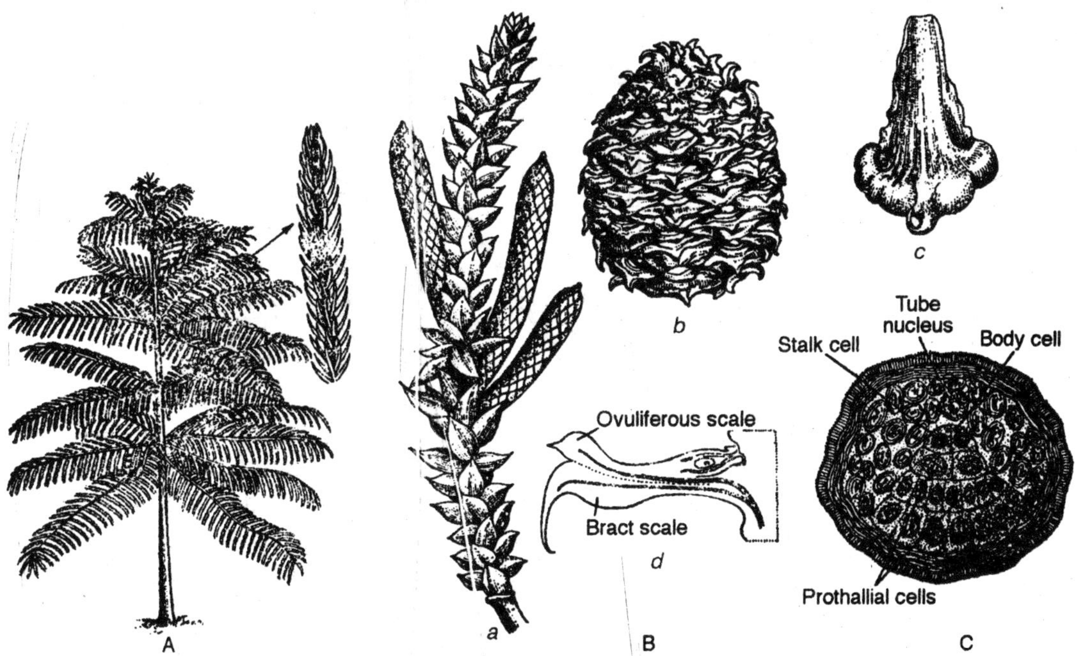

Fig. 48.28: *Araucaria*. A. Habit of *A. columnaris* (= *A. cookii*)—leafy shoot inset, B. *A. bidwilii*—a. Male cones on a shoot, b. A female cone, c. A scale with a seed, d. L.s. of c (the ovuliferous scale is called ligule by some), C. Male gametophyte of *A. cunninghamii* (after Chamberlain)

Tribe: *Araucarieae*

Genus: 2. *Araucaria* (about 15 spp.): New Guinea, East Australia, New Caledonia, South America. Timber as well as ornamental tree. Some species are grown in Indian parks, etc., viz., *A. columnaris* (= *A. cookii*), *A. bidwilii, A. cunninghamii* (Fig. 48.28).

The two following families are often associated with the Taxaceae (placed here in a separate order Taxales) into a group *Taxares* (as opposed to the *Pinares*) because of the absence of obvious female cones in most genera.

Family 10. PODOCARPACEAE

Shrubs or trees with linear or lanceolate leaves. In *Podocarpus* the leaf is fairly broad with parallel veins. Dwarf shoots transformed into phylloclades in *Phyllocladus*. Male strobilus of microsporophylls each with two microsporangia. Pollens winged. Definite female cones in some species, in others it is very much reduced with two or even one scale laterally bearing a single ovule. In addition, the ovule is covered by an outgrowth called *epimatium* (Figs. 48.29C and D) which is probably the ovuliferous scale though it has been supposed to be an aril or a second integument. When ripe the epimatium as well as the stalk of the ovule (*peduncle*) become fleshy while the integument turns stony. The male gametophyte (Fig. 48.29E) resembles *Araucaria* but with fewer prothallial cells.

This is an ancient family with fossils clearly known up to Upper Triassic even in India. They may even be related to the Triassic-Jurassic Cheirolepidaceae (c.f. page 1304) in which an epimatium is present.

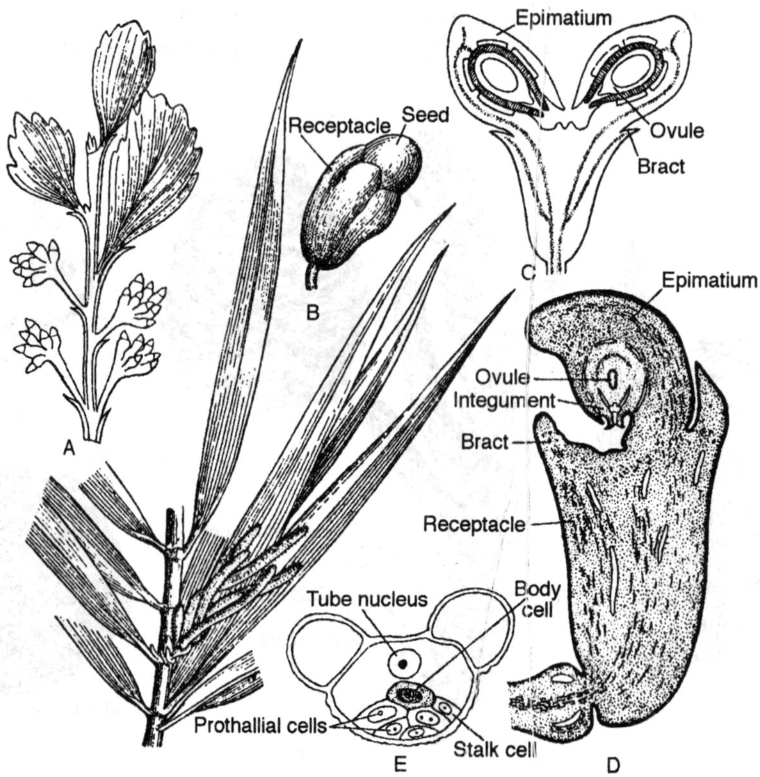

Fig. 48.29: A. *Phyllocladus* showing phylloclade dwarf shoots and female cones, B-E. *Podocarpus*, B. *P. nerifolius* shoot with male strobili and a seed inset, C. L.s. of *Podocarpus* female cone—usually only one ovule matures, D. L.s. of seed, E. Male gametophyte

Podocarpaceae is mainly confined to the Southern Hemisphere but *Podocarpus* extends to India, China and Japan. There are 7 genera with about 111 species.

Subfamily: *Pherosphaeroideae*

Genus: 1. *Pherosphaera* (2 spp.): Tasmania

Subfamily: *Phyllocladoideae*

Genus: 2. *Phyllocladus* (6 spp.): Borneo, Philippines, New Zealand, Tasmania, New Guinea (Fig. 48.29A)

Subfamily: *Podocarpoideae*

Genera: 3. *Saxegothaea* (1 sp.): Chili.

4. *Microcachrys* (1 sp.): Tasmania.

5. *Dacrydium* (20 spp.): New Zealand, Australia, Chili.

6. *Podocarpus* (about 80 spp.): As important in the Southern Hemisphere as *Pinus* in the North: Africa, India, Burma, Japan, East Indies, Australia, New Zealand, C. and S. America, West Indies (Fig. 48.29B to E).

7. Acmophyle (2 spp.): New Caledonia, Fiji.

INDIAN SPECIES OF PODOCARPACEAE

1. *Podocarpus nerifolius* D. Don.: Spreading from the plains (tropical evergreen climax forests) to 900m in the Eastern Himalaya, Sikkim eastwards to Khasia, Chittagong, Burma, Andamans. The timber is of value and is used for oars, paddles, spears, masts, etc. (Fig. 48.29B to E).
2. *P. wallichianus* C. Presl. (= *P. latifolia* Wall.): Assam, Burma, Western Ghats from Nilgiri southwards up to 1500 m, Nicobar Island. Slightly aromatic. This is the only natural conifer in South India.

Distribution of these species is shown in the map in Figure 48.27.

Family 11. CEPHALOTOXACEAE

The family comes even nearer to Taxaceae and is often included within it. They are trees with opposite or whorled branches and distichous (spiral on vertical stems), linear, persistent leaves with midribs. The leaves and

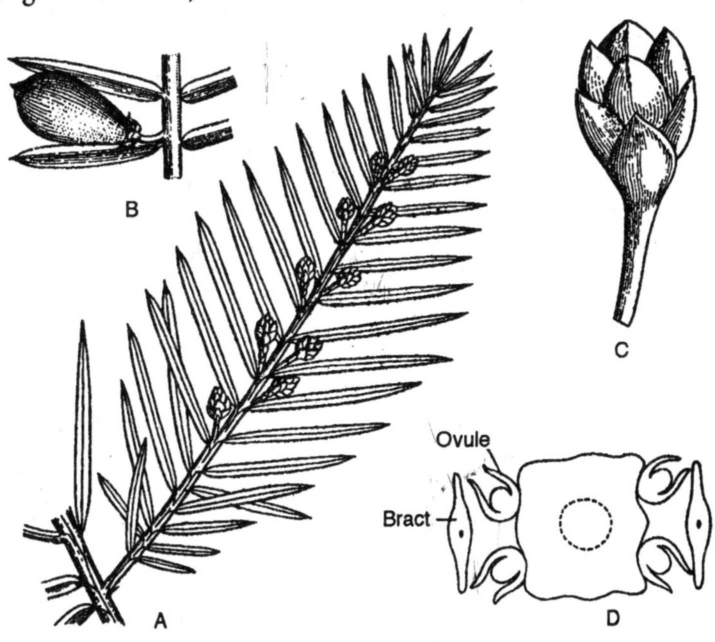

Fig. 48.30: *Cephalotaxus manii*. A. Shoot with male strobili, B. Shoot with mature seed, C. *Cephalotaxus* female cone, D. T.s. of C showing two bracts subtending two ovules each

branches are covered with a white powder on the underside. Male flowers in globose heads in leaf axils. Each microsporophyll with three microsporangia. Female cones with several pairs of bracts each subtending two erect ovules (Fig. 48.30D). Usually one or two ovules mature into stalked drupaceous fruits in each cone (Fig. 48.30B). *Cephalotaxus* has a mesarch stem which is unique among conifers and causes its separation from the Taxaceae. Fossils (*Cephalotaxopsis* and *Cephalotaxites*) are known since the Cretaceous. Only one genus is usually recognized:

Genus *Cephalotaxus* (6 spp.): Tropical Eastern Himalaya, Burma, Thailand, S. and M. China, Japan (Fig. 48.30).

INDIAN SPECIES OF CEPHALOTAXACEAE

1. *Cephalotaxus manii* Hook. f.: Khasia and Naga Hills, 1370-2600 m (Fig. 48.30).
2. *C. griffithii* Hook f.: Arunachal, Naga Hills, Manipur, Mishmi Hills, Burma, 1520-1820 m.

Both the species are small trees but the timber does not seem to be of much use. Their distribution is shown in the map in Figure 48.27.

Order III. **TAXALES**

The family Taxaceae is placed in a separate order as many authorities beginning from Sahni are of opinion that the Taxaceae represent a phylogenetic line quite different from that of the general Coniferales. They do not show any sign of a female 'strobilus' and the solitary ovules are placed in the tips of side dwarf shoots. Although their fossil history is known up to the Upper Triassic there is no indication of their origin from any strobilate plant.

Sahni (1920) expressed the opinion that the Taxaceae should be separated from the Coniferales. Buchholz (1933) proposed the name *Phanerostrobilares* for the general conifers with female cones and *Aphanerostrobilares* for Podocarpaceae and Taxaceae. Florin (1938-1941) also considered the Taxales as separate from the Coniferales. Podocarpaceae and Cephalotaxaceae are sometimes included within the Taxales by various authors. But, there is some dispute as to their non-strobilate nature and, at any rate, arguments for placing them in a separate order is not as unanimous as for Taxaceae. Only Taxaceae is, therefore, placed here within the Taxales. It may also be noted that Taxales has been raised to a separate class, Taxopsida in Engler's current Syllabus and some authorities have even suggested raising the Taxales into a separate division.

Family: TAXACEAE

The family Taxaceae comprise of evergreen trees with flat, linear, apparently distichous leaves and without resin canals. They are mostly confined to the Northern Hemisphere. Male flowers solitary or in small, axillary cones. Pollen grains wingless and do not cut off any prothallial cell during male gametophyte formation. Microsporophylls peltate or apically thickened, with 3-9 microsporangia. Ovules not in strobili, axillary, terminal on dwarf shoots, seated on a meristematic tissue whose megasporophyll nature is not clear. Seed erect, hard, covered by a fleshy, coloured *aril* which is the only part in the tree free from the alkaloid *taxin*. There are 5 genera with about 15 species:

Tribe: *Torreyeae*

Genera: 1. *Amentotaxus* (1 sp.): China. Known as Tertiary fossil.

2. *Torreyea* (5 spp.): China, Japan, California, Florida. Known as fossil since Upper Jurassic.

Tribe: *Taxeae*

Genera: 3. *Austrotaxus* (1 sp.): New Caledonia.

4. *Nothotaxus* (1 sp.): China.

5. *Taxus* (Yew—7 to 8 spp.): England, Continental Europe, Asia Minor, Himalaya, China, Japan, Philippines, Celebes, N. America, Mexico. Known as fossil since Jurassic.

Fossils named *Palaeotaxus* are known from Upper Triassic, as also *Taxoxylon* wood from Createaceous and *Taxites* twigs from Jurassic to recent. *Taxites* is known from Jurassic Indian Upper Gondwana in the Rajmahal series.

INDIAN SPECIES OF TAXACEAE

1. *Taxus baccara* L.: Spreading from England and Europe over to temperate Himalaya, Khasia Hills, Naga Hills, Manipur and Upper Burma. Up to 50 ft. tall and often very old. Hard timber. Used as fodder in spite of the alkaloid (Fig. 48.31). (Distribution shown on map in Figure 48.27).

T. wallichiana Zucc. with narrower leaves has sometimes been considered as a separate species but it is doubtful whether it is different form *T. baccata.*

Genus: TAXUS

The sporophyte plant

Taxus is an evergreen, densely branched tree of medium height which is known to be long living. Some of the British yew trees are claimed to be 2000 years old. About 7 or 8 species are distributed all over the Northern Hemisphere. *Taxus baccata* L. is the predominant species spreading from England to Japan over the Himalaya, to the Pacific Islands and to

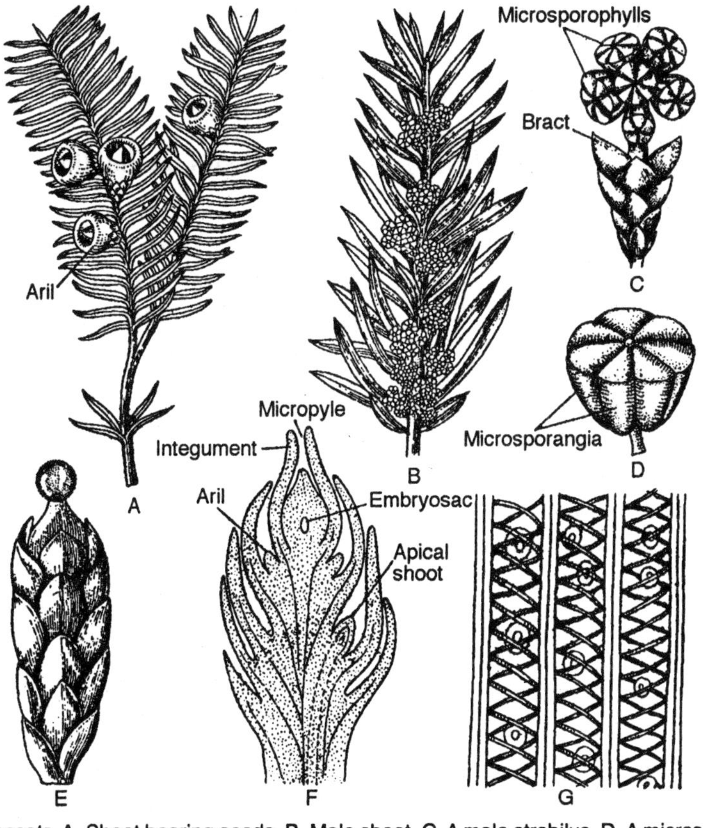

Fig. 48.31: *Taxus baccata.* A. Shoot bearing seeds, B. Male shoot, C. A male strobilus, D. A microsporophyll, E. A female flowering shoot, F. Median l.s. of E, G. L.s of *Taxus* wood showing tracheides with tertiary spiral thickening in addition to bordered pits

North America going as south as Mexico. The 'yew' tree is of historical importance in Europe. The trunk of the tree is quite thick and is usually formed by the fusion of several basal branches.

The leaves are dark-green, 2 to 4 cm long, more or less broadly linear (Figs. 48.31A and B) with a short stalk-like portion broadening in a decurrent base and ending in a spinous tip. There is a single median vein. They are spiral in arrangement but in the shoots they are twisted to give a distichous appearance. Although persistent, they fall off after some time leaving the leaf base as permanent ridges. All the leaves are of one type and the shoots also are uniform. The poisonous alkaloid *taxin* is present in the leaves and all parts of the plant excepting the aril. When present in large quantity it may be injurious to cattle. Scaly *winter buds* develop onshoots for overwintering.

The *anatomy* of the leaf and the stem generally resembles *Pinus* with certain differences. In the leaf the mesophyll is differentiated into palisade and spongy parenchymas and there is a single vascular bundle. Otherwise the leaf presents the same xerophytic structure. The stem anatomy differs in having no resin duct, in the medullary rays being simpler and always uniseriate, and in the cortex and the pith occupying a lesser area of the stem. The trancheides have uniseriate bordered pits and they show a tertiary spiral thickening (Fig. 48.31G) rendering the wood more elastic. This was the reason why yew was the favourite for construction of bows among the European medieval archers. The root structure is similar to that of *Pinus.*

Taxus is dioecious although semetimes male and female shoots are found on the same tree. The flowering shoots are borne on the underside of the shoots of the previous year at the axils of leaves (Figs. 48.31A and B). Flowering in India takes place in early spring (February-March).

The *male strobilus* or *flower* (Fig. 48.31C) is a solitary, stalked, subglobose, short catkin of 6 to 10 microsporophylls borne on an axis. There are a few scaly bracts at its base. Each microsporophyll (Fig. 48.31D) is peltate (as in *Equisetum*) with 5 to 9 microsporangia hanging below the peltate disc. Each microsporangium has a wall, a tapetum inside, and contains numerous microspores or pollens which are not winged. The microsporangia split and curl out when ripe to liberate the pollens.

The *female shoot* (Figs. 48.31E and F) arises in the axil of a leaf on the previous year's branch as a small bud. It bears a number of closely overlapping scale leaves and a growing apex on the axis which soon ceases to function after the development of a secondary lateral shoot in the axil of one of its top scales. This secondary shoot may be called the female flower as it bears a few scale leaves and ends in a single terminal ovule. The structure of the ovule is as in *Pinus* excepting the *aril* which is described below.

Gametophytes, Pollination, Fertilization

Pollination is anemophilous as in *Pinus* but the wingless pollens are shed in the uninucleate stage. The development of the *female gametophyte* also resembles *Pinus* excepting a few minor differences. The *archesporial cells* seem to be hypodermal in origin but the *megaspore mother cells* are found deep within the nucellus. In *Taxus,* cases are known where different megaspore mother cells have given rise to more than one female gametophyte. But in such cases only one gametophyte survives to the last. The megaspore mother cell gives rise to a *linear tetrad of megaspores* of which only the lowest one is functional. But, here again, cases are known in *Taxus* where more than one megaspore from the same linear tetrad start forming female gametophytes though only one of them persists to the end. As in *Pinus,* the female gametophyte develops several *archegonia.* An

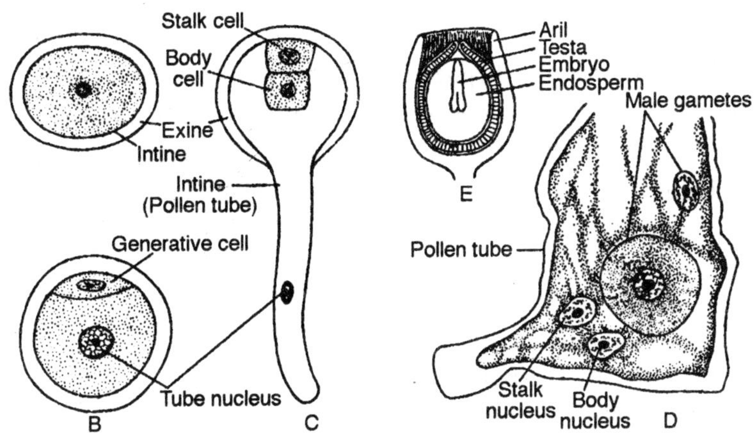

Fig. 48.32: *Taxus baccata*, A-D. Stages of development of the male gametophyte, E. Median l.s. of a mature seed

important point of difference is that no cell wall develops between the *ventral canal nucleus* and the *egg nucleus* before the former disintegrates so that there is no ventral canal 'cell' in *Taxus.*

There is a big difference in the development of the *male gametophyte* between *Pinus* and *Taxus.* The wingless pollen of *Taxus* (Fig. 48.32A), shed in the uninucleate stage, begins germination after it is lodged on the tip of the megasporangium nucellus. The first division of its nucleus results in a *generative cell* and a *tube cell* (Fig. 48.32B) so that the prothallial cells are completely eliminated. Thereafter it develops (Figs. 48.32C and D) as in *Pinus.* The two *male gametes* are always of unequal size. The pollen tube reaches the female gametophyte top piercing through the nucellus very quickly, usually two months before the archegonia are formed, but the actual bursting of the pollen tube and fusion of male gamete and the egg take place later. *Fertilization* results in the *zygote* or the *oospore.*

Sporophyte embryo, seed and germination of seed

Development of the embryo is similar to *Pinus* but the *proembryo* fills up the whole of the zygote cavity and its tiers are irregular. Usually there is a single embryo from a proembryo there being no cleavage polyembryony. But, several archegonia within the same gametophyte may send down embryos resulting in *simple polyembryony.* However, as in *Pinus*, a single embryo in a seed is the usual rule. The *embryo* shows two *cotyledons*, a *plumule* and a *radicle* which are surrounded by the *endosperm* (gametophytic tissue) in the young stage (Fig. 48.32E).

During the development of the *seed* (Fig. 48.32E) the integument, three-layered as in *Pinus* at the beginning, dries up into a hard *testa.* A peculiarity is the development of the aril. The aril is present in the ovule as a low collar surrounding its base (Fig. 48.31F). As the seed matures the aril grows as a fleshy envelope covering the seed except at the top (Figs. 48.31A and 48.32E) giving the seed a berry-like appearance. The aril, at first green, becomes red in the mature seed. It is edible, much liked by birds and is the only part of the *Taxus* plant devoid of the poisonous alkaloid *Taxin.* The seed itself, however, is poisonous. The morphological nature of the aril has been greatly discussed. It has been supposed by different authors to be a modified ovuliferous scale, an epimatium (as in *Podocarpus*), a delayed outer layer of the integument, a second integument, a cupule-like structure, a scale, etc. In the other genus *Torreya* the aril completely covers the seed.

The *germination* of the seed is hypogeal with the first two linear cotyledonary leaves piercing the soil. The primary root persists.

Economic use of Taxus

Taxus wood is very durable but is not much used in India because of its relative inaccessibility. It is locally used for poles, ploughs, cart axles, etc. Formerly it used to be of great demand for bows. As the poisonous alkaloid content is usually low it is often used as fodder. In Ladakh the bark is used as a sort of tea. The aril is eaten by birds.

AFFINITIES OF THE CONIFERALES AND THE TAXALES

That the Coniferales evolved out the Cordaitales is now accepted by most authorities. Florin (1951) has shown by the comparison of fossils that the female strobilus of the Pinaceae has evolved from the Cordaitales through Lebachiaceae and Voltziaceae. The reduction of the broader leaves of the cordaitales to the needle leaves of Pinaceae has also been taken for granted although fossil evidences are not quite so strong.

Within the Coniferales the two familes Pinaceae (Abietaceae) and Araucariaceae both seem to be very primitive, there ancestry being traceable perhaps down to Carboniferous. So, the question as to which of them is more primitive has been often argued. Arucariaceae no doubt resembles the Cordaitales in its secondary wood and foliage. Its male gametophyte and embryo are also more primitive than that of Pinaceae. A direct origin has been suggested on these grounds. But Jeffrey (1917) considers the wood of the Pinaceae more directly related to the Cordaitales and suggests that the similarity with Araucariaceae is a later development. He supposes (Fig. 48.33A) that the Araudariacdeae evolved out of the Pinaceae at a very early stage. The fusion of the ovuliferous and bract scales evolved at that stage led to the Araucariaceae. The other families arose later as shown in the diagram. Takhtajan (1953) more or less adheres to a similar view incorporating the findings of Florin (Fig. 48.33B).

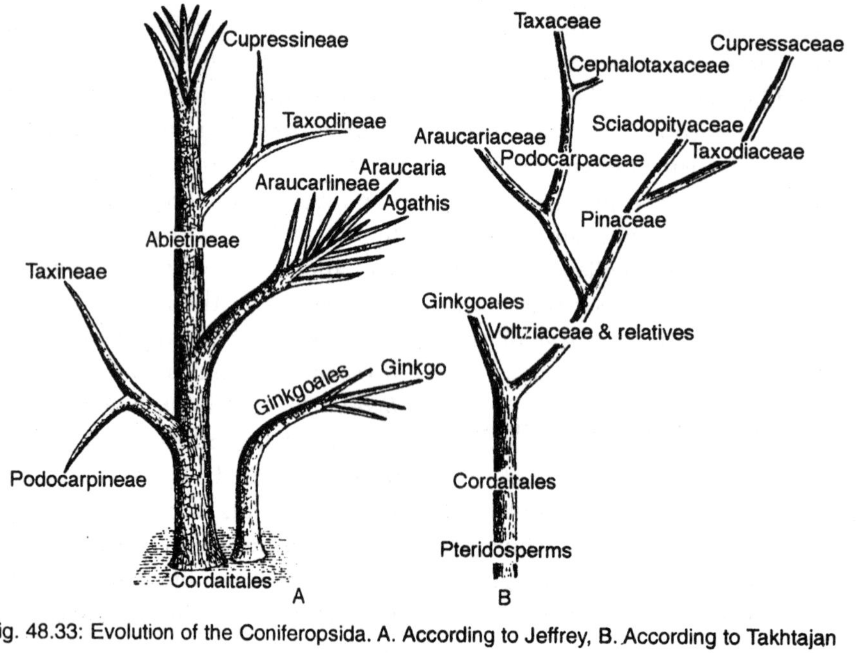

Fig. 48.33: Evolution of the Coniferopsida. A. According to Jeffrey, B. According to Takhtajan

The views of Seward (1905) about the Lycopodean origin of Araucariaceae and of Pennhallow (1904, 1907) that the Pinaceae originated from the the Taxaceae which came from the Cordaitales no longer attract much attention. Lam (1955) suggests the origin of the Coniferopsida from a separate Lycopsida stock. This view also does not seem to be very much supported.

The position of the Taxaceae (Taxales) has been even more problematic. Formerly included within the Coniferales, it was recognized that the family shows no sign of a strobiloid ancestor although it may be clearly traced to the Upper Triassic and probably even earlier. It has been attempted to derive the family out of the Podocarpaceae stock by Jeffrey as well as Takhtajan. This does not seem to be very satisfactory. So Sahni (1920) suggested the separation of the Taxaceae from the Coniferales. This has also been the view of Pilger in Engler's Syllabus (1916, 1954) and of Florin (1951). Another point of argument is whether Podocarpaceae and Cephalotaxaceae also should come within the new order. It is safest, how-ever, to consider Taxaceae alone. Pennhallow's view of deriving the Taxaceae directly out of the Cordaitales has been cited above. At the present state, it is better to consider the Taxales as a separate but problematic order.

The affinities of the Coniferopsida with the Pteridospermopsida and the Cycadopsida seem to be very remote. Any connection does not seem probable in the Spermaphytic level though earlier relationship is possible.

Chapter 49

Coniferopsida: Ginkgoales and Czekanowskiales

Order IV. GINKGOALES

The Ginkgoales are included within the Coniferopsida because of its coniferous type tree, wood and gametophytes. But, its distinctiveness has given rise to the argument as to whether it should better be placed in a separate class Ginkgopsida or even a separate division Ginkgophyta. The present-day monotypic *Ginkgo biloba* is a relic of a much more extensive group of plants which probably evolved in the Devonian, began to expand in the Triassic and attained its climax in the Jurassic after which it began to wane. Even the single species present today probably escaped extinction because of its cultivation in Chinese and Japanese temples.

FOSSIL GINKGOALES

Fossils which have been definitely ascribed to the Ginkgoales are known since the Permian. Besides these, there are several forms leading to the Upper Devonian which have been suspected to belong to this group. *Ginkgophyllum* and *Platyphyllum* leaves from the Upper Devonian look Ginkgoan but it is not definite whether they are not fern leaves. *Psygmophyllum* is a similar leaf fossil widely known from the Upper Devonian and Permian. In India, it is known from the Lower Gondwana Kashmir *Gongamopteries* beeds (Lower Permian). *Rhipidiopsis* from the Indian Lower Gondwana (Raniganj—Upper Permian and Barakar—Lower Permian) also has a similar position. *Dichophyllum* from Kansas Upper Carboniferous shows finely dissected branching (without clear leaves) and, with its seed, has been suspected to have Ginkgoan affinity.

Among the more definite early members ascribed to the Ginkgoales is *Trichopitys* (Fig. 49.1A) from French Lower Permian. In this the spirally arranged leaves are finely dissected dichotomously and in the axils of the leaves are borne branched systems bearing 4 to 6 (sometimes even 20) ovules. Florin (1949) suggests that such a structure could have given rise to the modern *Ginkgo* by evolving perennial dwarf shoots and reducing the number of ovules. *Sphenonbiera* is another genus spreading from Lower Permian to Lower Createaceous showing finely dichotomously dissected leaves but these were in short shoots as well as long shoots. The male shoot were also dwarf in

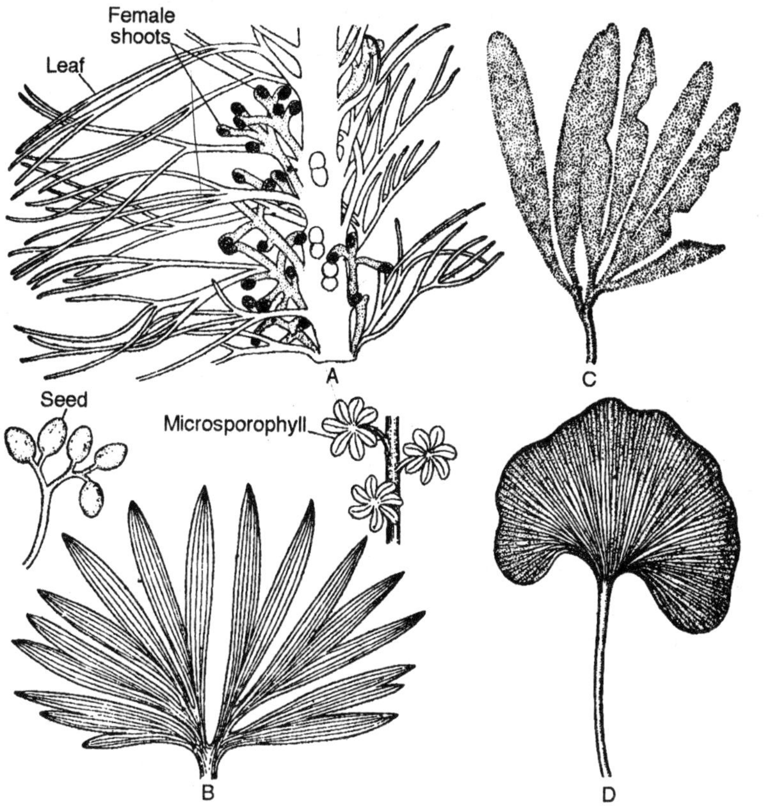

Fig. 49.1: A. *Trichopitys heteromorpha* (after Florin), B. *Baiera gracilis* leaf (after Renault) with male and female shoots of *B. muensteriana* (after Schimper and Schank) inset, C. *Ginkgoites hermelini* lead (after Harris), D. *Ginkgo adiantoides* leaf (after Brown)

which the central axis branched and then bifurcated, each bunch terminating in a cluster of 3 to 5 sporangia. There are also a few other fossil genera which probably belonged to the Ginkgoales, viz., *Saportaea* (Permian), *Phoenicopsis* (Jurassic-Createaceous), *Ginkgodium* (Jurassic), *Windwardia* (Createaceous).

Fossil genera which are considered to belong to the Ginkgoaceae itself are *Baiera, Ginkgoites* and those assigned to the genus *Ginkgo.* Besides the leaf shape they also show similar cuticle and stomata. *Baiera* (Fig. 49.1B), found from Lower Permian to Cretaceous, has deeply incised leaves (almost segmented) without petioles and showing distinctly dichotomous venation. Several seeds were borne on branched axes of the female shoot while the microsporangia were on male shoots consisting of several sporophylls each with six or more sporangia. The name *Ginkgoites* (Triassic to Tertiary) has been loosely applied to all *Ginkgo*-like leaf fossils but this should better be applied to highly incised leaves with petioles (Fig. 49.1C) only. Typical *Ginkgo* leaves showing different degrees of incision has been found from the Middle Jurassic onwards (Fig. 49.1D).

In the Indian Upper Gondwana *Ginkgo crassipes* is known in the Rajmahal series (Lower Jurassic) while Jabalpur rocks (Upper Jurassic) show *Ginkgoites lobata* as well as *Phoenicopsis* sp.

Family GINKGOACEAE: *GINKGO BILOBA* L.

At present the order Ginkgoales is reduced to a single family Ginkgoaceae containing the single genus *Ginkgo* with the solitary species *biloba* called the 'maiden-hair tree' for its leaves resembling the 'maiden-hair fern' (*Adiantum*). The name *Ginkgo* originated as a wrong form of the Chinese name *Ginkyo*. It was sometimes known as *Salisburia adiantifolia* Sm. and, until Hirase discovered multiciliated sperms in it, it was considered as an exceptional member of the Taxaceae. The tree genus is one of the oldest known as its representatives (mostly leaf fossils) can be traced definitely to the Lower Jurassic (Liassic) and possibly even to Upper Triassic when it had worldwide distribution. It has now shrunk to the single species even which has probably escaped extinction by cultivation in Chinese and Japanese temples. Its presence in the wild state has been disputed although Li (1956) claims its presence in the forests of South-East China. Because of such a history *Ginkgo* is called a 'living fossil'. It is now being cultivated throughout the world in temperate and subtropical regions. One or two plants are known at Darjeeling, Dehradun and Mussoorie and more are now being grown elsewhere in India.

The sporophyte plant

Young *Ginkgo* plants show the characteristic excurrent habit of the conifers but branching becomes irregular with age, the crown becoming spreading and broad. The trunk is of slow growth but, ultimately, attains a height of 30 metres and a girth of 1 metre. It bears two kinds of branches—the long branches of indefinite growth bearing short branches (*dwarf shoots* or *brachyplasts*) of limited growth (Fig. 49.2). Leaves are deciduous and spirally borne on the long shoot with *spurs* in axillary positions. Each spur (dwarf shoot) bears leaf scars, some scale leaves and a crown of leaves (Internodes being suppressed) resembling a miniature *Cycas* tree. Each spur shoot remains covered by bud scales when young. As stated above, the beautiful leaves resemble those of *Adiantum*. The venation is symmetrical and dichotomous. On the long shoots the leaves are lobed while those on spurs have merely wavy margins.

The *anatomy* of the *stem* is different in the long shoot (Fig. 49.3A) and in the spurs (Fig.

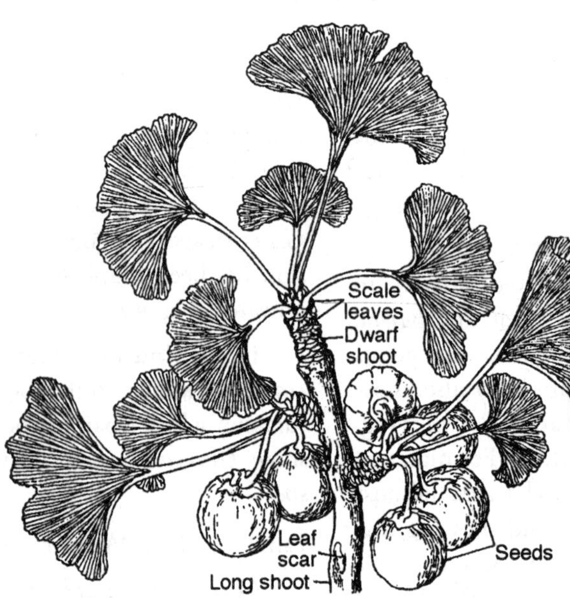

Fig. 49.2: A shoot of *Ginkgo biloba* bearing dwarf shoots and seeds

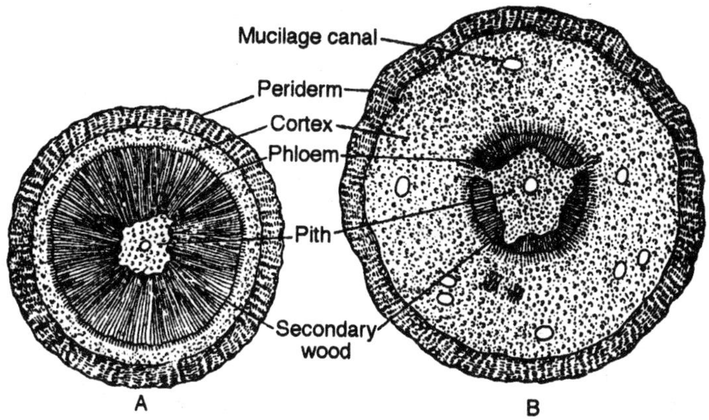

Fig. 49.3: *Ginkgoo biloba.* T.s. of long shoot stem (A) and of spur stem (B)

49.3B). The long shoot has a comparatively smaller pith and cortex, its wood is harder and there are not as many mucilage canals. *Mucilage canals* are found in the pith, primary cortex, root, petiole, leaf blade, seed with its peduncle and even the sporangium. The stem anatomy conforms to the coniferous type with its small *pith* and *cortex*, persistent *cambium*, large *secondary wood* and *primary endarch solenostele* of collateral bundles. *Protoxylem* tracheides are spiral. The tracheides of the secondary wood are with bordered pits occurring irregularly in one or two rows and showing 'bars of Sanio'. *Medullary rays* are characteristically short in vertical extent, many of them being only one cell in width and almost always one cell in height. There are no parenchyma cells in the wood excepting the medullary rays. The wood shows feeble *annual rings* which are not as well developed as in *Pinus*. There is the development of a distinct *periderm*.

In the seedling, the cotyledon shows a mesarch bundle, a cycadean character. This character is lost in the *leaves* where two exarch, collateral *vascular bundles* covered by a *sheath* enter each petiole and ultimately dichotomises in the *veins*. The spur leaves do not contain *palisade cells* but those on the long shoots (Fig. 49.4) have them. The upper *epidermis* is continuous while stomata are present in the lower. The lower *mesophyll parenchyma* cells are placed parallel to the leaf surface forming a *transfusion tissue*. The *vascular bundles* (veins) are closed collateral with the xylem on top. The mesophyll also shows prominent *mucilage canals*.

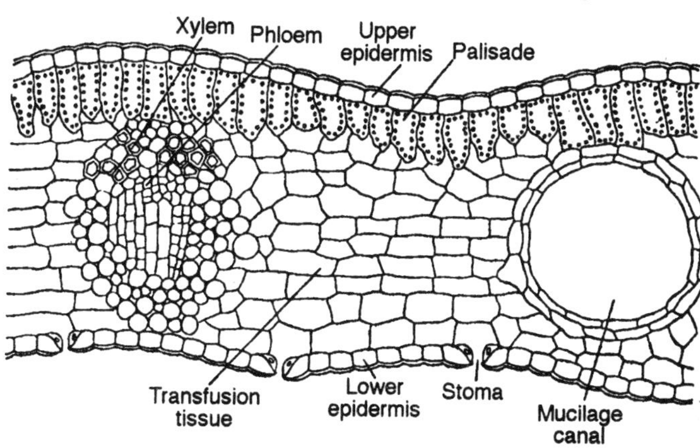

Fig. 49.4: T.s. of *Ginkgo biloba* long shoot leaf

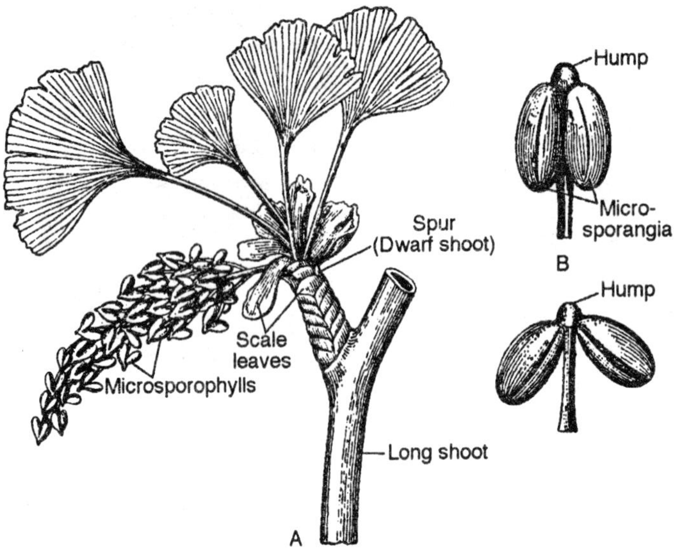

Fig. 49.5: *Ginkgo biloba*. A. A male dwarf shoot. B. Microsporophylls from A (magnified)

The *root* is diarch with radial xylem and phloem. The stele is surrounded by a pericycle, an endodermis and large cortex. Secondary growth takes place here also.

Ginkgo is dioecious, there being strictly male and female trees. Male and female strobili (Figs. 49.5A and 49.6A) are borne in the axils of the scale leaves in the crowns of the dwarf shoots forming clusters there.

The *male strobilus* (Fig. 49.5A) consists of an axis bearing spirally a large number of elaborate microsporophylls. The slender *microsporophyll* (Fig. 49.5B) is surmounted by a *hump* containing a large mucilage cavity and bears two pendent *microsporangia*. The hump is considered by some as a sterile third microsporangium capable of developing a fertile one. The microsporophyll, in that case, is peltate as in *Taxus*. Each microsporangium develops in the eusporangiate way, an *archesporial cell*

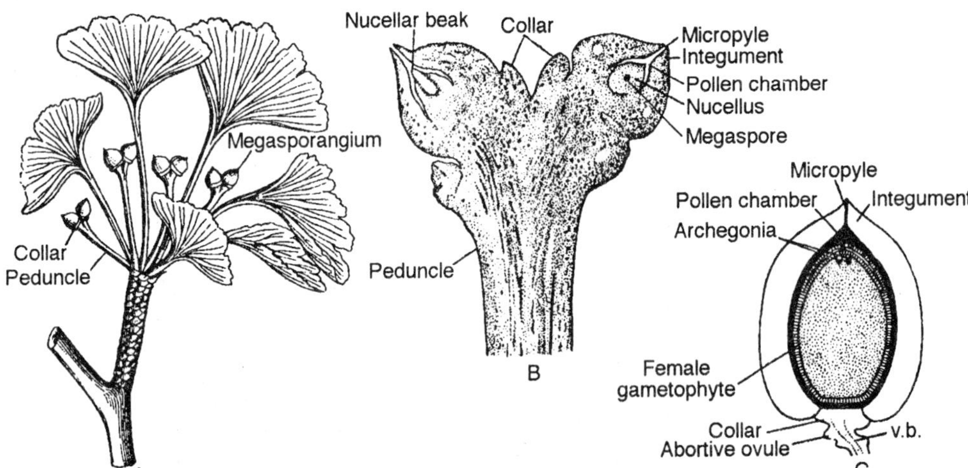

Fig. 49.6: *Ginkgo biloba*. A. A female dwarf shoot, B. Median l.s. of a female axis (strobilus) showing megasporangium (ovule) structure, C. Median l.s. of ovule. The integument shows three layers

giving rise to a *primary wall cell* and a *primary sporogenous cell*, the latter giving rise to a *sporogenous tissue*. The *spore mother cells* give rise to the *microspores* or pollens on reduction division. The *n* chromosome number is 8. Each *pollen* is covered by an intine and an exine but the latter is incomplete so that two lateral ears are seen protruding from the pollen (Fig. 49.7A). The microsporangium ruptures by a longitudinal split and the pollens are discharged.

In the *female dwarf shoot* the *peduncles* bearing the ovules are borne in good numbers on such a spur shoot as in the case of the male strobili (Fig. 49.6A). In the young stage a peduncle bears two *megasporangia* or *ovules* (Figs. 49.6A and B) but, usually one aborts and only one matures (Fig. 49.2). The young ovules become mature while still covered by the bud scales and then break through them. Each *ovule* is surrounded by a *collar*-like growth which is supposed by some to represent the sporophyll a structure otherwise missing in *Ginkgo*. As the peduncle bears four vascular traces, it is probable that it represents the axis of a *female strobilus* bearing two *megasporophylls*. The *ovule* (Fig. 49.6C) generally resembles *Cycas*. The prominent *nucellus* has a large *pollen chamber* surmounted by a beak which becomes hard. The nucellus is free from the *integument* at top. The single *integument* has an outer green fleshy layer, a thin hard middle part and a thin fleshy inner part which later becomes papery. The two *vascular strands* supplying the base of the integument do not branch. There is a *micropylar opening* on top of the nucellus. One or two *megaspore mother cells* are formed but only one of them undergoes reduction division forming a *linear tetrad* of *megaspores* of which only one matures and becomes functional.

The gametophytes

MALE GAMETOPHYTE: The *pollen* begins to germinate while still within the microsporangium. It divides to form a small *first prothallial cell* and a large *antheridial initial cell* (Fig. 49.7A). The antheridial initial cuts off a *second prothallial cell* which persists while the first prothallial cell disintegrates early (Fig. 49.7B). The antheridial cell now divides to form a *generative cell* (Fig. 49.7C). Usually the microsporangium dehisces and the pollens are shed at this 4-celled stage. The wind-carried pollens are deposited on the micropyle of the ovule whence they are sucked down along with the drying up of a drop of liquid (*pollination drop*) and ultimately deposited in the pollen chamber. Here the bulging intine (which part was never covered by the exine) on the pollen protrudes as a pollen tube and becomes

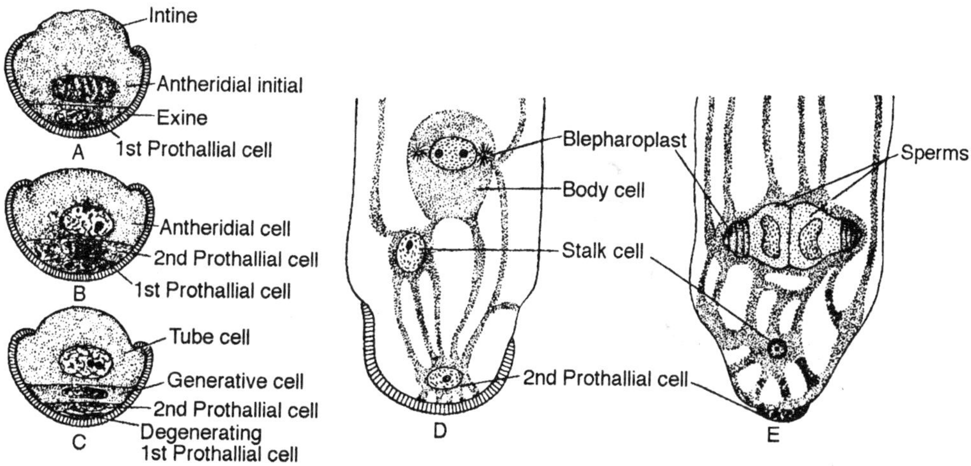

Fig. 49.7: *Ginkgo biloba*. A-E. Stages of development of male gametophyte

anchored in the tissue of the nucellus where, as in *Cycas*, it acts as a haustorium, the original function of a pollen tube. The sepermatogenous end of the pollen tube now advances towards the female gametophyte breaking down and absorbing the tissue that comes on the way. Ultimately the pollen chamber is enlarged so that nothing remains between the pollen tubes and the female gametophyte. Then the generative cell divides into a *stalk cell* and a *body cell* (Fig. 49.7D). The body cell divides to form two multiciliated *sperms* just as in *Cycas* and the *blepharoplasts* behave in the same way (Fig. 49.7E). The sperms are slightly more elongated than in *Cycas* and the spirally arranged cilia are more restricted to the apical region.

FEMALE GAMETOPHYTE: The functional megaspore (Fig. 49.6B) enlarges and the nucleus divides by rapid free nuclear division. Cell wall formation then begins at the periphery and gradually proceeds towards the centre. A spongy nutritive tissue is formed round the growing gametophyte as in *Cycas*. The gametophyte becomes covered by a thin membrane in addition to the megaspore membrane. The gametophyte is soon filled with an *endosperm* tissue which develops chlorophyll at a later stage as the integument is more or less transparent. *Archegonium* initials appear long before wall formation in the gametophyte has reached the centre. Usually two (rarely three) archegonia are formed in the top part of the gametophyte above the general endosperm. At first a two-celled *neck* and a rapidly increasing *central cell* are formed. The central cell nucleus divides to form a *ventral canal nucleus* and an *egg nucleus* between which a definite wall is seen (Fig. 49.8A). In this respect *Ginkgo* seems to be more primitive than *Cycas* where no such partition wall is formed. In *Pinus* there is an ephemeral partition wall. Further, instead of the cup-shaped *archegonial chamber* of the Cycads, *Ginkgo* has a somewhat cylindrical chamber surrounding a mass of solid gametophytic tissue upon which the nucellus wall rests like a tent on a pole (Fig. 49.8B).

Pollination and *fertilization* take place as in *Cycas*. The process is slow although not as slow as in *Pinus*. The strobili appear in February-March, pollination takes place about May and fertilization about the end of September.

The embryo sporophyte and seed

Fertilization is followed by free nuclear division which usually takes place eight times giving rise to 256 free nuclei evenly placed in the egg cytoplasm when cell walls are formed simultaneously all through. The embryonic tissue now faintly shows three zones: the

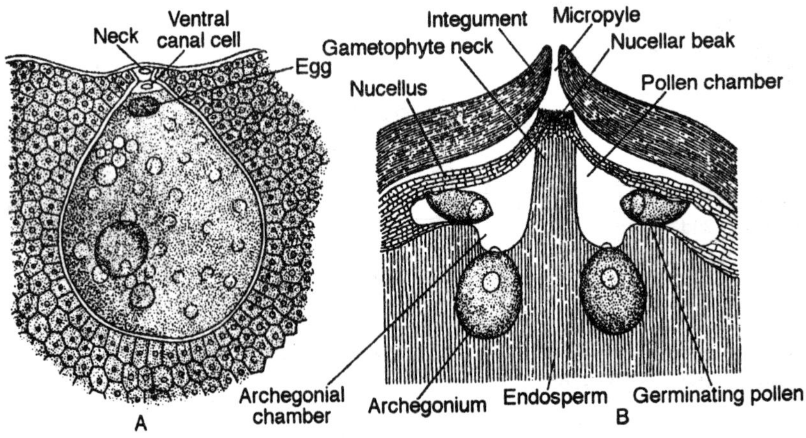

Fig. 49.8: *Ginkgo biloba*. A. An archegonium, B. Median l.s. of top of ovule during fertilization

micropylar cells are somewhat elongated, the middle cells remain large, the basal cells become smaller and more compact by rapid division. Subsequent growth takes place only at the base which rapidly forms two *cotyledons*, the *radicle* and the *plumule* containing rudiments of five leaves (Fig. 49.9). Normally there is only one embryo in a seed. The seeds (Fig. 49.2) are quite large (about 1 inch long) and the outer fleshy layer of the seed coat (mature outer layer of integument) contains lot of mucilage. The seed germinates quickly. Germination is hypogeal. The terminal parts of the cotyledons remain inside the seed until all the endosperm (gametophytic tissue) is absorbed. The cotyledons come above surface only after the radicle has developed the tap root.

Economic importance

The seed is edible and yields an oil. The timber is of commercial value. In temperate America it has been recommended as a hardy roadside tree.

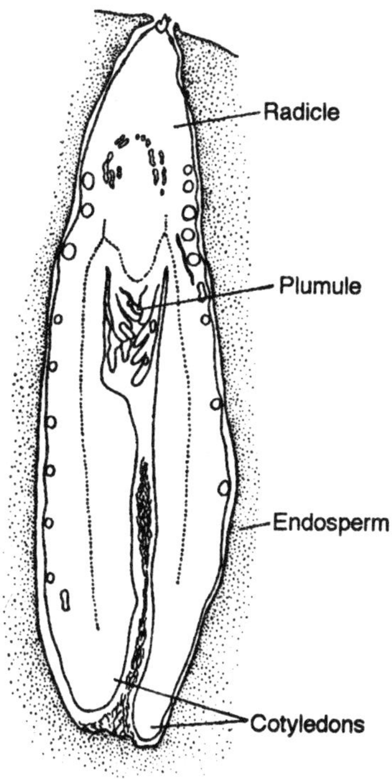

Fig. 49.9: L.s. of embryo in *Ginkgo biloba* seed

AFFINITIES OF THE GINKGOALES

The Ginkgoales are included within the Coniferopsida because of the coniferous habit of the tree, the coniferous wood and the gametophytic structures. But, it does not seem to be as close to the Coniferales as the Cordaitales although, possibly, the Cordaitales had motile spermatozoids in common with the Ginkgoales. Nevertheless, the Ginkgoales have been placed within the Cycadopsida in Engler's Syllabus (1954) as the ovule and the manner of fertilization resembles *Cycas* and because, if the collar of the ovule be regarded as a modified sporophyll, the ovules becomes phyllospermous. Coulter and Chamberlain (1910) were impres-sed by the cordaitean affinity of the Ginkgoales and suggested that either the Ginkgoales originated out of the Cordaitales or both the orders evolved from a common ancestor. Arnold (1947) suggested a Pteridospermous origin because of the phyllospermous ovule and the fern-like leaf, seeds, motile sperms and double leaf traces. But it is quite possible that the origin of the Ginkgoales lies even beyond them and still undiscovered, may be near the Pteridospermic group of Progymnosperms. Because of this uncertainty, although the majority opinion is in favour of a coniferous affinity, there are quite a number who think that *Ginkgo* should be in an independent division or class (Ginkgophyta or Ginkgopsida) representing a separate line of evolution of seed plants in parallel with the Cycads and the Conifers as *Ginkgo* shows some characters common to each of the two groups.

Order V. CZEKANOWSKIALES

Czekanowskia is a fossil of leafy shoots from the Jurassic and was supposed to belong to the Ginkgoales. It is a short shoot with scales on the lower parts and a cluster of about 10 leaves at the tip. The leaves are about 15 cm long, dichotomise 3 or 4 times and the ultimate segments are very narrow (only 1 mm

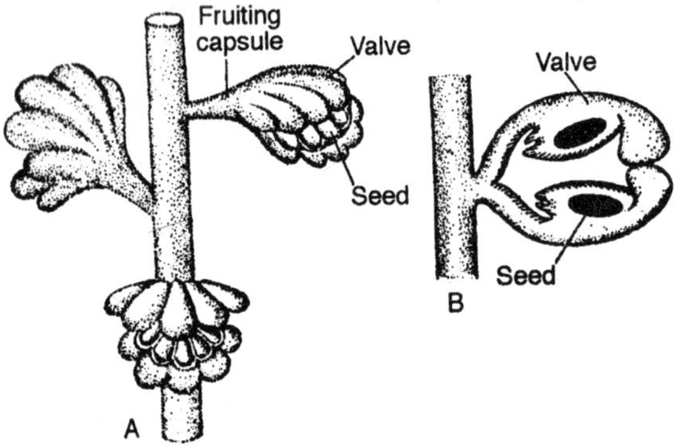

Fig. 49.10: *Leptostrobus longus* (fruiting shoot of *Czekanowskia*) (after Harris). A. Part of the shoot showing three fruiting capsules, B. T.s. through a fruiting capsule

broad) bearing single veins. While *Czekanowskia* was considered as a member of the Ginkgoales, Harris (1951) described *Leptostrobus* (Fig. 49.10), the seed bearing structure found in association with *Czekanowskia* leaves. This seems to be a fertile shoot with some scale leaves at the base and fruiting capsules arranged on the axis at an interval of 5 mm. Each such fruiting capsule shows a pair of lobed valves up to 5 mm long and 5 mm broad with 3 to 5 seeds borne on the inner face of each valve. Such a peculiar seed structure is so different from the Ginkgoales that these fossils seem to represent a quite different group of Gymnosperms and have been placed at least in a separate order, Czekanowskiales (with the single family Czekanowskiaceae) which some suppose to have the same relationship with the Ginkgoales at the Taxales has with the Coniferales.

Chapter 50

Chlamydospermopsida

AFFINITIES OF THE GROUP

The three genera *Ephedra, Welwitschia* and *Gnetum* present such advanced features among Gymnosperm, including the apparent presence of a *chlamydon* or perianth that they have been placed in a group, the Chlamydosperms, representing the most advanced Gymnosperms. The three genera show uniformity in (1) the compound male and female strobili representing inflorescences of flowers, (2) the two apparent integuments of which the inner forms a long micropylar tube, (3) the opposite decussate leaves, (4) the apparent vessels in the stems, (5) the absence of resin canals, and (6) the dicotyledonous embryos. But, otherwise, the three genera are so different that it is difficult to consider them as members of the same family although Chamberlain treated them as members of the single family Gnetaceae. The more general view, however, is to consider the three genera as three separate families—Ephedraceae, Welwitschiaceae and Genetaceae. Again, in certain other characters the diffe-rences between these three are so great that it has been proposed (viz., Eames 1952) to raise them to three orders: Ephedrales, Welwitschiales and Gnetales. Of these three, *Ephedra* shows certain characters which mark it out from the other two: (1) Its stomata are *haplocheilic* in common with the Pteridosperms, Cycadales, Cordaitales, Coniferales and Ginkgoales while those of *Welwitschia* and *Gnetum* are *syndetocheilic* (but haplocheilic in *Gnetum ula*, c.f. page 1123) as found in the Cycadeoidales and Pentoxylales. (2) According to Eames (1952) the *Ephedra* ovule is not cauline but appendicular while it is not so in the other two genera. (3) The male gametophyte development in *Ephedra* conforms to the other Gymnosperms while it is quite different in *Gnetum.* (4) *Ephedra* embryosac is monosporic as in other Gymnosperms but in *Gnetum* it is tetrasporic. (5) The archegonium and embryogeny in *Ephedra* is as in the other Gymnosperms while it is different in the other two genera consequent on the suppression of the archegonia. (6) Anatomical features like the stem apex and the leaf trace number are different in *Ephedra* and *Gnetum.* Such differences have lead Eames (1951) and Wardlaw (1955) to conclude that the Ephedrales are nearer to the Cordaitales and Coniferales while this cannot be said about the other two orders. Because of this, Pant (1957) has kept the Ephedrales within the Conifers while only Welwitschiales and Genetales are considered as Chlamydosperms. There are good reasons for coming to such a conclusion but Coulter and Chamberlain (1910) are of view that, however nearer to the true Gymnosperms may *Ephedra* be, it has so many features common with the other two genera that there is no justification in not considering the three genera to form one group. Provisionally, however, they are placed in different orders.

As to the origin of the Chlamydosperms the question is still open for lack of sufficient evidence. Fossil evidence is almost non-existent. Fossil *Gnetum* is now known. *Ephedra* and *Welwitschia* are known only from fossil pollens. *Ephedra* pollens are known from Createaceous and Tertiary while *Welwitschia* pollens are known from Tertiary. Recently Wilson (1959) has reported fossil pollens *Ephedripites* and *Vittatina* from Permian in Oklahoma and these are supposed to be related to *Ephedra* and *Welwitschia*. From morphological evidences as explained above it has been concluded that *Ephedra* arose out of the Coniferopsida stock although it is not possible to point out the exact origin. But the relationship between the other two genera and *Ephedra* is still not clear and the origin of *Gnetum* may be elsewhere. *Gnetum* differs from the true Gymnosperms in many characters. Lam (1955) suggested a *Proto-angiosperm* stock separating out of the Psilophytan stock earlier than the evolution of the Coniferopsida. This line, according to him, gave rise to both the Chlamydosperms and the Angiosperms. On the other hand, the similarities of the Chlamydosperms with the Cycadales have been stressed by Thoday and Takhtajan has supposed its origin from the Cycadopsida. On the whole, however, we may consider the origin of the Gnetales yet unsolved.

The Chlamydosperms show certain affinities with the Angiosperms. Arber and Parkin (1908) supposed that Chlamydosperms and Angiosperms arose from a common stock—the 'Hemiangiosperms'. Hagerup (1934) opined that the Piperaceae of Angiosperms arose directly from this group. Fagerlind (1946) also derives the view that the Chlamydosperms arose from an ancestral stock which also gave rise to 'Proangiosperms', the ancestors of the Angiosperms. The Chlamydosperm side line had two arms—one arm gave rise to *Ephedra* while the other arm gave rise to *Welwitschia* and *Gnetum* (Fig. 50.1). The affinities with the Angiosperms are more pronounced in the genus *Gnetum* and Thompson (1916) said that the ancestor of the Angiosperms was a relative of this genus. The Angiospermic affinities of the Chlamydosperms as a whole may be stated as follows:

(1) The leaves are opposite, scaly in *Ephedra*, parallel veined like monocots in *Welwitschia* (though with unusual, indefinite growth) and net-veined like dicot leaves in *Gnetum*.

(2) The growing stem apex of most species of *Ephedra* and *Gnetum* shows tunica and corpus as in the Angiosperms.

(3) True vessels are present in the secondary xylem along with tracheides. But the vessels of *Gnetum* and the vessels of Angiosperms, though similar when mature, are different in ontogeny. The vessels of Angiosperms (as well as those found in *Selaginella* and *Peridium*) originate first by the loss of membranes from transversely elongated bordered pits on the cross-walls of tracheides and then these slit-like perforations give rise to narrow scalariform perforations which finally lose the bars between the slits and become completely open. In the Chlamydosperms on the other hand round bordered pits on the cross-walls lose their membranes and coalesce to form single large perforations. Because of this some authorities doubt whether the presence of vessels in the Chlamydosperms may be considered as an

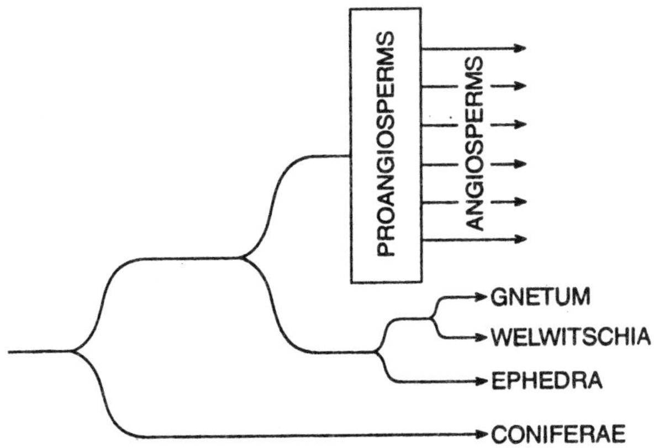

Fig. 50.1: Fagerlind's concept of the affinities of the Chlamydosperms

Angiospermic affinity or is merely a case of parallel evolution. But, it has been found that the vessels in Rosaceae, Ranunculaceae and Geraniales are of both *Gnetum* and true Angiosperm types. Thus, it may not be so important to distinguish between the two types of vessels.

(4) Staminate and ovulate strobili are compound and represent the inflorescences and flowers of Angiosperms.

(5) There is a 'perianth' structure associated with the flowers and there are two apparent integuments. The nature of the 'perianth' however may not be different from the bracts of other Gymnosperms and the integuments may not be really double. These points are later discussed in connections with each genus.

(6) A long micropylar tube is present in each genus. In *Gnetum* pollen grains are found lodged in it and it has been argued that this tube is an approach to the Angiosperm carpel structure developing a style.

(7) Male and female gametophyte development approaches the Angiosperms. In *Welwitschia* and *Gnetum* the archegonia are eliminated. The stalk cell in the male gametophyte is also eliminated in *Welwitschia* and *Gnetum*.

(8) In embryogeny the free nuclear stage, still persisting in *Ephedra*, is eliminated in *Welwitschia* and *Gnetum*.

(9) A dicotyledonous embryo is a regular feature in the Chlamydosperms but this feature is also present in some other Gymnosperms.

A general conclusion is that the Chlamydosperms, specially *Welwitschia* and *Gnetum*, do show some Angiosperm characters. But the character are not such as to allow one to suppose that the Angiosperms arose out of them. A more cautious and logical conclusion is that the Chlamydosperms arose from a stock of ancestral plants related to the Gymnosperms out of which the Angiosperms arose. Among the Chlamydosperms, Ephedrales is closer to the Gymnosperms whereas the Welwitschiales and Gnetales represent another arm of evolution in this group which shows more Angiosperm characters. Thus, it is easy to conclude that the Angiospermic characters in the Chlamydosperms arose as a result of parallel evolution and not direct relation.

There are reports that the chlamydospermopsida often designated as Gnetophyta with three distinct orders viz. Ephedrales, Welwitschiales and Gnetales.

Order I. **EPHEDRALES:** *EPHEDRA*

The Ephedrales consist of a single family Ephedraceae with a single genus *Ephedra* composed of about 42 species almost equally divided between the Old and the New Worlds in the temperature-subtropical regions. *Ephedra* is usually a low, profusely branching shrub of xerophytic habit growing in exposed situations on dry rocks and even sandy deserts (including saline tracts) in which latter situation it acts as a sandbinder. Sometimes the branches may be long trailers or climbers. In the Old World they spread from the Canary Island and Madeira in the Pacific on the west along the Mediterranean coasts up to the Black Sea occurring as far north as the Alps in Switzerland and western France in Europe. From the Mediterranean region it crosses into Asia over the dry tracts of Arabia and Iran to the north-west of the Indian subcontinent (Sind, Rajasthan, Punjab, Kashmir and Western Himalaya). It then spreads along the Himalaya along the drier ridges up to Sikkim. Across the Himalaya it passes over Western Tibet to China along the upper basins of Yang-tse-Kiang and Hoang Ho rivers finally reaching as far north as the Siberian Amur basin. In the New World it is found along the drier western coast. From California it spreads to Arizona, New Mexico (only one species spreading as far east as Taxas), Mexico then spreading along South American Andes to Bolivia, Paraguay, Argentina, Chile and Patagonia. No species is found to be common between the Eastern and the Western or the Northern and the Southern Hemispheres. Such world-wide distribution naturally points to great antiquity although fossil history is not at all clear. However, there are some pollen fossils. Pollens ascribable to *Ephedra* are known down to the Createaceous while one type of pollen fossil from the Permian, *Ephedripites*, was probably related to *Ephedra*.

Hooker recorded three species of *Ephedra* from India: *Ephedra vulgaris* from subalpine Himalaya spreading from Western Tibet (2-4000 m) to Sikkim (3500-5000 m); *E. pachyclada* from Chitral and Western Tibet in alpine regions up to 5000 m which is more robust than *E. vulgaris*; and *E. peduncularis* from Sind, Rajasthan and Punjab which is a tall climber often developing filiform leaves. These species are with sessile or subsessile anthers. The species have been re-examined and revised so that the following six Indian species are recognized at present: (1) *Ephedra foliata* Boiss. from Sind, Punjab and Rajasthan; (2) *E. intermedia* Schrenk var. *tibetica* Stapf. from Kashmir, Punjab, Himachal and Jaunsar; (3) *E. gerardiana* Wall. from dry alpine and temperate Himalaya spreading from Kashmir to Sikkim; (4) *E. saxatilis* Royle var. *sikkimensis* (Stapf.) Florin from Sikkim; (5) *E. nebrodensis* Tin. var. *procera* Stapf. from Kashmir; and (6) *E. regeliana* Florin (*E. major*) from Ladakh and Lahul.

Ephedra is the source of the valuable drug *Ephedrine* which is extracted from the Asiatic species *gerardiana, regeliana, equisetina* and *sinica.*

The sporophyte plant

The plant is a straggling woody shrub usually up to 2 metres high (may be much longer or shorter) with characteristic jointed, fistular and green stems. Branching begins early at the cotyledonary stage. Some of the axillary buds develop into rhizomes which may give rise to new plants if they come in contact with soil. The internodes grow by a meristematic zone at the base of each which dries up at the end of each growing season causing brittleness and shedding of branches. New axillary shoots develop in the next season. The apical buds of branches contain four or five undeveloped internodes covered by scale leaves.

The leaves are very small, scaly, and so rudimentary as to be of no use to the plant (Fig. 50.2A). They are cyclic, usually opposite decussate, but may be in whorls of three or, rarely, even four. Each scaly leaf shows a thick midrib with a thin border (which may scarcely be called a blade) which very rarely shows any stoma. Leaves of a whorl are fused at the thicker base and branches arise from the axils.

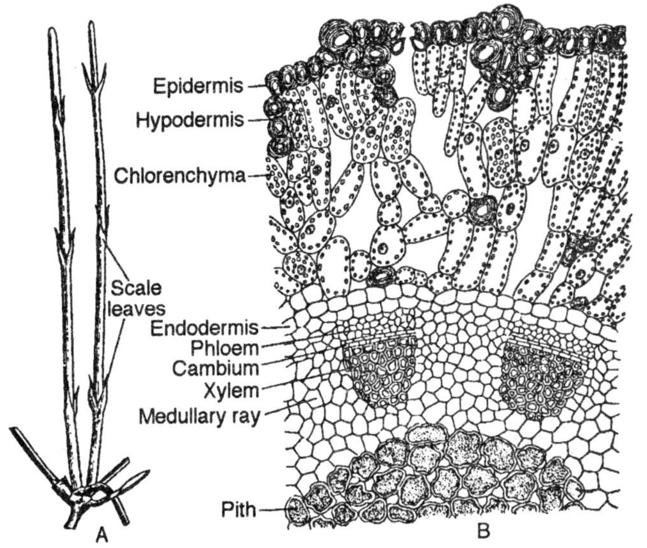

Fig. 50.2: *Ephedra*. A. A bit of sterile shoot showing scaly leaves, B. T.s of young stem (primary)

An early abscission layer causes shedding of the leaf leaving a broad scar. Every leaf is supplied at the base by a double leaf trace.

The stem is the photosynthetic organ and may be called a phylloclade. It is green when young and resembles *Equisetum* both externally (with ridges and furrows) and internally (with prominent chlorenchyma). A t.s. of the stem (Fig. 50.2B) shows an epidermis of thick-walled cells with a thick cuticle. A group of very thick-walled cells (fibrous) under each ridge form a *hypodermis*. Sunken stomata are distributed at the bottom or along the sides of the furrows between the ridges. A *cortex* of thinwalled chlorenchymatous cells, often radially elongated, contain abundant chloroplasts and intercellular spaces. This spreads up to the stele and is internally delimited by an *endodermis* which is not always clear. The *vascular cylinder* is an endarch siphonostele. In each *vascular bundle* there is an external phloem and an internal xylem with the protoxylem at the base and there is a cambial tissue between the xylem and the phloem. This primary cambium is persistent and gives rise to a thin *secondary wood*. The xylem has ordinary coniferous tracheides mixed with broad vessels formed as already explained. Most of the wood, however, consists of tracheides with bordered pits. *Medullary rays* are very long and wide. Annual rings are clear, no vessel being formed during the winter. The *pith* in the centre is a group of rather thick-walled cells with very dense contents (probably tannin). Resin canals are absent but cells with stellate calcium oxalate crystals are present. At the end of the growing season a diaphragm-like transverse abscission plate at the base of each internode adds to the brittleness of the branches.

Ephedra is normally dioecious though abnormal monoecious plants are known. Both male and famale flowers are borne in compound inflorescences.

The *staminate strobili* are borne in whorls at the axils of scale leaves at each node of the fertile branch (Fig. 50.3A). If the leaves are opposite decussate they are also so. Each strobilus (Fig. 50.3B) consists of a short axis bearing 2 to 8 opposite decussate pairs of bracts. Of the bracts the lower one or two pairs are sterile while all the upper ones bear solitary flowers at their axils. The staminate strobilus is, therefore,

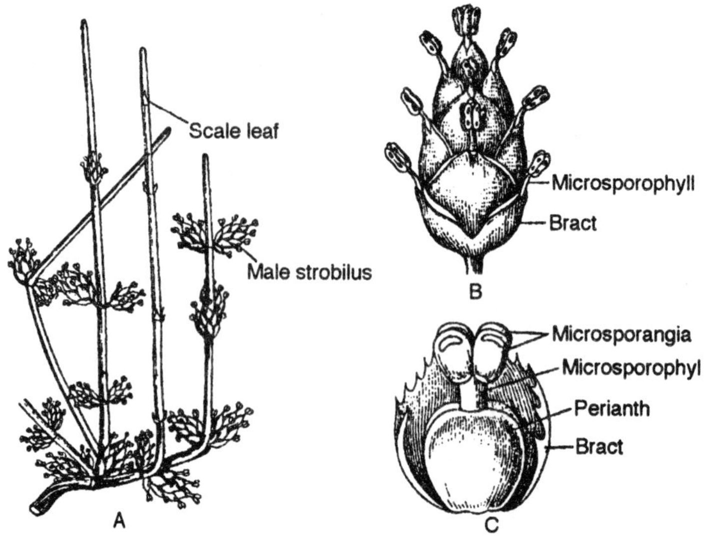

Fig. 50.3: *Ephedra*. A. A staminate shoot, B. A male strobilus, C. A male flower

a compound structure or an inflorescence. The axillary flower (Fig. 50.3C) has an axis which may be called a *sporangiophore*. The sporangiophore is clasped at the base by two anterior-posterior scaly bracteoles which are united at the base and possess no vascular tissue. These bracteoles have been termed the *perianth*. The sporangiophore bears at its tip two to eight 2-3 locular, sessile or stalked *microsporangia* or anthers. The sporangiophore has been supposed to be a branched *sporophyll* (the filaments of the anthers, if present, being the branches) or as an axis bearing several sporophylls (each anther with filament is then a sporophyll). Eames is of opinion that the sporangiophore is formed by the fusion of two sporophylls. The

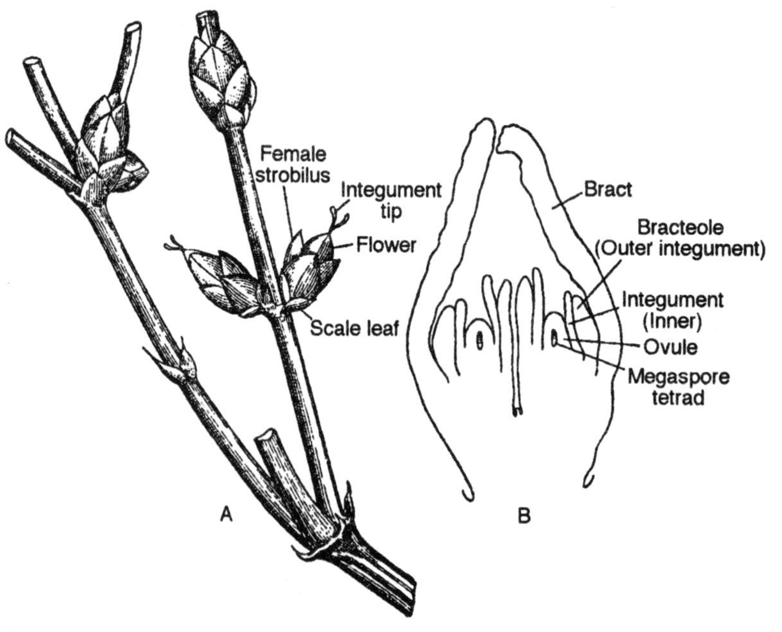

Fig. 50.4: *Ephedra*. A. An ovulate shoot, B. Median l.s. of the top of the ovulate shoot showing the ovules

microsporangia develop in the eusporangiate way. The microsporangial wall is thin and the tapetal cells are large. The pollens are round or elliptical. The microsporangium loculi dehisce by transverse or oblique slits.

The *ovulate strobili* (Fig. 50.4A) are similarly borne but they are elongated and pointed while the staminate ones are globular. In this strobilus most of the lower pairs of bracts (four or more pairs) are sterile there being two female flowers in the axils of the uppermost pair of bracts (Fig. 50.4B). Often one of the two flowers aborts and the other takes up a false terminal position. Each flower consists of an ovule surrounded, apparently, by an outer and an inner integument. But, the outer integument is formed of four coalescent segments with four rudimentary veins and it is considered as a fused pair of bracteoles forming a *perianth* as in the male flower. The inner integument also is formed by the fusion of a pair of segments showing two rudimentary veins. The tip of the inner integument elongates immensely by the time of pollination (Fig. 50.6A) forming a sort of micropylar tube which is often twisted spirally and exudes a pollination drop catching the pollens. The inner integument is free from the nucellus of the ovule for more than half the latter's length and is also completely free from the outer integument.

The gametophytes

MALE GAMETOPHYTE: The *microspore* or the pollen cell divides and cuts off the *first prothallial cell* (Fig. 50.5A). A second division cuts off a *second prothallial nucleus* (Fig. 50.5B) but no wall is cut off between this nucleus and the other which is the *antheridial initial.* The latter divides to form a *tube nucleus* and a *generative cell* (Fig. 50.5C). The generative cell next divides to form a *stalk nucleus* and a *body nucleus* which remain in a common mass of cytoplasm (Fig. 50.5D). The pollen is shed at this stage. On reaching the pollen chamber the

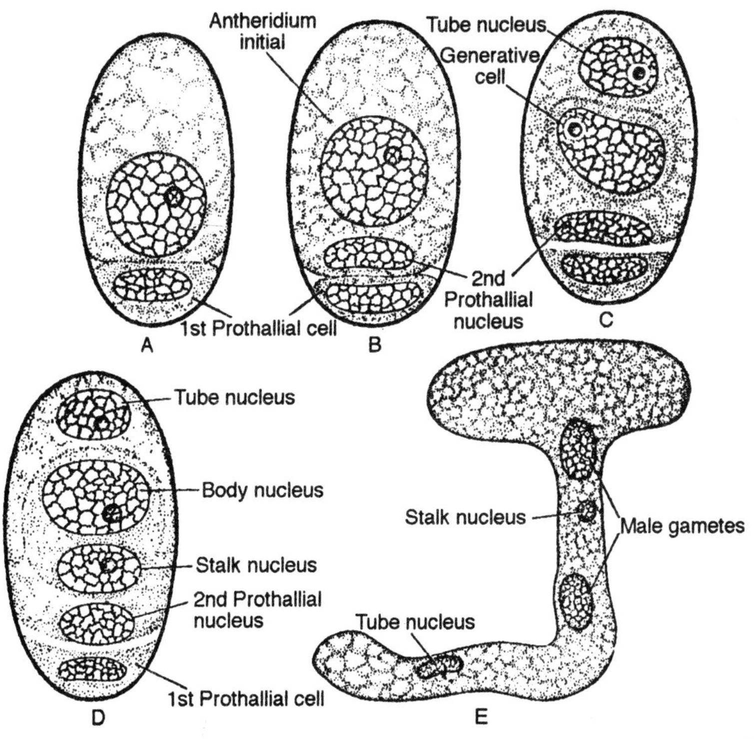

Fig. 50.5: *Ephedra*. A-E Stages of development of the male gametophyte (after Land)

exine of the pollen cracks off and the body nucleus divides to form two *male gametes.* The intine now forms a pollen tube from the base of the pollen and the two male nuclei lie in its general cytoplasm (Fig. 50.5E). The prothallial nuclei get disorganized by this time.

FEMALE GAMETOPHYTE: A single hypodermal *archesporial cell* in the nucellus of the ovule gives rise to a single *megaspore mother cell* (at which stage it takes up a more deep-seated position) which on reduction usually develops a linear tetrad of *megaspores.* The lowest megaspore is functional while the other three disintegrate. This megaspore gives rise to the female gametophyte. The germination of the megaspore starts with a stage of free nuclear division round a central vacuole ultimately giving rise to 256 or more nuclei is about 20 days when cell wall formation starts and proceeds from periphery to centre. The tissue of the female gametophyte is gradually differentiated into three zones—an upper reproductive zone of loose, very thin-walled cells where the archegonia are later formed; a lower storage zone (*endosperm*) of more compact cells full of food reserve; and a basal haustorial zone also of compact cells (Fig. 50.6A). In the reproductive or micropylar zone usually two (rarely one or three) superficial cells become the *archegonial initials.* The initial

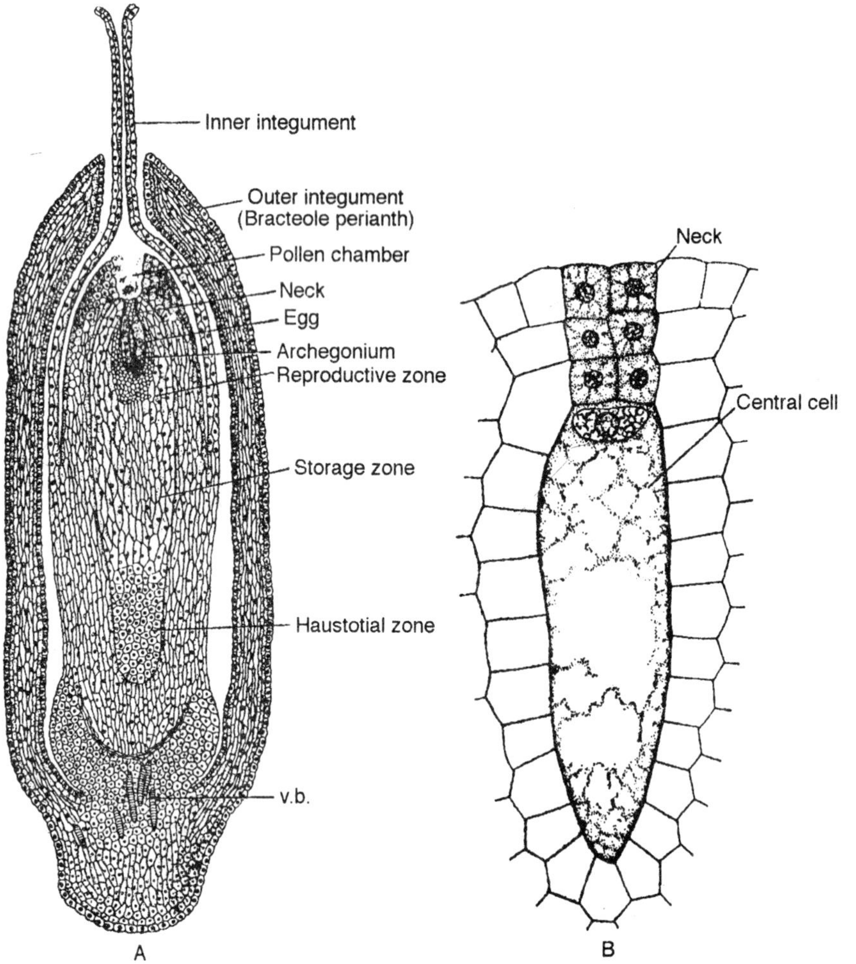

Fig. 50.6: *Ephedra.* A. Median l.s. of a mature ovule, B. L.S. of a young archegonium (after Land)

form a *primary neck cell* and a *central* ...ne primary neck cell divides to form ...nt tiers of neck cells (about 32 cells in all). The formation of the neck cells is very regular at first (Fig. 50.6B) but later becomes irregular so that the neck cells can no longer be distinguished from the surrounding gametophytic cells. The archegonial neck of *Ephedra* is the longest among the Gymnosperms and is associated with the deepest of pollen chambers. The central cell nucleus divides to form a *ventral canal nucleus* and an *egg*, there being no cell wall between the two. A zone of tapetal cells forms a jacket round the ventral region of the archegonium which now greatly elongated. Simultaneously with the development of the archegonium all the nucellar tissue above the gametophyte disintegrates developing a deep *pollen chamber*, the deepest among the Gymnosperms. The l.s of a mature ovule is shown in Figure 50.6A.

Pollination and fertilization

Pollination is carried on by wind in early spring when a sparkling pollinating drop is exuded by the long tubular tip (which is slit on one side) of the inner integument and the pollens get caught in this liquid. The pollens get sucked in with the drying of the liquid and soon rest at the bottom of the pollen chamber almost directly in contact with the archegonia when the pollen tubes are formed. The latter are necessary only for penetrating the neck. Fertilization takes place within 12 hours of pollination which is rather quick in comparison with the other Gymnosperms. The pollen tube discharges all the four (Fig. 50.5E) nuclei into the egg where both the ventral canal and the egg nuclei are present. The first male gamete fertilizes the egg forming the *zygote*. A very interesting feature has been observed in some cases (Herzfeld, 1922; Khan, 1943) where the second male gamete fuses with the ventral canal nucleus and a transient tissue is formed which disappears with the development of the embryo. This certainly recalls the 'double fertilization' and the 'endosperm' of the Angiosperms although it is not yet possible to be too definite about its evolutionary importance.

The new sporophyte

The zygote divides by free nuclear division to form eight nuclei which remain evenly distributed in the cytoplasm. Each nucleus then invests itself with cytoplasm and cell wall and becomes a *proembryo* (Fig. 50.7A) which is capable of developing an independent embryo. So, this is a distinctive type of polyembryony without a regular cleavage. Usually 3 to 6 embryos begin to develop (Fig. 50.7B). The proembryo cell develops a tubular *suspensor tube* inside which the nucleus divides. This nucleus may also divide earlier in other cases. One nucleus is cut off at the tip by a cell wall and becomes the *embryo initial* while the other remains in the suspensor cell (Fig. 50.7C). The embryo initial now begins division (Fig. 50.7D) ultimately giving rise to a multicellular *secondary suspensor* and a dicotyledonous *embryo* below. The suspensor and the secondary suspensor elongate and push down the growing embryo into the endosperm below. Although several embryos begins to devolop (which number may be up to 18 or 19 as observed by Khan and may even involve some cleavage polyembryony) usually only one attains maturity in one seed.

As the seed (Fig. 50.7E) matures, in addition to the integuments, the adjacent bracts of the female strobilus also become fleshy and become additional investments to the seed. Apparently there is no resting stage and the seed is capable of germination at once. The germination is hypogeous. The cotyledons persist as green photosynthetic organs for some time.

As stated before, the life history clearly shows strong gymnospermous characters with a few angiospermic features.

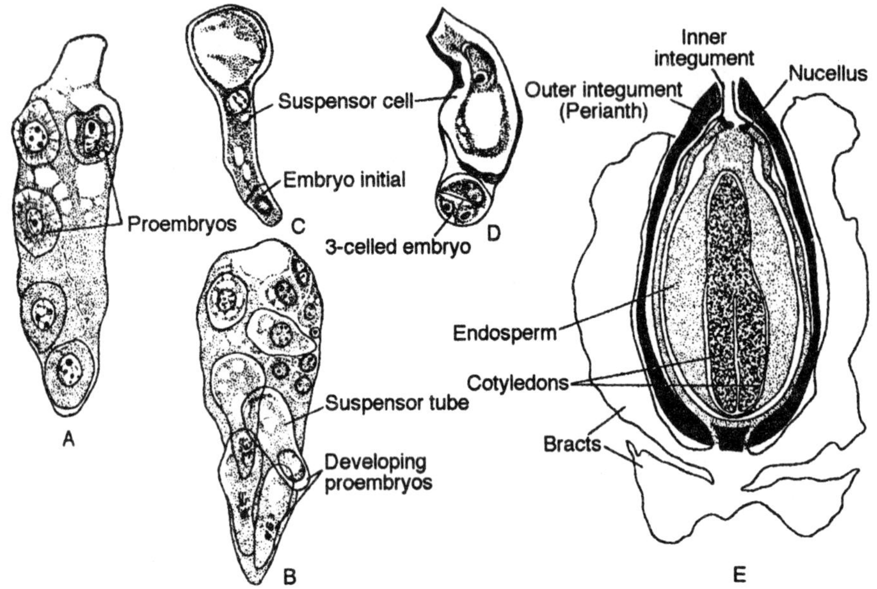

Fig. 50.7: *Ephedra foliata* (after Khan). A-D. Stages of development of embryo. A and B show whole embryosac, C and D show development of one proembryo, E. Median l.s. of seed

Order II. **WELWITSCHIALES:** *WELWITSCHIA*

The peculiar species *Welwitschia mirabilis* is the sole representative of the order Welwitschiales and is placed in the family Welwitschiaceae. As the plant is found only along the North and the South coasts of the Walvis Bay in Southwest Africa and in Angola, it is not so well known. The taxon was originally named as *Tumboa mirabilis* by its discoverer Welwitsch. Hooker, while describing it, renamed it as *Welwitschia mirabilis.* Although such renaming is against international rules, in this case Hooker's name has been internationally accepted (*nomina conservenda*).

Welwitschia mirabilis (Fig. 50.8) is an extreme xerophyte growing on rocky beds in areas with rainfall less than 1″ in a year. The stem has the shape of a transversely elliptical turnip which is bilobed, rarely more than 45 cm above ground and frequently almost covered except at the top. This stem may reach a diameter of more than 1 metre, suddenly tapers down and the extremely long tap root elongates to reach the water table below. Plants grow scattered on the dry sand or a few may grow in a cluster. The stem above the leaf is the 'crown' which has some semicircular ridges marked by scars of falling inflorescences. The complete stem is covered by a periderm which is the thickest at the crown. Thre is a single pair of opposite leaves which is the first and the last pair of leaves borne at right angles to the two cotyledons (seedling leaves). This pair persists throughout the life of the plant (more than 100 years). The leaves are entire until they are 15 to 20 cm long, after which they begin to become shredded longitudinally into ribbons and to dry and wear out at the tips while they are constantly renewed from the leaf bases which remain permanently meristematic. The leaves are thick and leathery, about 2 metres long and the venation is parallel with transverse connections here and there. They grow from the base of a groove which extends nearly halfway around, one on each side of the crown. A leaf shows a thick epidermis, sunken stomata in furrows, hypodermal sclerenchymatous patches, pallisade mesophyll, transfusion tissue and a

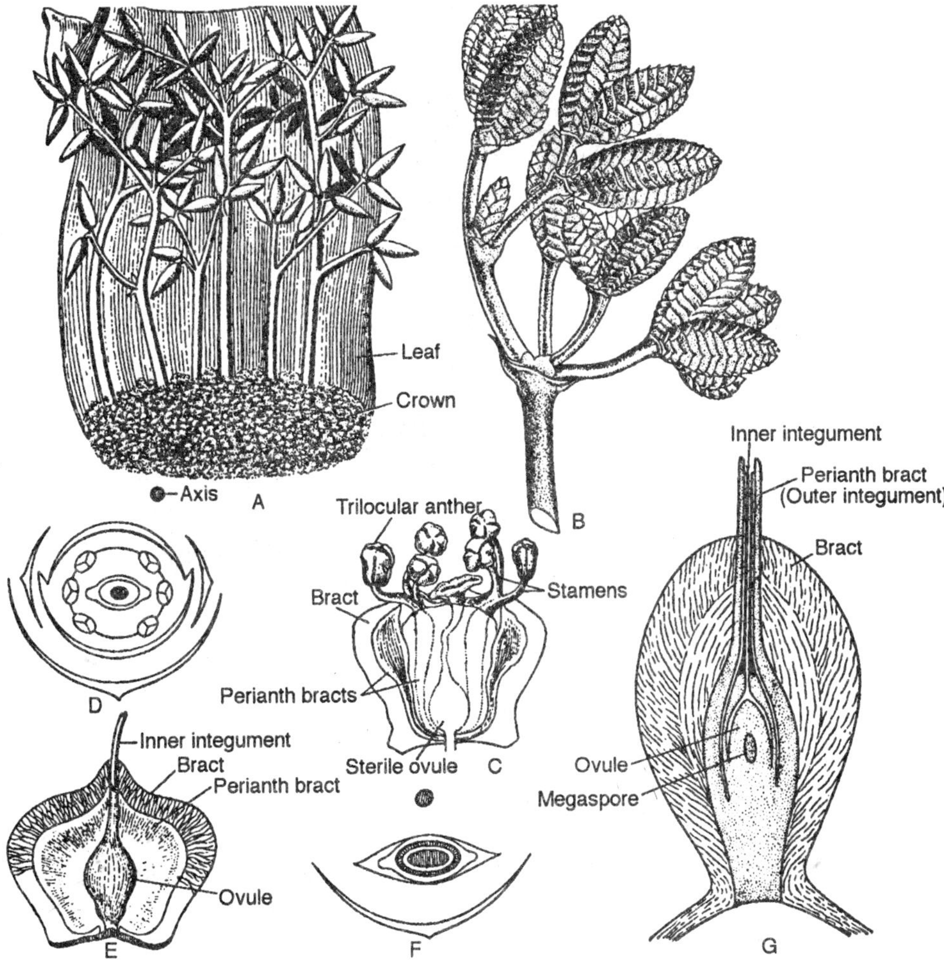

Fig. 50.9: *Welwitschia mirabilis*. A. Crown of plant showing male inflorescences, B. A. flower stalk with female inflorescences, C. A male flower, D. Floral diagram of male flower, E. A female flower, F. Floral diagram of female flower, G. Median l.s. of female flower

which develop broad but thin wings and closely invest the ovule even at maturity forming a samara-type fruit. These fused *perianth bracts* have been called the *outer integument* by certain authorities like Pearson. Inside this is the *integument* (or the *inner integument*) which is short at first but soon becomes much longer being drawn out into a long neck overtopping the perianth and with a notch at the apex (Fig. 50.9G). There is one hypodermal *archesporial cell* giving rise to a single *megaspore mother cell*. The megaspores develop as a linear tetrad inside the nucellus in the usual way and only the lowermost one is functional.

The gametophytes

It is in the gametophytes of *Welwitschia* that reduction towards Angiosperms is clearly noticed.

MALE GAMETOPHYTE: The microspore or the *pollen* develops only a single *prothallial nucleus* (not cell), a *tube nucleus* and a *generative nucleus* (Fig. 50.10A). The pollen is shed at this stage and on the top of the ovule develops a polen tube within which the generative cell divides to form two naked *male gametes*.

FEMALE GAMETOPHYTE: The megaspore nucleus divides by free cell formation to from

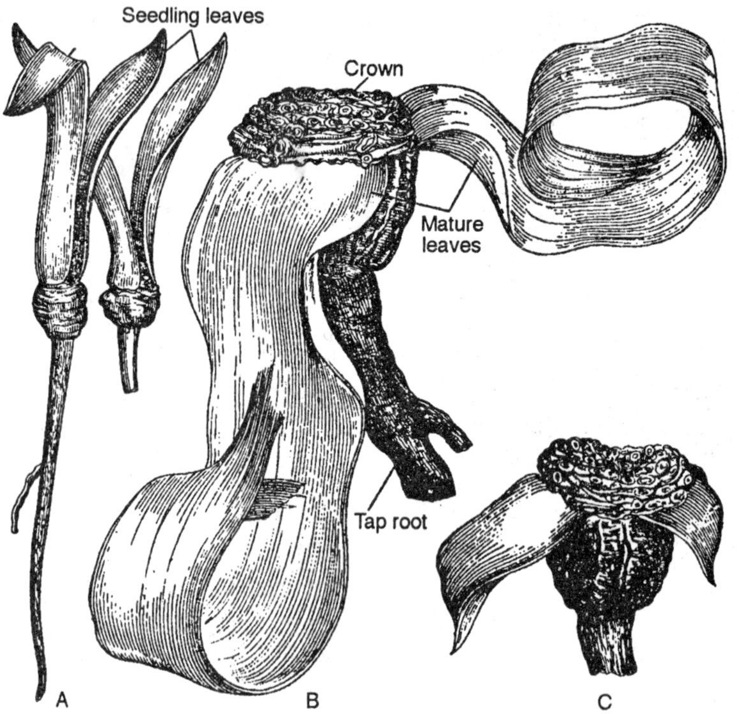

Fig. 50.8: *Welwitschia mirabilis*. Young (A) and mature (B and C) plants

water storage tissue in the middle. The stem stele shows endarch, open, collateral bundles. Annual rings are not present and the cambium is transitory. Growth ring-like marks may develop beacause of polycyclic secondary cambium as in *Cycas*. The secondary xylem shows tracheides with bordered pits (with reticulation between pits) and vessels. The phloem, as in other Gymnosperms, is devoid of companion cells. Thin-walled, branching 'spicular cells' incrusted with calcium oxalate crystals is a special feature of the stem, root and leaf.

Welwitschia is dioecious. Both the staminate (Fig. 50.9A) and the ovulate (Fig. 50.9B) strobili are compound and borne in clusters on branching shoots arising from the rim just above the leaf groove. The inflorescence stalk branches dichotomously and the arrangement of the members on it is cyclic. Each strobilus or inflorescence shows close opposite decussate bracts on the axis. All the upper bracts bear single flowers at the axils. All the bracts are bright red at maturity, almost as in *Poinsettia*, and there seems to be some nectar secretion pointing to insect pollination.

In a male flower (Figs. 50.9C and D) there are two opposite decussate pairs of *perianth bracts* of which the upper pair is smaller and united with one another at the base while the lower pair is bigger and free. These enclose six stamens, two against each one of the larger bracts and one against each of the smaller ones. The stamers or the *microsporphylls* are united below and each shows an another with three pollen sacs or *microsporangia*. The pollens (*microspores*) are elliptical with exine and intine of equal thickness. In the centre of the flower is a rudimentary sterile ovule with nucellus but no sporogenous tissue and covered by a single long integument. The male flower is, therefore, actually bisporangiate.

The female strobilus is very conspicuous because of the brilliant red colour of the bracts at maturity. Each female flower (Figs. 50.9E and F) has only one pair of fused *perianth bracts*

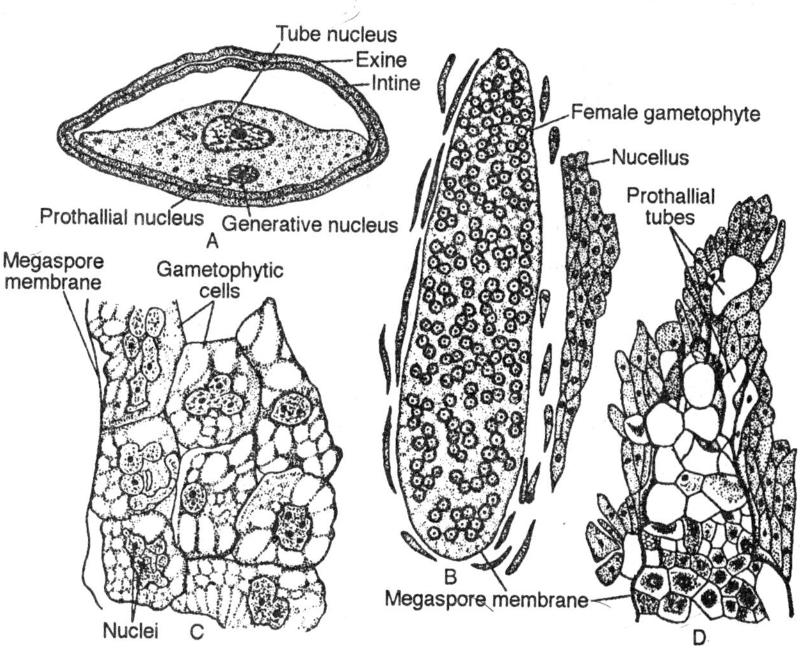

Fig. 50.10: *Welwitschia mirabilis*. A. 3-nucleate stage of pollen, B. Female gametophyte in free-nuclear condition, C. Female gametophyte cells undergoing fusion of the multinuclei in each cell, D. Top of female gametophyte showing prothallial tubes (after Pearson)

about 1024 nuclei which are evenly distributed in the cytoplasm without any central vacuole (Fig. 50.10B). Cell formation now sets in to result in a large number of multinucleate cells. The nuclei in each cell then fuse (Fig 50.10C) to form uninucleate cells of variable chromosome constitution. The most remarkable feature is that no archegonium is formed in the female gametophyte. Some large cells at the top of the female gametophyte, which may be regarded as archegonial initials, elongate breaking through the megaspore membrane and grow into the nucellus like pollen tubes. These are called *prothallial tubes* (Fig. 50.10D). The lower part of the gametophyte functions as the *endosperm.*

Pollination and fertilization

Welwitschia is insect pollinated (probably hemipteran) being favoured by the bright colour and nectar. No pollen chamber is formed in the ovule and the pollens, getting entrance through the micropylar tube where there is a pollinating drop by which it is draw in germinate into pollen tubes on the top of the nucellus. The pollen tubes grow down the nucellus and meet the prothallial tubes halfway down it. The end walls of the tubes fuse and fertilization takes place when the nucleus of the prothallial tube passes into the pollem tube and fuses with a male gamete forming a *zygote.*

The new sporophyte

The zygote cell elongates towards the top of the endosperm and divides into two, the upper cell forming the *primary suspensor* and the lower cell the *embryo initial* (Fig. 50.11A). The embryo initial then divides rapidly forming an *embryonic plate* of 8 cells while its upper cells develop *embryonal tubes* which encircle the suspensor as *inner* and *outer cortical rings* (Fig. 50.11B). The suspensor elongates quickly and becomes much coiled, pushing down the growing embryo into the endosperm. The embryo is dicotyledonous. As a number of embryos may develop, the seed is polyembryonic at the beginning but the

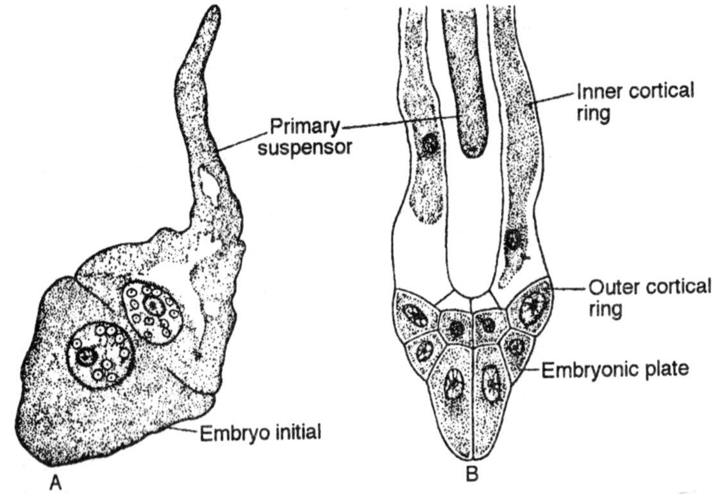

Fig. 50.11: *Welwitschia mirabilis*. A. First division of zygote, B. Later stage of embryo development (after Pearson)

mature seed usually shows only one embryo. The seed with the wings is about an inch wide and is wind dispersed. On germination, the two cotyledons, spread out into a pair of cotyledonary leaves which are later replaced by the pair of persistent leaves at right angles to the former.

Welwitschia is very peculiar and abnormal in its gametophytes and fertilization. Nevertheless, it points out to reduction in the angiospermic line.

Order III. **GNETALES:** *GNETUM*

Gnetum, which constitutes the family Gnetaceae and the order Gnetales is a genus formed predominantly of lianas in tropical forests in contrast to *Ephedra* and *Welwitschia* which are xerophytic. The plant, when not in fruit, may easily be taken as a dicot and had sometimes been actually described as a dicot. It shows the nearest approach to the Angiosperms and the approach is more natural than the very abnormal condition in *Welwitschia*. Apart form the lianas, only a few species are trees or shrubs.

The number of species included within *Gnetum* varies from 30 to 40 in literature. Markgraf (1930, 1951) listed 28 species distri-

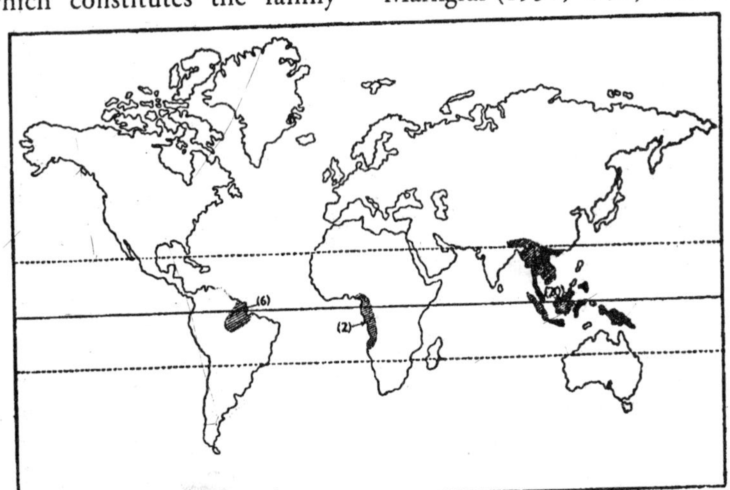

Fig. 50.12: Distribution of *Gnetum* species over the Earth with number of species in each are shown in brackets. (Modified from Markgraf)

buted over the Earth as shown in the map (Fig. 50.12). Of these, 20 species are recorded from tropical Southest Asia including India, Southeast China and spreading to East Indies Islands. There is no *Gnetum* in Australia, North America or Europe. 6 species are recorded by Markgraf in the Northeast of South America and West Indies although Pearson recognized 12 species in this region. 2 species are recorded from West Central Africa. No species is common between the Eastern and Western Hemispheres.

Bharadwaj (1957) recognized five species in India and a sixth in East Pakistan. Their distribution is shown on the map in Figure 50.13. These are:

1. *Gnetum gnemon* L.: Trees or erect shrubs, rarely scandent; showing 6 varieties occurring in Arunachal, Assam, Manipur, Burma, Malay, Moluccas, Sunda Islands, Celebes, Borneo, Philippines, New Guinea. Only 2 varieties, *brunonianum* (Griff.) Mgf. and *griffithii* Mgf. occur in Arunachal and Assam.
2. *G. ula* Brongn. (= *G. scandens* Brandis): A most extensive liana most common in India. Found in the tropical forests of South India and occurring as far north as Orissa. Also found in the Andaman Islands.
3. *G. montanum* Mgf. (= *G. scandens* Hook. f. and *G. edule* Kurz): A robust liana from Sikkim and Darjeeling, Khasia and Jaintian. Assam, Sylhet, Mayurbhanj, Burma, Siam, China and Indochina. There is a form *megalocarpum* Mgf. with larger seed from the Dafla Hill, Arunachal.
4. *G. contractum* Mgf. (= *G. scandens* Hook. f. in part): A scandent shrub from Kerela and Nilgiri Hills.

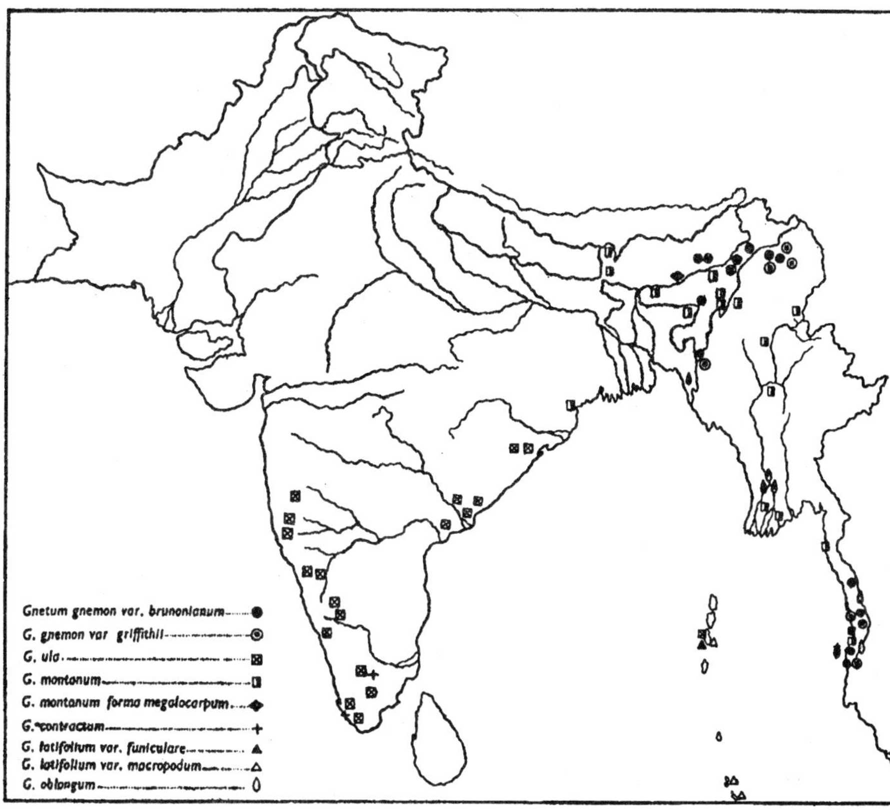

Fig. 50.13: Distribution of *Gnetum* species in the Indian subcontinent. (Modified from Bharadwaj)

5. *G. latifolium* Blume: A scandent shrub from the Andaman and Nicobar Islands in addition to Malay, Siam, Sumatra and Java. There are 5 varieties of which var. *macropodum* (Kurz) Mgf. is known in Andaman and Nicobar while var. *funiculare* (Blume) Mgf. is known from the South Andaman.

The sixth from Bangladesh is: *G. oblongum* Mgf. (= *G. funiculare* Kurz): A scandent shrub from Chittagong and South Burma.

A key to the five Indian species with their distinguishing features is given below:

1. Male spikes with long internodes so that axis is visible between bracts (*G. gnemon* var. *griffithii* shows variations with shorter internodes). Seed sessile. (Section *Gnemonomorphi*) — *G. gnemon* (Leaves yellow-green to brown when dry, up to 20 × 9 cm, oblong or elliptic. Seed smooth or papillose, 1-2.5 cm long, ellipsoid, in lax inflorescence in var. *brunonianum*, almost round in contracted branched inflorescence in var. *griffithii*).

1. Male spike more or less cylindrical, internodes short and contracted so that axis is not visible between bracts. Seed stalked. (Section *Cylindrostachys*).

 2. Bracts of the male spikes flat with margins recurved outwards so as to expose the flowers.

  3. Male inflorescence usually simple — *G. ula* (Leaves black on drying, 8-15 × 7.5 cm, ovate-elliptic. Female inflorescence simple. Seed rugose, ellipsoid 3-4 × 1.5-2 cm).

  3. Male inflorescence usually branched — *G. montanum* (Leaves black on drying, into 30 × 12 cm, ovate to oblong-ovate Female inflorescence also branched. Seed 1.5 × 1 cm, ellipsoid, longer than stalk).

 2'. Bracts of the male spikes cylindrical, margins not recurved outwards, flowers included.

  3'. Fertile parts of the male spike up to 1.5 cm long and 0.4 mm thick, inflorescence short. Leaf lamina 10 × 5 cm. — *G. contractum* (Leaves black on drying, elliptical or ovate. Male inflorescence one branched. Female inflorescence branched. Seed 3-3.5 × 1.5 cm, oblong-ovate, long acuminate, longer than stalk. Not distinct from *G. ula* in vegetative condition).

  3'. Fertile part of male spike up to 4 cm thick. Inflorescence lax up to 12 cm long. Leaf up to 25 × 12 cm. — *G. latifolium* (Leaves black to brown on drying, elliptic to broadly ovate. Male inflorescence branched only once. Female inflorescence once or twice branched. Seed 2-2.5 × 1-1.5 cm, broad or narrowly ellipsoid, shorter than stalk).

The Sporophyte plant

As is usual for tropical woody lianas, most of the species twine, climb or trail over other vegetation reaching the tops of the tallest trees. In an old plant there may be no leaf on the first 40 or 50 ft. The branches of unlimited growth ('long shoots') bear short 'dwarf shoots' of limited growth though in some species (e.g. the tree species *G. gnemon*) such difference is not marked. The nodes of the long shoots are articulated and swollen in most species. Leaves are borne on the dwarf shoots (Fig. 50.15A) and are remarkable in being large, oval, entire, leathery and pinnate reticulate veined exactly like dicot leaves so that a plant not in flower may be consi-

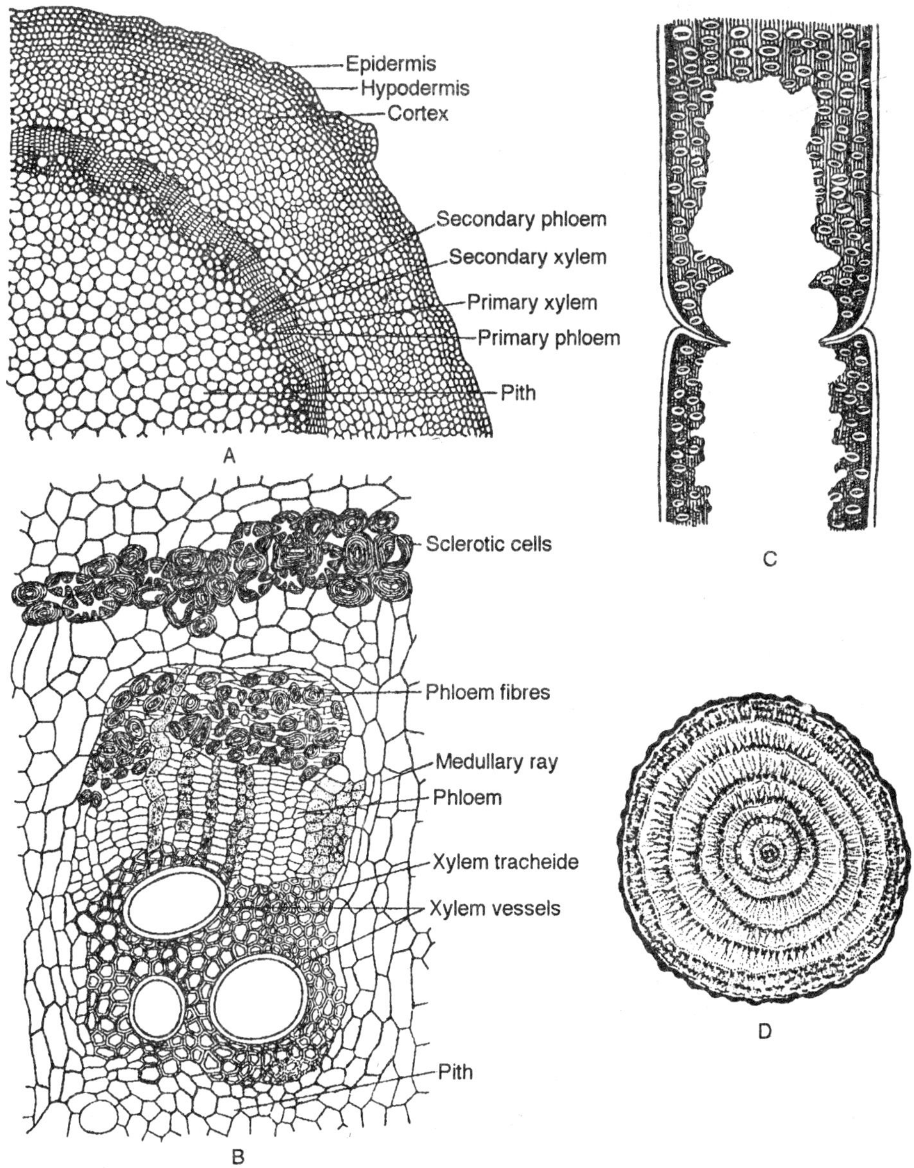

Fig. 50.14: A. T.s. of *Gnetum gnemon* stem, B. T.s. of part of *G. ula* stem, C. A vessel of *G. gnemon* showing a large pit perforating the cross-wall and numerous bordered pits on longitudinal walls, D. T.s. of a *Gnetum* stem showing polycyclic vascular rings. (Figs. A and B after Maheswari and Vasil, C after Chamberlain)

dered as an Angiosperm. These leaves are borne in opposite decussate pairs, are exstipulate and have short petioles. In the long shoots of climbers the leaves are reduced to scales.

The t.s. of the stem also resembles a dicot. It shows an endarch siphonostele formed of 20 to 24 open, conjoint, collateral bundles arranged in a ring (Fig. 50.14A). A section-shows from the periphery an *epidermis* with thick cuticle and sunken stomata and a *cortex* of thin-walled parenchymatous cells which shows a collenchymatous *hypodermis*. The outer cortex later develops suberized fibres in irregular patches. The endodermis and pericycle are not distinct. The *primary vascular bundles* are separated by wide *primary medullary rays*. Just outside the phloem is a patch of *spicular sclerotic cells* (Fig. 50.14B). The *phloem* of some bundles show fibres on the outside. The phloem proper is very regular in *G. ula* showing the sieve cells and the so-called 'companion cells' in radial rows. These 'companion cells', however, are different from those in Angiosperms as they are cut off independently from the cambium. The *cambium* itself is not distinct. The *xylem* is endarch but the protoxylem is very small. The *secondary xylem*, in addition to the normal tracheides, show a number of vessels of different sizes. These (Fig. 50.14C) are with multiseriate pits but the end walls show large single pits formed by the fusion or enlargement of circular pits and ultimate disappearance of end walls—a method more advanced than that in *Ephedra* but still different from that of most of the Angiosperms where the vessels are scalariform. *Secondary growth* is normal in the arborescent *G. gnemon*. But in the lianas (e.g., *G. ula*) several successive rings of cambia develop one after another giving rise to polycyclic rings of anomalous secondary wood (Fig. 750.14D). Sometimes the rings are incomplete and eccentric. *Periderm* is usually thin and irregular and lenticel is usually absent.

The root is diarch or tetrarch with scanty primary xylem. Secondary growth is normal. The vessels are larger than in stem and the tracheides show conspicuous 'bars of Sanio'.

The leaf anatomy resembles that of a dorsiventral dicotyledonous leaf. Maheshwari and Vasil noted the development of stomata in *G. ula* to be *haplocheilic* as in *Ephedra* while Florin and others found that most *Gnetum* species have *syndetocheilic* stomata. The stomata are on the lower epidermis only. *Mesophyll* shows palisade and spongy layers but *spicular sclerotic cells* are mixed with the latter.

Gnetum is generally dioecious but bisexual inflorescences are also known in a few cases. As in other genera, both male and female flowers are borne in compound strobili or inflorescences. Both types of inflorescences are usually panicles in the axils of leaves or in terminal positions on dwarf shoots.

MALE STROBILUS: Each individual strobilus (borne as panicles : Fig. 50.15A) develops at the axil of a bract or apically on a dwarf shoot. It shows a slender axis with a pair of concrescent bracts at its base and higher up each node of the axis bears opposite decussate bracts which are connate throughout their whole lengths making them look like cups. In the axils of these *connate bracts* (also called *collars*), all the way around the axis, the staminate flowers are borne in several rings (Fig. 50.15B). In some cases male strobili are found with female flowers in the upper rings (e.g., in *G. gnemon*) but the latter usually soon abort. Each male flower (Fig. 50.15C) consists of a stalk with two unilocular anthers or *microsporangia* and a sheathing *perianth* which is apparently formed by the fusion of two bracteoles. The flowers, in the young stage, remain hidden under the rim of the connate bracts. Then the perianths only peep out. Finally, the flowers open by a quick elongation of the stalks which pierce the perianth. Hairs are often associated

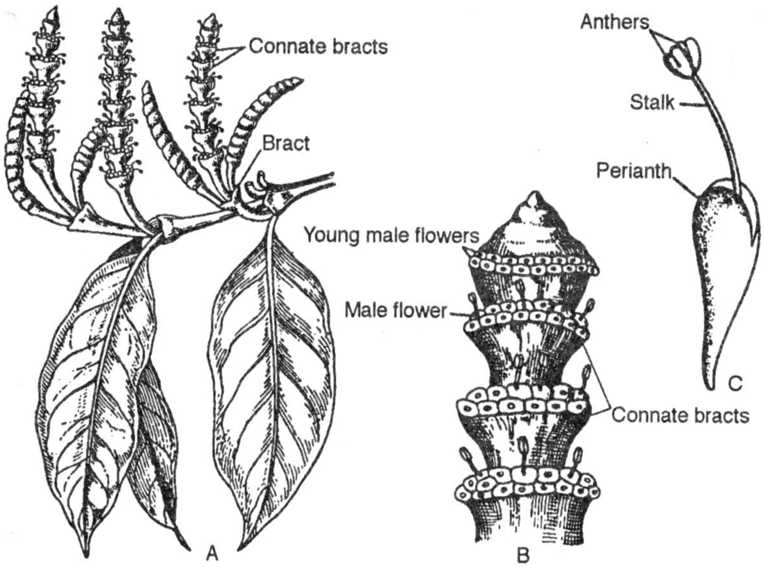

Fig. 50.15: *Gnetum gnemon*. A. A twig bearing male strobili, B. Part of a male spike, C. A male flower

with the male as well as the female flowers of *Gnetum*. In the development of a microsporangium the tapetum is formed from the sporogenous tissue and not from the wall cells. The *microspores* develop by reduction division of the *spore mother cells* as usually and at maturity all cells between the microspores or pollens and the epidermis disintegrate. The anthers dehisce by longitudinal slits to release the pollens.

FEMALE STROBILI: The female strobili or spikes also develop from positions as the male and have the same outward appearance. In the young stage (Fig. 50.16A) the flowers remain hidden under the connate bracts. At maturity the flowers, placed in one ring only, peep out (Fig. 50.16B). Only a few of the ovules in a spike mature into seeds (Fig. 50.16D). The structure of the female flower of *Gnetum* (Fig. 50.16C) is very problematic. There are *three*

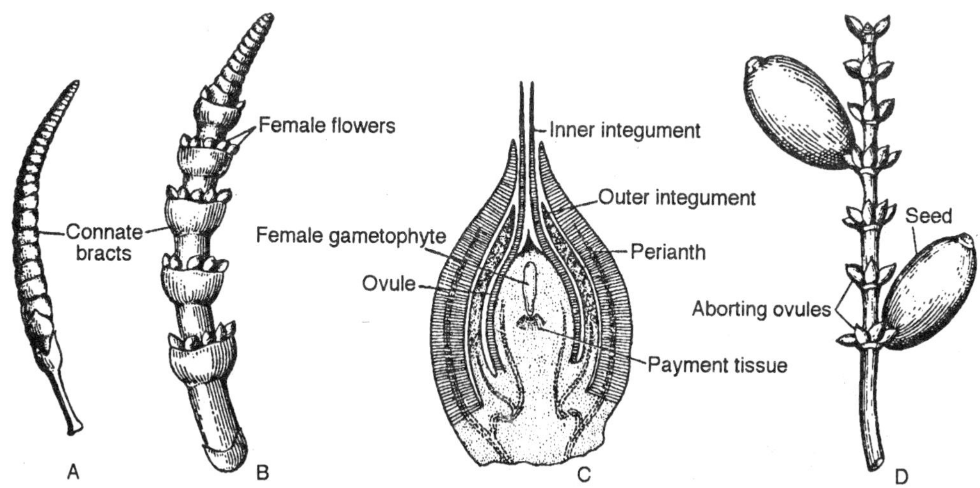

Fig. 50.16: A-B. Young an mature conditions of *Gnetum ula* female strobilus, C. Median l.s. of female flower (after Lotsy), D. A female strobilus with two mature seeds

investments round the ovule nucellus instead of the two of *Ephedra* and *Welwitschia.* Each investment, if formed of bracteoles, are completely fused. There are separate vascular supplies for each of the three investments. Chamberlain and Beccari consider the outermost investment as the *perianth* and the two inner ones as the *outer* and *inner integuments* so that at least *Gnetum* of the Chlamydosperms has got two real integuments as in Angiosperms. The inner integument shows the long micropylar tube as in *Ephedra* and *Welwitschia.* In the mature seed the inner integument is reduced to a thin *papery* layer while the outer integument is differentiated into an inner hard *stony* layer formed of sclerotic and fibrotic cells while its outside becomes *fleshy*, green in young stage, red when ripe and often edible. Thus, there are three coats of the seed covering as in *Cycas*, only the three coats are developed from two integuments instead of one. Strassburger and Celakovsky, however, considered the middle integument as merely the outer layer of the inner integument while Lotsy considered only the innermost investment as integument bringing it on the same level as *Ephedra* and *Welwitschia.* It has also been suggested (Thompson) that the inner integument is analogous to the Angiosperm ovary. There are several hypodermal *archesporial cells* and the *sporogenous cells* become deep-seated after cutting off the *wall cells* on top. Thompson found that reduction of one or more *megaspore mother cells* gave rise to linear tetrads of four *megaspores* of which the lowest became functional. More than one megaspore of even the same linear tetrad may be functional. But Maheswari and Vasil (1961) concluded that in *Gnetum* the four megaspores do not have partition walls between them and the whole structure (i.e., all the four megaspores) take part in developing the female gametophyte. Thus, the female gametophyte is *tetrasporic.* There is no nucellar beak in the female gametophyte and the tip is slightly disorganized to form a rather shallow pollen chamber.

The gametophytes

Male Gametophyte: The *pollen* has a spiny thick exine and an intine (Fig. 50.17A). Thompson (1916) decided that there is no prothallial cell or nucleus develops in the pollen. The pollen nucleus divides to form a *tube nucleus* and a *generative nucleus* which lettar divides to form a *stalk nucleus* and a *body nucleus.* The pollen is shed at this three-nucleate stage. The conclusion was that *Gnetum* was very near the Angiosperms in eliminating the prothallus. But later investigations, as discussed by Maheshwari and Vasil (1961), show that this is

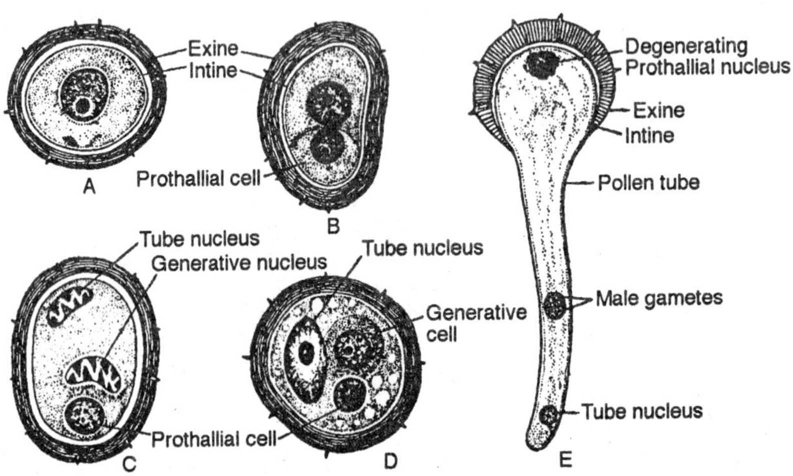

Fig. 50.17: A-E. Stages of germination of ***Gnetum*** pollen. (A-D after Madhulata)

not the true position. The pollen nucleus does divide to cut off a *prothallial cell* (Fig. 50.17B) whose partition wall may be transitory when the prothallial cell becomes naked (Fig. 50.17C). The other nucleus divides into a *tube nucleus* and a *generative nucleus* (Fig. 50.17C) the latter of which soon becomes a naked *generative cell* with its own cytoplasm (Fig. 50.17D). The pollen is actually shed at this three-nucleate stage. On reaching the shallow pollen chamber it germinates with the intine coming out as a *pollen tube* (Fig. 50.17E). The prothallus cell slowly disintegrates and the generative cell, preceded by the tube nucleus, divides to form two naked *male gametes.*

Female Gametophyte: The tetrasporic cell (megaspore mother cell with four megaspore nuclei but no partition wall) forms the female gametophyte. Several such female gemetophytes may start developing but only a few (not more than two or three) reach full maturity. The four nuclei inside divide by free nuclear division forming a large number of nuclei round a central vacuole the upper part of which latter is more swollen (Fig. 50.18A). Wall formation starts from the lower chalazal end, slowly proceeds upwards and is not complete at the micropylar end even at the time of fertilization. The tissue of the nucellus immediately below the gametophyte serves for its nutrition and becomes specially organized to form the 'pavement tissue' (Fig. 50.16C).

There is no archegonium initial so that *Gnetum* has reached the final stage of reduction of archegonium (a stage which is higher than in *Welwitschia* but not abnormal as in the latter) attaining the Angiosperm level. One or more nuclei out of the free nuclei on the top of the gametophyte become a little larger, accumulate some cytoplasm round them and are then identified as *eggs* (Fig. 50.18B) in *G. gnemon.* In *G. ula* it seems that groups of cells are formed on the top, a few out of which act as eggs (Fig. 50.18C).

Pollination and fertilization

The pollen, carried by wind, is caught by the pollination drop on the tip of the long micropylar tube of the inner integument when

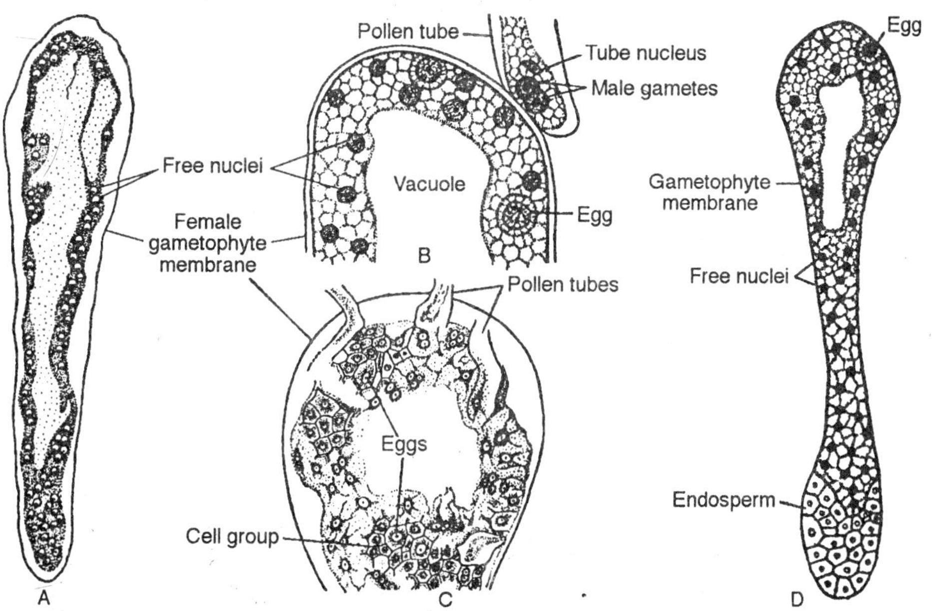

Fig. 50.18: A. Female gametophyte of *Gnetum ula* in free nuclear stage (after Vasil), B. Top of *G. gnemon* female gametophyte showing eggs (after Thompson), C. Top of *G. ula* female gametophyte showing eggs (modified from Vasil), D. Female gametophyte of *Gnetum* showing first part of endosperm at base

it gets sucked in by the drying up of this drop and gets placed in the shallow *pollen chamber* on the top of the nucellus. There it germinates and the pollen tube reaches the female gametophyte after penetrating the nucellus (Figs. 50.18B and C). It is doubtful whether even the lower part of the female gametophyte starts developing cell walls before the pollen tube has entered the gametophyte. At the time of fertilization a cellular tissue is formed at the base (Fig. 50.18D), which is the *endosperm* where there are multinucleate as well as uninucleate cells and the multinucleate cells also become uninucleate by fusion. One male gamete fertilizes an egg and the two gametes of a pollen tube may fertilize two eggs. A number of *zygotes* are formed in the female gametophyte by the fertilization of a number of eggs by different pollen tubes. The nuclei on the top of the gametophyte, which are in the free nuclear stage even at the time of fertilization, organize a cellular tissue forming part of the endosperm immediately after fertilization. Although there is no evidence of 'double fertilization', this post-fertilization development of the bulk of the endosperm has an angiospermic touch.

The new sporophyte and the seed

The germination of the *zygote* has been found to be very variable. It may swell and then give rise to two tubular growths (Fig. 50.19A). But, both the tubes do not survive, only the

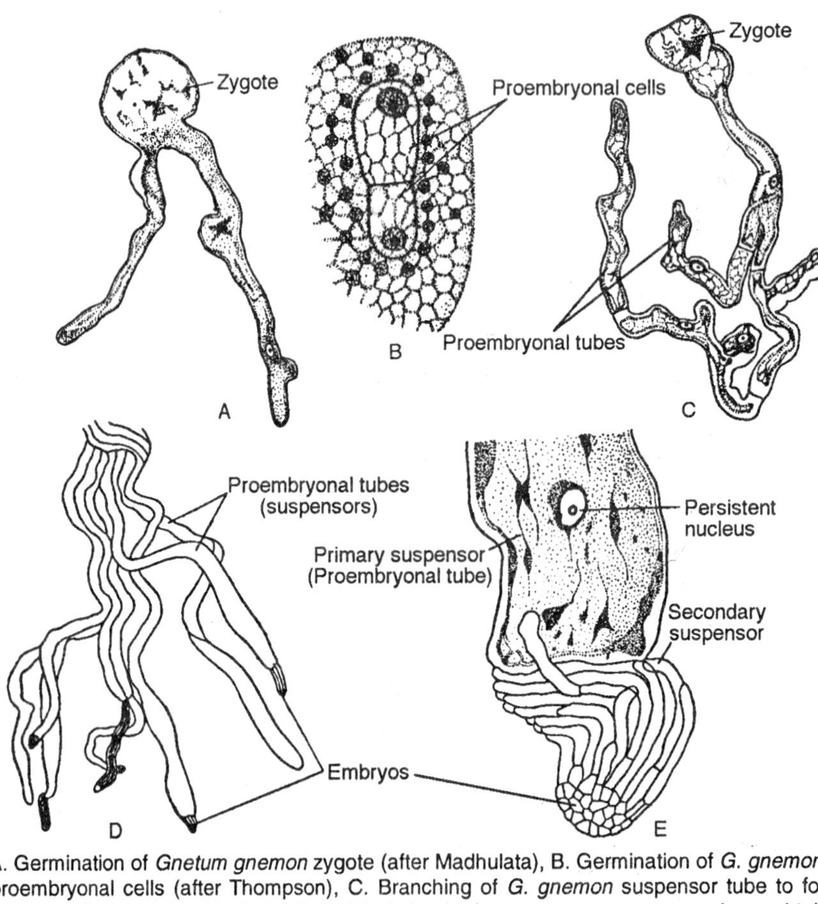

Fig. 50.19: A. Germination of *Gnetum gnemon* zygote (after Madhulata), B. Germination of *G. gnemon* zygote forming proembryonal cells (after Thompson), C. Branching of *G. gnemon* suspensor tube to form several *proembryonal tubes* (after Madhulata), D. Intertwined primary suspensor or proembryonal tubes and embryo initials in young seed of *G. ula* (after Vasil), E. Tip of proembryonal tube with growing embryo in *G. ula* (after Vasil)

one in which the zygote nucleus passes becomes functional. The zygote nucleus gets cut off at the tip of this tube by a cell wall. This cell now begins division and the *primary suspensor tube* is formed. The primary suspensor tube organizes the *embryo cells* at its tip. In some cases the zygote cell has been found to divide into two (Fig. 50.19B) and each of the two cells may form a separate primary suspensor tube or one of the two cells may get disorganized with only one suspensor tube developing. Sometimes, the primary suspensor tube branches to give rise to several branch suspensor tube or *proembryonal tubes* (Fig. 50.19C). Each proembryonal tube elongates greatly and these as well as others from other zygotes form an entangled mass within the endosperm (Fig. 50.19D). It, therefore, results in distinct 'polyembryony'. Further development of the embryo takes place only after the seed is formed and, often, after the detached seed is in soil. The tip nucleus of each primary suspensor or proembryonal tube divides into two. The upper remains in the tube as a persistent nucleus while the other is cut off by a wall and becomes the *embryo initial.* This embryo initial gives rise to a number of elongated *secondary suspensor cells* above and the *embryo* below (Fig. 50.19E).

The seed becomes very large with a *seedcoat* showing an outer fleshy, a middle stony and an inner papery covering as already stated. Seeds hang in bunches looking like fruits (Fig. 50.20A). Inside the seed is a copious *endosperm.* Although the seed is polyembryonic usually a single embryo lasts to the final stage. At this stage (Fig. 50.20B) the embryo inside shows a *dicotyledonous embryo.* A *feeder* is developed from the base of the hypocotyl and this absorbs nourishment from the endosperm. The *radicle* comes out first and germination is hypogeal. Germination is much delayed after the shedding of the seed as a considerable part of embryo development takes place after the seed is in the soil. They hypocotyl pushes out the two cotyledons which develop into the first pair of leaves.

The affinities of *Gnetum,* approaching the Angiosperms, has already been discussed.

Fig. 50.21 shows the life cycle of *Gnetum.*

Economic use

Gnetum gnemon is cultivated in Malay, Indonesia and neighbouring islands for its edible seed. The coat is discarded and the seed is roasted for consumption and also crushed into cakes, biscuits, etc. The young shoots and inflorescences are used as a vegetable. A fibre is obtained from its bark and is used in ropes and thread for fishing nets. Seed of *G. ula* is similarly used in South India and also yields an oil. *G. latifolium* seed and fibre are also of similar

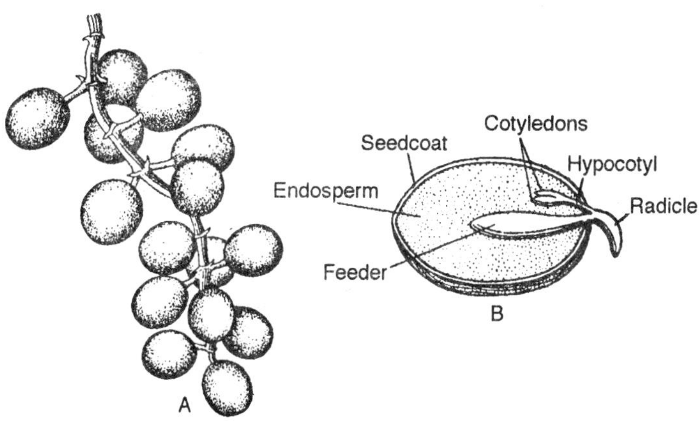

Fig. 50.20: A. Bunch of *Gnetum* seeds, B. L.s. of a seed of *Gnetum gnemon* about to germinate

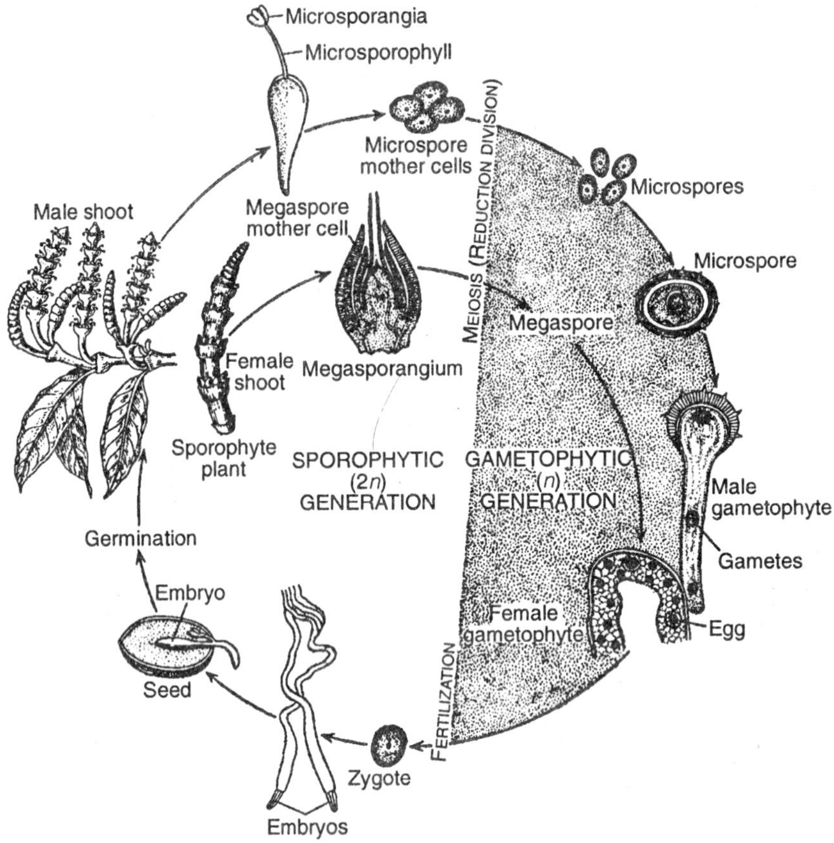

Fig. 50.21: Life cycle of *Gnetum*

use. *G. montanum* is supposed to be able to kill fish.

Gnetophytes, a Heterogeneous Group

We must emphasize that the gnetophytes represent, morphologically and phylogenetically, a paradoxical group of seed plants. On the one had, as Chamberlain (1935) maintained, the *combination* of certain characters—such as the compound nature of strobili and the presence of vessels in the xylem—is sufficient to keep the three genera together. On the other hand, Eames (1952) considered *Ephedra* to be closer to the conifers than to *Gnetum* or *Welwitschia* based upon differences in the organization of strobili. Details of the reproductive cycles also tend to support this belief (Table 50.1).

As can be seen in Table 50.1, *Gnetum* and *Welwitschia* share certain reproductive characters, but the methods of fertilization and embryo development are quite different and specialized in the two. The differences among all three genera suggest that the gnetophytes are a very heterogeneous assemblage of plants. This was the view of Martens (1971), who remarked that he was more struck by the contrasts among the three genera than by the traits they have in common. *Gnetum,* at various times, has been proposed as a possible progenitor of angiosperms, but the genetophytes as well as all extant gymnosperms lack "pollenkitt" (a sticky lipoidal substance on the surface of pollen grains). Pollenkitt is apparently restricted to angiosperms, leading one scientist to rule out the gnetophytes as progenitors of angiosperms (Hesse, 1984).

Table 50.1: Comparizon of certain reproductive characteristics in *Ephedra, Gnetum,* and *Welwitschia*

	Ephedra	Gnetum	Welwitschia
Sterile (stalk) cell in the pollen grain	Present	Absent	Absent
Female gametophyte monosporic in origin	Present	Absent (Tetrasporic)	Absent (Tetrasporic)
Alveolation in development of female gametophytic	Present	Absent	Absent
Archegonia	Present	Absent	Absent
A free nuclear period in embryogenesis	Present	Absent	Absent

Chapter 51

Progymnosperms and Origin of Gymnosperms

BACKGROUND

The affinities of the classes Pteridospermopsida, Cyacadopsida, Coniferopsida and Chlamydospermopsida including some problematic orders have been discussed in the preceding chapters and the discussions give an idea of the Gymnosperms as known today.

Jeffrey used the term 'Archigymnospermae' to those gymnosperms which show motile sperms. But this character is found or is assumed in very different taxa (Pteridospermopsida, Cycadopsida, Cordaitales and Ginkgoales of Coniferopsida) so that it is not possible to point out to any known taxon (extinct or extant) which may be marked out as the ancestor of all the Gymnosperms. It is possible that the origin was polyphyletic and the several origins are to be sought within some plexus of pteridophytic plants like the hypothetical 'Progymnosperms'. Even the fragmentary fossil history, which is far from complete, shows the presence of the Cycadofilicales and the Cordaitales and also probable fossils of Gnetales as early as the Devonian without any sign of convergence. Probably the different main taxa differentiated earlier than the Devonian at a time when the main pteridophytic taxa were also differentiating (c.f. Fig. 51.1).

Morphologically, the Gymnosperms culminated in the Angiosperms. But, here again, there is no Gymnosperm taxon which may be pointed out as the ancestor of the Angiosperms. The Caytoniales, the Cycadeoidales and the Glossopteridales show some angiosperm characters but that is not enough to be a direct ancestor. *Gnetum* and *Welwitschia* are quite near the Angiosperms but cannot be the ancestors. It may be assumed after Fagerlind (c.f. Fig. 50.1) that possibly the Chlamydosperms are related to the 'Proangiosems' which, again, probably came from the ancestors of the Conifers—related to the 'Progymnosperms' at the pteridophytic level.

A word need be added about the view that the term 'Gymnosperm' should also be considered obsolete along with the term 'Pteridophyta' as both these taxa are assemblages of some parallel lines of evolution. Such a step seems to be too drastic at the present state of our knowledge as both these taxa, in spite of their diversities and the uncertainty of origin, show certain uniformity in the patterns of life history which sufficiently keep them together.

Because the gymnosperms include such ancient lines of seed-bearing plants, understanding their origins depends on the analysis of the fragmentary fossil record. The formidable nature of the task that confronts paleobotanists is aggravated by the fact that many gymnosperm fossils—like the remains of other, more primitive vascular plants—are often isolated portions of leaves, strobili, sporangia, spores, stems, and roots. Many of these separate parts have been provisionally assigned generic names, but efforts to reconstruct a "complete plant" from the synthesis of such "organ genera" are often theoretical and subject to drastic revision as new paleobotanical discoveries are made. The most significant progress in understanding, however, is made when two or more "organ genera" can be demonstrated to be physicallay interconnected. One of the most remarkable discoveries of this sort was made by Beck (1960a, 1960b) who found two well-known organ genera, *Archaeopteris* (the "frond" of a presumed fern) and *Callixylon* (the stem of a gymnospermous tree) in organic connection. The plant represented by the interconnection of *Archaeopteris* and *Callixylon* was placed by Beck (1960b) in a new class of vascular plants, the Progymnospermopsida. As relatively advanced representatives of this group, *Archaeopteris* combined the anatomy typical of gymnosperms (i.e., secondary xylem tracheids with circular bordered pits) with fern-like fronds and so-called pteridophytic reproduction (i.e., reproduction by dispersed spores, rather than by seeds). According to Beck's (1970) interpretation, "the progymnosperms are of great significance because they seem to be the immediate ancestors of seed plants."

Following the original reconstruction of *Archaeopteris*, Beck (1962, 1964, 1970, 1971, 1976, 1981) and others (Carluccio *et al.*, 1966; Banks, 1968) have considerably advanced our knowledge of its morphology, anatomy, and phylogenetic relationships.

Historical Résumé

Prior to Beck's (1960b) investigations, *Callixylon* and *Archaeopteris* were two of the most widespread organ genera of the Upper Devonian. The generic name *Archaeopteris* was given to compressions of fernlike "leaves" which Arnold (1947) considered "the most widely distributed fern in the Upper Devonian." According to Beck (1960a), *Archaeopteris* is a common fossil in eastern North America and has been collected from Gaspé Peninsula, Quebec, to south-western Pennsylvania; Beck found the specimen that he used in his original reconstruction in beds of Upper Devonian age near Sidney, New York. *Callixylon*, on the other hand, is a name originally applied to petrifications of certain Devonian trees with gymnospermous secondary xylem. Stumps of *Callixylon* 1.5 meters in diameter have been reported, and longs 6 to 9 meters long or more have been uncovered in Kentucky and Texas.

The specimen, which revealed the organic connection between the frond of *Archaeopteris* and the stem of *Callixylon*, consisted of a small axis "to which are attached several fragments of bipinnately compound leaves, one of which bears both fertile and vegetative parts" (Beck 1960b). In accordance with the rule of priority, established by the International Code of Nomenclature, Beck used the earlier generic name *Archaeopteris* for the entire plant represented by the two organ genera (*Archaeopteris* and *Callixylon*); the closest affinity of his specimen was with *Archaeopteris macilenta.*

Obviously, the critical point in Beck's discovery was his demonstration that the axis of his specimen represents a portion of the stem of *Callixylon.* This was shown by the presence of well-defined *groups* of circular bordered pits on the radial walls of the tracheids of the secondary xylem, a feature which had long been considered distinctive of the wood of *Callixylon* (see Beck 1960b, 1970) and Beck *et al.*, 1982, for a detailed analysis of the secondary xylem of *Callixylon*).

Using data based on previous descriptions of the habit of *Callixylon* by Arnold (1931), *Archaeopteris* was considered by Beck (1962) to have been a large tree-attaining a height of 18 meters or more—with a crown of branches bearing large bipinnately compound leaves (Fig. 51.1). The leaves were regarded as megaphylls, comparable to the leaves of ferns and certain pteridosperms (extinct seed ferns).

A very different interpretation of the "frond" of *Archaeopteris* was subsequently proposed by Carluccio *et al.* (1966). They found that the so-called rachis of the frond has a radially symmetrical vascular system like the

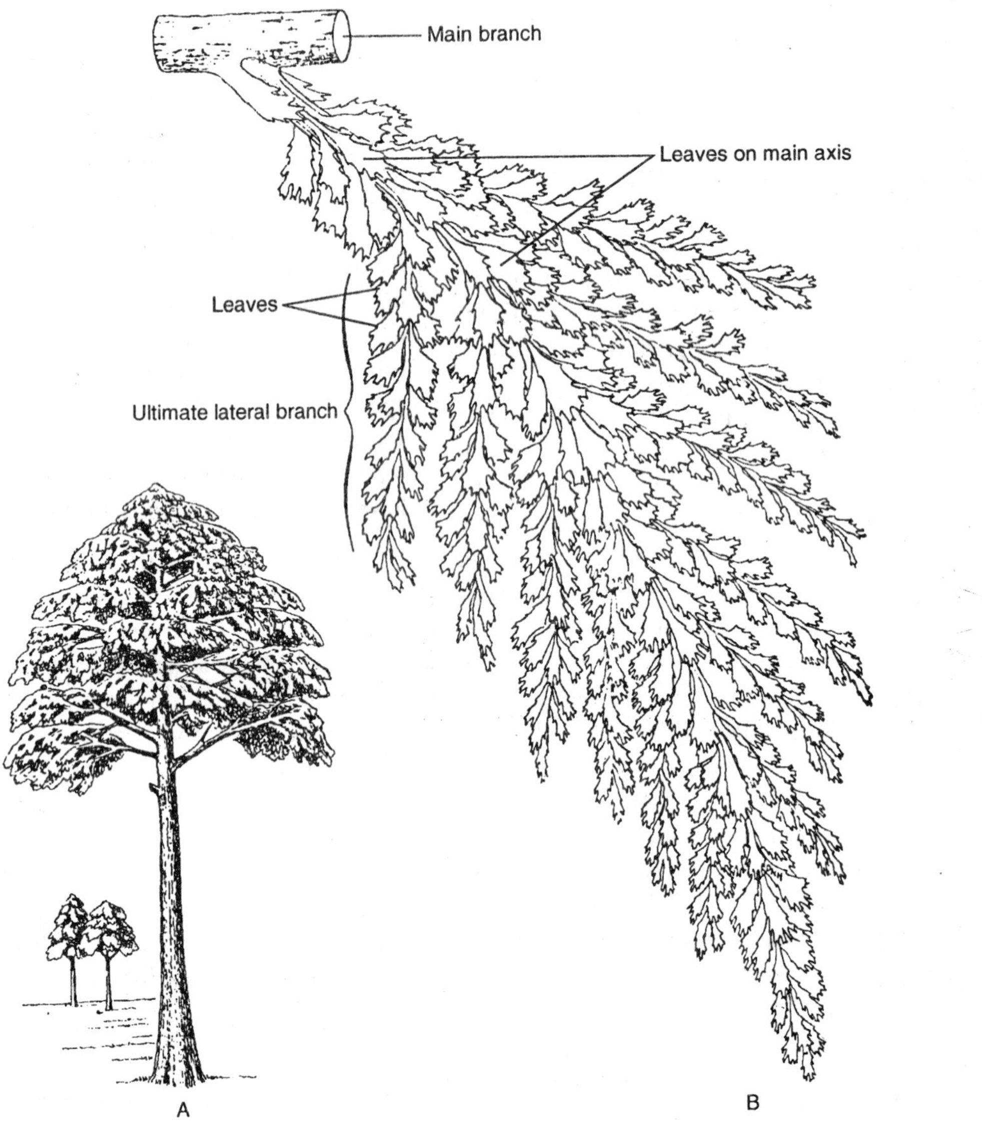

Fig. 51.1: Reconstruction of *Archaeopteris*. A. growth habit, B. reconstruction of a complete vegetative, lateral branch system attached to a larger axis

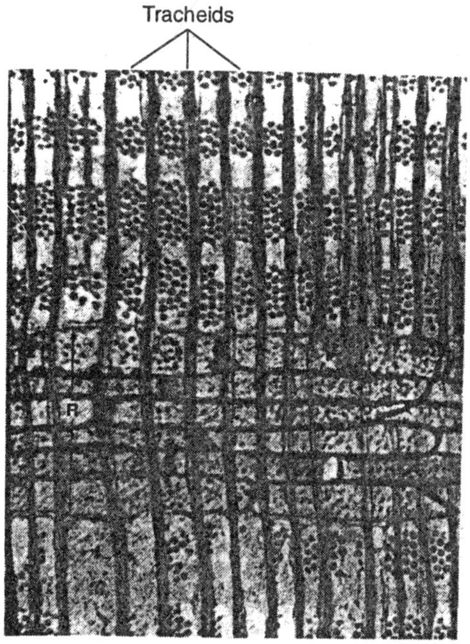

Fig. 51.2: Rdial longitudinal section, secondary wood of *Callixylon newberryi* showing groups of pits on tracheid walls

stem and that the axes of the lateral "pinnae" had a similar vasculature. Using these facts, Carluccio *et al.* concluded that the "frond" of *Archaeopteris* is not a pinnately compound leaf but a flattened leafy *branch system.* According to this interpretation the actual leaves, which in *Archaeopteris macilenta* are broadly laminate, are attached to the ultimate lateral branches (Fig. 51.1). Beck (1970) subsequently adopted this revised interpretation. In some species the leaves are more finely divided. Fertile, ultimate apendages are considered homologous with vegetative leaves and are laminate, often dichotomously branched, distally. The ultimate fertile shoots exhibit the characteristics of strobili (Beck 1981). Numerous sporangia, which dehisced longitudinally, occurred on the adaxial side of sporophylls (Fig. 51.3).

Vascular Anatomy

The conclusions reached by Beck (1971) from his study of the external morphology of compression-type specimens were supported by his investigations of stelar anatomy. A reconstruction of the vascular of a main axis is shown in Fig. 51.2. The primary xylem forms a cylinder of nine ribs, each of which is represented by an axial (or sympodial) vascular bundle; primary xylem is mesarch. Branch traces and leaf traces diverge from the axial bundles as follows: two of the nine stelar bundles give rise to traces that vascularize the members of the two distichous rows of lateral, determinate branches. The remaining seven "ribs," or axial bundles—three on one side of the stele and four on the opposite side—are the points of origin of the traces the extend into the helically arranged leaves of the main axis. Each leaf trace follows a steep, obliquely radial course through the cortex of the axis and then becomes divided into two strands as it enters the base of a leaf. The further branching of these two strands produces the dichotomous venation characteristic of *Archaeopteris* leaves. The stele is siphonostelic but there are no leaf gaps (parenchymatous regions) directly above a departing leaf trace. Secondary xylem (sx) was produced, of the type described earlier; secondary phloem also was formed, but more completely described for an earlier (older) progymnosperm in the order Aneurophytales (Wight and Beck, 1984). The production of secondary phloem and xylem is another unique feature among Devonian plants and serves to strengthen the case for a phylogenetic relationship

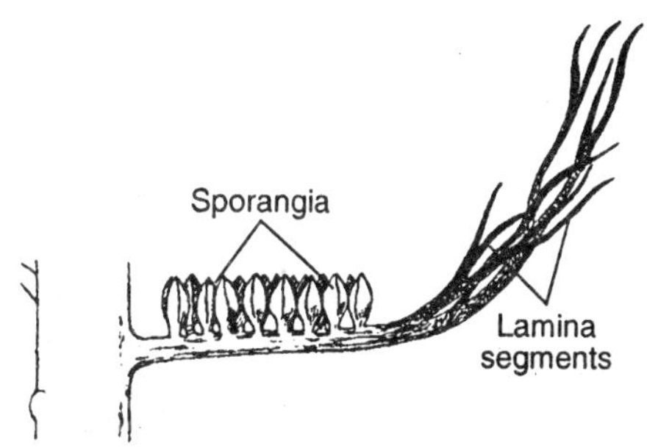

Fig. 51.3: Reconstruction of a fertile appendage of *Archaeopteris*. Note the two rows of adaxial sporangia

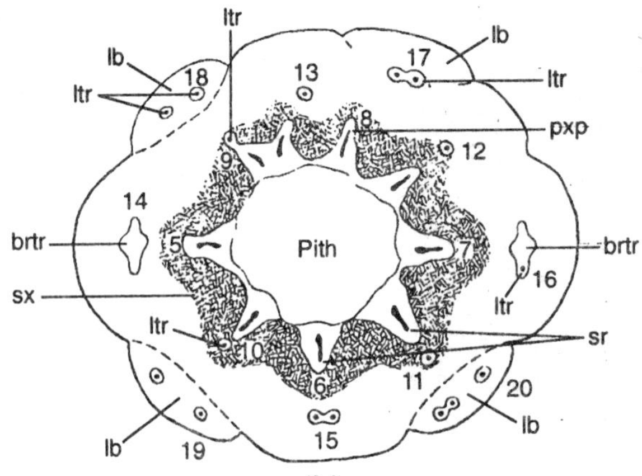

Fig. 51.4: Reconstruction of the vascular anatomy of the main axis of *Archaeopteris macilenta*. The stele consists of nine "ribs" (sr) and is enclosed by a cylinder of secondary xylem (sx) indicated by hatching. The helical sequence in the divergence of branch traces (brtr) and leaf traces (ltr) is indicated by numbers 5 to 20; higher numbers indicate older traces, lower numbers younger traces. lb, leaf base; pxp, protoxylem pole

between the progymnosperms and gymnosperms.

Phylogenetic significance of the Progymnosperms

The external morphology and vascular anatomy of *Archaeopteris* support the concept that the so-called frond of this progymnosperm represents a complex branch system to which leaves are attached. If this interpretation is correct, it seems doubtful that the *Archaeopteris* branch system typifies a stage in the evolution of the megaphyllous type of leaf characteristic of either the ferns or of such ancient gymnosperms as the seed ferns (pteridosperms). Beck (1971) noted the striking resemblance between the leafy branch systems of *Archaeopteris* and the frondlike branch systems of such living conifers as *Metasequoia* and *Taxodium*. Some of the Permian conifers, such as *Carpentieria*, also closely resembled *Archaeopteris* even in details of leaf morphology. Heterospory has been demonstrated in several species, including *A. macilenta* (see Beck, 1976, 1981). In another species, *Archaeopteris halliana*, the fertile leaves were interspersed among vegetative leaves on ultimate branches. Each fertile leaf was dichotomously branched with erect sporangia located on the adaxial surface. The sterile leaves were fan-shaped with dichotomous venation. This species also was heterosorous (Phillips *et al.*, 1972).

The Progymnospermophyta, as originally conceived by Beck (1960b), includes not only *Archaeopteris* but also other Devonian genera with secondary xylem and free-sporing reproduction. Genera included in the order Aneurophytales of the Progymnospermophyta. Branching in this order was mainly three dimensional, that of vegetative branches being either helical or decussate. Sporangia were borne terminally. The primary xylem was a ribbed protostele except in ultimate appendages where it was terete. It is particularly interesting to note that *Protopteridium*, now designated *Rellimia* (Bonamo, 1977), was formerly regarded as a primitive fern. It had a three-lobed protostele with mesarch protoxylem in each lobe, and sporangia were borne terminally on ultimate axes. In *Tetraxylopteris* branching was decussate, and sporangia were borne terminally on axes arranged in an irregularly pinnate manner (Bonamo and Banks, 1967). The protostele was a four-lobed action-stele, and secondary xylem has been described. The discovery of members of the Aneurophytales in the same fossil bed with the most ancient seeds led Rothwell and Erwin (1987) to propose that members of this order are more likely to be ancestral to the gymnosperms than the Archaeopteridales.

If *Archaeopteris* is neither a primitive fern nor the progenitor of the ferns, what evidence supports the possible evolution of coniferophytic gymnosperms from the progymnosperms? As stated earlier, *Archaeopteris* (including the form genus *Callixylon*) produced secondary xylem similar to extant conifers. Also, the investigations of Namboodiri and Beck (1968) have suggested that an understanding of evolutionary relationships lies in the correct interpretation of the structure and evolution of the stele in progymnosperms and gymnosperms.

However, before describing the investigations of Namboodiri and Beck, it is necessary to review briefly the concept of Jeffrey (1917). He believed that the first step in stelar evolution from the protostele to a tubular siphonostele was the enclosure of cortical tissue within the stele and the formation of parenchymatous regions, *leaf gaps*, above the point of departure of leaf traces. The tubular nature of the stele would be obscured if the leaf gaps overlapped as in many ferns. By this reasoning, Jeffrey concluded that the vasculature of seed plants was derived phylogenetically from ferns. The presence of the leaf gap was one of the definitive characters used by Jeffrey in establishing the "Pteropsida," a major taxon under which he grouped ferns, gymnosperms, and angiosperms. This system of classification has generally been abandoned by botanists (For details of his theory the reader is referred to Jeffrey's 1917 publication).

According to Namboodiri and Beck (1968) the evolution of the stele in progymnosperms and gymnosperms has never involved the formation of leaf gaps (in the sense of Jeffrey). Beck (1964) concluded that "stelar evolution in ferns and gymnosperms comprizes two entirely separate lines." (Also, see Rothwell, 1976.) Let us now examine the nature of the evidence supporting the conclusions of Namboodiri and Beck.

Figure 51.5 represents diagrammatically the main steps in the evolution of the primary vascular system of progymnosperms and gymnosperms. The most primitive stage was a protostele, the ribs of which represent the points of origin of the radially diverging traces to the "appendages". This kind of stele is illustrated by certain primitive progymnosperms such as *Aneurophyton* and *Tetraxylopteris.* The next stage in stelar evolution was characterized by the gradual dissection of the protostele into "longitudinal columns", a process accompanied by "medullation," i.e., the origin of a central pith region. Further evolutionary specialization led to a more definitive system of discrete *sympodial bundles,* each of which, at the node, divided *tangentially* into two *radially aligned* strands, an outer leaf trace and an inner "reparatory strand"; the latter continued its upward course, and a leaf gap was not formed. This is the type of primary vascular system that Beck (1971) found in the main axis of the "frond" of *Archaeopteris.* The next step is a change in the natue of the division of a sympodial bundle. A sympodial bundle divides radially to form two bundles, initially placed side by side along the same tangential plane. One bundle is the departing leaf trace and the other is the reparatory strand, continuing upward as the sympodial bundle. This mode of separation of leaf traces also occurred without the formation of leaf gaps. The spaces between sympodial bundles can be described as interfasicular regions. This type of stele is illustrated by the Carboniferous seed fern, *Lyginopteris.* According to Beck (1970) the stelar system of *Lyginopteris*

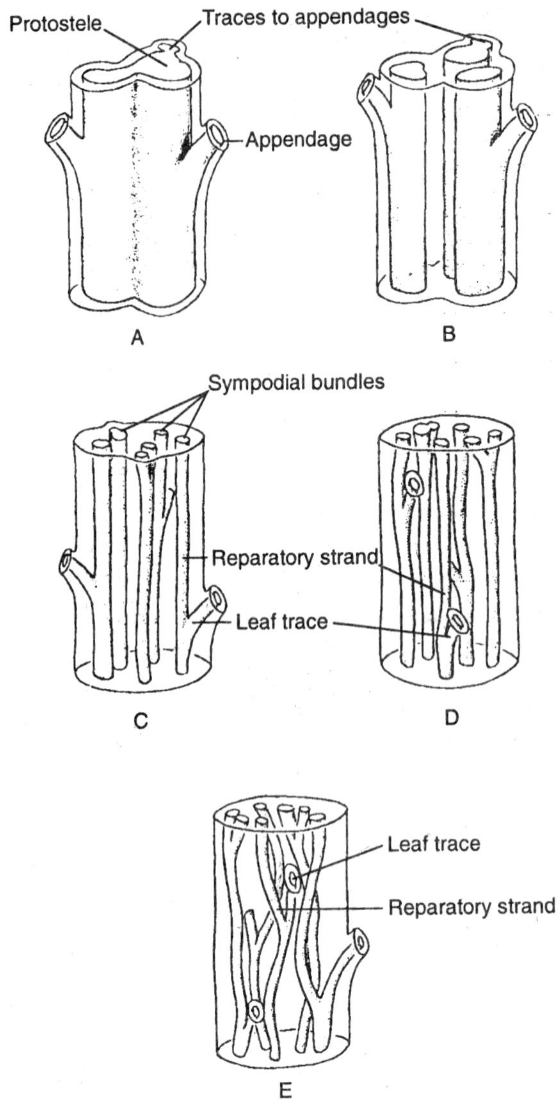

Fig. 51.5: Diagrams showing the probable evolutionary development of the primary vascular systems of progymnosperms and gymnosperms: A. the primitive protostelic condition, B. division of protostele into longitudinal columns, C. vascular system composed of discrete sympodial bundles, each of which, at a nodal level, divides tangetially into a preparatory strand and a leaf trace, as illustrated in *Archaeopteris*, D. vascular system in which the reparatory strand and leaf trace are formed by the radial division of the sympodial bundle, as illustrated in *Lyginopteris*, a seed fern, E. primary vascular system characteristic of living conifers with helical phyllotaxis. See text for further explanation (Redrawn from Namboodiri and Beck, *Amer. Jour. Bot.* 55:464, 1968.)

oldhamia, "in all of its major features, including the clear absence of a leaf gap, is identical with that of coniferns."

The salient feature of the concept of stelar evolution proposed by Namboodiri and Beck (1968c) concerns the similarity between the general mode of ontogenetic origin and divergence of leaf traces in progymnosperms and conifers. In both groups, leaf traces are described as arising directly from sympodial bundles of the eustele without the formation of **leaf gaps** (that is, the formation of leaf gaps or "perforations" in an otherwise tubular siphonostele as described by Jeffrey). In other words, the stele in gymnosperms evolved initially from a "dissected" protostele.

In stems of seed plants in which the primary vascular cylinder consists of sympodial bundles and departing leaf traces (eustele), the recognition of "leaf gaps" is rather uncertain because the parenchyma confronting the diverging leaf trace is confluent with adjacent interfascicular regions of the stele. Frequently it is only after some secondary growth has occurred that the "gap" becomes evident because secondary xylem formation is delayed in the interfascicular region confronting the departing leaf trace. In transectional view, the foliar gaps are then strictly correlated *in position* with the departing leaf traces. It is this aspect of anatomy that has made possible the recognition of various types of nodal anatomy characteristic of woody dicotyledons and conifers. From purely descriptive and practical standpoints, the term leaf gap is a useful one as long as its application is tempered with the realization that the steles in ferns and seed plants have probably evolved independently.

From the comprehensive studies on stelar anatomy, the origin of gymnosperms from ferns—as postulated by Jeffrey (1917)—no longer appears to be a useful hypothesis. Ancestors of the gymnosperms are to be found in the progymnosperms—one line leading to the seed ferns, the other to the coniferophyte assemblage. The ferns probably had their origin from some group of ancient plants of the *Psilophyton* type within the Trimerophytophyta.

Index

C

D

E

F

G

I

J

K

L

M

N

O

P

Q

R

S

T

T

T

U

U

V

W